The Geology of North America
Volume G-1

The Geology of Alaska

Edited by

George Plafker
U.S. Geological Survey
MS 904, 345 Middlefield Road
Menlo Park, California 94025

and

Henry C. Berg
115 Malvern Avenue
Fullerton, California 92632

1994

Acknowledgment

Publication of this volume, one of the synthesis volumes of *The Decade of North American Geology Project* series, has been made possible by members and friends of the Geological Society of America, corporations, and government agencies through contributions to the Decade of North American Geology fund of the Geological Society of America Foundation.

Following is a list of individuals, corporations, and government agencies giving and/or pledging more than $50,000 in support of the DNAG Project:

Amoco Production Company
ARCO Exploration Company
Chevron Corporation
Cities Service Oil and Gas Company
Conoco, Inc.
Diamond Shamrock Exploration Corporation
Exxon Production Research Company
Getty Oil Company
Gulf Oil Exploration and Production Company
Paul V. Hoovler
Kennecott Minerals Company
Kerr McGee Corporation
Marathon Oil Company
Maxus Energy Corporation
McMoRan Oil and Gas Company
Mobil Oil Corporation
Occidental Petroleum Corporation
Pennzoil Exploration and Production Company
Phillips Petroleum Company
Shell Oil Company
Caswell Silver
Standard Oil Production Company
Oryx Energy Company (formerly Sun Exploration and Production Company)
Superior Oil Company
Tenneco Oil Company
Texaco, Inc.
Union Oil Company of California
Union Pacific Corporation and its operating companies:
Union Pacific Resources Company
Union Pacific Railroad Company
Union Pacific Realty Company
U.S. Department of Energy

Published by The Geological Society of America, Inc.
3300 Penrose Place, P.O. Box 9140, Boulder, Colorado 80301

Printed in U.S.A.

Library of Congress Cataloging-in-Publication Data

The Geology of Alaska / edited by George Plafker and Henry C. Berg.
p. cm. — (The Geology of North America ; v. G-1)
Includes bibliographical references and index.
ISBN 0-8137-5219-1
1. Geology—Alaska. 2. Mines and mineral resources—Alaska.
I. Plafker, George, 1929– . II. Berg, Henry C. III. Series.
QE71.G48 1986 vol. G-1
[QE83]
557 s—dc20
[557.98] 94-18088
CIP

Cover Photo: Mt. Saint Elias and Tyndall Glacier looking north from over the Chaix Hills along the Gulf of Alaska. Bedded marine Tertiary rocks of the Yakutat terrane underlie the foothills in the foreground, and Late Cretaceous metamorphosed flysch and oceanic volcanic rocks of the Chugach terrane underlie snow-covered parts of Mt. Saint Elias and the ridge to the west. The Tertiary strata are tightly folded and imbricated as a consequence of post-Oligocene thrusting relatively beneath the Chugach terrane along the Chugach–Saint Elias fault system at the base of the mountains. Ongoing deformation is manifested by active seismicity, by 1,000 m of emergence of marine strata in the foreground since Pliocene time, and by uplift of coastal terraces at rates that average as much as 11 mm/yr. Mt. Saint Elias, the second highest mountain in Alaska, was the first landfall made by Vitus Bering's discovery expedition in July 1741. Vertical relief from the summit of Mt. Saint Elias to the tidal front of Tyndal Glacier at the head of Icy Bay is 5,489 m in a horizontal distance of 24 km, making it among the steepest in the world.

10 9 8 7 6 5 4 3 2 1

Note: Mt. Saint Elias, mentioned in the above caption, was cropped out of the cover photograph. Mt. Saint Elias would appear to the right.

Contents

PALEOMAGNETISM

QUATERNARY GEOLOGY

RESOURCES

SYNTHESIS

Plates
(in accompanying slipcase)

Preface

The Geology of North America series has been prepared to mark the Centennial of The Geological Society of America. It represents the cooperative efforts of more than 2,000 individuals from academia, state and federal agencies of many countries, and industry to prepare syntheses that are as current and authoritative as possible about the geology of the North American continent and adjacent oceanic regions.

This series is part of the Decade of North American Geology (DNAG) Project, which also includes six wall maps at a scale of 1:5,000,000 that summarize the geology, magnetic and gravity anomaly patterns, regional stress fields, thermal aspects, and seismicity of North America and its surroundings. Together, the synthesis volumes and maps are the first co-ordinated effort to integrate all available knowledge about the geology and geophysics of a crustal plate on a regional scale.

The products of the DNAG Project present the state of knowledge of the geology and geophysics of North America through the 1980s, and they point the way toward work to be done in the decades ahead.

In addition to the contributions from organizations and individuals acknowledged at the front of this book, major support for this volume has been provided by the U.S. Geological Survey.

A. R. Palmer
General Editor for the volumes
published by The Geological
Society of America

J. O. Wheeler
General Editor for the volumes
published by the Geological
Survey of Canada

Foreword

To those of you who are using this book for the first—or fiftieth—time, we greet you. Welcome to Alaska, geologically unique among the 50 states and a land of varied geology: active and passive plate margins; volcanoes, great earthquake faults, tsunami waves, landslides, glaciers, permafrost, and mineral and energy resources; and, we believe, a collage of disparate crustal terranes welded to each other and to the North American continent over a span of at least 500 million years.

The Geology of Alaska consists of 33 chapters and 13 plates that describe the geology and geophysics of each of Alaska's principal onshore and offshore regions and also cover a spectrum of topical subjects that include physiography, lithotectonic terranes, igneous and metamorphic petrology, geochronology, geophysics, geochemistry, sedimentary basins, mineral and energy resources, glaciation, permafrost, neotectonics, and tectonic evolution. Areal and topical coverage of the volume closely follows the original outline created during an organizational workshop in 1982, despite several authorship changes and two changes in editorship in the interim.

The geologic premise for most, but not all, of the areal chapters is that most of Alaska consists of displaced fault-bounded slivers, slices, and blocks of crust that were emplaced in their present positions relative to the craton by a variety of tectonic processes. Some of the authors, however, do not agree on the number, distribution, and configuration of these lithotectonic (or tectonostratigraphic) terranes, and others argue that Alaska consists mainly of crust having a local, not distant origin. As a consequence, interpretations of the same data may differ in some chapters in the volume. We regard such diversity of geologic opinion as healthy and a challenge to those who use this book as a basis for future geologic studies in Alaska.

Our aim in *The Geology of Alaska* was to bring together in one volume a summary and bibliography of virtually all that now is known about the geology of the state and its offshore margins, and to offer an interpretation of its tectonic evolution. Like Alaska, the book is unique: there has never been one like it, nor is there likely to be another for a long time. We offer it to the earth science community at large and especially to those willing to wrestle with its still-numerous geologic problems, and, we hope, to gather new data and offer their own new interpretations. For you who take up the challenge, it will be an exciting quest. Welcome!

The Geology of Alaska is the result of the dedicated efforts of more than 80 authors and scores of contributors and peer reviewers. We wish to express our deep appreciation to all of these individuals who selflessly diverted time from their personal research to make this volume possible. In our judgment, their efforts were certainly worthwhile.

We gratefully acknowledge David L. Jones for his lead role in conceiving and organizing the volume and in identifying and recruiting authors, and John P. Galloway, John S. Lull, Leslie Gergen, and James W. Laney, of the U.S. Geological Survey, for their outstanding technical support to the editors in drafting illustrations for the volume and in shepherding its myriad components along the long and complex path to publication. We also thank the Geologic Division of the U.S. Geological Survey and the Alaska Division of Geological and Geophysical Surveys, whose research scientists, ably supported by technical and secretarial staffs, authored or co-authored most of the chapters and plates; the Branch of Western Technical Reports of the U.S. Geological Survey, whose staff edited most of the chapters and prepared a majority of the illustrations for publication; illustrators of the Branch of Central Technical Reports who prepared Plate 13; and the geology departments of the University of Arizona, Cornell University, Johns Hopkins University, and the ARCO Alaska Company, whose teaching and research staffs made major contributions to the volume.

We dedicate this volume to the generations of Alaska field geologists whose work, often under trying conditions, provided the solid foundation on which the contributions in the book are built.

George Plafker
Henry C. Berg
February 1994

The Geology of North America
Vol. G-1, The Geology of Alaska
The Geological Society of America, 1994

Chapter 1

Introduction

George Plafker
U.S. Geological Survey, MS-904, 345 Middlefield Road, Menlo Park, California 94025
Henry C. Berg
115 Malvern Avenue, Fullerton, California 92632

PURPOSE AND SCOPE

The Geology of Alaska summarizes the onshore and offshore geology, tectonic evolution, and mineral resources of Alaska and the adjacent continental margin. The volume was prepared at a particularly appropriate time because it follows a period during which there has been an explosive increase in the amount, quality, and regional coverage of earth science data collected in Alaska and because the unifying concepts of plate tectonics and accretionary terranes have become available as a framework for interpreting the data. These new concepts have led to recognition that all of Alaska, except possibly for one area that underlies less than one percent of the state, consists of lithotectonic terranes (also referred to as "suspect" or "tectonostratigraphic" terranes) that have been added to, displaced from, and/or rotated to varying degrees relative to autochthonous parts of the continental margin (Silberling and others, this volume, Plate 3). Thus, a first-order division of the geology of Alaska can be made into (1) the small area of probable autochthonous rocks in east-central Alaska; (2) terranes underlain by known or probable pre-Late Proterozoic continental crust that were part of the North American miogeocline; and (3) terranes, flysch basins, and overlap assemblages that have either been added to, or built along, the south and west margins of the miogeoclinal assemblages in a belt 550 to 700 km wide (Fig. 1). Much of the geologic research in Alaska during the past 15 years has focused on defining these lithotectonic terranes on the basis of their biostratigraphic, magmatic, metamorphic, structural, and paleomagnetic histories. These data have been used to constrain the degree of allochthoneity and the timing of emplacement of individual lithotectonic terranes.

In this introductory chapter we first present brief overviews of the physiographic, tectonic, and geologic setting of Alaska. We then summarize major aspects of the history of geologic and geophysical research in Alaska. Finally, we outline the organization of the volume and the areal or topical coverage of the various contributions. Substantial overlap with the companion volume on the Canadian Cordillera (Gabrielse and Yorath, 1991; Monger, 1989) is inevitable because the political boundary bears no relation to geologic boundaries and because many important geologic relations of the northern Cordillera are within Canada. Excellent recent overviews of the western Cordillera concerning plate and terrane tectonics concepts include those of Coney (1989), Oldow and others (1989), and classic syntheses of the tectonics of North America, including Alaska, by King (1969a, 1969b).

The region covered in this volume includes all of Alaska and its offshore margins between the international boundaries with Russia and Canada. It spans the North American continent between the Arctic Ocean basin on the north and the Pacific Ocean basin on the south. The continent is about 1300 km wide at the border with Canada and it widens to as much as 2000 km in western Alaska near long 170°W. The intraoceanic part of the Aleutian arc ranges from 75 to 180 km wide. The east-west dimension of the region ranges from about 1100 km along the Arctic margin at lat 70°N to 3600 km along the Pacific Ocean margin at lat 58°N, where it includes the entire Aleutian arc. This area is about 3.4×10^6 km^2, of which 1.52×10^6 km^2 is onshore and the remainder is offshore (Fig. 1). For purposes of comparison, this total area is equal to ~40% of the onshore conterminous United States.

PHYSIOGRAPHIC AND GEOLOGIC SETTING

Onshore Alaska is divisible into a number of domains that reflect broadly both the underlying geologic complexity and ongoing neotectonic deformation and volcanism (Fig. 2). We describe briefly the major physiographic subdivisions to acquaint the reader with the most important geographic features in Alaska that are cited throughout this book, as well as their relations to major physiographic provinces elsewhere in the western Cordillera.

Arctic coastal plain

The Arctic coastal plain in northern Alaska is part of the Interior Plains physiographic division of North America that extends through Canada and the conterminous United States along

Plafker, G., and Berg, H. C., 1994, Introduction, *in* Plafker, G., and Berg, H. C., eds., The Geology of Alaska: Boulder, Colorado, Geological Society of America, The Geology of North America, v. G-1.

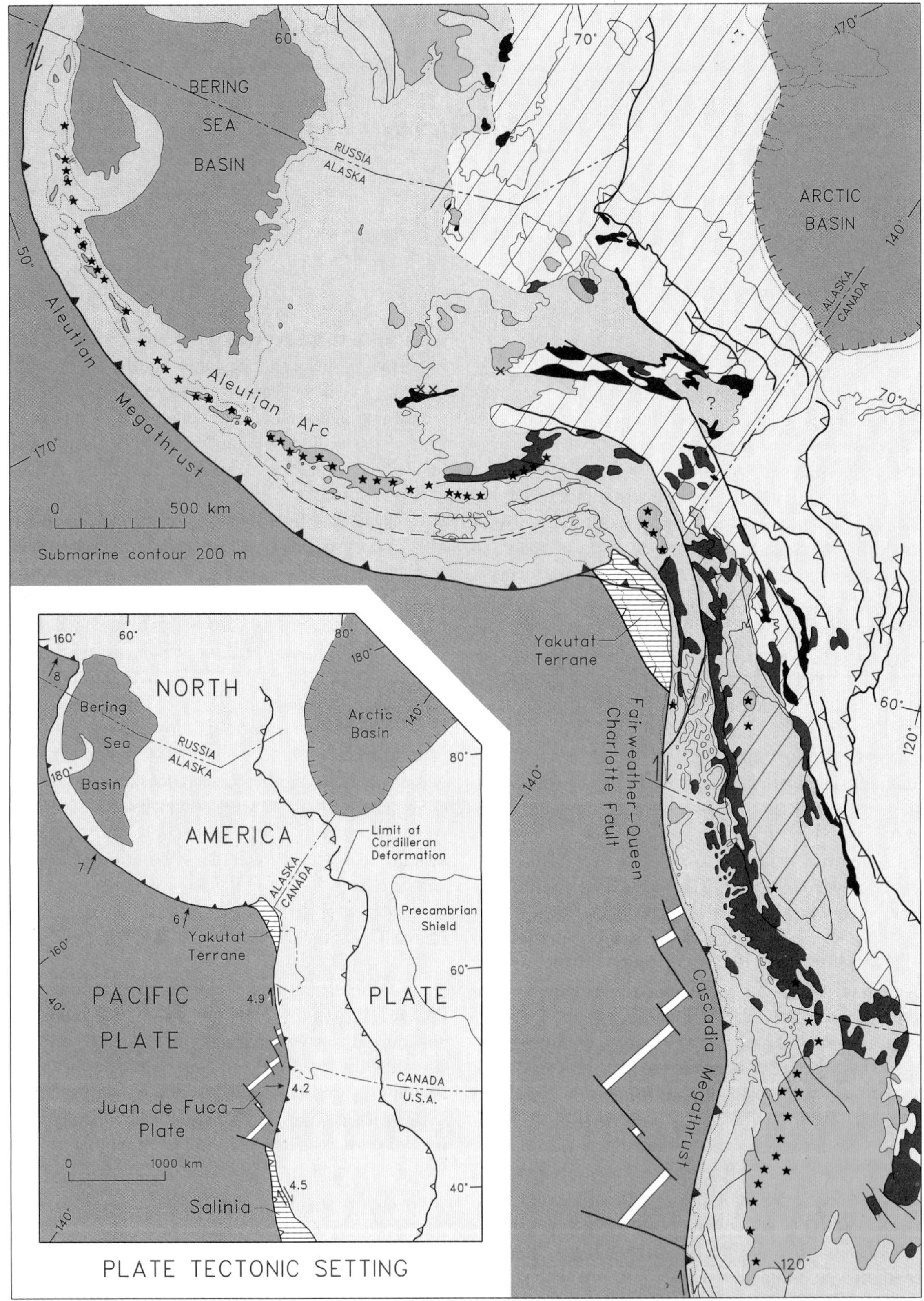

Figure 1. Generalized tectonic map showing the relation of major tectonic elements of Alaska to contiguous regions of Canada, the northern conterminous United States, and eastern Russia. Modified from King and Edmonston (1972), Silberling and others (1993), and unpublished data of Nokleberg and others. Explanation of patterns and symbols on facing page.

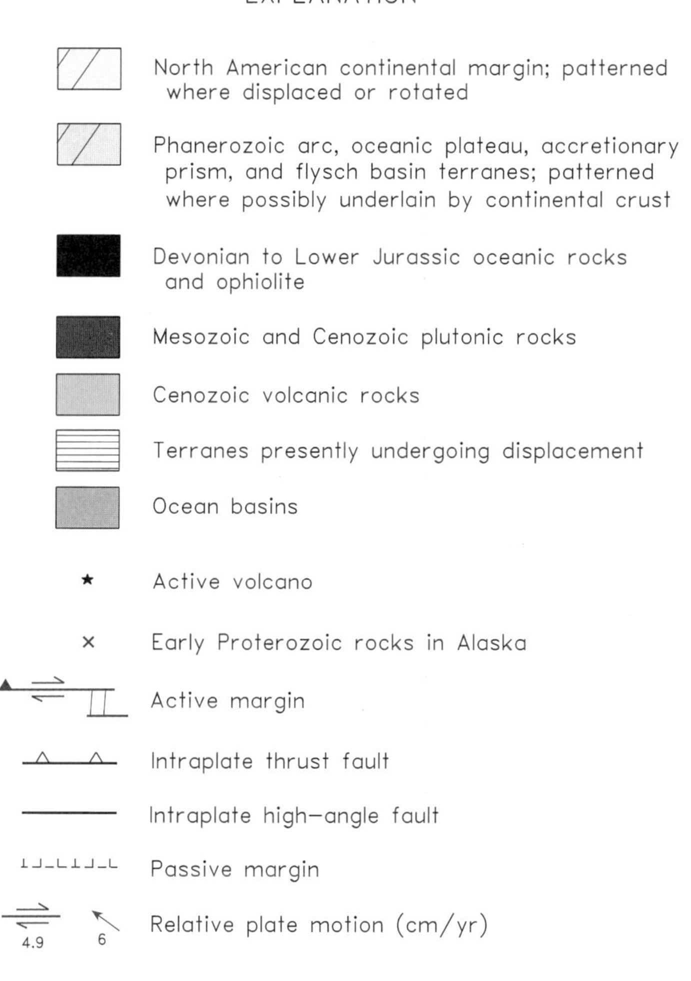

the east side of the Rocky Mountains. In Alaska, the plain is alluvium covered and slopes gently northward from maximum elevations of about 200 m along the southern margin of the province. It is underlain by at least 300 m of permafrost, and has many thaw lakes. A polygonal network of vertical ice wedges close to the surface underlies most of the plain away from water bodies. A dune field covers much of the coastal plain west of the Colville River. Subsurface exploration indicates that it is underlain by a sequence of Paleozoic and early Mesozoic miogeoclinal rocks which appear to have affinities with rocks of the North American miogeocline in the Ellesmere basin of Arctic Canada.

Cordillera of North America

All of Alaska south of the Arctic coastal plain lies within the Cordillera of North America (also variously referred to as the "Cordilleran orogenic belt" and the "Cordilleran orogen"). The Cordillera extends southeastward along the Pacific Ocean margin as far as Guatemala. The Cordillera in Alaska can be broadly subdivided into (1) a northern mountainous belt, (2) a central low-lying intermontane region, and (3) a southern mountainous belt. This part of the Cordillera in North America is unique in that the predominantly northwest trending topographic belts to the south bend abruptly into complex arcuate and linear east-west to southwest trends (Fig. 2B).

Northern Cordillera. The northern Cordillera in Alaska is equivalent to the northern part of the Rocky Mountain system of North America that extends southward through Canada and the northern part of the conterminous United States. In Alaska, these mountains are dominated by the east-west–trending Arctic Foothills, Delong Mountains, Baird Mountains, and Brooks Range, in which average summit elevations are between 1000 to 1500 m in the western part and rise to 2100 to 2400 m in the eastern part. During many of the Pleistocene glaciations, the central and eastern Brooks Range, as well as the Baird and Delong Mountains and Noatak lowland, were covered with a mountain ice cap that imposed a characteristic glacial topography, and nourished ice tongues that extended into lower lying areas to the north and south. Fission-track data and the presence of river terraces suggest that the mountains are young and rising.

The stratigraphy and structure of this region exhibit affinities with the geology of the North American miogeocline in both adjacent parts of Canada and the Ellesmere basin of Arctic Alaska.

Interior Alaska. Between the mountain barriers of northern and southern Alaska is an extensive region that is drained into the Bering Sea mainly by the Yukon and Kuskokwim river systems. This region is part of the intermontane Plateaus physiographic province that extends southward through Canada and into the conterminous United States, where it includes all of the Great Basin and Colorado Plateau. The entire region is commonly referred to simply as "interior" Alaska, even though it extends to the Bering Sea coast. Wide alluvium-covered lowlands border the Bering Sea and broad alluvium-floored interior lowlands and basins are distributed along the major drainages. The anomalous Yukon Flats are located about 800 km from the coast, but they are as flat as a coastal plain and are less than 200 m above sea level. Elsewhere the region consists of plateaus, hills, and rolling uplands in which average summit elevations are mainly between 600 and 1500 m. Numerous domes, ridges, and mountains rise 300 to 600 m above the general upland level. Interior Alaska was free of ice throughout the Pleistocene, except for ice caps over the Ahklun Mountains in southwestern Alaska and the southern mountains of the Seward Peninsula, and small cirque and valley glaciers in some of the highland areas elsewhere. Ice-free areas of interior Alaska served as part of the migration route for humans and other land animals from Eurasia to the Americas.

The geology in much of this region is poorly exposed because of an extensive cover of loess and vegetation, as well as the effects of mass wasting related to permafrost. Pre–mid-Cretaceous basement rocks in the region include displaced and rotated lithotectonic terranes of Late Proterozoic and Paleozoic age with affinity to the North American miogeocline, Devonian to Early Jurassic terranes of oceanic affinity, domi-

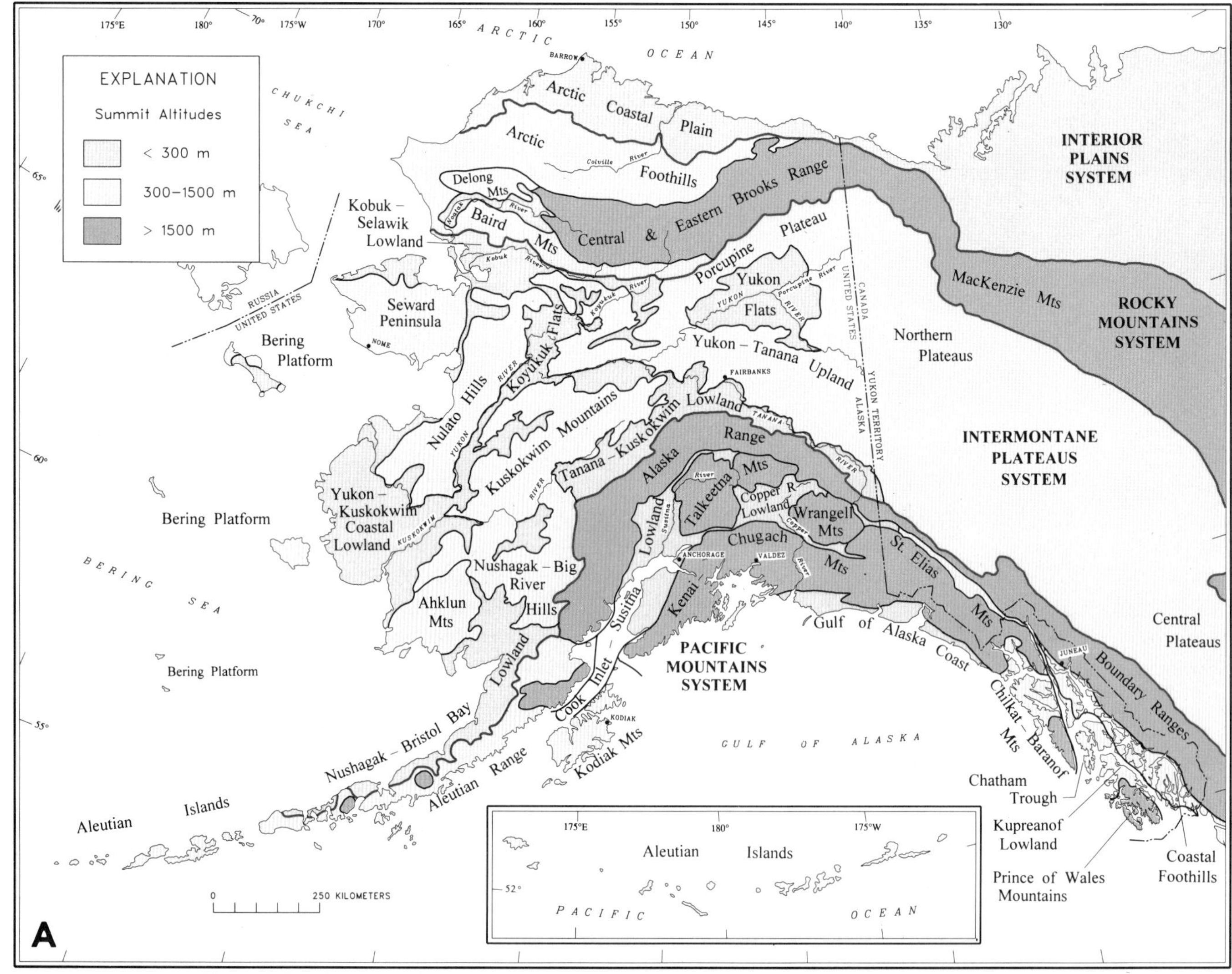

Figure 2. Physiography of Alaska and adjacent parts of Canada. A: Physiographic subdivisions of Alaska from Wahrhaftig (this volume, Plate 2). Heavy black lines bound Alaska physiographic divisions; heavy red lines bound major North American physiographic systems. B (on facing page): Shaded relief map is from Harrison (1969).

nantly Jurassic and Early Cretaceous intraoceanic arc terranes, and small terranes of uncertain affinity. Older rocks in much of this region are concealed by mid-Cretaceous and younger plutonic rocks, arc-related volcanic rocks, flysch basins, and alkalic basalt. Basement beneath a large part of western interior Alaska and the Bering Sea shelf is unknown; limited data suggest that it may be constructed mainly of mid-Cretaceous and younger magmatic and clastic rocks.

Southern Cordillera. The southern Cordillera is part of the Pacific Mountain system of North America that rims the Pacific Ocean margin from Alaska to Central America. Physiographically, the southern Cordillera in Alaska, like the Pacific Mountain system in Canada and the conterminous United States, is made up of two rugged mountain belts with an intervening discontinuous lowland belt (Fig. 2).

The northern mountain belt in mainland Alaska consists of the Alaska Range, which describes a convex-north bend roughly parallel to the Gulf of Alaska margin. To the west, the Alaska Range extends into the Aleutian Range and Aleutian Islands, which form a convex-south arc; the continuation of the Alaska Range into southeastern Alaska and Canada comprises the linear Boundary Ranges and the subparallel Coastal Foothills. The southern mountain belt in south-central Alaska is made up of the Saint Elias, Chugach, Kenai, and Kodiak mountains, and the Chilkat-Baranof and Prince of Wales mountains in southeastern Alaska. The lowland belt between the northern and southern mountains includes, from west to east, the Cook Inlet–Susitna and Copper River lowlands and the island-studded Kupreanof lowlands in southeastern Alaska. These lowlands are connected by narrow and discontinuous valleys along the southern margin

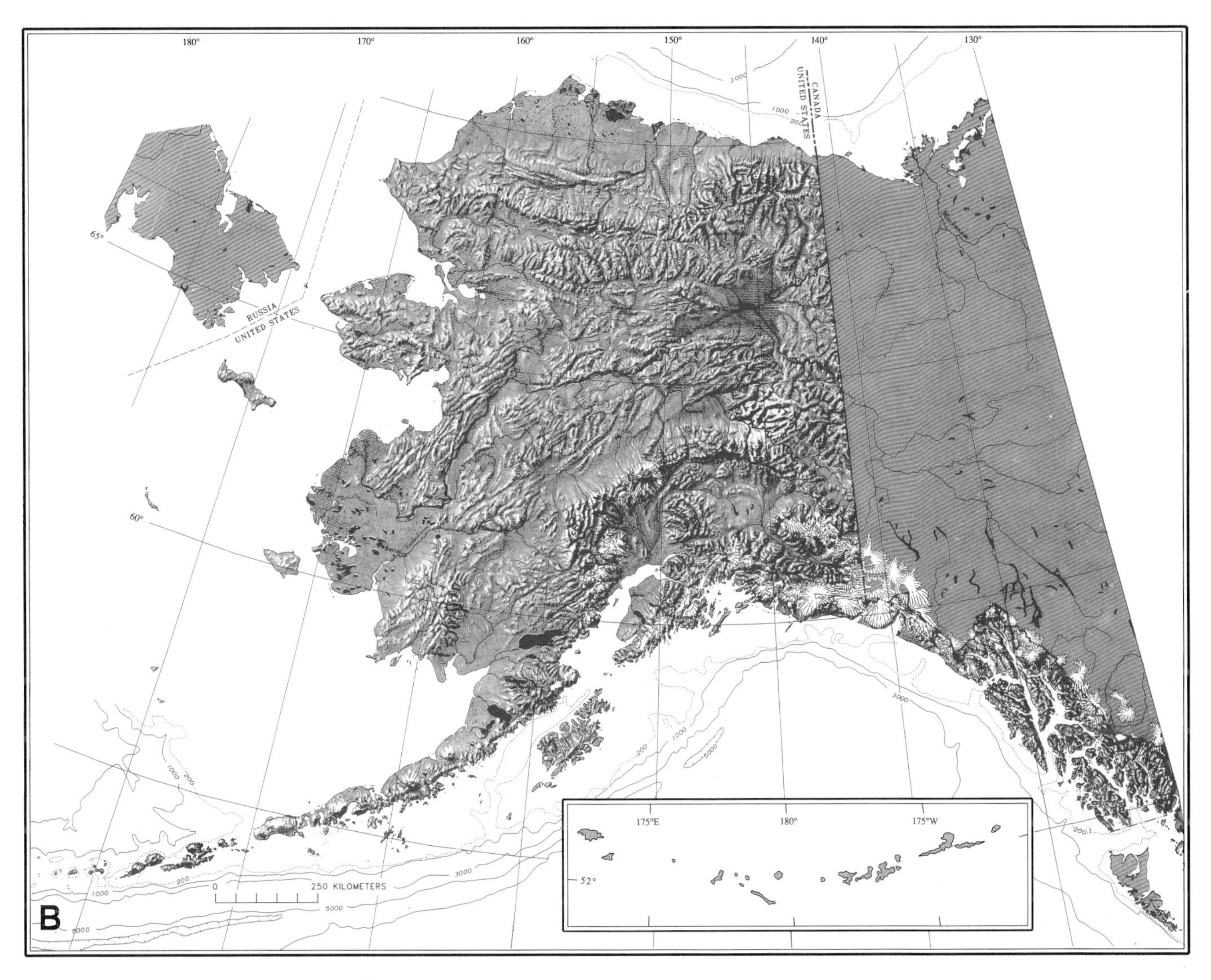

of the roughly circular Talkeetna Mountains and along the north side of the Wrangell and St. Elias mountains.

Except for the Saint Elias Mountains, most of the summit altitudes in the southern Cordillera are between 2000 and 3000 m. The Saint Elias Mountains include several extremely high mountains, including Mount Logan (5745 m) in Canada and Mount Saint Elias (5488 m) in Alaska. A compact group of mountains rising to 4000 m lies in the Chugach Range just north of Prince William Sound. Three isolated groups of mountains over 3000 m surmount the Alaska Range; the highest of these is the group that culminates in Mt. McKinley, at an altitude of 6180 m the highest peak in North America.

The entire southern Cordillera has been glaciated episodically during the Pleistocene, and lowlands were filled with broad ice fields fed from mountain ice caps or networks of giant valley glaciers. Depositional glacial landforms, including arcuate belts of moraines and stagnant-ice topography, abound in the lowlands, along with large lakes in ice-carved basins within and along the mountain fronts, and numerous lakes in ground moraines and stagnant-ice topography. Extensive networks of mountain glaciers and ice fields exist in parts of the Alaska, Aleutian, Kenai-Chugach, Talkeetna, Wrangell, Saint Elias, and Coast mountains. The glaciers invade the margins of the lowlands and, along the Gulf of Alaska, they form the largest piedmont glacier lobes in the world (Malaspina and Bering glaciers). Glaciers currently extend to tidewater at numerous bays and fiords and at one locality along the open Gulf of Alaska coast (La Perouse Glacier).

Many large rivers in the southern Cordillera cross the mountain ranges through deep canyons. Most notable of these are the Copper, Delta, Nenana, Susitna, and Alsek rivers in southern Alaska and the Stikine and Taku rivers in southeastern Alaska. They are probably antecedent streams that predate Neogene uplift of these mountain ranges.

All of the southern Cordillera is underlain by accreted intraoceanic arc and plateau terranes, arc-related accretionary prisms, and flysch basins that range in age from Proterozoic through Cenozoic; the terranes are extensively intruded by post-accretion plutons mainly of mid-Cretaceous and Paleogene age and are overlapped by Late Cretaceous and younger basinal strata and volcanic rocks. Much of the present mountainous topography is the result of Paleogene deformation and uplift related to ongoing accretion of the Yakutat terrane along the northern Gulf of Alaska (Fig. 1). In the Aleutian Islands, Alaska Range, and Wrangell Mountains much of the mountainous topography is constructional, consisting of stratovolcanoes related to the Aleutian magmatic arc. An isolated stratovolcano also occurs near Sitka in southeastern Alaska (Fig. 1).

TECTONIC SETTING

Mainland Alaska and its submerged extensions span the North American plate from the passive trailing margin along the Arctic Ocean basin to the tectonically active southern boundary with the Pacific plate; the intraoceanic part of the Aleutian arc (west of about long 170°W) lies along the relatively passive northern margin between the arc and the Bering Sea basin (Fig. 1). Most of Alaska south of the southern Brooks Range has been a locus of plate-tectonic activity intermittently since Late Proterozoic time and the entire region was dominated by large-scale convergence or oblique convergence since the Late Triassic. We infer from geologic and paleomagnetic data that the continental margin in Alaska had an original northwest to north trend (in present coordinates) and that the present east-west to southwest structural trends in central and western Alaska are the result of large-scale late Early Cretaceous to early Tertiary counterclockwise rotations. Plate reconstructions (Engebretsen and others, 1985) indicate dominantly orthogonal convergence during the subduction of the Farallon plate (100–85 Ma) and dominantly dextral-oblique convergence during the subduction of the Kula plate (85–55? Ma). After about 55 Ma, Pacific–North American plate relative motions were northwesterly with dextral to dextral-oblique convergence on the northwest-trending transform margin and orthogonal convergence on the northeast-trending Aleutian Arc margin. Total subduction beneath the south margin of Alaska is about 7000 km since formation of the Kula plate in mid-Cretaceous time, and the coast-parallel dextral component during this interval is about 2000 km (Engebretsen and others, 1985).

At present, the Pacific plate is moving northwest relative to Alaska at rates that range from 4.9 cm/yr along the Queen Charlotte–Fairweather transform fault system in the east to as much as 7.7 cm/yr at the western end of the Aleutian arc; the small Yakutat terrane is coupled tightly to the northwest-moving Pacific plate and is currently being accreted to the southern margin of Alaska (Fig. 1). Stress trajectories derived from neotectonic data suggest that late Cenozoic mountain-building deformation and seismicity throughout Alaska and adjacent parts of Canada are driven mainly or entirely by interaction between the Pacific plate (and Yakutat terrane) and the southern margin of the continent (Plafker and others, this volume, Plate 12). Manifestations of this plate interaction include the development of the Aleutian magmatic arc and trench, active dextral displacement on the Queen Charlotte–Fairweather transform fault and on northwest- to west-trending intraplate faults, basaltic volcanism throughout much of interior and western Alaska and on the Bering Sea shelf, extreme uplift and topographic relief in the coastal mountains and Alaska Range, intense ocean-verging compressional deformation within and adjacent to the Yakutat terrane, and by some of the most active seismicity in the world, including the largest earthquakes and measured horizontal and vertical coseismic deformation recorded in North America.

REGIONAL GEOLOGIC SETTING

Major crustal types known or inferred to underlie mainland Alaska and adjacent parts of Canada are (1) continental crust of the Cordilleran miogeocline, (2) heterogeneous crust consisting mainly of amalgamated magmatic arcs, oceanic plateaus, melange, and flysch belts in southern and western Alaska, and (3) a narrow intervening discontinuous belt of oceanic crustal rocks (including ophiolite) that in part overrides the adjacent rocks (Fig. 1). These original crustal types have been extensively modified and obscured by magmatism, metamorphism, and overlapping sedimentary deposits that are mainly of Cretaceous and Cenozoic age, and they have been disrupted by Late Cretaceous and Cenozoic faulting and large-scale rotations. Together these tectonic processes have produced the complex structural trends that characterize much of Alaska and that are reflected in the topography (Fig. 2B). The western half of the Cenozoic Aleutian magmatic arc is built on younger oceanic crust and the eastern half of the arc extends across older accreted crust that constitutes the basement along the southern margin of Alaska.

The early continental margin of North America, including what is now northern and eastern Alaska and northern Canada, was shaped by Late Proterozoic rifting beginning at about 850 Ma and by intermittent rifting that continued into early Paleozoic time. Rifting was followed by gradual subsidence of the continental margin and initial deposition of thick bedded sequences of Late Proterozoic and Paleozoic age that constitute the Cordilleran miogeocline, which extends inland to the Precambrian shield (inset map, Fig. 1). Early Proterozoic (2.1 Ga) metamorphic rocks are exposed only in small isolated areas of western Alaska where their structural relations to nearby miogeoclinal rocks are unknown (Fig. 1); their isotopic signatures are compatible with an origin in a continental-margin magmatic arc (Decker and others, this volume).

From the inception of the Cordilleran miogeocline in Late Proterozoic time, the continental margin has been affected repeatedly by plate-tectonic activity to form the present Cordilleran orogenic belt. Manifestations of this tectonism in Alaska and adjacent parts of Canada include early Paleozoic rift-related sedimentation and magmatism as well as Silurian(?) and Devonian to Early Mississippian continental margin magmatism and deformation that may be arc related.

Mainly during Jurassic and Cretaceous time, plate convergence along the continental margin resulted in formation of a collage of diverse intraoceanic arcs, arcs on possible rifted continental crust, arc-related accretionary prisms, flysch basins, oceanic plateaus, and vast slabs of oceanic crustal rocks (Fig. 1). The emplacement of these disparate terranes was accompanied by repeated episodes of deformation, magmatism, and metamorphism within the accreted assemblages and in adjacent parts of the miogeocline and by initial formation of the Cordilleran fold and thrust belt with associated foreland basins in northern Alaska and Canada (Fig. 1). The oceanic crustal rocks are the remnants of vast Devonian to Early Jurassic marginal sea basins that were closed between the continental margin and the intraoceanic terranes. Although they were greatly modified by subsequent tectonism, magmatism, and erosion, these remnants of deep-sea crust are a fundamental feature of the Cordilleran orogenic belt because they delineate the suture zones along which the autochthonous or parautochthonous rocks of the continental margin are juxtaposed against the outboard collage of predominantly noncontinental rocks (Fig. 1).

From about mid-Cretaceous to early Tertiary time an Andean arc system was developed along the continental margin of Alaska and British Columbia in response to rapid convergence between the Kula and North America plates. Voluminous arc-related volcanism, plutonism, and associated metamorphism in a belt as wide as 500 km welded the accreted terranes to one another and to the continental margin. Arc-related magmatic rocks and volcanogenic sediments probably built much of the crust that underlies the Bering Sea shelf and adjacent parts of western Alaska, and an arc-related accretionary prism as wide as 300 km developed along the seaward side of the arc.

The present complex structural configuration of mainland Alaska evolved mainly during this Cretaceous to Early Tertiary interval as a result of large-scale rotations and translations, some of which occurred simultaneously. During late Early Cretaceous time the Arctic Ocean basin formed by counterclockwise rotation of northern Alaska away from the Arctic Canada segment of the continental margin. Probably during this same time interval, continental fragments in central Alaska (Ruby terrane) were initially rotated clockwise away from the southern Brooks Range. During Late Cretaceous to early Tertiary time much of the Cordillera south of the Brooks Range was disrupted by hundreds of kilometers of offset along the Tintina and Denali faults and their major splays. In central and western Alaska, this faulting accompanied, or was immediately followed by, counterclockwise oroclinal rotation and associated displacements on preexisting transcurrent faults.

HISTORY OF GEOLOGIC AND GEOPHYSICAL RESEARCH

Geologic and geophysical investigations in Alaska have taken place in five general phases that overlap to varying degrees. They are (1) the period of Russian ownership from 1741 to 1867; (2) pioneering explorations and investigations of metallic minerals and fuels from 1867 to about 1906; (3) reconnaissance geologic mapping and expanded metallic minerals and fuels investigations from 1907 to about 1939; (4) systematic reconnaissance geologic mapping in the more accessible parts of Alaska and accelerated investigations of metallic minerals and fuels from 1940 to about the time of Alaska statehood in 1959; and (5) detailed geologic mapping, resource investigations, and topical studies throughout Alaska and contiguous offshore areas since 1960. Geologic concepts since 1960 can be subdivided into three subgroups: an early period dominated by geosynclinal theory; a second period beginning in the mid- to late 1960s during which the concepts of plate tectonics emerged; and a period beginning in the late 1970s in which it was recognized that most of Alaska consists of terranes that are displaced and rotated in varying degrees relative to the North America continental margin.

The following summary is based largely on two unpublished manuscripts by R. M. Chapman and D. J. Grybeck of the U.S. Geological Survey that provide detailed historical overviews of geologic and geophysical investigations in Alaska and the evolution of geologic concepts. Additional historical data and data on resource exploration and production are given in appropriate chapters in this volume. A helpful introduction to the geologic literature of Alaska was prepared by Grybeck (1976), and Hopkins (1983a, 1983b) compiled data on the status of geologic mapping in Alaska through 1982. We begin this section with a brief discussion of some of the unique aspects of field work in Alaska.

Note on the logistics of geologic field work in Alaska

The problems of working in Alaska are unique, and to a large extent they have shaped the progress of geologic research. Formidable obstacles to field investigations are its vast size, its remoteness from the conterminous United States, the scarcity of road and rail access, and barriers to ground travel posed by rugged mountain ranges, extensive regions of swampy tundra, swift icy rivers, and extensive glaciers.

Most of the early exploratory work in Alaska was along the coast and major drainages where boats could be used for travel. Elsewhere, traverses were primarily by foot except in some parts of southern Alaska where horses could be used. Geologic investigations during the early years in Alaska often had to be carried out concurrently with topographic surveys because of the absence of suitable base maps. Field seasons routinely were long, arduous, and hazardous with only sporadic communication with the outside world. As eloquently described by Smith (1939, p. 4), "Some of the conditions of field work in the early days would stagger any but those possessed of indomitable courage and perseverance as well as keen technical insight. Many of the best-founded present-day beliefs were first reached by those early geologists from some chance observation made during the intervals snatched from the racking grind of back-packing or other labors entailed in the grim necessity of self-preservation under the exigencies of exploratory work."

Except in the vicinity of the limited road and railroad network, geologic field work requiring overland or boat travel remained basically unchanged in much of Alaska until the late 1950s, when helicopters came into general use. However, access and logistics were significantly improved beginning in the late 1920s, when light fixed-wing aircraft became practical for moving and supplying personnel, for reconnoitering, and for taking aerial photographs. The availability of aerial photographs for mapping beginning in the late 1920s also did away with the need to conduct topographic surveys along with the geologic mapping. In the mid-1940s, tracked vehicles such as the U.S. Army "weasel" began to be used in winter geophysical surveys on the North Slope and in support of geological investigations in areas of low and moderate relief. The use of portable short-wave radios, beginning in the mid-1940s, ended the isolation of field parties in remote parts of Alaska and greatly expedited the logistics and safety of field work.

It was the increasingly routine use of helicopters for logistic support beginning in the mid- to late 1950s that revolutionized field work for geologists and geophysicists in Alaska. Helicopters have made all but the most rugged parts of Alaska accessible for investigation and they significantly improve the efficiency and quality of field work by reducing the time required in getting to and from study areas.

Russian Alaska: 1741 to 1867

The first recorded observations on the geology of Alaska were made in 1741 by the German naturalist Georg Wilhelm Steller at the time of Vitus Bering's brief discovery landing in the New World at Kayak Island in the Gulf of Alaska. During the ensuing period of Russian occupation of Alaska, the primary impetus for exploration was to exploit the fur trade, to consolidate Russian territorial claims, and to further missionary interests. Nevertheless, some prospecting and mining was carried out by the Russians. The first recorded mining venture in Alaska was on the Kenai Peninsula, where a placer gold mine was operated from 1848 to about 1850. The Russians also began producing coal in 1854 from deposits at Port Graham on the southern Kenai Peninsula for use as fuel for steamships, but the mine was abandoned after about 10 years.

Published knowledge of the geography of Alaska was scarce and largely confined to maps of the coast and major drainages. Geologic and geographic information about the interior was derived from a few reconnaissance expeditions by the Russians and by surveys for a telegraph route across Alaska that was to connect by submarine cable with Siberia.

From 1867 to 1906

In 1867 Alaska became a possession of the United States through purchase from Russia and territorial status was conferred in 1912. Following the change of ownership of Alaska, mining and exploration activity continued at a low level almost until the end of the century. In 1877 the first lode-gold mine began production in Alaska near the former Russian capital at Sitka, and by 1882 the Treadwell lode mine near Juneau began operation. The bonanza gold discoveries in the Canadian Klondike on the upper Yukon River in 1896 started a stampede to the north that quickly spread into Alaska. In 1898 the rush to Alaska began in earnest with discovery of rich gold placer deposits near Nome. Within a few years, gold prospecting and important discoveries spread throughout Alaska. Gold mining has made a major contribution to Alaska mineral production every year since 1877, except during World War II when a moratorium was imposed on gold mining.

Deposits of copper and associated precious metals were known in Alaska as early as 1867 and important copper deposits were mined in the Prince William Sound area from 1897 to 1930 and in the Ketchikan area from 1906 to 1941. The bonanza copper deposits near Kennicott were discovered in 1900 and the district produced large amounts of copper and silver from 1911 until the mines closed in 1938. Cinnabar deposits in the Kuskokwim River region have been mined since 1902 and are the source of all the mercury produced in Alaska. Coal was mined on a small scale from several deposits throughout Alaska in the period from 1880 to 1915. Seeps of oil and gas were long known to the native people along the Gulf of Alaska coast, on the Alaska Peninsula, and in northern Alaska. In 1902 the shallow Katalla oil field was developed in a seepage area and it produced small quantities of oil until the refinery burned and the field was abandoned in 1933.

A systematic effort spurred by the influx of gold prospectors was begun in 1898 by the U.S. Geological Survey (USGS) and U.S. Army to investigate Alaskan resources and to produce topographic and geologic maps. The geologic investigations focused mainly on areas having potential for gold, copper, and mineral fuels. Although most of the early publications on Alaskan geology emphasized mineral resources, regional geologic maps were routinely prepared along the routes of travel. In addition to the work of Federal agencies, the USGS and the American Geographical Society of New York jointly sponsored I. C. Russell's two expeditions in 1889 and 1891 to the Mount Saint Elias area and the studies of the effects of the great 1899 Yakutat Bay earthquakes that were carried out in 1906 by R. S. Tarr and G. C. Martin. A major scientific expedition to parts of the south coast of Alaska in 1899 that included geologic observations was privately funded by Averill Harriman. The earliest geophysical studies in Alaska were a few pendulum gravity measurements along the coast made by T. C. Mendenhall of the U.S. Coast and Geodetic Survey.

First synthesis of the geology of Alaska. In 1906 A. H. Brooks, who was then head of the Alaska Division of the USGS, published the first summary of the geology and physiography of Alaska (Brooks, 1906). At that time, less than 25% of Alaska had been visited by geologic parties, and the geology of the state could be depicted on a page-size map with 10 map units. Major geosynclinal and geanticlinal areas were delineated, but faults are absent from the map and structure sections. Nevertheless, the broad outlines of the geography and physiography were delineated and many of the stratigraphic units still in use in Alaska had already been defined. A noteworthy achievement was the recognition of the limits of Pleistocene glaciation and the fact that much of interior Alaska had not been covered by continental glaciers.

From 1907 to 1939

Geologic mapping and metallic mineral resource investigations continued throughout this period and they were expanded to the basinal areas in a search for coal and petroleum. Exploration for coal was carried out in the Controller Bay and other coal fields of southern Alaska by private companies and the USGS. Mining of coal in the Matanuska Valley from 1916 to 1970 and near Healy in the Alaska Range since 1918 was stimulated by completion of the Alaska Railroad in 1914. The Fairbanks Exploration Company experimented with seismic, electrical, and magnetic surveys as aids to delineating placer gold deposits in permafrost. Basinal areas were investigated by company and USGS parties throughout Alaska, and unsuccessful test wells were drilled by oil companies during the 1920s and 1930s along the Gulf of Alaska and on the Alaska Peninsula. Naval Petroleum Reserve No. 4 (Pet-4) was established in northern Alaska by Presidential Proclamation in 1923. Under an agreement with the U.S. Navy, the USGS conducted reconnaissance geologic and geographic surveys to evaluate the petroleum potential of Pet-4 from 1923 to 1926. The National Geographic Society funded glaciological studies in the Yakutat Bay area by R. S. Tarr and G. S. Martin in 1909 and 1910 as well as research on the 1912 Katmai eruption by C. N. Fenner and R. F. Griggs. Private funds also supported the outstanding geologic mapping carried out by E. deK. Leffingwell in the Canning River area of the eastern Brooks Range between 1906 and 1914. Nevertheless, by 1939, less than half of what was then the Territory of Alaska had been surveyed by geologists, even by minimum reconnaissance standards (Smith, 1939).

Second synthesis of the geology of Alaska. P. S. Smith, who became head of the USGS in Alaska in 1924, wrote the second comprehensive publication on the geology of Alaska (Smith, 1939). Smith's summary mainly described the general distribution, lithology, and relative ages of rock units and Quaternary deposits as well as metallic mineral deposits and coal resources. The accompanying map, at 1:2,500,000 scale, depicted the geology of about half of the territory with 16 map units; an accompanying table shows the age, lithology, and correlations of stratigraphic units. Structural data are not shown on the map and there is no section on structure in the text. The limits of Pleistocene glaciation, however, were well defined, the general distribution of permanently frozen ground (permafrost) and its importance in placer-gold mining were clearly recognized, and most of the active volcanoes in Alaska are shown. In 1941 Smith published the first in a series of reports on the petroleum potential of Alaska.

From 1940 to 1959

From about 1940 until 1959, there was a dramatic increase in the amount and diversity of earth science research in Alaska.

World War II and postwar investigations by USGS and other federal agencies. World War II provided the impetus for exploration by the USGS throughout Alaska to locate strategic mineral resources and petroleum. Construction related to military activities provided the incentive for USGS mapping of surficial deposits and permafrost that was to continue as an important program into the late 1950s. In northern Alaska, a modern oil-exploration program, including test drilling, was conducted by the U.S. Navy through Arctic Contractors in Pet-4 from 1944 to 1953, and resulted in the discovery of several noncommercial oil and gas deposits. In conjunction with exploration in Pet-4, the USGS continued geologic mapping in Pet-4 and other basins; this led to a summary publication on the geology and petroleum potential of Alaska (Gryc and others, 1951). Military needs spurred acquisition of aerial photo coverage for all of Alaska and these photographs were extensively utilized to speed up geologic mapping using photogeologic interpretation. In 1944 USGS geologists and geophysicists began a search for radioactive minerals in support of the secret atomic research program that led to development of one uranium mine in southeastern Alaska.

A joint topographic mapping program by the USGS, Army Corps of Engineers, and U.S. Coast and Geodetic Survey begun in 1948 produced reconnaissance 1:250,000 topographic map coverage of Alaska by 1953, and in 1989 modern topographic map coverage was completed at scales of 1:62,500, 1:63,360, and 1:250,000. At the request of the War Department, the USGS conducted a systematic reconnaissance of the geology of the Aleutian Islands from late 1945 through 1954.

The postwar years witnessed a significant expansion of topographic, geologic, and geophysical mapping, drilling, and earth science research. Exploration for minerals by industry remained active, and mines in Alaska produced gold, coal, silver, platinum, and mercury. Much of the mapping by the USGS and other federal agencies was conducted as part of the requirements for land classifications and management related to establishment of the State of Alaska in 1959. An innovation in the early 1950s was the first use by the USGS of geochemical analysis of plant and soil samples as a guide to metallic minerals exploration.

Petroleum exploration and Cook Inlet discoveries. Renewed postwar exploration and drilling activity for petroleum by oil companies was rewarded in 1957 by the discovery of a commercial oil deposit on the Kenai lowland, and in 1963 by discovery of a major oil field offshore in the adjacent Cook Inlet. Since 1957 almost two dozen oil and gas fields have been discovered onshore and offshore throughout the Cook Inlet basin. Oil production peaked in the Cook Inlet basin in 1970 and declining production is projected to the late 1990s. The Cook Inlet discoveries sparked a boom in leasing and exploratory drilling throughout southern Alaska that was accompanied by an emphasis on geologic mapping in basinal areas by the USGS.

Geophysical studies. A variety of geophysical methods began to be applied to investigations in Alaska during this period by the USGS. In the early 1940s magnetic and resistivity methods began to be used in mineral resource studies and in the mid-1950s aeromagnetic and radiation surveys were employed. Exploration for petroleum in Pet-4 and elsewhere in Alaska employed seismic reflection, aeromagnetic, and gravity surveys beginning in the mid-1940s. Offshore geophysical studies included submarine gravity measurements off the continental margin by the Lamont Geological Observatory, and explosion refraction experiments in the Skagway and Prince William Sound areas by the Carnegie Institution in 1956 and by the Scripps Institution of Oceanography in 1956–1957 at Dixon Entrance and at three sites near Kodiak, Unimak, and Adak islands. The University of Wisconsin and Woods Hole Oceanographic Institution released a preliminary list of Alaskan gravity stations in 1959, and in 1958 the USGS began a long-term gravity mapping program that provided reconnaissance coverage of much of Alaska by 1977.

Compilations of geologic mapping and research in Alaska 1957–1959. By 1959, reconnaissance geologic studies had been carried out in about 80% of the territory. However, geologic maps at scale of 1:250,000 or smaller were available for only about one-third of the territory and no more than a few percent was covered by more detailed larger scale maps. Although numerous areas remained essentially unknown, coverage was adequate for compilation of a significantly improved geologic map of Alaska at a scale of 1:2,500,000 that had 55 map units and showed many of the major faults for the first time (Dutro and Payne, 1957). The book *Landscapes of Alaska: Their geologic evolution* (Williams, 1958) summarized in layman's terms the physiography of Alaska and its geologic underpinnings, and a comprehensive treatment of the physiography of Alaska was written by Wahrhaftig (1965). A compilation by Payne (1955) depicted major Mesozoic and Cenozoic tectonic elements of Alaska at a 1:5,000,000 scale and discussed them in terms of discrete orogenies, and subparallel geosynclinal sedimentary basins and geanticlinal source regions. These concepts were to influence thinking on the tectonics of Alaska until the late 1960s, after which they began to be displaced by more mobilistic plate and accretionary tectonics models. The geology and stratigraphy of possible petroleum basins in Alaska was summarized by Miller and others (1959); in this synthesis discussions of tectonic framework of the basins followed Payne's geosynclinal concepts. This was followed by similar summaries that focused on the petroleum resources of Alaska but also provided excellent synopses of the geologic setting (Gates and Gryc, 1963; Gates and others, 1968).

Early concepts of crustal mobility in Alaska. Some of the first studies to suggest large-scale crustal mobility in Alaska include: (1) the recognition that seismic zones associated with volcanic arcs, including the Aleutian arc, occur along major complex reverse faults that dip from the trench beneath the magmatic arc (Benioff, 1954); (2) the interpretation of the Denali fault as a major San Andreas–type dextral strike-slip fault along which hundreds of kilometers of displacement occurred (Saint Amand, 1957); (3) studies of the effects of the 1958 Lituya Bay earthquake (M_S = 7.9) that documented that the Fairweather fault is a major strike-slip structure along which at least 4 m dextral displacement occurred during the earthquake (Tocher, 1960); and (4) interpretations by Carey (1958) that the Arctic- Ocean basin formed as a rotational rift of Alaska away from the Canadian Arctic islands and that western Alaska is a gigantic counterclockwise oroclinal bend about a pivot in Prince William Sound.

Post-1960 investigations

After Alaska became a state, there was a marked expansion in the number of organizations doing geologic work in the state and in the diversity of the studies. Contributions were increasingly made by the newly formed Alaska Geological Survey (now the Alaska Division of Geological and Geophysical Surveys [ADGGS]), by students and faculty members in the Geology Department of the University of Alaska, which began a graduate program in 1959, and by faculty and students from several universities outside the state.

Regional mapping and metallic minerals and land-use investigations. Since 1960, the USGS has continued to carry out statewide regional mapping and minerals evaluations. Geologic mapping and mineral resource investigations were initially undertaken by the USGS to fill important gaps in geologic knowledge of Alaska. Later, much of the work was done as part of national programs, including: (1) efforts to delineate gold, silver, and platinum resources (1965 to 1969); (2) land and resource evaluations required for land selections as part of the Alaska Native Claims Settlement Act of 1971 that still continue; (3) studies of areas designated or proposed for wilderness classification (1969 to the early 1980s); and (4) the Alaska Mineral Resources Assessment Program, which began systematic multidisciplinary geologic, geophysical, and geochemical studies of 1:250,000 scale quadrangles (1974 to present).

Exploration by industry for metallic minerals peaked in the early 1980s and declined thereafter due to unfavorable metals prices, but several deposits containing very large resources of molybdenum, copper, lead, zinc, gold, tin, tungsten, cobalt, fluorite, and beryllium have been delineated. Of these, production began from the Greens Creek zinc-lead-copper-silver-gold deposit in southeastern Alaska in 1989 and from the Red Dog zinc-lead-silver-barite deposit near Kotzebue in 1991, making these the first producing base-metal mines in Alaska since 1941.

1964 earthquake and geologic hazards studies. In 1964, the destruction wrought by the largest earthquake ever recorded in North America (M_W = 9.2) brought a new awareness of geologic hazards that stimulated geotechnical studies by government agencies, private companies, and academia. Comprehensive studies were made of the geologic effects of the earthquake and tsunami and the resulting damage to the works of man and ecosystems by the USGS, other federal and state agencies, and geotechnical companies. Among the notable new results of this research were (1) some of the earliest compelling evidence for crustal plate convergence and large-scale thrust faulting in an arc environment; (2) the first demonstration of a direct relation between coseismic vertical displacement of the sea floor and generation of a destructive tsunami; and (3) the earliest application of geologic techniques (now referred to as "paleoseismology") to determine the recurrence intervals of earthquakes (Plafker, 1965; Plafker and Rubin, 1967).

In 1974, a long-term program of geologic studies was undertaken by the USGS to identify and evaluate potentially active surface faults and other earthquake-related hazards throughout Alaska. Significant contributions to the geologic hazards assessment have also been made by geotechnical contractors to the oil companies, by State of Alaska personnel, and by researchers from universities.

At the time of the earthquake, two seismographs operated in Alaska. As a direct result of the disaster, between 1966 and 1972 the University of Alaska and the U.S. Coast and Geodetic Survey (now the National Oceanographic and Atmospheric Administration) established a regional seismograph network of nearly 40 stations throughout Alaska. Since 1971, the USGS has operated a regional seismic monitoring network of as many as 54 stations in southern Alaska in response to the need to develop seismic design criteria for the Trans-Alaska oil pipeline and for offshore petroleum drilling platforms in the Gulf of Alaska. In addition, the University of Alaska operated local networks of as many as 17 stations to study the details of seismicity in northeastern, northwestern, and central Alaska.

Petroleum exploration and the Prudhoe Bay discovery. Interest in petroleum increased dramatically after Alaska emerged as a potential world-class producer in 1968, when the supergiant Prudhoe Bay field on the North Slope was discovered. The field contains roughly 13 billion barrels of recoverable oil and nearly 30 trillion cubic feet of gas. Since production began in 1977, oil from Prudhoe Bay and other oil fields discovered nearby has far outstripped all other mineral production in Alaska in value, and the income from this production has been a mainstay of the economy of Alaska. After the Prudhoe Bay discovery, intensified geologic and geophysical investigations by industry quickly expanded throughout adjacent areas of the North Slope and most of the onshore and offshore basins, providing a vast amount of new information on the surface and subsurface geology.

In 1974 the U.S. Navy renewed its program to explore Pet-4. In 1976 Congress transferred Pet-4 to the Department of Interior and renamed it the National Petroleum Reserve in Alaska. The continuing exploration responsibility and the contract to Husky Oil NPR Operations, Inc., were assigned to the USGS in 1976. Under this program, 28 exploration holes were drilled, including the two deepest exploratory wells in Alaska, and additional geophysical and geologic studies were completed.

During the 1970s and early 1980s an explosive growth of information by the USGS and oil companies on the continental shelves of southern and western Alaska was stimulated by federal and state petroleum lease sales. Much new geophysical coverage of the continental shelves was provided by aeromagnetic surveys and by surface ship gravity, magnetic, and seismic reflection and refraction surveys by the USGS, U.S. Coast and Geodetic Survey, Oregon State University, Lamont-Doherty Geological Observatory, and private industry. Geologic information offshore was obtained from bottom samples by ships of the USGS and petroleum industry and from drilling of test wells by oil companies, including 6 in the Bering Sea, 19 in the Gulf of Alaska, and 9 in lower Cook Inlet and Shelikof Strait. Since the late 1980s exploration has extended offshore onto the Arctic Ocean continental shelf, where 65 exploratory wells were drilled and three commercial oil fields were discovered as of 1993.

Miscellaneous research programs. From 1968 to 1971 construction of the Trans-Alaska pipeline and other major facilities related to production of Prudhoe oil required intensive geotechnical investigations and surficial geologic mapping by engineering firms and the USGS, with special emphasis on the distribution and foundation properties of permafrost in unconsolidated deposits in and near the pipeline corridor and on heat flow. A landmark analysis by Lachenbruch (1970) of the thermal effects of a heated pipeline in permafrost had a major impact on decisions to elevate the pipeline above ground in most permafrost areas. In 1971 core drilling on the ocean floor and lower continental slope at 5 sites off Kodiak Island and at 10 sites both north and south of the Aleutian Ridge as part of the Deep Sea Drilling Program of the National Science Foundation provided the first deep-sea data on the stratigraphy and structure along the ocean-continent interface. A program of marine geologic and geophysical investigations has been carried out in the Beaufort and Chukchi seas intermittently since 1969 by the USGS from a USGS research vessel and U.S. Coast Guard icebreakers. From 1984 through 1992, the USGS carried out a multidisciplinary study (TACT) to determine the structure, composition, and evolution of the Alaskan crust along a north-south transect across Alaska, generally along the Trans-Alaska oil pipeline corridor, and offshore across the continental margin in the Gulf of Alaska. Among the new insights provided by the uniform and high-quality geophysical and geologic data are multiple underplating of oceanic crust beneath the southern continental margin, possible large-scale underthrusting of flysch beneath parts of interior Alaska, and the complex structural imbrication and detachments in the upper crust beneath parts of the Brooks Range. In 1989 the National Science Foundation sponsored a marine deep seismic reflection transect (EDGE) across the eastern Aleutian arc and trench between Kodiak Island and the mainland that provided new data on the configuration of the subducting oceanic plate and the internal structure of the arc accretionary prism and overlying forearc basin. The 1992 eruption of Mt. Spurr near Anchorage resulted in establishing the Alaska Volcano Observatory by the USGS to carry out volcanological and geophysical studies and to develop eruption prediction capabilities at Mt. Spurr and other active volcanoes in Alaska.

Status of geologic mapping. In 1980, the fourth and most recent geologic map of Alaska was published at 1:2,500,000 scale

(Beikman, 1980, this volume, Plate 1). By that time, an estimated 80% of Alaska was mapped by reconnaissance standards, about 30% was covered by geologic maps to modern standards at scales of 1:250,000 or larger, and about 2.5% was covered by more detailed larger scale maps (Hopkins, 1983a, 1983b). The 1980 geologic map is the first to show coverage for all of Alaska, although the geology in some areas is generalized. The map delineates 62 stratigraphic units, 5 metamorphic rock units, and 114 igneous rock units as well as most of the major faults. Ages of the stratigraphic units are considerably improved over earlier maps thanks to more refined dating by microfossils such as nannoplankton, radiolarians, conodonts, and palynomorphs. Ages and classifications of crystalline rocks were improved by isotopic dating methods and by a variety of high-precision geochemical and isotopic petrologic techniques. Important syntheses that contributed to the structural features shown on Beikman's geologic map include the study of major strike-slip faults in Alaska by Grantz (1966) and delineation of the principal tectonic elements of Alaska in the classic tectonic map and accompanying text by King (1969a, 1969b).

Considerable new geologic mapping has been completed since the publication of the 1980 map. As of 1993, about 50% of the state has been mapped by detailed reconnaissance or better standards at 1:250,000 scale, and almost 10% is covered by larger scale geologic maps. Much of the post-1980 map data have been incorporated in the various contributions in this volume.

Plate tectonic concepts in Alaska. During the mid-1960s the concepts of plate tectonics grew and developed largely outside Alaska, and by the late 1960s the global plate tectonic theory was fully developed (Isaacs and others, 1968). However, Coats (1962) published a remarkably insightful, but rarely cited, synthesis of the tectonics and magmatism of the Aleutian arc that predated by several years the plate tectonic explanation of magmatic arcs at convergent plate margins. Coats correctly interpreted (1) that the origin of the dipping zone of seismicity beneath the Aleutian arc occurs along a megathrust above underthrusting oceanic crust and its sedimentary cover; (2) the relation between the position of the active volcanoes and depth to the underthrust oceanic crust; and (3) the role of fluids derived from the downgoing slab and magmatic differentiation in determining compositions of erupted volcanic rocks. The time-sequential sections drawn by Coats (1962, Fig. 9) differ only in detail from many of the post-plate tectonic models for the Aleutian arc. Coats's interpretation of convergence at the Aleutian arc was dramatically confirmed by studies of the tectonic deformation and seismicity associated with the 1964 Alaska earthquake at the eastern end of the Aleutian arc (Plafker, 1965, 1969).

Most of the success of plate tectonics in interpreting geologic history is based on the concept that the interactions of a small number of rigid plates and the resulting tectonic response provide modern analogs for interpreting past tectonic regimes and orogenesis (Dewey and Bird, 1970). Thus, petrotectonic assemblages in the rock record can be used to deduce plate tectonic settings (Dickinson, 1972) from which paleogeographies and scenarios of plate tectonic evolution can be reconstructed. In Alaska, this approach is applicable to understanding the evolution of the Cretaceous and Tertiary arcs along the southern margin of Alaska. However, it soon became apparent that plate tectonic models had little relevance to understanding much of the geology of the remainder of Alaska or of parts of the Cordillera to the south. For example, stratigraphic and structural studies in northern Alaska provided support for Carey's (1958) hypothesis for rotation of Arctic Alaska away from the margin of Arctic Canada and suggested large-scale north-south shortening by thrusting (>240 km) with resultant juxtaposition of disparate facies, including sheets of oceanic crustal rocks along the southern margin of the Brooks Range (Tailleur and Brosgé, 1970). Similarly, in central Alaska, the rotated and geologically disparate belts of rocks could not be related to plate tectonics models.

Accretionary tectonics in Alaska. The importance of accretionary tectonics began to be recognized in the early 1970s, with the emergence of the general concept of the Cordillera as a collage of displaced terranes, each of which must be analyzed as a complex succession of possibly unique tectonic events, rather than as the result of Cordillera-wide orogenies (Helwig, 1974). Berg and others (1972) and Jones and others (1972) were the first to define and describe terranes in southeastern Alaska, and by 1978 the first terrane map of southeastern Alaska was published (Berg and others, 1978). The terrane concept as a method of regional tectonic analysis gained some credibility with publication of the landmark paper on the allochthonous Wrangellia terrane (Jones and others, 1977), in which paleomagnetic data provided quantitative evidence for post-Triassic northward displacement measured in thousands of kilometers. Subsequently, terrane recognition and analysis (Coney and others, 1980) have been major topics of study throughout the Cordillera and evidence for large-scale telescoping, delamination, and wedging of upper crustal rocks has accumulated from structural and geophysical studies (for example, Coney, 1989; Oldow and others, 1989; Fuis and Plafker, 1991; Grantz and others, 1991). In Alaska, about 50 terranes and subterranes have been defined (Silberling and others, this volume, and several earlier publications) and terrane concepts have played and will continue to play an important role in attempts to reconstruct the tectonic evolution and deep crustal structure of Alaska.

ORGANIZATION OF THIS VOLUME

This volume represents the third attempt to synthesize the geology of Alaska, following those of Smith in 1939 and Brooks in 1906. The volume consists of 33 chapters and 13 plates. Besides this introduction, chapters in this volume are divided into two sections: 12 chapters that summarize and interpret the geology of major regions of Alaska (Fig. 3) and 20 chapters that cover a spectrum of topical subjects. The chapters are supplemented by 13 plates, all of which are at 1:2,500,000 scale, except for the maps showing metallogenic provinces (Plate 11), which are at 1:5,000,000 scale.

The regional and topical chapters contain diverse and occasionally conflicting interpretations of the geology of Alaska. We feel that such diversity is inevitable given the complex geology and our imperfect knowledge of many key problems at the time this volume was written. Important areas for future research are highlighted by these differences in interpretation.

Plates

The first three plates provide coverage of the geology (Plate 1), physiography (Plate 2), and lithotectonic terranes (Plate 3) of Alaska. They are reproductions of previously published maps and depict the geology as known before 1980 (Beikman, 1980), the physiographic provinces as delineated by Wahrhaftig (1965), and the lithotectonic terrane subdivisions as of about 1987 (Silberling and others, 1993). Much of the geologic mapping and data relevant to terrane definitions and interpretations obtained since these maps were compiled have been incorporated in the other plates and chapters of this volume.

Plates 4, 5, and 13 are comprehensive maps of the distribution, lithology, and ages of metamorphic, plutonic, and volcanic rocks in Alaska. Also shown in Plate 13 are tables of geochemical data for many of the pre-Cenozoic volcanic belts. Plate 6 includes a new geologic map and regional structure sections that cover the Brooks Range and North Slope. Plate 7 depicts the geology and structure of sedimentary basins of Alaska and its continental margins, the locations of oil and gas fields, and many of the important wildcat wells drilled for petroleum. Plate 8 shows locations of published, and some unpublished, isotopically dated crystalline rocks in Alaska and it includes an extensive tabulation of data relevant to the dated samples. Plates 9 and 10 are isostatic gravity and aeromagnetic maps for Alaska and parts of the continental margin. Plate 11 shows the locations of major metallic mineral deposits of Alaska and their relations to lithotectonic terranes. Plate 12 comprises a map and tables depicting neotectonic features of Alaska including seismicity, distribution and relative ages of late Cenozoic faults, distribution and composition of young volcanic rocks, thermal springs, Holocene vertical and horizontal displacement rates, and horizontal stress trajectories. An accompanying text describes the tectonic setting and evaluates earthquake hazards.

Regional chapters

The regional chapters (2 to 13) present comprehensive summaries of the bedrock geology of nine major regions of onshore Alaska and three major offshore regions covering the Arctic Ocean margin, the Bering Sea margin, and the Aleutian arc (Fig. 3). Areas not covered by the onshore regional chapters consist primarily of Quaternary unconsolidated deposits or volcanic rocks and basinal sequences that are shown on the map of sedimentary basins of Alaska (Plate 7) or are discussed in topical chapters. The stratigraphy and geologic history in these chapters are discussed in the traditional oldest-first chronological order to facilitate comparisons among the various geologic elements in each region.

Interpretations of the geologic history in the regional chapters are based on summaries of the bedrock stratigraphy, crystalline rocks, and structure, together with relevant geochemical, isotopic, paleomagnetic and other geophysical data. The rocks of the onshore regions are so heterogeneous, varied in age, and discontinuous, that it is helpful to discuss the geology in terms of stratigraphic and structural assemblages, or terranes, rather than individual formations (see discussion in Chapter 33). However, the terrane concept is not accepted by all geologists working in Alaska, and in a few of these regions the authors have chosen to avoid the use of terrane terminology to describe disparate structurally juxtaposed geologic units.

Topical chapters

The topical chapters are summaries of geologic features and phenomena in Alaska that span geographic boundaries of the regional chapters. They include summaries, interpretations, and discussions of time-independent features of Alaska, including magmatic arcs and major belts of crystalline rocks, sedimentary basins, paleomagnetic data, glacial history and permafrost, and mineral and energy resources. Eight plates contain maps and tabular data that complement topical chapters.

Crystalline rocks. Chapters 15 to 25 highlight the distribution, petrology, chemistry, age, and evolution of igneous and metamorphic rocks throughout Alaska or in large regions of the state.

The Aleutian arc is the dominant tectonic feature of southern Alaska and its magmatic rocks and their evolution are the focus of Chapters 22 to 24; various aspects of the arc rocks are also included in regional and topical chapters that include the arc (Chapters 10, 11, and 18) and Plate 5. The petrology, structure, age, and evolution of crystalline rocks in Alaska are summarized in chapters describing magmatic belts (Chapters 16 to 20 and Plates 5 and 13), metamorphic assemblages (Chapter 15 and Plate 4), and ophiolitic rocks (Chapter 21). Isotopic compositions of intrusive rocks were used to interpret the general composition of the crust in much of Alaska in Chapter 25.

Sedimentary basins. The stratigraphy, structure, resource potential, and evolution of the basins of interior Alaska (exclusive of the Cook Inlet basin) are the subjects of Chapter 14. The petroliferous Cook Inlet basin and basins of the North Slope are discussed in Chapter 30, and coal-bearing areas are described in Chapter 31. Coastal and offshore basins along the Bering Sea and Gulf of Alaska are covered in regional chapters (Chapters 2, 8, 11, and 12). The sedimentary basins of Alaska and its shelves are shown in Plate 7.

Paleomagnetic data. Paleomagnetic data in Alaska have provided a major impetus for inferring that large regions of the state are allochthonous with respect to their surroundings and in formulating the concept of tectonostratigraphic terranes. The paleomagnetic data used in evaluating latitudinal displacements and rotations in Alaska are reviewed and summarized in Chapter 26.

Figure 3. Areas described in regional chapters in this volume. Large numbers indicate chapter numbers; intersecting hachure patterns indicate areas of overlap between chapters. Unpatterned areas are characterized by unconsolidated deposits, minor young volcanic rocks, and basinal strata that are described in topical chapters or are shown on the geologic map of Alaska (this volume, Plate 1), the map showing sedimentary basins of Alaska (this volume, Plate 7), and maps showing distribution of Late Cretaceous and Cenozoic volcanic rocks (this volume, Plates 7 and 12). Base map shows standard 2° and 3° quadrangles in Alaska.

Quaternary geology. Special aspects of Alaskan geology concern its Quaternary glacial history, periglacial phenomena, and neotectonic activity. Chapter 27 summarizes the glacial history with its implications for migrations of early humans across the Bering Strait. The distribution and nature of permafrost as well as its importance to engineering works in the Arctic are discussed in Chapter 28. Neotectonic data are compiled in Plate 12.

Mineral and energy resources. The mineral and energy resources of Alaska are of major economic importance to the state. The distribution, significance, and genesis of metallic minerals, petroleum, coal, and geothermal resources are the focus of Chapters 29 to 32.

Chapter 29 and Plate 11 summarize the distribution and the geologic and tectonic setting for major mineral deposits in Alaska and interpret their origin according to mineral-occurrence models. Chapter 30 and Plate 7 present data on the geology and models for petroleum generation and accumulation in the productive North Slope and Cook Inlet basins. Chapter 31 is a synthesis of the distribution, geologic occurrence, grade, and reserves of the enormous coal resources throughout Alaska and considers models for their origin. Chapter 32 evaluates Alaska's untapped geothermal resource potential in the context of distribution of thermal springs, heat flow, and young volcanic belts.

Tectonic overview and synthesis. The last chapter of the volume is a brief overview and synthesis of the tectonic evolution of Alaska. Chapter 33 is illustrated by figures showing lithotectonic terranes, composite terranes, magmatic belts, and data on rotations, translations, and plate motions. A set of eight figures shows an interpretation of the Phanerozoic evolution.

For the reader unfamiliar with the geology of Alaska, we suggest that one way to gain an appreciation of the location, physiography, geology, and tectonic evolution of the major elements that make up Alaska would be to begin with the first and last chapters of this volume.

REFERENCES CITED

Beikman, H. M., 1980, Geologic map of Alaska: U.S. Geological Survey, scale 1:2,500,000.

Benioff, H., 1954, Orogenesis and deep crustal structure—Additional evidence from seismology: Geological Society of America Bulletin, v. 65, p. 385–400.

Berg, H. C., Jones, D. L., and Richter, D. H., 1972, Gravina-Nutzotin belt—Tectonic significance of an upper Mesozoic sedimentary and volcanic sequence in southern and southeastern Alaska, *in* Geological Survey research 1972: U.S. Geological Survey Professional Paper 800-D, p. D1–D24.

Berg, H. C., Jones, D. L., and Coney, P. J., 1978, Pre-Cenozoic tectonostratigraphic terranes of southeastern Alaska and adjacent areas: U.S. Geological Survey Open-File Report 78-1085, 2 sheets, scale 1:1,000,000.

Brooks, A. H., 1906, The geography and geology of Alaska with a section on climate by Cleveland Abbe, Jr.: U.S. Geological Survey Professional Paper 45, 327 p.

Carey, S. W., 1958, A tectonic approach to continental drift, *in* Carey, S. W., ed., Continental drift: A symposium: Hobart, Tasmania University, p. 177–355.

Coats, R. R., 1962, Magma type and crustal structure in the Aleutian arc, *in* The crust of the Pacific basin: American Geophysical Union Geophysical Monograph 6, p. 92–109.

Coney, P. J., 1989, Structural aspects of suspect terranes and accretionary tectonics in western North America: Journal of Structural Geology, v. 11, p. 107–125.

Coney, P. J., Jones, D. L., and Monger, J.W.H., 1980, Cordilleran suspect terranes: Nature, v. 288, p. 329–333.

Dewey, J. F., and Bird, J. M., 1970, Mountain belts and the new global tectonics: Journal of Geophysical Research, v. 75, p. 2625–2647.

Dickinson, W. R., 1972, Evidence for plate-tectonic regimes in the rock record: American Journal of Science, v. 272, p. 551–576.

Dutro, J. T., Jr., and Payne, T. G., 1957, Geologic map of Alaska: U.S. Geological Survey, scale 1:2,500,000.

Engebretsen, D. C., Cox, A., and Gordon, R. G., 1985, Relative motions between oceanic and continental plates in the Pacific basin: Geological Society of America Special Paper 206, 59 p.

Fuis, G. S., and Plafker, G., 1991, Evolution of deep structure along the Trans-Alaska Crustal Transect, Chugach Mountains and Copper River Basin, southern Alaska: Journal of Geophysical Research, v. 96, p. 4229–4253,

Gabrielse, H., and Yorath, C. J., eds., 1991, Geology of the Cordilleran orogen in Canada: Geological Survey of Canada, Geology of Canada, no. 5 (also Geological Society of America, The Geology of North America, v. G-1), 844 p.

Gates, G. O., and Gryc, G., 1963, Structure and tectonic history of Alaska, *in* Childs, O. E., and Beebe, B. W., eds., The backbone of the Americas, a symposium: American Association of Petroleum Geologists Memoir 2, p. 264–277.

Gates, G. O., Grantz, A., and Patton, W. W., Jr., 1968, Geology and natural gas and oil resources of Alaska, *in* Natural gases of North America—Pt. 1, Natural gases in rocks of Cenozoic age: American Association of Petroleum Geologists Memoir 9, v. 1, p. 3–48.

Grantz, A., 1966, Strike-slip faults in Alaska: U.S. Geological Survey Open-File Report 267, 82 p.

Grantz, A., Moore, T. E., and Roeske, S. M., 1991, A-3 Gulf of Alaska to Arctic Ocean: Boulder, Colorado, Geological Society of America, Centennial Continent/Ocean Transect no. 15, scale 1:500,000, 3 sheets.

Grybeck, D. J., 1976, An introduction to the geologic literature of Alaska: U.S. Geological Survey Open-File Report 76-235, 23 p.

Gryc, G., Miller, D. J., and Payne, T. G., 1951, Alaska, *in* Ball, M. W., ed., Possible future petroleum provinces of North America: American Association of Petroleum Geologists Bulletin, v. 35, p. 151–168.

Harrison, R. E., 1969, Shaded relief [Alaska]: U.S. Geological Survey, National Atlas of the United States of America Sheet 58, scale 1:7,500,000.

Helwig, J., 1974, Eugeosynclinal basement and a collage concept of orogenic belts, *in* Dott, H., Jr., and Shaver, R. H., eds., Modern and ancient geosynclinal sedimentation; Problems of palinspastic restoration: Tulsa, Oklahoma, Society of Economic Paleontologists and Mineralogists Special Publication 19, p. 359–376.

Hopkins, B. A., 1983a, Geologic maps in Alaska published by the U.S. Geological Survey, post-1930: Scales 1:96,000 to 1:250,000: U.S. Geological Survey Open-File Report 83-577, 1 plate, scale 1:2,500,000, 19 p.

—— , 1983b, Geologic maps in Alaska published by the U.S. Geological Survey, post-1930: scales 1:20,000 to 1:63,360: U.S. Geological Survey Open-File Report 83-578, 1 plate, scale 1:2,500,000, 17 p.

Isaacs, B. L., Oliver, J. E., and Sykes, L. R., 1968, Seismology and the new global tectonics: Journal of Geophysical Research, v. 73, p. 5855–5899.

Jones, D. L., Irwin, W. P., and Ovenshine, A. T., 1972, Southeastern Alaska; a displaced continental fragment?: U.S. Geological Survey Professional Paper 800-B, p. B211–B217.

Jones, D. L., Silberling, N. J., and Hillhouse, J. W., 1977, Wrangellia—A displaced terrane in northwestern North America: Canadian Journal of Earth Sciences, v. 14, p. 2565–2577.

King, P. B., 1969a, The tectonics of North America—A discussion to accompany the tectonic map of North America, scale 1:5,000,000: U.S. Geological Survey Professional Paper 628, 94 p.

—— , compiler, 1969b, Tectonic map of North America: U.S. Geological Survey, scale 1:5,000,000.

King, P. B., and Edmonston, G. J., 1972, Generalized tectonic map of North America: U.S. Geological Survey Miscellaneous Geological Investigations Map I-688, scale 1:5,000,000.

Lachenbruch, A. H., 1970, Some estimates of the thermal effects of a heated pipeline in permafrost: U.S. Geological Survey Circular 632, 23 p.

Miller, D. J., Payne, T. G., and Gryc, G., 1959, Geology of possible petroleum provinces in Alaska, with annotated bibliography by E. H. Cobb: U.S. Geological Survey Bulletin 1094, 131 p.

Monger, J.W.H., 1989, Overview of Cordilleran geology, *in* Ricketts, B. D., ed., Western Canada sedimentary basin: A case history: Calgary, Canadian Society of Petroleum Geologists, p. 9–32.

Oldow, J. S., Bally, A. W., Avé Lallemant, H. G., and Leeman, W. P., 1989, Phanerozoic evolution of the North American Cordillera; United States and Canada, *in* Bally, A. W., and Palmer, A. R., eds., The Geology of North America—An overview: Boulder, Colorado, Geological Society of America, The Geology of North America, v. A, p. 139–232.

Payne, T. G., 1955, Mesozoic and Cenozoic tectonic elements of Alaska: U.S. Geological Survey Miscellaneous Geological Investigations Map OM-126, scale 1:1,000,000.

Plafker, G., 1965, Tectonic deformation associated with the 1964 Alaska earthquake: Science, v. 148, p. 1675–1687.

—— , 1969, Tectonics of the March 27, 1964, Alaska earthquake: U.S. Geological Survey Professional Paper 543-I, 74 p.

Plafker, G., and Rubin, M., 1967, Vertical tectonic displacements in south-central Alaska during and prior to the great 1964 earthquake: Osaka City University, Journal of Geosciences, v. 10, art. 1-7, p. 1–14.

Saint Amand, P., 1957, Geological and geophysical synthesis of the tectonics of portions of British Columbia, the Yukon Territory, and Alaska: Geological Society of America Bulletin, v. 68, p. 1343–1370.

Silberling, N. J., Jones, D. L., Monger, J.W.H., and Coney, P. J., 1993, Lithotectonic terrane map of the North American Cordillera: U.S. Geological Survey Miscellaneous Investigations Map I-2176, scale 1:5,000,000.

Smith, P. S., 1939, Areal geology of Alaska: U.S. Geological Survey Professional Paper 192, 100 p.

—— , 1941, Possible future oil provinces in Alaska *in* Levorsen, A. I., ed., Possible future oil provinces of the United States and Canada: American Association of Petroleum Geologists Bulletin, v. 25, p. 1440–1446.

Tailleur, I. L., and Brosgé, W. P., 1970, Tectonic history of northern Alaska, *in* Adkison, W. L., and Brosgé, M. M., eds., Proceedings, Geological seminar on the North Slope of Alaska: Los Angeles, American Association of Petroleum Geologists Pacific Section Meeting, p. E1–E20.

Tocher, D., 1960, The Alaska earthquake of July 10, 1958—Movement on the Fairweather fault and field investigation of southern epicentral region: Seismological Society of America Bulletin, v. 50, p. 267–292.

Wahrhaftig, C., 1965, The physiographic provinces of Alaska: U.S. Geological Survey Professional Paper 482, 52 p.

Williams, H., ed., 1958, Landscapes of Alaska: Their geologic evolution: Berkeley and Los Angeles, University of California Press, 148 p.

Manuscript Accepted by the Society March 16, 1994

ACKNOWLEDGMENTS

Our colleagues in the Branch of Alaskan Geology of the USGS provided much of the geologic and historical information contained in this Introduction. We thank D. F. Barnes, R. M. Chapman, Clyde Wahrhaftig, and Warren Yeend for informal reviews of all of parts of this chapter. We are grateful to George Gryc, D. G. Howell, and W. W. Patton for their constructive technical reviews of the entire chapter and to James Laney, who prepared the figures.

Printed in U.S.A.

The Geology of North America
Vol. G-1, The Geology of Alaska
The Geological Society of America, 1994

Chapter 2

Geology of the Arctic continental margin of Alaska[1]

Arthur Grantz, Steve D. May, and Patrick E. Hart
U.S. Geological Survey, 345 Middlefield Road, Menlo Park, California 94025

INTRODUCTION

Location and physiography

Alaska faces the Canada Basin of the Arctic Ocean along an arcuate continental margin, gently concave to the north, that stretches unbroken from the Mackenzie Delta, near 137°W to Northwind Ridge of the Chukchi Borderland near 162°W. (Marine geographic features mentioned below can be found on Plates 1 and 11 of Grantz and others, 1990a.) This margin, with an arc-length of about 1,050 km, marks one side of a continental rift along which the Canada Basin opened by rotation about a pole in the Mackenzie Delta region during middle Cretaceous time. The rift-margin structures, which lie beneath the inner shelf and coastal plain in the eastern Alaskan Beaufort Shelf and beneath the outer shelf in the western Beaufort and Chukchi Shelf, are now buried by a thick middle Lower Cretaceous to Holocene progradational continental terrace sedimentary prism.

We divide the Arctic continental margin of Alaska into three sectors of strongly contrasting geologic structure and physiographic expression. In the Barter Island sector (see Figs. 3 and 4) the structure is dominated by the effects of Eocene to Holocene convergence and uplift, and the continental slope is upwardly convex; in the Barrow sector the structure is dominated by the effects of middle Early Cretaceous rifting and continental breakup, and the continental slope is upwardly concave; and in the Chukchi sector the structure is controlled by an easterly trending middle Early Cretaceous rift, and the continental slope abuts the Chukchi Borderland.

Physiographically, the Alaska continental margin is expressed by the Alaska continental rise and slope (Fig. 1), which lies between the oceanic Canada Basin to the north and the flat and shallow continental shelves of the Beaufort and Chukchi Seas to the south. Outer shelf and slope morphology along the margin is dominated by gravity-driven slope failures. These include both surficial slumps and deeply penetrating slope failures related to listric normal faults that dip toward the modern and ancient free face of the continental slope. Some of the listric faults offset Quaternary deposits or the seabed (Grantz and others, 1983a).

A major influence on the physiography of the region is the relative size and age of the sediment sources that built the progradational sedimentary prisms of the narrow Alaska slope and rise and the much more extensive continental rise and abyssal plain of the Canada Basin. The great volume of clastic sediment that poured into the Canada Basin from the Mackenzie River drainage system and from the Canadian Shield via the Amundsen Gulf, in Cretaceous and especially Cenozoic time, overwhelmed that which built the Alaska slope and rise. The extensive Canada Basin fill is banked against the Alaska slope and rise at depths of a little more than 1,000 m at the Mackenzie Delta—the major sediment source for the basin—to almost 4,000 m at the abyssal plain at the foot of the Northwind Escarpment. The disparity in sediment supply is illustrated by the fact that the deepest part of the Canada abyssal plain lies near the foot of the narrow Alaska slope and rise (Grantz and others, 1990a, Plate 1).

Previous studies

Early ideas on the geology of the northern Alaska continental margin were extrapolated from studies of the surrounding landmasses. Carey (1958) originally suggested that the Beaufort Sea margin of Alaska was created by a rift in which northern Alaska was rotated away from the Canadian Arctic Islands by oroclinal bending about a pivot in the Gulf of Alaska. Regional geologic considerations led Tailleur (1969a, b, and 1973) also to postulate that rotational rifting played a crucial role in the tectonic evolution of the region. Based on data acquired during oil exploration, Rickwood (1970) concurred with the rotational rift hypothesis and proposed that rifting was crucial to creation of the trap that holds the supergiant hydrocarbon deposit at Prudhoe Bay.

In the past decade, interpretation of the geology of the Beaufort Shelf has relied heavily on multichannel seismic reflection data, but the first geological cross sections of the region were drawn from other types of geophysical data. Wold and others

[1]In order to provide coverge of the entire state of Alaska in one volume, the editors have reprinted this chapter from Grantz and others, 1990a where it appears as Chapter 16, pp. 257–288. The only differences between this reprinted text and the original are in the references to plates and chapters in Grantz and others (1990a).

Grantz, A., May, S. D., and Hart, P. E., 1994, Geology of the Arctic continental margin of Alaska, *in* Plafker, G., and Berg, H. C., eds., The Geology of Alaska: Boulder, Colorado, Geological Society of America, The Geology of North America, v. G-1.

(1970) published an interpretation of the Beaufort Sea margin based on gravity data collected from light airplane landings on sea ice. A more detailed survey, based on shipborne gravity meter readings, was presented and interpreted by Dehlinger (1980). To date, the most detailed published gravity map of the region, compiled from shipborne data gathered by the U.S. Geological Survey between 1972 and 1982, was published by May (1985). Extensive proprietary surveys also exist.

Two early seismic refraction studies of the Alaskan continental margin produced reconnaissance profiles. One, across the lower continental slope and upper rise north of Barrow by Milne (1966), is unreversed and of limited usefulness. The other, a long reversed profile from Barrow to the northern Chukchi Shelf (Hunkins, 1966), suggests that the shelf there is underlain by continental crust and that the top of basement deepens from about 1 km near Barrow to about 6 km near the shelf edge. Two studies of the seismic velocity structure of the Beaufort Sea and Chukchi Sea margins were based on sonobuoys. Houtz and oth-

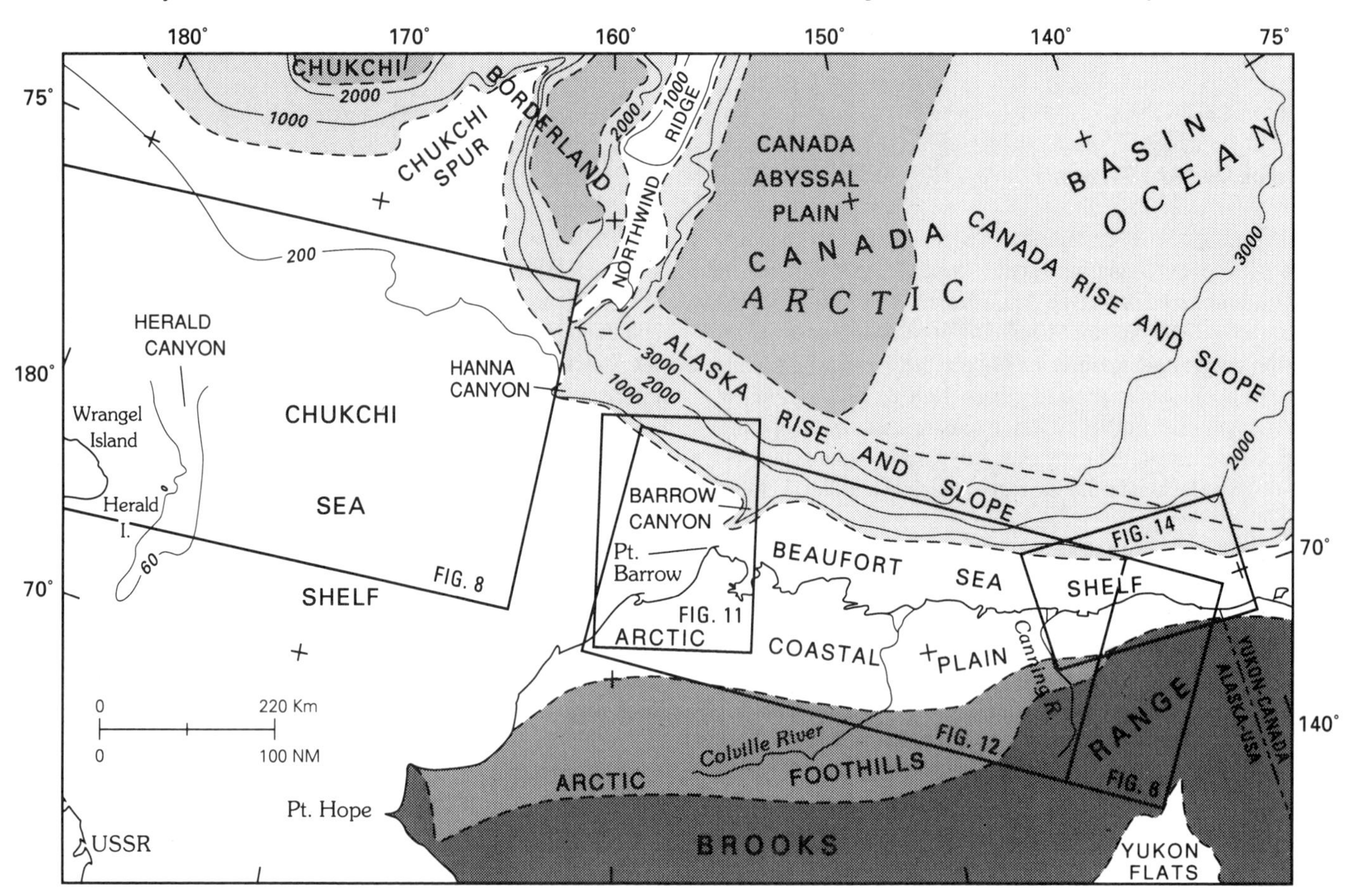

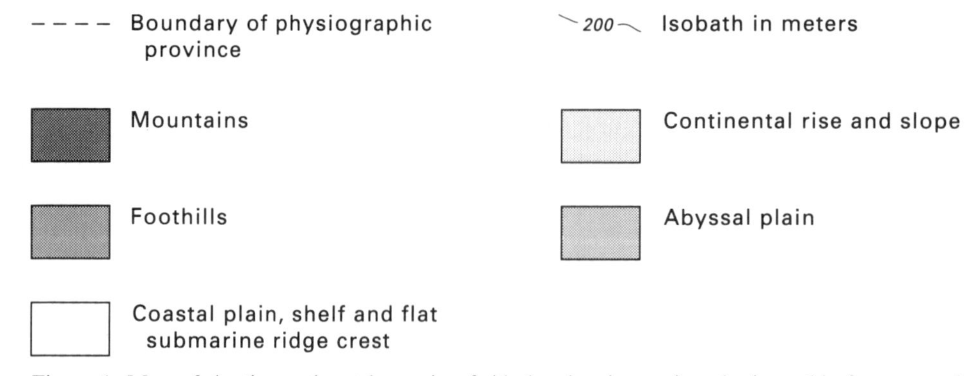

Figure 1. Map of Arctic continental margin of Alaska showing major physiographic features and location of Figures 6, 8, 11, 12, and 14.

ers (1981) used more than 100 sonobuoy records from the northern Alaskan shelves to interpret the general velocity structure of the region. A more detailed study of the velocity structure of the Beaufort Shelf between Prudhoe Bay and Cape Simpson was published by Bee and others (1984).

Published data on both present magnetic field anomalies and paleomagnetic field directions in the region are limited. Extensive proprietary aeromagnetic surveys have been conducted over the Beaufort Shelf, but published data consist primarily of a reconnaissance magnetic anomaly map (Cramer and others, 1986) of the area west of 155°W. Pioneering paleomagnetic studies of upper Paleozoic strata in the Brooks Range were interpreted to agree with the proposed counterclockwise rotation of Arctic Alaska away from the Canadian Arctic Islands (Newman and others, 1979), but Hillhouse and Grommé (1983) showed that the magnetic field direction was overprinted during the Cretaceous. More recently, Halgedahl and Jarrard (1987) have reported that oriented drill cores from the Kuparuk oil field, near the Beaufort Sea coast, apparently escaped overprinting. These cores indicate that Arctic Alaska was rotated counterclockwise with respect to North America since deposition of the cored rocks in middle Early Cretaceous time.

An early discussion of the regional geologic framework of the Beaufort and Chukchi Shelves, based primarily on analog single channel seismic reflection data, was published by Grantz and others (1975). In 1977, 1978, and 1980 the U.S. Geological Survey conducted multichannel seismic reflection surveys of the entire Alaska Beaufort Shelf and the central and eastern parts of the Chukchi Shelf. Regional geological interpretations of these data have been published in several reports (Grantz and May, 1983, 1987; Grantz and others, 1979, 1981, 1987a, b). Craig and others (1985) used proprietary subsurface and seismic reflection data, in conjunction with previously published reports, to summarize the regional geology, petroleum resources, and environmental geology of the Alaskan Beaufort Shelf. An interpretation of the stratigraphy and geologic history of the Beaufort Shelf, also based on proprietary data, was recently presented by Hubbard and others (1987). An important recent contribution is a discussion of the regional geology of the central and northern Chukchi Shelf by Thurston and Theiss (1987).

Time-to-depth conversion functions

The seismic reflection time-to-depth conversion functions used in this study were calculated from regionally averaged multichannel seismic reflection stacking velocities (Fig. 2). These functions enable the reader to determine the generalized depth of seismic horizons in Figures 7A, 7B, 8, 9, 11, 13, and 14. The curves for the Beaufort Shelf and North Chukchi Basin exhibit velocities which are common for Mesozoic and Cenozoic sedimentary strata, whereas the curve for the Chukchi Shelf shows a considerably higher velocity structure. The Chukchi Shelf velocities are higher because lower Mesozoic and Paleozoic strata lie at shallower depths beneath large areas of this shelf than they do beneath the Beaufort Shelf and North Chukchi Basin. Users of the

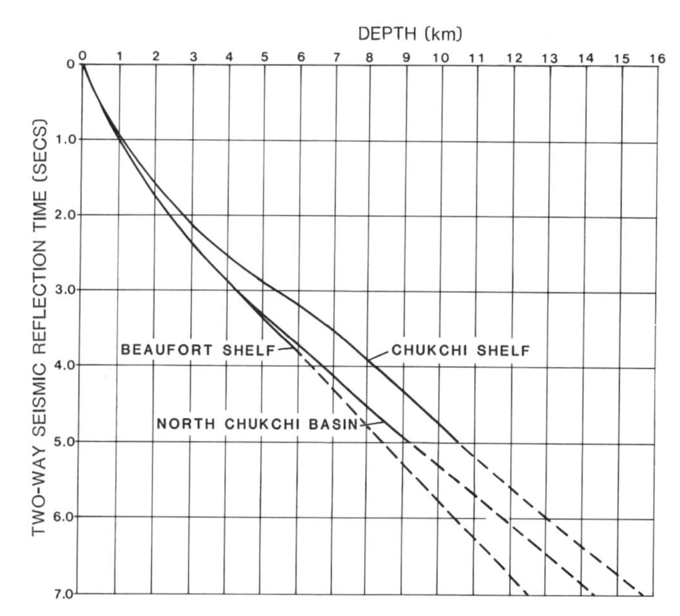

Figure 2. Depth as a function of regionally averaged seismic-reflection time for the Beaufort and Chukchi Shelves and the North Chukchi Basin, derived from seismic-stacking-velocity measurements. Dashed where inferred.

time-to-depth functions of Figure 2 should keep in mind that these are regional averages, and that they become increasingly uncertain with depth as a result of the increasing uncertainty of stacking velocities with depth.

GEOLOGIC FRAMEWORK AND REGIONAL STRATIGRAPHY

Major provinces

Three regional tectonic provinces meet at the continental margin north of Alaska: the Canada Basin of the Arctic Ocean, the Arctic Platform of the Arctic Alaska plate of northern Alaska and adjacent shelves, and the Chukchi Borderland of the Arctic Basin north of the Chukchi Shelf (Figs. 1, 3, and 4). The geologic character of the margin and the juxtaposition of these first-order features are the result of five distinct, but in part related, tectonic events. (1) Rifting beginning in Early Jurassic time separated the Arctic Platform from the North American craton and produced a series of rifts now located beneath the Alaska Beaufort Shelf and slope and Banks Island of the Canadian Beaufort margin (Fig. 12 in Grantz and others, 1990b). This extensional event was almost synchronous with the initiation of major convergence in the Brooks Range orogen in Middle Jurassic time. (2) Rotational rifting beginning in Hauterivian (middle Early Cretaceous) time led to continental breakup and drift of the Arctic Alaska plate away from the Canadian Arctic Islands about a pole of rotation near 68.5°N, 136°W (Fig. 3) and formed the Canada Basin in mid-Cretaceous time. The initiation of this event, in turn, was almost synchronous with the end of major convergence in the

Brooks Range orogen in early Albian time. (3 and 4) Poorly understood, but more localized rifting in the western Chukchi Shelf and Chukchi Borderland created the North Chukchi Basin during two events—Jurassic to Neocomian and late Early Cretaceous. An oversimplified model for these events shown in Figure 3 is speculative, but it illustrates one possible geometry for the proposed rifting after the added complication of segmentation of the borderland into north-trending ridges and basins by late Late Cretaceous and early Paleogene rifting. (For convenience, events of this general age will hereafter be called "Laramide" in this chapter.) (5) Eocene to Quaternary thrusting in the Brooks Range orogen in northeastern Alaska and the adjacent continental margin.

Arctic Platform

Regional structure. The Arctic platform slopes gently southward from the broad crest of the Barrow arch, near the sea coast, to beneath the northward-thrust nappes of the Brooks

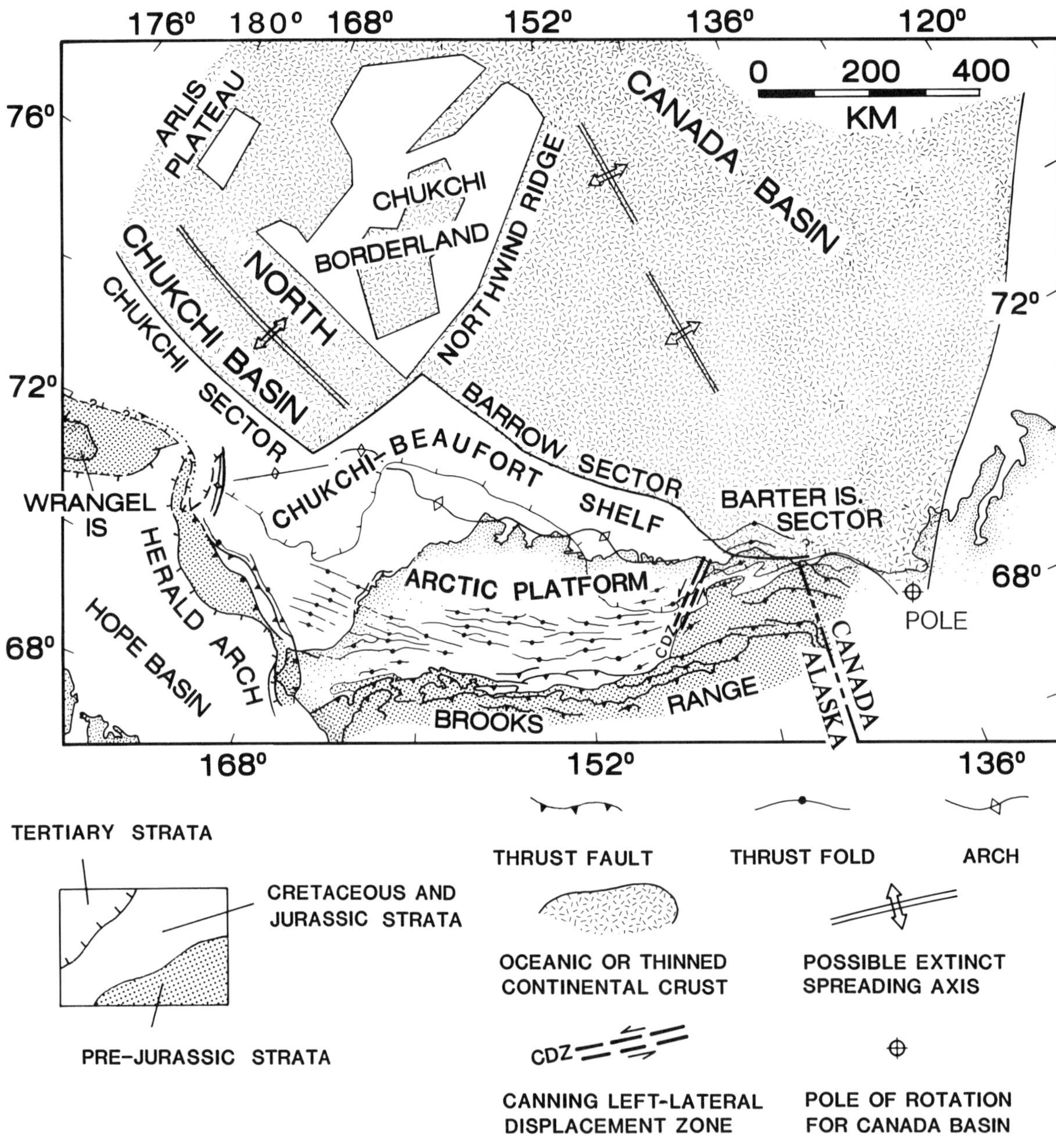

Figure 3. Tectonic model for Alaska segment of Arctic Ocean Basin showing major structural features, inferred distribution of continental versus oceanic or thinned continental crust, and the relation of the Chukchi, Barrow, and Barter Island sectors to the structural features that define them. Possible spreading axis in Canada Basin after Taylor and others (1981), in North Chukchi Basin after Grantz and others (1979).

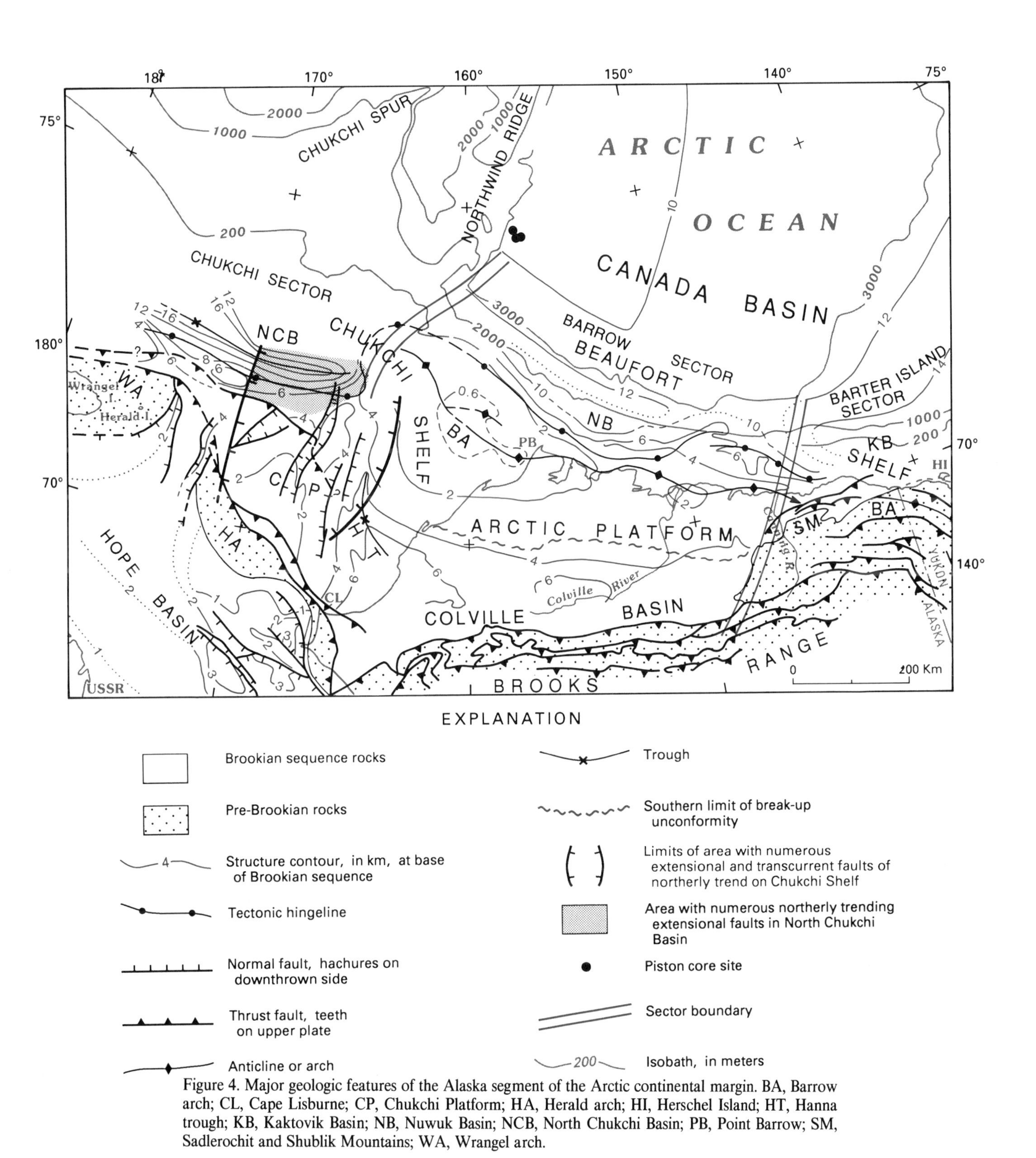

Figure 4. Major geologic features of the Alaska segment of the Arctic continental margin. BA, Barrow arch; CL, Cape Lisburne; CP, Chukchi Platform; HA, Herald arch; HI, Herschel Island; HT, Hanna trough; KB, Kaktovik Basin; NB, Nuwuk Basin; NCB, North Chukchi Basin; PB, Point Barrow; SM, Sadlerochit and Shublik Mountains; WA, Wrangel arch.

Range orogenic belt (Fig. 3). The platform gradient is about 1° beneath the Arctic Coastal Plain and about 2° to 4° beneath the Arctic (Northern) Foothills of the Brooks Range. This gradient developed in two stages. The initial gradient developed in Mississippian to Neocomian time, when the Arctic Platform was the site of clastic and carbonate deposition in a stable shelf environment. The sourceland for the clastic sediments (Barrovia of Tailleur, 1973) lay north of the present continental shelf, and the platform sloped generally southward to a rifted margin of Devonian age south of the present Brooks Range. The initial gradient was augmented in Middle or Late Jurassic and Early Cretaceous time by loading of the southern part of the Arctic Platform by multiple nappes from the Brooks Range orogen to the south and by the weight of sediment deposited in the Colville Basin, a foreland basin on the north side of the orogen.

Brooks Range orogen. Multiple, far-travelled nappes of the Brooks Range orogen override the Arctic Platform beneath the present Brooks Range across the width of Alaska, and related thrust faults and folds extend into the Cretaceous and Tertiary strata of the Colville foreland basin beneath the Arctic Foothills of the Brooks Range. (See Moore and others, this volume, Chapter 3), for a recent summary of Brooks Range tectonics.) Nappe emplacement was a response to convergence between the Paleo-Pacific Basin and the southern margin of Arctic Alaska. Convergence began in Middle Jurassic time, and the major displacements, which were northerly directed, were largely completed during Albian time. Mayfield and others (1988) estimate that crustal shortening across the western Brooks Range exceeds 700 to 800 km.

In the eastern Brooks Range, renewed thrusting in Eocene or earlier time was superimposed upon the Jurassic and Cretaceous compressional events. The Cenozoic thrusting is strongest east of a narrow, northeasterly striking zone of earthquake epicenters designated the Canning displacement (fault) zone (Fig. 4) by Grantz and others (1983a). This zone is a northern extension of Aleutian arc Benioff zone earthquakes from central Alaska to the Beaufort Sea. Some Tertiary thrust displacement also occurred in the foreland west of this zone. Counterclockwise rotation of fold axes, faults, and geologic contacts indicate that the Canning displacement zone was, and continues to be, the site of significant left-lateral deformation. Earthquake epicenters suggest that the zone has a counterpart to the east, along the north-trending Cordilleran front at the east face of the Richardson Mountains, which lie west of the lower Mackenzie River valley north of 66°N. The intervening region, which corresponds to the Barter Island sector of the continental margin, has been uplifted 300 to 500 m above the surrounding plains and plateaus to form the northeastern Brooks Range.

Elevation of the northeastern Brooks Range is related to thickening of the crust by thrusting. Depression of the crust to form the deep Kaktovik Basin (Fig. 4), which lies north of the northeast Brooks Range, resulted from loading by the northward-thrust nappes and by thick accumulations of Upper Cretaceous and Cenozoic detritus derived from accelerated erosion of the new topography generated by nappe emplacement. The amount of Cenozoic displacement of the nappes is uncertain, but total tectonic transport, including that which occurred in late Mesozoic time, is estimated by Rattey (1985) to exceed 400 km. On the basis of rotated fold axes and faults in Upper Cretaceous and lower(?) Tertiary rocks in and adjacent to the Canning displacement zone, Grantz and others (1983a) estimated that Cenozoic tectonic transport in the northeast Brooks Range was at least 25 to 50 km northward with respect to the Arctic Coastal Plain and Foothills to the west.

Barrow arch. A broad structural high known as the Barrow arch (Fig. 4) trends along the Arctic coast of Alaska from the northeastern Chukchi Shelf to Yukon Territory. This feature is the product of multiple Jurassic and Cretaceous events of regional influence rather than the product of a single episode of folding or upwarping. The south flank of the arch is the south-sloping Arctic Platform. The north flank is a collapsed continental margin formed by thermal subsidence and subsequent sedimentary loading adjacent to the rifts that separated northern Alaska from the Canadian Arctic Islands in Jurassic and Early Cretaceous time. The overall slope of the top of Paleozoic basement toward the Canada Basin, including the effects of faulting, is as steep as 12° and locally may exceed 16°.

The western part of the Barrow arch appears to have been uplifted in Tertiary time. Shale compaction studies by Ervin (1981) suggest that west of the Colville River the crest of the arch has been differentially uplifted with respect to areas south and east. The uplift increases westward from about 100 to 200 m near the Colville River to a maximum of 600 to 900 m near Barrow. The uplifted area presumably extends westward beneath the northeastern Chukchi Shelf, where the western limit of uplift is not defined by compaction data. Regional stratigraphy suggests that the uplift was post-Cretaceous in age. Structural relief on the Barrow arch at the meridian of Prudhoe Bay, east of the area of presumed Tertiary uplift, is about 5 km on the south flank and about 10 km on the north flank.

Characteristic stratigraphy permits the Barrow arch to be recognized even where the position of its crestline is obscured by structural complexity. The most distinctive stratigraphic feature is a Hauterivian breakup unconformity that overlies the crest and upper flanks of the arch. A second characteristic is the absence or patchy occurrence of the otherwise widespread stable shelf clastic and carbonate strata of the Mississippian to Neocomian Ellesmerian sequence from the crestal region and north flank of the arch. A third is the presence of failed rift deposits of probable Jurassic to Neocomian age seaward of the Barrow arch crestline. By these criteria, the Barrow arch can be traced from the northeast Chukchi Sea to Flaxman Island and projected southeast from Flaxman Island beneath the Arctic Coastal Plain to the front of the Brooks Range some 40 or 50 km west southwest of Demarcation Bay. An analogous feature can be traced across the north slope of the British Mountains in northern Yukon Territory. If the position of the arch is correctly projected beneath the Arctic Coastal Plain in northeastern Alaska, its counterpart in coastal Yukon has been offset some 20 or 30 km northward with respect to its position beneath the coastal plain on the frontal thrust faults of the Brooks Range (Fig. 4).

Stratigraphy. Stable shelf deposits of northern provenance and Early Mississippian to Neocomian age that lie beneath and north of the far-travelled nappes of the Brooks Range orogen define the Arctic Platform of northern Alaska. The rocks upon which the platform was developed, and those that were deposited upon it, have been divided into four distinct stratigraphic sequences whose lithologic character reflects their tectonic environment (Fig. 5). The pre-platform rocks, of Ordovician or older through Devonian age, are correlated with the lithologically similar main part of the Franklinian sequence of Lerand (1973). This sequence was named for the Upper Cambrian through Devonian eugeoclinal and miogeoclinal strata of the areally extensive Franklinian geosyncline and adjacent cratonic shelf of the Canadian Arctic Islands.

In Late Silurian to Early Devonian time the Franklinian rocks of northern Alaska were strongly deformed, mildly metamorphosed, and structurally consolidated (cratonized) by an early stage of the Ellesmerian orogeny. Large grabens and half grabens developed in the tectonized Franklinian rocks during probably two cycles of late- and early post–Ellesmerian orogeny extension. Basins of the older cycle were filled by Lower(?) or Middle Devonian nonmarine sediments, now strongly folded but unmetamorphosed, that are assigned to the upper part of the Franklinian sequence. Basins of the younger cycle, well observed on seismic reflection profiles (Plate 9.2 in Kirschner and Rycerski, 1988), were filled by Upper Devonian(?) and Lower Mississippian nonmarine clastic rocks of local derivation, which are here designated the Eo-Ellesmerian sequence. Deep erosion and extensional faulting associated with the formation of these basins constituted the final stage in the development of the Arctic Platform. A measure of the tectonic stability achieved by these Devonian events is the virtual lack of deformation of the 340-Ma Arctic Platform beneath the North Slope in spite of middle Early Cretaceous rifting, which severed the platform from the Canadian Arctic Islands, and Middle Jurassic to Quaternary convergence, which buried its southern part beneath the large nappes of the Brooks Range orogen. Gravity modelling along the Trans-Alaska Pipeline route south of Prudhoe Bay tied to refraction measurements in the Canada Basin suggests (Grantz and others, 1990) that the crust in the area of the Arctic platform is about 34 km thick, and thus of normal continental thickness.

On the Arctic Platform the Eo-Ellesmerian sequence is succeeded by the gently dipping stable-shelf clastic and platform carbonate rocks of the Early Mississippian to Neocomian Ellesmerian sequence, which is of northern provenance (Fig. 5). The term Ellesmerian sequence was proposed by Lerand (1973) for the carbonate and overlying clastic rocks of Mississippian through Jurassic age in the Sverdrup Basin of the Canadian Arctic Islands. In northern Alaska, platformal Neocomian beds of northern provenance have been added to the Ellesmerian sequence as defined by Lerand (1973). Failed rift deposits of inferred Jurassic and Neocomian age beneath the central and eastern Alaskan Beaufort Shelf, the Dinkum succession, are correlative with the upper part of the Ellesmerian sequence of the North Slope. The Ellesmerian and Dinkum beds are overlain by the Brookian sequence. This sequence consists of thick continental-deltaic and shelf clastic deposits of southern (Brooks Range) provenance that prograded across the Arctic Platform and the newly established continental margin in Aptian or Albian to Quaternary time. The Ellesmerian rocks wedge out toward an ancient sourceland (Barrovia) near the present Beaufort Sea coast, but the underlying Franklinian and overlying Brookian sequences extend to the outer shelf and slope. Principal sources for the discussion of the stratigraphy of the Arctic Platform that follows include Moore and Mull (1989), Moore and others (this volume, Chapter 3), Kirschner and Rycerski (1988), and Molenaar and others (1986).

Franklinian sequence. Outcrops in the northeast Brooks Range and test wells on the North Slope show that the Arctic Platform, and probably the adjacent continental shelf, is underlain by lithologically varied sedimentary and volcanic rocks of Proterozoic and early Paleozoic age. West of the Canning River, deep test wells encounter strongly deformed, mildly metamorphosed argillite, arenite, carbonate rocks, and chert containing graptolites and chitinozoans of Ordovician and Silurian age (Carter and Laufeld, 1975). Lithologically and faunally these beds resemble the Ordovician and Silurian flysch of the Franklinian geosyncline in the Canadian Arctic Islands (Trettin, 1972; Trettin and Balkwill, 1979), Ordovician and Silurian graptolite-bearing argillite in northern Yukon Territory about 50 km south of Herschel Island (Lane and Cecile, 1989), and graptolitic flysch and shale of Ordovician and Silurian age on the Lisburne Peninsula of the westernmost Brooks Range (Grantz and others, 1983b).

In test wells on Flaxman Island, near the mouth of the Canning River, and in outcrops in the Sadlerochit and Shublik Mountains east of the river (Fig. 4) are found Proterozoic to Devonian carbonate rocks with minor amounts of quartzite, argillite, and mafic volcanic rocks (Blodgett and others, 1986; Clough and others, 1988). Only the Cambrian to Devonian beds of this basin plain to carbonate platform assemblage correlate with the Franklinian sequence. Farther east, beneath the narrow Arctic Coastal Plain of northeastern Alaska, seismic reflection data suggest to Fisher and Bruns (1987) the presence of pre-Mississippian rocks there are more than 5 to 7 km thick and possess a simple structure.

Upper part of Franklinian sequence: More than 100 m of strongly folded, but unmetamorphosed nonmarine beds at the bottom of the Topagoruk test well, in the western part of the North Slope, are here placed in the upper part of the Franklinian sequence. These beds are chert-pebble conglomerate and dark gray shale with carbonaceous partings and plant fragments of Middle (possibly Early) Devonian age. As described by Collins (1958), these beds dip 35° to 60° and are overlain unconformably by gently inclined beds that we correlate with the Ellesmerian sequence. Lack of metamorphism in the Middle Devonian rocks indicates that they postdate the main orogeny, which regionally deformed and mildly metamorphosed the un-

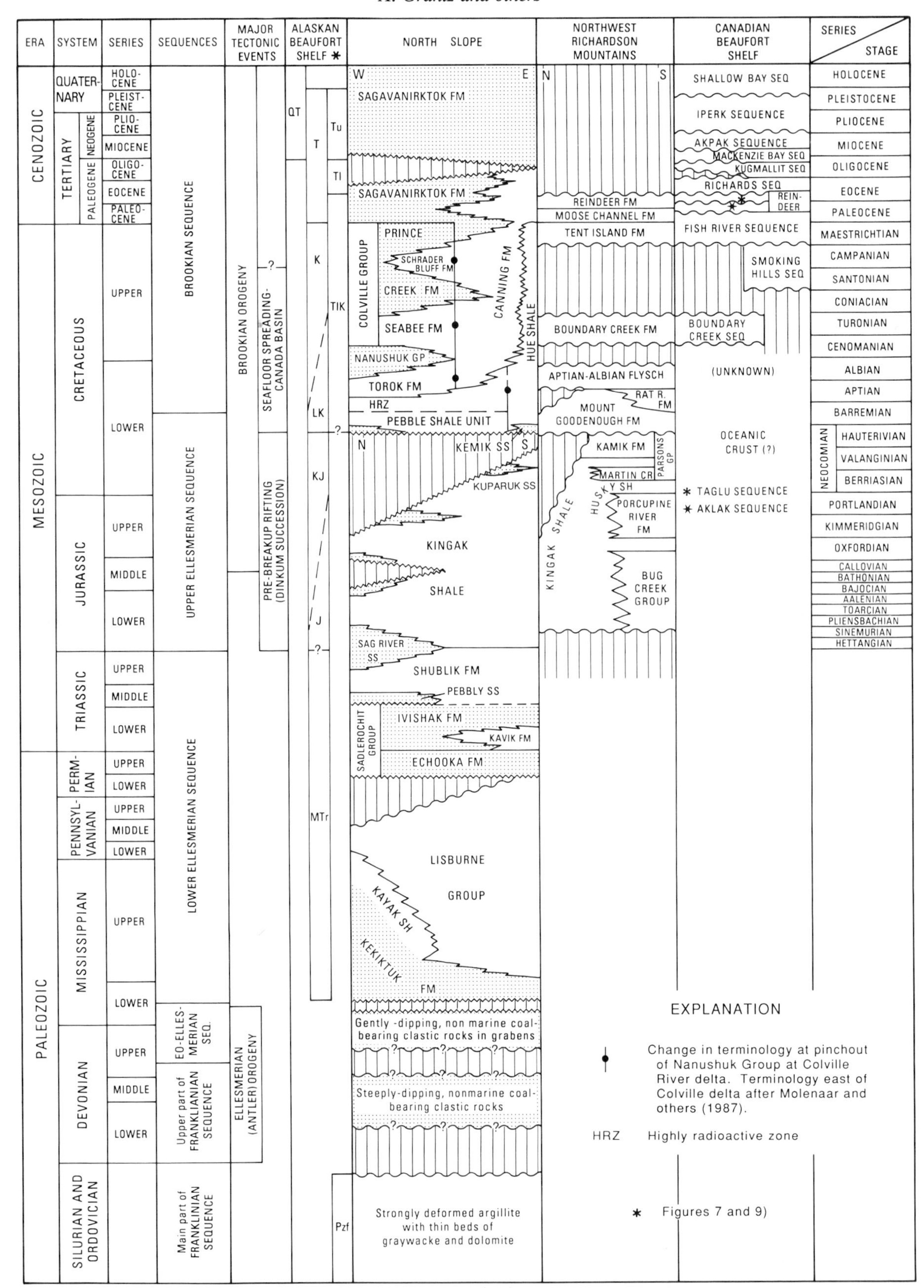

Figure 5. Stratigraphy of Arctic Alaska and adjacent continental shelves after Mickey and Haga (1987), Molenaar and others (1987), Dietrich and others (1989), J. Dixon (personal communication, 1989), and this chapter.

derlying Ordovician and Silurian argillite and graywacke. The moderately steep dip of the Lower(?) or Middle Devonian beds indicates that they were, in turn, folded and truncated by erosion before the low-dipping beds of the overlying Ellesmerian sequence were deposited. The Late Devonian structural event that tilted the Lower(?) or Middle Devonian beds correlates with the Late Devonian and Early Mississippian Ellesmerian orogeny of Lerand (1973). Together the stronger Early Devonian event and the weaker Late Devonian or Early Mississippian event constitute the main phases of the Ellesmerian orogeny on the North Slope.

Ellesmerian sequence. Eo-Ellesmerian sequence: Unmetamorphosed Upper Devonian(?) and Lower Mississippian coal-bearing nonmarine strata, the oldest beds of the Ellesmerian sequence on the Arctic Platform, rest nonconformably on Franklinian beds. Seismic reflection records (Kirschner and Rycerski, 1988, Plate 9.2) show that these strata occupy fault-bounded basins (grabens and half grabens) and range in thickness from a knife edge to more than 3,000 m. Kirschner and Rycerski (1988) place these strata in the Endicott Group, but their confinement to grabens isolates them from the Upper Devonian and Lower Mississippian type-Endicott strata of the Brooks Range allochthons, from which the coastal plain rocks also differ greatly in character of substrate, lithology, and depositional environment. In this chapter we refer to these beds as Eo-Ellesmerian in the sense that they are atypical beds in the earliest part of the Ellesmerian sequence that are transitional between the nonmarine beds of the upper Franklinian sequence below and the typical Ellesmerian marine shelfal strata above. The seismic reflection records of Kirschner and Rycerski (1988) show that an erosional unconformity with mild angular discordance overlies the Eo-Ellesmerian strata and that the main body of Ellesmerian rocks oversteps the areally more restricted graben deposits. This Lower Mississippian unconformity represents the waning, final phase of the Ellesmerian orogeny on the North Slope.

Lower part of the Ellesmerian sequence: Ellesmerian strata above the Eo-Ellesmerian graben deposits consist of four transgressive-regressive cycles whose clastic components were derived from the northern sourceland of Barrovia. The sequence as a whole, and many of its constituent units, thins northward to pinch-outs, mainly erosional truncations in paralic facies, near the Beaufort coast and beneath the northern Chukchi Shelf.

Well-bedded sedimentary rocks of shelf facies that in general produce strong seismic reflections characterize the lower part of the Ellesmerian sequence on the North Slope, where they have been divided into three major, partly unconformity-bounded transgressive-regressive sedimentary cycles (Moore and Mull, 1989). The lowest cycle consists of platform carbonate rocks of the Lower Mississippian to Lower Permian Lisburne Group and the partly underlying, partly time-equivalent Kayak Shale and subjacent basal Kekiktuk Conglomerate. The cycle ranges in thickness from a wedge edge near the coast to more than 1,700 m beneath the central part of the coastal plain. The middle cycle is a dominantly clastic deposit, the Lower Permian to Lower Triassic Sadlerochit Group, which overlies the Lisburne Group on a regional unconformity. The Sadlerochit grades from commonly coarse nonmarine facies near the coast to marine lutite and fine-grained sandstone in North Slope wells. Its thickness ranges from a wedge edge near the coast to more than 800 m beneath the central coastal plain. The highest cycle consists of Middle and Upper Triassic marine shelf deposits of the Shublik Formation and its proximal facies, the partly overlying Sag River Sandstone and the coeval Karen Creek Sandstone of the eastern North Slope. The Shublik is as much as 200 m thick and consists of phosphatic shale, siltstone, and coquinoid limestone.

Upper part of the Ellesmerian sequence and the Hauterivian breakup unconformity: Basinal marine lutite with a few thin sandstone bodies of the Jurassic and early Neocomian (Berriasian and Valanginian) Kingak Shale constitutes the lower stratigraphic unit of the upper part of the Ellesmerian sequence beneath the North Slope and the central Chukchi Shelf. Hauterivian and Barremian (Mickey and Haga, 1987) organic-rich marine shale of the informally named pebble shale unit (PSU in this chapter) constitutes the upper stratigraphic unit. An angular unconformity lies between these units on Barrow arch, but south of a middle Neocomian shelf break beneath the northern part of the Arctic Foothills the PSU rests conformably on lower Neocomian clinoforms at the top of the Kingak Shale (Molenaar, 1988).

Foreset beds interpreted from seismic reflection records show that the Kingak prograded south- or southeastward from the Barrow arch onto a subsiding shelf that was about 1 km deep beneath the Arctic Foothills. Neocomian erosion at the base of the PSU has in many places stripped the Kingak from the north flank and crest of the Barrow arch. Total thickness of the Kingak ranges from a knife edge along the crest of the arch to a maximum of 1,200 m in the northern part of the Arctic Foothills (Bird, 1988a).

The PSU consists mainly of highly organic marine shale characterized by floating grains and pebbles of frosted quartz and chert, elevated levels of gamma ray activity, and low seismic velocity. The unit is 60 to 150 m thick on the Barrow arch and 60 to 75 m thick beneath the Arctic Foothills (Bird, 1988a). At several places on the arch, as much as 15 m of locally and northerly sourced sandstone and pebble conglomerate occur at the base of the PSU. According to Witmer and others (1981) and Mickey and Haga (1987) foraminifer and palynomorphs indicate that below a highly radioactive zone (HRZ) in its uppermost 6 to 12 m, the PSU is Hauterivian and Barremian in age. The floating quartz and chert grains that characterize the PSU below the HRZ are thought to have a local source on Barrow arch or its northern flank, as did the more voluminous Kingak Shale.

The HRZ is a condensed organic shale that contains Barremian to possibly lower or middle Albian radiolarians and palynomorphs (Mickey and Haga, 1987), and it lacks the floating quartz and chert grains that characterize the PSU. Its scant thickness (6 to 12 m), modest time span, fine-grained texture, high organic content, and high gamma ray activity demonstrate that the HRZ is a pelagic or hemipelagic deposit largely isolated from sources of terrigenous clastic sediment. If the HRZ were a bottomset basinal deposit at the base of the prograding Brookian sequence, as suggested by Mickey and Haga (1987), it should contain distal turbidites from the adjacent continental terrace, and be thicker. Because the HRZ is a pelagic or hemipelagic deposit with neither Ellesmerian nor Brookian sedimentary contributions, it should be regarded as an independent transitional unit. Because of its thinness, however, and because it cannot ordinarily be separated from the PSU on seismic reflection records, the HRZ is in practice lumped with the PSU and the Ellesmerian sequence (Fig. 5).

The erosional unconformity at the base of the PSU is thought to be the breakup unconformity associated with initiation of continental drift in the Canada Basin. The thin, strongly reflective, highly organic PSU is considered to be the product of feeble, locally sourced, synbreakup or earliest post-breakup sedimentation on the rift-margin high. Like the HRZ, the PSU is a transitional, synrift unit between the Ellesmerian and Brookian sequences. The change from the feeble clastic sedimentation of the PSU to the condensed, pelagic or hemipelagic sedimentation of the HRZ records the early post-breakup subsidence of the rift-margin high below wave base. This subsidence ended the era of rift-margin sedimentation from local (Barrow arch) sources that had nourished the PSU and began the era of condensed sedimentation in a deeper environment, which was isolated from terrigenous sourcelands that produced the HRZ. Overtopping of the Barrow arch by detritus from the Brooks Range orogen terminated HRZ deposition, and initiated south-sourced Brookian sedimentation above the HRZ by Aptian or Albian time.

Dinkum succession (Failed rift deposits of the Beaufort Shelf and slope). On the central Beaufort Shelf north of Barrow arch, the PSU rests unconformably on strata of inferred Jurassic and early Neocomian age in and north of the Dinkum graben (Figs. 6, 7). We interpret these beds to have been deposited in a failed rift system along the continental margin. Hubbard and others (1987) have grouped the graben-filling beds of the Beaufort Shelf and the upper beds of the Ellesmerian sequence (for convenience, herein referred to as the upper Ellesmerian beds) of the North Slope in a new sequence, the Beaufortian, on the premise that both are rift related. In this chapter we place these deposits in separate sequences because, although largely coeval, they were deposited in different tectonic and sedimentary environments and have differ-

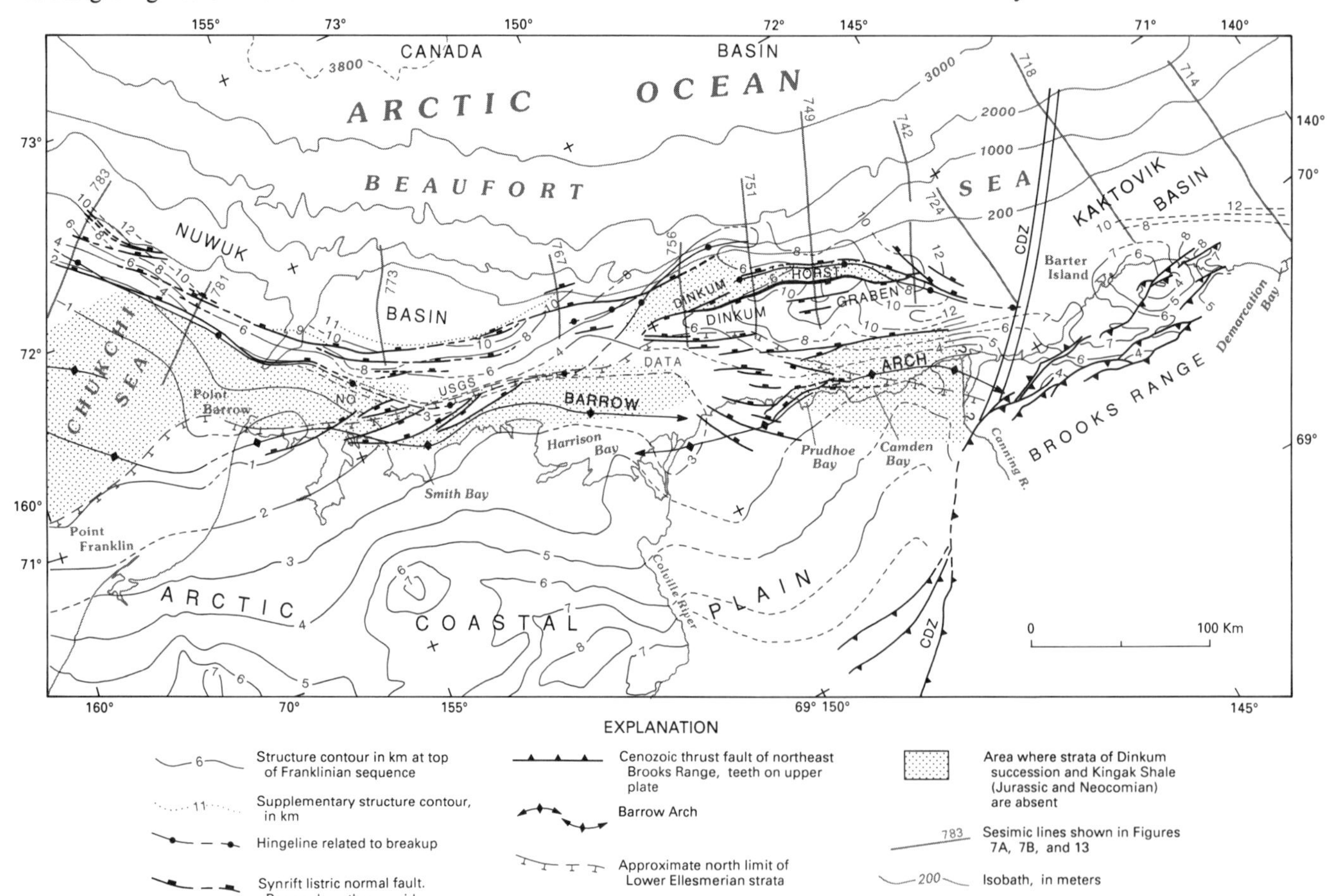

Figure 6. Map of Arctic Alaska continental margin showing depth in kilometers to top of Franklinian sequence, location of synrift structures, and area adjacent to rift stripped of Jurassic and early Neocomian strata in late Neocomian time. Onshore data from Bruynzeel and others (1982), Tailleur and others (1978), Bruns and others (1987), Bird (1988a, b), and Jamison and others (1980). Offshore data from profiles in Grantz and others (1982, 1986) and maps in Hubbard and others (1987), Craig and others (1985), and Pessel and others (1978). Lines dashed where projected or speculative.

ent stratigraphies. The upper Ellesmerian beds of the North Slope were deposited on an areally extensive stable shelf that subsided and tilted southward in response to tectonic overriding and loading by the early phases of the Brookian orogeny. In contrast the Dinkum graben fill, and the probably related deposits of the outer shelf and slope, were deposited in fault-bounded half grabens or grabens and have much higher ratios of length and thickness to width. We retain the term upper Ellesmerian for the areally extensive Jurassic and Neocomian marine deposits of the tectonically stable Arctic Platform and use Dinkum succession to designate the thick, but elongate and areally limited deposits of Dinkum graben and the tectonically related deposits of the continental margin.

Brookian sequence. The Brookian sequence of the North Slope was deposited in the Colville Basin, a foreland basin, and as a progradational continental terrace sedimentary prism along the continental margin. The stratigraphy of the sequence is complicated by east-west changes in facies and sediment thickness related to the location and timing of local domains of convergent tectonism and uplift in the Brooks Range.

South-sourced syntectonic flysch and molasse of the Aptian(?) and early Albian Fortress Mountain Formation lie at the base of the Colville Basin section in the southern North Slope. These sandstones, lutites, and conglomerates grade laterally northward into the lower part of the Torok Formation, which consists of foreset and bottomset turbidite and hemipelagic clastic deposits of Aptian(?) and Albian age that extend north to the continental shelf. The partly coeval, partly younger Torok Formation is about 6,000 m thick beneath the Arctic Foothills and less than 150 m thick on the Barrow arch. Beneath the central and western parts of the Arctic Coastal Plain the Torok downlaps onto the south-dipping upper surface of the HRZ.

The Fortress Mountain Formation is overlain by the post-tectonic (molassoid) Nanushuk Group—regressive shallow marine and deltaic sandstone, lutite, conglomerate, and coal of middle or late Albian to early Cenomanian age. The Nanushuk intertongues with foreset beds in the partly underlying, partly equivalent upper Torok Formation in the central and northern parts of the Colville Basin. By middle or late Albian time, post-rift subsidence of the rift margin permitted Torok sediments to overtop the Barrow arch in many places, and to begin the progradation of a continental terrace sedimentary prism along the then recently formed Beaufort Sea margin of the Canada Basin. The Nanushuk-Torok interval is several thousand meters thick west of the lower Colville River, but east of the river it thins drastically and is represented by condensed bottomset shale, which is no more than a few tens of meters thick on the Barrow arch east of Prudhoe Bay (Molenaar and others, 1986). The post-Nanushuk deposits of the Colville Basin rest on a middle Cenomanian unconformity that is distinct on logs and seismic sections according to Bruynzeel and others (1982), but correlated well sections by Molenaar and others (1986) suggest that at least in places this contact is a facies boundary. A widespread transgressive marine shale, the late Cenomanian and Turonian Seabee Formation of the Colville Group, rests on the unconformity and is presumably related in origin to the major long-term worldwide rise in sea level that culminated in late Cenomanian and early Turonian time (Haq and others, 1987). Molenaar and others (1986, 1987) apply the name "Hue Shale" to the highly organic and bentonitic basinal marine shale of Aptian(?) to Maastrichtian (and possibly Paleocene) age that constitutes the diachronous basal Brookian unit east of the Colville Delta. The unit is a condensed deposit, 300 m or less thick. It is overlain by prodelta or basin slope marine shale of Aptian or Albian to Eocene or Oligocene age, with turbidites in the lower beds, which Molenaar and others (1986, 1987) named the Canning Formation. The Canning is laterally equivalent to all of the shales of the Torok Formation, Colville Group, and lower Sagavanirktok Formation to the west of the Colville Delta. Between the Canning and Colville Rivers the Canning Formation is 1,500 to 3,000 m or more thick, but east of the Canning River where the formation is Campanian to Eocene or Oligocene in age, its thickness is unknown.

All of the nonmarine and marine deltaic and alluvial plain sandstones, lutites, and conglomerates above the Canning Formation east of the Colville River were placed in the Sagavanirktok Formation by Molenaar and others (1986, 1987). Near the Canning River the unit is 1,800 to 2,300 m thick and Late Eocene through Pliocene in age. The base of the unit becomes older to the west and near the Colville River the basal beds are Campanian. West of the Colville River, where the preexisting stratigraphic terminology is retained, the post-Nanushuk-Torok section is placed into two units: the Colville Group of middle Cenomanian through Maastrichtian age, and the Sagavanirktok Formation of Tertiary age. All of these units extend beneath the continental shelf. The overlying glacial and interglacial deposits, which are no more than a few tens of meters thick beneath the North Slope, thicken to 100 meters or more near the shelf break. These heterogeneous deposits are assigned to the upper Pliocene and Quaternary Gubik Formation, which is summarized in Dinter and others (1990).

Canada Basin

The Canada Basin is thought to have formed by rotational rifting during Hauterivian to early Late Cretaceous time. Available data (May and Grantz, 1990; Grantz and others, 1990b) suggest that oceanic crust (layers 2 and 3) beneath the southern Canada Basin is of near normal thickness (6 to 8 km) but that it is overlain by an unusually thick sedimentary layer (part 1). The layer 1 sediments are stratigraphically continuous with those of the post-rift sedimentary prism of the Barrow sector of the continental margin. An empirical relation between regional bathymetry and sediment thickness derived from seismic refraction data (Fig. 8 in Grantz and others, 1990c) suggests that oceanic crust lies less than 10 km below sea level north of the Chukchi Shelf to 12 km or more off the Mackenzie Delta, and that the overlying sedimentary layer may be a little more than 6 km thick in the former area and 12 km in the latter. The apex of the large

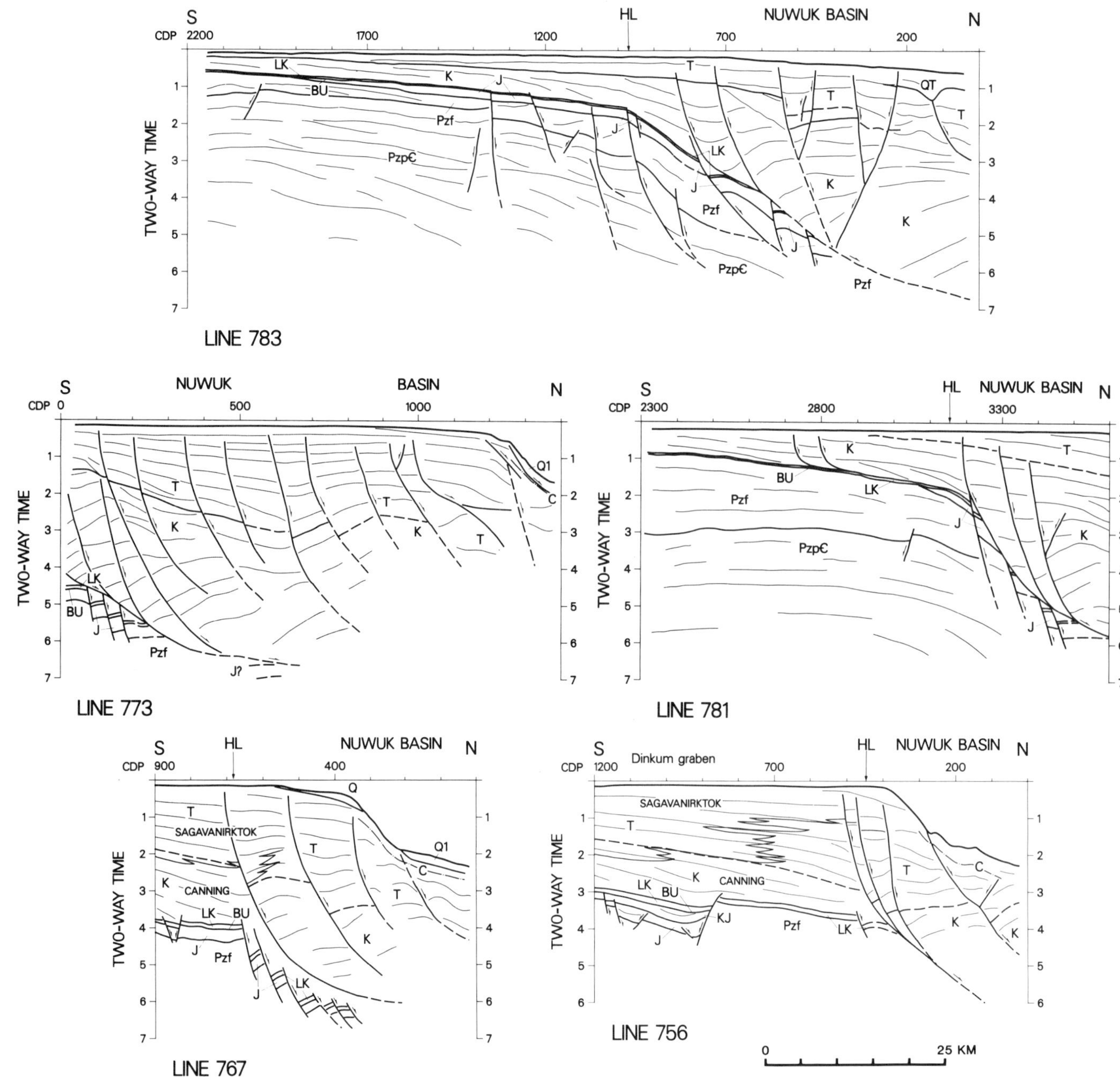

Figure 7. (This and facing page) Line drawings showing geologic structure and stratigraphy on seismic reflection profiles 756, 767, 773, 781, and 783 (shown on A) and 742, 749, and 751 (shown on B) across the Arctic Alaska continental margin. Time sections from Grantz and others (1982). Figure 2 shows approximate equivalence of seismic reflection time and depth, Figures 6 and 12 show location of profiles. Vertical exaggeration in water column is 7; in subsea sediments roughly 3.5, but highly variable. Q1, Quaternary submarine slide deposits; QT, Quaternary and late Tertiary marine sedimentary strata; T, Tertiary strata of the marine and nonmarine Gubik Formation, the dominantly nonmarine Sagavanirktok Formation, and the marine Canning Formation; Tu, upper Tertiary marine and nonmarine strata of the eastern Beaufort Shelf and slope; T1, post–middle Eocene lower Tertiary marine sedimentary strata of the eastern Beaufort Shelf and slope; T1K pre–middle Eocene to Cretaceous, and in places possibly Jurassic, marine sedimentary strata of the eastern Beaufort Shelf and slope; K, Upper to upper Lower Cretaceous strata of the marine and nonmarine Colville Group and Nanushuk Group and the marine Torok Formation and their equivalents beneath the continental slope and rise; LK, Hauterivian to early or middle Albian condensed marine shale with local lenses of sandstone and conglomerate of

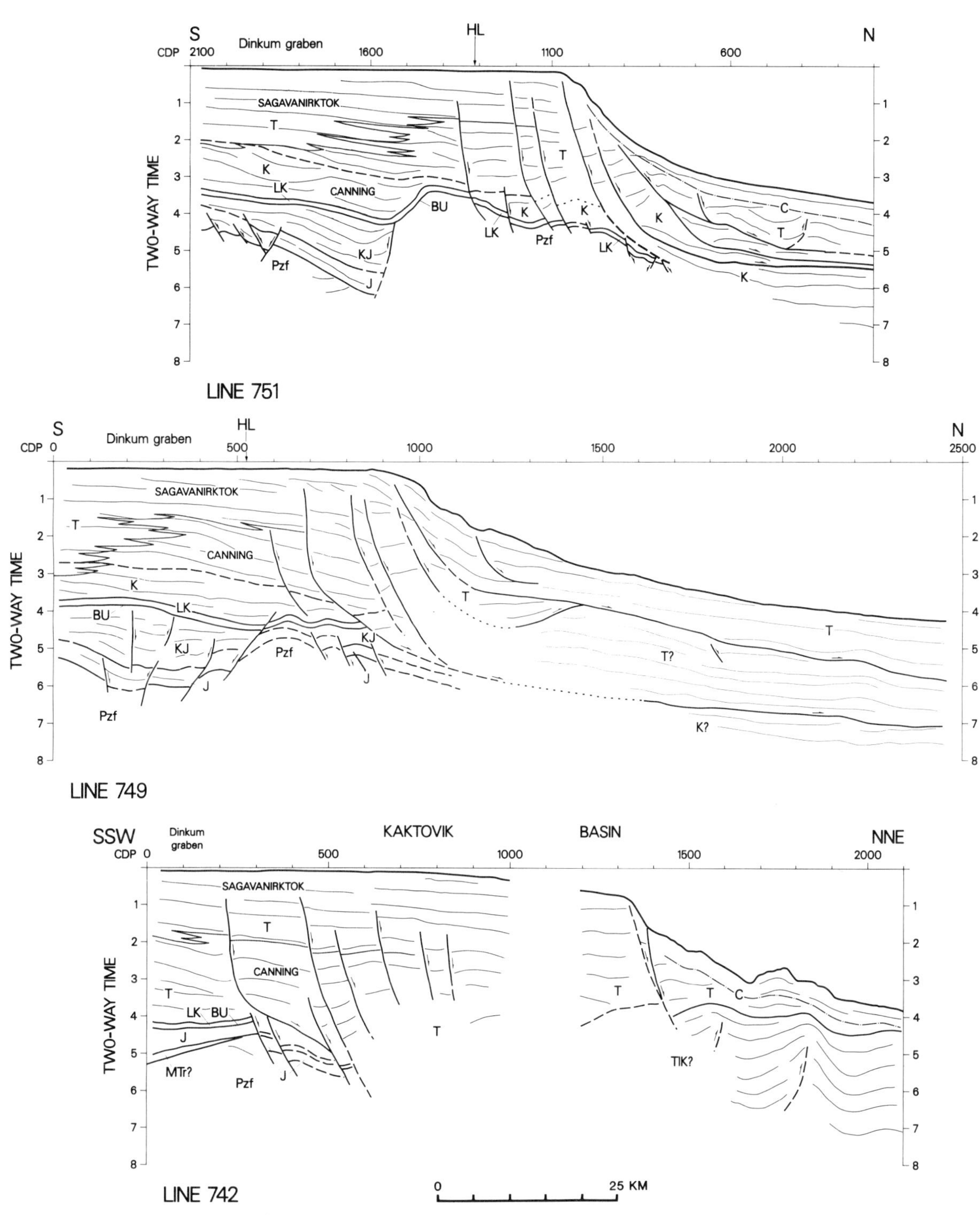

the "pebble shale unit" (PSU) on the western Beaufort Shelf and the Barremian to Maastrichtian Hue Shale on the eastern Beaufort Shelf; KJ, Jurassic and Neocomian synrift marine lutite of Dinkum succession; J, inferred sandstone and possibly conglomerate equivalent to the lower part of the Kingak Shale and possibly lower Ellesmerian sequence marine and paralic stable shelf deposits; MTR, lower Ellesmerian sequence marine and paralic stable shelf deposits; Pzf, lower Paleozoic (pre-Mississippian) eugeoclinal clastic and miogeoclinal carbonate and clastic strata; Pzp€, lower Paleozoic and Precambrian strata, inferred to include carbonate or possibly quartzite; BU, breakup unconformity; SU, syntectonic unconformity; SU-Teu, middle Eocene syntectonic unconformity; HL, tectonic hinge line related to rifting; D, diapiric core of thrust fold; C, base of solid gas hydrate.

continental rise sedimentary prism that fills the southern Canada Basin heads against the Mackenzie Delta and Amundsen Gulf (Grantz and others, 1990a, Plate 1), which indicates that most of the fill originated in the Mackenzie River valley and the glacial drainages entering Amundsen Gulf. Most of the geologic structure in the Canada Basin was imposed by stresses that originated within or south of the Beaufort Shelf. These structures include large submarine slides and, in the Barter Island sector, large thrust faults and thrust-folds related to detachment faults that root beneath the Brooks Range. Grantz and others (1990b) discuss the structure and stratigraphy of the Canada Basin.

Chukchi Borderland

A compact group of aseismic submarine ridges and plateaus and intervening deep basins, the Chukchi Borderland, intersects the continental shelf and slope west of 162°W (Fig. 1). The ridges and plateaus are high-standing, flat-topped features with a dominant trend of about N20°E, whose geologic character and mode of origin are poorly known. Present knowledge of Chukchi Borderland geology and geophysics is summarized by Hall (1990), who concludes that the borderland consists of three or more highstanding continental blocks separated by deep, graben-like basins formed by rifting. The physiography of the borderland suggests that the bounding faults of the continental blocks and grabens have a northerly trend.

Understanding of the character and origin of the borderland is a prerequisite for a complete understanding of the Alaskan Arctic continental margin because the Northwind Escarpment, which forms the eastern boundary of the borderland, is on strike with the eastern boundary of the North Chukchi Basin of the northwestern Chukchi Shelf. This juxtaposition suggests that these features may be genetically linked. Grantz and others (1979) and Vogt and others (1982) proposed that the highstanding ridges of the Chukchi Borderland were rifted from the continental shelf in the western Chukchi and East Siberian Seas, an area now occupied by the North Chukchi Basin.

Some data bearing on the origin of the borderland were collected over the southern part of Northwind Ridge in 1988. A two-channel seismic profile across the eastern half of the ridge showed that it is underlain by 400 to 600 m of flat-lying strata characterized by strong seismic reflectors. The flat-lying strata rest unconformably on moderately dipping to probably strongly deformed beds that are at least 2 km, and possibly 4 km or more thick. Three piston cores from the lower slopes of the Northwind Escarpment near 74°35′N (Fig. 3), at water depths of 3,200 m to 3,700 m, sampled the deformed beds. The lowest unit in the cores consists of bedrock or coarse rubble composed of yellowish brown to light gray, oxidized, foraminifer-bearing marine lutite and dark brown lutite containing abundant palynomorphs. The foraminifers, dated by W. V. Sliter of the U.S. Geological Survey as Albian (personal communication, 1989), resemble forms found in parts of the thick Nanushuk Group of paralic sedimentary rocks of the North Slope. The character and age of the flat-lying upper beds are not known, but their structural position, local deep erosion, and the presence of block faulting that is on strike with that of the eastern North Chukchi Basin suggest that they are no older than latest Cretaceous or Tertiary.

STRUCTURE AND STRATIGRAPHY OF THE ALASKAN MARGIN

Chukchi sector

North Chukchi Basin. The North Chukchi Basin and the Chukchi Borderland are the characterizing features of the Chukchi sector of the continental margin. Principal sources of data on the geologic character of the North Chukchi Basin are eight multichannel and a few single-channel U.S. Geological Survey seismic reflection profiles (Grantz and others, 1972a, b, and 1986) shown in Figures 8 and 9, and Thurston and Theiss (1987), which is based on unpublished proprietary data. We rely mostly on the sparser, but published, Geological Survey data.

The full extent of the North Chukchi Basin is not known. On the east the basin's Brookian strata thin toward the western extension of the Barrow arch (the North Chukchi High of Thurston and Theiss, 1987) beneath the north-central Chukchi Shelf (Fig. 8), and on the southeast they thin toward the Chukchi Platform. To the south the basin is bounded by a south-dipping Cenozoic thrust-fault system near 72°N that extends to the base of the Quaternary cover at the sea floor (Figs. 4 and 8). South of this fault system lies the Wrangel arch, which on Herald and Wrangel Islands (Fig. 8), exposes Proterozoic(?) to Middle Cambrian(?) metavolcanic and metasedimentary rocks and folded and thrust-faulted Paleozoic to Triassic sedimentary rocks (Cecile and Harrison, 1987). Between the Chukchi Platform and Wrangel arch, the south limit of the basin is obscured by thick sequences of Tertiary strata in northerly striking horst and graben structures (Jessup, 1985). Multichannel seismic reflection data extend the known area of the basin as far north as 73°10′N and as far west as 176°15′W, and single-channel data suggest that the basin extends north of 74°N. The basin axis is incompletely mapped, but it appears to lie near 72°30′ to 73°N and to trend N65° to 85°W.

Stratigraphy and internal structure. We recognize five major seismic-stratigraphic units in the North Chukchi Basin and tentatively correlate them with units in northern Alaska and the Beaufort Shelf. Line drawings in Figure 9 show these units plus pre-Cretaceous (unit pK) strata for which no correlation is apparent. The oldest unit (LE) is interpreted to consist of strata in the lower part of the Ellesmerian sequence that are poorly bedded in the lower part and well bedded in the upper part. These beds appear to lack clinoforms and may be shelf deposits. Unit LE is interpreted to be overlain by Jurassic and Neocomian marine clastic rocks of the upper Ellesmerian sequence (unit UE), which thickens from less than 1 km at the basin edge to more than 2.5 km within the basin 15 km to the west. It may have been a precursor of the North Chukchi Basin, just as the Dinkum succession of the Barrow sector was a precursor to the Hauterivian and younger rift basins of the Beaufort margin. The lower two-thirds or more of unit UE consists of weakly reflective beds thought to

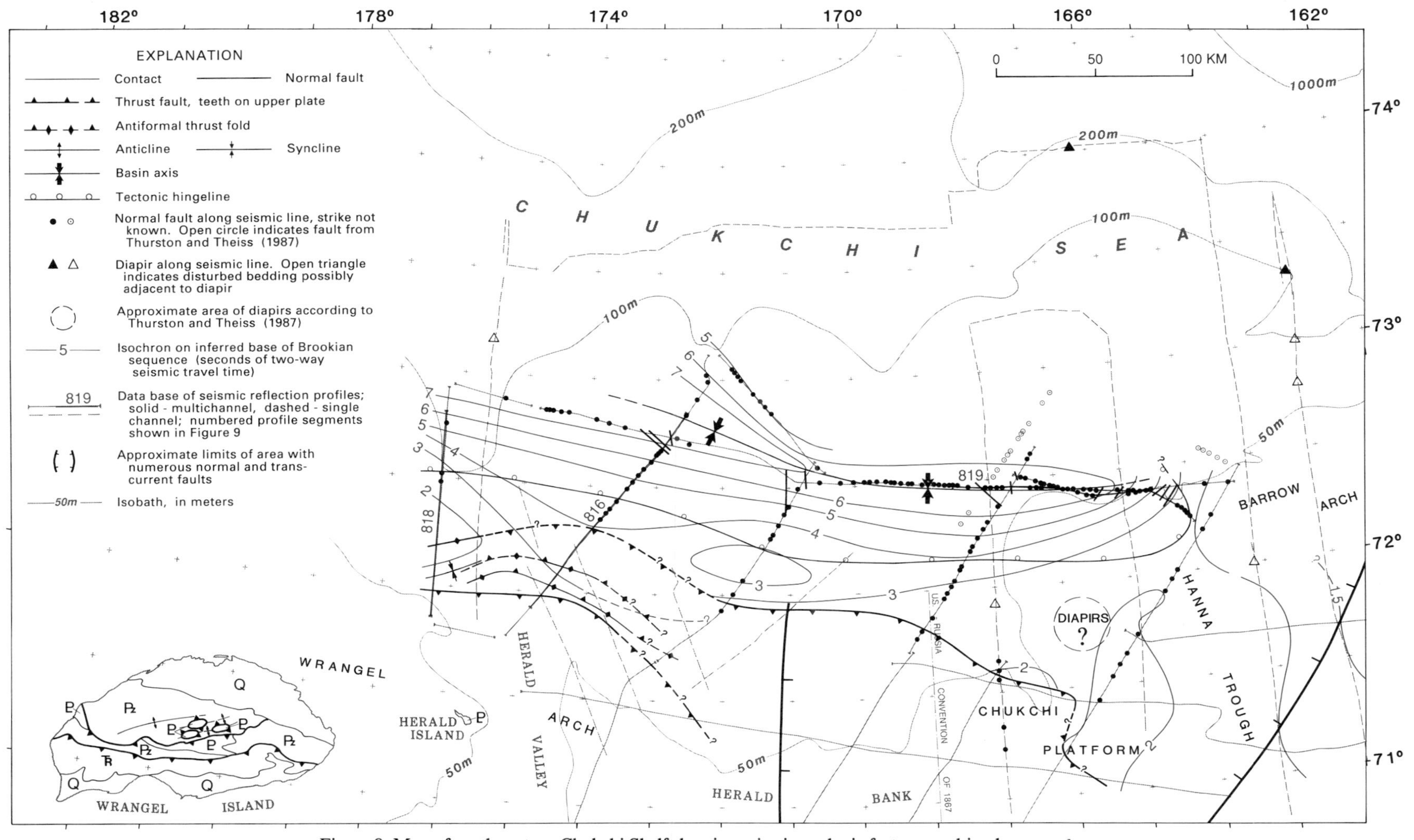

Figure 8. Map of northwestern Chukchi Shelf showing seismic-geologic features and isochrons on base of Brookian sequence in North Chukchi Basin, generalized geology of Wrangel Island after Cecile and Harrison (1987), and location of seismic reflection profiles shown in Figure 9. See Figure 2 for relationship between isochrons and depth. Q, Quaternary deposits; T̶R, Triassic flyschoid strata; Pz, Paleozoic fluvial to shelfal clastic and platform to slope carbonate rocks; P̶, Proterozoic(?) to Middle Cambrian(?) metavolcanic and metasedimentary rocks intruded by granitic sills and dikes (Gromov Complex). Contacts, faults, and axes are dashed where inferred, queried where speculative.

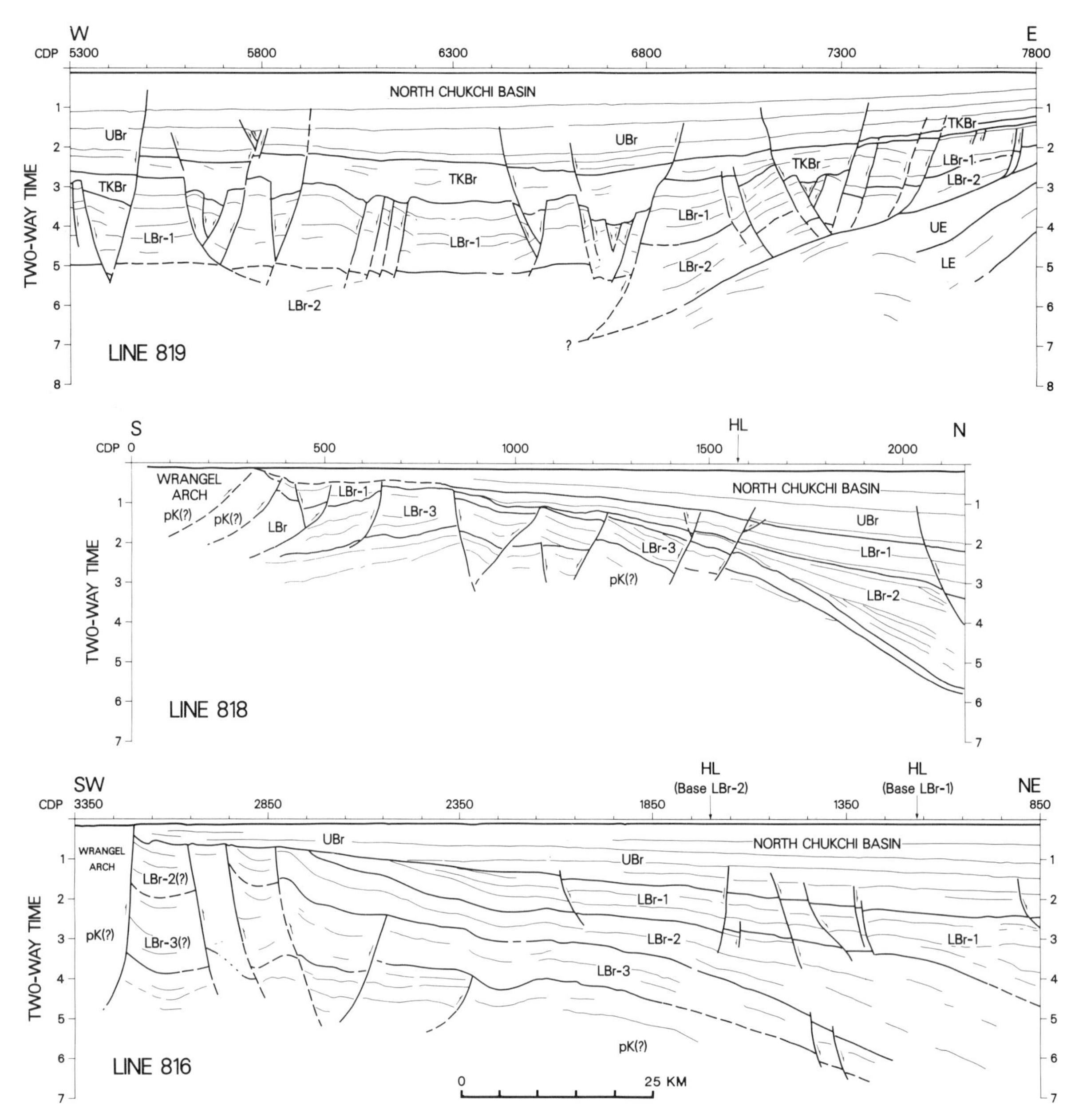

Figure 9. Line drawings showing inferred geologic structure and stratigraphy on seismic reflection profiles 816, 818, and 819 across North Chukchi Basin. Profiles from Grantz and others (1986). See Figure 2 for equivalence of reflection time and depth and Figure 8 for location of profiles. Vertical exaggeration in water column is 7; in subsea sediments roughly 3.5, but highly variable. UBr, upper Brookian (Tertiary) marine strata; TKBr, "Laramide" (uppermost Cretaceous and/or lower Paleogene) marine strata; LBr-1, LBr-2, and LBr-3, seismic-stratigraphic subdivisions of inferred lower Brookian (Cretaceous) marine strata; pK, inferred pre-Cretaceous strata; UE, inferred upper Ellesmerian (Jurassic) marine strata; LE, inferred lower Ellesmerian stable shelf marine strata; HL, tectonic hinge line.

be lutites, but the upper third or less on profile 819 (Fig. 9) also contains beds that produce stronger reflections and may include sandstone. These upper beds are gently folded and have been truncated by an angular unconformity at the base of the overlying Brookian sequence. Erosion has locally removed most of unit UE from the east flank of the basin.

A thick sequence of reflections, interpreted to represent clastic rocks in the lower (Cretaceous) part of the Brookian sequence (LBr), unconformably overlies units pK and UE. The internal stratigraphy of unit LBr is hard to define on the basis of our widely spaced profiles, but it may be divisible into three units. Unit LBr-3, the oldest, is shown on profiles 816 and 818 (Fig. 9), where it thins basinward from 1.8 to 3.2 km or more on the paleoshelf at the southern margin of the basin to less than 0.2 to

2.0 km where the unit fades out of our records in the axial deep to the north. The unit consists of weakly to moderately reflective beds, apparently topsets, that have been offset locally by mild folds and at least one thrust fault. These beds were erosionally truncated before unit LBr-2 was deposited upon them.

Units LBr-2 and -1 consist of moderately to strongly reflective, well-bedded rocks that are in topset facies south and east of the tectonic hinge line that separates the paleoshelf regions of the basin from its axial deep (Figs. 8 and 9). On the paleoshelf, unit LBr-2 ranges from 0 to at least 1.7 km, and unit LBr-1 from 0 to at least 1.4 km in thickness. North and west of the hinge line units LBr-2 and -1 increase dramatically in thickness—LBr-2 to more than 6.7 km and LBr-1 to more than 5.4 km. Reflective properties of the constituent beds also are more varied basinward of the hinge line, ranging from weak to strong, and with foreset as well as topset beds present, especially in the lower half of each unit.

The tectonic hinge line at the base of unit LBr-1 (Fig. 9) is about 20 to 30 km farther north than is the hinge line at the base of unit LBr-2, which suggests that subsidence was a multistage event that progressed basinward. On the westernmost profile (818 in Fig. 9), Unit LBr-2 was faulted, uplifted, and eroded on the paleoshelf before overstepping unit LBr-1 was deposited. On the eastern part of line 819 in Figure 9, which crosses the edge of the axial deep, unit LBr-2 can be seen to rest on a glide plane. In consequence, the relation of unit LBr-2 to older units on this profile is not known, and as drawn the unit may include part of unit LBr-3.

An extensional event block-faulted the LBr units in the eastern part of the basin (Fig. 8) and created a spectacular horst and graben paleotopography that was infilled and smoothed by unit TKBr (profile 819, Fig. 9). The extension produced fault scarps with local relief that in places exceeds 2.5 km. The extensional faulting that produced this relief was below wave base because the height of the fault scarps approximates the structural relief on the faults. The area affected is shown on Figures 4 and 8. Unit TKBr, of inferred "Laramide" age, consists of weakly to moderately reflective foreset and bottomset beds that were deposited in local sediment catchments, as well as some topset beds. The unit is as much as 3.2 km thick. A change to markedly better bedded, and more strongly reflective rocks in overlying unit UBr suggests that the deposition of unit TKBr brought the seabed close to wave base and largely filled the depression formed by the extensional faulting.

Unit TKBr thins eastward and southeastward toward a bordering paleoshelf beneath the present central Chukchi Shelf, and it thins over broad structural highs on the shelf such as the Chukchi Platform. It also infills low-lying areas such as the partly erosional, partly structural "Laramide" depression that overlies part of the Hanna trough (Figs. 4 and 8) between the Chukchi Platform and the western part of the Barrow arch. Unit TKBr is about 2 km thick in the trough, is about 2 km deep, and is connected with the main body of unit TKBr in the North Chukchi Basin by two very broad channels. These channels, like the "Laramide-age" depression in the Hanna trough, are partly structural, partly erosional. We postulate that a graded surface originally extended from the depression over the Hanna trough to the deeper North Chukchi Basin during unit TKBr deposition. This surface was uplifted after unit TKBr deposition in the area between the northern part of the Chukchi Platform and the western part of the Barrow arch, and the base of the broad channel now lies above the base of the "Laramide" depression. In the western part of the mapped area of the North Chukchi Basin, beyond the area affected by intense extensional faulting (see Figs. 4 and 8), unit TKBr assumes the geometry of the underlying and overlying formations and has been included in unit LBr-1.

Conformably above unit TKBr in the North Chukchi Basin lies a sequence of well-bedded, moderately to strongly reflective rocks, unit UBr, that is 3.7 km or more thick. The unit was deposited on a generally smooth surface with a long-wavelength structural sag in the eastern part of the basin (profile 819 in Fig. 9). Topset beds are dominant in unit UBr, although it contains some foreset and bottomset beds in the western part of the mapped area. The predominance of topset beds throughout this thick unit indicates that sedimentation generally kept pace with subsidence and that the sediment supply was abundant. Continuing or renewed activity on the larger listric faults that distended units LBr and TKBr has in places offset beds in the lower part of unit UBr (profile 819). Some of the faults that offset lower UBr beds are of moderate displacement and have similar thicknesses of TKBr in their hanging and footwalls. These appear to represent tectonic movement during the early stages of unit UBr deposition. Most of the faults that offset lower UBr beds, however, have displacements of only 100 to 200 m in the UBr, much larger displacements in underlying units, and markedly disparate thicknesses of unit TKBr in their hanging walls and footwalls. These small offsets are thought to have resulted from differential compaction.

Diapirs. Thurston and Theiss (1987) report that a north-trending graben on the northern part of the Chukchi Platform appears to be the source of diapirs, diapir mounds similar to "salt pillows," and associated collapse and withdrawal structures (Fig. 8). The diapirs originate in the upper part, and possibly in the lower part of the Ellesmerian sequence in the graben, and some of them pierce strata as young as upper Brookian (Tertiary).

Several diapirs, and structural features possibly associated with diapirs, were also observed in upper Brookian strata on Geological Survey seismic profiles in and near the North Chukchi Basin. Figure 10 shows two of these diapirs, piercement structures reaching to within 200 m of the seabed, that were identified on a single-channel seismic reflection profiles. A star pattern of seismic profiles across the diapir at 73°21′N shows that this feature is about 2 km in diameter, that its base lies more than 2.5 km below sea level, and that it may be surrounded by a withdrawal syncline. Morphologically these features resemble late-stage salt diapirs, but reconnaissance gravity and refraction measurements (Fig. 10) showed no associated negative density or velocity anomalies. Grantz and others (1975) therefore inferred that these diapirs may be shale cored.

Listric-normal fault province. The eastern part of North Chukchi Basin is underlain by a province of listric faults, horsts, and grabens illustrated in profile 819 (Fig. 9). The faults are mainly of latest Cretaceous or earliest Tertiary age because they disrupt unit LBr, inferred to be of middle Early to Late Cretaceous age, and are overlain by mainly undeformed unit TKBr, of inferred "Laramide" age. A few offset the lower beds of unit UBr of inferred Tertiary age. The extent of the fault province is shown in Figures 4 and 8. Correlation of the faults between U.S. Geological Survey seismic profiles and text figures in Thurston and Theiss (1987) indicate that the strike of the faults is variable, but mainly northerly. Maximum observed displacement on individual fault sets of the listric fault system is about 2.5 km. Extension along our seismic profiles in the listric fault area, as measured by offsets in unit LBr, ranges from 5 to 12 percent. The down-dip terminations of the listric faults in the axial deep lie below our deepest identifiable seismic reflections (6.5 seconds, about 13 km), but the faults can be seen on profile 819 to merge in a detachment, or sole, fault on the eastern slope of the basin Fig. 9). This detachment fault follows the top of unit UE, which underwent some fault truncation of its uppermost beds, from a depth of 3 km near the eastern end of profile 819 to more than 14 km at the east end of the axial deep, 55 km to the west.

The extensional fault terrane lies between a zone of north- to northeast-striking faults to the south and east, and the north-northeast–striking ridges and basins of the Chukchi Borderland to the north (Fig. 4). The faults to the south of the extensional fault terrane were mapped by Jessup (1985), who reported that they were normal faults that displace upper Brookian (Tertiary) strata. Those to the east and south were mapped by Thurston and Theiss (1987) as the Hanna wrench fault zone. Thurston and Theiss (1987) suggest that many of the faults in the Hanna wrench-fault zone formed in transtensional or transpressional stress fields, but the sense of displacement is not reported. Wrench-fault features are best developed in the eastern part of the Hanna fault zone and may give way to normal faults to the west and in the North Chukchi Basin. The wrench faults reportedly do not extend south of Herald arch and they can not be traced into the North Chukchi high at the west end of the Barrow arch.

Salient features of the listric normal fault province that bear on its origin are the north to northeast trend of its individual faults, the large structural relief on individual fault sets (2.5 km or more), its compactness (fault densities are high within the province and low beyond its boundaries), the significant extension produced by the fault system (5 to 12 percent), and the position of the province at the continental margin opposite the north- to northeast-trending aseismic ridges of the Chukchi Borderland. Restriction of the extensional fault system to only the eastern part of the North Chukchi Basin and the northerly trend of the fault is incompatible with a rift origin above the east-trending axial deep of the North Chukchi Basin or with sliding toward the west-northwesterly striking free face of the adjacent continental margin.

An origin for the listric fault province is suggested by its resemblance to the extensional fault system, which disrupts the Eurasian continental margin where it is impinged by the Arctic Mid-Ocean Ridge of the Eurasia Basin in the Laptev Sea (see Grantz and others, 1990a, Plate 11). Similarities include structural position at the continental margin, trend and character of faulting, and dimensions. Based on this resemblance, we suggest that entry of a "Laramide" spreading center from the Arctic Basin into the continent also created the north-northeast–trending ridge and basin topography of the eastern Chukchi Borderland and the listric normal faults of the eastern North Chukchi Basin. Insufficient data are available to constrain a geometric model, but bathymetry and the regional distribution of extensional features of broadly "Laramide" age in the Arctic region suggest that the normal faults of the western Chukchi Shelf may belong to a regional spreading ridge and transform fault system of latest Cretaceous–early Tertiary age. This system might extend from the Chukchi Shelf to the Mid-Atlantic Ridge via north-south trending structures within the Chukchi Borderland, Mendeleev Ridge,

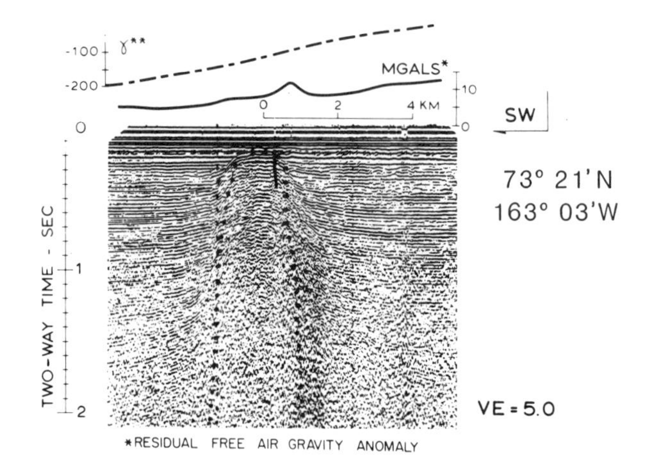

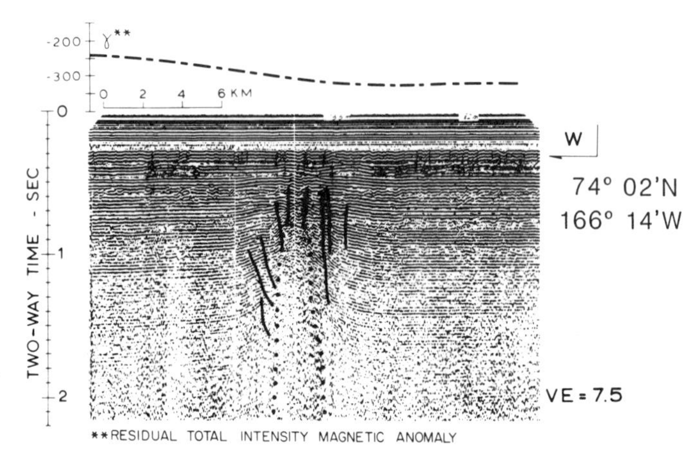

Figure 10. Single-channel air gun seismic reflection, gravity, and magnetic profiles across piercement diapirs in North Chukchi Basin. The upper profile is a central crossing of diapir at 73°21′N, 163°03′W; the lower profile is probably an oblique crossing of diapir at 74°02′N, 166°14′W. See Figure 8 for location.

Makarov Basin, and Baffin Bay, and transform faults such as Nares Strait and possibly others, as yet not identified, in the central Arctic Basin (Figs. 1, 3, 4; Grantz and others, 1990a, Plate 11).

Barrow sector

An arcuate continental margin of simple geometry faces the Canada Basin between Northwind Ridge on the west and the Canning displacement (fault) zone on the east (Figs. 4 and 6). This, the Barrow sector of the Arctic Alaska margin, is characterized by the Dinkum graben, a failed rift of Jurassic and Neocomian age, and the Hauterivian to early Late Cretaceous sea-floor spreading that created the Canada Basin. A syn- and post-rift progradational continental terrace sedimentary prism, the Nuwuk Basin, consists of post-Neocomian (Brookian) clastic sediment that buries the rifts. Cretaceous beds are the principal component of the basin on the west and Tertiary beds to the east. The principal structural features of the Barrow sector are shown in map Figures 3, 4, and 6, seismic reflection profiles 724 to 783 in Figures 7A, B, and 13, and seismic profiles 9D and E in Grantz and others, 1990a, Plate 9.

Stratigraphy. *Franklinian rocks.* Seismic reflection and refraction data indicate that in many areas the Ordovician and Silurian (Franklinian) argillite and graywacke encountered in many wells on the central and western North Slope extend offshore beneath the Beaufort and Chukchi Shelves. For example, sonobuoy refraction measurements by Bee and others (1984) from the inner Beaufort Shelf from west of Smith Bay to Prudhoe Bay found uppermost basement velocities of 4.24 to 6.08 km/s, which they interpret to represent Franklinian basement composed of argillite and phyllite. These workers, however, also found sonobuoy refraction velocities of 6.4 to 7.07 km/s from the top of basement in a smaller area of the inner shelf adjacent to 153°W, between Smith and Harrison Bays. The measurements suggest to these workers that the top of basement in this area, which is about 30 km in east-west dimension, consists of crystalline rock, probably silicic in composition. This high-velocity basement may be related to granite found immediately below the basal Ellesmerian unconformity at the nearby East Teshekput well (Bird and others, 1978), which lies near 153°W about 20 km west-southwest of Harrison Bay. The granite underlies Late Mississippian beds, and K/Ar dating and correlation with plutonic events in the northeast Brooks Range suggest that it may be Devonian. If a large negative gravity anomaly in the area of the well (Barnes, 1977) is created by the pluton, then the diameter of the pluton may be on the order of 50 km—commensurate in size with the area of the inner shelf underlain by high-velocity basement.

West of Point Barrow, beneath the northeastern Chukchi Shelf, Mississippian strata at the base of the Ellesmerian sequence are underlain by a broadly folded unit, Pzf, that is at least 7 km thick (profiles 781 and 783 in Fig. 7A and Grantz and others, 1990a, Plate 9) and has sonobuoy seismic velocities of 4.15 to 5.35 km/s (Houtz and others, 1981). These beds, inferred to consist of clastic sedimentary rocks, overlie about 5 km of well-bedded, strongly reflective beds with sonobuoy seismic velocities of 5.7 to 7.3 km/s in the upper part of unit Pzp€ (profiles 781 and 783). The lower part of unit Pzp€ consists of conformable, but less reflective strata that may be 6 km or more thick. Some seismic reflection profiles in the eastern part of this region show that a structural detachment zone in places lie between broadly folded areas of unit Pzf and the underlying, nonfolded, unit Pzp€. This detachment zone can be seen in Figure 18A of Haimela and others (1990) and Plate 3 of Craig and others (1985).

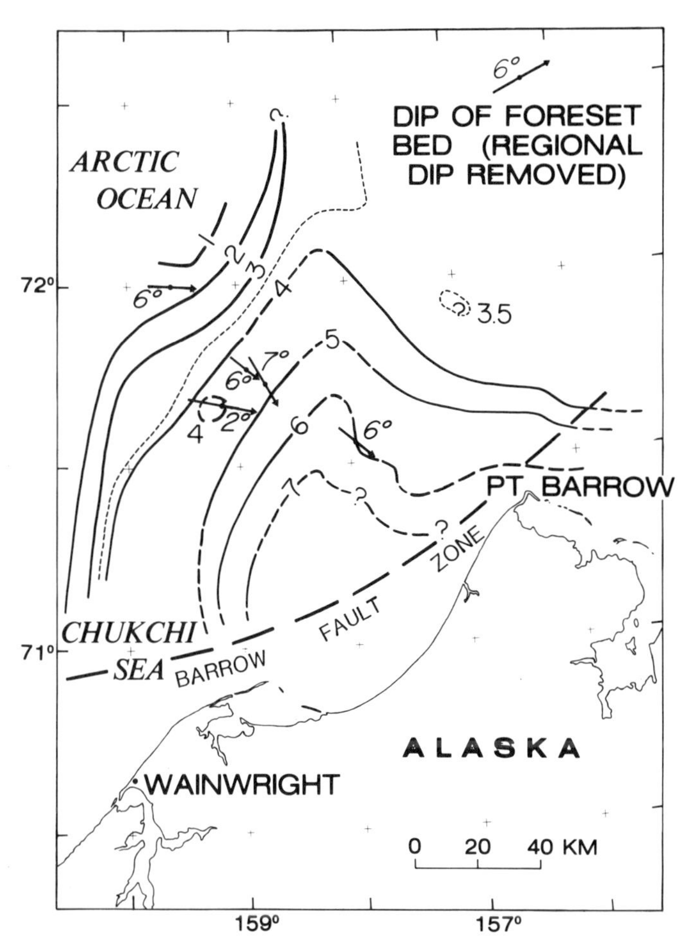

Figure 11. Isopach map (km) of unit Pzf in northeastern Chukchi Shelf showing dip of foreset beds and location of Barrow fault zone.

An isopach map of unit Pzf in the northeast Chukchi Shelf (Fig. 11) shows that the unit thins from more than 7 km near Point Barrow and Wainwright to less than 1 km about 160 km to the northwest. Seismic reflection profiles show that the unit consists of two contrasting facies and that the unit was eroded at an angular unconformity at the base of the PSU. Beyond 60 km west northwest of Point Barrow, where the base of the unit dips 6° or 7° east-southeast, the unit consists of southeasterly sloping interlayered strong to weak reflections, which we interpret to represent interbedded sandstone and lutite in foreset facies. Closer to the coast the dip flattens and the foresets grade into weak reflections, which we infer to represent basinal lutites in bottomset facies. The foreset unit thins from more than 5 km on the northwest to approximately 2.2 km where it grades into the basinal facies about 40 km offshore. This geometry indicates that the sourceland of this unit lay to the northwest, toward the sourceland of Barrovia (Tailleur, 1973).

Correlation of units Pzf and Pzp€ of the northeastern Chukchi Shelf with the North Slope section is uncertain due to the structural contrast between the mildly deformed pre-Mississippian units of the Chukchi Shelf and the strongly deformed and mildly metamorphosed pre-Mississippian strata of the western North Slope. A complicating factor is the presence of a northeast-striking structurally disturbed zone along the northwest coast of Alaska that lies between these areas of contrasting structural intensity. This feature, the Barrow fault zone, is postulated by Craig and others (1985) to be a major northeast-striking, northwest-dipping normal fault with about 10 km of stratigraphic displacement (Figs. 4 and 11).

Craig and others (1985) suggest conditionally that unit Pzf may be "age equivalent to the allochthonous Middle to Upper Devonian rocks of the Baird and Endicott Groups in the west-central Brooks Range." Underlying unit Pzp€ is thought by these workers and by Thurston and Theiss (1987, Fig. 11) to consist of Devonian carbonate and acoustic basement. Grantz and May (1983), on the other hand, suggest that unit Pzf is a mildly structured facies of the Ordovician and Silurian argillite and graywacke of the western North Slope and that the underlying, strongly reflective beds of unit Pzp€ may consist of upper Proterozoic or Cambrian carbonate or metamorphic rocks.

The stratigraphy proposed by Craig and others (1985) requires that the pre-Mississippian section of the northeast Chukchi Shelf be very different than that of the western North Slope. Thus, if offshore unit Pzf consists of Devonian clastic strata, as these authors propose, the correlative units beneath the North Slope would include the Devonian nonmarine clastic rocks encountered at the bottom of the Topagarok test well of the western North Slope (Collins, 1958). The Topogarok rocks can be inferred to rest unconformably on seismically incoherent Ordovician and Silurian argillite and graywacke, whereas unit Pzf rests on a thick section of strongly reflective beds (unit Pzp€) offshore. A contrast in stratigraphy of this magnitude across the Barrow fault zone would require large transcurrent or thrust displacement, and could not be explained by normal displacement, as proposed by Craig and others (1985).

The stratigraphy suggested by Grantz and May (1983) correlates offshore unit Pzf with the Ordovician and Silurian argillite and graywacke of the North Slope and requires that the mild metamorphism and strong deformation that characterize the North Slope rocks die out at the Barrow fault zone. This contrast in deformation could be explained if the Barrow fault zone is a major splay of a regional detachment fault that thrusts more strongly deformed argillite and graywacke of the North Slope against less strongly deformed argillite and graywacke beneath the northeast Chukchi Shelf. The detachment surface between units Pzf and Pzp€ noted above is thought to be the sole fault at which the Barrow fault zone roots. In support of this hypothesis, we note that seismic velocities of unit Pzf in the northeast Chukchi Sea (Vp = 4.15 to 6.0, average = 4.9 km/s; Houtz and others, 1981) are similar to velocities in the argillite and graywacke of the central North Slope (Vp = 4.3 to 4.9, average = 4.6 km/s; Fisher and Bruns, 1987) and the central Beaufort Shelf (Vp = 4.25 to 6.08 km/s; Bee and others, 1984), and that the argillite and graywacke unit is only mildly metamorphosed beneath the North Slope. In addition there is no subsurface or seismic reflection evidence that a well-bedded, high-velocity bedded section more than 7 km thick, such as unit Pzp€, lies between the argillite and graywacke and the unmetamorphosed Devonian and Mississippian nonmarine clastic rocks of the central and western North Slope. We consider the possibility that more than 7 km of unit Pzp€–like rocks once lay between these units on the central and western North Slope, but were subsequently completely removed by erosion, to be unlikely.

Ellesmerian and Brookian rocks. The character of Brookian offshore stratigraphic units in the Barrow sector is shown in profiles 742 to 783 (Figs. 7A, B). Ellesmerian rocks pinch out beneath the inner Beaufort Shelf and the northern Chukchi Shelf to the south of these profiles. Their maximum thickness on the Beaufort Shelf, about 1.3 km, occurs in the Colville Delta–Prudhoe Bay area.

A unit of northward-thickening, moderately strong seismic reflections that is 1.1 km or more thick lies at the base of the sedimentary fill in Dinkum graben (unit J in profiles 742 to 756; Figs. 7A, B). The tectonic setting of these beds at the base of a graben fill suggests that they consist of alternating fine- and coarse-grained clastic materials. The strength of the reflections may corroborate this inference, but other lithologies could also produce such reflections. Near Prudhoe Bay the unit appears to project into upper Ellesmerian strata of the North Slope. We infer that the basal unit of the Dinkum succession in Dinkum graben correlates with the upper Ellesmerian sequence of the North Slope, but it may include lower Ellesmerian beds in paralic facies adjacent to the northern sourceland (Grantz and May, 1983). Overlying the basal beds in Dinkum graben is a unit that produces mainly weak, but some moderate-amplitude seismic reflections. This unit, KJ in Figures 7A and B, is 3.5 km or more thick. Reflection character suggests that the unit consists of topset lutites and some sandstone. Bedding in unit KJ appears to be parallel to that in the underlying, better bedded unit J, and therefore, unit KJ may also consist of topset beds deposited in the subsiding graben. Possible south-dipping foreset beds in the unit on line 751 may indicate that some of these beds are deep-water deposits of northerly provenance. Northward thickening of the Dinkum succession and many of its subunits indicates deposition in a subsiding, north-tilting half graben.

A well-developed angular unconformity, which cuts down-section to the south, forms the upper contact of the Dinkum graben succession (profile 751, Fig. 7B). This is the breakup unconformity of Hauterivian age. The stratigraphic position of this unconformity indicates that the extension represented by the Dinkum succession and graben was an older event than the breakup unconformity and the extension that opened the Canada Basin. A Jurassic and Neocomian age is inferred for the Dinkum succession because it underlies the Hauterivian unconformity; because the succession can be projected into upper Ellesmerian rocks of the North Slope; and because Jurassic marine sedimentary rocks in an analogous stratigraphic and structural position crop out on strike in coastal northern Yukon Territory (Norris, 1984; Poulton, 1978).

A poorly observed, generally deep, northward-thickening seismic reflection unit (J) lies between unit Pzf and the breakup unconformity beneath the outer shelf and slope north of the Dinkum graben. Based on its stratigraphic position and similarities in reflection character, we correlate these beds with unit J in Dinkum graben (Figs. 7A, B). Unit J of the outer shelf produces mainly weak, but some moderately strong seismic reflections with apparent topset morphology. The unit is more uniform in thickness than unit KJ, and is more than 1.3 km thick northwest of Point Barrow and 1.9 km thick north of Prudhoe Bay. We speculate that unit J of the outer shelf was deposited on a subsiding shelf or basin floor in a graben that was tectonically related to, and coeval with, the Dinkum graben. We have, however, identified only north-dipping extensional faults, and the existence of the graben is only inferred from these faults and the position of unit J beneath the continental margin. Correlation of unit J of the outer shelf with that in the Dinkum graben is supported by the proximity of these beds across faults on the north side of the graben on profiles 742 and 749 (Fig. 7B). Unit J is an important marker for measuring the effects of down-to-the-basin rift faulting beneath the outer shelf, and it appears to be the preferred detachment surface for listric normal faults beneath the outer shelf and slope.

Units PSU and HRZ are lumped as seismic unit LK on profiles from the Beaufort and northern Chukchi Shelves (Figs. 7A and 7B). This unit thickens from an average of about 50 m (range 20 to 70 m) beneath the northern Chukchi Shelf (profile 783 in Fig. 7A) to between 400 and 500 m on profiles in the eastern part of the Barrow sector, where it consists of bottomset beds. The age of LK in the western part of the Barrow sector, where it is overlain by the Torok Formation, is Hauterivian to Aptian(?) and possibly early Albian. However, the top of the unit gets progressively younger as it thickens to the east, the direction in which the foreset facies of the Torok and younger formations also pass into bottomset facies. Near the Canning River, in the easternmost part of the Barrow sector, unit LK may include beds as young as Paleocene. Onshore, Molenaar and others (1986, 1987) assign post-PSU bottomset beds, which are equivalent to the upper part of unit LK, to the Hue Shale.

Units K and T in the seismic profiles of Figures 7A and B represent lower Brookian (Cretaceous) and upper Brookian (Tertiary) deposits of the Colville Basin on the North Slope and Chukchi Shelf and the prograding prodelta and intradelta rocks in the continental terrace sedimentary prism of the Nuwuk Basin. The boundary, which was roughly projected from the North Slope, is useful to illustrate structure but it is oblique to the lithologic and facies boundaries in these rocks. The subdivision was effected by roughly projecting the Cretaceous-Tertiary boundary offshore from the North Slope. In both units K and T, topset and foreset facies in the continental shelf pass into bottomset (basinal) facies in the Canada Basin to the north. The approximate position of the interfingering topset-foreset boundary, which is also the contact between the Canning and Sagavanirktok Formations of Molenaar and others (1986), is shown on profiles 742 to 767 (Figs. 7A, B).

On the western North Slope and central Chukchi Shelf, unit K consists of the Torok Formation and the overlying Nanushuk Group. The unit is oldest (Aptian?) in the southern part of the Colville Basin and becomes younger northward by downlap onto unit LK. It is youngest on the crest and north flank of the Barrow arch, where the basal beds are Albian. On the Beaufort Shelf and the northern Chukchi Shelf the uppermost part of unit K also includes topset and foreset beds of the Colville Group. East from Point Barrow, bottomset beds at the base of unit K increase in thickness, the foreset beds at the top of the unit decrease in thickness, and the boundary between these lithofacies becomes younger to the east. Beneath the inner shelf east of Prudhoe Bay the Cretaceous-Tertiary boundary passes from foreset to bottomset facies, and the thickness of unit K is reduced to only about 200 m. Between Prudhoe Bay and Flaxman Island the entire Canning Formation becomes Tertiary (unit T), and unit K is represented only by the condensed bottomset facies of the Hue Shale. The Canning is entirely Tertiary in age beneath the eastern Beaufort Sea.

Sedimentation in unit T of the upper Brookian sequence was a continuation of progressive progradational continental terrace sedimentation established at the beginning of Brookian time. As a result, foreset facies are predominant in unit K, and topset facies in unit T beneath the continental shelf of the Barrow sector, as suggested by bedding traces in Figures 7A and B. Unit T correlates with the Sagavanirktok Formation of the North Slope west of Prudhoe Bay, but only with its upper part to the east (Molenaar and others, 1986; Kirschner and Rycerski, 1988). The lower part of unit T east of Prudhoe Bay belongs to the Canning Formation, which encloses a thick intertongue of the Sagavanirktok Formation. The Gubik Formation, which caps the section on the Beaufort Shelf, is included in unit T.

Structure. Contours on the top of the Franklinian sequence (Fig. 6) and on the base of the Brookian sequence (Fig. 12) illustrate the regional geologic structure of the Barrow sector. The Franklinian surface slopes seaward with gradients of 30 to 100 m/km from a broad culmination at the crest of the Barrow arch to the tectonic hinge line of Early Cretaceous rifting beneath the outer shelf. At the hinge line the gradient increases abruptly to 250 to 500 m/km, which carries the surface to the lower limit of our seismic data (12 km below sea level) beneath the slope. In places, down-to-basin normal faults of modest displacement offset the Franklinian surface at the hinge line. These faults char-

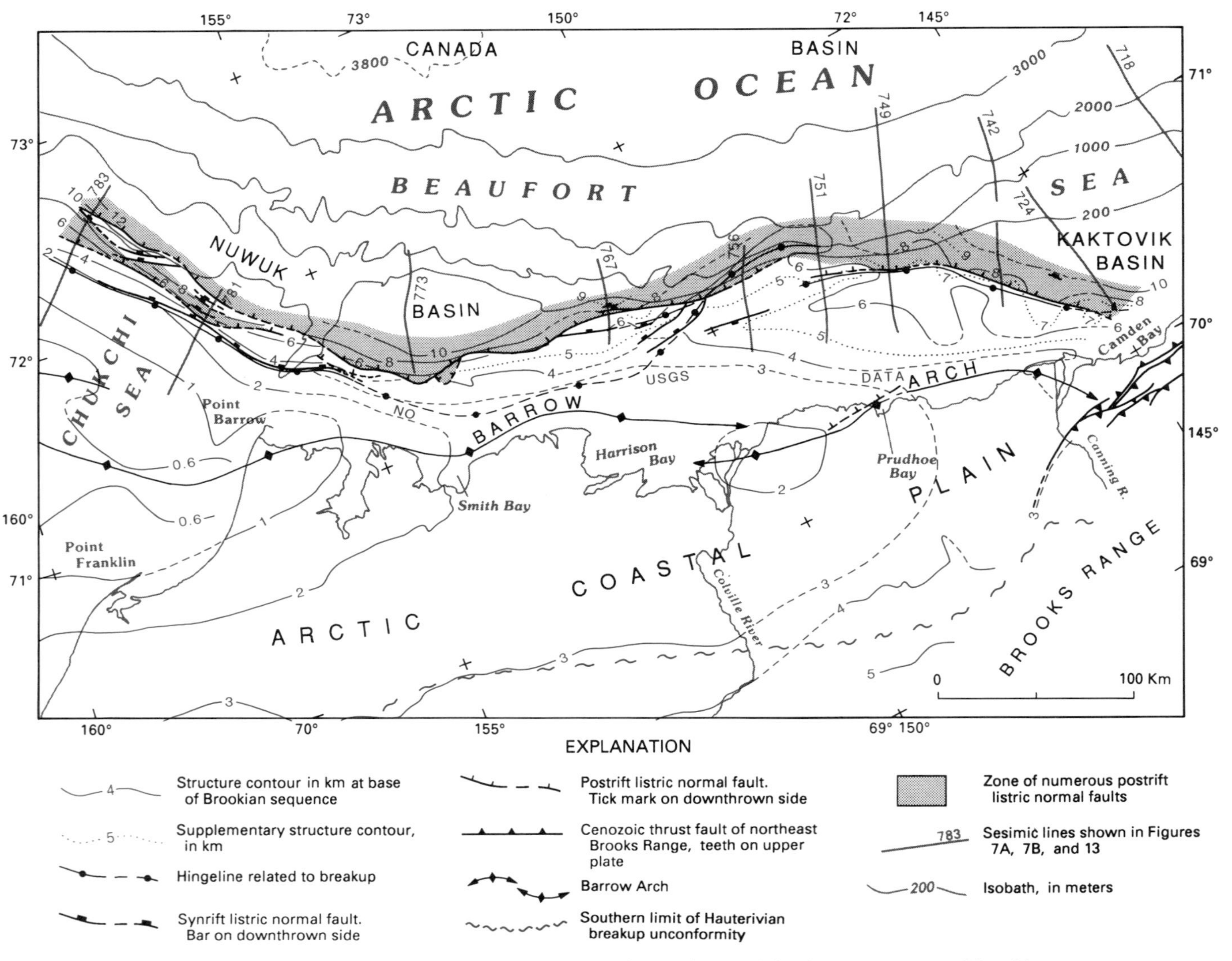

Figure 12. Map of Arctic Alaska continental margin showing depth in kilometers to base of Brookian sequence, which approximates position of Hauterivian breakup unconformity, and synrift and postrift structures. Data sources as for Figure 6. Lines dashed where projected or speculative.

acteristically are overstepped by the breakup unconformity or extend above it only a short distance into the basal Brookian strata. These faults are therefore related to the rifting events that immediately preceded and accompanied breakup. Seaward of the hinge line the Franklinian surface is offset by down-to-basin normal faults with displacements of 1.5 to 2 km. In places in the eastern part of the Barrow sector, the faults of this set have almost inverted the Dinkum graben by down-dropping rocks on the north flank of this structure below the position of their counterpart units within the graben to the south.

A 50- to 75-km northward excursion of the Early Cretaceous tectonic hinge line at the Colville River delta with respect to its position to the west (Figs. 4 and 6) reflects the influence of extension at the Dinkum graben. Extension at the graben dies out westward in the area of the excursion, suggesting that the excursion is a result of spreading at the graben. The Dinkum fault, which bounds the half graben on the north, has a vertical displacement exceeding 4 km. The principal displacement was prebreakup (pre–late Hauterivian) but the fault is also the locus of early post-breakup grabenward downflexing and minor faulting (seismic profiles 749 and 751 of Fig. 7B). Dinkum horst, which lies between the graben and the seaward-sloping Franklinian surface north of the hinge line, plunges gently east, as does the floor of the graben, at least as far east as the Canning River. East of the Canning, in the Barter Island sector, the horst and graben pass beneath the thick sedimentary section of the Kaktovik Basin and are lost on our seismic reflection records. The east-southeast trend of the graben would, if not deviated, carry its axis to the coastal plain near Barter Island. Smaller rift features of the same generation as the graben also offset the top Franklinian surface southwest and west of the Dinkum graben, in an area that is shoreward of our seismic coverage (Craig and others, 1985, Fig. 9).

The northward excursion of the hinge line trend (Fig. 4) at the west end of the Dinkum graben forms the east boundary of the middle Early Cretaceous to Tertiary Nuwuk Basin of the central and western parts of the Barrow sector. West of this excursion the width of the sedimentary prism between the hinge line and the shelf break increases from 5 km at profile 756 (Fig. 7A) near the west end of the Dinkum graben to more than 50 km at profile 783 (Fig. 7A), 80 km northwest of Point Barrow. The progradational continental terrace strata of the Nuwuk Basin are virtually undeformed and slope gently seaward south of the hinge line. North of the hinge line these strata are offset by numerous down-to-the-basin listric normal faults, and the structure is dominated by a broad hanging-wall rollover anticline with as much as 1.5 km of structural relief. The rollover formed above a system of north-dipping listric growth faults and some associated south-dipping normal faults that reach nearly to the seabed and sole out at or near the breakup unconformity (see profiles 756 to 783 in Fig. 7A). The growth faults die out within the continental rise sedimentary prism.

North of Dinkum graben, where the hinge-line/shelf-break interval is in places as narrow as 5 to 10 km, the hanging-wall rollover anticline is less well developed and is extensively dismembered by down-to-the-basin listric normal faults. Where the hinge-line/shelf-break interval widens again east of the graben, a broad, high-amplitude rollover anticline is again present beneath the outer shelf (vicinity of CDP 1,500, profile 724, Fig. 13).

Barter Island sector

Large thrust and detachment folds of Cenozoic age with northward vergence characterize the Cretaceous and Cenozoic (Colville Basin) strata of the Barter Island sector (Plates 9 and 11 and Fig. 14; Grantz and others, 1990b, Fig. 11). These structures root in thrusts superimposed on the Middle Jurassic to Albian convergent structures that regionally are the dominant features of the Brooks Range orogen. Historic earthquakes along the Camden anticline beneath the continental shelf, folding and faulting of Quaternary sediment beneath the shelf and the Arctic Coastal Plain, and warping of the seabed indicate that this compression is still active.

The thrust and detachment folds developed over a large, multistrand detachment fault system that is projected to lie between 5 and 10 km below sea level beneath the continental slope and that dies out in small folds and back thrusts beneath the continental rise about 170 km north of the coast (Fig. 3 in Grantz and others, 1990b; Reiser and others, 1980; Bruns and others, 1987). The western limit of the thrust and detachment folds and related faults, and of the Barter Island sector, is the seismically active Canning displacement zone. The eastern limit is the seismically active eastern front of the Richardson Mountains, in the lower Mackenzie River valley. Note that the Canning displacement zone is a northward extension of the Aleutian Benioff zone (Grantz and others, 1990a, Plate 11), which is of Cenozoic age. This extension and the orientation of the thrust folding and faulting suggest that Cenozoic compression in the Barter Island sector originated in convergence between the Pacific and the North American plates at the Aleutian megathrust. Cenozoic thrusting in the northeast Brooks Range is estimated to have carried older rocks now exposed in the range 25 to 50 km or more northward across the Colville Basin beds of the Barter Island sector. The basin may lie near its original position beneath the thrust sheets, but this has not been documented in seismic reflection profiles across the Arctic Coastal Plain and Foothills (see Bruns and others, 1987). Crustal loading by Cenozoic nappes in the northeast Brooks Range and accelerated erosion and clastic sedimentation from the newly uplifted nappes is also postulated to have depressed the foreland to create the Kaktovik Basin of the Barter Island sector continental margin (Fig. 4).

Stratigraphy. Platform carbonate and clastic strata more than 4.7 to 5.1 km thick, which crop out in south-dipping thrust-fault panels in the Sadlerochit and Shublik Mountains (Blodgett and others, 1986, 1988; Robinson and others, 1989), are the oldest rocks observed in the Barter Island sector. As discussed above under Stratigraphy, Arctic Platform, these beds are Late Proterozoic to Devonian in age. The platform rocks are inferred from outcrop data to rest on a large sole fault system of northern vergence that is rooted in the Brooks Range orogen.

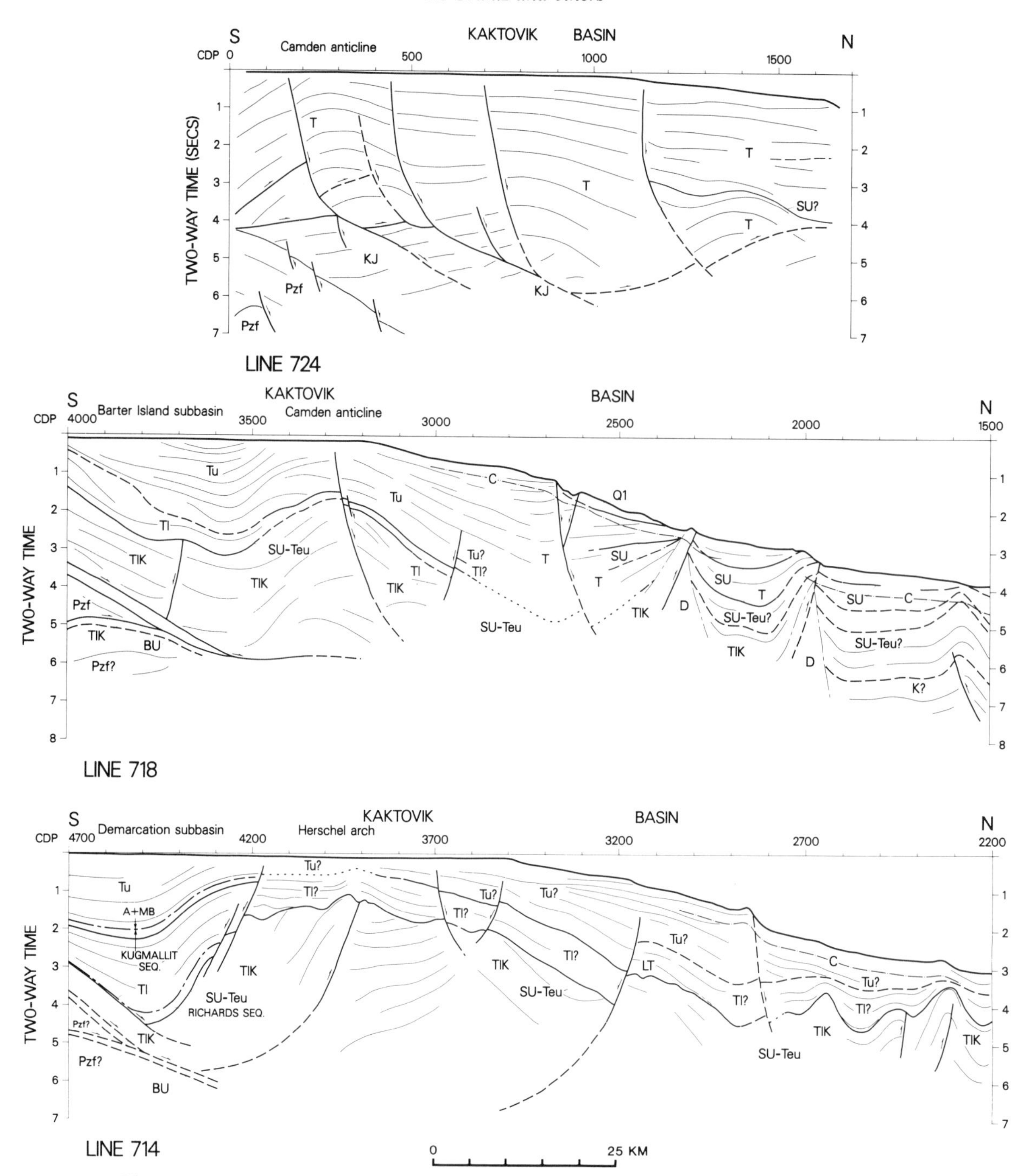

Figure 13. Line drawings showing geologic structures and stratigraphy on seismic reflection profiles 714, 718, and 724 across the Arctic Alaska continental margin. See Figure 7 for explanation and definitions. Sequence names on section 714 extrapolated from Dietrich and others (1989). Time sections from Grantz and others (1982).

East of the Sadlerochit and Shublik Mountains the oldest strata beneath the Arctic Coastal Plain are pre-Mississippian bedded rocks observed on seismic reflection records (Fisher and Bruns, 1987). These rocks have generally low dips, are 5 to 7 km thick, and are apparently disrupted by low-angle thrust faults. Offshore the pre-Mississippian bedded rocks have been tentatively identified at the south ends of profiles 714 and 718 (Fig. 13), where they have been designated unit Pzf (?). In thickness and reflection character this unit and the pre-Mississippian beds beneath the coastal plain resemble the well-bedded clastic rocks of inferred early Paleozoic age in the northeastern Chukchi Shelf (unit Pzf in profiles 781 and 783 in Fig. 7A). We suggest that all of these beds belong to the argillite and graywacke terrane of the central and western North Slope, and that they are a basinal

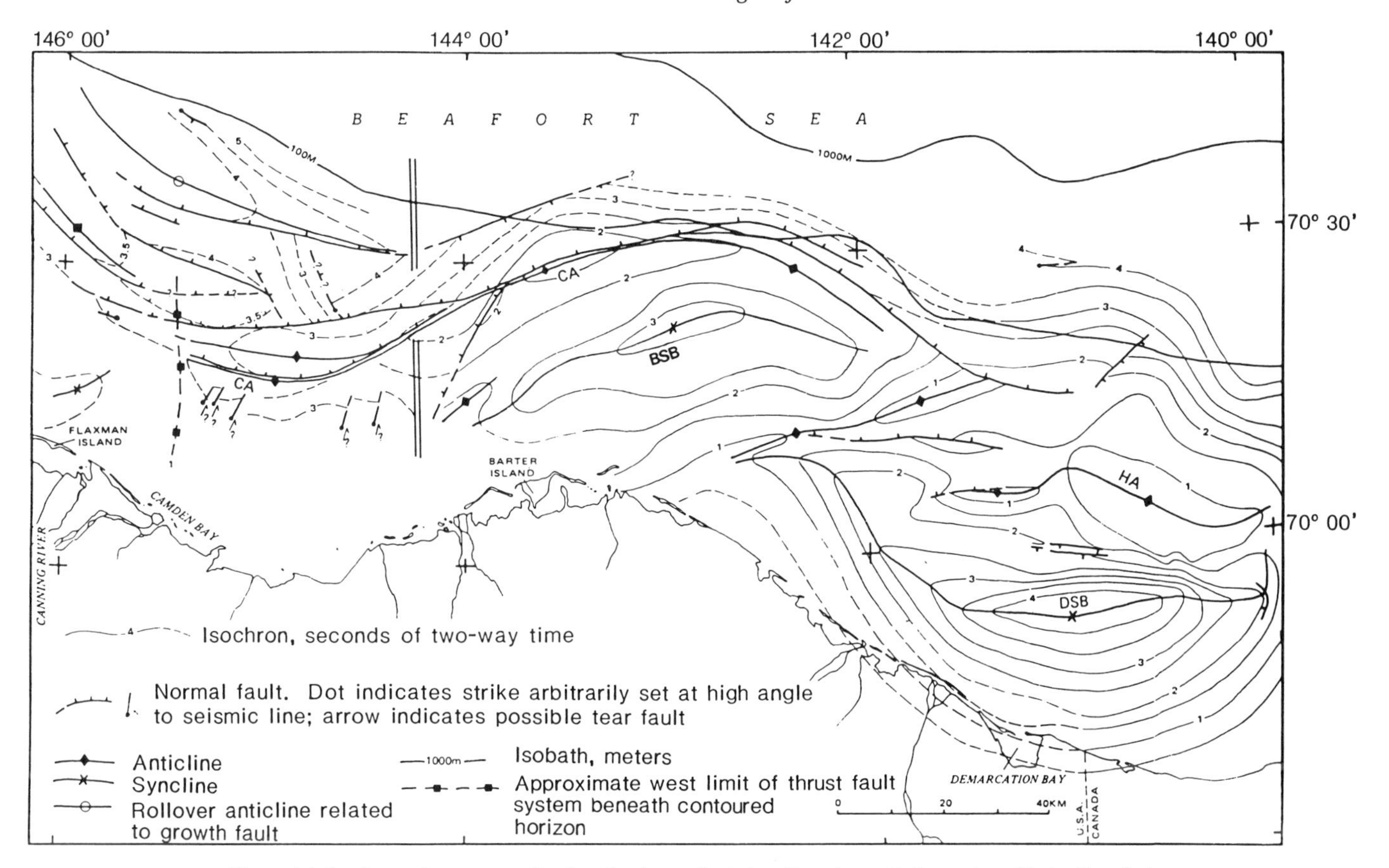

Figure 14. Isochrons (two-way reflection time) on selected surfaces beneath the eastern Alaska Beaufort Shelf; dashed where inferred. See Figure 2 for approximate equivalence of isochrons and depth. Contours are on middle Eocene unconformity east of 144°15′W and on a middle(?) Eocene horizon west of 144°15′W. Contours east of 141°W by James Dixon and James Dietrich (written communication, 1985). CA, Camden anticline; HA, Herschel arch; BSB, Barter Island subbasin; DSB, Demarcation subbasin.

facies of part of the Late Proterozoic to Devonian platformal succession of the Sadlerochit Mountains and vicinity. The north vergence of the platform rocks in the Sadlerochit and Shublik Mountains and their generally high structural position with respect to the argillite and graywacke terrane on either side suggest that the platform rocks have been thrust northward over the argillite and graywacke. The distribution of the platform rocks suggests that they, and any thrust faults that separate them from the argillite and graywacke terrane, extend beneath the continental shelf near Camden Bay and the Canning River.

The simple progradational depositional geometry of the Brookian sedimentary prism in the Barrow sector is replaced in the Barter Island sector by a complex of local Tertiary uplifts, erosional unconformities, and local sedimentary subbasins. Figure 14 shows the largest of these features. Camden anticline and Herschel arch are interpreted as large detachment folds, and the Barter Island and Demarcation subbasins as genetically related syndepositional basins. The cross-sectional geometry of these features is shown in profiles 724 to 714 in Figure 13. Growth of the detachment structures began in middle Eocene time, as dated by an erosional unconformity that formed in response to their initial growth. The unconformity, which lies between seismic-stratigraphic units TlK and Tl in profiles 714 and 718, was dated by correlation with a seismic-reflection event dated in test wells on the Canadian Beaufort Shelf (James Dixon and James Dietrich, 1990 and personal communication, 1986; Dietrich and others, 1989). It is the major stratigraphic break within the Brookian strata of the Barter Island sector. Above it lie stratigraphically simpler, but more complexly deformed units, and below it stratigraphically more complex, but less deformed units (see profiles 714 to 724 in Fig. 13, and profile 9C in Grantz and others, 1990a, Plate 9).

The character and age of the strata that lie between Franklinian basement and the middle Eocene unconformity (unit TlK in Fig. 13) are uncertain. Reflection character suggests that this unit consists of interbedded shale and sandstone or siltstone, and sonobuoy seismic refraction velocities in the range of 3.0 to 3.8 km/s suggest that the strata are Mesozoic or Paleogene, and not Paleozoic. Regional considerations and comparison with seismic profiles beneath the adjacent coastal plain (Bruns and others,

1987) suggest that near the coast the oldest strata of the unit are basal Brookian. Farther north, however, the oldest strata may, in places, belong to the Dinkum succession and contain Jurassic beds. The uppermost beds of unit TlK are middle Eocene. An unconformity is inferred to lie between unit TlK and the underlying Franklinian basement, but the unconformity is not clearly seen on the seismic profiles and its position in Figure 13 is in part speculative. The thickness of unit TlK may locally exceed 8.6 km, but its true thickness is masked by structural complexity.

Comparison with seismic-reflection profiles tied to test wells on the western Canadian Beaufort Shelf (Dixon and Dietrich, 1990; profile 8E, in Grantz and others, 1990a, Plate 8; and Dietrich and others, 1989) suggests that the upper part of unit TlK corresponds to Maastrichtian to middle Eocene strata of intertonguing intradelta and prodelta facies (Fish River and Reindeer depositional sequences on the Canadian shelf; Tent Island, Moose Channel, and Reindeer Formations of northern Yukon—see Fig. 5). The lower part of unit TlK correlates with marine shales of the Jurassic Kingak Formation (Canadian usage), a prodeltaic shale in the late Hauterivian to Aptian Mount Goodenough Formation, Albian flysch composed of detritus from the Brooks Range orogen, and organic-rich shale of the late Cenomanian to Turonian Boundary Creek Formation of northern Yukon. The Boundary Creek contains bentonite beds and ironstone concretions and correlates with lithologically similar beds in the Hue Shale and Seabee Formation of the North Slope.

Dixon and Dietrich (1990) report that upper Hauterivian strata at the base of the Mount Goodenough Formation rest on a regional unconformity that records a significant tectonic event. The character and age of the unconformity suggest to us that it correlates with the Hauterivian breakup unconformity of the Alaskan margin.

Comparison of unit TlK with the stratigraphic section in coastal wells in the eastern part of the Barrow sector indicate that it corresponds to the condensed shale of the PSU, the condensed distal shale of the Hue Shale, and the lower part of the Canning Formation. If the possible presence of Dinkum succession strata beneath the Beaufort Shelf and slope is excluded, unit TlK ranges in age from Hauterivian to middle Eocene.

Local depocenters and sedimentological complexity characterize the strata that overlie the middle Eocene unconformity in the Kaktovik Basin (units Tl and Tu in profiles 714 and 718 in Fig. 13; profile 9C in Grantz and others, 1990a, Plate 9). The largest local depocenters, the Barter Island and the Demarcation subbasins, lie on the south side of the Camden anticline and the Herschel arch detachment structures (Fig. 14). Narrower and shallower linear depocenters formed behind the ten or more long thrust folds that buckle the Cenozoic strata, and in some places the seabed, beneath the adjacent continental slope and rise. A map of these thrust folds is shown in Figure 11 of Grantz and others (1990b), and the folds are shown in cross section in profiles 714 and 718, Figure 13, and profile 9C, Grantz and others, 1990a, Plate 9.

Angular bedding discordances and thickness changes in the synclinal basins that lie upslope of Herschel arch, Camden anticline, and the thrust anticlines of the shelf and slope indicate that these basin are syntectonic with detachment and thrust folding. Total section in the Demarcation subbasin exceeds 7 km. Of this, 4 km is fill in the initial depression and 3 km is overburden. Thickening of strata into the subbasin, most strongly in the Kugmallit(?) sequence, exceeds 4 km across the south limb and 3 km across the north limb. Fill in the Barter Island subbasin is at least 1.4 km, and the overburden is more than 2.2 km thick. Fill in the basins upslope of the thrust folds of the continental slope is 0.1 to more than 1.0 km thick. The geostatic load created by the trapped sediment appears to have generated soft-sediment flow in underlying unit TlK, causing some of it to move from beneath the Demarcation subbasin, and perhaps the Barter Island subbasin, into the uplifted cores of Herschel arch and Camden anticline (profiles 714 and 718 in Fig. 13).

The basin fills in the Barter Island subbasin, and less clearly in the Demarcation subbasin, are progradational and consist of a lower unit dominated by foreset beds and an upper unit dominated by topset beds. The facies boundary between these units becomes younger to the north. We infer that this boundary, which can be observed on profile 718, Figure 13, corresponds to the similar boundary between the Canning and Sagavanirktok Formations of the Barrow sector. On our seismic records we are unsure whether the inferred Canning beds extend north of Camden anticline or Herschel arch, but locally the Sagavanirktok appears to rest directly on the middle Eocene unconformity over and north of these structures.

Correlation with data in Dixon and Dietrich (1990) and Dietrich and others (1989) indicates that the Demarcation subbasin contains the Richards and Kugmallit sequences and possibly the Mackenzie Bay, Akpak, and Iperk sequences of the Canadian western Beaufort–Mackenzie Basin (see profile 714 in Fig. 13). The Richards is middle to upper Eocene interbedded nearshore sandstone and mudstone less than 0.3 to more than 0.5 km thick; the Kugmallit about 0.8 to 1.0 km of lower and middle Oligocene mud-dominant prodelta and shelf sediment; the Mackenzie Bay as much as 2.0 km of upper Oligocene and lower Miocene prodelta sediment, and the Akpak middle and upper Miocene prodelta deposits that in places are close to 2.0 km thick. The Pliocene and Pleistocene Iperk sequence, unconsolidated gravel, sand and mud with abundant woody detritus, is about 0.2 km thick. In the Demarcation subbasin the Richards is well bedded, about 0.8 km thick, and appears to consist of topset (shelf) facies deposited on the middle Eocene unconformity. The overlying Kugmallit(?) beds consist of northward prograding topset and foreset beds that are 1.3 km thick about midslope on the south flank of the subbasin, about 2.4 km or less in thickness high on the north flank, and at least 3.8 km thick in the axial region. A correlation based on seismic reflection character suggests that an interval of strong reflections near 2.2 s in the axial region of the Demarcation subbasin (profile 714) may mark the base of the Mackenzie Bay sequence, and the base of an interval of weaker reflections near 1.05 s may mark the base of

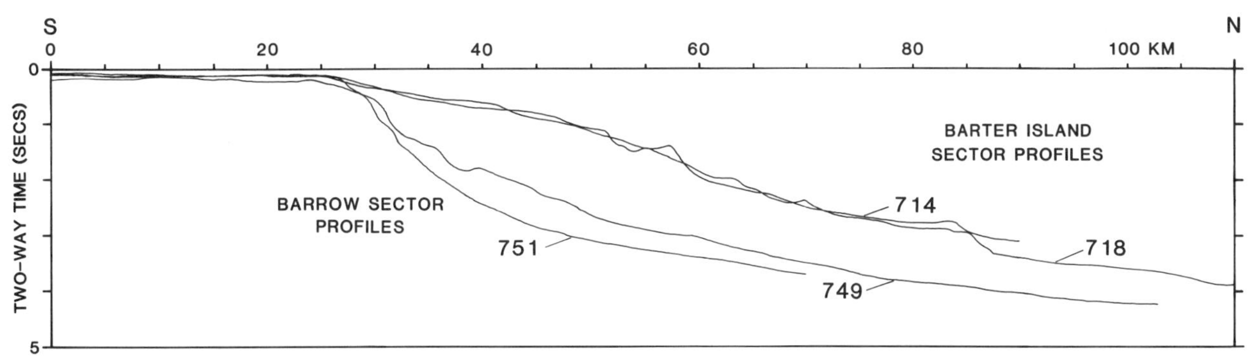

Figure 15. Comparison of seabed morphology across Arctic continental margin in Barter Island sector (profiles 714 and 718 of Fig. 13) and Barrow sector (profiles 749 and 751 of Fig. 7B). Area of prism between the two sets of profiles, equivalent to a tectonic welt almost 1 km high and 60 km wide, is a rough measure of minimum volume of Cenozoic tectonic transport across the convergent continental margin of the Barter Island sector with respect to the extensional (passive) margin of the Barrow sector.

the Akpak sequence. The Mackenzie Bay(?) interval is about 1.8 km thick and consists of mainly strongly bedded topset beds. The Akpak is about 0.7 km thick and also appears to consist of topset beds, but the reflections are obscured by multiples. If the Iperk sequence is present, it is obscured by multiples.

Structure. Convergent structures of Cenozoic age characterize the Barter Island sector and distinguish it from the Barrow sector, where convergent tectonics are absent. The convergence has also given these continental margin sectors contrasting morphologies, which are illustrated by the superimposed bathymetric profiles shown in Figure 15. The profiles across the slope of the Barrow sector are concave upward, recording the influence of extensional processes, whereas those across the Barter Island sector slope are convex upward, reflecting the influence of convergence. The cross-sectional area between the two sets of profiles extrapolated across the length of the Barter Island sector represent a bulge in the face of the continental terrace equivalent to a tectonic welt almost 1 km high, 60 km wide, and 500 km long. The average horizontal distance between the concave and convex profiles, about 30 km, is a rough measure of the minimum tectonic transport across the convergent Barter Island sector with respect to the extensional Barrow sector in Cenozoic time. An overview of the compressional structures of the Barter Island sector and their relation to regional structure is shown in Grantz and others (1990a, Plate 11.) Maps showing the character of the convergent structures are presented in Figure 14 of this chapter and Figure 11 of Grantz and others, 1990b.

Four types of Cenozoic structures, which lie in discrete, east-west–trending belts, are dominant in the upper crust of the Barter Island sector. On the south, beneath the inner and middle shelf, are the deep Barter Island and Demarcation subbasins (Fig. 14, profiles 714 to 724 in Fig. 13, and profile 9C in Plate 9). These basins are paired with broad, northward-convex, en echelon detachment folds, the Camden anticline and Herschel arch, that underlie the middle and outer shelf. North of the detachment folds, beneath the continental slope and rise, is a wide zone of slope-parallel thrust folds of large amplitude. Many of these folds buckle the sea floor and act as sediment traps. Beneath the outer shelf and upper slope are basinward-dipping listric normal faults, which are prominent only in the westernmost part of the sector (Figs. 6 and 12 and profile 724 in Fig. 13). The largest of these dip north and are related to deep slumping of Quaternary age near the present-day continental margin, but a few dip south.

Key to understanding the structure of the continental margin in the Barter Island sector is the character of Herschel arch and Camden anticline (profiles 714, 718, and 724 in Fig. 13, profile 9C in Grantz and others, 1990a, Plate 9). These folds are asymmetric, with a short south limb that dips landward at a large angle to the sea floor and a long north limb that dips seaward only a little more steeply than the sea floor. At the top of unit TlK their cross sections resemble south-facing monoclines at the short south limbs of the folds. The amplitudes of the monoclines at this horizon are 6.75 km for Herschel arch and 3.0 km or more for Camden anticline. The arch, which lies beneath the inner and central shelf, contains structural duplexes within unit TlK. Thrust-fault panels in the core of Herschel arch have a moderate south dip, which suggests that the structure was initiated in the regional, north-vergent compressional fault system of northeastern Alaska. Multiple structural culminations of the unconformity are present at the crest of the arch and must have been uplifted to wave base or the sea surface because they were the source of local sediment aprons. The culminations appear to represent small structural uplifts of the middle Eocene sea floor (see vicinity of CDP 3900, profile 714, Fig. 13). The culminations and the associated sedimentary aprons suggest that thrusting and uplift were penecontemporaneous with erosion and that the unconformity is a product of the early and middle Eocene convergence. The great disparity in thickness of unit TlK between Herschel arch and the basement for the Demarcation subbasin (Fig. 13) suggests that post-unconformity soft sediment flow in unit TlK from beneath the subbasin into the core of the arch may also have contributed to the large structural relief (6.75 km) across its south limb.

Camden anticline, in contrast to Herschel arch, appears to be a relatively simple buckle above a north-sloping detachment fault. In the central part (profile 718 in Fig. 13) the detachment fault has a relatively simple geometry and lies about 10 km below sea level at the fold axis. The amplitude of its south flank here is about 3 km. On the west (profile 724 in Fig. 13) the basal detachment fault system is a little more complex and it lies only about 6 km below sea level at the fold axis. The anticline here has an amplitude at the south flank of 4 km or more, but it is disrupted by several down-to-the-basin listric normal faults that merge with the detachment fault system at depth.

The thrust folds that lie north of Herschel arch and Camden anticline differ greatly from these structures in size and structural character. More than ten thrust folds have been identified beneath the continental slope and rise (profiles 714 and 718 in Fig. 13 and profile 9C in Grantz and others, 1990a, Plate 9; Fig. 11 in Grantz and others, 1990b). These structures are slope-parallel, range from a few tens of meters to at least 1.4 km in amplitude, and have wavelengths of 5 to 20 km. The folds are relatively narrow, elongate, and typically separated by broad, flat-bottomed synclines, in which respects they resemble detachment folds. Most are north vergent and appear to be cored by south-dipping thrusts, but a few are unfaulted or have north-dipping back thrusts in the core. Many of the folds have buckled the seabed, and some have acted as sediment traps in which hundreds of meters of sediment has accumulated. Additional discussion of these features can be found in Grantz and others (1990b).

Demarcation and Barter Island subbasins are compact, ovate extensional subbasins that lie en echelon on the landward side of Herschel arch and Camden anticline (Fig. 14). The Demarcation subbasin is 7.5 km deep, and the Barter Island subbasin a little more than 4 km deep (Fig. 13). Their north flanks, which lie against the arch and anticline, are steeper than their south flanks, and the north flank of the Demarcation subbasin is broken by south-dipping normal faults. Bedding is parallel to the subbasin floors or onlaps them in a southerly direction at low angles on the south flanks of the subbasins. In contrast, many beds onlap the floors of the north flanks of the subbasins and do so in a northerly direction and at steeper angles to the subbasin floors. In the Demarcation subbasin the lower part of the subbasin fill (upper Eocene and Oligocene beds of the Richards and Kugmallit sequences) thickens from less than 0.6 km over the south flank and less than 0.2 km over the north flank to 3.5 km at the subbasin axis. Some additional thickening occurred in higher beds, in the central part of the subbasin section, and part of the thinning on the north flank is due to normal faulting.

The regional geologic structure of northeastern Alaska, including the geologic structure of the Arctic Coastal Plain (Bruns and others, 1987), shows that the Cenozoic structures of the shelf in the Barter Island sector formed in a compressional environment. It therefore seems anomalous that the Demarcation and Barter Island subbasins are extensional features, that cross sections of Herschel arch and Camden anticline resemble south-facing monoclines, and that these large features lack the strongly compressional structural overprint of the thrust-fold belt of the adjacent continental slope and rise. Our seismic data do not penetrate deeply enough to fully resolve this anomaly, but the following hypothesis may explain our observations. We suggest that the growth of Herschel arch began in early or middle Eocene time as a compressional structure above a sole fault that extended from beneath the Brooks Range to the continental slope. As the sole fault passed from the Brooks Range to the Eocene shelf and continental margin, we propose that its dip changed from low south to low north. Compressionally thickened unit TlK was unstable where it overlay the northward-sloping portion of the sole fault beneath the continental margin, and we postulate that it began to glide slowly basinward under body forces. The rate of displacement was moderated by the resistance offered by sediment down-slope, in front of the gliding block. The space vacated by the block guide created the proto-Demarcation subbasin in late Eocene and Oligocene time. Thick and rapid Oligocene sedimentation in the subbasin may have mobilized the underlying unit TlK, facilitating the gliding and perhaps transferring some material to the glide block. An angular discordance at the base of the fill beneath the center and south flank of the basin (profile 714, Fig. 13) is interpreted as a detachment fault on which some of the gliding took place. Camden anticline and Barter Island subbasin are thought to be analogous to Herschel arch and the Demarcation subbasin; there was less thinning of unit TlK beneath the Barter Island subbasin, however, and its structure and stratigraphy indicate that this subbasin did not originate until late Cenozoic time.

Most of the thrust folds beneath the slope and rise are Neogene or Quaternary in age, but some were active during mid-Tertiary time. We suggest that these north-vergent compressional structures were formed in consequence of both middle and late Cenozoic movement on the Herschel arch block glide. Large-scale folding of the entire fill in both the Demarcation and Barter Island subbasins indicates that both subbasins were also affected by strong late Cenozoic tectonic movement. The geometry of this movement is not clear from our data. Presumably it would involve continued transmission of north-vergent compression to the rocks above the detachment faults of the Barter Island sector, moderately large displacements on the basal detachment faults where they dip north, and continued thickening and buckling of unit TlK above the detachment faults in both Herschel arch and Camden anticline. Distally, the displacement is transmitted to, and expended within the thrust folds that buckle the sea floor beneath the continental slope. The earthquake and aftershock swarm of 1968 on the Beaufort Shelf in the western part of the Barter Island sector suggest that this activity is continuing (Biswas and others, 1977).

The geometry of Herschel arch and Camden anticline suggest that these large structures formed under similar paleogeographic and structural conditions. If the Demarcation subbasin was created by block gliding, its 30 km width, is a measure of the minimum seaward displacement of the postulated Herschel arch glide block. Likewise, if the Barter Island subbasin was formed by block gliding, its width indicates that horizontal displacement of the Camden anticline glide block was also 30 km or more. The geometry and trend of the anticline indicate that its west end was fixed and that its initial position on the east was parallel to the coast about 10 or 20 km seaward of Barter Island.

TECTONIC HISTORY

The Arctic Ocean Basin is closely surrounded by the major continents of the Northern Hemisphere, which are characterized tectonically by numerous, large convergent orogens of Paleozoic, Mesozoic, and Cenozoic age. Yet, paradoxically, the Arctic Basin was formed by a series of extensional, or rifting events that began at least as early as late Paleozoic time. The best known of the extensional events that preceded formation of the Canada Basin and the Alaskan continental margin during middle Early Cretaceous time occurred in the Sverdrup basin of the western Canadian Arctic Islands. In the western part of this basin, Balkwill and Fox (1982) describe an incipient rift zone of Carboniferous to Late Cretaceous age marked by aligned normal faults, gabbro dikes, magnetic anomalies, and evaporite diapirs. The incipient rift also has stratigraphic expression in Carboniferous to Upper Cretaceous rocks. The rift strikes N40°E from Melville to northern Ellef Ringnes Island, about 25° counterclockwise from the trend of the adjacent present-day continental margin. Aligned dikes show the location of this feature on Plate 11 of Grantz and others (1990a). Balkwill and Fox (1982) state that the Sverdrup basin rift zone is likely to be the supracrustal expression of a crustal suture in Precambrian basement.

A second failed rift zone (Grantz and May, 1983) occurs along the continental margin from Banks Island to the continental slope west of Point Barrow. Stratigraphic relations in the northern slopes of the British Mountains in northern Yukon Territory indicate that this rift postdated thin, northerly sourced nearshore deposits of Late Triassic age and that the rift contains thick lutites as old as late Early Jurassic (Poulton, 1978). The presence of shale and siltstone of earliest Early to latest Middle Jurassic age in the Bonnet Lake area of the northern Richardson Mountains suggests that the oldest deposits in the rift may date back to the earliest Jurassic. Seismic stratigraphic relations along the Alaskan continental margin, detailed in previous sections of this chapter, indicate that the failed rift system contains as much as 4.7 km of Jurassic and Neocomian strata. The known and inferred distribution of the Jurassic failed rift deposits indicate that the rift was parallel to, and lay along or near the present continental margin along both the Canadian Arctic Islands and Alaskan sectors of the Canada Basin, and that the rift extended southward into the continent in the lower Mackenzie Valley. Restoring the Hauterivian to early Late Cretaceous counterclockwise rotation of Arctic Alaska away from the Canadian Arctic Islands about a pivot in the lower Mackenzie Valley (see Fig. 12 in Grantz and others, 1990b) would unite the now-separated fragments of the Jurassic failed rift deposits into a single rift system. This rift system would have lain within the peripheral platform of the Canadian Shield near, and subparallel to, the Carboniferous to Cretaceous failed rift of the Sverdrup basin.

The Jurassic rifting episode was a precursor to the late Neocomian rifting and Hauterivian breakup events that finally separated Arctic Alaska from the Canadian Arctic Islands. The newly formed Canada Basin was sufficiently well developed by Aptian time to receive more than 8.4 km of Albian progradational clastic deposits, and to serve as base level for development of submarine canyons with 1.4 km or more of erosional relief by Turonian time (Collins and Robinson, 1967). Firm evidence for the duration of rifting is not available, but the assumption of moderate rates for the spreading suggests that spreading was completed by middle Late Cretaceous time.

The geometry of the tectonic hinge line created by Hauterivian rifting and breakup (Figs. 4, 6, and 12) shows that this rifting event was the culmination of Jurassic rifting, and not a wholly independent event. Thus, the apparent position and trend of the Jurassic failed rift, as reconstructed from seismic and scattered outcrop data, is close to that of the successful Hauterivian and Barremian rift that produced the present continental margin. At a finer scale, the northward displacement of the Hauterivian hinge line at the mainly older Dinkum graben indicates that the breakup structures followed, rather than cut across, those of the precursor Jurassic rift system.

The southward excursion of the tectonic hinge line near 165°W, in the northwestern Chukchi shelf is associated with the genesis of the North Chukchi Basin. The great depth of this basin, its parallelism to the rifted margin to the east, and the apparent absence of a major transcurrent structure at its eastern margin, suggest that this offset segment of the continental margin also may have originated as a failed Jurassic rift. An intriguing unknown is whether the high-standing, northerly striking aseismic ridges of the adjacent Chukchi Borderland constitute fragments of the former northern margin of the North Chukchi Basin left behind by local complexities in the plate motions and geometry that created the Canada Basin.

A fourth episode of rifting may be represented by the wide zone of northerly trending "Laramide" extensional faults that disrupt the northwestern Chukchi Shelf and the eastern part of the adjacent North Chukchi Basin. This zone enters the continental margin from the north and apparently dies out within the Chukchi Platform of the central Chukchi Shelf. The position, trend, age, and deformational vigor of this event suggest that it may have been responsible for the dismemberment of a formerly compact Chukchi Borderland into its present array of northerly striking ridges and basins. These characteristics also suggest that the rift zone may have been connected, by a route which at present can only be conjectured, with rifting and sea-floor spreading of similar age in Baffin Bay and the Labrador Sea.

Convergent tectonics did not return to the Arctic Basin until

Cenozoic time when compression strongly deformed sediments of the continental margin in the Barter Island sector and adjacent parts of the southeastern Canada Basin. Distal effects of the deformation extend as far as 170 km north of the coast line, where water depths are in excess of 3,000 m. This episode of convergence probably began in Eocene time, and historic earthquakes and an abundance of deformed Quaternary sediments demonstrate that the convergence is still active. The tectonic position of the structures and earthquakes produced by this activity indicate that the compressional structures of the Barter Island sector in northeastern Alaska and northwestern Canada are distal effects of convergence between the Pacific Plate and North America at the Aleutian subduction zone.

REFERENCES CITED

Balkwill, H. R., and Fox, F. G., 1982, Incipient rift zone, western Sverdrup Basin, Arctic Canada, *in* Embry, A. F., and Balkwill, H. R., eds., Arctic geology and geophysics: Calgary, Canadian Society of Petroleum Geologists Memoir 8, p. 171–187.

Barnes, D. F., 1977, Bouguer gravity map of Alaska: U.S. Geological Survey Geophysical Investigations Map GP–913, 1 sheet, scale 1:2,500,000.

Bee, M., Johnson, S. H., and Chiburis, E. F., 1984, Marine seismic refraction study between Cape Simpson and Prudhoe Bay, Alaska: Journal of Geophysical Research, v. 89, no. B8, p. 6941–6960.

Bird, K. J., 1988a, Alaskan North Slope stratigraphic nomenclature and data summary for government-drilled wells, *in* Gryc, G., ed., Geology and exploration of the National Petroleum Reserve in Alaska, 1974 to 1982: U.S. Geological Survey Professional Paper 1399, p. 317–354.

—— , 1988b, Structure-contour and isopach maps of the National Petroleum Reserve in Alaska, *in* Gryc, G., ed., Geology and exploration of the National Petroleum Reserve in Alaska, 1974 to 1982: U.S. Geological Survey Professional Paper 1399, p. 355–380.

Bird, K. J., Connor, C. L., Tailleur, I. L., Silberman, M. L., and Christie, J. L., 1978, Granite on the Barrow Arch, northeast NPRA: U.S. Geological Survey Circular 772–B, p. 24–25.

Biswas, N. N., Gedney, L. D., and Huang, P., 1977, Seismicity studies in northeast Alaska by a localized seismographic network: University of Alaska Geophysical Institute Final Report No. 14–08–0001–15220, 22 p.

Blodgett, R. B., and 5 others, 1986, Age revisions for the Nanook Limestone and Katakturuk Dolomite, northeastern Brooks Range, Alaska: U.S. Geological Survey Circular 978, p. 5–10.

Blodgett, R. B., Rohr, D. M., Harris, A. G., and Rong, J., 1988, A major unconformity between Upper Ordovician and Lower Devonian strata in the Nanook Limestone, Shublik Mountains, northeastern Brooks Range, *in* Galloway, J. P., and Hamilton, T. D., eds., Geologic studies in Alaska by the U.S. Geological Survey during 1987: U.S. Geological Survey Circular 1016, p. 18–23.

Bruns, T. R., Fisher, M. A., Leinbach, W. J., and Miller, J. J., 1987, Regional structure of rocks beneath the coastal plain, *in* Bird, K. J., and Magoon, L. R., eds., Petroleum geology of the northern part of the Arctic National Wildlife Refuge, northeastern Alaska: U.S. Geological Survey Bulletin 1778, p. 249–254.

Bruynzeel, J. W., Guldenzopf, E. C., and Pickard, J. E., 1982, Petroleum exploration of NPRA, 1974–1981: Houston, Texas, Tetra-Tech, Inc., prepared for Husky Oil National Petroleum Reserve Operations, Inc., under contract to U.S. Geological Survey, 3 vols., 183 p.

Carey, S. W., 1958, The tectonic approach to continental drift, *in* Carey, S. W., ed., Continental drift; A symposium: Hobart, Tasmania University, p. 177–355.

Carter, C., and Laufield, S., 1975, Ordovician and Silurian fossils in well cores from North Slope of Alaska: American Association of Petroleum Geologists Bulletin, v. 59, p. 457–464.

Cecile, M. P., and Harrison, J. C., 1987, Review of the geology of Wrangel Island, Chukchi and east Siberian Sea, far northeastern Soviet Union: Geological Survey of Canada Open-File Report 1955, 109 p.

Clough, J. G., Blodgett, R. B., Imm, T. A., and Pavia, E. A., 1988, Depositional environments of Katakturuk Dolomite and Nanook Limestone, Arctic National Wildlife Refuge, Alaska [abs.]: American Association of Petroleum Geologists Bulletin, v. 72, p. 172.

Collins, F. R., 1958, Test wells, Topagoruk area, Alaska: U.S. Geological Survey Professional Paper 305–D, p. 265–316.

Collins, F. R., and Robinson, F. M., 1967, Subsurface stratigraphic, structural, and economic geology, northern Alaska: U.S. Geological Survey Open-File Report, 250 p.

Craig, J. D., Sherwood, K. W., and Johnson, P. P., 1985, Geologic report for the Beaufort Sea planning area, Alaska; Regional geology, petroleum geology, environmental geology: Anchorage, Alaska, U.S. Minerals Management Service OCS Report MMS 85–0111, 192 p.

Cramer, C. H., May, S. D., Hanna, W. F., Grantz, A., and Holmes, M. L., 1986, Magnetic anomaly map of the Chukchi Sea and adjacent northwest Alaska: U.S. Geological Survey Miscellaneous Investigations Map I–1182–F, scale 1:1,000,000.

Dehlinger, P., 1980, Gravity and crustal structure in the southern Beaufort Sea (north of Alaska): Final report to the Office of Naval Research on contract N 0014–75–C–0714 with the University of Connecticut, Storrs, September 1980, 31 p.

Dietrich, J. R., and 5 others, 1989, The geology, biostratigraphy, and organic geochemistry of the Natsek E–56 and Edlok N–56 wells, western Beaufort Sea, *in* Current Research, Part G: Geological Survey of Canada Paper 89–1G, p. 133–158.

Dinter, D. A., Carter, L. D., and Brigham-Grette, J., 1990, The late Neogene and Quaternary stratigraphy of the Canadian Beaufort continental shelf, *in* Grantz, A., Johnson, L., and Sweeney, J. F., eds., 1990a, The Arctic Ocean Region: Boulder, Colorado, Geological Society of America, The Geology of North America, V. L, p. 459–490.

Dixon, J., and Dietrich, J. R., 1990, Canadian Beaufort Sea and adjacent land areas, *in* Grantz, A., Johnson, L., and Sweeney, J. F., eds., 1990a, The Arctic Ocean Region: Boulder, Colorado, Geological Society of America, The Geology of North America, V. L, p. 239–256.

Ervin, L. D., 1981, The geologic significance of compaction gradient plots of wells in the National Petroleum Reserve; Alaska: Report prepared for Husky Oil NPR Operations, Inc., under contract to U.S. Geological Survey Office of National Petroleum Reserve, Alaska, 21 p.

Fisher, M. A., and Bruns, T. R., 1987, Structure of Pre-Mississippian rocks beneath the Coastal Plain, *in* Bird, K. J., and Magoon, L. B., eds., Petroleum geology of the northern part of the Arctic National Wildlife Refuge, northeastern Alaska: U.S. Geological Survey Bulletin 1778, p. 245–248.

Grantz, A., and May, S. D., 1983, Rifting history and structural development of the continental margin north of Alaska, *in* Watkins, J. S., and Drake, C. L., eds., Studies in continental margin geology: American Association of Petroleum Geologists Memoir 34, p. 77–100.

—— , 1987, Regional geology and petroleum potential of the United States Chukchi Shelf north of Point Hope, *in* Scholl, D. W., Grantz, A., and Vedder, J. G., eds., Geology and resource potential of the continental margin of western North America and adjacent ocean basins; Beaufort Sea to Baja California: Houston, Texas, Circum-Pacific Council for Energy and Mineral Resources, Earth Science Series 6, p. 37–58.

Grantz, A., Hanna, W. F., and Wallace, S. L., 1972a, Chukchi Sea seismic reflection and magnetic profiles, 1971, between northern Alaska and Herald Island: U.S. Geological Survey Open-File Report 72–137, 38 sheets.

Grantz, A., Holmes, M. L., Riley, D. C., and Wallace, S. L., 1972b, Seismic reflection profiles; Part 1, Seismic, magnetic, and gravity profiles; Chukchi Sea and adjacent Arctic Ocean, 1972: U.S. Geological Survey Open-File Report 72–138, 19 sheets.

Grantz, A., Holmes, M. L., and Kososki, B. A., 1975, Geologic framework of the Alaskan continental terrace in the Chukchi and Beaufort Seas, *in* Yorath, C. J., Parker, E. R., and Glass, D. J., eds., Canada's continental margins and offshore petroleum exploration: Canadian Society of Petroleum Geologists Memoir 4, p. 669–700.

Grantz, A., Eittreim, S. L., and Dinter, D. A., 1979, Geology and tectonic development of the continental margin north of Alaska, *in* Keen, C. E., ed., Crustal properties across passive margins: Tectonophysics, v. 59, p. 263–291.

Grantz, A., Eittreim, S. L., and Whitney, O. T., 1981, Geology and physiography of the continental margin north of Alaska and implications for the origin of the Canada Basin, *in* Nairn, A.E.M., Churkin, M., and Stehli, F. G., eds., The ocean basins and margins; Vol. 5, Geology of the Arctic Ocean Basin and its margins: New York, Plenum, p. 439–492.

Grantz, A., Mann, D. M., and May, S. D., 1982, Tracklines of multichannel seismic-reflection data collected by the U.S. Geological Survey in the Beaufort and Chukchi Seas in 1977 for which profiles and stack tapes are available: U.S. Geological Survey Open-File Report 82–735, 1 map sheet with text.

Grantz, A., Dinter, D. A., and Biswas, N. N., 1983a, Map, cross sections, and chart showing late Quaternary faults, folds, and earthquake epicenters on the Alaskan Beaufort Shelf: U.S. Geological Survey Miscellaneous Investigations Series Map I–1182–C, scale 1:500,000.

Grantz, A., Tailleur, I. L., and Carter, C., 1983b, Tectonic significance of Silurian and Ordovician graptolites, Lisburne Hills, northwest Alaska: Geological Society of America Abstracts with Programs, v. 15, p. 274.

Grantz, A., Mann, D. M., and May, S. D., 1986, Multichannel seismic-reflection data collected in 1978 in the eastern Chukchi Sea: U.S. Geological Survey Open-File Report 86–206, 3 p.

Grantz, A., Dinter, D. A., and Culotta, R. C., 1987a, Geology of the continental shelf north of the Arctic National Wildlife Refuge, *in* Tailleur, I. L., and Weimer, P., eds., Alaskan North Slope geology: Los Angeles, California, Society of Economic Paleontologists and Mineralogists, Pacific Section, and Anchorage, Alaska Geological Society, v. 2, p. 759–762.

Grantz, A., May, S. D., and Dinter, D. A., 1987b, Regional geology and petroleum potential of the United States Beaufort and northeasternmost Chukchi Seas, *in* Scholl, D. W., Grantz, A., and Vedder, J G., eds., Geology and resource potential of the continental margin of western North America and adjacent ocean basins; Beaufort Sea to Baja California: Houston, Texas, Circum-Pacific Council for Energy and Mineral Resources, Earth Science Series 6, p. 17–35.

Grantz, A., Johnson, L., and Sweeney, J. F., eds., 1990a, The Arctic Ocean Region: Boulder, Colorado, Geological Society of America, The Geology of North America, V. L, 644 p.

Grantz, A., May, S. D., and Hart, P. E., 1990b, Geology of the Arctic Continental Margin of Alaska, *in* Grantz, A., Johnson, L., and Sweeney, J. F., ed., 1990a, The Arctic Ocean Region: Boulder, Colorado, Geological Society of America, The Geology of North America, V. L, p. 257–288.

Grantz, A., May, S. D., Taylor, P. T., and Lawver, L. A., 1990c, Canada Basin, *in* Grantz, A., Johnson, L., and Sweeney, J. F., eds., 1990a, The Arctic Ocean Region: Boulder, Colorado, Geological Society of America, The Geology of North America, V. L, p. 379–402.

Halgedahl, S. L., and Jarrard, R. D., 1987, Paleomagnetism of the Kuparuk River Formation from oriented drill core; Evidence for rotation of the North Slope block, *in* Tailleur, I. L., and Weimer, P., eds., Alaskan North Slope geology: Los Angeles, California, Society of Economic Paleontologists and Mineralogists, Pacific Section, and Anchorage, Alaska Geological Society, v. 2, p. 581–617.

Haimela, N. E., Kirschner, C. E., Nassichuk, W. W., Ulmichek, G., and Procter, R. M., 1990, Sedimentary basins and petroleum resource potential of the Arctic Ocean region, *in* Grantz, A., Johnson, L., and Sweeney, J. F., eds., 1990a, The Arctic Ocean Region: Boulder, Colorado, Geological Society of America, The Geology of North America, V. L, p. 503–538.

Hall, J. K., 1990, Chukchi Borderland, *in* Grantz, A., Johnson, L., and Sweeney, J. F., eds., 1990a, The Arctic Ocean Region: Boulder, Colorado, Geological Society of America, The Geology of North America, V. L, p. 337–350.

Haq, B. U., Hardenbol, J., and Vail, P. R., 1987, Chronology of fluctuating sea levels since the Triassic: Science, v. 235, p. 1156–1167.

Hillhouse, J. W., and Grommé, S. C., 1983, Paleomagnetic studies and the hypothetical rotation of Arctic Alaska: Journal of the Alaska Geological Society, v. 2, p. 27–39.

Houtz, R. E., Eittreim, S., and Grantz, A., 1981, Acoustic properties of northern Alaska shelves in relation to the regional geology: Journal of Geophysical Research, v. 86, no. B5, p. 3935–3943.

Hubbard, R. J., Edrich, S. P., and Rattey, R. P., 1987, Geologic evolution and hydrocarbon habitat of the "Arctic Alaska Microplate": Marine and Petroleum Geology, v. 4, no. 1, p. 2–34.

Hunkins, K., 1966, The Arctic continental shelf north of Alaska, *in* Poole, W. H., ed., Continental margins and island arcs: Geological Survey of Canada Paper 66–15, p. 197–205.

Jamison, H. C., Brockett, L. B., and McIntosh, R. A., 1980, Prudhoe Bay; A 10-year perspective, *in* Halbouty, M. T., ed., Giant oil and gas fields of the decade 1968–1978: American Association of Petroleum Geologists Memoir 30, p. 289–314.

Jessup, D. D., 1985, Reconnaissance geology of the Chukchi platform; West-central Chukchi shelf, offshore Alaska [M.S. thesis]: East Lansing, Michigan State University, 105 p.

Kirschner, C. E., and Rycerski, B. A., 1988, Petroleum potential of representative stratigraphic and structural elements in the National Petroleum Reserve in Alaska, *in* Gryc, G., ed., Geology and exploration of the National Petroleum Reserve in Alaska, 1974 to 1982: U.S. Geological Survey Professional Paper 1399, p. 191–208.

Lane, L. S., and Cecile, M. P., 1989, Stratigraphy and structure of the Neruokpuk Formation, northern Yukon, *in* Current Research, Part G: Geological Survey of Canada Paper 89–19, p. 57–62.

Lerand, M., 1973, Beaufort Sea, *in* McCrossan, R. G., ed., The future petroleum provinces of Canada; Their geology and potential: Canadian Society of Petroleum Geologists Memoir 1, p. 315–386.

May, S. D., 1985, Free-air gravity anomaly map of the Chukchi and Alaskan Beaufort Seas, Arctic Ocean: U.S. Geological Survey Miscellaneous Investigations Series Map I–1182–E, scale 1:1,000,000.

May, S. D., and Grantz, A., 1990, Sediment thickness in the southern Canada Basin: Marine Geology, v. 93, p. 331–347.

Mayfield, C. F., Tailleur, I. L., and Ellersieck, I., 1988, Stratigraphy, structure, and palinspastic synthesis of the western Brooks Range, northwestern Alaska: U.S. Geological Survey Professional Paper 1399, p. 143–186.

Mickey, M. B., and Haga, H., 1987, Jurassic-Neocomian biostratigraphy, North Slope, Alaska, *in* Tailleur, I. L., and Weimer, P., eds., Alaskan North Slope geology: Los Angeles, California, Society of Economic Paleontologists and Mineralogists, Pacific Section, and Anchorage, Alaska Geological Society, v. 1, p. 397–404.

Milne, A. R., 1966, A seismic refraction measurement in the Beaufort Sea: Seismological Society of America Bulletin, v. 56, no. 3, p. 775–779.

Molenaar, C. M., 1988, Depositional history and seismic stratigraphy of Lower Cretaceous rocks in the National Petroleum Reserve in Alaska and adjacent areas, *in* Gryc, G., ed., Geology and exploration of the National Petroleum Reserve in Alaska, 1974 to 1982: U.S. Geological Survey Professional Paper 1399, p. 593–622.

Molenaar, C. M., Bird, K. J., and Collett, T. S., 1986, Regional correlation sections across the North Slope of Alaska: U.S. Geological Survey Miscellaneous Field Studies Map MF–1907, 1 sheet.

Molenaar, C. M., Bird, K. J., and Kirk, A. R., 1987 Cretaceous and Tertiary stratigraphy of northeastern Alaska, *in* Tailleur, I. L., and Weimer, P., eds., Alaskan North Slope geology: Los Angeles, California, Society of Economic Paleontologists and Mineralogists, Pacific Section, and Anchorage, Alaska Geological Society, v. 1, p. 513–528.

Moore, T. E., and Mull, C. G., 1989, Geology of the Brooks Range and North Slope, *in* Nokleberg, W. J., and Fisher, M. A., eds., Alaskan geological and geophysical transect, Valdez to Coldfoot, June 24–July 5, 1989, 28th International Geological Congress Field Trip Guidebook T104, Washington: American Geophysical Union, p. 107–131.

Newman, G. W., Mull, G. G., and Watkins, N. D., 1979, Northern Alaska paleomagnetism, plate rotation, and tectonics, *in* Sisson, A., ed., Relationship of plate tectonics to Alaskan geology and resources; Proceedings, Alaska Geological Society Symposium, Anchorage, Alaska, April, 1977: Alaska Geological Society, p. C-1 to C-7.

Norris, D. K., 1984, Geology of the Northern Yukon and Northwestern District of Mackenzie: Geological Survey of Canada Map 1581A, scale 1:500,000.

Pessel, G. H., Levorsen, J. A., and Tailleur, I. L., 1978, Generalized isopach map of Jurassic and possibly Lower Cretaceous shale, including Kingak Shale, eastern North Slope petroleum province, Alaska: U.S. Geological Survey Miscellaneous Field Studies Map MF–928K, 1 sheet, scale 1:500,000.

Poulton, T. P., 1978, Pre-Late Oxfordian Jurassic biostratigraphy of northern Yukon and adjacent Northwest Territories, *in* Stelck, C. R., and Chatterton, B.D.E., eds., Western and Arctic Canadian biostratigraphy: Geological Association of Canada Special Paper 18, p. 445–472.

Rattey, R. P., 1985, Northeastern Brooks Range, Alaska; New evidence for complex thin-skinned thrusting: American Association of Petroleum Geologists Bulletin, v. 69, p. 676–677.

Reiser, H. N., Brosgé, W. P., Dutro, J. T., Jr., and Detterman, R. L., 1980, Geologic map of the Demarcation Point Quadrangle, Alaska: U.S. Geological Survey Miscellaneous Investigations Series Map I–1133, scale 1:250,000.

Rickwood, F. K., 1970, The Prudhoe Bay field, *in* Adkison, W. L., and Brosgé, M. M., eds., Proceedings of the geological seminar on the North Slope of Alaska: Los Angeles, California, American Association of Petroleum Geologists, Pacific Section, p. L1–L11.

Robinson, M. S., and 6 others, 1989, Geology of the Sadlerochit and Shublik Mountains, Arctic National Wildlife Refuge, northeastern Alaska: Alaska Department of Natural Resources, Division of Geological and Geophysical Surveys Professional Report 100, 1 map sheet, scale 1:63,360.

Tailleur, I. L., 1969a, Speculations on North Slope geology: Oil and Gas Journal, v. 67, no. 38, p. 215–220, 225–226.

—— , 1969b, Rifting speculation on the geology of Alaska's North Slope: Oil and Gas Journal, v. 67, no. 39, p. 128–130.

—— , 1973, Probable rift origin of the Canada Basin, *in* Pitcher, M. G., ed., Arctic geology: American Association of Petroleum Geologists Memoir 19, p. 526–535.

Tailleur, I. L., Enwicht, S. E., Pessel, G. H., and Leverson, J. A., 1978, Maps showing land status and well locations and tables of well data, eastern North Slope petroleum province, Alaska: U.S. Geological Survey Miscellaneous Field Investigations Map MF–928–A, 5 sheets, scale 1:500,000.

Taylor, P. T., Kovacs, L. C., Vogt, P. R., and Johnson, G. L., 1981, Detailed aeromagnetic investigation of the Arctic Basin, 2: Journal of Geophysical Research, v. 86, p. 6323–6333.

Thurston, D. K., and Theiss, L. A., 1987, Geologic report for the Chukchi Sea planning area, Alaska; Regional geology, petroleum geology, and environmental geology: Anchorage, Alaska, U.S. Minerals Management Service OCS Report MMS 87–0046, 193 p.

Trettin, H. P., coordinator, 1972, The Innuitian Province, *in* Price, R. A., and Douglas, R.J.W., eds., Variations in tectonic styles in Canada: Canada Geological Association Special Paper 11, p. 83–180.

Trettin, H. P., and Balkwill, H. R., 1979, Contributions to the tectonic history of the Innuitian Province, Arctic Canada: Canadian Journal of Earth Sciences, v. 16, no. 3, p. 748–769.

Vogt, P. R., Taylor, P. T., Kovacs, L. C., and Johnson, G. L., 1982, The Canada Basin; Aeromagnetic constraints on structure and evolution: Tectonophysics, v. 89, p. 295–336.

Witmer, R. J., Haga, H., and Mickey, M. B., 1981, Biostratigraphic report of thirty-three wells drilled from 1975 to 1981 in National Petroleum Reserve in Alaska: U.S.. Geological Survey Open-File Report 81–1166, 47 p.

Wold, R. J., Woodzick, T. L., and Ostenso, N. A., 1970, Structure of the Beaufort Sea Continental Margin: Geophysics, v. 35, p. 849–861.

Manuscript Accepted by the Society January 19, 1990

Printed in U.S.A.

The Geology of North America
Vol. G-1, The Geology of Alaska
The Geological Society of America, 1994

Chapter 3

Geology of northern Alaska

Thomas E. Moore
U.S. Geological Survey, MS 904 Middlefield Road, Menlo Park, California 94025
Wesley K. Wallace
Department of Geology and Geophysics, University of Alaska, Fairbanks, Alaska 99775
Kenneth J. Bird
U.S. Geological Survey, MS 904, 345 Middlefield Road, Menlo Park, California 94025
Susan M. Karl
U.S. Geological Survey, 4200 University Drive, Anchorage, Alaska 99508
Charles G. Mull
Alaska Division of Geological and Geophysical Surveys, 794 University Avenue, Suite 200, Fairbanks, Alaska 99709
John T. Dillon*
Alaska Division of Geological and Geophysical Surveys, 794 University Avenue, Suite 200, Fairbanks, Alaska 99709

INTRODUCTION

This chapter describes the geology of northern Alaska, the largest geologic region of the state of Alaska. Lying entirely north of the Arctic Circle, this region covers an area of almost 400,000 km^2 and includes all or part of 36 1:250,000 scale quadrangles (Fig. 1). Northern Alaska is bordered to the west and north by the Chukchi and Beaufort seas, to the east by the Canadian border, and to the south by the Yukon Flats and Koyukuk basin. Geologically, it is notable because it encompasses the most extensive area of coherent stratigraphy in the state, and it contains the Brooks Range, the structural continuation in Alaska of the Rocky Mountain system. Northern Alaska also contains the largest oil field in North America at Prudhoe Bay, the world's second-largest zinc-lead-silver deposit (Red Dog), important copper-zinc resources, and about one-third of the potential coal resources of the United States (Kirschner, this volume; Magoon, this volume; Nokleberg and others, this volume, Chapter 10; Wahrhaftig and others, this volume).

Although geologic research in northern Alaska has resulted in many publications, including two symposium volumes (Adkison and Brosgé, 1970; Tailleur and Weimer, 1987), there exists no comprehensive geologic summary of the region. In this chapter we attempt to provide a summary by reviewing the essential stratigraphic and structural elements of the region, outlining the current hypotheses for the evolution of northern Alaska, and reporting problems that are currently the subject of controversy. Our objectives here are to provide a basic framework for the geology of the region and to establish a milepost from which the courses of future research can be planned. We have approached this complex topic by describing systematically the stratigraphy and structure in separate sections. These are then followed by an interpretive section that chronologically summarizes the geologic history of northern Alaska. We present the stratigraphy according to the existing tectonostratigraphic nomenclature; however, deformational features are discussed by structural province to highlight the orogenic features that developed after the formation of the major tectonostratigraphic boundaries. The data summarized here have been taken from published and unpublished geologic maps, open-file reports, circulars, abstracts, and field notes, as well as the more accessible geologic literature. Throughout the text we have related paleogeographic reconstructions to present geographic coordinates; however, many workers believe that the rocks of northern Alaska have been rotated and/or displaced significant distances from their sites of origin. An expanded version of this chapter is presented in Moore and others (1992).

*Deceased.

Geographic and geologic framework

Northern Alaska is divided into three major, parallel physiographic provinces: from south to north, these are the Arctic Mountains (Brooks Range), the Arctic Foothills, and the Arctic Coastal Plain (Wahrhaftig, 1965, and this volume). The Brooks Range consists of rugged, east-trending, linear mountain ranges, ridges, and hills that rise to more than 3000 m in the east but progressively decrease in elevation and relief toward the west. The Arctic Foothills consists of a series of rolling hills, mesas, and east-trending ridges that descend northward from more than 500 m to less than 300 m in elevation. From the Arctic Foothills,

Moore, T. E., Wallace, W. K., Bird, K. J., Karl, S. M., Mull, C. G., and Dillon, J. T., 1994, Geology of northern Alaska, *in* Plafker, G., and Berg, H. C., eds., The Geology of Alaska: Boulder, Colorado, Geological Society of America, The Geology of North America, v. G-1.

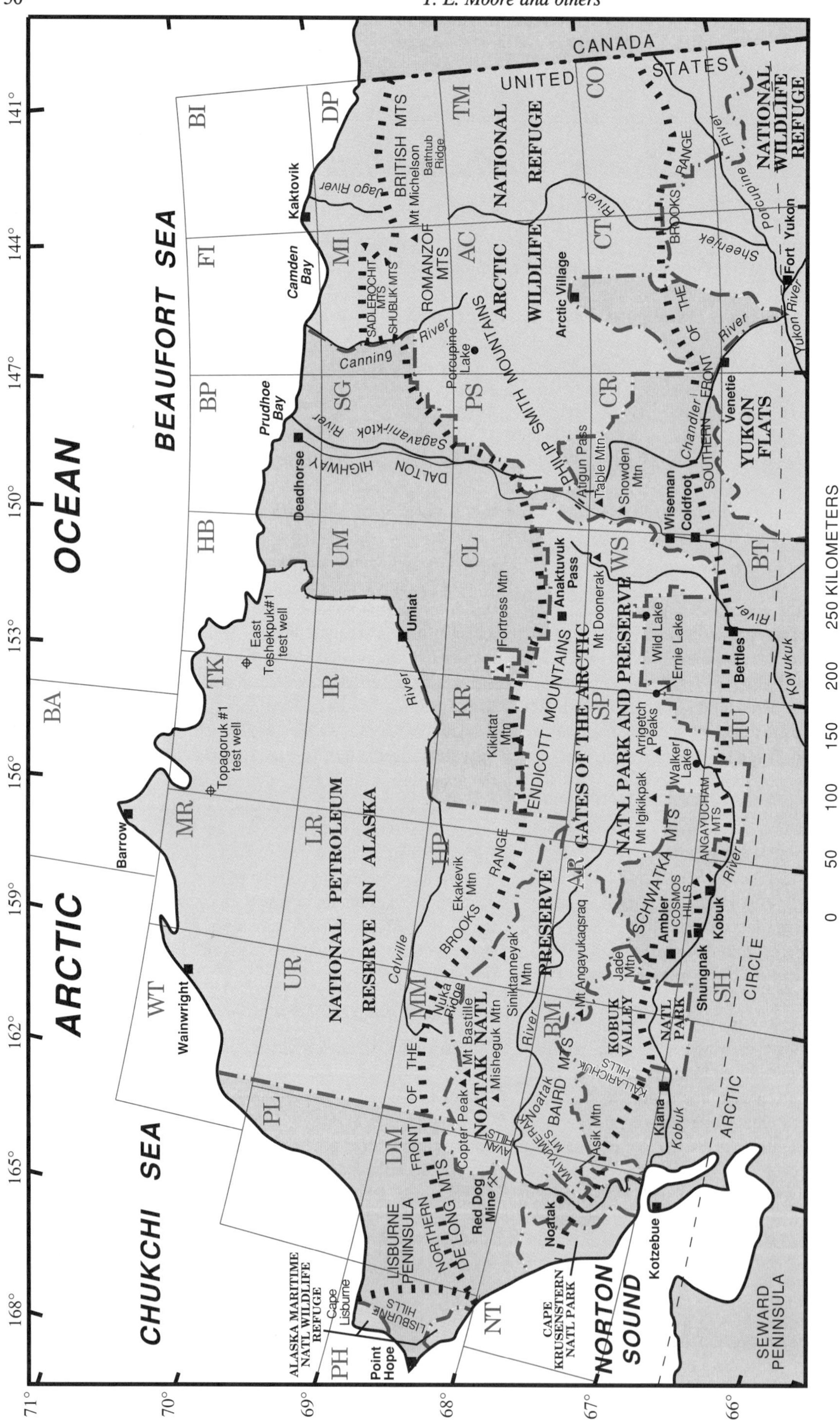

Figure 1. Selected geographic features of northern Alaska. Red shaded areas designate selected federal lands. Quadrangles (1:250,000 scale): AC, Arctic; AR, Ambler River; BA, Barrow; BI, Barter Island; BP, Beechy Point; BM, Baird Mountains; BT, Bettles; CL, Chandler Lake; CO, Coleen; CR, Chandalar; CT, Christian; DM, De Long Mountains; DP, Demarcation Point; FI, Flaxman Island; HB, Harrison Bay; HP, Howard Pass; HU, Hughes; IR, Ikpikpuk River; KR, Killik River; LR, Lookout Ridge; MI, Mount Michelson; MM, Misheguk Mountain; MR, Meade River; NT, Noatak; PH, Point Hope; PL, Point Lay; PS, Philip Smith Mountains; SG, Sagavanirktok; SH, Shungnak; SP, Survey Pass; TK, Teshekpuk; TM, Table Mountain; UM, Umiat; UR, Utukok River; WS, Wiseman; WT, Wainwright.

the marshy Arctic Coastal Plain slopes gradually northward to the Arctic Ocean. The latter two provinces, together composing the North Slope, narrow toward the east and are truncated on the west by the Chukchi seacoast (Wahrhaftig, this volume). The low, north-trending Lisburne Hills are situated along this coast.

Early investigations of the geology of northern Alaska showed that the North Slope is underlain by the large, west-trending Colville basin, a foreland basin of Cretaceous and Tertiary age (Fig. 2). The northern margin of the basin is delineated by the west-trending Barrow arch, a subsurface structural high composed of pre-Mississippian to Lower Cretaceous rocks. The northern limb of the Barrow arch lies beneath the Beaufort Sea, where it is overlain by Cenozoic rocks of the Beaufort Sea shelf. The southern part of the Colville basin is gently folded at the surface and is bordered by the west-trending disturbed belt (Brosgé and Tailleur, 1971), which is composed of mostly incompetent and structurally imbricated rocks along the northern front of the central and western Brooks Range. This belt marks the location of important north to south changes in the Paleozoic and lower Mesozoic stratigraphy. The Tigara uplift is a northwest-trending structural high that marks the southwestern limit of the Colville basin in the Lisburne Peninsula.

North of the disturbed belt in the eastern Brooks Range, the folded and faulted pre-Cretaceous rocks of the Colville basin are uplifted and exposed in the northeastern salient (Tailleur and Brosgé, 1970) of the Brooks Range. The disturbed belt is bounded on the south by the crestal belt of the Brooks Range. This belt is composed of folded and thrusted Devonian to Lower Cretaceous sedimentary rocks. South of the crest of the range, the central belt (Till and others, 1988) forms the geographic core of the Brooks Range. This belt consists of ductilely deformed Paleozoic slate, phyllite, schist, carbonate rocks, and orthogneiss that commonly retain primary textures. South of the central belt, in the southern Brooks Range, is the schist belt, which consists of

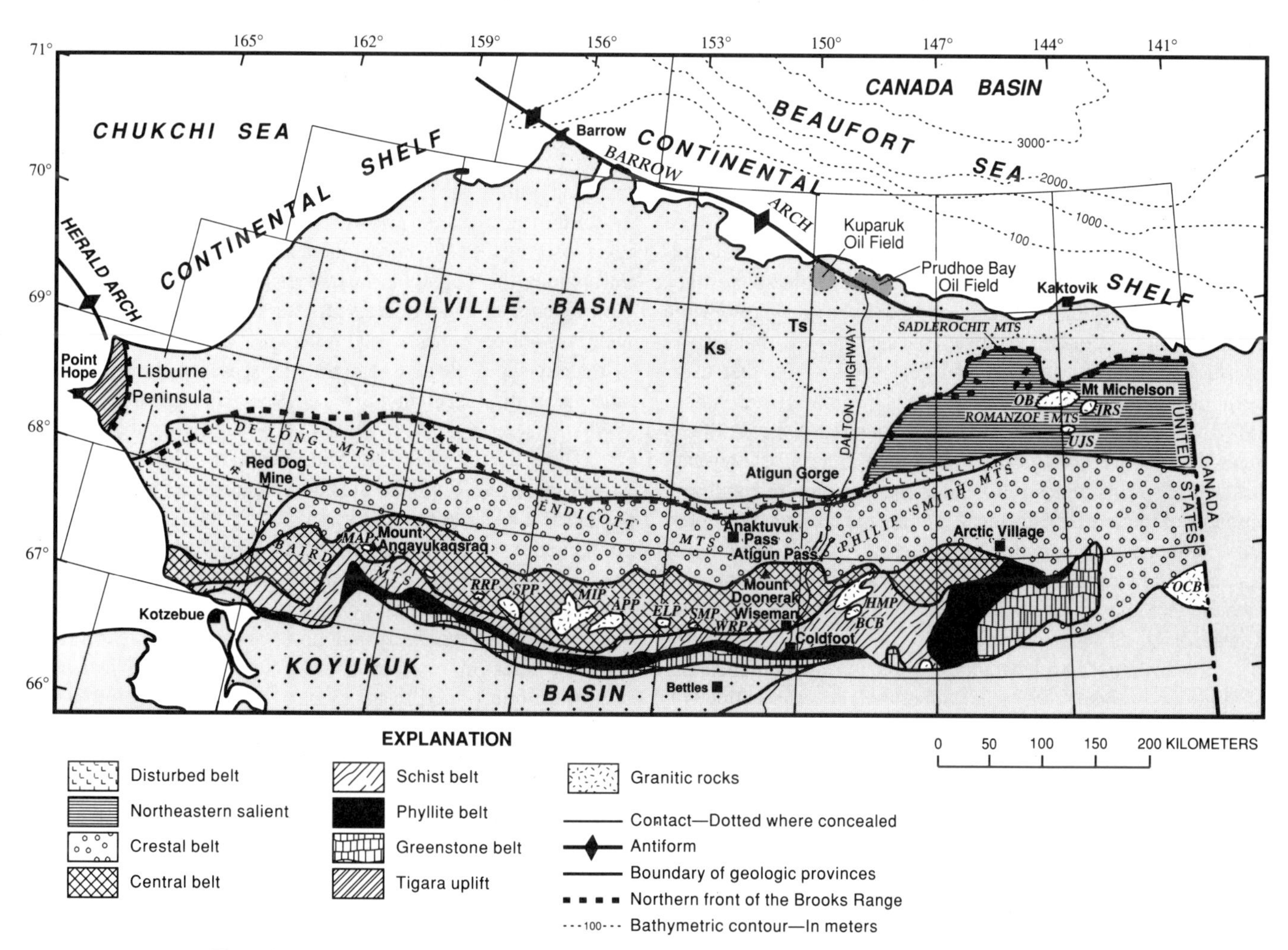

Figure 2. Geologic provinces of northern Alaska. Grid represents 1:250,000 scale quadrangle boundaries (see Fig. 1 for quadrangle names). APP, Arrigetch Peaks pluton; BCB, Baby Creek batholith; ELP, Ernie Lake pluton; HMP, Horace Mountain plutons; JRS, Jago River stock; Ks, Cretaceous sedimentary rocks; MAP, Mount Angayukaqsraq pluton; MIP, Mount Igikpak pluton; OB, Okpilak batholith; OCB, Old Crow batholith; RRP, Redstone River pluton; SMP, Sixtymile pluton; SPP, Shishakshinovik Pass pluton; Ts, Tertiary sedimentary rocks; UJS, Upper Jago River stock; WRP, Wild River pluton.

schistose metasedimentary and minor metavolcanic rocks (Brosgé, 1975; Turner and others, 1979). Two narrow belts lie south of the schist belt and underlie the southern foothills of the Brooks Range (the Ambler-Chandalar ridge and lowland section of the Arctic Mountains of Wahrhaftig, 1965, and this volume). The phyllite belt (Dillon, 1989), the more northerly of the two narrow belts, consists of recessive metasedimentary rocks, whereas the greenstone belt (Patton, 1973), the more southerly of the two, consists of resistant metabasalt and chert. The greenstone belt is commonly considered exotic with respect to most of the pre-Cretaceous rocks of northern Alaska, and its southern boundary delineates the northern margin of the Koyukuk basin (Cretaceous) to the south.

From the schist belt north to the Colville basin, exposed rocks of northern Alaska display a regional decrease in average age, grade of metamorphism, and deformational intensity. An apparent decrease in grade of metamorphism and deformation also extends southward from the schist belt into the Koyukuk basin. Although the various belts described above are commonly used to designate geologic provinces, their boundaries are not well defined, and their structural character and significance are controversial.

Lithotectonic terranes of northern Alaska

Jones and others (1987) and Silbering and others (this volume) have divided nearly all of Alaska into several tectonostratigraphic or lithotectonic terranes and subterranes. Terranes are defined as fault-bounded geologic packages of rock that display internal stratigraphic affinities and geologic histories that differ from neighboring terranes (Jones, 1983; Howell and others, 1985). We regard subterranes as fault-bounded divisions of terranes that can be geologically linked with adjacent subterranes but whose stratigraphies differ sufficiently to require significant amounts of relative displacement. We use the terrane and subterrane nomenclature herein to designate major structurally bounded stratigraphic or lithologic packages without inferring specific tectonic models of origin or distances of structural displacement.

For this report we have modified the terrane nomenclature of Jones and others (1987) and Silberling and others (this volume) for northern Alaska to incorporate new map and age data and to simplify discussion of the stratigraphy and tectonic history (Fig. 3). Principal revisions are (1) modification of the terrane-subterrane hierarchial nomenclature for northern Alaska on the basis of likely affinities among the various tectonostratigraphic units; (2) extension of the Coldfoot subterrane of the Arctic Alaska terrane into the western Brooks Range; (3) incorporation of both the southern part of the Coldfoot subterrane (that is, the phyllite belt) and the Brooks Range part of the Venetie terrane of Jones and others (1987) into the Slate Creek subterrane, a new subterrane of the Arctic Alaska terrane; (4) inclusion of the Sheenjek terrane of Jones and others (1987) into the De Long Mountains subterrane of the Arctic Alaska terrane; (5) inclusion of the Kagvik terrane of Jones and others (1987) in the Endicott Mountains subterrane of the Arctic Alaska terrane; and (6) inclusion of the Brooks Range part of the Tozitna terrane of Jones and others (1987) into the Angayucham terrane. Our revised terrane map is shown in Figure 3.

Northern Alaska consists of two lithotectonic terranes, the Arctic Alaska terrane and the Angayucham terrane (Silberling and others, this volume). The most extensive is the Arctic Alaska terrane (Newman and others, 1977; Fujita and Newberry, 1982; Mull, 1982), which underlies all of the North Slope and most of the Brooks Range. Rocks of this terrane span Proterozoic to Cenozoic time and are mostly of continental affinity. The Arctic Alaska terrane is bordered on the north by the Canada basin, which formed by sea-floor spreading in the Cretaceous (Grantz and May, 1983). The terrane extends into the Mackenzie delta region of northwest Canada, where it terminates beneath Cenozoic sedimentary cover. To the west, it may extend under the Chukchi Sea and into the Chukotsk Peninsula of the Russian Far East and terminate at the South Anyuy suture (Churkin and Trexler, 1981; Fujita and Newberry, 1982). At its southernmost exposure in Alaska, the Arctic Alaska terrane dips southward beneath the Angayucham terrane at a boundary called the Kobuk suture by Mull (1982) and Mull and others (1987c), and may continue southward beneath rocks of the Koyukuk basin. Along its southeastern margin, however, the Arctic Alaska terrane is juxtaposed against the Porcupine terrane of North American affinity along the Porcupine lineament (Grantz, 1966). Although the existence of a continuous structural break along this lineament has not been firmly established, it is commonly interpreted as a strike-slip fault of unknown age, sense of movement, and amount of displacement (Grantz, 1966; Churkin and Trexler, 1981).

The other terrane composing northern Alaska is the Angayucham terrane, which includes the greenstone belt of the southern Brooks Range and the structurally highest klippen in the crestal and disturbed belts. The Angayucham terrane is generally less than 5–10 km thick and consists largely of mafic and ultramafic rocks and siliceous pelagic rocks of Devonian through Jurassic age. Its structurally high position has led most workers to conclude that it represents the remnants of an extensive northward-transported thrust sheet of mostly oceanic rocks that overrode the Arctic Alaska terrane during Jurassic and Cretaceous time (Roeder and Mull, 1978).

On the basis of differing stratigraphy, facies, and structural position, the Arctic Alaska terrane has been divided into several subterranes and/or allochthons (Mull, 1982; Jones and others, 1987; Mayfield and others, 1988). The term "allochthon," as used by Mull and others (1987c) and Mayfield and others (1988), refers to rock packages of regional extent that are bounded by major thrust faults. These faults separate one package from adjacent ones of differing lithofacies but commonly of the same stratigraphic nomenclature. In this chapter, allochthon is used as a tectonostratigraphic subdivision of subterrane. The North Slope subterrane, the structurally lowest subterrane, consists of Cre-

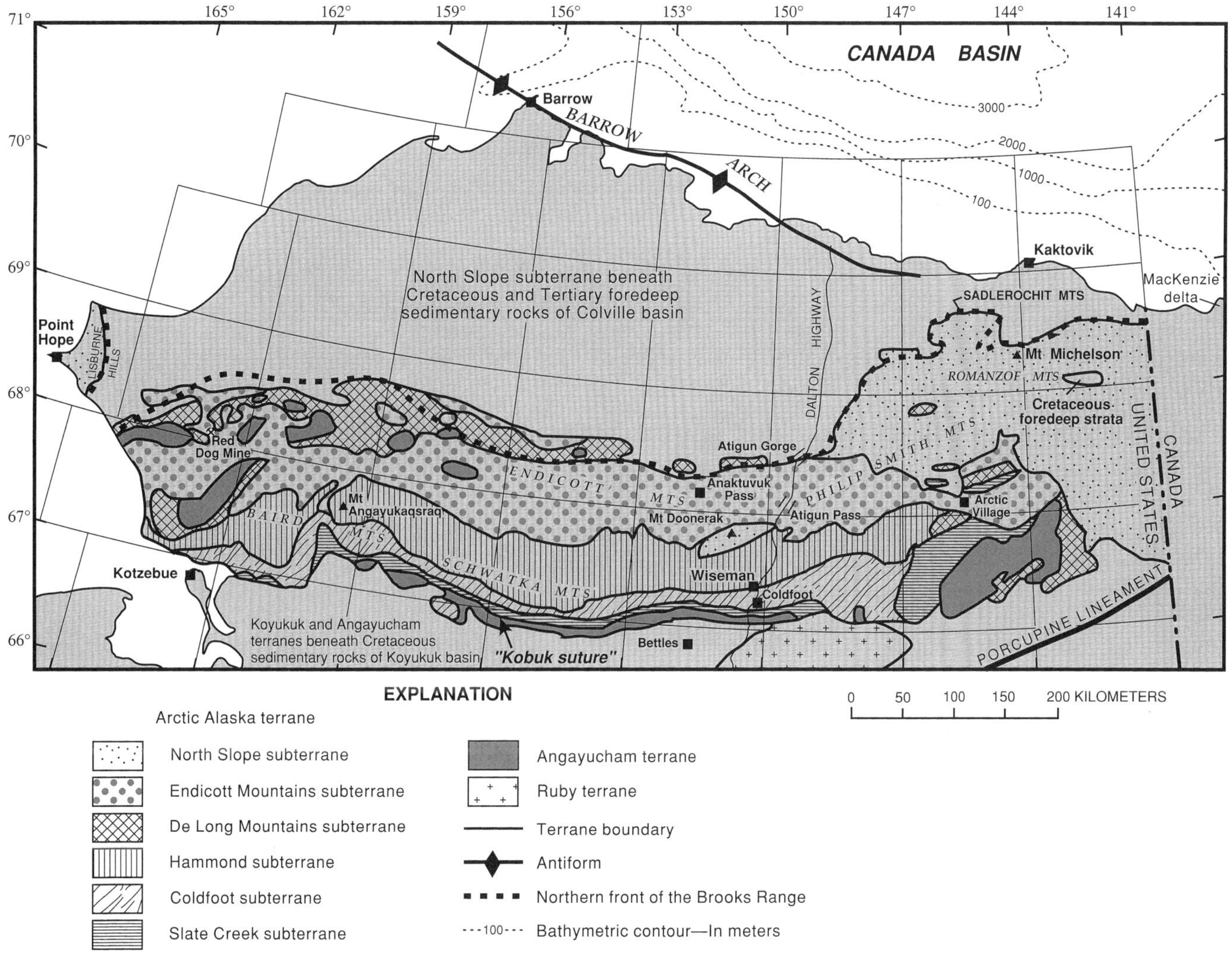

Figure 3. Terranes and subterranes of northern Alaska. Grid represents 1:250,000 scale quadrangle boundaries (see Fig. 1 for quadrangle names).

taceous and older rocks underlying the Colville basin and composing the Barrow arch, the northeastern salient of the Brooks Range, and the Lisburne Hills. These rocks were not extensively deformed by post-Mississippian orogenic activity in the subsurface of the North Slope, but they have been extensively deformed by Cretaceous and Cenozoic contractional tectonism in the northeastern salient of the Brooks Range and in the Lisburne Hills. Rocks of the North Slope subterrane are also exposed in a structural window, the Mt. Doonerak fenster, at Mt. Doonerak in the central Brooks Range (Fig. 3). This exposure is extremely significant because it indicates that the North Slope subterrane lies at depth beneath the other subterranes of the Arctic Alaska terrane. The Endicott Mountains subterrane consists of the Endicott Mountains (or Brooks Range) allochthon. This subterrane includes most of the imbricated sedimentary rocks of the crestal belt in the central Brooks Range. The presence of rocks of the North Slope subterrane both north and south (in the Mt. Doonerak fenster) of the Endicott Mountains subterrane has led most workers to conclude that the Endicott Mountains subterrane is a klippe of imbricated middle Paleozoic to Mesozoic rocks that overlie the North Slope subterrane. In various places along the length of the Brooks Range, the Endicott Mountains subterrane is structurally overlain by the De Long Mountains subterrane. Like the Endicott Mountains subterrane, the De Long Mountains subterrane consists of imbricated Paleozoic and Mesozoic sedimentary rocks but differs in important stratigraphic aspects from the North Slope and Endicott Mountains subterranes. The De Long Mountains subterrane is divided into four allochthonous successions: from base to top these are the Picnic Creek, Kelly River, Ipnavik River, and Nuka Ridge allochthons. These allochthons are the structurally highest and most displaced rocks of the Arctic Alaska terrane.

The stratigraphy of metamorphic rocks in the southern Brooks Range is not as well understood as that of the sedimentary rocks in the northern Brooks Range. For this reason, the subterranes of the southern Brooks Range consist of lithologic assemblages that are inferred, but cannot be proved, to constitute depositional successions. The Hammond subterrane includes most of the Proterozoic and lower Paleozoic mixed clastic and carbonate rocks of the central belt. These rocks display ductile deformational structures and low greenschist to local blueschist facies mineral assemblages, but relict igneous and sedimentary textures are commonly retained. The Coldfoot subterrane, which lies south of the Hammond subterrane, consists of quartz-mica schist, calc-schist, marble, and metavolcanic rocks of the schist belt. This subterrane displays Mesozoic blueschist facies metamorphic assemblages that are partly to completely overprinted by greenschist facies assemblages. The Slate Creek subterrane (also called the Slate Creek thrust panel of the Angayucham terrane by Patton and others, this volume, Chapter 7) consists of phyllite and subordinate metasandstone with rare Devonian fossils that compose the phyllite belt. The Coldfoot and Slate Creek subterranes are thought to belong to the Arctic Alaska terrane because of their quartzose composition and apparent contiguity with that terrane, but their precise relation to the other subterranes of the Arctic Alaska terrane is unknown.

The Koyukuk basin, Colville basin, and the Beaufort Sea shelf (Fig. 2) are regarded as post-tectonic basins by most workers, but they are partly overprinted by both shortening and extension that continued in the Cretaceous and Cenozoic (Tailleur and Brosgé, 1970). The Koyukuk basin is discussed by Patton and others (this volume, Chapter 7) and the Beaufort Sea continental-margin succession is discussed by Grantz and others (1990a). The Colville basin is the Early Cretaceous (Aptian[?] to Albian) and younger foredeep of the Brookian orogen (discussed in the following section). The strata of this basin rest almost entirely on the North Slope subterrane and therefore are described with that subterrane. Older Jurassic and Lower Cretaceous (Neocomian) foredeep deposits within the orogen compose the proto-Colville basin. Because the proto-Colville basin strata are preserved mainly in the allochthonous sequences of the Brooks Range, and palinspastic restoration of these rocks is uncertain, they are described below with the subterrane to which they have been assigned.

Orogenic events of northern Alaska

Geologic structures in northern Alaska are assigned to two major orogenic systems, the Ellesmerian and Brookian orogenies. The Ellesmerian orogeny affected units that lie unconformably beneath less-deformed Lower Mississippian strata in the northeastern Brooks Range and North Slope, but angular truncation by the sub-Mississippian unconformity elsewhere may indicate a greater extent for the orogen. The sub-Mississippian unconformity extends westward in the North Slope subsurface at least as far as the Lisburne Peninsula and is also exposed south of the crest of the Brooks Range at Mt. Doonerak and in the Schwatka Mountains (Fig. 1). Because Lerand (1973) inferred the sub-Mississippian structures of northern Alaska to be coeval with the Late Devonian and Early Mississippian Ellesmerian orogeny of the Canadian Arctic Islands, he extended the Ellesmerian orogen to include all early Paleozoic deformation along the Arctic Ocean margin from northern Greenland to Wrangell Island in the Russian Far East. However, recent work in the northeastern Brooks Range (Anderson and Wallace, 1990) suggests that the Ellesmerian orogen of northern Alaska is pre–Middle Devonian and is therefore older than the Ellesmerian orogen of the Canadian Arctic Islands.

The younger orogenic system is represented by the major east-trending, north-vergent fold-thrust belt that forms the Brooks Range. The orogenic event that produced these structures, the Brookian orogeny, can be divided into two major phases. The early Brookian phase is characterized by ductile deformation and metamorphism in the southern Brooks Range and by the emplacement of relatively far traveled thrust sheets in the northern Brooks Range during the Middle Jurassic and Early Cretaceous (Mull, 1982; Mayfield and others, 1988; Dillon, 1989). The thrust sheets advanced northward as far as the disturbed belt along the northern margin of the Brooks Range. The late Brookian orogenic phase deforms Albian and younger strata and is represented by structures that indicate at least three deformational episodes. The earliest episode produced gentle long-wavelength folds and north-directed thrust faults that display relatively small amounts of displacement in Albian and younger strata of the northern foothills of the Brooks Range. The Tigara uplift, exposed in the Lisburne Peninsula (Fig. 2; Payne, 1955), is an east-directed fold-thrust belt that continues northwestward under the Chukchi Sea along the Herald arch. This fold belt was regarded as Late Cretaceous or Tertiary by Grantz and others (1981), but Mull (1979, 1985) suggested that it was active in Albian (late Early Cretaceous) or older time. The youngest episode of deformation is the Romanzof uplift in the northeastern Brooks Range. This episode of deformation formed by north-directed folds and thrust faults that produced the northeastern salient of the Brooks Range and a middle Eocene unconformity that truncates deformed strata as young as early Eocene (Bird and Molenaar, 1987) in the subsurface of the Arctic Coastal Plain of Arctic National Wildlife Refuge (ANWR). Deformed Neogene and Quaternary strata also present beneath the Arctic Coastal Plain and on the adjacent Beaufort Sea continental terrace indicate that late Brookian deformation has continued into Quaternary time (Grantz and others, 1983a, 1987).

The relation between early and late Brookian structures is exposed at Ekakevik Mountain (Howard Pass quadrangle) (Figs. 1 and 4), where allochthonous Devonian to Neocomian strata of the De Long Mountains subterrane (Ipnavik River and Nuka Ridge allochthons) are unconformably overlain by gently folded conglomerate and sandstone of the Aptian(?) and Albian (upper Lower Cretaceous) Fortress Mountain Formation (Tailleur and others, 1966; Mull, 1985; Mull and others, 1987c; Mayfield and others, 1988; Molenaar and others, 1988). Brookian deformation was accompanied by regional metamorphism, as evidenced by K-Ar ages of 170 to 54 Ma, averaging 110 Ma (Turner and others, 1979), for most metamorphic rocks in the southern Brooks Range. Although the K-Ar data indicate that metamor-

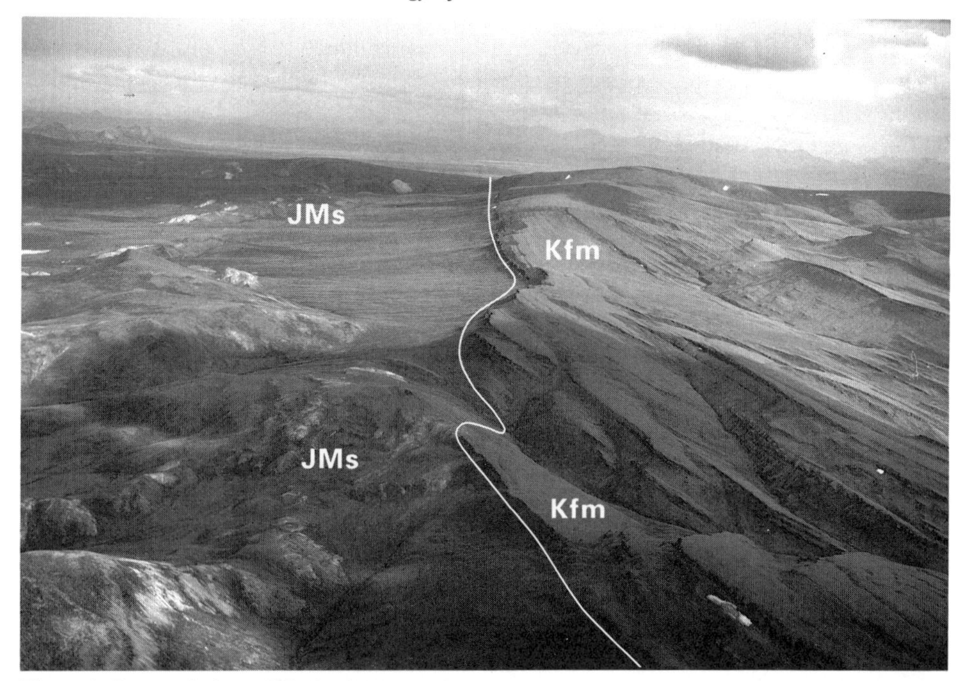

Figure 4. Eastward view of Ekakevik Mountain (Howard Pass quadrangle) depicting relation between early and late Brookian structures. Imbricate structures (early Brookian) that deform Paleozoic and Mesozoic sedimentary and diabasic rocks (JMs) of Ipnavik River and Nuka Ridge allochthons, De Long Mountains subterrane are unconformably truncated by Aptian(?) and Albian Fortress Mountain Formation (Kf), which displays gentle folds (late Brookian).

phism accompanied both Brookian orogenic phases, the data point out that regional uplift occurred mainly during the late Brookian phase.

STRATIGRAPHY OF THE ARCTIC ALASKA TERRANE

The description of the stratigraphy of the Arctic Alaska terrane is organized below by its six subterranes (North Slope, Endicott Mountains, De Long Mountains, Hammond, Coldfoot, and Slate Creek) so that the stratigraphy of these tectonic units can be compared easily. Of the six subterranes, the North Slope subterrane is the most widespread, the best studied, the least deformed, and represents the greatest amount of geologic time. For these reasons, and because of its major petroleum resources, the North Slope subterrane is the benchmark to which the others have been compared. The stratigraphic description of the Arctic Alaska terrane begins below with description of the North Slope subterrane and moves progressively southward, first to the sedimentary Endicott Mountains and De Long Mountains subterranes, which are exposed mainly in the northern Brooks Range, and finally to the metamorphic Hammond, Coldfoot, and Slate Creek subterranes, which are exposed in the southern Brooks Range.

North Slope subterrane

The North Slope subterrane, the northernmost and largest of the six subterranes that compose the Arctic Alaska terrane, is divided by structural features into four parts. The North Slope subsurface west of long 147°W consists of a relatively undisturbed post-Devonian succession with a laterally continuous, coherent stratigraphy that provides a detailed record of the depositional history of the subterrane. Information about these rocks comes primarily from wells tied to seismic-reflection records (Bird, 1982, 1988a; Bruynzeel and others, 1982; Grantz and May, 1983; Kirschner and others, 1983). To the east, the northeastern salient of the Brooks Range and adjacent North Slope expose a west-plunging cross-sectional view of the eastern Colville basin and its underlying geology that are complicated somewhat by Late Cretaceous and Cenozoic deformation. This region is important because it contains exposures of most of the units recognized in the subsurface of the North Slope. Stratigraphic information from the southern and southwestern parts of the subterrane comes from outcrop studies in the Mt. Doonerak fenster in the central Brooks Range and from the Lisburne Peninsula. The Mt. Doonerak fenster is an elongate, east-northeast–trending structural culmination, about 110 km long and 20 km wide, that lies in the central Brooks Range along the southern margin of the Endicott Mountains subterrane. The fenster is an antiform that exposes the North Slope subterrane beneath the Endicott Mountains and Hammond subterranes. The Lisburne Peninsula is a physiographic uplift and structural culmination separate from the Brooks Range, exposing a thrusted succession of strata that differs in some aspects from the North Slope subsurface. Because of these differences, Jones and others (1987) included rocks of the Lisburne Peninsula with the De Long Mountains subterrane, whereas Blome and others (1988) and Mayfield and others

(1988) have included them with the Endicott Mountains subterrane. However, an affinity of the Lisburne Peninsula succession to the North Slope subterrane is indicated by (1) the position of the region north of the northern limit of allochthonous strata in the disturbed belt of the Brooks Range, (2) the lithology of its Mississippian and older rocks, and (3) the presence of a prominent sub-Mississippian unconformity. We therefore include a description of the rocks of this region in our discussion of the North Slope subterrane.

The North Slope subterrane consists of a great thickness and variety of sedimentary rocks with a minor amount of igneous rocks (Fig. 5). Lerand (1973) grouped Phanerozoic rocks of the lands bordering the Beaufort Sea into three sequences on the basis of source area: the Franklinian sequence (northern source, Upper Cambrian through Devonian), Ellesmerian sequence (northern source, Mississippian to Lower Cretaceous), and Brookian sequence (southern source, Upper Jurassic or Lower Cretaceous to Holocene). Although drawn largely from his work in northern Canada, Lerand's scheme has provided a simplification of the complex stratigraphic nomenclature for rocks of the North Slope subterrane and is in widespread use. However, recent work has shown that the Franklinian is not a single, genetically related sequence of rocks because it includes rocks considerably older (that is, Proterozoic) and more diverse than originally defined, and the term is no longer used in Canada. Rocks assigned to the Franklinian sequence are therefore discussed under the heading pre-Mississippian rocks. This nomenclature reflects the truncation of most older rocks of the North Slope subterrane by the sub-Mississippian unconformity, which is the most distinctive feature of the subterrane. We also divide Lerand's Ellesmerian sequence into a lower carbonate and clastic succession, the lower Ellesmerian sequence, and an upper shale-rich succession, the upper Ellesmerian sequence, following the lead of other workers (Hubbard and others, 1987; Grantz and others, 1990a). The lower Ellesmerian sequence records deposition on a continental shelf that persisted from Mississippian to Triassic time. The upper Ellesmerian sequence reflects continued deposition on the continental shelf from the Jurassic through the Early Cretaceous and in addition documents the influence of rifting prior to the opening of the Canada basin.

Pre-Mississippian rocks. Pre-Mississippian rocks are known from various parts of the North Slope subterrane but for the most part have been mapped and studied only at the reconnaissance level. Some workers have inferred a regional stratigraphy for these rocks (Dutro, 1981), but existing descriptions and interpretations indicate that these rocks vary widely in age and record a number of depositional and tectonic environments. Because of the apparent absence of through-going stratigraphy and the common presence of faults bounding mapped units of pre-Mississippian rocks, we describe these rocks below by geographical area within the North Slope subterrane.

North Slope subsurface. Although pre-Mississippian rocks are buried to depths exceeding 10 km (Fig. 6), they have been penetrated by many wells along the Barrow arch. The well data show that these rocks consist mostly of steeply dipping and slightly metamorphosed, thin-bedded argillite, siltstone, and fine-grained quartzose sandstone. The argillite is locally interbedded with graywacke, limestone, dolomite, and chert, and has been dated as Early Silurian by graptolites and as Middle and Late Ordovician and Silurian by chitinozoans (Carter and Laufeld, 1975). More than 100 m of chert-pebble conglomerate and sandstone interstratified with siltstone, carbonaceous shale, and claystone was also found in the U.S. Navy Topagoruk #1 test well (Collins, 1958). Middle (or perhaps Early) Devonian plant fossils have been recovered from this nonmarine succession. Like the argillite, these rocks are truncated in angular unconformity by flat-lying Mississippian to Permian(?) strata of the lower Ellesmerian sequence, but they appear to be less metamorphosed and deformed than the argillite. This lack of significant deformation suggests that the Devonian clastic rocks in the Topagoruk well may have been deposited after the main phase of deformation of the older argillitic rocks but were tilted or folded prior to deposition of the overlying Mississippian rocks.

Northeastern salient of the Brooks Range. Pre-Mississippian rocks are widely exposed in the northeastern Brooks Range and have been divided into several assemblages or terranes (Dutro and others, 1972; Moore and others, 1985a). The best studied of these assemblages is a depositional succession of Proterozoic to Devonian carbonate strata exposed in the Shublik and Sadlerochit mountains (Fig. 1). These rocks structurally, and probably stratigraphically, overlie several hundred meters of quartzite, argillite, and tholeiitic basalt of probable continental affinity (Moore, 1987a). The Katakturuk Dolomite (Proterozoic), about 2400 m thick, forms the lowest part of the carbonate sequence and is overlain at a low-angle unconformity by the 1100-m-thick Nanook Limestone (Late Proterozoic?, Cambrian, and Ordovician) (Dutro, 1970; Blodgett and others, 1986, 1988, 1991; Robinson and others, 1989) and the 72-m-thick Mount Copleston Limestone (Lower Devonian) (Blodgett and others, 1991). Fossils in the upper 500 m of the Nanook are Cambrian and Ordovician; these ages imply that perhaps the lower part of the Nanook and presumably all of the Katakturuk are Proterozoic (R. B. Blodgett, 1991, oral commun.). A disconformity or very low angle unconformity separates Middle and/or Upper Ordovician limestone of the Nanook from Lower Devonian (Emsian) strata of the Mount Copleston Limestone.

The Katakturuk Dolomite grades upward from rocks of basin-plain to those of carbonate-platform depositional environments (Clough and others, 1988). The Nanook likewise grades upward from deep-marine carbonate turbidites at its base to shallow subtidal to intertidal deposits at its top. The Mount Copleston Limestone consists entirely of subtidal and intertidal carbonate rocks deposited on a partially restricted shallow-water carbonate platform (Clough and others, 1988; Blodgett and others, 1991).

In contrast to the carbonate succession of the Sadlerochit and Shublik mountains, the pre-Mississippian rocks of the nearby Romanzof Mountains are thick, lithologically diverse, and structurally complicated. The Neruokpuk Quartzite (Leffingwell,

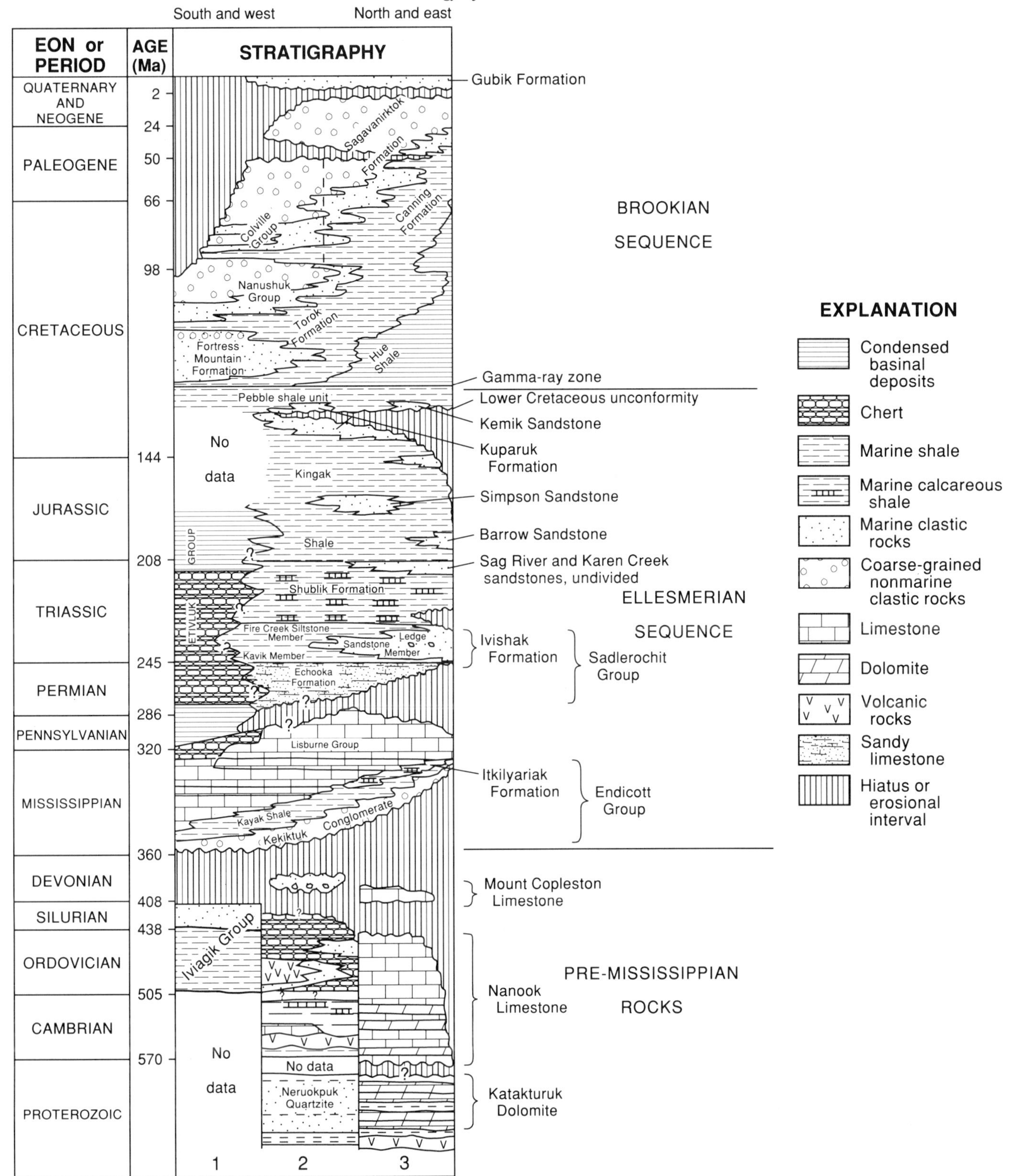

Figure 5. Generalized stratigraphic column of North Slope subterrane (Arctic Alaska terrane). Pre-Mississippian rocks are grouped according to region and are depicted in three subcolumns: 1, Lisburne Peninsula; 2, Romanzof Mountains; 3, Sadlerochit and Shublik mountains. Ordovician and Silurian Iviagik Group is that of Martin (1970). Jurassic Simpson and Barrow sandstones are of local usage. Brookian sequence depicts North Slope units only; less well known Brookian rocks in Lisburne Peninsula and northeastern Brooks Range are not shown. Absolute time scale (Palmer, 1983) is variable.

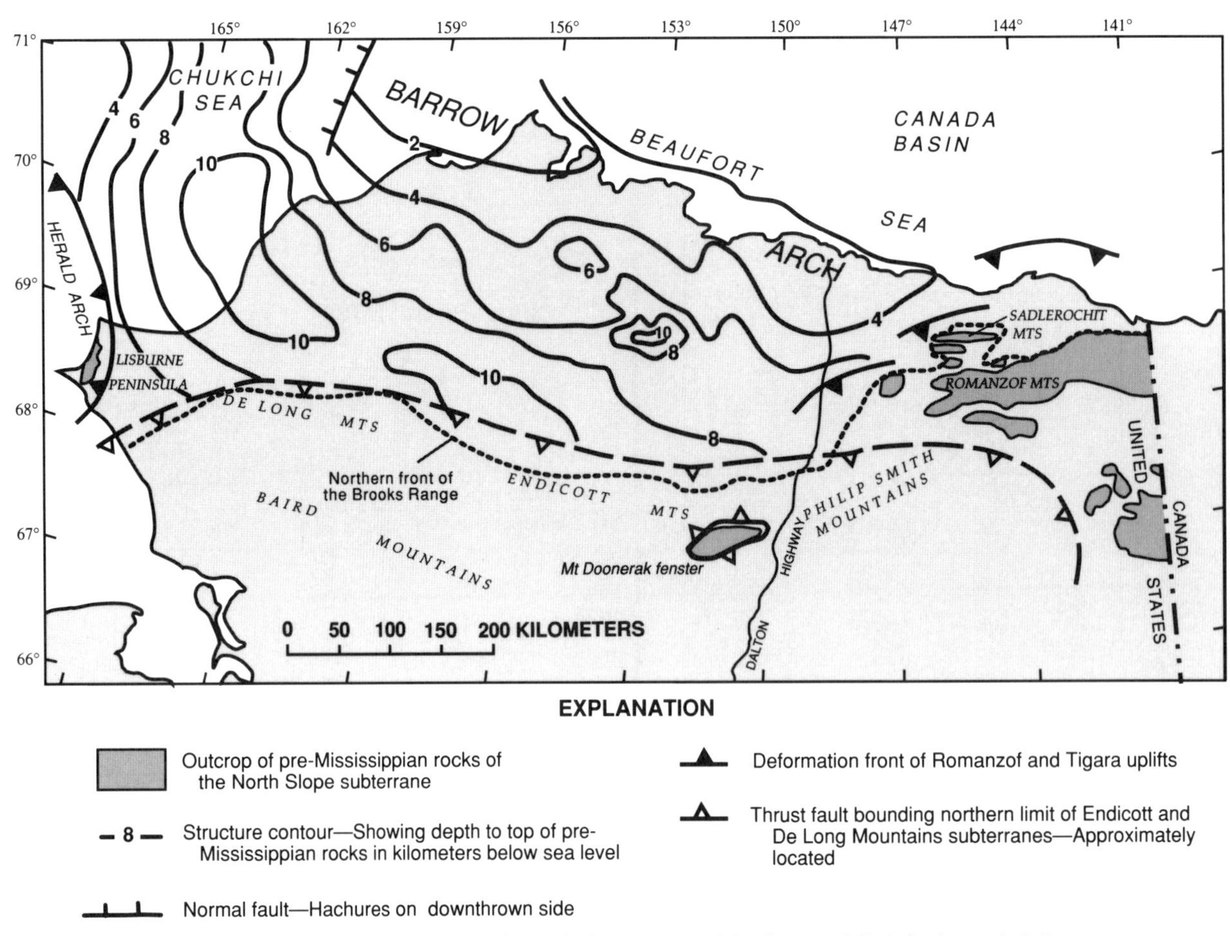

Figure 6. Distribution of pre-Mississippian rocks in outcrop and depth to pre-Mississippian rocks below sea level in North Slope subsurface and adjacent parts of Beaufort and Chukchi seas. Contours based on seismic (Gutman and others, 1982; Hubbard and others, 1987) and well data (Bird, 1982).

1919; Reiser and others, 1978, 1980) consists of metamorphosed quartzite turbidites, pebble conglomerate, phyllite, and quartzitic semischist, all of probable Proterozoic age and continental derivation. Other prominent lithologic packages of possible miogeoclinal affinity include recrystallized pelletoidal limestone and interbedded phyllite, quartzite, calcareous sandstone, and limestone of unknown age. South of these rocks lies an oceanic assemblage consisting of structurally incoherent mafic volcanic rocks, graywacke, radiolarian chert, and argillite that may structurally overlie the miogeoclinal strata. Trilobites of Cambrian age and North American affinity (Dutro and others, 1972; A. R. Palmer, 1988, oral commun.) are associated with the volcanic rocks in the oceanic assemblage, and chert of the assemblage contains Ordovician graptolites (Moore and Churkin, 1984). The oceanic assemblage is truncated by an angular unconformity, which is overlain by Middle Devonian chert arenite. The unconformity and overlying Devonian sandstone are in turn truncated at a low angle by the regional angular unconformity beneath the Mississippian rocks of the lower Ellesmerian sequence (Reiser and others, 1980; Dutro, 1981). Anderson and Wallace (1990) suggested that most deformation in this area occurred during pre–Middle Devonian time, but minor deformation may have continued into Late Devonian time. Stratigraphic relations and composition of the Devonian rocks are analogous to Devonian rocks of the Topagoruk well; the similarity suggests that pre–Middle Devonian deformation and subsequent local deposition of chert-rich clastic strata may have taken place over a wide area.

Mt. Doonerak fenster. Pre-Mississippian rocks in the Mt. Doonerak fenster consist of a structurally higher metasedimentary assemblage and a structurally lower metavolcanic assemblage. The metasedimentary assemblage consists of structurally disrupted dark argillite, phyllite, and slate with silty quartzitic laminae, black siliceous siltstone (metachert?), and lenticular limestone bodies. Middle Cambrian trilobites of Siberian affinity have been recovered from the limestone bodies (Dutro and others, 1984), and other fine-grained strata of the metasedimentary assemblage have yielded Ordovician and Early Silurian graptolites and conodonts (Repetski and others, 1987). The metavolcanic assemblage forms a series of thrust packages more than 2 km thick, consisting of pillow basalt and fragmental volcanic rocks of island-arc affinity (Moore, 1987a; Julian, 1989). Potassium-argon dating of the metavolcanic rocks indicates ages of about 470 Ma (Ordovician), although dikes intruding the assemblage yield ages of about 380 Ma (Devonian) (Dutro and others, 1976).

Lisburne Peninsula. Moderately deformed units of pre-Mississippian argillite and overlying lithic turbidites that are exposed along the western side of the Lisburne Hills were assigned to the Iviagik Group by Martin (1970). The argillite unit consists of more than 100 m of weakly metamorphosed siliceous shale and minor chert. The turbidite unit is at least several hundred meters thick and consists largely of thick-bedded, medium-grained graywacke and pebbly sandstone. A Middle Ordovician graptolite fauna has been recovered from both the argillite and turbidite units, although intercalated shale in the turbidite sequence has also yielded late Early Silurian (late Llandoverian) conodonts (A. G. Harris, 1982, written commun. to I. L. Tailleur; Grantz and others, 1983b).

Granitic rocks. A few scattered plutons intrude the pre-Mississippian rocks of the North Slope subterrane (Figs. 2 and 3). These plutons are generally elliptical in outline, range from 1 to 80 km in diameter, and consist largely of biotite granite and quartz monzonite. Porphyritic textures are common, and microcline megacrysts are reported from some plutons (Sable, 1977; Barker, 1982). Hornblende is abundant only in the small quartz monzonite stock at the headwaters of the Hulahula River in the Demarcation Point quadrangle, but it is also present in the Okpilak batholith. Dillon and others (1987b) reported a peraluminous composition for the Okpilak batholith (Demarcation Point and Mt. Michelson quadrangles) and a metaluminous composition for the nearby Jago stock (Demarcation Point quadrangle). Uranium, tin, and associated base metals are reported from the Old Crow (Coleen quadrangle) and Okpilak batholiths (Sable, 1977; Barker, 1982).

Isotopic ages for the plutons are sparse, but the available ages generally indicate crystallization in the Devonian. Devonian K-Ar, Pb-alpha, and U-Pb ages have been determined for the Okpilak batholith and nearby Jago stock. Granite found in the U.S. Navy East Teshekpuk #1 test well (Fig. 7) and the Old Crow batholith have yielded Mississippian K-Ar ages, which have been interpreted as minimum ages by Bird and others (1978) and Barker (1982). Mississippian strata rest nonconformably on both the Okpilak and Teshekpuk bodies and on hornfels associated with the Old Crow batholith (W. P. Brosgé, 1989, oral commun.), indicating a pre-Mississippian crystallization age for these plutons. In contrast, the small pluton in the headwaters of the Jago River, 15 km south of the Okpilak batholith, has yielded a Silurian K-Ar age. The distinct mineralogy and apparent age may indicate an earlier period of granitic intrusion (Silurian) than that of the other plutons in the North Slope subterrane.

Lower Ellesmerian sequence. The lower Ellesmerian sequence consists of marine carbonate rocks and quartz- and chert-rich marine and nonmarine clastic rocks that rest unconformably on pre-Mississippian rocks throughout the North Slope subterrane. Representing about 150 m.y. (Mississippian through Triassic) of sedimentation (Fig. 5), the sequence contains the most productive reservoirs of the Prudhoe Bay oil field. The lower Ellesmerian sequence averages 1–2 km thick but in local basins may exceed 5 km in thickness. It extends along the entire east-west length of the North Slope (Figs. 7, 8, and 9) but thins and fines southward beneath the foothills of the Brooks Range (Bird, 1985; Kirschner and Rycerski, 1988). The sequence also thins northward, because of onlap onto pre-Mississippian rocks, truncation by unconformities within the succession, and erosion in the Early Cretaceous along the Barrow arch. The extent of the lower Ellesmerian sequence, coupled with northward coarsening, erosional onlap, and progression to more shallow-marine and nonmarine facies, suggests that deposition occurred in shelf and platform environments along a slowly subsiding, south-facing continental margin (Bird and Molenaar, 1987). Three transgressive-regressive cycles are represented by this sequence: Mississippian to Early Permian, Early Permian to Early Triassic, and Early to Late Triassic. The lower and middle cycles are separated by a regional unconformity, whereas the middle and upper cycles are separated by a local basin-margin unconformity.

Endicott and Lisburne groups (Mississippian to Early Permian cycle). The first transgressive-regressive cycle is composed of nonmarine and shallow-marine clastic rocks of the Endicott Group (Upper Devonian to Permian?) (principally the Kekiktuk Conglomerate and Kayak Shale) and the overlying marine carbonate rocks of the Lisburne Group (Mississippian to Permian) (Figs. 5 and 9). These units generally become progressively younger to the north and northwest, and undivided rocks of the Lisburne Group and possibly Endicott Group are as young as Early Permian in the northern National Petroleum Reserve in Alaska (NPRA) (Bird, 1988a, 1988b). Together the Endicott and Lisburne groups compose a genetically related sequence that is bounded at the top and base by regional unconformities. A depositional model proposed by Armstrong and Bird (1976) suggests that nonmarine and nearshore-marine clastic sediments (Endicott Group) were deposited adjacent to, and north of, carbonate sediments (Lisburne Group) on a broad shallow-marine platform.

The Kekiktuk Conglomerate (Lower Mississippian) is a discontinuous, largely nonmarine unit that characteristically rests in angular unconformity on older rocks (Mull, 1982, p. 27). It is typically less than 500 m thick in the Prudhoe Bay area, less than 100 m thick in the northeastern Brooks Range, and only about 40 m thick in the Mt. Doonerak fenster. The Kekiktuk Conglomerate was defined by Brosgé and others (1962) in the Mt. Michelson quadrangle and consists of well-sorted, cross-bedded sandstone with lenses of conglomerate and interbeds of carbonaceous shale and coal. Conglomerate in the unit consists of various proportions of subangular to subrounded chert and quartz clasts and a small to locally large percentage of quartzite and argillite clasts (Brosgé and others, 1962; Nilsen and others, 1981). Maximum clast size, which is greater than 22 cm in the northeastern Brooks Range, decreases toward the west and south (Nilsen and others, 1981). In the Mt. Doonerak fenster, the unit is no younger than late Early Mississippian (Osagean) (Armstrong and others, 1976), whereas in the subsurface of the North Slope along the

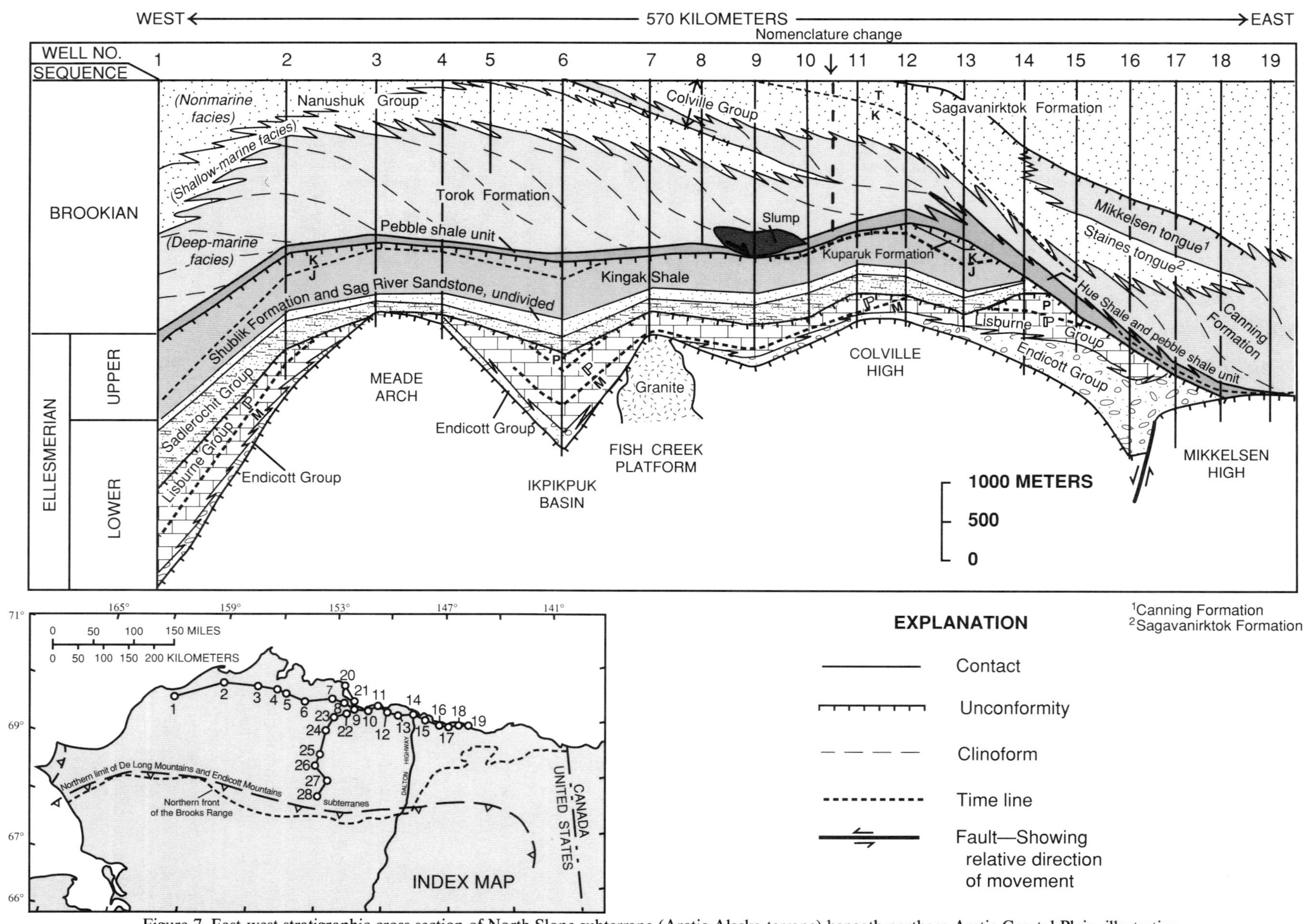

Figure 7. East-west stratigraphic cross section of North Slope subterrane (Arctic Alaska terrane) beneath northern Arctic Coastal Plain, illustrating major subsurface post-Devonian rock units of region and selected period boundaries (modified from Molenaar and others, 1986). Datum is mean sea level. Index map shows location of wells used to construct section: 1, Tunalik-1; 2, Kugrua-1; 3, South Meade-1; 4, Topagoruk-1; 5, East Topagoruk-1; 6, Ikpikpuk-1; 7, East Teshekpuk-1; 8, North Kalikpik-1; 9, South Harrison Bay-1; 10, Nechelik-1; 11, Colville Delta-1; 12, Kalubik Creek-1; 13, West Sak River-1; 14, Prudhoe Bay State-1; 15, Foggy Island Bay-1; 16, West Mikkelsen Bay State-1; 17, East Mikkelsen Bay State-1; 18, Point Thomson Unit-2; 19, Alaska State A-1.

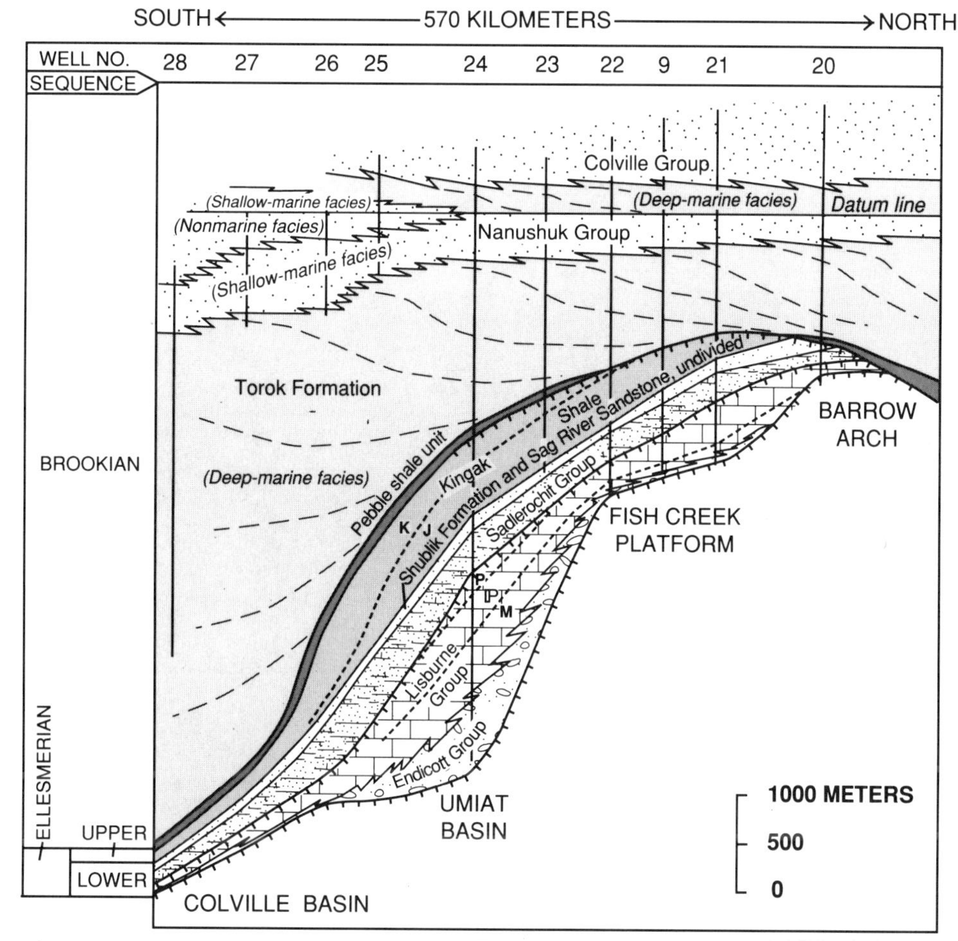

Figure 8. North-south stratigraphic cross section of North Slope subterrane (Arctic Alaska terrane) from Colville basin to Barrow arch, illustrating major subsurface post-Devonian rock units of region and selected period boundaries (modified from Molenaar and others, 1986). Datum is transgressive surface near or at top of Nanushuk Group. Numbers indicate wells used to construct diagram: 20, W.T. Foran-1; 21, Atigaru Point-1; 9, South Harrison Bay-1; 22, West Fish Creek-1; 23, North Inigok-1; 24, Inigok-1; 25, Square Lake-1; 26, Wolf Creek-3; 27, Little Twist-1; 28, East Kurupa-1. For location of wells and explanation of symbols, see Figure 7.

Barrow arch, the Kekiktuk may be as young as Late Mississippian (Meramecian), according to Woidneck and others (1987).

Seismic reflection lines show several basin-fill successions in the subsurface of the North Slope that lie conformably beneath Carboniferous strata of the Endicott Group (for example, the Umiat and Meade basins) (Fig. 9). These basins are as much as 4 km thick and postdate development of the regional middle Paleozoic unconformity. Although the basin-fill successions are not exposed and only their upper parts have been penetrated by deep wells, the basins are inferred to contain Lower Mississippian sedimentary rocks that are commonly assigned to the Kekiktuk Conglomerate and/or undivided Endicott Group (for example, Bird, 1985; Kirschner and Rycerski, 1988). Oldow and others (1987d) and Grantz and others (1990a), however, related the sedimentary fill of these basins to Devonian clastic rocks in the Topagoruk well and distinguished these two basin-fill successions from those of the thinner and more widespread strata of the Kekiktuk Conglomerate. The morphology of the basin-fill successions suggests deposition within a system of half grabens that developed during a period of regional extension (Oldow and others, 1987d; Kirschner and Rycerski, 1988; Grantz and others, 1990a).

The Kekiktuk Conglomerate and related coarse-grained clastic rocks are overlain gradationally by less than 400 m of gray to black, carbonaceous marine shale of the Kayak Shale (Mississippian), which was defined in the Endicott Mountains subterrane by Bowsher and Dutro (1957). The Kayak commonly contains various amounts of interbedded sandstone near its base and increasing amounts of fossiliferous, argillaceous limestone toward its top. Fossils from the unit indicate deposition in the Mississippian (Osagean and Meramecian?) in the Mt. Doonerak fenster (Armstrong and Mamet, 1978) and suggest deposition in the Late Mississippian (Meramecian) in the northeastern Brooks Range and North Slope subsurface (Armstrong and Bird, 1976). The Kayak interfingers laterally (northward) and vertically in the

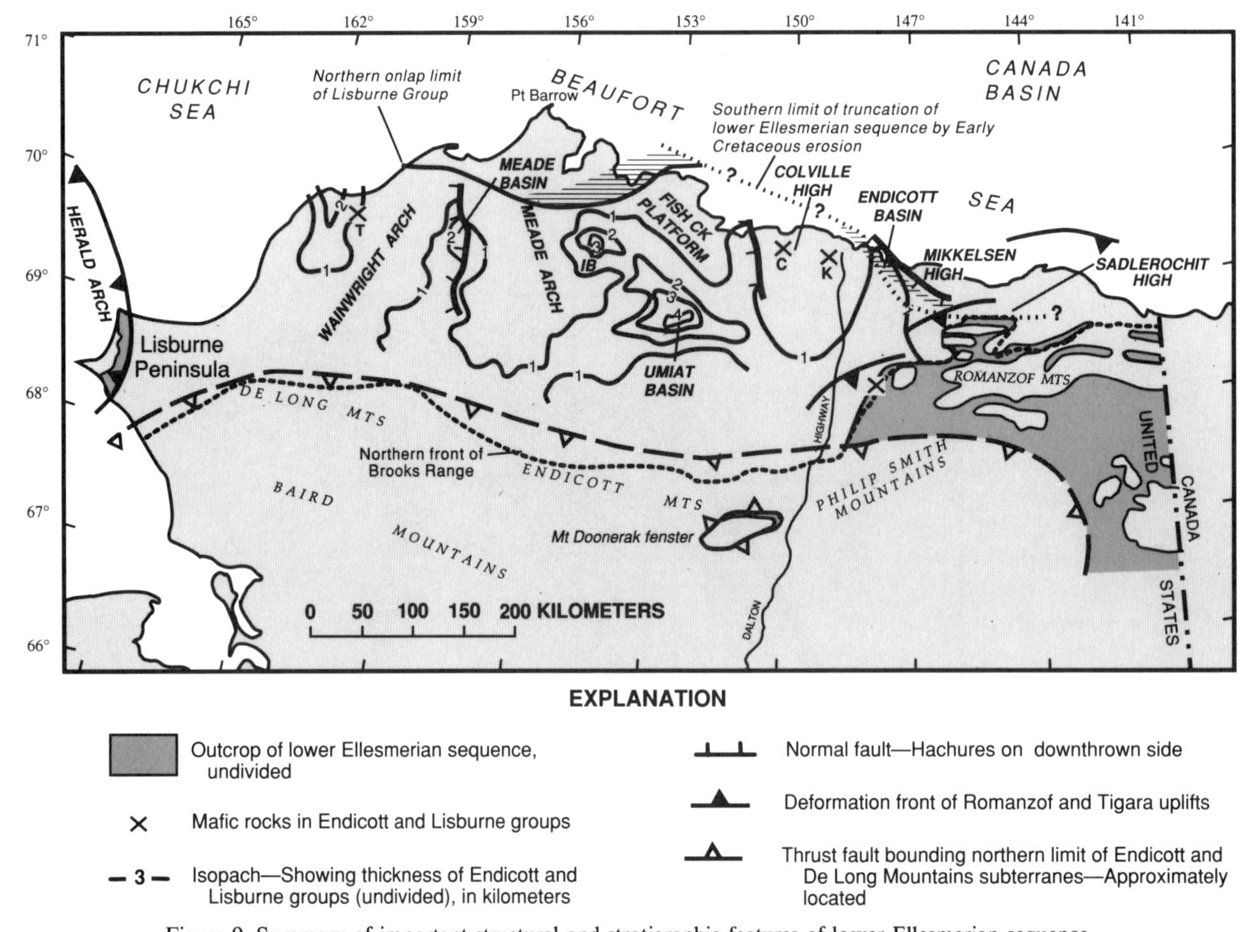

Figure 9. Summary of important structural and stratigraphic features of lower Ellesmerian sequence, North Slope subterrane (Arctic Alaska terrane), including isopachs of undivided Endicott and Lisburne groups (Gutman and others, 1982). Unusual thicknesses of clastic and carbonate rocks assigned to Endicott and Lisburne groups filled basins that developed as sags and half grabens in Mississippian or possibly Devonian time (see Figs. 7 and 8). Basin-bounding normal faults shown are generally overlapped by Lisburne Group or younger strata. Unconformity at top of Lisburne Group cut down into Endicott Group in a narrow band southeast of Point Barrow and between Mikkelsen and Colville highs (horizontal-line pattern). C, Colville State-1 well; IB, Ikpikpuk basin; K, Kuparuk State-1 well; T, Tunalik-1 well.

North Slope subsurface with the Itkilyariak Formation, a distinctive sequence of Upper Mississippian red, gray, and green shale, limestone, and sandstone (Mull and Mangus, 1972). The Kayak is a northward-transgressive unit and was deposited in brackish-water and shallow-marine environments, whereas the Itkilyariak Formation was deposited on an arid coastal plain that was flooded peridically by the sea (Bird and Jordan, 1977).

The Lisburne Group is present throughout most of the North Slope subterrane, where it consists of limestone and dolomite and various amounts of shale, sandstone, and nodular replacement chert. Its thickness is variable, averaging about 600 m and locally exceeding 1200 m, and it is thickest in basins established during deposition of the Endicott Group (Fig. 9). In most of the North Slope subterrane, the Lisburne Group consists of the Alapah Limestone (Upper Mississippian) and the Wahoo Limestone (Upper Mississippian to Middle Pennsylvanian), which was defined by Brosgé and others (1962) in the Mount Michelson quadrangle and is present only in the North Slope subterrane. The underlying Wachsmuth Limestone (Lower and Upper Mississippian) is present locally in the northeastern Brooks Range (Armstrong and Mamet, 1978). The Lisburne Group is young as Early Permian in the subsurface of the NPRA, but the Permian rocks have not yet been assigned to formations (Bird, 1988a). On the Lisburne Peninsula, for which Schrader (1902, 1904) named the group, the Lisburne consists, in ascending order, of the Nasorak Formation (Upper Mississippian), Kogruk Formation (Upper Mississippian and Lower Pennsylvanian), and Tupik Formation (Upper Mississippian and Lower Pennsylvanian?) (Campbell, 1967; Armstrong and others, 1971; Dutro, 1987; Bird, 1988a). Units of the Lisburne Group are lithologically similar, commonly having been distinguished only by age, and are difficult to map; therefore, most workers have treated the

Lisburne Group as an undifferentiated unit. Three environmental assemblages, however, are recognized within the Lisburne Group: a transgressive assemblage, a platform assemblage, and a deeper water assemblage (Armstrong, 1974; Armstrong and Bird, 1976) (Fig. 10).

The transgressive assemblage, represented primarily by the Nasorak Formation, the Wachsmuth Limestone, and the lower part of the Alapah Limestone in the North Slope subterrane, rests conformably on the Kayak Shale. This assemblage consists of spiculitic, argillaceous lime mudstone and overlying spiculitic, pelletoid crinoid-bryozoan wackestone and packstone. Dolomite, replacement chert, and shale are locally interbedded with the limestone. Upward in the assemblage, the carbonate rocks become slightly coarser grained, and the argillite content decreases, although grainstone is relatively uncommon, and well-developed oolite has not been observed. The presence of bryozoans, echinoderms, corals, brachiopods, foraminifers, and algae indicates open-marine conditions. Although this assemblage records many oscillations and, undoubtedly, hiatuses, it represents mainly open-marine carbonate environments and upward in the assemblage, the development of a carbonate platform.

The platform assemblage, which comprises the Kogruk Formation, the Wahoo Limestone, and the upper part of the Alapah Limestone in the North Slope subterrane, covers a full spectrum of carbonate lithologies from lime mudstone to grainstone. Color of the assemblage changes from gray and green in the south to red and gray in the north, near the paleoshoreline. Clastic content is also variable and increases north of the Brooks Range. The depositional model for the platform assemblage (Armstrong, 1974) is similar to those for most Phanerozoic carbonate platforms, except for a lack of reef-building organisms. Outcrop studies by Armstrong (1972) and Wood and Armstrong (1975) in the northeastern Brooks Range show that the platform assemblage is composed of many incomplete depositional cycles that indicate relatively rapid sea-level rises, protracted periods of stability, and prograding carbonate offlap. A complete cycle of carbonate deposition in the platform assemblage consists of crinoid-bryozoan packstone and wackestone overlain gradationally by ooid or crinoid packstone and grainstone that, in turn, is capped by packstone, wackestone, mudstone, and microdolomite. This succession represents progressive shallowing from open-marine through carbonate shoal, restricted platform, intertidal, and supratidal environments. Presence of algal mats, mud chips and cracks, and gypsum and anhydrite cements and replacements in beds deposited in the intertidal and supratidal environments suggests an arid climate.

In the North Slope subterrane, the deeper water assemblage is restricted to the Lisburne Peninsula, where it forms the uppermost part of the Lisburne Group (Tupik Formation). This third environmental assemblage consists of subequal amounts of dark, thin-bedded limestone, dolomite, chert, and siliceous shale, and contains relatively abundant sponge spicules, radiolarians, cephalopods, and phosphatic intervals. Limestone beds, commonly graded and laminated, consist of lime mudstone and carbonate turbidite. Chert in the assemblage is typically stratified, lenticular, and nodular, but locally crosscuts bedding, which indicates a replacement origin. The deeper water assemblage is typically thin compared to the other two environmental assemblages and represents deposition in deeper shelf, slope, and basinal environments at or near starved-basin conditions (Armstrong and Mamet, 1978).

The foraminiferal zonation of B. L. Mamet (Armstrong and others, 1970) has facilitated correlations among outcrop sections and well penetrations of the Lisburne Group (Armstrong, 1974; Armstrong and Mamet, 1977; Bird and Jordan, 1977; Bird, 1978; Witmer and others, 1981). These paleontologic correlations,

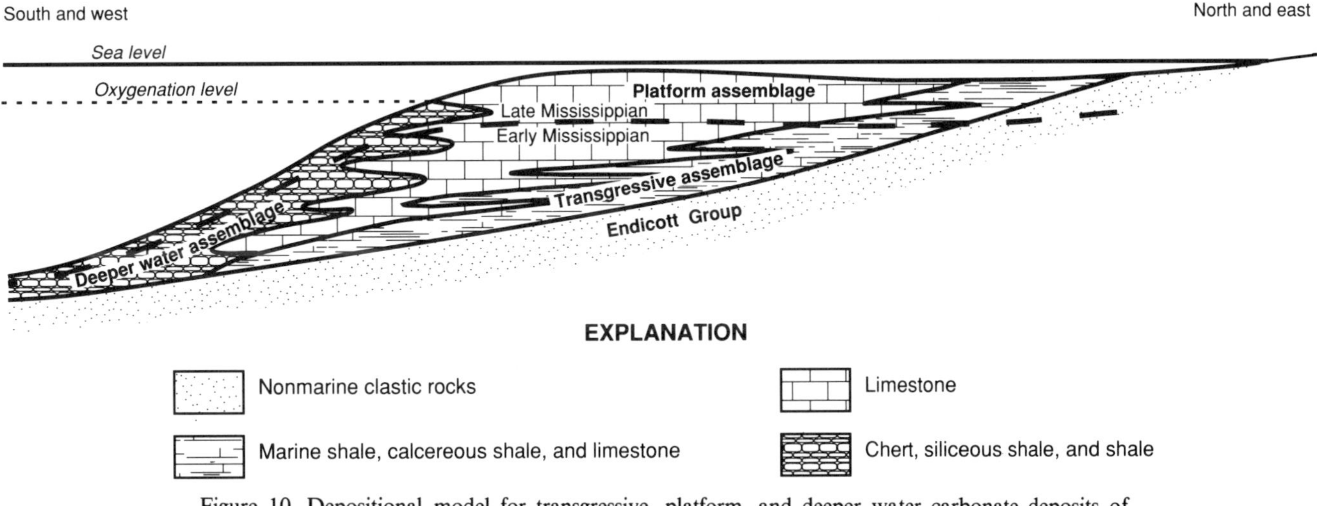

Figure 10. Depositional model for transgressive, platform, and deeper water carbonate deposits of Lisburne Group (modified from Armstrong and Bird, 1976). Distribution of facies was probably partly controlled by interplay of regional subsidence and oxygenation level in basin. Maximum thickness of carbonate strata of Lisburne Group is about 1000 m.

coupled with seismic data from the NPRA (Bruynzeel and others, 1982), show that east of the Meade arch (Fig. 9) the Lisburne Group transgressed about 100 km northward during the early Late Mississippian (Meramecian), whereas in the area of the Meade arch and the western NPRA, the Lisburne transgressed toward the northwest during Late Mississippian to Early Permian time. Gradual facies changes and wide areal distribution suggest that the sea floor upon which the Lisburne of the North Slope subterrane was deposited was a very low gradient ramp as much as 300 km wide. The areal extent of, and amount of erosion on, the regional unconformity at the top of the Lisburne are not well established and represent important unanswered questions.

Echooka Formation and lower and middle Ivishak Formation, Sadlerochit Group (Early Permian to Early Triassic cycle). Originally designated as the Sadlerochit Sandstone by Leffingwell (1919) for outcrops in the northeastern Brooks Range, the Sadlerochit is a widespread rock unit divided into two formations, the Echooka and Ivishak, and elevated to group rank by Detterman and others (1975). It is primarily a clastic, nonmarine to marine-shelf deposit of northern derivation that gradually thickens southward in the subsurface to more than 600 m (Fig. 11). The Sadlerochit overlies a regional unconformity above the Lisburne Group, which is marked by significant erosional relief but little discordance. West of long 154°W, the northern limit of the Sadlerochit is an onlap-pinchout against pre-Mississippian rocks, whereas to the east it is truncated by an Early Cretaceous unconformity.

The Echooka Formation (Permian) consists of about 100 m of calcareous mudstone, radiolarian chert, and bioclastic, glauconitic limestone of the Joe Creek Member and an overlying 100 m of quartzose sandstone and siltstone of the Ikiakpaurak Member (Detterman and others, 1975). A crudely channelized chert-pebble and chert-cobble conglomerate is present locally at the base of the formation in the Sadlerochit and Shublik mountains (Crowder, 1990). Deposits of the Echooka document the northward advance of the Sadlerochit sea across the eroded platform of

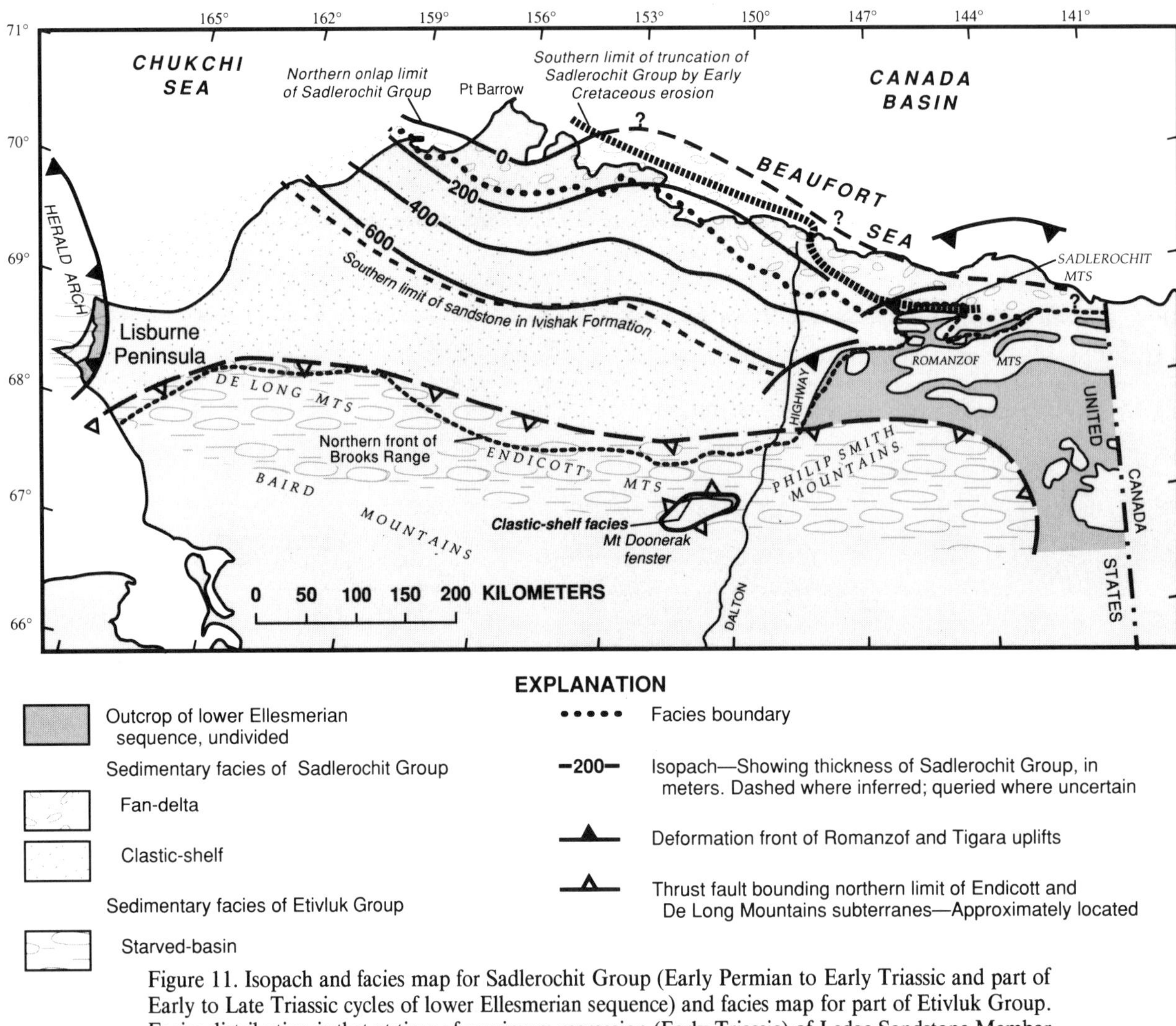

Figure 11. Isopach and facies map for Sadlerochit Group (Early Permian to Early Triassic and part of Early to Late Triassic cycles of lower Ellesmerian sequence) and facies map for part of Etivluk Group. Facies distribution is that at time of maximum regression (Early Triassic) of Ledge Sandstone Member of Ivishak Formation.

Lisburne carbonate rocks. Brachiopods indicate that the Joe Creek Member is Early and Late Permian (Sakmarian to Kazanian), and the Ikiakpaurak Member is Late Permian (Kazanian; late Guadalupian) (Detterman and others, 1975). The Echooka Formation is characterized by the trace fossil *Zoophycos.*

The Ivishak Formation (Triassic) (Keller and others, 1961; Detterman and others, 1975) consists of fine- to coarse-grained clastic rocks deposited in marine and nonmarine environments. Detterman and others (1975) divided the Ivishak into three members: in ascending order, these are the Kavik Member, the Ledge Sandstone Member, and the Fire Creek Siltstone Member. We assign the Kavik Member and Ledge Sandstone Member to the upper part of the Early Permian to Early Triassic depositional cycle, whereas the Fire Creek Member represents a younger transgressive episode and is assigned to the overlying Early to Late Triassic depositional cycle.

The Kavik Member of the Ivishak Formation (Triassic) abruptly overlies the Echooka Formation. This abrupt contact, thought to be a disconformity by Detterman and others (1975), is probably a surface of downlap by the southward-prograding Kavik Member. The Kavik consists of up to 213 m of dark-colored, laminated to thin-bedded silty shale and siltstone that thicken southward from the Barrow arch. These rocks represent prodelta deposits that grade upward into massive deltaic sandstones and conglomerates of the Ledge Sandstone Member. In outcrop, the Kavik is dated as Early Triassic (late Griesbachian, in part) by ammonites and pelecypods (Detterman and others, 1975), but the Kavik is Late Permian in the subsurface of the Prudhoe Bay area (Jones and Speers, 1976).

The Ledge Sandstone Member of the Ivishak Formation is as much as 200 m thick and consists of sandstone beds that thicken and coarsen northward, and contains thin siltstone and shale interbeds. Chert-pebble to chert-cobble conglomerate is present in its northernmost facies. Because the Ledge is the primary reservoir for the Prudhoe Bay oil field, it has been studied in considerable detail (Detterman, 1970; Eckelmann and others, 1976; Jones and Speers, 1976; Wadman and others, 1979; Jamison and others, 1980; Melvin and Knight, 1984; Lawton and others, 1987; Marinai, 1987; Payne, 1987; Atkinson and others, 1988). At Prudhoe Bay, the Ledge is a fluvial-deltaic complex, which can be divided into a lower progradational, upward-coarsening megacycle, ranging from prodelta siltstone to an alluvial-fan clast-supported conglomerate, and an upper upward-fining sandstone megacycle. Lawton and others (1987) suggested that the fluvial-deltaic facies of the Ledge was deposited on an elongate, relatively narrow coastal plain that was traversed by both braided and meandering streams. Marine sandstone of the Ledge extends southward for as far as 100 km in the subsurface (Fig. 11). A greater percentage of sandstone and conglomerate east of long 154°W suggests greater uplift in the nearby source highlands in this area than to the west.

Southward from the Barrow arch, the Sadlerochit Group becomes finer grained, more marine, and more difficult to subdivide. In the Romanzof Mountains, the Sadlerochit consists largely of a thick sequence of siliceous mudstone and locally thin bedded limestone and is overlain by dark shale and thin-bedded, ripple-marked, fine-grained sandstone that was deposited in a marine-shelf environment. In the Mt. Doonerak fenster, the Sadlerochit consists of a 55-m-thick lower unit of calcareous, very fine grained sandstone and siltstone and a 70-m-thick upper unit of black, phyllitic shale (Mull and others, 1987a). The lower unit contains the trace fossil *Zoophycos* and Early Permian (Wolfcampian) brachiopods. The sedimentary structures and fauna suggest that the Sadlerochit shelf extended southward at least as far as the Mt. Doonerak fenster.

Upper Ivishak Formation, Shublik Formation, Sag River Sandstone, and Karen Creek Sandstone (Early to Late Triassic cycle). The third transgressive-regressive cycle in the lower Ellesmerian sequence consists of the Fire Creek Siltstone Member of the upper Ivishak Formation (Sadlerochit Group), the Shublik Formation, and the Sag River and Karen Creek sandstones. The Fire Creek Siltstone Member of the Ivishak Formation is an upward-fining and northward-thinning unit composed of as much as 135 m of thin-bedded to massive, commonly laminated siltstone and argillaceous sandstone. The unit gradationally overlies the Ledge Sandstone Member and in its northernmost extent may either pinch out or be erosionally truncated by an unconformity at the base of the overlying Shublik Formation. Burrows and sparse ammonites and pelecypods indicate that the Fire Creek Siltstone Member is marine and represents a deepening of the sea and the initiation of the next transgressive-regressive cycle. Ammonites date the Fire Creek as Early Triassic (Smithian) (Detterman and others, 1975).

The Shublik Formation (Triassic) (Leffingwell, 1919; Mull and others, 1982) is a relatively thin, dark-colored, poorly exposed, heterogeneous assemblage of richly fossiliferous shale, mudstone, carbonate rocks (bioclastic limestone, dolomite, siderite), siltstone, and sandstone of Middle and Late Triassic age (Detterman and others, 1975). This unit onlaps pre-Mississippian rocks in the subsurface near Point Barrow and in outcrop just east of the international boundary bordering the northeastern Brooks Range. It rests disconformably on the Sadlerochit Group where the Fire Creek Siltstone Member of the Ivishak Formation is thin or absent, as at Prudhoe Bay and in the Sadlerochit Mountains. Along the axis of the Barrow arch and northward, the Shublik Formation is truncated by the extensive Lower Cretaceous unconformity (Fig. 12). The Shublik Formation is a blanket-like deposit that averages about 100 m thick in most areas but is nearly 200 m thick in the northeastern NPRA and in the northeastern Brooks Range, where it may have been deposited close to points of clastic-sediment influx (Bird, 1987).

The Shublik was deposited on a low-gradient, southward-sloping shelf that was inherited from the underlying Sadlerochit Group; it represents an important regional marine transgression that overstepped the northern depositional limit of the Sadlerochit Group. Parrish (1987) identified a north to south succession of facies within the Shublik that may represent regional upwelling of oceanic water from the south. The northernmost facies consists

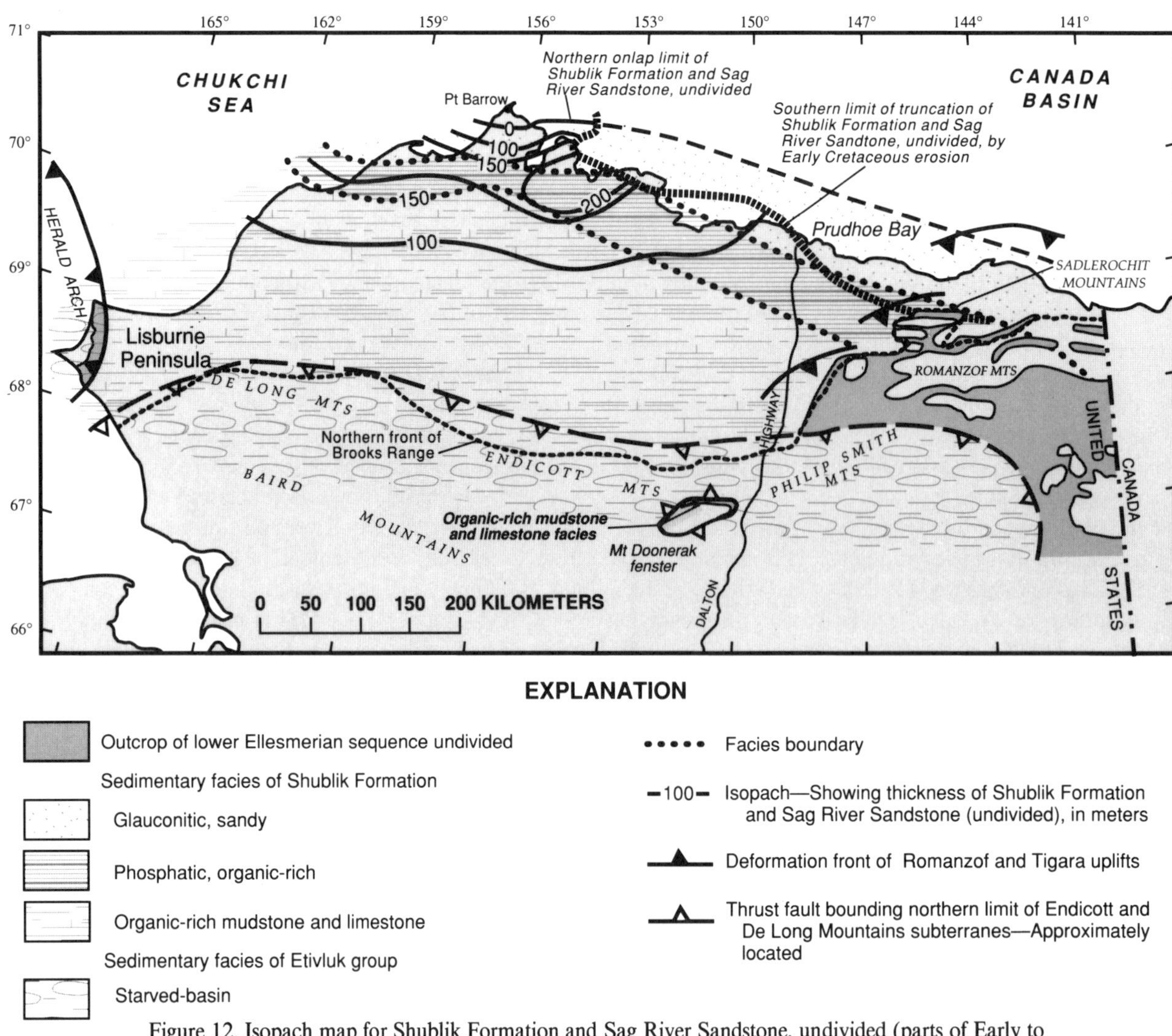

Figure 12. Isopach map for Shublik Formation and Sag River Sandstone, undivided (parts of Early to Late Triassic cycle of lower Ellesmerian sequence) and facies map for Shublik Formation (facies from Hubbard and others, 1987; Parrish, 1987), and coeval part of Etivluk Group.

of nearshore, fossiliferous sandstone and siltstone with variable amounts of glauconite. This facies grades southward into siltstone, calcareous mudstone, and limestone that contain phosphate nodules. The phosphate-bearing facies grades southward into the southernmost facies of black, organic-rich calcareous mudstone and fossiliferous limestone deposits, both of which contain abundant *Halobia* and *Monotis* bivalves. Shublik deposition was terminated by a minor regression, which deposited the overlying widespread, thin, shallow-marine sandstone sequence (Sag River and Karen Creek sandstones).

The Sag River Sandstone (Triassic) in the subsurface of the North Slope (North Slope Stratigraphic Committee, 1970) and its lithologic correlative in outcrop, the Karen Creek Sandstone (Detterman and others, 1975), are discontinuous, southward-thinning units. The maximum thickness of the Sag River Sandstone is about 100 m in the northeastern NPRA, and it thins rapidly southward (Fig. 12). The Sag River and Karen Creek sandstones comprise an intensely bioturbated succession of fine-grained to very fine grained, argillaceous, glauconitic sandstone and interbedded siltstone, and shale (Detterman and others, 1975; Barnes, 1987). Sedimentologic and stratigraphic relations of the Sag River are comparable to modern low-energy offshore-marine environments (Barnes, 1987). Late Norian bivalves from the base of the Karen Creek date the formation as Late Triassic (Detterman and others, 1975), whereas spores and pollen date the Sag River as Late Triassic to earliest Jurassic (Rhaetian to Hettangian) (Barnes, 1987). In the Prudhoe Bay area, the Sag River may become slightly older to the north, which suggests that the formation is time transgressive. Barnes (1987) interpreted this trend as evidence for Sag River deposition during culmination of a regionally significant marine regression in the North Slope that began in middle Shublik time.

Etivluk Group. Upper Paleozoic and lower Mesozoic rocks of the Lisburne Peninsula consist of 200 m of fine-grained clastic deposits and chert assigned to the Etivluk Group (Mull and others, 1982). The Etivluk of the Lisburne Peninsula is partly coeval with more proximal units of the Early Permian to Early Triassic and Early to Late Triassic cycles described above and is lithologically correlative with similar strata of the Etivluk Group in the Endicott Mountains and De Long Mountains subterranes described below. The lower part of the Etivluk succession on the Lisburne Peninsula is about 125 m thick and consists, in ascending order, of thoroughly bioturbated, gray, maroon, and green siliceous argillite, gray-green bedded chert, and argillaceous chert. The upper part is about 75 m thick and consists, in ascending order, of black siliceous shale, dark gray to black chert, thin-bedded, fossiliferous limestone, and gray chert and shale. Blome and others (1988) and Murchey and others (1988) reported that radiolarians collected from the middle part of the lower unit are Late Pennsylvanian or Early Permian, and those near the top of the lower unit are Permian; radiolarians and megafossils from the upper unit range from Middle (Ladinian) to Late (Norian) Triassic (Blome and others, 1988). The fine-grained, siliceous character of these strata, the type of faunal assemblages, and intense bioturbation indicate that the Etivluk Group rocks of the Lisburne Peninsula were deposited under starved-basin conditions in inner to outer shelf environments (Figs. 11 and 12).

Upper Ellesmerian sequence. Seismic stratigraphy shows that northern Alaska underwent a 100 m.y. interval (Jurassic to Early Cretaceous–Aptian) of extension, during which a failed rift episode in the Jurassic was followed by a successful rift episode in the Early Cretaceous (Hauterivian) (Grantz and May, 1983; Hubbard and others, 1987). This extension led to opening of the Canada basin and ultimate displacement of the Arctic Alaska terrane from the northern land mass that supplied quartz- and chert-rich sediments to the Ellesmerian sequence. The record of extension prior to opening of the Canada basin is contained in two rock sequences in northern Alaska, the Dinkum graben sequence on the Beaufort Sea shelf and the upper Ellesmerian sequence onshore.

Seismic stratigraphy of the Beaufort Sea shelf (Hubbard and others, 1987) reveals areally restricted clastic sedimentary units more than 3 km thick in the Dinkum graben and related half grabens (Fig. 13; Plate 13). These units, assumed to be coarse grained, represent rift-basin deposits (Grantz and May, 1983; Hubbard and others, 1987). Because they have been described by Grantz and others (1990a, and this volume), they are not discussed further here.

The upper Ellesmerian sequence, discussed in detail below, consists of areally extensive, fine-grained clastic strata deposited on a south-dipping shelf and slope beneath the present-day North Slope (Fig. 13) and south of the main axis of Jurassic and Early Cretaceous extension. The upper Ellesmerian sequence consists principally of the marine Kingak Shale, lower Kongakut Formation, and pebble shale unit, plus other sandstone units of local extent (Figs. 5, 7, and 8). The Kingak Shale was deposited over most of the North Slope subterrane in Jurassic and Early Cretaceous time but was uplifted along the incipient Arctic Ocean margin in the Early Cretaceous (Valanginian and Hauterivian) as part of a northwesterly elongate landmass about 250 km wide (Fig. 14). Later, the landmass was eroded to a surface of low relief and transgressed by the sea, resulting in deposition across a regional unconformity in later Neocomian time of a thin sequence of scattered sand bodies (upper part of Kuparuk Formation and Kemik Sandstone) and the blanket-like pebble shale unit. The regional unconformity, commonly referred to as the Lower Cretaceous unconformity, is restricted to the Barrow arch region, where it played an important role in the development of porosity and sealing of the North Slope petroleum reservoirs (Bird, 1987). To the south, the pebble shale and local sandstone units lie conformably on basinal, slope, and shelf deposits of the Kingak Shale. The upper Ellesmerian sequence totals more than 1.2 km thick in the NPRA but depositionally thins southward to less than half that thickness and thins northward because of erosional truncation in the Early Cretaceous (Fig. 13).

Kingak Shale. The Kingak Shale (Jurassic and Lower Cretaceous) consists predominantly of dark gray to black, micromicaceous, noncalcareous, pyritic shale and siltstone as thick as 1200 m (Detterman and others, 1975; Molenaar, 1983, 1988; Bird, 1987). The Kingak was considered Jurassic by Detterman and others (1975), but new information from the subsurface of the NPRA and reevaluation of outcrop data extended the Kingak to include Lower Cretaceous (Neocomian) black shale (Molenaar, 1983, 1988).

Seismic and outcrop data show that the Kingak Shale is composed of at least four southward-prograding, offlapping, and downlapping wedges of sedimentary rock (Bruynzeel and others, 1982; Kirschner and others, 1983; Bird, 1987; Hubbard and others, 1987; Molenaar, 1988). The clastic wedges consist of shelf and slope sequences that grade into basinal facies to the south (Molenaar, 1988). Molenaar (1988) calculated foreset angles of 1°–2° from clinoform reflectors and interpreted water depths of more than 400 to 1000 m for basinal Kingak strata. These cycles may represent local tectonism from the interplay between active rifting to the north and eustatic changes.

In vertical profiles from wells and outcrops, the sedimentary prisms are represented by gradual upward-coarsening cycles of shale and siltstone that are abruptly overlain by shale of the next cycle. The base of each cycle usually represents a downlap surface, characterized by very low rates of sedimentation (or nondeposition) and missing biozones. Fine-grained sandstone is present locally at the tops of the coarsening-upward cycles, particularly in the northern parts of the NPRA. Some of the sandstone units, such as the Lower Jurassic Simpson and Middle or Upper Jurassic Barrow sandstones (Bird, 1988a), are glauconitic and heavily bioturbated, suggesting offshore-bar deposition. The Simpson, and probably the Barrow, grade both northward and southward into finer grained marine facies of the Kingak Shale.

Kuparuk Formation. The Kuparuk Formation (Lower Cretaceous), a major oil-producing reservoir about 50 km west of Prudhoe Bay, consists of about 120 m of glauconitic sandstone

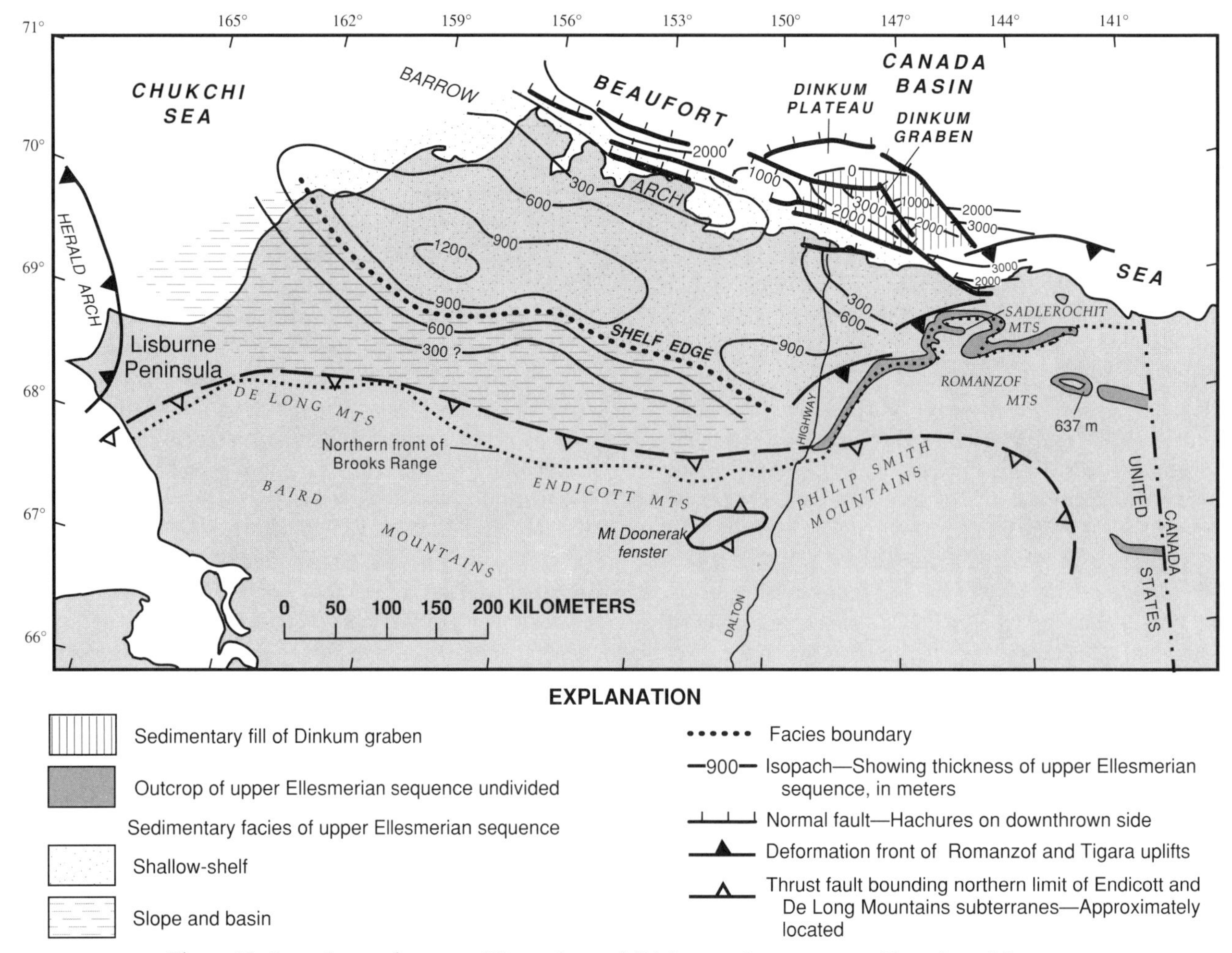

Figure 13. Isopach map for upper Ellesmerian and Dinkum graben sequences (Jurassic and Lower Cretaceous) and facies map for upper Ellesmerian sequence. Isopachs and rift-margin normal faults from Hubbard and others (1987). Abundance of normal faults on Beaufort Sea shelf suggests that main axis of Jurassic and Early Cretaceous extension was north of present coastline.

with interbedded siltstone and shale that gradationally overlies Lower Cretaceous marine shale of the Kingak (Carman and Hardwick, 1983; Molenaar and others, 1986; Masterson and Paris, 1987; Bird, 1988a; Gaynor and Scheihing, 1988). Masterson and Paris (1987) divided the Kuparuk (their Kuparuk River Formation) into two members. These are separated by the regional Lower Cretaceous unconformity. The lower member consists of six southeasterly prograding sandstone intervals, interpreted as storm deposits derived from a northern source and deposited on a marine shelf. Individual sand bodies in this member are as much as 24 m thick, 64 km long, and 24 km wide. Sandstone intervals in the upper member, as much as 15 m thick, were deposited on a marine shelf during an episode of extensional tectonism that produced local northwest-striking faults. Faulting influenced the thickness of the sandstone intervals and contributed to development of an intraformational unconformity that is probably related to the regional Lower Cretaceous unconformity. Stratigraphic thickening and rock-fragment composition suggest that an uplift near Prudhoe Bay was a source area for some of these sandstones. Dinoflagellates, palynomorphs, and pelecypods indicate that the lower part of the formation was deposited in the Berriasian(?) and Valanginian and that the upper part was deposited from the Hauterivian to the Barremian; therefore, erosional truncation occurred in late Valanginian and/or early Hauterivian time (Carman and Hardwick, 1983; Masterson and Paris, 1987).

Lower part of the Kongakut Formation. The Kongakut Formation (Lower Cretaceous) was defined by Detterman and others (1975) for a 637-m-thick sequence of shale and siltstone exposed at Bathtub Ridge in the eastern Brooks Range. They divided the Kongakut into four members: in ascending order, the clay shale, Kemik Sandstone, pebble shale, and siltstone members. In contrast to the two upper members of the Kongakut, which contain beds of feldspathic lithic sandstone and hence compose the lower part of the Brookian sequence (see below), the lower two members contain laminae and beds of quartzose sandstone and are therefore here assigned to the Upper Ellesmerian sequence.

The clay shale member consists of about 150 m of fissile,

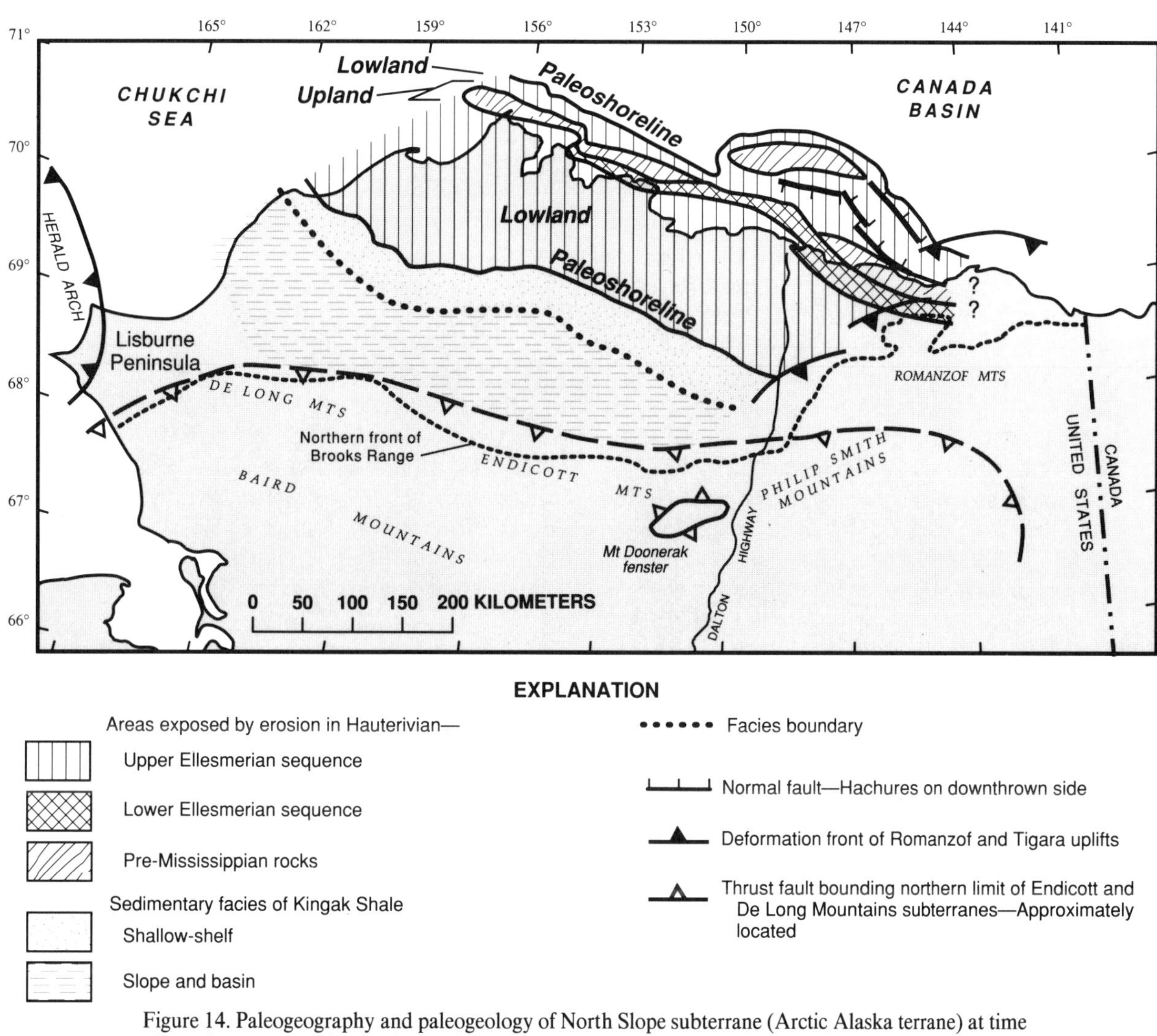

Figure 14. Paleogeography and paleogeology of North Slope subterrane (Arctic Alaska terrane) at time of maximum regression in Early Cretaceous (Hauterivian). Exposed area shows extent of Lower Cretaceous unconformity and underlying truncated rock units; shelf, slope, and basin facies of offshore areas are represented by Kingak Shale.

dark gray, marine shale that contains sparse beds of bioclastic limestone and coquinite. The overlying Kemik Sandstone Member consists of fine-grained to very fine grained sandstone turbidites and is less than 2 m thick (C. G. Mull, 1992, unpublished data), although Detterman and others (1975) reported a thickness of 80 m for this unit. *Buchia* fossils in the limestone of the clay shale member are Valanginian; the age and lithology of this member are correlative with at least part of the Kingak Shale. The Kemik Sandstone Member of the Kongakut has not been dated but is inferred to be a deep-marine equivalent of the Hauterivian, shallow-marine Kemik Sandstone, which is exposed farther north in the northeastern Brooks Range (see below).

Kemik Sandstone and related sandstone units. Many apparently discontinuous sandstone bodies lying above the Lower Cretaceous unconformity along the Barrow arch are stratigraphically equivalent to the upper part of the Kuparuk Formation. These sandstone bodies range in thickness from a few meters to as much as 100 m, have detrital compositions indicating nearby sources, and represent a variety of nearshore shallow-marine to offshore-bar environments. In the subsurface, most of the sandstone bodies are unnamed and generally are found in only one or two wells; a few are petroleum reservoirs. Two of the better-known sandstone bodies in the subsurface are the Put River Sandstone (Jamison and others, 1980) and the Thomson sand of local usage (Bird and others, 1987; Gautier, 1987). The Kemik Sandstone (Keller and others, 1961; Detterman and others, 1975; Molenaar, 1983; Molenaar and others, 1987; Mull, 1987), a lithologically similar unit exposed in the foothills of the northeastern Brooks Range, consists of as much as 40 m of sandstone and local pebble and cobble conglomerate that unconformably overlie Triassic, Jurassic, and lowermost Cretaceous rocks. The Kemik Sandstone locally contains abundant thick-shelled megafossils, grades laterally and interfingers with bioturbated pebbly siltstone and shale, and was deposited in lagoons, barrier islands, and offshore sand ridges on a shallow shelf (Knock, 1987; Mull, 1987). Ammonites indicate an Early Cretaceous (Hauterivian) age for the unit.

Pebble shale unit. The pebble shale unit (Robinson and others, 1956; Collins, 1958, 1961; Robinson, 1959) is thin (<160 m) but widespread in the subsurface and in outcrop of the North Slope subterrane. The pebble shale rests conformably on the discontinuous Kemik Sandstone beneath the coastal plain province and in outcrop along the mountain front, but where the Kemik is absent, the pebble shale lies unconformably on older strata. The pebble shale unit is characterized by black, organic-rich, fairly fissile marine shale of Hauterivian and Barremian age that contains sparse matrix-supported, polished pebbles of chert and quartz and well-rounded, frosted sand grains (Witmer and others, 1981; Mull, 1987).

The pebble shale unit is locally pyritiferous and glauconitic and contains minor sandstone and thin beds of greenish, possibly tuffaceous shale (Molenaar, 1983, 1988; Bird, 1987; Molenaar and others, 1987). The matrix-supported pebbles are generally less than a few centimeters in diameter, but cobbles and boulders as large as 25 cm are known (Molenaar and others, 1984); sand grains are fine to coarse grained. These clasts are thought to have been derived from the uplifted Early Cretaceous rift margin to the north, but the mechanism by which they were transported and deposited in the pebble shale unit is controversial (Mull, 1987; Molenaar, 1988).

Isopachs of the pebble shale unit (Bird, 1987) are irregular, ranging from 60 to 160 m. The area of greatest thickness of the pebble shale unit (>150 m) is near Barrow, where the shale contains interbedded sandstone and is closest to its clastic source that lay to the north (Blanchard and Tailleur, 1983). South of the coastal plain and in the northeastern Brooks Range, the pebble shale passes into shelf and slope settings, where no erosion took place during the Early Cretaceous.

Brookian sequence. The Brookian sequence consists of enormous quantities of sediment that were shed northward into the adjacent foredeep from the developing Brooks Range orogenic belt. Sandstones of the Brookian sequence reflect their orogenic provenance in that they contain significantly less quartz and more feldspar and labile rock fragments than sandstones of the Ellesmerian sequence. The Brookian sequence was deposited over at least 150 m.y. (Late Jurassic to the present), but deposition may have begun as much as 30 m.y. earlier, in the Middle Jurassic. The oldest and southernmost Brookian strata were probably deposited several hundred kilometers south of the present Brooks Range in the proto-Colville basin during the Jurassic and Early Cretaceous (Neocomian). These strata were transported northward with the allochthonous sequences of the Brooks Range and are now partially preserved as the Okpikruak Formation in the Endicott Mountains and De Long Mountains subterranes (Martin, 1970; Mull, 1982, 1985; Mayfield and others, 1988). Deposition of the allochthonous older rocks of the Brookian sequence was therefore coeval with deposition of the upper Ellesmerian sequence to the north. The younger rocks of the Brookian sequence, in contrast, were deposited after most of the northward migration of the Brooks Range thrust front. These strata rest mostly on older rocks of the North Slope subterrane, form the modern Colville basin, and are less deformed and more completely preserved than the older rocks of the Brookian sequence (Fig. 5).

The Brookian sequence can be subdivided into sedimentary packages or megacycles that grade from deep-marine deposits upward into nonmarine deposits (Mull, 1985). The oldest megacycles (Jurassic and Early Cretaceous—Berriasian and Valanginian) are represented by the Okpikruak Formation in the De Long Mountains and Endicott Mountains subterranes and by lithologically similar strata along the eastern and southern parts of the Lisburne Peninsula (the Ogotoruk, Telavirak, and Kisimilok formations; see Campbell, 1967). The older Brookian strata on the Lisburne Peninsula, like the Okpikruak Formation, may have been transported northward with the allochthonous sequences of the Brooks Range but were later faulted beneath rocks of the North Slope subterrane during the east-vergent thrusting event that produced the Tigara uplift in the Late Cretaceous or Tertiary.

Within the Colville basin, at least four sedimentary megacycles can be distinguished in rocks of the Brookian sequence: (1) the Aptian(?) to Albian megacycle, consisting of the Fortress Mountain Formation, upper part of the Kongakut Formation, and Bathtub Graywacke; (2) the Albian to Cenomanian megacycle, consisting of the Torok Formation and Nanushuk Group; (3) the Cenomanian to Eocene megacycle, consisting of the Colville Group and parts of the Hue Shale, Canning Formation, and Sagavanirktok Formation; and (4) the Eocene to Holocene megacycle, consisting of the upper parts of the Hue Shale, Canning Formation, and Sagavanirktok Formation, and the entire Gubik Formation. Rocks of the Aptian(?) to Albian megacycle are exposed in the north-central foothills of the Brooks Range, but the deposits of the younger three megasequences are shingled from west to east along the length of the Colville basin.

Seismic-reflection profiles of the Colville basin delineate a series of well-developed topset, foreset, and bottomset reflectors within each megacycle that mark, respectively, (1) fluvial, deltaic, and shelf deposits, (2) slope shale and turbidite deposits, and (3) basin-plain and turbidite deposits. The age and distribution of these megacycles, coupled with relevant paleocurrent and seismic data, show that the Colville basin was filled longitudinally as sediments prograded from the west toward the northeast in the late Early Cretaceous and onward into the eastern North Slope in the Late Cretaceous and Cenozoic (Chapman and Sable, 1960; Ahlbrandt and others, 1979; Molenaar, 1983, 1985, 1988; Huffman and others, 1985; Molenaar and others, 1986, 1987, 1988). The progressive northward and eastward infill of the basin may have resulted from the migration of the Brookian thrust front from the southwest to the northeast (Hubbard and others, 1987). Composition of Brookian sandstone ranges upsection from lithic and volcanic rich to chert and quartz rich. Mull (1985) has related this compositional change to progressive unroofing of the allochthonous sequences of the Brookian orogen.

Base of the Brookian sequence. Throughout the northern Colville basin and underlying parts of all the megacycles, the base of the Brookian sequence is marked by a widespread, 8–45-m-thick interval of laminated black shale and interbedded bentonite. Like the pebble shale, this unit contains isolated well-rounded, frosted sand grains and chert pebbles and has an average carbon content of greater than 3%. In contrast, however, the basal part of the Brookian sequence in the Colville basin is characterized by relatively high gamma radiation, which can be detected on gamma-ray well logs or by scintillometer in outcrop, and is therefore variously known as the gamma-ray zone (GRZ) or the highly radioactive zone (HRZ) (Carman and Hardwick, 1983; Bird, 1987; Molenaar and others, 1987) (Fig. 5). Dinoflagellates and radiolarians from the radioactive zone indicate that it was deposited during the Aptian and Albian (Carman and Hardwick, 1983; Molenaar and others, 1987) and perhaps in the Barremian (Mickey and Haga, 1983).

The GRZ may be the distal, condensed shale facies of the Brookian sequence, deposited on the north flank of the Colville basin, on the Barrow arch, and probably north of the arch. Its high carbon content and laminated character suggest an anoxic condition of deposition. The shale probably pinches out southward, where higher rates of Brookian sedimentation prevailed, but thickens in the northeastern Colville basin, where, in sections of the Hue Shale, condensed sedimentation spanned most of Late Cretaceous time (Molenaar and others, 1987).

Fortress Mountain Formation, upper part of the Kongakut Formation, and Bathtub Graywacke (Aptian? to Albian megacycle). Rocks of the Aptian(?) to Albian megacycle consist of shale, sandstone, and conglomerate exposed along the southern margin of the Colville basin. In the central and western Brooks Range, this megacycle is represented by the Fortress Mountain Formation (Aptian? and Albian) (Patton, 1956; Patton and Tailleur, 1964). This unit is as much as 3000 m thick and consists largely of coarse-grained graywacke turbidites (Fig. 5), although some of its southernmost and stratigraphically highest units may be nonmarine (Hunter and Fox, 1976; Crowder, 1987, 1989; Molenaar and others, 1988). In some places, the Fortress Mountain Formation conformably overlies fine-grained rocks of the Torok Formation, but elsewhere it rests unconformably on deformed rocks of the De Long Mountains subterrane (Tailleur and others, 1966; Mull, 1985) (Fig. 4). The unconformity represents either subaerial or submarine erosion (Molenaar and others, 1988). Facies change abruptly in the Fortress Mountain Formation, reflecting alluvial, fluvial, submarine-canyon, inner fan channel, outer fan, and basin-plain deposits (Crowder, 1987, 1989; Molenaar and others, 1988). The Fortress Mountain Formation may represent local coastal deltas or fan deltas that were shed toward the north from the ancestral Brooks Range. Regionally, the Fortress Mountain becomes thinner bedded and finer grained to the north and grades laterally into, and intertongues with, shale and siltstone turbidites of the lower Torok Formation (Mull, 1985; Molenaar and others, 1988). A Brooks Range provenance for the Fortress Mountain is supported by abundant clasts of chert and mafic igneous rocks that were derived from the De Long Mountains subterrane and Angayucham terrane. Abundant muscovite and carbonate detritus in the compositionally distinct Mount Kelly Graywacke Tongue of the Fortress Mountain Formation of the western Brooks Range suggests that the provenance in this area included rocks of the Hammond or Coldfoot subterranes (Mull, 1985). Ammonites and pelecypods, rare in the Fortress Mountain Formation, indicate that the unit is largely early Albian, but the undated lower part of the Fortress Mountain may be as old as Aptian (Molenaar and others, 1988).

In the eastern Brooks Range, the Aptian(?) to Albian megacycle consists of the upper part of the Kongakut Formation and the conformably overlying Bathtub Graywacke. The upper part of the Kongakut Formation (the pebble shale and siltstone members of Detterman and others, 1975) (Lower Cretaceous) consists of about 800 m of very thin bedded and fine-grained phosphatic, feldspathic lithic turbidites (C. G. Mull, 1992, unpublished data). The Bathtub Graywacke (Albian?) (Detterman and others, 1975) (Fig. 15) consists of 750 m of sandstone and shale turbidites that compose a submarine-fan sequence (Mull, 1985). The upper part of the Kongakut contains poorly preserved Aptian pelecypods, whereas the Bathtub Graywacke is inferred to be Albian and is at least partly equivalent to the Fortress Mountain Formation (Detterman and others, 1975). Because the upper part of the Kongakut Formation rests conformably on the upper Ellesmerian sequence (that is, the lower part of the Kongakut Formation), the Kongakut records continuous deposition from the upper Ellesmerian sequence into the Brookian sequence, and, with the Bathtub Graywacke, probably represents an uplifted remnant of the axial part of the eastern Colville basin.

Torok Formation and Nanushuk Group (Albian to Cenomanian megacycle). The Albian to Cenomanian megacycle consists primarily of the Nanushuk Group and laterally equivalent parts of the finer grained Torok Formation, which together compose the bulk of the Brookian sequence in the central and western Colville basin. The Torok Formation (Albian) (Gryc and others, 1951) consists of dark marine shale and sandstone that ranges in thickness from 6000 m near the Colville River to less than 100 m in its distal parts east of Prudhoe Bay. In the latter area, seismic-reflection data show the Torok Formation as a clastic wedge that onlaps northward onto the Barrow arch (Fig. 8). The upper part of the Torok grades into, and intertongues with, shallow-marine sandstone of the Nanushuk Group.

On seismic sections, the Torok Formation corresponds to bottomset (basinal) and foreset (slope and shelf) units (Molenaar, 1988). The bottomset units, each 150 to more than 700 m thick, consist of black, pyritic shale and siltstone with thin beds of basin-plain, fine-grained to very fine grained sandstone turbidites. These strata were deposited in water depths of 450 to 1000 m and were deposited on, and probably pass northward into, condensed radioactive shale (GRZ) at the base of the Brookian sequence. The foreset units of the Torok consist of slope and gradationally overlying shelf deposits of shale, siltstone, and minor thin-bedded sandstone. The slope deposits of the Torok are

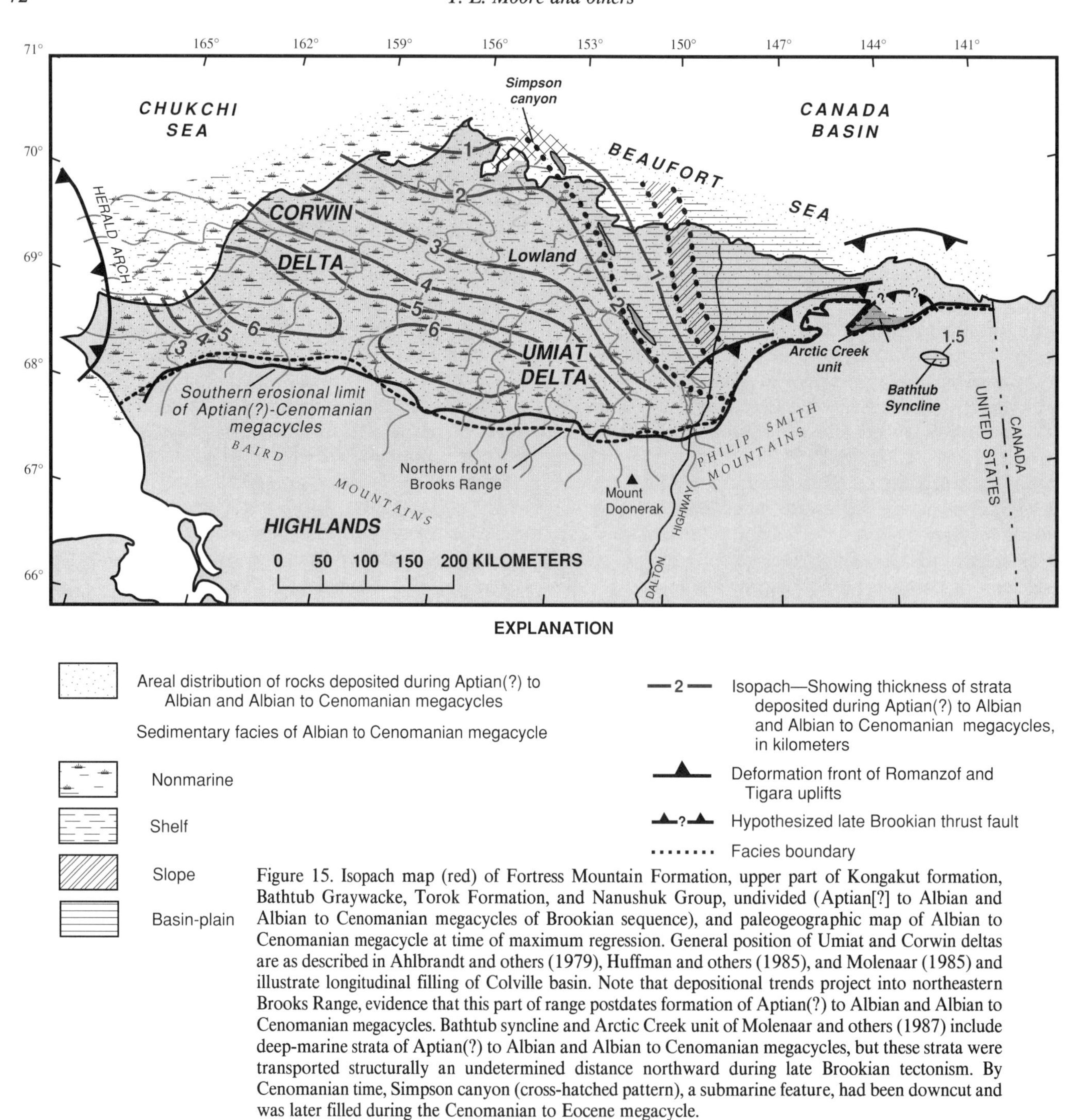

Figure 15. Isopach map (red) of Fortress Mountain Formation, upper part of Kongakut formation, Bathtub Graywacke, Torok Formation, and Nanushuk Group, undivided (Aptian[?] to Albian and Albian to Cenomanian megacycles of Brookian sequence), and paleogeographic map of Albian to Cenomanian megacycle at time of maximum regression. General position of Umiat and Corwin deltas are as described in Ahlbrandt and others (1979), Huffman and others (1985), and Molenaar (1985) and illustrate longitudinal filling of Colville basin. Note that depositional trends project into northeastern Brooks Range, evidence that this part of range postdates formation of Aptian(?) to Albian and Albian to Cenomanian megacycles. Bathtub syncline and Arctic Creek unit of Molenaar and others (1987) include deep-marine strata of Aptian(?) to Albian and Albian to Cenomanian megacycles, but these strata were transported structurally an undetermined distance northward during late Brookian tectonism. By Cenomanian time, Simpson canyon (cross-hatched pattern), a submarine feature, had been downcut and was later filled during the Cenomanian to Eocene megacycle.

450 to 1000 m thick, whereas the shelf deposits are a few meters to 335 m thick. The foreset reflectors are less distinct and less steep (<2°) in the western part of the NPRA than in the eastern part of the NPRA, where the foreset geometry is more distinct and dip angles are in the 4°–6° range. The steeper slope angle is equated with higher rates of progradation (Molenaar, 1988). The direction of progradation was northeastward, as determined from foreset directions in the Torok and paleocurrent directions and facies trends in the Nanushuk Group (Bird and Andrews, 1979).

The Nanushuk Group (Albian to Cenomanian) (Schrader, 1904; Gryc and others, 1951; Detterman and others, 1975) is a thick deltaic unit represented on seismic sections by topset reflectors that can be traced into the foreset and bottomset reflectors of the Torok Formation (Molenaar, 1985, 1988). The Fortress Mountain Formation may be a proximal equivalent of part of the Nanushuk Group (Kelley, 1988), but compositional differences and older, relatively rare megafossils (early and middle Albian) from the Fortress Mountain Formation suggest that it is in part older than the Nanushuk Group (middle Albian to Cenomanian) (Mull, 1985; Molenaar and others, 1988). The Nanushuk achieves a maximum thickness of more than 3000 m in the western North Slope. The lower part of the Nanushuk consists of a thick sequence of intertonguing shallow-marine sandstone and neritic shale and siltstone, whereas the upper part consists of dominantly nonmarine facies, including paludal shale and fluvial sandstone. The deltaic deposits contain huge, undeveloped resources of low-sulfur, low-ash bituminous to subbituminous coal in beds as thick as 6 m (Sable and Stricker, 1987; Wahrhaftig and others, this volume).

Two river-dominated delta systems, the Corwin and Umiat deltas, have been identified in strata of the Nanushuk Group (Ahlbrandt and others, 1979; Huffman and others, 1985) (Fig. 15). The Corwin delta was the larger of the two and prograded toward the northeast from a highland in the area of the Lisburne Peninsula, the present Chukchi Sea, or beyond. The width of the Corwin prodelta shelf ranged between 75 and 150 km. To the east, the smaller Umiat delta prograded northward from a source to the south and may represent a number of small deltas of rivers that once drained the ancestral central Brooks Range. By Cenomanian time, the Nanushuk deltas (principally the Corwin) had completely filled the western part of the Colville basin, prograded across the Barrow arch, and deposited sediment along the margin of the rapidly subsiding Canada basin. The Simpson canyon (Fig. 15), later filled with shale of the Upper Cretaceous Colville Group, is believed to have been cut at this time (Payne and others, 1951).

Colville Group and parts of the Hue Shale, Canning Formation, and Sagavanirktok Formation (Cenomanian to Eocene megacycle). A relative rise in sea level beginning in the Cenomanian ended the Nanushuk regression and initiated the third regressive megacycle of deposition in the Colville basin. Rocks of this megacycle include the Colville Group and parts of the Hue Shale, Canning Formation, and Sagavanirktok Formation (Fig. 5). These strata are lithologically similar to those of the Albian to Cenomanian megacycle, but rocks of the Cenomanian to Eocene megacycle characteristically contain thin beds of bentonite and tuff of mainly Late Cretaceous age. Deposition of the Cenomanian to Eocene megacycle began in the central North Slope and prograded northeastward into the eastern North Slope in the Late Cretaceous and Tertiary (Fig. 16). This progradation continued the regional northeastward shift of the main depocenter of the Brookian sequence (Molenaar, 1983; Bird and Molenaar, 1987; Molenaar and others, 1987).

In the foothills of the central North Slope, rocks of the Cenomanian to Eocene megacycle consist of the Colville Group (Schrader, 1902; Gryc and others, 1951; Brosgé and others, 1966; Detterman and others, 1975), which rests on shallow-marine to nonmarine rocks of the Nanushuk Group. The lower part of the Colville Group consists of about 500 m of marine shelf to marine basin shale, sandstone, bentonite, and tuff of the Seabee Formation (Cenomanian to Turonian). The Schrader Bluff Formation (Cenomanian to Campanian), which overlies these rocks, consists of about 800 m of shallow-marine sandstone and shale. These shallow-marine rocks intertongue with a 600-m-thick interval of nonmarine sandstone, conglomerate, shale, and coal of the uppermost Colville Group, the Prince Creek Formation (Santonian to Maastrichtian).

In the eastern Colville basin, beneath the foothills and coastal plain of the northeastern Brooks Range, the facies of the Colville basin are younger than those to the west and are represented primarily by the Hue Shale and Canning Formation (Molenaar and others, 1987). Because of deformation, poor exposure, and diachronous character of the Upper Cretaceous and Tertiary rocks of the Colville basin, Molenaar and others (1987) defined this change of nomenclature east of the eastern limit of the Nanushuk Group, about long 151°W (Fig. 7). The 300-m-thick Hue Shale (Aptian? to Campanian and probably Tertiary under the Beaufort Sea) is a condensed, basinal sequence consisting of shale, bentonite, and tuff. The basal 45 m of this unit contains the GRZ; the upper part of the unit consists of similar, but less radioactive, black shale. Because the Hue Shale forms the basal part of the Brookian sequence in the eastern Colville basin, this region must have been distal to the primary area of deposition of the earlier megacycles.

The first appearance in the eastern Colville basin of northeastward-prograding slope and shelf facies of the Cenomanian to Eocene megacycle is represented by the 1200-m-thick Canning Formation (Lower Cretaceous—Aptian to Tertiary). Although dominantly shale, the lower part of the Canning contains thin basin-plain sandstone turbidites; these pass upward into slope and shelf facies. South of Prudhoe Bay, the Canning Formation is largely Aptian to Cenomanian, whereas near the Canning River it is largely Campanian to Eocene; thus, the Canning is markedly diachronous (Molenaar and others, 1987).

The Sagavanirktok Formation (Campanian to Pliocene) (Gryc and others, 1951; Detterman and others, 1975; Molenaar and others, 1987) is a thick shallow-marine and nonmarine unit which overlies and intertongues with slope and shelf facies of the Canning Formation. It is as much as 2600 m thick and consists of sandstone, bentonitic shale, conglomerate, and coal, composing the regressive part of the megacycle. In northeastern Alaska, the Sagavanirktok Formation is primarily Paleocene and younger, but because of its diachronous nature, it may be as old as Campanian to the west, where it is stratigraphically equivalent to the Schrader Bluff and Prince Creek formations (Molenaar and others, 1987).

Gubik Formation and upper parts of the Hue Shale, Canning Formation, and Sagavanirktok Formation (Eocene to Holocene megacycle). The Eocene to Holocene megacycle consists of the Gubik Formation and the upper part of the Sagavanirktok Formation onshore, but it also includes parts of the Canning Formation and the Hue Shale offshore of the eastern North Slope (Fig. 5). Deposits of this megacycle reach a thickness of about 2 km under the coastal plain east of Prudhoe Bay (Fig. 16), but they are much thicker offshore (Grantz and others, 1990a). Wells west of the ANWR (Molenaar and others, 1986) penetrate deposits of the Sagavanirktok Formation that consist mostly of sandstone and conglomerate with about 30% interbedded siltstone and shale. These strata, representing a fluvial-deltaic environment, grade eastward and northward into finer grained shelf and slope deposits of the Canning Formation (Bird and Molenaar, 1987). The Gubik Formation (Pliocene and Pleistocene) (Schrader, 1902; Gryc and others, 1951; Detterman and others, 1975; Nelson and Carter, 1985) consists of unconsolidated, marine and nonmarine, poorly stratified to well-stratified gravel, sand, silt, and clay.

The base of the Eocene to Holocene megacycle is defined by a poorly dated erosional unconformity. This unconformity has been traced in wells along the coastline between the Colville River and the ANWR and may correlate with a late Eocene unconformity in the Mackenzie delta area (Bird and Molenaar, 1987; Molenaar and others, 1987). West of the ANWR, strata above and below the unconformity are disconformable, but in the ANWR, the unconformity separates more highly deformed rocks below from less-deformed rocks above (Bruns and others, 1987; Kelley and Foland, 1987). Development of the unconformity may be related to Tertiary thrusting and uplift in the northeastern Brooks Range, as indicated by apatite fission-track ages of 45 to 25 Ma (O'Sullivan, 1988; O'Sullivan and others, 1989).

Endicott Mountains subterrane

The Endicott Mountains subterrane is one of the three largest subterranes of the Arctic Alaska terrane, extending for about 900 km from the Chukchi Sea on the west to near the Canadian

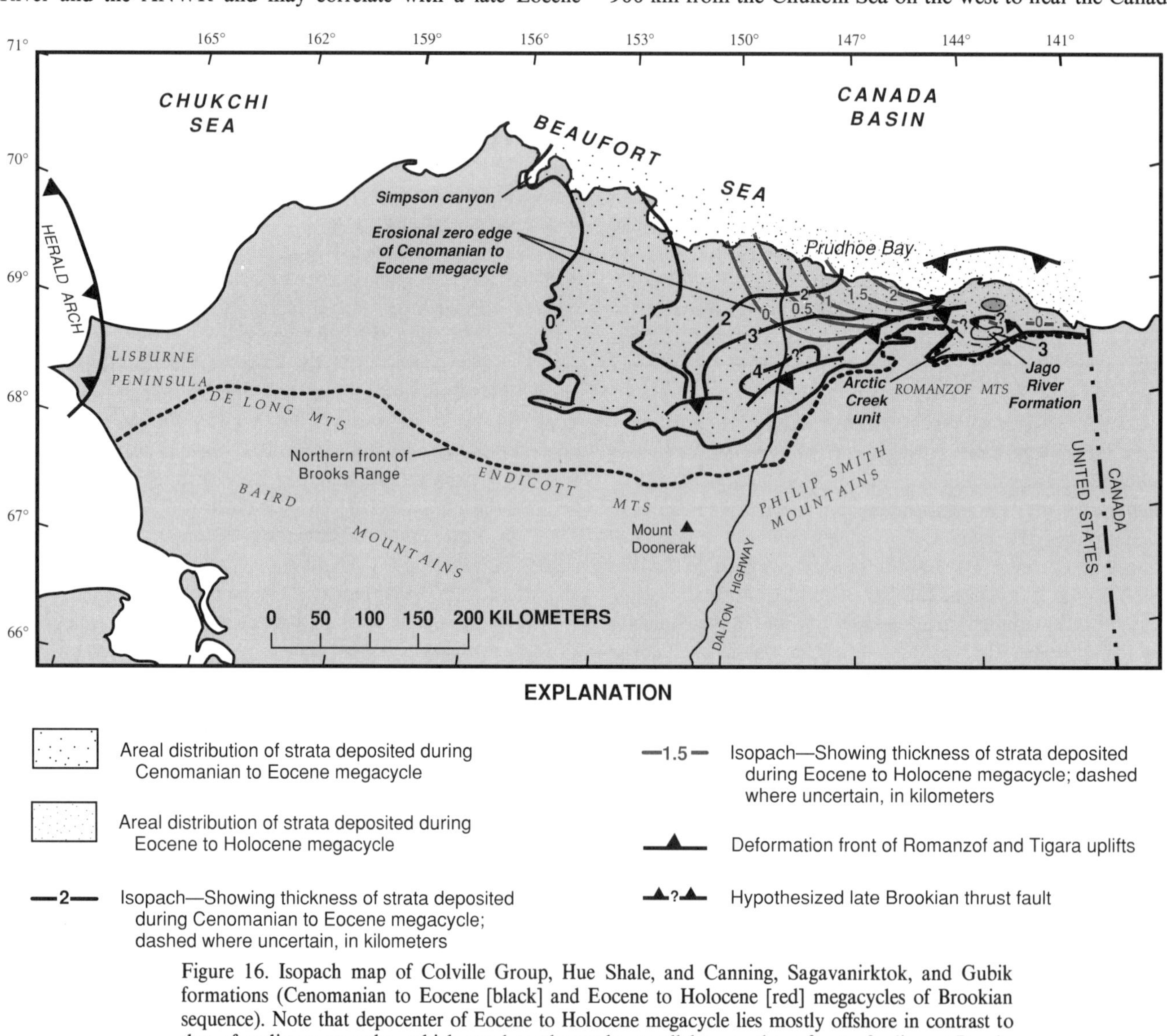

Figure 16. Isopach map of Colville Group, Hue Shale, and Canning, Sagavanirktok, and Gubik formations (Cenomanian to Eocene [black] and Eocene to Holocene [red] megacycles of Brookian sequence). Note that depocenter of Eocene to Holocene megacycle lies mostly offshore in contrast to that of earlier megacycles, which trend northeasterly, parallel to northern front of adjacent Brooks Range. Upper Cretaceous deep-water deposits of Arctic Creek unit (Molenaar and others, 1987) and Upper Cretaceous and Tertiary nonmarine deposits of Jago River Formation (Buckingham, 1987) are part of Cenomanian to Eocene megacycle, but these strata were transported structurally an undetermined distance northward during late Brookian tectonism.

border on the east (Fig. 3). The Endicott Mountains subterrane consists solely of the Endicott Mountains allochthon (Mull, 1982, 1985), which is the lowest of a stack of seven major allochthons in the Brooks Range that have been distinguished by Martin (1970) and Mayfield and others (1988).

Endicott Mountains allochthon. Although the stratigraphically lower part of the Endicott Mountains subterrane (allochthon) (Fig. 17) includes a transgressive succession analogous to that of the Endicott and Lisburne groups of the North Slope subterrane, it differs in that it has a faulted base, contains a regressive Upper Devonian sequence, lacks a sub-Mississippian unconformity, and has a much greater thickness of clastic rocks. The Permian to Lower Cretaceous part of the stratigraphic succession of the Endicott Mountains subterrane consists entirely of fine-grained rocks that represent shelf to basinal deposition in contrast to coeval shallower water deposits of much of the North Slope subterrane. Stratigraphic thickness of pre-Cretaceous rocks in this subterrane is 1500 m in the western Brooks Range and more than 6000 m in the central Brooks Range. The subterrane comprises the Beaucoup Formation, the Endicott Group, the Lisburne Group, the Etivluk Group, Ipewik unit, and the Okpikruak Formation.

Beaucoup Formation. The oldest rocks of the Endicott Mountains subterrane are those of the Beaucoup Formation (Upper Devonian) (Dutro and others, 1979) in the central Brooks Range and the correlative Nakolik River unit of Karl and others (1989) in the western Brooks Range. These units consist of a heterogeneous marine assemblage of phyllitic, calcareous siltstone and shale with lenticular limestone bodies. In its type area east of the Dalton Highway, the Beaucoup Formation forms a 545-m-thick depositional succession at the base of the Endicott Mountains subterrane conformably beneath the Hunt Fork Shale (Dutro and others, 1979). Elsewhere, however, the unit is extensively faulted and detached from the Endicott Mountains allochthon (Moore and others, 1991). Limestone bodies in the Beaucoup Formation and Nakolik River unit, although com-

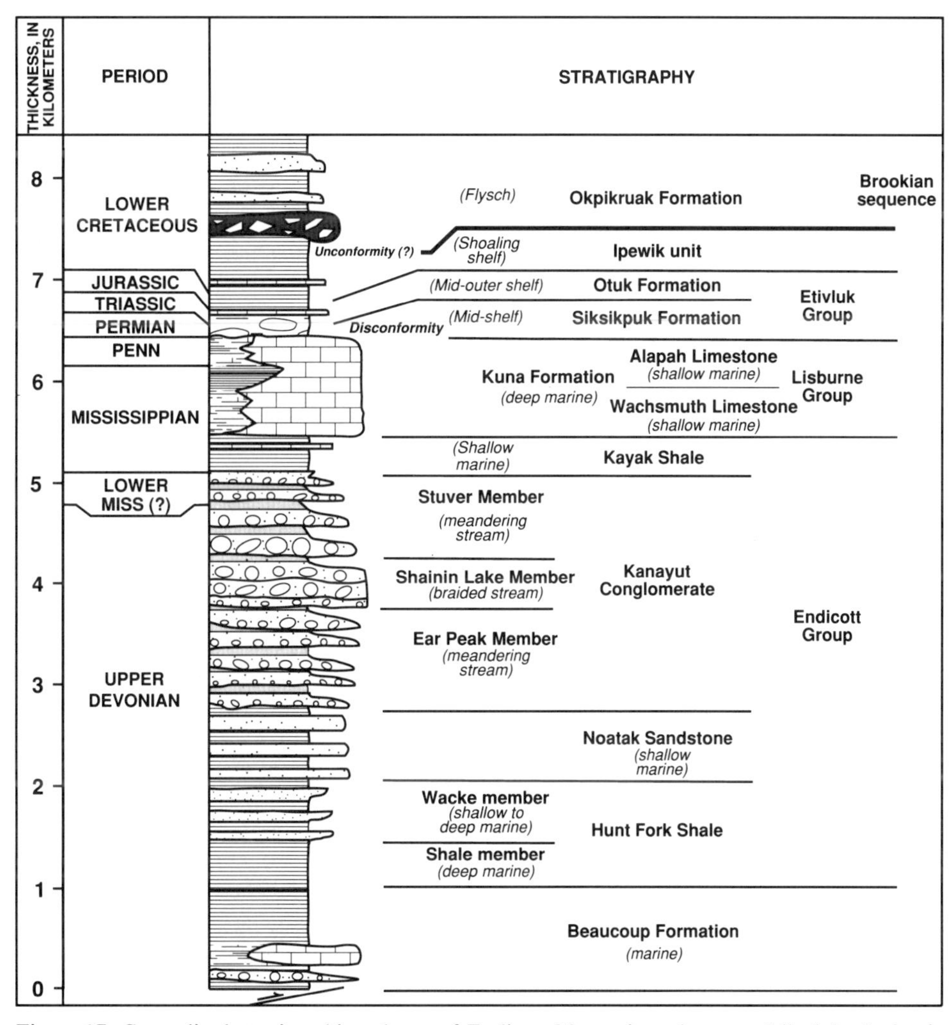

Figure 17. Generalized stratigraphic column of Endicott Mountains subterrane (allochthon), Arctic Alaska terrane. Thicknesses are maximum except that of Kuna Formation (<100 m), which could not be drawn to scale. Half arrow indicates relative northward thrust movement. For explanation of lithologic symbols, see Figure 21.

monly recrystallized, consist of bioclastic packstone and wackestone that contain early Late Devonian (Frasnian) megafossils. Dutro and others (1979) interpreted the bodies as stromatoporoid patch reefs.

Rock types not present in the type section of the Beaucoup Formation but associated with it in its type area include limestone-phyllite-pebble conglomerate, quartz-pebble conglomerate, maroon and green phyllite and argillite, mafic volcanic rocks, and silicic volcaniclastic rocks (Dutro and others, 1979). These rock types also have been mapped in the northern part of the Hammond subterrane along most of its length, leading some workers (Dillon and others, 1986; Dillon, 1989) to include many of the rocks of that subterrane in the Beaucoup Formation. However, this correlation may not be justified because the age and stratigraphic relations between Beaucoup rocks of the Hammond subterrane and those of the Endicott Mountains subterrane have not been established and because the Beaucoup would be an integral part of two discrete thrust-bounded packages of rock. Dutro and others (1979) originally envisioned the Beaucoup Formation as a link between the carbonate rocks of the Skajit Limestone (Hammond subterrane) and the clastic rocks of the Endicott Group (Endicott Mountains subterrane). At present, the Beaucoup Formation may be interpreted as (1) a tectonically disrupted stratigraphic interval that links the Endicott Mountains and Hammond subterranes; (2) two or more undifferentiated, but stratigraphically distinct, units; or (3) a detachment zone that separates the Endicott Mountains and Hammond subterranes and consists of rocks derived from both subterranes.

Endicott Group. In the Endicott Mountains subterrane, the Endicott Group, which is as much as 4500 m thick, consists in ascending order of the Hunt Fork Shale (marine), Noatak Sandstone (marine), Kanayut Conglomerate (nonmarine), and Kayak Shale (marine). This sequence represents a major fluvial-dominated deltaic clastic wedge shed southwestward during the Late Devonian and Early Mississippian from at least two major sources, one in the eastern Brooks Range and the other north of Anaktuvuk Pass in the central Brooks Range (Tailleur and others, 1967; Nilsen, 1981; Moore and Nilsen, 1984) (Fig. 18). Clasts in conglomerate of the sequence are largely chert, some containing radiolarian ghosts, and minor vein quartz, chert arenite, and chert-pebble conglomerate.

The Hunt Fork Shale (Upper Devonian) (Chapman and others, 1964) is a widespread sequence of thin-bedded, dark-gray shale, micaceous siltstone, and fine-grained quartzose sandstone more than 1000 m thick. Sedimentary structures preserved within the Hunt Fork indicate that the formation consists of thin-bedded turbidites and marginal-marine deposits that represent slope and prodelta depositional environments. Megafossils and conodonts from sparse, thin-bedded, bioclastic turbidites in the lower part of the unit are Frasnian (early Late Devonian), whereas fossils found higher in the unit, typically in shallow-marine sandstone, are Famennian (late Late Devonian) (Brosgé and others, 1979).

The Hunt Fork Shale grades upward into the Noatak Sandstone (Upper Devonian) (Dutro, 1952). The Noatak is 200 to 300 m thick and consists of coarsening-upward packages of fine- to medium-grained calcareous sandstone that are separated by units of dark siltstone and shale (Nilsen and Moore, 1984). Abundant trough cross-stratification, megafossils, and other features indicate that the Noatak was deposited in a marginal-marine environment in the Famennian (late Late Devonian).

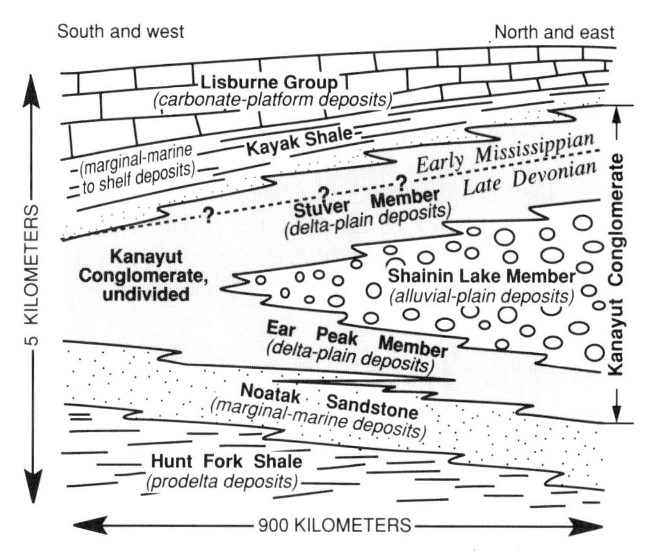

Figure 18. Diagrammatic cross section showing stratigraphic relations of Endicott Group, Endicott Mountains subterrane (allochthon) (modified from Nilsen and Moore, 1984).

The overlying Kanayut Conglomerate (Upper Devonian and Lower Mississippian?) consists of as much as 2600 m of interbedded sandstone, conglomerate, and shale, subdivided, in ascending order, into the Ear Peak, Shainin Lake, and Stuver members (Bowsher and Dutro, 1957; Nilsen and Moore, 1984). The Ear Peak and Stuver members are shale-bearing successions that were deposited by meandering streams on a delta plain. The Shainin Lake Member, in contrast, consists almost entirely of sandstone and conglomerate with clasts as large as 23 cm that were deposited by braided streams on an alluvial plain (Moore and Nilsen, 1984). The Kanayut thins and fines southward and westward from near Anaktuvuk Pass, and its three members cannot be distinguished west of the Killik River quadrangle. The Kanayut is dated by Late Devonian and Early Mississippian(?) plant fossils, sparse Famennian brachiopods, and its stratigraphic position (Nilsen and Moore, 1984).

A marine transgression following deposition of the Kanayut is recorded in rippled, thin-bedded, very fine grained quartzose sandstone in the basal part of the overlying Kayak Shale (Lower Mississippian). Above this sandstone unit (basal sandstone member of Bowsher and Dutro, 1957), the Kayak Shale consists of about 250 m of black shale and minor fossiliferous limestone-debris beds, especially in its upper part (Bowsher and Dutro, 1957). Abundant megafossils and microfossils in the intercalated limestone indicate that the basal Kayak in the Endicott Mountains subterrane is at least as old as late Kinderhookian (early Early Mississippian), and the uppermost part may be as young as

Osagean (late Early Mississippian) (Armstrong and Mamet, 1978). The Kayak records submergence of the Kanayut clastic wedge and offshore deposition of fine-grained sediment prior to northward progradation of the Lisburne carbonate platform.

Lisburne Group. Carbonate rocks of the Lisburne Group in the Endicott Mountains subterrane typically consist of about 700 m of cliff-forming Mississippian and Pennsylvanian echinoderm-bryozoan wackestone and packstone that commonly contain articulated crinoid stems and bryozoan fronds. Dolomitization, and chert nodules, veins, and layers with replacement textures, are common. As in the North Slope subterrane, transgressive, platform, and deeper water assemblages are recognized in the Lisburne Group, but in the Endicott Mountains subterrane, these assemblages are somewhat older, and the deeper water assemblage is more extensive. Bowsher and Dutro (1957) divided the Lisburne Group in the Endicott Mountains subterrane into the Wachsmuth Limestone and overlying Alapah Limestone, but regional mapping has shown that these units do not coincide with the upward change from transgressive to platform-carbonate facies and are otherwise difficult to distinguish.

In the north-central Brooks Range, the basal, transgressive part of the Lisburne Group consists of massive argillaceous limestone that grades upward into cherty, less argillaceous deposits of the platform facies (Armstrong and Mamet, 1978). To the south, the platform facies interfingers with unnamed black chert, radiolarian and spiculitic lime mudstone, and black shale that represent slope and starved-basin deposits. This deeper water assemblage increases in thickness and composes a greater proportion of the Lisburne Group in the southern and western parts of the subterrane. These facies relations suggest that deposition occurred on a slowly subsiding open-marine shelf that interfingered with an euxinic basin south or southwest of the main carbonate platform (Armstrong and Mamet, 1978).

Abundant foraminifers and conodonts indicate that deposition of the Lisburne Group limestones along the southern margin of the Endicott Mountains allochthon in the central Brooks Range began in the early Osagean (late Early Mississippian) (Armstrong and Mamet, 1978). Carbonate deposition spread northward during later Osagean time and continued throughout most of the Late Mississippian. Recently, Morrowan and Atokan fossils have been recovered from carbonate rocks near the top of the Lisburne Group, showing that carbonate deposition continued into Early Pennsylvanian time in the Endicott Mountains subterrane (Siok, 1985).

In the Killik River quadrangle, carbonate rocks of the Lisburne Group become thinner, more thinly bedded, and have a higher percentage of secondary chert, possibly as a result of a westward increase in the abundance of sponge spicules. These rocks are thought to grade laterally into the Kuna Formation (Mississippian and Pennsylvanian) (Mull and others, 1982), which is exposed in nearby thrust imbricates and in structural windows in the western Brooks Range. The Kuna Formation, a deep-water assemblage, consists of less than 100 m of sooty, phosphatic black shale and dolomite interbedded with lesser amounts of black, radiolarian- and spiculite-bearing chert. Interstratified platy micritic limestone, thin-bedded quartzose turbidites, and basaltic to rhyodacitic volcanic rocks are reported from a few places in the western Brooks Range (Nokleberg and Winkler, 1982; Moore and others, 1986). The Kuna Formation hosts the strata-bound zinc-lead-silver deposit of the Red Dog Mine in the De Long Mountains quadrangle (Moore and others, 1986).

Megafossil and microfossil data indicate that the Kuna Formation ranges from late Early Mississippian (Osagean) to Early or Middle Pennsylvanian (Mull and others, 1982) and occupies the same stratigraphic position as carbonate rocks of the Wachsmuth and Alapah limestones farther east. The Kuna Formation may represent sponge-rich mud deposited in a partly oxygenated starved-basin environment (Murchey and others, 1988) that lay southwest of the platform-carbonate facies of the Lisburne Group. This environment may have been in the same euxinic basin that Armstrong and Mamet (1978) inferred to exist south and west of the platform-carbonate facies (see above).

Etivluk Group. The Etivluk Group (Mull and others, 1982, 1987b) consists of the Permian Siksikpuk Formation (115 m) and the Triassic and Jurassic Otuk Formation (100 m). The Siksikpuk Formation was named by Patton (1957) for exposures mainly of shale and siltstone in the Chandler Lake quadrangle and was subsequently extended by Mull and others (1982) to include chert-rich sequences in the central and western Brooks Range. Mull and others (1987b) restricted the Siksikpuk to the shale- and siltstone-rich facies described by Patton (1957) and reassigned the chert-rich facies to the Imnaitchiak Chert (see Picnic Creek allochthon below). We herein agree with the restriction of the Siksikpuk Formation as proposed by Mull and others (1987b).

Four lithostratigraphic units have been recognized in the Siksikpuk Formation in the central Brooks Range by Siok (1985), Adams and Siok (1989), and Adams (1991). In ascending order, these are (1) yellow-orange–weathering, pyritic siltstone (2–17 m); (2) gray to greenish-gray and maroon mudstone and siltstone, containing nodules of barite and siderite (20–100 m); (3) wispy-laminated, greenish-gray silicified mudstone (frequently referred to as chert) (1–24 m); and (4) wispy-laminated, dark gray fissile shale and minor siltstone (1–40 m). Northeastward, the Siksikpuk Formation becomes progressively darker, thicker, coarser grained, and less siliceous, but more carbonate rich, features characteristic of the coeval Echooka Formation of the North Slope subterrane (Adams, 1991). The basal contact of the Siksikpuk Formation on the Lisburne Group may be a disconformity, because Upper Pennsylvanian strata have not been recognized in the Endicott Mountains subterrane (Patton, 1957). The Siksikpuk is a transgressive unit, representing mostly suspension sedimentation in an inner to middle neritic environment (Siok, 1985; Murchey and others, 1988; Adams, 1991).

Patton (1957) initially reported the age of the Siksikpuk Formation as Permian. Mull and others (1982) regarded their extended Siksikpuk as Pennsylvanian, Permian, and Early Triassic. Later, however, Mull and others (1987b) restated the age of their restricted Siksikpuk as Permian. Siok (1985) and Adams (1991) concluded from megafossil and microfossil evidence that the Siksikpuk Formation is largely Early Permian (Wolfcampian and Leonardian), except for its uppermost part, which extends into the early Late Permian (Guadalupian). We herein agree with an age assignment of Permian for the Siksikpuk Formation.

The overlying Otuk Formation (Mull and others, 1982) consists of four members. From base to top, these are (1) the shale member, consisting of dark gray to black, organic-rich shale with thin limestone beds (6–14 m); (2) the chert member, consisting mainly of green and black silicified mudstone and rhythmically interbedded black, calcareous shale (17–53 m); (3) the limestone member, consisting of yellow-brown–weathering limestone with minor shale (7–19 m); and (4) the Blankenship Member, consisting of black, fissile, bituminous shale with minor dark gray chert and dolomitic limestone (7 m). The chert and limestone members typically contain abundant pectinid pelecypods (commonly *Monotis* and *Halobia*). The lower part of the Otuk Formation is correlative with the Shublik Formation of the North Slope subterrane. This part of the Otuk is well dated on the basis of its abundant pectinid pelecypod fauna, conodonts, and radiolarians and ranges from Early Triassic (Scythian) in the shale member to as young as Late Triassic (late Norian) in the limestone member (Mull and others, 1982; Blome and others, 1988; Murchey and others, 1988). The Blankenship Member contains pelecypods and ammonites that indicate a middle Early Jurassic (Sinemurian) to early Middle Jurassic (Bajocian) age and is correlative with the lower part of the Kingak Shale of the North Slope subterrane (Mull and others, 1982; Bodnar, 1989). The Otuk represents condensed sedimentation in an open-marine, middle neritic to inner bathyal environment distant from a source of clastic detritus (Murchey and others, 1988; Bodnar, 1989).

Ipewik unit. In the western Brooks Range, the Ipewik unit (Jurassic and Lower Cretaceous) of Crane and Wiggins (1976) and Mayfield and others (1988) consists of about 100 m of poorly exposed soft, dark, maroon and gray clay shale, concretionary mudstone, fissile oil shale, and reddish coquinoid limestone containing highly compressed *Buchia sublaevis* fossils. Local intervals of resistant, fine- to medium-grained quartzose sandstone (the Tingmerkpuk subunit of Crane and Wiggins, 1976) are also present within the Ipewik of this area. Crane and Wiggins (1976) reported that the lower part of the unit contains Early and Late Jurassic megafossils and Middle Jurassic to Early Cretaceous dinoflagellate faunas, whereas the coquinoid limestone and Tingmerkpuk subunits of the upper part of the Ipewik contain abundant megafossils of Valanginian (Early Cretaceous) age. In the foothills of the central Brooks Range, the Ipewik (clay shale unit of Molenaar, 1988) is much thinner and consists of dark gray and black shale characterized by a distinctive interval, as much as 2 m thick, of reddish-weathering Valanginian coquinoid limestone beds and interbedded maroon shale. In this area, the Ipewik may rest unconformably on Middle Jurassic beds at the top of the Otuk Formation (Mull, 1989).

The Ipewik unit is a condensed section deposited in a quiescent marine basin. The widespread coquinoid limestone is commonly interpreted as a relatively shallow water unit deposited on an intrabasin medial sill or ridge (Jones and Grantz, 1964; Tailleur and Brosgé, 1970; Molenaar, 1988), although the limestone has also been thought of as deep-marine turbidites that consist of intrabasinal shallow-marine fossil debris (Molenaar, 1988). If the limestone were deposited on an intrabasin ridge, it is thought to have separated coeval early Brookian foredeep deposits to the south, represented by the Okpikruak Formation of the De Long Mountains subterrane, from the tectonically stable upper Ellesmerian shale basin to the north, represented by the upper part of the Kingak Shale of the North Slope subterrane.

Deposits of the Brookian sequence (Okpikruak Formation) in the Endicott Mountains subterrane. The youngest rocks of the Endicott Mountains subterrane, exposed mainly in the northern foothills of the Brooks Range, consist of gray, deep-marine mudstone and minor thin-bedded sandstone and conglomerate of the Okpikruak Formation (Upper Jurassic and Lower Cretaceous). At its type locality (herein assigned to the Endicott Mountains subterrane) in the Killik River quadrangle, the Okpikruak is at least 600 m thick (Gryc and others, 1951), and in the western Brooks Range, it is estimated to be more than 1000 m thick. Conglomerate is locally prominent and contains rounded cobbles and boulders of chert, limestone, granitic rocks, dacite, diabase, and gabbro. These rock types are like those of structurally higher allochthons (Mull and others, 1976; Crane, 1987), with the exception of the granitic clasts dated by K-Ar methods at 186–153 Ma (Jurassic), which are unlike any rock type mapped in the Brooks Range (Mayfield and others, 1978). *Buchia* pelecypods in the Okpikruak indicate a Valanginian age for the formation in the Endicott Mountains subterrane; however, Curtis and others (1990) reported Berriasian fossils from one exposure of the Okpikruak in the De Long Mountains quadrangle.

The Okpikruak largely represents turbidites and local olistostromes deposited either in a foredeep that migrated northward with the advancing Brooks Range thrust front (Mull, 1985; Crane, 1987; Mayfield and others, 1988) or possibly in a piggyback basin. Commonly, however, the Okpikruak is deformed, comprising broken formation or melange, the structural position of which is difficult to ascertain. The Okpikruak Formation rests conformably to unconformably on older rocks of the Endicott Mountains subterrane in a few places in the Killik River, Misheguk Mountain, and De Long Mountains quadrangles (Curtis and others, 1984, 1990; C. G. Mull, 1987, unpublished data).

De Long Mountains subterrane

The De Long Mountains subterrane, the structurally highest subterrane of the Arctic Alaska terrane, consists of four of the seven allochthons recognized by Tailleur and others (1966), Martin (1970), Mull (1985), and Mayfield and others (1988). In ascending order, these are the Picnic Creek, Kelly River, Ipnavik

River, and Nuka Ridge allochthons. Although these allochthons display overall stratigraphic similarity to the North Slope and Endicott Mountains subterranes, they differ primarily in aspects of their constituent Mississippian to Lower Cretaceous rocks (Fig. 19). The De Long Mountains subterrane is best exposed in the De Long Mountains of the western Brooks Range (Fig. 20). It also underlies much of the disturbed belt in the central Brooks Range and occurs as thrust imbricates in the eastern Brooks Range, where it has been called the Sheenjek terrane by Jones and others (1987). Not all of the four allochthons are present everywhere in the subterrane, but all of the allochthons present in any one area occur in the same vertical succession. The youngest strata present in all four allochthons are locally derived flysch of the Okpikruak Formation that is assumed to record northward progradation of the Brookian thrust front in the Late Jurassic and Early Cretaceous (Mull and others, 1976; Crane, 1987; Mayfield and others, 1988). Because the structural position of strata of the Okpikruak Formation has not been determined in many places, the Okpikruak strata of all four allochthons of the De Long Mountains subterrane are described together at the end of this section under the heading "Deposits of the Brookian sequence (Okpikruak Formation) in the De Long Mountains subterrane."

Picnic Creek allochthon. The Picnic Creek allochthon is named for exposures in the Picnic Creek fenster in the Misheguk Mountain quadrangle (Mayfield and others, 1984). However, the most detailed examination of the stratigraphy of the allochthon has been in the disturbed belt in the Killik River quadrangle in the central Brooks Range, where a stratigraphic nomenclature has been established by Mull and others (1987b; Fig. 19). The thickness of pre-Cretaceous strata of the Picnic Creek allochthon is less than 1000 m (Mayfield and others, 1988), but these rocks are imbricated so that the true thickness is uncertain. The allochthon comprises the Endicott Group, the Lisburne Group, the Etivluk Group, and the Okpikruak Formation.

Endicott Group. In the Killik River quadrangle, the base of the Picnic Creek allochthon consists of about 100 m of dark greenish-gray shale that grades upward into interbedded dark gray to black shale and thin sandstone beds. This unit, exposed in only a few places, is unfossiliferous, but Mull and others (1987b) correlated it with the Hunt Fork Shale (Upper Devonian) because of its fine-grained character and stratigraphic position below the Lower Mississippian Kurupa Sandstone. The disrupted and incompetent nature of this unit suggests that it has acted as a detachment surface along which the Picnic Creek allochthon was emplaced. In the western Brooks Range, the basal unit of the allochthon consists of about 50 m of calcareous, fine- to medium-grained sandstone with intercalated siltstone. These rocks contain brachiopods of Fammenian (late Late Devonian) age, which implies that the unit is at least partly correlative with the coeval marine Noatak Sandstone (Curtis and others, 1984; Ellersieck and others, 1990).

In the Killik River quadrangle, the Hunt Fork Shale grades up to about 40 m of quartzose sandstone, named the "Kurupa Sandstone" (Lower Mississippian) by Mull and others (1987b); we herein adopt this nomenclature. This relatively competent unit becomes more shale rich to the west and is not recognized in the western Brooks Range. The Kurupa Sandstone consists of thin- to medium-bedded, medium- to coarse-grained sandstone with siltstone and minor granule and pebble conglomerate. Mull and others (1987b) interpreted the Kurupa as turbidites that were deposited toward the southeast on a prodelta ramp. Abundant plant fossils near the top of the formation and a sparse brachiopod fauna suggest an Early Mississippian age and a Siberian affinity for the unit (Mull and others, 1987b).

The Kayak Shale (Lower Mississippian) is 15 to 40 m thick and, in the Killik River quadrangle, conformably overlies the Kurupa Sandstone. The Kayak consists of recessive siltstone and black clay shale with red-brown–weathering ironstone concretions and contains Mississippian megafossils (Mull and others, 1987b). In the western Brooks Range, the formation also contains local, thin, rusty- to buff-weathering bioclastic limestone beds. In a few places, the Kayak also contains as much as 30 m of fossiliferous, quartzose, fine- to medium-grained sandstone and sandy limestone. Megafossils and microfossils in the sandstone are Early Mississippian (Osagean) (Curtis and others, 1984; Ellersieck and others, 1984).

Lisburne Group. A distinctive unit in the Picnic Creek allochthon is the 87-m-thick sequence of thin-bedded, black, spicule- and radiolarian-bearing chert that rests conformably on the Kayak Shale. In the Killik River quadrangle, Mull and others (1987b) named this unit the "Akmalik Chert" (Upper Mississippian and Lower Pennsylvanian); we herein adopt this nomenclature. The Akmalik contains black shale partings and minor siliceous black mudstone and rare dolomitic limestone beds. Radiolarian and conodont assemblages and a plant fossil indicate a Late Mississippian (Meramecian and Chesterian) and Early Pennsylvanian (Morrowan) age for the unit. Mull and others (1987b) considered these rocks to be at least partly correlative with the Kuna Formation of the Endicott Mountains subterrane and to represent a deep-water, basinal equivalent of the platform carbonate sequence elsewhere in the Lisburne Group.

Etivluk Group. In the Picnic Creek allochthon, the Etivluk Group consists of gray, radiolarian chert with lesser amounts of brown, green, and maroon siliceous shale. These strata were assigned to the Siksikpuk Formation by Chapman and others (1964) and Mull and others (1982), but Mull and others (1987b) later restricted the Siksikpuk Formation to the dominantly shaly and silty beds typical of the stratotype and reassigned the more siliceous rocks of the Picnic Creek allochthon to the Imnaitchiak Chert (Lower Pennsylvanian to Jurassic?); we herein agree with this reassignment. In its type locality in the Killik River quadrangle, the Imnaitchiak Chert is 75 m thick and can be divided into six subunits (Siok, 1985). From base to top, these subunits are (1) greenish-gray glauconitic and phosphatic siltstone and sandstone and a local conglomerate bed that consists mostly of spherical oncoids, replaced by barite, and chert clasts (Siok and Mull, 1987) (<2 m); (2) bedded gray chert that contains an increasing amount of shale upsection (14–17 m); (3) green chert,

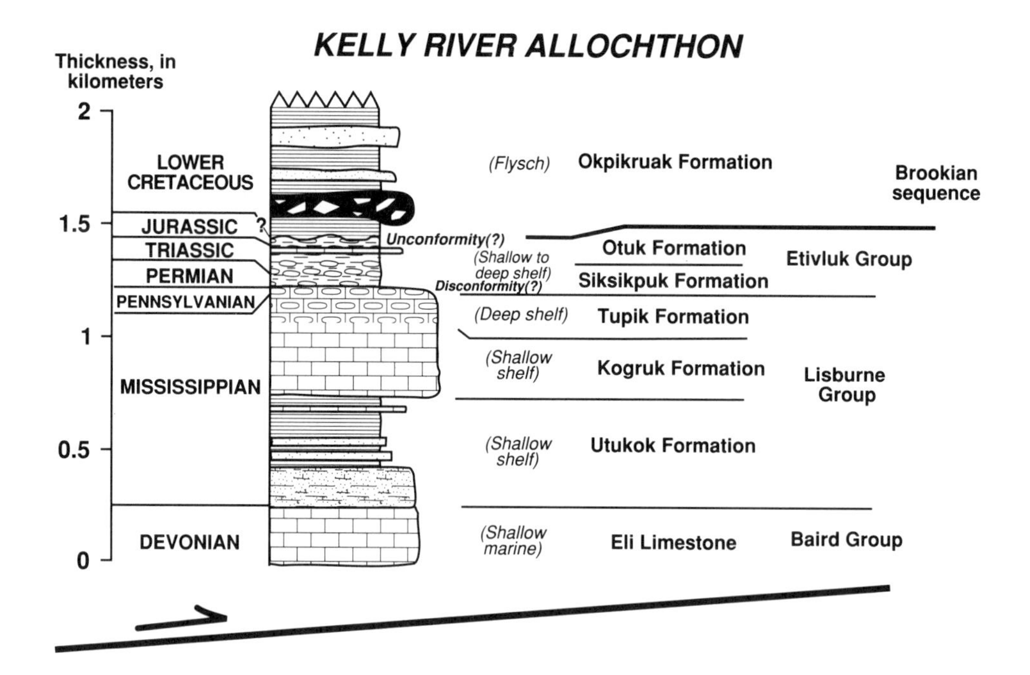

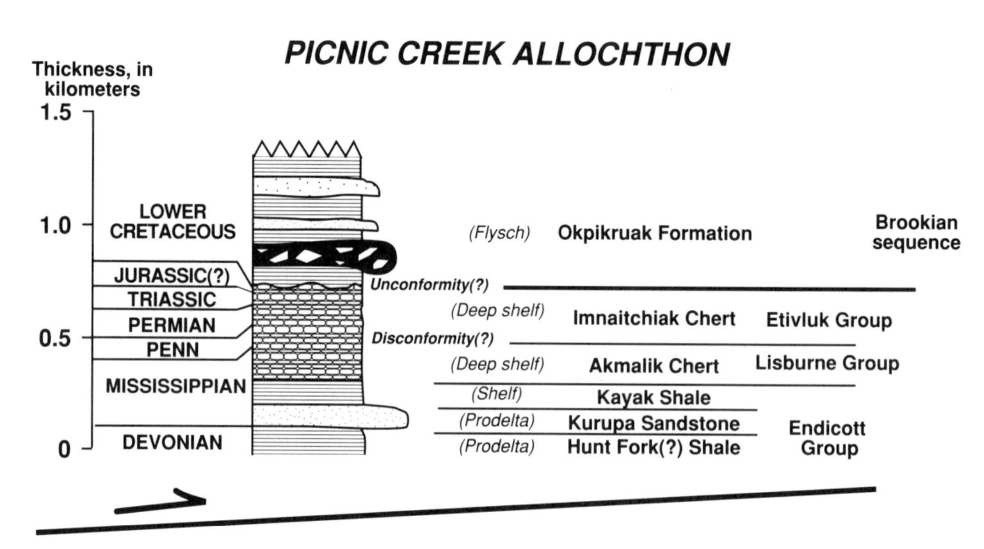

Figure 19. Generalized stratigraphic columns of allochthons of De Long Mountains subterrane, Arctic Alaska terrane. Thicknesses (approximate) from Mayfield and others (1988). Half arrows indicate relative northward thrust movement between allochthons. For explanation of lithologic symbols, see Figure 21.

siltstone, and yellow-orange claystone (8–10 m); (4) interbedded greenish-gray or red siltstone and shale (10–25 m); (5) rhythmically interbedded green and red chert that contains a decreasing amount of shale upsection (10–12 m); and (6) violet-gray chert and cherty siltstone that contain laminae of volcanogenic sandstone and grade upsection into silty shale (>8 m). The Imnaitchiak Chert contains conodonts and abundant radiolarians that indicate a Pennsylvanian and Permian(?) age for most of the unit (Mull and others, 1987b); however, the faunal assemblages may be as old as Late Mississippian (Meramecian to Chesterian) (Siok, 1985; Holdsworth and Murchey, 1988). Radiolarian assemblages at the top of the Imnaitchiak Chert indicate that the unit is as young as "Middle/Late" Triassic (Mull and others, 1987b, p. 650); an unconformity beneath Lower Cretaceous flysch at the top of the unit at the type locality hints at the possibility that elsewhere the Imnaitchiak may include Jurassic strata. The increased proportion and diversity of radiolarians relative to the underlying sponge-spicule–rich Akmalik Chert suggest deposition in relatively deep water in a subsiding, distal-platform environment (Murchey and others, 1988).

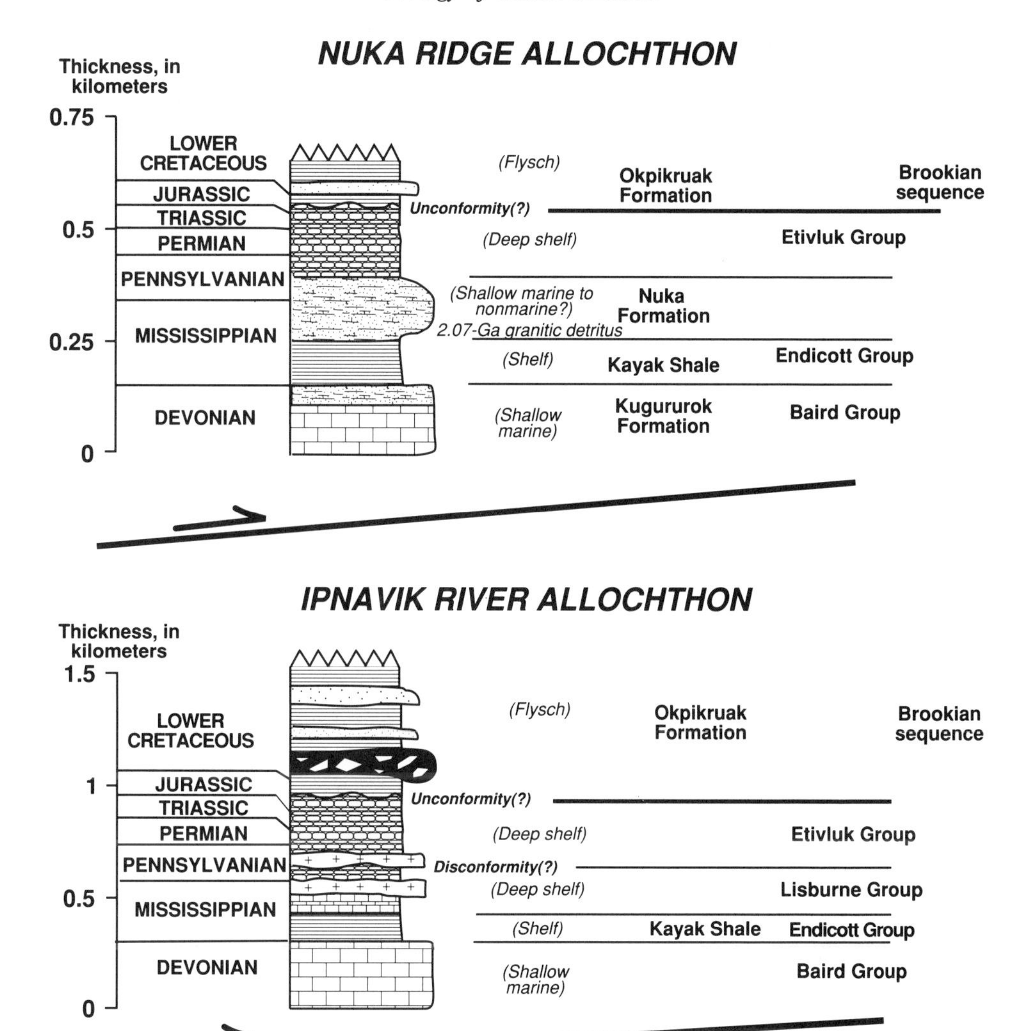

Kelly River allochthon. The Kelly River allochthon (Fig. 19), which is prominently exposed in the western Brooks Range, has not been identified east of the Misheguk Mountain quadrangle (Mayfield and others, 1988). The thickness of pre-Cretaceous strata of this allochthon prior to Brookian deformation is estimated to be less than 1500 m (Mayfield and others, 1988). The allochthon consists of strata assigned to the Baird Group, the Lisburne Group, the Etivluk Group, and the Okpikruak Formation.

Baird Group and related units. The term "Baird Group" includes a variety of Devonian and older carbonate units that are presumed to lie stratigraphically below the Lisburne Group (Tailleur and others, 1967). Recent workers, however, have questioned the usefulness of the Baird Group nomenclature and are currently reassessing the classification (see Hammond subterrane).

Thick, northeasterly-striking belts of rock assigned to the Baird Group compose the base of the Kelly River allochthon in the Baird Mountains and Misheguk Mountain quadrangles. In other areas of the Kelly River allochthon, carbonate rocks of the Baird Group are preserved as isolated fault-bounded slivers at the base of the allochthon. The carbonate rocks typically consist of massive to thick-bedded, light gray–weathering limestone and lesser amounts of dark gray–weathering limestone and dolomite; locally the unit contains thin yellowish-brown–weathering silty limestone. Foraminifers, conodonts, and brachiopods indicate Early, Middle, and late Late Devonian (Famennian) ages, but the fossil data may allow ages as old as Silurian and as young as Early Mississippian (Osagean) for various parts of the Baird Group of the Kelly River allochthon (Curtis and others, 1984, 1990; Ellersieck and others, 1984; Mayfield and others, 1984, 1987, 1988).

Recent work on the Baird Group of the Kelly River allochthon in the northwestern Baird Mountains quadrangle has shown that it consists of about 200 m of pelletoidal dolostone and rare metalimestone that contain Early and Middle Devonian (Emsian and Eifelian) conodonts (Karl and others, 1989). These strata are similar in lithofacies and biofacies to carbonate rocks of similar age in the Hammond subterrane (west-central Baird Mountains sequence), exposed immediately to the east, except that they lack

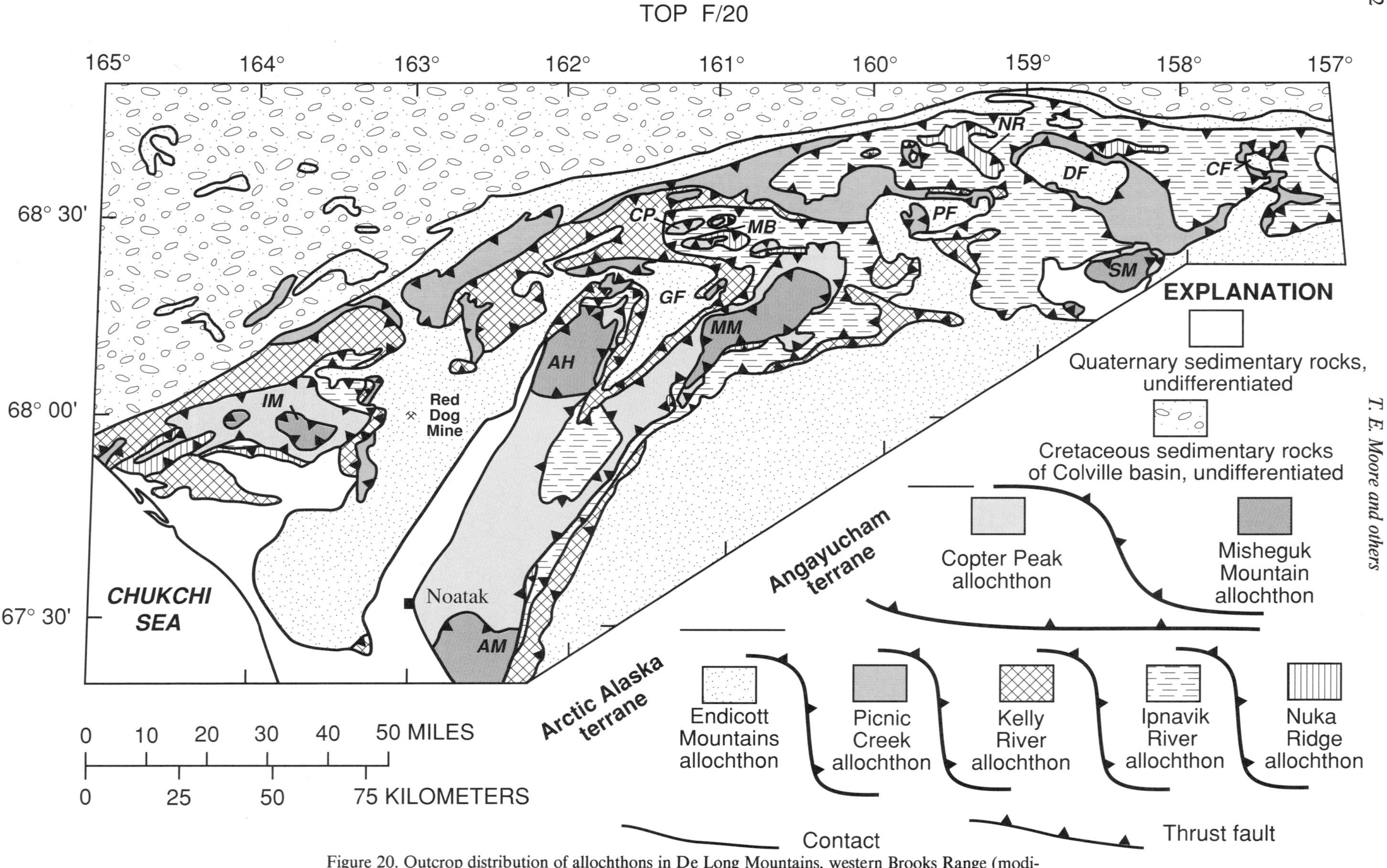

Figure 20. Outcrop distribution of allochthons in De Long Mountains, western Brooks Range (modified from Mayfield and others, 1988). AH, Avan Hills klippe; AM, Asik Mountain klippe; CF, Cutaway fenster; CP, Copter Peak klippe; DF, Drinkwater fenster; GF, Ginny fenster; IM, Iyikrok Mountain klippe; MB, Mount Bastille klippe; MM, Misheguk Mountain klippe; NR, Nuka Ridge klippe; PF, Picnic Creek fenster; SM, Siniktanneyak Mountain klippe.

evidence for blueschist facies metamorphism. These Early and Middle Devonian carbonate rocks of the Kelly River allochthon grade upward into 165 m of bioturbated, laminated, locally argillaceous shallow-marine carbonate rocks assigned to the Eli Limestone of the Baird Group by Tailleur and others (1967). Conodonts from the basal Eli Limestone are Middle Devonian (Givetian), whereas those from near its top are latest Devonian (Famennian) (Karl and Long, 1990).

Lisburne Group. The Lisburne Group in the Kelly River allochthon consists of three formations composed chiefly of carbonate rocks. In ascending order, these are the Utukok Formation, Kogruk Formation, and the Tupik Formation (Sable and Dutro, 1961). In contrast to exposures of the Lisburne Group in other allochthons, the Lisburne Group in the Kelly River allochthon rests conformably on carbonate strata of the Eli Formation rather than on terrigenous clastic rocks of the Endicott Group, which are not present in the Kelly River allochthon. However, the calcareous clastic rocks of the Utukok Formation may be correlative with the Endicott Group of other allochthons (Sable and Dutro, 1961). The Lisburne in the Kelly River allochthon represents progressive onlap onto a shallow platform that slowly subsided through Mississippian time (Armstrong, 1970).

The Utukok Formation (Mississippian), defined by Sable and Dutro (1961) in the Misheguk Mountain quadrangle, consists of interbedded and reddish-brown–weathering, fine-grained, quartzose sandstone, calcareous sandstone, sandy limestone, and gray, calcareous shale. The Utukok Formation is nearly 1000 m thick at its type locality, but it thins to as little as 10 m to the south in the Baird Mountains quadrangle (Sable and Dutro, 1961; J. A. Dumoulin, 1988, written commun.; Karl and others, 1989; Mayfield and others, 1990). At the type locality, sandstone is most abundant in the basal part of the unit, whereas limestone, sandy limestone, and shale are more abundant in the upper part of the unit. The sandstone is thin to medium bedded, consists almost entirely of well-sorted grains of quartz with minor chert and opaque-rich argillite, and is characterized by bioturbation, current-worked megafossils, abundant ripple marks, and small-scale cross-stratification. Megafossils and conodonts from the Utukok indicate an Early Mississippian (Kinderhookian and Osagean) age for the formation, whereas foraminifers indicate an Early and Late Mississippian (Osagean and Meramecian) age (Sable and Dutro, 1961; Curtis and others, 1984, 1990; Ellersieck and others, 1984, 1990; Mayfield and others, 1984, 1987, 1990; Karl and Long, 1990).

The Kogruk Formation (Mississippian and Pennsylvanian), also defined by Sable and Dutro (1961) in the Misheguk Mountain quadrangle, conformably overlies the Utukok Formation and consists of 650 m of light gray weathering, cliff-forming limestone with abundant nodular black chert and silicified zones. The limestone comprises medium-bedded to massive bryozoan-echinoderm packstone and wackestone and lesser amounts of peloidal lime mudstone and ooid grainstone and packstone (Armstrong, 1970). The lower 300 m of the unit is an oolite-bearing transgressive assemblage deposited in shoaling-water and tidal-channel environments. The upper 350 m is a carbonate-platform assemblage deposited in an open-marine to shoaling-water environment on a subsiding shelf on which subsidence and carbonate deposition were in near equilibrium (Armstrong, 1970). Corals, brachiopods, foraminifers, and conodonts indicate Early to Late Mississippian (late Osagean to early Chesterian) ages for the Kogruk in the Kelly River allochthon (Sable and Dutro, 1961; Armstrong, 1970; Curtis and others, 1984, 1990; Ellersieck and others, 1984, 1990; Mayfield and others, 1984, 1987, 1990; Dutro, 1987; J. A. Dumoulin, 1988, written communication), although the Kogruk Formation of the North Slope subterrane on the Lisburne Peninsula is as young as Pennsylvanian (Dutro, 1987).

The uppermost unit of the Lisburne Group in the Kelly River allochthon, the Tupik Formation (Upper Mississippian), consists of subequal amounts of thinly bedded, dark gray limestone and nodular and bedded, spiculitic, black chert that conformably overlie the Kogruk Formation. The Tupik is less than 30 m thick at its type section in the Misheguk Mountain quadrangle, but locally it may be as thick as 230 m (Sable and Dutro, 1961). The formation represents the deep-water assemblage of Armstrong and Bird (1976). Recovered foraminifers and a sparse megafauna indicate a Late Mississippian (Meramecian) age for the unit, but its undated upper part may include Pennsylvanian strata (Sable and Dutro, 1961; Curtis and others, 1984).

Etivluk Group. The Etivluk Group of the Kelly River allochthon consists of relatively nonresistant and poorly exposed, thinly interbedded silicified limestone and mudstone, chert, and shale of the Siksikpuk (20–40 m) and Otuk (20–40 m) formations. These units contain more shale than those of the Etivluk Group of the underlying Picnic Creek and overlying Ipnavik River allochthons, but they are similar to Etivluk Group units of the Endicott Mountains subterrane.

The Siksikpuk Formation (Pennsylvanian, Permian, and Triassic) of the Kelly River allochthon has not been studied in detail; therefore, its stratigraphy is not well known. The lower part of the formation, which may disconformably overlie the Tupik Formation, consists dominantly of chert, locally containing radiolarians, and silicified limestone and mudstone. Higher in the section, the Siksikpuk Formation consists of thinly interbedded, greenish-gray to gray silicified mudstone and nodular bedded chert that grade upward to maroon nodular chert and silicified mudstone with isolated barite crystals. Although the upper part of the Siksikpuk Formation is poorly exposed, it apparently consists of greenish-gray to maroon siltstone and mudstone. Few fossils have been described from Siksikpuk sections of the Kelly River allochthon, but because of the similarity of these sections with dated Siksikpuk sections in the Endicott Mountains subterrane, the Siksikpuk of the Kelly River allochthon is considered to be Pennsylvanian to Triassic (Curtis and others, 1984; Ellersieck and others, 1984; Mayfield and others, 1984).

The Otuk Formation (Triassic and Jurassic) rests conformably on the Siksikpuk Formation (Curtis and others, 1984; Ellersieck and others, 1984; Mayfield and others, 1984). Like the Otuk Formation in the Endicott Mountains subterrane, the Otuk Formation of the Kelly River allochthon can be divided into shale, chert, and limestone members; however, the uppermost member of the Otuk, the Blankenship, has not been recognized in the area of the Kelly River allochthon. The lower part of the chert member of the Otuk Formation consists dominantly of green to greenish-gray or maroon chert beds. Higher up, bedding is thinner and more regular, and yellowish-gray–weathering beds are more common. The lower part of the limestone member consists chiefly of banded, black- and yellowish-gray–weathering silicified limestone beds with abundant *Monotis* fauna. Maroon chert, similar to that in the Siksikpuk Formation, has been observed in the Otuk Formation at several localities in the Kelly River allochthon. Recovered radiolarians and the abundant pelecypod fauna are Late Triassic.

Ipnavik River allochthon. The Ipnavik River allochthon (Fig. 19) comprises the Baird Group, the Endicott Group, the Lisburne Group, the Etivluk Group, and the Okpikruak Formation. In addition, the allochthon is distinguished by many reddish-brown–weathering diabase sills that intrude black chert and limestone of the Lisburne and lower Etivluk groups. Reconstructed pre-Cretaceous strata of this allochthon total less than 900 m thick (Mayfield and others, 1988).

Baird Group. White- to light gray–weathering limestone assigned to the Baird Group is well exposed at the base of the Ipnavik River allochthon in the Picnic Creek fenster in the Misheguk Mountain quadrangle. At this locality, the Baird Group can be divided into a lower unit of limestone and interbedded gray shale (200 m) and an upper unit of massive to thin-bedded, coarse- to fine-grained limestone (500 m) (Mayfield and others, 1988). Sparse fossils, including stromatoporoids, corals, conodonts, brachiopods, and foraminifers, indicate an Early and Middle Devonian and early Late Devonian (Frasnian) age for the lower unit and a late Late Devonian (Famennian) age for the upper unit (Ellersieck and others, 1984; Mayfield and others, 1987, 1990).

Endicott Group. In the Ipnavik River allochthon, the Endicott Group consists only of the Kayak Shale (Mississippian). Here, the Kayak comprises 40 to 70 m of poorly exposed and sparsely fossiliferous, fissile black shale with interbedded orange-weathering limestone, siltstone, and ironstone concretions. Although the Kayak commonly crops out along faults at the base of the allochthon, the shale was originally deposited on carbonate rocks of the Baird Group (Ellersieck and others, 1984; Mayfield and others, 1990). Conodonts from the Kayak of the Ipnavik River allochthon are generally Early Mississippian (Kinderhookian) (Mayfield and others, 1990), but Late Mississippian conodonts were reported by Ellersieck and others (1984). Locally, the unit interfingers with as much as 200 m of buff limestone, sandstone, and gray shale that has been mapped as the Utukok Formation (Lisburne Group) by Mayfield and others (1990).

Lisburne Group. The Lisburne Group of the Ipnavik River allochthon is undivided and consists of as much as 250 m of interbedded black chert, black siliceous shale, and dark gray, laminated micritic limestone and fine-grained dolomite. These rocks overlie the Kayak Shale at a sharp, but conformable, contact (Ellersieck and others, 1984; Mayfield and others, 1984). Black chert, the dominant lithology of the Lisburne Group in this allochthon, is thin bedded and spiculitic and interfingers laterally with the carbonate rocks and siliceous shale of the Lisburne. The carbonate rocks locally compose more than 50% of the unit, are thin bedded, and typically display bioclastic textures. Foraminifers from the Lisburne Group of the Ipnavik River allochthon are Late Mississippian, whereas radiolarians from the unit are Mississippian to Early Pennsylvanian. The abundance of chert interstratified with thin-bedded, fine-grained limestone suggests that the Lisburne Group of the Ipnavik River allochthon is a deep-water assemblage deposited near the edge of a carbonate platform.

Etivluk Group. Although Mayfield and others (1984, 1990) suggested that rocks of the Etivluk Group of the Ipnavik River allochthon be assigned to the Siksikpuk and Otuk formations, the Etivluk Group of the Ipnavik River allochthon is more like the Imnaitchiak Chert of the Picnic Creek allochthon. The Etivluk of the Ipnavik River allochthon consists of a lower unit of gray to maroon chert with minor siliceous argillite and an upper unit of gray to greenish-gray chert with rare, *Monotis*-bearing siliceous limestone that together are 100 m thick (Blome and others, 1988). The base of the Etivluk Group is conformable with the underlying Lisburne Group. Radiolarians, foraminifers, and conodonts indicate a Pennsylvanian and Permian age for the lower part of the Etivluk Group and a Triassic to Early Jurassic (late Pliensbachian to Toarcian) age for the upper part (Mayfield and others, 1984, 1988, 1990; Blome and others, 1988).

Igneous rocks. Diabase sills that intrude the Lisburne Group and lower Etivluk Group of the Ipnavik River allochthon are up to 100 m thick and consist of microgabbro of tholeiitic or mildly alkaline composition (Ellersieck and others, 1984; Mayfield and others, 1984, 1988; Karl and Long, 1990). Several geochemical analyses from sills in the Misheguk Mountain quadrangle display nearly flat rare earth element patterns, suggesting that the sills were extruded at a mid-ocean ridge, continental margin, or ocean island (Moore, 1987b). No definitive radiometric dates have been obtained from these igneous rocks; therefore, their age, post–Early Mississippian, can be delimited only by stratigraphic relations.

Nuka Ridge allochthon. The Nuka Ridge allochthon, the highest of the allochthons of the De Long Mountains subterrane, is distinguished by an unusual, widespread arkosic limestone of Carboniferous age (Fig. 19). Although extensive exposures of the allochthon are limited to the Nuka Ridge and Mount Bastille klippen (Fig. 20) in the Misheguk Mountain quadrangle, the arkosic rocks are imbricated with Cretaceous strata in a number of widely scattered localities from the Chukchi Sea coast to the Chandler Lake quadrangle. Some of these exposures are clearly fault-bounded slivers, but many others are isolated blocks as much as a few tens of meters in maximum dimension; some of these may be olistoliths in the Upper Jurassic and Lower Cretaceous Okpikruak Formation. The maximum thickness of reconstructed pre-Cretaceous strata of the allochthon is probably less than 600 m thick, although the allochthon is as thick as 1.5 km due to structural repetition (Mayfield and others, 1988). The allochthon consists of strata assigned to the Baird Group, Endicott Group, Nuka Formation, Etivluk Group, and Okpikruak Formation.

Baird Group. In the Nuka Ridge allochthon, the Baird Group consists of the Kugururok Formation (Devonian) (Sable and Dutro, 1961). The lower part of the Kugururok as mapped by Ellersieck and others (1984) is about 300 m thick and consists of calcareous shale and minor hematite-bearing conglomeratic limestone overlain by massive to thin-bedded limestone and dolomite with sparse chert lenses. The upper part of the formation consists of 125 m of light colored, locally glauconitic, laminated to cross-bedded limestone, and toward the top of the formation, the limestone contains as much as 15% potassium feldspar grains (Sable and Dutro, 1961; Ellersieck and others, 1984). Megafossils and conodonts from the lower Kugururok indicate a Middle and Late Devonian (Givetian and Frasnian) age (Sable and Dutro, 1961; Ellersieck and others, 1984), but Ellersieck and others (1984) and Mayfield and others (1990) suggested that the feldspar-bearing uppermost part of the unit may be equivalent to the Upper Mississippian and Lower Pennsylvanian(?) Nuka Formation.

Endicott Group. The Endicott Group in the Nuka Ridge allochthon is represented only by the Kayak Shale (Mississippian). Here, the Kayak consists of as much as 350 m of poorly exposed, fissile black shale with minor interbedded reddish-brown–weathering bioclastic limestone and fine-grained sandstone that is locally feldspathic. Where exposed, the Kayak Shale is bounded at the base by a thrust fault, but it may stratigraphically pinch out within the allochthon. Mayfield and others (1984, 1987) reported that the Kayak Shale contains Mississippian foraminifers and brachiopods.

Nuka Formation. The Nuka Formation (Mississippian and Pennsylvanian?) consists of as much as 300 m of interbedded, light gray–weathering limestone to arkosic limestone, arkose, and quartzose sandstone that rest conformably on the Kayak Shale. In the area of Nuka Ridge in the Misheguk Mountain quadrangle (Tailleur and Sable, 1963; Tailleur and others, 1973), the Nuka Formation consists of massive to medium-bedded, fine-grained to very coarse grained sandstone and minor granule conglomerate that display parallel stratification and abundant tabular and trough cross-stratification. Strata of this area also exhibit many diagnostic features of a shallow-marine environment, including herringbone cross-stratification, inclined lamination, calcareous and bright green glauconitic zones, and beds rich in marine fossils. Elsewhere, however, Nuka sections display a prominent red coloration and sedimentary structures suggestive of nonmarine deposition or other sedimentary structures indicative of turbidite deposition. Sandstone and limestone of the Nuka Formation contain abundant, unweathered, coarse-grained, subangular potassium feldspar grains indicative of a nearby granitic source area. Uranium-lead isotopic ages of detrital zircons in the sandstone indicate that the granitic rocks from which the arkose was derived had an age of about 2.07 Ga (Hemming and others, 1989). This age is older than any known granitic source in northern Alaska. Although the abundant megafauna in the Nuka Formation was originally interpreted to range from Mississippian to Permian (Tailleur and Sable, 1963), foraminifers and conodonts indicate that the Nuka ranges only from Late Mississippian (late Meramecian) to Early Pennsylvanian(?) (early Morrowan?) (Mayfield and others, 1984, 1987, 1990).

Etivluk Group. The Etivluk Group of the Nuka Ridge allochthon consists of less than 150 m of gray and greenish-gray to maroon chert and minor shale to siliceous shale. The chert is typically well bedded and contains abundant Late Triassic radiolarians; the upper part of the sequence also contains rare *Monotis* fossils (Mayfield and others, 1984; T. E. Moore, 1985, unpublished data). Mayfield and others (1984) reported that the Etivluk Group rests conformably on the Nuka Formation and that, locally, chert at the base of the Etivluk Group contains feldspar grains, which supports their interpretation. Although this cherty succession of the Etivluk Group has been assigned to the Siksikpuk and Otuk formations (Mayfield and others, 1987), it is lithologically similar in most respects to the Imnaitchiak Chert of the Picnic Creek allochthon. On the basis of megafossils and the stratigraphic position, Mayfield and others (1984) considered the Etivluk of the Nuka Ridge allochthon to be Pennsylvanian to Jurassic.

Deposits of the Brookian sequence (Okpikruak Formation) in the De Long Mountains subterrane. The Upper Jurassic and Lower Cretaceous Okpikruak Formation (Gryc and others, 1951; Mayfield and others, 1988) consists of flysch that is widely exposed in the northern foothills of the central and western Brooks Range. The Okpikruak rests unconformably on the Etivluk Group of the Picnic Creek, Kelly River, Ipnavik River, and Nuka Ridge allochthons of the De Long Mountains subterrane, but the structural position of many other outcrops of the Okpikruak is difficult to determine because of poor exposure and structural complexity. Although detailed petrographic descriptions of the Okpikruak Formation of the Picnic Creek allochthon were provided by Wilbur and others (1987) and Siok (1989) for a few localities in the central Brooks Range, only generalized descriptions are available for the unit in most areas (for example, Patton and Tailleur, 1964).

In the De Long Mountains subterrane, the Okpikruak Formation consists of thin- to thick-bedded, fine- to coarse-grained lithic sandstone and gray, brown, or black shale and mudstone. Lenticular units of polymict pebble to boulder conglomerate and pebbly mudstone are locally present in beds as thick as 50 m. Clasts in the conglomerate are rounded to angular and consist of chert, limestone, granitic rocks, and felsic and mafic volcanic rocks. Thick intervals of fine-grained strata are common in the unit, although sandstone-to-shale ratios are as high as 5:1 for some intervals. Sedimentary structures in sandstone beds include flute casts, tool marks, ripple cross-lamination, graded bedding, shale rip-up clasts, starved ripples, and Bouma sequences. In the Okpikruak Formation of the Nuka Ridge allochthon, conglomerate and coarse-grained sandstone appear to be absent, but calcareous concretions are common. A maximum thickness of more than 1000 m is estimated for the Okpikruak in the Picnic Creek, Kelly River, and Ipnavik River allochthons, but a thickness of only 200 m has been reported for the unit in the Nuka Ridge allochthon (Curtis and others, 1984).

Sandstone of the Okpikruak Formation in the Picnic Creek allochthon in the Killik River and Chandler Lake quadrangles contains moderate to high proportions of lithic grains, including chert, volcanic, and clastic-sedimentary rocks. Although the lithic proportions of these rocks vary widely with location, the proportions indicate derivation from mixed magmatic-arc and recycled-orogen provenances (Wilbur and others, 1987; Siok, 1989). Siok (1989) concluded from sedimentologic evidence that these strata were deposited at bathyal to abyssal depths in the inner to middle, and possibly outer, regions of a submarine fan.

Olistostromes, locally present in the Okpikruak Formation in all except the Nuka Ridge allochthon, are the most widespread and have the most diverse olistolith compositions in the Ipnavik River allochthon. The olistoliths, some more than 10 m in diameter, consist of chert, mafic rocks, limestone, and arkosic limestone (Mull, 1979) that were derived from the Angayucham terrane and the Baird Group, Kogruk Formation, Nuka Formation, and Etivluk Group of the De Long Mountains subterrane (Mull, 1979; Crane, 1987; Curtis and others, 1990). Mull and others (1976) and Crane (1987) suggested that the olistostromes were deposited by debris flow and submarine gravity sliding adjacent to a tectonically active area.

Fossils are rare in the Okpikruak and consist of various species of *Buchia* pelecypods that indicate a mainly Neocomian age for the formation (Jones and Grantz, 1964; Patton and Tailleur, 1964). Where assigned to the Picnic Creek allochthon in the Chandler Lake, Killik River, and De Long Mountains quadrangles, and the Kelly River allochthon in the Misheguk Mountain quadrangle, the *Buchia* fossils indicate a Valanginian (Early Cretaceous) age for at least part of the unit (Curtis and others, 1984; Wilbur and others, 1987; Mayfield and others, 1988; Ellersieck and others, 1990). Both Berriasian and Valanginian species of *Buchia* have been recovered in Okpikruak strata of the Ipnavik River allochthon in the Misheguk Mountain quadrangle (Curtis and others, 1984); no fossils have been reported from Okpikruak strata of the Nuka Ridge allochthon. Okpikruak strata in the De Long Mountains quadrangle have yielded a *Buchia* of Late Jurassic age, but the outcrop from which the fossil was recovered may be assigned to either the Kelly River, Ipnavik River, or Nuka Ridge allochthons (Curtis and others, 1990). These data suggest that the Okpikruak Formation of structurally higher allochthons of the De Long Mountains subterrane was deposited earlier than the remainder of the formation of underlying allochtons (Mayfield and others, 1988).

Hammond subterrane

The Hammond subterrane extends for nearly 800 km along most of the length of the southern Brooks Range and includes most of the rocks of the central belt (Figs. 2 and 3). The Hammond subterrane consists of a structurally imbricated assemblage of thick carbonate and clastic units with phyllitic and schistose textures and greenschist and locally blueschist and amphibolite facies metamorphic mineral assemblages (Jones and others, 1987). Deformational fabrics, typically inhomogeneous, vary from penetrative to nonpenetrative; however, relict sedimentary and igneous textures are commonly preserved. Some local allochthonous sequences have been identified within the Hammond subterrane, but their regional extent is not fully known (for example, Kugrak River allochthon of Mull and others, 1987c; Skajit allochthon of Oldow and others, 1987d). Fossils, found primarily in carbonate rocks, are rare, but they indicate that much of the subterrane consists of lower Paleozoic rocks. Isotopic ages from sparse granitic rocks intruding the terrane yield Late Proterozoic and Devonian ages and indicate the presence of Precambrian basement rocks in the subterrane (Fig. 21).

Because of limited age control, metamorphism, and structural complications, the stratigraphy of the Hammond subterrane is poorly known; stratigraphic studies of the subterrane have been concentrated mainly in the Baird Mountains quadrangle and along the Dalton Highway. For the purpose of this discussion, we divide the rocks of the subterrane into several general lithologic assemblages: (1) Proterozoic and Proterozoic(?) metasedimentary and metavolcanic rocks; (2) thick units of pre-Mississippian carbonate rocks assigned to the Baird Group; (3) lower Paleozoic metasedimentary rocks; (4) metasedimentary rocks assigned to the Devonian Beaucoup Formation and the Hunt Fork Shale; (5) upper Paleozoic carbonate and metaclastic rocks assigned to the Endicott, Lisburne, and Sadlerochit(?) groups; and (6) metagranitic rocks. Stratigraphic relations within and between the various assemblages are controversial, and future work may indicate that the assemblage comprises two or more tectonostratigraphic units.

Figure 21. Generalized stratigraphic columns for Arctic Alaska and Angayucham terranes. Subterranes and terranes are shown in present relative positions; allochthons of De Long Mountains subterrane are shown in restored positions relative to one another (Mayfield and others, 1988). Shaded regions highlight selected rock units; dashed line indicates base of overlapping sedimentary rocks of Colville and Koyukuk basins. Subcolumn 1, undivided pre-Mississippian rocks of North Slope subterrane in northeastern salient of Brooks Range, except Sadlerochit and Shublik mountains; subcolumn 2, pre-Mississippian carbonate succession in Sadlerochit and Shublik mountains; subcolumn 3, undivided pre-Mississippian rocks of Hammond subterrane; subcolumn 4, pre-Mississippian rocks of Baird Group in Baird Mountains quadrangle. Abbreviations: BG, Baird Group; CLV, Brookian sequence in Colville basin; END, Endicott Group; ETIV, Etivluk Group; KING, Kingak Shale; KOY, sedimentary rocks of Koyukuk basin; LCU, Lower Cretaceous unconformity; LISB, Lisburne Group; NK, Nuka Formation; OKPK, Okpikruak Formation; SAD, Sadlerochit Group; SF, Shublik Formation. Absolute time scale from Palmer (1983).

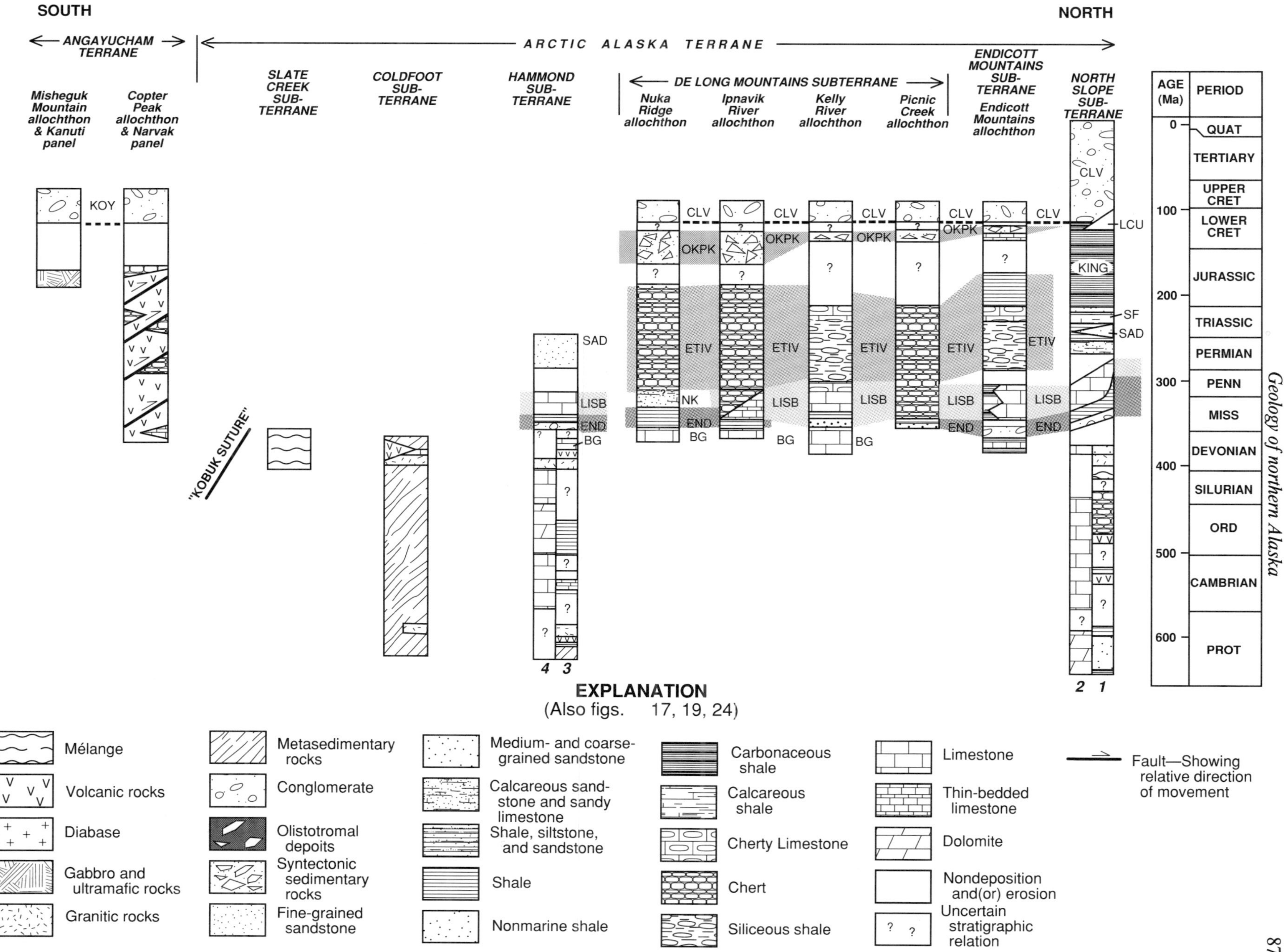

SOUTH
NORTH
ANGAYUCHAM TERRANE
ARCTIC ALASKA TERRANE
Misheguk Mountain allochthon & Kanuti panel
Copter Peak allochthon & Narvak panel
SLATE CREEK SUB-TERRANE
COLDFOOT SUB-TERRANE
HAMMOND SUB-TERRANE
DE LONG MOUNTAINS SUBTERRANE
Nuka Ridge allochthon
Ipnavik River allochthon
Kelly River allochthon
Picnic Creek allochthon
ENDICOTT MOUNTAINS SUB-TERRANE
Endicott Mountains allochthon
NORTH SLOPE SUB-TERRANE
AGE (Ma)
PERIOD
0
100
200
300
400
500
600
QUAT
TERTIARY
UPPER CRET
LOWER CRET
JURASSIC
TRIASSIC
PERMIAN
PENN
MISS
DEVONIAN
SILURIAN
ORD
CAMBRIAN
PROT
KOY
"KOBUK SUTURE"
CLV
OKPK
ETIV
LISB
NK
END
BG
SAD
SF
KING
LCU
4 3
2 1
EXPLANATION
(Also figs. 17, 19, 24)
Mélange
Volcanic rocks
Diabase
Gabbro and ultramafic rocks
Granitic rocks
Metasedimentary rocks
Conglomerate
Olistotromal depoits
Syntectonic sedimentary rocks
Fine-grained sandstone
Medium- and coarse-grained sandstone
Calcareous sandstone and sandy limestone
Shale, siltstone, and sandstone
Shale
Nonmarine shale
Carbonaceous shale
Calcareous shale
Cherty Limestone
Chert
Siliceous shale
Limestone
Thin-bedded limestone
Dolomite
Nondeposition and(or) erosion
Uncertain stratigraphic relation
Fault—Showing relative direction of movement

Proterozoic and Proterozoic(?) metasedimentary and metavolcanic rocks. Although Proterozoic rocks of the Hammond subterrane have been recognized only in structural culminations at Mount Angayukaqsraq in the northeastern Baird Mountains quadrangle and near Ernie Lake in the Wiseman and Survey Pass quadrangle, Proterozoic rocks may be more widespread throughout the subterrane (Dillon and others, 1986; Karl and others, 1989; Till, 1989). At Mount Angayukaqsraq, Proterozoic rocks consist of two imbricated lithologic assemblages. The structurally lower assemblage consists mainly of undated dolostone, metalimestone, and marble with subordinate intercalated phyllite, quartzite, carbonate-cobble metaconglomerate, and metabasite. The dolostone contains well-preserved stromatolitic mounds, fenestral fabrics, oolitic intraclasts, and grainstone that contains ooids, composite grains, and pisoids. Together these features suggest an intertidal to shallow subtidal depositional environment (Dumoulin, 1988). The intercalated metabasite displays massive, layered, pillowed, and pillow-breccia structures. The structurally higher assemblage contains interleaved metavolcanic and metasedimentary units with well-developed metamorphic fabrics and no relict sedimentary or volcanic structures. The metasedimentary unit consists of interlayered quartzite, micaceous quartzite, calc-schist, pelitic schist, and garnet amphibolite that are intruded by granitic rocks. The granitic rocks yielded a U-Pb zircon age of 750 ± 6 Ma (Karl and others, 1989). Hornblende and white mica K-Ar and Rb-Sr ages on rocks of the metasedimentary unit suggest that amphibolite facies metamorphism occurred at about 655 to 594 Ma (Turner and others, 1979; Mayfield and others, 1982; Armstrong and others, 1986). The metavolcanic unit, also intruded by Proterozoic granitic rocks, contains thick bodies of massive to crudely layered metabasite that are interlayered with thinner lenses of carbonate rock. Relict green hornblende found in the cores of blue amphibole of the metabasite (A. B. Till, oral commun., cited *in* Karl and others, 1989) indicate that these rocks were also affected by the Proterozoic amphibolite facies metamorphic event and were overprinted by blueschist facies assemblages in Mesozoic time (Karl and others, 1989; Till, 1989). The metavolcanic unit has yielded K-Ar hornblende ages of 729 ± 22 and 595 ± 30 Ma (Turner and others, 1979; Mayfield and others, 1982).

In the Ernie Lake area, Proterozoic(?) granitic gneiss intrudes interlayered quartz-mica schist, quartzite, metaconglomerate, marble, graphitic phyllite, calcareous schist, and metabasite (banded schist unit of Dillon and others, 1986). These rocks have not been investigated in detail; they are the structurally lowest rocks in the area and may extend eastward along the southern margin of the Hammond subterrane into the vicinity of the Proterozoic rocks at Mount Angayukaqsraq (W. P. Brosgé, 1988, oral commun.; Till, 1989).

Massive carbonate units of the Baird Group. The Hammond subterrane is characterized by widespread, prominent units of thick-bedded to massive, light gray limestone, dolostone, and marble as much as 1 km thick. These thick carbonate bodies extend over distances of 50 km or more and wedge out laterally into thinner carbonate units that are interlayered with metaclastic rocks. Mapping suggests that the lateral changes in thickness of these carbonate units may result from facies changes, as well as structural truncation along low-angle faults. Schrader (1902) gave the name Skajit Formation to one of these carbonate massifs, a 400-m-thick sequence of unfossiliferous marble (Henning, 1982) along the John River in the southern Wiseman quadrangle. Smith and Mertie (1930) later extended the formation to include most of the thick, massive carbonate units throughout the southern Brooks Range and renamed the formation the Skajit Limestone. Rare megafossils of Silurian and Devonian age were recovered from some of these limestone bodies in both the eastern and western Brooks Range, leading most workers to conclude that the Skajit Limestone is a regional stratigraphic marker unit of middle Paleozoic age (Brosgé and others, 1962, 1979; Tailleur and others, 1967; Oliver and others, 1975; Mayfield and Tailleur, 1978; Nelson and Grybeck, 1980; Dillon and others, 1986; Dillon, 1989).

To distinguish the Skajit Limestone and other older carbonate units in the southern Brooks Range from the upper Paleozoic Lisburne Group, Tailleur and others (1967) defined the Baird Group to consist of the Skajit Limestone, Kugururok Formation, and Eli Limestone (the latter two herein included in the De Long Mountains subterrane), and other unnamed carbonate units. They also suggested that the Baird Group represents remnants of a once-widespread carbonate-platform succession of largely Devonian age. However, subsequent recovery of Ordovician graptolites from the Baird Group (Carter and Tailleur, 1984) and the detailed micropaleontologic investigations of Dillon and others (1987a), Dumoulin and Harris (1987), and Dumoulin (1988) have revealed that the Baird Group consists of two or more successions of carbonate strata of mainly early Paleozoic age. These data have raised questions about the correlation and the stratigraphic usefulness of the various carbonate bodies assigned to the Skajit Limestone and the Baird Group as a whole.

The most detailed work on the carbonate rocks of the Baird Group is the conodont biostratigraphy and associated sedimentary facies studies of Dumoulin and Harris (1987), who have recognized two distinctive metacarbonate sequences, the northeastern carbonate sequence and the west-central carbonate sequence, in the Baird Mountains quadrangle in the western Brooks Range. Both sequences are tectonically disrupted, so they were reconstructed from incomplete, but overlapping, sections in different thrust sheets.

The northeastern carbonate sequence in the Baird Mountains quadrangle consists of at least 360 m of Middle Cambrian to Upper Silurian and Devonian(?) metalimestone and dolostone. The Middle and Upper Cambrian rocks grade upward from massive marble to thin-bedded metalimestone-dolostone couplets that contain mollusks, acrotretid brachiopods, and agnostid trilobites. These rocks were deposited under shallow-marine conditions (Dumoulin and Harris, 1987). The Cambrian rocks are overlain by Lower and Middle Ordovician metalimestone and graptolitic phyllite. Graptolites and conodonts indicate that the

Ordovician rocks were deposited in cool-water, mid-shelf to basinal conditions (Carter and Tailleur, 1984; Dumoulin and Harris, 1987). Upper Ordovician rocks of the sequence consist of bioturbated to laminated dolostone containing warm, shallow-marine conodonts, whereas Upper Silurian rocks consist of thinly laminated dolomitic mudstones that were deposited in a restricted shallow-marine environment. A few conodont species from the succession range into the Devonian, but no uniquely Devonian conodonts have been recovered. Although the geographic extent of this sequence is unknown, lithofacies similar to those of this sequence are found as far east as the Dalton Highway.

The 1200-m-thick west-central carbonate sequence of the Baird Mountains quadrangle consists largely of Ordovician metalimestone with Silurian dolostone and Devonian metalimestone (Dumoulin and Harris, 1987). The Lower Ordovician rocks consist of argillaceous metalimestone deposited in a normal-marine environment and fenestral dolostone deposited in a shallow-water, locally restricted platform environment. Middle Ordovician rocks consist partly of dolostone that was deposited in a warm, very restricted, shallow-water, innermost platform environment, whereas sparse Upper Ordovician rocks are metalimestone that was deposited in a cool and deep-water environment. Middle and Upper Silurian rocks consist of at least 100 m of shallow-water dolostone that may unconformably overlie the older rocks. Devonian rocks, widely exposed in the west-central sequence, conformably overlie the Silurian dolostone and consist of fossiliferous metalimestone deposited in a range of normal-marine to slightly restricted shelf environments. This sequence, which appears to be restricted to the southwestern Brooks Range, is found in a structurally higher position than the northeastern carbonate sequence. Lower and Middle Ordovician strata of the west-central carbonate sequences are not recognized outside the Baird Mountains quadrangle in the southern Brooks Range, but these strata show many similarities in biofacies and lithofacies to age-equivalent rocks of the York Mountains terrane on the Seward Peninsula. Harris and others (1988) reported that these Lower and Middle Ordovician strata contain a significant proportion of conodont species that are of Siberian provincial affinity.

Lower Paleozoic metasedimentary rocks. In the Baird Mountains quadrangle, lower Paleozoic metaclastic and metavolcanic rocks of the Tukpahlearik Creek unit of Karl and others (1989) are situated between thrust sheets containing the contrasting sequences of carbonate rocks of the Baird Group as described by Dumoulin and Harris (1987). The Tukpahlearik Creek unit consists of black carbonaceous quartzite and siliceous argillite with lenses of dolostone and marble, pelitic schist, chert-pebble metaconglomerate, calc-schist, thin-bedded micaceous marble, and metabasite. Karl and others (1989), who recovered Ordovician conodonts from a dolostone lens in the black quartzite, inferred a basinal depositional environment for these rocks.

Cambrian and Ordovician metaclastic rocks have been reported from near the Dalton Highway in the Hammond subterrane by Dillon and others (1987a). The Cambrian rocks are at least 100 m thick and consist of thin-bedded, red-weathering, phyllitic, calcareous siltstone and sandstone with intercalated black, carbonaceous phyllite. A thin but massive limestone unit, which forms the uppermost part of the sequence, contains Middle Cambrian phosphatic brachiopods and trilobites that Palmer and others (1984) considered similar to those in open-shelf facies of the Siberian platform. However, recent mapping shows that the trilobite-bearing limestone has ambiguous relations with both the metaclastic rocks and marble of the Skajit Limestone and may well be a fault slice of the Skajit Limestone (T. E. Moore, 1990, unpublished data). Dillon and others (1987a) inferred that the Cambrian rocks are overlain by at least 50 m of imbricated thin-bedded, black, carbonaceous phyllite and crinoidal limestone that are interstratified with thick-bedded marble and sparse thin-bedded, black quartzite and calcareous sandstone. Conodonts from the basal part of the black phyllite and crinoidal limestone unit are late Early Ordovician (middle Arenigian) and those in the uppermost part are Late Ordovician (Caradocian or younger) (Dillon and others, 1987a). These carbonaceous Ordovician rocks represent a basinal or off-platform depositional environment, but their relation to the assumed underlying rocks is uncertain because of extensive faulting.

Devonian metasedimentary rocks. Rocks exposed extensively in the Hammond subterrane, especially its northern part, compose a foliated, imbricated assemblage of black calcareous phyllite, calcareous chlorite phyllite, black siliceous phyllite, maroon and green phyllite, argillite- and limestone-pebble conglomerate, metagraywacke, mafic pillowed flows and volcaniclastic rocks, quartz- and chert-pebble conglomerate, quartzose sandstone, and lenticular limestone bodies. Stratigraphy of the assemblage is ambiguous due to variations in thickness, in proportion of rock types, and in stacking order. In the Wiseman quadrangle, Dillon and others (1986) divided the assemblage into three primary units: (1) a lower, unnamed unit of coarse-grained, siliceous, clastic metasedimentary rocks; (2) a middle, heterogenous unit locally assigned to the Beaucoup Formation; and (3) an upper unit of phyllite and slate assigned to the Hunt Fork Shale. Metamorphosed volcanic rocks and diabase are present locally in all three units.

The lowest unit consists of complexly interlayered and metamorphosed calcareous, chloritic siltstone, sandstone, and quartz conglomerate. Dillon (1989) reported that these rocks rest unconformably on Cambrian and Ordovician metasedimentary rocks and grade laterally and upward into the Skajit Limestone and associated metamorphosed silicic volcanic rocks and volcanogenic graywacke informally named the "Whiteface Mountain volcanics" (Dillon, 1989). Although no fossils have been recovered from the lowest unit, Dillon (1989) inferred a Middle Devonian age.

Where mapped as part of the Hammond subterrane in the Wiseman quadrangle, the Beaucoup Formation (Devonian) (Dutro and others,1979) consists of, from base to top, (1) black calcareous phyllite, siltstone, and other fine-grained rocks; (2) lenticular bodies of fossiliferous limestone or marble containing Middle and Late Devonian (Givetian and Frasnian) brachiopods and conodonts; and (3) calcareous, chloritic phyllite, metasandstone, and metaconglomerate (Dillon and others, 1986). These rocks have been correlated with the Beaucoup Formation of the Endicott Mountains subterrane because of their similarity in overall lithology and in age of enclosed carbonate strata and because of their structural proximity to the Endicott Mountains subterrane.

Dillon (1989) hypothesized that the Beaucoup Formation in the Hammond subterrane is characterized by complex facies changes. He suggested that the lenticular bodies of Devonian limestone of the middle Beaucoup grade laterally and downward into the carbonate massifs of the Skajit Limestone. To the west, where the Skajit is absent, the upper Beaucoup Formation grades laterally and downward into thick units of the Devonian metagraywacke and felsic metavolcanic rocks of Dillon's Whiteface Mountain volcanics. Dillon (1989) interpreted the metavolcanic rocks as the volcanic equivalents of Devonian plutonic rocks exposed elsewhere in the subterrane and correlative with the Ambler sequence of the Coldfoot subterrane. Associated rock units representing local facies of the clastic upper part of the Beaucoup Formation include quartzitic schist and metaconglomerate, calcareous schist and phyllite, chlorite-rich metasiltstone and phyllite, and quartz conglomerate (Dillon and others, 1987a). Units of black metachert and argillaceous, thin-bedded carbonate rocks were interpreted as facies of the lower black phyllite unit of the Beaucoup. All these associated units are unfossiliferous, and many are exposed over wide areas of the Hammond subterrane. Recent conodont biostratigraphic studies and new structural data (Moore and others, 1991), however, suggest that many of the rocks currently assigned to the Beaucoup Formation in the Hammond subterrane may be parts of allochthonous sequences that are unrelated, or only partly related, to each other and to the type section of the Beaucoup Formation in the Endicott Mountains subterrane.

Where mapped in the Hammond subterrane, rocks assigned to the Hunt Fork Shale consist of an undetermined thickness of noncalcareous, black and dark gray slate and phyllite that weather dark brown. Interstratified metasandstone beds are sparse but are locally concentrated at the base of the unit, along with stretched quartz- and chert-pebble conglomerate (Brosgé and Reiser, 1964; Dillon and others, 1986). Relict siltstone and very fine sandstone laminae indicate that the protolith of the slate and phyllite was mud-rich turbidites. Fossils are absent in these rocks; therefore, a Late Devonian age for the Hunt Fork of the Hammond subterrane is inferred on the basis of lithologic correlation to the fossiliferous Upper Devonian Hunt Fork Shale of the Endicott Mountains subterrane. Dillon (1989) concluded that in the Hammond subterrane the Hunt Fork Shale rests on a regional unconformity marked by local conglomeratic beds but locally grades downward into clastic rocks of the Beaucoup Formation.

Mississippian to Permian(?) rocks. Restricted exposures of the Endicott, Lisburne, and Sadlerochit(?) groups in the Schwatka Mountains (Fig. 1) (Ambler River and Survey Pass quadrangles) are the youngest strata included in the Hammond subterrane (Fig. 21). Mull and Tailleur (1977), Tailleur and others (1977), and Mull (1982) compared the Schwatka Mountains succession to the lower Ellesmerian sequence at the Mt. Doonerak fenster in the North Slope subterrane. However, the upper Paleozoic rocks of the Hammond subterrane differ from those in the Mt. Doonerak fenster in that (1) the former rest on granitic rocks and on carbonate rocks of the Baird Group instead of on clastic and mafic volcanic rocks as in the Mt. Doonerak fenster and (2) existing mapping indicates that the Schwatka Mountains succession does not occur in a structural window like that at Mt. Doonerak.

Endicott Group. The structurally lowest unit in the upper Paleozoic Schwatka Mountains succession is the Kekiktuk Conglomerate; it consists of 100 to 200 m of cross-stratified quartzite and stretched pebble to cobble conglomerate with intercalated red, green, and gray phyllite. These rocks unconformably overlie Devonian granitic rocks (Mull and Tailleur, 1977; Mull, 1982; Mull and others, 1987c) and older carbonate rocks of the Baird Group (Tailleur and others, 1977; Mayfield and Tailleur, 1978; Nelson and Grybeck, 1980), although Till and others (1988) interpreted the basal contact as a fault. The Kekiktuk Conglomerate grades upward into less than 300 m of Kayak Shale, including black carbonaceous phyllite, slate, and argillite with thin, red-weathering, fossiliferous limestone interbeds. These rocks are commonly tectonically thickened but may lie unconformably on older rocks in the Survey Pass and the Baird Mountains quadrangles (Nelson and Grybeck, 1980; Karl and others, 1989). Conodonts from the Kayak in the Ambler River quadrangle indicate an Early Mississippian (Osagean) age (Tailleur and others, 1977), whereas megafossil and microfossil assemblages from the Kayak in the Survey Pass quadrangle are Late Devonian (late Famennian) to Early Mississippian (Kinderhookian) and Late Mississippian (Nelson and Grybeck, 1980).

Lisburne Group. The Lisburne Group in the Schwatka Mountains consists of about 100 m of medium- to thick-bedded, gray to black limestone and white dolomite that are commonly replaced with nodular chert; thin, irregular chert beds are also present near the base of the unit. The Lisburne is deformed and locally metamorphosed to marble. Microfossils from the basal part of the group in this area are Early Mississippian (no younger than Osagean), whereas microfossils from the uppermost part of the group are Late Mississippian (late Meramecian to Chesterian) (I. L. Tailleur, 1988, oral commun.). Megafossils from the Lisburne in this area are also Early and Late Mississippian (Mayfield and others, 1978).

Sadlerochit(?) Group. Rocks mapped by Mayfield and Tailleur (1978) as the Sadlerochit(?) Group at Shishakshinovik Pass in the Schwatka Mountains consist of a schistose and reddish-brown–weathering succession of calcareous, fine-grained quartzose sandstone and interbedded siltstone that grades upward into black phyllitic shale. These rocks, less than 100 m thick, are well bedded and overlie the Lisburne Group on a probable disconformity. No fossils have been recovered from these rocks, but their lithology and stratigraphic position suggest a possible correlation with the clastic rocks of the Sadlerochit Group mapped elsewhere (Mull and Tailleur, 1977; Tailleur and others, 1977; Mayfield and Tailleur, 1978).

Metagranitic rocks. Metagranitic rocks of Proterozoic and Devonian age have been recognized in the Hammond subterrane (Plate 13). The Proterozoic metagranitic rocks comprise widely scattered, typically fault-bounded stocks and small plutons along the southern margin of the subterrane. The best studied of these are the intrusive rocks at Mount Angayukaqsraq (Baird Mountains quadrangle) (Dillon and others, 1987b; Karl and others, 1989), which consist of about 70% gabbro and leucogabbro and 30% granodiorite and alkali feldspar granite. These rocks, which yielded a U-Pb crystallization age of 750 Ma, intrude well-foliated metasedimentary and mafic volcanic rocks that were metamorphosed to amphibolite facies prior to intrusion in Proterozoic time (Karl and others, 1989). Granitic rocks of this intrusive complex are highly evolved and mildly peraluminous and are interpreted as "within-plate" magmas derived from a weakly fractionated source and emplaced in a non-arc, continental setting (Karl and others, 1989). Nelson and others (1989) reported Sm-Nd model ages based on an alkali-depleted mantle source for the crust of 1.3 to 1.5 Ga and ϵ_{Nd} values of 1.2 for the Proterozoic granitic rocks at Mount Angayukaqsraq. They inferred from those data that a major source for the granitic rocks was continental crust at least as old as Early Proterozoic. Granitic gneiss of the Ernie Lake (Survey Pass quadrangle) and nearby Sixtymile (Wiseman quadrangle) plutons has yielded highly discordant Late Proterozoic U-Pb ages that are broadly similar to ages of the Mount Angayukaqsraq rocks (Dillon and others, 1980, 1987b; Karl and others, 1989); however, the Ernie Lake and Sixtymile plutons show evidence of an older crustal component (1000–800 Ma) (Karl and others, 1989).

The younger granitic rocks comprise several large, metamorphosed plutons that define a west-trending belt in the Chandalar, Survey Pass, and Ambler River quadrangles. The plutons are elliptical, range from 5 to 50 km in length, and are commonly fault bounded. The largest are the Arrigetch Peak and Mount Igikpak plutons (Fig. 2) in the Survey Pass quadrangle. They consist mostly of peraluminous muscovite-biotite granite but range from alkali-feldspar granite to tonalite (Nelson and Grybeck, 1980; Newberry and others, 1986; Hudson, this volume). However, microprobe analysis indicates that muscovite in some of the Devonian plutons in the Survey Pass quadrangle may be partly or entirely metamorphic in origin (A. B. Till, 1990, oral commun.). Commonly associated with these plutons are augen gneiss, schistose orthogneiss, and aplitic and pegmatitic gneisses that all locally display isoclinal folds (Newberry and others, 1986). In contrast, the Horace Mountains plutons (Fig. 2) in the Chandalar quadrangle are composed largely of foliated metaluminous, porphyritic biotite ± hornblende granodiorite, porphyritic hornblende-biotite granodiorite, and leucogranite; these plutons also contain diorite, quartz monzonite, and tonalite (Newberry and others, 1986; Dillon, 1989). Discordant U-Pb zircon ages indicate crystallization of the Arrigetch Peak pluton at 366 ± 10 Ma (Late Devonian) and the Horace Mountain plutons at 402 Ma (Early Devonian) (Dillon, 1989). Initial Sr ratios for the Survey Pass plutons are high (about 0.715), and their granitic composition shows affinity with S-type granitoids, which suggests that the Survey Pass plutons were formed by melting of continental crust (Nelson and Grybeck, 1980; Newberry and others, 1986). The calculated Sm-Nd model ages for the Survey Pass plutons range from 0.7 to 1.6 Ga, and initial ϵ_{Nd} values for these plutons range from –6 to +3. These figures indicate varying involvement of older crust and younger material in the genesis of the Survey Pass plutons (Nelson and others, 1989). Compositional data from the Horace Mountain plutons, in contrast, indicate affinity with I-type granitoids (Newberry and others, 1986). Metamorphic aureoles surrounding the plutons locally contain noneconomic Sn-W skarns in the Survey Pass quadrangle and Cu-Ag and Pb-Zn-Ag skarns in the Chandalar quadrangle (Newberry and others, 1986).

Coldfoot subterrane

The Coldfoot subterrane consists largely of fine- to coarse-grained metasedimentary rocks that form the schist belt, a continuous 15–25-km-wide belt that extends for at least 600 km along the southern Brooks Range (Brosgé and Reiser, 1964; Mayfield and Tailleur, 1978; Nelson and Grybeck, 1980; Dillon and others, 1986; Dillon, 1989; Karl and others, 1989) (Figs. 2, 3, and 21). These metamorphic rocks are bounded to the south by the Slate Creek subterrane and to the north by the Hammond subterrane along boundaries that are difficult to identify in the field and are of uncertain structural significance. The Coldfoot subterrane is distinguished from the Hammond subterrane by the former's pervasive penetrative deformation, generally higher textural grade, smaller proportion and size of enclosed carbonate units, and near absence of relict sedimentary and igneous textures. The Coldfoot subterrane consists of four primary lithologic assemblages that have undetermined stratigraphic significance: (1) a lower unit of Proterozoic and lower Paleozoic metasedimentary rocks, exposed in structural windows; (2) the quartz-mica schist unit, consisting of various units of pelitic and semipelitic schist; (3) the Bornite carbonate sequence of Hitzman and others (1986), composed mainly of carbonate rocks that locally overlie the quartz-mica schist unit; and (4) the Ambler sequence of Hitzman and others (1982), composed of mixed metaclastic rocks, metavolcanic rocks, and marble that are enclosed by the quartz-mica schist unit. Like the Hammond subterrane, the Coldfoot subterrane contains granitic rocks of Late Proterozoic and Devonian age.

Lower unit of Proterozoic and lower Paleozoic metasedimentary rocks. The structurally lowest rocks in the Coldfoot subterrane are exposed along the northern margin of the subterrane and in structural windows through the quartz-mica schist unit. These rocks, mostly pelitic and calcareous schists, vary in lithology from place to place; this variability suggests that the protoliths represent either diverse sedimentary facies or different tectonostratigraphic units. Some of the schists are compositionally and temporally similar to rocks of the Hammond subterrane but differ in having structural features characteristic of the Coldfoot subterrane. The unit includes mixed schists in the Kallarichuk Hills (Baird Mountains quadrangle), the Kogoluktuk schist of Hitzman and others (1982), and unnamed units of calcareous schist and marble in the Chandalar and Wiseman quadrangles (Brosgé and Reiser, 1964; Dillon and others, 1986).

The undivided mixed schists of the Kallarichuk Hills unit (Karl and others, 1989) consist of silvery green quartz-mica schist with intercalated black quartzite and brown calcareous schist. Sparse lenses of gray marble, blue amphibole-bearing metabasite, felsic metavolcanic rocks, and rare metaconglomerate are also present in the unit. Conodonts indicate a Silurian to Mississippian age for one carbonate lens, whereas two-hole crinoids recovered in another area indicate a Devonian age. However, a Proterozoic age for part of the unit is suggested by a Proterozoic granodiorite body that intrudes marble of the unit.

The Kogoluktuk schist of Hitzman and others (1982) consists of schists that underlie the quartz-mica schist unit in a structural window in the Cosmos Hills. In the window, the Kogoluktuk consists of about 2500 m of interlayered pelitic schist, micaceous quartzite, feldspathic schist, metabasite, chlorite schist, chloritic dolomite, and marble that are distinguished from overlying units by coarser texture, relict epidote-amphibolite metamorphic assemblages, and more complex structural fabric. Hitzman and others (1982) correlated the Kogoluktuk with thinly layered micaceous quartzite and minor dolostone, marble, metaconglomerate, calc-schist, semipelitic schist, and metabasite along the northern margin of the subterrane in the Ambler River quadrangle. This northern assemblage can be traced westward into a structural window at the border of the Ambler River and Baird Mountains quadrangles. Marble in this structural window locally contains stromatolites, which suggest that it is correlative with the stromatolite-bearing Proterozoic rocks of the Hammond subterrane to the north (Karl and others, 1989). Hitzman and others (1982) inferred a Devonian or older age for the Kogoluktuk because it is intruded by granitic gneiss of probable Devonian age in the Cosmos Hills.

In a structural window west of the Dalton Highway in the Wiseman quadrangle, interlayered calcareous schist, pelitic schist, and marble underlie the quartz-mica schist unit along a tectonic contact. These rocks, at least 1000 m thick, contain prominent, 5–10-m-thick, laterally discontinuous, highly strained marble units that are structurally thickened by isoclinal folding or imbrication. Conodonts recovered from a dark argillaceous carbonate lens indicate an early Early Devonian (Lochkovian) age for at least some of these rocks (A. G. Harris, 1989, written commun.). Near the town of Wiseman, these rocks can be traced northward into a narrow belt of highly strained calcareous and pelitic schist that extends along strike for more than 150 km across much of the Wiseman and Chandalar quadrangles (Dillon and others, 1986).

Quartz-mica schist unit. The quartz-mica schist unit, exposed throughout the Coldfoot subterrane, consists generally of uniform pelitic and semipelitic quartz-mica schist and minor metabasite and metacarbonate rocks. This unit has an estimated structural thickness of 3–12 km (Hitzman and others, 1982; Dillon and others, 1986; Gottschalk, 1987) and is characterized by quartz stringers, boudins, and segregations that commonly define isoclinal folds. These rocks (Anirak and Maunelek schist units of Hitzman and others, 1982; knotty-mica schist unit of Dillon and others, 1986; Koyukuk schist unit of Gottschalk, 1987) typically consist of green-, gray-, or brown-weathering quartz + white mica + chlorite + albite ± chloritoid schist with graphitic and quartz-rich compositional layers generally less than a few meters thick. East of the Dalton Highway, the unit consists of about 95% interlayered pelitic schist, quartz-rich schist, and paragneiss, and about 5% metabasite, metagabbro, and albite-schist lenses (Gottschalk, 1987). The metabasite lenses contain greenschist facies assemblages (chlorite + albite + epidote + actinolite + sphene) but retain relict blueschist facies assemblages and pseudomorphs. Notably, one metabasite body near the Dalton Highway retains an eclogite (garnet + sodic clinopyroxene + rutile) assemblage (Gottschalk, 1990). The quartz-mica schist unit represents a thick, deformed, and metamorphosed succession of organic-rich shale and siliciclastic sedimentary rocks (Hitzman and others, 1982, 1986; Gottschalk, 1990); however, no relict primary structures have been preserved and possible earlier structural fabrics have been mostly transposed into a position parallel with the regional south-dipping foliation. The age of the protolith of the quartz-mica schist unit is unknown, but Hitzman and others (1986) considered it Devonian and older. Dillon and others (1986) and Dillon (1989) correlated the apparently less deformed parts of this unit with the Devonian Beaucoup Formation and Hunt Fork Shale of the Endicott Mountains subterrane, largely because of structural position and assumed comagmatic character of the interstratified metavolcanic rocks.

Bornite carbonate sequence. The Bornite carbonate sequence of Hitzman and others (1986) consists of about 1000 m of carbonate rocks in the Cosmos Hills that may conformably overlie rocks of the quartz-mica schist unit. The carbonate rocks grade upward from phyllitic marble to carbonate breccia, marble, massive fossiliferous dolostone, and massive encrinitic dolostone, and represent a carbonate mudbank complex or bioherm (Hitzman and others, 1986). Nearby lateral equivalents include marble, graphitic marble, carbonaceous phyllite, and fossiliferous, laminated dolostone, and represent back-reef lagoonal limestone and intratidal to supratidal deposits. A variety of Middle to Late Devonian or earliest Mississippian megafossils are preserved within the Bornite carbonate sequence (Patton and others, 1968), including well-preserved brachiopods of Middle Devonian age (R. B. Blodgett, 1990, oral commun.). Conodonts from one locality in the unit indicate a Silurian age (A. G. Harris and J. A. Dumoulin, 1987, unpublished data); thus, the Bornite carbonate sequence spans a large part of middle Paleozoic time. A 1-km-wide body of hydrothermal dolostone in the biohermal facies hosts the Ruby Creek copper deposit, which contains more than 100 million tons graded at 1.2% copper (Hitzman and others, 1986).

Ambler sequence. A well-known lithologic assemblage of the Coldfoot subterrane is the Ambler sequence of Hitzman and others (1982), named for the volcanogenic massive sulfide mineral district in the southern Ambler River and Survey Pass quadrangles. In this area, the sequence consists of a complexly interfingering and deformed, 700–1850-m-thick succession of foliated to massive metarhyolite, felsic schist, metabasite, marble, chloritic schist, calcschist, and graphitic schist that is inferred to wedge out to the south within the quartz-mica schist unit (Hitzman and others, 1982). Dillon and others (1986) extended the sequence eastward into the Wiseman quadrangle, where it forms thinner and laterally less extensive units and lenses.

Hitzman and others (1986) estimated the composition of the sequence in the Ambler River quadrangle to be 60% metavolcanic and volcaniclastic rocks, 25% marble, and 15% metasedimentary rocks. The metavolcanic rocks of the Ambler sequence, which include both felsic and basaltic lithologies, have been metamorphosed to blueschist and greenschist facies assemblages. The felsic rocks consist in part of porphyritic metarhyolite with relict potassium feldspar and resorbed bipyramidal quartz phenocrysts (the so-called button schist). The metabasalt is tholeiitic in composition and is massive and concordant, except locally where it contains pillow and breccia structures (Hitzman and others, 1986). Hitzman and others (1982) interpreted the metavolcanic rocks of the Ambler sequence as a compositionally bimodal succession of submarine mafic flows and felsic domes, ignimbrites, ash flows, pyroturbidites, and reworked clastic aprons of volcanic debris. Major copper, zinc, lead, and silver sulfide resources in the Ambler mineral district are distributed among many deposits associated with the felsic metavolvanic rocks of the Ambler sequence (Hitzman and others, 1986; Nokleberg and others, this volume, Chapter 29 and Plate 11).

Corals recovered from dolomitic lenses within the Ambler sequence are Late Devonian to Mississippian (Hitzman and others, 1986; Smith and others, 1978), but poor preservation casts doubt on their identification (R. B. Blodgett, 1990, oral commun.). Discordant zircon U-Pb and Pb-Pb ages of 373 to 327 Ma have been derived from felsic metavolcanic rocks in the sequence (Dillon and others, 1980); the ages indicate extrusion at 396 ± 20 Ma (Early Devonian) (Dillon and others, 1987b). On the basis of these ages, along with the reconstructed stratigraphy of the sequence and the bimodal composition of metavolcanic rocks, Hitzman and others (1982, 1986) proposed that the Ambler sequence was deposited in a continental-rift setting during Devonian time.

Metagranitic rocks. The oldest known intrusive rocks in the Coldfoot subterrane are small bodies of metamorphosed plutonic rocks of granitic to dioritic composition exposed in the Kallarichuk Hills (Baird Mountains quadrangle) (Fig. 1). Karl and others (1989) reported that the metagranitic rocks intrude marble of the Proterozoic and lower Paleozoic metasedimentary unit of the Coldfoot subterrane and yielded a Proterozoic U-Pb age of 705 ± 35 Ma.

A younger generation of metaplutonic rocks is represented by granitic gneiss in the Cosmos Hills (Ambler River quadrangle), by the Baby Creek batholith in the Chandalar quadrangle, and by several other small metagranitic bodies near Wild Lake and the village of Wiseman in the Wiseman quadrangle. The Baby Creek batholith is a metamorphosed S-type, peraluminous biotite-muscovite granite with a tentative initial Sr ratio of 0.707 (Newberry and others, 1986; Dillon, 1989). This batholith, which intrudes the quartz-mica schist unit, has yielded discordant U-Pb ages indicative of crystallization in the Early Devonian. The Baby Creek batholith is similar in most respects to most other porphyritic orthogneiss bodies of Devonian age in the Hammond and North Slope subterranes (Newberry and others, 1986) and is cogenetic with the felsic metavolcanic rocks of the Ambler sequence (Dillon and others, 1980; Dillon, 1989).

Slate Creek subterrane

Various parts of the 5–10-km-wide, topographically recessive phyllite belt along most of the southern Brooks Range have been included in the Venetie and Coldfoot terranes of Jones and others (1987), the Rosie Creek allochthon of Oldow and others (1987c, 1987d), and the Slate Creek thrust panel of the Angayucham terrane by Dillon (1989) and Patton and others (this volume, Chapter 7). Because of their quartzose character, chert-rich composition, and middle Paleozoic age, we combine these metaclastic rocks and discuss them here as the Slate Creek subterrane of the Arctic Alaska terrane (Fig. 3). Herein we limit the Slate Creek subterrane to rocks in the southern Brooks Range because these rocks are bounded to the south by prominent structures (South Fork–Malamute fault, Porcupine lineament). Rocks of similar lithologic character south of this area rest on rocks of the Ruby terrane (Patton and others, this volume, Chapter 7).

The Slate Creek subterrane consists of metamorphosed, thin-bedded, dark-colored shale, siltstone, and fine- to medium-grained sandstone, and local units of phyllonite and chloritic schist (Fig. 21). In a few areas, metagabbro and diabase dikes intrude the rocks (Nelson and Grybeck, 1980; Dillon and others, 1986; Dillon, 1989; Karl and others, 1989), and near Coldfoot, mafic intrusive rocks form lenses in phyllonite and argillite-matrix melange (Moore and others, 1991). Radiolarian chert of late Paleozoic age is associated with the metaclastic rocks along the southern margin of the subterrane, but a conformable relation between the chert and the metaclastic rocks has not been confirmed. No mappable internal stratigraphy has been recognized, and zones of tectonic melange and broken formation are prominent in the subterrane, especially along its southern margin. Zones of phyllonite and mylonite, pervasive along the northern margin of the subterrane, may represent an important zone of dislocation (Dillon, 1989). Elsewhere, Hitzman and others (1982) reported that the phyllitic rocks (their Beaver Creek phyllite unit) conformably grade downward into the quartz-mica schist unit of the Coldfoot subterrane (Dillon, 1989). Hitzman and others (1986) and Howell and others (1986, Fig. 2) estimated structural thicknesses of about 3 km for the Slate Creek subterrane in the western and eastern Brooks Range; Gottschalk (1987) estimated a thickness of about 1000 m for rocks of the subterrane near the Dalton Highway.

Where sedimentary structures are preserved, sandstone to shale ratios are typically less than 1:10 but in the eastern Brooks Range may be as high as 1:3. Sedimentary structures characteristically formed by turbidity currents (for example, graded beds, flute casts) have been reported by most workers (Hitzman and others, 1986; Murphy and Patton, 1988). Gottschalk (1987) also described shallow-marine sedimentary structures, including hummocky cross-stratification, in a local fault-bounded unit of the subterrane east of the Dalton Highway. The sandstone of the Slate Creek subterrane is largely composed of chert and quartz grains, but minor granitic, quartzose metamorphic, unfoliated sedimentary, and volcanic rock fragments are locally present. The quartzose composition of these strata suggests a continental provenance (Murphy and Patton, 1988).

Palynomorphs in the fault-bounded unit east of the Dalton Highway indicate an Early Devonian (Siegenian to Emsian) age (Gottschalk, 1987). In the eastern Brooks Range, palynomorphs indicate Early(?) Devonian and Middle or Late Devonian ages, and plant fossils indicate a Late(?) Devonian age (W. P. Brosgé, 1988, oral commun.). Murphy and Patton (1988) suggested that sedimentary rocks of the Slate Creek subterrane are the deep-marine equivalents of Upper Devonian rocks of the Endicott Group in the Endicott Mountains subterrane; Oldow and others (1987d), however, correlated the Slate Creek rocks with Devonian rocks of the Hammond subterrane.

DESCRIPTION OF THE ANGAYUCHAM TERRANE

The structurally highest units of the Brooks Range are allochthonous mafic and ultramafic rocks exposed in the greenstone belt in the southern foothills of the Brooks Range and to the north in a series of synclinal remnants of large thrust sheets in the crestal and disturbed belts of the western and eastern Brooks Range (Mull, 1982, p. 30; Figs. 2, 3, and 20). In both areas, the mafic rocks of the Angayucham terrane have been divided into two structural packages (Fig. 21). The structurally lower package consists principally of fault imbricates of metamorphosed mafic volcanic rocks. In the greenstone belt, this package has been called the Narvak thrust panel by Patton and Box (1989) (herein referred to as the Narvak panel), whereas in the crestal and disturbed belts it is called the Copter Peak allochthon by Mayfield and others (1988). The structurally higher package, which consists mainly of gabbroic and ultramafic rocks, is called the Kanuti thrust panel in the greenstone belt (Patton and Box, 1989) (herein referred to as the Kanuti panel) and the Misheguk Mountain allochthon in the crestal and disturbed belts (Mayfield and others, 1988). On the basis of similar lithology, age, and structural position, Roeder and Mull (1978) correlated the two packages of the greenstone belt with those of the crestal and disturbed belts and suggested that the former is the "root zone" for extensive thrust sheets now represented by the klippen in the crestal and disturbed belts. We organized rocks of the Angayucham terrane below by geographic location so that the evidence for this correlation, which is of fundamental importance to tectonic reconstructions, is apparent.

Greenstone belt

The Angayucham terrane in the greenstone belt is a narrow zone of slightly metamorphosed mafic igneous rocks, chert, and serpentinite that extends for more than 500 km along the southern margin of the Brooks Range. These rocks dip moderately to steeply southward and rest structurally on deformed and metamorphosed rocks of the Slate Creek subterrane of the Arctic Alaska terrane. Patton and Box (1989) have divided the igneous rocks into two lithotectonic units, herein referred to as the Narvak panel and Kanuti panel. These units are distinguished by the abundance of mafic volcanic rocks in the Narvak panel, in contrast with the large proportion of ultramafic rocks in the Kanuti panel.

Narvak panel. The Narvak panel consists of an imbricate stack of fault slabs composed of pillow basalt and subordinate diabase, basaltic tuff, argillite, limestone, and radiolarian chert that has a structural thickness of more than 10 km in the Angayucham Mountains and 6 km near the Dalton Highway (Dillon, 1989; Pallister and others, 1989) (Fig. 22). Bodies of amphibolite are locally present near the structural top of the unit in the Cosmos Hills and Angayucham Mountains (Hitzman and others, 1982; Pallister and others, 1989). The Narvak panel structurally overlies the Slate Creek subterrane along a zone of south-dipping faults, tectonic melange, cataclasite, and mylonite (Hitzman and others, 1986; Dillon, 1989). Fault slivers of diabase; Mississippian to Triassic chert and argillite; Devonian, Mississippian, Pennsylvanian, and Permian limestone; and calcareous arkose are locally present along this fault zone (Gottschalk, 1987; Pallister and Carlson, 1988; I. L. Tailleur, 1988, oral commun.; Dillon, 1989).

The volcanic rocks of the Narvak thrust panel consist largely of pillow basalt and pillow breccia. The basalts are very fine to fine-grained, nonvesicular to amygdaloidal, and commonly aphypric. Where porphyritic, they contain sparse microphenocrysts of plagioclase and/or clinopyroxene and, locally, titaniferous augite (Pallister and others, 1989). Microgabbro and rare cumulate layered gabbro have been reported locally from near the base of the Narvak (Barker and others, 1988). The rocks are metamorphosed to prehnite-pumpellyite or greenschist facies assemblages but have mainly static metamorphic textures (Gottschalk, 1987; Gottschalk and Oldow, 1988; Dillon, 1989; Pallister and others, 1989).

Associated sedimentary rocks consist largely of interpillow chert, thinly bedded radiolarian and tuffaceous chert, siliceous tuff breccia, argillite, and minor lenses of limestone and marble. The chert is gray, black, and red, and as much as 60 m thick (Gottschalk, 1987; Jones and others, 1988; Dillon, 1989; Pallister and others, 1989). Depositional contacts between the chert and basalt are present, but the bedded-chert sequences commonly

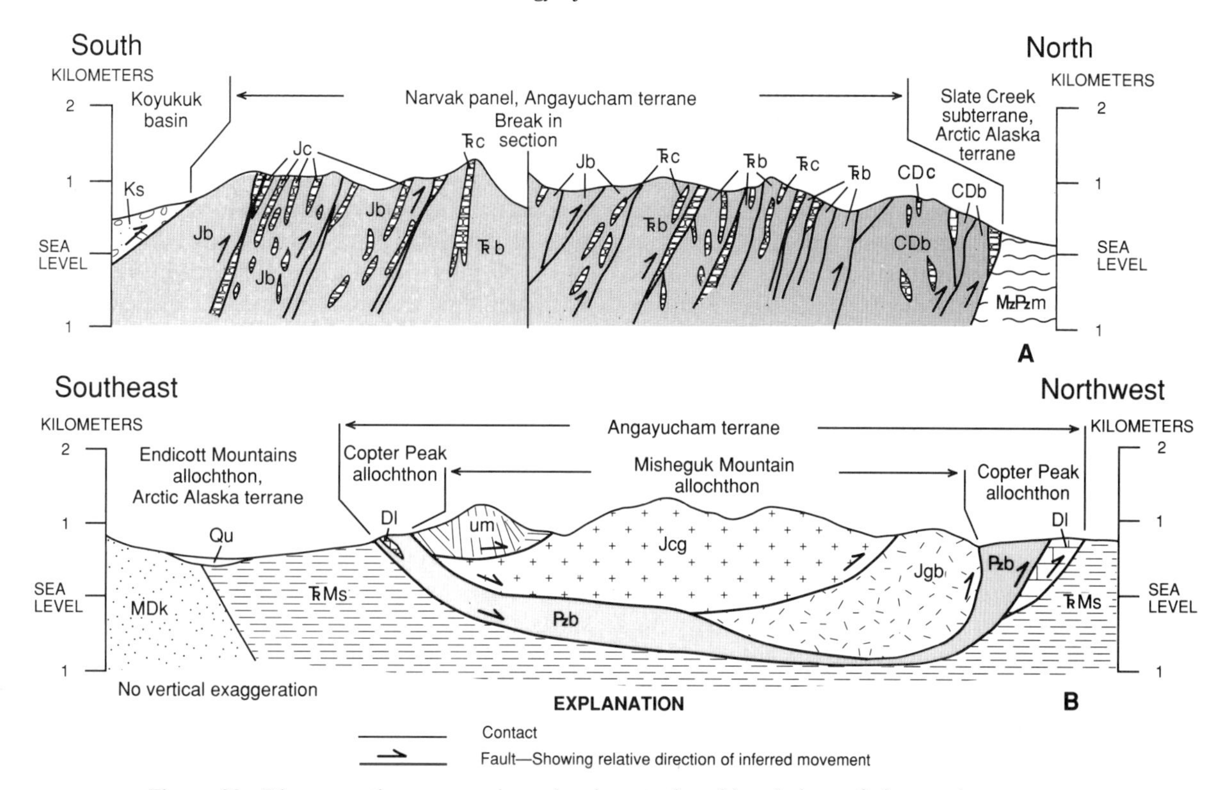

Figure 22. Diagrammatic cross sections showing stratigraphic relations of Angayucham terrane. A, Greenstone belt in Angayucham Mountains, Hughes quadrangle (modified from Pallister and others, 1989). Note imbricate structure and progressively younger age of rocks in thrust imbricates from structural base to top of Narvak panel. B, Crestal belt at Siniktanneyak Mountain, Howard Pass quadrangle (modified from Nelson and Nelson, 1982). Geologic units: Greenstone belt—Ks, Cretaceous sedimentary rocks; Jb, Jurassic pillow basalt, diabase, and microgabbro; Jc, Jurassic chert; ₮b, Triassic pillow basalt, diabase, and microgabbro; ₮c, Triassic chert; CDb, Devonian and Carboniferous pillow basalt, diabase, and microgabbro; CDc, Devonian and Carboniferous chert; MzPzm, Paleozoic and Mesozoic melange. Crestal belt—Qu, Quaternary deposits; Jcg, Jurassic(?) cumulate gabbro; Jgb, Jurassic(?) isotropic gabbro with plagiogranite dikes and stocks; ₮Ms, Mississippian to Triassic fine-grained sedimentary rocks of undivided Kayak Shale and Lisburne and Etivluk groups; MDk, Devonian and Mississippian(?) terrigenous clastic rocks of Kanayut Conglomerate; Dl, Devonian fossiliferous carbonate rocks; Pzb, Paleozoic diabase and pillowed volcanic rocks; um, dunite, harzburgite, and pyroxenite.

mark the position of faults bounding the imbricates that compose the Narvak panel.

Map relations and radiolarians from interpillow chert show that in the Angayucham Mountains, the Narvak panel consists of four to eight map-scale fault imbricates of restricted age (Pallister and Carlson, 1988). Radiolarian assemblages define an age progression, from structural base to top of the imbricate sequence, of (1) Late Devonian (Famennian), (2) Mississippian, (3) Pennsylvanian or Early Permian, (4) Triassic, and (5) Early Jurassic(?) (Murchey and Harris, 1985; B. L. Murchey, 1987, written commun.). In the vicinity of the Dalton Highway, various units of chert, including those of Late Devonian, Late Devonian to Early Mississippian(?), Mississippian, Late Mississippian(?) to Early Pennsylvanian, Early Permian, and Triassic to Early Jurassic indicate a similar spread of ages (Jones and others, 1988), although a consistent pattern is not apparent. Shallow-water fossils, abundant sponge spicules, and the finely laminated character of some of the chert units suggest deposition in intermediate (500 to 1500 m) water depths (Murchey and Harris, 1985), but chert containing abundant radiolarians is common and indicates deposition under deeper marine conditions (Karl, 1989).

Geochemical data from the volcanic rocks show that many are hypersthene-normative tholeiitic basalts (Barker and others, 1988; Pallister and others, 1989) and are transitional between normal and enriched mid-ocean ridge basalts (MORB). Rare earth element (REE) patterns vary from relatively flat to slightly depleted or moderately enriched in light REE (LREE). Considering the fine-grained, siliceous character of the associated sedimentary rocks and their wide range of ages, Barker and others (1988) and Pallister and others (1989) interpreted the geochemical data as evidence for extrusion of the igneous rocks on oceanic plateaus such as seamounts and ocean islands.

Kanuti panel. Ultramafic rocks in the southern Brooks Range foothills are exposed in the Jade Moutains and in the nearby Cosmos Hills (both in the Ambler River quadrangle) and as scattered fault slivers along the greenstone belt. In the Jade Mountains, serpentinite lies structurally on pillow basalt of the Narvak panel along a south-dipping fault contact and contains relict minerals and textures of a harzburgite protolith (Loney and Himmelberg, 1985). Prominent linear gravity and magnetic highs trend along the northern edge of the Koyukuk basin and pass through these exposures (Barnes, 1970; Cady, 1989), which suggests that the ultramafic rocks compose a regionally extensive sheet mostly buried beneath Cretaceous sedimentary rocks of the Koyukuk basin (Loney and Himmelberg, 1985; Dillon, 1989). Cady (1989), however, concluded that ultramafic rocks were rare and attributed the potential-field anomalies to mafic rocks of the Narvak panel.

Crestal and disturbed belts

The Angayucham terrane in the crestal and disturbed belts consists of klippen of slightly metamorphosed mafic volcanic rocks, diabase, gabbro, and ultramafic rocks that are exposed discontinuously along the length of the Brooks Range (Figs. 2 and 3). The klippen rest on unmetamorphosed sedimentary rocks of the De Long Mountains and Endicott Mountains subterranes of the Arctic Alaska terrane and consist of two lithotectonic units, the Misheguk Mountain allochthon and the subjacent Copter Peak allochthon (Mayfield and others, 1988). These units are distinguished by the abundance of mafic volcanic rocks and diabase in the Copter Peak allochthon in contrast with the largely ultramafic rocks and gabbro of the Misheguk Mountain allochthon.

Copter Peak allochthon. The structurally lower Copter Peak allochthon consists of imbricated units of basalt and diabase with subordinate basaltic tuff and breccia, microgabbro, siliceous tuff, radiolarian chert, and gray argillite. Although generally massive and highly fractured, pillow structures and lava flows are commonly reported, and columnar basalt is present locally (Moore, 1987b). The basalt is mostly very fine to fine grained, sparsely amygdaloidal and typically aphyric, although sparse plagioclase and/or clinopyroxene microphenocrysts are present in some rocks. These basalts have been partly to completely altered to assemblages of albite, green amphibole, chlorite, sphene, and calcite. Diabase dikes and sills intrude the lava flows and intercalated sedimentary rocks; they may compose a large part of some fault imbricates within the Copter Peak allochthon (Bird and others, 1985). In the klippen of the western Brooks Range, the Copter Peak allochthon is less than 3 km thick (Nelson and Nelson, 1982; Curtis and others, 1984; Ellersieck and others, 1984; Karl and Dickey, 1989) (Fig. 22). Geochemical data from the Kikiktak Mountain (Killik River quadrangle), Siniktanneyak Mountain (Howard Pass quadrangle), Copter Peak (Misheguk Mountain quadrangle), Asik Mountain (Noatak quadrangle), and Avan Hills (De Long Mountains and Misheguk Mountain quadrangles) klippen indicate that the mafic igneous rocks are tholeiites. REE patterns for most of the mafic rocks show moderate enrichment in LREEs, whereas others are flat. These data suggest that the mafic igneous rocks were extruded at a mid-ocean ridge or seamount (Moore, 1987b; Wirth and others, 1987). Data from the Maiyumerak Mountains klippen (Noatak and Baird Mountains quadrangles), however, indicate an oceanic-arc affinity for some of the rocks of the Copter Peak allochthon (Wirth and others, 1987; Karl and Dickey, 1989; Karl, 1991).

Interpillow chert is a minor, but ubiquitous, component throughout the Copter Peak allochthon; bedded chert is present locally. The chert is typically red or light colored, contains abundant radiolarians and little argillite, and represents slow sedimentation in an oxygenated, deep-water environment (Murchey and others, 1988). Limestone and marble, in places containing Devonian fossils, are present in the lower part of the Copter Peak allochthon. The carbonates are commonly interpreted as tectonic blocks or slivers that were incorporated along the basal thrust of the allochthon during its emplacement (Roeder and Mull, 1978; Mayfield and others, 1988). Nelson and Nelson (1982), however, reported that some fossiliferous carbonate units are interbedded with pillow lava and breccia of the Copter Peak allochthon; this evidence suggests extrusion of some of the basalts at relatively shallow depths.

Age and structural relations of the rocks of the Copter Peak allochthon are delimited mostly by fossils collected from intercalated chert and limestone. Interpillow chert in the Copter Peak allochthon in the western Brooks Range has yielded radiolarians that are largely Triassic (Ellersieck and others, 1984), but Mississippian and Pennsylvanian radiolarians have also been recovered from the Christian and Kikiktat mountain massifs (D. L. Jones, 1987, oral commun; B. L. Murchey, 1987, written commun.; C.G. Mull, 1987, unpublished data). Limestone interstratified with volcanic rocks near the base of the Siniktanneyak massif has yielded megafossils and conodonts of Late Devonian age (Nelson and Nelson, 1982).

Misheguk Mountain allochthon. The Misheguk Mountain allochthon consists of ultramafic tectonite and cumulate rocks, cumulate and isotropic gabbro, and diabase that have a reconstructed thickness of at least 6 km at Siniktanneyak Mountain (Fig. 22). The ultramafic rocks consists largely of dunite with chromitite layers and subordinate harzburgite, wehrlite, and pyroxenite (Zimmerman and Soustek, 1979; Bird and others, 1985). These are interlayered with, and pass upward into, layered cumulate gabbroic rocks, including troctolite, melagabbro, leucogabbro, and anorthosite; olivine, clinopyroxene, and plagioclase are the cumulate phases in these rocks (Zimmerman and Soustek, 1979; Nelson and Nelson, 1982; Bird and others, 1985). Noncumulate gabbro, locally intruded by small plagiogranite dikes and stocks, composes a large part of the Misheguk Mountain klippen and forms irregular intrusions in most other klippen in the western Brooks Range. This gabbro is ophitic, consisting of plagioclase, green hornblende, and uratilized clinopyroxene, and commonly displays miarolytic cavities. Diabase dikes locally intrude both the ultramafic and gabbroic rocks. In some of the klippen, dikes and stocks of potassium feldspar–bearing granitic rocks intrude the ultramafic rocks and gabbro and may represent a later plutonic episode (Zimmerman and others, 1981; Nelson and Nelson, 1982; Boak and others, 1987). Crystallization sequences and mineral chemistries of rocks in the Misheguk Mountain allochthon at Misheguk Mountain indicate crystallization in an arc, rather than a mid-ocean ridge setting (Harris, 1988).

Hornblende and biotite K-Ar ages from gabbro in the Siniktanneyak, Misheguk Mountain, and Christian klippen range from 172 to 147 Ma (Patton and others, 1977; Boak and others, 1987). Wirth and Bird (1992) reported ^{40}Ar-^{39}Ar incremental-heating ages of 187–184 Ma on hornblende from gabbro of the Asik Mountain klippe. These dates indicate that crystallization of the Misheguk Mountain allochthon occurred during the Middle Jurassic.

Metamorphic rocks. In klippen of the western Brooks Range, the contact between the Misheguk Mountain allochthon and the underlying Copter Peak allochthon is marked by pods and zones as much as a few tens of meters thick of low- to medium-grade, and locally high-grade, amphibolite. These rocks are schistose to gneissic and display cataclastic texture (Boak and others, 1987). Boak and others (1987) determined that the protolith for the amphibolite is volcanic and siliceous sedimentary rocks of the underlying Copter Peak allochthon. The metamorphic rocks have yielded K-Ar ages of 154 and 153 Ma and ^{40}Ar-^{39}Ar plateau ages of 171 to 163 Ma (Boak and others, 1987; Wirth and Bird, 1992) and are thought to have been metamorphosed at relatively low pressures (Zimmerman and Frank, 1982) and at metamorphic temperatures up to 560 °C (Boak and others, 1987).

Tectonic affinity of the Angayucham terrane

Rocks of the Angayucham terrane were originally thought to compose a dismembered ophiolite (Talleur, 1973b; Patton and others, 1977), but Roeder and Mull (1978) and Mayfield and others (1988) have shown that the structurally higher gabbroic and ultramafic assemblage is distinct from the underlying volcanic and diabase assemblage. Evidence indicating that these units are not cogenetic include the following. (1) The volcanic and diabase assemblage ranges from Late Devonian to Early Jurassic, whereas the gabbroic and ultramafic assemblage yields isotopic ages suggesting crystallization during the Middle Jurassic. (2) Trace element geochemistry of the volcanic and diabase assemblage indicates that it is composed largely of oceanic-plateau and seamount basalts (Moore, 1987b; Wirth and others, 1987; Barker and others, 1988; Pallister and others, 1989), whereas petrochemical data from the Misheguk Mountain allochthon suggest that it has an island-arc affinity (Harris, 1988). Most workers agree that the gabbroic and ultramafic rocks of the Kanuti panel and Misheguk Mountain allochthon may represent an incomplete ophiolite, but the age span of volcanic rocks and diabase in the Narvak panel and Copter Peak allochthon is much longer than that of known ophiolites; therefore, the volcanic rocks and diabase more likely represent an accreted assemblage of basaltic seamounts (Barker and others, 1988; Pallister and others, 1989). The dynamothermally metamorphosed rocks along the contact between the upper and lower assemblages were interpreted by Zimmerman and Frank (1982) and Boak and others (1987) as metamorphic aureoles that developed under the ultramafic rocks of the higher gabbroic and ultramafic assemblage during thrust emplacement. Metamorphism of the aureole rocks at relatively high temperatures (T) and low pressures (P) suggests that the ophiolite was obducted as a young, hot body in Middle Jurassic time, within 20 m.y. of crystallization (Zimmerman and Frank, 1982; Harris, 1988; Wirth and Bird, 1992).

The Copter Peak allochthon rests structurally on Valanginian (Lower Cretaceous) and older sedimentary rocks of the De Long Mountains and Endicott Mountains subterranes (Curtis and others, 1984; Ellersieck and others, 1984; C. G. Mull, 1987, unpublished data). Sedimentary debris derived from the Copter Peak and Misheguk Mountain allochthons is present in Jurassic and Neocomian (Lower Cretaceous) foredeep deposits of the Okpikruak Formation (Mull, 1985; Crane, 1987). These data indicate that the upper ophiolitic assemblage was emplaced onto the lower volcanic and diabase assemblage during the Jurassic, but emplacement of both assemblages onto the Arctic Alaska terrane was not completed until after Valanginian time (Boak and others, 1987; Mayfield and others, 1988). Final emplacement of some or all of the Angayucham terrane may have involved normal faulting along south-dipping detachment faults in the southern Brooks Range and along north-dipping faults in the northern Brooks Range (Miller, 1987; Gottschalk and Oldow, 1988; Harris, 1988).

STRUCTURAL GEOLOGY OF NORTHERN ALASKA

The northern and southern regions of northern Alaska have distinct structural characteristics. The northern region, encompassing most of the North Slope and continental shelf, is dominated by structures related to the Jurassic and Early Cretaceous rifting that formed the northern continental margin of Alaska. This rifting separated northern Alaska from a continent to the north, producing a structural high, the Barrow arch, that has played a continuing role in the structural and depositional history of the region. The northern flank of the Barrow arch has been dominated by passive-margin subsidence and sedimentation since formation of the continental margin in the Early Cretaceous (Grantz and May, 1983). The southern limb of the Barrow arch served as the continental foreland and the northern flank of the foredeep for the Brooks Range orogen to the south.

The southern region encompasses the Brooks Range, a major orogenic belt of more than 1000 km long and as much as 300 km wide. Like most orogens, the Brooks Range is an elongate belt that displays asymmetry both in the distribution and character of its major structural elements and in the dominant direction of tectonic transport. Throughout most of its extent, the Brooks Range displays east-striking, north-transported structures.

The deepest structural levels are exposed mainly to the south, in the internal part of the orogen, and are characterized by older rocks overprinted by metamorphism and ductile deformation. A fold and thrust belt has developed to the north in the external part of the orogen in mostly younger and unmetamorphosed, dominantly sedimentary rocks. The southern part of this fold and thrust belt consists of shortened preorogenic rocks, whereas its northern part includes deformed synorogenic foredeep strata. In the youngest part of the orogen to the east, the deformational front of the Brooks Range has migrated northward to the modern continental margin.

Although it displays many characteristics common to mountain belts throughout the world, the Brooks Range is unusual in a number of respects. Extensive preservation of the highest structural levels of the orogen (that is, the Angayucham terrane) and early synorogenic deep-marine sedimentation (Okpikruak Formation) indicate that structural relief was relatively low during the period of greatest contraction in the orogen. Unlike most other parts of the circum-Pacific region, Tethyan-type ophiolites (Coleman, 1984) are present in the Brooks Range (Kanuti panel and Misheguk Mountain allochthon of the Angayucham terrane), forming its structurally highest preserved elements. Furthermore, relatively high *P*/low *T* metamorphic rocks are exposed over large areas in the internal parts of the orogen, and relatively lower *P*/higher *T* metamorphism that overprinted these rocks did not reach particularly high temperatures, nor is there evidence for synorogenic to postorogenic magmatism. In most continental orogens, deformation proceeds over time toward the interior of the continent, whereas in the Brooks Range, deformation has migrated toward what is now the northern continental margin of Alaska. Major low-angle normal faulting along the south flank and elsewhere in the Brooks Range suggests that tectonic extension has played a major role in the unroofing and uplift of the internal part of the orogen.

For purposes of description, northern Alaska is here divided into six major structural provinces: the southern Brooks Range, the northern Brooks Range, the foothills, the Lisburne Peninsula, the northeastern Brooks Range, and the North Slope (Fig. 23). These provinces are defined by their structural characteristics (Table 1), but they coincide approximately with the physiographic (Plate 2) and geologic provinces of northern Alaska (Fig. 2) and with many, but not all, of its tectonostratigraphic subdivisions (Fig. 3).

Southern Brooks Range structural province

The southern Brooks Range structural province is the core or infrastructure of the Brooks Range orogenic belt, where relatively deep structural levels are exposed. Rocks in the northern part of the province dip to the north beneath the northern Brooks Range structural province and also plunge beneath it to the east and west (Fig. 23). To the south, rocks of the province dip to the south beneath the Kouykuk basin. The northern part of the province consists of the Hammond and Coldfoot subterranes, which display polymetamorphism and complex, largely penetrative polydeformation. Rocks of the North Slope subterrane in the Mt. Doonerak fenster also display some of these characteristics (Oldow and others, 1987d; Dillon, 1989) and so are included in this province. The effects of Brookian orogenesis are much less intense in the southern part of the province (Slate Creek subterrane and Angayucham terrane) than in the northern part of the province, although penetrative structures are locally present.

The structure of the southern Brooks Range province is dominated by a series of large, generally south-dipping, thrust-bounded packages (Plate 13, sections B-B′, C-C′). The structures within and bounding these packages are mostly north directed and east striking and include major and minor folds and thrust faults and associated penetrative fabrics. Rocks of the Coldfoot subterrane are thought to have been emplaced as a coherent package, whereas rocks of the Hammond subterrane may consist of a greater than 10-km-thick, imbricate stack of 1–3-km-thick thrust sheets (Oldow and others, 1987; Till and others, 1988; Karl and others, 1989). South-vergent folds and faults overprint north-vergent structures in many parts of the Hammond and Coldfoot subterranes, and east-vergent, north-trending structures are present in the western part of the province. In addition, south-dipping normal faults and east-striking right-lateral strike-slip faults are found throughout the province, especially along the southern flank of the range; displacements on the latter faults may have been quite large.

Dynamothermally metamorphosed rocks characterize most of the province. These comprise mainly greenschist facies mineral assemblages but locally retain blueschist, epidote-amphibolite, amphibolite, and eclogite facies assemblages. Prehnite-pumpellyite facies assemblages with static textures are also present, most notably in the Angayucham terrane along the southern edge of the province. Textural grade decreases gradually to the north across the Hammond subterrane and abruptly to the south across the Slate Creek subterrane; the texturally highest grade rocks are found in the quartz-mica schist unit of the Coldfoot subterrane.

Pre-Brookian structures and metamorphism. Although most structures in this province are assumed to be related to Jurassic and younger Brookian deformation, Proterozoic and Devonian and/or Mississippian deformational and metamorphic events can be inferred. These events may have been tectonically significant, but their record has been largely obscured or transposed by Brookian penetrative fabrics and metamorphism.

In the Hammond subterrane, Proterozoic metamorphism and deformation are indicated by Late Proterozoic amphibolite and metapelite in the Baird Mountains quadrangle and by Late Proterozoic plutons scattered throughout the subterrane. The metapelite and amphibolite have yielded minimum K-Ar ages of 729 ± 22 Ma (muscovite) and 594 ± 18 Ma (hornblende) (Mayfield and others, 1982). Isoclinal folds and lineations in the amphibolite facies rocks predate intrusion of the plutons, which have been dated at 750 ± 6 Ma (Karl and others, 1989). This relation indicates that the amphibolite facies rocks represent a regional metamorphic event (Karl and others, 1989; Till, 1989). Likewise,

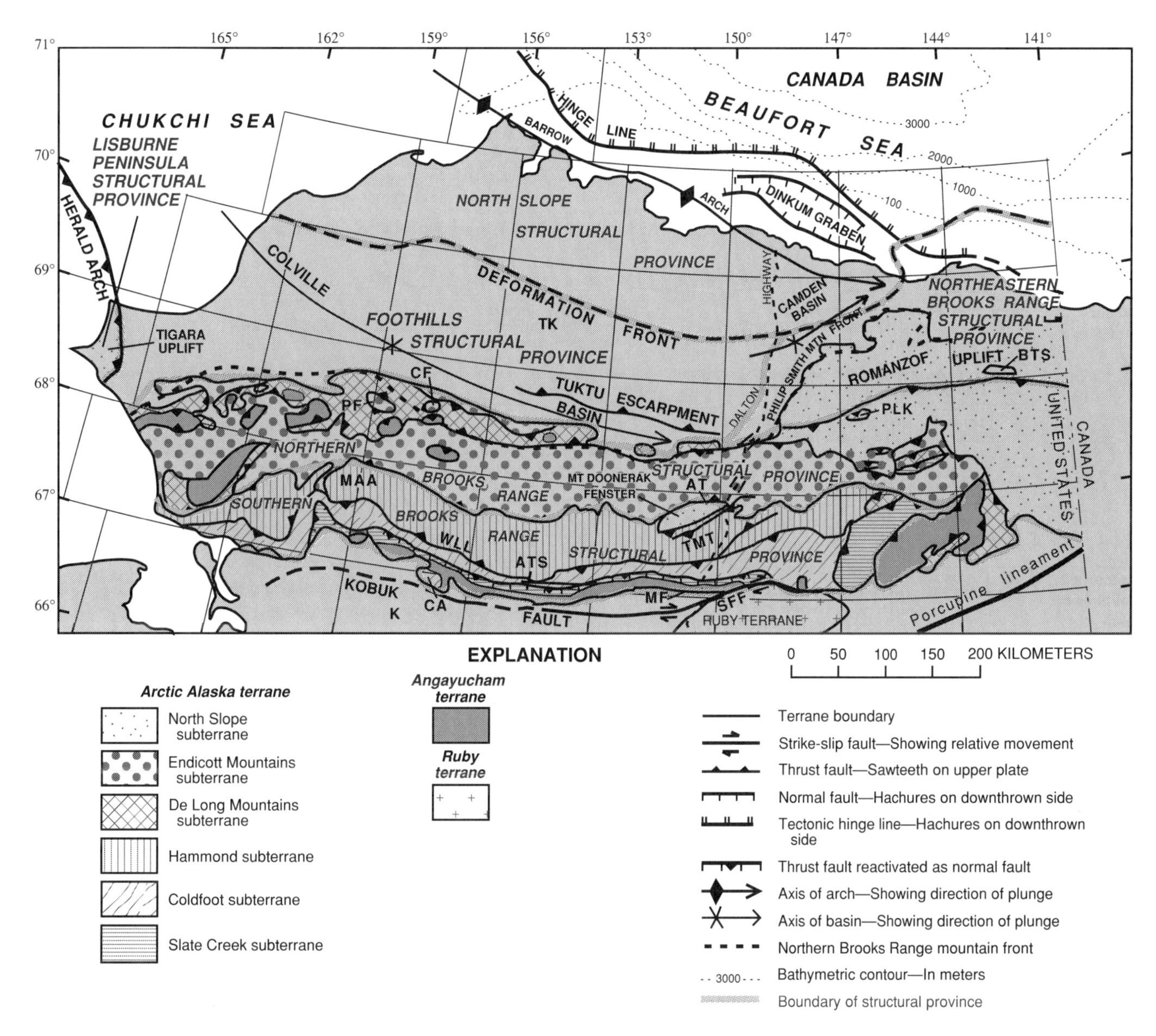

Figure 23. Structural provinces (red) and major tectonic elements of northern Alaska. AT, Amawk thrust; ATS, Angayucham "thrust" system; BTS, Bathtub syncline; CA, Cosmos arch; CF, Cutaway fenster; K, Cretaceous rocks of Koyukuk basin; MAA, Mount Angayukaqsraq antiform; MF, Malamute fault; PF, Picnic Creek fenster; PLK, Porcupine Lake klippe; SFF, South Fork fault; TK, Cretaceous and Tertiary rocks of Brookian sequence; TMT, Table Mountain thrust; WLL, Walker Lake lineament.

kyanite-bearing schist associated with Late Proterozoic gneiss of the Ernie Lake pluton (Survey Pass quadrangle) may represent remnants of a regional high-grade metamorphic belt (Nelson and Grybeck, 1980; Till, 1989).

In the Coldfoot subterrane, the earliest, coarse-grained fabric in the Proterozoic(?) Kogoluktuk Schist of Hitzman and others (1982) predates known Brookian structures and is associated with relict epidote-amphibolite facies assemblages (Hitzman and others, 1986). Also providing evidence for pre-Devonian orogenesis in this subterrane are fabrics in part of the quartz-mica schist unit that are not present in Devonian plutons (Dillon, 1989). Turner and others (1979) originally suggested that K-Ar data from the Coldfoot subterrane indicated a Precambrian episode of blueschist facies metamorphism, but Till and others (1988) considered the disparity of K-Ar ages to be the result of Late Proterozoic amphibolite facies assemblages overprinted by Mesozoic blueschist facies assemblages.

An Early or Middle Devonian intrusive event is indicated by the belt of orthogneiss bodies that intrudes rocks of both the Hammond and Coldfoot subterranes. Newberry and others (1988) suggested that narrow metamorphic aureoles around these plutons indicate emplacement at a high structural level. Late

TABLE 1. GEOLOGIC CHARACTERISTICS OF NORTHERN ALASKA STRUCTURAL PROVINCES
(Structural provinces shown in Fig. 23)

Geologic characteristics	Southern Brooks Range	Northern Brooks Range	Lisburne Peninsula
Stratigraphy	**Arctic Alaska terrane:** *North Slope subterrane:* In Mt. Doonerak fenster, lower Paleozoic argillite, volcanic rocks, and limestone unconformably overlain by Mississippian through Triassic, stratified clastic and carbonate rocks. *Hammond, Coldfoot, and Slate Creek subterranes:* Mainly Proterozoic to Upper Devonian and locally upper Paleozoic, metamorphosed clastic, carbonate, volcanic, and plutonic rocks. **Angayucham terrane:** Devonian to Jurassic basalt and subordinate chert, limestone, and argillite.	**Arctic Alaska terrane:** *North Slope subterrane:* Mainly Mississippian through Triassic, stratified clastic and carbonate rocks; pre-Mississippian rocks present locally. *Endicott Mountains, De Long Mountains, and Slate Creek subterranes:* Mainly Upper Devonian through Lower Cretaceous, stratified clastic and carbonate rocks. **Angayucham terrane:** Devonian to Triassic basalt and Jurassic(?) peridotite and gabbro.	**Arctic Alaska terrane:** *North Slope subterrane:* Pre-Mississippian clastic rocks unconformably overlain by Mississippian to Lower Cretaceous clastic and carbonate succession.
Province boundaries	**Northern, eastern, and western:** Thrust faults dipping beneath structurally higher northern Brooks Range province. **Southern:** Depositionally overlapped by Cretaceous rocks of Koyukuk basin. **Internal:** Thrust faults between North Slope and Hammond subterranes and between Hammond and Coldfoot subterranes. Thrust and/or normal faults between Coldfoot and Slate Creek subterranes and between Slate Creek subterrane and Angayucham terrane.	**Northern and western:** Depositionally overlapped by mid-Cretaceous terrigenous clastic foredeep deposits of the foothills province. Boundary commonly modified by late Brookian deformation. **Northeastern:** Zone of transition in structural style to northeastern Brooks Range province. **Southern:** North-dipping thrust fault structurally overlying southern Brooks Range province. **Internal:** Thrust faults underlying constituent allochthons.	**Northern and eastern:** Thrust over foredeep deposits of the foothills province. **Southern and western:** Depositionally over-lapped offshore by deposits of younger Hope basin. **Northwestern:** Offshore continuation (Herald arch) thrust over foredeep deposits of Colville basin.
Structural features of early Brookian deformation	**Arctic Alaska terrane:** *North Slope, Hammond, and Coldfoot subterranes:* Large, dominantly north vergent, thrust-bounded fault slices and associated major and minor, tight to isoclinal folds. Penetratively polydeformed with at least two generations of minor folds and associated foliation. Thrusting outlasted ductile deformation. *Slate Creek subterrane:* Penetratively polydeformed but monometamorphic. Contains major, south-dipping normal faults. **Angayucham terrane:** Brittle major and minor imbricate thrust faults.	**All units:** Dominantly north vergent folding and thrusting. Penetrative structures and metamorphic overprint present locally, mainly in association with major thrust faults. **Arctic Alaska terrane:** *North Slope subterrane:* Closely spaced north-vergent folds and imbricate thrust faults. *Endicott and De Long Mountains subterranes:* Five(?) regionally extensive, but commonly laterally discontinous, predictably stacked thrust packages or "allochthons." Allochthons contain folds and duplexes, but character and intensity of deformation varies according to lithology. **Angayucham terrane:** Comprises the two structurally highest allochthons of the province. These allochthons behaved as coherent thrust sheets with little internal deformation.	Not significantly affected by early Brookian deformation.
Structural features of late Brookian deformation	Broad, doubly plunging, ENE- to WSW-trending open folds and, mainly in Slate Creek subterrane and Angayucham terrane, south-dipping low-angle normal faults and east-trending high-angle faults, south side down, with probable major right-lateral strike-slip displacement. Local south-vergent folding and faulting in North Slope, Hammond, and Coldfoot subterranes.	**All units:** Regional west plunge exposes progressively deeper structural levels eastward. **Arctic Alaska terrane:** *North Slope subterrane:* Early and late Brookian structures indistinguishable. *Endicott and De Long Mountains subterranes:* Broad anticlines and synclines superimposed on early Brookian allochthons. Smaller late Brookian folds and faults probably present but difficult to distinguish from earlier deformational features. Range-front deformation, defined by abrupt northward decrease in structural relief, postdates emplacement of allochthons and mid-Cretaceous foredeep deposition. **Angayucham terrane:** Same as for Endicott and DeLong Mountains subterranes of Arctic Alaska terrane.	East-vergent thrust faults and associated folds. Related east-vergent structures probably superimposed on adjacent parts of foothills and northern Brooks Range provinces.
Structural features related to rifting of Canada basin	Not significantly affected by rifting in Canada basin.	Not significantly affected by rifting in Canada basin.	Not significantly affected by rifting in Canada basin.
Metamorphism	**Arctic Alaska terrane** *North Slope subterrane:* Prehnite-pumpellyite to lower greenschist facies. *Hammond and Coldfoot subterranes:* Greenschist and blueschist facies, mostly overprinted by lower greenschist facies. Grade increases southward. *Slate Creek subterrane:* Lower greenschist facies. **Angayucham terrane:** Prehnite-pumpellyite to lower greenschist facies.	Incipient metamorphism to lower greenschist facies. Conodont color alteration index values indicate metamorphism at <300 °C.	Little or no Brookian metamorphism.

TABLE 1 (continued)

Northeastern Brooks Range	Foothills	North Slope
Arctic Alaska terrane: *North Slope subterrane:* Upper Proterozoic and lower Paleozoic sedimentary and igneous rocks deformed in pre-Mississippian time unconformably overlain by Mississippian to Lower Cretaceous carbonate and clastic continental-margin succession. In Arctic Foothills and Coastal Plain, unconformably overlain by Albian and younger terrigenous clastic deposits.	**Arctic Alaska terrane:** Albian and younger terrigenous clastic rocks shed northward and eastward from Brooks Range into Colville basin foredeep. Deposited on gently south dipping, Mississippian to Lower Cretaceous clastic and carbonate continental-margin succession of North Slope subterrane.	**Arctic Alaska terrane:** *North Slope subterrane:* Mainly pre-Mississippian argillite unconformably overlain by south-facing, Mississippian to Lower Cretaceous, clastic and carbonate continental-margin succession. Albian and younger terrigenous clastic rocks shed from Brooks Range rest conformably to unconformably on these rocks and form a constructional continental-margin sequence along margin of Canada basin.
Northern and northeastern: Separated from Canada basin by northern front of compressional deformation. **Eastern:** Zone of transition to north-trending structural low in Canada. **Southern:** Zone of transition in structural style to northern Brooks Range province. **Western:** Zone of transition to similar, but probably older, structures of foothills province. **Northwestern:** Separated from North Slope province by northern front of compressional deformation.	**Northern:** Separated from North Slope province by northern front of compressional deformation. **Eastern:** Zone of transition to similar, but probably younger, structures of the northeastern Brooks Range province. **Southern:** Mid-Cretaceous foredeep deposits of foothills province depositionally overlap northern Brooks Range province. Boundary modified by late Brookian deformation. **Western:** Overthrust by Lisburne Peninsula province.	**Northern:** Separated from Canada basin by tectonic hinge line marked by downbowing of pre-Albian rocks. **Southern and southeastern:** Separated from northeastern Brooks Range and foothills provinces by northern front of compressional deformation. **Western:** Separated from rocks of North Chukchi basin by tectonic hinge line marked by zone of significant downbowing of pre-Albian rocks.
Not significantly affected by early Brookian deformation.	Not affected by early Brookian deformation.	Not affected by early Brookian deformation.
Regionwide: Northward-convex, north-vergent fold belt. **Mountains:** Broad, east-trending anticlinoria cored by pre-Mississippian rocks; overlying Mississippian to Lower Cretaceous rocks shortened by detachment folds or widely spaced thrust faults. **Arctic Foothills and Coastal Plain:** Sharp northward decrease in structural relief at mountain front, although similar structures present in mountains probably present in subsurface. To north, foredeep deposits display broad synclines and narrow anticlines; these structures probably underlain by complex north-vergent imbricate thrust faults at depth. Structural relief and intensity of deformation gradually decreases northward.	North-vergent thrusting and associated broad synclines and narrow anticlines mainly above detachment in Torok Formation; stratigraphically lower units locally involved in deformation to south. Structural relief and intensity of deformation decreases gradually northward.	Not affected by late Brookian deformation.
Jurassic to Cretaceous extensional faulting (Dinkum graben); post-Jurassic northward downbowing of pre-Cretaceous, south-dipping continental-margin deposits (Barrow arch); and local late Neocomian uplift and erosion along Barrow arch. All overprinted by late Brookian contractional structures.	Not significantly affected by rifting in Canada basin.	Jurassic to Cretaceous extensional faulting (Dinkum graben); post-Jurassic northward downbowing of pre-Cretaceous, south-dipping continental-margin deposits (Barrow arch); middle and Late Cretaceous listric faulting offshore and minor Jurassic and Cretaceous extensional faulting in subsurface; local late Neocomian uplift and erosion along Barrow arch.
Little or no Brookian metamorphism.	Not affected by Brookian metamorphism.	Not affected by Brookian metamorphism.

Devonian and Mississippian deformation in the province was interpreted by Hitzman and others (1986) as extensional on the basis of inferred down-to-basin (predominately southward) faulting in the Survey Pass quadrangle. Subsequent uplift and erosion during the Devonian and Early Mississippian may be indicated by the sub-Mississippian unconformity in the Schwatka Mountains (Hammond subterrane) and in the Mt. Doonerak fenster (North Slope subterrane). However, Oldow and others (1987d) argued that penetrative fabrics in the pre-Mississippian rocks in the Mt. Doonerak fenster were not formed by deformation in early Paleozoic time; rather, they are related to later Brookian orogenesis.

Early Brookian structures. The semipenetrative and penetrative fabrics that characterize most of the southern Brooks Range structural province are generally interpreted as contractional structures that were developed during early Brookian deformation. This interpretation is supported by the apparent stability of high-P/low-T mineral phases along foliation surfaces and ^{40}Ar-^{39}Ar loss spectra for white mica in glaucophane schist (A. B. Till, 1992, oral commun.). Although there is currently considerable discussion about the timing and significance of some of these fabrics, particularly in the Coldfoot and Slate Creek subterranes (for example, Miller and Hudson, 1991), we group them here as early Brookian structures and note points of controversy.

Hammond and Coldfoot subterranes. The Hammond and Coldfoot subterranes and the North Slope subterrane in the Mt. Doonerak fenster are characterized by pervasive polyphase deformation and metamorphism resulting from the Brookian orogeny. Two or more generations of Brookian fabrics are generally present, although assignment of specific structures to a particular deformational event is commonly difficult (Grybeck and Nelson, 1981; Hitzman and others, 1986; Oldow and others, 1987d; Dillon, 1989). Tight to isoclinal folds occur from thin-section to regional scale. Axes of the folds, most commonly subhorizontal, trend approximately eastward, though some earlier structures are more north trending. Fold axial surfaces, most axial-planar fabrics, foliations, and associated thrust faults are moderately to gently dipping, and they commonly have been folded during later deformational events.

Near the Dalton Highway, the area for which the most information is available, Dillon (1989) and Gottschalk (1990) recognized three major fabric elements. The earliest fabric element, locally preserved in rocks of the Coldfoot subterrane, is a penetrative schistosity associated with isoclinal folds and sheath folds that are mostly to completely transposed by the second schistosity. The second fabric element, the most prominent foliation in the Coldfoot and Hammond subterranes, is also present in the North Slope subterrane in the Mt. Doonerak fenster. It generally parallels lithologic layering and is associated with megascopic to mesoscopic, tight to isoclinal folds. These folds are commonly intrafolial and have fold axes that parallel mineral lineations. The third fabric element is a semipenetrative, axial-planar schistosity or cleavage that increases in intensity to the south. In the northern part of the Hammond subterrane and in the Mt. Doonerak fenster, this fabric element is a centimeter-spaced phyllitic cleavage that is locally intense near major faults, but in the southern part of the Hammond subterrane, it is a millimeter-spaced axial planar schistosity. The geometry of folds associated with the latest fabric is variable and includes asymmetric, upright, and kink folds that display both northward and southward vergence. The earliest, mostly transposed, structural fabric was attributed by Dillon (1989) to pre-Devonian deformation; however, Gottschalk (1990) attributed the earliest two sets of structures to progressive deformation and metamorphism of the Brookian orogeny under conditions of top-to-the-north ductile shear. The latest fabric element is interpreted as the result of top-to-the-south extensional deformation that was related to uplift in the orogenic belt (Dillon, 1989; Gottschalk, 1990).

It is currently unclear if structures of the Dalton Highway area are representative of structures along strike. A. B. Till (1990, written commun.) reported that Brookian deformation is more complex and associated metamorphic events less distinct in the Hammond subterrane in the Dalton Highway area than in rocks of the subterrane in the western Brooks Range. Zayatz (1987), however, reported that rocks in the western part of the Coldfoot subterrane (Kallarichuk Hills) had a structural history comparable to that of rocks of the Coldfoot subterrane in the Dalton Highway area.

The nature of the northern limit of the southern Brooks Range structural province (that is, the southern limit of the Endicott Mountains subterrane) is poorly understood. This contact has been mapped both as a conformable surface and a thrust fault, probably folded, above rocks here assigned to the Hammond subterrane, and the North Slope subterrane in the Mt. Doonerak fenster (Coney and Jones, 1985; Mull and others, 1987c; Karl and others, 1989) (Fig. 23). In the Ambler River and Survey Pass quadrangles (Fig. 1), the contact is a regional north-dipping thrust fault that places Devonian rocks of the Endicott Mountains subterrane on Mississippian and older rocks of the Hammond subterrane (Kugrak River allochthon of Mull and others, 1987c). In the Baird Mountains, Chandalar, and Philip Smith Mountains quadrangles, however, a north-dipping thrust requires younger rocks—Upper Devonian strata of the Endicott Mountains subterrane—to be thrust onto older rocks—Devonian and older units of the Hammond subterrane (Brosgé and Reiser, 1964; Dillon and others, 1986; Karl and others, 1989). Jones and Coney (1989), however, reported detailed biostratigraphic data in the Philip Smith Mountains quadrangle that show that this contact places older Upper Devonian strata on younger Upper Devonian rocks. Along the northern margin of the Mt. Doonerak fenster, the northern boundary of the southern Brooks Range structural province is the north-dipping Amawk thrust (Mull, 1982, p. 21; Plate 6, section B-B′), which places the Devonian Beaucoup Formation of the Endicott Mountains subterrane over the Triassic Shublik Formation and Karen Creek Sandstone of the North Slope subterrane (Mull, 1982; Mull and others, 1987a) (Fig. 23). South of the Mt. Doonerak fenster, however, the southern limit of the Endicott Mountains subterrane is the south-dipping Table Mountain thrust of Dillon (1987, 1989), which, at least locally, places older rocks of the Hammond subterrane over younger rocks of the Endicott Mountains subterrane (Plate 13, section B-B′). Oldow and others (1987d) interpreted the Table Mountain thrust as the primary contact between the Endicott Mountains subterrane and the Hammond subterrane (their Skajit allochthon) and, on the basis of existing mapping, extended it along much of the central Brooks Range. Grantz and others (1991), however, interpreted the Table Mountain thrust as a local out of sequence thrust fault that developed across an earlier, north-dipping thrust fault between the two subterranes.

The contact between the Hammond and Coldfoot subterranes is an important structural lineament (for example, in the Schwatka Mountains, the "Walker Lake lineament" of Fritts and others, 1971) that has been variously interpreted as a change in depositional facies, an unconformity, a metamorphic boundary, and a folded thrust fault. Oldow and others (1987d) and Till and others (1988) considered this boundary a thrust fault, although the nature of this contact is commonly ambiguous in the field. Oldow and others (1987d) suggested that the Coldfoot subterrane was deformed by ductile shear during high-pressure (>8 kbar) metamorphism in a crustal-scale, north-vergent duplex that was bounded above by a decollement. Above the decollement, which acted as the roof thrust of the duplex, rocks of the Hammond subterrane were imbricated under lower pressure (<5–6 kbar) conditions. The decollement was later breached by younger thrust faults, and Coldfoot subterrane rocks were thrust northward to a higher structural level onto rocks of the Hammond subterrane. Thus, the present contact between the Hammond and Coldfoot subterranes may be a compound structure that includes both north- and south-dipping thrust faults. Till (1988), in contrast, noted differences in K-Ar cooling ages and metamorphic assemblages between the Hammond and Coldfoot subterranes and, as a result, suggested that the contact is a major thrust system along which earlier metamorphosed rocks of the Coldfoot subterrane were uplifted and emplaced northward onto the Hammond subterrane while the latter rocks were still undergoing metamorphism.

Slate Creek subterrane and Angayucham terrane. Although rocks of the south-dipping Slate Creek subterrane are polydeformed, they display only a single lower greenschist facies metamorphic overprint, in contrast with the higher grade, polymetamorphosed rocks of the Coldfoot subterrane to the north (Hitzman and others, 1986; Dillon, 1989; Karl and others, 1989). Because cleavage in the Slate Creek subterrane is similar in appearance and orientation to the latest cleavage in the underlying Coldfoot subterrane, Dillon (1989) suggested that the two cleavages formed during the same metamorphic event. Gottschalk (1987) likewise reported that the Slate Creek subterrane displays several generations of north-vergent folds that correspond to those of the Coldfoot subterrane to the north. Miller and Hudson (1991), however, reported down-to-the-south sense of shear indicators in the Slate Creek subterrane and interpreted the pervasive south-dipping fabric as ductile deformation due to regional extension in mid-Cretaceous time.

The Angayucham terrane dips gently to moderately southward, and its base is defined by a complex fault zone, commonly a unit of melange (Pallister and Carlson, 1988; Dillon, 1989). Rocks of the Angayucham terrane are metamorphosed to prehnite-pumpellyite and greenschist facies but lack the penetrative, north-vergent structures characteristic of rocks structurally beneath them to the north (Hitzman and others, 1986; Dillon, 1989). However, the Angayucham terrane displays complex internal imbrication (Jones and others, 1988; Pallister and Carlson, 1988; Dillon, 1989; Pallister and others, 1989). For example, Pallister and others (1989) described multiple 1-km-thick fault slabs mostly of basalt, as well as a complex melange zone bordering the northern margin of the terrane. In addition, detailed biostratigraphic studies have shown that chert units within the terrane are highly imbricated (Jones and others, 1988; Dillon, 1989).

Brookian metamorphism. The regional metamorphic mineral assemblages that characterize the southern Brooks Range structural province developed during the Brookian orogeny. The earliest formed are the blueschist facies assemblages, preserved locally in the Coldfoot and Hammond subterranes in the Baird Mountains, Ambler River, Survey Pass, and Wiseman quadrangles (Turner and others, 1979; Nelson and Grybeck, 1981; Armstrong and others, 1986; Hitzman and others, 1986; Dusel-Bacon and others, 1989, and this volume). Elsewhere, early blueschist facies metamorphism is shown by pseudomorphs of glaucophane and pseudomorphs after lawsonite in garnet (Gottschalk, 1987, 1990; Till and others, 1988). Till (1988) reported that high-P/low-T assemblages of the Coldfoot subterrane consist of early lawsonite-bearing and later epidote-bearing blueschist facies assemblages, whereas the Hammond subterrane preserves crossite-bearing assemblages that are associated with greenschist facies assemblages.

Throughout much of the province, the earlier high-P/low-T assemblages are overprinted by pervasive retrograde chlorite-zone greenschist facies assemblages. The retrogradation represents a nearly isothermal drop in pressure during metamorphism (Hitzman and others, 1986). In the Baird Mountains quadrangle, the chlorite-zone retrograde assemblage is modified by late development of randomly oriented biotite at the expense of chlorite, which suggests that late prograde greenschist facies metamorphism was caused by an increase in temperature (Zayatz, 1987).

Estimates of the maximum temperature and pressure attained during metamorphism in the province are in the range of 400–500 °C and 6–11 kbar (Hitzman and others, 1986; Gottschalk and Oldow, 1988). Metamorphic zones extend over a distance of about 5 km from pumpellyite-actinolite facies in the Angayucham terrane to blueschist facies in the Coldfoot subterrane. These data suggest that peak metamorphism of the Coldfoot subterrane occurred at depths of more than 25 km, whereas metamorphism of the Angayucham terrane occurred above 10 km. Dusel-Bacon and others (1989) concluded that Brookian metamorphism of the Coldfoot subterrane followed a clockwise

P-T path that evolved from low-*T* to high-*T* subfacies of the blueschist facies followed by greenschist facies.

Although isotopic dating of the prograde high-*P*/low-*T* assemblages has proved to be a formidable problem because of the polymetamorphic history of the host rocks, the age of high-*P*/low-*T* metamorphism is generally regarded as Late Jurassic to Early Cretaceous (Armstrong and others, 1986; Hitzman and others, 1986). Recently obtained ^{40}Ar-^{39}Ar loss spectra for white mica in a glaucophane schist indicate a minimum age of 149 Ma for high-*P*/low-*T* metamorphism (A. B. Till, 1992, oral commun.). The high-*P*/low-*T* metamorphism is commonly interpreted to have been caused by southward subduction of the Arctic Alaska terrane beneath the Angayucham terrane, whereas the later greenschist facies metamorphism was probably due to later thermal recovery and decreasing pressure associated with uplift and unroofing of the Brooks Range orogen later in Cretaceous time (Dusel-Bacon and others, 1989).

Late Brookian structures. The onset of late Brookian deformation in the southern Brooks Range structural province is difficult to determine because late Brookian deformation is defined on the basis of stratigraphic relations not present in metamorphic rocks of the southern province. In the northern foothills of the Brooks Range, early Brookian structures are unconformably truncated by Aptian and Albian foredeep deposits; therefore, we consider Aptian and younger structural features in the Brooks Range to be part of late Brookian deformation. This timing corresponds roughly with the change from prograde to retrograde metamorphism. Late Brookian structures include uplift and late folding of the southern Brooks Range, extension and strike-slip faulting along its southern margin, and east-vergent deformation in part of the range.

Uplift of the southern Brooks Range. Potassium-argon and ^{39}Ar-^{40}Ar cooling ages of metamorphic minerals suggest that a major uplift and unroofing event occurred in the southern Brooks Range between 130 and 90 Ma and culminated at about 120 to 100 Ma (Turner and others, 1979; Mull, 1982; Dillon, 1989; Blythe and others, 1990; Miller and Hudson, 1991). Till (1988) reported that K-Ar cooling ages are 100–86 Ma for the Hammond subterrane and 130–100 Ma for the Coldfoot subterrane and concluded that uplift may have occurred somewhat earlier in the southern part of the southern Brooks Range province. This major uplift event was probably the result of (1) isostatic rebound following crustal thickening during early Brookian large-displacement thrust faulting; (2) continued shortening during late Brookian orogenesis; and (3) tectonic denudation resulting from crustal extension. Uplift in the core of the range was previously assumed to be the result of isostatic rebound or south-vergent folding and thrusting, and denudation was thought to be caused primarily by erosion during uplift (Mull, 1982, 1985; Dillon, 1989).

Extension in the southern Brooks Range. South-dipping faults, mylonite, and phyllonite in, and at the base of, the Slate Creek subterrane and the southern foothills belt of the Angayucham terrane have been interpreted as thrust faults by most workers (for example, Angayucham thrust system of Dillon, 1989; this volume, Plate 6). However, apparent younger over older relations, the abrupt upward decrease in metamorphic grade across some of the faults, and sense of shear indicators have led many workers (Carlson, 1985; Box, 1987; Miller, 1987; Oldow and others, 1987a, 1987c, 1987d; Gottschalk and Oldow, 1988; Miller and Hudson, 1991) to propose the existence of major south-dipping, low-angle normal faults along the southern margin of the Brooks Range. Gottschalk and Oldow (1988) described structures in the Wiseman quadrangle consistent with normal faulting and documented petrologic evidence for the omission of at least 10 km of structural section. Box (1987) and Miller and Hudson (1991) reported that kinematic indicators support down-to-the-south displacement on gently south dipping faults in the Ambler and Wiseman quadrangles. Miller and Hudson (1991) interpreted the mid-Cretaceous sedimentary deposits of the Koyukuk basin as detritus derived from the footwall and deposited on the hanging wall of a regional south-dipping normal fault.

Despite the possibility that the Angayucham terrane and the Slate Creek subterrane may now be underlain by normal faults, most workers agree that the Angayucham terrane and the Slate Creek subterrane were originally emplaced on north-vergent thrust faults. Subsequent extensional deformation is related to mid-Cretaceous uplift of the southern Brooks Range and filling of the Koyukuk basin, but extensional structures involve rocks as young as Late Cretaceous (Box, 1987), which indicates that the extension may have continued into Late Cretaceous or Tertiary time (Gottschalk and Oldow, 1988). It is unclear whether normal faulting occurred along preexisting or newly formed fault surfaces, whether it was brittle or ductile, and whether it occurred as a consequence of deformation in a contractional orogen or in association with a regional episode of extension.

Folding in the southern Brooks Range. Postmetamorphic broad, upright, open folds have been superimposed over early Brookian structures in the southern Brooks Range (Hitzman and others, 1986; Dillon, 1989). These folds are typically doubly plunging, are symmetric to slightly asymmetric, and have wavelengths of tens of kilometers. They trend east-northeast in the vicinity of the Dalton Highway (Dillon, 1989) but gradually change westward to a west-northwest orientation in the Ambler River quadrangle (Hitzman and others, 1986). This generation of folds accounts for many of the most conspicuous regional-scale folds within the southern Brooks Range, including the Cosmos arch (Hitzman and others, 1986), the Mt. Doonerak anticlinorium (Dillon, 1989), and the Mt. Angayukaqsraq anticlinorium (Till and others, 1988), and may also account for regional arching of the Coldfoot subterrane (for example, the Kalurivik arch of Hitzman and others, 1982). Although these folds generally cannot be related demonstrably to faults exposed at the surface, they probably are related to shortening above faults at depth. This can be best demonstrated for the Mt. Doonerak and Mt. Angayukaqsraq anticlinoria, both of which formed above an anticlinal stack of horses in a duplex (Oldow and others, 1987d; Till and others, 1988). The time of construction of the Mt. Angayukaqsraq anticlinorium is uncertain, but the Mt. Doonerak anticlinorium may have been constructed by thrusting related to shortening in the northeastern Brooks Range in the Late Cretaceous or Tertiary (Oldow and Avé Lallemant, 1989; Grantz and others, 1991). The Cosmos arch may also be a relatively late structural feature, as evidenced by deformed Upper Cretaceous rocks and antecedent drainages that cross the arch.

Strike-slip faults along the southern flank of the Brooks Range. Some east-striking high-angle faults with down-to-the-south and probable right-lateral strike-slip displacements have been observed or inferred in the southern Brooks Range and the adjacent Koyukuk basin (Dillon, 1989) (Fig. 23). The Kobuk fault (Grantz, 1966), which lies immediately south of the Brooks Range within Cretaceous deposits of the Koyukuk basin, was inferred to underlie a topographic depression occupied for much of its length by the Kobuk River. To the east, it merges with the Malamute fault to form the Malamute–South Fork fault system (Dillon, 1989). This fault system cuts Cretaceous deposits of the Koyukuk basin and rocks of the Angayucham terrane and juxtaposes the Brooks Range and Ruby geanticline. The Malamute fault is a more northerly strand of the Kobuk fault system that cuts rocks of the Angayucham terrane and the Slate Creek subterrane. Both the Malamute and South Fork faults are locally exposed as narrow zones of unrecrystallized, brittlely deformed rocks, commonly including breccia, gouge, and slickensides, and are marked by prominent steps in magnetic and gravity intensity. Dillon (1989) interpreted a minimum of 90 km of post–Early Cretaceous right separation for the system. Although the Malamute–South Fork fault system is broadly concordant with regional trends in the Brooks Range, it sharply truncates northeast-striking lithologic and structural trends within the Ruby geanticline to the south (Decker and Dillon, 1984; Coney and Jones, 1985; Dillon, 1989), suggesting that larger amounts of separation are possible. Grantz (1966) and Dillon (1989) suggested that the Kobuk and Malamute–South Fork fault systems may represent a westward continuation of the Tintina fault system, which was offset by right-lateral displacement on an inferred extension of the northeast-trending Kaltag fault system of western Alaska.

A major northeast-striking lineament, called the Porcupine lineament by Grantz (1966), parallels the Porcupine River southeast of the Brooks Range (Fig. 23). Exposures in this area are poor, but deformed rocks at least as young as Triassic are exposed over a broad area, along with overlying undeformed Miocene basalt (Plumley and Vance, 1988; Oldow and Avé Lallemant, 1989). Although no displacement of geologic features has been demonstrated across the Porcupine lineament, it separates rocks of northern Alaska from those of east-central Alaska and is thought to represent a regionally important strike-slip fault. Most workers have inferred a Cretaceous (post-Neocomian) age for the postulated fault but differ over the amount and direction of relative movement along the structure. Some have speculated that it represents as much as 2000 km of left slip (Dutro, 1981; Nilsen, 1981), whereas others have proposed 150–200 km of right slip (Churkin and Trexler, 1980, 1981; Jones, 1980, 1982b; Norris and Yorath, 1981; McWhae, 1986; Dillon, 1989), and at least one worker (Smith, 1987) suggested movement in both directions.

East-vergent deformation in the southwestern Brooks Range. The east-striking structures that dominate most of the Brooks Range give way to northeast- to north-striking structures in the southwestern Brooks Range (Fig. 23; Plate 13). This change in structural trend has been interpreted to represent an oroclinal bend of originally east-striking structures (Patton and Tailleur, 1977). However, on the basis of work in the Baird Mountains, Karl and Long (1987, 1990) suggested that the northeast- to north-striking structures have been superimposed over older east-striking structures. The younger structures consist of east-vergent folds and thrusts that decrease in intensity eastward but may extend as far east as the Dalton Highway (Gottschalk, 1990). Although the age of overprinting is uncertain in the Baird Mountains, it seems likely that this deformation was related to east- to northeast-directed deformation along the Tigara uplift and Herald arch.

Northern Brooks Range structural province

The northern Brooks Range structural province contains much of the preserved superstructure of the main axis of the Brooks Range orogen; it is a gently north sloping structural plateau between the structurally higher southern Brooks Range province to the south and the structurally lower foothills province to the north (this volume, Plate 6). Rocks are generally younger to the north (the "regional north dip" of Mull, 1982), but dip of bedding and thrust faults varies within the province. Throughout most of the northern Brooks Range province, dominantly north-vergent fold and thrust structures are spectacularly exposed; however, the rocks are only locally metamorphosed and penetratively deformed. All recognized structures in the province can be ascribed to Brookian orogenic events, although the presence of extensional faults of Devonian to Mississippian and Cretaceous age may be inferred from regional stratigraphic patterns and local structural features.

Allochthons and significance of the Mt. Doonerak fenster. The northern Brooks Range structural province is characterized by generally coeval, but distinctive, stratigraphic sequences that are structurally stacked in predictable succession over large areas. This observation, first recognized by Tailleur and others (1966), has led to the interpretation that the various stratigraphic sequences constitute regionally extensive thrust packages, or allochthons, stacked one on top of another (Martin, 1970; Ellersieck and others, 1979; Mull, 1982; Mayfield and others, 1988). On the basis of successions exposed in structural windows in the northwestern Brooks Range (Picnic Creek, Drinkwater, Ginny, and Cutaway fensters; Figs. 20 and 23), seven allochthonous sequences have been recognized. From base to top, these are (1) the Endicott Mountains (or Brooks Range) allochthon; (2) the Picnic Creek allochthon; (3) the Kelly River allochthon; (4) the Ipnavik River allochthon; (5) the Nuka Ridge allochthon; (6) the

Copter Peak allochthon; and (7) the Misheguk Mountain allochthon (Fig. 21). As described earlier, the rocks of the structurally lowest five allochthons belong to the Arctic Alaska terrane and are assigned to the Endicott Mountains subterrane (Endicott Mountains allochthon) and the De Long Mountains subterrane (Picnic Creek, Kelly River, Ipnavik River, and Nuka Ridge allochthons). The two structurally highest allochthons, the Copter Peak and Misheguk Mountain allochthons, constitute the Angayucham terrane in the crestal and disturbed belts.

Rocks thought to be relatively in place compared to the above units are those of the North Slope and the Hammond subterranes. However, with the exception of rocks of the North Slope subsurface, rocks of the North Slope and the Hammond subterranes have been transported northward above thrust faults and hence are allochthonous in the strict sense (Oldow and others, 1987d; Till and others, 1988). Mississippian to Triassic rocks (lower Ellesmerian sequence) of the North Slope subterrane structurally underlie the Endicott Mountains allochthon in the northeastern part of the northern Brooks Range province and in the Mt. Doonerak fenster along the southern margin of the province (Fig. 23; this volume, Plate 6). A Mississippian and younger sequence similar to that of the North Slope subterrane is also exposed discontinuously to the west in the Hammond subterrane (Schwatka Mountains) (Mull and Tailleur, 1977; Tailleur and others, 1977; Mayfield and others, 1988). The lithology of these Mississippian and younger stratigraphic sequences differ markedly from the coeval sequence in the structurally overlying Endicott Mountains allochthon (Fig. 21). Thus, the Mississippian and younger rocks exposed south of the Endicott Mountains allochthon probably underlie the Endicott Mountains allochthon and connect to the north with the lower Ellesmerian sequence of the North Slope subterrane on the North Slope and in the northeastern Brooks Range (Dutro and others, 1976; Mull and others, 1987a). Consequently, most workers agree that the Endicott Mountains and overlying allochthons must be restored to a position south of these rocks and the pre-Mississippian rocks that depositionally underlie them (Mull and others, 1987a, 1987c; Oldow and others, 1987d; Mayfield and others, 1988; Grantz and others, 1991). Mull and others (1987a) suggested a minimum northward displacement of the Endicott Mountains allochthon of 88 km, the distance from the northernmost exposures of the allochthon to the southern margin of the Mt. Doonerak fenster. This figure does not account for shortening within or below the allochthon. Oldow and others (1987d) suggested that the northern edge of the allochthon has been displaced about 200 km from its original position, assuming significant shortening within the subjacent North Slope subterrane. Oldow and others (1987d) also estimated an additional 35%–45% shortening to account for thrust imbrication and macroscopic folding within the allochthon.

Early Brookian structures. Although most completely and extensively preserved in the western Brooks Range, all but one of the allochthons (Kelly River) have been recognized along the length of the Brooks Range nearly to the Canadian border (Fig. 23). The structural stacking of coeval stratigraphic sequences over such a large area shows that deformation in the northern Brooks Range structural province is characterized by major shortening and indicates that the rocks of these allochthons are underlain by thrust faults of large displacement. Direction of tectonic transport, indicated by fold asymmetry and the tendency of thrust faults to cut stratigraphically upsection, is generally to the north, except in the westernmost part of the province, where the direction of tectonic transport is less certain. Most structures related to mapped thrusts probably formed during emplacement of the allochthons before Albian time, when foreland-basin deposits of the Aptian(?) and Albian Fortress Mountain Formation to the north unconformably overlapped the allochthons and their associated structures (Mull, 1982, 1985; Mayfield and others, 1988). The origin of many thrust-related structures, however, is unclear, and some may have developed in late Brookian time. Emplacement of the structurally lowest allochthon (Endicott Mountains allochthon) could not have occurred prior to deposition of its youngest strata in Early Cretaceous (Valanginian) time, but emplacement of the structurally higher allochthons in the De Long Mountains subterrane may have begun prior to the Early Cretaceous because foredeep strata of these allochthons are as old as Late Jurassic.

Although widespread, the allochthons commonly are not structurally intact and display internal folds, thrust faults, and fault-bounded changes in stratigraphic thickness and facies. In the northwestern Brooks Range, allochthons are laterally discontinuous, and one or more consecutive allochthons may be missing from the idealized structural sequence at any given location. Complete structural sequences consisting of all seven allochthons occur in only a few places, and even in many of those places, certain allochthons disappear laterally over distances of only a few kilometers. These features can be attributed to several factors, including (1) variation in original stratigraphic thickness within individual allochthons; (2) imbrication and development of duplexes within individual allochthons during thrusting, especially where relatively thin bedding and/or alternating competent and incompetent layers characterize all or part of the stratigraphy of an allochthon; (3) local extension of individual allochthons during thrusting; (4) breaching by out of sequence thrust faults; and (5) displacement along superimposed low-angle normal faults. The latter possibility is supported by the observation that faults at the base of allochthons locally cut downward through stratigraphic section to the north, in the inferred direction of tectonic transport (Roeder and Mull, 1978; Harris, 1988).

Angayucham terrane. Remnants of the Angayucham terrane in the northern Brooks Range structural province are locally preserved as klippen comprising the Copter Peak and Misheguk Mountain allochthons. These klippen range up to 150 km along strike and 20 km across strike (Patton and others, 1977; Roeder and Mull, 1978; Mayfield and others, 1988) and total less than 3 km thick. The thin dynamothermal aureole commonly marking the contact between the Copter Peak and Misheguk Mountain allochthons (Roeder and Mull, 1978; Boak and others, 1987;

Mayfield and others, 1988), and associated structural fabrics in adjacent parts of both allochthons, are related to original thrust emplacement of the Misheguk Mountain allochthon over the Copter Peak allochthon. Discontinuity of the aureoles and faulting of the metamorphic rocks within them indicate that the contact has been reactivated since its origin, probably along thrust faults that flatten upward beneath the Misheguk Mountain allochthon and later along down-to-the-north normal faults (Harris, 1988).

De Long Mountains subterrane. The structural thickness of the De Long Mountains subterrane is about 4 km, and constituent allochthons have respective structural thicknesses of no more than 3 km. These allochthons consist largely of structurally incompetent, thin-bedded rocks that formed multiple detachment horizons and complex and closely spaced (on the order of tens to hundreds of meters) folds and thrust faults. The thicker and more competent intervals, especially carbonate rocks of the Lisburne Group in the Kelly River allochthon, typically formed more extensive thrust sheets, broader folds, and more widely spaced thrust faults. Asymmetric to overturned folds in these competent intervals are as much as 1–2 km across. The stratigraphically lowest detachment horizons within the De Long Mountains subterrane are in fine-grained clastic rocks of the Endicott Group and older carbonate, or mixed carbonate and clastic, rocks.

Endicott Mountains subterrane. The Endicott Mountains subterrane (allochthon) has a structural thickness of at least 7 km in the central Brooks Range (Oldow and others, 1987d) and more than 10 km in the western Brooks Range (Mayfield and others, 1988). It is the stratigraphically and structurally thickest, most extensive, and most continuous of the allochthons. In the eastern part of the northern Brooks Range province, the Endicott Mountains allochthon comprises thick, structurally competent units, such as the Noatak Sandstone, Kanayut Conglomerate, and carbonate rocks of the Lisburne Group. In the southern part of the province, the basal detachment of the allochthon is developed in fine-grained clastic rocks of the Beaucoup Formation and Hunt Fork Shale, which display penetrative, dominantly north-vergent structures. These rocks acted as a shear zone and display an upward decrease in strain and in number of generations of small-scale folds (Handschy and Oldow, 1989). To the north in structurally higher levels of the allochthon, deformation is characterized by imbricate thrust sheets and large single-generation folds that detached within incompetent fine-grained rocks of the Hunt Fork Shale, Kayak Shale, and Etivluk Group (Kelley and others, 1985; Handschy and others, 1987; Kelley and Bohn, 1988; Handschy and Oldow, 1989). The shorter, steep to overturned limbs of anticlines face north in this area and display a strong sense of asymmetry; locally the folds are recumbent. In the western part of the province, the Endicott Mountains allochthon is composed largely of structurally incompetent units, especially the Kuna Formation, Etivluk Group, and Ipewik unit. Deformation in these rocks is characterized by complex and closely spaced folds and thrust faults; thus, fault spacing and fold wavelength regionally decrease to the west.

North Slope subterrane. Rocks assigned to the North Slope subterrane structurally underlie the Endicott Mountains subterrane and make up the eastern part of the northern Brooks Range structural province. North-vergent, asymmetric to overturned folds, between about 100 to 1000 m across, are the dominant structural element of these rocks. Thick and competent carbonate rocks of the Lisburne Group act as the rigid structural unit controlling the geometry of the folds. The folds, which are relatively closely spaced, are underlain and commonly separated by thrust faults rooted in the Kayak Shale. These structures probably formed during or after emplacement of the Endicott Mountains allochthon, but their absolute age is not determined.

Late Brookian structures. Exposures of the allochthons, particularly in the northwestern Brooks Range, are controlled primarily by folding. Structurally higher allochthons are preserved in broad synforms, and structurally lower allochthons are exposed in broad antiforms. These structural highs and lows, 15 to 30 km across, are gentle to open folds with gently to moderately dipping limbs. Local asymmetry of folds or associated thrust faults indicate tectonic transport to the north (in the western part of the province, to the northwest). The structures generally trend to the east, although there is a gradual change to northeast in the western part of the province. Both the anticlines and synclines tend to be doubly plunging, reflecting structural culminations and depressions along strike. The same regional pattern of folding affects all the allochthons. This pattern suggests that the folding occurred mainly after emplacement of the allochthons. As in the southern Brooks Range province, this style of large-scale folding may be directly related to postemplacement structural thickening by duplexing in underlying rocks.

Major features of the northern Brooks Range structural province. The northern Brooks Range structural province displays several major features that are the result of the accumulation of early and late Brookian deformation. These features are (1) the regional westward plunge of the orogen, (2) the disturbed belt, and (3) the range front of the western and central Brooks Range.

Regional westward plunge of the orogen. In the northern Brooks Range, progressively deeper structural levels are exposed from west to east due to regional westward plunge (Fig. 23). The structurally highest rocks, including the Angayucham terrane and De Long Mountains subterrane, are most extensively preserved in the northwestern Brooks Range. The central part of the province is underlain by the Endicott Mountains subterrane, indicating a relatively constant level of structural exposure for about 900 km in an east-west direction. Deformed rocks of the North Slope subterrane, the structurally lowest subterrane of the Arctic Alaska terrane, underlie the eastern part of the province. Increasing structural relief to the east probably resulted from greater depth to detachment but may also have resulted from greater shortening in the east and/or an oblique intersection of Brookian structures with regional Paleozoic and Mesozoic sedimentary facies patterns.

Disturbed belt. The northern part of the northern Brooks

Range structural province is characterized by complex folds and imbricate thrust faults and so is referred to as the "disturbed belt" (Brosgé and Tailleur, 1970, 1971; Tailleur and Brosgé, 1970) (Fig. 2). The disturbed belt consists of the northern part of the Endicott Mountains allochthon and remnants of other higher allochthons. Most of the allochthons are relatively thin and discontinuous in this region, probably because the original northern extent of the far-displaced allochthons corresponds roughly with the present northern boundary of the northern Brooks Range structural province.

The structural style of this belt has been strongly influenced by the dominance of thin-bedded and incompetent rock types. Fold wavelengths are short (meters to hundreds of meters), and faults are closely spaced in the thin-bedded rocks. Buckle folds are common where competent and incompetent rocks are interbedded. In dominantly incompetent (typically shale-rich) intervals, deformation produced complex, commonly incoherent, small-scale structures that are penetrative in many places. Where relatively thin competent layers make up a small percentage of a dominantly incompetent interval, broken formation is common. The incompetent intervals typically include flysch of the Okpikruak Formation that is interleaved with older, more competent allochthonous rocks. Where overprinted with a strong deformational fabric, it can be difficult to distinguish tectonically imbricated sections of the Okpikruak Formation from olistostromal units of the Okpikruak.

The disturbed belt has been mapped eastward across the mountain front and far into the eastern Brooks Range (Brosgé and Tailleur, 1970, 1971) (Fig. 2), where it is a major structural low that defines the transition between the northern and northeastern Brooks Range structural provinces (Wallace and Hanks, 1990) (Fig. 23). At its easternmost end, the disturbed belt consists entirely of the Endicott Mountains and North Slope subterranes, with the exception of an isolated klippe in the Porcupine Lake area (Arctic quadrangle) (Figs. 1 and 23). The klippe, composed of De Long Mountains subterrane and Angayucham terrane rocks, overlies rocks of the North Slope subterrane, confirming that, prior to erosion, highly allochthonous rocks once extended into the eastern Brooks Range at least as far north as the disturbed belt. To the northeast at Bathtub Ridge (Demarcation Point quadrangle (Fig. 1,), no allochthons are present; rather, autochthonous Lower Cretaceous deposits of the Colville basin are preserved in a structural low (the Bathtub syncline), conformably overlying rocks of the North Slope subterrane (Detterman and others, 1975; Reiser and others, 1980) (Figs. 3 and 23).

Range front of the western and central Brooks Range. An abrupt change in structural relief and elevation at the mountain front of the Brooks Range interrupts the progressive northward increase in level of structural exposure. The mountain front is most commonly marked by a sharp, down-to-the-north step of erosion-resistant carbonate rocks of the Lisburne Group. Where the carbonate rocks are stratigraphically thin or absent, the mountain front is marked by a step of Kanayut Conglomerate. In simplest terms, the range-front structures typically are down-to-the-north monoclines, in which the steep (north dipping) to overturned (south dipping) beds define the range front. The geometry of these monoclines suggests that they are underlain by north-directed thrust faults that generally are not exposed at the mountan front (Vann and others, 1986; Jamison, 1987). The east-trending range-front monoclines intersect older, presumably early Brookian, structures at an oblique angle (Crane and Mull, 1987), transect allochthon boundaries, and locally involve Albian and younger rocks. These observations suggest that the range-front structures are of late Brookian age.

Foothills structural province

The foothills structural province (Fig. 23) consists of deposits shed northward from the Brooks Range into its foredeep, the Colville basin, and later deformed during northward migration of the Brooks Range orogenic front (Mull, 1985). Deposition of Albian and younger clastic rocks of the Colville basin postdates the Late Jurassic to Neocomian emplacement of allochthons of the northern Brooks Range structural province; however, deformation of the clastic rocks indicates that contraction continued in the western and central Brooks Range to latest Cretaceous or earliest Tertiary time (Mull, 1985). Structures in the foothills structural province record shortening at least an order of magnitude less than that in the northern Brooks Range structural province. According to Kirschner and others (1983), only about 11 km of shortening (10%) has occurred in Brookian deposits of the foothills structural province; similarly, Oldow and others (1987d) suggested a figure of 15 km on the basis of a balanced cross section through the central Brooks Range.

The southern boundary of the province is the southern limit of exposure of the stratigraphically lowest deposits of the Colville basin, the Fortress Mountain and Torok formations of Albian age. There is a significant northward decrease in structural complexity across this boundary, in part because the allochthonous rocks to the south record the effects of large-scale thrust transport that has not affected the Colville basin deposits. The northern boundary of the province is the northern limit of structural thickening by thrust faulting and folding.

In vertical section, the foothills structural province is a northward-thinning wedge composed of deformed foredeep deposits (Plate 6, cross section B-B′). The wedge configuration is in large part the product of a southward increase in structural thickening, but it also reflects the original gentle south dip of the northern flank of the Colville basin. North of the deformation front, the top of the underlying upper Ellesmerian sequence dips about 1° south, as defined on seismic-reflection profiles by the reflector of the pebble-shale unit (Kirschner and others, 1983). South of the deformation front, the dip of this reflector increases to 3°, probably because of loading by both the Brookian thrusts and foredeep deposits. Reflectors in the Ellesmerian and subjacent Franklinian sequences can be traced southward at least as far as, and perhaps south of, the Brooks Range mountain front. These reflectors show little evidence of shortening over most of their length, but at least some evidence of thrusting and folding is visible to the south near the range front, despite the poor quality of data in this area (Kirschner and others, 1983; Mull and others, 1987c; Oldow and others, 1987d).

Outcrop, seismic, and well data indicate that the Torok and Fortress Mountain formations form a thick, northward-tapering, shale-rich wedge between the underlying homoclinal, south-dipping upper Ellesmerian sequence and overlying regionally north dipping deposits of the Nanushuk Group and younger units (Kirschner and others, 1983; Mull and others, 1987c; Molenaar, 1988). The Torok Formation is little deformed where it laps northward onto the northern flank of the Colville basin, but to the south it is imbricated and tectonically thickened, as is the Fortress Mountain Formation. Thickening in the wedge suggests that detachments exist within it and between it and the underlying relatively little deformed, gently south dipping rocks. Deformation within the wedge developed mainly by a combination of duplexing and detachment folding, though its precise character is difficult to define due to poor seismic data and lack of distinctive marker horizons. The sand-rich strata of the overlying Nanushuk Group are structurally more competent than the underlying shale of the Torok and Fortress Mountain formations and therefore have deformed more competently. The Nanushuk has been folded into sharp anticlines separated by broad synclines. The amplitude of the anticlines generally decreases and wavelength increases northward toward the deformation front. The anticlines typically are asymmetric, having steep limbs to the north, and are commonly breached by north-directed thrust faults.

In much of the central part of the foothills, a contrast in deformational geometry is marked by a prominent topographic feature, the Tuktu escarpment (Fig. 23), which delineates the southern limit of the north-dipping, more erosion-resistant sandstones of the Nanushuk Group. In the lowlands south of the Tuktu escarpment, small-scale, south-vergent folds in the Torok Formation are compatible with backthrusting near the top of the Torok and at the base of the Nanushuk Group, as hypothesized by Kelley (1988) in the Chandler Lake quadrangle (his Cobblestone fault). This geometry suggests that, as is typical in a triangle zone (Jones, 1982a), a thickened wedge of Torok and Fortress Mountain formations is overlain by a north-dipping, south-directed thrust fault at the Tuktu escarpment. Structurally lower and to the south, similar backthrusts separate synclinal remnants of competent sandstones and conglomerate of the Fortress Mountain Formation from underlying incompetent Okpikruak shale of the disturbed belt (Oldow and others, 1987d; Howell and others, 1992).

The Colville basin subsided by loading of the Brooks Range allochthons and sediments shed into the basin. However, analysis of gravity profiles across the Brooks Range and Colville basin suggests that an additional subsurface load is required to account for the total subsidence of the trough (Nunn and others, 1987). The nature of this load is unknown, but it may be due to (1) subduction of down-going lithosphere; (2) thinning of the dense lithospheric mantle beneath the crust prior to Brookian deformation in the lower plate of the orogen (that is, the southward continuation of the North Slope subterrane beneath the Brooks Range); and/or (3) obduction of a lithospheric block from the south (Angayucham terrane).

Lisburne Peninsula structural province

Pre-Cretaceous rocks of the Lisburne Peninsula are separated from coeval rocks of the Brooks Range proper by about 50 km of Cretaceous foredeep deposits of the Colville basin (Figs. 2 and 23) and display a northerly structural trend, in sharp contrast with the trend of the other structural provinces of northern Alaska. These pre-Cretaceous and overlying Lower Cretaceous deposits are deformed by west-dipping imbricate thrust faults and associated folds that characterize the structural style of the Lisburne Peninsula (Campbell, 1967). The southern part of the thrust front on the peninsula is marked by Mississippian and Pennsylvanian carbonate rocks of the Lisburne Group thrust over Neocomian clastic rocks of the Brookian sequence; the northern part exposes a regional-scale fold overturned to the northeast in Mississippian to Neocomian(?) rocks. These structures constitute an east-vergent fold and thrust belt that produced the Tigara uplift (Campbell, 1967), the onshore extension of the Herald arch of the Chukchi Sea (Grantz and others, 1970, 1981). Progressively older rocks are exposed to the west in the Lisburne Peninsula, probably reflecting progressively deeper basal detachment to the west.

The age of the Tigara uplift is not determined precisely. Campbell (1967) and Grantz and others (1970, 1981) reported that rocks as young as Albian are deformed, but Mull (1985) argued that the uplift already existed by Albian time. An Albian or older age for thrusting is inferred from paleocurrent data and distribution of Albian and younger sedimentary rocks in the Colville basin, which were deposited from west to east and were derived from a western source in the vicinity of the present Tigara uplift–Herald arch (Mull, 1985). The minimum age of thrusting is delimited only by undeformed Tertiary strata of the Hope basin that unconformably overlie the Tigara uplift south of Point Hope in the Chukchi Sea. This relation suggests that thrusting may be as young as early Tertiary (Grantz and others, 1981; Grantz and May, 1987). The Tigara uplift–Herald arch may represent a continuation of the Brooks Range, either formed originally along a different trend or later bent oroclinally (Patton and Tailleur, 1977). Alternatively, the Tigara uplift–Herald arch may have been formed in the Late Cretaceous or early Tertiary by tectonic processes unrelated to early Brookian orogenesis (Grantz and others, 1981).

Northeastern Brooks Range structural province

The northeastern Brooks Range structural province consists of the eastern part of the Arctic Coastal Plain and a prominent northward-convex arcuate topographic and structural salient, with respect to the northern Brooks Range structural province (Fig. 23). The topographically highest parts of the Brooks Range are found in the northeastern Brooks Range, and relatively deep structural levels of the North Slope subterrane are exposed extensively. Consequently, Proterozoic to lower Paleozoic rocks are widely exposed and display clear evidence of pre-Mississippian and younger deformational events. In structural style, the province is dominated at the surface by folding and lacks the closely spaced, large-displacement thrust faults characteristic of the northern Brooks Range structural province to the south (Mull, 1982; this volume, Plate 6; Wallace and Hanks, 1990; Howell and others, 1992). For this reason, the northern salient of the Brooks Range is thought to have escaped early Brookian deformation and was instead constructed mainly by late Brookian deformation, which extended north to the continental margin. Because the salient represents a younger deformational belt, it is known by the separate term, the Romanzof uplift (Fig. 23).

Pre-Mississippian structures. Pre-Mississippian rocks in the northeastern Brooks Range province display low metamorphic grades, and semipenetrative to penetrative structures that dip moderately to steeply with respect to the sub-Mississippian angular unconformity that characterizes the North Slope subterrane. The pre-Mississippian structures have been thought to have formed during a single Late Devonian to Early Mississippian event, the Ellesmerian orogeny. However, the presence of an angular unconformity beneath Middle Devonian strata not affected by the penetrative pre-Mississippian deformation indicates that major deformation preceded Middle Devonian deposition (Anderson and Wallace, 1990). East of the international border in the British Mountains, the youngest strongly deformed pre-Mississippian rocks are Early Silurian argillite (Lane and Cecile, 1989). This observation, coupled with the presence of undeformed Middle Devonian strata, indicates that the pre-Mississippian deformation occurred in the Silurian or Early Devonian in the North Slope subterrane. Polydeformational structures in Proterozoic to lower Paleozoic rocks of the northeastern Brooks Range suggest that an older deformation also may have affected some of these rocks (Anderson, 1991).

Geologic mapping by Reiser and others (1971, 1980) showed that faults displacing the pre-Carboniferous rocks in the northeastern Brooks Range dip mostly south. Reed (1968) and Reiser and others (1980) considered these faults evidence that structures formed during middle Paleozoic deformation were north directed and similar in orientation to younger Brookian structures. Oldow and others (1987b), however, interpreted these south-dipping faults and minor north-vergent structures as related to Brookian deformation and concluded from structural data that south-directed pre-Carboniferous penetrative deformation occurred in the northeastern Brooks Range.

Late Brookian structures. Although the northeastern Brooks Range province contains a fold and thrust belt, abundant evidence indicates that Brookian structures in the northeastern province were formed by late Brookian deformation. (1) The province lies north of the northern limit of the early Brookian allochthons. (2) The axis of the Colville basin strikes eastward into the province. (3) Albian foredeep deposits within the province have been uplifted and largely eroded, indicating significant late Brookian deformation (Mull, 1982, 1985). (4) Deformed Upper Cretaceous and Paleogene clastic rocks are locally preserved in the northern margin of the northeastern Brooks Range proper. (5) Neogene and Quaternary deposits are deformed above a middle Tertiary unconformity on the coastal plain and continental shelf (Craig and others, 1985; Bruns and others, 1987; Kelley and Foland, 1987). (6) Isotopic-cooling ages and apatite fission-track ages indicate uplift of the northeastern Brooks Range at about 60 Ma and uplift of the coastal plain to the north during later Tertiary time (Dillon and others, 1987b; O'Sullivan, 1988; O'Sullivan and others, 1989). (7) The province continues to be seismically active (Grantz and others, 1983a).

Romanzof uplift. The structure of the southern, mountainous part of the northeastern Brooks Range structural province is characterized by a series of east-trending anticlinoria, about 5–20 km wide, which expose pre-Mississippian rocks in their cores (Plate 6). These anticlinoria mark south-dipping horses in a duplex that is bounded by a floor thrust deep within the pre-Mississippian sequence and a roof thrust in the Mississippian Kayak Shale (Namson and Wallace, 1986; Wallace and Hanks, 1990). Although the overlying younger Mississippian through Triassic rocks conform to the structure of these anticlinoria, they also display shorter wavelength chevron folds above a major detachment horizon in the Kayak Shale. These detachment folds are hundreds of meters wide and do not display a strong or consistent sense of vergence. At structurally higher levels, detachment horizons occur in the Kingak Shale and in the pebble-shale unit. On the basis of a balanced cross section, Namson and Wallace (1986) estimated that about 40–45 km (27%–29%) of shortening occurred across the western part of the northeastern Brooks Range structural province, from its boundary with the northern Brooks Range province north to the range front.

North of the range front of the northeastern Brooks Range, rocks of the Arctic Coastal Plain are also deformed. Upper Cretaceous to Tertiary foredeep deposits are exposed at the surface and extend to considerable depth (Bader and Bird, 1986; Bruns and others, 1987; Kelley and Foland, 1987). These rocks define large antiforms that are currently thought to be the most prospective structures for petroleum exploration in the ANWR. Seismic reflection profiles suggest that these antiforms were probably formed by complex north-vergent imbrication beneath north-dipping roof thrusts. At deeper structural levels, large, north-tapering wedges are present above south-dipping thrust faults within pre-Mississippian rocks, similar to the horses in the pre-Mississippian rocks of the northeastern Brooks Range.

Offshore from Camden Bay eastward to Canada, Cretaceous and Tertiary clastic rocks have been deformed into an arcuate belt of folds (Grantz and May, 1983; Craig and others, 1985; Moore and others, 1985b; Grantz and others, 1987). Within this belt, structural relief and dip decrease northward toward the deformation front, which at its northernmost extent parallels the modern continental slope, about 170 km north of the landward limit of the northeastern Brooks Range structural province. On the coastal plain and offshore to the north, a major Eocene unconformity separates more highly deformed Paleogene deposits from underlying, less-deformed deposits. This unconformity suggests that a major deformational event occurred in Eocene time (Bruns and others, 1987; Kelley and Foland, 1987). Deformation has continued to the present, as indicated by exposures of steeply dipping Pliocene beds and offshore Quaternary structures, as well as active seismicity (Grantz and others, 1983a, 1987; Leiggi, 1987).

Late Cretaceous and Tertiary deformation in the northeastern Brooks Range structural province was influenced significantly by the Barrow arch and associated Lower Cretaceous unconformity and northward thinning of pre–Lower Cretaceous strata (Kelley and Foland, 1987; Wallace and Hanks, 1990). Because uplift of the Barrow arch occurred in Early Cretaceous time and prior to late Brookian deformation (see "North Slope structural province" below), the depth to Lower Cretaceous and older rocks was probably less in the northeastern Brooks Range province than anywhere else in the Brooks Range, and it decreased progressively northward toward the crest of the arch. Furthermore, the thickness of pre–Lower Cretaceous strata decreased northward because of onlap onto the northern highland that was the source of the Ellesmerian sequence and because of erosion during Early Cretaceous time. Consequently, the late Brookian deformation front prograded northward onto the northern flank of the Colville basin and southern flank of the Barrow arch and involved previously deformed upper Proterozoic(?) and Paleozoic rocks near the leading edge of the mountain belt.

Range front of the northeastern Brooks Range. The range front of the western part of the northeastern Brooks Range structural province trends northeasterly, diverging sharply from the easterly trend of the adjacent part of the front in the northern Brooks Range structural province. This northeast-trending segment is distinguished as the Philip Smith Mountains front (Fig. 23). The range front returns to a generally easterly trend to the northeast, where it is offset by a local salient defined by a series of east-trending front ranges, including the Sadlerochit and Shublik mountains (Figs. 1 and 23).

The range front of the northeastern Brooks Range province is probably defined by thrust-related folds, as in the northern Brooks Range province. However, the range front of the northeastern Brooks Range is younger than that of the northern Brooks Range province, having formed in response to the Cenozoic deformation that resulted in the Romanzof uplift. The origin of the arcuate trend of both the northeastern range front and the structures within the northeastern Brooks Range is uncertain. If tectonic transport was to the north-northwest, as structures in the central part of the arc suggest, then the northeast-trending Philip Smith Mountains front would mark an oblique ramp in a subsurface thrust fault (Rattey, 1985; Wallace and Hanks, 1990). Alternatively, the northeast-trending front may mark a zone of distributed left-lateral displacement that deformed earlier folds ("Canning displacement zone" of Grantz and May, 1983).

North Slope structural province

The North Slope structural province (Fig. 23) is characterized by nearly flat lying strata of Mississippian and younger age and by the absence of structures ascribed to the Brookian orogeny. Prominent structural features of this province, known from seismic reflection profiles and well data, are (1) pre-Mississippian structures of poorly known character truncated by a regional sub-Mississippian unconformity; (2) local basins of Devonian and/or Mississippian age; and (3) extensional structures related to formation of the northern continental margin of Alaska, the Barrow arch, and the regional Lower Cretaceous unconformity.

Pre-Mississippian structures. Structures in pre-Mississippian rocks of the North Slope province are poorly known, although existing data indicate that these rocks are penetratively deformed, weakly metamorphosed, and have a general easterly strike. Most of the pre-Mississippian rocks sampled by drill core are argillites or phyllites; they display slaty cleavage, small-scale isoclinal folds, or small-displacement faults. Drill cores and dipmeter logs indicate steep dips in the pre-Mississippian rocks, seismic data show local dipping and folded reflectors, and gravity and magnetic anomalies suggest the presence of major faults or dipping lithologic contacts. Except for the shallowest parts of the Barrow arch, there is little contrast in degree of induration between pre-Mississippian and immediately overlying Mississippian rocks. This observation suggests that the pre-Mississippian rocks were never buried to depths much greater than their present 3–5 km. However, the widespread presence of deformed Ordovician and Silurian argillitic rocks suggests that there is a great thickness of rocks of these ages, probably as a result of tectonic thickening.

Mississippian basins and faulting. By Mississippian time, following pre-Mississippian deformation, subsidence and deposition took place in several local basins (Meade, Umiat, Ikpikpuk, and Endicott) of the North Slope subterrane (Fig. 9). Seismic records indicate that these basins developed as sags or partly fault bounded basins (half grabens) and are filled with Mississippian and perhaps older strata. Well and seismic data show that bounding faults in the Endicott basin truncate Mississippian strata; in the Umiat basin, bounding faults truncate strata possibly as young as Pennsylvanian. Regionally, the bounding faults have north to northwest strikes and display as much as 700 m of throw. The origin of the basins has been attributed to extension and subsidence associated with formation of the passive continental margin of the Arctic Alaska terrane to the south during middle Paleozoic time (Kirschner and Rycerski, 1988; Grantz and others, 1991). Alternatively, Hubbard and others (1987) suggested that the basins represent local foredeeps and transtensional pull-apart basins associated with regional contraction during pre-Mississippian orogenesis.

Mesozoic rifting. A series of Jurassic and Early Cretaceous normal faults lie mostly beneath the continental shelf and strike subparallel to the northern continental margin of Alaska (Grantz and May, 1983; Craig and others, 1985; Hubbard and others,

1987). Seismic-sequence analysis indicates that faulting occurred over a span of about 60 m.y. (Bathonian to Aptian) (Hubbard and others, 1987; Grantz and others, 1990a, and this volume). An early episode of failed rifting, characterized by faults downthrown to the south, began in Middle Jurassic time and led to the development of sediment-filled grabens (for example, the Dinkum graben) (Fig. 23) under the modern Beaufort continental shelf. A later episode of successful rifting, characterized by faults downthrown to the north, resulted in continental breakup and opening of the oceanic Canada basin in Early Cretaceous (Hauterivian) time. Faulting and uplift associated with the continental breakup led to the formation of the north flank of the Barrow arch and truncation of its upper surface by the regional Lower Cretaceous unconformity, which developed in the Early Cretaceous (Valanginian to Hauterivian).

Barrow arch. The Barrow arch (also referred to as the "Barrow inflection" by Ehm and Tailleur [1985] and the "Beaufort sill" by Mull [1985] and Mull and others [1987c]) is a broad, west-northwest–trending structural high that underlies the coastal area of northern Alaska and separates the Colville basin to the south from the Canada basin to the north (Fig. 23). Flanks of the arch dip generally less than 2°, and its axis plunges eastward at about a half degree. At its crest near Barrow, pre-Mississippian rocks are at depths of only about 700 m, and structural relief across both flanks is about 10 km. Although recognized as a local structural high by Payne (1955), the full extent of the Barrow arch was first discussed by Rickwood (1970), who defined its crest by the inflection of dip in Ellesmerian strata. Recent mapping of the crest of the Barrow arch from seismic data has focused on the inflection of dip of either the Lower Cretaceous pebble-shale unit or the (erosional) structural top of the pre-Mississippian rocks that form the core of the Barrow arch. The southern flank of this structural high, with the associated Lower Cretaceous unconformity, forms the primary trap for the Prudhoe Bay oil field.

As pointed out by many workers, the Barrow arch did not form as a result of a single deformational event; instead, it is a composite structural high. Its southern flank was established by late Paleozoic time as the gently southward sloping continental margin of the Arctic Alaska terrane, and its dip was increased in the Early Cretaceous by tectonic and sedimentary loading related to emplacement of the early Brookian thrust sheets in the Brooks Range. The northern flank was initially developed as a discontinuous feature associated with failed rifting in the Middle Jurassic. It did not become a continuous structural feature, displaying a regional reversal of dip in Ellesmerian and older strata, until continental breakup and formation of the oceanic Canada basin in the Early Cretaceous (Hauterivian). Since Early Cretaceous time, the northern flank of the Barrow arch has been modified further by development of a thick prism of Cretaceous and Tertiary passive-margin deposits and continued southward instepping of normal faults in the Beaufort continental margin. The southern flank has been modified also by tectonic and sedimentary loading associated with late Brookian tectonism to the south. In the coastal plain adjacent to the northeastern Brooks Range, these modifying factors have converged, resulting in subsidence of the Barrow arch to depths exceeding 4 km and overshadowing of the structural high by late Brookian anticlinoria.

PALEOGEOGRAPHY AND TECTONIC HISTORY OF NORTHERN ALASKA

In the previous sections we have described the physical stratigraphy and structure of northern Alaska and related them to the major tectonic units in the region. In this section we discuss the depositional and tectonic implications of these data and use the tectonic units to construct speculative paleogeographic and tectonic models for northern Alaska. These models are discussed in chronological order but are considered in relation to the tectonic environment we have inferred for various intervals of time. Accordingly, the major subjects to be discussed are as follows: (1) depositional framework of a pre-Devonian continental margin, (2) early to middle Paleozoic orogenesis, (3) continental breakup along the southern margin of the Arctic Alaska terrane in Devonian time, (4) depositional framework of the latest Devonian to Jurassic passive continental margin, (5) Jurassic to Early Cretaceous (early Brookian) orogenesis, (6) origin of the present northern Alaska continental margin, (7) evolution of the Colville and Koyukuk basins, and (8) post-Neocomian (late Brookian) tectonism. We then discuss the relation of northern Alaska to the North America continent and consider the various models for its origin as part of the Arctic realm.

Pre-Devonian continental margin

Paleogeographic reconstruction of the pre-Devonian stratigraphy of the Arctic Alaska terrane is complicated not only by Brookian orogenesis during Mesozoic and Cenozoic time but also by one or more episodes of contractional and/or extensional deformation in pre-Mississippian time. The deformational style, vergence, and tectonic significance of the older orogenic episodes are poorly known, making pre-Devonian paleogeographic reconstructions speculative. For the purpose of discussion here, pre-Devonian rocks in the North Slope, Hammond, and Coldfoot subterranes are classified as carbonate-platform, continental-slope, and oceanic deposits. The carbonate-platform deposits in the North Slope and Hammond subterranes are thick; they span part of Proterozoic and much of early Paleozoic time. Continental-slope or distal continental-margin deposits in the North Slope subterrane are mostly fine-grained quartzose rocks, widespread in both the subsurface and surface; in the northeastern Brooks Range, these include quartzose turbidites (Neruokpuk Quartzite) that may be analogous to the Windemere Supergroup of the Canadian Cordillera. Other, typically fine-grained rocks that may be continental-margin or continental-slope deposits crop out in the Hammond subterrane and may make up much of the Coldfoot subterrane. Where dated, these fine-grained rocks are typically Ordovician and Silurian, but some of them were probably deposited during Proterozoic and Cambrian time. Rocks indicative of oceanic deposition include Cambrian, Ordovician, and Silurian radiolarian chert, argillite, graywacke turbidites, mafic volcanic rocks, and island-arc volcanic rocks. These rocks occur both as melange and coherent masses in the North Slope subterrane. They are abundant in the northeastern and south-central Brooks Range and Lisburne Peninsula and, on the basis of gravity and magnetic data (Grantz and others, 1991), are inferred to be in the subsurface of the North Slope. Volumetrically, however, oceanic deposits may compose only a small part of the pre-Devonian rocks of the Arctic Alaska terrane.

Norris (1985), Dillon and others (1987a), and Lane (1991) suggested that pre-Devonian rocks of the North Slope, Hammond, and Coldfoot subterranes formed a thick carbonate-shelf to deep-marine-slope succession marginal to North America in Late Proterozoic and early Paleozoic time. In such a reconstruction, the oldest of the pre-Devonian rocks represent lateral equivalents of the Tindir Group and related Proterozoic rocks, now 450 km to the south in the Canadian Cordillera and Kandik area of east-central Alaska, whereas the Cambrian, Ordovician, and Silurian oceanic rocks are interpreted as lateral equivalents of coeval fine-grained miogeoclinal rocks of the Selwyn basin in the central Yukon Territory. This reconstruction is supported by (1) the quartzose composition of most of the pre-Devonian siliciclastic rock of the Arctic Alaska terrane and their lateral equivalents, (2) the general stratigraphic similarities between the pre-Devonian rocks of the Arctic Alaska terrane and, as originally defined by Stewart (1976), the North American continental-margin succession of the Canadian Cordillera, (3) the North American affinity of most fauna in the northeastern Brooks Range, and (4) the general position of northern Alaska on depositional strike with the North American miogeocline. The continental margin represented by pre-Devonian rocks of the Arctic Alaska terrane may be the northward continuation of the Late Proterozoic and early Paleozoic passive margin of the Canadian Cordillera. Contemporaneous carbonate platforms may be represented by the Proterozoic to Devonian carbonate succession in the Sadlerochit and Shublik Mountains of the northeastern Brooks Range and the Baird Group in the southern Brooks Range, although their original positions relative to each other and the continental-margin deposits are unknown.

A passive-margin model alone does not explain the widespread evidence for pre-Mississippian deformation in the eastern part of the Arctic Alaska terrane. For this reason, and because of the presence of pre-Devonian oceanic and volcanic-arc deposits in the south-central and eastern Brooks Range, Moore and others (1985a), Moore (1986), and Grantz and others (1991) suggested that originally disparate tectonic elements (displaced terranes) may have been accreted to the pre-Devonian continental margin of North America along one or more sutures in the Brooks Range. They suggest that lower Paleozoic volcanic-arc rocks and lithic flysch in the North Slope subterrane record closure of an ocean basin outboard of the North American continent. Possible evidence of a closure event may be represented by faunas of different affinity in the Arctic Alaska terrane. In the North Slope subterrane in the northeastern Brooks Range, Cambrian trilobites are of North American affinity, whereas in the southern part of the North Slope subterrane at the Doonerak fenster and in the Hammond subterrane, Cambrian trilobites and Ordovician conodonts are of Siberian affinity (Dutro and others, 1984; Blodgett and others, 1986; Dillon and others, 1987a; Harris and others, 1988; A. R. Palmer, 1988, oral commun.). These paleontologic data suggest that the Siberian continent (or continental fragments related to it) was involved in the closure and that most or all of the lower Paleozoic rocks of the Hammond and Coldfoot subterranes are of peri-Siberian origin, whereas those of the North Slope subterrane north of the crest of the Brooks Range are of North American affinity.

Early to middle Paleozoic orogenesis

Brosgé and others (1962) were the first to suggest that the regional sub-Mississippian angular unconformity in the northeastern Brooks Range may be evidence for early to middle Paleozoic contractional or extensional deformation. They also noted the thick, widespread, coarse-grained Upper Devonian and Lower Mississippian(?) clastic rocks of the Kanayut Conglomerate (Endicott Group, Endicott Mountains subterrane) and suggested that these were derived by erosion from a mid-Paleozoic orogenic zone. Subsequent work has shown that the sub-Mississippian unconformity extends throughout the subsurface of the North Slope to the Lisburne Peninsula and also is present in the Mt. Doonerak fenster (North Slope subterrane) and in the Schwatka Mountains (Hammond subterrane) in the southern Brooks Range; the extent of this unconformity suggests that the hypothesized middle Paleozoic orogenic episode affected much of northern Alaska. Lerand (1973) inferred that this orogenic episode occurred in the Devonian, and he linked it to the Ellesmerian fold belt, which he traced from northern Greenland through the Canadian Arctic and northern Alaska to at least as far west as Wrangell Island.

In the northeastern Brooks Range, deformed rocks beneath the sub-Mississippian unconformity include highly strained rocks (Oldow and others, 1987b), the direction of structural transport of which is unknown or controversial. In the Sadlerochit and Shublik mountains, pre-Mississippian deformation is indicated by large-scale tilting. In the Hammond subterrane and in the North Slope subterrane in the Mt. Doonerak fenster, a regional pre-Mississippian orogenic episode has not been documented, even though Mississippian rocks and the sub-Mississippian unconformity are present; the significance of the unconformity is controversial. Pre-Mississippian penetrative deformation, however, has been suggested for some rocks of the Hammond subterrane in the southern Brooks Range (Dillon, 1989; Till, 1989), and uplift and tilting are likely for some of the others in the Hammond subterrane. In the Endicott Mountains and De Long Mountains subterranes, the sub-Mississippian unconformity is absent, and the Devonian to Mississippian stratigraphic section is conformable. Mississippian to Triassic strata of the North Slope subterrane onlapped northward across the sub-Mississippian unconformity onto older rocks presently in the subsurface of the North Slope. This relation indicates that a middle Paleozoic highland existed north of the Barrow arch. Southward sediment transport during deposition of the Upper Devonian to Lower Permian(?) Endicott Group also indicates a northern highland (Moore and Nilsen, 1984; Bird, 1988a; Mayfield and others, 1988). Taken together, this evidence suggests that the area of middle and late Paleozoic erosion extended south at least as far as the southern Brooks Range, but maximum uplift was located north of the present-day Barrow arch.

A minimum age for early to middle Paleozoic orogenesis is indicated by the Early Mississippian age of the Kekiktuk Conglomerate, which rests on the sub-Mississippian unconformity. Rocks as young as Middle Devonian are truncated at a low angle by the unconformity in the northeastern Brooks Range (Reiser ad others, 1971, 1980; Anderson and Wallace, 1990). A Devonian age for the orogenesis is also suggested by emplacement of large granitic plutons and batholiths yielding Devonian U-Pb crystallization ages in the North Slope, Hammond, and Coldfoot subterranes (Dillon and others, 1987b). In the northeastern Brooks Range, these plutons are truncated by the pre-Mississippian unconformity, indicating that uplift occurred between the time of crystallization in Devonian time and their erosional truncation in Early Mississippian time.

Brosgé and others (1962) reported several unconformities in pre-Mississippian strata of the northeastern Brooks Range and concluded that pre-Mississippian orogenesis in northern Alaska was diachronous or involved more than one event. In the southern Demarcation Point quadrangle, Reiser and others (1980) mapped Middle Devonian calcareous sandstone in angular unconformity with a highly deformed unit of Cambrian and Ordovician chert, argillite, mafic volcanic rocks, and lithic graywacke described by Dutro (1981) and Moore and Churkin (1984). The absence in the Middle Devonian rocks of complex structures present in the underlying, older rocks indicates that significant deformation took place in pre–Middle Devonian time (Anderson and Wallace, 1990). Lower Devonian (Emsian) limestone rests with angular unconformity on Upper Ordovician and older carbonate strata in the Sadlerochit and Shublik mountains (Blodgett and others, 1988). Dillon and others (1987a), who assumed a Devonian age for the Skajit Limestone, suggested that Devonian carbonate rests unconformably on older rocks in the central Brooks Range. On the basis of these observations and the apparent absence of Lower Devonian strata throughout most of the Hammond and North Slope subterranes, Dillon (1989) argued that a pre–Middle Devonian unconformity exists throughout the central and eastern Brooks Range and concluded that orogenesis occurred in Silurian or Early Devonian time.

The tectonic causes for early to middle Paleozoic orogenesis in northern Alaska are unclear. As discussed for the pre-Devonian continental margin, the presence of a deformed pre-Devonian oceanic assemblage in the northeastern Brooks Range, pre-Devonian volcanic-arc rocks in the Mt. Doonerak fenster, and exotic faunal affinities led Moore and others (1985a), Moore (1986), and Grantz and others (1991) to suggest convergent deformation between Ordovician and Early Devonian time. Although pre-Carboniferous ophiolitic assemblages have not been recognized in northern Alaska, a west-trending band of low-amplitude magnetic anomalies thought to originate in pre-Mississippian rocks in the southern North Slope may be interpreted as a suture marked by serpentinite or, alternatively, as a belt of intrusive or mafic extrusive rocks of oceanic or arc affinity (Grantz and others, 1991). Dillon and others (1980) suggested that Devonian granitic rocks of northern Alaska have an arc affinity and, on the basis of this interpretation, Hubbard and others (1987) concluded that convergent deformation continued into Early Devonian time. This conclusion may be supported by the predominance of radiolarian chert detritus in Upper Devonian clastic rocks of the Endicott Group in the Endicott Mountains subterrane (Moore and Nilsen, 1984), which implies uplift and exposure of pelagic deposits.

Continental breakup along the southern margin of the Arctic Alaska terrane

Following convergent deformation and tectonic juxtaposition of lower Paleozoic rocks of the Arctic Alaska terrane by Middle Devonian time, deposition of continental-shelf sediments of the Arctic Alaska terrane resumed. Depositional successions of Late Devonian to Jurassic age are characterized by overall deepening conditions that gradually evolved from nonmarine deposition in latest Devonian and earliest Mississippian time to carbonate-platform and platform-margin deposition in Carboniferous time and finally to fine-grained clastic-rock, siliceous-shale, pelagic-limestone, and chert deposition from Permian to Jurassic time. This succession suggests that the Arctic Alaska terrane was subjected to regional subsidence, particularly along its southern margin (Endicott Mountains and De Long Mountains subterranes) for more than 200 m.y. The presence of Devonian to Jurassic oceanic rocks of the Angayucham terrane resting on the southern margin of the Arctic Alaska terrane suggests that the Arctic Alaska terrane was bordered to the south by an oceanic region from which the Angayucham rocks were derived. These relations indicate that an ocean basin was opened, presumably by rifting, along the southern margin of the terrane in middle to late Paleozoic time and that all or part of the Upper Devonian to Jurassic stratigraphy of the Arctic Alaska terrane composed a south-facing passive-margin sequence.

The detailed history of rifting, continental breakup, and onset of passive-margin deposition is uncertain. Because a southern highland province (see below) is inferred from Early Proterozoic arkosic detritus in the Mississippian and Pennsylvanian(?) Nuka Formation (De Long Mountains subterrane), Mayfield and others (1988) proposed that continental breakup of the southern margin of the Arctic Alaska terrane and opening of the Angayucham ocean basin began in Early Pennsylvanian time. A rifting event of this age may be supported by the many undated mafic sills that intrude Carboniferous chert of the Ipnavik River allochthon and by rare mafic volcanic rocks within the Lisburne Group. Evidence is also provided by extensional structures of Carboniferous age (Moore and others, 1986) and the presence of Carboniferous evaporites and mineral deposits (Metz and others, 1982).

Alternatively, Einaudi and Hitzman (1986), Hitzman and others (1986), Schmidt (1987), and Dillon (1989) interpreted rhyolite-dominated bimodal volcanic rocks, associated massive sulfide deposits, and abrupt facies changes with related unconformities of Devonian age in the Hammond and Coldfoot subterranes as evidence of rift-related high-angle faulting and extensional tectonism of the southern Arctic Alaska terrane during Devonian time. Dillon and others (1987b) interpreted the more siliceous of the bimodal volcanic rocks as extrusive equivalents of Devonian plutonic rocks in the Hammond and Coldfoot subterranes and argued that isotopic data indicating a crustal source for the plutons provide evidence for their origin by partial melting of continental crust in an extensional setting. The complex facies relations of Middle and Late Devonian sedimentary rocks in the Hammond and North Slope subterranes, the multiple erosional episodes in the North Slope subterrane during Devonian and Early Mississippian time, and the regional Late Devonian to Jurassic subsidence of the Arctic Alaska terrane indicate that continental breakup took place in Middle to Late Devonian time (Grantz and others, 1991). Thick, but local, accumulations of Middle Devonian terrigenous clastic rocks in the northeastern Brooks Range (Anderson and Wallace, 1990) and in the North Slope subsurface (Collins, 1958) were tilted prior to the Early Mississippian, suggesting that extensional deformation related to rifting took place in Middle Devonian time. However, the consistent southwest-directed paleocurrent indicators and the uniform clast composition of the Upper Devonian to Lower Mississippian(?) chert-rich, fluvial-deltaic Kanayut clastic wedge of the Endicott Mountains subterrane indicate that latest Devonian sediments were deposited as a south-facing, constructional continental-margin succession rather than as a rift-basin succession. Thus, the Kanayut clastic wedge was probably shed from an uplifted area of older orogenic deposits along the northern shoulder of the rifted southern margin of the Arctic Alaska terrane and was deposited on the outboard, southern margin of the Arctic Alaska terrane after continental breakup in earlier Devonian time. Together, these relations suggest that Early Devonian or older convergent tectonism culminated in plutonism and was succeeded in Middle and Late Devonian time by a rifting event that resulted in continental breakup, formation of the Angayucham ocean basin, and establishment of a south-facing Atlantic-type continental margin by latest Devonian time. If continental drift began in Middle or Late Devonian time, as suggested here, then the inferred southern highland of the Arctic Alaska terrane of Mayfield and others (1988) may have been partly or wholly composed of a Proterozoic granitic terrane that had been accreted to the Arctic Alaska terrane in pre-Mississippian time and then subsided as a large continental block along the southern margin of the terrane during rifting and opening of the Angayucham ocean basin.

Latest Devonian to Jurassic passive continental margin

Two contrasting paleogeographic reconstructions of the Arctic Alaska terrane have been suggested for latest Devonian (Famennian) to Jurassic time. Mayfield and others (1988) hypothesized that the Brooks Range orogen consists of at least seven regional internally imbricated thrust sheets or allochthons characterized by distinct upper Paleozoic and Mesozoic stratigraphic sequences. They proposed a simple south-to-north thrust emplacement sequence for these allochthons. This model assumes that each allochthon restores to a position immediately south of the allochthon it overlies structurally, and that each allochthon was thrust into place prior to thrusting of the allochthon it overlies structurally. Thus, the structurally higher allochthons have been displaced farther than the structurally lower allochthons because the higher allochthons have been carried northward in piggyback fashion on top of the lower allochthons. Because of the presence of a lithologically equivalent upper Paleozoic to Mesozoic stratigraphic section in both the Hammond and North Slope subterranes, Mayfield and others (1988) believed the Hammond subterrane to be a deformed, southward continuation of the North Slope subterrane and that both subterranes are autochthonous or parautochthonous relative to the Endicott Mountains and De Long Mountains subterranes. This palinspastic reconstruction therefore restores the latter subterranes to positions south of the Hammond subterrane. The structural assumptions used in this reconstruction result in a relatively complex paleogeography of alternating basins and stable platforms, particularly during Mississippian time.

Churkin and others (1979), however, assumed a paleogeographic model featuring an uncomplicated south-facing passive continental margin throughout late Paleozoic and early Mesozoic time. They assumed that condensed, siliceous Mississippian to Triassic basinal sequences (their Kagvik sequence) were deposited on oceanic crust located south of the coeval thicker continental-shelf deposits of the North Slope and Endicott Mountains subterranes and the Kelly River and Nuka Ridge allochthons of the De Long Mountains subterrane. In contrast to the model of Mayfield and others (1988), this model requires a more complicated history of structural shortening during the Brookian orogeny, involving a currently undocumented interval of southward-vergent thrusting separating two intervals of northward-vergent thrusting. Mayfield (1980) and Mull (1980) pointed out several lines of evidence indicating that the deep-water siliceous successions were deposited partly in shelf and platform environments of the Endicott Mountains and De Long Mountains subterranes, and suggested a subsiding platform-margin and slope, rather than oceanic, site for deposition of the siliceous succession. The model of Mayfield and others (1988) is therefore generally preferred by most workers, although the Pennsylvanian or younger time of breakup called for in their model is not accepted by all workers (Hitzman and others, 1986; Grantz and others, 1991).

Figure 24 shows a series of block diagrams illustrating paleogeographic reconstructions for latest Devonian to Neocomian time. This model is modified from Mayfield and others (1988), who proposed a two-sided basin in northern Alaska in pre-Pennsylvanian time followed by a simple south-facing passive margin from Pennsylvanian to Jurassic time. The alternative model of Churkin and others (1979) requires restoration of the Kelly River and Nuka Ridge allochthons of the De Long Mountains subterrane to positions north of the Endicott Mountains subterrane but south of the combined Hammond and North Slope subterranes. Note that the position of the southern basin margin shown by Mayfield and others (1988) hinges on the palinspastic position of the inferred source area for clastic rocks of the Nuka Formation and has not been observed in outcrop.

The regional sub-Mississippian unconformity shows that during latest Devonian and earliest Mississippian time (Fig. 24A), much of northern Alaska (North Slope subterrane and at least part of the Hammond subterrane) was uplifted and exposed as extensive northern highlands. Quartz- and chert-rich detritus from these highlands was shed southward and westward along drainage courses through at least two major alluvial plains onto a broad delta plain, which together constituted the sites of deposition for fluvial strata of the Kanayut Conglomerate (Moore and Nilsen, 1984). These fluvial deposits (Kanayut Conglomerate) prograded southward across related shallow-marine sediments and prodelta shale (Noatak Sandstone and Hunt Fork Shale) in Famennian (late Late Devonian) time, resulting in construction of the thick, south-facing fluvial-deltaic clastic wedge (Brosgé and Tailleur, 1971; Nilsen, 1981; Moore and Nilsen, 1984) now contained in the Endicott Mountains subterrane. Distal, submarine parts of the fluvial-deltaic wedge may be represented by compositionally mature sandstone turbidites of the Slate Creek subterrane (Murphy and Patton, 1988).

By Early Mississippian time, erosion reduced the northern highlands source region (Hammond and North Slope subterranes) to a broad, southward-sloping platform. Marine transgression across the Kanayut clastic wedge and northward onto the erosional surface resulted in the unconformable infill of remaining broad drainage basins by thin sequences of fluvial strata of the Kekiktuk Conglomerate, followed by deposition of marine shale and fine-grained sandstone of the Kayak Shale. Grabenlike basins on seismic-reflection records and thick accumulations of the Kekiktuk Conglomerate (as much as 1500 m) in the subsurface of the North Slope suggest that local extensional basins modified the erosional surface and were infilled with locally derived nonmarine strata (Oldow and others, 1987d; Woidneck and others, 1987; Grantz and others, 1991) prior to the marine transgression. By the end of Visean (Late Mississippian) time, the paleostrandline had migrated northward across the eroded highlands to the vicinity of Prudhoe Bay (Melvin, 1987), resulting in deposition of the marine Kayak Shale over a large part of the Arctic Alaska terrane.

In contrast to the siliciclastic-dominated subterranes that restore to more northerly positions, the latest Devonian and earliest Mississippian parts of the more southerly Kelly River, Ipnavik River, and Nuka Ridge allochthons of the De Long Mountains subterrane consist largely of carbonate and siliceous strata with only local accumulations of siliciclastic strata. These carbonate and siliceous strata rest conformably on Devonian carbonate rocks that may represent a long-lived carbonate platform not influenced by coeval clastic sedimentation of the Kanayut Conglomerate and related formations of the Endicott Group. The accumulations of latest Mississippian and Pennsylvanian(?) arkosic sandstone in the Nuka Ridge allochthon of the De Long Mountains subterrane indicate a nearby granitic source for which there is no evidence in other parts of the Arctic Alaska terrane. Mayfield and others (1988) suggested that rocks of the De Long Mountains subterrane were deposited on a carbonate platform and slope that was marginal to a southern highland in Devonian and Mississippian time, because of the subterrane's structurally high position and inferred restoration to more southern positions. To distinguish the hypothetical southern highland from the demonstrated northern source region, Tailleur (1973a) and Mayfield and others (1988) termed the southern paleohighland "Nukaland" for the distinctive arkosic strata in the Nuka Formation, and the northern paleohighland "Barrovia" for the town of Barrow located in the northernmost part of the North Slope subterrane. The basinal area between the two highlands was named the Arctic Alaska basin and was interpreted as an epicontinental sea by Mayfield and others (1988) because of the presence of older carbonate-platform deposits beneath Upper Devonian rocks on both flanks of the basin.

Deposition of carbonate-platform rocks of the Lisburne Group began along both margins of the epicontinental basin in the Early Mississippian (Osagean) and migrated northward across the older clastic-wedge deposits (Endicott Group) of the northern margin following cessation of clastic deposition during the early Late Mississippian (Meramecian) (Fig. 24B). By middle Pennsylvanian time, carbonate-platform rocks of the Lisburne Group had been deposited as a diachronous sheet across most of the Endicott Mountains and North Slope subterranes and the intervening Hammond subterrane. Siliceous sediments deposited within the epicontinental basin in Mississippian and Early Pennsylvanian time consisted of radiolarian chert and siliceous shale (Akmalik Chert of the De Long Mountains subterrane and Kuna Formation of the Endicott Mountains subterrane). These units were deposited in basinal areas distant from the carbonate-platform deposits, but they migrated locally onto the southern and western margins of the carbonate platform (Wachsmuth and Alapah limestones, Endicott Mountains allochthon) as it gradually subsided. The Kelly River allochthon (De Long Mountains subterrane), the only allochthon of the De Long Mountains subterrane to contain a thick succession of Lisburne Group carbonate-platform rocks, is restricted to the western part of the Brooks Range. This restricted extent may indicate that the Kelly River allochthon was deposited on a local structural high, per-

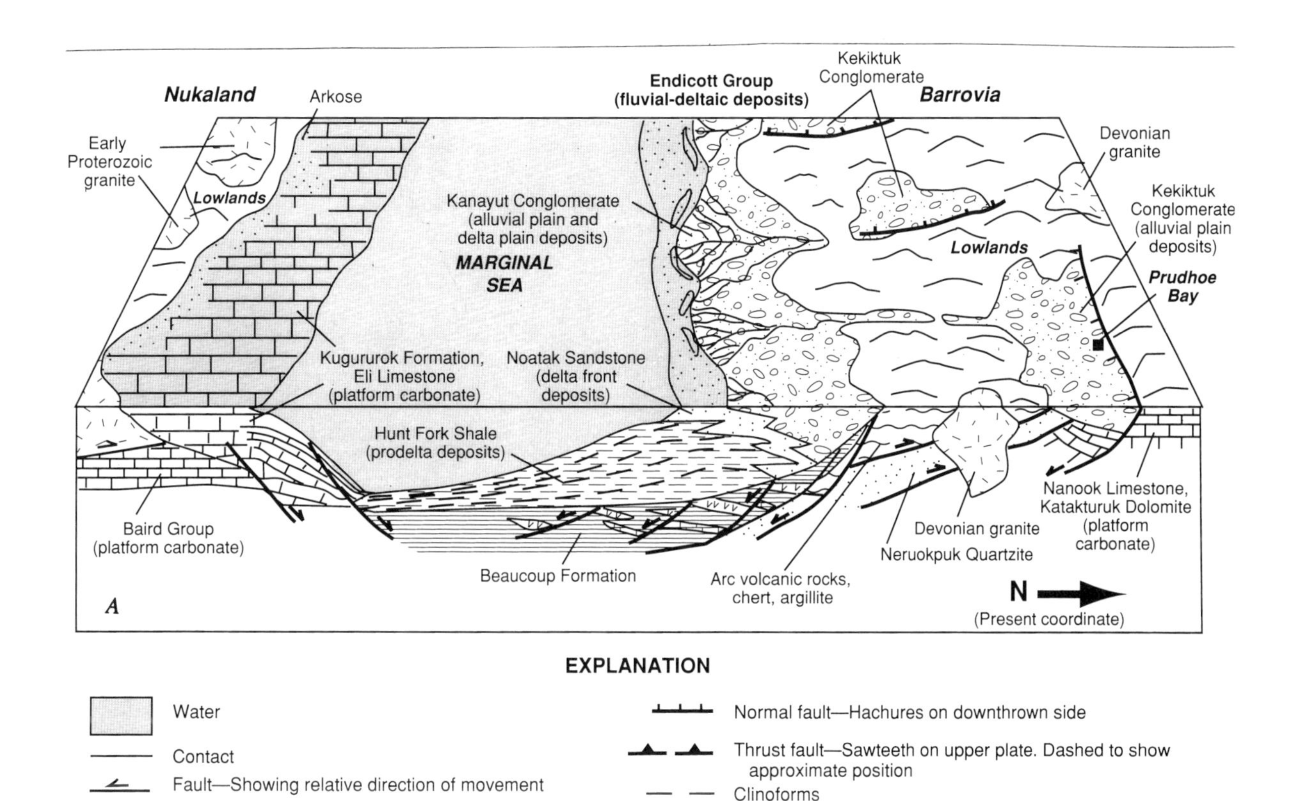

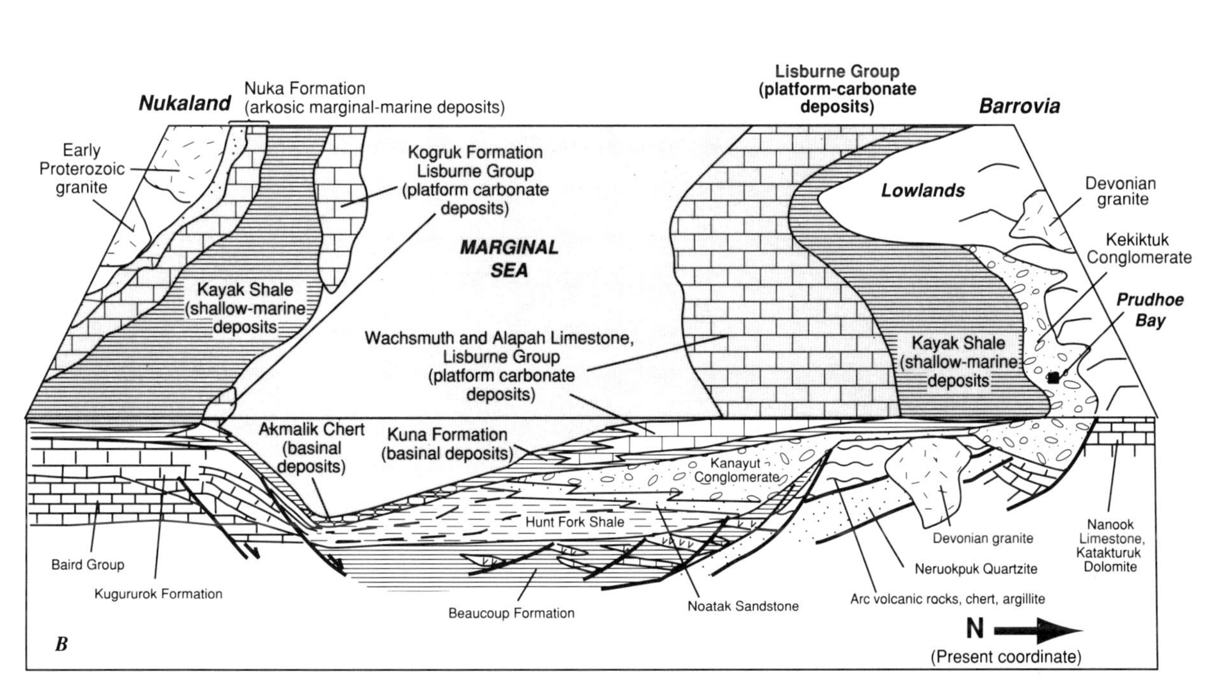

Figure 24 (on this and following two pages). Paleogeographic diagrams depicting depositional history of Arctic Alaska terrane for latest Devonian to Early Cretaceous. A, Latest Devonian (Fammenian) and earliest Mississippian. B, Early Late Mississippian (Meramecian). C, Late Permian. D, Early Triassic. E, Late Triassic. F, Early Cretaceous (Valanginian). For explanation of lithologic symbols, see Figure 21.

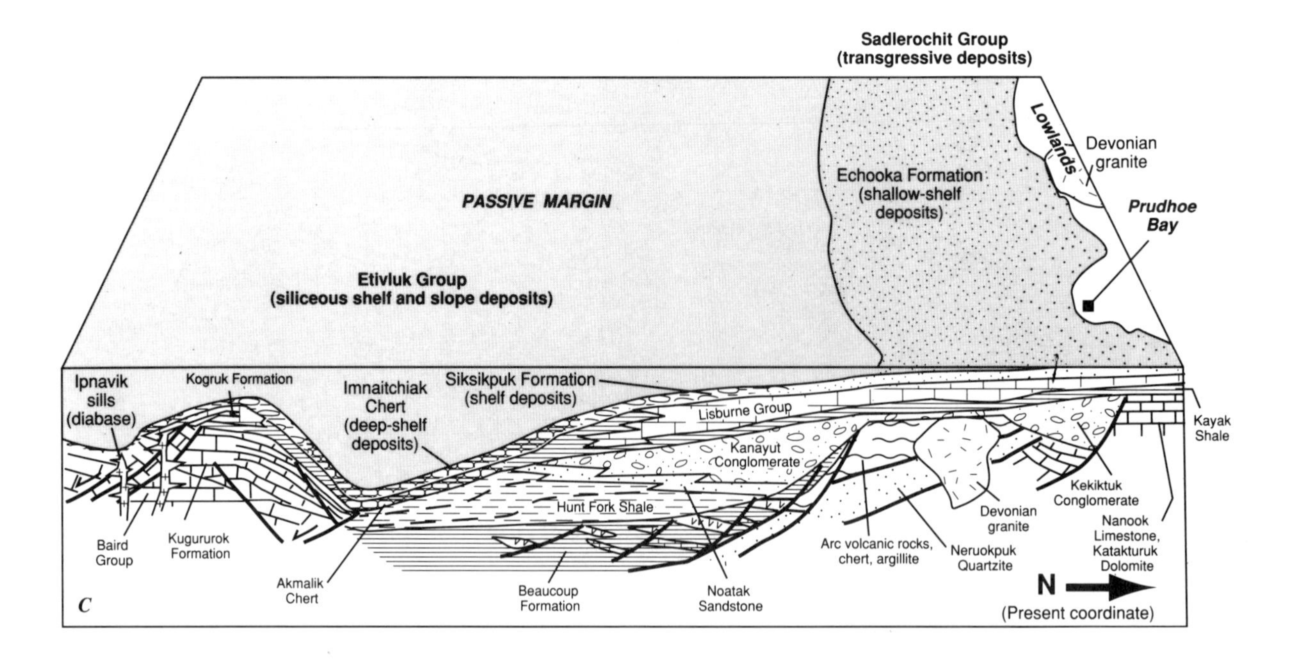

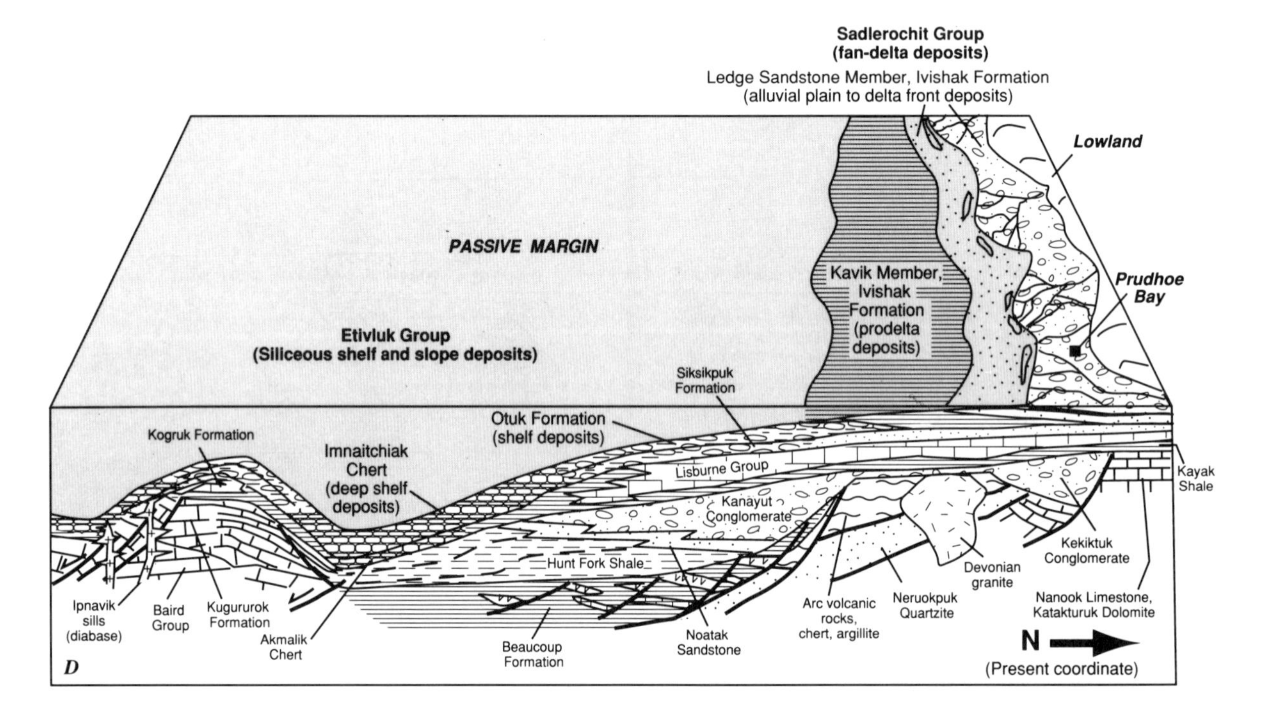

haps a fault block, within the epicontinental basin separating Barrovia and Nukaland.

By Permian time (Fig. 24C), deposition of carbonate-platform rocks in the North Slope subterrane gave way to deposition of transgressive siliciclastic rocks across a regional pre-Permian disconformity. These fine-grained siliciclastic rocks appear to fine and grade southward and westward into a shale-rich sequence in the Endicott Mountains subterrane (Siksikpuk Formation of Mull and others, 1987b), and into chert-rich sequences in most of the De Long Mountains subterrane (part of the Imnaitchiak Chert). The chert-rich sequences indicate that deposition in southern areas took place in basins remote from the influence of siliciclastic sedimentation. A regional increase with time in the ratio of radiolarians to sponge spicules in the siliceous rocks indicates gradual subsidence of the southern part of the Arctic Alaska terrane through the late Paleozoic (Murchey and others, 1988).

Uplift in the Early Triassic (Fig. 24D) in the region north of the present Beaufort Sea coastline is inferred to have caused erosion and deposition of coarse-grained marine and nonmarine

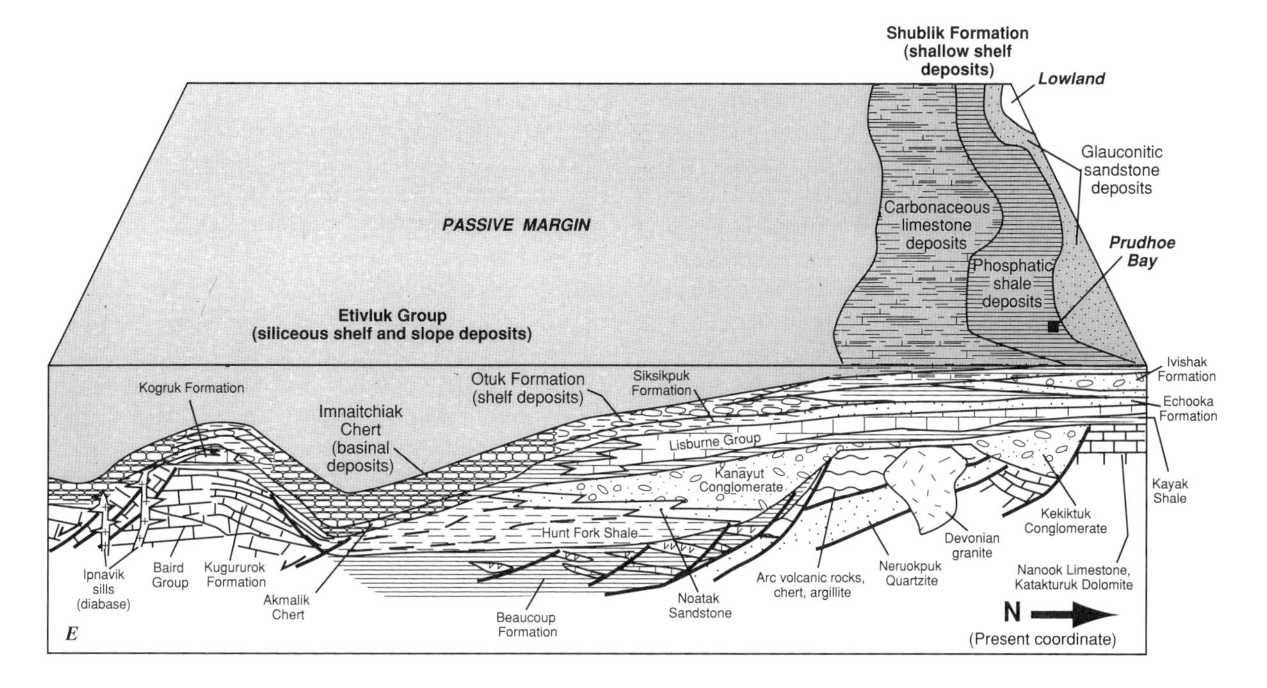

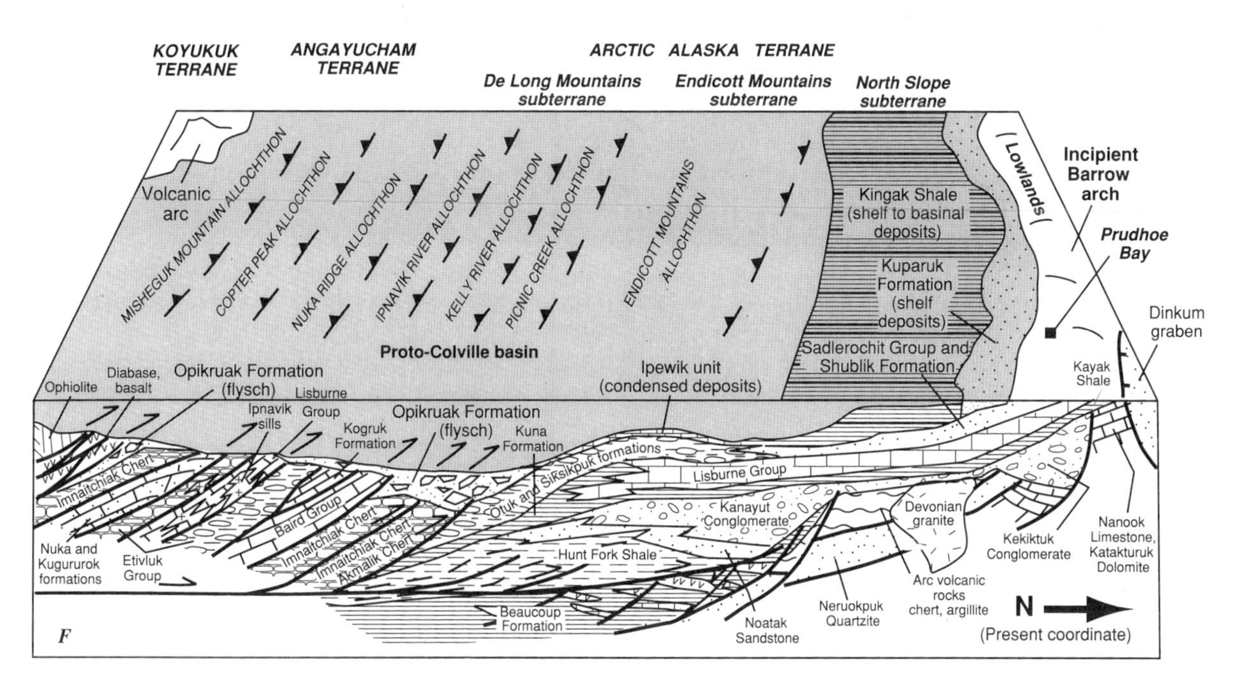

clastic detritus (Ledge Sandstone Member, Ivishak Formation, Sadlerochit Group). These deposits, inferred to represent a fan-delta (Melvin and Knight, 1984) or coastal-plain complex (Lawton and others, 1987) in the vicinity of the Prudhoe Bay oil field, were derived from a compositionally mature source area to the north and prograded southward across related prodelta deposits (Kavik Member of the Ivishak Formation) in the northern part of the North Slope subterrane. The thinness of the Sadlerochit Group (less than 700 m) indicates that the clastic wedge was constructed entirely across a shelf. Little is known about the cause of the hypothesized uplift in the northern highlands in Early Triassic time, but the alluvial-fan depositional environment favored by most workers for the northern and most proximal part of the Sadlerochit Group may suggest block faulting (I. L. Tailleur, 1988, oral commun.) that preceded Jurassic rifting of the Canada basin. Throughout the period of inferred tectonism, sedimentation in more southerly subterranes of the Arctic Alaska terrane consisted of siliceous shale and chert, indicating that Early Triassic tectonism was restricted to the northern margin of the terrane.

Subsequent marine transgression later in the Early Triassic through Late Triassic covered the coarse-grained marine and nonmarine detritus of the Sadlerochit Group with shallow-marine strata, including organic-rich shale, siltstone, and limestone of the Shublik Formation. General southward thinning and fining of the Shublik (Fig. 12) to siliceous shale, chert, and pelagic limestone of part of the Otuk Formation of the Etivluk Group in the Endicott Mountains and De Long Mountains subterranes (Fig. 24E) indicate that the Shublik Formation was deposited on a broad, southward-sloping, low-gradient shelf (Fraser and Clark, 1976). Furthermore, the phosphatic, glauconitic, and organic-rich character of the Shublik, as well as the abundance of the pelecypods *Monotis* and *Halobia* thought to represent mass kills, suggests that the shelf was subjected to a rising level, or episodic upwelling, of oxygen-depleted bottom water (Parrish, 1987). Ultimately, the Early Triassic to Late Triassic transgression may have been caused by cessation of tectonism and degradation of the source region (Melvin and Knight, 1984) or by a global or regional eustatic rise (Lawton and others, 1987).

During Jurassic time, marine sedimentation continued in the areas of the Endicott Mountains and North Slope subterranes south of the present Beaufort Sea coastline and lapped across the northern limit of the Shublik Formation and onto eroded pre-Mississippian rocks of the northern highlands region. Sedimentary rocks of the North Slope subterrane deposited during this time consist largely of dark, organic-rich, locally bioturbated shale and minor siltstone and sandstone that were derived from the north (part of the Kingak Shale). On seismic-reflection profiles, the Jurassic and Lower Cretaceous Kingak Shale displays a complex pattern of southward-prograding shallow-marine and slope deposits that downlap onto condensed basin deposits (Molenaar, 1988). These fine-grained progradational deposits provide a record of early uplift associated with rifting prior to continental breakup along the northern margin of the Arctic Alaska terrane in middle Early Cretaceous time. To the south in the Endicott Mountains subterrane, Jurassic strata consist of condensed basin deposits (Blankenship Member of the Otuk Formation); these deposits have also been inferred for the De Long Mountains subterrane by Mayfield and others (1988), but Jurassic rocks have yet to be documented. Although abundant in underlying upper Paleozoic and Triassic rocks in these subterranes, bedded chert is largely absent in the Jurassic rocks, probably reflecting a slight increase in the rate of clastic sedimentation related to rifting in the north and onset of Brookian tectonism in the south. By Late Jurassic time, flysch derived from the Brookian orogen was deposited in southernmost areas of the Arctic Alaska terrane, but condensed basin sedimentation continued until at least Valanginian (Early Cretaceous) time in more northerly areas (Fig. 24F). Near the top of the condensed basin deposits of the Endicott Mountains subterrane, the Valanginian coquinoid limestone (part of the Ipewik unit of Crane and Wiggins, 1976), composed of abundant pelecypods, is thought to have been deposited in situ and in relatively shallow water (Molenaar, 1988). This unit may represent an intrabasin structural high (or sill) (Jones and Grantz, 1964; Brosgé and Tailleur, 1971) that developed as a peripheral bulge in front of the northward-prograding Brookian orogenic belt. Alternatively, this unit may reflect initial uplift associated with onset of thrusting of the Endicott Mountains allochthon that was due to a northward shift of the main thrust front of the Brookian orogen in Valanginian time.

To the north, rift-margin uplift accompanying the successful opening of the Canada basin resulted in erosion of much of the crestal region of the Barrow arch in Early Cretaceous time. This erosional platform was composed, from south to north, of a broad southern coastal plain underlain by nonresistant rocks of the Kingak Shale, a narrow (50-km-wide) upland of relatively more resistant lower Ellesmerian and pre-Mississippian rocks, and a narrow northern coastal plain underlain by upper Ellesmerian rocks that were bordered on the north by the ancestral Arctic Ocean. Marine transgression in Hauterivian (Early Cretaceous) time resulted in local deposition of fine-grained, shallow-marine sand bodies (for example, Kemik Sandstone and upper part of the Kuparuk Formation) and in widespread deposition of the pebble shale unit onto shelf areas (Fig. 14).

Jurassic to Early Cretaceous (early Brookian) orogenesis

Northward emplacement of thrust sheets during the early phase of the Brookian orogeny in Jurassic and Early Cretaceous time is related by most workers to convergence between the oceanic Angayucham terrane and the continental Arctic Alaska terrane (Roeder and Mull, 1978; Mull, 1982; Box, 1985; Mayfield and others, 1988). The contact between the two terranes was called the Kobuk suture by Mull (1982) and the Angayucham thrust by Dillon (1989). Middle Jurassic to Early Cretaceous arc rocks deposited on the southern margin of the Angayucham terrane form the extensive Koyukuk terrane (Box, 1985; Box and Patton, 1989; Patton and Box, 1989; Patton and others, this volume, Chapter 7). The Angayucham and overlying Koyukuk terranes underlie the Albian (late Early Cretaceous) and younger Koyukuk basin, which borders the southern margin of the Brooks Range (Patton and Box, 1989). The Koyukuk and Angayucham terranes may have formed part of the paleo–Pacific Ocean basin during middle Mesozoic time (Box, 1985; Grantz and others, 1991).

Plate geometry during the early Brookian convergence is delimited by several important observations. (1) Peridotite and gabbro of the Misheguk Mountain allochthon (Angayucham terrane) form the structurally highest allochthon in the Brooks Range and are underlain by the pillowed basalts and diabase of the Copter Peak allochthon (Angayucham terrane), which are underlain by the subterranes of the Arctic Alaska terrane. (2) The structurally higher subterranes of the Arctic Alaska terrane (De Long Mountains, Endicott Mountains) were folded and faulted at shallow structural levels, whereas the structurally lower Hammond and Coldfoot subterranes were deformed ductilely at greater depths (Oldow and others, 1987d; Till and others, 1988). (3) Mineral assemblages in the ductilely deformed subterranes in the southern Brooks Range indicate metamorphism at high-*P*/low-*T* conditions in Jurassic and Cretaceous time followed by higher temperature metamorphism at somewhat lower pressures, whereas metamorphism of the Angayucham terrane and the structurally higher subterranes of the Arctic Alaska terrane occurred at relatively lower temperatures and pressures (Dusel-Bacon and others, 1989). (4) Plutonic and arc volcanic rocks of Jurassic and Early Cretaceous age are not present in the Arctic Alaska terrane, although sparse Middle Jurassic tuffaceous graywacke was reported from what is now mapped as the De Long Mountains subterrane (Jones and Grantz, 1964; Patton and Tailleur, 1964). (5) Middle Jurassic to Early Cretaceous plutonic and volcanogenic rocks, composing most of the Koyukuk terrane, are probably a remnant of an intraoceanic arc accreted to Alaska in Early Cretaceous time (Box, 1985; Patton and Box, 1989; Patton and others, this volume, Chapter 7). (6) Olistostromal blocks of volcanic and granitic detritus within the Okpikruak Formation of the De Long Mountains subterrane yield Jurassic K-Ar ages and may indicate that an arc terrane once formed the cover of the Misheguk Mountain allochthon (Angayucham terrane) prior to erosion of the orogenic belt (Mayfield and others, 1978). Roeder and Mull (1978), Mull (1982), Box (1985), Mayfield and others (1988), Patton and Box (1989), and Grantz and others (1991) interpreted these observations as evidence that early Brookian convergence between the Arctic Alaska terrane and the combined Angayucham and Koyukuk terranes involved south-dipping subduction of part of the Arctic Alaska terrane beneath the Angayucham and Koyukuk terranes. Box (1985) compared subduction of the continental Arctic Alaska terrane to Cenozoic subduction of Australia beneath Timor in the eastern Indian Ocean, and China beneath Taiwan in the western Pacific Ocean.

Stratigraphic evidence and isotopic-age data indicate that the early Brookian orogeny was diachronous. Onset of Brookian deformation is represented by selvages of amphibolite at the base of the ophiolitic Misheguk Mountain allochthon (Angayucham terrane). These amphibolite bodies, isotopically dated at 169 to 163 Ma, are inferred to have been metamorphosed during obduction of young, hot ocean crust (Wirth and Bird, 1992). The protoliths of the amphibolite, mafic volcanic rocks of the Copter Peak allochthon (Angayucham terrane), suggest that the Misheguk Mountain allochthon was emplaced onto the Copter Peak allochthon in the Middle or Late Jurassic. Sedimentary rocks of this age in the Arctic Alaska terrane, mostly condensed basinal sequences, suggest that the Arctic Alaska terrane was not involved in the initial phase of Brookian convergent deformation and that the earliest deformation involved only oceanic rocks of the Angayucham terrane.

Onset of involvement of the Arctic Alaska terrane in the Brookian orogeny is marked by the change from compositionally mature, northern source areas for uppermost Devonian to Jurassic strata to compositionally immature, southern source areas for Jurassic and younger strata of the proto-Colville basin. The oldest known compositionally immature strata in the Brooks Range are Upper Jurassic lithic turbidites of the Okpikruak Formation that were probably deposited on one of the structurally higher allochthons of the De Long Mountains subterrane (Curtis and others, 1984; Mayfield and others, 1988). The Okpikruak Formation is at least as old as Berriasian (earliest Early Cretaceous) in the Ipnavik River allochthon of the De Long Mountains subterrane but is no older than Valanginian in the Endicott Mountains allochthon. The Okpikruak Formation locally contains coarse-grained detritus derived from the Misheguk Mountain, Copter Peak, Ipnavik River, and Nuka Ridge allochthons, and it was deposited in, and carried with, the various allochthons of the De Long Mountains subterrane. The locally derived composition and olistostromal character of the Okpikruak Formation suggest syntectonic deposition into the foredeep along the northern limit of the early Brookian thrust front. Moreover, the apparent decrease in age of the Okpikruak structurally downward is interpreted as evidence that the thrust front advanced northward during Late Jurassic and Early Cretaceous time (Snelson and Tailleur, 1968; Mayfield and others, 1988).

The amount of Mesozoic to Cenozoic shortening across the Brookian orogen is poorly delimited, both because of the stratigraphic and paleogeographic uncertainties reviewed above and insufficient structural data. Assuming simple restorations, Mull (1982) estimated shortening ranging from a minimum of 96 km in the eastern Brooks Range to a minimum of 580 km in the western Brooks Range. Mayfield and others (1988), using a model for simple structural unstacking and rotation about a pivot point in the Mackenzie delta, estimated shortening of the Arctic Alaska and Angayucham terranes in the western Brooks Range to be 700 to 800 km or more. However, if the direction of structural transport during shortening was constant throughout the orogen, a minimum of only about 250 km of shortening is required for Mayfield and others' (1988) paleogeographic reconstruction, neglecting shortening across most intra-allochthon regional-scale folds and faults. Oldow and others (1987d) constructed balanced cross sections through the central Brooks Range and thereby made estimates of minimum shortening for the Endicott Mountains, Hammond, and Coldfoot subterranes of about 540 km. Although these estimates vary considerably in amount, they suggest that about 250 to 500 km of north-vergent shortening occurred during the Brookian orogeny. Such a large amount of shortening offers additional evidence that the Brookian orogeny is best explained as a consequence of subduction and ultimate closure of an ocean basin.

A tectonic model for Brookian tectonism modified from Mull (1982), Box (1985), Mayfield and others (1988), and Grantz and others (1991) is shown in Figure 25. During the Middle Jurassic, the Arctic Alaska terrane formed part of a passive continental margin adjoining an ocean basin that was underlain by late Paleozoic and early Mesozoic oceanic crust. Somewhere within the ocean basin, subduction of oceanic crust began during the Jurassic, and a continentward-facing intraoceanic arc (Koyukuk and related arc terranes) was established on oceanic crust (Kanuti panel, Misheguk Mountain allochthon) (Fig. 25A). Subduction of the oceanic crust flooring the basin resulted in con-

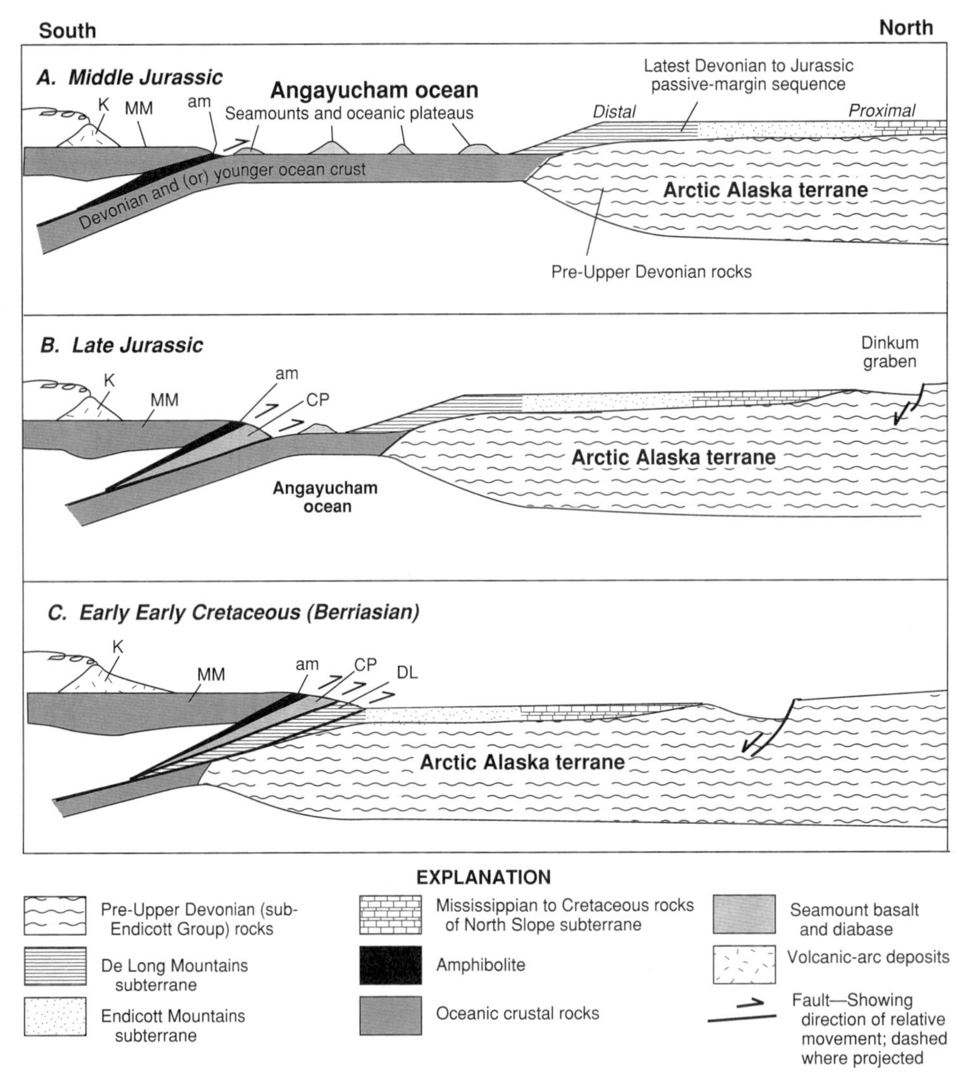

Figure 25. Tectonic model for Brookian orogeny. A, Initial subduction of late Paleozoic and early Mesozoic oceanic crust in Middle Jurassic. Underthrusting of older oceanic crust beneath young, hot oceanic crust (MM—Misheguk Mountain allochthon and Kanuti panel) and related arc terranes (K—Koyukuk terrane) produced amphibolite-grade metabasite (am). B, Subduction of oceanic crust is nearly complete; ocean basin is mostly closed by Late Jurassic. Subduction complex consists mostly of seamount and ocean-plateau fragments (CP—Copter Peak allochthon and Narvak panel) detached from subducted oceanic crust. Failed rifting associated with development of Dinkum graben in Jurassic time is shown schematically at right. C, Initial subduction of seaward part of Arctic Alaska terrane in latest Jurassic or earliest Cretaceous; outboard sedimentary cover (DL—De Long Mountains subterrane are detached from basement and imbricated. D, Subduction of continental substructure of Arctic Alaska terrane by middle Early Cretaceous to depth sufficient enough to produce blueschist facies metamorphism (bs); medial sedimentary cover of Arctic Alaska terrane (EN and SC—Endicott Mountains and Slate Creek subterranes) detached from basement, imbricated, and underthrust by inboard sedimentary cover and basement (HM and NS—Hammond and North Slope subterranes); rifting and continental breakup occurs along northern margin of Arctic Alaska terrane. E, Middle and lower crustal rocks containing blueschist facies assemblages (CF—Coldfoot subterrane) uplifted and emplaced at higher structural levels in late Early Cretaceous by large-scale out of sequence faulting, imbrication, duplexing, and thickening of lower crust; Canada basin opened and Barrow arch formed along northern margin of Arctic Alaska terrane. F, Continued contraction in latest Early Cretaceous thickened middle and lower crustal rocks; coeval crustal attenuation thinned imbricated crustal superstructure along down-to-south normal faults, particularly in southern Brooks Range. Resulting uplift and erosional unroofing shed huge volumes of clastic detritus southward into Koyukuk basin and northward into Colville basin.

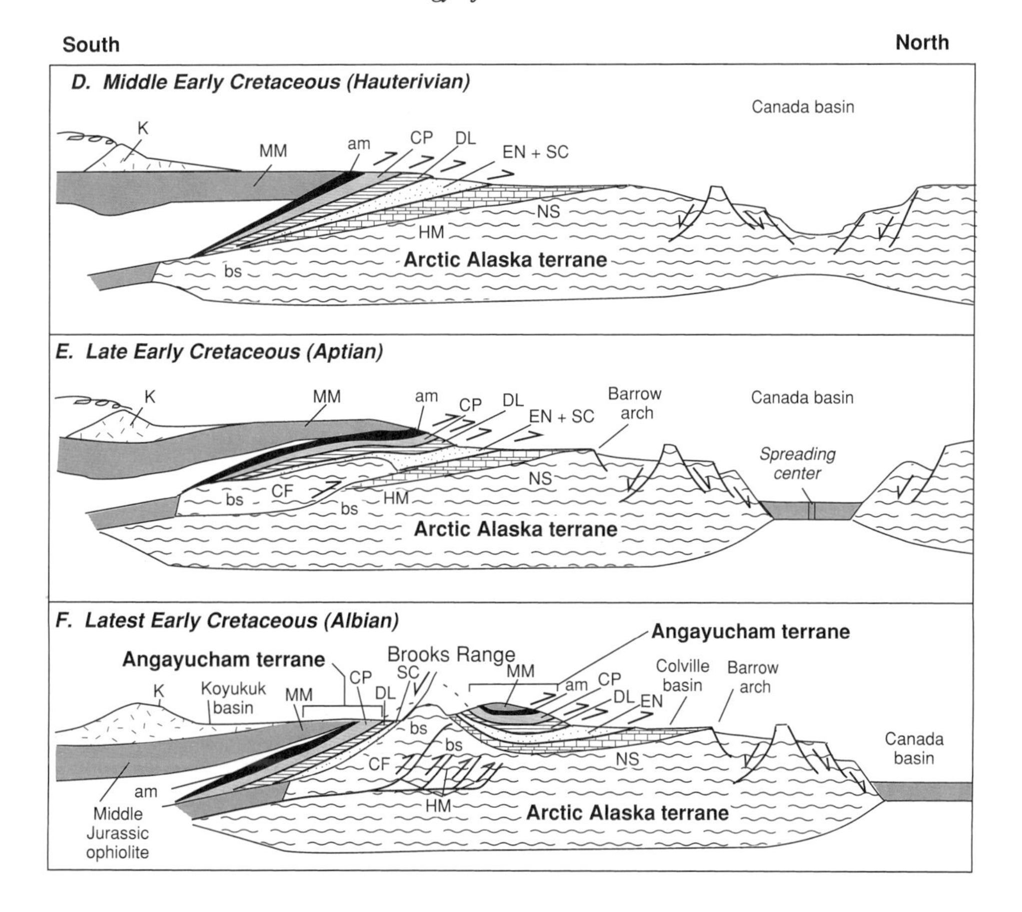

struction of a subduction complex (now represented by the Narvak panel and Copter Peak allochthon) in the forearc region by the progressive accretion of oceanic plateaus and seamounts of various ages (Fig. 25B). The earliest accreted oceanic supercrustal fragments were underplated beneath hot ophiolitic rocks of the Misheguk Mountain allochthon and underwent local dynamothermal metamorphism that reached amphibolite facies in Middle Jurassic time. The structurally lowest rocks of the Angayucham terrane, and hence latest accreted rocks, consist of Devonian volcanic rocks and locally interbedded limestone that may represent carbonate-covered volcanic atolls or, alternatively, carbonate-platform deposits drowned and covered by volcanic rocks during Devonian rifting (Mayfield and others, 1988; Patton and others, this volume, Chapter 7). In the latter case, accretion of these rocks would mark the onset of subduction of the distal continental margin of the Arctic Alaska terrane.

Stratigraphic evidence, described above, indicates that subduction of the Arctic Alaska terrane began by Late Jurassic time. Partial subduction during this time resulted in detachment of the outboard upper Paleozoic and lower Mesozoic sedimentary cover from underlying Proterozoic and lower Paleozoic crustal rocks of the terrane. The generally fine grained sedimentary rocks now forming the De Long Mountains subterrane were imbricated and underplated beneath the rocks of the Copter Peak allochthon, whereas their depositional basement was detached and carried to depth (Fig. 25C). Continued subduction resulted in detachment and imbrication of more landward parts of the sedimentary cover of the Arctic Alaska terrane (Endicott Mountains and Slate Creek subterranes), leading to their underplating beneath the De Long Mountains subterrane (Fig. 25D). The latest subducted, most landward parts of the sedimentary cover of the Arctic Alaska terrane (Hammond and North Slope subterranes) were thrust beneath the earlier emplaced, more seaward parts of the passive margin. The continental basement of the southern part of the Arctic Alaska terrane was subducted to depths of more than 25 km, where it underwent blueschist facies metamorphism.

Uplift of the high-pressure metamorphic rocks from their presumed site of metamorphism at depth in the subduction zone probably caused regional setting of K-Ar cooling ages (culminating at about 120–90 Ma) and the flood of clastic detritus from metamorphic source areas into the Colville and Koyukuk basins in Albian time (Till, 1988). This uplift may have resulted from thickening of the continental lithosphere of the Arctic Alaska terrane by isostatic resistance to subduction. Crustal thickening probably occurred in part by ductile deformation, imbrication, and duplexing at lower and mid-crustal levels. Oldow and others (1987d) and Grantz and others (1991) suggested that the structural high exposing the Coldfoot subterrane and other large antiforms in the southern Brooks Range were constructed as large-scale duplexes (Fig. 25, E and F). Thickening and uplift of the subducted continental lithosphere may have been accompanied or followed by attenuation of the overlying deformed sedimentary cover of the Arctic Alaska and Angayucham terranes (Fig. 25F), as suggested by missing structural section and major low-angle normal faulting along the southern margin of the Brooks Range (Miller, 1987; Gottschalk and Oldow, 1988; Miller and Hudson, 1991).

Origin of the northern Alaska continental margin

Depositional onlap relations, clast-size data, paleocurrent data, and compositional evidence show that a large continental highland (Barrovia) bordered the northern margin of the Arctic Alaska terrane and formed the source area for strata of the Ellesmerian sequence during late Paleozoic to late Mesozoic time. The Canada basin now occupies the site of this former highland, indicating that the Arctic Alaska terrane was severed from its northern source by creation of the basin. These considerations led Tailleur (1969, 1973a) and Rickwood (1970) to suggest that the modern continental margin of northern Alaska is an Atlantic-type (passive) continental margin formed at the end of deposition of the northerly derived strata of the upper Ellesmerian sequence in Early Cretaceous time. An Atlantic-type continental margin was confirmed by Grantz and May (1983), who presented evidence that the Canada basin was formed by sea-floor spreading in the Early Cretaceous.

Grantz and May (1983) and Hubbard and others (1987) inferred from stratigraphic relations in the Dinkum graben that crustal extension and subsidence related to rifting began at about 190–185 Ma (Early Jurassic). By Early Cretaceous time, extension and associated crustal warming along the northern margin of the Arctic Alaska terrane resulted in formation of a regional uplift, the Barrow arch, and subsequent erosional truncation of the underlying Kingak Shale and older rocks. The resulting regional Lower Cretaceous unconformity cuts progressively down stratigraphic section to the north into early Paleozoic rocks beneath the Beaufort Sea shelf, and it merges southward into a conformable contact between the Kingak Shale and the overlying Hauterivian pebble-shale unit (Molenaar, 1988). Grantz and May (1983) interpreted this as a breakup unconformity, which dates the age of initiation of sea-floor spreading in the Canada basin at about 125 Ma (Hauterivian). Subsequent cooling of the thinned continental-margin lithosphere caused rapid subsidence of the northern margin of the Arctic Alaska terrane, resulting in transgressive sedimentation (that is, overlap of the Kemik Sandstone and upper part of the Kuparuk Formation by the pebble-shale unit) across the unconformity.

Evolution of the Colville and Koyukuk basins

Brookian convergent deformation resulted in uplift of the Brooks Range and consequent shedding of clastic debris into the Colville and Koyukuk basins to the north and south of the orogen, respectively. In the Koyukuk basin, flysch of Albian age is overlain by molasse of Albian and Upper Cretaceous age (Patton and Box, 1989; Patton and others, this volume, Chapter 7). The clast composition of conglomerate in the Koyukuk basin along the southern margin of the Brooks Range varies with stratigraphic position (Dillon and Smiley, 1984; Dillon, 1989; Murphy and others, 1989). The oldest conglomerate units are enriched in chert and greenstone clasts, whereas the younger conglomerate units are enriched in quartzite and quartz clasts. On this basis, Dillon (1989) suggested that the conglomerate units record sequential erosional stripping of progressively lower structural units, which, from top to bottom in the southern Brooks Range, are the Angayucham terrane, the Slate Creek subterrane, and the Coldfoot subterrane. This sequence and mid-Cretaceous metamorphic-cooling ages in the orogen suggest that infilling of the northern Koyukuk basin resulted from incremental regional erosional unroofing of the internal part of the Brooks Range orogen. Because the uplifted region exposes the deepest structural levels of the orogen and is juxtaposed against Koyukuk basin deposits along south-dipping faults, deposition in the northern Koyukuk basin was probably controlled, at least in part, by extensional faulting. On the northern side of the orogen, strata of the Colville basin, as much as 10 km thick, define an asymmetric foredeep filled with Aptian(?) to Albian and younger clastic rocks. The northern flank of the basin originated as the gently south dipping continental shelf of the Barrovian passive margin and was steepened somewhat by subsidence of the basin and rift-related uplift of the Barrow arch in Early Cretaceous time. The southern flank of the basin is defined by uplifted and imbricated older rocks south of the Brookian thrust front.

The proto-Colville basin is represented by older foredeep deposits within the Brookian orogen. These strata are the allochthonous Upper Jurassic and Lower Cretaceous flysch and olistostromal rocks of the Okpikruak Formation now preserved in the De Long Mountains and Endicott Mountains subterranes. These foredeep deposits are coeval with the upper Ellesmerian sequence to the north, and thus indicate that shelf and platform sedimentation continued in more northerly parts of the Arctic Alaska terrane while foredeep deposition associated with early Brookian tectonism occurred to the south. Deposits of the proto-Colville basin collapsed and were imbricated along the southern flank of the basin and were then transported northward with the northward-migrating Brookian thrust front. Shortening along the basin's southern margin narrowed the basin and shifted its axis northward through early Neocomian time. Because no late Neocomian (Hauterivian and Barremian) foredeep deposits have been recognized in the Brooks Range, and the oldest deposits of the successor Colville basin are Aptian or Albian, late Neocomian deposits may be buried at depth, where they were probably overridden by the advancing early Brookian thrust front. Large-scale displacement of parts of the sedimentary cover of the Arctic Alaska terrane, which characterized early Brookian tectonism, was replaced by deeper seated tectonism and a reduction in northward displacement by Albian time. As a result, northward migration of the Brookian thrust front and the axis of the proto-Colville basin slowed, and the proto-Colville basin evolved into the Aptian(?) to Albian and younger Colville basin.

The filling of the Colville basin with enormous volumes of orogenic sediment began in Aptian or Albian time. The history of infilling is recorded by thick deposits of prodelta shale and thin-bedded turbidites of the upper part of the Torok Formation and overlying deltaic deposits of the Nanushuk Group in the western part of the basin, the Fortress Mountain and Torok formations in the central part of the basin, and the upper part of the Kongakut Formation and the Bathtub Graywacke in the eastern part of the basin. Although some of the deltaic deposits (for example, the Umiat delta of the Nanushuk Group) prograded to the north from the southern margin of the basin, the main basin-filling deltaic complex was the mud-rich, river-dominated Corwin delta (Nanushuk Group), which prograded from the west toward the east-northeast, almost parallel with the Colville basin axis. A large source area, probably an orogenic highland, is therefore postulated to have existed to the west or southwest in the area of the present Chukchi Sea or beyond (Fig. 15; Molenaar, 1985). Northward progradation of the Nanushuk deltas was initially constrained by the Barrow arch, which was a passive, but subsiding, high in Albian and later time. By the end of Albian time, however, deposits of the Corwin delta spilled northward across the western part of the Barrow arch into the newly formed Canada basin (Nuwuk basin of Grantz and May, 1983). The composition of the deltaic deposits suggests that fine-grained siliceous and carbonate-rich sequences like those of the De Long Mountains subterrane and Lisburne Peninsula formed the primary source region for the Corwin delta, whereas quartzose rocks of the Endicott Mountains subterrane formed the main source region for the Umiat delta (Bartsch-Winkler, 1979; Mull, 1985). Distinctive high-pressure metamorphic detritus in the Nanushuk Group suggests that the Coldfoot subterrane also contributed sediment to both deltaic systems (Till, 1988; A. B. Till, 1990, written commun.).

Following several significant marine transgressions in Cenomanian to Santonian time, deltaic deposits continued to prograde northeastward across the central North Slope, finally filling the eastern part of the basin in latest Cretaceous and Tertiary time. Because the eastern part of the Colville basin was the last to be filled, condensed basinal deposits in this area (Hue Shale) span Albian to Maastrichtian time; the delta-plain sediments (Sagavanirktok Formation) were not deposited until as late as Eocene time (Molenaar and others, 1987). By Paleocene time, slope-deltaic deposits of the Colville basin had prograded across the eastern part of the Barrow arch and had begun to form a thick constructional continental shelf and slope in the eastern Beaufort Sea (Kaktovik basin of Grantz and May, 1963). Later in Tertiary time, deltas built by sediment both transported along the axis of the Colville basin and shed directly from the rising Romanzof uplift to the south prograded into the eastern Beaufort Sea. Northward progradation was amplified further in Late Cretaceous and Cenozoic time by northward migration of the late Brookian foredeep with the northward advance of the late Brookian thrust front of the northeastern Brooks Range.

Post-Neocomian (late Brookian) tectonism

In post-Neocomian time, following emplacement of the far-traveled allochthons that characterize early Brookian tectonism, the nature of Brookian orogenic activity changed. Late Brookian deformational phases have a relatively restricted extent compared with early Brookian deformation, and late Brookian deformation involves distinct contractional, extensional, and translational events of varying orientation and age, which cannot be easily ascribed to a single tectonic episode.

The earliest and most widespread late Brookian tectonic event is significant uplift and unroofing along the main axis of the Brooks Range. This tectonism is recorded by the preponderance of K-Ar cooling ages at about 120–90 Ma (Turner and others, 1979) and by the flood of enormous volumes of clastic detritus into the Colville and Koyukuk basins during Albian and Cenomanian time. This uplift has been assumed to be the result of isostatic uplift following early Brookian convergence (Mull, 1982; Mayfield and others, 1988), but it may also reflect shortening and imbrication at deep structural levels and/or attenuation in the internal part of the orogen. Major duplexes in the internal part of the orogen, such as the ones at Mt. Angayukaqsraq and Mt. Doonerak, may have formed during this period. Displacement related to this uplift may have continued through Late Cretaceous time and may have been transmitted northward along deep detachments into the central and western North Slope, resulting in the long-wavelength folding of Albian and Upper Cretaceous sedimentary rocks of the Colville basin.

Post-Neocomian displacement along the southern flank of the Brooks Range on the Kobuk fault system may have controlled sedimentation in the Koyukuk basin. This relation was inferred by Dillon (1989) from the narrow basinward extent of Albian and Upper Cretaceous molasse, the presence of abundant debris-flow deposits within the basin, and the apparent association of the coarse-grained rocks with the Kobuk fault system. The amount of displacement on the Kobuk system is poorly delimited, but large-scale right-slip movement is possible (Grantz, 1966; Dillon, 1989). Large-scale post-Neocomian displacement has also been proposed for the Porcupine lineament along the southeast flank of the Brooks Range, but the sense and amount of displacement are controversial and have been based on regional tectonic models rather than field observations.

The southern flank of the Brooks Range was probably also the locus of extensional tectonism in post-Neocomian time. Miller and Hudson (1991) related development of the Koyukuk basin with its coarse-grained sedimentary fill and other features to a regional extensional event that began at about 115 Ma (Aptian), whereas Gottschalk and Oldow (1988), Pavlis (1989), and Gottschalk (1990) viewed extension as a byproduct of regional convergence that continued into Albian and/or younger time. Box (1987) noted that Upper Cretaceous rocks are penetratively deformed in the Cosmos Hills, and he suggested that, at least locally, extensional deformation took place in Late Cretaceous or younger time. Plumley and Vance (1988) reported that extensional tectonism to the east along the Porcupine lineament probably took place in Paleogene time. The change from convergence to extensional and translational tectonism has been related to regional right-slip motion along the Kula–North American plate boundary and to initiation of displacement along the Tintina fault system (Grantz, 1966; Pavlis, 1989; Grantz and others, 1991).

During the Late Cretaceous and/or Paleogene, contractional deformation took place in regions north of the main axis of the Brooks Range. The east-directed fold and thrust belt of the Herald arch and the Lisburne Peninsula, which may also be related to deformation in the western Brooks Range (Karl and Long, 1987), is probably the result of convergence between North America and Eurasia prior to the Neogene (Patton and Tailleur, 1977; Grantz and others, 1981). To the east in the northeastern Brooks Range and eastern North Slope, the Romanzof uplift resulted from Cenozoic north-vergent deformation. The Romanzof uplift produced the highest peaks and greatest relief in the Brooks Range, and it resulted in deposition of as much as 12 km of sediment along the Beaufort Sea continental margin. Significant shortening of these continental-margin deposits, coupled with modern seismicity (Grantz and others, 1983a), indicates that deformation has propagated northward and has continued into the Cenozoic with a corresponding uplift of pre-Mississippian, Ellesmerian, and Brookian rocks on the south flank of the Barrow arch and development of a Neogene foreland basin along the Beaufort Sea continental margin. Moore and others (1985b) and Grantz and others (1991) related the north-vergent tectonism to compressional stress imposed by Cenozoic convergence and accretion along the North American–Pacific plate boundary in southern Alaska. This compressional stress may have been transmitted to northeastern Alaska by displacement above deep crustal detachment surfaces across all of Alaska.

Where did the Arctic Alaska terrane originate?

The Arctic Alaska terrane is considered a suspect terrane (Coney and Jones, 1985) because its relation to the North American craton and neighboring terranes is uncertain. Much of this dilemma is a consequence of unresolved questions about the history of the rifting and the opening of the adjoining Canada basin. Paleomagnetic data cannot be used to determine the basin's origin because of the absence of clearly discernible magnetic lineations and the pervasive Cretaceous to Cenozoic remagnetization of northern Alaska (Van Alstine, 1986; Hillhouse and Grommé, 1988). Other less definitive lines of evidence have led to many plate reconstructions for the origin of the Canada basin, the Arctic Ocean as a whole, and adjacent plates. Although the Arctic region has been reconstructed in many ways, the Arctic Alaska terrane is typically restored to one of the four general positions shown in Figure 26 (A–D). Reviews of the various models have been given by Nilsen (1981) and Lawver and Scotese (1990).

The most widely accepted model for the origin of the Arctic Alaska terrane is rifting and counterclockwise rotation of the Arctic Alaska terrane away from northernmost Canada about a pole of rotation near the Mackenzie delta (Fig. 26D) (Carey, 1958; Tailleur, 1969, 1973a; Rickwood, 1970; Newman and others, 1977; Mull, 1982; Sweeney, 1982; Grantz and May, 1983; McWhae, 1986; Howell and Wiley, 1987; Ziegler, 1988; Grantz and others, 1990b). In this model, the Arctic Alaska terrane originated in a position contiguous with the Canadian Arctic Island segment of the Innuitian fold belt. The opening of the Canada basin in this model would require about 66° of counterclockwise rotation in Cretaceous time. Recent paleomagnetic results by Halgedahl and Jarrard (1987) on drill core from a single well in the North Slope support this amount of rotation; Lower Cretaceous and older strata in most other wells have been remagnetized in post–Early Cretaceous time (D. R. Van Alstine, 1988, oral commun.). Variants of the rotational origin for the Arctic Alaska terrane suggest that counterclockwise movement may have been accompanied by about 270 km of right slip along the northern Canadian continental margin in middle Mesozoic time (Grantz and others, 1990b) or preceeded by more than 1500 km of left slip in Paleozoic time (Sweeney, 1982). The rotational model (1) accounts for the inferred Cretaceous age of the Canada basin, (2) explains apparent similarities between the late Paleozoic and early Mesozoic geology of the Arctic Alaska terrane and the Innuitian fold belt, (3) allows for a northern source area for the Ellesmerian sequence, and (4) provides a reasonable fit for bathymetric and gravity data across the Canada basin. The hypothesized rotation of the Arctic Alaska terrane has been linked kinematically to convergence between the Arctic Alaska and Koyukuk arc terranes and the consequent formation of the Brooks Range orogen (Mayfield and others, 1988). However, such a linkage is unlikely because the amount of crustal shortening within the orogen does not decrease toward the pole of rotation as would be expected and because Brookian plate convergence began in Jurassic time, whereas sea-floor spreading in the Canada basin did not begin until Neocomian time (Rattey, 1985; Oldow and others, 1987b). Possible Cordilleran aspects of the terrane (for example, Devonian magmatic belt, overthrusting of oceanic rocks of the Angayucham terrane, general stratigraphic similarities of parts of the Arctic Alaska terrane with North American rocks of the Canadian Cordillera) may be accounted for by restoration to a position between the Innuitian orogen and the northern limit of the Cordilleran orogen. Problems with the

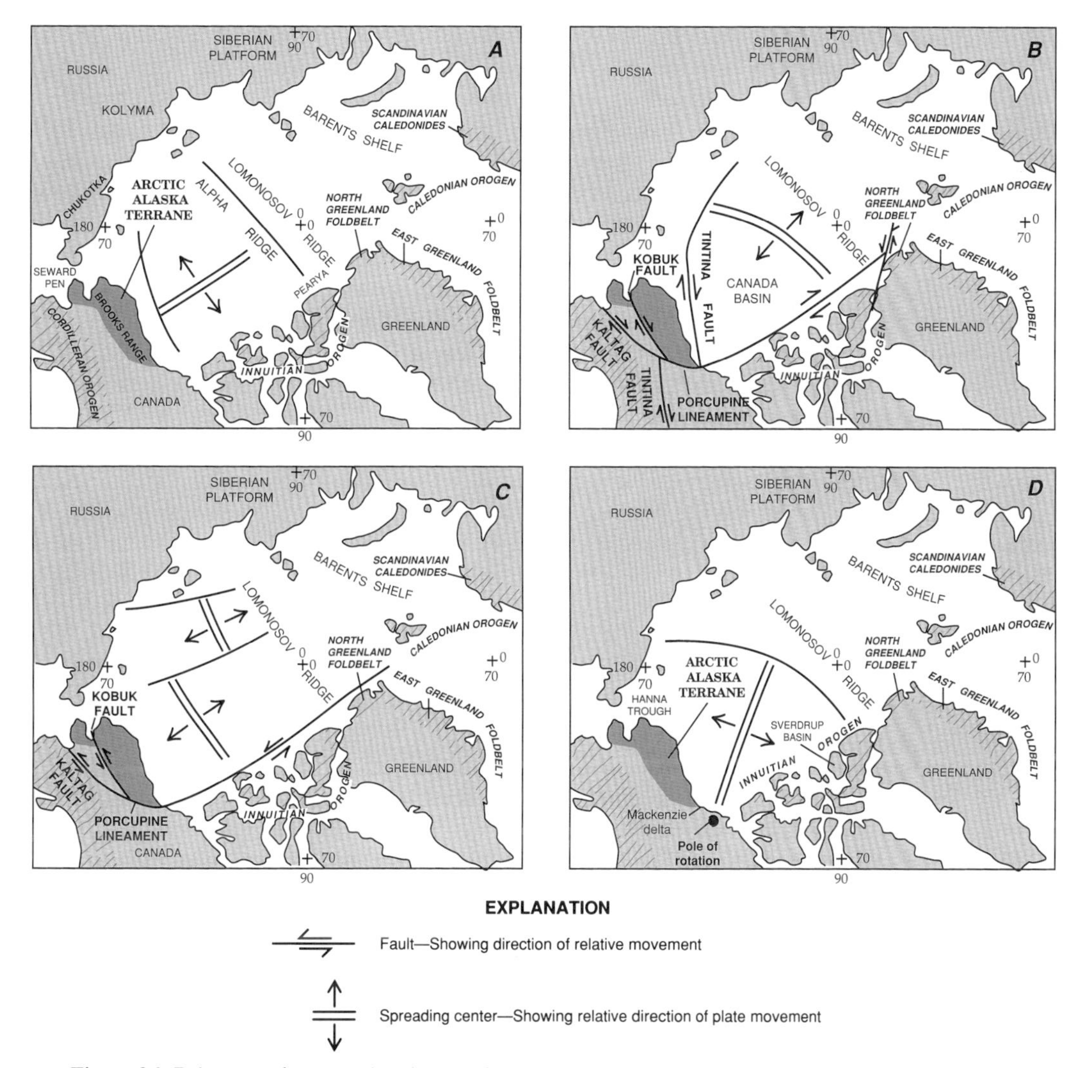

Figure 26. Paleotectonic maps showing possible restorations of Arctic Alaska terrane to positions in Cordilleran, Innuitian, and Caledonian orogens. A, In situ origin: Arctic Alaska terrane is not rotated nor translated (Bogdanov and Tilman, 1964; Herron and others, 1974; Churkin and Trexler, 1980, 1981). B, Yukon origin: Arctic Alaska terrane is translated by large-scale right slip along Tintina fault from position in northern Canadian part of Cordilleran orogen and is offset right laterally along Porcupine lineament (Jones, 1980, 1982b). C, Barents shelf origin: Arctic Alaska terrane is translated by large-scale left slip along transform fault parallel to Canadian Arctic Islands shelf edge from original position in northern Caledonian or Innuitian orogens (Dutro, 1981; Nilsen, 1981; Oldow and others, 1987b; Smith, 1987). D, Canadian Arctic Islands origin: Arctic Alaska terrane is rotated oroclinally from Innuitian orogen about a pole of rotation near present Mackenzie delta (Carey, 1958; Tailleur, 1969, 1973a; Rickwood, 1970; Newman and others, 1977; Mull, 1982; Sweeney, 1982; Grantz and May, 1983; McWhae, 1986; Howell and Wiley, 1987; Ziegler, 1988; Grantz and others, 1990b).

rotational hypothesis include structural and metamorphic differences between Devonian and older rocks across the restored boundary, opposition of middle Paleozoic sediment transport directions across this boundary, and the requirement that strike-slip displacements of several thousand kilometers must have occurred along the Lomonosov Ridge in the central Arctic Ocean as a consequence of rotation (Nilsen, 1981; Oldow and others, 1987b).

CONCLUSIONS

Northern Alaska consists of two principal tectonostratigraphic terranes: the Arctic Alaska and Angayucham terranes. The most extensive of the two is the Arctic Alaska terrane, which underlies the North Slope and most of the Brooks Range. Rocks of this terrane range from Proterozoic to Cenozoic and are divided into a structurally and stratigraphically complex pre-

Mississippian assemblage overlain by a once laterally continuous succession of Upper Devonian and Lower Mississippian to Lower Cretaceous, nonmarine to marine continental-margin deposits. These deposits are in turn overlain by upper Mesozoic and Cenozoic siliciclastic foredeep strata. The pre-Mississippian assemblage records early to middle Paleozoic convergent deformation and arc plutonism along the edge of North America and subsequent rifting in Devonian time. The Devonian rifting culminated in formation of an ocean basin and the development of a complex south-facing passive margin by Late Devonian time. Subsidence along the passive margin resulted in progressive northward onlap of fluvial-deltaic to continental-shelf and carbonate-platform deposits in the Late Devonian to Pennsylvanian and neritic, nonmarine, and bathyal deposits in the Permian to Early Cretaceous. Gradual deepening was accompanied by waning clastic input from the north, resulting in deposition of condensed basinal deposits of shale, siliceous limestone, and chert in more distal parts of the passive margin from Mississippian to Jurassic time.

The Angayucham terrane is a structurally thin succession of rocks that rests tectonically on the southern part of the Arctic Alaska terrane and consists of two assemblages, both of which originated in an ocean basin. The structurally lower assemblage comprises a structural collage of Devonian to Jurassic ocean-island basalts and pelagic sedimentary rocks that represent a subduction complex of Jurassic age. The upper assemblage comprises peridotite and gabbro that form an incomplete Middle Jurassic ophiolite of arc affinity. During Middle and Late Jurassic time, the lower assemblage was underplated by subduction to the upper ophiolitic assemblage. Sedimentary debris in Brookian foredeep deposits suggests that Jurassic granitic and volcanic rocks of magmatic-arc affinity may have once overlain the structurally higher ophiolitic assemblage, but later the volcanic rocks were eroded away.

The Brookian orogeny began in the Middle Jurassic with southward subduction of the ocean basin that lay south of the passive margin of the Arctic Alaska terrane. In Late Jurassic and early Neocomian time, progressively more inboard parts of the Arctic Alaska terrane were partially subducted beneath the oceanic forearc and earlier accreted oceanic rocks (Angayucham terrane). This convergence resulted in delamination and imbrication of the continental superstructure of the Arctic Alaska terrane, which consisted mainly of the passive-margin succession. As a consequence, more distal parts of the Arctic Alaska continental margin were progressively collapsed and thrust successively northward over more proximal parts. The axis of associated foredeep sedimentation in the proto-Colville basin migrated northward with the thrust front; later, the older foredeep deposits became involved in thrusting. The continental substructure of the Arctic Alaska terrane, meanwhile, was subducted to deeper structural levels, where it was tectonically thickened and subjected to high pressure–low temperature (blueschist facies) metamorphism.

In the northern part of the Arctic Alaska terrane, a failed episode of rifting occurred in the Early Jurassic and was followed by a successful episode of rifting in the Early Cretaceous (Hauterivian) that resulted in continental breakup and formation of the Canada basin. Northward subsidence along the newly rifted margin of the formerly south-dipping continental-margin sequence produced an inflection in dip of Lower Cretaceous and older rocks, thus forming the Barrow arch. The Barrow arch was uplifted and exposed above sea level at the time of continental breakup in the Early Cretaceous, producing a local erosional truncation of older rocks, but the arch has gradually subsided since late Early Cretaceous time.

The southern part of the Arctic Alaska terrane was rapidly uplifted and unroofed by the late Early Cretaceous (Albian) as plate convergence slowed, resulting in setting of isotopic cooling ages and in deposition of huge volumes of clastic detritus to the north in the Colville basin foredeep and to the south in the Koyukuk basin. Although sediments shed northward into the Colville basin built local northward-prograding deltas along much of the Brooks Range, the basin was filled largely by deposits of an eastward- to northeastward-prograding delta (Corwin delta) from Albian through Tertiary time. This delta eventually prograded northward across the western part of the Barrow arch, depositing an Albian and younger constructional continental-margin sequence along the margin of the Canada basin. Renewed north-vergent thrusting and uplift in Cenozoic time formed the northeastern salient of the eastern Brooks Range and caused northward migration of foredeep sedimentation across the Barrow arch and onto the adjacent part of the northern Alaska continental margin. Geologic structures and seismicity data indicate that thrusting has propagated northward across the continental margin in Neogene time and has deformed the eastern part of the Barrow arch and overlying sedimentary rocks.

Faunal affinities and broad stratigraphic similarities indicate that the Arctic Alaska terrane was part of North America by late Paleozoic time. Its exact site of origin, however, is controversial. The leading hypothesis suggests that, immediately following the culmination of early Brookian orogenesis in the Early Cretaceous, the northern margin of the Arctic Alaska terrane was rifted away from the Canadian Arctic Islands region and rotated clockwise about 67° to its present position, thus forming the Canada basin. Following rotation, east-vergent thrusting took place during the Cretaceous (post-Neocomian) along the western margin of northern Alaska (for example, the Lisburne Peninsula). This convergence probably resulted from local convergence between the Eurasian and North American plates. Likewise, convergent deformation during Cenozoic time in northeastern Alaska has been ascribed to the far-reaching effects of Pacific-North American plate interactions along the southern margin of Alaska.

REFERENCES CITED

Adams, K. E., 1991, Permian sedimentation in the north-central Brooks Range, Alaska: Implications for tectonic reconstructions [M.S. thesis]: Fairbanks, University of Alaska, 122 p.

Adams, K. E., and Siok, J. P., 1989, Permian stratigraphy in the Atigun Gorge area: A transition between the Echooka and Siksikpuk formations, *in* Mull, C. G., and Adams, K. E., eds., Dalton Highway, Yukon River to Prudhoe Bay, Alaska: Alaska Division of Geological and Geophysical Surveys Guidebook 7, v. 2, p. 267–276.

Adkison, W. L., and Brosgé, M. M., eds., 1970, Proceedings of the geological seminar on the North Slope of Alaska: Los Angeles, Pacific Section, American Association of Petroleum Geologists, 203 p.

Ahlbrandt, T. S., Huffman, A. C., Fox, J. E., and Pasternack, I., 1979, Depositional framework and reservoir-quality studies of selected Nanushuk Group outcrops, North Slope, Alaska, *in* Ahlbrandt, T. S., ed., Preliminary geologic, petrologic, and paleontologic results of the study of Nanushuk Group rocks, North Slope, Alaska: U.S. Geological Survey Circular 794, p. 14–31.

Anderson, A. V., 1991, Geologic map and cross sections, headwaters of the Kongakut and Aichilik rivers, Demarcation Point A-4 and Table Mountain D-4 quadrangles, eastern Brooks Range, Alaska: Alaska Division of Geological and Geophysical Surveys Public Data File 91-3, 24 p., 2 sheets, scale 1:25,000.

Anderson, A. V., and Wallace, W. K., 1990, Middle Devonian to Lower Mississippian clastic depositional cycles southwest of Bathtub Ridge, northeastern Brooks Range, Alaska: Geological Association of Canada/Mineralogical Association of Canada Program with Abstracts, v. 15, p. A2.

Armstrong, A. K., 1970, Carbonate facies and the lithostrotionoid corals of the Mississippian Kogruk Formation, De Long Mountains, northwestern Alaska: U.S. Geological Survey Professional Paper 664, 38 p.

—— , 1972, Pennsylvanian carbonates, paleoecology, and rugose colonial corals, north flank, eastern Brooks Range, Arctic Alaska: U.S. Geological Survey Professional Paper 747, 21 p.

—— , 1974, Carboniferous carbonate depositional models, preliminary lithofacies and paleotectonic maps, Arctic Alaska: American Association of Petroleum Geologists Bulletin, v. 58, p. 621–645.

Armstrong, A. K., and Bird, K. J., 1976, Carboniferous environments of deposition and facies, Arctic Alaska, *in* Miller, T. P., ed., Symposium on recent and ancient sedimentary environments in Alaska: Anchorage, Alaska Geological Society Symposium Proceedings, p. A1–A16.

Armstrong, A. K., and Mamet, B. L., 1977, Carboniferous microfacies, microfossils, and corals, Lisburne Group, Arctic Alaska: U.S. Geological Survey Professional Paper 849, 144 p.

—— , 1978, Microfacies of the Carboniferous Lisburne Group, Endicott Mountains, Arctic Alaska, *in* Stelck, C. R., and Chatterton, B.D.E., eds., Western and Arctic biostratigraphy: Geological Association of Canada Special Paper 18, p. 333–394.

Armstrong, A. K., Mamet, B. L., and Dutro, J. T., Jr., 1970, Foraminiferal zonation and carbonate facies of Carboniferous (Mississippian and Pennsylvanian) Lisburne Group, central and eastern Brooks Range, Arctic Alaska: American Association of Petroleum Geologists Bulletin, v. 54, p. 687–698.

—— , 1971, Lisburne Group, Cape Lewis–Niak Creek, northwestern Alaska, Chapter B, *in* Geological Survey research, 1971: U.S. Geological Survey Professional Paper 750-B, p. B23–B34.

Armstrong, A. K., Mamet, B. L., Brosgé, W. P., and Reiser, H. N., 1976, Carboniferous section and unconformity at Mount Doonerak, Brooks Range, Alaska: American Association of Petroleum Geologists Bulletin, v. 60, p. 962–972.

Armstrong, R. L., Harakal, J. E., Forbes, R. B., Evans, B. W., and Thurston, S. P., 1986, Rb-Sr and K-Ar study of metamorphic rocks of the Seward Peninsula and southern Brooks Range, *in* Evans, B. W., and Brown, E. H., eds., Blueschists and eclogites: Geological Society of America Memoir 164, p. 185–203.

Atkinson, C. D., Trumbly, P. N., and Kremer, M. C., 1988, Sedimentology and depositional environments of the Ivishak Sandstone, Prudhoe Bay field, North Slope, Alaska, *in* Lomando, A. J., and Harris, P. M., eds., Giant oil and gas fields, a core workshop: Tulsa, Oklahoma, Society of Economic Paleontologists and Mineralogists Core Workshop no. 12, p. 561–613.

Bader, J. W., and Bird, K. J., 1986, Geologic map of the Demarcation Point, Mount Michelson, Flaxman Island, and Barter Island quadrangles, northeastern Alaska: U.S. Geological Survey Miscellaneous Investigations Map I—1791, scale 1:250,000.

Barker, F., Jones, D. L., Budahn, J. R., and Coney, P. J., 1988, Ocean plateau-seamount origin of basaltic rocks, Angayucham terrane, central Alaska: Journal of Geology, v. 96, p. 368–374.

Barker, J. C., 1982, Reconnaissance of rare-metal occurrences associated with the Old Crow batholith, eastern Alaska–northwestern Canada: Short notes on Alaska Geology—1981: Alaska Division of Geological and Geophysical Surveys Geologic Report 73, p. 43–49.

Barnes, D. A., 1987, Reservoir quality in the Sag River Formation, Prudhoe Bay field, Alaska: Depositional environment and diagenesis, *in* Tailleur, I., and Weimer, P., eds., Alaskan North Slope geology: Bakersfield, California, Society of Economic Paleontologists and Mineralogists, Pacific Section, and Alaska Geological Society, Book 50, p. 85–94.

Barnes, D. F., 1970, Gravity and other regional geophysical data from northern Alaska, *in* Adkison, W. L., and Brosgé, M. M., eds., Proceedings of the geological seminar on the North Slope of Alaska: Los Angeles, American Association of Petroleum Geologists, p. I1–I19.

Bartsch-Winkler, S., 1979, Textural and mineralogical study of some surface and subsurface sandstones from the Nanushuk Group, western North Slope, Alaska, *in* Ahlbrandt, T. S., ed., Preliminary geologic, petrologic, and paleontologic results of the study of Nanushuk Group rocks, North Slope, Alaska: U.S. Geological Survey Circular 794, p. 61–76.

Bird, J. M., Wirth, K. R., Harding, D. J., and Shelton, D. H., 1985, Brooks Range ophiolites reconstructed: Eos (American Geophysical Union Transactions), v. 46, p. 1129.

Bird, K. J., 1978, New information on the Lisburne Group (Carboniferous and Permian) in the National Petroleum Reserve in Alaska: American Association of Petroleum Geologists Bulletin, v. 62, p. 880.

—— , 1982, Rock-unit reports of 228 wells drilled on the North Slope, Alaska: U.S. Geological Survey Open-File Report 82-278, 106 p.

—— , 1985, The framework geology of the North Slope of Alaska as related to oil-source rock correlations, *in* Magoon, L. B., and Claypool, G. E., eds., Alaskan North Slope oil/rock correlation study: American Association of Petroleum Geologists Studies in Geology no. 20, p. 3–29.

—— , 1987, The framework geology of the North Slope of Alaska as related to oil-source rock correlations, *in* Tailleur, I., and Weimer, P., eds., Alaskan North Slope geology: Bakersfield, California, Society of Economic Paleontologists and Mineralogists, Pacific Section, and Alaska Geological Society, Book 50, v. 1, p. 121–143.

—— , 1988a, Alaskan North Slope stratigraphic nomenclature and data summary for government-drilled wells, *in* Gryc, G., ed., Geology and exploration of the National Petroleum Reserve in Alaska, 1974 to 1982: U.S. Geological Survey Professional Paper 1399, p. 317–353.

—— , 1988b, Structure-contour and isopach maps of the National Petroleum Reserve in Alaska, *in* Gryc, G., ed., Geology and exploration of the National Petroleum Reserve in Alaska, 1974 to 1982: U.S. Geological Survey Professional Paper 1399, p. 355–377.

Bird, K. J., and Andrews, J., 1979, Subsurface studies of the Nanushuk Group, *in* Ahlbrandt, T. S., ed., Preliminary geologic, petrologic, and paleontologic results of the study of Nanushuk Group Rocks, North Slope, Alaska: U.S. Geological Survey Circular 794, p. 32–41.

Bird, K. J., and Jordan, C. F., 1977, Lisburne Group (Mississippian and Pennsylvanian), potential major hydrocarbon objective of Arctic Slope, Alaska: American Association of Petroleum Geologists Bulletin, v. 61, p. 1493–1512.

Bird, K. J., and Molenaar, C. M., 1987, Stratigraphy, *in* Bird, K. J., and Magoon, L. B., eds., Petroleum geology of the northern part of the Arctic National Wildlife Refuge, northeastern Alaska: U.S. Geological Survey Bulletin 1778, p. 37–59.

Bird, K. J., Connor, C. L., Tailleur, I. L., Silberman, M. L., and Christie, J. L., 1978, Granite on the Barrow Arch, northeast NPRA, *in* Johnson, K. M., ed., The United States Geological Survey in Alaska: Accomplishments during 1977: U.S. Geological Survey Circular 772-B, p. B24–B25.

Bird, K. J., Griscom, S. B., Bartsch-Winkler, S., and Giovannetti, D. M., 1987, Petroleum reservoir rocks, *in* Bird, K. J., and Magoon, L. B., eds., Petroleum geology of the northern part of the Arctic National Wildlife Refuge, northeastern Alaska: U.S. Geological Survey Bulletin 1778, p. 79–99.

Blanchard, D. C., and Tailleur, I. L., 1983, Pebble shale (Early Cretaceous) depositional environments in National Petroleum Reserve in Alaska (NPRA): American Association of Petroleum Geologists Bulletin, v. 67, p. 424–425.

Blodgett, R. B., Clough, J. G., Dutro, J. T., Jr., Ormiston, A. R., Palmer, A. R., and Taylor, M. E., 1986, Age revisions for the Nanook Limestone and Katakturuk Dolomite, northeastern Brooks Range, *in* Bartsch-Winkler, S., and Reed, K. M., eds., Geologic studies in Alaska by the U.S. Geological Survey during 1985: U.S. Geological Survey Circular 978, p. 5–10.

Blodgett, R. B., Rohr, D. M., Harris, A. G., and Jia-yu, R., 1988, A major unconformity between Upper Ordovician and Lower Devonian strata in the Nanook Limestone, Shublik Mountains, northeastern Brooks Range, *in* Hamilton, T. D., and Galloway, J. P., eds., Geologic studies in Alaska by the U.S. Geological Survey during 1987: U.S. Geological Survey Circular 1016, p. 18–23.

Blodgett, R. B., Clough, J. G., Harris, A. G., and Robinson, M. S., 1991, The Mount Copleston Limestone, a new Lower Devonian formation in the Shublik Mountains, northeastern Brooks Range, Alaska, *in* Bradley, D. C., and Ford, A., eds., Geologic studies in Alaska by the U.S. Geological Survey in 1990: U.S. Geological Survey Bulletin 1999, p. 1–5.

Blome, C. D., Reed, K. M., and Tailleur, I. L., 1988, Radiolarian biostratigraphy of the Otuk Formation in and near the National Petroleum Reserve in Alaska, *in* Gryc, G., ed., Geology and exploration of the National Petroleum Reserve in Alaska, 1974 to 1982: U.S. Geological Survey Professional Paper 1399, p. 725–751.

Blythe, A. E., Wirth, K. R., and Bird, J. M., 1990, Fission track and $^{40}Ar/^{39}Ar$ ages of metamorphism and uplift, Brooks Range, northern Alaska: Geological Association of Canada Program and Abstracts, v. 15, p. A-12.

Boak, J. M., Turner, D. L., Henry, D. J., Moore, T. E., and Wallace, W. K., 1987, Petrology and K-Ar ages of the Misheguk igneous sequence—An allochthonous mafic and ultramafic complex—and its metamorphic aureole, western Brooks Range, Alaska, *in* Tailleur, I., and Weimer, P., eds., Alaskan North Slope geology: Bakersfield, California, Society of Economic Paleontologists and Mineralogists, Pacific Section, and Alaska Geological Society, Book 50, p. 737–745.

Bodnar, D. A., 1989, Stratigraphy of the Otuk Formation and Cretaceous coquinoid limestone and shale unit, *in* Mull, C. G., and Adams, K. E., eds., Dalton Highway, Yukon River to Prudhoe Bay, Alaska: Alaska Division of Geological and Geophysical Surveys Guidebook 7, v. 2, p. 277–284.

Bogdanov, N. A., and Tilman, S. M., 1964, Similarities in the development of the Paleozoic structure of Wrangell Island and the western part of the Brooks Range (Alaska), *in* Soreshchanie po Problem Tektoniki: Moskva, Nauka, Skladchatye oblasti Evrazil, Materialy, p. 219–230.

Bowsher, A. L., and Dutro, J. T., Jr., 1957, The Paleozoic section in the Shainin Lake area, central Brooks Range, Alaska, *in* Part 3, Areal geology: Exploration of Naval Petroleum Reserve No. 4 and adjacent areas, northern Alaska, 1944–53: U.S. Geological Survey Professional Paper 303-A, p. 1–39.

Box, S. E., 1985, Early Cretaceous orogenic belt in northeastern Alaska: Internal organization, lateral extent, and tectonic interpretation, *in* Howell, D. G., ed., Tectonostratigraphic terranes of the circum-Pacific region: Circum-Pacific Council for Energy and Mineral Resources, Earth Science Series no. 1, p. 137–145.

—— , 1987, Late Cretaceous or younger SW-directed extensional faulting: Cosmos Hills, Brooks Range, Alaska: Geological Society of America Abstracts with Programs, v. 19, no. 6, p. 361.

Box, S. E., and Patton, W. W., 1989, Igneous history of the Koyukuk terrane, western Alaska: Constraints on the origin, evolution, and ultimate collision of an accreted island-arc terrane: Journal of Geophysical Research, v. 94, p. 15,843–15,867.

Brosgé, W. P., 1975, Metamorphic belts in southern Brooks Range, *in* Yount, M. E., ed., United States Geological Survey Program, 1975: U.S. Geological Survey Circular 722, p. 40–41.

Brosgé, W. P., and Reiser, H. N., 1964, Geologic map and section of the Chandalar quadrangle, Alaska: U.S. Geological Survey Miscellaneous Geologic Investigations Map I-375, scale 1:250,000.

Brosgé, W. P., and Tailleur, I. L., 1970, Depositional history of northern Alaska, *in* Adkison, W. L., and Brosgé, M. M., eds., Proceedings of the geological seminar on the North Slope of Alaska: Los Angeles, American Association of Petroleum Geologists, p. D1–D18.

—— , 1971, Northern Alaska petroleum province, *in* Cram, I. H., ed., Future petroleum provinces of the United States—Their geology and potential: American Association of Petroleum Geologists Memoir 15, p. 68–99.

Brosgé, W. P., Dutro, J. T., Jr., Mangus, M. D., and Reiser, H. N., 1962, Paleozoic sequence in eastern Brooks Range, Alaska: American Association Petroleum Geologists Bulletin, v. 46, p. 2,174–2,198.

Brosgé, W. P., Whittington, C. L., and Morris, R. H., 1966, Geology of the Umiat-Maybe Creek region, Alaska, *in* Part 3, Areal geology: Exploration of Naval Petroleum Reserve No. 4 and adjacent areas, northern Alaska, 1944–53: U.S. Geological Survey Professional Paper 303-H, p. 501–638.

Brosgé, W. P., Reiser, H. N., Dutro, J. T., Jr., and Detterman, R. L., 1979, Bedrock geologic map of the Philip Smith Mountains quadrangle, Alaska: U.S. Geological Survey Miscellaneous Field Studies Map MF-897B, scale 1:250,000.

Bruns, T. R., Fisher, M. A., Leinbach, W. J., Jr., and Miller, J. J., 1987, Regional structure of rocks beneath the coastal plain, *in* Bird, K. J., and Magoon, L. B., eds., Petroleum geology of the northern part of the Arctic National Wildlife Refuge, northeastern Alaska: U.S. Geological Survey Bulletin 1778, p. 249–254.

Bruynzeel, J. W., Guldenzopf, E. C., and Pickard, J. E., 1982, Petroleum exploration of NPRA, 1974–1981, Final report: Houston, Texas, Tetra Tech, 130 p.

Buckingham, M. L., 1987, Fluvio-deltaic sedimentation patterns of the Upper Cretaceous to lower Tertiary Jago River Formation, Arctic National Wildlife Refuge (ANWR), northeastern Alaska, *in* Tailleur, I., and Weimer, P., eds., Alaskan North Slope geology: Bakersfield, California, Society of Economic Paleontologists and Mineralogists, Pacific Section, and Alaska Geological Society, Book 50, p. 529–540.

Cady, J. W., 1989, Geologic implications of topographic, gravity, and aeromagnetic data in the northern Yukon–Koyukuk province and its borderlands, Alaska: Journal of Geophysical Research, v. 94, p. 15,821–15,841.

Campbell, R. H., 1967, Areal geology in the vicinity of the Chariot site, Lisburne Peninsula, northwest Alaska: U.S. Geological Survey Professional Paper 395, 71 p.

Carey, S. W., 1958, A tectonic approach to continental drift, *in* Carey, S. W., ed., Continental drift—A symposium: Hobart, Tasmania University, p. 177–355.

Carlson, C., 1985, Large-scale south-dipping, low-angle normal faults in the southern Brooks Range, Alaska: Eos (American Geophysical Union Transactions), v. 66, p. 1074.

Carman, G. J., and Hardwick, P., 1983, Geology and regional setting of the

Kuparuk oil field, Alaska: American Association of Petroleum Geologists Bulletin, v. 67, p. 1014–1031.
Carter, C., and Laufeld, S., 1975, Ordovician and Silurian fossils in well cores from North Slope of Alaska: American Association of Petroleum Geologists Bulletin, v. 59, p. 457–464.
Carter, C., and Tailleur, I. L., 1984, Ordovician graptolites from the Baird Mountains, western Brooks Range, Alaska: Journal of Paleontology, v. 58, p. 40–57.
Chapman, R. M., and Sable, E. G., 1960, Geology of the Utukok-Corwin region, northwestern Alaska: U.S. Geological Survey Professional Paper 303-C, p. 47–167.
Chapman, R. M., Detterman, R. L., and Mangus, M. D., 1964, Geology of the Killik-Etivluk rivers region, Alaska: U.S. Geological Survey Professional Paper 303-F, p. 325–407.
Churkin, M., Jr., 1975, Basement rocks of Barrow arch, Alaska, and circum-Arctic Paleozoic mobile belt: American Association of Petroleum Geologists Bulletin, v. 59, p. 451–456.
Churkin, M., Jr., and Trexler, J. H., Jr., 1980, Circum-Arctic plate accretion—Isolating part of a Pacific plate to form the nucleus of the Arctic basin: Earth and Planetary Science Letters, v. 49, p. 356–362.
—— , 1981, Continental plates and accreted oceanic terranes in the Arctic, *in* Nairn, A.E.M., Churkin, M., Jr., and Stehli, F. G., eds., The ocean basin and margins, Volume 5: The Arctic Ocean: New York, Plenum Press, p. 1–20.
Churkin, M., Jr., Nokleberg, W. J., and Huie, C., 1979, Collision-deformed Paleozoic continental margin, western Brooks Range, Alaska: Geology, v. 7, p. 379–383.
Clough, J. G., Blodgett, R. B., Imm, T. A., and Pavia, E. A., 1988, Depositional environments of Katakturuk Dolomite and Nanook Limestone, Arctic National Wildlife Refuge, Alaska: American Association of Petroleum Geologists Bulletin, v. 72, p. 172.
Coleman, R. G., 1984, The diversity of ophiolites: Geologie en Mijnbouw, v. 63, p. 141–150.
Collins, F. R., 1958, Test wells, Topagoruk area, Alaska, *in* Part 5, Subsurface geology and engineering data: Exploration of Naval Petroleum Reserve No. 4 and adjacent areas, northern Alaska, 1944–53: U.S. Geological Survey Professional Paper 305-D, p. 265–316.
—— , 1961, Core test and test wells, Barrow area, Alaska, *in* Part 5, Subsurface geology and engineering data: Exploration of Naval Petroleum Reserve No. 4 and adjacent areas, northern Alaska, 1944–53: U.S. Geological Survey Professional Paper 305-K, p. 569–644.
Coney, P. J., and Jones, D. L., 1985, Accretion tectonics and crustal structure in Alaska: Tectonophysics, v. 119, p. 265–283.
Craig, J. D., Sherwood, K. W., and Johnson, P. P., 1985, Geologic report for the Beaufort Sea planning area, Alaska: Minerals Management Service OCS (Outer Continental Shelf) Report MMS 85-0111, 192 p.
Crane, R. C., 1987, Cretaceous olistostrome model, Brooks Range, Alaska, *in* Tailleur, I., and Weimer, P., eds., Alaskan North Slope geology: Bakersfield, California, Society of Economic Paleontologists and Mineralogists, Pacific Section, and Alaska Geological Society, Book 50, p. 433–440.
Crane, R. C., and Mull, C. G., 1987, Structural style—Brooks Range mountain front, Alaska, *in* Tailleur, I., and Weimer, P., eds., Alaskan North Slope geology: Bakersfield, California, Society of Economic Paleontologists and Mineralogists, Pacific Section, and Alaska Geological Society, Book 50, p. 631–638.
Crane, R. C., and Wiggins, V. D., 1976, The Ipewik Formation, a significant Jurassic-Neocomian map unit in the northern Brooks Range fold belt: American Association of Petroleum Geologists, Pacific Section, 51st Annual Meeting, Program and Abstracts, p. 25–26.
Crowder, R. K., 1987, Cretaceous basin to shelf transition in northern Alaska: Deposition of the Fortress Mountain Formation, *in* Tailleur, I., and Weimer, P., eds., Alaskan North Slope geology: Bakersfield, California, Society of Economic Paleontologists and Mineralogists, Pacific Section, and Alaska Geological Society, Book 50, p. 449–458.
—— , 1989, Deposition of the Fortress Mountain Formation, *in* Mull, C. G., and Adams, K. E., eds., Dalton Highway, Yukon River to Prudhoe Bay, Alaska: Alaska Division of Geological and Geophysical Surveys Guidebook 7, v. 2, p. 293–301.
—— , 1990, Permian and Triassic sedimentation in the northeastern Brooks Range, Alaska: Deposition of the Sadlerochit Group: American Association of Petroleum Geologists Bulletin, v. 74, p. 1351–1370.
Curtis, S. M., Ellersieck, I., Mayfield, C. F., and Tailleur, I. L., 1984, Reconnaissance geologic map of southwestern Misheguk Mountain quadrangle, Alaska: U.S. Geological Survey Miscellaneous Investigations Series Map I-1502, scale 1:63,360.
—— , 1990, Reconnaissance geologic map of the De Long Mountains A-1 and B-1 quadrangles and part of the C-2 quadrangle, Alaska: U.S. Geological Survey Miscellaneous Investigations Series Map I-1930, scale 1:63,360.
Decker, J. E., and Dillon, J. T., 1984, Interpretation of regional aeromagnetic signatures along the southern margin of the Brooks Range, Alaska: Geological Society of America Abstracts with Programs, v. 16, p. 278.
Detterman, R. L., 1970, Sedimentary history of Sadlerochit and Shublik formations in northeastern Alaska, *in* Adkison, W. L., and Brosgé, M. M., eds., Proceedings of the geological seminar on the North Slope of Alaska: Los Angeles, California, American Association of Petroleum Geologists, Pacific Section, p. O1–O13.
Detterman, R. L., Reiser, H. N., Brosgé, W. P., and Dutro, J. T., Jr., 1975, Post-Carboniferous stratigraphy, northeastern Alaska: U.S. Geological Survey Professional Paper 886, 46 p.
Dillon, J. T., 1987, Root zone of the Endicott allochthon, Alaska: Geological Society of America Abstracts with Programs, v. 19, p. 372.
—— , 1989, Structure and stratigraphy of the southern Brooks Range and northern Koyukuk basin near the Dalton Highway, *in* Mull, C. G., and Adams, K. E., eds., Dalton Highway, Yukon River to Prudhoe Bay, Alaska: Alaska Division of Geological and Geophysical Surveys Guidebook 7, v. 2, p. 157–187.
Dillon, J. T., and Smiley, C. J., 1984, Clasts from the Early Cretaceous Brooks Range orogen in Albian to Cenomanian molasse deposits of the northern Koyukuk basin: Geological Society of America Abstracts with Programs, v. 16, p. 279.
Dillon, J. T., Pessel, G. H., Chen, J. A., and Veach, N. C., 1980, Middle Paleozoic magmatism and orogenesis in the Brooks Range, Alaska: Geology, v. 8, p. 338–343.
Dillon, J. T., Brosgé, W. P., and Dutro, J. T., Jr., 1986, Generalized geologic map of the Wiseman quadrangle, Alaska: U.S. Geological Survey Open-File Report OF 86-219, scale 1:250,000.
Dillon, J. T., Harris, A. G., and Dutro, J. T., Jr., 1987a, Preliminary description and correlation of lower Paleozoic fossil-bearing strata in the Snowden Mountain area of the south-central Brooks Range, Alaska, *in* Tailleur, I., and Weimer, P., eds., Alaskan North Slope geology: Bakersfield, California, Society of Economic Paleontologists and Mineralogists, Pacific Section, and Alaska Geological Society, Book 50, p. 337–345.
Dillon, J. T., Tilton, G. R., Decker, J., and Kelly, M. J., 1987b, Resource implications of magmatic and metamorphic ages for Devonian igneous rocks in the Brooks Range, *in* Tailleur, I., and Weimer, P., eds., Alaskan North Slope geology: Bakersfield, California, Society of Economic Paleontologists and Mineralogists, Pacific Section, and Alaska Geological Society, Book 50, p. 713–723.
Dumoulin, J. A., 1988, Stromatolite- and coated-grain-bearing carbonate rocks of the western Brooks Range, *in* Galloway, J. P., and Hamilton, T. D., eds., Geologic studies in Alaska by the U.S. Geological Survey in 1987: U.S. Geological Survey Circular 1016, p. 31–34.
Dumoulin, J. A., and Harris, A. G., 1987, Lower Paleozoic carbonate rocks of the Baird Mountains quadrangle, western Brooks Range, Alaska, *in* Tailleur, I., and Weimer, P., eds., Alaskan North Slope geology: Bakersfield, California, Society of Economic Paleontologists and Mineralogists, Pacific Section, and Alaska Geological Society, Book 50, p. 311–336.
Dusel-Bacon, C., Brosgé, W. P., Till, A. B., Doyle, E. O., Mayfield, C. F., Reiser, H. N., and Miller, T. P., 1989, Distribution, facies, ages, and proposed

tectonic associations of regionally metamorphosed rocks in northern Alaska: U.S. Geological Survey Professional Paper 1497-A, 44 p., 2 sheets, scale 1:1,000,000.

Dutro, J. T., Jr., 1952, Stratigraphy and paleontology of the Noatak and associate formations, Brooks Range, Alaska, *in* Naval Petroleum Reserve No. 4, Alaska: U.S. Geological Survey Geological Investigations, Special Report 33, 154 p.

——, 1970, Pre-Carboniferous carbonate rocks, *in* Adkison, W. L., and Brosgé, M. M., eds., Proceedings of the geological seminar on the North Slope of Alaska: Los Angeles, American Association of Petroleum Geologists, p. M1–M17.

——, 1981, Geology of Alaska bordering the Arctic Ocean, *in* Nairn, A.E.M., Churkin, M., Jr., and Stehli, F. G., eds., The ocean basins and margins, Volume 5: The Arctic Ocean: New York, Plenum Press, p. 21–36.

——, 1987, Revised megafossil biostratigraphic zonation for the Carboniferous of northern Alaska, *in* Tailleur, I., and Weimer, P., eds., Alaskan North Slope geology: Bakersfield, California, Society of Economic Paleontologists and Mineralogists, Pacific Section, and Alaska Geological Society, Book 50, p. 359–364.

Dutro, J. T., Jr., Brosgé, W. P., and Reiser, H. N., 1972, Significance of recently discovered Cambrian fossils and reinterpretation of Neruokpuk Formation, northeastern Alaska: American Association of Petroleum Geologists Bulletin, v. 56, p. 808–815.

Dutro, J. T., Jr., Brosgé, W. P., Lanphere, M. A., and Reiser, H. N., 1976, Geologic significance of Doonerak structural high, central Brooks Range, Alaska: American Association of Petroleum Geologists Bulletin, v. 60, p. 952–961.

Dutro, J. T., Jr., Brosgé, W. P., Reiser, H. N., and Detterman, R. L., 1979, Beaucoup Formation, a new Upper Devonian stratigraphic unit in the central Brooks Range, northern Alaska, *in* Sohl, N. F., and Wright, W. B., eds., Changes in stratigraphic nomenclature by the U.S. Geological Survey: U.S. Geological Survey Bulletin 1482-A, p. A63–A69.

Dutro, J. T., Jr., Palmer, A. R., Repetski, J. E., and Brosgé, W. P., 1984, Middle Cambrian fossils from the Doonerak anticlinorium, central Brooks Range, Alaska: Journal of Paleontology, v. 58, p. 1364–1371.

Eckelmann, W. R., Fisher, W. L., and DeWitt, R. J., 1976, Prediction of fluvial-deltaic reservoir, Prudhoe Bay field, Alaska, *in* Miller, T. P., ed., Symposium on recent and ancient sedimentary environments in Alaska: Anchorage, Alaska Geological Society, p. B1–B8.

Ehm, A., and Tailleur, I. L., 1985, Refined names for Brookian elements in northern Alaska: American Association of Petroleum Geologists Bulletin, v. 69, p. 664.

Einaudi, M. T., and Hitzman, M. W., 1986, Mineral deposits in northern Alaska: Introduction: Economic Geology, v. 81, p. 1583–1591.

Ellersieck, I., Mayfield, C. F., Tailleur, I. L., and Curtis, S. M., 1979, Thrust sequences in the Misheguk Mountain quadrangle, Brooks Range, Alaska: U.S. Geological Survey Circular 804-B, p. B8–B9.

Ellersieck, I., Curtis, S. M., Mayfield, C. F., and Tailleur, I. L., 1984, Reconnaissance geologic map of south-central Misheguk Mountain quadrangle: U.S. Geological Survey Miscellaneous Investigations Series Map I-1504, scale 1:63,360.

——, 1990, Reconnaissance geologic map of the De Long Mountains A-2 and B-2 quadrangles and part of the C-2 quadrangle, Alaska: U.S. Geological Survey Miscellaneous Investigations Series Map I-1931, scale 1:63,360.

Fraser, G. S., and Clark, R. H., 1976, Transgressive-regressive shelf deposition, Shublik Formation, Prudhoe Bay area, Alaska: American Association of Petroleum Geologists Bulletin, v. 60, p. 672.

Fritts, C. E., Eakins, G. R., and Garland, R. E., 1971, Geology and geochemistry near Walker Lake, southern Survey Pass quadrangle, Arctic Alaska: Alaska Division of Geological and Geophysical Surveys Annual Report, p. 19–27.

Fujita, K., and Newberry, J. T., 1982, Tectonic evolution of northeastern Siberia and adjacent regions: Tectonophysics, v. 89, p. 337–357.

Gautier, D. L., 1987, Petrology of Cretaceous and Tertiary reservoir sandstones in the Point Thompson area, *in* Bird, K. J., and Magoon, L. B., eds., Petroleum geology of the northern part of the Arctic National Wildlife Refuge, northeastern Alaska: U.S. Geological Survey Bulletin 1778, p. 117–122.

Gaynor, G. C., and Scheihing, M. H., 1988, Shelf depositional environments and reservoir characteristics of the Kuparuk River Formation (Lower Cretaceous), Kuparuk Field, North Slope, Alaska, *in* Lomando, A. J., and Harris, P. M., eds., Giant oil and gas fields, a core workshop: Tulsa, Oklahoma, Society of Economic Paleontologists and Mineralogists Core Workshop no. 12, p. 333–389.

Gottschalk, R. R., Jr., 1987, Structural and petrologic evolution of the southern Brooks Range near Wiseman, Alaska [Ph.D. thesis]: Houston, Texas, Rice University, 263 p.

——, 1990, Structural evolution of the schist belt, south-central Brooks Range fold and thrust belt, Alaska: Journal of Structural Geology, v. 12, p. 453–469.

Gottschalk, R. R., Jr., and Oldow, J. S., 1988, Low-angle normal faults in the south-central Brooks Range fold and thrust belt, Alaska: Geology, v. 16, p. 395–399.

Grantz, A., 1966, Strike-slip faults in Alaska: U.S. Geological Survey Open-File Report OFR-267, 82 p.

Grantz, A., and May, S. D., 1983, Rifting history and structural development of the continental margin north of Alaska, *in* Watkins, J. S., and Drake, C. L., eds., Studies in continental margin geology: American Association of Petroleum Geologists Memoir 34, p. 77–100.

——, 1987, Regional geology and petroleum potential of the United States Chukchi shelf north of Point Hope, *in* Scholl, D. W., Grantz, A., and Vedder, J. G., eds., Geology and resource potential of the continental margin of western North America and adjacent ocean basins—Beaufort Sea to Baja California: Circum-Pacific Council for Energy and Mineral Resources, Earth Science Series, v. 6, p. 37–58.

Grantz, A., Wolf, S. C., Breslau, L., Johnson, T. C., and Hanna, W. F., 1970, Reconnaissance geology of the Chukchi Sea as determined by acoustic and magnetic profiles, *in* Adkison, W. L., and Brosgé, M. M., eds., Proceedings of the geological seminar on the North Slope of Alaska: Los Angeles, American Association of Petroleum Geologists, p. F1–F28.

Grantz, A., Holmes, M. L., and Kososki, B. A., 1975, Geologic framework of the Alaskan continental terrace in the Chukchi and Beaufort Seas, *in* Yorath, C. J., Parker, E. R., and Glass, D. J., eds., Canada's continental margins and offshore petroleum exploration: Canadian Society of Petroleum Geologists Memoir 4, p. 669–700.

Grantz, A., Eittreim, S., and Whitney, O. T., 1981, Geology and physiography of the continental margin north of Alaska and implications for the origin of the Canada basin, *in* Nairn, A.E.M., Churkin, M., and Stehli, F. G., eds., The ocean basins and margins, Volume 5: The Arctic Ocean: New York, Plenum Press, p. 439–492.

Grantz, A., Dinter, D. A., and Biswas, N. N., 1983a, Holocene faulting, warping, and earthquake epicenters on the Beaufort Shelf north of Alaska: U.S. Geological Survey Miscellaneous Investigations Map I-1189C, 7 p., 3 sheets, scale 1:500,000.

Grantz, A., Tailleur, I. L., and Carter, C., 1983b, Tectonic significance of Silurian and Ordovician graptolites, Lisburne Hills, northwest Alaska: Geological Society of America Abstracts with Programs, v. 15, p. 274.

Grantz, A., Dinter, D. A., and Culotta, R. C., 1987, Structure of the continental shelf north of the Arctic National Wildlife Refuge, *in* Bird, K. J., and Magoon, L. B., eds., Petroleum geology of the northern part of the Arctic National Wildlife Refuge, northeastern Alaska: U.S. Geological Survey Bulletin 1778, p. 271–276.

Grantz, A., May, S. D., and Hart, P. E., 1990a, Geology of the Arctic continental margin of Alaska, *in* Grantz, A., Johnson, L., and Sweeney, J. F., eds., The Arctic Ocean region: Boulder, Colorado, Geological Society of America, The Geology of North America, v. L, p. 257–288.

Grantz, A., May, S. D., Taylor, P. T., and Lawver, L. A., 1990b, Canada basin, *in* Grantz, A., Johnson, L., and Sweeney, J. F., eds., The Arctic Ocean region: Boulder, Colorado, Geological Society of America, The Geology of North America, v. L, p. 379–402.

Grantz, A., Moore, T. E., and Roeske, S. M., 1991, A-3 Gulf of Alaska to Arctic Ocean: Boulder, Colorado, Geological Society of America Centennial Continent/Ocean Transect no. 15, 72 p., 3 sheets, scale 1:500,000.

Grybeck, D., and Nelson, S. W., 1981, Structure of the Survey Pass quadrangle, Brooks Range, Alaska: U.S. Geological Survey Miscellaneous Geologic Investigations Map MF—1176B, 8 p., scale 1:250,000.

Gryc, G., Patton, W. W., Jr., and Payne, T. G., 1951, Present Cretaceous stratigraphic nomenclature of northern Alaska: Washington Academy of Science Journal, v. 41, p. 159–167.

Gutman, S. I., Goldstein, A., and Guldenzopf, E. C., 1982, Gravity and magnetic investigations of the National Petroleum Reserve in Alaska: Houston, Texas, Tetra Tech, 88 p.

Halgedahl, S. L., and Jarrard, R. D., 1987, Paleomagnetism of the Kuparuk River Formation from oriented drill core: Evidence for rotation of the Arctic Alaska plate, *in* Tailleur, I., and Weimer, P., eds., Alaskan North Slope geology: Bakersfield, California, Society of Economic Paleontologists and Mineralogists, Pacific Section, and Alaska Geological Society, Book 50, p. 581–617.

Handschy, J. W., and Oldow, J. S., 1989, Strain variation in the Endicott Mountains allochthon, central Brooks Range, Alaska: Geological Society of America Abstracts with Program, v. 21, no. 5, p. 89.

Handschy, J. W., Oldow, J. S., and Avé Lallemant, H. G., 1987, Fold-thrust kinematics and strain distribution in the Endicott Mountains allochthon, Brooks Range, Alaska: Eos (American Geophysical Union Transactions), v. 68, p. 1457.

Harris, A. G., Repetski, J. E., and Dumoulin, J. A., 1988, Ordovician carbonate rocks and conodonts from northern Alaska, *in* Williams, S. H., and Barnes, C. R., eds., Program and Abstracts for the Fifth International Symposium on the Ordovician System, St. Johns, Newfoundland, Canada: Subcommission on Ordovician Stratigraphy (ICS/IUGS), IGCP Project 216—Global Bioevents, p. 38.

Harris, R. A., 1988, Origin, emplacement, and attenuation of the Misheguk Mountain allochthon, western Brooks Range, Alaska: Geological Society of America Abstracts with Programs, v. 20, no. 7, p. A112.

Hemming, S., Sharp, W. D., Moore, T. E., and Mezger, K., 1989, U/Pb dating of detrital zircons from the Carboniferous Nuka Formation, Brooks Range, Alaska: Evidence for a 2.07 Ga provenance: Geological Society of America Abstracts with Programs, v. 21, no. 6, p. A190.

Henning, M. W., 1982, Reconnaissance geology and stratigraphy of the Skajit Formation, Wiseman B-5 quadrangle, Alaska: Alaska Division of Geological and Geophysical Surveys Open-File Report AOF-147, scale 1:63,360.

Herron, E. M., Dewey, J. F., and Pitman, W. C., 1974, Plate tectonics model for the evolution of the Arctic: Geology, v. 2, p. 377–380.

Hillhouse, J. W., and Grommé, C. S., 1988, Cretaceous remagnetization of Paleozoic sedimentary rocks in the Brooks Range, Alaska, *in* Gryc, G., ed., Geology and exploration of the National Petroleum Reserve in Alaska, 1974 to 1982: U.S. Geological Survey Professional Paper 1399, p. 633–644.

Hitzman, M. W., Smith, T. E., and Proffett, J. M., Jr., 1982, Bedrock geology of the Ambler district, southwestern Brooks Range, Alaska: Alaska Division of Geological and Geophysical Surveys Geologic Report 75, 2 sheets, scale 1:125,000.

Hitzman, M. W., Proffett, J. M., Jr., Schmidt, J. M., and Smith, T. E., 1986, Geology and mineralization of the Ambler district, northwestern Alaska: Economic Geology, v. 81, p. 1592–1618.

Holdsworth, B. K., and Murchey, B. L., 1988, Paleozoic radiolarian biostratigraphy of the northern Brooks Range, Alaska, *in* Gryc, G., ed., Geology and exploration of the National Petroleum Reserve in Alaska, 1974 to 1982: U.S. Geological Survey Professional Paper 1399, p. 777–797.

Howell, D. G., and Wiley, T. J., 1987, Crustal evolution of northern Alaska inferred from sedimentology and structural relations of the Kandik area: Tectonics, v. 6, p. 619–631.

Howell, D. G., Jones, D. L., and Schermer, E. R., 1985, Tectonostratigraphic terranes of the circum-Pacific region, *in* Howell, D. G., ed., Tectonostratigraphic terranes of the circum-Pacific region: Houston, Texas, Circum-Pacific Council for Energy and Mineral Resources, Earth Sciences Series no. 1, p. 3–30.

Howell, D. G., Jones, D. L., and Coney, P. J., 1986, Convergent-margin geologic characterization for deep source hydrocarbon potential, *in* Deep source gas/gas hydrates: U.S. Geological Survey Quarterly Report (Annual Summary) for Department of Energy, Contract DE-AI21-83-MC20422, 11 p.

Howell, D. G., Bird, K. J, Huafu, L., and Johnsson, M. J., 1992, Tectonics and petroleum potential of the Brooks Range fold and thrust belt—A progress report, *in* Bradley, D. C., and Ford, A. B., Geologic studies in Alaska by the U.S. Geological Survey, 1990: U.S. Geological Survey Bulletin 1999, p. 112–126.

Hubbard, R. J., Edrich, S. P., and Rattey, R. P., 1987, Geologic evolution and hydrocarbon habitat of the Arctic Alaska microplate, *in* Tailleur, I., and Weimer, P., eds., Alaskan North Slope geology: Bakersfield, California, Society of Economic Paleontologists and Mineralogists, Pacific Section, and Alaska Geological Society, Book 50, p. 797–830.

Huffman, A. C., Jr., Ahlbrandt, T. S., Pasternac, I., Stricker, G. D., and Fox, J. E., 1985, Depositional and sedimentologic factors affecting the reservoir potential of the Cretaceous Nanushuk Group, central North Slope, Alaska, *in* Huffman, A. C., Jr., ed., Geology of the Nanushuk Group and related rocks, North Slope, Alaska: U.S. Geological Survey Bulletin 1614, p. 61–74.

Hunter, R. E., and Fox, J. E., 1976, Interpretation of depositional environments in the Fortress Mountain Formation, central Arctic Slope, *in* Cobb, E. H., ed. The United States Geological Survey in Alaska: Accomplishments during 1975: U.S. Geological Survey Circular 733, p. 30–31.

Jamison, H. C., Brockett, L. D., and McIntosh, R. A., 1980, Prudhoe Bay—A 10-year perspective, *in* Giant oil and gas fields of the decade 1968–1978: American Association of Petroleum Geologists Memoir 30, p. 289–314.

Jamison, W. R., 1987, Geometric analysis of fold development in overthrust terranes: Journal of Structural Geology, v. 9, p. 207–219.

Jones, D. L., 1983, Recognition, character, and analysis of tectonostratigraphic terranes in western North America, *in* Hashimoto, M., and Uyeda, S., eds., Accretion tectonics in the circum-Pacific regions: Boston, D. Reidel Publishing Co., p. 21–36.

Jones, D. L., and Coney, P., 1989, Regional decollement at base of Hunt Fork Shale, eastern Brooks Range, Alaska: Geological Society of America Abstracts with Programs, v. 21, no. 5, p. 99.

Jones, D. L., and Grantz, A., 1964, Stratigraphic and structural significance of Cretaceous fossils from Tiglukpuk Formation, northern Alaska: American Association of Petroleum Geologists Bulletin, v. 48, p. 1462–1474.

Jones, D. L., Silberling, N. J., Coney, P. J., and Plafker, G., 1987, Lithotectonic terrane map of Alaska: U.S. Geological Survey Miscellaneous Field Studies Map MF-1874A, scale 1:2,500,000.

Jones, D. L., Coney, P. J., Harms, T.A., and Dillon, J. T., 1988, Interpretive geologic map and supporting radiolarian data from the Angayucham terrane, Coldfoot area, southern Brooks Range, Alaska: U.S. Geological Survey Miscellaneous Field Studies Map MF-1993, scale 1:63,360.

Jones, H. P., and Speers, R. G., 1976, Permo-Triassic reservoirs of the Prudhoe Bay gas fields, *in* Braunstein, J., ed., North American oil and gas fields: American Association of Petroleum Geologists Memoir 24, p. 23–50.

Jones, P. B., 1980, Evidence from Canada and Alaska on plate tectonic evolution of the Arctic Ocean basin: Nature, v. 285, p. 215–217.

——, 1982a, Oil and gas beneath east-dipping underthrust faults in the Alberta foothills, *in* Powers, R. B., ed., Geologic studies of the Cordilleran thrust belt, Volume 1: Denver, Colorado, Rocky Mountain Association of Geologists, p. 61–74.

——, 1982b, Mesozoic rifting in the western Arctic Ocean basin and its relationship to Pacific seafloor spreading, *in* Embry, A. F., and Balkwill, H. R., eds., Arctic geology and geophysics: Canadian Society of Petroleum Geologists Memoir 8, p. 83–99.

Julian, F. E., 1989, Structure and stratigraphy of lower Paleozoic rocks, Doonerak window, central Brooks Range, Alaska [Ph.D. thesis]: Houston, Texas, Rice University, 127 p., scale 1:31,680.

Karl, S. M., 1989, Paleoenvironmental implication of Alaskan siliceous deposits, in Hein, J. R., and Obradovic, J., eds., Siliceous deposits of the Tethys and Pacific regions: New York, Springer-Verlag, p. 169–200.

——, 1991, Arc and extensional basin geochemical and tectonic affinities for the Maiyumerak basalts in the western Brooks Range, *in* Bradley, D. C., and Ford, A. B., eds., Geologic studies in Alaska by the U.S. Geological Survey, 1990: U.S. Geological Survey Bulletin 1999, p. 141–155.

Karl, S. M., and Dickey, C. F., 1989, Geology and geochemistry indicate belts of both ocean floor and arc basalt and gabbro in the Maiyumerak Mountains, northwestern Brooks Range, Alaska: Geological Society of America Abstracts with Programs, v. 21, no. 5, p. 100.

Karl, S. M., and Long, C. L., 1987, Evidence for tectonic truncation of regional east-west-trending structures in the central Baird Mountains quadrangle, western Brooks Range, Alaska: Geological Society of America Abstracts with Programs, v. 19, p. 392.

——, 1990, Folded Brookian thrust faults: Implications of three geologic/geophysical transects in the western Brooks Range: Journal of Geophysical Research, v. 95, p. 8581–8592.

Karl, S. M., Dumoulin, J. A., Ellersieck, I., Harris, A. G., and Schmidt, J. M., 1989, Preliminary geologic map of the Baird Mountains quadrangle, Alaska: U.S. Geological Survey Open-File Report 89-551, 65 p., scale 1:250,000.

Keller, A. S., Morris, R. E., and Detterman, R. L., 1961, Geology of the Shaviovik and Sagavanirktok rivers region, Alaska: U.S. Geological Survey Professional Paper 303-D, p. 169–219.

Kelley, J. S., 1988, Preliminary geologic map of the Chandler Lake quadrangle, Alaska: U.S. Geological Survey Open-File Report 88-42, 2 sheets, scale 1:125,000.

Kelley, J. S., and Bohn, D., 1988, Decollements in the Endicott Mountains allochthon, north-central Brooks Range, *in* Galloway, J. P., and Hamilton, T. D., eds., Geologic studies in Alaska by the U.S. Geological Survey during 1987: U.S. Geological Survey Circular 1016, p. 44–47.

Kelley, J. S., and Foland, R. L., 1987, Structural style and framework geology of the coastal plain and adjacent Brooks Range, *in* Bird, K. J., and Magoon, L. B., eds., Petroleum geology of the northern part of the Arctic National Wildlife Refuge, northeastern Alaska: U.S. Geological Survey Bulletin 1778, p. 255–270.

Kelley, J. S., Brosgé, W. P., and Reynolds, M. W., 1985, Fold-nappes and polyphase thrusting in the north-central Brooks Range, Alaska: American Association of Petroleum Geologists Bulletin, v. 65, p. 667.

Kirschner, C. E., and Rycerski, B. A., 1988, Petroleum potential of representative stratigraphic and structural elements in the National Petroleum Reserve in Alaska, *in* Gyrc, G., ed., Geology and exploration of the National Petroleum Reserve in Alaska, 1974 to 1982: U.S. Geological Survey Professional Paper 1399, p. 191–208.

Kirschner, C. E., Gryc, G., and Molenaar, C., 1983, Regional seismic lines in the National Petroleum Reserve in Alaska, *in* Bally, A. W., ed., Seismic expression of structural styles, Volume 1: American Association of Petroleum Geologists Studies in Geology Series 15, p. 1.2.5-1–1.2.5-14.

Knock, D. G., 1987, Depositional setting and provenance of upper Neocomian Kemik Sandstone, Arctic National Wildlife Refuge (ANWR), northeastern Alaska: Geological Society of America Abstracts with Programs, v. 19, p. 395.

Lane, L. S., 1991, The pre-Mississippian "Neruokpuk Formation," northeastern Alaska and northwestern Yukon: Review and new regional correlation: Canadian Journal of Earth Sciences, v. 28, p. 1521–1533.

Lane, L. S., and Cecile, M. P., 1989, Stratigraphy and structure of the Neruokpuk Formation, northern Yukon, *in* Current research, Part G: Geological Survey of Canada Paper 89-1G, p. 57–62.

Lawton, T. F., Geehan, G. W., and Voorhees, B. J., 1987, Lithofacies and depositional environments of the Ivishak Formation, Prudhoe Bay field, *in* Tailleur, I., and Weimer, P., eds., Alaskan North Slope geology: Bakersfield, California, Society of Economic Paleontologists and Mineralogists, Pacific Section, and Alaska Geological Society, Book 50, p. 61–76.

Lawver, L. A., and Scotese, C. R., 1990, A review of tectonic models for the evolution of the Canada basin, *in* Grantz, A., Johnson, L., and Sweeney, J. F., eds., The Arctic Ocean region: Boulder, Colorado, Geological Society of America, The Geology of North America, v. L, p. 593–618.

Leffingwell, E. de K., 1919, The Canning River region, northern Alaska: U.S. Geological Survey Professional Paper 109, 251 p.

Leiggi, P. A., 1987, Style and age of tectonism of the Sadlerochit Mountains to Franklin Mountains, Arctic National Wildlife Refuge, Alaska, *in* Tailleur, I., and Weimer, P., eds., Alaskan North Slope geology: Bakersfield, California, Society of Economic Paleontologists and Mineralogists, Pacific Section, and Alaska Geological Society, Book 50, p. 749–756.

Lerand, M., 1973, Beaufort Sea, *in* McCrossam, R. G., ed., The future petroleum provinces of Canada—Their geology and potential: Canadian Society of Petroleum Geology Memoir 1, p. 315–386.

Loney, R. A., and Himmelberg, G. R., 1985, Ophiolitic ultramafic rocks of the Jade Mountains–Cosmos Hills area, southwestern Brooks Range, *in* Bartsch-Winkler, S., ed., Geologic studies in Alaska by the U.S. Geological Survey during 1985: U.S. Geological Survey Circular 967, p. 13–15.

Marinai, R. K., 1987, Petrography and diagenesis of the Ledge Sandstone Member of the Triassic Ivishak Formation, *in* Bird, K. J., and Magoon, L. B., eds., Petroleum geology of the northern part of the Arctic National Wildlife Refuge, northeastern Alaska: U.S. Geological Survey Bulletin 1778, p. 101–115.

Martin, A. J., 1970, Structure and tectonic history of the western Brooks Range, De Long Mountains, and Lisburne Hills, northern Alaska: Geological Society of America Bulletin, v. 81, p. 3605–3622.

Masterson, W. D., and Paris, C. E., 1987, Depositional history and reservoir description of the Kuparuk River Formation, North Slope, Alaska, *in* Tailleur, I., and Weimer, P., eds., Alaskan North Slope geology: Bakersfield, California, Society of Economic Paleontologists and Mineralogists, Pacific Section, and Alaska Geological Society, Book 50, p. 95–107.

Mayfield, C. F., 1980, Comment *on* "Collision-deformed Paleozoic continental margin, western Brooks Range, Alaska": Geology, v. 8, p. 357–359.

Mayfield, C. F., and Tailleur, I. L., 1978, Bedrock geology map of the Ambler River quadrangle, Alaska: U.S. Geological Survey Open-File Map 78-120A, scale 1:250,000.

Mayfield, C. F., Tailleur, I. L., Mull, C. G., and Silberman, M. L., 1978, Granitic clasts from Upper Cretaceous conglomerate in the northwestern Brooks Range, *in* Johnson, K. M., ed., The United States Geological Survey in Alaska: Accomplishments during 1977: U.S. Geological Survey Circular 772-B, p. B11–B13.

Mayfield, C. F., Silberman, M. L., and Tailleur, I. L., 1982, Precambrian metamorphic rocks from the Hub Mountain terrane, Baird Mountains quadrangle, Alaska, *in* Coonrad, W. L., ed., The United States Geological Survey in Alaska: Accomplishments during 1980: U.S. Geological Survey Circular 844, p. 18–22.

Mayfield, C. F., Curtis, S. M., Ellersieck, I., and Tailleur, I. L., 1984, Reconnaissance geologic map of southeastern Misheguk Mountain quadrangle, Alaska: U.S. Geological Survey Miscellaneous Investigations Series Map I-1503, scale 1:63,360.

Mayfield, C. F., Ellersieck, I., and Tailleur, I. L., 1987, Reconnaissance geologic map of the Noatak C-5, D-5, D-6, and D-7 quadrangles, Alaska: U.S. Geological Survey Miscellaneous Investigation Series Map I-1814, scale 1:63,360.

Mayfield, C. F., Tailleur, I. L., and Ellersieck, I., 1988, Stratigraphy, structure, and palinspastic synthesis of the western Brooks Range, northwestern Alaska, *in* Gryc, G., ed., Geology and exploration of the National Petroleum Reserve in Alaska, 1974 to 1982: U.S. Geological Survey Professional Paper 1399, p. 143–186.

Mayfield, C. F., Curtis, S. M., Ellersieck, I., and Tailleur, I. L., 1990, Reconnaissance geologic map of the De Long Mountains A-3 and B-3 quadrangles and parts of the A-4 and B-4 quadrangles: U.S. Geological Survey Miscellaneous Investigations Series Map I-1929, scale 1,63,360.

McWhae, J. R., 1986, Tectonic history of northern Alaska, Canadian Arctic, and Spitsbergen regions since Early Cretaceous: American Association of Petroleum Geologists Bulletin, v. 70, p. 430–450.

Melvin, J., 1987, Fluvio-paludal deposits in the lower Kekiktuk Formation (Mississippian), Endicott Field, northeast Alaska, *in* Ethridge, F. G., Flores, R. M., and Harvey, M. D., eds., Recent developments in fluvial sedimentology: Contributions from the Third International Fluvial Sedimentology Conference: Society of Economic Paleontologists and Mineralogists Special Publication 39, p. 343–352.

Melvin, J., and Knight, A. S., 1984, Lithofacies, diagenesis and porosity of the Ivashak Formation, Prudhoe Bay area, Alaska, *in* McDonald, D. A., and Surdam, R. C., eds., Clastic diagenesis: American Association of Petroleum Geologists Memoir 37, p. 347–365.

Metz, P. A., Egan, A., and Johanson, O., 1982, Landsat linear features and incipient rift system model for origin of base-metal and petroleum resources in northern Alaska, *in* Embry, A. F., and Balkhill, H. R., eds., Arctic geology and geophysics: Canadian Society of Petroleum Geologists Memoir 8, p. 101–112.

Mickey, M. B., and Haga, H., 1983, Jurassic-Neocomian seismic stratigraphy, NPRA: Boulder, Colorado, National Geophysical Data Center, Item TGY—0220-BL, 133 p.

Miller, E. L., 1987, Dismemberment of the Brooks Range orogenic belt during middle Cretaceous extension: Geological Society of America Abstracts with Programs, v. 19, p. 432.

Miller, E. L., and Hudson, T. L., 1991, Mid-Cretaceous extensional fragmentation of a Jurassic–Early Cretaceous compressional orogen, Alaska: Tectonics, v. 10, p. 781–796.

Molenaar, C. M., 1983, Depositional relations of Cretaceous and lower Tertiary rocks, northeastern Alaska: American Association of Petroleum Geologists Bulletin, v. 67, p. 1066–1081.

——, 1985, Subsurface correlations and depositional history of the Nanshuk Group and related strata, North Slope, Alaska, *in* Huffman, A. C., Jr., ed., Geology of the Nanushuk Group and related rocks, North Slope, Alaska: U.S. Geological Survey Bulletin 1614, p. 37–59.

——, 1988, Depositional history and seismic stratigraphy of Lower Cretaceous rocks in the National Petroleum Reserve in Alaska and adjacent areas, *in* Gryc, G., ed., Geology and exploration of the National Petroleum Reserve in Alaska, 1974 to 1982: U.S. Geological Survey Professional Paper 1399, p. 593–621.

Molenaar, C. M., Kirk, A. R., Magoon, L. B., and Huffman, A. C., 1984, Twenty-two measured sections of Cretaceous–lower Tertiary rocks, eastern North Slope, Alaska: U.S. Geological Survey Open-File Report 84-695, 19 p.

Molenaar, C. M., Bird, K. J., and Collett, T. S., 1986, Regional correlation sections across the North Slope of Alaska: U.S. Geological Survey Miscellaneous Field Studies Map MF-1907.

Molenaar, C. M., Bird, K. J., and Kirk, A. R., 1987, Cretaceous and Tertiary stratigraphy of northeastern Alaska, *in* Tailleur, I., and Weimer, P., eds., Alaskan North Slope geology: Bakersfield, California, Society of Economic Paleontologists and Mineralogists, Pacific Section, and Alaska Geological Society, Book 50, p. 513–528.

Molenaar, C. M., Egbert, R. M., and Krystinik, L. F., 1988, Depositional facies, petrography, and reservoir potential of the Fortress Mountain Formation (Lower Cretaceous), central North Slope, Alaska, *in* Gryc, G., ed., Geology and exploration of the National Petroleum Reserve in Alaska, 1974 to 1982: U.S. Geological Survey Professional Paper 1399, p. 257–280.

Moore, D. W., Young, L. E., Modene, J. S., and Plahuta, J. T., 1986, Geologic setting and genesis of the Red Dog zinc-lead-silver deposit, western Brooks Range, Alaska: Economic Geology, v. 81, p. 1696–1727.

Moore, T. E., 1986, Stratigraphic framework and tectonic implications of pre-Mississippian rocks, northern Alaska: Geological Society of America Abstracts with Programs, v. 18, p. 159.

——, 1987a, Geochemistry and tectonic setting of some volcanic rocks of the Franklinian assemblage, central and eastern Brooks Range, *in* Tailleur, I., and Weimer, P., eds., Alaskan North Slope geology: Bakersfield, California, Society of Economic Paleontologists and Mineralogists, Pacific Section, and Alaska Geological Society, Book 50, p. 691–710.

——, 1987b, Geochemical and tectonic affinity of basalts from the Copter Peak and Ipnavik River allochthons, Brooks Range, Alaska: Geological Society of America Abstracts with Programs, v. 19, p. 434.

Moore, T. E., and Churkin, M., Jr., 1984, Ordovician and Silurian graptolite discoveries from the Neruokpuk Formation (sensu lato), northeastern and central Brooks Range, Alaska: Paleozoic Geology of Alaska and Northwestern Canada Newsletter, no. 1, p. 21–23.

Moore, T. E., and Nilsen, T. H., 1984, Regional sedimentological variations in the Upper Devonian and Lower Mississippian(?) Kanayut Conglomerate, Brooks Range, Alaska: Sedimentary Geology, v. 38, p. 464–498.

Moore, T. E., Brosgé, W. P., Churkin, M., Jr., and Wallace, W. K., 1985a, Pre-Mississippian accreted terranes of northeastern Brooks Range, Alaska [abs.]: American Association of Petroleum Geologists Bulletin, v. 69, p. 670.

Moore, T. E., Whitney, J. W., and Wallace, W. K., 1985b, Cenozoic north-vergent tectonism in northeastern Alaska: Indentor tectonics in Alaska [abs.]: Eos (American Geophysical Union Transactions), v. 66, p. 862.

Moore, T. E., Wallace, W. K., and Jones, D. L., 1991, TACT geologic studies in the Brooks Range: Preliminary results and implications for crustal structure: Geological Society of America Abstracts with Programs, v. 23, no. 2, p. 80.

Moore, T. E., Wallace, W. K., Bird, K. J., Karl, S. M., Mull, C. G., and Dillon, J. T., 1992, Stratigraphy, structure, and geologic synthesis of northern Alaska: U.S. Geological Survey Open-File Report OF 92-330, 183 p., scale 1:2,500,000.

Mull, C. G., 1979, Nanushuk Group deposition and the late Mesozoic structural evolution of the central and western Brooks Range and Arctic Slope, *in* Ahlbrandt, T. S., ed., Preliminary geologic, petrologic, and paleontologic results of the Nanushuk Group rocks, North Slope, Alaska: U.S. Geological Survey Circular 794, p. 5–13.

——, 1980, Comment *on* "Collision-deformed Paleozoic continental margin, western Brooks Range, Alaska": Geology, v. 8, p. 361–362.

——, 1982, The tectonic evolution and structural style of the Brooks Range, Alaska: An illustrated summary, *in* Powers, R. B., ed., Geological studies of the Cordilleran thrust belt, Volume 1: Denver, Colorado, Rocky Mountain Association of Geologists, p. 1–45.

——, 1985, Cretaceous tectonics, depositional cycles, and the Nanushuk Group, Brooks Range and Arctic Slope, Alaska, *in* Huffman, A. C., Jr., ed., Geology of the Nanushuk Group and related rocks, North Slope, Alaska: U.S. Geological Survey Bulletin 1614, p. 7–36.

——, 1987, Kemik Sandstone, Arctic National Wildlife Refuge, northeastern Alaska, *in* Tailleur, I., and Weimer, P., eds., Alaskan North Slope geology: Bakersfield, California, Society of Economic Paleontologists and Mineralogists, Pacific Section, and Alaska Geological Society, Book 50, p. 405–431.

——, 1989, Generalized stratigraphy and structure of the Brooks Range and Arctic Slope, *in* Mull, C. G., and Adams, K. E., eds., Dalton Highway, Yukon River to Prudhoe Bay, Alaska: Alaska Division of Geological and Geophysical Surveys Guidebook 7, v. 1, p. 31–46.

Mull, C. G., and Mangus, M. D., 1972, Itkilyariak Formation: New Mississippian formation of Endicott Group, Arctic Slope of Alaska: American Association of Petroleum Geologists Bulletin, v. 56, p. 1364–1369.

Mull, C. G., and Tailleur, I. L., 1977, Sadlerochit(?) Group in the Schwatka Mountains, south-central Brooks Range, *in* Blean, K. M., ed., The United States Geological Survey in Alaska: Accomplishments during 1976: U.S. Geological Survey Circular 751-B, p. B27–B29.

Mull, C. G., Tailleur, I. L., Mayfield, C. F., and Pessel, G. H., 1976, New structural and stratigraphic interpretations, central and western Brooks Range and Arctic Slope, *in* Cobb, E. H., ed., The United States Geological Survey in Alaska: Accomplishments during 1975: U.S. Geological Survey Circular 733, p. 24–26.

Mull, C. G., Tailleur, I. L., Mayfield, C. F., Ellersieck, I. F., and Curtis, S., 1982, New upper Paleozoic and lower Mesozoic stratigraphic units, central and western Brooks Range, Alaska: American Association of Petroleum Geologists Bulletin, v. 66, p. 348–362.

Mull, C. G., Adams, K. E., and Dillon, J. T., 1987a, Stratigraphy and structure of the Doonerak fenster and Endicott Mountains allochthon, central Brooks Range, Alaska, *in* Tailleur, I., and Weimer, P., eds., Alaskan North Slope geology: Bakersfield, California, Society of Economic Paleontologists and

Mineralogists, Pacific Section, and Alaska Geological Society, Book 50, p. 663–679.

Mull, C. G., Crowder, R. K., Adams, K. E., Siok, J. P., Bodnar, D. A., Harris, E. E., and Alexander, R. A., 1987b, Stratigraphy and structural setting of the Picnic Creek allochthon, Killik River quadrangle, central Brooks Range, Alaska: A summary, *in* Tailleur, I., and Weimer, P., eds., Alaskan North Slope geology: Bakersfield, California, Society of Economic Paleontologists and Mineralogists, Pacific Section, and Alaska Geological Society, Book 50, p. 649–662.

Mull, C. G., Roeder, D. H., Tailleur, I. L., Pessel, G. H., Grantz, A., and May, S. D., 1987c, Geologic sections and maps across Brooks Range and Arctic Slope to Beaufort Sea: Geological Society of America Map and Chart Series MC-28S, scale 1:500,000.

Murchey, B. L., and Harris, A. G., 1985, Devonian to Jurassic sedimentary rocks in the Angayucham Mountains of Alaska: Possible seamount or oceanic plateau deposits: Eos (American Geophysical Union Transactions), v. 66, p. 1102.

Murchey, B. L., Jones, D. L., Holdsworth, B. K., and Wardlaw, B. R., 1988, Distribution patterns of facies, radiolarians, and conodonts in the Mississippian to Jurassic siliceous rocks of the north Brooks Range, Alaska, *in* Gryc, G., ed., Geology and exploration of the National Petroleum Reserve in Alaska, 1974 to 1982: U.S. Geological Survey Professional Paper 1399, p. 697–724.

Murphy, J. M., and Patton, W. W., Jr., 1988, Geologic setting and petrography of the phyllite and metagraywacke thrust panel, north-central Alaska, *in* Galloway, J. P., and Hamilton, T. D., eds., Geologic studies in Alaska by the U.S. Geological Survey during 1987: U.S. Geological Survey Circular 1016, p. 104–108.

Murphy, J. M., Moore, T. E., Patton, W. W., Jr., and Saward, S. E., 1989, Stratigraphy of Cretaceous conglomerates, northeastern Koyukuk basin, Alaska: Unroofing of southeastern Brooks Range, Alaska: Geological Society of America Abstracts with Programs, v. 21, no. 5, p. 120.

Namson, J. S., and Wallace, W. K., 1986, A structural transect across the northeastern Brooks Range, Alaska: Geological Society of America Abstracts with Programs, v. 18, p. 163.

Nelson, B. K., Nelson, S. W., and Till, A. B., 1989, Isotopic evidence of an Early Proterozoic crustal source for granites of the Brooks Range, northern Alaska: Geological Society of America Abstracts with Programs, v. 21, p. A105.

Nelson, R. E., and Carter, L. D., 1985, Pollen analysis of a late Pliocene and early Pleistocene section from the Gubik Formation of Arctic Alaska: Quaternary Research, v. 24, p. 295–306.

Nelson, S. W., and Grybeck, D., 1980, Geologic map of the Survey Pass quadrangle, Brooks Range, Alaska: U.S. Geological Survey Miscellaneous Field Studies Map MF-1176A, scale 1:250,000.

—— , 1981, Metamorphic rocks of the Survey Pass quadrangle: U.S. Geological Survey Miscellaneous Field Studies Map MF-1176C, scale 1:250,000.

Nelson, S. W., and Nelson, W. H., 1982, Geology of the Siniktanneyak Mountain ophiolite, Howard Pass quadrangle, Alaska: U.S. Geological Survey Miscellaneous Field Studies Map MF-1441, scale 1:63,360.

Newberry, R. J., Dillon, J. T., and Adams, D. D., 1986, Regionally metamorphosed, calc-silicate–hosted deposits of the Brooks Range, northern Alaska: Economic Geology, v. 81, p. 1728–1752.

Newman, G. W., Mull, C. G., and Watkins, N. D., 1977, Northern Alaska paleomagnetism, plate rotation, and tectonics, *in* Sisson, A., ed., The relationship of plate tectonics to Alaskan geology and resources: Proceedings of the Sixth Alaska Geological Society Symposium: Anchorage, Alaska Geological Society, p. C1–C7.

Nilsen, T. H., 1981, Upper Devonian and Lower Mississippian redbeds, Brooks Range, Alaska, *in* Miall, A. D., ed., Sedimentation and tectonics in alluvial basins: Geological Association of Canada Special Paper 23, p. 187–219.

Nilsen, T. H., and Moore, T. E., 1984, Stratigraphic nomenclature for the Upper Devonian and Lower Mississippian(?) Kanayut Conglomerate, Brooks Range, Alaska: U.S. Geological Survey Bulletin 1529-A, p. A1–A64.

Nilsen, T. H., Moore, T. E., Brosgé, W. P., and Dutro, J. T., Jr., 1981, Sedimentology and stratigraphy of the Kanayut Conglomerate and associated units, central and eastern Brooks Range, Alaska—Report of the 1979 field season: U.S. Geological Survey Open-File Report 81-506, 39 p.

Nokleberg, W. J., and Winkler, G. R., 1982, Stratiform zinc-lead deposits in the Drenchwater Creek area, Howard Pass quadrangle, northwestern Brooks Range, Alaska: U.S. Geological Survey Professional Paper 1209, 22 p., 2 sheets, scale 1:19,800.

Norris, D. K., 1985, The Neruokpuk Formation, Yukon Territory and Alaska, *in* Current research, Part B: Geological Survey of Canada Paper 85-1B, p. 223–229.

Norris, D. K., and Yorath, C. J., 1981, The North American plate from the Arctic archipelago to the Romanzof Mountains, *in* Nairn, A.E.M., Churkin, M., Jr., and Stehli, F. G., eds., The ocean basins and margins, Volume 5: The Arctic Ocean: New York, Plenum Press, p. 37–103.

North Slope Stratigraphic Committee, 1970, The Sag River Sandstone and Kuparuk River sands, two important subsurface units in the Prudhoe Bay field, *in* Adkison, W. L., and Brosgé, M. M., eds., Proceedings of the geological seminar on the North Slope of Alaska: Los Angeles, California, American Association of Petroleum Geologists, Pacific Section, p. P1–P3.

Nunn, J. A., Czerniak, M., and Pilger, R. H., Jr., 1987, Constraints on the structure of Brooks Range and Colville basin, northern Alaska, from flexure and gravity analysis: Tectonics, v. 6, p. 603–617.

Oldow, J. S., and Avé Lallemant, H. G., 1989, Tectonic elements of eastern Arctic Alaska and northwestern Canada: Eos (American Geophysical Union Transactions), v. 70, p. 1337–1338.

Oldow, J. S., Avé Lallemant, H. G., and Gottschalk, R. R., 1987a, Large-scale extension within a contractional orogen: South-central Brooks Range, Alaska: Eos (American Geophysical Union Transactions), v. 68, p. 1457.

Oldow, J. S., Avé Lallemant, H. G., Julian, F. E., and Seidensticker, C. M., 1987b, Ellesmerian(?) and Brookian deformation in the Franklin Mountains, northeastern Brooks Range, Alaska, and its bearing on the origin of the Canada basin: Geology, v. 15, p. 37–41.

Oldow, J. S., Gottschalk, R. R., and Avé Lallemant, H. G., 1987c, Low-angle normal faults: Southern Brooks Range fold and thrust belt, northern Alaska: Geological Society of America Abstracts with Programs, v. 19, p. 438.

Oldow, J. S., Seidensticker, C. M., Phelps, J. C., Julian, F. E., Gottschalk, R. R., Boler, K. W., Handschy, J. W., and Avé Lallemant, H. G., 1987d, Balanced cross sections through the central Brooks Range and North Slope, Arctic Alaska: Tulsa, Oklahoma, American Association of Petroleum Geologists, 19 p., scale 1:200,000.

Oliver, W. A., Jr., Merriam, C. W., and Churkin, M., Jr., 1975, Ordovician, Silurian, and Devonian corals of Alaska: U.S. Geological Survey Professional Paper 823-B, p. 13–44.

O'Sullivan, P. O., 1988, Preliminary results of 42 apatite fission track analyses of samples from Arctic National Wildlife Refuge, northeastern Alaska: Alaska Division of Geological and Geophysical Surveys, Public Data File Report 88-25A, 57 p.

O'Sullivan, P. O., Decker, J., and Bergman, S. C., 1989, Apatite fission track thermal history of Permian to Tertiary sedimentary rocks in the Arctic National Wildlife Refuge, northeastern Alaska: Geological Society of America Abstracts with Programs, v. 21, no. 5, p. 126.

Pallister, J. S., and Carlson, C., 1988, Bedrock geologic map of the Angayucham Mountains, Alaska: U.S. Geological Survey Miscellaneous Field Studies Map MF-2024, scale 1:63,360.

Pallister, J. S., Budahn, J. R., and Murchey, B. L., 1989, Pillow basalts of the Angayucham terrane: Oceanic-plateau and island crust accreted to the Brooks Range: Journal of Geophysical Research, v. 94, p. 15,901–15,923.

Palmer, A. R., 1983, The Decade of North American Geology 1983 time scale: Geology, v. 11, p. 503–504.

Palmer, A. R., Dillon, J. T., and Dutro, J. T., Jr., 1984, Middle Cambrian trilobites with Siberian affinities from the central Brooks Range, northern Alaska: Geological Society of America Abstracts with Programs, v. 16, p. 327.

Parrish, J. T., 1987, Lithology, geochemistry, and depositional environment of the

Triassic Shublik Formation, northern Alaska, *in* Tailleur, I., and Weimer, P., eds., Alaskan North Slope geology: Bakersfield, California, Society of Economic Paleontologists and Mineralogists, Pacific Section, and Alaska Geological Society, Book 50, p. 391–396.

Patton, W. W., Jr., 1956, New and redefined formations of Early Cretaceous age, *in* Gryc, G., and others, eds., Mesozoic sequence in Colville River region, northern Alaska: American Association of Petroleum Geologists Bulletin, v. 40, p. 209–254.

—— , 1957, A new upper Paleozoic formation, central Brooks Range, Alaska, *in* Part 3, Areal geology: Exploration of Naval Petroleum Reserve No. 4 and adjacent areas, northern Alaska, 1944–53: U.S. Geological Survey Professional Paper 303-B, p. 41–45.

—— , 1973, Reconnaissance geology of the northern Yukon-Koyukuk province, Alaska, *in* Shorter contributions to general geology: U.S. Geological Survey Professional Paper 774-A, p. A1–A17.

Patton, W. W., Jr., and Box, S. E., 1989, Tectonic setting of the Yukon-Koyukuk basin and its borderlands, western Alaska: Journal of Geophysical Research, v. 94, p. 15,807–15,820.

Patton, W. W., Jr., and Tailleur, I. L., 1964, Geology of the Killik-Itkillik region, Alaska, *in* Part 3, Areal geology: Exploration of Naval Petroleum Reserve No. 4 and adjacent areas, northern Alaska, 1944–53: U.S. Geological Survey Professional Paper 303-G, p. 409–500.

—— , 1977, Evidence in the Bering Strait region for differential movement between North America and Eurasia: Geological Society of America Bulletin, v. 88, p. 1298–1304.

Patton, W. W., Jr., Miller, T. P., and Tailleur, I. L., 1968, Regional geologic map of the Shungnak and southern part of the Ambler River quadrangle, Alaska: U.S. Geological Survey Miscellaneous Geologic Investigations Map I-554, scale 1:250,000.

Patton, W. W., Tailleur, I. L., Brosgé, W. P., and Lanphere, M. A., 1977, Preliminary report on the ophiolites of northern and western Alaska: Oregon Department of Geology and Mineral Industries Bulletin 95, p. 51–57.

Pavlis, T. L., 1989, Middle Cretaceous orogenesis in the northern Cordillera: A Mediterranean analog of collision-related extensional tectonics: Geology, v. 17, p. 947–950.

Payne, J. H., 1987, Diagenetic variations in the Permo-Triassic Ivishak Sandstone in the Prudhoe Bay field and central-northeastern National Petroleum Reserve in Alaska, *in* Tailleur, I., and Weimer, P., eds., Alaskan North Slope geology: Bakersfield, California, Society of Economic Paleontologists and Mineralogists, Pacific Section, and Alaska Geological Society, Book 50, p. 77–83.

Payne, T. G., 1955, Mesozoic and Cenozoic tectonic elements of Alaska: U.S. Geological Survey Miscellaneous Geologic Investigations Map I-84, scale 1:5,000,000.

Payne, T. G., and 8 others, 1951, Geology of the North Slope of Alaska: U.S. Geological Survey Oil and Gas Investigations Map OM-126, 3 sheets, scale 1:1,000,000.

Plumley, P. W., and Vance, M., 1988, Porcupine River basalt field, northeastern Alaska: Age, paleomagnetism, and tectonic significance: Eos (American Geophysical Union Transactions), v. 44, p. 1458.

Rattey, R. P., 1985, Northeastern Brooks Range, Alaska—New evidence for complex thin-skinned thrusting: American Association of Petroleum Geologists Bulletin, v. 69, p. 676–677.

Reed, B. L., 1968, Geology of the Lake Peters area, northeastern Brooks Range, Alaska: U.S. Geological Survey Bulletin 1236, 132 p.

Reiser, H. N., Brosgé, W. P., Dutro, J. T., Jr., and Detterman, R. L., 1971, Preliminary geologic map of the Mount Michelson quadrangle, Alaska: U.S. Geological Survey Open-File Report 71-237, scale 1:250,000.

Reiser, H. N., Norris, D. K., Dutro, J. T., Jr., and Brosgé, W. P., 1978, Restriction and renaming of the Neruokpuk Formation, northeastern Alaska, *in* Sohl, N. F., and Wright, W. B., eds., Changes in stratigraphic nomenclature by the U.S. Geological Survey, 1977, Contributions to stratigraphy: U.S. Geological Survey Bulletin 1457-A, p. A106–A107.

Reiser, H. N., Brosgé, W. P., Dutro, J. T., Jr., and Detterman, R. L., 1980, Geologic map of the Demarcation Point quadrangle, Alaska: U.S. Geological Survey Miscellaneous Investigations Series Map I-1133, scale 1:250,000.

Repetski, J. E., Carter, C., Harris, A. G., and Dutro, J. T., Jr., 1987, Ordovician and Silurian fossils from the Doonerak anticlinorium, central Brooks Range, Alaska, *in* Hamilton, T. D., and Galloway, J. P., eds., Geologic studies in Alaska by the U.S. Geological Survey during 1986: U.S. Geological Survey Circular 998, p. 40–42.

Rickwood, F. K., 1970, The Prudhoe Bay field, *in* Adkison, W. L., and Brosgé, M. M., eds., Proceedings of the geological seminar on the North Slope of Alaska: American Association of Petroleum Geologists, Pacific Section, p. L1–L11.

Robinson, F. M., 1959, Core tests, Simpson area, Alaska, *in* Part 5, Subsurface geology and engineering data: Exploration of Naval Petroleum Reserve No. 4 and adjacent areas, northern Alaska, 1944–53: U.S. Geological Survey Professional Paper 305-L, p. 645–730.

Robinson, F. M., Rucker, F. P., and Bergquist, H. R., 1956, Two subsurface formations of Early Cretaceous age, *in* Gryc, G., and others, Mesozoic sequence in Colville River region, northern Alaska: American Association of Petroleum Geologists Bulletin, v. 40, p. 223–233.

Robinson, M. S., Decker, J., Clough, J. G., Reifenstuhl, R. R., Dillon, J. T., Combellick, R. A., and Rawlinson, S. E., 1989, Geology of the Sadlerochit and Shublik mountains, Arctic National Wildlife Refuge, northeastern Alaska: Alaska Division of Geological and Geophysical Surveys Professional Report 100, scale 1:63,360.

Roeder, D., and Mull, C. G., 1978, Tectonics of Brooks Range ophiolites, Alaska: American Association of Petroleum Geologists Bulletin, v. 62, p. 1696–1702.

Sable, E. G., 1977, Geology of the western Romanzof Mountains, Brooks Range, northeastern Alaska: U.S. Geological Survey Professional Paper 897, 84 p., scale 1:63,360.

Sable, E. G., and Dutro, J. T., Jr., 1961, New Devonian and Mississippian formations in De Long Mountains, northern Alaska: American Association of Petroleum Geologists Bulletin, v. 45, p. 585–593.

Sable, E. G., and Stricker, G. D., 1987, Coal in the National Petroleum Reserve in Alaska (NPRA): Framework geology and resources, *in* Tailleur, I., and Weimer, P., eds., Alaskan North Slope geology: Bakersfield, California, Society of Economic Paleontologists and Mineralogists, Pacific Section, and Alaska Geological Society, Book 50, p. 195–215.

Schmidt, J. M., 1987, Paleozoic extension of the western Brooks Range (WRB) continental margin—Evidence from mineral deposits, igneous rocks, and sedimentary facies: Geological Society of America Abstracts with Programs, v. 19, p. 447.

Schrader, F. C., 1902, Geological section of the Rocky Mountains in northern Alaska: Geological Society of America Bulletin, v. 13, p. 233–252.

—— , 1904, A reconnaissance in northern Alaska: U.S. Geological Survey Professional Paper 20, 139 p.

Siok, J. P., 1985, Geologic history of the Siksikpuk Formation on the Endicott Mountains and Picnic Creek allochthons, north-central Brooks Range, Alaska [M.S. thesis]: Fairbanks, University of Alaska, 253 p.

—— , 1989, Stratigraphy and petrology of the Okpikruak Formation at Cobblestone Creek, north-central Brooks Range, *in* Mull, C. G., and Adams, K. E., eds., Dalton Highway, Yukon River to Prudhoe Bay, Alaska: Alaska Division of Geological and Geophysical Surveys Guidebook 7, v. 2, p. 285–292.

Siok, J. P., and Mull, C. G., 1987, Glauconitic phosphatic sandstone and oncolite deposition at the base of the Etivluk Group (Carboniferous) Picnic Creek allochthon, north-central Brooks Range, Alaska, *in* Tailleur, I., and Weimer, P., eds., Alaskan North Slope geology: Bakersfield, California, Society of Economic Paleontologists and Mineralogists, Pacific Section, and Alaska Geological Society, Book 50, p. 367–370.

Smith, D. G., 1987, Late Paleozoic to Cenozoic reconstructions of the Arctic, *in* Tailleur, I., and Weimer, P., eds., Alaskan North Slope geology: Bakersfield, California, Society of Economic Paleontologists and Mineralogists, Pacific Section, and Alaska Geological Society, Book 50, p. 785–795.

Smith, P. S., and Mertie, J. B., Jr., 1930, Geology and mineral resources of

northwestern Alaska: U.S. Geological Survey Bulletin 815, 351 p.
Smith, T. E., Webster, G. D., Heatwole, D. A., Proffett, J. M., Kelsey, G., and Glavinovich, P. S., 1978, Evidence for mid-Paleozoic depositional age of volcanogenic base-metal massive sulfide occurrences and enclosing strata, Ambler district, northwest Alaska: Geological Society of America Abstracts with Programs, v. 10, p. 148.
Snelson, S., and Tailleur, I. L., 1968, Large-scale thrusting and migrating Cretaceous foredeeps in western Brooks Range and adjacent regions of northwestern Alaska: American Association of Petroleum Geologists Bulletin, v. 52, p. 567.
Stewart, J. H., 1976, Late Precambrian evolution of North America: Plate tectonics implications: Geology, v. 4, p. 11–15.
Sweeney, J. F., 1982, Mid-Paleozoic travels of Arctic Alaska: Nature, v. 298, p. 647–649.
Tailleur, I. L., 1969, Rifting speculation of the geology of Alaska's North Slope: Oil and Gas Journal, v. 67, no. 39, p. 128–130.
——, 1973a, Probable rift origin of Canada basin, *in* Pitcher, M. G., ed., Arctic geology: American Association of Petroleum Geologists Memoir 19, p. 526–535.
——, 1973b, Possible mantle-derived rocks in western Brooks Range, *in* U.S. Geological Survey research 1973: U.S. Geological Survey Professional Paper 850, p. 64–65.
Tailleur, I. L., and Brosgé, W. P., 1970, Tectonic history of northern Alaska, *in* Adkison, W. L., and Brosgé, M. M., eds., Proceedings of the geological seminar on the North Slope of Alaska: Los Angeles, American Association of Petroleum Geologists, p. E1–E19.
Tailleur, I. L., and Sable, E. G., 1963, Nuka Formation of Late Mississippian to Late Permian age, new formation in northern Alaska: American Association Petroleum Geologists Bulletin, v. 47, p. 632–642.
Tailleur, I., and Weimer, P., eds., 1987, Alaskan North Slope geology: Bakersfield, California, Society of Economic Paleontologists and Mineralogists, Pacific Section, and Alaska Geological Society, Book 50, 874 p.
Tailleur, I. L., Kent, B. H., and Reiser, H. N., 1966, Outcrop geologic maps of the Nuka-Etivluk region, northern Alaska: U.S. Geological Survey Open-File Report 66-128, 7 sheets, scale 1:63,360.
Tailleur, I. L., Brosgé, W. P., and Reiser, H. N., 1967, Palinspastic analysis of Devonian rocks in northwestern Alaska, *in* Oswald, D. H., ed., International symposium on the Devonian system, Volume 2: Calgary, Alberta Society of Petroleum Geologists, p. 1345–1361.
Tailleur, I. L., Mamet, B. L., and Dutro, J. T., Jr., 1973, Revised age and structural interpretation of Nuka formation at Nuka Ridge, northwestern Alaska: American Association of Petroleum Geologists Bulletin, v. 57, p. 1348–1352.
Tailleur, I. L., Mayfield, C. F., and Ellersieck, I. F., 1977, Late Paleozoic sedimentary sequence, southwestern Brooks Range, *in* Blean, K. M., ed., The United States Geological Survey in Alaska: Accomplishments during 1976: U.S. Geological Survey Circular 751-B, p. B25–B27.
Till, A. B., 1988, Evidence for two Mesozoic blueschist belts in the hinterland of the western Brooks Range fold and thrust belt: Geological Society of America Abstracts with Program, v. 21, no. 7, p. A112.
——, 1989, Proterozoic rocks of the western Brooks Range, *in* Dover, J. H., and Galloway, J. P., eds., Geologic studies in Alaska by the U.S. Geological Survey, 1988: U.S. Geological Survey Bulletin 1903, p. 20–25.
Till, A. B., Schmidt, J. M., and Nelson, S. W., 1988, Thrust involvement of metamorphic rocks, southwestern Brooks Range, Alaska: Geology, v. 10, p. 930–933.
Turner, D. L., Forbes, R. B., and Dillon, J. T., 1979, K-Ar geochronology of the southwestern Brooks Range, Alaska: Canadian Journal of Earth Sciences, v. 16, p. 1789–1804.
Turner, R. F., Martin, G. C., Risley, D. E., Steffy, D. A., Flett, T. O., and Lynch, M. B., 1986, Geologic report for the Norton Basin planning area, Bering Sea, Alaska: Minerals Management Service OCS (Outer Continental Shelf) Report MMS 86-0033, 179 p.
Van Alstine, D. R., 1986, Normal- and reversed-polarity synfolding CRM along the Brooks Range mountain front, northern Alaska: Eos (American Geophysical Union Transactions), v. 67, p. 269–270.
Vann, I. R., Graham, R. H., and Hayward, A. B., 1986, The structure of mountain fronts: Journal of Structural Geology, v. 8, p. 215–227.
Wadman, D. H., Lamprecht, D. E., and Mrosovsky, I., 1979, Joint geologic/engineering analysis of the Sadlerochit reservoir, Prudhoe Bay field: Journal of Petroleum Technology, v. 31, p. 933–940.
Wahrhaftig, C., 1965, Physiographic divisions of Alaska: U.S. Geological Survey Professional Paper 482, 52 p.
Wallace, W. K., and Hanks, C. L., 1990, Structural provinces of the northeastern Brooks Range, Arctic National Wildlife Refuge, Alaska: American Association of Petroleum Geologists Bulletin, v. 74, p. 1100–1118.
Wilbur, S., Siok, J. P., and Mull, C. G., 1987, A comparison of two petrographic suites of the Okpikruak Formation; a point count analysis, *in* Tailleur, I., and Weimer, P., eds., Alaskan North Slope geology: Bakersfield, California, Society of Economic Paleontologists and Mineralogists, Pacific Section, and Alaska Geological Society, Book 50, p. 441–447.
Wirth, K. R., and Bird, J. M., 1992, Chronology of ophiolite crystallization, detachment, and emplacement: Evidence from the Brooks Range, Alaska: Geology, v. 20, p. 75–78.
Wirth, K. R., Harding, D. J., and Bird, J. M., 1987, Basalt geochemistry, Brooks Range, Alaska: Geological Society of America Abstracts with Programs, v. 19, p. 464.
Witmer, R. J., Haga, H., and Mickey, M. B., 1981, Biostratigraphic report of thirty-three wells drilled from 1975 to 1981 in National Petroleum Reserve in Alaska: U.S. Geological Survey Open-File Report 81-1166, 47 p.
Woidneck, K., Behrman, P., Soule, C., and Wu, J., 1987, Reservoir description of the Endicott field, North Slope, Alaska, *in* Tailleur, I., and Weimer, P., eds., Alaskan North Slope geology: Bakersfield, California, Society of Economic Paleontologists and Mineralogists, Pacific Section, and Alaska Geological Society, Book 50, p. 43–59.
Wood, G. V., and Armstrong, A. K., 1975, Diagenesis and stratigraphy of the Lisburne Group limestones of the Sadlerochit Mountains and adjacent areas, northeastern Alaska: U.S. Geological Survey Professional Paper 857, 47 p.
Zayatz, M. R., 1987, Petrography of the Baird Mountains schistose lithologies, northwestern Alaska: U.S. Geological Survey Circular 998, p. 49–52.
Ziegler, P. A., 1988, Laurussia—The Old Red Continent, *in* McMillan, N. J., Embry, A. F., and Glass, D. J., eds., Devonian of the world, Volume I: Regional syntheses: Canadian Society of Petroleum Geologists Memoir 14, p. 15–48.
Zimmerman, J., and Frank, C. O., 1982, Possible obduction-related metamorphic rocks at the base of the ultramafic zone, Avan Hills complex, De Long Mountains, *in* Coonrad, W. L., ed., The United States Geological Survey in Alaska: Accomplishments during 1980: U.S. Geological Survey Circular 844, p. 27–28.
Zimmerman, J., and Soustek, P. G., 1979, The Avan Hills ultramafic complex, De Long Mountains, Alaska, *in* Johnson, K. M., and Williams, J. R., eds., The United States Geological Survey in Alaska: Accomplishments during 1978: U.S. Geological Survey Circular 804-B, p. B8–B10.
Zimmerman, J., Frank, C. O., and Bryn, S., 1981, Mafic rocks in the Avan Hills ultramafic complex, De Long Mountains, *in* Albert, N.R.D., and Hudson, T., eds., The United States Geological Survey in Alaska: Accomplishments during 1979: U.S. Geological Survey Circular 823-B, p. B14–B15.

Manuscript Received by the Society October 7, 1992

ACKNOWLEDGMENTS

This manuscript incorporates many published and unpublished ideas and data contributed by K. E. Adams, W. P. Brosgé, J. A. Dumoulin, A. Grantz, A. G. Harris, D. G. Howell, D. L. Jones, E. L. Miller, W. J. Nokleberg, J. S. Oldow, R. R. Reifenstuhl, I. L. Tailleur, and A. B. Till. We are grateful to reviewers D. C. Bradley and A. B. Till for their careful reading of the manuscript and thoughtful suggestions for its improvement. We also thank K. E. Adams, who provided a thorough and essential edit of the text.

Authors for various parts of the manuscript are as follows: T. E. Moore: Introduction; Endicott Mountains; DeLong Mountains; Hammond, Coldfoot, and Slate Creek subterranes of the Arctic Alaska terrane; Angayucham terrane; Paleogeography and tectonic models for northern Alaska; and Conclusion. W. K. Wallace: Structural geology: K. J. Bird: North Slope subterrane: S. M. Karl: Hammond, Coldfoot, and Slate Creek subterranes. C. G. Mull: Endicott Mountains and DeLong Mountains subterranes. Our colleague, John Dillon, who was killed in a light-plane accident while returning from field work in the Brooks Range during the initial stages of preparation of the manuscript in 1987, contributed data and ideas that were a great aid to our understanding of the geology of the Brooks Range.

NOTES ADDED IN PROOF

Since the final draft of the chapter on the bedrock geology of northern Alaska was prepared in 1992, it has become evident that descriptions of the major regional unconformities of the North Slope subterrane were not adequately covered in the text. These unconformities are significant to the stratigraphy, structure, tectonics, and economic geology of northern Alaska so will be summarized here. We also list below some significant new age and geologic data for northern Alaska.

Unconformities of the North Slope subterrane

Sub-Middle Devonian unconformity. Weakly metamorphosed, penetratively polydeformed Ordovician metasedimentary rocks are unconformably overlain by Middle Devonian (Eifelian) and younger(?) marine to nonmarine terrigenous clastic rocks in the eastern Brooks Range (Reiser and others, 1980) (Fig. 5, column 2). Although only locally preserved beneath the sub-Mississippian unconformity, the sub-Middle Devonian unconformity postdates major contractional deformation and is interpreted to mark the onset of rifting that led to formation of the late Paleozoic to early Mesozoic south-facing passive continental margin of northern Alaska (Anderson and others, 1994).

Sub-Mississippian unconformity. The regional sub-Mississippian unconformity, a characteristic feature of the North Slope subterrane, is present beneath Lower Mississippian rocks at the base of the Ellesmerian sequence (Fig. 5). The sub-Mississippian unconformity is weakly to prominently displayed on seismic-reflection records (Kirschner and others, 1983; Kirschner and Rycerski, 1988; Molenaar, 1988) and is confirmed by well penetrations along the Barrow arch (Bird, 1982, 1988a; Molenaar and others, 1986). The unconformity crops out in the northeastern Brooks Range (Brosgé and others, 1962), Mt. Doonerak fenster (Dutro and others, 1976; Mull and others, 1987a), and Lisburne Peninsula (Moore and others, 1984). In most areas, the unconformity is angular, although it may be a disconformity at Mt. Doonerak (Oldow and others, 1987d). Seismic-reflection records indicate that regionally the unconformity has low relief, although it is flexed downward in the Meade, Umiat, and other basins in the subsurface of the North Slope (Figs. 7, 8, and 9). Structures beneath the unconformity are variable, ranging from high-strain penetrative fabrics (Oldow and others, 1987b) to gentle folding in the northeastern Brooks Range (Robinson and others, 1989). Rocks below the unconformity are as young as late Early Devonian in the northeastern Brooks Range (Dillon and others, 1987b; Blodgett and others, 1991) and early Middle Devonian in the eastern Brooks Range (Anderson and others, 1993, 1994). The amount of erosion on the unconformity is unknown because of the poorly constrained stratigraphy of underlying rocks but may be several kilometers or more where the underlying rocks are penetratively deformed. The sub-Mississippian unconformity reflects northward transgression due to Early Mississippian continental breakup and thermal subsidence of the newly formed south-facing passive continental margin of northern Alaska (Fig. 24B).

Post-Lisburne disconformity. Throughout the North Slope subterrane, the Lisburne Group and overlying Echooka Formation are separated by a regional disconformity (Fig. 5). The disconformity is exposed in the northeastern Brooks Range (Detterman and others, 1975; Crowder, 1990) and confirmed by well penetrations in the North Slope (Bird, 1982, 1988a; Molenaar and others, 1986). To the south in the Endicott Mountains subterrane, the disconformity is overlain by the Siksikpuk Formation, a distal equivalent of the Echooka Formation (Siok, 1985; Adams, 1991). Over much of its area, the age of the disconformity is constrained by Early or Middle Pennsylvanian conodonts from the underlying Lisburne Group and by Early Permian brachiopods from the overlying Echooka Formation and coeval Siksikpuk Formation. In the subsurface of the North Slope south of Barrow, however, the Lisburne Group below the disconformity is as young as Early Permian (Bird, 1988a), which suggests that the disconformity is intra-Permian. The post-Lisburne disconformity was likely caused by regional uplift of the Lisburne carbonate platform or by a eustatic change in the Early Permian (Molenaar and others, 1986).

Lower Cretaceous unconformity (LCU). The regional Lower Cretaceous unconformity (LCU), or pebble shale unconformity, is present at the top of the upper Ellesmerian sequence in the northern part of the North Slope and northeastern Brooks Range (Fig. 5). The LCU is prominently displayed on seismic-reflection records (Kirschner and others, 1983; Kirschner and Rycerski, 1988; Molenaar, 1988), confirmed by well penetrations (Bird, 1982, 1988a; Molenaar and others, 1986), and observed in outcrop (Molenaar, 1983; Bird and others, 1987; Mull, 1987). In general, the greatest amount of erosional relief on the unconformity is in the north (more than 1 km on the Barrow arch) and the least amount is in the south (Tailleur and others, 1978; Bird, 1987, Figs. 4, 6, and 13; Bird and others, 1987, Fig. 7.1). Beneath the southern part of the coastal plain and northern part of the foothills, the LCU merges with a conformable stratal succession (Fig. 8). Near its southern extent, the age of the unconformity is intra-Hauterivian, where dated in wells such as Tunalik and Seabee in the NPRA (Haga and Mickey, 1983; Mickey and Haga, 1987) and in outcrops on the Echooka River in the foothills of the northeastern Brooks Range (Molenaar, 1983, Fig. 4). On the Barrow arch, where the LCU is contained within the Kuparuk Formation, detailed well correlations constrain its age as Valanginian and/or Hauterivian (Masterson and Paris, 1987).

The paleogeography and subcrop geology at the time of maximum development of the LCU (Fig. 14) show that an elongate landmass about 250 km wide was uplifted above sea level. The Jurassic and Lower Cretaceous Kingak Shale was exposed over most of the area south of the Barrow arch, whereas older rocks were exposed closer to the axis of the arch. Normal faulting occurred along the Barrow arch prior to, and during, development of the LCU (Hubbard and others, 1987; Masterson and Paris, 1987), resulting in variable amounts of erosion on different fault blocks.

The LCU is regarded as the breakup unconformity that marked the onset of subsidence of the rifted Arctic margin and a major marine transgression (Grantz and May, 1983). The LCU and overlying shale provide part of the trapping mechanism for many Prudhoe Bay area oil fields (Bird, 1991). The weathering of rocks exposed beneath the unconformity may be responsible for the enhanced porosity in oil field reservoirs (Shanmugam and Higgins, 1988; van de Kamp, 1988).

New geologic data

(1) Jurassic fossils have recently been recovered from a condensed shale interval within the uppermost part of the Etivluk Group on the Lisburne Peninsula (North Slope subterrane) (C. G. Mull, 1993, unpublished data), as shown diagrammatically in Figure 5.

(2) Detrital zircons from the quartz-mica schist unit of the Coldfoot subterrane near Wiseman have yielded U-Pb ages as young as 360 Ma. These ages

indicate that the protolith of the unit includes elements at least as young as latest Devonian (J. N. Aleinikoff, 1993, unpublished data).

(3) Plagiogranite from the Siniktanneyak Mountain ophiolitic klippe of the Misheguk Mountain allochthon (Angayucham terrane) has yielded a U-Pb zircon age of 170 ± 3 Ma (Moore and others, 1993). A potassium feldspar-bearing granitic dike in ultramafic rocks near the base of the ophiolite has yielded a U-Pb monazite age of 163 ± 3 Ma (J. N. Aleinikoff, 1993, unpublished data).

(4) Apatite fission track analysis of rocks from the northeastern Brooks Range indicates that the age of cooling, reflecting periods of rapid uplift and unroofing, becomes progressively younger northward from 62 ± 5 Ma at Bathtub Ridge in the south to 23 ± 3 Ma in the Sadlerochit Mountains in the north (Fig. 1) (O'Sullivan and others, 1993).

(5) Integrated seismic reflection and refraction data show that the Mt. Doonerak antiform in the central Brooks Range (Fig. 23) is a crustal-scale duplex (Levander and others, 1994). Apatite and zircon fission track analysis of rocks from the antiform indicates that cooling, reflecting construction of the antiform, occurred at 70–65 Ma and again at 24 ± 3 Ma (O'Sullivan and others, 1991).

(6) Integrated seismic reflection and refraction data show that along the route of the Dalton Highway the Moho lies at a depth of 33 km beneath the North Slope, 46 km beneath the crest of the Brooks Range, and 35 km at the southern edge of the range (Levander and others, 1994).

ADDITIONAL REFERENCES

Anderson, A. V., Mull, C. G., and Crowder, R. K., 1993, Mississippian terrigenous clastic and volcaniclastic rocks of the Ellesmerian sequence, upper Sheenjek River area, eastern Brooks Range, Alaska, *in* Solie, D. N., and Tannian, F., eds., Short notes on Alaskan geology 1993: Alaska Division of Geological and Geophysical Surveys Geologic Report 113, p. 1–6.

Anderson, A. V., Wallace, W. K., and Mull, C. G., 1994, Depositional record of a major tectonic transition in northern Alaska: Middle Devonian to Mississippian rift-basin margin deposits, upper Kongakut River region, eastern Brooks Range, Alaska, *in* Thurston, D., ed., Proceedings of the 1992 International Conference on Arctic Margins: Alaska Geological Society (in press).

Bird, K. J., 1991, North Slope of Alaska, *in* Gluskoter, H. J., Rice, D. D., and Taylor, R. B., eds., Economic Geology: U.S.: Boulder, Colorado, Geological Society of America, The Geology of North America, v. P-2, p. 447–462.

Haga, H., and Mickey, M. B., 1983, Jurassic-Neocomian seismic stratigraphy, NPRA: A report of work performed for the U.S. Geological Survey: San Diego, Biostratigraphics, unpublished report, 2 vol., 133 p., 14 plates.

Levander, A., Fuis, G. S., Wissinger, E. S., Lutter, W. J., Oldow, J. S., and Moore, T. E., 1994, Seismic images of the Brooks Range fold and thrust belt, Arctic Alaska, from an integrated seismic reflection/refraction experiment: Tectonophysics, v. 233 (in press).

Mickey, M. B., and Haga, H., 1987, Jurassic-Neocomian biostratigraphy, North Slope, Alaska, *in* Tailleur, I., and Weimer, P., eds., Alaskan North Slope geology: Bakersfield, California, Society of Economic Paleontologists and Mineralogists Pacific Section, and Alaska Geological Society Book 50, p. 397–404.

Moore, T. E., Nilsen, T. H., Grantz, A., and Tailleur, I. L., 1984, Parautochthonous Mississippian marine and non-marine strata, Lisburne Peninsula, Alaska, *in* Reed, K. M., and Bartsch-Winkler, S., eds., The United States Geological Survey in Alaska: Accomplishments during 1982: U.S. Geological Survey Circular 939, p. 17–21.

Moore, T. E., Aleinikoff, J. N., and Walter, M., 1993, Middle-Jurassic U-Pb crystallization age for Siniktanneyak Mountain ophiolite, Brooks Range, Alaska [abs.]: Geological Society of America Abstracts with Programs, v. 25, no. 5, p. 124.

O'Sullivan, P. B., Murphy, J. M., and Moore, T. E., 1991, Apatite fission track evidence for Tertiary uplift in the Doonerak Fenster region, central Brooks Range, Alaska [abs.]: Eos, American Geophysical Union 1991 Fall Meeting Programs and Abstracts, v. 19, no. 44, p. 299.

O'Sullivan, P. B., and 6 others, 1993, Multiple phases of Tertiary uplift and erosion in the Arctic National Wildlife Refuge, Alaska, revealed by apatite fission track analysis: American Association of Petroleum Geologists Bulletin, v. 77, p. 359–385.

Shanmugam, G., and Higgins, J. B., 1988, Porosity enhancement from chert dissolution beneath Neocomian unconformity: Ivishak Formation, North Slope, Alaska: American Association of Petroleum Geologists Bulletin, v. 72, p. 523–535.

Tailleur, I. L., Pessel, G. H., and Engwicht, S. E., 1978, Subcrop map at Lower Cretaceous unconformity and maps of Jurassic and Lower Cretaceous seismic horizons, eastern North Slope petroleum province, Alaska: U.S. Geological Survey Miscellaneous Field Studies Map MF-9281, scale 1:500,000.

van de Kamp, P. C., 1988, Stratigraphy and diagenetic alteration of Ellesmerian sequence siliciclastic rocks, North Slope, Alaska, *in* Gryc, G., eds., Geology and exploration of the National Petroleum Reserve in Alaska, 1974 to 1982: U.S. Geological Survey Professional Paper 1399, p. 833–854.

Printed in U.S.A.

The Geology of North America
Vol. G-1, The Geology of Alaska
The Geological Society of America, 1994

Chapter 4

Geology of Seward Peninsula and Saint Lawrence Island

Alison B. Till and Julie A. Dumoulin
U.S. Geological Survey, 4200 University Drive, Anchorage, Alaska 99508-4667

SEWARD PENINSULA: THE SEWARD AND YORK TERRANES

Seward Peninsula (Fig. 1) may be divided into two geologic terranes (Fig. 2) on the basis of stratigraphy, structure, and metamorphic history. The Seward terrane, an area 150 by 150 km in the central and eastern peninsula, is dominated by Precambrian(?) and early Paleozoic blueschist-, greenschist-, and amphibolite-facies schist and marble, and intruded by three suites of granitic rocks. The York terrane, roughly 100 by 75 km, occupies western Seward Peninsula and the Bering Straits region; it is composed of Ordovician, Silurian, Devonian, Mississippian, and possibly older limestone, argillaceous limestone, dolostone, and phyllite, which are cut by a suite of Late Cretaceous tin-bearing granites. The boundary between the Seward and York terranes is poorly exposed but is thought to be a major thrust fault because of its sinuous map trace, a discontinuity in metamorphic grade, and differences in stratigraphy across the boundary (Travis Hudson, oral communication, 1984). The boundary between the Seward terrane and the Yukon-Koyukuk province to the east is complicated by vertical faults (the Kugruk fault zone of Sainsbury, 1974) and obscured by Cretaceous and Tertiary cover.

The Seward Peninsula heretofore was thought to consist largely of rocks of Precambrian age (Sainsbury, 1972, 1974, 1975; Hudson, 1977). Microfossil data, however, indicate that many of the rocks considered to be Precambrian are early Paleozoic in age (Till and others, 1986; Dumoulin and Harris, 1984; Dumoulin and Till, 1985; Till and others, 1983; Vandervoort, 1985). It is likely that Precambrian rocks are a minor part of the stratigraphy of the Seward Peninsula.

Previous work

Regional mapping on Seward Peninsula began at the time of the Nome gold rush, with the work of Brooks, Richardson, Collier, and Mendenhall (1901), followed by the more detailed work of Knopf (1908), Smith (1910), Moffit (1913), and Steidtmann and Cathcart (1922). More recent mapping was initiated in the 1950s by Sainsbury (1969, 1972), who focused first on the stratigraphy and economic geology of the York Mountains in western Seward Peninsula and later on the reconnaissance-scale (1:250,000) mapping of the metamorphic rocks of the central peninsula (Sainsbury and others, 1972a, b; Sainsbury, 1974). Miller and others (1972) produced a preliminary geologic map of the southeastern peninsula at the same scale. Regional compilations have been made at 1:1,100,000 by Hudson (1977) and 1:500,000 by Robinson and Stevens (1984).

Detailed studies of the regionally extensive metamorphic rocks of the Seward terrane began with the description of blueschist-facies mineral assemblages of the Nome Group by Sainsbury and others (1970). Till and others (1986) formulated the stratigraphy of the metamorphic rocks of the Nome Group; and Thurston (1985), Forbes and others (1984), Evans and Patrick (1987), Patrick (1988), and Patrick and Evans (1989) conducted detailed mapping and petrogenetic studies of the blueschists. Detailed studies of the high-grade metamorphic rocks of the Kigluaik Mountains were undertaken by Till (1980), Sturnick (1984), and Patrick and Lieberman (1987).

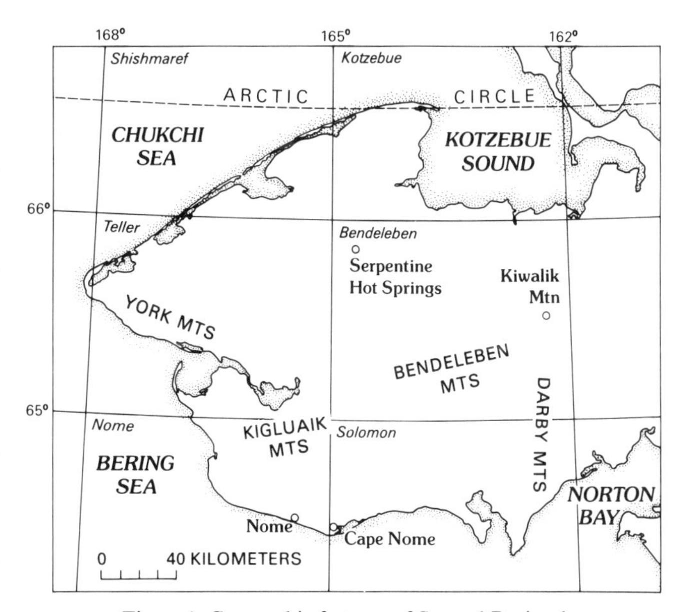

Figure 1. Geographic features of Seward Peninsula.

Till, A. B., and Dumoulin, J. A., 1994, Geology of Seward Peninsula and Saint Lawrence Island, *in* Plafker, G., and Berg, H. C., eds., The Geology of Alaska: Boulder, Colorado, Geological Society of America, The Geology of North America, v. G-1.

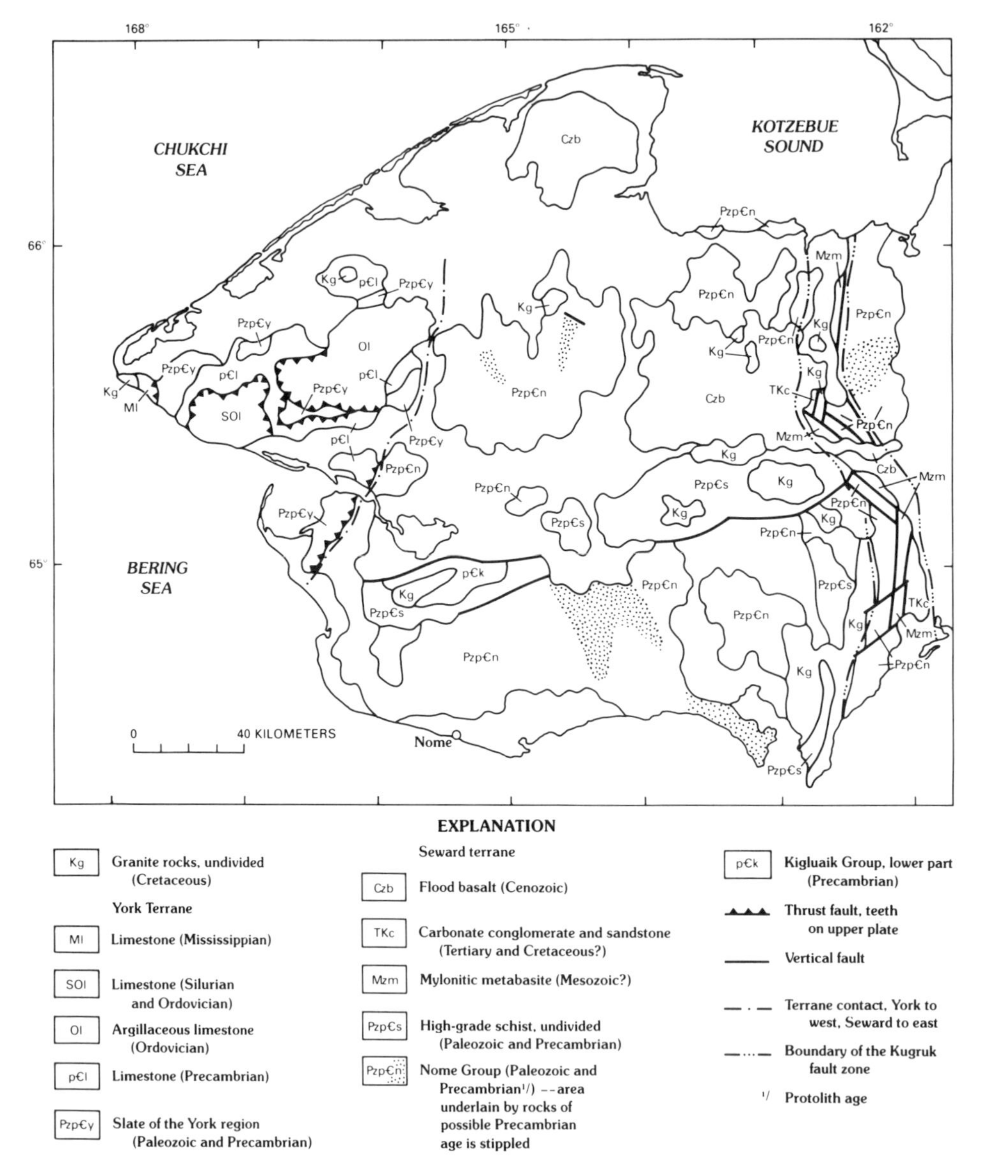

Figure 2. Simplified geologic map of Seward Peninsula, from Sainsbury (1972), Hudson (1977), and Till and others (1986).

Miller (1972) and Miller and Bunker (1976) conducted regional studies of the Cretaceous granitic rocks, and Hudson (1979) and Hudson and Arth (1983) studied geochemistry of the tin-granite suite in detail.

Tertiary deposits occur locally in the Seward terrane. Although they are poorly exposed, they have been studied in conjunction with flood basalts (Hopkins, 1963), coal deposits (Barnes, 1967), and uranium mineralization (Dickinson, 1984). Geothermal resources (hot springs), largely developed on plutonic contacts (Miller and others, 1975), have also drawn attention to Tertiary basins (Lockhart, 1984; Westcott and Turner, 1981). Studies of Quaternary deposits resulted from interest in the Bering land bridge and placer gold (Hopkins, 1967; Kaufman and Hopkins, 1986).

The Seward Peninsula is rich in mineral resources (Hudson and DeYoung, 1978; Puchner, 1986). Its gold deposits are well known, and it has substantial resources of tin and uranium. Numerous small base-metal vein deposits occur, and a stratabound Zn-Pb deposit has been postulated in the northern part of the peninsula (J. Briskey, written communication, 1985).

SEWARD TERRANE

The Seward terrane is dominated by low- and high-grade metamorphic rocks and at least three suites of granitic rocks (Fig. 2). Most common of the metamorphic rocks are the blueschist-facies schists of the Nome Group, which form the low rolling hills of the central peninsula. High-grade metamorphic rocks include the Kigluaik Group of the Kigluaik Mountains, large bodies of undivided schist and gneiss, and migmatite. These rocks, together with granitic bodies, are exposed in fault-bounded mountain ranges that transect the low-grade rocks from east to west (the Kigluaik and Bendeleben Ranges) and north to south (the Darby Range). Tertiary and Holocene basins filled with sedimentary and basaltic volcanic rocks are developed in the central and northern Seward terrane. Small amounts of basalt and coal-bearing sedimentary rocks occur in the eastern part of the terrane, commonly near the Kugruk fault zone. Within this north-south–trending fault zone, blocks of mylonitic metabasite, serpentinite, and Tertiary or Cretaceous(?) carbonate-clast conglomerate are juxtaposed with rocks of the Nome Group (Fig. 2).

Nome Group

The Nome Group, as defined by Moffit (1913), included two schist units with an interlayer of limestone. In his latest work, Sainsbury (1974) included all Nome Group rocks in the "slate of the York region." Recent mapping shows that the Nome Group is a metamorphic unit that includes two parts: (1) a coherent, mappable metamorphic stratigraphy; and (2) carbonate rocks, which have an indeterminate premetamorphic relation to that stratigraphy. Both parts are composed of early Paleozoic and possibly older protoliths that have undergone an episode of blueschist-facies metamorphism and deformation in the Jurassic (Forbes and others, 1984; Armstrong and others, 1986). the Nome Group, as used here, includes all lithologies that underwent blueschist-facies metamorphism and accompanying deformation. The Nome Group and related fossil data are discussed by Till and others (1986).

The stratigraphy of the metamorphic rocks includes four units mappable at a scale of 1:63,360: (1) a basal quartz-rich pelitic schist unit, (2) a mixed unit dominated by interlayered marble and quartz-graphite schist, (3) a mafic schist with calcareous components, and (4) an impure chlorite marble. These four units have a minimum combined thickness of 4.5 km; the schists share a common foliation and structural style and contain mineral assemblages that are stable under blueschist-facies conditions.

The pelitic schist forms tors several meters high and displays characteristic centimeter-thick layers and lenses of quartz that define at least two sets of isoclinal folds; the base is not exposed. These rocks were mapped as "tectonically metamorphosed York slate" by Sainsbury (1974).

The mixed unit is defined in part by its position between the more distinctive pelitic and mafic schist units, and also by the presence of quartz-graphite schist interlayered on a scale of 10 to 150 m with pure and impure marble and subordinate, thinner layers of pelitic and mafic schist. The quartz-graphite schist and marble thicken and thin significantly along strike on a scale of 5 to 10 km; pure marble dominates some exposure of the unit; quartz-graphite schist dominates others. The quartz-graphite schist of the mixed unit is equivalent to some of the rocks included by Sainsbury (1969) in the "slates of the York region." Mafic schist intercalated within the mixed unit is compositionally similar to parts of the overlying mafic schist unit. Metabasite boudins identical to those found in the mafic schist occur locally near the top of the mixed unit.

The mafic schist unit is composed of massive metabasite, mafic schist, calc-schist, chlorite-albite schist, and minor marble and quartzite. Massive, coarse-grained metabasite lenses and layers are conformably intercalated with schists of similar composition, as well as with quartz-poor calc-schist, pelite, semipelite and chlorite-albite schist. Glaucophane has been found in all major lithologies except pure marble and quartzite (Thurston, 1985). Repetitive interlayers of calc-schist and mafic schist occur locally in the unit. Two compositionally distinct types of metabasite have been recognized in the mafic schist: one exceptionally rich in iron and titanium, with abundant glaucophane and garnet; and one relatively rich in magnesium and aluminum, with actinolite and rare garnet (Forbes and others, 1984). The mafic schist unit is equivalent to the Casadepaga Schist of Smith (1910).

The impure chlorite marble directly overlies the mafic schist and consists of an orange-weathering, foliated marble, which characteristically includes lenses and thin layers of chlorite-albite schist. Blue amphibole forms inclusions in albite. Boudins of metabasite near the base of the impure marble are similar to the magnesium- and aluminum-rich metabasites in the mafic schist.

The presence of mafic material in the mixed unit, the mafic schist unit, and the impure chlorite marble suggests that the three lithologic units were part of the same sequence during the period of mafic igneous activity. Boudins in the mixed unit may represent feeder dikes to the volcanic protolith of the mafic schist, and boudins and lenses of mafic material at the base of the impure chlorite marble may have been produced at the waning stages of igneous activity.

Conodonts and radiolarians have been recovered from two of the upper three units of the metamorphic stratigraphy (Till and others, 1986). Radiolarians were observed in thin sections of finely laminated quartz-graphite schist from the mixed unit in the northern Darby Mountains. The age of the radiolarians has not been determined, but their presence indicates a Cambrian or younger age for that unit. Ordovician conodonts were found in marble near the top of the mixed unit; late Early to Middle Ordovician conodonts were recovered from the impure chlorite marble. Because these two units bracket the mafic schist, most of the metamorphic stratigraphy is Ordovician in age. The mafic igneous activity recorded in the mafic schist is therefore pre-late Early to Middle Ordovician in age. The mixed unit and mafic schist were intruded at Kiwalik Mountain by a granitic orthogneiss dated at 381 ± 2 Ma (J. Aleinikoff, written communication, 1983).

The original nature or age of the contact between the pelitic schist unit and the overlying unit is not known. No fossil or isotopic data constrain the age of the protolith of the pelitic schist unit. Gardner and Hudson (1984) indicate a Precambrian age for the unit on the basis of a structural analysis that suggested structures in the pelitic schist recorded at least one more deformational event than other units in the Nome Group. The pelitic schist unit may have been basement for Cambrian and Ordovician Nome Group rocks, or may have been juxtaposed with those rocks during the Jurassic.

The second part of the Nome Group, carbonate rocks with unknown relations to the premetamorphic stratigraphy, was divided by Dumoulin and Harris (1984) into several units on the basis of lithology and age. These rocks occur as isolated exposures and as fault slices; rarely, they are folded into the metamorphic stratigraphy. Till and others (1986) and Harris and Repetski (1987) provide fossil data and detailed lithologic descriptions of these rocks.

Cambrian, Ordovician, and Silurian dolostones, with sedimentary structures and fauna indicating shallow-water deposition, overlap in age with the metamorphic stratigraphy (Till and others, 1986); they may have formed in the same depositional basin. Deeper-water carbonate rocks, with sedimentary structures indicative of slope or basin environments, range in age from Cambrian to Devonian (Till and others, 1986; Ryherd and Paris, 1987).

Spatial relations suggest that Devonian rocks with sedimentary structures and fauna indicating shallow-water deposition may have been unconformably deposited on older Nome Group rocks (specifically the metamorphic stratigraphy and the deeper water Cambrian to Devonian carbonate rocks).

These lower Paleozoic units, along with carbonate rocks of unknown age, are included in the Nome Group on the basis of mineral assemblages, rock texture, and/or color alteration indices of conodonts. Depositional relations with other Nome Group rocks cannot be established, however, and they may not have formed in the same depositional basin as the carbonates in the metamorphic stratigraphy (principally in the mixed unit).

The flat-lying to gently dipping transposition foliation of the Nome Group is characterized by an axial planar schistosity, which is commonly parallel to lithologic layering and abundant intrafolial isoclinal folds. This foliation is the product of penetrative ductile deformation that presumably has significantly altered original geometric relations between lithostratigraphic units while leaving the stratigraphic succession largely intact. Lithologic contacts have been rotated parallel to the foliation in most exposures. Locally the transposition foliation crosses lithologic layering at high angles, presumably in the vicinity of the hinges of outcrop- to map-scale recumbent folds. In the northeast Seward Peninsula, in the vicinity of Kiwalik Mountain, the metamorphic stratigraphy is inverted relative to most exposures of the Nome Group; this may indicate that map-scale fold formation accompanied deformation. Stretching lineations (mineral aggregates, boudin axes) and isoclinal fold hinges that formed with the schistosity both have a north-south trend. All of the aforementioned structures were formed during regional blueschist-facies metamorphism, and indicate that the Nome Group schists were at a sufficient depth in the crust to undergo ductile deformation accompanied by significant extension along a north-south trend. Patrick (1988) used quartz petrofabrics and the geometry of rare sheath folds to determine that the deformation had northward vergence.

The metamorphic event occurred after the intrusion of the Kiwalik Mountain orthogneiss (381 ± 2 Ma) and before contact metamorphism of Nome Group rocks by the Kachauik and Windy Creek Plutons (100 to 110 Ma; Miller and others, 1972); that is, between the Late Devonian and middle Cretaceous. Armstrong and others (1986) obtained ages for the blueschist event ranging from 170 to 100 Ma using whole-rock-mica Rb-Sr isochrons. The same study obtained K-Ar mineral ages from rocks of the Nome Group ranging from 122 to 194 Ma, and suggested that a Jurassic metamorphic event was followed by partial resetting of the Rb-Sr isotopic system.

The timing of tectonic events studied in adjacent areas in northwestern Alaska probably can be related to this blueschist metamorphism event on the Seward Peninsula. In the western Brooks Range, sedimentation related to the Brooks Range orogeny began in the Middle Jurassic (Mayfield and others, 1983). Blueschist-facies rocks of the "schist belt," in the hinterland of the Brooks Range, have been related to the early phases of that orogenic episode (Gilbert and others, 1977; Hitzman and others, 1986). The predominantly lower Paleozoic platformal protolith sequence of the schist belt was metamorphosed in the Late Jurassic or Early Cretaceous (Armstrong and others, 1986). The similarities in protolith composition and metamorphic conditions of schist belt and Nome Group rocks have led to correlation of the two metamorphic terranes and the supposition that they formed in similar tectonic environments (Forbes and others, 1984; Patrick, 1988).

Kigluaik Group

Gneisses of the Kigluaik Group are the deepest crustal rocks exposed in northwestern Alaska and may be among the oldest. Moffit (1913) included in the group all schist and gneiss in the Kigluaik Mountains above greenschist facies in grade. These rocks can be divided into structurally (1) an upper section, amphibolite facies schists of the south flank of the Kigluaik Range (the Tigaraha Schist of Moffit); and (2) a lower section, granulite-facies gneiss of the Mt. Osborn area (the "heavily bedded limestone and basal gneiss" of Moffit). The upper section of the Kigluaik Group may be correlative with amphibolite-facies schists of the Bendeleben and Darby Ranges; no rocks correlative with the lower section of the group are known on the Seward Peninsula.

The upper section of the Kigluaik Group is composed of pelitic and calcareous schist, marble, and subordinate amphibolite and quartz-graphite schist. It is at least 3.5 km thick and locally includes foliated and nonfoliated granitic plutons. A massive pelitic and semipelitic biotite gneiss is the lowest lithologic unit in the upper section. It equilibrated above the second sillimanite isograd, and contains sillimanite and K-feldspar. The biotite gneiss is the only unit in the group from which an Rb-Sr age has been obtained (see below). Rocks from the upper part of the Kigluaik Group may be thermally upgraded equivalents of the Nome Group (Till, 1980; Thurston, 1985; Patrick and Lieberman, 1987). The contact between the Kigluaik and Nome Groups is commonly faulted, but may be a gradational metamorphic contact in the southwestern Kigluaik Mountains. Lithologic similarities and static metamorphic textures also provide possible evidence for this conclusion. High-grade equivalents of Nome Group units have been mapped in the Bedeleben and Darby Mountains (Till and others, 1986); these rocks may be correlative to the upper section of the Kigluaik Group.

The lower section of the Kigluaik Group consists of pure and impure marble, with interlayers of gneiss, mixed gneiss of pelitic, quartzo-feldspathic, mafic and ultramafic composition, and migmatite. Coarsely crystalline marble in the steepest faces of the cirques around Mt. Osborn ranges in composition from pure calcite marble to impure dolomitic marble containing olivine and diopside. Discontinuous layers and megaboudins of gneiss of varied composition are found below the marble-dominated section. Granulite-facies assemblages are common in these rocks, with two pyroxene-bearing assemblages in semipelitic and mafic gneiss and spinel peridotite in the ultramafic boudins. Mafic and ultramafic gneisses are most common at the base of the exposures and are interlayered with metasedimentary rocks. The only known alpine garnet lherzolite in North America occurs in these exposures and is partially recrystallized to spinel lherzolite. The migmatite is best exposed in a large cirque directly west of Mt. Osborn where it occupies the base of the exposed section of the Kigluaik Group. The upper contact of the migmatite locally cuts the foliation in the overlying gneiss. Large rafts of metasediment and peridotite derived from the mixed gneiss float in pegmatite and foliated granite.

The dominant structure in the Kigluaik Group is a transposition foliation parallel to lithologic layering. In the upper part of the Kigluaik Group, this is a schistosity; in the lower part of the group it is gneissic layering. Isoclinal introfolial folds with wavelengths of a few millimeters to tens of meters are present throughout the Kigluaik Group. Isoclinal fold hinges and stretching lineations are parallel and oriented along a north-south trend (Till, 1980). This foliation has been gently folded about an east-west axis to create a doubly plunging antiform with a culmination in the vicinity of Mt. Osborn.

Samples of pelitic and semipelitic gneiss from the biotite gneiss in the upper part of the Kigluaik Group yielded a whole-rock Rb/Sr isochron age of 735 Ma (Bunker and others, 1979). This age has been interpreted as the age of metamorphism for the Kigluaik Group; no confirmation of this interpretation has been obtained using mineral ages. This may represent the age of the protolith. The age of the migmatite is unknown. A geological argument can be made in favor of a Late Jurassic or Early Cretaceous age for the granulite-facies metamorphism (see "Tectonic history").

Biotite and hornblende separated from the upper Kigluaik Group have yielded K-Ar ages of 81 to 84 Ma (Westcott and Turner, 1981). Muscovite from pegmatite, which cuts the metamorphic fabric, yielded a K-Ar age of 84.0 ± 1.2 Ma (N. Shew, written communication, 1988). These ages represent minimum limits on the age of high-grade metamorphism.

High-grade metamorphic rocks of the Bendeleben and Darby Mountains

Amphibolite-, locally granulite-, and greenschist-facies rocks crop out in the Bendeleben and Darby Mountains. Lithologies include metapelite, marble, quartz-graphite schist, calc-schist, quartzo-feldspathic schist, amphibolite, and orthoamphibole-cordierite schist. These rocks may be in part correlative with the upper section of the Kigluaik Group; schist in the northern Darby Mountains is correlative with the Nome Group. Rocks in both mountain ranges have undergone both high- and low-pressure metamorphism.

In the eastern Bendeleben Mountains, sillimanite- and K-feldspar-bearing pelitic and quartzo-feldspathic schists occur in a large envelope of migmatite surrounding the Bendeleben Pluton; these schists record only one metamorphic event. In contrast, pelitic rocks in the western Bendeleben Mountains include an early high-pressure assemblage with kyanite and staurolite and late low-pressure assemblages with andalusite, sillimanite, and locally cordierite.

Regional mapping in the northern Darby Mountains has shown that Nome Group units can be recognized that have been upgraded from blueschist to greenschist facies. These units can be traced southward into the central Darby Range; the metamorphic grade of the rocks increases southward and reaches upper amphibolite or granulite facies in the vicinity of Mt. Arathlatuluk, where small stocks of anatectic granite crop out. A narrow contact aureole of andalusite-bearing schist is present around the Darby Pluton (95 Ma; Miller and Bunker, 1976) and overprints the upper amphibolite-facies metamorphism.

Protolithic ages of metasedimentary rocks in the Bendeleben and Darby Mountains are not known. However, where upgraded equivalents of Nome Group lithologies can be recognized, the protoliths are probably pre–Late Ordovician.

Biotite from an anatectic granitic stock in the southern Darby Mountains yields a K-Ar age of 102.6 ± 1.4 Ma (N. Shew, written communication, 1988), which is the minimum age limit for high-temperature metamorphism in the range. Geologic relations are consistent with this age, because relicts of the Late Jurassic to Early Cretaceous blueschist event were overprinted by the high-temperature event, and intrusion of the 95-Ma Darby Pluton occurred after the high-temperature metamorphism.

Ages of metamorphic events are difficult to discern in the Bendeleben Mountains. K-Ar mineral ages obtained from schist in the western Bendeleben Mountains and from the two plutons in the range fall in the range of 80 to 90 Ma (Westcott and Turner, 1981; Miller and Bunker, 1976). These ages record synchronous cooling of the intrusive bodies, their migmatitic carapaces, and the upper amphibolite-facies country rock.

Cretaceous magmatism

Large granitic plutons are associated with the high-grade metamorphic rocks of the Kigluaik, Bendeleben, and Darby Mountains (Fig. 2). These plutons range in composition from granodiorite to biotite granite but also include syenite and monzonite (Miller and Bunker, 1976; Miller, this volume, Chapter 16; Hudson, this volume); K-Ar ages of these plutons range from 100 to 80 Ma. Several alkaline subsilicic complexes and dike swarms occur in the southeastern Seward Peninsula; these are part of a belt of similar rocks of middle Cretaceous age, which extends some 1,200 km across western Alaska, the Bering Sea islands, and into eastern Siberia (Miller, 1972, 1989). Small late Cretaceous plutons and stocks of anorogenic origin are scattered across the northwestern Seward Peninsula and intrude both the York and Seward terranes. They consist of biotite granite and typically have tin deposits associated with them (Hudson and Arth, 1983).

Cenozoic sedimentary and basaltic rocks

Tertiary and Quaternary sedimentary rocks and basalt flows fill local basins and form large plateaus, respectively, on the Seward terrane. Sedimentary basins, tens of kilometers in diameter, locally may be very deep (Barnes and Hudson, 1977). Coal and uranium deposits occur in one or more of these basins (Dickinson, 1984). Geothermal activity at Pilgrim Springs, north of the Kigluaik Range, may be related to basin-forming faults (Westcott and Turner, 1981). Many of the Tertiary sedimentary rocks show signs of deformation, especially in the eastern part of the Seward terrane (Kugruk fault zone), where postdepositional deformation rotated coal-bearing strata to near-vertical dip. The maximum age of the strata is unknown, but pollen as old as Eocene have been collected from siltstone (T. Ager, written communication, 1985).

Alkalic and tholeiitic basalts form a plateau 46 by 25 km and an extensive maar and flow field in northern Seward Peninsula. Small flows occur in southern parts of the peninsula. Radiometric data indicate that flows as old as 29 Ma are present; the most recent flows are probably younger than 2,000 B.P. (Westcott and Turner, 1981; Hopkins, 1967, and unpublished data; Moll-Stalcup, this volume).

The Kugruk fault zone

A complex of north-south–trending vertical faults in the east part of the Seward terrane defines the Kugruk fault zone (Sainsbury, 1974; Fig. 2). Nome Group carbonate rocks from the most common slices in the fault zone. Blueschist-facies mylonitic metabasite, Tertiary or Cretaceous(?) carbonate- and mafic-clast conglomerate, altered tonalite, and rare serpentinite from mappable units in the zone. The presence of glaucophane and lawsonite in the mylonitic metabasite indicates that the zone may mark a significant tectonic boundary (Till, unpublished data). The carbonate- and mafic-clast conglomerate is a ridge-forming unit in the fault zone. It consists of material derived form bedrock units in the fault zone (principally the mylonitic metabasite and Paleozoic carbonate lithologies) but also contains material without analogues in the fault zone or in the surrounding Nome Group. The elongate Cretaceous Darby Pluton parallels the trace of the southern part of the fault zone, and it is possible that this is a zone of crustal weakness active from Cretaceous time to the present. Rocks of the Nome Group have been identified on both sides of the fault zone, so it cannot constitute the boundary between the Seward terrane and the Yukon-Koyukuk province (Dumoulin and Till, 1985).

YORK TERRANE

The York terrane consists mainly of Ordovician through Mississippian limestone, argillaceous limestone, dolostone, and fine-grained clastic rocks deposited in a shallow-water, normal marine to slightly restricted platform environment. Carbonate rocks of Ordovician age compose most of this sequence; some rocks are of unknown age and may be pre-Ordovician. York terrane carbonate rocks, especially those of Ordovician age, show many similarities in lithofacies and biofacies to the metacarbonate sequence of the western Baird Mountains Quadrangle in the southwestern Brooks Range (Dumoulin and Harris, 1987; Harris and Repetski, 1987). The color alteration indices (CAI) of most conodonts found in the York terrane range from 1.5 to 4.5 indicating that host rocks reached minimum temperatures of 60 to 300° C; locally higher values occur in rocks adjacent to plutonic intrusions (A. G. Harris, oral communication, 1985). Rocks of the York terrane have been deformed by thrust faulting, probably during the Brooks Range orogeny. Small stocks of 70- to 80-Ma biotite granite and mafic igneous rocks of unknown age locally intrude the sedimentary rocks. Tin deposits are associated with the granites; deposits at Lost River constitute the largest tin lode reserves in the United States (B. L. Reed, oral communication, 1985).

Ordovician rocks

Lower, Middle, and Upper Ordovician rocks occur in the York terrane. Lower Ordovician rocks are the most abundant and consist of a "shallow-water facies" and a "quiet-water facies" (Sainsbury, 1969). Rocks of the shallow water facies are interbedded quartz-carbonate sandstone and dolomitic lime mudstone, wackestone, and packstone. The sandstone is planar to cross-bedded, with locally well-developed oscillation and current ripples. The mudstones and wackestones are bioturbated, with bedding-plane feeding trails and subvertical burrows (Dumoulin, unpublished data). Other sedimentary features include intraclast conglomerates and oolites. This sequence may be as thick as 1,500 m (Sainsbury, 1969).

The sparse megafossil assemblage includes stromatolites, brachiopods, gastropods, and trilobite fragments (Sainsbury, 1969). Conodonts from a measured section at the mouth of Koteebue Creek indicate warm, shallow-water, normal marine conditions, and are of early Early Ordovician age (low Fauna D ofthe North American midcontinent faunal succession); they are very similar to correlative faunas from the Medfra Quadrangle, about 700 km southeast (A. G. Harris and J. Repetski, written communication, 1985).

Some rocks, lithologically similar to the Lower Ordovician shallow-water facies but considered by Sainsbury (1972) to be of Precambrian age, have recently been found to contain Early Ordovician conodonts. At one locality, bioclastic wackestones with 5 to 10 percent thin flasers of dolomitic mud are found; these rocks contain conodonts of early Early Ordovician age (low Fauna D) and indicate deposition in warm, shallow water (J. Repetski, written communication, 1985). Vandervoort (1985) reports additional collections of Early Ordovician conodonts from units previously thought to be Precambrian.

The other Lower Ordovician facies distinguished by Sainsbury (1969) consists of thick-bedded to massive micritic limestone with subordinate thinner interbeds of argillaceous limestone and locally abundant chert. As in the shallow-water facies, trace fossils are abundant. Unlike the shallow-water facies, these rocks lack ripple marks and sandstone interbeds, suggesting that deposition may have occurred in a quieter, perhaps somewhat deeper water environment with little input of terrigenous material (Sainsbury, 1969). Common rock types include bioclastic wackestones and pellet and intraclast grainstones (Dumoulin, unpublished data). The entire quiet-water sequence is thought to be about 2,200 m thick; the upper 70 m of this unit consists of a distinctive white to pinkish gray, blue-gray weathering, evenly bedded limestone with abundant trilobite fragments (Sainsbury, 1969).

Megafossils from this unit include cephalopods (Flower, 1941, 1946), trilobites, gastropods, brachiopods, and ostracodes (Sainsbury, 1969). These fossils indicate a Tremadocian or early Arenigian age for this unit (Dutro in Ross and others, 1982).

Conodonts collected from the quiet-water sequence support this age assignment. Collections from the middle and lower parts of the sequence contain conodonts of early Arenigian age (Fauna D) and indicate deposition on a shelf or platform chiefly under normal marine, although somewhat warm-water conditions. Collections from beds higher in the sequence contain conodonts of early middle Arenigian age diagnostic of slightly cooler water conditions (A. G. Harris and J. Repetski, written communication, 1985).

The entire Lower Ordovician sequence is characterized by 8- to 15-m-thick, shallowing-upward cycles (Vandervoort, 1985). The cycles appear to represent deposition in a range of subtidal to supratidal environments on a humid, pericontinental shelf.

Middle Ordovician outcrops are more limited in extent than those of the Lower Ordovician. The top of the quiet-water facies of the Lower Ordovician is succeeded, apparently conformably, by 7 to 30 m of fissile black shale with minor interbedded black limestone and dolomitic limestone. The shale sequence grades upward into flaggy, thin-bedded black limestone with shale partings. The Early/Middle Ordovician boundary seems to lie within this shale and limestone unit (Dutro in Ross and others, 1982). In the western part of the terrane, medium-bedded gray and brown limestone lithologically similar to older rocks also contains a Middle Ordovician fauna (Sainsbury, 1969; Dutro in Ross and others, 1982). Sainsbury (1972) considered the Middle Ordovician sequence to exceed 800 m in thickness.

Megafossils in Middle Ordovician rocks include cephalopods, graptolites, gastropods, and trilobites (Sainsbury, 1969; Dutro in Ross and others, 1982). Conodonts collected from the lower 50 m of the shale and limestone unit are of latest Arenigian to earliest Llanvirnian (earliest Middle Ordovician) age, and indicate cooler and deeper water conditions than those prevalent during deposition of the upper quiet water sequence (A. G. Harris and J. Repetski, written communication, 1985).

Upper Ordovician rocks crop out primarily along the Don River and are medium- to thick-bedded dark-gray limestone and dolomitic limestone, locally very fossiliferous. These rocks are wackestones and packstones, with pellets and skeletal hash in the micrite matrix (Dumoulin, unpublished data).

Megafossils from this unit include corals (Oliver and others, 1975), stromatoporoids, gastropods, ostracodes, trilobites, and brachiopods (Sainsbury, 1969). The trilobite fauna includes *Monorakis,* the only confirmed occurrence of this form outside of the northern Soviet Union, where it is widely distributed (Ormiston and Ross, 1979). Other evidence for a faunal tie between the Soviet Union and the area of the York terrane during the late Ordovician is provided by the coral and brachiopod faunas (Potter, 1984) and by conodont faunas from the Don River section. The conodont assemblage is of Late Ordovician age and is diagnostic of a warm, relatively shallow-water depositional environment; it includes forms thus far found only in Siberia and Alaska (A. G. Harris, written communication, 1988).

Silurian rocks

Silurian rocks are best exposed along the Don River where the Ordovician/Silurian boundary has been studied (Sainsbury and others, 1971). Light brown to dark gray limestones, dolomitic limestones, and dolostones contain local buildups of corals and stromatoporoids. The Silurian section consists chiefly of mudstones and bioclastic packstones and wackestones and contains more dolostones than does the underlying Ordovician section Some samples show well-developed fenestral fabric, suggesting deposition took place in very shallow water (Dumoulin, unpublished data). The Don River Silurian section may be as thick as 270 m (Sainsbury and others, 1971).

Megafossils from the Silurian portion of the Don River section include corals (Oliver and others, 1975), stromatoporoids, bryozoa, and brachiopods, interpreted to be of Middle and Late Silurian age by Sainsbury and others (1971). Biostratigraphically diagnostic conodonts from this locality are of Early and earliest Late Silurian age and indicate deposition in a warm, shallow-water environment (A. G. Harris, written communication, 1985).

Rocks of Silurian age also form a small klippe in the eastern part of the York terrane; these rocks were originally thought to be of Devonian age (Sainsbury, 1972). Megafossils at this locality indicate a Silurian age (W. Oliver, written communication, 1973). Conodonts from this locality are of middle Middle Silurian age and indicate deposition in a shallow-water environment (A. G. Harris, written communication, 1985).

Devonian? rocks

Only a single locality in the York terrane has produced fossils of possible Devonian age. Rocks in the north-central Teller quadrangle, originally shown by Sainsbury (1972) as part of the Lower Ordovician quiet-water facies, contain corals considered to be of probable Middle or early Late Devonian age by Oliver and others (1975). This age has been reevaluated, based on more recent worldwide collections, as Late Silurian (late Ludlovian)–early Late Devonian (Frasnian; A. Pedder, personal communication, 1988).

Mississippian rocks

Rocks of Mississippian age found near Cape Mountain, on the westernmost tip of the Seward Peninsula, consist of intensely deformed and recrystallized limestone intercalated with subordinate clastic rocks; the section is too deformed to allow an estimate of its thickness (Steidtmann and Cathcart, 1922; Sainsbury, 1972). A coral fauna of probable Late Mississippian age has been obtained from these rocks (Steidtmann and Cathcart, 1922).

Rocks of uncertain age

Two important units of uncertain age are found in the York terrane. There are extensive outcrops of slate, siltstone, graywacke, and subordinate carbonaceous limestone of the "slate of the York region" (Sainsbury, 1974) in the west, central, and south parts of the York terrane. This unit was thought by Sainsbury (1969, 1972) to be transitional upward into a locally strongly deformed, thin-bedded sequence of argillaceous and dolomitic limestone interbedded with silty claystone, called the Kanauguk Formation by Sainsbury (1974). Both the "slate of the York region" and the Kanauguk Formation were considered to be of Precambrian age because "the rocks are unfossiliferous and more deformed than the overlying Lower Ordovician limestones, contain quartz veins, and are intruded by numerous gabbros that do not intrude the Lower Ordovician limestones" (Sainsbury, 1972, p. 2).

The "slate of the York region" in the York terrane includes diverse lithologies, which may be grouped roughly into two sequences (Dumoulin and Till, unpublished data). One sequence consists of fine-grained schists and semischists of various, primarily pelitic and calcareous, composition; no fossils have yet been found to constrain their age. The other sequence is less metamorphosed and consists of locally calcareous, rhythmically intercalated mudstone, siltstone, and sandstone; clasts are mostly quartz and sedimentary lithic grains. Outcrop features include climbing ripples, cross beds, parallel laminae, convolute laminae, and graded beds, and suggest a turbidite origin for these rocks. A single conodont collection from this sequence is of middle Early through Late Ordovician age (A. G. Harris, written communication, 1985). Conodonts recovered from turbidite sequences may represent reworked older material, so it seems prudent to interpret this as a maximum age.

Rocks of the Kanauguk Formation are lithologically similar to rocks of the shallow-water facies of the Lower Ordovician; some limestones assigned a Precambrian age by Sainsbury (1972) have since been found to contain Lower Ordovician conodonts. Thus, much or all of Sainsbury's "Precambrian limestone unit" may be of Early Ordovician age and at least part of the "slate of the York region" is no older than middle Early Ordovician.

TECTONIC HISTORY

Lower Paleozoic rocks of the Seward Peninsula record a continental platform depositional environment punctuated by at least one episode of rifting. Iron- and titanium-rich metagabbros of the Nome Group are compositionally similar to modern ridge tholeiites (Forbes and others, 1984); the metagabbros intruded volcanogenic sediments and carbonate rocks during the Ordovician. Upper Paleozoic strata are not known on Seward Peninsula.

The Mesozoic history of the Seward Peninsula is dominated by convergent tectonics and plutonism. The Nome Group blueschist terrane is among the largest structurally coherent blueschist terranes in the world (Forbes and others, 1984). It is thought to have formed in or near a gently south-dipping subduction zone during collision of a volcanic arc with part of the North American continent in Late Jurassic to Early Cretaceous time (Box, 1985; Patrick, 1988).

The significance of the garnet lherzolite at the base of the Kigluaik Group is not well understood. The rare occurrence of alpine-type garnet lherzolite in crustal settings has caused many workers to infer that they are a relict of significant crustal upheaval (i.e., continental collision; Carswell and Gibb, 1980).

Upgraded equivalents of the Nome Group stratigraphy have been mapped in the Bendeleben and Darby Mountains (Till and others, 1986). Overprinting of higher-grade minerals directly onto blueschist-facies assemblages was noted by those workers in the northern Darby Mountains and by Patrick and Lieberman (1987) in the southern Kigluaik Mountains. The high-grade metamorphism in the Darby Mountains ended by 102 Ma (the age of small anatectic stocks that intruded the highest-grade rocks); there, the high-grade thermal overprint followed blueschist-facies metamorphism within 40 to 60 m.y. Patrick and Lieberman (1987) suggest that the high-grade rocks formed with restoration of a normal thermal gradient after subduction ceased, and they cite the metamorphic history of the Lepontine Alps as an analogue to the Seward Peninsula. Three periods of Late Cretaceous plutonism followed the crustal thickening episode.

In contrast to the deep crustal history of the Seward terrane, rocks of the York terrane show evidence of brittle deformation by thrust faulting (Sainsbury, 1969, 1972 and heating to relatively low temperatures (CAI 1.5 to 4.5; A. G. Harris, written communication, 1985). Sainsbury (1972) indicated a northward vergence for thrusting in the York terrane; structures with eastward vergence have also been observed (Till, unpublished data). (Sainsbury, 1972, also cited evidence for east-vergent folding in the Seward terrane.) More extensive microfossil dating, geologic mapping, and structural analysis are necessary before the age and nature of thrusting can be determined.

Confusion regarding the vergence of thrusting may be due to the occurrence of two separate events: one, synchronous with the development of the fold and thrust belt in the Brooks Range; and the other, in response to east-west compression centered in the Bering Straits area due to opening of the Atlantic Ocean. Folds with north-south–oriented axes, slightly overturned to the east, are found just east of Seward Peninsula in Albian and older sedimentary rocks, indicating that east-west compression occurred in post-Albian time (Patton and Tailleur, 1977). Patton and Tailleur (1977) postulated that an oroclinal bend formed between western Alaska and Chukotka (the Soviet Far East) in early Tertiary time, corresponding to the opening of the Atlantic Ocean. Their model would require a 90° counterclockwise rotation of Seward Peninsula relative to the western Brooks Range. Several lines of evidence argue against this model. Fold hinges and lineations formed during the Late Jurassic or Early Cretaceous in schist of the Seward terrane are parallel to their counterparts in the Brooks Range schist belt (Till, unpublished data). Quartz petrofabrics of Nome Group schist indicate that deformation on Seward Peninsula was northward-vergent (Patrick, 1988). Paleomagnetic studies (Plumley and Reusing, 1984) have been interpreted to preclude rotation of Seward Peninsula during the Tertiary. A major left-lateral strike-slip fault system trending northwest through Kotzebue Sound may instead be the major mechanism by which the Seward Peninsula moved into its present position.

Basin development and basaltic volcanism attest to the extensional nature of tectonics of the Seward Peninsula during Tertiary time.

SAINT LAWRENCE ISLAND

Saint Lawrence Island (Fig. 3) is situated in the shallow portion of the Bering Sea underlain by continental crust, 208 km from the Seward Peninsula and 64 km from the Chukotsk Peninsula of the U.S.S.R. The island is 5,000 km^2 in area, most of which is tundra-covered wave-cut platform. Streams are incised up to 50 m into the platform, and barren, rubble-covered mountains rise 300 to 600 m above it.

The island has been mapped at a scale of 1:250,000 (Patton and Csejtey, 1980), and topical studies have been completed on Pleistocene deposits (Hopkins and others, 1972) and plutonic rocks (Csejtey and Patton, 1974; Csejtey and others, 1971).

The bedrock of Saint Lawrence Island is composed of three lithologic belts, which are knit by mid-Cretaceous intrusives and overlain by a cover of Tertiary and Quaternary volcanic rocks.

The three lithologic belts include Paleozoic miogeoclinal sequence, a Permian to Triassic oceanic assemblage, and Lower Cretaceous andesitic rocks (Patton and Csejtey, 1980). The miogeoclinal assemblage, exposed in the eastern and central part of the island, is composed of a thick section of Devonian and Mississippian carbonate rocks and a condensed section of Triassic shale, limestone, and chert. The Permian to Triassic assemblage, in probable fault contact with the miogeoclinal rocks and exposed in the western and central part of the island, is composed of intensely deformed graywacke, grit, shale, and associated gabbro and diabase. Cretaceous volcanic rocks include flows, hypabyssal intrusives, and chiefly andesitic volcaniclastic rocks that range in composition from basalt to rhyolite. The Cretaceous rocks are also intensely folded and faulted (Patton and Csejtey, 1980). The three lithologic belts and a minor amount of undated but probable Precambrian or Paleozoic banded marly limestone were intruded

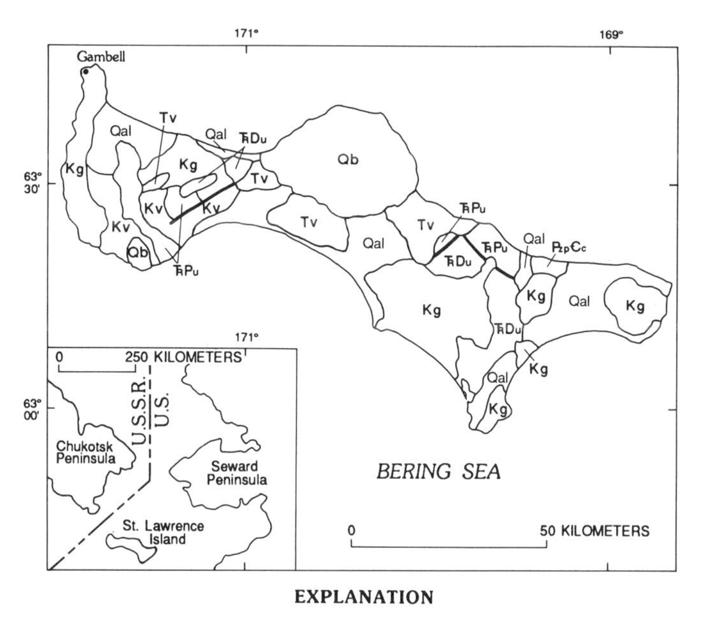

EXPLANATION

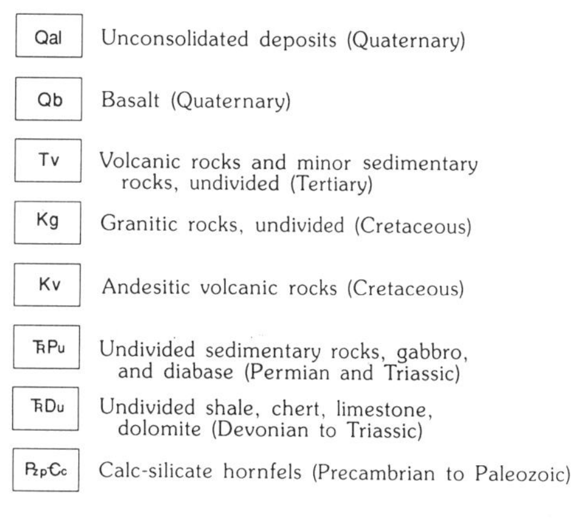

Figure 3. Geologic map of St. Lawrence Island, from Patton and Csejtey (1980).

during the Cretaceous by quartz monzonitic and lesser syenitic plutons, including small bodies of nepheline syenite that give K-Ar ages ranging from 93 to 108 Ma (Patton and Csejtey, 1980).

Tertiary volcanic rocks are of Paleocene and Oligocene age. Paleocene rocks are composed of sodarhyolite and basalt flows and hypabyssal intrusives with subordinate trachyandesite and andesite. Lignite float is found locally on these rocks. K-Ar ages of 62 to 64 Ma were obtained from the volcanic rocks (Patton and Csejtey, 1980). Oligocene rhyolitic and dacitic tuff with tuff-breccia and flows are intercalated with quartzose sandstone and conglomerate; plant fossils occur locally. These rocks are found on the eastern part of the island; correlative sedimentary rocks on the western part of the island include sandstone, coal, and tuff. A K-Ar age of 39.3 Ma was obtained from volcanic biotite; plant fossils of Oligocene age were collected from the sedimentary sequence (Patton and Csejtey, 1980). These Oligocene rocks are age-correlative to 37 to 43 Ma rhyolites and lesser basalt found in mainland western Alaska (Moll-Stalcup, personal communication, 1986; Miller and Lanphere, 1981).

Quaternary basalt cinder cones and flows form a large field in the north-central part of the island. These alkalic and theoleiitic basalts are part of the Bering Sea basalt field, which extends from the Seward Peninsula to the Pribilof Islands (Moll-Stalcup, personal communication, 1986; Moll-Stalcup, this volume).

Rocks of the Paleozoic to Mesozoic miogeoclinal sequence have been correlated with rocks in northwestern Alaska (Patton and Dutro, 1969). Devonian carbonate rocks with similarities to those found on St. Lawrence Island are found in the western and central Brooks Range. Shallow-water dolostones and marbles with particularly strong lithologic and faunal similarities occur in the Nome Group of the Seward terrane (Till and others, 1986). Mississippian and Triassic rocks of the island are correlative with the Alapah Limestone and Shublik Formation of the North Slope sequence (Patton and Dutro, 1969; Moore and others, this volume).

Saint Lawrence Island occupies a critical position in tectonic reconstructions of the Bering Straits region as a link between lithotectonic belts in Alaska and the northeast U.S.S.R. (Patton and Tailleur, 1977; Box, 1985).

REFERENCES CITED

Armstrong, R. L., Harakal, J. E., Forbes, R. B., Evans, B. W., and Thurston, S. P., 1986, Rb-Sr and K-Ar study of metamorphic rocks of Seward Peninsula and southern Brooks Range, Alaska, *in* Evans, B. W., and Brown, E. H., eds., Blueschists and eclogites: Geological Society of America Memoir 176, p. 185–203.

Barnes, D. F., and Hudson, T., 1977, Bouguer gravity map of Seward Peninsula, Alaska: U.S. Geological Survey Open-File Report 77–769C, scale 1:1,000,000.

Barnes, F. F., 1967, Coal resources of Alaska: U.S. Geological Survey Bulletin 1242-B, p. B28–B29.

Box, S. E., 1985, Early Cretaceous orogenic belt in northwestern Alaska; Internal organization, lateral extent, and tectonic interpretation, *in* Howell, D. G., ed., Tectonostratigraphic terranes of the Circum-Pacific region: Circum-Pacific Council for Energy and Mineral Resources, Earth Science Series, v. 1, p. 137–146.

Brooks, A. H., Richardson, G. B., Collier, A. J., and Mendenhall, W. C., 1901, Reconnaissances in the Cape Nome and Norton Bay regions, Alaska, in 1900: U.S. Geological Survey Special Publication, 185 p.

Bunker, C. M., Hedge, C. E., and Sainsbury, C. L., 1979, Radioelement concentrations and preliminary radiometric ages of rocks of the Kigluaik Mountains, Seward Peninsula, Alaska: U.S. Geological Survey Professional Paper 1129-C, 12 p.

Carswell, D. A., and Gibb, F.G.F., 1980, The equilibration conditions and petrogenesis of European crustal garnet lherzolites: Lithos, v. 13, p. 19–29.

Csejtey, B., Jr., and Patton, W. W., Jr., 1974, Petrology of the nepheline syenite of St. Lawrence Island, Alaska: U.S. Geological Survey Journal of Research, v. 2, p. 41–47.

Csejtey, B., Jr., Patton, W. W., Jr., and Miller, T. P., 1971, Cretaceous plutonic rocks of St. Lawrence Island, Alaska; A preliminary report, *in* Geological Survey Research, 1971: U.S. Geological Survey Professional Paper 750-D, p. D68–D76.

Dickinson, K. A., 1984, Death Valley, Alaska, Uranium deposit: Geological Society of America Abstracts with Programs, v. 16, p. 278.

Dumoulin, J. A., and Harris, A. G., 1984, Carbonate rocks of central Seward Peninsula, Alaska: Geological Society of America Abstracts with Programs, v. 16, p. 280.

——, 1987, Lower Paleozoic carbonate rocks of the Baird Mountains Quadrangle, western Brooks Range, Alaska, *in* Tailleur, I. L., and Weimer, P., eds., Alaskan North Slope geology: Pacific Section, Society of Economic Paleontologists and Mineralogists and Alaska Geological Society, v. 1, p. 311–336.

Dumoulin, J. A., and Till, A. B., 1985, Sea cliff exposures of metamorphosed carbonate and schist, northern Seward Peninsula, *in* Bartsch-Winkler, S., and Reed, K. M., eds., Accomplishments in Alaska during 1983: U.S. Geological Survey Circular 945, p. 18–22.

Evans, B. W., and Patrick, B. E., 1987, Phengite 3-T in high-pressure metamorphosed granitic orthogneisses, Seward Peninsula, Alaska: Canadian Mineralogist, v. 25, p. 141–158.

Flower, R. H., 1941, Cephalopods from the Seward Peninsula of Alaska: Bulletin of American Paleontology, v. 27, no. 102, 22 p.

——, 1946, *Alaskoceras* and the Plectoceratidae: Journal of Paleontology, v. 20, no. 6, p. 620–624.

Forbes, R. B., Evans, B. W., and Thurston, S. P., 1984, Regional progressive high-pressure metamorphism, Seward Peninsula, Alaska: Journal of Metamorphic Geology, v. 2, p. 43–54.

Gardner, M. C., and Hudson, T. L., 1984, Structural geology of Precambrian and Paleozoic metamorphic rocks, Seward terrane, Alaska: Geological Society of America Abstracts with Programs, v. 16, p. 285.

Gilbert, W. G., Wiltse, M. A., Carden, J. R., Forbes, R. B., and Hackett, S. W., 1977, Geology of Ruby Ridge, southwestern Brooks Range, Alaska: Alaska Division of Geological and Geophysical Surveys Geologic Report 58, 16 p.

Harris, A. G., and Repetski, J. E., 1987, Ordovician conodonts from northern Alaska: Geological Society of America Abstracts with Programs, v. 19, p. 169.

Hitzman, M. W., Proffett, J. M., Jr., Schmidt, J. M., and Smith, T. E., 1986, Geology and mineralization of the Ambler district, northwestern Alaska: Economic Geology, v. 81, no. 7, p. 1592–1618.

Hopkins, D. M., 1963, Geology of the Imuruk Lake area, Seward Peninsula, Alaska: U.S. Geological Survey Bulletin 1141-C, 101 p.
—— , ed., 1967, The Bering Land Bridge: Stanford, California, Stanford University Press, 485 p.
Hopkins, D. M., Rowland, R. W., and Patton, W. W., Jr., 1972, Middle Pleistocene mollusks from St. Lawrence Island and their significance for the paleoceanography of the Bering Sea: Quaternary Research, v. 2, p. 119–134.
Hudson, T., 1979, Igneous and metamorphic rocks of the Serpentine Hot Springs area, Seward Peninsula, Alaska: U.S. Geological Survey Professional Paper 1079, 27 p.
—— , 1977, Geologic map of the Seward Peninsula, Alaska: U.S. Geological Survey Open-File Report 77–796A, 1 sheet, scale 1:1,000,000.
Hudson, T., and Arth, J. G., 1983, Tin granites of Seward Peninsula, Alaska: Geological Society of America Bulletin, v. 94, p. 768–790.
Hudson, T., and DeYoung, J. R., Jr., 1978, Map and tables describing areas of mineral resource potential, Seward Peninsula, Alaska: U.S. Geological Survey Open-File Report 78–1–C, 62 p., 1 sheet, scale 1:1,000,000.
Kaufman, D. S., and Hopkins, D. M., 1986, Glacial history of Seward Peninsula, *in* Hamilton, T. D., Reed, K. M., and Thorson, R. M., eds., Glaciation in Alaska; The geologic record: Journal of the Alaska Geological Society, p. 51–77.
Knopf, A., 1908, Geology of the Seward Peninsula tin deposits, Alaska: U.S. Geological Survey Bulletin 358, 71 p.
Lockhart, A. B., 1984, Gravity survey of the central Seward Peninsula [M.S. thesis]: Fairbanks, University of Alaska, 83 p.
Mayfield, C. F., Tailleur, I. L., and Ellersieck, I., 1983, Stratigraphy, structure, and palinspastic synthesis of the western Brooks Range, northwestern Alaska: U.S. Geological Survey Open-File Report 83–779, 58 p.
Miller, T. P., 1972, Potassium-rich alkaline intrusive rocks of west-central Alaska: Geological Society of America Bulletin, v. 83, p. 2111–2128.
—— , 1989, Contrasting plutonic rock suites of the Yukon–Koyukuk basin and the Ruby geanticline, Alaska: Journal of Geophysical Research, v. 94, no. B11, p. 15969–15987.
Miller, T. P., and Bunker, C. M., 1976, A reconnaissance study of the uranium and thorium contents of plutonic rocks of southwestern Seward Peninsula, Alaska: U.S. Geological Survey Journal of Research, v. 4, p. 367–377.
Miller, T. P., and Lanphere, M. A., 1981, K-Ar age measurements on obsidian from the Little Indian River locality in interior Alaska, *in* Albert, N.R.D., and Hudson, T., eds., The U.S. Geological Survey in Alaska; Accomplishments during 1979: U.S. Geological Survey Circular 823-B, p. B39–B42.
Miller, T. P., Grybeck, D. L., Elliott, R. L., and Hudson, T., 1972, Preliminary geologic map of the eastern Solomon and southeastern Bendeleben Quadrangles, eastern Seward Peninsula, Alaska: U.S. Geological Survey Open-File Report 72–256, 11 p., 2 sheets, scale 1:250,000.
Miller, T. P., Barnes, I., and Patton, W. W., Jr., 1975, Geological setting and chemical characteristics of hot springs in west-central Alaska: U.S. Geological Survey Journal of Research, v. 3, no. 2, p. 149–162.
Moffit, F. H., 1913, Geology of the Nome and Grand Central Quadrangles, Alaska: U.S. Geological Survey Bulletin 533, 140 p.
Oliver, W. A., Merriam, C. W., and Churkin, M., 1975, Ordovician, Silurian, and Devonian corals of Alaska: U.S. Geological Survey Professional Paper 823-B, 31 p.
Ormiston, A. R., and Ross, R. J., Jr., 1979, *Monorakos* in the Ordovician of Alaska and its zoogeographic significance, *in* Gray, J., and Boucot, A. J., eds., Historical biogeography, plate tectonics, and the changing environment: Corvallis, Oregon State University Press, p. 53–59.
Patrick, B. E., 1988, Synmetamorphic structural evolution of the Seward Peninsula blueschist terrane, Alaska: Journal of Structural Geology, v. 10, no. 6, p. 555–565.
Patrick, B. E., and Evans, B. W., 1989, Metamorphic evolution of the Seward Peninsula blueschist terrane: Journal of Petrology, v. 30, p. 531–556.
Patrick, B. E., and Lieberman, J. E., 1987, Thermal overprint on the Seward Peninsula blueschist terrane; The Lepontine in Alaska: Geological Society of America Abstracts with Programs, v. 19, p. 800–801.
Patton, W. W., Jr., and Csejtey, B., Jr., 1980, Geologic map of St. Lawrence Island, Alaska: U.S. Geological Survey Miscellaneous Investigations Series Map I-1203, 1 sheet, scale 1:250,000.
Patton, W. W., Jr., and Dutro, J. T., Jr., 1969, Preliminary report on Paleozoic and Mesozoic sedimentary sequence on St. Lawrence Island, Alaska, *in* Geological Survey research 1969: U.S. Geological Survey Professional Paper 650-D, p. D138–D143.
Patton, W. W., Jr., and Tailleur, I. L., 1977, Evidence in the Bering Strait region for differential movement between North America and Eurasia: Geological Society of America Bulletin, v. 88, p. 1298–1304.
Plumley, P. W., and Reusing, S., 1984, Paleomagnetic investigation of Paleozoic carbonates, York Terrane, Seward Peninsula [abs.]: EOS Transactions of the American Geophysical Union, v. 65, no. 45, p. 869.
Potter, A. W., 1984, Paleobiogeographical relations of Late Ordovician brachiopods from the York and Nixon Fork terranes, Alaska: Geological Society of America Abstracts with Programs, v. 16, p. 626.
Puchner, C. C., 1986, Geology, alteration, and mineralization of the Kougarok Sn deposit, Seward Peninsula, Alaska: Economic Geology, v. 81, no. 7, p. 1775–1794.
Robinson, M. S., and Stevens, D. L., compilers, 1984, Geologic map of Seward Peninsula, Alaska: Alaska Division of Geological and Geophysical Surveys Report 34, 1 sheet, scale 1:500,000.
Ross, R. J., Adler, F. J., and 26 others, 1982, The Ordovician System in the United States; Correlation chart and explanatory notes: International Union of Geological Sciences Publication 12, 73 p.
Ryherd, T. J., and Paris, C. E., 1987, Ordovician through Silurian carbonate base-of-slope apron sequence, northern Seward Peninsula, Alaska [abs.], *in* Tailleur, I. L., and Weimer, P., eds., Alaskan North Slope geology: Pacific Section, Society of Economic Paleontologists and Mineralogists and Alaska Geological Society, v. 1, p. 347–348.
Sainsbury, C. L., 1969, Geology and ore deposits of the central York Mountains, Seward Peninsula, Alaska: U.S. Geological Survey Bulletin 1287, 101 p.
—— , 1972, Geologic map of the Teller Quadrangle, western Seward Peninsula, Alaska: U.S. Geological Survey Map I-685, 4 p., scale 1:250,000.
—— , 1974, Geologic map of the Bendeleben Quadrangle, Seward Peninsula, Alaska: Anchorage, Alaska, The Mapmakers, 31 p., scale 1:250,000.
—— , 1975, Geology, ore deposits, and mineral potential of the Seward Peninsula, Alaska: U.S. Bureau of Mines Open-File Report 73–75, 108 p.
Sainsbury, C. L., Coleman, R. G., and Kachadoorian, R., 1970, Blueschist and related greenschist facies rocks of the Seward Peninsula, Alaska, *in* Geological Survey research: U.S. Geological Survey Professional Paper 750-C, p. C52–C57.
Sainsbury, C. L., Dutro, J. T., Jr., and Churkin, M., 1971, The Ordovician–Silurian boundary in the York Mountains, western Seward Peninsula, Alaska: U.S. Geological Survey Professional Paper 750-C, p. C52–C57.
Sainsbury, C. L., Hummel, C. L., and Hudson, T., 1972a, Reconnaissance geologic map of the Nome Quadrangle, Seward Peninsula, Alaska: U.S. Geological Survey Open-File Report, 28 p., 1 sheet, scale 1:250,000.
Sainsbury, C. L., Hudson, T., Ewing, R., and Marsh, W. R., 1972b, Reconnaissance geologic map of the west half of the Solomon Quadrangle, Alaska: U.S. Geological Survey Open-File Report, 10 p., 1 sheet, scale 1:250,000.
Smith, P. S., 1910, Geology and mineral resources of the Solomon and Casadepaga Quadrangles, Seward Peninsula, Alaska: U.S. Geological Survey Bulletin 433, 234 p.
Steidtmann, E., and Cathcart, S. H., 1922, Geology of the York tin deposits, Alaska: U.S. Geological Survey Bulletin 733, 130 p.
Sturnick, M. A., 1984, Regional metamorphism of the eastern Kigluaik Mountains, Seward Peninsula, Alaska: Geological Society of America Abstracts with Programs, v. 16, p. 335.
Thurston, S. P., 1985, Structure, petrology, and metamorphic history of the Nome Group blueschist terrane, Salmon Lake area, Seward Peninsula, Alaska: Geological Society of America Bulletin, v. 96, p. 600–617.
Till, A. B., 1980, Crystalline rocks of the Kigluaik Mountains, Seward Peninsula, Alaska [M.S. thesis]: Seattle, University of Washington, 97 p.

Till, A. B., Dumoulin, J. A., Aleinikoff, J. N., Harris, A., and Carroll, P. I., 1983, Paleozoic rocks of the Seward Peninsula; New insight; New Developments in the Paleozoic Geology of Alaska and the Yukon: Alaska Geological Society Symposium Program and Abstracts, p. 25–26.

Till, A. B., Dumoulin, J. A., Gamble, B. M., Kaufman, D. S., and Carroll, P. I., 1986, Preliminary geologic map and fossil data, Solomon, Bendeleben, and southern Kotzebue Quadrangles, Seward Peninsula, Alaska: U.S. Geological Survey Open-File Report 86-276, 71 p., 3 sheets, scale 1:250,000.

Vandervoort, D. J., 1985, Stratigraphy, paleoenvironment, and diagenesis of the Lower Ordovician York Mountain carbonates, Seward Peninsula, Alaska [M.S. thesis]: Baton Rouge, Louisiana State University, 141 p.

Westcott, E., and Turner, D. L., eds., 1981, Geothermal reconnaissance survey of the central Seward Peninsula, Alaska: Fairbanks, University of Alaska Geophysical Institute Report UAG R-284, 123 p.

MANUSCRIPT RECEIVED BY THE SOCIETY DECEMBER 4, 1990

NOTES ADDED IN PROOF

Considerable advances have been made in understanding the metamorphic evolution of the Seward Peninsula. Conditions of the blueschist-facies metamorphism of the Nome Group were 460 ± 30°C and approximately 12 kilobars (Patrick and Evans, 1989); ductile fabrics formed during that event record high strain (Patrick, 1988). Amphibolite- to granulite-facies rocks of the Kigluaik Group reached peak metamorphic temperatures and pressures of 800°C and 8 kilobars (Lieberman, 1988). Magmatism was apparently synchronous with metamorphism and spanned the period 105–83 Ma (Amato and others, 1992). Interpretations of the tectonic setting of the high-grade metamorphism and the exhumation of the blueschists of the Nome Group remain controversial and have been assigned to both contractional (Patrick and Lieberman, 1988; Patrick, 1988) and extensional settings (Miller and Hudson, 1991; Miller and others, 1992).

Little work has been done on the protolith packages of the metamorphic units. However, a U-Pb zircon age from an orthogneiss body that intruded the lower part of the Nome Group indicates that the mixed unit is in part Proterozoic in age (B. Patrick, oral communication, 1993).

On Saint Lawrence Island, a new conodont collection shows that some of the Paleozoic carbonate rocks are at least as old as earliest Devonian and possibly older (J. Clough and A. G. Harris, oral communication, 1993).

ADDITIONAL REFERENCES

Amato, J. M., Wright, J. E., and Gans, P. B., 1992, The nature and age of Cretaceous magmatism and metamorphism on the Seward Peninsula, Alaska: Geological Society of America Abstracts with Programs, v. 24, p. 2.

Lieberman, J. E., 1988, Metamorphic and structural studies of the Kigluaik Mountains, western Alaska [unpublished Ph.D. thesis]: University of Washington, 189 p.

Miller, E. L., and Hudson, T. L., 1991, Mid-Cretaceous extensional fragmentation of a Jurassic–Early Cretaceous compressional orogen, Alaska: Tectonics, v. 10, p. 781–796.

Miller, E. L., Calvert, A. T., and Little, T. A., 1992, Strain-collapsed metamorphic isograds in a sillimanite gneiss dome, Seward Peninsula, Alaska: Geology, v. 20, p. 487–490.

Patrick, B. E., 1988, Synmetamorphic structural evolution of the Seward Peninsula blueschist terrane, Alaska: Journal of Structural Geology, v. 10, p. 555–565.

Patrick, B. E., and Evans, B. W., 1989, Metamorphic evolution of the Seward Peninsula blueschist terrane: Journal of Petrology, v. 30, p. 531–556.

Patrick, B. E., and Lieberman, J. E., 1988, Thermal overprint on blueschists of the Seward Peninsula; the Lepontine in Alaska: Geology, v. 16, p. 1100–1103.

Printed in U.S.A.

The Geology of North America
Vol. G-1, The Geology of Alaska
The Geological Society of America, 1994

Chapter 5

Geology of part of east-central Alaska

James H. Dover
U.S. Geological Survey, 4200 University Drive, Anchorage, Alaska 99508-4667

INTRODUCTION

The east-central Alaska region extends west from the Canadian border to the southeast edge of the Koyukuk basin, and north from the Yukon-Tanana upland to the southeast flank of the Brooks Range (Fig. 1). The region encompasses all or parts of 13 1:250,000 quadrangles and covers an area of about 140,000 km^2.

The physiography of the region (Wahrhaftig, this volume) is extremely diverse (Fig. 1). Steep, mountainous areas having high to moderate relief include parts of the Ogilvie Mountains, White Mountains, Yukon-Tanana upland, Ray Mountains, and southeastern Brooks Range. Areas of dissected plateaus and rolling hills with moderate to low relief include the Porcupine plateau and the Kokrines-Hodzana upland. Lowlands characterize the Yukon Flats basin and extend upstream into other parts of the Yukon River and Porcupine River drainage basins.

Bedrock exposures are severely limited in most of the east-central Alaska region. Mountainous areas are locally glaciated, but except in the most rugged parts of the Ogilvie Mountains and parts of the Brooks Range, bedrock is extensively mantled by surficial cover and the colluvial products of alpine weathering processes. Stream cuts commonly offer the best exposures in generally forested or tundra-covered areas of moderate relief. Lowland areas have few pre-Cenozoic bedrock exposures because of extensive surficial cover and tundra.

The pioneering reconnaissance studies by J. B. Mertie and his predecessors were carried out in the east-central Alaska region prior to 1940. A relatively small group of geologists conducted the basic geologic mapping during the 1960s and 1970s that defined the stratigraphic and structural framework of the region and that forms the foundation of more recent studies. Among these mappers, to whom I am especially indebted, are E. E. Brabb and Michael Churkin, Jr., in the Charley River and Black River quadrangles; H. L. Foster, in the Eagle quadrangle and the Yukon-Tanana upland; R. M. Chapman and F. R. Weber, in the Tanana and Livengood quadrangles; and W. P. Brosgè and H. N. Reiser, in the southeastern Brooks Range. Although the list of references accompanying this chapter attests to the number of contributors to the geology of the region, the fundamental contributions of these seven cannot be overestimated. Among the major geologic features of the east-central Alaska region that they helped define are the Charley River, Livengood, and southeastern Brooks Range fold-and-thrust belts; the low-grade metamorphic and plutonic complexes of the Ruby geanticline, Yukon-Tanana upland, and southern Brooks Range; the Tozitna mafic-igneous and sedimentary assemblage; and regionally prominent strike-slip fault systems (Fig. 2). More recent work has emphasized the wide range of stratigraphic facies and structural styles characterizing the east-central Alaska region. This work has led Churkin and others (1982) and Silberling and Jones (1984) to subdivide the region into numerous tectonostratigraphic (lithotectonic) terranes and subterranes (Fig. 2; Plate 3, this volume), the identity, origins, and significance of which are presently under debate. The application of terrane terminology and concepts to the east-central Alaska region, and their implications with regard to accretionary

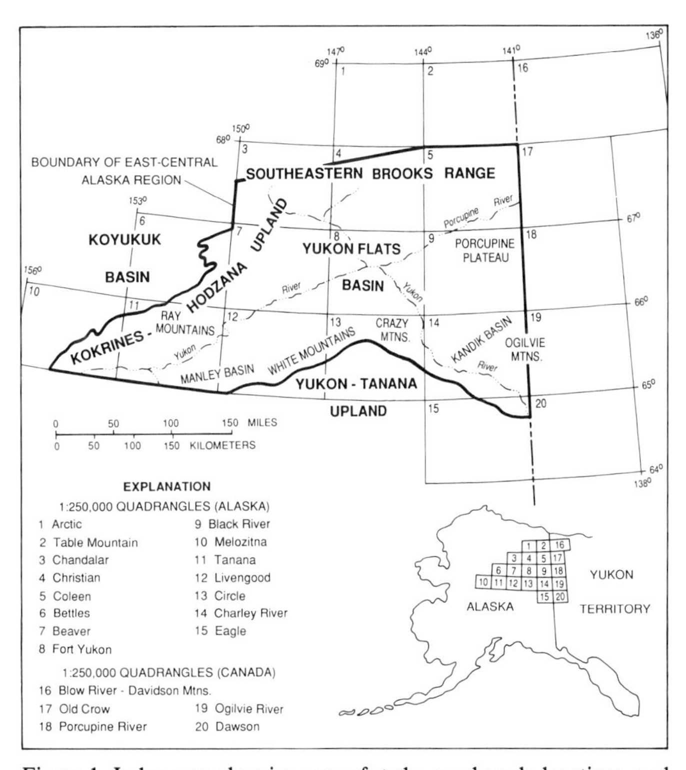

Figure 1. Index map showing area of study, quadrangle locations, and main physiographic features.

Dover, J. H., 1994, Geology of part of east-central Alaska, *in* Plafker, G., and Berg, H. C., eds., The Geology of Alaska: Boulder, Colorado, Geological Society of America, The Geology of North America, v. G-1.

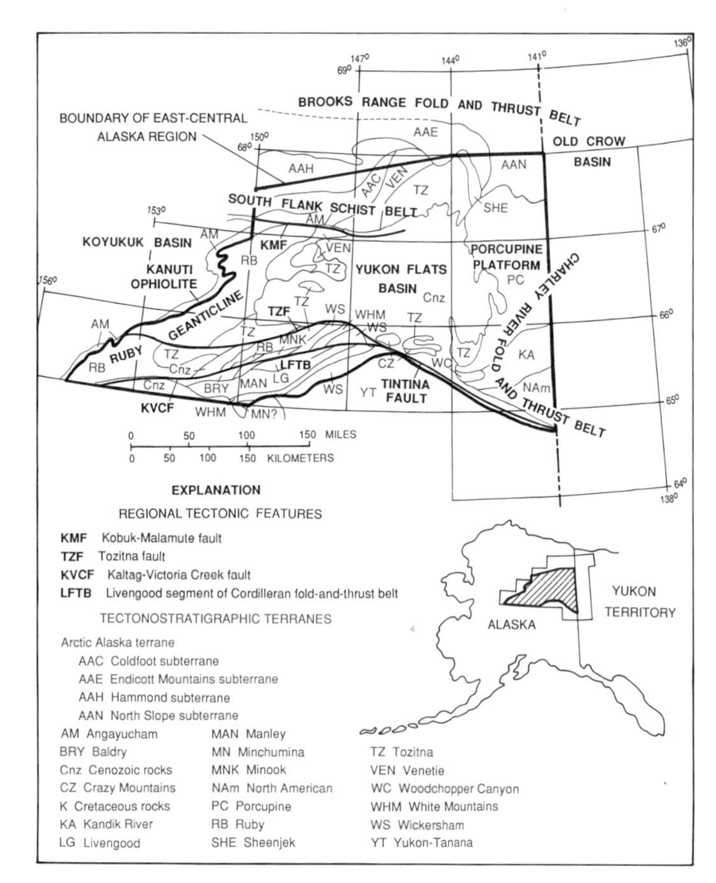

Figure 2. Major geotectonic features and tectonostratigraphic terranes. Terranes after Silberling and others (this volume), Silberling and Jones (1984), Churkin and others (1982).

tectonics and paleogeographic reconstruction in interior Alaska, have raised a series of fundamental questions. Among them are:

(1) To what degree are the stratigraphic sequences represented in east-central Alaskan terranes internally consistent and distinct from those of neighboring terranes, and can intermediate facies between terranes be reasonably interpreted or demonstrated?

(2) What is the character of terrane-bounding structures? Can the major suture zones required by accretionary tectonic models be identified in east-central Alaska? If so, how do they relate to other major structural features known in the region, such as fold-and-thrust belt structures or strike-slip faults?

(3) How and where did the ultramafic-bearing mafic volcanic/plutonic sequences of east-central Alaska originate, and how were they emplaced at their present sites?

(4) What is the significance of the granitic plutonism and metamorphic overprint that characterizes some terranes, and how distinct are the metamorphic differences between terranes?

(5) Can reasonable paleogeographic reconstructions be made for both accretionary and non-accretionary models of geologic development? How do they differ? Are they mutually incompatible, or is a combination of the two possible?

The approach taken in this chapter is first to briefly describe the geology of representative areas within the east-central Alaska region (Fig. 3), using newly compiled geologic maps for each of the areas. The areas considered are: (A) Charley River, (B) Livengood, (C) Ray Mountains (D) Beaver, and (E) Coleen. Each of these areas has undergone recent work or is critically situated with respect to regional geologic problems and lends itself to reevaluation in light of new studies elsewhere in the region. Together, these five areas encompass most of the bedrock exposed in the east-central Alaska region. Next, important topical problems of regional or interregional significance are discussed within the context of the areal geology as presently understood or interpreted, and the question of terrane accretion is considered. In a brief concluding section, a preferred paleogeographic and geotectonic model is suggested.

GEOLOGICAL SUMMARIES OF REPRESENTATIVE AREAS

Charley River area

As defined here, the Charley River area includes the parts of the Charley River and northeast Eagle 1:250,000 quadrangles lying north of the Yukon-Tanana upland (Fig. 3). Rocks of the Yukon-Tanana upland south of the Tintina zone are discussed by Foster and others (this volume).

Previous work. Geological observations were first made in the Charley River region by pioneering Alaskan geologists such

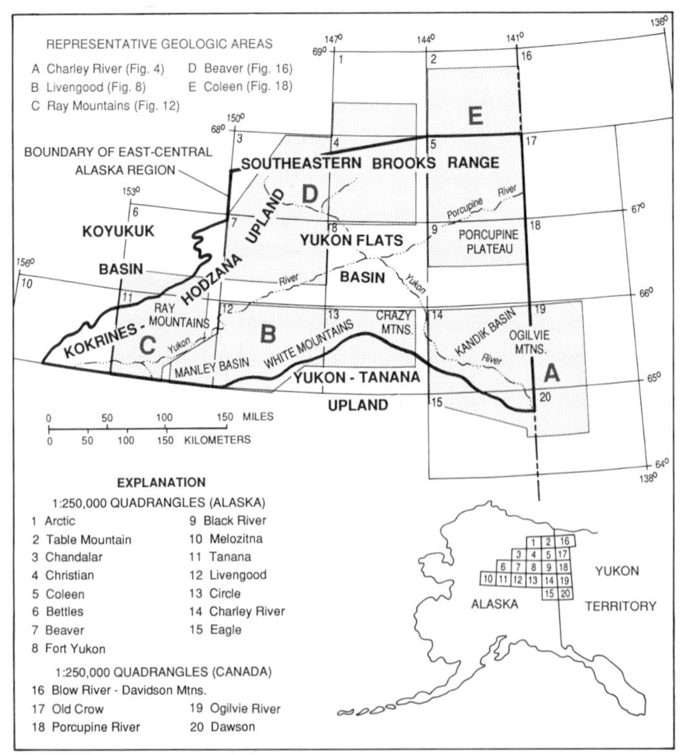

Figure 3. Location of representative geologic map areas discussed in text and illustrated in succeeding figures.

as Prindle (1905, 1906, 1913), Kindle (1908), Brooks and Kindle (1908), and Cairnes (1914), but the reconnaissance studies of Mertie (1930, 1932, 1937) laid the groundwork for subsequent geologic investigations in this part of east-central Alaska. For the past 20 years, the work of Brabb and Churkin has served as the geologic standard for the Charley River area. Current work relies largely on the distribution of rock units mapped by Brabb and Churkin (1969) and on the stratigraphic relations worked out during the course of their mapping (Brabb, 1967, 1969; Brabb and Churkin, 1967; Churkin and Brabb, 1965a, b). Reconnaissance petroleum and mineral exploration during the past 10 years has led to the realization that thrust faulting was an important deformational mechanism in the Charley River and adjacent areas (Gardner and others, 1984; unpublished data of British Petroleum Corporation and Louisiana Land and Exploration Corporation), and recent work suggests that the area lies within an extension of the northern Cordilleran fold-and-thrust belt of Canada (Dover, 1985a). Norris (1982, 1984) mapped a comparable style of folding and thrusting in adjacent Yukon Territory.

Stratigraphic and structural framework. The most recent map compilations of the Charley River area are those of Dover (1988) and Dover and Miyaoka (1988), shown in simplified form in Figure 4. The pattern of distribution of stratigraphic units on the map results from the tectonic repetition of a variety of stratigraphic sequences within a belt of folding and thrusting that extends across the entire Charley River area. The geology of parts of the Dawson and Ogilvie River quadrangles in adjacent Yukon Territory is compiled from Norris (1982) on Figure 4 in order to illustrate the regional continuity of stratigraphic and structural trends from the Charley River area.

The thrusted stratigraphic sequence in the eastern part of the Charley River area comprises a predominantly carbonate and marine clastic succession as much as 11,000 m thick of middle Proterozoic to Early Cretaceous age (Fig. 5). Detailed stratigraphic descriptions of the succession can be found in the reports by Brabb and Churkin already listed, and in reports by Clough and Blodgett (1984), Young (1982), and Laudon and others (1966). The three most prominent thrust repetitions involve two key stratigraphic units: (1) late Proterozoic and/or Lower Cambrian strata of the upper part of the Tindir Group, which includes *Oldhamia*-bearing beds in two of the repeated sections and basaltic volcanics and red beds in all three sections; and (2) the Permian Tahkandit Limestone and Triassic limestone and shale of the lower part of the Glenn Shale, also found in all three structurally repeated sections. The moderate to low dip of the bounding thrusts can be observed locally in bedrock exposures, but more commonly, it is interpreted from mapped thrust traces. The older-on-younger pattern of thrust repetition, required by the map patterns for the main thrust-bounded sequences in the southeastern part of the map, is illustrated in the five structure sections shown in Figure 6. Numerous imbricate fault splays complicate the internal structure of each of the thrust-bounded sequences, but the principal sole thrusts are indicated on both the map and the structure sections. The structure sections are constructed from surface geologic controls using classic fold-and-thrust belt styles and geometric rules (see Dover, 1992). Because of limited stratigraphic control, the sections should be considered conceptual rather than balanced.

Map patterns also demonstrate that at least some of the prominent faults in the western Charley River area are older-on-younger thrusts. The Snowy Peak and Threemile Creek thrusts, for example, bring late Paleozoic or older strata over Lower Cretaceous rocks of the Kandik Group (Fig. 4). However, in contrast to the eastern part of the Charley River area, the stratigraphic sequences involved in thrusting in the western part are mainly clastic rocks, bedded chert, and associated mafic igneous rocks of early, middle, and late Paleozoic, and Mesozoic age (Fig. 5). The drastic lithologic differences between these sequences and coeval ones in the eastern part of the Charley River area led Jones and others (1981) and Churkin and others (1982) to designate these western sequences as separate tectonostratigraphic terranes. Although some of these terranes are bounded by thrust faults comparable in style and magnitude to those elsewhere in the fold-and-thrust belt of the Charley River area, the character of the bounding structures of the rest of these terranes is equivocal because of poor exposures.

Stratigraphic variations in the Charley River fold-and-thrust belt. Although the stratigraphic succession involved in the eastern part of the Charley River fold-and-thrust belt is generally the one defined by Brabb and Churkin, numerous differences occur in the successions of various thrust plates (Fig. 7) as a result of facies variations and several major unconformities (see Fig. 5 and discussion by Dover, 1992). Variations occur along strike within individual thrust plates as well as between different thrust plates. For example, the predominant carbonate and quartzite-argillite components of the upper part of the Tindir Group, as well as subordinate interlayers of chert-clast conglomerate, diamictite, red beds, and basaltic volcanic and volcaniclastic rocks, interfinger within the Yukon and Three Castle Mountain thrust plates (Figs. 4 and 7). One example of facies telescoped across a thrust boundary involves the Jones Ridge Limestone and coeval but more argillaceous Ordovician to Cambrian strata across the Tatonduk thrust. Another example is the variations in limestone and chert-clast conglomerate components between the Permian Tahkandit Limestone and Step Conglomerate that occur in thrust slices of the Nation River thrust system (Figs. 4 and 7). Thrusting also juxtaposed sequences in which different amounts of section are missing beneath at least four major unconformities that punctuate the middle Proterozoic to Lower Cretaceous succession of the eastern part of the Charley River area (see Fig. 5). The unconformities are: (1) between the upper (Lower Cambrian and/or late Proterozoic) and lower (middle Proterozoic) parts of the Tindir Group, (2) below the Nation River Formation (Upper Devonian), (3) below the undivided Tahkandit Limestone and Step Conglomerate (Permian), and (4) between the upper (Lower Cretaceous and Jurassic) and lower (Middle and Upper Triassic) parts of the Glenn Shale. The most drastic and abrupt differences are in the amount of section missing beneath the

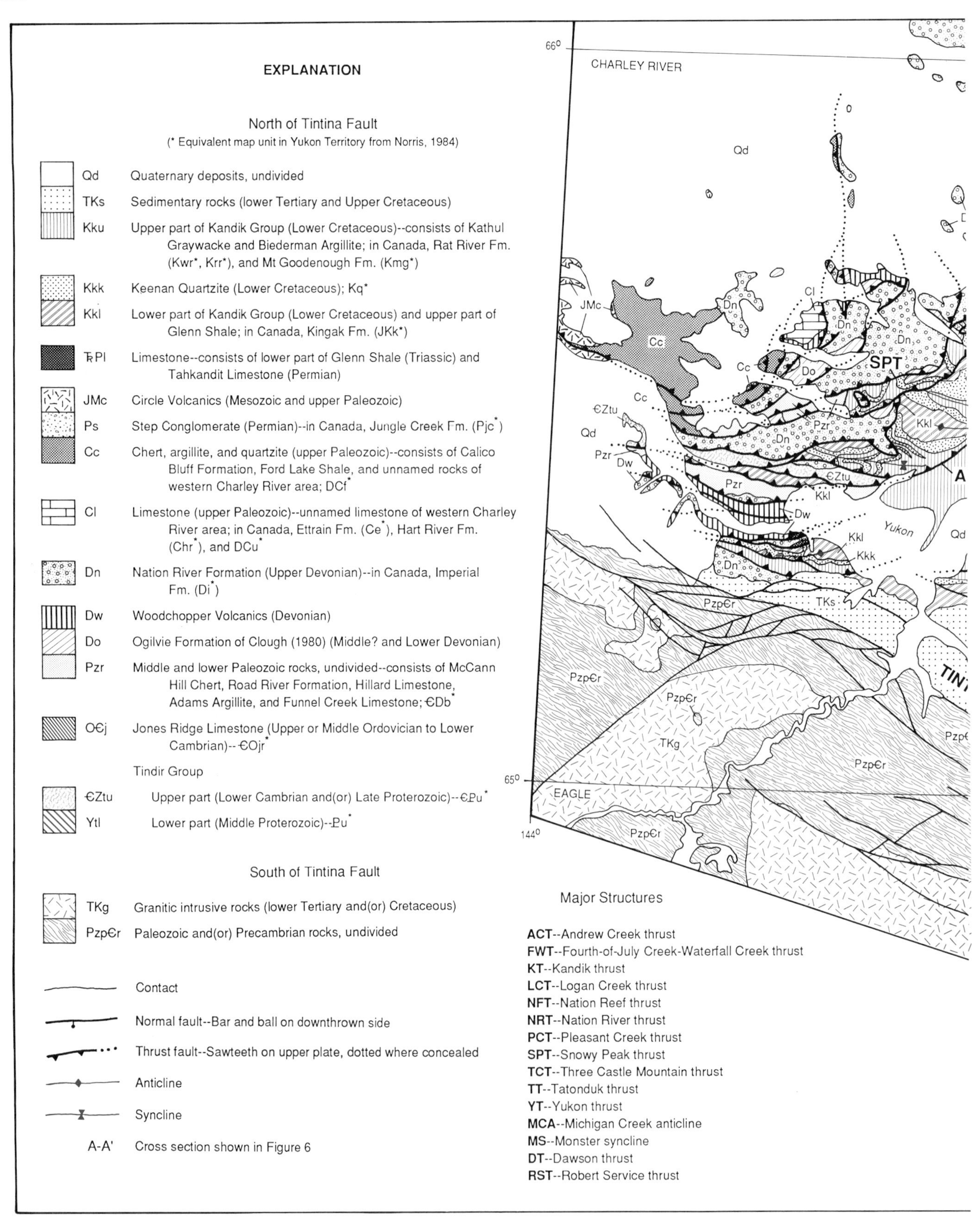

Figure 4. Geologic map of the Charley River area, and part of adjacent Yukon Territory. See Figure 3 for area location. Asterisk denotes correlative map unit in Yukon Territory from Norris (1984).

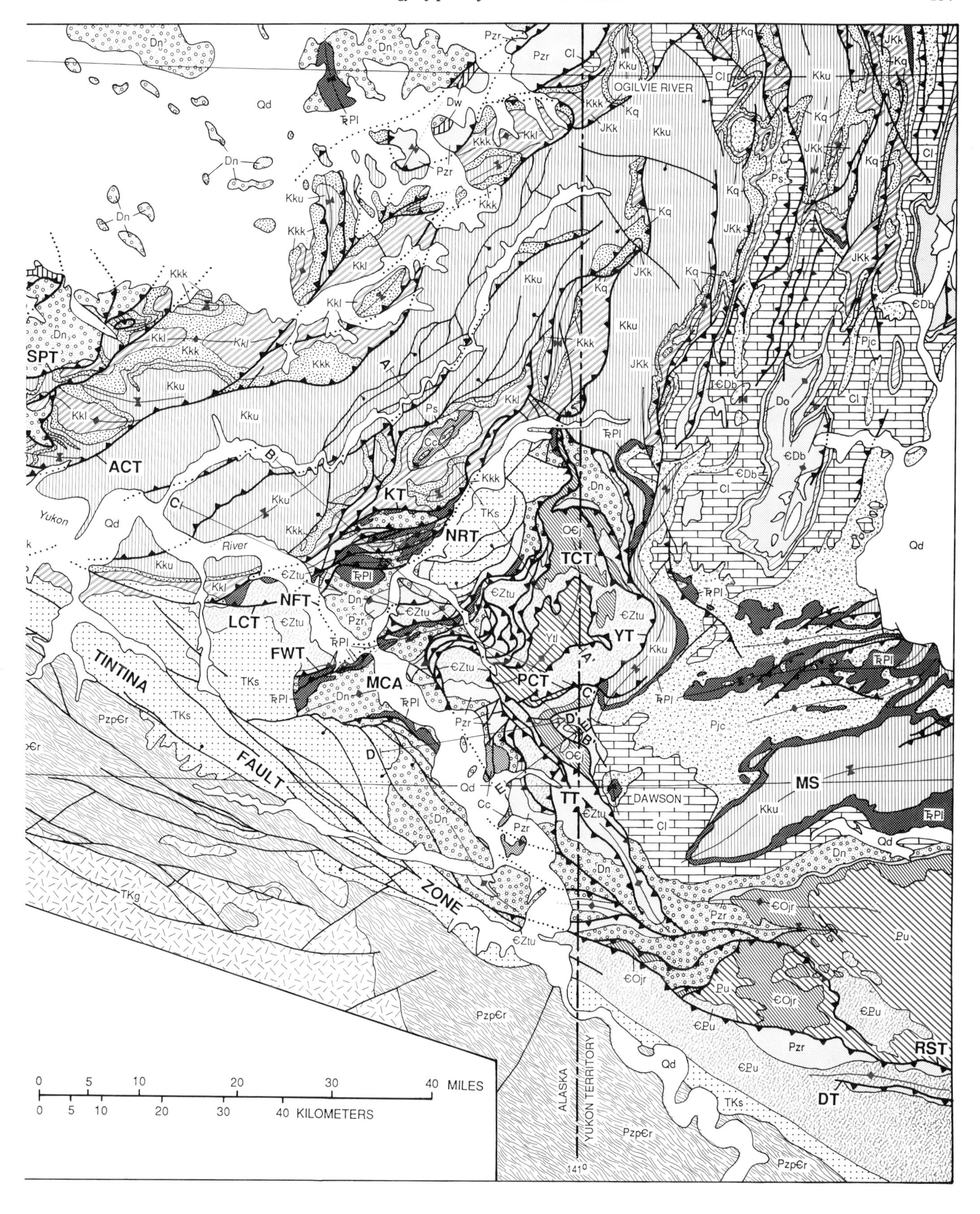
OGILVIE RIVER
Yukon
River
TINTINA
FAULT
ZONE
DAWSON
ALASKA
YUKON TERRITORY
141°
SPT
ACT
KT
NRT
TCT
NFT
LCT
FWT
MCA
YT
PCT
TT
MS
RST
DT
0 5 10 20 30 40 MILES
0 5 10 20 30 40 KILOMETERS

unconformity at the base of the Permian section, which rests directly on upper Tindir strata in the Nation River thrust complex (Figs. 4 and 7). Figure 7 represents a preliminary attempt to reconstruct facies patterns prior to thrusting across the southeastern part of the thrust belt. Little is known about facies variations and unconformities within the stratigraphic sequences of the western part of the Charley River fold-and-thrust belt.

Direction and amount of tectonic transport. Most thrusts in the Charley River quadrangle have northeast trends and underwent tectonic transport toward the southeast, based on the trends of thrust-generated folds, regional dips, and gross map patterns of older-on-younger stratigraphic repetition (see Fig. 4). Most major folds are upright, but a few are overturned in the direction of tectonic transport. A few northwest-directed back-thrusts are inferred in places, and some have been mapped by Norris (1982) in adjacent Yukon Territory. On the other hand, thrusts in the southeast corner of the Charley River quadrangle and in the northeast Eagle quadrangle trend predominantly northwest-southeast parallel to structures in the adjacent Dawson quadrangle of Yukon Territory, where the direction of tectonic transport is toward the northeast, based on the same kinds of map evidence mentioned above. The remarkable change of 90 to 110 degrees in orientation of the Charley River fold-and-thrust belt involves direction of transport as well as structural trend, and takes place at an abrupt "bend," around which individual structures can be mapped continuously from one trend to the other. The axis of the bend generally coincides with the Michigan Creek and other major anticlines in the Charley River map area, and projects eastward into a narrow zone separating belts of east-west and north-northeast structural trends in adjacent Yukon Territory (Fig. 4).

Modest amounts of thrust transport, generally of 5 km or less, are interpreted for most individual thrust plates from the geometric constraints of cross-section construction; total shortening distributed within some structural duplexes or imbricate thrust zones may be 10 km or more. Cumulative shortening of

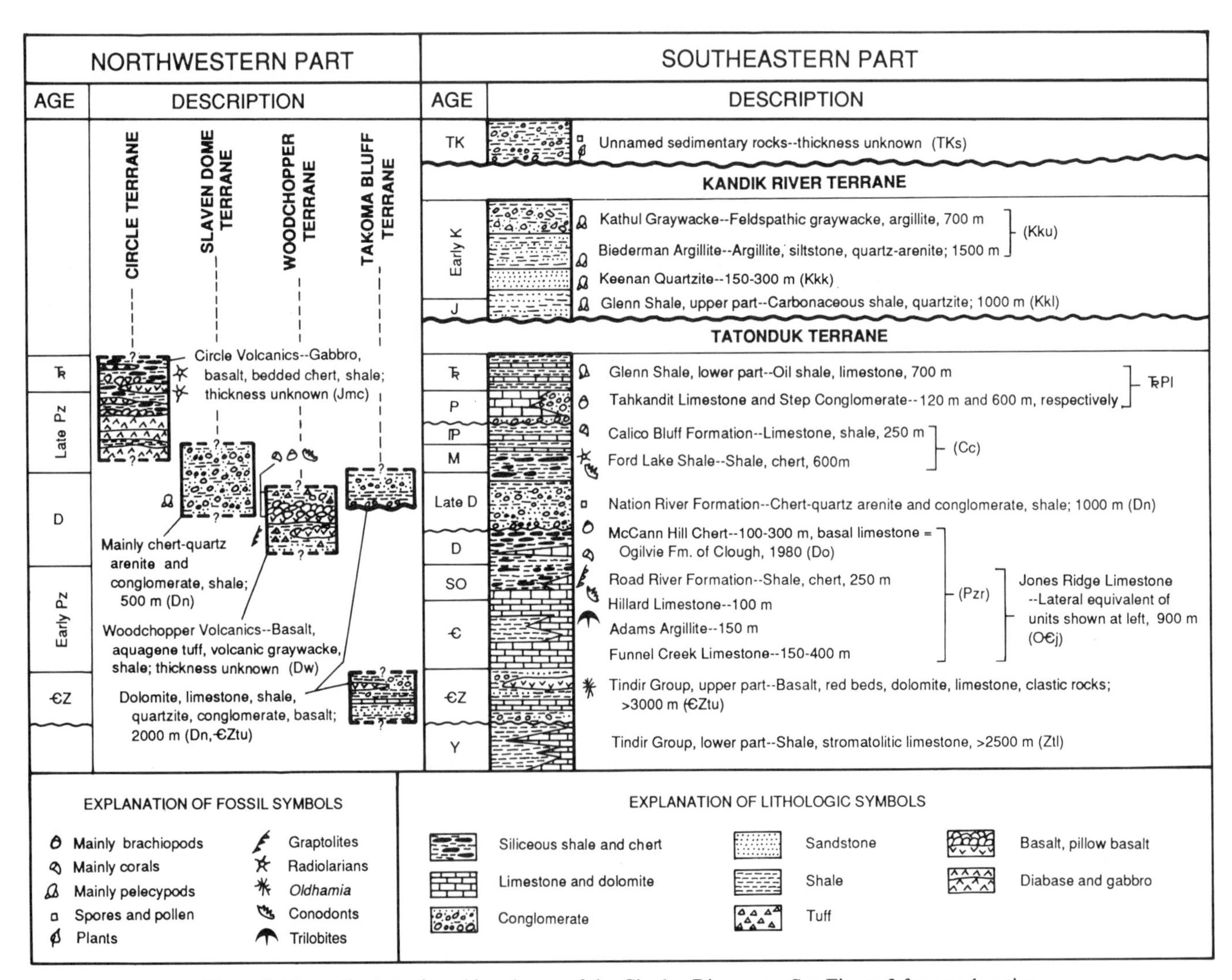

Figure 5. Generalized stratigraphic columns of the Charley River area. See Figure 3 for area location, and Figure 4 for terrane locations and explanation of map units (letter symbols in parentheses).

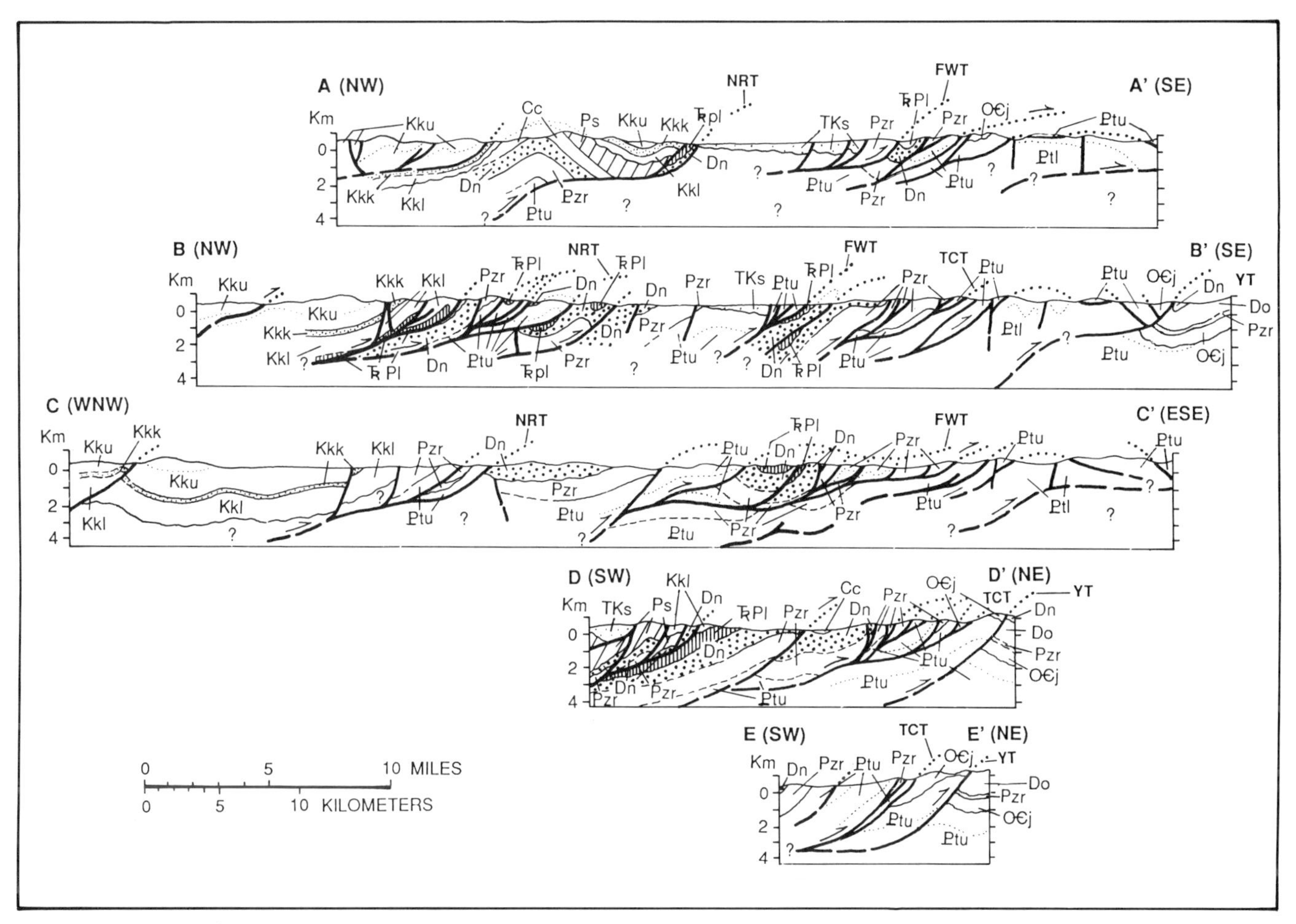

Figure 6. Structure sections of the eastern Charley River fold-and-thrust belt. See Figure 4 for section locations and explanation of map units and symbols.

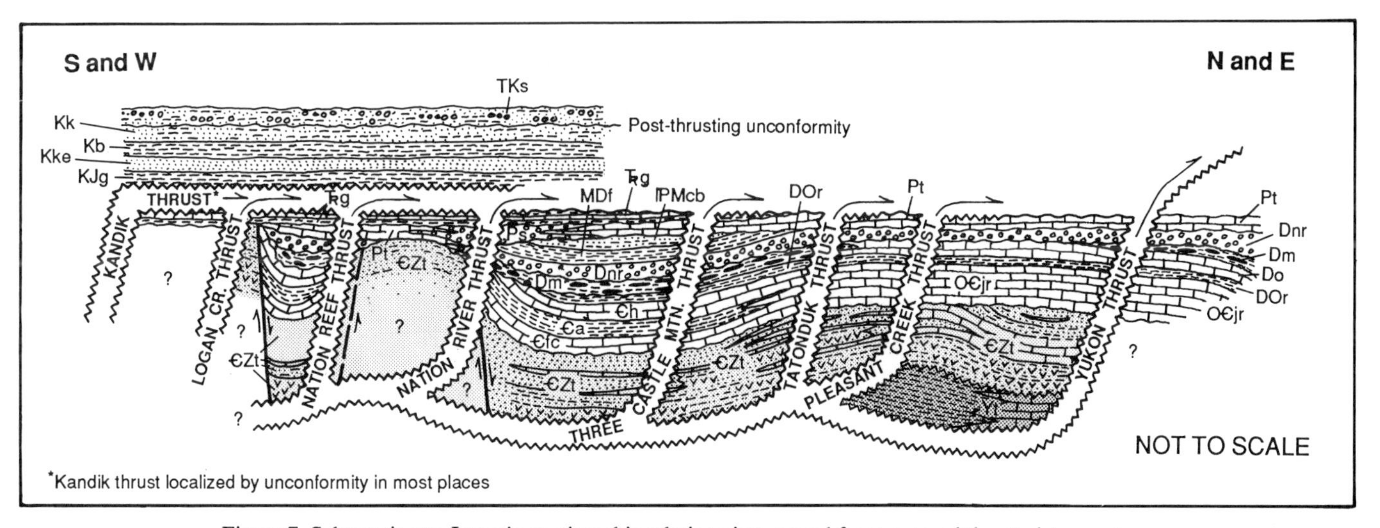

Figure 7. Schematic pre-Jurassic stratigraphic relations interpreted from restored thrust plates, eastern Charley River fold-and-thrust belt. See Figure 4 for explanation of map units and Figure 5 for explanation of lithologic symbols.

100 km is estimated for the segment of the Charley River fold-and-thrust belt in the Charley River area.

Numerous listric normal faults are shown on Figure 4. These faults generally involve Lower Cretaceous rocks of the Kandik Group and Upper Cretaceous to Tertiary rocks; in places the listric faults involve older rocks as well. The listric faults are interpreted to splay from relatively steep-dipping segments of thrust faults, where the thrusts are inferred to ramp upsection. As a corollary, ramps were suspected in cross-section construction below areas of abundant listric normal faulting (Fig. 6).

Timing of fold-and-thrust belt deformation. Lower Cretaceous rocks are involved in folding and thrusting throughout the Charley River region. There is no positive evidence to suggest compressional deformation prior to the Albian (Early Cretaceous) in the Charley River fold-and-thrust belt, but that possibility cannot be eliminated. Less deformed rocks of Late Cretaceous (Maastrichtian?) to mid-Paleocene age (unit TKs) unconformably overlap and post-date folding and thrusting (Fig. 4). In the two places where previous mapping suggested that Late Cretaceous- to Tertiary-age rocks were involved in thrusting, new evidence for an Early Cretaceous age of the deformed rocks has been found (Dover, 1992; Cushing and others, 1986). Therefore, at least the latest, and probably the main stage of fold-and-thrust belt development occurred after deposition of the Albian Kathul Graywacke and correlative rocks, and before unconformable deposition of unit TKs began in Maastrichtian(?) time. The same age of deformation is indicated for all parts of the Charley River fold-and-thrust belt, regardless of structural trend and direction of thrust transport.

Listric normal faulting involving Upper Cretaceous and Tertiary strata as young as Eocene(?) postdates compressional deformation.

Tintina fault zone. The Tintina fault zone cuts northwest-southeast across the southwest corner of the Charley River area. It separates weakly to nonmetamorphosed rocks of the Charley River fold-and-thrust belt, on the north, from igneous and metamorphic rocks of the Yukon-Tanana upland (see Foster and others, 1987, and this volume), to the south. The Tintina zone is expressed topographically as a linear trench, which contains Upper Cretaceous to lower Tertiary nonmarine clastic sedimentary rocks and/or Cenozoic sediments. It is a major strike-slip zone along which significant right-lateral strike separation occurred prior to Late Cretaceous and younger deposition; 450 km of right separation is documented in southern Yukon Territory (Tempelman-Kluit, 1979; Gordey, 1981). Although the magnitude of strike-slip that has occurred along the part of the Tintina fault zone in the Charley River area cannot be demonstrated locally, the Tintina zone's distinctly cross-cutting relation to the Charley River fold-and-thrust belt is evident from map patterns. Locally in the Charley River area, fold-and-thrust belt structures appear to be dextrally dragged as much as a few tens of kilometers along the Tintina zone.

Tintina trench-fill deposits are cut by an array of mainly dip-slip or oblique-slip faults, some with substantial displacements that accompanied uplift of the Yukon-Tanana block south of the Tintina zone. Some of these faults are still active (Foster and others, 1983). As much as 50 km of cumulative post-Eocene strike-separation is inferred for faults cutting Tintina trench-fill in Yukon Territory (see Hughes and Long, 1979) and in the Eagle quadrangle (Barker, 1986; Cushing and others, 1986), suggesting reactivation or continuation of Tintina strike-slip motion into post-Eocene time.

Livengood area

As defined here, the Livengood area includes most of the Livengood 1:250,000 quadrangle, as well as the part of the southeast Tanana quadrangle south of the Yukon River, and the Crazy Mountains and adjacent parts of the northern Circle quadrangle (Fig. 3). The area excludes the Yukon-Tanana upland, described by Foster and others (this volume).

Previous work. Geological investigations prior to 1935 in the Livengood area by Prindle, Brooks, and Eakin, are summarized and incorporated by Mertie (1937) in his geological description of the Yukon-Tanana region. Much of the impetus for those and subsequent geological studies in the area was the discovery of gold and the development of placer mines, many of which have operated intermittently to the present time. Regional (1:250,000-scale) map compilations by Chapman and others (1971, 1982) and Foster and others (1983) serve as a basis for current studies in the Livengood area. Reports of particular interest include those by Foster (1968), Brosgè and others (1969), Davies (1972), Chapman and others (1979), Foster and others (1982), Weber and Foster (1982), Albanese (1983), Cady and Weber (1983), Robinson (1983), Smith (1983), Cushing and Foster (1984), Weber and others (1985), Blodgett and others (1987), Wheeler-Crowder and others (1987), and Weber and others (1988).

Stratigraphic and structural framework. The degree of structural segmentation of the Livengood area and the complexity of its disrupted stratigraphy are clear from published geologic maps of the area, and are inherent in its subdivision into numerous tectonostratigraphic (lithotectonic) terranes by Churkin and others (1982), Silberling and Jones (1984), and Silberling and others (this volume). However, the actual character and regional significance of this geologic framework are not as obvious. In this summary, the geology of this critically situated area is updated in light of new data and work in progress.

Stratigraphic belts and terranes. The Livengood area contains 11 distinct stratigraphic belts[1] that are separated structurally from one another (Fig. 8). The bounding structures of the belts

[1]The term "stratigraphic belt" is used here to denote a typically elongate belt of rocks having characteristic stratigraphy or lithologic content that differs in some way from that of adjacent belts. This term is used in order to avoid any genetic connotation of "suspect" or accretionary origin that may be associated with the terms "tectonostratigraphic terrane" and "lithotectonic terrane," either by intent or by implication of common usage (for example, see usage in Silberling and Jones, 1984, p. A-2; and Bates and Jackson, 1987, p. 679).

are either strike-slip faults or thrust faults, but which of these types of fault is represented by individual structural boundaries has been in question. The geologic map of Figure 8 shows my preferred structural interpretation of the Livengood area. This interpretation differs in some important details from previously published map interpretations, reflecting the uncertainty of some field relations. However, there is general agreement on the character of the fault-bounded stratigraphic belts (Fig. 9).

(1) The *Wickersham stratigraphic belt,* called the Wickersham terrane by Silberling and Jones (1984) and the Beaver terrane by Churkin and others (1982), is the southeasternmost and structurally highest of the stratigraphic belts. It consists of the informally named Wickersham grit unit of late Proterozoic and Early Cambrian age (Weber and others, 1985). The Wickersham grit unit is subdivided by Chapman and others (1971) into a lower part of rhythmically interbedded grit, bimodal quartzite, slaty argillite, and subordinate chert and maroon and green slate, and an upper part of slaty argillite, maroon and green slate, quartzite, black limestone containing floating quartz grains, chert, and subordinate grit. The grit unit is overthrust from the southeast by greenschist-facies metamorphic rocks of the Yukon-Tanana upland, thought by Weber and others (1985) to be only modestly telescoped and slightly higher-grade equivalents of the grit unit.

(2) The *White Mountains stratigraphic belt,* called the White Mountains terrane by Silberling and Jones (1984) and Churkin and others (1982), is characterized by the Fossil Creek Volcanics (Ordovician) and the disconformably overlying Tolovana Limestone (Silurian). The Fossil Creek Volcanics is agglomeratic and has an alkalic basaltic composition. Recent mapping shows that (1) the sedimentary and volcanic divisions of the Fossil Creek recognized by Chapman and others (1971) are interfingering lateral facies rather than vertically stacked units, and (2) the Fossil Creek Volcanics were deposited on a basement of upper Wickersham grit unit that is identical in all respects to the grit unit in the Wickersham belt. Moreover, pebbles and cobbles of gritty quartzite resembling the Wickersham grit unit occur in the agglomeratic volcanic part of the Fossil Creek, indicating a nearby source of Wickersham rocks in Ordovician time. The character and distribution of Fossil Creek lithologies and their basaltic composition suggest a local volcanic center within an intracontinental depositional basin with locally steep, perhaps fault-controlled topography. From these lithologic relations and from map evidence for structural overturning and discordance, the boundary between the White Mountains and Wickersham/Beaver terranes is interpreted to be a thrusted unconformity of relatively small displacement.

Locally, chert- and quartzite-clast conglomerate and associated clastic rocks unconformably overlie the Fossil Creek–Tolovana section. The conglomerate is undated but is lithologically similar to that in the Nation River Formation (Devonian) and the Step Conglomerate (Permian) of the Charley River area.

Also included here in the White Mountains stratigraphic belt, but detached from the Fossil Creek–Tolovana section by another thrust fault, is a distinctive unit of gray, commonly bimodal, vitreous quartzite referred to as the Globe unit by Weber and others (1985). The Globe unit typically contains sheared argillite interbeds and abundant sills of hornblende-bearing quartz diorite or quartz gabbro. The age of this quartzite is unknown, and so far, the mafic intrusive rocks have proved too altered to date. However, the vitreous and bimodal character of the Globe unit, and its association with mafic sills, serve to distinguish it from other quartzites of the Livengood area. The Globe unit at the base of the White Mountains belt is in steep- to moderate-angle thrust contact on the Beaver Creek stratigraphic belt to the north.

(3) The *Beaver Creek stratigraphic belt* corresponds to the Manley terrane of Silberling and Jones (1984) and part of the Kandik terrane of Churkin and others (1982). The Beaver Creek belt contains mainly flyschoid rocks and chert-rich, locally derived polymictic conglomerate. There are three distinct flysch sequences in the Beaver Creek belt. The Vrain unit of Foster and others (1983) occurs in the eastern belt of the belt, the Wilber Creek unit of Weber and others (1985) has been mapped mainly in the central and western parts of the belt, and the Cascaden Ridge unit of Weber and others (1985) occurs in the central part of the belt along its north edge.

The Vrain unit, which is undated, consists mainly of carbonaceous, commonly iron-stained, rhythmically interlayered fine-grained to gritty clastic rocks, as well as tuff, and a prominent unit of typically stretched chert- and quartzite-pebble conglomerate. Minor amounts of chert and basalt also occur in rocks questionably assigned to the Vrain.

The Middle Devonian (Eifelian) Cascaden Ridge unit consists mainly of turbiditic shale, quartzite, and conglomerate. It has a locally developed basal conglomerate containing up to cobble- or boulder-sized clasts of mafic/ultramafic rocks, chert, dolomite, shale, and other locally occurring rock types (Weber and others, 1985). The similarity of some characteristic lithic components of the Cascaden Ridge and Vrain units, and their comparable stratigraphic position between the Wilber Creek unit and a persistent mafic/ultramafic band at the top of the underlying Livengood stratigraphic belt, raises the possibility that the Cascaden Ridge and Vrain units may be correlative, at least in part.

The Wilber Creek unit is Late Jurassic and Early Cretaceous in age, and contains shale, graywacke, quartzite, and polymictic conglomerate. It probably lies unconformable on both the Vrain and Cascaden Ridge units, but its basal contact is poorly constrained and questionably located on Figure 8 because of the difficulty in distinguishing between these three flysch units.

Polymictic conglomerates tentatively assigned to the Cascaden Ridge, Vrain, and Wilber Creek units of the Beaver Creek stratigraphic belt contain locally derived chert, argillite, quartzite, carbonate, mafic volcanic, and low-grade metamorphic clasts, and flat shale chips, as well as quartz, plagioclase, muscovite, chlorite, epidote, and opaque grains. All of these constituents are present in the various rock sequences of the Livengood area, suggesting the polymictic conglomerates of all these units were derived from local sources as early as Middle Devonian time. All three of the flysch units appear to be in faulted or faulted uncon-

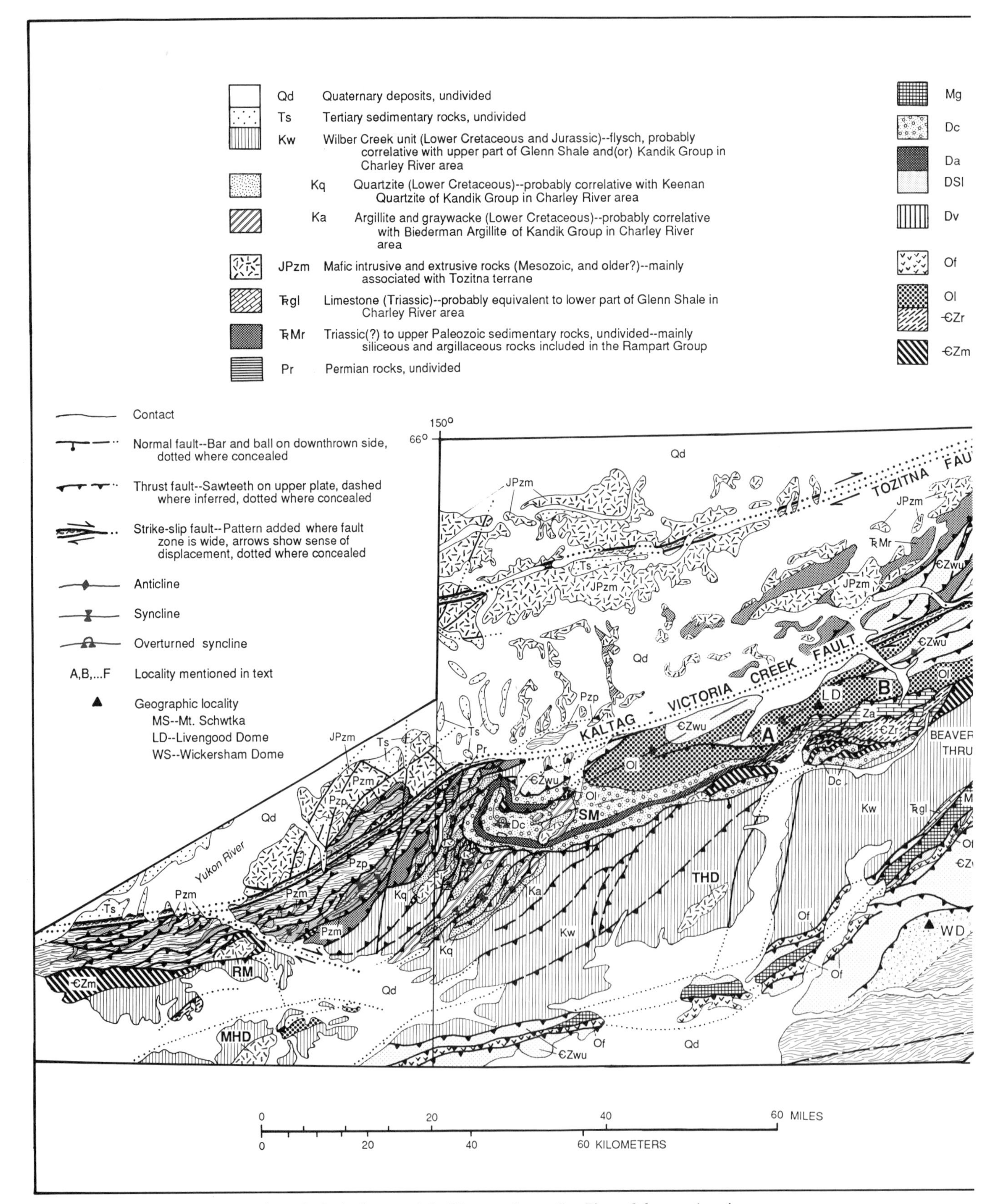

Figure 8. Geologic map of the Livengood area. See Figure 3 for area location.

EXPLANATION

Globe unit (Mississippian?)--mainly quartzite, but includes argillite interbeds and abundant gabbro in places

Clastic rocks (Devonian and(or) Mississippian?)--includes Cascaden Ridge and Vrain units, and other undated and unnamed rocks

Siliceous argillite (Devonian and(or) Mississippian?, or older)

Limestone (Devonian and Silurian)--includes Tolovana Limestone, the limestone of Lost Creek, and limestone in the Schwatka belt

Basalt and associated volcanoclastic and sedimentary rocks (Lower Devonian?)--probably correlative with Woodchopper Volcanics of Charley River area

Fossil Creek Volcanics (Ordovician?)--consists of volcanic and sedimentary faces, undivided

Livengood Dome Chert (Ordovician)

Heterogeneous sedimentary and mafic igneous rocks (Lower Cambrian? to Late Proterozoic?)

Mafic and serpentinized ultramafic rocks (Cambrian and(or) Late Proterozoic)

Za — Amy Creek (Late Proterozoic)--dolomite

Wickersham grit unit

- €Zwu — Upper part (Lower Cambrian and(or) Late Proterozoic)
- Zwl — Lower part (Late Proterozoic)

Low-grade metamorphic rocks

- Pzp — Pelitic and gritty metasedimentary rocks (Devonian and(or) older)
- Pzm — Marble (Paleozoic?)

Tg,Kg — Granitic intrusive rocks (lower Tertiary, Tg, and Cretaceous, Kg)

- **CM**--Cache Mountain pluton (lower Tertiary)
- **LP**--Lime Peak pluton (lower Tertiary)
- **MHD**--Manley Hot Springs Dome pluton (lower Tertiary)
- **RM**--Roughtop Mountain pluton (mid-Cretaceous)
- **SM**--Sawtooth Mountain pluton (mid-Cretaceous)
- **THD**--Tolovana Hot Springs Dome pluton (lower Tertiary)
- **VM**--Victoria Mountain pluton (lower Tertiary)

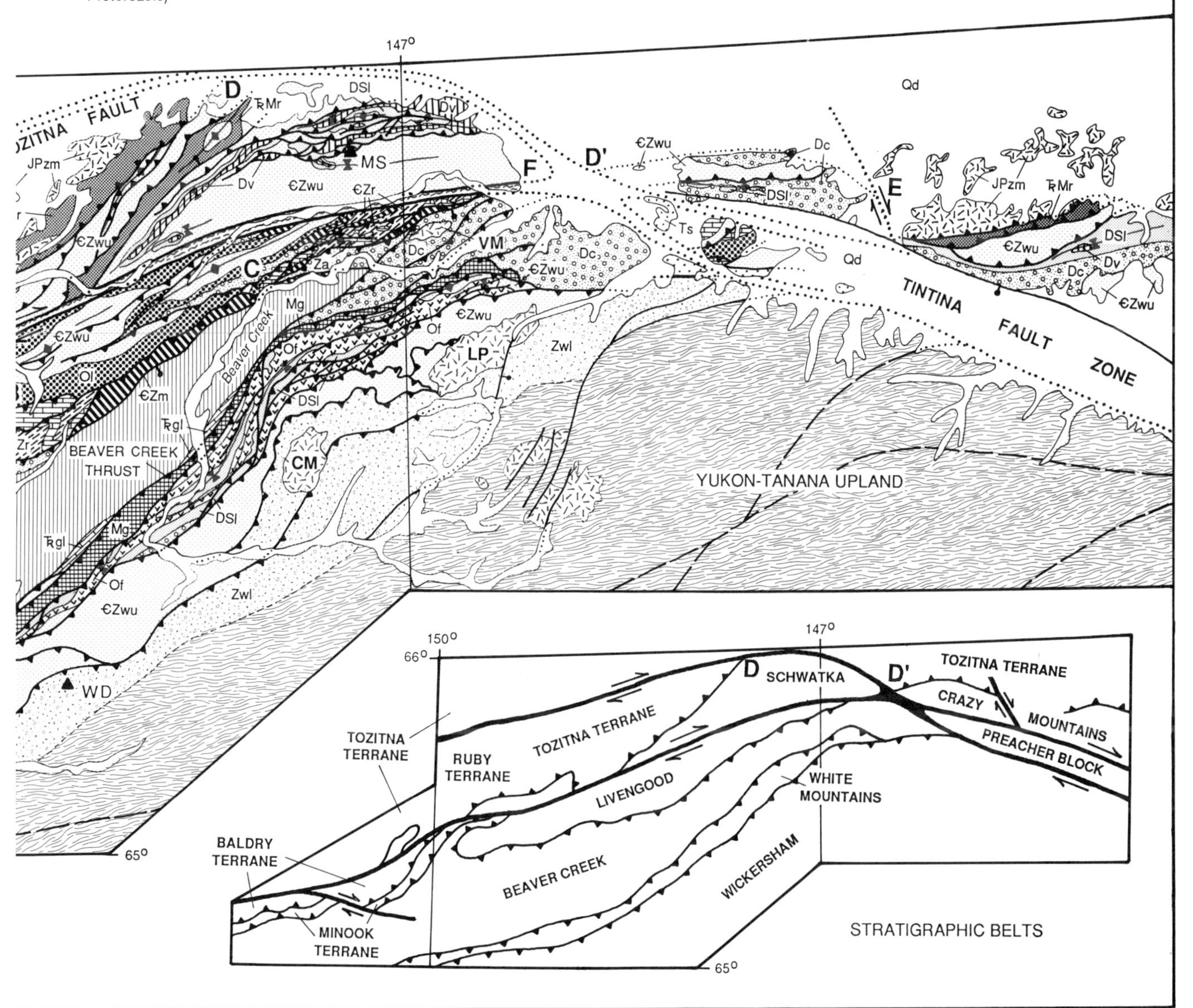

formable contact on the Livengood stratigraphic belt to the north.

Some of the clastic rocks formerly mapped as Wilber Creek unit in the Serpentine Ridge and Manley areas of the southeastern Tanana quadrangle (labeled SR and M, respectively, on Figure 8) are inferred here to be pre-Cretaceous based on lithologic associations and degree of induration.

An undated sequence of dark gray calcareous shale, silty limestone, and subordinate quartzite, tentatively correlated with the lithologically similar lower part of the Glenn Shale (Triassic) in the Charley River area, also occurs locally in the Beaver Creek stratigraphic belt.

(4) The *Livengood stratigraphic belt* generally corresponds with the Livengood terrane of Silberling and Jones (1984) and Churkin and others (1982). The Livengood belt is characterized by the Livengood Dome Chert (Ordovician), which in its type area consists mainly of intensely deformed black to varicolored chert and subordinate interbeds of graptolitic shale, tuffaceous rocks, and limestone. In some places, mafic sills or basalt interbeds occur in the Livengood Dome Chert. The rocks of the Livengood Dome Chert may comprise a less volcanic facies of the Ordovician Fossil Creek Volcanics, which contains dark chert interbeds in the White Mountains stratigraphic belt.

Many questions have arisen about the stratigraphic and structural relations of the Livengood Dome Chert with other units of the Livengood stratigraphic belt, because most of the units are poorly dated and contacts are seldom exposed. Uncertain identification of two key units is especially critical to map interpretation in the central and eastern parts of the Livengood belt. The two units are a distinctive dolomite, called the Amy Creek unit (Weber and others, 1985), and a clastic unit characterized by chert-pebble conglomerate (mapped as Za and Dc, respectively, on Figure 8).

The Amy Creek unit is a silicified, locally stromatolitic dolostone formed largely by dolomitization of pelletoidal lime mud. It contains thick black chert interbeds and lesser amounts of argillite, basalt, and volcaniclastic rocks. The Devonian to Silurian age originally assigned to the Amy Creek unit was based on fossils in a thin, presumably correlative limestone bed at a fossil locality on Lost Creek (locality A, Fig. 8). However, that correlation is now discounted, and the Amy Creek unit is currently undated (Robert Blodgett, oral communication, 1987). Based on its lithologic and petrologic character, and on its lack of fossils, the Amy Creek unit more closely resembles late Proterozoic dolomites of the east-central Alaska region and Yukon Territory

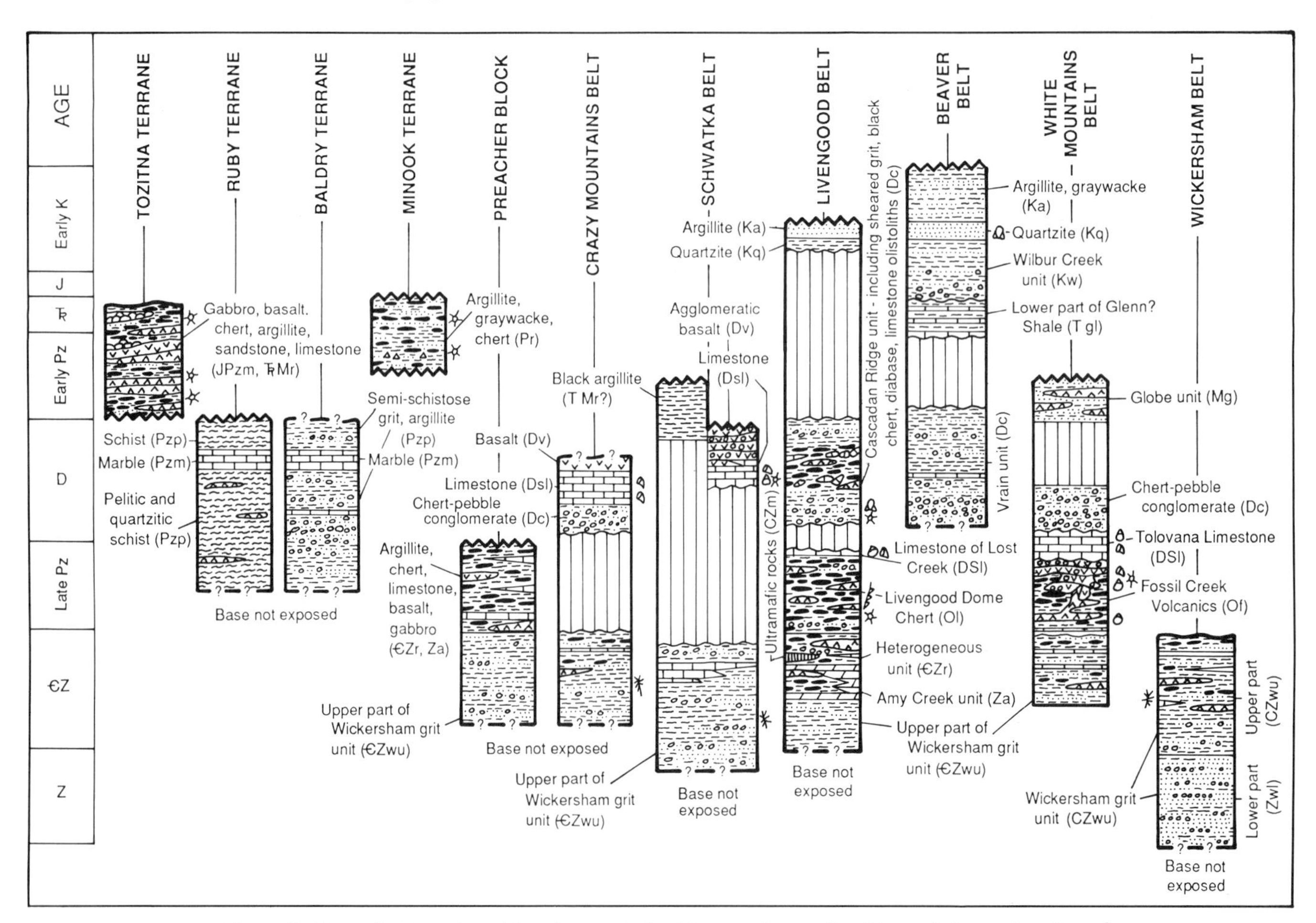

Figure 9. Generalized stratigraphic columns of the Livengood area. See Figure 5 for explanation of lithologic and fossil symbols, and Figure 8 for explanation of map units (letter symbols in parentheses).

than a middle Paleozoic carbonate. A late Proterozoic age is therefore tentatively adopted for the Amy Creek unit in this chapter for map and interpretive stratigraphic purposes. The fossiliferous Lost Creek beds are here designated as a separate unit of Late Silurian age, informally called the "limestone of Lost Creek" (see Figs. 8 and 9), that lies stratigraphically between the Livengood Dome Chert and a unit of chert-pebble conglomerate (Blodgett and others, 1988).

Chert-pebble conglomerate commonly occurs as isolated exposures between outcrops of the Livengood Dome Chert and maroon and green slate characteristic of the upper part of the Wickersham grit unit. At locality B, north of Livengood Dome (Fig. 8), where the conglomerate contains flat argillite chips and has an amorphous siliceous matrix, it has been interpreted as a basal conglomerate of the Livengood Dome Chert (Weber and others, 1985). At locality C (Fig. 8), chert-pebble conglomerate occupies a similar position between the Livengood Dome Chert and the maroon and green slate unit, but here the conglomerate also contains granular dolomite clasts, is associated with black argillite, graywacke, and other clastic rocks, and has a mylonitic quartz-chert arenite matrix. In the character of its matrix and associated rocks, this conglomerate more closely resembles conglomerates assigned to middle Paleozoic or younger clastic sequences in the Beaver Creek stratigraphic belt. The tentative interpretation, shown on Figure 8, is that the chert-pebble conglomerate at localities B and C is part of a Devonian clastic sequence lying in angular unconformity on the upper part of the Wickersham grit unit and overthrust by the Livengood Dome Chert.

East of locality C, the same conglomerate appears to lie in low-angle unconformity across a contact between maroon and green slate assigned to the upper part of the Wickersham grit unit, and a heterogeneous overlying unit of chert, siliceous argillite, greenstone, agglomeratic basalt, dolomite, dolomite-boulder conglomerate, siltite, quartzite, grit, and felsic volcanics(?), all injected by abundant, thick mafic sills. The sills range from quartz-bearing diorite and gabbro to sparse serpentinized cumulate ultramafic rocks thought to be differentiates of the sills.

The mafic and ultramafic rocks of the heterogeneous unit are similar in composition and petrologic character to those in a persistent but discontinuous northeast-trending band that marks the structural top of the Livengood stratigraphic belt. The rocks within this band include gabbro and diorite interlayered with serpentinized peridotite and dunite. K-Ar hornblende ages on three diorite or gabbro samples from this band near Amy Dome give Cambrian to late Proterozoic ages ranging from 518.3 ± 15.5 to 633 ± 19 Ma (D. L. Turner, unpublished data). Modeling of gravity and magnetic data across the mafic/ultramafic band confirms its layered character and the southward-decreasing dip of its structural base (Cady and Morin, 1990). The origin of these rocks is uncertain, but they are tentatively interpreted as a differentiated Cambrian or Precambrian sill complex by Cady and Morin (1990). Based on their lithologic associations and the petrologic similarities of their mafic and ultramafic components, the heterogeneous unit, the Amy Creek unit, and the mafic/ultramafic band of the Livengood stratigraphic belt may all represent thrust segments of a volcanic/plutonic center correlative with the upper part of the Tindir Group in the Charley River area, and the Mount Harper volcanic complex in the Dawson area of Yukon Territory.

(5) The *Schwatka stratigraphic belt* is separated from the Livengood belt by the Victoria Creek strike-slip fault. The Schwatka belt has been included in the Wickersham and White Mountains terranes by Silberling and Jones (1984), and in the corresponding Beaver and White Mountains terranes by Churkin and others (1982). *Oldhamia*-bearing Wickersham grit unit forms the basement of the Schwatka stratigraphic belt, but recent work indicates that overlying clastic, volcanic, and carbonate rocks differ from those of the White Mountains stratigraphic belt. Fossiliferous Lower (Emsian) and/or Middle Devonian (Eifelian) limestone in the Schwatka belt is distinctly younger than the Silurian Tolovana Limestone of the White Mountains, with which it was formerly correlated (Weber and others, 1988). If some of the limestone of the Schwatka belt is interbedded with the underlying unit of agglomeratic mafic volcanics, chert, and clastic rocks, as preliminary interpretation of recent mapping suggests, then at least part of the volcanic-bearing sequence may be as young as Early or Middle Devonian. If so, these volcanic rocks would be at least partly correlative with the Woodchopper Volcanics of the Charley River area, rather than with the Fossil Creek Volcanics of the White Mountains. This interpretation supports that originally made by Mertie (1937).

(6) The *Crazy Mountains belt* is separated from the rest of the Livengood area by strike-slip faults. Most of the rocks in the East and West Crazy Mountains are included in the Crazy Mountains terrane of Silberling and Jones (1984) and Churkin and others (1982). The upper part of the Wickersham grit unit forms the basement of the Crazy Mountains belt, where maroon and green slate, grit, and black limestone containing floating quartz grains are distinctive lithic components, and the section contains *Oldhamia*. Lithologies in the structurally(?) overlying sequence of chert, clastic rocks, mafic igneous rocks, and fossiliferous Lower Devonian limestone resemble those in the Schwatka stratigraphic belt. Overlying chert-pebble conglomerate is indistinguishable from middle to upper Paleozoic conglomerates in the White Mountains, Beaver Creek, and Livengood stratigraphic belts, and in the Charley River area.

(7) The *Preacher block* of Foster and others (1983) is bounded by splays of the Tintina strike-slip fault system. It contains thrust imbrications of lithologic components found mainly in the Livengood stratigraphic belt south of the Victoria Creek fault, including rocks of the upper part of the Wickersham grit unit, and probable correlatives of the Amy Creek unit and the heterogeneous mafic-bearing sequence.

Three additional terranes of small size are defined by Silberling and Jones (1984) at the west end of the Livengood area.

(8) The *Minook terrane* of Silberling and Jones (1984) contains upper Paleozoic flysch and chert-pebble conglomerate and

grit, lithologies resembling those in the Beaver Creek stratigraphic belt.

(9) The *Baldry terrane* has been described as a "structurally complex and polymetamorphosed assemblage of radiolarian chert, marble, greenschist, and mica schist" derived from probable lower to middle Paleozoic protoliths (Silberling and Jones, 1984). Although the low-grade metamorphic rocks of the Baldry terrane were tightly folded, thrusted, and sheared under low-grade ductile conditions, suitable protoliths occur in the White Mountains, Livengood, and Schwatka stratigraphic belts. Metamorphic recrystallization and ductile deformation in the Baldry terrane could be interpreted as a deeper-crustal or higher-temperature manifestation of the more brittle style of imbricate thrusting that characterizes the possible protolithic rocks.

(10) The *Ruby terrane* of Silberling and Jones (1984) also contains metamorphic rocks, but generally of higher grade than those in the Baldry terrane. The principal units are staurolite- and garnet-mica schist and marble. The rocks of the Ruby terrane are described more fully in the section on the Ray Mountains area, to follow. However, by analogy with similar rocks in the Ray Mountains area, where metamorphic facies gradations are abrupt and related to granitic intrusives, the rocks of the Ruby terrane are tentatively interpreted as locally higher-grade recrystallized equivalents of protoliths similar to those inferred for the Baldry terrane.

(11) The *Tozitna stratigraphic belt,* found along the north margin of the Livengood area, is included in the Tozitna terrane of Silberling and Jones (1984) and the Tozitna/Circle terranes of Churkin and others (1982). The Tozitna belt contains the Rampart Group (Brosgè and others, 1969) as well as voluminous mafic intrusive rocks and subordinate basaltic volcanics. The Rampart Group consists of intercalated chert, siliceous argillite, tuffaceous and volcaniclastic rocks, siltstone, sandstone, quartzite, and minor limestone. The mafic rocks, which may constitute 75 percent or more of the Tozitna belt, are dominantly diabase and gabbro, but range from diorite to basalt and pillow basalt. Cumulate ultramafic layers and lenses formed locally, most likely in differentiated sills. The gabbro has a K-Ar hornblende age of 210 $\pm$ 6 Ma (Triassic), according to Patton and others (1977). Cherts yield Late Mississippian, Pennsylvanian, and Triassic radiolarians (Foster and others, 1983; Jones and others, 1984), and a limestone bed near Rampart contains Permian fossils (Brosgè and others, 1969). The basal contact between the Tozitna belt and various underlying sequences, including those of the Schwatka and Crazy Mountains stratigraphic belts and the Ruby terrane, is interpreted to be a structural detachment. The content and contact relations of the Tozitna stratigraphic belt are considered further in the summary of the Ray Mountains area to follow.

Granitic intrusive rocks. Granitic plutons of mid-Cretaceous and early Tertiary age occur in the Livengood area. The oldest dated plutons have mid-Cretaceous K-Ar biotite and Pb-α zircon ages of about 90 Ma (Chapman and others, 1971, 1982). Two of these are the Roughtop and Sawtooth Mountain plutons, near the north edge of the Beaver Creek stratigraphic belt in the western part of the Livengood area. A third is a small syenitic plug that is enriched in thorium and rare earth elements and occurs within the Wickersham stratigraphic belt. All other dated plutons in the Livengood area give early Tertiary K-Ar biotite ages between 55 and 65 Ma (Chapman and others, 1971, 1982; Foster and others, 1983). Three of the early Tertiary plutons, the Manley Hot Springs Dome, Tolovana Hot Springs Dome, and Victoria Mountains plutons, intrude flyschoid rocks of the Beaver Creek stratigraphic belt. The two largest of the early Tertiary plutons, the Cache Mountain and Lime Peak plutons, are emplaced in grit of the Wickersham stratigraphic belt. The granitic plutons of the Livengood area are discussed in more detail by Weber and others (1988, p. 30–35) and by Burns and others (1987).

Structure. The keys to recognizing demonstrable strike-slip faults in the Livengood area are (1) their curvilinear topographic expression and continuity across diverse geologic belts; and (2) the width, intensity, and character of shearing of the fault zones. Two principal strike-slip faults are identified in the Livengood quadrangle: the Victoria Creek and Tozitna faults. These are the two main splays of the Tintina fault zone, which enters the Livengood area from the east. The Victoria Creek fault connects westward with the Kaltag fault of the Tanana quadrangle.

Exposures of strike-slip zones are rare because their sheared rocks are weak, easily eroded, and form linear topographic trenches containing valley fill materials. However, the Victoria Creek fault zone is well exposed along the lower reaches of Victoria Creek. Here, the main part of the fault zone is 1 to 1.5 km wide and contains disconnected tectonic lenses within an intensely and pervasively sheared, mylonitic matrix. Tectonic lenses range from nearly microscopic in size to blocks hundreds of meters long. Axes of crenulations and small folds vary but are usually steep to vertical; slickensides generally plunge at low angles. The continuity and linearity of the Victoria Creek strike-slip fault zone, and its low-angle truncation of fold-and-thrust trends, demonstrate its steep dip and post-thrusting age. Except for possible dextral drag of some preexisting thrusts, there is no direct evidence in the Livengood area alone for the sense and magnitude of slip on the Victoria Creek fault zone, but its character, its mappable continuity with both the Tintina and Kaltag faults, and its alignment between them mark the Victoria Creek fault as a fundamental link in a Tintina–Victoria Creek–Kaltag right-lateral strike-slip fault system.

The Tozitna fault zone, located along the north edge of the Livengood area, is not exposed in the Livengood area, but its physiographic expression and curvilinear trace indicate that it is a splay of the Tintina system. About 55 km of right separation is estimated for the Tozitna fault from the offset of the basal contact of the Tozitna stratigraphic belt from point D to D′ on Figure 8. The basal Tozitna contact also appears to be offset right-laterally 10 km or less between the East and West Crazy Mountains (locality E, Fig. 8), and between the East Crazy Mountains and the western Charley River area, possibly indicating two additional but minor splays of the Tintina fault system.

In contrast with the strike-slip zones, thrust contacts are

relatively narrow and sharply defined, typically are stratigraphically controlled, are associated with large-scale low-plunging folds, are not expressed by a topographic trench, and form systems of splaying individual thrusts, each of relatively limited extent and displacement. Most thrust faults are also poorly exposed, but they can be clearly identified in most places by stratigraphic truncations and older-on-younger stratigraphic repetition.

In most of the Livengood area south of the Victoria Creek fault, older-on-younger thrust relations, the mapped traces of thrust faults, rare overturned folds, and geophysical modeling (Cady and Morin, 1990; Long and Miyaoka, 1988) all indicate that the major northeast-trending thrusts dip to the southeast and had transport directions toward the northwest (Fig. 8). Based on preliminary resistivity cross sections by Long and Miyaoka (1988), the White Mountains stratigraphic belt was thrust about 10 km across the Beaver Creek belt along the largest of these thrusts, here named the Beaver Creek thrust (Fig. 8). Numerous northwest-directed older-on-younger thrusts also imbricate the Wickersham, White Mountains, and Livengood stratigraphic belts. Preliminary interpretation of geologic sections (Dover, 1988, Plate II-B) suggests most of these thrusts have small displacements of a few kilometers or less.

Near the west edge of the Livengood quadrangle, thrust trends wrap nearly 180 degrees around a major west-plunging anticline or anticlinal duplex, so that north- to northeast-trending thrusts in the western part of the Livengood area dip northwest and have southeast-directed movement (Fig. 8). This "bend" is analogous in character to that of the Charley River area. All major thrusts in the Schwatka, Crazy Mountains, and Tozitna stratigraphic belts north of the Tintina–Victoria Creek–Kaltag fault, also have southeastward-directed transport, including some with younger-on-older displacements that are interpreted as detached and thrusted stratigraphic contacts.

Structural timing. Dateable thrusting in the Livengood area involves rocks of the Wilber Creek unit as young as Early Cretaceous, and predates strike-slip movement on the Victoria Creek and Tozitna faults. At one locality (labeled F on Fig. 8), dolomite tentatively assigned to the Amy Creek unit is interpreted by Weber and others (1988) as having been thrust over the Victoria Creek fault, but this seems unlikely because the topographic trench representing the Victoria Creek zone appears to persist unbroken across and therefore postdate the alleged thrust. Pre-Cretaceous thrusting is possible but cannot be demonstrated.

Local evidence for the time of the principal strike-slip movement on the Tintina fault and its major splays in the Livengood area is that it cuts obliquely across and therefore postdates fold and thrust trends involving Wilber Creek rocks as young as Early Cretaceous, and its principal movement predates the deposition of poorly dated Tertiary(?) rocks. The idea that an earlier phase of major strike-slip and accompanying graben formation along an ancestral Tintina system controlled Jurassic to Lower Cretaceous (and possibly older) Wilber Creek deposition in the Beaver Creek stratigraphic belt (F. R. Weber, written communication, 1987) is incompatible with the regional distribution of lithologically equivalent rocks. All the rocks of the Beaver Creek belt appear to have lateral equivalents within regionally developed thrust-telescoped sequences of the Charley River and Dawson areas that are cut by the Tintina fault and extend far beyond the limits of the fault zone. The Wilber Creek unit, in particular, is correlated by Weber and others (1988, p. 36) with the Kathul Graywacke of the Kandik Group in the Charley River area. This unit represents a foreland basin sequence traceable into the foothills of the Brooks Range (Young, 1973), rather than a locally developed strike-slip graben deposit. Tertiary reactivation of the Tintina zone is indicated by numerous faults whose dip-slip components of movement are obvious, but which may have lateral-slip components as well. As much as 50 km of late Tertiary or Quaternary right-slip is postulated by Barker (1986) on the basis of geomorphic evidence. Modern fault scarps south of the Crazy Mountains (Foster and others, 1983) indicate that the Tintina fault zone is still active.

Although Tintina fault movement was of a magnitude and scale capable of affecting preexisting structures in adjacent wall rocks, such effects in the east-central Alaska region appear to have been relatively minor readjustments, such as the drag of pre-Tintina thrust faults recognized in the Charley River area, or possibly the tightening up of compressional structures—a process that would require closer control on the timing of thrust movements to demonstrate than is currently available. In this region at least, the interpretation most in accord with observed geologic relations is that Cordilleran fold-and-thrust belt development and Tintina strike-slip are unrelated events separated in time.

The Livengood area as a segment of the Charley River fold-and-thrust belt displaced by the Victoria Creek–Kaltag and Tozitna splays of the Tintina fault system. Right-lateral displacement on the east-central Alaskan segment of the main Tintina fault system was estimated by Chapman and others (1985) to be 160 to 180 km, based on: (1) estimates of displacement along the Kaltag fault farther west, assuming a Kaltag-Tintina connection; and (2) estimated offset between the Lower Cretaceous Kandik Group in the Charley River area and presumably correlative flyschoid rocks at the east end of the Beaver Creek stratigraphic belt in the Livengood area. This estimate differs considerably from the 450 km of right-lateral separation documented for the Tintina system in central Yukon Territory by Tempelman-Kluit (1979) and Gordey (1981). One possible explanation for this discrepancy is transpression, whereby the 270 km of excess Canadian strike-slip motion was absorbed in the Livengood area by compressive thrusting on splays that bend or horsetail southwestward from the main Tintina fault zone at the leading edge of the northwestward-driving upland. However, the geology of the Livengood and adjacent areas does not support a transpressive model because:

(1) The assumptions on which previous estimates of Tintina movement in east-central Alaska are based are invalid. Although a Kaltag-Tintina connection has been tracked through the Livengood area via the Victoria Creek fault by current mapping, other major faults with possibly large lateral components of movement

by which Tintina strike-slip could have been distributed and dissipated (Dover, 1985b) have now been recognized in southwestern Alaska (Patton and Moll, 1982). Thus, the discrepancy in amount of movement that the transpressive model was invoked to explain is in doubt.

(2) As presently mapped, the polymictic conglomerate-bearing Wilber Creek unit does not reach the easternmost end of the Beaver Creek stratigraphic belt, which has generally been restored against the Kandik basin sequence of the Charley River area in Tintina reconstructions used in support of the 180-km estimate of separation (Chapman and others, 1985). Even if it did, the polymictic lithology in the eastern part of the Beaver Creek belt does not match well with the Biederman Argillite and Kennan Quartzite, which are the units of the lower part of the Kandik Group exposed on the north side of the Tintina zone in the Charley River area. However, a section of folded and thrusted argillite and thick Keenan-like quartzite remarkably similar to that in the Kandik Group does characterize the western part of the Wilber Creek belt in the westernmost Livengood quadrangle, and the displacement required to juxtapose these western Livengood units with their lithologic counterparts in the Charley River area is about 400 km.

(3) Thrust faulting in the Livengood area appears to be comparable in style and magnitude, and involves some similar stratigraphic sequences as that in the Charley River fold-and-thrust belt, which predates and is clearly cut by the Tintina fault system. Even if all the thrusts were transpressional, the amount of shortening they could accommodate would be at most only a small fraction of the alleged 270-km discrepancy between previous estimates of Tintina fault displacement in central Yukon Territory and east-central Alaska. Nor is crustal thickening by brittle thrusting or ductile processes in the Yukon-Tanana upland (Foster and others, 1987, and this volume) known to be of appropriate Late Cretaceous to early Tertiary age and of sufficient magnitude to account for such drastic crustal shortening during Tintina time.

The main alternative to transpression is that Tintina fault displacement in east-central Alaska was comparable to that in Yukon Territory (Dover, 1985b). The concept of comparable displacement is supported by the remarkable similarity between the stratigraphic belts of the Livengood area and those of the Dawson area of west-central Yukon Territory, areas now separated by nearly 400 km, whose possible connection was recognized by Tempelman-Kluit (1971, 1984). In the southeast Dawson quadrangle, northeast-trending, northwestward-moving, thrust-bounded rock sequences described by Green (1972) and Thompson and Roots (1982) are truncated at a high angle by the Tintina fault zone (Fig. 10). These sequences, from structurally highest (on the southeast) to lowest (on the northwest), and their suggested Livengood-area equivalents, are:

(a) A belt of internally deformed and weakly metamorphosed Late Proterozoic to Lower Cambrian grit, maroon and green slate, and minor black limestone, overlain by the gabbro-bearing Ordovician and Silurian Road River Formation. Corresponding rocks of the Livengood area are the Wickersham grit unit and the Fossil Creek Volcanics in the Wickersham and White Mountains stratigraphic belts.

(b) The strongly folded and imbricately thrusted Keno Hill Quartzite of late Paleozoic age and associated gabbro (R. I. Thompson, written communication, 1986). Corresponding rocks of the Livengood area are the undated, but petrologically similar Globe quartzite and gabbro unit in the White Mountains stratigraphic belt.

(c) Carbonaceous phyllitic argillite, slate, and quartzite of a "lower schist" unit to which a Jurassic age was assigned by Green (1972), but which in the area remapped by Thompson and Roots (1982) contains at least some chert-pebble-rich polymictic conglomerate resembling that in the Lower Mississippian and Devonian Earn Group. Corresponding rocks of the Livengood area are the Vrain unit in the Beaver Creek stratigraphic belt. Also a discontinuous band of Triassic limestone and limy shale underlying the "lower schist unit" of the Dawson area corresponds to locally occurring but undated calcareous rocks tentatively assigned to the lower part of the Glenn Shale (Triassic) in the Beaver Creek belt.

(d) Chert-rich and tuffaceous(?) rocks in a second belt of the Ordovician and Silurian Road River Formation, which here is locally deposited on mafic sills. The corresponding unit of the Livengood area is the Livengood Dome Chert in the Livengood stratigraphic belt.

(e) A heterogeneous, late Proterozoic to Lower Cambrian unit of chert, clastic rocks, minor limestone, and abundant mafic intrusive and extrusive material marking local volcanic centers. Corresponding rocks of the Livengood area make up the heterogeneous unit in the Livengood stratigraphic belt.

(f) Late Proterozoic maroon and green slate, grit, and minor limestone similar to that in (a). Corresponding rocks of the Livengood area make up the Wickersham grit unit, which underlies the heterogeneous unit in the Livengood stratigraphic belt.

Predictably for a fold-and-thrust belt involving abrupt facies transitions, there are some important lithologic variations among the structural plates correlated between the Livengood and Dawson areas, especially in the relative abundances of lithic components. Furthermore, the Amy Creek unit of the Livengood stratigraphic belt (e) has no presently known counterpart in its corresponding sequence of the Dawson area. However, similar carbonates are common in the underlying structural plate at Dawson, below the Dawson thrust (Thompson and Roots, 1982). Considering the variability of stratigraphic details within the thrust belt, the remarkably high degree of correspondence of distinctive lithologic units and associations within individual thrust-bounded sequences between the two areas, as well as similarities in structural style and vergence, and in the distribution of low-grade metamorphic effects, all provide strong evidence that the two areas were directly connected before Tintina strike-slip fault movement. Therefore, I interpret the eastern Livengood area to be a segment of the northern Cordilleran fold-and-thrust belt displaced by the Tintina fault zone from the Dawson and Charley River areas.

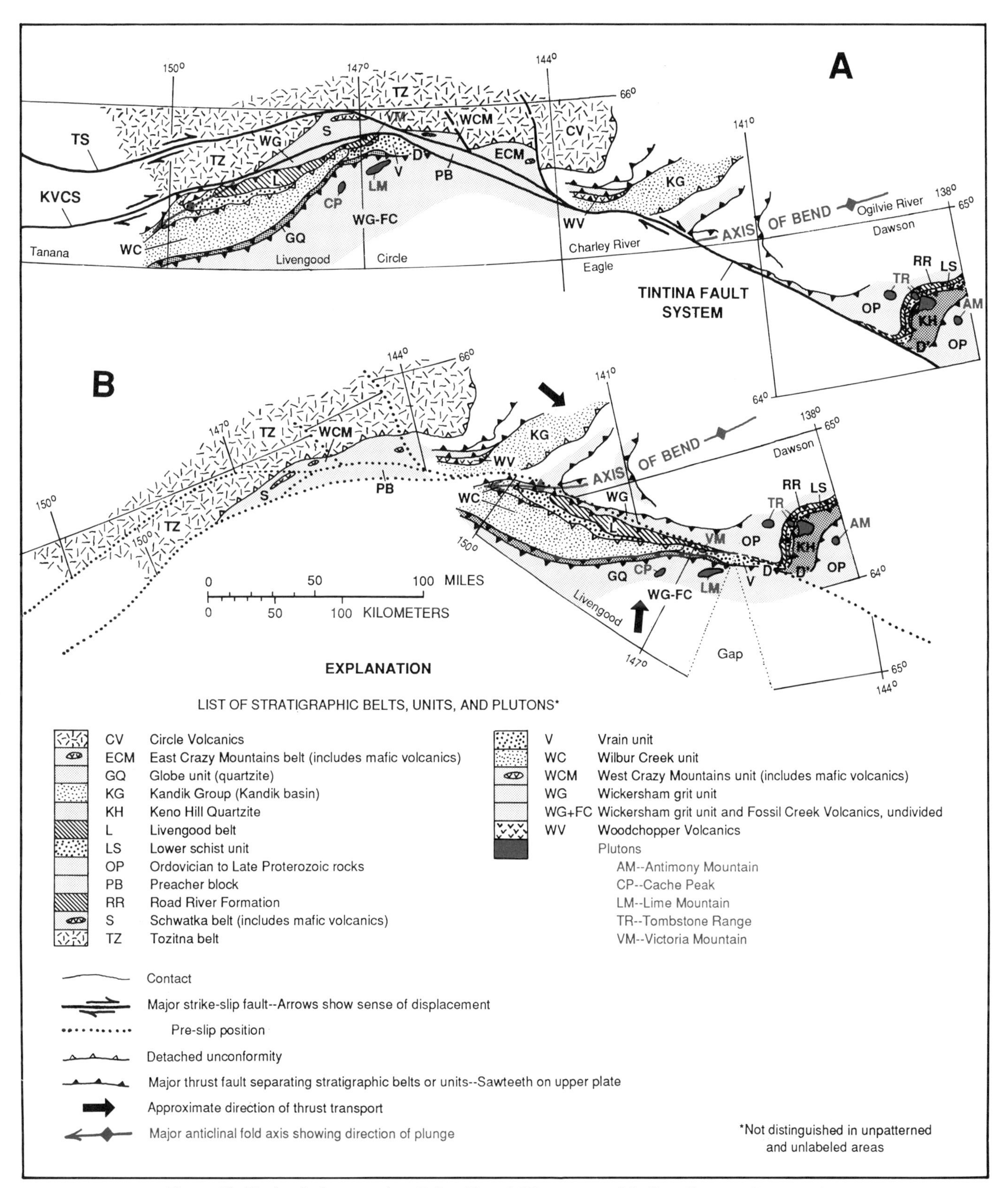

Figure 10. Tintina fault separation. Points D and D′ are two originally adjacent reference points displaced by Tintina fault movement. A. Present configuration. B. Restored pre-Tintina configuration.

Restoring the easternmost Livengood area to an original position opposite the Dawson area has two important consequences for correlations between the Livengood and Charley River areas. First, the Wilber Creek unit at the west end of the Beaver Creek stratigraphic belt, rather than that at the east end, is aligned with the Kandik basin (Figs. 8 and 10). This alignment produces a far more satisfactory connection between the lower units of the Kandik Group cut off by the Tintina fault in the Charley River area, and the part of the Wilber Creek unit containing a prominent Keenan-like quartzite unit. Second, the "bend" separating northwest- and southeast-directed thrusts in the Livengood area is aligned with the axis along which thrust trends and transport directions change in the same way in the Charley River area (Fig. 10).

Ray Mountains area

The Ray Mountains area, as defined here, includes the Ray Mountains proper, in the north half of the Tanana and southernmost part of the adjacent Bettles 1:250,000 quadrangles, and also the area between the Tozitna and Yukon Rivers in the south half of the Tanana quadrangle (Fig. 3). The area spans the Kokrines-Hodzana upland (also known as the Ruby geanticline), from the Koyukuk basin on the northwest (Patton and others, chapter 7, this volume) to the Yukon Flats basin. It is separated from the Livengood area on the southeast by the Kaltag–Victoria Creek splay of the Tintina fault system (Figs. 1 and 2).

Previous work. A brief reconnaissance study by Eakin (1916) covers part of the area considered here, but attention was first focused on the Ray Mountains area in a report by Patton and Miller (1970) on the Kanuti ultramafic belt. Preliminary geologic maps at a scale of 1:250,000 of the Bettles (Patton and Miller, 1973), Melozitna (Patton and others, 1978), and Tanana (Chapman and others, 1982) quadrangles now cover the entire Ray Mountains area, and a few preliminary topical reports related to that mapping or earlier work are also available. The most applicable of these to the Ray Mountains area deal with the Kaltag fault (Patton and Hoare, 1968), the age of the Rampart Group (Brosgè and others, 1969), and mapping at Sithylemenkat Lake (Herreid, 1969). The main impetus for more recent and on-going work in the Ray Mountains area was a multidisciplinary study along a transect across the Yukon-Koyukuk basin, from its southeast borderland in the Ray Mountains, to the south flank of the Brooks Range. Of the many short topical reports and abstracts generated so far by the Yukon-Koyukuk transect, mostly published since 1985, those of Dover and Miyaoka (1985a, b, c) in the Ray Mountains area are emphasized in this review, and others are cited where relevant.

Geologic framework. Dover and Miyaoka (1985a) separated the Ray Mountains segment of the Kokrines-Hodzana upland into three diverse and complex rock packages, and a fourth package is designated here. These are informally named: (1) the metamorphic suite of the Ray Mountains, in the central part of the range; (2) the Kanuti assemblage, on the northwest; (3) the Rampart assemblage, on the southeast; and (4) the Devonian metaclastic sequence separating packages 1 and 3 (Figs. 11 and 12). The metamorphic suite of the Ray Mountains is part of the Ruby terrane of Silberling and Jones (1984), and the structurally flanking Kanuti and Rampart assemblages correspond to their Angayucham and Tozitna terranes, respectively. The mid-Cretaceous Ray Mountains granitic batholith was intruded mainly into the metamorphic suite, but it locally cuts all of the other assemblages. The map distribution of these major units is shown on Figure 12.

Metamorphic suite of the Ray Mountains. Five informal units make up the metamorphic suite of the Ray Mountains. The structurally lowest two (units 1 and 2) may be autochthonous; the other three (units 3 to 5) are more structurally complex and are probably parautochthonous or possibly allochthonous.

Unit 1 is feldspathic quartzite containing 70 to 85 percent quartz, 10 to 20 percent intermediate plagioclase, and 2 to 10 percent mica. Most samples also contain a few percent of untwinned K-feldspar, and many have as much as 5 percent garnet and/or cordierite. The protolith was a slightly argillaceous, feldspathic, quartz-rich sediment of continental derivation but uncertain age.

Unit 2 is a mineralogically and texturally distinctive augen orthogneiss containing augen of igneous plagioclase, granulated quartz and plagioclase, or poikilitic simply twinned K-feldspar, in a cataclastic matrix of quartz, feldspar, and synkinematic micas. The augen orthogneiss has a Devonian U-Pb zircon age of 390 ± 12 Ma (Patton and others, 1987). Similar augen orthogneiss yields Middle Devonian to Early Mississippian Rb-Sr whole-rock and U-Pb zircon ages elsewhere in the metamorphic borderlands of the Yukon-Koyukuk basin (Dillon and others, 1979, 1980,

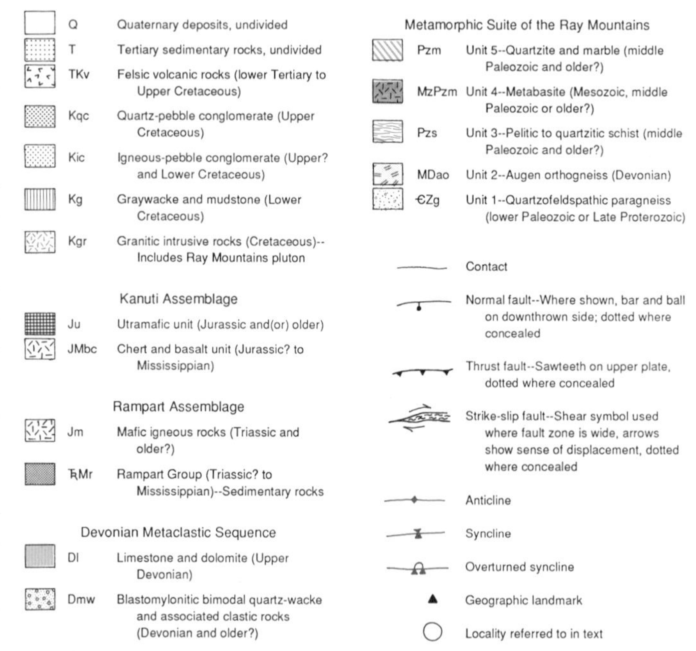

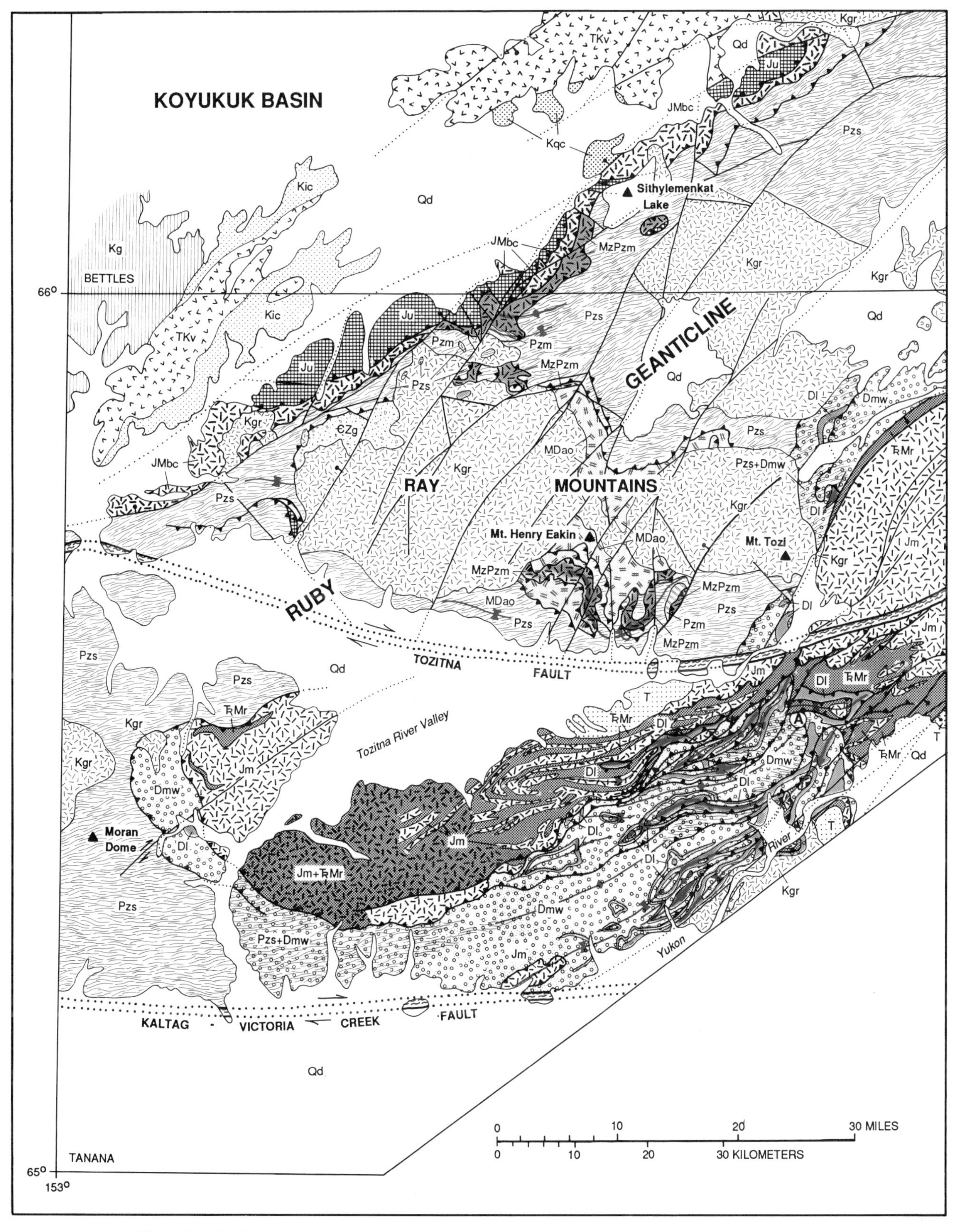

Figure 11. Geologic map of the Ray Mountains area. See Figure 3 for area location. Explanation on facing page.

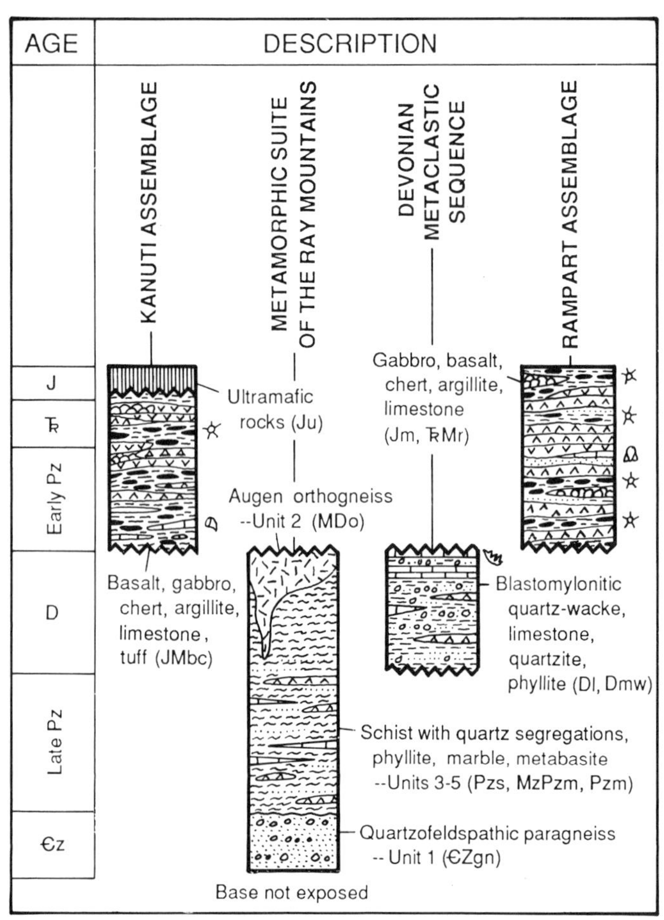

Figure 12. Generalized stratigraphic columns of the Ray Mountains area. See Figure 5 for explanation of lithologic symbols, and Figure 11 for explanation of map units.

1985). The protolith of the augen orthogneiss was granitic igneous rock that was probably intruded into the other four units.

Unit 3 is the most widely distributed and variable of the metamorphic units and consists mainly of pelitic to quartzitic schist. Subordinate interlayers are micaceous quartzofeldspathic schist, calc-schist or calc-silicate gneiss, amphibolitic schist, quartzite, marble, phyllonite, and at one locality, metaconglomerate containing metabasite clasts as large as cobble size. Quartz-segregation layering and blastomylonitic fabric are characteristic, and bedding is transposed along the main foliation, which parallels axial surfaces of isoclinal folds. The schist varies widely in mineralogy and texture because of the complex interaction of polyphase deformation and polymetamorphic recrystallization superimposed on its diverse lithologic components. The protolith of most of the schist was siliceous argillite and argillaceous feldspathic to calcareous quartz-rich sediment, which possibly originated as a turbidite having primary alternations of quartzofeldspathic and argillaceous layers.

Unit 4 consists of metamorphosed mafic igneous rocks, or metabasite, that occurs as large mappable bodies ranging from metagabbro and metadiabase to amphibolite and garnet-amphibolite. It also occurs as thin interlayers of amphibolitic schist and greenschist within unit 3. The mappable mafic rock bodies grade in fabric and degree of recrystallization from weakly sheared and incompletely recrystallized in their cores to strongly sheared and extensively recrystallized along their margins.

Unit 5 is composed of relatively pure quartzite and marble. Beds thick enough to map separately on Figure 11 are associated with metabasite bodies in the southern part of the Ray Mountains. A few thin interbeds of quartzite and marble also occur in unit 3.

Contacts between units within the metamorphic suite of the Ray Mountains are generally shallow-dipping, somewhat discordant blastomylonite zones in which ductile shear occurred under low-grade metamorphic conditions.

Preliminary analysis suggests that the mesoscopic fabric of the metamorphic suite of the Ray Mountains involves at least three deformational phases and three metamorphic episodes (Dover and Miyaoka, 1985b). The earliest folds (F1, Fig. 13) are isoclines to which the main schistosity (S1) is axial planar. F1 axes commonly parallel the axes of mapped folds. In thin section, F1 isoclines are associated with synkinematic amphibolite-facies minerals (M1). F1 isoclines are overprinted by second-generation folds (F2) of variable style that have axial plane cleavage (S2) at a low to moderate angle with S1 (Fig. 13). Most F2 folds are small chevron folds with tight or even isoclinal forms. Tight F2 folds increase in abundance toward cataclastic zones between major rock units and assemblages, where the cataclastic foliation is the dominant fabric element and appears to coincide with S2. These second-generation structures throughout the metamorphic suite of the Ray Mountains are invariably associated with greenschist-facies or locally glaucophanitic greenschist-facies mineral assemblages (M2). In rocks not strongly affected by the cataclastic event, M2 produced relatively minor retrogression of M1 minerals. In more strongly cataclastic rocks, many of phyllitic aspect, M1 minerals are locally preserved as granulated or pseudomorphic relics cut by cataclastic S2 foliation with synkinematic M2 mineral assemblages. Such rocks are retrogressive phyllonites. All the graphitic chlorite-muscovite-quartz schist of the metamorphic suite of the Ray Mountains is probably blastomylonite produced or affected by the M2 cataclastic event, and some of these rocks are demonstrably phyllonitic. In parts of the Ray Mountains not strongly affected by the cataclastic event, broad warps (F3) bend earlier structures in individual outcrops; F1 and F3 are coaxial in places. In strongly cataclasized zones, F3 folds are generally small crenulations that fold synkinematic M2 minerals and have only incipiently recrystallized axial-plane cleavage (S3) that cuts the cataclastic S2 foliation. Hornblende-hornfels facies mineral assemblages (M3) occur in a contact-metamorphic aureole several kilometers or less wide around the mid-Cretaceous Ray Mountains granitic batholith. M3 is generally post-kinematic relative to F2/S2 and earlier structures, and may represent a late post-kinematic phase of M_2. However, in the easternmost part of the metamorphic suite of the Ray Mountains, where it is in contact with the Devonian metaclastic unit and the

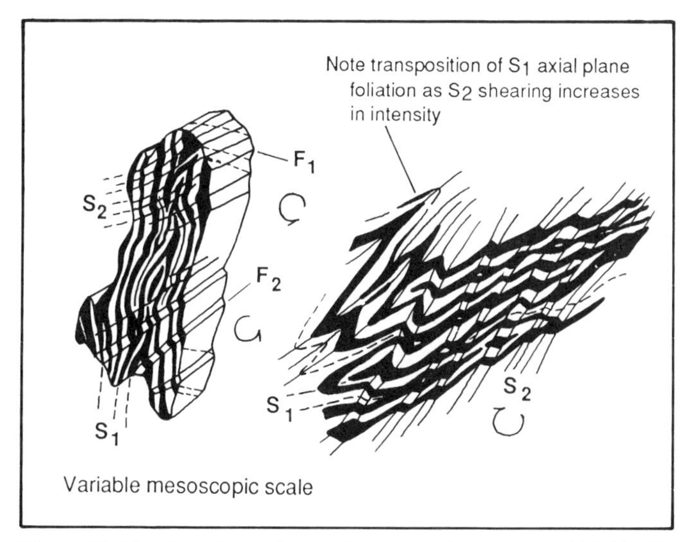

Figure 13. Sketches of superimposed mesoscopic structures. F, fold axis; S, axial plane foliation; subscripts indicate sequence of formation; arrows showing sense of rotation indicate asymmetry of folding.

Rampart assemblage, M3 garnet and albitic plagioclase porphyroblasts are rotated within a cataclastic foliation that is probably S3 but could possibly be S2.

Kanuti assemblage. Two tectonically juxtaposed units form the Kanuti assemblage, which lies on the northwest side of the metamorphic suite of the Ray Mountains. A lower chert-basalt unit contains discontinuous and largely unsheared blocks and lenses of incipiently recrystallized basalt, diabase, gabbro, and serpentinized ultramafic rocks in a mylonitic, melange-like matrix of low-grade metasedimentary rocks that include bedded chert of Triassic age, argillite, slate, Mississippian limestone, and volcaniclastic rocks (W. W. Patton, written communication, 1983). The structurally overriding ultramafic unit contains cryptically deformed, extensively serpentinized layered gabbro, peridotite, harzburgite, and dunite, with garnet-amphibolite possibly derived from eclogite at the base. K-Ar ages on hornblende from the garnet amphibolite are 138, 149, and 161 Ma (W. W. Patton, written communication, 1983). Primary igneous and sedimentary textures and mineralogy are typically preserved in the Kanuti assemblage, which was interpreted to be dismembered ophiolite by Patton and others (1977) and Loney and Himmelberg (1985a, b).

Although contacts between the Kanuti assemblage and the metamorphic suite of the Ray Mountains are poorly exposed, the Kanuti appears to lie in low- to moderate-angle thrust contact on the metamorphic suite because of the intensity of shearing in the lower chert-basalt unit, and the regional discordance of its basal contact with deformed units of the underlying metamorphic suite. The Kanuti ophiolite is interpreted to be a slice of the oceanic basement of the Koyukuk basin that was obducted southeastward onto the continentally derived metamorphic suite of the Ray Mountains (Patton and Box, 1985; Cady, 1986; Miller, 1985; Arth, 1985; Patton and others, chapter 16, this volume). If so, obduction appears to have been completed by mid-Cretaceous time, because emplacement of the Ray Mountains batholith locally produced contact metamorphic effects in the Kanuti ophiolite (W. W. Patton, written communication, 1983). Also, at one locality in the chert-basalt unit, bedded chert displays three phases of folding similar in style, orientation, and mutual angular relations to folds in the adjacent metamorphic suite (Dover and Miyaoka, 1985b). This similarity suggests that the Kanuti assemblage was in place or was close enough to the borderland in pre-mid-Cretaceous time to have participated in the same folding history as the metamorphic suite. The amount of obduction represented by the Kanuti belt is controversial and relates to the problem of the origin of the Tozitna terrane (Rampart assemblage) discussed below, but obduction may not have extended much beyond the present limits of Kanuti exposures.

Rampart assemblage. The term "Rampart assemblage," as used informally here, includes the Rampart Group of Mertie (1937), as well as voluminous associated mafic intrusive rocks. So defined, the Rampart assemblage corresponds to the Tozitna stratigraphic belt of the Livengood area. The Rampart Group contains basalt, pillow basalt, phyllite, argillite, slate, arkosic semischist, volcaniclastic rocks, conglomerate, Permian limestone, and radiolarian chert of Mississippian to Triassic(?) age (Jones and others, 1984). Intrusive rocks included in the Rampart assemblage are mainly diabase and gabbro, but subordinate diorite and minor serpentinized ultramafic rocks occur as well. A K-Ar age of 210 ± 6 Ma (Triassic) is reported by Brosgè and others (1969) on hornblende from gabbro near the town of Rampart.

Devonian metaclastic sequence. Also tentatively included in the Rampart assemblage by Dover and Miyaoka (1985a) were low-grade cataclastic rocks that lie structurally between the Rampart Group and the metamorphic suite of the Ray Mountains in the southeastern Ray Mountains. Blastomylonitic quartzwacke with bimodal quartz grain-size distribution is the dominant and most distinctive lithology, but associated rocks include contorted phyllite, platy arenaceous limestone, slightly argillaceous quartz-siltstone, poorly sorted limonite-spotted quartzite with limy cement, and pebbly mudstone with clasts as much as 2 cm in diameter of limestone, quartz, quartzite, chert, argillite, and feldspathic quartz-arenite. No local evidence was found in the southeastern Ray Mountains for the protolithic age of the bimodal quartz-wacke unit, but the Devonian(?) age (Dover and Miyaoka, 1985a) suggested by analogy with lithologicall similar rocks of the Beaver quadrangle to the north (Brosgè and others, 1973) is now confirmed by the recovery of latest Devonian (Famennian) conodonts (Anita Harris, written communication, 1985 and 1986) from platy limestone ("limestone and dolomite unit" of Fig. 11) associated with the bimodal unit in the area between the Tozitna and Yukon Rivers.

New lithologic, textural, and paleontological data on these cataclastic to blastomylonitic rocks in the Tozitna-Yukon Rivers area makes their designation as a separate Devonian metaclastic sequence lying between the Rampart assemblage and the metamorphic suite of the Ray Mountains more appropriate than assignment to the Rampart. The Devonian metaclastic sequence contains some rock types similar to those in the Rampart assemblage, but it differs in the predominance of quartz-rich, commonly bimodal clastic rocks, and in being slightly older than the oldest rocks documented n the Rampart assemblage. It contains some mafic intrusive material, but less than occurs in the Rampart assemblage.

The grade and intensity of metamorphic recrystallization of the metaclastic sequence appear to increase gradually westward (down section) from unequivocal Rampart rocks, toward the metamorphic suite of the Ray Mountains, with which its metamorphic textures merge and become indistinguishable. However, clearly recognizable primary sedimentary textures and clasts are typically preserved in the metaclastic sequence, and in this respect, it is less thoroughly tectonized and recrystallized in most places than is the metamorphic suite of the Ray Mountains. Metamorphic gradation was tentatively attributed to metamorphic convergence, caused by retrogressive phyllonitization in the metamorphic suite of the Ray Mountains, and progressive metamorphism in structurally detached rocks now assigned to the Devonian metaclastic sequence (Dover and Miyaoka, 1985b).

Fabric data from the Devonian metaclastic sequence in the southeastern Ray Mountains define broadly folded but generally southeast-dipping blastomylonitic foliation and include mineral lineations interpreted as indicators of northwest-southeast movement. Two lines of evidence were cited by Miyaoka and Dover (1985) for up-dip movement from southeast to northwest toward the metamorphic core of the Ray Mountains: (1) Five of the six asymmetrical folds observed in the metaclastic sequence verge to the northwest, and a major east-west–trending fold in the underlying metamorphic suite appears to be overturned northwestward. (2) In a preliminary study of shear sense, the preponderance of shear data also suggest northwestward-directed up-dip shear; movement indicators include s-c tectonites, asymmetrical augen, porphyroblast tails, and microfolds. In mylonitization was related to movement of the overlying Rampart assemblage, the microfabrics indicate it moved northwestward toward the crest of the Ruby geanticline. Recent additional fabric data confirm a consistent northwest-directed sense of shear for the southeastern flank of the Ray Mountains, but show numerous variations in trend and inconsistencies in sense of shear for the Ray Mountains area as a whole (Dover and Miyaoka, unpublished data).

Ray Mountains pluton. The Ray Mountains pluton is aligned east-west across the northeast-trending Ruby-Hodzana upland. A reconnaissance map by Puchner (1984) shows the Ray Mountains batholith to be a composite of at least four separate intrusive phases ranging from K-feldspar-rich porphyritic granite, biotite granite, and two-mica granite, to granodiorite. Several K-Ar, Rb-Sr whole-rock, and U-Pb ages from various phases of the Ray Mountains batholith range from 104 to 111.6 Ma (Puchner, 1984; Silberman and others, 1979a; Patton and others, 1987). The isotope chemistry of the Ray Mountains pluton and other granitic plutons of the Ruby geanticline indicates that they are S-type plutons derived from or contaminated by continental crust (Miller, chapter 19, this volume). Slight contact-metamorphic effects are present in the Kanuti assemblage at the west end of the Ray Mountains pluton, and an intrusive contact between a granitic pluton of similar age and the Kanuti assemblage was mapped by Patton and others (1978) in the Melozitna quadrangle, just southwest of the Ray Mountains. These relations establish the mid-Cretaceous as a younger limit for the time of obduction of the Kanuti assemblage. At the east end of the Ray Mountains batholith (Mt. Tozi lobe), contact-metamorphic (M3) garnet and albitic plagioclase porphyroblasts in the Devonian clastic sequence were rotated by mylonitization, indicating that at least the latest movement there postdates the mid-Cretaceous.

Brittle fracture and strike-slip faulting—Tozitna fault. A brittle fracture fabric and high-angle fault pattern are superimposed on the more ductile older fabrics of the Ray Mountains area (Fig. 11). The brittle structures are expressed mainly by northeast-trending topographic lineaments having generally small down-to-the-northwest dip-slip offsets. Geophysical modeling by Cady (1986) suggests that post-thrusting high-angle faults having similar northeast trend, and significant down-to-basin displacement cut the obducted northwest margin of the Ray Mountains just west of Kanuti ophiolite exposures.

The Tozitna fault (Fig. 11) in the southeastern Ray Mountains area is a strike-slip fault zone about 1 km wide containing steep kink-fold axes and pervasive low-plunging slickensides (Dover, 1985b). Its curvilinear trace can be followed for more than 250 km by aligned topographic valleys and low drainage divides or saddles across ridges from the margin of the Koyukuk basin, across the Tanana and Livengood quadrangles at the south edge of Yukon Flats basin, and into the Tintina fault system. It displaces the contact between the Rampart assemblage and the Devonian metaclastic sequence from its position in the southeastern Ray Mountains to that south of the Tozitna River about 55 km in a right-lateral sense, a separation comparable with that determined for the Tozitna fault in the Livengood ara.

Implications of new mapping between the Tozitna and Yukon Rivers. The best exposures of the contacts between the metamorphic suite of the Ray Mountains, the Devonian metaclastic sequence, and the Rampart assemblage are in the part of the Ray Mountains area lying between the Tozitna River and the Yukon River (Fig. 11). But here, as in the Mt. Tozi area of the southeastern Ray Mountains, the distinctions between these sequences are obscure in places. Thick diabase sills and associated sedimentary rocks typical of the Rampart assemblage are underlain by a bimodal quartz-wacke-bearing unit that is similar in most of its associated lithologies and petrologic characteristics to the metaclastic sequence of the southeastern Ray Mountains. The main differences are that in the Tozitna-Yukon Rivers area: (1) the carbonates are mainly dolomite or dolomitized limestone, and (2) low-grade metamorphic recrystallization of the quartz-wacke unit is locally more advanced than in the southeastern Ray Mountains. At locality A on Figure 11, an argillaceous part of the blastomylonitic quartz-wacke unit grades into garnet-muscovite schist, and the post-kinematic static growth of contact-metamorphic biotite, andalusite, and albite indicates a buried pluton nearby. Even so, primary sedimentary textures persist in most places within the bimodal quartz-wacke, and a few conodonts of latest Devonian (Famennian) age (A. G. Harris, written communication, 1984) are preserved in thin, platy arenaceous limestone beds associated with the dolomite.

Carbonate marker beds in the dominantly metaclastic sequence define large folds and thrusted folds. Folding and thrusting clearly involved rocks of the overlying Rampart assemblage. Locally, conodont-bearing Famennian limestone similar to but thicker than that typical of the metaclastic sequence appears to be infolded with only incipiently metamorphosed metachert, meta-argillite, and meta-diabase tectonite at the base of the Rampart assemblage. Also, metachert that yields Mississippian to Triassic radiolarians (Jones and others, 1984), limestone containing poorly preserved middle to late Paleozoic corals (Chapman, 1974), and thick diabase sills, all belonging to the Rampart assemblage, are structurally imbricated with the metaclastic sequence in places. Thrusts dip north, and where folds are asymmetrical, vergence is to the south or southeast. However, shear indicators in tectonites of the Devonian metaclastic unit generally indicate a top-to-the-north, -northwest, or -west sense of shear, suggesting a complex history of movement for the contact between the Rampart sequence and the Devonian metaclastic unit. The imbricated contact has a broadly anticlinal northeast-plunging form in the eastern part of the Tozitna-Yukon Rivers area. The area of highest metamorphic grade and contact-metamorphic effects, near the bend in Canyon Creek (locality A, Fig. 11), occurs along the anticlinal axis, suggesting that the buried pluton inferred there occurs in the core of the broad anticline.

Near Moran Dome (Fig. 11), at the west end of the Tozitna-Yukon Rivers area, low- to medium-grade schist with quartz segregations resembling schist in the metamorphic suite of the Ray Mountains, is intruded by probable mid-Cretaceous granite. Medium-grade mineral assemblages are restricted to the contact aureole of the pluton. The dominantly low-grade rocks of the Moran Dome area are similar in composition, mineralogy, and texture to the most extensively recrystallized parts of the metaclastic sequence in the Canyon Creek area to the east. However, continuity between the two areas is masked by poor exposures and the intervening occurrence of less completely recrystallized rocks of the metaclastic sequence and diabase of the Rampart assemblage. About 8 km northeast of Moran Dome, differences in metamorphic grade and degree of recrystallization between the Ray Mountains metamorphic suite and the Devonian metaclastic unit appear to be telescoped by east-dipping thrusts. Sense of shear determined from s-c tectonites in this area is variable, and blastomylonitic lineations participated in broad post-shear folding. However, the best-developed s-c fabrics indicate top-toward-the-west and -northwest up-dip movement (Dover and Miyaoka, unpublished data; Smith and Puchner, 1985), as they do in the southeastern Ray Mountains.

From relations mapped in the Tozitna-Yukon Rivers area, the contact between the Rampart assemblage and the Devonian metaclastic sequence is interpreted as a zone of ductile detachment, complicated by later folding and imbricate thrusting, separating two vertically stacked sequences that may originally have been in stratigraphic continuity. Detachment was probably controlled by fundamental differences in ductility between the two sequences—a more competent, mafic-rich, typically Rampart upper sequence that is latest Devonian and younger in age, and a less competent, quartz-wacke-rich lower sequence that is latest Devonian and older in age. Variable and locally divergent sense-of-movement data indicate a complex history of movement for this contact. The contact between the Devonian metaclastic sequence and the metamorphic suite of the Ray Mountains is also strongly sheared, and is further obscured by metamorphism that increases gradually in grade and degree of recrystallization with depth and toward granitic plutons. An original stratigraphic contact, possibly an unconformity, is possible, but cannot be proved. It is also possible that all of the multiphase deformation and polymetamorphism of the metamorphic suite of the Ray Mountains occurred in one prolonged and intermittently active orogenic event, the ductile and brittle effects of which were largely depth-controlled and overlapping in time. Early ductile fabrics formed moderately deep in the crust may have been overprinted by more brittle fabrics later in the orogenic cycle, as structural thickening of the crust drove uplift of the Ruby terrane.

Protoliths of metamorphic rocks in the Ray Mountains area. The protolith of the main body of the metamorphic suite of the Ray Mountains was quartz-rich, dominantly argillaceous, possibly turbiditic sediment, containing subordinate mafic igneous intercalations and a few interbeds of quartzite, calcareous sediments, volcaniclastic rocks, and limestone (units 3 to 5, Figs. 11 and 12); possible basement rocks are coarse-grained to gritty quartzitic to quartzofeldspathic rocks (unit 1, Figs. 11 and 12). The heterogeneous protolith of the Devonian metaclastic sequence (Figs. 11 and 12) includes bimodal quartz-wacke, siliceous argillite and chert, iron-rich and locally carbonate-cemented quartzite, polymictic conglomerate, mafic sills and dikes, and latest Devonian (Famennian) limestone. In lithologic character, these rock types resemble those in lower to middle Paleozoic assemblages deposited on gritty Wickersham (late Proterozoic) basement in the Livengood area, and in the outer part of the Cordilleran miogeocline in general. Appropriate protolithic models are the stratigraphic reconstructions described by Gordey (1981) for the Selwyn basin and Cassiar platform of south-central Yukon Territory (Fig. 14), which contain all of the lithic elements represented in the metamorphic suite of the Ray Mountains and the Devonian metaclastic unit.

Beaver area

As defined here, the Beaver area includes parts of the Beaver, Chandalar, Christian, and Arctic 1:250,000 quadrangles, and extends from the northeastern part of the Kokrines-Hodzana upland into the southeastern Brooks Range (Fig. 3).

Previous work. Parts of the Beaver area were included in the pioneering reconnaissance studies of Schrader (1900, 1904),

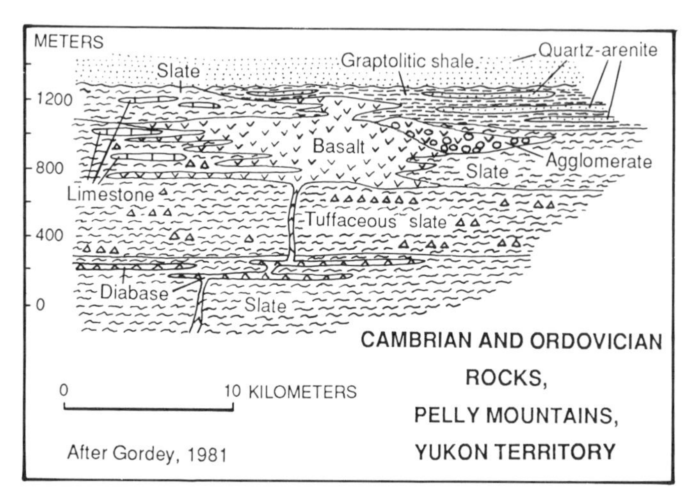

Figure 14. Possible protolith model for the metamorphic suite of the Ray Mountains.

Maddren (1913), and Mertie (1925, 1929). The most recent geologic maps of the Christian, Chandalar, and Arctic quadrangles, were completed by Brosgè and Reiser in 1962, 1964, and 1965, respectively; the Beaver quadrangle was mapped by Brosgè and others (1973). Topical studies, some of regional scope, include Brosgè (1960, 1975), Brosgè and others (1962), Reiser and others (1965), Chipp (1970), Brosgè and Dutro (1973), Holdsworth and Jones (1979), and Dillon and others (1979).

Stratigraphic and structural framework. The Beaver area has map relations that are critical to three major geologic problems: (1) stratigraphic correlation between the southeastern Brooks Range and the east-central Alaska region; (2) the relation of the mafic igneous complexes of the Beaver and Christian quadrangles to adjacent strata of the Ruby geanticline and the southeastern Brooks Range; and (3) the relation between faults inferred to have dominantly strike-slip movement of moderate to large displacement, and faults interpreted as thrusts. The brief summary presented here is based mainly on the maps of Brosgè and Reiser cited above. Part of the Beaver area is also discussed by Moore and others (this volume) in the context of the entire Brooks Range orogen.

Three principal rock packages are recognized in the Beaver area: (1) the mafic igneous complexes of the Beaver and Christian quadrangles, (2) low-grade metamorphic rocks of the Ruby geanticline and southeastern Brooks Range, and (3) an intervening belt of dominantly clastic, semi-schistose Devonian rocks (Figs. 15 and 16).

(1) The mafic igneous complexes of the Beaver and Christian quadrangles contain gabbro, diabase, diorite, and basalt of Jurassic and older(?) age (Reiser and others, 1965), intercalated with chert, argillite, tuff, and associated basinal rocks of Mississippian to Permian age (Holdsworth and Jones, 1979). These complexes appear from published descriptions to be equivalent in all respects—including lithologic content and age of basinal sedimentary rocks, composition and Mesozoic age of mafic igneous rocks, and stratigraphic and structural position with respect to adjacent rock sequences—to the Rampart assemblage of the Ray Mountains and Livengood areas, and the Circle Volcanics of the Charley River area. Regional correlation of these sequences is implied by the inclusion of all four in the Tozitna terrane of Silberling and Jones (1984).

(2) The rocks of the Ruby geanticline in the western part of the Beaver quadrangle include pelitic schist, marble, quartzite, calcareous schist, quartzo-feldspathic schist and gneiss, and greenstone; most are low-grade metamorphic rocks of the greenschist facies, but they have medium- to high-grade metamorphic assemblages near the Hodzana and Kanuti batholiths and other granitic plutons (Fig. 15). These metamorphic rocks closely resemble those in the metamorphic suite of the Ray Mountains, and they probably formed from the same protoliths. Structural and metamorphic details are not available for this part of the Ruby geanticline. Low-grade metamorphic rocks similar to those of the Ruby geanticline also occur in the "schist belt" along the part of the southern Brooks Range included in the Hammond and Coldfoot subterranes of the Arctic Alaska terrane (Fig. 2) of Silberling and Jones (1984). For example, generally low-grade metamorphic and blastomylonitic rocks exhibiting complex fabrics formed by multiple deformation and polyphase metamorphism have been described in the "schist belt" (Figs. 15 and 16) by Gottschalk (1987) and Grybeck and Nelson (1981). The Hammond subterrane differs lithologically from the Coldfoot subterrane and the Ruby geanticline in containing the locally thick Skajit Limestone (Devonian) and lower Paleozoic carbonate rocks. Locally, greenstone and greenschist are abundant and form large mappable masses in both the Hammond and Coldfoot subterranes, just as metabasites do in the Ray Mountains metamorphic suite.

(3) A belt of weakly metamorphosed Devonian clastic rocks composing the Venetie terrane of Silberling and Jones (1984) separates the mafic igneous complexes of the Beaver and Christian quadrangles from metamorphic rocks of the Ruby geanticline and the southern Brooks Range. The most characteristic features of these metaclastic rocks are their ferruginous quartz- and chert-rich wacke and lithic graywacke components and their pervasive but incompletely recrystallized blastomylonitic fabric, through which primary sedimentary features and rare invertebrate fossils are preserved. A unit of black phyllitic argillite with subordinate wacke interbeds appears to underlie the coarser clastic rocks. Available thin sections from rocks of this metaclastic sequence commonly contain more chert clasts than those from rocks of the Devonian metaclastic sequence in the Ray Mountains area, but some sections of quartz-rich siltstone and bimodal quartz-wacke are petrographically indistinguishable from those of the Ray Mountains. The quartz-wacke and graywacke units of the Beaver area (Brosgè and Reiser, 1962, 1964; Brosgè and others, 1973) appear to be mutually gradational and are most likely interbedded; these units are not separated in Figure 16. At least partly correlative rocks of the Hunt Fork Shale and Kanayut Conglomerate occur in the Endicott Mountains subterrane of the Arctic Alaska terrane of Silberling and Jones (1984).

The contacts between the three rock packages described above are shown on Figure 15 as thrust faults. In most places, these were mapped as stratigraphic contacts by Brosgè and Reiser (1962, 1964) and Brosgè and others (1973), but structural detachment is interpreted here for most of them, based on map discordances. If correct, analogy with the Ray Mountains area suggests that these detachments may range from thick zones of distributed ductile shear to brittle thrusts, or some complex combination of both. The age of ductile shearing in the Beaver area appears to be bracketed between the Jurassic age of the mafic igneous complexes of the Beaver and Christian quadrangles, which are locally involved, and the mid-Cretaceous age of the Hodzana pluton (Brosgè and others, 1973), which influenced the crystallization of blastomylonitic fabrics during or just after ductile deformation; if brittle thrusting occurred, it most likely postdates ductile deformation.

Two zones of right-lateral strike-slip faulting appear to cut east-west across the Beaver area. The southernmost zone, along the north edge of the Beaver quadrangle, contains two main fault strands that displace the east end of the Hodzana pluton, as well as the contacts between the three principal rock assemblages of the Beaver area, right-laterally about 17 km. The eastward continuation of this zone along the north edge of Yukon Flats basin is not exposed, but it may merge with the other major strike-slip zone, which parallels the Chandalar River in the southern Chandalar quadrangle. This zone, here called the Kobuk-Malamute fault zone, is poorly exposed, but it marks a prominent aeromagnetic lineament (Cady, 1978) and a topographic and geologic break between the Ruby geanticline and the schist belt of the southern Brooks Range, along which mafic and associated rocks (Angayucham terrane) at the margin of the Koyukuk basin appear to have been dragged and attenuated. The amount of right-lateral strike separation along the Kobuk-Malamute zone is undetermined. However, it could be 80 km or more if the more eastward position of the mafic-igneous complex of the Christian quadrangle, compared with that in the Beaver quadrangle, represents strike separation of an originally straighter belt of mafic complexes that was connected.

Discussion. Three observations by Brosgè and others (1973) in their descriptions of map units in the Beaver area suggest that the Devonian metaclastic sequence of the Beaver area shares or overlaps in metamorphic and lithologic characteristics with adjacent metamorphic sequences and with the mafic igneous complexes of the Beaver and Christian quadrangles:

(1) Although most exposures of the Devonian metaclastic sequence are weakly to strongly blastomylonitic, western exposures are more coarsely schistose and appear to grade into schistose rocks typical of the Ruby geanticline and the schist belt. This gradation is especially evident near the Hodzana and Kanuti batholiths in the Ruby geanticline, and west of the Chandalar River in the Chandalar and Christian quadrangles.

(2) The Devonian clastic rocks of the Endicott Mountains and Hammond subterranes and the Venetie terrane in the Chandalar and Christian quadrangles are similar to one another in their lithologic range and content of distinctive rock types. Based on the map distribution of these units, their variably recrystallized fabrics, and sparse fossil evidence, Brosgè and Reiser (1962, 1964) considered at least some of the metamorphic rocks of the schist belt to be more thoroughly and coarsely recrystallized, low- to medium-grade metamorphic equivalents of the Devonian clastic sequence. If correct, the terrane boundary between the Devonian clastic sequence and rocks of the adjacent schist belt may be primarily a metamorphic transition.

(3) Graywacke and quartzite like that of the Devonian clastic sequence locally intergrade with cherty rocks assigned to the mafic igneous complexes of the Beaver and Christian quadrangles. Brosgè and others (1973) also note places in the Christian quadrangle where these mafic rocks intrude and include Devonian metaclastic rocks.

These three observations heighten the similarity between the Beaver and Ray Mountains areas. They also support the possibility that the three principal rock packages here, as in the Ray Mountains area, may be structurally disrupted sequences that originally had stratigraphic continuity.

Coleen area

As defined here, the Coleen area includes the Coleen and most of the Table Mountain and Black River 1:250,000 quadrangles, and spans contiguous parts of the southeastern Brooks Range (Moore and others, this volume) and the part of the east-central Alaska region containing the Porcupine platform (Fig. 3).

Previous work. The earliest reconnaissance work in the Coleen area was by Kindle (1908), Cairnes (1914), and Maddren and Harrington (1955). Preliminary geologic maps at a scale of 1:250,000 were published for the Coleen quadrangle by Brosgè and Reiser (1969), for the Black River quadrangle by Brabb (1970), and for the Table Mountain quadrangle by Brosgè and others (1976). A detailed geologic traverse along the Porcupine River (Brosgè and others, 1966) supplements reconnaissance mapping in the Coleen quadrangle, and the geologic map of the Black River quadrangle (Brabb, 1970) incorporates unpublished mapping by geologists of the British Petroleum Exploration Company and Louisiana Land and Development Corporation. Numerous regional stratigraphic reviews discuss the stratigraphy of the Coleen area, including those by Laudon and others (1966), Churkin and Brabb (1968), Brosgè and others (1962), Brosgè and Dutro (1973), Dutro (1979), and Dutro and Jones (1984).

Stratigraphic framework. The Brooks Range and Porcupine platform parts of the Coleen area have important similarities in their stratigraphic and structural frameworks as well as the differences emphasized by terrane subdivision. For example, the Coleen quadrangle combines critical stratigraphic and structural relations of the Table Mountain quadrangle, representative of the Brooks Range orogen, with those of the Black River quadrangle, representative of the northern Canadian Cordillera. The stratigraphy of the Coleen quadrangle is divided by Brosgè and Reiser (1969) into four sequences, named the (1) Christian River, (2)

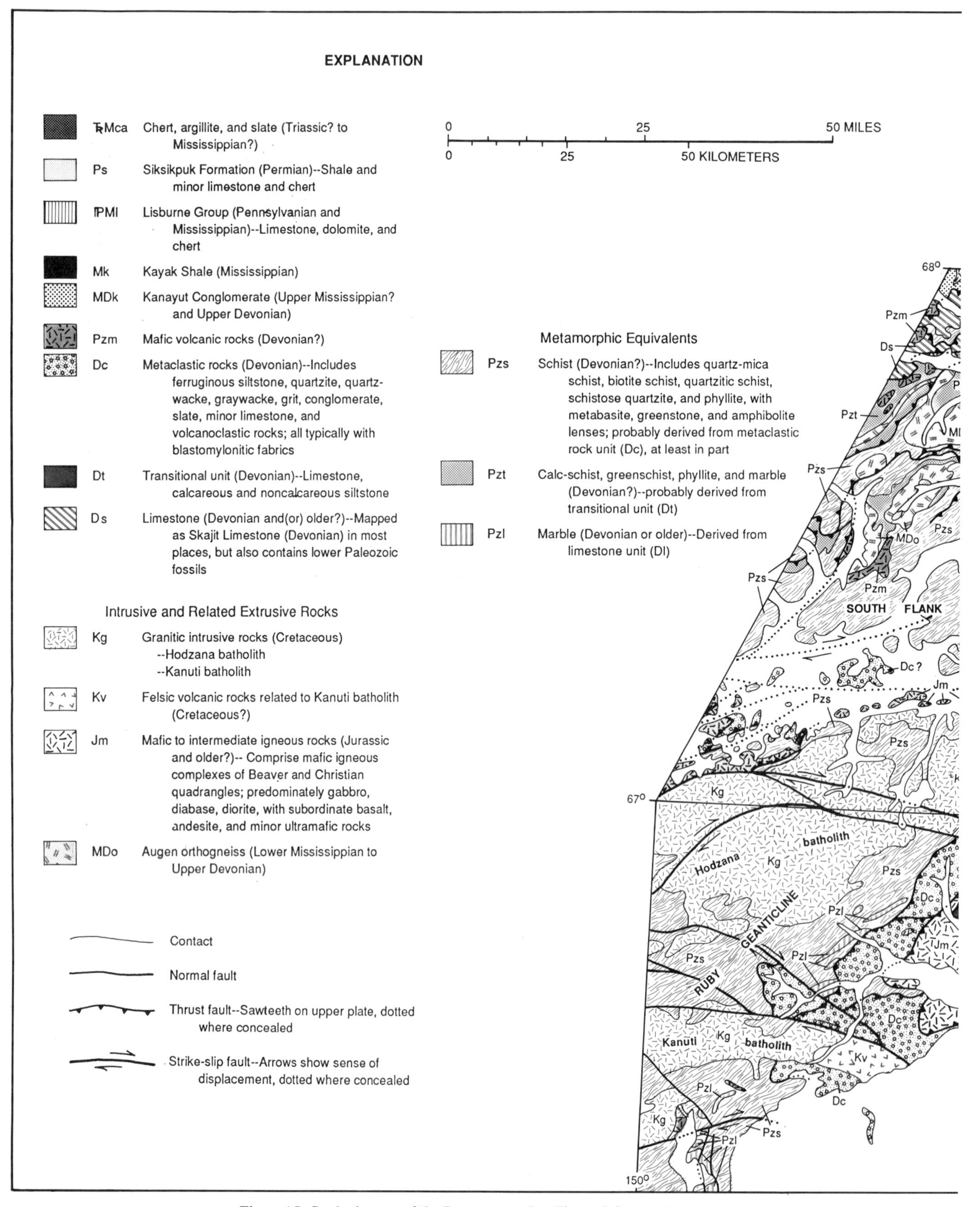

Figure 15. Geologic map of the Beaver area. See Figure 3 for area location.

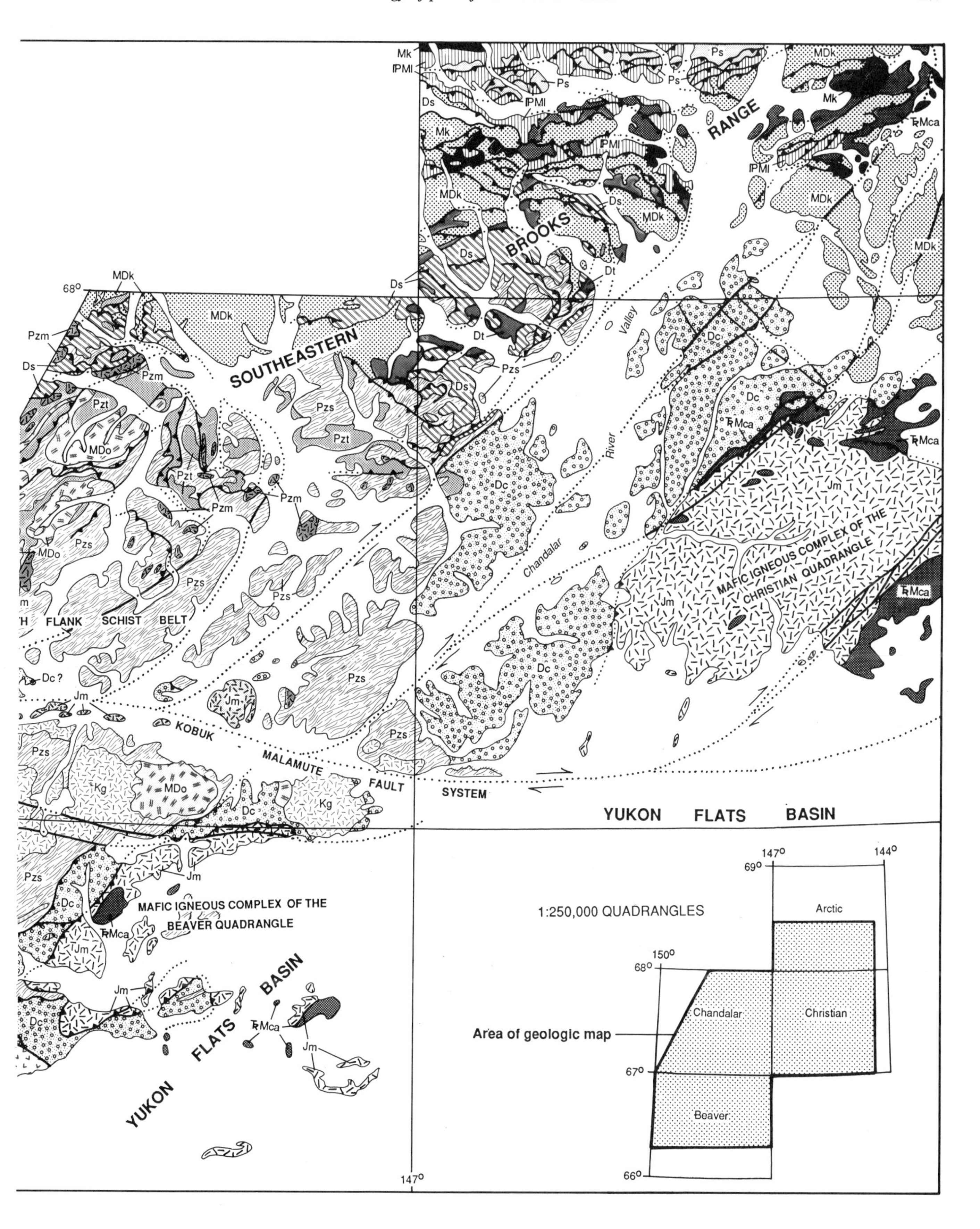
SOUTHEASTERN
BROOKS
RANGE
FLANK SCHIST BELT
KOBUK MALAMUTE FAULT SYSTEM
YUKON FLATS BASIN
MAFIC IGNEOUS COMPLEX OF THE CHRISTIAN QUADRANGLE
MAFIC IGNEOUS COMPLEX OF THE BEAVER QUADRANGLE
Chandalar River Valley
1:250,000 QUADRANGLES
Arctic
Chandalar
Christian
Beaver
Area of geologic map
68°
67°
66°
69°
147°
144°
150°
MDk
Mk
Ds
Dt
Pzs
Pzt
Pzm
MDo
Dc
Jm
Kg
Ps

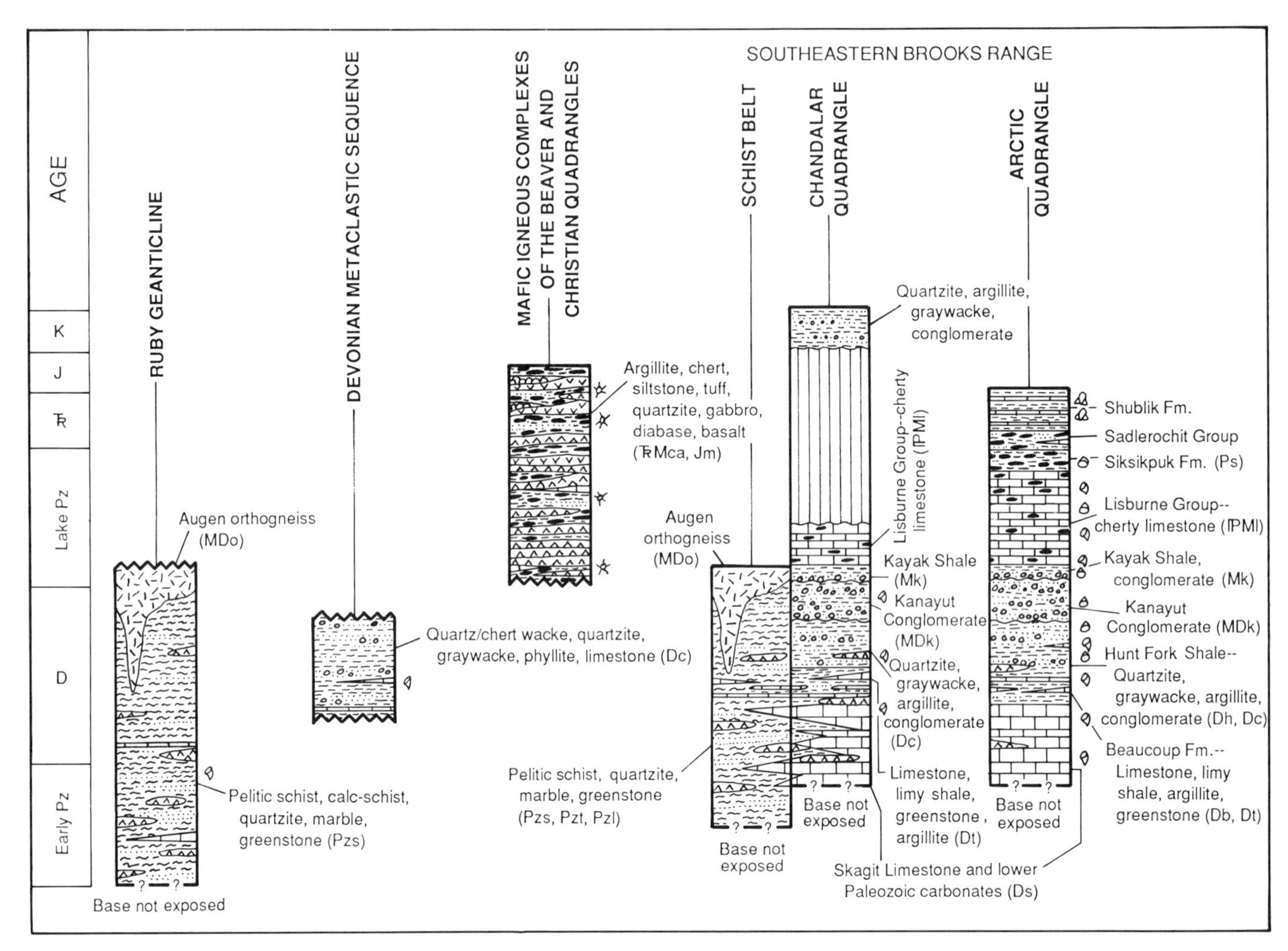

Figure 16. Generalized stratigraphic columns of the Beaver area. See Figure 5 for explanation of lithologic symbols, and Figure 15 for explanation of map units.

Brooks Range, (3) Strangle Woman Creek, and (4) Porcupine River sequences (Figs. 17 and 18).

(1) The Christian River sequence contains a basinal facies of black argillite, sandstone, and varicolored chert of Mississippian to Triassic(?) age, basalt and limestone of Mississippian age, all complexly injected by voluminous gabbro, diabase, and quartz diorite sills and dikes (Fig. 17). The Christian River sequence is contiguous with, and has a similar range of lithic content and age, as the mafic igneous complexes of the Christian and Beaver quadrangles in the Beaver area, and most of the sequence is assigned by Silberling and Jones (1984) to their Tozitna terrane. The lower part of the Christian River sequence differs from the other mafic igneous complexes in containing interbeds of thin, bioclastic, cherty, Mississippian limestone assigned by Brosgè and Reiser (1969) to the Lisburne Group. The basinal aspect of these limestone interbeds contrasts with the shallow-water platformal facies of the Lisburne that characterizes the adjacent (and underlying?) Brooks Range sequence (Endicott Mountains subterrane of Silberling and Jones, 1984). However, gradations between platformal and basinal facies occur in the Lisburne Group of the Brooks Range (Dutro, 1979; Dutro and Jones, 1984). The occurrence of the Mississippian carbonate rocks appears to have led Silberling and Jones (1984) to designate the lower part of the Christian River sequence as a separate Sheenjek terrane distinct from both the Tozitna terrane and the Endicott subterrane (Figs. 17 and 18). Alternatively, the interbedding of Lisburne limestone at the base of a mafic-injected and otherwise Tozitna-like sequence could be interpreted as lithologic intergradation between the Christian River and Brooks Range sequences (and between their equivalent terranes). Unconformably overlying the Mississippian limestone in the lower part of the Christian River sequence are Permian(?) to Triassic(?), calcareous, fine- to medium-grained clastic rocks and chert. In lithology and age, these rocks resemble chert-rich rocks in the Tozitna terrane, as well as chert-bearing rocks of the Siksikpuk Formation and Sadlerochit Group, which overlie the Lisburne Group in other sequences and terranes of the Coleen area.

(2) The Brooks Range sequence is equivalent to the Devo-

nian to Mississippian parts of sequences assigned throughout the Brooks Range by Silberling and Jones (1984) to their Endicott subterrane. The Brooks Range sequence consists of, from oldest to youngest, the Upper Devonian Hunt Fork Shale, Upper Devonian and Lower Mississippian(?) Kanayut Conglomerate, Mississippian Kekiktuk Conglomerate, Mississippian Kayak Shale, and Mississippian and Pennsylvanian Lisburne Group (Fig. 18). In the northwest corner of the Coleen quadrangle, and in places within the Table Mountain quadrangle to the north, a more heterogeneous succession of Devonian(?) limestone and volcanic rocks, and Paleozoic(?) semi-schistose ferruginous sandstone, conglomerate, and calacareous sandstone either underlies or interfingers with the Hunt Fork–Kanayut interval. In parts of the Table Mountain quadrangle, the typically ferruginous and semi-schistose clastic rocks are assigned to the schist unit (Pzp€s) of Brosgè and others (1976). Correlation of these clastic rocks with the late Proterozoic Neruokpuk Formation of Norris (1985), found in the easternmost Brooks Range northeast of the Coleen area, is possible; however, map descriptions of them in the Table Mountain quadrangle are most similar in lithologic character, petrologic details, and mylonitic to semi-schistose fabric to units assigned by Brosgè and Reiser (1962, 1964) and Brosgè and others (1973) to the Devonian metaclastic sequence in the Beaver area to the west.

In the Table Mountain quadrangle, the Lisburne Group of the Brooks Range sequence is overlain by calcareous, fine- to medium-grained clastic rocks like those of the Permian to Triassic Sadlerochit Group but yielding Pennsylvanian as well as Permian fossils. In parts of the Arctic quadrangle to the west, Permian and Triassic rocks of the Sadlerochit Group and Siksikpuk Formation overlying the Lisburne there contain bedded chert (Brosgè and Reiser, 1965).

(3) The Strangle Woman Creek sequence of the northeastern Coleen quadrangle is equivalent to the North Slope subterrane of Silberling and Jones (1984), and contains Mississippian limestone of the Lisburne Group and underlying conglomeratic sandstone (correlative with the Kekiktuk Conglomerate?) that lie discordantly on semi-schistose ferruginous quartzite and siltstone, and phyllitic argillite (Fig. 17); the contact was mapped by Brosgè and Reiser (1969) as an unconformity, but it also appears to mark a metamorphic break and may be a zone of structural detachment as well. The semi-schistose clastic rocks can be traced into those of the Brooks Range sequence in the Table Mountain quadrangle, but they are assigned to the Paleozoic here, based on the occurrence of fossil snails in a float block of the quartzite.

Unconformably overlying the Lisburne Group of the Strangle Woman Creek sequence in the Coleen area are questionably identified Jurassic quartzite and shale. Locally, chert and shale of Permian(?) (and Triassic?) age, and at one locality, calcareous shale, sandstone, and limestone assigned to the Triassic Shublik Formation, are mapped beneath the Jurassic rocks (Brosgè and Reiser, 1969). In the northern Table Mountain quadrangle, the Lisburne Group is overlain by lithologically similar, fine- to medium-grained clastic rocks and limestone of the Sadlerochit Group (Permian and Triassic) and Shublik Formation (Triassic) (Brosgè and others, 1976).

Jurassic(?) mafic rocks like those in the Christian River sequence occur locally within the Strangle Woman Creek sequence. They intrude carbonate rocks of the Lisburne Group in the northeastern part of the Coleen quadrangle, and Permian to Jurassic clastic rocks associated with the Shublik Formation near the contact with the Christian River sequence in the center of the quadrangle.

The Strangle Woman Creek sequence is intruded by the Old Crow batholith, which contains carboniferous granitic rocks yielding K-Ar biotite and muscovite ages of 295 ± 9 to 335 ± 10 Ma (Brosgè and Reiser, 1969), and associated felsic volcanics or metavolcanics. These suggest a middle to late Paleozoic igneous event corresponding to that recognized in the core of the Brooks Range orogen (Dillon and others, 1980), in the Ruby geanticline, and in the Yukon-Tanana upland. However, the Paleozoic age of the Old Crow batholith is in question, because the batholith: (a) appears to discordantly cut the Jurassic quartzite and shale unit; (b) cuts off faults that are interpreted from map patterns to be thrusts, which throughout the east-central Alaska and southeastern Brooks Range region are no older than late Mesozoic; and (c) appears to upgrade incompletely recrystallized semi-schistose rocks of the Strangle Woman Creek sequence to biotite and garnet schist near its contact, a relation reminiscent of the influence of Cretaceous plutons on the Devonian metaclastic sequence of the Beaver area. These contact relations raise the question of whether the Old Crow batholith may have been remobilized during regional heating in the Cretaceous, or whether the batholith might be a Cretaceous one in which the mid-Paleozoic age is inherited.

(4) The Porcupine River sequence is equivalent to the Porcupine terrane of Silberling and Jones (1984). It contains a dominantly carbonate succession in its lower and middle Paleozoic parts (Fig. 18) that is not present in sequences (1) to (3). This succession represents a carbonate platform that is best and most completely developed in the northeastern Black River quadrangle (Brabb, 1970). Sedimentologic and biostratigraphic aspects of the Porcupine platform are discussed by Churkin and Brabb (1968), Brosgè and Dutro (1973), Lane and Ormiston (1979), and Coleman (1985). Abundantly fossiliferous beds of the Lower or Middle Devonian Salmontrout Limestone occur in the lower part of the Porcupine platform sequence. Lower Devonian beds lithologically and biostratigraphically equivalent to part of the Salmontrout Limestone occur in the McCann Hill Chert (Churkin and Carter, 1970) and Ogilvie Formation of Clough (1980) in the Charley River area, and in the Schwatka stratigraphic assemblage of the Livengood area (R. B. Blodgett, written communication, 1987). Carbonate rocks coeval with older and younger parts of the Porcupine platform sequence are represented in the Silurian to Middle Devonian Tolovana Limestone of the White Mountains stratigraphic assemblage in the Livengood area, and in Middle to Upper Devonian and older carbonate beds assigned to the variably metamorphosed Skajit Limestone of the

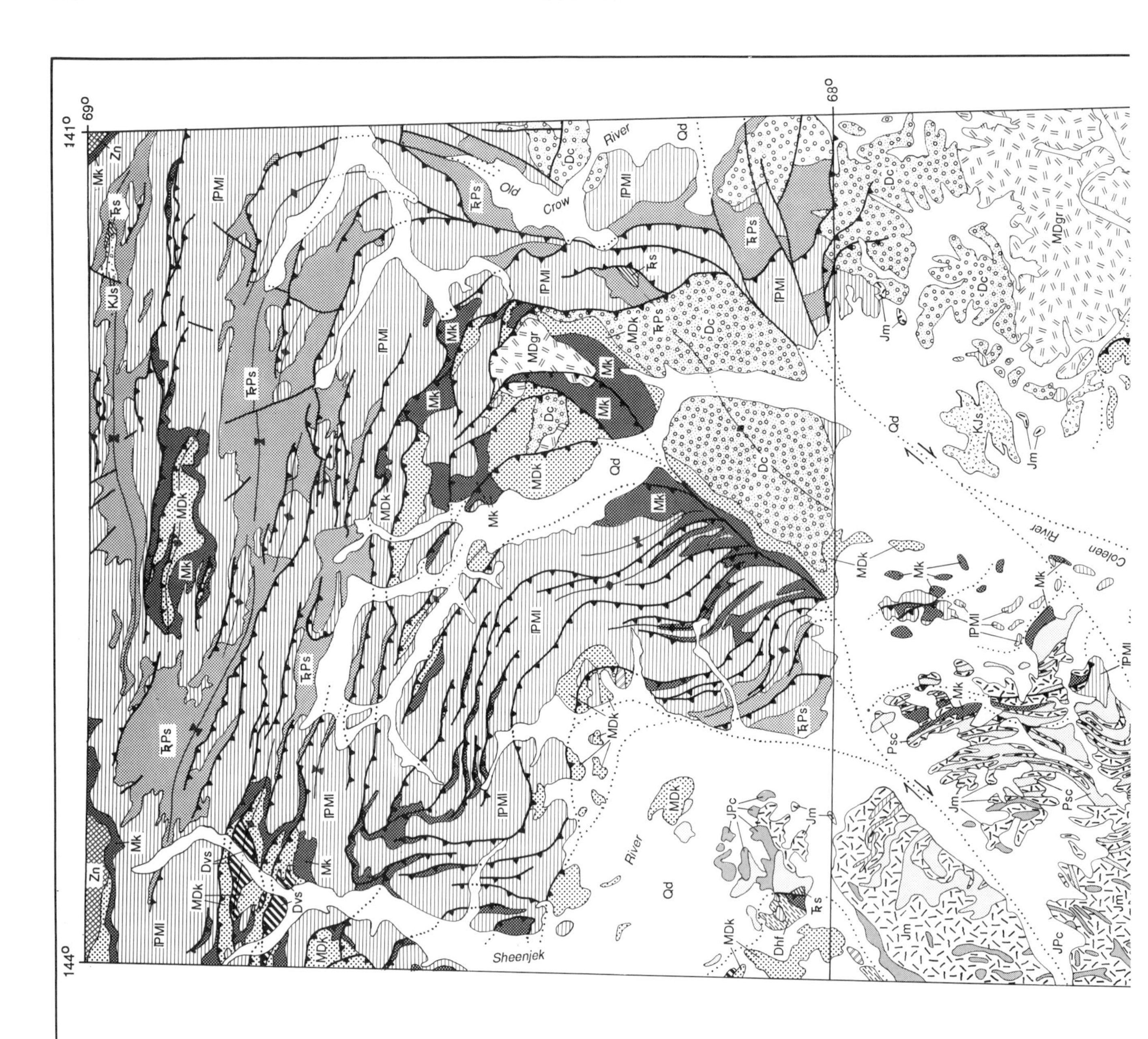
141°
69°
68°
144°
Old Crow River
Sheenjek River
Coleen River
Qd
Dc
MDgr
MDk
Mk
IPMl
TrPs
Trs
KJs
Jm
JPc
Psc
Dvs
Dhf
Zn

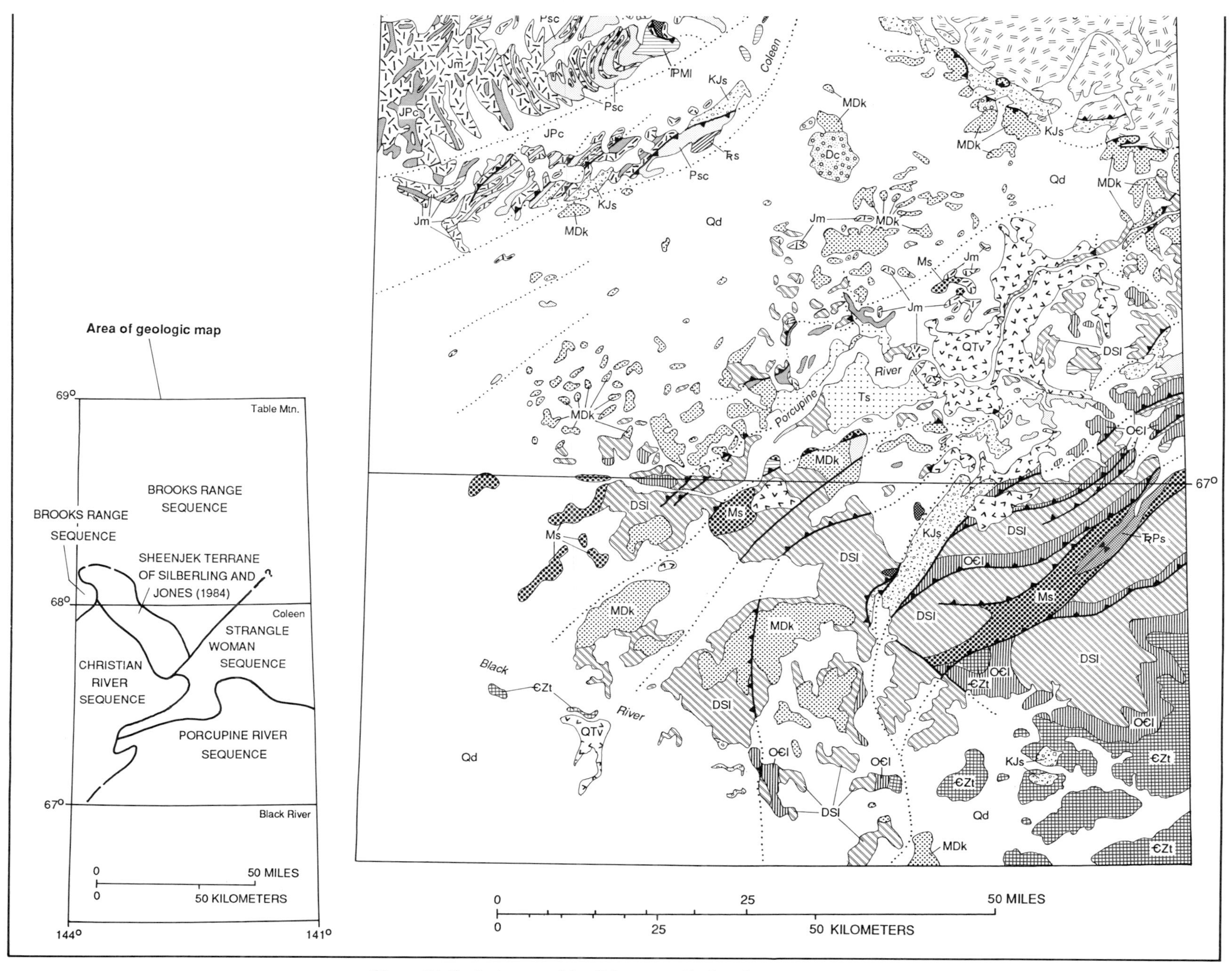

Figure 17. Geologic map of the Coleen area. Explanation on next page.

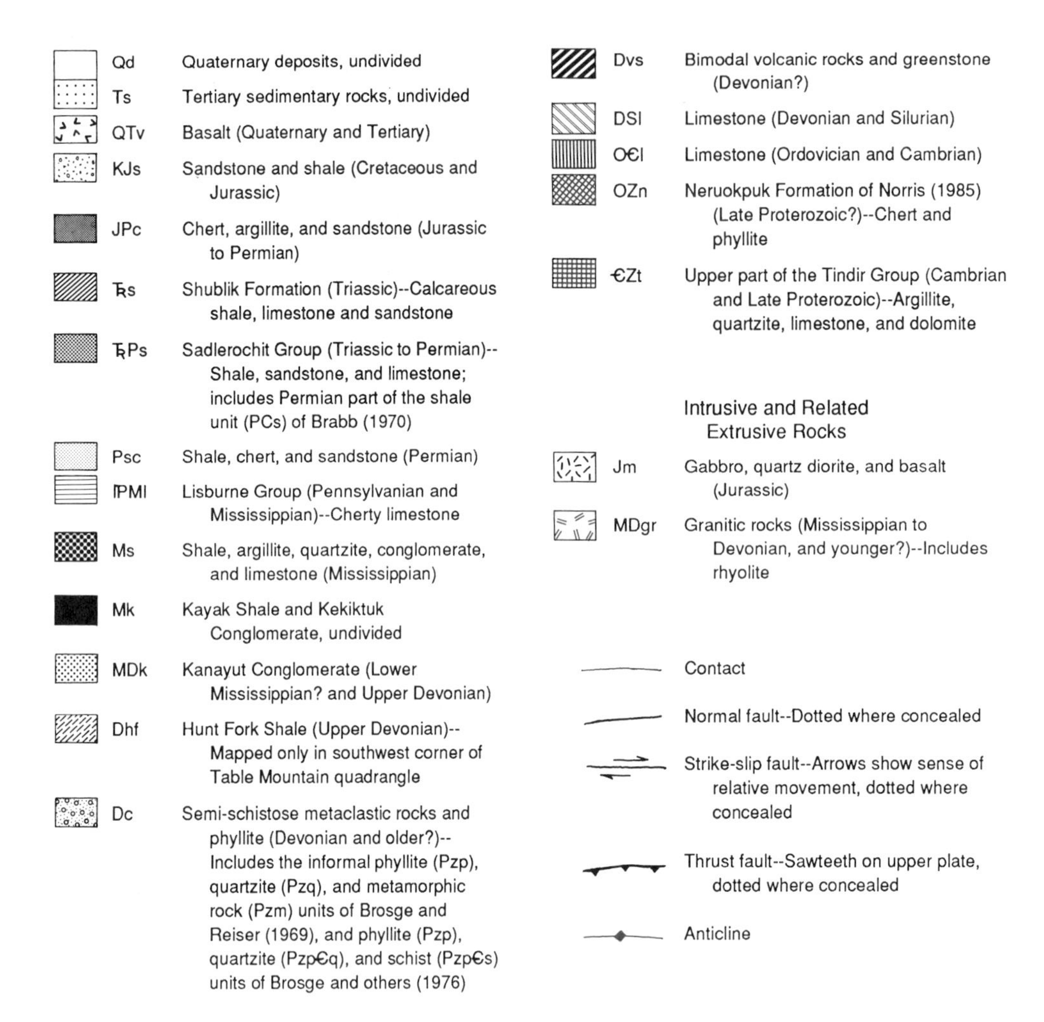

southern Brooks Range (Brosgè and Reiser, 1964; Nelson and Grybeck, 1980).

Above the Upper Ramparts of the Porcupine River, Brosgè and others (1966) traversed a succession of carbonate and clastic rocks originally considered to be Precambrian(?) in age, but which were later assigned to the Paleozoic(?) in the Coleen quadrangle (Brosgè and Reiser, 1969). The pattern of repetition of these units strongly suggests imbricate thrusting of the sequence. The clastic rocks consist of locally blastomylonitic ferruginous quartzite, sandstone, and shale, which track northwestward into a broad area of poor exposures that abuts the Strangle Woman and Christian River sequences. Laterally equivalent clastic rocks in adjacent Yukon Territory are questionably assigned to the late Proterozoic. However, the wide distribution of similar rocks that are locally fossiliferous in the southeastern Brooks Range raises the alternative possibility that these clastic rocks are part of a regional Devonian metaclastic package that unconformably overlies the Paleozoic carbonates and participated in older-on-younger thrusting in the upper Porcupine River area. Imbricate thrusting is also evident from the pattern of stratigraphic repetition in the western part of the Porcupine River traverse of Brosgè and others (1966), where upper Paleozoic units are involved in addition to the middle and lower Paleozoic parts of the section.

Carboniferous and younger strata of the Porcupine River sequence resemble those of sequences (1) to (3), except that the Lisburne Group and rocks of Triassic age are not identified. However, cherty limestone and shale like that characteristic of the Lisburne is present in two sections measured by Brosgè and others (1966); in neither case is the top exposed. The Carboniferous to Jurassic clastic rocks overlie Paleozoic strata of the Porcupine River sequence.

Jurassic(?) mafic rocks appear from map patterns to intrude Devonian carbonate and associated clastic rocks of both late and early Paleozoic age in the Porcupine River sequence. The mafic intrusive rocks are abundant in places, mainly north of the Porcupine River.

Structure. The boundaries of the four stratigraphic sequences of the Coleen area are poorly exposed, but some are shown as faults on existing maps (Brosgè and Reiser, 1969). The map evidence for sequence-bounding faults is equivocal.

The part of the Table Mountain quadrangle containing a thick Lisburne carbonate section and associated shaly units is characterized by a fold-and-thrust belt deformation style that is representative of the central and eastern Brooks Range. Most imbricate thrusts appear to splay from weak shaly zones of structural detachment, the main one being within the Mississippian Kayak Shale. Bundles of generally east-west–trending imbricate thrust faults bend southeastward toward, and apparently merge with, several northeast-trending topographic lineaments in the southern Table Mountain and northern Coleen quadrangles (Fig. 17). Neither the lineaments nor the thrust bundles can be traced beyond their points of mutual convergence. The lineaments appear to track southwestward into the Kobuk-Malamute fault zone of the Beaver area.

Faults interpreted as thrusts in the southeastern Coleen and northeastern Black River quadrangles have northeast trends paralleling those mapped by Norris (1984) in adjacent parts of northern Yukon Territory. The thrusts are cut by high-angle faults with more linear or curvilinear northeast trends and for which normal dip-slip (extensional) displacements are inferred. Among the most prominent of these are the Yukon and Kaltag faults of Norris (1984), which have mapped lengths of 100 to 200 km. Some of the high-angle faults may have a strike-slip component of movement, but none appear to be through-going features that produce any significant lateral offset of preexisting thrusts. They may, however, represent long-lived crustal fractures that intermittently influenced basin formation and sedimentation.

THE GEOLOGY OF EAST-CENTRAL ALASKA IN AN INTERREGIONAL CONTEXT

Most of the geologic features and problems identified in the foregoing summaries either involve the east-central Alaska region as a whole or are interregional in scope. In this section of the report, several key topics are discussed in an interregional context that draws on data and interpretations in adjacent regions of

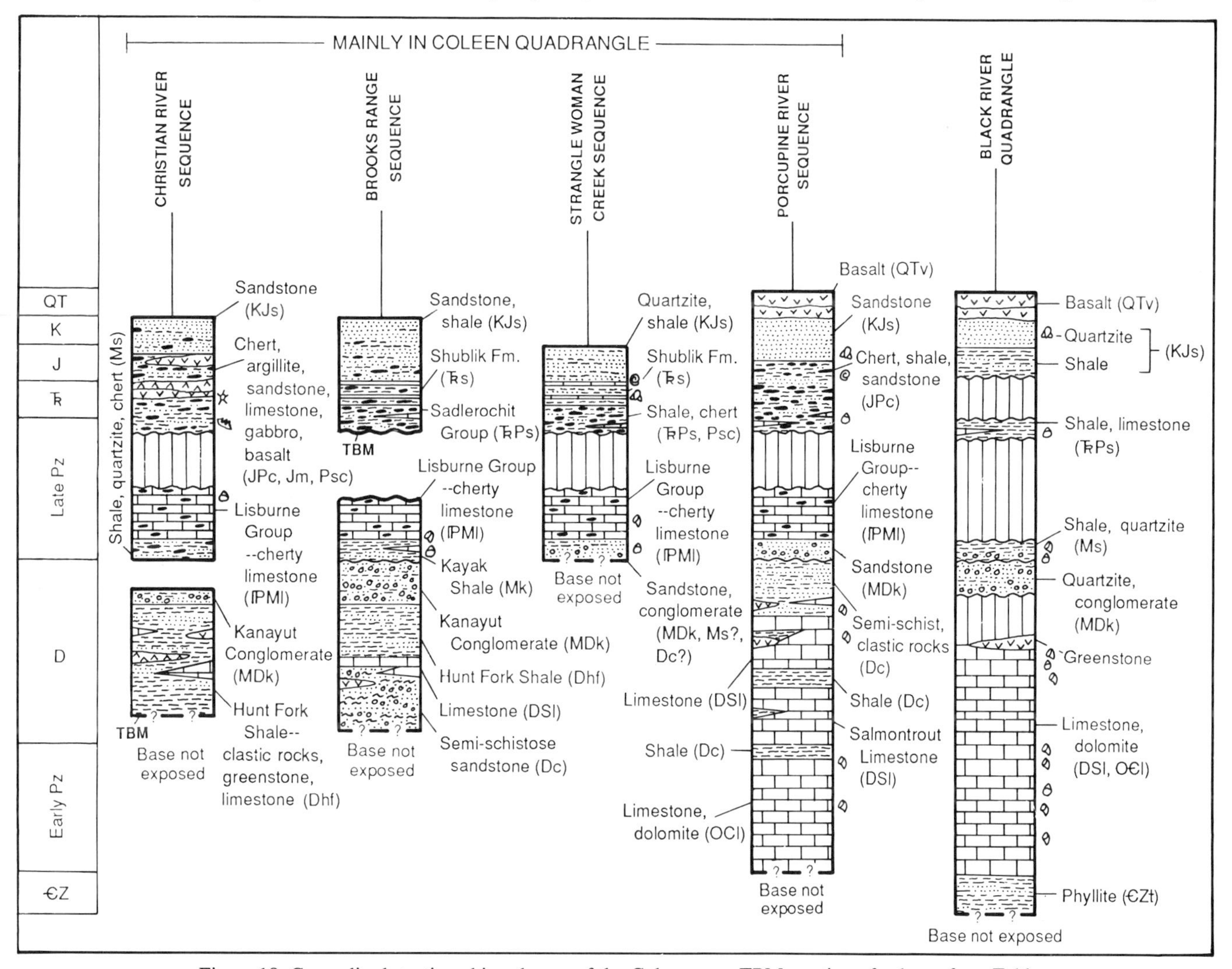

Figure 18. Generalized stratigraphic columns of the Coleen area. TBM, portion of column from Table Mountain quadrangle. See Figure 5 for explanation of lithologic symbols and Figure 17 for explanation of map units.

Alaska and Canada, as well as on the descriptive summaries of east-central Alaska geology.

Terrane subdivisions and paleogeographic setting—a regional stratigraphic perspective

The geology of east-central Alaska has been subdivided into numerous tectonostratigraphic (lithotectonic) terranes in recent publications (see for example, Coney and others, 1980; Jones and others, 1981, 1983, 1987; Dutro and Jones, 1984; Howell and others, 1987; Silberling and others, this volume). An accretionary[2] origin was proposed for some east-central Alaskan terranes by Churkin and others (1982). There is no question that east-central Alaska has undergone severe structural disruption. Most terrane boundaries here are either thrust faults or strike-slip faults. In evaluating the significance of these terranes, the most important questions are: (1) whether any of the terranes represent accretionary sequences that originated beyond the North American depositional margin, and whether any of the terrane-bounding structures represent fundamental sutures; or alternatively, (2) whether the terranes are compatible within a reasonably reconstructed North American paleostratigraphic framework.

In considering the stratigraphic framework of east-central Alaska, a continuity of generally coeval stratigraphic packages is more apparent for the region as a whole than is obvious from the study of stratigraphic details in local areas. Despite the stratigraphic differences emphasized by tectonostratigrahic terrane subdivisions, many threads of stratigraphic continuity and similarity are apparent from area to area and terrane to terrane in the foregoing geologic summaries. The similarities between many terranes are at least as impressive as the differences, and these similarities are of critical significance in evaluating relations among terranes. Furthermore, it seems clear that stratigraphic differences caused by facies variations and unconformities within some individual terranes are as great as those between terranes (e.g., see Dover, 1990; Brosgè and Dutro, 1973; Thompson and Roots, 1982; Dutro and Harris, 1987).

Correlation and North American affinity of rocks and terranes in the Charley River and Livengood areas. *Eastern Charley River area.* The lithologic content and sequence of stratigraphic units in the eastern part of the Charley River fold-and-thrust belt are remarkably similar to those of the middle Proterozoic to Triassic North American miogeoclinal succession exposed farther east in Canada. There are some obvious stratigraphic differences between the Charley River area and adjacent Yukon Territory (Jones and others, 1987), and facies variations are locally prominent, as should be expected for any depositional realm as extensive and complex as the Cordilleran miogeocline. Nevertheless, numerous key units of the eastern Charley River fold-and-thrust belt (Fig. 5) can be correlated with units in Yukon Territory of unquestioned North American origin (also see Tempelman-Kluit, 1984). Among these correlations are:

(1) interfingering shale and stromatolitic limestone of the middle Proterozoic lower part of the Tindir Group (Eberlein and Lanphere, 1988), with the upper part of the Wernecke Supergroup of east-central Yukon Territory (Delaney, 1981);

(2) basaltic volcanic and volcaniclastic rocks and associated rocks of the late Proterozoic and/or Lower Cambrian upper part of the Tindir Group, with the Mount Harper volcanic complex of Thompson and Roots (1982), Roots and Moore (1982), and Roots (1983, 1988) in the Dawson quadrangle;

(3) graptolitic shale and chert of the Lower Devonian to Lower Ordovician Road River Formation, with the Road River Formation and correlative rocks of Selwyn basin and northern Yukon Territory (see discussion in Cecile, 1982, p. 19–23);

(4) Ogilvie Formation of Clough (1980) and coeval Lower to Middle Devonian limestone, with the Ogilvie Limestone of the Dawson quadrangle (Clough and Blodgett, 1984);

(5) thick, quartzite- and chert-rich, coarse clastic rocks of the Nation River Formation (Upper Devonian), with the Earn Group, thought by Gordey and others (1987) to be derived from local uplifts of lower Paleozoic quartzite- and chert-bearing rocks within the outer part of the Cordilleran miogeocline, and with the Imperial Formation of northern Yukon Territory (Gordey and others, 1987);

(6) Ford Lake Shale (Upper Mississippian to Upper Devonian), with strata assigned to the same unit in central Yukon Territory (Norris, 1984);

(7) Calico Bluff Formation (Lower Pennsylvanian and Upper Mississippian), with the Hart River Formation (Bamber and Waterhouse, 1971);

(8) Permian Tahkandit Limestone and Step Conglomerate, with the Tahkandit Limestone and Jungle Creek Formation of central Yukon Territory (Norris, 1984); and

(9) Kandik Group (Lower Cretaceous) and upper part of the Glenn Shale (Jurassic), with coeval foreland basin deposits of northern Yukon Territory and the north slope of the Brooks Range (Young, 1973; Norris, 1984).

Western Charley River area. The regional correlation and paleogeographic significance of rocks in the terranes proposed by Silberling and Jones (1984) and Churkin and others (1982) for the western Charley River area (Fig. 5) are less obvious. Churkin and others (1982) interpreted the Woodchopper Canyon, Slaven Dome, and Circle terranes as oceanic fragments that were accreted to North America. Their interpretation was based on the basinal character of the middle to upper Paleozoic fine-grained siliciclastic rocks and cherts, and on the abundance of associated mafic volcanic and plutonic rocks. Alternatively, I suggest that all of these rocks originated in rifts or extensional basins within the outer part of the middle to late Paleozoic Cordilleran miogeocline and were imbricated with coeval carbonate and clastic strata by thrusting within the Charley River fold-and-thrust belt.

[2]The term "accretion" as used in this chapter means continental growth by the addition of "exotic" material that originated external to the craton and its pericratonal depositional prism, usually by convergent plate-tectonic processes. This usage excludes the juxtaposition of continental material tectonically dispersed from one part of a continental margin to another, either by strike-slip faulting or be telescoping on thrust faults.

The stratigraphic evidence supporting a North American miogeoclinal origin for the western Charley River terranes is that all of them, including those with oceanic-like rocks, contain units similar in lithology and age to units of North American affinity in the eastern part of the Charley River area and in Yukon Territory. The Takoma Bluff terrane of Churkin and others (1982) contains poorly dated late Proterozoic rocks similar to those in the upper part of the Tindir Group, and clastic rocks resembling the Upper Devonian Nation River Formation of the eastern Charley River sequence. Black carbonaceous to siliceous argillite and chert of late Paleozoic age in the Circle terrane of Churkin and others (1982) and the corresponding Tozitna terrane of Silberling and Jones (1984) are lithologically similar to parts of the Ford Lake Shale (Upper Mississippian to Upper Devonian) and Calico Bluff Formation (Lower Pennsylvanian and Upper Mississippian) of the eastern Charley River area, and to fine-grained parts of the Earn Group (Gordey and others, 1987). Additional correlations with rocks of central Yukon Territory suggested by Tempelman-Kluit (1984) are: (1) the Woodchopper Canyon terrane of Churkin and others (1982) with the Marmot Volcanics of Cecile (1982) in the Misty Creek embayment of Selwyn basin; and (2) chert-pebble conglomerate and associated clastic deposits of the Slaven Dome terrane of Churkin and others (1982) with the Upper Devonian Nation River Formation in the eastern Charley River fold-and-thrust belt, and with the Earn Group (Gordey and others, 1987). Major differences in the amount of local downcutting by unconformities and the abruptness of facies variations characterizing parts of the eastern Charley River stratigraphic succession, combined with the lack of evidence for compressional deformation of pre-Mesozoic age, indicate that late Proterozoic to Permian deposition was influenced by intermittently active pre-thrusting high-angle faults that controlled basin subsidence and block uplift of local source areas (Gordey and others, 1982, 1987; Dover, 1992). A late Proterozoic and Paleozoic extensional environment like that envisioned for the Canadian Cordillera by Thompson and Eisbacher (1984), Thompson and others (1987), Gabrielse and Yorath (1987), and Struik (1987) would be expected to produce the kinds of contrasting rock sequences that are structurally juxtaposed in the "terranes" of the western Charley River area. Rocks with oceanic affinity are inferred to have formed in the most dilated of the extensional basins.

The structural evidence against an accretionary origin for any of the terranes of the western Charley River area is that the style of the thrust faults that bound them, and the degree of internal deformation and shortening within them, are comparable to the style and intensity of deformation characterizing thrust-bounded sequences in the eastern part of the Charley River fold-and-thrust belt, where individual thrusts are interpreted to have displacements of 10 km or less. None of the western Charley River area thrusts, either individually or as a group, appears to be a suture in the sense required by the accretionary model. Conversely, the existence of the Charley River fold-and-thrust belt provides a ready explanation for facies telescoping without invoking accretionary tectonics.

Livengood area. As the geological summary of the Livengood area illustrates (Figs. 8 and 9), numerous lithologic similarities exist among the stratigraphic belts (and corresponding terranes designated by Silberling and Jones, 1984) recognized there. Six of the 11 stratigraphic belts have a basement composed of the Wickersham grit unit. These are the Wickersham, White Mountains, Livengood, Schwatka, and Crazy Mountains belts, and the Preacher block. At least five of the stratigraphic belts contain chert-rich, locally derived polymictic conglomerates of probable middle to late Paleozoic age. These are the White Mountains, Beaver, Livengood, and Crazy Mountains belts, and the Minook terrane. Basaltic volcanic and locally differentiated mafic intrusive rocks, thought to represent volcanic/plutonic centers of at least three distinct ages, occur in several of the stratigraphic belts of the Livengood area. The heterogeneous and mafic/ultramafic unit of the Livengood stratigraphic belt is interpreted as a late Proterozoic to Cambrian mafic volcanic/plutonic center comparable to those in extensional intracontinental North American settings farther east, such as the upper part of the Tindir Group and the Mount Harper complex (see Roots, 1988). Another center, represented by the Fossil Creek Volcanics of Ordovician age in the White Mountains stratigraphic belt, probably correlates with mafic volcanic rocks in unit II of Thompson and Roots (1982) in the Dawson quadrangle, and with the more sparsely volcanic Livengood Dome Chert in the Livengood stratigraphic belt (Fig. 9). The third mafic volcanic/plutonic center consists of the volcanics of probable Early to Middle Devonian age in the Schwatka stratigraphic belt. These most likely correlate, at least in part, with the Woodchopper Volcanics of the Charley River area.

No tectonic melange, relict accretionary prism, or other evidence of crustal suturing is evident in any of the stratigraphic belts of the Livengood area. On the contrary, the thrusts bounding these belts, and their styles of internal folding and imbrication, resemble fold-and-thrust belt styles in the Charley River area and adjacent Yukon Territory. Most of the stratigraphic belts (and equivalent "terranes") of the Livengood area seem best interpreted as thrust-bounded sequences of North American origin, whose lithologic differences resulted from facies telescoping of limited extent, analogous to that in the Charley River fold-and-thrust belt. The thrusted sequences were further segmented and dispersed by later strike-slip faulting. Ductility contrasts and differences in metamorphic character between some stratigraphic belts may have resulted from the juxtaposition of different crustal levels by fold-and-thrust belt telescoping and strike-slip movement.

Regional geologic ties between the Ray Mountains and Beaver areas, the southeastern Brooks Range, and adjacent areas. *Metamorphic rocks.* The protolithic model for the metasedimentary rocks and metabasites in the metamorphic suite of the Ray Mountains (Fig. 14) infers deposition and mafic intrusion in an intermittently extending, distal part of a depositional wedge like that characterizing the late Proterozoic through middle Paleozoic Cordilleran miogeocline. These metasedimentary rocks appear to be representative of the Ruby terrane of Silberling and Jones (1984) as a whole. Granitic plutons generated from continental crust were emplaced in the sedimentary protolith in Devonian time, prior to the culminating deformational, metamorphic, and plutonic events in Jurassic to mid-Cretaceous time.

The rocks of the Ruby terrane resemble those of the schist belt of the southern Brooks Range (and its terrane equivalents). A direct connection between the two has long been suspected but not proved (Patton and others, 1987). Rocks of the schist belt described by Dillon and others (1980, 1985, 1986), Nelson and Grybeck (1980), Karl and Long (1987), Gottschalk (1987), and Till and others (1987) resemble those in the metamorphic suite of the Ray Mountains in lithic content, protolith, fabric and structural sequence, and the occurrence of Upper Devonian to Lower Mississippian augen orthogneiss. An older episode of Precambrian metamorphism is recognized locally in the western Brooks Range (Turner and others, 1979); probable Precambrian metamorphic rocks also occur locally in the Ruby terrane southwest of the Ray Mountains (Silberman and others, 1979b). The Ruby terrane and schist belt differ in that the part of the Ruby terrane north of the Kaltag fault has a vastly greater volume of mid-Cretaceous granitic rocks than the schist belt, where Cretaceous plutonism has not been demonstrated (Miller, chapter 19, this volume); Cretaceous plutonism is also relatively sparse in the Ruby terrane south of the Kaltag fault.

All of the lithic components of the Ray Mountains metamorphic suite also have close counterparts within the Yukon-Tanana upland (Fig. 19; Foster and others, 1987, and this volume). Comparable blastomylonitic fabrics, deformational style and sequence, and paragenetic relations of polymetamorphic mineral assemblages have been described in the Yukon-Tanana upland by Laird and Foster (1984), Cushing (1984), Cushing and Foster (1984), Cushing and others (1982), Dusel-Bacon and Foster (1983), and Foster and others (1987, and this volume). Middle Devonian to Early Mississippian augen orthogneiss (Dusel-Bacon and Aleinikoff, 1985) and mid-Cretaceous granitic rocks of the Yukon-Tanana upland are indistinguishable from those in the Ray Mountains. Some siliciclastic rocks of the Yukon-Tanana upland that are as young as Permian and contain mafic/ultramafic rocks (Fig. 19) have no recognized counterparts in the metamorphic suite of the Ray Mountains, but less metamorphosed lithologic equivalents of that age occur in the Rampart assemblage.

Devonian metaclastic sequence. The variably metamorphosed Devonian clastic rocks along the southeast side of the Ruby geanticline represent a particularly important tie between the Ray Mountains area and parts of the southeastern Brooks Range in the Beaver and possibly the Coleen areas. These typically mylonitic to semi-schistose Devonian metaclastic rocks were originally correlated by Brosgè and Reiser (1962, 1964) throughout the area that is now subdivided into the Venetie, Coldfoot, and Hammond subterranes by Silberling and Jones (1984). The lithologic grounds for the original correlations are still valid, and metamorphic variations within the metaclastic unit are gradational in most places, and cross subterrane boundaries. Furthermore, the metaclastic sequence contains all the lithic components, including quartz- and chert-rich conglomerates and black argillaceous rocks, that characterize the middle Paleozoic part of the Brooks Range sequence in the Endicott Mountains subterrane of Silberling and Jones (1984), which contains the Upper Devonian Hunt Fork Shale, Upper Devonian and Lower Mississippian(?) Kanayut Conglomerate, Lower Mississippian Kekiktuk Conglomerate, and Mississippian Kayak Shale. The lithologically similar Earn Group of the northern Canadian Cordillera has a range of Devonian to Mississippian age encompass-

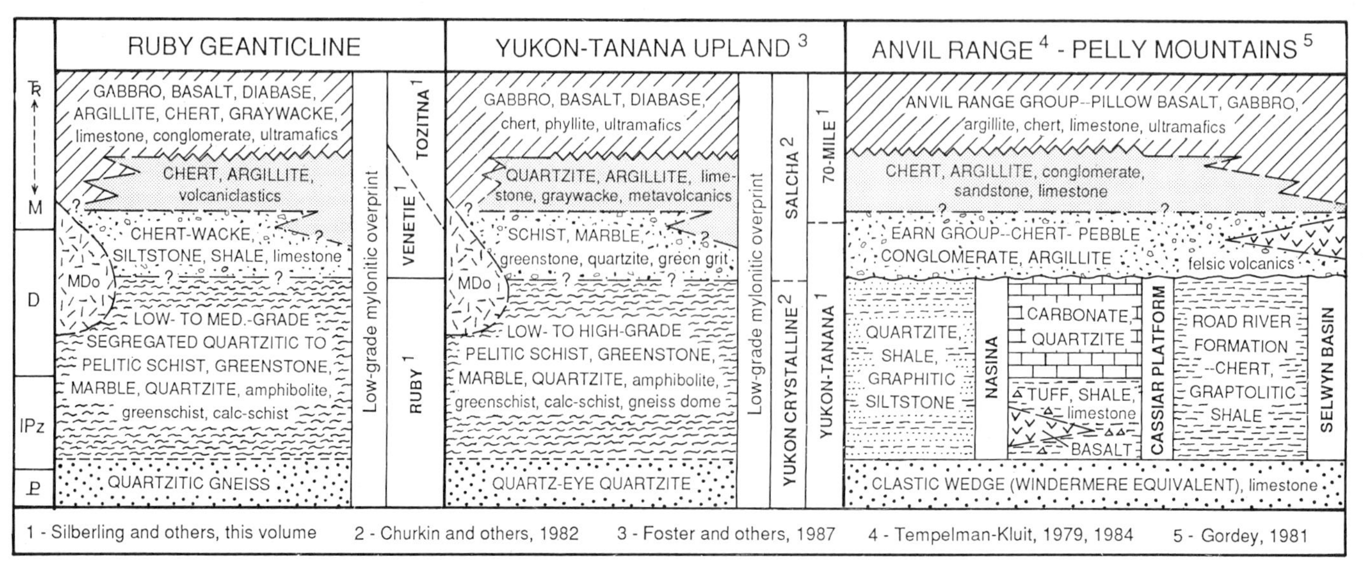

Figure 19. Comparison and correlation of stratigraphic sections of the Ruby geanticline and Yukon-Tanana upland with central Yukon Territory. MDg, Mississippian to Devonian augen orthogneiss.

ing that of the Devonian metaclastic sequence, as well as that of the Nation River Formation and the Cascaden Ridge unit in several sequences and terranes of the Charley River and Livengood areas. All of these clastic rocks can be considered as parts of a regionally developed but locally variable sequence of Devonian clastic rocks that overlapped the outer part of the Cordilleran miogeocline as it underwent intermittent fault-controlled basin subsidence and block uplift (Gordey and others, 1987).

If the semi-schistose clastic rocks of the Coleen area are indeed Devonian, then the stratigraphic similarities between the Brooks Range and Strangle Woman River sequences (and the equivalent terranes of Silberling and Jones, 1984) are compelling, and their differences can be attributed to minor facies variations and unconformities, just as comparable differences have been explained in the east-central Alaska region. The likelihood that the Sheenjek terrane of Silberling and Jones (1984) represents lithologic interfingering between basinal sediments of the mafic-rich Christian River sequence and a basinal facies of Lisburne limestone (Mississippian and Pennsylvanian) of the Brooks Range sequence bears on the origin of the Tozitna terrane, discussed below.

Other correlations involving units of the Coleen area. The Porcupine River sequence differs stratigraphically from the Christian River, Brooks Range, and Strangle Woman Creek sequences in some important ways, but it does not appear to be separated from them by a major structural discontinuity. Its boundary is irregular, and available mapping reveals no single through-going structural zone, such as the hypothetical "Porcupine lineament" (Churkin and Trexler, 1980, 1981), separating it from the others. The "Kaltag" fault of adjacent Yukon Territory (Norris, 1984) projects into the Porcupine sequence, and is not a bounding mega-structure along its north edge. Significantly, semi-schistose clastic rocks like those elsewhere in the Coleen area appear to overlie the lower Paleozoic carbonate section of the Porcupine sequence. If so, and if all of these clastic rocks are Devonian, then lower Paleozoic platform carbonates equivalent to those of the Porcupine sequence could be present in the subsurface under the clastic unit in the other sequences of the Coleen area.

Heterogeneous volcanic-bearing Devonian rocks in the Brooks Range sequence of the Coleen area appear to have lithologic equivalents in volcanic-bearing parts of the schist belt farther west, in the southern Wiseman and Chandalar quadrangles (Dillon and others, 1986; Brosgè and Reiser, 1964). These not only resemble rocks of the Ambler sequence of the southwestern Brooks Range, for which a rift origin is inferred (Hitzman and others, 1982, 1986), but also have lithologic counterparts farther north in the Brooks Range (Dutro and others, 1977), in the Woodchopper Volcanics and the Schwatka belt of the Charley River and Livengood areas, and in parts of the northern Cordilleran miogeocline in southern Yukon Territory (Mortensen and Godwin, 1982). The age and character of sediment-hosted volcanogenic mineral deposits in rift-controlled sequences of the Ambler district of the southwestern Brooks Range are also similar to deposits in comparable settings of the Selwyn basin in southern Yukon Territory (Einaudi and Hitzman, 1986).

Skajit carbonate rocks in the southern Brooks Range contain a few fossils, mainly of Devonian age, but rocks mapped as the Skajit Limestone locally contain fossils as old as Middle Ordovician (Nelson and Grybeck, 1980; J. T. Dutro, written communication, 1987). The thickness and distribution of the Skajit along the 660-km length of the southern Brooks Range suggest that it formed a regionally extensive but relatively narrow carbonate platform. Tectonic and erosional remnants of at least partly coeval and biostratigraphically similar carbonate platforms are represented in east-central Alaska by the Porcupine platform of the Coleen area, the Tatonduk sequence of the Charley River area, and the Nixon Fork platform (Blodgett, 1983) in southwestern Alaska. The Cassiar platform (Tempelman-Kluit, 1977) in southern Yukon Territory is analogous. Biostratigraphic similarities demonstrate the close paleogeographic ties among most of these terranes, and with the North American miogeocline (Blodgett, 1983; Savage and others, 1985). Furthermore, some much thinner, Lower to Middle Devonian carbonate and calcareous shale horizons containing similar faunas occur within more heterogeneous, clastic to volcanic-bearing, probably rift-controlled assemblages of east-central Alaska, suggesting that the platform carbonates may have interfingered with the rift assemblages (Brosgè and others, 1979). The best examples are the Cascaden Ridge unit and the Schwatka sequence of the Livengood area, and the Woodchopper Volcanics of the Charley River area (Lane and Ormiston, 1976; Blodgett, 1987).

The Mississippian Kekiktuk Conglomerate lies in regional unconformity on Devonian and older rocks. The unconformity is profoundly angular in some places, and a discontinuity elsewhere, suggesting an environment of localized Devonian and Mississippian tectonism comparable to that interpreted for the Earn Group in central Yukon Territory (Gordey and others, 1982, 1987). The Kekiktuk and the Kayak Shale are included within the Endicott Group, which is mapped within the northern Yukon extension of the Cordilleran fold-and-thrust belt.

Skeletal, micritic, and cherty limestone and dolomite of Lisburne age and character continue southeastward into the northern Yukon segment of the Cordilleran fold-and-thrust belt, where they are mapped as the Hart River and Ettrain Formations (Norris, 1984). Correlation of the Lisburne Group with the Hart River and Ettrain Formations would imply that late Paleozoic depositional environments were similar, if not continuous, between the eastern Brooks Range and the northern Cordilleran miogeocline. Subtle unconformities within the Lisburne have been related by Schoennagel (1977) to unconformities recording Antler orogenic pulses in the western United States.

Permian clastic lithologies of the Siksikpuk Formation and Sadlerochit Group resemble those of the Step Conglomerate and Tahkandit Limestone of the Charley River area, and the Jungle Creek Formation of northern Yukon Territory (Norris, 1984). The Triassic Shublik Formation is similar in lithology, age, and petroleum source-bed potential to the lower part of the Glenn Shale in the Charley River region (D. Morgridge, oral communication, 1987).

Origin of the Tozitna terrane. Mississippian to Triassic (or Jurassic) argillite, chert, and other sedimentary rocks in the Tozitna terrane of Silberling and Jones (1984) differ both in their basinal character and in their association with voluminous mafic igneous rocks from most of the coeval sequences in adjacent terranes of east-central Alaska (Fig. 20). Two fundamentally different interpretations have been proposed for the origin of the Tozitna terrane.

One interpretation is that the Tozitna originated as oceanic crust beyond the North American depositional margin, and was then transported and obducted onto its present substrate (Churkin and others, 1982; Patton and others, 1977; Coney, 1983; Coney and Jones, 1985). Opinions differ as to the direction of transport and the location of the root zone. Patton and others (1977) suggest that the terrane originated in the Koyukuk basin (Fig. 20) and was transported southeastward from a root zone now represented by the dismembered Kanuti ophiolite. Coney (1983) and Coney and Jones (1985) suggest an origin from south of the Yukon-Tanana upland (Fig. 20) involving generally northward transport. In either case, this accretionary model requires hundreds of kilometers of semi-coherent, unidirectional tectonic transport of a relatively thin and regionally extensive sheet across continental crust.

The other interpretation is that the Tozitna terrane originated locally as a parautochthonous intracratonic rift assemblage in which oceanic crust was generated (Gemuts and others, 1983). In this parautochthonous rift interpretation, collapse of the rift sequence and facies-telescoping at its margins during subsequent Mesozoic convergence produced ductilely sheared, bedding-controlled detachments and conventional fold-and-thrust belt structures that require much smaller distances of transport than does the accretionary model.

Geophysical data for the Tozitna terrane discussed by Cady (1987) are not diagnostic for either of these models, and paleomagnetic evidence bearing on the origin and amount of tectonic transport of the Tozitna terrane is sparse and inconclusive because of alteration, regional reheating, and complex structural relations.

The Tozitna problem is not a provincial one unique to east-central Alaska, but is interregional in scope. Rock sequences similar in lithic content, range of age, and deformational character to the Tozitna terrane are recognized the length of the Cordilleran orogen from Alaska to Nevada. Those sequences in east-central Alaska and the northern Canadian Cordillera for which accretionary origins have previously been suggested are the Seventymile (Foster and others, this volume); the Sylvester, Nina Creek, and Slide Mountain (Monger, 1977); and the Anvil allochthons (Tempelman-Kluit, 1979, 1984; Gordey, 1981). The equivalence of these sequences emphasizes the vast extent and continuity on an interregional scale of the transported sheet that would be required by any model of obductive accretion, and compounds the mechanical difficulty of transporting more or less coherently so extensive a sheet across hundreds of kilometers of continental crust.

Aside from the mechanical problem presented by such large-scale obduction, the geologic framework of east-central Alaska, as summarized in this chapter, best supports the collapsed intracratonic rift model to explain the origin of the Tozitna terrane. Evidence leading to this interpretation is:

(1) Tozitna rocks no older than Early Mississippian to latest Devonian occur interregionally on a substrate no younger than Late Devonian. Although the rocks of the Tozitna terrane and its substrate are typically imbricated with one another near their contact, the interregional extent and consistency of the stratigraphic stacking order of the Tozitna and its substrate are difficult to reconcile with any model of far-traveled accretion, and argue strongly for an originally depositional contact between the two (see Monger and Ross, 1979).

(2) Inconsistencies in the direction of tectonic transport, indicated by shear fabrics in mylonites at the base of the Tozitna terrane, are difficult to reconcile with unidirectional allochthonous transport of the magnitude required. Preliminary fabric studies in the Ray Mountains and southern Tanana quadrangle suggest movement generally upward and outward over crystalline terranes lying outboard from the Tozitna terrane (Fig. 20).

(3) Local stratigraphic ties exist between the Tozitna terrane and adjacent or underlying rock sequences. Possible stratigraphic interfingering is present in the Christian River sequence of the Coleen area, in the Yukon-Tozitna Rivers part of the Ray Mountains area, and in the western Charley River area. The interfingering is analogous to that already proposed between platform and rift-controlled sequences of the outer Cordilleran miogeocline that were active intermittently from late Proterozoic through Paleozoic time.

(4) Similarities in mafic igneous associations with other suspected rift sequences in the outer part of the Cordilleran miogeocline. The petrologic similarities of all of these occurrences suggest similar modes and sites of origin. The volcanic rocks of late Proterozoic to Early Cambrian(?) age in the upper part of the Tindir Group, the Devonian Woodchopper Volcanics, and the Tozitna terrane are lithologically indistinguishable from the Ordovician Fossil Creek Volcanics (Wheeler-Crowder and others, 1987), and from those in the Mount Harper complex (Roots, 1988), whose highly alkalic tholeiitic composition suggests an extensional, continental rift origin. The predominance of intrusive rocks in the Tozitna terrane may be a function of the degree of basin opening and tensional spreading (see Struik, 1987). Occurrences of mafic intrusive rocks resembling those of the Tozitna terrane are also mapped locally in adjacent rock sequences within the Ray Mountains, Beaver, and Coleen areas, suggesting that the mafic rocks are not "rootless" within a gigantic Tozitna allochthon, but also penetrated its substrate near the Tozitna basin margins (see Monger and Price, 1979).

(5) The possible North American affinity of faunas. The paleogeographic affinities of radiolarians and the sparse conodont

and shelly faunas characterizing the Tozitna terrane are not well known. However, in the northern Canadian Cordillera, terranes analogous to the Tozitna in lithic content and geologic setting contain a fusulinid assemblage comparable to that in the southwestern United States (Monger and Ross, 1971; Struik, 1981). This suggests fragmentation and dispersion of North American rocks by major strike-slip, but not an origin external to North America.

In combination, these factors strongly support the conclusion that the Tozitna terrane does not have a far-traveled, exotic, accretionary origin. On the contrary, the interregional continuity, consistency, and relative simplicity of the paleogeographic and paleotectonic framework inherent in the intracratonic rift model supports the conclusion that the Tozitna terrane represents the culmination of an extensional basin-forming process that operated intermittently from Proterozoic to early Mesozoic time within the outer part of the Cordilleran miogeocline. Struik (1987) reached the same conclusion for the equivalent terranes of the northern Canadian Cordillera.

I conclude that all the terranes of east-central Alaska, including the Tozitna terrane, are fragments of North American continental crust and intracontinental basins that were telescoped and dispersed by classical non-accretionary processes. Plate convergence is inferred to have been concentrated at the outboard edge of the Brooks Range–Ruby–Yukon/Tanana crystalline belt (Fig. 20), which most likely represents the outer limit of North American continental crust, and the inboard limit of terrane accretion.

Mesozoic flysch sequence. The Lower Cretaceous Kandik Group, Upper Jurassic and Lower Cretaceous Wilbur Creek unit, and upper part of the Glenn Shale in the southern part of the east-central Alaska region, and laterally equivalent rocks preserved locally in central Yukon Territory, all form part of a foreland depositional basin that can be traced continuously through northern Yukon Territory (Fig. 20) to the Blow trough north of the Brooks Range (Dixon, 1986; Young, 1973). The basin evolved from a shallow epicontinental marine basin to a deeper flysch trough. The continuity of the foreland basin and the similarity of its detritus on an interregional scale seem to preclude the possibility of diverse origins for local segments of the flysch sequence. The flysch was derived primarily from uplifted, orogenically active source areas containing mainly crystalline rocks and variably metamorphosed distal deposits of the Paleozoic Cordilleran miogeocline. The source areas were located outboard of the epicontinental basin, in what are now the hinterland portions of the northern Cordilleran and Brooks Range fold-and-thrust belts, and in the southern Brooks Range, Ruby, Yukon-Tanana, and Yukon crystalline terranes.

Crystalline terranes of interior Alaska—continental arc origin

The similarities in protoliths, structural and metamorphic histories, and granitic plutonism that lead to the correlation of the crystalline terranes of interior Alaska with one another, together with their regional distribution and relations to major tectonic features, suggest that all these terranes originally composed a more continuous crystalline belt that was tectonically segmented into the southern Brooks Range, Ruby, and Yukon-Tanana crustal blocks (Fig. 20). The crystalline belt is here interpreted to be a continental or Andean-type arc that formed along the outermost part of the Cordilleran miogeocline (Burchfiel and others, 1987). The arc underwent crustally contaminated granitic plutonism in mid-Paleozoic time (Dusel-Bacon and Aleinikoff, 1985), possibly generated in an ancestral continental arc- or aborted arc-forming event (Smith and Rubin, 1987). However, the mid-Paleozoic effects are largely masked by far more intense and widespread tectonic, metamorphic, and igneous arc-forming processes resulting from plate convergence in Jurassic to mid-Cretaceous time.

The stratigraphic framework on which the proposed arc was built included: (1) quartz-rich, continentally derived gritty sediments of late Proterozoic and Early Cambrian age similar to those of the Windermere Supergroup in Yukon Territory, through which older Precambrian blocks, interpreted to be rifted fragments of the North American craton, are locally preserved; (2) dominantly argillaceous distal deposits of the intermittently extending early and middle Paleozoic Cordilleran miogeocline, in which mafic intrusive materials and diverse sedimentary facies, as well as some regionally important carbonate platform deposits, had rift-controlled distributions; (3) quartz- and chert-rich conglomerates and related clastic deposits, with locally interfingering mafic igneous and carbonate buildups, derived mainly from locally active block uplifts during Devonian and Early Mississippian time; and (4) upper Paleozoic and lower Mesozoic rocks of the Tozitna terrane. As interpreted here, the Tozitna terrane developed mainly in a broad and regionally subsiding basin inboard of the axis of later Mesozoic arc development (Fig. 20), but Tozitna deposition may have overlapped the future site of the arc as well, possibly in smaller fault-controlled basins like those influencing earlier deposition. The position of the main Tozitna basin inboard from the axis of the Devonian and Mississippian arc also suggests an origin as an intracontinental back-arc basin (see Monger and Price, 1979).

A Canadian counterpart of the Alaska segment of the continental arc is the Yukon crystalline terrane of southern Yukon Territory, for which a displaced North American origin is suspected (Mortensen and Jilson, 1985; LeCouteur and Tempelman-Kluit, 1976; Gabrielse, 1985).

Distribution and significance of fold-and-thrust belts

Recognition of the Charley River fold-and-thrust belt, and of a displaced segment of it in the Livengood area, provides a critical link in Alaska between the fold-and-thrust belts of the northern Canadian Cordillera and the eastern Brooks Range (Fig. 20). This regionally continuous belt, as represented on numerous regional geologic and tectonic maps (King, 1969; Norris, 1984), forms the Alaskan arm of the bifurcated foreland fold-and-thrust

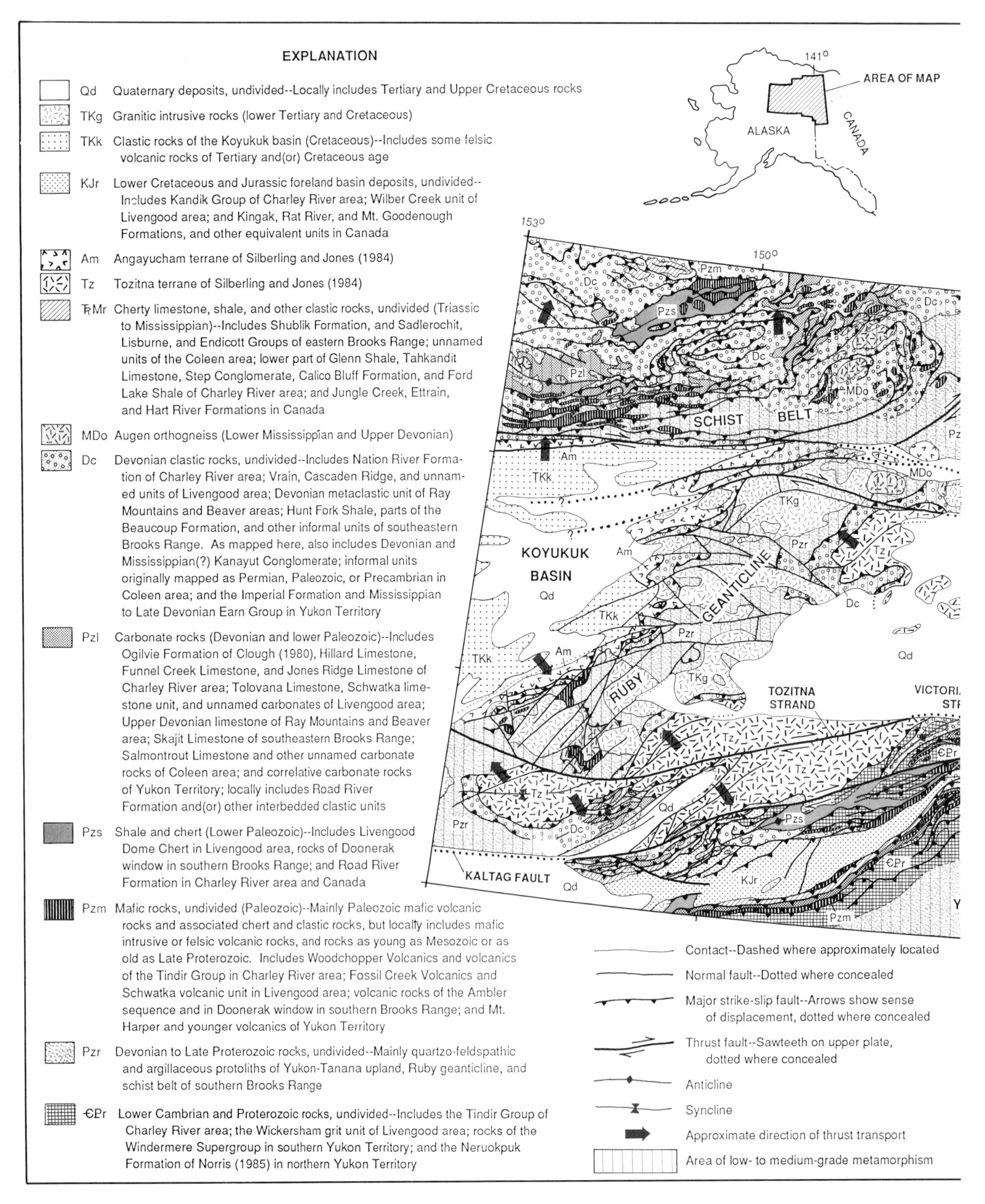

Figure 20. Regional tectonic map of east-central Alaska region.

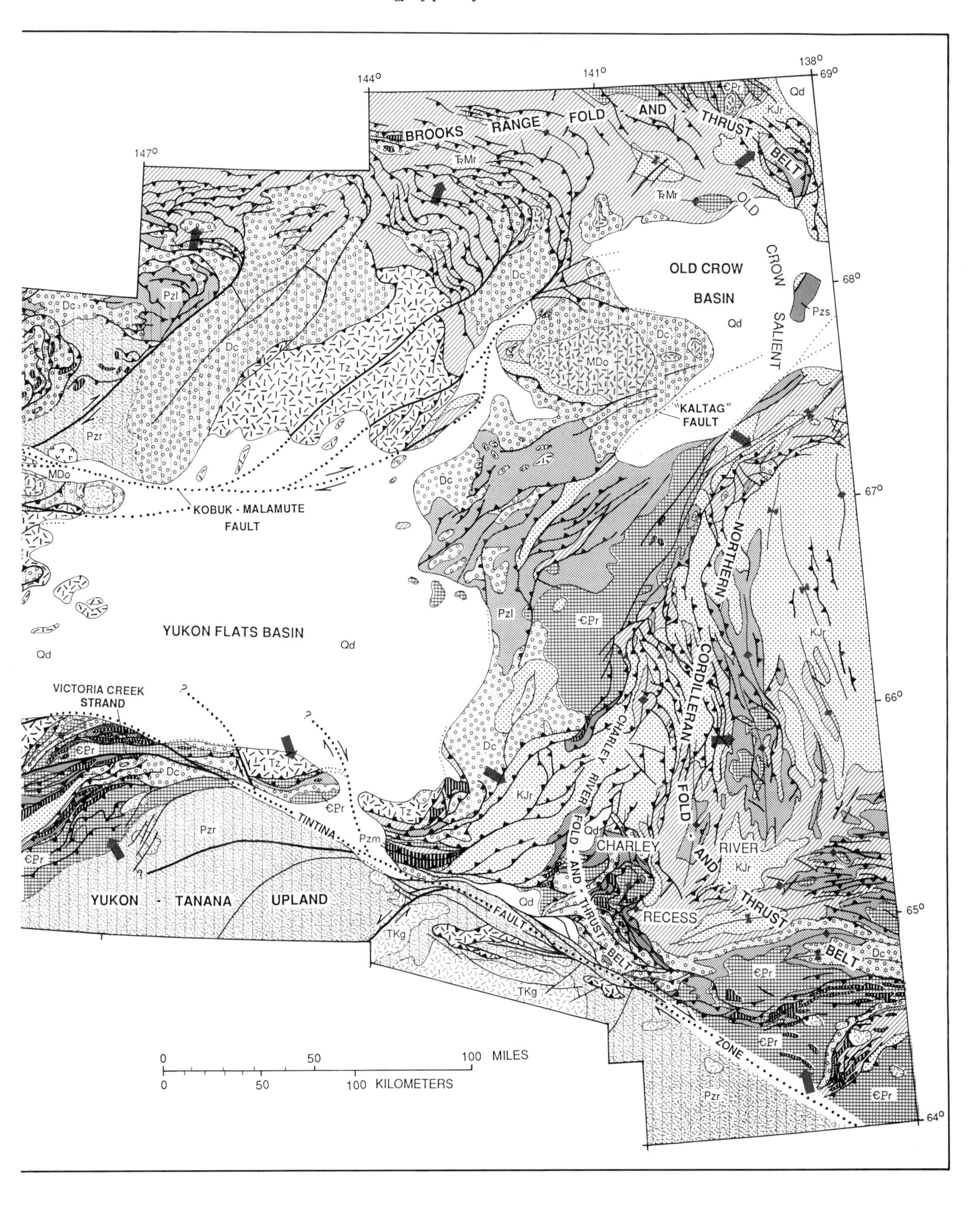

BROOKS RANGE FOLD AND THRUST BELT
OLD CROW BASIN
OLD CROW SALIENT
"KALTAG" FAULT
KOBUK - MALAMUTE FAULT
YUKON FLATS BASIN
VICTORIA CREEK STRAND
TINTINA FAULT ZONE
YUKON - TANANA UPLAND
NORTHERN CORDILLERAN FOLD AND THRUST BELT
CHARLEY RIVER FOLD AND THRUST BELT
CHARLEY RIVER RECESS
0 50 100 MILES
0 50 100 KILOMETERS

belt described by Norris (1987). The most prominent aspect of the belt is its Z-shaped configuration, consisting of a broad salient in northern Yukon Territory (Old Crow salient on Fig. 20) and a much tighter recess in the Charley River area (Charley River recess on Fig. 20). Equally striking is the systematic change in direction of tectonic transport around the "bends" of the fold-and-thrust belt so as to maintain a consistent "inboard" or fore-landward direction of transport, toward the North American craton, or in the case of the Brooks Range, toward an inferred former position of the craton (Fig. 20).

In east-central Alaska, datable fold-and-thrust belt deformation is bracketed between the Albian age of the youngest rocks involved and the Maastrichtian(?) age of the oldest unconformably overlapping rocks, and is therefore restricted to a relatively short span of about 20 m.y. in the mid-Cretaceous. If earlier compression occurred, its deformational effects cannot be identified from data currently available. However, flysch deposits of the upper part of the Glenn Shale and the Kandik Group, and their correlatives along the length of the fold-and-thrust belt, record uplift presumably associated with orogenesis in the more outboard or hinterland crystalline parts of the orogen in Jurassic and Early Cretaceous time. Farther east, toward the foreland, the Upper Cretaceous Eagle Plain Formation is involved in minor thrusting and broad folding indicating that compressional deformation there is either younger or lasted longer than in the Charley River fold-and-thrust belt. Moreover, major folds of the Eagle Plain Formation are truncated by high-angle faults that bound the southern part of the Old Crow salient of the fold-and-thrust belt (Fig. 20). If, as map patterns suggest, these faults merge with fold-and-thrust belt structures within the salient (Norris, 1984), then at least some thrust faults within that part of the salient must have had post–Late Cretaceous movement.

Jurassic to Lower Cretaceous foredeep deposits record the earliest pulses of uplift and orogeny in the Brooks Range, but major thrusting (Brookian) in the central Brooks Range also occurred in the mid-Cretaceous (Tailleur and Brosgè, 1970). Progressively younger thrusting occurred northward toward the foothills and the Arctic slope, where Tertiary thrusting predominates and compressive stresses remain active today (Kelley and Foland, 1987; Grantz and others, 1987). In the southeastern Brooks Range, several bundles of imbricate thrust faults appear to merge to the southeast with lineaments splaying northeastward from the Kobuk-Malamute fault zone (Fig. 20), suggesting that the thrust bundles are structures on which the predominantly right-slip motion of the lineaments is absorbed by imbricate thrusting. If, as map patterns suggest, the major strike-slip movement on the Kobuk-Malamute system postdates the main pulse of mid-Cretaceous Brookian thrusting, then concurrent movement on the imbricate thrust bundles of the southeastern Brooks Range would be of comparable late- to post-Brookian age.

Besides strengthening the structural link between the northern Cordilleran and Brooks Range orogens, recognition of the classical fold-and-thrust belt framework of much of the east-central Alaska region demonstrates a far greater degree of disruption of already complex stratigraphic patterns by typical fold-and-thrust belt processes than was previously suspected. Telescoping on conventional fold-and-thrust belt structures of small to moderate displacement provides a reasonable explanation for the juxtaposition of apparently disparate rock facies and sequences in east-central Alaska.

Major strike-slip fault systems—pattern, timing, significance, and restoration of offset

The two principal strike-slip fault systems of the east-central Alaska region are the Tintina and Kobuk-Malamute systems (Fig. 20). The continuity of the Tintina fault zone with its two main splays, the Kaltag–Victoria Creek and Tozitna faults, and the case for right-lateral separation in Alaska comparable to the 450 km recognized in central Yukon Territory, were discussed in the summary of the Livengood area.

There is no general agreement on the amount, timing, or even the existence of significant right-slip on the Kobuk-Malamute fault zone (Grantz, 1966). However, it follows a major aeromagnetic lineament (Cady, 1978) and is considered here to have produced at least 50 to 75 km of apparent right-separation of the Tozitna terrane and underlying rock sequences in the Beaver area. It cuts Upper Cretaceous conglomerate (Patton and Miller, 1973) derived from post-Brookian uplift of the Ruby-Hodzana upland the Brooks Range, and appears to laterally displace the east contact of the mid-Cretaceous Hodzana pluton (Brosgè and others, 1973). The possible genetic relation between the Kobuk-Malamute fault zone and imbricate thrust bundles in the eastern Brooks Range is based on the interpretation of prominent topographic lineaments and high-angle faults mapped discontinuously on published maps as predominantly right-slip splays of the Kobuk-Malamute system.

Other presumably fracture- or fault-controlled lineaments of the east-central Alaska region may also have a significant strike-slip component, judging from the extent of their linear or curvilinear topographic expression. Lineaments with lengths of 150 km or more cross the Ruby geanticline and the Koyukuk basin margin (Figs. 12 and 20). Though poorly documented and not systematically studied, these features locally coincide with mapped high-angle faults and the contacts of some large east-west–trending granitic plutons of mid-Cretaceous age.

Northeast-trending high-angle faults of comparable length are mapped in northern Yukon Territory near the south margin of the Old Crow salient of the northern Cordilleran fold-and-thrust belt (Fig. 20). A major component of right-lateral strike-slip has been inferred for these faults by Norris (1972a), but where they are inferred to cross the Old Crow salient and to cut and disrupt fold-and-thrust belt structures, none of them, including the "Kaltag" fault (Norris, 1984), displaces fold-and-thrust belt structures laterally to any significant degree. After crossing the Old Crow salient of the Cordilleran fold-and-thrust belt, some of the northeast-trending faults swing northward and merge with north-south–trending extensional faults (Norris, 1984) and/or thrust faults (Lane, 1988) of the Rapid depression. The limited and discontinuous extent of exposure in the critical area of structural intersection, combined with the sinuosity of some faults within the northeast-trending system, also seem to allow for the interpretation that at least some of the northeast-trending faults merge with fold-and-thrust belt structures of the Old Crow salient. If so, these could be additional examples of right-lateral slip transformed into thrusting within the Old Crow salient, like that inferred for the eastern Brooks Range.

The northeast-trending faults of northern Yukon Territory do not significantly displace middle to late Mesozoic stratigraphic and structural belts of the Old Crow salient, and available data indicate that they do not mark any fundamental pre-Cretaceous stratigraphic boundaries. Although they may have influenced pre-Cretaceous deposition to some extent, examples of apparent stratigraphic continuity across the zone of northeast-trending faults were summarized above for the Coleen area and northern Yukon Territory. These indicate at least intermittent stratigraphic continuity across this zone of faults. Furthermore, there is no geologic, geophysical, or physiographic evidence to suggest that any of the northeast-trending faults in northern Yukon Territory are major through-going features connected across the Yukons Flat basin with the strike-slip faults of the Livengood or Ray Mountains areas (Fig. 20). A connection between the "Kaltag" fault of northern Yukon Territory (Norris, 1984; Norris and Yorath, 1981, p. 85–87) and the Kaltag fault of west-central Alaska (Patton and Hoare, 1968) is denied not only by its lack of expression in the Coleen area and by the lack of supporting evidence across the intervening Yukon Flats basin, but also by new map evidence in the Livengood area for a connection between the Kaltag fault and the Tintina fault system. The "Kaltag" fault of Yukon Territory appears to be of no more individual importance than several other faults within the southern part of the Old Crow salient. Based on all the foregoing local and regional map interpretations, no major through-going "Porcupine lineament" (Churkin and Trexler, 1980, 1981), "Kaltag" megashear, or other fundamental discontinuity required by various megatectonic models (e.g., see Nilson, 1981; Jones, 1982) is recognized in east-central Alaska.

Paleogeographic and paleotectonic reconstruction of the complex stratigraphic and structural framework of east-central Alaska first requires the restoration of movement on the major strike-slip fault systems. All of the strike-slip motion documented or interpreted on these systems in east-central Alaska is right-lateral. Restoration of the 450 km of right-separation inferred for the Tintina fault system in the Livengood, Charley River, and Dawson areas is illustrated in Figure 10. Motion on the Kobuk-Malamute system and its possible splays in the Old Crow salient is more conjectural, but a maximum of about 100 km of right separation seems reasonable and is arbitrarily used here for palinspastic restoration. The restoration shown in Figure 21 for the east-central Alaska region and some adjacent areas is discussed by Dover (1985b).

The areal disposition of major dextral strike-slip fault systems with respect to the two "bends" in the Z-shaped configuration of the northern Cordilleran–Brooks Range fold-and-thrust belt suggests a genetic relation between the two—possibly a relation in map view analogous to duplex formation above a structural ramp in cross section (Fig. 22). However, the palinspastic restoration illustrated in Figure 21 shows that the Cordilleran–Brooks Range fold-and-thrust belt had a subdued but still pronounced Z-shaped configuration prior to strike-slip motion on the Tintina and Kobuk-Malamute fault systems. Therefore, although strike-slip attenuated and accentuated the "bends," it did not cause them.

Coincidence of the Tintina fault system along much of its length with the northwest trend of fault-controlled facies belts in the outer part of the late Proterozoic and Paleozoic northern Cordilleran miogeocline suggests that the Tintina trend may have been controlled by Paleozoic and older crustal rifting. The suggestion that the Tintina fault zone reactivates a Mesozoic suture zone (Churkin and others, 1982) is not supported by current interpretations based on the most recent mapping.

Oroclinal bending and crustal rotation

The Z-shaped pattern of stratigraphic and structural trends in east-central Alaska has significant implications for paleotectonic and paleographic reconstruction. The Z-shape is defined not only by the Cordilleran–Brooks Range fold-and-thrust belt, but also by all the major stratigraphic belts of east-central Alaska (Fig. 20), including: (1) the crystalline terranes; (2) Proterozoic and Paleozoic, chert- and mafic volcanic-bearing, outer miogeoclinal sequences; (3) carbonate platform sequences; (4) the Devonian and Mississippian metaclastic sequence; (5) the Tozitna terrane; and (6) Jurassic to Lower Cretaceous flysch (Fig. 21). This pattern is consistent for a strip of mainly continental crust with a minimum width of 250 km and a strike length of at least 2,000 km. The fact that the change of nearly 100 degrees in tectonic trend around the Charley River recess coincides with a comparable change in the direction of thrust transport, led Dover (1985a, 1992) to suggest that the recess might represent an oroclinal bend. If so, the means by which space is accommodated within the oroclinal core is not clear. Interpretation of the Old Crow salient as a second, complimentary orocline is inherent in the models of Canada basin opening and Brooks Range rotation proposed by Carey (1958), Tailleur and Brosgè (1970), Tailleur (1972), Grantz and others (1982), and Milazzo and others (1987). The model of Norris (1972b) relating paired en echelon belts of oppositely verging structures to continental margin irregularities does not explain the fundamental Z-shaped pattern considered here.

If the displacement history postulated here for the Tintina fault system is valid, then the abrupt 40-degree bend in the Tintina system in the Livengood area must have occurred after most of the strike-slip was completed. Otherwise, transpression would have been required on the part of the system now trending to the southwest. Counterclockwise bending of the Tintina system and its splays in southwestern Alaska is tentatively attributed by Dover (1985b) to a major Tertiary crustal rotation event now being documented in west-central Alaska (Coe and others, 1985; Thrupp and Coe, 1986; Coe and Thrupp, 1987).

Extensional origin of Yukon Flats basin

The position of the Yukon Flats basin (Kirschner, chapter 15, this volume) between the Tintina and Kobuk-Malamute strike-slip systems suggests that it could have originated as a pull-apart basin in a region of secondary tension generated by dextral shear between the Tintina and Kobuk-Malamute fault systems (Fig. 23A). Intermittent Tertiary and Quaternary movement accompanied by synorogenic deposition and local mafic volcanism, and continuing extension that maintains the low topography of the basin, would require that at least one of the bounding strike-slip systems is still active. Alternatively, because the Yukon Flats basin is on the convex side of the bend in the Tintina fault system, as are other extensional basins in south-

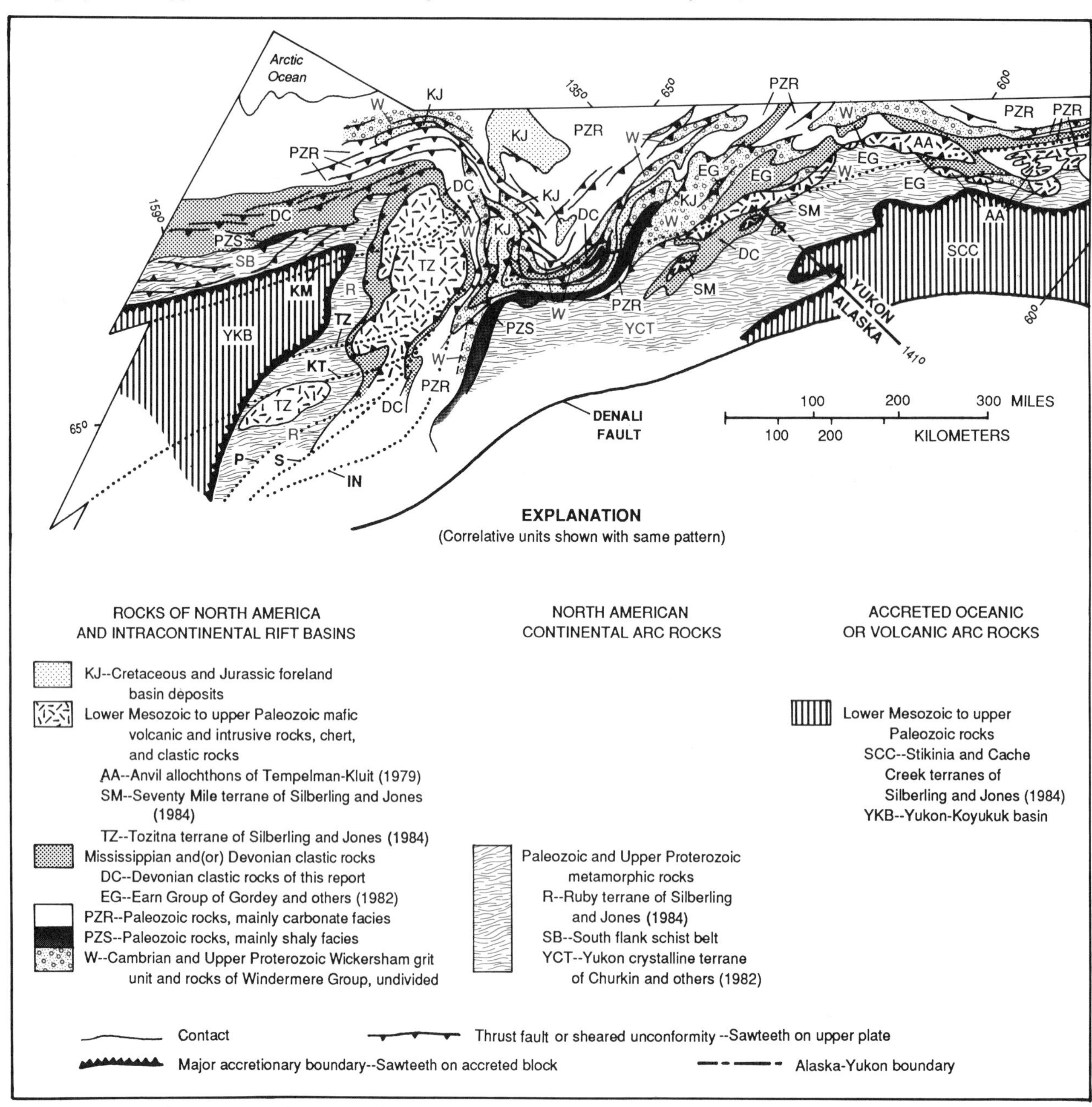

Figure 21. Pre-Tintina geologic configuration. 40 degrees of counter-clockwise bending and 425 km of cumulative right-slip removed from Tintina fault system, based on removal of 130 km from the Kaltag fault (KT), 20 km from unnamed faults inferred to cross Yukon Flats basin, 75 km from the Tozitna fault (TZ), 125 km on the Susulitna (S) and Poorman (P) faults, combined, and 75 km on the Iditarod–Nixon Fork fault. Also, 100 km removed from the Kobuk-Malamute fault (KM).

central Alaska relative to bends in the Denali and Border Ranges fault systems (Fig. 23B), all of these basins may be caused by tension induced by the Tertiary crustal rotation event, and localized within strike-slip-bounded crustal layers at the apices of the bends. If so, crustal rotation must still be active.

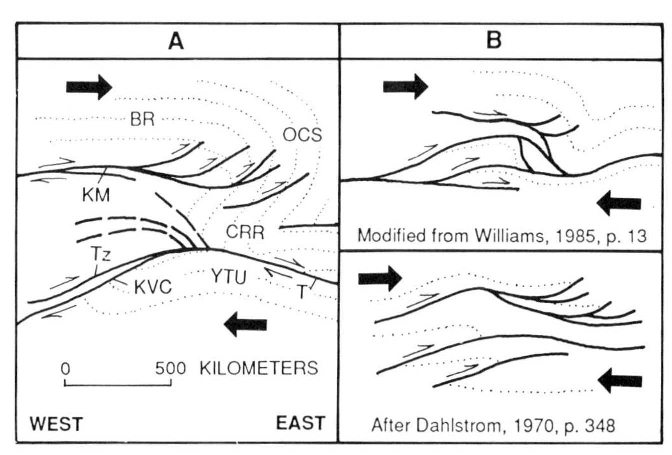

Figure 22. Comparison of fault patterns in map and cross-section views. A, Generalized strike-slip fault pattern of east-central Alaska in map view. B, Examples of thrust fault patterns in cross-section view. Explanation of symbols: BR, Brooks Range; OCS, Old Crow salient; KM, Kobuk-Malamute fault; CRR, Charley River recess; Tz, Tozitna fault; KVC, Kaltag–Victoria Creek fault; T, Tintina fault; YTU, Yukon-Tanana upland.

PALEOGEOGRAPHIC AND GEOTECTONIC MODEL

The paleogeographic and geotectonic model presented here is offered as a testable alternative to accretionary tectonic models implied in the tectonostratigraphic terrane subdivision of east-central and Arctic Alaska, and also as an alternative to geotectonic models requiring thousands of kilometers of lateral movement on megashears passing through east-central Alaska. The model that best fits the geologic framework of east-central Alaska, as presently understood and summarized in this chapter (Fig. 24), incorporates: (1) a passive and intermittently extending late Proterozoic to Triassic depositional prism that wrapped continuously from the Cordilleran miogeocline into the Innuitian miogeocline (Franklinian and Ellesmerian sequences) of the Canadian Arctic islands, along the ancient North American margin; (2) two periods of continental arc development generated by plate convergence along the outer edge of this margin, with possible back-arc spreading associated with the first, and a major fold-and-thrust belt developed during the second; (3) the bifurcating fold-and-thrust belt model of Norris (1987); and (4) concepts of rotational Canada basin opening held by Carey (1958), Tail-

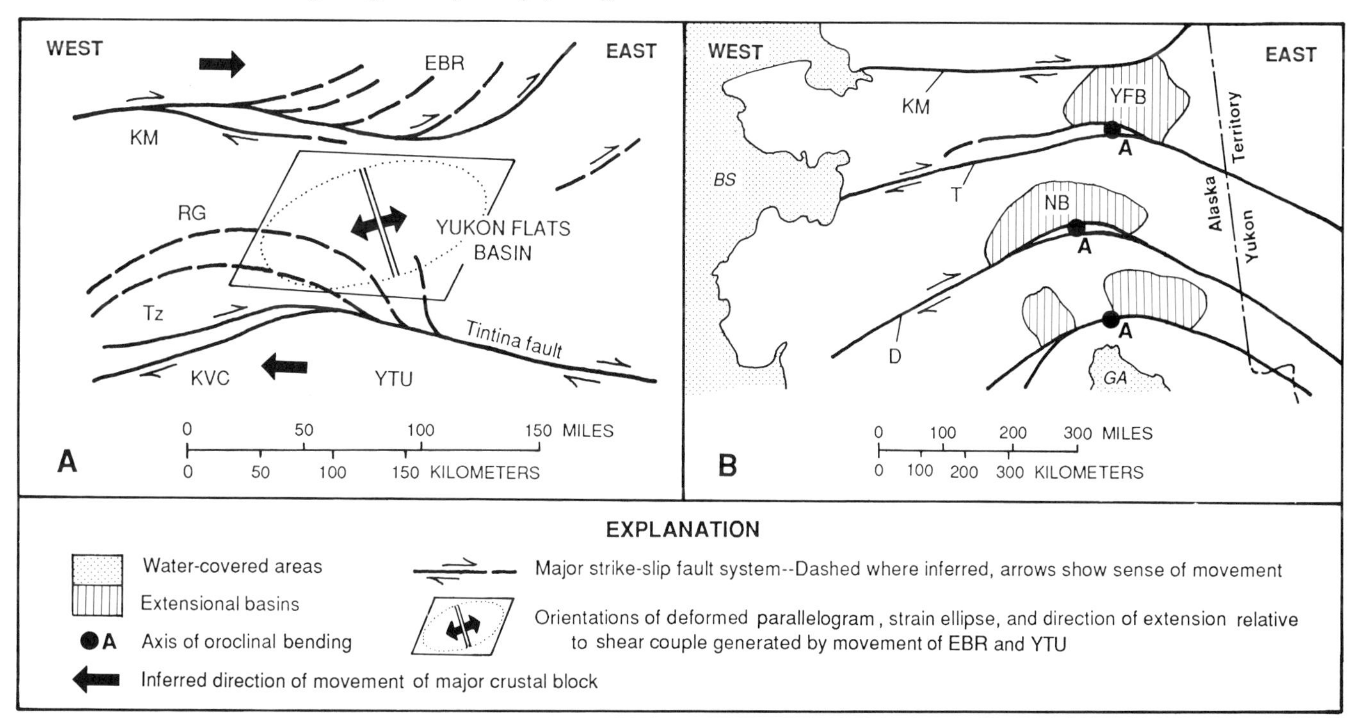

Figure 23. Possible origins of extensional basins. A, Extension by rotation between active strike-slip faults. B, Extension on convex sides of oroclinally bent strike-slip faults. EBR, Brooks Range; KM, Kobuk-Malamute fault; RB, Ruby geanticline; Tz, Tozitna fault; KVC, Kaltag–Victoria Creek fault; YTU, Yukon-Tanana upland; YFB, Yukon Flats basin; BS, Bering Sea; GA, Gulf of Alaska; T, Tintina fault; NB, Nenana basin; D, Denali fault; A, axis of oroclinal bending.

leur and Brosgé (1970), Tailleur (1972), Freeland and Dietz (1973), and Grantz and others (1982), and recently supported by paleomagnetic evidence from Arctic Alaska (Halgedfahl and Jarrard, 1987). In this model, the "bends" in the Z-pattern of stratigraphic and structural trends have two different origins (Fig. 24):

(1) The Charley River recess reflects an original swing in the trend of the ancient North American margin from predominantly northwest-southeast in the Canadian Cordillera to northeast-southwest in the Innuitian belt of Arctic Canada (represented by Norris's Innuitian arm of the bifurcated Cordilleran fold-and-thrust belt). The miogeoclinal prism flanking this originally curved continental margin had an intermittent history of extension from Proterozoic through early Mesozoic time, interrupted in the Devonian and Early Mississippian by an incipient continental arc-forming event that produced uplift and granitic plutonism along its outer edge but little documented metamorphism or compressional deformation. Rift basins within the miogeoclinal prism contain siliceous and siliciclastic rocks, and mafic igneous material; coarse clastic rocks were derived from accompanying block uplifts. Carbonate rocks like those of the Mackenzie, Cassiar, Nixon Fork, Porcupine, and Skajit platforms rimmed the basins, including the early Paleozoic Selwyn basin and Rich-

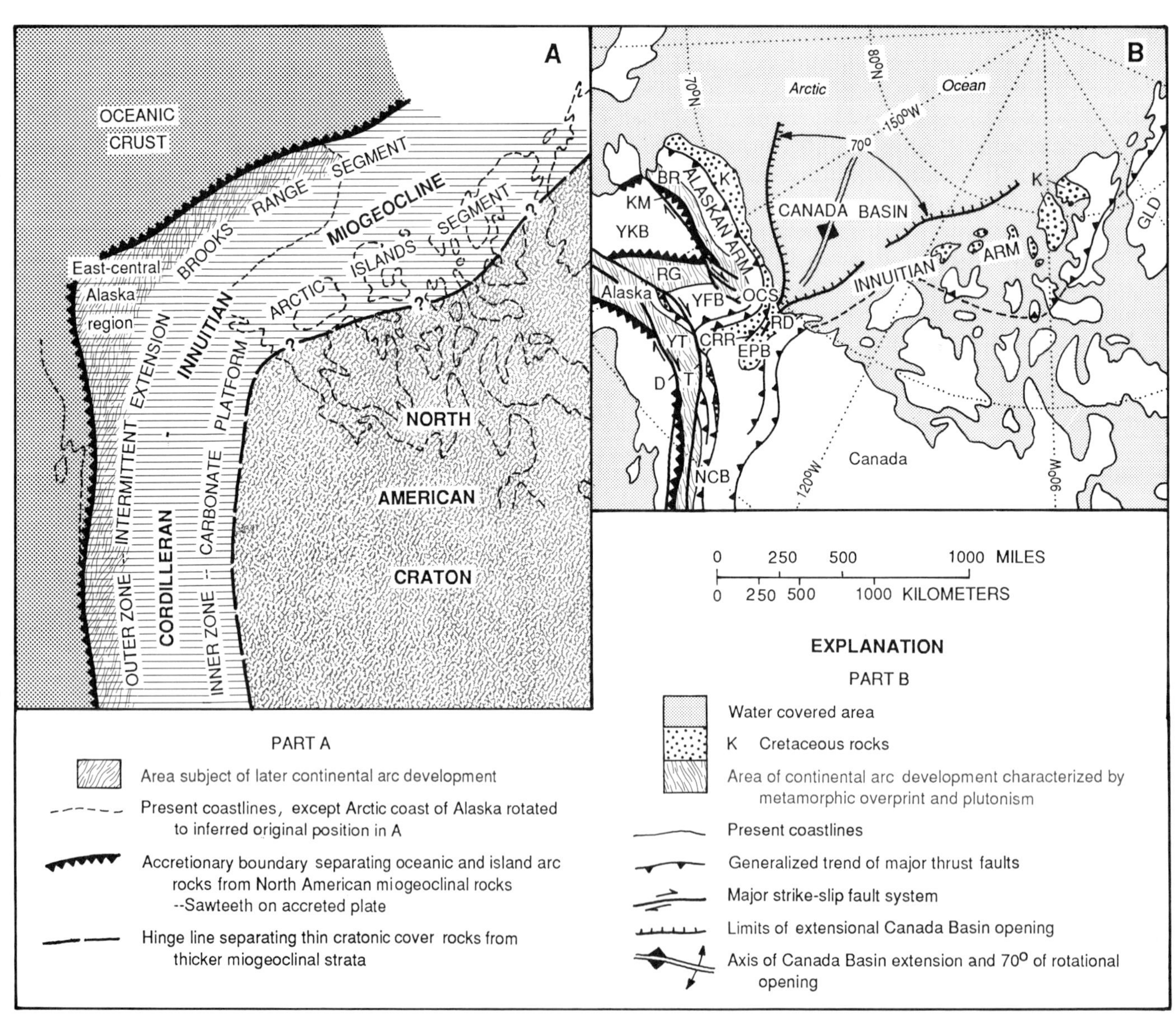

Figure 24. Regional geotectonic model. A, Inferred configuration of Cordilleran-Innuitan miogeocline prior to Canada basin opening. B, Present tectonic configuration, after Canada basin opening and associated oroclinal bending, and major strike-slip faulting (see also Norris, 1987). GLD, Greenland; AK, Alaska; KM, Kobuk-Malamute fault; BR, Brooks Range; YKB, Yukon-Koyukuk basin; YFB, Yukon Flats basin; RD, Rapid depression; OCS, Old Crow salient; CRR, Charley River recess; EPB, Eagle Plain basin; YT, Yukon-Tanana upland; D, Denali fault; T, Tintina fault; RG, Ruby geanticline; NCB, Northern Cordilleran fold-and-thrust-belt.

ardson trough, or capped intervening block uplifts. More advanced extension of the late Paleozoic to Triassic Tozitna basin, accompanied by regional subsidence, produced the most mafic rich of the extensional basins. The Tozitna is regarded as an aborted ocean-forming rift that may have originated as an intracratonic back-arc basin inboard of the aborted mid-Paleozoic continental arc. Beginning with metamorphism and plutonism in Jurassic time and continuing with major supracrustal fold-and-thrust belt development in the mid-Cretaceous, a full-fledged continental arc generated by plate convergence was superimposed along the outer edge of the North American miogeocline over the site of the aborted mid-Paleozoic one, and a foreland flysch basin developed along the length of the orogen. The outboard edge of the crystalline terranes of east-central Alaska, which formed the continental arc, is considered to be the inboard limit of accretionary terranes in Alaska.

(2) The Old Crow Salient, represented by Norris's Alaskan arm of the fold-and-thrust belt, is a true Cordilleran–Brooks Range orocline produced by rotation of Arctic Alaska from the Canadian Arctic islands during opening of the Canada basin (Fig. 24). The Rapid depression and probably the Eagle Plain depression to the south, which together separate the unrotated Innuitian and rotated Alaskan arms of the bifurcated Cordilleran fold-and-thrust belt, would be on-land intracratonic rift-basins representing the apex of Canada basin opening. Some rotation of Arctic Alaska may have been coeval with compressional development of the Cordilleran–Brooks Range fold-and-thrust belt.

The opposing "bends" of the Z-pattern were later attenuated by dextral strike-slip fault systems, possibly reflecting a change in plate motions at about 80 Ma along the Alaskan segment of the North American continental margin from direct to more oblique convergence. Alternatively, the period of major strike-slip may have coincided with east-west compression generated by convergence between North America and Eurasia.

REFERENCES CITED

Albanese, M. D., 1983, Bedrock geologic map of the Livengood B-4 Quadrangle, Alaska: Alaska Division of Geological and Geophysical Surveys Report of Investigations 83-3, scale 1:40,000.

Arth, J. G., 1985, Neodymium and strontium isotopic composition of Cretaceous plutons of the Yukon-Koyukuk basin, Ruby geanticline, and Seward Peninsula, Alaska [abs.]: EOS Transactions of the American Geophysical Union, v. 66, p. 1102.

Bamber, E. W., and Waterhouse, J. B., 1971, Carboniferous and Permian stratigraphy and paleontology, northern Yukon territory, Canada: Bulletin of Canadian Petroleum Geology, v. 19, no. 1, p. 29–250.

Barker, J. C., 1986, Placer gold deposits of the Eagle trough, upper Yukon River region, Alaska: Bureau of Mines Information Circular 9123, 20 p.

Bates, R. L., and Jackson, J. A., 1987, Glossary of Geology: Alexandria, Virginia, American Geological Institute, 3rd edition, 788 p.

Blodgett, R. B., 1983, Paleobiogeographic affinities of Devonian fossils from the Nixon Fork terrane, southwestern Alaska, *in* Stevens, C. H., ed., Pre-Jurassic rocks in western North American suspect terranes: Society of Economic Paleontologists and Mineralogists, p. 125–130.

—— , 1987, Taxonomy and paleobiogeographic affinities of an early Middle Devonian (Eifelian) gastropod faunule from the Livengood Quadrangle, east-central Alaska [Ph.D. thesis]: Corvallis, Oregon State University, 139 p.

Blodgett, R. B., Wheeler-Crowder, K. L., Rohr, D. M., Harris, A. G., and Weber, F. R., 1987, Late Ordovician fossils from the Fossil Creek volcanics of the Livengood Quadrangle; Significance for Late Ordovician glacio-eustasy: U.S. Geological Survey Circular 998, p. 54–58.

Blodgett, R. B., Zhang, N., Ormiston, A. R., and Weber, F. R., 1988, A Late Silurian age determination for the limestone of the "Lost Creek unit," Livengood C-4 Quadrangle, east-central Alaska: U.S. Geological Survey Circular 1016, p. 56–58.

Brabb, E. E., 1967, Stratigraphy of the Cambrian and Ordovician rocks of east-central Alaska: U.S. Geological Survey Professional Paper 559-A, 30 p.

—— , 1969, Six new Paleozoic and Mesozoic formations in east-central Alaska: U.S. Geological Survey Bulletin 1274-I, 26 p.

—— , 1970, Preliminary geologic map of the Black River Quadrangle, east-central Alaska: U.S. Geological Survey Miscellaneous Geologic Investigations Map I-601, scale 1:250,000.

Brabb, E. E., and Churkin, M., Jr., 1967, Stratigraphic evidence of the Late Devonian age of the Nation River formation, east-central Alaska, *in* Geological Survey Research 1967: U.S. Geological Survey Professional Paper 575-D, p. D4–D15.

—— , 1969, Geologic map of the Charley River Quadrangle, east-central Alaska: U.S. Geological Survey Miscellaneous Geologic Investigations Map I-573, scale 1:250,000.

Brooks, A. H., and Kindle, E. M., 1908, Paleozoic and associated rocks of the upper Yukon, Alaska: Geological Society of America Bulletin, v. 19, p. 255–314.

Brosgè, W. P., 1960, Metasedimentary rocks in the south-central Brooks Range, Alaska: U.S. Geological Survey Professional Paper 400-B, p. B351–B352.

—— , 1975, Metamorphic belts in the southern Brooks Range: U.S. Geological Survey Circular 722, p. 40.

Brosgè, W. P., and Dutro, J. T., Jr., 1973, Paleozoic rocks of northern and central Alaska: American Association of Petroleum Geologists Memoir 19, p. 361–375.

Brosgè, W. P., and Reiser, H. N., 1962, Preliminary geologic map of the Christian Quadrangle, Alaska: U.S. Geological Survey Open-File Map 62–15, scale 1:250,000.

—— , 1964, Geologic map and section of the Chandalar Quadrangle, Alaska: U.S. Geological Survey Miscellaneous Geologic Investigations Map I-375, scale 1:250,000.

—— , 1965, Preliminary geologic map of the Arctic Quadrangle, Alaska: U.S. Geological Survey Open-File Report 65–22, scale 1:250,000.

—— , 1969, Preliminary geologic map of the Coleen Quadrangle, Alaska: U.S. Geological Survey Open-File Map 69–25, scale 1:250,000.

Brosgè, W. P., Dutro, J. T., Jr., Mangus, M. D., and Reiser, H. N., 1962, Paleozoic sequence in eastern Brooks Range, Alaska: American Association of Petroleum Geologists Bulletin, v. 46, no. 12, p. 2174–2198.

Brosgè, W. P., Reiser, H. N., Dutro, J T., Jr., and Churkin, M., Jr., 1966, Geologic map and stratigraphic sections, Porcupine River Canyon, Alaska: U.S. Geological Survey Open-File Report 66–10, 4 sheets, scale 1:63,360.

Brosgè, W. P., Lanphere, M. A., Reiser, H. N., and Chapman, R. M., 1969, Probable Permian age of the Rampart Group, central Alaska: U.S. Geological Survey Bulletin 1294-B, 18 p.

Brosgè, W. P., Reiser, H. N., and Yeend, W., 1973, Reconnaissance geologic map of the Beaver Quadrangle, Alaska: U.S. Geological Survey Miscellaneous Field Studies Map MF-525, scale 1:250,000.

Brosgè, W. P., Reiser, H. N., Dutro, J. T., Jr., and Detterman, R. L., 1976, Reconnaissance geologic map of the Table Mountain Quadrangle, Alaska: U.S. Geological Survey Open-File Report 76–546, scale 1:200,000.

Brosgè, W. P., Reiser, H. N., and Dutro, J. T., Jr., 1979, Significance of Middle Devonian clastic rocks in the eastern Brooks Range, Alaska: U.S. Geological

Survey Circular 823-B, p. B24–B25.
Burchfiel, B. C., Eaton, G. P., Lipman, P. W., and Smith, R. B., 1987, The Cordilleran orogen: Conterminous U.S. sector, *in* Palmer, A. R., ed., Perspectives in regional geological synthesis: Geological Society of America DNAG Special Publication 1, p. 91–98.
Burns, L. E., Newberry, R. J., and Reifenstuhl, R. R., 1987, Intrusive rocks of the Lime Peak–Mt. Prindle area, *in* Smith, T E., Pessell, G. H., and Wiltse, M. A., 1987, Mineral assessment of the Lime Peak–Mt. Prindle area, Alaska: Alaska Division of Geological and Geophysical Surveys, p. 3-1 to 3-78.
Cady, J. W., 1978, Aeromagnetic map and interpretation, Chandalar Quadrangle, Alaska: U.S. Geological Survey Miscellaneous Field Studies Map MF-878C, scale 1:250,000.
——, 1986, Geophysics of the Yukon–Koyukuk province: U.S. Geological Survey Circular 978, p. 21–25.
——, 1987, Preliminary geophysical interpretation of the oceanic terranes of interior and western Alaska; Evidence of thick crust of intermediate density: American Geophysical Union Geodynamic Series v. 19, p. 301–305.
Cady, J. W., and Morin, R. L., 1990, Aeromagnetic and gravity data, *in* Weber, F. R., McCammon, R. B., Rinehart, C. D., and Light, T. D., Mineral resources of the White Mountain National Recreation Area, east-central Alaska: U.S. Geological Survey Open-File Report 88–284, in press.
Cady, J. W., and Weber, F. R., 1983, Aeromagnetic map and interpretation of magnetic and gravity data, Circle Quadrangle, Alaska: U.S. Geological Survey Open-File Report 83–170C, 29 p., scale 1:250,000.
Cairnes, D. D., 1914, The Yukon–Alaska international boundary, between Porcupine and Yukon Rivers: Geological Survey of Canada Memoir 67, 161 p.
Carey, S. W., 1958, A tectonic approach to continental drift, *in* Carey, S. W., ed., Continental drift, a symposium: Tasmania University, p. 177–355.
Cecile, M. P., 1982, The lower Paleozoic Misty Creek embayment, Selwyn Basin, Yukon and Northeast Territories, with a paleontologic index by W. H. Fritz, B. S. Norford, and R. S. Tipnis: Geological Survey of Canada Bulletin 335, 78 p., 1 map, scale 1:500,000.
Chapman, R. M., 1974, Metamorphic rock sequence between Rampart and Tanana dated: U.S. Geological Survey Circular 700, p. 42.
Chapman, R. M., Weber, F. R., and Taber, B., 1971, Preliminary geologic map of the Livengood Quadrangle, Alaska: U.S. Geological Survey Open-File Report 71–66, scale 1:250,000.
Chapman, R. M., Weber, F. R., Churkin, M., Jr., and Carter, C., 1979, The Livengood Dome chert, a new Ordovician formation in central Alaska, and its relevance to displacement on the Tintina fault: U.S. Geological Survey Professional Paper 1126-F, p. F1–F13.
Chapman, R. M., Yeend, W., Brosgè, W. P., and Reiser, H. N., 1982, Reconnaissance geologic map of the Tanana Quadrangle, Alaska: U.S. Geological Survey Open-File Report 82–734, 20 p., scale 1:250,000.
Chapman, R. M., Trexler, J. H., Jr., Churkin, M., Jr., an Weber, F. R., 1985, New concepts of the Mesozoic flysch belt in east-central Alaska: U.S. Geological Survey Circular 945, p. 29–32.
Chipp, E. R., 1970, Geology and geochemistry of the Chandalar area, Brooks Range, Alaska: Alaska Department of Natural Resources, Division of Mines and Geology Geologic Report 42, 39 p.
Churkin, M., Jr., and Brabb, E. E., 1965a, Ordovician, Silurian, and Devonian biostratigraphy of east-central Alaska: American Association of Petroleum Geologists Bulletin, v. 49, no. 2, p. 172–185.
——, 1965b, Occurrence and stratigraphic significance of Oldhamia, a Cambrian trace fossil, in east-central Alaska: U.S. Geological Survey Professional Paper 525-D, p. D120–D124.
——, 1968, Devonian rocks of the Yukon-Porcupine Rivers area and their tectonic relation to other Devonian sequences in Alaska, *in* Proceedings of the International Symposium on the Devonian System, vol. 2: Calgary, Alberta Society of Petroleum Geologists, p. 227–258.
Churkin, M., Jr., and Carter, C., 1970, Devonian tentaculitids of east-central Alaska; Systematics and biostratigraphic significance: Journal of Paleontology, v. 44, no. 1, p. 51–68.
Churkin, M., Jr., and Trexler, J. H., 1980, Circum-Arctic plate accretion; Isolating part of a Pacific plate to form the nucleus of the Arctic basin: Earth and Planetary Science Letters v. 48, p. 356–362.
——, 1981, Continental plates and accreted oceanic terranes in the Arctic, *in* Nairn, A. E., Churkin, M., Jr., and Stehli, F. G., eds., The ocean basins and margins; Volume 5, The Arctic ocean: New York, Plenum Press, p. 1–20.
Churkin, M., Jr., Foster, H. L., Chapman, R. M., and Weber, F. R., 1982, Terranes and suture zones in east-central Alaska: Journal of Geophysical Research, v. 87, no. B5, p. 3718–3730.
Clough, J. G., 1980, Fossil algae in Lower Devonian limestones, east-central Alaska, *in* Short notes on Alaskan geology, 1979–80: Alaska Division of Geological and Geophysical Surveys Geologic Report 63, p. 19–21.
Clough, J. G., and Blodgett, R. B., 1984, Lower Devonian basin to shelf carbonates in outcrop from the western Ogilvie Mountains, Alaska and Yukon Territory, *in* Eliuk, L. S., ed., Carbonates in subsurface and outcrop, 1984 Canadian Society of Petroleum Geologists Core Conference Manual: Calgary, Alberta, Canada, Canadian Society of Petroleum Geologists, p. 57–81.
Coe, R. S., and Thrupp, G. A., 1987, Tectonic implications of rotated paleomagnetic declinations in west-central Alaska: Geological Society of America Abstracts with Programs v. 19, p. 367.
Coe, R. S., Globerman, B. R., Plumley, P. W., and Thrupp, G. A., 1985, Paleomagnetic results from Alaska and their tectonic implications, *in* Howell, D. G., ed., Tectonostratigraphic terranes of the circum-Pacific region: Houston, Texas, Circum-Pacific Council for Energy and Mineral Resources, p. 85–108.
Coleman, D. A., 1985, Shelf to basin transition of Silurian-Devonian rocks, Porcupine River area, east-central Alaska: American Association of Petroleum Geologists Bulletin, v. 69, no. 4, p. 659.
Coney, P. J., 1983, Structural and tectonic aspects of accretion in Alaska, *in* Howell, D. G., Jones, D. L., Cox, A., and Nur, A., Proceedings of the Circum-Pacific Terrane Conference: Stanford, California, Stanford University Publications, p. 68–70.
Coney, P. J., and Jones, D. L., 1985, Accretion tectonics and crustal structure in Alaska: Tectonophysics, v. 119, p. 265–283.
Coney, P. J., Jones, D. L., and Monger, J.W.H., 1980, Cordilleran suspect terranes: Nature, v. 288, p. 329–333.
Cushing, G. W., 1984, Early Mesozoic tectonic history of the eastern Yukon–Tanana upland [M.S. thesis]: Albany, State University of New York, 255 p.
Cushing, G. W., and Foster, H. L., 1984, Structural observations in the Circle Quadrangle, Yukon–Tanana upland, Alaska: U.S. Geological Survey Circular 868, p. 64–65.
Cushing, G. W., Foster, H. L., Laird, J., and Burack, A. C., 1982, Description and preliminary interpretation of folds and faults in a small area in the Circle B-4 and B-5 Quadrangles, Alaska: U.S. Geological Survey Circular 844, p. 56–58.
Cushing, G. W., Meisling, K. E., Christopher, R. A., and Carr, T. R., 1986, The Cretaceous to Tertiary evolution of the Tintina fault zone, east-central Alaska: Geological Society of America Abstracts with Programs, v. 18, p. 98.
Dahlstrom, C.D.A., 1970, Structural geology in the eastern margin of the Canadian Rocky Mountains: Bulletin of Canadian Petroleum Geology v. 18, no. 3, p. 332–406.
Davies, W. E., 1972, The Tintina trench and its reflection in the structure of the Circle area, Yukon–Tanana upland, Alaska: Twenty-fourth International Geological Congress, Section 3, Tectonics, Montreal, Canada, p. 211–216.
Delaney, G. D., 1981, The mid-Proterozoic Wernecke Supergroup, Wernecke Mountains, Yukon Territory, *in* Campbell, F.H.A., ed., Proterozoic basins of Canada: Geological Survey of Canada Paper 81-10, p. 1–23.
Dillon, J. T., Pessel, G. H., Chen, J. H., and Veach, N. C., 1979, Tectonic and economic significance of Late Devonian and late Proterozoic U-Pb zircon ages from the Brooks Range, Alaska: Alaska Division of Geology and Geophysical Surveys Geologic Report 61, p. 36–43.
——, 1980, Middle Paleozoic magmatism and orogenesis in the Brooks Range, Alaska: Geology, v. 8, p 338–343.
Dillon, J. T., and 5 others, 1985, New radiometric evidence for the age and thermal history of the metamorphic rocks of the Ruby and Nixon Fork

terranes, west-central Alaska: U.S. Geological Survey Circular 945, p. 13–18.
Dillon, J. T., Brosgè, W. P., and Dutro, J. T., 1986, Generalized geologic map of the Wiseman Quadrangle, Alaska: U.S. Geological Survey Open-File Report 86–219, scale 1:250,000.
Dixon, J., 1986, Comments on the stratigraphy, sedimentology, and distribution of the Albian Sharp Mountain Formation, northern Yukon, *in* Current Research, Part B: Geological Survey of Canada Paper 86-1B, p. 375–381.
Dover, J. H., 1985a, Possible oroclinal bend in northern Cordilleran fold and thrust belt, east-central Alaska: Geological Society of America Abstracts with Programs, v. 17, p 352.
—— , 1985b, Dispersion of Tintina fault displacement in interior Alaska: Geological Society of America Abstracts with Programs, v. 17, p. 352.
—— , 1988, Geologic cross sections, *in* Weber, F. R., McCammon, R. B., Rinehart, C. D., Light, T. D., and Wheeler, K. L., Geology and mineral resources of the White Mountains National Recreation Area, east-central Alaska: U.S. Geological Survey Open-File Report 88–284, Plate II-B.
—— , 1992, Geologic map and fold-and-thrust-belt interpretation of the southeastern part of the Charley River Quadrangle, east-central Alaska: U.S. Geological Survey Miscellaneous Investigations Map I-1942, scale 1:100,000.
Dover, J. H., and Miyaoka, R. T., 1985a, Major rock packages of the Ray Mountains, Tanana and Bettles Quadrangles: U.S. Geological Survey Circular 945, p. 32–36.
—— , 1985b, Metamorphic rocks of the Ray Mountains; Preliminary structural analysis and regional tectonic implications: U.S. Geological Survey Circular 945, p. 36–38.
—— , 1985c, Metamorphic rocks and structure of the Ray Mountains, southeast borderland of Koyukuk basin, Alaska: EOS Transactions of the American Geophysical Union, v. 66, no. 46, p. 1101–1102.
—— , 1988, Reinterpreted geologic map and fossil data, Charley River Quadrangle, east-central Alaska: U.S. Geological Survey Miscellaneous Field Studies Map MF-2004, scale 1:250,000.
Dusel-Bacon, C., and Aleinikoff, J. N., 1985, Petrology and tectonic significance of augen gneiss from a belt of Mississippian granitoids in the Yukon-Tanana terrane, east-central Alaska: Geological Society of America Bulletin, v. 96, p. 411–425.
Dusel-Bacon, C., and Foster, H. L., 1983, A sillimanite gneiss dome in the Yukon crystalline terrane, east-central Alaska; Petrography and garnet-biotite geothermometry: U.S. Geological Survey Professional Paper 1170-E, p. E1–E25.
Dutro, J. T., Jr., 1979, Alaska, *in* The Mississippian and Pennsylvanian (Carboniferous) systems in the United States: U.S. Geological Survey Professional Paper 1110M-DD, p. DD1–DD16.
Dutro, J. T., Jr., and Harris, A. G., 1987, Some stratigraphic and paleontologic constraints on tectonic modelling of the Brooks Range, Alaska: Geological Society of America Abstracts with Programs, v. 19, p. 374.
Dutro, J. T., Jr., and Jones, D. L., 1984, Paleotectonic setting of the Carboniferous of Alaska: International Congress on Carboniferous Stratigraphy and geology, v. 9, no. 3, p. 229–234.
Dutro, J. T., Jr., Brosgè, W. P., and Reiser, H. N., 1977, Upper Devonian depositional history, central Brooks Range, Alaska: U.S. Geological Survey Circular 751-B, p. B16–B18.
Eakin, H. M., 1916, The Yukon-Koyukuk region, Alaska: U.S. Geological Survey Bulletin 631, 88 p.
Eberlein, D. E., and Lanphere, M. A., 1988, Precambrian rocks of Alaska; A review, *in* Harrison, J. E., and Peterman, Z. E., eds., Introduction to the correlation of Precambrian rock sequences: U.S. Geological Survey Professional Paper 1241B, 18 p.
Einaudi, M. T., and Hitzman, M. W., 1986, Mineral deposits in northern Alaska; Introduction: Economic Geology and the Bulletin of the Society of Economic Geologists v. 81, no. 7, p. 1583–1591.
Foster, H. L., and 6 others, 1982, Radiolaria indicate Carboniferous and Triassic ages for chert in the Circle Volcanics, Circle Quadrangle: U.S. Geological Survey Professional Paper 1375, p. 78.
Foster, H. L., Laird, J., Keith, T.E.C., Cushing, G. W., and Menzie, W. D., 1983, Preliminary geologic map of the Circle Quadrangle, Alaska: U.S. Geological Survey Open-File REport 83–170A, 32 p., 1 map, scale 1:250,000.
Foster, H. L., Keith, T.E.C., and Menzie, W. D., 1987, Geology of east-central Alaska: U.S. Geological Survey Open-File Report 87–188, 59 p.
Foster, R. L., 1968, Potential for lode deposits in the Livengood gold placer district, east-central Alaska: U.S. Geological Survey Circular 590, 18 p.
—— , 1969, Nickeliferous serpentinite near Beaver Creek, east-central Alaska: U.S. Geological Survey Circular 615, p. 2–4.
Freeland, G. L., and Dietz, R. S., 1973, Rotation history of Alaskan tectonic belts: Tectonophysics v. 18, p. 379–389.
Gabrielse, H., 1985, Major dextral transcurrent displacements along the northern Rocky Mountain trench and related lineaments in north-central British Columbia: Geological Society of America Bulletin, v. 96, no. 1, p. 1–14.
Gabrielse, H., and Yorath, C. J., 1987, The Cordilleran orogen; Canadian sector, *in* Palmer, A. R., ed., Perspectives in regional geological synthesis: Geological Society of America DNAG Special Publication 1, p. 81–89.
Gardner, M. C., Jarrard, R. D., and Mount, V., 1984, Style and origin of structures, Paleozoic and Precambrian sedimentary rocks of southern Tatonduk Terrane, Alaska: Geological Society of America Abstracts with Programs, v. 16, p. 285.
Gemuts, I., Puchner, C. C., and Steffel, C. I., 1983, Regional geology and tectonic history of western Alaska: Journal of the Alaska Geological Society v. 3, p. 67–85.
Gordey, S. P., 1981, Stratigraphy, structure, and tectonic evolution of southern Pelly Mountains in the Indigo Lake area, Yukon Territory: Geological Survey of Canada Bulletin 318, 44 p., 1 map, scale 1:60,000.
Gordey, S. P., Abbott, J. G., and Orchard, M. J., 1982, Devono-Mississippian (Earn Group) and younger strata in east-central Yukon, *in* Current Research, Part B: Geological Survey of Canada Paper 82–1B, p. 93–100.
Gordey, S. P., Abbott, J. G., Tempelman-Kluit, D. J., and Gabrielse, H., 1987, "Antler" clastics in the Canadian Cordillera: Geology, v. 15, p. 103–107.
Gottschalk, R. R., 1987, Tectonics of the schist belt metamorphic terrane near Wiseman, Alaska: Geological Society of America Abstracts with Programs v. 19, p. 383.
Grantz, A., 1966, Strike-slip faults in Alaska: U.S. Geological Survey Open-File Report 66–53, 82 p.
Grantz, A., and 9 others, 1982, The Arctic region, *in* Palmer, A. R., ed., Perspectives in regional geological synthesis; Planning for the Geology of North America: Geological Society of America DNAG Special Publication 1, p. 105–115.
Grantz, A., Dinter, D. A., and Culotta, R. C., 1987, Structure of the continental shelf north of the Arctic National Wildlife Refuge, *in* Bird, K. J., and Magoon, L. B., eds., Petroleum geology of the northern part of the Arctic National Wildlife Refuge, northeastern Alaska: U.S. Geological Survey Bulletin 1778, p. 271–276.
Green, L. H., 1972, Geology of Nash Creek, Larsen Creek, and Dawson map-areas, Yukon Territory: Geological Survey of Canada Memoir 364, 157 p., 3 maps, scale 1:250,000.
Grybeck, D., and Nelson, S. W., 1981, Structure of the Survey Pass Quadrangle, Brooks Range, Alaska: U.S. Geological Survey Miscellaneous Field Studies Map MF—1176-B, 7 p., 1 map, scale 1:250,000.
Halgedahl, S. L., and Jarrard, R. D., 1987, Paleomagnetism of the Kuparuk River Formation from oriented drill core; Evidence for rotation of the Arctic Alaska plate, *in* Tailleur, I., and Weimer, P., eds, Alaskan North Slope geology: Pacific Section, Society of Economic Paleontologists and Mineralogists and Alaska Geological Society, p. 581–617.
Herreid, G., 1969, Geology and geochemistry Sithylemenkat Lake ara, Bettles Quadrangle, Alaska: Alaska Department of Natural Resources, Division of Mines and Geology Geologic Report 35, 22 p.
Hitzman, M. W., Smith, T. E., and Proffett, J. M., Jr., 1982, Bedrock geology of the Ambler district, southwestern Brooks Range, Alaska: Alaska Division of Geological and Geophysical Surveys Geologic Report 75, scale 1:125,000.
Hitzman, M. W., Proffett, J. M., Jr., Schmidt, J. M., and Smith, T. E., 1986,

Geology and mineralization of the Ambler district, north-western Alaska: Economic Geology, v. 81, p. 1592–1618.
Holdsworth, B., and Jones, D. L., 1979, Late Paleozoic fossils in ophiolite, northeastern Alaska: U.S. Geological Survey Professional Paper 1150, p. 95.
Howell, D. G., Murray, R. W., Wiley, T. J., Boundy-Sanders, S., and Kaufman-Linam, L., 1987, Sedimentology and tectonics of the Devonian Nation River Formation, Alaska, part of yet another allocthonous terrane: American Association of Petroleum Geologists Abstracts with Programs, v. 71, p. 569.
Hughes, J. D., and Long, D.G.F., 1979, Geology and coal resource potential of Early Tertiary strata along Tintina trench, Yukon Territory: Geological Survey of Canada Paper 79-32, 21 p.
Jones, D. L., Silberling, N. J., Berg, H. C., and Plafker, G., 1981, Tectonostratigraphic terrane map of Alaska: U.S. Geological Survey Open-File Report 81–792, scale 1:250,000.
Jones, D. L., Howell, D. G., Coney, P. J., and Monger, J.W.H., 1983, Recognition, character, and analysis of tectono-stratigraphic terranes in western North America, *in* Hashimoto, M., and Uyeda, S., eds., Accretion tectonics in the Circum-Pacific regions: Tokyo, Japan, Terra Scientific Publishing Company, p. 21–35.
Jones, D. L., Silberling, N. J., Chapman, R. M., and Coney, P., 1984, New ages of radiolarian chert from the Rampart district, east-central Alaska: U.S. Geological Survey Circular 868, p 39–43.
Jones, D. L., Bounday-Sanders, S., Murray, R. L., Howell, D. G., and Wiley, T. J., 1987, Tectonic contacts of miogeoclinal strata in east-central Alaska: American Association of Petroleum Geologists Abstracts with Programs, v. 71, p. 573.
Jones, P. B., 1982, Mesozoic rifting in the western Arctic ocean basin and its relationship to Pacific seafloor spreading, *in* Embry, A. F., and Balkwill, H. R., eds., Arctic geology and geophysics; Proceedings of the Third International Symposium on Arctic Geology: Canadian Society of Petroleum Geologists Memoir 8, p. 83–99.
Karl, S., and Long, C. L., 1987, Evidence for tectonic truncation of regional east-west trending structures in the central Baird Mountains Quadrangle, western Brooks Range, Alaska: Geological Society of America Abstracts with Programs v. 19, p. 392.
Kelley, J. S., and Foland, R. L., 1987, Structural style and framework geology of the coastal plain and adjacent Brooks Range, *in* Bird, K. J., and Magoon, L. B., eds., Petroleum geology of the northern part of the Arctic National Wildlife Refuge, northeastern Alaska: U.S. Geological Survey Bulletin 1778, p. 255–270.
Kindle, E. M., 1908, Geologic reconnaissance of the Porcupine Valley, Alaska: Geological Society of America Bulletin, v. 19, p. 315–338.
King, P. B., compiler, 1969, Tectonic map of North America: U.S. Geological Survey Map, scale 1:5,000,000.
Laird, J., and Foster, H. L., 1984, Description and interpretation of a mylonitic foliated quartzite unit and feldspathic quartz wacke (grit) unit in the Circle Quadrangle, Alaska: U.S. Geological Survey Circular 939, p. 29–33.
Lane, L. S., 1988, The Rapid fault array; A foldbelt in Arctic Yukon, *in* Current Research, Part D: Geological Survey of Canada Paper 88-1D, p. 95–98.
Lane, H. R., and Ormiston, A. R., 1976, The age of the Woodchopper Limestone (Lower Devonian), Alaska: Geologica et Palaeontologica, v. 10, p. 101–108.
——, 1979, Siluro-Devonian biostratigraphy of the Salmontrout River area, east-central Alaska: Geologica et Palaeontologica, v. 13, p. 39–96.
Laudon, L. R., Hartwig, A. E., Morgridge, D. L., and Omernik, J. B., 1966, Middle and Late Paleozoic stratigraphy, Alaska-Yukon border area between Yukon and Porcupine Rivers: American Association of Petroleum Geologists Bulletin, v. 50, no. 9, p. 1868–1889.
LeCouteur, P. C., and Tempelman-Kluit, D. J., 1976, Rb/Sr ages and a profile of initial Sr^{87}/Sr^{86} ratios for plutonic rocks across the Yukon crystalline terrane: Canadian Journal of Earth Sciences, v. 13, p. 319–30.
Loney, R. A., and Himmelberg, G. R., 1985a, Distribution and character of the peridotite-layered gabbro complex of the southeastern Yukon-Koyukuk ophiolite belt: U.S. Geological Survey Circular 945, p. 46–48.
——, 1985b, Ophiolitic ultramafic rocks of the Jade Mountains–Cosmos Hills area, southwestern Brooks Range: U.S. Geological Survey Circular 967, p. 13–15.
Long, C. C., and Miyaoka, R., 1988, Resistivity cross-section, *in* Weber, F. R., McCammon, R. B., Rinehart, C. D., and Light, T. D., Mineral resources of the White Mountain National Recreation Area, east-central Alaska: U.S. Geological Survey Open-File Report 88–284, p. 73–76.
Maddren, A. G., 1913, The Koyukuk-Chandalar region, Alaska: U.S. Geological Survey Bulletin 532, 119 p.
Maddren, A. G., and Harrington, G. L., 1955, Geologic maps of the area along the Alaska-Canada boundary between the Porcupine River and Arctic Ocean: U.S. Geological Survey Open-File Report 55–105.
Mertie, J. B., 1925, Geology and gold placers of the Chandalar district, Alaska: U.S. Geological Survey Bulletin 773-E, p. 215–263.
——, 1929, The Chandalar-Sheenjek district, Alaska: U.S. Geological Survey Bulletin 810-B, p. 87–139.
——, 1930, Geology of the Eagle-Circle district, Alaska: U.S. Geological Survey Bulletin 816, 168 p.
——, 1932, The Tatonduk-Nation district, Alaska: U.S. Geological Survey Bulletin 836-E, 109 p.
——, 1937, The Yukon-Tanana region, Alaska: U.S. Geological Survey Bulletin 872, 276 p.
Milazzo, G., Plumley, P. W., and Whalen, M. T., 1987, Tertiary tectonics of the Porcupine terrane, NE Alaska: EOS Transactions of the American Geophysical Union, v. 68, no. 16, p. 292.
Miller, T. P., 1985, Petrologic character of the plutonic rocks of the Yukon–Koyukuk basin and its borderland [abs.]: EOS Transactions of the American Geophysical Union, v. 66, p. 1102.
Miyaoka, R. T., and Dover, J. H., 1985, Preliminary study of shear sense in mylonites, eastern Ray Mountains, Tanana Quadrangle: U.S. Geological Survey Circular 967, p. 29–32.
Monger, J.W.H., 1977, Upper Paleozoic rocks of the western Canadian Cordillera and their bearing on Cordilleran evolution: Canadian Journal of Earth Sciences, v. 14, p. 1832–1859.
Monger, J.W.H., and Price, R. A., 1979, Geodynamic evolution of the Canadian Cordillera; Progress and problems: Canadian Journal of Earth Sciences, v. 16, p. 770–791.
Monger, J.W.H., and Ross, C. A., 1971, Distribution of fusulinaceans in the western Canadian Cordillera: Canadian Journal of Earth Sciences, v. 8, p. 259–278.
——, 1979, Upper Paleozoic volcanosedimentary assemblages of the western North American Cordillera, *in* Neuvieme Congres International de Stratigraphie et de geologie du Carbonifere, v. 3: Carbondale, Southern Illinois University Press, p. 219–228.
Mortensen, J. K., and Godwin, C. I., 1982, Volcanogenic massive sulfide deposits associated with highly alkaline rift volcanics in the southeastern Yukon Territory: Economic Geology and the Bulletin of the Society of Economic Geologists, v. 77, no. 5, p. 1225–1230.
Mortensen, J. K., and Jilson, G. A., 1985, Evolution of the Yukon–Tanana terrane; Evidence from southeastern Yukon Territory: Geology v. 13, no. 11, p. 806–810.
Nelson, S. W., and Grybeck, D., 1980, Geologic map of the Survey Pass Quadrangle, Brooks Range, Alaska: U.S. Geological Survey Miscellaneous Field Studies Map MF-1176A, scale 1:250,000.
Nilsen, T. H., 1981, Upper Devonian and lower Mississippian redbeds, Brooks Range, Alaska, *in* Miall, A. D., ed., Sedimentation and tectonics in alluvial basins: Geological Association of Canada Special Paper 23, p. 187–219.
Norris, D. K., 1972a, Structural and stratigraphic studies in the tectonic complex of northern Yukon Territory, north of Porcupine River, *in* Report of Activities, Part B: Geological Survey of Canada Paper 72-1, p. 91–99.
——, 1972b, En echelon folding in the northern Cordillera of Canada: Bulletin of Canadian Petroleum Geology v. 20, no. 3, p. 634–642.
——, 1982, Geology of the Ogilvie River (116G and 116F) map area: Geological Survey of Canada Map 1526A, scale 1:250,000.
——, 1984, Geology of the northern Yukon and northwestern district of

Mackenzie: Geological Survey of Canada Map 1581A, scale 1:500,000.
——, 1985, The Neruokpuk Formation, Yukon Territory and Alaska, *in* Current Research, Part B: Geological Survey of Canada Paper 85-1B, p. 223–229.
——, 1987, Porcupine virgation; A key to the collapse of the Brooks Range orogen: Geological Society of America Abstracts with Programs, v. 19, p. 437.
Norris, D. K., and Yorath, C. J., 1981, The North American plate from the Arctic Archipelago to the Romanzof Mountains, *in* Nairn, A.E.M., Churkin, M., Jr., and Stehli, F. G., eds., The ocean basins and margins; Volume 5, The Arctic Ocean: New York, Plenum Press, p. 37–103.
Patton, W. W., Jr., and Box, S. E., 1985, Tectonic setting and history of the Yukon-Koyukuk basin, Alaska [abs.]: EOS Transactions of the Geophysical Union, v. 66, no. 46, p. 1101.
Patton, W. W., Jr., and Hoare, J. M., 1968, The Kaltag fault, west-central Alaska: U.S. Geological Survey Professional Paper 600-D, p. D147–D153.
Patton, W. W., Jr., and Miller, T. P., 1970, Preliminary geologic investigations in the Kanuti River region, Alaska: U.S. Geological Survey Bulletin 1312-J, 10 p.
——, 1973, Bedrock geologic map of Bettles and southern part of Wiseman Quadrangles, Alaska: U.S. Geological Survey Miscellaneous Field Studies Map MF-492, scale 1:250,000.
Patton, W. W., Jr., and Moll, E. J., 1982, Structural and stratigraphic sections along a transect between the Alaska Range and Norton Sound: U.S. Geological Survey Circular 844, p. 76–78.
Patton, W. W., Jr., Tailleur, I. L., Brosgè, W. P., and Lanphere, M. A., 1977, Preliminary report on the ophiolites of northern and western Alaska, *in* Coleman, R. G., ed., North American ophiolites: Oregon Department of Geology and Mineral Industries Bulletin 95, p. 51–57.
Patton, W. W., Jr., Miller, T. P., Chapman, R. M., and Yeend, W., 1978, Geologic map of the Melozitna Quadrangle, Alaska: U.S. Geological Survey Miscellaneous Investigations Series Map I-1071, scale 1:250,000.
Patton, W. W., Jr., Stern, T. W., Arth, J. G., and Carlson, C., 1987, New U/Pb ages from granite and granite gneiss in the Ruby geanticline and southern Brooks Range, Alaska: Journal of Geology, v. 95, p. 118–126.
Prindle, L. M., 1905, The gold placers of the Fortymile, Birch Creek, and Fairbanks regions, Alaska: U.S. Geological Survey Bulletin 251, 89 p.
——, 1906, The Yukon-Tanana region, Alaska; Description of Circle quadrangle: U.S. Geologic Survey Bulletin 295, 27 p., 1 pl.
——, 1913, A geologic reconnaissance of the Circle quadrangle, Alaska: U.S. Geological Survey Bulletin 538, 82 p.
Puchner, C. C., 1984, Intrusive geology of the Ray Mountains batholith: Geological Society of America Abstracts with Programs v. 16, p. 329.
Reiser, H. L., Lanphere, M. A., and Brosgè, W. P., 1965, Jurassic age of a mafic igneous complex, Christian quadrangle, Alaska: U.S. Geological Survey Professional Paper 525-C, p. C68–C71.
Robinson, M. S., 1983, Bedrock geologic map of the Livengood C-4 Quadrangle, Alaska: Alaska Division of Geological and Geophysical Surveys Report of Investigations 83-4, scale 1:40,000.
Roots, C. F., 1983, Mount Harper Complex, Yukon; Early Paleozoic volcanism at the margin of the Mackenzie platform, *in* Current Research, part A: Geological Survey of Canada Paper 83-1A, p. 423–427.
——, 1988, Cambo-Ordovician volcanic rocks in eastern Dawson map-area, Ogilvie Mountains, Yukon, *in* Yukon geology; Annual report of geology section: Department of Indian Affairs and Northern Development, v. 2, p. 81–87.
Roots, C. F., and Moore, J. M., Jr., 1982, Proterozoic and early Paleozoic volcanism in the Ogilvie Mountains, *in* Yukon exploration and geology 1982: Department of Indian and Northern Affairs, Exploration and Geological Services Division, p. 55–62.
Savage, N. M., Blodgett, R. B., and Jaeger, H., 1985, Conodonts and associated graptolites from the late Early Devonian of east-central Alaska and western Yukon Territory: Canadian Journal of Earth Sciences, v. 22, no. 12, p. 1880–1883.
Schoennagel, F. H., 1977, Mississippian unconformities in northern Alaska related to Antler tectonic pulses: American Association of Petroleum Geologists Bulletin, v. 61, no. 3, p. 435–442.
Schrader, F. C., 1900, Preliminary report on a reconnaissance along the Chandalar and Koyukuk Rivers, Alaska, in 1899: U.S. Geological Survey Twenty-first Annual Report, p. 441–486.
——, 1904, A reconnaissance in northern Alaska: U.S. Geological Survey Professional Paper 20, 139 p.
Silberling, N. J., and Jones, D. L., eds., 1984, Lithotectonic terrane maps of the North American Cordillera: U.S. Geological Survey Open-File Report 84–523, scale 1:2,500,000.
Silberman, M. L., Moll, E. J., Chapman, R. M., Patton, W. W., Jr., and Connor, C. L., 1979a, Potassium-argon age of granitic and volcanic rocks from the Ruby, Medfra, and adjacent quadrangles, west-central Alaska: U.S. Geological Survey Circular 804-B, p. B63–B66.
Silberman, M. L., Moll, E. J., Patton, W. W., Jr., Chapman, R. M., and Connor, C. L., 1979b, Precambrian age of metamorphic rocks from the Ruby province, Medfra and Ruby quadrangles; Preliminary evidence from radiometric age data: U.S. Geological Survey Circular 804-B, p. B66–B68.
Smith, G. M., and Puchner, C. M., 1985, Geology of the Ruby geanticline between Ruby and Poorman, Alaska, and the tectonic emplacement of the Ramparts Group: EOS Transactions of the American Geophysical Union, v. 46, p. 1102.
Smith, G. M., and Rubin, C. M., 1987, Devonian-Mississippian arc from the northern Sierras to the Seward Peninsula; Record of a protracted and diachronous Antler-age orogeny: Geological Society of America Abstracts with Programs v. 19, p. 849.
Smith, T. E., 1983, Bedrock geologic map of the Livengood C-3 Quadrangle, Alaska: Alaska Division of Geological and Geophysical Surveys Report of Investigations 83-5, scale 1:40,000.
Struik, L. C., 1981, A re-examination of the type area of the Devono-Mississippian Cariboo orogeny, central British Columbia: Canadian Journal of Earth Sciences, v. 18, p. 1767–1775.
——, 1987, The ancient western North American margin; An alpine rift model for the east-central Canadian Cordillera: Geologic Survey of Canada Paper 87-15, 19 p.
Tailleur, I. L., 1972, Probable rift origin of Canada Basin, Arctic Ocean, *in* Arctic geology: American Association of Petroleum Geologists Memoir 19, p. 526–535.
Tailleur, I. L., and Brosgè, W. P., 1970, Tectonic history of northern Alaska, *in* Adkinson, W. L., and Brosgè, M. M., eds., Proceedings of the Geological Seminar on the North Slope of Alaska: American Association of Petroleum Geologists, p. E1–E20.
Tempelman-Kluit, D. J., 1971, Stratigraphy and structue of the "Keno Hill Quartzite" in Tombstone River–Upper Klondike River map areas, Yukon Territory: Geological Survey of Canada Bulletin 180, 102 p.
——, 1977, Stratigraphic and structural relations between the Selwyn basin, Pelly-Cassiar platform, and Yukon crystalline terrane in Pelly Mountains, Yukon: Geological Survey of Canada Paper 77-1A, p. 223–227.
——, 1979, Transported cataclasite, ophiolite, and granodiorite in Yukon; Evidence of arc-continent collision: Geological Survey of Canada Paper 79-14, 27 p.
——, 1984, Counterparts of Alaska's terranes in Yukon [abs.], *in* Symposium; Cordilleran Geology and Mineral Exploration Status and Future Trends, Vancouver, B. C., Canada, Feb. 20–21, 1984: Cordilleran Section, Geological Association of Canada, p. 41–44.
Thompson, R. I., and Eisbacher, G. H., 1984, Late Proterozoic rift assemblages, northern Canadian Cordillera: Geological Society of America Abstracts with Programs, v. 16, p. 336.
Thompson, R. I., and Roots, C. F., 1982, Ogilvie Mountains project, Yukon; Part A, A new regional mapping program, *in* Current Research, part A: Geological Survey of Canada Paper 82-1A, p. 403–411.
Thompson, R. I., Mercier, E., and Roots, C. F., 1987, Extension and its influence on Canadian Cordilleran passive-margin evolution, *in* Coward, M. P., Dewey, J. F., and Hancock, P. L., eds., Continental extensional tectonics:

Geological Society of London Special Publication 28, p. 409–417.
Thrupp, G. A., and Coe, R. S., 1986, Paleomagnetic evidence for counterclockwise rotation of west-central Alaska since the Paleocene [abs.]: EOS Transactions of the American Geophysical Union, v. 67, no. 44, p. 921.
Till, A. B., Schmidt, J. M., and Nelson, S. W., 1987, Thrust-involvement of Proterozoic and Mesozoic metamorphic rocks, southwestern Brooks Range, Alaska: Geological Society of America Abstracts with Programs v. 19, p. 458.
Turner, D. L., Forbes, R. B., and Dillon, J. T., 1979, K-Ar geochronometry of the southwestern Brooks Range, Alaska: Canadian Journal of Earth Sciences, v. 16, p. 1789–1804.
Weber, F. R., and Foster, H. L., 1982, Tertiary(?) conglomerate and Quaternary faulting in the Circle Quadrangle, Alaska: U.S. Geological Survey Circular 844, p. 58–61.
Weber, F. R., Smith, T. E., Hall, M. H., and Forbes, R. B., 1985, Geologic guide to the Fairbanks-Livengood area, east-central Alaska: Anchorage, Alaska Geological Society, 44 p.
Weber, F. R., and 8 others, 1988, Geologic framework, *in* Weber, F. R., McCammon, R. B., Rinehart, C. D., and Light, T. D., Geology and mineral resources of the White Mountain National Recreation Area, east-central Alaska: U.S. Geological Survey Open-File Report 88–284, p. 4–58.
Wheeler-Crowder, K. L., Forbes, R. B., Weber, F. R., and Rinehart, C. D., 1987, Lithostratigraphy of the Ordovician Fossil Creek volcanics with petrology and geochemistry of included metabasalts, White Mountains, east-central Alaska: U.S. Geological Survey Circular 998, p. 70–73.
Williams, G. D., 1985, Thrust tectonics in the south-central Pyrenees: Journal of Structural Geology v. 7, no. 1, p. 11–17.
Young, F. G., 1973, Mesozoic epicontinental, flyschoid, and molassoid depositional phases of Yukon's north slope, *in* Aitken, J. D., and Glass, D. J., eds., Proceedings of the Symposium on the Geology of the Canadian Arctic: Geological Association of Canada and the Canadian Society of Petroleum Geologists, p. 181–202.
Young, G. M., 1982, The late Proterozoic Tindir Group, east-central Alaska; Evolution of a continental margin: Geological Society of America Bulletin v. 93, no. 8, p. 759–783.

Manuscript Accepted by the Society October 19, 1990

ACKNOWLEDGMENTS

Many colleagues contributed expertise as consultants, critics, or editors that significantly improved this presentation of the geology of east-central Alaska. The principal contributors and their main areas of interest are: Michael Churkin, stratigraphy of the Charley River area; Thomas Dutro, Paleozoic stratigraphy of Porcupine platform and southeastern Brooks Range; Florence Weber, geology of the Circle/Livengood area; John Cady, geophysics of east-central Alaska; Robert Chapman, geology of the Livengood/Tanana area; Steven Gordey, geology of southern Yukon Territory and the mid-Paleozoic Earn Group; and Robert Thompson, geology of the Dawson area. However, the author accepts full responsibility for the interpretations and conclusions presented here, which may not necessarily represent the views of the contributors.

Printed in U.S.A.

The Geology of North America
Vol. G-1, The Geology of Alaska
The Geological Society of America, 1994

Chapter 6

Geology of the Yukon-Tanana area of east-central Alaska

Helen L. Foster and Terry E. C. Keith
U.S. Geological Survey, 345 Middlefield Road, Menlo Park, California 94025
W. David Menzie
U.S. Geological Survey, Reston, Virginia 22092

INTRODUCTION

East-central Alaska as described in this volume (Fig. 1) is a physiographically diverse region that includes all or parts of the following physiographic divisions (Wahrhaftig, this volume): Northern Foothills (of the Alaska Range), Alaska Range (north of the northernmost strand of the Denali fault system), Tanana-Kuskokwim Lowland, Northway-Tanacross Lowland, and the Yukon-Tanana Upland. The Northern Foothills are largely rolling hills in Pleistocene glacial deposits and dissected Tertiary nonmarine sedimentary rocks. The included part of the Alaska Range is composed of highly dissected terranes of metamorphic rocks that have been intruded by Cretaceous and Tertiary igneous rocks. Mountain peaks reach altitudes as high as 4,000 m, and relief is commonly more than 1,000 m. Glaciers have carved a rugged topography. The Tanana-Kuskokwim Lowland is covered with thick glacial, alluvial, and wind-blown deposits. The Northway-Tanacross Lowland consists of three small basins mantled with outwash gravel, silt, sand, and morainal deposits. The Yukon-Tanana Upland, the largest of the physiographic divisions, consists of maturely dissected hills and mountains with altitudes as high as 1,994 m, and relief ranging from a few to hundreds of meters. Some of the highest areas supported small alpine glaciers during the Pleistocene, and rugged topography resulted locally.

With the exception of the Alaska Range, outcrops in east-central Alaska are commonly widely scattered and small, due to extensive surficial deposits and vegetation. The vegetation ranges from heavy spruce forests along large streams to tundra at elevations of approximately 1,000 m. The region is largely in the zone of discontinuous permafrost. Most of the low-lying areas, as well as the high mountain areas, are in the permafrost regime. Some areas, mostly intermediate in elevation, are permafrost free.

East-central Alaska is composed of a number of accreted terranes (Jones and others, 1984; Silberling and others, this volume) that have continental, oceanic, and possibly island-arc affinities. The largest terrane, the Yukon-Tanana, has mostly continental affinities. The small Seventymile terrane has oceanic affinities. The southern part of the Yukon-Tanana terrane, which includes the Lake George, Macomb, and Jarvis Creek Glacier terranes (or subterranes) of the Mount Hayes Quadrangle, may have island-arc affinities (Gilbert and Bundtzen, 1979; Nokleberg and Aleinikoff, 1985). The Hayes Glacier and Windy terranes in the Mount Hayes Quadrangle are also suggested to be tectonic slices of an island-arc or possibly of a submerged continental-margin arc (Nokleberg and Aleinikoff, 1985).

Terminology of igneous rocks used in this chapter follows that of Streckeisen (1976).

YUKON-TANANA TERRANE

The Yukon-Tanana terrane (YT) consists largely of the area lying between the Yukon and Tanana Rivers but, as defined by Silberling and others (this volume), it also includes some of the Alaska Range and its foothills north of the Denali fault system. Nokleberg and Aleinikoff (1985) describe the part of the Mount Hayes Quadrangle in the YT as the Lake George terrane, which lies north of the Tanana River, and the Macomb, Jarvis Creek Glacier, Hayes Glacier, and Windy terranes, which lie between the Tanana River and the Denali fault. For this chapter, we consider those terranes, with the exception of the Windy terrane, to be subterranes of the YT. In the Healy, Mount McKinley, and Kantishna Quadrangles, isolated, probably fault-bounded exposures of rocks are also included in the Yukon-Tanana terrane. For ease in discussing stratigraphy and structure, the Macomb, Jarvis Creek Glacier, Hayes Glacier, and Windy terranes and the areas to the southwest and southeast of them will be discussed separately from the part of the Yukon-Tanana terrane between the Yukon and Tanana Rivers. In this chapter, we refer to the part of the Yukon-Tanana terrane north of the Tanana River as the YTTN. The eastern YTTN also includes the small extension into Alaska of the Stikinia terrane (Silberling and others, this volume).

YUKON—TANANA TERRANE NORTH OF THE TANANA RIVER

The YTTN has been referred to as the "Yukon-Tanana region" (Mertie, 1937), the "Yukon-Tanana upland" (Foster and others, 1973), and the "Yukon Crystalline Terrane" (Tempelman-Kluit, 1976; Churkin and others, 1982). It is nearly coinci-

Foster, H. L., Keith, T.E.C., and Menzie, W. D., 1994, Geology of the Yukon–Tanana area of east-central Alaska, *in* Plafker, G., and Berg, H. C., eds., The Geology of Alaska: Boulder, Colorado, Geological Society of America, The Geology of North America, v. G-1.

dent with the previously described physiographic division called the Yukon-Tanana Upland (Wahrhaftig, this volume). The YTTN is primarily a terrane of quartzitic, pelitic, calcic, and mafic metasedimentary rocks with some mafic and felsic meta-igneous rocks that have been extensively intruded by Mesozoic and Cenozoic granitic rocks and minor amounts of intermediate and mafic rocks. Cretaceous and Cenozoic volcanic rocks are abundant in the eastern part. Late Cretaceous and Tertiary sedimentary rocks were deposited in small, widely separated nonmarine basins. The YTTN has been considered a composite terrane by Churkin and others (1982), and many problems related to its complex geologic history are as yet unresolved.

Metamorphic rocks

The geology of the YTTN was reported by Mertie (1937), who described the stratigraphy in some detail. Mertie included many of the metamorphic rocks in a formation called the Birch Creek Schist, of Precambrian age, and eventually most of the metamorphic rocks of east-central Alaska were included in this formation. Because usage of the name Birch Creek Schist became so broad, and because parts of the formation were found to be younger than Precambrian, the name lost usefulness. In 1973, Foster and others recommended that it be abandoned. In recent reconnaissance geologic mapping the metamorphic rocks have been divided into many units, but because of the lack of information on age and structural relations, few new formations have been formally named and described.

The metamorphic rocks within the YTTN vary in such aspects as composition and origin of protoliths, present lithology, structure, and metamorphic history. On the basis of these characteristics and other data, Churkin and others (1982) divided the YTTN into four subterranes (Y_1 to Y_4). Although structural

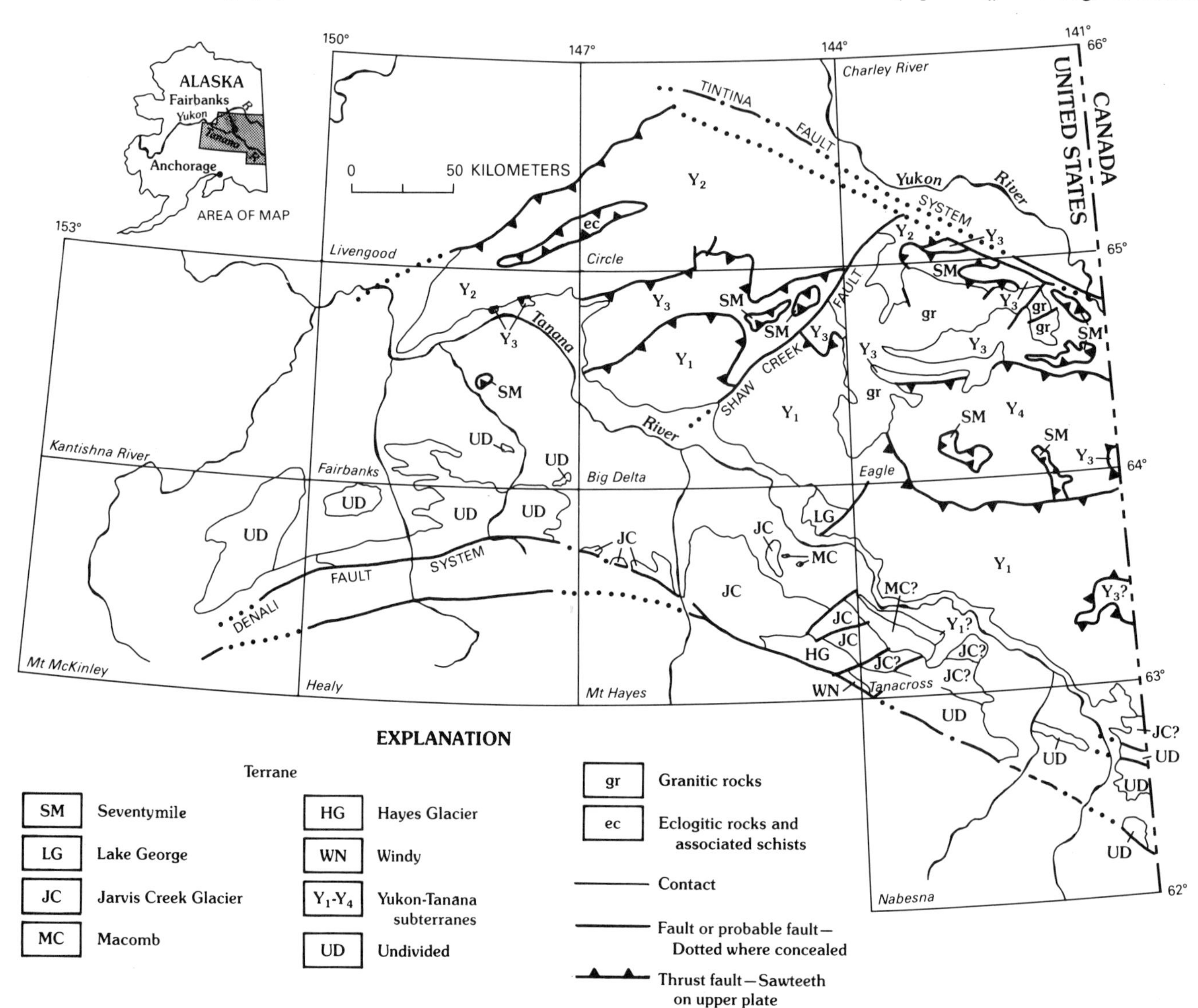

Figure 1. Map of east-central Alaska showing locations of quadrangles, terranes, and subterranes. Subterranes modified from Churkin and others (1982). Terranes of Mount Hayes Quadrangle from Nokleberg and Aleinikoff (1985).

details and stratigraphic relations are poorly known, these subdivisions are useful for descriptive purposes and for discussion of regional relations. We have used them, with minor modifications, as the framework for our discussion of the metamorphic rocks of the YTTN (Fig. 1; Table 1).

Subterrane Y_1 (northeastern Mount Hayes, southern Big Delta, southwestern Eagle and northern Tanacross Quadrangles). This is the largest and southernmost subterrane of the YTTN (Fig. 1). The part of this subterrane in the Mount Hayes Quadrangle has been termed the "Lake George terrane" (Nokleberg and others, 1983). The southern boundary of Y_1 in the Mount Hayes Quadrangle (the southern boundary of the Lake George terrane) is the largely concealed Tanana River fault (Nokleberg and Aleinikoff, 1985). An extension of this or adjoining faults probably forms the southern Y_1 boundary elsewhere. The northern boundary of Y_1 is probably a thrust fault. The rocks are all metamorphosed to the amphibolite facies, probably at intermediate pressures. Protoliths primarily were quartzitic and pelitic sedimentary rocks and felsic intrusive rocks with some intermediate and mafic intrusive and volcanic rocks. Calcareous rocks, rare to absent in the eastern part of Y_1, occur in very minor amounts in the western part of Y_1. Pelitic rocks are more abundant in the western part of Y_1, and quartzose rocks predominate in the eastern part of Y_1. The dominant rock types include quartz-biotite gneiss and schist, sillimanite gneiss, quartzite, amphibolite, and orthogneiss (unit ag, Fig. 2), including augen gneiss.

Augen gneiss, a widely distributed and characteristic rock type having a granitic composition and blastoporphyritic texture, occurs primarily east of a major high-angle fault, the Shaw Creek fault (Fig. 1; Dusel-Bacon and Aleinikoff, 1985). Large augen (megacrysts or porphyroblasts) of potassium feldspar range from 1 to 9 cm in longest dimension and have been modified into augen by mylonitization. These augen gneisses, occurring in the Big Delta, Eagle, Mount Hayes, and Tanacross Quadrangles, are considered to be a part of deformed and metamorphosed intrusions of porphyritic granite. Dusel-Bacon and Aleinikoff (1985) postulated that these and similar augen gneisses and other orthogneisses in the Yukon Territory are part of an intrusive belt that extends from the central part of the Big Delta Quadrangle in east-central Alaska into the Yukon Territory. Most of these augen gneisses were included in the Pelly Gneiss of McConnell (1905); Mertie (1937) used the term "Pelly Gneiss" for the augen gneisses in eastern Alaska but did not map them separately from the Birch Creek Schist. Other types of augen gneiss also occur in minor amounts in subterrane Y_1.

Sillimanite gneiss is a major rock type around the augen gneiss in the Big Delta Quadrangle and occurs in a large area that has been interpreted as a gneiss dome (Dusel-Bacon and Foster, 1983) on the northwest side of the Shaw Creek fault (Fig. 1). Metamorphic grade (second sillimanite isograd) is highest in the central part of the gneiss dome and decreases northward. Triple-point conditions for alumina silicate minerals are postulated for the schist on the north flank of the gneiss dome. The gneiss dome is mostly quartz-orthoclase-plagioclase-biotite-sillimanite ± muscovite gneiss. Cordierite occurs in some of the gneisses. To the north, muscovite gradually increases and K-feldspar decreases. North of the Salcha River, the pelitic rocks are interlayered with quartzitic schist, quartzite, marble, amphibole schist, and quartzofeldspathic schist.

The rocks of subterrane Y_1 are well foliated, and foliation is folded at least once. Gneissic banding is well developed in the augen gneiss and is concordant with the foliation in the surrounding metamorphic rocks (Dusel-Bacon and Aleinikoff, 1985). Foliation is plastically folded in the gneiss dome and some other gneisses. Most of the rocks show varying degrees of mylonitization and post-mylonitization recrystallization. In the Lake George terrane (Mount Hayes Quadrangle), Aleinikoff and Nokleberg (1985a) described small-scale isoclinal folds in pelitic schist; axial planes parallel foliation and compositional layering but fold an older foliation. A lineation formed by the intersection of the axes of small tight folds with foliations occurs locally in augen gneiss in the Big Delta Quadrangle and is especially well developed in the northeastern Tanacross Quadrangle.

The only information on the age of the protoliths of the metamorphic rocks in subterrane Y_1 comes from U-Pb dating of zircon. U-Pb analyses of zircon from medium-grained schistose granitic rock from the Mount Hayes Quadrangle indicate a Devonian intrusive age, about 360 Ma (Aleinikoff and Nokleberg, 1985a). Detailed study of augen gneiss from the Big Delta Quadrangle indicates an intrusive age of Mississippian, about 345 Ma (Dusel-Bacon and Aleinikoff, 1985). U-Pb analyses of zircon from augen gneiss in the Tanacross Quadrangle (Aleinikoff and others, 1986) and also of augen gneiss in the southeastern Yukon Territory (Mortensen, 1983) indicate a Mississippian intrusive age. Dusel-Bacon and Aleinikoff (1985) interpret the paragneisses, schists, and quartzites of subterrane Y_1 as wall rocks of the orthogneisses; therefore they are Mississippian or older in the Big Delta, Eagle, and Tanacross Quadrangles and Devonian or older in parts of the Mount Hayes Quadrangle. An early Paleozoic age seems most likely for most of them, but a Precambrian age for at least some cannot be ruled out. Both the augen gneiss and some quartzites are shown by U-Pb analyses of zircons to have an inherited early Proterozoic component (2.1 to 2.3 b.y. old) (Aleinikoff and others, 1986). The origin of this Proterozoic material is unknown, but a high initial $^{87}Sr/^{86}Sr$ ratio in the augen gneiss supports petrologic indications of involvement of continental crustal material in the formation of these rocks (Dusel-Bacon and Aleinikoff, 1985).

Deformed and recrystallized ultramafic rocks in isolated outcrops (too small to show on Fig. 2) are infolded with gneisses and schists of subterrane Y_1; most are concentrated in the south-central and southeastern Big Delta and southwestern Eagle Quadrangles. Locally preserved textures indicate that some of the ultramafic rocks were originally peridotite (harzburgite). Intense recrystallization has formed elongate-oriented olivine crystals with pods of granular magnetite as long as 5 cm. Other metamorphic minerals include hornblende, actinolite, serpentine, chlorite, talc, anthophyllite, and magnesite. The ultramafic rocks probably

TABLE 1. SUBTERRANES OF THE YUKON–TANANA TERRANE NORTH OF THE TANANA RIVER (YTTN)

Subterrane	Included map units (Fig. 2)	Lithology	Metamorphic facies	Distinctive characteristics	Age of protolith
Y_1 (Includes Lake George terrane)	ag	Gneiss, schist, amphibolite, quartzite	Amphibolite (moderate pressure)	Augen gneiss common; marble and other calcareous rocks rare or absent	Mississippian and pre-Mississippian
Y_2	qq	Quartzite and quartzitic schist with small amounts of pelitic schist, calc-silicate rocks, mafic schist, and rare marble	Greenschist (moderate pressure)	Quartzite and quartzitic schists that are commonly recrystallized mylonites with megacrysts of quartz and/or feldspar	Unknown, may be early Paleozoic
	ps	Pelitic schist, quartzite, marble, and amphibolite	Amphibolite (moderate pressure)	Sillimanite and kyanite-bearing pelitic schists	Unknown, may be early or middle Paleozoic
	ec	Eclogite, amphibolite pelitic schist, and mafic glaucophane-bearing schist	Amphibolite (high pressure)	Occurrence of eclogite and of glaucophane in mafic schists	Unknown
Y_3	ms	Mylonitic schist, semi-schist, quartz-white mica chlorite schist, quartzite, and minor phyllite, marble, and greenstone	Greenschist (moderate pressure)	Light greenish gray color characteristic. Abundant mylonitic schists	Unknown, may be Carboniferous or middle Paleozoic
	cp	Calcareous phyllite, phyllite marble, quartzite, and argillite	Low greenschist	Calcareous phyllite with thin, crumbly layers. Weathers leaving a lag gravel of white quartz on surface. Gray and dark gray quartzite and argillite overlie calcareous phyllite	Unknown, may be early or middle Paleozoic
	sm	Greenschist, calcareous greenschist, quartz-chlorite white mica schist, marble, greenstone, quartzite	Greenschist	Abundant layers and lenses of marble; some schists contain megacrysts of quartz and/or feldspar	Paleozoic; partly Mississippian
	ws	Quartz-chlorite white mica schist with minor quartzite, phyllite, and metavolcanic rocks	Greenschist	Marble layers rare to absent. Light green color characteristic. Some schists have megacrysts of quartz and/or feldspar	Unknown, presumably middle or late Paleozoic
	qs	Quartzite with minor phyllite, carbonaceous quartz schist and graphitic schist	Greenschist	Overall dark gray color characteristic; dark gray quartzite dominant rock type	Unknown, probably Paleozoic
Y_4	gs	Quartzitic and pelitic gneiss and schist, quartzite, marble, and amphibolite	Amphibolite	Thick masses and layers of coarsely crystalline marble. Quartz-biotite-hornblende gneiss a characteristic and common rock type	Probably Paleozoic

composed a thrust sheet over part of subterrane Y_1 early in the development of the subterrane and were metamorphosed and deformed with the rocks of subterrane Y_1.

The metamorphic history of subterrane Y_1 is not known in detail. Nokleberg and Aleinikoff (1985) interpreted the subterrane in the Mount Hayes Quadrangle (Lake George terrane) to be intensely deformed and regionally metamorphosed at least once, and more likely twice, at conditions of the middle amphibolite facies. Then, during a late stage of the second regional metamorphism and deformation, the Lake George terrane was intruded by Cretaceous granitic rocks, which were subsequently metamorphosed, along with the older wall rocks, under conditions of the lower greenschist facies. Interpretation of metamorphic history of subterrane Y_1 in the Big Delta, Eagle, and Tanacross Quadrangles is based largely on the study of the augen gneiss, adjacent wall rocks, and the gneiss dome. Dusel-Bacon and Aleinikoff (1985) suggested that major amphibolite-facies metamorphism and deformation may have closely followed intrusion of the augen gneiss, on the basis of structural characteristics of the augen gneiss and limited isotopic data on wall rocks.

An Early Cretaceous thermal event caused lead loss in zircon from the U-rich zircon fractions from sillimanite gneiss and quartzite and also affected Rb-Sr and K-Ar isotopic systems (Aleinikoff and others, 1986; Wilson and others, 1985). If temperatures reached only greenschist-facies levels in subterrane Y_1, they would have produced minor retrograde effects. However, if amphibolite-facies temperatures were reached, the effects might not be detected because the rocks were already metamorphosed to amphibolite facies. The only recognized petrologic changes that can be attributed to this Early Cretaceous metamorphism outside of the Mount Hayes Quadrangle are minor, and no younger metamorphic events have been identified.

Nokleberg and Aleinikoff (1985) interpreted the Lake George terrane and other terranes to the south as shallow to deep parts of a submarine igneous arc of Devonian age, and believed that it was either an island arc or submerged continental-margin arc. Dusel-Bacon and Aleinikoff (1985) suggested that a Mississippian belt of intrusions developed either below or inland from a continental arc, and that in the latter case, the belt of augen gneiss plutons could be analogous to belts of peraluminous plutons that occur inland from continental margins in some orogenic belts.

Subterrane Y_2 (Circle, northern Big Delta, southwestern Charley River, Livengood, and Fairbanks Quadrangles). The Y_2 subterrane is bounded on the north by the Tintina fault system and on the south and west by thrust faults; the eastern boundary is obscured by Mesozoic granitic plutons. The western fault contact is considered as the terrane boundary of the YT (Laird and Foster, 1984; Foster and others, 1983). Subterrane Y_2 is composed of three fairly distinct groups of rocks: one group (unit qq, Fig. 2) consists mostly of quartzites and quartz schists of greenschist to amphibolite facies and has been referred to informally as the Fairbanks schist unit in the Fairbanks Quadrangle (Bundtzen, 1982); the second group (unit ps, Fig. 2) consists of amphibolite- to epidote-amphibolite–facies schist, quartzite, marble, and amphibolites, some of which have been included in the Chena River sequence (Hall and others, 1984). These two groups of rocks are in thrust contact, as indicated by sharp lithologic changes and their map patterns, but thrusting occurred before major metamorphism because metamorphic isograds do not follow the unit contacts (Foster and others, 1983). A third group of thrust-bounded rocks (unit ec, Fig. 2) occurs in the southwestern Circle and southeastern Livengood Quadrangles and consists of eclogite associated with amphibolite, impure marble, pelitic schist, and rare glaucophane-bearing schist. These rocks, referred to as the Chatanika terrane by Bundtzen (1982), probably had a metamorphic and deformational history different from that of the remainder of subterrane Y_2 before they were thrust together.

Quartzite and quartzitic schists (unit qq). This group of rocks crops out in about half of the Circle Quadrangle and occurs in the Fairbanks and Livengood Quadrangles. Quartzite and quartzitic schist are the most abundant rock types in this unit, but minor amounts of pelitic schist, calc-silicate rocks, mafic schist, and rare marble are interlayered. In the Circle Quadrangle, quartzite and quartzitic schist are fine to coarse grained and equigranular or fine to coarse grained with rare to abundant megacrysts of quartz and less abundant feldspar, ranging from less than a millimeter to more than a centimeter in diameter. Megacrysts are clear, white, gray, blue-gray, or black, and may be strained monocrystalline or polycrystalline grains. The matrix is generally a mosaic of strained quartz, minor feldspar, and white mica. Locally, chlorite, biotite, and small garnets are present. Most of these rocks are mylonites (Wise and others, 1984); many show syntectonic recrystallization in quartz, especially near fault contacts (Foster and others, 1983).

Mylonitization is particularly evident along the northwestern margin of this unit. Foster and others (1983) and Laird and Foster (1984) interpreted this as a major zone of thrusting; however, some workers (Hall and others, 1984) considered the northwestern contact of this unit to be gradational with a grit unit of the Wickersham terrane of Jones and others (1984). A small area of "grit" and quartzite in the south-central part of the Circle Quadrangle was interpreted as a window in the thrust sheet of this unit (Foster and others, 1983).

Quartzitic rocks are interlayered with minor amounts of pelitic schist (quartz + plagioclase + muscovite + chlorite schist commonly with biotite ± garnet or with chloritoid ± garnet). Garnet is absent in the northern part of this unit. Rare, thin marble layers occur. Chlorite schist, locally magnetic, is interlayered and infolded with quartzite and pelitic schist. Interpretation of aeromagnetic data (Cady and Weber, 1983) suggests that magnetic chlorite schists, which are a poorly exposed unit, may be more abundant than is apparent from outcrops in the southwestern part of the Circle Quadrangle.

A mafic schist is the dominant rock type in a 190-km^2 area in the east-central part of the Circle Quadrangle. Green chlorite-quartz-carbonate schist, generally with abundant plagioclase porphyroblasts, is interlayered with amphibole (commonly actinolite) + chlorite + epidote + plagioclase + quartz + sphene + biotite

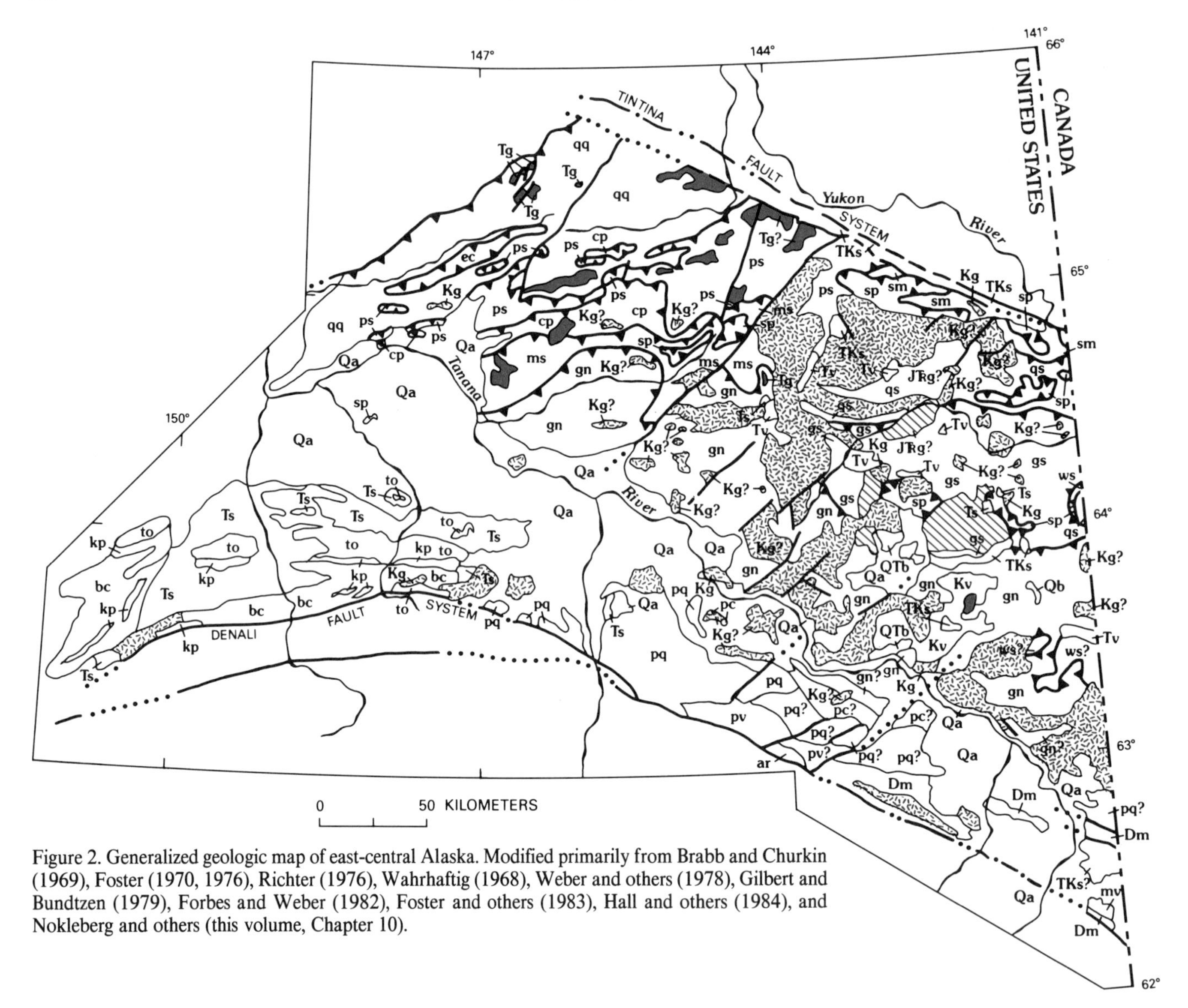

Figure 2. Generalized geologic map of east-central Alaska. Modified primarily from Brabb and Churkin (1969), Foster (1970, 1976), Richter (1976), Wahrhaftig (1968), Weber and others (1978), Gilbert and Bundtzen (1979), Forbes and Weber (1982), Foster and others (1983), Hall and others (1984), and Nokleberg and others (this volume, Chapter 10).

or white mica ± carbonate ± garnet schist. The protolith may have been, in part, mafic pyroclastic rocks. Minor marble, quartzite, and pelitic schist are interlayered with the mafic schist (Foster and others, 1983).

The westernmost metamorphic rocks of the Yukon-Tanana terrane form unit "qq" of the Fairbanks and Livengood Quadrangles (Bundtzen, 1982). Although poorly exposed, they have been examined in detail. In the Fairbanks Quadrangle, this unit is dominantly quartzite and muscovite-quartz schist ± garnet, biotite, and chlorite. It is estimated to be more than 1,000 m thick (Hall and others, 1984). Interstratified near the center of this group of rocks is a 130-m-thick sequence of interlensing felsic schist, micaceous quartzite, chloritic or actinolitic schist, graphitic schist, minor metabasite, metarhyolite, calc-silicate layers, banded gray marble, and quartzite, referred to informally as the Cleary sequence (Bundtzen, 1982). These rocks, interpreted to be largely of distal volcanic origin, host lode mineral occurrences in the Fairbanks mining district (Hall and others, 1984).

Four distinct deformational events are recognized in this unit in the Circle Quadrangle (Cushing and Foster, 1984). The first (D_1) produced a penetrative schistosity (S_1) parallel or subparallel to gently dipping axial planes of rarely observed tight to isoclinal recumbent folds. S_1 commonly parallels compositional layering and is everywhere prevalent in the metamorphic rocks. Folds associated with the second deformational event, D_2, are ubiquitous and range from tight to isoclinal with rounded and chevron fold hinges. Amplitudes and wavelengths range from microscopic to several meters. A second schistosity, S_2, developed locally owing to mechanical rotation of S_1. Folds of the third deformational event, D_3, are recumbent, tight to isoclinal. Because fold styles and orientations are similar, structural features of D_3 are difficult to distinguish from D_2. The fourth deformational event (D_4) is characterized by gentle and open folds that deform all previous structures. Wavelengths and amplitudes are generally less than 50 cm but may be as large as 5 m.

Unit "qq" was subjected to moderate-pressure greenschist-

EXPLANATION

Unconsolidated deposits

- **Qa** Quaternary alluvial deposits; primarily alluvium but includes colluvial, glacial, and eolian deposits

Sedimentary rocks

- **Ts** Tertiary sandstone, conglomerate, shale, coal, and tuffaceous rocks. Includes informally designated coal-bearing formation of local usage overlain by the Nenana Gravel
- **TKs** Cretaceous and/or Tertiary sedimentary rocks (conglomerate, sandstone, coal, shale, and tuffaceous rocks)

Igneous rocks

Volcanic

- **Qb** Quaternary alkali-olivine basalt
- **QTb** Tertiary and (or) Quaternary basalt and gabbro
- **Tv** Tertiary felsic to mafic volcanic rocks
- **Kv** Cretaceous volcanic and volcaniclastic rocks

Plutonic

- **Tg** Tertiary granitic rocks
- **Kg** Cretaceous granitic rocks
- **JTRg** Late Triassic and Early Jurassic granitic rocks

Metamorphic rocks

North of Tanana River

Rocks of higher metamorphic grade

- **ps** Pelitic schist, quartzite, marble, and amphibolite
- **gs** Gneiss, schist, marble, amphibolite, and quartzite
- **gn** Gneiss (including augen gneiss and sillimanite gneiss), schist, amphibolite, and quartzite
- **ec** Eclogite and associated rocks

Rocks of lower metamorphic grade

- **cp** Calc-phyllite and quartzite
- **ms** Mylonitic schist, semischist, quartzite, phyllite, marble, and greenstone
- **ws** Quartz-chlorite-white mica schist
- **qs** Quartzite and quartzitic schist
- **sm** Schist, including greenschist, marble, greenstone, and quartzite
- **qq** Quartzite, quartzitic schist, mafic schist, and calc-schist
- **sp** Serpentinized peridotite, greenstone, and associated metasedimentary rocks including chert of Mississippian, Permian, and Triassic age

South of Tanana River

- **gn** Gneiss (including augen gneiss), schist, amphibolite, and quartzite
- **pc** Pelitic schist, calc-schist, and quartz-feldspar schist
- **pq** Pelitic schist, quartzite, and schistose volcanic and plutonic rocks
- **pv** Phyllite and schistose volcanic rocks
- **ar** Argillite, limestone, conglomerate, and other metasedimentary rocks
- **mv** Metavolcanic and volcaniclastic rocks
- **Dm** Metasedimentary rocks of probable Devonian age intruded by diorite and gabbro
- **to** Totatlinika schist of possible Late Devonian to Mississippian age; mostly quartzitic schist and metavolcanic rocks
- **kp** Keevy Peak Formation of possible Ordovician to Devonian age; includes carbonaceous phyllite, quartzite, stretched conglomerate, and white mica-quartz schist
- **bc** Birch Creek schist of former usage; includes quartzitic, graphitic, chloritic, and calcareous schist, marble, and greenstone

Contact

Fault or probable fault—Dashed where approximately located; dotted where concealed

Thrust fault or postulated thrust fault—Dotted where concealed; sawteeth on upper plate facies regional metamorphism that was more intense in the southern part of the unit (Foster and others, 1983). Polymetamorphism of regional extent has not been identified. Rolled garnet and plagioclase grains are common but appear to be explained by one syntectonic growth event. Inclusions of chloritoid found within, but not outside of, garnet grains can be explained by progressive metamorphism through and above the conditions of chloritoid stability (Burack, 1983). Contact-metamorphic effects are superimposed upon the regional metamorphism around most Tertiary plutons; biotite and amphibole have developed across the foliation, and garnet is commonly all or partly chloritized.

The ages of the protoliths of this unit are unknown because no fossils have been found. U-Pb determinations on zircon from one quartzite indicate that the protolith included material from an early Proterozoic source of essentially the same age (2.1 to 2.3 Ga) as that of subterrane Y_1 (J. N. Aleinikoff, personal communication, 1983). The very large quartzitic component of these rocks and abundance of quartz megacrysts have led some workers to suggest that the protolith might have been a part of the Canadian Windermere Supergroup (F. R. Weber, personal communication, 1979).

Pelitic schist, quartzite, marble, and amphibolite (unit ps). These rocks are mostly medium- to coarse-grained pelitic schist and gneiss with minor interlayers of quartzite, quartzitic schist, marble, and amphibolite. Other rocks included in this unit are augen gneiss, calc-silicate, and ultramafic rocks. Regional metamorphism ranges from amphibolite to epidote-amphibolite facies (sillimanite + potassium feldspar to garnet grade in pelitic schist and gneiss) with the highest-grade rocks occurring in the southeastern part of the Circle Quadrangle; metamorphic grade decreases northward and westward. A characteristic mineral assemblage of the highest-grade rocks is quartz + plagioclase + white mica + biotite + sillimanite ± potassium feldspar ± garnet. Metamorphic grade seems to be close to the muscovite + quartz = sillimanite + potassium feldspar ± H_2O isograd. Other pelitic assemblages in the higher-grade part of the unit are: biotite + garnet + staurolite ± kyanite; biotite + garnet + kyanite; biotite + garnet + kyanite + sillimanite; all with quartz + white mica + plagioclase. Augen gneiss is mostly a biotite felsic gneiss containing augen-shaped potassium feldspar porphyroblasts. A characteristic mineral assemblage is potassium feldspar, commonly microcline + quartz + plagioclase + brown biotite + white mica. Augen are generally composed of two or more potassium feldspar crystals. Variations in the size of augen, relative proportions of major mineral constituents, and field relations suggest that the augen gneisses do not all have the same origin and that protoliths probably include both igneous and sedimentary rocks. Some augen gneiss occurrences may be folded and metamorphosed sills or dikes (F. R. Weber, personal communication, 1979), but other occurrences, especially those that cap high parts on ridges, could be thrust remnants of subterrane Y_1. A common pelitic assemblage to the north and west of the highest-grade rocks is quartz + white mica + biotite + garnet + chlorite + plagioclase. Mafic schist mineralogy is hornblende + plagioclase + quartz + epidote ± chlorite + biotite.

Small scattered outcrops of metamorphosed ultramafic rocks (too small to show on Fig. 2) consist mainly of actinolite, chlorite, serpentine, magnetite, chlorite, talc, and magnesite. Relict olivine, orthopyroxene, and clinopyroxene are found locally in rocks preserving textures of harzburgite. The ultramafic rocks appear to occur discontinuously at or near the edge of thrust plates composed of this unit.

Unit "ps" appears to have a metamorphic and deformational history similar to that described for unit "qq" in that isograds and folds are not related to contacts between the two units. As in unit "qq," all of the rocks of this unit are polydeformed, but polymetamorphism of regional extent has not been identified. However, contact metamorphism is indicated around some of the Tertiary plutons where pseudomorphs of white mica after staurolite and kyanite porphyroblasts occur.

The age of the protoliths of this unit is unknown. A single U-Pb age of 345 ± 5 Ma (Mississippian) (J. N. Aleinikoff, personal communication, 1985) was obtained on zircon from one augen orthogneiss, an age similar to that obtained on zircon from augen gneiss in subterrane Y_1 (Aleinikoff and others, 1986). Because this dated augen gneiss may be in thrust contact with the associated metasedimentary rocks rather than intrusive into them, its age may not provide an upper constraint on the protolith age of this unit. Although younger protolith ages cannot be ruled out, early and/or middle Paleozoic ages are reasonable possibilities for this unit.

Eclogite and associated rocks (unit ec). The only eclogitic rocks known in east-central Alaska occur in a small area in the southwestern part of subterrane Y_2, where bands and lenses of eclogite are intercalated with amphibolite, impure marble, pelitic schist, and mafic glaucophane-bearing schist (Fig. 2). They were first described by Prindle (1913) in what is now the southeastern part of the Livengood Quadrangle, but because of very limited exposure, were given little attention until rediscovered in the 1960s (Forbes and Brown, 1961). Swainbank and Forbes (1975) described their petrology. Eclogitic rocks also have been found in the southwestern part of the Circle Quadrangle (Foster and others, 1983).

The eclogitic rocks in the Livengood area consist of several combinations of garnet, omphacitic clinopyroxene, amphibole, calcite, phengitic mica, quartz, albite, epidote, sphene, and rutile (Swainbank *in* Hall and others, 1984). Bulk chemistry suggests that they may have been derived from marls and graywackes. Glaucophane and kyanite-staurolite-chloritoid–bearing assemblages also have been found (Brown and Forbes, 1984). Pyroxene-garnet, biotite-garnet, muscovite-paragonite, and aluminosilicate data suggest crystallization temperatures of 600° ± 50°C at pressures of 13 to 15 kb (Brown and Forbes, 1984). These eclogites are similar to eclogites from alpine-type orogenic terranes (Group C of Coleman and others, 1965). They occur in northwest-trending isoclinal recumbent folds that have been deformed by open or overturned folding along northeast-trending axes.

Eclogite in the Circle Quadrangle, which on the basis of garnet composition also falls within Group C of Coleman and others (1965), appears to be in a mafic layer within quartz + white mica + garnet (somewhat retrograded to chlorite) schist and quartzite. Where a contact is visible, foliation has the same orientation in both the mafic and pelitic layers. The mafic layer cuts across foliation and is more massive in the interior, which suggests that its protolith may have been a dike. A typical sample of the mafic layer consists of garnet, omphacite, quartz, clinoamphibole (barroisite to alumino-barroisite), clinozoisite, white mica, rutile, and sulfide (trace). Estimated conditions of metamorphism are 600° ± 50°C and 1.35 ± 0.15 GPa (Laird and others, 1984b).

Swainbank and Forbes (1975) recognized the probable fault relations of the eclogite unit es and suggested that it might be a window of older and more complexly metamorphosed rocks surrounded by an upper plate of younger, less metamorphosed rocks, or two terranes separated by a high-angle fault system. More recent examination of field relations and fabric orientations has led to the interpretation that this unit forms the upper plate of a folded thrust (Hall and others, 1984). In the Circle Quadrangle, Foster and others (1983) also show the eclogite in the upper plate of a thrust.

K-Ar determinations of age were made on several micas and amphiboles from the eclogite-bearing rocks of the Livengood area. A minimum age of 470 ± 35 Ma, determined from an amphibole in eclogite, is interpreted to indicate an early Paleozoic metamorphic event, probably associated with the early recumbent style of folding (Swainbank and Forbes, 1975). Several K-Ar ages of 103 to 115 Ma determined on mica in pelitic schist and garnet amphibolite are associated with a second metamorphic episode and with folding about northeast-trending axes (Swainbank *in* Hall and others, 1984). Because of the fault relations between the eclogite unit and the remainder of subterrane Y_2, the early Paleozoic radiometric age cannot be directly tied to events in other parts of subterrane Y_2 or YT. However, the late event (103 to 115 Ma) is most likely the same event that is widely recognized throughout much of the YT (see discussion of metamorphism in YTTN).

Subterrane Y_3 (Northern Big Delta, Fairbanks, southern Circle, Eagle, Charley River, and eastern Tanacross Quadrangles). Subterrane Y_3 was originally described only in the western part of the YTTN (Churkin and others, 1982), but we extend it to include the greenschist-facies rocks in the northeastern part of the YTTN. In the northern Big Delta, Fairbanks, and southern Circle Quadrangles, subterrane Y_3 separates subterrane Y_1 from Y_2. This part of subterrane Y_3 consists primarily of two distinct units of rocks that are in probable thrust contact with each other as well as with subterranes Y_1 and Y_2. The southernmost unit (unit ms, Fig. 2) consists mostly of greenish gray quartzose mylonitic schist; the more northerly unit (unit cp, Fig. 2) consists of gray calcareous phyllite and gray quartzite. In the eastern part of the Eagle Quadrangle, in the Tanacross Quadrangle, and in the south-central part of the Charley River Quadrangle, subterrane Y_3 includes three other units: one consists of fault blocks and slices of mylonitic schist, greenschist, quartzite, marble, greenstone, and phyllite (unit sm, Fig. 2); a second consists primarily of dark gray quartzite and gray quartz schist (unit qs, Fig. 2); and the third is characterized by light green quartz–chlorite–white mica schist (unit ws, Fig. 2). All three units are separated from one another and from subterrane Y_4 and Y_1 by faults.

Mylonitic schist (unit ms). The principal rock types of this unit (Fig. 2) are mylonitic schist, semischist, and quartz–white mica chlorite schist ± epidote, quartz sericite schist, and quartzite, with minor phyllite, marble, and greenstone.

Mylonitic textures, common throughout the unit, are more intense and more concentrated along the southern margin. Quartzite and quartzitic schist commonly have large gray, blue-gray, and clear glassy quartz grains, and some also have large feldspar grains (mostly microcline). The grains are commonly "eye-shaped" with "tails" of crushed recrystallized quartz. The marbles are fine to coarse grained and interlayered in schist and quartzite. Mafic greenschist and greenstone occur locally, particularly in the eastern exposures of the unit. Rocks of this unit are metamorphosed to the greenschist facies. Foliation, generally well developed, has been folded at least once.

The age of these rocks is unknown. Correlation has been suggested with the Totatlanika Schist in the northern Alaska Range, of probable Late Devonian to Mississippian age (Gilbert and Bundtzen, 1979), the Klondike Schist of the Yukon Territory (Weber and others, 1978), and the Macomb terrane of the Mount Hayes Quadrangle in the northern Alaska Range (W. J. Nokleberg, personal communication, 1985). Although there are some similarities in lithology among these units, the lack of data on age of protoliths and on their structural relations make such correlations speculative.

Calc phyllite–black quartzite (unit cp). A sequence of thin-layered calcareous phyllite, phyllite, and thin, crumbly carbonate layers is overlain by light to dark gray quartzite interlayered with dark gray to black argillite and phyllite. The quartzite is mostly medium grained, and thinly layered to massive. Foliation, incipient cleavage, and small isoclinal folds occur locally.

In the Circle Quadrangle, vitrinite reflectance studies (Laird and others, 1984a) indicate that most of these rocks were subjected to temperatures (180° ± 50°C) no higher than those normally associated with sediment diagenesis. One sample near a fault has been subjected to temperatures as high as 230° ±50°C. In the Big Delta Quadrangle, some of these rocks near Tertiary granitic intrusions have been contact metamorphosed. The grade of these rocks elsewhere in the Big Delta Quadrangle has not been determined but is probably mostly in the same range as those of the Circle Quadrangle.

No fossils have been found in these rocks; stratigraphic relations suggest that they may be of early or middle Paleozoic age. Contacts of this unit appear to be faults except where the contact is with Tertiary granite.

Schist, quartzite, and marble (unit sm). This is a varied unit of greenschist-facies rocks consisting of greenschist, calcareous greenschist, quartz–chlorite–white mica schist, marble, greenstone, quartzite including gray and dark gray quartzite, and schist with large quartz and/or feldspar grains. It crops out in the northeastern part of the Eagle Quadrangle south of the Tintina fault. Although this unit includes some rocks similar to those in units qs and ws (discussed below), it differs in having abundant layers and lenses of marble. The marble is mostly light gray, medium to coarsely recrystallized, and thin layered to massive; some is 100 m or more thick.

These rocks are commonly highly sheared and fractured and locally brecciated. In places they have a well-developed, northwesterly striking foliation. A strong, closely spaced (±10 cm apart) fold-axis lineation is locally conspicuous, especially in the more quartzose rocks. In places the resulting structure resembles mullions. Small faults having prominent gouge and breccia zones are abundant. Locally, the unit has the appearance of melange.

The stratigraphic and/or structural position of this unit is not clear, but the unit is probably overthrust by unit gs, making it the structurally lowest unit in the eastern part of the YTTN. The correlation of these rocks with rocks of the Yukon Territory is unknown. They may be a group of rocks not previously mapped in the western Yukon Territory or possibly a different facies or different part of the section of Green's (1972) unit B and/or unit A. We believe them to be of Paleozoic age on the basis of a few poorly preserved echinodermal fragments found in marble at several localities in the Eagle Quadrangle.

Quartz–chlorite–white mica schist group (unit ws). The most common rock type in this unit is light green or light greenish gray quartz–chlorite–white mica ± carbonate schist. Chlorite generally is minor. In the Tanacross Quadrangle, biotite locally may be present. Other interlayered rock types include quartzite, phyllite, and metavolcanic rocks of both mafic and felsic composition. Some schists have scattered coarse grains of gray, bluish gray, and glassy quartz, commonly with augen shape. Some quartz grains are single crystals; others are polycrystalline. Feldspar grains also occur in some schists. Minor actinolite and epidote locally may be present. Mylonite and partly recrystallized mylonitic texture occur.

Foliation is generally well developed. In the southeastern part of the Eagle Quadrangle, a strong lineation plunging westerly is formed by small, tight-to-isoclinal folds. These folds are refolded by generally northeast-trending folds that we believe may be related to thrusting. This unit probably also has been affected by late open folding that is not easily observed in the Eagle Quadrangle, but is evident in rocks adjacent to this unit in Canada.

In the Eagle Quadrangle, this unit is overthrust by, and in places imbricated with, amphibolite-facies rocks (Foster and others, 1985). In the Tanacross Quadrangle its contacts are poorly exposed. The unit is continuous to the east with similar rocks in the Yukon Territory, which are included in Green's (1972) unit B and were termed the "Klondike Schist" by McConnell (1905).

The age of protoliths of these rocks is unknown, and because all their contacts in Alaska are believed to be thrust faults, their stratigraphic position is also unknown. Middle or late Paleozoic protolith ages have been postulated (Tempelman-Kluit, 1976).

Quartzite and quartzitic schist (unit qs). Light gray to black quartzite, which ranges in grain size from very fine to medium, is the most common rock type in this unit (Fig. 2). Quartz grains are almost always highly strained. Outcrops are massive to schistose depending on the relative amounts of white mica and quartz. Gray phyllite, phyllitic and carbonaceous quartz schist, and graphitic schist layers occur locally.

These rocks generally have a well-developed foliation that regionally strikes approximately east-west, with both northward and southward dips ranging from 5 to 60°. Folds, commonly isoclinal, a few millimeters to several meters in wavelength and amplitude, deform foliation. Axes of these folds generally trend slightly north of west and have a shallow plunge (0 to 20°). Small (1 to 10 cm amplitude and wavelength), asymmetric, rootless folds defined by cream-colored fine-grained polycrystalline quartz may be an early phase of this deformation. An open folding, commonly having a wavelength of about 0.5 m and an amplitude averaging 10 cm, deforms early structures. The axial trend of the open folds is slightly west of north and plunges a few degrees to 25° northwest or southeast.

This unit crops out in a generally westerly trending belt in the east-central part of the Eagle Quadrangle and in a small area in the southeastern corner of that quadrangle. The unit continues eastward into Canada as part of Green's (1972) unit A and the Nasina Series of McConnell (1905). A Paleozoic age has been hypothesized (Foster, 1976), although no fossils have been found and contacts of the unit are probably all faults.

Subterrane Y_4 (Eagle and Tanacross Quadrangles). Subterrane Y_4 consists primarily of quartzitic and pelitic gneiss and schist, quartzite, marble, and amphibolite (unit gs, Fig. 2) metamorphosed to the amphibolite and epidote-amphibolite facies. On the north it is in thrust contact with greenschist-facies rocks of subterrane Y_3, on the west it has been extensively intruded by Cretaceous granitic plutons, and on the south a probable thrust fault separates it from subterrane Y_1. On the east it extends into the Yukon Territory of Canada. The unit typically consists of medium- to coarse-grained amphibolite, quartz–amphibole–biotite gneiss, quartz-biotite gneiss, marble, and quartzite. Because the composition of most of the gneiss and schist is quartzose, aluminum silicate minerals are rare; however, kyanite has been found in a few thin sections. White, light gray, or pinkish white marbles most commonly are coarse grained. Quartzite is mostly light gray and tan. In the northwestern part of subterrane Y_4 the unit consists of mostly schist, quartzite, and marble, all complexly deformed and intruded by dikes and plutonic rocks. The metamorphic grade locally may be as low as upper greenschist facies.

Foliation and compositional layering have been deformed into tight, asymmetric folds that are locally isoclinal. Wavelength and amplitude range from 0.5 m to several hundred meters. Fold axes have an average trend of N50°E and generally plunge less than 20° northwest or southeast. A second generation of folding, probably the same open-folding that has affected the rocks of adjacent subterranes, produced open folds with axes trending nearly north-south. The wavelength of these folds is about 0.5 m, and their average amplitude about 10 cm. Pre-metamorphic folding has not been definitely recognized, but may be indicated by rarely observed small, rootless folds. Thrust slices marked by zones of gouge, breccia, and small fault-related folds are common throughout subterrane Y_4.

The major regional metamorphism and accompanying deformation of these rocks is believed to have taken place during Late Triassic to Middle Jurassic time on the basis of ^{40}Ar-^{39}Ar incremental heating experiments. Granitic intrusion (Taylor Mountain, Mount Veta, and possibly other plutons) was probably synchronous with metamorphism and preceded thrusting of this unit over units ws and qs (Cushing and others, 1984a). The Early Cretaceous metamorphism that affected subterrane Y_1 also affected subterrane Y_4 but was of minor intensity (Cushing and others, 1984a).

The age of these rocks is poorly known, but they are considered Paleozoic on the basis of a few poorly preserved crinoid columnals (Foster, 1976) in marble. We know of no evidence for Precambrian protoliths in this unit.

Nonfoliated igneous rocks

Plutonic rocks. Nonfoliated igneous rocks occur throughout the YTTN and range in composition from ultramafic to felsic, but most are felsic. On the east side of the Shaw Creek fault (Fig. 2), some intrusions are of batholithic size, and small plutons, dikes, and sills are common throughout the YTTN. In the following discussion the terminology of Streckeisen (1976) is used.

Felsic granitic plutons. Three periods of Mesozoic and Cenozoic granitic intrusion are recognized in the YTTN. The oldest, Late Triassic and Early Jurassic (215 to 188 Ma) in age (unit JTrg, Fig. 2), occurs only in the eastern part of the YTTN (subterrane Y_4). The second, of Late Cretaceous age (95 to 90 Ma), occurs throughout the YTTN but is especially prominent in the central part (unit Kg, Fig. 2). The third, of Late Cretaceous and early Tertiary age (70 to 50 Ma), occurs throughout the YTTN but has particular significance in the northwestern part (unit Tg, Fig. 2).

The Late Triassic and Early Jurassic plutons include those of Taylor Mountain and Mount Veta, and possibly some plutons in the central and eastern parts of the Eagle Quadrangle. Locally, the margins of these plutons are sheared, crushed, and faulted. Taylor Mountain, in the south-central part of the Eagle and northern part of the Tanacross Quadrangles, is a batholith about 648 km^2 in area. It is mostly medium-grained, equigranular granite, but locally it is granodiorite to diorite. In places along the southern and eastern margins, its texture is slightly gneissic. Plagioclase (oligoclase to andesine), sericitized potassium feldspar, quartz, hornblende, and biotite are the major constituents; common accessory minerals are sphene, apatite, and opaque minerals. Plagioclase is generally zoned and sericitized. Quartz is strained, and the margins of some quartz grains are granulated. Hornblende is generally more abundant than biotite. Some biotite has altered to chlorite. The southern contact of the pluton is fairly sharp, and there is little evidence of thermal effects on the amphibolite-facies country rock. However, the country rock along the northeastern contact has been altered by the intrusion, and dikes and sills are numerous. Epidote is abundant in the contact zone. Conventional K-Ar dating has indicated an age of about 180 Ma for the Taylor Mountain batholith. ^{40}Ar-^{39}Ar determinations provide an integrated plateau age on hornblende of 209 $\pm$ 3 Ma (Cushing, 1984). A U-Pb determination on sphene gives a concordant age of 212 Ma (Aleinikoff and others, 1981). We thus believe that the batholith was emplaced in Late Triassic time.

Hornblende plagioclase porphyry is a characteristic intrusive rock in the vicinity of Mount Veta. Its mineralogy commonly is that of quartz monzodiorite, but more felsic compositions also occur. Some phases of the intrusion are equigranular. In some very coarse-grained porphyritic phases, a mineral alignment defines a weak foliation. Conventional K-Ar analysis of hornblende from this intrusion yielded an age of 177 $\pm$ 5 Ma (Foster, 1976), and an integrated plateau age of 188 $\pm$ 2 Ma was obtained by the ^{40}Ar-^{39}Ar method (Cushing, 1984). Some other hornblende-bearing plutons nearby have not been dated, but may be of similar age.

Figure 3 is a plot of fields of normative quartz, albite, and anorthite for representative granitic rocks in the YTTN. Two fields are plotted for Triassic and Jurassic plutons (Taylor Mountain and Mount Veta). These rocks are characteristically low in quartz, contain abundant hornblende, and in contrast to most granitic rocks in the YTTN, none is corundum normative. Mineralogical and common lead isotope data (J. N. Aleinikoff, personal communication, 1985) are consistent with a dominantly oceanic source for magmas that formed the Triassic and Jurassic plutons.

The Triassic and Jurassic plutons are similar in lithology and age to plutonic rocks of the Klotassin Suite in the Yukon Territory (Tempelman-Kluit, 1976). The Klotassin Suite has been interpreted as the roots of an island arc that formed on the margin of Stikinia (Tempelman-Kluit, 1979), a continental fragment that was joined to North America in Middle Jurassic time (Tempelman-Kluit, 1979). Tempelman-Kluit (1976) has suggested that the granitic rocks of Taylor Mountain may belong to the Klotassin Suite and thus would be a part of the Stikinia terrane (Silberling and others, this volume). However, the country rocks (subterrane Y_4) intruded by the pluton of Taylor Mountain are not typical of those that compose the Stikinia terrane in the parts of Canada where it is best known. Although subterrane Y_4 may not be a part of Stikinia, this does not preclude correla-

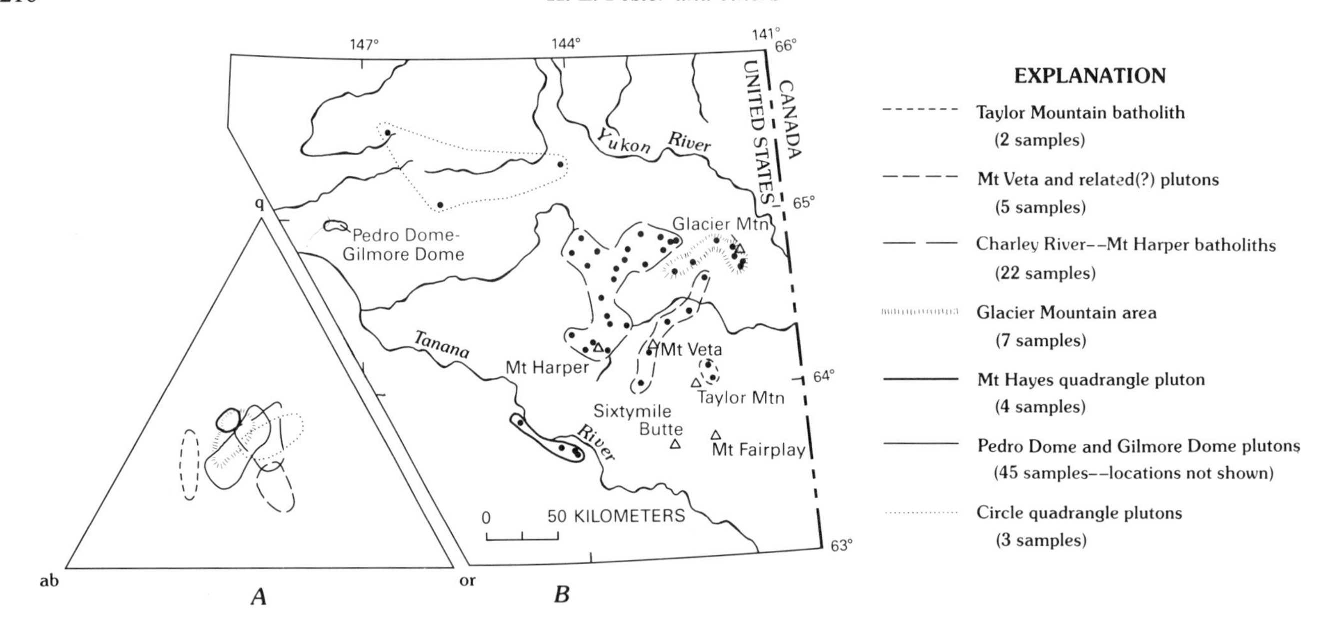

Figure 3. Normative mineralogical data for granitic plutons in the Yukon-Tanana terrane. A. Fields of normative mineralogy for samples of granitic plutons. B. Localities of samples (dots) used to define the mineralogical fields. Data primarily from Holmes and Foster (1968), Foster and others (1978a), Luthy and others (1981), Blum (1983), and Wilson and others (this volume).

tion of the granitic rocks of Taylor Mountain with the Klotassin Suite. Subterrane Y_4 may have been near or joined to the Stikinia terrane in Triassic or Early Jurassic time and involved in the same plutonic event.

Hornblende-bearing granitic rocks in the southeastern part of the Tanacross Quadrangle were included by Jones and others (1984) in the Stikinia terrane because they are adjacent to rocks of similar lithology to the east in the Yukon Territory that are included in the Klotassin Suite. However, these granitic rocks in Alaska are also adjacent, on the west, to lithologically similar Cretaceous granites. Neither the Alaskan nor the Canadian hornblende-bearing granites close to the U.S.-Canada border have been dated. More data are needed to determine the relations of these rocks and the extent of the Stikinia terrane.

Granitic rocks of Cretaceous age occur in bodies that range in size from less than 1 km^2 to plutons of batholithic proportions. Batholiths occur in the southeastern and northwestern parts of the Tanacross Quadrangle, the northeastern part of the Mount Hayes Quadrangle, and in the western part of the Eagle and eastern part of the Big Delta Quadrangle. The plutons range in composition from quartz monzonite to diorite but are dominantly granite and granodiorite. They are equigranular to porphyritic and are generally medium grained. The mafic minerals may be either hornblende or biotite, and most commonly both occur. Primary muscovite is rare and minor when present. Mylonitic textures are locally present, especially along major faults such as the Shaw Creek fault. Alteration of feldspars is slight to moderate, and biotite is commonly partly chloritized.

The age range of this group of intrusions, on the basis of conventional K-Ar analyses, is about 110 to 85 Ma, but most are about 95 to 90 Ma.

Nokleberg and others (1986) described some intrusive rocks of similar age in the northeastern part of the Mount Hayes Quadrangle as locally slightly to moderately schistose and slightly regionally metamorphosed, and suggested that those rocks were intruded during the waning stage of a major regional amphibolite-facies metamorphism in the mid- to Late Cretaceous. This Cretaceous metamorphism is not recognized as the major thermal event in other parts of the YTTN. We suggest that this event affected at least some other parts of the YTTN, but was of minor intensity. For instance, in subterrane Y_4 it has been detected by ^{40}Ar-^{39}Ar incremental heating experiments on minerals from amphibolite-facies metamorphic rocks (Cushing and others, 1984b).

Plots of normative quartz, albite, and orthoclase in samples from individual bodies of Cretaceous granitic rocks fall into distinct compositional fields. Figure 3 shows fields and locations for samples from the batholith(s) in the western part of the Eagle and eastern part of the Big Delta Quadrangle (Mount Harper–Charley River); from plutons in the northeastern part of the Mount Hayes Quadrangle; from small granitic bodies near Fairbanks (Pedro and Gilmore Domes); and from granitic bodies in the northeastern part of the Eagle Quadrangle (Glacier Mountain). The Cretaceous granitic rocks are richer in quartz than the Triassic and Jurassic granites, and most contain normative corundum. Blum (1983), in a detailed study of the small granitic bodies near Fairbanks, concluded, on the basis of petrography, major-oxide chemistry, and initial strontium isotopic ratios, that those granites formed either from a magma generated by partial melting of metamorphic rocks of continental affinities or from an initially mantle- or lower-crustal-derived melt with significant contamination from continental material. Common-lead ratios of Cretaceous granitic rocks from widely scattered localities in the YTTN also suggest a mixture of lead from two sources, one a continentally derived radiogenic source, and the other a more primitive oceanic source (J. N. Aleinikoff, personal communication, 1984).

In addition to granitic plutons, several small (generally less than 2 km^2) pyroxene diorite plutons are known in the Big Delta Quadrangle. These plutons range in age from about 93 to 90 Ma, based on conventional K-Ar analysis (Foster and others, 1979). Granitic rocks of this general age group are common in many parts of Alaska and the Yukon Territory. In the Yukon Territory they are included in the Coffee Creek Suite (Tempelman-Kluit, 1976).

The youngest intrusions, having Tertiary K-Ar ages of 70 to 50 Ma, are most prevalent in the northeastern part of the Big Delta and the Circle Quadrangle, although a few others are widely scattered throughout the YTTN. They are primarily granite in composition, but Mount Fairplay in the Tanacross Quadrangle ranges from hornblende–augite–biotite diorite through hornblende-augite syenite to hornblende-biotite quartz monzonite (Kerin, 1976). The Tertiary plutons are generally small (3 km^2 or less), medium to coarse grained, and equigranular to porphyritic. The mafic mineral is generally biotite. They are quartz-rich and corundum normative, characteristics of granites formed from magmas derived from continental crust. A field of normative quartz, albite, and orthoclase for samples of Tertiary granite from the Circle Quadrangle is shown in Figure 3.

Mafic and ultramafic differentiates. Bodies of coarse-grained (grains commonly 5 to 10 cm long) gabbro, hornblendite, and clinopyroxenite that appear to be mafic and ultramafic igneous differentiates occur in several widely scattered localities in subterranes Y_4, Y_2, and Y_1. Biotite-bearing hornblendite and clinopyroxenite (Type III of Foster and Keith, 1974) are most abundant, and in a few places, very coarse-grained gabbronorite and hornblende gabbro occur in association with coarse-grained ultramafic rocks. Epidote and garnet are locally abundant in the gabbros. The mafic and ultramafic igneous differentiates appear to have intruded regionally metamorphosed greenschist- to amphibolite-facies country rocks. Margins of some of the mafic and ultramafic igneous differentiates are foliated, but there is no foliation within the bodies. Felsic dikes intrude most of the mafic and ultramafic differentiates as well as the surrounding metamorphic country rock, but their genetic relations are not known. In the central part of the Eagle Quadrangle, K-Ar ages of biotite hornblendite are 175 ± 5.1 Ma on hornblende and 185 ± 5 Ma on biotite (recalculated from Foster, 1976). This ultramafic body could have a Late Triassic or Early Jurassic plutonic origin. In subterrane Y_1 in the Tanacross Quadrangle, biotite from olivine gabbro associated with a small ultramafic body has a K-Ar age of 66.6 ± 2 Ma (Wilson and others, 1985).

Ultramafic dikes or small plugs of fresh, nonmetamorphosed, cumulate-textured pyroxenite and olivine pyroxenite intrude greenschist-facies metamorphic rocks in the central Eagle Quadrangle. The dikes or plugs could either be differentiates of nearby granodiorite intrusions, or unrelated ultramafic intrusive rocks.

Volcanic rocks. Post-metamorphic volcanism occurred during the Cretaceous, Tertiary, and Quaternary and was most prevalent in the eastern YTTN. All of the nonmetamorphosed felsic volcanic rocks originally were considered to be of Tertiary age (Mertie, 1937), but K-Ar age dating has shown that welded tuffs and other felsic volcanic rocks (unit Kv, Fig. 2) covering a large area in the Tanacross Quadrangle are as old as Cretaceous (Bacon and others, 1985).

The Cretaceous volcanic rocks occur in and around three poorly exposed calderas in the central part of the Tanacross Quadrangle and consist of welded tuff, air-fall tuff, lava flows, ash-flow sheets, and small hypabyssal intrusions. Most are felsic and range from rhyolite to dacite in composition. Phenocrysts in rhyolite tuff, lava, and hypabyssal intrusions are quartz, sanidine, plagioclase, clinopyroxene, biotite, and iron and titanium oxides ± allanite. All phenocrysts except quartz, clinopyroxene, and allanite are generally altered. Common alteration minerals are sericite, illite, clinoptilolite, kaolinite, chlorite, quartz, and potassium feldspar (Bacon and others, 1985). Tuffaceous sedimentary rocks occur near the margin of one of the calderas.

Mafic and intermediate volcanic rocks are close to the felsic rocks, and we assume that some are also of Cretaceous age. However, in a few places where cross-cutting relations can be seen, the mafic rocks are younger than the felsic ones. Mafic rocks include potassic andesitic lava flows containing plagioclase, biotite, clinopyroxene, and in some places, hornblende phenocrysts, but detailed mineralogy of most of the mafic lavas is not known.

The K-Ar age of sanidine in welded tuff of the northernmost caldera is 93.6 ± 2 Ma, and of hornblende in rocks of the easternmost caldera is 90 ± 2.8 Ma (Bacon and others, 1985). These ages suggest that the volcanic rocks are related to the granitic plutons that were intruded during the 110- to 85-Ma intrusive episode and that they may be roof remnants of some of these plutons. Although the Mount Fairplay intrusive body in the north-central part of the Tanacross Quadrangle is partly surrounded by Cretaceous volcanic rocks, it appears to be unrelated to them and has a Tertiary K-Ar age (Foster and others, 1976).

Tertiary volcanic rocks (unit Tv, Fig. 2), as yet little studied, range from much-altered rhyolite to basalt in composition. They occur in the Tanacross, Eagle, and Big Delta Quadrangles and commonly are associated with hypabyssal dikes and small intrusions. They include lava flows, welded ash-flow tuff, and air-fall tuff. Locally, felsic tuff and possibly flows have been faulted, including thrust faulting, and small-scale deformation has resulted from the faulting. Sedimentary rocks, generally of limited extent, occur locally with volcanic rocks.

Sanidine in porphyritic rhyolite in the eastern part of the Big Delta Quadrangle was dated at 61.6 ± 2 Ma by the K-Ar method, and sanidine in welded tuff in the eastern part of the Tanacross Quadrangle yielded ages of 57.8 ± 2 and 56.4 ± 2 Ma (Foster and others, 1979).

We believe that some undated basaltic rocks (unit Qv, Fig. 2) in the Tanacross Quadrangle are of Quaternary age, largely on the basis of physiographic relations, but Prindle Volcano in the eastern part of the Tanacross Quadrangle leaves no doubt of its young age. Prindle Volcano is an isolated cone with a lava flow extending about 6.4 km downslope to the southeast from a breached crater. The cone and lava flow are composed of vesicular alkaline olivine basalt that contains abundant peridotite and granulite inclusions. The basalt consists of clinopyroxene, olivine, and opaque minerals in a fine-grained groundmass believed to contain occult nepheline and potassium feldspar (Foster and others, 1966). The peridotite inclusions range in size from xenocrysts to polycrystalline masses as much as 15 cm in diameter. Mineral assemblages include olivine, orthopyroxene, clinopyroxene, and spinel in at least five different combinations. The mineral assemblages of the inclusions are characterized by hypersthene and/or clinopyroxene and plagioclase, but also may include quartz and carbonate, along with such accessory minerals as apatite, zircon, magnetite, and rutile. The well-preserved cone suggests that the eruptive activity occurred during Quaternary time. Indirect evidence, which includes its possible correlation with a similar cone in the Yukon Territory, and the fact that its lava flow is overlain by white volcanic ash that is probably the White River Ash Bed, suggests that it is post–early Pleistocene, but older than 1,900 yr B.P. (Foster, 1981).

A number of unconsolidated volcanic ash deposits occur in the YTTN, and some are fairly well dated. Most, such as the White River Ash Bed, probably originated outside of the YTTN (see section on Quaternary geology: Volcanic ash).

Sedimentary rocks (YTTN). The unmetamorphosed sedimentary rocks of the YTTN (unit TKs, Fig. 2) are all of nonmarine origin, of Late Cretaceous and/or Tertiary age, and of limited areal extent. They appear to have been deposited in small, disconnected basins, at least some of which resulted from faulting. Some sedimentary rocks are closely associated with volcanic rocks, and most include considerable amounts of tuff. They have been deformed by folding (Foster and Cushing, 1985) and in some cases, by thrusting and high-angle faulting.

The largest area of sedimentary rocks is in the northern part of the Eagle and southern part of the Charley River Quadrangle both north and south of the Tintina fault. In some places, sedimentary rocks cover the faults of the Tintina system, but in other places they are cut by the faults. These rocks are dominantly conglomerate, but include sandstone, mudstone, shale, breccia, lignite, and coal. Most of the conglomerate consists of well-rounded white and tan quartz and black chert clasts 2 to 13 cm in diameter in a quartzose matrix. Where bedding can be detected, dips as steep as 60° occur, and most of the rocks dip at least 20°. The conglomerate and sandstone were not principally derived from the local metamorphic terrane but probably from more distant sources north of the Tintina fault. Pollen and poorly preserved plant fragments and impressions indicate that the rocks may range in age from Late Cretaceous to Pliocene (Foster, 1976).

Several small patches of Tertiary sedimentary rocks (unit Ts, Fig. 2) occur in the southeastern part of the Eagle Quadrangle. They are mostly conglomerate and sandstone, but near the town of Chicken also include coal seams, white tuff, and glassy tuff containing abundant plant fragments (Foster, 1976).

Folded conglomerate, sandstone, argillite, tuff, tuffaceous argillite, and sandstone with some lignite and carbonaceous layers form a discontinuous belt about 30 km long in the north-central Tanacross Quadrangle. Poorly preserved pollen indicates that deposition was in Late Cretaceous(?) time (Foster, 1967), but pollen in some of the deposits is as young as Neogene (Yaeko Igarashi, personal communication, 1985). In addition to Cretaceous and Tertiary pollen, these rocks also contain monosulcate pollen of Devonian age (Foster, 1967), suggesting that at the time of sediment deposition on the underlying metamorphic terrane, this terrane was located where pollen could be derived from nonmetamorphosed Devonian rocks (Foster, 1967).

Other sedimentary rocks appear to have been deposited in basins associated with a Late Cretaceous volcanic complex near Mount Fairplay and Sixtymile Butte (Foster, 1967).

In the northeastern part of the Tanacross Quadrangle, unconsolidated to poorly consolidated gravel and conglomerate resting unconformably on metamorphic rocks occur in a small area at about 1,266 m altitude. The gravel is composed mostly of yellowish white quartz pebbles and well-rounded, polished chert pebbles 1 to 15 cm in diameter. The chert pebbles are not locally derived. This deposit has been interpreted as most likely late Tertiary in age, but could be of early Pleistocene age (Foster, 1970).

Metamorphism

Regional metamorphism throughout the YTTN ranges from very low grade (about equivalent to burial metamorphism) to amphibolite facies of about the second sillimanite isograd. Changes in metamorphic grade across subterrane boundaries are commonly abrupt and can be attributed to juxtaposition of rocks of different metamorphic grade by faulting, particularly thrust faulting. Gradational changes in metamorphic grade are documented within subterranes Y_1 and Y_2. More than one period of regional metamorphism has not been recognized petrographically, except in the ecologitic rocks, where barroisite is rimmed by hornblende and omphacite is altered to cryptocrystalline material; but further work is needed, especially in consideration of possible evidence from radiometric age determinations of more than one period of regional metamorphism.

Pressures during metamorphism were probably mostly moderate, but the dominance of quartzitic over pelitic compositions makes determining the pressures (and temperatures) of metamorphism difficult. Kyanite is common in the southeastern part of subterrane Y_2 and northern part of subterrane Y_1 and occurs rarely in subterrane Y_4. In the Circle Quadrangle (subterrane

Y_2), mineral assemblages in pelitic rocks indicate progressive metamorphism along a P-T path similar to Barrovian metamorphism in Scotland (path A, Harte and Hudson, 1979). Medium-pressure metamorphism is also indicated by the amphibole composition in mafic schist (using the criteria summarized by Laird, 1982).

Andalusite + kyanite and andalusite + sillimanite occur in the southeastern part of subterrane Y_2, but it is not clear that the andalusite formed at the same time as kyanite and sillimanite, and it may instead be related to nearby Tertiary plutons. In subterrane Y_1, all three Al_2SiO_5 polymorphs have been identified along the Salcha River (Dusel-Bacon and Foster, 1983), suggesting metamorphism at about 0.4 GPa (using the data of Holdaway, 1971). Farther south, sillimanite + andalusite and sillimanite + cordierite (Dusel-Bacon and Foster, 1983, Fig. 2) may indicate low-pressure–facies series regional metamorphism.

Evidence of high-pressure metamorphism has been found in two areas, each fault-bounded. Glaucophane + epidote + chlorite + albite + white mica + sphene + carbonate ± garnet schist occurs in a fault slice in the northern part of the Eagle Quadrangle. Eclogite and rare glaucophane-bearing assemblages have been found in the southwestern part of the Circle Quadrangle and southeastern part of the Livengood Quadrangle. The estimated conditions of metamorphism in both areas of eclogitic rocks are about 600°C and 1.4 GPa (Laird and others, 1984b; Brown and Forbes, 1984).

The highest documented regional metamorphic grade is in the sillimanite gneiss dome of subterrane Y_1, where metamorphic grade reaches and probably surpasses the second sillimanite isograd, and partial melting may have occurred (Dusel-Bacon and Foster, 1983). Temperatures between 655° and 705° ± 30°C are indicated by garnet-biotite geothermometry. To the northeast, metamorphic grade decreases through the staurolite stability field to garnet grade, where the rocks are in contact with the weakly metamorphosed rocks of subterrane Y_3. In subterrane Y_2, metamorphic grade decreases gradually from sillimanite + muscovite and perhaps sillimanite + K feldspar in the southeast to staurolite + kyanite, garnet, and then biotite grade farther north and west.

The lowest metamorphic grade is in rocks in the southern part of the Circle Quadrangle (subterrane Y_3) and in the eastern part of the Eagle Quadrangle (Seventymile terrane). The quartzite and quartzitic schist (unit cp) in the southern Circle Quadrangle have been shown by vitrinite reflectance (Laird and others, 1984a) to be in the range of normal sediment diagenesis. We believe that these rocks are in thrust contact with garnet- and staurolite-grade rocks. In the Eagle Quadrangle, slightly metamorphosed sedimentary rocks occur with greenstone in thrust sheet remnants of the Seventymile terrane.

Contact metamorphism is primarily associated with Tertiary plutons, but in a few places may be peripheral to Cretaceous plutons. Around many Tertiary plutons, biotite has grown across the foliation of the regionally metamorphosed rocks. In the southern part of subterrane Y_2, contact metamorphism associated with Late Cretaceous and/or early Tertiary plutons have retrograded staurolite and kyanite porphyroblasts to white mica. In the central part of the Circle Quadrangle (subterrane Y_2), hornfelsic growth of biotite and amphibole is associated with small felsic intrusions and has overprinted probable garnet-grade regional metamorphism. Contact metamorphism up to sillimanite grade has overprinted staurolite + kyanite-grade regional metamorphism. In the south-central part of the Eagle Quadrangle, contact metamorphism is indicated by andalusite in schist near a small Cretaceous or Tertiary pluton.

Limited radiometric age data suggest that time(s) of major regional metamorphism(s) may have differed in the four subterranes of the YTTN. In subterrane Y_1, U-Pb zircon ages indicate that amphibolite-facies regional metamorphism was synchronous with or followed emplacement of a belt of felsic Mississippian plutons (now augen gneiss), but preceded intrusion of Cretaceous plutons (Dusel-Bacon and Aleinikoff, 1985). Metamorphism predates Tertiary and probable Cretaceous plutonism in subterranes Y_3 and Y_2. ^{40}Ar-^{39}Ar incremental heating experiments indicate that major regional amphibolite-facies metamorphism in subterrane Y_4 peaked about 213 ± 2 Ma, but the extent of the area affected by this event is not known (Cushing, 1984). Conventional K-Ar dating indicates a metamorphic event, which apparently was widespread in the YTTN, in the Early Cretaceous (125 to 110 Ma; Wilson and others, 1985). ^{40}Ar-^{39}Ar data indicate that in the eastern part of the Eagle Quadrangle this Early Cretaceous regional event was of low metamorphic grade and had relatively minor effects, although it could have been a major event elsewhere (Cushing and others, 1984a). Sufficient data are not yet available to delineate and compare the metamorphic events of the four subterranes.

Mineral resources

Gold, primarily in placer deposits, has been the most important mineral resource in the YTTN. It has also been produced from lodes in the Fairbanks district and in minor amounts elsewhere. Small amounts of antimony and tungsten have been produced from lode deposits in the Fairbanks district during periods of high prices. Exploration has identified occurrences of copper-molybdenum porphyry and tungsten skarn. Widespread geochemical anomalies, and scattered occurrences and prospects, suggest that the region may contain granite-related uranium deposits, lode tin deposits (most likely greisen), platinum deposits of mafic igneous association, and sedimentary exhalative zinc-lead deposits.

A little coal has been produced for local use, and unmeasured coal resources in Late Cretaceous and/or Tertiary sedimentary rocks may be significant. Geothermal springs are known at three localities (Fig. 4) in the YTTN, and are exploited for recreation and local use at two locations.

Figure 4 shows the distribution of selected deposits, prospects, and occurrences by probable deposit type. Entries were selected to show: (1) important deposits, (2) spatial patterns of occurrence, and (3) the different types of occurrence. Patterns in the distribution of particular types of deposits suggest that there is some relation of mineral occurrence to subterranes (Nokleberg and others, this volume, Chapter 29 and Plate 11). Although mining has taken place in the region since the 1890s, it is only in recent years that there has been systematic exploration for deposits other than gold lodes and placers. New types of mineral deposits have been found, and as the discovery of a diamond in a placer deposit in the Circle district (Fairbanks Daily News-Miner, Dec. 11, 1984) attests, unexpected types of deposits undoubtedly await discovery.

Gold placers and lode deposits. Gold has been, and continues to be, the most important mineral commodity in east-central Alaska. Most of the gold has been produced from placers; lode production has been significant only in the Fairbanks district, and in most districts the amount of gold in known lode sources is small relative to that in placers. Gold deposits are found in all of the subterranes of the YTTN, but the greatest production

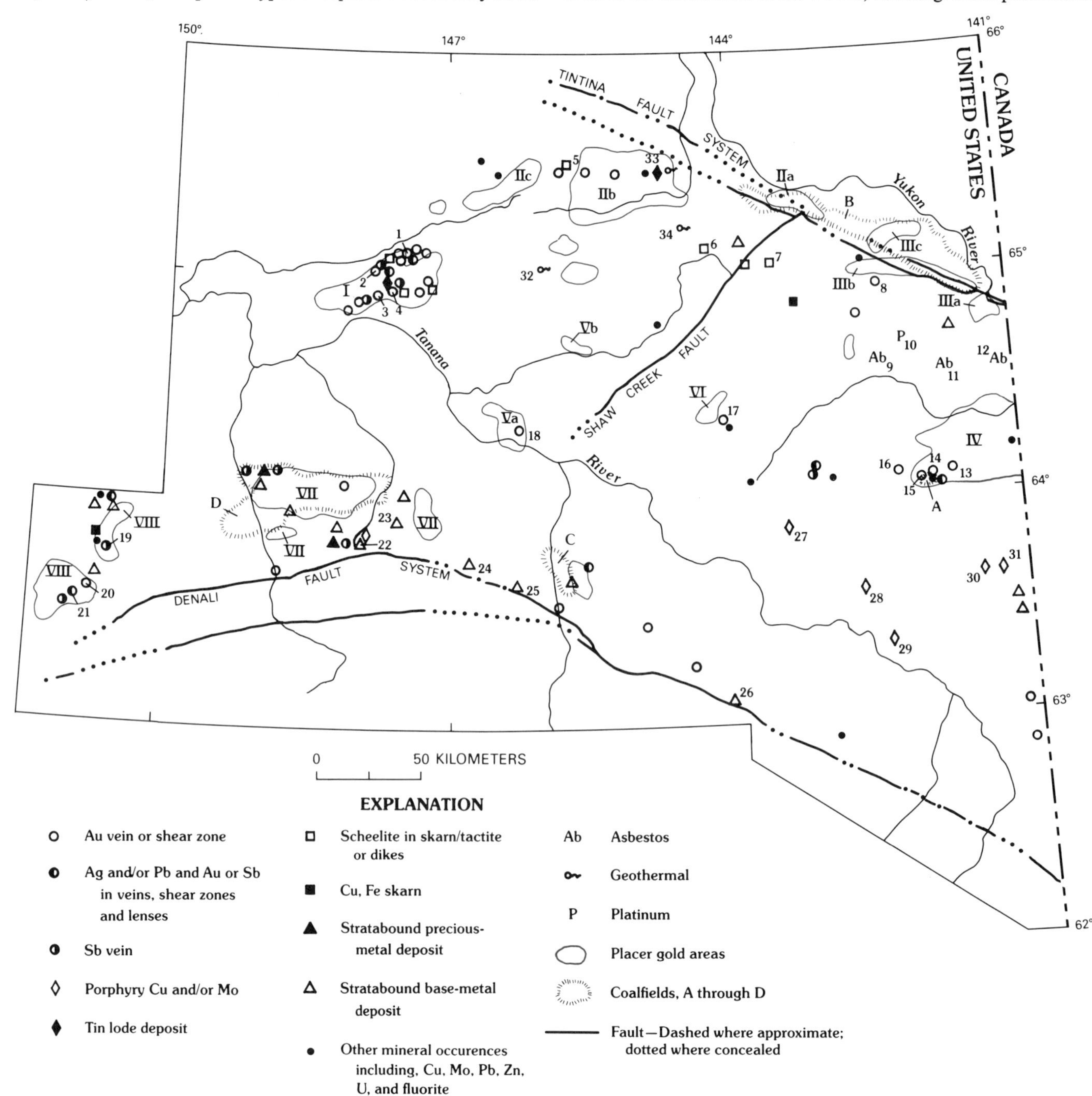

Figure 4. Map showing selected mineral deposits, occurrences, and prospects in east-central Alaska. Numbers and letters refer to localities mentioned in text. Data primarily from Berg and Cobb (1967), Cobb (1973), Singer and others (1976), Eberlein and others (1977), MacKevett and Holloway (1977), Barker (1978), Gilbert and Bundtzen (1979), and Nokleberg and others (this volume, Chapter 29 and Plate 11).

and most extensive placers are in subterrane Y_2. Because geologic characteristics and ore controls vary among the districts, these features are discussed separately for the largest districts.

Fairbanks district. Gold was first discovered in streams of the Fairbanks district (No. I, Fig. 4) in 1902 by Felix Pedro. This district, the largest producer in east-central Alaska, has yielded about 7,750,000 oz of gold from placers and approximately 250,000 oz of gold from lodes (Bundtzen and others, 1984). Production from lodes has been concentrated in four areas: Cleary Hill (No. 1, Fig. 4), Treasure-Vault Creek (No. 2, Fig. 4), Ester Dome (No. 3, Fig. 4), and Gilmore Dome (No. 4, Fig. 4). In each of those areas, placers are closely associated with lodes.

The placers drain, and the lodes occur in, the quartzite and quartzitic schist unit of subterrane Y_2 (particularly the Cleary sequence of the Fairbanks Schist of Smith and Metz, *in* Nokleberg and others, this volume, Chapter 29 and Plate 11). In the Fairbanks area, the quartzite (unit qq) has undergone early isoclinal folding and later east-northeast–trending broad open folding. Cretaceous felsic plutonic rocks (Blum, 1983) have intruded the Cleary sequence along the axes of broad anticlinal structures. Lodes, except on Ester Dome, occur primarily as east-west–trending auriferous veins in shears and crushed zones, but locally lodes lie parallel to foliation. Several types of lodes are recognized, including: (1) simple pyrite-arsenopyrite-gold quartz veins, (2) gold-sulfosalt-sulfide-quartz veins, and (3) late stibnite veins, and (4) gold-scheelite veins. Smith and Metz (*in* Nokleberg and others, this volume, Chapter 29 and Plate 11) present geochemical characteristics of some of the vein types. Early workers (Hill, 1933) stressed the structural control of lodes and their relation to granitic intrusions; recent workers (Smith and Metz *in* Nokleberg and others, this volume, Chapter 29 and Plate 11) have stressed the association of the lodes with the Cleary sequence and have suggested that, because of inherent geochemical enrichment, it is an important ore control.

Circle district. The Circle district, one of the oldest mining districts in interior Alaska, has produced at least 850,000 oz of gold (Bundtzen and others, 1984), all from placers. The deposits occur in three separate areas (Nos. IIa, IIb, and IIc, Fig. 4), and characteristics of the deposits and probable sources of gold vary with the area.

In the eastern area (most of which lies just outside of the YTTN) (No. IIa, Fig. 4) the principal productive streams (Woodchopper and Coal Creeks) have produced at least 20,000 oz of gold, mostly from stream, but including bench, placers. These streams drain a diverse group of rocks including schist and gneiss of subterrane Y_2; phyllite, argillite, and quartzite of subterrane Y_3; Tertiary sedimentary rocks; Cretaceous granite; and unmetamorphosed rocks north of the YTTN. Mertie (1938) believed that the proximal source of the placers was gold in the Tertiary sedimentary rocks.

Most of the district's gold has been produced from stream placers in the eastern half of the central part of the district (No. IIb, Fig. 4). Streams drain the quartzite and quartzitic schist unit of subterrane Y_2 and Tertiary granite. No large lode sources of gold have been discovered, but samples of small limonite-stained quartz veins, arsenopyrite-bearing shear zones, zones of silicified breccia, and felsic dikes contain traces of gold. Mertie (1938) suggested that sources of gold in the streams were auriferous fracture zones, veins, breccias, and felsic dikes that developed and were emplaced above the Circle pluton, and recent work (Menzie and others, 1983) supports this hypothesis.

The western part of the Circle district (No. IIc, Fig. 4) has been the least productive part. Only Nome Creek had substantial production. Considering the mining method, duration of operation, and limited production data, Nome Creek probably produced between 10,000 and 20,000 oz of gold. The streams in the western Circle district drain the quartzite and quartzitic schist unit of subterrane Y_2, a biotite granite pluton, and felsic hypabyssal rocks.

Eagle district. Gold was discovered in streams of the Eagle (Seventymile) district between 1895 and 1898. The district includes American Creek and its tributaries, Discovery Fork and Teddy's Fork (IIIa, Fig. 4); the Seventymile River and its tributaries, Flume, Alder, Barney, Broken Neck, Crooked, and Fox Creeks (IIIb, Fig. 4); and Fourth of July Creek and Washington Creek and their tributaries, which are mostly north of the YTTN (IIIc, Fig. 4). The district has produced at least 30,000 oz of gold and byproduct silver, all from stream and bench placers. Gold alloyed with platinum was found at Fourth of July Creek and the mouth of Broken Neck Creek, and a few platinum nuggets were reportedly recovered from a tributary (Lucky Gulch) to the Seventymile River (Cobb, 1973). Flume, Alder, and American Creeks drain mafic and ultramafic rocks of the Seventymile terrane and quartzite, marble, phyllite, and graphitic schist of subterrane Y_3 (Foster, 1976). Barney, Broken Neck, Fox, Crooked, Fourth of July, and Washington Creeks drain mostly sedimentary rocks of Cretaceous and/or Tertiary age (Foster, 1976; Brabb and Churkin, 1969).

Mertie (1938) believed that the sources of the placer gold in streams that drain the Seventymile terrane and subterrane Y_3 were quartz veins and mineralized zones that were genetically related to granitic rocks, and work by Clark and Foster (1971) supports this view. Clark and Foster reported anomalous values of gold in hydrothermally altered rocks, including silica-carbonate rocks, serpentinite and diorite, and quartz veins adjacent to a northwest-trending fault zone between Alder and Flume Creeks (No. 8, Fig. 4). They also detected arsenic in soil samples taken across a probable fault that strikes northwest across Teddy's Fork. Sources of the gold in streams that drain the Cretaceous and/or Tertiary sedimentary rocks may be paleoplacers (Mertie, 1938), but metamorphic rocks, which underlie the sedimentary rocks, and some dike rocks may also be sources.

Fortymile district. Gold was initially discovered on the Fortymile River in 1886 (Prindle, 1905) and on its tributaries between 1886 and 1895 (Mertie, 1938; IV, Fig. 4). Gold production of at least 417,000 oz (Cobb, 1973) has come mostly from stream and low bench placers; however, a few high bench placers have been worked occasionally. The streams of the Fortymile district drain a diverse group of rocks that include quartz-biotite gneiss and schist, quartzite, marble, and amphibolite of subterrane Y_4; quartzite, marble, phyllite, graphitic schist, and greenschist of subterrane Y_3; greenstone, serpentinite, and associated sedimentary rocks of the Seventymile terrane; granodiorite, quartz monzonite, quartz diorite, and diorite of the Taylor Mountain batholith; and various undivided granitic rocks whose age is thought to be Mesozoic or Tertiary (Foster, 1976).

At least three types of small lode sources of gold are present in the Fortymile district. One type, exemplified by the Purdy (Foster and Clark, 1970), Ingle, and Tweeden lodes (Nos. 14, 15, 16, Fig. 4), comprises gold-bearing quartz + calcite veins in metasedimentary and metavolcanic rocks of the Seventymile terrane and in intermediate-composition plutonic rocks that intrude the terrane. Gold has also been detected in altered diorite in the bedrock of Lost Chicken Creek (Foster and O'Leary, 1982) and in metatuff in an outcrop in the South Fork of the Fortymile River (No. 13, Fig. 4). The second type of lode gold occurrence is gold in quartz veins in subterrane Y_4. Mertie (1938) reported such an occurrence along Wade Creek and thought that the veins were related to a granite intrusion at depth. A third type of gold occurrence is gold in and adjacent to crushed zones and faults. Such zones are thought to be a source of gold in Dome Creek and Canyon Creek (Mertie, 1938). The recognition of major thrust faults within the Fortymile region (Foster and others, 1984) provides other possible sites of lode concentrations of gold.

Richardson district. The Richardson district includes scattered gold placers in the western part of the Big Delta Quadrangle. The two main areas of gold production are: (1) Tenderfoot Creek and adjacent streams, and (2) Caribou Creek and adjacent streams (areas Va and Vb, Fig. 4). Bundtzen and Reger (1977) estimated production at 95,000 oz of gold and 24,000 oz of silver; additional production from residual placers has taken place since 1979 (Eakins and others, 1983).

In the Tenderfoot Creek area, gold has been produced from stream and residual placers; Tenderfoot Creek is the main productive stream. The placers are buried beneath a deep cover of loess. The only known lode source of gold in this area is the Democrat lode (No. 18, Fig. 4), where gold occurs in altered and veined quartz porphyry. A sample of altered quartz porphyry yielded a K-Ar age of 89.1 ± 2.7 Ma (Wilson and others, 1985, after Bundtzen and Reger, 1977). Bundtzen and Reger regarded this age as the probable time of mineralization and noted that the distribution of the quartz porphyry and the gold placers in the Tenderfoot Creek area appears to be controlled by a northwest-trending lineament.

Gold production from the Caribou Creek (No. Vb, Fig. 4) area has come exclusively from stream placers, mostly along Caribou Creek. No lode sources of gold are known. However, the geochemistry of rock samples (Foster and others, 1978c) indicates that iron-stained areas adjacent to dikes, plutons, and local shear zones commonly contain anomalous amounts of arsenic, antimony, and zinc and in places contain silver, lead, or gold.

Tibbs Creek–Black Mountain area. The Tibbs Creek–Black Mountain area is one of the few areas that has produced lode gold, and this production probably exceeded that from the area's placers (VI, Fig. 4). The Blue Lead, Gray Lead, and Grizzly Bear (No. 17, Fig. 4), the principal mines, were operated mainly in the 1930s, although there was an attempt at further production in the late 1970s. About 32 oz of gold and 25 oz of silver were produced from quartz veins that cut metamorphic rocks and Cretaceous(?) granite. The veins are especially abundant and most productive near the contact of the metamorphic and granitic rocks. There are also minor occurrences of antimony and molybdenum nearby. Streams draining the area, such as Tibbs Creek, have had minor placer operations (Menzie and Foster, 1978).

Porphyry copper-molybdenum deposits. Porphyry copper-molybdenum occurrences in the Tanacross Quadrangle and in the adjacent Yukon Territory constitute the "interior porphyry belt" of Hollister and others (1975). A number of occurrences, including Mosquito, Paternie, Asarco, Bluff, and Taurus (Nos. 27, 28, 29, 30, and 31, Fig. 4), are known in the Tanacross Quadrangle (Singer and others, 1976), and ten occurrences are known in the Yukon Territory (Sinclair, 1978). The porphyry occurrences in Alaska are confined to subterrane Y_1 and are thought to be associated with Cretaceous and early Tertiary porphyritic, felsic subvolcanic stocks. These stocks were intruded into coeval volcanic rocks, Cretaceous plutonic rocks, and schist and gneiss. The porphyry deposits formed within and adjacent to subvolcanic stocks and breccia pipes. Although early Tertiary volcanic rocks and associated intrusive rocks occur in the eastern part of the Big Delta and western part of the Eagle Quadrangle, presently known porphyry occurrences are confined to the Tanacross Quadrangle.

The deposits of the interior belt display many characteristics typical of porphyry deposits elsewhere: (1) they are associated with felsic, subvolcanic intrusive rocks; (2) they are surrounded by large areas of hydrothermally altered rocks; (3) common hypogene minerals in the deposits are pyrite, chalcopyrite, molybdenite, and in some cases magnetite; and (4) supergene enrichment is an important control on the grade of mineralization. Deposits of the interior belt differ from other porphyry deposits in their reported lower grades and smaller tonnages (Singer and others, 1976). The inferred resources at Taurus are 50 million tons of 0.3 percent copper and 0.07 percent molybdenum (Chipp *in* Nokleberg and others, this volume). Taurus reportedly contains a considerable amount of supergene-enriched material. Hypogene ore in deposits of the interior belt is reported to be lower in grade than supergene ore by a factor of 1.5 to 2 (Godwin, 1976; Sawyer and Dickinson, 1976).

Tungsten. Tungsten, mainly in scheelite, occurs in the quartzite and quartzitic schist and the pelitic schist units (qq and ps) of subterrane Y_2; small amounts of tungsten have been produced from deposits hosted by the quartzite and quartzitic schist unit adjacent to Cretaceous granite or granodiorite intrusions northeast of Fairbanks. Production has been limited to the Fairbanks district, but exploration has identified a number of significant prospects elsewhere in subterrane Y_2. Near one prospect, Table Mountain (No. 5, Fig. 4), scheelite is present in sediments of streams that drain the quartzite and quartzitic schist unit, and a Tertiary(?) granite intrusion. There, tungsten occurs in thin marble layers in the quartzite and quartzitic schist unit, probably above a cupola of the intrusion. Scheelite is widely distributed in sediments of streams that drain the pelitic schist unit (Menzie and others, 1983). In recent years a number of occurrences have been discovered in this unit adjacent to Cretaceous(?) and Tertiary granitic intrusions in the southeastern part of the Circle Quadrangle (No. 6, Fig. 4), in the southwestern part of the Charley River Quadrangle (No. 7, Fig. 4), and in the northwestern part of the Eagle Quadrangle (Foley and Barker, 1981).

Published studies of the ore controls of the tungsten deposits are limited to the Fairbanks district. Byers (1957), who mapped many of the deposits and favored their contact-metasomatic origin, stated that the distribution of tungsten minerals was probably controlled by the occurrence of limestone within the contact zone of porphyritic granite; by local structural irregularities, such as drag folds, that localized ore deposition; and by tungsten-bearing quartz pegmatite that filled fractures in the rocks above the porphyritic granite. Metz and Robinson (1980), following the ideas of Maucher (1976), suggested, on the basis of amphibolite in the footwall of some of the Fairbanks deposits, that the deposits may be remobilized syngenetic deposits.

Petrologic and geochemical studies of the plutons related to the deposits of the Fairbanks district (Blum, 1983) suggest that they formed by remelting of Precambrian crustal material, but do not identify a specific source for the tungsten.

Tin. Cassiterite is widely distributed as an accessory mineral in stream sediments and placer concentrates of the northwest part of the YTTN; however, only a few lode tin occurrences have been identified, and little has been published on their characteristics. Nevertheless, the occurrence of cassiterite in stream sediments and the presence of both Cretaceous and Tertiary granites, which are petrologically and geochemically similar to granites in tin-bearing regions (Fig. 3), suggest that parts of the YTTN that contain such granites may also contain unidentified lode and/or associated placer tin deposits (Menzie and others, 1983). Lode deposits, if present, are likely to be large, low-grade greisen deposits, although skarn and vein deposits may also occur. A small amount of tin has been recovered from several creeks in the Circle District as a byproduct of gold mining (P. Jeffrey Burton, written communication, 1983).

Uranium. Anomalous levels of uranium have been detected in springs and stream sediments (Barker and Clautice, 1977; Menzie and others, 1983) in the northwestern part of the YTTN. The anomalies are spatially associated with biotite granites of early Tertiary age (Fig. 3).

Platinum. Anomalous amounts of platinum and palladium were detected in samples of a gabbro intrusion at one locality (No. 10, Fig. 4) in the Eagle Quadrangle (Foster, 1975), and smaller amounts of platinum were detected in similar mafic to ultramafic intrusive rocks elsewhere in that quadrangle (Foster and Keith, 1974). Such intrusive bodies could serve as sources for platinum in placer deposits.

Stratabound mineral occurrences. Although no stratabound lode deposits are known in the YTTN, the following general observations suggest that such deposits may be present: Paleozoic metasedimentary and metavolcanic rocks in subterranes Y_2, Y_3, and Y_4 are similar to those hosting such deposits in other parts of the northern Cordillera; geochemical anomalies in stream-sediment samples occur in several areas; and recent exploration has located areas favorable for prospecting for stratabound deposits in several subterranes of the YTTN.

In addition, galena has been reported as float and in stream sediments in subterrane Y_4, and possible stratabound deposits have been identified in the eastern parts of subterranes Y_3 and Y_4. In the western part of the YTTN, anomalous amounts of zinc, silver, and barium in sediment samples from streams draining the calc phyllite-black quartzite unit of subterrane Y_3 led Menzie and Foster (1978) to suggest that this unit may host stratabound deposits.

Coal. Coal deposits occur in Late Cretaceous and/or early Tertiary nonmarine sedimentary rocks in two areas of the YTTN. One area is near the village of Chicken, in the south-central part of the Eagle Quadrangle (A, Fig. 4), and the other is along the northeastern margin of the YTTN and probably extends north of the YTTN (B, Fig. 4). The characteristics of these coal fields are summarized in Table 2; their resources have not been estimated.

Geothermal resources. Three hot spring areas occur in the YTTN, all in the Circle Quadrangle (Waring, 1917). The hot springs are hydrothermal convective systems heated by deep circulation along faults associated with early Tertiary granitic plutons (Miller and others, 1975). The Chena Hot Springs (No. 32, Fig. 4) have a maximum measured surface temperature of 67°C and a discharge of approximately 800 l/min; maximum temperature at Circle Hot Springs (No. 33, Fig. 4) is 57°C, and discharge is approximately 500 l/min. Respective reservoir temperatures are calculated as 100° and 128°C; the springs thus are classified as intermediate-temperature hydrothermal convection systems (Brook and others, 1979). Springs in the third area (No. 34, Fig. 4) have a maximum measured surface temperature of 61°C (Keith and others, 1981b). Chena and Circle Hot Springs are used primarily for recreational purposes, although minor other uses of the hot water have been made. The third hot spring area is not developed because of its remote location.

Application of isotopic-dating techniques

Determining definitive ages of the rocks in the YTTN and the time(s) of their metamorphism, intrusion, and deformation is difficult owing to lack of fossils, poor exposures, and the region's complex metamorphic and deformational history. Isotopic techniques for dating rocks thus have become the principal means of

TABLE 2. COAL FIELDS OF YTTN*

Field	Size	Structure	Seam characteristics	Rank	Sulfur content	Development status
Eagle	130 by 3 to 16 km (80 by 2 to 10 miles)	Open folds	Near Washington Creek, five seams are at least 1.3 m thick	Subbituminous	Low	None
Chicken		Vertical beds	One seam is at least 6.8 m thick	Unknown	Unknown	Minor past production for local use

*Reference: Barnes (1967)

obtaining data on the ages of rocks and the thermal/metamorphic events.

The first isotopic data for rock samples from the YTTN came from lead-alpha (Pb-α) studies in the late 1950s (Matzko and others, 1958; Jaffe and others, 1959; Gottfried and others, 1959). In 1960, Stern (Holmes and Foster, 1968) obtained four Pb-α ages on plutonic rocks in the northern part of the Mount Hayes Quadrangle. K-Ar and Rb-Sr dating methods were applied soon after (Wasserburg and others, 1963). During the next 15 years, many K-Ar ages were determined on both igneous and metamorphic rocks, but mostly on granitic rocks. Ages determined by G. J. Wasserburg, M. A. Lanphere, D. L. Turner, J. G. Smith, F. H. Wilson, and others were compiled by Dadisman (1980). Little Rb-Sr work was done, partly because of the difficulty of obtaining unaltered material, but recently Blum (1983) reported Rb-Sr data for granitic rocks near Fairbanks. The K-Ar ages are most useful on nonmetamorphosed igneous rocks; ages obtained on metamorphic rocks are difficult to interpret. To better interpret the metamorphic rocks, McCulloch and Wasserburg (1978) applied a Nb-Sm method to Alaska Range rocks and Aleinikoff applied U-Pb methods (Aleinikoff and others, 1981) using zircons separated from an augen gneiss and other metamorphic rocks of the YTTN. Rb-Sr work was done to supplement the U-Pb work. Most recently, ^{40}Ar-^{39}Ar incremental heating methods have been used to help decipher the metamorphic history in the eastern part of the YTTN (Cushing and others, 1984a). Integrated studies of the various types of data for the YT have resulted in improved interpretations of the geologic history by making it possible to constrain the ages of many events more closely.

The following conclusions derive largely from K-Ar work based on about 142 age determinations (Wilson and others, 1985; this volume):

1. Three major periods of felsic plutonism occurred after major regional metamorphism, and volcanic rocks are associated with at least one and possibly two of these plutonic events. The oldest period of plutonism, during Late Triassic and Early Jurassic time, resulted in the emplacement of the granitic batholith at Taylor Mountain, and of other plutonic rocks (Mount Veta, Fig. 3B). Plutonic rocks of this age are presently known in eastern Alaska only in the southern part of the Eagle and northern part of the Tanacross Quadrangle.

The second period of plutonism occurred throughout most of the YT from about 105 to 85 Ma. The largest number of ages determined are between 95 and 90 Ma (Wilson and others, 1985; this volume). The largest and most numerous of these plutons are in the eastern part of the YTTN, mainly east of the Shaw Creek fault. The plutons mostly vary from quartz diorite to quartz monzonite. On the basis of K-Ar ages of 90 $\pm$ 2.8 Ma (hornblende) and 93.6 $\pm$ 2.1 Ma (sanidine) (Bacon and others, 1985), we interpret the extensive deposits of welded tuff in the Tanacross Quadrangle mainly as caldera-fill associated with this plutonic event.

The third period of felsic intrusions is indicated by 46 age determinations that range from 70 to 50 Ma (Wilson and others, 1985; this volume). These intrusions are generally small and commonly have contact aureoles. Most are located in the northwestern part of the YTTN northwest of the Shaw Creek fault, but one also occurs in the central part of the Tanacross Quadrangle (Mount Fairplay, Fig. 3). Comagmatic Tertiary volcanic rocks are documented by a K-Ar date of 57.8 $\pm$ 2 Ma in the eastern part of the Tanacross and one of 61.6 $\pm$ 2 Ma in the eastern Big Delta Quadrangles (Foster and others, 1976, 1979).

2. Although protolith ages and timing of metamorphism of metamorphic rocks are difficult to determine from K-Ar data, some important results have been obtained. Wilson and others (1985) studied 59 K-Ar age determinations on metamorphic rocks (43 from the YTTN) and identified two distinct clusters. Eighteen ages (7 from the YTTN) fall between 190 and 160 Ma; 24 ages (all from the YTTN) fall between 125 and 105 Ma. Although there is considerable scatter in the data, the Early Cretaceous cluster includes a number of concordant mineral pairs. Wilson and others concluded that this cluster probably reflects a metamorphic or thermal event that is distinct from the Cretaceous intrusive event and that the metamorphic ages generally are not reset by this later plutonism (F. H. Wilson, personal communication, 1984).

3. Although the number of determinations is limited and only a few are concordant mineral pairs, the K-Ar data strongly suggest a Jurassic metamorphic or thermal event (Wilson and others, 1985; this volume). As discussed below, ^{40}Ar–^{39}Ar data show such an event in the Eagle Quadrangle, but its areal extent and local intensities cannot be determined from the available data.

The U-Pb data from zircons indicate the following:

1. Plutonism occurred in subterrane Y_1 in the Mount Hayes Quadrangle (Lake George terrane) at about 360 Ma (Aleinikoff and Nokleberg, 1985a).

2. Extensive granitic intrusions occurred in subterrane Y_1 about 341 ± 3 Ma (lower intercept age) and are now represented by widely distributed augen gneisses (Dusel-Bacon and Aleinikoff, 1985).

3. The augen gneiss contains an inherited component of Early Proterozoic (2.1 to 2.3 Ga) zircons (Dusel-Bacon and Aleinikoff, 1985).

4. U-Pb ages of zircons from metamorphic rocks believed to have volcanic protoliths suggest that they were erupted 360 to 380 m.y. ago (Dusel-Bacon and Aleinikoff, 1985).

5. Time of metamorphism of the augen gneiss is not conclusively known.

6. Protolith age of the metamorphic rocks interpreted as wall rocks to the augen gneiss is not known, but they contain the Early Proterozoic inherited component.

The Rb-Sr whole-rock isochron obtained from widely separated outcrops of augen gneiss has an age of 333 ± 26 Ma, confirming the Mississippian intrusive age of the protolith obtained from the U-Pb determinations on zircon. Sm-Nd data also support the presence of an old crustal component in the augen gneiss (Aleinikoff and others, 1986).

Recent ^{40}Ar-^{39}Ar experiments on rock samples from subterrane Y_4 in the eastern part of the Eagle Quadrangle have established:

1. A cooling history for the granitic rocks of Taylor Mountain by analysis of hornblende, biotite, and K-feldspar. The granite was emplaced about 209 ± 3 Ma, and cooled from 500 to 175°C through a period of about 32 m.y. (Cushing and others, 1984b).

2. A major (amphibolite facies) metamorphic event that reached a peak about 213 ± 2 Ma, followed by igneous intrusions and cooling over a period of about 36 m.y. Amphibolite adjacent to the Taylor Mountain batholith has a Triassic integrated plateau age (213 ± 2 Ma) (Cushing and others, 1984b).

3. That thrusting occurred during cooling from the peak of metamorphism because greenschist-facies greenstone having a metamorphic age of 201 ± 2 Ma is thrust over the amphibolite, which is adjacent to the Taylor Mountain batholith. Other evidence of thrusting at about that time is based on the age of biotite crystallized in a thrust zone. The biotite has an integrated plateau age of 187 ± 2 Ma. The time of thrusting is also constrained by the age of a dike (integrated plateau age of 186 ± 2 Ma on muscovite) that cuts deformed metamorphic rocks in the thrust zone and is not deformed or significantly metamorphosed.

4. That the Cretaceous metamorphic or thermal event identified by the K-Ar work affected subterrane Y_4, as shown by minor plateaus in the ^{40}Ar-^{39}Ar data.

In summary, current radiometric age data indicate plutonism in the YTTN in Mississippian, Triassic, Cretaceous, and Tertiary time, and volcanism in the Mississippian, Cretaceous, and Tertiary. Major parts of the YTTN may have an Early Proterozoic basement or have received sediment from eroding Early Proterozoic sources. Metamorphism took place during Late Triassic and Early Jurassic time in the eastern part of YTTN (Y_4), but its extent is not known. A Cretaceous thermal event, of low grade at least in subterrane Y_4, was widespread in the YTTN. Deformation that included major thrusting occurred in subterrane Y_4 in Early Jurassic time.

YUKON–TANANA TERRANE SOUTH OF THE TANANA RIVER

Although exposure of bedrock in the YT south of the Tanana River is much better than in the YTTN, local cover by Tertiary rocks and by glacial deposits, restricted accessibility in the rugged Alaska Range, and a complex structural history make it difficult to relate this area to the YTTN and to other adjacent areas. Silberling and others (this volume) have grouped the rocks south of the Tanana River into several lithotectonic terranes, and Nokleberg and Aleinikoff (1985) have further divided the rocks in the Mount Hayes Quadrangle into several terranes or, in some cases, subterranes in the terminology used in this chapter. Our discussion of the area south of the Tanana River begins with terranes or subterranes in the Mount Hayes Quadrangle, continues with a description of the rocks of the Tanacross and Nabesna Quadrangles, and ends with the rocks in the Healy and Mount McKinley Quadrangles.

Metamorphic rocks

Mount Hayes Quadrangle. Rocks in the Mount Hayes Quadrangle south of the Tanana River and north of the Denali fault, most of which were included in the Yukon-Tanana terrane by Silberling and others (this volume), were divided by Nokleberg and others (1983) into the Macomb, Jarvis Creek Glacier, Hayes Glacier, and Windy terranes.

Macomb terrane. The Macomb terrane consists primarily of medium-grained mylonitic pelitic schist, calc-schist, and quartz-feldspar-biotite schist, intruded by quartz monzonite, granodiorite, quartz diorite, and diorite (unit pc, Fig. 2). The intrusive rocks have been almost completely recrystallized to mylonitic schist (Nokleberg and others, 1983). U-Pb analyses of zircons from the metamorphosed plutonic rocks indicate a Devonian (about 370 Ma) intrusion (Nokleberg and Aleinikoff, 1985; Aleinikoff and Nokleberg, 1985b). All of the rocks are polydeformed and metamorphosed under conditions of the lower amphibolite facies, but in places they have been retrograded to lower greenschist facies. Because no other age data are available, the age of protoliths of the intruded metasedimentary rocks cannot be determined more precisely than either Devonian or pre-

Devonian (pre-intrusion).

Jarvis Creek Glacier terrane. The Jarvis Creek Glacier terrane consists of fine-grained polydeformed schist (unit pq, Fig. 2) derived from sedimentary and volcanic rocks. Metasedimentary rocks are pelitic schist, quartzite, calc-schist, quartz-feldspar schist, and marble. Metavolcanic rocks are meta-andesite and metaquartz–keratophyre with some metadacite, metabasalt, and rare metarhyolite. All are cataclastically deformed, recrystallized, and metamorphosed under conditions of the greenschist facies. U-Pb analyses of zircons from metavolcanic rocks indicate a Devonian extrusive age of about 370 Ma (Nokleberg and Aleinikoff, 1985). The sedimentary protoliths are also considered to be of Devonian age, because they are interlayered with the metavolcanic schist (Nokleberg and Aleinikoff, 1985). Nokleberg and Lange (1985) suggested that the metavolcanic rock-rich part of Jarvis Creek Glacier terrane may be correlative with the Totatlanika Schist (unit to, Fig. 2) to the west in the Healy Quadrangle because both groups of rocks include abundant intermediate volcanic protoliths.

Hayes Glacier terrane. The Hayes Glacier terrane consists of two groups of phyllites: one is dominantly metasedimentary rocks with few to no metavolcanic rocks, and the other is mainly metavolcanic rocks with moderate to abundant amounts of metasedimentary rocks (unit pv, Fig. 2). Metasedimentary rock types are pelitic, quartzose, and quartz-feldspar phyllites, and minor calc-phyllite and marble. Metavolcanic rocks include meta-andesite, meta-quartz–keratophyre, and sparse metadacite and metabasalt. The rocks are cataclastically deformed and have been metamorphosed under conditions of the lower and middle greenschist facies. An early schistosity is folded into rarely seen, small-scale, isoclinal folds having axial planes parallel to schistosity. The dominant schistosity, which postdates the folding, dips moderately to steeply southward. Metamorphosed and deformed gabbro, diabase, and metagabbro dikes also occur and, on the basis of structural relations, are believed (W. J. Nokleberg, written communication, 1983; this volume, Chapter 10) to be middle or Late Cretaceous in age. Lamprophyre dikes and a small alkali-gabbro pluton were emplaced in early Tertiary(?) time (W. J. Nokleberg, written communication, 1983; this volume, Chapter 10).

Windy terrane. The Windy terrane consists predominantly of argillite, limestone, marl, quartz-pebble siltstone, quartz sandstone, metagraywacke, metaconglomerate, andesite, and dacite (unit ar, Fig. 2). Locally abundant megafossils and sparse conodonts indicate a Silurian(?) and Devonian age for these rocks. The rocks are generally slightly deformed and have poorly developed schistosity. Locally, deformation is intense, and phyllonite and protomylonite have formed in narrow zones. Rocks along the terrane's southern margin adjacent to the Denali fault are characterized by intense shearing, abundant fault gouge, and locally, by low-grade metamorphism (W. J. Nokleberg, written communication, 1985; this volume, Chapter 10).

Origin of terranes. On the basis of field relations and stratigraphic and structural data, Nokleberg and Aleinikoff (1985) interpret the Lake George, Macomb, Jarvis Creek Glacier, and Hayes Glacier terranes from north to south as successively shallower levels of a single, highly metamorphosed and deformed, Devonian submarine igneous arc. They suggest that the arc is either an island arc containing a slice of continental crust that contaminated the Devonian magmas, or a submerged continental-margin arc, with continental detritus being shed into a companion trench and subduction-zone system. Nokleberg (written communication, 1985; this volume, Chapter 10) interprets the Windy terrane as a surface-level slice of a Devonian island arc.

Tanacross and Nabesna Quadrangles. Amphibolite-facies gneiss and schist including augen gneiss (unit ag, Fig. 2) compose the northwestern part of the Alaska Range just south of the Tanana River in the Tanacross Quadrangle. The rocks are generally quartzose and commonly garnetiferous; calcareous rocks are rare. The lithology and metamorphic grade of the rocks, including the augen gneiss, are similar to those north of the Tanana River in subterrane Y_1. Quartz-mica schists in the foothills south of the Tanana River in the south-central part of the Tanacross Quadrangle (unit pc?, Fig. 2) have similarities in lithology and metamorphic grade to rocks in the Macomb terrane of the Mount Hayes Quadrangle.

The rocks in the Alaska Range in the Tanacross Quadrangle decrease in metamorphic grade to the south (Foster, 1970), and 10 to 20 km south of the Tanana River they are mostly greenschist-facies quartz–white mica schist ± chlorite, quartz-graphite schist, and quartzite (unit pq?, Fig. 2). All or part of these greenschist-facies rocks may be coextensive with the Jarvis Creek Glacier terrane of the Mount Hayes Quadrangle.

In the southeastern corner of the Tanacross Quadrangle, low-grade metamorphic rocks are largely light pink, light green, gray, and tan phyllite with discontinuous layers of marble and quartzite (unit pv?, Fig. 2). Greenstone also occurs. Because these rocks are adjacent to those of the Hayes Glacier terrane and have some similar lithologies, we tentatively correlate them. This group of rocks also appears to be coextensive with similar rocks in the Nabesna Quadrangle that have been considered of Devonian age.

In the Tanacross Quadrangle, the greenschist-facies schist and phyllite are intruded by dikes, sills, and lenses of altered diorite (not shown on map), which appear to be slightly metamorphosed (Foster, 1970).

Although there is considerable decrease in grade of metamorphism from amphibolite to greenschist facies from north to south in the Alaska Range within the Tanacross Quadrangle, there is little difference in the deformational characteristics. Foliation most commonly strikes northwest and dips predominantly southwest. Large-amplitude (several hundred meters) folds in layering and/or schistosity are visible in a few places, but small folds (amplitudes of 1 cm to more than 1 m) are common. S.H.B. Clark (written communication, 1972) recognized three generations of folds. The earliest is a set of small tight-to-isoclinal folds that fold the compositional layering and have well-developed axial plane schistosity. These folds are rarely preserved. A second set of folds deforms schistosity, are tight to isoclinal, and have axial-plane schistosity. The third-generation folds are kink folds and deform both previous generations of folds. Although major faults were not mapped between units in the Tanacross Quadrangle, Foster (1970) recognized the possible existence of such faults.

In the northern part of the Nabesna Quadrangle, a group of greenschist-facies rocks (unit pq?, Fig. 2) consists mostly of quartz-muscovite schist, quartz-muscovite-chlorite schist, graphitic schist, and minor calcareous mica schist. These schists may be coextensive with the Jarvis Creek Glacier terrane. South of the schists is a unit of slightly metamorphosed sedimentary and mafic volcanic rocks (unit Dm, Fig. 2). In the northwestern and north-central part of the quadrangle, this unit consists predominantly of dark gray phyllite, quartzite, porcellanite, quartz-mica schist, and marble. These rocks have been extensively intruded by mafic diorite and gabbro, which were emplaced after the main period of folding and metamorphism (Richter, 1976). Much of this area is included in the Pingston terrane of Silberling and others (this volume). Farther south, partly along the north side of the Denali fault, the rocks are chiefly phyllite and metaconglomerate with subordinate quartz-mica schist and quartzite. Scattered along strike are pinnacled outcrops of recrystallized limestone, a few of which contain rugose and tabulate corals of Middle Devonian age. In the east-central part of the Nabesna Quadrangle, probably bounded by faults, are weakly metamorphosed volcanic and volcaniclastic rocks (unit mv, Fig. 2). The western part of this volcanic unit consists mostly of andesite and basalt flows; the eastern part is dominantly volcanic sandstone, cherty argillite, quartzite, and tuff. Some of the east-central Nabesna Quadrangle is included in the Windy and McKinley(?) terranes of Silberling and others (this volume). Protoliths of the metamorphic rocks in the Nabesna Quadrangle are probably of Paleozoic age; the few fossils that have been found indicate that they may be largely Devonian.

Healy and Mount McKinley Quadrangles. Workers in the Healy and Mount McKinley Quadrangles have recognized three major groups of metamorphic rocks; none of these are continuous in outcrop with the metamorphic rocks in the Mount Hayes Quadrangle. The southernmost group, which is bounded on the south by the Hines Creek strand of the Denali fault system, formerly was called the Birch Creek Schist (unit bc, Fig. 2) (Wahrhaftig, 1968; Gilbert and Bundtzen, 1979; Bundtzen, 1981). North of this unit, a group of less crystallized rocks composes the Keevy Peak Formation (unit kp, Fig. 2); and a group of lithologically diverse rocks has been included in the Totalanika Schist (unit to, Fig. 2) (Wahrhaftig, 1968; Gilbert and Bundtzen, 1979). Differences in degree of metamorphism, lithology, and structural history suggest that these units are fault bounded (Wahrhaftig, 1968).

The southernmost group (bc, Fig. 2) consists predominantly of quartz–white mica schist, micaceous quartzite, and lesser amounts of graphitic schist, porphyroclastic quartz-feldspar schist, chlorite schist, greenstone, calcareous schist, and marble (Gilbert and Bundtzen, 1979). It was completely recrystallized during two or more periods of metamorphism: in the central Healy Quadrangle its metamorphic grade is greenschist facies, but to the west and north in the McKinley Quadrangle its grade is higher (Bundtzen, 1981; Wahrhaftig, 1968). In the McKinley Quadrangle, Bundtzen (1981) recognized an upper-greenschist-to-amphibolite–facies prograde event, followed after an unknown interval by lower-greenschist–facies retrograde metamorphism. Bundtzen suggested that differences in metamorphic grade of this unit from southeast to northwest in the McKinley Quadrangle may indicate different structural levels, with deepest levels to the northwest. The unit is complexly folded and faulted. Its age is unknown, but may be at least partly Paleozoic if rocks that contain echinodermal fragments belong to this unit (Gilbert and Bundtzen, 1979). In the Kantishna region of the McKinley Quadrangle, Bundtzen tentatively assigned a Precambrian age to this unit, but recognized that parts of it may be younger.

The Keavy Peak Formation consists of black or dark gray, carbonaceous phyllite; black quartzite; stretched-pebble conglomerate; gray, green, and purple slate; and white mica–quartz schist. Textures are commonly mylonitic, and some schists contain large, scattered bluish gray quartz grains, probably porphyroclasts. These rocks are less intensely deformed and recrystallized than those in the southernmost unit bc, but have been isoclinally folded. Because the Keevy Peak Formation is only slightly metamorphosed, original sedimentary features, such as graded bedding and cross-bedding, are preserved locally. Wahrhaftig (1968) indicated that the Keevy Peak Formation lies unconformably on unit bc. He also stated, "Several features suggest that the schist formations of the central Alaska Range have been cut by numerous unmapped thrusts and that many of the mapped lithologic contacts between schists of different units are, in fact, tectonic contacts whose original nature has been obscured by subsequent metamorphism." We believe that the contact of the Keevy Peak Formation probably is such a thrust; in the Kantishna Hills, Bundtzen (1981) also believed it to be a tectonic contact. Scarce fossils from the upper part of the Keevy Peak Formation are Middle and Late Devonian in age (Gilbert and Redman, 1977). Gilbert and Bundtzen (1979) suggested that the formation may range in age from Ordovician to Devonian.

The most northerly and apparently youngest metamorphosed formation in the Healy and Mount McKinley Quadrangles is the Totatlanika Schist, first defined by Capps (1912) and redefined and divided into five members by Wahrhaftig (1968). The characteristic lithology is quartz-orthoclase-sericite schist (and gneiss) (unit to, Fig. 2) that interfingers complexly with a large variety of lithologies in which felsic and mafic metavolcanic rocks predominate. Gilbert and Bundtzen (1979) described three main lithologies: metafelsite, metabasite, and metasedimentary rocks. The metafelsite consists primarily of porphyritic metarhyolite and felsic metatuff, now primarily quartz–orthoclase–white mica schist and gneiss. Wahrhaftig (1965) described augen of potassium feldspar 2.5 to 25 mm in diameter and smaller augen of quartz. Gilbert and Bundtzen (1979) interpreted the augen as relict phenocrysts and many of the rocks as probable mylonites (terminology of Wise and others, 1984). The metabasite primarily is probably calc-alkaline metabasalt, but there are also minor amounts of metavolcanic rocks of intermediate composition. Metasedimentary rocks predominate in the upper part of the Totatlanika Schist; their protoliths included sandstone, siltstone, and tuff. Locally, the rocks are calcareous, and carbonate layers occur. Relict sedimentary textures are visible in places. Black phyllite, indistinguishable from black phyllite in the Keevy Peak Formation, is interlayered with metavolcanic rocks throughout the Totatlanika Schist.

The Totatlanika Schist has undergone low-grade regional metamorphism, in most places probably no higher than low greenschist facies. A large component of the regional event has been dynamic rather than thermal, as evidenced by extensive development of mylonite (Bundtzen, 1981). Mica crenulations, cleavage, and isoclinal folding are common in the less competent layers of the unit.

A few fossils found in the Totatlanika Schist suggest that it probably ranges from Late Devonian to Mississippian in age. Gilbert and Bundtzen (1979) proposed that it may consist largely of volcanic-arc deposits formed above a subduction zone along the western margin of North America.

Mesozoic igneous rocks

In the Tanacross, Nabesna, and Mount Hayes Quadrangles, granitic rocks, probably mostly of Cretaceous age, intrude the metamorphic rocks. They range in composition from quartz diorite to quartz monzonite and are similar in composition and age to granitic rocks that cover large areas of the YT north of the Tanana River. In the Mount Hayes Quadrangle they occur in the Macomb, Jarvis Creek Glacier, and Hayes Glacier terranes. Nokleberg and others (1986) considered them to be slightly to moderately metamorphosed and suggested that they were intruded during the waning stage of an Early Cretaceous regional metamorphism.

Only a few small granitic bodies are known in the Tanacross Quadrangle south of the Tanana River, but in the Nabesna Quadrangle, three fairly large plutons occur. They are dominantly quartz monzonite, although they vary widely in composition. Most are foliated and have no xenoliths (Richter, 1976). The Gardiner Creek pluton (Richter, 1976) in the northeastern corner of the Nabesna Quadrangle is probably coextensive with the granitic plutons of the YTTN in the southeastern part of the Tanacross Quadrangle.

The Hayes Glacier terrane is intruded also by mafic dikes, commonly much deformed and metamorphosed. They are considered to be of mid- or Late Cretaceous age (W. J. Nokleberg, written communication, 1985; this volume, Chapter 10). In the central part of Jarvis Creek Glacier terrane, an intrusive suite of monzonite, alkali gabbro, lamprophyre, and quartz diorite, of early Tertiary(?) age, is partly surrounded by a ring dike of quartz monzonite. Locally extensive lamprophyre dikes and alkali gabbro are probably temporally associated with this suite. Foley (1982) described two dike swarms of nonmetamorphosed potassic alkali-igneous rocks, one near the West Fork of the Robertson River and the other to the east near the Tok River, and suggested that they are the youngest igneous rocks of the eastern Alaska Range (Late Cretaceous). They include biotite-lamprophyre dikes and sills, associated breccia dikes, and a stock of alkali gabbro and alkali diorite. Some of the mafic rocks of the Mount Hayes Quadrangle may be related to those in the southern part of the Tanacross Quadrangle (Foster, 1970) and/or Nabesna Quadrangle (Richter, 1976).

Mineral resources

In the southern part of the YT, gold has been produced from placer deposits, and gold, silver, antimony, and lead have been produced from several types of vein deposits. Other types of deposits present in the terrane include skarn or tactite deposits, copper vein deposits, and stratabound auriferous-sulfide bodies. Recent exploration for volcanogenic massive sulfide deposits has identified a number of significant prospects and occurrences. Perhaps the most important mineral resource of this region is coal, which occurs in Tertiary sedimentary rocks. Figure 4 shows the distribution of selected lode deposits, prospects, and occurrences in the southern YT.

Gold placers. Placer deposits in the Bonnifield and Kantishna districts (Nos. VII and VIII, Fig. 4) each yielded about 45,000 to 50,000 oz of gold between their discovery in 1903 and 1960 (Cobb, 1973). The deposits are mainly stream placers, but include bench placers. Sources of the gold are likely the various vein and stratabound lode deposits that occur in the districts.

Vein deposits. Most vein deposits in the southern YT belong to three types identified by Bundtzen (*in* Nokleberg and others, this volume, Chapter 29) in the Kantishna district: (1) auriferous quartz-arsenopyrite veins such as the Banjo (No. 20, Fig. 4); (2) galena-sphalerite-tetrahedrite-sulfosalt veins, such as Quigley Ridge (No. 21, Fig. 4), and (3) simple stibnite-quartz veins such as Stampede, Rambler, Glory Creeks, and Rock Creek (No. 19, Fig. 4).

Volcanogenic massive sulfide deposits. Most of the volcanogenic massive sulfide prospects and occurrences are located in the metavolcanic part of the Jarvis Creek Glacier terrane (Lange and Nokleberg, 1984). Important occurrences are known at Anderson Mountain (Freeman *in* Nokleberg and others, this volume, Chapter 29) (No. 22, Fig. 4); near Dry Creek (Gaard *in* Nokleberg and others, this volume, Chapter 29) (No. 23, Fig. 4); Miyaoka, Hayes Glacier, and McGinnis Glacier (Lang and Nokleberg *in* Nokleberg and others, this volume, Chapter 29) Nos. 24 and 25, Fig. 4); and in the Delta district (Nauman and Newkirk *in* Nokleberg and others, this volume, Chapter 29) (No. 26, Fig. 4). The deposits have many characteristics of deposits associated with felsic and intermediate volcanic rocks that form in island-arc settings.

Coal. Two coal fields, the Nenana and Jarvis Creek (D and C, Fig. 4), occur in Tertiary nonmarine sedimentary rocks that are described more fully in the next section. The Nenana field, which consists of several separate basins, has been a significant

TABLE 3. COAL FIELDS OF THE SOUTHERN PART OF THE YT*

Field	Size	Structure	Reserves/ resources†	Seam characteristics	Rank	Sulfur content	Development status
Nenana	Several basins 129 km by 16 to 48 km (80 by 10 to 30 miles)	Open folds and a few faults	780 x 10^6 5,400 x 10^6 7,900 x 10^6	Separate basins contain 8 to 9 seams that are at least 1.5 and up to 20 m thick	Subbituminous	Low	Produces about 800,000 tons/yr for local use. Export planned.
Jarvis Creek	40 km^2 (16 mi^2)	Open folds	0.3 x 10^6 12.5 x 10^6	Basin contains 30 seams that vary in thickness from 0.3 to 2.3 m	Subbituminous	Low	Some past production. Presently being developed for local use.

*References: Barnes (1967); Eakins and others (1983); Wahrhaftig and Hickcox (1955).
†tonnes proven, indicated, or inferred.

source of coal in Alaska, and both fields, whose characteristics are summarized in Table 3, contain substantial resources.

Sedimentary rocks

Tertiary nonmarine sedimentary rocks (unit Ts, Fig. 2) are fairly extensive in the northern part of the Healy Quadrangle and southern part of the Fairbanks Quadrangle and also occur in the southeastern Big Delta Quadrangle, northeastern and north-central Mount Hayes Quadrangle, and northeastern Mount McKinley Quadrangle. Two distinct units, shown as one composite unit on Figure 2, have been recognized: the coal-bearing formation and the overlying Nenana Gravel.

The coal-bearing formation is an informally designated sequence (based on local usage) consisting of interbedded lenses of poorly consolidated sandstone, siltstone, claystone, conglomerate, and lignitic and subbituminous coal (Wahrhaftig and Hickcox, 1955). The generally uncemented and poorly to moderately consolidated rocks erode readily. Both lithology and thickness vary greatly over short distances, and the range in thickness can be at least partly attributed to deposition on an uneven erosion surface of deeply weathered metamorphic rocks (Wahrhaftig and Hickcox, 1955). The total thickness of the coal-bearing formation reaches several hundreds of meters. Bedding is generally horizontal or has gentle dips. In places the formation has been warped and faulted. The number and thickness of coal beds is variable throughout the formation; in the Nenana coal field, however, there are a large number, and they range in thickness from a few centimeters to 20 m (Barnes, 1967). In most places only a few coal beds are thicker than 60 cm. The coal-bearing formation has long been considered of Tertiary age, but its position within the Tertiary is uncertain. A Miocene age is considered probable (Holmes and Foster, 1968).

The Nenana Gravel consists largely of poorly to moderately consolidated, poorly cemented, fairly well sorted conglomerate and sandstone (Wahrhaftig, 1958). Pebbles in the conglomerate are generally slightly weathered. The formation is more resistant to erosion than the underlying coal-bearing formation and commonly supports steep cliffs 15 to 30 m high (Wahrhaftig, 1958). It varies in thickness but is known to exceed 1,300 m in places. It generally has about the same attitude as the coal-bearing formation, and although a minor unconformity locally occurs between these units, the Nenana Gravel has also been warped and faulted since deposition. Patches of poorly consolidated gravel on the north flank of the Alaska Range in the Mount Hayes Quadrangle may be erosional remnants of the Nenana Gravel (Holmes and Foster, 1968). The exact age of the formation is uncertain. Pollen studies (Holmes and Foster, 1968) suggest that it probably is Pliocene, whereas Wolfe and Toshimasa (1980) consider it to be late Miocene and early Pliocene on the basis of Clamgulchian-stage fossils.

Geophysical data

Comparatively few geophysical studies for the YT have been published. Aeromagnetic maps are available for most of the quadrangles at scales of 1:250,000 and 1:63,360, and interpretations of the aeromagnetic maps have been made for the Nabesna (Griscom, 1975), Tanacross (Griscom, 1976), Big Delta (Griscom, 1979), and Circle (Cady and Weber, 1983) Quadrangles. No regional aeromagnetic interpretation that includes the YT has been made since a study based on widely spaced (10 mi) flight lines (Brosgé and others, 1970). Available gravity data for the YT are shown on a Bouguer gravity map of Alaska (Barnes, 1977; Barnes and others, this volume). Gravity maps have been published for the Nabesna and Circle Quadrangles at scales of 1:250,000 (Barnes and Morin, 1975; Cady and Barnes, 1983). The density of gravity stations and quality of data are variable throughout the YT. Geophysical methods have been used by private industry in exploration for asbestos, copper porphyry, and other types of deposits. Some of the geophysical work being done on the Trans-Alaska Crustal Transect (TACT) will include parts of the YT.

SEVENTYMILE TERRANE

The Seventymile terrane is a discontinuous belt of alpine-type (fault-bounded) ultramafic rocks and associated slightly metamorphosed mafic volcanic and sedimentary rocks that have been thrust upon and imbricated with rocks of subterranes Y_3 and Y_4 of the YTTN. Churkin and others (1982) referred to this belt of rocks as the Salcha terrane. The belt trends northwesterly from the Yukon Territory into the northern part of the Eagle Quadrangle; in the northeastern part of the Big Delta Quadrangle, the belt is displaced to the south along the northwest side of the Shaw Creek fault. From there it trends southwestward to the center of the Fairbanks Quadrangle (Fig. 5). Five large peridotite bodies, labeled 1 to 5 on Figure 5 (the peridotites of Boundary, American Creek, Mount Sorenson, Salcha River ["Nail" allochthon of Southworth, 1984], and Wood River Buttes) and three areas of massive greenstone bodies, labeled I to III on Figure 5 (the greenstones of Wolf Mountain, Chicken, and Ketchumstuk), make up the main part of the Seventymile terrane. Some of these rocks in the Eagle Quadrangle have been described as parts of a dismembered ophiolite (Foster and Keith, 1974; Keith and others, 1981a). Numerous sporadically distributed small lenses of serpentinized peridotite and serpentinite crop out south of the main belt of ultramafic rocks (Keith and Foster, 1973), especially in the Eagle Quadrangle. These small bodies may have detached from the sole of the thrust fault at the base of the Seventymile terrane.

Ultramafic rocks

The largest outcrops of the Seventymile terrane are composed mainly of alpine-type ultramafic rocks. The peridotite of Boundary is approximately 8 km^2 in area, the peridotite of American Creek approximately 31 km^2, the peridotite of Mount Sorenson approximately 41 km^2, the peridotite of Salcha River is 40 km long and is approximately 80 km^2 in area, and the peridotite of Wood River Buttes is approximately 6 km^2 in area.

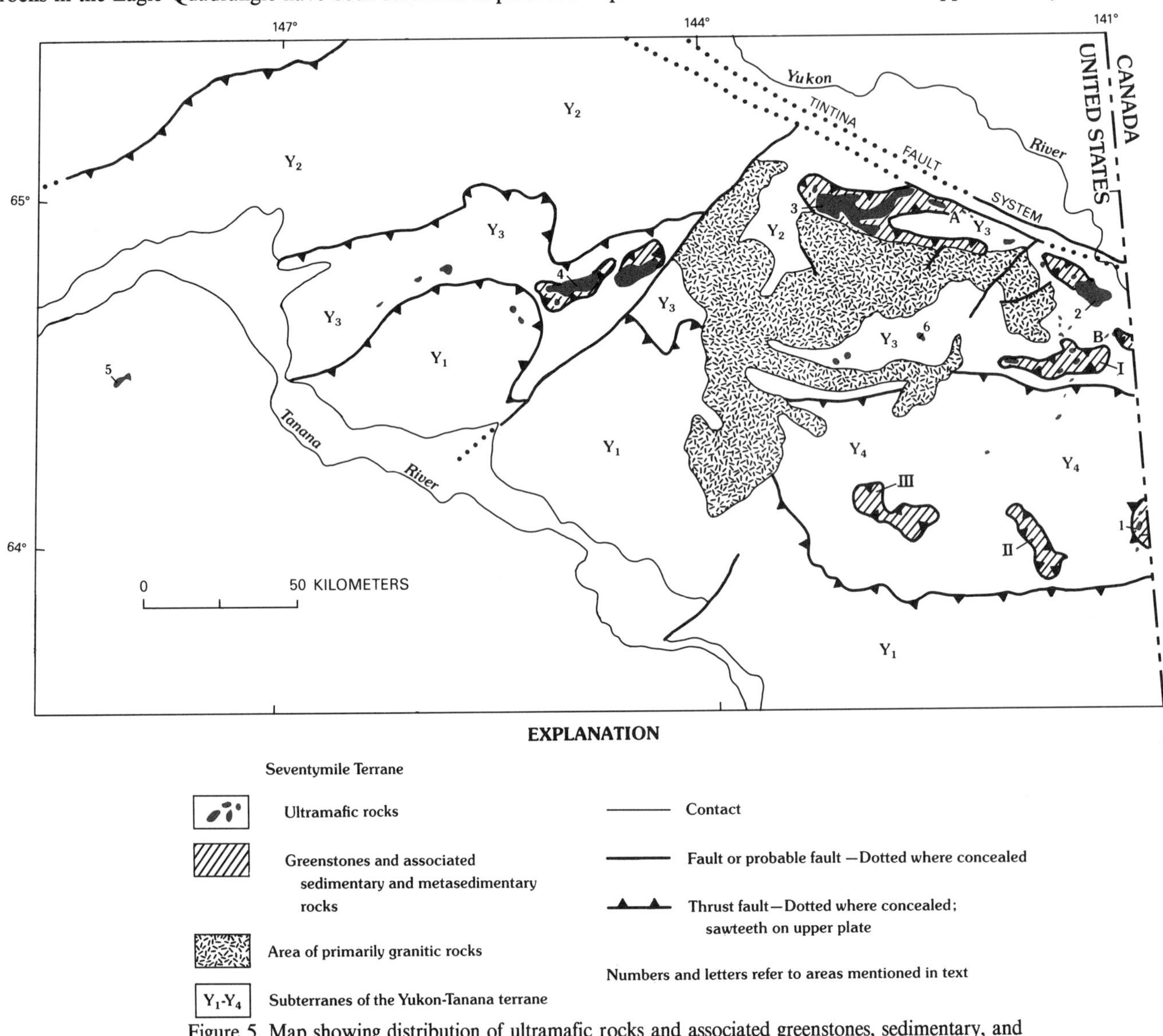

Figure 5. Map showing distribution of ultramafic rocks and associated greenstones, sedimentary, and metasedimentary rocks of the Seventymile terrane. Map after Foster (1976), Weber and others (1978), and Keith and others (1981a).

The rocks mainly are partly serpentinized harzburgite and dunite, and minor amounts of clinopyroxenite. Chromite is locally present, but nowhere abundant. Secondary magnetite that developed during serpentinization is common. Tectonic inclusions of rodingite occur in the large peridotite bodies. Bodies of coarse-grained cumulate gabbro are associated with the peridotite of Mount Sorenson. Silica-carbonate zones are well developed at the base of the peridotite of Salcha River and locally developed in the peridotite of Mount Sorenson.

Small serpentinite bodies, whose relict textures suggest mainly harzburgite protoliths, crop out as lenses or pods. Locally, at the contact with country rock, the small bodies have a rind of actinollite and/or chlorite, or hard slip-fiber serpentine, indicating a tectonic contact with the adjacent rocks. The distribution of the small outcrops is irregular, and they appear to be thrust over, and infolded and imbricated with, metamorphic rocks of subterranes Y_3 and Y_4. One of these small bodies, the serpentinite of Slate Creek (No. 6, Fig. 5), contains asbestos in potentially commercial quantities (Foster, 1969; Mullins and others, 1984).

Volcanic rocks

Greenstone bodies, originally basaltic pillow lavas and mafic lava flows, are in contact with the massive peridotite bodies at all the major outcrops. The greenstones are composed mainly of chlorite, actinolite, epidote, feldspar, magnetite, quartz, and calcite. Large masses of greenstone with minor associated serpentinite and silica-carbonate lenses also crop out as thrust remnants (Nos. I, II, and III, Fig. 5). The greenstones in the south-central part of the Eagle Quadrangle (No. II, Fig. 5), which lie upon subterrane Y_4, show thermal effects from the adjacent Taylor Mountain batholith.

Tuff that is metamorphosed to lower greenschist facies and which generally contains significant amounts of calcite and chlorite is associated with greenstone in many places.

Associated with the peridotite of Mount Sorenson are diabase dikes and plugs; some are metamorphosed and contain small amounts of pumpellyite and prehnite. Nonmetamorphosed basaltic pillow lavas and porphyritic silicic volcanic rocks crop out at the eastern end of the peridotite of Mount Sorenson and are in contact with cumulate gabbro.

At one locality (A, Fig. 5), in the north-central part of the Eagle Quadrangle, glaucophane, epidote, and sphene have formed from basaltic rock under blueschist metamorphic conditions (Foster and Keith, 1974).

Metasedimentary rocks

Low-grade metasedimentary rocks within the Seventymile terrane are chert, argillite, sandstone, conglomerate, graywacke, and fine-grained, dark gray limestone. Many were deposited alternately with submarine basaltic lava flows that now are greenstone; younger sedimentary strata were deposited in local basins. Probably all are submarine in origin and were deposited prior to thrusting of the Seventymile terrane.

Chert is interlayered with some of the greenstones and mafic volcanic rocks. Most is slightly recrystallized and contains abundant silica veins and veinlets. Radiolarians and conodonts from red chert in contact with the peridotite of Salcha River indicate an Early Permian age (Foster and others, 1978b). No fossils have been found in adjacent very low-grade metamorphosed graywacke sandstone and conglomerate adjacent to the peridotite. Slightly recrystallized chert interlayered with pillow basalts on the southeast side of the peridotite of Mount Sorenson has not yet yielded fossils, but Mississippian radiolarians occur in red and gray chert on the north side of the peridotite of Mount Sorenson (D. L. Jones, oral communication, 1984).

Slightly metamorphosed sedimentary rocks north of the Fortymile River (B, Fig. 5) include argillite, volcanic conglomerate, and fine-grained black limestone. An early Late Triassic age (early Early Norian) for a much fractured and deformed carbonaceous limestone is indicated by the conodont *Epigondolella primitia* Mosher (T. R. Carr, written communication, 1985). Some of these rocks are similar to those near the Clinton Creek asbestos deposit in the Yukon Territory that are considered by Abbott (1982) to be Late Triassic in age on the basis of a conodont. Metasedimentary rocks (tuff, argillite, and limestone) of slightly higher metamorphic grade occur near greenstones in the south-central part of the Eagle Quadrangle (C, Fig. 5).

Metamorphism

Much of the pervasive serpentinization of the ultramafic rocks is due to their hydration during emplacement into the crust. Large peridotite masses show no internal textural effects from regional metamorphism. However, many of the smaller ultramafic bodies that have been imbricated with parts of subterranes Y_3 and Y_4 have foliation and low-grade metamorphic mineral assemblages. Metamorphism of the mafic volcanic rocks may have occurred, in part, within an ocean basin prior to thrusting. However, subsequent thrusting and low-grade regional greenschist metamorphism has produced cross-cutting veinlets of serpentine (including cross-fiber asbestos), magnetite, chlorite (penninite), actinolite, anthophyllite, and magnesite. Closely associated tuff and sedimentary rocks show effects of low-grade regional metamorphism by a weak but distinct foliation and by chloritization of ferromagnesian minerals.

Structural and tectonic relations (also see "Geologic history")

The Seventymile terrane has been thrust into its present position with respect to the underlying metamorphic rocks of the Yukon-Tanana terrane. Thrust planes are nearly horizontal in most places. Imbricate thrusting of serpentinized peridotite with subterrane Y_4 and units ws and qs of subterrane Y_3 is prevalent in the central and extreme southeastern part of the Eagle Quadrangle.

The alpine-type ultramafic rocks, originally derived from the mantle, were emplaced into oceanic crust (Coleman, 1977; Patton and others, this volume, Chapter 21) in an ocean basin opening between subterranes Y_1 and Y_4, and also between subterranes Y_1 and Y_2. Later, as the ocean basin gradually closed and the tectonic components of Alaska moved northward into their present positions, some of the rocks of the ocean basin (Seventymile terrane) were obducted southward onto subterrane Y_4, but most were thrust northward as several slices onto subterrane Y_3. The large peridotite bodies were the leading edge of the main thrust sheet, and are structurally the highest part of the Seventymile terrane. Thrusting of the peridotite bodies was followed by thrusting of the greenstone and associated metasedimentary rocks. The trace of the former ocean basin is now indicated primarily by the distribution of the rocks of the Seventymile terrane lying in an arcuate belt upon subterrane Y_3 (Fig. 5).

Mineral resources

The Seventymile terrane contains large deposits of asbestos, minor occurrences of gold, and reported minor occurrences of nickel and chromium. The gold occurrences were discussed under the Seventymile district of the YTTN; because present information suggests that the nickel and chromium occurrences are small, they are not discussed further.

Asbestos. Asbestos occurs in two geologic settings within the Seventymile terrane: (1) in large, partially serpentinized ultramafic bodies, many of which are in contact with greenstone, and (2) in small serpentinite bodies (Keith and Foster, 1973). Significant deposits and prospects are confined to the second geologic setting and include the Slate Creek deposit (No. 9, Fig. 4), the Champion Creek prospect (No. 11, Fig. 4) (Mullins and others, 1984), and the Liberty Creek prospect (No. 12, Fig. 4). Large deposits in the Yukon Territory (Clinton Creek and Caley) occur in the same geologic setting (Abbott, 1982). The Slate Creek deposit has reported reserves of 60 million tons of ore containing 6.4 percent fiber of good quality for asbestos cement products (Mullins and others, 1984) and 67 million tons of indicated and possible ore. The deposits and prospects share a number of common characteristics. The asbestos occurs as cross-fiber chrysotile in 0.03- to 2-cm-thick, closely spaced veins that have been completely serpentinized in small bodies of serpentinized ultramafic rock (Foster, 1969; Hytoon, 1979). The margins of many of these bodies contain fibrous actinolite and slip-fiber serpentine, indicating that the bodies have been sheared.

The deposits and prospects are confined to a narrow zone along and approximately parallel to the contact between amphibolite-facies rocks and greenschist-facies rocks. This contact was interpreted as a thrust fault by Foster and others (1984). The thrust fault probably was formed in early Middle Jurassic time (Foster and others, 1984), and the small, fractured, completely serpentinized ultramafic bodies that host the asbestos deposits and prospects may be part of the thrust that soles the Seventymile terrane.

QUATERNARY GEOLOGY OF EAST—CENTRAL ALASKA

Much of east-central Alaska is covered by Quaternary sedimentary deposits. Almost all of the Fairbanks Quadrangle, for example, is covered by them, and the few natural outcrops occur mostly along rivers and streams and on the highest hilltops. The Quaternary deposits are unconsolidated, but many of them are perennially frozen. In the part of east-central Alaska described in this chapter, the largest area of Quaternary deposits is in the Tanana River valley, and widespread eolian deposits on the surrounding uplands also were derived from the Tanana valley. Quaternary deposits have had a major role in the economy of east-central Alaska in many ways, particularly as a source of rich gold placers, and of sand and gravel.

Ancient gravels

Several small areas of unconsolidated, poorly, or partly consolidated gravel are found at high elevations in parts of the northern Alaska Range and in the YTTN of the Tanacross Quadrangle. These gravels in the northern Alaska Range resemble the Tertiary Nenana Gravel and have generally been correlated with it. However, some of these gravels may be of early Pleistocene age. In the northern Alaska Range, they occur at elevations just under 1,000 to 1,200 m. They are generally only a few meters thick, deeply weathered, apparently flat-lying, and contain well-rounded clasts. They rest unconformably on metamorphic rocks or on coal-bearing sedimentary rocks (Foster, 1970). Small areas of residuum from metamorphic rocks occur on remnants of an old warped erosion surface that extends southeast from Mount Neuberger (in the Alaska Range 23 km southwest of the town of Tok). The surface is about 1,700 m in altitude near Mount Neuberger but is lower to the southeast. The residuum is at least 6 m thick in places. Its age is unknown, but it probably pre-dates the earliest glaciation in this area (Foster, 1970).

Glaciation

Most of east-central Alaska was not covered by glaciers during the Pleistocene. Continental ice sheets were not present, but alpine and piedmont glaciers were abundant in the Alaska Range, and small alpine glaciers occupied the upper reaches of many valleys in the highest parts of the YTTN. Glacial outwash was extensively deposited by the large streams, and morainal deposits are conspicuous in many valleys. Silt and sand derived from the dry exposed material in large outwash aprons that were widely deposited in the Tanana River valley were blown northward to form dunes at the northern edge of the valley and a loess mantle on adjacent hills. Material from the Yukon River valley and the alluvial fans of tributaries was blown southward; some of it mantled hills, terraces, and other landforms in the northernmost part of the area described in this chapter. Lacustrine deposits from ice-dammed lakes occur in the Tanana River valley and a few tributary valleys.

Old glaciations of pre-Wisconsin age have been recognized (Ten Brink, 1983; Weber, 1983) in both the northern Alaska Range and the YTTN. Ten Brink (1983) stated that a broad expanse of piedmont ice spread north from the Alaska Range to cover the ancestral foothills area, probably in late Tertiary time. Two other pre-Wisconsin to early Pleistocene glacial events are also recorded in the Alaska Range. The most extensive and best known Wisconsin-age glaciations are represented in many valleys by two major early Wisconsin advances and four late Wisconsin advances. At least six Holocene ice advances have been recorded, and small glaciers still exist (Table 4).

Weber (1983) recognized three pre-Wisconsin glacial episodes in the YTTN, the oldest being pre-Pleistocene in age, and the other two early(?) and middle(?) Pleistocene. Two probable early Wisconsin advances were followed by three or four advances in late Wisconsin time. Holocene glacial deposits generally only occur in deep north-facing cirques, but there is evidence of two short (less than 3 km) Holocene valley glaciers. No glaciers presently exist in the YTTN (Table 4).

Fluvial, eolian, and lacustrine deposits

The thickest and most extensive fluvial, eolian, and lacustrine deposits in east-central Alaska occur in the Tanana River valley. One of the most useful sections is on the northeast side of the Tanana River, 50 km southeast of the town of Tok, where 40 m of fluvial, eolian, and lacustrine deposits are exposed. Radiocarbon ages indicate that wood from near the base of the section has an age greater than 42,000 yr. An ash bed beneath the modern turf is about 1,400 yr old (Fernald, 1962; Carter and Galloway, 1984). Some interior valleys also have thick eolian and fluvial deposits, but exposures are limited. Commonly, loess is so thick (more than 60 m in places) that fluvial deposits beneath them are rarely exposed. Locally, there are thick carbonaceous silt and peat deposits that mostly are perennially frozen. Sand dunes occur north of the Tanana River in the Tanacross and Big Delta Quadrangles. The maximum thickness of sediments in the Tanana River valley is not known, but parts of the valley floor are below sea level (Péwé, 1974).

Placer gold occurs both in Holocene gravels and in gravels as old as early Pleistocene. Most of the gold-bearing gravels are buried beneath frozen silt and other sediments. Some compose terraces a few to many meters above present stream levels. High terraces are particularly well developed along the Fortymile and Seventymile Rivers and along the Yukon River.

Talus and landslide deposits, particularly abundant in the Alaska Range, also occur in the YTTN. Rock glacier deposits are locally significant in the Alaska Range (Holmes and Foster, 1968).

Frost polygons are well developed in lowland areas where permafrost is generally present and thaw lakes are common. Patterned ground, including stone polygons and stone stripes, is abundant in upland areas. Solifluction is a major mass-movement process on valley slopes. More than 300 open-system pingos are widely distributed mainly north of the Tanana River (Holmes and others, 1968). Most are on south- and southeast-facing slopes near the transition between valley-fill deposits and slope mantle. The pingos are composed primarily of silt, colluvium, and valley-fill material and range from 3 to 35 m in height.

Volcanic ash

Volcanic ash layers ranging in age from early Pleistocene to Holocene occur at several horizons in the Quaternary deposits of east-central Alaska. The most widely distributed is the White River Ash Bed, which originated about 1,400 yr ago at the east end of the Wrangell Mountains near the Alaska-Canada border (Lerbekmo and Campbell, 1969). It commonly occurs just beneath the turf but may be buried to a depth of several meters in active stream valleys or beneath recent eolian deposits. Numerous other ash deposits of unknown origin occur: they include some in the Fairbanks area, lower Delta River area, along the Chatinika River 40 km north of Fairbanks (Péwé, 1975), and in the central and southern Eagle Quadrangle (Weber and others, 1981). The Sheep Creek Tephra of Westgate (1984 *in* Porter, 1985), an ash of unknown source and older than 40,000 yr B.P., has recently been identified at four localities in Alaska and Yukon Territory: Eva Creek, near Fairbanks (Péwé, 1975); Canyon Creek in the Big Delta Quadrangle (Weber and others, 1981); Lost Chicken Creek in the Eagle Quadrangle (Lee Porter, written communication, 1985); and on the Stewart River in Yukon Territory. At Lost Chicken Creek, the ash is associated with in situ fossil mammal remains.

Vertebrate fossils

East-central Alaska is well known for the abundant remains of extinct Pleistocene mammals, found mostly in frozen deposits along rivers and streams. Most of the remains have been uncovered during placer gold mining, and the Fairbanks mining district has been especially productive of fossil vertebrates. At least 45 mammalian genera occupied parts of the area during the late Pleistocene, including the American lion, camels, giant beavers, and ground sloths, of which at least 16 genera have become extinct (Porter, 1985).

Vertebrate fossils were discovered (Weber and others, 1981) in 1974 along the Richardson Highway near the mouth of Canyon Creek, a small tributary to the Tanana River in the Big Delta Quadrangle. There, the Pleistocene fauna includes rodents, woolly mammoth, the Yukon wild ass, western camel, long-horned bison, mountain sheep, wolf, tundra hare, and caribou. Although the fauna is a standard Alaskan Pleistocene assemblage (Guthrie, 1968), it is one of the few central Alaskan mammalian collections that is stratigraphically controlled and radiometrically dated (40,000 yr old; Weber and others, 1981). The fossils from Lost Chicken Creek, a small tributary to the South Fork of the Fortymile River in the southern Eagle Quadrangle, have been intensively collected and studied. This locality has produced more than 1,000 fossils from 37 m of unconsolidated sediments, which range in age from 50,400 to 1,400 yr B.P. (Porter, 1985). The assemblage includes 16 vertebrate genera, among which are the unusual occurrence of gallinaceous birds, wolverine, the extinct American lion, collared lemmings, and saiga antelope. Human involvement may be indicated for some unknown time prior to

TABLE 4. GLACIAL ADVANCES IN EAST-CENTRAL ALASKA

	YTTN Glacial sequences (after Weber, 1983)			Northern Alaska Range Glacial advances (after Ten Brink, 1983)	
	Yukon-Tanana Upland	Mount Prindle area		Local names used and specific valleys by Péwé (1965, 1975) and Wahrhaftig (1958)	Regional informal nomenclature by Ten Brink and others (1983)
HOLOCENE	Ramshorn glaciation		HOLOCENE		Muldrow Foraker II Peters Yanert II Foraker I Yanert I
LATE WISCONSIN	Salcha glaciation	Convert glaciation	LATE WISCONSIN	Donnelly glaciation (Delta River; Péwé, 1965, 1975) Riley Creek glaciation (Nenana Valley; Wahrhaftig, 1958)	McKinley Park glaciation: McKinley Park stade IV McKinley Park stade III McKinley Park stade II McKinley Park stade I
EARLY(?) WISCONSIN	Eagle glaciation	American Creek glaciation	EARLY WISCONSIN	Delta glaciation (Delta River; Péwé, 1965, 1975) Healy glaciation (Nenana River; Wahrhaftig, 1958)	Early Wisconsin III(?) Early Wisconsin II Early Wisconsin I
MIDDLE(?) PLEISTOCENE	Mount Harper glaciation	Little Champion glaciation	PRE-WISCONSIN	Dry Creek glaciation (Nenana River; Wahrhaftig, 1958)	Pre-Wisconsin III
EARLY(?) PLEISTOCENE	Charley River glaciation	Prindle glaciation	PRE-WISCONSIN	Browne glaciation (Nenana River; Wahrhaftig, 1958)	Pre-Wisconsin II
PRE-PLEISTOCENE	Goodpaster glaciation		PRE-WISCONSIN	Nenana Gravel (upper part) (late Tertiary) (Nenana River; Wahrhaftig, 1958)	Pre-Wisconsin I

11,000 yr B.P. by broken and burned bones of mammoth, bison, horse, and caribou (Lee Porter, written communication, 1985). The Lost Chicken fauna represents a hardy, cold- and dry-adapted biologic community that lived during middle and late Wisconsin postglacial time.

The stratigraphy, flora, and fauna of east-central Alaska indicate that early Wisconsin time was cool and dry, that middle Wisconsin time was wetter, and that glacial climates terminated abruptly between 12,000 and 9,000 yr B.P. (Lee Porter, written communication, 1985). The Holocene had an early period (ca. 8,000 yr B.P.) of very warm temperatures, followed by a cool period (ca. 7,000 yr B.P.) (Lee Porter, written communication, 1985).

GEOLOGIC HISTORY

Determining the geologic history of east-central Alaska is hampered by a scarcity of fossils, high grades of metamorphism, and extensive cover by vegetation and surficial deposits. Because of these constraints, it is not surprising that geologists working in different parts of the YT have emphasized different data and arrived at different interpretations. The data of this chapter are presented within the framework of tectonostratigraphic terranes (Silberling and others, this volume). Interpretations are based largely on regional-scale geologic mapping, data from ^{40}Ar-^{39}Ar incremental heating experiments, and structural studies of rocks in the eastern part of the Eagle Quadrangle.

In the following discussion, we consider several tectonostratigraphic assemblages or "packages" of rocks, subterranes Y_1 through Y_4 (described above under "Stratigraphy"), in terms of their relations to each other and to the YTTN, and our conclusions are applied primarily to the YTTN. Nokleberg and others (1983) and Nokleberg and Aleinikoff (1985) have identified tectonostratigraphic assemblages and have proposed interpretations of the geologic relations of the rocks south of the YTTN, particularly in the Mount Hayes Quadrangle, and their conclusions are not discussed further in this chapter. We believe that many subterranes of the YT contain evidence of unique pre-Mesozoic histories, but at present, data are still insufficient to integrate their geologic evolution.

The geologic history recorded in rocks of the YTTN probably begins in latest Proterozoic or early Paleozoic time with deposition of continentally derived sediments in several different environments marginal to North America or to other continents. The deposits include the quartz-rich sediments of unit qq (subterrane Y_2), the pelitic schist, amphibolite, and marble of unit ps (subterrane Y_2), and the quartzose and pelitic sediments of subterrane Y_1. Probably beginning a little later, but perhaps overlapping in time deposition in subterranes Y_1 and Y_2, the protoliths of subterrane Y_4 were deposited. Other continental-margin sediments of Paleozoic age are those of units qs, ws, and cp of subterrane Y_3. Possible fore-arc sediments of Paleozoic age are those of units sm and ms of subterrane Y_3.

The first well-dated event in the YT is felsic intrusion in Devonian time (Aleinikoff and Nokleberg, 1985) in the southern part of subterrane Y_1, followed in Mississippian time by extensive porphyritic granitic intrusion (augen gneiss) throughout much of the northern part of subterrane Y_1. Dusel-Bacon and Aleinikoff (1985) postulated that metamorphism was synchronous with intrusion, although metamorphism may not have occurred until the Cretaceous (Dusel-Bacon, 1986). This continental fragment (subterrane Y_1) was probably a separate entity in mid–late Paleozoic time because the distribution of the augen gneiss is limited to subterrane Y_1, except for possible thrust remnants in the southern part of subterrane Y_2.

From late Paleozoic through Triassic time, basalt and minor amounts of sedimentary chert, graywacke, shale, and limestone accumulated in an ocean basin that separated the continental fragment comprising subterranes Y_2 and Y_3 from the one comprising subterranes Y_1 and Y_4. Mantle peridotite was tectonically emplaced into the oceanic crust and became a part of the ocean basin suite. Remnants of these mantle and oceanic rocks compose the Seventymile terrane. During this period, the continental fragments each had a poorly known but complex history of metamorphism and deformation that is implied by their heterogeneity, but the early histories of these subterranes are as yet undetermined.

The next well-dated events are represented in subterrane Y_4 and may have been limited to that subterrane. Paleozoic sedimentary rocks were metamorphosed to amphibolite facies in the Late Triassic (about 213 ± 2 Ma). Intrusion of granodiorite (Taylor Mountain batholith) occurred at 209 ± 3 Ma, shortly after the peak of metamorphism, and the pluton cooled with the regional geotherm.

During the cooling period, but before approximately 201 Ma, oceanic basalt and associated sediments, probably from the closing ocean basin to the north of subterrane Y_4, were emplaced tectonically. The oceanic strata were metamorphosed to greenschist facies during the waning stages of the regional metamorphism of subterrane Y_4. Deformation, probably related to northwestward movement of subterrane Y_4, produced major northeast-trending folds in unit qs when subterrane Y_4 and remnants of oceanic strata (Seventymile terrane) were thrust northward onto subterrane Y_3. Complex imbrication of subterrane Y_4 with units qs and ws of subterrane Y_3 occurred with the collision of these packages of rocks. The thrusting together of subterranes Y_3 and Y_4 occurred after regional metamorphism of subterrane Y_4, but before 187 ± 2 Ma, as indicated by ^{40}Ar–^{39}Ar dating of undeformed and unmetamorphosed dike rocks that cut subterrane Y_4. With continued northward movement, more imbrication, shearing, mylonitization, and thrusting occurred within the Seventymile terrane, and between it and subterrane Y_3.

In the western part of the YTTN, times of thrusting and metamorphism are not as well known as in the eastern part. However, within subterrane Y_2, thrusting of unit ps over unit qq occurred before regional metamorphism of subterrane Y_2, and the joining of subterranes Y_2 and Y_3 occurred after the regional metamorphism of subterrane Y_2. The Seventymile terrane was probably thrust over the western part of subterrane Y_3 in Jurassic time, at the same time it was thrust over the eastern part. Closing of the ocean basin that separated subterranes Y_4 and Y_1 from subterranes Y_2 and Y_3 is recorded only by the presence of thrust remnants of the Seventymile terrane on top of subterrane Y_3.

Differences in the sedimentary protoliths of subterranes Y_1 and Y_4, the restriction of augen gneiss to subterrane Y_1, and probable differences in ages of metamorphism of these subterranes suggest that they were not joined until after Early Jurassic time. The presence of a late Early Cretaceous metamorphic event in subterranes Y_1 and Y_4 provides an upper constraint to the time that subterrane Y_1 joined the rest of the YTTN. The subterranes were finally welded together by emplacement of the Cretaceous granitic plutons.

In the eastern part of the YTTN, volcanism associated with Cretaceous intrusion produced calderas and extensive deposits of welded tuff. At about the same time, and perhaps related either to volcanism or to further northwestward movement of the YT, the YT developed a northeasterly trending pattern of fractures. The best known of these fractures is the Shaw Creek fault, which appears to have considerable strike-slip displacement (Griscom, 1979). However, the southeast side seems to be upthrown, and on the basis of reconnaissance field data, M. C. Gardner has suggested (written communication, 1983) that the Shaw Creek fault may be a high-angle reverse fault. Some of the northeast-trending faults were reactivated at intervals in Tertiary and Quaternary time. The right-lateral strike-slip Tintina fault system marks the northern boundary of the YT. There is little direct evidence in Alaska of the time, kind, and extent of first and subsequent movements on the Tintina and its subsidiary faults; limited information suggests that, as might be expected, movement occurred at different times on various segments of the fault. In the Eagle Quadrangle, for instance, Upper Cretaceous and/or lower Tertiary conglomerates are broken and pulverized by fault movement at one locality but appear undisturbed at another. In Canada, Gabrielse (1985) has suggested that dextral displacement on faults such as the Tintina date from the mid-Cretaceous or earlier to late Eocene or Oligocene. In order to account for the difference in the amount of displacement on the Tintina from that on the northern Rocky Mountain Trench, he suggested that transcurrent displacement brackets the time of major regional thrusting and folding. If this were the case in Alaska, movement on the Tintina fault could possibly have begun as early as the Jurassic.

The last major plutonic event was the emplacement of felsic plutons primarily in the northwestern part of the YTTN from about 65 to 50 Ma. At about the same time, probably mafic and felsic volcanic rocks, and perhaps shallow felsic intrusions, were emplaced, primarily in the eastern part of the YTTN. A late Mesozoic and/or Tertiary folding event with some thrusting deformed Upper Cretaceous or Paleogene nonmarine sedimentary rocks and Neogene sedimentary and volcanic rocks. Open folding, possibly contemporaneous, affected the metamorphic rocks throughout the YTTN. Regional uplift and northward tilting (Mertie, 1937) followed, probably in Pleistocene time. The effects of this deformation are most evident in the northeastern part of the YTTN, where the Fortymile and Seventymile Rivers are deeply entrenched and are bordered by extensive high-level terraces.

Prindle Volcano, in the eastern part of the Tanacross Quadrangle, is a manifestation of Pleistocene or Holocene alkali-olivine basaltic volcanism. Because its lava contains spinel-bearing peridotite and granulite inclusions (Foster and others, 1966), it also suggests that extension is taking place at deep crustal and upper mantle levels. Prindle Volcano lies at the northern end of a belt of occurrences of alkali-olivine basalts that extends along the western continental margin of North America (Foster and others, 1966).

In general, our interpretation of the geologic history of the YTTN is similar to the one modeled by Tempelman-Kluit (1979) for part of the Yukon Territory. Some events may also correlate with those in other parts of the Canadian Cordillera (Monger and others, 1982). Following Tempelman-Kluit's (1979) model, the amphibolite-facies rocks of subterrane Y_4 might be considered as part of Stikinia (Stikinia terrane), and the Late Triassic Taylor Mountain granitic intrusion could be correlated with the Klotassin Suite of the Yukon Territory. In our view, however, it seems more likely that subterrane Y_4 was not a part of Stikinia, but instead was a discrete terrane that lay north of Stikinia in Late Triassic and Early Jurassic time. Later in Jurassic time, both Stikinia with the intruded Klotassin Suite and subterrane Y_4 with the Taylor Mountain and other Triassic and Jurassic intrusions were amalgamated with other terranes to form terrane I of Monger and others (1982). Subterrane Y_3 could correlate with Tempelman-Kluit's cataclastic unit, and the Seventymile terrane could have originated in the closing of the Anvil Ocean. Although tentative correlations can be made, some details in lithologies and timing of events differ. For instance, closing of the Anvil Ocean is postulated for Middle Jurassic time in Canada but may have been a little earlier in Alaska.

REFERENCES CITED

Abbott, G., 1982, Origin of the Clinton Creek asbestos deposit, *in* Yukon exploration and geology: Exploration and Geological Services Northern Affairs, Indian and Northern Affairs, Canada, p. 18–25.

Aleinikoff, J. N., and Nokleberg, W. J., 1985a, Age of intrusion and metamorphism of a granodiorite in the Lake George terrane, northeastern Mount Hayes Quadrangle, *in* Bartsch-Winkler, S., and Reed, K. M., eds., The United States Geological Survey in Alaska; Accomplishments during 1983: U.S. Geological Survey Circular 945, p. 62–65.

—— , 1985b, Age of Devonian igneous-arc terranes in the northern Mount Hayes Quadrangle, eastern Alaska Range, Alaska, *in* Bartsch-Winkler, S., ed., The United States Geological Survey in Alaska; Accomplishments during 1984: U.S. Geological Survey Circular 967, p. 44–49.

Aleinikoff, J. N., Dusel-Bacon, C., and Foster, H. L., 1981, Geochronologic studies in the Yukon-Tanana Upland, east-central Alaska, *in* Albert, N.R.D., and Hudson, T., eds., The United States Geological Survey in Alaska; Accomplishments during 1979: U.S. Geological Survey Circular 823-B, p. B34–B37.

—— , 1986, Geochronology of augen gneiss and related rocks, Yukon-Tanana terrane, east-central Alaska: Geological Society of America Bulletin, v. 97, p. 626–637.

Bacon, C. R., Foster, H. L., and Smith, J. G., 1985, Cretaceous calderas and rhyolitic welded tuffs in the Yukon-Tanana terrane, east-central Alaska: Geological Society of America Abstracts with Programs, v. 17, p. 339.

Barker, J. C., 1978, Mineral deposits of the Yukon-Tanana Uplands; A summary report: U.S. Bureau of Mines Open-File Report 88–78, 33 p.

Barker, J. C., and Clautice, K. H., 1977, Anomalous uranium concentrations in artesian springs and stream sediments in the Mount Prindle area, Alaska: U.S. Bureau of Mines Open-File Report 130–77, 18 p.

Barnes, D. F., 1977, Bouguer gravity map of Alaska: U.S. Geological Survey Geophysical Investigations Map GP-913, 1 sheet, scale 1:250,000.

Barnes, D. F., and Morin, R. L., 1975, Gravity map of the Nabesna Quadrangle, Alaska: U.S. Geological Survey Miscellaneous Field Studies Map MF-655-1, 1 sheet, scale 1:250,000.

Barnes, F. F., 1967, Coal resources of Alaska, *in* Contributions to economic geology: U.S. Geological Survey Bulletin 1242-B, p. B1–B85.

Berg, H. C., and Cobb, E. H., 1967, Metalliferous lode deposits of Alaska; U.S. Geological Survey Bulletin 1246, 254 p.

Blum, J. D., 1983, Petrology, geochemistry, and isotope geochronology of the Gilmore Dome and Pedro Dome plutons, Fairbanks district, Alaska: Alaska Division of Geological and Geophysical Surveys Report of Investigations 83-2, 59 p.

Brabb, E. E., and Churkin, M., Jr., 1969, Geologic map of the Charley River Quadrangle, east-central Alaska: U.S. Geological Survey Miscellaneous Geologic Investigations Map I-573, scale 1:250,000.

Brook, C. A., and 5 others, 1979, Assessment of geothermal resources of the United States–1978: U.S. Geological Survey Circular 790, p. 18–85.

Brosgé, W. P., Brabb, E. E., and King, E. R., 1970, Geologic interpretation of reconnaissance aeromagnetic survey of northeastern Alaska: U.S. Geological Survey Bulletin 1271-F, 14 p.

Brown, E. H., and Forbes, R. B., 1984, Paragenesis and regional significance of ecologitic rocks from the Fairbanks district, Alaska: Geological Society of America Abstracts with Programs, v. 16, p. 272.

Bundtzen, T. K., 1981, Geology and mineral deposits of the Kantishna Hills, Mount McKinley Quadrangle, Alaska [M.S. thesis]: Fairbanks, University of Alaska, 237 p.

—— , 1982, Bedrock geology of the Fairbanks mining district, western sector: Alaska Division of Geological and Geophysical Surveys Open-File Report 155, 2 plates.

Bundtzen, T. K., and Reger, R. D., 1977, The Richardson lineament; A structural control for gold deposits in the Richardson mining district, interior Alaska: Alaska Division of Geological and Geophysical Surveys Geologic Report 55, p. 29–34.

Bundtzen, T. K., and 6 others, 1984, Alaska's mineral industry, 1983: Alaska Division of Geological and Geophysical Surveys Special Report 33, 56 p.

Burack, A. C., 1983, Geology along the Pinnell Mountain Trail, Circle Quadrangle, Alaska [M.S. thesis]: Durham, University of New Hampshire, 98 p.

Byers, F. M., Jr., 1957, Tungsten deposits in the Fairbanks district, Alaska: U.S. Geological Survey Bulletin 1024-I, p. 179–215.

Cady, J. W., and Barnes, D. F., 1983, Complete Bouguer gravity map of Circle Quadrangle, Alaska; Folio of the Circle Quadrangle, Alaska: U.S. Geological Survey Open-File Report 83–170D, scale 1:250,000.

Cady, J. W., and Weber, F. R., 1983, Aeromagnetic map and interpretation of magnetic and gravity data, Circle Quadrangle, Alaska: U.S. Geological Survey Open-File Report 83–170–C, 29 p.

Capps, S. R., 1912, The Bonnifield region, Alaska: U.S. Geological Survey Bulletin 501, 64 p.

Carter, L. D., and Galloway, J. P., 1984, Lacustrine and eolian deposits of Wisconsin age at Riverside Bluff in the upper Tanana River valley, Alaska, *in* Coonrad, W. L., and Elliott, R. L., eds., The United States Geological Survey in Alaska; Accomplishments during 1981: U.S. Geological Survey Circular 868, p. 66–68.

Churkin, M., Jr., Foster, H. L., Chapman, R. M., and Weber, F. R., 1982, Terranes and suture zones in east-central Alaska: Journal of Geophysical Research, v. 87, no. B5, p. 3718–3730.

Clark, S.H.B., and Foster, H. L., 1971, Geochemical and geological reconnaissance in the Seventymile River area, Alaska: U.S. Geological Survey Bulletin 1315, 21 p.

Cobb, E. H., 1973, Placer deposits of Alaska: U.S. Geological Survey Bulletin 1374, 213 p.

Coleman, R. G., 1977, Ophiolites: New York, Springer-Verlag, 229 p.

Coleman, R. G., Lee, D. E., Beatty, L. B., and Brannock, W. W., 1965, Eclogites and eclogites; Their differences and similarities: Geological Society of America Bulletin, v. 76, p. 483–508.

Cushing, G. W., 1984, Early Mesozoic tectonic history of the eastern Yukon-Tanana Upland [M.S. thesis]: Albany, State University of New York, 235 p.

Cushing, G. W., and Foster, H. L., 1984, Structural observations in the Circle Quadrangle, Yukon-Tanana Upland, Alaska, *in* Coonrad, W. L., and Elliott, R. L., eds., The United States Geological Survey in Alaska; Accomplishments during 1981: U.S. Geological Survey Circular 868, p. 64–65.

Cushing, G. W., Foster, H. L., and Harrison, T. M., 1984a, Mesozoic age of metamorphism and thrusting in the eastern part of east-central Alaska: EOS Transactions of the American Geophysical Union, v. 65, no. 16, p. 290–291.

Cushing, G. W., Foster, H. L., Harrison, T. M., and Laird, J., 1984b, Possible Mesozoic accretion in the eastern Yukon-Tanana Upland, Alaska: Geological Society of America Abstracts with Programs, v. 16, p. 481.

Dadisman, S. V., 1980, Radiometric ages of rocks in south-central Alaska and western Yukon Territory: U.S. Geological Survey Open-File Report 80–183, 82 p., scale 1:1,000,000.

Dusel-Bacon, C., and Aleinikoff, J. N., 1985, Petrology and tectonic significance of augen gneiss from a belt of Mississippian granitoids in the Yukon-Tanana terrane, east-central Alaska: Geological Society of America Bulletin, v. 96, p. 411–425.

Dusel-Bacon, C., and Foster, H. L., 1983, A sillimanite gneiss dome in the Yukon crystalline terrane, east-central Alaska; Petrography and garnet-biotite geothermometry: U.S. Geological Survey Professional Paper 1170-E, 25 p.

Eakins, G. R., and 6 others, 1983, Alaska's mineral industry 1982: Alaska Division of Geological and Geophysical Surveys Special Report 31, 63 p.

Eberlein, G. D., Chapman, R. M., Foster, H. L., and Gassaway, J. S., 1977, Table describing known metalliferous and selected nonmetalliferous mineral deposits in central Alaska: U.S. Geological Survey Open-File Report 77–168D.

Fernald, A. T., 1962, Radiocarbon dates relating to a widespread volcanic ash deposit, eastern Alaska: U.S. Geological Survey Professional Paper 450-B, p. B29–B30.

Foley, J. Y., 1982, Alkaline igneous rocks in the eastern Alaska Range; Short notes on Alaskan geology, 1981: Alaska Division of Geological and Geophysical Surveys Geologic Report 73, p. 1–5.

Foley, J., and Barker, J. C., 1981, Tungsten investigations of the VABM Bend vicinity, Charley River and Eagle Quadrangles, eastern Alaska: U.S. Bureau of Mines Open-File Report 29–81, 22 p.

Forbes, R. B., and Brown, J. M., 1961, A preliminary map of the bedrock geology of the Fairbanks mining district, Alaska: Alaska Division of Mines and Minerals Mineral Investigations Report 194–1, scale 1:63,360.

Forbes, R. B., and Weber, F. R., 1982, Bedrock geologic map of the Fairbanks mining district, Alaska: State of Alaska, Division of Geological and Geophysical Surveys Open-File Report AOF–170, scale 1:63,360.

Foster, H. L., 1967, Geology of the Mount Fairplay area, Alaska: U.S. Geological Survey Bulletin 1241-B, p. B1–B18.

—— , 1969, Asbestos occurrence in the Eagle C-4 Quadrangle, Alaska: U.S. Geological Survey Circular 611, 7 p.

—— , 1970, Reconnaissance geologic map of the Tanacross Quadrangle, Alaska: U.S. Geological Survey Miscellaneous Geologic Investigations Map I-593, scale 1:250,000.

—— , 1975, Significant platinum values confirmed in ultramafic rocks of the Eagle C-3 Quadrangle, *in* Yount, M. E., ed., United States Geological Survey Alaska Program 1975: U.S. Geological Survey Circular 722, p. 42.

—— , 1976, Geologic map of the Eagle Quadrangle, Alaska: U.S. Geological Survey Miscellaneous Geologic Investigations Series Map I-922, scale 1:250,000.

—— , 1981, A minimum age for Prindle Volcano, Yukon-Tanana Upland, *in*

Albert, N.R.D., and Hudson, T., eds., The United States Geological Survey in Alaska; Accomplishments during 1979: U.S. Geological Survey Circular 823-B, p. B37–B38.

Foster, H. L., and Clark, S.H.B., 1970, Geochemical and geologic reconnaissance of a part of the Fortymile area, Alaska: U.S. Geological Survey Bulletin 1312-M, 29 p.

Foster, H. L., and Cushing, G. W., 1985, Tertiary(?) folding in the Tanacross Quadrangle, *in* Bartsch-Winkler, S., and Reed, K. M., eds., The United States Geological Survey in Alaska; Accomplishments during 1983: U.S. Geological Survey Circular 945, p. 38–40.

Foster, H. L., and Keith, T.E.C., 1974, Ultramafic rocks of the Eagle Quadrangle, east-central Alaska: U.S. Geological Survey Journal of Research, v. 2, no. 6, p. 657–669.

Foster, H. L., and O'Leary, R. M., 1982, Gold found in bedrock of Lost Chicken Creek, gold placer mine, Fortymile area, Alaska, *in* Coonrad, W. L., ed., The United States Geological Survey in Alaska; Accomplishments during 1980: U.S. Geological Survey Circular 844, p. 62–63.

Foster, H. L., Forbes, R. B., and Ragan, D. M., 1966, Granulite and peridotite inclusions from Prindle Volcano, Yukon-Tanana Upland, Alaska: U.S. Geological Survey Professional Paper 550-B, p. B115–B119.

Foster, H. L., Weber, F. R., Forbes, R. B., and Brabb, E. E., 1973, Regional geology of Yukon-Tanana Upland, Alaska, *in* Pitcher, M. G., ed., Arctic geology: American Association of Petroleum Geologists Memoir 19, p. 388–395.

Foster, H. L., and 6 others, 1976, The Alaskan Mineral Resource Assessment Program; Background information to accompany folio of geologic and mineral resource maps of the Tanacross Quadrangle, Alaska: U.S. Geological Survey Circular 734, 23 p.

Foster, H. L., Donato, M. L., and Yount, M. E., 1978a, Petrographic and chemical data on Mesozoic granitic rocks of the Eagle Quadrangle, Alaska: U.S. Geological Survey Open-File Report 78-253, 29 p., 2 maps, scale 1:250,000.

Foster, H. L., Jones, D. L., Keith, T.E.C., Wardlaw, B., and Weber, F. R., 1978b, Late Paleozoic radiolarians and conodonts found in chert of Big Delta Quadrangle, *in* Johnson, K. M., ed., United States Geological Survey in Alaska; Accomplishments during 1977: U.S. Geological Survey Circular 772-B, p. B34–B36.

Foster, H. L., O'Leary, R. M., McDanal, S. K., and Clark, A. L., 1978c, Analyses of rock samples from the Big Delta Quadrangle, Alaska: U.S. Geological Survey Open-File Report 78-469, 125 p., 1 map, scale 1:250,000.

Foster, H. L., and 6 others, 1979, The Alaskan Mineral Resource Assessment Program; Background information to accompany folio of geologic and mineral resource maps of the Big Delta Quadrangle, Alaska: U.S. Geological Survey Circular 783, 19 p.

Foster, H. L., Laird, J., Keith, T.E.C., Cushing, G. W., and Menzie, W. D., 1983, Preliminary geologic map of the Circle Quadrangle, Alaska: U.S. Geological Survey Open-File Report 83-170A, 29 p., scale 1:250,000.

Foster, H. L., Laird, J., and Cushing, G. W., 1984, Thrust faulting in the Eagle A-1 Quadrangle, Alaska, and its implications for the tectonic history of the Yukon-Tanana Upland [abs.]: EOS Transactions of the American Geophysical Union, v. 65, no. 16, p. 291.

Foster, H. L., Cushing, G. W., Keith, T.E.C., and Laird, J., 1985, Early Mesozoic tectonic history of the Boundary area, east-central Alaska: Geophysical Research Letters, v. 12, no. 9, p. 553–556.

Gabrielse, H., 1985, Major dextral transcurrent displacements along the northern Rocky Mountain Trench and related lineaments in north-central British Columbia: Geological Society of America Bulletin, v. 96, p. 1–14.

Gilbert, W. G., and Bundtzen, T. K., 1979, Mid-Paleozoic tectonics, volcanism and mineralization in north-central Alaska Range, *in* Sisson, A., ed., The relationship of plate tectonics to Alaskan geology and resources: Anchorage, Alaska Geological Society Symposium 6 Proceedings, 1977, p. F1–F22.

Gilbert, W. G., and Redman, E., 1977, Metamorphic rocks of Toklat-Teklanika Rivers area, Alaska: Alaska Division of Geological and Geophysical Surveys Geologic Report 50, 13 p.

Godwin, C. I., 1976, Casino, *in* Brown, S. C., ed., Porphyry deposits of the Canadian Cordillera: Canadian Institute of Mining and Metallurgy Special Volume 15, p. 344–354.

Gottfried, D., Jaffe, H. W., and Senftle, F. E., 1959, Evaluation of the lead-alpha (Larsen) method for determining ages of igneous rocks: U.S. Geological Survey Bulletin 1097-A, p. 1–63.

Green, L. H., 1972, Geology of Nash Creek, Larsen Creek, and Dawson map-areas, Yukon Territory: Geological Survey of Canada Memoir 364, p. 1–157.

Griscom, A., 1975, Aeromagnetic map and interpretation of the Nabesna Quadrangle, Alaska: U.S. Geological Survey Miscellaneous Field Studies Map MF-655H, 2 sheets, scale 1:250,000.

——, 1976, Aeromagnetic map and interpretation of the Tanacross Quadrangle, Alaska: U.S. Geological Survey Miscellaneous Field Studies Map MF-767A, 2 sheets, scale 1:250,000.

——, 1979, Aeromagnetic map and interpretation for the Big Delta Quadrangle, Alaska: U.S. Geological Survey Open-File Report 78–529–B, scale 1:250,000.

Guthrie, R. D., 1968, Paleoecology of the large mammal community in interior Alaska during the late Pleistocene: American Midland Naturalist, v. 79, p. 346–363.

Hall, M. H., Smith, T. E., and Weber, F. R., 1984, Geologic guide to the Fairbanks–Livengood area, east-central Alaska: Fairbanks, Alaska, Alaska Division of Geological and Geophysical Surveys, 30 p.

Harte, B., and Hudson, N.F.C., 1979, Pelite facies series and temperatures and pressures of Dalradian metamorphism in E. Scotland, *in* Harris, A. L., Holland, C. H., and Leake, B. E., eds., The Caledonides of the British Isles–Revisited: Scottish Academic Press, p. 323–337.

Hill, J. M., 1933, Lode deposits of the Fairbanks district, Alaska: U.S. Geological Survey Bulletin 849-B, p. 29–163.

Holdaway, M. J., 1971, Stability of andalusite and the aluminum silicate phase diagram: American Journal of Science, v. 271, p. 97–131.

Hollister, V. F., Anzalone, S. A., and Richter, D. H., 1975, Porphyry copper deposits of southern Alaska and contiguous Yukon Territory: Canadian Institute of Mining and Metallurgy Bulletin, v. 68, p. 104–112.

Holmes, G. W., and Foster, H. L., 1968, Geology of the Johnson River area, Alaska: U.S. Geological Survey Bulletin 1249, 49 p.

Holmes, G. W., Hopkins, D. M., and Foster, H. L., 1968, Pingos in central Alaska: U.S. Geological Survey Bulletin 1241-H, 40 p.

Hytoon, M., 1979, Geology of the Clinton Creek asbestos deposit, Yukon Territory [M.S. thesis]: Vancouver, University of British Columbia, 67 p.

Jaffe, H. W., Gottfried, D., Waring, C. L., and Worthing, H. W., 1959, Lead-alpha age determinations of accessory minerals of igneous rocks (1953–1957): U.S. Geological Survey Bulletin 1097-B, p. 65–148.

Jones, D. L., Silberling, N. J., Coney, P. J., and Plafker, G., 1984, Lithotectonic terrane map of Alaska (west of the 141st Meridian), Part A, *in* Silberling, N. J., and Jones, D. L., eds., Folio of the lithotectonic terrane maps of the North American Cordillera: U.S. Geological Survey Miscellaneous Field Studies Map MF-1874-A, 1 sheet, scale 1:2,500,000.

Keith, T.E.C., and Foster, H. L., 1973, Basic data on the ultramafic rocks of the Eagle Quadrangle, east-central Alaska: U.S. Geological Survey Open-File Report 73–140, 4 sheets.

Keith, T.E.C., Foster, H. L., Foster, R. L., Post, E. V., and Lehmbeck, W. L., 1981a, Geology of an alpine-type peridotite in the Mount Sorenson area, east-central Alaska, *in* Shorter contributions to general geology: U.S. Geological Survey Professional Paper 1170-A, p. A1–A9.

Keith, T.E.C., Presser, T. S., and Foster, H. L., 1981b, New chemical and isotope data for the hot springs along Big Windy Creek, Circle A-1 Quadrangle, Alaska, *in* Albert, N.R.D., and Hudson, T., eds., The United States Geological Survey in Alaska; Accomplishments during 1979: U.S. Geological Survey Circular 823-B, p. B25–B28.

Kerin, L. J., 1976, The reconnaissance petrology of the Mount Fairplay igneous complex [M.S. thesis]: Fairbanks, University of Alaska, 95 p.

Laird, J., 1982, Amphiboles in metamorphosed basaltic rocks; Greenschist to

amphibolite facies, *in* Veblen, D. R., and Ribbe, P. H., eds., Amphiboles; Petrology and experimental phase relations: Mineralogical Society of America Reviews in Mineralogy, v. 9B, p. 113–138.
Laird, J., and Foster, H. L., 1984, Description and interpretation of a mylonitic foliated quartzite unit and feldspathic quartz wacke (grit) unit in the Circle Quadrangle, Alaska, *in* Reed, K. M., and Bartsch-Winkler, S., eds., The United States Geological Survey in Alaska; Accomplishments during 1982: U.S. Geological Survey Circular 939, p. 29–33.
Laird, J., Biggs, D. L., and Foster, H. L., 1984a, A terrane boundary of the southeastern Circle Quadrangle, Alaska?; Evidence from petrologic data: EOS Transactions of the American Geophysical Union, v. 65, no. 16, p. 291.
Laird, J., Foster, H. L., and Weber, F. R., 1984b, Amphibole eclogite in Circle Quadrangle, Yukon-Tanana Upland, Alaska, *in* Coonrad, W. L., and Elliott, R. L., eds., The United States Geological Survey in Alaska; Accomplishments during 1981: U.S. Geological Survey Circular 868, p. 57–60.
Lange, I. M., and Nokleberg, W. J., 1984, Massive sulfide deposits of the Jarvis Creek Terrane, Mount Hayes Quadrangle, eastern Alaska Range: Geological Society of America Abstracts with Programs, v. 16, p. 294.
Lerbekmo, J. F., and Campbell, F. A., 1969, Distribution, composition, and source of the White River Ash, Yukon Territory: Canadian Journal of Earth Sciences, v. 6, p. 109–116.
Luthy, S. T., Foster, H. L., and Cushing, G. W., 1981, Petrographic and chemical data on Cretaceous granitic rocks of the Big Delta Quadrangle, Alaska: U.S. Geological Survey Open-File Report 81–398, 12 p.
MacKevett, E. M., Jr., and Holloway, C. D., 1977, Table describing metalliferous and selected nonmetalliferous mineral deposits in eastern southern Alaska: U.S. Geological Survey Open-File Report 77–169, 99 p.
Matzko, J. J., Jaffe, H. W., and Waring, C. L., 1958, Lead-alpha age determinations of granitic rocks from Alaska: American Journal of Science, v. 256, no. 8, p. 529–539.
Maucher, A., 1976, The strata-bound cinnabar-stibnite-scheelite deposits, *in* Wolf, K. H., ed., Handbook of strata-bound and stratiform ore deposits, Volume 7: Amsterdam, Elsevier Publishing Company, p. 477–503.
McConnell, R. G., 1905, Report on the Klondike gold fields: Geological Survey of Canada Annual Report, n.s., v. 14, pt. B, 71 p.
McCulloch, M. T., and Wasserburg, G. J., 1978, Sm-Nd and Rb-Sr chronology of continental crust formation: Science, v. 200, no. 4345, p. 1003–1011.
Menzie, W. D., and Foster, H. L., 1978, Metalliferous and selected nonmetalliferous mineral resource potential in the Big Delta Quadrangle, Alaska: U.S. Geological Survey Open-File Report 78–529D, 60 p.
Menzie, W. D., Foster, H. L., Tripp, R. B., and Yeend, W. E., 1983, Mineral resource assessment of the Circle Quadrangle, Alaska: U.S. Geological Survey Open-File Report 83–170B, 57 p.
Mertie, J. B., Jr., 1937, The Yukon-Tanana region, Alaska: U.S. Geological Survey Bulletin 872, 276 p.
—— , 1938, Gold placers of the Fortymile, Eagle, and Circle districts, Alaska: U.S. Geological Survey Bulletin 897-C, 261 p.
Metz, P. A., and Robinson, M. S., 1980, Investigation of mercury-antimony-tungsten metal provinces of Alaska, II: Fairbanks, University of Alaska, Mineral Research Industry Laboratory Open-File Report 80-8, 39 p.
Miller, T. P., Barnes, I., and Patton, W. W., Jr., 1975, Geologic setting and chemical characteristics of hot springs in west-central Alaska: U.S. Geological Survey Journal of Research, v. 3, no. 2, p. 149–162.
Monger, J.W.H., Price, R. A., and Tempelman-Kluit, D. J., 1982, Tectonic accretion and the origin of the two major metamorphic and plutonic welts in the Canadian Cordillera: Geology, v. 10, p. 70–75.
Mortensen, J. K., 1983, Age and evolution of the Yukon-Tanana terrane, southeastern Yukon Territory [Ph.D. thesis]: Santa Barbara, University of California, 155 p.
Mullins, W. J., McQuat, J. F., and Rogers, R. K., 1984, The Alaska asbestos project; Part 1, Project review: Industrial Minerals, no. 199, p. 41.
Nokleberg, W. J., and Aleinikoff, J. N., 1985, Summary of stratigraphy, structure, and metamorphism of Devonian igneous-arc terranes, northeastern Mount Hayes Quadrangle, eastern Alaska Range, Alaska, *in* Bartsch-Winkler, S., ed., The United States Geological Survey in Alaska; Accomplishments during 1984: U.S. Geological Survey Circular 967, p. 66–71.
Nokleberg, W. J., and Lange, I. M., 1985, Volcanogenic massive sulfide occurrences, Jarvis Creek Glacier terrane, western Mount Hayes Quadrangle, eastern Alaska Range, *in* Bartsch-Winkler, S., and Reed, K. M., eds., The United States Geological Survey in Alaska; Accomplishments during 1983: U.S. Geological Survey Circular 945, p. 77–80.
Nokleberg, W. J., Aleinikoff, J. N., and Lange, I. M., 1983, Origin and accretion of Andean-type and island arc terranes of Paleozoic age juxtaposed along the Hines Creek fault, Mount Hayes Quadrangle, eastern Alaska Range, Alaska: Geological Society of America Abstracts with Programs, v. 15, p. 427.
—— , 1986, Cretaceous deformation and metamorphism in the northeastern Mount Hayes Quadrangle, eastern Alaska Range, *in* Bartsch-Winkler, S., and Reed, K. M., eds., Geologic studies in Alaska by the U.S. Geological Survey during 1985: U.S. Geological Survey Circular 978, p. 64–69.
Péwé, T. L., 1965, Delta River area, Alaska Range, *in* Péwé, T. L., Ferrians, O. J., Jr., Nichols, D. R., and Karlstrom, T.N.V., eds., Guidebook for Field Conference F, Central and south-central Alaska; International Association for Quaternary Research, 17th Congress, U.S.A. 1965: Lincoln, Nebraska Academy of Science, p. 55–93.
—— , 1975, Quaternary geology of Alaska: U.S. Geological Survey Professional Paper 835, 145 p.
Porter, L., 1985, Late Pleistocene fauna of Lost Chicken Creek, Alaska [Ph.D. thesis]: Pullman, Washington State University, 173 p.
Prindle, L. M., 1905, Fortymile, Birch Creek, and Fairbanks regions, Alaska: U.S. Geological Survey Bulletin 251, 89 p.
—— , 1913, A geologic reconnaissance of the Fairbanks Quadrangle, Alaska: U.S. Geological Survey Bulletin 525, 220 p.
Richter, D. H., 1976, Geologic map of the Nabesna Quadrangle, Alaska: U.S. Geological Survey Miscellaneous Investigations Series Map I-932, scale 1:250,000.
Sawyer, J.P.B., and Dickinson, R. A., 1976, Mount Nansen, *in* Brown, S. A., ed., Porphyry deposits of the Canadian Cordillera: Canadian Institute of Mining and Metallurgy Special Volume 5, p. 336–343.
Sinclair, W. D., 1978, Porphyry occurrences of southern Yukon, *in* Current Research, Part A: Geological Survey of Canada Paper 78–1A, p. 283–286.
Singer, D. A., Curtin, G. C., and Foster, H. L., 1976, Mineral resources map of the Tanacross Quadrangle, Alaska: U.S. Geological Survey Miscellaneous Field Studies Map MF-767-E, scale 1:250,000.
Southworth, D. D., 1984, Geologic and geochemical investigations of the "Nail" allochthon, east-central Alaska: U.S. Bureau of Mines Open-File Report 176–84, 19 p.
Streckeisen, A. L., 1976, To each plutonic rock its proper name: Earth Science Reviews, v. 12, p. 1–33.
Swainbank, R. C., and Forbes, R. B., 1975, Petrology of eclogitic rocks from the Fairbanks district, Alaska, *in* Forbes, R. B., ed., Contributions to geology of the Bering Sea Basin and adjacent regions: Geological Society of America Special Paper 151, p. 77–123.
Tempelman-Kluit, D. J., 1976, The Yukon crystalline terrane; Enigma in the Canadian Cordillera: Geological Society of America Bulletin, v. 87, p. 1343–1357.
—— , 1979, Transported cataclasite, ophiolite, and granodiorite in Yukon; Evidence of arc-continent collision: Geological Survey of Canada Paper 79–14, 27 p.
Ten Brink, N. W., 1983, Glaciation of the northern Alaska Range, *in* Thorson, R. M., and Hamilton, T. D., eds., Glaciation in Alaska: Fairbanks, University of Alaska Museum Occasional Paper 2, p. 82–91.
Wahrhaftig, C., 1958, Quaternary geology of the Nenana River valley and adjacent parts of the Alaska Range, Part A *of* Quaternary and engineering geology in the central part of the Alaska Range: U.S. Geological Survey Professional Paper 293, p. 1–68.
—— , 1968, Schists of the central Alaska Range, *in* Contributions to stratigraphy: U.S. Geological Survey Bulletin 1254-E, p. E1–E22.
Wahrhaftig, C., and Hickcox, C. A., 1955, Geology and coal deposits, Jarvis

Creek coal field, Alaska: U.S. Geological Survey Bulletin 989-G, p. 353–367.

Waring, G. A., 1917, Mineral springs of Alaska: U.S. Geological Survey Water-Supply Paper 418, 118 p.

Wasserburg, G. J., Eberlein, G. D., and Lanphere, M. A., 1963, Age of Birch Creek Schist and some batholithic intrusions in Alaska: Geological Society of America Special Paper 73, p. 258–259.

Weber, F. R., 1983, Glacial geology of the Yukon–Tanana Upland; A progress report, *in* Thorson, R. M., and Hamilton, T. D., eds., Glaciation in Alaska: Fairbanks, University of Alaska Occasional Paper 2, p. 96–100.

Weber, F. R., Foster, H. L., Keith, T.E.C., and Dusel-Bacon, C., 1978, Preliminary geologic map of the Big Delta Quadrangle: U.S. Geological Survey Open-File Report 78–529-A, scale 1:250,000.

Weber, F. R., Hamilton, T. D., Hopkins, D. M., Repenning, C. A., and Haas, H., 1981, Canyon Creek; A late Pleistocene vertebrate locality in interior Alaska: Quaternary Research, v. 16, p. 167–180.

Wilson, F. H., Smith, J. G., and Shew, N., 1985, Review of radiometric data from the Yukon Crystalline Terrane, Alaska and Yukon Territory: Canadian Journal of Earth Sciences, v. 22, no. 4, p. 525–537.

Wise, D. U., and 7 others, 1984, Fault-related rocks; Suggestions for terminology: Geology, v. 12, p. 391–394.

Wolfe, J. A., and Toshimasa, T., 1980, The Miocene Seldovia Point flora from the Kenai Group, Alaska: U.S. Geological Survey Professional Paper 1105, 52 p.

MANUSCRIPT ACCEPTED BY THE SOCIETY MAY 8, 1990

ACKNOWLEDGMENTS

The authors were assisted throughout the preparation of this chapter by many fellow workers. The following colleagues and their contributions of unpublished data, written sections, and expertise are particularly acknowledged: Jo Laird, metamorphic stratigraphy and metamorphism north of the Tanana River, University of New Hampshire, Durham, New Hampshire 03824; G. W. Cushing, ^{40}Ar-^{39}Ar incremental heating experiments and structural geology, ARCO Oil and Gas Company, Plano, Texas 75075; F. H. Wilson and Nora Shew, K-Ar determinations, history of application of radiometric age dating methods, and results of K-Ar work in the Yukon-Tanana terrane, U.S. Geological Survey, Anchorage, Alaska 99508; J. N. Aleinikoff, U-Pb determinations on zircons and common lead studies, U.S. Geological Survey, Denver, Colorado 80225; W. J. Nokleberg, geology of the Mount Hayes Quadrangle and mineral resources of the Yukon-Tanana terrane, U.S. Geological Survey, Menlo Park, California 94025; Cynthia Dusel-Bacon, distribution and petrology of metamorphic rocks, U.S. Geological Survey, Menlo Park, California 94025; Lee Porter, vertebrate paleontology, Gig Harbor, Washington 98335; M. C. Gardner, the Shaw Creek fault, ARCO Oil and Gas Company, Plano, Texas 75075; T. R. Carr, identification of conodonts, ARCO Oil and Gas Company, Plano, Texas 75075; and P. J. Burton, general information on mineral resources, State of Alaska, Department of Natural Resources, Fairbanks, Alaska 99708. Assistance was also given by R. M. Chapman and Béla Csejtey, Jr., U.S. Geological Survey, Menlo Park, California 94025; J. T. Dutro, Jr., U.S. Geological Survey, Washington, D.C. 20560; F. R. Weber, U.S. Geological Survey, Fairbanks, Alaska 99708; John W. Cady, U.S. Geological Survey, Denver, Colorado 80225; T. K. Bundtzen and T. E. Smith, State of Alaska, Division of Mines and Geology, Fairbanks, Alaska, 99708; S. C. Bergman and Michael Churkin, Jr., ARCO Oil and Gas Company, Plano, Texas 75075; and other colleagues.

The able assistance of Christa Marting in drafting and other aspects of manuscript preparation and of W. M. Trollman in typing and word processing are also gratefully acknowledged.

NOTES ADDED IN PROOF

Major additions to the geologic and geophysical data for the western Yukon–Tanana region, made since this chapter was completed in 1986, have resulted largely from the Trans-Alaska Crustal Transect program (TACT) of the U.S. Geological Survey. The TACT data include new radiometric ages obtained both on metamorphic and igneous rocks (Aleinikoff and others, 1987, and Nokleberg and others, 1989). Data were also collected that were interpreted to indicate Cretaceous crustal extension in the western Yukon–Tanana region (Pavlis and others, 1993); these data provide additional information for interpreting the Mesozoic structural history of the region. Geophysical studies, particularly the use of deep-sensing, low-frequency magnetotelluric (MT) and shallow-sensing, high-frequency audiomagnetotelluric (AMT) soundings, along with a seismic reflection profile, provide information on the deep structure of the Yukon-Tanana terrane north of the Denali fault. One model for the Yukon-Tanana terrane suggested by these geophysical data is that of thin-skinned thrusting of the rigid units of the Yukon-Tanana terrane over Mesozoic flysch (Labson and Stanley, 1989; Beaudoin and others, 1992). Further testing of this model is taking place and alternative interpretations need more investigation.

In the eastern Yukon-Tanana region, recent data have been derived from largely U.S. Geological Survey-supported geothermobarometric studies of selected amphibolite-facies metamorphic rocks (Dusel-Bacon and Hansen, 1992; Dusel-Bacon and others, 1993). Other new information includes structural observations (Hansen, 1990), and new age dates both on igneous and metamorphic rocks (Hansen and others, 1991). Geothermobarometric calculations suggest that high-pressure amphibolite-facies metamorphic rocks are present both in the Y_4 subterrane (Taylor Mountain terrane of Dusel-Bacon and Hansen, 1992) and the Y_1 (Lake George subterrane of Dusel-Bacon and Hansen, 1992).

ADDITIONAL REFERENCES

Aleinikoff, J. N., Dusel-Bacon, C., Foster, H. L., and Nokleberg, W. J., 1987, Lead isotope fingerprinting of tectono-stratigraphic terranes, east-central Alaska: Canadian Journal of Earth Sciences, v. 24, p. 2089–2098.

Beaudoin, B. C., Fuis, G. S., Mooney, W. D., Nokleberg, W. J., and Christensen, N. I., 1992, Thin, low-velocity crust beneath the southern Yukon-Tanana terrane east-central Alaska: Results from Trans-Alaska Crustal Transect refraction/wide-angle reflection data: Journal of Geophysical Research, v. 97, no. B2, p. 1921–1942.

Dusel-Bacon, C., and Hansen, V. L., 1992, High-pressure amphibolite-facies metamorphism and deformation within the Yukon-Tanana and Taylor Mountain terranes eastern Alaska, *in* Bradley, D. W., and Dusel-Bacon, C., eds., Geologic studies in Alaska by the U.S. Geological Survey, 1991: Geological Survey Bulletin 2041, p. 140–158.

Dusel-Bacon, C., Csejtey, B., Jr., Foster, H. L., Doyle, E. O., Nokleberg, W. J., and Plafker, G., 1993, Distribution, facies, ages, and proposed tectonic associations of regionally metamorphosed rocks in east- and south-central Alaska: U.S. Geological Survey Professional Paper 1497-C, p. C1–C73.

Hansen, V. L., 1990, Yukon-Tanana terrane; A partial acquittal: Geology, v. 18, p. 365–369.

Hansen, V. L., Heizler, M. T., and Harrison, T. M., 1991, Mesozoic thermal evolution of the Yukon-Tanana composite terrane: New evidence from $^{40}Ar/^{39}Ar$ data: Tectonics, v. 110, p. 51–76.

Labson, V. F., and Stanley, W. D., 1989, Electrical resistivity structure beneath southern Alaska, *in* Nokleberg, W. J., and Fisher, M. A., eds., Alaskan Geological and Geophysical Transect, Field Trip Guidebook T104: Washington, D.C., American Geophysical Union, p. 75–78.

Nokleberg, W. J., Foster, H. L., and Aleinikoff, J. N., 1989, Geology of the northern Copper River Basin eastern Alaska Range and southern Yukon-Tanana Basin, southern and east-central Alaska, *in* Nokleberg, W. J., and Fisher, M. A., eds., Alaskan Geological and Geophysical Transect, Field Trip Guidebook T104: Washington, D.C., American Geophysical Union, p. 34–63.

Pavlis, T. L., Sisson, V. B., Foster, H. L., Nokleberg, W. J., and Plafker, G., 1993, Mid-Cretaceous extensional tectonics of the Yukon-Tanana terrane, Trans-Alaska Crustal Transect (TACT), east-central Alaska: Tectonics, v. 12, no. 1, p. 103–122.

Printed in U.S.A.

The Geology of North America
Vol. G-1, The Geology of Alaska
The Geological Society of America, 1994

Chapter 7

Geology of west-central Alaska

William W. Patton, Jr., Stephen E. Box*, and Elizabeth J. Moll-Stalcup*
U.S. Geological Survey, 345 Middlefield Road, Menlo Park, California 94025
Thomas P. Miller
U.S. Geological Survey, 4200 University Drive, Anchorage, Alaska 99508-4667

INTRODUCTION

West-central Alaska includes a broad area that stretches from the Bering and Chukchi seacoasts on the west to the upper Yukon-Tanana Rivers region on the east, and from the Brooks Range on the north to the Yukon-Kuskokwim delta on the south. It covers 275,000 km^2, nearly one-fifth of the entire state—and all or parts of 29 1:250,000 scale quadrangles (Fig. 1).

Rolling hills with summit altitudes between 300 and 1,000 m and isolated mountain ranges that rise to a maximum altitude of 1,500 m characterize the area (Wahrhaftig, this volume). The uplands are separated by broad alluviated coastal and interior lowlands that stand less than 200 m above sea level. Bedrock exposures are generally limited to elevations above 500 m and to cutbanks along the streams.

The bedrock underlying this huge area consists of six pre–mid-Cretaceous lithotectonic terranes, which were assembled by Early Cretaceous time and were subsequently overlapped by mid- and Upper Cretaceous terrigenous sediments (Figs. 2 and 3; Jones and others, 1987; Silberling and others, this volume). The bedrock in the east-central part is composed of lower Paleozoic sedimentary rocks and Precambrian metamorphic rocks that belong to the Nixon Fork and Minchumina terranes. A broad mid-Cretaceous uplift, the Ruby geanticline (Fig. 4), borders the Nixon Fork terrane on the northwest and extends diagonally across the area from the eastern Brooks Range to the lower Yukon River valley. The core of the geanticline consists of the Ruby terrane, an assemblage of Precambrian(?) and Paleozoic continental rocks that was metamorphosed to greenschist, blueschist, and amphibolite facies in late Mesozoic time. Structurally overlying the Ruby terrane are allochthonous masses of upper Paleozoic and lower Mesozoic oceanic rocks variously assigned to the Angayucham, Tozitna, and Innoko terranes. The Yukon-Koyukuk basin is a large wedge-shaped depression filled with mid- and Upper Cretaceous terrigenous sedimentary rocks that extends from the Ruby geanticline westward to the Seward Peninsula and Bering and Chukchi seacoasts (Fig. 4). Lower Cretaceous island-arc-type volcanic rocks of the Koyukuk terrane are exposed on structural highs within the basin. Large mid- and Late Cretaceous granitoid bodies intrude the northern part of the Yukon-Koyukuk basin and the Ruby geanticline; Upper Cretaceous and lower Tertiary calc-alkalic volcanic rocks are widely distributed over all but the northwestern part of the area. Continental flood basalts of late Cenozoic age overlap the west edge of the Yukon-Koyukuk basin along the boundary with the Seward Peninsula and the Bering and Chukchi seacoasts.

All pre-uppermost Cretaceous rocks in the area are tightly folded and broken by high-angle faults (Fig. 5). Uppermost Cretaceous and lowermost Tertiary rocks are broadly folded and cut by high-angle faults; those of Eocene and younger age are virtually undisturbed. The area is transected by three major east- and northeast-trending fault systems: the Kobuk, Kaltag, and Nixon Fork–Iditarod (Fig. 4). All three systems are believed to have had large-scale strike-slip movement in Late Cretaceous and early Tertiary time, but the amount and direction of movement are well documented only for the Kaltag fault (Patton and Tailleur, 1977; Patton and others, 1984).

Geographic localities mentioned in this chapter refer to the U.S. Geological Survey Alaska Topographic Series 1:250,000 scale maps shown in Figure 1. The reader will need to consult these topographic maps in order to locate specific geographic features.

PRE–MID-CRETACEOUS LITHOTECTONIC TERRANES

Minchumina terrane

Definition and distribution. The Minchumina terrane (Jones and others, 1987; Silberling and others, this volume) consists of a northeast-trending belt of thin-bedded limestone, chert, argillite, and quartzite of Precambrian(?) and early Paleozoic age that extends for a distance of nearly 300 km from the southern part of the Medfra Quadrangle to the northeastern part of the Kantishna River Quadrangle (Figs. 1, 2, and 3). This belt of sparsely exposed bedrock underlies the Tanana-Kuskokwim low-

Present addresses: Box, U.S. Geological Survey, Room 656, West 920 Riverside Avenue, Spokane, Washington 99201; Moll-Stalcup, U.S. Geological Survey, National Center, Reston, Virginia 22092.

Patton, W. W., Jr., Box, S. E., Moll-Stalcup, E. J., and Miller, T. P., 1994, Geology of west-central Alaska, *in* Plafker, G., and Berg, H. C., eds., The Geology of Alaska: Boulder, Colorado, Geological Society of America, The Geology of North America, v. G-1.

land and adjoining parts of the Kuskokwim Mountains (Wahrhaftig, this volume). Within the belt we recognize two subterranes—the Telida subterrane, which extends the entire length of the belt, and the East Fork subterrane, which is confined to the south-central part of the Medfra Quadrangle (Patton and others, 1980; Fig. 3).

Description. *Telida subterrane.* The stratigraphic sequence that makes up the Telida subterrane has been pieced together from a study of widely scattered exposures in the Medfra Quadrangle (Patton and others, 1980), the Kantishna River Quadrangle (Chapman and Yeend, 1981; Chapman and others, 1975), and the northwestern part of the Mt. McKinley Quadrangle (Chapman and others, 1981). Four separate lithologic units are recognized, but owing to lack of continuous exposures, their stratigraphic relations are uncertain. The following sequence, in ascending stratigraphic order, is based on regional structural relations and a few fossil collections.

1. The limestone and phyllite unit consists of impure limestone and dolomite, green-gray phyllite, and argillite, and is probably pre-Ordovician in age.

2. The argillite and quartzite unit consists of slaty argillite, quartzite, quartz and quartz-feldspar grit, calcareous argillite, and chert. It contains Ordovician corals and graptolites and may include pre-Ordovician rocks in the lower part.

3. The chert and argillite unit consists of dark gray to black chert, argillite, shaly limestone, and siliceous siltstone. It contains Ordovician graptolites and early Paleozoic radiolarians.

4. The limestone unit consists of reefal and algal limestone bodies containing Middle to Late Devonian corals.

East Fork subterrane. The East Fork subterrane (Fig. 3)

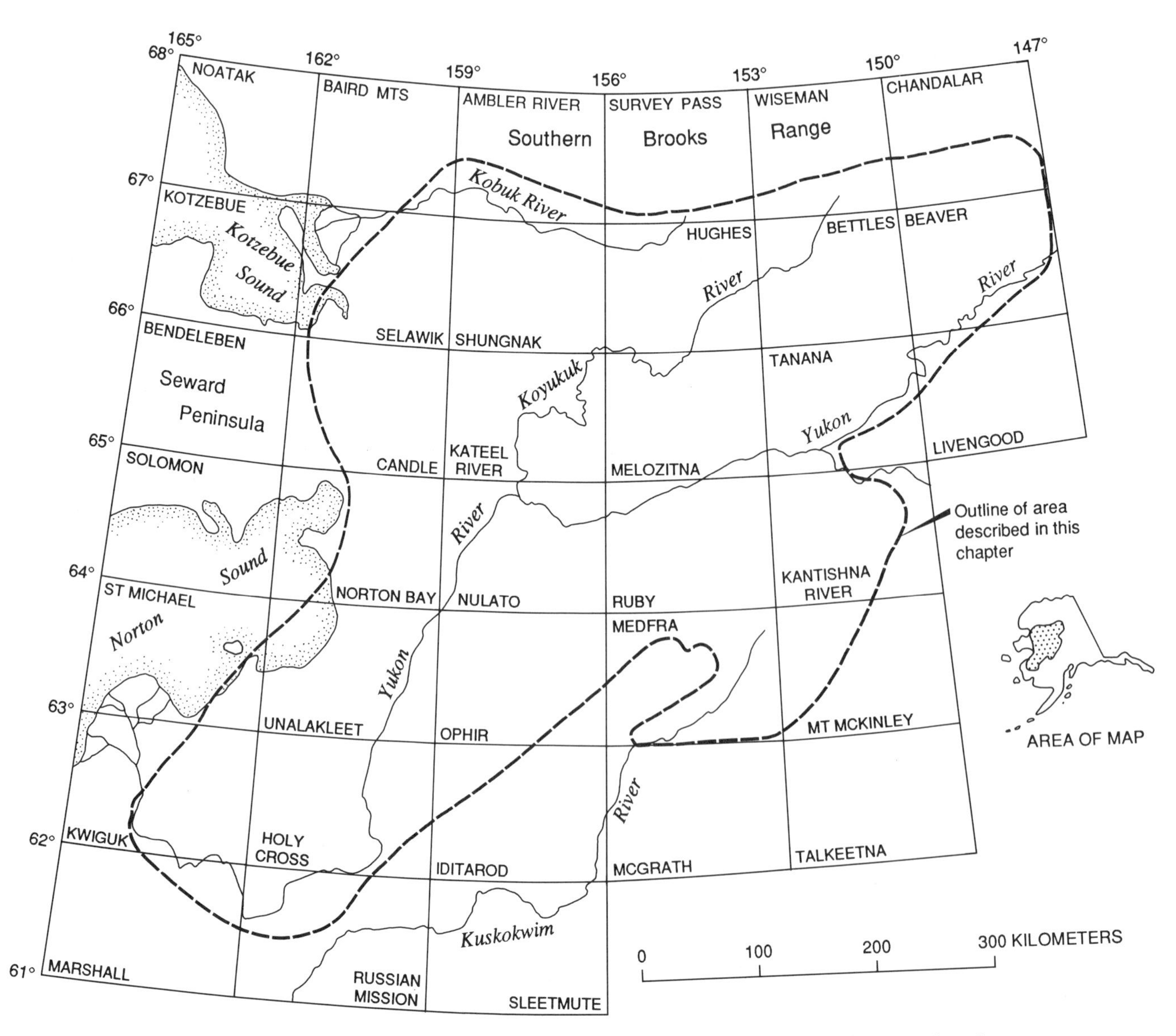

Figure 1. Index map of west-central Alaska showing location of quadrangles (1:250,000) and outline of area described in this chapter.

occupies a small wedge-shaped area in the Tanana-Kuskokwim lowland between the Telida subterrane and the Nixon Fork terrane in the south-central part of the Medfra Quadrangle. This subterrane consists entirely of the East Fork Hills Formation (Dutro and Patton, 1982), a succession of alternating thin beds of limestone and orange-weathering dolomite that is locally sheared and foliated. Laminated dolomite, dark chert, and siliceous siltstone occur in subordinate amounts. Owing to the lack of good exposures and to the structural complexity of this formation, no estimate of total thickness is possible. The formation has been assigned an Early Ordovician to Middle Devonian age based on scattered conodont collections (Dutro and Patton, 1982).

The East Fork subterrane is juxtaposed with the Nixon Fork terrane along a northeast-trending fault that appears to be a strand of the Nixon Fork–Iditarod fault system (Dutro and Patton, 1982; Fig. 5, D-D′). The East Fork subterrane is cut off on the southeast by the more north-trending Telida subterrane. The contact between the two subterranes is not exposed, but the divergence in their regional trends suggests that it is a fault.

Interpretation and correlation. The Minchumina terrane is composed chiefly of deep-water deposits, which we interpret to be a continental-margin facies that is coeval, at least in part, with the lower Paleozoic platform carbonate rocks of the Nixon Fork terrane. The reefy algal limestones of Middle and Late Devonian age (unit 4) that cap the Telida subterrane represent a southeastward progradation (relative to their present orientation) of the shallow-water beds across the older deep-water deposits.

The Minchumina terrane appears to be part of an extensive but discontinuous belt of deep-water Precambrian(?) and lower Paleozoic deposits that can be traced northeastward from the western part of the Alaska Range in the McGrath Quadrangle, through the Tanana-Kuskokwim lowland at least as far as the Livengood Quadrangle. This belt includes such deep-water assemblages as the Dillinger assemblage in the Alaska Range (Bundtzen and others, 1985), the Nilkoka Group along the lower Tanana River (Péwé and others, 1966), and the Livengood Dome Chert (Chapman and others, 1980) in the Livengood Quadrangle. In southwestern Alaska, Decker and others (this

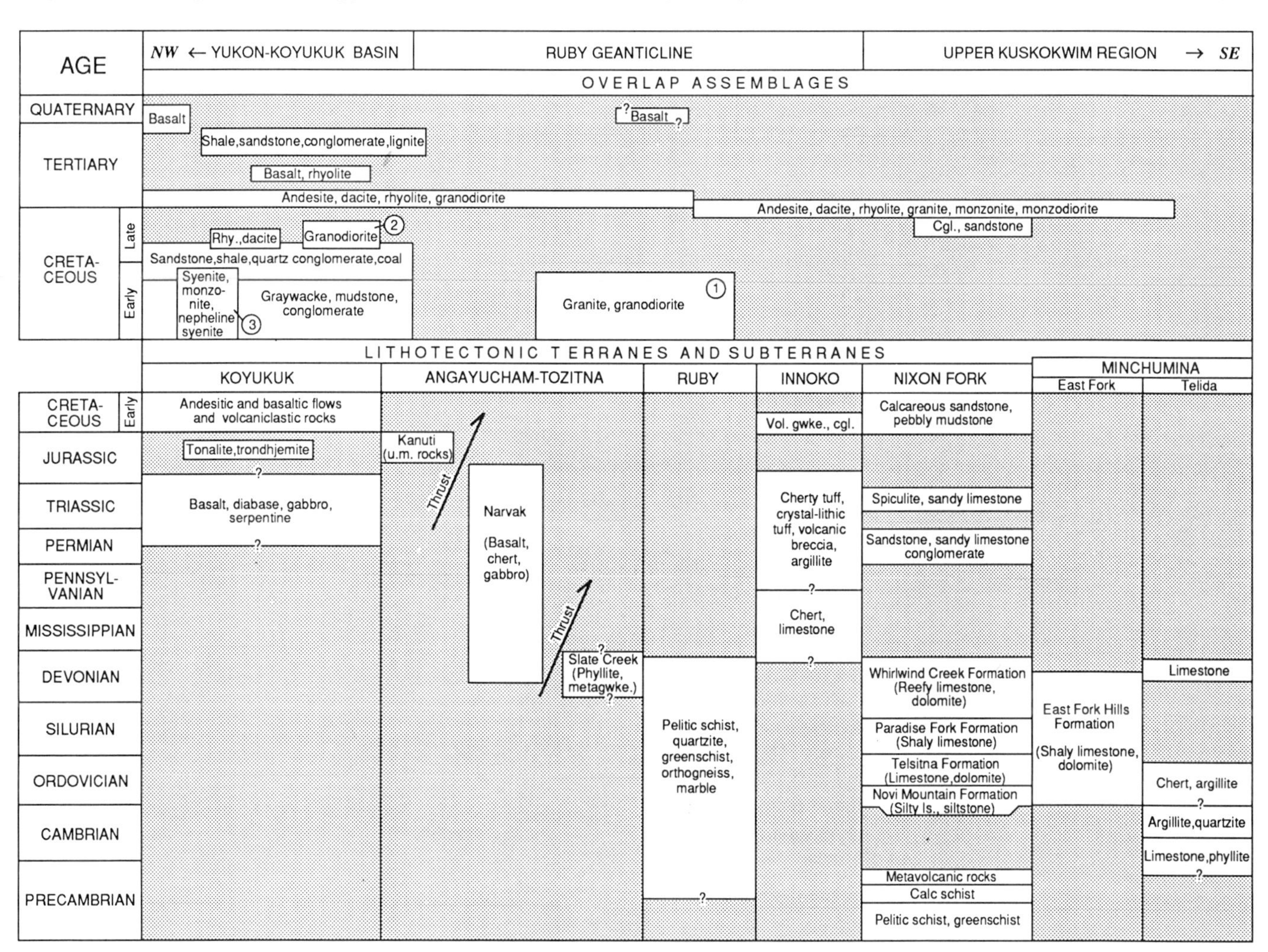

Figure 2. Correlation of pre–mid-Cretaceous lithotectonic terranes and mid-Cretaceous and younger overlap assemblages.

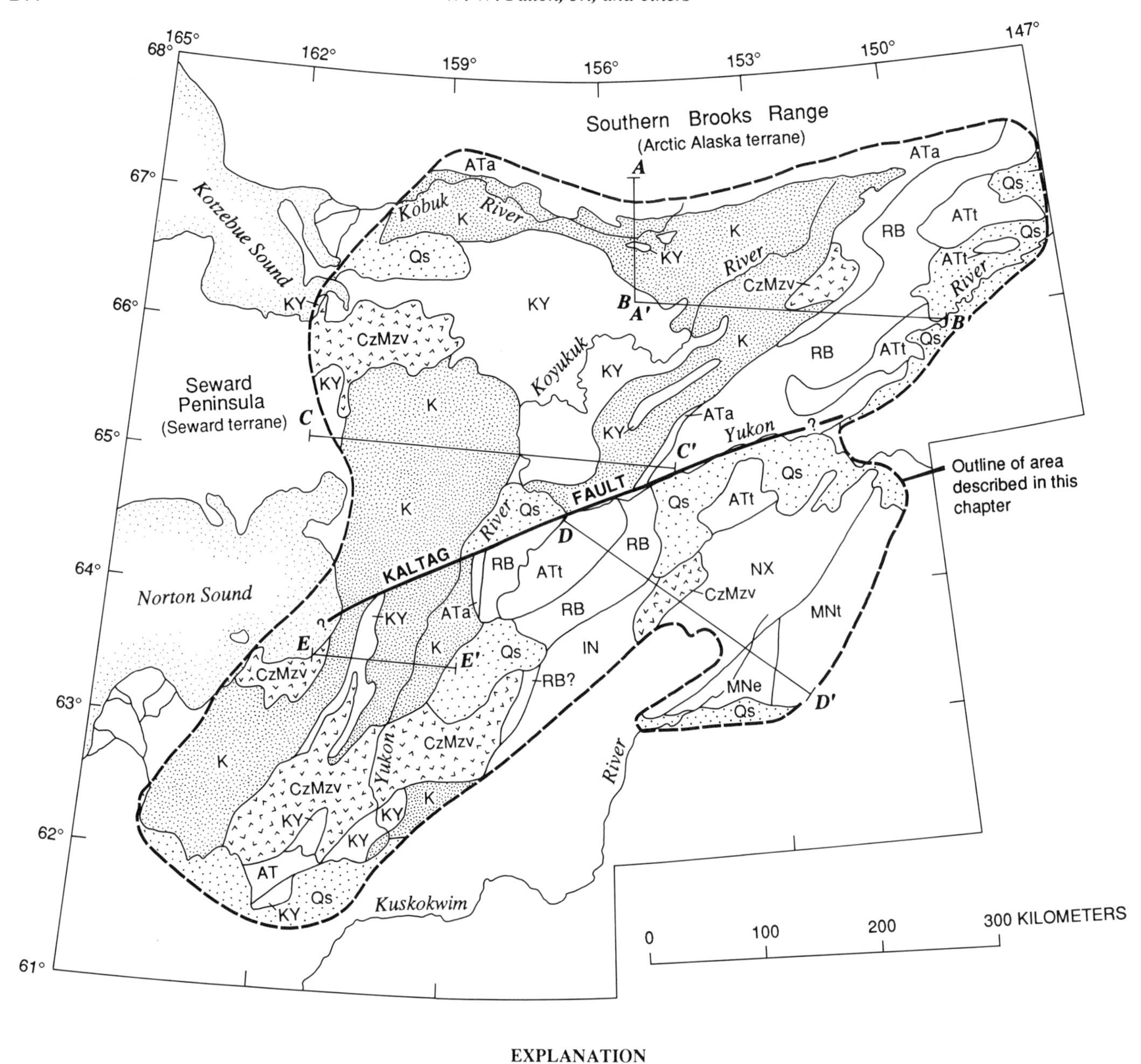

EXPLANATION

Overlap assemblages

Qs	Surficial deposits (Quaternary)
CzMzv	Volcanic rocks (Cenozoic and upper Mesozoic)
K	Terrigenous sedimentary rocks (mid and Upper Cretaceous)

Lithotectonic terranes

	Minchumina terrane
MNt	Telida subterrane
MNe	East Fork subterrane
NX	Nixon Fork terrane
IN	Innoko terrane
RB	Ruby terrane
AT	Angayucham-Tozitna (composite) terrane
ATa	Angayucham belt
ATt	Tozitna belt
KY	Koyukuk terrane
———	Contact

Figure 3. Pre–mid-Cretaceous lithotectonic terranes and mid-Cretaceous and younger overlap assemblages (modified from Jones and others, 1987; Silberling and others, this volume). Schematic cross sections along lines A-A′ to E-E′ shown in Figure 5.

volume) include the Minchumina terrane in the White Mountain sequence of the Farewell terrane.

Nixon Fork terrane

Definition and distribution. The Nixon Fork terrane (Jones and others, 1987; Silberling and others, this volume) is composed of three stratigraphic packages separated by major unconformities: a Precambrian metamorphic basement, a thick lower Paleozoic platform carbonate sequence, and a thin upper Paleozoic and Mesozoic terrigenous clastic sequence (Fig. 2; Patton and others, 1980). It forms a northeast-trending belt that extends for 500 km from the Holitna lowland in the Lime Hills Quadrangle of southwestern Alaska to the Nowitna lowland in the Ruby and Kantishna River Quadrangles of central Alaska. The area covered by this chapter (Fig. 3) includes the northern 300 km of this belt and some of the best exposures of the Nixon Fork terrane. The southwestern part of this belt is described by Decker and others (this volume). The Nixon Fork terrane is bounded on the northwest by the Innoko terrane along the Susulatna lineament—a conspicuous topographic alignment of stream valleys interpreted to be a major fault zone (Patton, 1978; Fig. 5, D–D′). On the southeast the Nixon Fork terrane adjoins the Minchumina terrane along a strand of the Nixon Fork–Iditarod fault system.

Description. *Precambrian metamorphic rocks.* The lowest stratigraphic package consists of a greenschist metamorphic facies assemblage of pelitic and calc schists, greenstone, and minor felsic metaplutonic rocks locally capped by felsic metavolcanic rocks (Patton and others, 1980). This assemblage is poorly exposed along the northwest side of the Nixon Fork terrane and dips southeastward beneath the lower Paleozoic carbonate rocks (Fig. 5, D–D′). Potassium-argon mineral-separate ages from this assemblage range from 921 to 296 Ma (Silberman and others, 1979; Dillon and others, 1985). Metaplutonic rocks from the lower part yielded a U-Pb zircon age of 1,265 ± 50 Ma, and metavolcanic rocks from the top yielded a U-Pb zircon age of 850 ± 30 Ma (Dillon and others, 1985).

Lower Paleozoic carbonate rocks. The middle stratigraphic package is composed of four platform carbonate formations that have an aggregate thickness of more than 5,000 m and an age range of Early Ordovician to Late Devonian (Dutro and Patton, 1982). The Novi Mountain Formation, the oldest of the four, consists of about 900 m of cyclically interbedded shallow-water to supratidal silty to micritic limestone, and calcareous siltstone of Early Ordovician age. This sequence is overlain by the Telsitna Formation, a 2,000-m-thick sequence of abundantly fossiliferous, shallow-water limestone and dolomite of Middle and Late Ordovician age. Unconformably(?) above the Telsitna Formation is the Paradise Fork Formation, a 1,000-m-thick deeper-water assemblage of graptolite-bearing, dark, thin-bedded limestone and shale of Early to Late Silurian age (Dutro and Patton, 1982). At the top is the Whirlwind Creek Formation, a Late Silurian to Late Devonian sequence of shallow-water limestone and dolomite, 1,000 to 1,500 m thick.

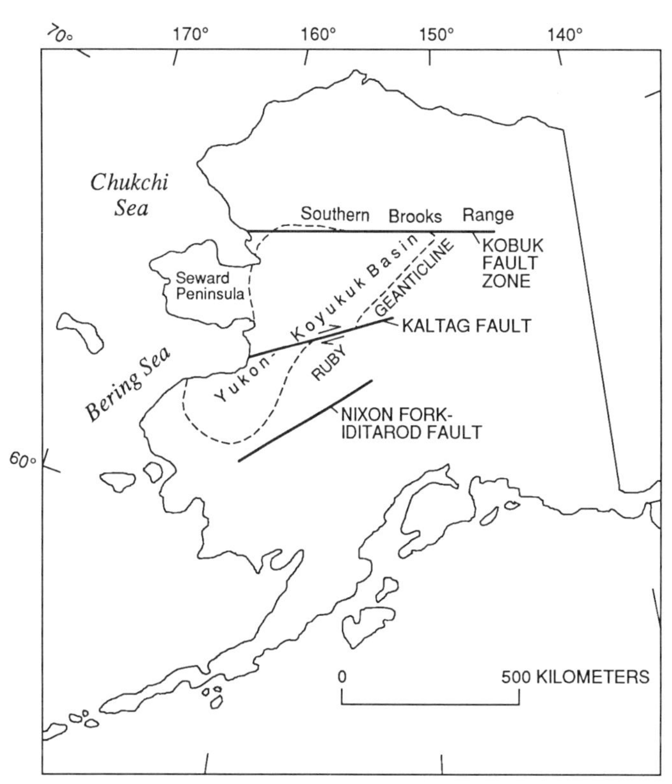

Figure 4. Index map showing location of major tectonic elements in west-central Alaska.

Upper Paleozoic and Mesozoic clastic rocks. The highest stratigraphic package consists of Permian to Lower Cretaceous, terrigenous, shallow-marine, sedimentary rocks, which unconformably overlie both the lower Paleozoic platform carbonate rocks and the Precambrian basement rocks. These beds are composed largely of quartz and carbonate debris eroded from the underlying carbonate and metamorphic assemblages. The total thickness of this succession probably does not exceed 1,000 m. The lowest strata are composed of quartz-carbonate sandstone and siltstone of early Late Permian age. Locally, where these strata rest directly on the metamorphic basement, they contain a coarse basal conglomerate consisting of large angular blocks of pelitic schist and greenstone set in a quartz-carbonate sandstone matrix (Patton and Dutro, 1979). The Permian strata are disconformably overlain by an Upper Triassic sequence composed of quartz-carbonate sandstone and conglomerate in the lower part, and dark spiculitic chert in the upper part. The Triassic beds, in turn, are disconformably overlain by Lower Cretaceous quartz-carbonate clastic rocks and pebbly mudstones that range in age from Valanginian at the base to Aptian at the top. No strata of Jurassic age have been identified in the Nixon Fork terrane.

Interpretation and correlation. The tectonic affinities and paleogeography of the Nixon Fork terrane are not clear. Most workers agree that the Minchumina terrane and other lower Paleozoic deep-water assemblages, such as the Dillinger and Liven-

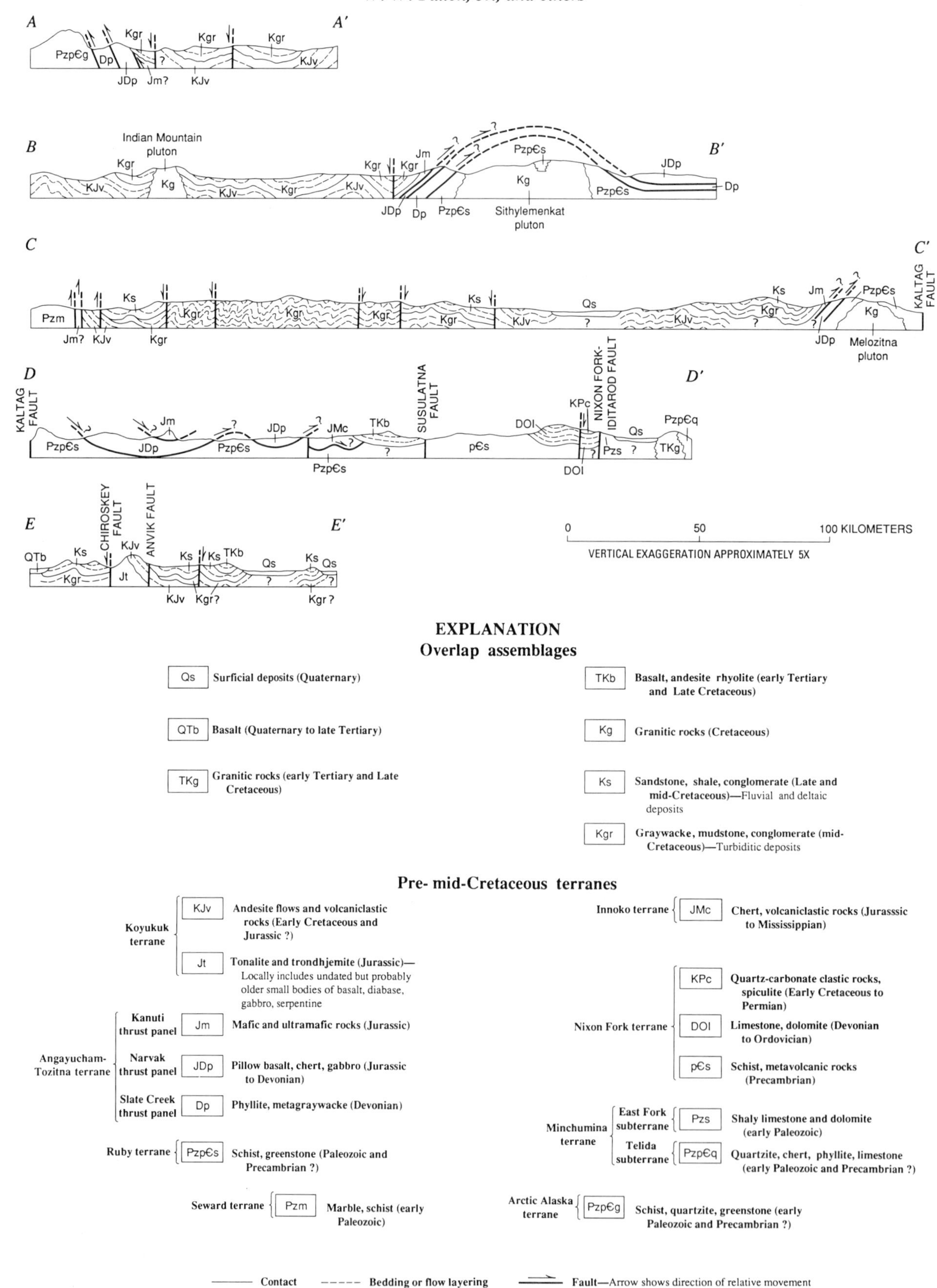

Figure 5. Schematic cross sections A-A′ to E-E′ illustrating gross structural features of west-central Alaska. Location of sections shown in Figure 3.

good terranes, represent a seaward "shale-out" of the platform carbonate rocks of the Nixon Fork terrane similar to that found along the early Paleozoic continental margin in the Canadian Cordillera (Churkin and Carter, 1979). However, the Nixon Fork terrane is not flanked on the opposite side by cratonic assemblages similar to those that lie along the inner margin of the Canadian Cordillera.

The paleontologic and paleomagnetic evidence bearing upon the biostratigraphic and tectonic affinities of the lower Paleozoic platform carbonate rocks of the Nixon Fork terrane is conflicting and controversial. Blodgett (1983) and Potter and others (1980), for example, argue that the Nixon Fork terrane has North American affinities, as indicated by the Devonian ostracod, brachiopod, gastropod, and trilobite fauna and the Middle and Late Ordovician brachiopod fauna. Palmer and others (1985), on the other hand, contend that the Nixon Fork terrane has Siberian platform affinities based on Middle Cambrian trilobites recently found in the Nixon Fork terrane southwest of the area covered by this chapter. Paleomagnetic studies by Plumley and others (1981) indicate that the Ordovician carbonate rocks of the Nixon Fork terrane have been displaced no more than 10° northward relative to the Ordovician pole position of the North American craton, but have been rotated clockwise about 70° prior to late Mesozoic time. Plumley and Coe (1982) proposed that the Nixon Fork terrane is "a section of the Canadian Cordillera" that rotated clockwise to its present position. A difficulty with this model, however, is that the present southeastward direction of shale-out of the Nixon Fork platform carbonate deposits is opposite that which would be expected if the west-facing, early Paleozoic continental margin of the Canadian Cordillera were rotated 70° clockwise. Churkin and others (1984) suggested the possibility that the Nixon Fork terrane represents a fragment of the Cassiar platform—an outboard ridge of lower Paleozoic platform carbonate rocks in the central Canadian Cordillera that shales out eastward into deeper-water deposits of the Selwyn basin.

Decker and others (this volume) favor a stablistic model for the origin of the Precambrian and lower Paleozoic part of the Nixon Fork and Minchumina terranes, which they lump together in the White Mountain sequence of the Farewell terrane. They argue that these terranes were "part of a coherent Paleozoic passive continental margin which lay upon a Precambrian continental crystalline basement to form a southwesterly-directed peninsular extension of the Paleozoic North America continent." In support of this model, they point out similarities between the major lithostratigraphic units of the Nixon Fork and Minchumina terranes with coeval rocks in northwestern Canada. A problem with this model, however, is that although there are many similarities between the lower Paleozoic platform carbonate rocks in the two areas, the underlying Precambrian rocks are quite different. Unlike the Nixon Fork terrane, the Canadian succession does not record a major regional metamorphic event in late Proterozoic or earliest Paleozoic time. Furthermore, the absence of a transitional facies between the fault-bounded blocks of the Nixon Fork terrane and the Telida and East Fork subterranes is an argument against their being a single coherent terrane.

We suggest that these terranes and subterranes represent an assemblage of small fault slices transported to their present positions from the North American (or Siberian?) continental margin along transform strike-slip faults. The paleomagnetic evidence for 70° of clockwise rotation of the Nixon Fork terrane may be the result of local rotation during tectonic transport of the fault slice that was sampled, rather than rotation of the entire continental margin.

Innoko terrane

Definition and distribution. The Innoko terrane (Jones and others, 1987; Silberling and others, this volume) consists of a poorly exposed sequence of radiolarian chert and andesitic volcaniclastic rocks ranging in age from Late Devonian to Early Cretaceous (Fig. 2). This assemblage of oceanic and volcanic-arc rocks extends from near Poorman in the Ruby Quadrangle, 250 km southwestward, to near Flat in the Iditarod Quadrangle (Fig. 3). The best exposures are in the northwestern part of the Medfra Quadrangle (Patton and others, 1980) and in the eastern part of the Ophir Quadrangle (Chapman and others, 1985), but even in those areas outcrops are limited to scattered small knobs and pinnacles along ridgetops and to a few stream banks. On the southeast side, the Innoko terrane is faulted against the Nixon Fork terrane and against the mid-Cretaceous sedimentary rocks of the Kuskokwim basin along the Susulatna lineament (Patton, 1978). On the northwest side, the contact with the Ruby and Tozitna terranes is generally covered by Quaternary alluvial deposits and Late Cretaceous to early Tertiary volcanic fields, but regional relations suggest that it is faulted. In the Iditarod Quadrangle the Innoko terrane narrows to less than 15 km in width, and at the southwest end near Flat, it appears to be overlapped by Upper Cretaceous and lower Tertiary volcanic rocks (M. L. Miller, personal communication, 1986).

Precambrian(?) and Paleozoic rocks, consisting chiefly of pelitic schist, carbonate rocks, and greenstone of greenschist metamorphic facies, are exposed in several structural windows beneath the Innoko terrane (Chapman and others, 1985). They are believed to be part of the Ruby terrane, which has been overridden by the Innoko along a low-angle thrust (Patton and Moll, 1982; Fig. 5, D-D′).

Large mafic-ultramafic complexes overlie the Innoko terrane at Mount Hurst in the southern Ophir Quadrangle and are on strike to the southwest along the upper Dishna River in the Iditarod Quadrangle (Patton and others, this volume; Miller and Angeloni, 1985). These ophiolite assemblages appear to be allochthonous with respect to the Innoko terrane and locally overlap onto the structural windows of the Ruby terrane. The origin of these bodies is uncertain; possibly they represent remnants of the mafic-ultramafic complex that makes up the Kanuti thrust panel of the composite Angayucham-Tozitna terrane (Fig. 2).

Description. Two different rock assemblages (not shown separately on Fig. 3) are recognizable in the Innoko terrane: an oceanic assemblage of radiolarian chert with minor carbonate rocks, basalt, and gabbro; and a volcanic arc-like assemblage of volcaniclastic rocks, cherty tuff, volcanic graywacke, argillite, and diabase and gabbro intrusive rocks. Cherts in the oceanic assemblage yield radiolarians of Late Devonian(?), Mississippian, Pennsylvanian, Permian, and Triassic age. The carbonate rocks, which occur as turbidites intercalated with the chert, contain reworked foraminifers and conodonts ranging in age from Late Devonian to Late Mississippian. The cherty tuffs in the arc assemblage have yielded Triassic and Early Jurassic(?) radiolarians and Triassic conodonts, and the volcanic graywacke in the upper part of the arc assemblage contains fragments of *Inoceramus* of Early Cretaceous(?) age. Field relations suggest that the arc assemblage overlies the oceanic assemblage, but the possibility that the two have been juxtaposed by thrust faulting cannot be ruled out.

Intrepretation and correlation. The Innoko terrane is treated here as a single coherent sequence but, in fact, may represent two unrelated assemblages that have been juxtaposed by thrust faulting. The lower radiolarian chert oceanic assemblage correlates closely in age and lithology with chert assemblages in the Narvak thrust panel of the Angayucham-Tozitna terrane and may be an eastward extension of a part of that terrane. The thick volcanic arc-like assemblage in the upper part of the Innoko terrane has no counterpart in the Angayucham-Tozitna terrane, but a similar Triassic to Lower Cretaceous arc assemblage has been described in the Togiak terrane, which lies roughly on strike to the southwest in the lower Kuskokwim River region (Jones and others, 1987; Silberling and others, this volume; Decker and others, this volume). It is possible that the Innoko and Togiak terranes are coextensive, and lie beneath a cover of overlapping Upper Cretaceous and Tertiary sedimentary and volcanic rocks in the southern Iditarod and northern Sleetmute Quadrangles (Fig. 1).

Ruby terrane

Definition and distribution. The Ruby terrane (Jones and others, 1987; Silberling and others, this volume) consists of a Precambrian(?) and Paleozoic assemblage of pelitic schist, quartzite, greenschist, orthogneiss, and marble (Fig. 2). It forms the metamorphic core of the Ruby geanticline—a pre-mid-Cretaceous uplift that trends diagonally across central Alaska along the southeast side of the Yukon-Koyukuk basin (Figs. 3 and 4). At its northeast end, the Ruby terrane is cut off by the Kobuk fault—a probable strike-slip fault that bounds the south edge of the Brooks Range. The Ruby terrane extends southwestward from the Brooks Range to the Yukon River where it is offset right laterally about 160 km by the Kaltag fault (Patton and others, 1984). South of the fault, it continues through the Kaiyuh Mountains in the Nulato Quadrangle and at least as far south as the Innoko River in the Ophir Quadrangle (Figs. 1 and 3). The extent of the Ruby terrane south of the Innoko River is uncertain because of the lack of bedrock exposure in the Innoko Lowlands. Recent mapping (Angeloni and Miller, 1985) shows that similar rocks of greenschist metamorphic facies extend at least as far south as the north-central part of the Iditarod Quadrangle. However, an assemblage of amphibolite-grade augen gneiss, amphibolite, and pelitic schist (assigned to the Idono sequence) is also found along the same structural trend but does not have a counterpart in the Ruby terrane (Miller and Bundtzen, 1985; Angeloni and Miller, 1985). These higher-grade rocks may correlate with the Kilbuck terrane of southwestern Alaska, which is described in this volume by Decker and others. A small area of metamorphic rocks on the lower Yukon River in the Russian Mission Quadrangle, shown as Ruby terrane on the lithotectonic terrane map by Jones and others (1987; Silberling and others, this volume), appears to be composed chiefly of metabasites and is interpreted by us to be Angayucham-Tozitna terrane (Fig. 3).

The Ruby terrane is bordered along the northwest side by a narrow band of rocks belonging to the Angayucham-Tozitna terrane, which rests in thrust contact on the Ruby terrane and dips northwestward beneath the mid-Cretaceous sedimentary rocks of the Yukon-Koyukuk basin (Fig. 5, B-B′, C-C′). On the southeast side, it is bordered by several large allochthonous masses of Angayucham-Tozitna and Innoko terranes and by the Quaternary deposits of the Yukon Flats and Nowitna Lowland (Fig. 3; Wahrhaftig, this volume).

Description. The Ruby terrane is characterized by regionally metamorphosed greenschist-facies metasedimentary rocks and metabasites of Precambrian(?) and Paleozoic age. Locally it also includes high-pressure greenschist-facies assemblages, which are distinguished by the presence of glaucophane; and amphibolite-facies assemblages, which are distinguished by the presence of sillimanite and kyanite. Typical rock types include quartz-mica schist, quartzite, calcareous schist, mafic greenschist, quartzofeldspathic schist and gneiss, metabasite, and marble. Potassium-argon dating and geologic relations indicate that regional metamorphism occurred in Late Jurassic to Early Cretaceous time, probably during tectonic emplacement of the allochthonous oceanic rocks of the Angayucham-Tozitna terrane (Turner, 1984; Patton and others, 1984). Subsequently, both the metamorphic rocks of the Ruby terrane and the Angayucham-Tozitna terrane were widely intruded by mid-Cretaceous granitic plutons and extensively altered to andalusite-cordierite hornfels, hornblende hornfels, and contact marbles.

Dusel-Bacon and others (1989) have summarized common mineral assemblages for the intermediate- and high-pressure greenschist facies and for the amphibolite facies in the Ruby terrane as follows:

Intermediate-pressure greenschist facies:

Pelitic schists: quartz+white mica+chloritoid+chlorite±epidote, calcite, and sphene. Quartz+white mica±epidote, calcite, and biotite.

Metabasites: Epidote+actinolite+chlorite±zoisite, biotite, sphene, and quartz.

High-pressure greenschist facies:

Pelitic schists: Glaucophane+white mica+garnet+chlorite+chloritoid

+quartz. Glaucophane+white mica+calcite.

Metabasites: Albite+chlorite+actinolite+epidote group minerals± calcite and sphene. Chlorite+epidote+sphene+plagioclase±white mica and glaucophane.

Amphibolite facies:

Pelitic schists: Quartz+muscovite+biotite+staurolite+garnet. Quartz +muscovite+biotite+garnet. Kyanite+sillimanite+biotite+quartz+potash feldspar.

In the Tanana Quadrangle, Dover and Miyaoka (1985) recognized polymetamorphism in which "low P/T amphibolite facies assemblages (M_1) formed during main-phase isoclinal folding (F_1); M_1 minerals occur as relics in rocks later subjected to a cataclastic (F_2), low-grade, retrogressive event (M_2)." Polymetamorphism has not been documented elsewhere, however, and as pointed out by Dusel-Bacon and others (1989), it is uncertain if it has affected all parts of the Ruby terrane. High-pressure assemblages, indicated by the presence of glaucophane, have been identified in the Kaiyuh Mountains of the Nulato Quadrangle (Patton and others, 1984; Forbes and others, 1971; Mertie, 1937), in the Kokrines Hills of the Melozitna and Tanana Quadrangles (Patton and others, 1978; Chapman and others, 1982), and at a single locality on the north side of the Ray Mountains in the Tanana Quadrangle (Dover and Miyaoka, 1985). Amphibolite-facies assemblages have been reported in the Melozitna and Ruby Quadrangles (Patton and others, 1978; Smith and Puchner, 1985), in the Tanana Quadrangle (Dover and Miyaoka, 1985), and in the Beaver Quadrangle (Brosgé and others, 1973).

Protolith age. The metamorphic rocks of the Ruby terrane are tentatively assigned an early and middle Paleozoic protolith age, as indicated by scattered fossil and radiometric ages and by the gross lithologic similarities of these rocks to the better-dated metamorphic assemblages in the southern Brooks Range (Arctic Alaska terrane). Fossils have been found in carbonate layers at three widely scattered localities: (1) early Middle Ordovician conodonts from near Illinois Creek in the Nulato Quadrangle (A. G. Harris, written communication, 1984), (2) Middle Devonian conodonts from near Wolf Creek in the Melozitna Quadrangle (A. G. Harris, written communication, 1983), and (3) a Devonian coral from near Yuki Mountain in the Ruby Quadrangle (Mertie and Harrington, 1924). A Devonian U/Pb zircon age was obtained from a granite gneiss in the Ray Mountains of the Tanana Quadrangle (Patton and others, 1987).

Precambrian rocks have not been identified in the Ruby terrane, but the possibility that they are present cannot be ruled out, particularly among the vast tracts of poorly exposed pelitic schists and quartzites. Precambrian rocks have been reported in similar assemblages in the southern Brooks Range (Mayfield and others, 1983; Dillon and others, 1980).

Age of metamorphism. Regional metamorphism both in the Ruby terrane and in the southern Brooks Range (Arctic Alaska terrane) is thought to be related to arc collision and overthrusting or "obduction" of the Angayucham-Tozitna and Innoko terranes in latest Jurassic to Early Cretaceous time (Turner, 1984; Patton, 1984). Dover and Miyaoka (1985) point out evidence for polymetamorphism in the Ray Mountains, and it is possible that part or all of the Ruby terrane also underwent an earlier metamorphic event. The regional metamorphism clearly pre-dates the widespread intrusion of Early Cretaceous (~110 Ma) granitic plutons and the deposition of upper Lower Cretaceous (Albian) marginal conglomerates of the Yukon-Koyukuk basin, which locally contain a significant component of metamorphic debris. Two metamorphic mineral K-Ar ages of 136 and 134 Ma from the Ruby terrane (Patton and others, 1984) and a large number of K-Ar ages ranging from 130 to 85 Ma from similar metamorphic rocks in the southern Brooks Range (Turner, 1984) probably represent cooling ages set during unroofing of the isostastically rebounding metamorphic terranes following overthrusting of the Angayucham-Tozitna terrane.

Interpretations and correlations. Early workers (Cady and others, 1955; Payne, 1955; Miller and others, 1959) viewed the Ruby terrane as a narrow mid-Cretaceous structural high or "geanticline" bordering the southeastern flank of the Yukon-Koyukuk basin. The metamorphosed Precambrian(?) and Paleozoic rocks of the Ruby terrane were presumed to extend beneath the basin and to connect with rocks of similar age and metamorphic grade exposed in the southern Brooks Range and on the Seward Peninsula. However, subsequent investigations around the margins of the Yukon-Koyukuk basin showed that the relations between the metamorphic borderlands and the basin were much more complex than previously supposed. Recognition of inward-dipping ophiolite assemblages along the margins of the basin, and the presence of a voluminous island-arc volcanic assemblage (Koyukuk terrane) within the basin, led to the suspicion that the basin has an ensimatic basement (Patton, 1970). Recent isotopic and petrologic studies of mid- and Upper Cretaceous granitic plutons and of Upper Cretaceous and lower Tertiary volcanic rocks in the Ruby terrane and adjacent parts of the basin (Arth, 1985, and this volume; Miller, 1985; Moll and Arth, 1985), and geophysical investigations along the margin of the Ruby terrane (Cady, 1985), suggest that the Precambrian(?) and Paleozoic ensialic rocks of the Ruby terrane do not extend beneath the basin.

The metamorphic assemblages of the Ruby terrane and the southern Brooks Range (Arctic Alaska terrane) are grossly similar and may be correlative. Protoliths of both terranes appear to have general lithologic similarities, including early to middle Paleozoic carbonate rocks dated by fossils and Devonian granitic rocks dated by zircons. The metamorphic grade, the K-Ar ages of the metamorphic minerals, and the general structural setting (i.e., overthrusting by the Angayucham-Tozitna terrane) are also similar. Both terranes contain abundant metabasalt and metadiabase, but felsic metavolcanic rocks of middle Paleozoic age, which locally are abundant in the southern Brooks Range, have not been identified in the Ruby terrane, and mid-Cretaceous granitic plutons, which are widespread in the Ruby terrane, are absent from the southern Brooks Range. The two terranes are juxtaposed at the northeast apex of the Yukon-Koyukuk basin along the Kobuk fault zone—an east-trending belt of strike-slip(?)

faulting of post–mid-Cretaceous age that extends along the south margin of the Brooks Range from Kotzebue Sound to the Yukon Flats (Patton, 1973; Fig. 4). The faults cut off the northeast trends of the Ruby terrane, and north of the fault zone there are no indications that regional trends in the Brooks Range bend around the apex of the Yukon-Koyukuk basin to conform to the northeast strike of the Ruby geanticline (Patton and Miller, 1973; Decker and Dillon, 1984).

The gross similarities between the Ruby terrane and southern Brooks Range have prompted workers to infer that both terranes were once parts of a single continuous belt. Several tectonic models have been offered to explain their present configuration and relation:

1. Tailleur (1973, 1980) proposed that the Ruby terrane is an eastward extension of the southern Brooks Range (Arctic Alaska terrane) and owes its present southwest trend to oroclinal bending related to east-west convergence between North America and Eurasia in Late Cretaceous and early Tertiary time.

2. Patton (1970) suggested that the Ruby terrane is a fragment of the southern Brooks Range, which was rifted away in late Paleozoic time and subsequently rotated counterclockwise to leave behind the V-shaped Yukon-Koyukuk basin. Subsequent Early Cretaceous collision of an oceanic island arc (Koyukuk terrane) with this V-shaped continental margin resulted in overprinting of the metamorphic fabric in roughly its present orientation (Patton, 1984; Box, 1985).

3. Churkin and Carter (1979) argued that the Ruby terrane represents an eastward extension of the southern Brooks Range that has been displaced right laterally (in Cretaceous? time) to its present position by a major southwest-trending strike-slip fault, which they labelled the "Porcupine lineament."

All three of these models are based on the assumption that the Ruby terrane and the southern Brooks Range (Arctic Alaska terrane) once formed a continuous belt. However, much additional study and detailed mapping of the Ruby terrane are needed in order to critically evaluate this assumption. At present we cannot rule out the possibility that the Ruby terrane is an exotic fragment unrelated to the southern Brooks Range that was plucked from some distant part of the Cordilleran or Siberian continental margin and rafted to its present position before the overthrusting of the Angayucham-Tozitna terrane.

Angayucham-Tozitna terrane

Definition and distribution. The (composite) Angayucham-Tozitna terrane, as defined in this chapter, is composed of an imbricated thrust assemblage of oceanic rocks that ranges in age from Devonian to Jurassic and occurs as allochthonous remnants of huge thrust sheets that overrode the Ruby geanticline and southern Brooks Range in Late Jurassic to Early Cretaceous time (Figs. 2, 3, and 5). Jones and others (1987) and Silberling and others (this volume) divide these oceanic assemblages into two separate terranes: the Angayucham and the Tozitna. However, our investigations show that these two terranes are composed of nearly identical thrust sequences that have been assembled in the same stacking order and show similar age ranges. Therefore, in this chapter, we treat the Angayucham and Tozitna terranes as separate belts of the same terrane. In addition, we include in our Angayucham-Tozitna terrane an assemblage of phyllite and metagraywacke, which underlies the oceanic rocks of both the Angayucham and Tozitna belts and was mapped separately by Jones and others as the Coldfoot and Venetie terranes. In our view, this assemblage has close tectonic affinities to the oceanic rocks of the Angayucham-Tozitna terrane, and we describe it in this chapter as the lowest of three thrust panels that compose the Angayucham-Tozitna terrane. We informally designate the three thrust panels, from lowest to highest, as: Slate Creek, Narvak, and Kanuti (Fig. 2).

The Angayucham belt of the Angayucham-Tozitna terrane forms a narrow but nearly continuous band for 500 km along the south edge of the Brooks Range from the lower Kobuk River to the northeast apex of the Yukon-Koyukuk basin, and along the northwest boundary of the Ruby terrane from the northeast apex of the basin to the Kaltag fault (Fig. 3). South of the fault the offset extension of the Angayucham belt can be traced, by scattered outcrops and aeromagnetic data, for an additional 50 km from near the village of Kaltag to the Innoko Lowland (Fig. 3).

The Tozitna belt of the Angayucham-Tozitna terrane consists of four large and several small allochthonous synformal masses that rest on Ruby terrane along the axis and southeastern flank of the Ruby geanticline. Two large masses lie south of the Kaltag fault, in the Ruby and Nulato Quadrangles, and two lie north of the fault, in the Tanana and Beaver Quadrangles (Figs. 1 and 3). The belt continues northeastward beyond the area covered by this chapter and includes a large synformal mass in the Christian Quadrangle (Brosgé and Reiser, 1962; Patton and others, 1977; Jones and others, 1987; Silberling and others, this volume).

A small area on the lower Yukon River in the Russian Mission Quadrangle, shown as Ruby terrane on the lithotectonic terrane map of Jones and others (1987), tentatively is assigned by us to the Angayucham-Tozitna terrane (Fig. 3). Sparse reconnaissance field data suggest that most of the rocks in this area are upper Paleozoic and lower Mesozoic metabasites, more like rocks of the Angayucham-Tozitna terrane than those of the Ruby terrane. The structural relation of the rocks in this area to the adjoining Middle Jurassic to Lower Cretaceous rocks of the Koyukuk terrane is uncertain.

Description. The Slate Creek, Narvak, and Kanuti thrust panels that make up the Angayucham-Tozitna terrane appear to represent a reversely stacked sequence that progresses from continental slope deposits in the Slate Creek or lowest panel, through oceanic rocks in the Narvak or middle panel, to cumulus and mantle peridotites and layered gabbro in the Kanuti or highest panel (Figs. 2 and 5, A-A′, B-B′, C-C′, D-D′). The contact between the Slate Creek and the Narvak thrust panels is generally blurred by tectonic shuffling, whereas the contact between the Narvak and Kanuti thrust panels is sharply defined and commonly marked by a thin layer of garnet amphibolite tectonite. A detailed description of each thrust panel follows:

The Slate Creek thrust panel is composed chiefly of phyllite and metagraywacke and minor amounts of carbonate rocks, basalt flows, and basalt breccias. The assemblage is easily eroded and typically forms a lowland between the more resistant basalt and chert of the Narvak thrust panel and the pelitic schist of the structurally underlying Ruby and Arctic Alaska terranes. The phyllite and metagraywacke are overprinted by a low-grade penetrative metamorphic fabric, but turbidite features, such as graded bedding and sole marks, are locally discernible. The metagraywacke is composed chiefly of quartz and chert clasts but, in places, contains a significant component of volcanic rock and feldspar clasts. Of particular interest is the presence in the upper part of the thrust panel of: (1) slices of exotic shallow-water Devonian carbonate rocks that commonly are covered or enveloped by basalt flows, and (2) debris-flow(?) breccias composed of angular blocks of vesicular basalt as much as 50 cm across set in a matrix of volcanic and carbonate rock debris. The possible tectonic significance of these will be discussed below.

The age of the phyllite and metagraywacke is controversial. Palynoflora recovered from the phyllite and metagraywacke in the Wiseman Quadrangle (Gottschalk, 1987) and in the Christian Quadrangle of the eastern Brooks Range (W. P. Brosgé, unpublished data, 1987) suggest a Devonian age. Dillon and others (1986), however, assign a Mississippian to Triassic age to this thrust panel on the basis of radiolarians from intercalated chert layers. We favor the Devonian age because, in our experience, the relation of the intercalated chert layers to the phyllite and metagraywacke is seldom clear, and the possibility that they are fault slices cannot be ruled out.

The Narvak thrust panel, the most widely exposed of the three panels, consists of multiple thrust slices of pillow basalt, chert, gabbro, and diabase and minor amounts of basaltic tuffs, volcanic breccia, and carbonate rocks. In the Angayucham Mountains (Hughes Quadrangle), where best exposed, the Narvak panel has an aggregate structural thickness of nearly 10 km (Pallister and Carlson, 1988). All of the rocks in the Narvak panel are weakly metamorphosed to prehnite-pumpellyite facies and show an overall increase in metamorphic grade downward in the panel. Greenschist-facies metamorphism and local high-pressure metamorphism, as indicated by the presence of glaucophane, occur at the base of the panel. Igneous textures are generally well preserved except in mylonite zones bordering thrust faults. The chert, which includes both interpillow and bedded types, ranges from pure radiolarite and spiculite to cherty tuffs. Sills and dikes of gabbro and diabase are common in both the Angayucham and Tozitna belts, but are especially abundant in the Tozitna.

Systematic detailed sampling for radiolaria from cherts in the well-studied section of the Narvak thrust panel in the Angayucham Mountains (Hughes Quadrangle) gives a range in age from Devonian at the base to Jurassic at the top (Murchey and Harris, 1985). A similar range of radiolarian ages has been reported from widely scattered localities elsewhere in this panel (Dillon and others, 1986; Patton and others, 1984; Jones and others, 1984). Carbonate rocks, which appear to be confined to the lower part of the Narvak panel, yield mixed conodont faunas ranging from Ordovician to Late Mississippian in age (A. G. Harris, written communication, 1985) and a few megafossils of Devonian, Mississippian(?), and Permian age. Some of the conodont collections clearly have been reworked from shallow-water sources.

The Kanuti thrust panel is composed of mafic-ultramafic complexes consisting of a tectonite mantle suite in the lower part and a cumulus plutonic suite in the upper part (Loney and Himmelberg, 1985). The composition and structural setting of these complexes are described in detail elsewhere (Patton and others, this volume) and will only be summarized here.

The mantle suite is composed of partly serpentinized harzburgite and dunite but it also contains minor clinopyroxenite and the cumulus suite of wehrlite, clinopyroxenite, and gabbro (Loney and Himmelberg, 1985). The base of the thrust panel commonly consists of a layer of amphibolite as much as 25 m thick composed of a highly tectonized aggregate of amphibole, plagioclase, and garnet. Amphibole dikes and gabbro in the cumulate plutonic suite yield K-Ar ages of 172 to 138 Ma, and metamorphic amphibole from amphibolite at the base of the panel yields K-Ar ages of 172 to 155 Ma (Patton and others, 1977; Patton and others, this volume). These are interpreted as cooling ages related to tectonic emplacement of the thrust panel.

Interpretation and correlation. The Angayucham-Tozitna terrane represents allochthonous remnants of huge thrust sheets of oceanic rocks that overrode the Precambrian and lower Paleozoic continental margin deposits of the Arctic Alaska and Ruby terranes in Late Jurassic to Early Cretaceous time. The Angayucham belt forms a narrow band of slab-like bodies that rim the north and southeast sides of the mid-Cretaceous Yukon-Koyukuk basin and dip inward beneath the basin. The Tozitna belt forms allochthonous synformal masses aligned along the axis and southeastern flank of the Ruby geanticline (Ruby terrane). The two belts are closely correlative in age and lithology, and we consider them to be remnants of the same overthrust sheets. Similar upper Paleozoic and lower Mesozoic oceanic assemblages appear to have overridden the early Paleozoic continental margin in Mesozoic time throughout the length of the North American Cordillera (Harms and others, 1984; Struik and Orchard, 1985). The Innoko terrane, which in its lower part is composed predominantly of radiolarian cherts that correlate in age with cherts in the Narvak thrust panel of the Angayucham-Tozitna terrane, may be an eastward extension of a part of the Angayucham-Tozitna terrane.

Several lines of evidence support a Late Jurassic to Early Cretaceous age of emplacement of the Angayucham-Tozitna terrane on the continental margin. Emplacement could not have begun earlier than Late Jurassic since K-Ar cooling ages of mafic rocks in the Kanuti thrust panel are Late Jurassic. The oldest flyschoid deposits derived from the Angayucham-Tozitna terrane in the western Brooks Range are Late Jurassic (Tithonian) in age

(Mayfield and others, 1983). The emplacement must have been completed by late Early Cretaceous (Albian) time because at that time large volumes of flyschoid sediments containing debris from both Angayucham-Tozitna terrane and from the metamorphic rocks of the Ruby and Arctic Alaska terranes were being deposited in the Yukon-Koyukuk basin. Furthermore, on the Ruby geanticline, the Ruby terrane and the Angayucham-Tozitna terranes are stitched together by granitic plutons that yield U/Pb and K-Ar ages of about 110 Ma (Patton and others, 1987).

We interpret the Slate Creek thrust panel to represent an assemblage of continental slope-and-rise deposits that accumulated along a middle Paleozoic rifted continental margin. The bulk of the assemblage is composed of fine-grained siliciclastic turbidites, but the presence of coarse volcanic-rich breccias and of exotic blocks of shallow-water carbonate rocks, which locally are enveloped in basalt flows, leads us to believe that the assemblage was deposited in an extensional environment. Bimodal volcanism along the south edge of the Brooks Range (Arctic Alasks terrane) provides additional evidence of extension in middle Paleozoic time (Hitzman, 1984).

The Narvak thrust panel is an imbricated assemblage of oceanic basalt and chert, which we suggest was accreted to the continental margin when an intraoceanic volcanic arc (Koyukuk terrane) collided with the margin in latest Jurassic to Early Cretaceous time. The lower (upper Paleozoic) part of this assemblage appears to have formed near the continental margin, probably in the continent-ocean transition zone. Intercalated carbonate rocks in the lower part of the panel contain redeposited shallow-water conodont faunas of Ordovician to Late Mississippian age. The wide age range and diversity of these faunas suggest that they were derived from platform carbonate rocks belonging to the continental margin rather than to a carbonate buildup in an intraoceanic setting. By contrast, the upper (lower Mesozoic) part or the Narvak thrust panel has no carbonate rocks and shows no evidence of having formed near a continental margin.

Geochemical data from basaltic rocks of the Narvak thrust panel are characteristic of intraplate volcanism (Barker and others, 1988; Pallister and others, 1989). Compositions of the major elements indicate that the basalts fall within the tholeiitic to alkalic rock series. Chondrite-normalized rare-earth element (REE) plots range from flat to strongly enriched in light rare-earth elements (Fig. 6A). These basalts lack the strongly depleted concentrations of Nb, Ta, and Ti that are characteristic of subduction-related arc volcanism. Instead they resemble basalts from intraplate seamounts (such as Hawaii) or plume-influenced spreading ridge islands (such as Iceland) (Thompson and others, 1984).

The Kanuti thrust panel is typical of the lower part of an ophiolite succession (Patton and others, this volume). However, its geologic setting and geochemistry indicate that the ophiolites may have formed as the roots of a volcanic arc rather than as the lower part of a crustal section generated along a mid-ocean ridge. No high-level oceanic crustal rocks have been found in this thrust panel, and structurally the next higher sequence exposed within the Yukon-Koyukuk basin is the Jurassic and Lower Cretaceous andesitic volcanic arc assemblage belonging to the Koyukuk terrane (Fig. 5, A-A′, B-B′, C-C′). The chemistry of chromian spinels, which Dick and Bullen (1984) believe is an indicator of the petrogenesis of the peridotites, suggests that the ultramafic cumulates in this thrust panel are not typical of peridotites dredged from modern mid-ocean ridges (Loney and Himmelberg, personal communication, 1985).

Metamorphic amphibole from amphibolite at the sole of the Kanuti thrust panel yields Middle and Late Jurassic K-Ar ages (Patton and others, this volume). These ages suggest that the Kanuti panel was thrust onto the Narvak panel before it collided with the continental margin in Late Jurassic to Early Cretaceous time (Patton and Box, 1985). A small tonalite-trondhjemite pluton of Middle and Late Jurassic age, exposed in the south-central part of the Yukon-Koyukuk basin (Fig. 5, E-E′), may represent an earlier phase of arc volcanism, a phase synchronous with emplacement of the Kanuti thrust panel onto the Narvak thrust panel (Patton, 1984).

We propose the following tectonic model to explain the evolution of the Angayucham-Tozitna terrane (Fig. 7):

1. Middle and Late Jurassic: an intraoceanic arc-trench system formed at some unknown distance from the continental margin. The trench lay between the arc and the continent, and the arc migrated toward the continent. The Kanuti thrust panel was generated within the arc-trench system and was underplated by oceanic seamounts, now constituting the upper part of the Narvak thrust panel, during underthrusting of the oceanic plate.

2. Late Jurassic and Early Cretaceous: oceanic crust between the arc and continent was consumed, and continental underthrusting began. The lower part of the Narvak thrust panel, at the continent-ocean boundary, was underplated beneath the upper part of the Narvak panel. This event was followed successively by underthrusting of sediments from the continental slope and rise, constituting the Slate Creek thrust panel, and finally by the underthrusting of the continental margin constituting the Arctic Alaska and Ruby terranes.

Although most workers agree that the oceanic rocks of the Angayucham-Tozitna terrane are allochthonous with respect to the metamorphic assemblages of the Ruby and Arctic Alaska terranes, not all agree that their emplacement was a product of subduction and terrane accretion. Gemuts and others (1983) argue that the mafic-ultramafic rocks of the Angayucham-Tozitna terrane were generated in local rift basins along the margin of the Yukon-Koyukuk basin and were emplaced on the metamorphic rocks during a later compressional event. Similarly, Dover (this volume) suggests that the mafic-ultramafic rocks were generated within the continental rocks of the Ruby geanticline and southern Brooks Range and were tectonically emplaced by "classical nonaccretionary deformational processes." In our view, however, the models of Gemuts and others and Dover are untenable, because they fail to account for: (1) the wide age range (Devonian to Jurassic) of the mafic-ultramafic rocks, (2) the characteristic reverse stacking order in which mantle and lower crustal mafic-ultramafic rocks overlie higher-level crustal mafic rocks, and (3) the presence of high-pressure blueschist mineral assemblages in the underlying metamorphic rocks of the Ruby and Arctic Alaska terranes.

Koyukuk terrane

Definition and distribution. The Koyukuk terrane, as defined by Jones and others (1987) and Silberling and others (this volume), lies wholly within the Yukon-Koyukuk basin. It is composed dominantly of Upper Jurassic(?) and Lower Cretaceous andesitic volcanic rocks, but south of the Kaltag fault it also includes Middle and Late Jurassic tonalite-trondhjemite plutonic rocks, which intrude an altered complex of mafic and ultramafic rocks of uncertain age and tectonic affinities (Figs. 2 and 3). The Koyukuk terrane is exposed on a broad arch that extends across the north-central part of the Yukon-Koyukuk basin from the Seward Peninsula to the Koyukuk River, and on smaller structural highs within the basin. The best outcrops are in cutbanks along the Koyukuk River in the Hughes and Melozitna Quadrangles and along the lower Yukon River in the Russian Mission Quadrangle (Fig. 1). In the interstream areas, exposures are confined largely to broad, thermally altered zones surrounding mid- and Late Cretaceous granitoid plutons. The Koyukuk terrane is presumed to underlie much of the basin, but the nature and geometry of the contacts with older rocks that rim the basin are generally obscured by overlapping mid-Cretaceous terrigenous sedimentary rocks and younger volcanic rocks (see, for example, Fig. 5, sections A-A′, B-B′, C-C′). It is possible that the Koyukuk terrane stratigraphically overlies the Angayucham-Tozitna terrane and therefore does not, in reality, constitute a separate terrane.

Description. The Koyukuk terrane is composed of two distinctly different assemblages: (1) Middle and Late Jurassic tonalite-trondhjemite plutonic rocks, and a complex of altered mafic volcanic rocks and mafic and ultramafic plutonic rocks of uncertain but probable late Paleozoic or early Mesozoic age; and (2) Upper Jurassic(?) and Lower Cretaceous andesitic volcaniclastic rocks and flows (Fig. 2). The andesitic volcanic and volcaniclastic rocks unconformably overlie the tonalite-trondhjemite plutonic rocks and mafic-ultramafic complex in the Unalakleet and Holy Cross Quadrangles and compose all of the exposed Koyukuk terrane elsewhere in the Yukon-Koyukuk basin.

The Middle and Late Jurassic tonalite-trondhjemite plutons occur in a linear fault-bounded complex extending along the Chiroskey fault from the Kaltag fault in the northern Unalakleet Quadrangle southward to the central part of the Holy Cross Quadrangle (Fig. 3; Fig. 5, section E-E′). These plutonic rocks are extensively sheared and fractured and are locally altered to pink granite by potassic metasomatism. K-Ar mineral ages range from 173 ± 9 to 154 ± 6 Ma. A representative spidergram of Middle Jurassic tonalite (Fig. 6B, sample 5) exhibits the characteristic signature of subduction-related magmatism, i.e., enrichment of LIL (large ion lithophile) elements and depletion of Nb-Ta relative to the light rare-earth elements (REE). Clasts from this plutonic suite occur in overlying Lower Cretaceous volcaniclastic rocks in the Unalakleet Quadrangle (Patton and Moll, 1985). In the Unalakleet Quadrangle, these plutonic rocks are faulted against, and locally appear to intrude, a poorly exposed assemblage of pillow basalt, diabase, gabbro, and serpentinized ultramafic rocks of uncertain tectonic affinities (Patton and Moll, 1985; unpublished data, 1986). Basalt from this assemblage shows a trace-element signature that is similar to that of basalts from the Angayucham-Tozitna terrane and is unlike that of basalts from the Upper Jurassic(?) and Lower Cretaceous part of the Koyukuk terrane. The trace-element characteristics of these pre–Middle Jurassic basalts are similar to "enriched" or "transitional" mid-ocean ridge basalts (E-MORB or T-MORB), to tholeiitic basalts from some ocean islands (tholeiitic OIB), and to some back-arc basalts (BABB) (Thompson and others, 1984; Saunders and Tarney, 1984). These basalts lack depletion of Nb and Ta relative to the light REE (such as La)—the primary diagnostic characteristic of volcanic arc magmas.

The Upper Jurassic(?) and Lower Cretaceous volcaniclastic rocks and flows, which make up the bulk of the exposed Koyukuk terrane, can be divided into a lower unit of Late Jurassic(?) and early Neocomian age, and an upper unit of late Neocomian and Aptian age. The lower unit is the most extensive regionally and records an episode of voluminous andesitic magmatism. All known fossils are early Neocomian (Berriasian and Valanginian) in age, but the lower part of the unit may be as old as Late Jurassic. Lava flows range from basalt to dacite and are characterized by abundant plagioclase and subordinate amounts of pyroxene and/or hornblende phenocrysts. Pyroclastic and epiclastic volcanic rocks are predominant. They include angular tuff breccias, pillow breccias, lapilli tuffs, crystal tuffs, lithic tuffs, hyaloclastites, and shallow- to deep-marine conglomerates, sandstones, shales, and tuffaceous cherts. Regional stratigraphic correlation is difficult because of the lack of distinct marker horizons. Widely scattered, fossiliferous coquinoid limestones of Valanginian age indicate widespread shallow-marine deposition at that time.

The geochemistry of the volcanic rocks of the lower unit is extremely variable (Fig. 6B, samples 2 to 4), and ranges from arc tholeiite to shoshonite. All rocks of this unit exhibit the chemical characteristics of subduction-related magmatism (such as enriched LIL elements and depleted Nb-Ta relative to the light REE). Chondrite-normalized REE patterns vary from flat (Fig. 6B, sample 4) to extremely enriched in light REE (Fig. 6B, sample 2). Although relative age control for samples within this unit is generally poor, samples of known Valanginian age are all extremely light REE–enriched (i.e., La content is more than 300 times the chondritic value). This suggests that the youngest samples of the lower unit of the Upper Jurassic(?) and Lower Cretaceous assemblage are the most enriched in LIL elements and in the light REE. However, neither the phenocryst mineralogy (plag±cpx±hbl) nor incompatible-element ratios (e.g., La/Th, K/Rb, Zr/Sm) vary significantly within this unit, which suggests

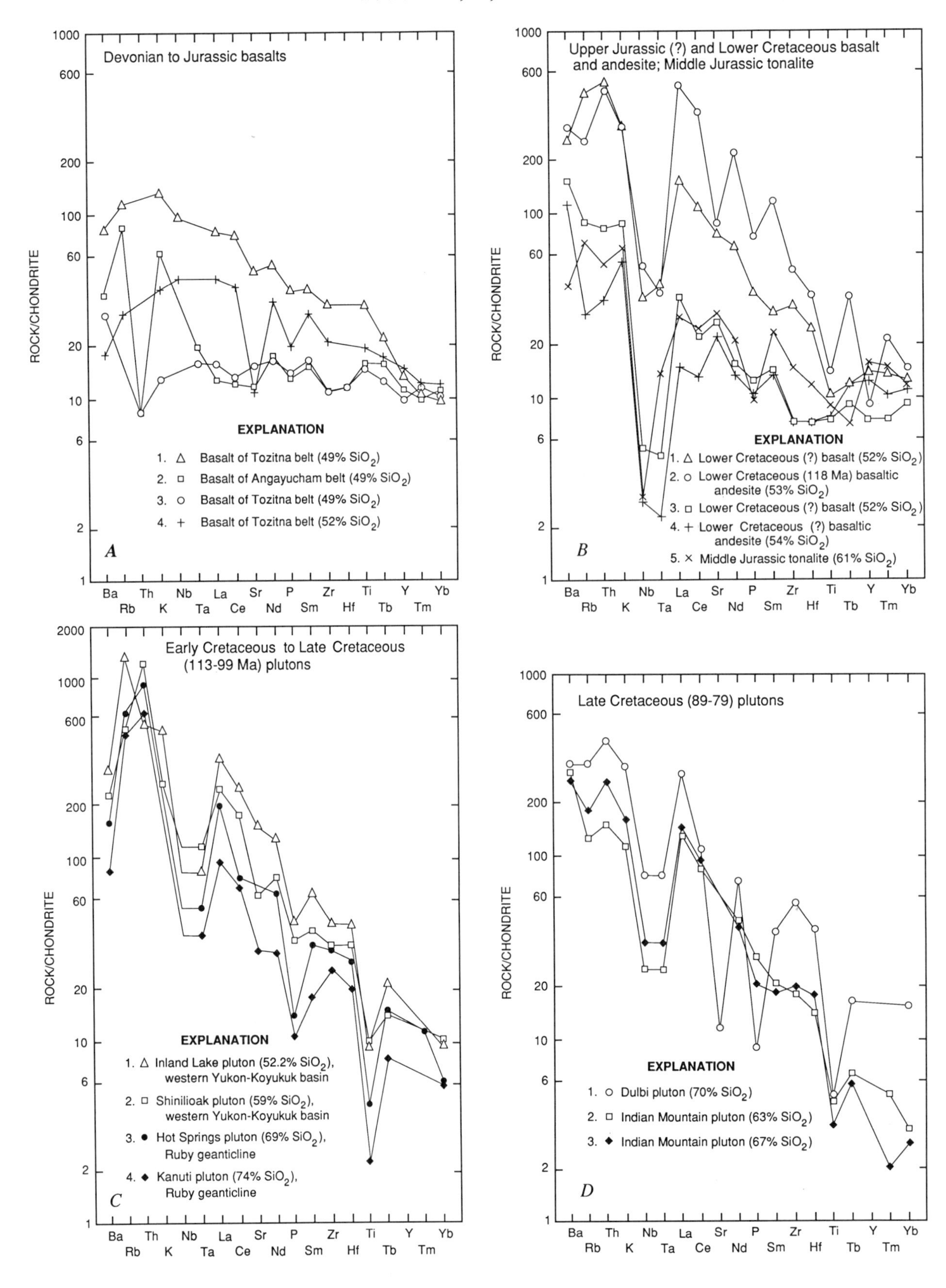
Devonian to Jurassic basalts
ROCK/CHONDRITE
EXPLANATION
1. △ Basalt of Tozitna belt (49% SiO2)
2. □ Basalt of Angayucham belt (49% SiO2)
3. ○ Basalt of Tozitna belt (49% SiO2)
4. + Basalt of Tozitna belt (52% SiO2)
A
Ba Rb Th K Nb Ta La Ce Sr Nd P Sm Zr Hf Ti Tb Y Tm Yb
Upper Jurassic (?) and Lower Cretaceous basalt and andesite; Middle Jurassic tonalite
EXPLANATION
1. △ Lower Cretaceous (?) basalt (52% SiO2)
2. ○ Lower Cretaceous (118 Ma) basaltic andesite (53% SiO2)
3. □ Lower Cretaceous (?) basalt (52% SiO2)
4. + Lower Cretaceous (?) basaltic andesite (54% SiO2)
5. × Middle Jurassic tonalite (61% SiO2)
B
Early Cretaceous to Late Cretaceous (113-99 Ma) plutons
EXPLANATION
1. △ Inland Lake pluton (52.2% SiO2), western Yukon-Koyukuk basin
2. □ Shinilioak pluton (59% SiO2), western Yukon-Koyukuk basin
3. ● Hot Springs pluton (69% SiO2), Ruby geanticline
4. ◆ Kanuti pluton (74% SiO2), Ruby geanticline
C
Late Cretaceous (89-79) plutons
EXPLANATION
1. ○ Dulbi pluton (70% SiO2)
2. □ Indian Mountain pluton (63% SiO2)
3. ◆ Indian Mountain pluton (67% SiO2)
D

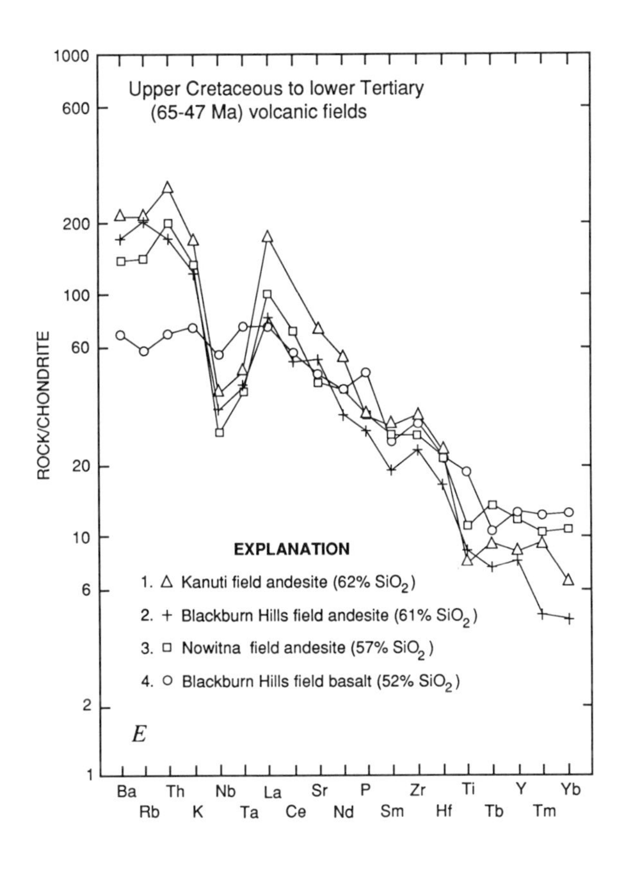

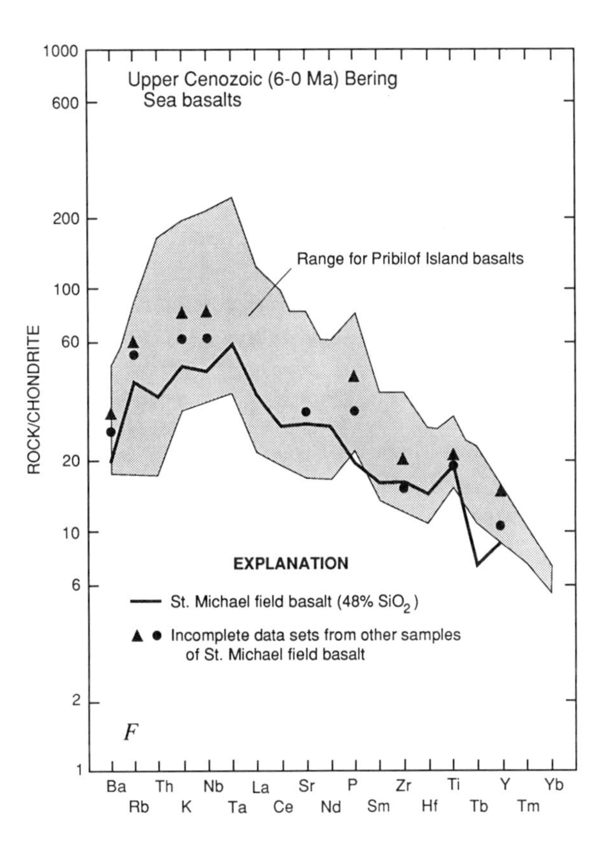

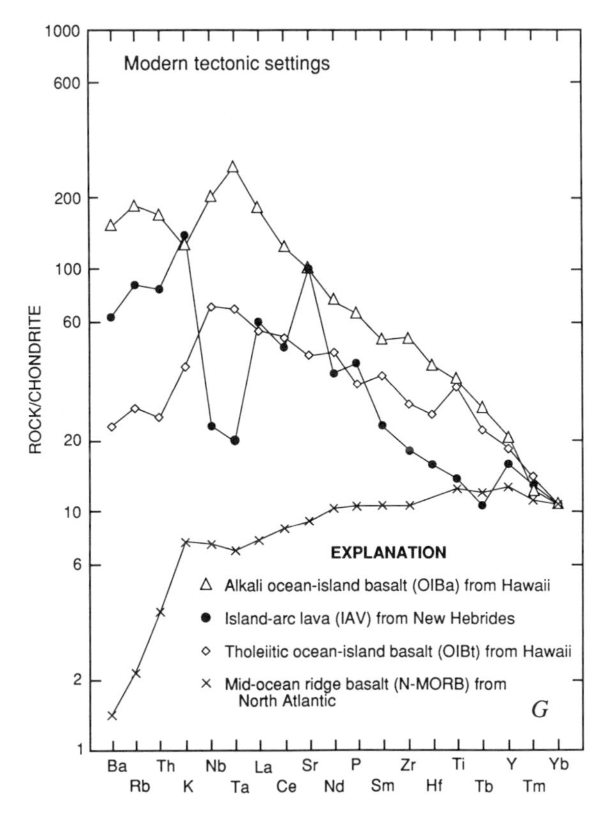

Figure 6. Chondrite-normalized extended rare-earth element diagrams (spidergrams using normalization factor from Thompson and others, 1984) for igneous rocks in western Alaska. *A,* Angayucham-Tozitna (composite) terrane (Box and Patton, unpublished data, 1987); *B,* Koyukuk terrane (Box and Patton, 1989); *C,* Ruby geanticline and western Yukon-Koyukuk basin (T. P. Miller and J. G. Arth, written communication, 1987); *D,* eastern Yukon-Koyukuk basin (T. P. Miller and J. G. Arth, written communication, 1987); *E,* Late Cretaceous-early Tertiary volcanic fields (Moll-Stalcup, this volume); *F,* upper Cenozoic Bering Sea basalts; Pribilof Island basalts from Lee-Wong and others (1979) and F. Lee-Wong (written communication, 1987); St. Michael basalts from E. J. Moll-Stalcup (unpublished data, 1987); *G,* Examples of chondrite-normalized diagrams from modern tectonic settings (Thompson and others, 1984).

that the late trace-element enrichment is due to decreasing degrees of partial melting of the same mantle source (Box and Patton, 1989).

The upper unit of the Upper Jurassic(?) and Lower Cretaceous assemblage is late Neocomian and Aptian in age and is gradational with the lower unit. It has been identified in the Hughes (Patton and Miller, 1966; Box and others, 1985), Unalakleet (Patton and Moll, 1985), Norton Bay (Patton and Bickel, 1956b) and Candle (Patton, 1967) Quadrangles. In the Hughes Quadrangle, where best studied (Box and others, 1985), the age ranges from Hauterivian to Aptian, based on mega- and microfossil identifications (J. W. Miller, written communication, 1983; N. Albert, written communication, 1985). K-Ar ages from the upper unit in the Unalakleet and Candle Quadrangles are 118 ± 3.5 and 124 ± 3 Ma, respectively (Patton and Moll, 1985; Patton, 1967). The upper unit has distinct sedimentologic, petrologic, and geochemical features that distinguish it from the lower unit. In each area the upper unit was deposited in a subsiding sedimentary environment. In the Norton Bay Quadrangle this volcanic and volcaniclastic unit was deposited in the transition between nonmarine and shallow-marine environments. In the Hughes and Unalakleet Quadrangles, the upper unit was deposited in a sub-wave-base (slope?) setting in the vertical transition between shallow-marine and basinal deep-marine environments. Lava flows are typically andesites but differ from the lower unit in containing sanidine, biotite, and/or melanite garnet phenocrysts.

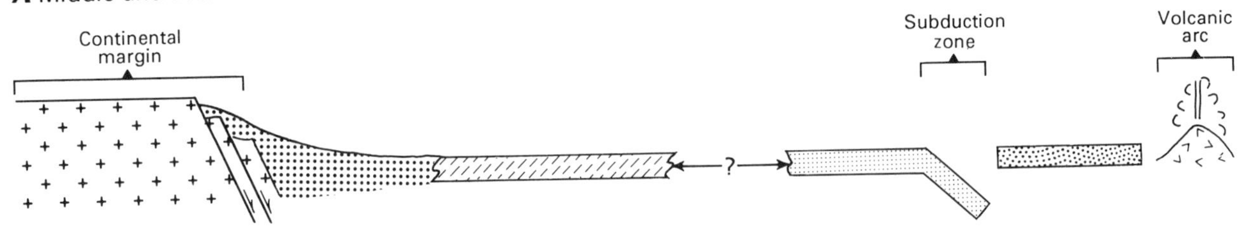

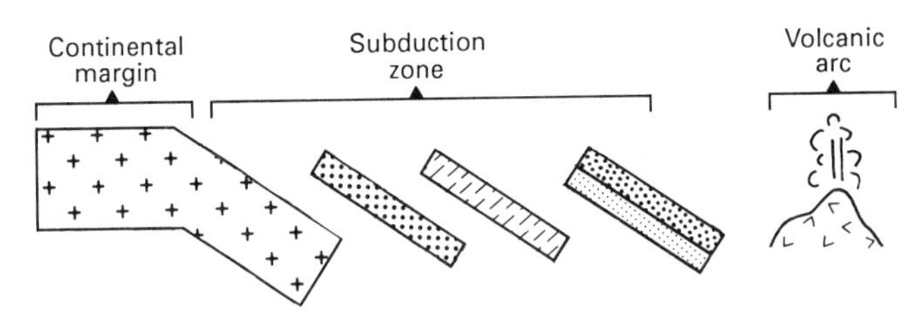

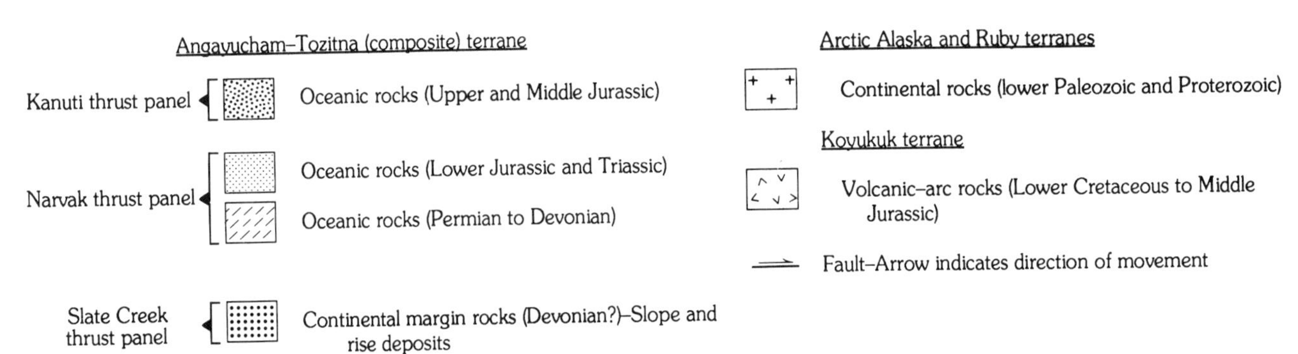

Figure 7. Tectonic model for evolution of the (composite) Angayucham-Tozitna terrane. *A,* In Middle and Late Jurassic time the upper part of the Narvak thrust panel of the Angayucham-Tozitna terrane is subducted beneath the Kanuti thrust panel at an interoceanic arc-trench system an unknown distance from the continental margin. *B,* By latest Jurassic and Early Cretaceous time the continental margin becomes involved in the arc-trench system with successive subduction of the lower part of the Narvak panel, the Slate Creek thrust panel, and finally the continental rocks of the Arctic Alaska and Ruby terranes.

Geochemically, they are similar to the Valanginian part of the lower unit in being extremely enriched in light REE (Fig. 6B, sample 2). However, they differ from flows of the lower unit in certain incompatible-element ratios (e.g., La/Th, K/Rb, Zr/Sm), which implies that they could not have been derived from the same source as the earlier flows (Box and Patton, 1989).

Interpretation. The Koyukuk terrane is believed to be a Mesozoic, intraoceanic, volcanic-arc complex that collided with continental North America in Late Jurassic to Early Cretaceous time (Roeder and Mull, 1978; Gealey, 1980; Box, 1985; Fig. 7). Both the Jurassic plutonic rocks and the Upper Jurassic(?) and Lower Cretaceous volcanic rocks of the Koyukuk terrane have the geochemical signature of subduction-related magmatism. Arc magmatic activity extended from 173 to 115 Ma, with a possible hiatus in the Late Jurassic.

Three lines of evidence suggest that the Koyukuk terrane developed in an intraoceanic setting. First, the lack of exposures of older continental basement rock within the terrane, or of continental detritus in clastic sections within the terrane, suggests the lack of continental basement. Second, the lower crustal and upper mantle part of an ophiolite sequence (namely, the Kanuti thrust panel of the Angayucham-Tozitna terrane) dips beneath the northern and southeastern flanks of the Koyukuk terrane. This sequence may constitute the basement for part or all of the Koyukuk terrane. And third, Sr, Nd, Pb, and O isotopic data from Upper Cretaceous plutons in the northeastern part of the terrane and from lower Tertiary volcanic rocks in the eastern part of the terrane indicate the lack of Paleozoic or older continental crustal contamination (Arth and others, 1984; Moll and Arth, 1985).

The position of the surface trace of the subduction zone (i.e., the trench) relative to the arc is suggested by several regional geologic features. The Angayucham-Tozitna terrane dips beneath the northern and southeastern flanks of the Koyukuk terrane and includes garnet amphibolites that record Middle to Late Jurassic oceanic thrusting. This suggests that the Middle to Late Jurassic intraoceanic subduction zone lay to the north and southeast of contemporaneous arc magmatism in the Koyukuk terrane, as related to present coordinates and to the present configuration of the Yukon-Koyukuk basin. Likewise, Early Cretaceous structural and metamorphic features of the Arctic Alaska and Ruby terranes (i.e., the outward-directed thrust belt and retrograded blueschist-facies metamorphism) imply partial underthrusting of the North American margin beneath the Koyukuk terrane from the north and southeast. Stratigraphic evidence from the western Brooks Range (Mayfield and others, 1983) suggests that continental underthrusting beneath the Koyukuk terrane began at the end of the Jurassic or the beginning of the Cretaceous (~144 Ma).

Continental underthrusting of the arc is apparently reflected in the Early Cretaceous volcanic geochemistry of the Koyukuk terrane. The flows of probable pre-Valanginian (i.e., pre-138 Ma) age are arc-tholeiitic and calc-alkaline rocks having moderate LIL and light REE enrichment typical of an intraoceanic volcanic arc. The flows of Valanginian (138 to 131 Ma) age are highly enriched in LIL and in light REE, and apparently were derived from the same source as the older flows but at sharply decreased degrees of partial melting. Presumably, this reflects decreasing convergence rates due to the difficulties of subducting continental crust. The Hauterivian to Aptian flows (130 to 115 Ma) are also highly enriched, but their different incompatible-element ratios suggest a change in the composition of the mantle source.

OVERLAP ASSEMBLAGES

Major convergent motion between the lithotectonic terranes of western Alaska ceased in Early Cretaceous time. In mid- and Late Cretaceous time these terranes were eroded and partly covered by terrigenous clastic rocks (see Decker and others, this volume, for a discussion of a similar relation in southwestern Alaska). Subsequently these mid- and Upper Cretaceous clastic rocks were themselves strongly deformed and displaced by large-scale strike-slip faulting. Both the lithotectonic terranes and the overlapping clastic rocks were subjected to several widespread magmatic events between mid-Cretaceous and Quaternary time.

The following section describes these overlap assemblages, which include: mid- and Upper Cretaceous terrigenous sedimentary rocks, mid- and Late Cretaceous plutonic rocks, Upper Cretaceous and lower Tertiary volcanic and plutonic rocks, and upper Tertiary and Quaternary flood basalts (Fig. 2).

Mid- and Upper Cretaceous terrigenous sedimentary rocks

Distribution. Mid- and Upper Cretaceous terrigenous clastic rocks underlie a large part of the Yukon-Koyukuk basin in stratigraphic sections as much as 5 to 8 km thick. They may be grossly subdivided into a lower flyschoid assemblage composed of graywacke and mudstone turbidites, and an upper molassoid assemblage of fluvial and shallow-marine conglomerate, sandstone, and shale.

Graywacke and mudstone turbidites. The turbidite flyschoid assemblage was deposited in two vaguely defined subbasins: the Lower Yukon, which extends in a broad band along the west edge of the basin from the latitude of Kotzebue Sound southward to the Yukon delta; and the Kobuk-Koyukuk, which occupies a V-shaped area along the northern and southeastern margins of the basin (Figs. 8 and 9). The two subbasins are separated by the Koyukuk terrane, a remnant volcanic arc that trends eastward from Kotzebue Sound to the Koyukuk River and then southward beneath the Koyukuk Flats (Wahrhaftig, this volume). Near the Kaltag fault the Koyukuk terrane narrows, and the two subbasins converge. South of the fault, the subbasins are offset 100 to 160 km to the southwest and are separated by only a narrow fault-bounded slice of the Koyukuk terrane (Fig. 3, section E-E′). We believe that the Yukon-Koyukuk basin and the two subbasins owe their present configuration in large measure to a strong east-west compressional event that occurred in western Alaska and the Bering Strait region in Late Cretaceous and early Tertiary time (Fig. 9; Patton and Tailleur, 1977; Tailleur, 1980).

The turbidite assemblage is sparsely exposed over about 60 percent of the Kobuk-Koyukuk subbasin and over about 50 percent of the Lower Yukon subbasin (Fig. 8). It includes midfan channel, midfan lobe, and outer fan to basin plain facies associations (Mutti and Ricci Lucchi, 1972), but lack of biostratigraphic control and poor exposures preclude delineation of these facies through time. Graywacke compositions are dominated by volcanic lithic fragments derived from the Angayucham-Tozitna (oceanic) terrane that rims the basin and from the Koyukuk volcanic arc terrane within the basin. Metamorphic debris from the Arctic Alaska, Seward, and Ruby terranes is present in variable but subordinate amounts.

In the Kobuk-Koyukuk subbasin, widely dispersed fossils in the turbidite assemblage all appear to be Albian in age. However, in the western Bettles Quadrangle, about 4 km of stratigraphic section are exposed beneath an early(?) Albian fossil occurrence (Box and others, 1985). The lower kilometer of this section has interbedded potassium-feldspar–bearing tuffs that are lithologically identical to, and presumably correlative with, tuffs of Barremian to Aptian age 70 km to the southwest in the Hughes Quadrangle. There, the possibly correlative Barremian to Aptian tuffaceous section is overlain by turbidites of early Albian age. This variation in the age of the base of the turbidites and the lack of conglomeratic facies bordering the Koyukuk terrane suggest that the Koyukuk terrane was progressively onlapped by turbiditic basin-fill sediments derived primarily from the northern and southeastern subbasin margins. The age of these turbidites is confined to the Barremian to Albian interval, although the age of the base of the turbidite assemblage in the central part of the Kobuk-Koyukuk subbasin is not known.

The turbidite assemblage in the Lower Yukon subbasin consists of a central belt of noncalcareous turbidites flanked both east and west by belts of calcareous turbidites. Field and aeromagnetic data suggest that these graywacke and mudstone beds have an aggregate thickness of more than 6,500 m along the Yukon River–Norton Sound divide (Gates and others, 1968). The central belt is composed of volcanic lithic graywackes with a very minor component of metamorphic detritus. It is divided by a fault into a western half of fine-grained turbidites with abundant carbonaceous material, and an eastern half of medium- to coarse-

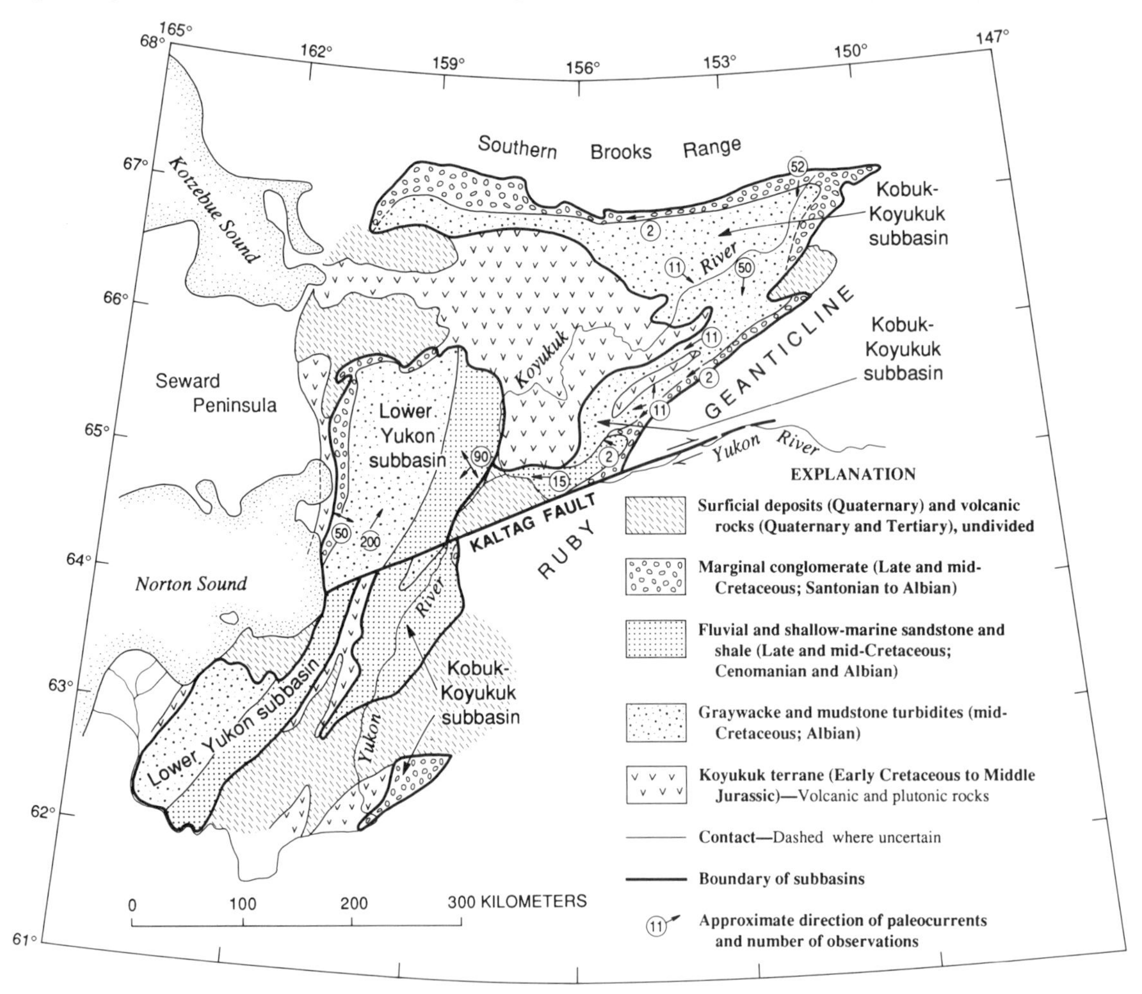

Figure 8. Map of Yukon-Koyukuk basin showing distribution of mid- and Upper Cretaceous terrigenous sedimentary rocks and of Middle Jurassic to Early Cretaceous Koyukuk terrane. Paleocurrent arrows indicate principal direction of sediment transport.

grained turbidites that have been partially replaced by secondary laumontite (Hoare and others, 1964). Paleocurrent directions are consistently to the northeast (Fig. 8). The flanking calcareous belts are significantly richer in metamorphic rock detritus and have a carbonate cement. They are typically fine-grained, and locally show reworking above storm base. Paleocurrent directions in these calcareous belts are scattered around the compass. Fossils in this Lower Yukon subbasin turbidite assemblage are confined to a few scattered ammonites of probable Albian age in the eastern calcareous belt.

Fluvial and shallow-marine conglomerate, sandstone, and shale. The molassoid assemblage of shallow-marine and nonmarine sedimentary rocks is composed of: (1) marginal conglomerates that rim the basin, and (2) deltaic deposits that extend from the southeastern margin across the southeastern limb of the Kobuk-Koyukuk subbasin and the east half of the Lower Yukon subbasin (Fig. 8).

Marginal conglomerates. Polymictic conglomerate and sandstone eroded from the borderlands rim the basin on all three sides in sections estimated to be locally as much as 2,500 m thick. On the borderland side they unconformably overlie the Angayucham-Tozitna and Koyukuk terranes, and on the basinward side they rest on, and in part may be laterally gradational with, the graywacke and mudstone turbidites. Along the northern and southeastern margins of the basin, the conglomerates are composed in the lower part of mafic rocks and chert derived from the Narvak and Kanuti thrust panels of the Angayucham-Tozitna terrane, and in the upper part of quartz and metamorphic rock clasts derived from the Arctic Alaska and Ruby terranes. Locally, an intermediate sequence consisting largely of metagraywacke clasts derived from the Slate Creek thrust panel of the Angayucham-Tozitna terrane occurs between the mafic-rich and quartz-rich conglomerates (Dillon and Smiley, 1984). This compositional progression in vertical section is thought to reflect erosional unroofing of the Arctic Alaska and Ruby terranes following overthrusting of the Angayucham-Tozitna terrane.

The mafic-rich conglomerates are restricted in outcrop to within 10 to 20 km of the northern and southeastern margins of the Kobuk-Koyukuk subbasin. They rest depositionally on Angayucham-Tozitna terrane and grade upward into the quartz-rich conglomerates. The mafic-rich conglomerates, which have an aggregate thickness of as much as 1,500 m, appear to have been deposited primarily in nonmarine alluvial fan and braided-stream environments. However, marine fossils recovered from these conglomerates in the Selawik Quadrangle suggest that some of these strata were deposited in a nearshore marine environment (Patton and Miller, 1968). Paleocurrents are generally directed either away from or parallel to the subbasin margins (Fig. 8). Sparse fossils indicate that these conglomerates are late Early Cretaceous (Albian) in age.

The quartz-rich conglomerates form a nearly continuous band along the northern and southeastern margins of the basin, where they rest depositionally on the mafic-rich conglomerates and overlap onto the Angayucham-Tozitna terrane. The quartz-rich conglomerates, which also include varying amounts of quartz sandstone, shale, and thin bituminous coal beds, have an aggregate thickness of as much as 1,000 m. They range in age from earliest Late Cretaceous (Cenomanian) to middle Late Cretaceous (Santonian) (Patton, 1973).

Marginal conglomerates also crop out along the western edge of the Yukon-Koyukuk basin (Fig. 8), where they are divisible into two stratigraphic units: a lower nonmarine unit composed of andesitic volcanic rocks derived primarily from the Koyukuk terrane, and an upper shallow-marine unit composed of carbonate-cemented volcanic and metamorphic rock clasts derived from both the Koyukuk and Seward terranes. The lower unit, which is in depositional contact with the Koyukuk terrane, consists of poorly sorted conglomeratic strata deposited on eastward-prograding alluvial fans. In the Candle Quadrangle these conglomerates contain granitic boulders petrographically similar to nearby granitic plutons dated at about 105 Ma (Patton,

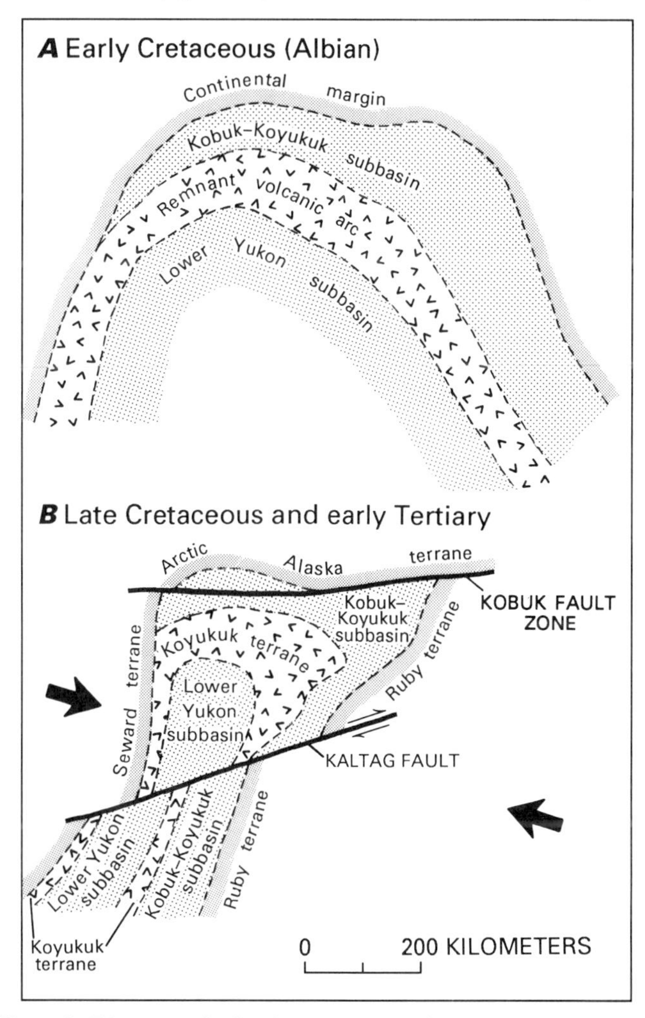

Figure 9. Diagrammatic sketches showing possible evolution of Kobuk-Koyukuk and Lower Yukon subbasins. *A,* Hypothetical configuration of subbasins during deposition of graywacke and mudstone turbidites in Early Cretaceous (Albian) time. *B,* Central and southern part of Yukon-Koyukuk basin strongly compressed (heavy arrows) and offset by Kaltag fault during Late Cretaceous and early Tertiary time.

1967; Miller and others, 1966), indicating that the conglomerates are no older than middle Albian. The upper unit was deposited in a nearshore environment, the currents being directed away from and parallel to the west subbasin margin (Nilsen and Patton, 1984). Sparse palynomorphs and foraminifers suggest that this unit is also Albian in age.

Fluvial and shallow-marine sandstone and shale. A westward-prograding assemblage of deltaic deposits, which includes quartzose sandstone, shale, and thin seams of bituminous coal, forms a broad belt along the southeast side of the Yukon-Koyukuk basin from the Melozitna Quadrangle southward to the Yukon delta (Fig. 8). This belt extends westward from the basin margin across the southeastern arm of the Kobuk-Koyukuk subbasin and the east third of the Lower Yukon subbasin. At the Kaltag fault the belt is offset from 100 to 160 km right laterally. The deltaic deposits rest on graywacke-mudstone turbidites in the subbasins and they grade laterally into the upper quartz-rich marginal conglomerates along the southeastern edge of the Yukon-Koyukuk basin. Along the narrow belt of Koyukuk terrane that separates the two subbasins, the graywacke-mudstone turbidite section thins or is missing, and locally the deltaic deposits rest directly on volcanic rocks of the Koyukuk terrane. The deltaic deposits grade from nonmarine debris-flow and braided-stream deposits on the east, through delta-plain meandering-stream deposits, to delta-front deposits on the west (Nilsen and Patton, 1984). The delta-front deposits are faulted against the graywacke-mudstone turbidites along their west edge. The deltaic deposits had their source in the Ruby terrane and are distinguished from the underlying graywacke-mudstone turbidites by better sorting and by the predominance of clasts of quartz and metamorphic rock over volcanic rock. Paleocurrent indicators show a wide array of sediment-transport directions from northwestward to southeastward.

The fluvial and shallow-marine deltaic deposits have been studied in greatest detail along the Yukon and lower Koyukuk River in the Nulato, Kateel River, and Ruby Quadrangles (Martin, 1926; Hollick, 1930; Patton and Bickel, 1956a; Patton, 1966; Nilsen and Patton, 1984; Harris, 1985). In this area they have an estimated thickness of 3,000 to 3,500 m and grade upward from shallow-marine beds with abundant mollusks of Albian age into nonmarine beds with plant fossils of Cenomanian and Turonian(?) age (R. A. Spicer, oral communication, 1987). To the south, in the Unalakleet Quadrangle, mollusks as young as Cenomanian have been found in the marine strata, suggesting that the contact with the overlying nonmarine beds is diachronous (Patton and Moll, 1985).

Interpretation. We believe that the Yukon-Koyukuk basin originated in early Albian or possibly Aptian time as two subbasins separated by a remnant volcanic arc (Koyukuk terrane), possibly in the configuration shown in Figure 9A. During early and middle Albian time, the subbasins were filled with voluminous flyschoid deposits derived from the continental margin and from the remnant arc. In late Albian and early Late Cretaceous time, a broad prograding-delta complex composed of sediments derived from the continental margin was built out across the southeastern limb of the Kobuk-Koyukuk subbasin, the adjoining part of the remnant arc, and the eastern part of the Lower Yukon subbasin. At approximately the same time, alluvial fans and narrow marine shelves composed of coarse detritus from the borderlands were built along the north and west sides of the basin. In Late Cretaceous to early Tertiary time, an east-west compressional event in western Alaska, probably related to convergence between the North American and Eurasian lithospheric plates, sharply constricted the central and southern parts of the Yukon-Koyukuk basin and offset the basin 100 to 160 km along the Kaltag fault (Fig. 9B; Patton and Tailleur, 1977). The foreshortening in the basin is reflected in the isoclinal folding of the graywacke-mudstone turbidites in the central part of the Lower Yukon subbasin (Fig. 5, C-C′) and in the juxtaposition of dissimilar sedimentary facies, probably by large-scale thrust faulting (Nilsen and Patton, 1984). Juxtaposition of divergent regional trends along the western margin of the basin suggests that large-scale thrusting also may have occurred there. The present structure of the basin and its margins is dominated by high-angle normal faults of late Cenozoic age, which tend to obscure the earlier pattern of low-angle thrust faults.

Mid- and Late Cretaceous plutonic rocks

Major plutonism in west-central Alaska took place from 113 to 99 Ma (late Early Cretaceous) with a volumetrically lesser episode from 79 to 89 Ma (Late Cretaceous). These plutonic rocks underlie about 5 percent of the region and can be grouped into three distinct suites on the basis of their composition, age, and distribution (Fig. 2); each suite apparently was derived from source material of different composition. Most of the variation within suites can be explained by fractional crystallization, variation in crustal melting, or local crustal contamination.

The largest of these suites intrudes the Ruby and Angayucham-Tozitna terranes along the Ruby geanticline. The remaining two suites occur entirely within the Yukon-Koyukuk basin and intrude both the Koyukuk terrane and the overlapping assemblage of mid- and Upper Cretaceous terrigenous sedimentary rocks. All three suites lack a metamorphic fabric and major deformation.

Ruby geanticline plutons. *Distribution and age.* A large suite of plutonic rocks in west-central Alaska forms a major part of the northeast-trending Ruby geanticline (Fig. 4) and is exposed over 8,000 km^2, primarily in the Ruby terrane. The plutonic rocks in this region are concentrated in the 400-km-long segment of the Ruby terrane north of the Kaltag fault (Figs. 3 and 4), where they are exposed over 40 percent of the terrane, as opposed to less than 1 percent south of the fault (150 km^2). North of the Kaltag fault, they are probably the single most voluminous rock type and constitute one of the major Mesozoic batholithic complexes of interior Alaska—second only in area to the plutons of the Yukon-Tanana Upland. The batholith, which consists of individual plutons surrounded by narrow thermal aureoles developed in regionally metamorphosed pelitic country rocks, covers about twice as much area as the other two suites of Cretaceous plutons in west-central Alaska. The plutons are somewhat elongated in an east-west direction, oblique to the northeast-striking trend of the Ruby geanticline.

The plutonic rocks intrude Precambrian(?) and Paleozoic crystalline rocks of the Ruby terrane and the Devonian to Jurassic rocks of the Angayucham-Tozitna terrane, and are therefore no older than Jurassic. Radiometric age data constrain the age of the Ruby geanticline plutons to the mid-Cretaceous. K-Ar ages, chiefly on biotite, range from 112 to 99 Ma (Miller, 1985; Miller, this volume). A mid-Cretaceous age for these plutonic rocks is corroborated by U/Pb zircon ages of 112 to 109 Ma from the Ray Mountains pluton in the Tanana Quadrangle (Patton and others, 1987) and a Rb-Sr age of 112 Ma (Blum and others, 1987) from the Jim River pluton in the Bettles Quadrangle.

Description. The Ruby geanticline plutons are composed chiefly (~80 percent) of leucocratic biotite granite and have lesser amounts of granodiorite and muscovite-biotite granite (Fig. 10); syenite and monzonite are rare but do occur at the extreme north end of the Ruby geanticline (Blum and others, 1987). The granites are typically coarse-grained, strongly porphyritic, and nonfoliated and form large plutons as much as 800 km^2 in area. Mineralogically, they are characterized by an abundance of quartz and K-feldspar and lesser amounts of albite and biotite; hornblende is rare. Primary muscovite occurs in some phases of the large southern plutons, but modal cordierite is lacking. Zircon, apatite, ilmenite, allanite, fluorite, and tourmaline are common accessory minerals. Some, and possibly all, of these plutons are composite bodies, but only the Ray Mountains pluton in the Tanana Quadrangle has been mapped in sufficient detail to document the separate phases (Puchner, 1984).

The plutons are highly evolved and are restricted in composition; the SiO_2 content generally ranges from 68 to 76 percent. They are K-rich, Na-depleted, and weakly to moderately peraluminous; normative corundum is usually greater than 0.8 percent (Miller, this volume). They appear to belong to the ilmenite (magnetite-free) series of granitic rocks (Ishihara, 1981) and show low Fe_2O_3/FeO and reduced magnetic susceptibility. Rb/Sr ratios are high. Initial $^{87}Sr/^{86}Sr$ ratios range from 0.7056 to 0.7294, show large internal variations, and have average values that decrease from southwest to northeast (Arth, 1985; Arth, this volume). Nd initial ratios (NIR) show a reverse relation to the $^{87}Sr/^{86}Sr$ ratios and increase to the northeast from 0.51158 to 0.51240 (Arth, 1985; Arth, this volume). Representative chondrite-normalized, extended rare-earth element diagrams from two Ruby geanticline plutons are shown in Fig. 6C, samples 3 and 4.

Mineral deposits of tin, tungsten, and uranium are associated with these highly evolved plutonic rocks (Nokleberg and others, this volume, both chapters). Such incompatible elements are characteristically associated with these types of granitic rocks.

Interpretation. The Ruby geanticline plutons make up a high-silica, K-rich, weakly to moderately peraluminous, Fe-reduced, compositionally restricted suite of granitoid rocks. Such modal and major-element characteristics are typical of granitic rocks thought to have been generated by melting of continental crust, and typify the S-type granites of Chappel and White (1974). Their high strontium initial ratios (SIR >0.7056) also indicate that significant amounts of Paleozoic or older continental crust were involved in the origin of the plutonic magmas (Arth, 1985). These characteristics are in direct contrast to those of neighboring granitic rocks of the eastern Yukon-Koyukuk basin.

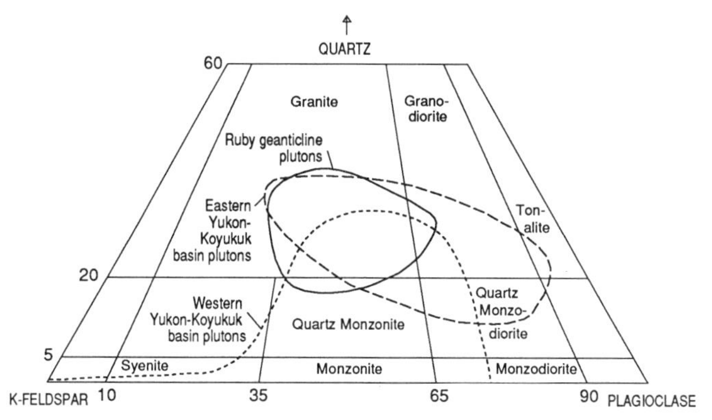

Figure 10. Modal plots of plutonic suites in the Ruby terrane, eastern Yukon-Koyukuk basin, and western Yukon-Koyukuk basin.

Eastern Yukon-Koyukuk basin plutons. *Distribution and age.* Plutons in the eastern Yukon-Koyukuk basin consist of several large bodies in the Hughes and Shungnak Quadrangles, and numerous small stocks in the eastern Melozitna Quadrangle. They intrude both the Lower Cretaceous volcanic rocks of the Koyukuk terrane and the overlap assemblage of mid- and Upper Cretaceous terrigenous sedimentary rocks (e.g., the Indian Mountain pluton shown in Fig. 5, B-B′). They yield Late Cretaceous K-Ar ages of 89 to 79 Ma, about 10 to 20 m.y. younger than the plutons in the adjoining Ruby geanticline and in the western Yukon-Koyukuk basin. Plutons belonging to this suite occur within 15 km of the plutons in the Ruby geanticline.

Description. The plutons in the eastern part of the basin consist of a compositionally expanded suite of tonalite to high-silica granite, the most typical rock type being a granodiorite (Fig. 10); gabbros and quartz diorites are lacking. Individual plutons commonly are compositionally zoned and show gradational internal contacts. The rocks are generally massive, leucocratic, medium-grained, hypidiomorphic, and equigranular, but locally are porphyritic. They are hornblende- and biotite-bearing, and contain abundant sphene, magnetite, apatite, zircon, and allanite. These eastern Yukon-Koyukuk basin rocks generally have SiO_2 contents of 62 to 73 percent, not including a 100-km^2 area of high-silica (76 to 78 percent) granite. They are relatively enriched in Na_2O (>3.2 percent) (Na_2O/K_2O >1) and CaO and have the high Fe_2O_3/FeO ratios of the magnetite series of Ishihara (1981). SIRs for this eastern suite of plutonic rocks range from 0.7038 to 0.7056 and show little internal variation (Arth, 1985; this volume); Rb/Sr ratios are low (Arth and others,

1984). Three representative diagrams of chondrite-normalized, extended rare-earth elements for this eastern suite are shown in Figure 6D, samples 1 through 3.

Coeval volcanic rocks. We believe that silicic volcanic rocks near the Shinilikrok River in the Shungnak Quadrangle (Patton and others, 1984) are coeval with the Late Cretaceous plutons of the eastern Yukon-Koyukuk basin. The volcanic complex, which underlies an area of about 170 km^2, consists of rhyolitic welded tuffs and dacitic flows, tuffs, and hypabyssal intrusive rocks and has yielded a K-Ar biotite age of 87 Ma (Patton and others, 1968). It rests unconformably on two western-basin mid-Cretaceous plutons and has been intruded and thermally metamorphosed by the Wheeler Creek pluton of the Late Cretaceous eastern-basin suite. The spatial distribution, field relations, tightly constrained age, and chemical trends of the volcanic rocks strongly suggest that they are comagmatic with the compositionally similar Late Cretaceous plutons and that the latter intruded their own ejecta.

Interpretation. The plutons of the eastern Yukon-Koyukuk basin constitute a compositionally expanded calc-alkaline magmatic suite ranging from tonalite to granite. This compositional trend, together with the relatively high Na_2O content, the high Na_2O/K_2O ratio, the oxidized state of Fe, the abundance of hornblende in addition to biotite, and the presence of mafic xenoliths are all characteristics similar to those proposed by Chappel and White (1974) for I-type granites, which are granites generated with no continental crustal component. SIRs and NIRs also suggest no involvement of Paleozoic or older crust in the generation of the plutonic magmas (Arth, 1985), but they are compatible with source areas that would include oceanic mantle and Mesozoic supracrustal rocks.

Western Yukon-Koyukuk basin plutons. *Distribution and age.* An east-west belt of 10 plutons and several small stocks, covering an area of about 1,600 km^2, extends from the Shungnak Quadrangle in the Yukon-Koyukuk basin to the margin of the Seward Peninsula in the Candle Quadrangle. The belt continues onto the southeastern Seward Peninsula and St. Lawrence Island (Miller and Bunker, 1976). These plutons intruded the Upper Jurassic(?) and Lower Cretaceous volcanic rocks of the Koyukuk terrane and contributed debris to the mid- and Upper Cretaceous terrigenous sedimentary rocks, suggesting an Albian age that is confirmed by numerous K-Ar ages ranging from 113 to 99 Ma (Miller, 1971, 1972).

Description. This suite (Fig. 10) can be subdivided into two distinct but related series: a potassic series (KS) and an ultrapotassic series (UKS). The KS series is represented by SiO_2-saturated to slightly oversaturated rocks, such as monzonite, syenite, and quartz syenite, that are characterized by low quartz and abundant K-feldspar contents. Hornblende and clinopyroxene are the principal varietal mafic minerals. The KS-series rocks average 4.5 percent K_2O; SiO_2 content ranges from 53.8 to 66.1 percent with a compositional gap to 73.9 percent. They have a lower abundance of incompatible elements and less radiogenic Sr than rocks of the UKS series. Representative chondrite-normalized, extended rare-earth element diagrams for the KS and for UKS series are shown in Figure 6C, sample 2, and in Figure 6C, sample 1, respectively.

The UKS series, which constitutes about 5 percent of the western Yukon-Koyukuk basin plutonic rocks, consists of single-feldspar, hypersolvus nepheline-bearing rock types, including malignite, foyaite, ijolite, biotite pyroxenite, and pseudoleucite porphyry. SiO_2 ranges from 44.5 to 58.4 percent, K_2O is as high as 16.6 percent, and the rocks are nepheline- and commonly leucite-normative (Miller, 1972). The UKS-series rocks define an ultrapotassic rock province consisting of at least 12 intrusive complexes and dike swarms that form a sinuous belt extending some 1,300 km from the western Yukon-Koyukuk basin westward through the southeastern Seward Peninsula and St. Lawrence Island (Csejtey and Patton, 1974) to the east tip of the Chukotsk Peninsula, USSR (Miller, 1972).

Interpretation. The identifying characteristics of the KS and UKS series of the western Yukon-Koyukuk basin plutonic rocks are strong enrichment in K_2O and depletion in SiO_2. Any model used to explain their petrogenesis must account for these features. The occurrence of the granitic plutons along a narrow linear belt in the Koyukuk terrane, despite marked differences in their composition, petrography, and age, suggests that the plutons are petrogenetically related. If, as we believe, the origin of the eastern Yukon-Koyukuk basin plutonic rocks is related to the melting of oceanic mantle and Mesozoic supracrustal rocks, possibly above a subduction zone, then the western Yukon-Koyukuk plutonic rocks may likewise have originated by melting of mantle material. This interpretation is supported by studies suggesting that ultrapotassic magmatic rocks generally originate in the mantle (Miller, 1972, 1985, this volume). The gradual but strong increase in K_2O from east to west in the Yukon-Koyukuk basin, however, and the corresponding increase in SIRs (Arth, this volume), suggest that continental material also was involved in the generation of the plutonic magmas. Such an origin remains compatible with our interpretation that the Koyukuk terrane developed in an intraoceanic setting having no continental basement, providing that the mantle beneath the western Koyukuk terrane is K-enriched subcontinental mantle, as has been suggested for the generation of K-rich magmas elsewhere (Varne, 1985).

Summary. The petrogenesis of the eastern Yukon-Koyukuk basin and the Ruby geanticline plutonic rocks appears relatively straightforward on the basis of mineralogical, chemical, and isotopic characteristics. The eastern Yukon-Koyukuk granitoids are confined to the basin and are typical of plutons derived from oceanic mantle or perhaps from Mesozoic supracrustal volcanic rocks. The Ruby geanticline granites are confined to the Ruby and overthrust Angayucham-Tozitna terranes and are typical of plutons generated in an anatectic continental environment. The boundary between these compositionally distinct granitic rocks is sharp—plutons of each suite are only about 15 km apart in the eastern Melozitna Quadrangle—and it almost certainly coincides with the tectonic boundary that separates the basin from the geanticline.

The western Yukon-Koyukuk basin plutonic rocks are more enigmatic in origin than the other two suites. The boundary between the plutonic rock suites of the western and eastern parts of the basin, for example, is more gradational, particularly in isotopic systematics; and whereas the western Yukon-Koyukuk granitoids also appear to have been derived, at least in part, from the mantle, their mantle source could have been continental rather than oceanic.

Late Cretaceous (Maastrichtian) and early Tertiary (early Eocene) volcanic and plutonic rocks

Distribution. Late Cretaceous and early Tertiary magmatic activity occurred in a vast region of western Alaska, extending from the Arctic Circle to Bristol Bay. Rocks of this age that occur within the area described by this chapter include parts of two northeast-trending belts: the Kuskokwim Mountains belt (Wallace and Engebretson, 1984), which is located in the northern Kuskokwim Mountains and extends from the south side of the Kaltag fault to beyond the southern edge of the area covered by this chapter; and the Yukon-Kanuti belt (Fig. 1 in Moll-Stalcup, this volume), which lies northwest of the Kuskokwim Mountains belt and extends from the Arctic Circle southwest to the Kaltag fault and continues south of the fault on the west side of the Yukon River. There is no clear boundary between the Kuskokwim Mountains and the Yukon-Kanuti belts in the region south of the Kaltag fault. We divide the Upper Cretaceous and lower Tertiary volcanic rocks into two belts because the rocks from the Yukon-Kanuti belt are younger (65 to 47 Ma) than those in the Kuskokwim Mountains belt (72 to 60 Ma), and because the Yukon-Kanuti belt lies within the Yukon-Koyukuk basin, whereas the Kuskokwim Mountains belt overlies Precambrian(?) and Paleozoic continental rocks of the Ruby, Nixon Fork, and Minchumina terranes. The volcanic fields in both belts are commonly preserved in broad open synclines with dips generally less than 30°. Most of the synclines are fault-bounded on at least one flank.

Description. *Kuskokwim Mountains belt.* The Kuskokwim Mountains belt consists of volcanic fields, volcanoplutonic complexes and small plutons, dikes, and sills, all of which have K-Ar ages between 72 and 60 Ma. The belt occurs south and east of the Yukon River within this study area (Fig. 1 in Moll-Stalcup, this volume) and extends into southwestern Alaska beyond the area covered by this study. The volcanic fields consist chiefly of andesite (Nowitna), dacite, and rhyolite (Dishna and Sischu) or of basalt, andesite, dacite, and rhyolite (Yetna). The Sischu volcanic field also includes at its base a 25-m-thick section of nonmarine conglomerate, sandstone, and lignite containing palynomorphs of latest Cretaceous (Campanian to Maastrichtian) age (Patton and others, 1980). We interpret the volcanoplutonic complexes as eroded volcanic centers, which now consist of circular outcrop areas of andesite flows and shallow hypabyssal rocks, intruded by small granitic stocks. Most of the volcanic rocks in the complexes are highly altered by the intrusions. Small dikes, sills, and stocks—many too small to be shown on published maps—occur throughout the Kuskowim Mountains belt. These dikes, sills, and plugs are compositionally similar to the volcanic rocks, and range from monzodiorite to granite.

Present exposures in the Kuskokwim Mountains belt suggest that andesite, followed by rhyolite, are the overwhelmingly dominant volcanic rock types. Basalt is relatively uncommon and rocks having less than 52 percent SiO_2 are rare. Most of the intrusive rocks are intermediate to felsic in composition. Major element data on volcanic and plutonic rocks show trends typical of calc-alkalic suites: MgO, FeO*, TiO_2, Al_2O_3, and CaO decrease, and K_2O and Na_2O increase, with increasing SiO_2. TiO_2 is low (1.75 percent) and Al_2O_3 is relatively high (12 to 17 percent). None of the suites shows Fe-enrichment. K_2O varies considerably from moderate (1.3 percent at 56 percent SiO_2) to very high values (4 percent at 56 percent SiO_2). Moderate- to high-K suites plot in the subalkaline field on a total alkali versus SiO_2 diagram. Very high-K suites plot in the alkalic field and are classified as shoshonitic (Morrison, 1980). In the northern Kuskokwim Mountains the high-K calc-alkalic and shoshonitic suites tend to be older (71 to 65 Ma) than the moderate-K suites (68 to 62 Ma), although there is considerable overlap.

Igneous rocks of the Kuskokwim Mountains belt are characterized by a high incompatible-element content. In general, K correlates with other incompatible elements (Rb, Ba, Th, Nb, Ra, U, Sr, and LREE), such that the shoshonitic suites are most enriched in incompatible elements and the moderate-K suites are the least enriched. Furthermore, the major- and trace-element data indicate that all of the volcanic and plutonic rocks are highly enriched in K, Rb, Ba, Th, U, and Sr, and depleted in Nb-Ta relative to La—chemical characteristics typical of subduction-related volcanic arc rocks (Fig. 6E; Perfit and others, 1980; Thompson and others, 1984). Trace-element ratios (Ba/Ta, Ba/La, La/Nb) of andesites in the northern Kuskokwim Mountains thus are similar to those in arc andesites (Gill, 1981), although these elements are somewhat more abundant in the high-K calc-alkalic and shoshonitic suites than in typical arc andesites.

Volcanic and plutonic rocks in the Kuskokwim Mountains belt have compositions that suggest they have undergone a significant amount of fractionation, and many show isotopic and geochemical evidence of having interacted with continental crust. In the northern Kuskokwim Mountains—where the basement is Precambrian and Paleozoic carbonate rocks and schist of the Nixon Fork, Minchumina, and Ruby terranes—andesites in the Nowitna volcanic field have initial Sr isotope ratios (SIR) of 0.7045 to 0.7053 and trace-element abundances that suggest that the magmas have assimilated small amounts of continental crust during crystal fractionation (Moll and Arth, 1985). Rhyolites in the Sischu volcanic field have a high Sn, Be, U, W, and F content and SIR greater than 0.7080 (Moll and Arth, 1985; Moll and Patton, 1983), which suggests that they either were contaminated by large amounts of continental crust or were partial melts of the crust.

Yukon-Kanuti belt. Only three areas in the Yukon-Kanuti

belt have been studied in detail (Moll-Stalcup, this volume; Moll-Stalcup and others, this volume): the Kanuti volcanic field in the Bettles Quadrangle, the Yukon River area in the Ophir and Unalakleet Quadrangles, and the Blackburn Hills field in the Unalakleet Quadrangle. Extensive volcanic rocks of probable Tertiary age also occur farther south in the Holy Cross Quadrangle, but no age, petrologic, or geochemical data are available for those rocks.

Available data indicate that the Yukon-Kanuti belt consists chiefly of volcanic rocks varying from basalt to rhyolite, and minor intrusive rocks. The Kanuti field is composed predominantly of dacite, ranging in age from 59 to 56 Ma; the Yukon River rocks consist of basalt, andesite, dacite, and rhyolite, ranging in age from 54 to 48 Ma; and the Blackburn Hills volcanic field consists of basalt, andesite, rhyolite, and granodiorite, ranging in age from 65 to 56 Ma.

Data on the Yukon-Kanuti belt also show that a transition in chemistry and mineralogy occurred at about 56 Ma. Rocks older than 56 Ma are calc-alkalic and show the characteristic geochemical enrichments and depletions of arc rocks. Rocks 56 Ma or younger occur in the Yukon River area and the Blackburn Hills and are a mixed assemblage of calc-alkalic and mildly alkalic suites. The calc-alkalic rocks in the younger suite are chemically and mineralogically similar to those in the older suite. The alkalic rocks have less Nb-Ta depletion, less alkali enrichment, and mildly alkalic mineral assemblages (rhyolites: anorthoclase+ hedenbergite; latites: anorthoclase+plagioclase+ biotite; and basalts: olivine+ plagioclase+clinopyroxene+ biotite). This chemical and mineralogical transition is not strongly reflected in the major-element data, which show a typical calc-alkaline affinity for all the rocks. Basalt, however, is restricted to the post–56 Ma assemblage, and three analyzed basalts have less Nb-Ta depletion and lower alkali/LREE ratios than basalts in typical arcs (Fig. 6E, sample 4) (Moll-Stalcup, this volume).

Comparison of the pre–56 Ma rocks in the Yukon-Kanuti belt with those in the Kuskokwim Mountains belt shows that the two belts are compositionally similar in most respects. Rocks in the Yukon-Kanuti belt are moderate- to high-K calc-alkalic and range in composition from basalt to rhyolite. Major- and trace-element trends for the Yukon-Kanuti belt are similar to the Kuskokwim Mountains belt. However, the Yukon-Kanuti belt has lower K_2O and incompatible-element content than the Kuskokwim Mountains belt. Some of the variation in K and in incompatible-element content may be due to the interaction of the Kuskokwim Mountains magmas with old continental crust of the Ruby, Nixon Fork, and Minchumina terranes. Basalts, andesites, dacites, and rhyolites in the Yukon-Kanuti belt have low SIR (0.7033 to 0.7053) and high NIR (0.51248 to 0.51290), which precludes significant interaction with old continental crust (Moll-Stalcup and Arth, 1989). Some of the variation in K_2O may be tectonically controlled, however, as suggested by the correlation of the K_2O content with age, and by the occurrence of moderate-K, high-K, and shoshonitic rocks in the northern Kuskokwim Mountains overlying old continental crust.

Interpretation and correlation. The widespread calc-alkalic magmatism in western and southern Alaska between 75 and 65 Ma has been interpreted by Moll-Stalcup (this volume) as representing a wide continental arc related to subduction of the Kula or Pacific plate beneath southern Alaska. She further interprets the transition to a mixed assemblage of mildly alkalic and calc-alkalic rocks at 56 Ma as marking the end of subduction-related magmatism in interior Alaska, and the transition to intraplate magmatism. The convergence angle (30°) between the present position of the Late Cretaceous and early Tertiary magmatic province and the plate-motion vector is close to the minimum required for arc magmatism (Wallace and Engebretson, 1984; Gill, 1981). Paleomagnetic data on a number of Late Cretaceous and early Tertiary volcanic fields, including the Nowitna and Blackburn Hills, indicate about 30 to 55° of counterclockwise rotation, but no major latitudinal displacement relative to North America since their formation (Hillhouse and Coe, this volume; Thrupp and Coe, 1986). These data indicate that the magmatic belt may have had a convergence angle of 55 to 80° before the rotation of western Alaska in the Eocene. The age and K_2O data also indicate that the magmatic arc was narrower and the K gradient across the arc steeper between 75 and 65 Ma, and that between 65 and 56 Ma, the arc broadened and the K-gradient was more gradual.

Upper Eocene volcanic rocks

Distribution. Small volumes of volcanic rocks were erupted in western Alaska at about 40 Ma. Occurrences of these volcanic rocks within the area described in this chapter include three volcanic bodies in the Melozitna Quadrangle: (1) a rhyolite field near the Indian River (41.6 and 39.9 Ma; Miller and Lanphere, 1981), (2) a basalt-rhyolite field in the Takhakhdona Hills (43.0 Ma; Harris, 1985), and (3) a rhyolite field near Dulbi Hot Springs (43.0 Ma; Patton and Moll, unpublished data, 1986). A fourth, little-known basalt field occurs at the boundary of the Unalakleet and Holy Cross Quadrangles along the Yukon River (42.7 Ma; Harris, 1985). All of these volcanic fields are situated along the southeastern margin of the Yukon-Koyukuk basin where the basin is relatively shallow.

Description. The upper Eocene volcanic rocks in western Alaska have not been well studied. The Indian River, Takhakhdona Hills, and Dulbi Hot Springs areas are described by Patton and others (1978) as consisting of rhyolite tuff, flows, and breccia and basalt flows. Obsidian occurs at the Indian Mountain locality. Four chemical analyses of basalt, andesite, dacite, and rhyolite obtained from the Takhakhdona Hills (Patton and Moll-Stalcup, unpublished data, 1987) suggest that the basalt is anorogenic and mildly alkalic. The andesite and dacite show "arc-like" enrichments and depletions similar to those in the nearby Late Cretaceous and early Tertiary Yukon-Kanuti belt. The other areas consist of basalt and rhyolite of uncertain affinity and may represent a bimodal suite related to movement along the many faults in western interior Alaska. The upper Eocene volcanic rocks are flat lying or only broadly folded, with dips generally less than 5°.

Tertiary nonmarine coal-bearing deposits

Small deposits of poorly consolidated nonmarine clay shale, sandstone, conglomerate, and lignite that contain palynomorphs ranging in age from Oligocene to Pliocene are scattered along or near the Kaltag fault in the Unalakleet (Patton and Moll, 1985), Norton Bay (Patton, unpublished map, 1987), Melozitna (Patton and others, 1978), and Tanana (J. P. Bradbury, written communication, 1979) Quadrangles. Float of lignite containing spores and pollen of Tertiary age has also been reported along the Tozitna river in the Tanana Quadrangle (Chapman and others, 1982) and on the Mangoak River and near Elephant Point in the Selawik Quadrangle (Patton, 1973). All the deposits appear to be of limited extent and confined to small structural or topographic basins.

Upper Cenozoic basalt

Distribution. Late Cenozoic volcanism was widespread along the westernmost margin of Alaska and on the adjacent Bering Sea shelf. Two large volcanic fields occur within the area covered by this chapter: one along the south shore of Norton Sound in the St. Michael and Unalakleet Quadrangles, and the other a less well-studied field southeast of Kotzebue Sound in the Candle, Selawik, and Kateel River Quadrangles (Figs. 1 and 3). Two small fields of olivine basalt of uncertain, but probable late Tertiary or Quaternary age, also have been mapped on the Ruby geanticline in the southeastern part of the Bettles Quadrangle (Patton and Miller, 1973).

Description. The St. Michael volcanic field, covering about 2,000 km^2, is located along the southern shore of Norton Sound and consists of tholeiite and alkali olivine basalt flows, basanite tuffs, cones, and maar craters. A young cone at Crater Mountain is composed of basanite that contains lherzolite nodules. The base of the volcanic field gives K-Ar ages of 3.25 and 2.80 Ma (D. L. Turner, written communication, 1987). Several steep-sided cones, which lack frost brecciation and lichen cover, are probably Holocene or Pleistocene in age.

The large field southeast of Kotzebue Sound covers about 4,500 km^2 and consists of vesicular olivine basalt flows and cones. Some of the cones contain peridotite nodules. None of the flows is dated, but they are presumed to be late Tertiary or Pleistocene in age and are correlative with the Imuruk Volcanics on the adjacent Seward Peninsula (Patton, 1967).

Chemical data on the St. Michael field and on a number of the volcanic fields located outside this study area (St. Lawrence Island, Nunivak Island, and the Seward Peninsula) indicate that all of the volcanic fields are compositionally similar, ranging from nephelinite through basanite through alkali olivine basalt to olivine tholeiite (Hoare, unpublished data, 1980; Swanson and Turner, unpublished data, 1987; Moll-Stalcup, this volume). The St. Michael field also has quartz tholeiite and hawaiite. Most of the volcanic rocks have Mg numbers (100 Mg/Mg + Fe^{2+}) greater than 65, which implies they are primary or near-primary melts of mantle peridotite. Alkali olivine basalt and tholeiite represent at least 95 percent of the volcanic rocks present in all the volcanic fields, and they form broad shield volcanoes. Basanite and nephelinite compose 2 to 3 percent of the rocks, and eruptions of those magmas both precede and postdate eruptions of the more voluminous, less alkalic basalts (Hoare, unpublished data, 1980). Basanite and nephelinite occur in steep cones, short viscous flows, and tephra deposits emanating from the maar crater. They commonly contain xenoliths of lherzolite, pyroxene granulite, dunite, harzburgite, chromite, gabbro, or Cretaceous sedimentary bedrock, and megacrysts of anorthoclase, clinopyroxene, and kaersutite.

Trace-element data from the St. Michael field (Fig. 6F) indicate that the rocks are LREE-enriched, and that the LREE content increases with alkalinity. All the rocks have positive Nb-Ta anomalies similar to oceanic island basalts (OIB). SIR on the St. Michael field is 0.7027 (Mark, 1971), similar to values for Nunivak Island and the Pribilof Islands, both of which plot in the field where OIB and MORB overlap on $^{143}Nd/^{144}Nd$-$^{87}Sr/^{86}Sr$ diagrams (Von Drach and others, 1986; Roden and others, 1984).

Interpretation and correlation. Bering Sea basalts are strikingly similar to Hawaiian lavas in composition, despite having formed in a different tectonic environment. The widespread Bering Sea basalts are not aligned along a hot-spot trace; rather they seem to be associated with extensional faulting. Young cones are aligned east-west, defining a fracture or fault in the St. Michael field, as well as on Nunivak and St. Lawrence Island. The volcanic fields south of the Selawik Hills are bounded on the north by several east-west–trending faults, and the basalts on the Seward Peninsula are associated with east-west–trending faults. Moll-Stalcup (this volume) believes that the Bering Sea basalts represent intraplate volcanism associated with regional north-south extension in the Bering Sea region in late Cenozoic time.

REFERENCES CITED (* = See Notes Added in Proof)

Angeloni, L. M., and Miller, M. L., 1985, Greenschist facies metamorphic rocks of north-central Iditarod Quadrangle, *in* Bartsch-Winkler, S., ed., U.S. Geological Survey in Alaska; Accomplishments during 1984: U.S. Geological Survey Circular 967, p. 19–21.

*Arth, J. G., 1985, Neodymium and strontium isotopic composition of Cretaceous calc-alkaline plutons of the Yukon–Koyukuk basin, Ruby geanticline, and Seward Peninsula, Alaska [abs.]: EOS Transactions of the American Geophysical Union, v. 66, no. 46, p. 1102.

*Arth, J. G., and 5 others, 1984, Crustal composition beneath the Yukon–Koyukuk basin and Ruby geanticline as reflected in the isotopic composition of Cretaceous plutons: Geological Society of America Abstracts with Programs, v. 16, p. 267.

Barker, F., Jones, D. L., Budahn, J. R., and Coney, P. J., 1988, Ocean plateau-seamont origin of basaltic rocks, Angayucham terrane, central Alaska: Journal of Geology, v. 96, p. 368–374.

Blodgett, R.B., 1983, Paleobiogeographic affinities of Devonian fossils from the

Nixon Fork terrane, southwestern Alaska, *in* Stevens, C. H., ed., Pre-Jurassic rocks in western North American suspect terranes: Pacific Section, Society of Economic Paleontologists and Mineralogists, p. 125–130.

Blum, J. D., Blum, A. E., Davis, T. E., and Dillon, J. T., 1987, Petrology of cogenetic silica-saturated and oversaturated plutonic rocks in the Ruby geanticline of north-central Alaska: Canadian Journal of Earth Sciences, v. 24, p. 159–169.

Box, S. E., 1985, Early Cretaceous orogenic belt in northwestern Alaska; Internal organization, lateral extent, and tectonic interpretation, *in* Howell, D. G., ed., Tectonostratigraphic terranes of the Circum-Pacific region: Circum-Pacific Council for Energy and Mineral Resources Earth Science Series, v. 1, p. 137–145.

Box, S. E., and Patton, W. W., Jr., 1989, Igneous history of the Koyukuk terrane, western Alaska; Constraints on the origin, evolution, and ultimate collision of an accreted island arc terrane: Journal of Geophysical Research, v. 94, no. B11, p. 15,843–15,867.

Box, S. E., Patton, W. W., Jr., and Carlson, C., 1985, Early Cretaceous evolution of the northeastern Yukon–Koyukuk basin, west-central Alaska, *in* Bartsch-Winkler, S., ed., The United States Geological Survey in Alaska; Accomplishments during 1984: U.S. Geological Survey Circular 967, p. 21–24.

Brosgé, W. P., and Reiser, H. N., 1962, Preliminary geologic map of the Christian Quadrangle, Alaska: U.S. Geological Survey Open-File Report 62-15, 2 sheets, scale 1:250,000.

Brosgé, W. P., Reiser, H. N., and Yeend, W., 1973, Reconnaissance geologic map of the Beaver Quadrangle, Alaska: U.S. Geological Survey Miscellaneous Field Studies Map MF-525, scale 1:250,000.

Bundtzen, T. K., Gilbert, W. G., Kline, J. T., and Solie, D. N., 1985, Summary of the Geologic Mapping Program by the Division of Geological and Geophysical Surveys (DGGS) in the McGrath Quadrangle, Alaska [abs.]: American Association of Petroleum Geologists Bulletin, v. 69, no. 4, p. 658.

*Cady, J. W., 1985, Geophysics of the Yukon–Koyukuk province, Alaska [abs.]: EOS Transactions of the American Geophysical Union, v. 66, no. 46, p. 1103.

Cady, W. M., Wallace, R. E., Hoare, J. M., and Weber, E. J., 1955, The central Kuskokwim region, Alaska: U.S. Geological Survey Professional Paper 268, 132 p.

Chapman, R. M., and Yeend, W., 1981, Geologic reconnaissance of east half of Kantishna River Quadrangle and adjacent areas, *in* Albert, N.R.D., and Hudson, T., eds., The United States Geological Survey in Alaska; Accomplishments during 1979: U.S. Geological Survey Circular 823B, p. B30–B32.

Chapman, R. M., Yeend, W. E., and Patton, W. W., Jr., 1975, Preliminary reconnaissance map of the western half of Kantishna River Quadrangle, Alaska: U.S. Geological Survey Open-File Map 75-351, scale 1:250,000.

Chapman, R. M., Weber, F. R., Churkin, M., Jr., and Carter, C., 1980, The Livengood Dome Chert; A new Ordovician formation in central Alaska and its relevance to displacement on the Tintina fault, *in* Shorter contributions to stratigraphy and structural geology, 1979: U.S. Geological Survey Professional Paper 1126-F, p. F1–F13.

Chapman, R. M., Churkin, M., Jr., Carter, C., and Trexler, J. H., Jr., 1981, Ordovician graptolites and early Paleozoic radiolarians in the Lake Minchumina area date a regional shale and chert, *in* Albert, N.R.D., and Hudson, T., eds., The United States Geological Survey in Alaska; Accomplishments during 1979: U.S. Geological Survey Circular 823B, p. B32–B33.

Chapman, R. M., Yeend, W., Brosgé, W. P., and Reiser, H. N., 1982, Reconnaissance geologic map of the Tanana Quadrangle, Alaska: U.S. Geological Survey Open-File Report 82-734, 18 p., scale 1:250,000.

Chapman, R. M., Patton, W. W., Jr., and Moll, E. J., 1985, Reconnaissance geologic map of the Ophir Quadrangle, Alaska: U.S. Geological Survey Open-File Report 85-203, scale 1:250,000.

Chappel, B. W., and White, A.J.R., 1974, Two contrasting granite types: Pacific Geology, no. 8, p. 173–174.

Churkin, M., Jr., and Carter, C., 1979, Collision-deformed Paleozoic continental margin in Alaska; A foundation for microplate accretion: Geological Society of America Abstracts with Programs, v. 11, p. 72.

Churkin, M., Jr., Wallace, W. K., Bundtzen, T. K., and Gilbert, W. G., 1984, Nixon Fork–Dillinger terranes; A dismembered Paleozoic craton margin in Alaska displaced from Yukon Territory: Geological Society of America Abstracts with Programs, v. 16, p. 275.

Csejtey, B., Jr., and Patton, W. W., Jr., 1974, Petrology of a nepheline syenite of St. Lawrence Island, Alaska: U.S. Geological Survey Journal of Research, v. 2, no. 1, p. 41–47.

Decker, J., and Dillon, J. T., 1984, Interpretation of regional aeromagnetic signatures along the southern margin of the Brooks Range, Alaska: Geological Society of America Abstracts with Programs, v. 16, p. 278.

Dick, H.J.B., and Bullen, T., 1984, Chromian spinel as a petrogenetic indicator of abyssal and alpine-type peridotites and spatially associated lavas: Contributions to Mineralogy and Petrology, v. 86, p. 54–76.

Dillon, J. T., and Smiley, C. J., 1984, Clasts from the Early Cretaceous Brooks Range orogen in Albian to Cenomanian molasse deposits of the northern Koyukuk basin, Alaska: Geological Society of America Abstracts with Programs, v. 16, p. 279.

Dillon, J. T., Pessel, G. H., Chen, J. H., and Veach, N. C., 1980, Middle Paleozoic magmatism and orogenesis in the Brooks Range, Alaska: Geology, v. 8, p. 338–343.

Dillon, J. T., and 5 others, 1985, New radiometric evidence for the age and thermal history of the metamorphic rocks of the Ruby and Nixon Fork terranes, west-central Alaska, *in* Bartsch-Winkler, S., and Reed, K. M., eds., The United States Geological Survey in Alaska; Accomplishments during 1983: U.S. Geological Survey Circular 945, p. 13–18.

Dillon, J. T., Brosgé, W. P., and Dutro, J. T., Jr., 1986, Generalized geologic map of the Wiseman Quadrangle, Alaska: U.S. Geological Survey Open-File Report 86-219, scale 1:250,000.

Dover, J. H., and Miyaoka, R. T., 1985, Major rock packages of the Ray Mountains, Tanana and Bettles Quadrangles, *in* Bartsch-Winkler, S., and Reed, K. M., eds., U.S. Geological Survey in Alaska; Accomplishments during 1983: U.S. Geological Survey Circular 945, p. 33–36.

Dusel-Bacon, C., and 6 others, 1989, Distribution, facies, ages, and proposed tectonic associations of regionally metamorphosed rocks in northern Alaska: U.S. Geological Survey Professional Paper 1497A, p. A1–A44.

Dutro, J. T., Jr., and Patton, W. W., Jr., 1982, New Paleozoic formations in the northern Kuskokwim Mountains, west-central Alaska, *in* Stratigraphic notes, 1980–1982: U.S. Geological Survey Bulletin 1529-H, p. H13–H22.

Forbes, R. B., Hamilton, T., Tailleur, I. L., Miller, T. P., and Patton, W. W., Jr., 1971, Tectonic implications of blueschists facies metamorphic terranes in Alaska: Nature; Physical Science, v. 234, p. 106–108.

Gates, G. O., Grantz, A., and Patton, W. W., Jr., 1968, Geology and natural gas and oil resources of Alaska, *in* Natural gases of North America: American Association of Petroleum Geologists Memoir 9, p. 3–48.

Gealey, W. K., 1980, Ophiolite obduction mechanism, *in* Panayiotow, A., ed., Ophiolites; Proceedings of an International Ophiolite Symposium: Cyprus Geological Survey Department, p. 228–243.

Gemuts, I., Puchner, C. C., and Steffel, C. I., 1983, Regional geology and tectonic history of western Alaska, *in* Proceedings of the 1982 Symposium on Western Alaska Geology and Resource Potential: Alaska Geological Society, v. 3, p. 67–85.

Gill, J. B., 1981, Orogenic andesites and plate tectonics: New York, Springer-Verlag, 390 p.

Gottschalk, R. R., Jr., 1987, Structural and petrologic evolution of the southern Brooks Range near Wiseman, Alaska [Ph.D. thesis]: Houston, Texas, Rice University, 263 p.

Harms, T. A., Coney, P. J., and Jones, D. L., 1984, The Sylvester allochthon, Slide Mountain terrane, British Columbia–Yukon; A correlative of oceanic terranes of northern Alaska: Geological Society of America Abstracts with Programs, v. 16, p. 288.

Harris, R. A., 1985, Paleomagnetism, geochronology, and paleotemperature of the Yukon–Koyukuk province, Alaska [M.S. thesis]: Anchorage, University of Alaska, 143 p.

Hitzman, M. W., 1984, Geology of the Cosmos Hills; Constraints for Yukon–

Koyukuk basin evolution: Geological Society of America Abstracts with Programs, v. 16, p. 290.
Hoare, J. M., Condon, W. H., and Patton, W. W., Jr., 1964, Occurrence and origin of laumontite in Cretaceous sedimentary rocks in western Alaska: U.S. Geological Survey Professional Paper 501-C, p. C74–C78.
Hollick, A., 1930, The Upper Cretaceous floras of Alaska: U.S. Geological Survey Professional Paper 159, 123 p.
Ishihara, S., 1981, The granitoid series and mineralization, *in* Skinner, B. J., ed., Economic Geology Seventy-fifth Anniversary Volume: Economic Geology Publishing Co., p. 458–484.
Jones, D. L., Silberling, N. J., Chapman, R. M., and Coney, P., 1984, New ages of radiolarian chert from the Rampart district, east-central Alaska, *in* Coonrad, W. L., and Elliott, R. L., eds., The U.S. Geological Survey in Alaska; Accomplishments during 1981: U.S. Geological Survey Circular 868, p. 39–43.
Jones, D. L., Silberling, N. J., Coney, P. J., and Plafker, G., 1987, Lithotectonic terrane map of Alaska, west of the 141st meridian; Folio of the lithotectonic terrane maps of the North American Cordillera: U.S. Geological Survey Miscellaneous Field Studies Map MF-1874-A, scale 1:2,500,000.
Lee-Wong, F., Vallier, T. L., Hopkins, D. M., and Silberman, M. L., 1979, Preliminary report on the petrography and geochemistry of basalts from the Pribilof Islands and vicinity, southern Bering Sea: U.S. Geological Survey Open-File Report 79-1556, 51 p.
*Loney, R. A., and Himmelberg, G. R., 1985, Distribution and character of the perioditite-layered gabbro complex of the southeastern Yukon–Koyukuk ophiolite belt, Alaska, *in* Bartsch-Winkler, S., and Reed, K. M., eds., U.S. Geological Survey in Alaska; Accomplishments during 1983: U.S. Geological Survey Circular 945, p. 46–48.
Mark, R. K., 1971, Strontium isotopic study of basalts from Nunivak Island, Alaska [Ph.D. thesis]: Stanford, California, Stanford University, 50 p.
Martin, G. C., 1926, Mesozoic stratigraphy of Alaska: U.S. Geological Survey Bulletin 776, 493 p.
*Mayfield, C. F., Tailleur, I. L., and Ellersieck, I., 1983, Stratigraphy, structure, and palinspastic synthesis of the western Brooks Range, northwestern Alaska: U.S. Geological Survey Open-File Report 83-779, 58 p.
Mertie, J. B., Jr., 1937, The Kaiyuh Hills, Alaska: U.S. Geological Survey Bulletin 868-D, p. 145–178.
Mertie, J. B., Jr., and Harrington, G. L., 1924, The Ruby–Kuskokwim region, Alaska: U.S. Geological Survey Bulletin 754, 129 p.
Miller, D. J., Payne, T. G., and Gryc, G., 1959, Geology of possible petroleum provinces in Alaska: U.S. Geological Survey Bulletin 1094, 131 p.
Miller, M. L., and Angeloni, L. M., 1985, Ophiolitic rocks of the Iditarod Quadrangle, west-central Alaska [abs.]: American Association of Petroleum Geologists Bulletin, v. 69, no. 4, p. 669–670.
Miller, M. L., and Bundtzen, T. K., 1985, Metamorphic rocks in the western Iditarod Quadrangle, west-central Alaska, *in* Bartsch-Winkler, S., and Reed, K. M., eds., U.S. Geological Survey in Alaska; Accomplishments during 1983: U.S. Geological Survey Circular 945, p. 24–28.
Miller, T. P., 1971, Petrology of the plutonic rocks of west-central Alaska: U.S. Geological Survey Open-File Report 454, 132 p.
——, 1972, Potassium-rich alkaline intrusive rocks of western Alaska: Geological Society of America Bulletin, v. 83, p. 2111–2127.
*——, 1985, Petrologic character of the plutonic rocks of the Yukon–Koyukuk basin and its borderlands [abs.]: EOS Transactions of the American Geo-Physical Union, v. 66, no. 46, p. 1102.
Miller, T. P., and Bunker, C. M., 1976, A reconnaissance study of the uranium and thorium contents of plutonic rocks of the southeastern Seward Peninsula, Alaska: U.S. Geological Survey Journal of Research, v. 4, p. 367–377.
Miller, T. P., and Lanphere, M. A., 1981, K-Ar age measurements on obsidian from the Little Indian River locality in interior Alaska, *in* Albert, N.R.D., and Hudson, T., eds., The United States Geological Survey in Alaska: Accomplishments during 1979: U.S. Geological Survey Circular 823-B, p. B39–B42.
Miller, T. P., Patton, W. W., Jr., and Lanphere, M. A., 1966, Preliminary report on a plutonic belt in west-central Alaska: U.S. Geological Survey Professional Paper 550-D, p. D158–D162.
Moll, E. J., and Arth, J. G., 1985, Sr and Nd isotopes from Late Cretaceous–early Tertiary volcanic fields in western Alaska; Evidence against old radiogenic continental crust under the Yukon–Koyukuk basin [abs.]: EOS Transactions of the American Geophysical Union, v. 66, no. 46, p. 1102.
Moll, E. J., and Patton, W. W., Jr., 1983, Late Cretaceous–early Tertiary calc-alkalic volcanic rocks of western Alaska: Geological Society of America Abstracts with Programs, v. 15, p. 406.
Moll-Stalcup, E. J., and Arth, J. G., 1989, The nature of the crust in the Yukon–Koyukuk province as inferred from the chemical and isotopic composition of five Late Cretaceous and early Tertiary volcanic fields in western Alaska: Journal of Geophysical Research, v. 94, no. B11, p. 15,989–16,020.
Morrison, G. W., 1980, Characteristics and tectonic setting of the shoshonite rock association: Lithos, v. 13, p. 97–108.
Murchey, B., and Harris, A. G., 1985, Devonian to Jurassic sedimentary rocks in the Angayucham Mountains of Alaska; Possible seamount or oceanic plateau deposits [abs.]: EOS Transactions of the American Geophysical Union, v. 66, no. 46, p. 1102.
Mutti, E., and Ricci Lucchi, F., 1972, Le torbiditi dell' Appennino setentrionale; Introduzione all'analisi de facies (Turbidites of the northern Apennines; Introduction to facies analysis): Societa Geologica Italiana Memoire, v. 11, p. 161–199.
Nilsen, T. H., and Patton, W. W., Jr., 1984, Cretaeous fluvial to deep-marine deposits of the central Yukon–Koyukuk basin, Alaska, *in* Reed, K. M., and Bartsch-Winkler, S., eds., U.S. Geological Survey in Alaska; Accomplishments during 1982: U.S. Geological Survey Circular 939, p. 37–40.
Pallister, J. S., Budahn, J. R., and Murchey, B. L., 1989, Pillow basalts of the Angayucham terrane; Oceanic-plateau and island crust accreted to the Brooks Range: Journal of Geophysical Research, v. 94, no. B11, p. 15,901–15,923.
Pallister, J. S., and Carlson, C., 1988, Bedrock geologic map of the Angayucham Mountains, Alaska: U.S. Geological Survey Miscellaneous Field Studies Map, MF-2024, scale 1:63,360.
Palmer, A. R., Egbert, R. M., Sullivan, R., and Knoth, J. S., 1985, Cambrian trilobites with Siberian affinities, southwestern Alaska [abs.]: American Association of Petroleum Geologists Bulletin, v. 69, no. 2, p. 295.
Patton, W. W., Jr., 1966, Regional geology of Kateel River Quadrangle, Alaska: U.S. Geological Survey Miscellaneous Geologic Investigations Map I-437, scale 1:250,000.
——, 1967, Regional geologic map of the Candle Quadrangle, Alaska: U.S. Geological Survey Miscellaneous Geologic Investigations Map I-492, scale 1:250,000.
——, 1970, A discussion of tectonic history of northern Alaska, *in* Adkison, W. L., and Brosgé, M. M., eds., Proceedings of the Geological Seminar on the North Slope of Alaska: Pacific Section, American Association of Petroleum Geologists, p. E20.
——, 1973, Reconnaissance geology of the northern Yukon-Koyukuk province, Alaska: U.S. Geological Survey Professional Paper 774-A, p. A1–A17.
——, 1978, Juxtaposed continental and oceanic-island arc terranes in the Medfra Quadrangle, west-central Alaska, *in* Johnson, K. M., ed., The United States Geological Survey in Alaska; Accomplishments during 1977: U.S. Geological Survey Circular 772-B, p. B38–B39.
——, 1984, Timing of arc collision and emplacement of oceanic crustal rocks on the margins of the Yukon–Koyukuk basin, western Alaska: Geological Society of America Abstracts with Programs, v. 16, p. 328.
Patton, W. W., Jr., and Bickel, R. S., 1956a, Geologic map and structure sections along part of the lower Yukon River, Alaska: U.S. Geological Survey Miscellaneous Geologic Investigations Map I-197, scale 1:200,000.
——, 1956b, Geologic map and structure sections of the Shaktolik River area, Alaska: U.S. Geological Survey Miscellaneous Geologic Investigations Map I-226, scale 1:80,000.
*Patton, W. W., Jr., and Box, S. E., 1985, Tectonic setting and history of the Yukon–Koyukuk basin, Alaska [abs.]: EOS Transactions of the American

Geophysical Union, v. 66, no. 46, p. 1101.
Patton, W. W., Jr., and Dutro, J. T., Jr., 1979, Age of the metamorphic complex in the northern Kuskokwim Mountains, west-central Alaska, *in* Johnson, K. M., and Williams, J. R., eds., The United States Geological Survey in Alaska; Accomplishments during 1978: U.S. Geological Survey Circular 804-B, B61–B63.
Patton, W. W., Jr., and Miller, T. P., 1966, Regional geologic map of the Hughes Quadrangle, Alaska: U.S. Geological Survey Miscellaneous Geologic Investigations Map I-459, scale 1:250,000.
——, 1968, Geologic map of the Selawik and southeastern Baird Mountains Quadrangles, Alaska: U.S. Geological Survey Miscellaneous Geologic Investigations Map I-530, scale 1:250,000.
——, 1973, Bedrock geologic map of Bettles and southern part of Wiseman Quadrangles, Alaska: U.S. Geological Survey Miscellaneous Field Studies Map MF-492, scale 1:250,000.
Patton, W. W., Jr., and Moll, E. J., 1982, Structural and stratigraphic sections along a transect between the Alaska Range and Norton Sound, *in* Coonrad, W. L., ed., U.S. Geological Survey in Alaska; Accomplishments during 1980: U.S. Geological Survey Circular 844, p. 76–78.
——, 1985, Geologic map of the northern and central parts of the Unalakleet Quadrangle, Alaska: U.S. Geological Survey Miscellaneous Field Studies Map MF-1749, scale 1:250,000.
Patton, W. W., Jr., and Tailleur, I. L., 1977, Evidence in the Bering Strait region for differential movement between North America and Eurasia: Geological Society of America Bulletin, v. 88, p. 1298–1304.
Patton, W. W., Jr., Miller, T. P., and Tailleur, I. L., 1968, Regional geologic map of the Shungnak and southern part of the Ambler River Quadrangles, Alaska: U.S. Geological Survey Miscellaneous Geologic Investigations Map I-554, scale 1:250,000.
Patton, W. W., Jr., Tailleur, I. L., Brosgé, W. P., and Lanphere, M. A., 1977, Preliminary report on ophiolites of northern and western Alaska, *in* Coleman, R. G., and Irwin, W. P., eds., North American ophiolites: Oregon Department of Geology and Mineral Industries Bulletin 95, p. 51–57.
Patton, W. W., Jr., Miller, T. P., Chapman, R. M., and Yeend, W., 1978, Geologic map of Melozitna Quadrangle, Alaska: U.S. Geological Survey Miscellaneous Geologic Investigations Map I-1071, scale 1:250,000.
Patton, W. W., Jr., Moll, E. J., Dutro, J. T., Jr., and Silberman, M. L., and Chapman, R. M., 1980, Preliminary geologic map of the Medfra Quadrangle, Alaska: U.S. Geological Survey Open-File Report 80-811A, scale 1:250,000.
Patton, W. W., Jr., Moll, E. J., Lanphere, M. A., and Jones, D. L., 1984, New age data for the Kaiyuh Mountains, west-central Alaska, *in* Coonrad, W. L., and Elliott, R. L., eds., U.S. Geological Survey in Alaska; Accomplishments during 1981: U.S. Geological Survey Circular 868, p. 30–32.
Patton, W. W., Jr., Stern, T. W., Arth, J. G., and Carlson, C., 1987, New U/Pb ages from granite and granite gneiss in the Ruby geanticline and southern Brooks Range, Alaska: Journal of Geology, v. 95, no. 1, p. 118–126.
Payne, T. G., 1955, Mesozoic and Cenozoic tectonic elements of Alaska: U.S. Geological Survey Miscellaneous Geologic Investigations Map I-84, scale 1:5,000,000.
Perfit, M. R., Gust, D. A., Bence, A. E., Arculus, R. J., and Taylor, S. R., 1980, Chemical characteristics of island-arc basalts; Implications for mantle sources: Chemical Geology, v. 30, p. 227–256.
Péwé, T. L., Wahrhaftig, C., and Weber, F., 1966, Geologic map of the Fairbanks Quadrangle, Alaska: U.S. Geological Survey Miscellaneous Geologic Investigations Map I-455, scale 1:250,000.
Plumley, P. W., and Coe, R. S., 1982, Paleomagnetic evidence for reorganization of the north Cordilleran borderland in late Paleozoic time: Geological Society of America Abstracts with Programs, v. 14, p. 224.
Plumley, P. W., Coe, R. S., Patton, W. W., Jr., and Moll, E. J., 1981, Paleomagnetic study of the Nixon Fork terrane, west-central Alaska: Geological Society of America Abstracts with Programs, v. 13, p. 530.
Potter, A. W., Gilbert, W. G., Ormiston, A. R., and Blodgett, R. B., 1980, Middle and Upper Ordovician brachiopods from Alaska and northern California and their paleogeographic implications: Geological Society of America Abstracts with Programs, v. 12, p. 147.
Puchner, C. C., 1984, Intrusive geology of the Ray Mountains batholith: Geological Society of America Abstracts with Programs, v. 16, p. 329.
Roden, M. F., Frey, F. A., and Francis, D. M., 1984, An example of consequent mantle metasomatism in peridotite inclusions from Nunivak Island, Alaska: Journal of Petrology, v. 25, p. 546–577.
Roeder, D., and Mull, C. G., 1978, Tectonics of Brooks Range ophiolites (Alaska): American Association of Petroleum Geologists Bulletin, v. 62, no. 9, p. 1696–1702.
Saunders, A. D., and Tarney, J., 1985, Geochemical characteristics of basaltic volcanism within back-arc basins, *in* Kolelaar, B. P., and Howells, M. F., eds, Marginal Basin Geology: Cambridge, Massachusetts, Shiva Publishing Ltd., p. 59–76.
Silberman, M. L., Moll, E. J., Patton, W. W., Jr., Chapman, R. M., and Connor, C. L., 1979, Precambrian age of metamorphic rocks from the Ruby province, Medfra and Ruby Quadrangles; Preliminary evidence from radiometric age data, *in* Johnson, K. M., and Williams, J. R., eds., The United States Geological Survey in Alaska; Accomplishments during 1978: U.S. Geological Survey Circular 804-B, p. B66–B67.
Smith, G. M., and Puchner, C. C., 1985, Geologic map and cross sections of the Ruby geanticline between Ruby and Poorman, Alaska: Geological Society of America Abstracts with Programs, v. 17, p. 720.
Struik, L. C., and Orchard, M. J., 1985, Late Paleozoic conodonts delineate imbricate thrusts within the Antler Formation of the Slide Mountain terrane central British Columbia: Geology, v. 13, p. 794–798.
Tailleur, I. L., 1973, Probable rift origin of Canada basin, Arctic Ocean, *in* Arctic geology: American Association of Petroleum Geologists Memoir 19, p. 526–535.
——, 1980, Rationalization of Koyukuk "Crunch" northern and central Alaska [abs.]: American Association of Petroleum Geologists Bulletin, v. 6415, p. 792.
Thompson, R. N., Morrison, M. A., Hendry, G. L., and Parry, S. J., 1984, An assessment of the relative roles of crust and mantle in magma genesis; An elemental approach: Philosophical Transactions of the Royal Society of London, series A, v. 310, p. 549–590.
Thrupp, G. A., and Coe, R. S., 1986, Early Tertiary paleomagnetic evidence and the displacement of southern Alaska: Geology, v. 14, p. 213–217.
Turner, D. L., 1984, Tectonic implications of widespread Cretaceous overprinting of K-Ar ages in Alaskan metamorphic terranes, Geological Society of America Abstracts with Programs, v. 16, p. 338.
Varne, R., 1985, Ancient subcontinental mantle; A source for K-rich orogenic volcanics: Geology, v. 13, p. 405–408.
Von Drach, V., Marsh, B. D., and Wasserburg, G. J., 1986, Nd and Sr isotopes in the Aleutians; Multicomponent parenthood of island-arc magmas: Contributions to Mineralogy and Petrology, v. 92, p. 13–34.
Wallace, W. K., and Engebretson, D. C., 1984, Relationship between plate motions and Late Cretaceous to Paleogene magmatism in southwestern Alaska: Tectonics, v. 3, p. 295–315.

MANUSCRIPT ACCEPTED BY THE SOCIETY MAY 9, 1990

ACKNOWLEDGMENTS

We thank Henry Berg, Marti Miller, and Bela Csejtey for their reviews and helpful suggestions for improving the manuscript.

NOTES ADDED IN PROOF

The following preliminary reports listed in 'References Cited' have been superseded by later reports.

Arth, J. G., 1985.
superseded by:
Arth, J. G., Criss, R. E., Zmuda, C. C., Foley, N. K., Patton, W. W., Jr., and Miller, T. P., 1989, Remarkable isotopic and trace element trends in potassic through sodic Cretaceous plutons of the Yukon-Koyukuk basin, Alaska, and the nature of the lithosphere beneath the Koyukuk terrane: Journal of Geophysical Research, v. 94, no. B11, p. 15,957–15,968.

Arth, J. G., and 5 others, 1984.
superseded by:
Arth, J. G., Zmuda, C. C., Foley, N. K., Criss, R. E., Patton, W. W., Jr., and Miller, T. P., 1989, Isotopic and trace element variations in the Ruby batholith, Alaska, and the nature of the deep crust beneath the Ruby and Angayucham terranes: Journal of Geophysical Research, v. 94, no. B11, p. 15,941–15,955.

Cady, J. W., 1985.
superseded by:
Cady, J. W., 1989, Geologic implications of topographic, gravity, and aeromagnetic data in the northern Yukon-Koyukuk province and its borderlands: Journal of Geophysical Research, v. 94, no. B11, p. 15,821–15,841.

Loney, R. A., and Himmelberg, G. R., 1985.
superseded by:
Loney, R. A., and Himmelberg, G. R., 1989, The Kanuti ophiolite, Alaska: Journal of Geophysical Research, v. 94, no. B11, p. 15,869–15,900.

Mayfield, C. F., Tailleur, I. L., and Ellersieck, I., 1983.
superseded by:
Mayfield, C. F., Tailleur, I. L., and Ellersieck, I., 1988, Stratigraphy, structure, and palinspastic synthesis of the western Brooks Range, northwestern Alaska, *in* Gryc, G., ed., Geology and exploration of the National Petroleum Reserve in Alaska, 1974 to 1982: U.S. Geological Survey Professional Paper 1399, p. 143–186.

Miller, T. P., 1985.
superseded by:
Miller, T. P., 1989, Contrasting plutonic suites of the Yukon-Koyukuk basin and the Ruby geanticline, Alaska: Journal of Geophysical Research, v. 94, no. B11, p. 15,969–15,987.

Patton, W. W., Jr., and Box, S. E., 1985.
superseded by:
Patton, W. W., Jr., and Box, S. E., 1989, Tectonic setting of the Yukon-Koyukuk basin and its borderlands, western Alaska: Journal of Geophysical Research, v. 94, no. B11, p. 15,807–15,820.

Printed in U.S.A.

The Geology of North America
Vol. G-1, The Geology of Alaska
The Geological Society of America, 1994

Chapter 8

Geology of the eastern Bering Sea continental shelf

Michael S. Marlow, Alan K. Cooper, and Michael A. Fisher
U.S. Geological Survey, 345 Middlefield Road, Menlo Park, California 94025

INTRODUCTION

The Bering Sea shelf south of the Bering Strait encompasses an area of 1,300,000 km^2, more than the combined area of California, Oregon, and Washington (840,000 km^2, Fig. 1). The shelf area lies between western Alaska and eastern Siberia. The outer shelf is underlain by three large basins, Bristol, St. George, and Navarin, filled with sedimentary rocks, as well as by three bedrock ridges that extend from the Alaska Peninsula to near Siberia (Figs. 1 and 2). The innermost part of the shelf, Norton Sound, is underlain by the large, sediment-filled Norton basin (Fig. 1; Fisher and others, 1982). A similar inner basin, Anadyr basin, underlies the Gulf of Anadyr along the western side of thee Bering shelf (Fig. 1).

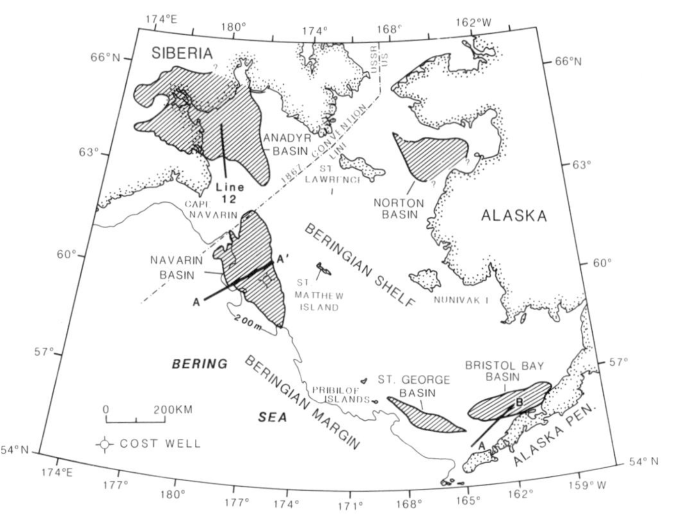

Figure 1. Outline and location of major basins beneath the Bering Sea shelf. See Table 1 for the sizes of the various basins. Location of seismic reflection profile across Navarin basin is shown by line A-A′ (see Fig. 10 for the profile). Albers equal-area projection.

Setting

The Bering shelf south of the Bering Strait is underlain by five large sedimentary basins that encompass 280,000 km^2 of the shelf. Three of them, the Norton, Anadyr, and Bristol Bay basins, are large structural sags in the Earth's crust, formed by block-faulting and down-dropping of the bedrock or basement beneath the basins. Two of the basins, Navarin and St. George, are grabens and half-grabens, which formed by extension and collapse of the outer Bering Sea margin. The latter basins are filled with 10 to 15 km of strata of mainly Cenozoic age.

Except for Anadyr, the shelf basins are mainly Cenozoic features that developed after the amalgamation of Alaska at the end of the Mesozoic. The oldest sedimentary units recovered from a well in Norton basin are Eocene in age. The oldest sedimentary bodies drilled in Anadyr basin are Cretaceous and those drilled in Navarin basin are Late Cretaceous. The bottom basin fill drilled along the flanks of St. George basin is Eocene in age.

Regional geologic studies onshore in western Alaska, along the Alaska Peninsula, and in eastern Siberia suggest that suitable source rocks for the generation of hydrocarbons exist offshore in the basins of the Bering Sea shelf. Also, the sedimentary fill in the offshore basins is sufficiently thick to favor the thermal development of hydrocarbons. Twenty onshore wells in Anadyr basin, nine onshore wells on the Alaska Peninsula, five offshore COST (Continental Offshore Stratigraphic Test) wells in the shelf basins, several offshore Deep Sea Drilling Project sites, and dredge samples from the Bering Sea continental margin all show that adequate reservoir beds of late Cenozoic age probably exist within the shelf basins. Seismic reflection data show large structural traps within the basin fills of the Bering subshelf. Such traps include anticlines, diapiric structures, growth-faults, stratigraphic pinch-outs against the bedrock flanks of the basins, discordant overlap of the basins' lower stratigraphic section by the overlying younger sequence, and drape structures formed by differential compaction of the basins' lower sequence over bedrock highs.

REGIONAL FRAMEWORK OF THE BERING SEA SHELF

Mesozoic structural trends

Large onshore geologic features in western Alaska strike toward the Bering Sea shelf, such as major strike-slip faults and tectonostratigraphic terranes. Two major strike-slip faults, the

Marlow, M. S., Cooper, A. K., and Fisher, M. A., 1994, Geology of the eastern Bering Sea continental shelf, *in* Plafker, G., and Berg, H. C., eds., The Geology of Alaska: Boulder, Colorado, Geological Society of America, The Geology of North America, v. G-1.

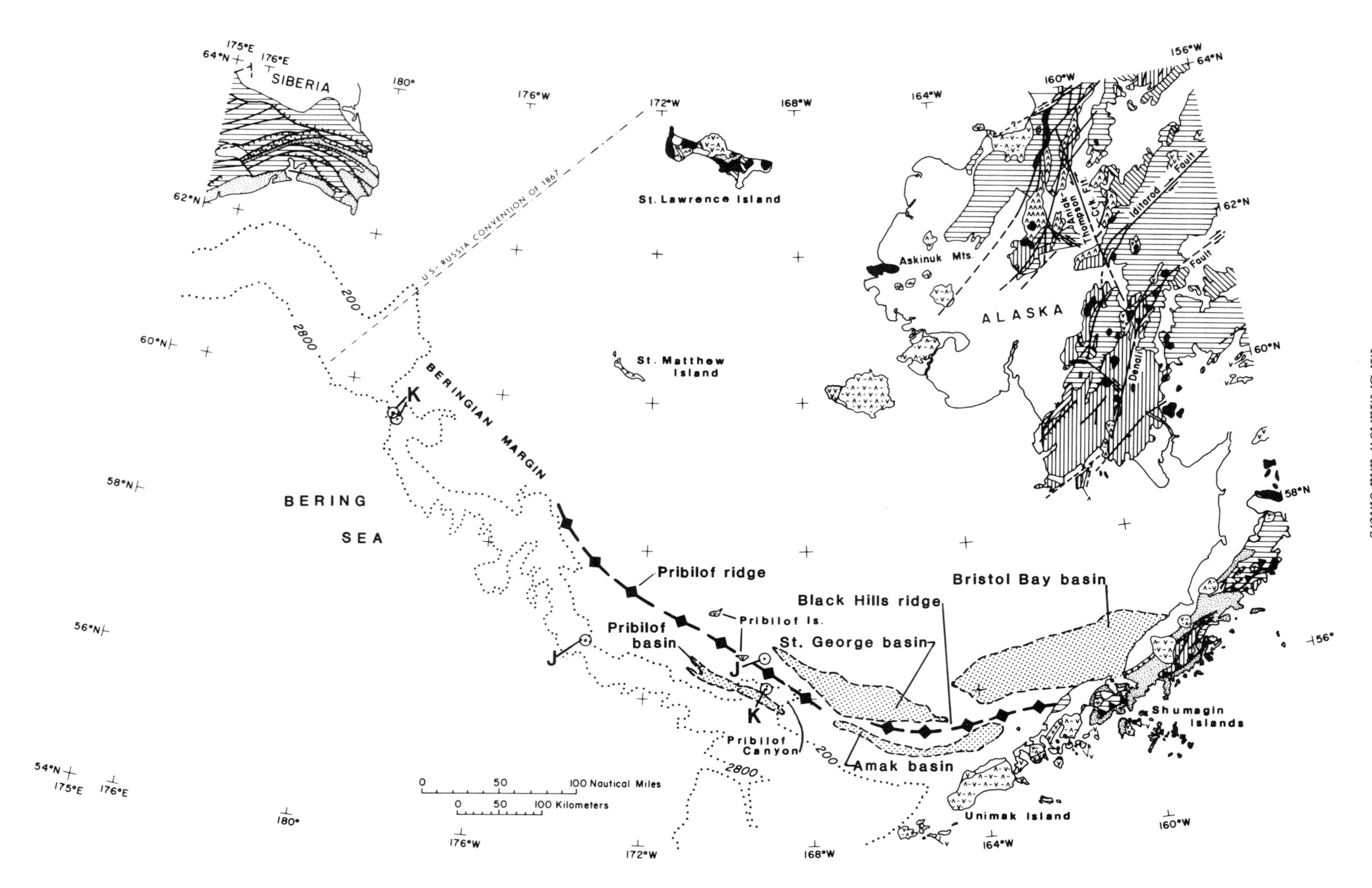
SIBERIA
ALASKA
BERING
SEA
BERINGIAN MARGIN
U.S.-RUSSIA CONVENTION OF 1867
St. Lawrence Island
St. Matthew Island
Askinuk Mts.
Aniak-Thompson Cr. Flt.
Iditarod Fault
Fault
Denali
Pribilof ridge
Pribilof Is.
Pribilof basin
St. George basin
Black Hills ridge
Bristol Bay basin
Amak basin
Pribilof Canyon
Shumagin Islands
Unimak Island
J
K
200
2800
0 50 100 Nautical Miles
0 50 100 Kilometers
175°E
176°E
180°
176°W
172°W
168°W
164°W
160°W
156°W
156°
64°N
62°N
60°N
58°N
56°N
54°N

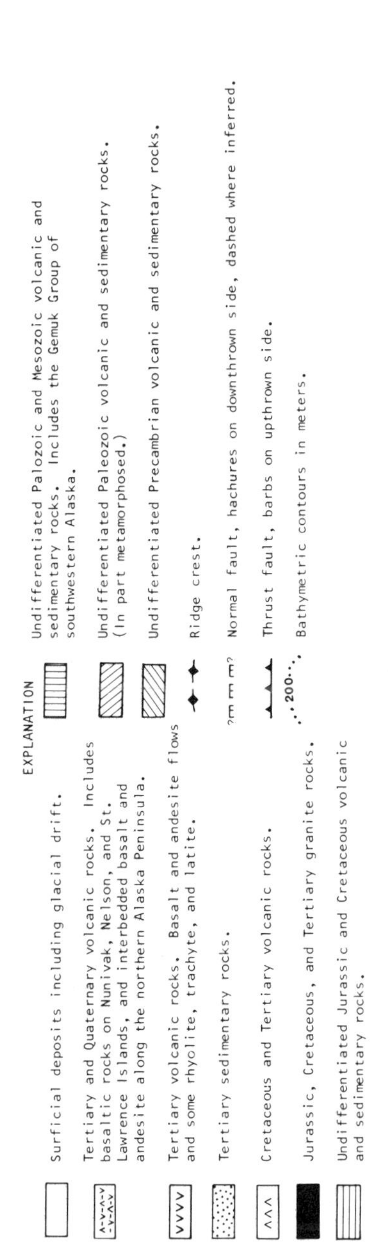

Figure 2. Generalized geology of western Alaska, Alaska Peninsula, and eastern Siberia. Onshore geology modified from Burk (1965), Yanshin (1966), and Beikman (1974; this volume). Dots enclosed by circles show sites where Jurassic (J) and Cretaceous (K) sedimentary rocks were dredged from the continental slope. Adapted from Marlow and Cooper (1980a). Albers equal-area projection.

Kaltag and the Denali faults, trend southwest from central and western Alaska toward the Bering Sea. The Goodnews and Togiak tectonostratigraphic terranes trend from western Alaska toward the Bering Sea (Jones and Silberling, 1979; Silberling and others, this volume, Plate 3); they continue offshore at least 50 to 100 km (Cooper and Marlow, 1983).

Gravity, magnetic, seismic reflection, dredge, and drill data suggest that the Upper Jurassic, shallow-marine rocks exposed in the Black Hills on the Alaska Peninsula extend west to northwest along the south flank of St. George basin and connect with Pribilof Ridge (Fig. 2; Marlow and Cooper, 1980a). In addition, Upper Cretaceous (Campanian) rocks were dredged from the southern flank of this ridge in nearby Pribilof Canyon (Fig. 2; Hopkins and others, 1969). Seismic reflection data show that the Cretaceous dredge samples may have been recovered from an isolated, intrabasement rock sequence that is folded into the flank of Pribilof Ridge (Marlow and Cooper, 1980b). Other dredge samples along the margin show that Mesozoic bedrock is unconformably overlain by shallow-water, diatomaceous mudstone of early Tertiary age (Jones and others, 1981).

The Upper Jurassic rocks that extend offshore beneath the Bering shelf are coeval with the Naknek Formation exposed along the Alaska Peninsula (Burk, 1965), and both Jurassic sections are part of the Peninsular terrane of Jones and Silberling (1979). The Peninsular terrane extends about 1,200 km from south-central Alaska to the western tip of the Alaska Peninsula (Fig. 3). Paleomagnetic studies of Mesozoic rocks from the Peninsular terrane suggest that the terrane originated at a paleolatitude south of the equator in Jurassic time (Packer and Stone, 1972; Stone and others, 1982). Geologic data suggest that the docking of the Peninsular terrane against the ancient North American continent was complete by the end of the Mesozoic (see Marlow and Cooper, 1983, for more detailed discussion of the docking time of the Peninsular terrane).

The offshore part of the Peninsular terrane can be traced by geophysical and dredge data about 1,000 km beneath the outer shelf of the Beringian margin (Figs. 2 and 3). Thus, docking of the Peninsular terrane in southern Alaska probably coincided in time with docking along the Beringian margin. Formation of the margin at the end of the Mesozoic presumably occurred about the same time that the large outer shelf basins, such as St. George and Navarin, began to form and fill with sediment.

We proposed in an earlier paper (Marlow and others, 1976) that a Jurassic, Cretaceous, and earliest Tertiary magmatic arc extended parallel to and inside (landward or toward Alaska) the outer shelf basins. This igneous belt is characterized by high-amplitude, high-frequency magnetic anomalies (Marlow and others, 1976). The magmatic arc is exposed as calc-alkalic volcanic and intrusive rocks of late Mesozoic and earliest Tertiary age on St. Matthew and St. Lawrence Islands on the Bering shelf and as similar rocks in southern and western Alaska and eastern Siberia (Reed and Lanphere, 1973; Patton and others, 1974, 1976; Scholl and others, 1975; Marlow and others, 1976). However, Cooper and Marlow (1983) suggest that the southern, inner

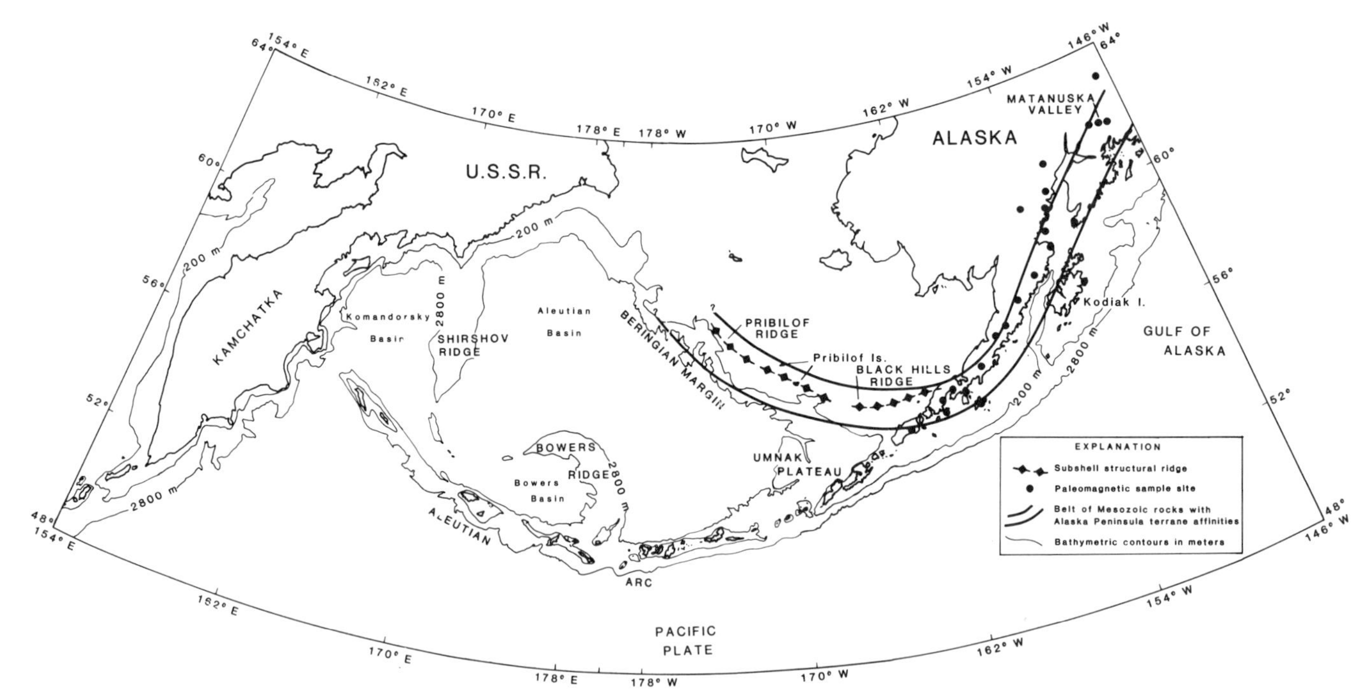

Figure 3. Locations of sample sites used in paleomagnetic studies of southern Alaska by Stone and others (1982). Offshore extension of the Peninsular terrane is shown by diamond pattern over bedrock ridges beneath the outer Bering Sea shelf (from Marlow and Cooper, 1980a). Adapted from Marlow and Cooper (1983).

Bering shelf is underlain in part by offshore extensions of allochthonous terranes of southern Alaska (Jones and Silberling, 1979; McGeary and Ben-Avraham, 1981). Whether these terranes merge with the Mesozoic igneous belt near St. Matthew Island is not known.

Cenozoic isolation and collapse

Plate motion along the Bering Sea margin apparently ceased with the formation of the Aleutian Island Arc in earliest Tertiary time, when the plate boundary shifted from the Bering Sea margin to a site near the present Aleutian Trench, thereby trapping a large section of oceanic plate (Kula?) within the abyssal Bering Sea (Scholl and others, 1975; Cooper and others, 1976a, b; Marlow and others, 1976; Marlow and Cooper, 1985). Cessation of the plate motion isolated and deactivated the Bering Sea margin. In early Tertiary time, the margin underwent extensional deformation and differential subsidence, which has continued throughout the Cenozoic. Elongate basins of great size and depth, such as the St. George and Navarin basins, formed along the modern outer Bering Sea shelf (Fig. 1; Marlow and others, 1976), while parts of the inner shelf also subsided, forming Bristol Bay and Norton basins. Extensional deformation of the folded rocks of the Mesozoic bedrock beneath the shelf has continued to the present, as evidenced by the normal faults flanking the outer shelf basins that commonly offset the sea floor. These normal faults are growth-type structures that typically rupture the entire Cenozoic basin fill. Collapse of the outer shelf and adjacent margin may have been aided by Cenozoic sediment-loading and continued subsidence of the adjacent oceanic crust (Kula? Plate) flooring the Aleutian Basin of the abyssal Bering Sea.

ST. GEORGE BASIN

Description

St. George basin is a long (300 km) and narrow (30 to 50 km) graben whose long axis strikes northwestward parallel to the trend of the Beringian margin (Figs. 1, 2, and 4, Table 1). The basin, near the Pribilof Islands and beneath the virtually featureless shelf, is filled with more than 10 km of sedimentary deposits.

A tracing of a 24-channel seismic reflection profile, 8B, across St. George basin, is shown in Figure 5. The flat acoustic basement or bedrock surface underlying the southwest end of

TABLE 1. SEDIMENTARY BASINS IN THE BERING SHELF

Basin Name	Length (km)	Width (km)	Area (km^2)
Bristol Bay	290	75	21,750
St. George	300	50	15,000
Navarin	400	90	76,000
Anadyr	400	300	120,000
Norton	250	180	45,000

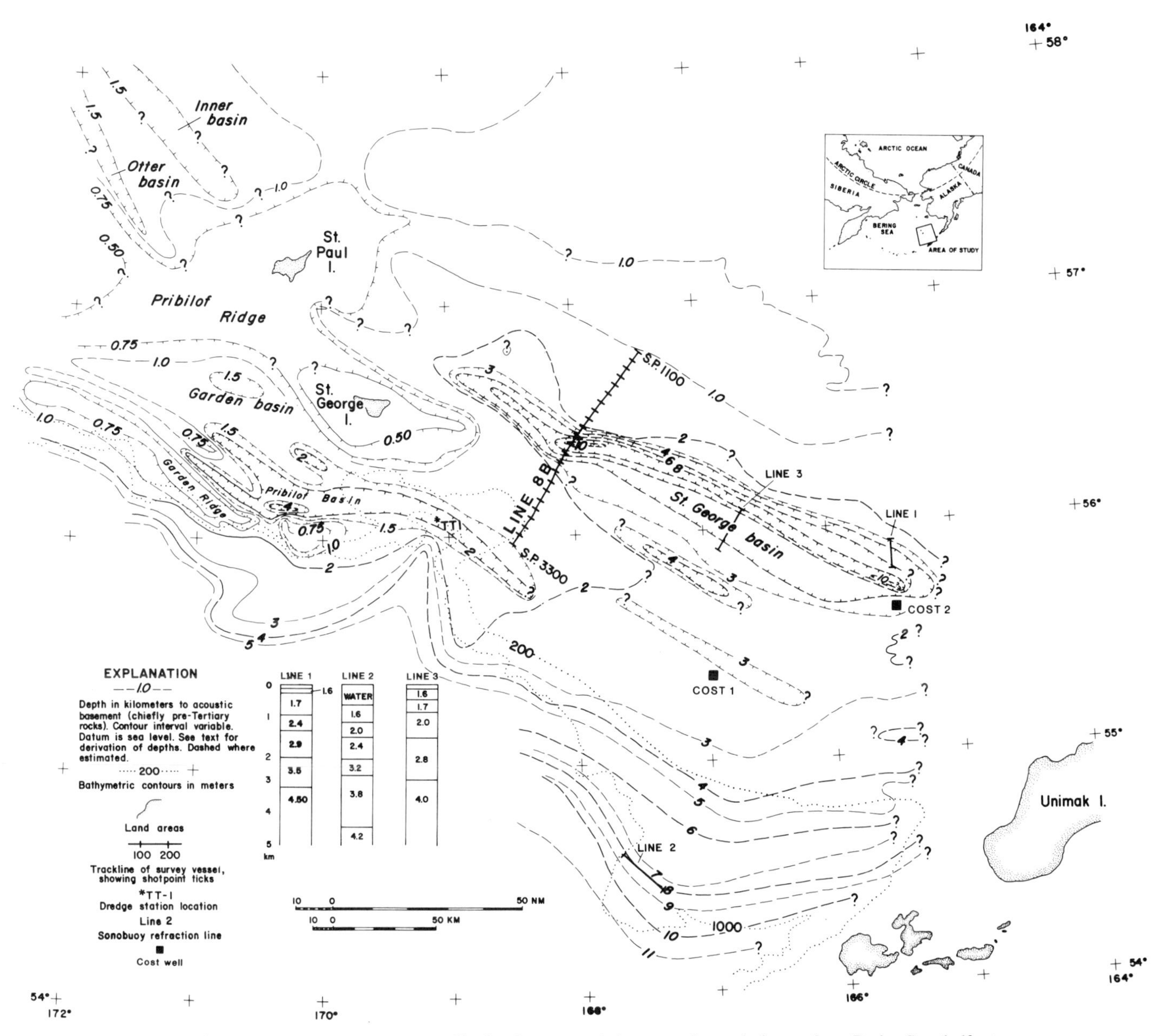

Figure 4. Structure-contour map of bedrock or acoustic basement beneath the southern Bering Sea shelf surrounding St. George basin. Line with shotpoint ticks and numbers (Line 8B) is profile 8B shown in Figure 5. Lines marked 1, 2, and 3 correspond to sonobuoy refraction stations shown in lower left of the figure (unnumbered top layer is sea water with an assumed velocity of sound of 1.5 km/s). Cretaceous dredge site in Pribilof Canyon is marked TT1 and corresponds to the site shown on Figure 2. Adapted from Marlow and others (1977).

profile 8B (Fig. 5; between shotpoints 3200 and 2700) is Pribilof ridge, which is overlain by an undisturbed, layered sequence 1.3 to 1.4 km thick (1.3 to 1.4 s). Beneath the flat-lying sequence are dipping reflectors within the bedrock of Pribilof ridge that suggest the ridge includes folded sedimentary beds (Jurassic Naknek[?] Formation). Farther north, between shotpoints 2700 and 2200, the basement or bedrock surface descends in a series of down-to-basin steps, plunging to a maximum subbottom depth of about 5.4 s (more than 10 km) beneath the axis of St. George basin. The overlying beds are broken by at least three major normal faults that dip toward the basin axis and appear to be related to offsets in the basin's bedrock. Within the basin fill, the offset along the faults increases with depth, implying that these are growth structures. Strata are synclinally deformed about the basin's structural axis. From shotpoints 2200 to about 1860, the bedrock surface rises rapidly to a minimum depth of 500 m (0.55 s; Fig. 5). Overlying strata are broken by normal faults that dip down to the basin axis.

TABLE 2. STRATIGRAPHIC SEQUENCES PENETRATED AT ST. GEORGE BASIN COST WELLS

Ate	Depth to Top (m)
COST WELL NO. 1	
Pliocene	487
Miocene	1,097
Oligocene	1,636
Eocene	2,563
Igneous	3,163
Basement (age unknown)	
COST WELL NO. 2	
Pliocene	445
Miocene	1,294
Oligocene	1,844
Eocene(?)	3,378
Early Cretaceous or Late Jurassic	3.822
Late Jurassic	4,075

Age and history

Two COST wells were drilled into the St. George basin, and data from those two wells are extensively described by Turner and others (1984a, b). We present below a short summary of each well taken from their report (Table 2). The first St. George COST well, 170 km southeast of St. George Island, Alaska, was drilled in 1976 to a total depth of 4,197 m in water 135 m deep. The second St. George COST well was drilled in 1982 to a total depth of 4,458 m in water 114 m deep, about 180 km northwest of Cold Bay, Alaska Peninsula.

The Cenozoic sedimentary section penetrated in the wells includes interbedded sandstone, siltstone, mudstone, diatomaceous mudstone, and conglomerate. The section was derived predominantly from volcanic source terranes, and the detrital material is mechanically and chemically unstable. Porosity and permeability in the section are reduced by ductile deformation of individual sediment grains, and by cementation and diagenesis of the section.

The Mesozoic sedimentary section at the bottom of the COST No. 2 well consists of shallow marine sandstone, siltstone, shale, conglomerate, and minor amounts of coal. Porosities and permeabilities in this section are also low.

BRISTOL BAY BASIN

Description

Underlying a large portion of the Bering Sea shelf, north of the Alaska Peninsula, is a sediment-filled structural depression known as Bristol Bay basin (Fig. 1, Table 1). The sedimentary deposits along the southern flank of the basin are exposed on the northern fringe of the Alaska Peninsula. The basin trends northeastward along the peninsula, about four-fifths of it lies offshore beneath the flat Bering Sea shelf. The basin is asymmetrical in cross section; the basement rocks of the northwest flank dip gently toward the basin axis, but the southeast flank of folded Jurassic rocks is steeply inclined, probably faulted, and crops out in the northern foothills of the Alaska Peninsula. The basin's sedimentary section, more than 6,000 m thick, is composed chiefly of Cenozoic deposits; they are thickest in the southeast part of the basin.

The northeast end of Bristol Bay basin is bounded by exposures of highly deformed, locally intruded, metamorphosed Paleozoic and Mesozoic rocks (Hatten, 1971). To the southwest, the basin is bordered by the offshore extension of the Black Hills, an anticlinal structure composed of Jurassic sedimentary rocks (Fig. 2; Marlow and Cooper, 1980a). South of the basin, the basement consists mainly of Mesozoic and Cenozoic volcanic and plutonic units that form the core of the Alaska Peninsula (McLean, 1977). To the north, beneath the Bering Sea shelf, the acoustic basement (as shown on geophysical records) flooring the basin probably consists of Mesozoic sedimentary, igneous, and metamorphic rocks.

The interpreted section of a 24-channel seismic reflection profile shown in Figure 6 illustrates the spatial relation of such large subshelf structures as the Amak basin, the Black Hills ridge, and the Bristol Bay basin (see Fig. 1 for location). Near the center of the profile, at a subbottom depth of about 1.0 s, moderately dipping strata are resolved within the core of a bedrock ridge, informally called the Black Hills ridge after similarly deformed rocks exposed nearby in the Black Hills of the Alaska Peninsula. Strata in the Black Hills ridge are cut by a major unconformity (Fig. 6). The age of this unconformity may be Cretaceous in age because middle Cretaceous rocks are missing on the nearby Alaska Peninsula owing to uplift and erosion (Burk, 1965). The unconformity can be traced to the east into Bristol Bay basin, where the surface of the unconformity appears to grade into a depositional contact having parallel reflectors above and below it (Fig. 6). If this horizon is indeed Cretaceous in age, then the layered strata beneath it could be the same age or older. Hence, the initial formation and filling of the basin could have begun by late Mesozoic time. The same argument can be made for the dipping strata near the bottom of the Amak basin along the western end of the same profile (Fig. 6). However, a major unconformity along the Bering Sea continental margin west of Bristol Bay basin is early Tertiary in age (Marlow and Cooper, 1985). If the unconformity along the margin is contemporaneous with the cutting of the Black Hills ridge, then the Bristol Bay and Amak basins may be no older than early Tertiary.

As shown by the northeastern half of profile A-B, Bristol Bay basin is a long structural sag containing more than 6.6 km (4.5 s) of moderately dipping to flat-lying reflectors (Fig. 6). The upper half of the fill (above the heavy line) is virtually flat lying and undeformed, and we interpret it to be late Cenozoic in age. The lower half of the basin fill contains discontinuous and dip-

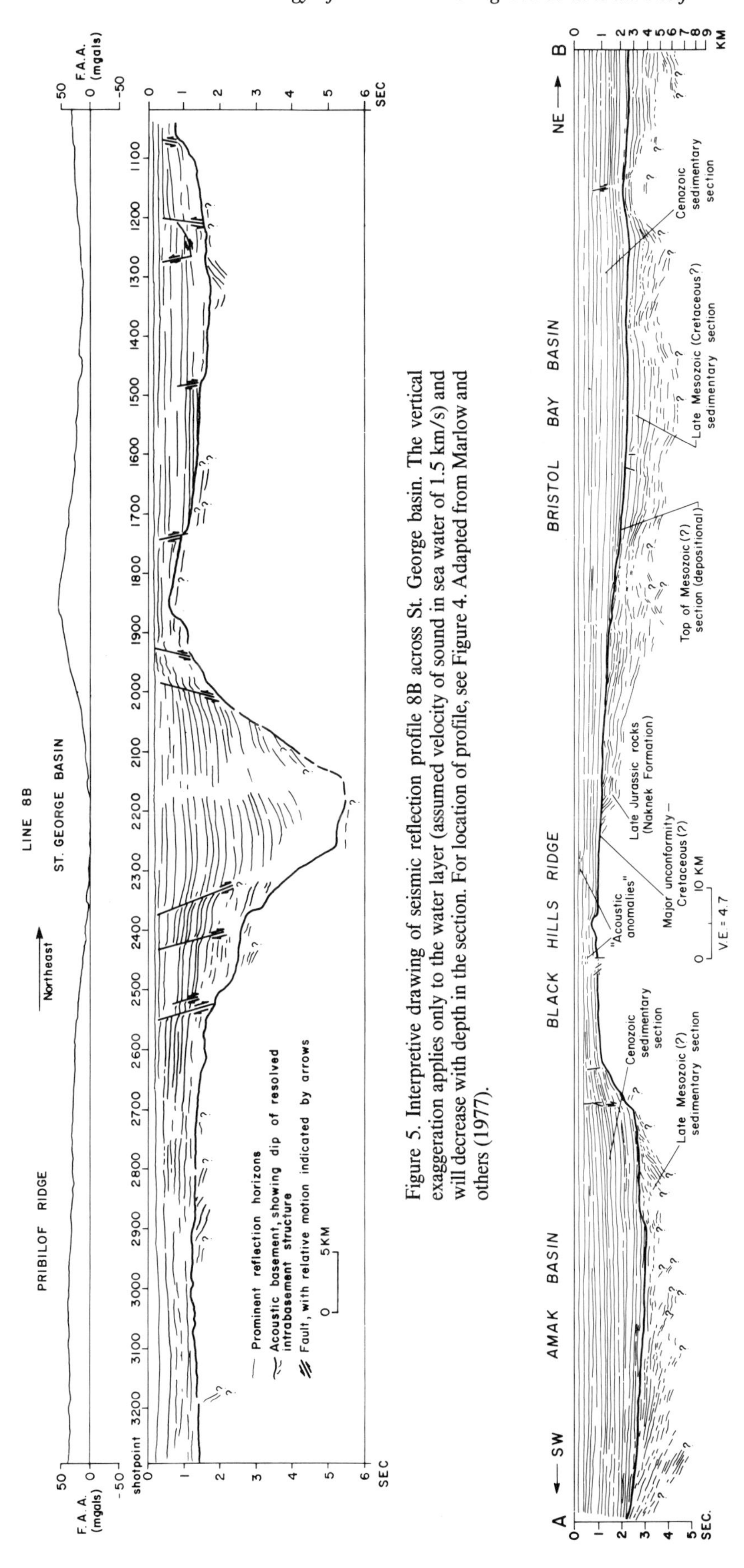

Figure 5. Interpretive drawing of seismic reflection profile 8B across St. George basin. The vertical exaggeration applies only to the water layer (assumed velocity of sound in sea water of 1.5 km/s) and will decrease with depth in the section. For location of profile, see Figure 4. Adapted from Marlow and others (1977).

Figure 6. Interpretive drawing of 24-channel seismic reflection profile A-B across the southern Bering Sea shelf and the western end of Bristol Bay basin. For location of profile see Figure 1. Travel time, in seconds, is two-way time. From Marlow and Cooper (1980a).

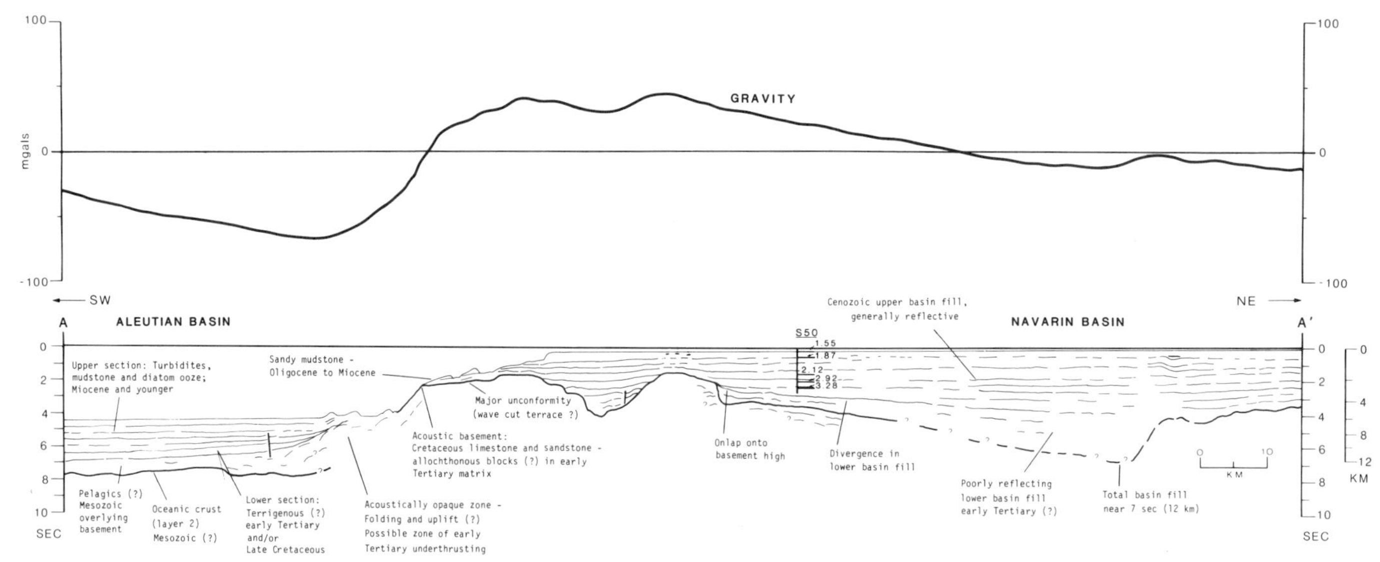

Figure 7. Interpretive drawing of seismic reflection and gravity profiles across the northern Beringian margin and Navarin basin. For location see A-A′ on Figure 1. The strong reflector beneath the Aleutian Basin can be traced beneath the base of the margin into an acoustically opaque zone. The thick sedimentary fill in Navarin basin extends to a subshelf depth of about 12 km. Adapted from Marlow and Cooper (1985).

ping reflectors, which suggest gentle folding and draping over older basement (bedrock) highs. Our profile was recorded to a subbottom depth of 5 s, which indicates that Bristol Bay basin is filled with more than 6 to 7 km of strata. The nearby Gulf Sandy River No. 1 well on the Alaska Peninsula bottomed in Oligocene sedimentary rocks of the Stepovak Formation at a total depth of about 4 km without reaching a basement or bedrock complex (Hatten, 1971; Brockway and others, 1975), indicating that the adjacent Bristol Bay basin may contain a much thicker Cenozoic sedimentary section.

Age and history

Only one COST well has been drilled in Bristol Bay basin, and data from that well are proprietary. Thus, little is publicly known about the stratigraphy and age of the fill in the offshore portions of the basin.

Bristol Bay basin, as a distinct structural trough, probably began to form in late Mesozoic or early Tertiary time, and the basin has continued to subside during most of Cenozoic time. The basin probably began to form after the docking of the Peninsular terrane in late Mesozoic time and during the formation of the ancestral Aleutian Island arc in earliest Tertiary time (Scholl and others, 1975, 1987; Marlow and others, 1976; Marlow and Cooper, 1983, 1985).

NAVARIN BASIN

Description

The Navarin basin province underlies the northwest corner of the Bering Sea shelf just south of Cape Navarin of eastern Siberia (Fig. 1). Isopach contours of sediment thickness derived from seismic reflection data define three basins within the province that contain 10 to 15 km of strata and underlie about 76,000 km^2 of the Bering Sea shelf (Table 1; Marlow, 1979).

A 160-km-long seismic reflection profile, A-A′ (see location on Fig. 1), transects the northwestern Beringian margin west of St. Matthew Island, and an interpretive drawing of the profile (Fig. 7) crosses on to the Aleutian Basin (southwestern end of profile) in water more than 3,000 m (4.0 s) deep. Underlying the basin floor are 4 to 5 km of undeformed strata that overlie a distinct acoustic basement. Magnetic and refraction-velocity data suggest that the basement is oceanic crust of Mesozoic age (Kula? plate; Cooper and others, 1976a, b).

Below the lower part of the continental slope, the rock sequence is characterized by scattered and discontinuous reflectors (Fig. 7). Oceanic basement was not resolved. In contrast, a strong basal reflector, a subshelf basement, can be traced beneath the continental slope to a basement outcrop. Dredged rocks from this exposure, and the flatness of the subslope basement surface, suggest that it is a wave-base unconformity cut across broadly folded rock sequences. These dredge and seismic reflection data attest that the subslope basement platform has subsided from a former shelf depth to at least 1,500 to 1,700 m since about Oligocene time (Marlow and Cooper, 1985).

Basement beneath the shelf can be followed to the northeast below the thick sedimentary section filling Navarin basin (Fig. 7). Near the northeastern end of the profile, strata in the basin are 12 km (7 s) thick. Reflections from the upper basin fill are strong, continuous, and flat. Apparent breaks in the continuity of these reflectors are associated with columnar "wipe-out" zones beneath short discontinuous reflectors in the upper few hundred meters of the section. The acoustic "wipe-out" zones may be produced by shallow accumulations of gas that mask the lateral continuity of deeper reflectors (Marlow and others, 1982). Dredge data originally indicated that the upper 3 to 4 km of the Navarin basin section included beds younger than Oligocene. Deeper strata are poorly reflective, diverge in dip from the overlying strata, and near the base of the section may be early Tertiary in age (Marlow and Cooper, 1985). In 1983, the basin's stratigraphic sequence was explored at a COST well.

Age and history

The Navarin basin province has been covered by many geophysical surveys, thus allowing the mapping of major basins and major structures capable of trapping hydrocarbons. The geologic history of the Navarin basin is poorly known because only one COST well has been drilled in the basin. The well, located on Figure 1, was drilled in 1983 to a total depth of 4,998 m in water 132 m deep (Table 3). Data from this well are described extensively in Turner and others (1984c).

The Cenozoic sedimentary section sampled includes silty and sandy mudstone, diatomaceous mudstone, muddy and fine-grained sandstone interbedded with sandy mudstone, and claystone. The Upper Cretaceous section consists of siltstone, fine-grained sandstone, mudstone, claystone, tuff, and coal. The sedimentary section below 335 m and above 3,895 m was deposited in a marine environment, whereas the section between 3,895 and 4,663 m was deposited in fluvial, flood-plain, and deltaic environments. The section from 3,901 to 4,587 m of Upper Cretaceous strata is intruded by numerous basaltic and diabasic sills. Minimum K-Ar dates on the igneous samples suggest that the emplacement of the sills was in late Oligocene or early Miocene time.

Dredge samples were recovered from the continental slope near the base of the section overlying the structural highs that flank Navarin basin. These samples are Paleocene to Oligocene in age (Jones and others, 1981). While it is difficult to trace this early Tertiary section into the subshelf basins in the Navarin basin, we suspect that a thick Cenozoic section fills the basins. The dipping strata observed in the lower fill of the basin are probably in large part early Tertiary in age. The flat-lying overlying section is probably Miocene and younger. In places, the upper section is broadly folded and truncated, especially in the northeastern end of Navarin basin. This period of deformation may be related to late Miocene and Pliocene uplift and folding in the nearby Koryak Mountains of eastern Siberia (Marlow, 1979; Marlow and others, 1983). Subsidence in the Navarin basin has apparently been continuous during the Cenozoic, possibly extending even into the present day.

TABLE 3. STRATIGRAPHIC SEQUENCES PENETRATED AT NAVARIN COST WELL

Age	Depth to Top (m)
Pliocene	468
Miocene	969
Oligocene	1,739
Eocene	3,743
Late Cretaceous (unconformity)	3,895

ANADYR BASIN

Description

Anadyr basin is a large structural depression that encompasses about 120,000 km^2 and is filled with Upper Cretaceous and Cenozoic sedimentary rocks (Table 1). Onshore, the basin underlies the Anadyr lowlands and is flanked on the south and west by folded Mesozoic rocks of the Koryak foldbelt (McLean, 1979). To the north and east the basin is flanked by the Okhotsk-Chukotsk volcanic belt, a broad bedrock high composed of plutonic and volcanic rocks that extends southeastward from eastern Siberia across the Beringian shelf at least to St. Matthew Island (Patton and others, 1974, 1976; Marlow and others, 1983). To the east and southeast, the basin extends offshore as far as 200 km (Fig. 1). The thickest part of the basin fill is more than 9 km (Marlow and others, 1983).

Seismic reflection and refraction data show that the Anadyr basin is separated from Navarin basin by Anadyr ridge, a northwest-trending bedrock high that is characterized by high-amplitude, short-wavelength magnetic anomalies (Figs. 1 and 8; Marlow and others, 1983). Anadyr ridge may be an offshore extension of a melange belt underlying the Koryak Range. Sonobuoy refraction data suggest that the velocity profile of strata in the offshore part of Anadyr basin is similar to that in Navarin basin. However, the basins are different structurally. Navarin basin is complex and contains both compressional and extensional elements, whereas Anadyr basin is a simple, broad crustal sag, semicircular in outline.

A smaller basement ridge underlies the south end of line 12 (Fig. 9) and is associated with a magnetic high (Fig. 10). Reflectors above horizon Beta (the presumed Paleogene-Mesozoic boundary; Marlow and others, 1983) are deformed in a diapir-like fashion toward the sea floor, suggesting Holocene(?) tectonism. Anadyr basin, in contrast, contains mainly undeformed fill, and appears in cross section as a saucer-shaped structural sag, undisturbed by faulting (Fig. 10).

Age and history

By 1973, 20 wells had been drilled into the onshore part of the basin; these wells are discussed by McLean (1979). Reconstructions of plate motions for the North Pacific suggest that during the Mesozoic, before the formation of the Aleutian arc, the Bering Sea continental margin was a zone of either oblique underthrusting or transform motion between the North American and Kula(?) Plates. Subduction of the Kula(?) Plate beneath Siberia presumably resulted in the formation of the Okhostsk-Chukotsk volcanic arc (Scholl and others, 1975; Marlow and others, 1976; Patton and others, 1976). South of the Okhotsk-

Chukotsk volcanic belt, the melange and olistostrome units, now exposed as discrete tectonic terranes in the Koryak Range (Aleksandrov and others, 1976), were formed by obduction of fragments scraped off the Kula(?) Plate (Scholl and others, 1975).

Uplift of the Okhotsk-Chukotsk volcanic belt as it formed, and uplift of the Koryak melange belt during the middle Cretaceous, presumably resulted in relative crustal downwarping of the intervening area and the initial formation of the forearc Anadyr basin. The bedrock sequence in Anadyr basin, as deduced from seismic reflection data and extrapolation from onshore geology, consists of strongly folded and broken tectonic blocks of Upper Jurassic and Lower Cretaceous rocks (Agapitov and others, 1973; Burlin and others, 1974; McLean, 1979; Marlow and others, 1983). This basement complex probably forms the lower reflecting sequence below the Beta horizon in our seismic reflection profile (Fig. 9) and is characterized by refraction velocities of 5 km/s or greater (Marlow and others, 1983). Offshore data further suggest that the pre-Cretaceous sequence below Anadyr basin is deformed into a broadly semicircular sag that trends southeast beneath the Gulf of Anadyr.

McLean (1979) compared the Anadyr basin and the flanking Okhotst-Chukotsk volcanic arc, respectively, to the Great Valley sequence in central California and its flanking Sierra Nevadan plutonic belt. The Great Valley sequence, which fills a forearc basin, is flanked on its seaward side by the Franciscan Complex or assemblage of coastal California. This sequence of tectonic elements is similar to that associated with Anadyr basin because Anadyr basin is bounded seaward by the Koryak assemblage of eastern Siberia.

NORTON BASIN

Description

The Norton basin underlies the northernmost part of the Bering Sea and Norton Sound and is made up of the St. Law-

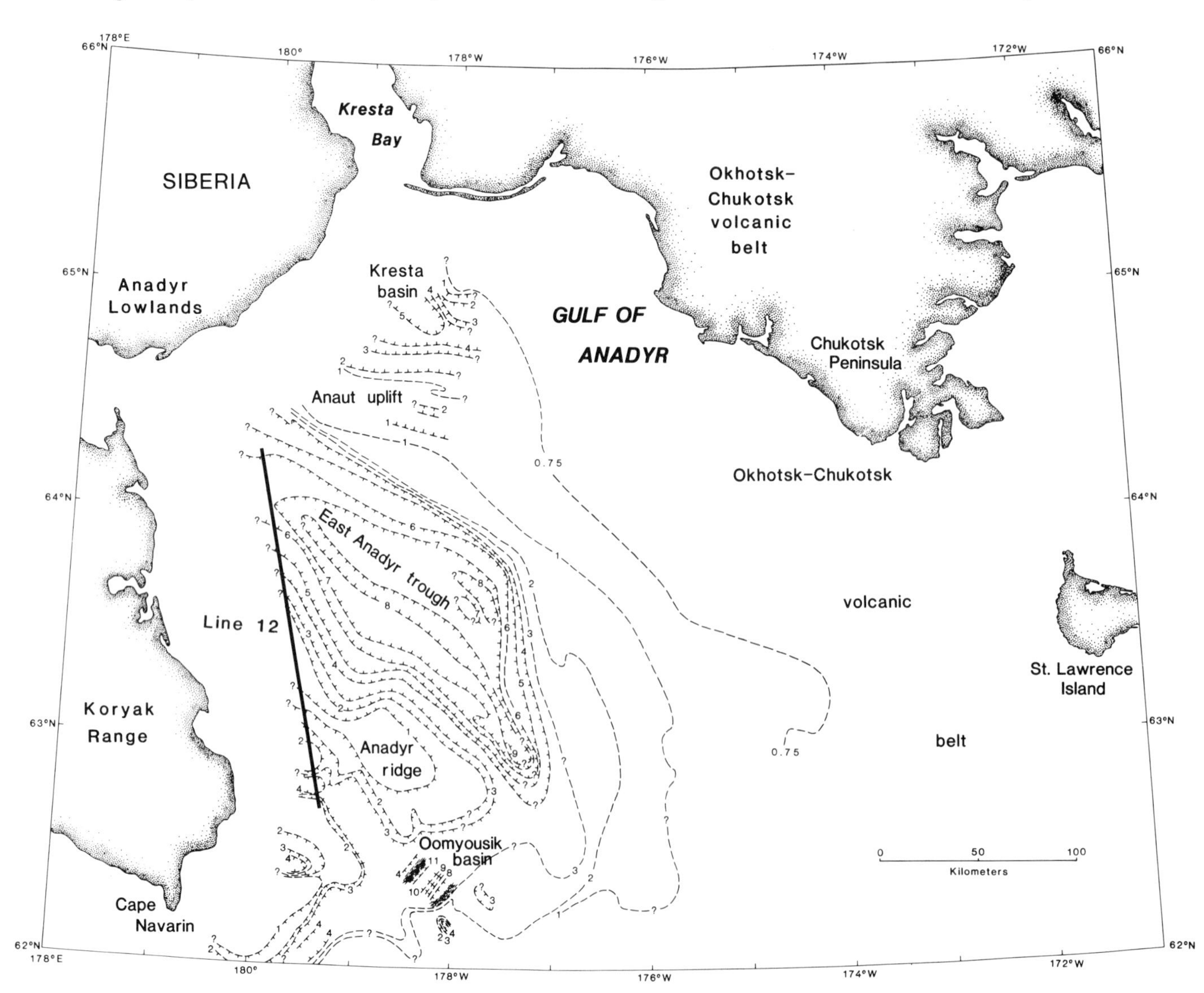

Figure 8. Structure-contour map of total sediment accumulation in the offshore portions of Anadyr basin. Albers equal-area projection. From Marlow and others (1983).

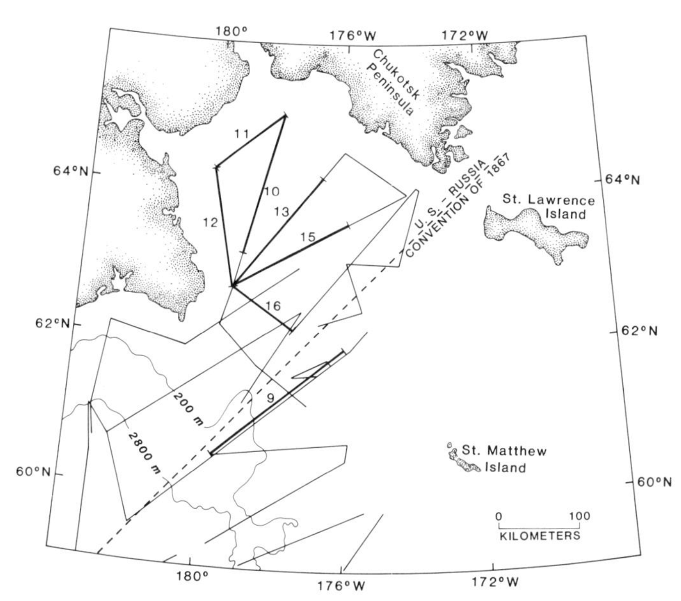

Figure 9. Trackline chart of multichannel seismic reflection surveys across the northern Bering Sea shelf and Anadyr basin. Heavy trackline number 12 is profile shown in Figure 10. Tracklines include surveys made in 1976, 1977, and 1980. Albers equal-area projection. From Marlow and others (1983).

rence and Stuart subbasins (Fisher and others, 1982). The St. Lawrence subbasin lies northeast of St. Lawrence Island and west of the Yukon horst, a structure that extends from the Yukon Delta toward Nome (Fig. 11). The St. Lawrence subbasin is surrounded by shallow basement that dips toward the basin from the Seward Peninsula and St. Lawrence Island, where Precambrian and Paleozoic rocks crop out. Although major normal faults have been active adjacent to the Yukon Delta, gravity data suggest that the Norton basin does not extend beneath the land areas covered by the delta (Fisher and others, 1982).

The Stuart subbasin lies west of the Yukon horst and north of Stuart Island. The main part of the Stuart subbasin is a subcircular depression and includes the deepest part of the Norton basin, where the basin fill is 6.5 km deep. Fault-bounded troughs, filled with basin rocks up to 3 to 4 km thick, extend northeastward and northwestward from the main basin deep. The Nome horst trends east, separating a narrow east-striking graben from the main part of the basin. Thus, the subbasins of the Norton basin differ both in structure and depth of basin fill. Large normal faults in the St. Lawrence subbasin strike northwest, and the maximum thickness of fill in this area is 5 km. In contrast, major normal faults in the Stuart subbasin strike east and northeast.

Stratigraphy

Two COST wells were drilled in the Norton basin by a consortium of oil companies (Table 4). Data from these wells are summarized by Turner and others (1983a, b).

The COST No. 1 well was drilled in the St. Lawrence subbasin, and the COST No. 2 well was drilled in the Stuart subbasin. The oldest rocks penetrated by the two COST wells are quartzite, phyllite, and marble of Precambrian or Paleozoic age. The only evidence for this age is that the rocks in the wells are similar to those of Precambrian and Paleozoic age exposed in the York Mountains on the western Seward Peninsula. These rocks form the acoustic basement and the economic basement for oil and gas. The deepest units of the basin fill penetrated by both wells are poor in fossils, but may be Eocene or older. Nonmarine sandstone, and some conglomerate and coal are 94 m and 537 m thick in wells 1 and 2, respectively. In the COST 2 well, these poorly dated rocks are separated by an unconformity from a 774-m-thick section of middle and upper Eocene siltstone, mudstone, conglomerate, and coal that was deposited in nonmarine and transitional environments. Rocks in the COST No. 1 well that may correlate with this Eocene section form a 776-m-thick unit of sandstone, mudstone, and siltstone dated as Oligocene or older that may have been deposited in a marine environment. Basalt that is anomalously low in alkalis was also encountered in this rock unit; the basalt yields a Miocene K/Ar age of 19.9 Ma. Because the sedimentary rocks that surround the basalt are probably older than Miocene, the basalt is intrusive. In the COST No. 1 well, Oligocene mudstone and shale are 1,584 m thick and

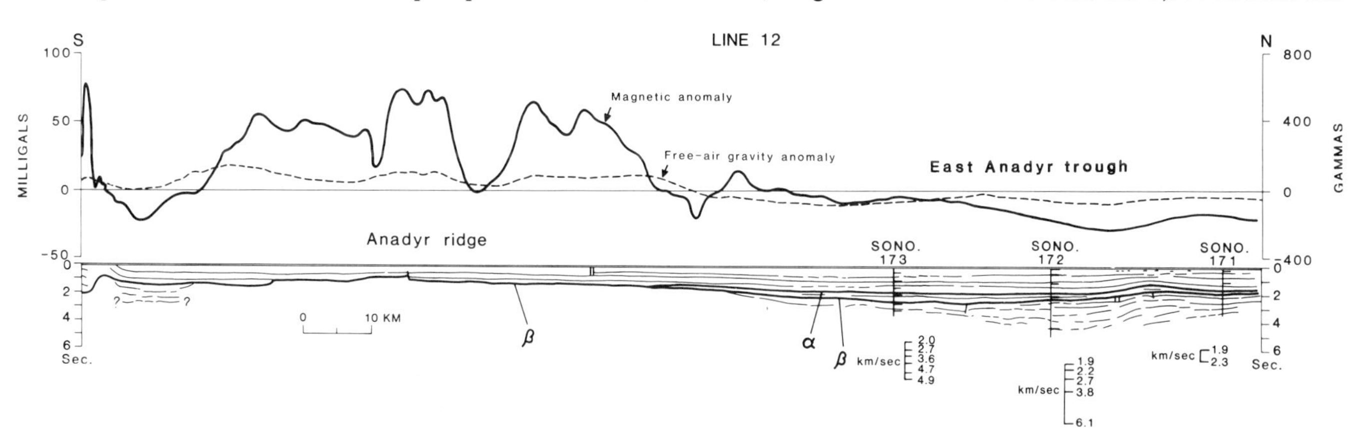

Figure 10. Gravity and magnetic profiles and interpretive drawing of 24-channel seismic reflection line 12 across Anadyr basin and ridge. For location of profile see Figure 9. Sono. 173 refers to sonobuoy refraction station 173 (Marlow and others, 1983).

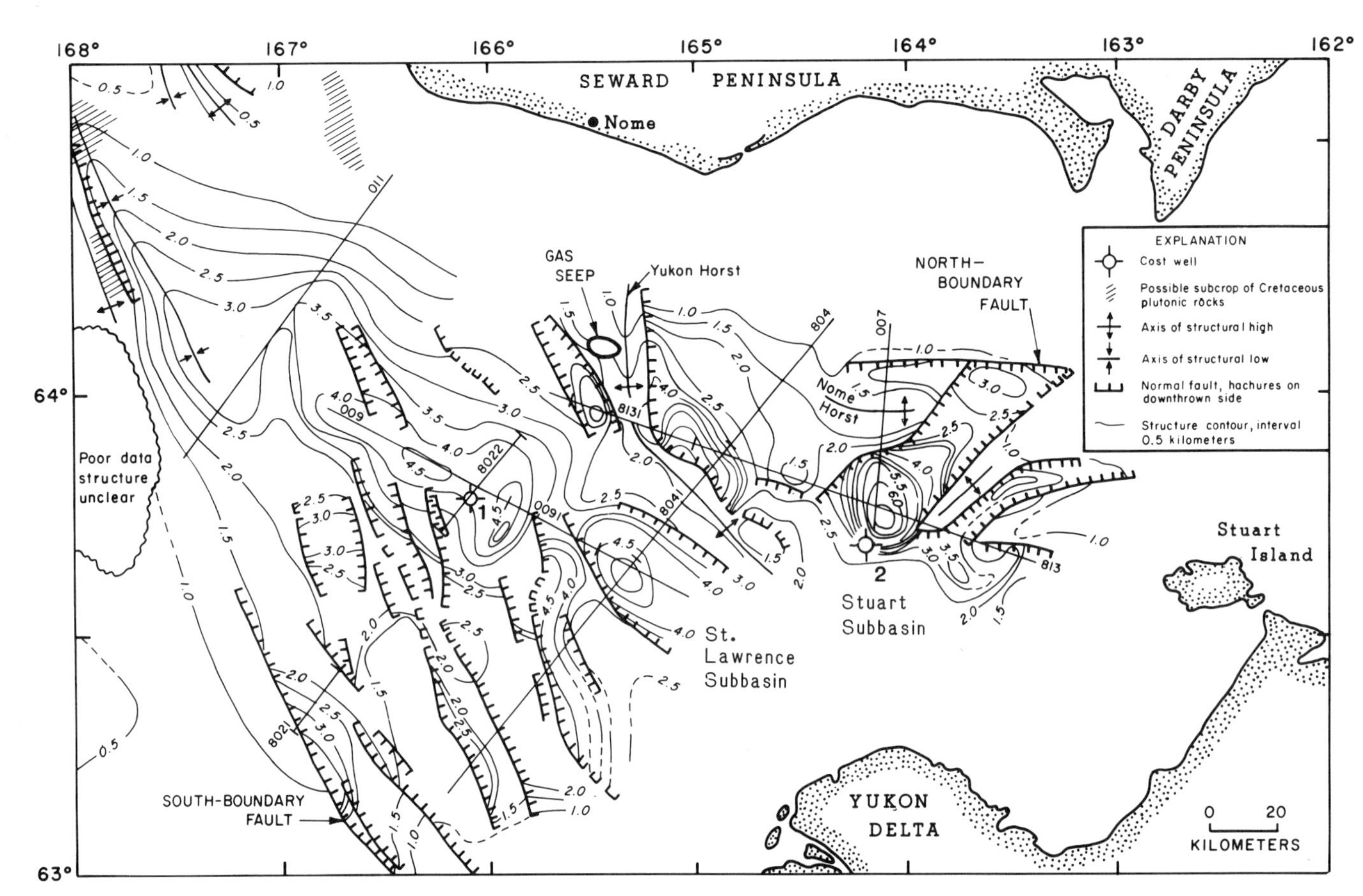

Figure 11. Structure contours on horizon A in Norton basin. Horizon A is an unconformity between probable Precambrian and Paleozoic rocks below and uppermost Cretaceous or lower Paleogene rocks above. From Fisher and others (1982).

were deposited in a middle neritic to upper bathyal environment. In the COST No. 2 well, the Oligocene section is 2,023 m thick. The lower half of the Oligocene section is composed of siltstone, mudstone, and some coal that were deposited in nonmarine or transitional environments, whereas the upper half of this section is composed of mudstone, siltstone, and sandstone that were deposited in transitional and neritic environments. Miocene neritic rocks are 566 m thick in the COST No. 1 well and they consist of mudstone, diatomite, and siltstone. In the COST No. 2 well, 288 m of Miocene siltstone, mudstone, diatomite, and some coal were deposited mainly in a neritic environment. Pliocene neritic diatomite, mudstone, and muddy sandstone in the COST No. 1 well are 402 m thick; neritic rocks of the same age and lithology in the COST No. 2 well are 384 m thick. Shallow-marine Pleistocene and Holocene sandstone, siltstone, and diatomaceous mudstone are 345 m thick in the COST No. 1 well and 355 m thick in the COST No. 2 well.

Age and history

Rocks exposed on the Seward Peninsula underwent a complex hsitory of deformation and intrusion. If basement rocks beneath the Norton basin record a similar history, then these rocks were deformed during the Late Jurassic and Early Cretaceous, during the orogeny that formed the Brooks Range of northern Alaska (Patton and Taileur, 1977; Patton and others, 1977; Roeder and Mull, 1978; Mull, 1979). During the Early and middle Cretaceous, after this orogeny, the area of Norton basin was exposed and eroded. Evidence for this erosion comes from the calcarenites preserved in the Yukon-Koyukuk province east of the Norton basin (Patton, 1973). These rocks were derived from the west and suggest that Paleozoic or older carbonate rocks were exposed where the basin now lies. During the Late Cretaceous, rocks on the Seward Peninsula and perhaps those in the basement under the Norton basin were metamorphosed by the intrusion of granitic rocks.

Fisher and others (1982) proposed, from seismic stratigraphy and onshore geology, that the Norton basin formed during the latest Cretaceous to the end of the Paleocene (70 to 55 Ma), contemporaneously with the formation of grabens in western Alaska that were filled with volcanic rocks. The COST wells penetrated basin fill of middle Eocene age or older. However, the age of the deepest basin fill penetrated is unknown, and nearly 2 km of unsampled strata underlie the basin adjacent to the deepest well penetration. Thus, the age of the initial basin formation is unknown, but is certainly before the middle Eocene. Extensional basins typically undergo rapid subsidence along major faults dur-

TABLE 4. STRATIGRAPHIC SEQUENCE PENETRATED AT NORTON COST WELLS

Age	Depth to Top (m)
COST WELL NO. 1	
Pleistocene	?
Pliocene	402
Miocene	804
Oligocene	1,369
Oligocene or older	2,953
Eocene(?) or older	3,729
Metamorphic basement:	
Paleozoic or older	3,824
COST WELL NO. 2	
Pleistocene	?
Pliocene	402
Miocene	786
Oligocene	1,074
Eocene	3,097
Eocene(?) or older	3,871
Metamorphic basement:	
Paleozoic or older	4,407

ing their early histories; consequently, the thick Oligocene section penetrated by both wells suggests that the basin was probably young and active during the Oligocene. Hence, Norton basin probably formed sometime during the early Cenozoic, from the Paleocene(?) to the middle Eocene. In the two COST wells the thickness of Oligocene rocks, uncorrected for compaction, makes up 35 to 45 percent of the basin fill. This thick Oligocene section suggests that during this epoch, and probably earlier, the basin underwent rapid subsidence. Seismic reflection data show that this subsidence occurred mainly along large normal faults. In the St. Lawrence subbasin these faults show the greatest throw near the Yukon Delta, suggesting that the Kaltag fault, a major strike-slip fault mapped east of Norton basin, may have contributed to basin formation.

Seismic reflection data show that alluvial fans were deposited along the flanks of some of the major basement highs as the basin underwent rapid subsidence during Paleocene(?) through Oligocene time. Subsidence was rapid enough in the St. Lawrence subbasin, so that during the Eocene or Oligocene, a marine incursion caused the environment to shift from nonmarine to outer neritic and upper bathyal. In contrast, in the Stuart subbasin, separated from the other subbasin by the high-standing Yukon horst, a nonmarine environment of deposition continued until well into the Oligocene. During Miocene, Pliocene, and Quaternary time, deposition of marine strata occurred in both subbasins. The Neogene and Quaternary rocks are considerably thinner (even without correction for compaction) than the Oligocene rocks, indicating that a general slowing of the rate of basin subsidence occurred during the past 22 m.y. This decreasing thickness of fill mirrors the decreasing throw of faults upward through the basin fill.

CONCLUSIONS

The major basins of the Beringian shelf are listed in Table 1. Detailed descriptions of the basins can be found in Marlow and Cooper (1980a), Fisher and others (1982), Marlow and others (1983), and Turner and others (1983a, b; 1984a, b, and c). Bristol Bay, Anadyr, Norton, and Navarin basins appear to have formed as large downwarps in the Earth's crust, and block-like structures formed during down warping. These structures are buried in places by sedimentary layers that may form traps for migrating hydrocarbons.

St. George basin and the southern half of Navarin basin are tensional features, formed by a pulling apart of the earth's crust. Continued subsidence of these basins through geologic time has formed fold structures associated with growth faults along the flanks of the basins. These fold structures may also trap oil and gas in volumes that are economically significant.

The northern half of Navarin basin is dominated by compressional structures such as anticlines and domes that formed when the basin fill may have been subjected to horizontal compression. Anticlines and domes are the classic geologic structures from which most of the world's oil has been produced.

Norton basin contains two subbasins that probably began to form in early Cenozoic (Paleocene[?] through middle Eocene) time. These subbasins are underlain by basement rocks of Precambrian or Paleozoic age. If significant amounts of hydrocarbons occur in Norton basin, then data from the COST wells suggest the hydrocarbons will be present mainly as condensate and gas.

The shelf basins are currently being explored for hydrocarbons by drilling, but their hydrocarbon potential remains unknown. Navarin basin is the second largest shelf basin, however, and contains many large geologic structures that produce geophysical signatures that may indicate hydrocarbons within the structures. The onshore portion of nearby Anadyr basin (Fig. 1) has been drilled by the Soviet Union (McLean, 1979). Those wells revealed shows of oil and gas that suggest that the offshore portion of the basin may also contain hydrocarbons.

REFERENCES CITED

Agapitov, D. I., Ivanov, V. V., and Krainov, V. G., 1973, New data on the geology and prospects in petroleum/gas-bearing Anadyr basin: Akademiya Nauk SSR North-East Group Institute of Interior Center Translations, v. 49, p. 23–29 (in Russian).

Aleksandrov, A. A., and 6 others, 1976, New data on the tectonics of the Koryak Highlands: Geotectonics, v. 9, p. 292–299.

Beikman, H., 1974, Preliminary geologic map of the southwest quadrant of Alaska: U.S. Geological Survey Miscellaneous Field Studies Map MF-611, 2 sheets, scale 1:200,000.

Brockway, R., and 7 others, 1975, Bristol Bay region, stratigraphic correlation section, southwest Alaska: Anchorage, Alaska Geological Society.

Burk, C. A., 1965, Geology of the Alaska Peninsula Island arc and continental margin: Geological Society of America Memoir 99, 250 p.

Burlin, Yu. K., Dontsov, V. V., and Pastukhova, T. N., 1974, Future direction of oil and gas exploration in northeastern U.S.S.R.: Geologiya Nefti Gaza, no. 5, p. 17–22 (in Russian) (1975 summary, *in* Petroleum Geology, v. 12, no. 5, p. 222–223).

Cooper, K. A., and Marlow, M. S., 1983, Preliminary results of geophysical and geological studies of the Bering Sea shelf during 1982: U.S. Geological Survey Open-File Report 82-322, 8 p.

Cooper, A. K., Marlow, M. S., and Scholl, D. W., 1976a, Mesozoic magnetic lineations in the Bering Sea marginal basin: Journal of Geophysical Research, v. 81, p. 1916–1934.

Cooper, A. K., Scholl, D. W., and Marlow, M. S., 1976b, Plate tectonic model for the evolution of the eastern Bering Sea basin: Geological Society of America Bulletin, v. 87, p. 1119–1126.

Fisher, M. A., Patton, W. W., Jr., and Holmes, M. L., 1982, Geology of Norton basin and continental shelf beneath northwestern Bering Sea, Alaska: American Association of Petroleum Geologists Bulletin, v. 66, p. 225–285.

Hatten, C. W., 1971, Petroleum potential of Briston Bay Basin Alaska: American Association of Petroleum Geologists Memoir 15, p. 105–108.

Hopkins, D. M., and 10 others, 1969, Cretaceous, Tertiary, and early Pleistocene rocks from the continental margin in the Bering Sea: Geological Society of America Bulletin, v. 80, p. 1471–1480.

Jones, D. L., and Silberling, N. J., 1979, Mesozoic stratigraphy; The key to tectonic analysis of southern and central Alaska: U.S. Geological Survey Open-File Report 79-1200, 37 p.

Jones, D. M., and 6 others, 1981, Age, mineralogy, physical properties, and geochemistry of dredge samples from the Bering Sea continental margin: U.S. Geological Survey Open-File Report 81-1297, 68 p.

Marlow, M. S., 1979, Hydrocarbon prospects in Navarin basin province, northwest Bering Sea shelf: Oil and Gas Journal, October, p. 190–196.

Marlow, M. S., and Cooper, A. K., 1980a, Mesozoic and Cenozoic structural trends under southern Bering Sea shelf: American Association of Petroleum Geologists Bulletin, v. 64, p. 2139–2155.

——, 1980b, Multichannel seismic-reflection profiles collected in 1976 in the southern Bering Sea shelf: U.S. Geological Survey Open-File Report 80-389, 2 p.

——, 1983, Wandering terranes in southern Alaska: The Aleutia microplate and implications for the Bering Sea: Journal of Geophysical Research, v. 88, p. 3439–3446.

——, 1985, Regional geology of the Beringian continental margin, *in* Nasu, N., and others, eds., Formation of active ocean margins: Tokyo, Terra Scientific Publishing Co., p. 497–515.

Marlow, M. S., Scholl, D. W., Cooper, A. K., and Buffington, E. C., 1976, Structure and evolution of Bering Sea shelf south of St. Lawrence Island: American Association of Petroleum Geologists Bulletin, v. 60, p. 161–183.

Marlow, M. S., Scholl, D. W., and Cooper, A. K., 1977, St. George basin Bering Sea shelf; A collapsed Mesozoic margin, *in* Talwani, M., and Pitman, W. C., III, eds., Island arcs, deep sea trenches, and back-arc basins: American Geophysical Union Maurice Ewing Series, v. 1, p. 211–220.

Marlow, M. S., Cooper, A. K., Scholl, D. W., Vallier, T. L., and McLean, H., 1982, Ancient plate boundaries in the Bering Sea region, *in* Leggett, J. K., ed., Trench-forearc geology; Sedimentation and tectonics on modern and ancient active plate margins: Geological Society of London Special Publication 10, p. 201–211.

Marlow, M. S., Cooper, A. K., and Childs, J. R., 1983, Tectonic evolution of Gulf of Anadyr and formation of Anadyr and Navarin basins: American Association of Petroleum Geologists Bulletin, v. 67, p. 646–665.

McGeary, S. E., and Ben-Avraham, S., 1981, Allochthonous terranes in Alaska; Implications for the structure and evolution of the Bering Sea shelf: Geology, v. 9, p. 608–613.

McLean, H., 1977, Organic geochemistry, lithology, and paleontology of Tertiary and Mesozoic rocks from wells on the Alaska Peninsula: U.S. Geological Survey Open-File Report 77-813, 63 p.

——, 1979, Review of petroleum geology of Anadyr and Khatyrka basins, USSR: American Association of Petroleum Geologists Bulletin, v. 63, p. 1467–1477.

Mull, C. G., 1979, Nanushuk Group deposition and the late Mesozoic structural evolution of the central and western Brooks Range and Arctic slope: U.S. Geological Survey Circular 794, p. 5–13.

Packer, D. R., and Stone, D. B., 1972, An Alaskan Jurassic paleomagnetic pole and the Alaskan orocline: Nature; Physical Science, v. 237, p. 25–26.

Patton, W. W., Jr., 1973, Reconnaissance geology of the northern Yukon-Koyukuk Province, Alaska: U.S. Geological Survey Professional Paper 774-A, 17 p.

Patton, W. W., Jr., and Tailleur, I. L., 1977, Evidence in the Bering Strait region for differential movement between North America and Eurasia: Geological Society of America Bulletin, v. 88, p. 1298–1304.

Patton, W. W., Jr., Lanphere, M. A., Miller, T. P., and Scott, R. A., 1974, Age and tectonic significance of volcanic rocks on St. Matthew Island, Bering Sea, Alaska: Geological Society of America Abstracts with Programs, v. 6, p. 905–906.

——, 1976, Age and tectonic significance of volcanic rocks on St. Matthew Island, Bering Sea, Alaska: U.S. Geological Survey Journal of Research, v. 4, p. 67–73.

Patton, W. W., Jr., Tailleur, I. L., Brosge, W. P., and Lanphere, M. A., 1977, Preliminary report on the ophiolites of northern and western Alaska, *in* Coleman, R. G., and Irwin, W. P., eds., North American ophiolites: Oregon Department of Geology and Mineral Industry Bulletin 95, p. 51–57.

Reed, B. L., and Lanphere, M. A., 1973, Alaska-Aleutian range batholith; Geochronology, chemistry, and relation to circum-Pacific plutonism: Geological Society of America Bulletin, v. 84, p. 2583–2610.

Roeder, D., and Mull, C. G., 1978, Tectonics of Brooks Range ophiolites: American Association of Petroleum Geologists Bulletin, v. 62, p. 1696–1713.

Scholl, D. W., Buffington, E. C., and Marlow, M. S.,1975, Plate tectonics and the structural evolution of the Aleutian-Bering sea region, *in* Forbes, R. B., ed., Contributions to the geology of the Bering Sea Basin and adjacent regions: Geological Society of America Special Paper 151, p. 1–32.

Scholl, D. W., Vallier, T. L., and Stevenson, A. J., 1987, Geologic evolution and petroleum geology of the Aleutian Ridge, *in* Scholl, D. W., Grantz, A., and Vedder, J. G., eds., Geology and resource potential of the continental margin of western North America and adjacent ocean basins; Beaufort Sea to Baja California: Houston, Texas, Circum-Pacific Council for Energy and Mineral Resources Series, v. 6, p. 123–155.

Stone, D. B., Panuska, B. C., and Packer, D. R., 1982, Paleolatitudes versus time for southern Alaska: Journal of Geophysical Research, v. 87, p. 3697–3707.

Turner, R. F., and 5 others, 1983a, Geological and operational summary Norton Sound COST No. 1 well Norton Sound, Alaska: U.S. Geological Survey Open-File Report 83-124, 164 p.

Turner, R. F., and 6 others, 1983b, Geological and operational summary Norton Sound COST No. 2 well Norton Sound, Alaska: U.S. Geological Survey Open-File Report 83-557, 154 p.

Turner, R. F., and 6 others, 1984a, Geological and operational summary St. George Basin COST No. 1 well Bering Sea, Alaska: Minerals Management Service OCS Report MMS 84-16, 105 p.

Turner, R. F., and 6 others, 1984b, Geological and operational summary St. George Basin COST No. 2 well Bering Sea, Alaska: Minerals Management Service OCS Report MMS 84-18, 100 p.

Turner, R. F., and 7 others, 1984c, Geological and operational summary Navarin Basin COST No. 1 well Bering Sea, Alaska: Minerals Management Service OCS Report MMS 84-31, 245 p.

Yanshin, A. L., 1966, Tectonic map of Eurasia: Moscow, Geological Institute, Academy of Sciences, USSR, scale 1:5,000,000. (in Russian)

Manuscript Accepted by the Society May 10, 1990

ACKNOWLEDGMENTS

We thank Paul Carlson and Mark Holmes for their useful and thoughtful reviews.

Printed in U.S.A.

The Geology of North America
Vol. G-1, The Geology of Alaska
The Geological Society of America, 1994

Chapter 9

Geology of southwestern Alaska

John Decker*
Alaska Division of Geological and Geophysical Surveys, 794 University Avenue, Fairbanks, Alaska 99709
Steven C. Bergman
ARCO Oil and Gas Company, 2300 West Plano Parkway, Plano, Texas 75075
Robert B. Blodgett*
Department of Geology, Oregon State University, Corvallis, Oregon 97331
Stephen E. Box*
Branch of Alaskan Geology, U.S. Geological Survey, 4200 University Drive, Anchorage, Alaska 99508
Thomas K. Bundtzen
Alaska Division of Geological and Geophysical Surveys, 3700 Airport Way, Fairbanks, Alaska 99709
James G. Clough
Alaska Division of Geological and Geophysical Surveys, 794 University Avenue, Fairbanks, Alaska 99709
Warren L. Coonrad
Branch of Alaskan Geology, U.S. Geological Survey, 345 Middlefield Road, Menlo Park, California 94025
Wyatt G. Gilbert
Alaska Division of Geological and Geophysical Surveys, 3601 C Street, Anchorage, Alaska 99510
Martha L. Miller and John M. Murphy*
Branch of Alaskan Geology, U.S. Geological Survey, 4200 University Drive, Anchorage, Alaska 99508
Mark S. Robinson
Alaska Division of Geological and Geophysical Surveys, 794 University Avenue, Fairbanks, Alaska 99709
Wesley K. Wallace*
ARCO Alaska, Inc., P.O. Box 100360, Anchorage, Alaska 99510

INTRODUCTION

Southwest Alaska lies between the Yukon-Koyukuk province to the north, and the Alaska Peninsula to the south (Wahrhaftig, this volume). It includes the southwestern Alaska Range, the Kuskokwim Mountains, the Ahklun Mountains, the Bristol Bay Lowland, and the Minchumina and Holitna basins. It is an area of approximately 175,000 km^2, and, with the exception of the rugged southwestern Alaska Range and Ahklun Mountains, consists mostly of low rolling hills.

The oldest rocks in the region are metamorphic rocks with Early Proterozoic protolith ages that occur as isolated exposures in the central Kuskokwim Mountains, and in fault contact with Mesozoic accretionary rocks of the Bristol Bay region. Precambrian metamorphic rocks also occur in the northern Kuskokwim Mountains and serve as depositional basement for Paleozoic shelf deposits. A nearly continuous sequence of Paleozoic continental margin rocks underlies much of the southwestern Alaska Range and northern Kuskokwim Mountains. The most extensive unit in southwest Alaska is the predominantly Upper Cretaceous Kuskokwim Group, which, in large part, rests unconformably on older rocks of the region. Volcanic rocks of Mesozoic age are common in the Bristol Bay region, and volcanic and plutonic rocks of latest Cretaceous and earliest Tertiary age are common throughout southwest Alaska.

Two major northeast-trending faults are known to traverse southwest Alaska, the Denali-Farewell fault system to the south, and the Iditarod–Nixon Fork fault to the north. Latest Cretaceous and Tertiary right-lateral offsets of less than 150 km characterize both faults. The Susulatna lineament (or Poorman fault), north of the Iditarod–Nixon Fork fault, juxtaposes contrasting Early Cretaceous and older rocks and may have had significant Cretaceous

*Present addresses: Decker, ARCO Alaska, Inc., P.O. Box 100360, Anchorage, Alaska 99510; Blodgett, Branch of Paleontology and Stratigraphy, U.S. Geological Survey, National Center MS 9709, Reston, Virginia 22092; Box, U.S. Geological Survey, Room 656, West 920 Riverside Avenue, Spokane, Washington 99201; Murphy, Department of Geology, University of Wyoming, Laramie, Wyoming 82071; Wallace, Department of Geology and Geophysics, University of Alaska, Fairbanks, Alaska 99775-0760.

Decker, J., Bergman, S. C., Blodgett, R. B., Box, S. E., Bundtzen, T. K., Clough, J. G., Coonrad, W. L., Gilbert, W. G., Miller, M. L., Murphy, J. M., Robinson, M. S., and Wallace, W. K., 1994, Geology of southwestern Alaska, *in* Plafker, G., and Berg, H. C., eds., The Geology of Alaska: Boulder, Colorado, Geological Society of America, The Geology of North America, v. G-1.

lateral displacement. The Mulchatna fault, south of the Denali-Farewell fault system roughly corresponds to a pronounced aeromagnetic discontinuity and is likely to be a major basement fault.

The rocks of southwest Alaska (Fig. 1) can be grouped into three contrasting assemblages: (1) the predominantly Paleozoic continental margin rocks of the southwestern Alaska Range and northern Kuskokwim Mountains (Farewell terrane, defined below), (2) the predominantly Mesozoic accretionary rocks of the

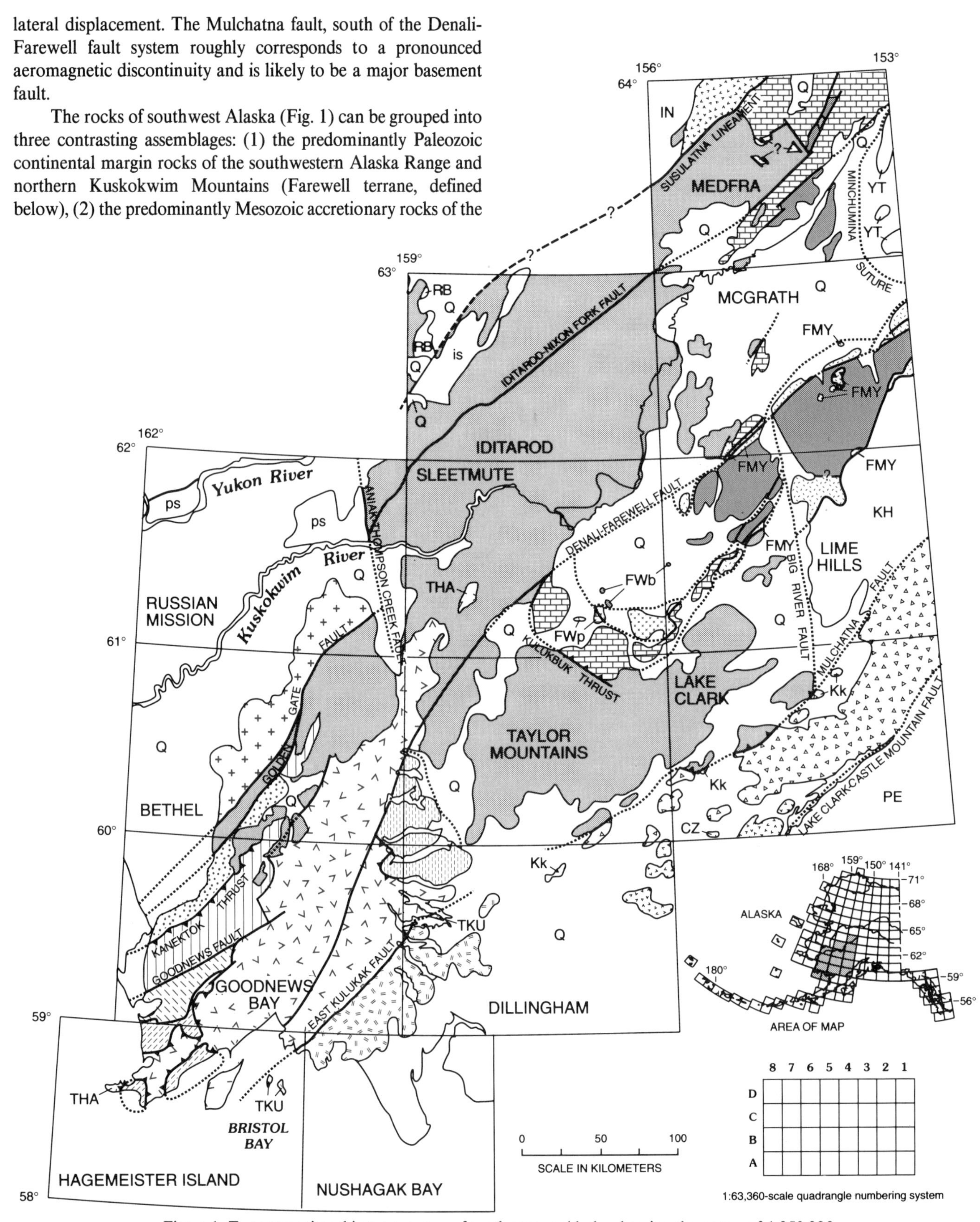

Figure 1. Tectonostratigraphic terrane map of southwestern Alaska showing the names of 1:250,000 scale quadrangles and the 1:63,360-scale quadrangle numbering system.

northern Bristol Bay region, and (3) postaccretionary clastic sedimentary rocks and mainly calc-alkaline volcanic and plutonic rocks that tie together assemblages 1 and 2. In order to establish a framework for discussion of these rocks, we have consolidated previously described tectonostratigraphic terranes (Jones and Silberling, 1979; Jones and others, 1981) in the southwestern Alaska Range and northern Kuskokwim Mountains, and subdivided the terranes of the Bristol Bay region. In the southwestern Alaska Range and northern Kuskokwim Mountains, sufficient stratigraphic information is now known to allow genetic relations to be established between previously defined terranes and to allow a more conventional stratigraphic approach to our discussion of the geology in this area. In the Bristol Bay region, however, many critical stratigraphic relations are unclear and our discussion of these rocks follows the tectonostratigraphic subterrane framework defined for this area by Box (1985). We believe that this somewhat different format for each area allows the best overall treatment of the geology of southwest Alaska.

EXPLANATION

OVERLAP ASSEMBLAGES

Q — Quaternary Surficial Deposits

K — Kuskokwim Group

Cz — Cenozoic Deposits

TERRANES OF SOUTHWEST ALASKA

Alaska Range and Kuskokwim Mountains

Farewell Terrane

FMY — Mystic Sequence

White Mountain Sequence

FWb — Basinal Facies

FWt — Transitional Facies

FWp — Platform Facies

Bristol Bay Region

NY — Nyack Terrane

Togiak Terrane

THA — Hagemeister Subterrane

TKU — Kulukak Subterrane

Goodnews Terrane

GNU — Nukluk Subterrane

GTC — Tikchik Subterrane

GPT — Platinum Subterrane

GCP — Cape Pierce Subterrane

KI — Kilbuck Terrane

ADJACENT TERRANES

KH — Northern Kahiltna Terrane

Kk — Southern Kahiltna Terrane

MC — McKinley Terrane

PE — Peninsular Terrane

IN — Innoko Terrane

PN — Pingston Terrane

RB — Ruby Terrane

YT — Yukon-Tanana Terrane

UNITS OF UNCERTAIN TERRANE AFFINITY

is — Idono Sequence

ps — Portage Sequence

——— Contact

—? -- Fault—Dashed where approximate, dotted where concealed, queried where uncertain.

▲▲····· Thrust fault—Dotted where concealed. Sawteeth on upper plate.

SOUTHWESTERN ALASKA RANGE AND NORTHERN KUSKOKWIM MOUNTAINS

The predominantly Paleozoic and Mesozoic rocks of the southwestern Alaska Range and northern Kuskokwim Mountains that are discussed in this chapter include parts of the Nixon Fork, Dillinger, and Mystic terranes, which Silberling and others (this volume) and Jones and Silberling (1979) originally defined as discrete, fault-bounded tectonostratigraphic terranes. Although these terranes have distinctive stratigraphies, several other authors have suggested that the Nixon Fork and Dillinger terranes consist instead of contrasting facies that were deposited within a single depositional basin (Bundtzen and Gilbert, 1983; Blodgett, 1983a; Blodgett and Gilbert, 1983; Gilbert and Bundtzen, 1983a; Blodgett and Clough, 1985), and that rocks of the Mystic terrane were deposited on rocks of the Dillinger terrane (Gilbert and Bundtzen, 1984). Recently, the identification of interfingering relationships and transitional or slope facies between the platform deposits of the Nixon Fork terrane, and the basinal deposits of the Dillinger terrane have been recognized in the White Mountain area (Gilbert, 1981; Blodgett and Gilbert, 1983), in the western Lime Hills and eastern Sleetmute Quadrangles (Blodgett and others, 1984; Clough and Blodgett, 1985; Smith and others, 1985), and in the Minchumina basin (M. W. Henning, written commun., 1985). Because genetic relationships exist among the Nixon Fork, Dillinger, and Mystic terranes, we herein redefine them as the Farewell terrane, comprising the White Mountain and Mystic sequences (Fig. 2).

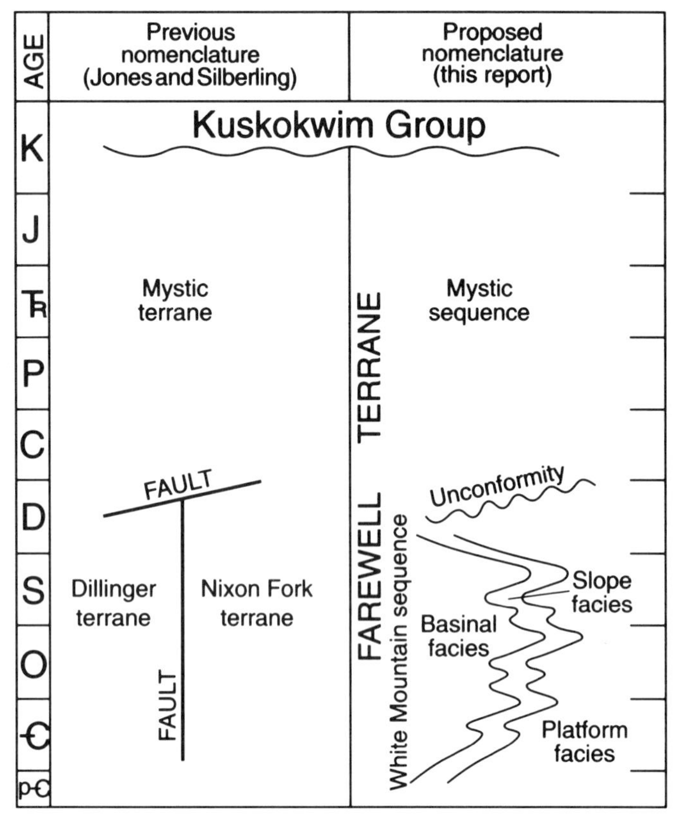

Figure 2. Farewell terrane terminology and relationships. This diagram compares the terminology proposed in this report with the terrane nomenclature of Jones and Silberling (1979). The figure schematically illustrates the genetic relations between the basinal facies (Dillinger terrane) and the platform facies (Nixon Fork terrane), and the unconformable relationship between the Mystic sequence and the White Mountain sequence. See text for discussion.

Farewell terrane

Definition. The Farewell terrane (Figs. 2 and 3) consists of a coherent but locally highly deformed lower Paleozoic through Lower Cretaceous continental margin sequence, deposited, in part, on Precambrian basement, and locally overlain by Cretaceous and Tertiary volcanic and sedimentary rocks. The Farewell terrane is divided into two distinctive sequences: (1) the White Mountain sequence, which consists of Middle Cambrian through Middle Devonian stable continental margin deposits; and (2) the overlying Mystic sequence, which consists of a highly variable assemblage of rocks of Late Devonian through Early Cretaceous age. We use the term "sequence" in roughly the same sense as "stratigraphic sequence" defined by Sloss (1963) and "depositional sequence" defined by Mitchum and others (1977). We define a sequence as a regionally extensive stratigraphic unit or body of rock bound above and below by unconformities or their correlative conformities.

White Mountain sequence. The White Mountain sequence is named for a well-exposed section near White Mountain in the McGrath A-4 Quadrangle. At White Mountain, three principal facies are recognized: a shallow-marine platform facies, a deep-marine basinal facies, and a transitional or slope facies. Historically, sections consisting predominantly of platform deposits have been called the Nixon Fork terrane (Patton, 1978; Jones and others, 1981) and sections consisting predominantly of basinal deposits have been called the Dillinger terrane (Jones and Silberling, 1979; Jones and others, 1982). In this chapter, we do not use the Nixon Fork and Dillinger terrane terminology; instead, we consider them together as comprising the White Mountain sequence.

Mystic sequence. The upper sequence is here designated the Mystic sequence. It includes rocks of the Mystic terrane of Jones and Silberling (1979) and age-equivalent rocks that overlie the White Mountain sequence. The Mystic sequence represents less stable tectonic conditions and more diverse local depositional environments than those in the White Mountain sequence. Facies relations between rock units within the Mystic sequence are poorly understood, but the upper and lower boundaries of the sequence are well defined. The Mystic sequence is stratigraphically bounded by the two most persistent units in southwest Alaska: the White Mountain sequence below and the Kuskokwim Group above. At most observed sites, the depositional contact between the Mystic and White Mountain sequences is an angular unconformity of Middle Devonian age, although apparently conformable relationships have been reported (Blodgett and Gilbert, 1983; Patton and others, 1984a). The upper boundary of the Mystic sequence with the Kuskokwim Group is a major angular unconformity of latest Early Cretaceous age that commonly cuts completely through the Mystic sequence into strata of the underlying White Mountain sequence. Further work may show that some of the wide variety of rocks we include within the Mystic sequence are tectonic slivers or independent terranes with no depositional link to rocks of the underlying White Mountain sequence. For this report, however, we include within the Mystic sequence all rock types of Late Devonian through Early Cretaceous age within the Farewell terrane that overlie the White Mountain sequence and/or underlie the Kuskokwim Group.

The main outcrop belt of the Mystic sequence (and type area of the Mystic terrane of Jones and Silberling, 1979) occurs in the Talkeetna Quadrangle (Nokleberg and others, this volume, chapter 10) where it is bounded on the north by the Denali-Farewell fault, and on the south by the Shellabarger fault (Jones and others, 1983). In this region, the Mystic sequence contains three fault-bounded assemblages (Jones and Silberling, 1979). The oldest, structurally lowest, and southernmost unit is a tectonic mixture (melange) composed of infolded and faulted blocks of typical White Mountain sequence lithologies, including Ordovician graptolitic shale, Silurian massive algal limestone, and pillow basalt (Bundtzen and Gilbert, unpublished 1983 field data). On the basis of stratigraphic and structural position, the melange most likely formed in post–Early Devonian time. To the north, structurally overlying these rocks, are strongly folded, partly coherent sanstone, gritstone, and limestone of Late Devonian age, phosphatic black radiolarian chert of latest Devonian and Mississippian age, and green chert and argillite of Pennsylvanian age (Jones and Silberling, 1979). The highest and most extensive unit is a thick, disrupted, flysch-like sequence containing minor chert, and thick, pebble to boulder conglomerate lenses containing clasts of limestone and black cherty argillite (Jones and Silberling, 1979). Permian plant fossils are known from one locality in the Talkeetna Quadrangle.

Description of key areas. *White Mountain area.* The White Mountain sequence (Fig. 3) is well exposed north of the Farewell fault in the Cheeneetnuk River area, near White Mountain in the McGrath A-4 and A-5 Quadrangles. There, Gilbert

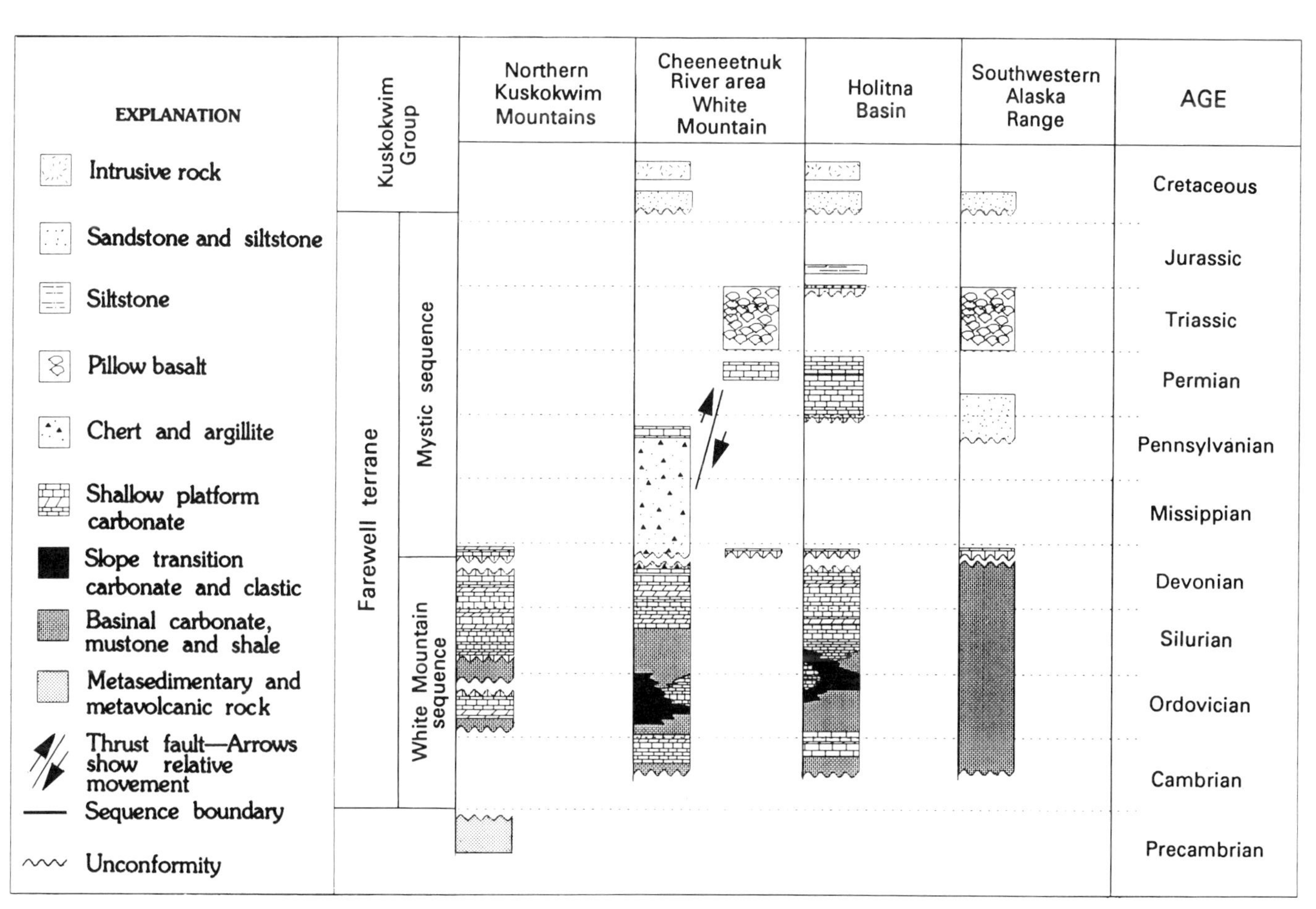

Figure 3. Generalized stratigraphic sections of the Farewell terrane showing the lithofacies variations within the White Mountain sequence. The Kuskokwim Group, the Mystic sequence, and the White Mountain sequence are separated by major interregional unconformities.

(1981) described a 6,000-m-thick sequence of Cambrian(?) through lower Middle Devonian rocks representative of marine shelf, slope, and basinal depositional environments. The oldest rocks in the Cheeneetnuk River area are of Cambrian(?) through Late Ordovician age and consist of oolitic limestone, banded mudstone, silty limestone turbidites, oncolitic algal limestone, and limestone breccia, representative of alternating deep-marine and shallow-marine depositional environments. The Upper Silurian through Middle Devonian strata consist of limestone and dolomite containing shallow-marine fauna, indicating relatively stable platform conditions. Fossils described from the region include Ordovician brachiopods (Potter and others, 1980) and Silurian conodonts (Savage and others, 1983) from White Mountain, and early Middle Devonian goniatites (House and Blodgett, 1982), sponges (Rigby and Blodgett, 1983), gastropods (Blodgett and Rohr, 1989), and the udotaecean alga *Lancicula sergaensis* Shuysky (Poncet and Blodgett, 1987) from the upper part of the Cheeneetnuk Limestone. Formal stratigraphic names remain to be applied in the White Mountain area, with the exception of the Cheeneetnuk Limestone (Blodgett and Gilbert, 1983), which is of Eifelian (early Middle Devonian) age in its upper part and forms the top of the platform carbonate succession in that area.

The White Mountain sequence in the White Mountain area is overlain by upper Paleozoic chert and argillite of the Mystic sequence. There the Mystic sequence consists of Middle(?) and Upper Devonian and Carboniferous deep-water argillite, siliceous shale, and chert, which depositionally overlie the platform carbonate rocks of the White Mountain sequence (Gilbert, 1981). Late Pennsylvanian trilobites occur in the McGrath A-5 Quadrangle and have been described by Hahn and others (1985) and Hahn and Hahn (1985). According to Blodgett and Gilbert (1983), the contact between the Cheeneetnuk Limestone at the top of the White Mountain sequence and the argillite and chert unit at the base of the Mystic sequence is abrupt, but apparently conformable.

Northern Kuskokwim Mountains. The White Mountain sequence in the northern Kuskokwim Mountains consists of a thick section of lower Paleozoic platform carbonate rocks that unconformably overlie metasedimentary and metavolcanic rocks of Precambrian age (Patton and others, 1984a; Silberman and others, 1979; Dillon and others, 1985). The metamorphic basement rocks consist of greenschist-facies pelitic schist, calc-schist, and metavolcanic rocks (Patton and others, 1984a). Dutro and Patton (1981) and Patton and others (1977), respectively, described stratigraphic sections for the lower and middle Paleozoic, and for the upper Paleozoic and Mesozoic in the White Mountain and Mystic sequences in the Medfra Quadrangle. Dutro and Patton (1982) later named four new formations with an aggregate thickness of more than 5,000 m (Novi Mountain, Telsitna, Paradise Fork, and Whirlwind Creek Formations) to encompass the Lower Ordovician through Devonian stratal sequence of the Medfra Quadrangle. As reported by Patton and others (1984a, p. 3–4). "Depositional environments range from mainly supratidal, characterized by laminated silty limestone in Lower Ordovician and Middle Devonian beds, to shallow marine, distinguished by a complex array of shallow-water carbonate facies that include reefoid bodies in the Upper Ordovician and Middle Devonian beds. Dark platy limestone and shale containing mid-Silurian graptolites indicate that deeper water paleoenvironments prevailed between Late Ordovician and Late Silurian time." Gastropods of late Early and early Middle Ordovician age (Rohr and Blodgett, 1988) and stromatoporoids of Middle or Late Ordovician age from the Medfra D-2 Quadrangle (Stock, 1981) support the shallow-water interpretation for rocks of these ages.

Rocks assigned to the Mystic sequence in the northern Kuskokwim Mountains include about 500 m of Permian, Triassic, and Lower Cretaceous (Valanginian to Albian) strata consisting of quartz-carbonate terrigenous sedimentary rocks and spiculitic chert beds that unconformably overlie metamorphic basement (Patton and others, 1984a). According to Patton and others (1984a, p. 4), "The relation of these terrigenous rocks [of the Mystic sequence] to the platform carbonate rocks [of the White Mountain sequence] is obscured by faulting. However, the debris which composes them clearly was derived, in large part, from erosion of the platform carbonate rocks."

Southwestern Alaska Range. The White Mountain sequence in the southwestern Alaska Range (Fig. 3) consists of a structurally complex assemblage of lower Paleozoic interbedded lime mudstone and shale, deep-water lime mudstone, and interbedded lithic sandstone and shale originally described by Capps (1926) along the Dillinger River in the eastern Talkeetna Quadrangle. In the southwestern Alaska Range, deep-marine strata correlative with the Dillinger River section occur south of the Denali-Farewell fault system in a wedge-shaped outcrop belt that broadens southwestward, where it grades into coeval slope facies, and platform carbonate rocks of the Holitna basin (Blodgett and Clough, 1985).

The following descriptions of stratigraphic relations within the White Mountain sequence in the southwestern Alaska Range are based on detailed field mapping by Gilbert (1981), Bundtzen and others (1982, 1985), Gilbert and others (1982), Gilbert and Solie (1983), and Kline and others (1986); and on measured sections by Armstrong and others (1977), Churkin and others (1977) and K. M. MacDonald (written commun., 1985). The presence of a rich graptolite succession allows for excellent biostratigraphic control.

The White Mountain sequence in the southwestern Alaska Range is a 1,400-m-thick section composed of five regionally persistent lithostratigraphic units: (1) Cambrian(?) and Ordovician rhythmically interbedded calcareous turbidites, shale, and minor greenstone; (2) Lower Ordovician to Lower Silurian graptolitic black shale and chert; (3) Lower and Middle Silurian laminated limestone and graptolitic black shale; (4) Middle and Upper Silurian lithofeldspathic sandstone turbidites and shale; and (5) Upper Silurian and Lower Devonian limestone, limestone breccia, sandstone, and shale (Bundtzen and Gilbert, 1983; Gilbert and Bundtzen, 1983a; K. M. MacDonald, written commun., 1985). Collectively, these units represent a shallowing-upward, deep basin to slope sequence (Bundtzen and Gilbert, 1983; Gilbert and Bundtzen, 1983a, 1984). Twenty-one graptolite zones have been identified by Michael Churkin and Claire Carter (U.S. Geological Survey) from units 2, 3, and 4, which represents one of the most complete graptolitic successions in North America.

Middle Devonian through Lower Cretaceous rocks of the Mystic sequence occur locally throughout the McGrath and northern Lime Hills Quadrangles. These rocks vary widely across the region but, in general, include (1) upper Middle Devonian thick-bedded Amphipora-bearing limestone and massive dolomite, which interfinger with quartzose sandstone, (2) Upper Devonian shallow-marine limestone, (3) post-Devonian plant-bearing sandstone and limestone conglomerate, (4) Upper Pennsylvanian and Lower Permian fusulinid-bearing shallow-water clastic rocks, (5) Mississippian or Triassic chert and pillow basalt, (6) post-Carboniferous mafic intrusive and extrusive rocks, and (7) Triassic and Lower Jurassic clastic and mafic volcanic rocks (Gilbert, 1981; Bundtzen and Gilbert, 1983; Gilbert and others, 1982). This heterogeneous suite of post-Middle Devonian rocks is, in part, depositional on underlying rocks of the White Mountain sequence, indicating that the White Mountain and Mystic sequences form a single depositional succession (Gilbert and Bundtzen, 1984).

The White Mountain sequence in the southwestern Alaska Range is deformed by isoclinal to open folds with axes trending N10 to 40°E and is locally overturned to the northwest (Bundtzen and Gilbert, 1983; Gilbert and Bundtzen, 1983b). Five major overturned isoclinal folds are responsible for at least 50 km of crustal shortening in the southeastern McGrath Quadrangle. Thrust faulting occurred concurrently with folding, but the amount of displacement along individual thrust planes is uncertain.

Holitna basin. The White Mountain sequence within the Holitna basin occurs in a broad, locally well-exposed outcrop belt in the Sleetmute, Taylor Mountains, and Lime Hills Quadrangles (Fig. 1). In the Sleetmute Quadrangle, it includes the Holitna Group (Cady and others, 1955), which is mapped in the Kulukbuk Hills and adjacent hills to the west and southwest. We estimate the White Mountains sequence to be at least 1,500 m and probably closer to 3,000 m thick. Cady and others (1955) recognized both Silurian and Devonian fossils in the upper part of the sequence, and suggested the undated lower part to be as old as Ordovician. Fossil collections made by the Alaska Division of Geological and Geophysical Surveys (ADGGS) during 1983 and 1984 establish the presence of Ordovician faunas in the Holitna basin (Blodgett and others, 1984). The oldest rocks in the basin are Middle Cambrian in age and are located along the axis of an anticline trending east-northeast in the southwestern corner of the Sleetmute A-2 quadrangle. Trilobites from these rocks were collected by Standard Oil Company and ADGGS geologists in 1984 from two closely spaced localities and are discussed in Palmer and others (1985). No Precambrian metamorphic basement rocks have been recognized in the Holtina basin.

A preliminary stratigraphic column for the White Mountain sequence of the Holitna basin is presented in Figure 3. In general, the oldest rocks (Middle Cambrian to Lower Ordovician) appear to be, at least in part, of deeper water character than the overlying Middle Ordovician to Middle Devonian carbonate succession. The deep-water nature of the older sequence is suggested by relatively thick intervals of graded carbonate debris flows, bedded chert, shale, and laminated lime mudstone lacking the abundant diverse benthic faunas typical of the overlying shallow-water platform carbonate sequence.

A Late Silurian to Early Devonian algal reef complex typically is developed along the outer edge of the carbonate platform (Blodgett and others, 1984; Blodgett and Clough, 1985; Clough and Blodgett, 1985). The reef complex is composed almost entirely of spongiostromate algae and locally is at least 500 m thick. Coeval cyclical lagoonal and peritidal environments were widely developed shoreward of the reef complex, and bear somewhat restricted faunas (Clough and others, 1984). The youngest Paleozoic fauna is of Eifelian (early Middle Devonian) age and is from the southeastern part of the Kulukbuk Hills (USGS collection 2688 mentioned in Cady and others, 1955, and recollected by the ADGGS). Fossils from this locality include brachiopod species also found in the upper part of the Cheeneetnuk Limestone of the McGrath A-4 and A-5 Quadrangles.

Along the Hoholitna River and in the vicinity of the Door Mountains, platform carbonate rocks appear to grade into coeval deeper water basinal rocks. This facies change is supported by (1) limestone debris flows that contain clasts apparently derived from the Late Silurian to Early Devonian algal reef complex, and (2) the close interdigitation of both shallow-water and deep-water facies near the interpreted outer edge of the carbonate platform.

Pennsylvanian and Permian brachiopods and Pennsylvanian foraminifera occur in coarse-grained fossil-rich detrital limestone interbedded with calcareous mudstone in the western Lime Hills Quadrangle (W. K. Wallace, unpublished data). These rocks are here considered part of the Mystic sequence. Late Triassic fossils have been collected from two localities in the Taylor Mountains Quadrangle, and are from units which most likely belong to the Mystic sequence. Rock units at these localities are silty lime mudstones bearing a diverse molluscan-rich fauna, accompanied by commonly occurring brachiopods (faunas identified by N. J. Silberling of the U.S. Geological Survey). Jurassic siliceous siltstone and volcanic conglomerate probably belonging to the Mystic sequence occur along the Stony and Swift Rivers in the western Lime Hills Quadrangle (Reed and others, 1985).

Minchumina basin. Within the Minchumina basin, a belt of the White Mountain sequence is exposed on several low rolling hills between the Iditarod–Nixon Fork and Denali-Farewell faults. Rocks within this belt consist mainly of laminated limestone and dolomitic limestone, dark gray chert, and siliceous siltstone, ranging in age from Ordovician to Middle Devonian (Patton and others, 1980). Platform carbonate rocks containing Late Ordovician (Ashgill) gastropods occur in the Lone Mountain region of the McGrath B-4 and C-4 Quadrangles (Rohr and Blodgett, 1985).

Poorly exposed sections of Mystic-like rocks cropout in low glacially scoured uplands north of the Farewell fault in the McGrath B-2, B-3, and C-1 Quadrangles. They consist of an undated clastic-carbonate sequence containing limestone breccia overlain by a fossiliferous Upper Devonian (Frasnian) limestone that is especially prominent at St. Johns Hill and immediately west of Farewell Mountain. The limestone units are overlain by bioclastic limestone of Permian age, then by chert-rich sandstone and ferruginous shale. The layered sequence is approximately 1,500 m thick, and is in probable fault contact with gabbro, pillow basalt, and chert, which yield both Mississippian and Triassic radiolaria (Kline and others, 1986).

Paleolatitude data

Sedimentological evidence from the White Mountain sequence of the Farewell terrane suggests that these rocks were deposited in warm, tropical latitudes. Evidence supporting this interpretation is primarily from the platform facies, and includes the presence of thick carbonate sections, the abundance of marine ooids in Ordovician and Silurian sedimentary rocks, and the presence of thick algal barrier reef complexes in Upper Silurian and Lower Devonian rocks. Supporting paleontological evidence includes the presence of green calcareous algae (Poncet and Blodgett, 1987) in shallow-water deposits. It also includes the paleobiogeographic affinities of the Alaskan Devonian faunas with those of the Cordilleran and Uralian regions of the Old World Realm (Blodgett, 1983a, 1983b; Blodgett and Gilbert, 1983), which are interpreted to have been located in equatorial climatic zones (Boucot and Gray, 1980, 1983). In addition, the Devonian gastropod are highly ornamented, as are present-day warm-water gastropod faunas, in contrast to the generally plain ornamentation of cold-water faunas.

Location of the Farewell terrane during Late Silurian and Early Devonian time in an equatorial humid belt, rather than in subequatorial arid to semiarid belts is suggested by the absence of evaporite deposits. The proposed low paleolatitudinal origin of the Farewell terrane is in agreement with similar positioning of the northwestern part of the North American continent based on Devonian lithofacies patterns (Heckel and Witzke, 1979; Boucot and Gray, 1980, 1983). However, the presence of distinctly Siberian faunas in the Middle Cambrian section of the Holitna basin (Palmer and others, 1985) suggests eastward longitudinal transport since Cambrian time, although the strongly Cordilleran affinities of the Lower and Middle Devonian faunas (Blodgett, 1983a) suggest that by that time, the Farewell terrane was affixed or closely adjacent to North America. This proximity during the Devonian is especially borne out by the Eifelian gastropod faunas of interior Alaska (known from rocks of the Farewell and Livengood terranes). These faunas belong to a single biogeographic entity, termed the Alaska-Yukon subprovince by Blodgett and others (1987), and are most closely allied with coeval faunas of western Canada and the Michigan basin.

Paleomagnetic data (Hillhouse and Coe, this volume) from Ordovician rocks of the White Mountain sequence have been discussed by Plumley and Coe (1982, 1983) and Vance-Plumley and others (1984). Their data suggest that the Farewell terrane had little or no major latitudinal displacement with respect to North America. This is concordant with our view, based on paleobiogeography and lithofacies trends. However, this does not discount the possibility of large-scale longitudinal transport.

Interpretation

Similarities between the major lithostratigraphic units of the Farewell terrane and rocks of the Selwyn basin and Mackenzie platform of northern Canada noted by Bundtzen and Gilbert (1983) suggest that the White Mountain sequence was part of a coherent Paleozoic passive continental margin that lay upon a Precambrian continental crystalline basement (Churkin and others, 1984; Gilbert and Bundtzen, 1983a) to form an extension of the Paleozoic North American continent (Blodgett and Clough, 1985).

Present-day lithofacies patterns have been complicated by the disruption of this Paleozoic continental margin by major right-lateral faults such as the Iditarod–Nixon Fork, Kaltag, and the Denali-Farewell faults (Fig. 4). Along the Denali-Farewell fault, right-lateral separation of 145 to 153 km (90 to 95 mi) is indicated by correlation of truncated trends of the Late Silurian and Early Devonian algal reef complex in the northeastern Kulukbuk Hills with exposures of the same complex in a similar paleogeographic setting (outer edge of carbonate platform) on the north side of the Denali-Farewell fault in the vicinity of White Mountain, McGrath A-4 Quadrangle (Blodgett and Clough, 1985). Bundtzen and Gilbert (1983) suggest right-lateral separation of about 60 km (37 mi) of an Upper Devonian (Frasnian) limestone and Triassic greenstone section along the Cheeneetnuk River south of the Denali-Farewell fault from a nearly identical section at Farewell Mountain to the northeast, north of the fault.

Rock sequences and the ages of major interregional unconformities within the Farewell terrane fit nicely the stratigraphic framework developed for northern Alaska and the Canadian Arctic by Lerand (1973). The White Mountain and Mystic sequences correspond respectively with Lerand's Franklinian and Ellesmerian sequences. The pre-Mississippian unconformity that separates the Franklinian and Ellesmerian sequences roughly correlates with the Middle Devonian unconformity which separates the White Mountain and Mystic sequences. In southwest Alaska, the Kuskokwim Group and related rock units occupy the same stratigraphic interval as the Brookian sequence in northern Alaska. The Early Cretaceous unconformity which separates the Ellesmerian and Brookian sequences corresponds to the Early Cretaceous unconformity between the Mystic sequence and the Kuskokwim Group.

In the southwestern Alaska Range, rocks of the Farewell terrane have undergone subisoclinal folding that postdates deposition of Triassic clastic rocks. Similar deformation, though with local variance in vergence, occurs in the adjacent Kahiltna terrane

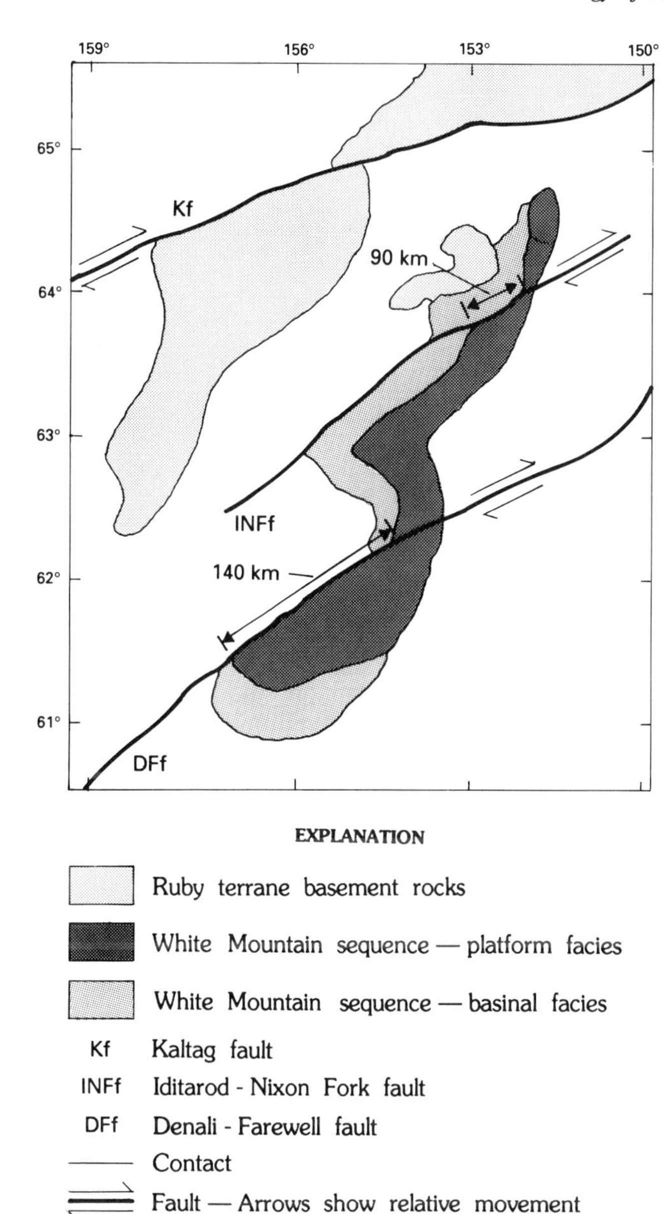

Figure 4. Generalized geologic map showing the offset of the White Mountain sequence along major faults. If the proposed offsets are removed, basement, shelf, and basin facies trends become contiguous and result in a consistent paleogeographic relationship.

to the southeast and apparently resulted from the same tectonic event. The timing of deformation in the Kahiltna terrane is bracketed in age by deformed Lower Cretaceous flysch and undeformed Late Cretaceous and Paleocene igneous rocks, which suggests that juxtaposition of the Farewell and Kahiltna terranes was completed by Late Cretaceous time. Although the contact between the Farewell and Kahiltna terranes is generally considered to be a fault, we cannot rule out the possibility that Jurassic and Lower Cretaceous flysch of the Kahiltna terrane was derived from and/or deposited on Farewell basement. In the McGrath A-1 quadrangle, facies relations within clastic rocks of the Kahiltna terrane suggest that the sediment was derived locally from a northern source (T. K. Bundtzen and W. G. Gilbert, unpublished field data). Conglomerate clasts, in part, resemble similar lithologies within the Farewell terrane. Granitic clasts also occur, however, and have no known local source.

TERRANES OF THE BRISTOL BAY REGION

The tectonostratigraphic terranes of the northern Bristol Bay region, named by Jones and Silberling (1979) include, from southeast to northwest, the Togiak, Goodnews, Kilbuck, and Nyack terranes (Fig. 5; Silberling and others, this volume). The Tikchik terrane of Jones and Silberling (1979) is considered here to be a subterrane of the Goodnews terrane. Extensive early mapping by Hoare and Coonrad (1959a, b, 1961a, b, 1978a) established the geologic framework of the region and provided the basis for the original terranes defined by Jones and Silberling (1979). Recent detailed mapping by Box (e.g., 1982, 1983b, 1985) along a coastal transect reveals a geologic history for these terranes reflecting episodic magmatism and accretion across an intraoceanic volcanic arc complex (Fig. 6), culminating in arc-continent collision and postcollisional strike-slip faulting. The terranes, and their respective subterranes, are described below in order of their interaction with the developing arc system. Unless otherwise noted, the subterrane nomenclature and descriptions used in this chapter for rocks of the Bristol Bay region are from Box (1985).

Togiak terrane

The Togiak terrane is composed of volcanic flows, coarse volcaniclastic breccias, tuffs, and associated epiclastic rocks of Late Triassic through Early Cretaceous age (Hoare and Coonrad, 1978a). Rocks of the Togiak terrane underwent only low-grade metamorphism (up to prehnite-pumpellyite or lower greenschist facies) and generally lack a penetrative metamorphic fabric. The Togiak terrane is divided into two northeast-trending subterranes, Hagemeister and Kulukak, that differ in sedimentary facies and structural style, but are linked by common provenance and stratigraphic history. Both subterranes record evidence for three deformational episodes accompanied by uplift and erosion: late Early Jurassic, latest Jurassic, and late Early Cretaceous. Volcanic episodes preceded each deformation event.

Hagemeister subterrane. *Definition.* The Hagemeister subterrane consists of Upper Triassic through Lower Cretaceous basaltic to dacitic volcanic and volcaniclastic rocks deposited on Upper Triassic ophiolitic rocks. The volcanic pile built upward, reaching sea level by Early Jurassic time. Rugged volcanic island topography resulted in a complex distribution of interfingering deep- and shallow-marine to subaerial facies.

Description. The Hagemeister subterrane includes three stratigraphic sequences separated by unconformities. The rocks within these sequences are generally not foliated, but record evi-

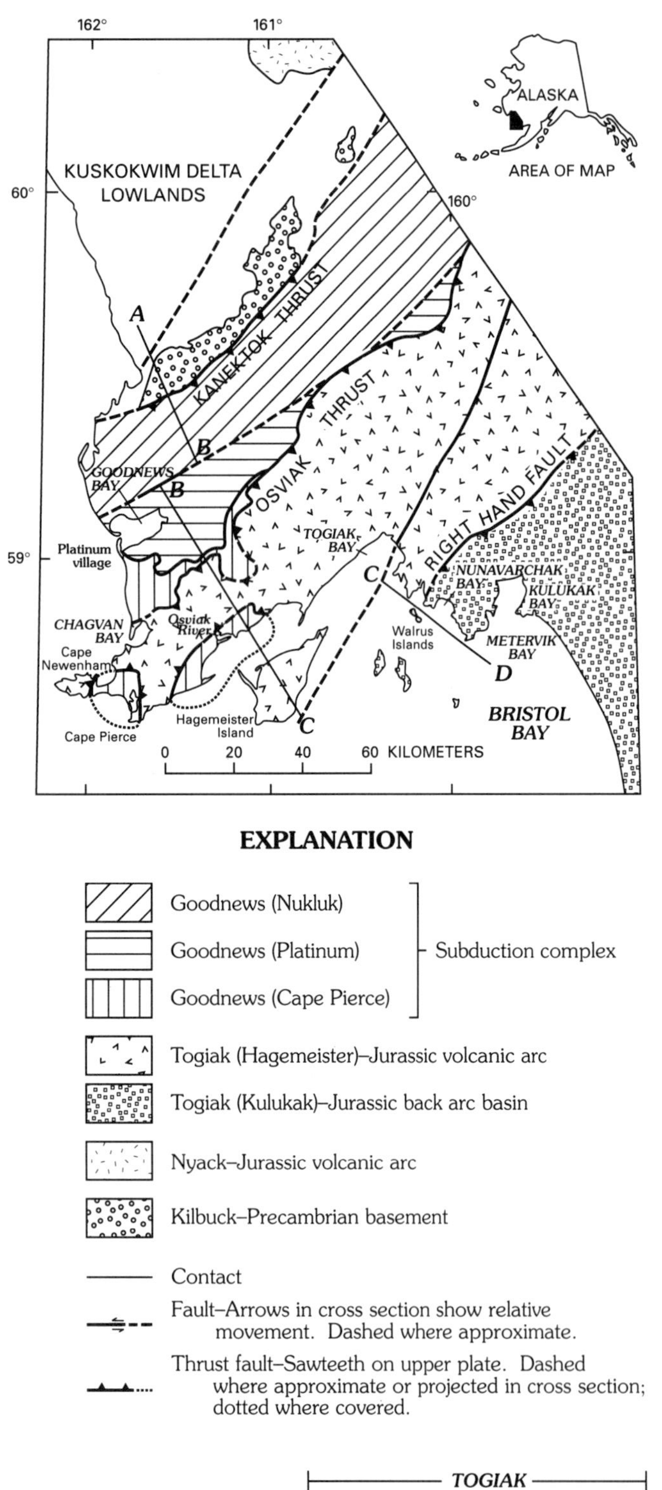

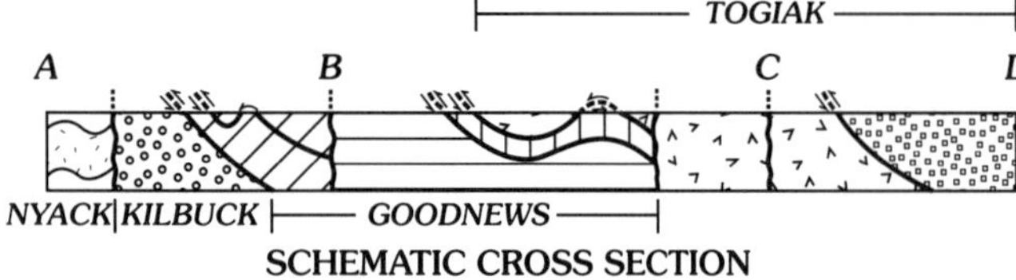

Figure 5. Terranes and subterranes of the Bristol Bay region showing the location of major faults and geographic features (Box, 1985).

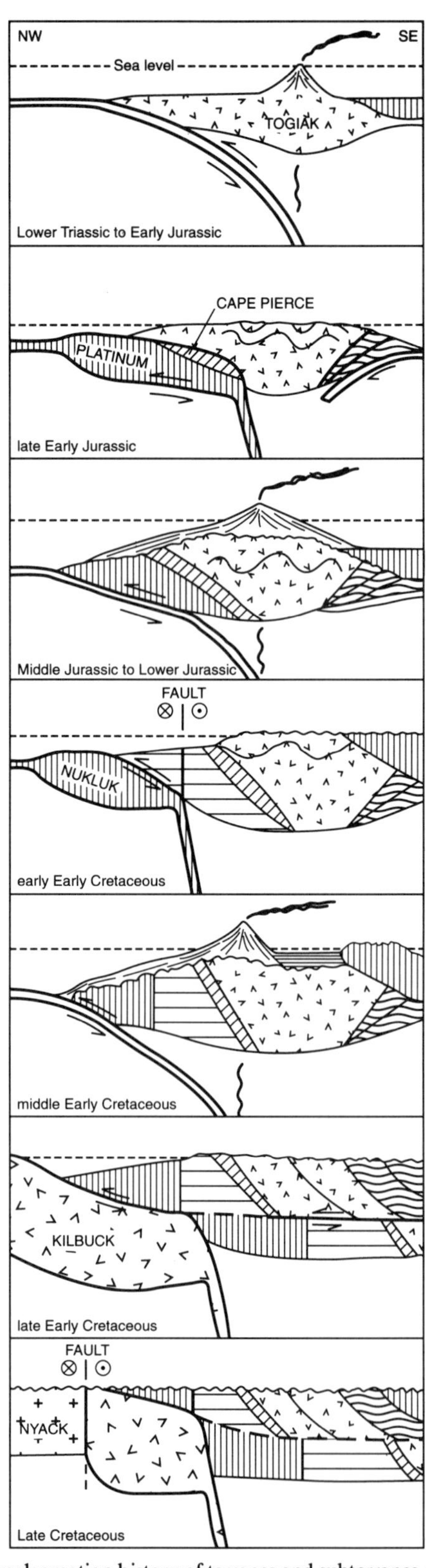

Figure 6. Amalgamation history of terranes and subterranes of the Bristol Bay region (Box, 1985). See Figure 5 for explanation of symbols. This diagram shows the successive accretion of terranes in the Bristol Bay region from Late Cretaceous to Late Triassic time. Arrow indicates direction of relative movement.

dence of up to prehnite-pumpellyite or lower greenschist-facies metamorphism, and moderate to severe deformation. The following description is based on coastal exposures of this subterrane (Box, 1985). Its application to the entire subterrane is untested.

The lowest sequence grades upward from an ophiolitic basement, containing Upper Triassic radiolarian chert, through thick volcanic breccia, into Lower Jurassic shallow-marine volcanogenic sandstone. This sequence records pre–Middle Jurassic deformation inferred by Box (1985) to relate to underthrusting by part of the Goodnews terrane from the northwest. The thrust contact (Osviak thrust) between the Goodnews and Togiak terranes is crosscut by Early to Middle Jurassic zoned mafic to ultramafic plutons, thus stitching these terranes by Middle Jurassic time. Elsewhere, a Middle Jurassic granitic pluton intrudes the lowest sequence.

The second sequence consists of Middle and Upper Jurassic (Bajocian to Tithonian) marine to nonmarine volcanic and volcaniclastic strata that, at least locally, rest with angular unconformity on the lowest sequence. Conglomerate from the second sequence locally contains clasts derived from the Middle Jurassic granitic pluton that intrudes the first sequence.

The third sequence consists of four widely separated belts of Lower Cretaceous (Valanginian to Albian) volcanic and sedimentary rocks (Hoare and Coonrad, 1983; Murphy, 1989) known in a few places to rest with angular unconformity on both the Togiak and Goodnews terranes. These belts record sedimentation in shallow- to deep-marine settings with locally derived, restricted provenances. Because no genetic relations are known to occur that link the four belts to a single depositional basin, we consider each belt to be part of the terrane it overlies.

Kulukak subterrane. *Definition.* The Kulukak subterrane consists predominantly of Jurassic volcaniclastic turbidites. In general, vertical changes in sandstone compositions and in turbidite facies record the partial unroofing of a volcanic terrane and the progradation of submarine fan environments. Although provenance links have been suggested by Box (1985), no direct depositional relationship with the Hagemeister subterrane has been observed.

Description. The Jurassic strata of the Kulukak subterrane consist of two informally named structural units, juxtaposed by a southeast-dipping thrust fault (Right Hand fault): the Nunavarchak structural unit, to the northwest; and the graywacke of Metervik Bay, to the southeast.

The Nunavarchak structural unit is characterized by structurally disrupted argillite displaying an anastomosing scaly fabric. The unit is assigned an Early(?) Jurassic age based on its structural and stratigraphic position, and on Jurassic or Cretaceous radiolaria (D. L. Jones, oral communication, *in* Box, 1985). Development of the structural fabric probably began in pre–Middle Jurassic time because angular clasts with distinct slaty cleavage that were most likely derived from the Nunavarchak structural unit occur in graywacke of the upper structural unit.

The graywacke of Metervik Bay consists of a basal unit composed of channelized volcaniclastic conglomerates (with minor angular slate clasts) and associated channel-margin facies, and three structurally higher units recording progradation of a submarine fan complex. Facies represent, from base to top, outer fan, midfan, and inner fan depositional environments. The fan complex ranges from Bajocian to Oxfordian in age, and shows a minor upsection increase in plutonic clasts, probably reflecting unroofing of an adjacent volcanic arc.

Goodnews terrane

The Goodnews terrane is a collage of variably metamorphosed blocks of laminated tuff, chert, basalt, graywacke, limestone, gabbro, and ultramafic rocks, in roughly that order of abundance. Devonian, Permian, and Upper Triassic limestones, and Mississippian, Upper Triassic, and Lower and Upper Jurassic cherts occur as fault-bounded blocks. Their original stratigraphic relations are uncertain. Box (1985) divided the Goodnews terrane into three subterranes with distinct stratigraphies and/or amalgamation histories. These subterranes, the Cape Peirce, Platinum, and Nukluk, were all linked with the Togiak terrane by earliest Cretaceous time; amalgamation of the Cape Pierce and Platinum subterranes with the Togiak terrane was completed by Middle Jurassic time. We interpret the Tikchik terrane of Jones and Silberling (1979) as a subterrane of the Goodnews terrane based on its similarities in age, lithology, and deformational style with rocks of the Platinum and Nukluk subterranes.

Cape Peirce subterrane. *Definition.* The Cape Peirce subterrane consists of foliated metamorphic rocks that are exposed in three separate outcrop belts in the Cape Newenham–Cape Peirce area (Fig. 1). These metamorphic rocks were probably derived from protoliths of probable Permian and Triassic ages (Box, 1985).

Description. The foliated metamorphic rocks of the Cape Peirce subterrane are in three nappes separated by low-angle faults. The structurally highest nappe is composed of mafic schist, the middle nappe is composed of metaclastic rocks, and the lowest nappe is composed of interbedded marble, slate, and mafic schist. Each nappe contains retrograded high-pressure mineral assemblages, and each low-angle fault zone contains structural blocks of nonfoliated mafic and ultramafic intrusive rocks. Protoliths of the highest nappe are similar to those of the structurally overlying, nonfoliated Hagemeister subterrane, and those of the lowest nappe are similar to the structurally underlying Platinum subterrane. Protoliths of the middle nappe are clastic rocks apparently derived from the Hagemeister subterrane.

The two lowest nappes record a premetamorphic deformation (D1) reflected by tight isoclinal folding, irregularly accompanied by axial-planar cleavage, or foliation. All three nappes record three later deformational events, including: (1) Late Triassic to Early Jurassic folding of the early isoclinal fold hinges and development of glaucophane- and lawsonite-bearing (Hoare and Coonrad, 1977, 1978b) high-pressure mineral assemblages (D2); (2) Late Jurassic(?) development of crenulation cleavage resulting from subhorizontal west- to northwest-directed stresses (D3);

and (3) post–Late Jurassic open folding of earlier fabrics along subhorizontal northeast-trending axes (D4).

Platinum subterrane. *Definition.* The Platinum subterrane consists of Lower and Middle Jurassic nonfoliated mafic flows, tuff, and volcaniclastic rocks metamorphosed to the prehnite-pumpellyite metamorphic facies. It occurs to the north of and structurally beneath the Cape Peirce subterrane in a wedge-shaped belt tapering to the northeast (Fig. 1).

Description. Rocks of the Platinum subterrane are interpreted by Box (1985) as the nonfoliated and less intensely deformed equivalent of the lowest schistose nappe of the Cape Peirce subterrane. Permian fossils occur in calcareous tuff interbedded with mafic volcanic rocks, and in limestone blocks within the fault zone between the Cape Peirce and Hagemeister subterranes. The Cape Peirce and Platinum subterranes both are intruded by zoned mafic-ultramafic plutons of early Middle Jurassic age (Southworth, 1984).

Platinum placer deposits at Goodnews Bay, America's largest source of this metal, are derived from one such pluton at Red Mountain. Hoare and Coonrad (1978a) suggested that the Red Mountain Pluton is part of a dismembered ophiolite belt of Jurassic age. However, the Red Mountain Pluton has a dunite core, pyroxenite border, and high-temperature contact zones, and may be analogous to the Alaskan-type of zoned ultramafic complex known in southeastern Alaska and in the Ural Mountains of Russia (Bird and Clark, 1976; Southworth, 1984).

Nukluk subterrane. *Definition.* The Nukluk subterrane is a structurally disrupted unit (mélange) containing blocks of diverse size and lithology ranging in age from Ordovician through Late Jurassic, locally set in a scaly argillaceous matrix. The blocks and matrix fabric of this mélange contrast with the more coherent nature of the Cape Peirce and Platinum subterranes. Deep-marine sedimentary rocks called the Eek Mountains belt by Hoare and Coonrad (1983) are of Early Cretaceous (Valanginian) age (Murphy, 1989) and rest with angular unconformity on older rocks of the Nukluk subterrane.

Description. Coherent limestone blocks of Ordovician, Devonian, and Permian age are distinctive constituents of the Nukluk subterrane, but constitute less than 5 percent of the outcrop area. The remaining area is underlain by radiolarian chert of Mississippian, Late Triassic, and Early and Late Jurassic age, laminated green or black mudstone, and basalt. At least two Devonian limestone blocks contain algal mounds indicative of a shallow water origin. Limestones containing Permian *Atomodesma* commonly are intercalated with amygdaloidal basalt flows; some are oolitic, indicating formation in a shallow-water environment. Massive limestones and basalts generally retain structural coherence, whereas laminated cherts, tuffs, and mudstones are more commonly highly tectonized.

A few elongate bands of serpentinite occur within this subterrane, at least one of which has an aeromagnetic signature indicative of a southeasterly dip (Griscom, 1978). Northwest-vergent overturned folds and stratigraphic facing directions in southeast-dipping blocks suggest imbrication along southeast-dipping thrust faults. Scattered occurrences of greenschist- to blueschist-facies schists along the northwestern edge of this terrane indicate high- to medium-pressure metamorphism of uncertain age in rocks adjacent to the fault boundary with the Precambrian Kilbuck terrane. Elsewhere, prehnite-pumpellyite facies metamorphic assemblages are sporadically developed. Penetrative deformation is reflected by the development of the scaly matrix fabric and the variable development of slaty cleavage within the more coherent blocks.

The northwestern boundary of the Nukluk subterrane is a southeast-dipping thrust fault that places the Nukluk subterrane on metamorphic rocks of the Kilbuck terrane (discussed below). Mid-Cretaceous (Albian) clastic rocks that overlie both assemblages are overturned along the fault. A latest Cretaceous pluton postdates deformation along this boundary. If the mid-Cretaceous clastic rocks are synorogenic deposits, deformation spanned the mid- and Late Cretaceous intervals.

The Eek Mountains belt (Hoare and Coonrad, 1983) is the westernmost of four coeval belts of Lower Cretaceous (Valanginian) marine clastic rocks. All four belts strike northeastward, reflecting underlying postdepositional structural trends, but not necessarily original basin geometries. The Eek Mountains belt is exposed along the crest of the doubly plunging, northeast-trending Eek Mountains anticline.

The Eek Mountains belt is a broken formation (Hsü, 1968) composed predominantly of siltstone and shale, and minor graywacke and conglomerate. In the northernmost Eek Mountains the belt is in fault contact with rocks of the Nukluk subterrane. A southeast-dipping structural fabric overprints all pre-Kuskokwim Group (pre-middle Albian) rocks but produced little metamorphic recrystallization of rocks within the Eek Mountains belt. The Kuskokwim Group overlaps older units with an angular unconformity preserving submarine canyon cut-and-fill (Murphy, 1989). Pre–Eek Mountains belt rocks have been metamorphosed to the prehnite and pumpellyite facies, while clastic rocks of the Eek Mountains belt and overlying Kuskokwim Group contain only detrital metamorphic minerals.

Petrographically, sedimentary rocks of the Eek Mountains belt are lithologically distinct from older rocks of the Nukluk subterrane, but are nearly identical to rocks of the Kuskokwim Group. Older rocks are entirely volcanogenic, while Eek Mountains belt rocks are polymictic. Clast components include metavolcanic and metasedimentary grains derived from the Nukluk subterrane, and, in the northern Eek Mountains, metamorphic clasts derived from the Kilbuck or distant Ruby terrane (Murphy and Decker, 1985). Previously, Kilbuck terrane detritus was presumed to be present only in Kuskokwim Group strata (Box, 1985). The presence of high-grade metamorphic clasts in sandstones of the Eek Mountains belt implies that the Kilbuck and/or Ruby terranes were in place relative to the Goodnews terrane by Valanginian time (Murphy, 1987).

Tikchik subterrane. *Definition.* The Tikchik [sub]terrane (Silberling and others, this volume) was named by Jones and Silberling (1979) and defined by Jones and others (1987). It occurs in the northern Tikchik Lakes area, southeast of the Togiak-Tikchik fault (Denali-Farewell fault system). The Tikchik subterrane is a structurally complex assemblage (melange) of clastic rocks, radiolarian chert of Paleozoic and Mesozoic age, Permian limestone and clastic rocks, Permian or Triassic pillow basalt and graywacke, and Upper Triassic clastic and mafic volcanic rocks. Ordovician radiolaria have also been identified from chert within the Tikchik subterrane (Hoare and Jones, 1981; D. L. Jones, J. M. Hoare and W. L. Coonrad, unpublished data), but the stratigraphic significance of the chert is unclear.

Description. Little more than what is described above is known about the Tikchik subterrane. The Tikchik subterrane is an apparently chaotic assemblage of blocks, particularly chert, of Ordovician, Triassic, Jurassic, and Early Cretaceous age, and of limestone of Permian age. The matrix is mainly graywacke, argillite, tuff, and mafic volcanic rock, although blocks of different ages commonly are juxtaposed without intervening matrix. Some of these blocks are quite large (kilometer-scale) and are structurally and stratigraphically coherent. The age(s) of deformation is uncertain, but predates deposition of the unconformably overlying clastic rocks of mid-Cretaceous age (W. K. Wallace, unpublished field data 1983).

Kilbuck terrane

Definition. The Kilbuck terrane, named by Jones and Silberling (1979), consists of multiply deformed, upper amphibolite- to granulite-facies metagranitic and metasedimentary rocks of Precambrian age (Hoare and Coonrad, 1979). These rocks appear to be folded into a large anticlinorium, with a high-grade gneissic core overlain by folded lower grade rocks.

Description. Hoare and others (1974), Hoare and Coonrad (1979), and D. L. Turner and others (unpublished data) provide the most detailed descriptions of this metamorphic complex. Their mapping indicates that the terrane is composed of a high-grade central zone flanked by lower grade schist. The central zone of layered biotite-hornblende gneiss with intercalated pyroxene granulite, garnet amphibolite, kyanite-garnet-mica schist, orthogneiss, and rare marble and quartzite record upper amphibolite- to lower granulite-facies metamorphism, partly retrograded under greenschist-facies conditions. In contrast, this high-grade core is flanked by schists recording greenschist-facies metamorphism. These rocks include chlorite schist, epidote-quartz-biotite schist, micaceous quartzite, marble, calc-phyllite, and metaconglomerate.

Geochronological studies by Turner and others (1983, and unpublished data) have yielded 2,050 Ma U-Pb and Pb-Pb zircon ages reflecting orthogneiss protolith crystallization, while coexisting sphene indicates a 1,800-Ma reset age recording the high-grade metamorphic event. A 1,800-Ma Rb-Sr whole-rock isochron also confirms this metamorphic event. K-Ar ages in both the high-grade core and low-grade carapace cluster in the range from 120 to 150 Ma, and may record uplift following the greenschist-facies metamorphic overprint. This metamorphic event is attributed to the underthrusting of the Kilbuck terrane beneath the northwest margin of the Goodnews terrane (Box, 1985). Postmetamorphic cataclasis in rocks along the southeastern boundary of the Kilbuck terrane record the return of this terrane to a shallower crustal level.

Boundaries between the Kilbuck and adjacent terranes are faults. Near the coast, the southeast-dipping Kanektok thrust (Fig. 5) places northwest-vergent greenschist- and blueschist-facies Goodnews terrane rocks over greenschist-facies rocks of the Kilbuck terrane (Box, 1982). Inland along strike, the Kanetok thrust places northwest vergent mid-Cretaceous clastic rocks and low-grade basalt and chert of the Goodnews terrane over amphibolite facies rocks of the Kilbuck terrane. The northwestern boundary of the weakly magnetic Kilbuck terrane may be defined by the sharp magnetic gradient along the southeastern flank of the Nyack terrane (Griscom, 1978). Where exposed, the Golden Gate fault (Fig. 5) juxtaposes the Kilbuck and Nyack terranes, but the sense of fault displacement is not clear.

Nyack terrane

Definition. The Nyack terrane (plate 3) is the least well known terrane in southwest Alaska. As originally defined by Jones and others (1981), the Nyack terrane consists of a Jurassic arc-related volcanic and volcaniclastic assemblage.

Description. Rocks of the Nyack terrane include andesitic, basaltic, and dacitic volcanic and volcaniclastic rocks interbedded with graywacke, siltstone, impure limestone, and conglomerate. It contains sparse marine fossils of Middle and Late Jurassic age (Hoare and Coonrad, 1959a, b). In the Nyack area, a granitic pluton intrudes the Jurassic volcanic rocks, and has yielded K-Ar ages of 108 to 120 Ma (Shew and Wilson, 1981; Decker and others, 1984a; Robinson and Decker, 1986). The rocks are unfoliated, but contain chlorite-epidote mineral assemblages of probable lower greenschist metamorphic facies (Hoare and Coonrad, 1959a, b).

The southeastern boundary of the Nyack terrane with the Kilbuck terrane is discussed above. The northwestern boundary is not exposed, but the irregular aeromagnetic pattern that characterizes the Nyack terrane is truncated along a linear, northeast-trending magnetic gradient that roughly follows the lower Kuskokwim river meander belt (Dempsey and others, 1957). In addition, the occurrence of rocks perhaps correlative with the Goodnews terrane (Portage sequence, discussed below) northwest of the Kuskokwim River suggests a concealed terrane boundary near there. The linearity of both the northwestern and southeastern boundaries of the Nyack terrane suggests that they are probably steep faults, but their amount and sense of displacement are unknown.

Enigmatic rocks

Two rock units of unknown terrane affinities occur in the Russian Mission and Iditarod Quadrangles (Fig. 1). Predominantly volcanic and sedimentary rocks exposed along the Kuskokwim River near Portage Mountain in the Russian Mission Quadrangle are here called the Portage sequence. Metamorphic rocks in the vicinity of VABM Idono in the Iditarod Quadrangle are called the Idono sequence. The southern Kahiltna terrane, exposed mainly in the Lake Clark Quadrangle, occurs between the Mulchatna and Lake Clark–Castle Mountain faults and constitutes a third enigmatic unit.

Portage sequence. Rocks of the Portage sequence occur in cutbank exposures of the Kuskokwim River near the village of Aniak, and in the vicinity of Portage Mountain in the Russian Mission Quadrangle (Fig. 1). These rocks were originally mapped by Hoare and Coonrad (1959b) as part of the undivided Gemuk Group. They were assigned to the Ruby and Innoko terranes (Silberling and others, this volume) by Jones and others (1981). Exposures along the river consist of volcanic flows, lahar deposits, green cherty tuff, volcanic breccia of probable andesitic and basaltic composition, altered diabase, graywacke, mudstone, calcareous conglomerate, and limestone (Hoare and Coonrad, 1959b; J. Decker and J. M. Hoare, unpublished field data, 1980). The unit is highly deformed and generally has a melange-like internal fabric. Nearest the projected location of the Aniak–Thompson Creek fault, river bluff exposures are highly shattered, moderately altered, and tightly to isoclinally folded. Fossils of Permian (Smith, 1939) and Triassic (U.S. Geological Survey, unpublished fossil report, 1963) age have been obtained from limestone within the Portage sequence. Hoare and Coonrad (1959b) suggest a Permian or Triassic age for these rocks based on correlation with similar rocks elsewhere in the region.

To the south, the Portage sequence may be correlative with rocks of the Nukluk subterrane; to the north and west, it may be correlative with the Innoko or Tozitna terranes.

Idono sequence. Gemuts and others (1983) first described a poorly exposed belt of augen gneiss and amphibolite in a 400-km^2 elliptical outcrop area in the central Iditarod Quadrangle (Fig. 1) and called them the Idono sequence. To be consistent with published nomenclature, we will also call these rocks the Idono sequence, although it does not strictly conform with our usage of the term "sequence" as defined earlier in this report. K-Ar ages of three samples led Miller and Bundtzen (1985) to tentatively correlate rocks of the Idono sequence with polymetamorphic rocks of the Ruby terrane exposed in the Kaiyuh Mountains (Patton and others, 1977b). However, Early Proterozoic U-Pb data and Precambrian K-Ar data subsequently obtained by Miller and Bundtzen (unpublished data), now suggest a closer similarity to rocks of the Kilbuck terrane.

Petrographic and field studies by Miller and Bundtzen (1985) indicate a variety of metamorphic rock-types in the Idono sequence including: (1) sphene-epidote-calcic oligoclase-quartz-hornblende-garnet amphibolite, (2) apatite-plagioclase-biotite-muscovite-quartz schist, and (3) andesine-biotite-muscovite quartzose schist. Unpublished 1984 and 1985 field studies by those workers also indicate metamorphosed augen gneiss, garnet amphibolite, meta-hornblendite, and foliated quartz diorite rock-units. The presence of hornblende and calcic oligoclase or andesine in the inferred mafic protoliths indicate that metamorphic conditions reached the lower amphibolite facies of regional metamorphism. K-Ar ages obtained from rocks at VABM Idono range from 126 to 134 Ma (Miller and Bundtzen, 1985), consistent with metamorphic ages obtained from greenschist- and locally blueschist-facies metamorphic rocks in the Kaiyuh Mountains (Patton and others, 1984b) about 120 km northeast of the Idono sequence. However, the varied lithologies in the Idono sequence also are similar to those described in the Kilbuck terrane discussed earlier in this chapter.

The Idono sequence has been multiply folded, and several foliation or cleavage surfaces are evident in outcrop. In most outcrops that we observed, the youngest foliation dips gently to moderately to the north-northwest, but the generally poor exposures severely limit structural interpretation.

Southern Kahiltna terrane. The southern Kahiltna terrane occupies a tectonically significant position between the Peninsular terrane to the east, and the terranes of southwestern Alaska to the west (Plate 3). The northern part of the Kahiltna terrane is discussed by Nokleberg and others (this volume, Chapter 10). We describe the southern part here because it is separated by younger plutonic and volcanic rocks from the rest of the terrane, and its relationship to the remainder of the terrane is uncertain. The southern Kahiltna terrane is best exposed in the Lake Clark Quadrangle, but small parts also occur in the Iliamna, Dillingham, Taylor Mountains, and Lime Hills Quadrangles (Fig. 1).

Wallace and others (1989) divided the southern Kahiltna terrane into two major stratigraphic units (Fig. 7): Chilikadrotna Greenstone of Bundtzen and others (1979), and the overlying Koksetna River sequence of Hanks and others (1985). The Chilikadrotna Greenstone is only locally exposed and consists predominantly of Upper Triassic and Jurassic(?) volcanic rocks. The Koksetna River sequence consists of extensive Upper Jurassic and Lower Cretaceous turbidite deposits.

The Chilikadrotna Greenstone is described by Wallace and others (1989) as consisting of a lower interval of basaltic pillow lavas and massive flows and an upper interval of andesitic flows and tuff breccia. These two intervals locally are separated by limestone pods that have yielded Late Triassic conodonts and brachiopods (Wallace and others, 1989). Silurian fossils previously reported from the same rocks (Bundtzen and others, 1979) has caused an age conflict that has not yet been resolved.

The Koksetna River sequence appears to consist entirely of turbidites composed of interbedded volcanogenic feldspathic-lithic graywacke, siltstone, and mudstone. Facies associations, from southeast to northwest, indicate deposition in slope, inner fan, middle fan, and outer fan submarine environments, suggesting a southeastern sediment source. A Late Jurassic and Early Cretaceous age for the Koksetna River sequence is based on sparse megafossils, mainly *Buchia,* in Lake Clark Quadrangle (Eakins and others, 1978; J. W. Miller, written communication to W. K. Wallace, 1983; Wallace and others, 1989). The nature of the original contact between the Chilikadrotna Greenstone and the Koksetna River sequence is uncertain because of poor expo-

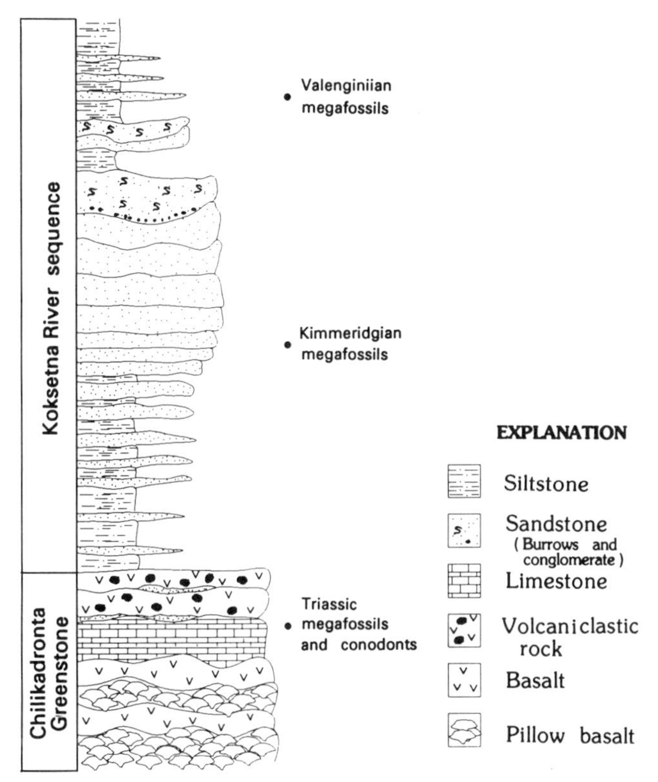

Figure 7. Composite stratigraphic column of the southern Kahiltna terrane showing points of fossil age control (Wallace and others, 1989).

sures and a strong deformational overprint. However, the Koksetna river sequence is younger, structurally higher, and locally contains distinctive detritus probably derived from the Chilikadrotna Greenstone, suggesting that the contact was probably originally depositional.

Rocks of the Koksetna River sequence superficially resemble those of the Kuksokwim Group to the west, and the two units thus were not distinguished in maps of the region published by Eakins and others (1978) and Nelson and others (1983). However, we believe that the two units can be distinguished on the basis of degree of induration, distribution of facies, available age data, and most diagnostically, clast composition (Hanks and others, 1985; Wallace and others, 1989). Sandstones of the Kuksokwim Group are generally less indurated and more carbonaceous than those of the Koksetna River sequence. Facies associations in the Kuskokwim Group indicate basin-plain to outer fan environments to the southeast, and middle- to inner fan environments to the northwest (Decker, 1984a; Hanks and others, 1985; Moore and Wallace, 1985; Wallace and others, 1989), suggesting a source to the northwest. This contrasts with the inferred southeastern source for the Koksetna River sequence. The quartzose lithic sandstones of the Kuskokwim Group contain higher percentages of sedimentary rock fragments and mica than sandstones of the Koksetna River sequence (Decker, 1984b; Hanks and others, 1985; Wallace and others, 1989).

The contact of the southern Kahiltna terrane with the Peninsular terrane probably is a fault, although it is now obscured, mainly by Late Cretaceous and Tertiary intrusions (Eakins and others, 1978; Nelson and others, 1983; Hanks and others, 1985; Wallace and others, 1989). The age and character of the Chilikadrotna Greenstone suggest that it may actually be part of the Peninsular terrane to the southeast; and parts of the sequence contain lithologies and structures similar to, but not diagnostic of, a subduction complex (Wallace and others, 1989). The Jurassic magmatic arc of the Peninsular terrane is interpreted by Reed and others (1983) to have formed above a southeast-dipping (present coordinates) subduction zone. If the Chilikadrotna Greenstone is part of the Peninsular terrane, it thus may have been located in a forearc setting adjacent to the convergent margin of the terrane. Following this assumption, the Koksetna River sequence may then have been shed from the axis of the Peninsular terrane into a basin to the northwest (present coordinates; Wallace and others, 1989). Therefore, on the basis of similarities in geologic history, and on the provenance and sediment source direction of the Koksetna River sequence, it is permissible to interpret the southern Kahiltna terrane as a northwestern extension of the Peninsular terrane (Hanks and others, 1985; Wallace and others, 1989). On the other hand, no genetic link has yet been established between the southern Kahiltna terrane and the Kuskokwim Group, and the two are probably separated by a fault (Wallace, 1984; Hanks and others, 1985; Wallace and others, 1989).

Paleolatitude data

Sedimentological evidence from lower and upper Paleozoic limestone blocks in the Goodnews terrane suggests that these rocks were deposited in tropical latitudes. Such evidence includes the occurrence of algal boundstone and oolitic limestone in the Nukluk subterrane (Hoare and Coonrad, 1978a; Box, 1985). Lower Jurassic rocks in the Hagemeister subterrane contain a faunal assemblage (characterized by the pelecypod *Weyla* sp.) that occurs well north of its northern range boundary elsewhere in North America (D. L. Jones, personal communication *in* Box, 1985), suggesting post–Early Jurassic poleward displacement.

Paleomagnetic analyses from the Tikchik subterrane by Karl and Hoare (1979) are inconclusive because the age and bedding attitudes could not be constrained. More detailed work in the same terrane (Engebretson and Wallace, 1985, and unpublished data) has confirmed a primary magnetic component in well-bedded Permian or Triassic basalts. Significant northward displacement (about $18 \pm 10°$) and counter-clockwise rotation (about $38 \pm 12°$) are indicated, although more precise age control would greatly decrease the uncertainty. If the basalts are Triassic in age, the data would suggest greater northward displacement, while a Permian age, which would be consistent with the reversed polarity of most of the samples, would suggest greater counter-clockwise rotation.

Paleomagnetic data (Globerman and Coe, 1984) from 81 volcanic flows in the post-accretionary upper sequence (72 to 65 Ma) of the Hagemeister subterrane show evidence for as much as

52° of counter-clockwise rotation since deposition, but no evidence for latitudinal displacement. Thus an unspecified amount of post–Early Jurassic to pre-Tertiary displacement is suggested by presently available data.

Interpretation

Many interpretations of relationships and correlations are possible among the terranes of the Bristol Bay region. However, the geology of the region is not known in sufficient detail to determine which of the possible relationships and correlations are correct, and many of the possibilities are not mutually exclusive. The most plausible interpretations, based on current data, are discussed below.

The Kilbuck terrane is unusual in that it consists of Proterozoic crystalline continental crust, yet is located between Mesozoic arc terranes. Although the Kilbuck terrane may be an allochthonous sliver of a distant continent, it seems more likely to us that it is a part of the North American continental backstop against which the Mesozoic arc terranes were accreted. Gemuts and others (1982) interpret the Kilbuck terrane to be continuous in the subsurface with the Idono sequence to the northeast. Similar Proterozoic protolith ages of rocks in the Kilbuck terrane and Idono sequence support their correlation. Instead of their being continuous in the subsurface, however, we offer the alternative model that the Kilbuck terrane has been offset from the Idono sequence along right-lateral strike-slip faults. The Idono sequence occurs north of the Iditarod-Nixon Fork fault, whereas the Kilbuck terrane occurs south of the Golden Gate fault. We believe that these two faults were once continuous but are now displaced about 43 km along the left-lateral Aniak–Thompson Creek fault (Fig. 1).

According to Box (1984b, 1985), the Togiak and Goodnews terranes formed as an intraoceanic arc-trench complex during Late Triassic to earliest Cretaceous time (Fig. 8A) that collided with North America during mid- to late Early Cretaceous time (Fig. 8B and 8C). Using present-day geography as a reference, southeastward-directed subduction occurred on the northwest flank of the volcanic arc (Hagemeister subterrane), creating an imbricately stacked subduction complex of accreted oceanic plate lithologies (Goodnews terrane) as shown in Figure 6. The Kulukak subterrane, southeast of the Hagemeister subterrane, consists of volcaniclastic turbidites deposited in a back-arc setting. Box (1983a, 1984a) suggests that the Nyack terrane is a slice of the Togiak terrane "doubled-up" by post-accretionary strike-slip faulting. According to this model, volcanic arc rocks of the Togiak and Nyack terranes continue northward into the Yukon-Koyukuk province where similar rocks of Jurassic and Cretaceous age occur (Fig. 8D). However, since the Togiak and Goodnews terranes trend southwestward into the Bering Sea, the continuation of these terranes into the Yukon-Koyukuk basin requires the arc-trench system to have been bent back upon itself analogous to the modern Banda arc (Hamilton, 1979), or to have had its trend reversed by some other means.

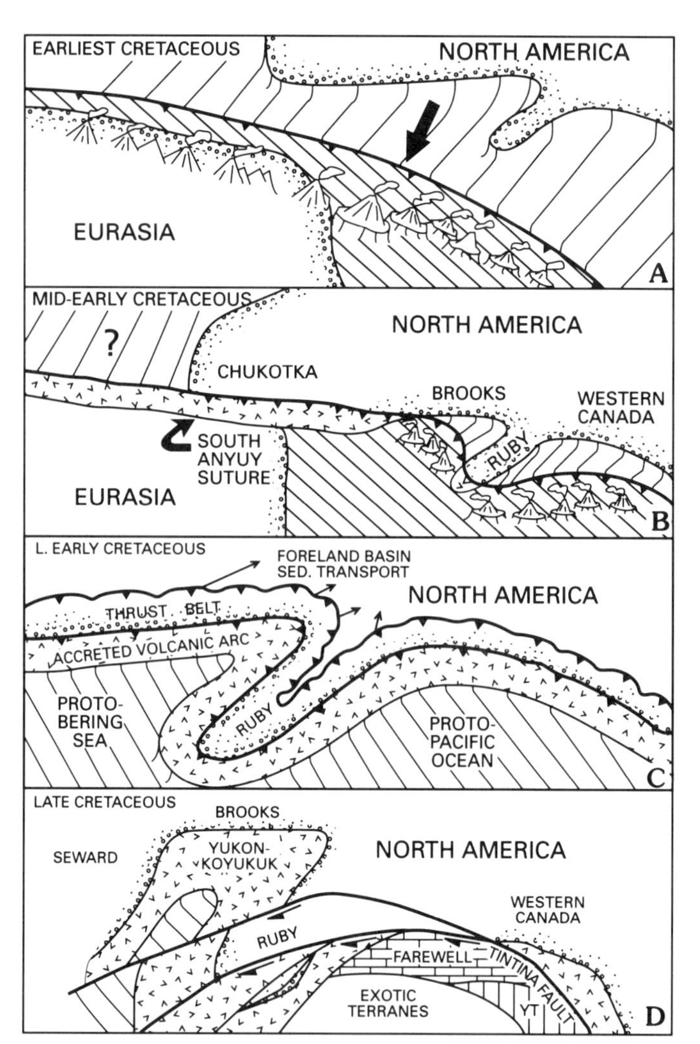

Figure 8. Schematic tectonic model of Box (1984a) showing correlation of Goodnews and Togiak terranes with terranes of Chukotka the Yukon-Koyukuk province. A, North America and Eurasia plate boundary (sawteeth on upper plate). Volcanic arc shown along plate boundary, large arrow indicates direction of relative movement. B, Development of the South Anyuy suture. Major mountain areas are labeled. Subduction boundary (large sawteeth on upper plate) and accreted volcanic arc complex (chevron pattern) along plate boundary. Uncertain terrane queried. C, Development of accreted volcanic arc and thrust belt along North American plate boundary. Arrows indicate direction of sediment transport. Sawteeth on upper plate. D, Major land areas in relation to accreted terranes. Arrows indicate direction of relative movement. Y-T = Yukon-Tanana terrane.

The southwestward projection of the Togiak and Goodnews terranes suggests that they may not be related to terranes onshore to the north and west, but instead continue offshore, possibly bending northwestward subparallel with the Bering Sea shelf margin (Marlow and others, 1976).

The Togiak and Goodnews terranes are also similar in many respects to the Peninsular and Kahiltna terranes to the south and east, suggesting a possible correlation (Wallace, 1983, 1984;

Decker and Murphy, 1985). However, paleomagnetic and geologic evidence (discussed below) indicates that any original relationship between the two pairs of terranes has been substantially modified by major displacement (Wallace and others, 1989). Although the two pairs of terranes probably formed in arc-trench settings, it is uncertain whether they were parts of a single continuous arc-trench system, or formed separately in the same tectonic setting. The latter relationship has been proposed by Wallace and others (1989) and is shown in Figure 9. This model assumes that the Yukon-Koyukuk arc, the Togiak arc, and the composite Peninsular-Wrangellia arc were separate but tectonically related magmatic arc complexes formed above a southwest-dipping subduction zone adjacent to North America in Middle and Late Jurassic time (Fig. 9A). During Late Jurassic and Early Cretaceous time, the volcanic arcs collided with North America, subduction stepped out beyond the arcs, and the subduction polarity reversed (Fig. 9B and 9C). Intermittent strike-slip faulting and subduction have modified the original relationships to the pattern that presently exists in southern Alaska (Fig. 9C to F).

The tectonic setting and paleogeography of the basin in which the Koksetna River sequence was deposited have important implications for the tectonic evolution of southwestern Alaska. Late Jurassic and Early Cretaceous turbidite-filled basins, including the Koksetna River sequence, the northern part of the Kahiltna terrane, and the Gravina-Nutzotin belt occur along the landward boundary of the Peninsular, Wrangellia, and Alexander terranes. It is generally agreed that these basins mark the suture along which the more outboard terranes of southern Alaska collided with North American. However, the basins are controversially interpreted to be precollisional (Jones and others, 1981, 1986; Csejtey and others, 1982, Coney and Jones, 1985) or postcollisional (Eisbacher, 1974; 1985; Decker and Murphy, 1985; Hanks and others, 1985; Wallace and others, 1989). The two interpretations differ with regard to basin configuration, timing and location of collision, and cause of deformation of the basinal rocks. If the basins were precollisional, they were probably relatively wide and floored by oceanic crust, and deformation of the basinal deposits resulted from postdepositional collision of the basins during or after Late Cretaceous time and near their present locations. If the basins were postcollisional, they were relatively narrow, were probably not floored by oceanic crust, and deformation of the basinal deposits resulted from late syntectonic compression followed by post-tectonic extension.

Paleomagnetic evidence (Stone and others, 1982; Hillhouse and Coe, this volume) suggests that there has been considerable northward translation of the Peninsular terrane since the Koksetna River sequence was deposited. Its depositional relationship with the Peninsular terrane thus requires significant tectonic transport of the southern Kahiltna terrane, whether the Koksetna River sequence was deposited in a pre- or postcollisional basin. On the other hand, the Farewell terrane and depositionally overlying Kuskokwim Group, which also rests depositionally upon the Goodnews and Togiak terranes (Wallace, 1983, 1984), does not appear to have been latitudinally displaced a significant amount since Paleozoic time (Plumley and others, 1981; Plumley and Coe, 1982, 1983). If the Koksetna River sequence was deposited on a part of the Peninsula terrane, the sequence therefore must be separated along its northwestern boundary from the Farewell, Goodnews, and Togiak terranes and from the Kuskokwim Group by a fault with considerable right-lateral strike-slip and/or thrust displacement. This contact is not exposed, but no pre–latest Cretaceous units have yet been correlated across it and field observations suggest that it is a major fault (Wallace, 1984; Hanks and others, 1985; Wallace and others, 1989).

POST-ACCRETIONARY ROCKS

Kuskokwim Group

The Lower and Upper Cretaceous (Albian to Coniacian) Kuskokwim Group consists of marine turbidites, and subordinate shallow-marine and fluvial strata, deposited in an elongate southwest-trending basin covering over 70,000 km^2 in southwestern Alaska (Bundtzen and Gilbert, 1983; Wallace, 1983; Decker, 1984b; Pacht and Wallace, 1984). The Kuskokwim Group is thickest (>10 km) and displays the deepest water facies in the central part of the basin between the villages of McGrath and Aniak. Mixed marine and nonmarine sections are relatively thin (<3 km) and are restricted to the basin margins and local basement highs. Sandstone and conglomerate clast compositions throughout the Kuskokwim Group suggest that the sediment was derived from local sources (Bundtzen and Gilbert, 1983; Wallace, 1983, 1984; Crowder and Decker, 1985; Murphy, 1989). Facies relations suggest rapid deposition within a regionally subsiding continental trough (Hoare, 1961; Decker and Hoare, 1982; Bundtzen and Gilbert, 1983). The basement upon which the Kuskokwim Group was deposited is known only locally near the margins of the basin. In the thick central part of the basin, especially between the Denali-Farewell and Iditarod–Nixon Fork faults, the underlying basement rocks are unknown (Decker and others, 1984b). The Kuskokwim Group is generally deformed into broad open folds, but locally, especially near its southeast margin, overturned or isoclinal folds and thrust faults predominate (Decker, 1984a).

The Iditarod–Nixon Fork and the Denali-Farewell fault systems divide the Kuskokwim Group into three geographic basins: the Iditarod basin to the northwest, the central Kuskokwim basin, and the Nushagak basin to the southeast. The geological relations among these three basins are unclear, but because they are defined by such postdepositional faults, they do not necessarily correspond with the original depositional basins. For example, there is little contrast in facies and clast composition across the Iditarod–Nixon Fork fault (Bundtzen and Laird, 1982, 1983a), whereas sandstone clast compositions are significantly different across the Denali-Farewell fault system (Crowder and Decker, 1985). Sandstones within the central Kuskokwim basin, north of the Denali-Farewell fault, are composed predominantly of metamorphic and volcanic rock fragments derived locally from

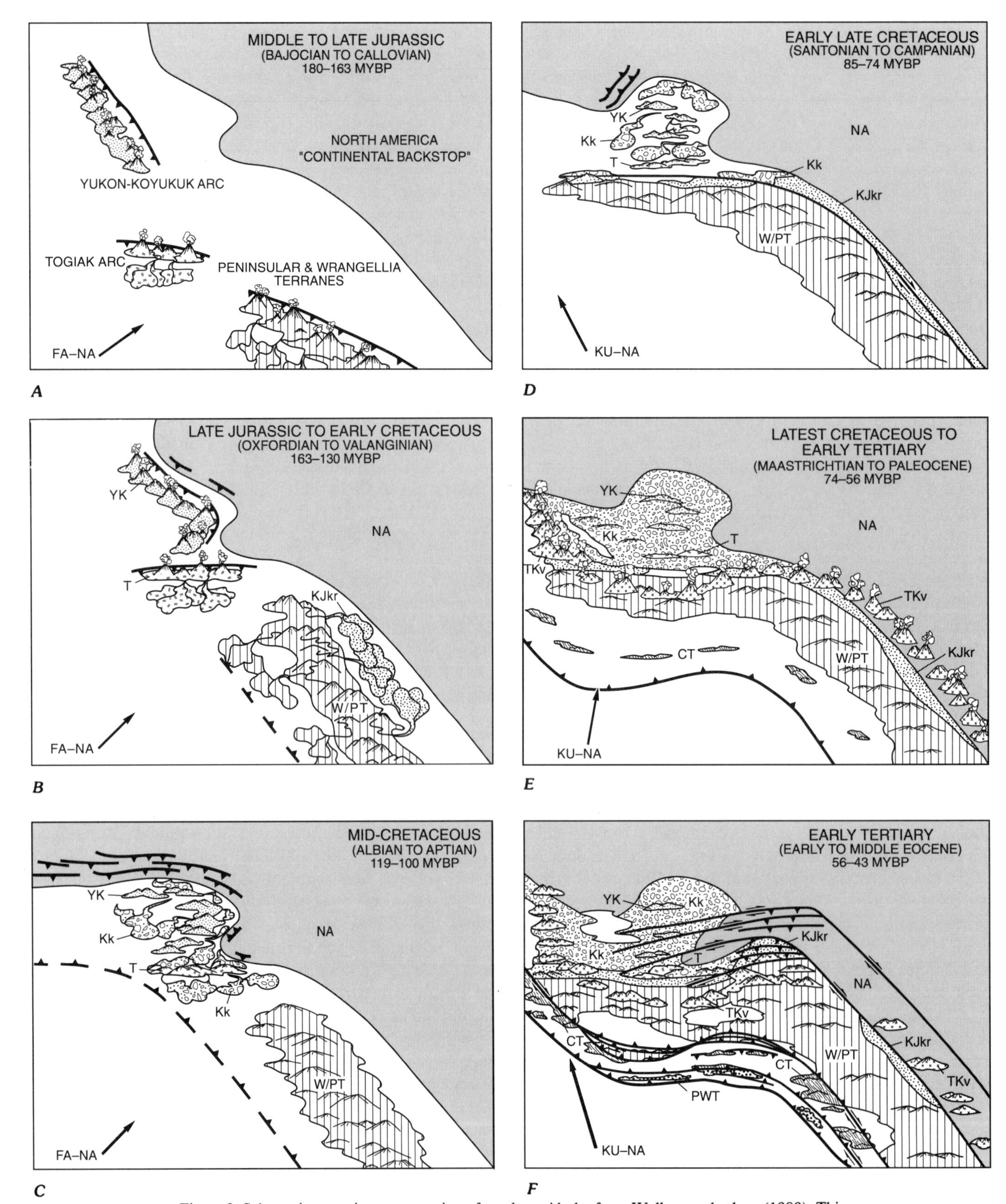

Figure 9. Schematic tectonic reconstruction of southern Alaska from Wallace and others (1989). This figure illustrates the possible paleogeographic affinities of the Goodnews and Togiak terranes with terranes of south-central Alaska. See text for discussion.

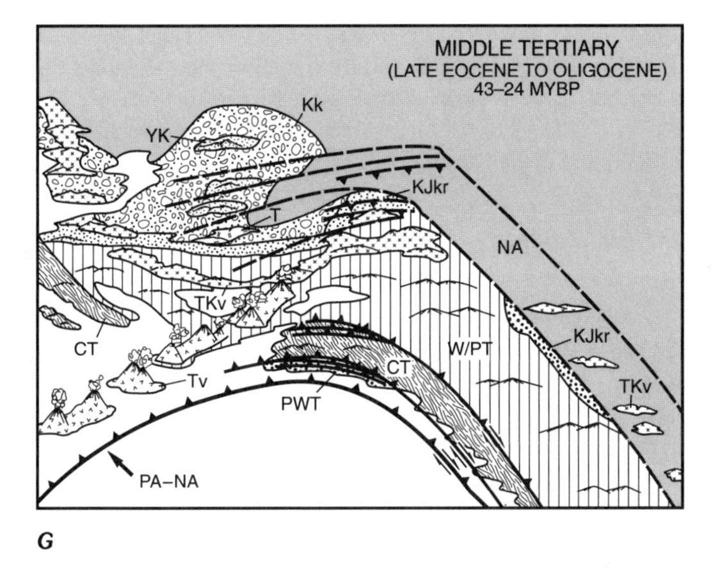

G

EXPLANATION

TECTONOSTRATIGRAPHIC TERRANES

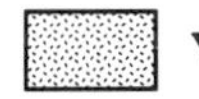
PWT–PRINCE WILLIAM TERRANE
Eocene accretionary rocks

CT–CHUGACH TERRANE
Early Cretaceous and Late Cretaceous accretionary rocks

W/PT–WRANGELLIA/PENINSULAR TERRANES
Paleozoic and Mesozoic sedimentary rocks and Early to Middle Jurassic island-arc rocks

T–TOGIAK ARC
Middle Jurassic to Early Cretaceous island-arc rocks

YK–YUKON KOYUKUK ARC
Middle Jurassic to Early Cretaceous island-arc rocks

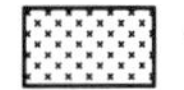
NA–NORTH AMERICA CONTINENTAL BACKSTOP

CRETACEOUS AND TERTIARY MAGMATIC ARCS

Tv–LATE EOCENE TO OLIGOCENE
Arc volcanic and intrusive rocks

TKv–LATE CRETACEOUS TO PALEOCENE
Arc volcanic and intrusive rocks

SUCCESSOR BASIN DEPOSITS

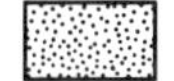
Kk–KUSKOKWIM GROUP
Late Early to Late Cretaceous clastic rocks

KJkr–JURA-CRETACEOUS FLYSCH
(Koksetna River sequence) Late Jurassic to Early Cretaceous deep-water clastic rocks

FAULTS–Solid during times of major displacement; dashed during times of minor displacement. Arrows indicate direction of relative movement.

SUBDUCTION ZONE–Dashed where existence is inferred from model and plate-motion vectors. Sawteeth on upper plate

Plate motion vector indicates motion of oceanic plate (FA–Farallon, KU–Kula, PA–Pacific) with respect to North America (NA). Length of plate motion vector proportional to convergence rate

various source terranes, but are noticeably poor in unmetamorphosed sedimentary rock fragments. Sandstones in the Nushagak basin, on the other hand, are characterized by sedimentary rock fragments.

Assuming that the Kuskokwim Group formed in one contiguous marine embayment, the Kuskokwim Group then stitches together most of the terranes of southern and western Alaska by Albian time. Kuskokwim strata of the Iditarod basin depositionally overlie rocks of the Ruby, Innoko, and Farewell terranes, and in the central Kuskokwim basin depositionally overlie rocks of the Farewell, Kilbuck, Goodnews, and Togiak (Hagemeister subterrane) terranes. Kuskokwim strata of the Nushagak basin depositionally overlie rocks of the Tikchik and Farewell terranes. The only terranes in or adjacent to southwestern Alaska not clearly overlain by the Kuskokwim Group are the Nyack, Kulukak subterrane of the Togiak, southern Kahiltna, and Peninsula terranes.

Igneous rocks

Igneous rocks of latest Cretaceous and earliest Tertiary age (45 to 80 Ma) are widespread throughout southwest Alaska (Wilson, 1977; Shew and Wilson, 1981; Robinson and Decker, 1986; Bergman and others, 1987; Wilson and others, this volume). These rocks have been divided by Wallace and Engebretson (1982, 1984) into two parallel northeast-trending belts, the Kuskokwim Mountains belt to the northwest, and the Alaska Range belt to the southeast. Only rocks within these two belts are summarized below; for additional information about Upper Cretaceous and Cenozoic volcanic rocks of southwest Alaska, see Moll-Stalcup (this volume).

The Kuskokwim Mountains belt consists of volcanic fields, isolated stocks, and volcano-plutonic complexes that vary widely in composition (Moll and others, 1981; Moll and Patton, 1982; Bundtzen and Laird, 1982, 1983a, b; Robinson and others, 1984a, b, 1986; Reifenstuhl and others, 1984; Robinson and Decker, 1986; Moll-Stalcup, 1987; Decker and others, 1990). Eleven volcano-plutonic complexes in the Iditarod and Ophir Quadrangles described by Bundtzen and Swanson (1984) chiefly consist of elliptical outcrop areas of andesite and basalt that overlie and flank alkali-gabbro to monzonite stocks, and are spatially associated with volumetrically minor garnet-bearing peraluminous rhyolite sills and chromium-enriched mafic dikes. Compositionally and temporally similar rocks occur to the north in the Medfra Quadrangle (Moll and others, 1980). A similar magma series is described by Robinson and others (1984a) in the southern Sleetmute Quadrangle, where granodiorite to quartz monzonite plutons intrude and are flanked by a bimodal suite of flows, tuff, lahar deposits, and hypabyssal rocks of basaltic andesite and rhyolite composition. A thick volcanic sequence consisting mainly of calc-alkaline basaltic andesite flows, tuff, and breccia are described on islands in the northern Bristol Bay region by Globerman and others (1984). Magmatic rocks of the Kuskokwim Mountains area intrude and/or overlie the Innoko, Ruby, Farewell, Kilbuck, Nyack, Goodnews, Togiak, and Tikchik terranes, and the Kuskokwim Group.

The Alaska Range belt consists of three longitudinal segments of magmatic rocks: a northeastern segment consisting of scattered batholiths and stocks with sharp discordant contacts, a central segment consisting of extensive elongate and concordant batholithic complexes with dynamothermally metamorphosed country rock, and a southwestern segment consisting of extensive volcanic fields and scattered hypabyssal plutons (Reed and Lanphere, 1973, 1974; Wallace and Engebretson, 1984). Intrusive rocks generally are tonalite, quartz-diorite, and granodiorite, and subordinate gabbro, syenite, quartz monzonite, and monzonite (Solie, 1983); volcanic rocks range from basalt to rhyolite and occur as flows and pyroclastic deposits (Reed and Lanphere, 1973, 1974; Solie and others, 1982; Wallace and Engebretson, 1984). The Alaska Range belt intrudes and/or overlies the Farewell, Kahiltna, and Peninsular terranes, and the Kuskokwim Group.

Moll and Patton (1982) and Bergman and Doherty (1986) combine the Kuskokwim Mountains belt and the Alaska Range belt into one province extending from the Alaska Range to the Yukon-Koyukuk Province. Bergman and Doherty (1986) and Bergman and others (1987) present age distribution and geochemical data on the 45- to 80-Ma igneous rocks in southern Alaska. They feel that the data for the 60- to 80-Ma rocks are consistent with subduction-related magmatism occurring in a 400- to 600-km-wide province. On the other hand, on the basis of trace element and isotopic data Bergman and others (1987) believe that the 45- to 60-Ma magmatism in the region resulted from post-subduction processes characterized mainly by crustal lithospheric melting (also see Hudson, this volume).

Upper Cretaceous and Tertiary sedimentary rocks

Nonmarine conglomerate, sandstone, shale, and coal occur in a series of small fault-bounded sedimentary basins that formed along the Denali-Farewell fault system in the McGrath Quadrangle (Gilbert, 1981; Dickey, 1982; Gilbert and others, 1982; Solie and Dickey, 1982; Kirschner, this volume). The rock assemblages in these basins are similar to those in the Usibelli Group in the Healy Quadrangle (Wahrhaftig and others, 1969; Wahrhaftig, 1987). In the McGrath Quadrangle, quartz and argillite clasts from the lowest unit were derived primarily from the polycrystalline basement rocks of the Yukon-Tanana terrane to the northeast (Dickey, 1984). Pollen ages for the quartz-argillite clast unit range from Eocene to middle Oligocene (unpublished company data cited in Dickey, 1984). Limestone-clast and felsite-clast units overlying the quartz-artillite unit are probably late Tertiary in age and were probably derived from the Farewell terrane in the Alaska Range to the south (Dickey, 1984). Coal in the Tertiary rocks in the McGrath Qaudrangle range in thickness from 1 to 20 m and in BTU rank from subbituminous C to A (Solie and Dickey, 1982).

Uppermost Cretaceous (Maestrichtian) nonmarine sedimentary rocks of the Summit Island Formation (Hoare and others, 1983) occur along coastal exposures in the Togiak Bay area. These rocks consist of thick-bedded conglomerate, sandstone, siltstone, carbonaceous mudstone, and sparse coal seams. The Summit Island Formation overlies with angular unconformity rocks of the Togiak terrane.

CONCLUSIONS

Southwest Alaska is composed of the predominantly Paleozoic continental margin rocks of the southwestern Alaska Range and northern Kuskokwim Mountains region (Farewell terrane), and of the predominantly Mesozoic accretionary rocks of the northern Bristol Bay region (Fig. 10). The Farewell terrane (together with the Kilbuck terrane and Idono sequence) probably formed a significant part of the North American continental backstop against which the Mesozoic terranes of southern Alaska were accreted. Although its contact with the Farewell terrane is not exposed, the Togiak-Goodnews arc-trench complex (here interpreted as a single terrane at least since Early Cretaceous time) has been amalgamated with the Farewell terrane at least since Albian time, when deposition of the Kuskokwim Group across the terranes in both regions began (Fig. 10; Cady and others, 1955; Hoare, 1961; Hoare and Coonrad, 1978a; Wallace, 1983). South of the Denali-Farewell fault, the suture zone between the Farewell terrane and the Togiak and Goodnews terranes probably is covered by the Kuskokwim Group and by younger deposits in the low-lying Nushagak Hills. North of the fault, the suture probably occurs within the centrak Kuskokwim Mountains, where it also is buried by the Kuskokwim Group and younger deposits.

The southern Kahiltna terrane has been juxtaposed against the Farewell terrane at least since latest Cretaceous to Paleocene time, when magmatism of the Alaska Range belt occurred within and northwest of the southern Kahiltna terrane (Wallace, 1983, 1984; Wallace and Engebretson, 1984; Wallace and others, 1989). The boundary between the southern Kahiltna terrane and the Farewell probably is a major suture zone (Fig. 10) along which the far-travelled terranes of southern Alaska (such as Wrangellia and the Peninsular terranes) were juxtaposed against North America. Considerable postcollisional northward translation of at least some of these terranes may have occurred by strike-slip displacement along this suture. The suture is generally buried by overlap deposits, intruded by post-accretion plutons, or modified by more recent faulting. The locations of the suture at depth may correspond with a pronounced aeromagnetic discontinuity (Decker and Karl, 1977) at the position of the Mulchatna fault at the surface. Considerable rearrangement of southwestern Alaska has occurred by strike-slip and thrust-faulting since deposition of the Albian and latest Cretaceous to Paleocene overlap strata. The disruptions, however, probably have been insignificant compared to the amount before amalgamation, as indicated by the fact that the Kuskokwim Group and the Alaska Range magmatic belt still are geologically relatively coherent.

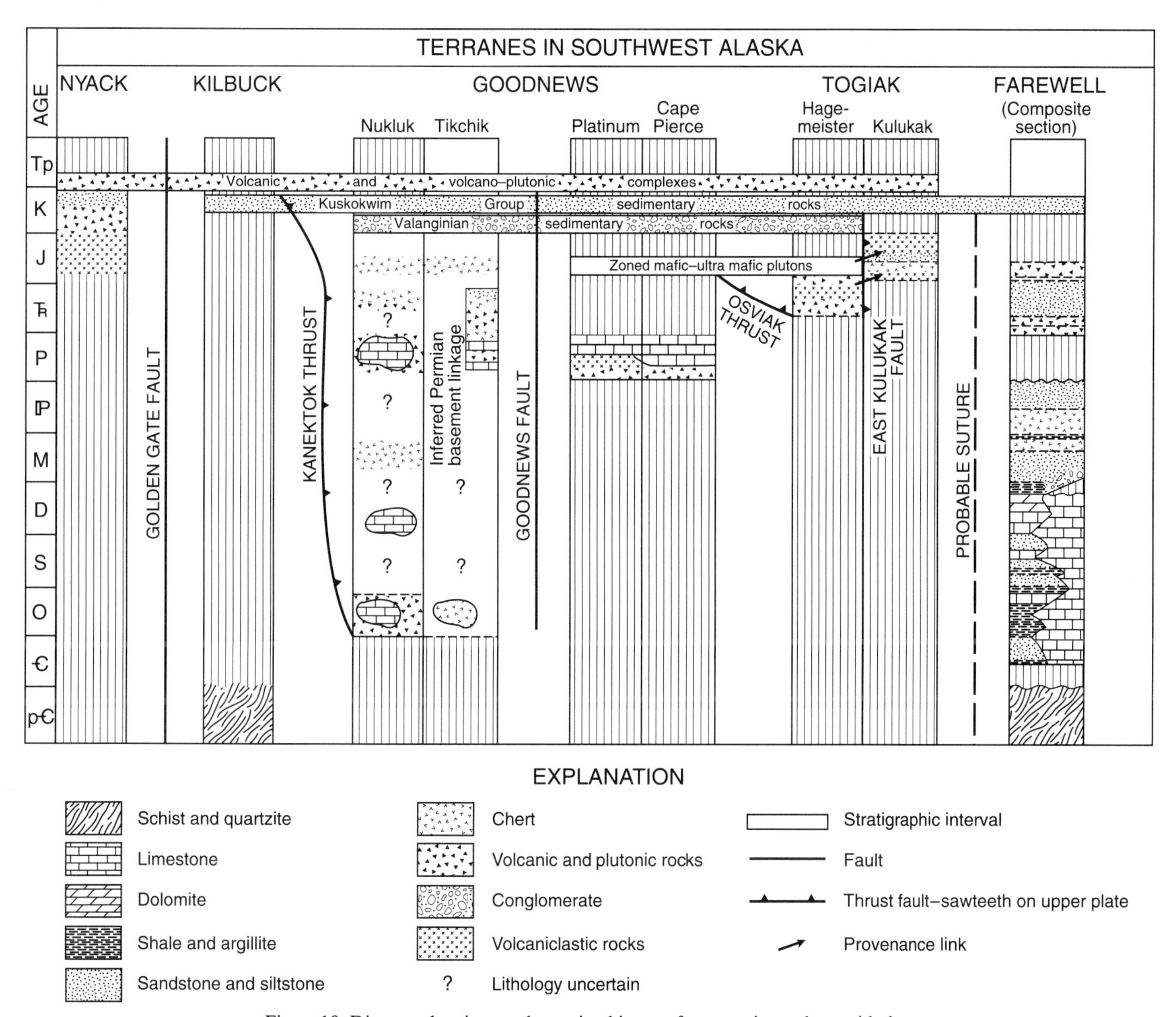

Figure 10. Diagram showing amalgamation history of terranes in southwest Alaska.

REFERENCES CITED

Armstrong, A. K., Harris, A. G., Reed, B. L., and Carter, C., 1977, Paleozoic sedimentary rocks in the northwest part of the Talkeetna Quadrangle, Alaska Range, Alaska, *in* Blean, K. M., ed., The United States Geological Survey in Alaska; Accomplishments during 1976: U.S. Geological Survey Circular 751-B, p. B61–B62.

Bergman, S. C., and Doherty, D. J., 1986, Nature and origin of 50–75 Ma volcanism and plutonism in W. and S. Alaska: Geological Society of America Abstracts with Programs, v. 18, p. 539.

Bergman, S. C., Hudson, T. L., and Doherty, D. J., 1987, Magmatic rock evidence for a Paleocene change in tectonic setting of Alaska: Geological Society of America Abstracts with Programs, v. 19, p. 586.

Bird, M. L., and Clark, A. L., 1976, Microprobe study of olivine chromites of the Goodnews Bay ultramafic complex, Alaska, and the occurrence of platinum: U.S. Geological Survey Journal of Research, v. 4, p. 717–725.

Blodgett, R. B., 1983a, Paleobiogeographic implications of Devonian fossils from the Nixon Fork terrane, southwestern Alaska, *in* Stevens, C. H., ed., Pre-Jurassic rocks in western North American suspect terranes: Pacific Section, Society of Exploration Petrologists and Mineralogists, p. 125–130.

——, 1983b, Paleobiogeographic implications of Devonian fossils from the Nixon Fork terrane, southwestern Alaska: Geological Society of America Abstracts with Programs, v. 15, p. 428.

Blodgett, R. B., and Clough, J. G., 1985, The Nixon Fork terrane; Part of an in place peninsular extension of the North American Paleozoic continent: Geological Society of America Abstracts with Programs, v. 17, p. 342.

Blodgett, R. B., and Gilbert, W. G., 1983, The Cheeneetnuk Limestone; A new Early(?) to Middle Devonian formation in the McGrath A-4 and A-5 Quadrangles, west-central Alaska: Alaska Division of Geological and Geophysical Surveys Professional Report 85, 6 p.

Blodgett, R. B., and Rohr, D. M., 1989, Two new Devonian spine-bearing pleurotomariacean gastropod genera from Alaska: Journal of Paleontology, v. 63, p. 47–52.

Blodgett, R. B., Clough, J. G., and Smith, T. N., 1984, Ordovician-Devonian paleogeography of the Holitna Basin, southwestern Alaska: Geological Society of America Abstracts with Programs, v. 16, p. 271.

Blodgett, R. B., Rohr, D. M., and Boucot, A. J., 1987, Early Middle Devonian gastropod biogeography of North America: Geological Society of America Abstracts with Programs, v. 15, p. 591.

Boucot, A. J., and Gray, J., 1980, A Cambro-Permian pangeaic model consistent with lithofacies and biogeographic data, *in* Strangeway, D. W., ed., The continental crust and its mineral deposits: Geological Association of Canada Special Paper 20, p. 389–419.

——, 1983, A Paleozoic Pangea: Science, v. 222, no. 4624, p. 571–581.

Box, S. E., 1982, Kanektok Suture, SW Alaska; Geometry, age, and relevance: EOS Transactions of the American Geophysical Union, v. 63, p. 915.

——, 1983a, Sinuous late Early Cretaceous arc-continent collisional belt in the NE USSR and NW Alaska: Geological Society of America Abstracts with Programs, v. 15, p. 531.

——, 1983b, Tectonic synthesis of Mesozoic histories of the Togiak and Goodnews terranes, SW Alaska: Geological Society of America Abstracts with Programs, v. 15, p. 406.

——, 1984a, Implications of a possible continuous 4,000 km long late Early Cretaceous arc-continent collisional belt in NE USSR and NW Alaska for the tectonic development of Alaska, *in* Howell, D. G., Jones, D. L., Cox, A., and Nur, A., eds., Proceedings of the Circum-Pacific Terrane Conference: Stanford, California, Stanford University Publications in the Geological Sciences, p. 33–35.

——, 1984b, Early Cretaceous arc-continent collision and subsequent strike slip displacement in southwest Alaska: Geological Society of America Abstracts with Programs, v. 16, p. 272.

——, 1985, Terrane analysis, northern Bristol Bay region, southwestern Alaska; Development of a Mesozoic intraoceanic arc and its collision with North America [Ph.D. thesis]: Santa Cruz, University of California, 163 p.

Bundtzen, T. K., and Gilbert, W. G., 1983, Outline of geology and mineral resources of the upper Kuskokwim region, Alaska, *in* Proceedings of the 1982 Symposium on Western Alaska Geology and Resource Potential: Alaska Geological Society Journal, v. 3, p. 101–119.

Bundtzen, T. K., and Laird, G. M., 1982, Geologic map of the Iditarod D-2 and eastern D-3 Quadrangles, Alaska: Alaska Division of Geological and Geophysical Surveys Geologic Report 72, 1 sheet, scale 1:63,360.

——, 1983a, Geologic map of the Iditarod D-1 Quadrangle, Alaska: Alaska Division of Geological and Geophysical Surveys Geologic Report 78, 1 sheet, scale 1:63,360.

——, 1983b Geologic map of the McGrath D-6 Quadrangle, Alaska: Alaska Division of Geological and Geophysical Surveys Professional Report 79, 1 sheet, scale 1:63,360.

Bundtzen, T. K., and Swanson, S. E., 1984, Geology and petrology of igneous rocks in Innoko River area, western Alaska: Geological Society of America Abstracts with Programs, v. 16, p. 273.

Bundtzen, T. K., Gilbert, W. G., and Blodgett, R. B., 1979, The Chilikadrotna Greenstone; An Upper Silurian metavolcanic sequence in the central Lake Clark Quadrangle, Alaska: Alaska Division of Geological and Geophysical Surveys Geologic Report 61, p. 31–35.

Bundtzen, T. K., Kline, J. T., and Clough, J. G., 1982, Preliminary geology of McGrath B-2 Quadrangle, Alaska: Alaska Division of Geological and Geophysical Surveys Open-File Report 149, 1 sheet, scale 1:40,000.

Bundtzen, T. K., Kline, J. T., Solie, D. N., and Clough, J. G., 1985, Geologic map of the McGrath B-2 Quadrangle, Alaska: Alaska Division of Geological and Geophysical Surveys Public Data File 85-14, 1 sheet, scale 1:40,000.

Cady, W. M., Wallace, R. E., Hoare, J. M., and Webber, E. J., 1955, The central Kuskokwim region, Alaska: U.S. Geological Survey Professional Paper 268, 132 p.

Capps, S. R., 1926, The Skwenta region: U.S. Geological Survey Bulletin 797, p. 67–98.

Churkin, M., Jr., Reed, B. L., Carter, C., and Winkler, G. R., 1977, Lower Paleozoic graptolitic section in the Terra Cotta Mountains, southern Alaska Range, Alaska, *in* Blean, K. M., ed., The United States Geological Survey in Alaska: Accomplishments during 1976: U.S. Geological Survey Circular 751-B, p. B37–B37.

Churkin, M., Jr., Wallace, W. K., Bundtzen, T. K., and Gilbert, W. G., 1984, Nixon Fork–Dillinger terranes; A dismembered Paleozoic craton margin in Alaska displaced from Yukon Territory: Geological Society of America Abstracts with Programs, v. 16, p. 275.

Clough , J. G., and Blodgett, R. B., 1985, Comparative study of the sedimentology and paleoecology of middle Paleozoic algal and coral-stromatoporoid reefs in Alaska: Papeete, Tahiti, 5th International Coral Reef Congress Proceedings, p. 593–598.

——, 1989, Silurian-Devonian algal reef mound complex of southwest Alaska, *in* Geldsetzer, H.H.J., James, N. P., and Tebbutt, G. E., eds., Reefs, Canada and adjacent areas: Canadian Society of Petroleum Geologists Memoir 13, p. 404–407.

Clough, J. G., Blodgett, R. B., and Smith, T. N., 1984, Middle Paleozoic subtidal to tidal flat carbonate sedimentation in southwest Alaska: Geological Society of America Abstracts with Programs, v. 16, p. 275.

Coney, P. J., and Jones, D. L., 1985, Accretion tectonics and crustal structure in Alaska: Tectonophysics, v. 119, p. 265–283.

Crowder, K. L., and Decker, J., 1985, Provenance of conglomerate clasts from the Upper Cretaceous Kuskokwim Group, southwest Alaska: 60th Annual Meeting, Pacific Section, American Association of Petroleum Geologists Program and Abstracts, p. 63.

Csejtey, B., Cox, D. P., Evarts, R. C., Stricker, G. D., and Foster, H. L., 1982, The Cenozoic Denali fault system and the Cretaceous accretionary development of southern Alaska: Journal of Geophysical Research, v. 87, p. 3741–3754.

Decker, J., 1984a, Geologic map of the Sparrevohn area: Alaska Division of Geological and Geophysical Surveys Public Data-File Report 84-42, 2 sheets, scale 1:40,000.

——, 1984b, The Kuskokwim Group; A post-accretionary successor basin in southwest Alaska: Geological Society of America Abstracts with Programs, v. 16, p. 277.

Decker, J., and Hoare, J. M., 1982, Sedimentology of the Cretaceous Kuskokwim Group, southwest Alaska, *in* Coonrad, W. L., ed., The United States Geological Survey in Alaska; Accomplishments during 1980: U.S. Geological Survey Circular 844, p. 81–83.

Decker, J., and Karl, S., 1977, Preliminary aeromagnetic map of the western part of southern Alaska: U.S. Geological Survey Open-File Report 77-169-J, 1 sheet, scale 1:1,000,000.

Decker, J., and Murphy, J. M., 1985, Large Upper Mesozoic clastic basins in Alaska, *in* 6th European Regional Meeting Abstracts: International Association of Sedimentologists, p. 551–553.

Decker, J., Reifenstuhl, R. R., and Coonrad, W. L., 1984a, Compilation of geologic data from the Russian Mission A-3 Quadrangle, Alaska: Alaska Division of Geological and Geophysical Surveys Report of Investigations 84-19, 1 sheet, scale 1:63,360.

Decker, J., Robinson, M. S., Murphy, J. M., Reifenstuhl, R. R., and Albanese, M. D., 1984b, Geologic map of the Sleetmute A-6 Quadrangle, Alaska: Alaska Division of Geological and Geophysical Surveys Report of Investigation 84-8, 1 sheet, scale 1:40,000.

Decker, J., Reifenstuhl, R., Robinson, M. S., and Waythomas, C. A., 1991, Geologic map of the Sleetmute A-5, A-6, B-5, and B-6 Quadrangles, southwest Alaska: Alaska Division of Geological and Geophysical Surveys Professional Report, 1 sheet, scale 1:63,360.

Dempsey, W. J., Meuschke, J. L., and Andreason, G. E., 1957, Total intensity aeromagnetic profiles of Bethel Basin, Alaska: U.S. Geological Survey Open-File Report 57–33, 3 sheets, scale 1:250,000.

Dickey, D. B., 1982, Tertiary sedimentary rocks and tectonic implications of the Farewell fault zone, McGrath Quadrangle, Alaska [M.S. thesis]: Fairbanks, University of Alaska, 54 p.

——, 1984, Cenozoic nonmarine sedimentary rocks of the Farewell fault zone, McGrath Quadrangle, Alaska: Sedimentary Geology, v. 38, p. 443–463.

Dillon, J. T., and 5 others, 1985, New radiometric evidence for the age and thermal history of the metamorphic rocks of the Ruby and Nixon Fork terranes, west-central Alaska, *in* Bartsch-Winkler, S., and Reed, K. M., ed., The United States Geological Survey in Alaska; Accomplishments during 1983: U.S. Geological Survey Circular 945, p. 13–18.

Dutro, J. T., Jr., and Patton, W. W., Jr., 1981, Lower Paleozoic platform sequence in the Medfra Quadrangle, west-central Alaska, *in* Albert, N.R.D., and Hudson, T., eds., The United States Geological Survey in Alaska; Accomplishments during 1979: U.S. Geological Survey Circular 823-B, p. B42–B44.

——, 1982, New Paleozoic formations in the northern Kuskokwim Mountains, west-central Alaska: U.S. Geological Survey Bulletin 1529-H, p. H13–H22.

Eakins, G. R., Gilbert, W. G., and Bundtzen, T. K., 1978, Preliminary bedrock geology and mineral resource potential of west-central Lake Clark Quadrangle, Alaska: Alaska Division of Geological and Geophysical Surveys Open-File Report 118, scale 1:125,000.

Eisbacher, G. H., 1974, Evolution of successor basins in the Canadian Cordillera, *in* Dott, R. H., Jr., and Shaver, R. H., eds., Modern and ancient geosynclinal sedimentation: Society of Economic Paleontologists and Mineralogists Special Publication 19, p. 274–291.

——, 1985, Pericollisional strike-slip faults and syn-orogenic basins, Canadian Cordillera, *in* Biddle, K. T., and Christie-Blick, N., eds., Strike-slip deformation, basin formation, and sedimentation: Society of Economic Paleontologists and Mineralogists Special Publication 37, p. 265–282.

Engebretson, D. C., and Wallace, W. K., 1985, Preliminary paleomagnetism and tectonic interpretation of the Tikchik terrane, southwest Alaska: Geological Society of America Abstracts with Programs, v. 17, p. 354.

Gemuts, I., Steefel, C. I., and Puchner, C. C., 1982, Geology and mineralization of western Alaska: Alaska Geological Society 1982 Symposium Program with Abstracts, p. 36–37.

Gemuts, I., Puchner, C. C., and Steefel, C. I., 1983, Regional geology and tectonic history of western Alaska, *in* Proceedings of the 1982 Symposium on Western Alaska Geology and Resource Potential: Alaska Geological Society, v. 3, p. 67–87.

Gilber, W. G., 1981, Preliminary geologic map of the Cheeneetnuk River area, Alaska: Alaska Division of Geological and Geophysical Surveys Open-File Report 153, scale 1:63,360.

Gilbert, W., G., and Bundtzen, T. K., 1983a, Paleozoic stratigraphy of Farewell area, southwest Alaska Range, Alaska, *in* Alaska Geological Society Symposium; New Developments in the Paleozoic Geology of Alaska and the Yukon, Anchorage, Alaska, 1983: Alaska Geological Society Program and Abstracts, p. 10–11.

——, 1983b, Two-stage amalgamation of west-central Alaska: Geological Society of America Abstracts with Programs, v. 15, p. 428.

——, 1984, Stratigraphic relationship between Dillinger and Mystic terranes, western Alaska Range, Alaska: Geological Society of America Abstracts with Programs, v. 16, p. 286.

Gilbert, W. G., and Solie, D. N., 1983, Preliminary bedrock geology of the McGrath A-3 Quadrangle, Alaska: Alaska Division of Geological and Geophysical Surveys Report of Investigations 83-7, scale 1:40,000.

Gilbert, W. G., Solie, D. N., and Dickey, D. B., 1982, Preliminary bedrock geology of the McGrath B-3 Quadrangle, Alaska: Alaska Division of Geological and Geophysical Surveys Open-File Report 148, scale 1:40,000.

Globerman, B. R., and Coe, R. S., 1984, Discordant declinations from "in place" Upper Cretaceous volcanics, SW Alaska; Evidence for oroclinal bending?: Geological Society of American Abstracts with Programs, v. 16, p. 286.

Globerman, B. R., Gill, J. B., and Batatian, D., 1984, Petrochemical data from Upper Cretaceous volcanic rocks; Hagemeister, Crooked, and Summit Islands, SW Alaska: Geological Society of America Abstracts with Programs, v. 16, p. 286.

Griscom, A., 1978, Aeromagnetic interpretation of the Goodnews and Hagemeister Island Quadrangles region, southwestern Alaska: U.S. Geological Survey Open-File Report 78-9C, 20 p.

Hahn, G., and Hahn, R., 1985, Trilobiten aus dem hohen Ober-Karbon oder Unter-Perm vol Alaska: Senckenbergiana Lethaea, v. 66, p. 445–485.

Hahn, G., Blodgett, R. B., and Gilbert, W. G., 1985, First recognition of the Gshelian (Upper Pennsylvanian) trilobite *Brachymetopus pseudometopina* Gairi and Ramovs in North America, and a description of accompanying trilobites from west-central Alaska: Journal of Paleontology, v. 59, no. 1, p. 27–31.

Hamilton, W., 1979, Tectonics of the Indonesian region: U.S. Geological Survey Professional Paper 1078, 345 p.

Hanks, C. L., Rogers, J. F., and Wallace, W. K., 1985, The western Alaska Range flysch terrane; What is it and where did it come from?: Geological Society of America Abstracts with Programs, v. 18, p. 359.

Heckel, P. H., and Witzke, B. J., 1979, Devonian world palaeogeography determined from distribution of carbonates and related lithic palaeoclimatic indicators, *in* House, M. R., Scrutton, C. T., and Bassett, M. G., eds., The Devonian System: London, Special Papers in Palaeontology 23, p. 99–123.

Hoare, J. M., 1961, Geology and tectonic setting of lower Kuskokwim Bristol Bay region, Alaska: American Association of Petroleum Geologists Bulletin, v. 45, p. 594–611.

Hoare, J. M., and Coonrad, W. L., 1959a, Geology of the Bethel Quadrangle, Alaska: U.S. Geological Survey Miscellaneous Geological Investigations Map I-285, scale 1:250,000.

——, 1959b, Geology of the Russian Mission Quadrangle, Alaska: U.S. Geological Survey Miscellaneous Geological Investigations Map I-292, scale 1:250,000.

——, 1961a, Geologic map of the Goodnews Quadrangle, Alaska: U.S. Geological Survey Miscellaneous Investigations Map I-339, scale 1:250,000.

——, 1961b, Geologic map of the Hagemeister Island Quadrangle, Alaska: U.S. Geological Survey Miscellaneous Investigations Map I-321, scale 1:250,000.

——, 1977, Blue amphibole occurrences in southwestern Alaska, *in* Blean, K. M., ed., The United States Geological Survey in Alaska; Accomplishments during 1976: U.S. Geological Survey Circular 751-B, p. B39.

——, 1978a, Geologic map of the Goodnews and Hagemeister Island Quadrangles region, southwestern Alaska: U.S. Geological Survey Open-File Report 78-9-B, scale 1:250,000.

——, 1978b, Lawsonite in southwest Alaska, *in* Johnson, K. M., ed., The United States Geological Survey in Alaska; Accomplishments during 1977: U.S. Geological Survey Circular 772-B, p. B55–B57.

——, 1979, The Kanektok metamorphic complex; A rootless belt of Precambrian rocks in southwestern Alaska, *in* Johnson, K. M., and Williams, J. R., eds., The United States Geological Survey in Alaska; Accomplishments during 1978: U.S. Geological Survey Circular 804-B, p. B72–B74.

——, 1983, Graywacke of Buchia Ridge and correlative Lower Cretaceous rocks in the Goodnews Bay and Bethel Quadrangles, southwestern Alaska: U.S. Geological Survey Bulletin 1529-C, 17 p.

Hoare, J. M., and Jones, D. L., 1981, Lower Paleozoic radiolarian chert and associated rocks in the Tikchik Lakes area, southwestern Alaska, *in* Albert, N.R.D., and Hudson, T., eds., The United States Geological Survey in Alaska; Accomplishments during 1979: U.S. Geological Survey Circular 823-B, p. B44–B45.

Hoare, J. M., Forbes, R. B., and Turner, D. L., 1974, Precambrian rocks in southwest Alaska, *in* Carter, C., ed., The United States Geological Survey in Alaska; Accomplishments during 1973: U.S. Geological Survey Circular 700, p. 46.

Hoare, J. M., Coonrad, W. L., and McCoy, S., 1983, Summit Island Formation; A new Upper Cretaceous formation in southwestern Alaska: U.S. Geological Survey Bulletin 1529-B, 18 p.

House, M. R., and Blodgett, R. B., 1982, The Devonian goniatite genera *Pinacites* and *Foordites* from Alaska: Canadian Journal of Earth Sciences, v. 19, no. 9, p. 1873–1876.

Hsü, K. J., 1968, The principles of mélanges and their bearing on the Franciscan-Knoxville paradox: Geological Society of America Bulletin, v. 79, p. 1063–1074.

Jones, D. L., and Silberling, N. J., 1979, Mesozoic stratigraphy; The key to tectonic analysis of southern and central Alaska: U.S. Geological Survey Open-File Report 79-1200, 41 p.

Jones, D. L., Silberling, N. J., Berg, H. C., and Plafker, G., 1981, Map showing tectonostratigraphic terranes of Alaska, columnar sections, and summary description of terranes: U.S. Geological Survey Open-File Report 81-792, scale 1:2,500,000.

Jones, D. L., Silberling, N. J., Gilbert, W. G., and Coney, P. J., 1982, Character, distribution, and tectonic significance of accretionary terranes in the central Alaska Range: Journal of Geophysical Research, v. 87, no. B5, p. 3709–3717.

—— , 1983, Tectonostratigraphic map and interpretive bedrock geologic map of the Mount McKinley region, Alaska: U.S. Geological Survey Open-File Report 83-11, scale 1:250,000.

Jones, D. L., Silberling, N. J., and Coney, P. J., 1986, Collision tectonics in the Cordillera of western North America; Examples from Alaska, *in* Coward, M. P., and Ries, A. C., eds., Collision tectonics: Geological Society of London Special Publication 19, p. 367–387.

Jones, D. L., Silberling, N. J., Coney, P. J., and Plafker, G., 1987, Lithotectonic terrane map of Alaska (west of the 141st Meridian): U.S. Geological Survey Miscellaneous Field Studies Map MF-1874-A, scale 1:2,500,000.

Karl, S., and Hoare, J. M., 1979, Results of a preliminary paleomagnetic study of volcanic rocks from Nuyakuk Lake, southwestern Alaska, *in* Johnson, K. M., and Williams, J. R., eds., The United States Geological Survey in Alaska; Accomplishments during 1978: U.S. Geological Survey Circular 804-B, p. B74–B78.

Kline, J. T., Gilbert, W. G., and Bundtzen, T. K., 1986, Preliminary geologic map of the McGrath C-1 Quadrangle, Alaska: Alaska Division of Geological and Geophysical Surveys Report of Investigations 86-25, 1 sheet, scale 1:63,360.

Lerand, M., 1973, Beaufort Sea, *in* McCrossan, R. G., ed., The future petroleum provinces of Canada: Canada Society of Petroleum Geologists Memoir 1, p. 315–386.

Marlow, M. S., Scholl, D. W., Cooper, A. K., and Buffington, E. C., 1976, Structure and evolution of the Bering Sea Shelf south of St. Lawrence Island: American Association of Petroleum Geologists Bulletin, v. 60, p. 161–183.

Miller, M. L., and Bundtzen, T. K., 1985, Metamorphic rocks in the western Idatarod Quadrangle, west-central Alaska, *in* Bartsch-Winkler, S., and Reed, K. M., eds., The United States Geological Survey in Alaska; Accomplishments during 1983: U.S. Geological Survey Circular 945, p. 24–28.

Mitchum, R. M., Jr., Vail, P. R., and Thompson, S., III, 1977, Seismic stratigraphy and global changes of sea level; Part 2, The depositional sequence as a basic unit for stratigraphic analysis, *in* Payton, C. E., ed., Seismic stratigraphy; Applications to hydrocarbon exploration: American Association of Petroleum Geologists Memoir 26, p. 53–62.

Moll, E. J., and Patton, W. W., Jr., 1982, Preliminary report on the Late Cretaceous and early Tertiary volcanic and related plutonic rocks in western Alaska, *in* Coonrad, W. L., ed., The United States Geological Survey in Alaska; Accomplishments during 1980: U.S. Geological Survey Circular 844, p. 73–76.

Moll, E. J. Silberman, M L., and Patton, W. W., Jr., 1980, Chemistry, mineralogy, and K-Ar ages of igneous and metamorphic rocks of the Medfra Quadrangle, Alaska: U.S. Geological Survey Open-File Report 80-811C, 2 sheets, scale: 1:250,000.

Moll-Stalcup, E. J., 1987, The petrology and Sr and Nd isotopic characteristics of five Late Cretaceous–early Tertiary volcanic fields in western Alaska [Ph.D. thesis]: Stanford, California, Stanford University, 310 p.

Moore, T. E., and Wallace, W. K., 1985, Submarine-fan facies of the Kuskokwim Group, Cairn Mountain area, southwestern Alaska: Geological Society of America Abstracts with Programs, v. 17, p. 371.

Murphy, J. M., 1987, Early Cretaceous cessation of terrane accretion, northern Eek Mountains, southwestern Alaska, *in* Hamilton, T. D., and Galloway, J. P., eds., Geologic studies in Alaska by the U.S. Geological Survey during 1986: U.S. Geological Survey Circular 998, p. 83–85.

—— , 1989, Geology, sedimentary petrology, and tectonic synthesis of Early Cretaceous submarine fan deposits, northern Eek Mountains, southwest Alaska [M.S. thesis]: Fairbanks, University of Alaska, 118 p.

Murphy, J. M., and Decker, J., 1985, The Goodnews terrane and Kuskokwim Group, Eek Mountains, southwest Alaska; Open marine to trench slope transition: American Association of Petroleum Geologists, Pacific Section 60th Annual Meeting, Program and Abstracts, p. 57.

Nelson, W. H., Carlson, C., and Case, J. E., 1983, Geologic map of the Lake Clark Quadrangle, Alaska: U.S. Geological Survey Miscellaneous Field Studies Map MF-1114A, scale 1:250,000.

Pacht, J. A., and Wallace, W. K., 1984, Depositional facies of a post-accretionary sequence; The Cretaceous Kuskokwim Group of southwest Alaska: Geological Society of America Abstracts with Programs, v. 16, p. 327.

Palmer, A. R., Egbert, R. M., Sullivan, R., and Knoth, J. S., 1985, Cambrian trilobites with Siberian affinities, southwestern Alaska: American Association of Petroleum Geologists Bulletin, v. 69, no. 2, p. 295.

Patton, W. W., Jr., 1978, Juxtaposed continental and ocean-island arc terranes in the Medfra Quadrangle, west-central Alaska, *in* Johnson, K. M., ed., The United States Geological Survey in Alaska; Accomplishments during 1977: U.S. Geological Survey Circular 772-B, p. B38–B39.

Patton, W. W., Jr., Dutro, J. T., Jr., and Chapman, R. M., 1977a, Late Paleozoic and Mesozoic stratigraphy of the Nixon Fork area, Medfra Quadrangle, Alaska, *in* Blean, K. M., ed., The United States Geological Survey in Alaska; Accomplishments during 1976: U.S. Geological Survey Circular 751-B, p. B38–B40.

Patton, W. W., Jr., Tailleur, I. L., Brosge, W. P., and Lanphere, M. A., 1977b, Preliminary report on the ophiolites of northern and western Alaska, *in* Colemen, R. G., and Irwin, W. P., eds., North American ophiolites: Oregon Department of Geology and Mineral Industries Bulletin 95, p. 51–57.

Patton, W. W., Jr., Moll, E. J., Dutro, J. T., Jr., Silberman, M. L., and Chapman, R. M., 1980, Preliminary geologic map of the Medfra Quadrangle, Alaska: U.S. Geological Survey Open-File Report 80-811A, 1 sheet, scale 1:250,000.

Patton, W. W., Jr., Moll, E. J., and King, H. H., 1984a, The Alaskan mineral resource assessment program; Guide to information contained in the folio of geologic and mineral resource maps of the Medfra Quadrangle, Alaska: U.S. Geological Survey Circular 923, 11 p.

Patton, W. W., Jr., Moll, E. J., Lanphere, M. A., and Jones, D. L., 1984b, New age data for the Kaiyuh Mountains, west-central Alaska, *in* Coonrad, W. L., and Elliott, R. L., eds., The United States Geological Survey in Alaska; Accomplishments during 1981: U.S. Geological Survey Circular 868, p. 30–32.

Plumley, P. W., and Coe, R. S., 1982, Paleomagnetic evidence for reorganization of the north Cordilleran borderland in Late Paleozoic time: Geological Society of America Abstracts with Programs, v. 14, p. 224.

—— , 1983, Paleomagnetic data from Paleozoic rocks of the Nixon Fork terrane, Alaska, and their tectonic implications: Geological Society of America Abstracts with Programs, v. 15, p. 428.

Plumley, P. W., Coe, R. S., Patton, W. W., and Moll, E. J., 1981, Paleomagnetic study of the Nixon Fork terrane, west-central Alaska: Geological Society of America Abstracts with Programs, v. 13, p. 530.

Poncet, J., and Blodgett, R. B., 1987, First recognition of the Devonian alga *Lancicula sergaensis* Shuysky in North America (west-central Alaska): Journal of Paleontology, v. 61, p. 1269–1273.

Potter, A. W., Gilbert, W. G., Ormiston, A. R., and Blodgett, R. B., 1980, Middle and Upper Ordovician brachiopods from Alaska and northern California and their paleogeographic implications: Geological Society of America Abstracts with Programs, v. 12, p. 147.

Reed, B. L., and Lanphere, M. A., 1973, Alaska–Aleutian Range batholith; Geochronology, chemistry, and relation to circum-Pacific plutonism: Geological Society of America Bulletin, v. 84, p. 2583–2609.

—— , 1974, Chemical variations across the Alaska-Aleutian Range batholith: U.S. Geological Survey Journal of Research, v. 2, p. 343–352.

Reed, B. L., Miesch, A. T., and Lanphere, M. A., 1983, Plutonic rocks of Jurassic age in the Alaska–Aleutian Range batholith; Chemical variations and polarity: Geological Society of America Bulletin, v. 94, p. 1232–1240.

Reed, K. M., Blome, C. D., Gilbert, W. G., and Solie, D. N., 1985, Jurassic radiolaria from the Lime Hills Quadrangle, *in* Bartsch-Winkler S., and Reed, K. M., eds., The United States Geological Survey in Alaska; Accomplishments during 1983: U.S. Geological Survey Circular 945, p. 53–54.

Reifenstuhl, R. R., Robinson, M. R., Smith, T. E., Albanese, M. D., and Allegro, G. A., 1984, Geologic map of the Sleetmute B-6 Quadrangle, Alaska: Alaska Division of Geological and Geophysical Surveys Report of Investigations 84-12, scale 1:40,000.

Rigby, J. K., and Blodgett, R. B., 1983, Early Middle Devonian sponges from the McGrath Quadrangle of west-central Alaska: Journal of Paleontology, v. 57, no. 4, p. 773–786.

Robinson, M. S., and Decker, J., 1986, Preliminary age dates and analytical data from the Sleetmute, Russian Mission, Taylor Mountains, and Bethel Quadrangles, southwestern Alaska: Alaska Division of Geological and Geophysical Surveys Public Data File Report 86-99, 8 p.

Robinson, M. S., Decker, J., Reifenstuhl, R. R., and Murphy, J. M., 1984a, Bedrock geology of the Chuilnuk and Kiokluk Mountains, southwest Alaska: Geological Society of America Abstracts with Programs, v. 16, p. 330.

Robinson, M. S., Decker, J., Reifenstuhl, R. R., Murphy, J. M., and Box, S. E., 1984b, Geologic map of the Sleetmute B-5 Quadrangle, Alaska: Alaska Division of Geological and Geophysical Surveys Report of Investigation 84-10, 1 sheet, scale 1:40,000.

Robinson, M. S., Decker, J., and Nye, C. J., 1986, Preliminary whole rock major oxide and trace element geochemistry of selected igneous rocks from the Sleetmute, Russian Mission, and Taylor Mountains Quadrangles, southwestern Alaska: Alaska Division of Geological and Geophysical Surveys Public Data File Report 86-98, 8 p.

Rohr, D. M., and Blodgett, R. B., 1985, Upper Ordovician Gastropoda from west-central Alaska: Journal of Paleontology, v. 59, p. 667–673.

—— , 1988, First occurrence of Helicotoma Salter (Gastropoda) from the Ordovician of Alaska: Journal of Paleontology, v. 62, p. 304–306.

Savage, N. M., Potter, A. W., and Gilbert, W. G., 1983, Silurian and Silurian to Early Devonian conodonts from west-central Alaska: Journal of Paleontology, v. 57, p. 873–875.

Shew, N., and Wilson, F. H., 1981, Map and table showing radiometric ages of rocks in southwestern Alaska: U.S. Geological Survey Open-File Report 81-866, 1 sheet, 26 p.

Silberman, M. L., Moll, E. J., Patton, W. W., Jr., Chapman, R. M., and Connor, C. L., 1979, Precambrian age of metamorphic rocks from the Ruby province, Medfra and Ruby Quadrangles; Preliminary evidence from radiometric age data: U.S. Geological Survey Circular 804-B, p. B66–B67.

Sloss, L. L., 1963, Sequences in the cratonic interior of North America: Geological Society of America Bulletin, v. 74, p. 93–114.

Smith, P. S., 1939, Areal geology of Alaska: U.S. Geological Survey Professional Paper 192, 100 p.

Smith, T. N., Clough, J. G., Meyer, J. F., and Blodgett, R. B., 1985, Petroleum potential and stratigraphy of Holitna Basin, Alaska: American Association of Petroleum Geologists Bulletin, v. 69, p. 308.

Solie, D. N., 1983, The Middle Fork plutonic complex, McGrath A-3 Quadrangle, southwest Alaska: Alaska Division of Geological and Geophysical Surveys Report of Investigations 83-16, 17 p.

Solie, D. N., and Dickey, D. B., 1982, Coal occurrences and analyses, Farewell–White Mountain area, Alaska: Alaska Division of Geological and Geophysical Surveys Open-File Report 160, 17 p.

Solie, D. N., Bundtzen, T. K., and Gilbert, W. G., 1982, Upper Cretaceous–Lower Tertiary volcanic rocks near Farewell, Alaska, *in* Sciences in the North: Fairbanks, Arctic Division, American Association for the Advancement of Science, 33rd Alaska Science Conference, p. 138.

Southworth, D. D., 1984, Red Mountain; A southeastern Alaska type ultramafic complex in southwestern Alaska: Geological Society of America Abstracts with Programs, v. 16, p. 334.

Stock, C. W., 1981, *Cliefdenella alaskanesis* n. sp. (Stromatoporoidea) from the Middle/Upper Ordovician of central Alaska: Journal of Paleontology, v. 55, no. 5, p. 998–1005.

Stone, D. B., Panuska, B. C., and Packer, D. R., 1982, Paleolatitudes versus time for southern Alaska: Journal of Geophysical Research, v. 87, no. B5, p. 3697–3707.

Turner, D. L., Forbes, R. B., Aleinikoff, J. N., Hedge, C. E., and McDougall, I., 1983, Geochronology of the Kilbuck terrane of southwestern Alaska: Geological Society of America Abstracts with Programs, v. 15, p. 407.

Vance-Plumley, P., Plumley, P. W., Coe, R. S., and Reid, J., 1984, Preliminary paleomagnetic results from three Ordovician carbonate sections in the McGrath Quadrangle, central Alaska: EOS Transactions of the American Geophysical Union, v. 65, no. 45, p. 866.

Wahrhaftig, C., 1987, The Cenozoic section at Suntrana, Alaska, *in* Hill, M. L., ed., Cordilleran Section of the Geological Society of America: Boulder, Colorado, Geological Society of America Centennial Field Guide, v. 1, p. 445–450.

Wahrhaftig, C., Wolfe, J. A., Leopold, E. B., and Lanphere, M. A., 1969, The coal-bearing group in the Nenana coal field, Alaska: U.S. Geological Survey Bulletin 1274-D, p. D1–D30.

Wallace, W. K., 1983, Major lithologic belts of southwestern Alaska and their tectonic implications: Geological Society of America Abstracts with Programs, v. 14, p. 406.

—— , 1984, Mesozoic and Paleogene tectonic evolution of the southwestern Alaska Range; Southern Kuskokwim Mountains region: Geological Society of America Abstracts with Programs, v. 16, p. 339.

Wallace, W. K. and Engebretson, D. C., 1982, Correlation between plate motions and Late Cretaceous to Paleocene magmatism in southwestern Alaska: EOS Transactions of the American Geophysical Union, v. 63, no. 45, p. 915.

—— , 1984, Relationships between plate motions and Late Cretaceous to Paleocene magmatism in southwestern Alaska: Tectonics, v. 3, p. 295–315.

Wallace, W. K., Hanks, C. L., and Rodgers, J. F., 1989, The southern Kahiltna terrane; Implications for the tectonic evolution of Alaska: Geological Society of America Bulletin, v. 101, p. 1389–1407.

Wilson, F. H., 1977, Some plutonic rocks of southwestern Alaska; A data compilation, 1977: U.S. Geological Survey Open-File Report 77-501, 9 p.

MANUSCRIPT ACCEPTED BY THE SOCIETY NOVEMBER 16, 1990

ACKNOWLEDGMENTS

Most of what is known about the geology of southwest Alaska is from systematic field mapping by J. M. Hoare, W. L. Coonrad, W. H. Condon, and S. E. Box in the Bristol Bay region, by W. M. Cady, J. Decker, J. M. Murphy, M. S. Robinson, and R. R. Reifenstuhl in the southern Kuskokwim Mountains, by W. W. Patton, Jr., T. K. Bundtzen, M. L. Miller, J. T. Dutro, E. J. Moll, and G. M. Laird in the northern Kuskokwim Mountains, and by B. L. Reed, T. K. Bundtzen, W. G. Gilbert, J. T. Kline, and D. N. Solie in the southwestern Alaska Range. Recent detailed mapping and topical studies, in large part by the authors of this chapter, and regional studies by D. L. Jones, N. J. Silberling, M. Churkin, and others, have contributed much to our current understanding of the geology of southwest Alaska. Discussions with geologists at the Alaska Division of Geological and Geophysical Surveys, in particular, D. B. Dickey, M. W. Henning, J. T. Kline, R. R. Reifenstuhl, T. E. Smith, and T. N. Smith have been very helpful, and we thank them for their cooperation.

This chapter was compiled by Decker, Murphy, and Robinson in 1984 and 1985, and was partially updated by Decker and Bergman in 1990 to include current references. The chapter was compiled from contributions on the southwestern Alaska Range and northern Kuskokwim Mountains by Blodgett, Bundtzen, Clough, Decker, Gilbert, and Miller; on the Holitna basin by Blodgett and Clough; on the Bristol Bay region and southern Kuskokwim Mountains by Box, Coonrad, Decker, Murphy, Robinson, and Wallace; on post-accretionary rocks by Bergman, Bundtzen, Decker, Robinson, and Wallace; and on the southern Kahiltna terrane by Wallace.

We dedicate this chapter to Joseph M. Hoare (1918 to 1981) for his pioneering contribution to the geology of southwest Alaska.

NOTES ADDED IN PROOF

Since the completion of the original draft of this chapter, the following relevant publications on paleontology and stratigraphy have appeared.

Babcock, L. E., and Blodgett, R. B., 1992, Biogeographic and paleogeographic significance of Middle Cambrian trilobites of Siberian aspect from southwestern Alaska: Geological Society of America Abstracts with Programs, v. 24, no. 5, p. 4.

Babcock, L. E., Blodgett, R. B., and St. John, J., 1993, Proterozoic and Cambrian stratigraphy and paleontology of the Nixon Fork terrane, southwestern Alaska: Proceedings of the First Circum-Pacific and Circum-Atlantic terrane conference, 5-22 November, 1993, Guanajuato, Mexico, p. 5–7.

Blodgett, R. B., 1993, *Dutrochus,* a new microdomatid (Gastropoda) genus from the Middle Devonian (Eifelian) of west-central Alaska: Journal of Paleontology, v. 67, p. 194–197.

Blodgett, R. B., and Gilbert, W. G., 1992, Upper Devonian shallow-marine siliciclastic strata and associated fauna and flora, Lime Hills D-4 quadrangle, southwest Alaska, *in* Bradley, D. C., and Dusel-Bacon, C., eds., Geologic Studies in Alaska by the U.S. Geological Survey, 1991: U.S. Geological Survey Bulletin 2041, p. 106–115.

Clough, J. G., and Blodgett, R. B., 1988, Silurian-Devonian algal reef mound complex of southwest Alaska, *in* Geldsetzer, H., and James, N. P., eds., Reefs, Canadian and adjacent areas: Canadian Society of Petroleum Geologists Memoir 13, p. 246–250.

Elder, W. P., and Box, S. E., 1992, Late Cretaceous inoceramid bivalves of the Kuskokwim Basin, southwestern Alaska, and their implications for basin evolution: Journal of Paleontology, Memoir 26, 39 p.

Johnson, J. G., and Blodgett, R. B., 1993, Russian Devonian brachiopod genera *Cvrtinoides* and *Komiella* in North America: Journal of Paleontology, v. 67, p. 952–958.

Mamet, B. L., and Plafker, G., 1982, A Late Devonian (Frasnian) Microbiota from the Farewell-Lyman Hills area, west-central Alaska: U.S. Geological Survey Professional Paper 1216-A, p. A1–A10.

Measures, E. A., Rohr, D. M., and Blodgett, R. B., 1992, Depositional environments and some aspects of the fauna of Middle Ordovician rocks of the Telsitna Formation, northern Kuskokwim Mountains, Alaska, *in* Bradley, D. C., and Dusel-Bacon, C., eds., Geologic Studies in Alaska by the U.S. Geological Survey, 1991: U.S. Geological Survey Bulletin 2041, p. 186–201.

Potter, A. W., and Blodgett, R. B., 1992, Paleobiogeographic relations of Ordovician brachiopods from the Nixon Fork terrane, west-central Alaska: Geological Society of America Abstracts with Programs, v. 24, no. 5, p. 76.

Rigby, J. K., Potter, A. W., and Blodgett, R. B., 1988, Ordovician sphinctozoan sponges of Alaska and Yukon Territory: Journal of Paleontology, v. 62, p. 731–746.

Rohr, D. M., Dutro, J. T., Jr., and Blodgett, R. B., Gastropods and brachiopods from the Ordovician Telsitna Formation, northern Kuskokwim Mountains, west-central Alaska, *in* Webby, B. D., and Laurie, J. R., eds., Global perspectives on Ordovician Geology, Proceedings of the Sixth International Symposium on the Ordovician System, Sydney, Australia: Balkema Press, p. 499–512.

Printed in U.S.A.

The Geology of North America
Vol. G-1, The Geology of Alaska
The Geological Society of America, 1994

Chapter 10

Geology of south-central Alaska

Warren J. Nokleberg and George Plafker
U.S. Geological Survey, MS-904, 345 Middlefield Road, Menlo Park, California 94025
Frederic H. Wilson
U.S. Geological Survey, 4200 University Drive, Anchorage, Alaska 99508

INTRODUCTION

Study area and major geologic units

South-central Alaska is defined as the region bounded by the Kuskokwim Mountains to the northwest, the basins north of the Alaska Range to the north, the Canadian border to the east, and the Chugach Mountains to the south (Fig. 1). This region, hereafter called the study area, includes the Alaska Range, the Wrangell, Nutzotin, and Talkeetna mountains, the Copper River and the Susitna basins, the northern flank of the Chugach Mountains, the Aleutian Range, and the Alaska Peninsula. This chapter describes and interprets the bedrock geology of the region, which consists mostly of a collage of Paleozoic and Mesozoic tectonostratigraphic terranes (hereafter referred to as terranes), Mesozoic flysch basin deposits, late Paleozoic and Mesozoic plutonic rocks, and younger late Mesozoic and Cenozoic sedimentary, volcanic, and plutonic rocks. Cited published sources and new data and interpretations of the authors are utilized for the descriptions and interpretations. The terranes, flysch basin deposits, and younger Mesozoic sedimentary, volcanic, and plutonic assemblages are described first in a general northwest to southeast order. Major faults or sutures are described second. Stratigraphic linkages and structural and tectonic relations between terranes are described last. Definitions of the various stratigraphic, structural, and tectonic terms are stated at the end of this introduction.

The three largest terranes in the study area are the Alexander, Peninsular, and Wrangellia terranes (Fig. 2) (Jones and others, 1981, 1984, 1987). Even though many boundaries between the three terranes are commonly faults, important stratigraphic linkages exist. These linkages suggest a common history since the Middle Pennsylvanian for the Alexander and Wrangellia terranes, and since at least the Late Triassic and possibly the late Paleozoic for the Peninsular and Wrangellia terranes (Plafker and others, 1989c; Plafker, 1990; Plafker and Berg, this volume, Chapter 33). As a result, this chapter interprets these three terranes as the Wrangellia composite terrane (WCT). An older name for the WCT is the Peninsular-Alexander-Wrangellia superterrane (Pavlis, 1983). The name of Wrangellia composite terrane is adopted in this chapter for conformity with Plafker and Berg (Chapter 33). A prior term, the Talkeetna superterrane, was used to describe the combined Wrangellia and Peninsular terranes by Csejtey and others (1982).

In addition to the major terranes in the study area, several small, narrow, highly deformed terranes occur adjacent to the Denali fault, south of the large masses of the Yukon-Tanana, Nixon Fork, and Dillinger terranes. These terranes are the Aurora Peak, Pingston, McKinley, Mystic, and Windy terranes, a terrane of ultramafic and associated rocks, and a fragment of the Dillinger terrane (Fig. 2). To the south, several other small terranes occur within highly deformed upper Mesozoic flysch assemblages, either as rootless nappes or as units bounded by steep faults (Jones and others, 1981, 1987). These terranes are the Broad Pass, Chulitna, Clearwater, Maclaren, Susitna, and West Fork terranes (Fig. 2). Two major late Mesozoic flysch basins, now tectonically collapsed, occur along the northern margin of the WCT (Fig. 2). To the northwest is the mainly Late Jurassic and Early Cretaceous flysch of the Kahiltna assemblage that probably originally depositionally overlapped the Peninsular terrane (Wallace and others, 1989). To the northeast is the partly coeval Late Jurassic and Early Cretaceous flysch and volcanic rocks of the Gravina-Nutzotin belt, which depositionally overlies the Wrangellia terrane in the study area and the Alexander terrane in southeastern Alaska (Berg and others, 1972).

Major faults or sutures are either known or are inferred to separate terranes. Along the northern boundary of the study area is the Denali fault and along the southern boundary is the Border Ranges fault (Fig. 2). Other important sutures are the Broxson Gulch thrust between the Maclaren and Wrangellia terranes, the Talkeetna thrust between the Peninsular and Wrangellia terranes, the West Fork fault between the southern Wrangellia and Peninsular terranes, and the Chitina Valley fault between the Wrangellia and southern Wrangellia terranes (Fig. 2). Major Cenozoic faults, such as the Castle Mountain, Hines Creek, and Totschunda faults, that occur mainly within terranes in the study area, are depicted and described by Plafker and others (Chapter 12).

Nokleberg, W. J., Plafker, G., and Wilson, F. H., 1994, Geology of south-central Alaska, *in* Plafker, G., and Berg, H. C., eds., The Geology of Alaska: Boulder, Colorado, Geological Society of America, The Geology of North America, v. G-1.

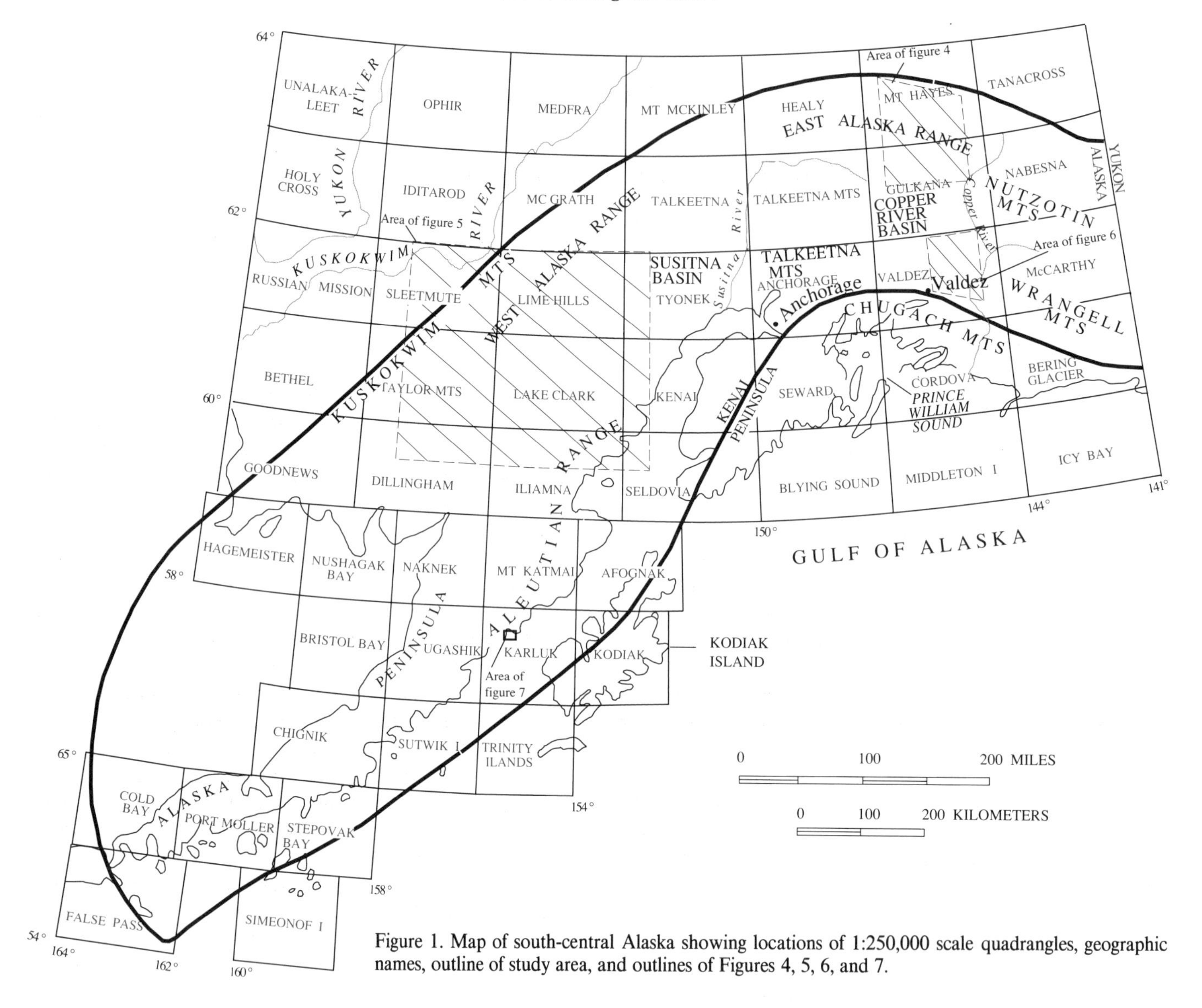

Figure 1. Map of south-central Alaska showing locations of 1:250,000 scale quadrangles, geographic names, outline of study area, and outlines of Figures 4, 5, 6, and 7.

Companion studies

Companion studies in this volume that describe the bedrock geology of areas adjacent to the study area are east-central Alaska by Foster and others (Chapter 6); southwestern Alaska by Decker and others (Chapter 9); southern Alaska continental margin by Plafker and others (Chapter 12); and the Aleutian arc by Vallier and others (Chapter 11). Other related topical studies in this volume that include data on the geology of the study area are accreted volcanic rocks by Barker and others (Chapter 17); metamorphic history of Alaska by Dusel-Bacon (Chapter 15); metallogenesis of Alaska by Nokleberg and others (Chapter 29); paleomagnetic data in Alaska by Hillhouse and Coe (Chapter 26); pre-Cenozoic plutonic history by Miller (Chapter 16); interior basins by Kirschner (Chapter 14); Late Cretaceous and Cenozoic magmatism in mainland Alaska by Moll-Stalcup (1990; Chapter 18); isotopic composition studies of igneous rocks of Alaska by Arth (Chapter 25); and volcanic rocks of the Aleutian and Wrangell arcs by Miller and Richter (Chapter 24).

Plates in this volume containing data relevant to the geology of the study area are geology of Alaska by Beikman; lithotectonic terrane map and columnar sections for Alaska by Silberling and others; metamorphic map of Alaska by Dusel-Bacon (Plate 4); map showing latest Cretaceous and Cenozoic magmatic rocks of Alaska by Moll-Stalcup and others; map showing selected pre-Cenozoic plutonic rocks and accreted volcanic rocks of Alaska by Barker and others; map showing sedimentary basins of Alaska by Kirschner (Plate 7); map and table showing isotopic age data in Alaska by Wilson and others; map showing locations of metalliferous lode deposits and placer districts of Alaska by Nokleberg and others (Plate 11); and neotectonic map of Alaska by Plafker and others (Plate 12).

Except where indicated, age designations used in this chapter are based on the Decade of North American Geology geologic time scale (Palmer, 1983); metamorphic facies are after Turner (1981); plutonic rock classifications are after Streckeisen (1976); volcanic rock classifications are after Streckeisen (1980); sandstone classifications are after Dickinson and Suczek (1979); deformed rock nomenclature is after Higgins (1971) and Wise and others (1984); terrane designations are modified from Jones and others (1987), Monger and Berg (1987), Silberling and others (this volume), Nokleberg and others (1994), and the authors; and older K-Ar isotopic ages are recalculated using the newer decay constants of Steiger and Jäger (1977).

Definitions

The following definitions are adapted from Coney and others (1980), Jones and others (1983), Howell and others (1985), Monger and Berg (1987), Wheeler and others (1988), and Nokleberg and others (1994).

Accretion. Tectonic juxtaposition of two or more terranes, or tectonic juxtaposition of a terrane(s) to a continental margin.

Accretionary wedge terrane. Fragment of a mildly to intensely deformed complex of turbidite deposits and lesser amounts of oceanic rocks. Formed adjacent to zones of thrusting and subduction along the margin of a continental or an island arc. Commonly associated with subduction zone terranes. May include large, fault-bounded units with coherent stratigraphy.

Assemblage. A group of stratigraphically or structurally related sedimentary and/or igneous rocks.

Belt. A regional linear group of geologic units.

Composite terrane. An aggregate of terranes that is interpreted to share either a similar stratigraphic kindred or affinity, or a common geologic history (Plafker, 1990; Plafker and Berg, this volume, Chapter 33). An approximate synonym is *superterrane* (Moore, 1992).

Continental margin arc terrane. Fragment of an igneous belt of coeval plutonic and volcanic rocks, and associated sedimentary rocks that formed above a subduction zone dipping beneath a continent. Inferred to possess a sialic basement.

Island (intraoceanic) arc terrane. Fragment of an igneous belt of plutonic rocks, coeval volcanic rocks, and associated sedimentary rocks that formed above an oceanic subduction zone. Inferred to possess a simatic basement.

Metamorphic terrane. Fragment of a highly metamorphosed and/or deformed assemblage of sedimentary, volcanic, and/or plutonic rocks that cannot be assigned to a single tectonic environment because original stratigraphy and structure are obscured. Includes highly deformed structural melanges that contain intensely deformed pieces of two or more terranes.

Overlap assemblage. A sequence of sedimentary and/or igneous rocks deposited on, or intruded into, two or more adjacent terranes. The sedimentary and volcanic parts depositionally overlie, or are interpreted to have originally depositionally overlain, two or more adjacent terranes, or terranes and the craton margin. Overlap plutonic rocks in some areas are coeval and genetically related to overlap volcanic rocks, and weld or stitch together adjacent terranes, or a terrane and a craton margin.

Oceanic crust, seamount, and ophiolite terrane. Fragment of part or all of a suite of eugeoclinal, deep-marine sedimentary rocks, pillow basalts, gabbros, and ultramafic rocks that are interpreted as oceanic sedimentary and volcanic rocks, and upper mantle. Includes both inferred offshore ocean and marginal ocean basin rocks. Includes minor volcaniclastic rocks of magmatic arc derivation. Mode of emplacement onto continental margin uncertain.

Postaccretion rock unit. Suite of sedimentary, volcanic, or plutonic rocks that formed in the late history of a terrane, after accretion. May occur also on an adjacent terrane(s) or on craton margin as an overlap assemblage unit. A relative term denoting rocks formed after juxtaposition of one terrane to an adjacent terrane.

Preaccretion rock unit. Suite of sedimentary, volcanic, or plutonic rocks that formed in the early history of a terrane, before accretion. Constitutes the stratigraphy inherent to a terrane. A relative term denoting rocks formed before juxtaposition of one terrane to an adjacent terrane.

Seamount and oceanic plateau. Major marine volcanic accumulations formed at a hot spot, fracture zone, or spreading center, or off the axis of a spreading center.

Subduction zone terrane. Fragment of variably to intensely deformed oceanic crust and overlying units, oceanic mantle, and lesser turbidite and continental margin rocks that were tectonically juxtaposed in a zone of major thrusting of one lithosphere plate beneath another. Many subduction zone terranes contain fragments of oceanic crust and associated rocks that exhibit a complex structural history, occur in a major thrust zone, and possess blueschist facies metamorphism. Commonly associated with accretionary wedge terranes. May include large, fault-bounded blocks with coherent stratigraphy.

Subterrane. Fault-bounded unit within a terrane that exhibits a similar, but not identical, geologic history to another fault-bounded unit in the same terrane.

Terrane. A fault-bounded assemblage or fragment that is characterized by a unique geologic history that differs markedly from that of adjacent terranes. Constitutes a physical entity, i.e., a stratigraphic succession bounded by faults, or an intensely deformed structural complex bounded by faults. Some terranes may be faulted facies equivalents of other terranes.

TERRANES ADJACENT TO DENALI FAULT

Several narrow terranes occur adjacent to the Denali fault in the central and eastern Alaska Range. These terranes are the Aurora Peak, McKinley, Mystic, Pingston, and Windy terranes, a terrane of ultramafic and associated rocks, and a fragment of the Dillinger terrane (Fig. 2). Some terranes, such as the Aurora Peak, Dillinger, McKinley, and Mystic terranes, extend along strike for only about 50 to 100 km, whereas others, such as the

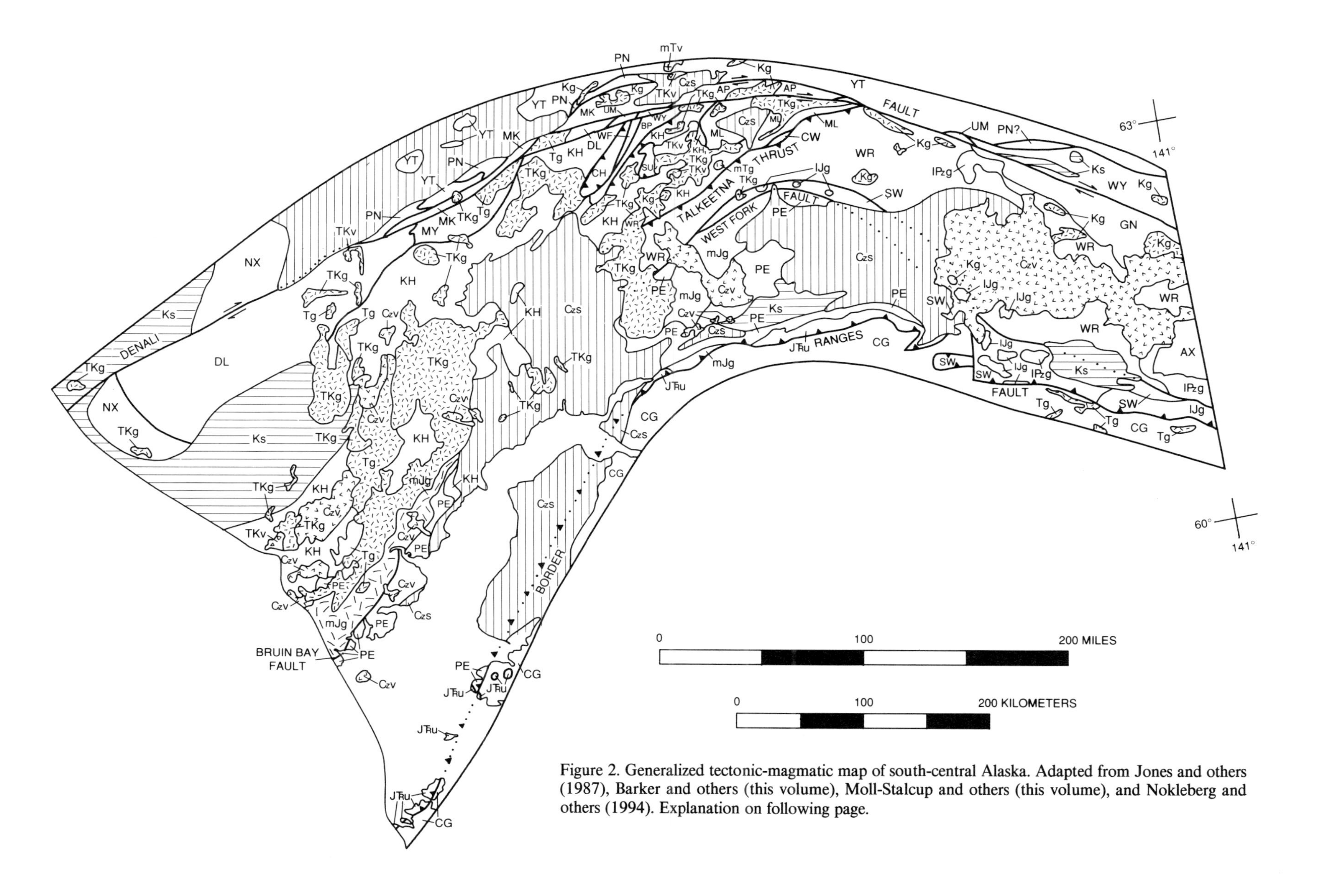

Figure 2. Generalized tectonic-magmatic map of south-central Alaska. Adapted from Jones and others (1987), Barker and others (this volume), Moll-Stalcup and others (this volume), and Nokleberg and others (1994). Explanation on following page.

EXPLANATION

CENOZOIC AND LATE CRETACEOUS SEDIMENTARY AND VOLCANIC ROCKS

Czs — Cenozoic sedimentary rocks and unconsolidated deposits

Czv — Cenozoic volcanic rocks

mTv — Middle Tertiary volcanic rocks

TKv — Late Cretaceous and early Tertiary volcanic rocks

Ks — Cretaceous sedimentary rocks

TERTIARY AND CRETACEOUS GRANITOID ROCKS

Tg — Early Tertiary granitoid rocks

mTg — Middle Tertiary granitoid rocks

TKg — Late Cretaceous and early Tertiary granitoid rocks

Kg — Cretaceous granitoid rocks

EARLY CRETACEOUS AND LATE JURASSIC SEDIMENTARY AND VOLCANIC ASSEMBLAGES

KH — Kahiltna assemblage

GN — Gravina-Nutzotin belt

LATE JURASSIC AND OLDER PLUTONIC ROCKS

lJg — Late Jurassic granitoid rocks: occur in Gravina-Nutzotin belt and Wrangellia terrane

mJg — Middle Jurassic granitoid rocks: occur in Peninsular terrane

JTRu — Late Triassic and Early Jurassic ultramafic. Mafic and granitoid rocks: occur in southern Peninsular terrane in Border Ranges ultramafic-mafic assemblage

lPzg — Late Paleozoic granitoid rocks: occur in Wrangellia and Alexander terranes

TECTONO-STRATIGRAPHIC TERRANES

AP — Aurora Peak Terrane

AX — Alexander terrane

BP — Broad Pass terrane

CG — Chugach terrane

CH — Chulitna terrane

CW — Clearwater terrane

DL — Dillinger terrane

MK — McKinley terrane

ML — Maclaren terrane

MY — Mystic terrane

NX — Nixon Fork terrane

PE — Peninsular terrane

PN — Pingston terrane

SW — Southern Wrangellia terrane

SU — Susitna terrane

TG — Togiak and Goodnews terranes, undivided

UM — Terrane of ultramafic and associated rocks

WF — West Fork terrane

WR — Wrangellia terrane

WY — Windy terrane

YT — Yukon-Tanana terrane

SYMBOLS

Contact

Fault—Dotted where concealed

Thrust fault—Dotted where concealed. Teeth on upper plate

Strike-slip fault—Dotted where concealed

Pingston and Windy terranes, and the terrane of ultramafic and associated rocks, occur discontinuously for several hundred kilometers. The terranes adjacent to the Denali fault are discussed in a general west to east order.

Pingston terrane

The Pingston terrane (Jones and others, 1981, 1982, 1983, 1984, 1987; Gilbert and others, 1984) occurs discontinuously along the Denali fault for several hundred kilometers north and northwest of the Dillinger, McKinley, and Mystic terranes in the Mount McKinley, Healy, and Mount Hayes quadrangles (Figs. 1, 2, and 3). The Pingston terrane consists of a weakly metamorphosed sequence of (1) Early Pennsylvanian and Permian phyllite, minor marble, and chert; (2) Late Triassic thin-bedded, laminated dark limestone, black sooty shale, calcareous sandstone, and minor quartzite; and (3) locally numerous bodies of gabbro, diabase, and diorite of Early Cretaceous(?) age (Reed and Nelson, 1980; Gilbert and others, 1984). The terrane is strongly folded and faulted, and displays a single slaty cleavage that parallels the axial planes of locally abundant isoclinal folds. A small lens of thin-bedded dark limestone, sooty shale, and minor quartzite, correlated with the Pingston terrane, occurs along the southern margin of the Windy terrane in the eastern Alaska Range (the unit is too thin to depict in Fig. 2) (W. J. Nokleberg, 1987, unpublished data). The Late Triassic stratified rocks are interpreted as a turbidite apron sequence deposited from deep-water turbidity currents that flowed from a cratonal source such as the Yukon-Tanana terrane to the north (Gilbert and others, 1984).

McKinley terrane

The McKinley terrane (Reed and Nelson, 1980; Jones and others, 1981, 1982, 1983, 1984, 1987; Gilbert and others, 1984) occurs adjacent to the Denali fault, north and northwest of the Mystic terrane, mainly in the McKinley quadrangle (Figs. 1, 2, and 3). The McKinley terrane consists mainly of (1) fine-grained Permian flysch, mainly graywacke, argillite, and minor chert; (2) Triassic chert; (3) a thick sequence of Late Triassic (Norian) pillow basalt; (4) Triassic(?) gabbro and diabase, interpreted as coeval with the pillow basalt; and (5) Late Jurassic(?) and Cretaceous flysch, mainly graywacke, argillite, minor conglomerate, and chert. Individual flows in the Late Triassic pillow basalt are commonly diabasic and locally have quenched margins; thickness ranges from 600 to 1,700 m. Discontinuous beds of fine-grained clastic sedimentary rocks are locally intercalated with the basalt. Also included in the McKinley terrane is Mississippian to Late Triassic chert that is thrust over, and folded with, the late Mesozoic flysch. The terrane is complex, thick, strongly folded and faulted, and weakly metamorphosed. On the basis of basalt whole-rock chemistry, the McKinley terrane was interpreted by Gilbert and others (1984) as a fragment of one or more Late Triassic seamounts possibly built on continental slope or rise crust.

Mystic and Dillinger terranes

The Mystic and Dillinger terranes (Jones and others, 1982, 1984, 1987) occur adjacent to the Denali fault, north and northwest of the Kahiltna assemblage, mainly in the McKinley and Healy quadrangles (Figs. 1, 2, and 3). The Dillinger terrane also forms a major unit beyond the western edge of the study area in southwestern Alaska (Decker and others, this volume).

The Mystic terrane (Figs. 2, and 3) (Reed and Nelson, 1980; Jones and others, 1982, 1983) consists of (1) Ordovician graptolitic shale and associated(?) pillow basalt; (2) massive Silurian limestone and Late Devonian sandstone, shale, conglomerate, and reefal limestone; (3) latest Devonian to Pennsylvanian radiolarian chert; (4) flysch, chert, argillite, and Permian conglomerate (locally plant bearing); and (5) associated Triassic(?) pillow basalt and gabbro. The terrane is a complexly deformed but partly coherent assemblage, and it is interpreted as a displaced fragment of the Paleozoic and early Mesozoic North American Cordillera continental margin (Jones and others, 1982; Decker and others, this volume).

The Dillinger terrane (Figs. 2 and 3) (Jones and others, 1982; Gilbert and Bundtzen, 1984; Patton and others, 1989; this volume, Chapter 7) consists chiefly of (1) Cambrian(?) and Ordovician calcareous turbidite, shale, and minor greenstone; (2) Early Ordovician and Early Silurian graptolitic black shale and chert; (3) Early and Middle Silurian laminated limestone and graptolitic black shale; (4) Middle to Late Silurian sandstone turbidites and shale; and (5) Late Silurian and Early Devonian limestone, breccia, sandstone, and shale that is complexly folded and faulted. The unit is interpreted as a displaced fragment of the Paleozoic and early Mesozoic North American Cordillera continental margin (Jones and others, 1982; Decker and others, this volume).

In southwest Alaska, the Dillinger, Nixon Fork, Mystic, and Minchumina terranes are interpreted by Decker and others (this volume) as various facies of the continental shelf and slope rocks of the Farewell terrane. In contrast, the related Minchumina and Nixon Fork terranes are interpreted by Patton and others (this volume, Chapter 7) as discrete fault-bounded units. Because of the distinctive, fault-bounded stratigraphy and structure in the study area, the Mystic and Dillinger rocks are herein interpreted as separate terranes within the study area. As originally interpreted by Jones and others (1982), the Dillinger, Nixon Fork, and Mystic terranes may have originally formed as facies of one another and were subsequently tectonically displaced several hundred kilometers from the northwestern part of the Canadian Cordillera.

Windy terrane

The Windy terrane (Figs. 2 and 3) (Jones and others, 1984, 1987; Nokleberg and others, 1985, 1989a, 1992a, 1992b) occurs in several narrow, discontinuous slivers within branches of the Denali fault, north of the Maclaren terrane, and south of the

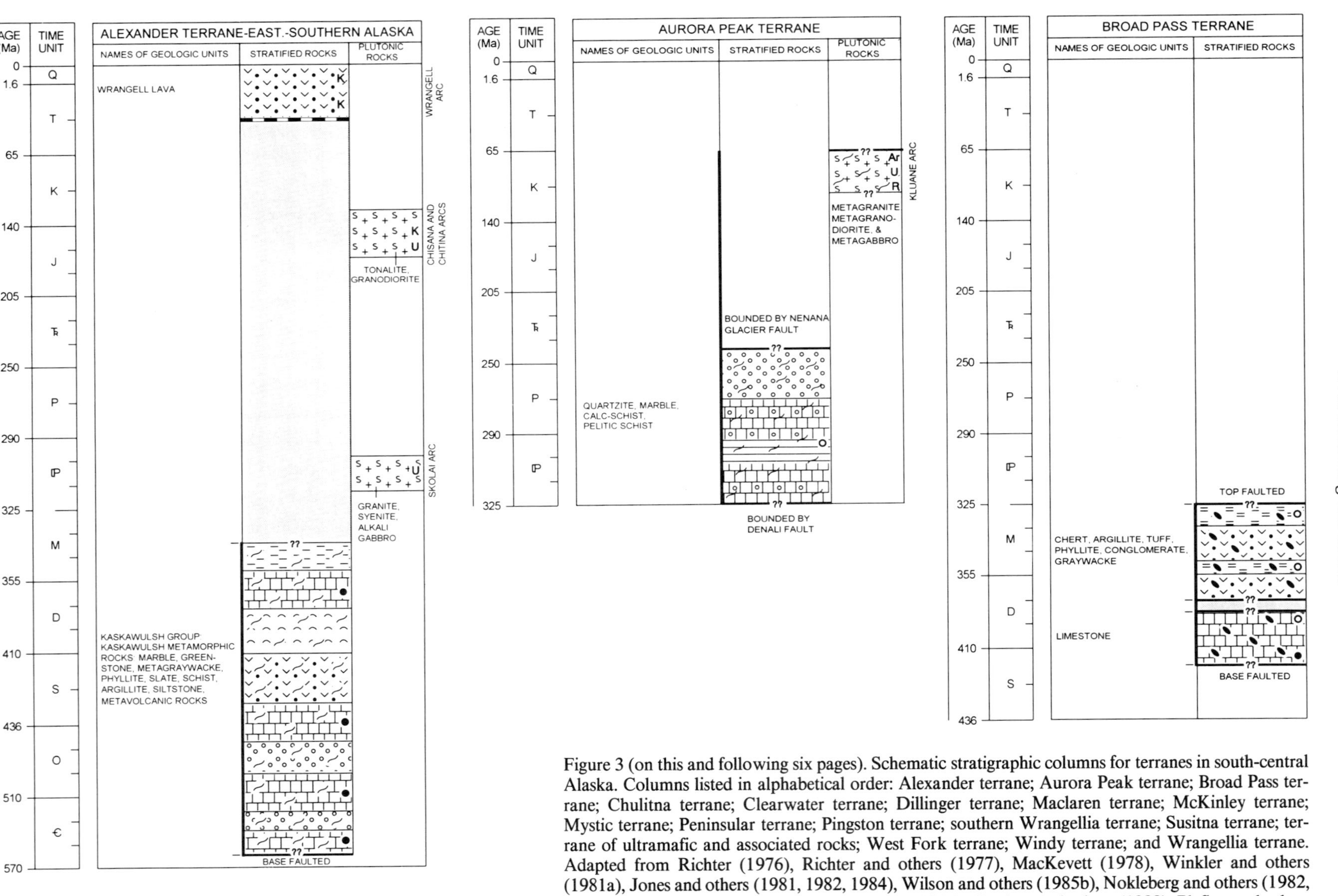

Figure 3 (on this and following six pages). Schematic stratigraphic columns for terranes in south-central Alaska. Columns listed in alphabetical order: Alexander terrane; Aurora Peak terrane; Broad Pass terrane; Chulitna terrane; Clearwater terrane; Dillinger terrane; Maclaren terrane; McKinley terrane; Mystic terrane; Peninsular terrane; Pingston terrane; southern Wrangellia terrane; Susitna terrane; terrane of ultramafic and associated rocks; West Fork terrane; Windy terrane; and Wrangellia terrane. Adapted from Richter (1976), Richter and others (1977), MacKevett (1978), Winkler and others (1981a), Jones and others (1981, 1982, 1984), Wilson and others (1985b), Nokleberg and others (1982, 1985, 1989a, 1992b, 1993), Csejtey and others (1986), Gardner and others (1988), Plafker and others (1989c), Wallace and others (1989), Winkler (1992), Silberling and others (this volume, Plate 3), and Wilson and others (1994). Geologic time scale from Decade of North American Geology geologic time scale (Palmer, 1983).

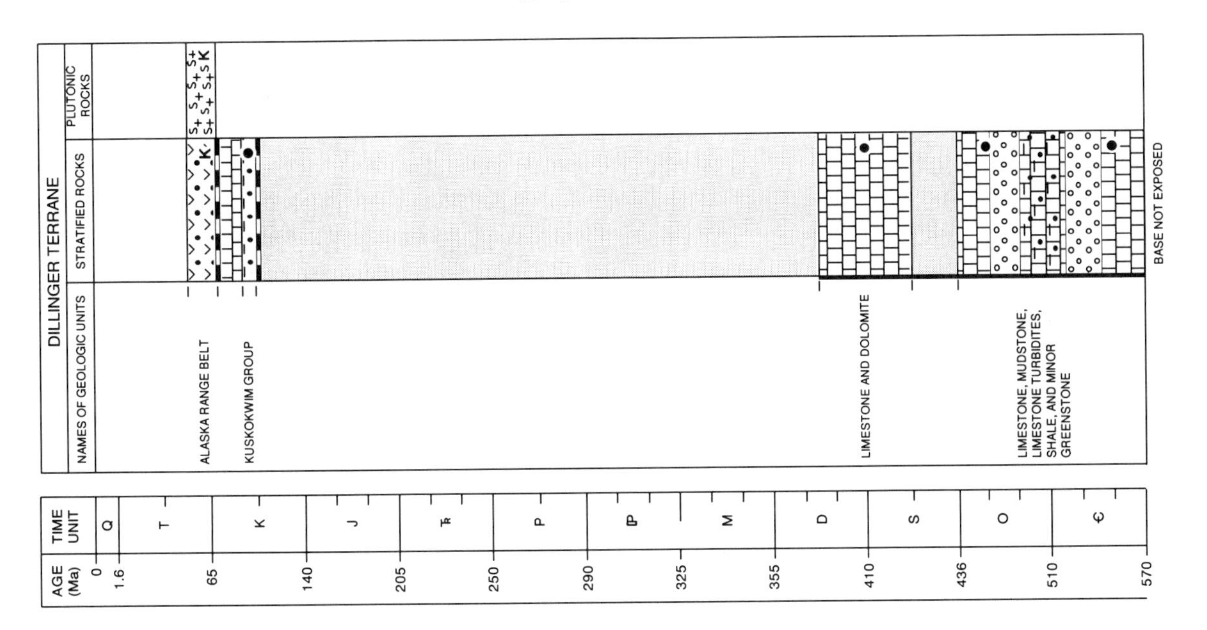
DILLINGER TERRANE
AGE (Ma)
TIME UNIT
NAMES OF GEOLOGIC UNITS
STRATIFIED ROCKS
PLUTONIC ROCKS
0
1.6
65
140
205
250
290
325
355
410
436
510
570
Q
T
K
J
P
M
D
S
O
ALASKA RANGE BELT
KUSKOKWIM GROUP
LIMESTONE AND DOLOMITE
LIMESTONE, MUDSTONE, LIMESTONE TURBIDITES, SHALE, AND MINOR GREENSTONE
BASE NOT EXPOSED

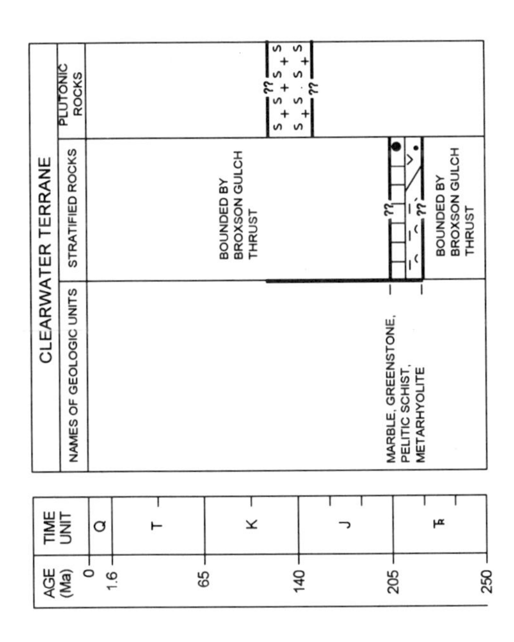
CLEARWATER TERRANE
AGE (Ma)
TIME UNIT
NAMES OF GEOLOGIC UNITS
STRATIFIED ROCKS
PLUTONIC ROCKS
0
1.6
65
140
205
250
Q
T
K
J
BOUNDED BY BROXSON GULCH THRUST
MARBLE, GREENSTONE, PELITIC SCHIST, METARHYOLITE
BOUNDED BY BROXSON GULCH THRUST

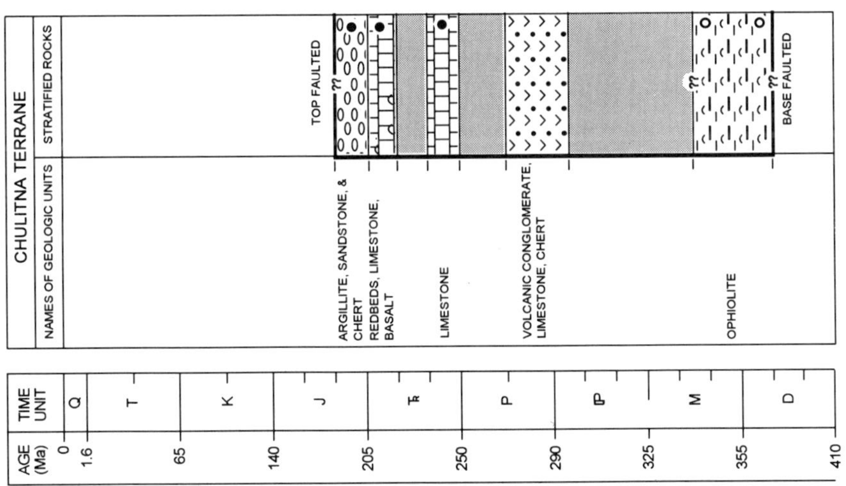
CHULITNA TERRANE
AGE (Ma)
TIME UNIT
NAMES OF GEOLOGIC UNITS
STRATIFIED ROCKS
0
1.6
65
140
205
250
290
325
355
410
Q
T
K
J
P
M
D
TOP FAULTED
ARGILLITE, SANDSTONE, & CHERT
REDBEDS, LIMESTONE, BASALT
LIMESTONE
VOLCANIC CONGLOMERATE, LIMESTONE, CHERT
OPHIOLITE
BASE FAULTED

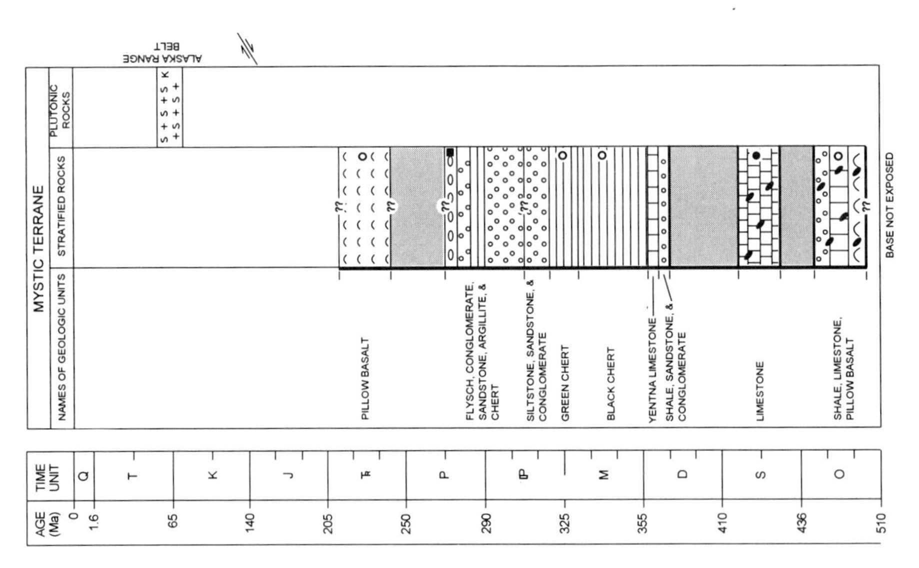

MYSTIC TERRANE
AGE (Ma)
TIME UNIT
NAMES OF GEOLOGIC UNITS
STRATIFIED ROCKS
PLUTONIC ROCKS
ALASKA RANGE BELT
PILLOW BASALT
FLYSCH, CONGLOMERATE, SANDSTONE, ARGILLITE, & CHERT
SILTSTONE, SANDSTONE, & CONGLOMERATE
GREEN CHERT
BLACK CHERT
YENTNA LIMESTONE
SHALE, SANDSTONE, & CONGLOMERATE
LIMESTONE
SHALE, LIMESTONE, PILLOW BASALT
BASE NOT EXPOSED

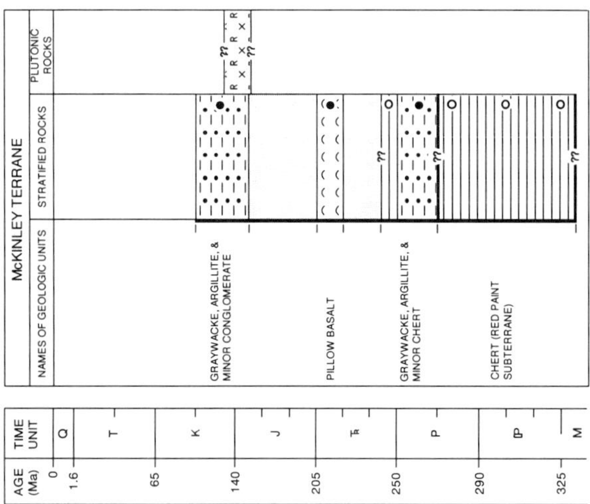

McKINLEY TERRANE
AGE (Ma)
TIME UNIT
NAMES OF GEOLOGIC UNITS
STRATIFIED ROCKS
PLUTONIC ROCKS
GRAYWACKE, ARGILLITE, & MINOR CONGLOMERATE
PILLOW BASALT
GRAYWACKE, ARGILLITE, & MINOR CHERT
CHERT (RED PAINT SUBTERRANE)

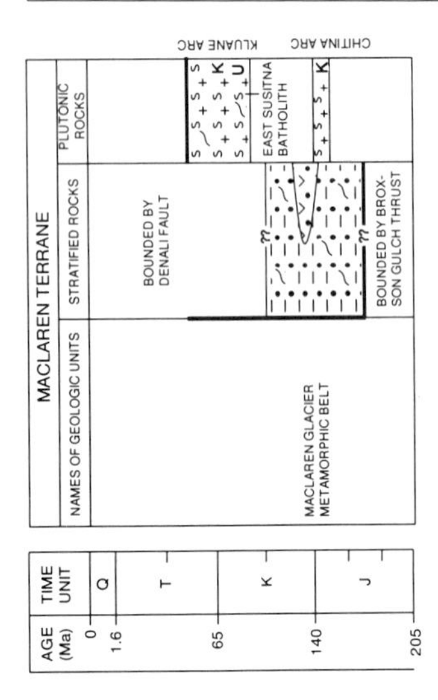

MACLAREN TERRANE
AGE (Ma)
TIME UNIT
NAMES OF GEOLOGIC UNITS
STRATIFIED ROCKS
PLUTONIC ROCKS
BOUNDED BY DENALI FAULT
MACLAREN GLACIER METAMORPHIC BELT
BOUNDED BY BROXSON GULCH THRUST
EAST SUSITNA BATHOLITH
KLUANE ARC
CHITINA ARC

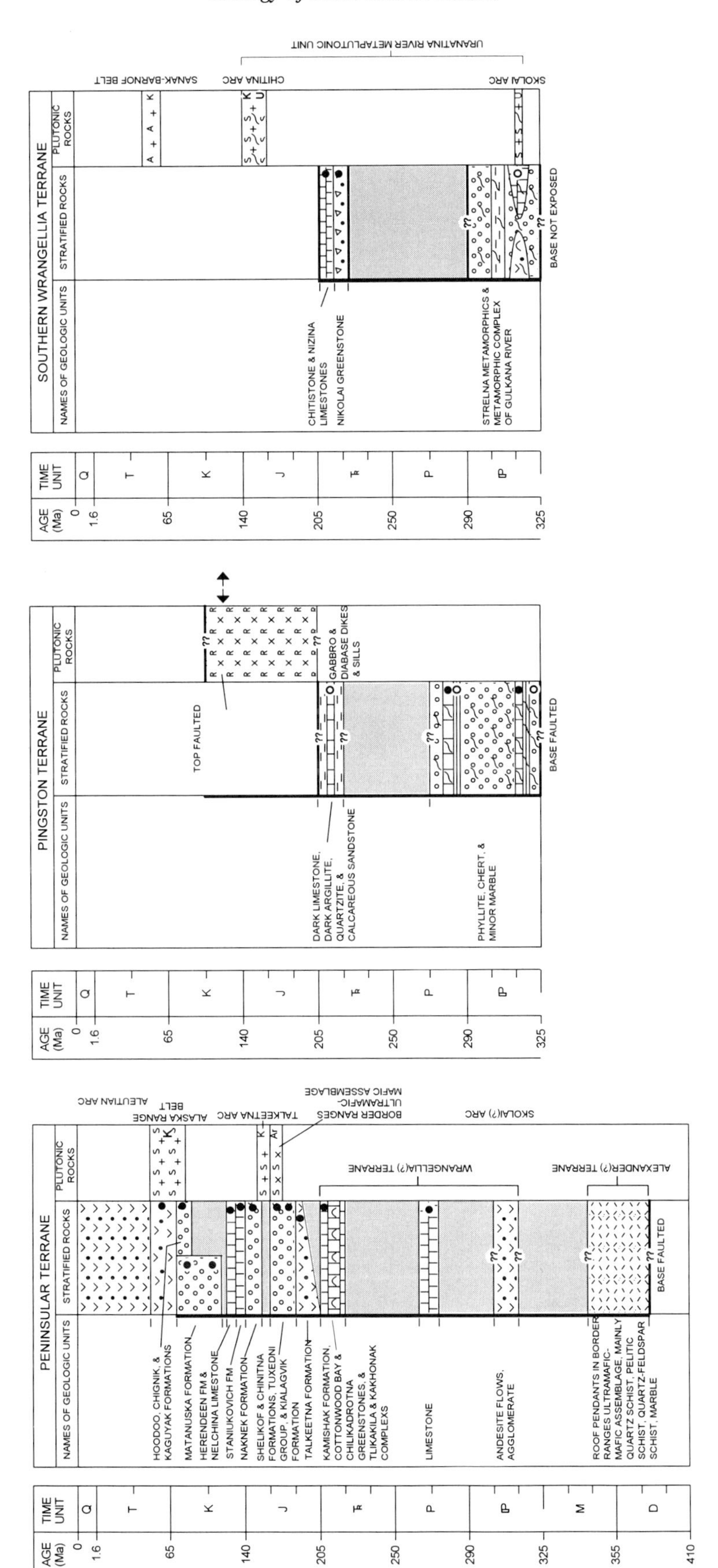
AGE (Ma)
TIME UNIT
0
1.6
65
140
205
250
290
325
355
410
Q
T
K
J
Ћ
P
IP
M
D
PENINSULAR TERRANE
NAMES OF GEOLOGIC UNITS
STRATIFIED ROCKS
PLUTONIC ROCKS
HOODOO, CHIGNIK, & KAGUYAK FORMATIONS
MATANUSKA FORMATION
HERENDEEN FM & NELCHINA LIMESTONE
STANIUKOVICH FM
NAKNEK FORMATION
SHELIKOF & CHINITNA FORMATIONS, TUXEDNI GROUP, & KIALAGVIK FORMATION
TALKEETNA FORMATION
KAMISHAK FORMATION, COTTONWOOD BAY & CHILIKADROTNA GREENSTONES, & TLIKAKILA & KAKHONAK COMPLEXS
LIMESTONE
ANDESITE FLOWS, AGGLOMERATE
ROOF PENDANTS IN BORDER RANGES ULTRAMAFIC-MAFIC ASSEMBLAGE, MAINLY QUARTZ SCHIST, PELITIC SCHIST, QUARTZ-FELDSPAR SCHIST, MARBLE
BASE FAULTED
WRANGELLIA(?) TERRANE
ALEXANDER(?) TERRANE
ALEUTIAN ARC
ALASKA RANGE BELT
TALKEETNA ARC
BORDER RANGES ULTRAMAFIC-MAFIC ASSEMBLAGE
SKOLAI(?) ARC
PINGSTON TERRANE
TOP FAULTED
DARK LIMESTONE, DARK ARGILLITE, QUARTZITE, & CALCAREOUS SANDSTONE
GABBRO & DIABASE DIKES & SILLS
PHYLLITE, CHERT, & MINOR MARBLE
SOUTHERN WRANGELLIA TERRANE
CHITISTONE & NIZINA LIMESTONES
NIKOLAI GREENSTONE
STRELNA METAMORPHICS & METAMORPHIC COMPLEX OF GULKANA RIVER
BASE NOT EXPOSED
SANAK-BARNOF BELT
CHITINA ARC
SKOLAI ARC
URANATINA RIVER METAPLUTONIC UNIT

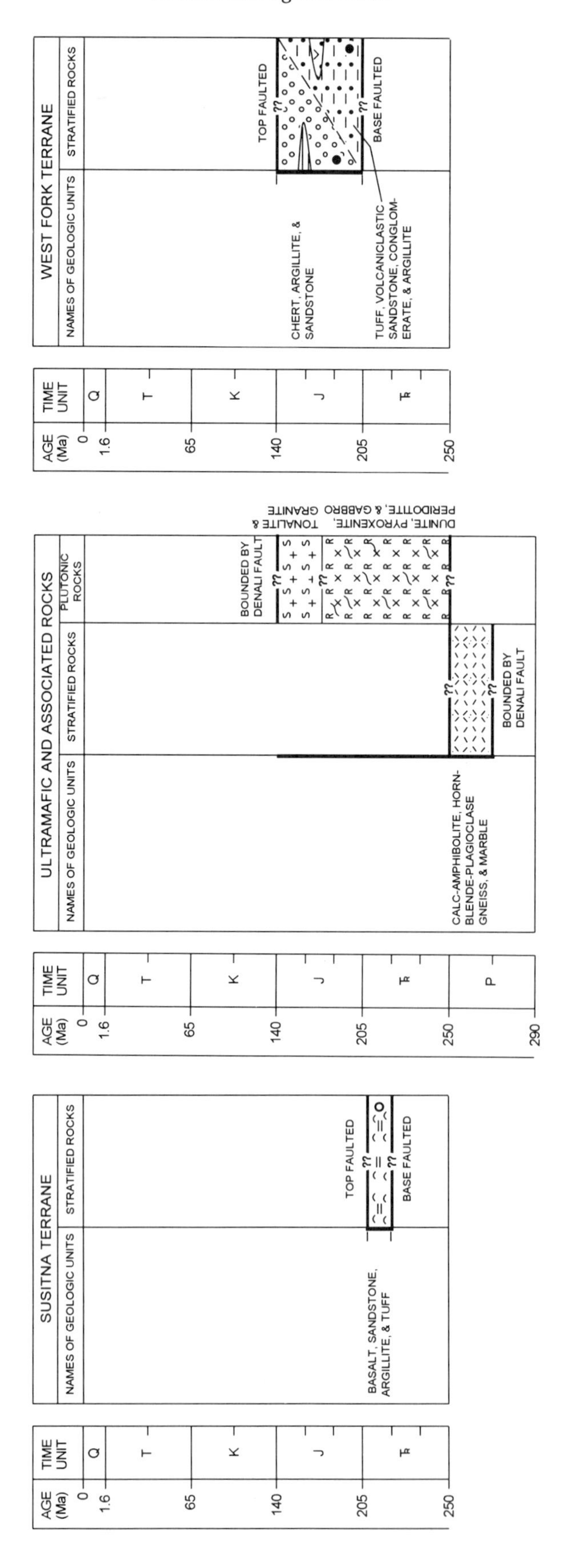
WEST FORK TERRANE
AGE (Ma)
TIME UNIT
NAMES OF GEOLOGIC UNITS
STRATIFIED ROCKS
0
1.6
65
140
205
250
Q
T
K
J
Ꞧ
TOP FAULTED
BASE FAULTED
CHERT, ARGILLITE, & SANDSTONE
TUFF, VOLCANICLASTIC SANDSTONE, CONGLOMERATE, & ARGILLITE
ULTRAMAFIC AND ASSOCIATED ROCKS
PLUTONIC ROCKS
290
P
BOUNDED BY DENALI FAULT
TONALITE & GRANITE
DUNITE, PYROXENITE, PERIDOTITE, & GABBRO
CALC-AMPHIBOLITE, HORNBLENDE-PLAGIOCLASE GNEISS, & MARBLE
SUSITNA TERRANE
BASALT, SANDSTONE, ARGILLITE, & TUFF

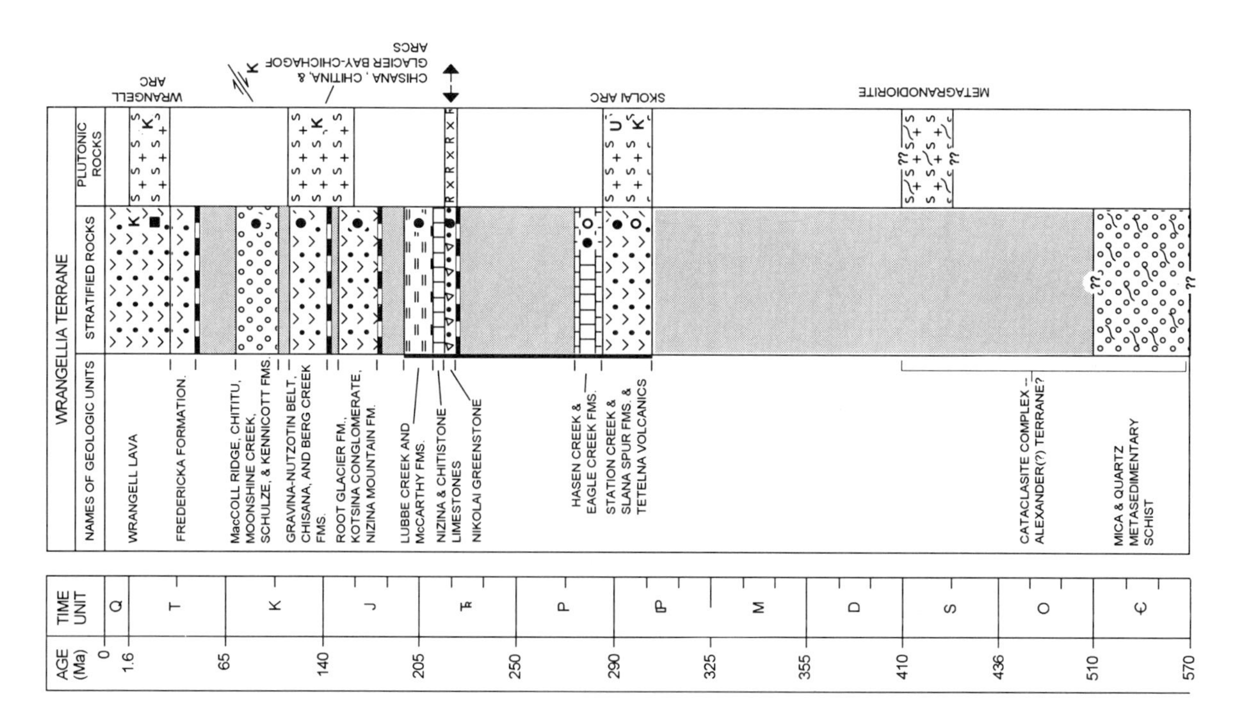

WRANGELLIA TERRANE
NAMES OF GEOLOGIC UNITS
STRATIFIED ROCKS
PLUTONIC ROCKS
WRANGELL LAVA
FREDERICKA FORMATION
MacCOLL RIDGE, CHITITU, MOONSHINE CREEK, SCHULZE, & KENNICOTT FMS.
GRAVINA-NUTZOTIN BELT, CHISANA, AND BERG CREEK FMS.
ROOT GLACIER FM., KOTSINA CONGLOMERATE, NIZINA MOUNTAIN FM.
LUBBE CREEK AND McCARTHY FMS.
NIZINA & CHITISTONE LIMESTONES
NIKOLAI GREENSTONE
HASEN CREEK & EAGLE CREEK FMS.
STATION CREEK & SLANA SPUR FMS. & TETELNA VOLCANICS
CATACLASITE COMPLEX -- ALEXANDER(?) TERRANE?
MICA & QUARTZ METASEDIMENTARY SCHIST
WRANGELL ARC
CHISANA, CHITINA, & GLACIER BAY-CHICHAGOF ARCS
SKOLAI ARC
METAGRANODIORITE
TIME UNIT
Q
T
K
J
P
M
D
S
O
AGE (Ma)
0
1.6
65
140
205
250
290
325
355
410
436
510
570

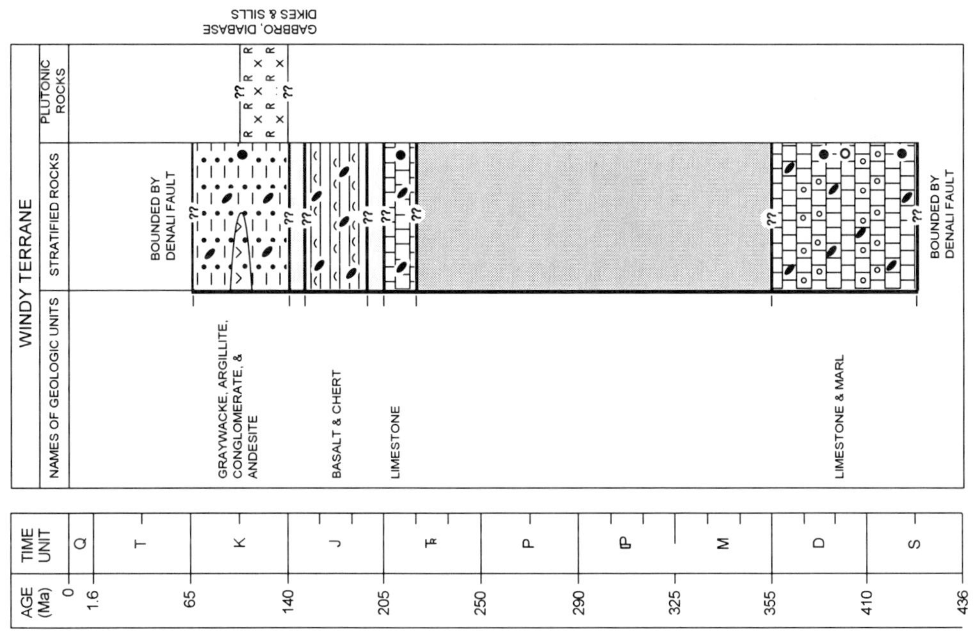

WINDY TERRANE
NAMES OF GEOLOGIC UNITS
STRATIFIED ROCKS
PLUTONIC ROCKS
BOUNDED BY DENALI FAULT
GRAYWACKE, ARGILLITE, CONGLOMERATE, & ANDESITE
BASALT & CHERT
LIMESTONE
LIMESTONE & MARL
GABBRO, DIABASE DIKES & SILLS
TIME UNIT
Q
T
K
J
P
M
D
S
AGE (Ma)
0
1.6
65
140
205
250
290
325
355
410
436

EXPLANATION FOR STRATIGRAPHIC COLUMNS

NON-MARINE CLASTIC DEPOSITS

SHALLOW-MARINE TERRIGENOUS DEPOSITS

SHALLOW-MARINE CARBONATE DEPOSITS

SHALLOW-MARINE SILICEOUS DEPOSITS

DEEP-MARINE HEMI-PELAGIC SEDIMENTARY AND VOLCANIC DEPOSITS

DEEP-MARINE PELAGIC AND SILICEOUS DEPOSITS

OCEANIC CRUST, SEAMOUNTS, AND OPHIOLITES

MAFIC VOLCANIC AND VOLCANICLASTIC ROCKS

TURBIDITE DEPOSITS

SUBDUCTION-RELATED VOLCANIC AND SEDIMENTARY ROCKS

RIFT-RELATED VOLCANIC AND SEDIMENTARY ROCKS

SUBDUCTION-RELATED GRANITIC ROCKS

SUBDUCTION-RELATED MAFIC AND ULTRAMAFIC PLUTONIC ROCKS

RIFT-RELATED MAFIC AND ULTRAMAFIC PLUTONIC ROCKS

INTENSELY METAMORPHOSED AND DEFORMED ROCKS OF INDETERMINATE ORIGIN

STRATIGRAPHIC HIATUS

NOTE: PATTERNS MAY BE COMBINED

TECTONIC OVERPRINTS:

- TECTONIC MELANGE
- PENETRATIVELY DEFORMED AND (OR) REGIONALLY METAMORPHOSED
- OLISTOSTROMAL DEPOSITS

MAJOR UNCONFORMITY

FAULT

TIME SPAN OF TERRANE

ISOTOPIC SYMBOLS:

K, K - Ar
Ar, Ar - Ar
R, Rb - Sr
U, U - Pb

AGE DIAGNOSTIC FOSSILS:

● MARINE MEGAFOSSIL
○ MARINE MICROFOSSIL
■ PLANT FOSSIL
?? AGE UNKNOWN

TECTONIC EVENTS:

- ACCRETIONARY EPISODE
- RIFTING EPISODE

CONTACTS:

- STRATIGRAPHIC OR INTRUSIVE
- TIME-TRANSGRESSIVE
- FACIES CHANGE

Aurora Peak and Yukon-Tanana terranes, mainly in the Healy and Mount Hayes quadrangles. The Windy terrane is a structural melange of diverse rock types that includes (Figs. 2, 3, and 4) (1) small to large fault-bounded lenses of limestone and marl of Silurian or Devonian age (Richter, 1976; Csejtey and others, 1986; Nokleberg and others, 1985, 1992a, 1992b); (2) Late Triassic limestone; (3) Jurassic basalt and chert; and (4) Cretaceous ammonite-bearing flysch and volcanic rocks composed mainly of argillite, quartz-pebble siltstone, quartz sandstone, metagraywacke, chert pebble and polymictic metaconglomerate, and lesser andesite and dacite (J. H. Stout, 1976, written commun.; Csejtey and others, 1986; Nokleberg and others, 1985, 1992a, 1992b).

Unlike adjacent terranes to the north and locally to the south, the Windy terrane exhibits mainly protolith textures and structures. Relict sedimentary structures include bedding, graded bedding, and crossbedding. The Windy terrane is intensely faulted and sheared and locally exhibits phyllonite and protomylonite with an intense schistosity formed at incipient lower greenschist facies metamorphism. The maximum structural thickness of the Windy terrane is about 5 km. The Windy terrane is interpreted as a structural melange that formed during tectonic mixing that occurred during Cenozoic dextral slip along the Denali fault. The Mesozoic flysch and associated volcanic rocks are interpreted as fragments of the Kahiltna assemblage and associated Chisana arc rocks (Stanley and others, 1990). The source for the Silurian or Devonian limestone and marl might be from the Mystic, Dillinger, and/or Nixon Fork terranes.

Aurora Peak terrane

The Aurora Peak terrane (Brewer, 1982; Nokleberg and others, 1985, 1989a, 1992b) occurs north of the Denali fault, west of the Richardson Highway, in the western Mount Hayes and eastern Healy quadrangles (Figs. 1, 2, and 3). The terrane consists of an older sequence of mainly metasedimentary rocks and a younger sequence of metaplutonic rocks (Fig. 4). Because of intense deformation, the stratigraphic thickness of the Aurora Peak terrane cannot be estimated. The maximum structural thickness is several thousand meters. Unless otherwise noted, the following description of the Aurora Peak terrane is after Aleinikoff (1984), Nokleberg (1985, 1989a, 1992), Brewer (1982), and Aleinikoff and others (1987).

The older sequence of metasedimentary rocks consists of mainly fine- to medium-grained and polydeformed calc-schist, marble, quartzite, and pelitic schist. One fragment of a conodont from marble indicates a Silurian to Triassic age. Protoliths for the metasedimentary rocks include marl, quartzite, and shale. The younger metaplutonic sequence consists of regionally metamorphosed and penetratively deformed, schistose quartz diorite, granodiorite, and granite, and sparse amphibolite derived from gabbro and diorite. U-Pb zircon isotopic analysis of a metamorphosed quartz diorite indicates an age of igneous intrusion of 71 Ma; isotopic analysis of lead from samples of metagranitic rocks suggests derivation from an ~1.2 Ga source.

The Aurora Peak terrane was twice metamorphosed and ductilely deformed. The core of the terrane exhibits an older, upper amphibolite facies metamorphism and associated mylonitic schist. Because of similar and parallel fabrics in the metasedimentary and metaplutonic rocks, the upper amphibolite facies metamorphism of both units is interpreted as having occurred during syntectonic intrusion of Late Cretaceous and early Tertiary granitic rocks. The margins of the terrane exhibit a younger middle greenschist facies metamorphism and formation of blastomylonite along an intense younger schistosity. A K-Ar biotite age of 27 Ma suggests that retrograde metamorphism occurred at least into the middle Tertiary. The younger, greenschist facies metamorphism is interpreted as having formed during dextral-slip transport of the terrane along the Nenana Glacier and Denali faults.

The Aurora Peak terrane and the Maclaren terrane to the south are interpreted as displaced continental margin arc fragments that were tectonically separated from the Kluane schist and the Ruby Range batholith that occur on the northeast side of the Denali fault some 400 km to the southeast in the Yukon Territory (Nokleberg and others, 1985; Plafker and others, 1989c). Between the Aurora Peak and Maclaren terranes is a structural melange of Paleozoic and Cretaceous sedimentary and volcanic rocks that constitutes the Windy terrane (Figs. 3 and 4). The Aurora Peak terrane is interpreted as having been tectonically transported to a position against the southern Yukon-Tanana terrane before transport of the Maclaren terrane and resultant formation of the structural melange of the Windy terrane between the Maclaren and Aurora Peak terranes.

Terrane of ultramafic and associated rocks

A terrane of ultramafic and associated rocks (Richter, 1976; Richter and others, 1977; Nokleberg and others, 1982, 1985, 1989a, 1992b) occurs in the eastern Alaska Range in the Mount Hayes and Healy quadrangles (Figs. 1 and 2). Generally the terrane occurs in narrow, fault-bounded lenses within branches of the Denali fault that are up to a few kilometers wide and several kilometers long. The terrane also occurs in a few small klippen south of the Denali fault in the eastern Mount Hayes quadrangle (Fig. 4).

The ultramafic rocks are mainly fine- to medium-grained pyroxenite and peridotite, and dunite, along with local hornblende gabbro (Fig. 3). The ultramafic rocks are largely altered to serpentinite and are generally highly sheared. The associated rocks are amphibolite, hornblende-plagioclase gneiss, and marble that host the ultramafic rocks. The associated rocks are interpreted as a suite of calcareous metasedimentary rocks metamorphosed to amphibolite facies. Sparse, small, elongate plutons of tonalite and granite locally intrude the ultramafic rocks and calcareous metasedimentary rocks. The ultramafic and associated rocks are moderately to intensely ductily deformed with a strong schistosity that is subparallel to contacts and enclosing faults. A weak schistosity occurs in the tonalite and granite that is subparallel to intrusive contacts. No isotopic age data exist for the terrane.

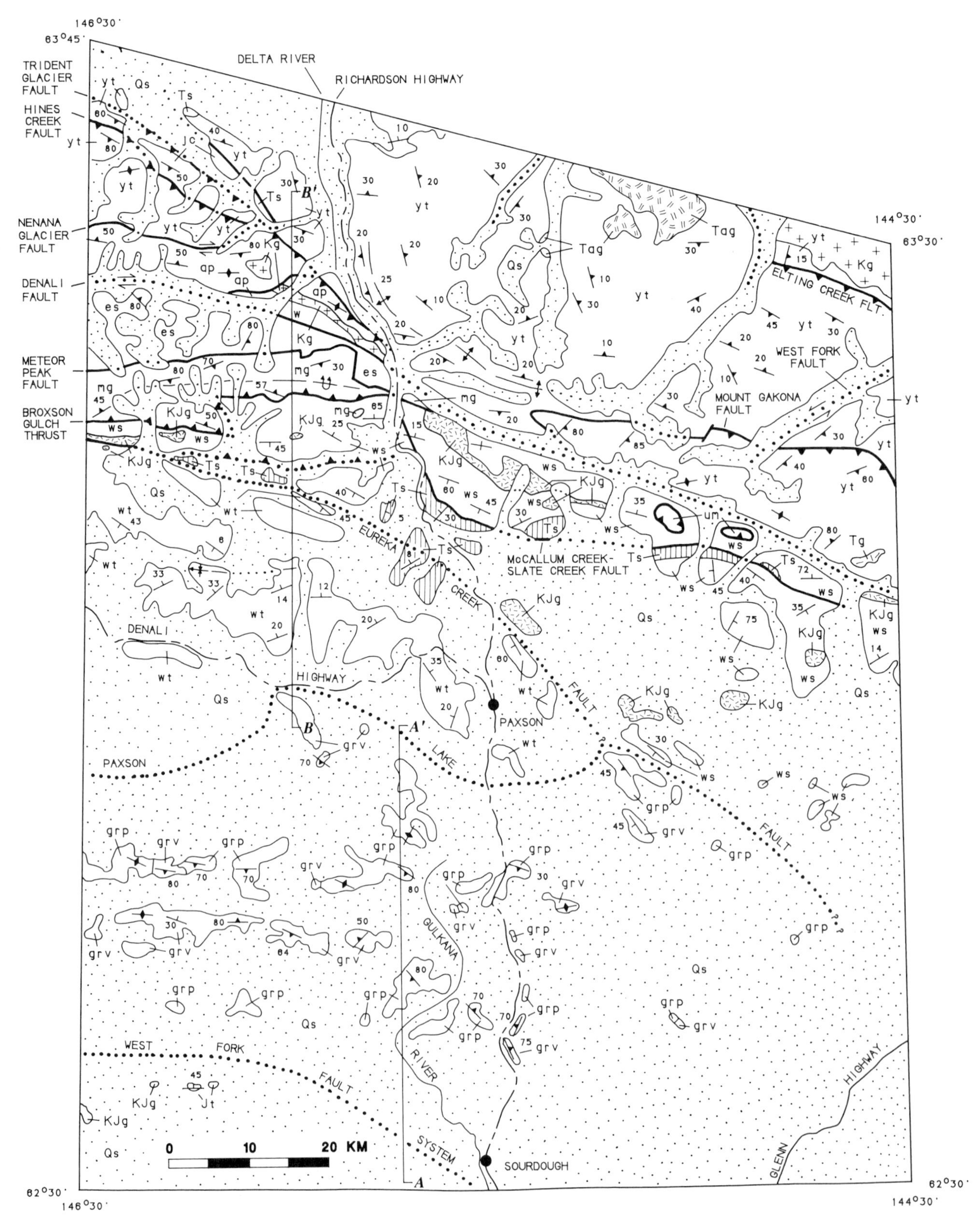

Figure 4. Generalized bedrock geologic map and cross sections of the northern Copper River basin and eastern Alaska Range along the Trans-Alaskan Crustal Transect (TACT). Refer to text for description of units. Adapted from Nokleberg and others (1985, 1989a, 1992b).

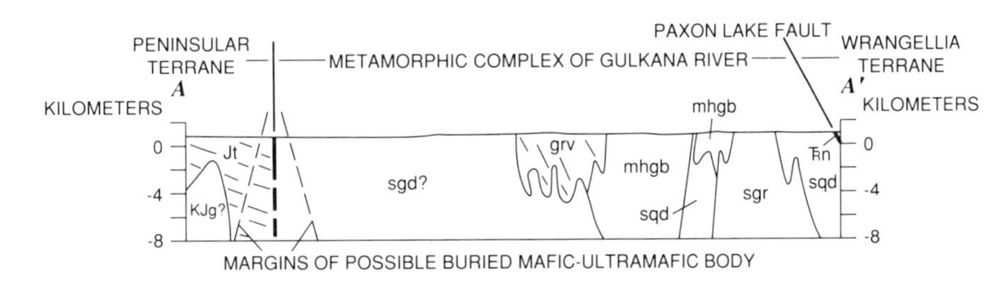

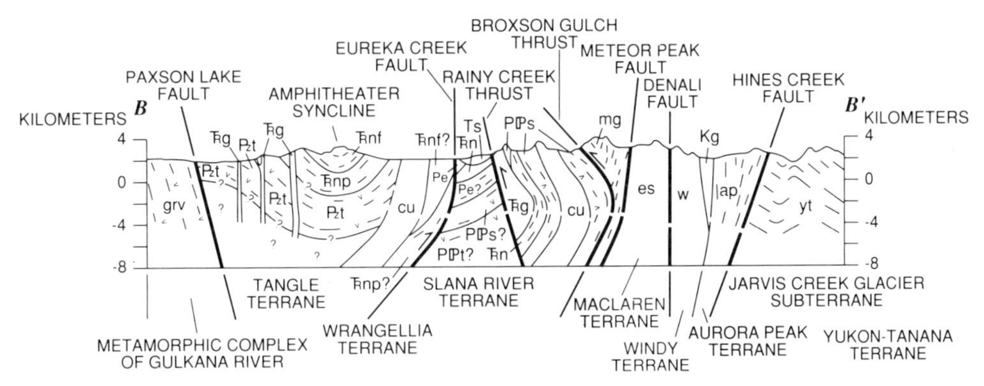

Notes: Surficial deposits omitted
No vertical exaggeration
All but one contact covered by surficial deposits in section *A-A'*

0 10 20 KILOMETERS

EXPLANATION

CENOZOIC SURFICIAL DEPOSITS AND SEDIMENTARY ROCKS

Qs

Quaternary surficial deposits

Tertiary sedimentary rocks

PLUTONIC ROCKS

Tg - Early Tertiary granitic rocks
Tag - Early Tertiary alkalic gabbro

Kg

Cretaceous granitic rocks

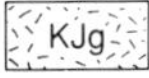

Early Cretaceous and Late Jurassic granitic rocks

YUKON-TANANA TERRANE

yt

AURORA PEAK TERRANE

ap

WINDY TERRANE

w

MACLAREN TERRANE

es

East Susitna batholith

mg

Maclaren Glacier metamorphic belt

TERRANE OF ULTRAMIFIC AND ASSOCIATED ROCKS

um

WRANGELLIA TERRANE

Slana River subterrane

ws: TRg, cu, TRn, Pe, PPs, PPt

ws - undifferentiated. On cross sections, divided into:
TRg - Gabbro and diabase;
cu - cumulate mafic & ultramafic rocks;
TRn - Nikolai Greenstone;
Pe - Eagle Creek Formation;
PPs - Slana Spur Formation; and
PPt - Tetelna Volcanics

Tangle subterrane

wt: TRg, cu, TRnf, TRnp, Pzt

wt - Undifferentiated. On cross sections, divided into:
TRg - Gabbro and diabase;
cu - cumulate mafic and ultramafic rocks;
Nikolai Greenstone—
TRnf - Subaerial basalt flow member and
TRnp - Pillow basalt flow member; and
Pzt - tuff, argillite, and chert

METAMORPHIC COMPLEX OF GULKANA RIVER

Metaplutonic rocks

grp: sgr, sgd, sqd, mhbd

grp - Undifferentiated. On cross sections, divided into:
sgr - Schistose granite;
sgd - Schistose granodiorite;
sqd - Schistose quartz diorite; and
mhgb - Metamorphosed hornblende gabbro

Metavolcanic and metasedimentary rocks

grv

PENINSULAR TERRANE

Jt

Talkeetna formation

SYMBOLS

Contact—Dotted where concealed
Fault—Dotted where concealed
Strike-slip
Thrust—Teeth on upper plate
Antiform—Dashed where approximately located; dotted where concealed
Anticline
Overturned anticline
Syncline—showing plunge
Strike and dip of beds
35 Inclined
Vertical
Strike and dip of schistosity and parallel compositional layer
75 Inclined
Vertical

The tonalite and granite are assumed to be Mesozoic in age (Richter and others, 1977); the other units of the terrane are therefore assumed to be Mesozoic or older.

The near-vertical lenses of the terrane along the Denali fault may be fragments emplaced from a source of depth during Late Cretaceous and Cenozoic dextral slip along the Denali fault. The two klippen of the terrane south of the Denali fault may represent remnants of amalgamation along the ancestral Denali fault, where substantial south-verging thrusting has been interpreted (Stanley and others, 1990). The terrane has been interpreted (1) as a crustal suture belt (Richter, 1976; Richter and others, 1977) or (2) as having uncertain, but possible, ophiolitic affinity (Patton and others, 1992; this volume, Chapter 21). Alternatively, because of the assemblage of high-grade metasedimentary rocks, metamorphosed ultramafic and mafic rocks, and metagranite plutons, the terrane might be a fragment of the deep levels of an igneous arc.

LATE JURASSIC AND EARLY CRETACEOUS FLYSCH BASINS

Two major tectonically collapsed flysch basins occur across the northern part of the study area, south of the Denali fault (Fig. 2) (Jones and others, 1982). To the northwest is the Late Jurassic and Early Cretaceous Kahiltna assemblage, and to the northeast is the Late Jurassic and Early Cretaceous Gravina-Nutzotin belt (Berg and others, 1972). These two assemblages of flysch and locally associated coeval volcanic and plutonic rocks form a major belt that is several thousand kilometers long and extends discontinuously from the Alaska Peninsula to the west, to the southern part of southeastern Alaska to the southeast (Berg and others, 1972; Rubin and Saleeby, 1991; Gehrels and Berg, this volume; Silberling and others, this volume). In the study area, the Kahiltna assemblage occurs discontinuously for more than 800 km along the northwestern margin of the Peninsular and Wrangellia terranes and is divided into southern and northern segments. The companion Gravina-Nutzotin belt occurs mainly along the northeastern margins of the Wrangellia and Alexander terranes for about 1,500 km in the study area.

Southern Kahiltna assemblage

The southern segment of the Kahiltna assemblage, hereafter called the southern Kahiltna assemblage for the sake of brevity, occurs along the northwestern margin of the Peninsular terrane in the northwestern Aleutian Range, mainly in the Lake Clark quadrangle (Figs. 1, 2, and 5). Almost everywhere in this area, the Kahiltna assemblage is separated from the main mass of the Peninsular terrane to the south by the Alaska-Aleutian Range batholith (Figs. 2 and 5) (Nelson and others, 1983; Wallace and others, 1989). In this area, the Kahiltna assemblage was defined as the informally named Koksetna River sequence and the Chilikadrotna Greenstone (Hanks and others, 1985; Wallace and others, 1989). The Koksetna River sequence is composed of widespread Late Jurassic and Early Cretaceous clastic rocks that are best exposed in the area drained by the Koksetna River. Conformably beneath the Koksetna River sequence is the Chilikadrotna Greenstone (Bundtzen and others, 1979; Wallace and others, 1989), which is interpreted below to be part of the Peninsular terrane. Unlike otherwise noted, the following description of the southern Kahiltna assemblage is from Eakins and others (1978) and Wallace and others (1989).

Koksetna River sequence. The Koksetna River sequence is generally poorly exposed, although a few excellent exposures occur in river gorges and the more rugged foothills west of the Alaska Range. The sequence consists mainly of complexly deformed volcanic-lithic turbidites that have no distinctive marker horizons. Only two localities with datable megafossils have been discovered, one each of Late Jurassic (Kimmeridgian) age and Early Cretaceous (Valanginian) age. Isolated outcrops of poorly exposed pebble and cobble conglomerate may occur in part of the sequence. Clast types in the conglomerate include intermediate to silicic volcanic and plutonic rocks, and subordinate chert and argillite. The depositional environments range from slope and inner fan (Mutti and Ricci Lucchi, 1978) in the southeast to middle fan and outer fan in the northwest; regional sediment transport was probably to the northwest or north. Clast compositions of the sandstones suggest derivation from a magmatic-arc provenance that was dominated by volcanic rocks, but with local exposures of plutonic rocks, fine-grained clastic rocks, regionally metamorphosed rocks, and contact-metamorphosed rocks.

Bedding generally dips steeply to the northwest. Upright and overturned beds indicate that the Koksetna River sequence has been isoclinally folded and faulted. The sequence is locally intruded and metamorphosed by latest Cretaceous to early Tertiary plutons of the Aleutian-Alaska Range batholith (Fig. 5). The sequence is progressively unconformably overlapped toward the southeast by latest Cretaceous to early Tertiary volcanic rocks preserved in a regional northeast-trending structural low. Contacts between the Koksetna River sequence and the stratigraphically subjacent Chilikadrotna Greenstone generally are not exposed. In the few places where exposed, the contact is intensely deformed. However, the Chilikadrotna Greenstone is exposed in the cores of a series of antiforms, indicating that it underlies the Koksetna River sequence. The Chilikadrotna Greenstone was probably both the depositional basement and a local sediment source for the overlying Koksetna River sequence.

Units adjacent to southern Kahiltna assemblage. The southern Kahiltna assemblage is bounded to the northwest by the sedimentary rocks ofthe Kuskokwim Group (Fig. 5), which consists of Late Cretaceous, deep-marine, shallow-marine, and local nonmarine sedimentary rocks (Patton and others, 1989; this volume, Chapter 7). Farther northwest, the Kuskokwim Group depositionally overlaps several major terranes, including the Nixon Fork, Dillinger, and Mystic terranes (Fig. 2). In the study area, the southern Kahiltna assemblage and the Kuskokwim Group are juxtaposed along the sharp Chilchitna fault (Fig. 5), which dips steeply northwest and is locally marked by a zone of phyllitic rocks. The fault may be a depositional contact that was structurally modified. However, more recent studies in this area indicate that the contact between the southern Kahiltna assemblage and the Kuskokwim Group may be gradational (D. C. Bradley, 1991, oral commun.). To the south, the southern Kahiltna assemblage may grade upward into the post-Turonian Kuskokwim Group. One locality within the southern Kahiltna assemblage in the Tyonek quadrangle has yielded the Late Cretaceous (Turonian) fossil *Inoceramus* (W. P. Elder, 1989, written commun. to D. C. Bradley).

Northern Kahiltna assemblage

The northern segment of the Kahiltna assemblage, hereafter called the northern Kahiltna assemblage for the sake of brevity, occurs in the northern Talkeetna Mountains and the central Alaska Range in the Talkeetna, Talkeetna Mountains, Healy, and adjacent quadrangles (Figs. 1 and 2) (Reed and Nelson, 1980; Csejtey and others, 1978, 1982, 1986). The assemblage consists of mainly monotonous, intensely deformed and locally highly metamorphosed flysch of Late Jurassic and Early Cretaceous age with a structural thickness of probably several thousand meters. The flysch consists of intercalated dark-colored argillite, phyllite, fine- to coarse-grained lithic graywacke, dark gray polymictic pebble conglomerate, subordinate chert-pebble conglomerate, a few thin beds of radiolarian chert, limy mudstone, and impure limestone. In the easternmost narrow lens of the northern Kahiltna assemblage in the eastern Healy and western Mount Hayes quadrangles, sparse andesite flows and volcanic metagraywacke are important components (Nokleberg and others, 1982, 1992b; Csejtey and others, 1986). The Late Jurassic and Early Cretaceous age of the northern Kahiltna assemblage is established by sparse microfossils and megafossils of Hauterivian to Barremian or Early Cretaceous age (Reed and Nelson, 1980; Csejtey and others, 1978, 1986). The flsych of the northern Kahiltna assemblage may be the protolith for the metasedimentary and metavolcanic rocks of the Maclaren terrane, discussed below. The flysch sequence is compressed into tight or isoclinal folds and is complexly faulted. Many areas are sheared and exhibit a pervasive axial plane cleavage. Because of the lithologically monotonous nature of the flysch sequence, faults are difficult to detect.

Gravina-Nutzotin belt

The Gravina-Nutzotin belt (Berg and others, 1972; Richter and Jones, 1973; Richter, 1976) occurs in the eastern Alaska Range and Nutzotin Mountains and is deposited on the Wrangellia terrane in the Mount Hayes, Nabesna, and McCarthy quadrangles (Figs. 1 and 2). The belt consists of more than 3,000 m of Late Jurassic and Early Cretaceous argillite, mudstone, and graywacke, and sparse conglomerate, limestone, volcanic flows, and volcaniclastic rocks and tuff deposited under shallow-marine to deep-marine conditions. Graded bedding is locally abundant and well developed, and turbidite deposits are common. The flysch sequence contains abundant intermediate-composition volcanic detritus (Berg and others, 1972; Richter, 1976; Nokleberg and others, 1982, 1985, 1992b). Rare fossils in the belt, mainly Buchia in the Nabesna quadrangle, indicate an Oxfordian through Barremian (Late Jurassic through Early Cretaceous) age. The rocks of the belt are locally highly deformed, particularly near the central part of the belt where isoclinal, overturned folds and companion thrust and reverse faults are common. The belt is faulted against the Yukon-Tanana terrane along the Denali fault to the north and east.

In the Nabesna quadrangle (Figs. 1 and 2), the Gravina-Nutzotin belt includes the Chisana Formation, which consists of marine and subaerial volcanic and volcaniclastic rocks, mainly andesite and basaltic andesite flows, and associated breccias, graywacke, and conglomerate that is more than 2,500 m thick (Richter, 1976). In this area, the Chisana Formation contains Early Cretaceous megafossils and locally grades downward through a tuffaceous unit into the stratigraphically subjacent part of the Gravina-Nutzotin belt (Richter, 1976). In the eastern Mount Hayes quadrangle, andesitic volcanic rocks are interlayered with flysch that contains Early Cretaceous megafossils (Nokleberg and others, 1992a, 1992b). The Chisana Formation is locally unconformably overlain by as much as 90 m of Late Cretaceous continental sedimentary rocks (Richter, 1976).

The Chisana Formation, the non-Chisana part of the Gravina-Nutzotin belt, and the Kahiltna assemblage are locally intruded by weakly deformed to nondeformed Late Jurassic and Early Cretaceous granitic plutons (Fig. 2) (Berg and others, 1972; Miller, this volume, Chapter 16). The source of the andesite flows and related rocks of the Chisana Formation, the volcanic detritus in the Gravina-Nutzotin belt, and the granitic plutons intruding these two units and the Kahiltna assemblage is interpreted as an extensive, coeval igneous arc flanked by the volcanic-derived flysch. The igneous arc was first recognized by Berg and others (1972) as the Gravina-Nutzotin basinal arc, and was named the Chisana arc by Plafker and others (1989c). Rare earth element (REE) whole-rock analyses suggest a transitional tholeiitic and calc-alkaline island-arc origin for the volcanic rocks (Barker and others, this volume; see also microfiche).

This island arc and companion flysch basin deposits of the Kahiltna assemblage are interpreted as having formed along the northern edge of the WCT prior to emplacement along the Alaskan continental margin (Berg and others, 1972; Plafker and others, 1989c; Barker, this volume). The main evidence for deposition of the flysch and associated rocks on the accreting margin of the WCT is the derivation of clasts in conglomerate from both the Wrangellia terrane and from continental sources (Berg and others, 1972; Richter, 1976).

TERRANES WITHIN KAHILTNA ASSEMBLAGE

Tectonically intermixed with the Kahiltna assemblage in the area south of the Denali fault are several small terranes, including the Chulitna, Susitna, West Fork, Broad Pass, Maclaren, and

Clearwater terranes (Fig. 2) (Jones and Silberling, 1979; Silberling and others, this volume).

Chulitna terrane

The Chulitna terrane (Nichols and Silberling, 1979; Jones and Silberling, 1979; Jones and others, 1980) occurs in the central Alaska Range (Figs. 1, 2, and 3). The terrane contains a number of thrust slivers, folded into a large southeastward-overturned syncline, that are tectonically underlain by flysch. A wide variety of rock sequences of different ages and depositional environments occurs in the terrane, which consists mainly of (1) a tectonically dismembered Late Devonian ophiolite composed of serpentinite, gabbro, pillow basalt, and red radiolarian chert; (2) Pennsylvanian chert and Permian limestone, argillite, and

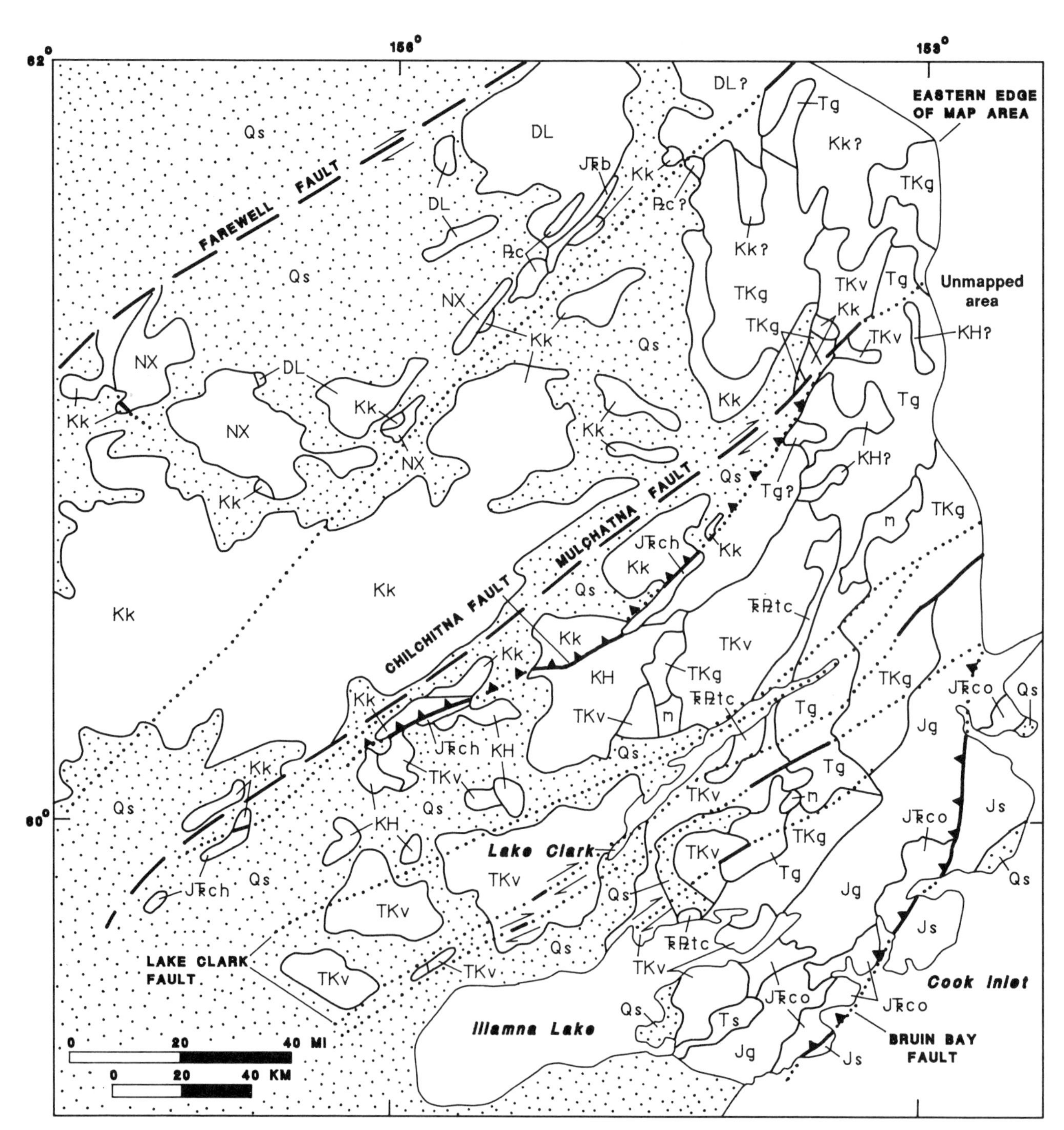

Figure 5. Generalized bedrock geologic map of the southern Kahiltna terrane in the Lake Clark region. Refer to text for description of units. Adapted and revised from Detterman and Reed (1980), Nelson and others (1983), Wallace and others (1989), and Beikman (this volume).

volcaniclastic rocks; (3) Early Triassic limestone; and (4) Late Triassic redbeds with minor interbedded limestone. Depositional contacts and/or reworked clastic detritus from underlying rocks indicate a stratigraphic continuity for this diverse package. For example, abundant Devonian ophiolite clasts occur with polycrystalline quartz pebbles in the Late Triassic redbeds. The ophiolitic clasts consist of serpentinite, basalt, and red radiolarian chert that contains the same distinctive radiolarians that are found in the main belt of Devonian ophiolitic rocks. Likewise, clasts of Mississippian chert occur in the basal Permian conglomerate, and clasts of Permian limestone are found in the Triassic redbeds. None of these sequences is known to occur elsewhere in Alaska. The terrane is highly folded and thrust faulted, but displays an internally coherent stratigraphy.

The geologic history of the Chulitna terrane commenced with formation of oceanic crust in Late Devonian time (Jones and others, 1980). Deposition of pelagic chert in an oceanic setting continued into the Mississippian; in late Paleozoic time, coarse volcanic conglomerates and flows covered the older cherts and incorporated ripped-up chert clasts in their basal beds. These late Paleozoic rocks are heterogeneous in character, and their internal stratigraphic relations are poorly known, mainly because of complex structure and lack of fossils throughout the section. The distinctive Triassic redbeds overlie ammonite-bearing cherty limestone of Early Triassic age (Nichols and Silberling, 1979). Fossils from this limestone show strong affinities with faunas from California, Nevada, and Idaho, and differ from assemblages known farther north in Canada. This relation implies a lower latitude position for the Chulitna terrane in the Early Triassic. The Late Triassic redbeds formed along a continental margin, as indicated by abundant polycrystalline quartz in some conglomeratic beds; however, no comparable sequence has been located along the western margin of North America that would indicate the position of the Chulitna terrane in Late Triassic time.

Susitna terrane

The Susitna terrane (Jones and others, 1980, 1982) occurs in the northern part of the Talkeetna Mountains (Figs. 1, 2, and 3) and consists of thick piles of pillow basalt, deep-marine tuffaceous sedimentary rocks, sandstone, and tuff. The fossils *Monotis subcircularis* and *Heterastridium* sp. are locally abundant in argilite interbedded with the volcanic rocks and indicate a Late Triassic (Norian) age. The Susitna terrane is a rootless nappe engulfed in the highly deformed Mesozoic flysch of the Kahiltna assemblage. The upper contact of basalt with flysch originally may have been depositional, but relations are now obscured by subsequent shearing along the contact. The Susitna terrane is herein interpreted as a possible fragment of an oceanic seamount and/or a fragment of the Peninsular terrane, possibly equivalent to either the Cottonwood Bay Greenstone or the Chilikadrotna Greenstone, that was tectonically decoupled from its basement and faulted into the flysch of the Kahiltna assemblage during the middle or Late Cretaceous. This interpretation would require at least moderate tectonic displacement from the core of the Peninsular terrane.

EXPLANATION

CENOZOIC AND LATE CRETACEOUS SEDIMENTARY ROCKS

Qs — Quaternary surficial deposits

Ts — Early Tertiary sedimentary rocks

Kk — Late Cretaceous Kuskokwim Group

CENOZOIC AND LATE CRETACEOUS GRANITOID ROCKS

Tg — Early Tertiary granitoid rocks

TKv — Early Tertiary and Late Cretaceous volcanic rocks

TKg — Early Tertiary and Late Cretaceous granitoid rocks

EARLY CRETACEOUS AND LATE JURASSIC SEDIMENTARY ROCKS

KH — Kahiltna assemblage

TECTONO-STRATIGRAPHIC TERRANES

Mystic terrane

JTRb — Pillow basalt, minor clastic rocks

Pzc — Clastic rocks, carbonate rocks, and chert

Dillinger terrane

DL

Nixon Fork terrane

NX

Peninsular terrane

Js — Sedimentary rocks, Tuxedni Group, Chinitna Formation, and Naknek Formation

Jg — Middle Jurassic granitic rocks

JTRch — Chilikadrotna Greenstone

JTRco — Cottonwood Bay Greenstone, Kamishak Formation, and Talkeetna Formation

TRPztc — Tlikakila complex. Mainly metamorphosed and tectonically disrupted sedimentary rocks and mafic and ultramafic rocks of uncertain age and local Triassic limestone

m — Metamorphosed rocks in roof pendants

SYMBOLS

Contact

Fault—Dotted where concealed

Thrust fault—Dashed where approximately located, dotted where concealed

Strike-slip fault—Dashed where approximately located, dotted where concealed

West Fork and Broad Pass terranes

The West Fork and Broad Pass terranes occur in the central Alaska Range structurally below the Chulitna terrane along its southeastern margin (Figs. 1, 2, and 3). The West Fork terrane consists of two separate lithologic units (Jones and others, 1982; Csejtey and others, 1986). The upper unit is chert, argillite, and sandstone ranging in age from Early to Late Jurassic. The lower unit is crystal tuff and volcaniclastic sandstone and argillite. The lower unit is mainly undated, but one occurrence of fossiliferous sandy conglomerate of Early Jurassic age is known. The West Fork terrane is interpreted as a fragment of a turbidite basin formed adjacent to the volcanic part of a Jurassic island arc (Jones and others, 1982).

The Broad Pass terrane (Jones and others, 1982; Csejtey and others, 1986) occurs in the central Alaska Range and is a poorly exposed structural melange of (1) chert, argillite, tuffaceous rocks, phyllite, conglomerate, and graywacke with Late Devonian to middle Mississippian radiolarians in chert; (2) blocks of Silurian and Devonian limestone; and (3) very minor serpentinite. No internal stratigraphy is known, and the basic structure is chaotic. The terrane extends northeastward beyond the known limits of both the Chulitna and West Fork terranes and terminates against the Denali fault (Fig. 2). The terrane is interpreted to be a structural mixture of diverse sedimentary rocks.

Maclaren terrane

The Maclaren terrane occurs in the central and eastern Alaska Range north of the Broxson Gulch thrust and south of the Denali fault (Figs. 2, 3, and 4) (Nokleberg and others, 1982, 1985, 1992b). The terrane consists of the Maclaren Glacier metamorphic belt to the south and the East Susitna batholith to the north. Included with the Maclaren terrane is the Nenana terrane of Jones and others (1984, 1987). Unless otherwise noted, the following description of the Maclaren terrane is after Smith and Lanphere (1971), Turner and Smith (1974), Smith (1981), Nokleberg and others (1982, 1985, 1989a, 1992a, 1992b), and Csejtey and others (1986).

Maclaren Glacier metamorphic belt. The Maclaren Glacier metamorphic belt is an inverted, prograde, Barrovian-type metamorphic belt (Smith, 1981; Nokleberg and others, 1985; Dusel-Bacon, 1991; this volume, Chapter 15; Dusel-Bacon and others, 1994) (Figs. 3 and 4). From south to north and structurally from bottom to top, the major fault-bounded units are pre–Late Jurassic argillite and metagraywacke, phyllite, and schist and amphibolite units. The argillite and metagraywacke unit consists predominantly of volcanic graywacke and siltstone, sparse andesite and basalt, and lesser calcareous and quartz siltstone metamorphosed at lower greenschist facies. The phyllite unit consists of interfoliated phyllite, metagraywacke, meta-andesite, and sparse marble metamorphosed at upper greenschist facies. The schist and amphibolite unit consists of interfoliated garnet amphibolite, garnet schist, amphibolite, mica schist, and calc-schist metamorphosed at amphibolite facies. The minimum structural thickness for the belt in the western Mount Hayes quadrangle is estimated to be several thousand meters.

A flysch protolith is interpreted for the argillite and metagraywacke unit. The only nearby data for the age of the argillite and metagraywacke unit are a 146 Ma K-Ar hornblende age and a 133 Ma K-Ar biotite age from an alkali gabbro stock that intrudes the argillite in the central Alaska Range. The alkali gabbro stock exhibits the same degree of lower greenschist facies metamorphism as the enclosing argillite and metagraywacke unit, indicating that both were metamorphosed together. This relation and the discordant ages for hornblende and biotite indicate that this part of the argillite and metagraywacke is Late Jurassic or older in age. This age, and a flysch protolith for the Maclaren Glacier metamorphic belt, suggests a correlation with the flysch of the Kahiltna assemblage to the west, which is faulted against the Maclaren terrane.

A general increase in metamorphic grade occurs from the lower greenschist facies argillite and metagraywacke unit, the lowest structural unit to the south, to the amphibolite facies schist and amphibolite unit, the highest structural unit to the north. The highest grade part of the metamorphic belt occurs in the Healy quadrangle to the west and is defined by the occurrence of sillimanite and kyanite. A Late Cretaceous and early Tertiary age of regional metamorphism for the schist and amphibolite unit of the Maclaren Glacier metamorphic belt is indicated by K-Ar hornblende and biotite ages ranging from 64.1 to 28.5 Ma.

East Susitna batholith. The East Susitna batholith (Fig. 4) consists of a suite of regionally metamorphosed and deformed, small to large plutons of gabbro, quartz diorite, granodiorite, and lesser quartz monzonite (Nokleberg and others, 1982, 1985, 1992b). Locally abundant migmatite, migmatitic schist, and schist and amphibolite, derived from older gabbro and diorite, also occur in the batholith. The schistose granitic rocks locally grade over a distance of a few centimeters to meters into migmatite, migmatitic schist, and schist and amphibolite.

The East Susitna batholith is ductilely deformed into mylonitic gneiss and schist, and regionally metamorphosed at middle amphibolite facies; there is local retrograde metamorphism to lower greenschist facies (Nokleberg and others, 1985). The batholith is separated from the Maclaren Glacier metamorphic belt by the Meteor Peak fault (Fig. 4). Isotopic studies indicate syntectonic intrusion of the East Susitna batholith from the middle Cretaceous through the early Tertiary. A sample of schistose quartz diorite from the batholith yields a U-Pb zircon age of 70 Ma, and various, abundnt K-Ar ages for metamorphic biotite and hornblende range from 87.4 to 29.8 Ma.

Coeval metamorphism of the East Susitna batholith and the Maclaren Glacier metamorphic belt (Nokleberg and others, 1985, 1989a) is indicated by (1) a regional relation of prograde units to the north, toward the East Susitna batholith; (2) a parallel fabric formed at amphibolite facies between the higher grade parts of the belt and the batholith; and (3) coeval K-Ar and U-Pb sphene isotopic ages for metamorphic minerals in both units. The increase of metamorphic grade in the Maclaren Glacier metamorphic belt toward the East Susitna batholith is interpreted as metamorphism that occurred in response to emplacement of the igneous magmas forming the batholith structurally above the metamorphic belt, thereby creating the inverted metamorphic sequence. A similar and subsequent interpretation is that the Maclaren terrane is an exhumed part of a deep crustal shear zone, where hot upper amphibolite facies rocks of the East Susitna batholith were emplaced over cooler lower grade rocks of the Maclaren Glacier metamorphic belt (Davidson and others, 1992). These relations are very similar to those observed for correlative units in the Gravina-Nutzotin belt and adjacent terranes, west of the foliated tonalite sill in the informally named Coast plutonic-metamorphic complex of Brew and Ford (1984) (Crawford and others, 1987; Gehrels and others, 1991).

The East Susitna batholith and companion Maclaren Glacier metamorphic belt are interpreted as having formed in a continental margin arc setting on the basis of (1) being a composite batholith intruding a quartz-rich flysch sequence; and (2) isotopic analysis of feldspar lead from samples of the East Susitna batholith, indicating derivation from a cratonal source of about 1.2 Ga (Aleinikoff and others, 1987). The Maclaren terrane is truncated to zero thickness where the Broxson Gulch thrust to the south abuts against the Denali fault to the north (Figs. 2 and 4). These relations strongly suggest that the Maclaren terrane is a displaced fragment, and that the rest of the Maclaren terrane may occur on the opposite side of the Denali fault (Forbes and others, 1973; Smith and others, 1974; Eisbacher, 1976; Stout and Chase, 1980; Smith, 1981; Nokleberg and others, 1985). The nearest rocks correlative with the Maclaren terrane are about 400 km away to the southeast in the Ruby Range batholith and associated metamorphic rocks of the Kluane Lake–Ruby Range area in the southwestern Yukon Territory, part of the Coast plutonic-metamorphic complex, Gravina-Nutzotin belt, and adjacent terranes.

Clearwater terranes

The Clearwater terrane (Jones and others, 1984) occurs in the eastern Alaska Range as a narrow, fault-bounded lens along the Broxson Gulch thrust (Figs. 1, 2, and 3). The terrane is a structurally complex assemblage of argillite, greenstone (metapillow basalt), shallow-water limestone and marble containing Late Triassic (late Norian) fossils, and sparse metarhyolite (Fig. 3) (Nokleberg and others, 1985, 1992b; Csejtey and others, 1986). Each lithology is fault bounded. The terrane is weakly metamorphosed and penetratively deformed at lower greenschist facies and locally contains a single intense schistosity (Nokleberg and others, 1982, 1985, 1992b; Dusel-Bacon and others, 1993). The Clearwater terrane is unique because it contains Late Triassic sedimentary rocks and presumably coeval basalt and rhyolite. Because it contains a bimodal suite of greenstone or metabasalt and metarhyolite, shallow-marine sedimentary rocks, and possibly a coeval granitic pluton, the Clearwater terrane is herein interpreted as a fragment of an island arc.

WRANGELLIA TERRANE

The Wrangellia terrane (Jones and others, 1977, 1981, 1987; Nokleberg and others, 1982, 1985, 1992b; Plafker and others, 1989c) occurs in the northern and eastern parts of the study area in the Gulkana, Healy, McCarthy, Mount Hayes, Nabesna, and Talkeetna Mountains quadrangles (Figs. 1 and 2). The terrane consists of several sequences (Richter, 1976; MacKevett, 1978; Winkler and others, 1981c; Nokleberg and others, 1982, 1985, 1992b; Plafker and others, 1989c): (1) a pre–late Paleozoic assemblage of metasedimentary and metagranitic rocks; (2) Pennsylvanian and Permian marine volcanic rocks (Tetelna volcanics); (3) interlayered Pennsylvanian and Early Permian marine volcanic and sedimentary rocks (Slana Spur and Station Creek formations); (4) Permian nonvolcanogenic limestone and argillite (Eagle Creek and Hasen Creek formations); (5) sparse hypabyssal to deep-seated late Paleozoic plutonic rocks; (6) a thick sequence of Late Triassic submarine and subaerial basalt (Nikolai Greenstone) and associated mafic and ultramafic intrusive rocks; (7) disconformably overlying Late Triassic and Early Jurassic shallow- and deep-water calcareous sedimentary rocks (Nizina and Chitistone limestones, and McCarthy and Lubbe Creek formations); (8) unconformably overlying Middle Jurassic through Early Cretaceous volcaniclastic and clastic rocks (Root Glacier and Nizina Mountain formations, Kotsina Conglomerate, and Berg Creek and Chisana formations) and flysch (Gravina-Nutzotin belt); and (9) younger Cretaceous marine basin deposits (Chititu, Moonshine Creek, Schulze, and Kennicott formations). Overlapping units are the middle Tertiary through Holocene subaerial volcanic and associated clastic rocks of the Wrangell Mountains volcanic field and coeval hypabyssal intrusive rocks of the Wrangell continental margin arc (Miller and Richter, this volume). In contrast to the McCarthy quadrangle to the southeast, the Late Jurassic through Cretaceous strata in the Nabesna and and Mount Hayes quadrangles to the northwest consist mainly of the Gravina-Nutzotin belt and the Chisana Formation (Richter, 1976; Nokleberg and others, 1982, 1992b). Unless noted, the following descriptions of the stratified rocks of the Wrangellia terrane in the eastern Alaska Range and Nutzotin Mountains in the Mount Hayes and Nabesna quadrangles are from Richter and Dutro (1975), Richter (1976), Richter and others (1977), and Nokleberg and others (1982, 1985, 1992b); descriptions from the Wrangell Mountains in the Nabesna quadrangle are from MacKevett (1969, 1971, 1978) and Smith and MacKevett (1970).

Pre–late Paleozoic units of Wrangellia terrane

In the Wrangellia terrane in the Nutzotin Mountains are roof pendants of pre–middle Pennsylvanian age. These roof pendants occur in the late Paleozoic granitic plutons of the Skolai arc and constitute a cataclasite unit (Richter, 1976) of mica and quartz schist, derived from clastic sedimentary rocks, and relatively younger, gneissose metagranitic rocks (Fig. 3) (W. J. Nokleberg and D. H. Richter, 1986, unpublished data). No fossil or isotopic data exist for the ages of the sedimentary or granitic rocks. Both parts of the cataclasite unit contain a regional upper greenschist facies schistose fabric that is also relatively older than the late Paleozoic plutons. This metasedimentary and metagranitic basement might be part of the mainly early and middle Paleozoic Alexander terrane which, to the southeast in the eastern Wrangell Mountains, is stitched to the Wrangellia terrane by a Pennsylvanian granitic pluton that intrudes the contact between the Alexander and Wrangellia terranes (see discussion below, and Gardner and others, 1988).

The basement of the Wrangellia terrane may have a variable nature. In some areas, an island-arc origin is indicated for the late Paleozoic Skolai arc by isotopic and chemical data (Aleinikoff and others, 1987; Barker, this volume), whereas in other areas, pre–late Paleozoic arc rocks occur either in roof pendants in the late Paleozoic plutonic rocks or, in the case of the northwestern Alexander terrane, are welded to the Wrangellia terrane by the late Paleozoic plutonic rocks. Analysis of seismic refraction data for the central part of the Wrangellia terrane in the Copper River Basin in the Gulkana quadrangle (Fig. 1) indicates a 35-km-thick, intermediate composition crust (Goodwin and others, 1989; Fuis and Plafker, 1991). Comparison of laboratory measurements of compression seismic velocities with modeled velocities indicates a seismically homogeneous middle crust and a thick and possibly more heterogeneous lower crust that together are likely composed of intermediate composition igneous rock and/or quartz-mica schist. Alternative interpretations for formation of the thick, intermediate composition crust are (Goodwin and others, 1989) (1) internal tectonic imbrication during accretion; (2) magmatic underplating after accretion of the Wrangellia terrane, possibly during formation of the late Cenozoic Wrangell Mountains volcanic field; or (3) buildup of at least the eastern part of the Wrangellia terrane on intermediate composition crust, possibly the Alexander terrane.

Late Paleozoic units of Wrangellia terrane

Tetelna volcanics, Slana Spur Formation, and Eagle Creek Formation. In the eastern Alaska Range and Nutzotin Mountains in the Mount Hayes and Nabesna quadrangles, the late Paleozoic stratified rocks of the Wrangellia terrane are the Tetelna volcanics, the Slana Spur Formation, and the Eagle Creek Formation (Figs. 3 and 4). The mainly Pennsylvanian or older Tetelna volcanics (Mendenhall, 1905; Richter, 1976; Richter and others, 1977; Nokleberg and others, 1982, 1992b) are dominantly andesite and lesser basalt flows, mud and debris avalanche deposits, and tuffs interbedded with fine- to coarse-grained volcaniclastic rocks that are more than 1,000 m thick. In one area in the Nabesna quadrangle (Fig. 1), Permian fossils and megafossils occur in part of the Tetelna volcanics (Richter, 1976). The base of the Tetelna volcanics is either not exposed or faulted. Conformably overlying the Tetelna volcanics is the middle Pennsylvanian to Early Permian Slana Spur Formation, which is mainly a thick sequence of marine calcareous and noncalcareous volcaniclastic rocks, including volcanic graywacke, volcanic breccia, and intermediate composition tuff, with subordinate limestone and argillite that are about 1,000 m thick. Conformably overlying the Slana Spur Formation is the Early Permian Eagle Creek Formation, which is mainly alternating units of shallow-marine argillite and limestone about 900 m thick.

Locally extensive hypabyssal dacite stocks, sills, and dikes and several large granitic plutons, including the Ahtell Creek pluton and a diorite complex (Richter and others, 1975; Barker and Stern, 1986; Beard and Barker, 1989; Barker, this volume), intrude only the Slana Spur Formation and the Tetelna volcanics. The hypabyssal and plutonic rocks are interpreted as being comagmatic with the volcanic rocks of the Slana Spur Formation, and together with this unit constitute part of the Skolai arc. U-Pb zircon isotopic analyses of the granitic rocks yield Pennsylvanian ages of 290 to 320 Ma.

Station Creek and Hasen Creek formations. In the Wrangell Mountains in the McCarthy quadrangle, the late Paleozoic stratified rocks of the Wrangellia terrane are the Station Creek and Hasen Creek formations (Figs. 2 and 3). The Pennsylvanian and Permian Station Creek Formation consists of altered andesite, basalt, volcanic breccia,and graywacke that is approximately 2,000 m thick. The base of the unit is either faulted or not exposed. The lower volcanic member of this unit is lithologically correlated with the Tetelna volcanics, and the upper volcaniclastic member is lithologically correlated with the Slana Spur Formation in the eastern Alaska Range. The Station Creek Formation, along with local crosscutting granitoids, constitutes the Skolai arc in the Wrangell Mountains (Richter and others, 1975; Barker and Stern, 1986; Beard and Barker, 1989; Barker, this volume; Miller, this volume, Chapter 16).

The Early Permian Hasen Creek Formation also occurs in the McCarthy quadrangle and is mainly fossiliferous limestone and argillite and lesser sandstone, chert, and conglomerate that is ~200 m thick. The unit is correlated with the Eagle Creek Formation in the eastern Alaska Range.

Mesozoic units of Wrangellia terrane

Nikolai Greenstone. In the eastern Alaska Range and Nutzotin Mountains, the Late Triassic Nikolai Greenstone, first described by Rohn (1900), consists mainly of massive, subaerial, amygdaloidal basalt flows, lesser pillow-basalt flows, and thin beds of argillite, chert, and mafic volcaniclastic rocks that are up to 4,350 m thick. The flows are predominantly intermixed aa and pahoehoe; individual units range from 5 cm to more than 15 m thick. Locally extensive gabbro dikes and cumulate mafic and ultramafic sills intrude the Nikolai Greenstone and older rocks in the subterrane; these dikes and sills are probably comagmatic with the basalt that formed the Nikolai Greenstone. Locally underlying the Nikolai is a unit of Permian through Middle Triassic shale, limestone, and chert that is up to 600 m thick (Silberling and others, 1981).

In the Wrangell Mountains, the Late Triassic Nikolai Greenstone is mainly altered, amygdaloidal, tholeiitic basalt ~3,000 m thick. Pillow basalt and argillite occur locally near the base. The unit occurs in intermixed pahoehoe and aa flows in units between 15 cm and 15 m thick. The basal part of the Nilolai is generally a volcanic conglomerate as much as 70 m thick.

The Nikolai Greenstone consists throughout of clinopyroxene and former calcic plagioclase. The unit is pervasively metamorphosed to a granoblastic suite of lower greenschist facies minerals, mainly chlorite, epidote, albite, zeolite, and prehnite. Amygdules are generally filled by calcite, chlorite, quartz, and epidote; some amygdules contain zeolites, prehnite, native copper, or Cu-sulfides.

Chitistone and Nizina limestones, McCarthy Formation, and Lubbe Creek Formation. The Late Triassic Chitistone and Mizina limestones occur mainly in the McCarthy quadrangle and consist of limestone, dolomite, algal-mat chips, and lesser chert that are ~1,100 m thick. In addition, the Nizina Limestone contains sparse tuffaceous detritus, mainly plagioclase and pyroxene crystals, and andesite or basalt fragments M. Mullen, 1991, written commun.). Similar unnamed limestone units occur stratigraphically above the Nikolai Greenstone in the Nutzotin Mountains and eastern Alaska Range. These dolomites and limestones are interpreted to have formed in supratidal settings; some beds have diagenetic features characteristic of sabkha facies (Armstrong and MacKevett, 1982).

The McCarthy and Lubbe Creek formations occur mainly in the McCarthy quadrangle and are primarily impure chert and limestone and radiolarian-rich siliceous shale that are as thick as 300 m. As with the Nizina Limestone, the McCarthy Formation contains sparse tuffaceous detritus (M. Mullen, 1991, written commun.). These deposits are interpreted as having formed in an open marine environment that gradually evolved upward to basinal limestone and siltstone (McCarthy Formation of Late Triassic and Early Jurassic ages) that was rich in pelagic mollusks and siliceous organisms. The tuffaceous material suggests proximity of an active volcanic arc in the Late Triassic and Early Jurassic, possibly the Talkeetna arc of the Peninsular terrane.

Nizina Mountain Formation, Kotsina Conglomerate, and Root Glacier, Berg Creek, Chisana, Kennecott, Chititu, Moonshine Creek, Schulze, and MacColl Ridge formations. The Jurassic and Cretaceous Nizina Mountain Formation, Kotsina Conglomerate, and Berg Creek, Chititu, Lubbe Creek, Kennecott, MacColl Ridge, Moonshine Creek, Root Glacier, and Schulze formations occur mainly in the McCarthy quadrangle and are primarily shallow-marine clastic rocks that compose a section more than 2,000 m thick. The dominant lithologies in each unit are spiculite and minor coquina (Lubbe Creek); graywacke (Nizina Mountain); conglomerate (Kotsina); siltstone, sandstone, shale, and conglomerate (Root Glacier); conglomerate, sandstone, siltstone, and calcarenite (Berg Creek); andesite flows, tuffs, and volcanic flysch (Chisina); sandstone, siltstone, and minor conglomerate (Kennecott); mudstone, shale, and subordinate porcellanite (Chititu); siltstone, sandstone, and minor conglomerate (Moonshine Creek); porcellanite with minor sandstone and conglomerate (Schulze); and coarse sandstone and minor conglomerate (MacColl Ridge). The Early Jurassic strata are basinal organic-rich siliceous limestone, cherty argillite, and shale, whereas the Middle and Late Jurassic strata are dominantly fine- to medium-grained clastic rocks; conglomerate is locally abundant in the upper part of the section. Locally abundant fossils throughout this sequence indicate deposition in water depths ranging from moderately deep (about 500–1,000 m) to shelfal (about 100–200 m). These Middle Jurassic to Late Cretaceous basinal strata contain many interformational disconformities and local intraformational hiatuses, and they vary markedly in thickness and distribution. An episode of strong north-verging folding and faulting occurred after deposition of the Late Jurassic strata; Early Cretaceous strata were deposited unconformably on these deformed strata.

Late Paleozoic and Mesozoic plutons and igneous arcs, and regional metamorphism of the Wrangellia terrane

The Wrangellia terrane contains remnants of four epochs of igneous activity: the late Paleozoic Skolai arc, Late Triassic mafic magmatism, the Late Jurassic Chitina arc, and the Early Cretaceous Chisana arc and Late Cretaceous Kluane arc. In addition, a period of middle Cretaceous regional metamorphism occurred throughout most of the terrane and the superjacent Gravina-Nutzotin belt.

Late Paleozoic Skolai arc. The Skolai arc forms a lithologically variable belt that is discontinuously exposed in the Wrangellia and Alexander terranes (Fig. 2), and in adjacent parts of Canada to the east. In the study area, the plutonic part of this belt consists of compositionally diverse early to middle Pennsylvanian plutons (unit lPzg, Fig. 2) that are mainly (1) the Ahtell pluton (diorite-gabbro-tonalite-anorthosite suite and diorite complex) in the eastern Alaska Range (Richter and others, 1975) with U-Pb ages of 290 to 316 Ma (Barker and Stern, 1986); (2) metagranite and metagranodiorite with U-Pb zircon ages of 308 to 310 Ma (Aleinikoff and others, 1988) in the Uranatina River metaplutonic unit of the southern Wrangellia terrane, and correlative units in the Chitina Valley, Chugach Mountains, and western Wrangell Mountains (Plafker and others, 1989c); and (3) the Bernard Glacier pluton with a U-Pb zircon age of 308 Ma (Gardner and others, 1988) intruding both the Wrangellia and Alexander terranes in the southeastern Wrangell Mountains. Possibly correlative rocks to the east in Canada include the alkalic Iceland Ranges plutonic suite of Pennsylvanian age in adjacent parts of Canada (Campbell and Dodds, 1982a, 1982b; Dodds and Campbell, 1988). This arc constitutes the lowermost widespread unit of the Wrangellia terrane. A south-facing arc is suggested by the increase in carbonate rock and chert contents in bedded sequences from the Wrangell Mountains southward across the Chitina Valley.

The origin of the Skolai arc was first discussed in two early studies. (1) In the Nabesna quadrangle, Richter and Jones (1973) interpreted the late Paleozoic volcanic and sedimentary rocks of the Wrangellia terrane as forming in an island arc developed on oceanic crust; and (2) to the west in the Mount Hayes quadrangle, Bond (1973, 1976) interpreted the late Paleozoic volcanic and sedimentary rocks of the Wrangellia terrane as forming in a marine volcanic chain on continental crust above a subduction zone. The main pieces of evidence for a marine origin for the Skolai arc are (1) submarine deposition of the volcanic flows, tuff, and breccia, and associated volcanic graywacke and argillite; and (2) locally abundant features that indicate deposition of sedimentary and volcanic debris from turbidity currents to form volcanic graywacke. The main supporting observations for an (oceanic) island-arc origin are (1) the absence of abundant continental crustal detritus in late Paleozoic stratified rocks; (2) little or no quartz in the volcanic rocks and associated shallow intrusive bodies; (3) common lead isotopic compositions for late Paleozoic granitic rocks, indicating low radiogenic lead values and derivation from a mixture of oceanic mantle and pelagic sediment leads, without an older continental component (Aleinikoff and others, 1987); and (4) Rb-Sr isotopic data and REE volcanic and plutonic whole-rock chemical analyses and petrologic data indicating an intra-oceanic island-arc origin (Barker and Stern, 1986; Beard and Barker, 1989; Arth, this volume; Barker, this volume; Miller, this volume, Chapter 16).

Late Triassic mafic magmatism and formation of Nikolai Greenstone and related rocks. A voluminous but short-lived period of mafic magmatism occurred in the Late Triassic and resulted in eruption of the widespread basalt of the Nikolai Greenstone and correlative units, such as the Karmutsen Formation in British Columbia, during a 7–8 m.y. interval (Richter and Jones, 1973; MacKevett, 1978; Jones and others, 1977; Silberman and others, 1981; Winkler and others, 1981a; Nokleberg and others, 1985; Plafker and others, 1989c). Coeval with the eruption was intrusion of gabbro and diabase dikes and sills, and of cumulate mafic and ultramafic sills.

The mafic magmatism resulting in the Nikolai Greenstone was first interpreted as forming in a rift setting (Jones and others, 1977; Nokleberg and others, 1985; Barker and others, 1989; Barker, this volume). Evidence cited for a rift setting is (1) a vast extent, occurrence in a linear belt, and a great thickness throughout the Wrangellia terrane; (2) a relatively constant igneous texture, petrology, and average chemical composition, approximating a typical high-Al tholeiite; and (3) REE analyses compatible with (back-arc) spreading.

The mafic magmatism resulting in the Nikolai Greenstone has been interpreted as forming in a mantle plume setting (Richards and others, 1991). Evidence cited for a plume origin consists of (1) the lack of any recognized sheeted dikes, rift facies, and rift structures; and (2) no indication of large amounts of crustal extension and graben formation usually associated with normal faulting. According to this interpretation, a buoyant rise of a large plume head would cause rapid dynamic uplift that preceded volcanism. Partial melting of the plume beneath oceanic lithosphere would result in an enormous volume of basalt being erupted quickly throughout the area centered on the head of the plume. The basaltic eruption would be a much shorter period compared with known rifted arcs. A rapid change would occur from deep water to shallow, near sea-level conditions immediately before basalt eruption.

Several features of the observed geology fit the plume model (Richards and others, 1991). In the submerged parts of the Skolai arc, thermal expansion and uplift resulted in surfacing of the Wrangellia terrane and formation of a vast thickness of marine and subaerial basalt with Nd and Sr isotopic compositions characteristic of oceanic plume basalt, up to 6,000 m thick, within about 7 to 8 m.y. in the Late Triassic. This upwelling also explains the lack of rift facies beneath the Nikolai Greenstone, as well as the absence of linear dike swarms that generally occur in rifted crust. Instead, widespread emplacement of gabbroic sills and cumulate mafic and ultramafic rocks occurred. Afterward, cooling and thermal subsidence controlled the post-basalt sedimentation in the stratigraphic sequence overlying the Nikolai, with a change from shallow-water carbonate facies of the Chitistone Limestone to the basinal deposits of the McCarthy Formation. This change may reflect thermal contraction induced by cooling of the underlying basaltic pile.

However, some problems exist with the plume origin. First, the Wrangellia terrane differs from modern oceanic plateaus in that the basalt of the Late Triassic Nilolai Greenstone was built on an inactive, late Paleozoic (Skolai) island arc and was locally succeeded by Early Jurassic arc volcanism that occurred no more than 10–15 m.y. later. Modern-day plateaus are not similar. Second, a serious problem may exist with the size and shape of the Nikolai Greenstone and correlative units in the Wrangellia terrane versus the size and shape of mantle plumes. The Wrangellia terrane is about 2,500 km long and less than 200 km wide. In contrast, plumes tend to be circular, and a very large one is only about 1,000 km in diameter (Richards and others, 1991). Additional studies are needed to resolve the origin of the basalts of the Nikolai Greenstone and coeval mafic and ultramafic plutonic rocks.

Late Jurassic Chitina arc. Late Jurassic granitic plutons of the Chitina arc (unit lJg, Fig. 2) intrude the older rocks of the Wrangellia and southern Wrangellia terranes in the southern Wrangell Mountains and Chitina Valley (Hudson, 1983; Miller, this volume, Chapter 16). The granitic rocks are mainly strongly foliated tonalite and granodiorite, and lesser granite and quartz diorite (MacKevett, 1978). The plutons extend from the southwestern Wrangell Mountains southeastward to Chichagof Island in southeastern Alaska and are interpreted as the roots of the Chitina arc (Plafker and others, 1989c).

The Chitina arc is interpreted as an island arc that formed during the Late Jurassic stage of subduction of the McHugh Complex and correlative units of the Chugach terrane to the south along the southern margin of the WCT (Plafker and others, 1989c). North to northeast subduction for the Chitina arc and the Chisana arc (described below) is inferred from the presence of coeval volcanic detritus of Late Jurassic to Early Cretaceous age in the adjacent McHugh Complex of the Chugach terrane (Plafker and others, 1989c). The axis of the Chitina arc is parallel to, and about 100 km south of, the Early Cretaceous Chisana arc.

The Late Jurassic plutons in the Chitina Valley were intensely penetratively deformed during a major regional orogeny that began in the Late Jurassic (MacKevett, 1978). In addition, the position of the Jurassic plutons along the southern margins of the Wrangellia terrane and the southern Wrangellia terrane, virtually at the Border Ranges fault system, requires that a substantial segment of the southern Wrangellia terrane margin has been tectonically removed since the Jurassic (MacKevett, 1978; Plafker and others, 1989c). Sinstral displacement may have offset the missing segment of the terrane margin that may be the part of the Wrangellia terrane in British Columbia (Plafker and others, 1989c).

Early Cretaceous Chisana arc and Late Cretaceous Kluane arc. The Chisana arc consists of a belt of Early Cretaceous volcanic and plutonic rocks in the Wrangell and Nutzotin Mountains and along the southern flank of the eastern Alaska Range (Barker, 1987; Plafker and others, 1989c). In the eastern Alaska Range, the volcanic unit is the Chisana Formation that overlies the Gravina-Nutzotin belt. Local andesite flows and volcaniclastic rocks occur within the Gravina-Nutzotin belt (Nokleberg and others, 1982, 1992b). In the Wrangell Mountains in the McCarthy quadrangle, the volcanic-detritus–rich unit is the Berg Creek Formation. These and subjacent units in the Wrangellia terrane are also intruded by coeval Early Cretaceous granitic plutons (older, Cretaceous part of unit TKg, Fig. 2). The Chisana arc is interpreted as an island arc that formed on the leading edge of the WCT during the Early Cretaceous stage of subduction of the McHugh Complex and correlative units of the Chugach terrane to the south along the southern margin of the WCT (Plafker and others, 1989s).

The Kluane arc consists of a widely spaced belt of Late Cretaceous granitic plutons that occurs mainly in the Gravina-Nutzotin belt and correlative units in Canada in the Wrangellia and Maclaren terranes (older, Late Cretaceous part of unit TKg, Fig. 2) (Plafker and others, 1989c). The younger Maastrichtian volcaniclastic sedimentary rocks of the MacColl Ridge Formation in the Wrangell Mountains (Fig. 3) are interpreted as a fore-arc basin sequence that was in part coeval with the younger Late Cretaceous Kluane arc. The Kluane arc is interpreted as a continental margin arc that formed during latest Cretaceous accretion of the Valdez Group and correlative units of the Chugach terrane to the south along the southern margin of the WCT (Plafker and others, 1989c). Regional geologic data indicate that the Kluane arc formed after accretion of the WCT to North America during the middle Cretaceous (Plafker and others, 1989c; Plafker, 1990).

Succeeding the Kluane arc is the Late Cretaceous and early Tertiary part of the Alaska-Aleutian Range batholith and coeval volcanic rocks (Miller, this volume, Chapter 16; Moll-Stalcup, 1990, and this volume. This vast array of voluminous plutons and volcanic rocks occurs in a belt 700 km long and 130 km wide that extends from the northwestern part of the Alaska Peninsula through the Talkeetna Mountains to the central Alaska Range (unit TKg, Fig. 2). This plutonic belt is interpreted as having formed during the early stages of rapid northward subduction in the latest Cretaceous and early Tertiary (Miller, this volume, Chapter 16; Moll-Stalcup, 1990, and this volume).

Middle(?)-Cretaceous regional metamorphism. The Wrangellia terrane, including the Gravina-Nutzotin belt, generally exhibits weak, prehnite-pumpellyite to lower greenschist facies metamorphism (Nokleberg and others, 1985; Dusel-Bacon and others, 1994). The texture is generally granoblastic, but local incipient development of cleavage occurs. Abundant relict igneous or sedimentary minerals and textures remain. Locally asymmetric folds and companion axial plane faults accompany the regional metamorphism (Csejtey and others, 1982). The regional deformation and metamorphism is interpreted as having occurred in the middle(?) Cretaceous because (1) the structural fabric and metamorphic minerals generally occur in Early Cretaceous and older units (Dusel-Bacon and others, 1994); (2) sparse middle Cretaceous K-Ar whole-rock metamorphic ages are determined for the Wrangellia terrane (Silberman and others, 1980, 1981; Nokleberg and others, 1985, 1992a; Dusel-Bacon and others, 1994); and (3) relatively undeformed Late Cretaceous intrusive rocks of the Kluane arc, discussed above, locally intrude the highly deformed Late Jurassic and Early Cretaceous flysch and older bedrock of the Wrangellia terrane (Csejtey and others, 1982; Nokleberg and others, 1985; Plafker and others, 1989c). This regional deformation and metamorphism is interpreted as having occurred during the accretion of the WCT to the North American continental margin (Csejtey and others, 1982; Nokleberg and others, 1985; Plafker and others, 1989c).

Late Paleozoic and Mesozoic regional differences in Wrangellia terrane

Regional variations occur in the late Paleozoic and Mesozoic stratigraphy of the Wrangellia terrane. Two subterranes in the eastern Alaska Range, the Slana River and Tangle subterranes, exhibit significant differences in late Paleozoic and Late Triassic strata (Nokleberg and others, 1985). In addition, in the eastern Alaska Range and Wrangell Mountains, Late Jurassic and Early Cretaceous strata exhibit significant differences (Richter, 1976; MacKevett, 1978).

Late Paleozoic and Triassic regional differences—Slana River and Tangle subterranes. In the eastern Alaska Range, the Wrangellia terrane is divided into northern Slana River and southern Tangle subterranes that are juxtaposed along the intervening Eureka Creek fault (Fig. 4) (Nokleberg and others, 1982, 1985, 1992b). Unless noted, the following descriptions and interpretations of the two subterranes are after Richter (1976), Stout (1976), Richter and others (1977), MacKevett (1978), Nokleberg and others (1982, 1985, 1992b).

The Slana River subterrane occurs south of the Denali fault

along the southern flank of the eastern Alaska Range. The Slana River subterrane consists (relative to the Tangle subterrane) mainly of (Fig. 4) (1) a thick sequence of Pennsylvanian and Permian island-arc volcanic and associated sedimentary rocks (Tetelna volcanics, Slana Spur Formation, Eagle Creek Formation), part of the Skolai arc; (2) a thin, 1,500-m-thick sequence of disconformably overlying massive basalt flows of the Late Triassic Nikolai Greenstone and coeval gabbro and diabase dikes and sills; (3) Late Triassic limestone; and (4) Late Jurassic and Early Cretaceous flysch of the Gravina-Nutzotin belt. The Slana River subterrane is the variant of the Wrangellia terrane in the Nutzotin Mountains and Wrangell Mountains.

The Tangle subterrane occurs south of the Slana River subterrane mainly in the southern foothills of the eastern and central Alaska Range. The Tangle subterrane consists (relative to the Slana River subterrane) mainly of (Fig. 4) (1) a relatively thin, lower sequence of upper Paleozoic and Lower Triassic sedimentary and tuffaceous rocks; (2) a relatively thick, disconformably overlying section of the Nikolai Greenstone, about 4,500 m thick, that is locally intruded by extensive cumulate mafic and ultramafic rocks, and gabbro and diabase dikes and sills; and (3) locally a thin unit of Late Triassic limestone. The late Paleozoic and Early Triassic sedimentary rocks consist mostly of aquagene tuff, dark gray argillite, minor andesite tuff and flows, and sparse light gray limestone estimated to be a few hundred meters thick. Sparse late Paleozoic and Early Triassic megafossils and radiolarians occur in this sequence. The mafic volcanic and associated rocks of the Nikolai Greenstone constitute the upper part of the Amphitheater Group of Stout (1976). The flysch of the Gravina-Nutzotin belt is missing in the Tangle subterrane.

The differences in the upper Paleozoic and lower Mesozoic parts of the Tangle and Slana River subterranes indicate that the two units (1) represent distal and proximal parts, respectively, of the same late Paleozoic Skolai arc; (2) represent proximal and distal parts, respectively, of the same Late Triassic mafic magmatism system; and (3) have been considerably shortened tectonically and juxtaposed during terrane migration and accretion. A stratigraphy somewhat similar to the Tangle subterrane occurs in the Wrangellia terrane on Vancouver Island in Canada (Muller and others, 1974; Muller, 1977; Nokleberg and others, 1985). If the Tangle subterrane originally formed close to the variant of the Wrangellia terrane on Vancouver Island, then considerable tectonic dismemberment of the Wrangellia terrane has occurred since deposition of the Gravina-Nutzotin belt in the Late Jurassic and Early Cretaceous.

Late Jurassic and Cretaceous regional differences—eastern Alaska Range and Wrangell Mountains. Jurassic and Cretaceous strata also exhibit significant differences between the eastern Alaska Range and Nutzotin Mountains to the northwest (Richter, 1976; Nokleberg and others, 1982, 1992b) and the southern flank of the Wrangell Mountains to the southeast (MacKevett, 1978). Three main differences occur. (1) The eastern Alaska Range and Nutzotin Mountains do not contain Middle Jurassic strata, whereas the southern Wrangell Mountains contain the Middle Jurassic Nizina Mountain Formation and Kotsina Conglomerate. (2) The eastern Alaska Range and Nutzotin Mountains contain a thick sequence of the flysch and volcanic rocks of the Late Jurassic and Early Cretaceous Gravina-Nutzotin belt, whereas the southern Wrangell Mountains contain the progradational clastic rocks of the Berg Creek and Chisana Formations. (3) Units similar to the younger marine clastic Cretaceous rocks of the southern Wrangell Mountains (Chititu, Moonshine Creek, Schulze, and MacColl Ridge formations) do not exist in the eastern Alaska Range and Nutzotin Mountains to the north. These stratigraphic differences, which persisted from the Middle Jurassic into the Cretaceous, suggest that vertical movements oscillated between the northern and southern areas of the Wrangellia terrane, with subsidence in one area being matched by uplift in the other.

Origin of Wrangellia terrane. The allochthonous origin of the Wrangellia terrane was first recognized during paleomagnetic investigations of the Late Triassic Nikolai Greenstone, which displays shallow paleomagnetic pole inclinations indicating eruption of the basalt near the Triassic paleoequatorial (Hillhouse, 1977; Hillhouse and Grommé, 1984; Jones and others, 1977). Field relations discussed below indicate that the Wrangellia terrane was attached to the Alexander terrane in the southeastern Wrangell Mountains by the middle Pennsylvanian (Gardner and others, 1988). Paleomagnetic and stratigraphic data suggest that the Wrangellia terrane in Alaska and the Peninsular terrane may have shared a common geologic history since the Late Triassic, and possibly since the late Paleozoic.

Five major tectonic events characterize the Wrangellia terrane, according to Richter and Jones (1973), Csejtey and others (1982), Nokleberg and others (1985), and Plafker and others (1989c): (1) eruption of volcanic rocks and intrusion of coeval hypabyssal and plutonic rocks of the late Paleozoic Skolai arc (Bond, 1973, 1976; Jones and others, 1977; Nokleberg and others, 1985; Jones and Silberling, 1979; Barker, this volume; Miller, this volume, Chapter 16); (2) extrusion of the basalt of Late Triassic Nikolai Greenstone and intrusion of coeval mafic intrusive rocks in either a rift or plume environment in a near-equator setting (Jones and others, 1977; Nokleberg and others, 1985; Plafker and others, 1989c; Richards and others, 1991); (3) formation of a major Late Jurassic and Early Cretaceous flysch basin on the northern or leading edge of the WCT during migration toward the North American continental margin, and formation of the coeval Chisana arc (Nokleberg and others, 1985; Plafker and others, 1989c); (4) accretion, deformation, and low-grade regional metamorphism in the middle Cretaceous (Nokleberg and others, 1985; Plafker and others, 1989c); and (5) Cenozoic dextral-slip movement along the Denali fault and internal dismemberment.

SOUTHERN WRANGELLIA TERRANE

The southern Wrangellia terrane (Plafker and others, 1989c) occurs south of the Wrangellia terrane in the northern Copper River basin, southern Wrangell Mountains, and northern

Chugach Mountains in the Gulkana, McCarthy, Talkeetna Mountains, and Valdez quadrangles (Figs. 1 and 2). The terrane is mainly a suite of generally highly deformed and metamorphosed late Paleozoic or older sedimentary and volcanic rocks, Late Triassic basalt and limestone, and Pennsylvanian and Late Jurassic metaplutonic rocks. The southern Wrangellia terrane contains some of the most structurally and petrologically complex rocks in the study area. In the northern Copper River basin, the terrane consists of the metamorphic complex of Gulkana River and is bounded to the north by the Paxson Lake fault and to the south by the West Fork fault and Peninsular terrane (Fig. 2). In the southern Wrangell Mountains and northern Chugach Mountains, the terrane consists of the Strelna metamorphics of Plafker and others (1989c) and the Urantina River metaplutonic unit, and is bounded to the north by the Chitina Valley fault and to the south by various splays of the Border Ranges fault and the Chugach terrane (Fig. 2). Unless noted, the following descriptions of the southern Wrangellia terrane in the Gulkana and Talkeetna Mountains quadrangles are from Csejtey and others (1978) and Nokleberg and others (1986, 1989a); descriptions from the Valdez and McCarthy quadrangles are from MacKevett (1978), Winkler and others (1981a), Nokleberg and others (1989b), and Plafker and others (1989c).

Metamorphic complex of Gulkana River

The metamorphic complex of Gulkana River along the southern flank of the central and eastern Alaska Range consists of three rock sequences (Fig. 4) (Nokleberg and others, 1986; Kline and others, 1990). The oldest sequence is an unfossiliferous unit of (1) chlorite schist derived from generally massive, locally pillowed hornblende andesite, lesser clinopyroxene basalt, and agglomerate; (2) metamorphosed felsic tuff and metarhyolite to metadacite flows; (3) sparse medium- to thin-foliated calc-schist, pelitic schist, and metachert; and (4) rare calcite and dolomite marble. The mafic metavolcanic rocks are interpreted to be comagmatic with the relatively younger metamorphosed mafic plutonic rocks, discussed below, which are interpreted to be late Paleozoic in age. The metamorphic complex of Gulkana River was first recognized by Csejtey and others (1978) in the northern Talkeetna Mountains as an unnamed unit consisting mainly of amphibolite, greenstone, and locally schistose granitic rocks.

The next younger sequence consists of schistose hornblende diorite and lesser schistose gabbro that occur in dikes, sills, and small plutons. K-Ar hornblende ages for these metamorphosed mafic rocks are 130, 131, 233, 282, 295, 306, and 1,369 Ma (W. J. Nokleberg, T. E. Smith, and D. L. Turner, 1986, unpublished data). The oldest age of 1,369 Ma may represent a younger plutonic rock containing excess argon. The intermediate ages of 282 to 306 Ma may be a late Paleozoic age of intrusion. The younger ages of 130 and 131 Ma are interpreted as minimum ages of regional metamorphism, as discussed below.

The youngest sequence consists of schistose granitic plutons, mainly biotite granodiorite and granite, with lesser biotite-muscovite trondhjemite and quartz diorite. Locally abundant relict igneous biotite and muscovite occur in some of the metagranitic plutons. K-Ar metamorphic biotite and white mica ages, discussed below, are interpreted as the age of synkinematic intrusion of the plutons in the Late Jurassic. The structural thickness of the metamorphic complex is estimated as several kilometers, although the base is not exposed.

The dominant major structure in the metamorphic complex of Gulkana River is an east-west–striking, steeply dipping to vertical structural homocline of metavolcanic and metasedimentary rocks (Fig. 4). The dominant minor structure is a locally intensely developed mylonitic schistosity in all three units that strikes generally east-west and dips steeply to vertically, parallel to the regional strike of major units and bounding faults (Fig. 4). Locally intense ductile deformation occurs along the schistosity, and there is formation of mylonitic schist in metasedimentary and metavolcanic rocks and mylonitic gneiss in metaplutonic rocks. Lower greenschist facies minerals, mainly chlorite, actinolite, epidote, albite, and white mica, occur along the schistosity away from the metamorphosed granitic plutons of the metamorphic complex. In more highly metamorphosed and deformed areas, hornblende, clinopyroxene, calcic plagioclase, and biotite in the metaigneous rocks are mostly to completely replaced by metamorphic minerals, whereas in lesser deformed areas, hornblende, clinopyroxene, calcic plagioclase, and biotite are only partly replaced by metamorphic minerals.

The grade of regional metamorphism of the country rocks adjacent to the metamorphosed plutons increases to lower amphibolite facies with the occurrence of hornblende, garnet, calcic plagioclase, and biotite in place of lower grade minerals farther away. In these areas, the schistose fabric of the country rocks continues into the metagranitic rocks. These relations are interpreted as regional metamorphism occurring during syntectonic intrusion of the metagranitic plutons. Schistose biotite-muscovite granodiorite and quartz diorite from three metagranitic plutons have K-Ar metamorphic biotite ages of 139, 142, 143, 146, and 149 Ma, and metamorphic white mica ages of 146 and 148 Ma (W. J. Nokleberg, T. E. Smith, and D. L. Turner, 1986, unpublished data). These K-Ar ages are interpreted as a Late Jurassic age of regional metamorphism, deformation, and syntectonic intrusion of the metamorphosed plutonic rocks. The Late Jurassic ages also provide a minimum age for the wall rocks.

Strelna metamorphics and Uranatina River metaplutonic unit

In the northern Chugach and southern Wrangell Mountains, the southern Wrangellia terrane is composed of the Strelna metamorphics of Plafker and others (1989c) and the Uranatina River metaplutonic unit. The Strelna metamorphics include the metasedimentary rocks of the Dadina River area that occur to the northwest in the southwestern Wrangell Mountains, and the Uranatina River metaplutonic unit includes the metagranitic rocks of the Dadina River area. In the area west of the Copper

River and south of Tonsina, the term "Haley Creek metamorphic assemblage" is adapted from Wallace (1981a) to include both the metamorphosed stratified and plutonic rocks of the southern Wrangellia terrane (Fig. 6). For a discussion of previous names for these units, refer to Plafker and others (1989c).

Most rocks assigned to the Strelna metamorphics and the Uranatina River metaplutonic unit occur in an east-west–striking belt of moderately to steeply dipping, highly deformed lithologies. However, west of the Copper River and south of Tonsina, the Haley Creek metamorphic assemblage occurs in a large, folded thrust sheet and small outlying klippen in an area up to 10 km wide in a north-south direction and more than 20 km long in an east-west direction (Fig. 6). The base of this thrust sheet is the part of the Border Ranges fault, which defines the boundary of the Chugach terrane (Fig. 2). Locally the thrust sheet is eroded to expose the underlying Valdez Group of the Chugach terrane in windows (Fig. 6) (Nokleberg and others, 1989b; Plafker and others, 1989b, 1989c). This maximum structural thickness of the thrust sheet is estimated to be 1 to 2 km.

Strelna metamorphics. The Strelna metamorphics of Plafker and others (1989c) comprise five main lithologies: greenschist, marble, schistose marble, quartzofeldspathic mica schist, and micaceous quartz schist. One body of carbonate rocks contains recrystallized crinoids and brachiopods (Moffit, 1938), indicating deposition in a relatively shallow marine environment. Intercalated marbles have yielded conodonts of early Pennsylvanian (Bashkirian) age (Plafker and others, 1985). The quartzofeldspathic mica schist is probably derived from a quartzofeldspathic sandstone with minor intercalated argillaceous sediment. The high silica content of the quartz schist and its intimate intercalation with micaceous layers indicate a probable protolith of banded argillaceous and tuffaceous chert. Field relations indicate that the greenschist is derived from massive flow units, pillow lavas, and thinner bedded units that were probably tuff or breccia intercalated with sedimentary rocks. The metavolcanic rocks are probably the southernmost extrusive part of the Skolai arc.

Intense deformation of the Strelna metamorphics has destroyed most primary textures except for primary layering, relict graded bedding, and sparse relict clastic textures. Metamorphic grade is mainly lower greenschist facies, but increases to amphibolite facies adjacent to Late Jurassic metaplutonic rocks of the Uranatina River metaplutonic unit (Nokleberg and others, 1989b). In these areas, the schistose fabric of the country rocks continues into the metaplutonic rocks. The Strelna is interpreted, on the basis of lithology and sparse megafauna and conodont microfauna, to represent a former marine sequence composed dominantly of quartzofeldspathic, pelitic, calcareous, and cherty rocks with subordinate volcanic rocks that were deposited in a shelf or upper slope basin.

Uranatina River metaplutonic unit. The metasedimentary rocks of the Strelna metamorphics of Plafker and others (1989c) are extensively intruded by compositionally diverse, strongly schistose to blastomylonitic metaplutonic rocks of the Pennsylvanian and Jurassic Uranatina River metaplutonic unit (Fig. 3) (Plafker and others, 1989c). This unit was defined for the area southeast of Tonsina in the northern Chugach Mountains (Fig. 6) by Plafker and others (1989c), but it is herein extended to include all late Paleozoic and early and middle Mesozoic plutonic rocks in the southern Wrangellia terrane. The Uranatina River unit is exposed in the southwestern Wrangell Mountains and northern Chugach Mountains (Winkler and others, 1981a), and it may correlate with metamorphic rocks to the west in the Anchorage quadrangle in the informally named Knik River terrane of Pavlis (1983). Wallace (1981a, 1981b) referred to these rocks informally in the vicinity of the Richardson Highway as the Uranatina River complex.

The Uranatina River metaplutonic unit consists of volumetrically minor amounts of Pennsylvanian metagranodiorite, metagranite, metagabbro, and interlayered gabbro and orthogneiss. U-Pb zircon isotopic ages for the Pennsylvanian metaplutonic rocks range from 309 to 310 Ma (Aleinikoff and others, 1988). The Pennsylvanian plutonic rocks are interpreted as the deeper levels of the southern margin of the Skolai arc (Plafker and others, 1989c). The relatively more abundant Jurassic metaplutonic rocks are compositionally variable, ranging from ultramafic to trondhjemitic, but are dominantly gneissic hornblende diorite, tonalite, and amphibolite, and they constitute the southern margin of the Late Jurassic Chitina arc (Plafker and others, 1989c). One U-Pb zircon age is 153 Ma, and K-Ar isotopic ages range from 138 to 157 Ma (MacKevett, 1978; Winkler and others, 1981a; Plafker and others, 1989c). The Jurassic metaplutonic rocks of the Uranatina River metaplutonic unit are correlated with the Chitina Valley batholith east of the Taral fault and in the Wrangellia terrane on the basis of similar ages, lithologies, geochemistry, and structural style. Sparse Early Cretaceous K-Ar ages from the Uranatina River metaplutonic unit (Winkler and others, 1981a) are tentatively interpreted as having been reset by a Late Cretaceous thermal event and by early Tertiary plutonism.

Correlation of metamorphic complex of Gulkana River with the Strelna metamorphics and Uranatina River metaplutonic unit

The metamorphic complex of Gulkana River is correlated with both the Strelna metamorphics of Plafker and others (1989c) and the Uranatina River metaplutonic unit because of (1) similar stratified protoliths; (2) similar suites of Late Jurassic metamorphosed plutonic rocks; and (3) a similar strongly mylonitic fabric that is interpreted in both units as forming during the late stages of, or immediately subsequent to, syntectonic plutonism (Nokleberg and others, 1986, 1989b; Pavlis and Crouse, 1989). The major differences between the metamorphic complex of Gulkana River and the units to the southeast are the occurrence of more abundant metasedimentary rocks and sparse Pennsylvanian plutonic rocks in the latter.

Origin of the southern Wrangellia terrane

The southern Wrangellia terrane was interpreted by Plafker and others (1989c) as a deeper and more metamorphosed equivalent of the type Wrangellia terrane that has been exposed by uplift along the Chitina fault system. The lithologic differences between the units may be explained by sedimentary facies changes and/or differences in level of exposure. Several relations appear to link the southern Wrangellia terrane with the Wrangellia terrane. (1) Similar protolith suites of upper Paleozoic sedimentary and intermediate composition volcanic rocks (Plafker and others, 1989c) and similar late Paleozoic plutonic rocks occur in both units (Aleinikoff and others, 1988; Gardner and others, 1988). (2) Spatial association of the southern Wrangellia and Wrangellia terranes for several hundred kilometers suggests a genetic link. (3) The southern Wrangellia terrane appears to contain Late Triassic Nikolai Greenstone and limestone, distinctive units of the Wrangellia terrane (Winkler and others, 1981a; Plafker and others, 1989c). (4) The Late Jurassic plutons of the Chitina Valley batholith occur in both the Wrangellia terrane and southern Wrangellia terrane close to the Chitina fault, but they are more abundant in the uplifted southern Wrangellia terrane. (5) Similar common lead crustal isotopic compositions for feldspar occur in both units (Aleinikoff and others, 1987).

Alternatively, D. L. Jones (1990, oral commun.), Grantz and others (1991), and Nokleberg (this chapter) interpret the southern Wrangellia terrane as a separate unit. Important differences occur between the southern Wrangellia terrane to the south and the Wrangellia terrane to the north. (1) The southern Wrangellia terrane contains a penetrative fabric and an amphibolite facies grade of metamorphism that is generally lacking in the Wrangellia terrane. (2) The southern Wrangellia terrane contains suites of syntectonic, two-mica granitic rocks of mainly Late Jurassic age that generally do not occur in adjacent parts of the Wrangellia terrane. (3) The late Paleozoic or younger stratified rocks of the Wrangellia terrane are not observed to stratigraphically overlie either the metamorphic complex of Gulkana River or the Strelna metamorphics. (4) The Late Triassic carbonate rocks and associated undated greenstone, mapped in the Strelna metamorphics (Plafker and others, 1989c), locally occur in fault-bounded blocks that are interpreted to have originally stratigraphically overlain the older, late Paleozoic stratified rocks of the Strelna. Because of these differences, the southern Wrangellia terrane in this chapter is designated as a unique unit. For similar reasons, this unit was also designated as a separate terrane, the Strelna terrane, by Grantz and others (1991). Additional studies are needed to resolve these differing interpretations.

ALEXANDER TERRANE

The Alexander terrane occurs southeast of the Wrangellia terrane in the southeastern Wrangell Mountains in the McCarthy and Bering Glacier quadrangles near the border between Alaska and the Yukon Territory of Canada (Figures 1, 2). The terrane was originally defined in, and comprises much of, southeastern Alaska (Berg and others, 1972; Gehrels and Saleeby, 1987; Jones and others, 1987; Monger and Berg, 1987; Gehrels and Berg, this volume).

In the study area, the Alexander terrane consists mainly of rocks informally referred to as the Kaskawulsh metamorphic rocks by Gardner and others (1988), which are equivalent to the Kaskawulsh Group that occurs in adjacent parts of Canada (Figure 3) (Kindle, 1953; MacKevett, 1978; Campbell and Dodds, 1982a, 1982b; Gardner and others, 1988). The early to middle Paleozoic Kaskawulsh metamorphic rocks are an intensely multiply folded and schistose sequence of marble, greenstone, metagraywacke, phyllite, slate, schist, argillite, metasiltstone, mafic volcanic rocks, and volcaniclastic rocks that are probably a few thousand meters thick. Numerous Devonian and older megafossils occur in the unit on structural strike in the Yukon Territory to the southeast (Campbell and Dodds, 1982a, 1982b). Metamorphic grade ranges from upper greenschist to amphibolite facies. To the east in Canada, the Alexander terrane also contains younger, late Paleozoic and early Mesozoic rocks that are correlated with the Wrangellia terrane (Gardner and others, 1988).

In the southeastern Wrangell Mountains, the Middle Pennsylvanian Barnard Glacier pluton stitches the Alexander and Wrangellia terranes (Gardner and others, 1988). This pluton intrudes both the Kaskawulsh metamorphic rocks of the Alexander terrane and the adjacent Pennsylvanian and Lower Permian Station Creek Formation of the Wrangellia terrane. The nature of the contact between the Pennsylvanian and Lower Permian Station Creek Formation and the Kaskawulsh metamorphic rocks is not exposed. Gardner and others (1988) suggested that the Station Creek Formation was deposited unconformably on the metamorphic rocks.

Intrusion of the Middle Pennsylvanian Bernard Creek pluton into both terranes requires juxtaposition by the Early Pennsylvanian and suggests that the Kaskawulsh metamorphic rocks of the northwestern Alexander terrane either may be basement for part of the Wrangellia terrane in the study area or may have been faulted against adjacent parts of the Wrangellia terrane, a suggestion originally made by Muller (1977). The roof pendants of pre–late Paleozoic metasedimentary and metagranitic rocks in the Nutzotin Mountains, described above, may be part of this basement.

The Alexander terrane in the study area and in southeastern Alaska is interpreted as a fragment of a long-lived, early and middle Paleozoic island arc (Gehrels and Saleeby, 1987; Gehrels and Berg, this volume). The above field relations indicate that this arc either was tectonically juxtaposed against the Wrangellia terrane by the middle Pennsylvanian or was the stratigraphic basement to at least part of the Wrangellia terrane. Similarly, the unit of unnamed Paleozoic(?) metamorphic rocks along the southern margin of the Peninsular terrane, discussed below, may be part of an Alexander terrane basement for part of the Peninsular terrane.

PENINSULAR TERRANE

The Peninsular terrane, originally defined by Jones and others (1977) and Jones and Silberling (1979, 1982), occurs along the southern edge of the study area, from the Alaska Peninsula to

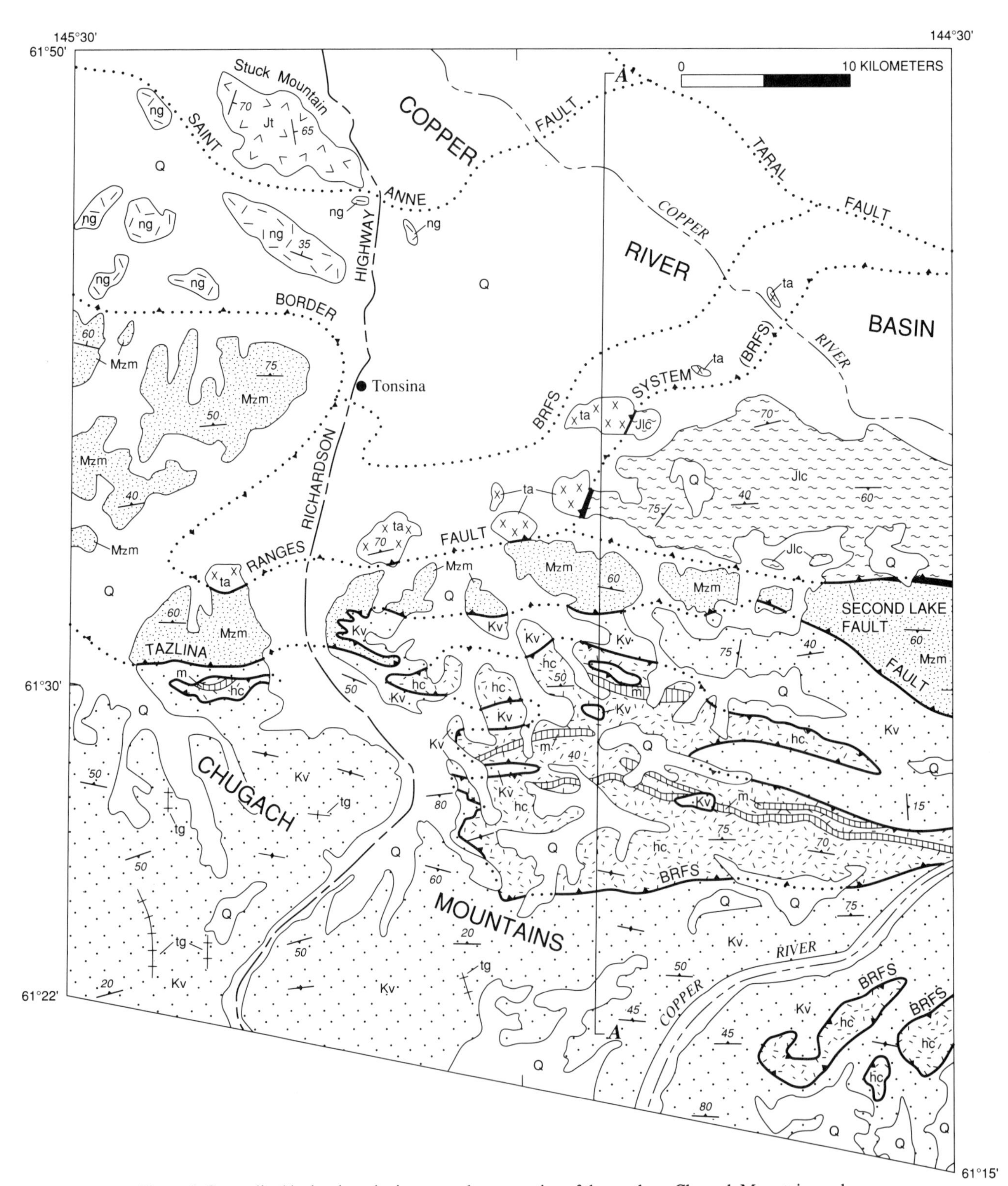

Figure 6. Generalized bedrock geologic map and cross section of the northern Chugach Mountains and southwestern Copper River basin along the Trans-Alaskan Crustal Transect (TACT). Refer to text for description of units. Adapted from Winkler and others (1981a), Nokleberg and others (1989b), and Plafker and others (1989c).

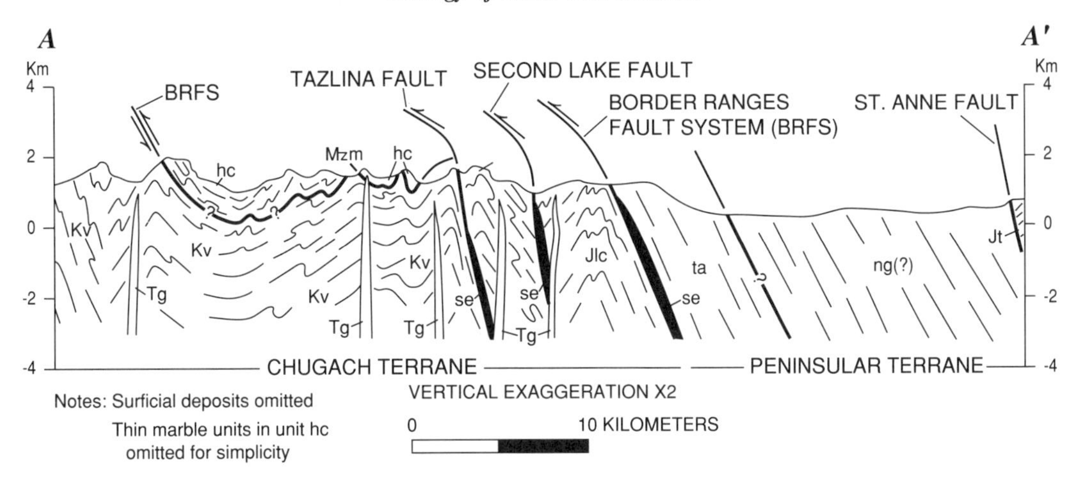

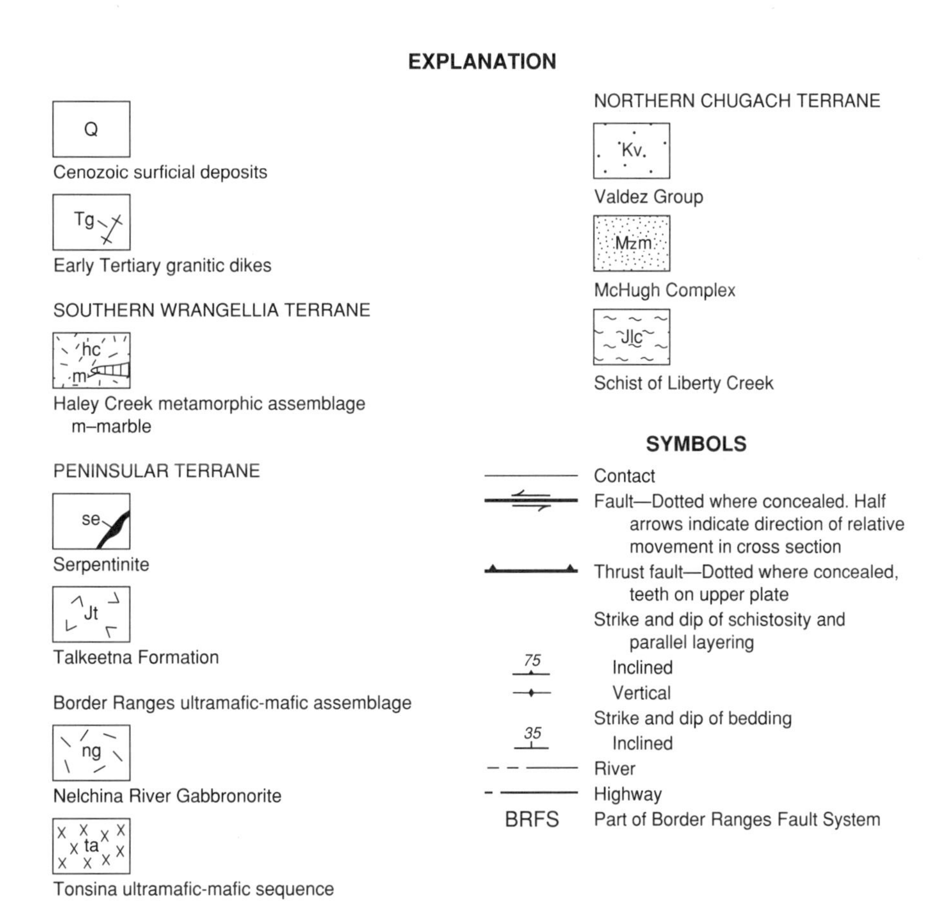

the southwest to the southeastern Copper River Basin to the northeast (Figs. 1 and 2). The terrane occurs south of the Wrangellia terrane, north of the Chugach terrane, and is bounded by the Talkeetna thrust and West Fork fault to the north and by the Border Ranges fault to the south (Fig. 2). The principal syntheses of geologic studies of the rocks and structures of the terrane are by Burk (1965), Wilson and others (1985b, 1994), Plafker and others (1989c), Detterman and others (1994), and Vallier and others (this volume). As defined by Plafker and others (1989c), the Peninsular terrane consists of (1) unnamed Paleozoic(?) metamorphic rocks that occur mainly as roof pendants in the Border Ranges ultramafic-mafic assemblage; (2) sparse late Paleozoic limestone and volcanic rocks; (3) Late Triassic limestone and basalt; (4) the Late Triassic to Middle(?) Jurassic Border Ranges ultramafic-mafic assemblage; (5) Early Jurassic marine andesite flows and related rocks; (6) Middle Jurassic granitic rocks that compose the older part of the Alaska-Aleutian Range batholith; (7) Middle Jurassic through Cretaceous fossiliferous clastic, volcaniclastic, and calcareous rocks; and (8) metamorphosed Permian to Late Jurassic stratified assemblages that occur either as roof pendants in, or occur adjacent to, the Alaska-Aleutian Range batholith. The original definition of the Peninsular terrane by Jones and others (1977) was expanded by Plafker and others (1989c) to include the Border Ranges ultramafic-mafic assemblage and the unnamed metamorphic rocks that are discontinuously exposed along the southern margin of the terrane. Large parts of the Peninsular terrane are also intruded by the Late Cretaceous through Tertiary part of the Alaska-Aleutian Range batholith, and the plutonic part of the Late Cretaceous and early Tertiary Alaska Range–Talkeetna Mountains igneous belt (unit TKg, Fig. 2) (Miller, this volume, Chapter 16; Moll-Stalcup, 1990, and this volume).

Alaska Peninsula terrane

Recent studies of the Mesozoic strata of the Alaska Peninsula lead to the definition of the Alaska Peninsula terrane (Wilson and others, 1985b, 1994). The Alaska Peninsula consists of (1) late Paleozoic through Late Triassic sedimentary and volcanic rocks that are correlated with the Wrangellia terrane; (2) Jurassic and Cretaceous sedimentary and volcanic rocks; and (3) Jurassic plutonic rocks of the Alaska-Aleutian Range batholith. The definitions of the Alaska Peninsula and Peninsular terranes differ principally in that the former excludes the unit of unnamed metamorphic rocks and the Border Ranges ultramafic-mafic assemblage. These latter two units were interpreted by Wilson and others (1985a, 1994) as a separate terrane.

The Alaska Peninsula terrane is composed of two distinct but related subterranes, the Chignik and Iliamna subterranes. These two subterranes share a partial common geologic history in that the Iliamna subterrane to the northwest has served at most times as a source area for the Chignik subterrane to the southeast. However, some rock units occur in both subterranes. The bedrock of the Alaska Peninsula terrane is overlapped by the upper Eocene to lower Miocene volcanic and volcaniclastic rocks of the Meshik arc (Wilson, 1985) and by the late Tertiary and Quaternary volcanic and associated sedimentary rocks of the Aleutian arc (Moll-Stalcup, 1990, and this volume; Vallier and others, this volume).

Iliamna subterrane. The Iliamna subterrane occurs northwest of the Bruin Bay fault and is named after exposures in the Iliamna quadrangle (Fig. 1). The unit is composed mainly of the Kakhonak Complex, the Late Triassic Cottonwood Bay Greenstone, the Upper Triassic Kamishak Formation, the Early Jurassic Talkeetna Formation, and the Middle Jurassic part of the Alaska-Aleutian Range batholith (Reed and Lanphere, 1969, 1972, 1973, 1974a, 1974b; Detterman and Reed, 1980). Each of these units is correlative with parts of the Chignik subterrane or, in the case of the Kamishak and Talkeetna formations, are actually part of both subterranes. The Iliamna subterrane is characterized by metamorphism up to amphibolite facies grade and folding that is most intense nearest the Alaska-Aleutian Range batholith.

The southernmost exposure of rocks of the Iliamna subterrane occurs in the vicinity of Becharof Lake, where Jurassic quartz diorite (Reed and Lanphere, 1972) is exposed on a small island on the south side of the lake and in the hills north of the lake in the Naknek quadrangle. Southwest of Becharof Lake, the Iliamna subterrane is covered by younger rocks, although its presence may be inferred from geophysical and drill data (Wilson and others, 1994). Interpretation of aeromagnetic data (Case and others, 1981) suggests that the Iliamna subterrane may continue at least as far south as Port Heiden. Pratt and others (1972) suggested continuation of the batholith into southern Bristol Bay, on the basis of interpretation of shipborne magnetic data. However, the nature of the data precludes determination with certainty that the anomalies can be specifically tied to the Jurassic part of the batholith.

Chignik subterrane. The Chignik subterrane occurs southwest of the Bruin Bay fault and is named after exposures in the Chignik quadrangle (Fig. 1). The unit is composed mainly of upper Paleozoic andesite flows, agglomerate, and limestone, the Late Triassic Kamishak Formation, the Early Jurassic Talkeetna Formation, and a long-lived succession of Middle Jurassic through Cretaceous clastic, volcaniclastic, and calcareous units. In much of the Chignik subterrane, the structural style is dominated by en echelon anticlines, normal faulting, and thrust and high-angle reverse faults that have minor displacement in a northwest to southeast direction. Structures are typically aligned subparallel to the general northeast-southwest trend of the peninsula. In general, compressional features are more common to the southwest in the Chignik subterrane. Metamorphism in the Chignik subterrane occurs only in narrow contact-metamorphic zones around plutons. Most of the Mesozoic rocks of the Alaska Peninsula are part of the Chignik subterrane.

Unnamed Paleozoic(?) metamorphic rocks along the southern margin of the Peninsular terrane

Various unnamed Paleozoic(?) metamorphic rocks are discontinuously exposed along the southern margin of the Peninsular terrane, mainly northeast of Anchorage (Clark, 1972; Carden and Decker, 1977; Pavlis, 1982, 1983, 1986; Burns and others, 1983, 1991; Pavlis and others, 1988; Winkler, 1992), possibly at Seldovia (Martin and others, 1915), and on the northwest side of the Kodiak Islands (Roeske and others, 1989). The metamorphic rocks, too small to depict in Figure 2, occur mainly along, and immediately north of, the Border Ranges fault system.

The metamorphic rocks include two units of the so-called Knik River schist terrane (Carden and Decker, 1977; Pavlis, 1983). The first unit consists mainly of highly metamorphosed and deformed quartz schist and lesser mica-quartz schist, garnet-mica schist, hornblende-biotite-quartz schist, calc-amphibolite, siliceous marble, and pelitic marble. Protoliths for this metasedimentary and metavolcanic unit (Pavlis, 1982; Burns and others, 1991) are shale, chert, tuffaceous arenite, quartz-rich graywacke, marl, shaley limestone, and possible minor basalt. Local fault-bounded lenses of limestone contain Permian fusulinids (Clark, 1972). The metamorphic grade ranges from amphibolite to greenschist facies; the lower grade may be retrogressive. Similar quartz-muscovite-albite-chlorite pelitic schist also occurs as roof pendants in granitic rocks in the southwestern Talkeetna Mountains (Csejtey and others, 1978), and may be related.

A second unit consists of Jurassic gabbro, diorite, quartz diorite, tonalite, and amphibolite that are intensely mixed by intrusion and faulting (Pavlis, 1983; Pavlis and others, 1988; Burns and others, 1991). This unit is interpreted to be part of the Border Ranges ultramafic-mafic assemblage. Both units are cut by steeply dipping zones of cataclasite composed of chlorite-rich, highly deformed and altered mafic and ultramafic plutonic and volcanic rocks. The zones may represent interconnected strands of the Border Ranges fault system, along which ductile (Pavlis, 1982) and Tertiary brittle (Little and Naeser, 1989) offset was concentrated.

The metasedimentary and metavolcanic rocks of the unnamed Paleozoic(?) metamorphic unit may correlate with the late Paleozoic metasedimentary and metavolcanic rocks in the southern Wrangellia terrane (Burns, 1985; Plafker and others, 1989c; Winkler, 1992) and are herein interpreted as representing the deeper structural levels of the Peninsular terrane upon which part of the Late Triassic(?) and Jurassic Talkeetna igneous arc was built. Alternatively, the unit of unnamed metamorphic rocks may constitute basement for a pre-Jurassic subduction complex (Pavlis, 1982, 1983; Pavlis and others, 1988). The origin of the unnamed Paleozoic(?) metamorphic rocks unit is a major problem because its age, lithologies, and structural fabric are dissimilar to the younger stratified and intrusive rocks that compose most of the terrane.

Late Paleozoic strata of Peninsular terrane—unnamed Permian limestone and Permian(?) volcanic rocks

The oldest dated rocks in the Peninsular terrane are unnamed fossiliferous, late middle Permian limestones and Permian(?) volcanic agglomerates and flows (Figs. 3 and 7). The limestone is a 40-m-thick succession of thin- to thick-bedded, medium-grained, crystalline tan to gray limestone with thin interbeds of chert that occurs on a small unnamed islet (100 by 200 m) at the entrance to Puale Bay. A late middle Permian (early

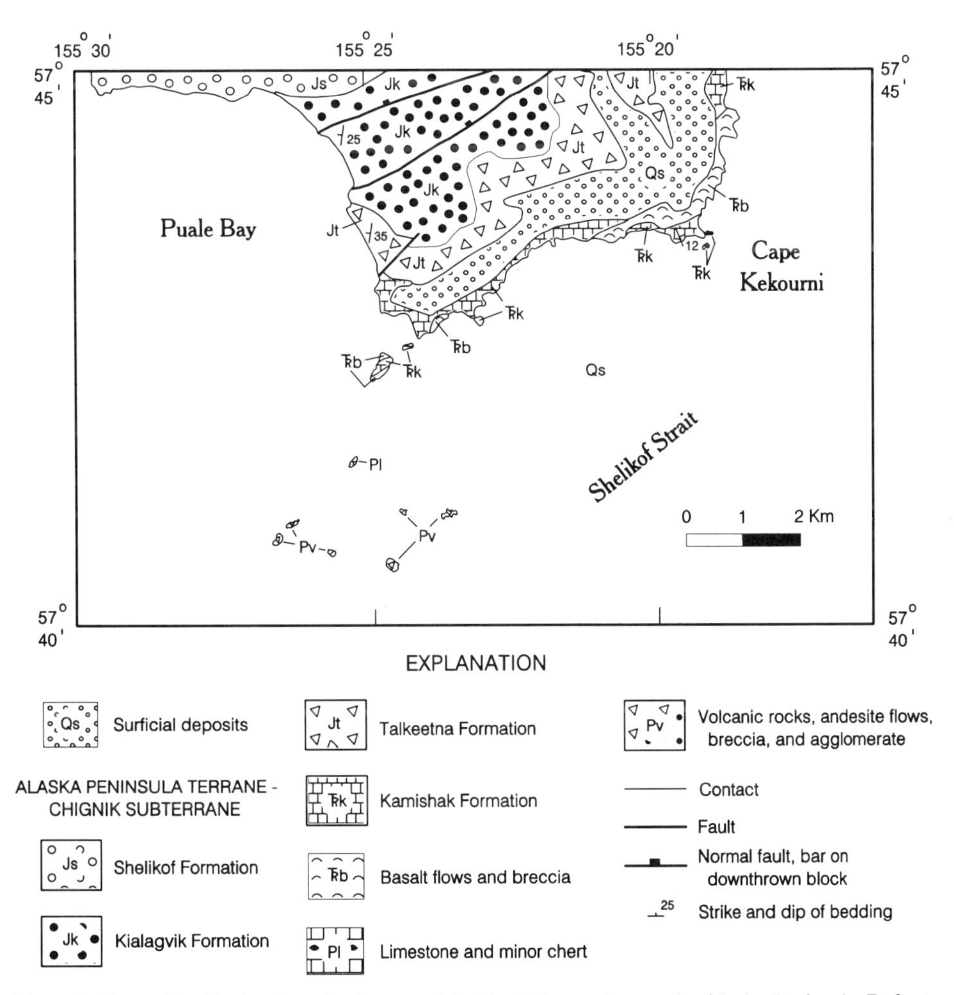

Figure 7. Generalized bedrock geologic map of the Puale Bay region on the Alaska Peninsula. Refer to text for description of units. Adapted from Wilson and others (1994).

Guadalupian) age is based on poorly preserved and silicified coral, brachiopod, and foraminifer fossils (Hanson, 1957). No contacts are exposed, although the highly contorted beds dip about 40° northwest, which places them apparently structurally beneath Triassic rocks on other islands about 1 km to the north.

Massive dark green to black volcanic breccia, agglomerate, and andesitic flows of unknown age and thickness are exposed on small islets southeast of Puale Bay and appear to be structurally beneath the Permian limestones (Figs. 3 and 7) (Hanson, 1957; Hill, 1979). Barring major faults between the islands, this volcanic sequence is assigned a Permian(?) age. Limited chemical data (Hill, 1979) suggest that these volcanic rocks are calc-alkaline. The Permian(?) volcanic rocks and Permian limestones may partially correlate with the Pennsylvanian and Early Permian volcanic and associated rocks of the Skolai arc, and overlying Permian limestone of the Wrangellia terrane.

Mesozoic strata of Peninsular terrane

Kakhonak and Tlikalika complexes. The Kakhonak Complex (Detterman and Reed, 1980) (Fig. 3) is a heterogeneous assemblage of metamorphic rocks of Permian(?) to Jurassic age exposed in the Iliamna quadrangle. The complex, unique to the Iliamna subterrane, is exposed mainly as roof pendants in the Alaska-Aleutian Range batholith, but it also crops out in areas surrounded by surficial deposits. The complex contains marble, quartzite, greenstone and other metavolcanic rocks, and schist. Most rocks are weakly metamorphosed to greenschist facies; relict bedding is often discernible. The maximum age of the enclosing batholith is about 155 Ma, placing a minimum age of early Late Jurassic on the metamorphic rocks. Lithologies indicate that the protoliths were similar to the Permian limestones and the overlying Cottonwood Bay Greenstone, Kamishak Formation, and Talkeetna Formation.

The informally named Tlikakila complex of Wallace and others (1989) occurs along the northwestern edge of the Alaska-Aleutian Range batholith in a narrow, northeast-trending, faulted belt along the southeastern margin of the Kahiltna assemblage (Fig. 5) (Wallace and others, 1989). The Tlikakila complex consists of a narrow belt of variably metamorphosed and highly deformed metasedimentary and mafic metavolcanic rocks, and mafic and ultramafic rocks. Sparse Late Triassic fossils occur in the complex. This unit is correlated with the Kakhonak Complex (Wilson and others, 1993).

Cottonwood Bay Greenstone. The Cottonwood Bay Greenstone (Detterman and Reed, 1980) (Fig. 3) is a sequence of metavolcanic rocks exposed on the west side of Cook Inlet that is unique to the Iliamna subterrane. The thickness may exceed 600 m; however, contact relations with other stratigraphic units are obscure. Bedding is locally discernible; in most areas, the rocks have been altered to hornfels or low-grade, schistose metamorphic rock. The least altered areas consist of massive, dark-green, amygdaloidal basalt flows. No isotopic age determinations exist for the greenstone, but a close association with the Kamishak Formation, described below, suggests a Late Triassic (Norian?) age. The lower pillowed greenstone member of the Shuyak Formation on northwest Kodiak Island was correlated by Moore and Connelly (1977) with the Cottonwood Bay Greenstone. A possible correlation of the Cottonwood Bay Greenstone with the Late Triassic Nikolai Greenstone of the Wrangellia terrane was suggested by Detterman and Reed (1980). On the basis of lithology, the Cottonwood Bay Greenstone has been interpreted as of oceanic affinity, possibly an ocean-island basalt (J. R. Riehle, 1991, oral commun.). On the basis of REE chemistry, the lower member of the Shuyak Formation is interpreted as an island-arc tholeiite (Hill, 1979).

Chilikadrotna Greenstone. The Chilikadrotna Greenstone occurs in the Lake Clark quadrangle in the southwestern Alaska Range (Figs. 3 and 5) (Eakins and others, 1978; Bundtzen and others, 1979; Wallace and others, 1989). The Chilikadrotna is poorly exposed in a series of discontinuous, northeast-trending antiformal exposures near the northwestern boundary of the southern Kahiltna assemblage, in low hills adjacent to the Chilikadrotna River (Eakins and others, 1978). The unit is composed principally of massive altered basalt that commonly is amygdaloidal, locally pillowed, and contains rare interpillow limestone and chert. Sparse pyroxene-bearing andesitic flows, tuff breccia, and crystal-lithic lapilli tuff also occur in the sequence. The volcanic rocks of the Chilikadrotna Greenstone commonly are metamorphosed to prehnite-pumpellynite facies. Penetrative deformational fabrics are developed only locally, but fault surfaces and shear zones are abundant.

Diverse metasedimentary rocks locally are spatially associated with the volcanic rocks (Wallace and others, 1989), although the stratigraphic relations with the volcanic rocks are obscure. The metasedimentary rocks include limestone, chert, cherty shale and tuff, and various metaclastic rocks, and they occur mainly in narrow, structurally concordant zones adjacent to the volcanic rocks. Conodonts from the limestone are Late Triassic (Norian) in age (T. R. Carr, 1983, written commun.; N. M. Savage, 1983, written commun.), and brachiopods from the limestone are also of probable Norian age (P. R. Hoover, 1984, written commun.). The metasedimentary rocks exhibit a weak slaty cleavage that dips steeply and mainly to the northwest. Locally, highly schistose zones occur. Much of the unit is highly imbricated.

The Chilikadrotna Greenstone was grouped with the Koksetna River sequence into the southern Kahiltna terrane by Wallace and others (1989). However, on the basis of lithology and association with the Norian sedimentary rocks, the Chilikadrotna Greenstone is herein interpreted as part of the Peninsular terrane that is depositionally overlain by the Koksetna River sequence that composes the southern Kahiltna assemblage. The Chilikadrotna Greenstone was correlated with the Cottonwood Bay Greenstone by Wallace and others (1989).

Kamishak Formation. The Kamishak Formation (Figs. 3 and 7), common to both the Iliamna and Chignik subterranes, consists of 1,200 m of thin-bedded to massive limestone, chert, minor calcareous siltstone and mudstone, and sparse tuff (Fig. 7). The depositional environment of the unit was shallow water and high energy intervals of the unit include both reefs and biohermal buildups. Fossils yield a Late Triassic (Norian) age (Detterman and Reed, 1980; C. D. Blome, U.S. Geological Survey, 1981, oral commun.).

The carbonate rocks of the Kamishak Formation and interbedded volcanic rocks are interpreted to overlie the Permian limestone and volcanic rocks that are exposed on the islands of Puale Bay (Fig. 7) (Detterman and others, 1994). Triassic rocks similar to the Kamishak Formation also occur in the subsurface in the Cathedral River area (McLean, 1977) at the southwest end of the Alaska Peninsula. Moore and Connelly (1977) suggested that the upper member of the Shuyak Formation, mainly tuff, volcaniclastic turbidite, massive sandstone, volcanic conglomerate, and mudstone (Connelly, 1978; Connelly and Moore, 1979), on northwestern Kodiak Island is a possible correlative of the Triassic volcanic and volcaniclastic rocks at Puale Bay. Limestone and volcanic rocks of the Kamishak Formation are herein interpreted as having formed in an island-arc environment (Moore and Connelly, 1977; Hill, 1979; Wang and others, 1988) at low latitude (Detterman, 1988; Hillhouse and Coe, this volume).

At Puale Bay, the upper contact of the Kamishak Formation is conformable and gradational with the overlying Talkeetna Formation (Figs. 3 and 7); Detterman and others (1994) arbitrarily placed the contact as the point where clastic sedimentary rocks replace limestone as the major constituent of the rock sequence. The Kamishak Formation is lithologically and faunally equivalent to the Nizina Limestone and the McCarthy Formation in the Wrangellia terrane (Detterman and Reed, 1980). The Port Graham sequence in the Seldovia area (Kelley, 1984) is also coeval with, and lithologically similar to the Kamishak Formation.

Talkeetna Formation. The Late Triassic(?) and Early Jurassic Talkeetna Formation (Fig. 3), common to both the Iliamna and Chignik subterranes, consists mainly of andesitic, basaltic, and volcaniclastic rocks (Paige and Knopf, 1907; Martin, 1926; Detterman and Hartsock, 1966; Csejtey and others, 1978; Winkler and others, 1981a; Detterman and others, 1983; Winkler, 1992). The unit crops out discontinuously from the southern Copper River basin, through the Talkeetna Mountains, to Puale Bay on the Alaska Peninsula. Stratigraphically and lithologically equivalent rocks are encountered in a drillhole at the southwest end of the Alaska Peninsula.

In the Talkeetna Mountains (the type area), and in the southern Copper River basin and northern Chugach Mountains, the Talkeetna Formation is mainly bedded andesitic volcaniclastic sandstone and tuff, ignimbrite, breccia, and agglomerate; andesite and lesser rhyolite and basalt flows; and shale (Csejtey and others, 1978; Plafker and others, 1989c; Burns and others, 1991; Winkler, 1992). On the Alaska Peninsula, the Talkeetna is mainly gray-green, coarse-grained tuffaceous sandstone; lesser green to red, massive coarse-grained tuff; and minor brownish-gray siltstone and gray to gray-brown limestone (Detterman and others, 1994). The thickness is variable throughout the area of outcrop, but is typically between 400 and 1,200 m. Apparently partly coeval shallow intrusive rocks (mainly dikes, sills, and plugs of andesite composition) and local compositionally diverse diorite, quartz diorite, and granodiorite plutons of Late Jurassic, Late Cretaceous, and early Tertiary age intrude the unit.

The Talkeetna Formation undergoes a gradual lithologic transition from predominantly volcanic lithologies in the northeast to predominantly sedimentary lithologies to the southwest. This transition suggests that the rocks of the unit become more distal from the arc volcanism to the southwest. Diagnostic Early Jurassic marine megafossils occur throughout, and Late Triassic fossils locally occur in the lower part of the unit on the southern Kenai Peninsula (Martin and others, 1915; Kelley, 1984; Detterman and others, 1994). The Talkeetna Formation was correlated by Plafker and others (1989c) with the lithologically and temporally equivalent andesitic arc sequence that includes the Bonanza Group in the Vancouver Island segment of the Wrangellia terrane and the Maude and Yakoun formations in the Queen Charlotte Island segments of the Wrangellia terrane (Sutherland-Brown, 1968; Jeletzky, 1976; Cameron and Tipper, 1985).

A shallow-marine to subaerial deposition for much of the Talkeetna Formation is indicated by fossil content and sedimentary facies (Grantz, 1960, 1960b; Imlay and Detterman, 1973; Detterman and Reed, 1980; Detterman and others, 1994). The Early Jurassic age (Hettangian and early Sinemurian) is based on an abundant megafauna. However, this megafauna is present in great abundance in only a few horizons and may represent mass kills as a result of volcanic eruptions (Detterman and others, 1994). At Puale Bay, the formation records an inner neritic to sublittorial environment (Detterman and others, 1994). The contact of the Talkeetna with the overlying Kialagvik Formation is disconformable, because rocks of the late Sinemurian, Pliensbachian, and most of the Toarcian stages are missing.

Kialagvik Formation and Tuxedni Group. The Early and Middle Jurassic (Callovian to Toarcian) Kialagvik Formation (Fig. 3) (Capps, 1923) overlies the Talkeetna Formation and is mainly cross-bedded sandstone, rhythmically bedded siltstone and sandstone, shale, conglomerate, and sparse limestone, and is about 400 m thick (Detterman and others, 1994). The Kialagvik Formation is exposed only in a limited area between Wide and Puale bays, and is interpreted as having formed in a gradually subsiding depositional basin. It records erosion of the volcanic part of the Talkeetna arc and incipient unroofing of the Jurassic plutonic root of the Talkeetna arc that forms the Jurassic part of the Alaska-Aleutian Range batholith (Reed and others, 1983).

The Middle Jurassic Tuxedni Group crops out from Iniskin Bay to the Talkeetna Mountains and is a 3,000-m-thick sequence of graywacke, sandstone, conglomerate, siltstone, and shale (Detterman and Hartsock, 1966; Imlay and Detterman, 1973; Imlay, 1984). The conglomerate of the Tuxedni is mainly composed of volcanic rocks in a graywacke matrix and, like the Kialagvik Formation, records erosion of the volcanic part of the Talkeetna arc. The lower part of the Tuxedni Group is correlated with the

Kialagvik Formation; the Tuxedni Group is a more complete sequence that ranges up to Bathonian(?) in age (Detterman and others, 1994).

Shelikof and Chinitna formations. The Middle Jurassic (Callovian) Shelikof and Chinitna formations (Fig. 3) (Capps, 1923) overlie the Kialagvik Formation and are composed of 1,400 m of volcaniclastic graywacke, conglomerate, and siltstone containing calcareous sandstone clasts. Deposition of these units reflects continued erosion of the volcanic part of the Talkeetna arc to the northwest (present-day coordinates). The sedimentary rocks record a shoaling depositional environment from bottom to top.

In the Cook Inlet area and in the Talkeetna Mountains, rocks similar in age to the Shelikof Formation are mapped as part of the Middle Jurassic (Callovian) Chinitna Formation (Detterman and Hartsock, 1966; Grantz, 1960a, 1960b). The Chinitna is divided into two members: a lower unit, the Tonnie Siltstone Member, and an upper unit, the Paveloff Siltstone Member. The Chinitna Formation consists dominantly of siltstone and concretionary siltstone. The upper part of the Chinitna is correlative with the Shelikof (Detterman and Hartsock, 1966). The Chinitna contains much less coarse volcanic debris than the Shelikof Formation. Both formations were deposited in similar quiescent marine depositional environments.

Naknek Formation. The Late Jurassic (Tithonian to Oxfordian) Naknek Formation (Fig. 3) consists of up to 3,200 m of arkosic sandstone, conglomerate, siltstone, and sparse limestone beds that occur within siltstone intervals. The unit was originally described by Spurr (1900) and was more thoroughly described and divided into five members by Detterman and others (1994). The Naknek is the most persistent and widespread unit on the Alaska Peninsula and possibly in southern Alaska. It crops out from the Talkeetna Mountains to the southwestern end of the Alaska Peninsula. The Naknek was deposited in a shallow-water shelf and nonmarine environment. Granitic lithic and mineral fragments are the dominant component and indicate unroofing and erosion of the Alaska-Aleutian Range batholith. Boulder conglomerate is developed in the Naknek adjacent to the Bruin Bay fault (Burk, 1965). The Naknek correlates with the Root Glacier Formation (MacKevett, 1969) of the Wrangellia terrane in the southern Wrangell Mountains.

Staniukovich and Herendeen formations and Nelchina Limestone. The Early Cretaceous (Valanginian and Berriasian) Staniukovich Formation (Fig. 3) (Detterman and others, 1994) conformably overlies the Naknek Formation in many areas. Lithologically, the Staniukovich is similar to the Naknek, although finer grained. The Staniukovich consists of siltstone, shale, and minor sandstone deposited in shallow-marine and nonmarine environments (Wilson and others, 1994). The sediment was derived from a source terrane composed in part of the Alaska-Aleutian Range batholith and in part from older sedimentary units.

The Early Cretaceous (Barremian and Hauterivian) Herendeen Formation (Fig. 3) (Detterman and others, 1994) conformably overlies the Staniukovich Formation; it is a thin, well-sorted calc-arenite that locally contains abundant *Inoceramus* prisms, and angular to subrounded quartz and feldspar grains. North of the Mount Katmai area, rocks of Early Cretaceous age are generally not known in the Chignik subterrane, although a thin (100 m thick) Early Cretaceous clastic unit that includes, in part, the Nelchina Limestone was mapped by Csejtey and others (1978) and Grantz (1960a, 1960b) in the Talkeetna Mountains quadrangle. The Nelchina Limestone is a lithologic and stratigraphic equivalent of the Herendeen Formation.

Uplift in middle Cretaceous time resulted in removal of much of the Herendeen and Staniukovich formations and parts of the Naknek Formation. The uplift appears to have been the most pronounced in the Puale Bay to Lake Iliamna area. However, in the Mount Katmai area, small patches of Albian rocks (Pedmar Formation) have been found that indicate deposition during Aptian to Santonian time.

Matanuska Formation. The stratigraphically uppermost part of the Chignik subterrane consists of the Matanuska, Chignik, Hoodoo, and Kaguyak formations, which were deposited in a short-lived marine transgression in the Early and Late Cretaceous (Fig. 3).

The Early and Late Cretaceous (Albian to Maastrichtian) Matanuska Formation (Martin, 1926; Grantz, 1961) was originally defined in the Matanuska Valley and in this area is the youngest part of the Chignik subterrane. The Matanuska also occurs in the Talkeetna Mountains and in the southern Copper River basin, mainly in the Anchorage, Valdez, and Talkeetna Mountains quadrangles (Figs. 1 and 3). The Matanuska is a flysch sequence of dark gray marine siltstone, sandstone, claystone, and minor conglomerate (Grantz and Jones, 1960; Grantz, 1964; Detterman and others, 1994). Several faunal gaps and unconformities divide the formation. The much longer depositional history of the Matanuska Formation, compared with older Late Cretaceous formations, indicates a return to a marine depositional basin sooner in the northeastern part of the subterrane than in more southwestern parts. However, the disconformities and unconformities within the Matanuska Formation indicate that the basin was not structurally stable. Conglomerate and coal throughout most of the underlying Chignik Formation indicate a nearshore to nonmarine environment, whereas conglomerate and coal seams are known only locally in the Matanuska Formation. The unit is interpreted as being deposited in a forearc apron adjacent to the Late Cretaceous and early Tertiary part of the Alaska-Aleutian Range batholith (Moore and Connelly, 1977; Winkler, 1992).

Chignik, Hoodoo, and Kaguyak formations. The Chignik and Hoodoo formations (Fig. 3) are of Late Cretaceous (Campanian and Maastrichtian) age and are exposed southwest of Puale Bay and Becharof Lake. The Chignik Formation consists of about 500 to 650 m of fluvial to shallow-marine sandstone, siltstone, and lesser coal and carbonaceous shale. The Hoodoo Formation consists of more than 600 m of rhythmically bedded sandstone and shale, and coarsens upward. The Kaguyak Formation is in the Mount Katmai area of the Alaska Peninsula, and is very similar lithologically to the Hoodoo, but is more than 1,200 m thick. The Hoodoo and Kaguyak formations were deposited in prograding submarine fans. The Chignik, Hoodoo, and Kaguyak formations are apparently in large part derived from the reworking of earlier, predominantly sedimentary strata of the area, although a volcanic-rock component was possibly added.

Younger Cenozoic assemblages. The Mesozoic rocks of the Peninsular terrane are overlain by more 6,000 m of Cenozoic volcanic and sedimentary rocks of the Aleutian arc on the Alaska Peninsula and to the northeast in the Talkeetna Mountains (Figs. 2 and 3) (Csejtey and others, 1978; Silberman and Grantz, 1984; Winkler, 1992; Detterman and others, 1994; Vallier and others, this volume; Miller and Richter, this volume) and the Meshik arc (Wilson, 1985). Thick, basinal coal-bearing rocks also occur in the Matanuska Valley, southern Talkeetna Mountains, Cook Inlet, Kenai Peninsula, and Copper River basin (Kirschner, this volume, Chapter 14).

Border Ranges ultramafic-mafic assemblage

A discontinuous belt of plutonic rocks, informally named the Border Ranges ultramafic-mafic assemblage (Plafker and others, 1989c), is up to 20 km wide and extends for more than 1,600 km along the southern margin of the Peninsular terrane from the Taral fault near the Copper River to southwest of the Kodiak Islands (unit JTru, Fig. 2; Figs. 3 and 6) (Winkler and others, 1981b; Hudson, 1983; Burns, 1985, 1994; Burns and others, 1991). The assemblage is marked by a remarkably continuous, high positive aeromagnetic anomaly throughout its length (Grantz and others, 1963; Andreason and others, 1974; Fisher, 1981; Case and others, 1986). The Border Ranges ultramafic-mafic assemblage is the southernmost unit in the Peninsular terrane and is bounded to the south by the Border Ranges fault system, which forms the southern margin of the Peninsular terrane (Figs. 2 and 6). Smaller scattered ultramafic bodies occur for about 100 km east of Tonsina, and other ultramafic bodies, locally with associated layered gabbros, occur discontinuously for several hundred kilometers to the southwest, and include the Wolverine Complex of Carden and Decker (1977) and the Eklutna complex of Burns (1985) near Anchorage, bodies on the Kenai Peninsula and on western Kodiak Island (Burns, 1985, 1993; Plafker and others, 1989c; Burns and others, 1991). Locally, klippen of the Border Ranges ultramafic-mafic assemblage, such as at Klanelneechena Creek in the western Valdez quadrangle and at Red Mountain in the Seldovia quadrangle (Fig. 1), occur south of the Border Range fault and structurally above the Chugach terrane to the south (Fig. 2). The klippen are interpreted as erosional remnants of the structurally higher Border Ranges ultramafic-mafic assemblage (Burns, 1985).

The Border Ranges ultramafic-mafic assemblage is best exposed in the eastern Peninsular terrane, along the northern Chugach Mountains, where it consists predominantly of layered gabbro with minor ultramafic rocks (Fig. 6). In the central and western parts of the assemblage, tonalite and quartz diorite are also abundant and are complexly intermixed with minor mafic and ultramafic variants (Burns and others, 1991; Winkler, 1992). The ultramafic-mafic rocks may be as thick as 14 km. Contact relations between the Border Ranges ultramafic-mafic assemblage and rocks to the north are also best exposed in the northern Chugach Mountains, where it forms a north-dipping sequence structurally overlain mainly by Mesozoic bedded volcaniclastic rocks of the Talkeetna Formation and granitic rocks along the St. Anne fault (Figs. 2 and 6) (Nokleberg and others, 1989b; Plafker and others, 1989c). The fault displays late Cenozoic normal displacement and is interpreted as having formed along a former intrusive contact. Similar Cenozoic normal displacement is interpreted for the youngest period of movement along the Border Ranges fault system to the south and southwest (Little and Naeser, 1989; Nokleberg and others, 1989b; Plafker and others, 1989c).

Tonsina ultramafic-mafic sequence and Nelchina River Gabbronorite. Detailed studies of the Border Ranges ultramafic-mafic assemblage in the northern Chugach Mountains recognize two major subdivisions: the deeper level Tonsina ultramafic-mafic sequence, and the shallower level Nelchina River Gabbronorite (Burns, 1985, 1994; Burns and others, 1991). The Tonsina ultramafic-mafic sequence is discontinuously exposed in a belt 40 km long and up to 4 km wide west of the Copper River (Fig. 6). It is juxtaposed against the northern margin of the Chugach terrane along the Border Ranges fault system and dips generally north to northwest toward a concealed contact with the Nelchina River Gabbronorite. The Tosina ultramafic-mafic sequence and correlative sequences consist of distinctive high-pressure and high-temperature cumulate sequences of variably folded and faulted layered ultramafic and mafic rocks. The principal rock types are dunite and harzburgite, websterite, and an upper unit of predominantly spinel- and garnet-bearing gabbro. The transition from the ultramafic rocks to the garnet gabbros may represent a fossil Moho.

The Nelchina River Gabbronorite (Burns and others, 1991; Burns, 1994) extends westward along the southern margin of the Peninsular terrane as a continuous belt up to 12 km wide for at least 120 km along the northern margin of the Chugach Mountains. The Nelchina River Gabbronorite consists of gabbronorite, two-pyroxene gabbro, magnetite gabbronorite, anorthositic gabbronorite, norite, and lesser hornblende gabbro (Burns and others, 1991). The gabbroic rocks are either massive with little or no directional fabric, or occur in layers up to 20 m thick that are defined by modal variations in pyroxene and plagioclase. Similar units of mafic cumulate rocks occur to the southwest and constitute most of the Border Ranges ultramafic-mafic assemblage (Burns, 1985).

In the northern part of the Chugach Mountains near Tonsina, ^{40}Ar-^{39}Ar isotopic analysis of hornblende indicates an early Middle Jurassic age of 171–181 Ma for the Border Ranges ultramafic-mafic assemblage (Winkler and others, 1981a; Plafker and others, 1989c; Onstott and others, 1989). In the central

Chugach Mountains, K-Ar, ^{40}Ar-^{39}Ar, and U-Pb isotopic analyses of zircon, hornblende, and biotite from these rocks range from about 165 to 195 Ma (Winkler, 1992). To the southwest on Kodiak Island, U-Pb zircon isotopic analysis of the Afognak pluton yields an age of 217 ± 10 Ma or latest Triassic (Roeske and others, 1989). K-Ar hornblende ages throughout the Border Ranges ultramafic-mafic assemblage yield an age range mainly from the Late Triassic to Middle Jurassic (Roeske and others, 1989; Plafker and others, 1989c; Winkler and others, 1981a; Burns and others, 1991; Winkler, 1992).

Origin of the Border Ranges ultramafic-mafic assemblage. The structural position, continuity, and lithology of the Border Ranges ultramafic-mafic assemblage clearly tie it to the Peninsular terrane (Plafker and others, 1989c). Although generally separated by faults, two areas link the Border Ranges ultramafic-mafic assemblage to the rest of the Peninsular terrane to the north. (1) Locally on northwest Kodiak Island, the latest Triassic Afognak pluton intrudes the volcanic rocks in the upper member of the Upper Triassic Shuyak Formation (Roeske and others, 1989), which is correlated with the Kamishak Formation on the Alaska Peninsula (Moore and Connelly, 1977; Connelly, 1978). (2) To the northwest near Tonsina in the northern Chugach Mountains, the Nelchina River Gabbronorite contains one inclusion of contact-metamorphosed volcanogenic sandstone interpreted as derived from the Talkeetna Formation (Plafker and others, 1989c). These relations indicate that the Border Ranges ultramafic-mafic assemblage locally intruded either the volcanic part of the arc or the stratigraphic underpinnings of the arc.

The Border Ranges ultramafic-mafic assemblage has been interpreted as originating by fractionation of basaltic magma in the roots of the Talkeetna arc, on the basis of its composition and spatial relation to the andesitic flows and related volcaniclastic rocks of the Late Triassic(?) and Early Jurassic Talkeetna Formation (Burns, 1985; DeBari and Coleman, 1989; Plafker and others, 1989c; Burns and others, 1991). Emplacement of the ultramafic part of the Border Ranges ultramafic-mafic assemblage in an active orogenic belt is suggested by high-temperature granulite facies metamorphism and textures characteristic of high-grade, synkinematic plastic deformation (DeBari and Coleman, 1989; Nokleberg and others, 1989b). Mineral assemblages in the Tosina ultramafic-mafic sequence, together with olivine, pyroxene, and spinel compositions, indicate crystallization of the lower level gabbro at about 9.5–11 kbar (28–33 km), temperatures in the range 1100–1150 °C; the higher level gabbronorite crystallized at temperatures greater than 1150 °C at unconstrained pressures above 2 kbar (6 km) (DeBari and Coleman, 1989).

The origin of the hornblende-rich gabbronorite in the upper parts of the Border Ranges ultramafic-mafic assemblage is interpreted as volatile-rich magmas that further fractionated to calc-alkalic diorites and trondhjemites that erupted to form the subparallel Talkeetna volcanic arc (Burns, 1985). Available data for the Nelchina River Gabbronorite indicate lower pressure crystallization of a hydrous tholeiitic magma (Coleman and Burns, 1985; DeBari and Coleman, 1989). The metamorphic grade of schists and hornfels intruded by the gabbronorite suggests intermediate- to shallow-level emplacement. Plastic deformation both of gabbroic rocks and of the quartz diorite, trondhjemite, and tonalite suite suggests intrusion while the gabbroic rocks were still hot enough to deform by flow (Burns, 1985, 1993; DeBari and Coleman, 1989). Major element and trace element mass-balance calculations of the Border Ranges ultramafic-mafic assemblage and of the overlying Talkeetna Formation yield a high-Mg basaltic bulk and low-Al bulk composition that is compatible with derivation by partial melting of a mantle wedge source, and that is too low to have been derived by partial melting of subducted oceanic crust (DeBari and Sleep, 1991).

An alternative interpretation (proposed by F. H. Wilson) is that the Border Ranges ultramafic-mafic assemblage is not genetically linked to the Talkeetna arc. The correlation and assumed genetic association of the Border Ranges ultramafic-mafic assemblage with the Talkeetna arc on the basis of chemistry is reasonable at the eastern end of the arc. However, the increasing spatial separation of the Border Ranges ultramafic-mafic assemblage, the Talkeetna Formation, and the Jurassic part of the Alaska-Aleutian Range batholith southwest of the Kenai Peninsula argues against a close genetic association. Given the great depth of emplacement and presumably slow cooling of the ultramafic and mafic rocks of the Border Ranges ultramafic-mafic assemblage (Burns, 1985), it is unlikely that the mainly Middle Jurassic ages reported for the Border Ranges ultramafic-mafic assemblage (Onstott and others, 1989; Plafker and others, 1989c; Winkler and others, 1981b) reflect emplacement. The Late Triassic(?) and Early Jurassic age of the Talkeetna Formation and the Middle Jurassic cooling age of the Jurassic part of the Alaska-Aleutian Range batholith are both well established. The similarity of these ages and those for the Border Ranges ultramafic-mafic assemblage also argues against a genetic association. It is unlikely that the deepest roots of the arc would yield older cooling ages than the middle or upper parts of the arc. Uplift and emplacement of the Border Ranges ultramafic-mafic assemblage adjacent to its own volcanic arc, nearly contemporaneous with arc activity, require an unlikely series of tectonic events to occur within a convergent margin.

Middle Jurassic plutonic rocks on Alaska Peninsula

Middle Jurassic plutonic rocks, the older part of the Alaska-Aleutian Range batholith, form a spectacular north-northwest–striking magmatic belt about 740 km long and from 15 to 35 km wide (unit mJg, Fig. 2) (Lanphere and Reed, 1973; Reed and Lanphere, 1969, 1973, 1974a; Reed and others, 1983; Miller, this volume, Chapter 16). The plutons intrude the Late Triassic(?) and Early Jurassic Talkeetna Formation and consist of a calc-alkaline suite with relatively low initial Sr ratios (Reed and others, 1983). The dominant lithologies are diorite, quartz diorite, and tonalite. Some studies of chemical variations across the batholith suggest emplacement above a southeast-dipping subduction zone (Reed and Lanphere, 1974a; Reed and others, 1983), whereas other studies suggest emplacement above a northwest-dipping subduction zone (Moore and Connelly, 1977). These characteristics and the spatial association with the mainly Early Jurassic Talkeetna Formation were used by Reed and others (1983) to define the Talkeetna arc.

Talkeetna arc and origin of Peninsular terrane

The major tectonic event characterizing the Peninsular terrane is the early Mesozoic Talkeetna arc, represented by the deep-level igneous rocks of the Border Ranges ultramafic-mafic assemblage, the Late Triassic(?) and Early Jurassic Talkeetna Formation, and the Middle Jurassic part of the Alaska-Aleutian Range batholith. The Talkeetna arc is a major Mesozoic feature in the study area that extends for more than 1,500 km along the length of the Alaska Peninsula and northeastward to the eastern Copper River basin (Fig. 2). The arc was originally defined by Plafker and others (1989c) as an Early Jurassic feature composed of the deep-level ultramafic and mafic rocks of the Border Ranges ultramafic-mafic assemblage that formed the roots to the volcanic and related rocks of the Talkeetna Formation. Burns (1985) interpreted the deep-level ultramafic and mafic rocks of the Border Ranges ultramafic-mafic assemblage as the source of the andesitic volcanic and associated rocks of the Talkeetna Formation. Miller (this volume, Chapter 16) places the Talkeetna Formation and Middle Jurassic granitic rocks of the Alaska-Aleutian Range batholith into the Talkeetna arc and mentions the possibility of the Border Ranges ultramafic-mafic assemblage also being part of the arc.

Moderately different ages exist for the various components of the Talkeetna arc. Plafker and others (1989c) interpreted that the mainly Middle Jurassic isotopic ages for the Border Ranges ultramafic-mafic assemblage are cooling ages from Early Jurassic igneous activity. However, a wider age range may exist for the arc. Volcanism associated with the Talkeetna arc may have begun in the Late Triassic, as indicated by the occurrence of *Halobia* and *Pseudomonotis subcircularis* in tuffaceous strata 1,500 m thick at the base of the section in the Port Graham area of the southern Kenai Peninsula (Martin and others, 1915). A latest Triassic isotopic age (217 Ma) was reported for the Afognak pluton within the Border Ranges ultramafic-mafic assemblage on Kodiak Island (Roeske and others, 1989). With this range of fossil and isotopic ages and the above field relations, the Talkeetna arc is herein interpreted to consist of (1) Late Triassic tuffaceous strata and the Early Jurassic Talkeetna Formation; (2) the Early and/or Middle Jurassic Border Ranges ultramafic-mafic assemblage; and (3) the Middle Jurassic part of the Alaska-Aleutian Range batholith.

An island-arc origin for the Talkeetna arc is indicated by geochemical data. Andesite and basalt from the Talkeetna Formation in the Talkeetna Mountains show affinities with a medium-K, tholeiitic type of orogenic andesite; these data are compatible with an island-arc origin (Barker and Grantz, 1982; Barker, this volume, see microfiche). Compatible geochemical data were also obtained at the easternmost outcrops of the Talkeetna Formation along the Richardson Highway (Plafker and others, 1989c). The Middle Jurassic plutonic rocks exhibit relatively low initial Sr ratios (Reed and Lanphere, 1973; Reed and others, 1983; Barker, this volume, see microfiche; Miller, this volume, Chapter 16). The bulk composition of the Talkeetna arc was calculated by DeBari and Sleep (1991) assuming an igneous-arc linkage for the Border Ranges ultramafic-mafic assemblage and the Talkeetna Formation.

The polarity of the arc is interpreted as southeast on the basis of the occurrence of northward-dipping, coeval subduction-zone assemblages in the northern Chugach terrane that occur across the Border Ranges fault to the south (Plafker and others, 1989c; this volume, Chapter 12). Units constituting these subduction-zone assemblages are mainly (1) discontinuous outcrops of blueschist and locally interlayered greenschist that are exposed along the northern margin of the Chugach terrane from Kodiak Island to the eastern Chugach Mountains with metamorphic ages that range mainly from 204 to 154 Ma (see summaries in Roeske, 1986; Roeske and others, 1989; Plafker and others, 1989c; this volume, Chapter 12); and (2) the older parts of the subduction-zone melange of the McHugh Complex, which contains Late Triassic through middle Cretaceous radiolarian chert (Nelson and others, 1986; Winkler and others, 1981c; Plafker and others, 1989c; this volume, Chapter 12) that may have been accreted during the existence of Talkeetna arc. These relations indicate that the latest Triassic through Middle(?) Jurassic Talkeetna arc to the north and the Late Triassic through Middle Jurassic parts of the accretionary assemblages to the south are the remnants of a paired igneous arc and subduction-zone complex, now juxtaposed along the Border Ranges fault (Plafker and others, 1989c; this volume, Chapter 12). However, these structural relations require tectonic erosion of the intervening part of the arc (Pavlis, 1983; Nokleberg and others, 1989b; Plafker and others, 1989c; Roeske and others, 1989).

An alternative interpretation (proposed by F. H. Wilson) is that the arc faced northwest and the subduction zone dipped southeastward (using present-day coordinates). This interpretation is based largely on whole-rock chemical differentiation trends for Jurassic plutonic rocks on the Alaska Peninsula (Reed and others, 1983). Included with this interpretation is the Kahiltna assemblage as a coeval accretionary prism. Problems with this interpretation are (1) the Kahiltna assemblage is younger than the Talkeetna Formation and the Middle Jurassic plutonic rocks; (2) the Kahiltna assemblage does not include any oceanic rocks typical of arc-related accretionary prisms; (3) the occurrence of the coeval blueschists and subduction-zone melange of the northern Chugach terrane cannot be explained readily by a southeast-dipping subduction zone; and (4) some studies of chemical variation across the Middle Jurassic part of the Alaska-Aleutian Range batholith suggest emplacement above a northwest-dipping subduction zone (Moore and Connelly, 1977).

In the northern Chugach Mountains, the upward transition from cumulate ultramafic and mafic rocks of the Border Ranges ultramafic-mafic assemblage to supracrustal volcaniclastic rocks of the Talkeetna Formation within a sequence no more than 6 to 7 km thick clearly suggests crustal thinning. It is possible that this thinning occurred during post–Middle Jurassic, low-angle, listric movement along detachment faults between the major units (Nokleberg and others, 1989b).

MAJOR FAULTS

Denali faults

The Denali fault (Figs. 2 and 4), one of the major tectonic boundaries in North America, extends from northern southeastern Alaska through the Alaska Range to the Bering Sea, a distance of more than 2,000 km. The fault has been studied by Sainsbury and Twenhofel (1954), St. Amand (1954, 1957), Twenhofel and Sainsbury (1958), Grantz (1966), Richter and Matson (1971), Stout and others (1973), Brogan and others (1975), Wahrhaftig and others (1975), Packer and others (1975), Plafker and others (1977; 1989c; this volume, Plate 12), Lanphere (1978), Stout and Chase (1980), Csejtey and others (1982), and Nokleberg and others (1985). In this chapter, the Denali fault is restricted to the main trace or the McKinley strand of Csejtey and others (1982). The Hines Creek splay to the north, which was included as part of the Denali fault system by Wahrhaftig and others (1975) and Csejtey and others (1982), was interpreted by Nokleberg and others (1989a, 1992b) as a separate structure that is not a terrane boundary, but instead is a younger structure mostly within the southern Yukon-Tanana terrane.

Evidence for the extent of displacement on the Denali fault consists of high-quality data for smaller amounts of movement during the Quaternary and lower quality data for greater amounts of movement during the late Mesozoic and Tertiary. Evidence for Quaternary movement on the Denali fault in the eastern and central Alaska Range was summarized by Richter and Matson (1971), Stout and others (1973), and Plafker and others (1977). Estimates of Quaternary movement on the fault since early Wisconsin or Illinoisan time range from 1 to 6.5 km (Stout and others, 1973); probable rates of Holocene movement average about 1.5 cm/yr (Plafker and others, this volume, Plate 12). The sense of movement in all places is right lateral. Evidence for earlier Tertiary dextral displacement along the Denali fault, based on offset of the Foraker and McGonagal plutons near Mount McKinley, is about 38 km in 38 m.y. (Reed and Lanphere, 1974b).

Evidence for substantial Late Cretaceous and early Tertiary dextral displacement along the Denali fault consists of (1) a minimum of 300 km of offset of flysch of the Late Jurassic and Early Cretaceous Gravina-Nutzotin belt between the Nutzotin Mountains in the Wrangell Mountains and similar rocks to the southeast in the Dezadeash Basin near the Ruby Mountains (Eisbacher, 1976); (2) offset of the Maclaren and Aurora Peak terranes, which have early Tertiary granitic rocks and metamorphic fabrics, in the eastern Alaska Range, from similar rocks about 400 km to the southeast in the southwestern Yukon Territory (Forbes and others, 1973; Smith and others, 1974; Nokleberg and others, 1985, 1989a); (3) offset of the Broxson Gulch thrust, which offsets the Maclaren and Wrangellia terranes, from the same area (Eisbacher, 1976; Nokleberg and others, 1985); and (4) post-Triassic dextral offset of part of the Wrangellia terrane in the eastern Wrangell Mountains to a position at least 350 km to the southeast on the Chilkat Peninsula in northern southeastern Alaska (Plafker and others, 1989a).

In addition to offset units, the very distinct structural melange of the Paleozoic and Mesozoic Windy terrane occurs along the Denali fault between the Aurora Peak terrane and the Late Cretaceous and early Tertiary East Susitna batholith of the Maclaren terrane (Figs. 2 and 4). The occurrence of the Windy terrane for several hundred kilometers along the Denali fault indicates substantial offset between highly disparate bedrock units. Also in the central and eastern Alaska Range, significantly different terranes occur on opposite sides of the Denali fault (Hickman and Craddock, 1976; Richter, 1976; Sherwood and Craddock, 1979; Jones and others, 1987; Brewer, 1982; Sherwood and others, 1984; Nokleberg and others, 1985, 1992b). All of the above data are incompatible with the interpretations of Csejtey and others (1982), who indicated only a few kilometers offset along the Denali fault during the Cenozoic.

Nenana Glacier fault

The Nenana Glacier fault (Brewer, 1982; Nokleberg and others, 1985, 1989a, 1992b) occurs a few kilometers north of the Denali fault in the central and eastern Alaska Range in the Mount Hayes and Healy quadrangles (Figs. 1 and 2). The fault juxtaposes the Aurora Peak terrane to the south from the Yukon-Tanana terrane to the north (Figs. 2 and 4). The chief evidence for the fault includes (1) a narrow zone of ductile deformation and intense shearing with locally abundant retrogressive metamorphism of the high-grade rocks of the Aurora Peak terrane; (2) local formation of structural melange between the Aurora Peak and Yukon-Tanana terranes; and (3) juxtaposition of the amphibolite facies Cretaceous rocks of the Aurora Peak terrane against the low-grade Devonian rocks of the Hayes Glacier subterrane of the southern Yukon-Tanana terrane. The Nenana Glacier fault is interpreted as a slightly older Late Cretaceous or early Tertiary branch of the Denali fault (Brewer, 1982; Nokleberg and others, 1985, 1989a, 1992b).

Ancestral Denali fault

Geologic and geophysical data suggest that Mesozoic flysch along the northern margin of the Wrangellia terrane was thrust under the southern margin of the multiply metamorphosed and deformed Paleozoic Yukon-Tanana terrane in the middle to Late Cretaceous along the ancestral Denali fault. Magnetotelluric and seismic reflection data for the southern Yukon-Tanana terrane are interpreted as indicating that large volumes of conductive rocks exist beneath the southern Yukon-Tanana terrane in the eastern and central Alaska Range (Labson and others, 1988; Stanley and others, 1990). The highly conductive rocks are interpreted to be carbonaceous units that appear to dip northward under the highly resistive rocks of the southern Yukon-Tanana terrane. The carbonaceous units are probably a combination of (1) a formerly extensive, but now collapsed, Mesozoic flysch basin now partly preserved in the Kahiltna assemblage and the Gravina-Nutzotin belt; and (2) other carbonaceous sedimentary rocks in Paleozoic sedimentary terranes now preserved as remnants along the Denali fault. This flysch basin, described above, is interpreted as initially having formed on the leading or northern edge of the WCT (Nokleberg and others, 1985; Plafker and others, 1989c; Rubin and Saleeby, 1991). The docking of this terrane may have caused large-scale oblique underthrusting, folding, and metamorphism in the flysch basin, within which the Cenozoic strike-slip Denali fault subsequently developed.

Two geologic observations support a period of Cretaceous oblique-slip underthrusting along the ancestral Denali fault. A zone of intense retrograde metamorphism, with K-Ar mica ages of 100 to 110 Ma (Nokleberg and others, 1986), associated ductile deformation, and south-verging structures occurs along the southern margin of the Yukon-Tanana terrane. This zone occurs for several hundred kilometers adjacent to the Denali fault and is up to 20 km wide (Nokleberg and Aleinikoff, 1985; Nokleberg and others, 1986). The intensity of retrograde metamorphism and ductile deformation increases toward the fault, and lowest greenschist facies rocks and intensely deformed, mylonitic schist occur adjacent to the fault. South-verging, tightly appressed to isoclinal minor folds with axial plane zones of mylonite are also observed in the zone of mylonitic schist. This fabric indicates that retrograde metamorphism occurred during thrusting of the southern Yukon-Tanana terrane onto units to the south. The retrograde metamorphism was interpreted by Stanley and others (1990) as occurring during the underthrusting of cool, wet flysch of the Kahiltna assemblage. Dark colored argillite and metagraywacke of Cretaceous age form a major part of the melange of the Windy terrane that occurs for several hundred kilometers along the Denali fault in the central and eastern Alaska Range (Fig. 2). These lenses are interpreted as relicts of the flysch of the Kahiltna assemblage that remained after oblique underthrusting along the ancestral Denali fault and subsequent Cenozoic dextral strike-slip movement along the fault.

Talkeetna thrust

The Talkeetna thrust dips moderately to steeply southeastward along the northwestern margin of the Peninsular terrane in the central and western Alaska Range, mainly in the Healy, Mount McKinley, and Talkeetna quadrangles (Figs. 1 and 2) (Csejtey and others, 1978, 1982, 1986; Csejtey and St. Aubin, 1981). The thrust occurs between the Kahiltna assemblage to the north and the older bedrock of the Peninsular and Wrangellia terranes to the south (Csejtey and others, 1982, 1986). Toward the east in the western Mount Hayes quadrangle, the Talkeetna thrust terminates against the younger, north-dipping Broxson Gulch thrust (Nokleberg and others, 1985, 1989a).

There are complicated structural relations along the Talkeetna thrust (Csejtey and others, 1982). Subsidiary, southeastward-dipping imbricate faults along the thrust consistently bring older rocks of the Wrangellia terrane to the south on top of younger rocks of the Kahiltna assemblage to the north. Folds associated with the thrust are generally tight or isoclinal, and they range in size from small secondary folds to major folds having an amplitude of several kilometers. A northwestward tectonic transport is also indicated by drag folds in the upper plates of the subsidiary thrusts and by northwestward, overturned folds in the flysch north of the Talkeetna thrust. Series of complex and imbricate thrust faults with local reversals of structural vergence adjacent to the Chulitna terrane are interpreted as the result of local stress reversal during late orogenic folding. In other areas, major folds with amplitudes of at least 10 km are associated with the thrust.

Movement along the Talkeetna thrust and the intense deformation of the Peninsular and Wrangellia terranes and the Kahiltna assemblage, both above and below the thrust, are interpreted as occurring during the middle and Late Cretaceous emplacement of the WCT to Alaska (Csejtey and others, 1982; Nokleberg and others, 1985; Plafker and others, 1989c). The age of the orogeny is bracketed by the youngest deformed rocks (the Late Jurassic and Early Cretaceous flysch) and by the ages (Late Cretaceous and early Tertiary) of the oldest undeformed intrusive rocks (Csejtey and others, 1978, 1982). The orogeny involved complex thrusting and folding, low- to medium-grade regional metamorphism, subordinate high-angle faulting, and late tectonic plutonism. This deformation produced a prominent, generally northeast trending structural grain.

Broxson Gulch thrust

The Broxson Gulch thrust (Stout, 1976; Nokleberg and others, 1982, 1985, 1992b) occurs in the eastern Alaska Range between the Wrangellia terrane to the south and the Maclaren terrane to the north (Figs. 2 and 4). The thrust extends to the west from an acute termination against the Denali fault in the eastern Alaska Range. To the west, the thrust truncates the relatively older Talkeetna thrust (Csejtey and others, 1986). The main evidence for the Broxson Gulch thrust is (1) truncation of various units in both terranes; (2) locally abundant, highly sheared rocks; and (3) juxtaposition of units of diverse stratigraphy, age, and structure.

Like the Talkeetna thrust, the Broxson Gulch displays a complex history of movement (Nokleberg and others, 1985, 1989a). The first period of movement consisted of thrusting of the Wrangellia terrane onto the Maclaren terrane after deposition of the Late Jurassic and Early Cretaceous strata of the Gravina-Nutzotin belt. The thrust is interpreted as having dipped southward (oceanward) during this period. The second period of movement consisted of substantial strike-slip offset, which resulted in dislocation of a small fault-bounded wedge of the Triassic and Jurassic McCarthy Formation from the eastern Wrangell Mountains to the central Alaska Range. Similar strike-slip movement may have occurred along the Talkeetna thrust to the west (Csejtey and others, 1982). The third period of movement consisted of overturning of the fault to a north dip and reactivation with south-verging movement. South-vergent movement is indicated by (Nokleberg and others, 1982, 1985, 1989a, 1992b) (1) juxtaposition of older bedrock of the Wrangellia terrane over Tertiary sedimentary and volcanic rocks, and over Wisconsin glacial deposits along various north-dipping branches of the fault; and (2) refolding of schistosity and compositional layering in the southern part of the Maclaren terrane to north dips near the thrust.

Paxson Lake and Chitina Valley faults

The Wrangellia terrane is bounded on the south by the Paxson Lake fault in the northern Copper River Basin (Fig. 2) (Nokleberg and others, 1986) and by the Chitina Valley fault along the southern margin of the Wrangell Mountains (Fig. 2) (Plafker and others, 1989c). To the south of these faults is the southern Wrangellia terrane (Fig. 2).

The Paxson Lake fault occurs in the southern flank of the Alaska Range and in the northern flank of the Talkeetna Mountains. The fault is not exposed in the southern Alaska Range, where it strikes mainly east-west; a steep to vertical dip is suggested by a straight trace (Figs. 2 and 4) (Nokleberg and others, 1986, 1989a). Evidence for the fault consists of (1) the intensely deformed and steeply dipping rocks of the metamorphic complex of Gulkana River adjacent to the nonpenetratively deformed and gently dipping units of the Wrangellia terrane; and (2) the presence of thick and massive hornblende andesite and two-mica granitic plutons in the metamorphic complex of Gulkana River compared with few, if any, similar rocks in the Wrangellia terrane. The Paxson Lake fault is most narrowly constrained on the southwest flank of Paxson Mountain to a colluvium-covered zone about 500 m wide. The only unit covering the fault is composed of Quaternary glacial deposits. The Paxson Lake fault continues to the west in the Talkeetna Mountains quadrangle, where it is mapped in bedrock either as (1) a southeast-dipping zone of intense shearing that occurs along the southern boundary of the Wrangellia terrane (Csejtey and others, 1978); or (2) a northwest-dipping, faulted disconformity (Kline and others, 1990).

The Chitina Valley fault occurs in the southern Wrangell Mountains (MacKevett, 1978; Gardner and others, 1988; Plafker and others, 1989c). The fault strikes east-southeast and dips steeply southwest. It juxtaposes late Paleozoic metamorphic rocks, Triassic greenstone and marble, and Late Jurassic plutons of the southern Wrangellia terrane to the south against the weakly metamorphosed and less-intensely deformed sedimentary, volcanic, and plutonic rocks of the Wrangellia terrane to the north. Pre–Early Cretaceous deformation along the fault is indicated by strata of Albian age deposited unconformably across the fault. Structures in rocks north of the fault are north-verging, whereas south of the fault, the upper Paleozoic rocks of the southern Wrangellia terrane are deformed into south-verging, nearly recumbent folds that are cut by intensely deformed Late Jurassic granitoids of the Chitina Valley batholith (MacKevett, 1978). These relations indicate that at least two periods of penetrative deformation have affected the rocks along the southern margin of the Wrangellia terrane.

West Fork fault

The West Fork fault occurs between the southern Wrangellia terrane to the north and the Peninsular terrane to the south in the northern Copper River basin (Fig. 2). The fault is everywhere covered by surficial deposits, mainly glaciofluvial or glacial lake deposits, but generally strikes east-west to southeast, and appears to be near vertical based on a straight trace (Fig. 3) (Nokleberg and others, 1986). Geologic evidence for the fault consists of the extensive, nonmetamorphosed, gently dipping Late Triassic(?) and Early Jurassic Talkeetna Formation to the south occurring adjacent to locally highly metamorphosed and deformed, late Paleozoic(?) metamorphosed volcanic, sedimentary, and plutonic rocks of the metamorphic complex of Gulkana River to the north. Geophysical evidence for the fault consists primarily of the major east-west–striking Sourdough magnetic high that occurs along the fault (Andreasen and others, 1974; Campbell and Nokleberg, 1986). This magnetic high is interpreted as a near-vertical body of mafic or ultramafic rocks between the Peninsular terrane and the metamorphic complex of Gulkana River. The West Fork fault trends southeastward and may connect with the Taral fault underneath the unconsolidated Quaternary deposits of the Copper River basin and the Tertiary volcanic rocks of the Wrangell arc (Winkler and others, 1981a), where a similar magnetic high marks the concealed inferred trace of the fault (Case and others, 1986).

Bruin Bay fault

The Bruin Bay fault separates the Iliamna and Chignik subterranes of the Peninsular terrane on the Alaska Peninsula. The fault occurs between the mainly Late Triassic and Early Jurassic strata of the Iliamna subterrane to the northwest from the mainly Middle and Late Jurassic strata of the Chignik subterrane on the southeast (Fig. 2). The fault strikes northeast, dips northwest, and displays high-angle reverse (northwest side up) movements with a total stratigraphic throw of as much as 3 km (Detterman and Reed, 1980). A component of left-lateral strike-slip motion is possible; however, the evidence is tenuous (Detterman and Reed, 1980, p. 69). Preferential development of conglomerate in the Naknek Formation adjacent to the Bruin Bay fault suggests that the fault may have been active in Late Jurassic (Oxfordian) time (Burk, 1965; Detterman and Reed, 1980). Displacement predates emplacement of Oligocene plutons across the fault. The Bruin

Bay fault has been inferred to extend south of Becharof Lake (Jones and Silberling, 1979; von Huene and others, 1985; Lewis and others, 1988). However, the available data are not conclusive because the faults along strike south off Becharof Lake dip in the opposite sense from the Bruin Bay fault and have the opposite sense of slip, i.e., northwest side downthrown.

LINKAGES BETWEEN WRANGELLIA, ALEXANDER, AND PENINSULAR TERRANES

As defined by Jones and others (1981), the terranes constituting the WCT were interpreted as structurally bound blocks with unique geologic and displacement histories. This interpretation was strongly influenced by early paleomagnetic data of variable reliability that suggested large relative motions between the Alexander, Peninsular, and Wrangellia terranes (Hillhouse, 1977; Hillhouse and Grommé, 1980, 1984; Stone and Packer, 1979; Stone and others, 1982). Geologic, paleontologic, and more recent paleomagnetic studies favor the alternative that these terranes have been a single terrane throughout much, if not all, of their decipherable histories, and that differences between them can be explained by differences in exposed structural level and/or facies variations. These available data suggest the following. (1) The Alexander terrane was amalgamated with the Wrangellia terrane and was the basement beneath at least part of the Wrangellia terrane by at least early Pennsylvanian time. (2) The northern part of the former Taku terrane of southeastern Alaska is a displaced fragment of Wrangellia. (3) The combined Wrangellia-Alexander terranes were amalgamated with the Peninsular terrane at least by Late Triassic time and possibly by the late Paleozoic. The various types of geologic linkages between sections of the Wrangellia, Alexander, Peninsular, and former northern Taku terranes are shown schematically in Figure 8 and are summarized briefly below.

Paleozoic linkages

A possible early Paleozoic linkage between the Wrangellia terrane in British Columbia and the Alexander terrane in southeastern Alaska is suggested by extensive arc magmatism in both terranes, followed by sporadic volcanism and deposition of thick carbonate and chert sequences in the late Paleozoic (Fig. 8, linkage A). Metamorphosed arc volcanic and plutonic rocks of the Sicker Group on Vancouver Island are dated at 370–420 Ma, although the age of the base of this sequence is not known (Brandon and others, 1986). In the southern Alexander terrane, the possibly coeval metamorphic arc complex is mainly of Early Ordovician to Early Devonian age; associated plutonic rocks range in age from 430–405 Ma (Gehrels and Saleeby, 1987). A difference between the two regions is that the coarse clastic rocks of the Early Devonian Klakas orogeny in the Alexander terrane (Gehrels and Saleeby, 1987) are not recognized on Vancouver Island. Paleomagnetic data suggest that the Alexander terrane was about 15° south of its present latitude relative to North America in Pennsylvanian time, assuming an origin in the Northern Hemisphere (Van der Voo and others, 1980); comparable paleomagnetic data are unavailable for the Wrangellia terrane in Alaska, where the abundant oldest known rocks are Pennsylvanian.

The Wrangellia terrane in Alaska and the northwestern part of the Alexander terrane in the southeastern Wrangell Mountains are stitched together by (1) a belt of gabbroic, granitic, and alkalic plutonic rocks with middle Pennsylvanian zircon ages of 315–305 Ma (Fig. 8, linkage B) (Gardner and others, 1988; Aleinikoff and others, 1988); and (2) volcanic rocks of the associated Pennsylvanian and Early Permian Skolai arc (Fig. 8, linkage C; Muller, 1967; Gardner and others, 1988). The southeastern limit of the late Paleozoic igneous complex is unknown; coeval andesitic volcanic rocks and syenitic intrusive rocks with a single K-Ar biotite minimum age of 277 Ma occur locally in the southern Alexander terrane in southeastern Alaska and could represent part of this arc (Fig. 8; Gehrels and Berg, 1992; this volume). Possibly equivalent schistose metavolcanic and metasedimentary rocks, locally associated with mainly Early Permian carbonate rocks, occur in the Peninsular terrane along the northern margin of the Chugach Mountains in the Seldovia area and on the Alaska Peninsula (Figs. 3 and 7) (Plafker and others, 1989c; Wallace and others, 1989).

Permian carbonate rocks also provide possible stratigraphic linkages between the Alexander, Peninsular, and Wrangellia terranes in the study area and the northern former Taku terrane (Fig. 8, linkage D) (Jones and others, 1987; Plafker and others, 1989a). The carbonate rocks are mainly Early Permian in age, except on the Alaska Peninsula, where a single Late Permian occurrence is known. Pennsylvanian and Permian megafaunas from these terranes at the species level show close ties among the terrane faunas and also between the terrane faunas and North America (Mackenzie Gordon, Jr. and J. T. Dutro, Jr., 1991, written commun.).

Middle and Late Triassic linkages

The Alexander, Peninsular, Wrangellia, and former northern Taku terranes are linked by distinctive thick Middle to Late Triassic eruptive sequences and overlying shallow-marine strata (Fig. 8; linkages E and F). The eruptive rocks in these terranes are dominantly basaltic in sequences up to several kilometers thick; those in the Alexander terrane are either entirely basaltic or bimodal basalt and felsic volcanic rocks that are interbedded with marine sedimentary rocks in sequences up to several hundred meters thick (Jones and others, 1977; Detterman and Reed, 1980; Gehrels and others, 1986; Plafker and others, 1989c; Winkler, 1992).

The Wrangellia terrane in Alaska is linked with the Wrangellia terrane in British Columbia by thick sequences of Ladinian to Carnian basalt (Nikolai Greenstone and Karmutsen Formation) and overlying Late Triassic marine strata (Fig. 8, linkages E and F) (Jones and others, 1977). A comparable sequence of Carnian and older(?) greenstone, overlain by thin carbonate rocks

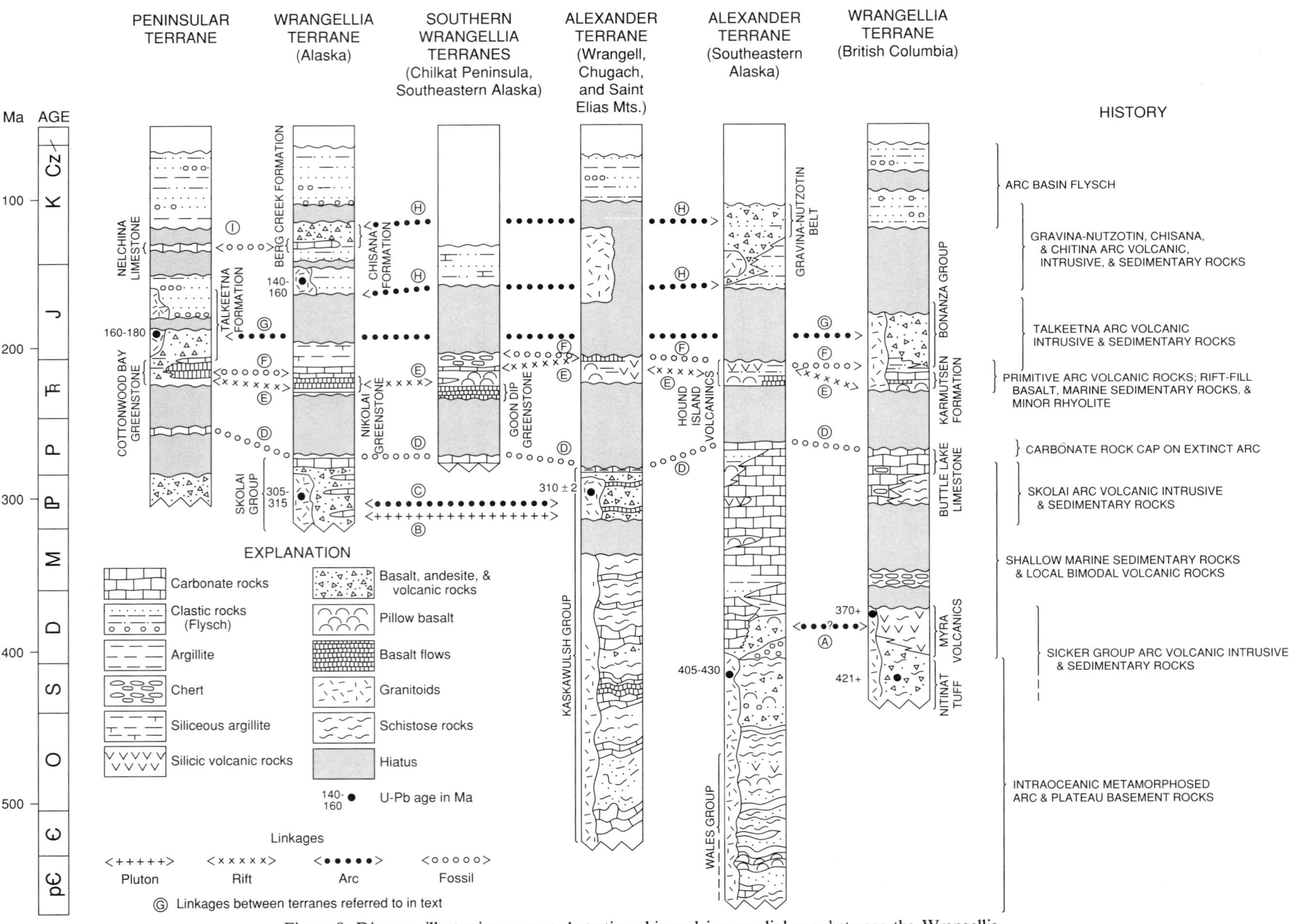

Figure 8. Diagram illustrating suggested stratigraphic and igneous linkages between the Wrangellia, southern Wrangellia, Peninsular, Alexander, and northern Taku (fragment of Wrangellia) terranes, south-central Alaska. Refer to text for discussion.

and chert of latest Carnian to late Norian age, occurs on the Chilkat Peninsula of southern Alaska in the Wrangellia terrane (former northern Taku terrane) (Plafker and others, 1989a). A similar, but slightly younger, sequence of basalt overlain by shallow-marine sedimentary rocks is also identified in the Peninsular terrane on the Alaska Peninsula, where both the Norian Cottonwood Bay Greenstone and the Kamishak Formation (limestone) have been correlated with the Nikolai Greenstone and Nizina Limestone, respectively, in the type area of the Wrangellia terrane (Detterman and Reed, 1980; Wilson and others, 1994). Undated greenstone and limestone in the southern Talkeetna Mountains area of the Peninsular terrane are tentatively correlated with the lithologically similar Nikolai Greenstone and overlying Chitistone Limestone of the Wrangellia terrane in the Wrangell Mountains (Winkler, 1992).

Reconstructions of paleogeographic distribution of the terranes based on studies of species endemism of Triassic bivalves suggest an ocean-island chain origin for the Wrangellia and Peninsular terranes (Newton, 1983), a scenario that is compatible with the paleomagnetically determined paleolatitude of about 30° south relative to the North American craton for the Alexander, Peninsular, and Wrangellia terranes (Hillhouse and Gromme, 1984; Haeussler and others, 1989; P. W. Plumley, 1990, oral commun.).

Late Triassic and Jurassic linkages

The Peninsular terrane and Wrangellia terrane of British Columbia are linked by major arc sequences, mainly of Early Jurassic age, but ranging in age from Late Triassic through early Middle Jurassic (Fig. 8, linkage G). These sequences are inferred (Plafker and others, 1989c; this volume, Chapter 12) to be a previously continuous arc that was disrupted by more than 600 km of sinistral strike-slip displacement. This correlation is based on the striking similarity in age and depositional environment of the two sequences, as well as the abrupt termination of continuous belts of the Talkeetna Formation and related intrusive rocks at the eastern end of the Peninsular terrane in Alaska (Jones and others, 1987) and the presence of coeval parts of the Bonanza Group and related units in the Wrangellia terrane of British Columbia (Muller and others, 1974; Cameron and Tipper, 1985). The Late Triassic and Early Jurassic section in the type area of the Wrangellia terrane in the Wrangell Mountains of Alaska is dominantly carbonate, but coeval arc volcanism in the region is inferred from the occurrence of lithic fragments of microlitic tuff throughout the Norian to Pleinsbachian Nizina Limestone and McCarthy Formation (M. Mullen, 1990, oral commun.). Faunal assemblages for both terranes are compatible with the paleomagnetic data in that they suggest deposition in subequatorial paleolatitudes (Imlay, 1984; Cameron and Tipper, 1985).

Late Jurassic and Cretaceous linkages

The Alaska segment of the Wrangellia terrane and the southern Alexander terrane are linked by the Chitina and Chisana magmatic arcs and associated sedimentary rocks during the latest Jurassic through Early Cretaceous (Fig. 8, linkage H). These sequences consist of distinctive arc-related volcanic, volcaniclastic, and intrusive rocks that compose the Gravina-Nutzotin belt (Berg and others, 1972; Gehrels and Berg, this volume) and probably the correlative Late Jurassic Tonsina-Chichagof and Early Cretaceous Nutzotin-Chichagof plutonic belts (Hudson, 1983; Miller, this volume, Chapter 16).

The Peninsular and Wrangellia terranes in Alaska are linked during the Valanginian to Hauterivian (early Early Cretaceous) by the distinctive fossiliferous Nelchina Limestone and Berg Creek Formation that extend discontinuously from the Alaska Peninsula to the southern Wrangell Mountains (Fig. 8, linkage I) (Plafker and others, 1989c). The two terranes may also be linked by the coeval and possibly originally connected sedimentary and lesser volcanic rocks of the Late Jurassic and Early Cretaceous Kahiltna assemblage and Gravina-Nutzotin belt.

WRANGELLIA COMPOSITE TERRANE

Stratigraphic relations and structural imbrication

The Alexander, Wrangellia, and Peninsular terranes exhibit stratigraphic, volcanic, and plutonic ties from the late Paleozoic through the Cretaceous. Because of these relations and linkages, the three terranes are interpreted as constituting the WCT (Plafker and others, 1989c). The chief differences between the three terranes are notable contrasts in the relative abundances of coeval strata. For example, in the study area, the Alexander terrane is mainly an early and middle Paleozoic volcanic arc terrane with only a thin expression of the late Paleozoic and early Mesozoic units of the Wrangellia terrane. Similarly, the Wrangellia terrane is mainly a late Paleozoic island-arc sequence and an early Mesozoic mafic magmatism sequence. Pre–late Paleozoic basement rocks, possibly of the Alexander terrane, which are interpreted to underlie the Wrangellia terrane or to have been juxtaposed against the Wrangellia terrane by the middle Pennsylvanian, are found locally in the eastern part of the study area in the Nutzotin Mountains and in the southeastern Wrangell Mountains. Similarly, the Peninsular terrane consists mainly of an Early and Middle Jurassic island-arc sequence. Only locally does a small part of the Peninsular terrane exhibit either (1) the volcanic and sedimentary rocks of the late Paleozoic Skolai island arc and the Late Triassic rift basalt that form the major part of the Wrangellia terrane or (2) possible fragments of metamorphosed Paleozoic or older metasedimentary and metavolcanic rocks that might be the early and middle Paleozoic part of the Alexander terrane.

The above relations indicate that, from east to west, the Alexander, Wrangellia, and Peninsular terranes exhibit successively higher levels of a structural-stratigraphic succession. The oldest part of the succession is mainly to the east in the predominantly early and middle Paleozoic Alexander terrane; the middle part of the succession in the central part of the predominately late Paleozoic and early Mesozoic Wrangellia terrane; and the youngest part of the succession is to the west in the predominantly middle and late Mesozoic Peninsular terrane.

Despite these linkages, substantial tectonic imbrication has occurred within the WCT (Plafker and others, 1989c). The abrupt termination of the Talkeetna arc and associated plutonic rocks in the eastern Peninsular terrane structurally against the very minor occurrence of this arc in the Wrangellia terrane requires extensive post–Middle Jurassic tectonic displacement of the Wrangellia terrane relative to the Peninsular terrane. This post-Jurassic displacement resulted in juxtaposition of different facies, and, therefore, in the original designation of the two units as separate terranes.

The timing of the suggested strike-slip displacement between the Peninsular and Wrangellia terranes cannot be determined more closely than the Late Jurassic to Late Cretaceous. The deformation affects Late Jurassic plutons and may have been synchronous with Late Jurassic Chitina arc magmatism. The displacement occurred prior to (mid[?]-Cretaceous) accretion of the younger parts of the melange of the McHugh Complex of Early Cretaceous (Berriasian to Valanginian) age to the southern Wrangellia terrane, because these rocks do not appear to be deformed by the strike-slip event. Along the northern margins of the Wrangellia and Peninsular terranes, geologic data indicate that the combined terranes were sutured to each other and to North America by the middle Cretaceous in mainland Alaska (Csejtey and St. Aubin, 1981; Nokleberg and others, 1985) and British Columbia (Monger and others, 1982).

Similar extensive tectonic imbrication has occurred within parts of the Wrangellia terrane. The tectonically juxtaposed Slana River and Tangle subterranes exhibit notable contrasts in the relative abundances and thicknesses from the late Paleozoic through the late Mesozoic. If, as suggested above, the Tangle subterrane formed originally close to the variant of the Wrangellia terrane on Vancouver Island, then considerable tectonic dismemberment and recombining of various elements of the Wrangellia terrane occurred in the late Mesozoic and Cenozoic.

Paleomagnetic studies indicate that the WCT was about 30° south of its present latitude relative to North America during the Late Triassic (Haeussler and others, 1989; Hillhouse and Coe, this volume). The WCT was juxtaposed against North America by the middle Cretaceous (Csejtey and others, 1982; Nokleberg and others, 1985; Plafker and others, 1989c; Grantz and others, 1991). The terrane is interpreted to have arrived at its present position between about the Cenomanian and the Eocene (Hillhouse and Coe, this volume). These studies suggest a long history of tectonic migration with ample opportunity for internal dismemberment of the WCT.

Structural and tectonic relations across the eastern Wrangellia composite terrane

The structural and tectonic relations across the eastern WCT, parallel to the Trans-Alaskan Crustal Transect (TACT) and the Richardson Highway, are displayed in three cross sections in Figures 4 and 6. In the following discussion, the three cross sections are described and interpreted from south to north.

Cross section A-A′ (Fig. 6) displays the important structural and tectonic relations between the northern Chugach, southern Peninsular, and southern Wrangellia terranes. To the south, the moderate-grade metamorphic rocks of the southern Wrangellia terrane, interpreted as a fragment of an island arc, occur in a major klippe that is floored by a splay of the Border Ranges fault. The klippe rests on the low-grade Late Cretaceous flysch of the Valdez Group in the northern Chugach terrane. Juxtaposition of the southern Wrangellia and Chugach terranes occurred in the early Cenozoic after accretion and underthrusting of the Valdez Group beneath the relatively older subduction-zone complex of the McHugh Complex (Nokleberg and others, 1989b; Plafker and others, 1989c). Farther north, the Border Ranges ultramafic-mafic assemblage to the north is thrust over the blueschist and greenschist unit of the schist of Liberty Creek along another splay of the north-dipping Border Ranges fault. If the deep-level island-arc rocks of the Border Ranges ultramafic-mafic assemblage are tectonically paired to the blueschists along the northern margin of the Chugach terrane, then the entire fore-arc basin that originally existed between the two units has been tectonically removed. Farther north, the Late Triassic(?) and Early Jurassic Talkeetna Formation is normally faulted along the St. Anne fault onto the Late Triassic through Middle(?) Jurassic Border Ranges ultramafic-mafic assemblage to the south. If both units are part of the Talkeetna arc, then juxtaposition of these two units requires moderate disappearance of stratigraphic section. Locally, one inclusion of volcaniclastic sandstone, interpreted as a part of the Talkeetna, is observed in the Nelchina River Gabbronorite (Plafker and others, 1989c).

Cross section A-A′ (Fig. 4) displays the important structural and tectonic relations between the northern Peninsular, southern Wrangellia, and Wrangellia terranes. To the south, the Talkeetna Formation of the Peninsular terrane is juxtaposed along the strike-slip (?) West Fork fault against the southern Wrangellia terrane to the north. The unmetamorphosed Jurassic or Cretaceous sedimentary, volcanic, and plutonic rocks of the Peninsular terrane contrast sharply with the regionally deformed and metamorphosed sedimentary, volcanic, and plutonic rocks of the southern Wrangellia terrane (Nokleberg and others, 1986, 1989a). Farther north, the southern Wrangellia terrane is juxtaposed along the strike-slip(?) Paxson Lake fault against the Wrangellia terrane to the north. In contrast to the southern Wrangellia terrane, the Wrangellia terrane exhibits a weak granoblastic, lower greenschist facies fabric. Strike-slip movement along both faults is poorly constrained because of poor exposure and extensive glacial deposits.

Cross section B-B′ (Fig. 4) displays the important structural and tectonic relations between the northern Wrangellia, Maclaren, Windy, Aurora Peak, and southern Yukon-Tanana terranes. To the south, the island-arc Wrangellia terrane is thrust under the continental margin arc Maclaren terrane along the north-dipping Broxson Gulch thrust. This fault is interpreted as the locus of middle Cretaceous accretion of the Wrangellia terrane onto the North American continental margin, and originally dipped south but was subsequently rotated to a north dip (Nokleberg and others, 1985). The low-grade metamorphism and broad folding of the Wrangellia terrane contrasts sharply with the high-grade and multiply deformed metamorphic rocks of the Maclaren terrane. Farther north, the high-grade metaplutonic rocks of the East Susitna batholith of the Maclaren terrane are juxtaposed against the low-grade metasedimentary rocks in the structural melange of the Paleozoic and Mesozoic sedimentary and volcanic rocks of the Windy terrane. Late Cretaceous and early Tertiary metamorphic ages for the Maclaren terrane require early to middle Tertiary juxtaposition against the Windy terrane. This relation and the occurrence of the Windy terrane for several hundred kilometers along the Denali fault are among several lines of evidence for substantial Cenozoic strike-slip displacement along the Denali fault. Farther north, the Windy terrane is juxtaposed against the continental margin arc Aurora Peak terrane which, like the Maclaren terrane, is interpreted as a fragment of the Mesozoic North American continental margin that migrated along the Denali and Nenana Glacier faults (west of the plane of section) from a site to the southeast (Nokleberg and others, 1985). Farther north, the Aurora Peak terrane is juxtaposed against the Paleozoic Yukon-Tanana terrane, which is interpreted as an older displaced fragment of the Paleozoic North American continental margin (Foster and others, 1987, and this volume; Nokleberg and others, 1989a). The Yukon-Tanana terrane occurs in east-central Alaska and discontinuously along the eastern side of southeastern Alaska to near Vancouver, Canada (Foster and others, 1987, and this volume; Gehrels and Berg, this volume).

CONCLUSIONS

The interpretations presented herein are compatible with most geologic data in the study area, paleomagnetic data for the Wrangellia terrane, and major aspects of plate reconstructions for the northeast Pacific region. These interpretations differ from previous ones (Coney and others, 1980; Jones and Silberling, 1982; Monger and others, 1982; Saleeby, 1983; Coney and Jones, 1985; Howell and others, 1985; Umhoefer, 1987) in that they require tectonic juxtaposition, by late Paleozoic time, of the Wrangellia and Alexander terranes, and by at least the Late Triassic of the Wrangellia and Peninsular terranes. These three terranes, interpreted as the WCT, may have been close to each other since the late Paleozoic, with younger and structurally higher units exposed from east to west. The mainly early and middle Paleozoic Alexander terrane to the east is interpreted as the stratigraphic basement for the mainly late Paleozoic and early Mesozoic Wrangellia terrane to the west. Similarly, the Wrangellia terrane is interpreted as the stratigraphic basement for the predominantly Mesozoic Peninsular terrane.

In addition to the WCT, a series of narrow and generally highly deformed terranes occur as discontinuous, fault-bounded lenses along the Denali fault, south of the Yukon-Tanana, Nixon Fork, and Dillinger terranes. These units are the Aurora Peak, Pingston, McKinley, Mystic, and Windy terranes; a terrane of ultramafic and associated rocks; and a fragment of the Dillinger terrane. Several small terranes also occur within highly deformed upper Mesozoic flysch assemblages, either as rootless nappes or as units bounded by steep faults. These units are the Chulitna, West Fork, Broad Pass, Susitna, Maclaren, and Clearwater terranes (Fig. 2). Diverse origins are interpreted for many of these units, including origins as displaced fragments of the Paleozoic and early Mesozoic North American continental margin, continental margin arcs, island arcs, ophiolites, seamounts, flysch basins, and structural melanges.

Two major late Mesozoic flysch basins, now tectonically collapsed, occur along the northern margin of the WCT. To the northwest is the Kahiltna assemblage, which probably originally depositionally overlapped the Peninsular terrane; to the northeast is the Gravina-Nutzotin belt, of mainly coeval flysch and volcanic rocks, which depositionally overlies the Wrangellia terrane. These major Late Jurassic and Early Cretaceous flysch basins, as well as the coeval Chitina and Chisana igneous arcs, extend for several thousand kilometers through southern and southeastern Alaska; they are interpreted as forming on the leading edge of the WCT during migration towards North America (Nokleberg and others, 1985; Plafker and others, 1989c).

Major faults or sutures are either known or are inferred to separate most terranes in the study area. Along the northern boundary is the Denali fault, and along the southern boundary is the Border Ranges fault (Fig. 2). Other important sutures are the Broxson Gulch thrust between the Maclaren and Wrangellia terranes, the Talkeetna thrust between the Kahiltna assemblage to the north and the Wrangellia terrane to the south, the West Fork fault between the southern Wrangellia and Peninsular terranes, and the Chitina Valley fault between the Wrangellia and southern Wrangellia terranes (Fig. 2). These faults are both the loci of accretion of adjacent terranes and the loci of tectonic erosion of the margins or marginal facies of terranes that existed before accretion.

FUTURE STUDIES

Many important questions remain for future studies.

1. What are the origins of the small terranes along the Denali fault and within the Kahiltna assemblage? The tectonic environments need further substantiation, because, if the interpretations are correct, they suggest migration from widely divergent loci.

2. What is the structural and tectonic setting of the Kahiltna assemblage? The occurrence of several small terranes, some possibly exotic, within the assemblage suggests substantial tectonic dismemberment. Is the assemblage a collage of various Mesozoic flysch units that formed in widely separated sites and were subsequently accreted together?

3. Significant discrepancies exist between geologic and paleomagnetic data indicating pre-Triassic amalgamation of the Wrangellia and Alexander terranes (Van der Voo and others,

1980; Yole and Irving, 1980; Hillhouse and Grommé, 1984; Haeussler and others, 1989; Gardner and others, 1988; Plafker and others, 1989c; P. W. Plumley, 1991, written commun.) and paleontologic data that indicate differing pre-Triassic displacement histories for these terranes (Newton, 1983; Silberling, 1985; Detterman, 1988).

4. An apparent contradiction exists between geologic data that indicate late Early to early Late Cretaceous (Albian to Cenomanian) docking of the WCT in about its present position relative to inboard terranes, and plate reconstructions (Engebretsen and others, 1985) indicating that northward displacement and docking of the WCT could not have occurred prior to the Campanian.

5. What was the nature of accretion of the island-arc assemblages of the WCT to the south against the displaced continental margin terranes of the Yukon-Tanana, Dillinger, Mystic, and Nixon Fork terranes to the north? Were substantial parts of the Mesozoic flysch and associated island-arc volcanic rocks along the northern margin of the WCT thrust under the southern margin of these displaced continental margin terranes to the north along the ancestral Denali fault (Stanley and others, 1990)? Were substantial parts of the WCT to the south and the Yukon-Tanana and other continental margin terranes to the north tectonically eroded during accretion?

6. The Paleozoic and early Mesozoic stratigraphic and igneous linkages between the various parts of the WCT rely mainly on comparative studies between generally widely separated occurrences of sedimentary and volcanic rocks, and on temporal, petrographic, and geochemical similarities between widely separated plutons in proposed arcs. Most of the definition of the WCT relies on these similarities. Currently, only one pluton, the Barnard Glacier pluton in the southeastern Wrangell Mountains, is observed to stitch together the Wrangellia and Alexander terranes. Abundant new data, particularly on the isotopic compositions of the volcanic and plutonic parts of arcs, are needed to clearly demonstrate the continuity of the interpreted arcs. Similarly, detailed stratigraphic and paleontologic studies are needed for the correlations of sedimentary rocks between the various parts of the WCT.

7. There is considerable debate on the origin of the mafic magmas that formed the Late Triassic basalts of the Nikolai Greenstone and coeval mafic and ultramafic plutonic rocks in the WCT. The available data appear to contradict formation in either a rift or plume setting.

8. Are the southern Wrangellia and Wrangellia terranes linked by stratigraphy and by igneous arcs? One major linkage between the two units is the occurrence of the Late Triassic limestone and undated greenstone in the southern Wrangellia terrane. However, these two units occur in fault-bounded fragments near the Chitina Valley fault and may be displaced from the Wrangellia terrane to the north. The igneous arc linkages rely mainly on petrologic and isotopic similarities of widely separated plutons.

9. Are parts of the Peninsular and Wrangellia terranes stratigraphically underlain by pre–late Paleozoic metasedimentary and metagranitic rocks of the Alexander terrane? Two possible occurrences of the Alexander terrane are the unit of unnamed Paleozoic(?) metamorphic rocks along the southern margin of the Peninsular terrane, and the pre–late Paleozoic metasedimentary and metagranitic rocks that occur as roof pendants in late Paleozoic granitic rocks in the Nutzotin Mountains in the Wrangellia terrane. Both areas need additional study.

10. Are the early Paleozoic Kaskawulsh metamorphic rocks equivalent to the Kaskawulsh Group of Canada, and are both part of the highly deformed and faulted Alexander terrane? More detailed studies, principally in western Canada, might establish this unit as a separate terrane. In that case, the Wrangellia and Alexander terranes would not be stitched together by late Paleozoic metagranitic rocks.

11. What was the nature of tectonic juxtaposition of the deep-level, low-temperature and high-pressure glaucophane blueschist and greenschist along the northern margin of the Chugach terrane, south of the Border Ranges fault, against the deep-level, high-temperature and high-pressure ultramafic rocks of the Border Ranges ultramafic-mafic assemblage to the north? These structural relations require substantial tectonic erosion of the intervening part of the arc.

12. Does a genetic linkage exist between the various parts of the Talkeetna island arc, the Late Triassic through Middle(?) Jurassic Border Ranges ultramafic-mafic assemblage, the Late Triassic(?) and Lower Jurassic Talkeetna Formation, and the Middle Jurassic part of the Alaska-Aleutian Range batholith? The present inferred linkage of the three units is based primarily on temporal data, level of emplacement of the three igneous units, and the apparent intrusion of the Border Ranges ultramafic-mafic assemblage into the Late Triassic and Early Jurassic stratified rocks of the Peninsular terrane in two widely separated areas. Additional isotopic data and field observations are needed to support this linkage.

REFERENCES CITED

Aleinikoff, J. N., 1984, Age and origin of metaigneous rocks from terranes north and south of the Denali fault, Mt. Hayes quadrangle, east-central Alaska: Geological Society of America Abstracts with Programs, v. 16, p. 266.

Aleinikoff, J. N., Dusel-Bacon, C., Foster, H. L., and Nokleberg, W. J., 1987, Pb-isotope fingerprinting of tectonostratigraphic terranes, east-central Alaska: Canadian Journal of Earth Sciences, v. 24, p. 2089–2098.

Aleinikoff, J. N., Plafker, G., and Nokleberg, W. J., 1988, Middle Pennsylvanian plutonic rocks along the southern margin of Wrangellia, *in* Hamilton, T. D., and Galloway, J. P., eds., The United States Geological Survey in Alaska: Accomplishments during 1987: U.S. Geological Survey Circular 1016, p. 110–113.

Andreasen, G. E., Grantz, A., Zeitz, I., and Barnes, D. G., 1974, Geologic interpretation of magnetic and gravity data in the Copper River Basin, Alaska: U.S. Geological Survey Professional Paper 316-H, p. 135–153.

Armstrong, A. K., and MacKevett, E. M., Jr., 1982, Stratigraphy and diagenetic history of the lower part of the Triassic Limestone, Alaska: U.S. Geological Survey Professional Paper 1212-A, p. A1–A26.

Barker, F., 1987, Cretaceous Chisana island arc of Wrangellia, eastern Alaska: Geological Society of America Abstracts with Programs, v. 19, p. 580.

Barker, F., and Grantz, A., 1982, Talkeetna Formation in the southeastern Talkeetna Mountains, southern Alaska: An Early Jurassic andesitic island arc: Geological Society of America Abstracts with Programs, v. 14, p. 147.

Barker, F., and Stern, T. W., 1986, An arc-root complex of Wrangellia, eastern Alaska Range: Geological Society of America Abstracts with Programs, v. 18, p. 534.

Barker, F., Sutherland Brown, A., Budahn, J. R., and Plafker, G., 1989, Back-arc with frontal arc component origin of Triassic Karmutsen basalt, British Columbia, Canada: Chemical Geology, v. 75, p. 81–102.

Beard, J. S., and Barker, F., 1989, Petrology and tectonic significance of gabbros, tonalites, shoshonites, and anorthosites in a late Paleozoic arc-root complex in the Wrangellia terrane, southern Alaska: Journal of Geology, v. 97, p. 667–683.

Berg, H. C., Jones, D. L., and Richter, D. H., 1972, Gravina-Nutzotin belt-tectonic significance of an upper Mesozoic sedimentary and volcanic sequence in southern and southeastern Alaska, *in* Geological Survey research 1972: U.S. Geological Survey Professional Paper 800-D, p. D1–D24.

Bond, G. C., 1973, A late Paleozoic volcanic arc in the eastern Alaska Range, Alaska: Journal of Geology, v. 81, p. 557–575.

—— , 1976, Geology of the Rainbow Mountain–Gulkana Glacier area, eastern Alaska Range, with emphasis on upper Paleozoic strata: Alaska Division of Geological and Geophysical Surveys Geologic Report 45, 47 p.

Brandon, M. T., Orchard, M. J., Parrish, R. R., Sutherland Brown, A., and Yorath, C. J., 1986, Fossil ages and isotopic dates from Paleozoic Sicker Group and associated intrusive rocks, Vancouver Island, British Columbia, *in* Current research, Part A: Geological Survey of Canada Paper 86-1A, p. 683–696.

Brew, D. A., and Ford, A. B., 1984, The northern Coast plutonic-metamorphic complex, southeastern Alaska and northwestern British Columbia, *in* Coonrad, W. C., and Elliott, R. L., eds., The United States Geological Survey in Alaska: Accomplishments during 1981: U.S. Geological Survey Circular 868, p. 120–124.

Brewer, W. M., 1982, Stratigraphy, structure, and metamorphism of the Mount Deborah area, central Alaska Range, Alaska [Ph.D. thesis]: Madison, University of Wisconsin, 318 p.

Brogan, G. E., Cluff, L. S., Korringa, M. K., and Slemmons, D. B., 1975, Active faults of Alaska: Tectonophysics, v. 29, p. 73–85.

Bundtzen, T. K., Gilbert, W. G., and Blodgett, R. B., 1979, The Chilikadrotna Greenstone, an Upper Silurian metavolcanic sequence in the central Lake Clark quadrangle, Alaska: Alaska Division of Geological and Geophysical Surveys Geologic Report 61, p. 31–35.

Burk, C. A., 1965, Geology of the Alaska Peninsula—Island arc and continental margin: Geological Society of America Memoir 90, 250 p.

Burns, L. E., 1985, The Border Ranges ultramafic and mafic complex, south-central Alaska: Cumulate fractionates of island-arc volcanics: Canadian Journal of Earth Sciences, v. 22, p. 1020–1038.

—— , 1994, Geology of part of the Nelchina River Gabbronorite and associated rocks: U.S. Geological Survey Bulletin 2058 (in press).

Burns, L. E., Little, T. A., Newberry, R. J., Decker, J. E., and Pessel, G. E., 1983, Preliminary geologic map of parts of the Anchorage C-2, C-3, D-2, D-3 quadrangles, Alaska: Alaska Division of Geological and Geophysical Surveys Report of Investigations 83-10, 3 sheets, scale 1:25,000.

Burns, L. E., Pessel, G. H., Little, T. A., Pavlis, T. L., Newberry, R. J., Winkler, G. R., and Decker, J., 1991, Geology of the northern Chugach Mountains, south-central Alaska: Alaska Division of Geological and Geophysical Surveys Professional Report 94, 63 p.

Cameron, B.E.B., and Tipper, H. W., 1985, Jurassic stratigraphy of the Queen Charlotte Islands, British Columbia: Geological Survey of Canada Bulletin 365, 49 p.

Campbell, D. L. and Nokleberg, W. J., 1986, Magnetic profile and model across northern Copper River basin, northwestern Gulkana quadrangle, Alaska, *in* Bartsch-Winkler, S., and Reed, K., eds., Geologic studies in Alaska by the U.S. Geological Survey during 1985: U.S. Geological Survey Circular 978, p. 35–38.

Campbell, R. B., and Dodds, C. J., 1982a, Geology of s.w. Kluane Lake map area, Yukon Territory: Geological Survey of Canada Maps 115F and 115G, Open File 829, scale 1:250,000.

—— , 1982b, Geology of the Mount St. Elias map area, Yukon Territory: Geological Survey of Canada Maps 115B and 115C, Open File 830, scale 1:250,000.

Capps, S. R., 1923, Recent investigations of petroleum in Alaska; the Cold Bay district: U.S. Geological Survey Bulletin 739-C, p. C77–C116.

Carden, J. R., and Decker, J. E., 1977, Tectonic significance of the Knik River schist terrane, south-central Alaska: Alaska Division of Geological and Geophysical Surveys Geologic Report 55, p. 7–9.

Case, J. E., Cox, D. P., Detra, D., Detterman, R. L., and Wilson, F. H., 1981, Geologic interpretation of aeromagnetic map of the Chignik and Sutwik Island quadrangles, Alaska: U.S. Geological Survey Miscellaneous Field Studies Map MF-1053-J, scale 1:250,000.

Case, J. E., Burns, L. E., and Winkler, G. R., 1986, Maps showing aeromagnetic survey and geologic interpretation of the Valdez quadrangle, Alaska: U.S. Geological Survey Miscellaneous Field Studies Map MF-1714, 2 sheets, scale 1:250,000.

Clark, S.H.B., 1972, Reconnaissance bedrock geologic map of the Chugach Mountains near Anchorage, Alaska: U.S. Geological Survey Field Studies Map MF-350, 70 p., 1 sheet, scale 1:250,000.

Coleman, R. G., and Burns, L. E., 1985, The Tonsina high-pressure mafic-ultramafic cumulate sequence, Chugach Mountains, Alaska: Geological Society of America Abstracts with Programs, v. 17, p. 348.

Coney, P. J., and Jones, D. L., 1985, Accretion tectonics and crustal structure in Alaska: Tectonophysics, v. 119, p. 265–283.

Coney, P. J., Jones, D. L., and Monger, J.W.H., 1980, Cordilleran suspect terranes: Nature, v. 288, p. 329–333.

Connelly, W., 1978, Uyak Complex, Kodiak Islands, Alaska: A Cretaceous subduction complex: Geological Society of America Bulletin, v. 89, p. 755–769.

Connelly, W., and Moore, J. C., 1979, Geologic map of the northwest side of the Kodiak and adjacent islands, Alaska: U.S. Geological Survey Miscellaneous Field Studies Map MF-1057, 2 sheets, scale 1:250,000.

Crawford, M. L., Hollister, L. S., and Woodsworth, G. J., 1987, Crustal deformation and regional metamorphism across a terrane boundary, Coast plutonic complex, British Columbia: Tectonics, v. 6, p. 343–361.

Csejtey, B., Jr., and St. Aubin, D. R., 1981, Evidence for northwestward thrusting of the Talkeetna superterrane, and its regional significance, *in* Albert, N.R.D., and Hudson, T., eds., 1981, The United States Geological Survey in Alaska: Accomplishments during 1979: U.S. Geological Survey Circular 823-B, p. B49–B51.

Csejtey, B., Jr., and 8 others, 1978, Reconnaissance geologic map and geochronology, Talkeetna Mountains quadrangle, northern part of Anchorage quadrangle, and southwest corner of Healy quadrangle, Alaska: U.S. Geological Survey Open-File Report 78-558-A, 60 p., scale 1:250,000.

Csejtey, B., Jr., Cox, D. P., Evarts, R. C., Stricker, G. D., and Foster, H. L., 1982, The Cenozoic Denali fault system and the Cretaceous accretionary development of southern Alaska: Journal of Geophysical Research, v. 87, p. 3741–3754.

Csejtey, B., Jr., and 13 others, 1986, Geology and geochronology of the Healy quadrangle, Alaska: U.S. Geological Survey Open-File Report 86-396, 92 p.

Davidson, C., Hollister, S. M., and Schmid, S. M., 1992, Role of melt in the formation of a deep-crustal compressive shear zone: The Maclaren Glacier metamorphic belt, south-central Alaska: Tectonics, v. 11, p. 348–359.

DeBari, S. M., and Coleman, R. G., 1989, Examination of the deep levels of an island arc: Evidence from the Tonsina ultramafic-mafic assemblage, Tonsina, Alaska: Journal of Geophysical Research, v. 94, p. 4373–4391.

DeBari, S. M., and Sleep, N. H., 1991, High-Mg, low-Al bulk composition of the Talkeetna arc, Alaska: Implications for primary magmas and the nature of arc crust: Geological Society of America Bulletin, v. 103, p. 37–47.

Detterman, R. L., 1988, Mesozoic biogeography of southern Alaska with implications for the paleogeography: U.S. Geological Survey Open-File Report 88-662, 27 p.

——, 1990, Stratigraphic correlation and interpretation of exploratory wells, Alaska Peninsula: U.S. Geological Survey Open-File Report 90-279, 51 p.
Detterman, R. L., and Hartsock, J. K., 1966, Geology of the Iniskin-Tuxedni region, Alaska: U.S. Geological Survey Professional Paper 512, 78 p.
Detterman, R. L., and Reed, B. L., 1980, Stratigraphy, structure, and economic geology of the Iliamna quadrangle, Alaska: U.S. Geological Survey Bulletin 1368-B, 86 p.
Detterman, R. L., Case, J. E., Wilson, F. H., Yount, M. E., and Allaway, W. H. Jr., 1983, Generalized geologic map of the Ugashik, Bristol Bay, and part of Karluk quadrangles, Alaska: U.S. Geological Survey Miscellaneous Field Studies Map MF-1539-A, scale 1:250,000, 1 sheet.
Detterman, R. L., Case, J. E., Miller, J. W., Wilson, F. H., and Yount, M. E., 1994, Stratigraphic framework of the Alaska Peninsula: U.S. Geological Survey Bulletin 1969-A (in press).
Dickinson, W. R., and Suczek, C. A., 1979, Plate tectonics and sandstone compositions: American Association of Petroleum Geologists Bulletin, v. 63, p. 2164–2182.
Dodds, C. J., and Campbell, R. B., 1988, Potassium-argon ages of mainly intrusive rocks in the Saint Elias Mountains, Yukon and British Columbia: Geological Society of Canada Paper 87-16, 43 p.
Dusel-Bacon, C., 1991, Metamorphic history of Alaska: U.S. Geological Survey Open-File Report 91-556, 48 p.
Dusel-Bacon, C., Csejtey, B., Foster, H. L., Doyle, E. O., Nokleberg, W. J., and Plafker, G., 1994, Distribution, facies, ages, and proposed tectonic associations of regionally metamorphosed rocks in east- and south-central Alaska: U.S. Geological Survey Professional Paper 1497-C, 59 p.
Eakins, G. R., Gilbert, W. R., and Bundtzen, J. K., 1978, Preliminary bedrock geology and mineral resource potential of west-central Lake Clark quadrangle, Alaska: Alaska Division of Geological and Geophysical Surveys Open-File Report 118, 15 p.
Eisbacher, G. H., 1976, Sedimentology of the Dezadeash flysch and its implications for strike-slip faulting along the Denali fault, Yukon Territory and Alaska: Canadian Journal of Earth Sciences, v. 13, p. 1495–1513.
Engebretsen, D. C., Cox, A., and Gordon, R. G., 1985, Relative motions between oceanic and continental plates in the Pacific basin: Geological Society of America Special Paper 206, 59 p.
Fisher, M. A., 1981, Location of the Border Ranges fault southwest of Kodiak Island, Alaska: Geological Society of America Bulletin, v. 92, p. 19–30.
Forbes, R. B., Turner, D. L., Stout, J. H., and Smith, T. E., 1973, Cenozoic offset along the Denali fault, Alaska [abs.]: Eos (Transactions, American Geophysical Union), v. 54, p. 495.
Foster, H. L., Keith, T.E.C., and Menzie, W. D., 1987, Geology of east-central Alaska: U.S. Geological Survey Open-File Report 87-188, 59 p.
Fuis, G. S., and Plafker, G., 1991, Evolution of deep structure along the Trans-Alaska Crustal Transect, Chugach Mountains and Copper River Basin, southern Alaska: Journal of Geophysical Research, v. 96, p. 4229–4253.
Gardner, M. C., Bergman, S. C., MacKevett, E. M., Jr., Plafker, G., Campbell, R. C., Cushing, G. W., Dodds, C. J., and McClelland, W. D., 1988, Middle Pennsylvanian pluton stitching of Wrangellia and the Alexander terrane, Wrangell Mountains, Alaska: Geology, v. 16, p. 967–971.
Gehrels, G. E., and Berg, H. C., 1992, Geologic map of southeastern Alaska: U.S. Geological Survey Miscellaneous Geologic Investigations Map I-1867, 1 sheet, scale 1:600,000.
Gehrels, G. E., and Saleeby, J. B., 1987, Geologic framework, tectonic evolution, and displacement history of the Alexander terrane: Tectonics, v. 6, p. 151–173.
Gehrels, G. E., Dodds, C. J., and Campbell, R. B., 1986, Upper Triassic rocks of the Alexander terrane, southeast Alaska and the Saint Elias Mountains of British Columbia and Yukon: Geological Society of America Abstracts with Programs, v. 18, p. 109.
Gehrels, G. E., McClelland, W. C., Samson, S. D., and Patchett, P. J., 1991, U-Pb geochronology of Late Cretaceous and early Tertiary plutons in the northern Coast Mountains batholith: Canadian Journal of Earth Sciences, v. 28, p. 899–911.
Gilbert, W. G., and Bundtzen, T. K., 1984, Stratigraphic relationships between Dillinger and Mystic terranes, western Alaska Range, Alaska: Geological Society of America Abstracts with Programs, v. 16, p. 286.
Gilbert, W. G., Nye, C. J., and Sherwood, K. W., 1984, Stratigraphy, petrology, and geochemistry of Upper Triassic rocks from the Pingston and McKinley terranes, central Alaska Range: Alaska Division of Geological and Geophysical Surveys Report of Investigations 84-30, 14 p.
Goodwin, E. B., Fuis, G. S., Nokleberg, W. J., and Ambos, E. L., 1989, The crustal structure of the Wrangellia terrane along the east Glenn Highway, eastern-southern Alaska: Journal of Geophysical Research, v. 94, p. 16,037–16,057.
Grantz, A., 1960a, Geologic map of Talkeetna Mountains (A-2) quadrangle, Alaska, and the contiguous area to the north and northwest: U.S. Geological Survey Miscellaneous Geologic Investigations Map I-313, scale 1:48,000.
——, 1960b, Geologic map of the Talkeetna Mountains (A-1) quadrangle and the south third of Talkeetna Mountains (B-1) quadrangle, Alaska: U.S. Geological Survey Miscellaneous Geologic Investigations Map I-314, scale 1:48,000.
——, 1961, Geologic map and cross sections of the Anchorage (D-2) quadrangle and northeastern most part of the Anchorage (D-3) quadrangle, Alaska: U.S. Geological Survey Miscellaneous Investigations Series Map I-0342, scale 1:48,000.
——, 1964, Stratigraphic reconnaissance of the Matanuska Formation in the Matanuska Valley, Alaska: U.S. Geological Survey Bulletin 1181-I, 33 p.
——, 1966, Strike-slip faults in Alaska: U.S. Geological Survey Open-File Report, 82 p.
Grantz, A., and Jones, D. L., 1960, Stratigraphy and age of the Matanuska Formation, south-central Alaska: U.S. Geological Survey Professional Paper 400-B, p. B347–B350.
Grantz, A., Zietz, I., and Andreasen, G. E., 1963, An aeromagnetic reconnaissance of the Cook Inlet area, Alaska: U.S. Geological Survey Professional Paper 316-G, p. 117–134.
Grantz, A., Moore, T. E., and Roeske, S. M., compilers, 1991, A-3 Gulf of Alaska to Arctic Ocean: Boulder, Colorado, Geological Society of America, Centennial Continental/Ocean Transect no. 15, 3 sheets with text, scale 1:500,000.
Haeussler, P. J., Coe, R. S., Onstott, T. C., and Renne, P., 1989, A second look at the paleomagnetism of the Late Triassic Hound Island Volcanics of the Alexander terrane [abs.]: Eos (Transactions, American Geophysical Union), v. 70, p. 1068.
Hanks, C. L., Rogers, J. F., and Wallace, W. K., 1985, The Western Alaska Range Flysch terrane: What is it and where did it come from?: Geological Society of America Abstracts with Programs, v. 17, p. 359.
Hanson, B. M., 1957, Middle Permian limestone on Pacific side of Alaska Peninsular: American Association of Petroleum Geologists Bulletin, v. 41, p. 2376–2378.
Hickman, R. G., and Craddock, C., 1976, Geologic map of central Healy quadrangle, Alaska: Alaska Division of Geological and Geophysical Surveys Open-File Report AOF-95, scale 1:63,360.
Higgins, M. W., 1971, Cataclastic rocks: U.S. Geological Survey Professional Paper 687, 97 p.
Hill, M. D., 1979, Volcanic and plutonic rocks of the Kodiak-Shumagin shelf, Alaska: Subduction deposits and near-trench magmatism [Ph.D. thesis]: Santa Cruz, University of California, 274 p.
Hillhouse, J. W., 1977, Paleomagnetism of the Triassic Nikolai Greenstone, McCarthy quadrangle, Alaska: Canadian Journal of Earth Sciences, v. 14, p. 2578–2592.
Hillhouse, J. W., and Grommé, C. S., 1980, Paleomagnetism of the Hound Island volcanics, Alexander terrane, southeastern Alaska: Journal of Geophysical Research, v. 85, p. 2594–2602.
——, 1984, Northward displacement and accretion of Wrangellia: New paleomagnetic evidence from Alaska: Journal of Geophysical Research, v. 89, p. 4461–4467.
Howell, D. G., Jones, D. L., and Schermer, E. R., 1985, Tectonostratigraphic

terranes of the circum-Pacific region, *in* Howell, D. G., ed., Tectonostratigraphic terranes of the circum-Pacific region: Circum-Pacific Council for Energy and Mineral Resources Earth Science Series, no. 1, p. 3–30.

Hudson, T., 1983, Calc-alkaline plutonism along the Pacific rim of southern Alaska, *in* Roddick, J. A., ed., Circum-Pacific plutonic terranes: Geological Society of America Memoir 159, p. 159–169.

Imlay, R. W., 1984, Early and middle Bajocian (Middle Jurassic) ammonites from southern Alaska: U.S. Geological Survey Professional Paper 1322, 38 p.

Imlay, R. W., and Detterman, R. L., 1973, Jurassic paleobiogeography of Alaska: U.S. Geological Survey Professional Paper 801, 34 p.

Jeletzky, J. A., 1976, Mesozoic and ?Tertiary rocks of the Quatsino Sound, Vancouver Island, British Columbia: Geological Survey of Canada Bulletin 242, 243 p.

Jones, D. L., and Silberling, N. J., 1979, Mesozoic stratigraphy—The key to tectonic analysis of southern and central Alaska: U.S. Geological Survey Open-File Report 79-1200, 37 p.

—— , 1982, Mesozoic stratigraphy key to tectonic analysis of southern Alaska and central Alaska, *in* Leviton, A. E., ed., Frontiers of geological exploration of western North America: San Francisco, California, Pacific Division, American Association of Petroleum Geologists, p. 139–153.

Jones, D. L., Silberling, N. J., and Hillhouse, J., 1977, Wrangellia—A displaced terrane in northwestern North America: Canadian Journal of Earth Sciences, v. 14, p. 2565–2577.

Jones, D. L., Silberling, N. J., Csejtey, B., Jr., Nelson, W. H., and Blome, C. D., 1980, Age and structural significance of ophiolite and adjoining rocks in the Upper Chulitna district, south-central Alaska: U.S. Geological Survey Professional Paper 1121-A, 21 p.

Jones, D. L., Silberling, N. J., Berg, H. C., and Plafker, G., 1981, Map showing tectonostratigraphic terranes of Alaska, columnar sections, and summary description of terranes: U.S. Geological Survey Open-File Report 81-792, 20 p., 2 sheets, scale 1:2,500,000.

Jones, D. L., Silberling, N. J., Gilbert, W., and Coney, P., 1982, Character, distribution, and tectonic significance of accretionary terranes in the central Alaska Range: Journal of Geophysical Research, v. 87, p. 3709–3717.

Jones, D. L., Silberling, N. J., and Coney, P. J., 1983, Tectono-stratigraphic map and interpretative bedrock geologic map of the Mount McKinley region, Alaska: U.S. Geologial Survey Open-File Report 83-11, 2 sheets, scale 1:250,000.

Jones, D. L., Silberling, N. J., Coney, P. J., and Plafker, G., 1984, Lithotectonic terrane map of Alaska (west of the 141st meridian), *in* Silberling, N. J., and Jones, D. L., eds., Lithotectonic terrane maps of the North American Cordillera: U.S. Geological Survey Open-File Report 84-523, scale 1:2,500,000.

—— , 1987, Lithotectonic terrane map of Alaska (west of the 141st meridian): U.S. Geological Survey Miscellaneous Field Studies Map MF-1874, scale 1:2,500,000.

Kelley, J. S., 1984, Geologic map and sections of the southwestern Kenai Peninsula west of the Port Graham fault, Alaska: U.S. Geological Survey Open-File Report 84-0152, 1 sheet, scale 1:63,360.

Kindle, E. D., 1953, Dezadeash map area, Yukon Territory: Canadian Geological Survey Memoir 268, 68 p.

Kline, J. T., Bundtzen, T. K., and Smith, T. E., 1990, Preliminary bedrock geologic map of the Talkeetna Mountains D-2 quadrangle, Alaska: Alaska Division of Geological and Geophysical Surveys Public-Data File 90-24, 13 p., 1 sheet, scale 1:63,360.

Labson, V. F., Fisher, M. A., and Nokleberg, W. J., 1988, An integrated study of the Denali fault from magnetotelluric sounding, seismic reflection, and geologic mapping: Eos (Transactions, American Geophysical Union), v. 69, p. 1457.

Lanphere, M. A., 1978, Displacement history of the Denali fault system, Alaska and Canada: Canadian Journal of Earth Sciences, v. 15, p. 817–822.

Lanphere, M. A., and Reed, B. L., 1973, Timing of Mesozoic and Cenozoic plutonic events in Circum-Pacific North America: Geological Society of America Bulletin, v. 84, p. 3773–3782.

Lewis, S. D., Ladd, J. W., and Bruns, T. R., 1988, Structural development of an accretionary prism by thrust and strike-slip faulting: Shumagin region, Aleutian trench: Geological Society of America Bulletin, v. 100, p. 767–782.

Little, T. A., and Naeser, C. W., 1989, Tertiary tectonics of the Border Ranges fault system, Chugach Mountains, Alaska: Deformation and uplift in a forearc setting: Journal of Geophysical Research, v. 94, p. 4333–4359.

MacKevett, E. M., Jr., 1969, Three newly named Jurassic Formations in the McCarthy C-5 quadrangle, Alaska, *in* Cohee, G. V., Bates, R. G., and Wright, W. B., eds., Changes in stratigraphic nomenclature by the U.S. Geological Survey 1967: U.S. Geological Survey Bulletin 1274-A, p. A35–A49.

—— , 1971, Stratigraphy and general geology of the McCarthy C-5 quadrangle, Alaska: U.S. Geological Survey Bulletin 1323, 35 p.

—— , 1978, Geologic map of the McCarthy quadrangle, Alaska: U.S. Geological Survey Miscellaneous Investigations Series Map I-1032, 1 sheet, scale 1:250,000.

Martin, G. C., 1926, The Mesozoic stratigraphy of Alaska: U.S. Geological Survey Bulletin 776, 493 p.

Martin, G. C., Johnson, B. L., and Grant, U. S., 1915, Geology and mineral resources of the Kenai Peninsula, Alaska: U.S. Geological Survey Bulletin 587, 243 p.

McLean, H., 1977, Organic geochemistry, lithology, paleontology of Tertiary and Mesozoic from wells on the Alaska Peninsula: U.S. Geological Survey Open-File Report 77-813, 63 p.

Mendenhall, W. C., 1905, Geology of the central Copper River region, Alaska: U.S. Geological Survey Professional Paper 41, 133 p.

Moffit, F. H., 1938, Geology of the Chitina Valley and adjacent area, Alaska: U.S. Geological Survey Bulletin 894, 137 p.

Moll-Stalcup, E., 1990, Latest Cretaceous and Cenozoic magmatism in mainland Alaska: U.S. Geological Survey Open-File Report 90-84, 80 p.

Monger, J.W.H., and Berg, H. C., 1987, Lithotectonic terrane map of western Canada and southeastern Alaska: U.S. Geological Survey Miscellaneous Field Studies Map MF-1874-B, 12 p., 1 sheet, scale 1:2,500,000.

Monger, J.W.H., Price, R. A., and Tempelman-Kluit, D. J., 1982, Tectonic accretion and the origin of the two major metamorphic and plutonic welts in the Canadian Cordillera: Geology, v. 10, p. 70–75.

Moore, J. C., and Connelly, W., 1977, Tectonic history of the continental margin of southwestern Alaska: Late Triassic to earliest Tertiary, *in* Sisson, A., ed., The relationship of plate tectonics to Alaskan geology and resources: Anchorage, Alaska Geological Society, 6th Symposium Proceedings, p. H1–H29.

Moore, T. E., 1992, The Arctic Alaska superterrane, *in* Bradley, D. C., and Dusel-Bacon, C., eds., Geologic studies in Alaska by the U.S. Geological Survey, 1991: U.S. Geological Survey Bulletin 2041, p. 238–244.

Muller, J. E., 1967, Kluane Lake area, Yukon Territory: Geological Survey of Canada Memoir 340, 137 p.

—— , 1977, Geology of Vancouver Island: Geological Survey of Canada Open-File Map 463, 1 sheet, scale 1:250,000.

Muller, J. E., Northcote, K. E., and Carlisle, D., 1974, Geology and mineral deposits of Alert–Cape Scott map area, Vancouver Island, British Columbia: Canadian Geological Survey Paper 74-8, 77 p.

Mutti, E., and Ricci Lucchi, F. T., 1978, Turbidites of the northern Apennines—Introduction to facies analysis: International Geology Review, v. 20, p. 125–166.

Nelson, S. W., Blome, C. D., Harris, A. G., Reed, K. M., and Wilson, F. H., 1986, Late Paleozoic and Early Jurassic fossil ages from the McHugh Complex, *in* Bartsch-Winkler, S., ed., The United States Geological Survey in Alaska: Accomplishments during 1984: U.S. Geological Survey of Circular 978, p. 60–69.

Nelson, W. H., Carlson, C., and Case, J. E., 1983, Geologic map of the Lake Clark quadrangle, Alaska: U.S. Geological Survey Miscellaneous Field Studies Map MF-1114-A, scale 1:250,000.

Newton, C. R., 1983, Paleozoogeographic affinities of Norian bivalves from the Wrangellian, Peninsular, and Alexander terranes, western North America, *in*

Stevens, C. H., ed., Pre-Jurassic rocks in western North American suspect terranes: Los Angeles, California, Pacific Section, Society of Economic Paleontologists and Mineralogists, p. 37–48.
Nichols, K. M., and Silberling, N. J., 1979, Early Triassic (Smithian) ammonites of paleoequatorial affinity from the Chulitna terrane, south-central Alaska: U.S. Geological Survey Professional Paper 1121-B, 5 p.
Nokleberg, W. J., and 10 others, 1982, Geologic map of the southern part of the Mount Hayes quadrangle, Alaska: U.S. Geological Survey Open-File Report 82-52, 26 p., 1 sheet, scale 1:250,000.
Nokleberg, W. J., Jones, D. L., and Silberling, N. J., 1985, Origin and tectonic evolution of the Maclaren and Wrangellia terranes, eastern Alaska Range, Alaska: Geological Society of America Bulletin, v. 96, p. 1251–1270.
Nokleberg, W. J., Wade, W. M., Lange, I. M., and Plafker, G., 1986, Summary of geology of the Peninsular terrane, metamorphic complex of Gulkana River, and Wrangellia terrane, north-central and northwestern Gulkana quadrangle, *in* Bartsch-Winkler, S., and Reed, K., eds., Geologic studies in Alaska by the U.S. Geological Survey during 1985: U.S. Geological Survey Circular 978, p. 69–74.
Nokleberg, W. J., Foster, H. L., and Aleinikoff, J. N., 1989a, Geology of the northern Copper River Basin, eastern Alaska Range, and southern Yukon-Tanana Basin, southern and east-central Alaska, *in* Nokleberg, W. J., and Fisher, M. A., eds., Alaskan geological and geophysical transect: International Geological Congress, 27th, Guidebook T104, p. 34–64.
Nokleberg, W. J., Plafker, G., Lull, J. S., Wallace, W. K., and Winkler, G. W., 1989b, Structural analysis of the southern Peninsular, southern Wrangellia, and northern Chugach terranes along the Trans-Alaskan Crustal Transect (TACT), northern Chugach Mountains, Alaska: Journal of Geophysical Research, v. 94, p. 4297–4320.
Nokleberg, W. J., Aleinikoff, J. N., Dutro, J. T., Jr., Lanphere, M. A., Silberling, N. J., Silva, S. R., Smith, T. E., and Turner, D. L., 1992a, Map, tables, and summary of fossil and isotopic age data, Mount Hayes quadrangle, eastern Alaska Range, Alaska: U.S. Geological Survey Miscellaneous Field Studies Map 1996-D, 86 p., 1 sheet, scale 1:250,000.
Nokleberg, W. J., Aleinikoff, J. N., Lange, I. M., Silva, S. R., Miyaoka, R. T., Schwab, C. E., and Zehner, R. E., 1992b, Preliminary geologic map of the Mount Hayes quadrangle, eastern Alaska Range, Alaska: U.S. Geological Survey Open-File Report 92-594, 39 p., 1 sheet, scale 1:250,000.
Nokleberg, W. J., Moll-Stalcup, E. J., Miller, T. P., Brew, D. A., Grantz, A., Plafker, G., Moore, T. E., and Patton, W. W., Jr., 1994, Tectonostratigraphic terrane and overlap assemblage map of Alaska: U.S. Geological Survey Open-File Report 94-194, 84 p., 1 sheet, scale 1:2,500,000.
Onstott, T. C., Sisson, V. B., and Turner, D. L., 1989, Initial argon in amphiboles from the Chugach Mountains, southern Alaska: Journal of Geophysical Research, v. 94, p. 4361–4372.
Packer, D. R., Brogan, G. E., and Stone, D. B., 1975, New data on plate tectonics of Alaska: Tectonophysics, v. 29, p. 87–102.
Paige, S., and Knopf, A., 1907, Stratigraphic succession in the region northeast of Cook Inlet, Alaska: Geological Society of America Bulletin, v. 18, p. 327–328.
Palmer, A. R., compiler, 1983, The Decade of North American Geology 1983 geologic time scale: Geology, v. 11, p. 503–504.
Patton, W. W., Jr., Box, S. E., Moll-Stalcup, E. J., and Miller, T. P., 1989, Geology of west-central Alaska: U.S. Geological Survey Open-File Report 89-554, 53 p.
Patton, W. W., Jr., Murphy, J. M., Burns, L. E., Nelson, S. W., and Box, S. E., 1992, Geologic map of ophiolitic and associated volcanic arc and metamorphic terranes of Alaska (west of the 141st meridian): U.S. Geological Survey Open-File Report 92-20A, 1 sheet, scale 1:2,500,000.
Pavlis, T. L., 1982, Origin and age of the Border Ranges fault of southern Alaska and its bearing on the late Mesozoic tectonic evolution of Alaska: Tectonics, v. 1, p. 343–368.
——, 1983, Pre-Cretaceous crystalline rocks of the western Chugach Mountains, Alaska: Nature of the basement of the Jurassic Peninsular terrane: Geological Society of America Bulletin, v. 94, p. 1329–1344.
——, 1986, Geologic map of the Anchorage C-5 quadrangle, Alaska: Alaska Division of Geological and Geophysical Surveys Public-Data File 86-7, 44 p., 1 sheet, scale 1:63,360.
Pavlis, T. L., and Crouse, G. W., 1989, Late Mesozoic strike slip movement on the Border Ranges fault system in the eastern Chugach Mountains, southern Alaska: Journal of Geophysical Research, v. 94, p. 4321–4332.
Pavlis, T. L., Monteverde, D. H., Bowman, J. R., Rubenstone, J. L., and Reason, M. D., 1988, Early Cretaceous near-trench plutonism in southern Alaska: A tonalite-trondhjemite intrusive complex injected during ductile thrusting along the Border Ranges fault system: Tectonics, v. 7, p. 1179–1199.
Plafker, G., 1990, Regional geology and tectonic evolution of Alaska and adjacent parts of the northeast Pacific Ocean margin: Proceedings of the Pacific Rim Congress 90: Queensland, Australia, Australasian Institute of minig and Metallurgy, p. 841–853.
Plafker, G., Hudson, T., and Richter, D. H., 1977, Preliminary observations on late Cenozoic displacements along the Totschunda and Denali fault systems, *in* Blean, K. M., ed., The United States Geological Survey in Alaska—Accomplishments during 1976: U.S. Geological Survey Circular 751-B, p. B67–B69.
Plafker, G., Harris, A. G., and Reed, K. M., 1985, Early Pennsylvanian conodonts from the Strelna Formation, Chitina Valley area, *in* Bartsch-Winkler, S., ed., The U.S. Geological Survey in Alaska: Accomplishments during 1984: U.S. Geological Survey Circular 967, p. 71–74.
Plafker, G., Blome, C. D., and Silberling, N. J., 1989a, Reinterpretation of lower Mesozoic rocks on the Chiklat Peninsula, Alaska, as a displaced fragment of Wrangellia: Geology, v. 17, p. 3–6.
Plafker, G., Lull, J. S., Nokleberg, W. J., Pessel, G. H., Wallace, W. K., and Winkler, G. R., 1989b, Geologic map of the Valdez A-4, B-3, B-4, C-3, C-4, and D-4 quadrangles, northern Chugach Mountains and southern Copper River Basin, Alaska: U.S. Geological Survey Open-File Report 89-569, 1 sheet, scale 1:125,000.
Plafker, G., Nokleberg, W. J., and Lull, J. S., 1989c, Bedrock geology and tectonic evolution of the Wrangellia, Peninsular, and Chugach terranes along the Trans-Alaskan Crustal Transect in the northern Chugach Mountains and southern Copper River basin, Alaska: Journal of Geophysical Research, v. 94, p. 4255–4295.
Pratt, R. M., Rutstein, M. S., Walton, F. W., and Buschur, J. A., 1972, Extension of Alaskan structural trends beneath Bristol Bay, Bering Shelf, Alaska: Journal of Geophysical Research, v. 77, p. 4994–4999.
Reed, B. L., and Lanphere, M. A., 1969, Age and chemistry of Mesozoic and Tertiary plutonic rocks in south-central Alaska: Geological Society of America Bulletin, v. 80, p. 23–44.
——, 1972, Generalized geologic map of the Alaska-Aleutian Range batholith showing potassium-argon ages of the plutonic rocks: U.S. Geological Survey Miscellaneous Field Studies Map MF–372, scale 1:1,000,000.
——, 1973, Alaska-Aleutian Range batholith: Geochronology, chemistry, and relations to circum-Pacific plutonism: Geological Society of America Bulletin, v. 84, p. 2583–2610.
——, 1974a, Chemical variations across the Alaska-Aleutian Range batholith: U.S. Geological Survey Journal of Research, v. 2, p. 343–352.
——, 1974b, Offset plutons and history of movement along the McKinley segment of the Denali fault system, Alaska: Geological Society of America Bulletin, v. 85, p. 1883–1892.
Reed, B. L., and Nelson, S. W., 1980, Geologic map of the Talkeetna quadrangle, Alaska: U.S. Geological Survey Miscellaneous Investigations Map I-1174A, 15 p., scale 1:250,000.
Reed, B. L., Miesch, A. T., and Lanphere, M. A., 1983, Plutonic rocks of Jurassic age in the Alaska-Aleutian Range batholith: Chemical variations and polarity: Geological Society of America Bulletin, v. 94, p. 1232–1240.
Richards, M. A., Jones, D. L., Duncan, T. A., and DePaolo, D. J., 1991, A mantle plume initiation model for the formation of Wrangellia and other oceanic flood basalt plateaus: Science, v. 254, p. 263–267.
Richter, D. H., 1976, Geologic map of the Nabesna quadrangle, Alaska: U.S. Geological Survey Miscellaneous Geological Investigations Series Map

I-932, scale 1:250,000.
Richter, D. H., and Dutro, J. T., Jr., 1975, Revision of the type Mankommen Formation (Pennsylvanian and Permian), Eagle Creek area, eastern Alaska Range, Alaska: U.S. Geological Survey Bulletin 1395-B, p. B1–B25.
Richter, D. H., and Jones, D. L., 1973, Structure and stratigraphy of the eastern Alaska Range, Alaska, *in* Arctic geology: American Association of Petroleum Geologists Memoir 19, p. 408–420.
Richter, D. H., and Matson, N. A., Jr., 1971, Quaternary faulting in the eastern Alaska Range: Geological Society of America Bulletin, v. 82, p. 1529–1540.
Richter, D. H., Lanphere, M. A., and Matson, N. A., Jr., 1975, Granitic plutonism and metamorphism, eastern Alaska Range: Geological Society of America Bulletin, v. 86, p. 819–829.
Richter, D. H., Sharp, W. N., Dutro, J. T., Jr., and Hamilton, W. B., 1977, Geologic map of parts of the Mount Hayes A-1 and A-2 quadrangles, Alaska: U.S. Geological Survey Miscellaneous Investigations Series Map I-1031, 1 sheet, scale 1:63,360.
Roeske, S. M., 1986, Field relations and metamorphism of the Raspberry Schist, Kodiak Islands, Alaska, *in* Evans, B. W., and Brown, E. H., eds., Blueschists and eclogites: Geological Society of America Memoir 164, p. 169–184.
Roeske, S. M., Mattinson, J. M., and Armstrong, R. L., 1989, Isotopic ages of glaucophane schists on the Kodiak Islands, southern Alaska, and their implications for the Mesozoic tectonic history of the Border Ranges fault system: Geological Society of America Bulletin, v. 101, p. 1021–1037.
Rohn, O., 1900, A reconnaissance of the Chitina River and the Skolai Mountains, Alaska: U.S. Geological Survey 21st Annual Report, part 2, p. 399–440.
Rubin, C. M., and Saleeby, J. B., 1991, The Gravina sequence: Remnants of a mid-Mesozoic oceanic arc in southern southeast Alaska: Journal of Geophysical Research, v. 96, p. 14,551–14,568.
Sainsbury, C. L., and Twenhofel, W. S., 1954, Fault patterns in southeastern Alaska: Geological Society of America Bulletin, v. 65, p. 1300.
St. Amand, P., 1954, Tectonics of Alaska as deduced from seismic data: Geological Society of America Bulletin, v. 65, p. 1350.
——, 1957, Geological and geophysical synthesis of the tectonics of portions of British Columbia, the Yukon Territory, and Alaska: Geological Society of America Bulletin, v. 68, p. 1343–1370.
Saleeby, J. B., 1983, Accretionary tectonics of the North American Cordillera: Annual Review of Earth and Planetary Sciences, v. 11, p. 45–73.
Sherwood, K. W., and Craddock, C., 1979, General geology of the central Alaska Range between the Nenana River and Mount Deborah: Alaska Division of Geological and Geophysical Surveys Open-File Report 116, 22 p., 2 plates, scale 1:63,360.
Sherwood, K. W., Brewer, W. M., Craddock, C., Umhoefer, P. J., and Hickman, R. G., 1984, The Denali fault system and terranes of the central Alaska Range region, Alaska: Geological Society of America Abstracts with Programs, v. 16, p. 332.
Silberling, N. J., 1985, Paleogeographic significance of the Upper Triassic bivalve *Monotis* in Circum-Pacific accreted terranes, *in* Howell, D. G., ed., Tectonostratigraphic terranes of the circum-Pacific region: American Association of Petroleum Geologist Circum-Pacific Earth Science Series, no. 1, p. 63–70.
Silberling, N. J., Richter, D. H., Jones, D. L., and Coney, P. J., 1981, Geologic map of the bedrock part of the Healy A-1 quadrangle south of the Talkeetna–Broxson Gulch fault system, Clearwater Mountains, Alaska: U.S. Geological Survey Open-File Report 81-1288, 1 sheet, scale 1:63,360.
Silberman, M. L., and Grantz, A., 1984, Paleogene volcanic rocks of the Matanuska Valley area and the displacement history of the Castle Mountain fault, *in* Coonrad, W. L., and Elliott, R. L., eds., The United States Geological Survey: Accomplishments during 1981: U.S. Geological Survey Circular 868, p. 82–86.
Silberman, M. L., MacKevett, E. M., Jr., Connor, C. L., and Matthews, A., 1980, Metallogenic and tectonic significance of oxygen isotope data and whole-rock potassium-argon ages of the Nikolai Greenstone, McCarthy quadrangle, Alaska: U.S. Geological Survey Open-File Report 80-2019, 29 p.
Silberman, M. L., MacKevett, E. M., Jr., Connor, C. L., Klock, P. R., and Kalechitz, G., 1981, K-Ar ages of the Nikokai Greenstone from the McCarthy quadrangle, Alaska—The "docking" of Wrangellia, *in* Albert, N.R.D., and Hudson, T., eds., The United States Geological Survey in Alaska: Accomplishments during 1979: U.S. Geological Survey Circular 823-B, p. B61–B63.
Smith, J. G., and MacKevett, E. M., Jr., 1970, The Skokai Group in the McCarthy B-4, C-4, and C-5 quadrangles, Wrangell Mountains, Alaska: U.S. Geological Survey Bulletin 1274-Q, p. Q1–Q26.
Smith, T. E., 1981, Geology of the Clearwater Mountains, south-central Alaska: Alaska Division of Geological and Geophysical Surveys Geologic Report 60, 72 p.
Smith, T. E., and Lanphere, M. A., 1971, Age of the sedimentation, plutonism and regional metamorphism in the Clearwater Mountains region, central Alaska: Isochron/West, no. 2, p. 17–20.
Smith, T. E., Forbes, R. B., and Turner, D. L., 1974, A solution to the Denali fault offset problem: Alaska Division of Geological and Geophysical Surveys Annual Report, 1973, p. 25–27.
Spurr, J. E., 1900, A reconnaissance in southwestern Alaska in 1898: U.S. Geological Survey 20th Annual Report, part 7, p. 31–264.
Stanley, W. D., Labson, V. F., Nokleberg, W. J., Csejtey, B., Jr., and Fisher, M. A., 1990, The Denali fault system and Alaska Range of Alaska: Evidence for suturing and thin-skinned tectonics from magnetotellurics: Geological Society of America Bulletin, v. 102, p. 160–173.
Steiger, R. H., and Jäger, E., 1977, Subcommission on geochronology: Convention on the use of decay constants in geo- and cosmochronology: Earth and Planetary Science Letters, v. 36, p. 359–362.
Stone, D. B., and Packer, D. R., 1979, Paleomagnetic data from the Alaska Peninsula: Geological Society of America Bulletin, v. 90, p. 545–560.
Stone, D. B., Panuska, B. C., and Packer, D. R., 1982, Paleolatitudes vs. time for southern Alaska: Journal of Geophysical Research, v. 87, p. 3697–3708.
Stout, J. H., 1976, Geology of the Eureka Creek area, east-central Alaska Range: Alaska Division of Geological and Geophysical Surveys Geologic Report 46, 32 p.
Stout, J. H., and Chase, C. G., 1980, Plate kinematics of the Denali fault system: Canadian Journal of Earth Sciences, v. 17, p. 1527–1537.
Stout, J. H., Brady, J. B., Weber, F. R., and Page, R. A., 1973, Evidence for Quaternary movement on the McKinley strand of the Denali fault in the Delta River area, Alaska: Geological Society of America Bulletin, v. 84, p. 939–947.
Streckeisen, A., 1976, To each plutonic rock its proper name: Earth Science Reviews, v. 12, p. 1–33.
——, 1980, Classification and nomenclature of volcanic rocks, lamprophyres, carbonatites, and melilitic rocks: IUGS subcommission on the systematic of igneous rocks: Geology, v. 7, p. 331–335.
Sutherland-Brown, A., 1968, Geology of the Queen Charlotte Islands, British Columbia: British Columbia Department of Mines and Petroleum Resources Bulletin 54, 225 p.
Turner, D. L., and Smith, T. E., 1974, Geochronology and generalized geology of the central Alaska Range, Clearwater Mountains and northern Talkeetna Mountains: Alaska Division of Geological and Geophysical Surveys Open-File Report AOF-72, 10 p.
Turner, F. J., 1981, Metamorphic petrology (second edition): New York, McGraw Hill, 524 p.
Twenhofel, W. S., and Sainsbury, C. S., 1958, Fault patterns in southeastern Alaska: Geological Society of America Bulletin, v. 69, p. 1431–1442.
Umhoefer, P. J., 1987, Northward translation of "Baja British Columbia" along the Late Cretaceous to Paleocene margin of western North America: Tectonics, v. 6, p. 377–394.
Van der Voo, R., Jones, M., Grommé, C. S., Eberlein, G. D., and Churkin, M., Jr., 1980, Paleozoic paleomagnetism and northward drift of the Alexander terrane, southeastern Alaska: Journal of Geophysical Research, v. 85, p. 5281–5296.
von Huene, R., Keller, G., Bruns, T. R., and McDougall, K., 1985, Cenozoic migration of Alaskan terranes indicated by paleontologic study, *in* Howell, D. G., ed., Tectonostratigraphic terranes of the circum-Pacific region: Amer-

ican Association of Petroleum Geologists Circum-Pacific Earth Science Series, no. 1, p. 121–136.

Wahrhaftig, C., Turner, D. L., Weber, F. R., and Smith, T. E., 1975, Nature and timing of movement on the Hines Creek strand of the Denali fault system, Alaska: Geology, v. 3, p. 463–466.

Wallace, W. K., 1981a, Structure and petrology of a portion of a regional thrust zone in the central Chugach Mountains, Alaska [Ph.D. thesis]: Seattle, University of Washington, 253 p.

——, 1981b, Structure and petrology of a regional thrust zone in the central Chugach Mountains, Alaska: Dissertation Abstracts International, v. 42, p. 1364-B.

Wallace, W. K., Hanks, C. L., and Rogers, J. F., 1989, The southern Kahiltna terrane: Implications for the tectonic evolution of southwestern Alaska: Geological Society of America Bulletin, v. 101, p. 1389–1407.

Wang, J., Newton, C. R., and Dunne, L., 1988, Late Triassic transition from biogenic to arc sedimentation on the Peninsular terrane: Puale Bay, Alaska Peninsula: Geological Society of America Bulletin, v. 100, p. 1466–1478.

Wheeler, J. O., Brookfield, A. J., Gabrielse, H., Monger, J.W.H., Tipper, H. W., and Woodsworth, G. J., 1988, Terrane map of the Canadian Cordillera: Geological Survey of Canada Open File Report 1894, 9 p., scale 1:2,000,000.

Wilson, F. H., 1985, The Meshik arc—An Eocene to earliest Miocene magmatic arc on the Alaska Peninsula: Alaska Division of Geological and Geophysical Surveys Professional Report no. 88, 14 p.

Wilson, F. H., Case, J. E., and Detterman, R. L., 1985a, Preliminary description of a Miocene zone of structural complexity in the Port Moller and Stepovak Bay quadrangles, Alaska, *in* Bartsch-Winkler, S., and Reed, K. M., eds., The United States Geological Survey in Alaska: Accomplishments during 1983: U.S. Geological Survey Circular 945, p. 54–56.

Wilson, F. H., Detterman, R. L., and Case, J. E., 1985b, The Alaska Peninsula terrane: A definition: U.S. Geological Survey Open-File Report 85-450, 19 p.

Wilson, F. H., Detterman, R. L., and DuBois, G. D., 1994, Geologic framework of the Alaska Peninsula, southwest Alaska, and the Alaska Peninsula terrane: U.S. Geological Survey Bulletin 1969-B (in press).

Winkler, G. R., 1992, Geologic map and summary geochronology of the Anchorage 1° × 3° quadrangle, southern Alaska: U.S. Geological Survey Miscellaneous Investigations Series Map I-2283, 1 sheet, scale 1:250,000.

Winkler, G. R., Silberman, M. L., Grantz, A., Miller, R. J., and MacKevett, E. M., Jr., 1981a, Geologic map and summary geochronology of the Valdez quadrangle, southern Alaska: U.S. Geological Survey Open-File Report 80-892-A, 2 sheets, scale 1:250,000.

Winkler, G. R., Miller, R. J., Silberman, M. L., Grantz, A., Case, J. E., and Pickthorn, W. J., 1981b, Layered gabbroic belt of regional extent in the Valdez quadrangle, *in* Albert, N.R.D., and Hudson, T., eds., The United States Geological Survey in Alaska: Accomplishments during 1979: U.S. Geological Survey Circular 823-B, p. B74–B76.

Winkler, G. R., Miller, R. J., and Case, J. E., 1981c, Blocks and belts of blueschist and greenschist in the northwestern Valdez quadrangle, *in* Albert, N.R.D., and Hudson, T., eds., The United States Geological Survey in Alaska: Accomplishments during 1979: U.S. Geological Survey Circular 823-B, p. B72–B74.

Wise, D. U., Dunn, D. E., Engelder, J. T., Geiser, P. A., Hatcher, R. D., Kish, S. A., Odum, A. L., and Schamel, W., 1984, Fault-related rocks: Suggestions for terminology: Geology, v. 12, p. 391–394.

Yole, R. W., and Irving, E., 1980, Displacement of Vancouver Island: Paleomagnetic evidence from the Karmutsen Formation: Canadian Journal of Earth Sciences, v. 17, p. 1210–1288.

Manuscript Accepted by the Society May 28, 1993

ACKNOWLEDGMENTS

We gratefully acknowledge invaluable discussions of the geology of south-central Alaska over the years with J. N. Aleinikoff, Fred Barker, H. C. Berg, G. C. Bond, T. K. Bundtzen, L. E. Burns, R. G. Coleman, P. J. Coney, Bela Cséjtey, Jr., S. M. DeBari, R. L. Detterman, Cynthia Dusel-Bacon, W. G. Gilbert, Arthur Grantz, C. S. Grommé, J. W. Hillhouse, Travis Hudson, D. L. Jones, M. A. Lanphere, E. M. MacKevett, Jr., T. E. Moore, W. W. Patton, Jr., T. L. Pavlis, S. M. Roeske, D. H. Richter, N. J. Silberling, V. B. Sisson, T. E. Smith, D. B. Stone, J. H. Stout, and G. R. Winkler. We appreciate constructive reviews of the manuscript by David W. Scholl and G. R. Winkler.

Printed in U.S.A.

The Geology of North America
Vol. G-1, The Geology of Alaska
The Geological Society of America, 1994

Chapter 11

Geologic framework of the Aleutian arc, Alaska

Tracy L. Vallier, David W. Scholl, Michael A. Fisher, and Terry R. Bruns
U.S. Geological Survey, 345 Middlefield Road, Menlo Park, California 94025
Frederic H. Wilson
U.S. Geological Survey, 4200 University Drive, Anchorage, Alaska 99508
Roland von Huene* and Andrew J. Stevenson
U.S. Geological Survey, 345 Middlefield Road, Menlo Park, California 94025

INTRODUCTION

The Aleutian arc is the arcuate arrangement of mountain ranges and flanking submerged margins that forms the northern rim of the Pacific Basin from the Kamchatka Peninsula (Russia) eastward more than 3,000 km to Cook Inlet (Fig. 1). It consists of two very different segments that meet near Unimak Pass: the Aleutian Ridge segment to the west and the Alaska Peninsula–Kodiak Island segment to the east. The Aleutian Ridge segment is a massive, mostly submerged cordillera that includes both the islands and the submerged pedestal from which they protrude. The Alaska Peninsula–Kodiak Island segment is composed of the Alaska Peninsula, its adjacent islands, and their continental and insular margins. The Bering Sea margin north of the Alaska Peninsula consists mostly of a wide continental shelf, some of which is underlain by rocks correlative with those on the Alaska Peninsula.

There is no pre-Eocene record in rocks of the Aleutian Ridge segment, whereas rare fragments of Paleozoic rocks and extensive outcrops of Mesozoic rocks occur on the Alaska Peninsula. Since the late Eocene, and possibly since the early Eocene, the two segments have evolved somewhat similarly. Major plutonic and volcanic episodes, however, are not synchronous. Furthermore, uplift of the Alaska Peninsula–Kodiak Island segment in late Cenozoic time was more extensive than uplift of the Aleutian Ridge segment. It is probable that tectonic regimes along the Aleutian arc varied during the Tertiary in response to such factors as the directions and rates of convergence, to bathymetry and age of the subducting Pacific Plate, and to the volume of sediment in the Aleutian Trench.

The Pacific and North American lithospheric plates converge along the inner wall of the Aleutian trench at about 85 to 90 mm/yr. Convergence is nearly at right angles along the Alaska Peninsula, but because of the arcuate shape of the Aleutian Ridge relative to the location of the plates' poles of rotation, the angle of convergence lessens to the west (Minster and Jordan, 1978). Along the central Aleutian Ridge, underthrusting is about 30° from normal to the volcanic axis. Motion between plates is approximately parallel along the western Aleutian Ridge.

In this paper we briefly describe and interpret the Cenozoic evolution of the Aleutian arc by focusing on the onshore and offshore geologic frameworks in four of its sectors, two sectors each from the Aleutian Ridge and Alaska Peninsula–Kodiak Island segments (Fig. 1). We compare the geologic evolution of the segments and comment on the implications of some new, previously unpublished data.

Sector 1, the Komandorsky (Russia)–Near Islands sector of the Aleutian Ridge segment, is an area along that part of the Pacific–North American plate boundary which has been virtually strike-slip in character since at least late middle Eocene time, when plate-direction and plate-velocity changes relative to the Hawaiian hot spot created the Emperor-Hawaiian seamount bend at about 43 Ma (Dalrymple and others, 1977; Clague, 1981). Sector 2, the Adak-Amlia sector, includes the large islands of Adak, Atka, and Amlia in the central Aleutians. Here, plate convergence is oblique. The third and fourth sectors constitute the Alaska Peninsula and its bordering islands. The Shumagin sector of the Alaska Peninsula, the third sector, is underthrusted by the Pacific Plate nearly normal to the arc's axis. The fourth, or Kodiak Island, sector on the east has been similarly underthrusted by the Pacific Plate.

Unpublished data incorporated in this chapter are mostly from the western part of the arc, where offshore studies in 1981 provided seismic-reflection profiles and other geophysical data, and onshore work in 1981 and 1983 yielded rock samples for stratigraphic and petrologic studies. In addition, GLORIA (Geological Long-Range Inclined Asdic) side-scan images were collected within the Exclusive Economic Zone (EEZ) of the entire arc during cruises in 1987 and 1988. Also collected during those latter cruises were single-channel seismic-reflection and bathymetric profiles, magnetics, and gravity data. Seismic-reflection

*Present address: GEOMAR, Christian-Albrechts University, Wischhofstrasse 1-3, 2300 Kiel 14, Germany.

Vallier, T. L., Scholl, D. W., Fisher, M. A., Bruns, T. R., Wilson, F. H., von Huene, R., and Stevenson, A. J., 1994, Geologic framework of the Aleutian arc, Alaska, *in* Plafker, G., and Berg, H. C., eds., The Geology of Alaska: Boulder, Colorado, Geological Society of America, The Geology of North America, v. G-1.

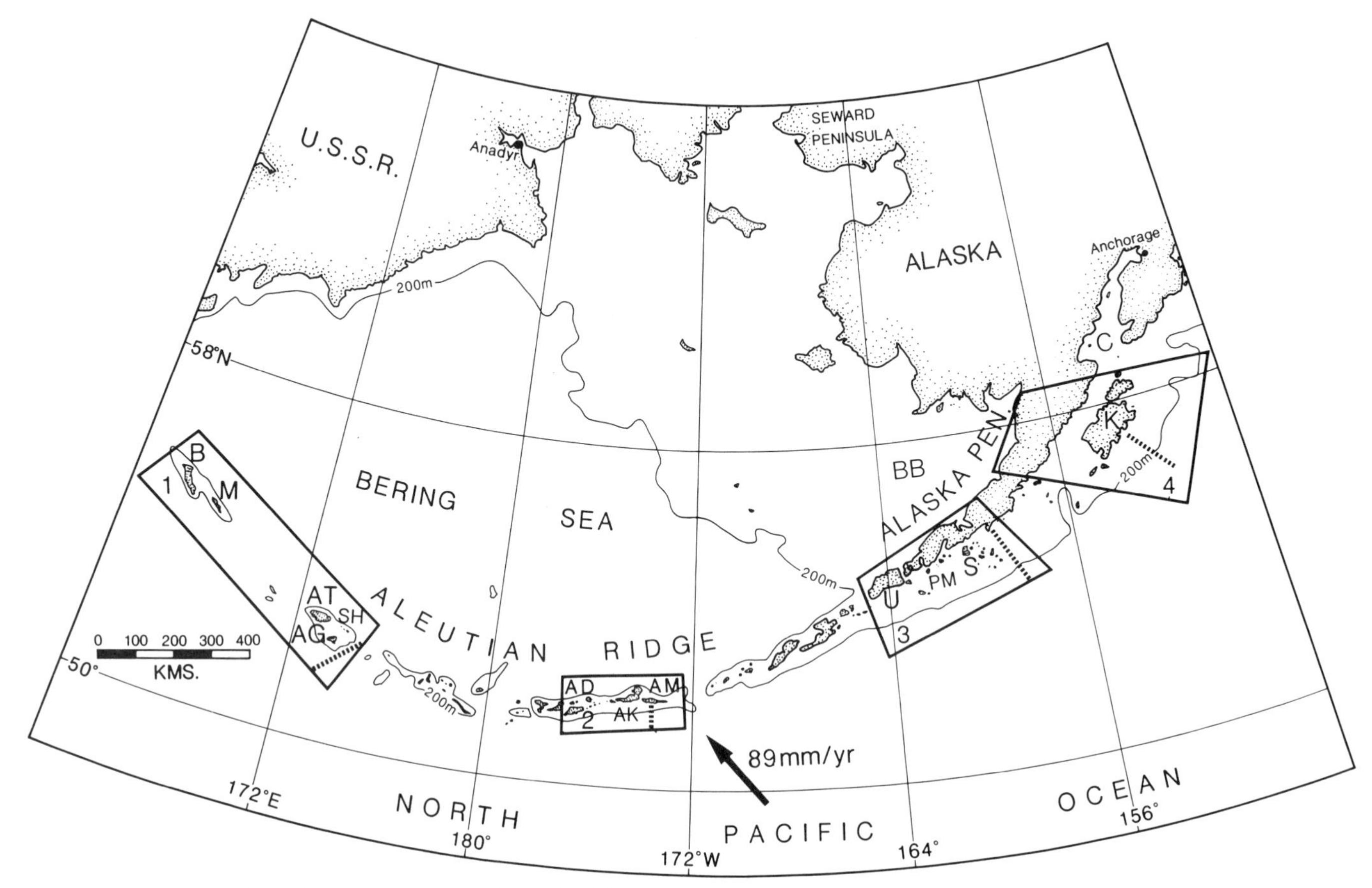

Figure 1. Index map of the North Pacific Ocean, Bering Sea, and Aleutian arc, with convergence direction (arrow) and rate shown for the central Aleutian Ridge segment (Engebretson and others, 1986). Outlined sectors: 1, Komandorsky–Near Islands sector; 2, Adak-Amlia sector; 3, Shumagin sector; and 4, Kodiak Island sector. The abbreviations (left to right) are B, Bering Island; M, Medny Island; AT, Attu Island; AG, Agattu Island; SH, Shemya Island; AD, Adak Island; AK, Atka Island; AM, Amlia Island; U, Unimak Island (Unimak Pass lies to the west of the island); BB, Bristol Bay; PM, Port Moller; S, Shumagin Islands; K, Kodiak Island; and C, Cook Inlet. Approximate locations of seismic-reflection profiles (Figs. 4, 6, 8, and 9) are shown as dashed lines within the outlined sectors.

profiles displayed in this paper were reduced photographically from migrated 24-channel records. We use the geologic time scale of Palmer (1983) throughout this chapter. Some previously unpublished stratigraphic and age data from the Alaska Peninsula also are incorporated in this chapter.

PREVIOUS STUDIES

The U.S. Geological Survey (USGS) Alaska Volcano Project focused attention on the Aleutian arc beginning in the late 1940s and continuing through the 1960s; results of those studies are published in USGS Bulletins (974 and 1028 series). Additional geologic studies preceded the nuclear tests on Amchitka Island in 1969 and 1971. In the Komandorsky Islands region, geological investigations of the major islands, Medny and Bering, were reported by Shmidt (1978) and Borsuk and Tsvetkov (1980). A regional geological and geophysical synthesis of onshore and offshore studies of the Bering Sea and Aleutian Ridge areas was written by Stone (1988).

The monumental work by Burk (1965) provided the first coherent summary of regional stratigraphy and structure of the Alaska Peninsula. Onshore studies (Fig. 2) of the Alaska Peninsula from the late 1960s to the present have been carried out in large part by several USGS geologists (e.g., Reed and Lanphere, 1969, 1973; McLean, 1979; Reed and others, 1983; Detterman and others, 1983, 1985; Detterman and Miller, 1985; Detterman, 1985, 1986; Detterman and others, 1990; Wilson, 1980, 1982, 1985; Wilson and others, 1981, 1985a, b). Stone and Packer (1977) provided important paleomagnetic data. Bordering islands, including the Kodiak and Sanak Islands, were studied by J. C. Moore and his colleagues at the University of California, Santa Cruz, and George Moore, Tor Nilsen, and others at the US Geological Survey (e.g., Moore, 1974a, b; Moore and others, 1983; Connelly, 1978; Moore, 1969; Nilsen and Moore, 1979; Nilsen and Zuffa, 1982). Particularly important to USGS studies on the Alaska Peninsula is the Alaska Mineral Resource Assessment Program (AMRAP), which began in 1974. Aleutian volcanoes have been studied during the past decade mostly by Robert Kay and Suzanne Kay and their associates at Cornell University (e.g., Kay and Kay, this volume), Thomas P. Miller at the USGS (Miller and Richter, this volume), and Bruce Marsh and his associates at Johns Hopkins University. James Myers at the University of Wyoming and James Brophy at the University of Indiana currently are studying several of the volcanoes.

Offshore geologic framework investigations along the Aleutian Ridge segment began in the early 1960s (Shor, 1964), continued with studies by Grow and Atwater (1970), the Deep Sea Drilling Project (Leg 19: Creager and others, 1973), and reached their most intense phase in the late 1970s and early 1980s when the USGS gathered multichannel seismic-reflection, seismic-refraction, gravity, and magnetic data, and rock samples (Scholl and others, 1975, 1983a, b, 1986, 1987; McCarthy and others, 1984; McCarthy and Scholl, 1985; Harbert and others, 1986; Geist and others, 1988; Ryan and Scholl, 1989). Lonsdale (1988) reported on the offshore region south of the Near Islands. Newly acquired GLORIA images and the associated geophysical data are increasing our knowledge and understanding of the Aleutian arc. Field studies complemented the offshore studies (Hein and McLean, 1980; McLean and others, 1983; Hein and others, 1984; Vallier and others, 1984).

Alaska Peninsula and Kodiak Island offshore geologic framework studies also began in the 1960s, were greatly enhanced by results from Leg 18 of the Deep Sea Drilling Project (Kulm and others, 1973), and reached their peak during the late 1970s and early 1980s. GLORIA images and associated geophysical data, collected in 1988 and 1989, are currently being studied. Interpretations of the offshore geology near Kodiak Island were given by Fisher (1980), Fisher and von Huene (1980, 1982, 1984), Fisher and others (1981, 1987), and von Huene and others (1979, 1985, 1987). Geological interpretations of the more westerly parts of the Alaska Peninsula offshore region near the Shumagin Islands were provided by Bruns and von Huene (1977), Bruns and others (1985, 1987a, b, c), and Lewis and others (1984, 1988).

GEOLOGIC FRAMEWORK OF THE ALEUTIAN RIDGE SEGMENT

Age and correlation

The Aleutian Ridge has been part of an active volcanic arc for at least 55 m.y. (Scholl and others, 1983a, b, 1986, 1987; Lonsdale, 1988; Ryan and Scholl, 1989). However, tectonic and igneous responses to subduction, as well as subduction rates and the process itself, have varied with time and location. Accordingly, large variations in lithofacies and structures are recorded in the rocks. Structural data from the islands are somewhat inconsistent because the ridge is segmented into discrete tectonic blocks, some of which have rotated (Harbert and Cox, 1984; Geist and others, 1988) in response to obliquely directed stresses that result from convergence of the Pacific and North American Plates.

A major problem in interpreting the Cenozoic evolution of the Aleutian Ridge is the lack of reliable age data. Although older rocks are typically more altered than younger rocks, compositions and stratigraphies are not reliable for correlation purposes; similar lithologies and stratigraphic sequences occur in rocks of all ages. Some specific factors that contribute to the correlation problems are (1) an abundance of unfossiliferous, coarse-grained volcaniclastic debris; (2) abrupt lithofacies changes; (3) a long time range over which many of the flora and fauna lived, thereby making the fossils of limited value for age determinations; and (4) the arc's complex igneous history, which thermally altered preexisting rocks, thereby destroying all but the most robust fossils and making K-Ar radiometric ages suspect, especially near plutons (Delong and McDowell, 1975).

Despite age and lithologic correlation problems, rock-stratigraphic units have been mapped on some islands of the Aleutian Ridge. Gates and others (1971), for example, described rock units on Attu and Agattu Islands. Shmidt (1978) mapped units on the Komandkorsky Islands and correlated them with lithologic units on other islands of the Aleutian Ridge.

In order to correlate onshore and offshore units, we follow the example of Marlow and others (1973) and Scholl and others (1975) who informally divided the rocks into chronostratigraphic units (early, middle, and late series), which subsequently were changed (Scholl and others, 1983a, b) to lower (LS), middle (MS), and upper (US) series (Fig. 2). Rocks of these series accumulated during time intervals (early, middle, and late series time) that are associated with the ridge's evolutionary stages; they are not related directly to lithology. By dividing the rocks into chronostratigraphic units, we can examine all the rocks deposited and magmatically emplaced within a specific time interval and thereby relate them to the evolution of the ridge. A major disadvantage in using chronostratigraphic units is that they are mappable only if their ages are known (as already noted, clues are not easily found). A notable advantage is that temporally related onshore and offshore rocks can be discussed together despite large lithologic differences.

The use of chronostratigraphic units (LS, MS, US) is amenable to both the islands and the offshore in the Aleutian Ridge segment, although the smaller offshore data base has produced reliable rock ages only in certain areas (Scholl and others, 1983a, 1987). Our chronostratigraphic division for the entire ridge is based mainly on the recognition of geophysically mapped offshore rock and sedimentary units in the Adak-Amlia sector of the Aleutian Ridge. The offshore lower, middle, and upper series are dated and correlated, in part, by tracing reflecting horizons on seismic-reflection profiles to onshore outcrops, and, in part, by data obtained from submarine outcrops and drilling (Scholl and others, 1983a). Onshore units, where undated, are separated by their relative amounts of alteration and deformation; LS rocks generally are more altered and deformed than younger rocks.

We assigned an age older than 37 Ma to LS rocks along the Aleutian Ridge. The 37-Ma age is based in part on our interpretation that the major building stage of the Aleutian Ridge was completed and voluminous and widespread igneous activity had greatly decreased by about that time (Scholl and others, 1987; Ryan and Scholl, 1989). Use of the 37-Ma age is strengthened by fossils in rocks from a dredge haul located at the base of MS rocks in the Adak-Amlia sector that have an age about the same as the Eocene-Oligocene boundary (Scholl and others, 1987). Deposition of these MS beds on older igneous masses of the LS occurred after igneous activity was more restricted along the ridge crest, thereby signaling the onset of the dominance of sedimentary processes over igneous processes on the ridge's crest and sloping flanks.

On the basis of island outcrops, the LS consists dominantly of flow, hypabyssal plutonic, and volcaniclastic rocks; rare are deep-seated plutonic rocks. Included in the LS, however, are rocks that may constitute a "basement" to the stratified rocks, such as the metamorphosed mafic plutonic suite that is exposed on Attu Island (Vallier and others, 1983). LS rocks are generally metamorphosed to greenschist (or albite-epidote hornfels), prehnite-pumpellyite, and zeolite facies. In places, the grade of metamorphism in LS rocks is directly related to the proximity of younger plutons.

Rocks from the LS offshore are generally characterized by coarse and laterally discontinuous acoustic layering. In many areas, a strong and irregular upper reflector is overlain by middle series strata that exhibit laterally continuous acoustic layering. In the Aleutian Ridge back-arc region, LS rocks merge acoustically with, or overlie, probable Cretaceous igneous oceanic crust; the boundary cannot be resolved on available seismic-reflection profiles.

MS rocks are assigned an age range of 37 to 5.3 Ma (Scholl and others, 1987). We base the younger age on physical evidence for accelerated deformation and igneous activity along the present ridge axis beginning in about the early Pliocene (or possibly the latest Miocene). Mostly for convenience we place the base of the US at the Miocene-Pliocene boundary, which is 5.3 Ma according to the time scale of Palmer (1983). Island exposures of MS strata include abundant volcanic flow and volcaniclastic rocks and common plutonic rocks. Many sills, dikes, and plugs and a

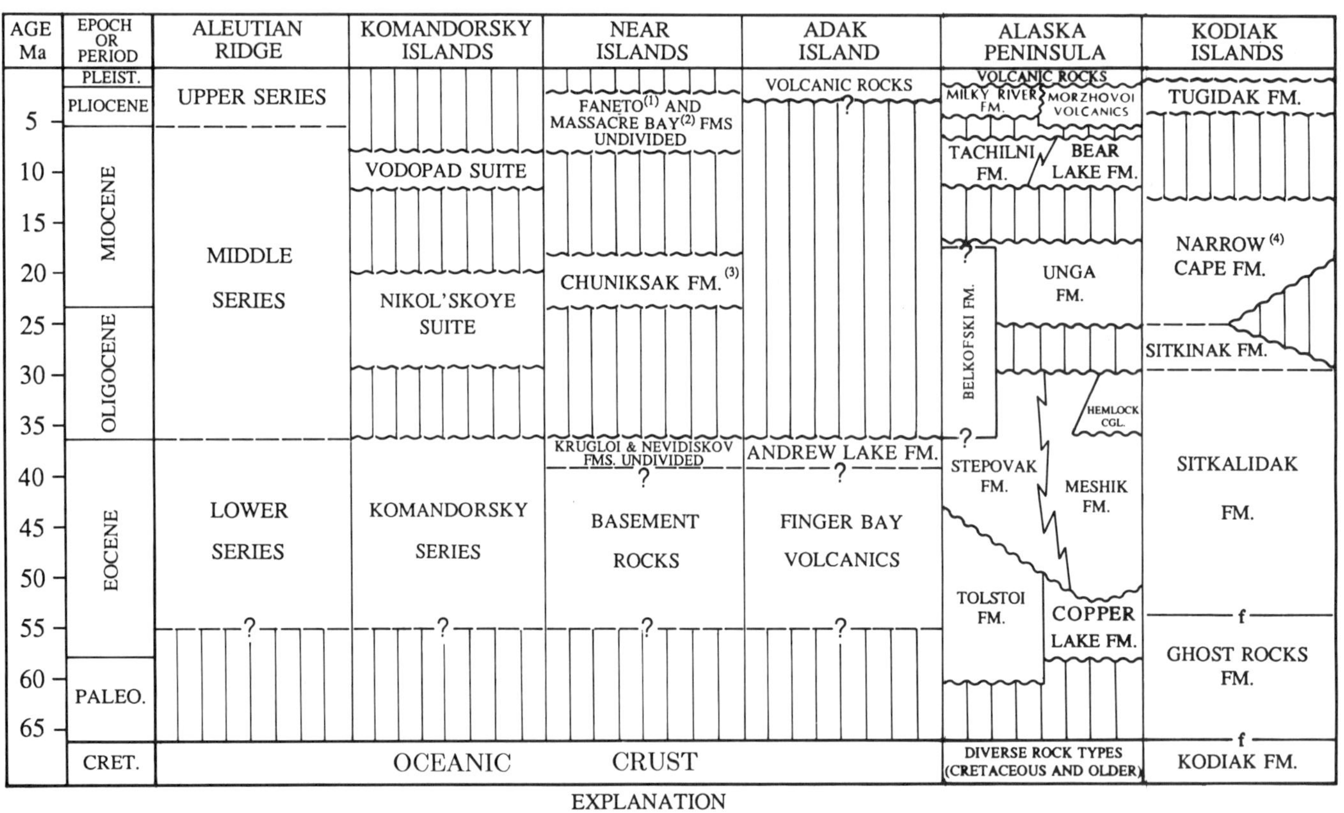

Figure 2. Correlation chart of Aleutian Ridge chronostratigraphic units and Upper Cretaceous and Tertiary rocks of the Alaska Peninsula and Kodiak Islands (compiled from Shmidt, 1978; Gates and others, 1971; Coats, 1956; Fraser and Snyder, 1959; Hein and McLean, 1980; Rubenstone, 1984; Detterman and others, 1990; G. Moore, 1969, and written communication, 1986). Time scale from Palmer (1983). Numbers signify the following: (1) age is late Tertiary or early Pleistocene; (2) age is Miocene; (3) age is Miocene (?); (4) age is late Oligocene(?) to middle Miocene.

relatively large pluton (Hidden Bay pluton on Adak Island) are of Oligocene (36 to 28 Ma) age and constitute an older plutonic phase of MS rocks (Citron and others, 1980; Pickthorn and Vallier, 1991). Large quartz diorite and granodiorite plutons, such as those exposed on Atka Island (Hein and others, 1984), constitute a younger plutonic phase and are of middle and early late Miocene (20 to 9 Ma) age (Pickthorn and others, 1984; Pickthorn and Vallier, 1991).

Middle series (MS) strata form an extensive sediment blanket on both fore-arc and back-arc slopes. Except in certain areas, the MS strata do not vary much in thickness; this indicates that slope basins were not subsiding rapidly and that, in general, local structural relief was not particularly great during the time of MS sediment deposition. Dip-slope sequences of MS strata are generally truncated along the outer edge of a very pronounced wave-planed surface of the summit platform.

Upper series (US) rocks and sediment accumulations formed approximately during the past 5.3 m.y. and include material not only from the latest stage of volcanism, but also most of the strata that fill offshore basins and mantle insular and continental slopes. These strata were deposited during a period of extensive ridge planation. Offshore, US accumulations fill fore-arc and summit basins, thinly mantle parts of the summit platform and sloping flanks of the ridge, and constitute most of the accretionary complex that underlies the landward slope of the trench. US sediments fill the Aleutian Trench.

Western Aleutian Ridge: Komandorsky–Near Islands sector

The geology of the Komandorsky Islands was reviewed by Shmidt (1978), and that of the Near Islands by Gates and others (1971), Delong and McDowell (1975), and Rubenstone (1984). Data from the corresponding offshore regions include single-channel seismic-reflection profiles and rock samples collected on a cruise of the R/V *Bartlett* (Buffington, 1973) and on Soviet cruises (e.g., Gnibidenko and others, 1980); on multichannel and single-channel seismic-reflection profiles, sonobuoy, magnetic, and gravity data collected during a 1981 cruise of the R/V *S.P. Lee* (Scholl and others, 1983a, b, 1987; McCarthy and Scholl, 1985; Ryan and Scholl, 1989); and on GLORIA side-scan images, single-channel seismic-reflection profiles, plus gravity and magnetic data collected on a cruise of the M/V *Farnella* in 1987 (unpublished).

Less than 100 km south of Attu Island the extinct Kula-Pacific Ridge intersects the Aleutian Trench (Lonsdale, 1988); a small fragment of the Kula Plate remains west of the Kula-Pacific Ridge and south of the trench. The south and west sides of the Kula Plate are terminated by Stalemate Ridge, a major transform fault that had separated Cretaceous crust of the Pacific Plate from early Tertiary crust of the Kula Plate during spreading along the Kula-Pacific Ridge. The axis of the Aleutian Trench is uplifted more than 1 km at the intersection of Stalemate Ridge and the trench. Stalemate Ridge has been partly responsible for extensive deformation along the inner trench wall of the western Aleutian Ridge as it moved westward along the arc during oblique convergence (Vallier and others, 1987).

Lower series. Rocks of the lower series (greater than 37 Ma) are well exposed on Bering and Medny Islands of the Komandorsky Islands group and on Agattu and Attu Islands of the Near Islands group (Figs. 2 and 3). The oldest unit on the Komandorsky Islands, the Komandorsky Series, is subdivided into four subunits (suites) of probable Eocene age (Shmidt, 1978). The Komandorsky Series is composed dominantly of volcaniclastic rocks and contains a relatively wide range of fossil marine faunas, including planktonic foraminifers. Radiometric ages of 35 to 28 Ma (K-Ar minimum ages of altered rocks) were obtained on samples of flows and small plutons in the Komandorsky Series (Borsuk and Tsvetkov, 1980; Pickthorn and Vallier, 1991).

On Attu Island (Fig. 3), metamorphosed mafic plutons (Vallier and others, 1983), volcanic and sedimentary rocks (the "basement rocks" unit of Gates and others, 1971), and the Chirikof and Nevidiskov Formations (Gates and others, 1971) are assigned to the lower series. LS rocks on Agattu Island consist of a "basement rocks" unit and the Krugloi Formation (Gates and others, 1971).

The mafic metaplutonic unit on Attu Island (Vallier and others, 1983), where mapped, is in fault contact with both the stratified "basement rock" unit of Gates and others (1971) and younger rocks. It consists of small stocks, sills, and dikes of metamorphosed gabbro, diorite, diabase, and basalt. It most likely is the pedestal upon which oldest stratified rocks were magmatically emplaced and deposited. Major minerals are plagioclase, blue-green amphibole, epidote, and iron-oxide minerals. The rocks have some trace-element characteristics of the island-arc tholeiite (IAT) magma series, with low TiO_2 values, flat rare-earth patterns (normalized to chondrite-element abundances), and low values of Ni, Co, and Cr. A high $^{143}Nd/^{144}Nd$ ratio, however, suggests a mid-ocean ridge basalt (MORB) or back-arc basin basalt (BABB) affinity of these rocks (James Rubenstone, written communication, 1988). Some of the metaplutonic rocks near faults have undergone variable penetrative deformation, and are now amphibolite, actinolite-plagioclase-quartz schist, and mylonite; most rocks, however, have retained their primary igneous textures. A K-Ar radiometric age of 41.5 ± 2 Ma was determined on mineral separates of blue-green amphibole (Vallier and others, 1983; Pickthorn and Vallier, 1991). That late Eocene age marks the time of metamorphism and not of igneous crystallization.

Rocks within the oldest stratified unit on Attu Island (the "basement rocks" unit of Gates and others, 1971), include volcanic flow rocks, volcaniclastic rocks, limestone, chert, and small hypabyssal plutonic rocks. Chemically analyzed volcanic rocks range in composition from basalt to rhyolite and have trace-element signatures of island arc tholeiite (IAT) magmas. Fossil nannoplankton specimens yield a "late Eocene–early Oligocene age" (David Bukry, written communication, 1983). The Chirikof Formation (Gates and others, 1971), based on our field mapping, may be a fine-grained facies of the "basement rocks" unit. The

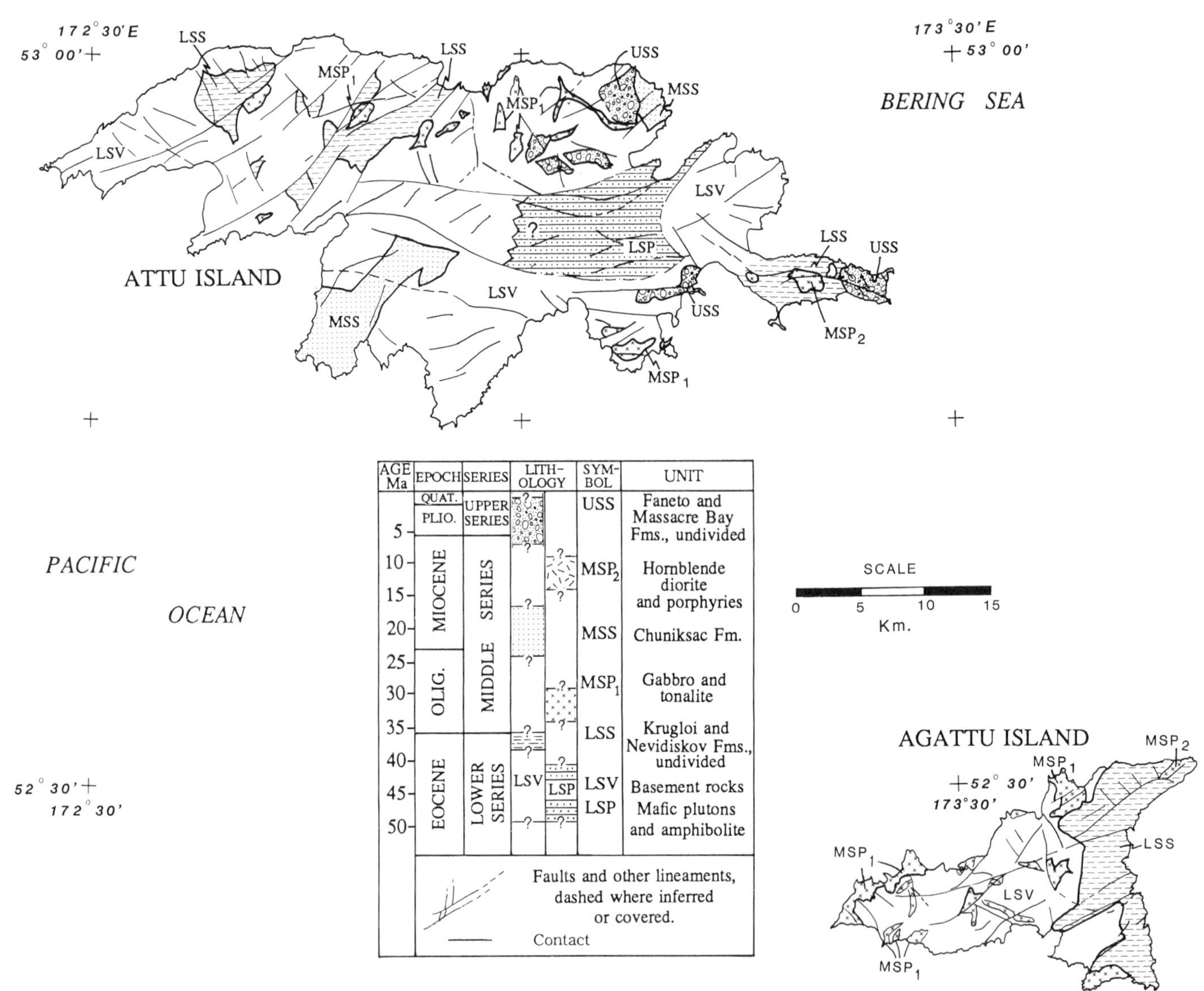

Figure 3. Generalized geologic map and stratigraphic column of Attu and Agattu Islands of the Near Islands group, sector 1 (modified from Gates and others, 1971).

Krugloi Formation on Agattu Island is well stratified and consists of flysch-like beds of sandstone, argillite, and limestone. Limestone samples within the unit have an age range of "middle Eocene to late Oligocene" based on very sparse fossil nannoplankton (Bukry, written communication, 1982). A 34-Ma K-Ar radiometric age (Pickthorn and Vallier, 1991), from altered quartz diorite that cuts the Krugloi Formation on Agattu Island, indicates that the Krugloi Formation is older than 34 Ma. The similarity between the Krugloi Formation on Agattu Island and Nevidiskov Formation (Gates and others, 1971) on Attu Island suggests that those two units are correlative.

Offshore, poorly stratified rocks of the LS underlie a prominent acoustic horizon that can be traced on a multichannel seismic profile from the summit platform seaward to an accretionary complex at the base of the inner trench wall (Fig. 4). LS rocks are truncated by the wave-cut surface of the summit platform, and can be acoustically traced northward across the insular slope, where they both merge with and overlie igneous oceanic crust.

Middle series. The middle series (Oligocene and Miocene; 37 to 5.3 Ma) rocks occur as flows, hypabyssal plutons, and volcaniclastic strata on both the Komandorsky and Near Islands groups. Offshore, on the Pacific flank of the Aleutian Ridge, the MS is characterized by relatively well-layered acoustic units (Fig. 4).

Middle series rocks on the Komandorsky Islands (Fig. 2) are the Nikol'skoye Suite, a unit that crops out mostly on Bering Island, and the Vodopad Suite of Medny Island (Shmidt, 1978). The Nikol'skoye Suite, of late Oligocene and early Miocene age, consists of flows and associated pyroclastic deposits, related volcaniclastic sedimentary rocks, conglomerate, and dolomite. Fossils include terrestrial plants and marine diatoms, mollusks, and foraminifers. Volcanic rocks of the Nikol'skoye Suite are alkalic

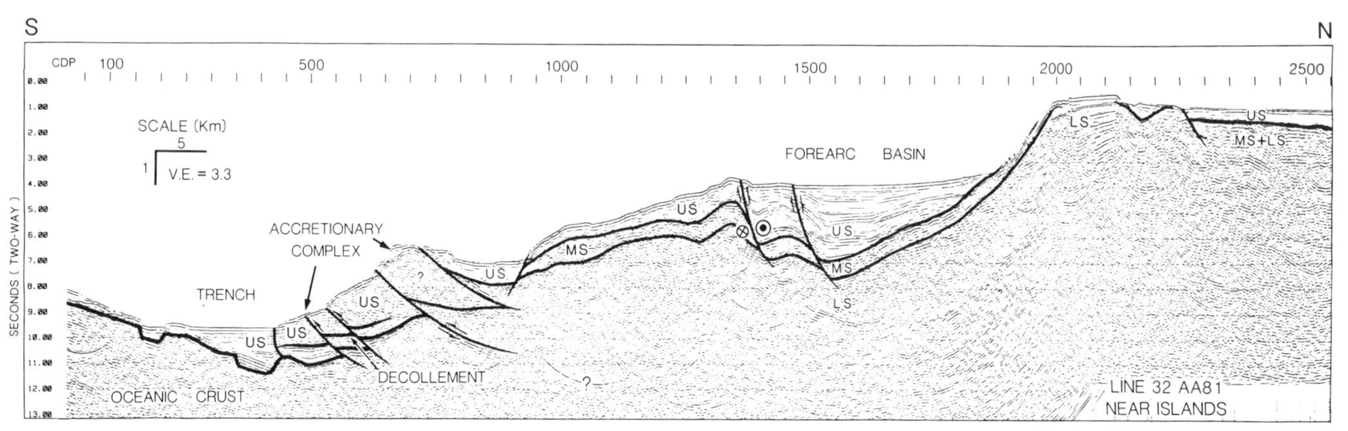

Figure 4. Migrated 24-channel seismic-reflection profile of Line 31 from the Near Islands region (see Fig. 1 for approximate location). Distance between common depth points (CDP) is about 50 m. The chronostratigraphic units are LS, lower series; MS, middle series; and US, upper series. Note the relatively small accretionary complex and the chaotic structure along the inner wall of the trench. Movement along some faults may be in part strike-slip; the fault that cuts the fore-arc basin shows suspected right-lateral displacement with circled dot (movement toward observer) and circled x (movement away from observer).

(Borsuk and Tsvetkov, 1980), as indicated by abundant ultramafic xenoliths in the basalt flows, mineralogy (titanoaugite, titanomagnetite, and barkevikite in the modes), and chemistry (relatively high Na_2O, K_2O, TiO_2, Ba, and Sr). Deeply eroded subaerial volcanoes on Medny Island make up the Vodopad Suite. The Vodopad Suite consists of lava flows and volcaniclastic rocks of basalt, andesite, and dacite compositions. Potassium-argon radiometric ages of Vodopad Suite rocks range from about 12 to 8 Ma (Borsuk and Tsvetkov, 1980; Pickthorn and Vallier, 1991).

Middle series strata on Attu Island are the Chuniksak and Massacre Bay Formations (Figs. 2 and 3). The Chuniksak Formation (Gates and others, 1971) consists of Miocene(?) siliceous fine-grained sedimentary rocks. The late Miocene (8 to 6 Ma) Massacre Bay Formation has andesitic and dacitic lavas, associated volcaniclastic sedimentary rocks, and a wide assortment of hypabyssal intrusive rocks (Gates and others, 1971; Delong and McDowell, 1975). Two major phases of MS plutonism are recognized on the Near Islands. The older phase, exposed both on Agattu and Attu Islands (Fig. 3), includes hypabyssal plutons of moderately altered gabbro, diorite, quartz diorite, and diabase. Whole-rock K-Ar radiometric ages range from about 34 to 28 Ma (Pickthorn and Vallier, 1991), indicating a probable Oligocene age of emplacement. Younger MS plutons, ranging in age from 15 to 11 Ma (Pickthorn and Vallier, 1991), are exposed on Shemya Island and along the northeastern coast of Agattu Island.

Offshore Attu and Agattu Islands, MS strata form a coherently layered and laterally continuous unit (Fig. 4). MS strata apparently mantled both the summit and the slopes of the western Aleutian Ridge before the formation of a fore-arc basin. Strata assigned to the MS are recognized beneath US strata in the fore-arc basin; they can be traced seaward on seismic-reflection profiles to the accretionary wedge. In general, acoustic layering of MS strata is not so regular and coherent as in the overlying US.

Upper series. Upper series (Pliocene and Quaternary; less than 5.3 Ma) rocks are not exposed in any large bodies on the Komandorsky Islands, but the Faneto Formation on Attu Island probably can be assigned to the upper series. The Faneto Formation is a thick sedimentary rock unit that was deposited in a fluvial environment. Gates and others (1971) stated that the age was uncertain because of an absence of fossils. However, because of the composition of the Faneto Formation, they assigned it to the late Tertiary or early Pleistocene and either syn– or post–Massacre Bay Formation in age. Based on our field work, we conclude that it is younger than the Massacre Bay Formation and assign it to the upper series (Fig. 3).

Upper series strata offshore the Near Islands are in places very thick and fill a series of discontinuous fore-arc basins (Karl and others, 1987; Vallier and others, 1987). Equivalent strata fill the Aleutian Trench (Fig. 4) and thinly mantle large parts of the summit platform. They also constitute most of the accretionary wedge that underlies the landward slope of the trench. US sediment packages are as thick as 3 km (2.8 sec) in the fore-arc basins and may be several kilometers thick in the structurally stacked accretionary wedge.

Central Aleutian Ridge: Adak-Amlia sector

Reviews of the geology of the central Aleutian Ridge in the Adak-Amlia sector are given by Marlow and others (1973), Hein and McLean (1980), Kay and others (1982), McLean and others (1983), Scholl and others (1983a, b; 1987), Hein and others (1984), Rubenstone (1984), McCarthy and others (1984), McCarthy and Scholl (1985), Geist and others (1988), Ryan and Scholl (1989), Fournelle and others (this volume), and Kay and Kay (this volume). Our offshore work in the central Aleutians was concentrated east of Adak Island near Amlia and Atka Islands (Fig. 1). New interpretations of the geologic evolution of the central Aleutian Ridge include the role of strike-slip faulting in the development of both summit and forearc basins and the tectonic rotation of discrete blocks along the ridge (Geist and others, 1988; Ryan and Scholl, 1989). Rocks on Adak Island (Fig. 5) are probably representative of the central Aleutian Ridge rock framework. All chronostratigraphic units are exposed on Adak Island.

Lower series (LS). The Finger Bay Volcanics, Finger Bay pluton, and Andrew Lake Formation on Adak Island (Coats, 1950, 1952, 1956; Fraser and Snyder, 1959; Scholl and others, 1970; Hein and McLean, 1980; Kay and others, 1982; Kay, 1983; Rubenstone, 1984) are all part of the lower series (Figs. 2 and 5). LS units constitute most of the rocks on Amlia Island and a large part of the flow and volcaniclastic rocks on southern Atka

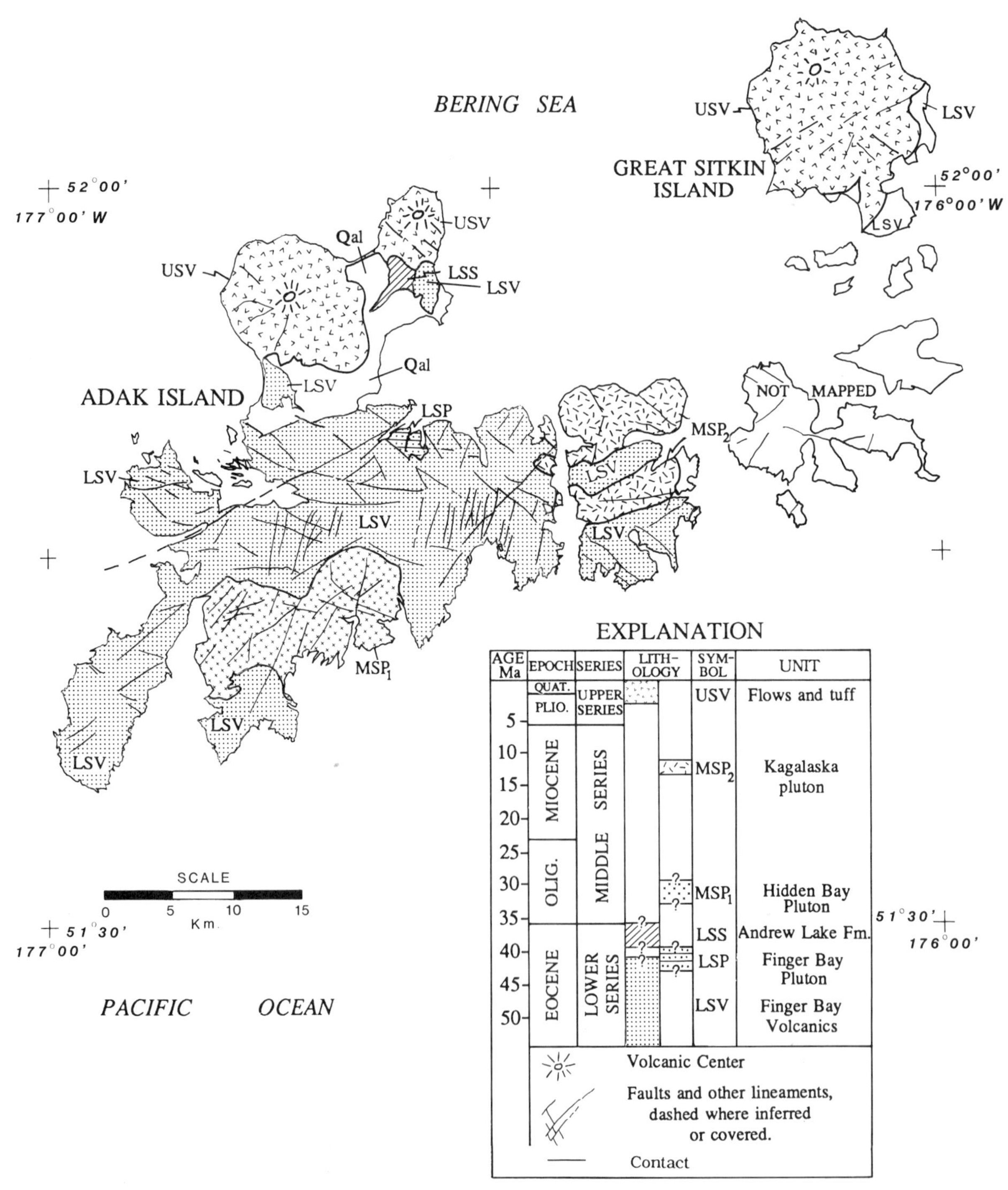

Figure 5. Generalized geologic map and stratigraphic column of Adak, Kagalaska, and Great Sitkin Islands, sector 2 (modified from Coats, 1950, 1956; Fraser and Snyder, 1959; Rubenstone, 1984).

Island (Hein and McLean, 1980; McLean and others, 1983; Hein and others, 1984).

Robust weather- and thermal-resistant fossil nannoplankton floras from calcareous rocks on Amlia Island have age ranges that extend from the "middle Eocene into the early Oligocene" (David Bukry, written communication, 1982). Somewhat altered (weathered and/or metamorphosed to the zeolite and prehnite-pumpellyite facies) gabbros and flow rocks have K-Ar radiometric ages of late Eocene (39.8 ± 1.2 Ma) and early Oligocene (32 ± 1 Ma) respectively; an altered rhyolite was dated at 24.5 ± 0.7 Ma (McLean and others, 1983). Based on the degree of alteration of radiometrically dated igneous rocks and ages of fossils in associated sedimentary rocks, we suspect that some K-Ar ages are too young and that the rocks are late Eocene and earliest(?) Oligocene in age. A $^{40}Ar/^{39}Ar$ radiometric age of about 50 Ma was determined on an altered flow rock from the Finger Bay Volcanics (Rubenstone, 1984), which suggests that rocks older than late Eocene may constitute parts of the lower series of the central Aleutian Ridge. In addition, a late Eocene faunal age is well established for the overlying(?) Andrew Lake Formation (Hein and McLean, 1980).

LS igneous rocks on islands of the Adak-Amlia sector have a broad range of compositions that are characteristic of an island-arc setting. Flow rocks of the Finger Bay Volcanics on Adak Island are calc-alkaline basalt and basaltic andesite (Rubenstone, 1984), whereas on Atka and Amlia Islands, compositions of flow rocks and hypabyssal plutonic rocks are tholeiitic, transitional, and calc-alkaline (Hein and others, 1984). Plutonic rocks of the lower series on Adak, Atka, and Amlia Islands are relatively small hypabyssal bodies of gabbro, diorite, diabase, monzodiorite, quartz diorite, and finer-grained porphyries of the same compositions (Kay and others, 1982; McLean and others, 1983; Hein and others, 1984). The Finger Bay pluton on Adak Island is the most thoroughly studied of the LS plutons (Kay, 1983; Kay and others, 1983).

LS sedimentary rocks on islands of the central Aleutians are dominantly coarse-grained volcaniclastic rocks. Finer-grained volcanic sandstone and siltstone are common components of the strata. Tuff and pyroclastic breccia are abundant locally, such as along the northeastern coast of Amlia Island, where a thick section of metamorphosed (prehnite-pumpellyite facies) silicic pyroclastic rocks is exposed.

Lower series strata offshore generally form a poorly reflective acoustic unit beneath an irregular, but persistent, strong reflector (Fig. 6; Scholl and others, 1983a, b, 1987; Ryan and Scholl, 1989). These massive accumulations form dip-slope sequences beneath the flanks of the arc. LS rocks are mostly submarine flows, hypabyssal plutons, and coarse-grained volcaniclastic rocks on islands near the ridge crest, but their lithologic characteristics are mostly unknown seaward. Based on the island outcrops and acoustic signatures of the oldest strata offshore, however, we suspect that similar rock types occur beneath the flanks of the arc.

Middle series (MS). MS rocks that crop out on large islands of the Adak-Amlia sector consist of plutons on Kagalaska and Adak Islands (Citron and others, 1980) and plutons, flows, and sedimentary rocks on Atka Island (Hein and others, 1984). On Adak and Kagalaska Islands (Fig. 5), both the Hidden Bay and Kagalaska plutons belong to the MS; K-Ar radiometric ages of about 36 to 31 Ma were determined on samples of the Hidden Bay pluton and about 14 to 12 Ma on samples from the Kagalaska pluton (Citron and others, 1980). Radiometric (K-Ar and U—Pb) ages of Tertiary plutons on Atka Island range from 24 to 9 Ma, with most dates falling in the 17- to 11-Ma interval; two flow rocks were dated at 12 to 11 Ma in the fossil-tree-bearing upper Miocene Martin Harbor section (Hein and others, 1984).

MS strata offshore form an extensive sediment blanket that drapes over the arc's Pacific (Fig. 6) and Bering Sea insular slopes. These strata unconformably overlie LS rocks in many areas and generally exhibit more coherent acoustic layering than the LS strata. MS strata are more deformed than US strata in most areas. The oldest MS sample dredged from offshore subma-

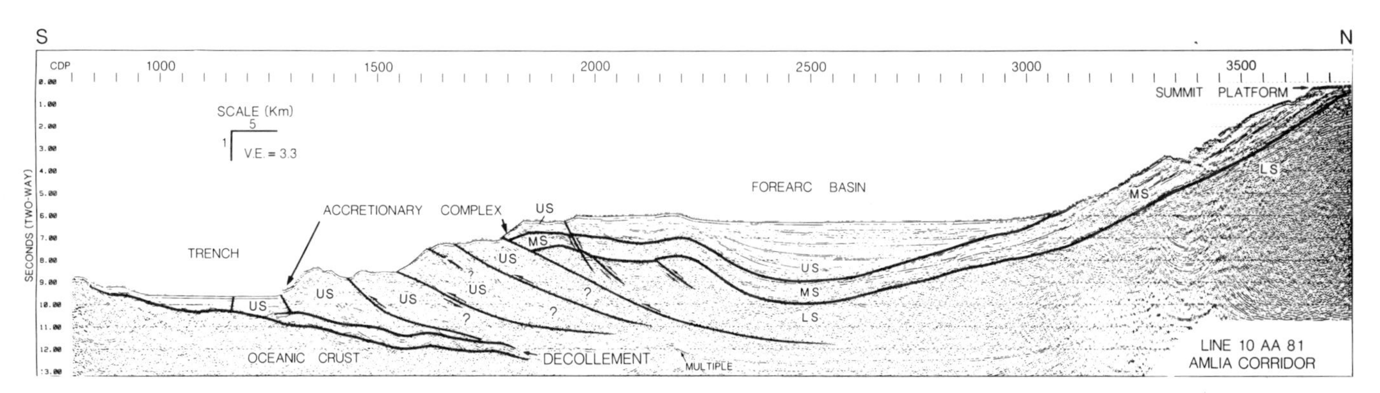

Figure 6. Migrated 24-channel seismic-reflection profile of Line 10 from the Adak-Amlia sector in the central Aleutians (see Fig. 1 for approximate location). See Ryan and Scholl (1989) for additional details of this seismic profile. Note the wide accretionary complex and the uplifted fore-arc basin sediments. Ryan and Scholl (1989) suspect that there has been some right-lateral strike-slip displacement (not shown on this profile) along a "backstop thrust" that separates the arc framework from the accretionary wedge.

rine outcrops is early Oligocene volcanic sandstone (Scholl and others, 1983a, b, 1987). The strata maintain a roughly uniform thickness beneath the fore-arc terrace (about 1.2 sec), and they can be traced seaward to the structural high that borders the seaward edge of the terrace.

Upper series (US). The Atka volcanic center (Marsh, 1976, and written communication, 1985) forms a large part of exposed US rocks on Atka Island. On Adak Island (Figs. 2 and 5), rocks of Pliocene and Quaternary age of the US include the volcanic flow and volcaniclastic rocks of Moffett and Adagdak volcanoes. Other active and dormant volcanoes in the Amlia-Adak sector (e.g., Seguam volcanic cones, Koniuji, Kasatochi, Great Sitkin, Kanaga, and Tanaga) are also part of the US.

Offshore, US strata are characterized acoustically by coherent and laterally continuous reflectors (Fig. 6). The strata typically are thin over the higher parts of the summit platform, but fill summit basins (i.e., Amukta and Amlia basins) to a thickness of 2 to 5 km (Scholl and others, 1983a; Geist and others, 1988). US strata also fill the fore-arc basin (Atka basin) beneath the Aleutian Terrace (Fig. 6) to a thickness of about 4 km and are uplifted across Hawley ridge, a prominent fore-arc structural antiform (Scholl and others, 1983a; Ryan and Scholl, 1989). US beds in Atka basin unconformably onlap south-dipping MS strata that underlie the arc's southern insular slope. In addition, equivalent sediment fills the Aleutian Trench to an average thickness of about 1.5 km (Fig. 6), but the trench fill is as thick as 3 to 4 km a short distance away near the intersection of the trench and the Amlia Fracture Zone at about 173° W (Scholl and others, 1982). US strata constitute the bulk of the accretionary complex, which is 6 to 8 km thick (4.6 sec.) along Line 10 (Fig. 6); the age of these offscraped deposits is poorly known, but based on regional relations, they are interpreted to be latest Miocene or Pliocene and younger (McCarthy and Scholl, 1985; Scholl and others, 1987; Ryan and Scholl, 1989).

GEOLOGIC FRAMEWORK OF THE ALASKA PENINSULA-KODIAK ISLAND SEGMENT

Age and correlation

The Alaska Peninsula and adjacent islands are composed predominantly of northeast-trending rock masses ranging in age from Permian to Holocene (Burk, 1965; Moore, 1974b; von Huene and others, 1979, 1987; Detterman and Reed, 1980; Detterman and others, 1981, 1983; Detterman, 1985; Wilson and others, 1985a; Fisher and others, 1987; Bruns and others, 1987b; Detterman and others, 1994). A complete discussion of the pre-Cenozoic rocks is beyond the purpose of this chapter; the most up-to-date review is given by Detterman and others (1994). Exposed rocks on the Alaska Peninsula, northern part of Kodiak Island, and inner Shumagin Islands include mostly arc-related Paleozoic to Holocene volcanic and plutonic rocks as well as shallow marine and continental deposits rich in volcanic and plutonic detritus (Fig. 7). In contrast, most rocks on the islands of Sanak, outer Shumagin, and the southern part of Kodiak are thick, flysch-like deposits of Late Cretaceous and Paleogene ages. The Cretaceous strata are intruded by Paleocene (about 60 Ma) plutons.

An understanding of the geologic evolution of the Alaska Peninsula and companion island groups was enhanced by the recognition of discrete, fault-bounded tectonostratigraphic terranes (e.g., Stone and Packer, 1977; Berg and others, 1978; Jones and others, 1981; Plumley and others, 1983; Silberling and others, this volume). In the study area, these terranes—initially named the Peninsular, Chugach, and Prince William terranes by Jones and Silberling (1979)—are believed to be allochthonous to southern Alaska (Hillhouse, 1987; Hillhouse and Coe, this volume). The Peninsular terrane was renamed Alaska Peninsula terrane by Wilson and others (1985a) who subdivided it into Iliamna and Chignik subterranes (Wilson and others, 1985a).

In this paper we refer to the Chugach and Prince William terranes south of the Alaska Peninsula as the Chugach-Prince William terrane. The contact between the mostly shallow-marine and continental rocks of the Alaska Peninsula terrane and the deep-water Shumagin and Kodiak Formations of the Chugach-Prince William terrane is the Border Ranges fault (Plafker and others, 1977; Plafker and others, this volume, Chapter 12; Fisher and von Huene, 1984). This structure is inferred (Fig. 7) to strike southwest and to lie between the inner and outer Shumagin Islands. Wilson and others (1985b) suggested that the Border Ranges fault in the Port Moller area may lie instead along the south coast of the Alaska Peninsula.

Western Alaska Peninsula: Shumagin sector

Onshore Shumagin sector. Stratigraphy of Cenozoic rock units on the Alaska Peninsula was discussed by Burk (1965), McLean (1979), Wilson (1980, 1985), Wilson and others (1981, 1985b), Detterman (1985), Detterman and Reed (1980), Detterman and others (1981, 1983, 1985), and Detterman and Miller (1985). Revisions in stratigraphic nomenclature, ages, and geologic interpretations will continue as work progresses. Detterman (1985) and his co-workers (Detterman and others, 1994), for example, have modified some stratigraphic and age assignments, and we use them in this short review (Fig. 2).

Five major stratigraphic packages of Cenozoic age are recognized on the Alaska Peninsula. The oldest consists of the Paleocene and Eocene Tolstoi and Copper Lake Formations, which unconformably overlie Upper Jurassic to Upper Cretaceous rocks; these early Tertiary formations consist predominantly of nonmarine, volcaniclastic, and carbonaceous sedimentary rocks. In places, the Tolstoi Formation is as thick as 1,500 m.

Unconformably overlying the Tolstoi and Copper Lake Formations, and making up the second package, are the Stepovak Formation and Meshik Volcanics of late Eocene and early Oligocene age and correlative volcanic rocks in the Katmai region and the informally defined Popof volcanic rocks of the Shumagin Islands (Fig. 2). These constitute most of the Meshik arc as defined by Wilson (1985). The Stepovak Formation consists mostly of arc-related sedimentary rocks, whereas the Meshik Volcanics are composed of a full range of volcanic rocks, including abun-

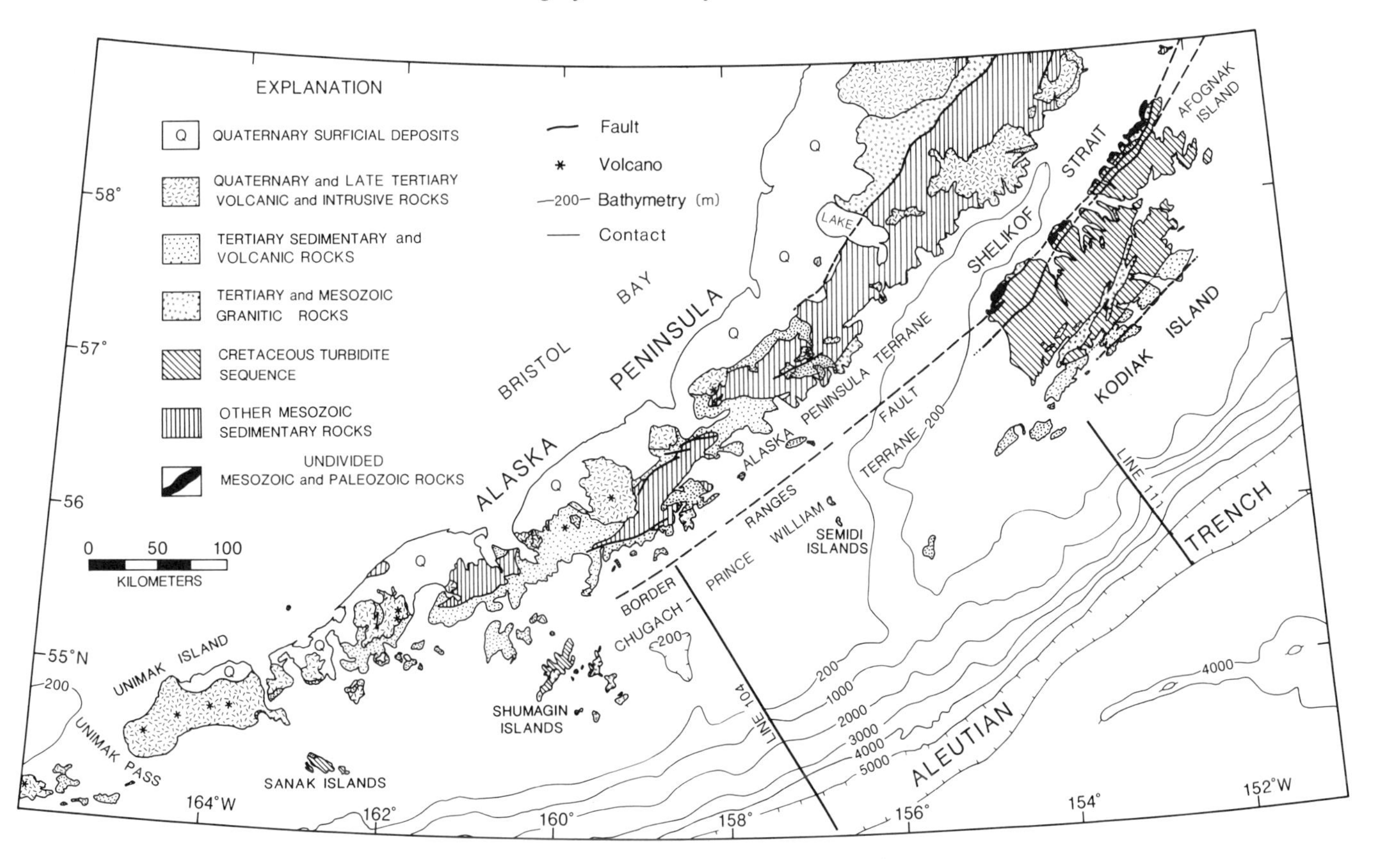

Figure 7 Generalized geologic map of the Alaska Peninsula, Kodiak Island, and adjacent island groups, sectors 3 and 4 (modified from Beikman, this volume; Bruns and others, 1985). Approximate locations of Lines 104 (Fig. 8) and 111 (Fig. 9) are shown.

dant andesite and dacite, along with their associated volcaniclastic rocks. A third package is the Oligocene and lower Miocene Belkofski Formation, the early Oligocene Hemlock Conglomerate, and the late Oligocene and early Miocene Unga Formation. The preferred interpretation is that the Unga and Belkofski Formations are at least in part correlative. The fourth package is the unconformably overlying upper Miocene Tachilni and Bear Lake Formations. These two formations were shown by Marincovich (1983) to be correlative. They consist of shallow-water marine and nonmarine units including sandstone, conglomerate, siltstone, shale, and thin coal beds. The Pliocene Milky River Formation and Morzhovoi Volcanics unconformably overlie the Bear Lake and Tachilni Formations. These units, as well as the unnamed Pliocene and Quaternary rocks and sediments, are the youngest in the region and constitute the fifth stratigraphic package of Cenozoic age.

The Upper Cretaceous Shumagin Formation is exposed on the outer islands of the Shumagin sector and is the major stratigraphic unit of the Chugach–Prince William terrane. These deep-water sedimentary rocks, and those of the partly equivalent Kodiak Formation, were probably deposited in slope basins and/or a trench (Moore, 1974a; Nilsen and Moore, 1979; Nilsen and Zuffa, 1982). Upper Cretaceous flysch-like rocks mapped on the Alaska Peninsula, the Kaguyak Formation in the Katmai region, and the Hoodoo Formation in the Port Moller area, are equivalent in age and lithologically similar to the Shumagin and Kodiak Formations (Detterman and Miller, 1985).

Igneous activity on the Alaska Peninsula was discussed by Reed and Lanphere (1969, 1973) for the plutons east of the Shumagin sector and by Wilson and others (1981) and Wilson (1985) for the volcanic rocks. Reed and Lanphere (1973) concluded that Jurassic plutons, ranging in age from about 175 to 155 Ma, compose most of the Alaska-Aleutian Range batholith. In addition, another period of plutonism occurred nearly continuously beginning in the Late Cretaceous, about 74 Ma, and ending in the early Tertiary, about 58 Ma and possibly as late as 55 Ma. In the Meshik arc (Wilson, 1985), ages of the volcanic flows and associated hypabyssal plutons range from 48 to 22 Ma; most fall within the interval from 40 to 30 Ma. Ages of volcanic rocks that make up the present regime on the Alaska Peninsula range in age from about 10 Ma (possibly as old as 15 Ma) to the present (Wilson and others, 1981; Wilson, 1982; Wilson and others, this volume).

Offshore Shumagin sector. The Shumagin margin (Bruns and others, 1985, 1987a, b, c; Lewis and others, 1988) has five major structural elements: (1) Shumagin basin, (2) Sanak basin,

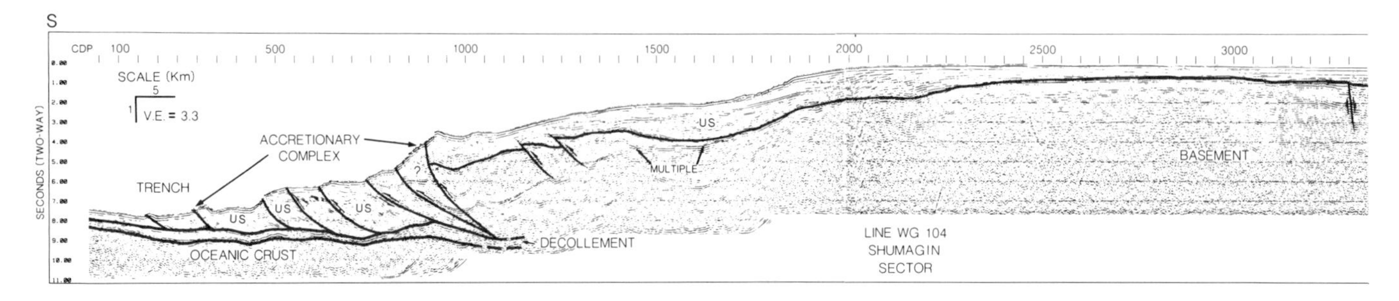

Figure 8. Migrated 24-channel seismic-reflection profile of Line 104 of the Shumagin sector. See Bruns and others (1987a, b) for additional interpretations on the region. The upper series (US) is the only chronostratigraphic unit shown; older units are difficult to correlate with Aleutian Ridge units. Note the thick upper series sequences and the wide accretionary complex.

(3) shelf edge and upper slope sedimentary wedges, (4) a mid-slope structural trend that includes Unimak ridge, and (5) a lower-slope accretionary complex. A multichannel seismic-reflection profile that crosses the margin near the Shumagin Islands, reveals a 35-km-wide accretionary complex beneath the landward trench slope and a sediment-filled Shumagin basin beneath the shelf (Fig. 8).

The difficulty in correlating chronostratigraphic units of the Aleutian Ridge with units of the Alaska Peninsula and Kodiak Island is their very different pre-Eocene histories. Furthermore, very few submarine outcrops have been sampled for age determinations. Therefore, only the upper series and "basement" designations are used for offshore regions of the Alaska Peninsula–Kodiak Island segment in the figures, where "basement" refers to acoustic basement and may be of any age older than late Cenozoic (e.g., probably late Miocene or Pliocene).

Shumagin basin (Fig. 8) has a relatively thin (about 2.5 km) sediment fill, whereas the sediment fill of nearby Sanak basin is as thick as 8 km (Bruns and others, 1985, 1987a, b). These basin-filling sedimentary units possibly include rocks as old as Oligocene, but most are probably of latest Miocene to Holocene age. They overlie an acoustic basement that probably consists of the same rock assemblages that compose older parts of the Chugach–Prince William terrane onshore. The sedimentary cover that drapes the shelf edge and upper slope ranges in thickness from 2 to 4 km; seismic-refraction data suggest that the thickness locally may be as great as 6 km.

An accretionary complex on the lower slope is nearly 35 km wide and several kilometers thick (Fig. 8). A decollement at the base of the accretionary complex can be traced for the entire width of the accretionary complex and, in places, for an additional 10 km beneath the margin's bedrock framework. We believe that most sediment in the accretionary complex is of Pliocene and Quaternary age.

Geophysical and regional geological studies indicate that the offshore structural fabric swings northwestward from the Sanak Islands region and merges with the outer part of the Beringian margin (Pratt and others, 1972; Lewis and others, 1984, 1988). This structural continuity strengthens the hypothesis that there was a tectonic link between the Alaska Peninsula and the Beringian margin before the growth of the Aleutian Ridge.

Eastern Alaska Peninsula: Kodiak Island sector

Onshore Kodiak Island sectors. The Kodiak Island sector (Fig. 7) includes parts of both the Alaska Peninsula and Chugach–Prince William terranes and has a relatively well known, though complex, geologic history (Moore, 1969; Reed and Lanphere, 1973; Connelly, 1978; Reed and others, 1983; Wilson and others, 1985a, von Huene and others, 1985, 1987; Fisher and others, 1987; Detterman and others, 1994). The Alaska Peninsula terrane portion of the sector is divided into the Chignik and the Iliamna subterranes by the Bruin Bay fault. On Kodiak Island the Alaska Peninsula and Chugach–Prince William terranes are separated by the Border Ranges fault.

Isolated exposures of possibly early Mesozoic and Paleozoic(?) quartzite, greenstone and other metavolcanic rocks, schist, and Triassic(?) limestone that occur as roof pendants in the Alaska-Aleutian Range batholith (Reed and Lanphere, 1973) in the Iliamna subterrane may be the oldest rocks of the Alaska Peninsula terrane. In the Chignik subterrane, Triassic volcanic rocks and a sequence of Permian to Late Cretaceous arc-derived sedimentary units occur in the Kodiak Island sector. A small part of the Alaska Peninsula terrane is also exposed on the northwest coasts of the Kodiak Island group; it consists of a complex assemblage of Late Triassic volcanic rocks, Early Jurassic schist, and Early Jurassic diorite and quartz diorite plutons. A plausible explanation for geologic relations in the Kodiak Islands is that the Chignik subterrane and rocks of the Alaska Peninsula terrane represent a back-arc region of the Iliamna subterrane Jurassic magmatic arc. This arc may have been constructed on the Wrangellia terrane (Jones and others, 1977); the Triassic and Permian rocks of the Chignik subterrane correlate well with Wrangellia rocks of the same age.

Tertiary rocks overlying the Alaska Peninsula terrane in the Kodiak Island sector are much thinner than in the Shumagin sector. The Copper Lake Formation in the Katmai area has a total thickness of about 1,000 m and consists of thick conglomer-

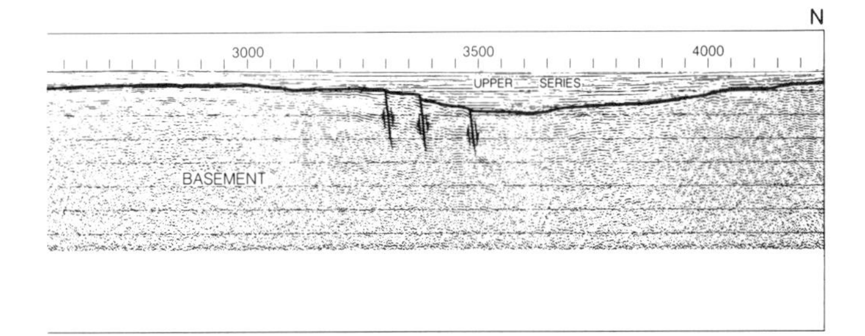

ate at the top and bottom and sandstone and siltstone with considerable carbonaceous debris and minor coal in the middle. Fossil megaflora suggest an age of early Eocene for the Copper Lake Formation, approximately the same as the Tolstoi Formation on the southern part of the Alaska Peninsula. The Copper Lake Formation, like the Tolstoi Formation, is dominantly nonmarine with some nearshore marine beds. The overlying Hemlock Conglomerate is a fluvial sandstone and conglomerate with interbedded siltstone, shale, and coal of presumably early Oligocene age on the basis of megaflora. The Hemlock Conglomerate was largely derived from erosion of Mesozoic sedimentary and granitic rocks; the proportion of volcanic debris in the unit increases southward toward the Katmai area.

Southwest of the Alaska Peninsula terrane, the combined Chugach–Prince William terrane rims the entire south side of the Alaska Peninsula. It is composed of Late Cretaceous and early Tertiary subduction complexes. The dominant geologic units are the Kodiak (Moore, 1969) and Shumagin (Moore, 1974b) Formations that consist of Cretaceous flysch. The Uyak Complex (Connelly, 1978), bounded on the northwest by the Border Ranges fault, structurally overlies the flysch unit on Kodiak, Afognak, and nearby islands. The Uyak Complex is a chaotic melange-like unit consisting of blocks of sandstone, greenstone, radiolarian chert, and limestone in a chert and argillite matrix. Early Tertiary granitic plutons (63 to 58 Ma; Wilson and others, this volume) intrude the flysch.

Nearer the Aleutian trench, the Ghost Rocks and Sitkalidak Formations (Moore, 1969; Nilsen and Moore, 1979) of Paleocene to Oligocene age may form a similar subduction complex, which is equivalent to the Orca Group of Prince William Sound. The Ghost Rocks Formation is an isoclinally folded, faulted, and sheared unit that is distinguished by its pillow basalt or greenstone, hard and black claystone, thin limestone beds, and locally prominent zeolite-bearing tuffaceous sandstone. It is lithologically similar to the Uyak Complex more so than to the structurally adjacent Kodiak Formation. The Sitalidak Formation, according to Moore (1969, p. 32), is ". . . a rather uniform sequence of sandstone and siltstone graded beds about 3,000 m thick . . ." and has many lithologic characteristics similar to the Kodiak Formation.

Overlap sequences of the Chugach–Prince William terrane are the Sitkinak, Narrow Cape, and Tugidak Formations of the Kodiak Islands. The Sitkinak Formation, of middle or late Oligocene age, stratigraphically overlies the Sitkalidak Formation. The Sitkinak Formation consists of 1,500 m of dominantly nonmarine coal-bearing siltstone, sandstone, and conglomerate and lesser amounts of marine sandstone and siltstone. The late Oligocene and Miocene Narrow Cape Formation is a richly fossiliferous marine sandstone and siltstone unit 700 m thick that conformably overlies the Sitkinak Formation. Its rocks record a quiet shallow marine environment. The youngest rock unit in the Kodiak Island sector is the Tugidak Formation. It consists of 1,500 m of richly fossiliferous Pliocene sandstone and siltstone with randomly distributed pebbles and cobbles. Correlation with the Yakataga Formation by Moore (1969) suggests by inference that the Tugidak Formation was deposited in a glaciomarine environment (Molnia, 1986).

Offshore Kodiak Island sector. The offshore deep-water (more than 1 sec water depth) framework of the Kodiak Island sector, in part shown in Figure 9, is well documented by several profiles shown in von Huene and others (1987). The older rocks have been deformed, and large structures (e.g., Albatross bank; not shown on Fig. 9) underlie extensive bathymetric highs.

Few rock samples have been collected from the submerged parts of the margin, although upper Miocene through Quaternary rocks and sediments were recovered from the top of Albatross bank. From studies of the seismic stratigraphy and drill hole data, the fill in Albatross Basin (a shelf basin) is considered to be late Miocene and younger in age (Fisher and von Huene, 1980; von Huene and others, 1987; Fisher and others, 1987) and to overlie a regional unconformity similar to that in the Shumagin sector. The nature of the underlying rocks can be interpreted from outcrops on Kodiak Island and from drill-hole samples. A drill hole near the shelf edge north of the Kodiak Island sector, near Middleton Island, which is located about 150 km northeast of Kodiak Island, penetrated this regional unconformity and bottomed in lower Eocene rocks (Rau and others, 1977; Keller and others, 1984); a drill hole near the edge of the shelf on the flank of Albatross bank bottomed in rocks of Eocene age (Herrera, 1978). The Middleton Island drill hole has sedimentary rocks and sediment of late middle Eocene (42 to 40 Ma), late Eocene and early Oligocene (38 to 34 Ma), early Miocene (23 to 22 Ma), and late Miocene to Pleistocene (younger than 6 Ma) ages (Keller and others, 1984). Structural variations within the Chugach–Prince Williams terrane and the distance between the Middleton Island well and the Kodiak Island sector suggest, however, that correlations between the two areas remain tentative.

The lower slope of the margin is underlain by an accretionary complex that is divided into two parts along a prominent decollement. The upper part, mostly trench and abyssal-plain sequences, was accreted along steeply dipping thrust faults; the lower half is being subducted (Fig. 9). On the Pacific Plate, the sediment sequence is probably similar to that penetrated at DSDP Site 178 (Kulm and others, 1973) where 730 m of Miocene to Quaternary hemipelagic and terrigenous sedimentary rocks and sediments overlie older pelagic sedimentary rocks and Eocene basalt.

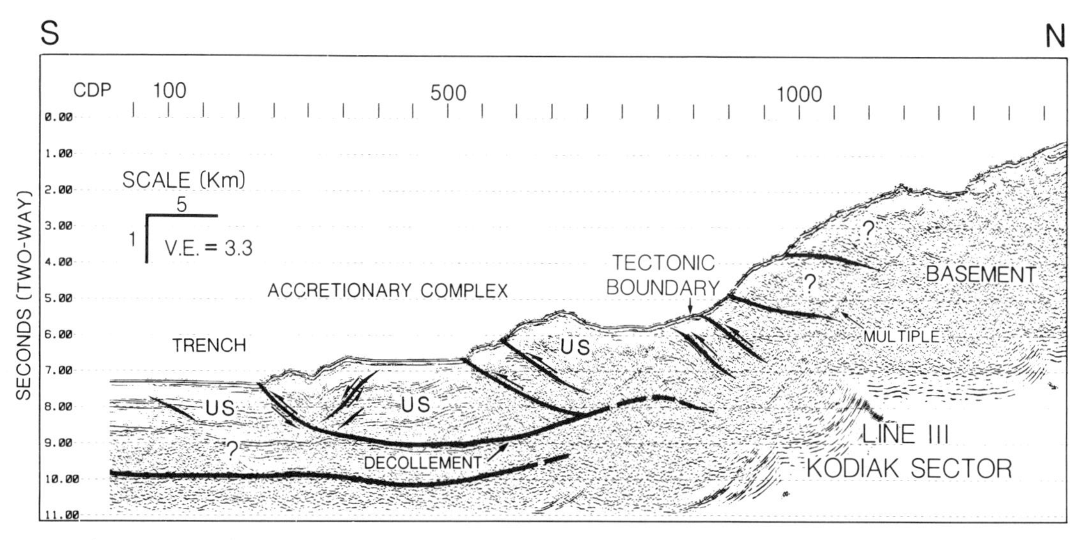

Figure 9. Migrated 24-channel seismic-reflection profile of the deeper parts in the Kodiak Island sector. Approximate location shown in Figure 1. See von Huene and others (1987) for a more complete interpretation of the Kodiak Islands sector. Similar to Figure 8, only upper series strata are designated. Note the relatively thick accumulation of trench sediments and thick section underlying the prominent decollement.

COMPARISON OF FORE-ARC SEISMIC REFLECTION PROFILES

Comparisons of seismic-reflection profiles along the Pacific margin of the Aleutian arc show distinct differences among the four sectors (Table 1). In the Komandorsky–Near Islands sector southeast of Agattu Island (Fig. 4), a relatively narrow fore arc is about 70 km wide between common depth points (CDP) 400 and 2000 (calculated by considering the 30° angle between the line of profile and the present axis of the ridge crest). Although the vector sum of Pacific and North American Plate motions is nearly parallel to the plate boundary, earthquake analyses indicate some reverse faulting beneath the ridge (Newberry, 1984; Newberry and others, 1986); compressional stress is confirmed by several high-angle reverse faults on the seismic profile. These faults, in places, offset the sea floor. We strongly suspect that movement along some of these faults is predominantly oblique-slip. The accretionary complex is narrow and thin if only the deformed sedimentary pile is considered. However, oceanic crust probably is involved, which may not be the case along other profiles (Figs. 6, 8, and 9). A poorly developed decollement can be traced on the profile (Fig. 4) for a maximum of only about 8 to 10 km. LS rocks, which make up the lithologic basement of a large part of the Aleutian Ridge, merge acoustically with the Pacific oceanic crust beneath the outer structural high that separates the accretionary complex from the arc's framework. The fore-arc basin is well developed; the sedimentary section is thick (the US alone has a thickness of about 3 to 4 km), and the basin is narrow (about 20 km).

The Adak-Amlia sector has the greatest number of multichannel seismic-reflection profiles in the Aleutian Ridge segment (e.g., Scholl and others, 1983a, b, 1987; McCarthy and others, 1984; McCarthy and Scholl, 1985; Harbert and others, 1986; Geist and others, 1988; Ryan and Scholl, 1989). There are some major differences between seismic-reflection profiles from the Komandorsky–Near Islands and Adak-Amlia sectors (Table 1). In the Amlia region (Fig. 6), for example, the accretionary complex, fore-arc basin, and decollement are about twice as wide as they are along the Near Islands profile (Fig. 4), and the accretionary complex is much thicker. However, sediment thicknesses are about the same in the trench, fore-arc basin, and subdecollement sequences. Not far from Line 10 (Fig. 6), near the intersection of the Amlia Fracture Zone and the Aleutian Ridge, the thickness of trench sediments is 3 to 4 km (Scholl and others, 1982). An antiform (Hawley Ridge; Ryan and Scholl, 1989) upwarped the sediment sequence along the seaward side of the fore-arc basin (at about CDP 2200 on Fig. 6).

In contrast to fore-arc regions of the Aleutian Ridge, the Pacific margin of the Alaska Peninsula in the Shumagin sector (Line 104, Fig. 8; Table 1) has no well-developed fore-arc terrace. Instead, the sea floor descends gradually from shelf depths to the Aleutian Trench. The depth to the trench floor is about 1,400 m shallower along Line 104 compared to depths in the Amlia corridor and Near Islands regions. A major thrust fault (at about CDP 1800, Fig. 8) structurally separates an accretionary complex from the margin's framework. Shelf basins (shallow fore-arc basins) have formed (Bruns and others, 1985, 1987a, b; Lewis and others, 1988); particularly well developed is Sanak basin. Thicknesses of the trench sediments, the accretionary complex, and the subdecollement sequence are about the same in the Shumagin sector as in the Amlia corridor region (Table 1).

Several seismic-reflection profiles, in addition to that shown here (Fig. 9) are displayed and discussed in detail by Fisher and von Huene (1980) and von Huene and others (1987). For example, an adjacent profile across the Kodiak Island margin (line 509; von Huene and others, 1987) shows a well-developed shelf basin (Albatross basin) that contains a thick (about 4 km) sequence of relatively young (late Miocene and younger, mostly correlative to upper series) sediments. Line 111 (Fig. 9) shows significant differences from the other profiles we selected for this chapter (Table 1; Figs. 4, 6, and 8). Particularly evident is the shallower depth of the trench floor and the greater volume of sediments in the trench. The subdecollement sedimentary sequence also is much thicker than in any of the other profiles (Figs. 4, 6, and 8).

GEOLOGIC EVOLUTION OF THE ALEUTIAN ARC

General discussion

The Aleutian arc formed along zones of convergence between the North American Plate and various oceanic plates, including the modern Pacific Plate and extinct Kula Plate. As a structural entity, the Aleutian arc has been an elongate structural feature, rimming the North Pacific Ocean, since at least the early Eocene. The geologic evolution of the western (Aleutian Ridge) segment was very different from that of the eastern (Alaska Peninsula–Kodiak Island) segment during the Mesozoic and early Tertiary with regards to dominant geologic processes. Since the early Eocene, however, the entire arc has had a similar history, although igneous episodes have not been synchronous and we believe that the western part of the arc is being translated westward along major strike-slip faults.

The Alaska Peninsula–Kodiak Island segment evolved along a zone of plate convergence since at least the Jurassic. Arc polarity may have changed after Jurassic plutonism (Reed and others, 1983), but the igneous rocks are clearly characteristic of a convergent, subduction-related margin. Parts of the Alaska Peninsula and Chugach–Prince William terranes apparently originated far to the south of their present position (Stone and Packer, 1977; Plumley and others, 1983; Hillhouse, 1987; Hillhouse and Coe, this volume) and were accreted onto nuclear Alaska during the Late Cretaceous and earliest Tertiary. Paleomagnetic data from the southern part of the Chugach–Prince William terrane indicate that some latitudinal displacement occurred until at least the Oligocene (Plumley and others, 1983).

Aleutian Ridge segment

Some broad, general interpretations can be made regarding the geological evolution of the Aleutian Ridge: (1) the major volumetric growth of the ridge occurred after a rotational change in Kula Plate motion (about 56 to 55 Ma; Lonsdale, 1988) and before the time of the Emperor-Hawaiian seamount bend (about 43 Ma; Dalrymple and others, 1977; Clague, 1981); (2) major arc-building volcanism had waned, but had not entirely ceased, by about 37 Ma; (3) the ridge subsequently was irregularly uplifted and fragmented by shearing and block rotation, eroded, and affected by volcanism during the ensuing approximately 30 m.y., and was intruded by plutons in episodes that occurred at about 36 to 28 Ma and 20 to 9 Ma; (4) the ridge has been rejuvenated volcanically since about 5 or 6 Ma; (5) the fore-arc and modern shelf (or summit) basins and the subduction complex have developed mostly since the Pliocene; and (6) the ridge is being translated westward along major strike-slip faults in response to oblique convergence between the Pacific and North American Plates.

Lower series rocks, erupted and deposited from about 55 to 37 Ma, are dominantly arc-type igneous rocks and derivative coarse-grained volcaniclastic rocks. Arc volcanism most likely began at the start of oceanic intraplate subduction, probably along a transform fault that resulted from the buckling of the extinct Kula Plate (now mostly subducted), with subsequent formation of the Aleutian subduction zone (Scholl and others, 1987; Lonsdale, 1988). Underthrusting during the arc's early stage of rapid growth was roughly due north, possibly at a relatively high rate (Engebretson and others, 1986). Resultant intraplate subduction and volcanism caused rapid growth of the ridge's igneous framework across a width of 150 to 200 km and along the entire length of the arc. At about 43 Ma (age of the Emperor-Hawaiian

TABLE 1. DEPTHS TO AND DIMENSIONS OF MAJOR FEATURES ALONG THE PACIFIC MARGIN OF THE ALEUTIAN ARC*

Line No.		Trench		Accretionary Complex		Decollement†		Forearc Basin				Shelf Basin	
		D	T	W	T	W	T	W	D	T(MS)	T(US)	D	T
32	(Fig. 4)	9.6	1.4	12	2.4	10	0.5	20	4.0	1.0	2.8	1.0	0.8
10	(Fig. 6)	9.6	1.1	26	5.0	25	0.6	50	6.2	1.2	2.0	....	....
104	(Fig. 8)	7.8	1.2	35	5.0	30	0.5	....	....	....	....	0.2	1.6
111	(Fig. 9)	7.2	2.8	32	4.5	27	1.1	....	....	....	....	0.2	3.0§

*Locations of lines and seismic-reflection profiles are shown in Figure 1. Widths are in kilometers; depths and thicknesses are two-way travel time in seconds. D = depth; T = thickness; W = width; MS = middle series; US = upper series.
†Thickness is measured between the decollement upper surface and oceanic crust.
§Data are from adjoining Line 509; R. von Huene's files.

seamount bend), the rate of convergence decreased and the oceanic plate rotated counterclockwise to travel in a west-northwest direction, thereby generally oblique to the trend of the curvilinear Aleutian Ridge. This change in plate vectors led to widespread extension along the arc and to strike-slip faulting that, with time, fragmented the arc into blocks that rotated in a clockwise sense and initiated extensive westward dispersion of the fore-arc regions, particularly in the central and western sectors.

The end of lower series time marks the waning of voluminous arc volcanism and the beginning dominance of tectonic erosion and sedimentation over volcanic processes. However, volcanic and plutonic episodes continued sporadically along the magmatic axis of the ridge. The lower slope (trench wall) of the ridge probably was sediment starved during the early growth of the ridge, similar to the insular lower slopes of the present-day Tonga and Mariana volcanic arcs.

The mafic metaplutonic suite on Attu Island is a significant LS unit because its metamorphic age indicates a thermal event at about 42 Ma (near the time of the Emperor-Hawaiian seamount bend and the shutdown of the Kula-Pacific Ridge). The metamorphic amphibole age (41.5 $\pm$ 2 Ma) indicates that the magmatic crystallization age of the igneous rocks is older; that age has not yet been determined. These metamorphosed mafic plutons compose the only large exposure of high-grade (amphibolite facies) metamorphic rocks known on the Aleutian Ridge. Some trace-element contents suggest island-arc tholeiite chemical affinities. A high $^{143}Nd/^{144}Nd$ ratio, however, suggests oceanic (or back-arc basin) MORB affinities (J. Rubenstone, written communication, 1988).

Other LS igneous rocks from the Aleutian Ridge are typical of island arcs and have calc-alkaline, tholeiitic, and transitional chemical affinities. No progression from tholeiitic to calc-alkaline affinities with time and/or space is apparent in LS rocks.

Lower series sedimentary rocks accumulated mostly as debris and turbidite flows (unsorted, mostly coarse-grained volcaniclastic rocks and size-graded breccia and sandstone associations). Finer-grained sedimentary rocks of the Krugloi (Agattu Island), Nevidiskov (Attu Island), and Andrew Lake (Adak Island) Formations probably are lithofacies of the coarser-grained units. Lower series rocks identified offshore of the islands can be traced acoustically on many seismic-reflection profiles to the outer structural high that borders the accretionary complex.

The Oligocene through Miocene middle series (37 to 5.3 Ma) had a beginning that was marked by virtual cessation of igneous activity on the ridge flanks and the beginning of their burial by sediments. Igneous activity did continue along the crestal area where plutons were emplaced during the intervals from 36 to 28 and 20 to 9 Ma. During MS time, terrigenous debris eroded from the arc covered LS rocks with a sedimentary blanket as thick as 2 to 3 km.

Most volcanism along the Aleutian Ridge during middle series time was of island-arc character; calc-alkaline volcanic rocks are prevalent. The Nikol'skoye Suite on Bering Island (the westernmost island in the arc) is the only known pre-Quaternary volcanic rock unit of alkalic character on the Aleutian Ridge. These late Oligocene and early Miocene alkalic rocks erupted in a tectonic regime that may have been different from the rest of the Aleutian Ridge at that time.

The apparent episodic plutonism and change in chemical affinities (tholeiitic to calc-alkaline) of plutonic rocks may be related to significant changes in either tectonic patterns or tectonic processes. After the buildup of LS rocks, for example, the first phase of plutonism occurred during the Oligocene (36 to 28 Ma). On the Agattu, Amatignak (small island southwest of Adak), and Umnak Islands, this phase is characterized by large (up to 100 m thick) sills and dikes that have island-arc tholeiite affinities on the basis of high FeO/MgO ratios, flat rare earth patterns, and low contents of compatible trace elements. Only on Adak Island has a relatively large calc-alkaline pluton (Hidden Bay pluton) of that age been recognized. The second phase (20 to 9 Ma; mostly 17 to 11 Ma) is characterized by large calc-alkaline plutons. Above about 52 percent SiO_2, these plutonic rocks have relatively constant FeO/MgO ratios and rare earth element patterns that are enriched in lighter elements and depleted in heavier elements. The pulses or phases of plutonism may be related to tectonic events such as greater extension of the ridge, changes of plate convergent vectors, crustal thickening related to underplating of subducted sediment, and the subduction of hotter (younger) sea floor. Much more information is needed, however, before pluton episodes and compositions can be related to tectonic patterns and processes in volcanic arcs.

Near the end of the Miocene the last major plutonic phase waned. During the past 5 to 6 m.y. the arc's summit area began to undergo significant erosion, synchronous with the formation of large offshore fore-arc and summit basins and to the beveling of the summit platform. Renewed volcanism has not been sufficient to offset the combined effects of subsidence and erosion. Since the Eocene, the Aleutian Ridge segment apparently has not traveled far latitudinally (Harbert and Cox, 1984).

Alaska Peninsula–Kodiak Island segment

The Alaska Peninsula–Kodiak Island segment of the Aleutian arc records a longer geologic history than the Aleutian Ridge segment, and its Cenozoic history more generally reflects a continental margin, rather than an island arc, response to the subduction environment. The Mesozoic history reflects the migration of terranes northward since the Triassic from near equatorial latitudes to near the present latitude. The Paleozoic history of the Alaska Peninsula is poorly known and poorly constrained by present data.

In early Mesozoic time, the Alaska Peninsula may have been an island arc located far to the south of its present position (Stone and Packer, 1977; Hillhouse and Coe, this volume). The backbone of the Alaska Peninsula, the Alaska-Aleutian Range batholith, was emplaced during Jurassic time, and magmatically associated volcanic rocks were erupted. Reed and others (1983) suggested that the Jurassic arc had a reverse polarity to the present arc. The Jurassic and Early Cretaceous sedimentary sequence in the Chignik subterrane records erosion of this arc and unroofing of the batholith.

Early Late Cretaceous rocks are generally unknown on the Alaska Peninsula; however, Late Cretaceous rocks of the Peninsula terrane and, in part, the Chugach–Prince William terrane, reflect continued erosion of the Jurassic arc and sedimentary recycling of earlier Mesozoic sedimentary rocks. Transport of some debris into the paleo-Aleutian trench, which was the locus of deposition for sediments of the Kodiak and Shumagin Formations, is shown by thin section analysis of the rocks from the Shumagin Islands (John Decker, oral communication to F. H. Wilson, 1985). Late Cretaceous plutons in the northwestern part of the Alaska-Aleutian Range batholith may indicate the presence of a Late Cretaceous magmatic arc. However, rocks of the Alaska Peninsula contain no evidence for significant Late Cretaceous volcanism.

The Border Ranges fault, the present boundary between the Chugach–Prince William and Alaska Peninsula terranes, was formed in Late Cretaceous or early Tertiary time. Foreshortening of the continental margin and accretion of the Chugach–Prince William terrane occurred soon afterward. In the Shumagin Islands region, Eocene and Oligocene volcanic and volcaniclastic sedimentary rocks may overlap the possible extension of the Border Ranges fault.

In contrast to the three recognized rock series (LS, MS, and US) of the Aleutian Ridge segment, Cenozoic rocks of the Alaska Peninsula–Kodiak Islands segment are not separated into chronostratigraphic units (Fig. 2). If they were, however, then the Tolstoi and Copper Lake Formations, and possibly parts of the Stepovak Formation and Meshik Volcanics, are correlative with the LS. The Belkofski, Unga, Tachilni, and Bear Lake Formations, plus the Hemlock Conglomerate, would be correlated with the MS. Furthermore, the Milky River Formation, Morzhovoi Volcanics, and younger volcanic rocks and sediments would be assigned to the US.

The first clear evidence of post-Jurassic volcanism on the Alaska Peninsula occurs in rocks of the Meshik arc. In contrast to the Aleutian Ridge segment, after the time of the Emperor-Hawaiian seamount bend, volcanism increased significantly on the Alaska Peninsula. The vast majority of these volcanic rocks yield K-Ar ages in the range from 42 to 30 Ma. However, at the southwest end of the Alaska Peninsula, ages of 54.8 ± 1.8 and 51.7 ± 5.5 Ma (Wilson and others, this volume) have been determined on basaltic volcanic rocks that crop out in a limited area on the mainland just northwest of the Shumagin Islands. Volcanic rocks of similar age are not known elsewhere on the Alaska Peninsula. These rocks may either be the earliest evidence for the formation of the Aleutian arc or be part of the ancient Beringian margin arc (Davis and others, 1989). On the north end of the Alaska Peninsula, plutons of late Eocene and Oligocene age overlap the age range of plutons in the lower series (e.g., the Finger Bay pluton on Adak Island) and the oldest of the middle series plutons (Hidden Bay pluton on Adak Island and sills on Attu, Agattu, and Umnak Islands). Farther south on the Alaska Peninsula, coeval volcanic rocks of the Meshik arc are exposed. The youngest rocks that are clearly part of the Meshik arc are sparsely distributed late Oligocene and earliest Miocene volcanic rocks. However, hypabyssal plutons and volcanic rocks correlative with the second phase of the middle series (20 to 9 Ma) are known from the inner Shumagin Islands and offshore islands west of the Shumagin Islands at the southwest end of the Alaska Peninsula.

A second episode of magmatism on the Pacific coast of the Alaska Peninsula was apparently initiated in late Miocene time (about 15 Ma). Since that time, the active volcanic front has migrated to the northwest, and rapid uplift and erosion have exposed large plutons as young as Pliocene. Late Miocene sedimentary rocks on the northwest side of the Peninsula (the Bear Lake Formation) record the continued erosion and sedimentary recycling of debris from the Mesozoic magmatic arc and are conspicuous because of the small portion of volcanic clasts compared to most other Tertiary sedimentary rocks on the Alaska Peninsula. In contrast, on the Pacific Ocean side of the Peninsula, the age equivalent late Miocene Tachilni Formation is composed dominantly of clasts derived from a volcanic terrane. By Pliocene time, volcanic rocks and volcanic debris were the major components of the section. The Quaternary rocks record a complex interaction between glacial, volcanic, and shallow-marine sedimentary processes. The glacial accumulation zone on the southwest half of the Alaska Peninsula was on the continental shelf and glaciers traveled across the peninsula to the Bering Sea, leaving well-developed morainal deposits that record successive ice advances across the Alaska Peninsula (Weber, 1985; Detterman, 1986).

Outboard of the Alaska Peninsula on Kodiak Island, Cenozoic rocks record an early Tertiary subduction event; Miocene sedimentary units record a progressive tectonic uplift and a progressive decrease in the proportion of volcanic lithic fragments (Nilsen and Moore, 1979). This decrease in volcanic lithic fragments is in stark contrast to the development of the modern Aleutian volcanic arc farther inland. Pliocene and later rocks on the shelf and offshore islands show an increasing dominance of glaciomarine processes with time.

In the Aleutian magmatic arc (including the Paleogene Meshik arc) of the Alaska Peninsula, calc-alkaline andesite and dacite are the dominant volcanic products. However, rocks of tholeiitic affinity are common, particularly on the south half of the Alaska Peninsula. Available data suggests no apparent relation between age and chemical affinity. In the modern (Holocene) volcanic arc, only some of the westernmost volcanic centers are tholeiitic (Miller and Richter, this volume).

Rocks as young as Oligocene are involved in folding that produced a series of northeast-trending en echelon anticlines on the central Alaska Peninsula, and rocks as young as late Miocene are overthrust by Mesozoic rocks at the south end of the Peninsula.

SUMMARY AND CONCLUSIONS

The Aleutian arc forms most of the northern boundary of the Pacific Ocean. It is separated near Unimak Pass into an Aleutian Ridge segment on the west and an Alaska Peninsula–Kodiak Island segment on the east. The Aleutian Ridge segment is an early(?) Eocene to Holocene intraoceanic island arc, whereas the Alaska Peninsula–Kodiak segment is composed of tectonostratigraphic terranes of Mesozoic and Cenozoic age that incorporate fragments of even older rocks, upon which a Cenozoic volcanic arc has evolved. The curvature of the Aleutian structural arc is such that it is underthrust at about 90° by the Pacific Plate in the east; in the far western part, the North American and Pacific Plate motions are nearly parallel.

Rocks of the Aleutian Ridge segment are divided into three chronostratigraphic units as a convenient method for correlating offshore and onshore rocks. These units are informally called the lower (middle early through late Eocene rocks, older than 37 Ma), middle (early Oligocene through late Miocene rocks, 37 to 5.3 Ma), and upper series (early Pliocene through Quaternary rocks, younger than 5.3 Ma).

The Aleutian Ridge segment is also divided into two geographic sectors: the Komandorsky–Near Islands sector in the extreme western part of the arc and the Adak-Amlia sector in the central part. The Komandorsky–Near Islands sector shows a wide range of strike-slip and extensional structures and arc-type rock compositions; the Pacific margin is narrow, but a well-developed fore-arc basin and a relatively small accretionary complex occur in some areas. A large mafic metaplutonic suite on Attu Island was metamorphosed to the amphibolite facies about 42 Ma. In the central part of the structural arc, the Adak-Amlia sector, plate convergence is about 30° west of a perpendicular line drawn to the axis of the arc. Here, the arc has a relatively wide fore-arc basin and accretionary complex.

The Aleutian Ridge developed on Early(?) Cretaceous oceanic crust, apparently beginning in the early Eocene about 55 Ma. Growth of the ridge waned in the 40 to 37 Ma interval, which corresponds to the end of lower series time. During middle series time (37 to 5.3 Ma) volcanism continued, but at a diminished rate; plutonic activity was particularly dominant during the intervals from 36 to 28 Ma and 20 to 9 Ma. A major summit-cutting erosional event took place in the latest Miocene and early Pliocene interval. During the past approximately 5.3 m.y. (upper series time), offshore fore-arc and summit basins formed, the Aleutian Trench filled with sediments, a broad accretionary complex formed, and the latest phase of arc magmatic activity commenced.

The Alaska Peninsula–Kodiak Island segment is converging with the Pacific Plate at about 90°. Two discrete tectonostratigraphic terranes, recognized on land, are separated by the Border Ranges fault: the Alaska Peninsula terrane on the north and the Chugach–Prince William terrane on the south. Most rocks in the Alaska Peninsula terrane are Middle Jurassic to Late Cretaceous in age. The Chugach–Prince William terrane contains mostly Late Cretaceous and younger rocks, many of which were deposited in trench and other deep marine basin settings. The terranes were juxtaposed during the Late Cretaceous and major activity on the Border Ranges fault ceased sometime in the Late Cretaceous and early Tertiary interval prior to 60 Ma. Early Tertiary to Holocene volcanic rocks, their associated plutonic rocks, and marine and nonmarine sedimentary rocks constitute the Cenozoic geologic column on the Alaska Peninsula.

Offshore along the Alaska Peninsula–Kodiak Island segment, Cenozoic sediments were deposited in shelf and slope basins and in the Aleutian Trench. A regional offshore unconformity of either early or middle Miocene age separates the older framework of the margin from acoustically well-stratified basinal sequences. A thick accretionary complex beneath the lower slope is late Miocene(?) to Holocene in age. Offshore structures and regional geologic data indicate that the structural fabric of the Alaska Peninsula–Kodiak Island segment changes from a southwest to a northwest trend near Unimak Pass and forms part of the margin of the Bering Sea.

REFERENCES CITED

Berg, H. C., Jones, D. L., and Coney, P. J., 1978, Map showing pre-Cenozoic tectonostratigraphic terranes of southeastern Alaska and adjacent areas: U.S. Geological Survey Open-File Report 78-1085, 2 sheets, scale 1:1,000,000.

Borsuk, A. M., and Tsvetkov, A. A., 1980, Magmatic associations of the western part of the Aleutian Island arc: International Geology Review, v. 24, p. 317–329.

Bruns, T. R., and von Huene, R., 1977, Sedimentary basins on the Shumagin shelf, western Gulf of Alaska: Houston, American Institute of Mining, Metallurgical, and Petroleum Engineers, Inc., 9th Annual Offshore Technology Conference Proceedings, v. 1, p. 41–50.

Bruns, T. R., von Huene, R., Culotta, R. D., and Lewis, S. D., 1985, Summary geologic report for the Shumagin Outer Continental Shelf (OCS) planning area, Alaska: U.S. Geological Survey Open-File Report 85-32, 58 p.

Bruns, T. R., von Huene, R., and Culotta, R. D., 1987a, Aleutian Trench, Shumagin segment, seismic section 104, *in* von Huene, R., ed., Seismic images of modern convergent margin tectonic structure: American Association of Petroleum Geologists Studies in Geology, no. 26, p. 14–21.

Bruns, T. R., von Huene, R., Culotta, R. C., Lewis, S. D., and Ladd, J. W., 1987b, Geology and petroleum potential of the Shumagin margin, Alaska, *in* Scholl, D. W., Grantz, A., and Vedder, J., eds., Geology and resource potential of the continental margin of western North America and adjacent ocean basins; Beaufort Sea to Baja California: Circum-Pacific Council for Energy and Mineral Resources Earth Science Series, v. 6, p. 157–190.

Bruns, T. R., Vallier, T. L., Pickthorn, L. B., and von Huene, R., 1987c, Early Miocene calc-alkaline basalt dredged from the Shumagin margin, Alaska, *in* Hamilton, T. D., and Galloway, J. P., eds., The United States Geological Survey in Alaska; Accomplishments during 1986: U.S. Geological Survey Circular 998, p. 143–146.

Buffington, E. C., 1973, The Aleutian-Kamchatka trench convergence and investigations of lithospheric plate interactions in the light of modern geotectonic

theories [Ph.D. thesis]: Los Angeles, University of Southern California, 350 p.
Burk, C. A., 1965, Geology of the Alaska Peninsula; Island arc and continental margin: Geological Society of America Memoir 99, 250 p.
Citron, G. P., Kay, R. W., Kay, S. M., Snee, L. W., and Sutter, J. F., 1980, Tectonic significance of early Oligocene plutonism on Adak Island, central Aleutian Islands, Alaska: Geology, v. 8, p. 375–379.
Clague, D. A., 1981, Linear island and seamount chains, aseismic ridges, and intraplate volcanism; Results from DSDP: Society of Economic Paleontologists and Mineralogists Special Publication 32, p. 7–22.
Coats, R. R., 1950, Volcanic activity in the Aleutian arc: U.S. Geological Survey Bulletin 974-B, p. 35–49.
——, 1952, Magmatic differentiation in Tertiary and Quaternary volcanic rocks from Adak and Kanaga Islands, Aleutian Islands, Alaska: Geological Society of America Bulletin, v. 63, p. 485–514.
——, 1956, Geology of northern Adak Island, Alaska: U.S. Geological Survey Bulletin 1028-C, p. 45–67.
Connelly, W., 1978, Uyak Complex, Kodiak Islands, Alaska; A Cretaceous subduction complex: Geological Society of America Bulletin, v. 89, p. 755–769.
Creager, J. S., Scholl, D. W., and others, 1973, Initial Reports of the Deep Sea Drilling Project: Washington, D.C., U.S. Government Printing Office, v. 19, 913 p.
Dalrymple, G. B., Clague, D. A., and Lanphere, M. A., 1977, Revised age for Midway volcanism, Hawaiian volcanic chain: Earth and Planetary Science Letters, v. 37, p. 107–116.
Davis, A. S., Pickthorn, L. G., Vallier, T. L., and Marlow, M. S., 1989, Petrology, geochemistry, and ages of volcanic-arc rocks from the continental margin of the Bering Sea; Implications for relocation of the plate margin: Canadian Journal of Earth Sciences, v. 26, p. 1474–1490.
Delong, S. E., and McDowell, F. W., 1975, K-Ar ages from the Near Islands, western Aleutian Islands, Alaska; Indication of a mid-Oligocene thermal event: Geology, v. 3, p. 691–694.
Detterman, R. L., 1985, The Paleogene sequence on the Alaska Peninsula [abs.]: American Association of Petroleum Geologists Bulletin, v. 69, p. 661.
——, 1986, Glaciation of the Alaska Peninsula, *in* Hamilton, T. D., Reed, K. M., and Thorson, R. M., Glaciation in Alaska; The geologic record: Alaska Geological Society, p. 151–170.
Detterman, R. L., and Miller, J. W., 1985, Kaguyak Fm; An Upper Cretaceous flysch deposit, *in* Bartsch-Winkler, S., and Reed, K. M., eds., The United States Geological Survey in Alaska; Accomplishments during 1983: U.S. Geological Survey Circular 945, p. 49–51.
Detterman, R. L., and Reed, B. L., 1980, Stratigraphy, structure, and economic geology of the Iliamna Quadrangle, Alaska: U.S. Geological Survey Bulletin 1368-b, 86 p.
Detterman, R. L., Miller, T. P., Yount, M. E., and Wilson, F. H., 1981, Geologic map of the Chignik and Sutwik Island Quadrangles, Alaska: U.S. Geological Survey Miscellaneous Investigations Series Map I-1229, scale 1:250,000.
Detterman, R. L., Case, J. E., Wilson, F. H., Yount, M. E., and Allaway, W. H., Jr., 1983, Generalized geologic map of the Ugashik, Bristol Bay, and part of the Karluk Quadrangles, Alaska: U.S. Geological Survey Miscellaneous Field Studies Map MF-1539-A, scale 1:250,000.
Detterman, R. L., Miller, T. P., Wilson, F. H., and Yount, M. E., 1985, Geologic map of the Ugashik and western Karluk Quadrangles, Alaska: U.S. Geological Survey Miscellaneous Geological Investigations Map I-1685, scale 1:250,000.
Detterman, R. L., Case, J. E., Miller, J. W., Wilson, F. H., and Yount, M. E., 1994, Stratigraphic framework of the Alaska Peninsula: U.S. Geological Survey Bulletin 1969-A (in press).
Engebretson, D. C., Cox, A., and Gordon, R. G., 1986, Relative motions between oceanic plates and continental plates in the Pacific basin: Geological Society of America Special Paper 206, 59 p.
Fisher, M. A., 1980, Petroleum geology of Kodiak Shelf, Alaska: American Association of Petroleum Geologists Bulletin, v. 64, p. 1140–1157.
Fisher, M. A., and von Huene, R., 1980, Structure of Upper Cenozoic strata beneath Kodiak Shelf, Alaska: American Association of Petroleum Geologists Bulletin, v. 64, p. 1014–1033.
——, 1982, Geologic structure of the continental shelf southeast and southwest of Kodiak Island, Alaska, from 24-fold seismic data: U.S. Geological Survey Miscellaneous Field Studies Map MF-1460, 1 sheet, scale 1:500,000.
——, 1984, Geophysical investigation of a suture zone; The Border Ranges fault of southern Alaska: Journal of Geophysical Research, v. 89, p. 11333–11351.
Fisher, M. A., Bruns, T. R., and von Huene, R., 1981, Transverse tectonic boundaries near Kodiak Island, Alaska; Geological Society of America Bulletin, v. 91, p. 218–224.
Fisher, M. A., Detterman, R. L., and Magoon, L. B., 1987, Tectonics and petroleum geology of the Cook-Shelikof Basin, southern Alaska, *in* Scholl, D. W., Grantz, A., and Vedder, J. G., eds., Geology and resource potential of the continental margin of western North America and adjacent ocean basins; Beaufort Sea to Baja California: Circum-Pacific Council for Energy and Mineral Resources Earth Science Series, v. 6, p. 213–228.
Fraser, G. D., and Snyder, G. L., 1959, Geology of southern Adak Island and Kagalaska Island, Alaska: U.S. Geological Survey Bulletin 1028-M, p. 371–407.
Gates, G. O., Powers, H. A., and Wilcox, R. E., 1971, Geology of the Near Islands, Alaska: U.S. Geological Survey Bulletin 1028-U, p. 709–822.
Geist, E. L., Childs, J. R., and Scholl, D. W., 1988, Evolution and petroleum geology of Amlia and Amukta intra-arc summit basins, Aleutian Ridge: Marine and Petroleum Geology, v. 4, p. 334–352.
Gnibidenko, G. S., and 5 others, 1980, Tectonics of the Kuril-Kamchatka deepwater shelf: Moscow, Nauka, 179 p. (in Russian)
Grow, J. S., and Atwater, T., 1970, Mid-Tertiary tectonic transition in the Aleutian Arc: Geological Society of America Bulletin, v. 81, p. 3715–3722.
Harbert, W. P., and Cox, A., 1984, Preliminary paleomagnetic results from Umnak Island, Aleutian Ridge: Geological Society of America Abstracts with Programs, v. 16, p. 288.
Harbert, W. P., Scholl, D. W., Vallier, T. L., Stevenson, A. J., and Mann, D. M., 1986, Major evolutionary phases of a forearc basin of the Aleutian Terrace; Relation to North Pacific tectonic events and the formation of the Aleutian subduction complex: Geology, v. 14, p. 757–761.
Hein, J. R. and McLean, H., 1980, Paleogene sedimentary and volcanogenic rocks from Adak Island, central Aleutian Islands, Alaska: U.S. Geological Survey Professional Paper 1126-E, 16 p.
Hein, J. R., McLean, H., and Vallier, T. L., 1984, Reconnaissance geology of southern Atka Island, Aleutian Islands, Alaska: U.S. Geological Survey Bulletin 1069, 19 p.
Herrera, R. C., 1978, Developments in Alaska in 1977: American Association of Petroleum Geologists Bulletin, v. 62, p. 1311–1321.
Hillhouse, J. W., 1987, Accretion of southern Alaska: Tectonophysics, v. 139, p. 107–122.
Jones, D. L., and Silberling, N. J., 1979, Mesozoic stratigraphy; The key to tectonic analysis of southern and central Alaska: U.S. Geological Survey Open-File Report 79–1200, 37 p.
Jones, D. L., Silberling, N. J., and Hillhouse, J. W., 1977, Wrangellia; A displaced terrane in northwestern North America: Canadian Journal of Earth Sciences, v. 14, p. 2565–2577.
Jones, D. L., Silberling, N. J., Berg, H. C., and Plafker, G., 1981, Map showing tectonostratigraphic terranes of Alaska, columnar sections, and summary description of terranes: U.S. Geological Survey Open-File Report 81–792, 20 p.
Karl, H. A., Vallier, T. L., Masson, D., and Underwood, M. B., 1987, Long-range side-scan sonar mosaic of the western Aleutian arc and trench and the adjacent North Pacific seafloor [abs.]: EOS Transactions of the American Geophysical Union, v. 68, p. 1485–1486.
Kay, S. M., 1983, Metamorphism in the Aleutian arc; The Finger Bay pluton, Adak, Alaska: Canadian Mineralogist, v. 21, p. 665–681.
Kay, S. M., Kay, R. W., and Citron, G. P., 1982, Tectonic controls on tholeiitic and calc-alkaline magmatism in the Aleutian arc: Journal of Geophysical

Research, v. 87, p. 4051–4072.
Kay, S. M., Kay, R. W., Brueckner, H. K., and Rubenstone, J. L., 1983, Tholeiitic Aleutian arc plutonism; The Finger Bay pluton, Adak, Alaska: Contributions to Mineralogy and Petrology, v. 82, p. 99–116.
Keller, G., von Huene, R., McDougall, K., and Bruns, T. R., 1984, Paleoclimatic evidence for Cenozoic migration of Alaskan terranes: Tectonics, v. 3, p. 473–495.
Kulm, L. D., von Huene, R., and others, 1973, Initial reports of the Deep Sea Drilling Project: Washington, D.C., U.S. Government Printing Office, v. 18, 1077 p.
Lewis, S. D., Ladd, J. W., Bruns, T. R., Hayes, D., and von Huene, R., 1984, Growth patterns of submarine canyons and slope basins, eastern Aleutian Trench, Alaska [abs.]: EOS Transactions of the American Geophysical Union, v. 65, p. 1104.
Lewis, S. D., Ladd, J. W., and Bruns, T. R., 1988, Structural development of an accretionary prism by thrust and strike-slip faulting; Shumagin region, Aleutian Trench: Geological Society of America Bulletin, v. 100, p. 767–782.
Lonsdale, P., 1988, Paleogene history of the Kula Plate; Offshore evidence and onshore implications: Geological Society of America Bulletin, v. 100, p. 733–754.
Marincovich, L. N., Jr., 1983, Molluscan paleontology, paleoecology, and North Pacific correlation of Miocene Tachilni Formation, Alaska Peninsula: Bulletin of American Paleontology, v. 84, p. 155.
Marlow, M. S., Scholl, D. W., and Buffington, E. C., 1973, Tectonic history of the central Aleutian arc: Geological Society of America Bulletin, v. 84, p. 1555–1574.
Marsh, B. D., 1976, Some Aleutian andesites; Their nature and source: Journal of Geology, v. 87, p. 27–45.
McCarthy, J., and Scholl, D. W., 1985, Mechanisms of subduction accretion along the central Aleutian trench: Geological Society of America Bulletin, v. 96, p. 691–701.
McCarthy, J., Stevenson, A. J., Scholl, D. W., and Vallier, T. L., 1984, Perspective remarks concerning the petroleum geology of the Aleutian accretionary prism: Petroleum and Marine Geology, v. 1, p. 151–167.
McLean, H., 1979, Tertiary stratigraphy and petroleum potential of Cold Bay-False Pass area, Alaska Peninsula: American Association of Petroleum Geologists Bulletin, v. 63, p. 1522–1526.
McLean, H., Hein, J. R., and Vallier, T. L., 1983, Reconnaissance geology of Amlia Island, Aleutian Islands, Alaska: Geological Society of America Bulletin, v. 94, p. 1020–1027.
Minster, B. J. and Jordan, T. H., 1978, Present-day plate motions: Journal of Geophysical Research, v. 83, p. 5331–5354.
Molnia, B. F., 1986, Glacial history of the northeastern Gulf of Alaska; A synthesis, *in* Hamilton, T. D., Reed, K. M., and Thompson, R. M., eds., Glaciation in Alaska; The geologic record: Alaska Geological Society, p. 219–236.
Moore, G., 1969, New formations on Kodiak Island and vicinity, Alaska: U.S. Geological Survey Bulletin 1274-A, p. 27–35.
Moore, J. C., 1974a, The ancient continental margin of Alaska, *in* Burk, C. S., and Drake, C. L., eds., The geology of continental margins: New York, Springer-Verlag, p. 811–816.
——, 1974b, Geologic and structural map of the Outer Shumagin Islands, southwestern Alaska: U.S. Geological Survey Miscellaneous Investigations Map I-815, scale 1:63,360.
Moore, J. C., Byrne, T., Plumley, P. W., Reid, M., Gibbons, H., and Coe, R. S., 1983, Paleogene evolution of the Kodiak Islands, Alaska; Consequences of ridge-trench interaction in a more southerly latitude: Tectonics, v. 2, p. 265–293.
Newberry, J. T., 1984, Seismicity and tectonics of the far western Aleutian islands [M.S. thesis]: East Lansing, Michigan State University, 93 p.
Newberry, J. T., Laclair, D. L., and Fujita, K., 1986, Seismicity and tectonics of the far western Aleutian Islands: Journal of Geodynamics, v. 6, p. 13–32.
Nilsen, T. H., and Moore, G. W., 1979, Reconnaissance study of Upper Cretaceous to Miocene stratigraphic units and sedimentary facies, Kodiak and adjacent islands, Alaska, with a section on sedimentary petrology by G. R. Winkler: U.S. Geological Survey Professional Paper 1093, 34 p.
Nilsen, T. H., and Zuffa, G. G., 1982, The Chugach terrane, a Cretaceous trench-fill deposit, southern Alaska, *in* Leggett, J. K., ed., Trench-forearc geology: Geological Society of London Special Publication 10, p. 213–227.
Palmer, A. R., 1983, The Decade of North America Geology 1983 geologic time scale: Geology, v. 11, p. 503–504.
Pickthorn, L. B. and Vallier, T. L., 1991, Radiometric age data of rocks from the Aleutian Ridge, Alaska: U.S. Geological Survey Open-File Report (unpublished).
Pickthorn, L. B., Vallier, T. L., and Scholl, D. W., 1984, Geochronology of igneous rocks from the Aleutian Island arc: Geological Society of America Abstracts with Programs, v. 16, p. 328–329.
Plafker, G., Jones, D. L., and Pessagno, E. A., Jr., 1977, A Cretaceous accretionary flysch and melange terrane along the Gulf of Alaska margin, *in* Blean, K. M., ed., The United States Geological Survey in Alaska; Accomplishments during 1976: U.S. Geological Survey Circular 751-B, p. 41–43.
Plumley, P. W., Coe, R. S., and Byrne, T., 1983, Paleomagnetism of the Paleocene Ghost Rocks Formation, Prince William terrane, Alaska: Tectonics, v. 2, p. 295–314.
Pratt, R. M., Rutstein, M. S., Walton, F. W., and Buschur, J. A., 1972, Extension of Alaskan structural trends beneath Bristol Bay, Bering Shelf, Alaska: Journal of Geophysical Research, v. 77, p. 4993–4999.
Rau, W. W., Plafker, G., and Winkler, G. R., 1977, Preliminary foraminiferal biostratigraphy and correlation of selected stratigraphic sections and wells in the Gulf of Alaska Tertiary province: U.S. Geological Survey Open-File Report 77–747, 54 p.
Reed, B. L., and Lanphere, M. A., 1969, Age and chemistry of Mesozoic and Tertiary plutonic rocks in south-central Alaska: Geological Society of America Bulletin, v. 80, p. 23–44.
——, 1973, Alaska-Aleutian Range batholith; Geochronology, chemistry, and relation to circum-Pacific plutonism: Geological Society of America Bulletin, v. 84, p. 2583–2610.
Reed, B. L., Miesch, A. T., and Lanphere, M. A., 1983, Plutonic rocks of Jurassic age in the Alaska-Aleutian range batholith; Chemical variations and polarity: Geological Society of America Bulletin, v. 94, p. 1232–1240.
Rubenstone, J. L., 1984, Composition of Paleogene submarine volcanic rocks of the central and western Aleutians and their bearing on the evolution of the Aleutian arc [Ph.D. thesis]: Ithaca, Cornell University, 350 p.
Ryan, H., and Scholl, D. W., 1989, The evolution of forearc structures along an oblique convergent margin, central Aleutian arc: Tectonics, v. 8, p. 497–516.
Scholl, D. W., Greene, H. G., and Marlow, M. S., 1970, Eocene age of the Adak Paleozoic(?) rocks, Aleutian Islands, Alaska: Geological Society of America Bulletin, v. 81, p. 3583–3592.
Scholl, D. W., Buffington, E. C., and Marlow, M. S., 1975, Plate tectonics and the structural evolution of the Aleutian-Bering Sea region, *in* Forbes, R., ed., Contributions to the geology of the Bering Sea Basin and adjacent regions: Geological Society of America Special Paper 131, p. 1–31.
Scholl, D. W., Vallier, T. L., and Stevenson, A. J., 1982, Sedimentation and deformation in the Amlia fracture zone sector of the Aleutian trench: Marine Geology, v. 48, p. 105–134.
——, 1983a, Arc, forearc, and trench sedimentation and tectonics; Amlia corridor of the Aleutian Ridge, *in* Watkins, J., and Drake, C. L., eds., Studies in continental margin geology: American Association of Petroleum Geologists Memoir 34, p. 413–439.
——, 1983b, Geologic evolution of the Aleutian Ridge; Implications for petroleum resources: Journal of the Alaska Geological Society, v. 3, p. 33–46.
——, 1986, Terrane accretion, production, and continental growth; A perspective based on the origin and tectonic fate of the Aleutian-Bering Sea region: Geology, v. 14, p. 43–47.
——, 1987, Geologic evolution and petroleum potential of the Aleutian Ridge, *in* Scholl, D. W., Grantz, A., and Vedder, J. G., eds., Geology and resource potential of the continental margin of western North America and adjacent ocean basins; Beaufort Sea to Baja California: Circum-Pacific Council for Energy and Mineral Resources Earth Science Series, v. 6, p. 123–156.

Shmidt, O. A., 1978, The tectonics of the Komandorsky Islands and the structure of the Aleutian Ridge: Geological Institute, Academy of Science U.S.S.R., v. 320, Moscow, Nauka, 100 p. (in Russian).

Shor, G. G., Jr., 1964, Structure of the Bering Sea and the Aleutian Ridge: Marine Geology, v. 1, p. 213–219.

Stone, D. B., 1988, Bering Sea-Aleutian arc, Alaska, *in* Nairn, A.E.M., Stehli, F. G., and Uyeda, S., The ocean basins and margins: v. 7B, p. 1–84.

Stone, D. B. and Packer, D. R., 1977, Tectonic implications of Alaska Peninsula paleomagnetic data: Tectonophysics, v. 37, p. 183–201.

Vallier, T. L., O'Connor, R., and Harbert, W. P., 1983, Gabbro and amphibolite from Attu Island, Aleutian island arc, Alaska [abs.]: EOS Transactions of the American Geophysical Union, v. 64, p. 870.

Vallier, T. L., McCarthy, J., Scholl, D. W., Stevenson, A. J., and O'Connor, R., 1984, Offshore structures in the Near Islands region, western Aleutian island arc: Geological Society of America Abstracts with Programs, v. 16, p. 338.

Vallier, T. L., Karl, H. A., and Underwood, M. B., 1987, Geologic framework of the western Aleutian arc and adjacent North Pacific seafloor [abs.]: EOS Transactions of the American Geophysical Union, v. 68, p. 1486.

von Huene, R., Moore, G. W., Moore, J. C., and Stephen, S. D., 1979, Cross section, Alaska Peninsula-Kodiak Island-Aleutian Trench; Summary: Geological Society of America Bulletin, v. 90, p. 417–430.

von Huene, R., Fisher, M. A., McClellan, P. H., Box, S., and Ryan, H., 1985, North Pacific to Kuskokwim Mountains: Boulder, Colorado, Geological Society of America Continent-Ocean Transect A2, scale 1:500,000.

von Huene, R., Fisher, M. A., and Bruns, T. R., 1987, Geology and evolution of the Kodiak margin, Gulf of Alaska, *in* Scholl, D. W., Grantz, A., and Vedder, J. G., eds., Geology and resource potential of the continental margin of western North America and adjacent ocean basins; Beaufort Sea to Baja California: Circum-Pacific Council for Energy and Mineral Resources Earth Science Series, v. 6, p. 191–212.

Weber, F. R., 1985, Late Quaternary glaciation of the Pavlof Bay and Port Moller areas, Alaska Peninsula, *in* Bartsch-Winkler, S., ed., The United States Geological Survey in Alaska; Accomplishments during 1984: U.S. Geological Survey Circular 967, p. 42–44.

Wilson, F. H., 1980, Late Mesozoic and Cenozoic tectonics and the age of porphyry copper prospects, Chignik and Sutwik Island Quadrangles, Alaska Peninsula: U.S. Geological Survey Open-File Report 80–543, 94 p.

——, 1982, Map and table showing preliminary results of K-Ar studies in the Ugashik Quadrangle, Alaska Peninsula: U.S. Geological Survey Open-File Report 82–140, scale 1:250,000.

——, 1985, The Meshik arc; An Eocene to earliest Miocene magmatic arc on the Alaska Peninsula: Alaska Division of Geological and Geophysical Surveys Professional Report 88, 14 p.

Wilson, F. H., Gaum, W. C., and Herzon, P. L., 1981, Map and tables showing geochronology and whole-rock geochemistry, Chignik and Sutwik Island Quadrangles, Alaska: U.S. Geological Survey Map MF-1053-M, 3 sheets, scale 1:250,000.

Wilson, F. H., Detterman, R. L., and Case, J. E., 1985a, The Alaska Peninsula terrane; A definition: U.S. Geological Survey Open-File Report 85-470, 20 p.

Wilson, F. H., Case, J. E., and Detterman, R. L., 1985b, Preliminary description of a Miocene zone of structural complexity, Port Moller and Stepovak Bay Quadrangles, *in* Bartsch-Winkler, S., and Reed, K. M., eds., The United States Geological Survey in Alaska; Accomplishments during 1983: U.S. Geological Survey Circular 945, p. 55–56.

Manuscript Accepted by the Society May 9, 1990

ACKNOWLEDGMENTS

We greatly appreciate the critical reviews of Hugh McLean, Henry Berg, and John Armentrout. In addition, Michael Garrison, Sean Stone, and James R. Le Compte gave valuable technical help with the manuscript. We thank George Moore for assistance with the stratigraphic column of Kodiak Island. Andre Tsvetkov sent samples from the Komandorsky Islands. Jill McCarthy, Dennis Mann, Jon Childs, and John Miller processed the seismic-reflection profiles. Robin Frisch helped map lineaments on Attu, Agattu, and Adak Islands. Leda Beth Pickthorn determined K-Ar radiometric ages of several samples. William Harbert and Robert O'Connor provided stimulating discussions while mapping parts of Attu, Shemya, Amlia, and Atka Islands during the 1983 field season. Hugh McLean and James Hein worked with us on Amlia Island in 1979.

NOTES ADDED IN PROOF

Significant studies have been completed since this manuscript was accepted for publication. Some interpretations have changed, particularly in the Alaska Peninsula segment, and new data have been collected. Suggested changes in the stratigraphy of the Komandorsky Islands are summarized by Tsvetkov and others (1990). In the Gulf of Alaska and Kodiak regions, the Trans-Alaska Crustal Transect (TACT) and EDGE projects (Fisher and others, 1989; Fuis and others, 1991; Moore and others, 1991; and Page and others, 1989) added new data for interpretations of the deep crust beneath the Alaskan margin and of the crustal structure of accreted terranes in southern Alaska. Preliminary data from section balancing of EDGE and other seismic profiles in the Aleutian trench region east of the Kodiak Islands, in fact, suggest that the accretionary wedge is younger than previously thought, possibly no more than 1 m.y. old.

GLORIA (Geological LOng-Range Inclined Asdic) side-scan and complementary geophysical surveys in 1986, 1987, and 1988, along both sides of the Aleutian arc and as far west as the international border between the United States and Russia, contributed a large amount of marine data within 200 nautical miles of the coastlines. Those cruises gathered side-scan images, single-channel seismic profiles, and magnetic and gravity data. Information from the Bering Sea (north) side of the Aleutian arc is published in an atlas (EEZ-SCAN Scientific Staff, 1991) and data from the Pacific Ocean (south) side of the arc will be published in a similar atlas in late 1995 or 1996.

Interpretations of Aleutian GLORIA surveys, to be published in the near future, deal with sedimentation in the Aleutian fore-arc region (M. R. Dobson, written communications, 1993) and with a small remnant of Kula plate that occurs offshore the Near Islands that was discovered by Lonsdale (1988). Extreme obliquity of collision between the western Aleutian arc and the oceanic plate resulted in the formation of extensive wrench faults that have fragmented the Near Islands and Komandorsky Islands regions into elongate slivers. The Aleutian arc is moving west along those wrench faults and is headed for a rendezvous with the Kamchatka Peninsula (E. Geist, written communications, 1993). One of the wrench faults, the Agattu fault, cuts the fore-arc region from south of Amchitka Island to west of the Near Islands. At the intersections of the Agattu fault with northeast-trending wrench faults, deep sediment basins have developed along the narrow Aleutian terrace.

The present-day Kula plate is a small fragment of a major oceanic plate that has been mostly consumed beneath the Aleutian arc. The plate is bounded on the south and west by Stalemate ridge and on the north by the Aleutian trench. The Kula-Pacific ridge, a relict spreading center, trends nearly north-south and intersects the Aleutian trench at about long 172°E. Spreading along that ridge ended ~42 m.y. ago, close to the same time as the bend of the Hawaii-Emperor seamount chains.

The Alaska Peninsula–Kodiak Island segment has a coherent and nearly complete stratigraphic sequence dating from Permian time. We now recognize that past reports giving Silurian and Devonian ages to some carbonate rocks on the Alaska Peninsula are in error (Detterman and others, 1994); those carbonates are Triassic in age. Some of the early rocks of the Alaska Peninsula record the rapid development and deep erosion of a Jurassic island arc developed on a carbonate and greenstone terrane that has Wrangellia affinities. Recent analyses (Wilson and others, 1993) suggest that the oldest rocks in the Kodiak Islands, which lie adjacent to the Border Ranges fault system, are not part of the Alaska Peninsula terrane. Their lithology and geologic history are distinctly different from rocks of equivalent age on the Alaska mainland. They in part represent a metamorphosed early Mesozoic accretionary complex (Roeske and others, 1989) and may be a sliver of an undefined terrane. These rocks are unknown in outcrop southwest of the Kodiak Islands; however, Fisher (1981) correlated a series of magnetic anomalies thought to be associated with plutonic and ultramafic rocks in this complex to near Sutwik Island, which does lie southwest of the Kodiak Islands. Late Cretaceous rocks consist of a subduction complex that includes fluvial and flyschoid deposits on the mainland and extensive flyschoid trench deposits on the offshore islands. The coherent Late Cretaceous stratigraphy and facies relationships, as originally reported by Mancini and others (1978), across the Alaska Peninsula and Chugach terrane suggest that at the southwest end of this segment rocks of the Chugach terrane were deposited adjacent to the Alaska Peninsula terrane and not rafted in from elsewhere.

Alaska Peninsula Tertiary rocks include the deposits of the Meshik arc, age-equivalent to the early part of the middle series in the Aleutian arc, and deposits from late Miocene to Holocene magmatic activity along the arc, age-equivalent to late middle series and upper series rocks of the Aleutian Ridge segment. The only rocks on the Alaska Peninsula that are now believed to be age-equivalent to lower series rocks in the Aleutian arc are the Tolstoi and Copper Lake formations. However, neither of these units provides evidence for the initiation of volcanism along the Aleutian arc; both have batholithic and sedimentary rock provenances. In the Kodiak Islands, Tertiary rocks include very early Tertiary plutons emplaced into Late Cretaceous trench deposits. In the eastern part of the islands, a Paleogene subduction complex (Ghost Rocks) equivalent to the Orca Group of Prince William Sound is overlapped by nonmarine and shallow-water marine clastic rocks as young as Pliocene. Southwest of the Kodiak Islands, rocks neither equivalent to the Ghost Rocks nor to the clastics are known onshore.

Migration of the Alaska Peninsula terrane as part of the larger Peninsula-Alexander-Wrangellia (PAW) composite terrane (see Nokleberg and others, this volume, Chapter 10) from southerly latitudes was largely complete by the end of Mesozoic time. The outboard composite Chugach and Prince William terranes are in part depositional on the PAW terrane and in part juxtaposed against it. The magmatic arc, subdivided into two components in this segment, is emplaced through the older rocks. The older component, called the Meshik arc, is primarily exposed on the southern Alaska Peninsula. The younger component extends the length of the segment and shows clear evidence for migration of the volcanic front to the northwest from the Pacific coast with time.

ADDITIONAL REFERENCES

EEZ-SCAN Scientific Staff, 1991, Atlas of the U.S. Exclusive Economic Zone, Bering Sea: U.S. Geological Survey Miscellaneous Investigations Series 1-2053, 145 p.

Fisher, M. A., 1981, Location of the Border Ranges fault southwest of Kodiak Island, Alaska: Geological Society of America Bulletin, v. 92, p. 19–30.

Fisher, M. A., Brocher, T. M., Nokleberg, W. J., Plafker, G., and Smith, G. L., 1989, Seismic reflection images of the crust of the northern part of the Chugach terrane, Alaska: Results of a survey for the Trans-Alaska Crustal Transect (TACT): Journal of Geophysical Research, v. 94, p. 4424–4440.

Fuis, G. S., Ambos, E. L., Mooney, W. D., Christensen, N. I., and Geist, E. L., 1991, Crustal structure of accreted terranes in southern Alaska, Chugach Mountains, and Copper River Basin, from seismic refraction results: Journal of Geophysical Research, v. 96, p. 4187–4227.

Lonsdale, P., 1988, Paleogene history of the Kula Plate: Offshore evidence and onshore implications: Geological Society of America Bulletin, v. 100, p. 733–754.

Mancini, E. A., Deeter, T. M., and Wingate, F. H., 1978, Upper Cretaceous arc-trench gap sedimentation on the Alaska Peninsula: Geology, v. 6, p. 437–439.

Moore, J. C., and 11 others, 1991, EDGE deep seismic reflection transect of the eastern Aleutian arc-trench layered lower crust reveals underplating and continental growth, Geology, v. 19, p. 420–424.

Page, R. A., Stephens, C. D., and Lahr, J. C., 1989, Seismicity of the Wrangell and Aleutian Wadati-Benioff zones and the North American plate along the Trans-Alaska crustal transect, Chugach Mountains and Copper River Basin, southern Alaska: Journal of Geophysical Research, v. 94, p. 16059–16082.

Roeske, S. M., Mattinson, J. M., and Armstrong, R. L., 1989, Isotopic ages of glaucophane schists on the Kodiak Islands, southern Alaska, and their implications for Mesozoic tectonic history of the Border Ranges fault system: Geological Society of America Bulletin, v. 101, p. 1021–1037.

Tsvetkov, A. A., Fedorchuk, A. V., and Gladenkov, A. Yu., 1990, Geology and magmatic evolution of Bering Island: International Geology Review, v. 32, p. 1202–1217.

Wilson, F. H., Detterman, R. L., and DuBois, G. D., 1993, Geologic framework of the Alaska Peninsula, southwest Alaska and the Alaska Peninsula terrane: U.S. Geological Survey Bulletin 1969-B, scale 1:500,000 (in press).

Printed in U.S.A.

The Geology of North America
Vol. G-1, The Geology of Alaska
The Geological Society of America, 1994

Chapter 12

Geology of the southern Alaska margin

George Plafker
U.S. Geological Survey, MS-904, 345 Middlefield Road, Menlo Park, California 94025
J. Casey Moore
Earth Sciences Board, University of California, Santa Cruz, California 95064
Gary R. Winkler
U.S. Geological Survey, Box 25046, MS-905, Federal Center, Denver, Colorado 80225

INTRODUCTION

This chapter summarizes the tectonic setting, geology, and tectonic evolution of the southern Alaska margin south of the Border Ranges fault system, which extends 2100 km from the Sanak Islands on the west to Chatham Strait on the east and seaward to the base of the continental slope (Fig. 1). Mesozoic and Cenozoic rocks that make up the southern Alaska continental margin record a complex history of subduction-related underplating, offscraping, and metamorphism, as well as transform-related large-scale strike-slip displacements. The region discussed in this chapter has an area of about 328,000 km^2, of which almost 30% is onshore. The land area includes parts or all of 26 1:250,000 scale quadrangles.

The mainland along the northern Gulf of Alaska margin consists of alluvium- and glacier-covered coastal lowlands, 0 to 40 km wide, backed by a belt as wide as 40 km of rugged foothills that rise to elevations of about 2000 m (Wahrhaftig, this volume). The foothills are bordered to the north by the exceedingly rugged Kenai, Chugach, and Saint Elias mountains. Average summit elevations are over 2000 m, and numerous peaks are over 5000 m; the highest peaks are Mt. Saint Elias in Alaska (5488 m) and nearby Mt Logan (5745 m) in Canada. All major drainages in the coastal mountains are occupied by glaciers except for the Alsek River, which drains across the Saint Elias Mountains from Canada, and the Copper River, which drains across the Chugach Mountains from the interior of Alaska. The Kodiak Islands group, the islands of Prince William Sound, and the islands within the study area between Cross Sound and Chatham Strait in the Alexander Archipelago of southeastern Alaska have moderately rugged mountains with summit altitudes between 400 and 1500 m, and generally irregular, drowned, cliffed shorelines.

The width of the continental shelf and slope averages about 110 km; it ranges from as much as 250 km off the southwestern tip of the Kenai Mountains to as little as 40 km at the southeast end of the area. Most of the topography of the shelf and slope is gently undulating, except where it is broken by seven major submarine valleys and a number of smaller valleys that were filled by glaciers to the edge of the continental shelf during the Pleistocene. East of Kayak Island, a relatively smooth and steep continental slope descends to a gentle continental rise at water depths of 2000 to 4000 m. West of Kayak Island, the slope makes up the more irregular inner wall of the eastern Aleutian Trench. The trench is as deep as 4500 m at its northeastern end and gradually deepens westward to about 6400 m off the south end of the Alaska Peninsula.

Unless otherwise indicated, age designations are based on the Decade of North American Geology geologic time scale (Palmer, 1983); metamorphic facies follow the classification of Turner (1981); plutonic rock classifications are after Streckeisen (1976); sandstone classifications are after Dickinson and Suczek (1979); terrane nomenclature follows that of Howell and others (1985); and terrane boundaries and descriptions are after Silberling and others (this volume)

TECTONIC SETTING

Included within the southern Alaska margin is a complex of highly deformed, offscraped, and accreted deep-sea rocks (Chugach, Saint Elias, Ghost Rocks, and Prince William terranes of Silberling and others [this volume]) and an allochthonous fragment of the continental margin (Yakutat terrane) (Fig. 2). The Southern Margin composite terrane includes the four accreted terranes south of the Border Ranges fault system, and the Wrangellia composite terrane includes the three terranes to the north (Peninsular, Wrangellia, and Alexander) (Plafker, 1990; Plafker and Berg, this volume, Chapter 33). The Southern Margin composite terrane boundary with the Pacific plate is the Aleutian megathrust fault system, and the Yakutat terrane boundaries with the Pacific plate are defined by the Transition and Fairweather fault systems. Boundaries between the Yakutat and Chugach terranes are the Kayak Island zone, the Chugach–Saint Elias fault system, and the Fairweather fault. In this chapter, the Saint Elias

Plafker, G., Moore, J. C., and Winkler, G. R., 1994, Geology of the southern Alaska margin, *in* Plafker, G., and Berg, H. C., eds., The Geology of Alaska: Boulder, Colorado, Geological Society of America, The Geology of North America, v. G-1.

terrane is considered to be part of the Chugach terrane and the Ghost Rocks terrane is part of the Prince William terrane, for reasons discussed below. Present relative plate motions between the Pacific plate and the continental margin are shown in Figure 2.

Structural relations between terranes, rocks that intrude them, and overlap assemblages in the study area are shown diagrammatically in Figure 3, which also shows the inferred configuration of the subducted Pacific plate. In this overview, relations between the terranes of the study area and the major fault boundaries are discussed from north to south.

Border Ranges fault system

The Border Ranges fault system is a suture along which accreted rocks of the Chugach terrane have been juxtaposed against and beneath terranes to the north (MacKevett and Plafker, 1974). The fault system has a history of Early Jurassic through Late Cretaceous subduction manifested by ductile structures in the accreted rocks of the Chugach terrane (Nokleberg and others, 1989, Fig. 2), which were underthrust a horizontal distance of at least 40 km beneath the Wrangellia composite terrane (Plafker and others, 1989). In some areas, most notably in the northern Chugach Mountains and in the Saint Elias Mountains, the Border Ranges fault has been modified by younger, brittle and ductile strike-slip displacements along or near the fault trace (e.g., Pavlis, 1982; Pavlis and Crouse, 1989; Roeske and others, 1991); it is locally offset by Paleogene normal faults in the Matanuska Valley area east of Anchorage (Little, 1990).

Southern Margin composite terrane

The deep-marine rocks south of the Border Ranges fault system constitute one of the largest subduction-related accretionary complexes in the world and record intermittent offscraping and underplating that has occurred probably from the Late Triassic to the present. Together, these rocks make up the Southern Margin composite terrane of Plafker (1990). The accreted rocks are overlain locally by basinal deposits and are intruded extensively by Tertiary granitoid plutons. The various assemblages of the Chugach terrane were successively accreted in the interval from latest Triassic to earliest Tertiary time, but mainly from latest Cretaceous to early Paleocene time. It is believed that throughout most or all of this time, the motion of the oceanic plates relative to the North American continental margin was orthogonal to moderately transpressive (see section on tectonic evolution, Fig. 15, A–C).

The Ghost Rocks Formation was thrust relatively northward beneath the Chugach terrane along the Resurrection fault prior to about 62 Ma, and the Prince William terrane was thrust relatively northward beneath the Ghost Rocks and the Chugach terrane along the Contact fault system (Fig. 3, A–A′, B–B′) prior to about 50 Ma. Counterclockwise oroclinal bending of what is now western Alaska in the early Tertiary formed the present concave-southward margin (Coe and others, 1985), and northwestward movement of the Pacific plate relative to the continental margin since the middle Eocene has resulted in oblique to orthogonal convergence along the present northern and western Gulf of Alaska.

The deep configuration of the underthrusting oceanic crust between the Aleutian Trench and the present magmatic arc of the Aleutian Islands, Alaska Peninsula, and Wrangell Mountains (Fig. 3, C–C′, D–D′) is constrained by earthquake foci and deep seismic reflection and refraction data (Plafker and others, 1982, 1989; Plate 12, this volume; von Huene and others, 1987; Fisher and others, 1989; Page and others, 1989; Fuis and others, 1991; Fuis and Plafker, 1991; Moore and others, 1991). Along much of

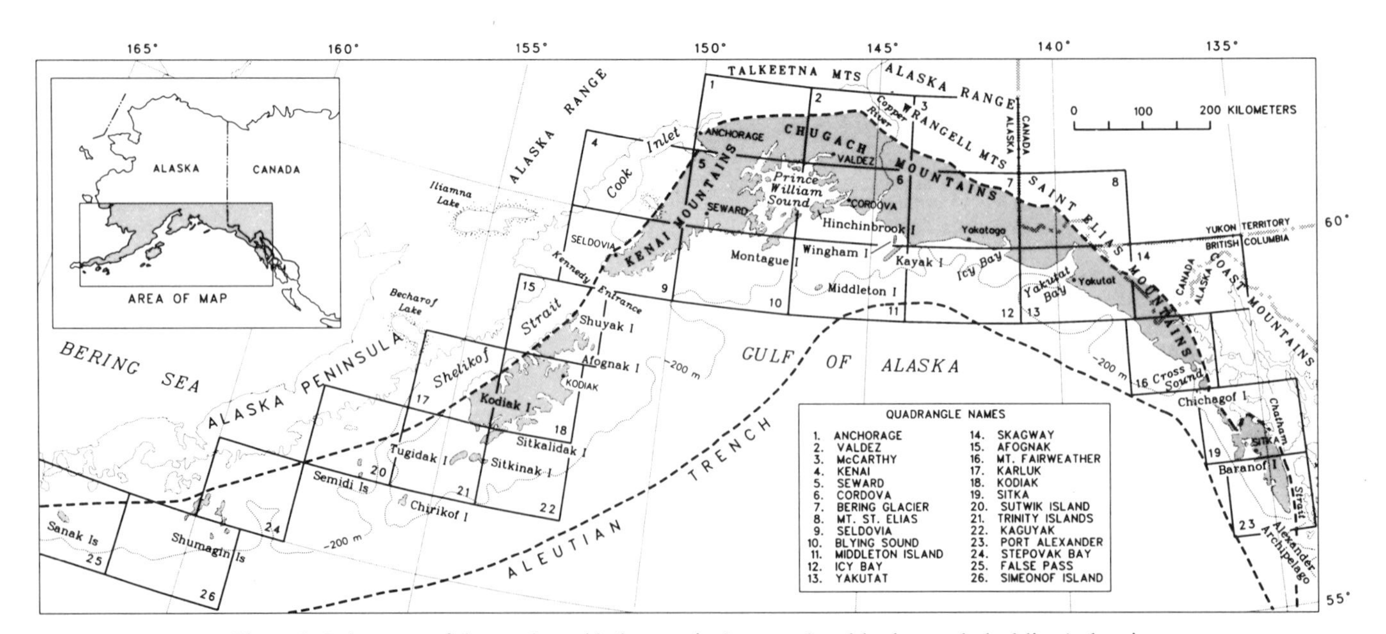

Figure 1. Index map of the southern Alaska margin (area enclosed by heavy dashed lines) showing major geographic features and areas of 1:250,000 scale quadrangles noted in text.

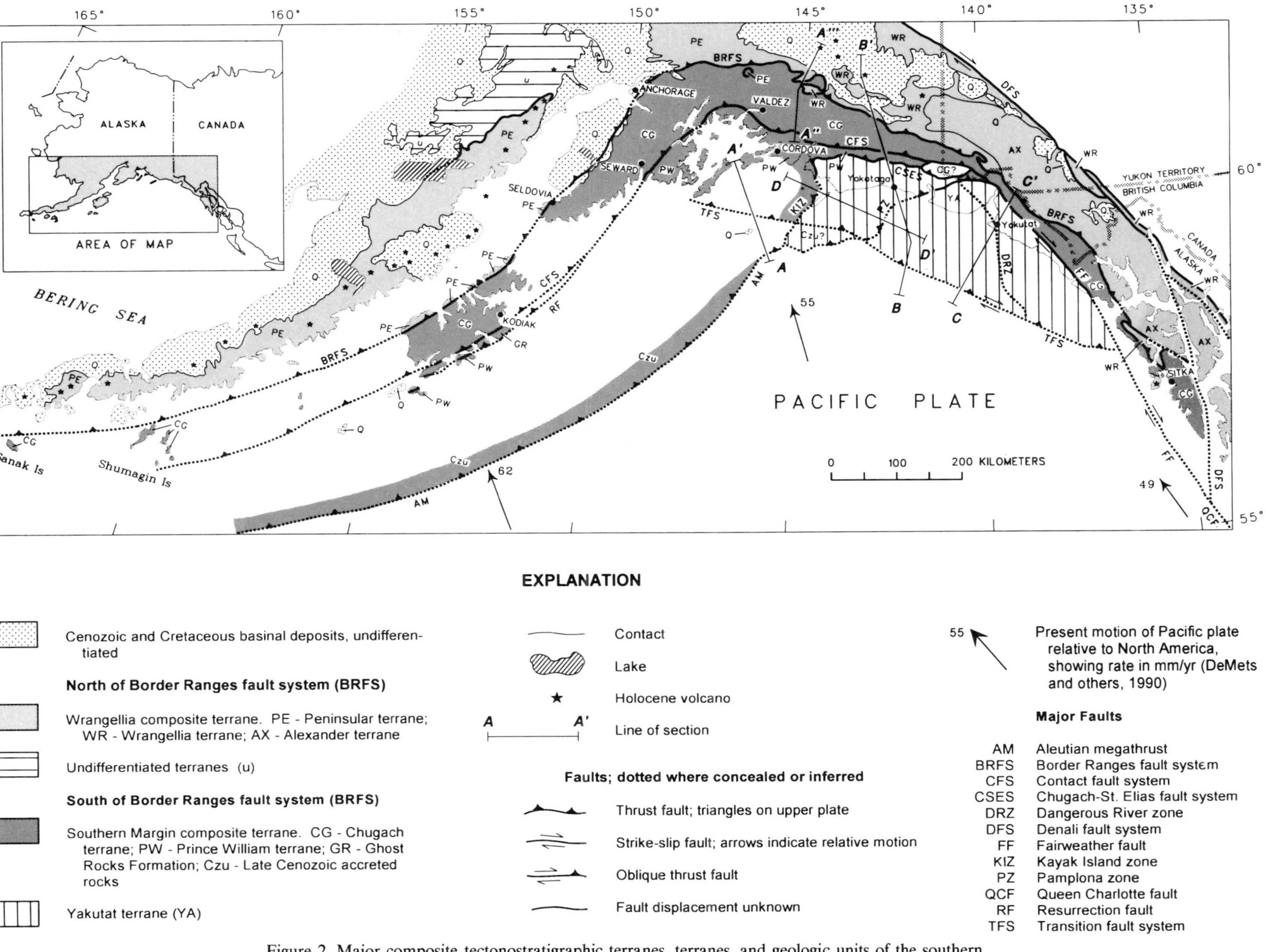

Figure 2. Major composite tectonostratigraphic terranes, terranes, and geologic units of the southern Alaska margin and adjacent areas. Modified from Silberling and others (this volume) and Plafker and Berg (this volume, Chapter 33).

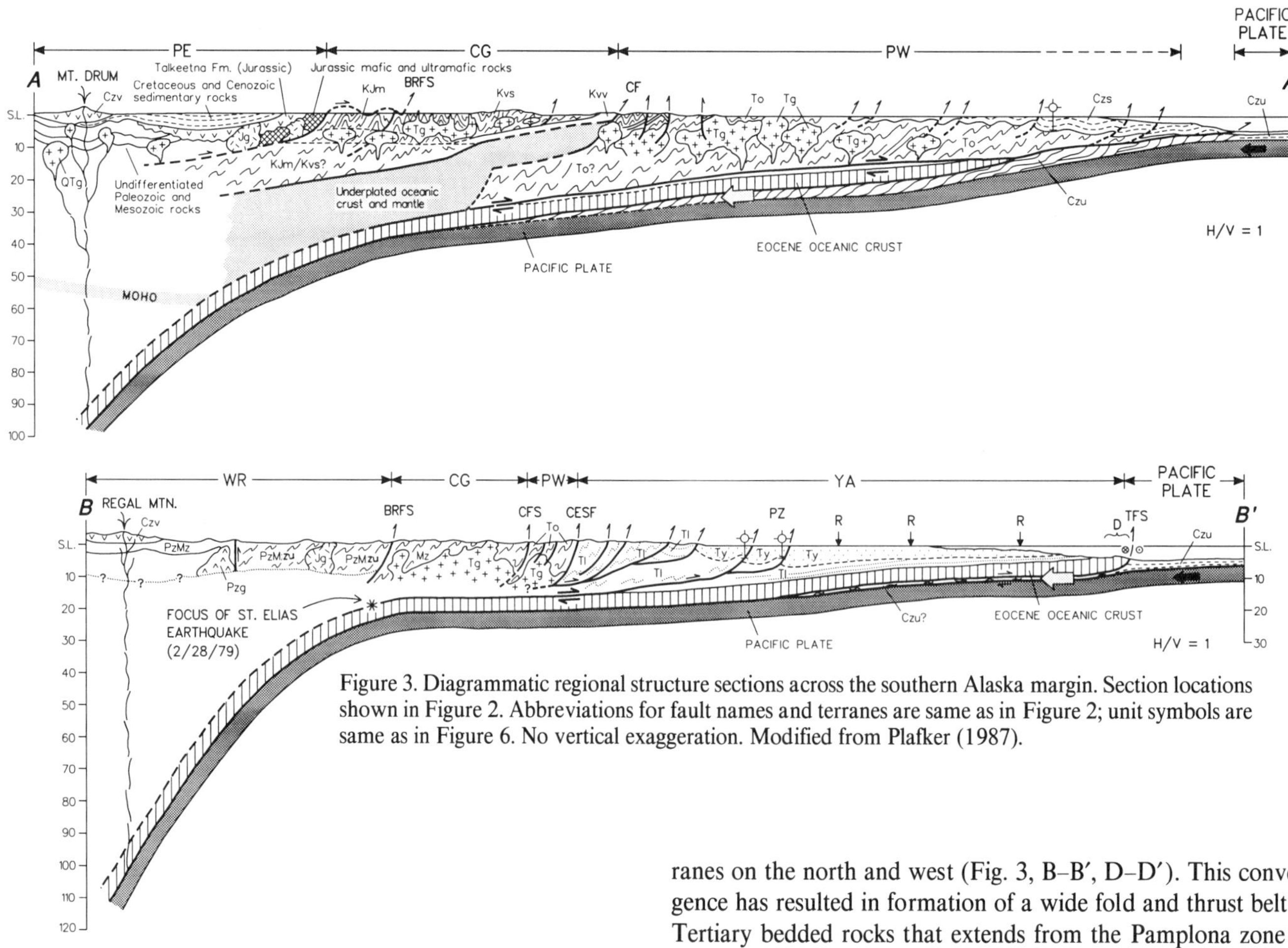

Figure 3. Diagrammatic regional structure sections across the southern Alaska margin. Section locations shown in Figure 2. Abbreviations for fault names and terranes are same as in Figure 2; unit symbols are same as in Figure 6. No vertical exaggeration. Modified from Plafker (1987).

the inner wall of the Aleutian Trench, marine seismic reflection data and deep ocean drilling suggest formation of a late Cenozoic accretionary wedge (von Huene and others, 1987; Moore and others, 1991). An apparent exception is at the eastern end of the trench, where the oceanic crust and 2.5 km of sedimentary cover have apparently been subducted beneath the continental margin and there is no evidence of a post-Eocene accretionary prism in the seismic reflection data (Plafker and others, 1982).

Yakutat terrane

The Yakutat terrane consists of a thick sequence of Cenozoic clastic sedimentary rocks and minor coal underlain in part by an offset fragment of the Chugach terrane and in part by Paleogene oceanic crust (Plafker, 1987). It was displaced about 600 km northwest along the Queen Charlotte–Fairweather transform fault during the late Cenozoic (Fig. 2). Along its northwest boundary, the terrane has been underthrust at shallow depth (<15 km) for at least 200 km beneath the Prince William terrane (Fig. 3, A–A′, D–D′; Griscom and Sauer, 1990). The south boundary of the Yakutat terrane is the dextral oblique Transition fault system along which the Pacific plate has underthrust the terrane at a low angle and is carrying it northwestward relative to adjacent terranes on the north and west (Fig. 3, B–B′, D–D′). This convergence has resulted in formation of a wide fold and thrust belt in Tertiary bedded rocks that extends from the Pamplona zone to the north and west margins of the terrane.

Wrangellia composite terrane north of the Border Ranges fault system

Terranes north of the Border Ranges fault system make up the backstop against and beneath which the Mesozoic accretionary complex was emplaced (Fig. 2). These are the dominantly intraoceanic Paleozoic and Mesozoic magmatic-arc assemblages of the Peninsular, Wrangellia, and Alexander lithotectonic terranes (Gehrels and Berg, this volume; Nokleberg and others, this volume, Chapter 10; Silberling and others, this volume). Together with a displaced fragment of the Wrangellia terrane (northern part of the Taku terrane of Silberling and others, this volume), these terranes constitute the Wrangellia composite terrane as defined by Plafker (1990). The Wrangellia composite terrane is approximately the equivalent of Terrane II of Monger and others (1982) and the Wrangellia superterrane of Saleeby (1983). The northern part of the composite terrane, consisting of the Alaska part of the Wrangellia terrane and the Peninsular terrane, was referred to as the Talkeetna superterrane by Csejtey and others (1982). Recent data favor the interpretation that, since Paleozoic time, the Wrangellia composite terrane has been a single large disrupted terrane, rather than an amalgamation of several distinct,

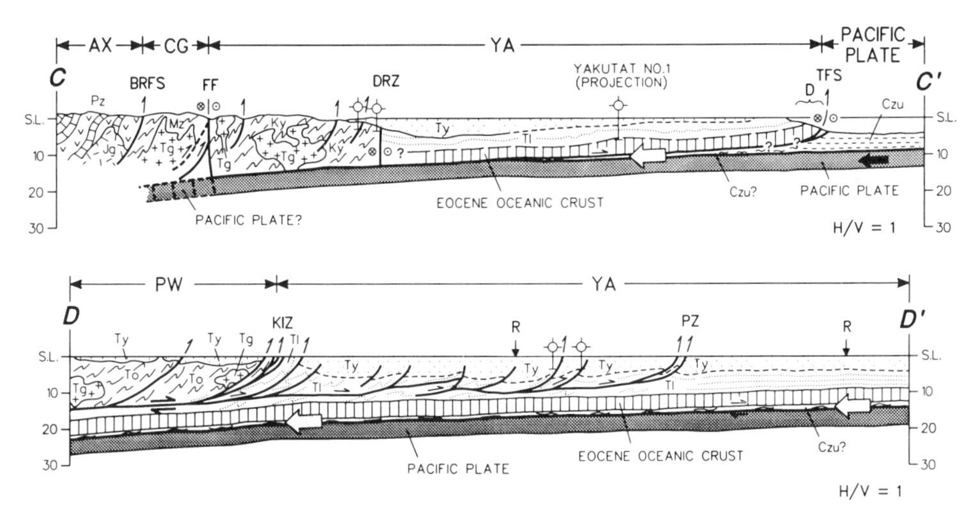

EXPLANATION

QTg	Quaternary and Tertiary intrusive rocks
Czv	Late Cenozoic Wrangell Volcanics
Czu	Late Cenozoic deep marine sedimentary rocks
Czs	Cenozoic marine sedimentary rocks
Tg	Tertiary granitic rocks
To	Paleocene and Eocene Orca Group
Ty	Neogene Yakataga Fm.
Tl	Paleogene sedimentary rocks
Ky	Cretaceous Yakutat Group
Kvv	Cretaceous Valdez Group volcanic rocks
Kvs	Cretaceous Valdez Group flysch
KJm	Cretaceous and Jurassic melange and blueschist
Mz	Cretaceous metaflysch and metamelange
Jg	Jurassic granitic rocks
Pz	Paleozoic rocks, undifferentiated
Pzg	Paleozoic plutonic rocks
PzMzu	Paleozoic and Mesozoic schistose rocks
PzMz	Paleozoic and Mesozoic sedimentary and volcanic rocks
⊗	Movement away from reader
⊙	Movement toward reader
R	Refraction line

structurally bound terranes at least since the end of the Paleozoic (Plafker, 1990; Plafker, 1990; Plafker and Berg, this volume, Chapter 33; Nokleberg and others, this volume, Chapter 10).

Paleomagnetic inclination anomalies indicate that the Peninsular, Wrangellia, and Alexander terranes were about 30° south of their present latitude relative to cratonic North America during the Late Triassic (Haeussler and others, 1989; P. W. Plumley, 1990, written commun.; Hillhouse and Coe, this volume). In addition, the Alexander terrane may have undergone about 15° of northward movement relative to North America since the Pennsylvanian, assuming an origin in the Northern Hemisphere (Van der Voo and others, 1980). The composite terrane was emplaced against the North America continental margin by mid-Cretaceous time (Csejtey and others, 1982; Gehrels and Berg, this volume; Nokleberg and others, this volume, Chapter 10) and possibly as early as the Middle Jurassic (McClelland and others, 1992); it was at about its present position between Cenomanian and Eocene time (Hillhouse and Coe, this volume).

Magmatic arcs

The history of arc magmatism in the Wrangellia composite terrane and regions to the north of it is relevant for interpretation of the Southern Margin composite terrane accretionary complex because the magmatic arcs reflect periods of relative convergence during which rocks may have been subducted, accreted, or offscraped, and because the arcs provided much of the sediment that makes up the accretionary complex.

Rocks interpreted as arc related extend from just north of the Border Ranges fault into interior Alaska and the intermontane region of Canada (Armstrong, 1988; Dodds and Campbell, 1988; Barker, this volume; Barker and others, this volume; Brew, this volume; Miller, this volume; Miller and Richter, this volume; Moll-Stalcup, this volume; Plafker and Berg, this volume, Chapter 33). These arcs appear to have been built on the Wrangellia composite terrane (Fig. 4) prior to about mid-Cretaceous time. After emplacement of the Wrangellia com-

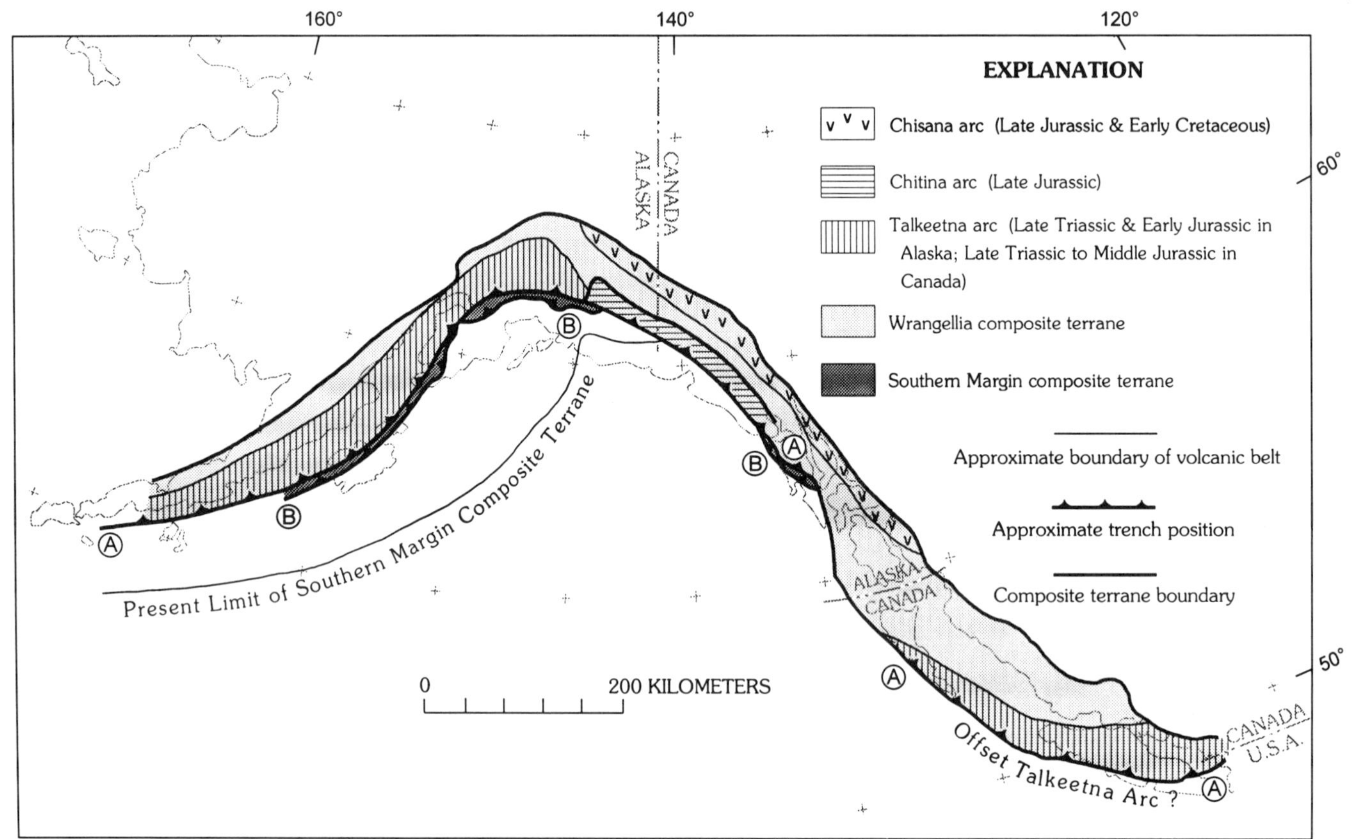

Figure 4. Relation of the Southern Margin composite terrane to Late Triassic to mid-Cretaceous magmatic arcs in the Wrangellia composite terrane. Circled letters indicate approximate positions of the associated trench at the beginning of Talkeetna arc activity (A) and at the end of Chisana arc activity (B). Trench A corresponds with present position of Border Ranges fault and trench B is at the present position of the Uganic, Chugach Bay, Eagle River, Tazlina, and related faults.

posite terrane along the southern Alaska margin, the Late Cretaceous and Paleogene arcs were built on the Wrangellia composite terrane and on terranes farther inboard in Alaska and Canada (Fig. 5). The Cenozoic Aleutian-Wrangell magmatic arc is entirely within the Wrangellia composite terrane in the Alaska Peninsula and Wrangell Mountains, and it is intraoceanic west of the Alaska Peninsula in the Aleutian Islands (Miller and Richter, this volume; Vallier and others, this volume).

Three Mesozoic arc-related magmatic sequences occur within the Wrangellia composite terrane (Fig. 4). They are (1) the latest Triassic and Early Jurassic Talkeetna arc of the Peninsular terrane and possibly the largely coeval latest Triassic to earliest Middle Jurassic arc rocks of the Bonanza Group in the British Columbia part of the Wrangellia terrane; (2) a belt of calc-alkalic plutonic rocks of Middle and Late Jurassic age (Tonsina-Chichagof belt of Hudson, 1983), which includes the Late Jurassic Chitina arc in the Wrangellia and Alexander terranes; and (3) the Late Jurassic and Early Cretaceous Chisana arc (Gravina-Nutzotin belt of Berg and others, 1972).

Three major known and inferred post–mid-Cretaceous arc sequences occur within and north of the Wrangellia composite terrane (Fig. 5). (1) The latest Cretaceous and Paleogene (mainly 77 to about 55 Ma, but locally may be as old as 93 Ma and as young as 45 Ma) Kluane arc consists of variably metamorphosed intermediate-composition plutonic rocks along and north of the northeastern margin of the Wrangellia composite terrane in the Coast Mountains of southeastern Alaska and British Columbia (Brew, this volume; Armstrong, 1988) and correlative plutonic and volcanic rocks within adjacent parts of British Columbia and Yukon Territory northeast of the Coast Mountains (Armstrong, 1988), and a broad belt of volcanic and plutonic rocks in western, central, and east-central Alaska that includes much or all of the Yukon-Tanana, Yukon-Kanuti, and Kuskokwim Mountains–Talkeetna Mountain belts of Moll-Stalcup (this volume). (2) Middle Tertiary (43 to 33 Ma) magmatism associated with the Aleutian arc forms a well-defined belt in Alaska that trends along the Alaska Peninsula and extends into central Alaska (Moll-Stalcup, this volume). (3) Neogene (<30 Ma) magmatism of the Aleutian-Wrangell magmatic arc coincides approximately with the middle Tertiary arc except for an extension eastward into the Wrangell Mountains and adjacent parts of Canada (Miller and Richter, this volume).

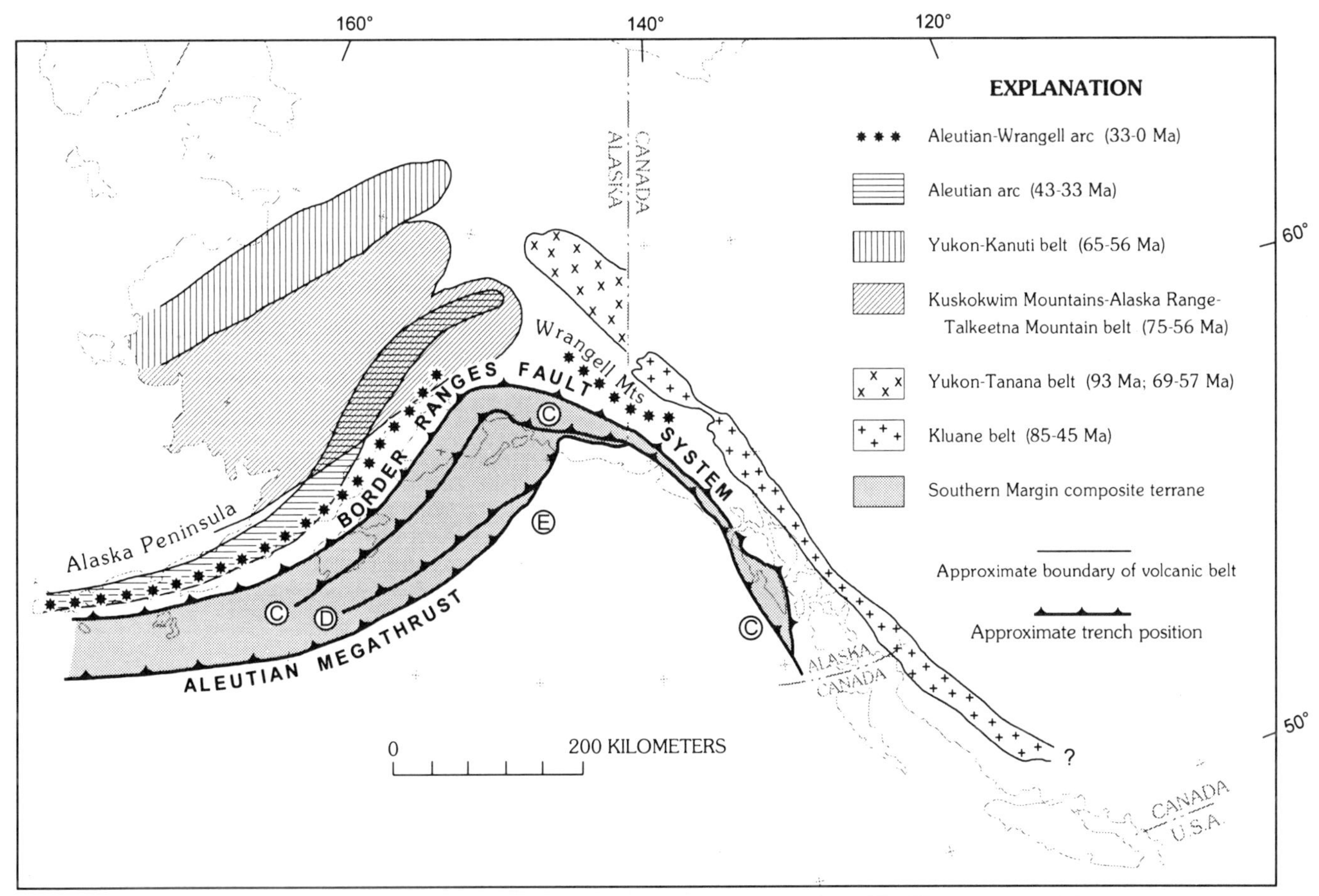

Figure 5. Relations of the Southern Margin composite terrane to Late Cretaceous and Cenozoic magmatic arcs (and belts) in Alaska. Circled letters indicate approximate positions of the associated trench at the end of the Cretaceous (C), during the middle Eocene and Oligocene(?) (D), and at present (E). Trench C corresponds with present position of the Contact fault, trench D corresponds with Chugach–Saint-Elias and Pamplona zone fault systems, and trench E corresponds with Aleutian megathrust. Magmatic belts after Moll-Stalcup (this volume) and Brew (this volume).

A mid-Cretaceous thermal event resulted in widespread plutonism and metamorphism throughout the Wrangellia composite terrane and in much of Alaska as far north as the southern Brooks Range (Miller, this volume; Dusel-Bacon, this volume). There is no evidence in Alaska that the plutonism was accompanied by arc volcanism in the Wrangellia composite terrane. During this time interval, widespread plutonism accompanied by arc volcanism was recorded in the intermontane region of Canada that extended into the eastern Yukon-Tanana belt of Alaska (Armstrong, 1988; Bacon and others, 1990).

Pacific plate

The Pacific plate oceanic crust is progressively younger from west to east along the southern Alaska margin. Its age ranges from chron 5C (about 17 Ma) off Chatham Strait, to chron 15 (about 37 Ma) at the east end of the Aleutian Trench, to chron 24 (about 58 Ma) off the Sanak Islands at the west end of the region (Atwater and Severinghaus, 1989).

The present motion of the Pacific plate relative to the North American plate is northwestward at about 66 mm/yr off the Sanak Islands, and decreases to about 50 mm/yr off southeastern Alaska (Fig. 2) (DeMets and others, 1990). Northwesterly relative motions of the Pacific and Kula plates at moderate rates have prevailed since about 55 Ma (Lonsdale, 1988). This motion is accommodated mainly by orthogonal underthrusting along the Aleutian megathrust system and the Kayak Island and Pamplona zones to the northeast, and by dextral to dextral oblique strike-slip along the Queen Charlotte–Fairweather transform fault system; lesser amounts of slip occur on faults along the south and north boundaries of the Yakutat terrane and within interior Alaska (Lahr and Plafker, 1980; Plafker and others, this volume, Plate 12).

Prior to counterclockwise rotation of what is now western Alaska in latest Cretaceous to earliest Tertiary time, the continental margin probably had a general northwest trend comparable to that of the present continental margins of British Columbia and the conterminous United States (Moore and Connelly, 1979; Plafker, 1987; Plafker and others, 1989; Hillhouse and Coe, this volume). From about 85 to 55 Ma, motion of the Kula plate relative to the continent was north-northeast at high rates (120 to 210 mm/yr), resulting in dextral transpression along the continental margin. From about 180 to 85 Ma, motion of the Farallon plate was probably east to northeast at moderate to high rates (70 to 120 mm/yr), and there was consequent orthogonal convergence or sinistral transpression along the continental margin (Engebretsen and others, 1985; Lonsdale, 1988).

SOUTHERN MARGIN COMPOSITE TERRANE

The minimum total onshore and offshore area of the Southern Margin composite terrane includes almost equal areas of Mesozoic and Cenozoic rocks covering close to 270,000 km^2 (Table 1). In general, the older Mesozoic assemblages of the Chugach terrane differ significantly from each other in age, lithology, and mode and time of emplacement, whereas differences between units in the Late Cretaceous part of the Chugach terrane and in the Paleogene Prince William terrane and related units are more subtle. These differences largely reflect the relative age spans for these rocks: greater than 100 m.y. for the Mesozoic units versus 20 m.y. for the Cenozoic units of the accretionary prism that are exposed at the surface. The more subtle differences in the younger, more voluminous sequences probably also reflect their proximity to active continental margin arcs.

TABLE 1. MESOZOIC AND PALEOGENE ACCRETIONARY UNITS OF THE SOUTHERN MARGIN COMPOSITE TERRANE

Terrane	Units	Area (km^2)*	Volume (km^3)†
	Ghost Rocks Formation	6100	61,000
Prince William	Orca Group (Includes Sitkalidak Formation and Resurrection Peninsula sequence)	139,300	1,393,000
	Subtotal Cenozoic	145,400	1,454,000
Chugach	Glaucophanic Greenschist Assemblage (Includes schists of Raspberry Strait, Seldovia, Iceberg Lake, Liberty Creek, and Pinta Bay)	1600	24,000
	Melange Assemblage (Includes Uyak Complex, Kachemak Bay sequence, McHugh Complex, Yakutat Group melange, and Kelp Bay Group)	13,200	198,000
	Flysch and Basalt Assemblage (Includes Shumagin Formation, Kodiak Formation, Valdez Group and its metamorphic equivalents in the Chugach Mountains, the metamorphic complex of the Saint Elias area, Sitka Graywacke, and possibly the informally named Ford Arm Formation)	108,600	1,629,000
	Subtotal Mesozoic	123,400	1,851,000
Total		268,800	3,305,000

*Includes inferred offshore areas.
†Average thicknesses based on geophysical data are 15 km for Mesozoic and 10 km for Cenozoic units. Does not include eroded material, much of which may be recycled into younger parts of the accretionary sequence.

Chugach terrane

Accreted rocks of Mesozoic age make up the Chugach terrane (Berg and others, 1972; Plafker and others, 1977). The terrane is 60 to 100 km wide and extends 2100 km along the Gulf of Alaska margin from the Sanak Islands to Chatham Strait (Fig. 2). The total onshore and offshore area underlain by these rocks is about 124,000 km^2. The Chugach terrane is composed of three major fault-bounded assemblages that are progressively younger from north to south and constitute about 1%, 10%, and 89%, respectively, of the total terrane area. These are (1) local coherent slabs of Late Triassic(?) and Jurassic(?) greenschist with local blueschist facies minerals closely associated with melange along the north margin of the terrane, (2) a discontinuous landward belt of Late Triassic to Early Cretaceous melange, and (3) a southern continuous belt of Late Cretaceous volcaniclastic flysch and oceanic basaltic rocks that make up most of the terrane. These three major assemblages of the Chugach terrane are referred to informally in this chapter as the glaucophanic greenschist, melange, and flysch and basalt assemblages. As thus defined, the glaucophanic greenschist assemblage includes the schist of Raspberry Strait of Roeske (1986) on the Kodiak Islands, the melange assemblage includes the Kachemak terrane of Jones and others (1987) on the southern Kenai Peninsula, and the flysch and basalt assemblage includes the Saint Elias terrane of Jones and others (1987) in the Saint Elias Mountains. Assemblages of Mesozoic melange, flysch, and mafic volcanic rocks that underlie the eastern part of the Yakutat terrane are inferred to be displaced fragments of the Chugach terrane (Plafker, 1987). The distribution of the Mesozoic accretionary assemblages is shown in Figure 6, and their lithologies, ages, and correlations are summarized in Figure 7.

The Chugach terrane is bounded by major faults, both landward and seaward throughout its extent, and its three constituent assemblages also are fault bounded (Fig. 6). Along its north boundary it is faulted against and beneath the Wrangellia composite terrane. Along the south margin younger Paleogene accreted sequences of the Ghost Rocks Formation and Prince

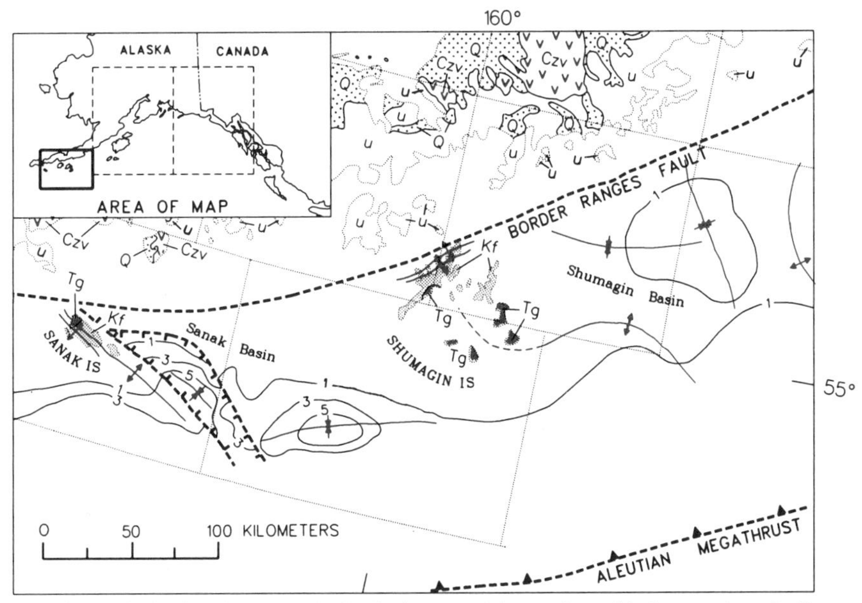

Figure 6 (on this and following three pages). Geologic map of the southern Alaska margin and adjacent regions. Geology modified from Beikman (this volume), Plafker (1987), Plafker and others (1989), and Kirschner (this volume, Plate 7).

William terrane are underthrust along the Contact fault system. In the eastern part of the Yakutat terrane (east of long 141°W), it is juxtaposed along the Fairweather dextral strike-slip fault against rocks inferred to be displaced fragments of the Chugach terrane. South of the Chatham Strait fault in southeastern Alaska, the Chugach terrane is in contact with probable oceanic crust of the Pacific plate along the Queen Charlotte dextral strike-slip fault.

We recognize that the flysch and basalt, melange, and glaucophanic greenschist assemblages could be interpreted either as subterranes of the Chugach terrane or as separate terranes. However, we prefer to retain the name "Chugach terrane" because of its extensive usage in the literature. In doing so, we have extended the original definition by Berg and others (1972) to include all Mesozoic accreted deep-sea rocks south of the Border Ranges fault. We have not attempted to define subterranes because uncertainties in dating and correlation, as well as structural complexity and metamorphism, preclude precise definition and delineation of the constituent units or of their time of accretion.

Glaucophanic greenschist assemblage. Blueschist-bearing metamorphic rocks have been recognized along and near the Border Ranges fault system from the Kodiak Islands on the west to Chatham Strait in southeastern Alaska on the east, and they underlie a total area of about 1500 km^2. Units large enough to show in Figure 6 are (1) the schist of Raspberry Strait, which comprises a discontinuous belt 2 to 8 km wide in the northern part of the Kodiak Islands (Roeske, 1986); (2) the schist of Seldovia, exposed in a small area near Seldovia in the southern Kenai Mountains (Carden and others, 1977; Seldovia schist terrane of Cowan and Boss, 1978); (3) the schist of Iceberg Lake, which occurs as a coherent tectonic slice 40 km long and up to 4 km wide as well as numerous smaller klippen that are surrounded by melange facies rocks in the western Valdez quadrangle of the northern Chugach Mountains (Winkler and others, 1981); and (4) the schist of Liberty Creek, which crops out over an area 28 km long and up to 13 km wide in the eastern Valdez quadrangle of the northern Chugach Mountains (Winkler and others, 1981; Plafker and others, 1989). The variably metamorphosed schist of Pinta Bay (Decker, 1980a) in the Chichagof Island area includes irregular fault slices containing blueschist minerals in a discontinuous belt 30 km long and 4 to 5 km wide. The presence of abundant blocks of glaucophanic greenschist in glacial moraines of the Yakutat Bay area in the Saint Elias Mountains suggests that blueschist facies rocks are present within the Chugach terrane in the rugged Saint Elias Mountains. Details of the metamorphism of the glaucophanic greenschist and other metamorphic rocks of the Chugach terrane are given by Dusel-Bacon (this volume).

Lithology. The rocks of the glaucophanic greenschist assemblage (Fig. 7) are chiefly greenschist and blueschist, but they include muscovite and actinolite schist, siliceous schist, metachert, and graphitic schist. The protolith was dominantly basaltic pillow flows, tuffs, tuff breccias, and volcaniclastic rocks with minor chert, carbonate, and argillaceous rocks. At Seldovia the schist occurs as isolated tectonic blocks of quartz-sericite schist, greenschist, and crossite-epidote blueschist immersed in a poorly exposed, completely sheared matrix of argillite (Cowan and Boss, 1978). A single large body of marble occurs with the schist of Seldovia (Forbes and Lanphere, 1973), and foliated calcareous rocks occur in the schist of Iceberg Lake (Winkler and others, 1981). Pillow structures are locally well preserved in the schist of Iceberg Lake, and faint primary structures suggestive of breccia and possible pillow breccia occur in the schist of Liberty Creek (Plafker and others, 1989).

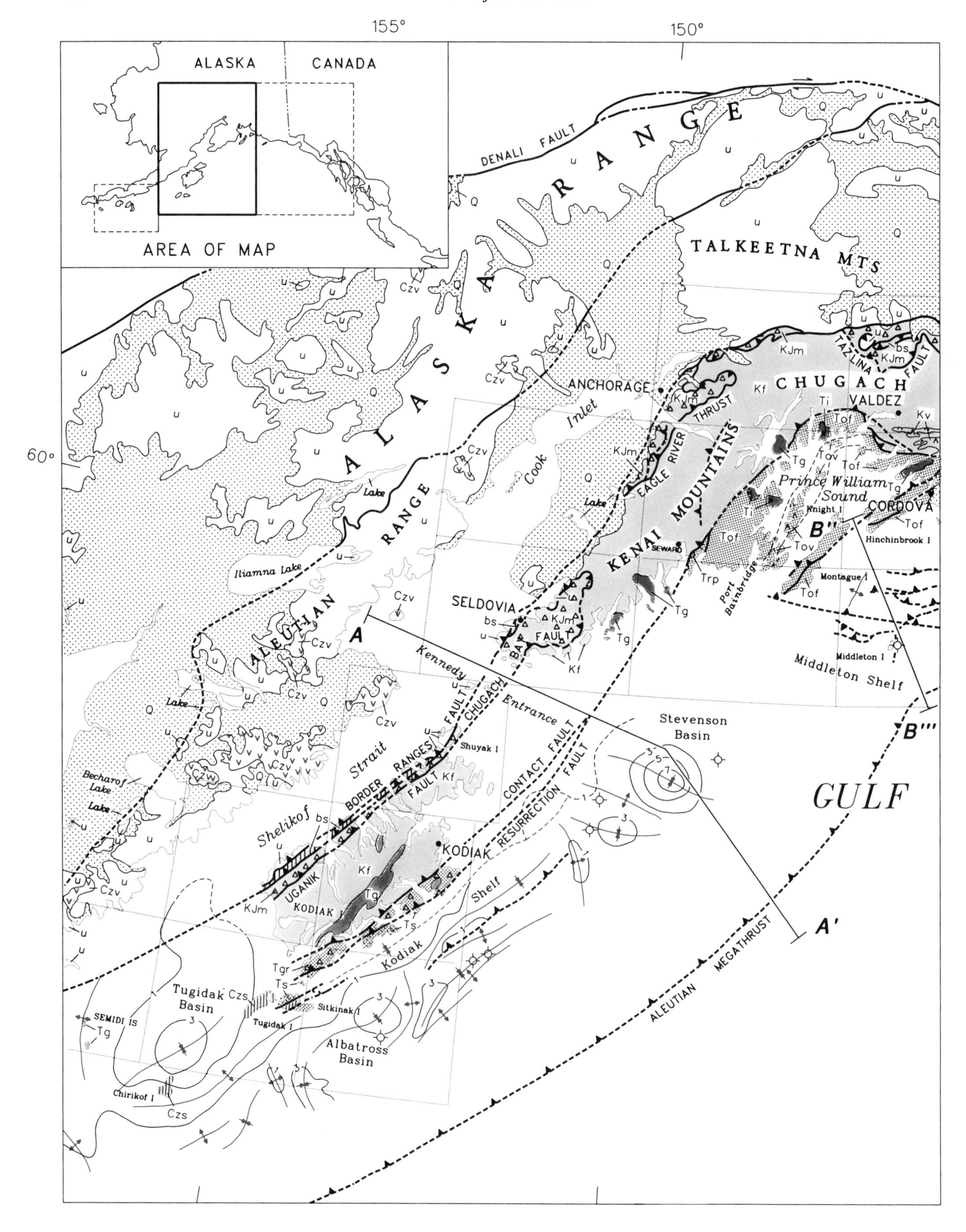
155°
150°
60°
ALASKA
CANADA
AREA OF MAP
DENALI FAULT
ALASKA RANGE
TALKEETNA MTS
ANCHORAGE
Cook Inlet
CHUGACH
TAZLINA FAULT
VALDEZ
THRUST
EAGLE RIVER
KENAI MOUNTAINS
Prince William Sound
Knight I
CORDOVA
Hinchinbrook I
SEWARD
Port Bainbridge
Montague I
Middleton I
Middleton Shelf
ALEUTIAN RANGE
Iliamna Lake
SELDOVIA
BAY FAULT
Kennedy Entrance
CHUGACH FAULT
BORDER RANGES FAULT
Shuyak I
Shelikof Strait
CONTACT FAULT
RESURRECTION FAULT
Stevenson Basin
GULF
Becharof Lake
KODIAK
UGANIK
KODIAK I
Kodiak Shelf
ALEUTIAN MEGATHRUST
Tugidak Basin
SEMIDI IS
Tugidak I
Sitkinak I
Albatross Basin
Chirikof I
A
A'
B''
B'''

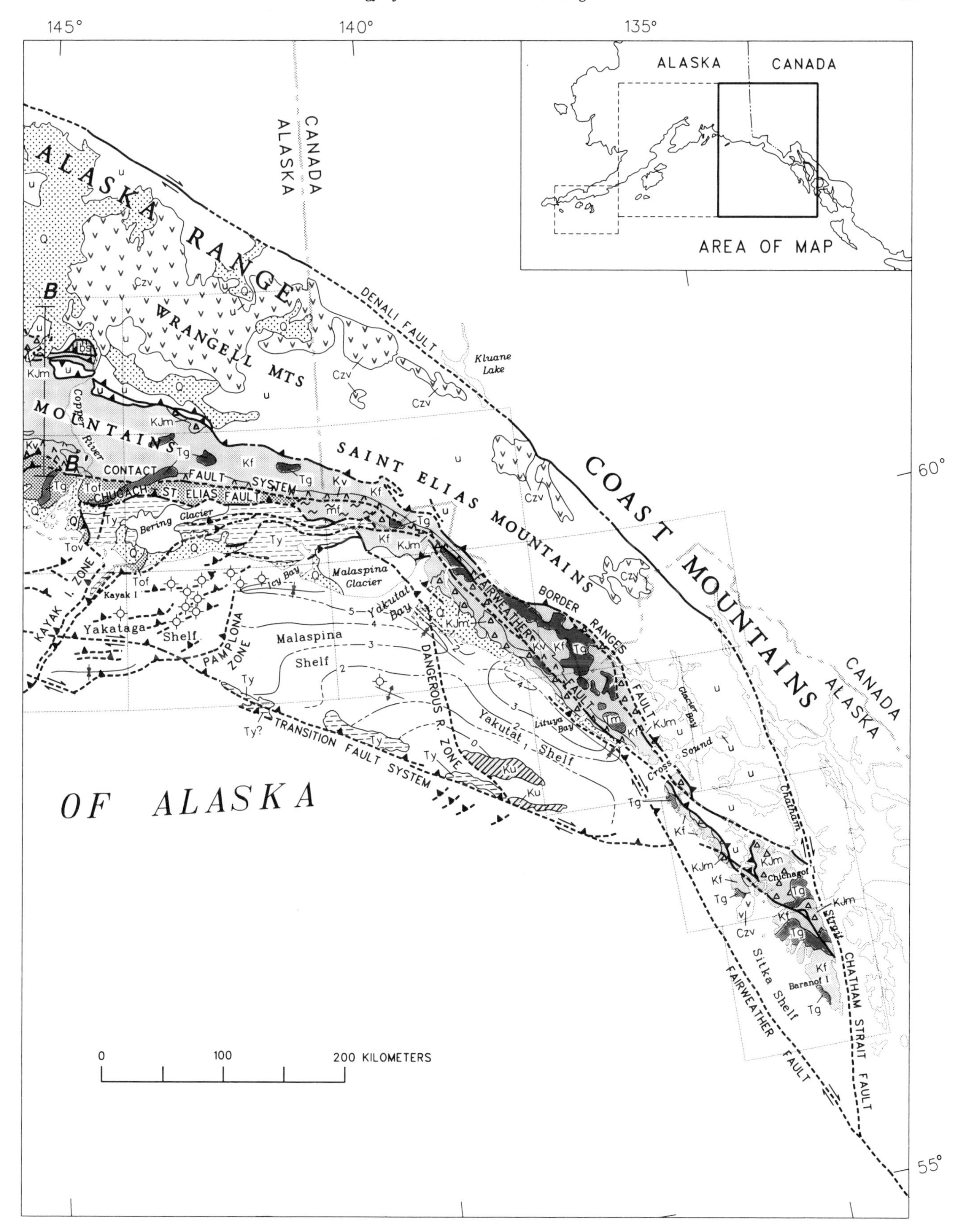
ALASKA
CANADA
AREA OF MAP
ALASKA RANGE
WRANGELL MTS
DENALI FAULT
Kluane Lake
COAST MOUNTAINS
SAINT ELIAS MOUNTAINS
CONTACT FAULT SYSTEM
CHUGACH-ST. ELIAS FAULT
Bering Glacier
Malaspina Glacier
Icy Bay
Kayak I
KAYAK I. ZONE
Yakataga Shelf
PAMPLONA ZONE
Malaspina Shelf
Yakutat Bay
DANGEROUS R. ZONE
FAIRWEATHER FAULT
BORDER RANGES FAULT
Yakutat Shelf
Lituya Bay
Glacier Bay
Cross Sound
Chatham Strait
Chichagof
Baranof I
Sitka Shelf
FAIRWEATHER FAULT
CHATHAM STRAIT FAULT
TRANSITION FAULT SYSTEM
OF ALASKA
0 100 200 KILOMETERS
145°
140°
135°
60°
55°

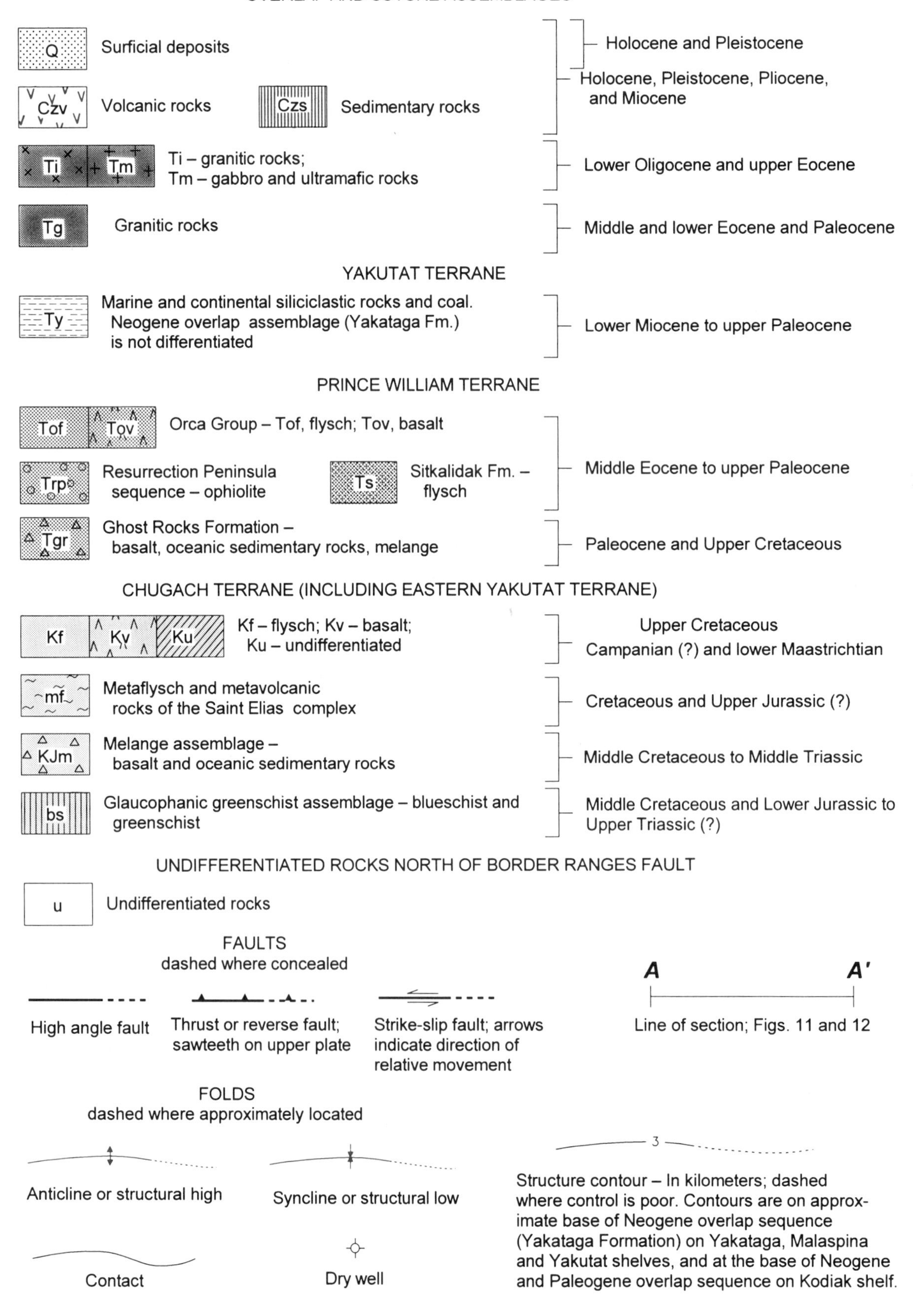
EXPLANATION
OVERLAP AND SUTURE ASSEMBLAGES
Q
Surficial deposits
Holocene and Pleistocene
Czv
Volcanic rocks
Czs
Sedimentary rocks
Holocene, Pleistocene, Pliocene, and Miocene
Ti
Tm
Ti – granitic rocks;
Tm – gabbro and ultramafic rocks
Lower Oligocene and upper Eocene
Tg
Granitic rocks
Middle and lower Eocene and Paleocene
YAKUTAT TERRANE
Ty
Marine and continental siliciclastic rocks and coal. Neogene overlap assemblage (Yakataga Fm.) is not differentiated
Lower Miocene to upper Paleocene
PRINCE WILLIAM TERRANE
Tof
Tov
Orca Group – Tof, flysch; Tov, basalt
Trp
Resurrection Peninsula sequence – ophiolite
Ts
Sitkalidak Fm. – flysch
Middle Eocene to upper Paleocene
Tgr
Ghost Rocks Formation – basalt, oceanic sedimentary rocks, melange
Paleocene and Upper Cretaceous
CHUGACH TERRANE (INCLUDING EASTERN YAKUTAT TERRANE)
Kf
Kv
Ku
Kf – flysch; Kv – basalt; Ku – undifferentiated
Upper Cretaceous
Campanian (?) and lower Maastrichtian
mf
Metaflysch and metavolcanic rocks of the Saint Elias complex
Cretaceous and Upper Jurassic (?)
KJm
Melange assemblage – basalt and oceanic sedimentary rocks
Middle Cretaceous to Middle Triassic
bs
Glaucophanic greenschist assemblage – blueschist and greenschist
Middle Cretaceous and Lower Jurassic to Upper Triassic (?)
UNDIFFERENTIATED ROCKS NORTH OF BORDER RANGES FAULT
u
Undifferentiated rocks
FAULTS
dashed where concealed
High angle fault
Thrust or reverse fault; sawteeth on upper plate
Strike-slip fault; arrows indicate direction of relative movement
A
A'
Line of section; Figs. 11 and 12
FOLDS
dashed where approximately located
Anticline or structural high
Syncline or structural low
3
Structure contour – In kilometers; dashed where control is poor. Contours are on approximate base of Neogene overlap sequence (Yakataga Formation) on Yakataga, Malaspina and Yakutat shelves, and at the base of Neogene and Paleogene overlap sequence on Kodiak shelf.
Contact
Dry well

Geochemically analyzed metabasalts from the schists of Liberty Creek and Iceberg Lake have variable major element composition, but stable minor element and rare earth element (REE) compositions that closely match average normal mid-ocean ridge basalt (N-type MORB) at about 7 to 19 times chondrites with slight depletion of light REEs (LREE) over heavy REEs (HREE) (Plafker and others, 1989).

Metamorphic grade is transitional greenschist-blueschist facies in most areas. However, it ranges from prehnite-pumpellyite and lawsonite-albite through lower greenschist and blueschist, to transitional blueschist-epidote-amphibolite facies. The characteristic blue amphibole in the rocks is crossite; glaucophane has been identified only in the schist of Raspberry Strait (Roeske, 1986). Lawsonite occurs in the schist of Raspberry Strait (Roeske, 1986) and rarely in the schists of Iceberg Lake and Liberty Creek (Winkler and others, 1981). Peak metamorphism for the schist of Raspberry Strait was 4.5 to 8 kbar and 350 to 500 °C, and for the schists of Iceberg Lake and Liberty Creek it was 6.2 ± 2 kbar and 350 to 375 °C (Roeske, 1986; Sisson and Onstott, 1986).

Structure. The schist is deformed into chevron and isoclinal folds that commonly verge south. All rocks are pervasively sheared and locally mylonitic. The schist of Raspberry Strait has undergone at least three regional deformations: one or two ductile deformations associated with the high-pressure metamorphism; a brittle crenulation; and a late brittle-shearing event that locally juxtaposes rocks of differing metamorphic facies (Roeske, 1986). This latest deformation formed a pervasive system of anastomosing faults ranging in width from a few centimeters to >100 m. In the schist of Liberty Creek, at least two generations of regional ductile deformations include an earlier south-verging set of folds that is overprinted by younger, north-verging folds (Fig. 8; Nokleberg and others, 1989) and a later brittle-shearing event of mid-Cretaceous or younger age.

Age. Isotopic dating of metamorphic minerals in the glaucophanic greenschist assemblage indicates Late Triassic to mid-Cretaceous crystallization ages (Fig. 7); crystallization presumably occurred during deep subduction of the rocks. The data indicate at least two episodes of subduction-related transitional blueschist facies metamorphism along the north margin of the Chugach terrane. The early episode is Early to Middle(?) Jurassic in age in the Kodiak Islands, southern Kenai Mountains, and Chugach Mountains. The younger episode in southeastern Alaska is considered by Decker and others (1980) to be mid-Cretaceous (91 to 106 Ma), on the basis of ^{40}Ar ages on white mica and actinolite in rocks associated with the blueschist minerals. The possibility of postemplacement resetting of these mineral ages by shearing and/or extensive Tertiary plutonism cannot be discounted.

The protolith age of the schist is conjectural. Its position in the accretionary complex indicates only that it is probably as old as, or older than, the less metamorphosed Late Triassic to mid-Cretaceous McHugh Complex, which borders the schist on the south. In the western Valdez quadrangle (Winkler and others, 1981) and in the Seldovia area (Forbes and Lanphere, 1973; Carden and others, 1977; Jones and others, 1987), glaucophanic greenschist rocks having Early to Middle Jurassic metamorphic ages occur in close proximity to Late Triassic oceanic rocks, which include paleontologically dated radiolarian chert. This association suggests that the glaucophanic greenschist in those areas may be a more deeply subducted part of the Late Triassic oceanic assemblage (Plafker and others, 1989; Bradley and Kusky, 1992). Emplacement age within the accretionary complex is considered to be close to the recrystallization age of the schist, or Early Jurassic for all but the schist of Pinta Bay in southeastern Alaska, which may have been emplaced in the mid-Cretaceous. Glaucophanic blueschist that occurs as erratic blocks in the Yakutat terrane is undated.

Interpretation. The combination of lithology of the sequence and the geochemistry and relict textures of the metavolcanic rocks suggests that the glaucophanic greenschist is an oceanic assemblage of basalt and associated marine sedimentary rocks. Metamorphism to glaucophanic greenschist indicates underthrusting to a depth of 12 to 25 km beneath a convergent margin followed by rapid uplift and retrograde metamorphism. Metamorphic ages indicate subduction during the Early Jurassic in the Kodiak Islands and the Chugach Mountains and during the mid-Cretaceous in southeastern Alaska. Because convergence and deep underthrusting probably occurred during much of the Mesozoic Era (see section on tectonic evolution), blueschist facies metamorphic rocks with ages other than Early to Middle(?) Jurassic and mid-Cretaceous will probably be discovered within the Chugach terrane.

Glaucophanic greenschist and the Talkeetna magmatic arc. Glaucophanic greenschist along and near the Border Ranges fault system occurs in close proximity to almost coeval Late Triassic to early Middle Jurassic volcanic and plutonic rocks along the south margin of the Peninsular terrane, including the informally named Border Ranges ultramafic and mafic complex of Burns (1985), the Kodiak-Kenai plutonic belt of Hudson (1983), and the Talkeetna Formation. Plafker and others (1989) interpreted this relation to indicate that both the glaucophanic blueschist of the Chugach terrane and the magmatic rocks of the Peninsular terrane are related to the Talkeetna arc (Fig. 4) and that together they record a Late Triassic, Early Jurassic, and possibly early Middle Jurassic history of northward to eastward subduction relative to the Wrangellia composite terrane. The close proximity of the high-pressure–low-temperature schists to high-temperature plutonic rocks suggests major post–Early Jurassic structural disruption of the Talkeetna arc margin, most probably by tectonic shortening and subsequent modification by extension (Plafker and others, 1989; Fuis and Plafker, 1991).

Melange assemblage. The melange facies of the Chugach terrane occurs in an area of about 13,000 km^2, mainly in a discontinuous belt 0 to 20 km wide along the north margin of the Chugach terrane and locally as slices that are structurally interleaved with the blueschist and flysch facies (Fig. 6). The melange facies includes (1) the Uyak Complex and terrane of Cape Current in the Kodiak Islands, (2) the McHugh Complex in the

Kenai and Chugach mountains, and (3) the Kelp Bay Group on Chichagof Island and rocks that extend northwestward through Cross Sound to Glacier Bay to the Tarr Inlet area along the northern Saint Elias Mountains (Fig. 7). The melange unit of the Yakutat Group of the eastern Yakutat terrane is inferred to correlate with the melange of the Chugach terrane and is discussed in this section.

On its north margin, the melange assemblage is faulted against rocks of the glaucophanic blueschist assemblage along a strand of the Border Ranges fault system in the Kodiak Islands (Roeske, 1986), along the Port Graham fault in the southern Kenai Mountains (Kelley, 1985), and along the near-vertical Second Lake fault zone in the Chugach Mountains (Fig. 8). Elsewhere in the Kenai and Chugach mountains, the melange is

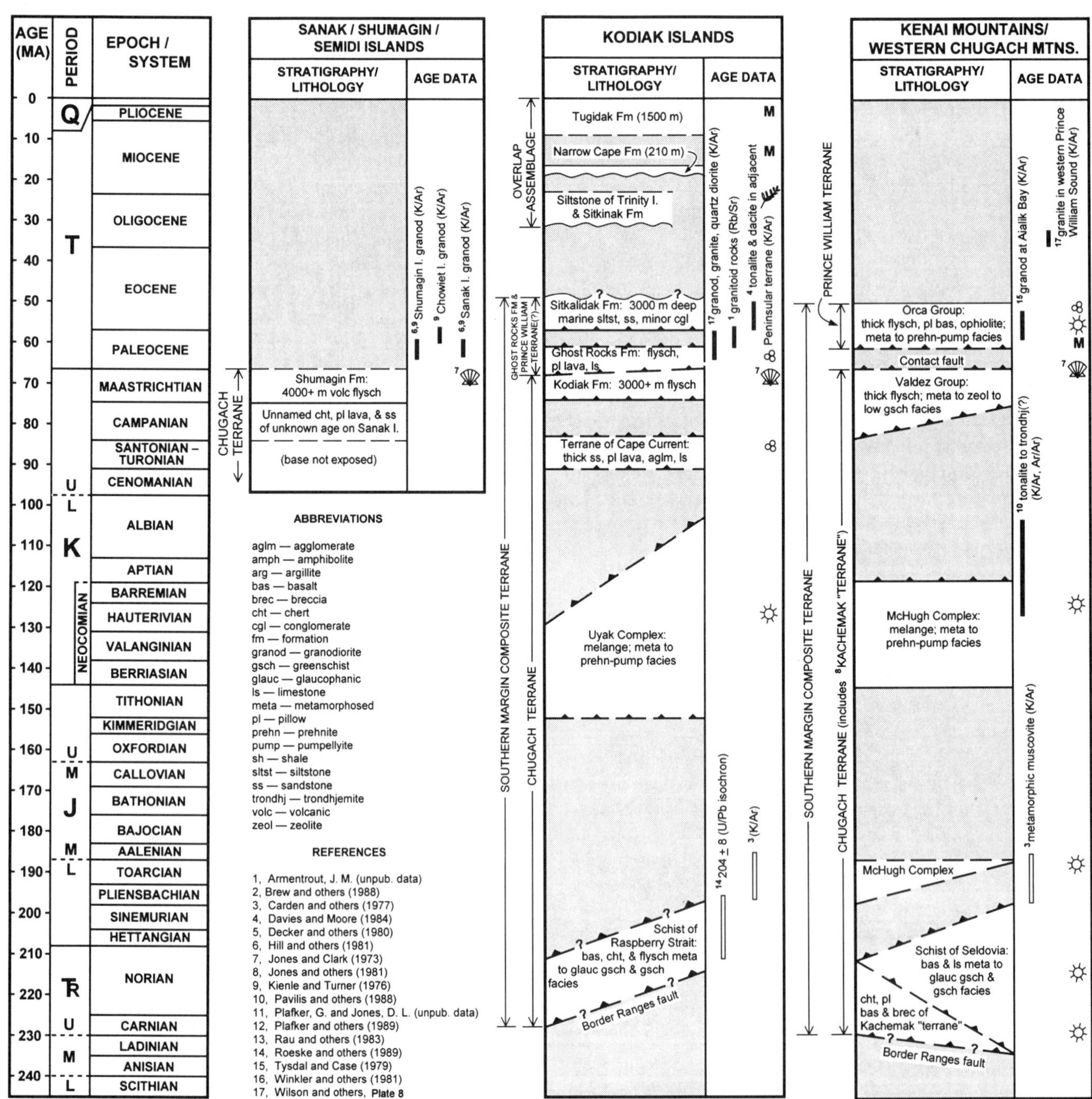

Figure 7. Representative columnar sections for selected areas of the Southern Margin composite terrane and Yakutat terrane. Selected data sources keyed by numbers to references; see text for unit descriptions. More detailed stratigraphic columns for Cenozoic basinal strata and Cenozoic strata of Yakutat terrane are shown in Figures 10 and 13, respectively.

juxtaposed against rocks of the Peninsular or Wrangellia terranes along the Border Ranges fault system. These faults are commonly marked by lenses of serpentinized ultramafic rocks and broad shear zones. In southeastern Alaska and in the Yakutat terrane, contact relations are commonly obscured by pervasive postemplacement disruption along strike-slip faults. In those areas, the melange may be intercalated with rocks of the glaucophanic schist and turbidite facies or they may be juxtaposed against the Wrangellia and Alexander terranes along the Border Ranges fault system (Plafker and Campbell, 1979; Decker and Plafker, 1982; Karl and others, 1982; Roeske and others, 1991).

Lithology. The melange typically consists of pervasively disrupted and variably metamorphosed broken formations that originally consisted mainly of dark greenish-black tholeiitic pillowed basalt and related fragmental volcanic rocks, subordinate amounts of black argillite, green tuff, tuffaceous argillite, and siliceous argillite, and minor lenticular bodies of radiolarian ribbon chert and gray carbonate rocks. Characteristic lithologic associations include (1) massive to weakly foliated green calcareous and glassy-appearing tuff and metavolcanic rocks in crudely discontinuous layers from a few millimeters to several tens of meters thick, together with volcaniclastic graywacke and rare dark-gray

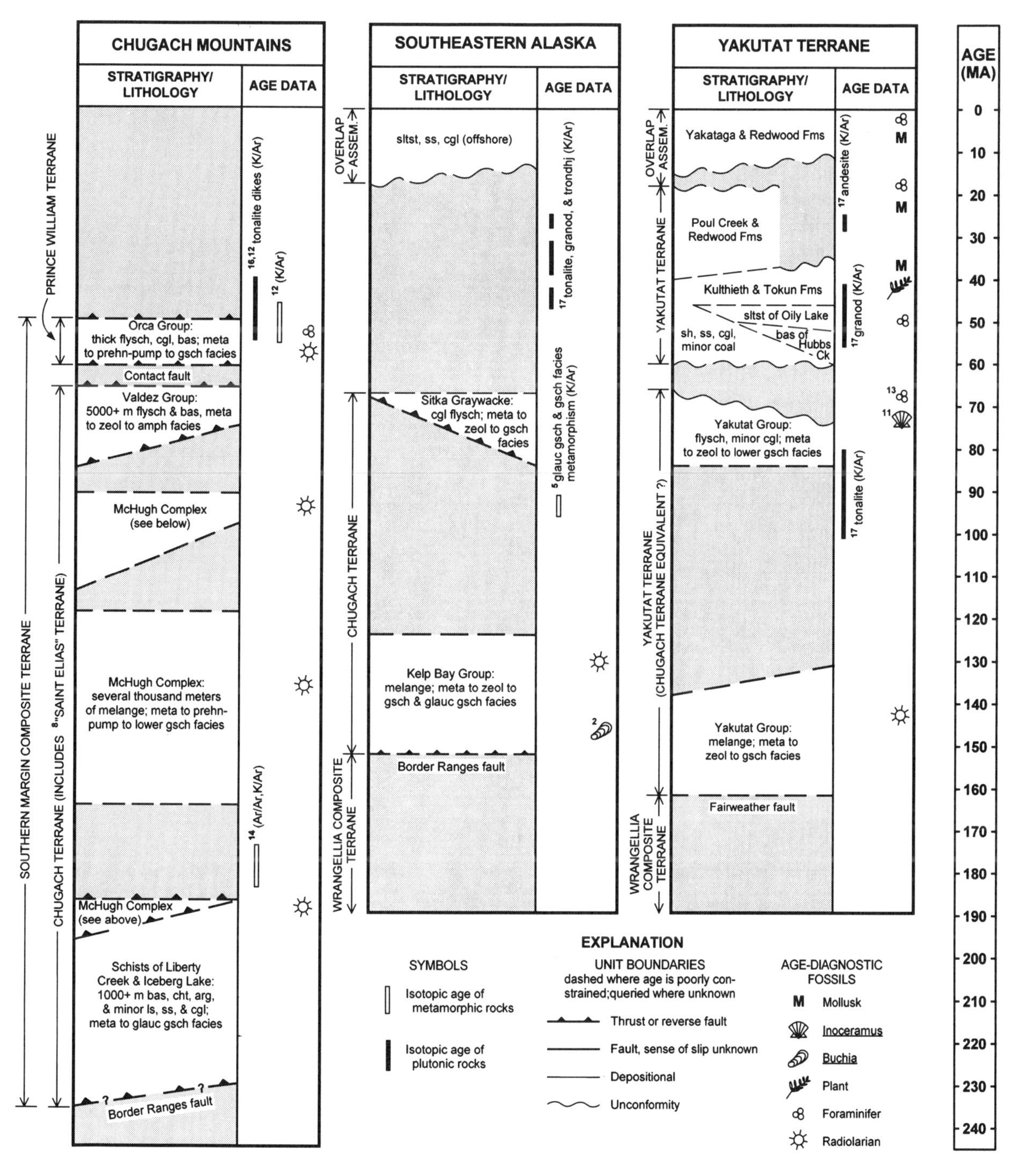

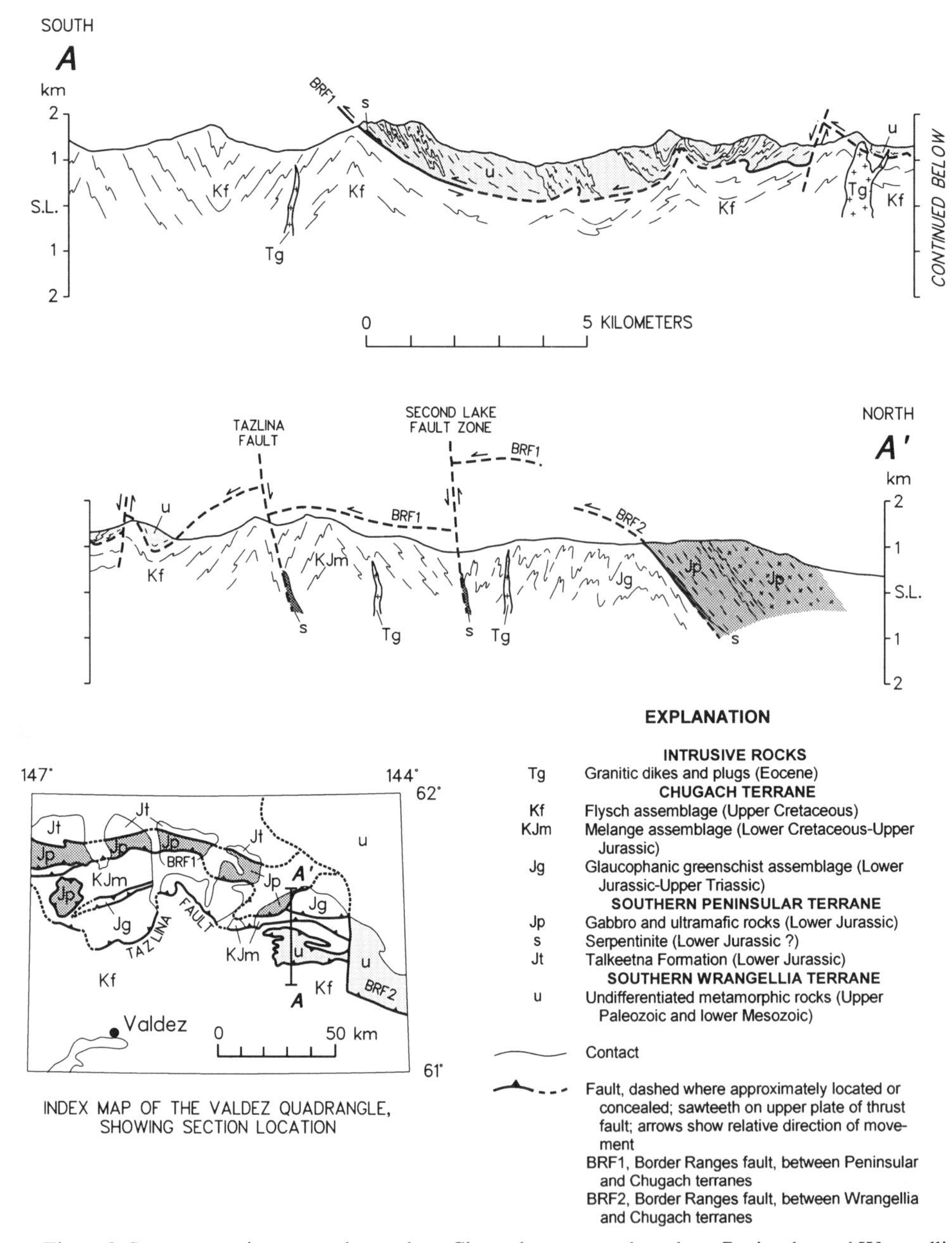

Figure 8. Structure section across the northern Chugach terrane and southern Peninsular and Wrangellia terranes along the Trans-Alaska Crustal Transect (modified from Plafker and others, 1989).

argillite; (2) thinly interlaminated green metatuff and recrystallized dark gray argillite with light green radiolarian chert; (3) greenish-gray volcaniclastic graywacke, with rare mudstone rip-ups, and lenses of pebble-cobble conglomerate, pebbly mudstone, and green chert; (4) chaotically intermixed green tuff and dark gray to black argillite that enclose blocks of bedded chert, graywacke, massive green metavolcanic rocks, recrystallized carbonate rocks, argillite, slaty argillite, and phyllitic green metatuff, which is locally interlayered with recrystallized gray phyllitic argillite; and (5) massive, red-weathering, green and greenish-gray metavolcanic rocks enclosing blocks of red ribbon chert, and associated with greenish-gray graywacke.

The ratio of argillite to tuff in the matrix commonly varies from 1:10 to 10:1, but locally may be nearly pure argillite or nearly pure tuff. Contacts among these rock types are rarely depositional, but more commonly have been disrupted by brittle shear fractures subparallel to bedding. Bedding is generally completely disrupted where argillite is most abundant.

The brittle rocks occur as angular phacoids or lenses from millimeters to kilometers in maximum dimension; in thin section, they show cataclastic fabrics overprinted by solution seams. They may be either enclosed in sheared argillite or streaky tuff and tuffaceous argillite, or juxtaposed against other phacoids. Some inclusions display pinch-and-swell structure, necking, and boudinage. All are oriented parallel to the foliation in the argillite matrix. These blocks and slabs include both (1) disrupted brittle basalt, sandstone, and chert beds within the original stratigraphic sequence and (2) exotic olistostromal blocks of variable age and lithology derived mainly from the Wrangellia composite terrane to the north. The proportion of matrix and blocks varies markedly along regional strike of the unit. In the Anchorage area, Clark (1973) delineated both predominantly metaclastic sequences of metamorphosed graywacke, argillite, and conglomerate and predominantly metavolcanic sequences of metamorphosed greenstone, including pillow basalt, radiolarian chert, and argillite.

The intraformational basaltic rocks characteristically occur in masses from a few meters to several hundred meters thick with variable amounts of chaotically intermixed maroon ribbon chert, green tuff, and black argillite. They are thoroughly altered to greenstone but locally retain textures and structures indicating derivation from massive flows, pillow lavas, and pillow breccia. Geochemically, the metavolcanic rocks are basalts with fairly consistent major and minor element compositions; minor variations are probably due to alteration and the presence of fine veinlets of quartz and calcite (Nelson and Blome, 1991; Barker, this volume). The rocks are predominantly N-MORB on the basis of their REE contents and most discrimination plots, but they include some samples from the Yakutat Group melange that may represent ocean-island or plateau basalt types. Metavolcanic rocks of the McHugh Complex are geochemically indistinguishable from those of the glaucophanic blueschist assemblage, except that average REE contents are slightly higher.

Greenish-gray metasandstone and metasiltstone occur in units as thick as a few tens of meters, mainly in association with the argillite and metatuff. Chaotic structures attributed to local soft-sediment deformation are common. Rarely, massive bodies of graywacke more than 1 km thick are surrounded by incipiently to pervasively sheared rocks. The graywacke is fine to medium grained and is massive to weakly schistose; its bedding-plane foliation is defined by flattened lithic grains. The metasiltstone, which may exhibit a spaced cleavage, occurs as laminae in argillite and as couplets graded with metasandstone. The metasandstone is mainly volcanogenic metagraywacke consisting of plagioclase and mafic to intermediate volcanic grains in variable amounts of carbonate matrix. The metasandstone commonly contains metamorphic pumpellyite and prehnite; veins and patches of prehnite, calcite, and quartz are abundant.

In the Kodiak Islands, large fault-bounded slabs of lithofeldspathic sandstone with minor pillow lava, pillow breccia, and pelagic limestone are incorporated along a fault that separates the melange from the flysch and basalt assemblage. These slabs, named the terrane of Cape Current by Connelly and Moore (1979), are discussed here because they are similar to the melange facies in lithology and structural style. The Cape Current rocks, however, differ in that they contain Turonian to Santonian foraminifers that are one or two stages younger than any other dated rocks of the melange assemblage.

Slabs and blocks as large as a few kilometers in maximum dimension occur within the melange; many are of probable or suspected extraformational origin. These include a wide variety of gabbroic, ultramafic, and schistose rocks, Paleozoic and Triassic marble, some pillowed and massive greenstone that is in part Late Triassic, Early Jurassic shallow-water sandstone, and Jurassic granitoid rocks of intermediate composition (Clark, 1973; Connelly, 1978; Decker, 1980a; Winkler and others, 1984; Plafker, 1987). At outcrop scale these slabs and blocks, together with those of intraformational origin, are commonly lensoid in shape and form a macroscopic melange foliation that strikes parallel to the faults that bound the assemblage. Fault-bounded pods and lenses of schistose to massive serpentinite as much as 250 m wide by 1600 m long occur locally in the Yakutat melange both east and west of Yakutat Bay (Plafker, unpublished data) and within the Kelp Bay Group at 28 or more localities on Baranof Island in southeastern Alaska (Loney and others, 1975).

Metamorphic grade of the melange assemblage is primarily prehnite-pumpellyite facies in the region west of the Richardson Highway, indicating that it was subducted to depths of about 10 to 22 km at temperatures below 400 °C (Dusel-Bacon, this volume). The pervasively sheared, partly mylonitic character of the melange assemblage makes it difficult to establish a clear chronology between metamorphism and deformation. In the Uyak Complex, the first-generation folds and associated axial-planar foliation formed before the peak prehnite-pumpellyite metamorphism of the melange (Moore and Wheeler, 1978). In the Chugach and Kenai mountains, prehnite occurs both in deformed patches and in veinlets that cut the foliation in the enclosing rocks, indicating that metamorphism may have been both syntectonic and post-tectonic (Winkler and others, 1984). In the western Valdez quadrangle in the northern Chugach Mountains and in the Yakutat Group in the Saint Elias Mountains, the melange has been overprinted by Paleogene greenschist facies metamorphism (Plafker and others, 1989). In southeastern Alaska the metamorphic grade commonly ranges from greenschist to amphibolite facies near Eocene plutons (Loney and others, 1975).

Structure. Structural style in this commonly massive-appearing unit consists of sparse south-verging structures (Fig. 8) and numerous, closely spaced zones of intense cataclasis. Within the assemblage, broad zones of intense shearing lack any stratal continuity and innumerable shears of unknown offset have juxtaposed contrasting rocks. The penetrative fabric of these chaotic rocks clearly records an overall ductile behavior during deformation. The occurrence of disrupted brittle phacoids at all scales in a foliated and sheared ductile argillite and tuff matrix imparts the characteristic appearance of a blocks-in-argillite melange to the unit.

Detailed fabric analysis of the Uyak Complex in the Kodiak Islands by Moore and Wheeler (1978) indicates that the melange is geometrically coherent, that it shows two generations of folds and related axial-plane foliations formed during emplacement into the hanging wall of a subduction zone, and that the tectonic transport direction is approximately perpendicular to the trend of the outcrop belt (N 38° ± 11° W). In most areas, the outcrop-scale planar fabric defined by slabs and blocks in the melange parallels a penetrative foliation in the rocks. In some areas, such as the northern Chugach Mountains, the fabric is locally chaotic or mylonitic (Nokleberg and others, 1989). The thickness of the melange assemblage is unknown because of the prevailing structural complexity and an absence of stratigraphic marker horizons; maximum structural thickness is estimated to be about 20 km in the Chugach Mountains (Winkler and others, 1981; Plafker and others, 1989).

Age. The most reliable depositional age for this melange of oceanic basalt and overlying sediment is from radiolarians in chert, which were presumably deposited as pelagic ooze on the sea floor. Radiolarians from numerous chert samples include Middle to Late Triassic (Ladinian, Carnian, Norian), Early Jurassic (Hettangian, Pleinsbachian, Toarcian), Late(?) Jurassic, Late Jurassic to Early Cretaceous (Tithonian to Albian), and early Late Cretaceous (Cenomanian) assemblages (Fig. 7; Yao and others, 1982; Winkler and others, 1981; Nelson and others, 1987; C. D. Blome, unpublished data). Where the radiolarian data are adequate to determine age progression, strata in the melange are generally younger toward the south despite their prevailing structural complexity (Plafker and others, 1989).

In the northern Kodiak Islands, the terrane of Cape Current, which occurs along the south margin of the melange unit, contains planktonic foraminifers of Turonian to Santonian age that appear to have been deposited on a seamount of pillow basalt. Early Cretaceous foraminifers occur in the Yakutat Group melange in the subsurface near Yakutat (Yakutat #3 well; Rau and others, 1983) but it is not known whether they are from pelagic sediment or exotic blocks.

The only age-diagnostic megafossils from sedimentary rocks of the melange assemblage are Tithonian *Buchia* from two localities in the Kelp Bay Group, including *Buchia fischeriana* (D'Orbigny) of late Tithonian age (Connelly, *in* Decker, 1980a; Brew and others, 1988). In addition, *Buchia* were collected from several other Kelp Bay Group sites where they occur in blocks several meters to tens of meters wide consisting of medium-grained calcareous sandstone with shallow-marine trace fossils. These blocks which are surrounded by distal turbidites and pelagic mudstone with deep-marine trace fossils, are clearly foreign to the deep-marine deposits and were probably redeposited by submarine landsliding (Decker, 1980a). The blocks include *Buchia* cf. *B. fischeriana* and *B. piochii*(?) of Tithonian age and *B. subokensis* and *B. okensis* of Berriasian age (Loney and others, 1975; D. L. Jones, 1978, written commun., cited *in* Decker, 1980a).

Interpretation. The melange assemblage is interpreted as oceanic crustal rocks and pelagic chert, carbonate rocks, and shale that were mixed with arc-derived sediments and fragments of older rocks at the south margin of the Wrangellia composite terrane. Volcanogenic sandstone and local andesitic green tuff in the sedimentary units indicate that much of the sediment was probably derived from an active arc source. The abundance of slump structures in the siliciclastic rocks reflects deposition on relatively steep slopes such as the inner wall of a trench or a continental margin. The structural style of the assemblage indicates that oceanic rocks were probably accreted, disrupted, and mixed with terrigenous sediments at a convergent plate margin (Clark, 1973; Connelly, 1978; Moore and Connelly, 1979). The serpentinite could represent altered fragments of ultramafic rocks incorporated in the melange from deep levels of the oceanic crust; suitable sources in adjacent terranes to the north are not known.

Fisher and Byrne (1987) interpreted the melange in the Kodiak Islands as having formed as shear zones at the top of a deeply subducted layer immediately below a major decollement in zones of concentrated fluid flow and fracture-dominated permeability. Evidence of relative landward underthrusting and dominantly prehnite-pumpellyite facies metamorphism argue for underplating and substantial burial of this assemblage (Moore and Wheeler, 1978).

Pavlis (1982) suggested that mid-Cretaceous trondhjemitic plutonic rocks in the McHugh Complex in the Anchorage quadrangle were emplaced during a single episode of accelerated accretion. A comparable hornblende K-Ar age (96 ± 4.5 Ma) was obtained for lithologically similar tonalite in coeval melange of the Yakutat Group in the Yakutat quadrangle (Wilson and others, this volume). Decker and others (1980) postulated that a minimum age for the correlative Kelp Bay Group on Chichagof Island is given by mid-Cretaceous K-Ar ages (109 to 91 Ma) from low-temperature–high-pressure regional metamorphic mineral assemblages, which they suggested formed during subduction and accretion of the group.

Data indicating that plate convergence occurred along the southwest boundary of the Wrangellia composite terrane during much or all of the Jurassic and Cretaceous (Engebretsen and others, 1985), together with the apparent southward decrease in age, suggest that the melange was probably accreted over a long time span and that the mid-Cretaceous accretion differed only in the local occurrence of near-trench plutonism. Accretion of melange facies rocks may have begun with inception of the Talkeetna magmatic arc in the Late Triassic and it may have been simultaneous with glaucophanic-greenschist facies metamorphism. The main phase of melange accretion postdates Early to early Middle Jurassic glaucophane-greenschist facies metamorphism, and the youngest accretion was coeval with, or postdated, the early Late Cretaceous rocks of the terrane of Cape Current.

Structural relations and paleontologic data suggest that the melange assemblage predates deposition and accretion of the adjacent flysch and basalt assemblage of the Chugach terrane. In areas affected by postemplacement structural interleaving along strike-slip faults, such as the Yakutat terrane and southeastern Alaska, we recognize that some turbidites of the flysch and basalt assemblage may be indistinguishable from comparable facies in the melange.

Provenance of exotic blocks. The age and lithology of distinctive exotic blocks and conglomerate clasts in the melange assemblage indicate a dominant provenance from the adjacent Peninsular and Wrangellia terranes. The distribution of most of these exotic rocks suggests modest relative strike-slip displacement following deposition; however, the Yakutat Group melange is interpreted to be displaced with the Yakutat terrane about 600 km from the continental margin south of Chatham Strait (see section on Yakutat terrane displacement history). At least six kinds of exotic clasts are included in these blocks. (1) Limestone phacoids in the Uyak Complex contain Permian Tethyan fusulinids (Connelly, 1978) and remnants of the probable Late Triassic hydrozoan Spongiomorpha, both of which are known from adjacent areas of the western Peninsular terrane (N. J. Silberling, cited *in* Connelly, 1978). (2) A carbonate cobble in the McHugh Complex near Anchorage contains a distinctive Early Pennsylvanian conodont fauna (Nelson and others, 1986) that is identical in age and lithology to carbonate rocks from the western part of the Strelna metamorphics of Plafker and others (1985a) near the south margin of the Wrangellia terrane. (3) Limestone blocks in the McHugh Complex near Anchorage contain Early Permian Tethyan fusulinids of the type found in nearby parts of the Peninsular terrane (Clark, 1972). (4) Abundant large blocks of Middle Jurassic quartz diorite (165 to 175 Ma) in the McHugh Complex in the Anchorage and western Valdez quadrangles were probably derived from nearby parts of the Peninsular terrane (G. R. Winkler, 1988, unpublished data). (5) Exotic blocks in the correlative Yakutat Group melange (and probably the Kelp Bay Group), include Permian marble, Late Triassic carbonate rocks with associated greenstone, Sinemurian ammonite-bearing volcanogenic sandstone, and Middle Jurassic quartz diorite; suitable sources for these rocks are present in the Wrangellia terrane south of Chatham Strait (see discussion of Yakutat terrane displacement history). (6) Blocks of Late Triassic marble with associated greenstone in the Kelp Bay Group in southeastern Alaska are best correlated with the Whitestripe Marble and Goon Dip Greenstone in nearby outcrops of the Wrangellia terrane (Karl and others, 1990). In addition, kilometer-scale slabs and blocks of mafic and ultramafic plutonic rocks near the north margins of the Uyak and McHugh complexes (Moore and Wheeler, 1978; Winkler and others, 1981; Kelley, 1985) are lithologically similar to rocks of the informally named Border Ranges ultramafic and mafic complex of Burns (1985) along the south margin of the Peninsular terrane. These larger exotic bodies are interpreted as klippen of the Peninsular terrane that were structurally interleaved with the Chugach terrane during Late Cretaceous or younger folding and/or displacement along steeply dipping faults (Plafker and others, 1989).

Melange assemblage and the Chitina and Chisana arcs. The volcanogenic sandstone and andesitic tuff that make up much of the melange assemblage are arc derived and appear to be mainly of Late Jurassic and Early Cretaceous age (Tithonian to Valanginian). The most likely sediment sources for the volcanogenic deposits are the Late Jurassic Chitina arc and Early Cretaceous Chisana arc in the Wrangellia terrane (Plafker and others, 1989) and their possible westward continuations in the Peninsular terrane (Fig. 4). It is likely that at least part of the volcanogenic sediment may also have been derived from the Late Triassic and Early Jurassic Talkeetna arc, which is widely distributed in the Peninsular terrane part of the Wrangellia composite terrane in Alaska and its inferred offset continuation in the British Columbia segment of the Wrangellia terrane (Plafker and others, 1989; Plafker, 1990).

Flysch and basalt assemblage. The flysch and basalt assemblage is an extremely thick, lithologically monotonous sequence that crops out in a continuous belt as wide as 80 km, extending the full length of the Chugach terrane (Fig. 6), and that underlies about 110,000 km^2. Volcanic rocks make up only a small percentage of the assemblage west of the Chugach Mountains, but they increase in relative abundance and thickness in the eastern Chugach Mountains and Saint Elias Mountains from Prince William Sound to Cross Sound. The assemblage consists of the Shumagin Formation (Moore, 1973, 1974), the Kodiak Formation (Moore, 1969; Moore, 1973), the Valdez Group (Tysdal and Plafker, 1978), the Sitka Graywacke (Decker and Plafker, 1982), and the flysch facies of the Yakutat Group (Fig. 7).

Lithology of flyschoid rocks. The flysch consists of slope, fan, and basin-plain turbidites, with volumetrically minor intercalations of volcanic tholeiite. Primary sedimentary features are commonly well preserved in the Sanak, Shumagin, and Kodiak islands. In most areas east of the Kodiak Islands, however, primary sedimentary features other than graded bedding are commonly obliterated by shearing along bedding planes and by development of a penetrative foliation. The rocks include gradational sequences of argillaceous mudstone containing sporadic thin beds of sandstone, sequences of rhythmically alternating thin beds of argillite and fine-grained sandstone, and medium- to thick-bedded sandstone with minor thin beds of argillite. Graded beds are common, and thicker sandstone beds locally contain small chips of phyllite and thin layers of granule-sized clasts. The sandstone is generally quartzofeldspathic, and grains are typically surrounded by a very fine grained matrix of recrystallized phyllosilicate minerals.

Petrographic studies indicate that the sandstone is dominantly volcaniclastic graywacke but contains sedimentary and metamorphic lithic grains from a broad spectrum of rock types, including some that may have been derived from both inside and outside the Chugach terrane (Decker, 1980a; Zuffa and others, 1980; Winkler and Plafker, 1981a; Winkler and others, 1981, 1984; Lull and Plafker, 1985; Dumoulin, 1987). Although most sandstone of the flysch and basalt assemblage is lithic rich, Connelly (1978) reported that samples from the Kodiak Formation are more feldspathic than those from correlative rocks to the southwest and northeast. Most sandstone is compositionally graywacke (percent lithic grains > percent feldspar grains) but is not texturally graywacke (matrix less than 15%). Lithic clasts are dominantly volcanic in origin, but the ratio of volcanic to total lithics (Lv/Lt) seems, in general, to decrease eastward across the Chugach terrane (Zuffa and others, 1980). The volcanic detritus is primarily andesitic (Moore, 1973; Winkler, *in* Nilsen and Moore, 1979; Zuffa and others, 1980), although Mitchell (1979) and Decker (1980a) suggested that the volcanic suite in southeastern Alaska is bimodal, with rhyolitic and dacitic compositions dominant over basaltic. A subordinate plutonic source for these sedimentary rocks is indicated by the occurrence of sericitized feldspars, microcline, micas, and plutonic rock fragments. Thus, the primary source for the sedimentary rocks of the Chugach terrane appears to have been an evolved arc with an increasing degree of dissection to the east (Zuffa and others, 1980). In Prince William Sound, the Valdez Group compositions that change systematically from west to east (and probably from older to younger) indicate derivation from progressively deeper levels of a magmatic arc (Dumoulin, 1987).

Turbidite facies associations have been delineated for several areas in flyschoid rocks of the Chugach terrane—particularly in the Shumagin Islands (Moore, 1973), the Kodiak Islands (Nilsen and Bouma, 1977; Nilsen and Moore, 1979), the Kenai Peninsula (Budnik, 1974; Mitchell, 1979), Chichagof Island (Decker, 1980a), and the Yakutat terrane (Nilsen and others, 1984). Nilsen and Zuffa (1982) summarized the regional facies associations as inboard slope facies and outboard inner fan and basin-plain facies, the more proximal facies being to the southeast. However, many, if not all, of the slope facies rocks discussed by Nilsen and Zuffa (1982) are probably part of the melange assemblage and predate deposition of the flysch and basalt assemblage. The facies relations suggest a prograding fan system in a narrow trough (probably a trench) with a principal source at the southeast end (Nilsen and Zuffa, 1982). This interpretation is consistent with general coarsening of the flysch from west to east along the length of the belt.

Goechemical analyses of seven samples of metagraywacke (67.3% to 70.5% SiO_2, 0.8% to 2.1% K_2O) and eight samples of metapelite (58.2% to 72.0% SiO_2, 1.6% to 3.6% K_2O) from the Valdez Group in the western Chugach Mountains are remarkably uniform in major and minor element concentrations (Plafker and others, 1989). LREEs are moderately enriched relative to HREEs, and REE concentrations are comparable with that of Andean-type andesite. Discrimination plots using the immobile trace elements La, Th, and Sc suggest a continental margin arc provenance, in agreement with the data from sandstone provenance studies elsewhere in the Chugach terrane.

Lithology of volcanic rocks. Volcanic rocks are present along the south margin of the flysch and basalt assemblage, mainly in the region east of Prince William Sound (Fig. 6). They are dominantly tholeiitic basalt, pillowed basalt, and basalt breccia with minor associated diabase intrusive rocks. The volcanic rocks occur in a belt as wide as 8 km that can be traced discontinuously from the Prince William Sound area to southeastern Alaska for a distance of 600 km (Lull and Plafker, 1990; Barker and others, this volume). In much of this region their positions correlate closely with highs on aeromagnetic maps (Godson, this volume) and in a general way with a discontinuous 60 to 100 mGal isostatic gravity anomaly (Barnes and others, this volume). Minor green basaltic to andesitic metatuff in lenticular beds from a few centimeters to about 15 m thick and up to 4 km long is locally interbedded with metasedimentary rocks overlying the massive basalt unit. Decker's (1980a) correlation of the metavolcanic rocks of the Valdez Group with the metavolcanic rocks near the westernmost exposures of the Ford Arm Formation of Decker (1980b) on Chichagof Island was based on the continuity of the magnetic anomaly associated with these rocks and on their similarity in structural position, sandstone petrography, and turbidite facies.

The mafic igneous rocks of the flysch and basalt assemblage are geochemically similar to LREE-depleted basalt (Lull and Plafker, 1990). They plot as tholeiites with an iron enrichment trend on the AFL diagram, and they straddle the tholeiite–calc-alkalic compositional boundary on a plot of FeO^*/MgO vs. SiO_2. On discrimination diagrams for Th-Hf-Ta, Ti-Zr-Y, and Ti-Zr, the metabasalts and the comagmatic basaltic tuff within the Valdez Group plot mostly within island-arc tholeiite fields. REE patterns have moderate negative slopes and primitive chondritic compositions and show moderate depletion of LREEs relative to HREEs (La 1 to 7 × chondrite, Lu 9 to 20 × chondrite). REE abundances of the intrusive rocks are systematically lower than those of the greenstones. Geochemical data suggest that the Valdez Group metabasalts could be transitional between island-arc tholeiite and MORB, but with a strong island-arc tholeiite signature. Together with the field relations, however, the data favor an island-arc petrogenetic setting, but close enough to the continental margin to be within reach of deep-sea fan deposits (Lull and Plafker, 1990).

Metamorphism. The flysch and basalt assemblage is variably metamorphosed to zeolite facies west of the Chugach Mountains, to zeolite and low-greenschist facies in the western Chugach Mountains, to greenschist and amphibolite facies in the eastern Chugach Mountains and Saint Elias Mountains, and to zeolite and low-greenschist facies in southeastern Alaska and in the Yakutat Group; local areas show contact metamorphism (Dusel-Bacon, this volume).

Detailed studies in the Kodiak Islands indicate that the Kodiak Formation was underthrust to significant depths and then underplated to the Chugach terrane (Sample and Moore, 1987). Both mesoscopic and regional structural trends indicate that underplating probably occurred by duplex accretion. Metamorphic mineralogy, vitrinite-reflectance data (Sample and Moore, 1987), and fluid-inclusion data (Vrolijk and others, 1988) indicate that deformation related to underthrusting occurred at a depth of 10 to 12 km at temperatures of 200 to 250 °C.

In the Chugach Mountains east of the Copper River, metamorphic grade increases along strike, culminating in the eastern

Chugach Mountains in a schist and gneiss complex with sillimanite-grade migmatitic rocks (the informally named Chugach metamorphic complex of Hudson and Plafker, 1982). Epidote-amphibolite and greenschist facies metamorphism characterizes the sequence throughout the Saint Elias Mountains to Cross Sound. The superimposition of progressive metamorphism across regional structural trends and the typical granoblastic textures indicate that metamorphism occurred after the Valdez Group part of the flysch and basalt assemblage was deformed against and amalgamated with the melange assemblage to the north. Rocks of the Chugach metamorphic complex record a regional Paleogene thermal event characterized by very high temperature at low pressure (Sisson and others, 1989). A minimum age for metamorphism of the group is given by K-Ar whole-rock and mineral ages that range from 53.5 to 47.9 ± 2 Ma in the Chugach and Saint Elias Mountains (Fig. 7). This age of metamorphism overlaps the age of anatectic plutonism in the region (Hudson and others, 1979; Hudson, this volume; Barker and others, 1992), which culminated about 50 m.y. ago (Fig. 7). Numerous felsic dikes, sills, and plugs of this age that intrude the Valdez Group throughout the full width of its outcrop belt in the central Chugach Mountains were formed during the anatectic melting event (Plafker and others, 1989).

Thickness. Estimated minimum stratigraphic thicknesses from outcrop data are 3 to 4 km for the Shumagin Formation (Moore, 1973) and 3 km or more for the Kodiak Formation (Nilsen and Moore, 1979). Deep-marine reflection data indicate a vertical structural thickness of as much as 30 km for the Kodiak Formation and thick, probably younger, underplated sedimentary sequences or intercalations of oceanic crust and sedimentary rocks (Moore and others, 1991). The vertical thickness of the Valdez Group, based on geophysical data, is between 10 and 20 km beneath the central Chugach Mountains (Fuis and Plafker, 1991); the outcrop thickness of the basaltic sequence is 3 to 5 km along the southern margin of the Chugach terrane near the Contact fault system east of Prince William Sound.

Structure. In the Kodiak Islands, and the Kenai, Chugach, and Saint Elias mountains, the flysch and basalt assemblage is faulted against and relatively beneath either the melange assemblage (along the Uganik, Chugach Bay, Eagle River, Tazlina, and related thrust faults), or it is juxtaposed against the Wrangellia composite terrane along the Border Ranges fault system (Fig. 6).

Detailed study of the structures and fabric of the Kodiak Formation (Fig. 9) indicates that (1) the unit can be divided structurally into a northwestern belt of moderate strain with steeply dipping structures, a high-strain central belt with shal-

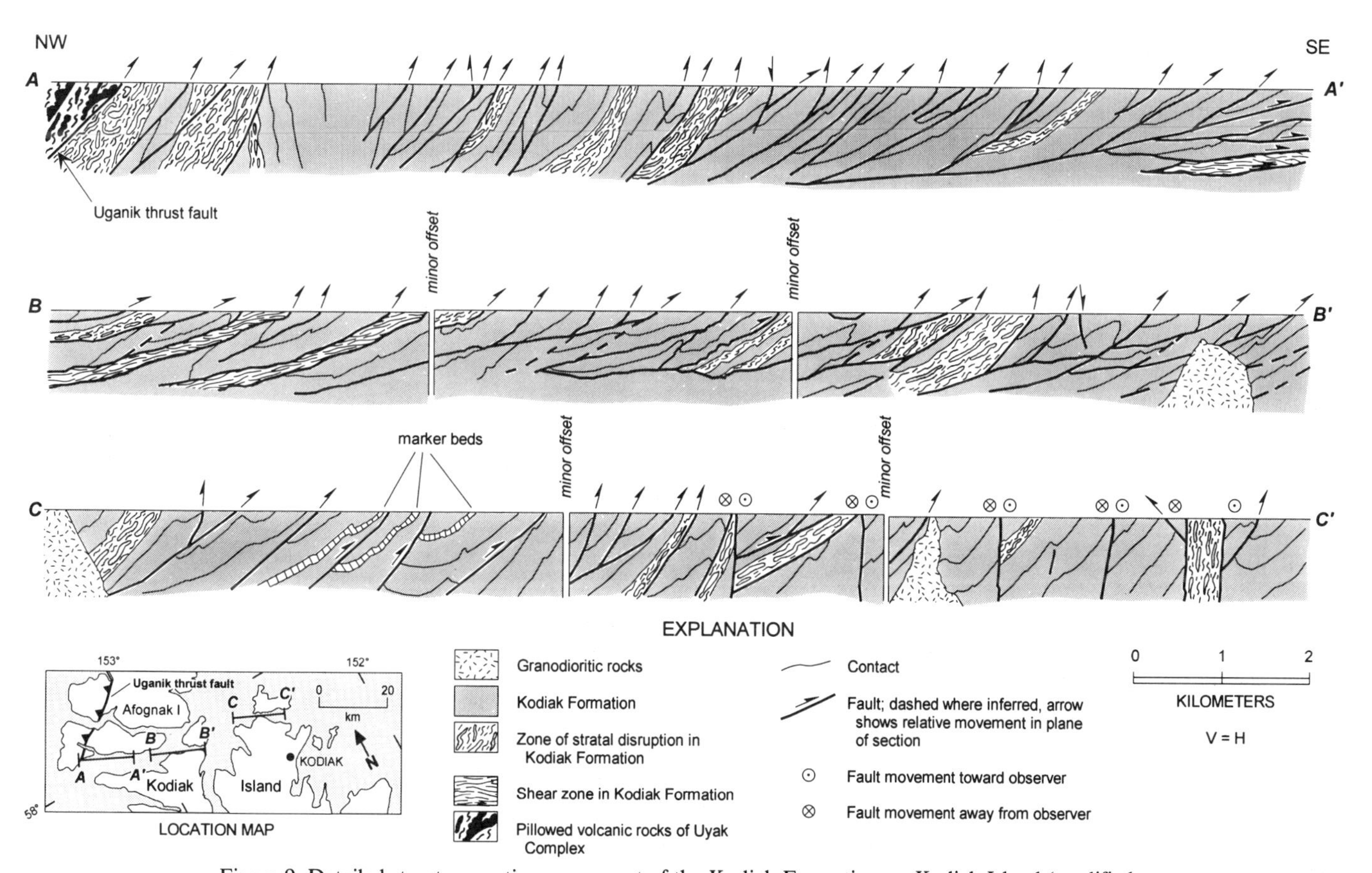

Figure 9. Detailed structure section across part of the Kodiak Formation on Kodiak Island (modified from Sample and Moore, 1987).

lowly dipping structures, and a low-strain southeastern belt with steeply dipping structures; and (2) fault geometries locally resemble duplex fault zones in which high-angle listric reverse faults stop at subhorizontal detachment faults below and above in strata that were previously tilted.

In the Chugach Mountains along the Trans-Alaska Crustal Transect, the flysch and basalt assemblage is thrust relatively northward beneath both the Wrangellia composite terrane along the Border Ranges fault and the melange assemblage along the Tazlina fault (Fig. 8). In this area, the Border Ranges fault (BRF1 in Fig. 8) is a folded low-angle thrust along which the Wrangellia terrane has been displaced as much as 40 km southward. The contact is marked by a zone of mylonitic rocks and by local blocks of variably serpentinized ultramafic rocks (Plafker and others, 1989). The lower plate sequence is complexly deformed by tight, chevron to isoclinal, major and minor folds and thrust faults that commonly verge south. Coaxial north-verging folds are superposed on south-verging structures over a large segment of the northern Chugach Mountains, including the area shown in Figure 8 (Nokleberg and others, 1989). The regional parallelism of axial planes of folds, faults, and foliation or cleavage is striking.

In the Kenai and western Chugach mountains, the fault contact between the flysch and basalt assemblage and the melange assemblage in many places is a well-defined thrust that dips northwestward at low to moderate angles and truncates tight folds in the underlying flysch (Cowan and Boss, 1978; Winkler and others, 1984; Plafker and others, 1989; D. C. Bradley, 1991, oral commun.; Winkler 1992). Elsewhere, the fault is a complex imbricated zone in which slivers of the two units may be juxtaposed. Deformation in the flysch and basalt assemblage is especially prounouced near the fault contact with numerous surfaces of intensive shearing in the Valdez Group that are defined by boudins of sandstone oriented subparallel to the main fault surface. Pervasively sheared rocks consistently grade structurally downward into well-bedded sandstone and phyllite from 100 m to 2 km south and east of the fault. Rocks of the overriding melange assemblage do not show a similar increase in intensity of deformation as they approach the fault. Similar fabric contrasts occur across the correlative Uganik thrust in the Kodiak Islands (Fig. 9), where a slip vector of 334° ± 7° was determined for the fault by Moore and Wheeler (1978).

The Eagle River fault in the northern Kenai and western Chugach mountains is cut by numerous high-angle faults, which generally are upthrown to the northwest. Map patterns of the fault near Anchorage (Clark, 1972; Winkler, 1992) indicate that the Eagle River fault may be folded into a broad open syncline as well. The age of major offset on the Eagle River fault is well constrained regionally by paleontologic and radiometric information. The fault truncates rocks of the Valdez Group that contain Late Cretaceous megafossils, and it is intruded by numerous felsic to intermediate hypabyssal dikes that are similar to intrusions dated as early to middle Eocene in nearby areas (Winkler and others, 1984). In the Seldovia area in the southern Kenai Mountains, Bradley and Kusky (1992) mapped folds that have wavelengths of about 1 km that deform the Chugach Bay fault (Fig. 6).

Age. The Late Cretaceous (early Maastrichtian and late Campanian) age of the flysch and basalt assemblage is based on age-diagnostic Inoceramids and foraminifers. Early Maastrichtian *Inoceramus,* including *I. kusiroensis, I. balticus,* and *I. concentrica* occur at 11 localities in the Shumagin Formation, 9 localities in the Kodiak Formation, and 15 localities in the Valdez Group in the northern Kenai and western Chugach mountains (Jones and Clark, 1973; Budnick, 1974; J. W. Miller, 1984, 1985, written communs. to J. C. Sample; J. W. Miller, 1986, written commun. to D. Fisher; J. W. Miller, *in* Winkler, 1992). One occurrence of poorly preserved Campanian or Maastrichtian Inoceramids was found in the Valdez Group in the southern Kenai Mountains (D. L. Jones, *in* Tysdal and Plafker, 1978), and a single specimen of the Campanian *I. schmidti* came from the flysch facies of the Yakutat Group (Jones and Clark, 1973). In addition, Late Cretaceous (Campanian to Maastrichtian) foraminifers have been recovered from well cores in the Yakutat Group flysch (W. V. Sliter, *in* Rau and others, 1983). No fossils have been recovered from the Sitka Graywacke, which has been tentatively correlated with the flysch and basalt assemblage on lithology and its structural position within the Chugach terrane (Plafker and others, 1977).

Deformation and accretion of the flysch and basalt assemblage was completed prior to emplacement of post-tectonic plutons of middle to late Paleocene age in the Shumagin, Sanak, and Kodiak islands and of early to early middle Eocene age on the mainland to the east (Fig. 7). Accretion, however, may have been essentially synchronous with deposition in an active subduction environment and could have begun as early as Campanian time.

Interpretation. Metabasalt in the Valdez Group is inferred to be the remnant of a primitive intraoceanic island arc that formed within the Kula plate during Campanian to early Maastrichtian time and migrated toward an Andean-type continental margin (Lull and Plafker, 1990). Deep-sea fan deposits of basaltic sediments and tuff from the intraoceanic arc were mixed with volcaniclastic sediments and tuff from a continent-margin arc as the island arc approached the continent. Arc-continent collision in latest Cretaceous to early Paleocene time resulted in successive offscraping and accretion to the continental margin of the flysch, mixed flysch and basaltic tuff, and basalt that make up the flysch and basalt assemblage of the Chugach terrane.

Petrography of the flysch and sparse interbedded metaandesite tuff indicate derivation from an evolving magmatic arc as sediments were derived from increasingly deeper levels from west to east in the accretionary prism. Turbidite facies associations and paleocurrent determinations indicate dominantly westward transport of sediment. Deposition was largely confined to a trench, probably as narrow fans that became elongated in the direction of the trench axis but possibly also as minor infilling from the north by slumps or through channels (Nilsen and Zuffa, 1982). Sedimentological and isotopic data indicate that a major part of this sediment came from the southeastern Alaska and British Columbia region and was transported northwestward along the continental margin (Nilsen and Moore, 1979; Nilsen and Zuffa, 1982; Decker and Plafker, 1983).

Structures and metamorphism in the Kodiak Formation have been interpreted to indicate underthrusting and underplating by accretion of duplex structures at a minimum depth of 10 km with attendant stratal disruption and formation of slaty cleavage, thrust faults, and folds (Sample and Fisher, 1986; Sample and Moore, 1987). Crenulation cleavage, secondary folds, and thrust faults were subsequently formed by postaccretion intrawedge shortening. Duplex structures have not yet been recognized in outcrop other than in the Kodiak Islands, and the extent to which this mechanism applies to accretion elsewhere in the Chugach terrane is not known.

Flysch and basalt assemblage and the Late Cretaceous Kluane arc. The turbidites of the flysch and basalt assemblage are inferred to have been deposited in a trench associated with a southward- to seaward-facing continental margin (Andean type) arc that extended from British Columbia to the Alaska Peninsula (Fig. 5). The continuity of the flysch and basalt assemblage suggests that it was accreted along a relatively linear margin before counterclockwise bending of western Alaska to form the Gulf of Alaska. We infer that this magmatic arc, informally named the Kluane arc by Plafker and others (1989), was the source of most of the detritus. Plutonic rocks of Late Cretaceous and younger age, having appropriate compositions to be the roots of the inferred Kluane arc, occur discontinuously in two belts that appear to be dextrally offset some 400 km along the Denali fault (Nokleberg and others, 1985; this volume, Chapter 10). The northern part of this belt extends northward from the central Alaska Range through the Talkeetna Mountains to the Denali fault (Alaska Range–Talkeetna Mountains belt of Hudson, 1983); the southern segment is inferred to extend from the Kluane Lake area in Canada southward through the Coast Mountains of southeastern Alaska and British Columbia (Monger and others, 1982; Brew and Ford, 1984). The belt consists of 85 to 45 Ma calc-alkalic plutonic rocks in British Columbia (Armstrong, 1988), 72 to 55 Ma rocks in southeastern Alaska (Barker and others, 1986; Loney and Brew, 1987; Brew, this volume), 74 to 50 Ma rocks in the Aleutian Range, and 77 to 65 Ma rocks in the Talkeetna Mountains (Hudson, 1983; Moll-Stalcup, this volume). Although rocks of Late Cretaceous age appear to compose a relatively small part of the southern segment (Brew, this volume), their scarcity may be due to masking by the voluminous Paleogene magmatism in the region.

Hollister (1979) and Crawford and Hollister (1982) have shown that the southern segment of the belt in the Coast Mountains of British Columbia was uplifted rapidly in two major pulses: a younger one between 62 and 48 Ma (early Tertiary), and an earlier one, probably Late Cretaceous in age. This crystalline belt includes plutonic rocks of appropriate composition for a magmatic arc and is interpreted as a deeply eroded segment of an extensive Late Cretaceous and early Tertiary magmatic arc that traverses much of southern Alaska (the Kluane arc of Plafker and others, 1989; Fig. 5). The Coast Mountains segment of the arc is inferred to be the primary sediment source for the flysch sequence; additional sediments were contributed from segments of the arc in central and western Alaska. Alternative sediment sources other than the Kluane arc are possible, but they all require highly improbable scenarios of longer sediment-transport distances from source regions and/or large postemplacement strike slip of the Chugach terrane away from its source region.

Metamorphic complex of the Saint Elias area. Metamorphosed flysch and basalt constitute an east-west–trending, structurally bound block 80 km long and as wide as 15 km that underlies the highest part of the rugged Saint Elias Mountains in Alaska, including Mt. Saint Elias (Saint Elias terrane of Jones and others, 1987, and Silberling and others, this volume). On the south the complex is emplaced against unmetamorphosed Cretaceous and Tertiary rocks of the Yakutat terrane along the Chugach–Saint Elias fault system; on the north it is emplaced against lower grade metamorphic rocks of the Chugach terrane along an inferred extension of the Fairweather fault system beneath glaciers of the Saint Elias Mountains (Fig. 6).

Interleaved and penetratively deformed epidote-amphibolite-grade metabasalt and metaflysch make up most of the metamorphosed complex of the Saint Elias area. The schistose rocks are intruded by small 50 Ma granitic plutons and by at least one undated layered gabbro that is known from float on glacial moraines derived from the south face of Mount Saint Elias (Plafker, unpublished data).

Rocks that make up the metamorphic complex of the Saint Elias area are similar to the metamorphosed flysch and basalt assemblage of the Chugach terrane in the southern Saint Elias Mountains in metamorphic grade, structure, and lithology, and in the age of the intruding Tertiary plutons. Hence, these rocks are here considered to be a displaced block of the Chugach terrane rather than a separate terrane.

Structural and lithologic data suggest that the metamorphosed complex of the Saint Elias area was probably sliced off the Chugach terrane and displaced to its present position by dextral movement along the Fairweather fault. The original position of the comlex is uncertain. However, a good match can be made with lithology and structure of the Chugach terrane south of the Alsek River, about 50 km southeast of Yakutat; this position requires some 200 km offset along the Fairweather fault.

Paleomagnetic data. The only paleomagnetic data for igneous rocks of the Chugach terrane, from the Crillon–La Perouse layered gabbro of Loney and Himmelberg (1983), have been interpreted as indicating that the Chugach terrane moved 25° to 26° northward and rotated 90° counterclockwise relative to the North America craton (Grommé and Hillhouse, 1981). Since the paleomagnetic work was completed, the gabbro has been dated as middle Tertiary in age, probably 28 ± 8 Ma (Loney and

Himmelberg, 1983). The young age of the gabbro indicates that the paleolatitude determination is incorrect because it requires that the pluton be translated northward at twice the displacement rate of the Pacific plate (Plafker, 1984). The assumption of initial horizontality for the gabbro layers may be incorrect and/or the pluton may have been affected by a postemplacement northeastward tilt to give the apparent inclination anomaly of 19° ± 8°. A prominent decrease in metamorphic grade of country rocks from southwest to northeast across the gabbro body is compatible with tilting, but the apparent large rotation remains unexplained.

Paleomagnetic studies of Cretaceous sedimentary rocks of the Chugach terrane were carried out at two sites in the Shumagin Formation (Stone and Packer, 1979) and one in the Kodiak Formation (Stone and others, 1982). Although these studies yield mean paleolatitudes from 43° to 69° south of the predicted North American reference, the magnetizations are possibly contaminated by a postdeformational component, so the Cretaceous paleolatitude determinations may be invalid (Hillhouse and Coe, this volume).

Latest Cretaceous and Paleogene accretionary assemblages

Most of the latest Cretaceous and Paleogene accretionary assemblages of the Southern Margin composite terrane are juxtaposed against the western Chugach terrane along the Contact fault system and related structures (Figs. 2 and 6). From oldest to youngest these accreted units are (1) the Late Cretaceous and early Paleocene Ghost Rocks Formation in the Kodiak Islands area, (2) the late Paleocene through early middle Eocene Orca Group in the Prince William Sound region, (3) the unnamed sequence of probable early Eocene age on the Resurrection Peninsula, and (4) the Eocene Sitkalidak Formation of the Kodiak Islands (Fig. 7). The Ghost Rocks Formation has also been referred to by some workers as the Ghost Rocks terrane (Plafker, 1987; Jones and others, 1987; Silberling and others, this volume), and the Prince William terrane consists principally of the Orca Group. Limited available age control suggests that both the Resurrection Peninsula sequence and the Sitkalidak Formation may be in part or entirely coeval with the Orca Group. However, because of uncertainties in the age of these units, they are described separately from it in the sections that follow.

The Contact fault system (Plafker and others, 1977) extends from just east of Mount Saint Elias (long 140°W) to the Kodiak Islands. In northern and eastern Prince William Sound, it consists of steep to gently north dipping reverse faults that separate foliated flysch and greenstone of the Valdez Group on the north from strongly deformed flysch and tholeiite of the Orca Group on the south (Winkler and Plafker, 1981b; Winkler and others, 1981). In this area, a profound change in deep crustal structure indicates that the fault is a fundamental structural boundary (Fuis and others, 1991; Fuis and Plafker, 1991; Wolf and others, 1991). Successive accretionary wedges within the Orca Group may be bounded by faults that splay southwestward from the Contact fault system (Winkler and Plafker, 1981b). Minor structures along faults in eastern Prince William Sound indicate that predominantly dextral strike-slip displacements have been superimposed on structures formed in a thrust or oblique thrust stress regime (Winkler and Plafker, 1981b; Plafker and others, 1986). Although some post–early Eocene shear has affected plutons south of the Contact fault, strike-slip displacement must be minor because plutons that intrude across the fault in the region east of Prince William Sound have no measurable offset (Fig. 6).

In much of western Prince William Sound, the Contact fault is not well defined due to a combination of pore exposure and similarity in the structural trends, metamorphic grade, and lithology of the juxtaposed units (Tysdal and Case, 1979; Bol and others, 1992; Dumoulin, 1987). Age relations suggest that the Contact fault should lie west of the Resurrection Peninsula sequence (Fig. 6).

In the Kodiak Islands, the Contact fault marks the boundary between flysch of the Chugach terrane and the Ghost Rocks Formation to the southeast. The Ghost Rocks Formation includes kinematic indicators that indicate underthrusting beneath the adjacent Kodiak Formation. The Kodiak Formation shows a late phase of right-lateral strike-slip faulting within 20 km of the Contact fault (Sample and Moore, 1987) that is not similarly developed in the Ghost Rocks Formation. This strike-slip faulting must have developed prior to emplacement of the Ghost Rocks Formation. The current exposure of the Contact fault is steep and may not represent the structure along which the Ghost Rocks Formation was underthrust. Seismic-reflection lines northeast of the Kodiak Islands show no obvious structural boundary along the projection of the Contact fault (Moore and others, 1991).

Ghost Rocks Formation. The Ghost Rocks Formation is exposed in a belt about 10 km wide along the southeast side of the Kodiak Islands and underlies a total onshore and offshore area of about 6000 km^2 (Fig. 6). In the Kodiak area the formation is juxtaposed against the Late Cretaceous Kodiak Formation on the northwest along the Contact fault. On the southeast, the Ghost Rocks Formation is faulted against younger Tertiary rocks along the Resurrection fault. Correlative rocks are not known elsewhere in the accretionary complex. The Resurrection Peninsula sequence, which has been tentatively correlated with the Ghost Rocks Formation (Jones and others, 1987), is now known to be younger than that sequence and is considered separately in a following section.

Lithology. The Ghost Rocks Formation comprises a complexly deformed assemblage of sandstone and mudstone with interbedded volcanic rocks and minor pelagic limestone that locally is devoid of any stratal continuity (Moore, 1969; Moore and others, 1983; Byrne, 1984). The clastic rocks include turbiditic sandstone-rich and argillite-rich units with local channelized cobble-boulder conglomerates containing clasts of chert, sandstone, limestone, and greenstone. Primary structures and depositional contacts are locally well preserved. The argillite-rich units include interbedded pillow basalt, tuff, chert-rich pebble conglomerate, and minor limestone containing pelagic foraminifers. Sandstone compositions are dominated by lithic fragments of volcanic origin; quartz and feldspar in about equal proportions are less abundant (G. R. Winkler, *in* Nilsen and Moore, 1979; Bryne, 1984). These compositions are compatible with derivation from a volcanic arc or slightly dissected volcanic-arc source region.

The volcanic rocks in the Ghost Rocks Formation include both LREE-depleted tholeiitic basalt and LREE-enriched calc-alkaline basaltic andesite to andesite (Hill and others, 1981; Barker, this volume). Andesite flows as thick as 10 m are interbedded with the coarse clastic rocks, mainly in the sandstone-rich units. The major and minor element composition of the basalt is most consistent with an oceanic source, most probably a spreading ridge. In contrast, the andesitic rocks are interpreted as having been derived from a hybrid magma that most probably formed by assimilation of flysch sediments of the accretionary prism in a MORB-like basalt. Hypabyssal plutons of gabbro to quartz diorite that cut the Ghost Rocks Formation and units to the northwest are inferred to be part of this same magmatic system (Hill and others, 1981; Barker, this volume).

The Ghost Rocks Formation has been metamorphosed to prehnite-pumpellyite facies. The combined metamorphic mineral assemblages, vitrinite reflectance data, and fluid-inclusion results suggest a maximum temperature of 250 to 260 °C and burial depths of 10 to 12 km (Moore and others, 1963; Vrolijk and others, 1988).

Structure. Foliation and axial traces of folds in the Ghost Rocks Formation generally trend northeast, parallel to the outcrop pattern. Coherent units show a broadly coaxial deformation succeeded by small-scale conjugate folds and the development of spaced cleavage. Melange units show a layer-parallel, noncoaxial deformation with pervasive small-scale cataclastic shear zones and a deformed, gently plunging lineation (Moore and others, 1983; Bryne, 1982, 1984; Fisher and Bryne, 1987). In some places igneous rocks are deformed by these structures and in others they postdate the structures; igneous activity was apparently contemporaneous with northwest-directed shortening (Bryne, 1982, 1984).

Age. The age of the Ghost Rocks Formation is uncertain but is considered in this chapter to be Late Cretaceous and Paleocene. The unit postdates accretion of the Campanian and early Maastrichtian Kodiak Formation to the northwest and predates emplacement of late Paleocene intrusive rocks. Limestone at four localities contain rare, poorly preserved planktonic foraminifers of both Late Cretaceous (middle to late Maastrichtian) age and Paleocene age (Table 2; Moore and others, 1983). Volcanic rocks in the unit are undated, although at one locality limestone of possible Paleocene age is in contact with pillow basalt (Nilsen and Moore, 1979) and at another locality Maastrichtian limestone occurs as lenses in basalt (Table 2). Andesitic rocks in the unit are inferred to be comagmatic with hypabyssal plutonic rocks in the Kodiak Islands that have K-Ar radiometric ages of about 62 Ma (Hill and others, 1981; Moore and others, 1983; Moll-Stalcup, this volume; Wilson and others, this volume). The hypabyssal intrusions, which cut across the deformed Ghost Rocks Formation, indicate an early Paleocene minimum age for deposition and accretion of the unit.

Paleomagnetic data. Paleomagnetic inclination anomalies from volcanic rocks in the Ghost Rocks Formation fall into two general populations that assume a Paleocene age. (1) Andesitic rocks from the inboard part of the unit at Alitak Bay are 16° ± 9° north of their Paleocene paleolatitude. (2) Basalt from the outboard part of the unit at Kiliuda Bay is 31° ± 9° north of its Paleocene paleolatitude (Plumley and others, 1983; Hillhouse and Coe, this volume). At Kiliuda Bay rotation was counterclockwise at about 45°, and at Alitak Bay it was clockwise at about 90°.

Interpretation. We interpret the Ghost Rocks Formation as a trench or trench-slope deposit with a significant component of ocean-floor basalt and andesite related to ridge-trench interaction (Hill and others, 1981; Moore and others, 1983; Barker, this volume). The argillite unit and associated basalt were most probably deposited in a hemipelagic environment. They may be the distal equivalent of the sandstone-rich unit with its associated andesitic rocks that probably were deposited adjacent to a subduction zone characterized by near-trench volcanism (Moore and others, 1983). Conglomerate-clast lithology in the sandstone-rich unit is compatible with derivation from nearby parts of the Chugach and Peninsular terranes. The deformation, metamorphism, and igneous activity associated with these units are appropriate for emplacement in a subduction zone along the seaward margin of the Chugach terrane. Deposition, major deformation, metamorphism, and intrusion occurred in the brief interval from the middle Maastrichtian to the early Paleocene.

The paleomagnetic data indicate latitudinal displacement of about 16° relative to the craton since the early Paleocene. The large inclination anomaly for the Kiliuda Bay basalt samples (31°) suggests that these rocks are exotic oceanic crustal fragments, possibly as old as Late Cretaceous, and that they do not necessarily indicate the paleolatitude of the enclosing accretionary prism (Plafker, 1990). This interpretation differs from that of Coe and others (1985), who inferred that the paleomagnetic data represent a single population with an averaged Paleocene paleolatitude 25° ± 7° farther south. Comagmatic granitic intrusive rocks that occur in close proximity across the boundaries of the Chugach, Peninsular, and Ghost Rocks terranes suggest that these terranes were in about their present relative positions by early Paleocene time (Davies and Moore, 1984). Paleomagnetic data suggest that the northeast margin of the Wrangellia composite terrane (which includes the Peninsular terrane) was in place relative to North America by about 55 Ma (Hillhouse and others, 1985; Hillhouse and Coe, this volume). If so, the Ghost Rocks Formation has not moved appreciably northward since that time, unless there are very large undetected dextral strike-slip displacements on suitably oriented faults inboard of these units. The 16° of relative motion must have occurred in the interval from early Paleocene to 55 Ma.

Orca Group. The Orca Group is a deep-sea-fan flysch complex interbedded with subordinate oceanic volcanic rocks

TABLE 2. FORAMINIFERA FROM THE GHOST ROCKS FORMATION, ALASKA*

Sample No. Collector	Latitude Longitude	Age	Lithology	Fauna	Paleontologist or Comments
AMe-76-21† MF-3421§ G. W. Moore	57°13.1'N 153°02.9'W	Probable Paleocene	Limestone	? *Subbotina* spp., *S.* sp. cf. *S. triangularis*, *S. triloculinoides* group	R. Z. Poore, U.S.G.S., written commun. to G. W. Moore, Oct. 7, 1976.
AMc-76-56† MF-3422§ G. W. Moore	57°23'N 152°36.1'W	Probable Paleocene	Limestone overlying pillow lava	*Globigerina pseudobulloides* Plummer (or a closely related taxon), *Planorotalites* sp.?	Age designation based on correct identification of *Planorotalites* sp.?
				?*Subbotina* spp., *S.* sp. cf. *S. triangularis*, *S. triloculinoides* group	R. Z. Poore, U.S.G.S., written commun. to G. W. Moore, Oct. 7, 1976; Oct. 22, 1976.
HZe† W. Connelly	57°13'N 153°00'W	Paleocene	Limestone block	*Globigerina* spp., *G.* sp. cf, *G. yeguaensis*, Wienzierl and Applin, abundant	W. V. Sliter, U.S.G.S., written commun. to G. Plafker, March 1992.
78-KN-C8-A† T. Byrne	56°56.5'N 153°38.8'W	Late Cretceous, middle to upper Maastrichtian	Limestone lenses in pillow basalt	*Contusotruncana patelliformis* (Gandolfi), *Globotruncana arca* (Cushman), *G.* sp. cf. *G. hilli* (Pessagno), *G. linneiana* (d'Orbigny), *Globotruncanella havanensis* (Voorwijk), *G.* sp. cf. *G. petaloidea* (Gandolfi), *Globotruncanita stuarti* (de Lapparent), *G.* sp. cf. *G. stuartiformis* (Dalbiez), *Hedbergella* spp., *Heterohelix* spp., *Planoglobulina acervulinoides* (Egger), *Pseudoguembelina* sp. cf. *P. costulata* (Cushman), *Pseudotextularia elegans* (Rzehak), *Rugoglobigerina rugosa* (Plummer), *R.* sp. cf. *R. hexacamerata* (Brönnimann)	W. V. Sliter, U.S.G.S., written commun. to G. Plafker, Feb. 8, 1993.

Note: U.S.G.S is United States Geological Survey.
*Identified in thin section
†Field number.
§U.S. Geological Survey collection number.

and minor hemipelagic mudstone (Winkler, 1976; Winkler and Plafker, 1981b; Helwig and Emmet, 1981; Nelson and others, 1985) that forms an outcrop belt more than 100 km wide that extends from the western Chugach Mountains beneath Prince William Sound and southwestward beneath the continental shelf and slope (Fig. 6). The Orca Group makes up most or all of the Prince William terrane as defined by Jones and others (1987). Possibly correlative units are the Sitkalidak Formation (Fig. 7) and the Resurrection Peninsula sequence, described below. The onshore area is roughly 21,000 km^2 and the total area, including the inferred offshore area underlain by the Orca Group, is at least 140,000 km^2.

The Orca Group is bounded on the north by the Contact fault system, on the south in the Chugach Mountains by the Chugach–Saint Elias fault system, and on the southeast by the offshore Kayak Island structural zone and its onshore extension east of the Copper River. The south and west boundaries offshore are not well known, although seismic reflection data and a characteristic magnetic signature of the volcanic rocks in the unit suggest that the Orca Group probably extends beneath much or all of the continental shelf. An exploratory well 3600 m deep near Middleton Island bottomed in deformed strata that are possibly correlative with the Orca Group, although the paleontological data do not preclude the alternative interpretation that they are as young as late Eocene (Plafker and others, 1982; Rau and others, 1983; Plafker, 1987).

Clastic rocks. Monotonous sequences of thin- to thick-bedded turbidites consisting of sandstone, siltstone, and mudstone are the dominant clastic components. Abundant primary sedimentary features indicate deposition from sediment gravity flows, chiefly turbidity currents.

The sandstone is feldspathic to lithofeldspathic. Lenticular, matrix-supported pebbly mudstone and sandstone also are widespread, and at one locality, along Valdez Arm north of Galena Bay, calcareous concretions in debris-flow deposits contain abundant fossil crabs and rare pelecypods and gastropods. The pebbly mudstone, sandstone, and debris flows apparently mark distributary channels that fed mid-fan depositional lobes.

Sandstone compositions are variable but on the average comprise roughly equal amounts of quartz, feldspar, and rock fragments (Winkler, 1976; Dumoulin, 1987; Gergen and Plafker, 1988). About 30% to 50% of the rock fragments are volcanic and as much as 20% are metamorphic; plagioclase makes up 80% to more than 90% of the total feldspar. These compositions are compatible with derivation from a mixed dissected magmatic arc and high-grade metamorphic provenance.

A minor but ubiquitous component of the unit is massive, clast-supported, well-rounded pebble, cobble, and boulder conglomerate. Typical clast assemblages include greenstone, sandstone, argillite, and limestone; less common are felsic porphyry and tuff, granitoid rocks, sandstone, limestone, and orthoquartzite. At one locality just east of the mouth of the Copper River, the conglomerate consists of rhyolitic welded tuff and porphyry cobbles and boulders and less than 10% sandstone clasts. The massive conglomerates probably represent inner fan channel-fill deposits that mark entry points of major feeder channels onto the Orca submarine-fan complex. The presence of cobble- to boulder-size siliceous welded tuff and orthoquartzite clasts suggests proximity to a continental source for part of the sediment.

Volcanic rocks. Mafic volcanic rocks, which constitute 15% to 20% of the Orca Group, crop out in three major discontinuous belts: (1) along the northern margin of Prince William Sound, (2) from Hinchinbrook Island eastward to the east side of the Copper River, and (3) offshore on the continental shelf. The largest of these volcanic sequences are shown in Figure 6.

The landward belt of mafic volcanic rocks is 150 km long, as much as 12 km wide, and extends in a broad arc along the western, northern, and northeastern Prince William Sound. The belt is marked by pronounced positive residual gravity anomalies (Case and others, 1966) and aeromagnetic anomalies (Case and Tysdal, 1979). Large changes in total amplitude of the anomalies along strike, however, indicate that in places the volcanic sequences thin markedly, either stratigraphically or structurally. The rocks are dominantly tabular bodies of altered basalt, consisting chiefly of pillowed, massive, or crudely columnar flows, pillow breccia, and aquagene tuff with variable amounts of associated gabbro and diabase as sills or sheeted dikes. At Knight Island in western Prince William Sound, an ophiolite complex consists of a core of peridotite, sheared ultramafic rocks, and thousands of meters (structural thickness) of sheeted dikes that are surmounted by more than 1500 m of pillow-basalt breccia and massive flows (Richter, 1965; Tysdal and Case, 1979; Nelson and others, 1985). Small bodies of gabbro intrude both the sheeted dike complex and the sedimentary section that is interbedded with and encloses the volcanic sequence. Southwest of Knight Island the proportion of argillaceous rocks interbedded with, or intruded by, the mafic igneous sequence increases and the stratigraphy of the volcanic rocks is disrupted. This area of strongly deformed rocks was referred to as the Bainbridge melange by Helwig and Emmet (1981). Along regional strike to the east there is no equivalent belt of melange.

A second belt of mafic volcanic rocks extends discontinuously from Hinchinbrook Island in southeastern Prince William Sound to the Cordova area and east of the Copper River mouth along the south and east margin of the Prince William terrane; thinner volcanic sequences occur sporadically in eastern Prince William Sound north of Cordova (Fig. 6; Winkler and Plafker, 1981b). Volcanic rocks in the southern belt typically have pronounced aeromagnetic anomalies but no associated gravity anomalies, suggesting that the bodies are thin or rootless. The volcanic sequences are dominantly massive flows, pillow basalt, pillow breccia, and aquagene tuff as thick as 2500 m. Sheeted dikes are not present in any of these sequences, and, in general, they contain a greater proportion of interbedded sedimentary rocks than those of the northern belt. Red or green argillite or varicolored chert is locally associated with the amygdaloidal volcanic rocks, pillow breccia, or aquagene tuff, and in a few places upper surfaces of volcanic rocks are mantled with thin beds of bioclastic limestone. Fragmental bioclastic limestone with abundant shallow-water fossils and pelagic limestone containing foraminifers indicate deposition above the carbonate compensation depth in shallower water than normal flysch (Winkler, 1976). In some parts of the belt, and as far southwest as Montague Island, red or green argillite, limestone, and tuffaceous fine-grained sedimentary rocks are interbedded with typical flysch but other volcanic rocks are lacking (Helwig and Emmet, 1981; Dumoulin and Miller, 1984).

The third belt of volcanic rocks occurs within the Orca Group on the continental shelf in the Gulf of Alaska where it is defined by large positive anomalies in the aeromagnetic data (Winkler and Plafker, 1981b; J. E. Case, G. Plafker, and G. R. Winkler, unpublished data). By analogy with the onshore geology and geophysics, they are interpreted as discontinuous volcanic sequences at or near the sea floor that are probably similar to the volcanic rocks of the southern Prince William Sound belt. Geochemical data show that the volcanic and related intrusive rocks of the Orca Group have diverse compositions that indicate several possible interpreted origins (Barker, this volume, Table 3). Of 19 samples analyzed, 12 are basalt, 5 are diabase, and 2 are basaltic andesite. On an AFL diagram, samples plot in both the tholeiitic and calc-alkalic fields. High CaO (13% to 14%) in three samples is probably due to abundance of calcite as veinlets and amygdule filings. On the basis of minor element geochemistry, these rocks fall into four distinct suites, as follows: (1) moderately LREE-enriched samples from east of the Copper River (at Ragged Mountain) in a setting that is compatible with enriched MORB; (2) samples from the northern Prince William Sound belt that have flat REE trends, slight negative Eu anomalies, mainly N-MORB compositions, and a slight calc-alkaline affinity; (3) samples from the southern belt near Cordova and east of the Copper River at Ragged Mountain, and one from the northern belt in northeastern Prince William Sound that have low REE abundances and strong depletion of LREE compared to HREE that could indicate low-K tholeiite; and (4) one basaltic metaandesite from the northern belt in western Prince William

Sound that has a composition suggestive of contaminated basaltic andesite (similar to basaltic andesite of the Ghost Rocks Formation).

Thickness. The stratigraphic thickness of the Orca Group in outcrop is estimated as many thousands of meters (possibly 6 to 10 km), but accurate determinations of thickness are precluded by the pervasive tight folding and imbricate faulting together with poor biostratigraphic control (Winkler and Plafker, 1981b; Tysdal and Case, 1979; Helwig and Emmet, 1981). Refraction and reflection seismic data indicate a vertical thickness of about 14 to 18 km beneath the continental shelf and 18 to 20 km in Prince William Sound (Brocher and others, 1991).

Metamorphism. The entire Orca Group was converted to transitional continental crust during a major thermal event that was accompanied by widespread plutonism at about 50 Ma (Barker and others, 1992; Moll-Stalcup, this volume). During this event the rocks underwent dominantly laumontite and prehnite-pumpellyite facies metamorphism and local greenschist facies metamorphism (to biotite grade) near the larger plutons east of Prince William Sound (Fig. 6). Widespread thermal metamorphism of the Orca Group accompanied emplacement of early Oligocene sills, stocks, and small plutons in northern and western Prince William Sound that range in composition from gabbro to granite (Tysdal and Case, 1979; Nelson and others, 1985; G. Plafker and G. B. Dalrymple, unpublished data).

Structure. The Orca Group was deformed in two major and distinct episodes. The first episode preceded intrusion of the granitoid rocks and probably was contemporaneous with accretion. It is characterized by complex folding and faulting that produced a penetrative slaty cleavage. Fold axial planes and faults are vertical to seaward vergent; a notable exception is a zone of landward-vergent structures up to 15 km wide that extends from Montague Island to Hinchinbrook Island (Plafker, 1967; Winkler, 1976; Helwig and Emmett, 1981; Plafker and others, 1986). Regional strike is generally northeast in most of the area, but in a zone extending eastward from northeastern Prince William Sound, it is highly variable and even north to northwest (Fig. 6; Plafker and others, 1986). In eastern Prince William Sound the axial planes, cleavage or schistosity, and compositional layering have vertical to steep north dips, and fold axes and lineations plunge gently northeast or southwest. Parallelism of subhorizontal major and minor fold axes and lineations, where they are not affected by later folding, indicates dominantly orthogonal convergence and local areas of oblique convergence during accretion. Within a few hundred meters of the Contact fault system, intense folding and faulting in both upper and lower plate rocks commonly resulted in general parallelism of structures on both sides of the fault. An exception is the area just east of Prince William Sound, where major northeast-trending units and structures in the Orca Group extend obliquely into a dominantly east trending segment of the Contact fault system (Fig. 6).

Upper contacts of the volcanic rocks are generally concordant with the enclosing flysch, and lower contacts may be either concordant or faulted, where exposed. These relations suggest that the volcanic sequences did not function as semirigid blocks that were decoupled from stresses imposed on the flyschoid sequences. Instead, volcanic sequences generally share the same pattern of folds and faults that typifies the complexly deformed flysch. Although exposed volcanic sequences southwest of Knight Island were described as blocks in melange by Helwig and Emmet (1981), many of the "blocks" exhibit intrusive and depositional contacts with the enclosing rocks, indicating that they are not structurally disrupted (Tysdal and Case, 1979). Superimposed on the accretionary fabric of the Orca Group are younger open folds and related structures that are associated with broad regional warping in northeastern Prince William Sound and in the area to the east (Nokleberg and others, 1989). This warping is best displayed in the arcuate trace of the Contact fault system and by planar structures within the Orca Group near the fault. The transition from east-west to northeast trends defines part of the axis of the Alaskan orocline of Carey (1958). North-northeast strikes and near-vertical dips of the axial planes of the open folds indicate mainly east-southeast to west-northwest compression. Local zones of right-lateral shear within the Orca Group west of this axis may reflect late-stage counterclockwise bending of the west limb of the orocline. Additional northwest-directed late Cenozoic compressional deformation, as manifested by slip on minor conjugate strike-slip faults, may be related to emplacement of the Yakutat terrane against and beneath the Orca Group.

Age. Paleontologic data indicate that Orca Group strata range in age from late Paleocene through middle Eocene. Age-diagnostic fossil assemblages include foraminifers (29), diatoms and silicoflagellates (1), radiolarians (8), coccoliths (1), dinoflagellates (1), and one megafauna of mollusks and crabs (Plafker and others, 1985b).

The lower age range of the unit is uncertain but is presumably late Paleocene on the basis of age-diagnostic planktonic foraminifers at nine localities. Radiometric data constrain the younger age limit for much, if not all, of the Orca Group to the early middle Eocene (51 to 53 $\pm$ 1.6 Ma). This age limit is taken as the emplacement age of the early middle Eocene plutons that intruded and metamorphosed the sequence.

Paleomagnetic data. Paleomagnetic studies of pillow basalt in the Orca Group have yielded low-quality and contradictory results. Data from the Knight Island area in western Prince William Sound suggest little or no northward displacement (Hillhouse and Grommé, 1977), whereas preliminary analysis of data from the northeastern part of Prince William Sound suggests an origin possibly 40° to the south (Plumley and Plafker, 1985). However, these data are suspect because of the possibility that widespread thermal events of probable early to middle Eocene and early Oligocene age have reset the magnetization. Attempts to obtain paleomagnetic data from sedimentary rocks in the Orca Group have been unsuccessful because the magnetization has been overprinted by younger events (Hillhouse and Grommé, 1985). Geologic data indicate that the Orca Group was accreted to the Chugach terrane by 51 ± 3 Ma and that postaccretion horizontal displacement along the Contact fault suture between these terranes is no more than a few tens of kilometers (Winkler and Plafker, 1981b).

Interpretation. The bulk of the siliciclastic rocks that constitute the Orca Group was deposited as a deep-sea fan complex with paleocurrents and distinctive associations of turbidite facies indicating sediment transport and deposition on westward-sloping fans (Winkler, 1976). The combined sandstone compositions and sediment transport directions are compatible with a provenance mainly in the Kluane arc (Fig. 5). Sedimentation coincides with the main phase of uplift of the Coast Mountains of British Columbia and Alaska between 62 and 48 Ma (Hollister, 1979, 1982; Crawford and Hollister, 1982) and with the counterclockwise oroclinal bending of western Alaska (Plafker, 1987; Hillhouse and Coe, this volume). Minor sediment contributions to the deep-sea fans came from the adjacent Chugach terrane (Winkler, 1976).

Coeval submarine volcanism resulted in intercalation of local basalt masses within prisms of terrigenous sediment. Rapid northeast to north movement of the Kula plate resulted in progressive offscraping of these deposits, together with far-traveled material carried into the area by the Kula plate, against previously accreted rocks along the continental margin of what is now the northern and western Gulf of Alaska (Helwig and Emmett, 1981). Deformation of the Orca Group began prior to complete dewatering of the sedimentary rocks (Winkler, 1976) and soon was followed by intrusions of granodiorite, granite, and tonalite plutons, ranging in age from 53.5 to 50.5 Ma (± 1.6 Ma) in eastern Prince William Sound and the area to the east (Plafker and others, 1989).

The mafic volcanic rocks and ophiolite that make up the northern belt are probably a relatively coherent slice of oceanic crust, on the basis of the composition, distribution, and associated positive magnetic and gravity anomalies. An origin at the Kula-Farallon ridge is compatible with the occurrence of ophiolite and MORB-type lavas. Volcanic rocks in the southern belt and offshore are more likely rootless fragments of submarine ridges or seamount chains, on the basis of their variable composition, lenticularity along strike, the high proportion of glassy, vesiculated, or fragmental textures, their local association with nonflyschoid shallow-water sediments, and the absence of ophiolite sequences or strong positive gravity anomalies. Possible sites for eruption of the basalts are leaky transform faults along the continental margin (Tysdal and Case, 1979) or primitive intraoceanic arcs—environments that give rise to more diverse magma types and the formation of both LREE-enriched and LREE-depleted lavas (Lull and Plafker, 1990).

Resurrection Peninsula sequence. The Resurrection Peninsula sequence occupies an outcrop belt 21 km long and as much as 6 km wide on the mainland south of Seward (Fig. 6). The sequence is juxtaposed against rocks of the Chugach terrane to the northwest along a structure that may be a continuation of the Contact fault system, although its position in this area has not been well determined. The nature of the southern contact beyond the Resurrection Peninsula is unknown.

This sequence is along structural strike with the Ghost Rocks Formation, and it was tentatively included within the Ghost Rocks terrane on the basis of structural continuity and paleomagnetism (Plafker, 1987). Subsequently, the Resurrection Peninsula sequence has been shown to be in part or entirely younger than the Ghost Rocks Formation (Nelson and others, 1989), and the available age data suggest that it may be correlative with the Orca Group in Prince William Sound. However, because it is not known to be continuous in outcrop with the Orca Group, and because of the possibility that the ages of dated igneous rocks are minimum ages, the sequence is described separately here from the Orca Group.

Lithology. The Resurrection Peninsula and adjacent islands are underlain by an ophiolitic sequence that grades from mainly gabbro (in part cumulate) at the base, through a sheeted mafic dike complex, to pillow basalt at the top (Tysdal and Case, 1979; M. L. Miller, *in* Winkler and others, 1984). Serpentinized peridotite and pyroxenite occur as pods in the gabbro and as fault-bounded slices in adjacent flysch; dikes and plugs of plagiogranite intrude the upper parts of the gabbro. Siltstone is locally interbedded with the pillow basalt and interbedded tuff, and siltstone locally appears to stratigraphically overlie the ophiolitic rocks.

All rocks in the sequence are affected by low-greenschist facies metamorphism with a later prehnite facies overprint (Dusel-Bacon, this volume). The metamorphism, together with serpentinization of the ultramafic rocks and a lack of penetrative fabric, indicates mainly thermal and hydrothermal metamorphism related to hot circulating water on the sea floor, rather than the regional greenschist metamorphism (to biotite grade) that has affected adjacent Valdez Group rocks. The prehnite, however, could also be the result of a younger prehnite-pumpellyite-grade metamorphic event related to emplacement of these rocks in an accretionary sequence.

Structure. The ophiolitic rocks occur in two fault-bounded, apparently coherent slices; the main body on Resurrection Peninsula forms a homoclinal slab that dips steeply to moderately west (M. L. Miller, *in* Winkler and others, 1984). Structural data indicate polyphase deformation that included folding about a vertical structural axis during one phase (Bol and others, 1992). All contacts of the Resurrection Peninsula sequence with the adjacent rocks of the Valdez Group are faults, and the contact is a thrust fault at least locally (Nelson and others, 1985; M. L. Miller, *in* Winkler and others, 1984; Bol and others, 1992).

Age. The best crystallization age for part of the Resurrection Peninsula sequence is a U-Pb zircon age of 57 ± 1 Ma from a trondhjemite intrusion (Nelson and others, 1989). Four samples of altered basaltic greenstone from the Resurrection Peninsula and one from the nearby Renard Island yield K-Ar whole-rock ages ranging from 54.4 ± 2.7 to 45.7 ± 2.3 Ma (Winkler and others, 1984, Table 5) that are probably minimum ages and are compatible with the U-Pb age. These data suggest a probable age close to that of the Paleocene-Eocene boundary and a correlation with the younger part of the Orca Group rather than with the Ghost Rocks Formation, as previously inferred by Plafker (1987).

Paleomagnetic data. An excellent data set for basalts and sheeted dikes of the Resurrection Peninsula sequence indicates a paleomagnetic discrepancy of $13° \pm 9°$, assuming an early Eocene age; they also indicate counterclockwise rotations of about 90° (Bol and others, 1992).

Interpretation. The Resurrection Peninsula sequence is a relatively coherent fragment of oceanic crust and upper mantle that was offscraped and accreted with flysch deposits, presumably at the inner wall of a trench. Emplacement age of the Resurrection Peninsula rocks is not known; it is inferred to postdate accretion of the Late Cretaceous Valdez Group and to be coeval with, or to slightly predate, accretion of the parts of the Orca Group that are outboard of the sequence.

Rocks of the Resurrection Peninsula sequence were tentatively correlated with the Ghost Rocks Formation by Plafker (1987) on the basis of similarities in paleomagnetic signature and structural position. However, this correlation is now considered unlikely because at least part of the sequence is 5 m.y. younger than the Ghost Rocks Formation. The early Eocene isotopic age reported by Nelson and others (1989) favors correlation with oceanic rocks in the Orca Group; a suggestion made by some of the earliest workers in the area (Grant and Higgins, 1910). The ophiolitic sequence is very similar to the thick undated pillow-basalt and sheeted-dike sequences of the Orca Group that make up most of Knight Island in western Prince William Sound (Tysdal and Case, 1979).

The paleomagnetic data indicate northward displacement of between 4° and 22° since crystallization of the igneous rocks. If the age of the sampled rocks is about 57 Ma, as suggested by isotopic data, only the minimal displacements are compatible with evidence indicating that the amalgamated Wrangellia and Southern Margin composite terranes were at about their present paleolatitude at 55 Ma (Hillhouse and Coe, this volume). Alternatively, if the 57 Ma age is a minimum emplacement age for these rocks, earlier relative poleward displacement of the sequence is required and could have occurred by a combination of northward movement relative to the Chugach terrane and by dextral slip along faults between these terranes and the craton, as proposed by Bol and others (1992). A major difficulty with this alternative, however, is that cumulative dextral slip on all known early Tertiary faults between the Resurrection Peninsula sequence and the craton can account for no more than about 8° of northward displacement (Plafker, 1987).

Sitkalidak Formation. The Sitkalidak Formation (Moore, 1969) crops out along the southeastern coast of Kodiak Island and on adjacent offshore islands, where it forms a belt 165 km long and up to 25 km wide (Fig. 6). It is separated from the older Ghost Rocks Formation to the northwest by the Resurrection fault and is conformably to unconformably overlain by younger Tertiary units (Moore, 1969; Moore and Allwardt, 1980). Although the age range is uncertain, available data suggest that the sequence is in part or entirely correlative with the Orca Group of the Prince William terrane (Fig. 7).

The Sitkalidak Formation and correlative deformed accreted rocks form the acoustic basement offshore from the Kodiak Islands. Seismic reflection data and stratigraphy in holes drilled for petroleum exploration suggest that comparable rocks underlie much or all of the continental shelf and possibly the upper slope off the Kodiak Islands (Fisher and others, 1987; Hoose, 1987; Olson and others, 1987).

Lithology. The Sitkalidak Formation consists of massive sandstone, interbedded sandstone and siltstone, mudstone, and conglomerate deposited by turbidity currents. Although the sandstone-siltstone ratio is highly variable, the overall ratio is about 50:50. The sandstone beds are medium to coarse grained, poorly graded, and average about 1.0 to 1.5 m in thickness. Intervening beds are well-graded fissile siltstone and fine-grained sandstone beds that vary in thickness from a few centimeters to 50 cm. Sandstone modes of the Sitkalidak Formation are characterized by quartz and feldspar, in about equal proportions, that together constitute aobut 50% to 60% of the total detrital grains (Stewart, 1976; Lyle and others, 1977; G. R. Winkler, *in* Nilsen and Moore, 1979; Moore and others, 1983). Lithic fragments of dominantly volcanic and sedimentary origin constitute most of the remainder, and minor constituents are epidote, pyroxene, biotite, and muscovite. Sandstone composition suggests a mixed provenance with a dominant magmatic-arc component. Stratigraphic thickness of the unit is about 3000 m (Moore, 1969).

The Sitkalidak Formation is metamorphosed to the zeolite facies; peak metamorphic temperatures were 100 to 125 °C and maximum burial depths were 2.4 to 3.9 km (Moore and Allwardt, 1980).

Structure. Two general structural styles that suggest different modes of accretion have been recognized within the Sitkalidak Formation (Moore and Allwardt, 1980). One characterizes an inboard unit of seaward-verging strata that are tightly folded but not disrupted; the other characterizes a more deformed outboard unit of landward-verging strata and commonly exhibits stratal disruption along cataclastic shear zones with web structures (Fig. 9; Sample and Moore, 1987).

Age. The Sitkalidak Formation is younger than the Late Cretaceous and early Paleocene Ghost Rocks Formation and older than the overlying Oligocene strata of the Sitkinak Formation. Although Moore (1969) and subsequent workers (e.g., Armentrout, 1975; Olson and others, 1987) assigned an Oligocene age to the formation, that assignment is based on a very sparse and poorly preserved fossil fauna. The megafauna consists of a single long-ranging fossil clam of unknown age significance (a new genus of Vesicomyidae) and small fragments of a fossil crab (*Calianassa* aff. *C. porterensis*) of a type that occurs in late Eocene rocks of western Washington but has an uncertain age range. Ten samples from the formation have yielded a poorly preserved predominantly agglutinated bathyal foraminiferal assemblage consisting of long-ranging forms (Lyle and others, 1977; Clendenen and others, 1992). This assemblage includes *Ammosphaeroidinia*? sp., *Bathysiphon alexanderi, B. sanctaecrucis, B. eocenica, Haplophragmoides* sp., *H. obliquicameratus* Marks, *H.* cf. *deformes, Nodosarella atlantisae hispidula* (Cushman), *Nodosaria arundinea* Schwager, *Praeglobobulimina ovata cowlitzensis* (Beck), *Reophax* sp., *Rhabdammina eocenica* Cushman and Hanna, and *Trochammina globegeriniformis* (Parker and Jones). The presence of *P. ovata cowlitzensis* constrains the depositional age to the Eocene (Clendenen and others, 1992). Possibly correlative strata penetrated in three stratigraphic test wells on the continental shelf, beneath a regional unconformity, contain Eocene microfossils; one well (KSSD No. 1; Fig. 10) includes strata of early and middle Eocene age, on the basis of dinocyst fossils and to a lesser extent on foraminifers (Olson and others, 1987).

Interpretation. The facies sequences, paleocurrents, and variable pebble-clast compositions suggest deposition as a series of coalescing submarine fans, probably on a trench floor or in a trench-slope basin (Nilsen and Moore, 1979). The unit consists mainly of outer fan turbidites that locally include some basin-plain and inner fan deposits; paleocurrents indicate general transport of sediments both downslope and parallel to the slope with a major feeder system in the vicinity of Sitkalidak Island (Nilsen and Moore, 1979; Moore and Allwardt, 1980).

The combined sedimentological and structural data suggest that the more deformed part of the Sitkalidak Formation was emplaced as an obductively offscraped trench-fill sequence and that the less-deformed inboard part probably accumulated within a slope basin (Moore and Allwardt, 1980). At least the older part of the unit may be equivalent to the Orca Group on the basis of its age, lithology, structure, and position within the accretionary complex. Sandstone compositions are compatible with a common dissected continental-margin magmatic-arc source for these units.

The difference in metamorphism between the Sitkalidak and

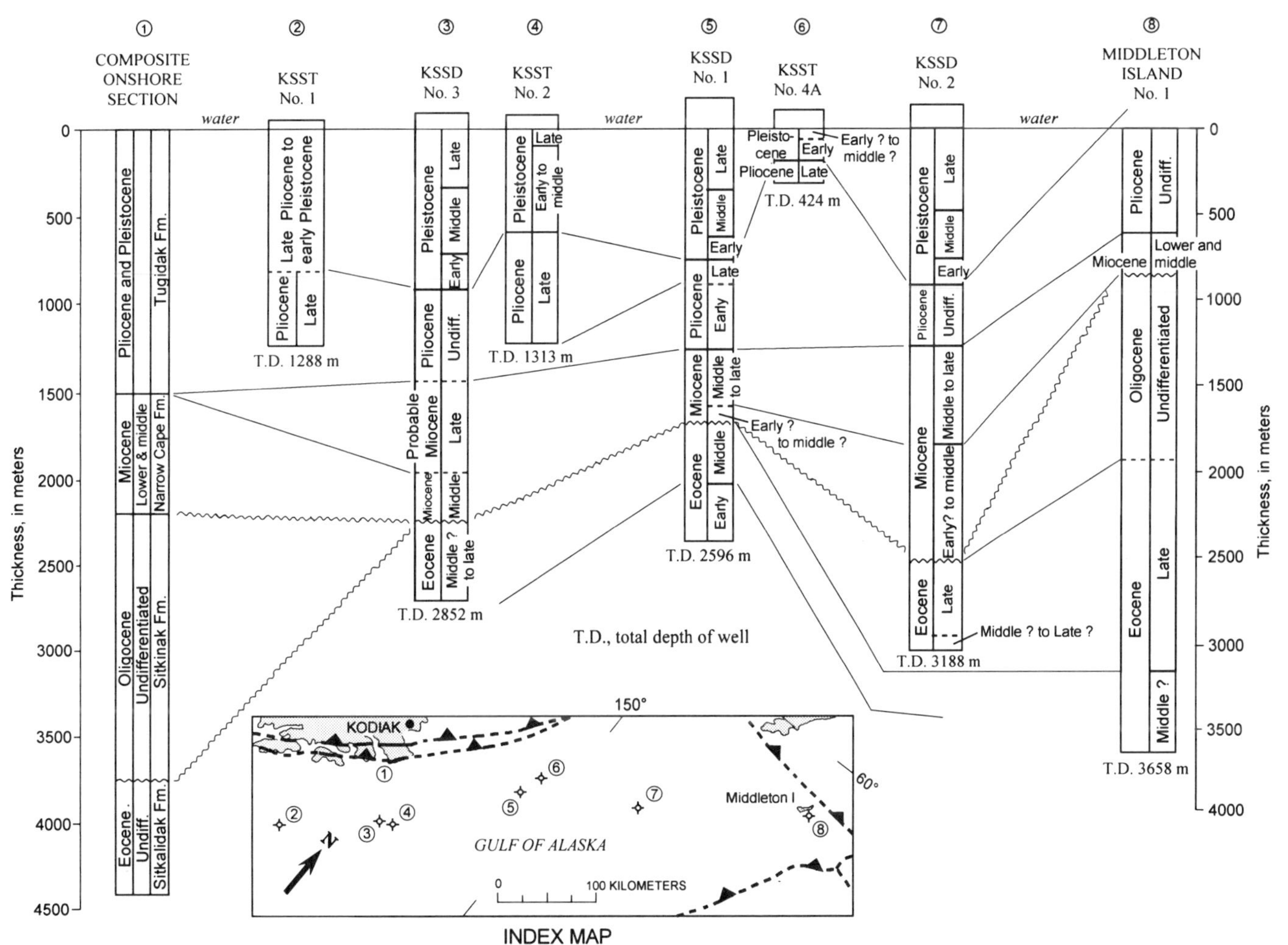

Figure 10. Correlation chart for Tertiary overlap assemblages from outcrops on the Kodiak Islands and from wells on the Kodiak and Middleton shelves. Adapted from North American Commission on Stratigraphic Nomenclature (1983); Rau and others (1983); Olson and others (1987); Schaff and Gilbert (1987).

Ghost Rocks formations indicates ~8 km of vertical displacement along the Resurrection fault that separates them. However, marine seismic-reflection data show no obvious structural boundary along the projection of the Resurrection fault to the northeast (Moore and others, 1991).

Late Cenozoic accretionary sequences associated with the Aleutian Arc and megathrust

Accreted post-Eocene strata are inferred to underlie large areas of the continental margin west of Middleton Island, and they occur at the base of the slope in an area about 100 km long by 40 km wide that is bounded by the easternmost Aleutian megathrust, the Transition fault system, and the Kayak Island fault zone (Fig. 6).

Sediments on the present oceanic crust consist of 3 to 5 km of deep-sea fan and trench-fill turbidites mainly of Pliocene and Quaternary age that overlie a thin layer of Tertiary abyssal sediments of Eocene to Miocene age (von Huene and others, 1987; Drummond, 1986). The age of the basal abyssal sediments and underlying basalt crust ranges from chron 15 (about 37 Ma) at the eastern end of the Aleutian Trench to chron 24 (about 56 Ma) off the Sanak Islands (Atwater and Severinghaus, 1989). The sedimentary rocks and oceanic crust are being accreted to, and underthrust beneath, the continental slope with resultant seaward progradation of the accretionary prism.

In the segment south of Middleton Island, accreted post-Eocene strata are inferred to underlie much of the continental margin in a belt that is close to 100 km wide perpendicular to the trench (von Huene and others, 1987; Moore and others, 1991). A decollement at the base of the slope typically separates the section into a deformed and accreted upper part and a relatively undeformed subducted lower part (von Huene and others, 1987). The deformed upper part, which underlies much of the continental slope, includes both seaward- and landward-verging thrust faults and associated folds (von Huene and others, 1987; Moore and others, 1991). The inner margin of these accreted rocks is marked, at least locally, by an out-of-sequence thrust fault that juxtaposes deformed and indurated Paleogene strata against Pleistocene slope sediments (see section on deep crustal structure, below). Where identified on deep seismic reflection profiles near Stevenson Basin (Fig. 11), this boundary is located close to the continental shelf edge. The sequence below the basal decollement is progressively underplated at depth beneath the accretionary prism along the entire 300 km width that it can be seen seismically (Moore and others, 1991).

At the eastern end of the Aleutian megathrust, offscraped oceanic sedimentary rocks underlie the lower 1 to 1.5 km of the

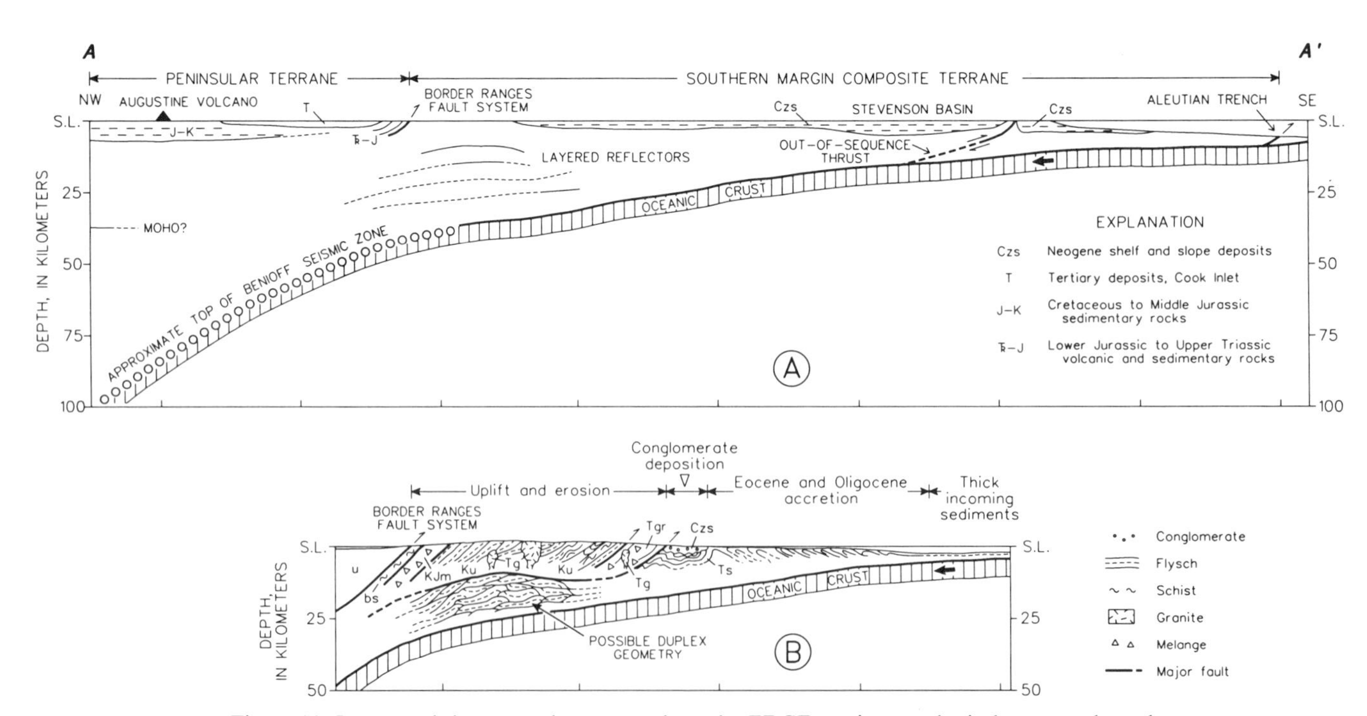

Figure 11. Interpreted deep crustal structure along the EDGE marine geophysical transect through Kennedy Entrance from the Pacific plate to the Aleutian arc; location of section shown in Figure 6. A: Section based on seismic-reflection data; base of oceanic crust projects into zone of intense seismicity of downgoing slab. Velocities for depth conversions obtained from wells, from velocity models used to locate earthquakes, and from preliminary results of unreversed piggyback ocean-bottom seismometer studies. B: Inferred structure of the Mesozoic and Paleogene accretionary prisms showing emplacement of layered reflectors by underplating in response to seaward growth of margin during late Eocene or Oligocene time; see Figure 6 for explanation of unit symbols (after Moore and others, 1991).

continental slope south of the Transition fault system (unit Czu, Fig. 6). These accreted strata form a large faulted anticlinal ridge and a subparallel basin on the landward side that contains young tilted sediments (Bruns and Schwab, 1983, section 922). The inner margin of accreted late Cenozoic strata cannot be resolved with available seismic reflection data. The margin shown in Figure 6 coincides with the south limits of the Yakutat terrane as deduced by Griscom and Sauer (1990) from magnetic data. Samples dredged from the south flank of the anticlinal ridge and the continental slope to the north are slightly indurated to well-indurated siliciclastic sediments and calcareous sediments that contain a microfauna of Pliocene through Pleistocene age and locally abundant reworked Paleogene microfossils (Plafker, 1987). The dredge samples appear to be accreted trench deposits that were subjected to unusually rapid lithification, possibly due to tectonic deformation related to plate convergence. Pelagic sediments have not been recovered, and the microfossils indicate a terrigenous sediment source. This suggests that the accretionary prism in this area consists dominantly of fan and trench-fill turbidites without a significant component of abyssal sediments.

Structures within and below the accretionary prism appear to be highly variable along strike. A significant gap in the late Cenozoic accretionary prism apparently occurs along this margin at the inner trench wall off Middleton Island. In that area the oceanic crust and its blanket of about 3.5 km of sediment is apparently thrust at least 14 km beneath the continental slope (Fig. 3, A–A‴) and there is no evidence for a Cenozoic accretionary prism comparable to those that occur both to the southwest and northeast.

Overlap assemblages and basinal sequences

Arc-related rocks of the Southern Margin composite terrane are in fault contact with, or are unconformably overlain by, relatively undeformed overlap assemblages that are exposed locally along the southeast margin of the Kodiak Islands, and underlie Chirikof, Tugidak, Sitkinak, and Middleton islands as well as much of the contiguous continental shelf and slope (Fig. 6). Thickness and lithology of the sequences has been studied in offshore areas mainly by regional seismic-reflection profiles, six deep stratigraphic test wells on the Kodiak shelf, and one well drilled for petroleum on the Middleton shelf near Middleton Island. Stratigraphic correlations between onshore sections in the Kodiak Islands and offshore wells of the Kodiak and Middleton shelves are shown in Figure 10. The following summary of these deposits is based mainly on data in Wiley (1986), Bruns and Carlson (1987), Bruns and others (1987), Plafker (1987), Schaff and Gilbert (1987), Turner (1987), von Huene and others (1987), and Vallier and others (this volume).

Basinal sequences east of the Middleton shelf are discussed in a following section on the Yakutat terrane.

Sanak and Shumagin basins. The Sanak and Shumagin basins (Fig. 6) are large offshore sedimentary basins that overlie the Mesozoic accretionary prism in the vicinity of the Sanak Islands and the Shumagin Islands (von Huene and others, 1987). The Sanak basin is a complex of fault-bounded troughs containing sediments more than 8 km thick, whereas the Shumagin basin is a relatively simple depression with as much as 2.5 km of fill (Fig. 6). The basin fills are inferred to be Miocene and younger clastic and volcaniclastic(?) strata. Acoustic basement probably consists of the same sequence of highly deformed Late Cretaceous turbidites and older melange that crops out on the nearby islands (Fig. 7). Additional detail on the stratigraphy and structure of these basins was given by Bruns and others (1987) and Vallier and others (this volume).

Kodiak Islands and Kodiak shelf. Basinal and overlap sequences of late Oligocene and younger age are intermittently exposed in outcrop from Kodiak to Chirikof Island and are inferred from seismic reflection and well data to underlie much of the contiguous continental shelf (Fig. 6).

The lower part of the overlap sequence in outcrop includes about 1500 m of fossiliferous sandstone, siltstone, conglomerate, and minor coal of the Sitkinak Formation and 700 m of sandy marine siltstone of the Narrow Cape Formation. Both units were deposited in a nearshore environment; the Sitkinak Formation is Oligocene in age (Moore, 1969), and the Narrow Cape is middle Miocene (Clendenen and others, 1992). Sandstone compositions suggest a local source with progressive exposure and recycling of older sediments and unroofing of granitic plutons (Fisher and von Huene, 1980). The upper part of the section consists of about 1500 m of Pliocene, fossiliferous, interbedded sandstone, siltstone, and conglomeratic sandy mudstone of the Tugidak Formation, much of which was deposited in a shallow glaciomarine environment. Strata penetrated in the offshore wells are generally similar in lithology to the Tugidak and Narrow Cape formations (Turner, 1987); Oligocene strata equivalent in age to the onshore Sitkinak Formation are known to be present offshore only in the Middleton Island No. 1 well on the Middleton shelf (Fig. 10).

Major continental margin basins in this region are the Tugidak and Albatross basins (each filled with about 5 km of strata) and the Stevenson basin (7 km of strata). The basin sequences are of middle Miocene or younger age (Fig. 10) and all contain folds and faults that are at least partly contemporaneous with basin filling (von Huene and others, 1987; Hoose, 1987; Olson and others, 1987). On the Kodiak shelf, local basin subsidence is inferred to result from downward flexure of the underlying lithosphere as a consequence of vertical thickening of the adjacent accretionary prism near the shelf edge (Moore and others, 1991; Clendenen and others, 1992).

Middleton shelf. Seismic data indicate that the late Cenozoic section beneath the Middleton shelf landward from Middleton Island is variable, with a maximum thickness of 5 km (Bruns, 1979). Anticlines are characteristically tight and extensively faulted and have irregular but predominantly east-west to northeast-southwest trends. Multiple minor unconformities and growth features within the late Cenozoic section indicate that local deformation was contemporaneous with deposition.

The section beneath the Middleton shelf is known primarily

from the Middleton Island No. 1 well (Fig. 10) that penetrated 3700 m of an uninterrupted sequence of Eocene and younger clastic sedimentary strata (Rau and others, 1983; Keller and others, 1984). Samples of Quaternary glaciomarine rocks were dredged from the adjacent continental slope, and about 1100 m of late Pliocene and early Pleistocene glaciomarine strata are exposed on Middleton Island (Plafker, 1987). The lower 1700 m in the Middleton Island well consists of Eocene siltstone and sandstone that were deposited in progressively shallower marine environments, from bathyal at the base to middle or upper bathyal at the top of the section, as indicated by benthonic foraminifera assemblages (Rau and others, 1983). The age of the base of this interval is uncertain; it is most likely late early to early middle Eocene, on the basis of benthonic foraminifers (Rau and others, 1983), late middle Eocene, on the basis of coccoliths (J. D. Bukry, *in* Keller and others, 1984), or earliest late Eocene, on the basis of sparse planktonic foraminifers (Keller and others, 1984). The Eocene section is overlain by about 1000 m of Oligocene and 300 m of early and middle Miocene siltstone, claystone, and minor sandstone deposited at middle to upper bathyal depths, as indicated by the benthonic foraminifera fauna (Rau and others, 1983). The upper 625 m of the well penetrates a Pliocene sequence of siltstone, mudstone, and conglomeratic sandy mudstone with a fauna that indicates deposition in a cold, shallow-marine environment, probably partly of glaciomarine origin, comparable to the depositional environment of the section exposed on Middleton Island.

Middleton Island is on a shelf-edge structural high. This is the only part of the basin where a thick, gently dipping section (Fig. 3, A–A‴) can be observed on marine seismic reflection profiles, and the reflectors can be traced from the vicinity of the well southward to the continental slope. Middleton Island has undergone Holocene northwestward tilting of as much as 28°, faulting, and earthquake-related tectonic uplift at a rate that averages close to 1 cm/yr and continues to the present time (Plafker and Rubin, 1978). However, paleontologic and dipmeter data suggest that the section in the nearby well is not complicated by steep dips or faulting, except possibly close to the bottom (Fig. 3, A–A‴). These relations suggest that there may be a fault between Middleton Island and the well.

The geology of the shelf near Middleton Island differs from that of the Kodiak shelf in that its thick, relatively undeformed Oligocene and Eocene marine section underlies the regional sub-Miocene unconformity (Fig. 10) and that the turbidite sequence of the Eocene Sitkalidak Formation is not encountered in the well. Stratigraphic relations require that the accretionary prism beneath the outer Middleton shelf is middle Eocene or older in age, and it is most probably the Orca Group (Plafker, 1987). The absence of deep reflectors in seismic profiles elsewhere on the Middleton shelf indicates that the Eocene and Oligocene parts of the basinal sequence occur mainly on the outer shelf and slope and that these strata may thin landward onto the Orca Group close to the shelf edge, as depicted schematically in Figure 3, section A-A‴. Paleontologic data from Eocene and younger Tertiary strata in the Middleton Island well favor the interpretation that they were deposited at about their present paleolatitude relative to the North America continental margin (Keller and others, 1984).

Edgecumbe volcanics and the Sitka shelf. In the Sitka area of southeastern Alaska, the Edgecumbe Volcanics unconformably overlap rocks of the Chugach terrane over an area of 260 km^2 (Fig. 6). The volcanic rocks comprise a bimodal sequence of tholeiitic basalt and calc-alkalic flow and pyroclastic rocks at least in part of Holocene age (Brew, this volume). Structural control for the location of this volcanic field is enigmatic; Brew (this volume) suggests that the field is localized along an unidentified zone of extension that is related to the offshore Fairweather strike-slip fault.

Rocks of the Chugach terrane beneath the continental shelf of southeastern Alaska west of Chatham Strait are unconformably overlain by a cover of acoustically transparent strata of probable Neogene age that locally fill small elongate basins as much as 3 km thick (Bruns and Carlson, 1987). The bedded strata are gently to moderately folded, and they are locally displaced along active strands of the Queen Charlotte fault.

Plutonic rocks

The Southern Margin composite terrane is extensively intruded by plutonic rocks of gabbroic to granitic composition and mid-Cretaceous to middle Cenozoic age. Widespread metamorphism of the accreted rocks is related to emplacement of these plutons, especially the numerous calc-alkaline plutons of Paleocene and Eocene age (Fig. 6). The geology of the plutonic rocks is briefly summarized here. Additional data on their occurrence, mineralogy, geochemistry, and age are given in other chapters of this volume, especially those by Brew, Hudson, Miller, and Moll-Stalcup, and in other references cited.

Cretaceous granitic rocks. Sparse, weakly foliated to nonfoliated Cretaceous granitic plutons less than a few square kilometers in area (too small to show in Fig. 6) are locally emplaced into melange along the north margin of the Chugach terrane in the western Chugach Mountains (Pavlis, 1982) and in the Tarr Inlet area of the southeastern Saint Elias Mountains (Yakutat-Chichagof belt of Hudson, 1983). Similar granitic plutonic rocks are emplaced into the Yakutat Group melange in the eastern part of the Yakutat terrane. One such pluton in the Mount Fairweather quadrangle north of Lituya Bay is 20 to 30 km long and averages 2 km in width (Fig. 6). A Cretaceous pluton in the western Chugach Mountains that is too small to show in Figure 6 reportedly is emplaced across the Border Ranges fault (Pavlis, 1982).

Most of the plutonic rocks are highly altered and locally foliated trondhjemite, tonalite, and diorite. Alteration consists of extensive saussuritization, sericitization, and replacement of biotite by prehnite and chlorite. Plutons in this suite have limited distribution and are dated mainly by K-Ar analyses. In the Chugach terrane, isotopic ages of these plutons range from 124 to 100 Ma in the western Chugach Mountains (Pavlis, 1982; Pavlis and others, 1988; Winkler, 1992) and about 119 to 97 Ma in the Tarr Inlet area (Decker and Plafker, 1982); the only dated rock of this suite in the Yakutat terrane is 96 ± 4.5 Ma (Fig. 7).

Pavlis (1982) interpreted granitic rocks of the Cretaceous suite in the western Chugach Mountains to be products of near-trench magmatism related to docking of the Wrangellia composite terrane against North America. Emplacement ages correspond to a widespread thermal event in Alaska that resulted in resetting of K-Ar ages in granitic rocks (Plafker and others, 1989) and greenstone (Silberman and others, 1981) in the southern part of the Wrangellia terrane. At about the same time or shortly thereafter, subduction was taking place along the south margin of the Chugach terrane as indicated by local blueschist facies metamorphism dated at 106 to 91 Ma on Chichagof Island (Decker and others, 1980). Thus, the distance between the mid-Cretaceous trench and the plutons was approximately the width of the Chugach terrane, or between 50 and 100 km. As with the better-studied Paleogene plutons described below, the Cretaceous plutons may represent anatectic melts above leaky or unusually hot segments of the subducting Kula plate.

Paleocene and Eocene granitic rocks. Granitic plutons of Paleocene and Eocene age (Sanak-Baranof belt of Hudson, 1983; and this volume) are by far the most abundant intrusive rocks in the southern margin accretionary complex. They are discontinuously exposed for 2100 km along the Gulf of Alaska margin from the Sanak Islands to Chatham Strait. They commonly underlie 5% to 10% of the outcrop area of the accretionary assemblage except in some of the more deeply eroded parts of the Chugach and Saint Elias mountains, where the plutons and associated migmatite make up as much as 20% to 25% of the outcrop over large areas (Fig. 6). These plutons appear to have been intruded in two major episodes, although this impression may be a function of the limited number of dated samples, most of which are by K-Ar or Rb-Sr methods.

During the first episode in the Paleocene (64 to 60 Ma), a western group of plutons was emplaced into the Shumagin, Kodiak, and Ghost Rocks formations from the Sanak Islands to the Kodiak Islands. The second episode was during the Eocene (about 57 to 42 Ma), when the eastern group of plutons was emplaced into flysch and melange of the McHugh Complex and into the Valdez, Kelp Bay, and Yakutat groups and their metamorphosed equivalents from the Kenai Peninsula to Chatham Strait (Fig. 7). The eastern group of plutons shows a general tendency to be younger from west to east, although the change is not uniform across the region; for example, plutonic rocks with biotite K-Ar ages as old as 49 Ma occur in the southern part of this belt (Brew, this volume). Coeval felsic to intermediate hypabyssal dikes, sills, and small stocks commonly are associated with the plutons, but coeval hypabyssal intrusions that are not obviously connected to plutons are also widely distributed throughout the Prince William and Chugach terranes. Similar shallow intrusive bodies also occur in the Wrangellia and Peninsular terranes north of the Border Ranges fault system.

Most plutons are 50 to 100 km^2 in area, but they range from less than 1 km^2 to about 600 km^2. They tend to be elongate parallel to the structural trend of the wall rocks, except in the eastern Chugach Mountains, where the plutons trend east-northeast obliquely across the prevailing east-west structures (Fig. 6). Compositionally, the plutonic rocks are calc-alkaline and dominantly granodiorite, tonalite, and quartz diorite; granite, diorite, and gabbro are less common. Contacts are generally gradational and migmatitic at deep structural levels in much of the eastern Chugach and Saint Elias mountains, but sharply crosscutting at the higher structural levels exposed elsewhere. In the Ghost Rocks Formation on the Kodiak Islands and in the Orca Group in eastern Prince William Sound, field relations, ages, and geochemistry indicate that small bodies of gabbro and diorite probably are comagmatic with nearby Paleogene siliceous plutonic rocks (Hill and others, 1981; Barker and others, 1992). Zoned hornfels aureoles are present around many of these bodies. On southern Baranof Island, for example, flysch in a 1200 km^2 area is metamorphosed to albite-epidote hornfels in an outer aureole and to hornblende hornfels, amphibolite, and possibly pyroxene hornfels in an inner aureole (Loney and Brew, 1987; Dusel-Bacon, this volume).

The Paleocene and Eocene belt of granitoid rocks is anomalous in that the rocks intrude the accretionary prism and are located more than 100 km oceanward from rocks of the coeval magmatic arc. The field relations and a variety of geochemical data suggest an origin largely by anatectic melting of cold flyschoid graywacke, argillite, and basalt of the accretionary prism (Hudson and others, 1979; Hudson, this volume; Hill and others, 1981; Farmer and others, 1987; Barker and others, 1992). Sustained magmatic activity is required to produce the plutons in a 2100-km-long belt over a time span of 10 to 15 m.y. Barker and others (1992) reviewed a variety of mechanisms that have been proposed for melting the flysch, but because of thermal considerations they favor underplating by basaltic liquid. Although the source of the basaltic magma is uncertain, the observed magmatic complexity could involve subduction of one or more ridge-transform systems of the Kula and Farallon plates (Marshak and Karig, 1977; Barker and others, 1992), possibly with opening of mantle windows beneath parts of the accretionary prism as a consequence of subduction of the landward limb of the Kula-Farallon ridge (Plafker and others, 1989).

Hypabyssal phases of these intrusive rocks in adjacent areas of the Peninsular, Chugach, and Ghost Rocks terranes in the Kodiak Islands suggest that there has been no large relative movement between these terranes since the early Paleocene (Davies and Moore, 1984). Similarly, hypabyssal intrusive rocks that cross the boundaries between the Peninsular, Wrangellia, Chugach, and Prince William terranes in the western Chugach Mountains indicate that these terranes were in about their present relative positions by the middle Eocene (Plafker and others, 1989). Paleomagnetic data place the Wrangellia composite terrane and the attached terranes of the Southern Margin composite terrane at about their present latitude by at least 55 Ma (Hillhouse and others, 1985; Hillhouse and Coe, this volume).

Oligocene and younger granitic rocks. Post-Eocene intrusive rocks in the Southern Margin composite terrane consist of a suite of small intermediate to felsic plutons that are similar in composition, mineralogy, and occurrence to the Paleogene plutons except that most of them are not foliated.

Oligocene plutons with roughly circular outcrop areas as large as 270 km^2 intrude both the Valdez and Orca groups in western Prince William Sound (Fig. 6; Tysdal and Case, 1979; Nelson and others, 1985). The plutons are dominantly granite; more mafic marginal and early intrusive phases range in composition from granodiorite to two-pyroxene gabbro. K-Ar ages for plutonic rocks of this suite range from 38 to 34 Ma.

Intermediate-composition rocks younger than Oligocene are mainly small elongate hypabyssal bodies in the Yakutat and Saint Elias quadrangles of the Saint Elias Mountains (Hudson and others, 1977) and on Baranof Island (Loney and Brew, 1987); these plutons are too small to show in Figure 6. The rocks are dominantly muscovite-biotite granodiorite, commonly with more mafic quartz diorite and tonalite in the border zones. Three plutons dated by K-Ar within the Chugach terrane in the Yakutat quadrangle yield ages of about 30 to 20 Ma (Hudson and others, 1977), and one yielded a U-Pb age on zircon of 35 Ma (G. Plafker, 1981, unpublished data).

In general, the available data indicate that the Oligocene and younger granitoid rocks are grossly similar in composition and mineralogy to those of the Eocene suite. Without detailed studies, we suspect that they also may have formed by anatectic melting of the Mesozoic accretionary prism that makes up the wall rocks.

Late Eocene(?) and younger gabbroic rocks. A belt of layered gabbro bodies extends from the Fairweather Range to northern Chichagof Island in southeastern Alaska, the largest of which are shown in Figure 6 (see also Brew, this volume; Patton and others, this volume, Chapter 21). These plutons are of special interest because they contain potentially valuable occurrences of nickel, copper, and platinum-group minerals (Nokleberg and others, this volume, Chapter 10). In the mountainous Fairweather Range, four plutons of layered gabbro are known, two of which contain ultramafic cumulates. The largest is the La Perouse pluton, which is 27 km long, is as wide as 12 km, and underlies an area of 260 km^2. It consists of as much as 6000 m of layered gabbro, gabbronorite, and norite with a basal zone of ultramafic rocks at least 680 m thick; the rocks were emplaced at about 1055 °C and 5.4 kbar (Loney and Himmelberg, 1983; Brew, this volume). The age of the layered gabbro is between 41 and 19 Ma, most probably 28 ± 8 Ma (Hudson and Plafker, 1981; Loney and Himmelberg, 1983).

Small Tertiary gabbro or norite bodies and associated tonalite plutons just south of Cross Sound on Yakobi Island and northern Chichagof Island may be related to the layered gabbro plutons in the Fairweather Range, although available age data from the tonalite suggest that their earlier emplacement at about 43 to 34 Ma (Himmelberg and others, 1987; Brew, this volume).

The tectonic setting for emplacement of the mafic-ultramafic rocks is poorly understood. The rocks are considerably younger than the youngest rocks of the accretionary complex they intrude, and they appear to be too close to the continental margin to be arc related. Furthermore, the age data suggest emplacement during a time of dominantly dextral strike-slip motion along the southeastern Alaska margin (Engebretsen and others, 1985). The most likely possibility is that the mafic-ultramafic rocks represent deep magma chambers and feeders to volcanoes that formed above local areas of extension along the middle Tertiary continental margin. In this respect they may be analogous to the present Edgecumbe volcanic field, which is just inboard of the transform margin in southeastern Alaska (Fig. 6).

Deep crustal structure

Seismic refraction-reflection studies, together with earthquake seismology and potential-field geophysical data, provide insights into the deep crustal structure across the Southern Margin composite terrane and more landward parts of the Aleutian arc (Figs. 11 and 12). Much of the data are from two major transects: (1) a marine transect (EDGE transect) that extends from seaward of the Aleutian Trench through Kennedy Entrance into the volcanic arc near the Alaska Peninsula (Fisher and others, 1987; Moore and others, 1991); and (2) the Trans-Alaska Crustal Transect (TACT), which runs from the Pacific plate near the eastern end of the Aleutian Trench into Prince William Sound and from there onshore across all of Alaska, generally following the Richardson Highway and the trans-Alaska oil pipeline (Page and others, 1989; Griscom and Sauer, 1990; Brocher and others, 1991; Fuis and Plafker, 1991; Fuis and others, 1991).

EDGE transect. Along the EDGE transect, Moore and others (1991) showed gently dipping oceanic crust and the underlying Moho extending from the Aleutian Trench northwestward for more than 200 km beneath the accreted rocks of the Southern Margin composite terrane to a depth of 30 km or more (Fig. 11). North of the accretionary complex, earthquake seismologic data show a marked steepening of the subduction zone to about 45° at a depth of about 100 km beneath the Aleutian magmatic arc on the Alaska Peninsula (Fig. 11). Thus, the EDGE transect provides a virtually complete image of the accretionary prism.

Along the projected axis of the Kodiak Islands and Kenai Mountains, the seismic reflection data show a series of layered reflectors that extend from about 10 to 35 km below the surface and through the bulk of the crust (Fisher and others, 1987; Moore and others, 1991). Surface exposures and wells along the seaward margin of the Kodiak Islands and across the Kodiak shelf indicate that a large volume of rock was accreted during Eocene and Oligocene(?) time. In order to maintain critical taper, the accretionary prism had to become thicker as well as wider. Thus the layered reflectors are interpreted as underplated deposits (Byrne, 1986), perhaps nappes and crustal-scale duplex structures, that caused the prism to thicken (Moore and others, 1991). These reflectors have also been interpreted as detached pieces of oceanic crust (Fisher and others, 1987). If the layered reflectors really represent underplated Eocene rocks, then the Prince William

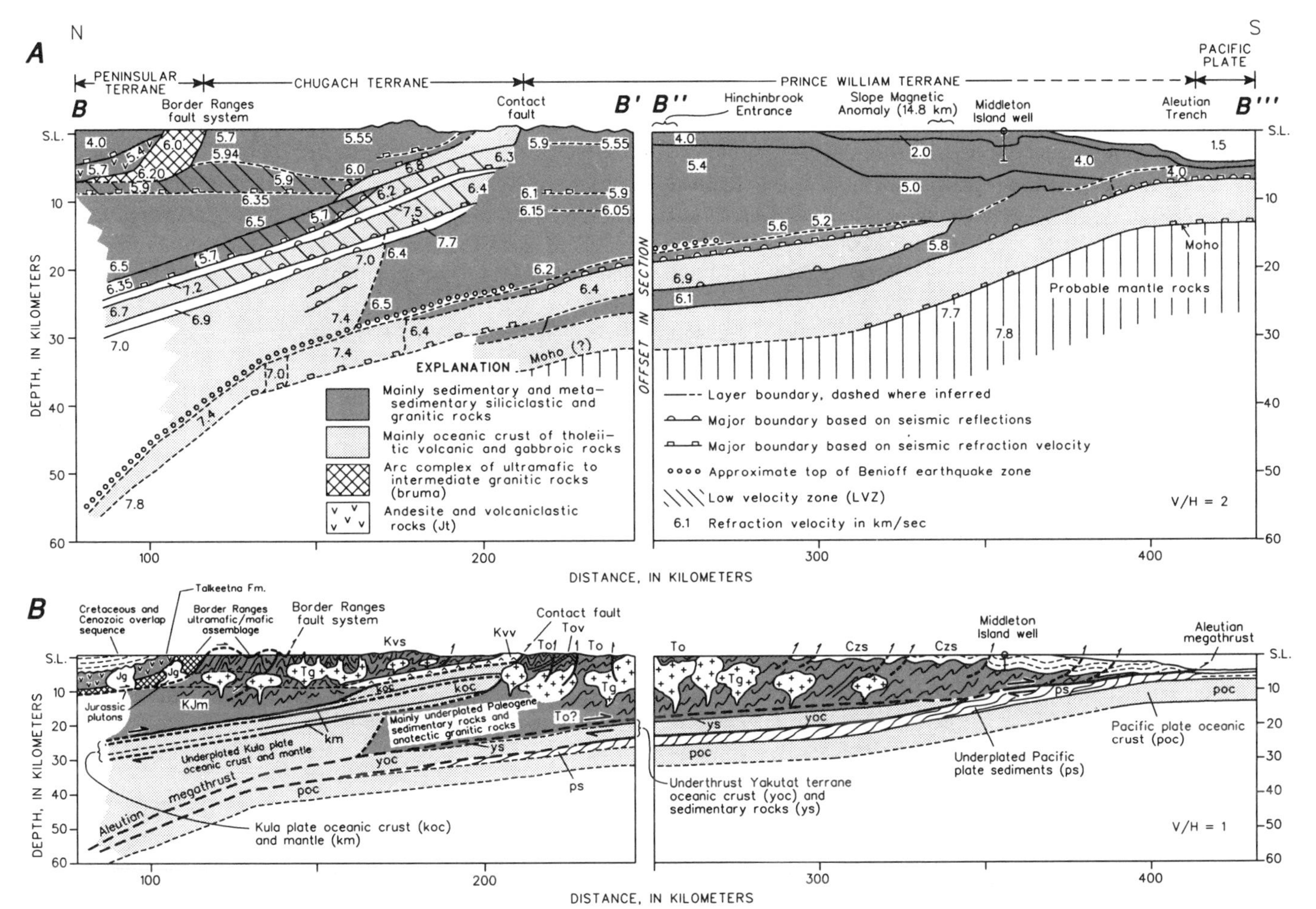

Figure 12. Interpreted deep crustal structure along the offshore and onshore Trans-Alaska Crustal Transect (TACT); location of section shown in Figure 6. A: Geophysical model based on refraction data (Fuis and others, 1991, and unpublished data; Brocher and others, 1991), seismic-reflection data offshore and along part of the transect onshore (Fisher and others, 1989 and unpublished data), earthquake seismic data (Page and others, 1989), offshore magnetic data (Griscom and Sauer, 1990) and surface and well geology (Plafker, 1987; Plafker and others, 1989). B: Interpreted geologic section showing underthrust Pacific plate and Pacific plate with overlying Yakutat terrane crust in seaward part of the section, the compressed, underthrust, and intruded, Mesozoic and Cenozoic accretionary complexes, and extensive underplating beneath the Mesozoic accretionary complex north of the Contact fault by oceanic crust and sedimentary rocks of the Kula plate. Modified from Plafker and others (1989), scenario B of Fuis and Plafker (1991), and Brocher and others (1991 and unpublished data). See Figure 6 for explanation of unit symbols.

terrane would directly underlie the Chugach terrane, indicating that the tectonostratigraphic terranes here are at least in part crustal flakes.

The EDGE seismic-reflection data (Fig. 11) and information from wells (Fig. 10) indicate that the shelf edge is marked by a prominent out-of-sequence thrust fault that separates the Paleogene and Neogene accretionary sequences. Such thrusts are one means whereby an accretionary prism may thicken and maintain critical taper. This out-of-sequence thrust separates accretionary packages of contrasting ages and may be an analog of the faults bounding the principal accretionary units in the Kodiak Islands.

Trans-Alaska crustal transect. Along the TACT route, seismic reflection data and earthquake seismicity indicate that the Benioff zone dips northward at an average angle of about 9° from the Aleutian Trench to 20 to 25 km depth beneath the Contact fault about 175 km arcward from the trench, that the zone is about 35 to 40 km deep beneath the Border Ranges fault, and that its maximum depth is 90 to 100 km beneath Mount Drum in the Wrangell Mountains (Page and others, 1989). The underthrusting slab consists of the Pacific plate and its sedimentary cover within about 85 km of the trench; to the north the Pacific plate is overlain by a platelet of subducted oceanic crust that may have originally been part of the Pacific plate but is now basement to the Yakutat terrane. The top of the subducted oceanic crust has been traced by seismic reflection data to depths of about 20 km beneath the Prince William terrane in northern Prince William

Sound (Fig. 12; Brocher and others, 1991). Geologic relations and earthquake data suggest that the main decollement (Aleutian megathrust) is near the top of the subducted Pacific plate seaward of the southern edge of the Yakutat terrane as defined by the slope magnetic anomaly (Griscom and Sauer, 1990) and that it is near the top of the Yakutat terrane oceanic crust and overlying sedimentary rocks to the north. The earthquake data suggest little or no relative motion at the present time between the Yakutat terrane and Pacific plate (Lahr and Plafker, 1980; Plafker, 1987; Page and others, 1989).

From south to north, salient features of the crustal structure shown in Figure 12 above the underthrusting Pacific plate and Yakutat terrane are as follows. (1) Reflection and refraction data reveal that Pacific plate oceanic crust with a cover of about 3.5 km of pelagic and hemipelagic sediment is underthrust along a subhorizontal decollement beneath the Prince William terrane (Orca Group and granitic rocks) and the overlying basinal sediments for about 40 km landward from the Aleutian Trench (Plafker and others, 1982) and that the subducted oceanic crust and mantle can be delineated downdip for almost 100 km (Brocher and others, unpublished). North-south–trending magnetic anomalies on the subducted Pacific plate can be mapped for as much as 150 km beneath the slope and shelf west of the transect (Griscom and Sauer, 1990). (2) The southern edge of the subducted Yakutat terrane basement, which is marked by a strong positive magnetic anomaly (slope magnetic anomaly), is about 15 km deep where it extends beneath the transect 75 km north of the Aleutian Trench (Fig. 6; Griscom and Sauer, 1990). (3) From the slope magnetic anomaly, the top of the underthrust Yakutat terrane oceanic crust (compressional wave velocities are 6.4 to 6.9 km/s) can be traced beneath the Orca Group of the Prince William terrane as a gently warped, highly reflective layer that deepens northwestward from 14 km or less beneath the outer continental shelf to about 20 km in northern Prince William Sound close to the Contact fault (Brocher and others, 1991). (4) The upper plate (Prince William terrane) consists dominantly of flysch and subordinate mafic volcanic rocks of the Orca Group and flysch-melt granitic intrusive rocks (4.0 km/s to 6.2 km/s). These rocks thicken northward from 4 to 6 km beneath the continental slope to as much as 22 km at the Contact fault and they also extend beneath the Chugach terrane as a thick wedge (average velocity to 6.5 km/s) that is as much as 27 km deep about 50 km north of the Contact fault. The overlying basinal sequence of late Eocene and younger clastic sedimentary rocks has velocities of 4.0 km/s or less and is probably no more than 6 km thick. (5) The Chugach terrane north of the Contact fault consists of an accreted primitive intraoceanic arc (average velocity of 6.8 km/s; Lull and Plafker, 1990) overlain by a Mesozoic accretionary prism of flysch and melange with associated flysch-melt granitic rocks (average velocity 5.5 to 6.5 km/s). This assemblage is interpreted in Figure 12 as a northward-thickening wedge that extends from its outcrop at the Contact fault to a depth of about 17 km beneath the Border Ranges fault (alternative B of Fuis and Plafker, 1991); an alternative interpretation is that the prism is truncated beneath the north half of the terrane by a subhorizontal decollement at a depth of about 9 km (marked by low-velocity layer, 5.9 km/s) that extends northward beneath the Border Ranges fault and much of the adjacent Peninsular terrane (alternative A of Fuis and Plafker, 1991). (6) The central and northern Chugach terrane is underplated by a northward-dipping and thickening sequence interpreted as tectonically underplated Kula plate and mantle (6.2 to 7.8 km/s). The upper layer of this sequence is a coherent slab that dips gently northward from a few kilometers below the surface at the Contact fault to 20 km or more beneath the Border Ranges fault and the Peninsular terrane (Fisher and others, 1989; Fuis and Plafker, 1991; Fuis and others, 1991). (7) Although the Border Ranges fault at the surface is marked by a major lithologic and structural contrast between accreted flysch and melange of the Chugach terrane against Jurassic layered gabbro and ultramafic rocks of the Peninsular terrane as well as by gravity and magnetic anomalies, upper crustal compressional wave velocities (between 5.7 and to 6.0 km/s) apparently do not reflect the lithologic change across the terrane boundary. This lack of velocity contrast suggests that the seismic velocities at depth in the Peninsular terrane have been significantly reduced by emplacement of Middle Jurassic and younger granitic rocks along the refraction line (Plafker and others, 1989; Fuis and Plafker, 1991; Nokleberg and others, this volume, Chapter 10).

YAKUTAT TERRANE

The Yakutat terrane lies outboard of the Southern Margin composite terrane in the northern Gulf of Alaska. It is 600 km long, a maximum of 200 km wide, and has an area of about 58,000 km^2. The terrane is bounded on the east and north by the Fairweather and Chugach–Saint Elias faults, on the west by the Kayak Island structural zone, and on the south by the Transition fault system (Figs. 2 and 6). Two other structural zones, the Pamplona and Dangerous River zones, divide the Yakutat terrane into three parts that have distinct lithologic and structural features (Fig. 6). Cenozoic rocks on opposite sides of the Pamplona zone have undergone marked differences in deformation. Basement of the terrane differs on opposite sides of the Dangerous River zone and the two parts of the basement are linked by overlapping terrigenous clastic rocks of Paleogene age (Figs. 7 and 13). Except as otherwise noted, the following summary of the geology of the Yakutat terrane is after Plafker (1987).

Basement rocks and the Dangerous River zone

The north- to northwest-trending Dangerous River zone marks an abrupt change in the basement geology and the thickness of the overlying stratigraphic section. Changes in the geology across this zone are documented by dredge samples across it on the continental slope and by seismic-reflection profiles that show abrupt westward thickening across it. On land, possibly correlative structural features are inferred from marked differences in the depth to basement in wells near Yakutat and from juxtaposed

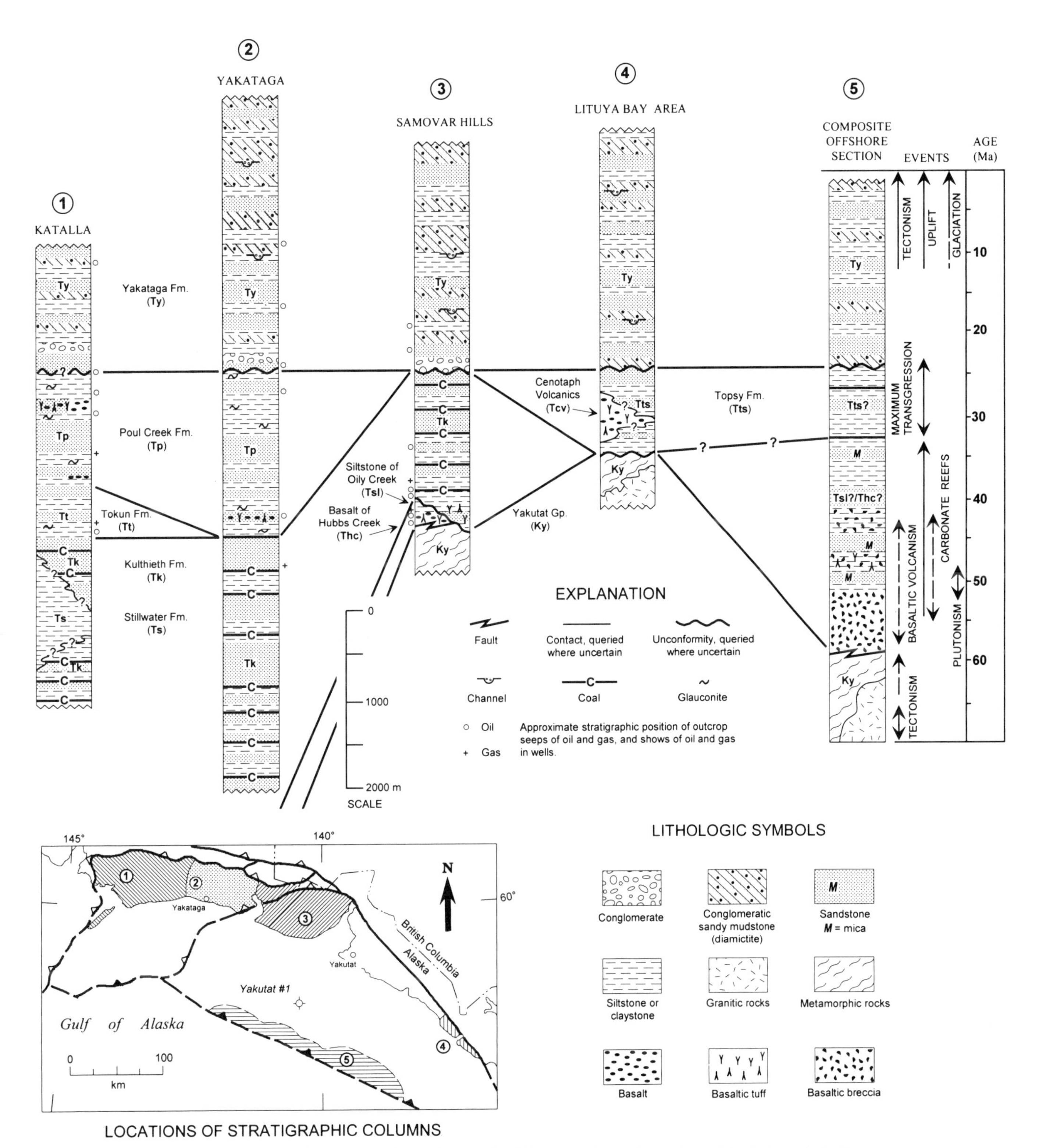

Figure 13. Generalized composite onshore and offshore stratigraphic sections of basinal strata in the Yakutat terrane. Modified from Plafker (1987). Circled numbers show locations of stratigraphic sections.

Cretaceous and Tertiary units along down-to-the-west faults in a nunatak north of the Malaspina Glacier (Fig. 13; Plafker, 1987, Fig. 1A). The Yakutat Group is believed to terminate at the Dangerous River zone, because it does not crop out to the west despite exposure of deep structural levels in areas of extreme topographic relief.

East of the Dangerous River zone. East of the Dangerous River zone, basement rocks are upper Mesozoic flysch and melange of the Yakutat Group, minor serpentinite, and a variety of Tertiary intrusive rocks, most of which are similar in composition and age to intrusive rocks emplaced in the southern part of the Chugach terrane. Metamorphic grade is dominantly prehnite-pumpellyite facies except near some of the plutons, where it is as high as lower greenschist facies. The geology of the Yakutat Group, interpreted as a displaced segment of the Chugach terrane, has already been described as part of the melange and flysch assemblage of the Chugach terrane. In outcrops in the Samovar Hills (a large nunatak north of Malaspina Glacier), the Yakutat Group is locally unconformably overlain by Eocene marine siltstone and sandstone with minor coal, and it is in fault contact with early Eocene bedded rocks (Fig. 13).

Structural trends, metamorphic isograds, and lithologic units in the Yakutat Group trend northwest roughly parallel to the trend of the Fairweather fault and are truncated obliquely on the west by the Dangerous River zone. The offshore Fairweather Ground is inferred to be underlain by the Yakutat Group, as indicated by samples dredged from the continental slope and shelf (Plafker, 1987) and by aeromagnetic anomalies interpreted to result from plutons within the sequence (Taylor and O'Neill, 1974).

West of the Dangerous River zone. West of the Dangerous River zone, the basement is Paleocene(?), early Eocene, and possibly Paleocene oceanic basalt beneath the continental shelf (determined from dredge samples and seismic refraction data). The basement onshore is unknown but is inferred to be comparable to that offshore, except that the sedimentary section near the north margin of the terrane may be deposited unconformably on older strata, such as equivalents of the Orca Group. This part of the terrane consists of a relatively undeformed segment between the Dangerous River and Pamplona zones and a highly deformed segment from the Pamplona zone to the Kayak Island zone (Fig. 6). Stratigraphically overlying the basement is an early Eocene through Oligocene bedded sequence of clastic rocks with minor coal and volcanic rocks having a composite thickness of almost 6 km. A clastic overlap sequence of middle Miocene and younger age that is as thick as 4 km unconformably overlies the Paleogene section.

The configuration and lithology of the basement rocks on which the Cenozoic bedded sequence was deposited (Fig. 3, B–B′, C–C′, D–D′) are known from (1) explosion-refraction data on the continental shelf (Bayer and others, 1978); (2) wide-angle seismic recordings obtained from a large airgun source during marine multichannel-reflection profiling in the northern Gulf of Alaska (Brocher and others, 1991); (3) dredge samples along the continental slope; (4) the Yakutat #1 well, which penetrated basalt at a depth of 5350 m and bottomed at 5430 m; (5) calculated depths of magnetic basement beneath the continental slope and shelf along two lines perpendicular to the Transition fault (Griscom and Sauer, 1990); and (6) outcrops of basalt in the Malaspina Glacier area described in the preceding section. The basement across much of the onshore area is concealed. It is likely that at least part of the Tertiary sequence along the north margin of the terrane was deposited on a basement of rocks such as the Orca Group, which is exposed just north of the Chugach–Saint Elias fault (Fig. 3, B–B′).

The basaltic basement rocks dredged from the continental slope consist of tholeiites some of which are enriched in large ion lithophile elements (LILE) and some of which are depleted in LILEs. These rocks were interpreted as normal mid-ocean ridge and oceanic island basalt, respectively, by Davis and Plafker (1986).

Cenozoic stratigraphy

Cenozoic sedimentary rocks are marine siliciclastic sediments, with minor amounts of volcanic rocks and coal, having a composite maximum thickness of 9500 m (Fig. 13). Each epoch from Paleocene (latest part only) through the Holocene is represented. Three major subdivisions of Cenozoic rocks are recognized onshore and offshore on the basis of fossils and gross lithologic characteristics: (1) late Paleocene and Eocene continental to marine clastic coal-bearing strata and early to middle Eocene basalt, (2) Oligocene and early Miocene marine clastic, glauconitic, and basaltic strata, and (3) an overlap assemblage of middle Miocene to Holocene, predominantly marine and glaciomarine clastic strata. Changes in depositional environment are characteristically gradational and appear to transgress time in different parts of the basin. Important exceptions are the regional unconformities that occur above the Yakutat Group and between the Paleogene strata, and overlap sequences of Miocene and younger age.

Late Paleocene, Eocene, and early Oligocene(?). The onshore sequence is at least 3000 m thick, and reflects marine regression, warm seas, and a subtropical to paratropical climate during the late Paleocene, Eocene, and possibly earliest Oligocene. Major stratigraphic units deposited during this interval are a basal sequence of deep marine volcanic and volcaniclastic strata (basalt of Hubbs Creek and siltstone of Oily Lake) and a thick sequence of slope and shelf clastic deposits (Tokun and Stillwater formations) that grade upward and landward into shallow-marine and nonmarine strata (Kulthieth Formation). These units crop out in a belt extending from the west side of Yakutat Bay to the western margin of the Yakutat terrane (Figs. 6 and 13).

The oldest paleontologically dated rocks in the Yakutat terrane are Paleocene coal-bearing nearshore marine strata that unconformably overlie Yakutat Group flysch at one small isolated nunatak in the Malaspina Glacier area, east of the Dangerous River zone (Samovar Hills section, Fig. 13). Elsewhere, the oldest dated strata above pre-Tertiary basement rocks are early to middle Eocene marine mafic volcanic rocks and fine-grained clastic rocks that were deposited in a shallow to moderately deep marine environment. Strata of this age are known to be exposed onshore only in a section in the Samovar Hills section north of Malaspina Glacier (Fig. 13).

The basalt of Hubbs Creek occurs as a large fault-bounded block and in small isolated outcrops. Its distribution suggests that it may conformably overlie the Yakutat Group and is in turn overlain by the siltstone of Oily Lake. The basalt unit consists of several hundred meters of ~15% basaltic flows. 75% agglomerate, and 10% tuff, and has been cut by numerous diabasic dikes and sills. Shell fragments and algal and bryozoan remains within the volcaniclastic units are nondiagnostic as to age; they indicate a shallow subtropical marine environment of formation, but were probably deposited in slope to bathyal water depths. Whole-rock K-Ar ages on basalt fragments yield apparent minimum ages of 40.0 ±4.8 and 50.0 ± 3.9 Ma (analyses by Krueger Enterprises, Inc., 1980). The older age probably is closest to the crystallization age of the rock, because it is similar to the paleontologically dated overlying marine unit. The basaltic unit is correlated with basaltic breccia on the continental slope (Fig. 13) and at the bottom of a petroleum exploration well (Yakutat #1 in Fig. 6) on the adjacent continental shelf because of similarities in lithology, age, and stratigraphic occurrence. The basalt of Hubbs Creek is of special interest because large seepages of oil and gas are associated with its outcrop distribution (Plafker, 1987, Fig. 1, inset A).

The siltstone of Oily Lake consists of 100 to 200 m of thick-bedded tuffaceous siltstone with a small percentage of basaltic lapilli tuff, vitric tuff, and very fine grained, calcareous, locally graded sandstone in beds 10 to 30 cm thick. A sparse fauna of foraminifers indicates a middle Eocene (Ulatisian) age and suggests outer shelf to upper slope conditions of deposition. The siltstone unit onshore correlates closely in age and lithology with the oldest siltstones dredged on the continental slope (Fig. 13). The siltstone units onshore and offshore are particularly rich in mature organic material and numerous seepages of oil and gas occur within the outcrop area of the siltstone of Oily Lake (Plafker, 1987, Fig. 1, inset A).

The Kulthieth Formation on the mainland consists of a thick sequence of coal-bearing, alluvial-plain, delta-plain, barrier-beach, and shallow-marine deposits. These units intertongue offshore and on Kayak and Wingham islands into deeper marine deltaic and prodeltaic strata of the Tokun Formation; farther west the Kulthieth Formation intertongues with the prodeltaic strata of the Stillwater Formation. The Kulthieth Formation and correlative units contain an abundant marine mollusk fauna, a sparse marine microfauna, and a leaf flora of early to middle Eocene, late Eocene, and possible early Oligocene age. These strata correlate in lithology and age with the lower middle part of the composite sections on the continental slope (Fig. 13).

Late Eocene(?), Oligocene, and early Miocene. Rocks of possible late Eocene, Oligocene, and possible early Miocene age are exposed intermittently along the coastal belt from Icy Bay to Ragged Mountain and on Kayak Island, where they are as much as 1800 m thick (Fig. 13). They also were penetrated in wells drilled for petroleum along the coast and in at least three wells drilled on the adjacent outer continental shelf. Predominantly argillaceous sediment—in part glauconitic, organic rich, and intercalated with intrabasinal water-laid alkalic basaltic tuff, breccia, and pillow lava—accumulated to form the Poul Creek Formation and perhaps the lower sandy part of the Redwood Formation. The strata were deposited during a general marine transgression, and much of the sequence reflects neritic to bathyal environments within the oxygen-minimum zone. Poorly dated volcaniclastic strata of the Cenotaph Volcanics and the intercalated siltstone and sandstone of the Topsy Formation, which overlie pre-Tertiary basement in a relatively small basin between Lituya Bay and Cross Sound in the eastern part of the terrane (Fig. 13), may be partly equivalent in age to the Poul Creek Formation. The Poul Creek Formation is equivalent in age to the upper middle part of the composite continental slope section.

Paleogene sandstone compositions. The Paleogene sandstones are predominantly lithofeldspathic but include rare feldspatholithic and feldspathic compositions (average Q_{37-40} F_{34-41} L_{5-27}). Lithic fragments are dominantly volcanic (67% to 92%), but include moderately abundant metamorphic (5% to 27) and sparse sedimentary (3% to 6%) components (Winkler and others, 1976). Polycrystalline quartzose grains average close to 1% of total quartz grains. Heavy minerals and micas make up an average of 0.8% to 2.2%, and generally increase upward in the sequence. Composition of the sandstone suggests that the source is a complex volcanic-arc terrane where progressively deeper granitic and metamorphic sources were exposed from the early Eocene through the Oligocene.

Neogene overlap assemblages. Since the middle Miocene, and possibly as early as latest Oligocene, an enormous volume of clastic sediment that makes up the Yakataga Formation and most or all of the Redwood Formation was shed from the Chugach and Saint Elias mountains onto the adjacent continental margin (Fig. 13). The formation unconformably overlaps older units and the boundaries of the Yakutat terrane with the Prince William terrane to the west and the Chugach terrane to the southeast. Uplift of the Chugach and Saint Elias mountains during this interval resulted from collision of the Yakutat terrane with, and from partial underthrusting of it beneath, adjacent terranes to the north.

The Yakataga Formation is characterized by rapidly deposited siliciclastic sediments that include an abundance of glacially derived material from the mountains to the north, whereas the Redwood Formation consists of nonglacial siliciclastic sediments with abundant conglomerate. The sediment was deposited predominantly in a shelf to upper bathyal environment, but locally in nearshore or even continental environments onshore in the Malaspina Glacier area, and in deeper bathyal environments offshore. The Yakataga Formation is widely distributed onshore and underlies much of the continental shelf and slope. It is as much as 4600 m thick onshore (Fig. 13) and more than 5000 m thick offshore (see isopach contours, Fig. 6). The Redwood Formation is about 1370 m thick and crops out only onshore in the western part of the Yakutat terrane (Katalla area of the eastern Cordova quadrangle). Sandstone compositions are similar to those of the Paleogene strata, which were the source for much of the Neogene sediment (Winkler and others, 1976).

The abundant microfauna and megafauna indicate that the age of the Yakataga Formation increases from east to west across the region and that the most likely age for the base is early to middle Miocene at Kayak Island and early middle Miocene in the Yakataga section (see a review of paleontological data in Marincovich, 1990). Yakataga Formation sedimentation continues to the present in structurally low areas both onshore and offshore.

Structure

Structural style within the Yakutat terrane differs significantly in its three major segments: east of the Dangerous River zone, between the Dangerous River and Pamplona zones, and between the Pamplona and Kayak Island zones (Fig. 6). These differences mainly reflect the type of basement, thickness of the Cenozoic sequence, and degree of coupling with the Pacific plate (Bruns, 1979, 1983; Bruns and Schwab, 1983; Plafker, 1987).

East of the Dangerous River zone. East of the Dangerous River zone, the structural style of the Yakutat Group is generally similar to that of the melange and flysch units within the Chugach terrane north of the Fairweather fault. A significant difference is that the rocks are also deformed and interleaved in a broad distributive zone of late Cenozoic dextral shear related to the Fairweather transform fault system, which forms the northeast boundary of the terrane.

The only fold known within the Cenozoic strata in this segment is a tight anticline that lies between the Fairweather fault and an inferred offshore fault along the northeast margin of the Yakutat basin (Fig. 6). The structure consists of a gently dipping north flank and a vertical to slightly overturned south limb. Approximate parallelism of the fold and the Fairweather fault reflects a significant component of compression in a broad zone across the northeast margin of the Yakutat terrane. Uplifted Holocene marine terraces on the south limb of this structure indicate continuing active growth (Hudson and others, 1976).

Between the Dangerous River and Pamplona zones. The Dangerous River and Pamplona zones bound the roughly triangular central part of the Yakutat terrane on the northeast and northwest, respectively. This segment of the terrane consists of Paleogene oceanic crust overlain by Cenozoic strata that thicken northward from about 6 km beneath the middle of the continental slope to almost 10 km beneath the inner continental shelf (Bayer and others, 1978). This segment of the Yakutat terrane has been carried passively on the underthrust Pacific plate, and the Cenozoic strata are virtually undeformed, except where they are bowed downward and fill deep Neogene basins or where they are draped over structural highs in the basement (Figs. 3, B–B′, C–C′, D–D′, and 6).

West of the Pamplona zone. In the western and northwestern parts of the Yakutat terrane, late Cenozoic underthrusting of the Prince William terrane that continues to the present has resulted in formation of a fold and thrust belt that is as wide as 120 km; the Pamplona zone is the present deformational front (Figs. 3, B–B′, C–C′, and 6). This deformed belt is a northeastward continuation of the convergence associated with the Aleutian megathrust, and it occupies the entire western part of the terrane west of the Pamplona zone and its onshore equivalents to the northeast. Its southern margin is obscure because the Pamplona zone extends into the Transition fault system along the base of the continental slope. The terrane boundary is arbitrarily assumed to follow the inferred westward projection of the Transition fault as indicated by the slope magnetic anomaly. Magnetic and seismic data indicate that the oceanic crust of the western Yukatat terrane is about 10 to 12 km deep within and near the Kayak Island zone (Bayer and others, 1978; Brocher and others, 1991; Griscom and Sauer, 1990).

The onshore and offshore structure of most of the western segment is dominated by broad synclines and tight anticlines and faulted anticlines that trend east to northeast; the largest of these faults are delineated in Figure 6. Many of the anticlinal structures were targets for unsuccessful exploratory drilling onshore in the coastal belt from 1954 to 1963 and offshore from 1969 to 1983. Individual structures range in width from about 4 to 10 km, and their closure on the lower Yakataga horizon extends 15 to 40 km along strike. Anticlines are generally asymmetric and doubly plunging and are commonly bounded on the seaward side by high-angle thrust faults. Dips on the flanks of the anticlines commonly range from 5° to 45° on the landward side to vertical or overturned on the seaward side. Sharp but unfaulted flexures on the south limb of many folds suggest deformation above structural ramps at depth.

Along the north and west margins of the western segment of the Yakutat terrane, the intensity of folding and magnitude of fault displacements increases significantly (Figs. 3, B–B′, D–D′, and 6). This complexity is attributed to deformation of the bedded sequence of the Yakutat terrane against the relatively resistant Prince William terrane. On Kayak Island, the entire sequence is nearly vertical or overturned and is repeated along several major thrust faults due to deformation after the middle Miocene and prior to emplacement of a dacite plug (Plafker, 1987) dated at 6.2 Ma by Nelson and others (1985). Rogers (1977) noted that seismic profiles show sequential folds that formed as a front of deformation migrated southeastward from the Kayak Island area to the Pamplona zone during the development of this late Cenozoic structural belt. Additional details on the pattern and development of the offshore structures were given by Bruns and Schwab (1983).

Structural shortening in the vicinity of Kayak Island is conservatively calculated to exceed 20 km across a horizontal distance of 10 km (Plafker, 1974). In the outcrop belt to the northeast, shortening normal to the regional strike is estimated to be about 45% in the Paleogene sequence and less than 25% in the

Neogene units at surface levels (Plafker, 1974). Abnormally high fluid pressures encountered at depth in all offshore holes in this belt, together with bore-hole deformation, reflect northwest-southeast–directed lateral stresses (Hottman and others, 1979). High fluid pressures probably facilitate large-scale beding-plane detachment thrusts above the underthrusting basaltic basement of the Yakutat terrane, as depicted schematically in Figure 3, section D–D′. Wide-angle seismic data confirm this model of underthrusting of the Yakutat terrane (Brocher and others, 1991).

Ongoing deformation of structures within this segment of the terrane in indicated by (1) earthquake-related uplift of late Holocene marine terraces along the south flank of a coastal anticline (Sullivan anticline) west of Icy Bay at an average rate of 10.5 mm/yr (Plafker and others, Plate 12, this volume), (2) geodetically detectable horizontal deformation in the same region (Savage and Lisowski, 1988), (3) deformation of well bore holes (Hottman and others, 1979), and (4) seismic activity (Plafker and others, Plate 12, this volume).

Intrusive rocks

East of the Dangerous River zone, intrusive rocks within the Yakutat Group are generally comparable in age and composition to intrusive rocks within the accretionary complex of the Southern Margin composite terrane. These rocks include trondhjemite and diorite plutons of known or probable Early Cretacoeus age (largest about 50 km^2) and tonalite and granodiorite of early Eocene age (largest about 40 km^2) and middle to late Eocene age (Fig. 6). Less common are small tonalite-granodiorite plutons and diabase sills and plugs of Oligocene age and one middle Miocene hypabyssal dacite porphyry body at least 12 km long and as wide as 3 km that has a whole-rock K-Ar age of 13 ± 06 Ma.

West of the Dangerous River zone, intrusive rocks are scarce and too small to show in Figure 6. The largest is a unique elongated dacite plug 4 km long and 1.5 km wide that forms a prominent landmark 500 m high at the south end of Kayak Island (Winkler and Plafker, 1981b; Plafker, 1987). The plug has a micrograntic and porphyritic texture and is emplaced into nearly vertical late Oligocene and early Miocene marine clastic sedimentary strata that are thermally metamorphosed within 100 m of the contact. It has a whole-rock K-Ar age of 6.2 ± 0.3 Ma (Nelson and others, 1985). This intrusion is of special interest because it is situated within a structurally complex belt just east of the suture between the Yakutat and Prince William terranes (Fig. 3, D–D′). Emplacement of the plug postdates the main phase of subduction-related deformation at Kayak Island and suggests that near-trench magmatism continued beneath the underthrusting Yakutat terrane as recently as latest Miocene time.

Small altered alkalic to diabasic hypabyssal intrusive rocks of unknown age are locally exposed on the mainland in the Katalla area. Diabase to gabbroic dikes and sills are sparsely distributed in outcrop areas of Paleogene strata and are probably feeders to early and late Oligocene volcanic rocks in the Poul Creek Formation and Cenotaph Volcanics.

Boundary faults

Displacement of the Yakutat terrane relative to the continental margin and the Pacific plate has been driven largely by late Cenozoic motion of the Pacific plate. The present motion of this plate relative to the North America plate in the northern Gulf of Alaska at long 145°W is deduced as N17°W at about 55 mm/yr (DeMets and others, 1990). It averaged about N16°W at 63 mm/yr for the past 5 m.y. and from 5 to 28 Ma it was N24°–34°W at an average of 45 mm/yr (Engebretsen and others, 1985). The degree of coupling between the Pacific plate and Yakutat terrane has undoubtedly varied through time, depending upon how displacement was shared between the Fairweather and Transition fault systems. At present, the Yakutat terrane is moving mainly with the Pacific plate at about 50 mm/yr relative to North America and the relative velocity between the Pacific plate and Yakutat terrane is estimated at only about 4 mm/yr (Plafker and others, Plate 12, this volume).

Fairweather fault. The dextral strike-slip Fairweather fault, which forms the north boundary of the Yakutat terrane, is part of a ridge-trench transform system that includes the Queen Charlotte fault off the coast of British Columbia. The surface trace of the Fairweather fault is readily visible on land for about 280 km from Cross Sound on the southeast to its juncture with the Chugach–Saint Elias fault in the vicinity of Yakutat Bay, and it has been delineated offshore across the continental shelf to Chatham Strait (Fig. 6) by seismic-reflection and earthquake data. It is inferred to connect with the Queen Charlotte fault in the vicinity of Chatham Strait.

Onshore, the trace of the Fairweather fault is marked by broad shear zones in bedrock and by a topographic trench as much as 1 km wide. Holocene scarps and dextrally offset drainages occur along the trace at numerous localities, and coseismic dextral displacements of at least 3.5 m and vertical displacements of 1 m occurred during the M_s 7.9 Lituya Bay earthquake of July 10, 1958 (Plafker and others, 1978). For most of its length, the fault juxtaposes rocks interpreted as different facies of the Chugach terrane. These are mainly zeolite facies Mesozoic flysch and melange of the Yakutat Group, with associated Paleogene granitoid rocks on the southwest against schistose mafic and pelitic rocks of epidote-amphibolite and amphibolite grade and associated Paleogene plutons on the northeast. The pre-Holocene displacement history along the fault is conjectural. As discussed below, correlation of the Dangerous River zone with the Chatham Strait fault requires about 600 km of Neogene dextral slip.

Transition fault system. The boundary of the Pacific plate with the eastern and central segments of the Yakutat terrane is the Transition fault system (Figs. 2 and 6), which extends along the base of the continental slope. At its northwest end it extends into the maze of faults at the juncture of the Aleutian megathrust and Pamplona zone; however, its associated slope magnetic anomaly can be traced about 200 km to the northwest beneath the Prince William terrane (Griscom and Sauer, 1990). The east end of the fault is poorly constrained but most likely intersects the Fairweather–Queen Charlotte fault system south of Cross Sound (Fig. 6).

East of the Dangerous River zone, the Transition fault system truncates the pre-Tertiary basement; west of this zone, it truncates the Paleogene oceanic crust and overlying clastic sequence. This truncation, together with numerous slickensided rocks in dredge samples from the lower continental slope, suggests large-scale displacement along the Transition fault system—displacement that has affected dredged samples of Cretaceous(?), Eocene, Oligocene, and late Pliocene or younger age.

Truncation of the Paleogene basaltic basement west of the Dangerous River zone is the likely source of the prominent linear positive anomaly informally referred to as the "slope magnetic anomaly" along much of the continental slope inboard of the Transition fault (Griscom and Sauer, 1990; Godson, this volume). Apparent northwestward extension of the slope magnetic anomaly beneath the Prince William terrane suggests a minimum of 200 km of subduction parallel to the Transition fault system. This amount of subduction represents about 7.3 m.y. of displacement parallel to the Transition fault system at present convergence rates and directions as given by DeMets and others (1990), or about 5.7 m.y. using the displacement rates for this time interval given by Engebretsen and others (1985). The magnetic anomaly beneath the Prince William terrane south of Montague Island has been modeled as a basalt slab, the top of which is about 15 km deep; it is ~3 km thick at its south margin (Fig. 3, A–A′; Griscom and Sauer, 1990).

Dextral oblique slip on the Transition fault system during the past 5 m.y. is suggested by the present orientation of the Transition fault system relative to Pacific plate motion, by the orientation of the folds and faults adjacent to the fault system southwest of the Fairweather Ground structural high (Fig. 6), and by the focal mechanisms for the Cross Sound earthquakes (Page, 1975) along the fault trace east of the Fairweather Ground.

A complex fold and thrust belt associated with convergence along the Aleutian megathrust intersects the Transition fault at its western end. From this intersection, the fold and thrust belt extends northeastward across the continental slope between the Kayak Island and Pamplona zones (Fig. 6). It then extends onshore as a poorly exposed zone of deformation that ultimately connects with the Fairweather transform fault in the vicinity of Yakutat Bay (Fig. 6). The entire Yakutat terrane west of the Pamplona zone and much of the adjacent Prince William terrane have been deformed during the late Cenozoic as a result of this convergence.

Chugach–Saint Elias fault system. The Chugach–Saint Elias fault system along the north boundary of the Yakutat terrane (Fig. 6) marks the inner margin of a broad zone along which unmetamorphosed Late Cretaceous and Paleogene strata of the Yakutat terrane are thrust relatively beneath variably metamorphosed rocks of the Valdez and Orca groups. Dips are in the range of 30° to 45° N at the few localities where the fault plane is exposed. At its west end it terminates at an inferred intersection with the onshore extension of the Kayak Island zone (Ragged Mountain fault), and on the east it terminates against the Fairweather fault.

The Kayak Island zone. The west margin of the Yakutat terrane onshore is an irregular series of structures that include (1) west-dipping thrust faults on the mainland and on Kayak Island (Ragged Mountain and Kayak Island faults, respectively) and (2) the offshore Kayak Island zone, which is characterized by complex structures that appear as seismic basement on marine seismic reflection profiles. At its southwest end, the Kayak Island zone is inferred to merge with the east end of the Aleutian megathrust, although it cannot be traced across the contitental slope with available geophysical data.

Displacement history of the Yakutat terrane

The Yakutat terrane is clearly allochthonous with respect to adjacent terranes to the north on the basis of differences in lithology and metamorphism of the late Mesozoic rocks, the provenance of Paleogene siliciclastic rocks, the age and lithology of Paleogene basaltic rocks, paleomagnetic data, and much of the paleontologic data. Together, these data suggest that the Yakutat terrane was transported to its present position during late Cenozoic time by dextral strike slip along the Fairweather–Queen Charlotte transform fault system from a site along the continental margin north of Puget Sound off southeastern Alaska and British Columbia (Plafker, 1987). An alternative model by Keller and others (1984) that is based on limited foraminiferal data requires much greater northward terrane transport from a region off northern California. That model, however, appears to be incompatible with most other critical data for the terrane source, as discussed below.

Correlation of the Dangerous River zone with the Chatham Strait fault. The Chugach terrane is abruptly truncated on the east at the Chatham Strait fault (Fig. 2), where the belt is about 100 km wide across structural trend and consists of melange with minor serpentinite (Kelp Bay Group) and flysch (Sitka Graywacke), into which are emplaced abundant Paleogene granitic plutons. To the southeast, a conspicuous gap in the Mesozoic accretionary belt extends at least as far south as Vancouver Island, Canada. Similarly, in the Yakutat terrane, melange (with minor serpentinite) and flysch of the Yakutat Group and associated Paleogene plutons constitute a belt about 100 km wide that is abruptly terminated on the west at the Dangerous River zone (Fig. 6). The Kelp Bay Group and Sitka Graywacke in southeastern Alaska correlate with the Yakutat Group on the basis of similarities in age, lithology, structure, flysch source directions, and the plutons emplaced in these units (Plafker, 1987; Fig. 7). According to this interpretation, the Yakutat Group was originally southeast of Chatham Strait (Fig. 14), and the Dangerous River zone is the offset equivalent of the Chatham Strait fault. This correlation requires ~600 km of dextral strike slip of the Yakutat Group and Dangerous River zone relative to the Chugach terrane and Chatham Strait fault along the Fairweather fault (see section on tectonic evolution).

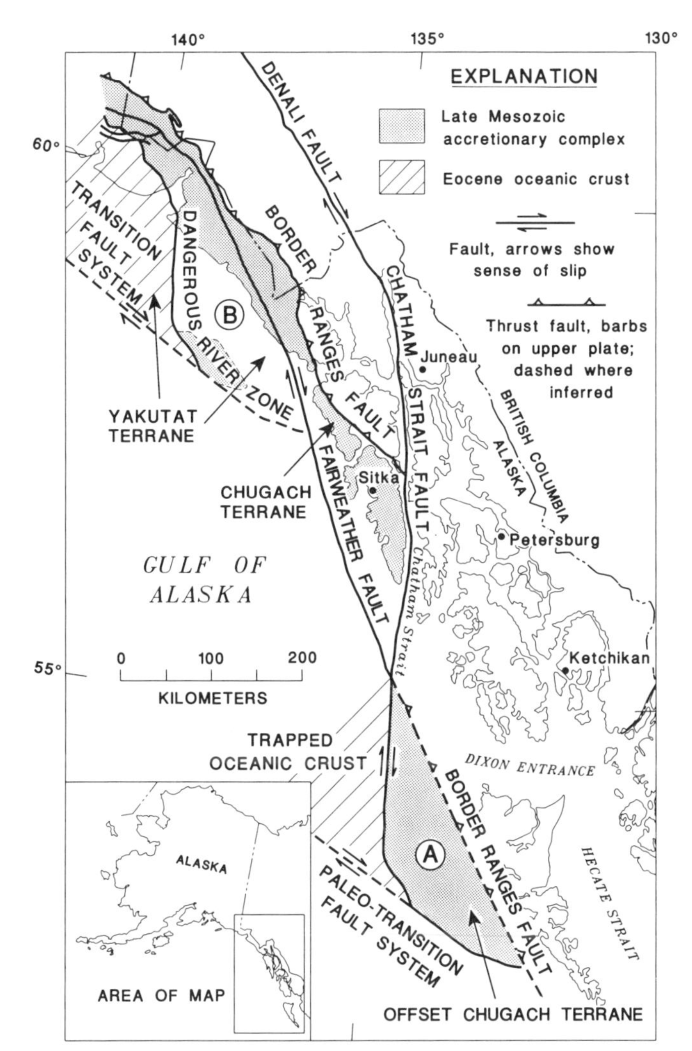

Figure 14. Diagram showing preferred origin of Yakutat Group rocks in the eastern Yakutat terrane as a displaced fragment of the late Mesozoic accretionary complex that comprises the Chugach terrane. A: Early(?) Eocene reconstructed position of the late Mesozoic Chugach terrane after about 180 km dextral offset along the Chatham Strait fault segment of the Denali fault system (Hudson and others, 1982). B: Present position of displaced late Mesozoic Chugach terrane and position after about 600 km late Cenozoic northward displacement along the Fairweather fault. Note that according to this reconstruction, the Dangerous River zone is an offset segment of the Chatham Strait fault and the Transition fault system is an offset segment of a Paleogene transform fault. See Figure 7 for Mesozoic columnar sections in southeastern Alaska and in the Yakutat terrane.

Wrangellia terrane source of exotic blocks in melange. Exotic olistostromal blocks within the Yakutat Group melange, some of which are kilometer-scale in size, require a nearby Wrangellia terrane source during the Late Jurassic and Early Cretaceous. The most suitable source is the segment of Wrangellia terrane that is now exposed in the Queen Charlotte Islands and Vancouver Island; similar rocks may also be present on the continental margin off British Columbia and as far north as Chatham Strait (Fig. 14; Table 3). No other source terrane for this combination of exotic blocks is known south of Vancouver Island. Thus, the exotic blocks constrain the allowable maximum relative motion between the Wrangellia terrane and the Yakutat terrane to about 800 km, the probable exposure length of the Wrangellia terrane along the continental margin south of Chatham Strait (Monger and Berg, 1987).

Provenance of clastic sedimentary rocks. Systematic regional compositional changes in the sandstone of the Late Cretaceous flysch of the Chugach and Yakutat terranes (Zuffa and others, 1980; Winkler and others, 1976) indicate a provenance from a dissected magmatic arc. The inferred arc is most likely the southern part of the Kluane arc (Fig. 5), the roots of which are now represented by the plutonic belt that makes up much of the Coast Mountains of British Columbia and southeastern Alaska (Plafker and others, 1989). Further evidence of a provenance along the continental margin to the southeast comes from (1) regional stratigraphic studies of Paleogene sandstones that tie sandstones in the Yakutat terrane with coeval sandstones in continental basins of southeastern Alaska (Chisholm, 1985); (2) isotopic studies that tie the flysch in the Chugach Mountains to a source in the Coast Mountains (Farmer and others, 1987); and (3) evidence for two major pulses of uplift and deep erosion in the Coast Mountains during the Paleogene and probable Late Cretaceous that could have supplied the enormous quantity of sediment in the latest Cretaceous and Paleogene sequences (Hollister, 1979, 1982). In addition, isotopic and trace element analysis of a single suite of detrital muscovite from Eocene sandstone of the Yakutat terrane suggests a source in high-grade crystalline rocks of the types that occur in the Omineca crystalline belt of southern British Columbia (Heller and others, 1992).

Paleomagnetic data. Reliable paleomagnetic data on the displacement history of the Yakutat terrane have not been obtained onshore (Hillhouse and Grommé, 1985). Van Alstine and others (1985) reported $13° \pm 3°$ of post–early Eocene displacement relative to cratonal North America, with no more than 20° of rotation, on the basis of preliminary data from the offshore Yakutat #1 well (Fig. 6). If correct, these data suggest that the sampled strata were deposited somewhere between what is now Puget Sound, Washington, and Chatham Strait during the early Eocene; a more exact position depends on the amount of post–early Eocene dextral slip on faults between the craton and the Yakutat terrane (Plafker, 1987).

TECTONIC EVOLUTION—A MODEL

Our data indicate a complex Mesozoic to Holocene tectonic evolution for the southern Alaska continental margin. This interpretation is focused on evolution of the accreted terranes that constitute the Southern Margin composite terrane and the Yakutat terrane, and is based mainly on the geologic and geophysical data discussed in this chapter. Relevant regional geologic data for adjacent regions are incorporated where necessary to provide a broad perspective. We emphasize, however, that no single model has yet been devised that satisfactorily explains all the geologic,

TABLE 3. EXOTIC BLOCKS OF THE YAKUTAT GROUP MELANGE AND THEIR POSSIBLE SOURCE UNITS IN THE WRANGELLIA TERRANE

Melange Block				Possible Source	
Lithology	Maximum dimension	Age	Reference	Unit	Reference
Tonalite and diorite	2 km	Middle and Late Jurassic (178 ± 10 Ma and 160 ± 3.5 Ma)	Hudson and others, 1977	Island intrusions; Queen Charlotte Island and Vancouver Island	Sutherland-Brown, 1968; Muller, 1977; Armstrong, 1988
Volcanogenic sandstone	Float (about 20 cm)	Early Jurassic (Sinemurian based on the ammonite *Crucilobiceras* sp.)	G. Plafker, unpublished data; fossil identification by R. W. Imlay, written commun., Nov. 5, 1968	Kunga Formation, Queen Charlotte Island; Bonanza Volcanics, Vancouver Island	Sutherland-Brown, 1968; Muller, 1977
Limestone	800 m	Possible Late Triassic (silicified gastropods and thin-walled pelecypods)	G. Plafker, unpublished data; fossils identified by N. J. Silberling, written commun., 1968	Kunga Formation, Queen Charlotte Island; Quatsino Limestone, Vancouver Island	Sutherland-Brown, 1968; Muller, 1977
Pillow basalt (greenstone)	5 km	Possible Late Triassic (close association with fossiliferous limestone)	G. Plafker, unpublished data	Karmutsen Formation, Queen Charlotte Island and Vancouver Island	Sutherland-Brown, 1968; Muller, 1977
Shallow-water limestone	5 m	Early Permian (conodonts: *Hindeodus* sp., *Neogondolella* sp., *Sweetoganthus whitei* (Rhodes)	S. M. Karl, written commun., Oct. 1980; fossils identified by B. R. Wardlaw, written commun., Oct. 9, 1980	Buttle Lake Limestone, Vancouver Island	Brandon and others, 1986

paleontologic, geophysical, and paleomagnetic data, because some of these data are mutually contradictory and/or internally inconsistent.

Our model for the tectonic evolution of terranes in and adjacent to the study area is shown diagrammatically in plan view (Fig. 15) for eight time periods and in cross sections (Fig. 16) for five time periods from the Late Triassic to the present.

Relative plate motions are after Engebretsen and others (1985), except that the cessation of Kula-Pacific plate spreading is assumed to predate by a few million years the onset of arc volcanism in the Alaska Peninsula at about 50 ± 2 Ma (Plafker, 1987) and to coincide approximately with the early to middle Eocene widespread thermal event along the east margin of the Gulf of Alaska. The change in relative plate motions probably resulted from welding of the Kula and Pacific plates together, rather than from subduction of the entire Kula plate, as suggested by Lonsdale (1988). The model assumes that counterclockwise oroclinal bending of western Alaska, including the Southern Margin composite terrane, occurred mainly in the Paleogene for reasons summarized by Plafker (1987). The present coordinate grid and coastal outline of western North America shown in Figure 15 (C–H) are for reference only; they do not imply knowledge of the actual position of the terranes of southern Alaska relative to the North American craton before early Tertiary time.

Paleozoic and Triassic; pre ~215 Ma

The tectonic model begins with an amalgamated Wrangellia composite terrane having an intraoceanic volcanic, plutonic, and sedimentary basement as old as Precambrian at the south end (Alexander terrane) and late Paleozoic at the north end (Wrangellia and Peninsular terranes) (Gehrels and Berg, this volume; Nokleberg and others, this volume, Chapter 10). Pennsylvanian and Early Permian volcanism of the Skolai arc was followed by widespread shallow-marine conditions and carbonate deposition in many parts of the composite terrane. In the Middle and Late Triassic (Ladinian to Norian), widespread marine to subaerial basaltic volcanism and local bimodal volcanism throughout much of the composite terrane reflect a major regional event variously attributed to rifting, back-arc spreading, or mantle-plume activity (see discussions in Barker and others, 1989; Barker, this volume; Nokleberg and others, this volume, Chapter 10). Volcanism was followed by shallow-marine sedimentation. Biostratigraphic data have been interpreted as indicating that parts of the Wrangellia composite terrane were probably isolated by a seaway from terranes to the east until accretion to North America in the middle to Late Cretaceous (Jones and Silberling, 1982). Paleomagnetic data from Alaska (Hillhouse and Coe, this volume) suggest a Late Triassic paleolatitude about 3100 ± 600 km to the south, at about the present latitude of southern California (Fig. 15A, inset), assuming a Northern Hemisphere origin.

Late Triassic to Late Jurassic; ~215 to 160 Ma

By Late Triassic time, the Wrangellia composite terrane was assembled into approximately its present configuration except that the northern and British Columbia parts of the Wrangellia terrane probably had not been separated by sinistral strike-slip offset (Plafker and others, 1989). The Late Triassic (late Norian) and Early Jurassic to early Middle Jurassic ocean-facing Talkeetna magmatic arc developed across the Peninsular terrane and across the Queen Charlotte Islands and Vancouver Island parts of the Wrangellia terrane during left-oblique subduction of the Farallon plate. Accretion and metamorphism of the glaucophanic greenschist assemblage of the Chugach terrane along the seaward margin of the Talkeetna arc during latest Triassic to early Middle Jurassic time indicate a northward-dipping subduction zone (Fig. 16A). Although the time of accretion of the relatively low grade metamorphic rocks of the melange assemblage is unknown, some of this unit may have been emplaced during the Jurassic.

The location of the Wrangellia composite terrane and associated Southern Margin composite terrane during the existence of the Talkeetna arc is speculative; the position shown in Figure 15A (inset) is based on (1) inferred continuity of the Talkeetna arc with coeval arcs that are now distributed along the continental margin from central California to northern British Columbia and (2) the requirement from paleomagnetic data that the composite terrane was about 3100 km south of its present position in the Late Triassic.

During the Middle and Late(?) Jurassic orogenic deformation resulted in tectonic attenuation of the seaward margin of the Talkeetna arc with collapse of any forearc that may have been present, so that parts of the glaucophanic greenschist assemblage were juxtaposed against deep-level plutonic rocks of the adjacent Peninsular terrane (Fig. 16A).

Late Jurassic to Early Cretaceous; 160 to 120 Ma

In Late Jurassic time the Chitina magmatic arc, now represented by a linear belt of tonalitic plutons (about 153 ± 4 Ma), developed along the seaward margin of the Alexander and Wrangellia terranes. The southeastward continuation of this magmatic belt may be the younger plutons of the Island intrusives on Queen Charlotte and Vancouver islands (Armstrong, 1988). During Valanginian to Barremian (about 135 to 120 Ma) time magmatism shifted inland to the Chisana andesitic arc in the Alexander and northern Wrangellia terranes, possibly due to a decrease in dip of the subduction zone (Fig. 16B).

Flysch and volcaniclastic rocks from these arcs were deposited on both sides of the Wrangellia composite terrane during this interval. Within the Chugach terrane, abundant arc-derived volcaniclastic rocks were deposited on the ocean floor and intertongued with oceanic pelagic sediments and basalt. These deposits were subsequently emplaced into the accretionary complex as the melange assemblage. Olistostromal blocks in the melange were derived from the adjacent Peninsular and Wrangellia terranes and indicate limited postemplacement strike-slip movement between the composite terrane and the Chugach terrane. Although the time of accretion of the melange assemblage is not well constrained, at least part of it probably continued to be accreted during this interval above a north-dipping subduction zone (Fig. 16B).

Truncation and intense shearing of the southern part of the Wrangellia composite terrane are inferred to have occurred during 600 to 1000 km of sinistral displacement of the British Columbia segment of the Wrangellia terrane relative to the Alaska segment (Plafker and others, 1989; Nokleberg and others, this volume, Chapter 10). The inferred displacement postdates Late Jurassic plutonism and predates Late Cretaceous accretion of the flysch and basalt assemblage of the Chugach terrane.

Early Cretaceous to late Paleocene; 120 to 62 Ma

During the interval from the Aptian to Campanian (about 120 to 84 Ma) the Kula plate formed and moved northeast with increased velocity relative to the North American craton and terranes along the continental margin (Fig. 15C). During this interval of dextral-oblique transpression along the continental margin, we infer that the combined Southern Margin and Wrangellia composite terranes were displaced rapidly northward to approximately their present position relative to inboard terranes. Docking of the Wrangellia composite terrane against inboard terranes in mainland Alaska during this interval is indicated along and near the suture zone by termination of marine sedimentation after Cenomanian time, by deformation of flysch basins, and by widespread magmatism and metamorphism (Nokleberg and others, this volume, Chapter 10). Data from southeastern Alaska, however, suggest that the Wrangellia composite terrane may have been emplaced in about its present position relative to inboard terranes as early as Middle Jurassic to middle Cretaceous time (Gehrels and Berg, this volume).

The Campanian and early Maastrichtian interval was characterized by a major outpouring of volcanogenic sediment from the Kluane arc, most of which was deposited on the Kula plate. Simultaneously, a primitive island arc developed within the Kula plate but within reach of deep-sea fans from the Kluane arc. The flysch and basaltic island-arc rocks were successively accreted to older rocks of the accretionary complex above a northeast-dipping subduction zone to form the bulk of the Mesozoic Chugach accretionary prism (Fig. 16C).

The early and middle(?) Paleocene was an interval of relatively slow sedimentation during which the predominantly oceanic Late Cretaceous and early Paleocene Ghost Rocks Formation was accreted to the southern margin of the Chugach terrane in what is now the Kodiak Islands. Accretion was followed by a major mid-Paleocene (64 to 62 Ma) thermal event that resulted in plutonism throughout much of the Southern Margin composite terrane from the Kodiak Islands area westward; local near-trench andesitic volcanism contributed to the Ghost Rocks Formation. Although coeval rocks have not been found east of the Kodiak Islands, they may be present at depth within the accretionary complex.

Late Paleocene to middle Eocene; 62 to 48 Ma

Paleomagnetic data indicate that the Wrangellia composite terrane and the attached Southern Margin composite terrane were at their present latitude by at least 55 Ma. The Paleocene and Eocene interval was characterized by rapid northward underthrusting of the Kula plate beneath the continental margin and by counterclockwise oroclinal bending of what is now western Alaska, probably beginning in the Late Cretaceous or early Paleocene.

Oblique to orthogonal underthrusting resulted in plutonism and uplift that formed the ancestral Coast Mountains of British Columbia and Alaska and were related to late-stage activity of the Kluane magmatic arc and at least two subparallel coeval magmatic belts that developed in western Alaska (Fig. 15D). During this interval, multiple slices of Kula plate crust and sediments were emplaced beneath the older units of the accretionary complex (Fig. 16D).

From about late Paleocene through middle Eocene time (62 to 48 Ma), an enormous flood of sediment derived dominantly from renewed uplift of the what is now the Coast Mountains was deposited mainly onto the Kula plate as deep-sea fans. These fans were subsequently carried rapidly northward on the Kula plate and, together with compositionally variable basaltic oceanic rocks, were progressively emplaced in a trench along the northern and western parts of the Southern Margin composite terrane to form the Sitkalidak Formation, Resurrection Peninsula sequence, and Orca Group of the Prince William terrane (Fig. 15, D and E).

Cumulative dextral displacement of about 400 km on the Denali fault (Nokleberg and others, 1985) and 600 to 1000 km on the subparallel Tintina fault system (Gabrielse, 1985) may have occurred in part or entirely during this interval. However, the timing for major displacements on these interior faults cannot be bracketed more closely than latest Cretaceous to late Eocene; displacement at low rates has continued throughout the Cenozoic

Figure 15. Diagrams illustrating the inferred Mesozoic and Cenozoic evolution of the geology of the southern Alaska margin (modified from Plafker, 1987; Plafker and others, 1989). Tectonostratigraphic terranes are after Silberling and others (this volume), composite terranes are after Plafker and Berg (this volume, Chapter 33), and vectors for relative motions of oceanic plates are after Engebretsen and others (1985), except for present relative motions, which are after DeMets and others (1990). Abbreviations as in Figures 2, 3, and 6. **A:** Late Triassic and Early to Middle Jurassic configuration of composite terranes showing major magmatic arcs and glaucophanic greenschist assemblage rocks of the Chugach terrane accreted to southwest margin of Peninsular and Wrangellia terranes. Inset diagram shows a possible reconstruction of the position of the terranes relative to North America on the basis of paleomagnetic data suggesting about 28° of post-Triassic northward displacement of the Wrangellia composite terrane (WCT) relative to the craton and inferred continuity of Stikine and Talkeetna arcs. **B:** Late Jurassic and Early Cretaceous locations of the Chitina and Chisana magmatic arcs on the Wrangellia composite terrane and partly coeval melange facies rocks of the Chugach terrane accreted to southwestern margin of the Wrangellia composite terrane. During this interval sinistral strike slip may have moved the British Columbia part of the Wrangellia terrane as much as 600 km into its present position relative to the northern part of the Wrangellia terrane. Middle to Late Jurassic deformation and plutonism along the east margin of the Wrangellia composite terrane may possibly be due to collisions with continental terranes to the east. **C:** Initiation of Kula plate relative motion with docking of the Wrangellia composite terrane against inboard terranes in the interval from late Early to early Late Cretaceous. The Late Cretaceous Kluane magmatic arc developed in part over crushed flysch basins along the suture and was the main source of the Campanian and Maastrichtian flysch facies rocks of the Chugach terrane that were accreted to the southwestern margin of the Wrangellia composite terrane. Onset of counterclockwise oroclinal bending of western Alaska. Accretion of the Ghost Rocks terrane and exotic far-traveled fragments of oceanic rocks to the southern margin of the Chugach terrane by 62 Ma and late Paleocene anatectic plutonism from the Sanak Islands to the Kodiak Islands. **D:** About 8° northward translation of the Chugach and Ghost Rocks terranes relative to craton by dextral displacement on the Tintina and Denali fault systems. Counterclockwise oroclinal bending of western Alaska with formation of magmatic arcs in the western interior. Main episode of dextral offset on northwest-trending intracontinental faults (Tintina and Denali). Plutonism and uplift of the Coast Mountains, deep-sea fan sedimentation on the Kula plate, and accretion to the northern and western limbs of the Alaska orocline. **E:** Accretion of the Prince William and related terranes completed by the early middle Eocene. Eocene anatectic plutonism and associated metamorphism inferred to be related to subduction of the Kula plate. Wrangellia composite terrane at or close to its present position relative to North America by about 55 Ma. Offset of continental margin and Chugach terrane by 180–150 km displacement along the Chatham Strait fault (CSF) to form the proto-Dangerous River zone (DRZ); local accumulation of sediments, including carbonate reef detritus along continental margin. Probable continued dextral offset on intracontinental faults. **F:** Initiation of northwestward relative motion of the Pacific plate; beginning of arc volcanism on the Alaska Peninsula. Formation of the Transition fault system (TFS) as a transform fault outboard of the proto-Yakutat terrane; continued rapid sedimentation along continental margin and on ocean floor; possible accumulation of Zodiac fan deposits on oceanic crust off what is now southern British Columbia. **G:** Landward stepping of transform boundary to the Queen Charlotte–Fairweather fault system (QCF/FF) to form the Yakutat terrane. Onset of Wrangell Mountains volcanism after northwestward subduction of the Yakutat terrane to about 100 km depth. Continued slow deep-marine sedimentation on Yakutat terrane. **H:** Continued northward displacement totaling 600 km for the Yakutat terrane, possibly with about 20° counterclockwise rotation. Extreme uplift due to transpression along north and northeast margins of the Yakutat terrane accompanied by major pulse of clastic sedimentation on continental margins and dep-sea floor, culminating in Pliocene and Quaternary time.

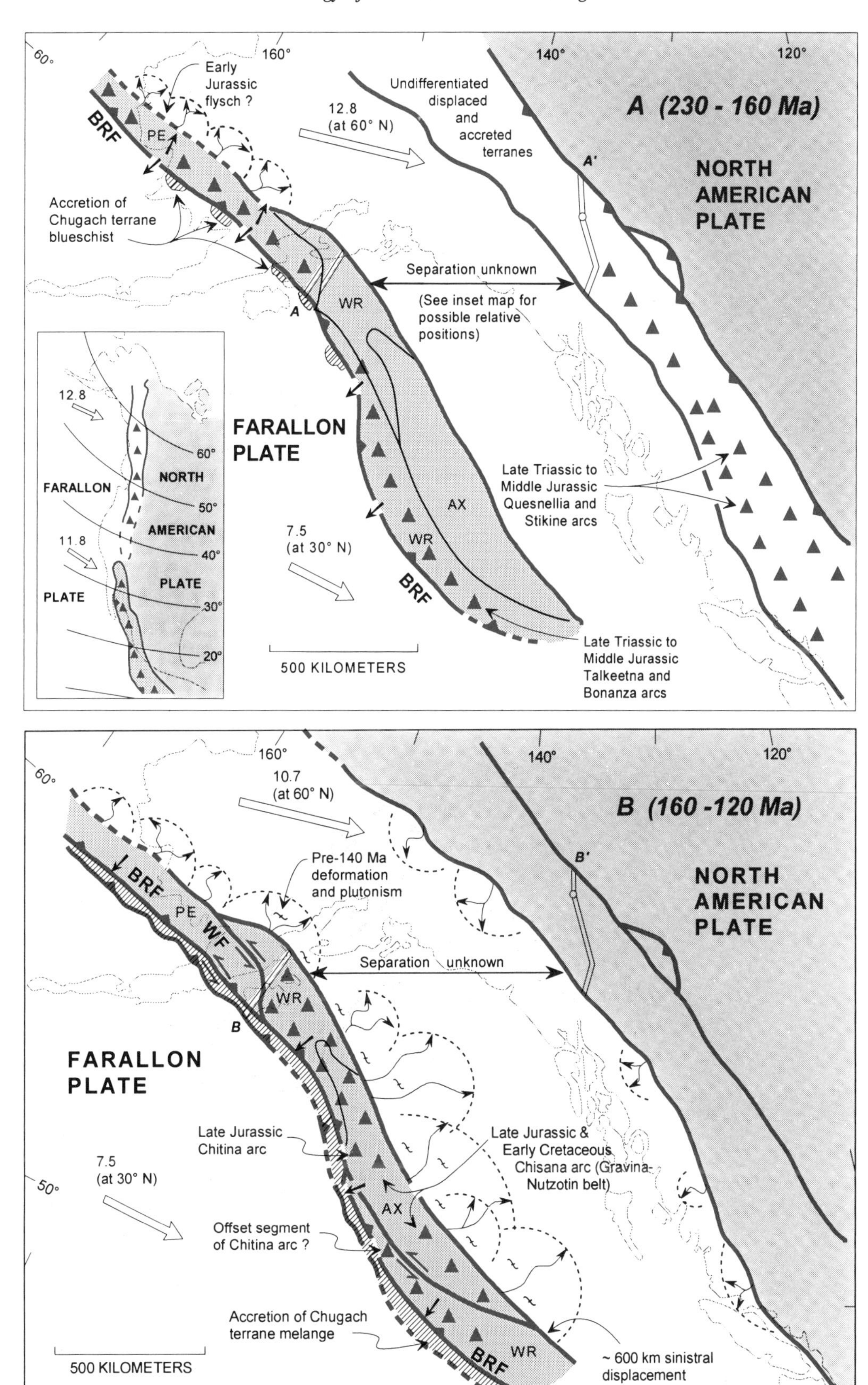
A (230 - 160 Ma)
160°
140°
120°
Early Jurassic flysch ?
Undifferentiated displaced and accreted terranes
12.8 (at 60° N)
BRF
PE
A'
NORTH AMERICAN PLATE
Accretion of Chugach terrane blueschist
Separation unknown
(See inset map for possible relative positions)
WR
A
FARALLON PLATE
Late Triassic to Middle Jurassic Quesnellia and Stikine arcs
AX
7.5 (at 30° N)
WR
BRF
500 KILOMETERS
Late Triassic to Middle Jurassic Talkeetna and Bonanza arcs
12.8
11.8
60°
50°
40°
30°
20°
NORTH
AMERICAN
PLATE
FARALLON
PLATE
B (160 -120 Ma)
10.7 (at 60° N)
Pre-140 Ma deformation and plutonism
BRF
PE
WF
B'
Separation unknown
WR
B
FARALLON PLATE
Late Jurassic Chitina arc
Late Jurassic & Early Cretaceous Chisana arc (Gravina-Nutzotin belt)
7.5 (at 30° N)
50°
AX
Offset segment of Chitina arc ?
Accretion of Chugach terrane melange
WR
BRF
500 KILOMETERS
~ 600 km sinistral displacement

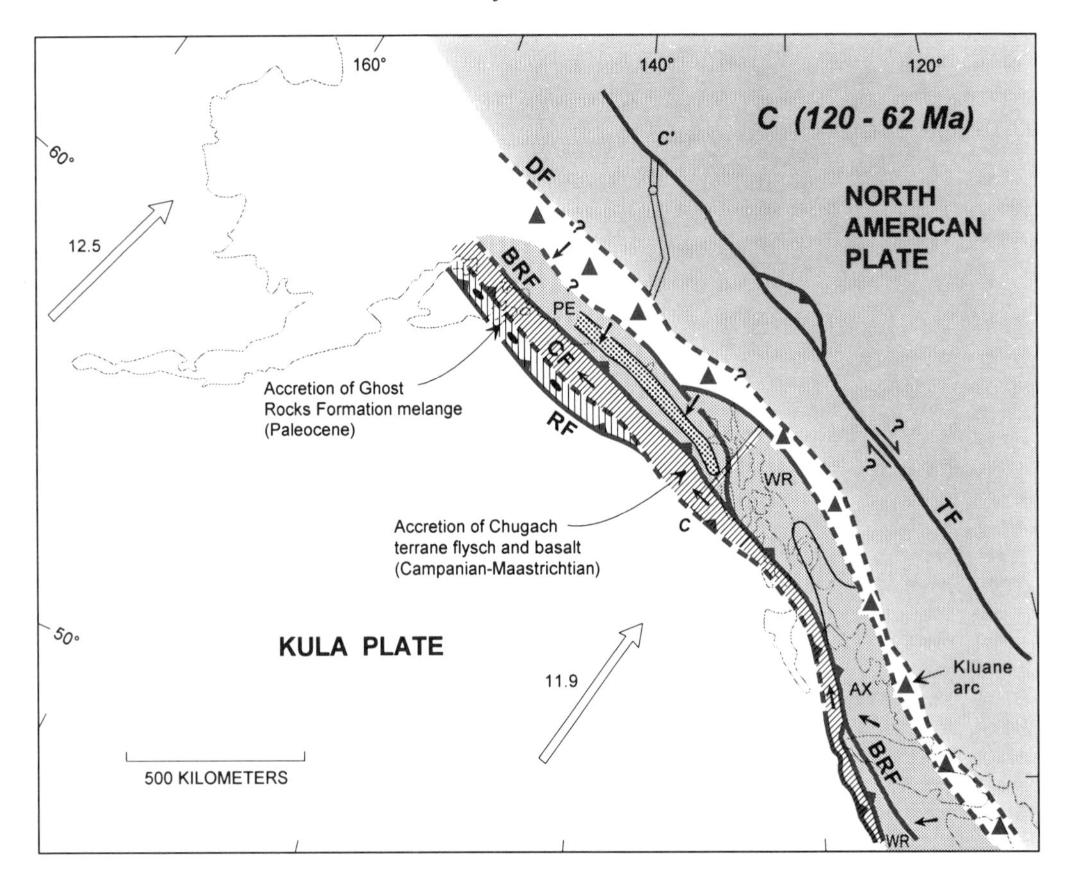
160°
140°
120°
C (120 - 62 Ma)
C'
60°
DF
NORTH
AMERICAN
PLATE
12.5
BRF
PE
CF
Accretion of Ghost
Rocks Formation melange
(Paleocene)
RF
WR
TF
Accretion of Chugach
terrane flysch and basalt
(Campanian-Maastrichtian)
C
50°
KULA PLATE
Kluane
arc
AX
11.9
BRF
500 KILOMETERS
WR

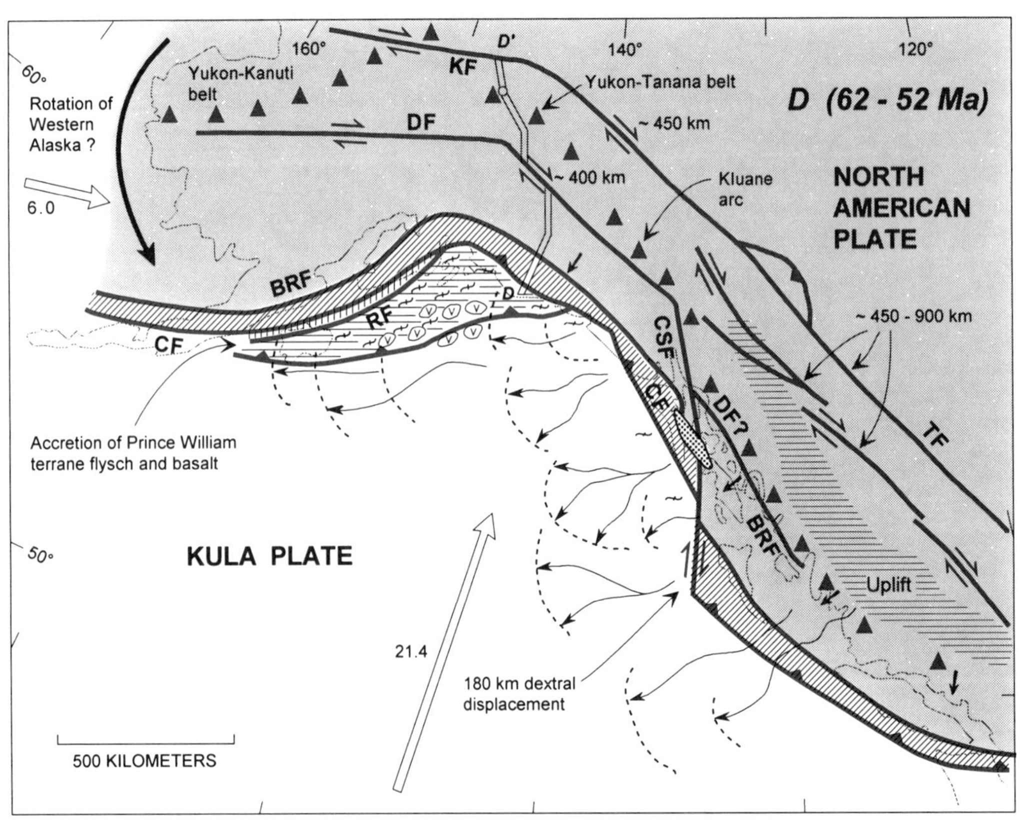
160°
D'
140°
120°
60°
Yukon-Kanuti
belt
KF
Yukon-Tanana belt
D (62 - 52 Ma)
Rotation of
Western
Alaska ?
DF
~ 450 km
~ 400 km
Kluane
arc
NORTH
AMERICAN
PLATE
6.0
BRF
RF
D
CSF
~ 450 - 900 km
CF
CF
DF?
Accretion of Prince William
terrane flysch and basalt
TF
50°
KULA PLATE
BRF
Uplift
21.4
180 km dextral
displacement
500 KILOMETERS

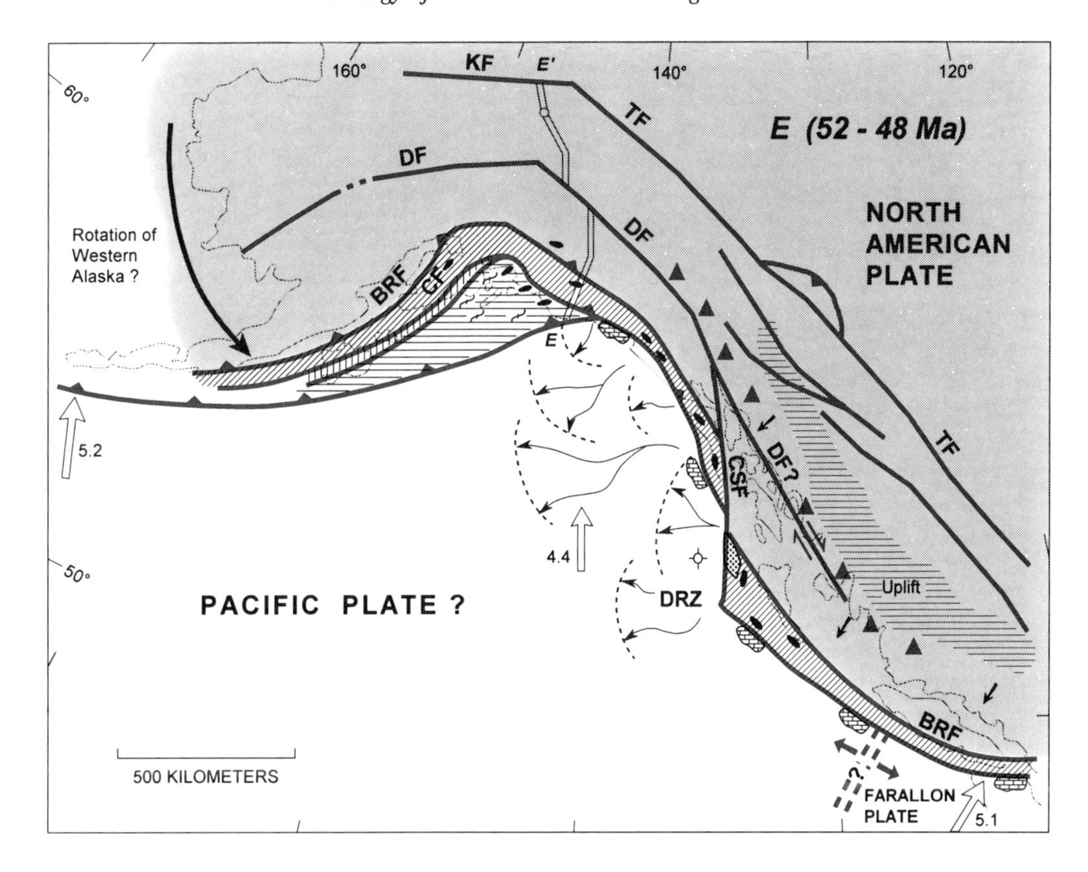
160°
140°
120°
KF
E'
TF
E (52 - 48 Ma)
DF
NORTH
AMERICAN
PLATE
Rotation of
Western
Alaska ?
BRF
CF
E
5.2
CSF
DF?
4.4
DRZ
Uplift
PACIFIC PLATE ?
BRF
500 KILOMETERS
FARALLON
PLATE
5.1

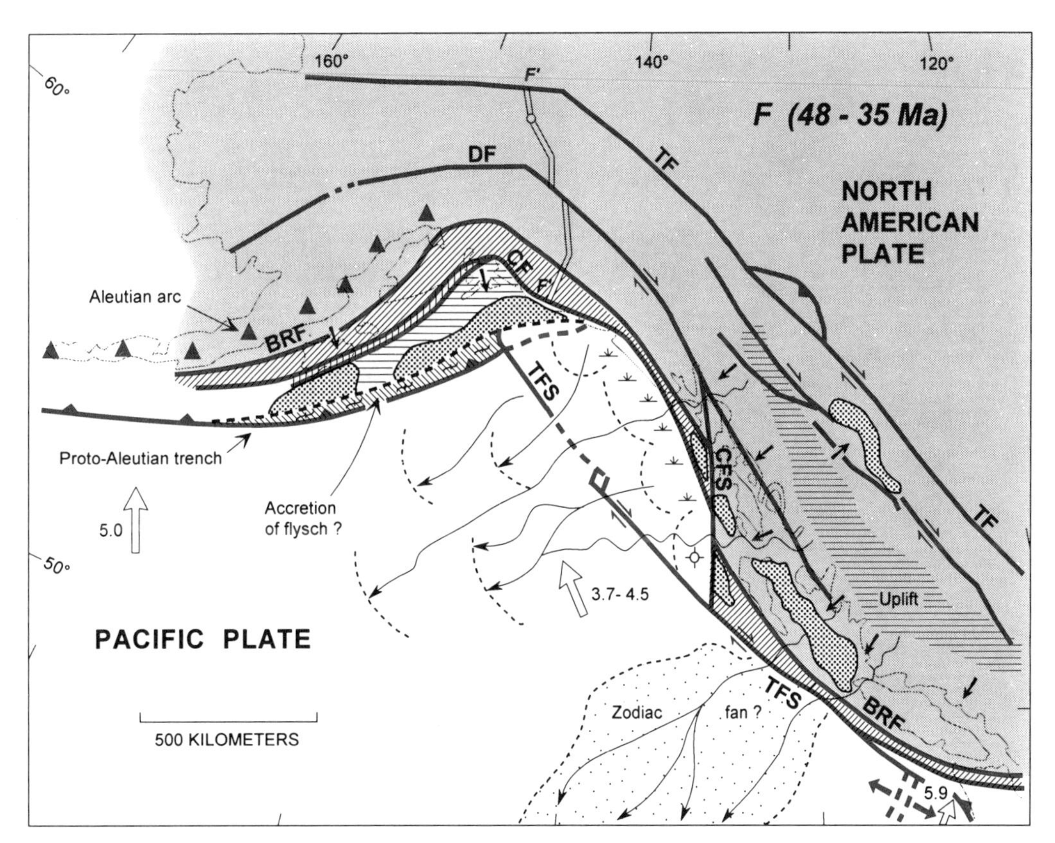
160°
140°
120°
F'
F (48 - 35 Ma)
DF
TF
NORTH
AMERICAN
PLATE
Aleutian arc
BRF
CF
F
TFS
Proto-Aleutian trench
CFS
5.0
Accretion
of flysch ?
3.7- 4.5
Uplift
PACIFIC PLATE
TF
Zodiac
fan ?
TFS
BRF
500 KILOMETERS
5.9

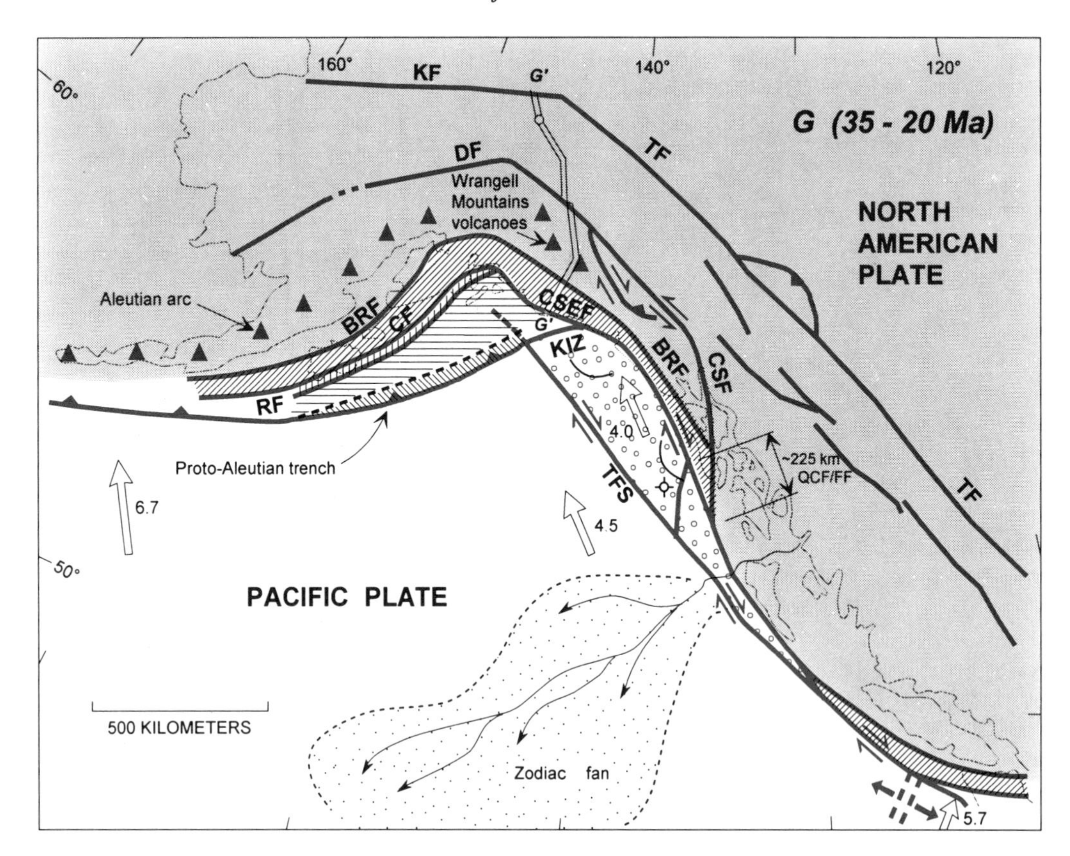
G (35 - 20 Ma)
160°
140°
120°
60°
50°
KF
G'
TF
DF
Wrangell
Mountains
volcanoes
NORTH
AMERICAN
PLATE
Aleutian arc
BRF
CF
CSEF
KIZ
CSF
RF
4.0
~225 km
QCF/FF
Proto-Aleutian trench
6.7
TFS
4.5
PACIFIC PLATE
500 KILOMETERS
Zodiac fan
5.7

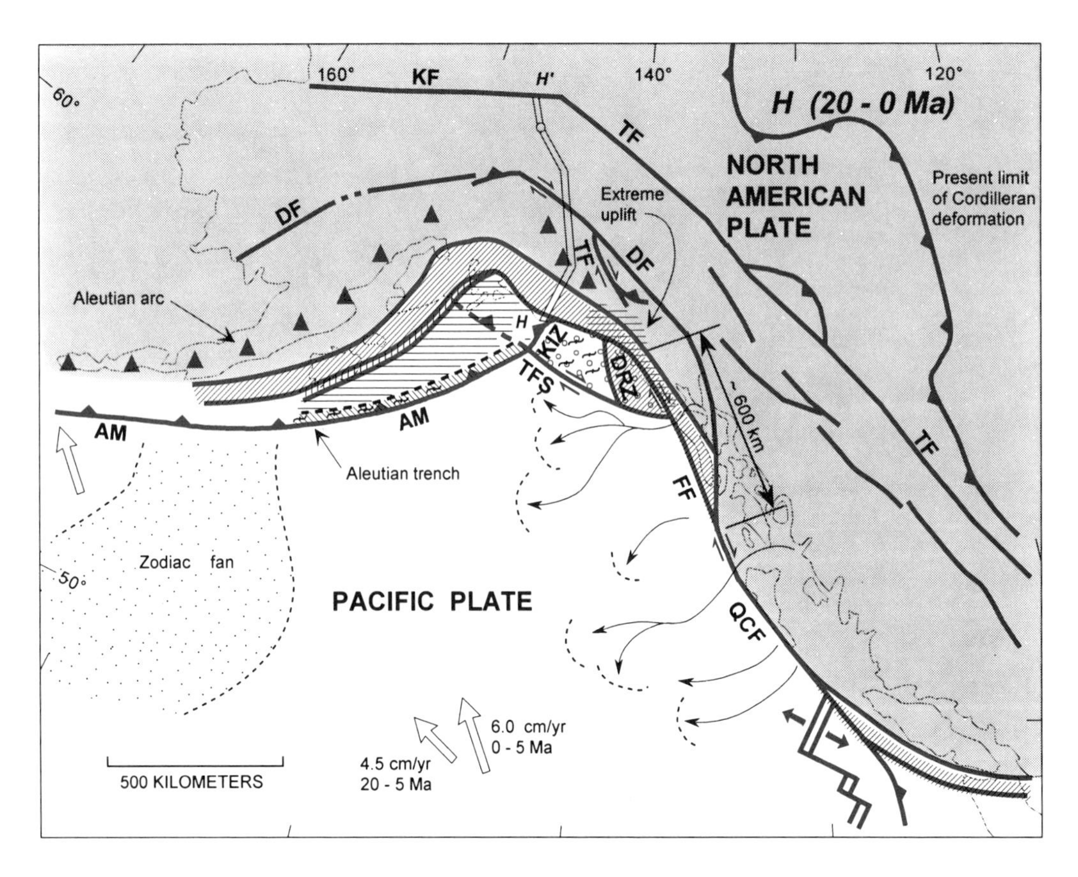
H (20 - 0 Ma)
160°
140°
120°
60°
50°
KF
H'
TF
NORTH
AMERICAN
PLATE
Present limit
of Cordilleran
deformation
DF
Extreme
uplift
Aleutian arc
H
KIZ
DRZ
TFS
~ 600 km
AM
Aleutian trench
FF
Zodiac fan
PACIFIC PLATE
QCF
6.0 cm/yr
0 - 5 Ma
4.5 cm/yr
20 - 5 Ma
500 KILOMETERS

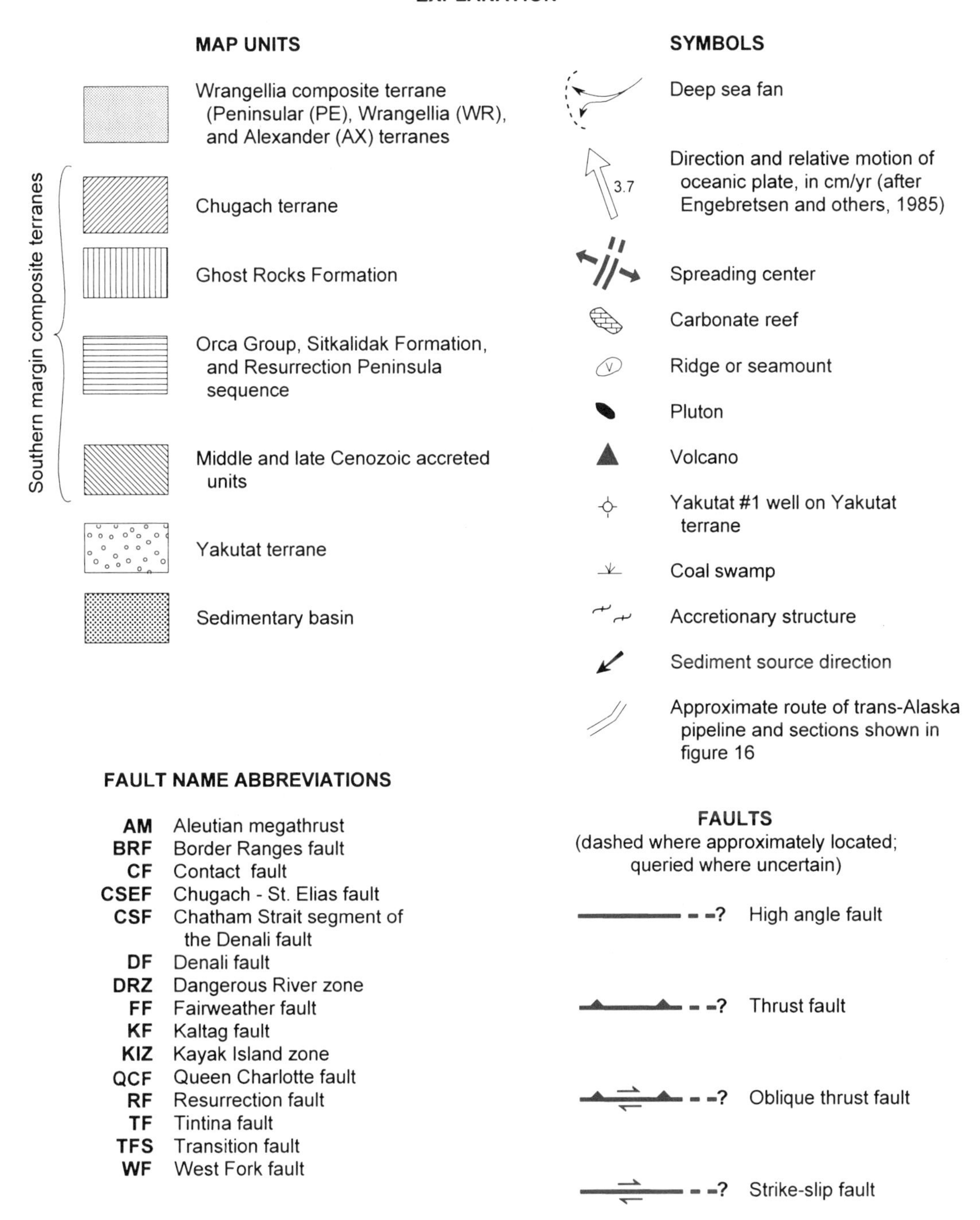

on some of these structures, most notably on the Denali fault system.

In the northern and eastern Gulf of Alaska, the middle Eocene interval (52 to 48 Ma) was marked by north to northwest subduction and the demise of the Kula-Farallon ridge, an event that probably mimics the one that began as much as 10 m.y. earlier to the west. A major coeval thermal event resulted in plutonism and low-pressure high-temperature metamorphism from about 54 to 48 Ma in the Chugach, Prince William, and eastern Yakutat terranes east of the Kodiak Islands area. These plutons are emplaced across the suture between the Southern Margin and Wrangellia composite terranes, indicating that the terranes were in about their present relative positions at that time. The thermal event may have resulted from juxtaposition of mantle rocks directly beneath the accretionary complex as the landward limb of the Kula plate was subducted (as depicted in Figure 16D); other mechanisms such as leakage of basaltic magma through a fragmented Kula plate, or subduction of very young and hot oceanic crust may have caused, or contributed to, the heating.

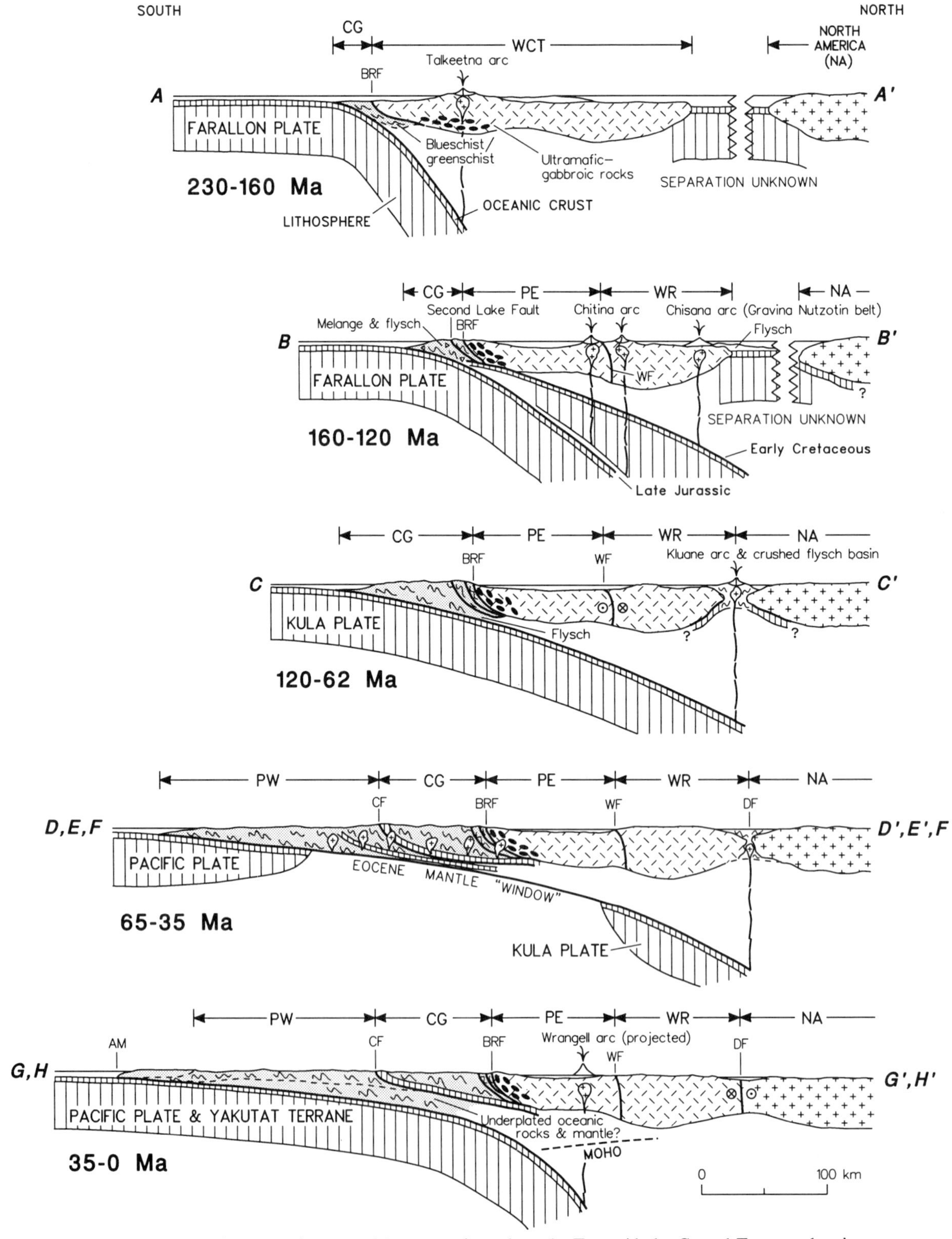

Figure 16. Diagrammatic sequential cross sections along the Trans-Alaska Crustal Transect showing inferred deep-crustal configuration and relative positions of tectonostratigraphic terranes for five of the time intervals depicted in Figure 15. Letters on cross sections correspond to maps of Figure 15. Abbreviations for terrane and fault names are the same as in Figure 15. Modified from Plafker and others (1989). Abbreviations as in Figures 2, 3, and 6.

Minor oroclinal bending continued that refolded structures in the Prince William terrane and possibly in the Chugach terrane. Dextral slip of ~180 km on the Chatham Strait fault is inferred to have truncated the Chugach terrane toward the beginning of this time interval or possibly a few million years earlier (Figs. 14 and 15E). During and immediately after faulting, early to middle Eocene marine clastic deposits from a crystalline-complex source, carbonate-reef detritus, and minor coal were deposited along the margins of the offset fragment of Chugach terrane and on oceanic crust to the west.

Middle Eocene to early Oligocene; 48 to 35 Ma

Andesitic volcanism associated with the Aleutian arc began on the Alaska Peninsula at about 50 Ma, a few million years after the onset of northwestward motion of the Pacific plate. Subduction beneath the Kodiak shelf was probably accompanied by accretion and underplating of deep-sea sediments and oceanic crust outboard of the Orca Group and equivalent rocks.

The Transition fault system developed across the northeastern Gulf of Alaska, and the main deposition of the Paleogene alluvial-fan-delta sequence took place on trapped oceanic crust northeast of the Transition fault system. Early Oligocene felsic plutonism in western Prince William Sound and small felsic and mafic to ultramafic intrusive bodies to the east may be related to leakage of basalt through the crust along this paleotransform-fault system. A locally derived Paleogene basinal sequence overlying the Yakutat Group was deposited in the eastern part of the Yakutat terrane. Sediments of interior and shelf-margin basins, and possibly the Zodiac deep-sea fan, were derived mainly from erosion of the uplifted crystalline rocks of what are now the Coast Mountains.

Late Oligocene to early Miocene; 35 to 20 Ma

Continued northwestward subduction of the Pacific plate was accompanied by andesitic volcanism on the Alaska Peninsula and Aleutian Islands, by sedimentation in shelf basins, and possibly by accretion and underplating of trench deposits and oceanic crust to the seaward part of the accretionary complex.

At about 30 Ma, the transform-fault boundary stepped inboard from the Transition fault system to the Queen Charlotte–Fairweather fault system to form the Yakutat terrane; motion of the Pacific and North American plates was shared between the Queen Charlotte–Fairweather system and the Transition fault system. Andesitic arc volcanism began in what are now the Wrangell Mountains at about 25 Ma following about 225 km subduction of the Yakutat terrane (the amount required to reach about 100 km depth beneath the Wrangell Mountains). Scattered small felsic intrusions emplaced near the Fairweather fault may be related to local areas of high heat flow along zones of extension. Fold and thrust belts were initiated along the Kayak Island zone and Chugach–Saint Elias fault system as a consequence of convergence and underthrusting of the Yakutat terrane beneath adjacent terranes to the north and west.

Early Miocene to present; 20 to 0 Ma

Continued northwestward subduction of the Pacific plate and Yakutat terrane was accompanied by andesitic volcanism in the Alaska Peninsula and Wrangell Mountains segments of the ancestral Aleutian arc, by continued sedimentation in shelf basins, and by subduction and probable underplating beneath the accretionary complex. About 1000 km of northwestward motion of the Pacific plate relative to the continental margin was shared approximately equally between the Queen Charlotte–Fairweather transform and the Transition fault systems. Transpression along the north margin of the Yakutat terrane resulted in extreme uplift of the Chugach Mountains and Fairweather Range and concurrent deposition of thick, locally derived late Cenozoic clastic sequences, including abundant marine glacial deposits (Yakataga and Redwood formations) in continental margin basins on the continental shelves and on the deep-sea floor. Folding and thrusting along the north and west margins of the Yakutat terrane progressed southward and eastward with time, so the deformational front is now at the Pamplona zone. Underthrusting and large-scale subduction of ocean crust and the overlying sediments took place along both the Aleutian megathrust and the Kayak Island zone; compressional deformation within the eastern Prince William terrane was relatively minor. Paleomagnetic data from Eocene sedimentary rocks suggest that the Yakutat terrane may have rotated counterclockwise ~20° as it moved into the bend of the northern Gulf of Alaska.

As a consequence of ~600 km of northwestward displacement and subduction, less than half of the Yakutat terrane remains. Subduction of this remaining part of the terrane, which is caught in the junction between the Aleutian megathrust and the Queen Charlotte–Fairweather transform-fault system, becomes increasingly difficult because it involves the relatively thick eastern segment of the Yakutat terrane with its Mesozoic transitional continental crust. The complex and extremely active deformation that characterizes the late Cenozoic history of the northern Gulf of Alaska will continue until the remaining part of the Yakutat terrane is either completely subducted or is accreted to the continental margin.

Discussion

The tectonic model presented here is compatible with most geologic, seismic, potential field, and paleomagnetic data, and with major aspects of plate reconstructions for the northeast Pacific region.

Nevertheless, many important questions can only be addressed by additional high-quality geologic mapping of critical areas and by detailed topical studies. Unresolved problems due to conflicting or inadequate data have been noted throughout this chapter. Some of the more pressing subjects for future research are listed here. (1) The mechanism for juxtaposition of the deep-level glaucophanic greenschist and plutonic rocks of the Talkeetna magmatic arc along the Border Ranges fault and the disposition of the sediment derived from subsequent uplift and erosion that exposed these rocks. (2) The timing of accretion of the Mesozoic melange assemblage of the Chugach terrane and the provenance of the included blocks. (3) Discrepancies between some paleomagnetic data for the Ghost Rocks Formation, Orca Group, and Resurrection Peninsula sequence that indicate deposition of volcanic rocks in the unit 16° to 31° south of their present latitudes, and geologic evidence and other paleomagnetic data that suggest accretion at about their present latitudes by 55 Ma with limited (about 8°) northward displacement of these terranes relative to the craton by dextral slip on major transcurrent faults. (4) The sources of channel-fill conglomerate in the Orca Group that includes sparse orthoquartzite and abundant metarhyolite tuff derived from a probable craton. (5) Apparent major differences between the Eocene and younger basinal strata penetrated in the Middleton Island well and the deformed Eocene and Oligocene turbidites penetrated in the Kodiak shelf stratigraphic test wells. (6) The discrepancy between estimated subduction of about 600 km of Yakutat terrane crust along the Kayak Island zone and the absence of either a greatly thickened sequence within the fold-thrust belt between the Kayak Island and Pamplona zones, as would be expected for large-scale offscraping and imbrication, or geophysical evidence west of the Pamplona zone that might indicate a volume of underplated low-density sediment commensurate with the amount of underthrusting. (7) Controls for post-Eocene magmatism along the dominantly transform margin of the eastern Gulf of Alaska. (8) The apparent contrast in structural style along the eastern Aleutian Trench between accretion off the Kodiak area and underplating off the Middleton Island area.

ACKNOWLEDGMENTS

We gratefully acknowledge helpful technical reviews of parts or all of this manuscript by our colleagues Fred Barker, H. C. Berg, D. C. Bradley, G. S. Fuis, and Andrew Griscom. W. V. Sliter and R. Z. Poore kindly contributed the new paleontological data for the Ghost Rocks Formation (in Table 2) and J. W. Laney prepared digital versions of the illustrations. This manuscript has benefitted from a thorough edit by M. L. Callas.

REFERENCES CITED

Armentrout, J. M., 1975, Molluscan biostratigraphy of the Lincoln Creek Formation, southwest Washington, *in* Weaver, D. W., Hornaday, G. R., and Tipton, A., eds., Paleogene symposium and selected technical papers on future energy horizons of the Pacific Coast: Long Beach, California, American Association of Petroleum Geologists, Society of Economic Paleontologists and Mineralogists, Society of Economic Geologists, Pacific sections, p. 14–48.

Armstrong, R. L., 1988, Mesozoic and early Cenozoic magmatic evolution of the Canadian Cordillera, *in* Clark, S. P., Jr., Burchfiel, B. C., and Suppe, J., eds., Processes in continental lithospheric deformation: Geological Society of America Special Paper 218, p. 55–91.

Atwater, T., and Severinghaus, J., 1989, Tectonic maps of the northeast Pacific, *in* Winterer, E. L., Hussong, D. M., and Decker, R. W., eds., The eastern Pacific Ocean and Hawaii: Boulder, Colorado, The Geology of North America, v. N, p. 15–20.

Bacon, C. R., Foster, H. L., and Smith, J. G., 1990, Rhyolitic calderas of the Yukon-Tanana terrane, east central Alaska: Volcanic remnants of a mid-Cretaceous magmatic arc: Journal of Geophysical Research, v. 95, p. 21,451–21,461.

Barker, F., Arth, J. G., and Stern, T. W., 1986, Evolution of the Coast batholith along the Skagway traverse, Alaska and British Columbia: American Mineralogist, v. 71, p. 632–643.

Barker, F., Sutherland Brown, A., Budahn, J. R., and Plafker, G., 1989, Back-arc with frontal-arc component origin of Triassic Karmutsen basalt, British Columbia, Canada: Chemical Geology, v. 75, p. 81–102.

Barker, F., Farmer, G. L., Ayuso, R. A., Plafker, G., and Lull, J. S., 1992, The 50-Ma granodiorite of the eastern Gulf of Alaska: Melting in an accretionary prism in the forearc: Journal of Geophysical Research, v. 97, p. 6757–6778.

Bayer, K. C., Mattick, R. E., Plafker, G., and Bruns, T. R., 1978, Refraction studies between Icy Bay and Kayak Island, eastern Gulf of Alaska: U.S. Geological Survey Journal of Research, v. 6, p. 625–636.

Berg, H. C., Jones, D. L., and Richter, D. H., 1972, Gravina-Nutzotin belt—Tectonic significance of an upper Mesozoic sedimentary and volcanic sequence in southern and southeastern Alaska, *in* Geological Survey research 1972: U.S. Geological Survey Profession Paper 800-D, p. D1–D24.

Bol, A. J., Coe, R. S., Grommé, C. S., and Hillhouse, J. W., 1992, Paleomagnetism of the Resurrection Peninsula, Alaska: Implications for the tectonics of southern Alaska and the Kula-Farallon ridge: Journal of Geophysical Research, v. 97, p. 17,213–17,232.

Bradley, D. C., and Kusky, T. M., 1992, Deformation history of the McHugh Complex, Seldovia quadrangle, south-central Alaska, *in* Bradley, D. C., and Ford, A. B., eds., Geologic studies in Alaska by the U.S. Geological Survey, 1990: U.S. Geological Survey Bulletin 1999, p. 17–32.

Brandon, M. T., Orchard, M. J., Parrish, R. R., Sutherland Brown, A., and Yorath, C. J., 1986, Fossil ages and isotopic dates from Paleozoic Sicker Group and associated intrusive rocks, Vancouver Island, British Columbia, *in* Current research, Part A: Geological Survey of Canada Paper 86-1A, p. 683–696.

Brew, D. A., and Ford, A. B., 1984, Tectonostratigraphic terranes in the Coast plutonic-metamorphic complex, southeastern Alaska, *in* Reed, K. M., and Bartsch-Winkler, S., eds., The United States Geological Survey in Alaska: Accomplishments during 1982: U.S. Geological Survey Circular 939, p. 90–93.

Brew, D. A., Karl, S. M., and Miller, J. W., 1988, Megafossils (*Buchia*) indicate Late Jurassic age for part of the Kelp Bay Group on Baranof Island, southeastern Alaska, *in* Galloway, J. P., and Hamilton, T. D., eds., Geological studies in Alaska by the United States Geological Survey during 1987: U.S. Geological Survey Circular 1016, p. 147–149.

Brocher, T. M., Fisher, M. A., Geist, E. L., and Christensen, N. I., 1989, A high-resolution seismic reflection/refraction study of the Chugach-Peninsular terrane boundary, southern Alaska: Journal of Geophysical Research, v. 94, p. 4441–4455.

Brocher, T. M., Moses, M. A., Fisher, M. A., Stephens, C. D., and Geist, E. L., 1991, Images of the plate boundary beneath southern Alaska, *in* Meissner, R., and others, eds., Continental lithosphere; deep seismic reflections: American Geophysical Union Geodynamic Series, v. 22, p. 241–246.

Bruns, T. R., 1979, Late Cenozoic structure of the continental margin, northern Gulf of Alaska, *in* Sisson, A., ed., The relationship of plate tectonics to Alaskan geology and resources (Proceedings, Alaska Geological Society Symposium, 6th, 1977, Anchorage): Anchorage, Alaska Geological Society, p. I1–I30.

—— , 1983, Structure and petroleum potential of the Yakutat segment of the northern Gulf of Alaska continental margin: U.S. Geological Survey Miscellaneous Field Studies Map MF-1480, 22 p., 3 sheets, scale 1:500,000.

Bruns, T. R., and Carlson, P. R., 1987, Geology and petroleum potential of the southeast Alaska continental margin, 1987, *in* Scholl, D. W., Grantz, A., and Vedder, J. G., eds., Geology and petroleum potential of the continental margin of western North America and adjacent ocean basins—Beaufort Sea to Baja California: Houston, Texas, Circum-Pacific Council for Energy and Mineral Resources, Earth Science Series, v. 6, p. 269–282.

Bruns, T. R., and Schwab, W. C., 1983, Structure maps and seismic stratigraphy of the Yakataga segment of the continental margin, northern Gulf of Alaska: U.S. Geological Survey Miscellaneous Field Studies Map MF-1424, 20 p., 4 sheets, scale 1:250,000.

Bruns, T. R., von Huene, R., Culotta, R. C., Lewis, S. D., and Ladd, J. W., 1987, Geology and petroleum potential of the Shumagin margin, Alaska, *in* Scholl, D. W., Grantz, A., and Vedder, J. G., eds., Geology and petroleum potential of the continental margin of western North America and adjacent ocean basins—Beaufort Sea to Baja California: Houston, Texas, Circum-Pacific Council for Energy and Mineral Resources, Earth Science Series, v. 6, p. 157–189.

Budnick, R. T., 1974, The geologic history of the Valdez Group, Kenai Peninsula, Alaska—Deposition and deformation at a Late Cretaceous consumptive plate margin [Ph.D. thesis]: Los Angeles, University of California, 139 p.

Burns, L. E., 1985, The Border Ranges ultramafic and mafic complex, south-central Alaska: Cumulate fractionates of island-arc volcanics: Canadian Journal of Earth Sciences, v. 22, p. 1020–1038.

Byrne, T., 1982, Structural evolution of coherent terranes in the Ghost Rocks Formation, Kodiak Island, Alaska, *in* Leggett, J. K., ed., Trench and forearc sedimentation and tectonics in modern and ancient subduction zones: Geological Society of London Special Publication 10, p. 229–242.

—— , 1984, Early deformation in melange terranes of the Ghost Rocks Formation, Kodiak Islands, Alaska, *in* Raymond, L. A., ed., Melanges: Their nature, origin, and significance: Geological Society of America Special Paper 198, p. 21–51.

—— , 1986, Eocene underplating along the Kodiak shelf, Alaska: Implications and regional correlations: Tectonics, v. 5, p. 403–421.

Carden, J. R., Connelly, W., Forbes, R. B., and Turner, D. L., 1977, Blueschists of the Kodiak Islands, Alaska: An extension of the Seldovia schist terrane: Geology, v. 5, p. 529–533.

Carey, S. W., 1958, The tectonic approach to continental drift, *in* Carey, S. W., convenor, Continental drift—A symposium (1956): Hobart, Australia, Tasmania University Geology Department, p. 177–355.

Case, J. E., and Tysdal, R. G., 1979, Geologic interpretation of aeromagnetic map of the Seward and Blying Sound quadrangles, Alaska: U.S. Geological Survey Miscellaneous Field Studies Map MF-880D, 2 sheets, scale 1:250,000.

Case, J. E., Barnes, D. F., Plafker, G., and Robbins, S. L., 1966, Regional gravity survey of Prince William Sound: U.S. Geological Survey Professional Paper 543-C, 12 p.

Chisholm, W. A., 1985, Comment *on* "Model for the origin of the Yakutat block, an accreting terrane in the northern Gulf of Alaska": Geology, v. 13, p. 87.

Clark, S.H.B, 1972, Reconnaissance bedrock geologic map of the Chugach Mountains near Anchorage, Alaska: U.S. Geological Survey Miscellaneous Field Studies Map MF-350, scale 1:250,000.

—— , 1973, The McHugh Complex of south-central Alaska: U.S. Geological Survey Bulletin 1372-D, p. D1–D11.

Clendenen, W. S., Sliter, W. V., and Byrne, T., 1992, Tectonic implications of the Albatross sedimentary sequence, Sitkinak Island, Alaska, *in* Bradley, D. C., and Ford, A. B., eds., Geologic studies in Alaska by the U.S. Geological Survey, 1990: U.S. Geological Survey Bulletin 1999, p. 52–70.

Coe, R. S., Globerman, B. R., Plumley, P. W., and Thrupp, G. A., 1985, Paleomagnetic results from Alaska and their tectonic implications, *in* Howell, D. G., ed., Tectonostratigraphic terranes of the circum-Pacific region: Houston, Texas, Circum-Pacific Council for Energy and Mineral Resources, Earth Sciences Series, v. 1, p. 85–108.

Connelly, W., 1978, Uyak Complex, Kodiak Islands, Alaska: A Cretaceous subduction complex: Geological Society of America Bulletin, v. 89, p. 755–769.

Connelly, W., and Moore, J. C., 1979, Geologic map of the northwest side of the Kodiak and adjacent islands, Alaska: U.S. Geological Survey Miscellaneous Field Studies Map MF-1057, 2 sheets, scale 1:250,000.

Cowan, D. S., and Boss, R. F., 1978, Tectonic framework of the southwestern Kenai Peninsula, Alaska: Geological Society of America Bulletin, v. 89, p. 155–158.

Crawford, M. L., and Hollister, L. S., 1982, Contrast of metamorphic and structural histories across the Work Channel lineament, Coast Plutonic Complex, British Columbia: Journal of Geophysical Research, v. 87, p. 3849–3860.

Csejtey, B., Jr., Cox, D. P., Evarts, R. C., Stricker, G. D., and Foster, H. L., 1982, The Cenozoic Denali fault system and the Cretaceous accretionary development of southern Alaska: Journal of Geophysical Research, v. 87, p. 3741–3754.

Davies, D. L., and Moore, J. C., 1984, 60 m.y. intrusive rocks from the Kodiak Islands link the Peninsular, Chugach and Prince William terranes: Geological Society of America Abstracts with Programs, v. 16, p. 277.

Davis, A., and Plafker, G., 1986, Eocene basalts from the Yakutat terrane: Evidence for the origin of an accreting terrane in southern Alaska: Geology, v. 14, p. 963–966.

Decker, J. E., 1980a, Geology of a Cretaceous subduction complex, western Chichagof Island, southeastern Alaska [Ph.D. thesis]: Stanford, California, Stanford University, 135 p.

—— , 1980b, Geology of a Cretaceous subduction complex, western Chichagof Island, southeastern Alaska: Ann Arbor, Michigan, Dissertation Abstracts International, v. 41, p. 4040B–4041B.

Decker, J. E., and Plafker, G., 1982, Correlation of rocks in the Tarr Inlet suture zone with the Kelp Bay Group, *in* Coonrad, W. L., ed., The United States Geological Survey in Alaska: Accomplishments during 1980: U.S. Geological Survey Circular 844, p. 119–123.

—— , 1983, The Chugach terrane—A Cretaceous subduction complex in southern Alaska: Geological Society of America Abstracts with Programs, v. 15, p. 386.

Decker, J. E., Wilson, F. H., and Turner, D. L., 1980, Mid-Cretaceous subduction event in southeastern Alaska: Geological Society of America Abstracts with Programs, v. 12, p. 103.

DeMets, C., Gordon, R. G., Argus, D. F., and Stein, S., 1990, Current plate motions: Geophysical Journal International, v. 101, p. 425–478.

Dickinson, W. R., and Suczek, C. A., 1979, Plate tectonics and sandstone compositions: American Association of Petroleum Geologists Bulletin, v. 63, p. 2164–2182.

Dodds, C. J., and Campbell, R. B., 1988, Potassium-argon ages of mainly intrusive rocks in the Saint Elias Mountains, Yukon and British Columbia: Geological Survey of Canada Paper 87-16, 43 p.

Drummond, K. J., 1986, Plate-tectonic map of the circum-Pacific region, northeast quadrant: Houston, Texas, Circum-Pacific Council for Energy and Mineral Resources, scale: 1:10,000,000.

Dumoulin, J. A., 1987, Sandstone composition of the Valdez and Orca groups, Prince William Sound, Alaska: U.S. Geological Survey Bulletin 1774, 37 p.

Dumoulin, J. A., and Miller, M. L., 1984, The Jeanie Point complex revisited, *in* Coonrad, W. L., and Elliott, R. L., eds., The United States Geological Survey in Alaska: Accomplishments during 1981: U.S. Geological Survey Circular 868, p. 75–77.

Engebretsen, D. C., Cox, A., and Gordon, R. G., 1985, Relative motions between oceanic and continental plates in the Pacific basin: Geological Society of America Special Paper 206, 59 p.

Farmer, G. L., Barker, F., and Plafker, G., 1987, A Nd and Sr isotopic study of Mesozoic and early Tertiary granitic rocks in south-central Alaska: Geological Society of America Abstracts with Programs, v. 19, no. 6, p. 376.

Fisher, D., and Byrne, T., 1987, Structural evolution of underthrusted sediments, Kodiak Islands, Alaska: Tectonics, v. 6, p. 775–793.

Fisher, M. A., and von Huene, R., 1980, Structure of the upper Cenozoic strata beneath Kodiak Shelf, Alaska: American Association of Petroleum Geologists Bulletin, v. 64, p. 1014–1033.

Fisher, M. A., von Huene, R., and Smith, G. L., 1987, Reflections from midcrustal rocks within the Mesozoic subduction complex near the eastern Aleutian Trench: Journal of Geophysical Research, v. 92, p. 7907–7915.

Fisher, M. A., Brocher, T. M., Nokleberg, W. J., Plafker, G., and Smith, G. L., 1989, Seismic-reflection images of the crust of the northern part of the

Chugach terrane, Alaska: Results of a survey for the Trans-Alaska Crustal Transect (TACT): Journal of Geophysical Research, v. 94, p. 4424–4440.

Forbes, R. B., and Lanphere, M. A., 1973, Tectonic significance of mineral ages of blueschists near Seldovia, Alaska: Journal Geophysical Research, v. 78, p. 1383–1386.

Fuis, G. S., and Plafker, G., 1991, Evolution of deep structure along the Trans-Alaska Crustal Transect, Chugach Mountains and Copper River basin, southern Alaska: Journal of Geophysical Research, v. 96, p. 4229–4253.

Fuis, G. S., Ambos, E. L., Mooney, W. D., Christensen, N. I., and Geist, E., 1991, Crustal structure of accreted terranes in southern Alaska, Chugach Mountains and Copper River basin, from seismic refraction results: Journal of Geophysical Research, v. 96, p. 4187–4227.

Gabrielse, H., 1985, Major dextral transcurrent displacements along the Northern Rocky Mountain Trench and related lineaments in north-central British Columbia: Geological Society of America Bulletin, v. 96, p. 1–14.

Gergen, L. D., and Plafker, G., 1988, Petrography of sandstones of the Orca Group from the southern Trans-Alaskan Crustal Transect (TACT) route and Montague Island, *in* Galloway, J. P., and Hamilton, T. D., eds., Geologic studies in Alaska by the U.S. Geological Survey during 1987: U.S. Geological Survey Circular 1016, p. 156–159.

Grant, U. S., and Higgins, D. F., 1910, Preliminary report on the mineral resources of the southern part of Kenai Peninsula, *in* Brooks, A. H., and others, compilers, Mineral resources of Alaska, report on progress of investigations, 1909: U.S. Geological Survey Bulletin 442, p. 166–178.

Griscom, A., and Sauer, P. E., 1990, Interpretation of magnetic maps of the northern Gulf of Alaska, with emphasis on the source of the slope anomaly: U.S. Geological Survey Open-File Report 90-348, 18 p.

Grommé, C. S., and Hillhouse, J. W., 1981, Paleomagnetic evidence for northward movement of the Chugach terrane, southern and southeastern Alaska, *in* Albert, N.R.D., and Hudson, T., eds., The United States Geological Survey in Alaska: Accomplishments during 1979: U.S. Geological Survey Circular 823-B, p. B70–B72.

Haeussler, P. J., Coe, R. S., Onstott, T. C., and Renne, P., 1989, A second look at the paleomagnetism of the Late Triassic Hound Island Volcanics of the Alexander terrane: Eos (Transactions, American Geophysical Union), v. 70, p. 1068.

Heller, P. L., Tabor, R. W., O'Neil, J. R., Pevear, D. R., Shafiqullah, M., and Winslow, N. S., 1992, Isotopic provenance of Paleogene sandstones from the accretionary core of the Olympic Mountains, Washington: Geological Society of America Bulletin, v. 104, p. 140–153.

Helwig, J., and Emmet, P., 1981, Structure of the early Tertiary Orca Group in Prince William Sound and some implications for the plate tectonic history of southern Alaska: Alaska Geological Society Journal, v. 1, p. 12–35.

Hill, M. D., Morris, J., and Whelan, J., 1981, Hybrid granodiorites intruding the accretionary prism, Kodiak, Shumagin, and Sanak islands, southwest Alaska: Journal of Geophysical Research, v. 86, p. 10,569–10,590.

Hillhouse, J. W., and Grommé, C. S., 1977, Paleomagnetic poles from sheeted dikes and pillow basalt of the Valdez(?) and Orca groups, southern Alaska: Eos (Transactions, American Geophysical Union), v. 58, p. 1127.

——, 1985, Paleomagnetism of sedimentary rocks, Prince William Sound and Yakutat terranes, *in* Bartsch-Winkler, S., ed., The United States Geological Survey in Alaska: Accomplishments during 1984: U.S. Geological Survey Circular 967, p. 60–62.

Hillhouse, J. W., Grommé, C. S., and Csejtey, B., Jr., 1985, Tectonic implications of paleomagnetic poles from lower Tertiary volcanic rocks, south-central Alaska: Journal of Geophysical Research, v. 90, p. 12,523–12,535.

Himmelberg, G. R., Loney, R. A., and Nabelek, P. I., 1987, Petrogenesis of gabbronorite at Yakobi and northwest Chichagof Islands, Alaska: Geological Society of America Bulletin, v. 98, p. 265–279.

Hollister, L. S., 1979, Metamorphism and crustal displacements: New insights: Episodes, v. 1979, no. 3, p. 3–8.

——, 1982, Metamorphic evidence for rapid (2 mm/yr) uplift of a portion of the Central Gneiss Complex, Coast Mountains, B.C.: Canadian Mineralogist, v. 20, p. 319–332.

Hoose, P. J., 1987, Seismic stratigraphy and tectonic evolution of the Kodiak shelf, *in* Turner, R. F., ed., Geological and operational summary, Kodiak shelf stratigraphic test wells, western Gulf of Alaska: Anchorage, Alaska, Minerals Management Service Outer Continental Shelf Report MMS 87-0109, p. 71–101.

Hottman, C. E., Smith, J. H., and Purcell, W. R., 1979, Relationship among earth stresses, pore pressure, and drilling problems offshore Gulf of Alaska: Journal of Petroleum Technology, v. 31, p. 1477–1484.

Howell, D. G., Jones, D. L., and Schermer, E. R., 1985, Tectonostratigraphic terranes of the circum-Pacific region, *in* Howell, D. G., ed., Tectonostratigraphic terranes of the circum-Pacific region: Houston, Texas, Circum-Pacific Council for Energy and Mineral Resources, Earth Science Series, v. 1, p. 3–30.

Hudson, T., 1983, Calc-alkaline plutonism along the Pacific rim of southern Alaska, *in* Roddick, J. A., ed., Circum-Pacific plutonic terranes: Geological Society of America Memoir 159, p. 159–169.

Hudson, T., and Plafker, G., 1981, Emplacement age of the Crillon–La Perouse pluton, Fairweather Range, *in* Albert, N.R.D., and Hudson, T., eds., The United States Geological Survey in Alaska: Accomplishments during 1979: U.S. Geological Survey Circular 823-B, p. B90–B94.

——, 1982, Paleogene metamorphism of an accretionary flysch terrane, eastern Gulf of Alaska: Geological Society of America Bulletin, v. 93, p. 1280–1290.

Hudson, T., Plafker, G., and Rubin, M., 1976, Uplift rates of marine terrace sequences in the Gulf of Alaska, *in* Cobb, E. H., ed., The United States Geological Survey in Alaska: Accomplishments during 1975: U.S. Geological Survey Circular 733, p. 11–13.

Hudson, T., Plafker, G., and Lanphere, M. A., 1977, Intrusive rocks of the Yakutat–Saint Elias area, south-central Alaska: U.S. Geological Survey Journal of Research, v. 5, p. 155–172.

Hudson, T., Plafker, G., and Peterman, Z. E., 1979, Paleogene anatexis along the Gulf of Alaska marine: Geology, v. 7, p. 573–577.

Hudson, T., Plafker, G., and Dixon, K., 1982, Horizontal offset history of the Chatham Strait fault, *in* Coonrad, W. L., ed., The United States Geological Survey in Alaska: Accomplishments during 1980: U.S. Geological Survey Circular 844, p. 128–132.

Jones, D. L., and Clark, S.H.B., 1973, Upper Cretaceous (Maestrichtian) fossils from the Kenai-Chugach mountains, Kodiak and Shumagin islands, southern Alaska: U.S. Geological Survey Journal of Research, v. 1, p. 125–136.

Jones, D. L., and Silberling, N. J., 1982, Mesozoic stratigraphy: The key to tectonic analysis of southern and central Alaska, *in* Leviton, A. E., and others, eds., Fronters of geological exploration of western North America (American Association for the Advancement of Science, Pacific Section, Annual Meeting, 60th Moscow, Idaho, 1979): American San Francisco, California, American Association for the Advancement of Science, p. 139–153.

Jones, D. L., Silberling, N. J., Berg, H. C., and Plafker, G., 1981, Map showing tectonostratigraphic terranes columnar sections, and summary description of terranes: U.S. Geological Survey Open-File Report 81-792, scale 1:2,500,000.

Jones, D. L., Silberling, N. J., Coney, P. J., and Plafker, G., 1987, Lithotectonic terrane map of Alaska (west of the 41st meridian): U.S. Geological Survey Miscellaneous Field Studies Map MF-1874-A, scale 1:2,500,000.

Karl, S. M., Decker, J. E., and Johnson, B. R., 1982, Discrimination of Wrangellia and the Chugach terrane in the Kelp Bay Group on Chichagof and Baranof islands, southeastern Alaska, *in* Coonrad, W. L., ed., The United States Geological Survey in Alaska: Accomplishments during 1980: U.S. Geological Survey Circular 844, p. 124–128.

Karl, S. M., Brew, D. A., and Wardlaw, B. R., 1990, Significance of Triassic marble from Nakwasina Sound, southeastern Alaska, *in* Dover, J. H., and Galloway, J. P., eds., Geologic studies in Alaska by the U.S. Geological Survey, 1989: U.S. Geological Survey Bulletin 1946, p. 21–28.

Keller, G., von Huene, R., McDougall, K. R., and Bruns, T. R., 1984, Paleoclimatic evidence for Cenozoic migration of Alaskan terranes: Tectonics, v. 3, p. 473–495.

Kelley, J. S., 1985, Geologic setting of the Kenai Peninsula and Cook Inlet Tertiary basin, south-central Alaska, *in* Sisson, A., ed., Guide to the geology of the Kenai Peninsula, Alaska: Anchorage, Alaska Geological Society, p. 3–19.

Kienle, J., and Turner, D. L., 1976, The Shumagin-Kodiak batholith—A Paleocene magmatic arc?, *in* Short notes on Alaskan geology—1976: Alaska Division of Geological and Geophysical Surveys Geologic Report 51, p. 9–11.

Lahr, J. C., and Plafker, G., 1980, Holocene Pacific–North American plate interaction in southern Alaska: Implications for the Yakataga seismic gap: Geology, v. 8, p. 483–486.

Little, T. A., 1990, Kinematics of wrench and divergent-wrench deformation along a central part of the Border Ranges fault system, northern Chugach Mountains, Alaska: Tectonics, v. 9, p. 585–611.

Loney, R. A., and Brew, D. A., 1987, Regional thermal metamorphism and deformation of the Sitka Graywacke, southern Baranof Island, southeastern Alaska: U.S. Geological Survey Bulletin 1779, 17 p.

Loney, R. A., and Himmelberg, G. R., 1983, Structure and petrology of the La Perouse gabbro intrusion, Fairweather Range, southeastern Alaska: Journal of Petrology, v. 24, p. 377–423.

Loney, R. A., Brew, D. A., Muffler, L.J.P., and Pomeroy, J. S., 1975, Reconnaissance geology of Chichagof, Baranof, and Kruzof islands, southeastern Alaska: U.S. Geological Survey Professional Paper 792, 105 p.

Lonsdale, P., 1988, Paleogene history of the Kula plate: Offshore evidence and onshore implications: Geological Society of America Bulletin, v. 100, p. 733–754.

Lull, J. S., and Plafker, G., 1985, Petrography of sandstone from the Yakutat Group, Malaspina district, southern Alaska, *in* Bartsch-Winkler, S., and Reed, K. M., eds., The U.S. Geological Survey in Alaska: Accomplishments during 1983: U.S. Geological Survey Circular 945, p. 73–77.

—— , 1990, Geochemistry and paleotectonic implications of metabasaltic rocks in the Valdez Group, southern Alaska, *in* Dover, J. H., and Galloway, J. P., eds., Geologic studies in Alaska by the U.S. Geological Survey, 1989: U.S. Geological Survey Bulletin 1946, p. 29–38.

Lyle, W., Morehouse, J., Palmer, I. F., Jr., Bolm, J. G., Moore, G. W., and Nilsen, T. H., 1977, Tertiary formations in the Kodiak Island area, Alaska, and their petroleum reservoir and source-rock potential: Fairbanks, Alaska Division of Geological and Geophysical Surveys Open-File Report 114, 48 p.

MacKevett, E. M., Jr., and Plafker, G., 1974, The Border Ranges fault in south-central Alaska: U.S. Geological Survey Journal of Research, v. 2, p. 323–329.

Marincovich, L., Jr., 1990, Molluscan evidence for early middle Miocene marine glaciation in southern Alaska: Geological Society of America Bulletin, v. 102, p. 1591–1599.

Marshak, R. S., and Karig, D. E., 1977, Triple junctions as a cause for anomalously near-trench igneous activity between the trench and volcanic arc: Geology, v. 5, p. 233–236.

McClelland, W. C., Gehrels, G. E., and Saleeby, J. B., 1992, Upper Jurassic–Lower Cretaceous basinal strata along the Cordilleran margin: Implications for the accretionary history of the Alexander-Wrangellia-Peninsular terrane: Tectonics, v. 11, p. 823–835.

Mitchell, P. A., 1979, Geology of the Hope-Sunrise (gold) mining district, north-central Kenai Peninsula, Alaska [M.S. thesis]: Stanford, California, Stanford University, 123 p.

Monger, J.W.H., and Berg, H. C., 1987, Lithotectonic terrane map of western Canada and southeastern Alaska: U.S. Geological Survey Miscellaneous Field Studies Map MF-1874-B, 12 p., scale 1:2,500,000.

Monger, J.W.H., Price, R. A., and Tempelman-Kluit, D. J., 1982, Tectonic accretion and the origin of the two major metamorphic and plutonic welts in the Canadian Cordillera: Geology, v. 10, p. 70–75.

Moore, G. W., 1969, New Formations on Kodiak and adjacent islands, Alaska, *in* Cohee, G. V., Bates, R. G., and Wright, W. B., compilers, Changes in stratigraphic nomenclature by the U.S. Geological Survey, 1967: U.S. Geological Survey Bulletin 1274-A, p. A27–A35.

Moore, J. C., 1973, Cretaceous continental margin sedimentation, southwestern Alaska: Geological Society of America Bulletin, v. 84, p. 595–613.

—— , 1974, Geologic and structural map of part of the outer Shumagin Islands, southwestern Alaska: U.S. Geological Survey Miscellaneous Investigation Series Map I-815, scale 1:63,360.

Moore, J. C., and Allwardt, A., 1980, Progressive deformation of a Tertiary trench slope, Kodiak Islands, Alaska: Journal of Geophysical Research, v. 85, p. 4741–4756.

Moore, J. C., and Connelly, W., 1979, Tectonic history of the continental margin of southwestern Alaska: Late Triassic to earliest Tertiary, *in* Sisson, A., ed., The relationship of plate tectonics to Alaskan geology and resources: (Proceedings, Alaska Geological Society Symposium, 6th, 1977, Anchorage): Anchorage, Alaska Geological Society, p. H1–H29.

Moore, J. C., and Wheeler, R. L., 1978, Structural fabric of a melange, Kodiak Islands, Alaska: American Journal of Science, v. 278, p. 739–765.

Moore, J. C., Byrne, T., Plumley, P. W., Reid, M., Gibbons, H., and Coe, R. S., 1983, Paleogene evolution of the Kodiak Islands, Alaska: Consequences of ridge-trench interaction in a more southerly latitude: Tectonics, v. 2, p. 265–293.

Moore, J. C., and 11 others, 1991, EDGE deep seismic reflection transect of the eastern Aleutian arc-trench layered lower crust reveals underplating and continental growth: Geology, v. 19, p. 420–424.

Muller, J. E., 1977, Geology of Vancouver Island: Geological Survey of Canada Open-File Map 463, scale 1:250,000.

Nelson, S. W., and Blome, C. D., 1991, Preliminary geochemistry of volcanic rocks from the McHugh Complex and Kachemak terrane, southern Alaska: U.S. Geological Survey Open-File Report 91-134, 14 p.

Nelson, S. W., Dumoulin, J. A., and Miller, M. L., 1985, Geologic map of the Chugach National Forest, Alaska: U.S. Geological Survey Miscellaneous Field Studies Map MF-1645-B, 16 p., scale 1:250,000.

Nelson, S. W., Blome, C. D., Harris, A. G., Reed, K. M., and Wilson, F. H., 1986, Late Paleozoic and Early Jurassic fossil ages from the McHugh Complex, *in* Bartsch-Winkler, S., and Reed, K. M., eds., Geological studies studies in Alaska by the U.S. Geological Survey during 1985: U.S. Geological Survey Circular 978, p. 60–64.

Nilsen, T. H., and Bouma, A. H., 1977, Turbidite sedimentology and depositional framework of the Upper Cretaceous Kodiak Formation and related stratigraphic units, southern Alaska: Geological Society of America Abstracts with Programs, v. 9, p. 1115.

Nilsen, T. H., and Moore, G. W., 1979, Reconnaissance study of Upper Cretaceous to Miocene stratigraphic units and sedimentary facies, Kodiak and adjacent islands, Alaska: U.S. Geological Survey Professional Paper 1093, 34 p.

Nilsen, T. H., and Zuffa, G. G., 1982, The Chugach terrane, a Cretaceous trench-fill deposit, southern Alaska, *in* Leggett, J. K., ed., Trench-forearc geology; sedimentation and tectonics on modern and ancient active plate margins, conference: Geological Society of London Special Publication 10, p. 213–227.

Nilsen, T. H., Plafker, G., Atwood, D. E., and Hill, E. R., 1984, Sedimentology of flysch of the upper Mesozoic Yakutat Group, Malaspina district, Alaska, *in* Reed, K. M., and Bartsch-Winkler, S., eds., The United States Geological Survey in Alaska: Accomplishments during 1982: U.S. Geological Survey Circular 939, p. 57–60.

Nokleberg, W. J., Jones, D. L., and Silberling, N. J., 1985, Origin and tectonic evolution of the Maclaren and Wrangellia terranes, eastern Alaska Range, Alaska: Geological Society of America Bulletin, v. 96, p. 1251–1270.

Nokleberg, W. J., Plafker, G., Lull, J. S., Wallace, W. K., and Winkler, G. R., 1989, Structural analysis of the southern Peninsular, southern Wrangellia, and northern Chugach terranes along the Trans-Alaska Crustal Transect, northern Chugach Mountains, Alaska: Journal of Geophysical Research, v. 94, p. 4297–4320.

North American Commission on Stratigraphic Nomenclature, 1983, North American stratigraphic code: American Association of Petroleum Geologists Bulletin, v. 67, p. 841–875.

Olson, D. L., Larson, J. A., and Turner, R. F., 1987, Biostratigraphy, *in* Turner, R. F., ed., Geological and operational summary, Kodiak shelf stratigraphic test wells, western Gulf of Alaska: Anchorage, Alaska, Minerals Management Service, Outer Continental Shelf Report MMS 87-0109, p. 139–191.

Page, R. A., 1975, Evaluation of seismicity and earthquake shaking at offshore sites: Houston, Texas, Annual Offshore Technology Conference, 7th, Proceedings, v. 3, p. 179–190.

Page, R. A., Stephens, C. D., and Lahr, J. C., 1989, Seismicity of the Wrangell and Aleutian Wadati-Benioff zones and the North American plate along the Trans-Alaska Crustal Transect, Chugach Mountains and Copper River basin, southern Alaska: Journal of Geophysical Research, p. 16,059–16,082.

Palmer, A. R., 1983, The Decade of North American Geology 1983 geologic time scale: Geology, v. 11, p. 503–504.

Pavlis, T. L., 1982, Origin and age of the Border Ranges fault of southern Alaska and its bearing on the late Mesozoic evolution of Alaska: Tectonics, v. 1, p. 343–368.

Pavlis, T. L., and Crouse, G. W., 1989, Late Mesozoic strike-slip movement on the Border Ranges fault system in the eastern Chugach Mountains, southern Alaska: Journal of Geophysical Research, v. 94, p. 4321–4332.

Pavlis, T. L., Monteverde, D. H., Bowman, J. R., Rubenstone, J. L., and Reason, M. D., 1988, Early Cretaceous near-trench plutonism in southern Alaska: A tonalite-trondhjemite intrusive complex injected during ductile thrusting along the Border Ranges fault system: Tectonics, v. 7, p. 1179–1199.

Plafker, G., 1967, Surface faults on Montague Island associated with the 1964 Alaska earthquake: U.S. Geological Survey Professional Paper 543-G, 42 p.

—— , 1974, Gulf of Alaska coastal zone, *in* Spencer, A. M., ed., Mesozoic-Cenozoic orogenic belts; data for orogenic studies: Geological Society of London Special Publication 4, p. 573–576.

—— , 1984, Comment *on* "Model for the origin of the Yakutat block, an accreting terrane in the northern Gulf of Alaska": Geology, v. 12, p. 563.

—— , 1987, Regional geology and petroleum potential of the northern Gulf of Alaska continental margin, *in* Scholl, D. W., Grantz, A., and Vedder, J. G., eds., Geology and resource potential of the continental margin of western North America and adjacent ocean basins—Beaufort Sea to Baja California: Houston, Texas, Circum-Pacific Council for Energy and Mineral Resources, Earth Science Series, v. 6, p. 229–268.

—— , 1990, Regional geology and tectonic evolution of Alaska and adjacent parts of the northeast Pacific Ocean margin: Proceedings of the Pacific Rim Congress 90: Queensland, Australia, Australasian Institute of Mining and Metallurgy, p. 841–853.

Plafker, G., and Campbell, R. B., 1979, The Border Ranges fault in the Saint Elias Mountains, *in* Johnson, K. M., and Williams, J. R., eds., The United States Geological Survey in Alaska: Accomplishments during 1978: U.S. Geological Survey Circular 804-B, p. B102–B104.

Plafker, G., and Rubin, M., 1978, Uplift history and earthquake recurrence as deduced from marine terraces on Middleton Island, Alaska, *in* Proceedings of Conference VI, Methology for identifying seismic gaps and soon-to-break gaps: U.S. Geological Survey Open-File Report 78-943, p. 687–721.

Plafker, G., Jones, D. L., and Pessagno, E. A., Jr., 1977, A Cretaceous accretionary flysch and melange terrane along the Gulf of Alaska margin, *in* Blean, K. M., ed., The United States Geological Survey in Alaska: Accomplishments during 1976: U.S. Geological Survey Circular 751-B, p. B41–B43.

Plafker, G., Hudson, T., Bruns, T. R., and Rubin, M., 1978, Late Quaternary offsets along the Fairweather fault and crustal plate interactions in southern Alaska: Canadian Journal of Earth Sciences, v. 15, p. 805–816.

Plafker, G., Bruns, T. R., Winkler, G. R., and Tysdal, R. G., 1982, Cross section of the eastern Aleutian arc, from Mount Spurr to the Aleutian Trench near Middleton Island, Alaska: Geological Society of America Map and Chart Series MC-28P, scale 1:250,000.

Plafker, G., Harris, A. G., and Reed, K. M., 1985a, Early Pennsylvanian conodonts from the Strelna Formation, Chitina Valley, *in* Bartsch-Winkler, S., ed., The U.S. Geological Survey in Alaska: Accomplishments during 1984: U.S. Geological Survey Circular 967, p. 71–74.

Plafker, G., Keller, G., Barron, J. A., and Blueford, J. R., 1985b, Paleontologic data on the age of the Orca Group, Alaska: U.S. Geological Survey Open-File Report 85-429, 24 p.

Plafker, G., Nokleberg, W. J., Lull, J. S., Roeske, S. M., and Winkler, G. R., 1986, Nature and timing of deformation along the Contact fault system in the Cordova, Bering Glacier, and Valdez quadrangles, *in* Bartsch-Winkler, S., and Reed, K. M., eds., Geologic studies in Alaska by the U.S. Geological Survey during 1985: U.S. Geological Survey Circular 978, p. 74–77.

Plafker, G., Nokleberg, W. J., and Lull, J. S., 1989, Bedrock geology and tectonic evolution of the Wrangellia, Peninsular, and Chugach terranes along the Trans-Alaska Crustal Transect in the Chugach Mountains and southern Copper River basin, Alaska: Journal of Geophysical Research, v. 94, p. 4255–4295.

Plumley, P. W., and Plafker, G., 1985, Additional estimate of paleolatitude for the Paleocene/Eocene(?) Prince William terrane–Orca volcanics: Geological Society of America Abstracts with Programs, v. 17, p. 401.

Plumley, P. W., Coe, R. S., and Byrne, T., 1983, Paleomagnetism of the Paleocene Ghost Rocks Formation, Prince William terrane, Alaska: Tectonics, v. 2, p. 295–314.

Rau, W. W., Plafker, G., and Winkler, G. R., 1983, Foraminiferal biostratigraphy and correlations in the Gulf of Alaska Tertiary province: U.S. Geological Survey Oil and Gas Investigations Chart OC-120, 11 p., 3 sheets.

Richter, D. H., 1965, Geology and mineral deposits of central Knight Island, Prince William Sound, Alaska: Alaska Division of Mines and Minerals Geologic Report 16, 37 p.

Roeske, S. M., 1986, Field relations and metamorphism of the Raspberry Schist, Kodiak Islands, Alaska, *in* Evans, B. W., and Brown, E. H., eds., Blueschists and eclogites: Geological Society of America Memoir 164, p. 169–184.

Roeske, S. M., Mattinson, J. M., and Armstrong, R. L., 1989, Isotopic ages of glaucophane schists on the Kodiak Islands, southern Alaska, and their implications for the Mesozoic tectonic history of the Border Ranges fault system: Geological Society of America Bulletin, v. 101, p. 1021–1037.

Roeske, S. M., Snee, L. W., and Pavlis, T. L., 1991, Strike-slip and accretion events along the southern Alaska plate margin in the Cretaceous and early Tertiary: Geological Society of America Abstracts with Programs, v. 23, p. 428–429.

Rogers, J. F., 1977, Implications of plate tectonics for offshore Gulf of Alaska petroleum exploration: Houston, Texas, Annual Offshore Technology Conference, 9th, Proceedings, v. 1, p. 11–16.

Saleeby, J. B., 1983, Accretionary tectonics of the North American Cordillera: Annual Review of Earth and Planetary Sciences, v. 11, p. 45–73.

Sample, J. C., and Fisher, D. M., 1986, Duplex accretion and underplating in an ancient accretionary complex, Kodiak Islands, Alaska: Geology, v. 14, p. 160–163.

Sample, J. C., and Moore, J. C., 1987, Structural style and kinematics of an underplated slate belt, Kodiak and adjacent islands, Alaska: Geological Society of America Bulletin, v. 99, p. 7–20.

Savage, J. C., and Lisowski, M., 1988, Deformation in the Yakataga seismic gap, southern Alaska, 1980–1986: Journal of Geophysical Research, v. 93, p. 4731–4744.

Schaff, R. G., and Gilbert, W. G., 1987, Southern Alaska region correlation chart, *in* Correlation of stratigraphic units of North America (COSUNA) project: Tulsa, Olkahoma, American Association of Petroleum Geologists.

Silberman, M. L., MacKevett, E. M., Jr., Connor, C. L., Klock, P. R., and Kalechitz, G., 1981, K-Ar ages of the Nilolai Greenstone from the McCarthy quadrangle, Alaska—The "docking" of Wrangellia, *in* Albert, N.R.D., and Hudson, T., eds., The United States Geological Survey in Alaska: Accomplishments during 1979: U.S. Geological Survey Circular 823-B, p. B61–B63.

Sisson, V. B., and Onstott, T. C., 1986, Dating blueschist metamorphism: A combined $^{49}Ar/^{39}Ar$ and electron microprobe approach: Geochimica et Cosmochimica Acta, v. 50, p. 2111–2117.

Sisson, V. B., Hollister, L. S., and Onstott, T. C., 1989, Petrologic and age constraints on the origin of a low-pressure/high-temperature metamorphic complex, southern Alaska: Journal of Geophysical Research, v. 94,

p. 4392–4410.

Stewart, R. J., 1976, Turbidites of the Aleutian abyssal plain: Mineralogy, provenance, and constraints for Cenozoic motion of the Pacific plate: Geological Society of America Bulletin, v. 87, p. 793–808.

Stone, D. B., and Packer, D. R., 1979, Paleomagnetic data from the Alaska Peninsula: Geological Society of America Bulletin, v. 90, p. 545–560.

Stone, D. B., Panuska, B. C., and Packer, D. R., 1982, Paleolatitudes versus time for southern Alaska: Journal of Geophysical Research, v. 87, p. 3697–3707.

Streckeisen, A., 1976, To each plutonic rock its proper name: Earth Science Reviews, v. 12, p. 1–33.

Sutherland-Brown, A., 1968, Geology of the Queen Charlotte Islands, British Columbia: British Columbia Department of Mines and Petroleum Resources Bulletin 54, 225 p.

Taylor, P. T., and O'Neill, N. J., 1974, Results of an aeromagnetic survey of the Gulf of Alaska: Journal of Geophysical Research, v. 79, p. 719–723.

Turner, F. J., 1981, Metamorphic petrlogy (second edition): New York, McGraw Hill, 524 p.

Turner, R. F., ed., 1987, Geological and operational summary, Kodiak shelf stratigraphic test wells, western Gulf of Alaska: Anchorage, Alaska, Minerals Management Service, Outer Continental Shelf Report MMS 87-0109, 341 p.

Tysdal, R. G., and Case, J. E., 1979, Geologic map of the Seward and Blying Sound quadrangles, Alaska: U.S. Geological Survey Miscellaneous Investigations Map I-1150, 12 p., scale 1:250,000.

Tysdal, R. G., and Plafker, G., 1978, Age and continuity of the Valdez Group, southern Alaska, *in* Sohl, N. F., and Wright, W. B., compilers, Changes in stratigraphic nomenclature by the U.S. Geological Survey, 1977: U.S. Geological Survey Bulletin 1457-A, p. 120–131.

Van Alstine, D. R., Bazard, D. R., and Whitney, J. W., 1985, Paleomagnetism of cores from the Yakutat well, Gulf of Alaska: Eos (Transactions, American Geophysical Union), v. 66, p. 865.

Van der Voo, R., Jones, M., Grommé, C. S., Eberlein, G. D., and Churkin, M., Jr., 1980, Paleozoic paleomagnetism and northward drift of the Alexander terrane, southeastern Alaska: Journal of Geophysical Research, v. 85, p. 5281–5296.

von Huene, R., Fisher, M. A., and Bruns, T. R., 1987, Geology and evolution of the Kodiak margin, Gulf of Alaska, *in* Scholl, D. W., Grantz, A., and Vedder, J. G., eds., Geology and resource potential of the continental margin of western North America and adjacent ocean basins—Beaufort Sea to Baja California: Houston, Texas, Circum-Pacific Council for Energy and Mineral Resources, Earth Science Series, v. 6, p. 191–212.

Vrolijk, P., Myers, G., and Moore, J. C., 1988, Warm fluid migration along tectonic melanges in the Kodiak accretionary complex, Alaska: Journal of Geophysical Research, v. 93, p. 10,313–10,324.

Wiley, T. J., 1986, Sedimentary basins of offshore Alaska and adjacent regions: U.S. Geological Survey Open-File Report OF 86-35, 118 p.

Winkler, G. R., 1976, Deep-sea fan deposition of the lower Tertiary Orca Group, eastern Prince William Sound, Alaska, *in* Miller, T. P., ed., Recent and ancient sedimentary environments in Alaska (Symposium Proceedings): Anchorage, Alaska Geological Society, p. R1–R20.

——, 1992, Geologic map, cross sections, and summary geochronology of the Anchorage quadrangle, southern Alaska: U.S. Geological Survey Miscellaneous Investigations Map I-2283, scale 1:250,000.

Winkler, G. R., and Plafker, G., 1981a, Tectonic implications of framework grain mineralogy of sandstone from the Yakutat Group, *in* Albert, N.R.D., and Hudson, T., eds., The United States Geological Survey in Alaska: Accomplishments during 1979: U.S. Geological Survey Circular 823-B, p. B68–B70.

——, 1981b, Geologic map and cross sections of the Cordova and Middleton Island quadrangles, southern Alaska: U.S. Geological Survey Open-File Report 81-1164, scale 1:250,000.

Winkler, G. R., McLean, H., and Plafker, G., 1976, Textural and mineralogical study of sandstones from the onshore Gulf of Alaska Tertiary province, southern Alaska: U.S. Geological Survey Open-File Report 76-198, 48 p.

Winkler, G. R., Silberman, M. L., Grantz, A., Miller, R. J., and MacKevett, E. M., Jr., 1981, Geologic map and summary geochronology of the Valdez quadrangle, southern Alaska: U.S. Geological Survey Open-File Report 80-892-A, scale 1:250,000, 2 sheets.

Winkler, G. R., Miller, M. L., Hoekzema, R. B., and Dumoulin, J. A., 1984, Guide to the bedrock geology of a traverse of the Chugach Mountains from Anchorage to Cape Resurrection: Anchorage, Alaska Geological Society, 40 p.

Wolf, L. W., Stone, D. B., and Davies, J. N., 1991, Crustal structure of the active margin, south-central Alaska: An interpretation of seismic refraction data from the Trans-Alaska Crustal Transect: Journal of Geophysical Research, v. 96, p. 16,455–16,469.

Yao, A., Matsuoka, A., and Nakatani, T., 1982, Triassic and Jurassic radiolarian assemblages in southwest Japan: News of Osaka Micropaleontologists, Special Volume 5, p. 27–43.

Zuffa, G. G., Nilsen, T. H., and Winkler, G. R., 1980, Rock-fragment petrography of the Upper Cretaceous Chugach terrane, southern Alaska: U.S. Geological Survey Open-File Report 80-713, 28 p.

MANUSCRIPT ACCEPTED BY THE SOCIETY FEBRUARY 24, 1993

Printed in U.S.A.

The Geology of North America
Vol. G-1, The Geology of Alaska
The Geological Society of America, 1994

Chapter 13

Geology of southeastern Alaska

George E. Gehrels
Department of Geosciences, University of Arizona, Tucson, Arizona 85721
Henry C. Berg
115 Malvern Avenue, Fullerton, California 92632

INTRODUCTION

Southeastern Alaska, an archipelago also known as the "panhandle" of Alaska, is an approximately 52,000-mi^2 area of intensely glaciated and heavily forested mountains that rise abruptly from a complex system of deep fiords and inland marine waterways. This area is underlain by a complex and heterogeneous assemblage of rocks, and is cut by an intricate network of thrust, normal, and strike-slip faults (Buddington and Chapin, 1929; Gehrels and Berg, 1992).

Rocks in the panhandle record a long and complete geologic history beginning in the Proterozoic, representing every Phanerozoic period, and continuing into the Holocene. These rocks are herein subdivided into ten tectonic assemblages (Figs. 1, 2, and 3), five of which are terranes that apparently contain distinct geologic records, and five of which are lithic assemblages that are in depositional, intrusive, or unknown contact with the terranes.

This chapter begins with a summary of the regional geology of southeastern Alaska derived primarily from the compilation of Gehrels and Berg (1992) and from more recent studies by us and many others. Next, we discuss the components and characteristics of each of the primary tectonic assemblages that make up southeastern Alaska and then discuss constraints and speculations on the relations between the terranes. We then present a general overview of the tectonic evolution of the area.

SUMMARY OF REGIONAL GEOLOGY

Stratified rocks

Pre-Jurassic stratified rocks in southeast Alaska constitute a series of northwest-elongate belts that are of various depositional ages and degrees of deformation and metamorphism. These belts occur along the east and west flanks of the Coast Mountains batholith (known as the Coast Plutonic Complex in Canada and referred to as the Coast Range batholith in Alaska) and on the islands to the west (Fig. 2). Rocks east of the batholith include a relatively narrow belt of Proterozoic(?) or Lower Paleozoic(?) schist, gneiss, and marble (Werner, 1977, 1978; Monger and Berg, 1987) and Devonian, Carboniferous, Permian, and Triassic sedimentary and volcanic rocks that extend eastward into the interior of British Columbia. Rocks along the west flank of the batholith include a narrow and poorly understood assemblage of moderately to strongly deformed Permian and Triassic metasedimentary and metavolcanic rocks. The most extensive assemblage of pre-Jurassic strata occurs on the islands to the west, where a surprisingly complete section of sedimentary and volcanic rocks ranges in age from the latest Proterozoic(?) and Cambrian through the Late Triassic.

Upper Mesozoic strata include Jurassic and Cretaceous graywacke and mafic-intermediate volcanic rocks along both the east and west flanks of the Coast Mountains batholith. Strata along the west flank overlap both belts of pre-Jurassic rocks and grade from relatively nondeformed on the west to high-grade schist and gneiss toward the east. A third assemblage occurs along the west coast of northern southeastern Alaska and consists of strongly deformed and disrupted Cretaceous graywacke and volcanic rocks. Cenozoic strata are widespread on both sides of the batholith. They range from Paleocene to Holocene in age, and some volcanic rocks erupted as recently as 360 ± 60 yr ago (Elliott and others, 1981).

Within the Coast Mountains batholith, metastratified rocks consist primarily of amphibolite- to granulite-facies schist, gneiss, and marble derived from pre-Tertiary protoliths. Some of these rocks may be correlative with the Proterozoic(?) and lower Paleozoic(?) metamorphic rocks along the eastern flank of the batholith, some were probably derived from the Cretaceous and older stratified rocks to the east and west, and some protoliths may be unique to the batholith.

Intrusive rocks

Intrusive rocks range in age from Cambrian to middle Tertiary, but most are Cretaceous or early Tertiary (Gehrels and others, 1984; Brew, this volume; Miller, this volume). Paleozoic plutons occur on several of the islands west of the Coast Mountains batholith and include: (1) small bodies of Cambrian metagranodiorite and metadiorite, (2) Ordovician and Early Silurian

Gehrels, G. E., and Berg, H. C., 1994, Geology of southeastern Alaska, *in* Plafker, G., and Berg, H. C., eds., The Geology of Alaska: Boulder, Colorado, Geological Society of America, The Geology of North America, v. G-1.

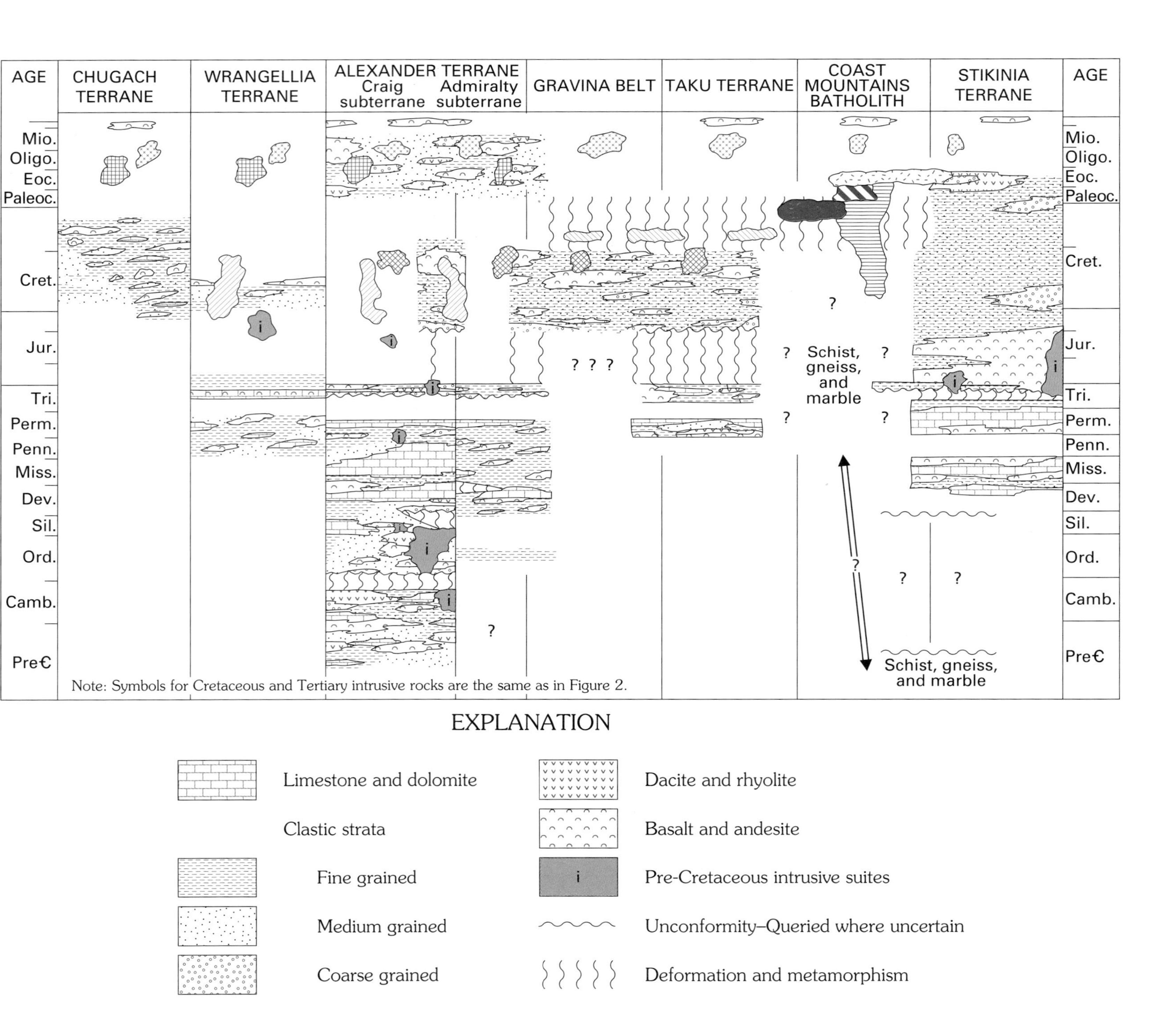
AGE
CHUGACH TERRANE
WRANGELLIA TERRANE
ALEXANDER TERRANE
Craig subterrane
Admiralty subterrane
GRAVINA BELT
TAKU TERRANE
COAST MOUNTAINS BATHOLITH
STIKINIA TERRANE
AGE
Mio.
Oligo.
Eoc.
Paleoc.
Cret.
Jur.
Tri.
Perm.
Penn.
Miss.
Dev.
Sil.
Ord.
Camb.
PreЄ
Schist, gneiss, and marble
Schist, gneiss, and marble
Note: Symbols for Cretaceous and Tertiary intrusive rocks are the same as in Figure 2.
EXPLANATION
Limestone and dolomite
Clastic strata
Fine grained
Medium grained
Coarse grained
Dacite and rhyolite
Basalt and andesite
Pre-Cretaceous intrusive suites
Unconformity–Queried where uncertain
Deformation and metamorphism

calc-alkaline granitoids, (3) Middle Silurian to Early Devonian trondhjemite and leucodiorite, and (4) Pennsylvanian or Permian syenite and diorite. Triassic plutons include a pyroxene gabbro body on Duke Island and a large granodiorite body along the east flank of the Coast Mountains. Jurassic plutons occur in a belt from Baranof Island to the west side of Glacier Bay, on southern Prince of Wales Island, and perhaps in close association with Jurassic and Cretaceous strata along the west flank of the Coast Mountains batholith. All of the pre-Cretaceous intrusive rocks, except the Pennsylvanian or Permian syenitic-dioritic plutons and the Jurassic plutons on Baranof and Prince of Wales Islands, appear to be cogenetic with neighboring volcanic rocks.

Cretaceous and early Tertiary plutons include: (1) granitoids of Early Cretaceous and early Tertiary age on islands west of the Coast Mountains batholith and along its eastern flank; (2) belts of mid-Cretaceous ultramafic bodies and Late Cretaceous granodioritic plutons that occur along the west flank of the batholith; (3) Cretaceous and early Tertiary plutons in the Coast Mountains, which are primary constituents of the Coast Mountains batholith; and (4) Oligocene and Miocene granitoids that trend west-northwest across the batholith and belts to the west.

Regional metamorphism and deformation

Several phases of deformation and/or metamorphism have punctuated the evolution of southeastern Alaska (Dusel-Bacon, this volume). The most significant and widespread event occurred during Late Cretaceous and early Tertiary time, when stratified protoliths now in the Coast Mountains batholith were regionally metamorphosed to amphibolite or granulite facies, strongly deformed, and intruded by a variety of plutonic suites. Rocks east of and along the eastern flank of the batholith were metamorphosed originally before middle Paleozoic time and uplifted, eroded, and perhaps deformed during the Middle Triassic Tahltanian orogeny (Southern, 1971; Monger, 1977). Rocks along the west flank of the batholith were strongly deformed and metamorphosed primarily during formation of the batholith in Late Cretaceous and early Tertiary time. Toward the west, these rocks decrease in metamorphic grade and degree of deformation, and older events are distinguishable. An unconformity at the base of the Jurassic and Cretaceous section indicates that the Permian and Triassic strata to the east and the Triassic and older strata to the west were uplifted and eroded, and at least locally deformed and metamorphosed, between Late Triassic and Late Jurassic time (Gehrels and Berg, 1993; McClelland and Gehrels, 1987b). West of the belt of Jurassic and Cretaceous strata, Paleozoic rocks record deformational and metamorphic events during Middle Cambrian to Early Ordovician and Middle Silurian to earliest Devonian time, and an uplift and erosional event during the Late Permian(?) and Triassic.

Figure 1. Schematic stratigraphic columns from southwest to northeast for the terranes and lithic assemblages of southeastern Alaska. Cretaceous-Tertiary intrusive suites are shown in red and explained in Figure 2. Areas queried are uncertain. Double-headed arrow indicates possible correlation.

Faults

The most conspicuous structural features in southeast Alaska are regional strike-slip fault zones that cut the bedrock into a great jigsaw pattern (Twenhofel and Sainsbury, 1958). On the west, the panhandle is truncated at the North American continental margin by the Queen Charlotte–Fairweather fault system. Faults of this system are known, from geologic mapping, earthquake seismology, and marine and onshore geophysical studies, to be active right-lateral structures with considerable displacement (Plafker and others, this volume, Chapter 12 and Plate 12). To the north, they splay into a set of complex thrust faults (Plafker, 1987). The second major strike-slip system is the Chatham Strait fault, which offsets rocks as young as middle Tertiary by as much as 150 km (Lathram, 1964; Hudson and others, 1982a). This fault is apparently truncated to the southwest by the Fairweather–Queen Charlotte fault system and connects to the north with the Denali fault system. The difference in displacement between the Chatham Strait fault (150 km) and the Denali fault (350? km: Lanphere, 1978; Nokleberg and others, 1985) is a major unresolved problem. The third major strike-slip system in southeastern Alaska is the Clarence Strait fault, which coincides with a major topographic lineament but has only approximately 15 km of dextral displacement (Gehrels and others, 1987).

Thrust, low-angle normal, and steep dip-slip faults are also common in southeastern Alaska. The oldest known of these are southwest-vergent thrusts in the southern panhandle, which moved during Middle Silurian to earliest Devonian time. Their movement was followed by normal movement on gently dipping faults on southern Prince of Wales Island (Keete Inlet fault of Redman, 1981) probably during Late Permian(?) and Triassic time (Gehrels and Saleeby, 1987a). Southwest-vergent thrust faults have been mapped only locally along the west flank of the Coast Mountains batholith (Berg and others, 1988; Rubin and Saleeby, 1987a, b; Gehrels and McClelland, 1988a; McClelland and Gehrels, 1988) but probably are much more widespread. Such faults along the west coast of British Columbia, just south of the panhandle, regionally juxtaposed high-grade metamorphic rocks southwestward over lower-grade rocks during Late Cretaceous time (Crawford and others, 1987). Faults of similar age and style but steeper orientation have also been recognized along the east flank of the batholith (Berg and others, 1978; Crawford and others, 1987). Soon after movement on these thrust faults, rocks of the batholith were uplifted along high-angle faults, one of which is referred to as the Coast Range megalineament (Brew and Ford, 1978).

TECTONIC ASSEMBLAGES

Rocks of southeast Alaska were initially divided into regional geologic belts or assemblages by Buddington and Chapin (1929). Schuchert (1923) noted that rocks in one of these belts

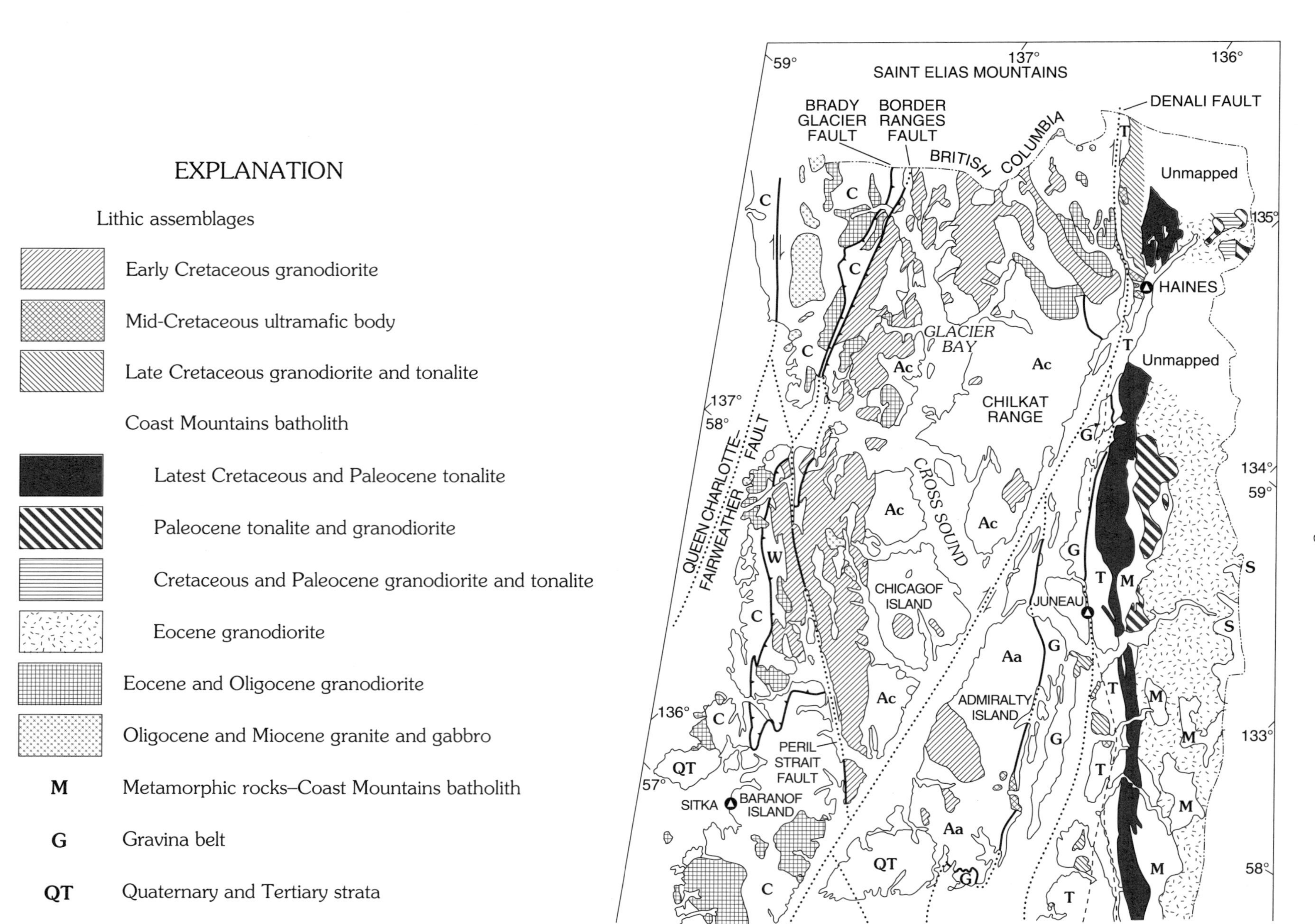
EXPLANATION
Lithic assemblages
Early Cretaceous granodiorite
Mid-Cretaceous ultramafic body
Late Cretaceous granodiorite and tonalite
Coast Mountains batholith
Latest Cretaceous and Paleocene tonalite
Paleocene tonalite and granodiorite
Cretaceous and Paleocene granodiorite and tonalite
Eocene granodiorite
Eocene and Oligocene granodiorite
Oligocene and Miocene granite and gabbro
M Metamorphic rocks–Coast Mountains batholith
G Gravina belt
QT Quaternary and Tertiary strata
SAINT ELIAS MOUNTAINS
DENALI FAULT
BRADY GLACIER FAULT
BORDER RANGES FAULT
BRITISH COLUMBIA
QUEEN CHARLOTTE–FAIRWEATHER FAULT
PERIL STRAIT FAULT
HAINES
JUNEAU
SITKA
Unmapped
GLACIER BAY
CHILKAT RANGE
CROSS SOUND
CHICAGOF ISLAND
ADMIRALTY ISLAND
BARANOF ISLAND
59°
137°
136°
135°
134°
133°
58°
57°

Terranes

Alexander terrane

Aa Admiralty subterrane

Ac Craig subterrane

C Chugach terrane

S Stikinia terrane

T Taku terrane

W Wrangellia terrane

Contact

Coast Range megalineament

Fault and (or) tectonic boundary–Dashed where approximately located, dotted where concealed or projected, queried where uncertain. Arrows indicate direction of relative movement. Values in parenthesis indicate relative distance of displacement. Sawteeth on upper plate of thrust fault

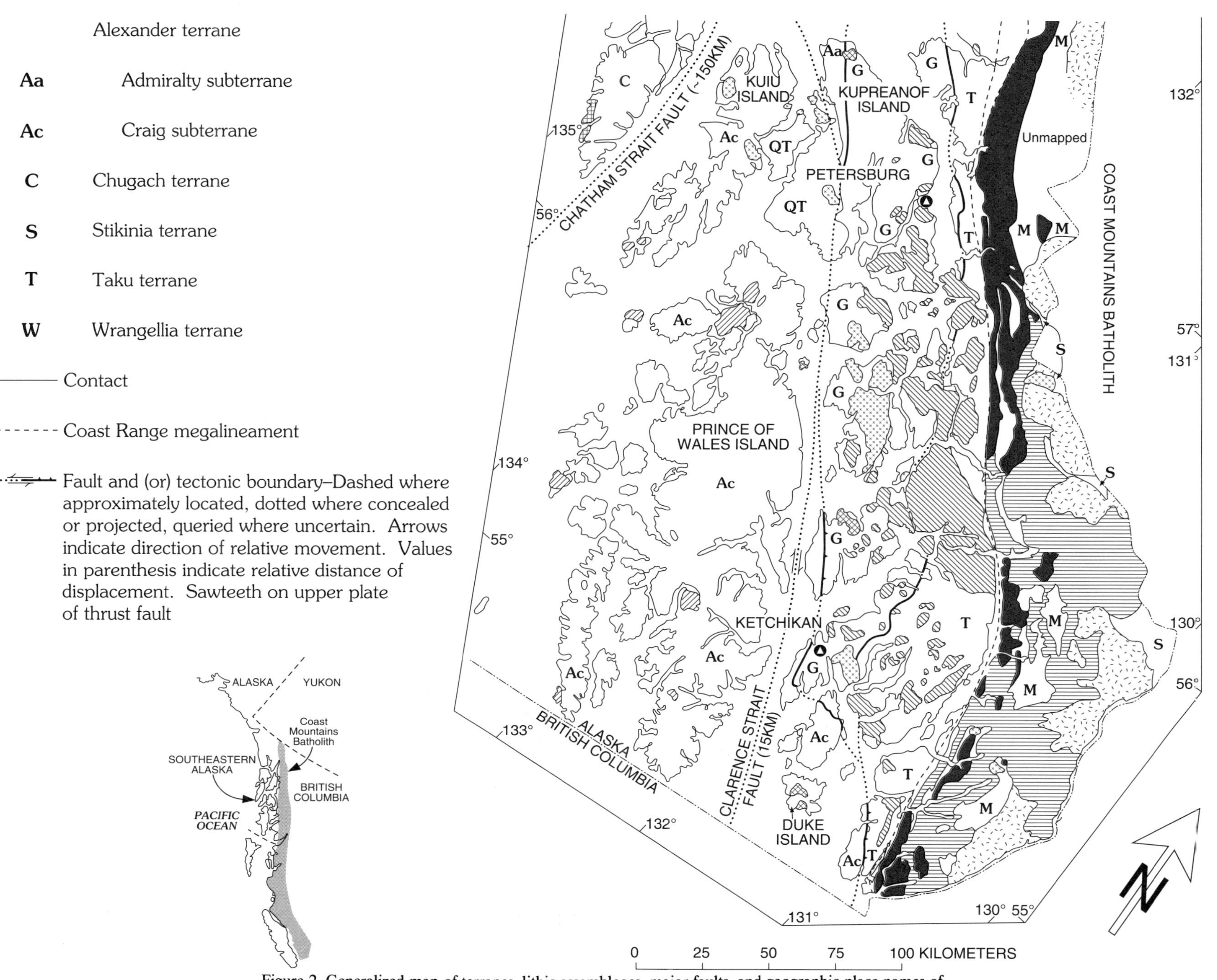

Figure 2. Generalized map of terranes, lithic assemblages, major faults, and geographic place names of southeastern Alaska (modified from Gehrels and Berg, 1992, and Monger and Berg, 1987).

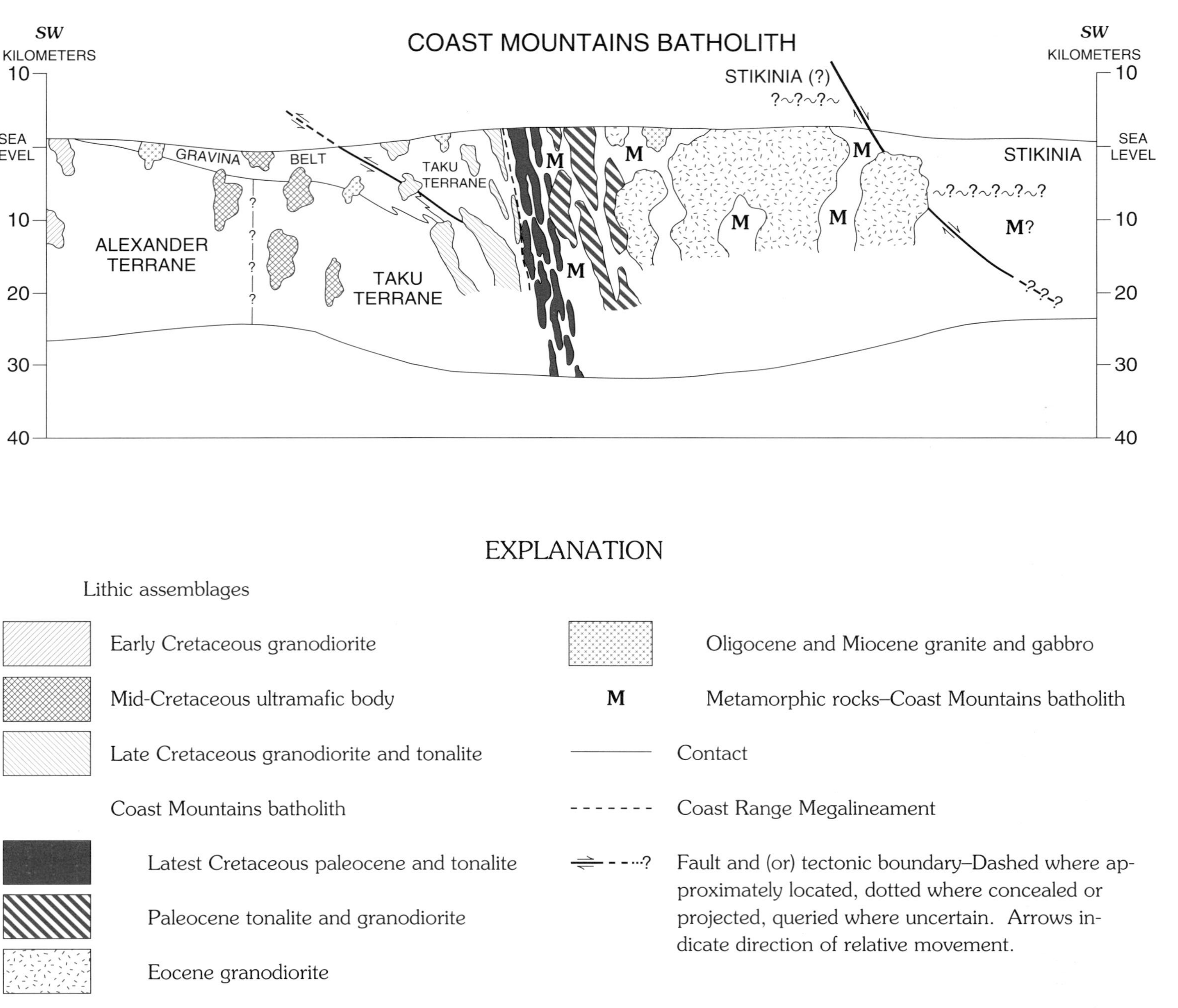

Figure 3. Highly interpretive northeast–southwest cross section that displays known and inferred tectonic and intrusive relations within and adjacent to the Coast Mountains batholith. Symbols for intrusive suites are the same as Figures 1 and 2.

record a geologic history that is significantly different from other regions of the northern Cordillera—this assemblage he referred to as the "Alexandrian embayment" within the Cordilleran geosyncline. Geosynclinal theory dominated syntheses of southeastern Alaska geology until: (1) Wilson (1968) recognized that Paleozoic rocks of the panhandle occur outboard of coeval miogeoclinal strata and must therefore have been accreted to North America; (2) Monger and Ross (1971) documented the Tethyan or equatorial affinity of Permian fusulinid faunas inboard of southeastern Alaska; and (3) Berg and others (1972) recognized that the pre-Jurassic rocks of southeastern Alaska belong to several distinct tectonic fragments called "terranes" that have disparate geologic records. The terrane concept was applied in a more comprehensive fashion by Berg and others (1978), wherein they divided all rocks of southeast Alaska into: (1) fundamentally distinct, fault-bounded tectonic fragments called "tectonostratigraphic terranes" and (2) assemblages that were emplaced into or deposited on more than one terrane and are accordingly interpreted to have formed after adjacent fragments were juxtaposed. Except as otherwise noted, we follow the terrane designations of Monger and Berg (1987) and Silberling and others (this volume).

In this chapter, we subdivide the rocks of southeast Alaska into ten tectonic assemblages (Figs. 1, 2, and 3). Five of these are identified as terranes because they apparently have distinct geologic records. These include the Alexander, Chugach, Stikinia, Taku, and Wrangellia terranes. The other five are lithic assemblages that consist of metamorphic rocks of unknown tectonic affinity or of rocks that are known or reasonably interpreted to be in depositional or intrusive contact with the terranes. Such lithic assemblages include: (1) Jurassic and Cretaceous strata of the Gravina belt (part of the Gravina-Nutzotin belt of Berg and others, 1972) along the west flank of the Coast Mountains batholith; (2) metamorphic pendants and screens of unknown tectonic affinity within the batholith; (3) plutonic rocks within the batholith; (4) Cretaceous, Eocene and Oligocene, and Oligocene and Miocene plutons west of the batholith; and (5) Tertiary and Quaternary strata that are widespread throughout southeast Alaska.

Alexander terrane

The Alexander terrane (Ac and Aa on Fig. 2) comprises a variety of stratified, metamorphic, and plutonic rocks of latest Precambrian(?) to Cambrian through Middle(?) Jurassic age that underlie much of the Alaskan panhandle and continue northward into the Saint Elias Mountains region of British Columbia and the Yukon (Berg and others, 1972; Churkin and Eberlein, 1977; Gehrels and Saleeby, 1987a) and then westward into the Wrangell Mountains of Alaska (MacKevett, 1978). Volcaniclastic turbidites, shallow-marine carbonate rocks, and subordinate conglomerate of Silurian age are the most widespread units in the terrane. Pre-Silurian rocks occur primarily in the southern part of southeast Alaska, upper Paleozoic rocks occur in relatively restricted areas, and Upper Triassic strata crop out in a fairly narrow belt along the eastern margin of the terrane.

The oldest rocks recognized are arc-type metasedimentary and metavolcanic rocks (Wales Group) that were metamorphosed and deformed during the Middle Cambrian to Early Ordovician Wales orogeny (Gehrels and Saleeby, 1984, 1987a). These rocks form the depositional and intrusive basement for an arc-type volcanic-plutonic-sedimentary complex of Early Ordovician to Early Silurian age, which underlies much of the southern part of the terrane. Ordovician chert and argillite on Admiralty Island and marine clastic strata and carbonate rocks in northernmost southeast Alaska and the Saint Elias Mountains region are interpreted to have formed in a deep- to shallow-marine basin behind this arc system (Gehrels and Saleeby, 1987a). This phase of arc-type activity ceased with onset of the Middle Silurian to earliest Devonian Klakas orogeny, which is manifest in the southern part of the terrane as: (1) southwest-vergent thrusting, (2) regional metamorphism and deformation in some areas, (3) uplift (locally greater than 5 km) and erosion of the arc complex, and (4) generation of anatectic(?) trondhjemite and leucodiorite bodies.

Upper Paleozoic strata in much of the Alexander terrane consist primarily of shallow-marine carbonate rocks, clastic strata, and subordinate mafic-intermediate volcanic rocks. These strata now occur in restricted erosional remnants but probably were much more widespread originally. The predominance of shallow-marine limestone, the lack of regionally significant unconformities or thick conglomerate in the section, and restriction of volcanic rocks to the Middle and Upper Devonian and the Lower Permian parts of the section suggest that the terrane evolved in a tectonically stable environment when compared to the early Paleozoic orogenic and magmatic activity.

Upper Triassic strata overlie the Permian and older rocks on a regional unconformity. In the southern part of the terrane, the section generally consists, from bottom to top, of basal conglomerate and sedimentary breccia, rhyolite and rhyolitic tuff, massive limestone, calcareous argillite, and basaltic-andesitic pillow flows and breccia. Toward the northwest, the amount of rhyolite and conglomerate decreases, and the proportion of mafic-intermediate volcanic rocks increases. These strata and their subjacent unconformity are interpreted to have formed in a rift environment on the basis of: (1) the bimodal (basalt-rhyolite) composition of the volcanic rocks, (2) occurrence of the section in a relatively narrow belt along the eastern margin of the terrane, (3) stratigraphic evidence for syndepositional faulting, and (4) evidence for Late Permian(?) and Triassic uplift and erosion without accompanying deformation and metamorphism (Gehrels and others, 1986).

The youngest component of the terrane is the Bokan Mountain Granite, which is a Middle(?) Jurassic peralkaline ring-dike complex on southern Prince of Wales Island (Thompson and others, 1982; Saint-Andre and others, 1983; Armstrong, 1985). The tectonic significance of this body is as yet unknown.

The Alexander terrane was subdivided by Berg and others (1978) and Monger and Berg (1987) into the Craig, Annette, and Admiralty subterranes on the basis of regional variations in stratigraphy and in degree and age of metamorphism and deformation. Gehrels and others (1987) suggest that the Annette and Craig subterranes should not be differentiated, however, because they share similar pre–Middle Devonian and Triassic lithic assemblages, record the same early Paleozoic tectonic histories, and both lack upper Paleozoic strata. We accordingly include rocks of the Annette subterrane with rocks of the Craig subterrane.

Relations between the Admiralty (Aa on Fig. 2) and Craig (Ac on Fig. 2) subterranes are difficult to assess because rocks of the Admiralty subterrane have only locally been studied in detail, have yielded few fossils, and are regionally metamorphosed and deformed in most areas. However, studies to date on Kupreanof (Muffler, 1967; Brew and others, 1984; McClelland and Gehrels, 1987a, b) and Admiralty (Lathram and others, 1965) Islands and in the Chilkat Range (Lathram and others, 1959; Brew and others, 1985) support the interpretation that Paleozoic strata of the Admiralty subterrane record a different history from coeval rocks in the rest of the terrane. In general, the Admiralty subterrane consists of Ordovician and Devonian to Triassic basinal clastic strata, mafic-intermediate volcanic rocks, and subordinate limestone (Fig. 1). A Late Devonian and Early Mississippian metamorphic and deformational event has recently been recognized in Admiralty subterrane rocks northwest of Haines (Forbes and others, 1987). The timing of this event and the basinal and volcanic-rich nature of these strata contrast with the history of the Craig subterrane and support their distinction as two subterranes.

The earliest reliable link between the two assemblages occurs in the Permian, when clasts of chert from the Admiralty assemblage were deposited on the Craig subterrane and dolomite of the Pybus Formation was deposited over both assemblages (Muffler, 1967; Berg and others, 1978; Jones and others, 1981). Triassic strata of the two subterranes are apparently correlative. Thus, prior to Permian time, the Admiralty subterrane may have been a basinal, volcanic-rich facies adjacent to a tectonically stable, dominantly shallow-marine facies, or it may have been a distinct tectonic fragment.

The displacement history of the Alexander terrane prior to its Mesozoic accretion is poorly constrained and is a subject of considerable speculation. Apparently reliable constraints on its paleoposition include: (1) Nd-Sr isotopic data that indicate the terrane contains juvenile crustal materials and was not in proximity to a large continental landmass prior to Jurassic time (Samson and others, 1987, 1988); (2) paleomagnetic data that suggest the terrane evolved near the paleoequator from Ordovician through Pennsylvanian time (Van der Voo and others, 1980); (3) biogeographic indications of a late Paleozoic position between eastern and western Pacific faunal realms (Mamet and Pinard, 1985); (4) the low paleolatitude and eastern Pacific affinity of Triassic bivalves of the terrane (Tozer, 1982; Newton, 1983; Silberling, 1985); and (5) occurrence of the terrane outboard of the Cache Creek terrane, which contains Permian fusulinid faunas of Tethyan (equatorial) affinity (Monger and Ross, 1971) and apparently remained in an oceanic setting into Jurassic time (Cordey and others, 1987). Accretion against the western margin of inboard terranes probably began during Early or Middle Jurassic time (McClelland and Gehrels, 1987b; Gehrels and Saleeby, 1985) and was completed by the early Tertiary.

Chugach terrane

The Chugach terrane (C on Fig. 2) has two structural components (Plafker and others, 1977; Johnson and Karl, 1985; Gehrels and Berg, 1992): (1) a strongly deformed but coherent assemblage of flyschoidal graywacke, argillite, and slate of Late Cretaceous age; and (2) a deformed and disrupted assemblage (melange) composed of blocks of basic volcanic rocks, radiolarian chert, ultramafic rocks, limestone, and plutonic rocks in a matrix of cherty, tuffaceous argillite. The composition and age of some blocks suggest that they probably were derived from upper Paleozoic and Triassic strata of Wrangellia. Greenschist- to amphibolite-facies regional metamorphism overprints local remnants of blueschist-facies metamorphism. Radiolarians in chert in the matrix of the melange are generally Late Jurassic and Early Cretaceous in age. Upper Jurassic *Buchia* occur in blocks and possibly in situ in the matrix (Brew and others, 1988). The two assemblages are structurally interleaved in many places, but in general, the disrupted assemblage tends to lie structurally above and east of the coherent flysch assemblage. Most workers ascribe formation of the Chugach terrane to plate convergence along the Pacific margin of the Alexander and Wrangellia terranes.

Juxtaposition of the terrane against inboard assemblages is thought to have occurred during middle Cretaceous time (Decker and others, 1980) on the basis of the age of blueschist-greenschist metamorphism of inboard units. Although Cowan (1982) hypothesized that the terrane was displaced northward from an original position near Vancouver Island after 40 Ma, there are no known faults along which the postulated offset could have occurred (Plafker, 1987, p. 257).

Stikinia terrane

The most significant components of the Stikinia terrane (S on Fig. 22) in and adjacent to southeastern Alaska include Devonian carbonate rocks, Carboniferous arc-type volcanic and sedimentary rocks and widespread carbonate rocks, Lower Permian basinal strata, Lower and Upper Permian platformal limestone, and Upper Triassic to Middle Jurassic arc-type volcanic, plutonic, and clastic sedimentary rocks (Fig. 2; Monger, 1977; Gehrels and Berg, 1992; Robert G. Anderson, oral communication, 1987). Most of the terrane in British Columbia is overlain by Jurassic and Cretaceous sedimentary and volcanic rocks; Triassic and older strata crop out primarily along the eastern flank of the Coast Mountains. The boundary between these strata and high-grade metamorphic rocks in the Coast Mountains batholith is in most areas obliterated by plutons of the batholith. However, in some regions: (1) an east-dipping fault (shown speculatively as a normal fault in Fig. 3) juxtaposes high-grade rocks against strata of the Stikinia terrane, (2) strata belonging to the Stikinia terrane grade with increasing metamorphism into the high-grade rocks, and (3) Upper Triassic strata apparently overlie high-grade metamorphic rocks (Souther, 1971; Berg and others, 1978; Werner, 1977, 1978; Bultman, 1979; Monger and Berg, 1987; Brew and others, 1985; Hill and others, 1985; Crawford and others, 1987).

Rocks of the Stikinia terrane are known to have been displaced because they occur outboard of Cache Creek rocks containing Permian fusulinid faunas of Tethyan or equatorial affinity (Monger and Ross, 1971), and their primitive Nd-Sr isotopic signature precludes primary relations with North America (Samson and others, 1987). Their accretionary history remains enigmatic, however, because (1) paleomagnetic data from Permian and Triassic rocks of the Stikinia terrane do not record significant latitudinal transport relative to North America (May and Butler, 1986; Irving and Monger, 1987), (2) the Cache Creek terrane apparently remained in an oceanic setting through Early to Middle Jurassic time (Cordey and others, 1987), and (3) paleomagnetic data from Cretaceous strata and from older rocks with interpreted Cretaceous magnetic signatures record large-scale northward transport in Cretaceous to early Tertiary time (Irving and others, 1985; Marquis and Globerman, 1987).

Taku terrane

The Taku terrane (T on Fig. 2) is a poorly understood assemblage of deformed and metamorphosed strata of Early Permian, Middle and Late Triassic, and perhaps pre-Permian age (Silberling and others, 1982; Brew and Grybeck, 1984). As mapped on Figure 2, the terrane also contains a significant proportion of Upper Jurassic to mid-Cretaceous strata of the Gravina belt (Gehrels and Berg, 1992; Rubin and Saleeby, 1987a, b) and lower Paleozoic rocks of the Alexander terrane (Saleeby, 1987). Regionally significant pre-Jurassic components include Permian crinoidal marble intercalated with pelitic phyllite and felsic metatuff; Permian(?) basaltic metatuff, agglomerate, and pillow flows; Middle and Upper Triassic basalt, pillow basalt, basaltic breccia, carbonaceous limestone, slate, and phyllite; Jurassic and Cretaceous(?) calcareous flysch; undated quartzite and quartzofeldspathic gneiss presumably derived from felsic volcanic rocks; and metaconglomerate containing clasts of granitic rocks, quartzite, and fine-grained clastic strata. Rocks of Jurassic(?) and Cretaceous(?) age include greenschist- and amphibolite-facies metagraywacke, meta-argillite, metabasaltic pillow flows and breccia, and metaconglomerate.

Rocks of the Taku terrane can be subdivided into a variety of assemblages, the relations between which are as yet unknown (Monger and Berg, 1987). Near Haines (Fig. 2), Upper Triassic basalt and overlying Triassic and Jurassic(?) sedimentary rocks are reported to be geochemically and biostratigraphically similar to Wrangellian strata in western British Columbia and southern Alaska (Plafker and Hudson, 1980; Davis and Plafker, 1985; Plafker and others, 1989). This sequence is overlain conformably by undated calcareous flysch. Northwest of Juneau, Permian(?) and Triassic metabasaltic pillow flows predominate and are overlain unconformably by less deformed Jurassic and Cretaceous flysch of the Gravina-Nutzotin belt (Redman, 1984; Gehrels and Berg, 1993). In southern southeastern Alaska, Triassic metasedimentary rocks predominate, and contact relations with Jurassic and Cretaceous strata are obscured by Late Cretaceous deformation and metamorphism. Rubin and Saleeby (1987a, b) and Saleeby (1987) report that much of what has been mapped as the southern Taku terrane consists of lower Paleozoic metasedimentary and metavolcanic rocks of the Alexander terrane. In central southeastern Alaska, the terrane includes Permian and Triassic metasedimentary and metavolcanic rocks and a thick sequence of metarhyolite, metabasalt, and quartz-rich metaturbidites of unknown age and affinity (Gehrels and McClelland, 1988a). Hence, the Taku terrane, as shown on Figure 2, includes rocks that belong to the Wrangellia terrane to the north, and to the Gravina belt and Alexander terrane to the south. Permian and Triassic rocks of the Taku terrane in the central and southern part of the panhandle may be southern continuations of the Wrangellia terrane rocks to the north, or they may constitute a distinct tectonic assemblage.

Contact relations with adjacent terranes are poorly understood. To the southwest, the terrane is generally thrust southwestward over Jurassic and Cretaceous strata of the Gravina belt. In the Haines area and adjacent parts of Canada, the Denali fault marks the southwestern contact. To the northeast, rocks of the terrane increase in metamorphic grade to amphibolite and perhaps locally granulite facies and are intruded by plutons of the Coast Mountains batholith. We tentatively draw the eastern boundary within tonalitic bodies in the western part of the batholith, although metamorphic rocks within the batholith are lithically indistinguishable from some high-grade members of the Taku terrane.

Wrangellia terrane

A coherent sequence of unfossiliferous strata on Chichagof and Baranof Islands is intepreted to be a fragment of the Wrangellia terrane (W on Fig. 2) on the basis of similarities in lithic types and age (Plafker and others, 1976; Jones and others, 1977; Berg and others, 1978). The sequence is distinguished by: (1) thick, mainly subaerial basalt flows (Goon Dip Greenstone) similar to those of the Middle and/or Upper Triassic Nikolai Greenstone in southern Alaska (Jones and others, 1977); (2) shallow- to deep-marine carbonate rocks (Whitestripe Marble) that are similar to the Upper Triassic Chitistone Limestone; and (3) pelitic sedimentary rocks similar to the Upper Triassic and Lower Jurassic McCarthy Formation. Jurassic tonalitic plutons are the youngest components of the terrane. The sequence apparently overlies a heterogeneous assemblage of upper(?) Paleozoic mafic volcanic rocks, pyroclastic rocks, clastic sedimentary rocks, and minor chert and marble ranging in metamorphic grade from greenschist to amphibolite facies.

The contacts between these rocks and adjacent terranes are poorly understood. Along much of the east boundary the rocks of the Wrangellia terrane are apparently juxtaposed against rocks of the Alexander terrane along a Tertiary right-lateral fault (Peril Strait fault)—the original boundary between the two is difficult to identify because of emplacement of abundant Jurassic(?) and Cretaceous plutons and widespread metamorphism and deformation along the boundary. Elsewhere on Baranof Island and western and northern Chichagof Island, the Wrangellia terrane is juxtaposed against rocks of the Chugach terrane along the Border Ranges fault, which is interpreted as a west-vergent thrust (Plafker and others, 1976).

If correlations with rocks in other parts of the Wrangellia terrane are correct, then the rocks in southeastern Alaska must also have been transported considerable distances northward since Late Triassic time (Jones and others, 1977; Hillhouse and Grommé, 1984; Tozer, 1982).

Gravina belt

The Gravina belt (G on Fig. 2) comprises Upper Jurassic to mid-Cretaceous marine argillite and graywacke, interbedded andesitic to basaltic volcanic and volcaniclastic rocks, subordinate polymictic conglomerate, and perhaps plutons ranging from quartz diorite to dunite and peridotite (Berg and others, 1972, 1978; Gehrels and Berg, 1992). These strata occur in a narrow belt separating the Alexander and Taku terranes and record the transition from lower-grade rocks on the west to higher-grade rocks along the flank of the Coast Mountains. In general, the metamorphic grade increases from greenschist or subgreenschist facies to the west to amphibolite facies toward the east. Contact relations with adjacent terranes are uncertain. Although strata of the Gravina belt are interpreted to depositionally overlie rocks of the Alexander terrane (Berg and others, 1972), depositional contacts between the two assemblages are apparently preserved only on Gravina Island (Berg, 1973) and Kupreanof Island (McClelland and Gehrels, 1987a). In contrast, the Gravina rocks depositionally overlie metamorphosed and deformed rocks of the Taku terrane northwest of Juneau (Redman, 1992) and possibly on Chilkat Peninsula (Plafker and others, 1989). Similar relations may also occur near Ketchikan (Gehrels and Berg, 1992; McClelland and Gehrels, 1987b; C. Rubin, oral communication, 1987; C. Rubin, *in* Barker, this volume).

The eastern margin of the Gravina belt is in most areas difficult to identify. In southern southeastern Alaska, high-grade metamorphic rocks assigned to the Taku terrane were probably derived in part from Gravina belt protoliths. To the north, the eastern margin may occur along strike-slip or thrust faults along the west flank of the Coast Mountains batholith or within tonalitic bodies in the western part of the batholith.

Regional relations suggest that the Gravina belt is part of a basinal assemblage that accumulated along the eastern margin of the previously juxtaposed Alexander and Wrangellia terranes and continued eastward across the Taku terrane (Berg and others, 1972, 1978). The continuation or correlatives of Gravina belt strata east of the Taku terrane have not yet been identified.

Metamorphic rocks of the Coast Mountains batholith

Metasedimentary and metavolcanic rocks make up approximately 20 percent of the Coast Mountains batholith (M on Fig. 2) and consist primarily of pelitic, semi-pelitic, and quartzofeldspathic schist and gneiss and subordinate amphibolite, quartzite, marble, and calc-silicate rocks. These rocks have been referred to as components of the Central Gneiss Complex by previous workers. Protoliths are generally interpreted to have been argillaceous marine strata, limestone and/or dolomite, chert, and subordinate mafic to felsic volcanic rocks. Some rocks may also have been derived from plutonic protoliths. Protolith ages of Proterozoic(?), early Paleozoic(?), Carboniferous(?), Permian(?), Triassic, Jurassic(?), and Cretaceous(?) are indicated by: (1) relations north and east of Juneau, which suggest that the metamorphic rocks locally grade eastward into Triassic and older strata of the Stikinia terrane and are at least locally overlain by the Triassic strata (Souther, 1971; Werner, 1977, 1978; Bultman, 1979; Brew and others, 1985); (2) a preliminary Rb-Sr isochron of Proterozoic apparent age determined on high-grade metamorphic rocks along the Alaska–British Columbia border north of Juneau (Werner and Armstrong *in* Monger and Berg, 1987; Werner, 1977, 1978); and (3) regional relations that suggest that metasedimentary rocks in the central part of the batholith southeast of southeastern Alaska were derived from Jurassic and Cretaceous strata (Douglas, 1986), from the Jurassic Bowser Lake Group (Woodsworth and others, 1983), and from strata of Permian(?) (Hill, 1985) and pre-Permian(?) (Hutchison, 1982) age (Hill and others, 1985).

Regional amphibolite- and locally granulite-facies metamorphism occurred primarily during Late Cretaceous and early Tertiary time, although metamorphism prior to deposition of Jurassic and Cretaceous strata of the Gravina belt, prior to Triassic time and between Proterozoic and Carboniferous time, may have previously affected various parts of the Coast Mountains batholith.

Intrusive suites of the Coast Mountains batholith

Plutons of the Coast Mountains batholith generally belong to three distinct suites (shown separately on Fig. 2) that become progressively younger toward the east and to a large unit of undivided granodioritic rocks ranging in age from Early(?) Cretaceous to Paleocene. The oldest suite is composed of narrow but very long sheetlike masses of hornblende-dominant tonalite and quartz diorite that extend in a linear fashion along the western margin of the batholith. These tabular bodies (commonly referred to as "tonalite sills") are interpreted by some workers as marking a fundamental tectonic boundary within the batholith (Berg and others, 1978; Brew and Ford, 1981; Gehrels and Berg, 1992). Strong foliation and lineation within the bodies and their contact relations with country rocks suggest that they were intruded during the later stages of regional metamorphism and deformation within the Coast Mountains. U-Pb (zircon) ages on individual bodies become younger to the south: from near 70 Ma in northern Alaska (Barker and others, 1986), through 67 to 64 Ma in central southeastern Alaska (Gehrels and others, 1984), to 55 to 60 Ma in southernmost southeastern Alaska and adjacent parts of British Columbia (Arth and others, 1988; Armstrong and Runkle, 1979).

East of the tonalitic bodies are discrete plutons to large batholithic complexes of Paleocene granodiorite. These granodioritic bodies are commonly elongate but appear to have been emplaced after most of the deformation in the batholith. In southern southeastern Alaska, these bodies apparently engulf the tonalitic bodies and are interpreted to make up most of the batholith. The youngest and volumetrically most significant suite in the batholith contains huge bodies of biotite-dominant granodiorite of Eocene age (Gehrels and others, 1984; Gehrels and Berg, 1993). These rocks were generally emplaced at shallow crustal levels, as their volcanic cover (Sloko Group) is locally preserved adjacent to the plutons.

Barker and Arth (1984), Barker and others (1986), and Arth and others (1988) conclude that plutons in the batholith are components of an Andean-type or continental margin arc formed in response to subduction of oceanic crust. In contrast, Monger and others (1982), Kenah and Hollister (1983), and Crawford and others (1987) indicate that some components may be anatectic melts generated during the main phase of deformation and metamorphism in the Coast Mountains.

Intrusive suites west of the Coast Mountains batholith

Intrusive bodies west of the batholith belong to several suites that include the following from oldest to youngest:

1. Large, generally isolated plutons, predominantly of granodiorite composition, that intrude the Alexander and Wrangellia terranes. K-Ar ages on these bodies are generally Early Cretaceous, but some bodies may be coeval with similar plutons in the Saint Elias Mountains region that yield Late Jurassic K-Ar ages (Dodds and Campbell, 1988). These intrusive rocks are tentatively interpreted to be genetically related to Upper Jurassic to Lower Cretaceous volcanic rocks of the Gravina belt.

2. Zoned ultramafic complexes ranging in composition from dunite, commonly in the centers of the complexes, to clinopyroxenite (Taylor, 1967; Irvine, 1967, 1974). These bodies yield K-Ar ages from Early to mid-Cretaceous (Lanphere and Eberlein, 1966). The ultramafic bodies intrude Triassic and older rocks of the Alexander terrane, Permian(?) and Triassic(?) rocks of the Taku terrane, and probably Jurassic and Cretaceous strata of the Gravina belt. Irvine (1973, 1974) has argued that these bodies are subvolcanic to mafic, clinopyroxene-bearing flows in the Gravina belt.

3. Granodioritic, tonalitic, and subordinate quartz monzonite to quartz dioritie bodies on the west flank of the Coast Mountains that intrude strata of the Taku terrane and Gravina belt. Most intrusive bodies contain biotite and/or hornblende; many contain garnet, muscovite, and primary epidote; and some are pyroxene bearing. K-Ar, $^{40}Ar/^{39}Ar$, and U-Pb ages indicate emplacement primarily during mid-Cretaceous time. Geobarometric studies of epidote and garnet in these bodies suggest that they crystallized at mid-crustal to lower crustal levels (Zen and Hammarstrom, 1984a, b). Arth and others (1988) conclude that these plutons formed in a subduction-related magmatic arc on the basis of geochemical and isotopic analyses in the Ketchikan area.

4. Large granodioritic bodies of Eocene and Oligocene age that intrude Chugach, Wrangellia, and Alexander terranes in northwestern southeastern Alaska. The rocks range from muscovite and locally garnet-bearing granodiorite, granite, and tonalite in the Baranof Island–Glacier Bay area, to biotite- and hornblende-bearing quartz diorite and granodiorite in the Chilkat Range.

5. Stocks of biotite-, hornblende-, and pyroxene-bearing granite; alkali granite; quartz monzonite; granodiorite; diorite; and layered and locally zoned bodies of gabbro, quartz gabbro, and other mafic-ultramafic intrusives. These stocks occur in two distinct regions, one extending from the Coast Mountains east of Ketchikan northwestward to the northern tip of Kuiu Island and the other in northern southeastern Alaska on Chichagof Island and in the Glacier Bay area. K-Ar ages on these bodies are generally Oligocene and Miocene. The bodies in southern and central southeastern Alaska intrude the Coast Mountains batholith, Gravina belt, and the Taku terranes, whereas the northern assemblage intrudes the Alexander, Wrangellia, and Chugach terranes.

Quaternary and Tertiary strata

Tertiary and Quaternary strata (QT on Fig. 2) underlie large regions of Kupreanof, southern Admiralty, and western Baranof Islands and occur in many other more restricted areas throughout southeastern Alaska. Figure 2 shows the distribution of these strata only where they cover large regions. In central southeastern Alaska, these strata include the lower and middle Tertiary Kootznahoo Formation (nonmarine sandstone, shale, and conglomerate) and Admiralty Island Volcanics (basalt and andesite), and younger basaltic to rhyolitic volcanic rocks and associated sedimentary rocks. West of Glacier Bay, Oligocene(?) and Miocene strata (not shown on Fig. 2) belong to the Cenotaph Volcanics (basalt) and the Topsy Formation (marine calcareous sandstone and siltstone). Tertiary and Quaternary basaltic to rhyolitic volcanic rocks and subordinate sedimentary rocks also occur at Mount Edgecumbe (west of Sitka), in the Coast Mountains east of Ketchikan and Petersburg, in the western Prince of Wales Island region, on islands in Cross Sound, and in many other areas of southeastern Alaska.

RELATIONS AMONG TERRANES

The primary and present-day relations among the terranes of southeastern Alaska are controversial. Uncertainties about their similarities and differences and about the existence and characteristics of boundaries between them arise from the fact that the geology of much of the panhandle has been studied only in reconnaissance fashion, and because many critical relations within and between terranes are obscured by Cretaceous and early Tertiary metamorphism, deformation, and/or plutonism. To date, two fundamentally different interpretations of the tectonic framework of southeastern Alaska have been proposed. Berg and others (1978) and Monger and Berg (1987) believe that the Alexander, Chugach, Stikinia, Taku, and Wrangellia terranes and the metamorphic rocks of the Coast Mountains batholith each have distinct lithic components and tectonic histories and that each is (or was) fault bounded. In contrast, Brew and Ford (1983, 1984) believe that the differences between most of these terranes result from facies changes within a single crustal fragment. Specifically, Brew and Ford (1984) suggest that Permian and Triassic rocks of the Taku and Stikinia terranes are facies equivalents of the upper parts of the Alexander terrane, and that the metamorphic rocks of the Coast Mountains batholith and the older rocks of the Stikinia terrane are facies equivalents of the lower parts of the Alexander terrane.

In the following sections, we assess what is known about the similarities and differences among the Alexander, Taku, Wrangellia, and Stikinia terranes and the metamorphic rocks in the Coast Mountains batholith and offer some tentative interpretations about primary relations among the terranes.

Alexander-Wrangellia relations

The Alexander and Wrangellia terranes were originally interpreted as separate tectonic entities prior to their juxtaposition during Jurassic time (Berg and others, 1978; Coney and others, 1980). This interpretation needs to be modified because Pennsylvanian dioritic and syenitic intrusive bodies are now known to intrude both terranes and their boundary in southern Alaska and southwestern Yukon (MacKevett and others, 1986; Gardner and others, 1988). These relations indicate that the Alexander and Wrangellia terranes have been in proximity since at least Pennsylvanian time. In addition, it is likely that Upper Triassic rocks of the Alexander terrane are facies equivalents of the Upper Triassic rift assemblage of the Wrangellia terrane (Gehrels and others, 1986).

Pre-Pennsylvanian relations between the two terranes, however, are as yet uncertain. The Devonian through Permian volcanic and basinal sedimentary assemblages that characterize the Wrangellia terrane in British Columbia (Jones and others, 1977; Brandon and others, 1986) are different from the carbonate-dominated upper Paleozoic rocks of much of the Alexander terrane (Craig subterrane). It is possible that the Wrangellia terrane rocks correlate with distal sedimentary and volcanogenic components of the Admiralty subterrane, but a rigorous comparison of these two assemblages must await more detailed studies of the Admiralty subterrane.

Alexander-Taku relations

Rigorous comparisons between the Taku terrane and adjacent assemblages are hindered by a lack of age constraints on most protoliths of the Taku terrane. Recent studies have also shown that in many areas, rocks previously assigned to the Taku terrane belong to other assemblages. In the Ketchikan area, for example, Saleeby (1987) and Rubin and Saleeby (1987a, b) conclude that the Taku terrane of Monger and Berg (1987) consists of lower Paleozoic rocks of the Alexander terrane and Jurassic and Cretaceous strata of the Gravina belt, as well as the Permian and Triassic metasedimentary and metavolcanic rocks that are characteristic of the Taku terrane. In contrast, in northern southeastern Alaska, Davis and Plafker (1985) and Plafker and others (1989) argue on the basis of geochemical and biostratigraphic similarities that Triassic strata of the northern Taku terrane are correlative with Wrangellia terrane basalts and overlying sedimentary rocks in the Alaska Range and in British Columbia. These relations, the proximity of Permian and Triassic Taku rocks to older Alexander terrane strata near Ketchikan, and the pre-Triassic linkage of the Alexander and Wrangellia terranes are consistent with a scenario in which the Triassic rocks in the Alexander, Wrangellia, and Taku terranes are parts of a once-contiguous rift assemblage. The differences in Upper Triassic rock types among the three terranes may reflect varying positions in the extensional environment: bimodal volcanic rocks and coarse conglomerate of the Alexander terrane may have formed on thicker, more evolved crust near the basin margin; flood basalts of the Wrangellia terrane may be the result of extension within ensimatic or less evolved crust; and basalt and fine-grained clastic strata of the Taku terrane may have formed within an entirely basinal regime.

The principal arguments against primary links between the Alexander and Taku terranes are that: (1) Permian rocks are known in only two restricted regions of the Alexander terrane in the panhandle but apparently constitute much of the Taku terrane; (2) in spite of their present-day close proximity, Triassic lithic types and stratigraphic relations in the two terranes are quite different; and (3) zircon populations in U-Pb samples from the Alexander terrane do not show evidence of inheritance (Gehrels and Saleeby, 1987b; Gehrels and others, 1987), whereas Cretaceous intrusive bodies in the Taku terrane have inherited significant Precambrian zircon components (Rubin and Saleeby, 1987a).

Correlation of metamorphic rocks of the Coast Mountains batholith

The high metamorphic grade, penetrative deformation, and lack of protolith age control limit arguments concerning the regional tectonic affinity of metamorphic rocks of the Coast Mountains batholith. Most relations indicate, however, that rocks along the eastern margin of the batholith either are metamorphic equivalents of the Stikinia terrane strata or, as shown schematically on Figure 3, belong to a metamorphic complex that may be overlain and intruded by Triassic rocks of the Stikinia terrane. These metamorphic rocks are apparently indistinguishable from rocks in the western part of the batholith and in some eastern, high-grade part of the Taku terrane. In addition, inherited Precambrian zircon components also occur in intrusive bodies in both the Taku terrane (Rubin and Saleeby, 1987a, b) and the Coast Mountains batholith (Gehrels and others, 1984). Thus, although the suite of tonalitic bodies in the batholith and/or the Coast Range megalineament may mark the primary boundary between the batholith and the Taku terrane (Berg and others, 1978; Brew and Ford, 1978, 1984; Gehrels and Berg, 1992; Arth and others, 1988), it is not yet possible to document a significant change in protolith content across either one. We draw the boundary within the belt of tonalitic bodies because we do not view the Coast Range megalineament as a significant tectonic boundary.

Brew and Ford (1984) have suggested that metamorphic rocks in the Coast Mountains batholith are the metamorphic equivalents of strata in the lower part of the Alexander terrane. Beyond a general comparison of proportions of rocktypes, this possibility is difficult to test geologically because so little is known about the protolith age of the metamorphic rocks. Isotopically, however, the two can be distinguished on the basis of: (1) the presence of inherited zircon components in the batholith (Gehrels and others, 1984) but not in the Alexander terrane (Gehrels and Saleeby, 1987b; Gehrels and others, 1987), and (2) significantly more primitive $^{87}Sr/^{86}Sr$ and $^{143}Nd/^{144}Nd$ initial ratios in the Alexander terrane (Samson and others, 1987, 1988) than in intrusive bodies of the Coast Mountains batholith (Barker and others, 1986; Arth and others, 1988).

TECTONIC HISTORY OF SOUTHEASTERN ALASKA

The currently decipherable tectonic history of southeastern Alaska begins during latest Proterozoic(?) to Early Cambrian time with the formation of an intraoceanic arc-type basement for the Alexander terrane. Arc-type activity, punctuated by Late Cambrian to Early Ordovician and Middle Silurian to earliest Devonian orogenic events, continued through Silurian and perhaps into Devonian time. The early Paleozoic paleoposition of the terrane within this tectonically active, intraoceanic regime is problematic. Gehrels and Saleeby (1984, 1987a) argued that the terrane bears tectonic similarities to orogenic systems that formed along the paleo-Pacific margins of Australia, Antarctica, and crustal fragments now residing in Asia; they hypothesized that the terrane may have formed within the western part of the paleo-Pacific Ocean basin. As noted by Savage (1987), however, early paleozoic faunas of the Alexander terrane are different from faunas found in eastern Australia and more closely resemble North American forms. In apparent contrast to both comparisons, Nd-Sr isotopic data indicate that the terrane is constructed of juvenile crustal materials and that it was not near any continental landmasses during early Paleozoic time (Samson and others, 1987, 1988).

Proterozoic(?) and lower Paleozoic(?) rocks along the eastern margin of the Coast Mountains batholith differ from those in the Alexander terrane because they are dominated by quartz-rich clastic strata and yield Rb-Sr isotopic data consistent with an age of approximately 900 Ma (Werner and Armstrong *in* Monger and Berg, 1987). These rocks may have formed in a continental margin environment and are reported by Werner (1977, 1978) and Bultman (1979) to form the depositional basement to Triassic rocks of the northern Stikinia terrane. The primitive Nd-Sr isotopic signature of rocks of the central Stikinia terrane suggests, however, that this older metamorphic basement does not extend beneath the central part of the Stikinia terrane.

Beginning in Devonian time, the Wrangellia and Stikinia terranes and the Admiralty subterrane of the Alexander terrane all evolved in an environment characterized by intraoceanic arc-type volcanic rocks, basinal marine clastic sediments, and subordinate carbonate rocks. In contrast, upper Paleozoic rocks of the Craig subterrane record tectonic stability through at least Late Pennsylvanian time and perhaps through the mid-Permian. The Wrangellia terrane and the Craig and Admiralty subterranes were probably in close proximity during most of this time (MacKevett and others, 1986; Gardner and others, 1988; Muffler, 1967; Berg and others, 1978; Jones and others, 1981). These relations, combined with Nd-Sr isotopic data from upper Paleozoic rocks of the Craig subterrane and Stikinia terrane, require the large tectonic fragments in and adjacent to the panhandle to have evolved in an intraoceanic realm through late Paleozoic time. Similarities between Alexander terrane faunas and both North American and Tethyan forms (Mamet and Pinard, 1985; Ross and Ross, 1983, 1985) are consistent with an intraoceanic setting within the paleo-Pacific basin.

Triassic rocks of the Alexander and Wrangellia terranes are interpreted to have formed in a rift environment (Jones and others, 1977; Gehrels and others, 1986). In the Wrangellia terrane, huge volumes of tholeiitic flood basalt covered the terrane, whereas in the Alexander terrane a bimodal volcanic suite was erupted along the eastern margin of the terrane. Triassic basalt and andesite of the Stikinia terrane are interpreted to have erupted within a volcanic arc environment, presumably related to subduction along the eastern (inboard) margin of the terrane (Monger and Ross, 1971). The tectonic environment of Triassic rocks of the southern part of the Taku terrane is unknown. The northern part is interpreted by Plafker and others (1989) as a rift-fill sequence that developed on and along the northeastern margin of the composite Alexander and Wrangellia terranes.

The tectonic history of southeastern Alaska after Triassic time is dominated by accretion of the Alexander and Wrangellia terranes against the Stikinia or other inboard terranes. An unconformity separating Upper Jurassic to mid-Cretaceous strata of the Gravina belt from underlying Triassic and older rocks is the first evidence of this accretionary activity. Prior to Late Jurassic time, Permian and Triassic rocks of the Taku terrane were deformed and regionally metamorphosed (Gehrels and Berg, 1992), and rocks of the eastern part of the Alexander terrane were deformed and disrupted along the Duncan Canal shear zone on Kupreanof Island (McClelland and Gehrels, 1987a). This deformation is interpreted to record either the northward movement of the composite Alexander-Wrangellia terrane along the California-Washington continental margin (Gehrels and Saleeby, 1985) or perhaps to the initial juxtaposition against the western margin of the Stikinia terrane (McClelland and Gehrels, 1987b).

Rocks of the Gravina belt accumulated along the western margin of a marine basin of unknown width. As proposed by Berg and others (1972), volcanism within this basin was probably distally related to granitic plutonism in the Alexander terrane, to formation of part of the Chugach accretionary complex, and to plate convergence along the outboard margin of the Alexander and Wrangellia terranes. The Gravina belt as a depositional basin may have formed as: (1) an extensional structure within or behind a west-facing arc (Berg and others, 1972; Brew and Ford, 1983), (2) a collapsing sedimentary basin that records closure of the suture between the Alexander and Stikinia terranes (Pavlis, 1982), (3) a western continuation of the Jurassic and Cretaceous marine basin that formed on the Stikinia terrane (Muller, 1977), (4) a pull-apart structure in a right-lateral transform system along which the outboard terranes were transported northward (Gehrels and Saleeby, 1985), or (5) a fore-arc basin with respect to a northeast-facing arc constructed along the inboard margin of the composite Alexander-Wrangellia terranes (Fred Barker, written communication, 1988). Without additional information, all of these scenarios apparently remain viable.

Structural accretion of the Alexander terrane against the western margin of the Stikinia terrane began soon after deposition of mid-Cretaceous strata of the Gravina belt (Berg and others, 1972, 1978; Coney and others, 1980; Monger and others, 1982; Sutter and Crawford, 1985). This accretionary event is recognized as movement on west-vergent thrust faults, widespread deformation and regional metamorphism of rocks within the Coast Mountains batholith and along its western flank, and anatectic and/or subduction-generated plutonism within the suture zone separating the Alexander and Stikinia terranes (Monger and others, 1982; Arth and others, 1988). These events apparently culminated between approximately 95 and 65 Ma and were followed soon after by rapid uplift of the Coast Mountains batholith. Thermobarometric studies within the batholith indicate that uplift rates of 2 mm/yr were achieved at about 55 Ma, bringing rocks that formed at more than 20 km depth to the surface (Hollister, 1982; Crawford and others, 1987). In Figure 3, we follow Gehrels and McClelland (1988b) in speculating that much of the early Tertiary uplift of the batholith occurred along east-dipping, west-side-up normal faults that may form the present boundary between the batholith and strata of the Stikinia terrane to the east.

Intrusive bodies of Oligocene and Miocene gabbro and granite having low initial $^{87}Sr/^{86}Sr$ and of swarms of lamprophyre and quartz porphyry dikes suggests still younger post-accretionary extensional(?) tectonism, possibly tapping mantle sources. These intrusive rocks trend west-northwesterly, across the regional northwest trends of the Coast Mountains batholith, Taku terrane, Gravina belt, and Alexander terrane.

Southeastern Alaska continues to be tectonically active, as shown by Holocene faulting and uplift (Hudson and others, 1982b) and the eruption of lava flows as recently as 360 ± 60 yr ago (Elliott and others, 1981).

REFERENCES CITED

Armstrong, R. L., 1985, Rb/Sr dating of the Bokan Mountain granite complex and its country rocks: Canadian Journal of Earth Sciences, v. 22, p. 1233–1236.

Armstrong, R. L., and Runkle, D., 1979, Rb-Sr geochronometry of the Ecstall, Kitkiata, and Quottoon plutons and their country rocks, Prince Rupert region, Coast Plutonic complex, British Columbia: Canadian Journal of Earth Sciences, v. 16, p. 387–399.

Arth, J. G., Barker, F., and Stern, T. W., 1988, Coast batholith and Taku plutons near Ketchikan, Alaska; Petrography, geochronology, geochemistry, and isotopic character: American Journal of Science, v. 288-A, p. 461–489.

Barker, F., and Arth, J. G., 1984, Preliminary results, Central Gneiss Complex of the Coast Range batholith, southeastern Alaska; The roots of a high-K calc-alkaline arc?: Physics of the Earth and Planetary Interiors, v. 35, p. 191–198.

Barker, F., Arth, J. G., and Stern, T. W., 1986, Evolution of the Coast batholith along the Skagway traverse, Alaska and British Columbia: American Mineralogist, v. 71, p. 632–643.

Berg, H. C., 1973, Geology of Gravina Island, Alaska: U.S. Geological Survey Bulletin 1373, 41 p.

Berg, H. C., Jones, D. L., and Richter, D. H., 1972, Gravina–Nutzotin belt; Tectonic significance of an upper Mesozoic sedimentary and volcanic sequence in southern and southeastern Alaska: U.S. Geological Survey Professional Paper 800-D, p. D1-D24.

Berg, H. C., Jones, D. L., and Coney, P. J., 1978, Map showing pre-Cenozoic tectonostratigraphic terranes of southeastern Alaska and adjacent areas: U.S. Geological Survey Open-File Report 78-1085, scale 1:1,000,000.

Berg, H. C., Elliott, R. L., and Koch, R. D., 1988, Geologic map of the Ketchikan and Prince Rupert Quadrangles, southeastern Alaska: U.S. Geological Survey Miscellaneous Investigations Series Map I-1807, scale 1:250,000.

Brandon, M. T., Orchard, M. J., Parrish, R. R., Sutherland-Brown, A., and Yorath, C. J., 1986, Fossil ages and isotopic dates from the Paleozoic Sicker Group and associated intrusive rocks, Vancouver Island, British Columbia: Geological Survey of Canada Paper 86-1A, p. 683–696.

Brew, D. A., and Ford, A. B., 1978, Megalineament in southeastern Alaska marks southwest edge of Coast Range batholithic complex: Canadian Journal of Earth Sciences, v. 15, p. 1763–1772.

——, 1981, The Coast plutonic complex sill, southeastern Alaska, *in* Albert, N.R.D., and Hudson, T. L., eds., The United States Geological Survey in Alaska; Accomplishments during 1980: U.S. Geological Survey Circular 823-B, p. B96-B98.

——, 1983, Comment on 'Tectonic accretion and the origin of the two major metamorphic and plutonic welts in the Canadian Cordillera': Geology, v. 11, p. 427–429.

——, 1984, Tectonostratigraphic terranes in the Coast plutonic–metamorphic complex, southeastern Alaska, *in* Bartsch-Winkler, S., and Reed, K. M., eds., The United States Geological Survey in Alaska; Accomplishments during 1982: U.S. Geological Survey Circular 939, p. 90–93.

Brew, D. A., and Grybeck, D., 1984, Geology of the Tracy Arm–Fords Terror wilderness study area and vicinity, Alaska: U.S. Geological Survey Bulletin 1525-A, p. 21–52.

Brew, D. A., Overshine, A. T., Karl, S. M., and Hunt, S. J., 1984, Preliminary reconnaissance geologic map of the Petersburg and parts of the Port Alexander and Sumdum 1:250,000 Quadrangles, southeastern Alaska: U.S. Geological Survey Open-File Report 84-405.

Brew, D. A., Ford, A. B., and Garwin, S. L., 1985, Fossiliferous Middle and(or) Upper Triassic rocks within the Coast plutonic–metamophic complex southeast of Skagway, *in* Bartsch-Winkler, S., ed., The U.S. Geological Survey in Alaska; Accomplishments during 1984: U.S. Geological Survey Circular 967, p. 86–89.

Brew, D. A., Karl, S. M., and Miller, J. W., 1988, Megafossils (*Buchia*) indicate Late Jurassic age for part of the Kelp Bay Group on Baranof Island, southeastern Alaska, *in* Galloway, J. P., and Hamilton, T. D., eds., Geologic studies in Alaska by the U.S. Geological Survey during 1987: U.S. Geological Survey Circular 1016, p. 147–149.

Buddington, A. S., and Chapin, T., 1929, Geology and mineral deposits of southeastern Alaska: U.S. Geological Survey Bulletin 800, 398 p.

Bultman, T. R., 1979, Geology and tectonic history of the Whitehorse trough west of Atlin, British Columbia [Ph.D. thesis]: New Haven, Connecticut, Yale University, 284 p.

Churkin, M., Jr., and Eberlein, G. D., 1977, Ancient borderland terranes of the North American Cordillera; Correlation and microplate tectonics: Geological Society of America Bulletin, v. 88, p. 769–786.

Coney, P. J., Jones, D. L., and Monger, J.W.H., 1980, Cordilleran suspect terranes: Nature, v. 288, p. 329–333.

Cordey, F., Mortimer, N., DeWever, P., and Monger, J.W.H., 1987, Significance of Jurassic radiolarians from the Cache Creek terrane, British Columbia: Geology, v. 15, p. 1151–1154.

Cowan, D. S., 1982, Geological evidence for post-40 m.y. B.P. large-scale northwestward displacement of part of southeastern Alaska: Geology, v. 10, p. 309–313.

Crawford, M. L., Hollister, L. S., and Woodsworth, G. J., 1987, Crustal deformation and regional metamorphism across a terrane boundary, Coast Plutonic Complex, British Columbia: Tectonics, v. 6, p. 343–361.

Davis, A., and Plafker, G., 1985, Comparative geochemistry of Triassic basaltic rocks from the Taku terrane on the Chilkat Peninsula and Wrangellia: Canadian Journal of Earth Sciences, v. 22, p. 183–194.

Decker, J. E., Wilson, F. H., and Turner, D. L., 1980, Mid-Cretaceous subduction event in southeastern Alaska: Geological Society of America Abstracts with Programs, v. 12, p. 103.

Dodds, C. J., and Campbell, R. B., 1988, Potassium-argon ages of mainly intrusive rocks in the Saint Elias Mountains, Yukon and British Columbia: Geological Survey of Canada Paper 87-16, 43 p.

Douglas, B. J., 1986, Deformational history of an outlier of metasedimentary rocks, Coast Plutonic Complex, British Columbia, Canada: Canadian Journal of Earth Sciences, v. 23, p. 813–826.

Elliott, R. L., Koch, R. D., and Robinson, S. W., 1981, Age of basalt flows in the Blue River valley, Bradfield Canal Quadrangle, *in* Albert N.R.D., and Hudson, T. L., eds., The United States Geological Survey in Alaska; Accomplishments during 1979: U.S. Geological Survey Circular 832-B, p. B115–B116.

Forbes, R. B., Gilbert, W. G., and Redman, E. C., 1987, The Four Winds complex; A newly recognized Paleozoic metamorphic complex in southeastern Alaska: Geological Society of America Abstracts with Programs, v. 19, p. 378.

Gardner, M. C., and 8 others, 1988, Pennsylvanian pluton stitching of Wrangellia and the Alexander terrane, Wrangell Mountains, Alaska: Geology, v. 16, p. 967–971.

Gehrels, G. E., and Berg, H. C., 1992, Geologic map of southeastern Alaska: U.S. Geological Survey Miscellaneous Investigations Series Map I-1867, scale 1:600,000.

Gehrels, G. E., and McClelland, W. C., 1988a, Outline of the Taku terrane and Gravina belt in the Cape Fanshaw–Windham Bay region of central southeastern Alaska: Geological Society of America Abstracts with Programs, v. 20, p. 163.

——, 1988b, Early Tertiary uplift of the Coast Range batholith along west-side-up extensional shear zones: Geological Society of America Abstracts with Programs, v. 20, p. A111.

Gehrels, G. E., and Saleeby, J. B., 1984, Paleozoic geologic history of the Alexander terrane, and comparisons with other orogenic belts: Geological Society of America Abstracts with Programs, v. 16, p. 516.

——, 1985, Constraints and speculations on the displacement and accretionary history of the Alexander–Wrangellia–Peninsular superterrane: Geological Society of America Abstracts with Programs, v. 17, p. 356.

——, 1987a, Geologic framework, tectonic evolution, and displacement history of the Alexander terrane: Tectonics, v. 6, p. 151–173.

——, 1987b, Geology of southern Prince of Wales Island, southeastern Alaska: Geological Society of America Bulletin, v. 98, p. 123–137.

Gehrels, G. E., Brew, D. A., and Saleeby, J. B., 1984, Progress report on U/Pb (zircon) geochronologic studies in the Coast Plutonic–metamorphic complex east of Juneau, southeastern Alaska, *in* Bartsch-Winkler, S., and Reed, K. M., eds., The United States Geological Survey in Alaska; Accomplishments during 1982: U.S. Geological Survey Circular 939, p. 100–102.

Gehrels, G. E., Dodds, C. J., and Campbell, R. B., 1986, Upper Triassic rocks of the Alexander terrane, SE Alaska, and the Saint Elias Mountains of B.C. and Yukon: Geological Society of America Abstracts with Programs, v. 18, p. 109.

Gehrels, G. E., Saleeby, J. B., and Berg, H. C., 1987, Geology of Annette, Gravina, and Duke Islands, southeastern Alaska: Canadian Journal of Earth Sciences, v. 24, p. 866–881.

Hill, M. L., 1985, Remarkable fossil locality; Crinoid stems from migmatite of the Coast Plutonic Complex, British Columbia: Geology, v. 13, p. 825–826.

Hill, M. L., Woodsworth, G. J., and van der Heyden, P., 1985, The Coast Plutonic Complex near Terrace, B.C.; A metamorphosed western extension of Stikinia: Geological Society of America Abstracts with Programs, v. 17, p. 362.

Hillhouse, J. W., and Gromme, C. S., 1984, Northward displacement and accretion of Wrangellia; New paleomagnetic evidence from Alaska: Journal of Geophysical Research, v. 89, p. 4461–4477.

Hollister, L. S., 1982, Metamorphic evidence for rapid (2 mm/yr) uplift of a portion of the Central Gneiss Complex, Coast Mountains, B.C.: Canadian Mineralogist, v. 20, p. 319–332.

Hudson, T., Plafker, G., and Dixon, K., 1982a, Horizontal offset history of the Chatham Strait fault, *in* Coonrad, W. L., ed., The United States Geological Survey in Alaska; Accomplishments during 1980: U.S. Geological Survey Circular 844, p. 128–132.

Hudson, T., Dixon, K., and Plafker, G., 1982b, Regional uplift in southeastern Alaska, *in* Coonrad, W. L., ed., The United States Geological Survey in Alaska; Accomplishments during 1980: U.S. Geological Survey Circular 844, p. 132–135.

Hutchison, W. W., 1982, Geology of the Prince Rupert–Skeena map area, British Columbia: Geological Survey of Canada Memoir 394, 116 p.

Irvine, T. N., 1967, The Duke Island ultramafic complex, southeastern Alaska, *in* Wyllie, P. J., ed., Ultramafic and related rocks: New York, John Wiley and Sons, p. 84–97.

——, 1973, Bridget Cove volcanics, Juneau area, Alaska; Possible parental magma of Alaskan-type ultramafic complexes: Carnegie Institute of Washington Yearbook, v. 72, p. 478–491.

——, 1974, Petrology of the Duke Island ultramafic complex, southeastern

Alaska: Geological Society of America Memoir 138, 240 p.

Irving, E., and Monger, J.W.H., 1987, Preliminary paleomagnetic results form the Permian Asitka Group, British Columbia: Canadian Journal of Earth Sciences, v. 24, p. 1490–1497.

Irving, E., Woodsworth, G. J., Wynne, P. J., and Morrison, A., 1985, Paleomagnetic evidence for displacement from the south of the Coast Plutonic Complex, British Columbia: Canadian Journal of Earth Sciences, v. 22, p. 584–598.

Johnson, B. R., and Karl, S. M., 1985, Geologic map of western Chichagof and Yakobi Islands, southeastern Alaska: U.S. Geological Survey Miscellaneous Investigations Series Map I-1506, scale 1:250,000.

Jones, D. L., Silberling, N. J., and Hillhouse, J., 1977, Wrangellia; A displaced terrane in northwestern North America: Canadian Journal of Earth Sciences, v. 14, p. 2565–2577.

Jones, D. L., Berg, H. C., Coney, P., and Harris, A., 1981, Structural and stratigraphic significance of Upper Devonian and Mississippian fossils from the Cannery Formation, Kupreanof Island, southeastern Alaska, *in* Albert, N.R.D., and Hudson, T., eds., The United States Geological Survey in Alaska; Accomplishments during 1979: U.S. Geological Survey Circular 823-B, p. B109–B112.

Kenah, C., and Hollister, L. S., 1983, Anatexis in the Central Gneiss Complex, British Columbia, *in* Atherton, M. P., and Gribble, C. D., eds., Migmatites, melting, and metamorphism: Cheshire, United Kingdom, Shiva Publishing Ltd., p. 142–162.

Lanphere, M. A., 1978, Displacement history of the Denali fault system, Alaska and Canada: Canadian Journal of Earth Sciences, v. 15, p. 817–822.

Lanphere, M. A., and Eberlein, G. D., 1966, Potassium-argon ages of magnetite-bearing ultramafic complexes in southeastern Alaska: Geological Society of America Special Paper 87, p. 94.

Lathram, E. H., 1964, Apparent right-lateral separation on Chatham Strait fault, southeastern Alaska: Geological Society of America Bulletin, v. 75, p. 249–252.

Lathram, E. H., Loney, R. A., Condon, W. H., and Berg, H. C., 1959, Progress map of the geology of the Juneau Quadrangle, Alaska: U.S. Geological Survey Miscellaneous Geologic Investigations Map I-303, scale 1:250,000.

Lathram, E. H., Pomeroy, J. S., Berg, H. C., and Loney, R. A., 1965, Reconnaissance geology of Admiralty Island, Alaska: U.S. Geological Survey Bulletin 1181-R, p. R1-R48.

MacKevett, E. M., Jr., 1978, Geologic map of the MacCarthy Quadrangle, Alaska: U.S. Geological Survey Miscellaneous Investigations Series Map I-1032, scale 1:250,000.

MacKevett, E. M., Gardner, M. C., Bergman, S. C., Cushing, G., and McClelland, W. C., 1986, Geologic evidence for Late Pennsylvanian juxtaposition of Wrangellia and the Alexander terrane, Alaska: Geological Society of America Abstracts with Programs, v. 18, p. 128.

Mamet, B. L., and Pinard, S., 1985, 9 Carboniferous algae from the Peratrovich Formation, southeastern Alaska, *in* Toomey, D. F., and Nitecki, M. H., eds., Paleoalgology; Contemporary research and applications: New York, Springer-Verlag, p. 91–100.

Marquis, G., and Globerman, B. R., 1987, Paleomagnetism of the Upper Cretaceous Carmacks Group, west of Tintina Trench fault, Yukon and British Columbia: EOS Transactions of the American Geophysical Union, v. 68, p. 1254.

May, S. R., and Butler, R. F., 1986, North American Jurassic apparent polar wander; Implications for plate motion, paleogeography, and Cordilleran tectonics: Journal of Geophysical Research, v. 91, p. 11519–11544.

McClelland, W. C., and Gehrels, G. E., 1987a, Analysis of a major shear zone in Duncan Canal, Kupreanof Island, southeastern Alaska: Geological Society of America Abstracts with Programs, v. 19, p. 430.

——, 1987b, Evidence for early-Middle Jurassic deformation and metamorphism along the inboard margin of the Alexander terrane, SE Alaska: Geological Society of America Abstracts with Programs, v. 19, p. 764.

——, 1988, Characteristics of the Taku terrane (TT) and Gravina belt (GB) in the Petersburg region, central southeastern Alaska: Geological Society of America Abstracts with Programs, v. 20, p. 211.

Monger, J.W.H., 1977, Upper Paleozoic rocks of the western Canadian Cordillera and their bearing on Cordilleran evolution: Canadian Journal of Earth Sciences, v. 14, p. 1832–1859.

Monger, J.W.H., and Berg, H. C., 1987, Lithotectonic terrane map of western Canada and southeastern Alaska: U.S. Geological Survey Miscellaneous Field Studies Map MF-1874-B, scale 1:2,500,000.

Monger, J.W.H., and Ross, C. A., 1971, Distribution of Fusulinaceans in the western Canadian Cordillera: Canadian Journal of Earth Sciences, v. 8, p. 259–278.

Monger, J.W.H., Price, R. A., and Templeman-Kluit, D. J., 1982, Tectonic accretion and the origin of the two major metamorphic welts in the Canadian Cordillera: Geology, v. 10, p. 70–75.

Muffler, L.J.P., 1967, Stratigraphy of the Keku Islets and neighboring parts of Kuiu and Kupreanof Islands, southeastern Alaska: U.S. Geological Survey Bulletin 1241-C, 52 p.

Muller, J. E., 1977, Evolution of the Pacific margin, Vancouver Island, and adjacent regions: Canadian Journal of Earth Sciences, v. 14, p. 2062–2085.

Newton, C. R., 1983, Paleozoogeographic affinities of Norian bivalves from the Wrangellian, Peninsular, and Alexander terranes, *in* Stevens, C. H., ed., Pre-Jurassic rocks in western North American suspect terranes: Pacific Section, Society of Economic Paleontologists and Mineralogists, p. 37–48.

Nokelberg, W. J., Jones, D. L., and Silberling, N. J., 1985, Origin and tectonic evolution of the Maclaren and Wrangellia terranes, eastern Alaska Range, Alaska: Geological Society of America Bulletin, v. 96, p. 1251–1270.

Pavlis, T. L., 1982, Origin and age of the Border Ranges fault of southern Alaska and its bearing on the Late Mesozoic evolution of Alaska: Tectonics, v. 1, p. 343–368.

Plafker, G., 1987, Regional geology and petroleum potential of the northern Gulf of Alaska continental margin, *in* Scholl, D. W., Grantz, A., and Vedder, J., eds., Geology and resource potential of the continental margin of western North America and adjacent ocean basins; Beaufort Sea to Baja California: Circum-Pacific Council for Energy and Mineral Resources Earth Science Series, v. 6, p. 229–268.

Plafker, G., and Hudson, T., 1980, Regional implications of Upper Triassic metavolcanic and metasedimentary rocks on the Chilkat Peninsula, southeastern Alaska: Canadian Journal of Earth Sciences, v. 17, p. 681–689.

Plafker, G., Jones, D. L., Hudson, T., and Berg, H. C., 1976, The Border Ranges fault system in the Saint Elias Mountains and Alexander Archipelago, *in* Cobb, E. H., ed., The United States Geological Survey in Alaska; Accomplishments during 1975: U.S. Geological Survey Circular 733, p. 14–16.

Plafker, G., Jones, D. L., and Pessagno, E. A., Jr., 1977, A Cretaceous accretionary flysch and mélange terrane along the Gulf of Alaska margin, *in* Blean, K. M., ed., The United States Geological Survey in Alaska; Accomplishments during 1976: U.S. Geological Survey Circular 751-B, P. B41–B43.

Plafker, G., Blome, C. D., and Silberling, N. J., 1989, Reinterpretation of lower Mesozoic rocks on the Chilkat Peninsula, Alaska; Closer links with Wrangellia: Geology, v. 17, p. 3–6.

Redman, E., 1981, The Keete Inlet thrust fault, Prince of Wales Island: Alaska Division of Geological and Geophysical Surveys Report 73, p. 17–18.

——, 1984, An unconformity associated with conglomeratic sediments in the Berners Bay area of southeastern Alaska: Alaska Division of Geological and Geophysical Surveys Professional Report 86, p. 1–4.

Ross, C. A., and Ross, J.R.P., 1983, Late Paleozoic accreted terranes of western North America, *in* Stevens, C. H., ed., Pre-Jurassic rocks in western North American suspect terranes: Pacific Section, Society of Economic Paleontologists and Mineralogists, p. 7–22.

——, 1985, Carboniferous and Early Permian biogeography: Geology, v. 13, p. 27–30.

Rubin, C. M., and Saleeby, J. B., 1987a, The inner boundary zone of the Alexander terrane, southern SE Alaska; A newly discovered thrust belt: Geological Society of America Abstracts with Programs, v. 19, p. 445.

——, 1987b, The inner boundary zone of the Alexander terrane in southern SE Alaska; Part 1, Cleveland Peninsula to southern Revillagigedo Island: Geo-

logical Society of America Abstracts with Programs, v. 19, p. 826.
Saint-Andre, B. de, Lancelot, J. R., and Collot, B., 1983, U-Pb geochronology of the Bokan Mountain peralkaline granite, southeastern Alaska: Canadian Journal of Earth Sciences, v. 20, p. 236–245.
Saleeby, J. B., 1987, The inner boundary zone of the Alexander terrane in southern SE Alaska; Part 2, Southern Revillagigedo Island (RI) to Cape Fox (CF): Geological Society of America Abstracts with Programs, v. 19, p. 828.
Samson, S. D., McClelland, W. C., Gehrels, G. E., Patchett, P. J., and Anderson, R. G., 1987, Nd isotopes and the origin of the accreted Alexander and Stikine terranes in the Canadian Cordillera: EOS Transactions of the American Geophysical Union, v. 68, p. 1548.
Samson, S. D., McClelland, W. C., Gehrels, G. E., and Patchett, P. J., 1988, The Alexander terrane, Nd and Sr isotopic evidence for a primitive magmatic history: Geological Society of America Abstracts with Programs, v. 20, p. 227.
Savage, N. M., 1987, Eastern Australia is an unlikely source terrane for the Alexander terrane of southeastern Alaska: Geological Society of America Abstracts with Programs, v. 19, p. 830.
Schuchert, C., 1923, Sites and nature of the North American geosynclines: Geological Society of America Bulletin, v. 34, p. 151–230.
Silberling, N. J., 1985, Biogeographic significance of the Upper Triassic bivalve *Monotis* in Circum-Pacific suspect terranes, *in* Howell, D. G., ed., Tectonostratigraphic terranes of the circum-Pacific region: Circum-Pacific Council for Energy and Mineral Resources, p. 63–70.
Silberling, N. J., Wardlaw, B. R., and Berg, H. C., 1982, New paleontologic age determinations from the Taku terrane, Ketchikan area, southeastern Alaska, *in* Coonrad, W. L., ed., The United States Geological Survey in Alaska; Accomplishments during 1980: U.S. Geological Survey Circular 844, p. 117–119.
Souther, J. G., 1971, Geology and mineral deposits of the Tulsequah map area, British Columbia: Geological Survey of Canada Memoir 362, 84 p.
Sutter, J. F., and Crawford, M. L., 1985, Timing of metamorphism and uplift in the vicinity of Prince Rupert, British Columbia and Ketchikan, Alaska: Geological Society of America Abstracts with Programs, v. 17, p. 411.
Taylor, H. P., Jr., 1967, The zoned ultramafic complexes of southeastern Alaska, *in* Wyllie, P. J., ed., Ultramafic and related rocks: New York, John Wiley and Sons, p. 96–118.
Thompson, T. B., Pierson, J. R., and Lyttle, T., 1982, Petrology and petrogenesis of the Bokan Mountain Granite complex, southeastern Alaska: Geological Society of America Bulletin, v. 93, p. 898–908.
Tozer, E. T., 1982, Marine Triassic faunas of North America; Their significance for assessing plate and terrane movements: Geologische Rundschau, v. 71, p. 1077–1104.
Twenhofel, W. S., and Sainsbury, C. L., 1958, Fault patterns in southeastern Alaska: Geological Society of America Bulletin, v. 69, p. 1431–1442.
Van der Voo, R., Jones, M., Gromme, C. S., Eberlein, G. D., and Churkin, M., Jr., 1980, Paleozoic paleomagnetism and northward drift of the Alexander terrane, southeastern Alaska: Journal of Geophysical Research, v. 85, p. 5281–5296.
Werner, L. J., 1977, Metamorphic terrane, northern Coast Mountains west of Atlin Lake, British Columbia: Geological Survey of Canada Paper 77-1A, p. 267–269.
—— , 1978, Metamorphic terrane, northern Coast Mountains west of Atlin Lake, British Columbia: Geological Survey of Canada Paper 78-1A, p. 69–70.
Wilson, J. T., 1968, Static or moble earth, the current scientific revolution: Proceedings of the American Philosophical Society, v. 112, p. 309–320.
Woodsworth, G. J., Crawford, M. L., and Hollister, L. S., 1983, Metamorphism and structure of the Coast Plutonic Complex and adjacent belts, Prince Rupert and Terrace areas, British Columbia: Geological Association of Canada Field Trip Guide 14, 62 p.
Zen, E-an, and Hammarstrom, J. M., 1984a, Mineralogy and petrogenetic model for the tonalite at Bushy Point, Revillagigedo Island, Ketchikan 1° by 2° Quadrangle, southeastern Alaska, *in* Bartsch-Winkler, S., and Reed, K. M., eds., The United States Geological Survey in Alaska; Accomplishments during 1982: U.S. Geological Survey Circular 939, p. 118–123.
—— , 1984b, Magmatic epidote and its petrologic significance: Geology, v. 12, p. 515–518.

Manuscript Accepted by the Society May 9, 1990

ACKNOWLEDGMENTS

We thank George Plafker, Fred Barker, and Christopher Dodds for helpful reviews of the manuscript and Jason Saleeby, William McClelland, and Jonathan Patchett for extensive discussions concerning the geology and tectonics of southeastern Alaska. Many workers contributed information and insights to our compilation geologic map of southeastern Alaska, which serves as the basis for this overview. Research by G. E. Gehrels and co-workers in and adjacent to the Coast Mountains has been supported by the National Science Foundation (EAR-8616473 and EAR-8706749). Acknowledgment is made to the Donors of The Petroleum Research Fund, administered by the American Chemical Society, for partial support of our recent studies of the Alexander terrane.

Printed in U.S.A.

The Geology of North America
Vol. G-1, The Geology of Alaska
The Geological Society of America, 1994

Chapter 14

Interior basins of Alaska

Charles E. Kirschner
P.O. Box 154, Union, Washington 98592

INTRODUCTION

Interior Alaska has historically been considered the onshore geographic area of Alaska between the Brooks Range and the Seward Peninsula on the north and west respectively, and the Alaska Range on the south (Fig. 1). It is an area of more than 600,000 km^2 that covers the central one-third of Alaska and has been variously referred to as the "intermontane plateaus" (Wahrhaftig, this volume) and the "central plateaus" (Raisz, 1948). The Yukon River, the largest river in Alaska, approximately bisects the province. The topography of the region is generally subdued. Broad alluviated lowland areas are underlain by Cenozoic nonmarine sedimentary rocks. Extensive areas in the western and southwestern parts of the province are characterized by ridge and valley topography cut in complexly deformed Jurassic and Cretaceous flysch rocks. The Brooks Range and the Alaska Range were extensively glaciated during the Pleistocene; however, only a few very small glaciers were present in the higher mountains of the interior province. Muskeg and tundra at the lower elevations, willow and alder brush in the stream valleys, and spruce and birch forests up to the tree line, at about 750 to 900 m elevation, form a thick cover of vegetation so that bedrock exposures are generally limited to ridge tops above the tree line and river-cut bank exposures.

REGIONAL FRAMEWORK

In onshore interior Alaska there are three types of basins: Cenozoic nonmarine basins; Mesozoic, mildly to complexly deformed flysch basins; and the Kandik basin, a hinterland segment of the Cordilleran fold and thrust belt (Fig. 2; Table 1). The structural and stratigraphic development of the interior basins will be discussed in the context of their petroleum potential. Only seven exploratory wells have been drilled in all of these basins: two in Paleozoic rocks of the Kandik basin, two shallow wells in Cenozoic nonmarine rocks of the Nenana basin, one each in Mesozoic flysch rocks of the Kandik and the Yukon-Koyukuk belts, and one in the Bethel basin that penetrated Mesozoic flysch beneath a thin Cenozoic section. Many of these basins also contain coal deposits that are described in more detail by Wahrhaftig and others (this volume).

Some common characteristics of the Cenozoic basins are: (1) The sedimentary fill is less dense than the pre-Tertiary rocks so the basins form distinct gravity lows. (2) The fill is mainly nonmarine fluvial and coal-bearing sedimentary rocks deposited in cyclic fining-upward sequences. (3) A pattern of three cycles of Tertiary sedimentation appears to be characteristic, an early cycle of Paleocene to early Eocene age, a middle cycle of late Eocene to late Miocene age, and a late cycle of late Miocene and Pliocene age. (4) The depocenter for each younger cycle is commonly displaced from the preceding cycle as a result of deformation and uplift that produces a local to regional unconformity between the cycles; therefore, it cannot be predicted that early-cycle rocks will necessarily be present at depth beneath mid-cycle or late-cycle rocks. (5) Structure is commonly extensional, but folding related to thrust faulting, high-angle reverse faulting, or transpression by dextral faulting is also recognized. This style of structural development of "pull-apart" basins or "rhomb grabens and horsts," along major strike-slip fault systems, has been aptly described by Aydin and Nur (1982), Chinnery (1965), Wilcox and others (1973), and many others.

Mesozoic basins, characterized as flysch belts and flysch terranes, cover extensive areas of western and southwestern interior Alaska, are also present in the Alaska Range south of the Denali fault zone, and arc around the Gulf of Alaska from Bristol Bay on the southwest to southeastern Alaska (Fig. 3). Small, fault-bounded flysch terranes are present in central and east-central interior Alaska and may represent dismembered segments of formerly larger co-extensive basins. The common characteristic of the flysch belts of interior and southern Alaska is that they represent volcano-plutonic arc-related basin deposits. They will be discussed briefly in the context of their structural and stratigraphic development and petroleum potential.

The area generally referred to as the Kandik basin in east-central Alaska is a small hinterland segment of the Cordilleran fold and thrust belt. The Paleozoic to early Mesozoic stratigraphic sequence in the Kandik basin includes organic shales and platform carbonate and clastic rocks, but its petroleum potential appears to be very low due to strong structural deformation.

A conspicuous element of interior Alaska is the suite of relatively small, northeast-trending dextral strike-slip faults on the west limb of the Alaska "orocline," and the major southeast-

Kirschner, C. E., 1994, Interior basins of Alaska, *in* Plafker, G., and Berg, H. C., eds., The Geology of Alaska: Boulder, Colorado, Geological Society of America, The Geology of North America, v. G-1.

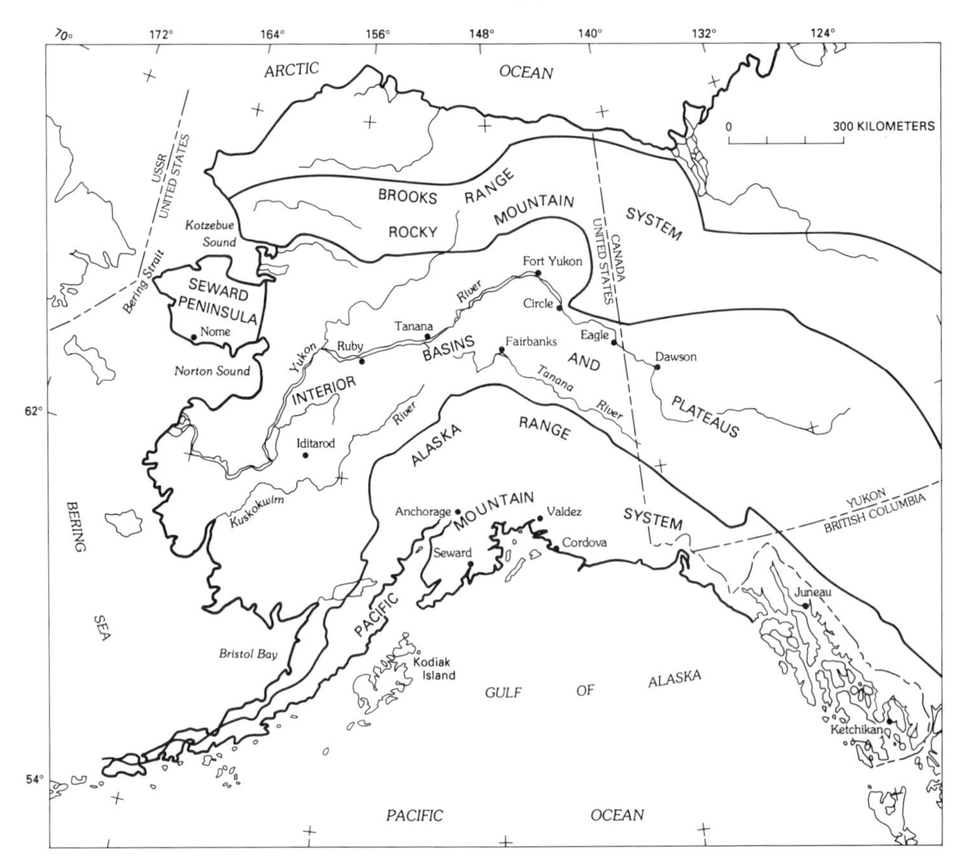

Figure 1. Map showing major North American physiographic divisions in Alaska.

trending dextral rift zones of the Tintina and Denali faults on the east limb of the Alaska "orocline" (Fig. 3; Grantz, 1966).

The northeast-trending faults in the suite are each about 400 to 500 km in length and have dextral offsets in the range of 100 km (e.g., Patton and Hoare, 1968). These faults are confined to the west limb of the Alaska "orocline" and thus are probably related to its development. The major Tintina and Denali fault zones, by comparison, are 1,500 to more than 2,500 km in length, with dextral offsets in the range of hundreds of kilometers (e.g., Gabrielse, 1985). Both the Denali and Tintina fault zones appear to end at the apex of the Alaska "orocline" in highly complex imbricate thrust systems or collision zones that must have accommodated much of the strike-slip motion on these major fault systems. The Denali fault terminates in the Alaska Range suture zone (Jones and others, 1963). The Tintina fault terminates in a similar suture zone referred to as the Beaver Creek suture (Churkin and others, 1982; Figs. 3 and 10).

Another conspicuous element of north-central interior Alaska and the southern Brooks Range, which probably indicates structural deformation on a major scale, consists of ophiolite assemblage terranes (Fig. 3; Patton and others, 1977). Around the northeastern margin of the Yukon-Koyukuk-Kobuk (YKK) flysch basin, the ophiolite assemblages are slab-like bodies that dip beneath the flysch basin. They are interpreted to represent the oceanic floor of the basin and the root zone of obducted allochthonous sheets that once extended as much as 300 km over continental crust.

Lithotectonic (tectonostratigraphic) terranes referred to in this chapter are described by Silberling and others (this volume).

TERTIARY BASINS

Bethel basin

Introduction. The Bethel basin is a large lowland area bordered on the south and west by the Bering Sea, and on the north by metamorphic and sedimentary rock uplands north of the Yukon River and east of the Kuskowim River. The area thus defined includes about 50,000 km^2. It is a lake-dotted marshy plain rising 30 to 90 m above sea level. The marshy ground is underlain by patchy permafrost and is about 30 percent lake surface. Numerous basalt flows and cinder cones are present in the west-central area of the plain (Fig. 4). The basin has seen a very low level of petroleum industry interest due to evidence of a thin Tertiary section and poor source and reservoir potential in the Cretaceous and older(?) sedimentary rocks. One deep exploratory well has been drilled in the basin: the Pan American, Napatuk Creek No. 1, with a total depth of 4,541 m (14,890 ft).

TABLE 1— *Cenozoic basins of interior and southern Alaska*

[Note: Well symbols follow usage in AGI data sheets (Dietrich and others, 1982). COST, Continental offshore stratigraphic test]

BASIN	MAXIMUM SQ KM	PERCENT >1 KM FILL	PERCENT >3 KM FILL	GRAVITY LOW MG	STRUCTURAL STYLE		EXPLORATORY WELL DATA, COMMENTS
BENDELEBEN	600	*40G	0	20	Graben or half-graben	--	Probable Cenozoic fill.
BETHEL	50,000	0	0	10	Lowlands		One exploratory well. Thin Cenozoic fill over Cretaceous flysch. No shows.
BRISTOL BAY	70,000	90	35	40	Half-graben Tertiary folding		Oil and gas shows in exploratory wells. One COST well offshore.
COOK INLET	28,000	65	50	100+	Half-graben Tertiary folding		Over 200 exploratory wells. Several producing oil and gas fields. Oil seeps.
COPPER RIVER	4,500	<1	0	30	Lowlands		Several exploratory wells. Only minor oil and gas shows in Cretaceous flysch. Thin tertiary fill.
GALENA	9,000	<1G	0	40	Lowlands half-graben	--	Lowland area with thin fill except for half-graben(?) low along Kaltag fault zone.
HOLITNA	5,000	1G	0	40	Gaben or half-graben trough	--	Lowland area with thin fill except for graben trough(?) along Farewell fault zone.
IMURUK	600	2G	0	20	Half-graben?	--	Probable Cenozoic fill.
INNOKO	6,000	1G	0	20	Half-graben or graben trough?	--	Probable Cenozoic fill localized by undefined dextral fault zone?
KOBUK	2,000	3G	0	20	Half-graben or graben trough?	--	Probable Upper Cretaceous and Cenozoic fill.
KOTZEBUE	40,000	30	<10	30	Half-graben		Two exploratory wells, no shows. Nonmarine Tertiary section immature.
MINCHUMINA	20,000	15G	0	30	Graben and half-graben complex	--	Large shallow basin with local small graben(?) and half-graben(?) basins.
MIDDLE TANANA	22,000	20G	<1	50	Half-graben		Large shallow basin. One small deep low. Two wells, no shows. Reported oil seeps unconfirmed.
NOATAK	1,600	15G	<1	50	Graben?	--	Probable Cenozoic fill.
NORTHWAY	3,000	0G	0	0	Lowlands		Thin Pleistocene fill. Shallow wells encountered marsh gas trapped by permafrost.
NORTON SOUND	50,000	60	10	30	Extensional graben complex		Two COST wells and six industry wells, oil and gas shows. Commercial production not established.
RUBY-RAMPART	6,000	25G	0	20	Half-graben trough	--	Trough along trend Kaltag fault zone. Upper Cretaceous and Cenozoic fill.
SELAWIK	2,300	>50G	0	30	Graben trough	--	Long narrow deep trough with Cenozoic fill.
SUSITNA	10,000	25	10	40	Graben and half-graben complex		Two exploratory dry holes. Up to four km Cenozoic fill.
TINTINA	2,500	<50G	<10	40	Graben trough	--	Trough along trend of Tintina fault zone. Thick Upper Cretaceous and Cenozoic fill.
YUKON FLATS	22,000	>25G	<1	40	Half-graben or graben complex?	--	Probable Cenozoic fill.

*G indicates thickness of fill based on gravity data

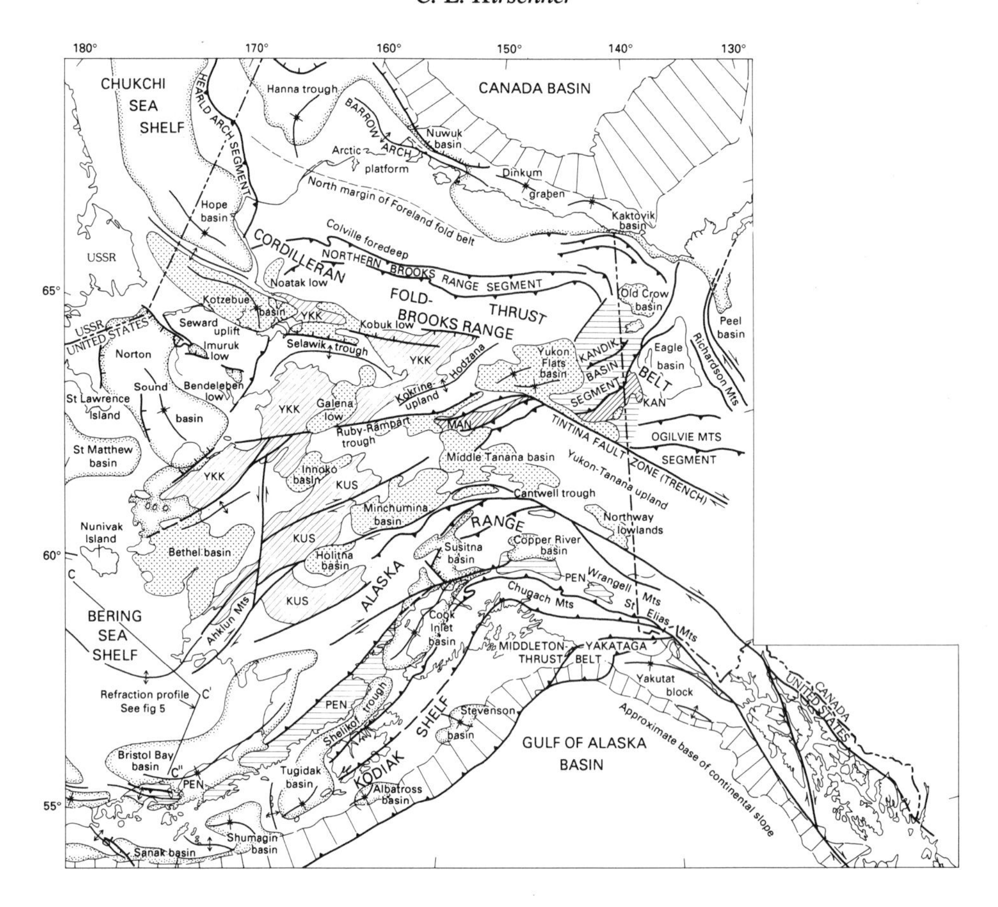

EXPLANATION

Cenozoic basins discussed in this report that contain mainly Tertiary nonmarine strata

Mesozoic basins

Belts of flysch and molasse strata, mainly of volcanogenic origin

YKK **Yukon-Koyuk-Kobuk basin**

KUS **Kuskokwim basin**

Belts of argillite and arenite, mainly of continental origin

KAN **Kandik River terrane**

MAN **Manley terrane**

PEN **Peninsular terrane**

Kandik basin—A segment of the Cordilleran fold and thrust belt

Contact—Approximately located

Fault—Sense of throw unknown. Dashed where speculative

Strike-slip fault—Arrows show relative movement. Dashed where speculative.

Thrust fault—Sawteeth on upper plate. Dashed where speculative

Normal fault—Hachures on downthrown side. Dashed where speculative

Trend of major uplift or anticlinoria

Trend of major downwarp or synclinoria

Figure 2. Map showing major basins, terranes, and structural features of Alaska.

Stratigraphy. Basement rock terranes north of the basin consist of Precambrian schist, gneiss, and migmatite of the Ruby terrane, and an Early Cretaceous volcaniclastic andesitic arc assemblage, the Koyukuk terrane (Jones and others, 1987; Silberling and others, this volume). Southeast of Bethel in the Ahklun Mountains, a similar Jurassic volcanic and volcaniclastic andesitic arc assemblage is widely exposed. North of Cape Newenham, smaller disparate terranes include an early Mesozoic to Paleozoic oceanic assemblage in a tectonic melange, ultramafic rocks of the Goodnews terrane, and Precambrian gneiss and phyllite of the Kilbuck terrane.

Cretaceous flysch and molasse of the Yukon-Koyukuk-Kobuk (YKK) basin are extensively exposed north of the Bethel lowlands (Hoare and Condon, 1966). Similar rocks of the Kuskokwim (KUS) basin are exposed east and northeast of the lowlands (Hoare and Coonrad, 1959; Decker, 1984). The flysch basins characteristically contain very thick marine and nonmarine graywacke and siltstone sequences derived largely from andesitic arc provenance terranes, and deposited in both fore-arc and back-arc settings. The rocks have been strongly deformed, altered, and locally metamorphosed. Hoare and Condon (1966) and Hoare and others (1964) have described extensive calcareous and laumontite diagenesis in the YKK sequence. Regional structural trends, supported in part by gravity, magnetic, and seismic data, suggest that similar flysch sequences underlie a thin Tertiary cover in parts of the Bethel basin and may extend southwesterly beneath the offshore waters of Kuskokwim Bay and the Bering Sea shelf. The Pan American Napatuk Creek No. 1 well penetrated about 3,900 m of sedimentary rocks at least in part coeval and lithologically similar to the flysch sequences (Fig. 5).

Tertiary sedimentary rocks in the Bethel basin are known only from the Napatuk Creek well, which penetrated 440 m of Miocene and Pliocene diatomaceous clay deposited in a near-shore marine environment. A shallow gravity low surrounds the Napatuk Creek well location (Fig. 4; Barnes, 1977). Similar lows are present northeast and southwest of the well, and indicate that Tertiary fill in the basin probably does not greatly exceed 610 m and may be less over most of the basin. Refraction data in Kuskokwim Bay indicate less than 1 m of low velocity (1.9 km/s) rocks that probably represent Cenozoic cover over Cretaceous flysch (Fig. 6).

Structure. Regional structural trends are southwesterly in the uplands around the basin, and gravity and magnetic data indicate that this trend is present beneath most of the Bethel basin. However, structural trends are westerly to northwesterly at Cape Romanzof, Cape Vancouver, and Cape Newenham, suggesting an oroclinal(?) flexure with an axial trace trending NNW. Structure in the Cretaceous flysch of the YKK and KUS basins is extremely complex, with sharp, commonly faulted anticlines and isoclinal folds. Northeast- and northwest-trending fault sets cut both the Cretaceous flysch and older basement rocks. Extensional rift basins, which are the locus of late Tertiary and Quaternary basalt flows, are associated with some of the larger fault systems, as for example, the Hagemeister-Togiak-Tikchik system (Hoare and Coonrad, 1978). Seismic maps on the Shell and Pan American development contracts define simple southwest-trending anticline-syncline pairs in shallow Tertiary rocks, but report incoherent deeper data. It is presumed that the incoherent data reflect complex structure in Cretaceous flysch, similar to that in outcrop.

Petroleum potential. Thermal maturity and visual kerogen analysis of the Napatuk Creek well indicate that the Tertiary rocks are immature, the Cretaceous (Campanian to Turonian) rocks are mature, and the Cretaceous(?) and older(?) rocks below 1,555 m (5,100 ft) are overmature. All samples are dominated by cellulosic gas-type kerogens. Analysis of outcrop samples from localities around the basin are similar to the Napatuk Creek well data and additionally indicate low organic carbon content and very low porosities and permeabilities (Lyle and others, 1982). These data suggest the potential for dry gas generation in the Late Cretaceous flysch, but complex structure would appear to preclude significant volumes of accumulation.

Kotzebue basin and Selawik trough

Introduction. The Kotzebue basin in northwestern Alaska covers an area of more than 20,000 km^2 beneath Kotzebue Sound, the northern lowland margin of the Seward Peninsula, and the Kobuk River delta east of Kotzebue (Fig. 7). A shallow arm of the basin extends westerly south of the Kotzebue arch beneath the southern Chukchi Sea, beyond the map area shown in Figure 7. If this area were included, the total area of the basin would approximately double (Fig. 2). The basin is a Cenozoic graben or half-graben downwarped across the metamorphic terranes of the southern Brooks Range and the northern Seward Peninsula. Associated gravity lows are the Selawik trough and the Kugarak-Kobuk low east of the basin, and the Noatak low north of the basin.

Stratigraphy. Basement rocks of the basin are probably Paleozoic to Precambrian schist and carbonate rocks like those exposed in the Brooks Range and on the Seward Peninsula. Similar rocks were drilled beneath the Tertiary fill of the basin in both of the Chevron wells. East of the basin, Upper Cretaceous conglomerate with carbonaceous shale, coal, and tuffaceous strata, and Lower Cretaceous flysch are exposed along the Kobuk River and in the Waring Mountains uplift.

The composite Cenozoic section in the basin appears to include three cycles of deposition; age control is not precise, however, and the stratigraphic correlations shown in Figure 9, which are based on a combination of lithology, palynology, and vitrinite reflectance data, are tentative. The early(?) Tertiary cycle is represented by 915 m of Eocene volcanic and volcaniclastic rocks in the Cape Espenberg well. The middle Tertiary cycle is represented by about 1,200 m of Pliocene to Oligocene conglomeratic sandstone, carbonaceous shale, and coal. The late Tertiary cycle is represented by about 600 m of Pliocene(?) and Pleistocene marine sandstone and shale in the Nimiuk Point well. The late Tertiary cycle is clearly represented on segment A-B of

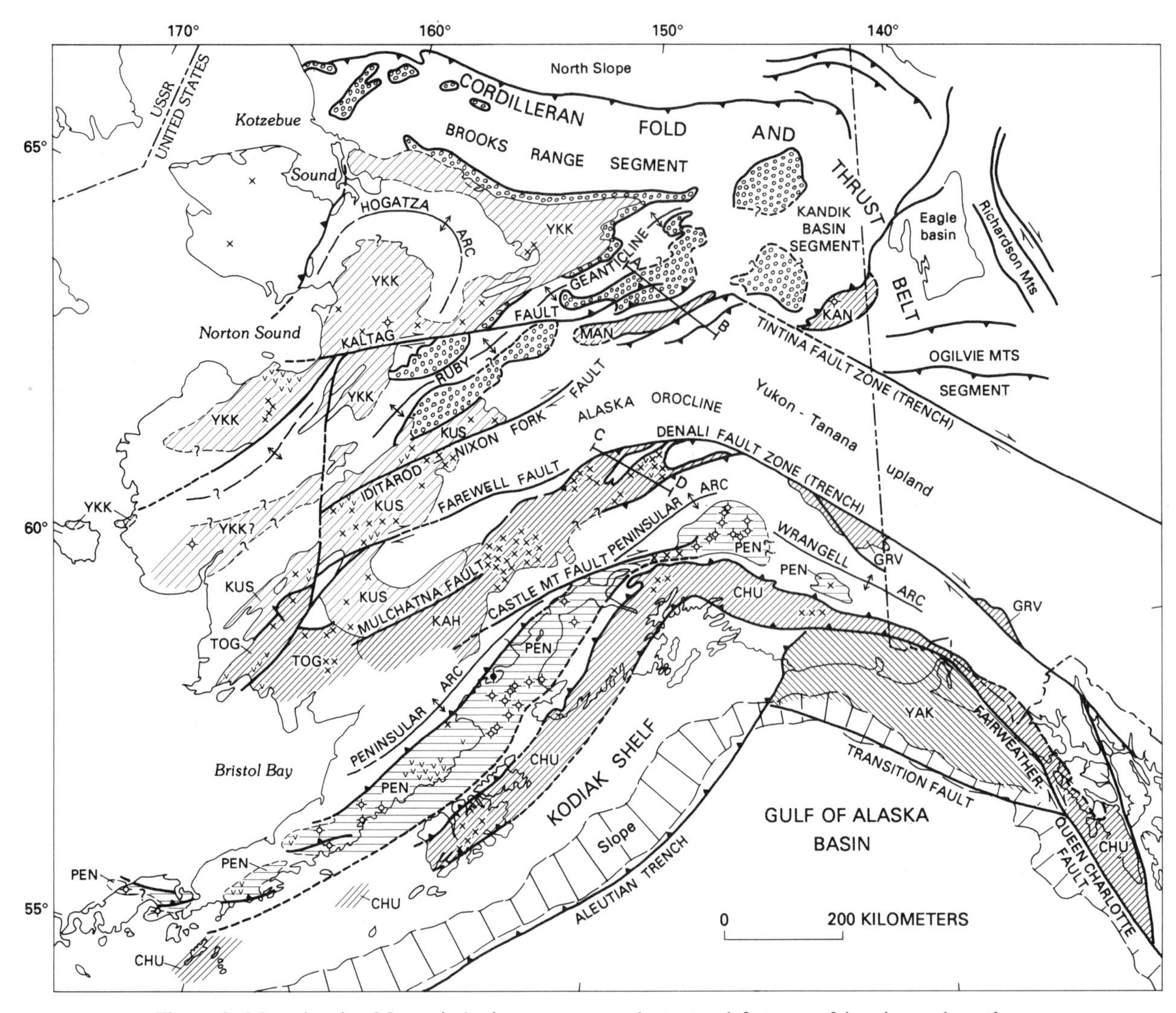

Figure 3. Map showing Mesozoic basins, terranes, and structural features of interior and southern Alaska.

the seismic line (Fig. 8) by faint, essentially horizontal reflectors to a maximum depth of about 1 sec. The middle Tertiary cycle is represented by strong distinctive reflectors from 1 to 2+ sec, which show gentle folding, minor faulting, and irregularity in continuity probably related to the nonmarine deltaic nature of the strata.

The western end of the Selawik trough is shown in segment B-C (Fig. 8), which defines a rift-graben complex with a little more than 2 sec or about 2,700 m of Cenozoic fill. Barnes and Tailleur (1970) have modeled the 50-milligal Noatak gravity low and interpret about 3 km of less dense fill. Ellersieck and others (1979) report an Upper Cretaceous or Tertiary outcrop on the Noatak River at the location of the gravity low. Patton (1973) reports small deposits of coal-bearing strata containing early Tertiary pollen at Elephant Point and on the Mangoak River near the south margin of the Selawik trough.

Structure. The broad structural pattern of the Kotzebue basin is a complex Tertiary extensional half-graben. The Selawik trough is a graben complex whose linearity, as defined by gravity, suggests a rift-graben. The en echelon Kobuk low is probably a similar feature. The Kobuk fault zone (Patton, 1973) trends easterly about 320 km along the south margin of the Brooks Range, from the Kobuk gravity low. Throughout its length the fault zone localizes Late Cretaceous nonmarine trough deposits. Comparison of Precambrian–Lower Paleozoic structural trends in the southern Brooks Range and the Seward Peninsula suggests a component of dextral offset. The Kotzebue basin appears to represent extension at the trailing edge of the north block, analogous to the Norton Sound basin at the trailing edge of the north block on the dextral Kaltag fault (Fig. 2; Fischer and others, 1982).

Petroleum potential. No pre-Tertiary source rocks can be predicted for the basin, and so the only known generative rocks

EXPLANATION

Mildly to moderately deformed rocks of the Peninsular terrane, locally petroliferous, common laumontite grade diagenesis

Moderately to complexly and isoclinally folded and faulted rocks. Pervasive laumontite and local higher grade metamorphism

- YKK **Yukon-Koyukuk-Kobuk basin**
- KUS **Kuskowim basin**

Pervasively deformed and structurally disrupted rocks; laumontite to amphibolite grade metamorphism

- CHU **Chugach terrane**
- GRV **Gravina-Nutzotin belt**
- KAH **Kahiltna terrane**
- TOG **Togiak terrane**
- KAN **Kandik River terrane**
- MAN **Manley terrane**
- YAK **Yakutat terrane**

Andesitic or basaltic rocks of all ages

Intrusive rocks of all ages

Ophiolite assemblage rocks

Contact—Approximately located. Dashed where speculative

Fault—Sense of throw not defined. Dashed where speculative

Strike-slip fault—Arrows show relative movement. Dashed where speculative

Thrust fault—Sawteeth on upper plate. Dashed where speculative

Normal fault—Hachures on downthrown side. Dashed where speculative

Trend of (volcanic-plutonic) arc and/or geanticline

Abandoned exploratory well

Abandoned exploratory well that produced a small volume of high, 50°API, oil

A B **Beaver creek suture zone**

C D **Central Alaska Range megasuture zone**

AGE AND CORRELATION OF MAP UNITS

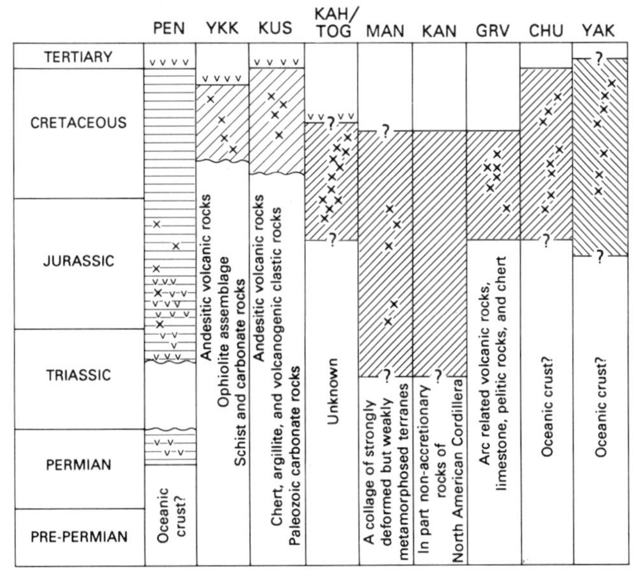

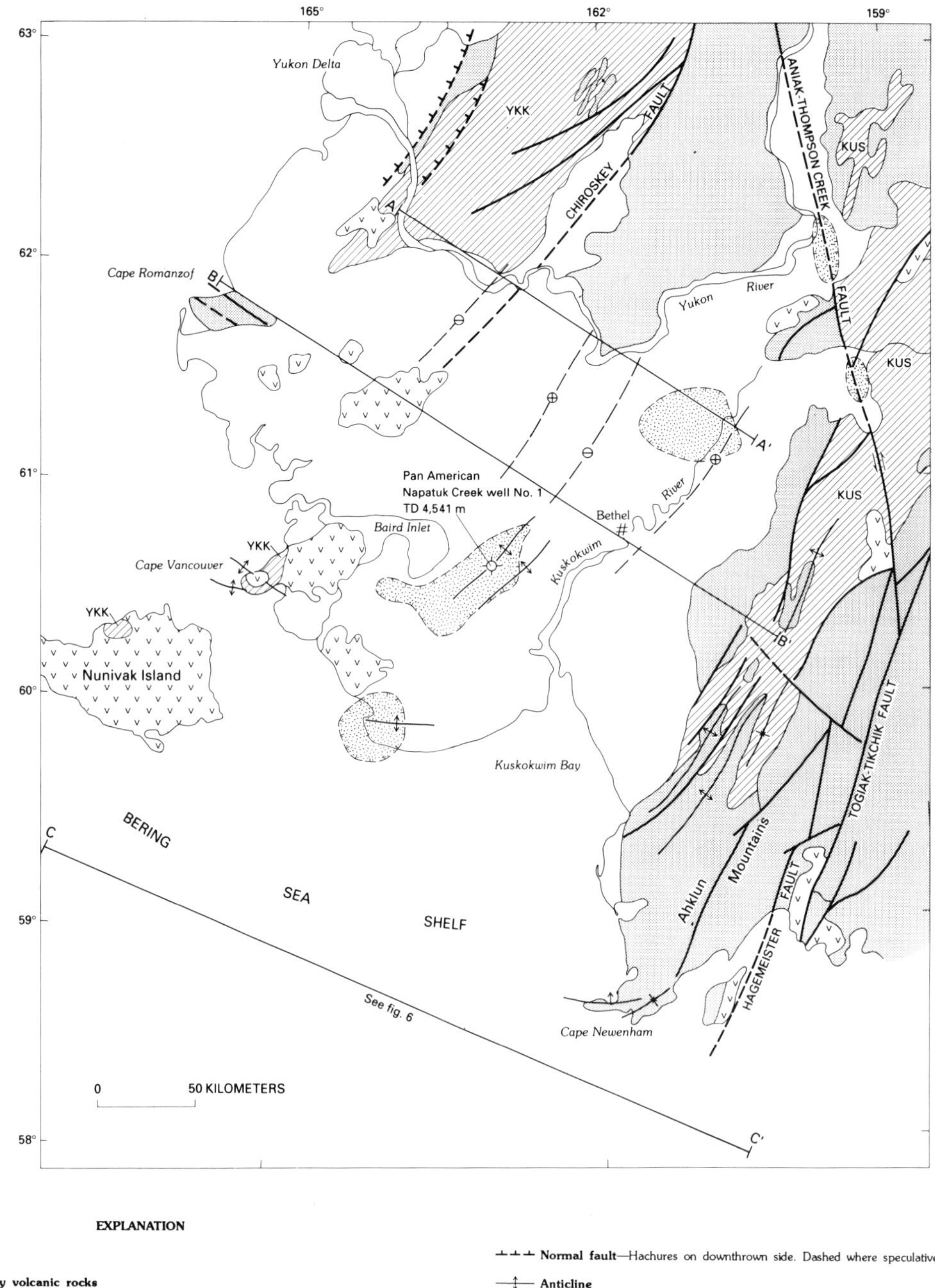

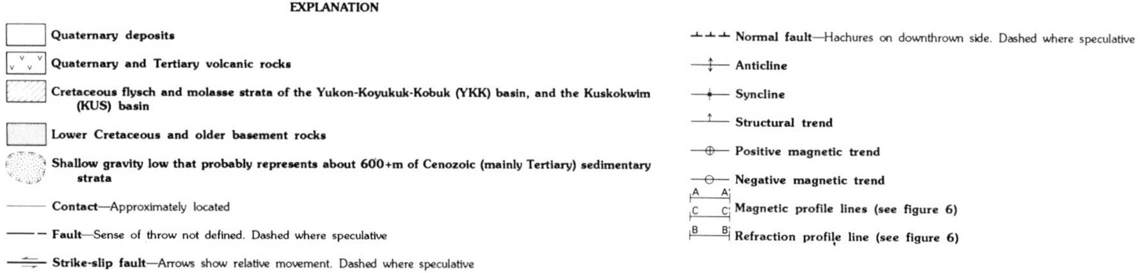

Figure 4. Generalized geologic map of Bethel basin showing gravity lows and locations of refraction and magnetic profiles.

would have to be the Tertiary basin fill and coal-bearing strata of Miocene to Oligocene age. Vitrinite reflectance data (Fig. 9) indicate that the Tertiary rocks are immature, and so the most likely potential in the Kotzebue basin is for dry gas.

Middle Tanana basin, Ruby-Rampart trough, Cantwell trough, and Northway lowlands

Introduction. The Middle Tanana basin of central Alaska is an alluvial and swampy lowland area of about 22,000 km^2 north of the central Alaska Range and south and west of the city of Fairbanks and the Yukon-Tanana upland (Fig. 10). The basin is drained by the Tanana River, which collects a large outflow of glacial meltwater from rivers flowing north out of the Alaska Range.

The eastern part of the basin between Nenana and Big Delta is believed to have a thin Cenozoic section on the basis of the local outcrop of basement monadnocks and of a shallow magnetic signature (Andreason and others, 1964). North of Nenana, the Minto gravity low in excess of 50 milligals suggests about 3 km of late Cenozoic fill. At the south margin of the basin near Healy, about 700 m of the Usibelli Group (in Nenana coal field) and as much as 1,200 m of the overlying Nenana Gravel are present in outcrop. The Ruby-Rampart trough northwest of the basin and the Cantwell trough south of the basin contain significant thicknesses of early Tertiary rocks.

The Northway lowlands on the Tanana River is a lowland area of about 3,000 km^2 near the Canada-Alaska international boundary, referred to by Miller and others (1959) as the upper Tanana basin (Fig. 10). Methane trapped by permafrost was encountered in a well drilled for water at a depth of about 60 m. A second well drilled by Alaska Propane Co., Inc. about 5 km northwest of the first well also encountered gas at about the same depth.

Stratigraphy. Basement pre-Tertiary rocks north and south of the Middle Tanana basin are thoroughly metamorphosed rocks of the Yukon-Tanana terrane. The Union Nenana No. 1 well bottomed in schist that is probably part of this terrane. Northwest

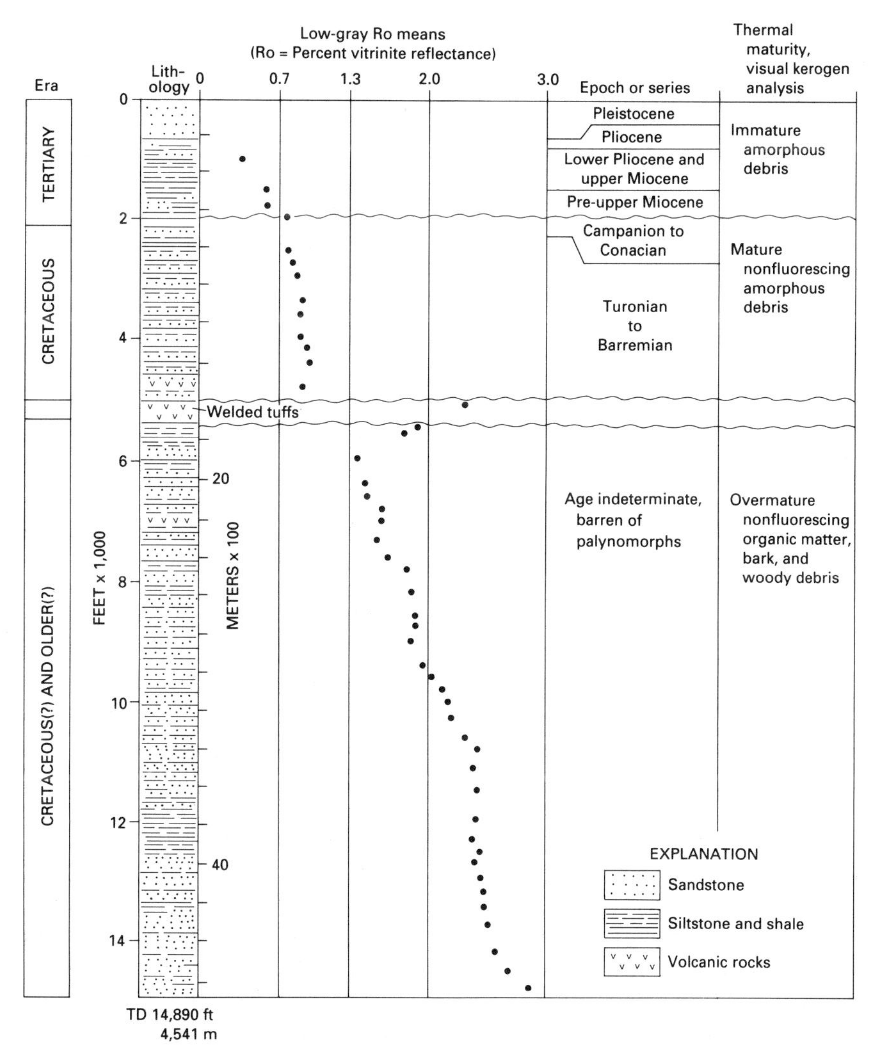

Figure 5. Lithologic log, stratigraphic column, and vitrinite reflectance of Pan American–Napatuk Creek No. 1 well, located southwest of Bethel (Fig. 4, this chapter).

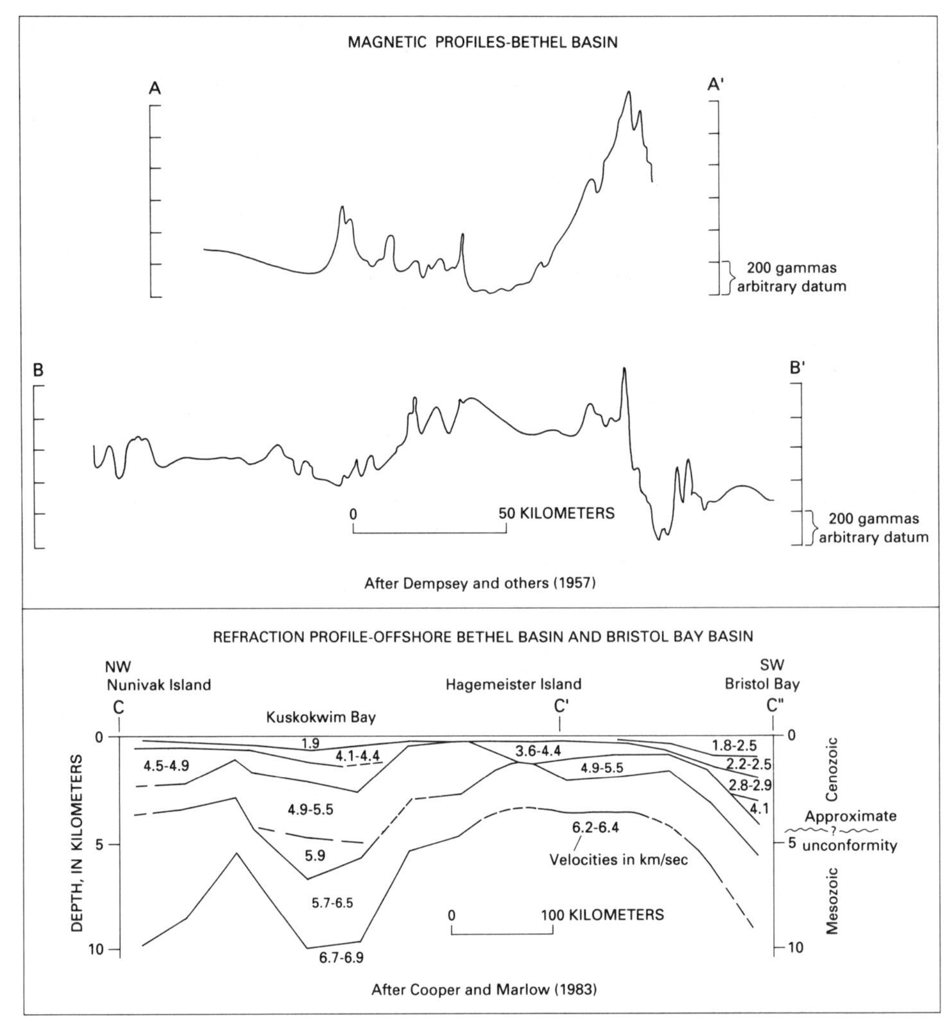

Figure 6. Magnetic (A–A′ and B–B′) and seismic refraction (C–C′–C″) profiles of Bethel and Bristol Bay basins (see Figs. 2 and 4).

of the basin, a collage of disparate Mesozoic and Paleozoic terranes, referred to by Churkin and others (1982) as the Beaver Creek suture zone, crops out in the Livengood and Tanana Quadrangles.

Nonmarine sedimentary fill in the Middle Tanana basin and adjacent areas represents three cycles of Tertiary sedimentation (Fig. 11). The early cycle is represented by outcrops in the Ruby-Rampart trough and the Cantwell trough. The mid-Tertiary cycle is represented by the Usibelli (coal-bearing) Group of the Healy coal basin and the coal-bearing section in the Union Nenana No. 1 well. The late Tertiary cycle is represented by the Nenana Gravel. The depositional cycles are punctuated by orogenic episodes in late Eocene and late Miocene time. Following each orogenic episode the depocenter for the succeeding cycle of deposition shifted, so it cannot be predicted that the deepest part of the Middle Tanana basin will contain early-cycle sedimentary deposits, and it is likely the coal-bearing beds of the Nenana coal field and the Union Nenana No. 1 well were deposited in separate basins (Wahrhaftig and others, this volume).

Early-cycle rocks of the Cantwell Formation average 600 to 1,500 m but are locally as much as 3,000 m in thickness. The formation consists of nonmarine, coal-bearing, clastic and volcanic rocks in the upper part of the formation (Wolfe and Wahrhaftig, 1970). The volcanic rocks have been designated the Teklanika Formation (Gilbert and others, 1976). The formation is complexly folded and is well indurated. Approximately equivalent-age rocks (early cycle) along the Yukon River in the Ruby-Rampart trough consist of about 900 to 1,500 m of conglomeratic sandstone, shale, and coal in thick, fluvial, fining-upward sequences (Page, 1959). Early Tertiary volcanic and volcaniclastic rocks are also recognized in the Ruby-Rampart trough (Chapman and others, 1971, 1982).

Mid-Tertiary cycle rocks of the Usibelli Group near Healy are subdivided into five formations (Fig. 11; Wahrhaftig and

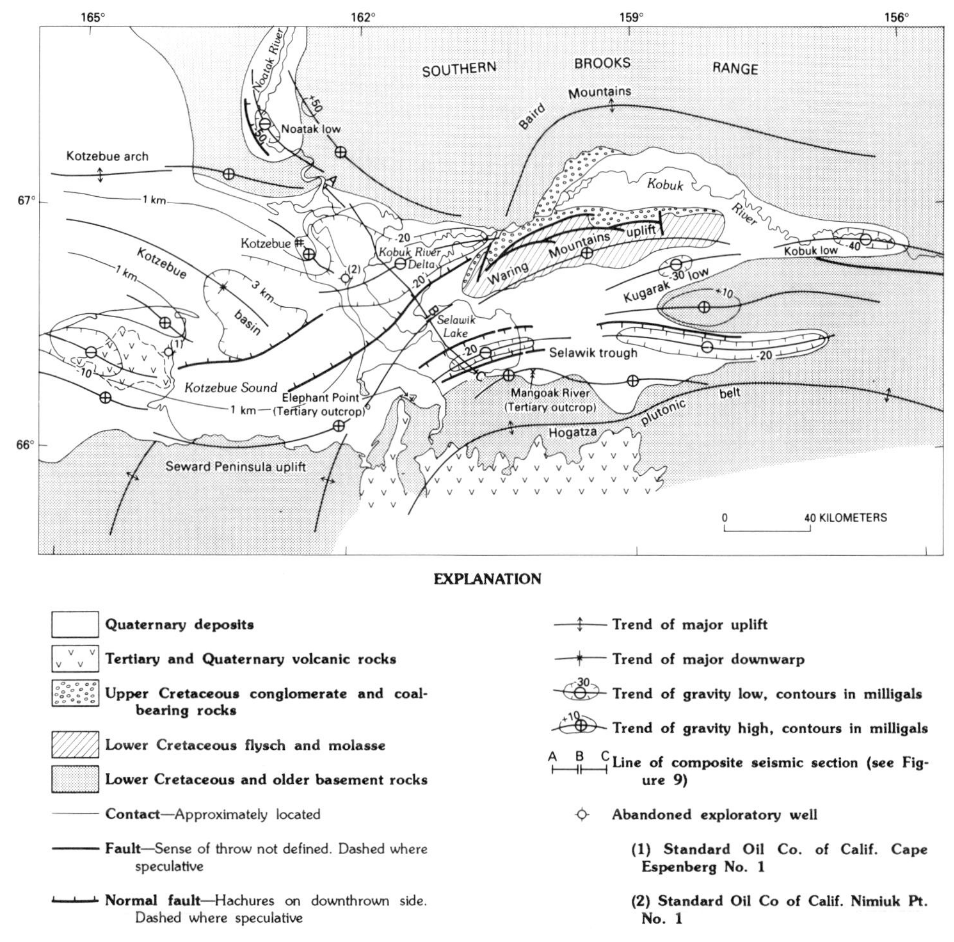

Figure 7. Generalized geologic map of Kotzebue basin, Selawik trough, and adjacent areas showing gravity trends and isopachs of Cenozoic basin fill at 1 and 3 km thickness.

others, 1969, this volume). The group is about 700 m thick and consists primarily of interbedded conglomeratic sandstone, shale, and coal in thick, fluvial, fining-upward sequences. Coal rank ranges from lignite to sub-bituminous B. Two formations, the Sanctuary Formation in the lower part of the group and the Grubstake Formation at the top of the group, are lacustrine shales that total about 60 m in thickness. Paleocurrent data indicate that a northerly provenance terrane, probably the Yukon-Tanana uplands, supplied sediment to the coal-bearing group in the Healy area during the Miocene (Wahrhaftig and others, 1969).

Late Tertiary–cycle rocks are represented by the Nenana Gravel, which is as much as 640 m thick in outcrop along the north front of the Alaska Range and at least 450 m thick in the Union, Nenana No. 1 well. The formation consists primarily of thick conglomerate and conglomeratic sandstone beds with minor lenticular interbeds of shale and lignite. The rocks were derived from the rising Alaska Range and deposited in large alluvial fans along the north flank of the range, and are regionally unconformable on the underlying coal-bearing group.

Structure. The structure of the Minto gravity low north of Nenana may be an extensional half-graben or graben complex (Fig. 12). Structure to the south in the coal-bearing group of the Healy area is a series of northeast- to east-trending folds and minor normal faults of latest Miocene and Pliocene age. Structure in the early Tertiary rocks in the Cantwell trough and the Ruby-Rampart trough is complex and reflects a period of strong folding and volcanism of Eocene age. The early Tertiary basins are strongly deformed and eroded to a fraction of their former depositional extent.

Petroleum potential. Miller and others (1959, p. 84–86) report an oil seep on Totatlanika Creek and provide an analysis of oil from an oily sand and gravel sample from that locality. An oil seep near the mouth of the Nenana River and oil-saturated tundra on the Wood River have also been reported but never confirmed. The most likely source rocks for oil or gas are coal beds and lacustrine shales. Coal rank ranges from lignite to sub-bituminous B in the Healy coal fields, so that methane is the most likely hydrocarbon product that the coals could generate. The

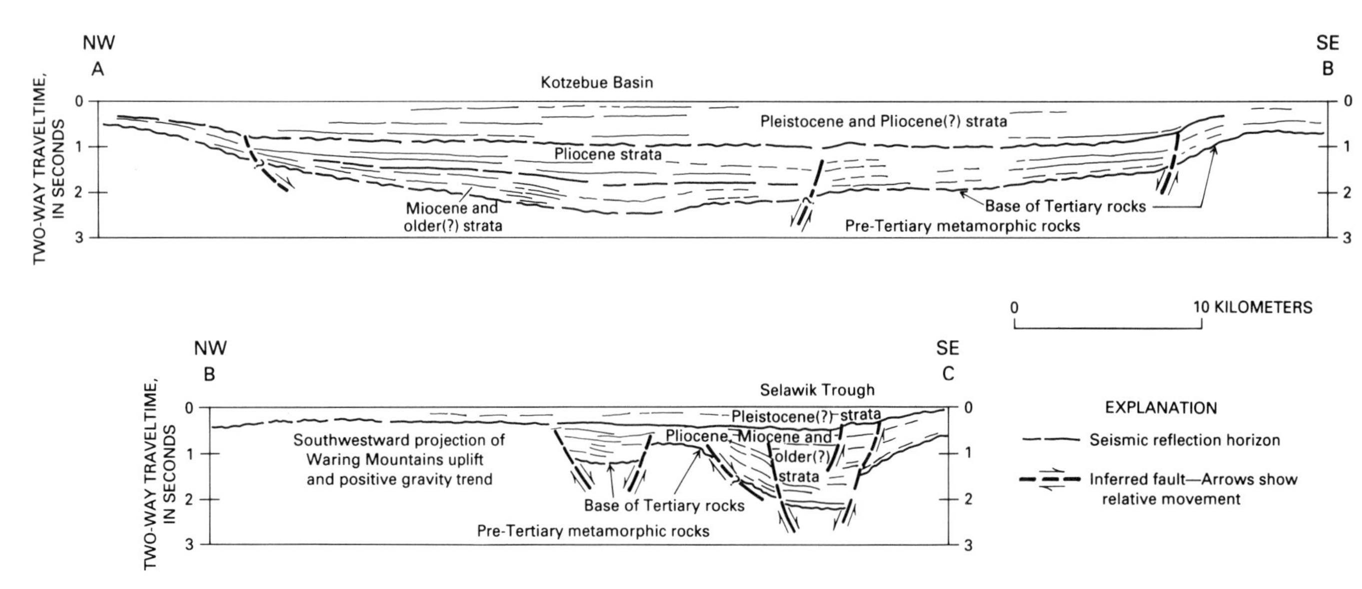

Figure 8. Interpretation of regional composite seismic line A–B–C (line location shown on Fig. 7) across eastern Kotzebue basin and western Selawik trough.

lacustrine beds in outcrop are relatively thin and have rather low (as much as about 2 percent) organic carbon content, but could be thicker and have a higher organic carbon content in the unexplored subsurface. An optimistic evaluation is that gas reserves of economic importance for local consumption could be present.

Minchumina, Holitna, and Innoko basins

Introduction. The Minchumina basin is a large Cenozoic basin of about 21,000 km^2 northwest of the central Alaska Range and Mt. McKinley, and southeast of the Kuskokwim Mountains (Fig. 13). It merges with the middle Tanana basin on the northeast and the Holitna basin on the southwest. The Holitna basin is a small Cenozoic basin of about 5,000 km^2 astride the Farewell fault zone (Fig. 14). The Minchumina basin has topographic and geologic similarities to the Nenana basin, having local basement metamorphic rock monadnocks and local sharp gravity lows that suggest a few kilometers of Cenozoic fill in small extensional basins. The most conspicuous feature of the Holitna basin is a long, narrow gravity trough localized along the trend of the Farewell fault zone, which suggests a rift-graben with a few kilometers of Cenozoic fill. Small outcrops of middle(?) and late Tertiary nonmarine coal-bearing strata are present locally along the Farewell fault zone, on trend with the Holitna basin. The Innoko basin is a lowland area of about 6,000 km^2 in the Kuskokwim Mountains, 150 km west of the Minchumina basin (Fig. 2). It localizes an elongate gravity low having about 20 milligals of relief that could represent as much as 2 km of Cenozoic fill. Surrounding terranes are metamorphic and volcanic rocks (Chapman and others, 1985).

Stratigraphy. The northeastern part of the Minchumina basin is underlain by metamorphic rocks of the Yukon-Tanana

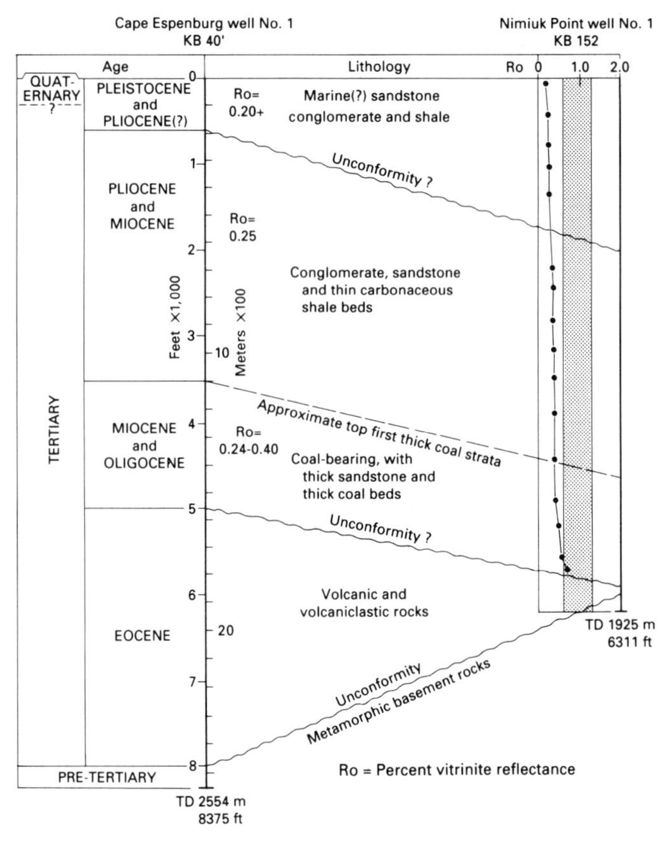

Figure 9. Well correlation section of Standard Oil Company of California Cape Espenburg No. 1 and Nimiuk Point No. 1 wells in the Kotzebue basin (Fig. 7 shows well locations).

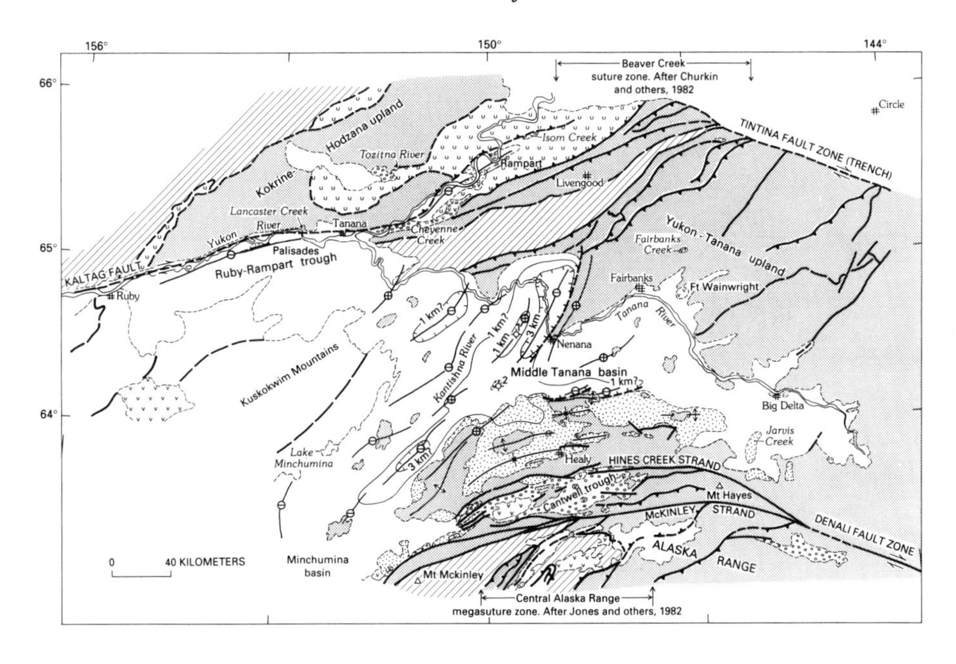

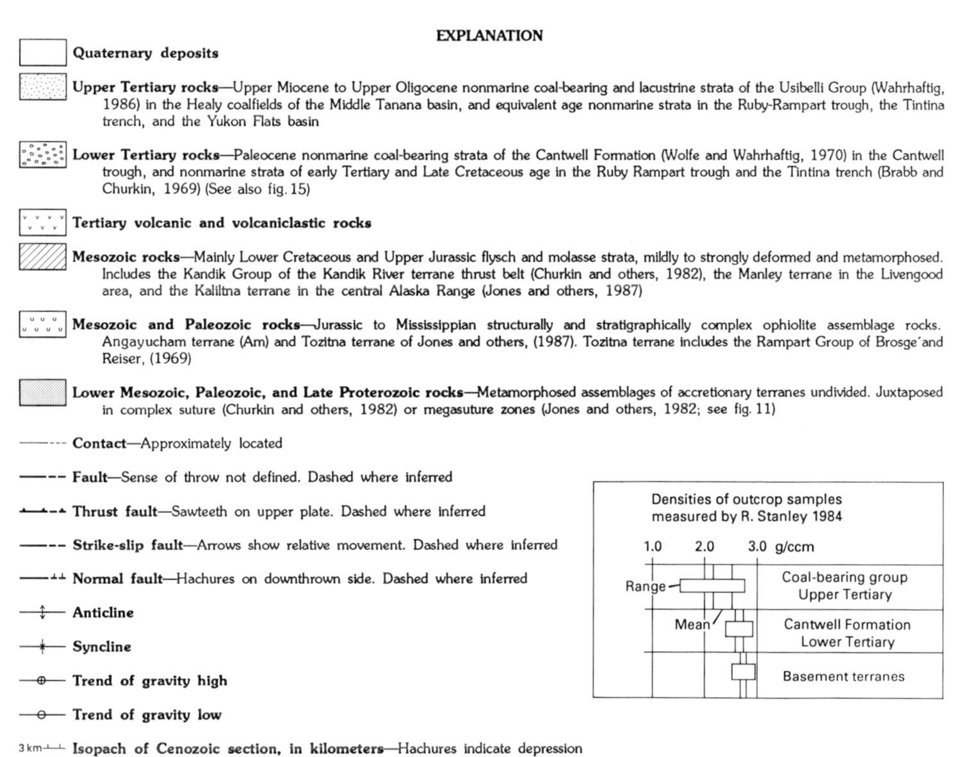

Figure 10. Generalized geologic map showing gravity and structural trends of the Middle Tanana basin, Ruby-Rampart trough, and adjacent upland terranes.

terrane that crop out in monadnocks surrounded by Quaternary alluvial deposits. Southwest of the Minchumina suture zone (Bundtzen and Gilbert, 1983), the Nixon Fork and Dillinger terranes probably underlie much of the Cenozoic fill in the basin, where several gravity lows suggest thick Tertiary fill. The Nixon Fork terrane represents a Paleozoic carbonate platform sequence, and the Dillinger terrane a basinal turbidite shale-out facies (Bundtzen and Gilbert, 1983; Churkin and others, 1984). Henning and others (1984) concluded that these rocks are overmature and have poor reservoir characteristics.

Flysch of the Cretaceous Kuskokwim Group may underlie parts of the Holitna and Minchumina basins. Generally, this flysch is strongly deformed and extensively intruded. Overall, these Cretaceous flysch rocks are unlikely to have oil source or reservoir rocks, or to provide a petroleum source for the overlying Tertiary fill.

Tertiary rocks of the Minchumina and Holitna basins appear to represent three cycles of deposition. Paleocene rocks of the Cantwell Formation are known only in the Cantwell trough northeast of the basin and represent the early cycle. Although not recognized farther southwest along the Farewell fault zone, it would not be surprising to find early-cycle rocks in the rift graben of the Holitna basin trough. A middle Tertiary cycle is represented by nonmarine deposits of conglomerate, sandstone, siltstone, and lignite (Dickey and others, 1982). The middle(?) Tertiary rocks were derived from northerly metamorphic provenance terranes and deposited by southerly flowing braided streams. The late Tertiary cycle consists of conglomerate that represents alluvial fan deposition from the rising Alaska Range on the southeast. The middle(?) and late Tertiary rocks are about 1,800 m thick along the Farewell fault zone near Farewell. Smith and others (1985) interpret as much as 4,500 m of Tertiary rocks in the Holitna basin trough, which has a maximum of about 40 milligals negative relief. Several gravity lows in the Minchumina

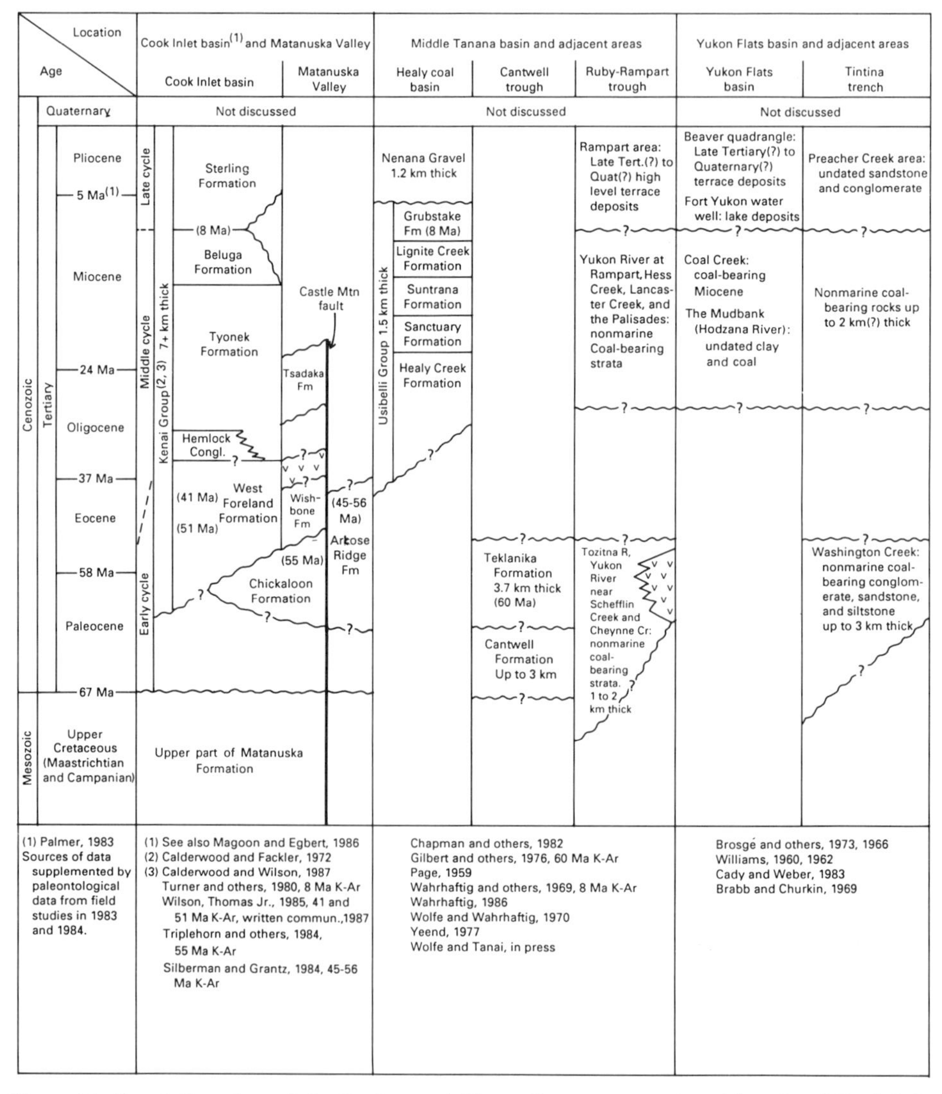

Figure 11. Correlation chart of the Tertiary and Upper Cretaceous stratigraphic units of Cook Inlet basin, Middle Tanana basin, Yukon Flats basin, and adjacent areas.

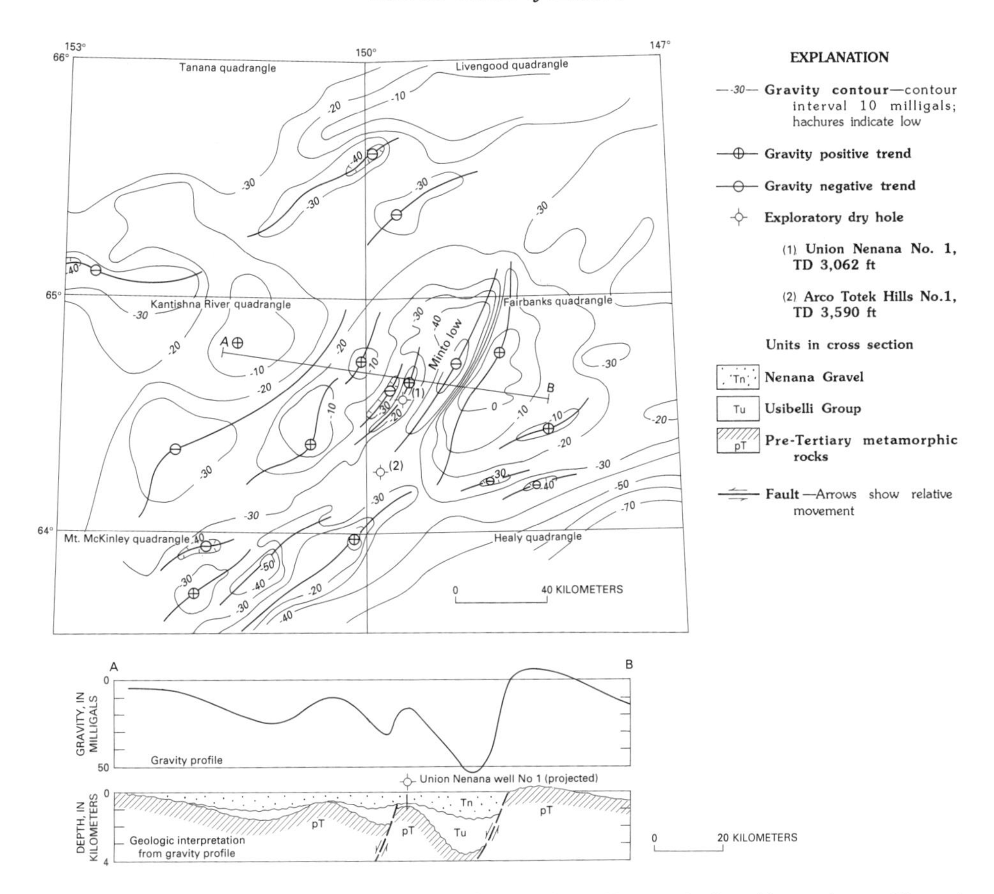

Figure 12. Gravity map of the western part of the Middle Tanana basin with gravity profile and cross-section interpretation of the Minto gravity low.

basin have 20 to 30 milligals negative relief, and so by comparison, may be expected to have 1,800 to 3,000 m of middle to late Tertiary fill.

Structure. Regionally, the Minchumina basin lies between the dextral Iditarod–Nixon Fork fault zone on the northwest and the dextral Farewell fault zone–Denali fault complex on the southeast. Henning and others (1984) have noted that steep gravity gradients associated with basement highs suggest large-displacement, high-angle block faulting and folding in the subsurface, possibly an extensional horst-and-graben complex having north- to northeast-trending structures imposed by dextral strain.

Petroleum potential. The areas of either the Minchumina or the Holitna basin that could have 3 km or more of nonmarine Tertiary fill are less than 1 percent of the total basin area. Although coal-bearing beds in these basins could generate gas, and fluvial sandstones are likely reservoirs, the size of any accumulation would probably be small; it is concluded that the Minchumina basin potential, at best, is limited to small gas prospects.

Yukon Flats basin and Tintina trench

Introduction. The Yukon Flats basin of east-central Alaska is an alluvial and marshy, lake-dotted lowland of more than 22,000 km^2, south of the southern Brooks Range and north of the Yukon-Tanana upland (Fig. 15). The basin is confined on the west by the Kokrine-Hodzana highlands and on the east by the Kandik thrust belt, a hinterland segment of the Cordilleran fold and thrust belt of Northwest Territories, Canada. On the basis of gravity modeling (Fig. 16), it is suspected the Yukon Flats basin may have as much as 3 km of Cenozoic fill. Hite and Nakayama (1980) report that seismic data along portions of the Yukon River indicate that the Cenozoic section locally may be as much as 4.5 km thick. The Tintina fault system and trench trends southeasterly from the southern margin of the basin and from the northern edge of the Beaver Creek suture zone (Churkin and others, 1982). The Tintina trench contains both late Tertiary and early Tertiary to late Cretaceous (Maastrichtian) nonmarine coal-bearing clastic rocks that are about 1.0 km thick in outcrop (Brabb and Churkin, 1969) and may be as thick as 3.0 km in the Circle Hot Springs gravity low (Cady and Weber, 1983).

Stratigraphy. Basement terranes north, west, and south of the Yukon Flats basin are low- to high-grade metamorphic terranes and mostly cannot be considered potential source beds for overlying Tertiary reservoir rocks. One unlikely exception is the presence of tasmanite (TA) in the Tozitna terrane (TZ) near Christian quadrangle (Fig. 15). Tasmanite is an oil shale that may yield a significantly higher volume of oil per unit volume than normal oil shales. The occurrence near Christian has been known for many years and is recorded in Mertie (1927). Apparently, the outcrops are extremely small and limited in extent; follow-up effort by geologists of the U.S. Geological Survey (Tailleur and others, 1967) and the U.S. Bureau of Mines (Donald W. Braggs and Donald P. Blasko, oral communication, 1986) have not been able to delineate additional deposits. The occurrence of the tasmanite in rocks of the Tozitna terrane is of interest because this terrane is present in outcrop around the north, west, and south margins of the basin, and it has a distinctive magnetic signature that can be correlated with the outcrop pattern and traced beneath the Tertiary fill over part of the basin (Fig. 17). Consequently, if larger occurrences of tasmanite were present in the Tozitna terrane in the subsurface, there is a remote potential for pre-Tertiary source beds in part of the basin.

The Porcupine terrane (PC) in the Kandik segment of the Cordilleran fold and thrust belt east of the basin includes Precambrian metamorphic rocks overlain by a thick, structurally and stratigraphically complex assemblage of limestone, dolomite, and shale of Cambrian to late Devonian age. Pennsylvanian and Permian rocks include shale, argillite, limestone, quartzite, and conglomerate. Two exploratory test wells have been drilled in rocks of the Porcupine terrane: the Louisiana Land and Exploration Company Doyon Nos. 2 and No. 3, which encountered Devo-

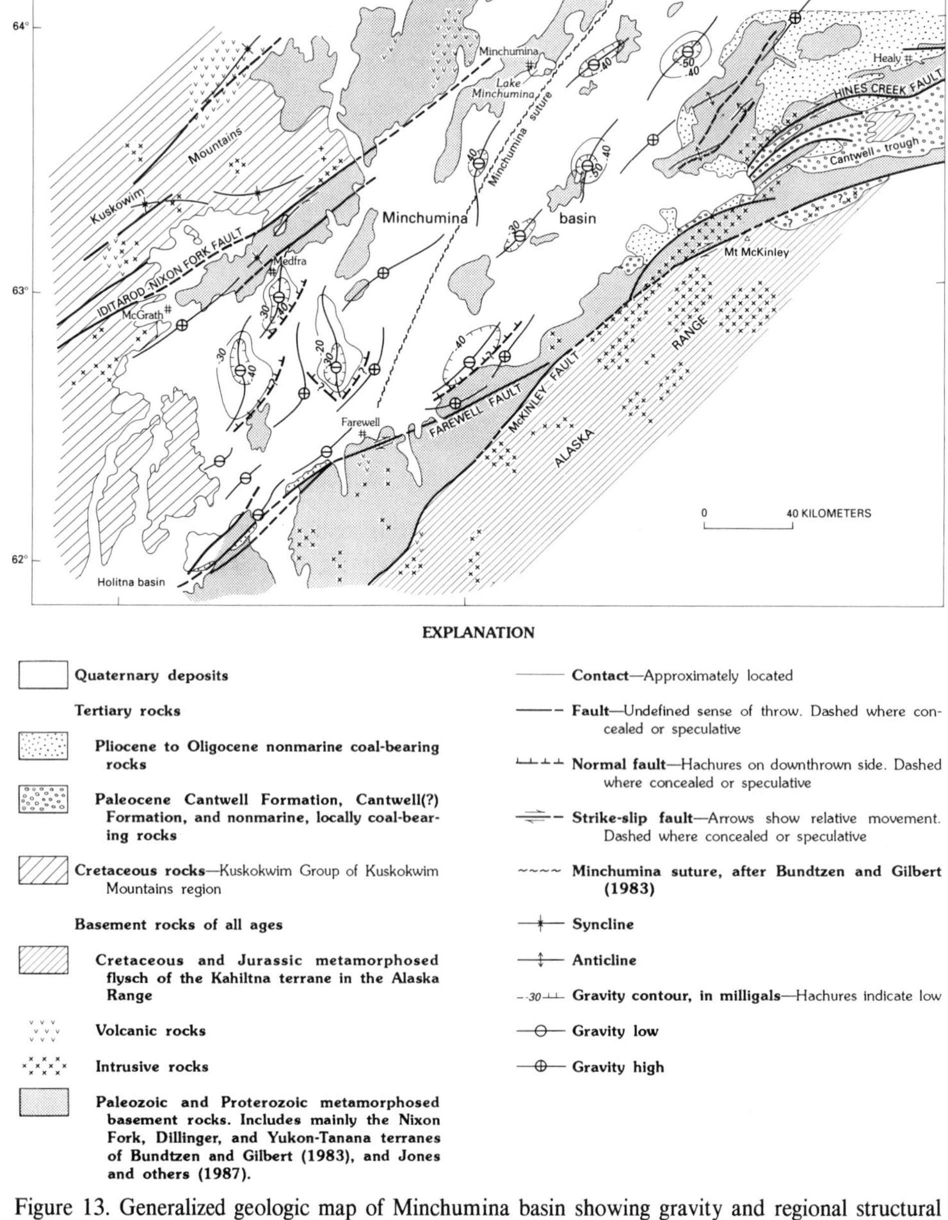

Figure 13. Generalized geologic map of Minchumina basin showing gravity and regional structural trends in adjacent upland terranes.

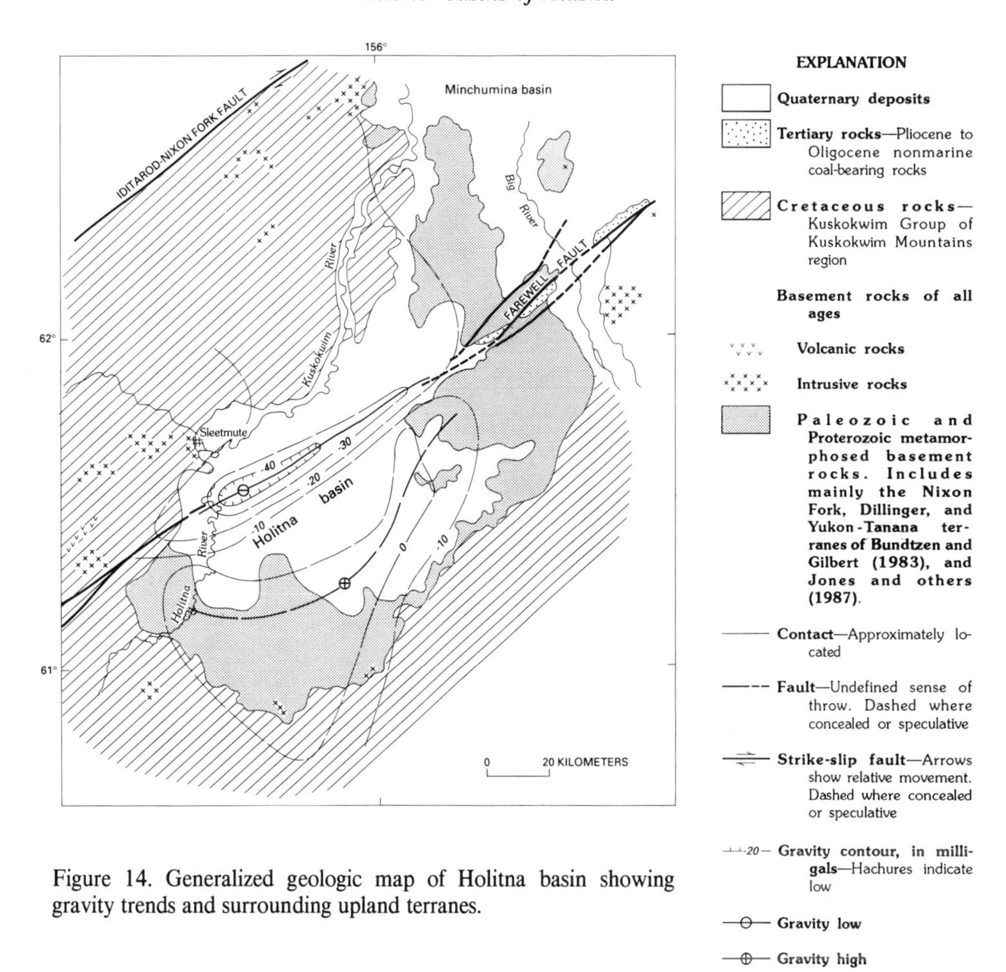

Figure 14. Generalized geologic map of Holitna basin showing gravity trends and surrounding upland terranes.

nian sandstone and shale, and Devonian, Silurian, Ordovician, and older(?) limestone and dolomite. A minor indication of dead oil is reported in the cuttings from the Doyon No. 2 well.

Nonmarine Tertiary sediments are locally present in small, widely scattered outcrops around the margin of the Yukon Flats basin, and in the Tintina trench. As in the Nenana basin, three cycles of sedimentation are suspected. An early cycle of latest Cretaceous (Maastrichtian) and early Tertiary age is represented by at least 1.0 km of conglomerate, sandstone, shale, and minor coal beds in the Tintina trench (Brabb and Churkin, 1969). These rocks are strongly folded and moderately indurated so that their average densities are comparable to densities of the Cantwell Formation (Fig. 10). A middle Tertiary cycle of Miocene age may be represented by coal-bearing beds at Coal and Mudbank Creeks in the Beaver Quadrangle, and in outcrops in the Tintina trench and in the Coleen Quadrangle northeast of Fort Yukon. The rocks consist primarily of conglomerate, sandstone, coal-bearing siltstone or shale, and lacustrine silt and clay. In the Coleen Quadrangle, lacustrine beds contain Miocene(?) clams (Brosgé and Reiser, 1969). As much as 2.0 km of section has been recognized in the Tintina trench southwest of the Preacher Creek fault. A late Tertiary cycle is probably represented by high-level sand and gravel deposits of Tertiary(?) and Quaternary(?) age in the Beaver Quadrangle (Brosgé and others, 1973) and by late Tertiary lake deposits in a water well at Fort Yukon (Williams, 1960).

Structure. The Yukon Flats basin appears to represent an extensional graben complex at the northwesterly terminus of the dextral Tintina fault system. It is interpreted to be a typical pull-apart basin or rhomb-graben. Sharp topographic breaks associated with steep gravity gradients on the northwest and southeast flanks of the regional gravity low are interpreted to represent normal faults (Fig. 16). Beneath the Tertiary fill of the basin there are divergent magnetic fabrics (Fig. 17). Beneath the central and eastern parts of the basin there is a distinctive northwesterly trending magnetic fabric that may be in part Tozitna(?) terrane and in part other, unknown, terrane(s). In the Kandik thrust belt the magnetic fabric trends northeast. The magnetic fabric in the basin and in the Kandik thrust belt suggests that the Porcupine terrane of the Kandik thrust belt may not project southwesterly beneath the eastern part of the Yukon Flats basin. The magnetic trends also do not support the extension of the Kaltag fault in Canada (Norris, 1985) beneath the basin to join the Kaltag fault of west-central Alaska (Patton and Hoare, 1968), as has been proposed by numerous authors (see, for example, McWhae, 1968). In contrast, the Tintina fault system has a strong gravity and magnetic signature (Cady and Weber, 1983) that ends in the northwest corner of the Circle Quadrangle, where it appears to merge with

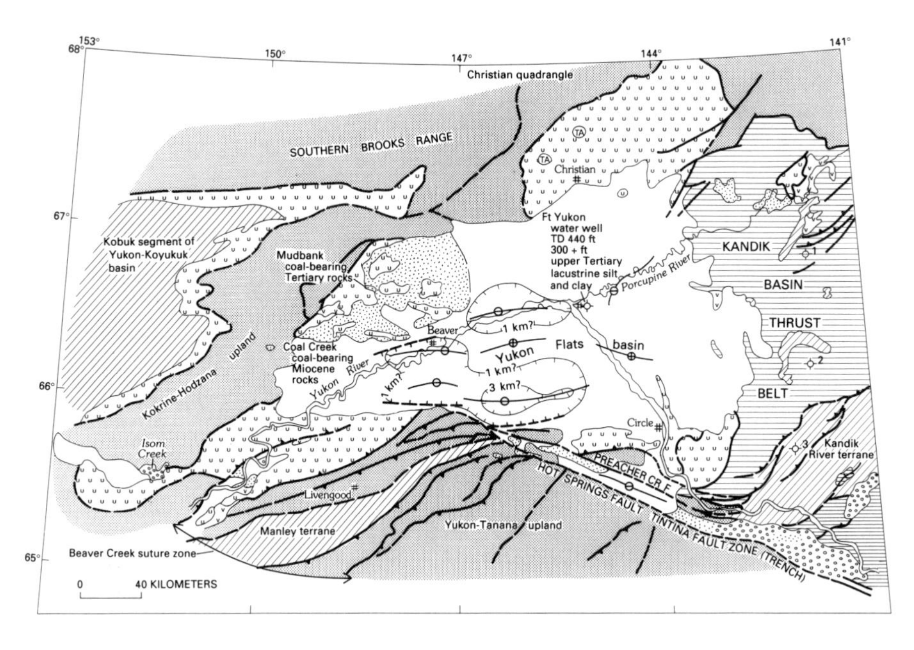

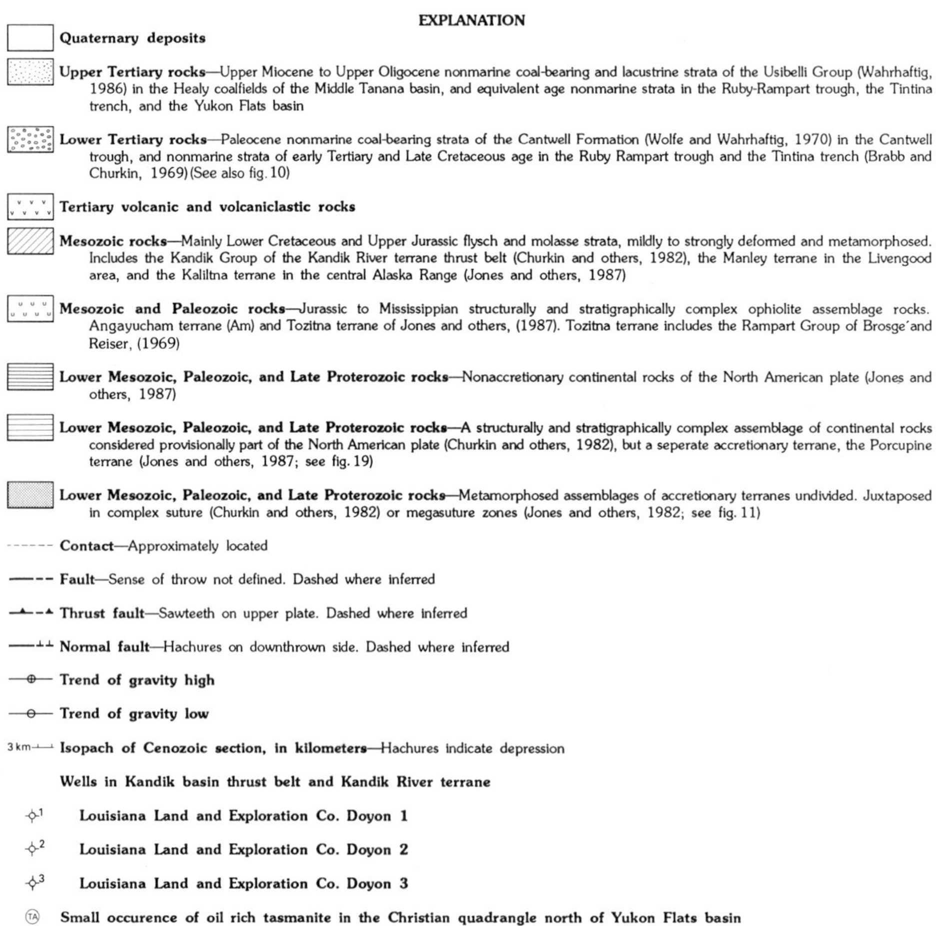

Figure 15. Generalized geologic map of Yukon Flats basin, Kandik basin thrust belt, and adjacent areas.

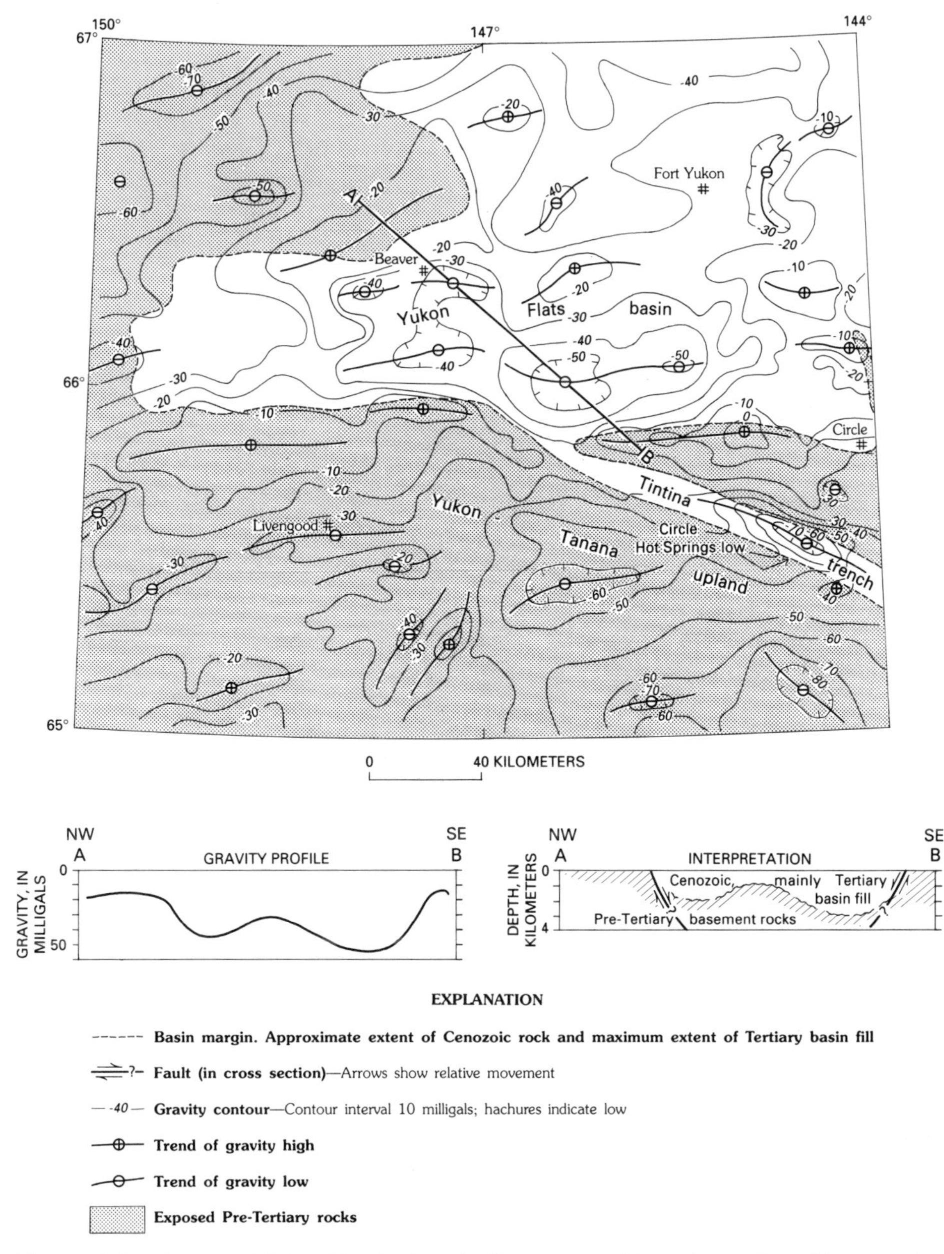

Figure 16. Gravity map of Yukon Flats basin and adjacent areas with gravity profile and interpretation of Cenozoic fill in the basin.

the Beaver Creek suture zone. These geophysical data suggest that the Beaver Creek suture zone accommodates much of the several hundred kilometers of Paleozoic to Tertiary dextral displacement on the Tintina fault.

Petroleum potential. No direct evidence for hydrocarbons has been reported in the Yukon Flats basin. The prospect for pre-Tertiary tasmanite source rocks in the Tozitna terrane appears to be extremely remote. Paleozoic and Triassic organic shales of North American plate affinity (NA) south of the Kandik flysch belt (KAN) cannot logically be projected beneath the basin, and two test wells in the Porcupine (PC) terrane east of the basin indicate that these rocks do not have significant source or reservoir characteristics. Thus, if there is petroleum potential in the basin, source beds most probably are nonmarine Tertiary lacustrine or coal beds. As in the Nenana basin, an optimistic evaluation is that gas reserves of economic importance for local consumption could be present.

MESOZOIC BASINS

The Mesozoic basins, characterized here as flysch belts, cover extensive areas of western and southwestern interior Alaska, and south-central Alaska. The sedimentary fill of these

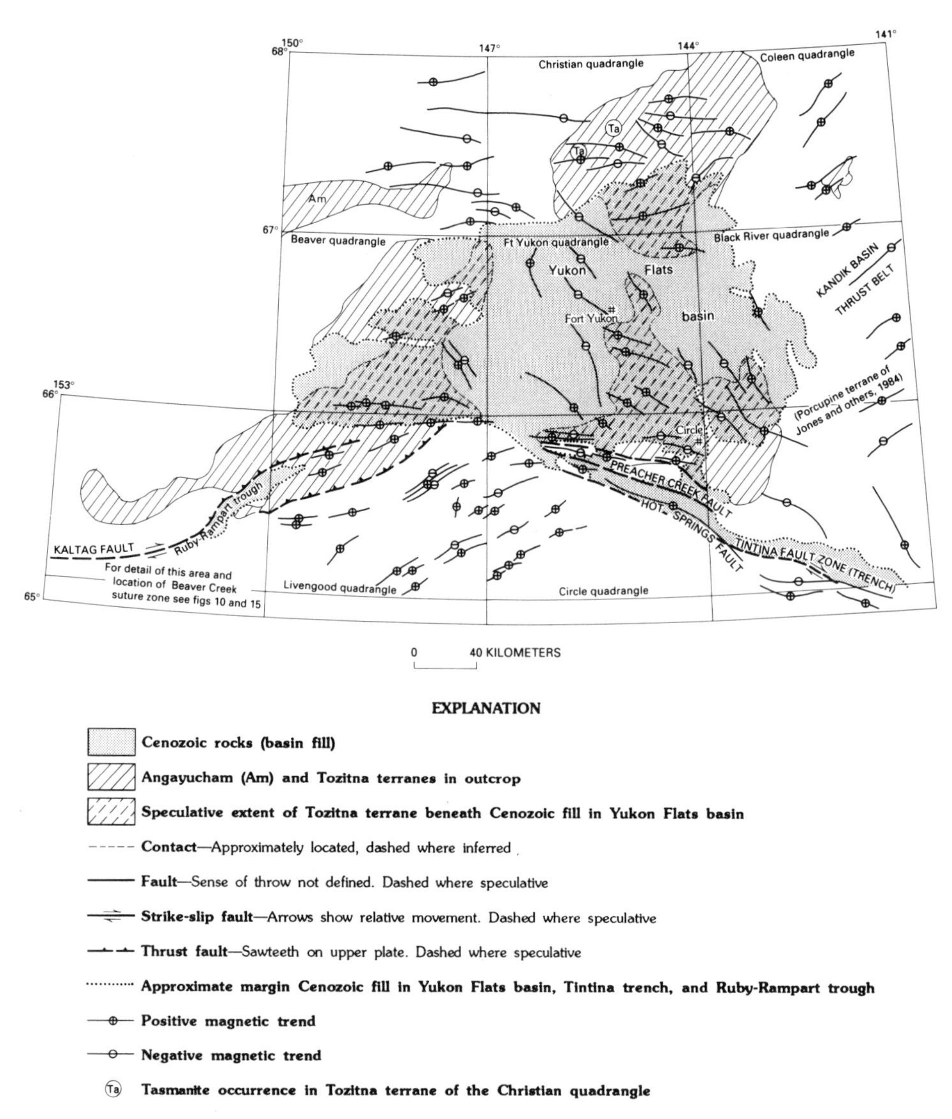

Figure 17. Map showing Angayucham and Tozitna terranes in outcrop, speculative extent of Tozitna terrane beneath Tertiary fill, trends of magnetic anomalies, and tasmanite occurrences in the Yukon Flats basin.

basins includes thick sequences of Mesozoic marine volcaniclastic graywacke and mudstone turbidites, and thick, coal-bearing paralic and marine shelf deposits in the regressive late-stage fill of the basins.

Basin fill was derived largely from volcanic-plutonic arc complexes and in part from adjacent continental metamorphic highlands. Deposition is interpreted to have been in fore-arc (mainly oceanic?) and back-arc (mainly continental?) basins synchronous with volcanism. The extent of the Mesozoic flysch basins is shown in Figure 3. In general, the data permit the interpretation that the fore-arc basins are the more strongly deformed and have no oil and gas potential, whereas the back-arc(?) basins are less deformed and may have oil and gas potential. Three exploratory wells have penetrated flysch rocks in the interior basins, one in the Kandik (KAN) basin flysch, one in the YKK flysch, and one in YKK(?) flysch in the Bethel lowlands beneath a thin Tertiary section. The Peninsular terrane (PEN) has been extensively explored without success.

Peninsular terrane (PEN)

The Peninsular terrane extends northeasterly the length of the Alaska Peninsula province and beneath the Tertiary fill of the Cook Inlet basin and the Copper River basin. It is interpreted that the basinal rocks were deposited in a back-arc setting, as tentatively proposed by Reed and others (1983), and that the pervasively deformed and metamorphosed flysch rocks of the Togiak (TOG) and Kahiltna (KAH) terranes on the northwest flank of the Peninsular arc represent partly synchronous fore-arc deposits. The stratigraphy of the Peninsular terrane includes Permian limestone; Upper Triassic petroliferous limestone, argillite, and volcanic rocks; Lower Jurassic andesitic volcanic rocks and volcaniclastic siltstone and sandstone; Middle Jurassic through Cretaceous fossiliferous clastic rocks (including Middle Jurassic organic shale), and minor bioclastic limestone and quartzite (Jones and others, 1987). Locally, oil seepages appear to be associated with faults, and oil and gas shows have been logged in exploratory tests. Structurally, the rocks are moderately to mildly deformed and faulted. They were once deeply buried, have high densities, and show the effects of diagenetic alteration, including laumontite and locally higher-grade metamorphism. About 40 exploratory tests with Cretaceous and Upper to Middle Jurassic objectives have been drilled in the Copper River and Cook Inlet basins and on the Alaska Peninsula. Only minor oil shows have been encountered in Cretaceous sandstone and Middle Jurassic fractured shale. These discouraging results may not eliminate the potential for future discoveries, but the lack of discoveries to date indicates that the resource potential is limited.

Yukon-Koyukuk-Kobuk (YKK) terrane

The Yukon-Koyukuk segment of the YKK terrane includes a belt of strongly deformed and mildly to strongly metamorphosed flysch rocks trending northeasterly along the eastern margin of Norton Sound in west-central Alaska. These rocks are interpreted to represent fore-arc deposits of the Hogatza volcano-plutonic arc. The Kobuk segment of this terrane trends easterly north of the Hogatza arc and south of the Brooks Range. The Kobuk segment probably represents back-arc molasse deposits derived from both the Hogatza arc and the Brooks Range and Kokrine-Hodzana upland. Stratigraphically, the terrane comprises a Lower and Upper Cretaceous volcaniclastic sequence, including turbidite, prodelta, and deltaic coal-bearing facies (Patton, 1973; Patton and others, this volume, chapter 7). Basement rocks are andesitic volcanics of early Cretaceous metamorphic terranes. The thickness of the stratigraphic sequence is poorly constrained. Patton (1973) suggests that the deepest part of the basin may be more than 7.0 km deep on the basis of magnetic data. Structure of the Yukon-Koyukuk segment is complex. Isoclinal folding is characteristic. Locally, broad synclines are flanked by sharply folded and faulted anticlines. Structure in the Kobuk back-arc(?) segment is less complex, and a few large anticlinal structures have been defined. For this reason it has been suggested that the Kobuk segment of the province could have petroleum potential (Patton, 1978; Hite and Nakayama, 1980); however, no direct evidence of oil and gas in the form of seepage is known from the region. Speculative potential petroleum resources based on a cubic mile of sediment in this type of basin may be misleading (Hite and Nakayama, 1980). Compared with the Peninsular terrane, which has numerous oil seepages and oil and gas shows in wells, even the most favorable areas of the Yukon-Koyukuk-Kobuk province could be expected to have, at best, minor gas reserves.

Kuskokwim terrane (KUS)

The Kuskokwim terrane covers about 60,000 km^2 in southwestern Alaska. The stratigraphy of the basin is similar to that of the Yukon-Koyukuk-Kobuk basin and includes several thousand meters of Lower to Upper Cretaceous quartzose lithic conglomerate, and sandstone and siltstone turbidites. Basement rocks include early Mesozoic andesitic volcanic rocks, chert, argillite, volcanogenic clastic rocks, and Paleozoic carbonate rocks, amalgamated prior to the deposition of the Cretaceous flysch (Decker and others, this volume). Structure of the province includes both open folds and tight chevron folds. Numerous high-angle faults and large strike-slip faults segment the terrane and localize many intrusive dikes and plutons. The intrusive bodies host numerous mineral occurrences. Petroleum potential of the province is believed to be precluded by structural complexity, intrusive bodies, and mineralization.

Kandik (KAN) and Manley (MAN) terranes

The Kandik and Manley terranes incorporate thick, highly deformed flyschoid rocks of Triassic to Early Cretaceous age (Dover, this volume). The thicknesses of the stratigraphic sections are poorly constrained, but they probably include at least 3.0 to 4.5 km of strata. The stratigraphy of the two terranes is similar enough to suspect that the Manley terrane is a dismembered segment of the Kandik terrane, displaced about 170 km by the Tintina fault zone since Early Cretaceous time. Churkin and Brabb (1969) have summarized the oil potential of the Kandik terrane as negligible, owing to complex structure and low-grade metamorphism. The Manley terrane is similar and has no petroleum potential.

PALEOZOIC AND MESOZOIC BASIN

Kandik basin

Introduction. The Kandik basin of east-central Alaska, as discussed here, is confined approximately by the Porcupine River on the north, the Tintina fault system on the south, the Alaska-Canada boundary on the east, and the Yukon Flats Cenozoic basin on the west (Fig. 18). As thus defined it covers an area of more than 20,000 km^2. Along the Canadian border, elevations reach 600 to 1,200 m in relatively rugged terrain but drop away to the west to the low elevations of the Yukon Flats. The basin has long been of interest to geologists and petroleum companies because it has the most complete Paleozoic to early Mesozoic stratigraphic section in Alaska, and has organic shales and oil shows at several stratigraphic levels from the Ordovician to the Jurassic (Fig. 19).

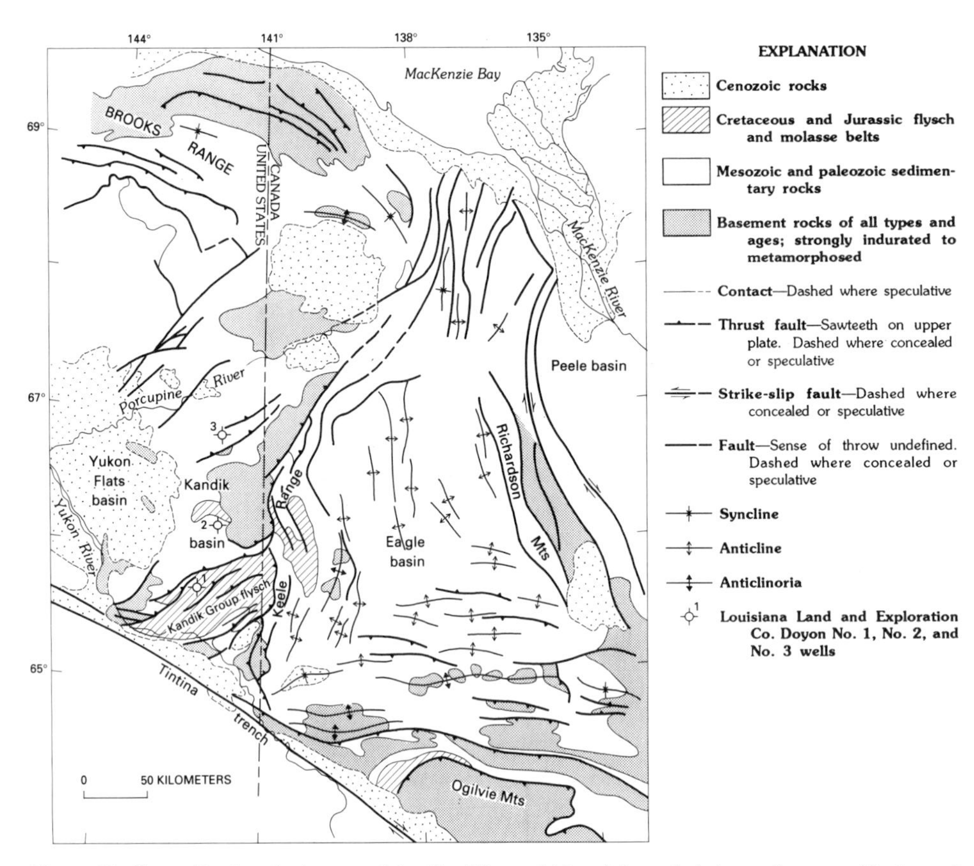

Figure 18. Generalized geologic map of the Cordilleran fold and thrust belt in northeastern Alaska and northwestern Canada, showing the Kandik basin and its geographic location to other basins and structural trends in the Cordillera.

Stratigraphy. The stratigraphic succession in the Kandik basin includes metamorphosed late Precambrian rocks, overlain by a thick Paleozoic continental-margin section deposited in shelf to open-marine environments, and a thick Jurassic-Cretaceous (Kandik Group) flysch sequence. The flysch sequence separates the Precambrian and Paleozoic sequences in Alaska into a small triangular area south of the flysch belt and a larger northern area from the flysch belt to the Porcupine River. South of the flysch belt, geologists agree that the terrane is part of the North American plate (unit NAm of Jones and others, 1987; Silberling and others, this volume; unit T of Churkin and others, 1982), but differ as to whether the rocks north of the flysch belt are different from those of NAm (Porcupine [unit PC] of Jones and others, 1987, and Silberling and others, this volume), or (provisionally) the same (Tatonduk terrane [unit T?] of Churkin and others, 1982). Rocks of the Porcupine (unit PC) or Tatonduk (unit T?) terranes are not as well known as those of NAm. The missing strata may be due to facies changes, to structural complication, or simply to inadequate data. A notable difference between the PC and T(?) terranes is the apparent absence of highly organic shales, petroliferous limestones, or oil shows from the PC terrane. The Doyon No. 2 well drilled 2,783 m of Devonian sandstone and Devonian to Cambrian(?) dolomite. A minor dead oil show was reported in dolomite in the lower part of the well. The Doyon No. 3 well drilled 4,128 m of Devonian to Ordovician dolomite and limestones. Thrust faulting is suspected, so that stratigraphic thickness of the well sections is uncertain.

The intervening Kandik unit of Jones and others (KA or K terrane) includes about 4.5 km of strongly deformed, low-grade metamorphic flysch rocks that are probably regionally unconformable or in fault contact with underlying Paleozoic rocks (Brabb and Churkin, 1969). The Doyon No. 1 well drilled 3,368 m in this terrane, probably in a thrust-repeated section. Dead oil reportedly was found in the cuttings.

Structure. Structure of the Kandik basin is complex and not mapped in detail. However, the regional structural setting of the basin, in the hinterland of the North America Cordilleran fold and thrust belt, offers some clues to its structure by analogy to other parts of the Cordillera. Dover (1985, this volume) interprets the structural framework as resembling the classical fold and thrust belt structure of the Canadian Cordillera. I agree in general with this concept and suggest an analogy to the Main Ranges or Front Ranges of the southern Canadian Cordillera as shown in Bally and others (1966, Plate 12). Reports of recent field studies (Gardner and others, 1984) tend to confirm this general analogy for at least the NAm (or T) terrane segment of the basin. Thrust

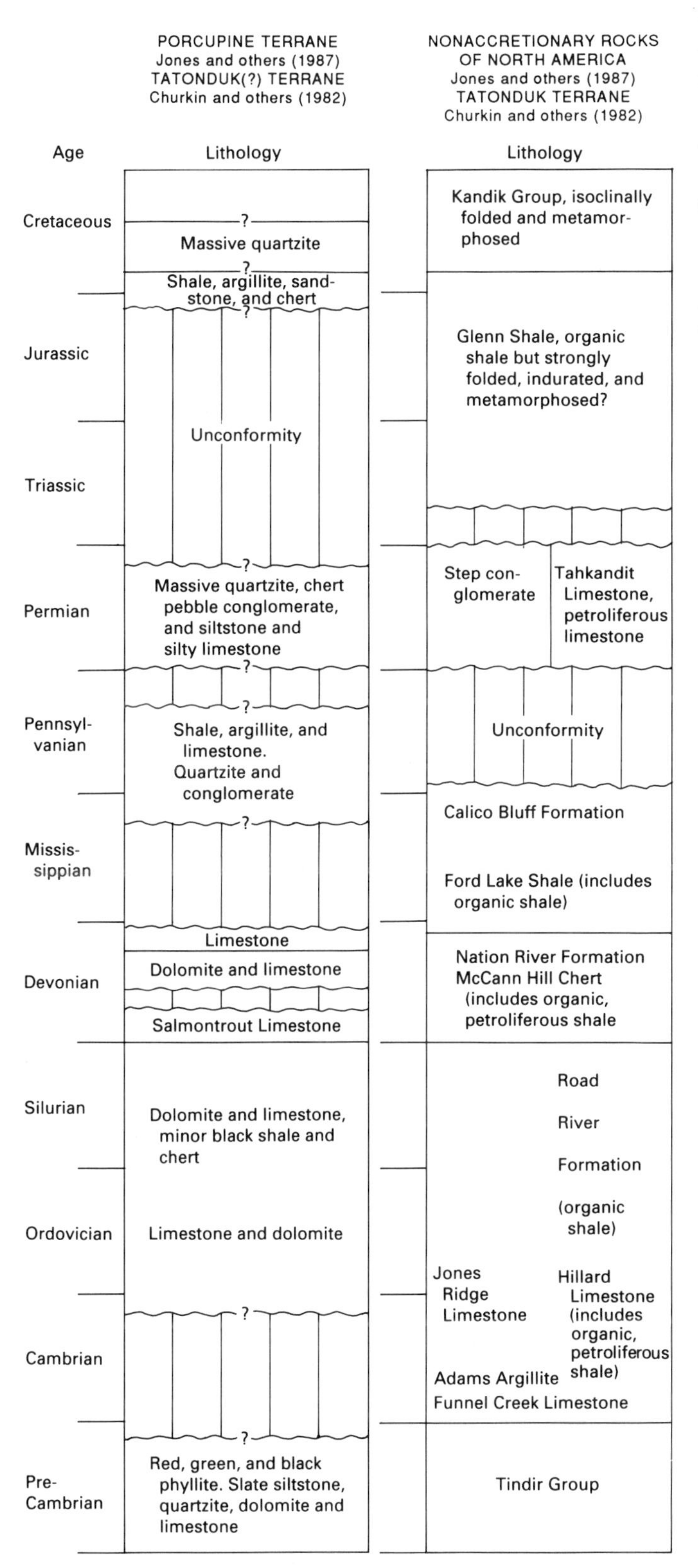

Figure 19. Stratigraphy of the Porcupine terrane compared to that of the nonaccretionary rocks of the North American plate in Kandik basin.

shortening in underlying Paleozoic rocks is apparently accommodated in the overlying Mesozoic flysch by isoclinal folding and accompanying low-grade metamorphism. Alternatively, the flysch belt could conceal a terrane boundary between the PC or (T?) terrane to the north and the NAm (or T) terrane to the south and east.

Petroleum potential. In spite of the abundant evidence for source rocks and the possible presence of reservoir rocks in the Paleozoic section, the extreme structural deformation, low-grade metamorphism, and discouraging results of three dry holes suggest little potential for significant or economic petroleum resources in the Kandik basin.

REFERENCES CITED

Andreason, G. E., Wahrhaftig, C., and Zietz, I., 1964, Aeromagnetic reconnaissance of the east-central Tanana lowland, Alaska: U.S. Geological Survey Geophysical Investigations Map GP-447, scale 1:125,000.

Aydin, A., and Nur, A., 1982, Evolution of pull-apart basins and their scale dependence: Tectonics, v. 1, no. 1, p. 71–105.

Bally, A. W., Gordy, P. L., and Stewart, G. A., 1966, Structure, seismic data, and orogenic evolution of southern Canadian Rocky Mountains: Bulletin of Canadian Petroleum Geology, v. 14, no. 3, p. 337–381.

Barnes, D. F., 1977, Bouguer gravity map of Alaska: U.S. Geological Survey Geophysical Investigations Map GP-913, 1 sheet, scale 1:250,000.

Barnes, D. F., and Tailleur, I. L., 1970, Preliminary interpretation of geophysical data from the Lower Noatak River basin, Alaska: U.S. Geological Survey Open-File Report 70–18, scale 1:250,000.

Brabb, E. E., and Churkin, M., Jr., 1969, Geological map of the Charley River Quadrangle, east-central Alaska: U.S. Geological Survey Miscellaneous Geological Investigations Map I-573, scale 1:250,000.

Brosgé, W. P., and Reiser, H. N., 1969, Preliminary geologic map of the Coleen Quadrangle, Alaska: U.S. Geological Survey Open-File Report 69-370, scale 1:250,000.

Brosgé, W. P., Reiser, H. N., Dutro, J. T., Jr., and Churkin, M., Jr., 1966, Geologic map and stratigraphic sections, Porcupine River Canyon, Alaska: U.S. Geological Survey Open-File Report 66–10, 4 sheets, scale 1:63,360.

Brosgé, W. P., Reiser, H. N., and Yeend, W., 1973, Reconnaissance geologic map of the Beaver Quadrangle, Alaska: U.S. Geologic Survey Miscellaneous Field Studies Map MF-525, scale 1:250,000.

Bundtzen, T. K., and Gilbert, W. G., 1983, Outline of geology and mineral resources of upper Kuskokwin region, Alaska, *in* Proceedings of the 1982 Symposium, Western Alaska Geology and Resource Potential, Anchorage, Alaska: Journal of the Alaska Geological Society, p. 101–117.

Cady, J. W., and Weber, F. R., 1983, Aeromagnetic map and interpretation of magnetic and gravity data, Circle Quadrangle, Alaska: U.S. Geological Survey Open-File Report 83–170C, scale 1:250,000.

Calderwood, K. W., and Fackler, W. C., 1972, Proposed stratigraphic nomenclature for Kenai Group, Cook Inlet Basin, Alaska: American Association of Petroleum Geologists, v. 56, p. 739–754.

Calderwood, K. W., and Wilson, T., 1987, Southern Alaska region: Tulsa, Oklahoma, The American Association of Petroleum Geologists, Correlation of stratigraphic units of North America (COSUNA) project.

Chapman, R. M., Weber, F. R., and Taber, B., 1971, Preliminary geologic map of Livengood Quadrangle, Alaska: U.S. Geological Survey Open-File Report 71–66, scale 1:250,000.

Chapman, R. M., Yeend, W., and Brosgé, W. P., 1982, Reconnaissance geologic map of the Tanana Quadrangle, Alaska: U.S. Geological Survey Open-File Report 82–734, scale 1:250,000.

Chapman, R. M., Patton, W. W., and Moll, E. J., 1985, Reconnaissance geologic map of the Ophir Quadrangle, Alaska: U.S. Geological Survey Open-File Report 85–203, scale 1:250,000.

Chinnery, M. A., 1965, The vertical displacements associated with transcurrent faulting: Journal of Geophysical Research, v. 70, no. 18, p. 4627–4632.

Churkin, M., Jr., and Brabb, E. E., 1969, Prudhoe Bay discovery forces a look at other petroliferous areas: Oil and Gas Journal, v. 67, no. 5, p. 104–110.

Churkin, M., Jr., Foster, H. L., and Chapman, R. M., 1982, Terranes and suture zones in east-central Alaska: Journal of Geophysical Research, v. 87, p. 3718–3730.

Churkin, M., Jr., Wallace, W. K., and Wesley, K., 1984, Nixon Fork-Dillinger terranes; A dismembered Paleozoic craton margin in Alaska displaced from Yukon Territory: Geological Society of America Abstracts with Programs, v. 16, p. 275.

Cooper, A. K., and Marlow, M. S., 1983, Preliminary results of geophysical and geological studies of the Bering Sea shelf during 1982: U.S. Geological Survey Open-File Report 83–322, 5 p.

Decker, J., 1984, The Kuskokwim Group; A post-accretionary successor basin in southwest Alaska: Geological Society of America Abstracts with Programs, v. 16, p. 277.

Dempsey, W. J., Menchke, J. L., and Andreasen, G. E., 1957, Total intensity aeromagnetic profiles of Bethel basin, Alaska: U.S. Geological Survey Open-file Report 57-33, 3 sheets.

Dickey, D. B., Gilbert, W. G., and Kline, J. T., 1982, Cenozoic nonmarine sedimentary rocks and structure of southwest Alaska: Geological Society of America Abstracts with Programs, v. 14, p. 160.

Dietrich, R. V., Dutro, J. T., Jr., and Fose, R. M., 1982, AGI data sheets for geology in the field, laboratory, and office: Falls Church, Virginia, American Geological Institute.

Dover, J. H., 1985, Possible oroclinal bend in northern cordilleran fold and thrust belt, east-central Alaska: Geological Society of America Abstracts with Programs, v. 17, p. 352.

Ellersieck, I., Tailleur, I. L., Mayfield, C. F., and Curtis, S. M., 1979, A new find of Upper Cretaceous or Tertiary sedimentary rocks in the Noatak Valley, *in* Johnson, K. M., and Williams, J. R., eds., The United States Geological Survey in Alaska; Accomplishments during 1978: U.S. Geological Survey Circular 804-B, p. B13.

Fisher, M. A., Patton, W. W., Jr., and Holmes, M. L., 1982, Geology of Norton basin and continental shelf beneath northwestern Bering Sea, Alaska: The American Association of Petroleum Geologists Bulletin, v. 66, no. 3, p. 255–285.

Gabrielse, H., 1985, Major dextral transcurrent displacements along the northern Rocky Mountain trench and related lineaments in north-central British Columbia: Geological Society of America Bulletin, v. 96, p. 1–14.

Gardner, M. C., Jarrard, R. D., and Mount, V., 1984, Style and origin of structures, Paleozoic and Precambrian sedimentary rocks of southern Tatonduk terrane, Alaska: Geological Society of America Abstracts with Programs, v. 16, p. 285.

Gilbert, W. G., Ferrell, V. M., and Turner, D. L., 1976, The Teklanika Formation; A new Paleocene volcanic formation in the central Alaska Range: Alaska Division of Geological and Geophysical Surveys Geologic Report 47, 16 p.

Grantz, A., 1966, Strike-slip faults in Alaska: U.S. Geological Survey Open-File Report 66-267, 82 p.

Henning, M. W., Meyer, J., Kombrath, R., and Krouskop, D., 1984, Geology and gravity evaluation of oil and gas potential of the Minchumina basin, Alaska: Geological Society of America Abstracts with Programs, v. 16, p. 289.

Hite, D. M., and Nakayama, E. M., 1980, Present and potential petroleum basins of Alaska, *in* Landwehr, M. L., ed., New ideas, new methods, new developments: Exploration and Economics of the Petroleum Industry, v. 18, p. 511–560.

Hoare, J. M., and Condon, W. H., 1966, Geologic map of the Kwiguk and Black Quadrangles, western Alaska: U.S. Geological Survey Miscellaneous Geologic Investigations Map I-469, scale 1:250,000.

Hoare, J. M., and Coonrad, W. L., 1959, Geology of the Russian Mission Quadrangle, Alaska: U.S. Geological Survey Miscellaneous Geological Investigations Map I-292, scale 1:250,000.

——, 1978, New geologic map of the Goodnews and Hagemeister Island Quadrangles, Alaska, *in* Johnson, K. M., ed., The United States Geological Survey in Alaska; Accomplishments during 1977: U.S. Geological Survey Circular 772-B, p. B50–B55, B112–B113.

Hoare, J. M., Condon, W. H., and Patton, W. W., Jr., 1964, Occurrence and origin of laumontite in Cretaceous sedimentary rocks in western Alaska: U.S. Geological Survey Professional Paper 501-C, p. C74–78.

Jones, D. L., Silberling, N. J., Berg, H. C., and Plafker, G., 1981, Tectonostratigraphic terrane map of Alaska: U.S. Geological Survey Open-file Report 81-792, 2 sheets.

Jones, D. L., Silberling, N. J., Gilbert, W. G., and Coney, P. J., 1982, Character, distribution, and tectonic significance of accretionary terranes in the central Alaska Range: Journal of Geophysical Research, v. 87, p. 3709–3717.

Jones, D. L., Silberling, N. J., Coney, P., and Gilbert, W. G., 1983, Interpretive bedrock geologic map of the McKinley region and tectonostratigraphic map of the Mount McKinley region: U.S. Geological Survey Open-File Report 83–11, 2 sheets, scale 1:250,000.

Jones, D. L., Silberling, N. J., Coney, P. J., and Plafker, G., 1987, Lithotectonic terrane map of Alaska (west of 141st Meridian); Folio of the lithotectonic terrane maps of the North America Cordillera: U.S. Geological Survey Miscellaneous Field-Studies Map MF-1874-A, scale 1:2,500,000.

Klein, R. M., Lyle, W. M., Dobey, P. L., and O'Conner, K. M., 1974, Energy and mineral resources of Alaska and the impact of federal land policies on their availability: State of Alaska, Department of Natural Resources, Division of Geological and Geophysical Surveys Open-File Report 50.

Lanphere, M. A., 1978, Displacement history of the Denali fault system, Alaska and Canada: Canadian Journal of Earth Sciences, v. 15, p. 817–822.

Lyle, W. M., Palmer, I. F., Bolm, J. G., and Flett, T. O., 1982, Hydrocarbon reservoir and source rock characteristics from selected areas of southwestern Alaska: Alaska Division of Geological and Geophysical Surveys Professional Report 77, 35 p.

Magoon, L. B., and Egbert, R. M., 1986, Framework geology and sandstone composition, *in* Magoon, L. B., ed., Geologic studies of the lower Cook Inlet COST No. 1 well, Alaska Outer Continental Shelf: U.S. Geological Survey Bulletin 1596, p. 65–90.

McWhae, J. R., 1986, Tectonic history of northern Alaska, Canadian Arctic, and Spitzbergen regions since Early Cretaceous: American Association of Petroleum Geologists Bulletin, v. 70, no. 4, p. 430–450.

Mertie, J. B., Jr., 1927, The Chandalar-Sheenjek district, Alaska: U.S. Geological Survey Bulletin 810, p. 87–139.

Miller, D. J., Payne, T. E., and Gryc, G., 1959, Geology of possible petroleum provinces in Alaska: U.S. Geological Survey Bulletin 1094, 131 p.

Norris, D. K., 1985, Geology of northern Yukon and Northwestern District of Mackenzie, Canada: Geological Survey of Canada Map 1581A, scale 1:500,000.

Page, R. A., 1959, Tertiary geology of the Cheyenne Creek area, Alaska [M.S. thesis]: Seattle, University of Washington, 66 p.

Patton, W. W., Jr., 1973, Reconnaissance geology of the northern Yukon-Koyukuk province Alaska: U.S. Geological Survey Professional Paper 774-A, p. A1–A17.

——, 1978, Maps and table describing areas of interest for oil and gas in central Alaska: U.S. Geological Survey Open-File Report 78-1-F, 2 p.

Patton, W. W., Jr., and Hoare, J. M., 1968, The Kaltag fault, west-central Alaska, *in* Geological Survey research 1968: U.S. Geological Survey Professional Paper 600-D, p. D147–153.

Patton, W. W., Jr., Miller, T. P., 1970, Preliminary geologic investigations in the Kanuti River region, Alaska: U.S. Geological Survey Bulletin 1312-J, p. J1–J10.

Patton, W. W., Jr., Miller, T. P., Tailleur, I. L., Brosgé, W. P., and Lanphere, M. A., 1977, Preliminary report on the ophiolites of northern and western Alaska, *in* Coleman, R. G., and Irwin, W. P., eds., North American ophio-

lites: Oregon Department of Geology and Mineral Industries Bulletin 95, p. 51–57.
Raisz, E., 1948, Land form map of Alaska: Cambridge, Massachusetts, Harvard University Institute of Geographical Exploration, 1 sheet, scale approx. 1:3,000,000.
Reed, B. L., Miesch, A. T., and Lanphere, M. A., 1983, Plutonic rocks of Jurassic age in the Alaska-Aleutian Range batholith; Chemical variations and polarity: Geological Society of America Bulletin, v. 94, p. 1232–1240.
Silberman, M. L., and Grantz, A., 1984, Paleogene volcanic rocks of the Matanuska Valley area and the displacement history of the Castle Mountain fault, *in* Coonrad, W. L., and Elliott, R. L., eds., The U.S. Geological Survey in Alaska, Accomplishments during 1981: U.S. Geological Survey Circular 868, p. 82.
Smith, T. N., Clough, J. G., Meyer, J. F., and Blodgett, R. B., 1985, Petroleum potential and stratigraphy of Holitna basin, Alaska [abs.]: American Association of Petroleum Geologists Bulletin, v. 69, no. 2, p. 308.
Tailleur, I. L., Brosgé, W. P., and Reiser, H. N., 1967, Oil shale in Christian region of northeastern Alaska: U.S. Geological Survey Professional Paper 575, p. A12.
Triplehorn, D. M., Turner, D. L., and Naeser, C. W., 1984, Radiometric age of the Chickaloon Formation of south-central Alaska; Location of the Paleocene–Eocene boundary: Geological Society of America Bulletin, v. 95, p. 740–742.
Turner, D. L., Triplehorn, D. M., Naesser, C. W., and Wolfe, J. A., 1980, Radiometric dating of ash partings in Alaska coal beds and upper Tertiary paleobotanical stages: Geology, v. 8, p. 92–96.
Wahrhaftig, C., 1986, The Cenozoic section in Suntrana, Alaska, *in* Hill, M. L., ed., Cordilleran section: Boulder, Colorado, Geological Society of America Centennial Field Guide, v. 1, p. 445–450.
Wahrhaftig, C., Wolfe, J. A., Leopold, E. B., and Lanphere, M. A., 1969, The coal-bearing group in the Nenana coal field, Alaska: U.S. Geological Survey Bulletin 1274-2, 30 p.
Wilcox, R. E., Harding, T. P., and Seely, D. R., 1973, Basic wrench tectonics: American Association of Petroleum Geologists Bulletin, v. 57, no. 1, p. 74–96.
Williams, J. R., 1960, Cenozoic sediments beneath the central Yukon Flats, Alaska, *in* Geological Survey research 1960: U.S. Geological Survey Professional Paper 400-B, p. B329.
——, 1962, Geologic reconnaissance of the Yukon Flats district, Alaska: U.S. Geological Survey Bulletin 1111-H, p. 289–331.
Wolfe, J. A., and Tanai, T., 1987, Systematics, phylogeny, and distribution of *Acer* (Maples) in the Cenozoic of Western North America: Journal of the Faculty of Science of Hokkaido University, Series 4, v. 22, p. 1–246.
Wolfe, J. A., and Wahrhaftig, C., 1970, The Cantwell Formation of the central Alaska Range: U.S. Geological Survey Bulletin 1294-A, p. A41–A46.
Yeend, W. E., 1977, Tertiary and Quaternary deposits at the Palisades, central Alaska: U.S. Geological Survey Journal of Research, v. 5, no. 6, p. 747–752.

Manuscript Accepted by the Society October 19, 1990

ACKNOWLEDGMENTS

This chapter has been significantly improved by helpful discussion and technical reviews by Clyde Wahrhaftig, William W. Patton, George Plafker, Leslie B. Magoon III, and Kenneth J. Bird.

Printed in U.S.A.

The Geology of North America
Vol. G-1, The Geology of Alaska
The Geological Society of America, 1994

Chapter 15

Metamorphic history of Alaska

Cynthia Dusel-Bacon
U.S. Geological Survey, 345 Middlefield Road, Menlo Park, California 94025

INTRODUCTION

This chapter presents a summary of the major, regionally developed, metamorphic episodes that affected Alaska throughout the evolution and accretion of its many lithotectonic terranes. Plate 4 (map and table showing metamorphic rocks of Alaska, 2 sheets, 1:2,500,000 scale) accompanies this chapter. The metamorphic scheme (Zwart and others, 1967) used for the map (Fig. 1, Table 1) is based on the occurrence of pressure- and temperature-sensitive metamorphic minerals. Regionally metamorphosed rocks are divided into four facies groups, each of which reflects a different grade of metamorphism. In order of increasing temperatures of crystallization, they are: (1) lamontite and prehnite-pumpellyite facies (LPP), shown on Plates 4A and 4B in shades of gray and tan; (2) greenschist facies (GNS), shown in shades of green; (3) epidote-amphibolite and amphibolite facies (AMP), shown in shades of orange and yellow; and (4) two-pyroxene (granulite) facies (2PX), which occurs only on the Seward Peninsula, shown in reddish brown. Where possible, the greenschist-facies and the epidote-amphibolite- and amphibolite-facies groups are further divided on the basis of pressure of crystallization into three facies series: high-, intermediate-, or low-pressure series. These facies series are indicated by an H, I, or L in place of the final letter in the symbol used for the facies group. High-pressure greenschist-(blueschist) facies rocks, and rocks metamorphosed under blueschist-facies conditions that evolved to intermediate- or low-pressure greenschist-facies conditions during a single episode, are shown in shades of blue. The metamorphic facies symbol for each episode is followed by a symbol showing the age of metamorphism or its minimum and maximum age limits. Subscripts are used to differentiate units that have the same metamorphic grade and age but that have different protoliths and are believed to have different metamorphic histories.

Plate 4B gives summary information for each map unit, including the number by which the unit is referred to in this chapter; the facies and age designation of the unit and its color and pattern on the map; the lithotectonic terrane(s) in which the unit occurs; the lithology and age range of the protoliths; the metamorphic rock types and minerals; and the known, minimum, and/or maximum age of metamorphism and the evidence for the three types of ages. Most, but not all, of the units shown on Plates 4A and B are discussed in this chapter. The reader is referred to four more detailed, regional metamorphic reports and accompanying 1:1,000,000-scale metamorphic facies maps on Alaska (Fig. 2) for sources of information and a more complete discussion and listing of references for all the units shown on Plates 4A and B.

Where sufficient data are available, the possible tectonic origin of a given metamorphic episode is discussed. Unless otherwise defined, all lithotectonic terranes are those of Jones and others (1987) west of the 141st meridian and of Monger and Berg (1987) east of it. All radiometric ages cited have been calculated or recalculated using the decay constants of Steiger and Jäger (1977). The Decade of North American Geology time scale (Palmer, 1983) is adopted in relating radiometric ages to geologic time. Abbreviations used in this chapter are explained in Table 2.

DESCRIPTION AND ORIGIN OF METAMORPHIC EPISODES

Brooks Range

Late Proterozoic amphibolite-facies metamorphism. A sequence of polymetamorphosed amphibolite-facies rocks recrystallized to greenschist- and blueschist-facies assemblages (unit 1) crops out in the Baird Mountains of the southwestern Brooks Range. The sequence includes pelitic schist, minor amounts of interlayered quartzite, marble, and metabasite, and crosscutting intermediate to mafic metaplutonic rocks. It makes up the Hub Mountain terrane of Mayfield and others (1982). Mineral assemblages formed duing M_1 include gt, hb, and pl in metabasite, and bt and gt in pelitic schist (A. B. Till, written communication, 1987). A Late Proterozoic age for M_1 is indicated by K-Ar ages on mu and hb between 729 ± 22 and 594 ± 18 Ma (Turner and others, 1979; Mayfield and others, 1982), and by an Rb-Sr mineral-whole rock isochron age of 686 ± 116 Ma (Armstrong and others, 1986). This metamorphic episode is the oldest recorded anywhere in the Brooks Range and is the only documented evidence of Proterozoic metamorphism in the region.

Dusel-Bacon, C., 1994, Metamorphic history of Alaska, *in* Plafker, G., and Berg, H. C., eds., The Geology of Alaska: Boulder, Colorado, Geological Society of America, The Geology of North America, v. G-1.

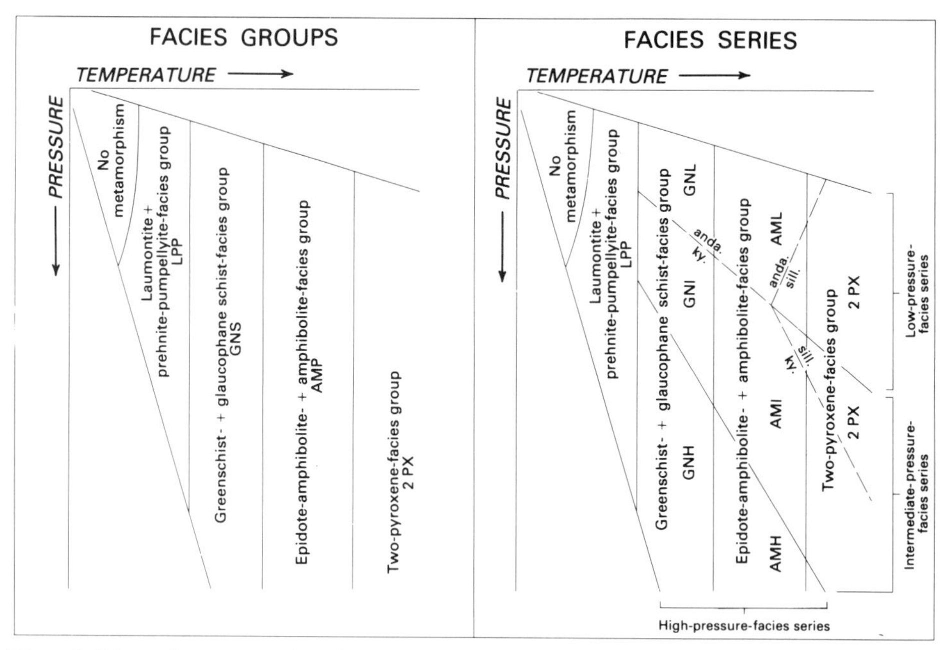

Figure 1. Schematic representation of metamorphic-facies groups and series in P-T space and their letter symbols used on Plates 4A and B (modified from Zwart and others, 1967). Stability fields of Al_2SiO_5 polymorphs andalusite (anda.), kyanite (ky.), and sillimanite (sill.) shown by dashed lines.

Most M_1 assemblages have been partially or, locally, totally recrystallized to greenschist- and blueschist-facies assemblages during subsequent Mesozoic metamorphism (M_2). Common M_2 minerals are ch, ep, wm, ab, sp, bar amph, ac, and bl amph (A. B. Till, written communication, 1987). Recent unpublished mapping indicates that blueschist-facies assemblages are most prevalent in rocks that lie structurally above and below thrust slices of the amphibolite-facies rocks (A. B. Till, written communication, 1987). M_2 is attributed to a high-P evolving to low-P, low-T metamorphic episode that affected the entire southern Brooks Range (unit 3 and M_2 of unit 2) between Middle Jurassic and mid-Cretaceous time (discussed in a later section).

Possible Proterozoic to middle Paleozoic epidote-amphibolite-facies metamorphism. Areas of epidote-amphibolite–facies rocks partly recrystallized to lower grade assemblages crop out in the central Brooks Range (unit 2). By analogy with unit 1, they may also record pre-Mesozoic metamorphism. These areas consist of polymetamorphosed pelitic, feldspathic, and graphitic schist, quartzite, marble, orthogneiss, and metabasite. Protoliths are older than the Middle Devonian granitoids and, between longitudes 151 and 153°, are older than the Proterozoic(?) or pre-middle Paleozoic granitoids (Dillon and others, 1980) that intrude them. It is unclear, however, whether the granitoids, shown with the pattern that denotes a metamorphosed pluton, were metamorphosed prior to the widespread Mesozoic episode that affected the entire southern Brooks Range.

In general, the metamorphic rocks of this unit are distinguished from contiguous rocks of unit 3 by having a coarser crystallinity, relict epidote-amphibolite–facies mineral assemblages, and a more complex structural fabric (Hitzman and others, 1982; Dillon and others, 1987). The structural fabric includes a penetrative fabric that predates that developed in unit 3, as well as one or more younger penetrative fabrics also present in unit 3.

Middle Paleozoic prehnite-pumpellyite- to greenschist-facies metamorphism. Low- to medium-grade metasedimentary, metavolcanic, and metacarbonate rocks of Proterozoic to Middle(?) and Late Devonian protolith age crop out in the Romanzof and Davidson Mountains in the eastern Brooks Range. In the Romanzof Mountains, metamorphic grade increases southward from prehnite-pumpellyite facies (unit 11) to greenschist facies (unit 12). The metamorphic contact between these units was probably gradational, although it has been subsequently modified by thrust faulting (Dusel-Bacon and others, 1989). Metamorphism predates the Mississippian age of unmetamorphosed rocks that unconformably overlie these units and postdates the Middle(?) Devonian protolith age of the youngest metasedimentary rocks (Sable, 1977; Dusel-Bacon and others, 1989).

The indicated range in metamorphic age is similar to the Devonian intrusive age (380 ± 10-Ma U-Pb upper intercept age of zircon; Dillon and Bakke, 1987) of a peraluminous batholith that intrudes units 11 and 12, suggesting that metamorphism and plutonism may have been products of the same tectonic regime. Parts of the batholith are gneissic and mylonitically or cataclastically deformed (Sable, 1977), but the age of this deformation is uncertain. Structural data collected from greenschist-facies rocks in the western end of unit 12 suggest that the pre-Mississippian rocks in that area were transported southeastward during Middle Devonian (Ellesmerian?) thrusting (Oldow and others, 1987). Additional studies are needed to determine the nature and extent

TABLE 1. SCHEME FOR DETERMINING METAMORPHIC FACIES*

Facies symbol	Diagnostic minerals and assemblages	Forbidden minerals and assemblages	Common minerals and assemblages	Remarks
		LAUMONTITE AND PREHNITE-PUMPELLYITE FACIES		
LPP	Laumontite + quartz, prehnite + pumpellyite.	Pyrophyllite, analcime + quartz, heulandite.	"Chlorite," saponite, dolomite + quartz, ankerite + quartz, kaolinite, montmorillonite, albite, K-feldspar, "white mica."	Epidote, actinolite, and "sphene" possible in prehnite-pumpellyite facies.
		GREENSCHIST FACIES		
GNS		Staurolite, andalusite, cordierite, plagioclase (An>10), laumontite + quartz, prehnite + pumpellyite.	Epidote, chlorite, chloritoid, albite, muscovite, calcite, dolomite, actinolite, talc.	
		Low- and intermediate-pressure greenschist facies		
GNL and GNI		Hornblende, glaucophane, crossite, lawsonite, jadeite + quartz, aragonite.		Biotite and manganiferous garnet possible; stilpnomelane mainly restricted to intermediate-pressure greenschist facies.
		High-pressure greenschist (blueschist) facies		
GNH	Glaucophane, crossite, aragonite, jadeite + quartz.		Almandine, paragonite, stilpnomelane.	Subcalcic hornblende (barroisite) may occur in highest temperature part of this facies.
		Low-temperature subfacies of high-pressure greenshist facies		
GNH (with stipple, Plate 4)	Above minerals plus pumpellyite and/or lawsonite.			
		EPIDOTE-AMPHIBOLITE AND AMPHIBOLITE FACIES		
AMP	Staurolite.	Orthopyroxene + clinopyroxene, actinolite + calcic plagioclase + quartz, glaucophane.	Hornblende, plagioclase, garnet, biotite, muscovite, diopside, K-feldspar, rutile, calcite, dolomite, scapolite.	
		Low-pressure amphibolite facies		
AML	Andalusite + staurolite, cordierite + orthoamphibole.	Kyanite.	Cordierite, sillimanite, cummingtonite.	Pyralspite garnet rare in lowest possible pressure part of this facies.
		Intermediate- and high-pressure amphibolite facies		
AMI and AMH	Kyanite + staurolite.	Andalusite.		Sillimanite mainly restricted to intermediate-pressure amphibolite facies.
		TWO-PYROXENE FACIES		
2PX	Orthopyroxene + clinopyroxene.	Staurolite, orthoamphibole, muscovite, epidote, zoisite.	Hypersthene, clinopyroxene, garnet, cordierite, anorthite, K-feldspar, sillimanite,biotite, scapolite, calcite,dolomite, rutile.	Hornblende possible. Kyanite may occur in higher pressure part of this facies and periclase and wollastonite in low-pressure part.

*Modified from Zwart and others (1967).

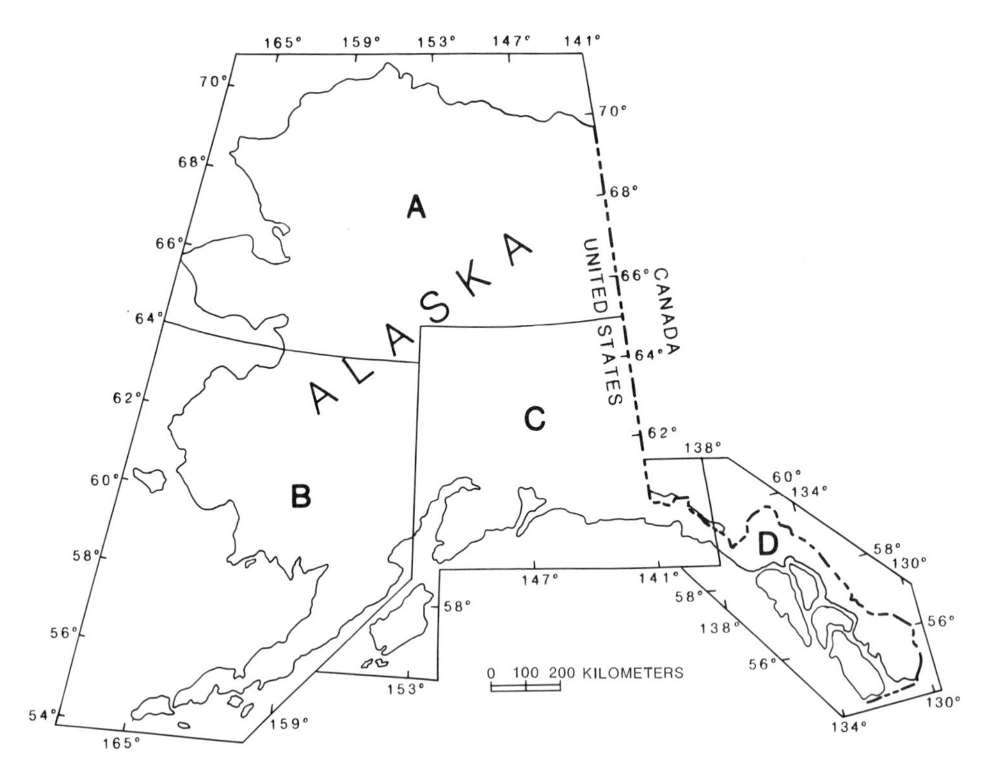

Figure 2. Map showing areas of study in the series of metamorphic facies reports of Alaska. A, Dusel-Bacon and others (1989); B, Dusel-Bacon and others (1991c); C, Dusel-Bacon and others (1991b); D, Dusel-Bacon and others (1991a).

of the proposed southeastward-thrusting event and its temporal relation to the metamorphism.

Isotopic and structural data (Sable, 1977; Dillon, 1987; Oldow and others, 1987) suggest that the rocks of units 11 and 12 and perhaps some parts of the adjacent unmetamorphosed rock unit in the Romanzof Mountains area were subsequently involved in Mesozoic and perhaps early Cenozoic deformation and metamorphism that was part of the widespread Jurassic and Early Cretaceous orogeny that affected the schist belt of the southern Brooks Range (unit 3 and related units).

In the Davidson Mountains, metamorphism of greenschist-facies rocks of unit 13 predated, but was associated with, intrusion of Late Devonian granitoids (Dusel-Bacon and others, 1989). This relation is suggested by an increase in the metamorphic grade from the ch- to the bt-gt-zone with decreasing distance from the crosscutting Late Devonian plutons, and by the closeness in age between the maximum metamorphic age provided by the youngest protolith age (Late Devonian) and the intrusive age of the undeformed crosscutting granitoids.

Jurassic to mid-Cretaceous high evolving to low-P, low-T metamorphism. *Schist belt.* A sequence of polydeformed blueschist- and greenschist-facies rocks (unit 3) crops out across almost the entire width of the southern Brooks Range. This metamorphic sequence, informally called the schist belt, consists of Devonian and older calcareous, pelitic, and graphitic metasedimentary rocks with volumetrically minor metacarbonate rocks, metarhyolite, metabasite, and granitoid orthogneiss (Dillon and others, 1980; Hitzman and others, 1982), and a subordinate amount of upper Paleozoic and locally Triassic metapelite and metacarbonate rocks along its northern margin. The rocks were metamorphosed during a single, prolonged, polyfacial episode and followed a clockwise P-T path that evolved from blueschist- to greenschist-facies conditions. This path reflects tectonic loading followed by decompression. Two phases of penetrative deformation are recognized. Both are characterized by isoclinal folding, and their relation to each other suggests refolding of early formed isoclines during decompression (Gottschalk, 1987). In the Wiseman area, which is probably typical of much of the southern schist belt, lineations and fold axes plunge to the south, and rocks have undergone north-vergent ductile shear deformation concurrent with metamorphism (Gottschalk, 1987).

Most of the rocks in which the high-P minerals gl, lw, and jdpx have been identified occur in a zone within the southern part of this unit (Plate 4A). The restricted occurrence of these minerals may be due to compositional controls (most rocks whose composition favors the development of these minerals are restricted to this zone) or, in part, to structural controls, as proposed by Hitzman (1980). He observed that blueschist assemblages occurred within large nappe-like folds. Glaucophane is by far the most commonly developed high-P mineral and occurs in metabasite, iron-rich metasedimentary rocks, and metatuff (Dusel-Bacon and others, 1989). The assemblage ky+ctd also occurs locally in iron-

TABLE 2. ABBREVIATIONS USED IN TEXT

CAI	conodont alteration index of Epstein and others (1977)	amph	amphibole	kf	potassium feldspar
		anda	andalusite	ky	kyanite
M_1	first metamorphic episode of polymetamorphic unit	bar amph	barroisitic amphibole	lw	lawsonite
		bt	biotite	mu	muscovite
M_2	second metamorphic episode of polymetamorphic unit	ch	chlorite	pa	paragonite
		co	cordierite	pl	plagioclase
S_1	fabric formed during first metamorphic episode	cpx	clinopyroxene	pr	prehnite
		cr	crossite	pu	pumpellyite
S_2	fabric formed during second metamorphic episode	ctd	chloritoid	qz	quartz
		ep	epidote group mineral	sil	sillimanite
P	pressure	gl	glaucophane	sp	spinel
T	temperature	gt	garnet	sph	sphene
		hb	hornblende	st	staurolite
ab	albite	jd	jadeite	tr	tremolite
ac	actinolite	jdpx	jadeitic pyroxene	wm	white mica

rich metasedimentary rocks, but it is not known whether the ky formed during the Jurassic to Cretaceous blueschist-to-greenschist episode or during the possible pre-Devonian episode discussed in the previous section.

The metamorphic grade and overall degree of deformation within unit 3 decreases to the north. The northern limit of this unit is defined, in part, on the basis of a CAI isotherm that delineates the first occurrence of CAI values of less than 5 (corresponding to a temperature of less than 300°C) for Ordovician through Triassic rocks (A. G. Harris, written communication, 1984). It is possible that the northern part of this unit never experienced the high-P episode recorded in the southern part, but available metamorphic and structural data collected by W. P. Brosgé and C. F. Mayfield during reconnaissance mapping of the region did not indicate a discrete change in the metamorphic history across the east-west strike of this unit. Future studies will undoubtedly define a more precise northern limit to the area that experienced early high-P metamorphism.

The high-P phase of the P-T path began in the low-T gl-lw stability field and evolved into the higher T ep-gt-gl stability field. Evidence for this increase in T with time consists of inclusions composed of pseudomorphs of pa and ep after lw within gt from metabasite at several localities across unit 3 (A. B. Till, oral communication, 1987). Jadeite+quartz has been identified in the Ambler River Quadrangle (Gilbert and others, 1977; Turner and others, 1979), and jdpx occurs in eclogite about 200 km to the east near Wiseman (Gottschalk, 1987). Jadeite+quartz probably formed locally during the low-T phase of the high-P metamorphism. Such formation is compatible with mineral assemblage data from the area near the eclogite, which suggest that the earliest phase of metamorphism occurred at $P > 8$ kb and $T \approx 450$°C and subsequently continued under blueschist- evolving to greenschist-facies conditions at $P < 8$ kb and $T \approx 480$°C (Gottschalk, 1987).

If much of at least the southern half of unit 3 was originally metamorphosed under blueschist-facies conditions (as suggested by the distribution of gl), then the degree of recrystallization under greenschist-facies conditions is variable, ranging from very little to almost total. Core-to-rim zoning in amphiboles from gl to ac to bar amph (Gottschalk, 1987; A. B. Till, oral communication, 1987) best records the decrease in P and increase in T experienced during this episode. The latest phase of the greenschist-facies part of the metamorphic episode produced a semipenetrative cleavage defined by the presence of aligned flakes of mu and ch and by dislocations in the earlier formed foliation; this was followed by growth of largely postkinematic helicitic ab porphyroblasts and randomly oriented bt, partial to total replacement of early formed gt by ch, and local formation of new idioblastic and unaltered gt porphyroblasts (Gilbert and others, 1977).

Low-grade rocks of the Doonerak window. Prehnite-pumpellyite–facies rocks (unit 4) and greenschist-facies rocks (unit 5), exposed in a structural window in the Mount Doonerak area of the Endicott Mountains, were also metamorphosed, but to a much lesser degree, during the widespread metamorphic episode that affected the schist belt between Middle Jurassic and mid-Cretaceous time (Dusel-Bacon and others, 1989). Both of these units consist of metasedimentary and metavolcanic rocks of Cambrian through Silurian age, and metabasite sills of Ordovician and Devonian age; unit 5 also contains an unconformably overlying sequence of Mississippian through Triassic metacarbonate and metasedimentary rocks along its northern and northeastern border (Dillon and others, 1986).

The angular unconformity between the Mississippian rocks and the underlying Devonian and older rocks in the window has been affected locally by normal faulting in the central part (Dillon and others, 1986) and thrust faulting in the eastern part (Julian and others, 1984). Structural analysis indicates that the schistosity and deformational structures in the Ordovician through Devonian rocks correspond with those in the overlying Mississippian through Triassic rocks (Julian and others, 1984). Thus, rocks both above and below the unconformity were metamorphosed during the same metamorphic episode.

The rocks of the Doonerak window are considered to be an exposure of the basement of the northern Brooks Range, and both areas are included in the North Slope terrane. Units 4 and 5 appear to have the same structures as those in the structurally overlying blueschist- and greenschist-facies rocks of unit 3 outside the window (H. G. Avé Lallemant, oral communication, 1987), suggesting that metamorphism of the basement rocks of the window probably also occurred during Middle Jurassic to mid-Cretaceous time as a result of northward overthrusting. Rocks in the eastern part of the window appear to be part of a duplex structure that formed after the earliest and most pervasive metamorphic foliation and during formation of a second generation of structures, dominated by asymmetric kink folds with northwest-dipping axial planes (Julian and others, 1984). The anomalously low (prehnite-pumpellyite facies) metamorphic grade of the structurally lowest rocks (unit 4), together with the lack of high-P minerals in units 4 and 5, are in accordance with the structural observations of Julian and her coworkers and suggest that the rocks of the window were metamorphosed under low-T and moderate- to low-P conditions late in the metamorphic episode. They were then overthrust, probably from the south, by the more deeply buried blueschist-facies rocks.

Age and tectonic origin of metamorphism. Metamorphism of unit 3 and related units of the southern Brooks Range clearly postdates the Triassic age of the youngest protoliths and probably took place in Jurassic to mid-Cretaceous time as a result of north-vergent tectonic loading. The spatial association between unit 3 and obducted oceanic rocks of the Angayucham terrane (unit 8) along the southern margin of the Brooks Range and Jurassic ultramafic and mafic rocks (Ju) in the northwestern and eastern ends of the range has been used as evidence to suggest that the rocks of the southern Brooks Range were tectonically loaded by north-directed overthrusting of oceanic rocks of Mississippian to Jurassic age along a south-dipping subduction zone (Patton and others, 1977; Roeder and Mull, 1978; Turner and others, 1979). Prior to, and probably closely preceding, its emplacement onto the continental rocks of the schist belt, the oceanic sequence was internally imbricated, and ultramafic rocks were emplaced on top of mafic rocks, becoming the structurally highest part of the sequence (Patton and others, 1977). A Middle to Late Jurassic time for this tectonic mixing is provided by K-Ar ages on hb of about 172 to 154 Ma from gt amphibolite, presumably formed during thrusting, which occurs at the base of the ultramafic sheets (Patton and others, 1977; Boak and others, 1985). Stratigraphic evidence in the western Brooks Range suggests that overthrusting of the oceanic sequence onto the continental rocks in that area began in the Middle Jurassic (Tailleur and Brosgé, 1970; Mayfield and others, 1983).

Dynamothermal metamorphism clearly had ceased by mid-Cretaceous time because Albian and Cenomanian conglomeratic rocks in the Yukon-Koyukuk basin record the uplift and progressive erosional stripping of the oceanic rocks and the underlying metamorphosed continental rocks of unit 3 (Patton, 1973; Dillon and Smiley, 1984). This timing is consistent with Early to mid-Cretaceous (120 to 90 Ma) K-Ar cooling ages on mica from unit 3 (Turner and others, 1979; Turner, 1984; Dillon and Smiley, 1984).

Obduction of the oceanic rocks apparently occurred in response to counterclockwise rotation and oroclinal bending of the lower plate continental rocks of the Arctic Alaska Plate (including the schist belt and Doonerak window) driven by rifting in the Arctic region between Early Jurassic and Early Cretaceous time (Tailleur, 1969; Mayfield and others, 1983; Grantz and May, 1983). Collision of a Jurassic and Cretaceous intraoceanic arc terrane with the southward-facing continental margin adjacent to the Yukon-Koyukuk basin has been proposed as an additional cause of the northward obduction (Box, 1985a).

Structural and metamorphic relations between the high-P schist belt and the structurally overlying, lower T and P, greenschist-facies continental rocks (unit 9) and prehnite-pumpellyite–facies oceanic rocks (unit 8) to the south suggests that post- or late-metamorphic down-to-the-south low-angle extensional faulting has dismembered the upper plate, removing much of the section that originally buried the blueschists. This late extensional phase of the orogeny has been postulated by several workers on the basis of map patterns (Carlson, 1985; Miller, 1987), field observations near Wiseman (Gottschalk and Oldow, 1988), and field and kinematic data from unit 10 in the Cosmos Hills east of Ambler (Box, 1987). Additional evidence in support of late- or post-metamorphic extensional movement between upper and lower plate rocks is found in the Cosmos Hills, where the allochthonous oceanic rocks of unit 8 cut across the metamorphic mineral zones in unit 3 (Hitzman, 1984). Continuation of extensional tectonism into Late Cretaceous, and perhaps early Tertiary, time is recorded in the deformational and metamorphic history of unit 10.

Jurassic to mid-Cretaceous low-grade metamorphism of oceanic thrust sheets. Weakly metamorphosed sedimentary and volcanic oceanic rocks crop out in a narrow V-shaped belt around the margins of the Yukon-Koyukuk basin along the southern margin of the Brooks Range and the northwestern margin of the Ruby geanticline, outboard of metamorphosed continental rocks to the north and southeast, respectively. These rocks make up the Angayucham terrane and consist of an inner and structurally lowest thrust sheet of ocean-continent transition-zone (Patton and others, this volume; W. W. Patton, Jr., and J. M. Murphy, oral communication, 1988) greenschist-facies metagraywacke and phyllite, with minor amounts of metalimestone and metachert, of Devonian to Triassic protolith age (unit 9); and an outer (basinward) overlying sheet of oceanic prehnite-pumpellyite–facies metabasite, metachert, metatuff, metalimestone, and argillite of Mississippian to Jurassic protolith age (unit 8). The structurally highest thrust sheet is of peridotite and gabbro (Ju).

These thrust sheets have been interpreted as components of an allochthonous oceanic complex, rooted in the Yukon-Koyukuk basin, that was thrust onto the Proterozoic and early Paleozoic continental margin during Middle Jurassic to mid-Cretaceous time, causing blueschist- and greenschist-facies metamorphism in the underlying rocks in the southern Brooks Range and possibly along the western margin of the Ruby geanticline (unit 28, discussed below in the section entitled Central Alaska)

(Patton and others, 1977, this volume, Chapter 7; Turner and others, 1979; Patton and Box, 1985). Some or all of the prehnite-pumpellyite–facies metamorphism of unit 8 may have accompanied thrust emplacement (Hitzman, 1983; Dusel-Bacon and others, 1989), as suggested by the following observations. (1) In the Cosmos Hills, prehnite-pumpellyite–facies metamorphism is most intense adjacent to the thrust surface between unit 8 and underlying unit 3 (Hitzman, 1983). (2) In the Ruby geanticline, gl occurs locally in metabasalt of unit 8 near the base of its tectonic contact with underlying unit 28 (Patton and Moll, 1982). The occurrence of gl indicates that some of these rocks were metamorphosed under high-P, low-T conditions and may have been tectonically intermixed with other rocks of this unit, either during the internal imbrication (emplacement of peridotite and gabbro on top of the basaltic thrust sheets) that preceded the obduction of the oceanic complex, or during the obduction itself (Patton and Moll, 1982).

The metamorphic history of unit 9 is less certain. In the Wiseman area of the southern Brooks Range, Dillon and others (1987) consider that the rocks of this unit have a single semipenetrative cleavage, in sharp contrast with the more complexly metamorphosed and deformed rocks of unit 3 to the north. However, mapping in this same sequence of rocks by Gottschalk (1987) in the Wiseman area, and Hitzman and others (1982) in the Cosmos Hills area, suggests that the rocks of this unit shared a common metamorphic history with unit 3.

Seward Peninsula

Jurassic to Early Cretaceous blueschist-evolving to greenschist-facies metamorphism. High-P blueschist-facies rocks that were partly recrystallized under intermediate-P greenschist-facies conditions crop out over a large area, 125 by 150 km (shown as unit 16), across the central Seward Peninsula. These rocks, referred to as the Nome Group, consist of pelitic schist, quartzite, marble, metabasite, mafic schist, and orthogneiss and are thought to have originated in a continental platform environment (Sainsbury and others, 1970; Till, 1983; Forbes and others, 1984). Protoliths of metasedimentary and metavolcanic rocks are Proterozoic and early Paleozoic (largely Ordovician) in age, and crosscutting orthogneiss has a Devonian intrusive age (Till and others, 1986; Armstrong and others, 1986; Till and Dumoulin, this volume).

Metamorphic minerals in the metabasite are gl, ep, gt, ab, sp, ac, ch, wm; pseudomorphs of ep, pa, and locally ab occur after lw (Forbes and others, 1984; Thurston, 1985). Glaucophane is also present in pelitic and mafic schist, and impure marble. Glaucophane-bearing eclogitic rocks have been found just east of the Nome River (Thurston, 1985). Local stabilization of eclogitic rocks is attributed to either outcrop-scale metamorphic conditions marginally different (perhaps lower P_{H_2O}) from those in the surrounding rocks (Thurston, 1985), or to small but complex variations in bulk composition (Evans and others, 1987).

Petrographic, structural, and phase-equilibrium data indicate that crystallization of the blueschist- and greenschist-facies assemblages occurred during a single episode of high-P evolving to intermediate-P, low-T metamorphism, similar to that recorded in the schist belt of the southern Brooks Range. Metamorphic conditions during the high-P phase of this monocyclic polyfacial episode started in the gl-lw-jdpx stability field and, with increasing T, evolved into the ep-gt-gl stability field (Thurston, 1985). The P-T path passed through approximately 9 to 11 kb at 400 to 450°C during the highest P phase of the prograde episode and passed into the ab-ep-bar amph field during initial stages of decompression and thermal relaxation (Forbes and others, 1984). Static retrograde alteration under greenschist-facies conditions occurred during subsequent rapid uplift. The final stages of this metamorphic episode are inferred to have taken place under intermediate-P conditions, because subsequent Cretaceous amphibolite-facies metamorphism (of unit 18), discussed below, that overprints the blueschist-facies fabric began in the ky stability field.

Metamorphism was synkinematic with penetrative ductile deformation, mesoscopic intrafolial isoclinal folding, and development of flat-lying to gently dipping transposition foliation (Thurston, 1985). Stretching lineations and isoclinal fold axes have a north-south trend (Till and others, 1986). The ubiquitous parallelism between stretching lineations and fold axes suggests a highly noncoaxial deformation, during which fold axes rotated toward the stretching direction, as noted by Patrick (1986). Quartz petrofabrics indicate northward vergence during deformation (Patrick, 1986).

Cretaceous amphibolite- and granulite-facies metamorphism. Amphibolite-facies pelitic, calcareous, and quartzo-feldspathic schist, marble, quartzite, and amphibolite (unit 18) crop out within the Kigluaik, Bendeleben, and Darby Mountains of the southern Seward Peninsula. Protoliths probably include upgraded lithologic equivalents of the Nome Group—the same protoliths as those of unit 16 (Till and others, 1986). Intermediate-P conditions are indicated by the assemblage ky-st-sil in pelitic schist; T conditions range from those of the bt zone to the sil+kf zone.

On the south flank of the Kigluaik Mountains, isograds that define a northward-increasing prograde metamorphic sequence are closely spaced (Till, 1980; Thurston, 1985), indicating a fairly steep geothermal gradient within the ky stability field. Thurston (1985) proposed that, in this area, the intermediate-P amphibolite-facies minerals were statically superimposed on pelitic rocks whose foliation developed during the widespread Jurassic blueschist-facies metamorphic episode (of unit 16) and that the ky-bearing assemblages were produced during intermediate-P thermal metamorphism associated with Cretaceous plutonism. Work by Patrick and Lieberman (1987) also indicates structural and metamorphic continuity across the contact between units 16 and 18 and supports the hypothesis that the amphibolite-facies assemblages (as well as granulite-facies assemblages, discussed below) were superimposed on (and mostly obliterated) the earlier formed blueschist-facies assemblages.

In the Bendeleben and Darby Mountains, amphibolite-facies

metamorphism was more dynamothermal in character and produced ky-bearing assemblages that define a penetrative fabric. These assemblages are overprinted by static, low-P, high-T assemblages that apparently formed as a result of thermal metamorphism associated with Late Cretaceous (80 Ma) plutonism (Till and others, 1986).

In one area of the Kigluiak Mountains, high-T rocks, whose metamorphism is inferred to have been associated with the thermal episode that culminated in the intrusion of 80-Ma plutons (A. B. Till, oral communication, 1987), form an area large enough to show as a separate unit (unit 19). This unit consists of upper amphibolite-facies bt gneiss; granulite-facies marble, pelitic, quartzofeldspathic, and mafic gneiss, and migmatite; and gt lherzolite partially recrystallized to sp lherzolite (Till, 1980, 1983). Protoliths are assumed to include upgraded lithologic equivalents of the Nome Group (Till and others, 1986). Relict ky inclusions within gt formed during granulite-facies metamorphism indicate that granulite-facies metamorphism postdated the intermediate-P amphibolite-facies metamorphism (A. B. Till, oral communication, 1987).

Age and tectonic origin of blueschist-, greenschist-, amphibolite-, and granulite-facies metamorphism. Blueschist-facies metamorphism was apparently caused by rapid tectonic loading of a continental plate (Forbes and others, 1984). Rb-Sr whole-rock-mica isochron ages and K-Ar mineral ages suggest that the high-P metamorphic cycle took place during the Middle or Late Jurassic, before about 160 Ma, followed by decompression and partial reequilibration between about 160 and 100 Ma (Armstrong and others, 1986). Similarities in protoliths, metamorphic grade, structural style, and apparent metamorphic age suggest a correlation between the high-P, low-T metamorphic and tectonic history of unit 16 and the schist belt of the southern Brooks Range (Armstrong and others, 1986; Patrick, 1986; Dusel-Bacon and others, 1989). By analogy with the proposed history of the schist belt, multiple thrust sheets of oceanic rocks (Angayucham terrane) may have once covered the Seward Peninsula (Till, 1983; Forbes and others, 1984). The nearest possible remnant of that oceanic terrane is a sliver of north-south–trending blueschist-facies rocks (unit 17, discussed below) that crops out on the eastern Seward Peninsula; other possible remnants have been proposed by Box (1985a).

A major difference in the subsequent metamorphic histories of the two areas, however, is the subsequent occurrence of moderate- to high-T metamorphism and plutonism on the Seward Peninsula and the absence of these thermal episodes in the schist belt. The rapid change from blueschist metamorphism to intermediate-P amphibolite-facies metamorphism and plutonism is similar to that described in the southern Aegean by Lister and others (1984). Intermediate-P metamorphism in the Aegean was synkinematic with extensional deformation. It was probably driven by gravitational spreading, following the compressional (blueschist) phase of tectonism. An alternative comparison is made by Patrick and Lieberman (1987) who compare the sequence of metamorphic and plutonic episodes on the Seward Peninsula to that observed in the Central Alps. They attribute the thermal overprinting to relaxation of isotherms following subduction, leading to the onset of crustal anatexis. Because no evidence of extensional faulting has been identified on the Seward Peninsula, the tectonic history of the Central Alps appears to be a better analog than does that of the southern Aegean.

The environment of crystallization of the gt lherzolite of unit 19 is unknown. Textural relations indicate that gt was stable in the lherzolite prior to granulite-facies metamorphism, during which time the sp-bearing assemblages apparently formed (Till, 1980; Lieberman and Till, 1987). The gt-bearing assemblage may be a relict of an upper mantle environment or may have been metamorphosed in a deep crustal setting. In the latter case, formation of the gt either occurred during a pre-Mesozoic event, or during the early phases of the Jurassic and Cretaceous metamorphic episode of the schist belt and Seward Peninsula, simultaneous with formation (at shallower levels) of the extensive blueschist-facies terranes (Lieberman and Till, 1987).

Probable Jurassic to Early Cretaceous blueschist-facies metamorphism. Blueschist-facies mylonitic metabasite and minor amounts of serpentinite crop out in a narrow fault-bounded belt (unit 17) along the east side of the Darby Mountains (Till and others, 1986). Protoliths are considered to range in age from middle Paleozoic to Jurassic on the basis of a tentative correlation with the low-grade oceanic rocks of the Angayucham terrane (unit 8) that are present around the margins of the Yukon-Koyukuk basin. Mylonitic metabasite in the northern part of the belt contains the assemblage cr-lw-pu, which is indicative of the low-T subdivision of the blueschist facies; mylonitic metabasite in the southern part of the belt contains the assemblage cr-ep-ac, indicative of the high-T (epidote-bearing) subdivision of the blueschist facies (subdivisions of Taylor and Coleman, 1968; and Evans and Brown, 1987). Relict igneous pyroxene grains are common in mylonitic metabasite in both areas (Till, 1983). The presence of cr and lw in this unit indicates that pressures probably occurred within the lower part of the range of P conditions in the nearby and more extensive unit 16. Somewhat different crystallization and deformational histories for the two units are indicated, however, by the incomplete recrystallization and brittle deformation of this unit compared with the complete recrystallization and ductile deformation of unit 16 (Till and others, 1986).

A middle Jurassic to mid-Cretaceous metamorphic age is assigned to unit 17 on the basis of correlation of its metamorphic history with that of the Angayuchum terrane. Arguing against this correlation is the widespread development of high-P minerals in unit 17 and the general absence of high-P minerals in the Angayucham terrane.

Central Alaska

Pre-middle Paleozoic greenschist- and amphibolite-facies metamorphism. Metamorphic units 22 and 23 crop out southeast of the Susulatna fault and were metamorphosed sometime during Proterozoic to middle Paleozoic time. The more extensive unit (22) consists of greenschist-facies Late Proterozoic felsic metavolcanic rocks (Dillon and others, 1985) and pre-Ordovician schist, quartzite, phyllite, argillite, marble, and mafic metavolcanic rocks (Silberman and others, 1979; Patton and others, 1980 and this volume, Chapter 7). Metamorphic grade is mostly of the ch and bt zones, but locally reaches the gt zone. Pre-Ordovician metamorphic, as well as protolith, ages are indicated for this unit because it is overlain by virtually unmetamorphosed Ordovician through Devonian strata that yield conodont-alteration indices that correspond with very low temperatures—generally less than 200°C (A. G. Harris, written communication, 1984). A minimum metamorphic age of 514 Ma is provided by the oldest of three K-Ar ages on mica from qz-mu-ch schist within this unit (Silberman and others, 1979). K-Ar and U-Pb data suggest that these rocks were not affected by the Late Jurassic and/or Early Cretaceous metamorphic episode that occurred northwest of the Susulatna fault in the Ruby geanticline (Dillon and others, 1985).

The other pre-middle Paleozoic metamorphic unit (23) is limited in area. It consists of polymetamorphosed and locally mylonitic pre-Ordovician schist, sheared grit, quartzite, phyllite, mafic and felsic metavolcanic rocks, and schistose metaplutonic rocks (Dusel-Bacon and others, 1989). Metaigneous rocks give Middle Proterozoic protolith ages (Silberman and others, 1979; Dillon and others, 1985). Polymetamorphism is suggested by replacement textures in pelitic rocks. The M_1 episode (or alternatively, the maximum-T phase of a P-T loop) occurred under amphibolite-facies conditions and produced the assemblage bt + gt ± st ± co in qz-mica schist. Subsequent retrograde metamorphism (M_2, or alternatively, a late phase of M_1) resulted in the almost complete replacement of st by ctd, gt by ch, and co by wm and ch. Textural evidence suggests that retrogressive greenschist-facies metamorphism was accompanied by shearing (Dusel-Bacon and others, 1989).

A maximum metamorphic age for both postulated metamorphic episodes is indicated by the Middle Proterozoic protolith ages; a middle Paleozoic minimum metamorphic age for the episodes is tentatively provided by the U-Pb lower intercept age on zircon (390 ± 40 Ma) from the metavolcanic rocks (Dillon and others, 1985) and by the virtually unmetamorphosed overlying Ordovician through Devonian strata. A Late Proterozoic (663 Ma) K-Ar age on mu from recrystallized mylonite along the border of metaquartz diorite (Silberman and others, 1979) may date the time of uplift and cooling, following retrogressive metamorphism of the country rocks and metamorphism and shearing of the plutonic body.

Pre-Early Cretaceous greenschist- and amphibolite-facies metamorphism. Pre-Early Cretaceous metamorphism affected four monometamorphic units (24, 25, 28, 30) and two polymetamorphic units (26, 27) in or near the Ruby geanticline. All of the units are considered to be part of the Ruby terrane and contain continental sedimentary, volcanic, and plutonic protoliths of Proterozoic and/or Paleozoic age. Intermediate- to high-P metamorphism of unit 28, discussed in a later section, is interpreted (Dusel-Bacon and others, 1989) to have taken place during the Mesozoic obduction of oceanic thrust sheets onto the continental margin. Timing of metamorphism(s) in the other units is more uncertain, and in most areas is known only to predate the intrusion of regionally extensive plutons that have yielded Early Cretaceous (about 110 Ma) K-Ar ages (Silberman and others, 1979; Patton and others, 1987).

Little is known about the metamorphic history of monometamorphic greenschist- and epidote-amphibolite-facies unit 24 and the higher grade areas of amphibolite-facies rocks (unit 25) within it. In the northeastern exposure of unit 25, some of the amphibolite-facies minerals may have been produced by thermal metamorphism caused by the adjacent Early Cretaceous plutons. In the southwestern exposure of unit 25, however, foliation of the amphibolite-facies rocks trends northwestward, subperpendicular to regional trends of thrust fault traces and plutons; metamorphism in this area clearly predates, and is unrelated to, the intrusion of Cretaceous or early Tertiary plutons, most of which are too small to be shown on Plate 4A (G. M. Smith, written communication, 1986).

The third monometamorphic unit (30) consists of lower amphibolite-facies rocks that crop out in a small area north of the Iditarod–Nixon Fork fault. These rocks include amphibolite, orthogneiss, pelitic schist, and quartzite (Miller and Bundtzen, 1985) and were informally designated as the Idono sequence by Gemuts and others (1983). U-Pb data on zircon from orthogneiss indicate an Early Proterozoic age for their granitoid protolith (M. L. Miller and T. W. Stern, unpublished data, 1987). K-Ar dates on hb from amphibolites include both Middle Proterozoic (1.22 and 1.08 Ga) and Early Cretaceous (126 Ma) ages; K-Ar dates on bt from amphibolite are about 324 Ma and 133 Ma (Miller and Bundtzen, 1985; M. L. Miller, written communication, 1986).

Polydeformed and polymetamorphosed metasedimentary and metaigneous rocks (unit 26) crop out in the Ray Mountains (Dover and Miyaoka, 1985a, b). This unit consists of quartzofeldspathic paragneiss, augen gneiss (shown as a metamorphosed pluton), schist, gneiss, marble, quartzite, phyllonite, metabasite, and amphibolite. Only the protolith age of the augen gneiss (Devonian; Patton and others, 1987) is known. Dover and Miyaoka (1985a, b) proposed that the unit experienced at least three deformational episodes and two major metamorphic episodes.

M_1 occurred primarily under amphibolite-facies conditions. It was synkinematic with ductile deformation, and produced blastomylonitic fabrics that are axial planar to isoclinal folds. Isoclines produced during the M_1 episode are overprinted by a second generation of folds that have an axial planar cleavage (S_2) at a low to moderate angle to the older schistosity. S_2 folds are tight and increase in abundance toward cataclastic zones in which a cataclastic foliation is the dominant fabric; this fabric appears to coincide with the S_2 cleavage. S_2 structures are invariably associated with greenschist-facies minerals attributed to a retrogressive metamorphic episode (M_2). Within intensely cataclasized phyllonite zones, M_2 minerals replace gt, bt, and st that were produced during M_1 metamorphism; M_2 minerals also grew synkinematically along shear surfaces. Glaucophane occurs with ctd in an M_2 mineral assemblage at one locality within a phyllonite zone (Dover and Miyaoka, 1985b; Plate 4A).

A similar sequence of polydeformed, polymetamorphosed, and moderately to strongly mylonitized schist, gneiss, quartzite, marble, and amphibolite, and singly metamorphosed granitoid gneiss, including augen gneiss (unit 27), crops out in the Kokrines Hills (Patton and others, 1978). The protolith age of the granitoid gneiss is unknown, but on the basis of lithologic similarity with dated augen gneiss in the Ray Mountains (unit 26) and in the southern Brooks Range (unit 2), a Devonian age is likely. I propose a tentative polymetamorphic history for this unit on the basis of the following field observations and interpretations made by J. T. Dillon (written communication, 1983): (1) metamorphic foliation (S_1) in quartzite and associated schist and gneiss is truncated by the granitoid gneiss whose foliation (S_2) is approximately perpendicular to the intrusive contact and locally to S_1; (2) S_1 foliation formed during an earlier metamorphic episode (M_1), and S_2 foliation and cleavage formed during a later metamorphic episode (M_2); and (3) M_1 produced bt, gt, sil, and locally ky and kf in pelitic rocks, and M_2 produced gt, mu, and bt in granitoid gneiss. Broad structural relations suggest that this unit may form an east-northeast–trending gneiss dome (Patton and others, 1978), but the age and origin of doming or remobilization are unknown.

Age and tectonic origin of pre-Early Cretaceous metamorphic episodes. Proterozoic and/or Paleozoic metamorphism may have occurred in several of the areas. The Idono sequence (unit 30) yields both Middle Proterozoic and Early Cretaceous mineral ages, which may indicate a correlation between its metamorphic history and that proposed for polymetamorphic unit 31 in southwestern Alaska. As discussed in a later section, metamorphism of unit 31 apparently took place during both Early Proterozoic and Jurassic to Cretaceous time. Amphibolite-facies metamorphism (M_1 of unit 27) in the Kokrines Hills may have predated the probable Devonian intrusive age of granitic gneiss, whose presumed metamorphic fabric is reported to crosscut the S_1 fabric of the rest of the unit. Similar orthogneiss, forming the structurally lowest thrust sheets in the Ray Mountains, may also have experienced a pre- or syn-Devonian metamorphic episode (M_1 of unit 26).

Obduction of mafic-ultramafic oceanic rocks onto the Proterozoic and middle Paleozoic continental margin (area of the southern Brooks Range and Ruby geanticline) during Middle Jurassic to mid-Cretaceous time is the most likely cause of the greenschist- and epidote-amphibolite–facies metamorphism of unit 24, and the greenschist-facies M_2 metamorphism of unit 26. The occurrence of gl in an M_2 retrogressive greenschist-facies assemblage, and the increased development of cataclastic fabrics and retrogressive metamorphism in shear zones in the Ray Mountains (unit 26), supports the overthrust origin proposed above. This tectonic model is the same one that is more clearly indicated for unit 28 (discussed below).

Mesozoic low-grade, locally high-P metamorphism. Greenschist-facies continental rocks (unit 28) and prehnite-pumpellyite–facies oceanic rocks (unit 29) were metamorphosed during an Early Cretaceous or older Mesozoic episode that occurred locally under high-P (blueschist-facies) conditions. These units are exposed within and east of both the Kokrines Hills and the Kaiyuh Mountains. The units in both areas are correlative and have been offset by approximately 160 km of right-lateral movement along the Kaltag fault (Patton and others, 1984). The greenschist-facies rocks (unit 28) make up a basement assemblage that consists of schist, quartzite, phyllite, slate, and mafic metavolcanic rocks of Proterozoic(?) and Paleozoic age and recrystallized limestone, dolomite, and chert of early to middle Paleozoic age. This unit is defined by the local presence of gl in pelitic schist (+ wm + qz ± gt ± ch ± ctd) and metabasite (+ ch + ab + ep ± ac). It, like the similar undifferentiated greenschist- and epidote-amphibolite–facies unit (24) that is devoid of gl, is included in the Ruby terrane.

The prehnite-pumpellyite–facies oceanic rocks (unit 29) occur as large thrust sheets (Patton and others, 1977 and this volume, Chapter 21) composed of metabasite, metachert, metasedimentary rocks, metalimestone, and metatuff. Protoliths range in age from Late Devonian to Late Triassic. The northwesternmost thrust sheets are included in the Tozitna terrane, and the southeasternmost thrust sheets in the Innoko terrane. The degree of low-T metamorphism varies considerably within the unit and appears to be a function of the structural position within the thrust sheet (W. W. Patton, Jr., and S. E. Box, oral communication, 1985; Patton and others, this volume, Chapter 21). Glaucophane occurs locally in metabasite near the structural base of the Tozitna terrane (Patton and Moll, 1982) where localized higher P conditions may have existed.

The intermediate- to locally high-P metamorphism of units 28 and 29 resulted presumably from tectonic loading accompanying the obduction of thrust sheets of oceanic rocks onto the Proterozoic and early Paleozoic continental margin. The primary evidence for this model is the occurrence of gl at the base of the oceanic thrust sheets (the northwestern exposures of unit 29, assigned to the Tozitna terrane, and unit 8, assigned to the Angayucham terrane and discussed previously), as well as in the underlying continental greenschist-facies rocks of unit 29. Patton and others (this volume, Chapter 21) present stratigraphic evidence that the obducted oceanic thrust sheets assigned by Jones and others (1987) to the Tozitna and the Angayucham terranes are parts of a single terrane.

The direction from which these oceanic terranes were thrust, and therefore correlation of their metamorphic histories, is a matter of some debate. According to one hypothesis (discussed earlier), the thrust sheets of a (composite) Angayucham-Tozitna terrane were rooted in the Yukon-Koyukuk basin and thrust southeastward over the continentally derived rocks of the Ruby geanticline (Patton and others, 1977 and this volume, Chapter 21; Patton and Moll, 1982; Turner, 1984). According to an alternative hypothesis, based on structural analysis of S-C fabrics (non-coaxial schistosity and shear surfaces) and the sense of rotation of large-scale nappe-like folds (Miyaoka and Dover, 1985; Smith and Puchner, 1985; G. M. Smith, written communication, 1986), the Tozitna terrane was thrust in the opposite direction—from the southeast toward the northwest. Dover and Miyaoka (1985b) attribute the development of cataclastic fabrics and accompanying retrogressive metamorphism within a part of the lower plate rocks (M_2 of unit 26) to the northwestward obduction of the Tozitna terrane that lies to the south.

Middle Jurassic and mid-Cretaceous maximum and minimum metamorphic age constraints for metamorphism caused by southeastward thrusting out of the Yukon-Koyukuk basin are discussed in the Brooks Range section. Metamorphism, if caused by northwestward obduction of the Tozitna terrane, would have to postdate the Triassic age of the youngest protoliths of that terrane. Metamorphism of units 28 and 29, prior to late Early Cretaceous time, is also indicated by K-Ar ages of 134 and 136 Ma on metamorphic mu from gl-bearing schist (unit 28) in the Kaiyuh Mountains (Patton and others, 1984), and by the 111-Ma age of a granitoid pluton that intrudes both the Ruby and Angayucham terranes in the Kokrines Hills (Patton and others, 1977, 1978).

Southwestern Alaska

Early Proterozoic and amphibolite-facies metamorphism. The oldest dated metamorphic episode in Alaska is recorded in a narrow, northeast-trending, fault-bounded belt of continentally derived amphibolite-facies rocks east of Kuskokwim Bay. These rocks, shown as unit 31, form the antiformal (informal) Kanektok metamorphic complex of the Kilbuck terrane and are composed of bt + hb + gt ± cpx gneiss, gt-mica schist, orthogneiss, gt amphibolite, and rare marble (Hoare and Coonrad, 1979; Turner and others, 1983; Decker and others, this volume). Kyanite, indicative of intermediate- to high-P conditions, occurs in gt-mica schist at one locality. Protoliths are Early Proterozoic sedimentary, mafic volcanic, and granitic rocks (Turner and others, 1983). Metamorphic mineral grains generally define a strong lineation and a foliation that is parallel to compositional layering. All of these structural features strike consistently to the northeast, roughly parallel to the trend of the complex (Hoare and Coonrad, 1979; D. L. Turner, written communication, 1982). On the basis of aeromagnetic, gravity, and field data, the structural setting of this metamorphic complex (Kilbuck terrane) has been interpreted as a rootless subhorizontal klippe (Hoare and Coonrad, 1979) or, alternatively, as a block extending to an unknown depth between two southeast-dipping thrust faults (Box, 1985c).

A 1.77-Ga (Early Proterozoic) metamorphic age for amphibolite-facies metamorphism is indicated by a U-Pb age on sph from orthogneiss, by the oldest of five Proterozoic K-Ar hb ages from amphibolite, and possibly also by a whole-rock Rb-Sr "isochron" (Turner and others, 1983). A minimum age for this metamorphic episode is provided by a 1.2-Ga age from $^{40}Ar/^{39}Ar$ incremental heating studies on hb from gt amphibolite (Turner and others, 1983).

On the basis of isotopic data, subsequent greenschist-facies retrogressive metamorphism (M_2) is proposed to have affected the amphibolite-facies rocks during Late Jurassic to Early Cretaceous time. Nearly all of the 58 dated rocks collected throughout the metamorphic complex show a total or partial resetting of K-Ar hb and bt ages and fall in the range of 150 to 120 Ma (D. L. Turner, written communication, 1982; Turner and others, 1983). Because only a limited and cursory petrographic study of the amphibolite-facies rocks has been made, it is not known to what degree, and under what T and P conditions, recrystallization accompanied the Mesozoic thermal episode that is documented by the isotopic dating. A latest Early Cretaceous minimum age for M_2 is indicated because: (1) overlying unmetamorphosed Albian conglomerates contain Kanektok components, and (2) unmetamorphosed Valanginian sediments to the south contain metamorphic gt and ep thought to be derived from the metamorphic complex (Murphy, 1987).

Unit 31 is similar in its lithology, internal structure, structural relation to adjacent mafic complexes, and Mesozoic K-Ar ages to the metamorphic belt that occurs along the southeastern borderland of the Yukon-Koyukuk basin to the north (Ruby terrane) discussed in the previous section (Box, 1985a; Patton and others, this volume, Chapter 7). The part of this belt closest to unit 31 is the informally designated Idono sequence (unit 30) 250 km to the northeast (Miller and Bundtzen, 1985)—a sequence that also gives Proterozoic and Early Cretaceous metamorphic-mineral ages. Several linear northeast-trending faults separate the two metamorphic units. One of these, the Iditarod–Nixon Fork fault, shows evidence of about 110 km of post-Cretaceous right-lateral offset (Grantz, 1966). If the other faults show similar senses of displacement, units 30 and 31 may prove to be right-lateral offset equivalents, making both part of the Ruby terrane (Dusel-Bacon and others, 1994c).

Low-grade, locally high-P Mesozoic metamorphism of oceanic rocks. The predominant period of metamorphism in southwestern Alaska occurred under low-grade conditions during Mesozoic time. Most of the rocks that were metamorphosed are of oceanic affinity; the continentally derived Early Proterozoic metamorphic rocks of unit 31 apparently were retrograded during the Mesozoic. Blueschist-facies rocks occur in two areas (units 33, 34) of the low-grade rocks, suggesting a subduction-related origin for most of the low-grade metamorphism.

One of the areas of blueschist-facies rocks (unit 33) is part of a nappe complex of predominantly schistose metavolcanic and metasedimentary rocks; other rocks in the nappe complex were metamorphosed under pumpellyite-actinolite–facies (facies terminology of Turner, 1981) to greenschist-facies conditions (unit 32). Protoliths are thought to be of Permian and Late Triassic age (Box, 1985b). Recrystallization is generally incomplete and primary textures and minerals are partly preserved. Diagnostic high-P minerals gl (+ ep + ac ± pu) and lw are sparse and poorly developed in unit 33 (Box, 1985b).

Metamorphism of units 32 and 33 is bracketed between the Late Triassic age of the youngest protolith and the Middle Jurassic age of postmetamorphic mafic and ultramafic plutons (Hoare and Coonrad, 1978). A 231 ± 7-Ma K-Ar age on amph (Box, 1985b) from schist of blueschist-facies unit 33 suggests that metamorphism of the nappe complex may have begun during Late Triassic time.

The other area of greenschist- and locally blueschist-facies rocks, shown as unit 34, consists of mafic schist interlayered with metachert, metalimestone, phyllite, and minor amounts of quartzose and calcareous schist (Hoare and Coonrad, 1959). Protolith ages are unknown, but limestone is probably Ordovician, Devonian, or Permian in age, and chert may be as young as latest Jurassic (Tithonian; Box, 1985b). Rocks of this unit have been affected by postmetamorphic, northwest-vergent imbrication (Box, 1985c). Greenschist-facies mafic schist is characterized by ch, ep, and ac, and blueschist-facies mafic schist by gl, ch, ac, and ep.

Actinolite from mafic schist of the northernmost exposure of unit 34 gives a K-Ar age of 146 ± 15 Ma (Box and Murphy, 1987), suggesting a Late Jurassic or Early Cretaceous minimum metamorphic age. The Triassic maximum age of metamorphism of units 32 and 33 is tentatively considered to apply also to unit 34.

The most extensive low-grade metamorphic unit (35) forms a diverse assemblage of prehnite-pumpellyite–facies metabasalt, metagabbro, metavolcaniclastic and metasedimentary rocks, metachert, and metalimestone. Protolith ages range from early or middle Paleozoic to Cretaceous (Silberling and others, this volume; Jones and others, 1987; Dusel-Bacon and others, 1994c). Primary igneous or depositional fabrics are generally well preserved, but locally the rocks are slaty, schistose, or highly sheared and disrupted. The lack of structural fabric, the disrupted character, and the very low grade of this unit make it difficult both to determine which rocks have been metamorphosed and to assess the relation between the timing of metamorphism and the intrusion of crosscutting igneous bodies. A pre-Early Cretaceous age of metamorphism is inferred for all areas of unit 35 because unmetamorphosed Valanginian andesitic volcanic rocks unconformably overlie the northernmost exposure of this unit (Patton and Moll, 1985), and unmetamorphosed Valanginian sedimentary rocks in the southern exposure of the unit contain pr-pu-bearing metavolcanic clasts (Murphy, 1987).

Mesozoic metamorphism is attributed to progressive underthrusting of a composite subduction complex (Goodnews terrane) beneath the northwestern margin of an oceanic arc (Togiak terrane), followed by underthrusting of the Early Proterozoic continental metamorphic complex (Kilbuck terrane) beneath the northwestern margin of Goodnews terrane (Fig. 3) (Box, 1985b, c).

According to this tectonic model, metamorphism of blueschist- and greenschist-facies units 33 and 32, respectively, occurred during the first episode of underthrusting. These two units make up the Cape Peirce subterrane of the Goodnews terrane of Box (1985b, c). Box believes this subterrane structurally underlies the prehnite-pumpellyite–facies rocks of the Togiak terrane and overlies the prehnite-pumpellyite–facies rocks of the Platinum subterrane of the Goodnews terrane (terranes and subterranes are those of Box, 1985c; also see Decker and others, this volume) along low-angle southeast-dipping faults (Fig. 3). Both areas of prehnite-pumpellyite–facies rocks are included in unit 35. Metamorphism of the Cape Peirce subterrane is presumed to have occurred during collision and partial subduction of an oceanic plateau (Platinum subterrane of the Goodnews terrane) beneath an overriding intraoceanic subduction-related volcanic arc (Togiak terrane). Lithologic similarities between the protoliths of the schistose blueschist- and greenschist-facies rocks of the Cape Peirce subterrane and those of the relatively undeformed low-grade overlying and underlying subterranes, suggest that the rocks of the Cape Peirce subterrane are the more tectonized equivalents of the adjacent two subterranes (Box, 1985b). Mafic and ultramafic plutons that intrude the Cape Peirce subterrane, the overlying Togiak terrane, and the underlying Platinum subterrane, provide a Middle Jurassic minimum age for amalgamation of the three subterranes.

Structural data suggest that the overriding arc of the Togiak terrane was originally thrust to the north-northeast over the Goodnews terrane (Box, 1985b). However, the low-angle fault mapped between the upper plate Togiak terrane and the underlying Cape Peirce terrane juxtaposes lower T and P rocks over higher T and P rocks, suggesting that the fault is low-angle normal fault rather than a thrust fault (Box, 1985b). As suggested by Box, a good explanation for the present relation between the plates is that early north-northeastward compressional faulting was followed by extensional (detachment) faulting. This same fault relation (lower grade rocks above higher grade rocks) occurs in the southern Brooks Range and Ruby geanticline (Plate 4A); faulting in all these areas may have the same origin (extensional reactivation of earlier compressional structures).

The greenschist- and blueschist-facies rocks of unit 34 were probably metamorphosed during the second episode of underthrusting. These rocks crop out along the northwestern margin of the Nukluk subterrane of the Goodnews terrane of Box (1985c). Late Jurassic to Early Cretaceous metamorphism of unit 34, and retrograde metamorphism of unit 31, probably took place as the continental Kilbuck terrane (unit 31) was partially thrust beneath the accretionary fore arc (Goodnews terrane) of an intraoceanic volcanic arc (Togiak terrane) (Box, 1985c). The following evidence supports this interpretation of the metamorphic history: (1) unit 34 occurs along the tectonic boundary between the Kilbuck and Goodnews terranes, and (2) the K-Ar age on ac from unit 34 falls in the same range as the Mesozoic K-Ar ages from the Kilbuck terrane.

Yukon-Tanana Upland and Alaska Range north of the McKinley and Denali faults

Overview. The age and origin of metamorphism of many of the units in the Yukon-Tanana upland and northern Alaska Range are poorly known. Metamorphism throughout the Yukon-Tanana upland predates the widespread intrusion of undeformed mid-Cretaceous granitoids. Mylonitic and blastomylonitic textures are common in most rocks, reflecting a history of dynamic metamorphism, followed by varying degrees of more static re-

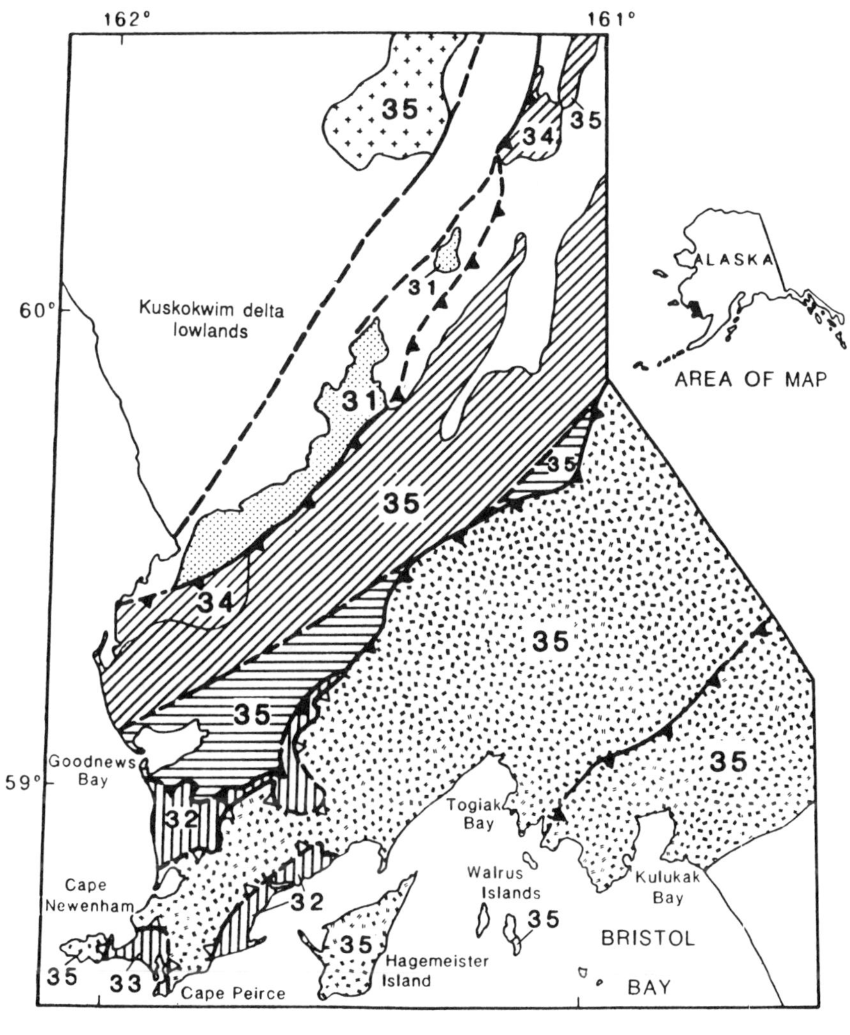

EXPLANATION

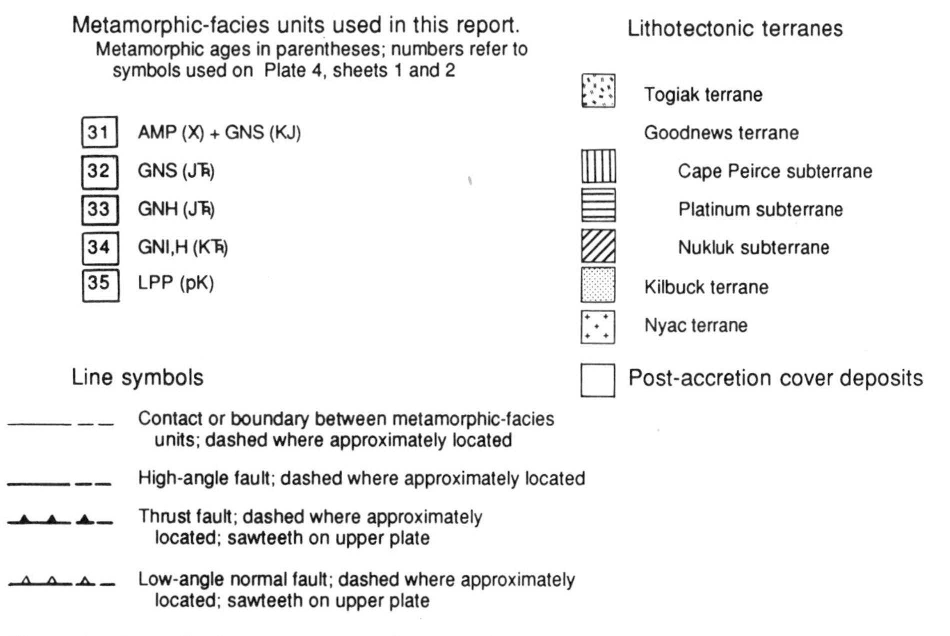

Figure 3. Generalized map showing lithotectonic terranes (modified from Box, 1985c) and metamorphic-facies units in southwestern Alaska.

crystallization. Many metamorphic unit boundaries are also terrane or subterrane boundaries defined by low-angle faults. Metamorphic grade changes abruptly across many of the faults, indicating that major metamorphism predated final emplacement of the fault-bounded units. Some of the low-angle faults place higher grade over lower grade rocks (the relation expected of compressional faulting), whereas others place lower grade over higher grade rocks (the relation expected of extensional faulting), suggesting a complex syn- or post-metamorphic structural evolution. In their regional synthesis, Foster and her co-workers (this volume) interpret the low-angle faults as being south-dipping thrusts. Thrust sheets near the Canadian border were emplaced in Early Jurassic time. Thrusting of other sheets may have occurred throughout a compressional episode that included crustal thickening and metamorphism in either or both Jurassic and/or Cretaceous time. In one area, however, kinematic data indicate that the most recent movement along the low-angle fault that separates greenschist-facies unit 45 of the hanging wall from amphibolite-facies unit 46 of the footwall was extensional, and that the hanging wall moved to the east-southeast (Pavlis and others, 1988b).

Early Paleozoic(?) high-P, high-T metamorphism. The oldest metamorphic episode inferred to have taken place in east-central Alaska produced eclogite and interlayered bands and lenses of mafic, calc-magnesian, quartzitic, pelitic schist, and impure marble. These rocks, shown as unit 36, are thought to be two klippen that form the upper plate of a folded thrust (Brown and Forbes, 1984; Foster and others, 1983). Protoliths are Proterozoic or early Paleozoic in age. Bulk chemistry suggests sedimentary protoliths for the eclogites of the western klippe (Swainback and Forbes, 1975), whereas discordant contacts exhibited by eclogitic rocks in the eastern klippen suggest that they originated as mafic dikes (Laird and others, 1984).

Eclogite within the eastern klippe consists of various combinations of gt, omphacitic cpx, barroisite, phengitic mu, qz, and rutile (Laird and others, 1984). Eclogite within the western klippe consists of combinations of these minerals plus gl, tr, ab, ep, and sph; ky + st + gt occurs in interlayered pelitic schist (Brown and Forbes, 1986). Phase equilibria (excluding that proposed for gl) and thermobarometry suggest P-T conditions of about 15 ± 2 kb and 600 ± 25°C (Laird and others, 1984; Brown and Forbes, 1986). Such temperatures exceed the experimentally determined maximum stability limit of gl (Maresch, 1977) by about 50°C. One explanation for this discrepancy is that the gl formed after the main phase of eclogite crystallization, as was proposed for an eclogitic block along the Tintina fault in Yukon Territory by Erdmer and Helmstaedt (1983). However, textural relations in the gl-bearing rocks of the western klippe show no evidence of such a progression, and the discrepancy is thus far unresolved.

A polymetamorphic history for this unit is suggested, because some eclogitic rocks in both klippen are overprinted by greenschist and epidote-amphibolite assemblages, which are characterized by hb, ab, ep, bt, and ch (Brown and Forbes, 1986; Foster and others, this volume).

An early Paleozoic metamorphic age for the first, and dominant, episode is tentatively suggested by a 470 ± 35-Ma K-Ar age on amph from eclogite in the western klippe (Swainbank and Forbes, 1975). Early isoclinal, recumbent folds about a northwest-trending axis are attributed to this episode. This high-P, high-T metamorphism only affected this group of thrust-bounded rocks and thus predated the time of their emplacement. An Early Cretaceous age for subsequent retrograde metamorphism is proposed on the basis of 115- to 103-Ma K-Ar ages on mica that were determined for associated rocks; the development of folds about a northeast-trending axis is attributed to this metamorphic episode (Swainbank and Forbes, 1975). Early Cretaceous metamorphism, discussed below, was widespread throughout east-central Alaska.

Mineral chemistry and the occurrence of gl indicate that the eclogites are similar to those from alpine-type orogenic environments (Group C of Coleman and others, 1965). The tectonic and metamorphic history of the eclogites may be similar to that of other isolated eclogite occurrences on strike along the Tintina fault in the Yukon Territory, as proposed by Erdmer and Helmstaedt (1983) and Brown and Forbes (1986). Eclogites in central Yukon Territory experienced a subduction-cycle P-T trajectory that included eclogite metamorphism, uplift through the stability field of gl, and finally greenschist-facies metamorphism (Erdmer and Helmstaedt, 1983). The present distribution and geologic position of the eclogite bodies in east-central Alaska and Yukon Territory suggest that rocks in both regions were emplaced as thrust sheets against or onto the cratonic margin of western North America (Erdmer and Helmstaedt, 1983). However, the timing of metamorphism and tectonic emplacement of these eclogite-bearing terranes are not well enough constrained to allow more than a tentative correlation. The early Paleozoic metamorphic age suggested for the Alaskan eclogitic rocks, if valid, would argue against correlation with the Canadian eclogites, which are believed to have middle Paleozoic protoliths and to have been metamorphosed during the late Paleozoic (Erdmer and Armstrong, 1988).

Late Triassic to Early Jurassic metamorphism. Metamorphism during this period was part of an orogenic episode that consisted of metamorphism, plutonism, folding, and thrusting. This episode resulted from the closing of an ocean basin, represented by the weakly metamorphosed rocks of unit 38 and associated ultramafic rocks (MzPzu). Accretion of amphibolite-facies rocks of unit 37 was related to this episode (Foster and others, 1985, this volume).

The earliest phase of this orogenic episode produced intermediate-P amphibolite- and epidote-amphibolite–facies bt gneiss and schist, amphibolite, marble, quartzite, and pelitic schist of unit 37. Rocks are well foliated and multiply deformed; at least some protoliths are of Paleozoic age (Foster and others, 1985). This unit is intruded by latest Triassic to earliest Jurassic granitoids, similar to those in the Stikinia terrane of Yukon Territory, Canada (Tempelman-Kluit, 1976). Unit 37 is in thrust contact with adjacent rocks. It is probably part of the Stikinia terrane or a comparable, but different, part of composite terrane I of Monger and others (1982) that includes the Stikinia terrane and that was accreted to the margin of North America in Jurassic time (Foster and others, this volume).

Amphibolite-facies metamorphism of unit 37 reached its peak about 213 ± 2 Ma ($^{40}Ar/^{39}Ar$ integrated plateau age on amph), followed by synmetamorphic intrusion of the Taylor Mountain batholith (shown as unit JTRg) at about 209 ± 3 Ma (Cushing and others, 1984). Northward thrusting of the amphibolite-facies rocks and low-grade oceanic rocks (unit 38) took place during cooling and was completed by about 185 Ma (Foster and others, this volume). The Early Cretaceous thermal event that strongly affected the adjacent augen gneiss-bearing amphibolite-facies unit 46 to the south (described below) was of only minor intensity in unit 37, as indicated by the Early Cretaceous apparent ages of the low-temperature gas fractions in most of the $^{40}Ar/^{39}Ar$ age spectra (Foster and others, this volume).

Basalt and related oceanic protoliths of unit 38 were metamorphosed under prehnite-pumpellyite– and lower greenschist-facies conditions during the waning stages of the regional metamorphism that accompanied the closing of the ocean basin (Foster and others, this volume). This low-grade oceanic unit makes up the Seventymile terrane and consists of massive and locally pillowed greenstone, argillite, metatuff, qz-wm schist, qz-ac schist, quartzite, metalimestone, metachert, and metagraywacke. Most protolith ages are unknown, but conodonts and radiolarians of Permian age occur in weakly metamorphosed chert in the Big Delta Quadrangle, and radiolarians of Mississippian age, brachiopods of Permian age, and conodonts of Triassic age occur in the northern Eagle Quadrangle (Foster and others, this volume). These rocks are associated with ultramafic and gabbroic rocks. In at least one area, this package of rocks is part of a dismembered ophiolitic assemblage (Keith and others, 1981). Unit 38 consists of a number of isolated thrust remnants, which are themselves broken by internal thrust faults; metamorphic grade differs between individual thrust remnants. Glaucophane (+ep±gt) occurs in one such thrust remnant in a small exposure of metabasalt just south of the Tintina fault (Keith and others, 1981; Plate 4A).

An Early Jurassic age for low-grade metamorphism of unit 38 is indicated by a 201 ± 5-Ma $^{40}Ar/^{39}Ar$ integrated plateau age on ac from greenstone in southeastern Eagle Quadrangle (G. W. Cushing, unpublished data, 1984). Northwestward accretion of the amphibolite-facies unit resulted in thrusting of remnants of the telescoped ocean basin, including this unit and the associated ultramafic rocks, northward onto greenschist-facies unit 44 and southward onto amphibolite-facies unit 37 (Foster and others, this volume). This followed, or was synchronous with, the low-grade metamorphism. The outcrop of gl-bearing greenstone may be part of a fault sliver that was dragged to a greater depth in a subduction zone or a transpressive boundary along the convergent margin.

Metamorphism of greenschist-facies unit 39, included in the Yukon-Tanana terrane, was probably also part of the Late Triassic through Early Jurassic orogenic episode. This unit probably formed part of the continental margin north of the ocean basin onto which the previous two metamorphic units were accreted. Common rock types are qz-wm (±ch±ac) schist, quartzite, phyllite, and metavolcanic rocks; protoliths are probably Paleozoic in age. Foliation is well developed, and rocks are multiply folded and commonly are lineated. This unit is correlated with the Klondike Schist of McConnell (1905) that crops out across the Canadian border (Foster and others, 1985). A Late Triassic to Early Jurassic metamorphic age is suggested by a 175 ± 14-Ma K-Ar age on mu (Tempelman-Kluit and Wanless, 1975) and by a 202 ± 11-Ma Rb-Sr whole-rock age (Metcalfe and Clark, 1983) determined for Klondike Schist in Canada.

Paleozoic or Mesozoic metamorphism of uncertain age and origin. The metamorphic ages of many units in east-central Alaska (units 40 to 45 in the Yukon-Tanana upland, and units 53 and 54, and M_1 of units 47 and 48 in the Alaska Range; Plate 4B) are unknown. Metamorphic ages are bracketed between the known, or probable, Paleozoic age of the youngest protoliths and the Early to mid-Cretaceous age of widespread plutonism that postdates regional metamorphism. Scattered late Early Cretaceous K-Ar mineral and whole-rock ages on schist and a U-Pb age on monazite from orthogneiss (ages given on Plate 4B) tentatively suggest an Early Cretaceous age for latest metamorphism. Unit 47 and its higher grade equivalent, unit 48, crop out south of the Tanana River and consist of metasedimentary, metavolcanic, and metaplutonic rocks of Devonian and older protolith age (Aleinikoff and Nokleberg, 1985; T. K. Bundtzen, oral communication, 1988). M_1 records greenschist-facies conditions in unit 47 but increases in intensity northward; it grades into the amphibolite-facies M_1 of unit 48 in the west (Bundtzen, 1981). M_1 metamorphism in units 47 and 48 postdates the Devonian age of the youngest known protoliths and predates the Early Cretaceous age of the episode that is thought to have caused lower greenschist-facies retrograde metamorphism (M_2) in units 47 and 48 as well as the monometamorphism of unit 49 (discussed in a later section). In unit 48 and the western part of unit 47, Bundtzen and Turner (1979) proposed an Early Jurassic minimum age for M_1 on the basis of the oldest of four K-Ar ages (195, 144, 123, and 104 Ma) on hb from gt amphibolite in unit 48.

Unit 46, characterized by augen gneiss, gives Early Cretaceous ages from a number of isotopic systems and on this basis, its latest, and probably its only, metamorphism is believed to be Early Cretaceous in age. However, limited evidence from U-Pb zircon data from quartzite (Aleinikoff and others, 1984b, 1986) and sillimanite gneiss from unit 46 (Aleinikoff and others, 1984a), and from structural relations between augen gneiss and wall rocks in east-central Alaska and Yukon Territory, suggest that an earlier metamorphic episode may have accompanied the Mississippian intrusion of batholithic sheets of what is now augen gneiss (Dusel-Bacon and Aleinikoff, 1985). These structural relations are: (1) that some areas of augen gneiss coincide with large structural and metamorphic culminations (Mortensen, 1983); and (2) that the concordant contacts of the augen gneiss bodies, and parallelism between lithologic contacts and the gently dipping regional penetrative fabric, suggest the augen gneiss bodies were intruded synkinematically into ductile crust (Dusel-Bacon and Aleinikoff, 1985).

Early Cretaceous metamorphism of augen gneiss-bearing unit. Amphibolite-facies unit 46 is characterized by large bodies of augen gneiss that form a discontinuous belt of metamorphosed Mississippian plutons, believed to have developed beneath, or inland from, a continental magmatic arc of Late Devonian to Early Mississippian age (Dusel-Bacon and Aleinikoff, 1985). Other rock types, interpreted as having been wall rocks to the augen gneiss protolith, are amphibolite-facies qz-mica schist, bt and bt-hb gneiss, quartzite, amphibolite, sillimanite gneiss, and minor amounts of marble, calc-schist, and felsic gneiss. The Mississippian protolith age for the augen gneiss (about 340 Ma, on the basis of a U-Pb lower intercept age of zircon and a Rb-Sr whole-rock isochron; Aleinikoff and others, 1986) establishes a minimum protolith age for the adjacent wall rocks. Protoliths of some metavolcanic rocks are Devonian in age (Aleinikoff and others, 1986).

Transitional low- to intermediate-P conditions are indicated for unit 46 by the local occurrence of ky and/or anda (Plate 4A) in qz-mica (± gt ± st ± sil) schist. All the rocks of this unit are well foliated; commonly the foliation is multiply folded into isoclinal folds on various scales. Many rocks exhibit a well-developed stretching lineation, and most show some degree of mylonitization followed by varying degrees of recrystallization.

A 600-km^2 body of sillimanite gneiss and flanking pelitic schist crops out in the western part of this unit (shown on Plate 4A by the sillimanite isograd) and has been interpreted as a gneiss dome by Dusel-Bacon and Foster (1983). Metamorphic grade increases across the pelitic schist on the flanks of the dome, where P-T conditions locally were near the Al_2SiO_5 triple point (approximately 3.8 kb and 500°C; Holdaway, 1971), to the gneissic core of the dome, where P-T conditions were near those of the second sillimanite isograd. Garnet-biotite geothermometry (calibration of Ferry and Spear, 1978) indicates an equilibration T of 535 to 600 ± 30°C for pelitic schist north of the dome and 655 to 705 ± 30°C for sillimanite gneiss in the core (Dusel-Bacon and Foster, 1983).

Petrographic evidence of a regional retrogressive metamorphic episode is minimal in most of this unit and consists of local, and minor, chloritization of bt and gt, sericitization of pl, kf, and co, and the development of ac from hb. In the southwestern part of unit 46, however, Nokleberg and others (1986a) note that amphibolite-facies rocks are consistently retrograded to greenschist-facies assemblages, and that the degree of retrogression increases to the south. This retrogressive metamorphism is shown on the map as the second greenschist-facies episode in adjacent unit 47 to the south.

Metamorphism of unit 46 postdates the Mississippian intrusive age of the augen gneiss protolith and predates the intrusion of Early and mid-Cretaceous plutons (generally with 105- to 85-Ma K-Ar cooling ages; Wilson and others, 1985). A 115-Ma U-Pb age on zircon from an unmetamorphosed (late metamorphic?) pluton that intrudes the sillimanite gneiss dome (Aleinikoff and others, 1984a) provides the most reliable minimum age of metamorphism. Abundant isotopic data from the metamorphic rocks of this unit suggest that metamorphism occurred between about 135 and 115 Ma: most conventional K-Ar mineral ages fall in the range of about 125 to 110 Ma; a Rb-Sr mineral isochron for augen gneiss is 115 Ma; sph from augen gneiss gives a concordant U-Pb age of 134 Ma; U-rich zircon fractions from sillimanite gneiss and quartzite show Early Cretaceous lead loss (Aleinikoff and others, 1986); and hb from augen gneiss gives a $^{40}Ar/^{39}Ar$ incremental-heating plateau age of 119 Ma (T. M. Harrison, written communication, 1987).

Early to Late Cretaceous metamorphism of other units. Effects of Early Cretaceous metamorphism are believed to be widespread across much of east-central Alaska. The following limited isotopic data suggest an Early Cretaceous age for metamorphism in units 41 and 42 in the northwestern Yukon-Tanana upland: three K-Ar mica ages and one whole-rock age from greenschist-facies unit 41 range from 138 to 100 Ma (Forbes and Weber, 1982), and monazite from orthogneiss of amphibolite-facies unit 42 gives a concordant U-Pb age of 115 Ma (J. N. Aleinikoff, written communication, 1987). As mentioned earlier, retrograde effects on the eclogite-bearing klippen (unit 36) that overlie unit 41 are attributed to this same episode.

Farther south, in the Alaska Range, retrograde metamorphism in polymetamorphic units 47 and 48 is also interpreted as having taken place in the Early Cretaceous. In that area, M_2 metamorphism occurred under ch-grade conditions; its effects are most evident where the metamorphic grade during M_1 was highest, namely in retrograded amphibolite-facies unit 48 and in the northern part of polymetamorphic greenschist-facies unit 47. M_2 assemblages in unit 48 and the adjacent area of unit 47 define a weak metamorphic foliation that is axial planar to broad northeast-trending folds (Bundtzen, 1981). Characteristics of M_2 metamorphism and accompanying deformation in the other parts of unit 47 vary widely, and thus correlation of M_2 episodes throughout the unit is tentative (Dusel-Bacon and others, 1991b). A late Early Cretaceous metamorphic age for M_2 is suggested for unit 48 and the adjacent area of unit 47 by the oldest K-Ar mica age (100 Ma) determined for those rocks (Bundtzen and Turner, 1979). A similar late Early Cretaceous age is suggested by a 107-Ma K-Ar age on mu from phyllite of unit 47 in the central Alaska Range (Sherwood, 1979). Nokleberg and others (1986a; this volume, Chapter 10) report that mid-Cretaceous plutons in the northern part of unit 47 in the Mt. Hayes Quadrangle appear to have been weakly metamorphosed under lower greenschist-facies conditions together with their polymetamorphosed wall rocks, suggesting that metamorphism in that area continued longer, perhaps into Late Cretaceous time.

Low-grade regional metamorphism of monometamorphic unit 49 was synkinematic with the development of northeast-trending folds and has been correlated with the M_2 of adjacent units 47 and 48 (Bundtzen and Turner, 1979). Mylonitic textures are common throughout this unit and indicate a large dynamic component to the regional metamorphic episode. An Early Cretaceous age for this deformational and metamorphic episode is proposed on the basis of a whole-rock K-Ar age of 108 ± 3 Ma on metafelsite (Bundtzen and Turner, 1979).

An eastward-increasing metamorphic sequence developed in Early(?) to Late Cretaceous time within units 50 (at least in that part east of longitude 151°), 51, and 52—units that are bounded to the south by the McKinley fault. Low- to intermediate-P conditions are indicated for the amphibolite-facies part of the sequence (unit 52) and are inferred for the lower grade part of the sequence to the west. Evidence for this P range consists of the sparse occurrence of relict anda, indicating low P, and the presence of gt in both metabasic and metapelitic rocks, suggesting intermediate P. Metamorphism may have begun earlier in the highest grade, eastern, part of the sequence. Geologic relations in the Healy Quadrangle indicate that metamorphism preceded, and continued during, intrusion of an Early Cretaceous pluton (105-Ma $^{40}Ar/^{39}Ar$ hb incremental-heating plateau age) into unit 52. The pluton generally crosscuts the metamorphic fabric but, locally, igneous contacts are migmatitic, and in places the pluton is foliated (Csejtey and others, 1986). Weak metamorphic effects also have been noted in plutonic rocks of Late Cretaceous (70 Ma) age farther east within unit 52 (Nokleberg and others, 1991). Metamorphism of this sequence may have been part of the M_2 metamorphic episode that affected unit 47 to the north. A similar eastward increase in metamorphic grade occurs in M_2 assemblages within the adjacent part of unit 47 south of the Hines Creek fault (Dusel-Bacon and others, 1991b).

Problematic tectonic origin of Cretaceous metamorphism and an alternative interpretation of Early Cretaceous isotopic ages. The assumption that much of east-central Alaska was metamorphosed during Cretaceous time is based on the interpretation that isotopic ages from throughout the region record uplift and cooling (to blocking temperatures of about 500°C in hb and about 300°C in bt) that followed (by not more than 20 to 40 m.y.) an initial episode of crustal thickening and metamorphism. Northward migration and accretion of the Wrangellia terrane against the North American margin, which would have included the rocks of east-central Alaska north of the McKinley and Denali faults, might be a possible cause of Cretaceous crustal thickening, but the timing of the accretion appears to be too late to explain the widespread Early Cretaceous (110 to 135 Ma) isotopic ages. As discussed in a subsequent section, accretion of the Wrangellia terrane is believed to have postdated the early Late Cretaceous (Cenomanian: 98 to 91 Ma) age of the youngest flysch in a basin that separated the Wrangellia terrane from North America, and to have resulted in intermediate-P metamorphism and synkinematic plutonism within the flysch basin at about 70 to 56 Ma. The mid- to Late Cretaceous metamorphism of units 50 to 52 that crop out just north of the McKinley fault probably overlapped with accretion of the Wrangellia terrane.

Given the problem of identifying the cause of a compressional episode in Early Cretaceous time, an alternative possibility is that the Early Cretaceous ages may not date the time of crustal thickening and heating but instead date uplift and cooling, perhaps brought about by extension, that followed the latest Triassic to Early Jurassic compressional episode (discussed above) that affected units 37 and 39 near the United States–Canada boundary. The original western extent of this episode is unknown. It is possible that metamorphism of other units in the Yukon-Tanana upland, whose metamorphic age is either given as Early Cretaceous or is bracketed between Paleozoic and Early Cretaceous time, may have been initially heated during the latest Triassic to Early Jurassic episode.

An argument in favor of this interpretation is the fact that oceanic rocks of unit 38 and associated ultramafic rocks (MzPz u), both of which are interpreted as having been part of the ocean basin that separated North America from a composite terrane that included the Stikinia terrane (Foster and others, this volume), occur far west of units 37 and 39, suggesting a greater original extent of the accreted composite terrane than is now recognized. Moreover, extensional faulting recently has been proposed for several areas in the Yukon-Tanana upland (Pavlis and others, 1988b; Duke and others, 1988; Hansen, 1989). Structural data collected near the fault contact between the flanks of the sillimanite gneiss dome of unit 46 and the structurally overlying mylonitic greenschist-facies rocks of unit 45 indicate that the latest fault movement was extensional (east to southeast movement of the hanging wall; Pavlis and others, 1988b). Other low-angle faults that place lower grade rocks over higher grade rocks (such as the fault that separates units 41 and 42 from units 43 and 44), the relation common in extended terranes, are possible candidates for extensional faulting. A correspondence between the age of extension and the Early Cretaceous isotopic ages remains to be proven, but it appears to be a reasonable supposition because augen gneiss and associated rocks of unit 46 are wall rocks to 90-Ma calderas (Bacon and others, 1990) and therefore were virtually at the surface by that time.

Arguing against latest Triassic to Early Jurassic accretion-related metamorphism throughout much of the Yukon-Tanana terrane time is the fact that latest Triassic to Early Jurassic plutons only occur in association with unit 37 (of possible Stikinia terrane affinity) in the eastern part of the Yukon-Tanana upland, and isotopic ages from upper greenschist- and amphibolite-facies metamorphic rocks in the rest of the upland and adjacent parts of the Alaska Range are predominantly late Early Cretaceous.

Area of southern Alaska between the McKinley and Denali faults and the Border Ranges fault system

Jurassic metamorphism in the Peninsular and Wrangellia terranes. Several areas of amphibolite- and greenschist-facies metaigneous rocks and associated metasedimentary rocks crop out in the Alaska Peninsula, the Talkeetna Mountains, the Gulkana area northeast of the Talkeetna Mountains, and the eastern Chugach Mountains (Nokleberg and others, this volume,

Chapter 10; Plafker and others, this volume, Chapter 12). Lithologic assemblages in all areas (units 55 to 60) are similar and include varying amounts of most of the following rock types: amphibolite and other amphibolite-facies rocks including mafic, calcareous, and pelitic schist, bt gneiss, marble, and quartzite or metachert; greenschist-facies rocks including greenschist, greenstone, metavolcaniclastic rocks, phyllite, argillite, and slate; and admixtures of the above rock types with massive to schistose intermediate to mafic plutonic rocks that are variably altered, sheared, and foliated. The association of protoliths (mafic to intermediate extrusive and intrusive rocks, siliciclastic rocks, calcareous rocks, and chert) suggests an oceanic affinity for most rocks. Unit 55 (Kakhonak Complex; see Detterman and Reed, 1980), and unnamed wall rocks to Jurassic plutons mapped in the Talkeetna Mountains (see Csejtey and others, 1978), and unit 56 (retrograded schist at the southern edge of the Talkeetna Mountains; Csejtey and others, 1978) are included in the Peninsular terrane. Unit 57 (informally designated metamorphic complex of Gulkana River; see Nokleberg and others, 1986b) crops out along the contact between the Peninsular and Wrangellia terranes. Unit 58 (informally designated Haley Creek metamorphic assemblage; see Plafker and others, 1989b), unit 59 (part of the Strelna metamorphics of Plafker and others, 1989b), and unit 60 (informally designated Dadina River metamorphic assemblage; see Winkler and others, 1981; Aleinikoff and others, 1988; Plafker and others, 1989b) are included by the cited workers in the Wrangellia terrane.

I interpret metamorphism in all the areas to have been part of the same metamorphic episode that was early tectonic or syntectonic with the intrusion of Jurassic batholiths. Assumed protolith ages in the Peninsular terrane differ from those known or assumed in the Wrangellia terrane (Plate 4B), however, and the widespread metamorphic episode probably was imposed on different protolith assemblages. Evidence for a Jurassic metamorphic age differs in the various areas and is summarized in Plate 4B; the reader is referred to the appropriate regional report (Fig. 2) for a detailed discussion of these units.

Following the model of Plafker and others (1989b), metamorphism and synkinematic plutonism probably took place within a magmatic arc(s) that developed as a result of left-oblique subduction of the Farallon Plate beneath a composite terrane composed of the Peninsular, Wrangellia, and Alexander terranes.

Possible correlatives of Jurassic metamorphism of the Peninsular and Wrangellia terranes in southern and southeastern Alaska. Unit 61 consists of a diverse sequence of greenschist- and epidote-amphibolite-facies tectonized metaplutonic, metasedimentary, and metavolcanic oceanic rocks. They have been informally called the "Knik River schist" by Carden and Decker (1977) and are described in detail by Pavlis (1983) and Pavlis and others (1988a). This unit is spatially associated with both Jurassic mafic and ultramafic plutons that form part of the Border Ranges ultramafic and mafic complex of Burns (1985; shown on Plate 4A as unit Jmu) and with Early Cretaceous trondhjemite. Pavlis and others (1988a) suggest that metamorphism of this unit accompanied intrusion of the Early Cretaceous (135 to 110 Ma) tonalitic to trondhjemitic plutons. K-Ar ages on rocks from this unit include a 177-Ma age on ac (Carden and Decker, 1977) and three younger ages on hb of 135, 121, and 107 Ma (Pavlis, 1983). These ages are compatible with either a model in which metamorphism took place during Jurassic time (in which case the younger ages were partly or totally reset during Early Cretaceous intrusion), or in which some or all the rocks were metamorphosed twice, each time in association with nearby plutonism.

The metamorphic history of unit 62 is difficult to assess because this heterogeneous assemblage of metaplutonic and metasedimentary rocks, referred to informally as the Cottonwood Creek complex of Richter and others (1975) and Richter (1976), occurs as a narrow slice within the Denali fault zone that is widely separated from the rest of the complex (probably the Alexander or Wrangellia terrane) from which it was derived.

Unit 63, informally referred to as the metamorphic complex of Tlikakila River, crops out as a northeast-trending belt within the Late Cretaceous and early Tertiary plutons of the Alaska–Aleutian Range batholith (Carlson and Wallace, 1983; Nelson and others, 1983). Metamorphism is known to postdate the Late Triassic protolith age of the youngest rocks. The spatial association between the metamorphic rocks of this unit and the adjacent Late Cretaceous and Tertiary plutons suggests that metamorphism may have been related to either one or both of the plutonic episodes. However, similarities in metamorphic rock types between this unit and unit 55 (discussed above), suggest that at least some of the rocks of the metamorphic complex of Tlikakila River may have been metamorphosed during the widespread episode of metamorphism and tectonism that slightly preceded or accompanied Middle to Late Jurassic plutonism in the Alaska-Aleutian Range and Talkeetna Mountains area.

Unit 94 crops out on Chichagof and Baranof Islands in southeastern Alaska and is also included in the Wrangellia terrane. This unit consists of amphibolite-facies mafic metavolcanic rocks and marine metasedimentary rocks whose protoliths predate the intrusion of Middle to Late Jurassic (168- to 155-Ma K-Ar ages on bt and hb) diorite plutons (Loney and others, 1967). R. A. Loney (oral communication, 1985) considers that the metamorphism of this unit is unrelated to the intrusion of the Jurassic plutons, because the direction of increase in metamorphic grade bears no relation to the distance from the plutons. However, Johnson and Karl (1985) report that rocks of this unit grade into the dioritic rocks of Jurassic, and Jurassic or Cretaceous, age and become more migmatitic close to the plutons, implying a genetic connection between plutonism and metamorphism. Because of the uncertainty in the metamorphic history of this unit, the age of metamorphism is widely bracketed between Paleozoic and Early Cretaceous time, allowing for the possibilities that metamorphism occurred long before plutonism, assuming the oldest possible protolith age for the rocks; or that metamorphism was associated with plutonism, as appears to be the case in other parts of the Wrangellia terrane basement, as described above.

Medium-grade metamorphism in the adjacent northern Alexander terrane. The greenschist-facies marble, phyllite, greenschist, mica schist, and weakly metamorphosed late Paleozoic plutons (shown with a diagonal dash overprint) of unit 64 crop out in the Wrangell and Saint Elias Mountains area (MacKevett, 1978; Hudson and others, 1977b; Campbell and Dodds, 1978; Miller, this volume). Marble is, in part, Devonian in age and may be as old as Cambrian (Gardner and others, 1988).

In the Saint Elias Mountains, the late Paleozoic plutons have been altered to greenschist-facies assemblages. Relations between the metamorphic rocks and the late Paleozoic plutons may indicate that metamorphism in that area was synkinematic with plutonism. These relations, reported by Hudson and others (1977a, b) consist of: (1) limited data that suggest that slightly higher grade metamorphic mineral assemblages are developed adjacent to the plutons; and (2) the fact that the plutons are dominantly foliate and commonly altered, and that contact relations between plutons and country rocks are locally complex, sometimes being sharp and crosscutting and sometimes gradational.

In the southeastern McCarthy Quadrangle, a pluton of Middle Pennsylvanian age (shown as a weakly metamorphosed pluton of unit 65, discussed below) intrudes both the Alexander and Wrangellia terranes, thereby indicating that the two terranes have been sutured since late Paleozoic time (MacKevett and others, 1986; Gardner and others, 1986). Because of this Middle Pennsylvanian minimum age for the juxtaposition of the two terranes, low-grade metamorphism in much of the Alexander terrane (this unit), like the slightly higher grade M_1 metamorphism in the Strelna Metamorphics of the Wrangellia terrane (unit 59) may have been associated with the intrusion of Late Jurassic plutons. Although Late Jurassic plutons have not been mapped within the Alexander terrane in Alaska, Late Jurassic to Early Cretaceous (160 to 130 Ma) plutons make up a major intrusive suite within the continuation of these terranes in Canada (Dodds and Campbell, 1988). Geologic evidence from this unit in Canada suggests that regional metamorphism and deformation may have occurred during latest Jurassic to earliest Cretaceous time (about 160 to 130 Ma) and been associated with plutonism of that age (R. B. Campbell and C. J. Dodds, written communication, 1986). Evidence on which Campbell and Dodds base this interpretation consists of: (1) metamorphism and deformation appear to postdate the deposition of Upper Triassic strata that probably rest unconformably on Paleozoic rocks but nevertheless appear to be equally deformed and metamorphosed; (2) the younger plutons within the belt of 160- to 130-Ma plutons seem clearly to postdate the metamorphism and deformation in the northeast where they produce distinct contact metamorphic aureoles; and (3) the older plutons of this group to the southwest may have been intruded during the metamorphism and deformation, because they are commonly elongate parallel to the regional structural grain, but they clearly have local crosscutting contacts and probably in part postdate those events.

The late Paleozoic to Early Cretaceous age constraints given for this unit reflect the possibility that one or both metamorphic episodes discussed above may have affected different parts of this unit.

Post–Early Jurassic to pre–mid-Cretaceous(?) low-grade metamorphism of the Wrangellia terrane. Weakly metamorphosed oceanic oceanic rocks of the Wrangellia terrane crop out south of the Talkeetna thrust and Denali fault, and north of faults that separate them from the belt of greenschist- and amphibolite-facies rocks that constitute the basement of the Peninsular and Wrangellia terranes (Nokleberg and others, this volume, Chapter 10; Plafker and others, this volume, Chapter 12). These weakly metamorphosed rocks, shown as unit 65, include upper Paleozoic arc-related metavolcanic and metaplutonic rocks, metalimestone, argillite, and metachert; overlying Middle Triassic argillite, Upper Triassic metabasalt (Nikolai Greenstone), and Upper Triassic and Lower Jurassic metasedimentary rocks of the McCarthy Formation; and, north of latitude 62°, small areas of overlying Triassic metalimestone, and Mesozoic marine metasedimentary rocks.

Most rocks have not been penetratively deformed (except near major faults) and exhibit well-preserved volcanic, sedimentary, or plutonic textures (MacKevett, 1978; Richter, 1976; W. J. Nokleberg, oral communication, 1984; Beard and Barker, 1989). Locally, however, in the general area that is proposed to be the leading edge along which the Wrangellia terrane was accreted (Csejtey and others, 1982), rocks are weakly phyllitic or schistose in the central Alaska Range (Smith, 1981; Nokleberg and others, 1985) and intensely folded and sheared in the Talkeetna Mountains (Csejtey and others, 1978).

Triassic greenstone in most areas contains the assemblage ch-pr-pu-ab-ep, indicating prehnite-pumpellyite–facies conditions (Richter, 1976; MacKevett, 1978; Csejtey and others, 1978; Smith, 1981; Nokleberg and others, 1985). CAI values of 5.5 to 6.0 from Upper Triassic conodonts collected the McCarthy Quadrangle from the Nizina and Chitistone Limestones and the lower part of the McCarthy Formation suggest metamorphic temperatures of about 350 to 450°C (M. W. Mullen, oral communication, 1989). Locally, in the more deformed northern areas of this unit, greenstone may contain the assemblage ac-ch-ep-ab, indicating lower greenschist-facies conditions.

The metamorphic grade of these late Paleozoic metavolcanic and metaplutonic rocks is generally comparable to that of the low-grade Mesozoic rocks. In the northeastern Gulkana Quadrangle, the Ahtell pluton contains metamorphic wm, ch, ab, and qz (W. J. Nokleberg, written communication, 1988). Metamorphic minerals developed in the correlative diorite complex of Richter (1976) consist of ac, ch, ep, and wm (Barker and Stern, 1986). Late Paleozoic metavolcanic rocks in the northwestern Nabesna Quadrangle generally contain the low-grade assemblage ch-ep-pu, except near the diorite complex where it has been crystallized to massive, fine-grained assemblages of hb, ep, ch, and feldspar (Richter, 1976). In the southeastern McCarthy Quadrangle, the (informal) Barnard Glacier pluton of Middle Pennsylvanian age (309 ± 5-Ma U-Pb age on zircon; Gardner and others, 1988) is generally unfoliated, but locally, it is cataclastically deformed (MacKevett, 1978). It is this pluton, mentioned in the previous section, that is intrusive into both the low-grade rocks of unit 65 to the west, included in the Wrangellia terrane, and the greenschist-facies rocks of unit 64 to the east, included in the Alexander terrane, thereby indicating that the two terranes were sutured together by Middle Pennsylvanian time (MacKevett and others, 1986; Gardner and others, 1988). Although the data are inconclusive, the pluton is tentatively shown on Plate 4A to have been weakly metamorphosed along with the Wrangellia terrane wall rocks on the west, although numerous other possible correlations of its metamorphic history are possible given the uncertainty of the timing of metamorphism in this region.

The age and cause of metamorphism may be different in other parts of unit 65. Because the metamorphic grade is so low, it is difficult to determine with certainty whether some associated rocks have been metamorphosed. In the southern part of this unit, the CAI values from the Late Triassic conodonts indicate that metamorphism is post–Late Triassic in age. The minimum metamorphic age in that area is uncertain, but I also assume that it postdates the Lower Jurassic part of the McCarthy Formation. A tentative Middle or Late Jurassic minimum age of metamorphism for at least some areas of this unit is suggested by the apparent lack of metamorphism in the Upper Jurassic and Lower Cretaceous Nutzotin Mountains sequence in the Nabesna Quadrangle (Richter, 1976) and in the Jurassic and Cretaceous sedimentary rocks that overlie the McCarthy Formation in the Valdez and McCarthy Quadrangles (Winkler and others, 1981; MacKevett, 1978).

Hornblende and biotite from the diorite complex in the northeastern part of unit 65 yield Early Jurassic K-Ar ages (Richter and others, 1975), but it is uncertain whether these ages (1) date the predominant period of metamorphism in that area, (2) provide a minimum age for a late Paleozoic or early Mesozoic episode, or (3) represent a partial resetting of the Pennsylvanian protolith age by a possible Cretaceous metamorphic episode, described below.

Metamorphism in some areas along the southern margin of unit 65 may have been associated with the intrusion of the Jurassic plutons that I propose occurred during the greenschist- to amphibolite-facies metamorphism of the Strelna Metamorphics (unit 59) to the south. Although the two units that may grade into each other are separated by a thrust fault of Cretaceous age (Chitina fault: Gardner and others, 1986), an overall proximity of the units during the Late Jurassic intrusive and metamorphic episode is suggested by the facts that (1) a Late Jurassic pluton also intrudes unit 65 in one area in the western McCarthy Quadrangle, (2) higher grade unit 59 includes a minor component of Upper Triassic rocks that are correlated with those of unit 65, and (3) metamorphic temperatures, as determined from Late Triassic conodonts, are comparable across the Chitina fault (that is, about 350 to 450°C to the north and about 350°C to the south: M. W. Mullen, written communication, 1989).

Some data suggest an alternative or, perhaps, additional, mid-Cretaceous metamorphic age for at least the northern part of unit 65. Field and petrographic observations in the Talkeetna Mountains and the eastern Alaska Range suggest that Jurassic and Cretaceous rocks that overlie the Triassic Nikolai Greenstone also have undergone low-grade metamorphism (B. Csejtey, Jr., written communication, 1984; Nokleberg and others, 1985). Three K-Ar whole-rock ages from samples of the Nikolai Greenstone from the southern part of this unit in the central McCarthy Quadrangle fall on a 112 ± 11-Ma isochron (Silberman and others, 1981).

Silberman and his co-workers (1981) propose that the late Early Cretaceous K-Ar ages from the McCarthy Quadrangle date an episode of low-grade metamorphism that was caused by frictional heating that accompanied the accretion of the Wrangellia terrane to the North American margin. Arguing against this hypothesis for at least the McCarthy Quadrangle, however, is the fact that the Upper Triassic rocks show no sign of being deformed (although they have been heated to about 350 to 450°C since the Late Triassic). Assuming that the late Early Cretaceous K-Ar ages do in fact date the timing of low-grade metamorphism in the central McCarthy Quadrangle, an alternative interpretation is that metamorphism may have been coeval with the northeast-directed, Early Cretaceous movement along the nearby Chitina fault (Gardner and others, 1986) that placed the medium-grade metamorphic rocks (unit 59) and synkinematic Jurassic plutonic rocks of the Wrangellia terrane on top of the low-grade rocks of unit 65.

The localized development of a penetrative fabric and higher-T ac-bearing mineral assemblages in the central Alaska Range and the Talkeetna Mountains suggests that metamorphism, at least in these areas, occurred during late Mesozoic accretion. Both areas were near the leading edge along which this terrane is thought to have been accreted in mid-Cretaceous time (Csejtey and others, 1982; Nokleberg and others, 1985).

Similar low-grade oceanic rocks crop out in southeastern Alaska where they are shown as units 91 and 99. Unit 91 is included in the northern part of the Taku terrane by Monger and Berg (1987), but subsequently has been reinterpreted as part of the Wrangellia terrane by Plafker and others (1989a); unit 99 is included in the Wrangellia terrane by Monger and Berg (1987). The age and tectonic origin of their metamorphism are also uncertain.

Mid-Cretaceous to early Tertiary metamorphism. *Low-grade metamorphism within a compressed flysch basin.* Very weakly metamorphosed and highly deformed flyschoid rocks, primarily metagraywacke, semischist, and argillite, and rocks from several tectonically interleaved fragments within the flysch, crop out as a northeastward-tapering wedge (shown as unit 67) in the central Alaska Range between the low-grade rocks of the Wrangellia terrane to the southeast, and the unmetamorphosed to moderately metamorphosed rocks of the continental margin to the northwest. Flyschoid rocks are Late Jurassic to mid-Cretaceous (Cenomanian) in protolith age (Csejtey and others, 1986). Tectonic fragments include protoliths that range in age from Late Devonian to Late Jurassic age and include a variety of sedimentary, volcanic, and ophiolitic rocks (Jones and others, 1980; Silberling and others, this volume). The tectonic juxtaposition of the disparate fragments (terranes) included within this unit is considered to have taken place during mid-Cretaceous time (Csejtey and others, 1978, 1982; Jones and others, 1980). According to Csejtey and others (1982), the flyschoid rocks were deposited in a basin between the North American craton and the approaching Talkeetna superterrane (equivalent to the composite Peninsular-Wrangellia terrane) to the south, and the small terranes within the flysch were transported in front of the superterrane by northward plate movement. Because the Wrangellia and Alexander terranes are stitched together by a Pennsylvanian pluton, as mentioned in a previous section, the superterrane also must have included the Alexander terrane as its southeasternmost component (Plafker and others, this volume, Chapter 12).

The dominant structural style of this metamorphic unit is compression and attendant thrust faulting that has juxtaposed fragments of what were parts of extensive coherent terranes (Csejtey and others, 1978; Jones and others, 1980). Deformation and recrystallization within the flysch terrane is most intense along zones of concentrated shear; rocks in these zones are commonly phyllitic, semischistose, or protomylonitic. The degree of metamorphism may vary within unit 67, but this aspect of the terranes has not been studied in detail. Metamorphic minerals developed in flyschoid rocks indicate metamorphic conditions characteristic of the prehnite-pumpellyite facies (Dusel-Bacon and others, 1994b).

Metamorphism is bracketed between the mid-Cretaceous age of the youngest metamorphosed rocks and the latest Paleocene and Eocene age of the overlying unmetamorphosed sedimentary and volcanic rocks, and the early Tertiary age of postmetamorphic granitoids that intrude the flyschoid rocks (Csejtey and others, 1982, 1986). The apparent increase in metamorphic grade toward zones of shearing, together with age brackets for accretion that are approximately the same as those for low-grade metamorphism, suggest that much of the metamorphism probably accompanied northward migration and accretion. Low-grade metamorphism of some of the elements of the tectonic fragments may have occurred even earlier.

Intermediate-P metamorphism of the Maclaren metamorphic belt. The Maclaren metamorphic belt consists of a largely fault-bounded, 140-km-long, roughly symmetrical, intermediate-P (Barrovian) sequence of: (1) prehnite-pumpellyite–facies metasedimentary rocks and metagabbro (unit 68); (2) greenschist-facies phyllite, metagraywacke, marble, quartzite, metapelite, and greenstone (unit 69); and (3) amphibolite-facies schist, gneiss, and amphibolite (unit 70) intruded by foliated, synkinematic plutons of intermediate composition, shown as unit TKg with a cross pattern (Smith, 1981; Csejtey and others, 1982, 1986; Nokleberg and others, 1985). Protoliths are Triassic to Cretaceous in age.

Intermediate-P conditions for the sequence are indicated by the presence of ky (+ sil + gt + st) in pelitic schist and gneiss of unit 70 (Smith, 1981), and sil pseudomorphs after ky (L. S. Hollister, written communication, 1985). Kyanite is also reported to occur in amphibolite (Smith, 1981)—further evidence of intermediate- or even high-P conditions. Andalusite has been reported from two localities, one near the western and one near the eastern boundary of unit 70 (Dusel-Bacon and others, 1994b). These reported occurrences of anda warrant further investigation, but they may simply indicate that higher structural levels are exposed on the ends of the metamorphic belt, relative to deeper level exposure in the central part of it.

Where not affected by tectonic shortening, metamorphic grade increases gradually from the flanks of the sequence toward its core. Along the southern limb of the belt, lower grade rocks dip northward under higher grade rocks to form an inverted metamorphic sequence (Smith, 1981). East of longitude 147°, there is an abrupt, rather than gradational, contact between greenschist- and amphibolite-facies rocks due to tectonic shortening along steep north-dipping faults (Nokleberg and others, 1985). A similarly sharp change in metamorphic grade occurs on either side of a north-dipping thrust (overturned to the south along its eastern end) that forms the southern margin of the foliated plutonic body, referred to as the East Susitna batholith (Nokleberg and others, 1985; Smith, 1981). Deeper level rocks are exposed north of the thrust, as indicated by the first appearance of sil and, with one exception, ky in upper plate rocks adjacent to the thrust (Smith, 1981).

Metamorphic recrystallization occurred during and after two dynamic phases of a prolonged metamorphic episode (Smith, 1981). The concordancy between intrusive contacts and metamorphic foliations in the granitoids and in the metamorphosed wall rocks, together with an increase in metamorphic grade toward the East Susitna batholith, indicates that the foliated granitoids intruded during the early part of the metamorphic episode (Smith, 1981; Nokleberg and others, 1985). This may have been toward the end of an early shearing phase or interkinematically before a final phase of shearing (Smith, 1981). U-Pb data on zircon from schistose quartz diorite indicate a 70 ± 7-Ma intrusive age of the East Susitna batholith (Aleinikoff and others, 1982). U-Pb data on sphene and K-Ar data on biotite from the same rock indicate a 56-Ma metamorphic age (Aleinikoff and others, 1982), which apparently marks the end of the prolonged Late Cretaceous to early Tertiary metamorphic episode. Biotite from pelitic schist gives a similar K-Ar age of 57 Ma (Smith and Lanphere, 1971). Rapid uplift and cooling during metamorphism is indicated by the fact that approximately the same age is given by sphene, whose closure temperature is greater than 600°C (Mattinson, 1978), and biotite, whose closure temperature is about 280°C (Harrison and others, 1985).

Metamorphism and tectonic shortening of the Maclaren metamorphic belt apparently resulted from the accretion of the previously amalgamated Peninsular and Wrangellia terranes to the Yukon-Tanana and Nixon Fork terranes of the ancient North American continent (Csejtey and others, 1982) and the synorogenic intrusion of the East Susitna batholith (Nokleberg and others, 1985). The flyschoid protoliths of the Maclaren metamorphic belt, as well as those of unit 67, were deposited mostly in the narrowing and subsequently collapsed ocean basin between the converging terranes (Csejtey and others, 1982; Nokleberg and others, 1985).

The location in which the convergence, deformation, plutonism, and metamorphism took place is disputed, however. Multiple lines of field and isotopic evidence suggest that the Maclaren metamorphic belt and East Susitna batholith are the offset equivalents of the Kluane Schist and Ruby Range batholith in Yukon Territory, displaced 400 km by right-lateral Cenozoic movement along the Denali and McKinley faults (summary of evidence given in Nokleberg and others, 1985; see also Aleinikoff and others, 1987). Nokleberg and his co-workers (1985) postulate that intense deformation and prograde metamorphism of the belt began during mid- to Late Cretaceous time as a result of the accretion of the Wrangellia terrane onto the North American margin further to the south and continued during early Tertiary time as a result of the northward migration of the flyschoid (Maclaren) terrane and the Wrangellia terrane along the North American margin. Csejtey and others (1982) dispute the correlation between the two metamorphic-plutonic complexes and propose instead that regional metamorphism of the Maclaren metamorphic belt occurred in place, extends across the McKinley fault, and is only slightly offset by it. Although there is an apparent similarity in metamorphic grade and a similar eastward increase in metamorphic grade on either side of the McKinley fault in the east-central part of the Healy Quadrangle, metamorphic data are insufficient to document continuity of metamorphic history across the fault. Arguing against continuity of metamorphic history on either side of the McKinley fault is the occurrence of a fault-bounded block of unit 54 (Windy terrane) between the areas of amphibolite-facies rocks that Csejtey would correlate across the fault (units 52 and 70).

Area of southern and southeastern Alaska that lies south of the Border Ranges fault system

Jurassic blueschist- to greenschist-facies metamorphism. A belt of transitional, and tectonically intermixed high-P blueschist-facies to intermediate-P greenschist-facies metabasalt, metachert, mica schist, marble, and fine-grained clastic rocks, derived from oceanic protoliths, crops out immediately south of the Border Ranges fault system, at the northern margin of the Chugach terrane (Plafker and others, this volume, Chapter 12). The belt, shown as unit 71, extends discontinuously for about 750 km from Kodiak Island on the west to the Copper River on the east. This unit consists of fault-bounded, commonly internally imbricated blocks, and includes, from southwest to northeast, the Raspberry Schist of Roeske (1986) on Kodiak and Afognak Islands, the informally designated schist of Seldovia on the Kenai Peninsula (Forbes and Lanphere, 1973; Carden and others, 1977), the informally designated schist of Iceberg Lake near Tazlina Glacier (Winkler and others, 1981; Sisson and Onstott, 1986), and the informally designated schist of Liberty Creek just west of the Copper River (Metz, 1976; Winkler and others, 1981; Plafker and others, 1989b). Protolith ages are unknown.

In most areas, greenschists that contain ch + ac commonly are finely intercalated with blue-amph–bearing schists that contain cr + ep (Forbes and Lanphere, 1973; Carden and others, 1977; Carden, 1978; Winkler and others, 1981). Glaucophane (+ ep) has been identified only in the Raspberry Schist on Afognak Island, and the assemblage gt + cr + ep is present in the schist of Iceberg Lake near Tazlina Glacier (Winkler and others, 1981). Lawsonite coexists with blue amph at scattered localities along the belt (Plate 4A).

The coexistence of blue amphibole with ep, and in one area with gt, is indicative of the high-T subdivision of the blueschist facies (Taylor and Coleman, 1968; Evans and Brown, 1987). However, the sporadic occurrence of lw, which is diagnostic of the low-T subdivision of the blueschist-facies, indicates that temperatures during metamorphism were probably near the boundary bedtween the two subdivisions. Phase equilibria that involve the breakdown of pu to form ep (Nitsch, 1971), and the breakdown of lw to form zo (Franz and Althaus, 1977), suggest temperatures between about 350 and 400°C. Phase equilibria and cr composition indicate crystallization at about 6 ± 2 kb for the schists of Iceberg Lake and Liberty Creek (Sisson and Onstott, 1986). This P-T range is consistent with the hypothesis of Carden and others (1977) that the finely developed intercalation of ac-ch-bearing layers and cr-bearing layers in the Raspberry Schist (equivalent to the Kodiak schist unit of Carden and others, 1977) and the schist of Seldovia are probably due to minor variations in original chemistry of layers that were metamorphosed under conditions close to the boundary between the greenschist and blueschist facies (Turner, 1981).

Detailed mapping in the area of Kodiak Island indicates that postmetamorphic faults separate blocks ranging from meters to hundreds of meters wide, and that overall the metamorphic grade of the blocks increases from southeast to northwest (Roeske, 1986). To the north, the schist of Seldovia also occurs as fault-bounded blocks of varying metamorphic grade (S. M. Roeske, oral communication, 1984).

The schist of Iceberg Lake makes up an elongate, 40- by 4-km, fault-bounded belt enclosed by the low-grade McHugh Complex (unit 72) near the Tazlina Glacier. Several small elongate blocks of this unit (too small to show on Plate 4A) also occur in melange along the Border Ranges fault system to the north (Winkler and others, 1981). Similar metamorphic mineral assemblages (primarily those of the greenschist-facies) are developed in the schist of Liberty Creek and the schist of Iceberg Lake, but rocks are noticeably finer grained (generally less than 3 mm) in the former than in the latter (Plafker and others, 1989b). Crossite-bearing rocks in the schist of Iceberg Lake locally contain gt and those in the schist of Liberty Creek, hematite.

Data for several isotopic systems suggest a Jurassic (primarily Early to early Middle Jurassic) age for the intermediate- to high-P greenschist-facies episode. K-Ar ages on wm and cr (as well as on ac from the schist of Seldovia) from unsheared rocks in all units except the schist of Liberty Creek range from Early to Late Jurassic, from 190 to about 152 Ma (Forbes and Lanphere, 1973; Carden and others, 1977; Winkler and others, 1981). Crossite-bearing rocks from the Raspberry Schist give an Early

Jurassic age of 196 Ma for a Rb-Sr whole rock-ph isochron and 204 ± 8 Ma for a U-Pb isochron of sph, wm, ab, and amph (Roeske and Mattinson, 1986). Near the eastern end of the belt, cr and ph from unsheared rocks of the schist of Iceberg Lake yield $^{40}Ar/^{39}Ar$ plateau ages of about 185 Ma (Sisson and Onstott, 1986).

Strongly sheared rocks from the schist of Iceberg Lake yield Early Cretaceous (138 to 113 Ma) K-Ar mineral ages (Winkler and others, 1981), indicating partial resetting subsequent to the well-dated Jurassic metamorphic episode. Resetting of the isotopic ages may have taken place during the emplacement of the schistose rocks of the McHugh Complex, the adjacent seaward subduction complex (unit 72) that was primarily accreted between Early Cretaceous and early Tertiary time (Winkler and others, 1981).

Early Cretaceous (123 to 107 Ma) K-Ar whole-rock ages have been determined for three samples of sheared rock from the schist of Liberty Creek (Plafker and others, 1989b), but interpretation of these ages is uncertain. Because of the very fine grain size of the unit, no minerals suitable for isotopic dating have been successfully separated. If the schist of Liberty Creek, like the rest of the belt, was originally metamorphosed during a Jurassic subduction-related transitional greenschist- to blueschist-facies episode, then its Early Cretaceous ages probably represent the same resetting event, attributed to emplacement of the McHugh Complex, that was proposed for the nearby schist of Iceberg Lake. Alternatively, the Early Cretaceous whole-rock ages may in fact date the timing of subduction-related metamorphism of the schist of Liberty Creek. In this case, a better analog for its metamorphic history would be that of the sparse blue-amph–bearing schist within the prehnite-pumpellyite- to greenschist-facies melange of unit 73 (Kelp Bay Group) of the Chugach terrane on Chichagof Island, over 600 km to the southeast (discussed below). Arguing against the analogy with the rocks on Chichagof Island is the fact that blue amphibole has only been found at one locality on Chichagof Island, whereas it occurs across a much larger area and in more abundance in the schist of Liberty Creek.

Similarities in lithology, mineralogy, and isotopic ages suggest that all parts of this unit, with the possible exception of the schist of Liberty Creek, are segments of a formerly continuous belt of accreted, subduction-related rocks. Metamorphism of the Raspberry Schist and schist of Seldovia is thought to have occurred as a result of northwest-directed subduction, which was coeval with magmatism in the nearby Alaska–Aleutian Range during Early Jurassic time (Carden and others, 1977; Connelly, 1978). Plafker and others (1989b) concur with this general hypothesis and propose that northward to eastward subduction beneath the (composite) Peninsular-Wrangellia terrane began in Late Triassic time and continued to Middle(?) Jurassic time as a result of left-oblique subduction of the Farallon plate. As pointed out by Plafker and his co-workers, the juxtaposition across the Border Ranges fault system of the low-T, high-P rocks of unit 71 with the approximately coeval high-T plutonic and volcanic rocks of the Triassic to Jurassic arc (including unit Jmu and the Talkeetna Formation that forms part of unit N) implies structural disruption of the seaward margin of the arc. On the basis of the separation between the inner margins of accretionary prisms and magmatic belts in modern arcs, Plafker and others (1989b) proposed that the observed juxtaposition indicates relative underthrusting, on the order of 50 km, of the high- to intermediate-P rocks beneath the plutonic rocks.

Jurassic to early Tertiary low-grade metamorphism of Chugach and Yakutat terranes melange. Tectonic melange, included in the melange facies of the Chugach terrane (Plafker and others, 1977 and this volume, Chapter 12), occurs immediately outboard of the Border Ranges fault or, in a few areas in southern Alaska, separated from it by the high- to intermediate-P subduction complex discussed above. The melange makes up all of unit 72 and the inboard parts of units 73 to 75. It consists of disrupted, deformed, and weakly metamorphosed blocks, ranging from tens of meters to several kilometers in longest dimension, of greenstone, mafic schist, metatuff, metagraywacke, metaargillite, metachert, metalimestone, phyllite, quartzite, serpentinite, and mafic plutonic rocks. Blocks are aligned in a sheared matrix of argillite, metatuff, and metachert.

Unit 72 crops out around the Gulf of Alaska and records primarily prehnite-pumpellyite–facies conditions; exotic metamorphic blocks in the melange locally record blueschist-facies conditions. This unit makes up the McHugh Complex in the Chugach Mountains (Clark, 1973; Tysdal and Case, 1979; Winkler and others, 1981), the Seldovia Bay complex in the Kenai Mountains (Cowan and Boss, 1978), and the Uyak Complex in the Kodiak Island area (Connelly, 1978). Near Seldovia, a small area of this unit is included in the Kachemak terrane by Jones and others (1987). Paleontologic ages of radiolarians from the melange matrix of this unit range from Late Triassic to mid-Cretaceous (Albian to Cenomanian), the bulk of the fossil ages being Late Jurassic to Early Cretaceous (Plafker and others, 1977; Connelly, 1978; Nelson and others, 1987). Fossils from blocks within the melange are as old as late Paleozoic.

Unit 73 crops out in southeastern Alaska and is made up primarily of melange (revised Kelp Bay Group of Johnson and Karl, 1985) that records prehnite-pumpellyite–facies, lower greenschist-facies, and rare blueschist-facies conditions (Decker, 1980). Along its western margin, unit 73 also includes a less extensive sequence of moderately deformed and disrupted bedded turbiditic metasedimentary and metavolcanic rocks (Sitka Graywacke) metamorphosed under prehnite-pumpellyite– and greenschist-facies conditions; these rocks are included in the flysch facies of the Chugach terrane (Plafker and others, 1977). Blocks within the melange are Triassic or Jurassic, Late Jurassic (Tithonian), and Early Cretaceous (Valanginian) in age (Loney and others, 1975; Decker and others, 1980; Johnson and Karl, 1985; Brew and others, 1988). Deposition of the melange matrix took place, in part, during the Late Jurassic (Tithonian; Brew and others, 1988) and presumably continued during at least the Early Cretaceous (age of the youngest blocks; Decker, 1980; Johnson and Karl, 1985; Brew and others, 1988). The depositional age of the bedded rocks is unknown but is considered to be Cretaceous on the basis of correlation with lithologically similar rocks in the Valdez Group and Yakutat Group to the northwest (Plafker and others, 1977; Brew and Morrell, 1979).

Around the Gulf of Alaska, rocks of the flysch facies of the Chugach terrane (units 77 and 78, discussed below) have been shown to have a younger depositional, and presumably, metamorphic age than that of the inboard melange facies of the Chugach terrane (unit 72). In southeastern Alaska, however, the depositional age of the flysch facies is unknown, and separate metamorphic histories have not, as yet, been demonstrated for the two tectonic facies. For this reason, both of these tectonic facies are included in unit 73 and in two related polymetamorphic units (units 74 and 75) in which the early low-grade metamorphic episode was overprinted by regionally extensive, low-P, thermal metamorphism associated with Eocene plutonism (Loney and others, 1975; Loney and Brew, 1987).

The melange of both areas is considered to be a subduction complex consisting of oceanic sedimentary and igneous rocks, offscraped fragments of continental margin, or older subduction assemblages (Clark, 1973; Moore and Connelly, 1979; Plafker and others, 1977; Cowan and Boss, 1978; Decker, 1980; Winkler and others, 1981). The relation of crystallization to deformation observed on Kodiak and Afognak Islands suggests that metamorphism occurred during active underthrusting, continued after accretion of the subduction complex onto the overthrust plate, and was followed by late fracturing and cataclasis during uplift of the complex (Connelly, 1978). This evolution is a reasonable hypothesis for the other areas of melange as well. As pointed out by Connelly (1978), a similar progression of deformation has been proposed for the Franciscan Complex (subduction complex; Glassley and Cowan, 1975).

Accretion of the melange facies may have taken place over a long time span that extended throughout the Jurassic and Cretaceous. This prolonged period of accretion is suggested by Plafker and others (1989b) because of an apparent southward decrease in age from Late Triassic to mid-Cretaceous noted in the western Valdez Quadrangle (Winkler and others, 1981), and the probable convergent plate motion that is indicated along the southern margin of the (composite) Peninsula-Wrangellia-Alexander terrane during much or all of Jurassic and Cretaceous time (Engebretsen and others, 1985). As suggested by Plafker and others (1989b), the earliest accretion of Chugach terrane rocks is probably represented by the Jurassic intermediate- to high-P greenschist-facies metamorphism of unit 71.

Although the innermost parts of the melange complex of the Chugach terrane may have been accreted as early as Jurassic time, accretion and metamorphism of much of units 72 to 75 probably occurred in Early to mid-Cretaceous time, following accumulation of the youngest matrix material. Low-grade metamorphic minerals that were interpreted by Decker and his coworkers (1980) to have formed during subduction and accretion of unit 73 on Chichagof Island give Cretaceous K-Ar ages (106 to 91 Ma; Decker and others, 1980). Early Cretaceous (135 to 110 Ma) plutons that intrude unit 72 northeast of Anchorage are interpreted to be the result of near-trench plutonism that occurred during underthrusting of the melange complex (Pavlis, 1982; Pavlis and others, 1988a). An early Tertiary minimum metamorphic age is indicated by the facts that: (1) emplacement and metamorphism of the melange complex preceded the probable latest Cretaceous or early Tertiary emplacement and metamorphism of the tectonically underthrust flysch facies of the Chugach terrane that occurs outboard of unit 72; and (2) metamorphic assemblages within flysch and melange on Baranof Island in southeastern Alaska (units 74 and 75) are overprinted by thermal metamorphism associated with Eocene plutons (Loney and others, 1975).

Lithologically similar melange composed of structurally disrupted lenses of Upper Jurassic to Lower Cretaceous chert, argillite, conglomerate, mafic volcanic rocks, and rare blocks of exotic lithologies (Hudson and others, 1977b) also occurs within units 79 to 81 of the Yakutat terrane. The Yakutat terrane lies outboard of the Chugach terrane and has been correlated with it by Plafker and others (1977, 1989b). Melange of the Yakutat terrane occurs tectonically interleaved with flysch that makes up the dominant part of units 79 to 81 (discussed below); tectonic mixing is on a scale too small to allow delineation of the melange. As with the Chugach terrane melange, metamorphism of parts of the Yakutat terrane melange may have occurred in the Jurassic or Early Cretaceous. At one locality within unit 79, deformed melange is crosscut by a tonalite pluton (unit Kg) that gives discordant K-Ar ages of 96 Ma on hb and 84 Ma on bt (G. Plafker, unpublished data, 1978). It is also likely that, as is the case with the Chugach terrane melange, blueschist-facies metamorphism affected some of these rocks because glacial erratics of crossite-bearing metabasalt occur locally along Russell Fiord and Yakutat Bay (G. Plafker, unpublished data, 1987). The lithology and occurrence of these erratics are most compatible with a source in the ice-covered parts of the Yakutat terrane.

Late Cretaceous to early Tertiary low-P metamorphism of Chugach and Yakutat terranes flysch. A low-P facies series of low- to medium- (and locally high-) T metamorphosed flysch of the Chugach terrane (units 76 to 78) and Yakutat terrane (units 79 to 81) forms an arcuate belt of rocks that extends from the Sanak and Shumagin Islands off the southeast coast of the Alaska Peninsula in the west to the Saint Elias Mountains area south of Yakutat Bay in the east. Protoliths of the Chugach terrane (western and central parts of the belt) consist of a steeply dipping Upper Cretaceous (Maastrichtian) turbidite sequence of graywacke, slate, and locally intercalated conglomerate and volcanic rocks (Moore, 1973; Jones and Clark, 1973; Nilsen and Moore, 1979). Protoliths of the Yakutat terrane (eastern part of the belt) include these rock types, in addition to small amounts of the structurally interleaved melange (undifferentiated on Plate 4A) discussed above.

The flysch of the Chugach terrane forms a north-dipping accretionary prism that was underthrust beneath either the melange facies of that terrane or beneath the combined Peninsular and Wrangellia terranes to the north. Metamorphic grade within the accretionary prism increases progressively to the east around the Gulf of Alaska where it culminates in the polymetamorphic rocks of the (informal) Chugach metamorphic complex (unit 78) of Hudson and Plafker (1982), which crops out in the eastern Chugach Mountains and the Saint Elias Mountains. Flysch on Sanak and Shumagin Islands underwent laumontite-facies metamorphism that was ascribed by Moore (1973) to burial. Within the rest of the prism, flysch and related rocks underwent prehnite-pumpellyite–facies (unit 76) to lower greenschist-facies (unit 77 and M_1 of unit 78) low-P metamorphism that probably accompanied north-directed underthrusting of the Chugach terrane beneath the (composite) Peninsular-Wrangellia terrane, and the development of south-vergent folds, during latest Cretaceous to early Tertiary time (Moore and others, 1983; Sample and Moore, 1987; Nokleberg and others, 1989). Metamorphism during this low-grade episode postdates the Maastrichtian age of the protoliths and predates the intrusion of crosscutting tonalitic plutons that give K-Ar ages of 63 Ma on Kodiak Island (Byrne, 1982) and approximately 60 to 50 Ma (data summarized by Plafker and others, 1989b; and Sisson and others, 1989) in the Chugach Mountains.

The Chugach metamorphic complex (unit 78), forms an elongate, 200-km-long and less than 50-km-wide, east-west–trending belt made up of anda- and co-bearing schist and gneiss, and a core zone of sil-bearing migmatite. These rocks represent the deepest parts of the accretionary prism that makes up the Chugach terrane. They were initially metamorphosed during the greenschist-facies episode that affected the adjacent rocks of unit 77 and further heated under low-pressure amphibolite-facies conditions that developed during the widespread Eocene intrusion of tonalitic plutons. Metamorphic grade generally increases from the edges toward the elongate core of the metamorphic complex. This overall increase is independent of the exposed distribution of major plutons, and as Sisson and others (1989) point out, this progression is not solely a product of contact-metamorphic effects. In addition to the overall distribution of metamorphic grades, local contact metamorphism has produced high-grade rocks near the contacts of large felsic intrusions (Sisson and Hollister, 1988). Amphibolite-facies metamorphism and partial melting in the core of the metamorphic complex overlapped in time with the development of second-generation north-vergent folds, steeply dipping cleavage and schistosity, and near horizontal east-west–trending fold axes and lineations (Sisson and Hollister, 1988).

Intrusion and M_2 metamorphism of unit 78 postdated the accretion of the upper Paleocene to middle Eocene Orca Group of the Prince William terrane against the southern margin of the Chugach terrane (south of the Contact fault). This relation is indicated by the fact that an elongate 51-Ma tonalite pluton (Winkler and Plafker, 1981) and the metamorphic effects associated with it (M_2 of unit 85; Miller and others, 1984; Sisson and others, 1989) crosscut the Chugach/Prince William terrane boundary near Miles Glacier.

Metamorphic P throughout the metamorphic complex (unit 78) was between 2 and 3 kb (about 10 km) and T between 500°C near the edge of the complex to about 650°C in its migmatitic core (Sisson and others, 1989). These P-T estimates, together with those from the adjacent greenschist-facies unit 77, suggest a nearly isobaric P-T-time path in which the rocks of unit 78, already heated to greenschist-facies conditions at a depth of about 10 km during M_1, were further heated to amphibolite-facies conditions during M_2.

The tectonic origin of the heat that produced both the amphibolite-facies Chugach metamorphic complex and the widespread belt of early Tertiary plutons that crops out within the Chugach terrane and the outboard Prince William terrane is problematic. Marshak and Karig (1977) pointed out that in a normal subduction setting, the temperatures within an accretionary prism, such as the Chugach and Prince William terranes, are much too low to cause partial melting. They proposed that the anomalous near-trench plutonism was the result of eastward migration of a ridge-trench-trench triple junction along the continental margin, and subduction of the Kula-Farallon spreading ridge. A second but related hypothesis proposed by Plafker and others (1989b; see also Plafker and Berg, this volume) suggests that the anomalous heating during the early Tertiary metamorphic and plutonic episode resulted from the opening of a high-T oceanic-slab–free mantle window beneath the continental margin, as the subducting Kula Plate pulled away from the Farallon-Pacific Plate, in a manner analogous to one described along the San Andreas transform fault system by Dickinson and Snyder (1979). According to a third hypothesis, the high temperatures that produced the Chugach metamorphic complex resulted from a combination of heat introduced by extensive horizontal, as well as vertical, transport of fluids (beginning during initial greenschist-facies metamorphism) followed by felsic melts, both of which were generated from downdip in the subduction zone and involved either subduction of young hot oceanic crust at a high rate and a low angle, or subduction of a spreading ridge (Sisson and Hollister, 1988; Sisson and others, 1989).

A low-P facies series, similar to and probably correlative with the series that developed within the Chugach terrane flysch, also was formed within Late Cretaceous flysch and the structurally interleaved melange of the Yakutat terrane (Hudson and others, 1977b). The flysch and melange of the Yakutat Group are correlative with the flysch and melange of the Chugach terrane (Plafker and others, 1989b). The Yakutat Group, like the Chugach terrane in south-central and southeastern Alaska, is extensively intruded by Eocene granitoid plutons. Yakutat terrane rocks affected by the low-P metamorphic episode include: (1) laumontite- and prehnite-pumpellyite–facies rocks of unit 79; (2) a narrow fault-bounded sliver of greenschist-facies rocks of unit 80; and (3) epidote-amphibolite- and amphibolite-facies rocks of unit 81 (Hudson and others, 1977b). A steep thermal gradient is indicated by a rapid progressive increase in metamorphic grade from prehnite-pumpellyite–facies rocks to an area (oval in plan view) of amphibolite-facies rocks, with only a narrow interval of intervening greenschist-facies rocks (not shown on Plate 4A). Within the oval outcrop area of amphibolite-facies rocks, anda porphyroblasts are developed locally, and rocks are characterized by semigranoblastic textures, providing evidence that metamorphism of these rocks was dominantly thermal in nature. The oval outline suggests a buried pluton.

The rest of epidote-amphibolite– and amphibolite-facies unit 81 occurs as an elongate fault-bounded sliver that is separated from the amphibolite-facies rocks of the Chugach terrane (unit 83, described below) by the Fairweather fault to the northeast, and from the prehnite-pumpellyite–facies rocks of the Yakutat terrane by another major fault to the southwest. Tectonic shortening, by thrust faulting or strike-slip faulting with a significant dip-slip component, is suggested by the juxtaposition of high- and low-grade rocks along the southwestern fault contact, and by the absence of intervening greenschist-facies rocks.

Major metamorphism is considered to be latest Cretaceous to early Tertiary in age, based on: (1) the latest Cretaceous protolith age of the youngest rocks; (2) K-Ar ages on hb of about 65 Ma, determined for amphibolites (Hudson and others, 1977b), and (3) the interpretation that this unit shared a common metamorphic history with that of the adjacent Chugach terrane discussed above.

Late Cretaceous to mid-Tertiary intermediate(?)-P metamorphism of Chugach terrane flysch. A sequence of transitional greenschist- to amphibolite-facies rocks (unit 82) and amphibolite-facies rocks (unit 83) crops out along the eastern margin of the Gulf of Alaska in the Saint Elias Mountains and Fairweather Range (Brew, 1978; Hudson and others, 1977b; Hudson and Plafker, 1982). Protoliths are interpreted as being turbidites and tholeiite of Late Cretaceous age because of lithologic similarity with the flysch facies of the Chugach terrane (Plafker and others, 1977 and this volume, Chapter 12; Barker and others, 1985). Most boundaries of the sequences are faults. The two fault blocks that make up the higher grade part of the sequence are presumed to have shared the same metamorphic history and to have subsequently been separated by right-lateral displacement along the Fairweather fault (Dusel-Bacon and others, 1994b).

The timing and number of metamorphic episodes that affected units 82 and 83 are unknown. A Cretaceous maximum age of metamorphism is proposed on the basis of the probable age of the protoliths. At least some of the metamorphism is known to have occurred prior to the intrusion of crosscutting plutons of intermediate composition that have K-Ar ages on hb of 61 ± 2 Ma in the southeastern Yakutat Quadrangle (Hudson and others, 1977a) and K-Ar ages of about 52 Ma in the southwestern Skagway Quadrangle (George Plafker, unpublished data, 1978). A K-Ar age on hb of 67 Ma from amphibolite in the Nunatak Fiord area, 55 km northeast of Yakutat, suggests a latest Cretaceous metamorphic age (Barker and others, 1985). In the same area, however, K-Ar ages on hb and bt from metamorphic rocks between Nunatak Fiord and the southwestern Skagway Quadrangle range from about 23 to 19 Ma, which suggests an additional, or alternative, Miocene metamorphic age; these K-Ar ages fall close to or within the 37- to 21-Ma range of K-Ar ages from widespread felsic intrusive rocks (Hudson and others, 1977b; George Plafker, unpublished data, 1978). Farther to the south, in the area northwest of Cross Sound, metamorphism appears to predate and to be unrelated to the intrusion of Oligocene (28 ± 8-Ma $^{40}Ar/^{39}Ar$ age on hb; Loney and Himmelberg, 1983) gabbroic plutons (Dusel-Bacon and others, 1994a).

The origin of the metamorphic episode(s) that affected units 82 and 83 is unknown. In the northern and central part of the area made up of these units, the spatial relation of the higher grade metamorphic rocks and Tertiary plutons suggests a genetic link. Unlike the Chugach metamorphic complex (unit 78, and adjacent parts of unit 77), described above, the thermal history of at least some parts of units 82 and 83 appears to be complicated by the occurrence of multiple periods of plutonism (and perhaps metamorphism) in the Tertiary and also probably by a much younger uplift history, as indicated by the exposure of the Miocene plutonic rocks and discordant biotite-hornblende pairs. Another difference between these two metamorphic sequences is the fact that ky, indicative of intermediate-P conditions, occurs at one locality in unit 83, whereas anda, indicative of low-pressure conditions, is present in unit 78.

Retrograde effects in overlying units corresponding with metamorphism and underthrusting of Chugach terrane flsych. North-directed underthrusting of Chugach terrane flysch beneath the southern margin of the (composite) Peninsular-Wrangellia-Alexander terrane, which is thought to have occurred during latest Cretaceous to early Tertiary time, may be responsible for (1) greenschist-facies retrograde metamorphism (M_2) of overlying unit 58 (Haley Creek metamorphic assemblage; Wallace, 1981, 1984; Nokleberg and others, 1989); (2) prehnite-pumpellyite–facies retrograde metamorphism (M_2) of overlying unit 59 (part of the Strelna metamorphics of Plafker and others, 1989b); and (3) a low-grade overprint in the Jurassic plutons that intrude unit 59 (Dusel-Bacon and others, 1994b). Characteristic M_2 minerals in unit 59 are pr (which occurs most commonly as lenses that bow apart the cleavage planes of metamorphic and igneous bt, and less commonly as veins), ep, ch, and rare pu or lau (Dusel-Bacon and others, 1994b). Calcium-rich fluids which formed veins of pr, and probably played a part in crystallization of M_2 phases, may have been derived from the underlying graywackes of the Valdez Group (Chugach terrane flysch). Lower plate rocks belonging to the Valdez Group may underlie unit 59 at a fairly shallow level, as is the case with the correlative Haley Creek metamorphic assemblage (unit 58), where the Valdez Group occurs at a depth of only 1 km (Page and others, 1986). Fracturing and fluid migration probably occurred during extension of upper plate rocks belonging to the Strelna metamorphics as they were underthrust by rocks of the Valdez Group. A possible analog to the proposed formation of pr in this unit is provided by the study of pr in plutonic and metamorphic rocks of the Salinian block of California. Ross (1976) proposed that hydrous solutions, derived from a "substratum" of Franciscan(?) graywackes, migrated through fractured rocks of the tectonically thinned margin of the Salinian block near the Sur fault zone, causing widespread crystallization of Ca-Al silicates.

Early Tertiary low-grade metamorphism of Prince William and Ghost Rocks terranes. A weakly metamorphosed and strongly deformed subduction complex of flysch and tholeiite, shown as prehnite-pumpellyite–facies unit 84 and related polymetamorphosed unit 85, crops out seaward of the slightly older Chugach terrane subduction complex. This complex is separated from the Chugach terrane by the landward-dipping Contact fault system (Plafker and others, this volume, Chapter 12). In the northern Gulf of Alaska, units 84 and 85 make up the Orca Group of the Prince William terrane (Tysdal and Case, 1979; Winkler, 1976; Winkler and Plafker, 1981; Helwig and Emmet, 1981). On Kodiak Island, unit 84 constitutes the Ghost Rocks Formation of Byrne (1982) of the Ghost Rocks terrane (Connelly, 1978; Byrne, 1986). The depositional age of the Orca Group is late Paleocene to middle Eocene (Plafker and others, 1985). Eocene fossils are reported from the Ghost Rocks Formation (Connelly, 1978), but the bulk of the formation accumulated during earliest Paleocene time with melange units including some Late Cretaceous material (Moore and others, 1983). Units 84 and 85 are generally isoclinally folded and include metagraywacke, argillite, metalimestone, greenstone, and mafic schist. Intense localized shearing has produced many zones of tectonic melange on Kodiak Island (Byrne, 1984) and in the southwestern part of Prince William Sound. In general, rocks of the Orca Group of unit 84 show a gradual increase in metamorphic grade from south to north (Goldfarb and others, 1986).

Low-grade metamorphism and deformation within both the Ghost Rocks and Prince William terranes took place during early Tertiary time and predated, perhaps by very little, the intrusion of the early Tertiary plutons that stitch them to the Chugach terrane. On Kodiak Island, these plutons, dated at 62 Ma, contact metamorphosed the Upper Cretaceous and Paleocene Ghost Rocks Formation and cut its structural fabric (Moore and others, 1983). Eocene and younger rocks in the Kodiak Islands are only slightly altered—further supporting a Paleocene metamorphic age for that area (Moore and others, 1983). Intrusion and metamorphism of the upper Paleocene to middle Eocene Orca Group occurred slightly later than did metamorphism of the Ghost Rocks Formation. From Prince William Sound to the east, 53- to 48-Ma plutons crosscut and contact metamorphosed already deformed and weakly metamorphosed rocks of the Orca Group (Winkler and Plafker, 1981; Miller and others, 1984). If the timing of metamorphism parallels that of plutonism, metamorphism began earlier in the west than in the east.

Low-grade metamorphism was probably associated with accretion of the terranes. Asymmetric north-dipping south-verging folds, which developed within unit 84 and the adjacent greenschist-facies rocks of the Chugach terrane (unit 77) near the Contact fault zone west of the Copper River, are interpreted as being synaccretionary structures, developed during oblique thrust convergence (Plafker and others, 1986). Similarly, Byrne (1986) proposed that the development of conjugate folds and spaced cleavage within the Ghost Rocks Formation occurred as a result of subhorizontal shortening of the accretionary complex during underthrusting.

East of the Copper River within unit 85, the low-grade mineral assemblages and associated structures, presumed to have formed during accretion, are overprinted by gt-co-bearing, low-P, amphibolite-facies assemblages. These higher grade assemblages were produced during the regionally extensive, predominantly thermal, metamorphism that accompanied the intrusion of Eocene plutons in this area of the Prince William and adjacent Chugach terranes (Miller and others, 1984; Sisson and others, 1989). The metamorphic effects of this episode extend across the Contact fault zone and correspond to the M_2 episode of unit 78 (Chugach metamorphic complex), as discussed in a previous section.

Southeastern Alaska

Middle Cambrian to Early Ordovician metamorphism. The oldest metamorphic episode in southeastern Alaska occurred under dominantly greenschist-facies conditions (M_1 of unit 86) on southern Prince of Wales Island (Gehrels and Saleeby, 1987b), in small areas (not shown on Plate 4A) on the southern tip of Gravina Island and adjacent islands (Gehrels and others, 1987), and under amphibolite-facies conditions (M_1 of unit 87) on southern Dall Island to the west (G. E. Gehrels, oral communication, 1987; Gehrels and Berg, this volume). Both of these units consist of metavolcanic and metasedimentary rocks. They have been described as the Wales Group (Buddington and Chapin, 1929; Herreid and others, 1978; Eberlein and others, 1983) and most recently as the informally designated Wales metamorphic suite (Gehrels and Saleeby, 1987b). Unit 86 consists of greenschist-facies mafic schist, greenstone, pelitic schist, phyllite, and marble, and small areas of amphibolite-facies schist and metaplutonic rocks. Unit 87 contains amphibolite-facies equivalents of unit 86. Preliminary U-Pb ages on zircon indicate Middle and Late Cambrian protolith ages for interlayered metaplutonic bodies, and thus Late Proterozoic and/or Cambrian protolith ages for the associated metasedimentary and metavolcanic rocks of these two units (Gehrels and Saleeby, 1987b).

In most areas, protolith features are obscured by penetrative metamorphic recrystallization and a high degree of flattening (Gehrels and Saleeby, 1987b). Mineral lineations are common in the amphibolite-facies rocks, and are present locally in mafic schist (Herreid and others, 1978; Eberlein and others, 1983).

A Late Cambrian and Early Ordovician age is indicated for M_1 metamorphism because: (1) metaplutonic rocks of Middle and/or Late Cambrian age are metamorphosed and deformed (Gehrels and Saleeby, 1987b), but uppermost Lower and Middle Ordovician strata of unit 88 that occur nearby and probably overlie unit 86 are only weakly metamorphosed and lack the penetrative metamorphic fabric characteristic of unit 86 (Eberlein and others, 1983; Gehrels and Saleeby, 1987b); and (2) rocks from unit 86 yield K-Ar ages of about 483 Ma (Turner and others, 1977). This metamorphic episode is part of the Wales orogeny of Gehrels and Saleeby (1987a, b). Although retrograde metamorphic effects have not been reported for units 86 and 87, geologic relations indicate that they were probably affected by the same low-grade metamorphic episode during Silurian and earliest Devonian time (shown as M_2) that is recorded in adjacent unit 88.

Silurian to earliest Devonian low- to medium-grade metamorphism. A weakly to moderately developed, Silurian to earliest Devonian metamorphic episode is recorded in the prehnite-pumpellyite–facies rocks of unit 88; in the lower greenschist-facies rocks of unit 89; and in the M_1 episode of the polymetamorphosed greenschist- and locally epidote-amphibolite-facies rocks of unit 90. Included in these units are basaltic to rhyolitic metavolcanic rocks, metasedimentary rocks, metachert, and metalimestone of late Early Ordovician to Early Silurian protolith age, and quartz dioritic plutons of Middle Ordovician to Early Silurian age (Eberlein and others, 1983; Gehrels and Berg, 1992; this volume).

The metamorphic grade is lowest on southern Prince of Wales Island and increases westward and eastward. Unit 88 is not penetratively deformed, and relict sedimentary and volcanic textures are widespread. Crosscutting plutonic rocks, also assumed to have been weakly metamorphosed, are locally brecciated. Metamorphism in the westernmost exposure of unit 88 increases southward into semischistose rocks of unit 89. East of Prince of Wales Island, the polymetamorphosed rocks of unit 90 generally are cataclastically deformed and show no pronounced foliation; locally rocks are schistose (Berg and others, 1988; Gehrels and others, 1983). Minor retrogressive metamorphism, apparent in the higher grade rocks of unit 90, is believed to have occurred during one or more of the Cretaceous metamorphic episodes, described below.

A Silurian to earliest Devonian metamorphic age is indicated for the following reasons: (1) strata of Early Silurian age are metamorphosed, but overlying strata of middle Early Devonian age and younger are either unmetamorphosed, as on Prince of Wales Island, or only affected by the Cretaceous metamorphic episode, as on the islands to the east (Gehrels and others, 1983); (2) metamorphosed Silurian rocks of unit 87 are cut by an undeformed latest Silurian to earliest Devonian (408 ± 10 Ma) pyroxenite (Eberlein and others, 1983; G. E. Gehrels, written communication, 1984); and (3) Late Silurian trondhjemite dikes crosscut the foliation of metadioritic rocks of Late Ordovician to Early Silurian protolith age, but have the same post-metamorphic deformational features as their wall rocks, indicating that the trondhjemite dikes and plutons were intruded before the final deformation that occurred during the latter stages of the episode (Gehrels and others, 1983; G. E. Gehrels, oral communication, 1985). Metamorphism of this unit is considered to have been part of an orogenic event referred to as the Klakas orogeny (Gehrels and others, 1983; Gehrels and Saleeby, 1987b).

Early Cretaceous metamorphism. Metamorphism in units 95 to 98 was apparently associated with the intrusion of elongate bodies of highly foliated tonalite and diorite of Early Cretaceous age (120 to 110 Ma; Loney and others, 1967; Decker and Plafker, 1982; Dusel-Bacon and others, 1994a; Brew, this volume). These units are included in the Alexander terrane.

Unit 95 crops out near Glacier Bay and on Chicagof Island and consists of a diverse assemblage of amphibolite-facies and hornblende-hornfels facies pelitic and semipelitic schist and gneiss, marble, and amphibolite, and minor amounts of lower grade greenstone and greenschist; protoliths are sedimentary and volcanic rocks of Silurian to Devonian age (Loney and others, 1975; Brew, 1978). On Chichagof Island, rocks are intensely folded, and there is a complete gradation in metamorphic textures between hornfels and foliated rocks (Loney and others, 1975). Structural trends in metamorphic rocks parallel those of the Cretaceous plutons. The general parallelism between the foliate fabric of the plutons, pluton/wall-rock contacts, and structures in the wall rocks, suggests that plutonism, folding, and thermal and dynamothermal metamorphism all took place as part of a continuum that occurred under roughly the same stress conditions.

Units 96 to 98 crop out on Admiralty Island and the adjacent mainland and form a sequence of metasedimentary, metavolcanic, and metaplutonic rocks that range in grade from prehnite-pumpellyite facies (unit 96), to greenschist (or albite-epidote-hornfels) facies (unit 97), and finally to undifferentiated greenschist (or albite-epidote-hornfels) facies and amphibolite (or hornblende-hornfels) facies (unit 98). Protoliths range in age from Ordovician to Early Cretaceous (references given in Dusel-Bacon and others, 1994a). Most medium and higher grade rocks are penetratively deformed. Intrusive rocks of the largest batholith on Admiralty Island are poorly to well foliated, but the trend of the foliation relative to that of the country rocks has not been studied in detail (Lathram and others, 1965). Evidence that metamorphism was associated with late Early Cretaceous plutonism consists of an apparent progressive increase in metamorphic grade toward the plutons, and a merging of contact aureoles with large areas of dynamothermally metamorphosed phyllite, schist, and gneiss of unit 98 (Loney and others, 1967).

The age and origin of metamorphism of greenschist- and, very locally, amphibolite-facies rocks of unit 92 is unknown. This unit crops out northeast of Glacier Bay and is bounded on the east by the Denali fault. Protoliths include mafic volcanic rocks, sedimentary rocks, and limestone, and have been correlated with rocks of Silurian to Permian age (MacKevett and others, 1974). Unit 92 also is intruded by a pluton of the belt of 120- to 110-Ma plutons that are thought to have been associated with metamorphism of units 95 to 98 to the south; therefore, metamorphism of unit 92 may have had a similar origin. An alternative and slightly older metamorphic history is suggested by geologic evidence from the apparent continuation of this unit about 100 km to the northwest in Canada. In that area, regional metamorphism and deformation appear to have occurred between Late Triassic and Early Cretaceous time and may have been associated with latest Jurassic to earliest Cretaceous (150 to 130 Ma) plutonism (R. B. Campbell and C. J. Dodds, written communication, 1986). This possible metamorphic episode is analogous to that discussed for unit 64 in the previous section (area of southern Alaska between the McKinley and Denali faults and the Border Ranges fault system).

Mid-Cretaceous low-grade metamorphism. Prehnite-pumpellyite– to lower greenschist-facies metasedimentary rocks, intermediate to mafic metavolcanic rocks, metalimestone, and metachert (unit 100 and related polymetamorphic unit 102) crop out in a 150-km-long southeast-trending belt from Kupreanof Island to Cleveland Peninsula. Protoliths of rocks correlated with the protoliths range in age from Late Triassic to mid-Cretaceous–Albian or Cenomanian (Berg and others, 1972; Brew and others, 1984; Gehrels and Berg, 1992). Rocks have been weakly metamorphosed (ch-, ac-, and rarely bt-zone assemblages) and locally are intensely folded and faulted. Metasedimentary rocks are generally poorly foliated, but fine-grained variants have good cleavage. Greenschist and greenstone locally contain abundant relict pyroxene phenocrysts (Brew and others, 1984).

Regional low-grade metamorphism is known to predate the intrusion of Alaskan-type mafic-ultramafic bodies that have yielded K-Ar ages of 110 to 100 Ma (Lanphere and Eberlein, 1966; Clark and Greenwood, 1972; Brew and others, 1984; Douglass and Brew, 1985). The late Early Cretaceous minimum metamorphic age indicated by these dates is close to the Albian or Cenomanian protolith age of the youngest rocks included in this unit. The geographic limits of the area affected by this episode are not known with certainty; they may have extended into the area shown as unit 101, discussed below.

Early Late Cretaceous intermediate-P metamorphism associated with the intrusion of 90-Ma plutons. Amphibolite-facies pelitic schist, quartzofeldspathic schist and gneiss, amph schist and gneiss, and minor amounts of marble, calc-schist, migmatite, and metaplutonic rocks (Berg and others, 1988; Brew and others, 1984) of unit 103 crop out from near Wrangell to Revillagigedo Island adjacent to and extending some distance from 90-Ma plutons. Protoliths of unit 103 are considered to include Jurassic and/or Cretaceous flysch, Permian and Triassic limestone, and intrusive rocks of probable Jurassic to Cretaceous age. Rocks are sufficiently recrystallized so that neither the original textures nor the original structures remain.

Kyanite, indicative of intermediate-P metamorphic conditions, is common in the st + gt ± sil–bearing pelitic schist of unit 103 (Berg and others, 1988; Douglass and Brew, 1985) and in aureoles developed around the 90-Ma plutons that intrude unit 102. In the northern part of unit 103 and within adjacent unit 102, relict anda also has been observed in pelitic schist from the aureoles of 90-Ma plutons (Plate 4A). In these areas, relict porphyroblasts of statically formed anda have been replaced by static (radial) ky or in some locations by mineral aggregates of intergrown ky and st (Dusel-Bacon and others, 1994a). This crystallization sequence of the Al_2SiO_5 polymorphs appears to indicate an increase from low- to intermediate-P conditions in the northern part of this unit during intrusion and metamorphism.

Most of the 90-Ma plutons referred to above are of intermediate composition and contain primary gt and epidote; they are part of a plutonic belt that extends from southern Revillagigedo Island north to the vicinity of Haines (Zen and Hammarstrom, 1984a; Brew, this volume). Sillimanite and ky isograds are located around the large 90-Ma plutons in the areas of Wrangell and northern Revillagigedo Islands, and metamorphic grade increases toward the plutons (Dusel-Bacon and others, 1994b), providing evidence that metamorphism was associated with plutonism. These plutons (shown as Kg with a "+" overprint on Plate 4A) are interpreted as having been emplaced during the waning stages of metamorphism and deformation (Brew and others, 1984; Douglass and Brew, 1985; Berg and others, 1988).

Geothermometric and geobarometric data from two samples of gt-ky schist in the southern part of unit 103 indicate a final equilibration T and P of 600°C, 7.5 to 8.5 kb, and 575 to 600°C, 8.5 to 9.2 kb for mineral rims (M. L. Crawford, written communication, 1983; Dusel-Bacon and others, 1994a). A similar, moderately high, P of final crystallization has been proposed for the primary gt- and epidote-bearing 90-Ma plutons that intruded this unit late in, or immediately following, the metamorphic episode. Zen and Hammarstrom (1984b), citing experimental data on the composition of magmatic gt and on the P required to crystallize magmatic epidote, propose that the magma began to crystallize at a minimum P of 13 to 15 kb (about 40 to 50 km) and finally crystallized at about 6 to 10 kb (about 20 to 30 km). The combination of the high- to intermediate-P magmatic and crystallization history inferred for the plutons, and the occurrence of relict anda, indicative of low-P conditions (less than 3.8 kb; Holdaway, 1971), in their aureoles in the Wrangell Island area, is indeed problematic.

Greenschist-facies unit 101 crops out south of amphibolite-facies unit 103 in the area of Revillagigedo Island and consists of metasedimentary and metavolcanic rocks, and minor amounts of marble and metaplutonic rocks (Berg and others, 1988). Protolith ages range from Devonian to Early Cretaceous (Berg and others, 1988; Gehrels and others, 1987). On Revillagigedo Island and the peninsula to the northwest, metamorphic foliation dips to the northeast, and the metamorphic sequence is cut by southwest-vergent thrust faults (Berg and others, 1988; Rubin and Saleeby, 1987).

Metamorphism of unit 101 also may have been part of the same thermal episode that culminated in the intrusion of early Late Cretaceous (approximately 90 Ma) plutons, as was the case for unit 103. This relation is suggested by the observation made by Berg and others (1988) that metamorphic mineral assemblages show an apparent gradational increase in grade from the southwest to the northeast, beginning in the greenschist-facies rocks of unit 101 and continuing into the amphibolite-facies unit 103. An argument against this interpretation is the recent detailed mapping by M. L. Crawford (oral communication, 1988), which indicates an abrupt, rather than gradational, increase in metamorphic grade at the boundary between units 101 and 103. An alternative, but not necessarily mutually exclusive, interpretation is that regional greenschist-facies metamorphism is a higher grade equivalent of the mid-Cretaceous episode that affected unit 100

(Dusel-Bacon and others, 1991a). This hypothesis is based on an extension of the interpretation of the metamorphic history of unit 100 to the northwest (Brew and others, 1984; Douglass and Brew, 1985) and on similarities noted during a reconnaissance of northern Revillagigedo Island by D. A. Brew (unpublished data, 1983).

K-Ar age determinations on amphibolite- and greenschist-facies rocks on Revillagigedo Island show a decrease in maximum apparent ages northward and eastward (Smith and Diggles, 1981; Berg and others, 1988). $^{40}Ar/^{39}Ar$ plateau ages on hb from plutonic and metamorphic rocks on Revillagigedo Island and from similar rocks to the south in British Columbia, show a similar age pattern, ranging from greater than 90 Ma in the west to about 56 Ma near the eastern boundary of unit 103 (Sutter and Crawford, 1985). Uplift of this block of rocks was greatest and occurred latest in the eastern part of the block. The eastern boundary of the block is spatially related to the Coast Range megalineament (Brew and Ford, 1978)—a topographic, structural, and geophysical feature that appears to have been near the western limit of large-scale regional uplift of southeastern Alaska and adjacent parts of British Columbia beginning in early Tertiary time (Crawford and Hollister, 1982, 1983).

Late Cretaceous to early Tertiary synkinematic metamorphism and plutonism. An intermediate-P (Barrovian) metamorphic sequence consisting of prehnite-pumpellyite–facies rocks (unit 105), greenschist-facies rocks (unit 106), and amphibolite-facies rocks (unit 107) crops out as an elongate, northwest-trending belt along the mainland of southeastern Alaska from Skagway to the area east of Wrangell. The Barrovian sequence increases in metamorphic grade to the northeast (Dusel-Bacon and others, 1994a, and references contained therein). Protoliths are thought to be clastic sedimentary rocks, mafic to intermediate volcanic, intrusive, and volcanogenic sedimentary rocks, limestone, and chert. Few fossils have been found in these rocks, but protolith ages are considered to be Permian, Triassic, and Jurassic to Cretaceous (Brew and Ford, 1984).

An early foliation, presumably formed during either or both of the low-grade episodes recorded in adjacent units 93 and 96, is locally detectable in unit 105 and the lowest grade part of unit 106 (Brew and others, 1984). With increasing metamorphic grade, rocks develop well-defined crenulation cleavage and transposition layering. Higher grade rocks in unit 106 and all those in unit 107 are well foliated and lineated. Foliation in gneissic rocks is locally anastamosing or lenticular. Mineral isograds marking the first appearance of bt, gt, st, ky, and sil trend north-northwest, generally parallel with elongate quartz dioritic plutons referred to as a tonalite sill complex by Brew and Ford (1981; also see Brew, this volume). It is shown on Plate 4A as the 600-km-long synkinematic intrusive unit TKg, just east of the Coast Range megalineament. Isogradic surfaces dip moderately to steeply northeast (Ford and Brew, 1973, 1977a; Brew and Ford, 1977), and hence are inverted. In the area east of Kupreanof Island, isogradic surfaces appear to steepen northeastward toward the Coast Range megalineament (Brew and others, 1984).

Garnet-biotite geothermometry for the sil-zone rocks of unit 107 (Himmelberg and others, 1984) indicates an equilibration T of about 690°C (calibration of Thompson, 1976) or 750°C (calibration of Ferry and Spear, 1978). The absence of anda and the abundance of ky indicates a minimum equilibration P of about 3.8 kb (Holdaway, 1971). Preliminary sphalerite geobarometry of three massive-sulfide deposits within the megalineament zone on the mainland east of central Admiralty Island indicates a general P range that is consistent with values of 3.8 to 4.5 kb at 575°C, calculated from silicate mineral equilibria of st-zone rocks in the same general area (Stowell, 1985). Geobarometry calculated by several methods indicates a P of about 9 kb for rocks to the north, but the data exhibit large scatter (G. R. Himmelberg, unpublished data, 1987; Brew and others, 1987).

Amphibolite-facies rocks on the mainland east and southeast of Wrangell Island (unit 108) contain lithologies similar to those in the amphibolite-facies part of the Barrovian metamorphic sequence (unit 107) to the north. Although they occur on strike, they are differentiated on the basis of possible P differences during final equilibration of the rocks. Unit 108 is a heterogeneous complex of migmatite, massive to foliated or gneissic batholiths, and smaller plutons that enclose metamorphic screens and roof pendants of paragneiss (Berg and others, 1988). Protolith ages are not generally known, but those of paragneiss are probably Paleozoic or Mesozoic (Brew and Ford, 1984; Berg and others, 1988), and at least some of those of orthogneiss are Early Cretaceous (Barker and Arth, 1984; Hill, 1984).

Garnet and sil are common constituents of the paragneiss and pelitic schist of unit 108; co occurs locally in the pelitic schist. Intermediate- to low-P conditions are suggested by mineral assemblages, and by thermobarometric data that indicate P-T conditions of 3.5 to 4.5 kb and 650°C (calibration of Ferry and Spear, 1978) for a sample of sil-gt-bt-qz-pl schist in northeastern Ketchikan Quadrangle (M. L. Crawford, written communication, 1983; Dusel-Bacon and others, 1994a). In many areas, paragneiss grades downward and laterally into gneissic granodiorite; elsewhere it is in sharp contact with plutonic rocks, or passes gradually into them through a zone of migmatite (Berg and others, 1988). Quartz diorite of the tonalite sill is commonly at least weakly foliated, and in the extreme case, is gneissic. Its foliation generally strikes north or northwest and is parallel to the outcrop trend and to the internal structure of the adjoining metamorphic rocks of unit 108. Much of the quartz diorite also has mylonitic or cataclastic textures, such as undulose quartz and granulated grain boundaries (Berg and others, 1988).

Metamorphism of units 105 to 108 (and M_2 of unit 104) is considered to be slightly pre- and synkinematic with the latest Cretaceous and early Tertiary mesozonal intrusion of the tonalite sill. Intrusion of the sill has been dated by U-Pb zircon methods of about 69 and 62 Ma in the north (Gehrels and others, 1984) and at about 58 to 55 Ma in the south (Berg and others, 1988). Evidence for the association of metamorphism and plutonism consists of the increase in metamorphic grade toward the sill; general parallelism between the sill and isograds that define the

Barrovian metamorphic sequence; and parallelism of foliation, contacts, and locally developed lineation in the sill with structural elements in the adjacent metamorphic rocks. In the Juneau area, truncation of metamorphic isograds by the sill, and parallelism between foliation in the sill and that in the metamorphosed wall rocks, suggest that intrusion of the sill accompanied a late stage of the regional metamorphism—a stage occurring after the thermal maximum but before the end of penetrative deformation (Ford and Brew, 1977b). Epizonal plutons (Tg) intruded the eastern part of the amphibolite-facies units during Eocene time. These Eocene plutons are surrounded by low-P high-T metamorphic rocks and by migmatites. The original eastern limit of the Barrovian metamorphism is obscured by the Eocene intrusions, but limited evidence indicates that the low-P metamorphism around the Eocene plutons was superimposed over the previous intermediate-P metamorphism.

A P-T-time path has been determined for the plutonic and metamorphic sequence that crops out near Prince Rupert, British Columbia, across the international boundary from unit 108. Many aspects of that path may also apply to the metamorphic history of unit 108 and perhaps also of units 105 to 107. Near Prince Rupert, metamorphic reactions, thermobarometric data, and isotopic data indicate that rocks correlative with those of unit 108 were uplifted and eroded at a rate of about 1 mm/yr between about 60 and 48 Ma, beginning at a depth of about 20 km and terminating at about 5 km (Hollister, 1982; revised in Crawford and others, 1987). The emplacement of the elongate 60-Ma Quotoon pluton, which is the continuation of the Alaskan tonalite sill, apparently occurred at deep levels during the early stages of uplift. Emplacement of intermediate and felsic plutons along the eastern margin of the complex occurred at high levels during the end stages of uplift. According to the Canadian work, metamorphism continued throughout the period of uplift under evolving P-T conditions (Hollister, 1982; Crawford and Hollister, 1982, 1983). Because of similarities in style and conditions of metamorphism between rocks west of the megalineament in British Columbia (correlative with unit 101) and early metamorphic relicts found in rocks east of the megalineament (correlative with unit 108), Crawford and Hollister (1982, 1983) suggest that high-grade rocks east of the megalineament were also metamorphosed during the episode associated with the intrusion of the 90-Ma plutons, discussed above. The high-grade crustal block east of the megalineament is thought to have remained at depth until it was displaced by rapid vertical uplift as a result of the weakening of the crust by anatexis and the development of melt-lubricated shear zones—particularly the one represented by the tonalite sill, along which rapid vertical movement was concentrated (Hollister and Crawford, 1986; Crawford and others, 1987).

The tectonic environment of the widespread plutonometamorphic episode that occurred along the western edge of the Coast Mountains in early Late Cretaceous and early Tertiary time was dominated by crustal thickening due to the accretion of an outboard terrane to the west (Plafker and Berg, this volume, Chapter 33). Monger and others (1982, 1983) proposed that the plutonometamorphic belt of the Coast Mountains of southeastern Alaska and British Columbia developed as a welt resulting from the accretion of the amalgamated Wrangellia and Alexander terranes against the previously accreted Stikinia and other terranes in Cretaceous and Tertiary time. Brew and Ford (1983) interpret the stratigraphic and paleomagnetic evidence to suggest that the Alexander and Stikinia terranes are one and the same and that a rift developed in the megaterrane as it migrated northward. According to their model, that rift was filled with the flysch and volcanic rocks of the Gravina belt (Berg and others, 1972) during Late Jurassic and Early Cretaceous time. They propose that the plutonometamorphic belt formed as a result of the closure of the rift and the resultant crustal thickening during accretion of the Chugach terrane—a terrane that lies to the west and southwest of the Alexander and Wrangellia terranes.

Workers in the southern extension of this metamorphic belt near Prince Rupert, British Columbia, concur with the terrane accretion model of Monger and others (1982). They propose that the crustal thickening resulted from west-directed tectonic stacking of crustal slabs along east-dipping thrusts. In places the thrusts were possibly lubricated by the intrusion of melt (parent magma of intermediate epidote-bearing plutons and sills) generated at the base of the crust (Hollister and Crawford, 1986; Crawford and others, 1987). These thrust faults, which were synchronous with 100- to 90-Ma plutonism near Prince Rupert, may be correlative with thrusts identified on Revillagigedo Island in Alaska (M. L. Crawford, oral communication, 1987).

REFERENCES CITED

Aleinikoff, J. N., and Nokleberg, W. J., 1985, Age of Devonian igneous-arc terranes in the northern Mount Hayes Quadrangle, eastern Alaska Range, Alaska, *in* Bartsch-Winkler, S., ed., The United States Geological Survey in Alaska: Accomplishments during 1984: U.S. Geological Survey Circular 967, p. 44–49.

Aleinikoff, J. N., Nokleberg, W. J., and Herzon, P. L., 1982, Age of intrusion and metamorphism of the East Susitna batholith, Mount Hayes B-6 Quadrangle, eastern Alaska Range, Alaska, *in* Coonrad, W. L., ed., The United States Geological Survey in Alaska; Accomplishments during 1980: U.S. Geological Survey Circular 844, p. 97–100.

Aleinikoff, J. N., Dusel-Bacon, C., and Foster, H. L., 1984a, Uranium-lead isotopic ages of zircon from sillimanite gneiss and implications for Paleozoic metamorphism, Big Delta Quadrangle, east-central Alaska, *in* Coonrad, W. L., and Elliott, R. L., eds., The United States Geological Survey in Alaska; Accomplishments during 1981: U.S. Geological Survey Circular 868, p. 45–48.

Aleinikoff, J. N., Foster, H. L., Nokleberg, W. J., and Dusel-Bacon, C., 1984b, Isotopic evidence from detrital zircons for Early Proterozoic crustal material, east-central Alaska, *in* Coonrad, W. L., and Elliott, R. L., eds., The United States Geological Survey in Alaska; Accomplishments during 1981: U.S. Geological Survey Circular 868, p. 43–45.

Aleinkoff, J. N., Dusel-Bacon, C., and Foster, H. L., 1986, Geochronology of augen gneiss and related rocks, Yukon-Tanana terrane, east-central Alaska: Geological Society of America Bulletin, v. 97, p. 626–637.

——, 1987, Lead isotopic fingerprinting of tectono-stratigraphic terranes, east-central Alaska: Canadian Journal of Earth Sciences, v. 24, p. 2089–2098.

Aleinikoff, J. N., Plafker, G., and Nokleberg, W. J., 1988, Middle Pennsylvanian plutonic rocks along the southern margin of the Wrangellia terrane, *in* Galloway, J. P., and Hamilton, T. D., eds., Geologic studies in Alaska by the

U.S. Geological Survey during 1987: U.S. Geological Survey Circular 1016, p. 110–113.

Armstrong, R. L., Harakal, J. E., Forbes, R. B., Evans, B. W., and Thurston, S. P., 1986, Rb-Sr and K-Ar study of metamorphic rocks of the Seward Peninsula and southern Brooks Range, Alaska, *in* Evans, B. W., and Brown, E. H., eds., Blueschists and eclogites: Geological Society of America Memoir 164, p. 185–203.

Bacon, C. R., Foster, H. L., and Smith, J. G., 1990, Rhyolitic calderas of the Yukon-Tanana terrane, east central Alaska; Volcanic remnants of a mid-Cretaceous magmatic arc: Journal of Geophysical Research, v. 95, no. B13, p. 21,451–21,461.

Barker, F., and Arth, J. G., 1984, Preliminary results, Central Gneiss Complex of the Coast Range batholith, southeastern Alaska; The roots of a high-K, calc-alkaline arc?: Physics of the Earth and Planetary Interiors, v. 35, p. 191–198.

Barker, F., and Stern, T. W., 1986, An arc-root complex of Wrangellia, eastern Alaska Range: Geological Society of America Abstracts with Programs, v. 18, p. 534.

Barker, F., McLellan, E. L., and Plafker, G., 1985, Partial melting of amphibolite to trondhjemite at Nunatak fiord, St. Elias Mountains, Alaska: Geological Society of America Abstracts with Programs, v. 17, p. 518–519.

Beard, J. S., and Barker, F., 1989, Petrology and tectonic significance of gabbros, tonalites, shoshonites, and anorthosites in a late Paleozoic arc-root complex in the Wrangellia terrane, southern Alaska: Journal of Geology, v. 97, no. 6, p. 667–683.

Berg, H. C., Jones, D. L., and Richter, D. H., 1972, Gravina-Nutzotin belt; Tectonic significance of an upper Mesozoic sedimentary and volcanic sequence in southern and southeastern Alaska, *in* Geological Survey research 1972: U.S. Geological Survey Professional Paper 800-D, p. D1–D24.

Berg, H. C., Elliott, R. L., and Koch, R. D., 1988, Geologic map of the Ketchikan and Prince Rupert Quadrangles, southeastern Alaska: U.S. Geological Survey Miscellaneous Investigations Series Map I-1807, scale 1:250,000.

Boak, J. L., Turner, D. L., Wallace, W. K., and Moore, T. E., 1985, K-Ar ages of allochthonous mafic and ultramafic complexes and their metamorphic aureoles, western Brooks Range, Alaska: American Association of Petroleum Geologists Bulletin, v. 69, no. 4, p. 656–657.

Box, S. E., 1985a, Early Cretaceous orogenic belt in northwestern Alaska; Internal organization, lateral extent, and tectonic interpretation, *in* Howell, D. G., ed., Tectonostratigraphic terranes of the Circum-Pacific region: Circum-Pacific Council for Energy and Mineral Resources Earth Science Series 1, p. 137–145.

——, 1985b, Geologic setting of high-pressure metamorphic rocks, Cape Newenham area, southwestern Alaska, *in* Bartsch-Winkler, S., ed., The United States Geological Survey in Alaska; Accomplishments during 1984: U.S. Geological Survey Circular 967, p. 37–42.

——, 1985c, Terrane analysis, northern Bristol Bay region, southwestern Alaska, *in* Bartsch-Winkler, S., ed., The United States Geological Survey in Alaska; Accomplishments during 1984: U.S. Geological Survey Circular 967, p. 32–37.

——, 1987, Late Cretaceous or younger southwest-directed extensional faulting, Cosmos Hills, Brooks Range, Alaska: Geological Society of America Abstracts with Programs, v. 19, p. 361.

Box, S. E., and Murphy, J. M., 1987, Late Mesozoic structural and stratigraphic framework, eastern Bethel Quadrangle, southwestern Alaska, *in* Hamilton, T. D., and Galloway, J. P., eds., Geologic studies in Alaska by the U.S. Geological Survey during 1986: U.S. Geological Survey Circular 998, p. 78–82.

Brew, D. A., 1978, Geology, *in* Brew, D. A., and 7 others, eds., Mineral resources of the Glacier Bay National Monument Wilderness Study area, Alaska: U.S. Geological Survey Open-File Report 78-494, p. B1–B21.

Brew, D. A., and Ford, A. B., 1977, Preliminary geologic and metamorphic-isograd map of the Juneau B-1 Quadrangle, Alaska: U.S. Geological Survey Miscellaneous Field Studies Map MF-846, scale 1:31,680.

——, 1978, Megalineament in southeastern Alaska marks southwest edge of Coast Range batholithic complex: Canadian Journal of Earth Sciences, v. 15, no. 11, p. 1763–1772.

——, 1981, The Coast plutonic complex sill, southeastern Alaska, *in* Albert, N.R.D., and Hudson, T., eds., The United States Geological Survey in Alaska; Accomplishments during 1979: U.S. Geological Survey Circular 823-B, p. B96–B99.

——, 1983, Comment *on* 'Tectonic accretion and the origin of the two major metamorphic and plutonic welts in the Canadian Cordillera': Geology, v. 11, p. 427–428.

——, 1984, The northern Coast plutonic-metamorphic complex, southeastern Alaska and northwestern British Columbia, *in* Coonrad, W. L., and Elliott, R. L., eds., The United States Geological Survey in Alaska; Accomplishments during 1981: U.S. Geological Survey Circular 868, p. 120–124.

Brew, D. A., and Morrell, R. P., 1979, Correlation of the Sitka graywacke, unnamed rocks in the Fairweather Range, and Valdez Group, southeastern and south-central Alaska, *in* Johnson, K. M., and Williams, J. R., eds., The United States Geological Survey in Alaska; Accomplishments during 1978: U.S. Geological Survey Circular 804-B, p. B123–B125.

Brew, D. A., Ovenshine, A. T., Karl, S. M., and Hunt, S. J., 1984, Preliminary reconnaissance geologic map of the Petersburg and parts of the Port Alexander and Sumdum Quadrangles, southeastern Alaska: U.S. Geological Survey Open-File Report 84-405, 43 p., scale 1:250,000.

Brew, D. A., Ford, A. B., and Himmelberg, G. R., 1987, Late Cretaceous sedimentation, volcanism, plutonism, metamorphism and deformation in the northern Cordillera, southeastern Alaska: Geological Society of America, Abstracts with Programs, v. 19, no. 7, p. 600.

Brew, D. A., Karl, S. M., and Miller, J. W., 1988, Megafossils (Buchia) indicate Late Jurassic age for part of Kelp Bay Group on Baranof Island, southeastern Alaska, *in* Galloway, J. P., and Hamilton, T. D., eds., Geologic studies in Alaska by the U.S. Geological Survey during 1987: U.S. Geological Survey Circular 1016, p. 147–149.

Brown, E. H., and Forbes, R. B., 1984, Paragenesis and regional significance of eclogitic rocks from the Fairbanks District, Alaska: Geological Society of America Abstracts with Programs, v. 16, p. 272.

——, 1986, Phase petrology of eclogitic rocks in the Fairbanks District, Alaska, *in* Evans, B. W., and Brown, E. H., eds., Blueschists and eclogites: Geological Society of America Memoir 164, p. 155–167.

Buddington, A. F., and Chapin, T., 1929, Geology and mineral deposits of southeastern Alaska: U.S. Geological Survey Bulletin 800, 398 p.

Bundtzen, T. K., 1981, Geology and mineral deposits of the Kantishna Hills, Mt. McKinley Quadrangle, Alaska [M.S. thesis]: Fairbanks, University of Alaska, 237 p.

Bundtzen, T. K., and Turner, D. L., 1979, Geochronology of metamorphic and igneous rocks in the Kantishna Hills, Mount McKinley Quadrangle, Alaska, *in* Short notes on Alaskan Geology, 1978: Alaska Division of Geological and Geophysical Surveys Geologic Report 61, p. 25–30.

Burns, L. E., 1985, The Border Ranges ultramafic and mafic complex, south-central Alaska; Cumulate fractionates of island-arc volcanics: Canadian Journal of Earth Sciences, v. 22, p. 1020–1038.

Byrne, T., 1982, Structural geology of coherent terranes in the Ghost Rocks Formation, Kodiak Islands, Alaska, *in* Leggett, J. K., ed., Trench and forearc sedimentation and tectonics: Geological Society of London Special Publication 10, p. 229–242.

——, 1984, Early deformation in mélange terranes of the Ghost Rocks Formation, Kodiak Islands, Alaska, *in* Raymond, L. A., ed., Mélanges; Their nature, origin and significance: Geological Society of America Special Paper 198, p. 21–51.

——, 1986, Eocene underplating along the Kodiak Shelf, Alaska; Implications and regional correlations: Tectonics, v. 5, p. 403–421.

Campbell, R. B., and Dodds, C. J., 1978, Operation Saint Elias, Yukon Territory, *in* Current Research, part A: Geological Survey of Canada Paper 78-1A, p. 35–41.

Carden, J. R., 1978, The comparative petrology of blueschists and greenschists in the Brooks Range and Kodiak-Seldovia schist belts [Ph.D. thesis]: Fairbanks,

University of Alaska, 242 p.
Carden, J. R., and Decker, J. E., 1977, Tectonic significance of the Knik River schist terrane, south-central Alaska: Alaska Division of Geological and Geophysical Surveys Geologic Report 55, p. 7–9.
Carden, J. R., Connelly, W., Forbes, R. B., and Turner, D. L., 1977, Blueschists of the Kodiak Islands, Alaska; An extension of the Seldovia schist terrane: Geology, v. 5, p. 529–533.
Carlson, C., 1985, Large-scale south-dipping, low-angle normal faults in the southern Brooks Range, Alaska: EOS Transactions of the American Geophysical Union, v. 66, no. 46, p. 1074.
Carlson, C., and Wallace, W. K., 1983, The Tlikakila complex; A disrupted terrane in the southwestern Alaska Range: Geological Society of America Abstracts with Programs, v. 15, p. 406.
Clark, A. L., and Greenwood, W. R., 1972, Petrographic evidence of volume increase related to serpentinization, Union Bay, Alaska: U.S. Geological Survey Professional Paper 800-C, p. C21–C27.
Clark, S.H.B., 1973, The McHugh Complex of south-central Alaska: U.S. Geological Survey Bulletin 1372-D, p. D1–D11.
Coleman, R. G., Lee, D. E., Beatty, L. B., and Brannock, W. W., 1965, Eclogites and eclogites; Their differences and similarities: Geological Society of America Bulletin, v. 76, p. 483–508.
Connelly, W., 1978, Uyak Complex, Kodiak Islands, Alaska; A Cretaceous subduction complex: Geological Society of America Bulletin, v. 89, p. 755–769.
Cowan, D. S., and Boss, R. F., 1978, Tectonic framework of the southwestern Kenai Peninsula, Alaska: Geological Society of America Bulletin, v. 89, p. 155–158.
Crawford, M. L., and Hollister, L. S., 1982, Contrast of metamorphic and structural histories across the Work Channel lineament, Coast Plutonic Complex, British Columbia: Journal of Geophysical Research, v. 87, no. B5, p. 3849–3860.
——, 1983, Correction *to* 'Contrast of metamorphic and structural histories across the Work Channel lineament, Coast Plutonic Complex, British Columbia': Journal of Geophysical Research, v. 88, no. B12, p. 10645–10646.
Crawford, M. L., Hollister, L. S., and Woodsworth, G. J., 1987, Crustal deformation and regional metamorphism across a terrane boundary, Coast Plutonic Complex, British Columbia: Tectonics, v. 6, no. 3, p. 343–361.
Csejtey, B., Jr., and 8 others, 1978, Reconnaissance geologic map and geochronology, Talkeetna Mountains Quadrangle, northern part of Anchorage Quadrangle, and southwest corner of Healy Quadrangle, Alaska: U.S. Geological Survey Open-File Report 78-558-A, 60 p., scale 1:250,000.
Csejtey, B., Jr., Cox, D. P., and Evarts, R. C., 1982, The Cenozoic Denali fault system and the Cretaceous accretionary development of southern Alaska: Journal of Geophysical Research, v. 87, no. B5, p. 3741–3754.
Csejtey, B., Jr., and 14 others, 1986, Geology and geochronology of the Healy Quadrangle, Alaska: U.S. Geological Survey Open-File Report 86-396, 96 p., scale 1:250,000.
Cushing, G. W., Foster, H. L., Harrison, T. M., and Laird, J., 1984, Possible Mesozoic accretion in the eastern Yukon-Tanana Upland, Alaska: Geological Society of America Abstracts with Programs, v. 16, p. 481.
Decker, J. E., Jr., 1980, Geology of a Cretaceous subduction complex, western Chichagof Island, southeastern Alaska [Ph.D. thesis]: Stanford, California, Stanford University, 135 p.
Decker, J. E., and Plafker, G., 1982, Correlation of rocks in the Tarr Inlet suture zone with the Kelp Bay Group, *in* Coonrad, W. L., ed., The United States Geological Survey in Alaska; Accomplishments during 1980: U.S. Geological Survey Circular 844, p. 119–123.
Decker, J. E., Wilson, F. H., and Turner, D. L., 1980, Mid-Cretaceous subduction event in southeastern Alaska: Geological Society of America Abstracts with Programs, v. 12, p. 103.
Detterman, R. L., and Reed, B. L., 1980, Stratigraphy, structure, and economic geology of the Iliamna Quadrangle, Alaska: U.S. Geological Survey Bulletin 1368-B, p. B28–B32, scale 1:250,000.
Dickinson, W. R., and Snyder, W. S., 1979, Geometry of subducted slabs related to the San Andreas transform: Journal of Geology, v. 87, p. 609–627.
Dillon, J. T., 1987, Latest Cretaceous–earliest Tertiary metamorphism in the northeastern Brooks Range, Alaska: Geological Society of America Abstracts with Programs, v. 19, p. 373.
Dillon, J. T., and Bakke, A. A., 1987, Evidence for Devonian age of the Okpilak batholith, northeastern Brooks Range, Alaska: Geological Society of America Abstracts with Programs, v. 19, p. 373.
Dillon, J. T., and Smiley, C. J., 1984, Clasts from the Early Cretaceous Brooks Range orogen in Albian to Cenomanian molasse deposits of the northern Koyukuk basin, Alaska: Geological Society of America Abstracts with Programs, v. 16, p. 279.
Dillon, J. T., Pessel, G. H., Chen, J. H., and Veach, N. C., 1980, Middle Paleozoic magmatism and orogenesis in the Brooks Range, Alaska: Geology, v. 8, p. 338–343.
Dillon, J. T., and 5 others, 1985, New radiometric evidence for the age and thermal history of the metamorphic rocks of the Ruby and Nixon Fork terranes, west-central Alaska, *in* Bartsch-Winkler, S., and Reed, K. M., eds., The United States Geological Survey in Alaska: Accomplishments during 1983: U.S. Geological Survey Circular 945, p. 13–18.
Dillon, J. T., Brosgé, W. P., and Dutro, J. T., Jr., 1986, Generalized geologic map of the Wiseman Quadrangle, Alaska: U.S. Geological Survey Open-File Report 86-219, scale 1:250,000.
Dillon, J. T., Pessel, G. H., Lueck, L., and Hamilton, W. B., 1987, Geologic map of the Wiseman A-4 Quadrangle: Alaska Division of Geological and Geophysical Surveys Professional Report 87, scale 1:63,360.
Dodds, C. J., and Campbell, R. B., 1988, Potassium-argon ages of mainly intrusive rocks in the Saint Elias Mountains, Yukon and British Columbia: Geological Survey of Canada Paper 87-16, 43 p.
Douglass, S. L., and Brew, D. A., 1985, Polymetamorphism in the eastern part of the Petersburg map area, southeastern Alaska, *in* Bartsch-Winkler, S., ed., The United States Geological Survey in Alaska; Accomplishments during 1985: U.S. Geological Survey Circular 967, p. 89–92.
Dover, J. H., and Miyaoka, R. T., 1985a, Major rock packages of the Ray Mountains, Tanana and Bettles Quadrangles, *in* Bartsch-Winkler, S., and Reed, K. M., eds., The United States Geological Survey in Alaska; Accomplishments during 1983: U.S. Geological Survey Circular 945, p. 32–36.
——, 1985b, Metamorphic rocks of the Ray Mountains; Preliminary structural analysis and regional tectonic implications, *in* Bartsch-Winkler, S., and Reed, K. M., eds., The United States Geological Survey in Alaska: Accomplishments during 1983: U.S. Geological Survey Circular 945, p. 36–38.
Duke, N. A., Nauman, C. R., and Newkirk, S. R., 1988, Evidence for Late Cretaceous extensional uplift of the Yukon-Tanana crystalline terrane, eastern Alaska Range: Geological Society of America Abstracts with Programs, v. 20, p. A112.
Dusel-Bacon, C., and Aleinikoff, J. N., 1985, Petrology and tectonic significance of augen gneiss from a belt of Mississippian granitoids in the Yukon-Tanana terrane, east-central Alaska: Geological Society of America Bulletin, v. 96, p. 411–425.
Dusel-Bacon, C., and Foster, H. L., 1983, A sillimanite gneiss dome in the Yukon crystalline terrane, east-central Alaska; Petrography and garnet-biotite geothermometry: U.S. Geological Survey Professional Paper 1170E, 25 p.
Dusel-Bacon, C., and 6 others, 1989, Distribution, facies, ages, and proposed tectonic associations of regionally metamorphosed rocks in northern Alaska: U.S. Geological Survey Professional Paper 1497-A, 44 p.
Dusel-Bacon, C., Brew, D. A., and Douglass, S. L., 1994a, Metamorphic facies map of southeastern Alaska; Distribution, facies, and ages of regionally metamorphosed rocks: U.S. Geological Survey Professional Paper 1497-D (in press).
Dusel-Bacon, C., and 5 others, 1994b, Distribution, facies, ages, and proposed tectonic associations of regionally metamorphosed rocks in east and south-central Alaska: U.S. Geological Survey Professional Paper 1497-C, 59 p.
Dusel-Bacon, C., Doyle, E. O., and Box, S. E., 1994c, Distribution, facies, ages, and proposed tectonic associations of regionally metamorphosed rocks in southwestern Alaska and the Alaska Peninsula: U.S. Geological Survey Professional Paper 1497-B (in press).

Eberlein, G. D., Churkin, M., Jr., Carter, C., Berg, H. C., and Ovenshine, A. T., 1983, Geology of the Craig Quadrangle, Alaska: U.S. Geological Survey Open-File Report 83-91, 26 p., scale 1:250,000.

Engebretsen, D. C., Cox, A., and Gordon, R. G., 1985, Relative motions between oceanic and continental plates in the Pacific basin: Geological Society of America Special Paper 206, 59 p.

Epstein, A. G., Epstein, J. B., and Harris, L. D., 1977, Conodont color alteration; An index to organic metamorphism: U.S. Geological Survey Professional Paper 995, 27 p.

Erdmer, P., and Armstrong, R. L., 1988, Permo-Triassic isotopic dates for blueschist, Ross River area, Yukon: Canada, Exploration and Geological Services Division, Yukon Indian and Northern Affairs, Yukon Geology, p. 33–36.

Erdmer, P., and Helmstaedt, H., 1983, Eclogite from central Yukon; A record of subduction at the western margin of ancient North America: Canadian Journal of Earth Sciences, v. 20, p. 1389–1408.

Evans, B. W., and Brown, E. H., 1987, Reply *on* 'Blueschists and eclogites': Geology, v. 15, p. 773–775.

Evans, B. W., Patrick, B. E., and Irving, A. J., 1987, Compositional control of blueschist/greenschist and genesis of Seward Peninsula metabasites: Geological Society of America Abstracts with Programs, v. 19, p. 375.

Ferry, J. M., and Spear, F. S., 1978, Experimental calibration of the partitioning of Fe and Mg between biotite and garnet: Contributions to Mineralogy and Petrology, v. 66, no. 2, p. 113–117.

Forbes, R. B., and Lanphere, M. A., 1973, Tectonic significance of mineral ages of blueschists near Seldovia, Alaska: Journal of Geophysical Research, v. 78, no. 8, p. 1383–1386.

Forbes, R. B., and Weber, F. R., 1982, Bedrock geologic map of the Fairbanks mining district, Alaska: Alaska Division of Geological and Geophysical Surveys Open-File Report AOF-170, scale 1:63,360.

Forbes, R. B., Evans, B. W., and Thurston, S. P., 1984, Regional progressive high-pressure metamorphism, Seward Peninsula, Alaska: Journal of Metamorphic Geology, v. 2, p. 43–54.

Ford, A. B., and Brew, D. A., 1973, Preliminary geologic and metamorphic-isograd map of the Juneau B-2 Quadrangle, Alaska: U.S. Geological Survey Miscellaneous Field Studies Map MF-527, scale 1:31,680.

——, 1977a, Preliminary geologic and metamorphic-isograd map of the northern parts of the Juneau A-1 and A-2 Quadrangles, Alaska: U.S. Geological Survey Miscellaneous Field Studies Map MF-847, scale 1:31,680.

——, 1977b, Truncation of regional metamorphic zonation pattern of the Juneau, Alaska, area by the Coast Range batholith, *in* Johnson, K. M., ed., The United States Geological Survey in Alaska; Accomplishments during 1976: U.S. Geological Survey Circular 751-B, p. B85–B87.

Foster, H. L., Cushing, G. W., Keith, T.E.C., and Laird, J., 1985, Early Mesozoic tectonic history of the Boundary area, east-central Alaska: Geophysical Research Letters, v. 12, no. 9, p. 553–556.

Foster, H. L., Laird, J., Keith, T.E.C., Cushing, G. W., and Menzie, W. D., 1983, Preliminary geologic map of the Circle Quadrangle, Alaska: U.S. Geological Survey Open-File Report 83-170-A, 30 p., scale 1:250,000.

Franz, G., and Althaus, E. K., 1977, The stability relations of paragenesis paragonite-zoisite-quartz: Stuttgart, Federal Republic of Germany, Neues Jahrbuch für Mineralogie Abhandlungen, v. 130, p. 159–167.

Gardner, M. C., MacKevett, E. M., Jr., and McClelland, W. D., 1986, The Chitina fault system of southern Alaska; An Early Cretaceous collisional suture zone: Geological Society of America Abstracts with Programs, v. 18, p. 108.

Gardner, M. C., and 6 others, 1988, Pennsylvanian pluton stitching of Wrangellia and the Alexander terrane, Wrangell Mountains, Alaska: Geology, v. 16, p. 967–971.

Gehrels, G. E., and Berg, H. C., 1992, Geologic map of southeastern Alaska: U.S. Geological Survey Miscellaneous Investigations Series Map I-1867, scale 1:600,000.

Gehrels, G. E., and Saleeby, J. B., 1987a, Geologic framework, tectonic evolution, and displacement history of the Alexander terrane: Tectonics, v. 6, no. 2, p. 151–173.

——, 1987b, Geology of southern Prince of Wales Island, southeastern Alaska: Geological Society of America Bulletin, v. 98, p. 123–137.

Gehrels, G. E., Saleeby, J. B., and Berg, H. C., 1983, Preliminary description of the Klakas orogeny in the southern Alexander terrane, southeastern Alaska, *in* Stevens, C. H., eds., Pre-Jurassic rocks in western North American suspect terranes: Pacific Section, Society of Economic Paleontologists and Mineralogists, p. 131–141.

Gehrels, G. E., Brew, D.A., and Saleeby, J. B., 1984, Progress report on U/Pb (zircon) geochronologic studies in the Coast plutonic-metamorphic complex, east of Juneau, southeastern Alaska, *in* Reed, K. M., and Bartsch-Winkler, S., eds., The United States Geological Survey in Alaska; Accomplishments during 1982: U.S. Geological Survey Circular 939, p. 100–102.

Gehrels, G. E., Saleeby, J. B., and Berg, H. C., 1987, Geology of Annette, Gravina, and Duke Islands, southeastern Alaska: Canadian Journal of Earth Sciences, v. 24, p. 866–881.

Gemuts, I., Puchner, C. C., and Steffel, C. I., 1983, Regional geology and tectonic history of western Alaska: Journal of the Alaska Geological Society, v. 3, p. 67–86.

Gilbert, W. G., Wiltse, M. A., Carden, J. R., Forbes, R. B., and Hackett, S. W., 1977, Geology of Ruby Ridge, southwestern Brooks Range, Alaska: Alaska Division of Geological and Geophysical Surveys Geologic Report 58, 16 p.

Glassley, W. F., and Cowan, D. S., 1975, Metamorphic history of Franciscan greenstone, blueschist, and eclogite: EOS Transactions of the American Geophysical Union, v. 56, no. 12, p. 1081.

Goldfarb, R. J., Leach, D. L., Miller, M. L., and Pickthron, W. J., 1986, Geology, metamorphic setting, and genetic constraints of epigenetic lode-gold mineralization within the Cretaceous Valdez Group, south-central Alaska, *in* Keppie, J. D., Boyle, R. W., and Haynes, S. J., eds., Turbidite-hosted gold deposits: Geological Association of Canada Special Paper 32, p. 93–113.

Gottschalk, R. R., 1987, Structural and petrologic evolution of the southern Brooks Range near Wiseman, Alaska [Ph.D. thesis]: Houston, Texas, Rice University, 263 p.

Gottschalk, R. R., and Oldow, J. S., 1988, Low-angle normal faults in the south-central Brooks Range fold and thrust belt, Alaska: Geology, v. 16, p. 395–399.

Grantz, A., 1966, Strike-slip faults of Alaska: U.S. Geological Survey Open-File Report, 82 p.

Grantz, A., and May, S. D., 1983, Rifting history and structural development of the continental margin north of Alaska, *in* Watkins, J. S., and Drake, C. L., eds., Studies in continental margin geology: American Association of Petroleum Geologists Memoir 34, p. 77–100.

Hansen, V. L., 1989, Mesozoic evolution of the Yukon-Tanana terrane: Geological Society of America Abstracts with Programs, v. 21, p. 90.

Harrison, T. M., Duncan, I., and McDougall, I., 1985, Diffusion of ^{40}Ar in biotite: Temperature, pressure, and compositional effects: Geochimica et Cosmochimica Acta, v. 49, no. 11, p. 2461–2468.

Helwig, J., and Emmet, P., 1981, Structure of the Early Tertiary Orca Group in Prince William Sound and some implications for the plate tectonic history of southern Alaska: Journal of the Alaska Geological Society, v. 1, p. 12–345.

Herreid, G., Bundtzen, T. K., and Turner, D. L., 1978, Geology and geochemistry of the Craig A-2 Quadrangle and vicinity, Prince of Wales Island, southeastern Alaska: Alaska Division of Geological and Geophysical Surveys Geologic Report 48, 49 p.

Hill, M. L., 1984, Geology of the Redcap Mountain area, Coast Plutonic Complex, British Columbia [Ph.D. thesis]: Princeton, New Jersey, Princeton University, 216 p.

Himmelberg, G. R., Ford, A. B., and Brew, D. A., 1984, Progressive metamorphism of pelitic rocks in the Juneau area, southeastern Alaska, *in* Coonrad, W. L., and Elliott, R. L., eds., The United States Geological Survey in Alaska; Accomplishments during 1981: U.S. Geological Survey 868, p. 131–134.

Hitzman, M. W., 1980, Devonian to recent tectonics of the southwestern Brooks

Range, Alaska: Geological Society of America Abstracts with Programs, v. 12, p. 447.

—— , 1983, Geology of the Cosmos Hills and its relationship to the Ruby Creek copper-cobalt deposit [Ph.D. thesis]: Stanford, California, Stanford University, 266 p.

—— , 1984, Geology of the Cosmos Hills; Constraints for Yukon-Koyukuk basin evolution: Geological Society of America Abstracts with Programs, v. 16, p. 290.

Hitzman, M. W., Smith, T. E., and Proffett, J. M., 1982, Bedrock geology of the Ambler district, southwestern Brooks Range, Alaska: Alaska Division of Geological and Geophysical Surveys Geologic Report 75, 2 sheets, scale 1:125,000.

Hoare, J. M., and Coonrad, W. L., 1959, Geology of the Bethel Quadrangle, Alaska: U.S. Geological Survey Miscellaneous Geologic Investigations Series Map I-285, scale 1:250,000.

—— , 1978, Geologic map of the Goodnews and Hagemeister Island Quadrangles region, southwestern Alaska: U.S. Geological Survey Open-File Report 78-9-B, scale 1:250,000.

—— , 1979, The Kanektok metamorphic complex; A rootless belt of Precambrian rocks in southwestern Alaska, *in* Johnson, K. M., and Williams, J. R., eds., The United States Geological Survey in Alaska; Accomplishments during 1978: U.S. Geological Survey Circular 804-B, p. B72–B74.

Holdaway, M. J., 1971, Stability of andalusite and the aluminum silicate phase diagram: American Journal of Science, v. 271, no. 2, p. 97–131.

Hollister, L. S., 1982, Metamorphic evidence for rapid (2 mm/yr) uplift of a portion of the Central Gneiss Complex, Coast Mountains, B.C.: Canadian Mineralogist, v. 20, p. 319–332.

Hollister, L. S., and Crawford, M. L., 1986, Melt-enhanced deformation: A major tectonic process: Geology, v. 14, p. 558–561.

Hudson, T., and Plafker, G., 1982, Paleogene metamorphism of an accretionary flysch terrane, eastern Gulf of Alaska: Geological Society of America Bulletin, v. 93, p. 1280–1290.

Hudson, T., Plafker, G., and Lanphere, M. A., 1977a, Intrusive rocks of the Yakutat–St. Elias area, south-central Alaska: U.S. Geological Survey Journal of Research, v. 5, no. 2, p. 155–172.

Hudson, T., Plafker, G., and Turner, D. L., 1977b, Metamorphic rocks of the Yakutat–St. Elias area, south-central Alaska: U.S. Geological Survey Journal of Research, v. 5, no. 2, p. 173–184.

Johnson, B. R., and Karl, S. M., 1985, Geologic map of the western Chichagof and Yakobi Islands, southeastern Alaska: U.S. Geological Survey Miscellaneous Investigations Series Map I-1506, 15 p., scale 1:125,000.

Jones, D. L., and Clark, S.H.B., 1973, Upper Cretaceous (Maestrichtian) fossils from the Kenai-Chugach Mountains, Kodiak and Shumagin Islands, southern Alaska: U.S. Geological Survey Journal of Research, v. 1, no. 2, p. 125–136.

Jones, D. L., Silberling, N. J., Csejtey, B., Jr., Nelson, W. H., and Blome, C. D., 1980, Age and structural significance of ophiolite and adjoining rocks in the Upper Chulitna district, south-central Alaska: U.S. Geological Survey Professional Paper 1121-A, p. A1–A21.

Jones, D. L., Silberling, N. J., Coney, P. J., and Plafker, G., 1987, Lithotectonic terrane map of Alaska (west of the 41st meridian): U.S. Geological Survey Miscellaneous Field Studies Map MF-1874-A, scale 1:2,500,000.

Julian, F. E., Phelps, J. S., Seidensticker, C. M., Oldow, J. S., and Avé Lallemant, H. G., 1984, Structural history of the Doonerak window, central Brooks Range, Alaska: Geological Society of America Abstracts with Programs, v. 16, p. 291.

Keith, T.E.C., Foster, H. L., Foster, R. L., Post, E. V., and Lehmbeck, W. L., 1981, Geology of the alpine-type peridotite in the Mount Sorenson area, east-central Alaska: U.S. Geological Survey Professional Paper 1170-A, 9 p.

Laird, J., Foster, H. L., and Weber, F. R., 1984, Amphibole eclogite in the Circle Quadrangle, Yukon-Tanana Upland, Alaska, *in* Coonrad, W. L., and Elliott, R. L., eds., The United States Geological Survey in Alaska; Accomplishments during 1981: U.S. Geological Survey Circular 868, p. 57–60.

Lanphere, M. A., and Eberlein, G. D., 1966, Potassium-argon ages of magnetite-bearing ultramafic complexes in southeastern Alaska, *in* Abstracts for 1965: Geological Society of America Special Paper 87, p. 94.

Lathram, E. H., Pomeroy, J. S., Berg, H. C., and Loney, R. A., 1965, Reconnaissance geology of Admiralty Island, Alaska: U.S. Geological Survey Bulletin 1181-R, p. R1–R48, scale 1:250,000.

Lieberman, J. E., and Till, A. B., 1987, Possible crustal origin of garnet lherzolite: Evidence from the Kigluaik Mountains, Alaska: Geological Society of America Abstracts with Programs, v. 19, p. 746.

Lister, G. S., Banga, G., and Feenstra, A., 1984, Metamorphic core complexes of Cordilleran type in the Cyclades, Aegean Sea, Greece: Geology, v. 12, p. 221–225.

Loney, R. A., and Brew, D. A., 1987, Regional thermal metamorphism and deformation of the Sitka Graywacke, southern Baranof Island, southeastern Alaska: U.S. Geological Survey Bulletin 1779, 17 p.

Loney, R. A., and Himmelberg, G. R., 1983, Structure and petrology of the La Perouse gabbro intrusion, Fairweather Range, southeastern Alaska: Journal of Petrology, v. 24, p. 377–423.

Loney, R. A., Brew, D. A., and Lanphere, M. A., 1967, Post-Paleozoic radiometric ages and their relevance to fault movements, northern southeastern Alaska: Geological Society of America Bulletin, v. 78, p. 511–526.

Loney, R. A., Brew, D. A., Muffler, L.P.J., and Pomeroy, J. S., 1975, Reconnaissance geology of Chichagof, Baranof, and Kruzof Islands, southeastern Alaska: U.S. Geological Survey Professional Paper 792, 105 p., scale 1:250,000.

MacKevett, E. M., Jr., 1978, Geologic map of the McCarthy Quadrangle, Alaska: U.S. Geological Survey Miscellaneous Investigations Series Map I-1032, 1 sheet, scale 1:250,000.

MacKevett, E. M., Jr., Robertson, E. C., and Winkler, G. R., 1974, Geology of the Skagway B-3 and B-4 Quadrangles, southeastern Alaska: U.S. Geological Survey Professional Paper 832, 33 p., scale 1:63,360.

MacKevett, E. M., Gardner, M. C., Bergman, S. C., Cushing, G., and McClelland, W. D., 1986, Geological evidence for Late Pennsylvanian juxtaposition of Wrangellia and the Alexander terrane, Alaska: Geological Society of America Abstracts with Programs, v. 18, p. 128.

Maresch, W. V., 1977, Experimental studies on glaucophane; An analysis of present knowledge: Tectonophysics, v. 43, p. 109–125.

Marshak, R. S., and Karig, D. E., 1977, Triple junctions as a cause for anomalously near-trench igneous activity between the trench and volcanic arc: Geology, v. 5, p. 233–236.

Mattinson, J., 1978, Age, origin, and thermal histories of some plutonic rocks from the Salinian block: Contributions to Mineralogy and Petrology, v. 67, p. 233–245.

Mayfield, C. F., Silberman, M. L., and Tailleur, I. L., 1982, Precambrian metamorphic rocks from the Hub Mountain terrane, Baird Mountains Quadrangle, Alaska, *in* Coonrad, W. L., ed., The United States Geological Survey in Alaska; Accomplishments during 1980: U.S. Geological Survey Circular 844, p. 18–22.

Mayfield, C. F., Tailleur, I. L., and Ellersieck, I., 1983, Stratigraphy, structure, and palinspastic synthesis of the western Brooks Range, northwestern Alaska: U.S. Geological Survey Open-File Report 83-779, 58 p.

McConnell, R. G., 1905, Report on the Klondike gold fields: Geological Survey of Canada Annual Report, n.s., v. 14, part B, 71 p. (Reprinted, 1957, *in* Bostock, H. S., ed., Yukon Territory; Selected field reports of the Geological Survey of Canada, 1898 to 1933: Geological Survey of Canada Memoir 284, p. 43–46).

Metcalfe, P., and Clark, G. S., 1983, Rb-Sr whole-rock age of the Klondike Schist, Yukon Territory: Canadian Journal of Earth Sciences, v. 20, no. 5, p. 886–891.

Metz, P. A., 1976, Occurrences of sodic amphibole-bearing rocks in the Valdez C-2 Quadrangle: Alaska Division of Geological and Geophysical Surveys Geologic Report 51, p. 27–28.

Miller, E. L., 1987, Dismemberment of the Brooks Range orogenic belt during middle Cretaceous extension: Geological Society of America Abstracts with Programs, v. 19, p. 432.

Miller, M. L., and Bundtzen, T. K., 1985, Metamorphic rocks in the western Iditarod Quadrangle, west-central Alaska, *in* Bartsch-Winkler, S., and Reed, K. M., eds., The United States Geological Survey in Alaska; Accomplishments during 1983: U.S. Geological Survey Circular 945, p. 24–28.

Miller, M. L., Dumoulin, J. A., and Nelson, S. W., 1984, A transect of metamorphic rocks along the Copper River, Cordova and Valdez Quadrangles, Alaska, *in* Reed, K. M., and Bartsch-Winkler, S., eds., The United States Geological Survey in Alaska; Accomplishments during 1982: U.S. Geological Survey Circular 939, p. 52–57.

Miyaoka, R. T., and Dover, J. H., 1985, Preliminary study of shear sense in mylonites, eastern Ray Mountains, Tanana Quadrangle, *in* Bartsch-Winkler, S., ed., The United States Geological Survey in Alaska; Accomplishments during 1984: U.S. Geological Survey Circular 967, p. 29–32.

Monger, J.W.H., and Berg, H. C., 1987, Lithotectonic terrane map of western Canada and southeastern Alaska: U.S. Geological Survey Miscellaneous Field Studies Map MF-1874-B, 12 p., scale 1:2,500,000.

Monger, J.W.H., Price, R. A., and Tempelman-Kluit, D. J., 1982, Tectonic accretion and the origin of the two major metamorphic and plutonic welts in the Canadian Cordillera: Geology, v. 10, p. 70–75.

——, Reply *to* Comment *on* 'Tectonic accretion and the origin of the two major metamorphic and plutonic welts in the Canadian Cordillera': Geology, v. 11, p. 428–429.

Moore, J. C., 1973, Cretaceous continental margin sedimentation, southwestern Alaska: Geological Society of America Bulletin, v. 84, p. 595–614.

Moore, J. C., and Connelly, W., 1979, Tectonic history of the continental margin of southwestern Alaska; Late Triassic to earliest Tertiary, *in* Sisson, A., ed., The relationship of plate tectonics to Alaskan geology and resources; Proceedings of the 6th Alaska Geological Society Symposium, 1977: Anchorage, Alaska Geological Society, p. H1–H29.

Moore, J. C., and 5 others, 1983, Paleogene evolution of the Kodiak Islands, Alaska; Consequences of ridge-trench interaction in a more southerly latitude: Tectonics, v. 2, no. 3, p. 265–293.

Mortensen, J. K., 1983, Age and evolution of the Yukon-Tanana terrane, southeastern Yu;kon Territory [Ph.D. thesis]: Santa Barbara, University of California, 155 p.

Murphy, J. M., 1987, Early Cretaceous cessation of terrane accretion, northern Eek Mountains, southwestern Alaska, *in* Hamilton, T. D., and Galloway, J. P., eds., Geologic studies in Alaska by the U.S. Geological Survey during 1986: U.S. Geological Survey Circular 998, p. 83–85.

Nelson, S. W., Blome, C. D., and Karl, S. M., 1987, Late Triassic and Early Cretaceous fossil ages from the McHugh Complex, southern Alaska, *in* Hamilton, T. D., and Galloway, J. P., eds., Geologic studies in Alaska by the U.S. Geological Survey during 1986: U.S. Geological Survey Circular 998, p. 96–98.

Nelson, W. H., Carlson, C., and Case, J. E., 1983, Geologic map of the Lake Clark Quadrangle, Alaska: U.S. Geological Survey Miscellaneous Field Studies Map MF-1114-A, scale 1:250,000.

Nilsen, T. H., and Moore, G. W., 1979, Reconnaissance study of Upper Cretaceous to Miocene stratigraphic units and sedimentary facies, Kodiak and adjacent islands, Alaska: U.S. Geological Survey Professional Paper 1093, 34 p.

Nitsch, K. H., 1971, Stabilitatsbeziehungen von prehnit- und pumpellyit-haltigen paragenesen: Contributions to Mineralogy and Petrology, v. 30, p. 240–260.

Nokleberg, W. J., Jones, D. J., and Silberling, N. J., 1985, Origin and tectonic evolution of the Maclaren and Wrangellia terranes, eastern Alaska Range, Alaska: Geological Society of America Bulletin, v. 96, p. 1251–1270.

Nokleberg, W. J., Aleinikoff, J. N., and Lange, I. M., 1986a, Cretaceous deformation and metamorphism in the northeastern Mount Hayes Quadrangle, eastern Alaska Range, *in* Bartsch-Winkler, S., ed., The United States Geological Survey in Alaska; Accomplishments during 1985: U.S. Geological Survey Circular 978, p. 64–69.

Nokleberg, W. J., Wade, W. M., Lange, I. M., and Plafker, G., 1986b, Summary of geology of the Peninsular terrane, metamorphic complex of Gulkana River, and Wrangellia terrane, north-central and northwestern Gulkana Quadrangle, *in* Bartsch-Winkler, S., ed., The United States Geological Survey in Alaska; Accomplishments during 1985: U.S. Geological Survey Circular 978, p. 69–74.

Nokleberg, W. J., Plafker, G., Lull, J. S., Wallace, W. K., and Winkler, G. R., 1989, Structural analysis of the southern Peninsular, southern Wrangellia, and northern Chugach terranes along the Trans-Alaska Crustal Transect, northern Chugach Mountains, Alaska: Journal of Geophysical Research, v. 94, no. B4, p. 4297–4320.

Nokleberg, W. J., Aleinikoff, J. N., Lange, I. M., Silva, S. R., Miyaoka, R. T., Schwab, C. E., and Zehner, R. E., 1992, Preliminary geologic map of the Mount Hayes quadrangle, eastern Alaska Range, Alaska: U.S. Geological Survey Open-File Report 92-594, 39 p., 1 sheet, scale 1:250,000.

Oldow, J. S., Avé Lallemant, H. G., Julian, F. E., and Seidensticker, C. M., 1987, Ellesmerian(?) and Brookian deformation in the Franklin Mountains, northeastern Brooks Range, Alaska, and its bearing on the origin of the Canada Basin: Geology, v. 15, p. 37–41.

Page, R. A., and 6 others, 1986, Accretion and subduction tectonics in the Chugach Mountains and Copper River Basin, Alaska; Initial results of the Trans-Alaska Crustal Transect: Geology, v. 14, p. 501–505.

Palmer, A. R., 1983, The Decade of North American Geology 1983 geologic time scale: Geology, v. 11, p. 503–504.

Patrick, B. E., 1986, Relationship between the Seward Peninsula blueschists and the Brooks Range orogeny; Evidence from regionally consistent stretching lineations: Geological Society of America Abstracts with Programs, v. 18, p. 169.

Patrick, B. E., and Lieberman, J. E., 1987, Thermal overprint on the Seward Peninsula blueschist terrane; The Lepontine in Alaska: Geological Society of America Abstracts with Programs, v. 19, p. 800–801.

Patton, W. W., Jr., 1973, Reconnaissance geology of the northern Yukon-Koyukuk province, Alaska: U.S. Geological Survey Professional Paper 774-A, p. A1–A17.

Patton, W. W., Jr., and Box, S. E., 1985, Tectonic setting and history of the Yukon-Koyukuk basin, Alaska: EOS Transactions of the American Geophysical Union, v. 66, no. 46, p. 1101.

Patton, W. W., Jr., and Moll, E. J., 1982, Structural and stratigraphic sections along a transect between the Alaska Range and Norton Sound, *in* Coonrad, W. L., ed., The United States Geological Survey in Alaska; Accomplishments during 1980: U.S. Geological Survey Circular 844, p. 76–78.

——, 1985, Geologic map of the northern and central parts of the Unalakleet Quadrangle, Alaska: U.S. Geological Survey Miscellaneous Field Studies Map MF-1749, scale 1:250,000.

Patton, W. W., Jr., Tailleur, I. L., Brosgé, W. P., and Lanphere, M. A., 1977, Preliminary report on the ophiolites of northern and western Alaska, *in* Coleman, R. G., and Irwin, W. P., eds., North American ophiolites: Oregon Department of Geology and Mineral Industries Bulletin 95, p. 51–58.

Patton, W. W., Jr., Miller, T. P., Chapman, R. M., and Yeend, W., 1978, Geologic map of the Melozitna Quadrangle, Alaska: U.S. Geological Survey Miscellaneous Investigations Series Map I-1071, scale 1:250,000.

Patton, W. W., Jr., Moll, E. J., Dutro, J. T., Jr., Silberman, M. L., and Chapman, R. M., 1980, Preliminary map of the Medfra Quadrangle, Alaska: U.S. Geological Survey Open-File Report 80-811-A, scale 1:250,000.

Patton, W. W., Jr., Moll, E. J., Lanphere, M. A., and Jones, D. L., 1984, New age data for the Kaiyuh Mountains, *in* Coonrad, W. L., and Elliot, R. L., eds., The United States Geological Survey in Alaska; Accomplishments during 1981: U.S. Geological Survey Circular 868, p. 30–32.

Patton, W. W., Jr., Stern, T. W., Arth, J. G., and Carlson, C., 1987, New U/Pb ages from granite and granite gneiss in the Ruby geanticline and southern Brooks Range, Alaska: Journal of Geology, v. 95, p. 118–126.

Pavlis, T. L., 1982, Origin and age of the Border Ranges fault of southern Alaska and its bearing on the late Mesozoic evolution of Alaska: Tectonics, v. 1, p. 343–368.

——, 1983, Pre-Cretaceous crystalline rocks of the western Chugach Mountains, Alaska: Nature of the basement of the Jurassic Peninsular terrane: Geological Society of America Bulletin, v. 94, p. 1329–1344.

Pavlis, T. L., Monteverde, D. H., Bowman, J. R., Rubenstone, J. L., and Reason, M. D., 1988a, Early Cretaceous near-trench plutonism in southern Alaska: A tonalite-trondhjemite intrusive complex injected during ductile thrusting along the Border Ranges fault system: Tectonics, v. 7, no. 6, p. 1179–1199.

Pavlis, T. L., Sisson, V. B., Nokleberg, W. J., Plafker, G., and Foster, H., 1988b, Evidence for Cretaceous crustal extension in the Yukon Crystalline terrane, east-central Alaska: EOS Transactions of the American Geophysical Union, v. 69, no. 44, p. 1453.

Plafker, G., Jones, D. L., and Pessagno, E. A., Jr., 1977, A Cretaceous accretionary flysch and mélange terrane along the Gulf of Alaska margin, *in* Blean, K. M., ed., The United States Geological Survey in Alaska; Accomplishments during 1976: U.S. Geological Survey Circular 751-B, p. B41–B42.

Plafker, G., Keller, G., Barron, J. A., and Blueford, J. R., 1985, Paleontologic data on the age of the Orca Group Alaska: U.S. Geological Survey Open-File Report 85-429, 24 p.

Plafker, G., Nokleberg, W. J., Lull, J. S., Roeske, S. M., and Winkler, G. R., 1986, Nature and timing of deformation along the Contact fault system in the Cordova, Bering Glacier, and Valdez Quadrangles, *in* Bartsch-Winkler, S., ed., The United States Geological Survey in Alaska; Accomplishments during 1985: U.S. Geological Survey Circular 978, p. 74–77.

Plafker, G., Blome, C. D., and Silberling, N. J., 1989a, Reinterpretation of lower Mesozoic rocks on the Chilkat Peninsula, Alaska, as a displaced fragment of Wrangellia: Geology, v. 17, p. 3–6.

Plafker, G., Nokleberg, W. J., and Lull, J. S., 1989b, Bedrock geology and tectonic evolution of the Wrangellia, Peninsular, and Chugach terranes along the Trans-Alaskan crustal transect in the Chugach Mountains and southern Copper River basin, Alaska: Journal of Geophysical Research, v. 94, no. B4, p. 4255–4295.

Richter, D. H., 1976, Geologic map of the Nabesna Quadrangle, Alaska: U.S. Geological Survey Miscellaneous Investigations Series Map I-932, scale 1:250,000.

Richter, D. H., Lanphere, M. A., and Matson, N. A., Jr., 1975, Granitic plutonism and metamorphism, eastern Alaska Range, Alaska: Geological Society of America Bulletin, v. 86, p. 819–829.

Roeder, D., and Mull, C. G., 1978, Tectonics of Brooks Range ophiolites, Alaska: American Association of Petroleum Geologists Bulletin, v. 62, p. 1696–1702.

Roeske, S. M., 1986, Field relations and metamorphism of the Raspberry Schist, Kodiak Islands, Alaska, *in* Evans, B. W., and Brown, E. H., eds., Blueschists and eclogites: Geological Society of America Memoir 164, p. 169–184.

Roeske, S. M., and Mattinson, J. M., 1986, Early Jurassic blueschists and island arc volcanism in southern Alaska; A paired metamorphic belt?: Geological Society of America Abstracts with Programs, v. 18, p. 178.

Ross, D. C., 1976, Prehnite in plutonic and metamorphic rocks of the northern Santa Lucia Range, Salinian block, California: U.S. Geological Survey Journal of Research, v. 4, no. 5, p. 561–568.

Rubin, C. M., and Saleeby, J. B., 1987, The inner boundary zone of the Alexander terrane, southern SE Alaska; A newly discovered thrust belt: Geological Society of America Abstracts with Programs, v. 19, p. 445.

Sable, E. G., 1977, Geology of the western Romanzof Mountains, Brooks Range, northeastern Alaska: U.S. Geological Survey Professional Paper 897, 84 p.

Sainsbury, C. L., Coleman, R. G., and Kachadoorian, R., 1970, Blueschist and related greenschist facies rocks of the Seward Peninsula, Alaska: U.S. Geological Survey Professional Paper 700-B, p. B33–B42.

Sample, J. C., and Moore, J. C., 1987, Structural style and kinematics of an underplated slate belt, Kodiak and adjacent islands, Alaska: Geological Society of America Bulletin, v. 99, p. 7–20.

Sherwood, K. W., 1979, Stratigraphy, metamorphic geology, and structural geology of the central Alaska Range, Alaska [Ph.D. thesis]: Madison, University of Wisconsin, 692 p.

Silberman, M. L., Moll, E. J., Patton, W. W., Jr., Chapman, R. M., and Connor, C. L., 1979, Precambrian age of metamorphic rocks from the Ruby province, Medfra and Ruby Quadrangles; Preliminary evidence from radiometric age data, *in* Johnson, K. M., and Williams, J. R., eds., The United States Geological Survey in Alaska; Accomplishments during 1978: U.S. Geological Survey Circular 804-B, p. B66–B68.

Silberman, M. L., MacKevett, E. M., Jr., Connor, C. L., Klock, P. R., and Kalechitz, G., 1981, K-Ar ages of the Nikolai Greenstone from the McCarthy Quadrangle, Alaska; The "docking" of Wrangellia, *in* Albert, N.R.D., and Hudson, T., eds., The United States Geological Survey in Alaska; Accomplishments during 1979: U.S. Geological Survey Circular 823-B, p. B61–B63.

Sisson, V. B., and Hollister, L. S., 1988, Low-pressure facies series metamorphism in an accretionary sedimentary prism, southern Alaska: Geology, v. 16, p. 358–361.

Sisson, V. B., and Onstott, T. C., 1986, Dating blueschist metamorphism; A combined $^{40}Ar/^{39}Ar$ and electron microprobe approach: Geochimica et Cosmochimica Acta, v. 50, p. 2111–2117.

Sisson, V. B., Hollister, L. S., and Onstott, T. C., 1989, Petrologic and age constraints on the origin of a low-pressure/high-temperature metamorphic complex, southern Alaska: Journal of Geophysical Research, v. 94, no. B4, p. 4392–4410.

Smith, G. M., and Puchner, C. C., 1985, Geology of the Ruby geanticline between Ruby and Poorman, Alaska, and the tectonic emplacement of the Rampart Group: EOS Transactions of the American Geophysical Union, v. 66, no. 46, p. 1102.

Smith, J. G., and Diggles, M. F., 1981, Potassium-argon determinations in the Ketchikan and Prince Rupert Quadrangles, southeastern Alaska: U.S. Geological Survey Open-File Report 78-73-N, 16 p., scale 1:250,000.

Smith, T. E., 1981, Geology of the Clearwater Mountains, south-central Alaska: Alaska Division of Geological and Geophysical Surveys Geologic Report 60, 72 p.

Smith, T. E., and Lanphere, M. A., 1971, Age of the sedimentation, plutonism, and regional metamorphism in the Clearwater Mountains region, central Alaska: Isochron/West, no. 2, p. 17–20.

Steiger, R. H., and Jäger, E., 1977, Subcommission on geochronology; Convention on the use of decay constants in geo- and cosmochronology: Earth and Planetary Science Letters, v. 36, no. 3, p. 359–362.

Stowell, H. H., 1985, Metamorphic pressures based on sphalerite geobarometry for the Coast Range megalineament zone, southeast Alaska: Geological Society of America Abstracts with Programs, v. 17, p. 411.

Sutter, J. F., and Crawford, M. L., 1985, Timing of metamorphism and uplift in the vicinity of Prince Rupert, British Columbia, and Ketchikan, Alaska: Geological Society of America Abstracts with Programs, v. 17, p. 411.

Swainbank, R. C., and Forbes, R. B., 1975, Petrology of eclogitic rocks from the Fairbanks district, Alaska, *in* Forbes, R. B., ed., Contributions to geology of the Bering Sea Basin and adjacent regions: Geological Society of America Special Paper 151, p. 77–123.

Tailleur, I. L., 1969, Rifting speculation on the geology of Alaska's North Slope: Oil and Gas Journal, v. 67, p. 128–130.

Tailleur, I. L., and Brosgé, W. P., 1970, Tectonic history of northern Alaska, *in* Adkison, W. L., and Brosgé, M. M., eds., Proceedings of a Geological Seminar on the North Slope of Alaska: Pacific Section, American Association of Petroleum Geologists, p. E1–E19.

Taylor, H. P., and Coleman, R. G., 1968, O^{18}/O^{16} ratios of co-existing minerals in glaucophane-bearing metamorphic rocks: Geological Society of America Bulletin, v. 79, p. 1727–1756.

Tempelman-Kluit, D. J., 1976, The Yukon Crystaline terrane; Enigma in the Canadian Cordillera: Geological Society of America Bulletin, v. 87, p. 1343–1357.

Tempelman-Kluit, D. J., and Wanless, R. K., 1975, Potassium-argon age determinations of metamorphic and plutonic rocks in the Yukon Crystalline terrane: Canadian Journal of Earth Sciences, v. 12, no. 11, p. 1895–1909.

Thompson, A. B., 1976, Mineral reactions in pelitic rocks; 2, Calculation of some P-T-X(Fe-Mg) phase relations: American Journal of Science, v. 276, no. 4, p. 425–454.

Thurston, S. P., 1985, Structure, petrology, and metamorphic history of the Nome

Group blueschist terrane, Salmon Lake area, Seward Peninsula, Alaska: Geological Society of America Bulletin, v. 96, p. 600–617.
Till, A. B., 1980, Crystalline rocks of the Kigluiak Mountains, Seward Peninsula, Alaska [M.S. thesis]: Seattle, University of Washington, 97 p.
—— , 1983, Granulite, peridotite, and blueschist; Precambrian to Mesozoic history of Seward Peninsula, *in* Western Alaska geology and resource potential; Proceedings of the 1982 Symposium: Journal of the Alaska Geological Society, p. 59–66.
Till, A. B., Dumoulin, J. A., Gamble, B. M., Kaufman, D. S., and Carroll, P. I., 1986, Preliminary geologic map and fossil data, Solomon, Bendeleben, and southern Kotzebue Quadrangles, Seward Peninsula, Alaska: U.S. Geological Survey Open-File Report 86-276, 69 p., scale 1:250,000.
Turner, D. L., 1984, Tectonic implications of widespread Cretaceous overprinting of K-Ar ages in Alaskan metamorphic terranes: Geological Society of America Abstracts with Programs, v. 16, p. 338.
Turner, D. L., Herreid, G., and Bundtzen, T. K., 1977, Geochronology of southern Prince of Wales Island, Alaska: Division of Geological and Geophysical Surveys Geologic Report 55, 47 p.
Turner, D. L., Forbes, R. B., and Dillon, J. T., 1979, K-Ar geochronology of the southwestern Brooks Range, Alaska: Canadian Journal of Earth Sciences, v. 16, p. 1789–1804.
Turner, D. L., Forbes, R. B., Aleinikoff, J. N., Hedge, C. E., and McDougall, I., 1983, Geochronology of the Kilbuck terrane of southwestern Alaska: Geological Society of America Abstracts with Programs, v. 15, p. 407.
Turner, F. J., 1981, Metamorphic petrology; Mineralogical, field, and tectonic aspects, 2nd edition: New York, McGraw-Hill, 524 p.
Tysdal, R. G., and Case, J. E., 1979, Geologic map of the Seward and Blying Sound Quadrangles, Alaska: U.S. Geological Survey Miscellaneous Investigations Series Map I-1150, 12 p., scale 1:250,000.
Wallace, W. K., 1981, Structure and petrology of a portion of a regional thrust zone in the central Chugach Mountains, Alaska [Ph.D. thesis]: Seattle, University of Washington, 253 p.
—— , 1984, Deformation and metamorphism in a convergent margin setting, northern Chugach Mountains, Alaska: Geological Society of America Abstracts with Programs, v. 16, p. 339.
Wilson, F. H., Smith, J. G., and Shew, N., 1985, Review of radiometric data from the Yukon Crystalline terrane, Alaska and Yukon Territory: Canadian Journal of Earth Sciences, v. 22, p. 525–537.
Winkler, G. R., 1976, Deep-sea fan deposition of the lower Tertiary Orca Group, eastern Prince William Sound, Alaska, *in* Miller, T. P., ed., Recent and ancient sedimentary environments in Alaska: Alaska Geological Society, p. R1–R20.
Winkler, G. R., and Plafker, G., 1981, Geologic map and cross sections of the Cordova and Middleton Islands Quadrangles, southern Alaska: U.S. Geological Survey Open-File Report 81-1161, 26 p., scale 1:250,000.
Winkler, G. R., Silberman, M. L., Grantz, A., Miller, R. J., and MacKevett, E. M., Jr., 1981, Geologic map and summary geochronology of the Valdez Quadrangle, Alaska: U.S. Geological Survey Open-File Report 80-892-A, 2 sheets, scale 1:250,000.
Zen, E-an, and Hammarstrom, J. M., 1984a, Magmatic epidote and its petrologic significance: Geology, v. 12, p. 515–518.
—— , 1984b, Mineralogy and a petrogenetic model for the tonalite pluton at Bushy Point, Revillagigedo Island, Ketchikan 1° by 2° Quadrangle, southeastern Alaska, *in* Reed, K. M., and Bartsch-Winkler, S., eds., The United States Geological Survey in Alaska; Accomplishments during 1982: U.S. Geological Survey Circular 939, p. 118–123.
Zwart, H. J., and 7 others, 1967, A scheme of metamorphic facies for the cartographic representation of regional metamorphic belts: International Union of Geological Sciences Geological Newsletter, v. 1967, p. 57–72.

MANUSCRIPT COMPLETED JUNE 1989

MANUSCRIPT ACCEPTED BY THE SOCIETY MAY 18, 1990

ACKNOWLEDGMENTS

This metamorphic compilation and summary would not have been possible had it not been for the willingness of numerous colleagues from the U.S. Geological Survey, the State of Alaska's Division of Geological and Geophysical Surveys, and several universities to contribute their time, thoughts, and unpublished data. I am particularly indebted to S. E. Box, D. A. Brew, H. L. Foster, G. Plafker, and A. B. Till. Drafting and technical assistance were provided by S. L. Douglass and E. O. Doyle. J. Y. Bradshaw and S. M. Roeske made valuable suggestions to improve the original manuscript. Finally, I especially thank Jan Detterman for her invaluable and exceedingly patient editing of the various, and evolving, formats of the metamorphic facies studies of Alaska.

NOTES ADDED IN PROOF

Significant additions to our understanding of the metamorphic history of Alaska have been made since this chapter was written five years ago. More up-to-date coverage of metamorphic data for much of Alaska is presented in the regional-scale treatments of the topic in U.S. Geological Survey Professional Papers 1497-C, published in 1993, and 1497-B and 1497-D planned for publication in 1994. Detailed metamorphic studies in three areas—the Yukon-Tanana upland, the Seward Peninsula, and the southern Brooks Range—conducted since this chapter was written have yielded important new constraints, which are outlined briefly below.

Recent geothermobarometric data from the Yukon-Tanana Upland of east-central Alaska (Dusel-Bacon and Hansen, 1992) indicate that high- to intermediate-pressure/medium-temperature metamorphism affected both the ductilely deformed amphibolite-facies rocks of unit 37 (Taylor Mountain terrane of Dusel-Bacon and Hansen [1992]) and those of the adjacent, eastern part of unit 46 (Lake George subterrane of the Yukon-Tanana terrane of Dusel-Bacon and Hansen [1992]). Rocks of unit 37 give latest Triassic to Middle Jurassic metamorphic cooling ages, whereas those of unit 46 give Early Cretaceous metamorphic cooling ages (Hansen and others, 1991). On the basis of geothermometric, kinematic, thermochronologic, and lithologic data, Hansen and others (1991) and Dusel-Bacon and Hansen (1992) proposed that high-pressure metamorphism of both units 37 and 46 took place during different phases of the latest Paleozoic through early Mesozoic shortening episode, resulting from closure of an ocean basin now represented by klippen of the Seventymile–Slide Mountain terrane (units 38 and MzPz u). According to their model, high-pressure metamorphism of unit 37 took place within a southwest-dipping (present-day coordinates) subduction system, whereas high-pressure metamorphism of unit 46 occurred during continentward overthrusting of the Taylor Mountain and Seventymile–Slide Mountain terranes and imbrication of the continental margin in Jurassic time. Hansen and others (1991) and Dusel-Bacon and Hansen (1992) propose that the difference in metamorphic cooling ages between the Taylor Mountain terrane and adjacent parts of the Lake George subterrane is best explained by Early Cretaceous unroofing of the Lake George subterrane caused by crustal extension, recorded in a younger top-to-the-southeast fabric. Unequivocal evidence for widespread mid-Cretaceous extension within the Yukon-Tanana terrane, discussed briefly in this chapter under the subheading "Problematic tectonic origin of Cretaceous metamorphism . . . ," is presented in detail by Pavlis and others (1993) in connection with their important kinematic study of mylonites within the footwall of the sillimanite gneiss dome southeast of Fairbanks, shown on Plate 4.

Important geophysical constraints on the tectonic history of the Yukon-Tanana terrane of east-central Alaska have been provided by seismic refraction/wide-angle reflection and magnetotelluric experiments conducted as part of the Trans-Alaska Crustal Transect program (TACT). Major crustal features identified by these studies are an anomalously thin crust, approximately 30 km thick (Beaudoin and others, 1992), a highly conductive zone in the middle crust interpreted as underthrust Mesozoic flysch (Stanley and others, 1990), and a basal crustal section that Beaudoin and others (1992) interpreted as the underthrust basement of the composite Peninsular-Alexander-Wrangellia superterrane. Largely on the basis of the above interpretations of the present makeup of the crust along the TACT line, Nokleberg and others (1991) interpret the Early Cretaceous metamorphic cooling ages in unit 46 to record regional metamorphism and deformation associated with underplating of Mesozoic flysch and the Gravina arc and oblique collision of the Peninsular-Alexander-Wrangellia superterrane with the Yukon-Tanana terrane, rather than to record cooling related to metamorphism during an extensional event. However, as mentioned in this chapter under the subheading "Problematic tectonic origin of Cretaceous metamorphism . . . ," problems remain with the relative timing of events inherent in the underplating model.

Additional data on the evolution of the blueschist belt on the Seward Peninsula (unit 16, Plate 4) published since this chapter was written are presented in papers by Evans and Patrick (1987) and Patrick and Evans (1989). Recent work in the Kigluaik Mountains in the southwestern Seward Peninsula has shown that Cretaceous magmatism was synchronous with amphibolite- to granulite-facies metamorphism (of units 18 and 19, Plate 4) (Amato and others, 1992). Two different interpretations of the tectonic setting of the high-grade metamorphic episode have been made. Patrick (1988) and Patrick and Lieberman (1988) proposed that the high-grade episode overprinted the earlier widespread blueschist-facies metamorphism (of unit 16, Plate 4) and was a natural consequence of crustal telescoping followed by thermal relaxation and anatexis. According to Amato and others (1992) and Miller and others (1992), however, high-grade metamorphism probably occurred as a result of mafic magmatism associated with regional extension within the Cretaceous magmatic belt.

Important additional metamorphic, structural, and thermochronologic studies have also been made that relate to the evolution of the widespread blueschist facies belt in the southern Brooks Range (units 1-3, Plate 4). Till and others (1988) delineated two temporally and structurally distinct high-pressure/low-temperature metamorphic belts in the southern Brooks Range: the schist belt, penetratively and ductilely deformed in the Jurassic, and the central belt, inhomogeneously deformed during mid-Cretaceous time. An inverted metamorphic gradient overprinted schist-belt blueschists in the Walker Lake area at 110 Ma (Patrick and others, 1991; Till and Patrick, 1991). The schist belt was exhumed by mid-Albian, based on deposition of detritus from the belt in the foreland basin at that time (Till, 1992).

Structural fabrics in the schist belt have been attributed to a ductile event related to contractional tectonism and a later less penetrative event related to extension (Gottschalk and Oldow, 1988; Gottschalk, 1990) or to a ductile event related to extension (Miller and others, 1990a, b). The ductile extension proposed by Miller and her coworkers for the schist belt is just one of many lines of evidence used by Miller and Hudson (1991) to postulate that the crust of northern and central Alaska was thinned by lithospheric-scale extension during the mid-Cretaceous. This theory remains controversial and has been discussed by Till and others (1993) and Miller and Hudson (1993).

ADDITIONAL REFERENCES

Amato, J. M., Wright, J. E., and Gans, P. B., 1992, The nature and age of Cretaceous magmatism and metamorphism on the Seward Peninsula, Alaska: Geological Society of America Abstracts with Programs, v. 24, p. 2.

Beaudoin, B. C., Fuis, G. S., Mooney, W. D., Nokleberg, W. J., and Christensen, N. I., 1992, Thin, low-velocity crust beneath the southern Yukon-Tanana terrane, east-central Alaska: Results from the Trans-Alaskan Crustal Transect refraction/wide-angle reflection data: Journal of Geophysical Research, v. 97, no. 1, p. 103–122.

Dusel-Bacon, C., and Hansen, V. L., 1992, High-pressure amphibolite-facies metamorphism and deformation within the Yukon-Tanana and Taylor Mountain terranes, eastern Alaska, *in* Bradley, D. C., and Dusel-Bacon, C., eds., Geologic Studies in Alaska by the U.S. Geological Survey, 1991: U.S. Geological Survey Bulletin 2041, p. 140–159.

Evans, B. W., and Patrick, B. E., 1987, Phengite 3-T in high-pressure metamorphosed granitic orthogneisses, Seward Peninsula, Alaska: Canadian Mineralogist, v. 25, p. 141–158.

Gottschalk, R. R., 1990, Structural evolution of the schist belt, south-central Brooks Range fold and thrust belt, Alaska: Journal of Structural Geology, v. 12, p. 453–470.

Gottschalk, R. R., and Oldow, J. S., 1988, Low-angle normal faults in the south-central Brooks Range fold and thrust belt, Alaska: Geology, v. 16, p. 395–399.

Hansen, V. L., Heizler, M. T., and Harrison, T. M., 1991, Mesozoic thermal evolution of the Yukon-Tanana composite terrane: New evidence from $^{40}Ar/^{39}Ar$ data: Tectonics, v. 10, no. B2, p. 1921–1942.

Miller, E. L., and Hudson, T., 1991, Mid-Cretaceous extensional fragmentation of a Jurassic–Early Cretaceous compressional orogen, Alaska: Tectonics, v. 10, p. 781–796.

——, 1993, Reply to comment on "Mid-Cretaceous extensional fragmentation of a Jurassic–Early Cretaceous compressional orogen, Alaska": Tectonics, v. 12, p. 1082–1086.

Miller, E. L., Christiansen, P. P., and Little, T. A., 1990a, Structural studies in the southernmost Brooks Range, Alaska: Geological Association of Canada, Mineralogical Association of Canada Annual Meeting, Programs with Abstracts, v. 15, p. A89.

Miller, E. L., Law, R. D., and Little, T. A., 1990b, Evidence for extensional deformation on the southern flank of the Brooks Range in the Florence Creek area, Wiseman A-3 quadrangle, Alaska: Geological Society of America Abstracts with Programs, v. 22, p. A183.

Miller, E. L., Calvert, A. T., and Little, T. A., 1992, Strain-collapsed metamorphic isograds in a sillimanite gneiss dome, Seward Peninsula, Alaska: Geology, v. 20, p. 487–490.

Nokleberg, W. J., Foster, H. F., Lanphere, M. A., Aleinikoff, J. N., and Pavlis, T. L., 1991, Structure and tectonics of the Yukon-Tanana, Wickersham, Seventymile, and Stikinia terranes along the Trans-Alaskan Crustal Transect (TACT), east-central Alaska: Geological Society of America Abstracts with Programs, v. 23, no. 2, p. 84.

Patrick, B. E., 1988, Synmetamorphic structural evolution of the Seward Peninsula blueschist terrane, Alaska: Journal of Structural Geology, v. 10, p. 555–565.

Patrick, B. E., and Evans, B. W., 1989, Metamorphic evolution of the Seward Peninsula blueschist terrane: Journal of Petrology, v. 30, p. 531–556.

Patrick, B. E., and Lieberman, J. E., 1988, Thermal overprint on the Seward Peninsula blueschist terrane: the Lepontine in Alaska: Geology, v. 16, p. 1100–1103.

Patrick, B. E., Dinklage, W. S., and Till, A. B., 1991, Metamorphism and progressive deformation in the Walker Lake region of the southern Brooks Range, Alaska: Geological Society of America Abstracts with Programs, v. 23, p. 87.

Pavlis, T. L., Sisson, V. B., Foster H. L., Nokleberg, W. J., and Plafker, G., 1993, Mid-Cretaceous extensional tectonics of the Yukon-Tanana terrane, Trans-Alaskan Crustal Transect (TACT), east-central Alaska: Tectonics, v. 12, no. 1, p. 103–122.

Stanley, W. D., Labson, V. F., Nokleberg, W. J., Csejtey, B., Jr., and Fisher, M. A., 1990, The Denali fault system and Alaska Range of Alaska: Evidence for underplated Mesozoic flysch from magnetotelluric surveys: Geological Society of America Bulletin, v. 102, no. 2, p. 160–173.

Till, A. B., 1992, Detrital blueschist-facies metamorphic mineral assemblages in Early Cretaceous sediments of the foreland basin of the Brooks Range, Alaska, and implications for orogenic evolution: Tectonics, v. 11, p. 1207–1223.

Till, A. B., and Patrick, B. E., 1991, Ar-Ar evidence for a 110–105 Ma amphibolite-facies overprint on blueschist in the south-central Brooks Range, Alaska: Geological Society of America Abstracts with Programs, v. 23, p. A436.

Till, A. B., Schmidt, J. M., and Nelson, S. W., 1988, Thrust involvement of metamorphic rocks, southwestern Brooks Range, Alaska: Geology, v. 16, p. 930–933.

Till, A. B., Box, S. E., Roeske, S. M., and Patton, W. W., Jr., 1993, Comment on "Mid-Cretaceous extensional fragmentation of a Jurassic–Early Cretaceous compressional orogen, Alaska" by E. L. Miller and T. Hudson: Tectonics, v. 12, p. 1076–1081.

Printed in U.S.A.

The Geology of North America
Vol. G-1, The Geology of Alaska
The Geological Society of America, 1994

Chapter 16

Pre-Cenozoic plutonic rocks in mainland Alaska

Thomas P. Miller
U.S. Geological Survey, 4200 University Drive, Anchorage, Alaska 99508-4667

INTRODUCTION

Studies during the past decade have revealed that much of Alaska consists of a collection of generally far-traveled tectonostratigraphic terranes, most of which were transported to their present locations and accreted to North America in late Mesozoic to early Tertiary time (Jones and others, 1984; Silberling and others, this volume). This collage results in the exceedingly complex geologic and tectonic framework that constitutes much of present-day Alaska.

Magmatic activity in Alaska was influenced by a host of factors, including many involving plate interactions, such as the rate of subduction, the angle of dip, the motion of individual plates, and the composition, thickness, and age of material that was subducted beneath or collided with Alaska. The identification and interpretation of magmatic patterns as reflected by time of intrusion, areal distribution, and composition can therefore contribute to an understanding of the tectonic history of Alaska.

The following overview focuses on plutonic rocks in mainland (excludes southeastern) Alaska emplaced from the Proterozoic into the earliest Tertiary. Plutonic rocks and belts emplaced during a specific time frame (e.g., the Early Cretaceous) are not necessarily everywhere related in terms of genesis or tectonic setting, and some related rocks may be shown in different temporal episodes. Postplutonic terrane movement and Cenozoic strike-slip faulting with large displacements have further complicated the identification and interpretation of plutonic events and patterns. Rocks assigned to specific plutonic belts are assumed to be cogenetic regardless of the mechanism of formation. The informal name given to a temporal episode of intrusive activity (e.g., Early to Middle Jurassic) may in some cases have been chosen to reflect more accurately the epoch or epochs where the major part of the activity occurred rather than to adhere strictly to absolute ages. Additional data on the latest Cretaceous and Cenozoic plutonism in Alaska are in the chapters by Moll-Stalcup (this volume) and Brew (this volume); and on Plate 13 (Barker and others, this volume).

Overviews of plutonic events in southern Alaska by Hudson (1983) and in southwestern Alaska by Wallace and Engebretson (1984) and Wallace and others (1989) were particularly helpful in preparing this compilation, as was that of Armstrong (1988) for adjacent parts of Canada. This discussion of individual plutonic belts and provinces that crop out over 1,000,000 km^2 is of necessity brief, and an attempt has been made to list pertinent references to guide the reader who wishes to delve further into specific regions.

The age designations used in this chapter are based on the Decade of North American Geology geologic time scale (Palmer, 1983), and plutonic rock classifications are after Streckeisen (1976). The pre-Cenozoic plutonic rocks of Alaska discussed in this chapter are shown in more detail on 1:2,500,000-scale maps in Barker and others (this volume) and Moll and others (this volume). Geographic and physiographic names follow those of Wahrhaftig (this volume).

PROTEROZOIC PLUTONIC ROCKS

Most of the few occurrences of known Precambrian rocks in mainland Alaska (Eberlein and Lanphere, 1988) are in terranes north of the Denali fault (Plate 13), and all those dated are of Proterozoic age (2,500 to 570 Ma). Proterozoic plutonic rocks, generally orthogneiss or otherwise deformed, have been reported at widely separated localities in the Brooks Range, Seward Peninsula, and in a northeast-trending belt in southwestern Alaska (Fig. 1). Compositional data for these plutonic rocks are sparse. Protolith ages are also relatively few but range between 2,050 and 750 Ma; all of these plutons are in accreted terranes that have been subjected to one or more Phanerozoic thermal events.

Brooks Range

The Brooks Range has been included in the Arctic Alaska terrane (Jones and others, 1984), which in turn is subdivided into several subterranes. The structural core of the central and eastern Brooks Range consists of a complex assemblage of slightly to moderately metamorphosed Proterozoic and early Paleozoic rocks that is unconformably overlain by Mississippian strata. Rocks composing the assemblage are largely metasedimentary units of quartzite, quartz arenite, carbonate rocks, phyllite, argillite, chert, and graywacke and intrastratified metavolcanic rocks. These units were deformed during middle Paleozoic, middle Mesozoic, and Tertiary orogenic events and are now exposed in

Miller, T. P., 1994, Pre-Cenozoic plutonic rocks in mainland Alaska, *in* Plafker, G., and Berg, H. C., eds., The Geology of Alaska: Boulder, Colorado, Geological Society of America, The Geology of North America, v. G-1.

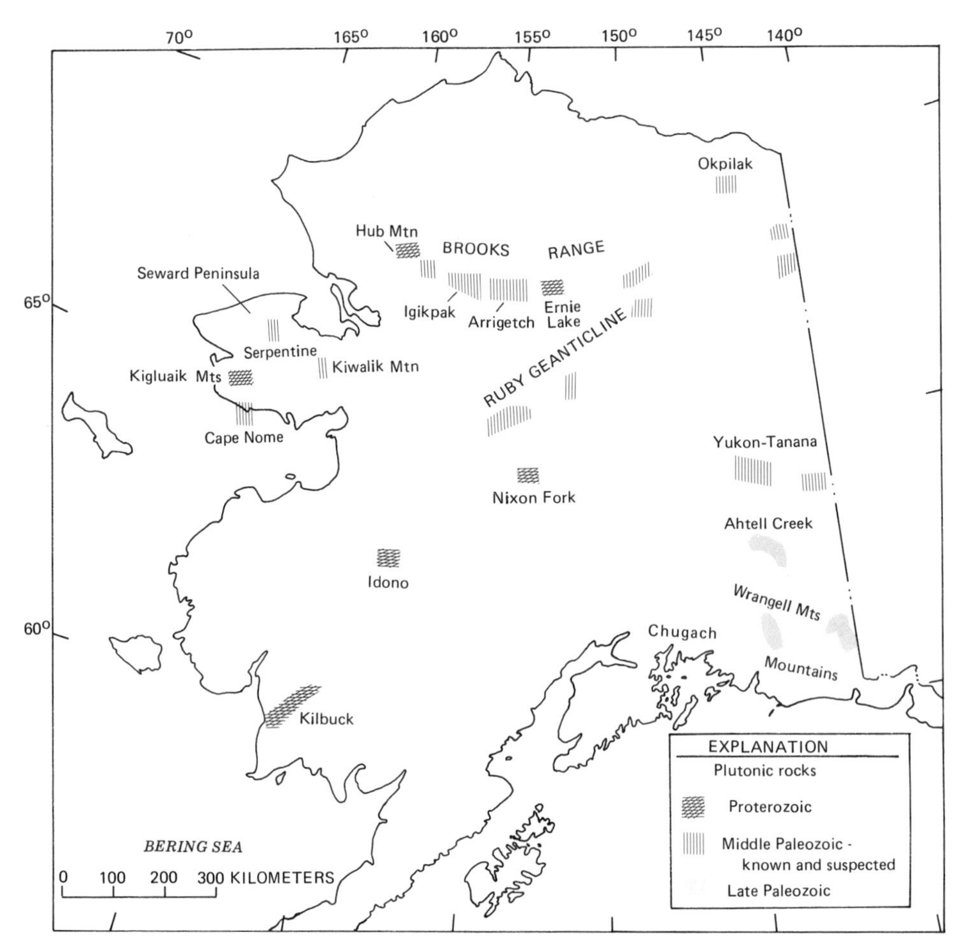

Figure 1. Distribution of Proterozoic and Paleozoic plutons in mainland Alaska.

imbricate thrust sequences and sole complexes. The Late Jurassic to Early Cretaceous thrust faulting is the result of collision between the Arctic Alaska and Angayucham terranes. Metamorphism related to this plate collision affected rocks throughout the southern Arctic Alaska terrane (Dusel-Bacon, this volume). The assemblage is locally intruded by granitic plutons of Proterozoic and middle Paleozoic age (Dillon and others, 1980).

Most felsic meta-igneous rocks in the Brooks Range yield Cretaceous K-Ar isotopic apparent ages (Turner and others, 1979). Dillon and others (1980, 1987), however, obtained preliminary U-Pb zircon ages on metagranitic rocks of the western Brooks Range metamorphic complex at Mount Angayukaqsraq, also known as Hub Mountain (Fig. 1), and of the central Brooks Range. U-Pb ratios from the Ernie Lake orthogneiss pluton (and the adjacent Sixtymile pluton) in the central Brooks Range (Fig. 1) yielded Late Proterozoic U-Pb and Pb-Pb ages (Dillon and others, 1980); however, because the data do not define a well-constrained chord, the time of intrusion and crystallization cannot be determined. The metamorphic country rocks for these bodies are similar in lithology and metamorphic grade to those at Mount Angayukaqsraq (Dillon and others, 1987). These orthogneiss bodies may therefore be similar in age to those at Mount Angayukaqsraq and may represent windows into related Late Proterozoic basement. Further isotopic age studies are needed to resolve this question.

The plutonic rocks at Mount Angayukaqsraq, first described by Mayfield and others (1983), were found by Karl and others (1989) to underlie an area of only about 4 km^2 and to consist chiefly (about 70 percent) of gabbro and leucogabbro intruded by granodiorite and alkali feldspar granite (about 30 percent). The intrusive rocks are typically massive, nonfoliated, medium grained, and possess relict igneous textures. The granite is highly evolved (more than 75 percent SiO_2) and slightly peraluminous. U-Pb zircon dating done on granitic samples from Mount Angayukaqsraq provides a 750 $\pm$ 6 Ma intrusive age (Karl and others, 1989) and provides the best measurement yet of the timing of late Proterozoic magmatism in the Brooks Range.

The metamorphic complex, in which the plutonic rocks are included, is part of an antiformal structure at the western end of the structural core of the Brooks Range. The stratigraphically oldest rocks in the metamorphic complex (late Proterozoic to early Paleozoic age) extend discontinuously nearly 800 km to the east (Till and others, 1988) and are part of the "central metamorphic belt" of Moore and others (this volume).

Seward Peninsula

The Seward Peninsula, a structurally and stratigraphically complex region of about 26,000 km ^{2} including the Seward and York terranes, is bounded on the east by the Yukon-Koyukuk basin and on the north by the Brooks Range fold-and-thrust belt. The Seward terrane, forming the east two-thirds of the peninsula, is dominated by blueschist- and greenschist-facies schists of the Nome Group (Till and Dumoulin, this volume). High-grade metamorphic rocks (amphibolite and granulite facies) are exposed in the fault-bounded Kigluaik, Bendeleben, and Darby Mountains (Fig. 1). These metamorphic rocks are composed of continental crustal material of Proterozoic to middle Paleozoic age and were subjected to crustal imbrication and thickening in middle Mesozoic time and widespread plutonism in mid-Cretaceous to Late Cretaceous time. Latest Jurassic to earliest Cretaceous compressional deformation of the Brooks Range–Seward Peninsula continental margin resulted in the formation of a north-directed fold-and-thrust belt with local blueschist metamorphism in its southern part (Forbes and others, 1984; Thurston, 1985; Einaudi and Hitzman, 1986). Late to post-tectonic plutonism occurred in mid-Cretaceous to Late Cretaceous time.

Granitic orthogneiss of possible late Proterozoic(?) to middle Paleozoic age crops out in widely scattered areas on the Seward Peninsula (Fig. 1), only the largest of which are shown on Plate 6. Compositional data are lacking for most of these bodies, and they are generally defined as metamorphosed granite, granodiorite, or tonalite. They range from layers structurally conformable to surrounding rock units and only a few meters thick to masses several kilometers in breadth. These metaplutonic bodies intruded or are intercalated with the late Proterozoic through Devonian miogeoclinal carbonate rocks, pelite, quartzite, and volcanogenic metasedimentary rocks that constitute most of the Seward Peninsula (Sainsbury, 1975; Till and others, 1986). They were subjected to blueschist-facies metamorphism during Middle Jurassic to Early Cretaceous burial, followed by overprinting to greenschist facies during decompression (Armstrong and others, 1986; Evans and Patrick, 1987; Patrick, 1988) and finally to postkinematic intrusion of Cretaceous granitic rocks (Till and others, 1986). The nature of the protoliths is uncertain, but Till and others (1986) suggest that the sediments were deposited on the shallow platform, shelf, and slope of a rifted continental margin.

Although the granitic orthogneiss has not been mapped or studied in detail, these rocks have been assigned ages ranging from Proterozoic (Sainsbury, 1975; Bunker and others, 1979) to Devonian (Till and others, 1986) on the basis of field relations and preliminary radiometric dating. Bunker and others (1979), for example, regarded all orthogneiss bodies on the Seward Peninsula as being Proterozoic in age. Their interpretation was based on Rb-Sr whole-rock ages of orthogneiss and associated schist in the Kigluaik Mountains, which indicated a 735-Ma age for the metamorphism. Till and others (1986), however, reported a U-Pb zircon age of 381 $\pm$ 2 Ma from the metagranite at Kiwalik Mountain in the eastern Seward Peninsula and suggested that similar orthogneiss bodies in the northern and southern parts of the peninsula may also be Devonian in age.

The stratigraphic position and structural setting of some orthogneiss bodies (e.g., in the Kigluaik Mountains and at Serpentine Hot Springs), indicate a possible Proterozoic age (Gardner and Hudson, 1984; Armstrong and others, 1986). Armstrong and others (1986), however, in a detailed Rb-Sr and K-Ar study of the Seward Peninsula, found no compelling evidence for Proterozoic metamorphism and therefore no definitive isotopic evidence for a Proterozoic protolith age for the orthogneiss. Their whole-rock Rb-Sr plots of metamorphic rocks and orthogneiss defined a fan bounded by isochrons of about 720 and 360 Ma. The available isotopic data therefore allow but do not confirm a Proterozoic age of intrusion. At present, the existence of Proterozoic plutonic rocks on the Seward Peninsula must be regarded as problematical.

Southwest and central Alaska

Proterozoic plutonic rocks have been identified in three widely separated localities in the Kilbuck, Idono, and Nixon Fork terranes, which are a group of terranes composed chiefly of Proterozoic(?) and Paleozoic metamorphic rocks extending from southwest to central Alaska (Fig. 1). The distribution and compositions of these plutonic rocks are poorly known. The isotopic ages are sufficient to identify these three localities as Precambrian but do not give the relations of one to another.

Kilbuck terrane. The Kilbuck terrane, which was called the "Kanektok metamorphic complex" by Hoare and Coonrad (1979), consists of a narrow, 110-km-long sliver of continental crust, most of which appears to be plutonic in origin. The metamorphic complex consists chiefly of quartz diorite, granodiorite gneiss, granite gneiss, and lesser amounts of greenschist- to amphibolite-facies metasedimentary rocks and marble. Chemical and modal data from these plutonic rocks are lacking, and the compositional nature of the protoliths is therefore uncertain. Box (1985a) has suggested that the continental Kilbuck terrane was partially thrust beneath the accretionary fore arc and intraoceanic volcanic arcs of the Goodnews and Togiak terranes, respectively, in Early Cretaceous time.

Turner and others (1983) interpret U-Th-Pb analyses and Rb-Sr whole-rock plots from granitic orthogneiss as indicating that the plutonic rocks crystallized at about 2,050 Ma and that a later metamorphic event occurred at about 1,770 Ma. The oldest 58 K-Ar mineral ages, most of which have been reset by a Mesozoic thermal event, also suggest a 1,770-Ma metamorphic event.

Idono sequence. Gemuts and others (1983) correlated an area of augen gneiss and amphibolite east of the Yukon River (Fig. 1; Plate 6) (which they called the Idono sequence) on the basis of lithology and geologic setting with similar rocks of Proterozoic(?) and Paleozoic age in the Ruby terrane. M. L. Miller and T. K. Bundtzen (unpublished data, cited in Decker and others, this volume) report preliminary U-Pb isotopic data on whole populations of zircons as indicating a late Proterozoic age of crystallization.

Nixon Fork terrane. The Nixon Fork terrane in central Alaska consists of metasedimentary and felsic metavolcanic and metaplutonic rocks that are overlain by unmetamorphosed Ordovician through Devonian shelf carbonate rocks and Permian to mid-Cretaceous terrigenous sedimentary rocks. Late Cretaceous and early Tertiary volcanism and plutonism were widespread throughout the terrane. Dillon and others (1985) have speculated that the Nixon Fork terrane may be an amalgamation of several Proterozoic terranes.

A sheared porphyritic quartz diorite from one of the metaplutonic units of the Nixon Fork terrane is intrusive into quartz-mica schist thought to be pre-Ordovician in age and has yielded a 921 ± 25 Ma K-Ar age (Silberman and others, 1979a, b). Dillon and others (1985) report a variety of discordant U-Pb zircon ages from associated foliated quartz porphyry metavolcanic rocks. Emplacement ages range from 1,265 ± 50 to 850 ± 30 Ma, with lead-loss events in middle Paleozoic and Late Cretaceous time. The available stratigraphic and radiometric age data indicate that Proterozoic magmatism has occurred, although they are not sufficiently precise to constrain the age of individual events.

PALEOZOIC PLUTONIC ROCKS

Paleozoic plutonic rocks, although relatively dispersed and underlying small areas, are important to the tectonic history of Alaska. Outside of southeastern Alaska, two principal episodes of Paleozoic plutonism are represented. The older and more areally extensive episode is composed of more than 30 middle Paleozoic granitic orthogneiss plutons that extend in a sinuous band from east-central Alaska across much of the Brooks Range and into the Seward Peninsula (Plate 13). A younger, late Paleozoic suite of plutonic rocks that occurs in the eastern Alaska Range and along the south flank of the Wrangell Mountains in southern east-central Alaska has important implications relating to terrane linkages.

Middle Paleozoic

The plutons consist chiefly of peraluminous biotite granite (Fig. 2) ranging in age from Middle Devonian to Early Mississippian. Individual plutons range in area from 5 to 500 km^2, and all have been slightly to moderately recrystallized to give a gneissic fabric. The plutons are intrusive into, and thus link or stitch, the Yukon-Tanana and Ruby terranes in east-central Alaska and Yukon Territory; the Hammond, Endicott, and North Slope terranes of the Brooks Range and Arctic Slope; and the Seward terrane of the Seward Peninsula (Plate 13). They are contiguous with bodies of similar age and composition in the Slide Mountain terrane of southeastern Yukon Territory (Hansen, 1988) and in the British Mountains of northernmost Yukon Territory. The existence of similar rocks in eastern Siberia is uncertain, although early Paleozoic plutons are reported on Wrangell Island in the Arctic Ocean, some 500 km due west of the Brooks Range (Kosygin and Popeko, 1987).

Yukon-Tanana Upland. Metamorphosed granitic rocks were first noted in the Yukon-Tanana Upland (the Yukon-Tanana terrane of Coney and others, 1980; see Foster and others, this volume) by Prindle (1909) and Mertie (1937). The plutons have been mapped by Foster (1970, 1976) and Weber and others (1978). Their petrology, tectonic setting, and distribution have been discussed by Dusel-Bacon and Aleinikoff (1985) and Dusel-Bacon and others (1989).

These studies, particularly that of Dusel-Bacon and Aleinikoff (1985), indicate that a belt of porphyritic peraluminous granitic rocks (Fig. 1), now deformed to augen gneiss, extends 200 km in an east-west direction across the Alaskan part of the Yukon-Tanana terrane and into Yukon Territory. The plutons intrude amphibolite-grade quartz-mica schist, hornblende-bearing schist and gneiss, biotite gneiss, leucocratic gneiss, and quartzite of probable early to middle Paleozoic age.

Early Mississippian intrusive ages are indicated by a 341 ± 3 Ma U-Pb lower intercept age of zircon, a 333 ± 26 Ma Rb-Sr whole-rock isochron for augen gneiss (Aleinikoff, 1984; Dusel-Bacon and Aleinikoff, 1985), and a 342 ± 5 Ma Rb-Sr whole-rock isochron for correlative augen gneiss from Yukon Territory (Mortensen, 1983).

The metagranite characteristically has an augen gneiss character and well-developed fluxion structure. Gneissic banding is generally concordant with layering in the surrounding metamorphic country rocks. Wall-rock contacts are sharp, however, and the plutons are demonstrably intrusive into the country rocks.

The metagranite ranges in SiO_2 content from 71.2 to 77.1 percent (water-free), is peraluminous (normative corundum 1.3 to 3.9 percent), and belongs to the ilmenite series of granitic rocks (Ishihara, 1981). These characteristics plus high initial $^{87}Sr/^{86}Sr$ ratios of 0.728 and 0.719 (Mortensen, 1983; Aleinikoff, 1984), the possibility of primary garnet and muscovite, and several other lines of evidence cited by Dusel-Bacon and Aleinikoff (1985) indicate that the parent magma of the granitoid protoliths was derived from, or more likely, contaminated by, metasedimentary crustal sources. A U-Pb upper intercept age on zircon fractions and a Sm-Nd whole-rock model age suggest an early Proterozoic (2.2 to 2.0 Ga) age for the crustal component (Aleinikoff and others, 1981, 1986).

Ruby geanticline. Middle Paleozoic granitic gneiss is found in the Ray Mountains near the center of the Ruby geanticline (Fig. 1; Plate 13), in the northern part of the Ruby terrane (Patton and others, 1987) in central Alaska. Possibly similar granite gneiss bodies crop out at either end of the Ruby geanticline and, although they have not been dated, are tentatively correlated with the gneiss in the Ray Mountains (Brosgé and Reiser, 1964; Patton and others, 1978).

The Ruby geanticline is a northeast-trending, uplifted area

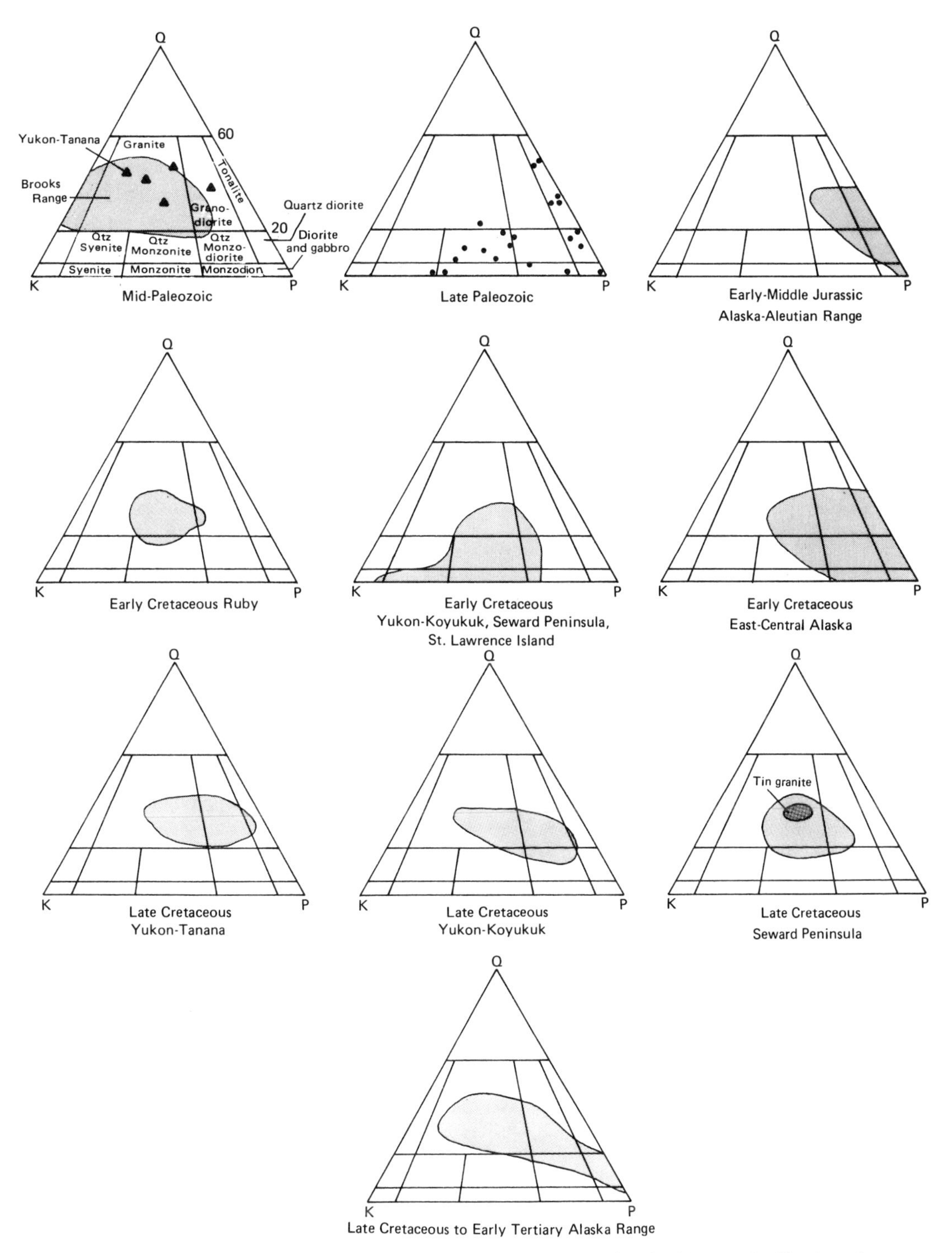

Figure 2. Representative modal compositions of plutonic rock suites in Alaska (classification after Streckeisen, 1976). Q, quartz; K, potassic feldspar; P, plagioclase. See text for references.

composed of a core of Proterozoic(?) and Paleozoic metamorphic rocks flanked by overthrust oceanic rocks of Jurassic age, all of which have been intruded by Early Cretaceous granitic plutons. It is bounded on the northwest by the Cretaceous Yukon-Koyukuk basin and on the southeast by Tertiary and Quaternary rocks of the Yukon Flats. The metamorphic country rocks consist of chlorite-quartz-mica schist, quartzite, greenstone, carbonate rocks, and minor quartzofeldspathic schist and gneiss. These rocks are generally greenschist-facies grade and locally reach almandine-amphibolite facies.

Strongly foliated quartzofeldspathic gneiss crops out on the north and south margin of the Ray Mountains pluton (Plate 13). The gneiss consists of a coarse- to medium-grained, porphyroclastic biotite + quartz + feldspar + muscovite mylonite gneiss with resistant augen of K-feldspar in a finer grained groundmass of biotite + quartz + feldspar + muscovite. The fabric of the gneiss is similar to that of other metamorphic rocks with which it appears to be interlayered. Although detailed compositional data are lacking, field observations suggest that the gneiss has a relatively restricted granitic composition (Patton and others, 1987). Contacts between the gneiss and the crosscutting Cretaceous granite are sharp.

U-Pb zircon ages were obtained from four whole-zircon populations from samples of gneiss that were collected from the Ray Mountains pluton (Patton and others, 1987). The four zircon ages are discordant, but chords (or a chord) indicate that the age of the granite gneiss is 390 ± 25 Ma (Middle Devonian) and that the apparent age of lead loss during metamorphism is about 110 Ma (Early Cretaceous)—the approximate age of the Ray Mountains pluton (Miller, 1989).

Two other masses of poorly mapped granitic gneiss occur at either end of the Ruby geanticline (Fig. 1; Plate 13) in contact with Cretaceous granitic rocks. Brosgé and Reiser (1964) describe the migmatitic rocks at the north end of the Ruby geanticline as consisting of granitic rocks intercalated with biotite schist. A Devonian(?) age is assigned by Brosgé and Reiser (1964) based on field relations with intercalated rocks of supposed Devonian age.

Patton and others (1978) describe the orthogneiss in the southern area as consisting of quartz-feldspar-biotite gneiss. Wall rocks are commonly garnetiferous and locally contain sillimanite. Patton and others (1978) suggest that the orthogneiss and associated wall rocks may be part of a gneiss dome and assign this body a Paleozoic age on the basis of field relations with intercalated Paleozoic rocks. Features such as the foliated character of the gneiss, the intrusive nature of the Early Cretaceous plutons, and the field relations between the gneiss and the other metamorphic rocks suggest that these gneiss bodies may be relatively old. An attempt was made to date the southernmost body of U-Pb analysis of zircon (Dillon and others, 1985). The data from two samples of the orthogneiss were insufficient to determine the protolith age but did indicate that the orthogneiss contained a Proterozoic or Paleozoic component, and that the zircon had lost lead during Cretaceous time. No radiometric age studies have been done on the northernmost gneiss body.

The Middle Devonian age of the orthogneiss near the Ray Mountains pluton is well constrained, but the lack of detailed mapping, petrologic studies, and radiometric age dating makes the tentative correlation of the other two localities subject to considerable uncertainty.

Brooks Range. Middle Paleozoic plutons are exposed along a 900-km-long belt across the length of the Brooks Range (Fig. 1; Plate 13) into the British Mountains of northernmost Yukon Territory (Dawson, 1988). The plutons underlie an aggregate area of about 4,500 km^2, including six bodies whose areal extents range from 200 to 2,000 km^2 and over a dozen smaller bodies of less than 50 km^2 area. Few, if any, plutons have been mapped in detail (1:63,360 scale or larger). The available data base is sparse and consists of approximately 100 major-element chemical analyses and about 120 modal analyses (Fig. 2), chiefly from the five largest plutons; published trace-element data are not available.

The plutonic rocks as described by Sable (1977), Nelson and Grybeck (1980), Dillon and others (1980, 1987), and Newberry and others (1986), consist chiefly (more than 95 percent) of highly evolved, peraluminous biotite and two-mica granite (Fig. 2). Hornblende-bearing metaluminous phases are found only in a few localities, generally as border phases of larger peraluminous plutons. The cataclastic and recrystallized nature of plutonic rocks suggests the possibility of mobilization of alkalis, which may account for the modal scatter reported by Nelson and Grybeck (1980). Most plutons probably belong in the ilmenite series because magnetite is sparse and Fe_2O_3/FeO ratios are usually less than 1. Primary aluminosilicate minerals such as sillimanite are lacking, although most analyzed samples record normative corundum. Many of these highly evolved rocks are "tin" granite, commonly with associated Sn-W skarns and, locally, greisens (Newberry and others, 1986).

The presence of albite-chlorite-muscovite-epidote assemblages in the country rocks indicates greenschist-facies metamorphism. Garnet and biotite isograds surround and extend south of the Arrigetch Peak and Mount Igikpak plutons; elsewhere, mid-greenschist- to epidote-amphibolite-facies assemblages occur in the vicinity of the plutons.

The plutons in the central Brooks Range contain metamorphic fabrics with orientations subparallel to those in the surrounding country rocks, and most of the plutons also contain large blastoporphyritic aggregates of quartz and K-feldspar (Dillon and others, 1980). Many plutons in the central and eastern Brooks Range have been dismembered by thrusting (Till and others, 1988; Hanks and Wallace, 1990). Dynamothermal metamorphism of many of the plutons throughout the area is indicated by strained cataclastic quartz phenocrysts; development of near isotropic, very fine-grained mylonitic zones; granulation and rotation of phenocrysts; the presence of folded dikes; and the development of schistosity and crenulation surfaces.

The plutons intrude Proterozoic(?) to early Paleozoic quartzofeldspathic schist, quartzose and pelitic metasedimentary basement rocks, and Devonian marble. Although contacts are commonly concordant, sheared, or faulted because of subsequent metamorphic transposition, granitic dikes and apophyses are present in the country rocks, and country-rock inclusions are present in the granitic rocks. Early Mississippian, coarse, clastic sedimentary rocks unconformably overlie the Okpilak pluton in the northeastern Brooks Range. Field evidence thus suggests a post-Silurian to pre–Early Mississippian age for at least some of the Brooks Range plutons (Sable, 1977; Dillon and others, 1987).

U-Pb and Rb-Sr age measurements that support a Devonian

crystallization age for individual plutons range from 400 to 370 Ma. U-Pb analyses were obtained on 31 zircon fractions from the middle Paleozoic plutons in the central Brooks Range (Dillon and others, 1987) and yielded an interpreted age of 390 ± 20 Ma. Lead-loss ages (lower intercepts on concordia) from multiple fractions of individual plutons range from 150 to 60 Ma. Three samples each from the Mount Igikpak and Arrigetch Peak plutons in the central Brooks Range (Fig. 1) give an $^{87}Sr/^{86}Sr$ $^{87}Rb/^{86}Sr$ age of 373 ± 25 Ma (similar to the U-Pb data) and an intercept (SIR) of 0.714 ± 0.003 (Silberman and others, 1979a). K-Ar studies (Turner and others, 1979) of muscovite and biotite from both metamorphic and plutonic rocks in the western and central Brooks Range indicate a Cretaceous age for cooling following the final metamorphic event but do not rule out earlier metamorphic events. A Tertiary resetting of K-Ar ages appears to have occurred in the northeastern Brooks Range plutons (Dillon, 1987).

These Devonian plutons are probably coeval (based on similar U-Pb zircon ages) and comagmatic (based on similar bulk chemistry) with a 250-km-long belt of felsic metavolcanic and hypabyssal intrusive rocks (Ambler sequence of Schmidt, 1986) that lies 10 to 30 km south of the Devonian plutons of the central Brooks Range. The Ambler (volcanic) sequence is 700 to 3,000 m thick, has a strike length of about 120 km, and is the host for several large volcanogenic massive sulfide deposits.

Seward Peninsula. The aforementioned granitic orthogneiss that crops out in several widely scattered areas on the Seward Peninsula (Fig. 1) occurs as bodies of metamorphosed granite, granodiorite, and tonalite intruded into metasedimentary rocks. The orthogneiss units range from structurally comformable layers only a few meters thick to masses several kilometers across such as at Kiwalik Mountain and Cape Nome (Fig. 1; Plate 13).

Till and others (1986) and Till and Dumoulin (this volume) report a U-Pb age on zircon fractions of 381 ± 2 Ma from the metagranite at Kiwalik Mountain in the eastern Seward Peninsula, an age supported by the crosscutting relations with the surrounding early Paleozoic metamorphic rocks. Armstrong and others (1986) noted that the very radiogenic character of Sr in most orthogneiss samples (SIR range for Kiwalik Mountain is 0.7081 to 0.7088) and the distinctive range of whole-rock Rb-Sr ages occurring between 720 and 360 Ma are characteristics similar to those reported for the middle Paleozoic plutons in the Yukon-Tanana terrane and suggest a common origin.

Discussion. The Middle Devonian to Early Mississippian metaplutonic rocks of northern and east-central Alaska apepar to be cogenetic members of a middle Paleozoic plutonic belt composed chiefly of highly evolved, peraluminous granite. These characteristics, together with high $^{87}Sr/^{86}Sr$ ratios, a geologic setting in miogeoclinal continental rocks, and in the case of the Brooks Range plutons, the associated tin mineralization, suggest that the plutons have an inherited crustal component. Dusel-Bacon and Aleinikoff (1985) propose that these plutons originated as granitic magmas that assimilated early Proterozoic crust or metasedimentary rocks derived from a provenance of that age, and that the Early Mississippian plutonic belt formed by these magmas thus represents a middle Paleozoic continental magmatic arc that developed near the edge of the Proterozoic craton. The actual location of the arc is uncertain because the plutons were probably translated northward to their present location during Cretaceous and Tertiary time. Monger and others (1982), Gabrielse and others (1982), and Aleinikoff and others (1986) have all noted the existence of a belt of Devonian and Mississippian granitic rocks extending from southern British Columbia to the Seward Peninsula. Rubin and others (1989) suggest that magmatic suites of this age in the western North America Cordillera define a middle Paleozoic East Pacific fringing arc system that developed across a host of terranes as far south as the northern Sierra and eastern Klamath terranes.

Late Paleozoic

The existence of late Paleozoic plutonic rocks in east-central Alaska was first confirmed by Richter and others (1975), who obtained Late Pennsylvanian K-Ar ages of 285 and 282 Ma on an assemblage of monzonite, granite, and syenodiorite at the Ahtell Creek pluton in the eastern Alaska Range (Fig. 1). Regional mapping and topical studies in the general area (Nokleberg and others, 1986; Gardner and others, 1988; Plafker and others, 1989; Beard and Barker, 1989) have confirmed several more Late Pennsylvanian plutons of similar composition in the Wrangell and northern Chugach Mountains (Fig. 1; Plate 13). Probable coeval plutonic rocks have been mapped in southernmost Yukon Territory and northern British Columbia (Dodds and Campbell, 1988).

The most extensive areas of Pennsylvanian plutonic rocks are near the Wrangell Mountains in east-central Alaska (Fig. 1; Plate 13). In the eastern Wrangell Mountains, several large plutons, partially fault bounded, underlie a total area of about 1,400 km^2 (MacKevett, 1978; Gardner and others, 1988) and intrude Pennsylvanian rocks of the Station Creek Formation of the Wrangellia terrane and an early to middle Paleozoic unit referred to by Gardner and others as the Kaskawulsh metamorphic rocks of the Alexander terrane. Gardner and others (1988) describe one of these intrusions, the Barnard Glacier pluton, as a composite body, with numerous apophyses extending into the country rocks, and surrounded by a contact aureole. The main body consists chiefly of nonfoliated, medium- to coarse-grained, equigranular to porphyritic quartz monzonite along with subordinate quartz syenite, alkali granite, and monzodiorite. U-Pb zircon analyses indicate that the pluton was emplaced at 309 ± 5 Ma (Late Pennsylvanian), although K-Ar hornblende ages range from 312 to 279 Ma, or into the Early Permian.

Other small bodies of metaplutonic rocks of Late Pennsylvanian age have recently been described (Nokleberg and others, 1986; Aleinikoff and others, 1988; Plafker and others, 1989) along the south and west flanks of the Wrangell Mountains and northern Chugach Mountains (Plate 13). U-Pb zircon ages of 310 ± 29 and 309 ± 11 Ma were obtained from the Dadina pluton at the west end of the Wrangell Mountains and from the Uranatina River pluton 60 km to the south in the northern Chugach Mountains, respectively. Both plutons consist of deformed metagranodiorite with strong enrichment of LREE versus HREE (Plafker and others, 1989). The close similarity in age, composition, and deformational style among all these plutons suggests that they are part of the same suite (Plafker and others, 1989).

Beard and Barker (1989) recently completed a detailed study of the Ahtell Creek and associated plutons along the north flank of the Wrangell Mountains (Fig. 1; Plate 13). They considered these rocks to be part of the late Paleozoic Skolai island arc volcanic system and identified three groups of plutonic rocks that they felt recorded a temporal progression from typical arc magmatism (gabbro-diorite and silicic intrusions, including tonalite, granodiorite, and granite) related to the Skolai island-arc system, to shoshonitic magmatism (monzonite-syenite). U-Pb ages obtained on zircon fractions from the silicic rocks indicate Early to Late Pennsylvanian crystallization ages between 320 and 290 Ma.

Gardner and others (1988) point out that the timing of the Late Pennsylvanian plutonism in east-central Alaska has important tectonic implications. Intrusion of the Barnard Glacier pluton into both the Alexander and Wrangellia terranes means that these terranes were stitched together by Late Pennsylvanian time. Beard and Barker (1989) suggest that the shoshonitic magmatism, commonly thought to occur as a result of tectonic instability, may have resulted from the collision of the Wrangellia and Alexander terranes, and that the age of pluton emplacement is also the age of Wrangellia-Alexander amalgamation.

MESOZOIC PLUTONIC ROCKS

Mesozoic plutonic rocks are widespread and voluminous in Alaska and fall into two principal age groups separated by 40 to 50 m.y. The earlier group spans most of the Jurassic, is confined to the half of Alaska south of the Kaltag and Tintina faults, and is generally concentrated near the south coast of Alaska. The younger and more widespread group ranges in age from Early Cretaceous to early Tertiary. The two principal groups have each been arbitrarily subdivided into three subgroups on the basis of similarities in age, composition, and distribution (Plate 13).

Late Triassic to Middle Jurassic

The earliest reported Mesozoic salic plutonism in Alaska occurs in the eastern Yukon-Tanana Upland of east-central Alaska (Fig. 3; Plate 13). The Taylor Mountain batholith, the largest of several Late Triassic to Early Jurassic plutons in the area, underlies an area of about 650 km^2 and consists of medium-grained granite and subordinate granodiorite to diorite (Foster and others, this volume). Locally, the margins of the pluton are shared and faulted. Other nearby plutons include quartz monzodiorite. Characteristically, the rocks contain abundant hornblende and minor quartz and are not corundum normative (Fig. 2).

Radiometric ages obtained from the Taylor Mountain and associated plutons range from latest Triassic to Middle Jurassic. The Taylor Mountain pluton itself yielded a U-Pb sphene age of 212 Ma (Aleinikoff and others, 1981), $^{40}Ar/^{39}Ar$ age determinations gave an integrated plateau age on hornblende of 209 $\pm$ 3 Ma, and K-Ar ages range from 198 to 180 Ma (Wilson and others, 1985). An associated pluton has provided an $^{40}Ar/^{39}Ar$ age determination of 188 $\pm$ 2 Ma and a K-Ar hornblende age of 177 $\pm$ 5 Ma (Foster and others, this volume).

These plutons are similar in composition and age to a suite of plutonic rocks in neighboring Yukon Territory. Some parts of the Klotassin batholith 75 km east of the Alaska border, for example, are considered Early Jurassic in age bsed on a 192 Ma U-Pb zircon age (Tempelman-Kluit and Wanless, 1975). In addition, Armstrong (1988) and Mortensen (1988) discuss widespread episodes of magmatic activity in western British Columbia that range from latest Late Triassic to Early Jurassic (214 to 200 and 214 to 190 Ma, respectively). Mortensen (1988) suggests that these plutons represent the deeper parts of the continental Stikine volcanic arc, which now borders the Yukon-Tanana terrane on the north and south.

Small (less than 20 km^2) calc-alkaline intermediate-composition plutons of Early to Middle Jurassic age crop out near the south coast of Alaska (Fig. 3; Plate 13). These plutonic rocks are in a complex structural and intrusive relation with the (informal) Border Ranges ultramafic-mafic complex of Burns (1985). The ultramafic to intermediate-composition plutons of this complex form a discontinuous belt that extends 1,000 km across southern Alaska (Fig. 3; Plate 13) from Kodiak Island through the Kenai Peninsula and northern Chugach Mountains to the Copper River. They are thought (Burns, 1985) to represent the lower crust and uppermost mantle of an Early to Middle Jurassic interoceanic island arc that lies on the north side of, and adjacent to, the Border Ranges fault and entirely within the Peninsular terrane.

The plutons are elongate in plan, parallel to regional trends, and locally fault bounded. Intrusive contacts are sharp to migmatitic, and diking and thermal metamorphism of nearby country rocks are locally well developed. Petrologic data are scarce, but the intermediate-composition plutons appear to consist chiefly of quartz diorite, tonalite, and minor granodiorite and make up another 40 percent of the plutonic rocks in the arc. Although the intermediate-composition plutons intrude the mafic plutonic rocks, they are considered by Burns (1985) to be, at least in part, broadly contemporaneous with them.

The emplacement age of the plutons is uncertain, as the available radiometric ages show considerable scatter. Roeske and others (1989) suggest a crystallization age of 217 $\pm$ 10 Ma (Late Triassic) for the intermediate-composition Afognak pluton on Afognak Island based on nearly identical $^{207}Pb^*/^{206}Pb^*$ ages on coarse and fine zircon fractions. K-Ar ages from this pluton tend to be younger, 197 $\pm$ 11 to 188 $\pm$ 11 Ma (Carden and others, 1977; Hudson, 1985), and according to Roeske and others (1989), may record post-crystallization cooling, a major later thermal event, or a lengthy emplacement period. K-Ar ages from

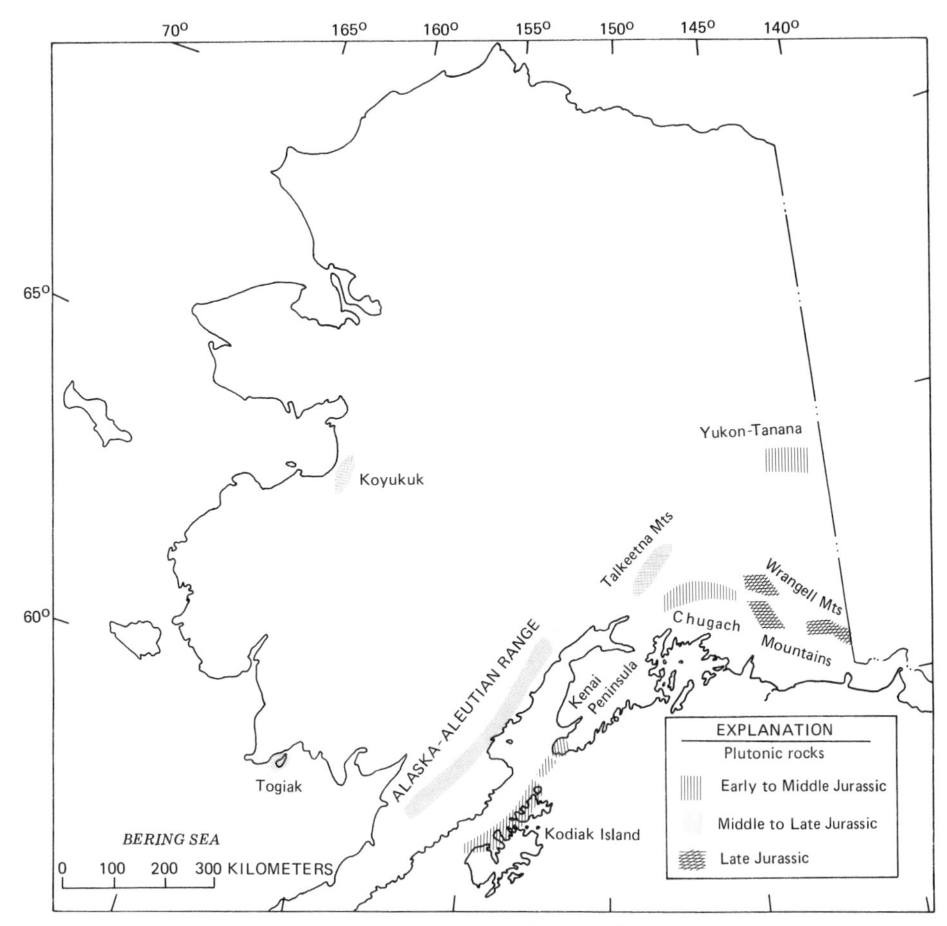

Figure 3. Distribution of Jurassic plutons in mainland Alaska.

the mainland part of the arc range from 194 to 163 Ma (Early to Middle Jurassic).

Middle to Late Jurassic

Plutonic rocks of Middle to Late Jurassic age occur in three widely separated parts of southern and central Alaska (Fig. 3; Plate 13) and represent the roots of the Peninsular, Togiak, and Koyukuk intraoceanic arc terranes. The plutonic rocks typically compositionally expanded calc-alkaline differentiation trends. The relations among these terranes, if any, is unknown, and the amount of detailed petrologic and geochronologic study varies greatly between areas. Box (1985b) has suggested that the Togiak and Koyukuk terranes are dismembered fragments of the same arc, whereas Wallace and others (1989) consider it more likely that these are separate arcs formed in a similar tectonic setting.

Koyukuk terrane. Plutonic rocks are exposed in a narrow, 80-km-long fault-bounded block in west-central Alaska south of the Kaltag fault (Plate 13) and adjacent to the Chirosky fault. These rocks are considered to be the oldest known remnants of an intraoceanic arc that collided with continental North America in latest Jurassic to Early Cretaceous time (Box and Patton, 1989). The plutonic rocks consist of tonalite and trondhjemite that locally intrude an assemblage of late Paleozoic to early Mesozoic volcanic rocks of probable non-arc affinity and appear to underlie the same volcanic arc rocks that make up the major part of the Koyukuk terrane in the north-central part of the Yukon-Koyukuk basin (Box and Patton, 1989). Both intrusive and country rocks are extensively sheared, fractured, and altered by potassic metasomatism. Patton and Moll (1984) report K-Ar ages ranging from 173 to 154 Ma. The plutonic rocks exhibit enrichment of LIL elements and the depletion of Nb-Ta relative to light REE characteristic of subduction-related magmatism (Box and Patton, 1989).

Togiak terrane. Several Middle Jurassic plutons ranging in composition from dunite to hornblende gabbro-diorite to biotite-hornblende granodiorite (only the latter is large enough to be shown on Plate 13) occur in the Goodnews Bay–Hagemeister Island region of southwest Alaska. Although these plutons were originally mapped by Hoare and Coonrad (1979) as Cretaceous to Tertiary in age, K-Ar dating by Box (1985a) on several different rock types yielded ages ranging from 186 to 162 Ma, with the granodiorite from Hagemeister Island giving a hornblende age of 183 ± 7 Ma. Box (1985b) described this granodiorite as medium grained, slightly altered, and intrusive into rocks of the Togiak terrane, a Mesozoic andesitic volcanic and volcaniclastic terrane of Late Triassic through Early Cretaceous age. Clasts of similar granodiorite occur in adjacent early Middle Jurassic (Bajocian) conglomerate.

Few analytical data are available on the granodiorite and associated mafic and ultramafic rocks. Box (1985a, b), however, considers that these ultramafic-mafic-intermediate-composition plutonic rocks represent the roots of a mostly Middle Jurassic magmatic arc that may be correlative with the Koyukuk arc to the north.

Peninsular terrane. One of the more spectacular magmatic belts in Alaska is the Alaska-Aleutian Range composite batholith and its northeastern extension into the Talkeetna Mountains (Fig. 3; Plate 13). This 15- to 35-km-wide, 740-km-long northeast-trending belt of plutonic rocks intrudes rocks of the intraoceanic Talkeetna arc that compose the Peninsular terrane (Jones and others, 1984; Silberling and others, this volume). The plutonic rocks were formed during three distinct magmatic episodes: (1) the Middle to Late Jurassic (174 to 158 Ma) emplacement of chiefly quartz diorite and tonalite; (2) the Late Cretaceous to Paleocene (83 to 58 Ma) emplacement of quartz diorite, tonalite, and granodiorite; and (3) an Oligocene (38 to 26 Ma) event consisting of a varied assemblage of rock types including quartz diorite, peraluminous granite, and alkali granite (Reed and Lanphere, 1969, 1973; Reed and others, 1983). The age, spatial distribution, and broad compositional trends of these plutons are well known following the work of Reed and Lanphere (1969, 1973), Turner and Smith (1974), Csejtey and others (1978), and Reed and others (1983). Published trace-element and isotopic data, however, are sparse.

The Jurassic plutonic rocks in the Alaska-Aleutian Range batholith occur along its southeast side, west of Cook Inlet, and extend into the northern Alaska Peninsula. The plutons intrude the Early Jurassic Talkeetna Formation, an assemblage of basalt, andesitic and dacitic flows, and volcaniclastic rocks with subordinate shale and graywacke (see Barker, this volume). Reed and Lanphere (1969, 1973) showed the plutonic rocks to define a calc-alkaline magmatic suite that ranges from hornblende gabbro through hornblende-biotite diorite, hornblende-biotite quartz diorite (Fig. 2), tonalite and rare granodiorite and quartz monzonite (Reed and others, 1983). Mafic phases appear to have been emplaced first although intrusive relations are complex. SiO_2 content ranges from 45.6 to 67.6 percent (mean = 58.3 percent) and Na_2O content is much greater than K_2O content (Reed and others, 1983).

The calc-alkaline character of the Alaska-Aleutian Range batholith, its emplacement into slightly older andesitic volcanic rocks, and low $^{87}Sr/^{86}Sr$ initial ratios of 0.7033 to 0.7037 (M. A. Lanphere and B. L. Reed, unpublished data, in Reed and others, 1983) led Reed and others (1983) to suggest that the Alaska-Aleutian Range batholith represents the root of the Talkeetna arc.

The relation of the Jurassic plutonic rocks of the Alaska-Aleutian Range batholith and the Talkeetna Mountains to the adjacent Early Jurassic to Middle Jurassic Kodiak-Kenai-Chugach magmatic arc is uncertain. The youngest radiometric ages of the latter belt overlap with some of the oldest ages of the Alsaka-Aleutian Range batholith. Reed and others (1983) thought that the two were unrelated and that the Kodiak-Kenai-Chugach belt might be a slightly older and discrete accreted arc. Alternatively, they suggested that the latter arc might have split parallel to its length and rifted, with the rift opening now occupied by the Cook Inlet basin. The northwest margin of this basin then became the site of renewed calc-alkaline magmatism reflected in the Alaska-Aleutian Range batholith. Hudson (1983) also initially considered the belts as separate but later (Hudson, 1985) raised the possibilitty that Early Jurassic plutonic rocks on Kodiak Island were mafic precursors to the Alaska-Aleutian Range batholith. The actual age range of these adjoining plutonic belts is as yet poorly constrained owing in part to widespread Cretaceous and Tertiary magmatism. If the belts are related, however, and the presently available ages are taken at face value, then, as Roeske (1986) has pointed out, volcanism and associated plutonism covered a long span of about 60 m.y. Since the lack of detailed geochronologic and petrologic studies prevents resolution of this question, the plutonic belts in this compilation are regarded as representing separate magmatic arcs.

Although Reed and Lanphere (1969, 1973) initially suggested that the Alaska-Aleutian Range batholith represented magmatism related to a northwest-dipping subduction zone, more detailed study (Reed and others, 1983) of chemical variation across the batholith led them to suggest that the subduction was directed toward the southeast. Plafker and others (1989) consider this possibility less likely because of the lack of a coeval (with the Talkeetna arc) accretionary prism along the north margin of the Peninsular terrane. A south-dipping subduction zone, however, would be similar in polarity to that postulated for the Togiak and Koyukuk terranes (Box, 1985b; Wallace and others, 1989).

Late Jurassic

Plutonism in Late Jurassic time in Alaska is confined to a narrow belt extending from southern east-central Alaska (Fig. 3; Plate 13) through southwestern Yukon Territory and northwestern British Columbia to Chichagof Island in southeastern Alaska (Dodds and Campbell, 1988; Dusel-Bacon, this volume, Chapter 15 and Plate 4). The plutons are commonly elongate in plan, parallel to regional trends, and typically foliated and mylonitic. They intrude the Strelna Metamorphics, defined and considered by Plafker and others (1989) to be in part as old as Early Pennsylvanian, and are unconformably overlain by Early Cretaceous sedimentary rocks. Preliminary ages are about 164 to 140 Ma, or Late Jurassic to earliest Cretaceous.

Hudson (1983) describes a characteristic compositional range that includes biotite-hornblende quartz diorite, tonalite, and granodiorite. Plafker and others (1989) report the Uranatina River pluton just west of the Copper River as being composed chiefly of diorite with lesser amounts of tonalite and quartz diorite. Sparse major- and trace-element analytical data from metatonalite suggest a calc-alkaline Alk-F-M differentiation trend and strong to moderate enrichment of LREE relative to HREE.

Hudson (1983) referred to these plutons as the Tonsina-Chichagof belt and noted that they were emplaced in the upper plate of the Border Ranges fault. He considered them to be the roots of a Late Jurassic magmatic arc. This arc, referred to as the Chitina arc by Plafker and others (1989), was built along the south margin of the Wrangellia terrane in Alaska. The polarity of the arc is uncertain because an associated accretionary prism or shelf basin has not been found, although northerly subduction is suggested by the presence of coeval volcanic detritus in the adjacent Chugach terrane (Plafker and others, 1989).

Early Cretaceous

Plutonic rocks of Early Cretaceous age occur in widely separated parts of mainland Alaska but are concentrated north of the Kaltag fault and south of the Brooks Range (Fig. 4; Plate 13) in an area that includes the Ruby, Koyukuk, and Seward terranes. Early Cretaceous plutons are voluminous in the part of the Ruby terrane north of the Kaltag fault but sparse and widely scattered in the Ruby and adjacent terranes south of the Kaltag fault. They are abundant in a belt extending more than 1,000 km across the northern Koyukuk basin through the southeastern Seward Peninsula and St. Lawrence Island into the Chukotkan Peninsula of Siberia. Short belts of Early Cretaceous plutonic rocks occur north of the Wrangell Mountains in east-central Alaska and in the western Chugach Mountains. Early Cretaceous plutons also are scattered across the Yukon-Tanana Upland within an extensive mid-Cretaceous (approximately 110 to 90 Ma) granitoid belt; they will be discussed in a later section.

Early Cretaceous plutons in Alaska show a great diversity in composition, including ultrapotassic nepheline syenite, tonalite through granodiorite, and anatectic biotite and two-mica granite. This compositional range is far greater than that seen for any other group of pre-Cenozoic plutons in Alaska.

West-central Alaska. Major plutonism in west-central Alaska took place from 115 to 98 Ma (Early Cretaceous), and a volumetrically lesser episode occurred at 89 to 78 Ma (Late Cretaceous). These plutonic rocks underlie about 5 percent of the region and can be grouped into distinct suites on the basis of their composition, age, and distribution.

Ruby and Angayucham-Tozitna terranes. Cretaceous plutonic rocks underlie a major part of the Ruby geanticline where they constitute a batholith larger than 8,000 km^2. The Ruby

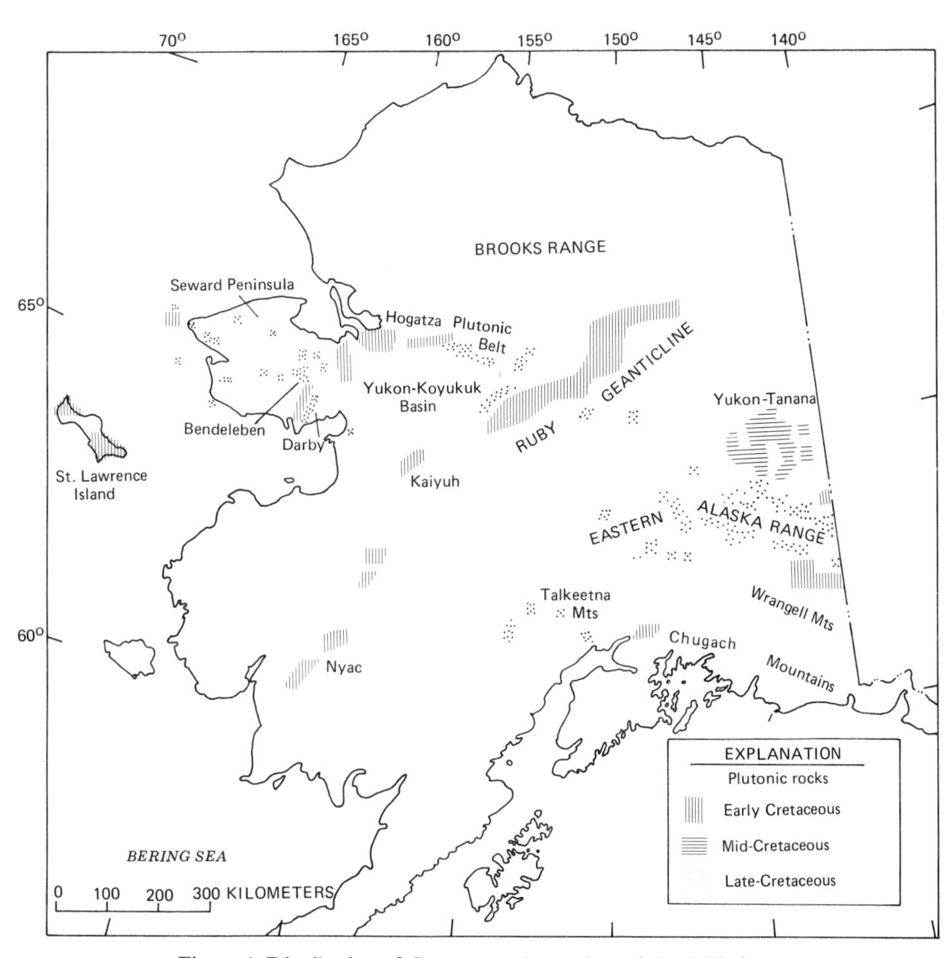

Figure 4. Distribution of Cretaceous plutons in mainland Alaska.

geanticline is a pre–mid-Cretaceous uplift that extends 400 km diagonally southwestward from the Brooks Range to the Yukon River, where it is offset right laterally about 160 km by the Kaltag fault (Patton and others, 1984). The core of the Ruby geanticline composes the northern part of the Ruby terrane (Jones and others, 1984; Silberling and others, this volume) and is truncated at its northeast end by the Brooks Range.

The Ruby terrane is primarily made up of greenschist-facies metasedimentary rocks and metabasite of Proterozoic(?) and Paleozoic age intruded by voluminous mid-Cretaceous plutons. High-pressure glaucophane-bearing greenschist-facies and intermediate- to high-pressure sillimanite-kyanite amphibolite-facies assemblages are present locally. Typical rock types include quartz-mica schist, quartzite, calcareous schist, mafic greenschist, quartzo-feldspathic schist and gneiss, and marble. Regional metamorphism predates the widespread Early Cretaceous plutonism and is thought to be Late Jurassic to Early Cretaceous in age (Turner, 1984; Patton and others, 1984).

Gross similarities between the metamorphic rocks of the Ruby terrane and the so-called schist belt of the Arctic terrane of the southern Brooks Range have led to suggestions that both terranes were once part of a single continuous belt. The Ruby terrane may also be an exotic fragment rafted to its present position prior to overthrusting of the Angayucham and Tozitna terranes (Patton and Box, 1989).

Although the lithologies and mineral assemblages of Brooks Range and Ruby terrane metamorphic rocks suggest similar protoliths, the Cretaceous granitic rocks that are the single most abundant rock type in the Ruby terrane (close to 50 percent of the exposed outcrop) are absent from the Brooks Range. When the voluminous amounts of granite are considered in proportion to the metamorphic country rock, the Ruby terrane more closely resembles the Yukon-Tanana terrane (Jones and others, 1984) to the southeast than the Brooks Range to the north.

The plutonic rocks also intrude the Jurassic mafic volcanic rocks of the Angayucham terrane that overlie the Ruby terrane and in turn are overlain by calc-alkaline volcanic rocks of Paleocene age (59 to 56 Ma; Moll and Patton, 1983). Radiometric dating has more closely constrained the age of the Ruby plutons as Early Cretaceous. K-Ar ages chiefly on biotite from seven different plutons range from 112 to 99 Ma (Miller, 1989). An Early Cretaceous age for these plutonic rocks is strengthened by U-Pb zircon ages of 112 to 109 Ma from the Ray Mountains pluton near the south end of the batholith (Patton and others, 1987) and a Rb-Sr age of 112 Ma (Blum and others, 1987) from the Jim River pluton at the northeast end.

The plutons of the Ruby geanticline are composed chiefly (more than 80 percent) of leucocratic biotite granite (Fig. 2), with lesser amounts of granodiorite and muscovite-biotite granite; syenite and monzonite are rare but occur at the extreme north end of the Ruby geanticline (Blum and others, 1987). The granitic rocks are typically coarse grained, strongly porphyritic, non-foliated and occur in plutons up to 800 km^2 in area. The individual plutons that compose the Ruby batholith are surrounded by narrow thermal aureoles. Contact wall rocks are thermally altered to andalusite-cordierite hornfels, hornblende hornfels, and marble. The plutons are somewhat elongated in an east-west direction and lie oblique to the northeast-striking trend of the Ruby geanticline.

The plutons are highly evolved and generally have SiO_2 contents of 68 to 76 percent. They are K-rich, Na-depleted, and weakly to moderately peraluminous, with normative corundum usually greater than 0.8 percent (Miller, 1989). Primary muscovite occurs in some phases of the large southern plutons, but modal cordierite is lacking. The plutons appear to belong to the ilmenite (magnetite-free) series of granitic rocks (Ishihara, 1981) with low Fe_2O_3/FeO ratios and reduced magnetic susceptibility. Initial Sr^{87}/Sr^{86} ratios are in the range 0.7056 to 0.7294, show large internal variations, and have average values that decrease from southwest to northeast (Arth and others, 1989a). Initial Nd^{143}/Nd^{144} ratios show a reverse relation and increase to the northeast, ranging from 0.51158 to 0.51240 (Arth and others, 1989a).

These modal and major-element characteristics are typical of granitic rocks thought to have been generated by melting, or contamination, of continental crust (Miller, 1989). High SIRs (>0.7056) also indicate the involvement of significant amounts of Paleozoic or older continental crust in the origin of the plutonic magmas (Arth and others, 1989a).

Ruby and Nyack terranes south of Kaltag fault. The difference in the amount of Early Cretaceous plutonic rocks on either side of the Kaltag fault is striking. Plutonic rocks underlie more than 40 percent of the Ruby terrane north of the Kaltag fault, where they are the single most voluminous rock type and constitute one of the major batholithic complexes of interior Alaska, second in area only to the plutons of the Yukon-Tanana Upland.

South of the Kaltag fault, however, Early Cretaceous plutons, though present, are relative sparse and constitute less than 1 percent of the terrane. The plutons occur as several small (less than 200 km^2) bodies scattered along a northeast-trending, 500-km-long belt extending from the Kaltag fault almost to the Bering Sea (Fig. 4; Plate 13). The plutons are intrusive into the Proterozoic(?) to Paleozoic crystalline rocks of the Ruby terrane and into Middle to Late Jurassic accreted volcanogenic rocks of the Nyack terrane. Their ages, however, are poorly constrained, since only the Kaiyuh Mountains pluton at the northeast end of the belt and the Nyack pluton near the southwest end have been dated. Patton and others (1984) report a biotite K-Ar age of 112 ± 3.4 Ma for the Kaiyuh Mountains pluton and speculate that it may be a faulted extension of the Early Cretaceous Melozitna pluton in the Ruby geanticline to the east. The Nyack pluton has yielded K-Ar ages ranging from 117 ± 3.3 Ma (Wilson, 1977) to 101.1 ± 3.0 Ma (Frost and others, 1988). Age designations for the remaining plutons in the belt are based on similarities in composition and geologic setting and are tentative (S. E. Box, oral communication, 1989). Few modal or chemical analyses are available, but the plutons have been mapped as granodiorite, granite, and quartz monzonite.

Koyukuk Basin–Seward Peninsula–St. Lawrence Island. Early Cretaceous plutons form an east-west trending belt that extends from the central part of the northern Koyukuk basin into the southeastern Seward Peninsula and across the Bering Sea shelf (Fig. 4; Plate 13) through St. Lawrence Island (Miller, 1971; Miller and Bunker, 1976; Csejtey and others, 1971; Miller and others, 1966) to the Chukotkan Peninsula (Kosygin and Popeko, 1987). These plutons intrude Neocomian andesitic volcanic rocks of the Yukon-Koyukuk basin and are overlain by, and shed debris into, Albian sedimentary rocks (Patton, 1973). The Early Cretaceous age suggested by their stratigraphic setting is confirmed by numerous K-Ar ages ranging from 113 to 99 Ma (Miller, 1989).

These plutons consist of two distinct but related groups: a potassic series (KS) and an ultra potassic series (UKS; Miller, 1989). The KS series is represented by SiO_2-saturated to slightly oversaturated monzonite, syenite, and quartz syenite characterized by low quartz and abundant K-feldspar (Fig. 2); hornblende and clinopyroxene are the principal varietal mafic minerals. K_2O content is greater than 4.5 percent, and SiO_2 content ranges from 53.8 to 66.1 percent, with a compositional gap from 66.1 to 73.9 percent. The KS rocks have a lower abundance of incompatible elements and radiogenic Sr than the UKS.

The UKS series constitutes about 5 percent of the western Koyukuk terrane plutonic rocks and consists of single-feldspar, hypersolvus nepheline-bearing rocks such as malignite, foyaite, ijolite, biotite pyroxenite, and pseudoleucite porphyry. SiO_2 content ranges from 44.5 to 58.4 percent, K_2O content is as high as 16.6 percent, and the rocks are nepheline and commonly leucite normative. These UKS rocks define an ultrapotassic rock province consisting of 12 known intrusive suites and dike swarms that form a sinuous belt extending some 1,500 km across the northwestern Koyukuk terrane (Plate 13) westward through the southeastern Seward Peninsula and St. Lawrence Island (Csejtey and Patton, 1974) to the east tip of Siberia (Miller, 1972). The trend of this alkaline rock belt is roughly parallel to the contact between the Koyukuk and Seward terranes.

The Early Cretaceous plutons of west-central Alaska are thus divisible into two contrasting compositional trends whose origins are enigmatic: (1) the voluminous, highly evolved, compositionally restricted suite of predominantly biotite and two-mica granite with elevated SIRs confined to the Ruby geanticline; and (2) the more than 1,000-km-long sinuous belt of silica-saturated to undersaturated granite, monzonite, and syenite characterized by strong enrichment in potassium (including an associated ultrapotassic rock province) and intrusive into two very different geologic provinces and terranes.

The compositional characteristics of the Ruby geanticline plutons suggest that they were generated by the melting, or contamination, of continental crust either in response to thickening of that crust following collision of the Koyukuk terrane with the Ruby terrane or perhaps as a result of underthrusting of the Koyukuk terrane beneath the Ruby terrane (Miller, 1989; Arth and others, 1989b). The long sinuous belt of Early Cretaceous plutons in the northern Koyukuk basin and its association over much of its length with the tectonic boundary between the Koyukuk and Seward terranes suggest that these plutons may be subduction related. The coeval Ohkotsk-Chukotsk magmatic belt of the northeastern USSR is considered (Khrenov and Bukharov, 1973; Parfenov and Natal'in, 1986) to result from northward underthrusting of Pacific Ocean crust from the south. The surface trace of such a subduction zone is unknown but could lie concealed beneath younger sedimentary and volcanic rocks to the south.

Ultrapotassic rocks such as those in west-central Alaska are generally considered to have a mantle origin (Bergman, 1987). A gradual increase in K_2O content and $^{87}Sr/^{86}Sr$ initial ratios from east to west across the Yukon-Koyukuk basin reported by Miller (1989) and Arth and others (1989b) suggests that continental crust or a keel of K-rich subcontinental mantle may underlie the west half of the basin.

East-central Alaska. A belt of Early Cretaceous plutons occurs in the eastern Alaska Range along the north flank of the Wrangellia terrane and south of the Denali fault (Fig. 4; Plate 13). The plutonic belt, first described in Alaska by Richter and others (1975) and Richter (1976), extends across southwestern Yukon Territory and northern British Columbia (Dodds and Campbell, 1988) into southeastern Alaska and was called the Nutzotin-Chichagof belt by Hudson (1983). In east-central Alaska, the plutonic belt consists of two relatively large bodies—the Nabesna pluton (250 km^2) and the Klein Creek pluton (more than 305 km^2)—and six smaller (less than 30 km^2) stocks (Richter and others, 1975).

K-Ar ages have been obtained on hornblende and biotite from five of the eastern Alaska Range plutons (Richter and others, 1975) and range from 117 to 105 Ma. The plutons intrude, and are assumed to be cogenetic with, a thick sequence of predominantly andesitic volcanic and volcaniclastic rocks of the Early Cretaceous Chisana Formation.

The plutonic rocks are massive, nonfoliated, and generally equigranular. They are strongly discordant to the country rock and locally contain abundant xenoliths; these features and the association with coeval volcanic rocks indicate that the plutonic rocks are epizonal (Richter and others, 1975).

Individual plutons range from diorite to granite (rare) and include syenodiorite, monzonite, and trondhjemite (Fig. 2). Melanocratic diorite and granodiorite are perhaps the most common lithologies. Chemical data, sparse relative to those available for other Early Cretaceous suites in Alaska, show a narrower range of SiO_2 content (48 to 64 percent) than might be expected from the modal plots. This narrow range is probably a reflection of the high mafic content of the suite.

Western Chugach Mountains. Early Cretaceous plutonic rocks have been reported in an unusual tectonic environment in the western Chugach Mountains (Fig. 4; Plate 13). Pavlis and others (1988) have described a group of tonalite-trondhjemite plutons as recording a period of Early Cretaceous near-trench plutonism along the paleo-Aleutian subduction zone. Radiometric dating of the plutons (summarized in Pavlis and others, 1988) yielded K-Ar ages of 135 to 110 Ma, $^{40}Ar/^{39}Ar$ plateau ages of 129 to 114 Ma, two Rb-Sr mineral isochrons of 133 and 130 Ma, and a U-Pb zircon age of 103 Ma. Pavlis and others (1988) believe that these data are consistent with pluton emplacement in the middle Early Cretaceous at about 135 to 125 Ma.

Most of the plutonic rocks are confined to five elongate stocks ranging from 2 to 12 km^2 in area. They consist chiefly of leucocratic biotite tonalite, with gradations to relatively mafic hornblende-biotite tonalite and biotite trondhjemite. These gradations are thought (Pavlis and others, 1988) to reflect crystal fractionation from a tonalitic magma rather than separate intrusions. Chemically, the plutons define a low-K series with the SiO_2 content of the main plutonic phases ranging from about 60 to 73 percent. MgO, CaO, FeO (total iron), and Al_2O_3 show systematic linear decreases with increased SiO_2, while Na_2O increases slightly and K_2O remains relatively constant; K_2O values are somewhat higher than trondhjemite associated with ophiolite sequences (Pavlis and others, 1988). On ternary K-Na-Ca plots, the rocks display a gabbro-trondhjemite rather than calc-alkaline trend.

The Early Cretaceous plutons were emplaced along the tectonic join between the older Jurassic arc and the younger Cretaceous melange terrane that was accreted by subduction beneath the old arc (Pavlis and others, 1988). The plutonic rocks are interpreted by Pavlis and others (1988) to have been injected during a major thrusting event that placed upper-mantle(?) ultramafic rocks atop the Cretaceous subduction assemblages. The generation of plutonic rocks in the generally cool, near-trench environment is atypical. Pavlis and others (1988) suggest shallow melting of amphibolite or metagraywacke along a young subduction zone as the most reasonable explanation, although a model involving a ridge-trench encounter is also allowable.

Mid-Cretaceous and Late Cretaceous

Mid-Cretaceous and particularly Late Cretaceous plutonic rocks are the most widespread of all plutonic rocks in mainland Alaska (Fig. 4; Plate 13) and record a variety of magmatic events. The age range of these plutonic suites and their tectonic setting, however, suggest that they can be subdivided (for the purposes of this compilation) into those rocks emplaced approximately between 97 and 74 Ma (discussed here) and those whose emplacement spanned the Cretaceous/Tertiary boundary between about 74 and 55 Ma (discussed later). This latter subgroup overlaps, for the sake of completeness, the discussion of Cenozoic magmatic rocks by Moll-Stalcup (this volume).

Mid-Cretaceous plutonic rocks in the age range 115 to 89 Ma are concentrated in the Yukon-Tanana Upland of east-central Alaska; Late Cretaceous plutonic rocks are moderately abundant in south-central Alaska and scattered across west-central Alaska (Fig. 4; Plate 13).

Yukon-Tanana Upland. Plutonic rocks of mid-Cretaceous age are the most widespread of all the voluminous plutonic rocks in the Yukon-Tanana Upland and underlie an area of more than 10,000 km^2; individual bodies range in area from smaller than 1 to larger than 300 km^2. The plutons have a highly irregular map pattern and are post-tectonic. They intrude several large middle Paleozoic and Jurassic plutons and their metamorphosed host rocks and contain numerous roof pendants, screens, and enclaves of the metamorphic country rock. Locally, they are covered by Cenozoic sedimentary and volcanic rocks. Northeast-trending high-angle faults have further disrupted many of the plutons, contributing somewhat to their irregular plan.

This period of plutonism in the Yukon-Tanana Upland began in the late Early Cretaceous and extended into the Late Cretaceous. Wilson and others (1985), in a summary report on 138 K-Ar mineral age determinations from plutonic rocks in the Yukon-Tanana terrane, pointed out that the ages fall into a bimodal cluster—110 and 50 Ma with a sharp maximum at 95 to 90 Ma and a spread of ages from 70 to 50 Ma. The oldest age reported is a 115-Ma U-Pb age on zircon from a post-metamorphic intrusion into a gneiss dome (Aleinikoff and others, 1984). Bacon and others (1985) obtained K-Ar ages of 94 and 90 Ma on welded tuffs in and around three large calderas spatially associated with the Late Cretaceous plutons. A Rb-Sr whole-rock isochron for a pluton in the western Yukon-Tanana terrane near Fairbanks yielded an age of 90.0 ± 0.9 Ma (Blum, 1985), which is similar to K-Ar ages of 95.3 ± 5.0 and 93.0 ± 5.1 Ma from the same group of plutons.

Foster and others (1978, 1987; see also Luthy and others, 1981) report that the plutons consist predominantly of granite and granodiorite and show a compositional range from quartz monzonite to diorite (Fig. 2). Biotite and hornblende are the typical mafic minerals; primary muscovite is rare. The rocks are generally medium grained and equigranular to porphyritic.

Cretaceous plutons of similar age and composition (the Omineca Belt; Armstrong, 1988) in the Yukon Territory and western Canada have distinctive compositional patterns exhibiting some S-type characteristics and enrichment in Al, K, Sr^{87}/Sr^{86} (SIRs commonly are greater than 0.7100), and O^{18}. As pointed out by Armstrong (1988), most of these plutons represent a mixture of mantle-derived magma and continental crust that was later modified by fractionation. The plutons may be subduction-related, but the nature and extent of the magmatic arc are uncertain.

Western Alaska–Seward Peninsula. Late Cretaceous plutonic rocks ranging in age from 94 to 78 Ma occur in a number of short-lived magmatic belts of limited extent in the Yukon-Koyukuk basin and the Seward Peninsula (Fig. 4; Plate 13). Late Cretaceous plutons in the eastern Koyukuk terrane consist of several large bodies in the east half of the Hogatza plutonic belt (Miller and others, 1966) and numerous small stocks (Fig. 4) to the south. They intrude both Neocomian andesitic volcanic rocks of the Koyukuk terrane and overlying Albian graywacke and mudstone, indicating that they can be no older than Albian in age. The plutonic rocks have yielded Late Cretaceous K-Ar ages of 89 to 79 Ma (Miller, 1989), some 10 to 20 m.y. younger than the Early Cretaceous plutons of the western Koyukuk terrane and the adjacent Ruby geanticline.

The plutons range from tonalite to high-silica granite, with granodiorite the most typical lithology; gabbro and quartz diorite are absent (Fig. 2). Individual plutons are commonly compositionally zoned, and they have sharp country-rock contacts. The rocks are generally leucocratic, medium-grained, massive, hypidiomorphic, and generally epigranular; locally they are porphyritic.

SiO_2 content generally ranges from 62 to 73 percent, but a 100-km^2 area of high-silica (76 to 78 percent) granite occurs near the west end of this belt. The rocks are relatively enriched in Na_2O (greater than 3.2 percent; Na_2O/K_2O greater than 1) and CaO; they have high Fe_2O_3/FeO ratios and belong to the magnetite series of Ishihara (1981). SIRs for this eastern suite of plutonic rocks range from 0.7038 to 0.7056 and show little internal variation (Arth and others, 1989b; Arth, this volume).

Mesozoic silicic magmatism on the Seward Peninsula occurred from about 108 to 69 Ma, or from about the middle to the end of the Cretaceous. The plutonic rocks can be grouped into three suites, each of which has a distinctive age range, distribution, and composition. The suites are chiefly of moderately to highly evolved silicic granite (Fig. 2) and consist of (1) the 96- to 91-Ma Darby pluton (Miller and Bunker, 1976; Till and others, 1986) and associated small bodies to the north (Plate 13); (2) scattered granite bodies ranging in area from less than 2 km^2 to over 200 km^2 and in age from 87 to 81 Ma and forming the core of the Bendeleben and Kigluaik Mountains (Miller and Bunker, 1976; Till and others, 1986), and (3) a group of small (0.2 to 70 km^2) tin granite stocks in the northwestern Seward Peninsula that range in age from 80 to 69 Ma (Hudson and Arth, 1983; Swanson and others, 1988). No particular area appears to have been the focus of repeated or episodic plutonism. The oldest plutonic rocks are in the southeastern part of the peninsula; the youngest are in the northwest.

Although the plutons are composed chiefly of granite, compositional differences exist between the suites. The adjacent Darby and Bendeleben plutons are chiefly granite (Fig. 4; Plate 13) and have similar modal compositional ranges, but the Darby pluton has higher SiO_2 contents, a much higher Fe_2O_3/FeO ratio, and higher U and Th contents (Miller and Bunker, 1976). The tin granite of the northwestern Seward Peninsula consists chiefly of highly evolved, compositionally restricted biotite granite (Fig. 2) with SiO_2 content of 72.6 to 77.1 percent (Hudson and Arth, 1983). Arth (1987) reported SIRs of 0.708 to 0.711 for the Darby and Bendeleben plutonic suites and higher SIRs of 0.708 to 0.720 for the northwestern tin granite.

The calc-alkaline trend of the Yukon-Koyukuk basin plutons, together with their relatively high Na_2O and Na_2/K_2O ratio, the oxidized state of Fe (magnetite series), the abundance of hornblende in addition to biotite, and the presence of mafic xenoliths, suggest that the plutons were not derived from sialic crust (Miller, 1989). SIRs and NIRs also show no evidence for the involvement of Paleozoic or older crust in the generation of the plutonic magmas (Arth and others, 1989b) but are compatible with source regions that would include oceanic mantle and Mesozoic supracrustal rocks. These compositional characteristics and the linear configuration of the belt suggest that the Yukon-Koyukuk basin plutons resulted from a short-lived period of subduction. In contrast, the composition and geologic setting of the Late Cretaceous plutons in the Seward Peninsula, particularly the high SIRs, suggest instead that they were derived from melting of the continental Proterozoic and Paleozoic metamorphic rocks.

Late Cretaceous to Early Tertiary

Plutonic rocks of Late Cretaceous to early Tertiary (74 to 55 Ma) age are widespread in the south half of Alaska (Fig. 5; Plate 13), where they form two extensive northeast-trending belts with a probable common origin. Although this suite of plutonic rocks spans the Cretaceous/Tertiary boundary, much of the magmatic activity appears to have been concentrated during the early (Late Cretaceous) part of the magmatic episode (Wallace and Engebretson, 1984). Moll-Stalcup (this volume) discusses this and associated suites of magmatic rocks in additional detail.

The more voluminous of the two belts of Late Cretaceous plutonic rocks occurs in the Alaska Range as a 700-km-long by up to 130-km-wide belt extending from the northern Alaska Peninsula through the Talkeetna Mountains to the central Alaska Range (Fig. 5). The belt is associated with, but on the landward side of, the Middle Jurassic plutonic belt of south-central Alaska and forms a major part of the Alaska-Aleutian Range batholith of Reed and Lanphere (1973). The belt has been termed the "Alaska Range–Talkeetna Mountains belt" by Hudson (1983) and the "Alaska Range belt" by Wallace and Engebretson (1984); farther east in Yukon Territory, British Columbia, and southeastern Alaska, plutonic rocks thought to be correlative have been informally referred to as belonging to the Kluane magmatic arc (Plafker and others, 1989).

The plutons intrude a number of terranes, including the Wrangellia and Peninsular terranes and the Early and Late Cretaceous (Albian and Cenomanian) Kuskokwim Group. Plutonic emplacement within this belt is well constrained by K-Ar dating in the Alaska-Aleutian Range batholith by Reed and Lanphere (1969, 1973) and in the Talkeetna Mountains and Alaska Range by Turner and Smith (1974), Csejtey and others (1978), and Smith (1981). These ages suggest a near continuum of plutonism (Reed and Lanphere, 1973) extending from about 74 to 55 Ma. The older rocks (older than 62 Ma) chiefly lie along the southeast margin of the belt (Hudson, 1983; Wallace and Engebretson, 1984).

The Alaska-Aleutian Range part of the belt consists of an elongate, concordant batholith intrusive into Paleozoic and Mesozoic metamorphic rocks and Mesozoic plutonic rocks. In the Talkeetna Mountains and Alaska Range, the belt consists of scattered plutons bounded by sharp, discordant contacts. The composition appears to vary across the trend of the belt, with biotite-hornblende tonalite, quartz diorite, and granodiorite occurring along the southeast margin and more silicic biotite grano-

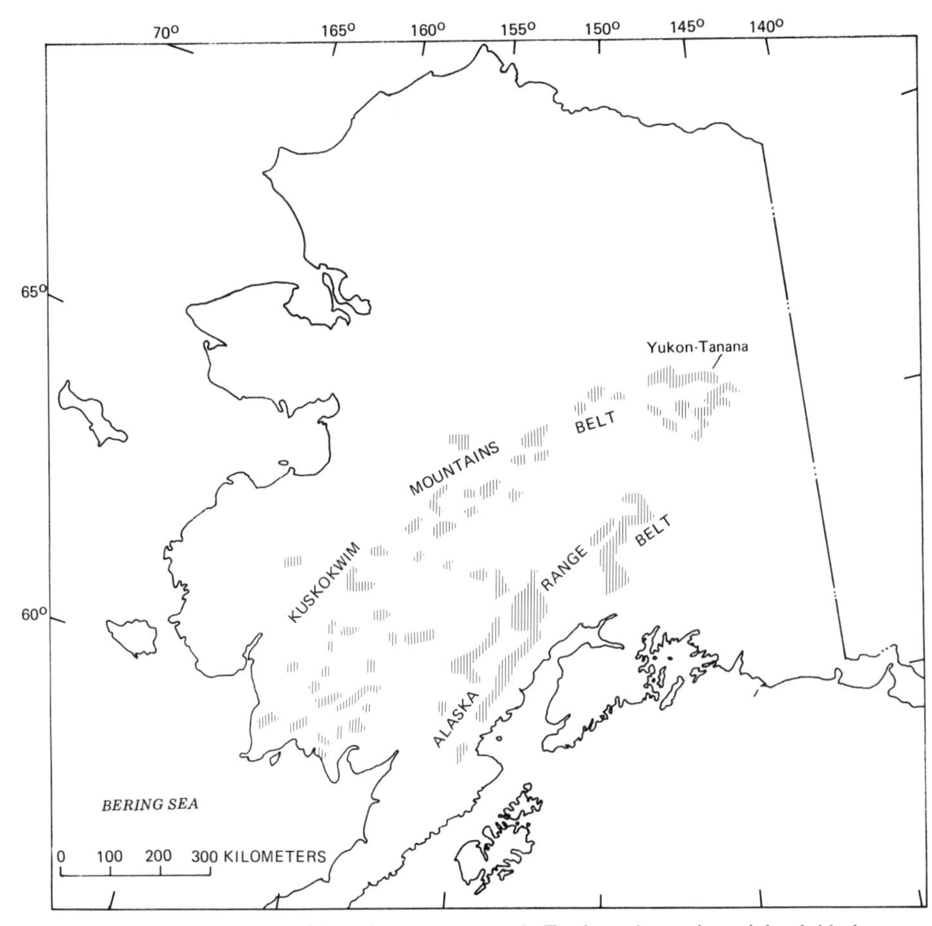

Figure 5. Distributions of Late Cretaceous to early Tertiary plutons in mainland Alaska.

diorite and granite with subordinate quartz monzonite and monzonite along the northwest margin (Hudson, 1983; Wallace and Engebretson, 1984). SiO_2 content ranges from about 60 to 76 percent, and K_2O content from about 2.8 to 5.6 percent (Reed and Lanphere, 1973). The plutonic rocks of this belt define a calc-alkaline trend (Fig. 2) and show a general increase in SiO_2 and K_2O from southeast to northwest.

The second belt of Late Cretaceous to early Tertiary plutonic rocks, though smaller in aggregate area than the Alaska Range belt, is much more widespread. This belt extends in an arcuate northeast direction from the Bering Sea in southwestern Alaska to the western Yukon–Tanana Upland (Fig. 5; Plate 13). More than 100 small (less than 200 km^2) plutons and volcano-plutonic complexes closely associated with a variety of volcanic rocks are more or less equally distributed along this 1,200-km-long by 200-km-wide trend, which Wallace and Engebretson (1984) referred to as the "Kuskokwim Mountains belt." The plutons are circular to irregular in plan, clearly discordant, and surrounded by thermal aureoles. The volcano-plutonic complexes are circular to elliptical piles of volcanic flows with intrusive cores.

A large number of K-Ar ages (Moll and Patton, 1982; Moll-Stalcup, this volume) show that these epizonal plutons and associated volcanic rocks were emplaced between about 72 and 60 Ma. They intruded the Innoko, Yukon-Tanana, Ruby, Nixon Fork, Nyack, Kilbuck, and (composite) Goodnews-Tikchik-Togiak terranes as well as the Early and Late Cretaceous Kuskokwim Group.

The plutons have an intermediate to felsic compositional range (Wallace and Engebretson, 1984; Moll-Stalcup, this volume) from gabbro and monzogabbro to granite. Monzonite, quartz monzonite, and granite predominate to the northeast, whereas tonalite, quartz diorite, and quartz monzonite are more common to the southwest. Major-element data show trends typical of calc-alkaline suites, although a more high-K, subalkalic character occurs to the northeast; some of the volcanic suites have a shoshonitic character (Moll-Stalcup, this volume). SIRs of associated volcanic rocks range from 0.7045 to 0.7053. High Sn, Be, U, W, and F contents, and possibly the SIRs as well, suggest that at least some of the magmas that formed the plutons either were contaminated by continental crust or were partial melts of the crust (Bergman and Doherty, 1986; Moll-Stalcup, this volume).

The Alaska Range and Kuskokwim Mountains plutonic belts are considered part of a more extensive plutonic belt that includes equivalent rocks in southwestern Yukon Territory, British Columbia, and southeastern Alaska (Plafker and others, 1989; Wallace and others, 1989). These rocks all appear to be arc-related and represent northward and eastward subduction during a period of rapid northward movement of the Kula Plate from 74 to 56 Ma (Wallace and Engebretson, 1984; Bergman and Doherty, 1986; Moll-Stalcup and others, this volume). The relation between the Alaska Range and the Kuskokwim Mountains belts is still uncertain, but they may represent a single arc that widened with time in response to a shallow dip of the Kula Plate (Bergman and Doherty, 1986).

SUMMARY

The preceding discussion has illustrated the episodic nature of plutonism in mainland Alaska by delineating a minimum of nine predominantly pre-Cenozoic plutonic episodes scattered about the state. Plutonic rocks defining an individual episode are commonly widely separated. In many cases, they are probably not comagmatic and almost certainly were formed by different processes. Some episodes are quite sharply defined by composition, age, and field relations, whereas others are much less well established. Terrane movements and "post-docking" faulting have further served to complicate the recognition and definition of episodes.

Proterozoic plutonism is so poorly defined in terms of composition and absolute age that it is presently impossible to ascertain the number of intrusive episodes that are represented by the widely separated occurrences of these rocks. All that can be said is that Proterozoic plutonic rocks of diverse composition and ranging in age from 2,050 to 750 Ma have been identified in several terranes of continental affinity in western and northern Alaska.

A belt of middle Paleozoic plutons that extends in sinuous but discontinuous fashion across the north half of Alaska is well defined by existing data. The plutons are continuous with rocks of similar age and composition in western Canada. Additional isotopic studies may narrow the presently rather broad Middle Devonian to Early Mississippian age range or show true spatial age differences along the belt. The plutons are thought to reflect a middle Paleozoic arc system that developed along the west edge of the North American craton. The highly evolved peraluminous composition of these plutons, their high SIRs, and the upper intercept ages of zircon discordia that compose most of the plutons suggest that considerable melting of Proterozoic continental crust was involved in their origin.

The scattered late Paleozoic (Pennsylvanian) plutons of shoshonitic affinities that flank the Wrangell Mountains are considered to be part of the Skolai island arc system (Beard and Barker, 1989) and may well have resulted from, and effectively date, the collision and amalgamation of the Wrangellia and Alexander terranes.

Plutonic rocks of early Mesozoic age are widespread throughout mainland Alaska and form extensive magmatic belts. Most of these pre-Cretaceous plutonic belts reflect the development and migration of a series of magmatic arcs that formed as various terranes impinged against the North American continental backstop (Wallace and others, 1989). The earliest Mesozoic plutonic rocks are associated with the Early to Middle Jurassic development of a north-dipping (present direction) subduction zone beneath an intraoceanic island arc that extended for over 1,000 km across the Peninsular terrane of southern Alaska. Felsic plutonic rocks (chiefly quartz diorite and tonalite) are confined to small bodies along the north side of Kodiak Island, on the tip of the Kenai Peninsula, and in the northern Chugach Mountains. Associated ultramafic and mafic rocks of the (informal) Border Ranges ultramafic-mafic complex of Burns (1985) occur over much of the intervening areas. These intermediate-mafic-ultramafic assemblages are assumed to form the core of the arc.

Several intermediate-composition isolated plutons in the heart of the eastern Yukon-Tanana Upland also were emplaced in latest Triassic to earliest Jurassic time. These plutons are assumed to be correlative with plutons of similar composition and age, including part of the Klotassin batholith, in adjacent Yukon Territory and elsewhere in western Canada and may represent part of a magmatic arc (the Stikine arc) that developed outboard of the North American continent.

Most Jurassic plutonic rocks, however, are Middle to Late Jurassic in age and are represented by three widely separated intraoceanic arc terranes. The most extensive of these arcs resulted in the formation of most of the Alaska-Aleutian Range batholith of south-central Alaska. The relations among these arcs, all of which may have had a south-dipping polarity, is uncertain. Also uncertain is the relation, if any, between the Alaska-Aleutian Range batholith and the seaward Early to Middle Jurassic Kodiak-Kenai-Chugach plutonic belt.

Cretaceous plutonism in mainland Alaska began in late Early Cretaceous time and continued into the early Tertiary. It exceeds that of the Jurassic in terms of area and distribution. Plutonic belts of Cretaceous age occur throughout the state and in some cases clearly represent the root zones of island and continental arc systems that collided with Alaska. In a lesser number of cases, however, subduction-related assemblages are lacking or concealed and the nature and polarity of the postulated arcs are uncertain. In the Ruby geanticline and perhaps in the Yukon-Tanana Upland, large areas of plutonic rocks appear to have formed within thickened continental crust, probably as a response to terrane collision and underthrusting.

Mesozoic plutonism in mainland Alaska ended with the widespread emplacement of Late Cretaceous to early Tertiary plutons throughout much of the south half of the state. This period of magmatic activity is subduction related and apparently resulted from continued underthrusting of the Kula Plate beneath an assemblage of terranes accreted to the North American continent that defined the North American Plate.

REFERENCES CITED

Aleinikoff, J. N., 1984, Age and origin of metaigneous rocks from terranes north and south of the Denali fault, Mt. Hayes Quadrangle, east-central Alaska: Geological Society of America Abstracts with Programs, v. 16, p. 266.

Aleinikoff, J. N., Dusel-Bacon, C., Foster, H. L., and Futa, K., 1981, Proterozoic zircon from augen gneiss, Yukon-Tanana Upland, east-central Alaska: Geology, v. 9, p. 469–473.

——, 1984, Uranium-lead isotopic ages of zircon from sillimanite gneiss and implications for Paleozoic metamorphism, Big Delta Quadrangle, east-central Alaska, *in* Coonrad, W. L., and Elliott, R. L., eds., The U.S. Geological Survey in Alaska; Accomplishments during 1981: U.S. Geological Survey Circular 868, p. 45–48.

Aleinikoff, J. N., Dusel-Bacon, C., and Foster, H. L., 1986, Geochronology of augen gneiss and related rocks, Yukon-Tanana terrane, east-central Alaska: Geological Society of America Bulletin, v. 97, p. 626–637.

Aleinikoff, J. N., Plafker, G., and Nokleberg, W. J., 1988, Middle Pennsylvanian plutonic rocks along the southern margin of Wrangellia, *in* Hamilton, T. D., and Galloway, J. P., eds., The U.S. Geological Survey in Alaska; Accomplishments during 1987: U.S. Geological Survey Circular 1016, p. 110–113.

Armstrong, R. L., 1988, Mesozoic and early Cenozoic magmatic evolution of the Canadian Cordillera, *in* Clark, S. P., Jr., ed., Processes in continental lithospheric deformation: Geological Society of America Special Paper 218, p. 55–91.

Armstrong, R. L., Harkal, J. E., Forbes, R. B., Evans, B. W., and Thurston, S. P., 1986, Rb-Sr and K-Ar study of metamorphic rocks of the Seward Peninsula and southern Brooks Range, Alaska, *in* Evans, B. W., and Brown, E. H., eds., Blueschists and eclogites: Geological Society of America Memoir 164, p. 185–203.

Arth, J. G., 1987, Regional isotopic variations in the Cretaceous plutons of northern Alaska: Geological Society of America Abstracts with Programs, v. 19, p. 355.

Arth, J. G., and 5 others, 1989a, Isotopic and trace-element variations in the Ruby batholith, Alaska, and the nature of the deep crust beneath the Ruby and Angayucham terranes: Journal of Geophysical Research, v. 94, p. 15941–15956.

——, 1989b, Remarkable isotopic and trace element trends in Cretaceous plutons of the Yukon-Koyukuk Basin, Alaska, and the nature of crustal lithosphere beneath the Koyukuk terrane: Journal of Geophysical Research, v. 94, p. 15957–15968.

Bacon, C. R, Foster, H. L., and Smith, J. G., 1985, Cretaceous calderas and rhyolitic welded tuffs in the Yukon-Tanana terrane, east-central Alaska: Geological Society of America Abstracts with Programs, v. 17, p. 339.

Beard, J. S., and Barker, F., 1989, Petrology and tectonic significance of gabbros, tonalites, shoshonites, and anorthosites in a late Paleozoic arc-root complex in the Wrangellia terrane, southern Alaska: Journal of Geology, v. 97, p. 667–683.

Bergman, S. C., 1987, Lamproites and other potassium-rich igneous rocks; A review of their occurrence, mineralogy, and geochemistry, *in* Fitton, J. G. and Upton, B. G., eds., Alakaline igneous rocks: Oxford, Blackwell Scientific Publications, p. 103–190.

Bergman, S. C., and Doherty, D. J., 1986, Nature and origin of 50–75 Ma volcanism and plutonism in W. and S. Alaska: Geological Society of America Abstracts with Programs, v. 18, p 539.

Blum, J. D., 1985, A petrologic and Rb-Sr isotopic study of intrusive rocks near Fairbanks, Alaska: Canadian Journal of Earth Sciences, v. 22, p. 1314–1321.

Blum, J. D., Blum, A. E., Davis, T. E., and Dillon, J. T., 1987, Petrology of cogenetic silica-saturated and -oversaturated plutonic rocks in the Ruby geanticline of north-central Alaska: Canadian Journal of Earth Sciences, v. 24, p. 159–169.

Box, S. E., 1985a, Mesozoic tectonic evolution of the northern Bristol Bay region, southwestern Alaska [Ph.D. thesis]: Santa Cruz, University of California, 163 p.

——, 1985b, Early Cretaceous orogenic belt in northwestern Alaska; Internal organization, lateral extent, and tectonic interpretation, *in* Howell, D. G., ed., Tectonostratigraphic terranes of the Circum-Pacific region: Houston, Texas, Circum-Pacific Council for Energy and Mineral Resources Earth Science Series 1, p. 137–145.

Box, S. E., and Patton, W. W., Jr., 1989, Igneous history of the Koyukuk terrane, western Alaska; Constraints on the origin, evolution, and ultimate collision of an accreted island arc terrane: Journal of Geophysical Research, v. 94, no. B11, p. 15843–15867.

Brosgé, W. P., and Reiser, H. N., 1964, Geologic map of the Chandalar Quadrangle, Alaska: U.S. Geological Survey Miscellaneous Geologic Investigations Map I-375, scale 1:250,000.

Bunker, C. M., Hedge, C. E., and Sainsbury, C. L., 1979, Radioelement concentrations and preliminary radiometric ages of rocks of the Kigluaik Mountains, Seward Peninsula, Alaska: U.S. Geological Survey Professional Paper 1129-C, p. 1–12.

Burns, L. E., 1985, The Border Ranges ultramafic and mafic complex, south-central Alaska; Cumulate fractionates of island-arc volcanics: Canadian Journal of Earth Sciences, v. 22, p. 1020–1038.

Carden, J. R., Connelly, W., Forbes, R. B., and Turner, D. L., 1977, Blueschists of the Kodiak Islands, Alaska; An extension of the Seldovia schist terrane: Geology, v. 5, p. 529–533.

Coney, P. J., Jones, D. L., and Monger, J.W.H., 1980, Cordilleran suspect terranes: Nature, v. 288, p. 329–333.

Csejtey, B., Jr., and Patton, W. W., Jr., 1974, Petrology of the nepheline syenite of St. Lawrence Island, Alaska: U.S. Geological Survey Journal of Research, v. 2, p. 41–47.

Csejtey, B., Jr., Patton, W. W., Jr., and Miller, T. P., 1971, Cretaceous plutonic rocks on St. Lawrence Island, Alaska; A preliminary report: U.S. Geological Survey Professional Paper 750-D, p. D68–D76.

Csejtey, B., Jr., and 8 others, 1978, Reconnaissance geologic map and geochronology, Talkeetna Mountains Quadrangle, northern part of Anchorage Quadrangle, and southwest corner of Healy Quadrangle, Alaska: U.S. Geological Survey Open-File Report 78–558A, 62 p.

Dawson, K. M., 1988, Mineral deposits and principal mineral occurrences of the Canadian Cordillera: Geological Survey of Canada Map 1513A, scale 1:2,000,000.

Dillon, J. T., 1987, Latest Cretaceous–earliest Tertiary metamorphism in the northeastern Brooks Range, Alaska: Geological Society of America Abstracts with Programs, v. 19, p. 373.

Dillon, J. T., Pessel, G. H., Chen, J. H., and Veach, N. C., 1980, Middle Paleozoic magmatism and orogenesis in the Brooks Range, Alaska: Geology, v. 8, p. 338–343.

Dillon, J. T., and 5 others, 1985, New radiometric evidence for the age and thermal history of the metamorphic rocks of the Ruby and Nixon Fork teranes, *in* Bartsch-Winkler, S., and Reed, K. M., eds., U.S. Geological Survey in Alaska; Accomplishments during 1983: U.S. Geological Survey Circular 945, p. 13–18.

Dillon, J. T., Tilton, G. R., Decker, J., and Kelly, M. J., 1987, Resource implications of magmatic and metamorphic ages for Devonian igneous rocks in the Brooks Range, *in* Tailleur, I. L., and Weimer, P., eds., Alaskan North Slope geology: Bakersfield, California, Pacific Section, Society of Economic Paleontologists and Mineralogists and Alaska Geological Society, Book 50, p. 713–723.

Dodds, C. J., and Campbell, R. B., 1988, Potassium-argon ages of mainly intrusive rocks in the St. Elias Mountains, Yukon and British Columbia: Geological Survey of Canada Paper 87–16, 43 p.

Dusel-Bacon, C., and Aleinikoff, J. N., 1985, Petrology and tectonic significance of augen gneiss from a belt of Mississippian granitoids in the Yukon-Tanana terrane, east-central Alaska: Geological Society of America Bulletin, v. 96, p. 411–425.

Dusel-Bacon, C., and 6 others, 1989, Distribution, facies, ages, and proposed tectonic associations of regionally metamorphosed rocks of northern Alaska: U.S. Geological Survey Professional Paper 1497-A, 44 p.

Eberlein, G. D., and Lanphere, M. A., 1988, Precambrian rocks of Alaska: U.S. Geological Survey Professional Paper 1241-B, p. B1–B18.

Einaudi, M. T., and Hitzman, M. W., 1986, Mineral deposits in northern Alaska; Introduction: Economic Geology, v. 81, no. 7, p. 1583–1591.

Evans, B. W., and Patrick, B. E., 1987, Phengite (3T) in high-pressure metamorphosed granitic orthogneisses, Seward Peninsula, Alaska: Canadian Mineralogist, v. 25, p. 141–158.

Forbes, R. B., Evans, B. W., and Thurston, S. P., 1984, Regional progressive high-pressure metamorphism, Seward Peninsula, Alaska: Journal of Metamorphic Geology, v. 2, p. 43–54.

Foster, H. L., 1970, Reconnaissance geologic map of the Tanacross Quadrangle, Alaska: U.S. Geological Survey Miscellaneous Geologic Investigations Map I-593, scale 1:250,000.

——, 1976, Geologic map of the Eagle Quadrangle, Alaska: U.S. Geological Survey Miscellaneous Geologic Investigations Map I-922, scale 1:250,000.

Foster, H. L., Donato, M. M., and Yount, M. E., 1978, Petrographic and chemical data on Mesozoic granitic rocks of the Eagle Quadrangle, Alaska: U.S. Geological Survey Open-File Report 78–253, 29 p.

Foster, H. L., Keith, T.E.C., and Menzie, W. D., 1987, Geology of east-central Alaska: U.S. Geological Survey Open-File Report 87–188, 59 p.

Frost, T. P., Calzia, J. P., Kistler, R. W., and Davison, V. V., 1988, Petrogenesis of the Crooked Mountains pluton, Bethel Quadrangle; A preliminary report, *in* Galloway, J. P., and Hamilton, T. D., eds., Geologic studies in Alaska: U.S. Geological Survey Circular 1016, p. 126–131.

Gabrielse, H., Loveridge, W. E., Sullivan, R. W., and Stebens, R. D., 1982, U-Pb measurements on zircon indicate middle Paleozoic plutonism in the Omineca Crystalline Belt, north-central British Columbia, *in* Current research, part C: Geological Survey of Canada Paper 82-1C, p. 139–146.

Gardner, J. C., and 8 others, 1988, Pennsylvanian pluton stitching of Wrangellia and the Alexander terrane, Wrangell Mountains, Alaska: Geology, v. 16, p. 967–971.

Gardner, M. C., and Hudson, T. L., 1984, Structural geology of Precambrian and Paleozoic metamorphic rocks, Seward Terrane, Alaska: Geological Society of America Abstracts with Programs, v. 16, p. 285.

Gemuts, I., Puchner, C. C., and Steffel, C. I., 1983, Regional geology and tectonic history of western Alaska: Journal of the Alaska Geological Society, v. 3, p. 67–85.

Hanks, C. L., and Wallace, W. K., 1990, Cenozoic thrust emplacement of a Devonian batholith, northeastern Brooks Range; Involvement of crystalline rocks in a foreland fold-and-thrust belt: Geology, v. 18, p. 395–398.

Hansen, V. L., 1988, A model for terrane accretion; Yukon-Tanana and Slide Mountain terranes, northwest North America: Tectonics, v. 7, no. 6, p. 1167–1178.

Hoare, J. M., and Coonrad, W. L., 1979, The Kanektok metamorphic complex, a rootless belt of Precambrian rocks in southwestern Alaska, *in* Johnson, K. M., and Williams, J. R., eds., The U.S. Geological Survey in Alaska; Accomplishments during 1978: U.S. Geological Survey Circular 804-B, p. B72–B74.

Hudson, T. L., 1983, Calc-alkaline plutonism along the Pacific rim of southern Alaska, *in* Roddick, J. A., ed., Circum-Pacific plutonic terranes: Geological Society of America Memoir 159, p. 159–170.

——, 1985, Jurassic plutonism along the Gulf of Alaska: Geological Society of America Abstracts with Programs, v. 17, p. 362.

Hudson, T. L., and Arth, J. G., 1983, Tin granites of Seward Peninsula, Alaska: Geological Society of America Bulletin, v. 94, p. 768–790.

Ishihara, S., 1981, The granitoid series and mineralization: Economic Geology 75th Anniversary Volume, p. 458–484.

Jones, D. L., Silberling, N. J., Coney, P. J., and Plafker, G., 1984, Lithotectonic terrane map of Alaska, *in* Silberling, N. J., and Jones, D. L., eds., Lithotectonic terrane maps of the North American Cordillera: U.S. Geological Survey Open-File Report 84–523, 12 p.

Karl, S. M., Aleinikoff, J. N., Dickey, C. F., and Dillon, J. T., 1989, Age and chemical composition of Proterozoic intrusive rocks at Mt. Angayukaqsraq, western Brooks Range, Alaska, *in* Dover, J. H., and Galloway, J. P., eds., 1988 Geologic studies in Alaska: U.S. Geological Survey Bulletin 1903, p. 10–19.

Khrenov, P. M., and Bukharov, A. A., 1973, Marginal volcano-plutonic belts in the North Asian craton: International Geology Review, v. 15, p. 688–697.

Kosygin, U. A., and Popeko, V. A., 1987, Map of magmatic formations of the Far Eastern USSR: Institute of Tectonics and Geophysics, Far East Branch of the USSR Academy of Sciences, scale 1:2,000,000 (in Russian).

Luthy, S. T., Foster, H. L., and Cushing, G. W., 1981, Petrographic and chemical data on Cretaceous granitic rocks of the Big Delta Quadrangle, Alaska: U.S. Geological Survey Open-File Report 81–398, 12 p.

MacKevett, E. M., Jr., 1978, Geologic map of the McCarthy Quadrangle, Alaska: U.S. Geological Survey Miscellaneous Geologic Investigation Map I-1032, scale 1:250,000.

Mayfield, C. F., Tailleur, I. L., and Ellersieck, I., 1983, Stratigraphy, structure and palinspastic synthesis of the western Brooks Range, northwestern Alaska: U.S. Geological Survey Open-File Report 83–779, 58 p.

Mertie, J. B., Jr., 1937, The Yukon-Tanana region, Alaska: U.S. Geological Survey Bulletin 872, 276 p.

Miller, T. P., 1971, Petrology of the plutonic rocks of west-central Alaska: U.S. Geological Survey Open-File Report 71–210, 136 p.

——, 1972, Potassium-rich alkaline intrusive rocks in western Alaska: Geological Society of America Bulletin, v. 83, p. 2111–2128.

——, 1989, Contrasting plutonic rock suites of the Yukon-Koyukuk basin and the Ruby geanticline, Alaska: Journal of Geophysical Research, v. 94, p. 15969–15989.

Miller, T. P., and Bunker, C. M., 1976, A reconnaissance study of the uranium and thorium content of plutonic rocks of the southeastern Seward Peninsula: U.S. Geological Survey Journal of Research, v. 4, no. 1, p. 57–74.

Miller, T. P., Patton, W. W., Jr., and Lanphere, M. A., 1966, Preliminary report on a plutonic belt in west-central Alaska: U.S. Geological Survey Professional Paper 550-D, p. D158–D162.

Moll, E. J., and Patton, W. W., Jr., 1982, Preliminary report on the Late Cretaceous and early Tertiary volcanic and related plutonic rocks in western Alaska, *in* Coonrad, W. L., ed., The U.S. Geological Survey in Alaska; Accomplishments during 1980: U.S. Geological Survey Circular 844, p. 73–76.

——, 1983, Late Cretaceous–early Tertiary calc-alkaline volcanic rocks of western Alaska: Geological Society of America Abstracts with Programs, v. 15, p. 406.

Monger, J.W.H., Price, R. A., and Tempelman-Kluit, D. J., 1982, Tectonic accretion and the origin of the two major metamorphic and plutonic belts in the Canadian Cordillera: Geology, v. 10, p. 70–75.

Mortensen, J. K., 1983, Age and evolution of the Yukon-Tanana terrane, southeastern Yukon Territory [Ph.D. thesis]: Santa Barbara, University of California, 155 p.

——, 1988, Significance of episodic magmatism in Yukon-Tanana terrane, Yukon and Alaska: Geological Society of America Abstracts with Programs, v. 20, p. A111.

Nelson, S. W., and Grybeck, D., 1980, Geologic map of the Survey Pass Quadrangle, Brooks Range, Alaska: U.S. Geological Survey Map Miscellaneous Field Studies MF-1176-A, scale 1:250,000.

Newberry, R. J., Dillon, J. T., and Adams, D. D., 1986, Regionally metamorphosed, calc-silicate-hosted deposits of the Brooks Range, northern Alaska: Economic Geology, v. 81, p. 1728–1752.

Nokleberg, W. J., Wade, W. M., Lange, I. M., and Plafker, G., 1986, Summary of geology of the Peninsular terrane, metamorphic complex of Gulkana River, and Wrangellia terrane, north-central and northwestern Gulkana Quadrangle, *in* Bartsch-Winkler, S., and Reed, K. M., eds., Geologic studies in Alaska: U.S. Geological Survey Circular 978, p. 69–74.

Palmer, A. R., 1983, The Decade of North American Geology 1983 geologic time scale: Geology, v. 11, p. 503–504.

Parfenov, L. M., and Natal'in, B. A., 1986, Mesozoic tectonic evolution of northeastern Asia: Tectonophysics, v. 127, p. 291–304.

Patrick, B. E., 1988, Synmetamorphic structural evolution of the Seward Peninsula blueschist terrane, Alaska: Journal of Structural Geology, v. 10, no. 6, p. 555–565.

Patton, W. W., Jr., 1973, Reconnaissance geology of the northern Yukon-Koyukuk province, Alaska: U.S. Geological Survey Professional Paper 774-A, 17 p.

Patton, W. W., Jr., and Box, S. E., 1989, Tectonic setting of the Yukon-Koyukuk basin and its borderlands: Journal of Geophysical Research, v. 94, no. B11, p. 15807–15820.

Patton, W. W., Jr., and Moll, E. J., 1984, Reconnaissance geology of the northern part of the Unalakleet Quadrangle, *in* Coonrad, W. L., and Elliott, R. L., eds., U.S. Geological Survey in Alaska; Accomplishments during 1981: U.S. Geological Survey Circular 868, p. 24–27.

Patton, W. W., Jr., Miller, T. P., Chapman, R. M., and Yeend, W., 1978, Geology of the Melozitna Quadrangle, Alaska: U.S. Geological Survey Miscellaneous Geologic Investigations Map I-1071, scale 1:250,000.

Patton, W. W., Jr., Moll, E. J., Lanphere, M. A., and Jones, D. L., 1984, New age data for the Kaiyuh Mountains, west-central Alaska, *in* Coonrad, W. L., and Elliott, R. L., eds., U.S. Geological Survey in Alaska; Accomplishments during 1981: U.S. Geological Survey Circular 868, p. 30–32.

Patton, W. W., Jr., Stern, T. W., Arth, J. G., and Carlson, C., 1987, New U/Pb ages from granite and granite gneiss in the Ruby geanticline and southern Brooks Range, Alaska: Journal of Geology, v. 95, p. 118–126.

Pavlis, T. L., Monteverde, D. H., Bowman, J. R., Rubenstone, J. L., and Reason, M. D., 1988, Early Cretaceous near-trench plutonism in southern Alaska; A tonalite-trondhjemite intrusive complex injected during ductile thrusting along the Border Ranges fault system: Tectonics, v. 7, no. 6, p. 1179–1200.

Plafker, G., Nokleberg, W. J., and Lull, J. S., 1989, Bedrock geology and tectonic evolution of the Wrangellia, Peninsular, and Chugach terranes along the Trans-Alaska crustal transect in the Chugach Mountains and southern Copper River Basin, Alaska: Journal of Geophysical Research, v. 94, no. B4, p. 4255–4295.

Prindle, L. M., 1909, The Fortymile Quadrangle, Yukon-Tanana region, Alaska: U.S. Geological Survey Bulletin 375, 52 p.

Reed, B. L., and Lanphere, M. A., 1969, Age and chemistry of Mesozoic and Tertiary plutonic rocks in south-central Alaska: Geological Society of America Bulletin, v. 80, p. 23–44.

——, 1973, Alaska-Aleutian Range batholith; Geochronology, chemistry, and relation to circum-Pacific plutonism: Geological Society of America Bulletin, v. 84, p. 2583–2610.

Reed, B. L., Miesch, A. T., and Lanphere, M. A., 1983, Plutonic rocks of Jurassic age in the Alaska-Aleutian Range batholith; Chemical variations and polarity: Geological Society of America Bulletin, v. 94, p. 1232–1240.

Richter, D. H., 1976, Geologic map of the Nabesna Quadrangle, Alaska: U.S. Geological Survey Miscellaneous Geologic Investigations Map I-932, scale 1:250,000.

Richter, D. H., Lanphere, M. A., and Matson, N. A., 1975, Granitic plutonism and metamorphism, eastern Alaska Range, Alaska: Geological Society of America Bulletin, v. 86, p. 819–829.

Roeske, S. M., 1986, Field relations and metamorphism of the Raspberry Schist, Kodiak Islands, Alaska, *in* Evans, B. W., and Brown, E. H., eds., Blueschists and eclogites: Geological Society of America Memoir 164, p. 169–184.

Roeske, S. M., Mattinson, J. M., and Armstrong, R. L., 1989, Isotopic ages of glaucophane schists on the Kodiak Islands, southern Alaska, and their implications for the Mesozoic tectonic history of the Border Ranges fault system: Geological Society of America Bulletin, v. 101, p. 1021–1037.

Rubin, C. M., Smith, G., and Miller, M. M., 1989, The geologic evolution and tectonic setting of mid-Paleozoic arc basement in the North America Cordillera: Geological Society of America Abstracts with Programs, v. 21, p. 137.

Sable, E. G., 1977, Geology of the western Romanzof Mountains, Brooks Range, northeastern Alaska: U.S. Geological Survey Professional Paper 897, 84 p.

Sainsbury, C. L., 1975, Geology, ore deposits, and mineral potential of the Seward Peninsula, Alaska, U.S. Bureau of Mines Open-File Report 73–75, 108 p.

Schmidt, J. M., 1986, Stratigraphic setting and mineralogy of the Arctic volcanogenic massive sulfide prospect, Ambler District, Alaska: Economic Geology, v. 81, no. 7, p. 1619–1643.

Silberman, M. L., Brookins, D. G., Nelson, S. W., and Grybeck, D., 1979a, Rubidium-strontium and potassium-argon dating of emplacement and metamorphism of the Arrigetch Peaks and Mount Igikpak plutons, Survey Pass Quadrangle, Alaska, *in* Johnson, K. M., and Williams, J. R., eds., The U.S. Geological Survey in Alaska; Accomplishments during 1978: U.S. Geological Survey Circular 804-B, p. B18–B19.

Silberman, M. L., Moll, E. J., Patton, W. W., Jr., Chapman, R. M., and Connor, C. L., 1979b, Precambrian age of metamorphic rocks from the Ruby province, Medfra and Ruby Quadrangles; Preliminary evidence from radiometric age data, *in* Johnson, K. M., and Williams, J. R., eds., The U.S. Geological Survey in Alaska; Accomplishments during 1978: U.S. Geological Survey Circular 804-B, p. B66–B68.

Smith, T. E., 1981, Geology of the Clearwater Mountains, south-central Alaska: Alaska division of Geological and Geophysical Surveys Geologic Report 60, 72 p.

Streckeisen, A., 1976, To each plutonic rock its proper name: Earth Science Reviews, v. 12, p. 1–33.

Swanson, S. E., Bond, J. F., and Newberry, R. J., 1988, Petrogenesis of the Ear Mountain Tin Granite, Seward Peninsula, Alaska: Economic Geology, v. 83, no. 1, p. 46–61.

Templeman-Kluit, D. J., and Wanless, R. K., 1975, Potassium-argon age determinations of metamorphic and plutonic rocks in the Yukon Crystalline Terrane: Canadian Journal of Earth Sciences, v. 12, p. 1895–1909.

Thurston, S. P., 1985, Structure, petrology, and metamorphic history of the Nome Group blueschist terrane, Salmon Lake area, Seward Peninsula, Alaska: Geological Society of America Bulletin, v. 96, p. 600–617.

Till, A. B., Dumoulin, J. A., Gamble, B. M., Kaufman, D. S., and Carroll, P. I., 1986, Preliminary geologic map and fossil data, Solomon, Bendeleben, and southern Kotzebue Quadrangles, Seward Peninsula, Alaska: U.S. Geological Survey Open-File Report 86–276, 18 p., scale 1:250,000.

Till, A. B., Schmidt, J. M., and Nelson, S. W., 1988, Thrust involvement of metamorphic rocks, southwestern Brooks Range, Alaska: Geology, v. 16, p. 930–933.

Turner, D. L., 1984, Tectonic implications of widespread Cretaceous overprinting of K-Ar ages in Alaskan metamorphic terranes: Geological Society of America Abstracts with Programs, v. 16, p. 338.

Turner, D. L., and Smith, T. E., 1974, Geochronology and generalized geology of the central Alaska Range, Clearwater Mountains, and northern Talkeetna Mountains: Alaska Division of Geological and Geophysical Surveys Open-File Report 72, 11 p.

Turner, D. L., Forbes, R. B., and Dillon, J. T., 1979, K-Ar geochemistry of the southwestern Brooks Range, Alaska: Canadian Journal of Earth Sciences, v. 16, p. 1789–1804.

Turner, D. L., Forbes, R. B., Aleinikoff, J. N., Hedge, C. E., and McDougall, I., 1983, Geochronology of the Kilbuck terrane of southwestern Alaska: Geological Society of America Abstracts with Programs, v. 15, p. 407.

Wallace, W. K., and Engebretson, D. C., 1984, Relationships between plate motions and Late Cretaceous to Paleogene magmatism in southwestern Alaska: Tectonics, v. 3, no. 2, p. 295–315.

Wallace, W. K., Hanks, C. L., and Rogers, J. F., 1989, The southern Kahiltna terrane; Implications for the tectonic evolution of southwestern Alaska: Geological Society of America Bulletin, v. 101, p. 1389–1407.

Weber, F. R., Foster, H. L., Keith, T.E.C., and Dusel-Bacon, C., 1978, Preliminary geologic map of the Big Delta Quadrangle, Alaska: U.S. Geological Survey Open-File Report 78–529A, scale 1:250,000.

Wilson, F. H., 1977, Some plutonic rocks of southwestern Alaska, a data compilation: U.S. Geological Survey Open-File Report 77–501, 9 p.

Wilson, F. H., Smith, J. G., and Shew, N., 1985, Review of radiometric data from the Yukon crystalline terrane, Alaska and Yukon Territory: Canadian Journal of Earth Sciences, v. 22, no. 4, p. 525–537.

MANUSCRIPT ACCEPTED BY THE SOCIETY FEBRUARY 12, 1991

Printed in U.S.A.

The Geology of North America
Vol. G-1, The Geology of Alaska
The Geological Society of America, 1994

Chapter 17

Some accreted volcanic rocks of Alaska and their elemental abundances

Fred Barker
U.S. Geological Survey, Box 25046, Denver Federal Center, Denver, Colorado 80225

With contributions by:
John N. Aleinikoff
U.S. Geological Survey, Box 25046, Denver Federal Center, Denver, Colorado 80225
Stephen E. Box
U.S. Geological Survey, Room 656, West 920 Riverside Avenue, Spokane, Washington 99201
Bernard W. Evans
Department of Geological Sciences, University of Washington, Seattle, Washington 98195
George E. Gehrels
Department of Geosciences, University of Arizona, Tucson, Arizona 85721
Malcolm D. Hill
Department of Geology, Northeastern University, Boston, Massachusetts 02115
Anthony J. Irving
Department of Geological Sciences, University of Washington, Seattle, Washington 98195
John S. Kelley
U.S. Geological Survey, P.O. Box 53, Anchorage, Alaska 99513
William P. Leeman
National Science Foundation, Washington, D.C. 20550
John S. Lull and Warren J. Nokleberg
U.S. Geological Survey, 345 Middlefield Road, Menlo Park, California 94025
John S. Pallister
U.S. Geological Survey, Box 25046, Denver Federal Center, Denver, Colorado 80225
Brian E. Patrick
Department of Geological Sciences, University of Washington, Seattle, Washington 98195
George Plafker
U.S. Geological Survey, 345 Middlefield Road, Menlo Park, California 94025
Charles M. Rubin
Division of Geological Sciences, California Institute of Technology, Pasadena, California 91125

INTRODUCTION

Fred Barker

This chapter describes and gives elemental abundances of many of the accreted volcanic rocks and of a few hypabyssal rocks of Alaska. These rocks range from early Paleozoic (or perhaps late Precambrian) to Eocene age. All formed prior to accretion of the terrane containing them and thus were generated either as primary features in the ancestral Pacific Ocean or on terranes or superterranes carried by plates underlying that ocean. Most formed in intraoceanic island arcs and related rifts; one oceanic plateau has been identified; basalts of mid-ocean-ridge (MORB) and of seamount types also are described; other basalts may have been extruded from leaky transforms.

These accreted volcanic rocks are important in terms of continental growth by accretion of oceanic rocks. Various workers have asserted that such growth is by accretion of intraoceanic island arcs. This assertion, however, must be appreciably modified for the ca. 400,000-km^2 region of southern and central Alaska that is underlain by accreted rocks. Though these rocks

Barker, F., 1994, Some accreted volcanic rocks of Alaska and their elemental abundances, *in* Plafker, G., and Berg, H. C., eds., The Geology of Alaska: Boulder, Colorado, Geological Society of America, The Geology of North America, v. G-1.

are not known in sufficient detail to yield a precise figure, I estimate that no more than 70 to 75 percent of this newly formed crust consists of former island arcs and arc-derived epiclastic sedimentary rocks.

Most of the tectonostratigraphic (lithotectonic) terranes of Alaska have minor exposures of volcanic rocks. Accounts of local and regional geology of the state contain cursory to extensive descriptions of such rocks. However, a catalog of such occurrences is not considered appropriate for this volume, and we discuss here only rocks studied by modern methods. The particular terranes containing these rocks are shown on Plate 13 (Barker and others, this volume), whereas all tectonostratigraphic terranes of Alaska are shown on Plate 3 (Silberling and others, this volume).

Though virtually all of the rocks considered in this chapter have been metamorphosed to prehnite-pumpellyite facies or higher grade, the emphasis is on the nature and environments of the protoliths. We also point out that most of these rocks have been subjected to alteration by seawater, by percolating fluids, or by metamorphism during or after accretion. Such processes affect the abundances of such "mobile" elements as K, Rb, Ba, Sr, and others in minor to major ways and obscures their original, magmatic abundances. Interpretations of environments of emplacement by using magmatic compositions thus devolve upon the relatively "immobile" elements such as Ti, Nb, Ta, rare-earth elements (REEs), and others. In addition, the uncertainties of using the various discriminant plots of elemental abundances or ratios as indicators of environments of generation now are well known. The authors of this chapter use geological information in conjunction with such plots where possible.

Analyses of more than 270 accreted volcanic and hypabyssal rocks are given in Tables 1 through 17 (on microfiche, back of this volume). Locations of these rocks are shown on Plate 6 (Barker and others, this volume). Place names not given on the plate may be found on Plate 2 (Wahrhaftig, this volume). The analysis in Tables 1 through 17 (on microfiche) were obtained by the following methods: major elements and Ba, Nb, Rb, Sr, Y, and Zr by x-ray fluorescence; FeO, H_2O, and CO_2 by wet methods; Sc, Cr, Co, and Ni by induction-coupled plasma optical emission spectrometry; rare-earth elements and other remaining elements by instrumental neutron activation. Precision of analyses cannot be given, but is within accepted standards.

SOME PRE-LATE CRETACEOUS VOLCANIC ROCKS OF SOUTHEASTERN ALASKA

G. E. Gehrels

Volcanic rocks of southeastern Alaska range in age from Late Proterozoic or Cambrian to Holocene, and include each intervening geologic period except the Pennsylvanian (Gehrels and Berg, 1992). Despite this surprisingly long and complete record, little is known about the tectonic environments in which the volcanic activity occurred. Hence, in the following discussion I give a general overview of all pre-Late Cretaceous volcanic and metavolcanic rocks in southeastern Alaska, and outine the scant geochemical and (or) petrologic data available. I describe Mesozoic (post-Triassic) rocks of Chichagof and Baranof Islands in the section on the Chugach terrane. The latest Mesozoic and Cenozoic volcanic rocks of southeastern Alaska are discussed by Brew (this volume).

TECTONIC SETTING OF SOUTHEASTERN ALASKA

As described in the chapter on the geology of southeastern Alaska (Gehrels and Berg, this volume), pre-Late Cretaceous rocks of the Alaska Panhandle belong to four terranes. From west to east, these terranes are: the Chugach terrane (mélange and other rocks of Cretaceous sedimentary and rocks of ocean-floor affinity); the Wrangellia terrane (hereafter informally designated Wrangellia; Triassic? basalt and marble); the Alexander terrane (Late Proterozoic or Cambrian through Jurassic sedimentary, volcanic, and intrusive rocks); and the Taku terrane (Permian and Triassic sedimentary and volcanic rocks, and undivided Jurassic and Cretaceous strata of the Gravina-Nutzotin belt).

Most workers agree that all of these terranes were firmly attached to the North American landmass by the end of Cretaceous time, but the timing and processes involved in their displacement, assembly, and accretion remain controversial. The only other relatively certain relationship between these terranes is that the Alexander terrane, Wrangellia, and the Taku terrane are all overlain by Upper Jurassic and Lower Cretaceous strata of the Gravina-Nutzotin belt (Berg and others, 1972, 1978; Barker, this chapter; Rubin, this chapter). The pre-Late Jurassic positions of the terranes relative to each other, and to North America, are as yet poorly constrained. Thus, the pre-Late Jurassic volcanic rocks described in this section, indeed most of the rocks in southeastern Alaska, may well have accumulated far from where they are found today.

The following discussion outlines the volcanic and metavolcanic rocks of southeastern Alaska from oldest to youngest.

LATE PROTEROZOIC OR CAMBRIAN

The oldest volcanic rocks known in southeastern Alaska belong to a greenschist-amphibolite-facies metamorphic complex (Wales Group) that forms the apparent basement to the Alexander terrane (Gehrels and Saleeby, 1987a). These rocks occur in a small area of southern southeast Alaska (Barker and others, this volume, Plate 13) and have been studied only in reconnaissance fashion. Protoliths of the metavolcanic rocks include basaltic-andesitic pillow flows, pillow breccia, and tuff breccia, and rhyolitic tuff and tuff breccia. Small dioritic and granodioritic intrusive bodies of Middle or Late Cambrian age are interpreted to be subvolcanic (Gehrels and Saleeby, 1987b).

ORDOVICIAN TO EARLY SILURIAN

Much of the southern Alexander terrane is underlain by a volcanic-plutonic-sedimentary complex of Ordovician to Early Silurian age, which apparently intrudes and overlies the older metamorphic basement. Volcanic rocks of this complex belong to the Descon Formation. They range from basaltic-andesitic pillow flows and breccia through dacitic breccia, tuff breccia, and tuff, to rhyolitic tuff, pyroclastic breccia, and extrusive dome complexes (Eberlein and others, 1983; Gehrels and Saleeby, 1987b; Gehrels and others, 1987). Coeval plutons range from diorite and subordinate gabbro, through quartz diorite, tonalite, and granodiorite, to granite, quartz monzonite, and quartz syenite. Geochemical analyses have been obtained from various compositional members of the plutonic suite, a basalt-andesite pillow flow, and a rhyolite breccia (Gehrels and Saleeby, 1987b). Major-, minor-, trace-, and rare-earth-element compositions of these rocks are indistinguishable from those of present-day island-arc systems and suggest that the southern Alexander terrane evolved in a convergent-margin environment during Ordovician to Early Silurian time.

DEVONIAN

Devonian volcanic rocks occur in widely scattered localities in the Alexander terrane, but nowhere are they laterally continuous or of great thickness. Lower Devonian volcanic rocks are heterogeneous in composition. Feldspar-porphyritic dacite breccia and basaltic pillow flows and breccia occur on southern Prince of Wales Island (Gehrels and Saleeby, 1987b) and on southern Annette Island (Gehrels and others, 1987); rhyolite flows occur in a very restricted region of east-central Prince of Wales Island (Eberlein and others, 1983); and plagioclase-phyric andesite, possibly correlative with that to the south, occurs on northern Kuiu Island (Muffler, 1967). Middle Devonian rocks occur only in a restricted area of west-central Prince of Wales Island, where they comprise basaltic pillow flows, breccia, and tuff of the Coronados Volcanics (Eberlein and others, 1983). In contrast, volcanic rocks of Late Devonian age occur as thick sections of basaltic-andesitic pillow flows and breccia on northern Chichagof Island (Freshwater Bay Formation, Loney and others, 1975) and western Prince of Wales Island (Port Refugio Formation, Eberlein and others, 1983).

Volcanic and metavolcanic rocks of known, probable, and possible Devonian age occur in several other areas of southeastern Alaska. Such rocks include basaltic flows and breccia of the St. Joseph Island Volcanics (western Prince of Wales Island, Eberlein and others, 1983); unnamed basaltic flows, agglomerate, and tuff of Silurian or Devonian age in the Chilkat Range (Brew and Ford, 1985); and strongly deformed and metamorphosed greenstone and greenschist (Retreat Group and Gambier Bay Formation) of Middle(?) Devonian age on Admiralty and Kupreanof Islands (Lathram and others, 1965; McClelland and Gehrels, 1987).

MISSISSIPPIAN

Mississippian andesitic rocks have been recognized in a small area of northern Kupreanof Island but may be more widespread on Amdiralty Island. These rocks occur on Kupreanof Island as layers and lenses in a disrupted chert-argillite-limestone complex (part of Cannery Formation, see Muffler, 1967).

PERMIAN

Permian volcanic rocks are known in two separate areas of the Alexander terrane, and probably constitute a significant part of the Taku terrane. In the Alexander terrane, they occur on northern Kuiu Island as olivine-rich basalt pillow flows and breccia interbedded with Lower Permian limestone and clastic strata of the Halleck Formation (Muffler, 1967). A similar but less well-known sequence of basaltic(?) volcanic rocks, limestone (marble), and clastic strata occurs in the Chilkat Range and north of Glacier Bay (Brew and Ford, 1985), and may extend northward into the area northwest of Haines (Gehrels and Berg, 1992).

Volcanic rocks of the Taku terrane are predominantly basaltic to andesitic, and range from well-preserved flows and breccia, through moderately deformed greenstone and greenschist, to amphibolite-facies schist and gneiss of uncertain volcanic origin (Berg and others, 1972; Gehrels and Berg, 1992). Fossils have not been found in these strata; their age is constrained only by stratigraphic relations with clastic rocks and limestone-bearing Permian and (or) Triassic fossils. Thus, although such volcanic rocks constitute a significant percentage of the Taku terrane, their age has nowhere been documented. North of Petersburg, these rocks are in unknown contact with a thick(?), but areally restricted, sequence of quartzo-feldspathic schist and gneiss that may have been derived from rhyolite and rhyolitic tuff and could be of Permian age (Monger and Berg, 1987).

TRIASSIC

Triassic volcanic rocks are widespread in southeastern Alaska and occur in the Alexander terrane, Wrangellia terrane, and probably the Taku terrane. The sequence in the Alexander terrane is distinctive in that it consists of a bimodal basalt-rhyolite suite, generally with rhyolite below basalt. These rocks can be traced in a nearly continuous band along the eastern margin of the Alexander terrane and separate subjacent pre-Triassic rocks from overlying Jurassic and Cretaceous strata of the Gravina-Nutzotin belt. A regionally extensive angular unconformity at the base of the Triassic section, a basal conglomerate or sedimentary breccia that displays rapid lateral thickness changes, and facies relations upsection all indicate deposition during a phase of uplift, erosion, and faulting along the eastern (inboard) margin of the Alexander terrane. The bimodal composition of the volcanic rocks and the occurrence of the strata in a narrow band along the eastern margin of the terrane indicate a probable rift origin (Gehrels and others, 1986). However, this hypothesis has not yet been tested by geochemical or petrologic analyses. The Triassic(?) Goon Dip Greenstone and the overlying Triassic(?) Whitestripe Marble (Johnson and Karl, 1985) form a small fragment of the Wrangellia terrane on western Chichagof and northern Baranof Islands. Decker (1980) has indicated that about a 2-km-thick section of the Goon Dip Greenstone is preserved—its base not being exposed. His three analyses show the Goon Dip to lie in the range of major and minor elements given by the Nikolai Greenstone (Barker and others, this chapter) and basalts of the Karmtusen Formation (Barker and others, 1989). Thus, the Goon Dip is presumed to have formed in a back-arc basin, as did the Nikolai and Karmtusen.

Volcanic rocks of the Taku terrane have been described under Permian rocks. However, some of the basaltic rocks within the northern Taku terrane are of Triassic age and are probably correlative with the tholeiitic basalts of Wrangellia. Plafker and Hudson (1980) and Davis and Plafker (1985) report that well-preserved Triassic basalts about 3 km thick on Chilkat Peninsula (northern southeast Alaska) are similar in age and chemistry to the Nikolai Greenstone of southern Alaska. Furthermore, megafossils of the associated limestone are like those of the correlative limestone of Wrangellia. Plafker and others (1989a) recently suggested that northern parts of the "Taku terrane," indeed, are a part of Wrangellia. Because the basaltic rocks on Chilkat Peninsula apparently are correlative with the volcanic and metavolcanic rocks in the Taku terrane to the southeast, it appears likely that the Taku terrane and Wrangellia terrane were contiguous during Late Triassic time (Davis and Plafker, 1985).

The occurrence of similar and coeval rift assemblages in Wrangellia, the Taku terrane, and the Alexander terrane raises the possibility that these three tectonic fragments have been in close proximity since at least Late Triassic time. Gardner and others (1988) note that syenitic intrusions of 309-Ma age stitch together the Alexander and Wrangellia terranes at their mutual contact in the Wrangell Mountains.

METAMORPHOSED MAFIC IGNEOUS ROCKS IN THE SEWARD PENINSULA

B. W. Evans, A. J. Irving, and B. E. Patrick

INTRODUCTION

Low-grade metamorphic rocks of the Nome Group (Till and Dumoulin, this volume) constitute a major part of the bedrock in the Seward Peninsula. Recent fossils finds (Dumoulin and Harris, 1984; Till and others, 1986) have shown that the stratigraphic age of the Nome Group extends from at least Cambrian through Devonian. Whole-rock Rb-Sr isotope data on pelitic schist and granitic orthogneiss are consistent with this age span (Armstrong and others, 1986). The metasedimentary rocks of the Nome Group (pelitic schist, calcareous schist, calcitic and dolomitic marble, graphitic quartzite, and chloritic schist) were deposited in shallow water, apparently in basins and flanking platforms on the rifted lower Paleozoic continental margin of North America. Metagabbros, metadiabases, and probable basaltic lava flow and pyroclastic rocks are particularly abundant in the chlorite schist unit (= Casadepaga Schist) of the Nome Group (Till and Dumoulin, this volume).

Most of the analyzed metabasites are from this unit, which is bounded above and below by marbles of Ordovician age (Till and others, 1986). A few samples are from poor exposures whose structural and stratigraphic relationships are unclear. Thus, the probable age of most of the samples is Ordovician, although no attempt has yet been made to date them radiometrically. In areas of limited deformation, such as south of Teller and near American River (Fig. 1), some metabasites occur in sill-like bodies, and contacts in a few places are seen to crosscut the layering of the metasediments. In these locations, and in the central parts of more massive layers and boudins elsewhere, igneous textures (blastophitic texture, former FeTi-oxide domains, unrecrystallized apatite) are still visible microscopically. In the Teller area, igneous augite, plagioclase, and hornblende survived the greenschist-facies metamorphism. Although contrasting lithologies (more versus less differentiated) occur together in some frost-heaved outcrops, no clear evidence of any primary-layered structure within the former sills has been found.

The rocks of the Nome Group underwent epidote-blueschist and greenschist facies metamorphism in the Jurassic and Early Cretaceous (Forbes and others, 1984; Thurston, 1985; Armstrong and others, 1986). As a result, the metabasites are now composed of various combinations of glaucophane, actinolite, epidote, almandine, paragonite, phengite, chlorite, albite, calcite, quartz, titanite, apatite, and rare sodic pyroxene or chloritoid, with lesser rutile, magnetite, zircon, and pyrite, and rare baddelyite. The growth and preservation of sodic amphibole was a function of both whole-rock composition and metamorphic history.

IGNEOUS ROCK TYPES

Major- and trace-element analyses have been conducted by ICP emission spectrometry on 41 samples of metabasite from the Nome Group rocks on the Seward Peninsula. Major-element

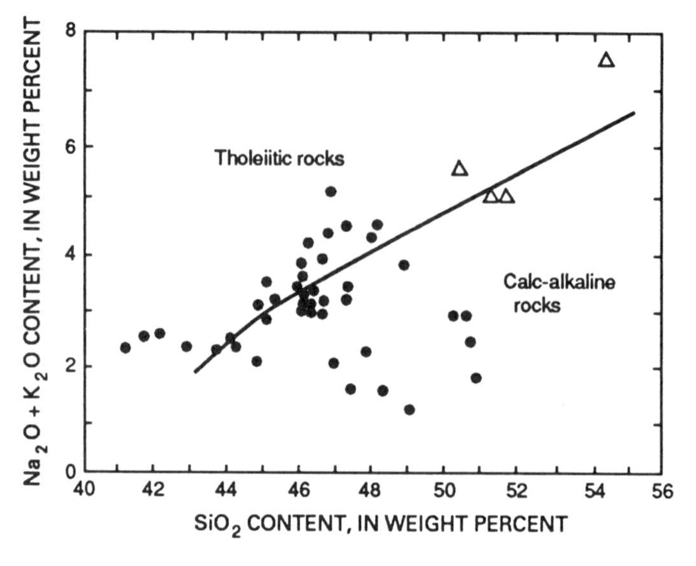

Figure 1. Plot of $Na_2O + K_2O$ versus SiO_2 (weight percent) for Seward Peninsula metabasites. Triangles signify Na-rich, Ca-poor samples (see text). Line separates tholeiitic and alkalic fields.

data on a further six samples were published by Thurston (1985). In addition, data for rare-earth and other trace elements were obtained for 14 of these samples by instrumental neutron activation analysis. The complete data for these 14 are given in Table 1 (on microfiche).

The majority of samples are hypersthene-normative, and can be classified as olivine tholeiite, quartz tholeiite, and olivine basalt (transitional to alkalic basalt); only two samples have more than 0.5 percent normative nepheline. Despite their essentially tholeiitic character many of the samples plot in the mildly alkalic field on an alkalies-SiO_2 plot (Fig. 1). Four samples (designated separately on Figs. 1 and 2) have notably high Na_2O (4.8 to 6.7 weight percent), low CaO (1.3 to 6.9 weight percent), high SiO_2, and relatively high Al_2O_3. These rocks may have experienced spilitic alteration or else may represent plagioclase-enriched igneous protoliths; one of these samples is probably a mafic metasediment. Except for several of these four, all samples plot outside the calc-alkaline field on an AFM diagram and in the tholeiitic field on a SiO_2-FeO^*/MgO plot (Fig. 2).

Mg-values for the Seward Peninsula metabasites range from 0.65 to 0.30 (Fig. 3), and thus, if the protoliths represent magmatic liquids, they show a range of evolved magma compositions relative to typical primary magmas. Many of the more Fe-rich rocks have very high TiO_2 contents (the most Ti-rich sample, now eclogite, contains 7.7 weight percent TiO_2, but more than half of the analyzed samples contain more than 3.5 weight percent TiO_2). Nevertheless, the most evolved (i.e., Fe-rich) compositions also include examples with lower TiO_2 contents (see Fig. 3). TiO_2 correlates fairly well with La/Yb (Fig. 4) and, to a lesser degree, with P_2O_5 and Zr, and shows a broad negative correlation with SiO_2. Relic textures make it clear that many of the samples were rich in igneous FeTi-oxides; the high TiO_2 contents are therefore believed to be an original property of the suite. Low K_2O contents (less than 0.2 weight percent in half the samples) could, on the other hand, reflect loss of potassium prior to or during the metamorphism, or alternatively may imply that some of the protoliths were cumulates.

When normalized to primitive mantle abundances (Fig. 5), the compositions of most Seward Peninsula metabasites are more enriched than "typical" MORB, but have similarities to "enriched" MORB and ocean-island tholeiites. Relative depletions in K, Rb, Ba, Zr, and Hf are also evident from this diagram. Both

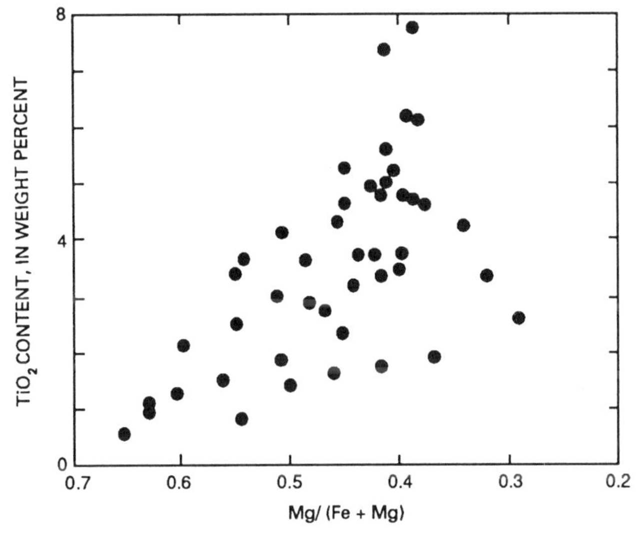

Figure 3. Plot of TiO_2 versus mg (Mg/[Mg+Fe]) for Seward Peninsula metabasites.

Figure 2. Plot of SiO_2 versus FeO^*/MgO (where $FeO^* = FeO + 0.9\ Fe_2O_3$) for Seward Peninsula metabasites. Symbols as in Fig. 1. Line separates calc-alkalic (left) and tholeiitic fields (after Miyashiro, 1974).

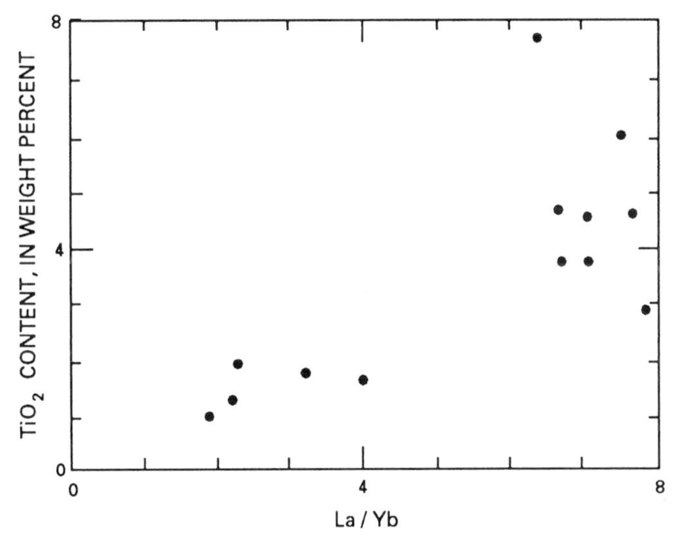

Figure 4. Plot of TiO_2 versus La/Yb for 14 Seward Peninsula metabasites.

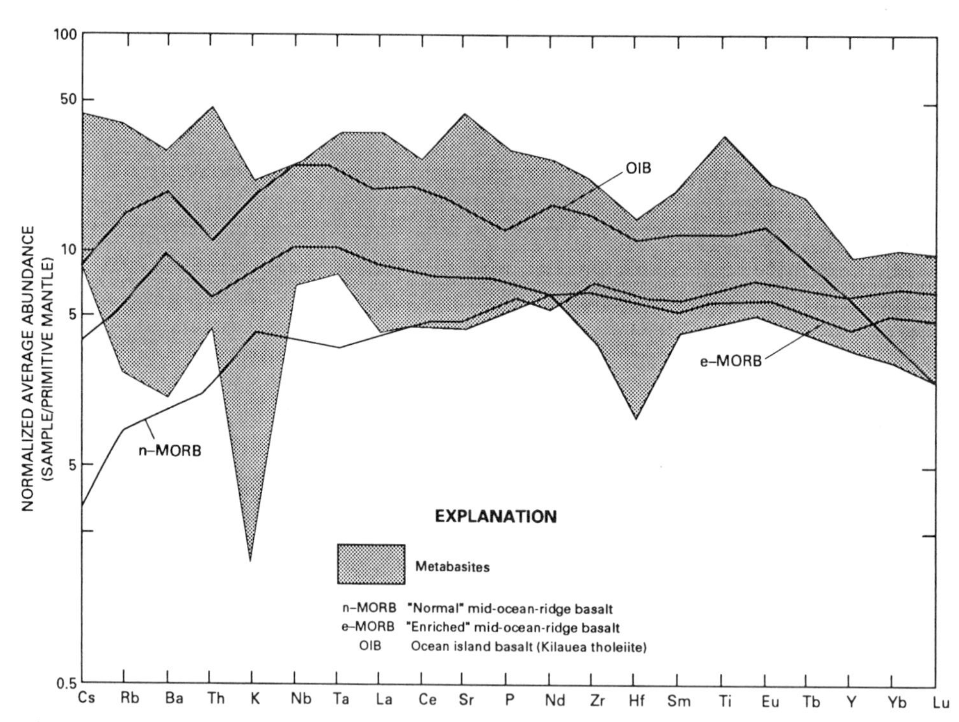

Figure 5. Primitive-mantle-normalized abundance diagram showing range for 14 Seward Peninsula samples and typical patterns for "normal" MORB, "enriched" MORB, and Kilauea tholeiite (OIB). After Sun and Nesbitt (1977), Pearce (1983), and BVSP (1981). Normalizing values from Sun and McDonough (1989).

enrichments and depletions in Sr and Ti are found, and some samples have small Eu anomalies. One interpretation of these chemical characteristics is that the protoliths formed from a variety of liquids and cumulates which were interrelated by addition or subraction of plagioclase, Fe-Ti oxides, and augite. The average composition of relict igneous augite in metagabbros in the Teller area yields a formula $(Ca_{.81}Na_{.02}Fe,Mg_{.17})$ $(Fe,Mg_{.95}Al_{.02}Ti_{.03})$ $(Si_{.92}Al_{.08})O_6$, with Mg/(Mg+Fe) = 0.70. Overall the data are consistent with relatively low-pressure differentiation processes such as would be expected in a subvolcanic dike and sill complex.

Most conventional trace-element discriminant diagrams for these rocks show wide scatter, presumably as a result of both cumulus processes and subsequent alteration. The rocks apparently can be distinguished from island-arc tholeiites on the basis of Ti/V ratios (Fig. 6) and their lack of depletion in Nb, Ta, and Ti (Fig. 5). Consideration of all the geochemical characteristics of the Seward Peninsula metabasites as well as their geological setting leads us to conclude that they represent basaltic magmatism associated with a rifted continental margin or shallow oceanic plateau.

ANGAYUCHAM TERRANE

J. S. Pallister and F. Barker

GENERAL DESCRIPTION

The Angayucham terrane derives its name from the Angayucham Mountains of the south-central Brooks Range, where as much as 8 to 10 km (composite thickness) of structurally interleaved pillow basalt, diabase, basaltic tuff, and chert are exposed (Pallister and Carlson, 1988). Basalt-chert sequences are exposed in a V-shaped outcrop belt that extends from the Angayucham Mountains to the east, past Coldfoot, then southwest through the Kanuti ophiolite to the Ruby Mountains (Barker and others, this volume, Plate 13). Similar rocks are exposed 20 to 100 km to the south and east in the Tozitna terrane (Jones and others, 1987; and see Patton and others, this volume, Chapter 21).

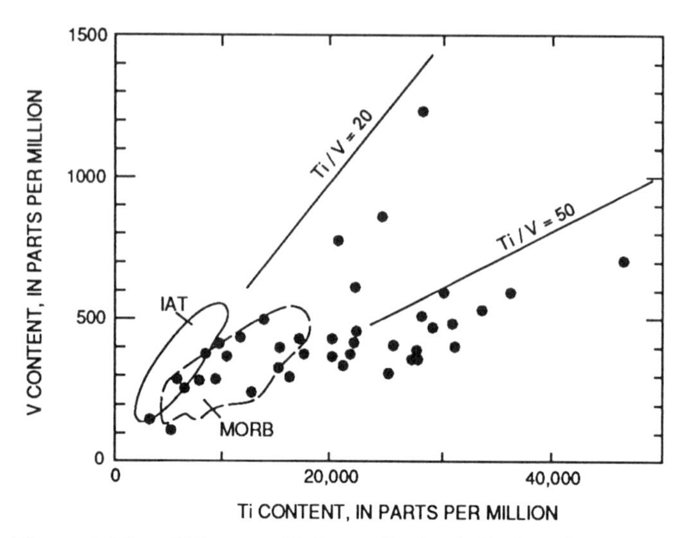

Figure 6. Plot of V versus Ti (normalized volatile free) for Seward Peninsula metabasites with fields after Shervais (1982).

These basaltic-chert terranes are near the top of the sequence of thrust sheets in much of western and northern Alaska and western Canada. Emplacement is generally related to Late Jurassic to Early Cretaceous accretion of the oceanic rocks onto the Brooks Range continental margin (Roeder and Mull, 1978; Jones and others, 1986; Pallister and others, 1989; Patton and others, this volume, Chapter 21). Pillow basalt and chert form the lower and more widely exposed of two lithologic assemblages; the upper assemblage is composed of basal ophiolite rocks (amphibolite, peridotite, and gabbro) of Jurassic age (Patton and others, 1977). Viewed together, the two assemblages appear to comprise an inverted ophiolite sequence; however, they are separated by thrust faults, and the lower basalt-chert assemblage contains fault slices that yield a variety of ages. Sections constructed across fault slices of the basalt-chert assemblage in the central Angayucham Mountains show a telescoped age progression from highly disrupted block mélange of Paleozoic and Mesozoic rocks on the north, through Triassic to Jurassic basalt-chert sequences to the south (Pallister and Budahn, 1989).

BASALT-CHERT ASSEMBLAGE

Thick piles of pillow basalt flows containing sills and dikes of diabase form the bulk of the Angayucham terrane. Intercalated basaltic tuff, red argillite, and lenses of radiolarian chert are common; fault-bounded sequences of chert and tuff are as thick as several hundred meters. The pillow basalts are nonvesicular to amygdaloidal; pillows are typically 30 cm to more than a meter in diameter; and pillow breccias are locally abundant. Two areas within the basalt-chert assemblage have been studied in detail. The central Angayucham Mountains were mapped by Hitzman and others (1982) and by Pallister and Carlson (1991), and the basalts were the subject of a petrologic and geochemical study by Pallister and others (1989). The basalt-chert sequence near Coldfoot was studied by Barker and others (1988).

The central Angayucham Mountains are underlain by prehnite-pumpeyllite facies pillow basalt, subordinate diabase and basaltic tuff, and minor radiolarian chert. The basalts are separated from higher grade metamorphic and crystalline rocks of the Brooks Range successively by a narrow belt of tectonic block–mélange along the northern flank of the Angayucham Mountains and then by a broad, low-relief belt of Paleozoic phyllite, greenschist, and minor blueschist (Patton and Miller, 1966, 1973; Hitzman and others, 1982).

Major-element analysis indicate that many of the basalts are hypersthene-normative olivine tholeiites. Classification based on immobile trace elements confirms the tholeiitic character of most of the basalts and suggests some had primary compositions that were transitional to alkali-basalt (Table 2, on microfiche). Although field and petrographic features of the basalts are similar, trace-element characteristics allow definition of geographically distinct suites. A central outcrop belt along the crest of the mountains is made up of basalt with relatively flat rare-earth element (REE) patterns. This belt is flanked to the north and south by basalts enriched in light rare earth elements (LREE). Radiolarian and conodont ages from interpillow and interlayered chert and limestone indicate that the central belt of basalts is Triassic in age, the southern belt is Jurassic in age, and the northern belt contains rocks of both Paleozoic and Mesozoic ages (Pallister and Carlson, 1988).

The abundance and thickness of pillow basalt and interlayered pelagic sedimentary rocks provide clear evidence that the Angayucham terrane formed in an oceanic setting. None of the basalts have trace-element characteristics of island-arc basalt. Data for most of the basalts cluster in the "within-plate basalt" fields of trace-element discriminant diagrams and lack the Nb and Ta depletions that characterize modern arc basalts (Fig. 7). The Triassic and Jurassic basalts are geochemically most akin to modern oceanic-plateau and island basalts (Pallister and others, 1989).

Field evidence also favors an oceanic plateau or island setting. The great composite thickness of pillow basalt probably resulted from obduction faulting, but the lack of fault slabs of gabbro or peridotite suggests that obduction faults did not penetrate below oceanic layer 2, a likely occurrence if layer 2 was anomalously thick, as in the vicinity of an oceanic island. The presence of basaltic tuff interbeds indicates proximity to an explosive basaltic eruptive center.

The juxtaposition of submarine basalts of differing chemical affinity and age adjacent to higher grade Paleozoic metamorphic rocks of the Brooks Range to the north may be explained by obduction of internally complex (thickened) oceanic crust formed in an ocean plateau and island setting. Emplacement and rotation of thrust plates to steep attitudes occurred during the Late Jurassic and Early Cretaceous accretion of the Brooks Range passive margin. Earlier workers, however, suggested that this juxtaposition was a result of extensional faulting.

Pillow basalt, diabase, chert, and argillite are thrust and folded into duplex structures near Coldfoot (Barker and others, 1988). Peak metamorphism was to epidote-amphibolite facies; originally glassy basalts were most reactive and were converted to blue-green amphibole-sodic plagioclase-epidote-magnetite assemblages. Intrusive and stratigraphic relations and dated radiolarian cherts suggest that the mafic rocks range from Mississippian to latest Triassic or Jurassic in age. Diabase sills and massive to cumulus-banded gabbro occur at or near the base of several large pillow lava sequences in the Coldfoot area. Intrusive relations indicate that these hypabyssal and plutonic rocks are stratiform bodies emplaced at shallow levels within preexisting oceanic crust; contact relations and geochemical affinities suggest that the intrusive rocks represent subvolcanic feeders for overlying pillow lavas.

The Coldfoot basalts are similar to those of the Angayucham Mountains in having the trace-element characteristics of modern within-plate basalts. Barker and others (1988) suggest that they formed in an oceanic intraplate setting (oceanic plateau or seamount) by the mixing of parent magmas having trace-element characteristics of both "normal" mid-ocean-ridge basalt (MORB) and within-plate or "plume-type" MORB.

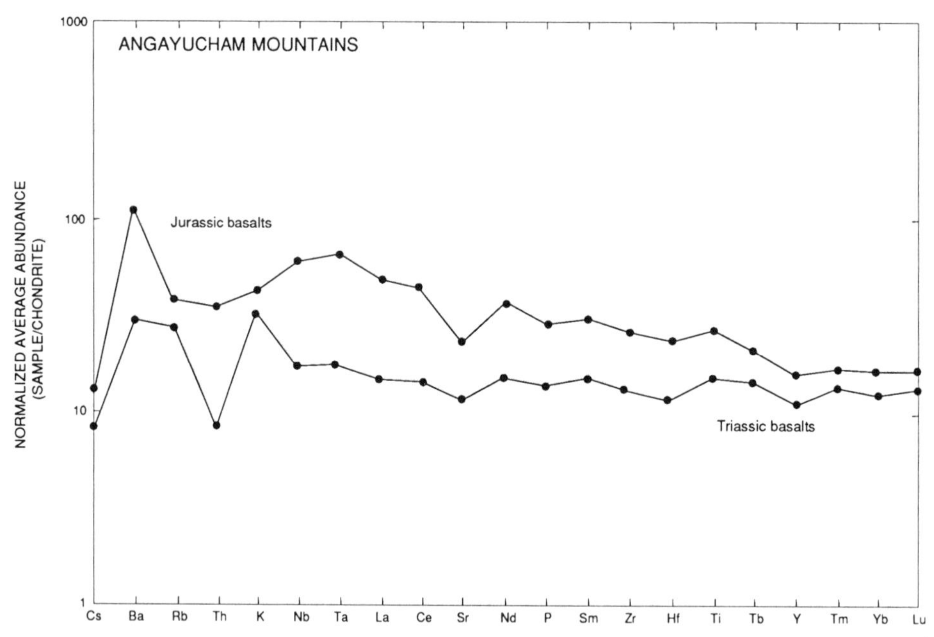

Figure 7. Spidergram of Triassic and Jurassic basalts of western Angayucham terrane (after Thompson and others, 1983) showing chondrite-normalized abundances of selected elements arranged in order of decreasing compatibility and mobility—to yield a smooth pattern for "normal" mid-ocean-ridge basalts (N-MORB).

SOUTHERN YUKON-TANANA TERRANE

W. J. Nokleberg and J. N. Aleinikoff

The southern margin of the Yukon-Tanana terrane in the eastern Alaska Range comprises, from north to south, the Lake George, Macomb, and Jarvis Creek Glacier subterranes (Jones and others, 1987; Aleinikoff and Nokleberg, 1985; Nokleberg and Aleinikoff, 1985). These terranes are interpreted as forming various stratigraphic/structural levels of a Devonian and Mississippian continental-margin arc (Aleinikoff and others, 1981, 1986) that formed along the western margin of North America (Nokleberg and Aleinikoff, 1985). To the north, in the deeper levels of the terrane, the Lake George and Macomb subterranes consist of a sequence of medium- to coarse-grained, multiply deformed, pelitic, mylonitic schists intruded by Devonian schistose quartz diorite, granodiorite, and granite (Aleinikoff and Nokleberg, 1985). To the south, in the higher levels of the terrane, the Jarvis Creek Glacier subterrane consists of Devonian metavolcanic and metasedimentary mylonitic schist and minor phyllonite (Nokleberg and Aleinikoff, 1985). Metavolcanic schists are derived mainly from andesite and quartz keratophyre, and from lesser amounts of dacite and basalt. Metasedimentary schists are derived from fine-grained clastic, calcareous, and volcanogenic sediments. In some areas, volcanogenic massive sulfide deposits occur extensively in the terrane (Nauman and others, 1980; Nokleberg and Lange, 1985). A submarine origin is indicated for these deposits by the interlayering of metavolcanic rocks and sulfide lenses and pods with fine-grained thinly layered metasedimentary rocks. Both suites of metavolcanic and metasedimentary rocks are multiply deformed and metamorphosed, pervasive, younger, middle to upper greenschist facies minerals to the south, and a few areas of older, relict lower amphibolite–facies minerals to the north.

Farther south, the Hayes Glacier subterrane (Nokleberg and Aleinikoff, 1985), consisting of Devonian metasedimentary and metavolcanic phyllonites, is derived from many of the same protoliths as the Jarvis Creek Glacier subterrane. The Hayes Glacier subterrane of the Yukon-Tanana terrane differs from the Jarvis Creek Glacier subterrane in having more black to dark gray carbonaceous pelitic rocks, sparse small volcanogenic massive sulfide deposits, and few metavolcanic and volcanically derived rocks. Stratigraphic and structural relations indicate that these subterranes represent, from north to south, successively higher levels of a single, now highly metamorphosed and deformed, Devonian submarine igneous arc (Nokleberg and Aleinikoff, 1985).

Thin metamorphosed volcanic flows and tuffs of the southern margin of the Yukon-Tanana terrane occur in layers as much as a few meters thick, intercalated mostly with quartz schist and pelitic schist. Estimate of the original thickness is precluded because of intense multiple deformation. These rocks are light to dark gray-green and consist of plagioclase and some quartz microphenocrysts set in a groundmass of thoroughly metamorphosed aggregates of plagioclase, quartz, actinolite, chlorite, epidote, and opaque minerals. The samples chosen for geochemical study were picked on the basis of relict igneous features in thin section, because in outcrop, protolith features are obscured by multiple metamorphism and deformation. Abundant relict igneous features in thin section consist of microphenocrysts of plagioclase with normal and oscillatory zoning, phenocryst outlines for plagioclase, complicated twinning in plagioclase, embayed (resorbed) outlines for quartz phenocrysts, and a massive character to the unit, indicating a probable flow origin. These criteria, however, do not totally exclude the possibility that some samples contain a minor intercalated clastic component.

Major-element whole-rock chemistry of 14 samples of metamorphosed volcanic flows and tuffs in the southern part of the Yukon-Tanana terrane is listed in Table 3 (on microfiche) and displayed in the plot (Fig. 8) of FeO^*/MgO versus SiO_2 (where $FeO^* = FeO + 0.9\ Fe_2O_3$). A wide range of major and minor oxides occurs. SiO_2 ranges from 53.2 to 79.7 percent with an average of 70.1 percent. Two samples are andesite (as defined by SiO_2 = 53 to 63 percent), four are dacite (SiO_2 = 63 to 69 percent), and eight are rhyolite (SiO_2 = 69 percent or more). On a plot of FeO^*/MgO versus SiO_2 (Fig. 8), the suite exhibits limited iron enrichment for low SiO_2-content samples. The suite is about equally divided between the tholeiitic and calc-alkaline fields. Silica variation diagrams, not published here, show two important features. First, for the "immobile" oxides—CaO, FeO^*, MgO, and TiO_2—relatively smooth variations approximate a calc-alkaline trend. Second, there is an almost random pattern for the more "mobile" elements—K_2O and Na_2O—indicating extensive metasomatism either during submarine volcanism and (or) metamorphism.

Origin as a Devonian and Mississippian submerged continental-margin arc is suggested by the marine character, calc-alkaline trend, and abundant quartz detritus in the associated metasedimentary rocks (Nokleberg and Aleinikoff, 1985; this chapter). In addition, preliminary common lead-isotopic studies on sulfide samples from the volcanogenic massive sulfide deposits (LeHuray and others, 1985), and similar studies on feldspars from the metaplutonic and metavolcanic rocks (Aleinikoff and others, 1987), indicate a continental component derived from an Early Proterozoic source. A continental margin setting also is indicated by the abundant quartz-rich schist containing detrital zircons derived from an Early Proterozoic source(s) (Dusel-Bacon and Aleinikoff, 1985; Aleinikoff and others, 1986).

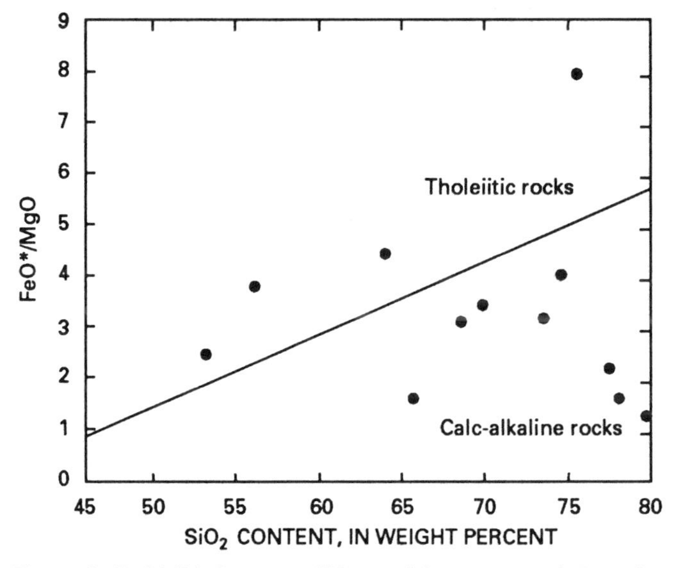

Figure 8. FeO^*/MgO versus SiO_2 (weight percent) of Devonian metavolcanic rocks from the Jarvis Creek Glacier and Hayes Glacier subterranes of the southern Yukon–Tanana terrane, eastern Alaska Range. Tholeiitic-calc-alkaline boundary after Miyashiro (1974).

WRANGELLIA TERRANE

F. Barker, W. J. Nokleberg, G. Plafker, and W. P. Leeman

The Wrangellia terrane, also known as "Wrangellia," was defined by Jones and others (1977). It is exposed in the Wrangell Mountains, eastern Alaska Range, St. Elias Mountains, southeastern Alaska at Chichagof and Baranof Islands, and near Haines, Queen Charlotte Islands, and Vancouver Island. This terrane consists largely of Devonian (Vancouver Island) and late Paleozoic (Alaska) intraoceanic island arcs, Permian limestone, Triassic back-arc-basin basalt, and Triassic limestone (see, e.g., Richter, 1976; MacKevett, 1978; Nokleberg and others, 1985; Brandon and others, 1986; Barker and others, 1989). In the Mount Hayes Quadrangle, Nokleberg and others (1985) have divided Wrangellia into the Slana River subterrane, which is like the type-Wrangellia terrane of the Wrangell Mountains and Nabesna Quadrangle; and into the Tangle subterrane, which is a deeper water facies of type-Wrangellia terrane.

Jurassic and (or) Cretaceous flysch and Early Cretaceous island-arc volcanic rocks formed on the northeast flank of Wrangellia prior to emplacement onto North America. These rocks are termed the Gravina-Nutzotin belt (Berg and others, 1972) and their volcanic rocks are described below. Oceanic volcanic and sedimentary rocks that accreted to the southwestern margin of Wrangellia are discussed in the section (this chapter) below on the Chugach terrane. We next consider the Paleozoic island arc, termed the Skolai arc; and the Triassic arc-rift basalts, the Nikolai Greenstone.

THE SKOLAI ISLAND ARC

The oldest exposed rocks of the Wrangellia terrane in the Wrangell Mountains, eastern Alaska Range, and Kluane Range, Yukon, are the remnants of an intraoceanic island arc of Late Mississippian or Early Pennsylvanian to Early Permian age. This island arc, first recognized by Bond (1973) and Richter and Jones (1973a), is informally termed the Skolai island arc. It consists of the Station Creek Formation of the Skolai Group of the Wrangell Mountains (Smith and MacKevett, 1970; MacKevett, 1978); the Tetelna Volcanics, and the Slana Spur Formation of the eastern Alaska Range (Bond, 1973, 1976; Richter, 1976; Jones and others, 1977; Nokleberg and others, 1985), and the Strelna Metamorphics of Plafker and others (1989b) found in the Chitina Valley region and along the southern margin of this part of

Wrangellia. The stratigraphic relations of these sparsely fossiliferous formations are not entirely clear—as one might expect of island-arc rocks. The Station Creek Formation probably is correlative with most of the Tetelna Volcanics; the Tetelna is overlain by the Slana Spur Formation in western Nabesna Quadrangle (Richter, 1976); but in the Mount Hayes Quadrangle to the west relations of the Tetelna Volcanics and Slana Spur Formation are poorly known (Nokleberg and others, 1985). The Strelna Metamorphics of Plafker and others (1989b) is described below.

Rocks of the Skolai island arc are the oldest exposed in this part of Wrangellia, with the possible exceptions of the quartz-feldspar schist of the Nabesna Quadrangle (Richter, 1976) and the orthogneiss and gabbro of the eastern Wrangell Mountains (MacKevett, 1978). The diorite-gabbro-tonalite complex of Richter (1976) that intruded the Tetelna Volcanics yielded four $^{206}Pb/^{238}U$ ages on zircon of 290 to 316 Ma (Barker and Stern, 1986; Beard and Barker, 1989), indicating that Tetelna volcanism, at least, may have commenced in latest Mississippian (e.g., prior to 330 Ma) or earliest Pennsylvanian time.

Like the axial regions of other island arcs, the lower part of the Station Creek Formation and the Tetelna Volcanics consists largely of massive and pillowed to brecciated basalt and andesite flows 3 m to more than 100 m thick. Intercalated mud flows (lahars), pyroclastic breccias, fine- to coarse-grained volcaniclastic to limy sedimentary rocks, and lapilli tuffs also are present. Exposed thicknesses range from approximately 420 to 1,400 m (Bond, 1973, 1976; Richter, 1976; MacKevett, 1978; Nokleberg and others, 1985). The upper part of the Station Creek and the Slana Spur Formations consist of interlayered silty to conglomeratic volcaniclastic to calcareous sedimentary rocks, lapilli tuff, pyroclastic breccias, and basaltic to dacitic flows (Nokleberg and others, 1982). Thickness of these strata ranges from less than 300 to 2,300 m (Bond, 1973; Richter, 1976). They were metamorphosed, probably during accretion in mid-Cretaceous time, to prehnite-pumpellyite or greenschist facies, and locally to hornfels or amphibolite facies. These rocks, however, were penetratively deformed only near large faults.

Tables 4 and 5 (on microfiche) list 52 new analyses of various volcanic and hypabyssal rocks from the McCarthy, Nabesna, and Mount Hayes 1:250,000-scale Quadrangles and from southwestern Yukon. Ten of these are basalt (SiO_2 <53 percent), 20 are andesite (SiO_2 = 53 to 63 percent), seven are dacite (SiO_2 = 63 to 69 percent), and seven are low-K rhyolite. These analyses show much scatter in K_2O, Na_2O, CO_2, Rb, Sr, and other elements—even though samples free of weathered crusts, calcite or epidote veinlets, and amygdules were collected—that resulted from exposure to seawater and (or) metamorphic fluids. On an FeO^*/MgO versus SiO_2 plot (Fig. 9) samples of the arc from the McCarthy and Nabesna Quadrangles show little iron enrichment and are essentially calc-alkaline. Most samples from the Mount Hayes Quadrangle also show a calc-alkaline signature (Fig. 10). These rocks show moderate to pronounced light-REE enrichment, having La abundances 20 to 90 times chondrites. About half the samples show small negative Eu anomalies. REEs show no correlation with SiO_2. A spidergram of average Station Creek and Tetelna basalts compares moderately well with that of average calc-alkaline arc basalt of Pearce (1982), as shown in Figure 11—except that Nb and Ta of these Alaskan rocks show greater abundances. We judge that these Skolai island-arc rocks probably are of Gill's (1981) medium-K orogenic andesite class.

Strelna Metamorphics

Background. The term Strelna Metamorphics of Plafker and others (1989b), refers to the metamorphosed rocks formerly called the Strelna Formation by Moffit (1938), that are exposed along the southern margin of Wrangellia within and adjacent to the Chitina Valley and to the nonplutonic part of the Haley Creek terrane (Winkler and others, 1981) between the Copper River and Richardson Highway in the northern Chugach Mountains.

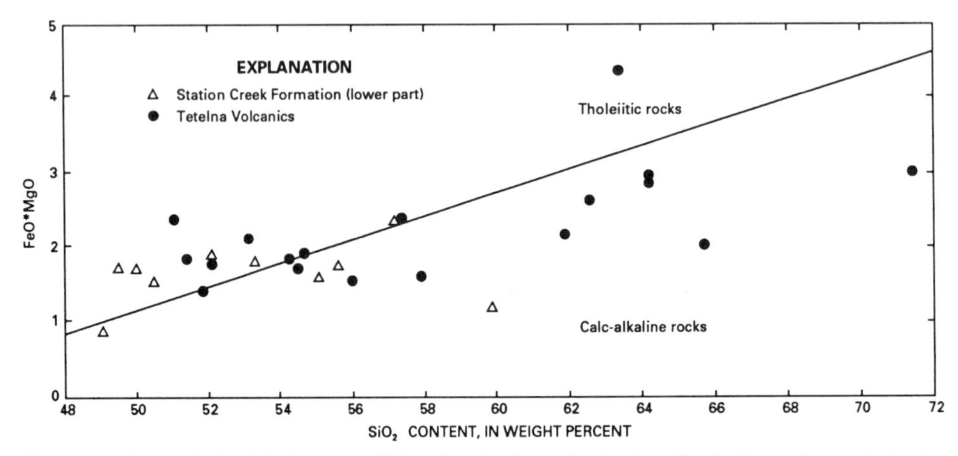

Figure 9. Plot of FeO*/MgO versus SiO_2 of rocks from the Station Creek Formation and Tetelna Volcanics, Wrangellia terrane. Tholeiitic-calc-alkaline boundary after Miyashiro (1974).

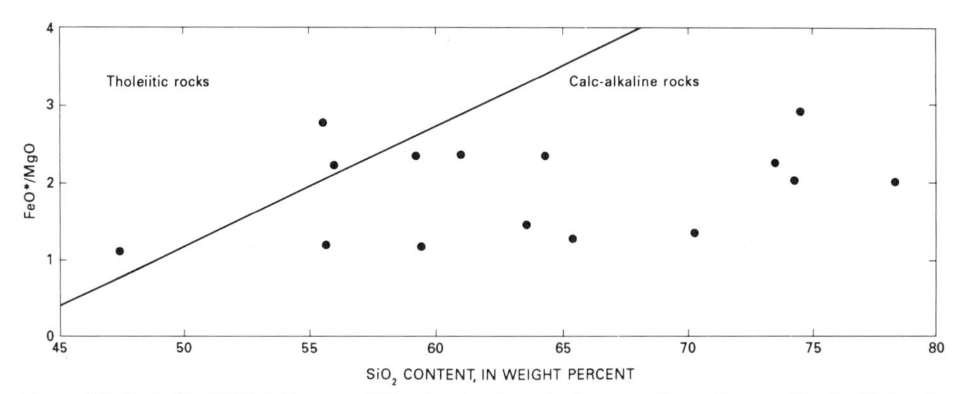

Figure 10. Plot of FeO*/MgO versus SiO_2 of volcanic rocks from the Pennsylvanian Tetelna Volcanics and Pennsylvanian and Permian Slana Spur Formation in the Slana River subterrane of the Wrangellia terrane, eastern Alaska Range. Tholeiitic-calc-alkaline boundary after Miyashiro (1974).

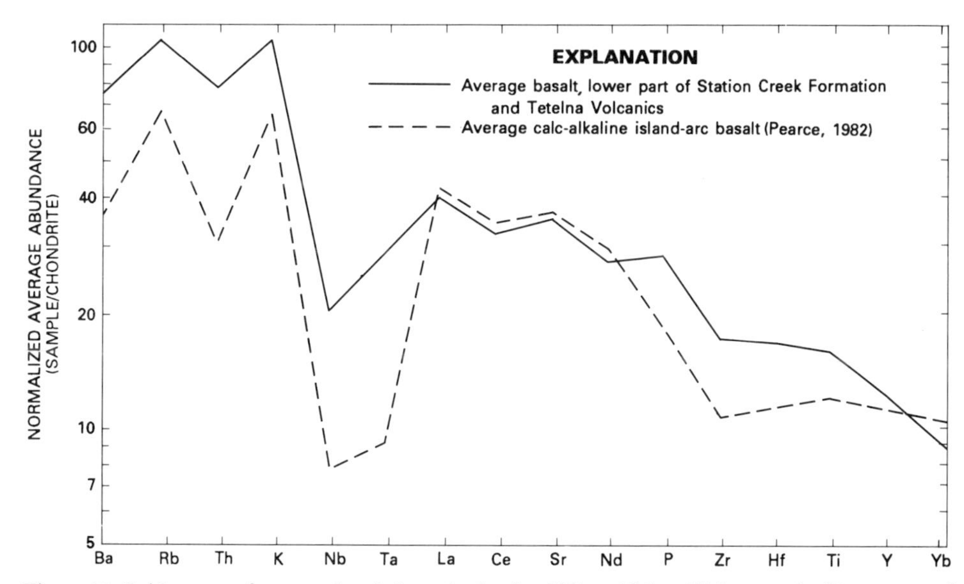

Figure 11. Spidergram of average basalt (samples having SiO_2 = 49.1 to 52.1 percent) of lower part of the Station Creek Formation and the Tetelna Volcanics, Wrangellia terrane, and spidergram of average calc-alkaline island-arc basalt (Pearce, 1982). Normalizing values from Thompson and others (1983).

Correlative rocks extend in a nearly continuous belt southeastward through adjacent parts of Canada (Campbell and Dodds, 1985), and possibly into the Yakutat Bay region (Barker and others, this volume, Plate 13). Within this belt, uplift has exposed deeper levels of basement rocks that consist of penetratively deformed greenschist- and amphibolite-facies metaandesite, metagraywacke, metachert, and many linear units of marble and schistose marble that are at least in part of Early Pennsylvanian age. The metamorphic rocks are intruded by abundant foliated plutonic rocks of the Late Jurassic Chitina Valley batholith (MacKevett, 1978) and an older suite of gneissic plutonic rocks, some of which are of Middle Pennsylvanian age (Aleinikoff and others, 1988). Geochemical data are available for only two samples of metabasite from the Strelna Metamorphics form the vicinity of the Copper River in the northern Chugach Mountains and two from the inferred correlative rocks near Yakutat Bay (Table 6, on microfiche).

Copper River area. Two analyzed samples of greenschist derived from volcanic rocks consist mainly of albite, quartz, epidote, and chlorite ± actinolite (samples 84APR213 and 84ANK155). Their mode of occurrence indicates that the volcanic rocks included massive flow units and thinner bedded units that were probably tuff or breccia intercalated with sedimentary rocks. Geochemically, they are tholeiites (50.1 to 52.1 percent SiO_2, 14.0 to 15.3 percent Al_2O_3, 5.3 to 8.7 percent MgO, 1.51 to 1.88 percent TiO_2) with a slight negative REE slope, La 9 to

12 × chondrites, and Lu 15 to 19 × chondrites. Though not plotted here, both samples lie within the tholeiite field on a plot of FeO^*/MgO versus SiO_2; they plot on a Th-Hf/3-Ta diagram as N-type MORB. On a Ti-Zr-Y diagram and similar discrimination diagrams, these samples plot in low-K tholeiite to ocean-floor basalt fields.

Yakutat area. In the Yakutat area our two samples of the Strelna(?) Metamorphics are tectonized, banded amphibolites that are cut by two generations of in-situ partial melt of trondhjemite. Though both samples show relatively high Al_2O_3 (17.5 to 18.0 percent), one is magnesian and low in incompatible elements (MgO = 8.25 percent, La = 5.3 ppm, Nb < 5 ppm, Zr = 107 ppm), and the other is of typical MgO and is enriched in incompatible elements (MgO = 5.65 percent, La = 31 ppm, Nb = 28 ppm, Zr = 217 ppm).

Interpretation. The Strelna Metamorphics are interpreted as representing a probable marine sequence of mixed quartzofeldspathic, pelitic, calcareous, and cherty rocks and variable amounts of mafic volcanic rocks that accumulated along the oceanward flank of a volcanic arc (Plafker and others, 1989b). These rocks are tentatively correlated with the less-metamorphosed and dominantly andesitic sequence that comprises the lower part of the Skolai arc (Station Creek Formation), although direct linkage with that unit cannot be demonstrated because of intervening faults.

NIKOLAI GREENSTONE: AN ARC-RIFT THOLEIITE

The Middle and (or) Upper Triassic Nikolai Greenstone is found in the Wrangell Mountains, eastern Alaska Range, and the Kluane Ranges of Yukon. Two occurrences of probable correlatives in southeastern Alaska are mentioned by Gehrels (this chapter). The Nikolai is a prominent formation of the Wrangellia section. It consists of 1,000 to 3,000 m of ferrotholeiite and local, intercalated volcaniclastic sedimentary rocks. The Nikolai unconformably overlies rocks of the Skolai arc; in many areas it conformably overlies approximately 100 m of unnamed Middle Triassic (Ladinian) chert, siltstone, and shale that bear a distinctive *Daonella* bivalve fauna; and it is disconformably overlain by the upper Karnian Chitistone Limestone (see summary of Jones and others, 1977; and Nokleberg and others, this volume, Chapter 10) or by younger rocks. The Nikolai typically shows a basal volcanic conglomerate approximately 70 m thick; but locally, in the McCarthy and Nabesna Quadrangles, interlayered pillow basalt and argillite are found (Richter, 1976; MacKevett, 1978). In these two quadrangles, this formation is largely aa and pahoehoe flows 5 cm to 20 m thick, which show amygdaloidal and typically oxidized tops. The Nikolai Greenstone of the Tangle subterrane, however, consists of a lower member of largely pillow basalt and an upper member of subaerial flows. Sills, dikes, and plugs of cogenetic gabbro and diabase are emplaced in or cut Nikolai flows; furthermore, Nokleberg and others (1982, 1985) also report that mafic to ultramafic sills containing abundant cumulus phases are found in, or subjacent to, the Nikolai Greenstone in the Mount Hayes Quadrangle. The Nikolai is correlative with the Karmutsen Formation of Vancouver Island and Queen Charlotte Islands (summary description by Barker and others, 1989).

Basalts of this formation typically are dark green-gray where fresh and brown, purple, or reddish where oxidized. They are massive and of ophitic texture; do not show penetrative deformation except near major faults; typically contain saussuritized labradorite phenocrysts and fresh or altered clinopyroxene phenocrysts; show serpentinized relicts of olivine; have altered matrices of chlorite, epidote, serpentine, Fe-Ti oxides, actinolite, and other minerals; and contain amygdules of quartz, calcite, chlorite, epidote, pumpellyite, prehnite, zeolites, and copper. Metamorphism mostly attained greenschist facies, but many samples show preprinting or overprinting of prehnite-pumpellyite or (and) zeolite facies (Richter, 1976; MacKevett, 1978; Nokleberg and others, 1982, 1985).

We present nine new major-minor-element analysis of the Nikolai Greenstone from the McCarthy and Nabesna Quadrangles and Kluane Ranges in Table 7 (on microfiche), and 18 from the Tangle subterrane of the Mount Hayes Quadrangle in Table 8 (on microfiche). These basalts, like those of other accreted oceanic rocks, show perturbed abundances of K_2O, Rb, Sr, Ba, and other mobile elements, but REEs (except Ce in some submarine rocks), high-field-strength elements (Sc, Ti, Y, Zr, and Nb), and transition elements (Cr, Mn, Fe, Co, and Ni) were largely or wholly immobile. Basalts of the Nikolai are ferruginous, most show FeO^*/MgO ratios greater than 1.5, and so are tholeiitic (see Miyashiro's discriminant, 1974). These FeO^*/MgO ratios are so high, except perhaps those of samples 82-25 and 78ANK155A, as to preclude origin directly from the mantle—that is, fractionation occurred prior to emplacement (see, e.g., Crawford and others, 1987). However, these basalts are not chemically homogeneous: like the correlative Karmutsen basalts of southern Wrangellia (Barker and others, 1989), they are of three types (Tables 7 and 8, on microfiche): (1) light-REE-depleted arc-type basalt, showing relatively low abundances of Ti, P, and Zr, and high abundances of Mg and Ni, as in sample 82-25; (2) a heavy-REE-depleted type, also low in Zr, as in samples 82-10 and 82-24; and (3) abundant, light-REE-enriched basalt, showing La at 20 to 30 times chondrites and heavy REEs at 10 to 19 times chondrites, and relatively high Ti, P, and Zr—compositions like those of back-arc-basin basalt (BABB) or mid-ocean-ridge tholeiite (MORB).

Three samples of the Nikolai Greenstone (Table 7, on microfiche) were analyzed for Sr and Nd isotopic ratios. These results, as calculated for a crystallization age of 220 Ma, are as follows:

Sample No.	$^{87}Sr/^{86}Sr$	$^{143}Nd/^{144}Nd$	ϵNd_o
82-9	0.70412	0.51264	+5.5
82-21	0.70369	0.51262	+5.2
82-25	0.70401	0.51255	+3.9

The $^{87}Sr/^{86}Sr$ ratios may be slightly higher than the primary or magmatic values, owing to interaction of these rocks with seawater or metamorphic fluids containing Sr of higher 87–86 ratios. Because Nd is relatively immobile in postmagmatic processes, the ϵNd_0 values probably are magmatic. These Sr and Nd initial ratios are close to those of the mantle array. We note that ϵNd_0 of +3.9 of sample 82-25, which is of the light-REE-depleted type of basalt, differs significantly from ϵNd_0 of the other two samples—indicating that 82-25 came from different mantle than did the other, light-REE-enriched samples.

By analogy with the results of Barker and others (1989) on the origin of basalts of the Karmutsen Formation, we suggest that the Nikolai Greenstone originated by: (1) mixing of residual island-arc tholeiitic (IAT) magma, as exemplified by sample 82-25 (Table 7, on microfiche), and having ϵNd_0 of +3.9 (as mentioned above), and a rift magma, as exemplified by the melts that were parental to samples 82-10 and 82-24 (Table 7, on microfiche), to produce a hybrid liquid; an (2) major fractionation of olivine, pyroxene(s), and plagioclase from this hybrid liquid to give the relatively abundant light-REE-enriched basalt (of ϵNd_0 of +5.2 and 5.5 in two samples). Tectonic events causing extrusion of the 1- to 3-km-thick Nikolai pile, by further analogy with the Karmutsen Formation, were (1) rifting of the Skolai arc, (2) formation of a new frontal arc, (3) formation of a back-arc basin or rift and mixing and emplacement of arc-type and rift-type magmas as suggested above. Vallier (1986) has suggested the possibility that the frontal arc of this model, coeval with Nikolai and Karmutsen basalts, is found in the island-arc volcanic rocks of the Wild Sheep Creek and Doyle Creek Formations of the Wallowa terrane, Idaho and Oregon.

GRAVINA-NUTZOTIN BELT

C. M. Rubin and F. Barker

DEFINITION

The Gravina-Nutzotin belt, named and defined by Berg and others (1972), is a distinctive lithostratigraphic package of rocks that formed on the margins of the Alexander and Wrangellia terranes in Late Jurassic and Early Cretaceous time. The Gravina-Nutzotin belt exends from Ketchikan through southeastern Alaska to the vicinity of Haines, intermittently through the southwest Yukon Territory (Kindle, 1953; Muller, 1967), and into the easternmost Alaska Range (Richter and Jones, 1973b; Richter, 1976; MacKevett, 1978).

GRAVINA-NUTZOTIN BELT IN SOUTHEASTERN ALASKA

C. M. Rubin

Background

In southeastern Alaska the Gravina-Nutzotin belt consists of marine basinal volcanic and sedimentary strata that range in age from Oxfordian (Late Jurassic) to early Albian (Early Cretaceous). Age constraints, however, are scant. On Annette and Gravina Islands (southern southeastern Alaska), Gravina-Nutzotin belt volcanic and hypabyssal rocks locally intrude and appear to unconformably overlie Triassic strata of the Alexander terrane (Berg, 1972; Berg and others, 1988; Rubin and Saleeby, 1987b; Gehrels and Berg, this volume), and on Kupreanof and southern Admiralty Islands the Gravina-Nutzotin belt appears to lie disconformably on Triassic rocks of the Alexander terrane (Muffler, 1967; Loney, 1964). On Chilkat Peninsula, possible Gravina-Nutzotin belt flysch disconformably overlies upper Norian (Upper Triassic) chert of the Taku terrane (Plafker and others, 1989a). Elsewhere the boundary is marked by either a high-angle fault or, as on the Cleveland Peninsula (Ketchikan area), a northeast-dipping thrust fault (Rubin and Saleeby, 1987a). As the eastern boundary of the Gravina-Nutzotin belt is approached, there is a general increase in deformation and in grade of metamorphism; however, the nature of the boundary between the Gravina-Nutzotin belt and most parts of the Taku terrane is not well understood. At Berners Bay (55 km north of Juneau), Gravina-Nutzotin belt strata lie unconformably on metamorphic rocks that have been assigned to the Taku terrane (Redman, 1986) because they are on strike with the Taku rocks of Chilkat Peninsula to the north. However, these rocks have not been dated, and the correlations should be considered tentative.

Ketchikan area

The volcanic part of the Gravina-Nutzotin belt consists of phyric to aphyric, massive to pillowed, basalt and basaltic andesite lavas and silicic flows, which are locally known as the Gravina Island Formation (Berg, 1973; Berg and others, 1988; Rubin and Saleeby, 1987b). Metamorphic grade ranges from greenschist to amphibolite facies; deformation and metamorphic grade increase toward the east and northeast. The mafic volcanic rocks consist of flows, pillow breccia, mud-flow deposits or lahars, tuff, and pyroclastic breccia. This submarine volcanic sequence has a structural thickness of approximately 8 km. The mafic flows typically are dark gray and contain abundant phenocrysts of clinopyroxene and minor ones of hornblende and plagioclase. These rocks show minor to pervasive alteration of both the clinopyroxene and hornblende phenocrysts and the fine-grained matrix to chlorite and white mica. Augite porphyry dikes and sills intrude the volcanic section and are probably cogenetic with the extrusive rocks. Hornblendite xenoliths are common in the volcanic flows and augite porphyries. Elongate bodies of porphyritic diorite and quartz diorite of varied texture intrude the volcanic rocks and probably formed as subvolcanic feeders. The silicic volcanic rocks consist of quartz-phyric dacite flows and tuff. The average of selected elemental abundances of six basalts of the Gravina Island Formation gives SiO_2 at 49.1 percent, Al_2O_3 at 16.2 percent, total Fe as Fe_2O_3 at 9.6 percent, TiO_2 at 0.59 percent, P_2O_5 at 0.12 percent, La at 5.7 ppm, Zr at 51 ppm, Hf at 1.02 ppm, and Nb at 10 ppm.

The sedimentary part of the Gravina-Nutzotin belt consists of interbedded tuffaceous turbidites, argillaceous turbidites, and minor impure carbonate, conglomerate, lithic sandstone, and breccia. The turbidites are typically dark to pale gray fining-upward sequences of tuffaceous argillite, lithic sandstone, and mudstone. Clasts from the channel-fill deposits consist of leucoquartz diorite to leucodiorite, argillite, and volcanic clasts in an argillaceous matrix.

Admiralty Island–Lynn Canal area

To the north, the volcanic parts of the Gravina-Nutzotin belt belong to the Douglas Island Volcanics and Brothers Volcanics and the correlative Bridget Cove Volcanics of Irvine (1973; Barker, 1957; Ford and Brew, 1973, 1977; Knopf, 1911; Lathram and others, 1965). These rocks consist of massive to pillowed basalt and andesite lavas, breccia, tuff, and volcanic mudflows. The mafic rocks are typically dark greenish gray and contain abundant phenocrysts of diopsidic augite and minor phenocrysts of plagioclase and magnetite (Irvine, 1973). The existence of olivine phenocrysts is inferred from chlorite pseudomorphs (Irvine, 1973). These rocks show minor to pervasive alteration of both clinopyroxene and matrix to chlorite, albite, calcite, and epidote. The exposed structural thickness of this section is approximately 9 km (Lathram and others, 1965; Loney, 1964). The sedimentary section of the Gravina-Nutzotin belt in this area consists of argillite and massive wacke with minor conglomerate; its estimated structural thickness is approximately 900 to 2,400 m (Loney, 1964).

The average elemental abundances of mafic metavolcanic strata of the Gravina-Nutzotin belt exposed on Douglas Island (immediately west of Juneau) are SiO_2 at 46.9 percent, Al_2O_3 at 13.1 percent, total Fe as Fe_2O_3 at 10.7 percent, TiO_2 at 0.74 percent, and P_2O_5 at 0.32 percent (Ford and Brew, 1987). Correlative metavolcanic strata, exposed at Berners Bay (Lynn Canal), have similar elemental abundances: SiO_2 at 48.2 percent, Al_2O_3 at 13.1 percent, total Fe as Fe_2O_3 at 10.7 percent, TiO_2 at 0.7 percent, and P_2O_5 at 0.41 percent (Ford and Brew, 1987). These geochemical data indicate a basalt to a high K-basalt protolith for the mafic metavolcanic rocks exposed on Douglas Island, Glass Peninsula, and Berners Bay.

The Gravina-Nutzotin belt volcanic-sedimentary package probably represents marine pyroclastic and volcaniclastic deposition within an arc-related basinal environment. The volcanic arc complex was constructed upon the Alexander terrane, and perhaps also on the Taku terrane during Late Jurassic to Early Cretaceous(?) time. Deformation probably occurred between early Albian time and 91 Ma, because of the presence of Albian ammonites in Gravina-Nutzotin belt strata on Etolin Island and crosscutting epidote-bearing plutons that yield Pb-U zircon ages of 91 to 94 Ma (Berg and others, 1972; Rubin and Saleeby, 1987b).

GRAVINA-NUTZOTIN BELT OF EASTERN ALASKA RANGE

F. Barker

In the Nutzotin Mountains of the eastern Alaska Range and in the adjacent Wrangell Mountains (Richter and Jones, 1973b; Richter, 1976) the lower, unnamed, sedimentary part of the Gravina-Nutzotin belt comprises predominant, shallow- to deep-water marine argillite, graywacke, and mudstone, and of minor conglomerate, impure limestone, local nonmarine sandstone, and other rocks. This section, whose base is not exposed and which lies in fault contact with older rocks, ranges in age from Tithonian (Late Jurassic) to Valanginian (Early Cretaceous) and is more than 3,000 m thick. This turbidite package, derived from Wrangellia (as known from paleocurrent directions, D. L. Jones, oral communication, 1983) and deposited on the margin of that oceanic complex, grades upward into the Chisana Formation (Richter and Jones, 1973b). The lowermost 600 m of the Chisana section consists of interbedded submarine lahars, basalt and andesite flows, tuff, volcaniclastic sedimentary rocks, and marine argillite, graywacke, and conglomerate. The upper part of the Chisana Formation, which is more than 2,500 m thick, consists of interlayered basalt and andesite flows, lahars, pyroclastic breccia, tuff, and volcaniclastic graywacke and conglomerate. The rocks are partly marine and partly continental. The Chisana is unconformably overlain by Upper Cretaceous continental sedimentary rocks (Richter and Jones, 1973b).

The flows, sills, and dikes in the Chisana Formation typically are dark greenish gray. They contain abundant phenocrysts of plagioclase (mostly labradorite) and minor ones of clinopyroxene and hornblende. These rocks show minor to pervasive alteration of both phenocrysts and their fine-grained matrices to albite, chlorite, calcite, and smectite. Lahars in the lowermost 600 m of the section are as thick as 100 m, contain blocks as large as 20 m, and are proximal to their volcanic source. Sedimentary xenoliths are common in the extrusive rocks. Pillow lava also is found in the lowermost 600 m of the marine section. A stock of hornblende diorite, heavily altered to albite, calcite, and chlorite, cuts the Chisana volcanic rocks and is believed to be cogenetic.

Table 9 (on microfiche) gives new major-minor-element analyses of ten basalts and six andesites of the Chisana Formation and of the associated diorite. The Chisana samples are Al_2O_3-rich, averaging 18.6 percent; even the three samples showing 20.6 to 22.1 Al_2O_3 have Eu/Eu^* ratios of about 1.07 to 1.2, so accumulation of plagioclase in these is only moderate. Because of disturbance of mobile elements (e.g., K_2O ranges from 0.13 to 1.82 percent), we use the FeO^*/MgO versus SiO_2 diagram (Miyashiro, 1974) to classify these rocks as tholeiitic transitional to calc-alkaline (Fig. 12). In Figure 13, spidergrams show average basalt of the Chisana plots—except for high Nb, P, and Sr—between Pearce's (1982) average tholeiitic and calc-alkaline island-arc basalts. The Chisana suite shows mild to pronounced light-REE enrichment, with La at 13 to 75 times chondrites. Lu is 6.5 to 12 times chondrites; nine samples show small to moderate positive Eu anomalies; and REE abundances do not correlate with SiO_2.

These basalts and andesites of the Chisana Formation form what Barker (1987) called the Chisana arc, a short-lived feature that formed on the northeast margin of the so-called (composite) Wrangellia-Alexander superterrane. This arc apparently formed

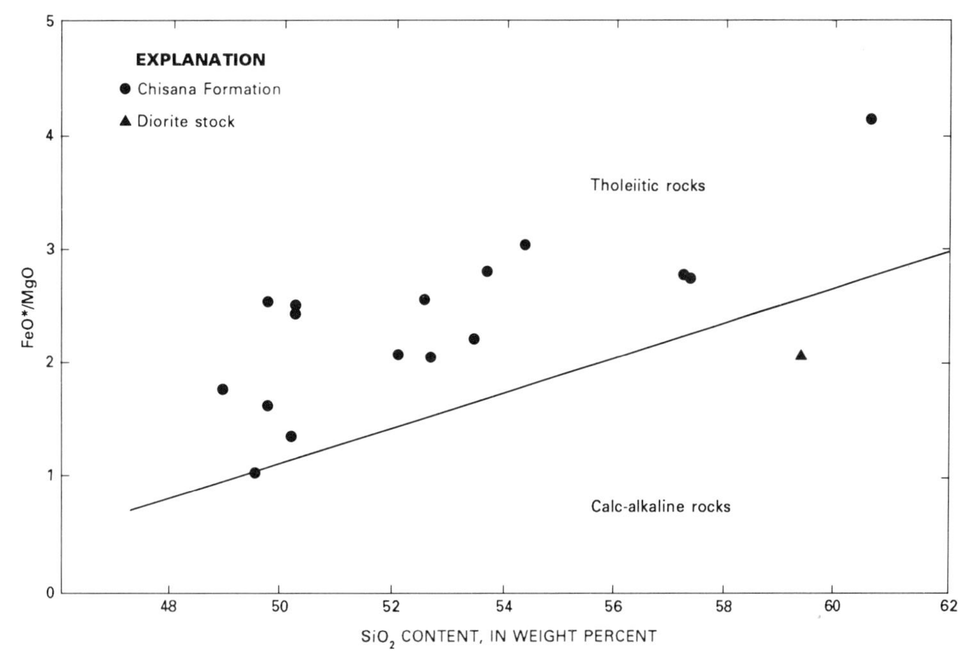

Figure 12. Plot of FeO*/MgO versus SiO_2 of rocks of the Chisana Formation and associated diorite stock, Gravina-Nutzotin belt, Nutzotin Mountains. Tholeiitic-calc-alkaline reference line from Miyashiro (1974).

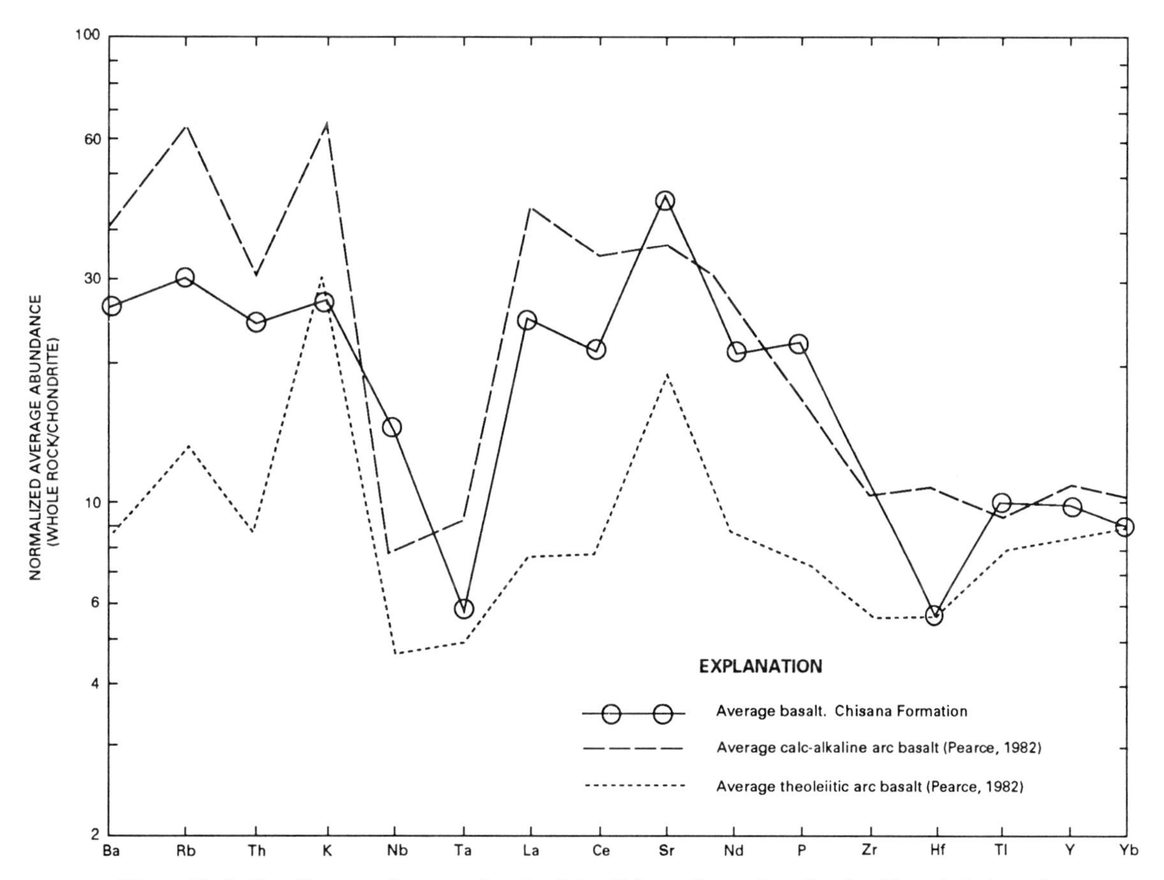

Figure 13. Spider diagram of average basalt of the Chisana Formation, Gravina-Nutzotin belt, and average tholeiitic basalt and calc-alkaline basalt of Pearce (1982).

in response to subduction, as this superterrane moved toward, and finally impinged on North America.

PRE–LATE CRETACEOUS VOLCANIC ROCKS OF THE BRISTOL BAY REGION, SOUTHWESTERN ALASKA

S. E. Box

Several terranes underlying broad areas of western Alaska have a major component of volcanic rocks of pre–Late Cretaceous age. They are the Koyukuk, Tozitna, Innoko, Togiak, Goodnews, Tikchik, and Nyack terranes (Jones and others, 1987; Silberling and others, this volume; and Patton and others, this volume, Chapter 7). None of these terranes shows evidence of a Precambrian continental basement or shows any associated quartzofeldspathic sedimentary rocks of continental derivation until late Early Cretaceous time. Their individual geologic histories suggest that these terranes originated by igneous processes in an oceanic-plate setting and were structurally thickened both during oceanic amalgamation events and during Early Cretaceous accretion against the older continental rocks of interior Alaska (Box, 1985a, b; Patton and others, this volume, Chapter 7; Decker and others, this volume). The geology and geochemistry of volcanic rocks from the northern part of this region (Koyukuk and Tozitna terranes) are discussed by Patton and others in this volume and will not be discussed further here. In this report the petrology, geochemistry, and stratigraphic context of the volcanic rocks of the southern part of western Alaska (Togiak, Goodnews, Tikchik, and Nyack terranes) are summarized by terrane. A broader interpretation of the tectonic histories of these latter terranes is given in Box (1985a, b) and Decker and others (this volume).

The Togiak terrane can be broadly divided into a lower unit of Late Triassic ophiolitic rocks depositionally overlain by an upper unit of Lower Jurassic through Lower Cretaceous volcanic rocks. The partly disrupted ophiolitic section consists of a 2-km-thick pillow basalt sequence that grades downward by increasing intrusion into a sheeted diabase sill complex. The sill complex is faulted against unlayered gabbroic intrusive rocks that grade in some places into layered gabbros and are faulted over serpentinized harzburgites and dunites. The pillow basalts are interbedded with red radiolarian cherts of Late Triassic age. The pillow basalts grade upward through a thick sequence of basaltic breccia and epiclastic strata into the upper unit of Lower Jurassic through Lower Cretaceous volcanic, pyroclastic, and epiclastic strata, which includes indications of shallow-marine deposition. Important unconformities of late Early Jurassic and latest Jurassic age occur within the upper unit.

The petrology and geochemistry of the two broad units of the Togiak terrane are distinct. Amygdaloidal pillow lavas and diabase sills of the lower unit are aphyric to sparsely porphyritic with a subophitic texture of intergrown plagioclase and clinopyroxene. Rocks are generally basaltic but range upward in SiO_2 content to basaltic andesites. TiO_2 contents are generally above 1.5 percent and K_2O contents are low (0.04 to 0.73 percent), and the rocks plot in the tholeiitic field on an AFM diagram. The rocks generally plot in a MORB-OIB field on Ba versus Nb and K_2O versus Nb plots (Fig. 13). Because of their relatively high TiO_2 and Nb contents, the basalts of the lower unit are more like lavas of ocean islands, seamounts, or elevated parts of mid-ocean ridges ("E-MORB") than typical lavas of the deeper parts of the mid-ocean ridges ("N-MORB"; Sun and others, 1979; BVSP, 1981; Thompson and others, 1984). In contrast, the lavas of the upper unit are strongly plagioclase-phyric and contain minor clinopyroxene, hornblende, and orthopyroxene, and rare biotite phenocrysts. They range from basalts to high-silica andesites; andesites are predominant. TiO_2 contents are moderate (0.65 to 1.6 percent), K_2O contents are moderate (0.5 to 2.2 percent), and the rocks fall in both the tholeiitic and calc-alkaline fields of the AFM diagram. The rocks show strong enrichment of alkaline elements relative to high-field-strength elements (e.g., Ba versus Nb and K_2O versus Nb plots in Fig. 14), the primary distinguishing characteristic of convergent-margin volcanism (Gill, 1981; BVSP, 1981; Thompson and others, 1984). The igneous history of the Togiak terrane records a Late Triassic (and older?) episode of nonarc, ocean-island volcanism succeeded by an Early Jurassic to Early Cretaceous period of island-arc volcanism.

The Goodnews terrane is a structurally complex assemblage of pillowed basalt flows, mafic and ultramafic rocks, cherts, limestones, and fine tuffaceous sedimentary rocks ranging from Late Devonian to Late Jurassic in age. The terrane was amalgamated against the Togiak terrane in two stages (late Early Jurassic and latest Jurassic), prior to Early Cretaceous accretion against the Precambrian continental rocks of the Kilbuck terrane (Box, 1985a, b; Decker and others, this volume). Volcanic and plutonic rocks of the Togiak terrane overlapped rocks of Goodnews terrane rocks after each amalgamation event. Preamalgamation volcanic flows of Permian age are widespread in the Goodnews terrane (Hoare and Coonrad, 1978); however, the bulk of these volcanic rocks are only known to be of pre–Early Cretaceous age.

Our geochemical database for lavas of the Goodnews terrane consists of major-element analyses from six samples, and trace-element data (XRF analyses only) from four samples. Flows are generally pillowed, and in some places are vesicular. They are generally aphyric or weakly prophyritic and have intergrown plagioclase and clinopyroxene. All flows are basalts, have moderate TiO_2 contents (0.9 to 1.5 percent), low K_2O contents (0.01 to 0.17), and plot in the tholeiitic field on an AFM diagram. On Ba versus Nb and K_2O versus Nb diagrams (Fig. 14), these rocks plot in the MORB-OIB field. Given the limited database, I feel confident only in coming to the general conclusion that these rocks were erupted in a nonarc oceanic-plate setting.

No geochemical data are presently available for volcanic rocks from the Nyack or Tikchik terranes. Petrographic and lithologic data from the Nyack terrane indicate that it is generally similar to the Togiak terrane in the predominance of plagioclase phyric flows and abundance of interbedded volcaniclastic strata of Middle and Late Jurassic age. The general lithologic assemblage of the Tikchik terrane—a structurally complex assemblage of basalt, chert, and limestone—resembles that of the Goodnews terrane. Tikchik flows consist of scattered occurrences as well as a thick section of mafic flows and volcanic breccias that are bracketed in time by Permian and Triassic calcareous strata. This thick section includes both aphyric and plagioclase and (or) clinopyroxene porphyries. We infer that Tikchik lavas represent nonarc, oceanic-plate volcanism by analogy with flows of the Goodnews terrane.

PENINSULAR TERRANE

TALKEETNA FORMATION

F. Barker and J. S. Kelley

The Peninsular terrane of southern Alaska contains a remnant of an intraoceanic island arc (A. Grantz, unpublished data, 1961; Moore and Connelly, 1977; Barker and Grantz, 1982)—termed the Talkeetna island arc—that now is exposed as

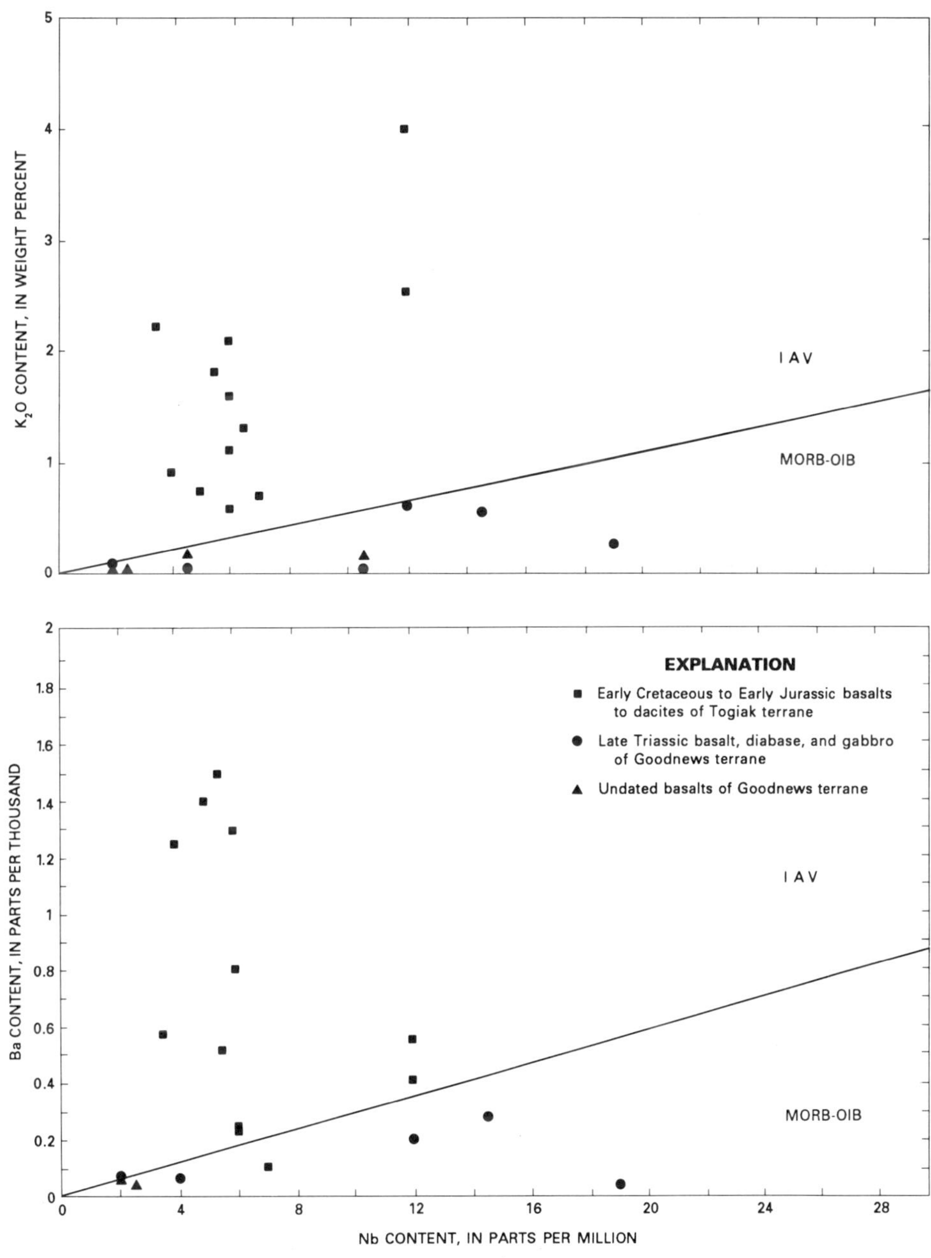

Figure 14. Plots of K_2O and Ba against Nb for samples from the Togiak and Goodnews terranes of southwestern Alaska. Line separating island-arc volcanic rocks (IAV) from mid-ocean-ridge basalt (MORB) and ocean-island basalt (OIB) from Thompson and others (1984) for the K_2O versus Nb, and from Gill (1981) for Ba versus Nb.

the Talkeetna Formation and related plutonic rocks. The Talkeetna Formation (see, e.g., Martin, 1926; Grantz, 1960a, b; Imlay and Detterman, 1973; Detterman and Hartsock, 1966; Detterman and Reed, 1980; Winkler and others, 1981) is exposed in the Talkeetna Mountains, the northern Chugach Mountains, and in the Alaska-Aleutian Range west of Cook Inlet; is of Early Jurassic age (early Sinemurian or older to late Toarcian in the type area); is at least 2,000 m and probably more than 3,000 m thick; and consists of marine and nonmarine pyroclastic breccia, lahars, tuff, silty to cobbly graywacke, flows and sills of basalt and andesite, and other rocks. Older rocks have not been recognized in the Talkeetna Mountains but may be present.

Older parts of the Talkeetna island arc, however, are found well to the southwest of the type area. Martin and others (1915) and Kelley (1980, 1984, 1985) report that 4,000 to 5,300 m of lower Hettangian to lower Sinemurian (earliest Jurassic) rocks are found on Kenai Peninsula between Port Graham and Seldovia. Deep-water marine andesitic to dacitic pyroclastic flows, volcaniclastic turbidites, tuffs, and other rocks range upward to shallow marine and nonmarine volcaniclastic sedimentary rocks and thin beds of coquina and coal. Informally termed the Pogibshi sequence by Kelley (1985), these rocks consist of lower offlap and upper onlap deposits. They overlie Norian (Upper Triassic) to Early Jurassic(?) argillite, limestone, and volcaniclastic rocks. Cowan and Boss (1978) report that 600 m of Talkeetna island-arc rocks are exposed southwest of the Kenai Peninsula in the Barren Islands.

The oldest known evidence of activity in the Talkeetna Island arc is found in the Norian rocks of Afognak and Kodiak Islands and the Alaska Peninsula (Moore and Connelly, 1977; Hill, 1979; Detterman and Reed, 1980). In the latter locality Detterman and Reed (1980) named the Cottonwood Bay Greenstone and the Kamishak Formation.

We present ten new analyses of rocks of the Talkeetna Formation from the well-exposed 1,000 m section of Glacial Fan Creek–Gunsight Mountain, eastern Talkeetna Mountains, and four analyses elsewhere in that range (Table 10, on microfiche). The Glacial Fan–Gunsight section is a heterogeneous one consisting of approximately 40 percent massive mud-flow deposits and pyroclastic breccias, approximately 40 percent airfall and water-laid lapilli tuff and fine-grained breccia, approximately 10 percent flows and sills of basalt and andesite, and the remainder of buff to reddish purple silty to cobbly graywacke. Most clasts of the fragmental rocks are plagioclase-phenocrystic basalt and andesite. Original magmatic phases, except for the relict clinopyroxene, have been altered to albite, epidote, chlorite, calcite, serpentine, zeolites, smectite, Fe-Ti oxides, and other phases. The tuffs contain 20 to 50 percent albitized plagioclase phenocrysts and minor clinopyroxene and (or) embayed quartz set in a matrix of highly altered glass and minerals. The flows are also rich in relict phenocrysts of plagioclase. A plug of 125 m diameter at Glacial Fan Creek also was analyzed (no. B80A-19, Table 10, on microfiche). The two most siliceous clasts (Table 10, on microfiche) contain embayed quartz crystals as well as abundant plagioclase.

The samples comprise three dikes, three flows, three water-laid tuffs, three clasts of breccias, and a sill, as well as the plug—so the number of extrusive cycles represented is not known. Abundances of Na, K, Rb, Sr, and Ba show much variability or mobility, as do Ca and Fe in samples B81A-37 and 79AGx-9, resulting from exposure to seawater and (or) metamorphic fluids. Though classification by alkalies cannot be used, the FeO^*/MgO versus SiO_2 (where $FeO^* = FeO+0.9Fe_2O_3$) plot (Fig. 15) of Miyashiro (1974) indicates that the basalt and andesite are tholeiitic, and the more siliceous rocks are calc-alkaline. On a spidergram of 16 elements (Fig. 16), of which 12 are relatively immobile in alteration processes, the average Talkeetna basalt plots between Pearce's (1982) average island-arc tholeiite and calc-alkaline basalt—a position that agrees with the near-transitional result of Figure 15. The low Nb and Ta, and the generally high Sr abundances are typical of island-arc basalt. Rare-earth elements (REEs) show La at approximately 10 to 28 times, and Lu at 7 to 18 times chondritic values. REE abundances do not correlate with SiO_2, which we interpret to mean that these samples are not related by simple crystal-liquid fractionation of pyroxene, plagioclase, and olivine. The moderately light-REE enrichment may mean that the Talkeetna arc is of medium-K type, as defined by Gill (1981). Average abundances of Zr at 66 ppm, Hf at 1.8 ppm, Ni at 22 ppm, and Co at 34 ppm, as well as least-disturbed(?) K_2O contents of 0.78 to 1.17 percent at 52.4 to 56.5 percent SiO_2 (Table 10, on microfiche), fit the medium-K class better than any other.

SHUYAK FORMATION

M. D. Hill

The oldest known evidence of activity of the Talkeetna island arc of the Peninsula terrane is found in the Norian (Upper Triassic) Shuyak Formation of the Afognak and Shuyak Islands (with possible correlatives at Middle Cape on Kodiak Island); at Port Graham on Kenai Peninsula; in unnamed units at Puale Bay on the Alaska Peninsula (Moore and Connelly, 1977; Connelly and Moore, 1977; Connelly, 1978); and in the Cottonwood Bay Greenstone of the Iliamna region, which is overlain by Norian limestone of the Kamishak Formation (Detterman and Reed, 1980). In the Shuyak Formation, *Halobia*-bearing (early to middle Norian) arc-derived volcaniclastic turbidites intruded by diabase sills structurally overlie a thick sequence of pillow basalts, some with interpillow limestone matrix (Connelly, 1978; Hill, 1979).

At Puale Bay, basaltic flows, agglomerate and conglomerate are interbedded with *Monotis*-bearing limestone and siltstone (Moore, 1967; Imlay and Detterman, 1977; Newton, 1983). In contrast, basaltic andesite to andesite cobbles predominate in conglomerates of uncertain age (probably older than Norian) that are exposed on the small islets 6 to 8 km southwest of Cape Kerkurnoi (Hill, 1979). The undated Middle Cape sequence on Kodiak Island tentatively has been correlated with the Shuyak

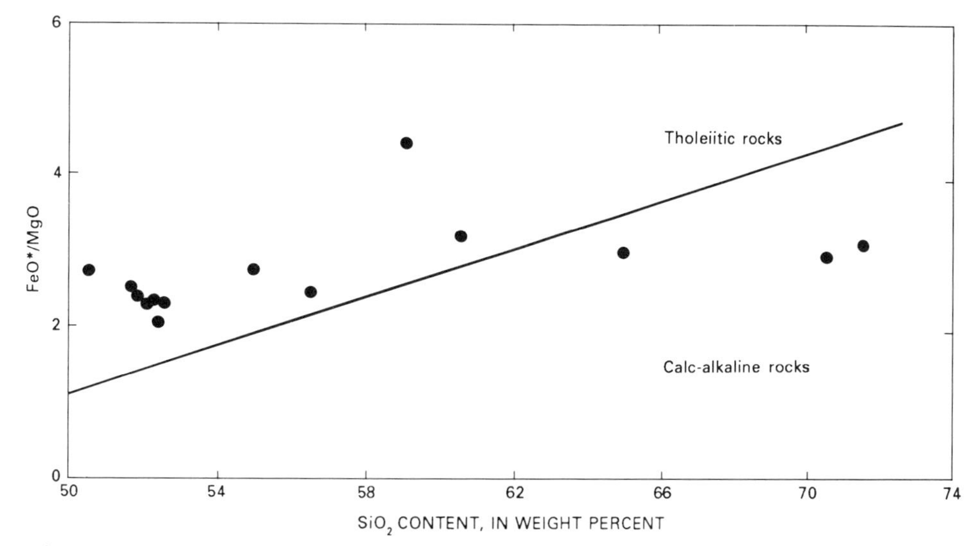

Figure 15. Plot of FeO*/MgO versus SiO_2 of rocks of the Talkeetna Formation, Talkeetna Mountains, Peninsula terrane. Boundary separating tholeiitic rocks from calc-alkaline rocks from Miyashiro (1974).

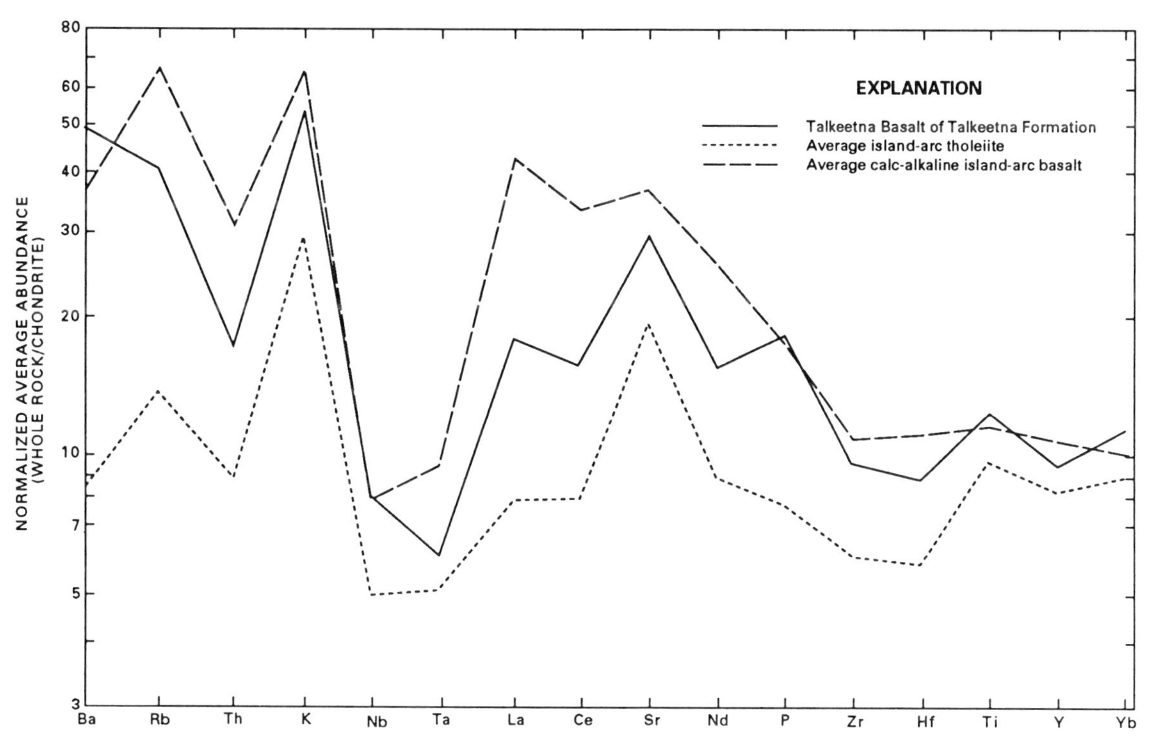

Figure 16. Spidergram of average basalt (samples having SiO_2 = 50.5 to 52.6 percent) of the Talkeetna Formation, Peninsula terrane, and spidergrams of average island-arc tholeiite and calc-alkaline basalt from Pearce (1982).

Formation because of its stratigraphic position (Connelly and Moore, 1977; Connelly, 1978).

Volcanic Geochemistry

The greenstones of the Shuyak Formation are prehnite-pumpellyite–grade metabasalts. Twenty-two analyses are given in Table 11 (on microfiche, with four samples from the Middle Cape sequence on Kodiak Island and six from the unnamed Norian (and older?) strata at Puale Bay. There are two distinct varieties of basalt in the Shuyak Formation—a low-Ti group (0.4 to 1.0 percent TiO_2) and a high-Ti group (1.7 to 3.1 percent TiO_2). The two are not clearly separated in the field, although the high-Ti basalts are more common in the lower part of the sequence. Both groups are subalkaline (high Y/Nb = 2 to 18, and high Zr/Nb = 8 to 29, Table 11A, on microfiche), and cannot be related by low-pressure fractional crystallization (Hill, 1979). The low-Ti suite is LREE—depleted and enriched in Sc, Cr, Co, and Ni (Table 11A, on microfiche); the high-Ti group is LREE-enriched and contains less Sc, Cr, Co, and Ni. On a spider diagram, neither group displays significant depletion of Ta-Nb or Zr-Hf, which is commonly seen in many arc basalts. Hill (1979) described equivocal results for the Shuyak Formation when Ti, Zr, Y, and Ni were used in a discriminant-analysis comparison against basalts from known tectonic environments. This phase of Talkeetna island arc volcanism is more enriched in Fe-Cr-Ni than is common among modern arc tholeiites.

In contrast, the section at Middle Cape contains calc-alkaline dacite to rhyodacite (Table 11B, on microfiche; and Fig. 17) The basalts on the mainland at Puale Bay ("P2-xx" samples in Table 11B, on microfiche and the basaltic-andesite to andesite cobbles of the offshore islets ("P3-xx" samples) are transitional to calc-alkaline in nature (Fig. 17). They include high-alumina compositions, and the low Zr and high Sr contents are typical of arc volcanic rocks.

CHUGACH TERRANE

G. Plafker, M. D. Hill, J. Lull, and F. Barker

TECTONIC SETTING

The Chugach terrane consists of accreted deep-sea sedimentary rocks and oceanic crustal rocks that form a continuous belt approximately 2,000 km long and 60 to 100 km wide in the coastal mountains of southern Alaska (Plafker and others, 1977; Plafker and others, this volume, Chapter 12). Volcanic rocks occur in three main units in the Chugach terrane (Barker and others, this volume, Plate 13). From north to south these units are: (1) Early Jurassic or older coherent blocks and bands of glaucophanic greenschist that include the schists of Raspberry Strait, Seldovia, Iceberg Lake, and Liberty Creek; (2) Late Triassic to mid-Cretaceous mélange of the Uyak Complex, McHugh Complex, Kelp Bay Group, and Yakutat Group; and (3) Upper Cretaceous basaltic volcanic rocks and minor andesitic rocks of the Valdez Group. The volcanic rocks of units (1) and (2) may represent different metamorphic facies of the same mélange protolith.

Basement rocks of the eastern Yakutat terrane are included in this discussion because they are interpreted to be an allochthonous fragment of the Chugach terrane that has been displaced to its present position from the continental margin of southeastern Alaska and (or) British Columbia by movement on the Fairweather fault (Plafker, 1987; Plafker and others, this volume, Chapter 12). Undated pillowed greenstone and related igneous rocks on the Resurrection Peninsula near Seward, previously considered to be correlative with the Valdez Group, have been tentatively assigned to the Paleocene Ghost Rocks terrane south of the Chugach terrane, because of lithologic and paleomagnetic similarities with the Ghost Rocks Formation of Plafker and others (this volume, Chapter 12) of Kodiak Island (Plafker, 1987; and see Hill, this chapter).

Rocks of the Chugach terrane range in age from Late Triassic through Late Cretaceous. The rocks are generally moderately to strongly deformed and locally are multiply deformed. They are variably metamorphosed to glaucophanic greenschist facies and zeolite to amphibolite facies (Dusel-Bacon, this volume, Chapter 15), and they are extensively intruded and locally thermally metamorphosed by Paleocene and Eocene plutons and dikes (Moll-Stalcup, this volume). Terrane boundaries are the Border Ranges fault system on the north and the Contact, Chugach–Saint Elias, and Fairweather faults to the south (Plafker and others, this volume, Chapter 12).

BLUESCHIST-GREENSCHIST UNIT

Glaucophane-bearing blueschist and greenschist of Early Jurassic age and mid-Cretaceous age are exposed intermittently

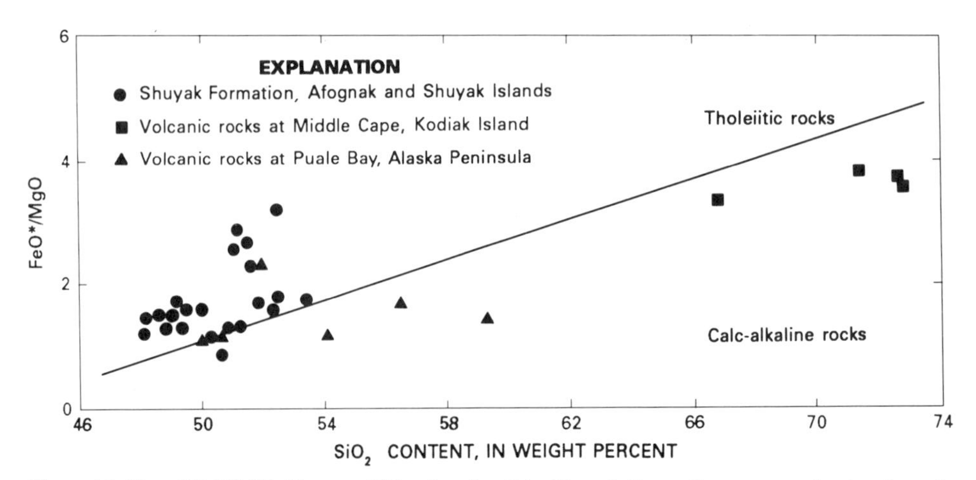

Figure 17. Plot of FeO^*/MgO versus SiO_2 of rocks of the Shuyak Formation, unnamed volcanic rocks at Puale Bay and at Middle Cape, Peninsula terrane. Tholeiitic-calc-alkaline boundary from Miyashiro (1974).

along and near the northern margin of the Chugach terrane (Barker and others, this volume, Plate 13), as summarized by Roeske (1986). Although Roeske included the fault-bounded blueschist and greenschist of the Kodiak Islands area as Peninsular terrane, we here consider it to be part of the Chugach terrane, because of its similarities in age and lithology to comparable schists on strike with those of the Chugach terrane, and because of its structural position relative to the Peninsular terrane. The unit consists predominantly of metabasites, including pillows, tuffs, and tuff breccias, and subordinate metachert, metagraywacke, and metapelite. Geochemical analyses are available for the schists of Liberty Creek (three samples) and Iceberg Lake (one sample), which are described below.

Schists of Liberty Creek and Iceberg Lake

Background. The schists of Liberty Creek (Metz, 1976; Plafker and others, this volume, Chapter 12) and Iceberg Lake (Winkler and others, 1981) form elongated, fault-bounded belts of about 280 km^2 and 120 km^2, respectively, in the northern Chugach Mountains.

The schist of Liberty Creek consists predominantly of regionally metamorphosed and multiply deformed, mafic volcanic rocks and minor pelitic rocks. Millimeter-scale crenulated lamination is characteristic. Rarely, the more massive greenschist exhibits faint primary structures suggestive of breccia and possible pillow breccia. Greenschist-facies mineral assemblages (epidote-rich actinolite-albite schist ± chlorite, quartz, calcite, and white mica) are intercalated with sparse bands and lenses from a few millimeters to a few centimeters wide containing variable amounts of very fine-grained blue amphibole (<0.1 mm) that is mainly crossite. Lawsonite has been found at only one locality in the schist of Liberty Creek, but in many places in the schist of Iceberg Lake (Winkler and others, 1981). Highly deformed carbonaceous (graphitic?) phyllite occurs locally as pods and irregular anastomosing layers as much as a few meters thick within the metavolcanic rocks. The lithology and mineralogy of the schist of Iceberg Lake are similar to the schist of Liberty Creek except that the schist of Iceberg Lake contains more chert and pelitic rocks, is coarser grained (0.1 to 0.7 mm), and the blueschist locally contains garnet.

Volcanic Geochemistry. Major- and minor-element analyses of three samples of greenschist from the schist of Liberty Creek and one from blueschist at Iceberg Lake are available (Table 12, on microfiche; and Figs. 18 through 20). The Liberty Creek rocks are basalts but show wide variations in the abundances of silica, CO_2, and other mobile elements (49.7 to 58.5 percent SiO_2; 0.08 to 2.3 percent K_2O; 0.63 to 2.85 percent NaO_2) as a result of mineral alteration and the presence of fine veinlets of quartz and calcite. The high-field-strength elements and transition elements were largely or wholly immobile (e.g., 1.03 to 1.93 percent TiO_2, and FeO^*/MgO is 1.8 to 3.3). Two of the samples plot in the tholeiitic field and the high-silica sample (84ANK152) plots in the calc-alkaline field on an AFM diagram (Fig. 18). All three samples have nearly identical REE patterns that closely match the pattern for average N-type MORB at about 7 to 19 × chondrite with slight depletion of LREE over HREE (Fig. 19). All samples plot in the field of N-type MORB (Wood, 1980) on the Th-Hf-Ta discrimination diagram (Fig. 20).

The sample from the schist of Iceberg Lake (84APr136A, Table 12, on microfiche) is a blue amphibole-bearing pillow basalt (49.0 percent SiO_2, 1.0 percent K_2O) that is geochemically similar to the least altered samples from the schist of Liberty Creek, except for a slightly higher Na_2O content (3.24 percent versus 0.63 to 2.85 percent).

Interpretation. The lithology and geochemistry of the metavolcanic rocks suggest that the schists of Liberty Creek and Iceberg Lake are oceanic tholeiites that were subducted with dominantly pelagic sediments at a convergent plate margin. The blueschist of the Iceberg Lake unit has a metamorphic age of 186 ± 1.5 Ma based on Ar^{40}/Ar^{39} on phengite (Sisson and Onstott, 1986) and K-Ar ages on crossite as old as 175 ± 5 Ma (Winkler and others, 1981)—a date that probably approximates the time of subduction. This age is close to the K-Ar age of 192 Ma reported by Roeske (1986) for glaucophane-bearing blueschist from Kodiak Island. The metamorphic age of the schist of Liberty Creek is assumed to be similar, but has not yet been determined.

The protolith age of the schists of Liberty Creek and Iceberg Lake is not known; their position in the accretionary complex indicates only that they are probably older than the less-metamorphosed, paleontologically dated Upper Triassic to mid-Cretaceous part of the McHugh Complex to the south (Winkler and others, 1981). Transitional blueschist-facies rocks of Early to Middle Jurassic metamorphic ages occur in close association with Upper Triassic oceanic rocks that include paleontologically dated radiolarian chert along the south side of the Border Ranges fault system (Jones and others, 1987). This association suggests the likelihood that the glaucophanic greenschist is a more deeply subducted part of the oldest part of the mélange unit described below.

MÉLANGE UNIT

The mélange unit of the Chugach terrane crops out as a nearly continuous fault-bounded belt as wide as 20 km that extends almost 500 km through the Kodiak Islands (Uyak Complex) and the northern margin of the Kenai and Chugach Mountains (McHugh Complex). Lithologically and, at least partially, temporally equivalent rocks are also exposed in a belt extending from the Tarr Inlet area of Glacier Bay to Chichagof and Baranof Islands (Kelp Bay Group), and as part of the allochthonous Yakutat terrane in the Yakutat Bay area (mélange of the Yakutat Group). In those areas, however, the contact relationships of the mélange have been obscured by structural interleaving with flysch and metasedimentary rocks along strike-slip faults related to the Fairweather fault system. The mélange unit consists of mixed oceanic basalt and sedimentary rocks, arc-derived volcanic and sedimentary rocks, and variable numbers of

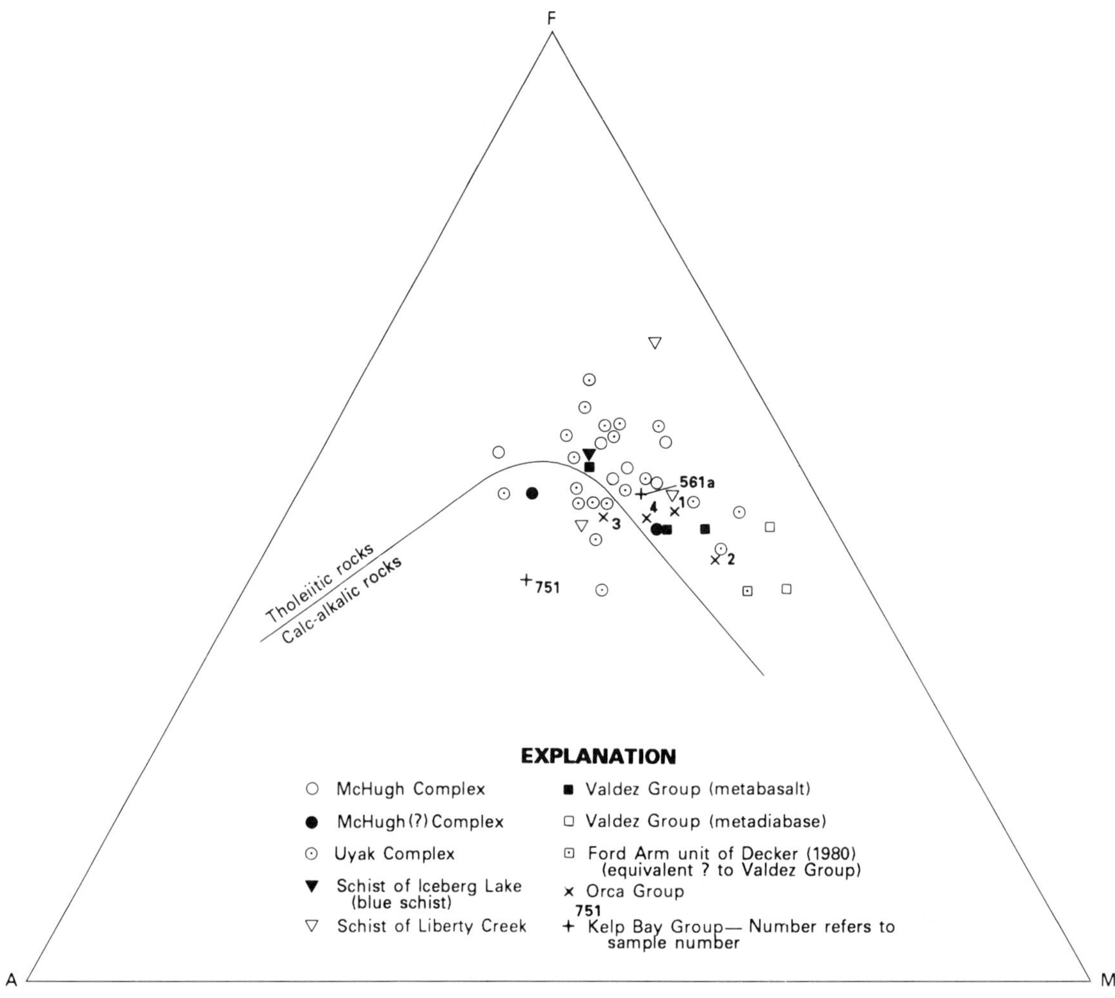

Figure 18. AFM diagram of mafic volcanic and shallow intrusive rocks of the Chugach terrane.

olistostromal blocks (Plafker and others, 1977; Plafker and others, this volume, Chapter 12).

Uyak Complex

Background. The Uyak Complex is a prehnite-pumpellyite–grade tectonic mélange that contains fossils ranging in age from mid-Permian to middle Early Cretaceous. It is exposed on Kodiak, Afognak, and Shuyak Islands (Connelly and Moore, 1977). It contains blocks of chert, greenstone, gabbro and ultramafic rocks, limestone, tuff, argillite, and graywacke (Connelly, 1978). The Uyak Complex is in contact with the Jurassic blueschist-facies Raspberry Schist of Roeske (1986) to the northwest along an unnamed fault that was correlated with the Border Ranges fault by Roeske (1986). Plafker and others (this volume, Chapter 12) believe the contact lies south of the Border Ranges fault. The Uganik thrust separates the Uyak Complex from the Upper Cretaceous (Maastrichtian) turbidites of the Kodiak Formation to the southeast. Connelly (1978) noted that a gently deformed sequence of Upper Cretaceous volcanic and sedimentary rocks (Cape Current terrane) occurs along the Uganik thrust on Shuyak Island (Connelly and Moore, 1977). Similar rocks, much less deformed than the neighboring Uyak Complex or Kodiak Formation to either side, are exposed along the west shore of Raspberry Strait on Kodiak Island, west of Dolphin Point. Two samples in Table 13 (on microfiche) are from that sequence.

Volcanic Geochemistry. The greenstones of the Uyak Complex are prehnite-pumpellyite–grade metabasalts (Table 13, on microfiche). They are subalkaline, and have high Y/Nb (2 to 15) and Zr/Nb (>6). Most samples plot in the tholeiitic field on Figure 21. Hill (1979) found that a discriminant analysis using Ti, Zr, Y, and Ni indicates that 94 percent of the basalts of the Uyak Complex are similar to ocean-ridge basalts. On a MORB-normalized spider diagram (not shown: data in Table 13, on microfiche), these samples display horizontal patterns clustering around 1.0, typical of MORB, for all elements except the highly mobile alkalies and alkaline earths.

The younger Upper Cretaceous(?) basalts within the Uganik thrust belt (V2-lw and V7-lz) are indistinguishable from the basalts of the Uyak Complex (Table 13, on microfiche; Barker

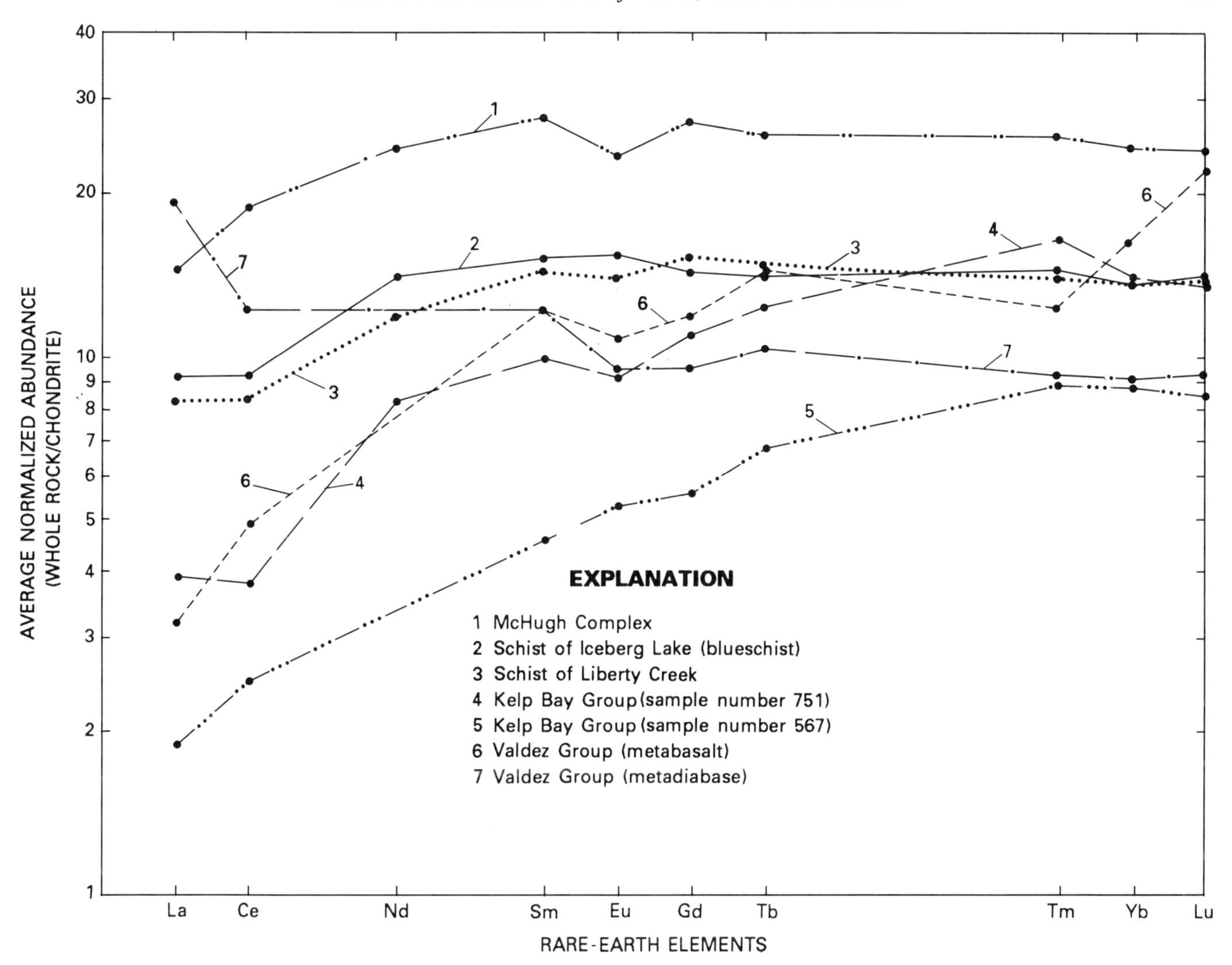

Figure 19. Chondrite-normalized REE distributions for mafic volcanic and shallow intrusive rocks of the Chugach terrane.

and others, this volume, Plate 13). If they formed within topographic lows along the Uganik thrust, they may be more closely related to the Cretaceous(?) and Paleocene Ghost Rocks Formation (discussed below) than to the Uyak Complex.

McHugh Complex

Background. The McHugh Complex as defined by Clark (1973), crops out discontinuously along the northern margin of the Chugach terrane in the Chugach and Kenai Mountains (Plafker and others, 1977). As used here, it also includes the (informal) Seldovia Bay Complex of Cowan and Boss (1978). Along its northern margin, the McHugh Complex is in contact with the Peninsular terrane, Wrangellia terrane, and the Raspberry blueschist unit along vertical, to steeply north-dipping, faults. It is juxtaposed against the Valdez Group on the south along thrust faults that dip moderately to gently northward. Structural style in this commonly massive unit consists of sparse south-verging structures, and an intense penetrative ductile fabric exhibited by pinch-and-swell structure and boudinage, and numerous closely spaced zones of intense cataclasis (Clark, 1973; Cowan and Boss, 1978; Winkler and others, 1981; Nokleberg and others, 1989). The common occurrence of disrupted brittle phacoids at all scales in a sheared argillite-and-tuff matrix imparts a characteristic blocks-in-argillite mélange appearance to parts of the unit. Metamorphism is commonly prehnite-pumpellyite facies except where locally it reaches greenschist facies near its eastern limit of exposure.

The metabasites of the McHugh Complex are dark green to dark greenish gray and characteristically form rough reddish-weathering outcrops. They occur in masses from a few meters to several hundred meters thick, together with variable amounts of chaotically intermixed ribbon chert, tuff, and argillite. The rocks are thoroughly altered to "greenstone," but locally they retain

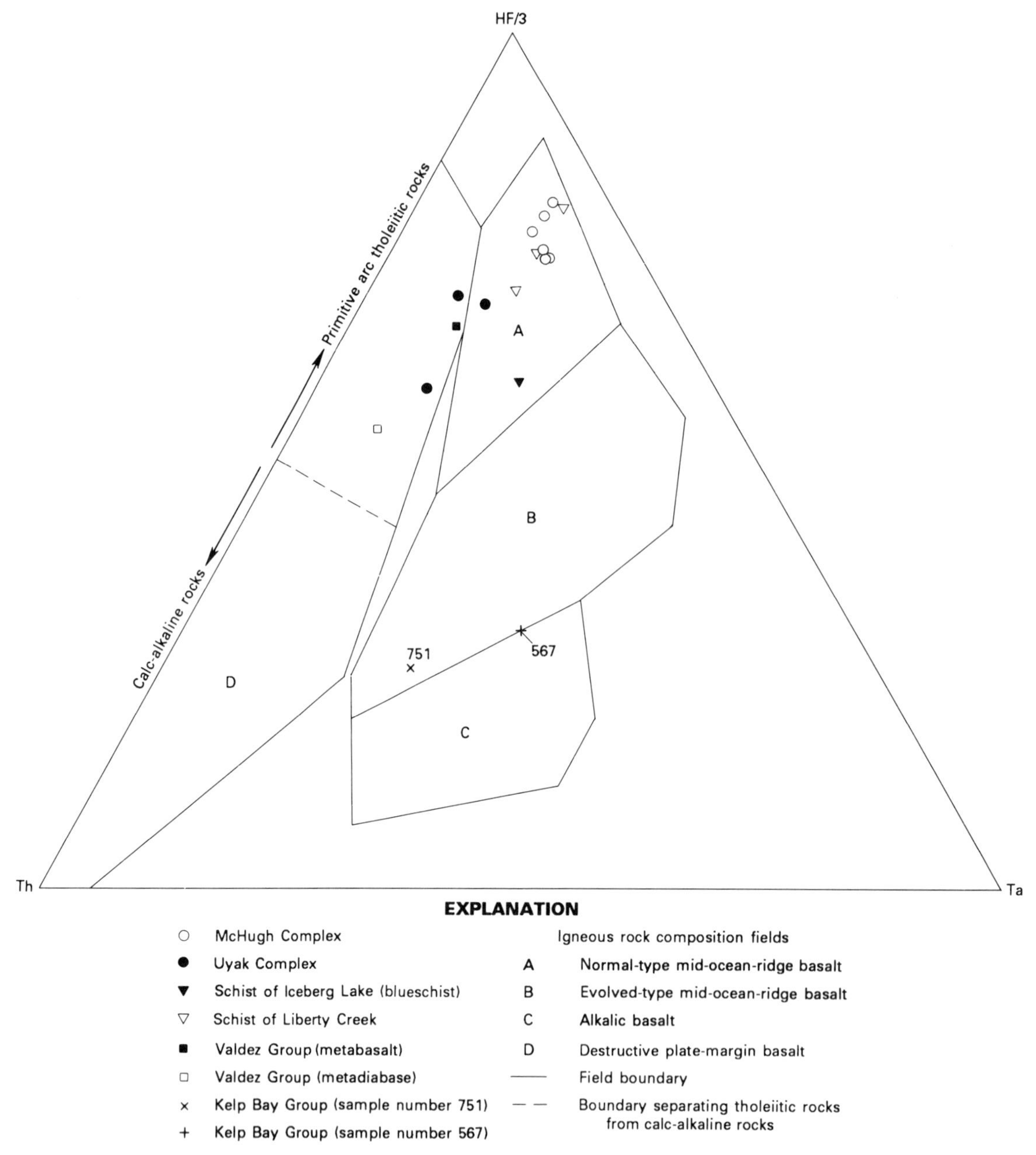

Figure 20. Ternary diagram of (Hf/3)-Th-Ta for selected mafic volcanic and shallow intrusive rock samples of Chugach terrane, showing predominantly normal-type mid-ocean-ridge basalt distribution.

textures and structures indicating that they originally were massive flows, pillow lavas, and pillow breccia. Textures are mainly microporphyritic, intergranular, and microbrecciated. Recognizable minerals in thin section are relict plagioclase, clinopyroxene (commonly as euhedral microphenocrysts), chlorite, sparse epidote, and quartz in a murky optically irresolvable matrix. Crosscutting veinlets consisting of quartz, calcite, serpentine, chlorite, and rare stilpnomelane are numerous.

The McHugh Complex was deposited from Late Triassic to mid-Cretaceous (Albian to Cenomanian) time, as indicated by the age of radiolarian assemblages from chert (Winkler and others, 1981; Nelson and others, 1986; Jones and others, 1987). Olistostromal blocks and clasts within the mélange range in age from Early Pennsylvanian to Jurassic (Nelson and others, 1986).

Volcanic geochemistry. Geochemical data for six analyzed samples of greenstone from the McHugh Complex are summarized in Table 14 (on microfiche) and on Figures 18 through 20. The rocks sampled are basalts that have fairly consistent major- and minor-element compositions (45.5 to 52.2 percent SiO_2, 1.6 to 3.3 ratio of FeO^*/MgO). There are minor variations, however,

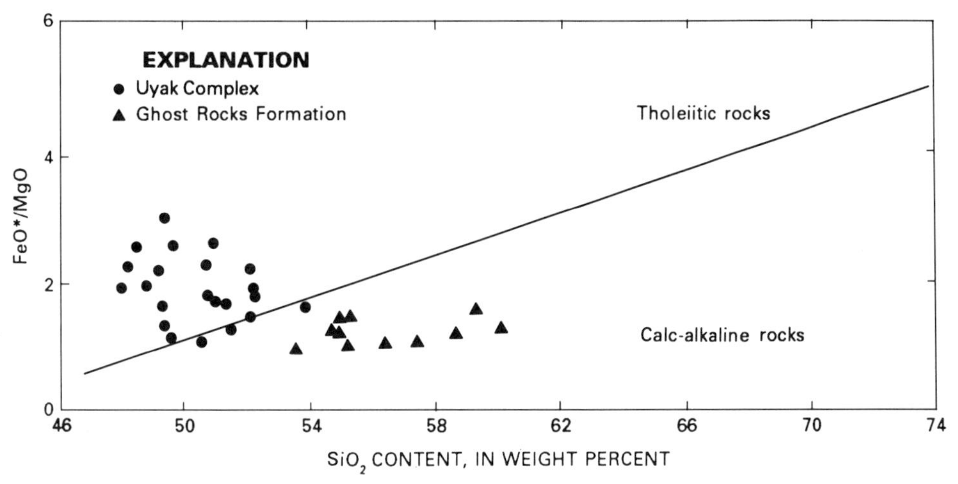

Figure 21. Plot of FeO*/MgO versus SiO_2 of the Uyak Complex and Ghost Rocks Formation, Chugach terrane.

in the alkali elements, probably due to alteration, and in silica and lime contents because of the ubiquitous presence of fine veinlets of quartz and calcite. The samples have low Al_2O_3 (12.82 to 13.64 percent) and moderate to high TiO_2 (1.89 to 2.89 percent). They plot in the tholeiitic field on an AFM diagram (Fig. 18) and on similar discrimination diagrams. The REE pattern is relatively flat at (La 13 to 19 × chondrite and Lu 20 to 28 × chondrite) with a slight depletion of LREE over HREE ($[La/Sm]_n$ = 0.5 to 0.6) and a slight negative Eu anomaly (Fig. 19). All samples plot in the field of N-type ocean-ridge basalt (N-MORB) (Wood, 1980) on the Th-Hf-Ta discrimination diagram (Fig. 20) and as ocean-floor basalts on Ti-Zr-Y and Ti-Zr discrimination plots (not shown) of Pearce and Cann (1973). The TiO_2 and REE abundances, however, are relatively high for MORB. The enrichment in HREE relative to LREE suggests derivation from a somewhat depleted mantle source. Geochemically, the McHugh Complex metavolcanic rocks are similar to the metavolcanic rocks in the schists of Liberty Creek and Iceberg Lake, except that their average REE contents are slightly higher.

Kelp Bay Group

Background. Metabasite constitutes an estimated 50 percent of the Kelp Bay Group on Baranof and Chichagof Islands (Loney and others, 1975; Decker, 1980; Johnson and Karl, 1985), and this belt of rocks is inferred to extend northwestward onto the mainland in the Tarr Inlet region of Glacier Bay (Decker and Plafker, 1981). The mélange of the Kelp Bay Group is generally similar to the McHugh Complex in lithology and structural style. It differs in that the Kelp Bay mélange is faulted into a collage of kilometer-scale blocks of lithologically heterogeneous rocks, includes well-foliated rocks ranging from prehnite-pumpellyite facies to greenschist and glaucophanic greenschist facies, and contains kilometer-scale blocks of carbonate and greenstone—some of which are olistostromal in origin (Loney and others, 1975; Decker, 1980; Johnson and Karl, 1985). The age of the matrix appears to be restricted to the Late Jurassic and Early Cretaceous (Decker, 1980; E. Pessagno, written communication, 1987). K-Ar ages of 109 to 91 Ma on sericite and actinolite from the Kelp Bay Group are interpreted by Decker (1980) as dating the time of accretion and initial metamorphism of the unit.

Volcanic geochemistry. Major- and minor-element geochemical analyses are available for two samples (751, 561a) of metabasite from the Kelp Bay Group (Table 14, on microfiche; and Figs. 18 through 20) and 15 additional samples have been analyzed for major and selected minor elements (Decker, 1980). The samples analyzed from the Kelp Bay Group include both intermixed green tuff and massive or pillowed blocks from within the mélange, as well as rocks of all textural grades from massive to schistose. Of the 17 samples analyzed, 15 greenstones are tholeiite (44.6 to 50.6 percent SiO_2, 0.8 to 2.77 percent TiO_2), and two samples of metatuff and metabreccia (661a, 661b) are andesite (57.6 to 59.3 percent SiO_2 and 0.26 to 0.31 percent TiO_2). The samples for which complete analyses are available have FeO^*/MgO ratios of 1.5 to 1.6. On a plot of FeO^*/MgO versus SiO_2 they are in the tholeiitic field (not shown), although on an AFM diagram one plots in the tholeiitic field and one is well into the calc-alkalic field (Fig. 18). The REE pattern is relatively flat (La 3 to 20 × chondrite and Lu 9 to 22 × chondrite); one sample (no. 751) shows significant LREE depletion (Fig. 18).

Yakutat Group

Background. The mélange of the Yakutat Group is similar to that of the Kelp Bay Group in lithology, structure, and matrix age, and it includes abundant lithologically diverse kilometer-scale olistostromal blocks suggestive of a Wrangellian provenance (Plafker, 1987). Most of the unit is typically subgreenschist facies; higher grade rocks occur in schistose fault-bounded slivers near the Fairweather fault, and abundant float of glaucophane greenschist, possibly derived from Yakutat Group mélange, occurs in glacial moraines in the area (G. Plafker, unpublished data).

Volcanic geochemistry. Four samples of tholeiitic metabasalt form the mélange of the Yakutat Group (Table 15, on microfiche) are of two general types. Greenstone from typical mélange northeast of Yakutat (samples 80Apr-43 and 80-APr-44) shows low Al_2O_3 (13.5 to 13.9 percent), moderate TiO_2 (1.4 to 1.9 percent), and mild light-REE enrichment ($[La/Sm]_n \sim 1.2$ to 1.5). Sheared greenstone from west of the Alsek River, samples 78APr-17 and 78APr-18A, are more aluminous (15.6 to 17 percent Al_2O_3), titaniferous (2.3 to 2.4 percent TiO_2), and light-REE enriched ($[La/Sm]_n = 1.3$ to 2.9). Sample 78APr-17, having 0.56 percent P_2O_5, 39 ppm Nb, 2.6 ppm Ta, and 216 ppm Zr, is subalkaline, of enriched MORB composition, and possibly a fragment of seamount basalt. The other three tholeiites appear to be transitional between N-MORB and E-MORB.

Interpretation

Characteristics of the basaltic rocks of the Chugach terrane mélange unit—whether massive, pillowed, tuffaceous, or schistose—suggest predominantly ocean-floor volcanism. Most indicate an origin as normal MORB or as seamounts, an origin that is compatible with their ubiquitous association with pelagic sediments. The compositions of two samples of breccia and tuff from the Kelp Bay Group, which commonly occur intimately interbedded with pelagic sediment within the mélange matrix, suggest possible derivation from a volcanic arc. The Late Jurassic Chitina arc and (or) the Early Cretaceous Chisana arc (see description, this chapter) are possible sources (Plafker and others, 1989b; this volume, Chapter 12). However, G. Winkler (oral communication, 1988) suggested an origin as aquagene tuff. Structural and lithologic data indicate that the complex was probably accreted, disrupted, and mixed with terrigenous sediments at a convergent plate margin (Clark, 1973; Moore and Connelly, 1977). The age distribution of dated radiolarian chert in the western Valdez Quadrangle (Winkler and others, 1981) suggests progressive accretion of slices of the mélange complex ranging from Late Triassic time along the northern margin of the complex to mid-Cretaceous time at the southern margin.

VALDEZ GROUP

Setting

Metavolcanic rocks occur in a discontinuous belt 2 to 8 km wide and 600 km long that defines the southern margin of the Valdez Group and correlative units from the Prince William Sound area to southeastern Alaska (Plafker and others, this volume, Chapter 12). To the north, the metavolcanic rocks are intercalated with structurally higher volcaniclastic flysch that comprises most of the Valdez Group; to the south, the volcanic rocks are in fault contact with rocks of the Prince William and Yakutat terranes. The rocks are polydeformed and polymetamorphosed to metamorphic grades that range from low greenschist facies in the western part of the belt to high greenschist facies, epidote amphibolite facies, and amphibolite facies in the eastern Chugach Mountains and Saint Elias Mountains. The age of the Valdez Group metavolcanic rocks is Late Cretaceous (Maastrichtian and Campanian?), as indicated by fossils in the overlying flysch unit (Tysdal and Plafker, 1978).

Data on the geochemistry of the volcanic rocks were obtained primarily from near the type area of the Valdez Group between Prince William Sound and the Copper River (five analyses), from higher grade schists in the area between Yakutat Bay and the Alsek River (four analyses), and from rocks on Chichagof Island (one analysis) that are tentatively correlated with the Valdez Group. These three areas are referred to below, respectively, as the "western," "Yakutat," and "Chichagof" areas.

Western area

Background. Structural thickness of the metavolcanic rocks is at least 4 km near the western end of the outcrop belt, where it has been studied in detail (Plafker and others, 1986). The rocks consist of green basaltic metatuff that occurs in lenticular beds ranging from a few centimeters to about 15 m thick and 4 km in length, interbedded with metasedimentary rocks, dark green to black metabasalt, amygdaloidal pillow flows, breccia ("greenstone"), and diabase intrusives. Flattened pillow structures are recognizable in some of the more massive volcanic units. Schistose dikes and sills of metadiabase and metagabbro associated with the thicker volcanic units commonly show conspicuous rusty brown alteration zones in outcrop.

The metatuff mainly consists of very fine-grained, sheared chlorite, fibrous actinolite, and epidote ± quartz, pyrite, white mica, and biotite. It commonly is microfolded and may be intimately interlayered with metapelite on a millimeter to centimeter scale. The basaltic rocks are mineralogically similar to the metatuff except that chloritized amygdules are locally abundant. The metadiabase and metagabbor are slightly foliated, equigranular, fine- to medium-grained rocks of diabasic or allotriomorphic granular texture. They consist of pale green actinolite and completely saussuritized plagioclase and as much as 20 percent each of fine- to very fine-grained epidote and chlorite. The diabase in places is cut by millimeter-scale epidote veinlets.

Volcanic geochemistry. Analyses are available for one sample of metatuff, two samples of schistose basalt (greenstone), and two samples of comagmatic metadiabase and metagabbro from the western part of the Valdez Group (Plate 6; Table 15, on microfiche). The samples are LREE—depleted tholeiite (48.5 to 53.7 percent SiO_2 and 0.35 to 1.17 percent TiO_2, $FeO^*/MgO = 0.8$ to 2.1) with low Ba, Sr, and K_2O/Na_2O. They plot as tholeiites with an "iron enrichment" trend on the AFM diagram (Fig. 18), and they straddle the tholeiite-calc-alkaline compositional boundary on a plot of FeO^*/MgO versus SiO_2 (not shown). Two samples with Th contents above detection levels plot in the field of primitive-arc tholeiites (Wood, 1980) on a Th-Hf-Ta discrimination diagram (Fig. 20). On discrimination diagrams of Ti-Zr-Y and Ti—Zr (not shown), the samples plot either in or just outside the low-K arc tholeiite field. REE abundances are low (Fig. 19) and show moderate depletion of LREE relative to HREE (La 1 to 6 × chondrites, Lu 9 to 15 × chondrites, and $[La/Sm]_n = 0.3$ to 0.5); REE contents of the intrusive rocks are systematically lower than for the greenstones. The geochemical data for these rocks are inconclusive in regard to implications about their petrotectonic setting. They could represent contaminated MORB or primitive-island-arc tholeiite.

Yakutat area

The four samples of the Valdez(?) Group are three metabasalts and one meta-andesite. The metabasalts, hornblende-plagioclase amphibolite, and actinolite-albite-chlorite schist are of light-REE-depleted tholeiitic compositions (Table 15, on microfiche). Because their Ta abundances of 0.5 to 0.6 ppm are higher than those of arc tholeiites, we suggest that these rocks probably are offscraped remnants of oceanic crust or MORB. The meta-andesite (sample 78-APr-7), now epidote-actinolite-chlorite-albite-calcite schist, contains 61.5 percent SiO_2, shows moderate light-REE enrichment (La approximately 40 × chondrites), and plots as calc-alkaline on an FeO^*/MgO-SiO_2 diagram (not shown). This rock probably originated as an ash fall from a distant island arc.

Chichagof area

The Ford Arm unit of Decker (1980) is a fault-bounded, undated sequence of flysch locally interbedded with schistose basaltic tuff, flow breccia, and massive flows. The unit is correlated by Decker with the Valdez Group on the mainland, because of its apparent structural continuity and lithology. The one sample (no. 690) for which a complete chemical analysis is available is an actinolite-chlorite-epidote schist that is tholeiitic on AFM (Fig. 18) and FeO^*/MgO versus SiO_2 (not shown) plots, and has a flat REE pattern with La at about 10 × chondrite (Table 15, on microfiche). If this one sample is representative of the Ford Arm unit, the geochemical data (especially the absence of strong LREE depletion) are more compatible if the unit is correlated with MORB tholeiites of the mélange unit rather than with the Valdez Group.

Interpretation

Characteristics of the sedimentary rocks of the Valdez Group suggest that they were deposited mainly in an ocean-floor or trench environment and were derived from a probable Andean type volcano-plutonic arc source (Zuffa and others, 1980; Decker and Plafker, 1981; Nilson and Zuffa, 1982; Plafker and others, this volume, Chapter 11). The basalt and related fragmental volcanic rocks of the Valdez Group are interpreted as the upper levels of oceanic crust upon which the clastic rocks were deposited. The apparent absence of pelagic sediments at the interface and the occurrence of mafic aquagene tuff intercalated on a fine scale with volcanogenic sediments suggest that the oceanic crustal rocks were, at least in part, erupted relatively close to the sediment source.

GHOST ROCKS TERRANE

M. D. Hill and G. Plafker

The Ghost Rocks terrane occurs as a 10-km-wide belt of mafic low-grade metamorphic volcanic rocks and mélange along the southern margin of Kodiak Island and as a thick accumulation of pillow basalt and associated intrusive rocks on the Resurrection Peninsula on the mainland near Seward (Barker and others, this volume, Plate 13; Plafker, 1987; Plafker and others, this volume, Chapter 12). The terrane is juxtaposed against the Chugach terrane to the north along the Contact fault, and against the Prince William terrane to the south along the Resurrection fault. Geochemical data are available only for the Ghost Rocks Formation of Kodiak Island, as discussed below.

GHOST ROCKS FORMATION

The Cretaceous(?) and Paleocene Ghost Rocks Formation (Plafker and others, this volume, Chapter 12) on Kodiak Island consists of pillowed and massive volcanic flows, tuff, sandstone, argillite, diabase, limestone, and conglomerate. Reid (in Moore and others, 1983) showed that the volcanic rocks include both LREE-depleted tholeiitic basalt and LREE-enriched calc-alkaline basaltic andesite to andesite. The data in Table 16, (on microfiche) and on Figure 21, from Hill (1979), are entirely from the latter group, sampled along the north shore of the Aliulik Peninsula on Kodiak Island. The Nd isotope data summarized by Moore and others (1983) provide convincing evidence that the calc-alkaline units of the Ghost Rocks Formation represent hybrid magmas derived by contamination of tholeiitic, MORB-like basalt with sediments of the accretionary prism. Demonstration of this mechanism within the volcanic rocks of the Ghost Rocks Formation and the occurrence of gabbro to quartz diorite plutons within the Ghost Rocks provide further support for the contention made by Hill and others (1981) that Ghost Rocks magma invaded the accretionary prism and was responsible for generating the approximately 60-Ma batholiths on Kodiak, Shumagin, and Sanak Islands. The basalt volcanism, preserved in relatively undeformed rocks of the Upper Cretaceous units within the Uganik thrust, may represent an early state of this Ghost Rocks volcanic activity.

PRINCE WILLIAM TERRANE

G. Plafker and J. S. Lull

INTRODUCTION

The Prince William terrane forms an outcrop belt more than 100 km wide that extends from the Copper River and Prince William Sound areas southwestward beneath the continental shelf and slopes off the Kodiak Island (Jones and others, 1987; Nokleberg and others, this volume, Chapter 10). The terrane basement consists of the upper Paleocene through middle Eocene Orca Group, overlain in the offshore areas by thick, younger Tertiary sedimentary sequences. The terrane is in fault contact with the Ghost Rocks, Chugach, and Yakutat terranes on the north and east; the offshore contact relationships to the south are unknown.

ORCA GROUP

Background

The Orca Group is a structurally complex deep-sea-fan flysch complex interbedded with subordinate oceanic volcanic and minor pelagic sedimentary rocks (Winkler, 1976; Winkler and others, 1981; Nelson and others, 1986). The sedimentary rocks were derived from a dominantly plutonic and high-grade metamorphic source and had a minor contribution from a probable cratonal source. Outcrops of predominantly mafic volcanic rocks as much as several kilometers thick occur in discontinuous belts as large as 150 km long and 12 km wide (Barker and others, this volume, Plate 13). They consist of basalt, including pillowed basalt, and basaltic breccia, tuff, and basaltic volcanogenic sedimentary rocks that locally contain pelagic microfossils and are interlayered with the flysch.

Volcanic geochemistry

Chemical analyses are available for 15 samples of basaltic rocks and four samples of diabase from the Prince William Sound area and the Ragged Mountains area (Table 17, on microfiche): 48.0 to 55.4 percent SiO_2, 11.8 to 20.6 percent Al_2O_3, O.4 to 2.4 percent TiO_2, FeO^*/MgO = 0.9 to 2.0). On an AFM diagram, most samples plot as tholeiitic or close to the field boundary with four samples in the calc-alkalic field (Fig. 22). On plots of FeO^*/MgO versus SiO_2 (not shown), the samples straddle the tholeiite-calc-alkalic boundary, but most plot in the tholeiite field. High CaO (13 to 14 percent) in three samples is probably due to veinlets and amygdules of calcite.

Based on minor-element geochemistry, four distinct types of mafic volcanic rocks can be recognized in the Orca Group. Type one (six samples) is best characterized as enriched (transitional) MORB (La 23 to 52 × chondrites, Lu 9 to 14 < chondrites, $[La/Sm]_n$ = 1.4 to 2.2) and the samples plot as E-MORB and ocean-island basalts on the Th-Hf-Ta and Ti-Zr-Y discrimination diagrams (not shown). Type two (three samples of basalt, four samples of diabase) is normal MORB. REE patterns for the basalt show a flat trend (La 17 × chondrites, Lu 17 to 18 × chondrites, $[La/Sm]_n$ = 1.0) and a slight negative Eu anomaly. REE abundances for the diabase are slightly lower than for the basalt (Fig. 23), show slight LREE depletion (La 9 to 12 ×

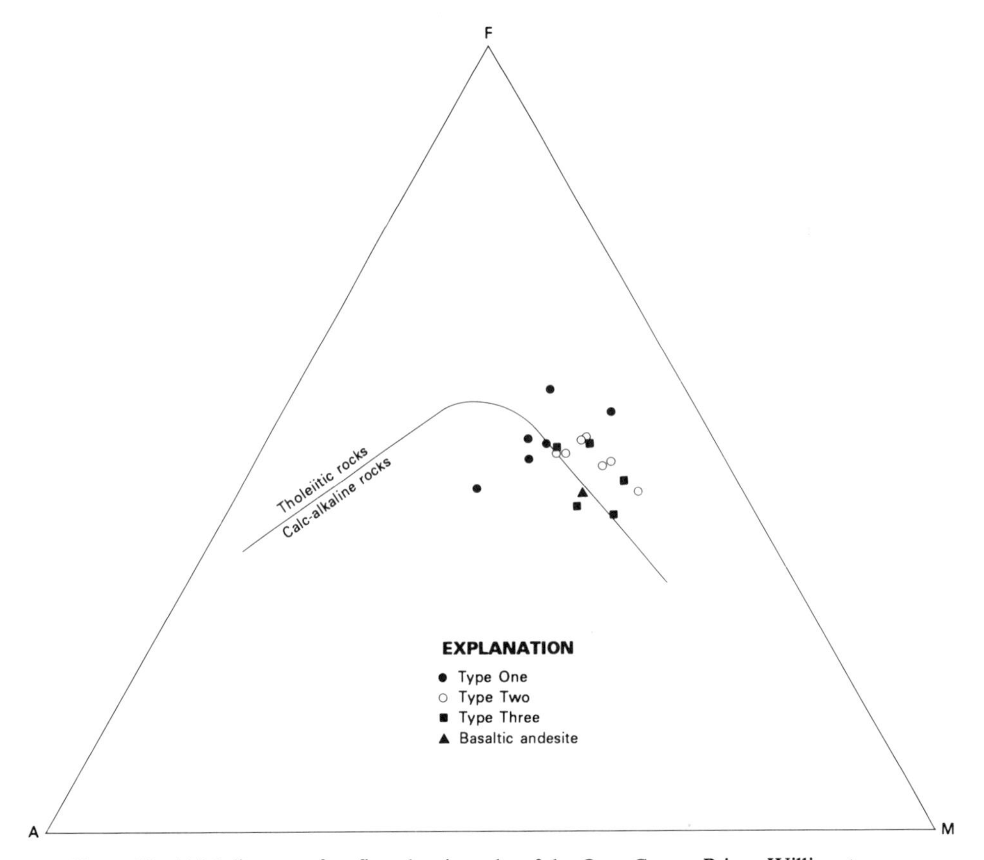

Figure 22. AFM diagram of mafic volcanic rocks of the Orca Group, Prince William terrane.

chondrites, Lu 13 to 16 × chondrites, $[La/Sm]_n$ = 0.7 to 0.9), and no Eu anomaly. This trend is nearly identical to the average MORB trend of Sun and others (1979). Type two basalts and diabase plot as ocean-floor basalt on a Ti-Zr-Y discrimination diagram and as N-MORB on a La/Ta plot (not shown). Type three basalts (five samples) are low-Ti (TiO_2 <1 percent) and low-Zr (<40 ppm) basalts similar to arc tholeiites. REE abundances are low and depleted in LREE (La 2 to 7 × chondrites, Lu 6 to 18 × chondrites, $[La/Sm]_n$ = 0.3 to 0.4), although one sample shows LREE enrichment ($[La/Sm]_n$ = 1.8). On a Ti-Sr-Y diagram, type three basalts plot within and just outside the low-K arc tholeiite field. Type four is represented by one sample of "arc" calc-alkalic andesite from southwest Prince William Sound. The REE pattern shows slight LREE enrichment (La 20 × chondrites, Lu 13 × chondrites, $[La/Sm]_n$ = 1.4) and a slight negative Eu anomaly. This sample plots in the field for calc-alkalic basalt on both the Ti-Zr-Y and the Hf-Ta discrimination diagrams (not shown).

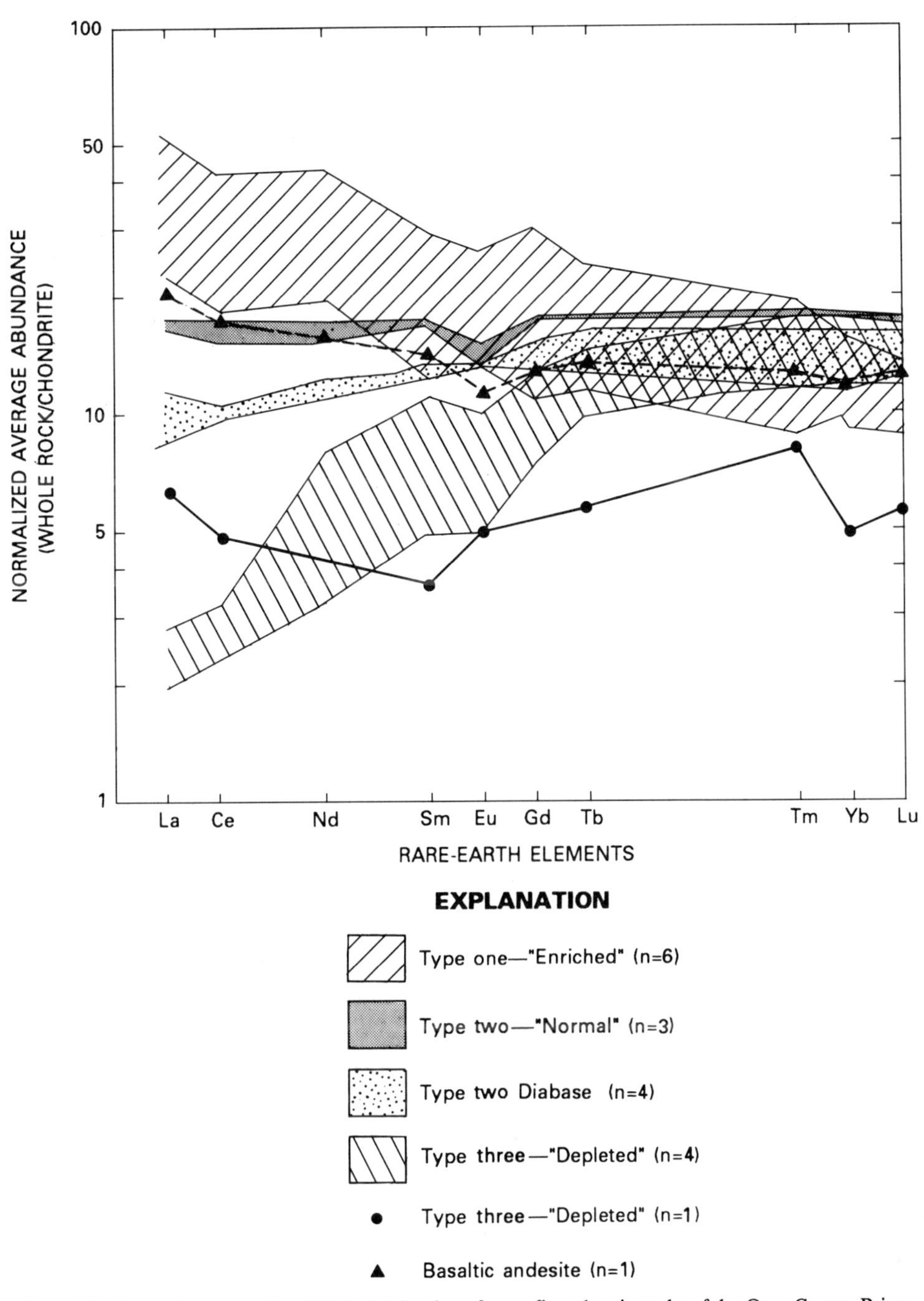

Figure 23. Chondrite-normalized REE distributions for mafic volcanic rocks of the Orca Group, Prince William terrane.

Interpretation

The geochemical data, in conjunction with geological factors, suggest eruption as submarine ridges or volcano chains within reach of submarine fans built out from the continental margin. The basalts may have erupted from leaky transforms along the continental margin (Tysdal and Case, 1979) or from the Kula-Farallon ridge that was rapidly approaching the continental margin during Orca time (Plafker and others, this volume, Chapter 12). The thermal and mechanical regime within fracture zones and near ridge-transform intersections favors ineffective magma mixing and less total melt production, giving rise to more diverse magma types and the formation of both LREE-enriched and LREE-depleted lavas (Batiza and Vanko, 1984). Such a regime gives one possible explanation for the chemical diversity of basalts of the Orca Group.

PALEOGENE BASALT OF THE YAKUTAT TERRANE

G. Plafker and F. Barker

BACKGROUND

The Yakutat terrane is a composite allochthonous terrane 600 km long and 200 km wide along the northern Gulf of Alaska margin. It is bounded on the east and north by the Fairweather and Chugach–Saint Elias faults, on the west by the Kayak Island zone, and on the south by the Transition fault system (Plafker, 1987). Oceanic crust of early Eocene and possibly Paleocene age underlies the western two-thirds of the block. A dominantly clastic sequence of early Eocene to Quaternary age overlies the oceanic basement. Dredge samples of basalt found in this clastic sequence have been described by Davis and Plafker (1985). We summarize in the following paragraph.

VOLCANIC GEOCHEMISTRY

Eight samples of this continental-slope basalt were collected just north of the Transition fault (Davis and Plafker, 1985). These rocks are altered flows, hyaloclastites, and pillow breccias that are aphyric to moderately phyric (>10 percent phenocrysts). Plagioclase is the common phenocryst phase, diopsidic augite is minor in amount, and olivine is rare. These basalts are of two chemical types: one shows depletion of light REE, Ti, Nb, and Zr, and is of MORB type; the other shows enrichment of light REEs and LILE and is of ocean-island or seamount type. Davis and Plafker infer that these basalts originated at and near the Kula-Farallon spreading center, were accreted to the continental margin at British Columbia or (and) Washington at ca. 48 Ma, and were carried northwestward by transform faulting.

REFERENCES CITED

Aleinikoff, J. N., and Nokleberg, W. J., 1985, Age of Devonian igneous-arc terranes in the northern Mount Hayes Quadrangle, eastern Alaska range, Alaska, *in* Bartsch-Winkler, S., ed., The United States Geological Survey in Alaska; Accomplishments during 1984: U.S. Geological Survey Circular 967, p. 44–45.

Aleinikoff, J. N., Dusel-Bacon, C., Foster, H. L., and Futa, K., 1981, Proterozoic zircon from augen gneiss, Yukon-Tana Upland, east-central Alaska: Geology, v. 9, p. 469–463.

Aleinikoff, J. N., Dusel-Bacon, C., and Foster, H. L., 1986, Geochronology of orthoaugen gneiss and related rocks, Yukon-Tanana terrane, east-central Alaska: Geological Society of America Bulletin, v. 97, p. 626–637.

Aleinikoff, J. N., Dusel-Bacon, C., Foster, H. L., and Nokleberg, W. J., 1987, Pb-isotope fingerprinting of tectonostratigraphic terranes, east-central Alaska: Canadian Journal of Earth Sciences, v. 24, p. 2089–2098.

Aleinikoff, J. N., Plafker, G., and Nokleberg, W. J., 1988, Middle Pennsylvanian plutonic rocks along the southern margin of Wrangellia, *in* Hamilton, T. D., and Galloway, J. P., eds., U.S. Geological Survey in Alaska; Accomplishments during 1987: U.S. Geological Survey Circular 1016, p. 110–113.

Armstrong, R. L., Harakal, J. E., Forbes, R. B., Evans, B. W., and Thurston, S. P., 1986, Rb-Sr and K-Ar study of metamorphic rocks of the Seward Peninsula and southern Brooks Range, Alaska, *in* Evans, B. W., and Brown, E. H., eds., Blueschists and eclogites: Geological Society of America Memoir 164, p. 185–204.

Barker, F., 1957, Geology of the Juneau B-3 Quadrangle, Alaska: U.S. Geological Survey Geologic Quadrangle Map GQ-100, 1 sheet, scale 1:63,360.

——, 1987, Cretaceous Chisana Island arc of Wrangellia, eastern Alaska: Geological Society of America Abstracts with Programs, v. 19, p. 580.

Barker, F., and Grantz, A., 1982, Talkeetna Formation in the southeastern Talkeetna Mountains, southern Alaska; An early Jurassic andesitic intraoceanic island arc: Geological Society of America Abstracts with Programs, v. 14, p. 147.

Barker, F., and Stern, T. W., 1986, An arc-root complex of Wrangellia, east Alaska Range: Geological Society of America Abstracts with Programs, v. 18, p. 534.

Barker, F., Jones, D. L., Budahn, J. R., and Coney, P. J., 1988, Seamount origin for basaltic rocks of the Angayucham terrane, central Alaska: Journal of Geology, v. 96, p. 368–374.

Barker, F., Sutherland Brown, A., Budahn, J. R., and Plafker, G., 1989, Back-arc with frontal-arc component origin of Triassic Karmutsen basalt, British Columbia: Chemical Geology, v. 75, p. 81–102.

Batiza, R., and Vanko, D., 1984, Petrology of young Pacific seamounts: Journal of Geophysical Research, v. 89, no. B13, p. 11235–11260.

Beard, J. S., and Barker, F., 1989, Petrology and tectonic significance of gabbros, tonalites, shoshonites, and anorthosites in a late Paleozoic arc-root complex in the Wrangellia terrane, southern Alaska: Journal of Geology, v. 97, p. 667–683.

Berg, H. C., 1972, Geologic map of Annette Island, Alaska: U.S. Geological Survey Miscellaneous Geologic Investigations Map I-684, 1 sheet, scale 1:63,360.

——, 1973, Geology of Gravina Island, Alaska: U.S. Geological Survey Bulletin 1373, 41 p.

Berg, H. C., Jones, D. L., and Richter, D. H., 1972, Gravina-Nutzotin belt; Tectonic significance of an upper Mesozoic sedimentary and volcanic sequence in southern and southeastern Alaska: U.S. Geological Survey Professional Paper 800-D, p. D1–D24.

Berg, H. C., Jones, D. L., and Coney, P. J., 1978, Map showing pre-Cenozoic tectonostratigraphic terranes of southeastern Alaska and adjacent areas: U.S. Geological Survey Open-File Report 78-1085, 2 sheets, scale 1:1,000,000.

Berg, H. C., Elliott, R. L., and Koch, R. D., 1988, Geologic map of the Ketchikan and Prince Rupert Quadrangles, southeastern Alaska: U.S. Geological Survey Miscellaneous Geologic Investigations Map I-1807, 27 p., scale 1:250,000.

Bond, G. C., 1973, A late Paleozoic volcanic arc in the eastern Alaska Range, Alaska: Journal of Geology, v. 81, p. 557–575.

——, 1976, Geology of the Rainbow Mountain–Gulkana Glacier area, eastern Alaska Range, with emphasis on upper Paleozoic strata: Alaska Division of Geological and Geophysical Surveys Geologic Report 45, 47 p.

Box, S. E., 1985a, Terranes of the northern Bristol Bay region, *in* Bartsch-Winkler, S., ed., The United States Geological Survey in Alaska; Accomplishments during 1984: U.S. Geological Survey Circular 967, p. 32–37.

——, 1985b, Early Cretaceous orogenic belt in northwestern Alaska; Internal organization, lateral extent, and tectonic interpretation, *in* Howell, D. G., ed., Tectonostratigraphic terranes of the Circum-Pacific region: Houston, Texas, Circum-Pacific Council of Energy and Mineral Resources Earth Sciences Series 1, p. 137–147.

Brandon, M. T., Orchard, M. J., Parrish, R. R., Sutherland Brown, A., and Yorath, C. J., 1986, Fossil ages and isotopic ages from the Paleozoic Sicker Group and associated intrusive rocks, Vancouver Island, British Columbia: Geological Survey of Canada Current Research, Part A, Paper 86-1A, p. 683–696.

Brew, D. A., and Ford, A. B., 1985, Preliminary reconnaissance geologic map of the Juneau, Taku River, Atlin, and part of the Skagway 1:250,000 Quadrangles, southeastern Alaska: U.S. Geological Survey Open-File Report 85–395, 23 p., 2 sheets, scale 1:250,000.

BVSP, 1981, Basaltic volcanism on the terrestrial planets: New York, Pergamon Press, 1,286 p.

Campbell, R. B., and Dodds, C. J., 1985, Geology of the Mount St. Elias map area (115B and C [E1/2]): Canadian Geological Survey Open-File Report 830, scale 1:250,000.

Clark, S.H.B., 1973, The McHugh Complex of south-central Alaska: U.S. Geological Survey Bulletin 1372-D, p. D1–D11.

Connelly, W., 1978, Uyak Complex, Kodiak Islands, Alaska; A Cretaceous subduction complex: Geological Society of America Bulletin, v. 89, p. 755–769.

Connelly, W., and Moore, J. C., 1977, Geologic map of the northwest side of the Kodiak Islands, Alaska: U.S. Geological Survey Open-File Map 77-382, 1 sheet, scale 1:250,000.

Cowan, D. S., and Boss, R. F., 1978, Tectonic framework of the southwestern Kenai Peninsula, Alaska: Geological Society of America Bulletin, v. 89, p. 155–158.

Crawford, A. J., Falloon, T. J., and Eggins, S., 1987, The origin of island arc high-alumina basalts: Contributions to Mineralogy and Petrology, v. 97, p. 417–430.

Davis, A., and Plafker, G., 1985, Comparative geochemistry of Triassic basaltic rocks from the Taku terrane on the Chilkat Peninsula and Wrangellia: Canadian Journal of Earth Sciences, v. 22, p. 183–194.

Decker, J. E., 1980, Geology of a Cretaceous subduction complex, western Chichagof Island, southeastern Alaska [Ph.D. thesis]: Stanford, California, Stanford University, 135 p.

Decker, J., and Plafker, G., 1981, Correlation of rocks in the Tarr Inlet suture zone with the Kelp Bay Group, *in* Coonrad, W. L., ed., The United States Geological Survey in Alaska; Accomplishments during 1980: U.S. Geological Survey Circular 844, p. 119–123.

Detterman, R. L., and Hartsock, J. K., 1966, Geology of the Iniskin-Tuxedni region, Alaska: U.S. Geological Survey Professional Paper 512, 78 p.

Detterman, R. L., and Reed, B. L., 1980, Stratigraphy, structure, and economic geology of the Iliamna Quadrangle, Alaska: U.S. Geological Survey Bulletin 1368-B, 86 p.

Dumoulin, J. A., and Harris, A., 1984, Carbonate rocks of central Seward Peninsula, Alaska: Geological Society of America Abstracts with Programs, v. 16, p. 280.

Dusel-Bacon, C., and Aleinikoff, J. N., 1985, Petrology and significance of augen gneiss from a belt of Mississippian granitoids in the Yukon-Tanana terrane: Geological Society of America Bulletin, v. 96, p. 411–425.

Eberlein, G. D., Churkin, M., Jr., Carter, C., Berg, H. C., and Ovenshine, A. T., 1983, Geology of the Craig Quadrangle, Alaska: U.S. Geological Survey Open-File Report 83-91, 28 p., 2 sheets, scale 1:250,000.

Forbes, R. B., Evans, B. W., and Thurston, S. P., 1984, Regional progressive high-pressure metamorphism, Seward Peninsula, Alaska: Journal of Metamorphic Geology, v. 2, p. 43–54.

Ford, A. B., and Brew, D. A., 1973, Preliminary geologic and metamorphic-isograd map of the Juneau B-2 Quadrangle, Alaska: U.S. Geological Survey Miscellaneous Field Investigations Map MF-527, 1 sheet, scale 1:31,680.

——, 1977, Preliminary geologic and metamorphic-isograd map of the northern parts of the Juneau A-1 and A-2 Quadrangles, Alaska: U.S. Geological Survey Miscellaneous Field Investigations Map MF-847, 1 sheet, scale 1:31,680.

——, 1987, Major-element chemistry of metabasalts of the Juneau-Haines region, southeastern Alaska: U.S. Geological Survey Circular 1016, p. 150–155.

Gardner, M. C., and 8 others, 1988, Pennsylvanian pluton stitching of Wrangellia and Alexander terranes, Wrangell Mountains, Alaska: Geology, v. 16, p. 967–971.

Gehrels, G. E., and Berg, H. C., 1992, Geologic map of southeastern Alaska: U.S. Geological Survey Miscellaneous Geologic Investigations Map I-1867, 1 sheet, scale approximately 1:600,000.

Gehrels, G. E., and Saleeby, J. B., 1987a, Geologic framework, tectonic evolution, and displacement history of the Alexander terrane: Tectonics, v. 6, p. 151–173.

——, 1987b, Geology of southern Prince of Wales Island, southeastern Alaska: Geological Society of America Bulletin, v. 98, p. 123–137.

Gehrels, G. E., Dodds, C. J., and Campbell, R. B., 1986, Upper Triassic rocks of the Alexander terrane, SE Alaska and the Saint Elias Mountains of B.C. and Yukon: Geological Society of America Abstracts with Programs, v. 18, p. 109.

Gehrels, G. E., Saleeby, J. B., and Berg, H. C., 1987, Geology of Annette, Gravina, and Duke Islands, southeastern Alaska: Canadian Journal of Earth Sciences, v. 24, p. 866–881.

Gill, J. B., 1981, Orogenic andesites and plate tectonics: New York, Springer-Verlag, 390 p.

Grantz, A., 1960a, Geologic map of Talkeetna Mountains (A-2) Quadrangle, Alaska, and the contiguous area to the north and northwest: U.S. Geological Survey Miscellaneous Geologic Investigations Map I-313, 1 sheet, scale 1:48,000.

——, 1960b, Geologic map of Talkeetna Mountains (A-1) Quadrangle, and the south third of Talkeetna Mountains (B-1) Quadrangle, Alaska: U.S. Geological Survey Miscellaneous Geologic Investigations Map I-314, 1 sheet, scale 1:48,000.

Hill, M. D., 1979, Volcanic and plutonic rocks of the Kodiak-Shumagin Shelf, Alaska; Subduction deposits and near-trench magmatism [Ph.D. thesis]: Santa Cruz, University of California, 259 p.

Hill, M. D., Morris, J., and Whelan, J., 1981, Hybrid grandiorites intruding the accretionary prism, Kodiak, Shumagin, and Sanak Islands, southwest Alaska: Journal of Geophysical Research, v. 86, p. 10569–10590.

Hitzman, M. W., Smith, T. E., and Profett, J. M., 1982, Bedrock geology of the Ambler district, southwestern Brooks Range, Alaska: Alaska division of Geological and Geophysical Surveys Geologic Report 75, 2 sheets, scale 1:250,000.

Hoare, J. M., and Coonrad, W. L., 1978, Geologic map of the Goodnews and Hagemeister Islands region, southwestern Alaska: U.S. Geological Survey Open-File Report 78-9B, 2 sheets, scale 1:250,000.

Imlay, R. W., and Detterman, R. L., 1973, Jurassic paleobiogeography of Alaska: U.S. Geological Survey Professional paper 801, 34 p.

——, 1977, Some Lower and Middle Jurassic beds in Puale Bay–Alinchak Bay area, Alaska Peninsula: American Association of Petroleum Geologists Bulletin, v. 61, p. 607–611.

Irvine, T. N., 1973, Bridget Cove volcanics, Juneau area; Possible parental magma of Alaska-type ultramafic complexes: Carnegie Institute of Washington Yearbook 72, p. 478–491.

Johnson, B. R., and Karl, S. M., 1985, Geologic map of western Chicagof and Yakobi Islands, southeastern Alaska: U.S. Geological Survey Miscellaneous Investigations Series Map I-1506, 15 p., 1 sheet, scale 1:125,000.

Jones, D. L., Silberling, N. J., and Hillhouse, J., 1977, Wrangellia; A displaced terrane in northwestern North America: Canadian Journal of Earth Sciences, v. 14, p. 2565–2577.

Jones, D. L., Silberling, N. J., and Coney, P. J., 1986, Collision tectonics in the Cordillera of western North America; Examples from Alaska, *in* Coward, M. P., and Ries, A. C., eds., Collision tectonics: Geological Society of London Special Publication 19, p. 367–387.

Jones, D. L., Silberling, N. J., Coney, P. J., and Plafker, G., 1987, Lithotectonic terrane map of Alaska (west of the 141st Meridian, *in* Silberling, N. J., and Jones, D. L., eds., Lithotectonic terrane maps of the North American Cordillera: U.S. Geological Survey Miscellaneous Field Studies Map MF-1874A, 1 sheet, scale 1:2,500,000.

Kelley, J. S., 1980, Environments of deposition and petrography of Lower Jurassic volcaniclastic rocks, southwestern Kenai Peninsula, Alska [Ph.D. thesis]: Davis, University of California, 304 p.

——, 1984, Geologic map and sections of the southwestern Kenai Peninsula west of the Port Graham fault, Alaska: U.S. Geological Survey Open-File Report 84-152, 1 sheet, scale 1:63,360.

——, 1985, Geology of the southwestern tip of the Kenai Peninsula, *in* Sisson, A., ed., Guide to the geology of the Kenai Peninsula: Alaska Geological Society, p. 50–68.

Kindle, E. D., 1953, Dezadeash map area, Yukon Territory: Canada Geological Survey Memoir 268, 68 p.

Knopf, A., 1911, Geology of the Berners Bay region, Alaska: U.S. Geological Survey Bulletin 446, 58 p.

Lathram, E. H., Pomeroy, J. S., Berg, H. C., and Loney, R. A., 1965, Reconnaissance geology of Admiralty Island, Alaska: U.S. Geological Survey Bulletin 1181-R, 48 p.

LeHuray, A. P., Church, S. E., and Nokleberg, W. J., 1985, Lead isotopes in massive sulfide deposits from the Jarvis Creek Glacier and Wrangellia terranes, Mount Hayes Quadrangle, eastern Alaska Range, Alaska, *in* Bartsch-Winkler, S., and Reed, M. M., eds., The United States Geological Survey in Alaska; Accomplishments in 1983: U.S. Geological Survey Circular 945, p. 72–73.

Loney, R. A., 1964, Stratigraphy and petrography of the Pybus-Gambier area, Admiralty Island, Alaska: U.S. Geological Survey Bulletin 1176, 103 p.

Loney, R. A., Brew, D. A., Muffler, L.J.P., and Pomeroy, J. S., 1975, Reconnaissance geology of Chichagof, Baranof, and Kruzof Islands, southeastern Alaska: U.S. Geological Survey Professional Paper 792, 105 p.

MacKevett, E. M., Jr., 1978, Geologic map of the McCarthy Quadrangle, Alaska: U.S. Geological Survey Miscellaneous Investigations Series Map I-1032, scale 1:250,000.

Martin, G. C., 1926, The Mesozoic stratigraphy of Alaska: U.S. Geological Survey Bulletin 776, 493 p.

Martin, G. C., Johnson, B. L., and Grant, U.S., 1915, Geology and mineral resources of [the] Kenai Peninsula, Alaska: U.S. Geological Survey Bulletin 587, 243 p.

McClelland, W. C., and Gehrels, G. E., 1987, Analysis of a major shear zone in Duncan Canal, Kupreanof Island, southeastern Alaska: Geological Society of America Abstracts with Programs, v. 19, p. 430.

Metz, P. A., 1976, Occurrences of sodic amphibole-bearing rocks in the Valdez C-1 Quadrangle: Alaska Division of Geological and Geophysical Surveys Report 51, p. 27–28.

Miyashiro, A., 1974, Volcanic rock series in island arcs and active continental margins: American Journal of Science, v. 275, p. 321–355.

Moffit, F. H., 1938, Geology of the Chitina Valley and adjacent area, Alaska: U.S. Geological Survey Bulletin 894, 137 p.

Monger, J.W.H., and Berg, H. C., 1987, Lithotectonic terrane map of western Canada and southeastern Alaska: U.S. Geological Survey Miscellaneous Field Studies Map MF-1874B, 1 sheet, scale 1:2,500,000.

Moore, G. W., 1967, Preliminary geologic map of the Kodiak Islands and vicinity, Alaska: U.S. Geological Survey Open-File Report 67-271, 1 sheet, scale 1:250,000.

Moore, J. C., and Connelly, W., 1977, Mesozoic tectonics of the southern Alaska margin, *in* Taiwani, M., and Pitman, W. C., III, eds., Island arcs, deep sea trenches, and back-arc basins: American Geophysical Union Maurine Ewing Series, v. 1, p. 71–82.

Moore, J. C., and 5 others, 1983, Paleogene evolution of the Kodiak Islands, Alaska; Consequences of ridge-trench interaction in a more southerly latitude: Tectonics, v. 2, p. 265–293.

Muffler, L.J.P., 1967, Stratigraphy of the Keku Islets and neighboring parts of Kuiu and Kupreanof Islands, southeastern Alaska: U.S. Geological Survey Bulletin 1241-C, p. C1–C52.

Muller, J. E., 1967, Kluane Lake map area, Yukon Territory: Geological Survey of Canada Memoir 340, 137 p.

Nauman, C. R., Blakestad, R. A., Chipp, E. R., and Hoffman, B. L., 1980, The north flank of the Alaska Range; A newly discovered volcanogenic massive sulfide belt: Geological Association of Canada Program with Abstracts, v. 5, p. 73.

Nelson, S. W., Blome, C. D., and Karl, S. M., 1986, Late Triassic and Early Cretaceous fossil ages from the McHugh Complex, southern Alaska: U.S. Geological Survey Circular 998, p. 96–98.

Newton, C. R., 1983, Paleozoogeographic affinities of Norian bivalves from the Wrangellian, Peninsular, and Alexander terranes, northwestern North America, *in* Stevens, C. H., ed., Pre-Jurassic rocks in western North American suspect terranes: Pacific Section, Society of Economic Paleontologists and Mineralogists, p. 37–48,

Nilsen, T. H., and Zuffa, G. G., 1982, The Chugach terrane; A Cretaceous trench-fill deposit, southern Alaska, *in* Leggett, J. K., ed., Trench-forearc geology; Sedimentation and tectonics on modern and ancient active plate margins: London, Blackwells, p. 213–227.

Nokleberg, W. J., and Aleinikoff, J. N., 1985, Summary of stratigraphy, structure, and metamorphism of Devonian igneous arc terranes, northeastern Mount Hayes Quadrangle, eastern Alaska Range, *in* Bartsch-Winkler, S., ed., The United States Geological Survey in Alaska; Accomplishments during 1984: U.S. Geological Survey Circular 967, p. 66–71.

Nokleberg, W. J., and Lange, I. M., 1985, Volcanogenic massive sulfide occurrences, Jarvis Creek Glacier terrane, western Mount Hayes Quadrangle, eastern Alaska Range, Alaska, *in* Bartsch-Winkler, S., and Reed, K. M., eds., The United States Geological Survey in Alaska; Accomplishments during 1983: U.S. Geological Survey Circular 945, p. 77–80.

Nokleberg, W. J., and 10 others, 1982, Geologic map of the southern Mount Hayes Quadrangle, Alaska: U.S. Geological Survey Open-File Report 82–52, 27 p., 1 sheet, scale 1:250,000.

Nokleberg, W. J., Jones, D. L., and Silberling, N. J., 1985, Origin and tectonic evolution of the Maclaren and Wrangellia terranes, eastern Alaska Range, Alaska: Geological Society of America Bulletin, v. 96, p. 1251–1270.

Nokleberg, W. J., Plafker, G., Wallace, W. K., and Winkler, G. R., 1989, Structural analysis of the southern Peninsular and northern Chugach terranes along the Trans-Alaskan Crustal Transect (TACT), eastern Chugach Mountains, Alaska: Journal of Geophysical Research, v. 94, p. 4297–4320.

Pallister, J. S., and Carlson, C., 1988, Bedrock geology map of the Angayucham Mountains: U.S. Geological Survey Miscellaneous Field Studies Map, MF-2024, scale 1:250,000.

Pallister, J. S., Budahn, J. R., and Murchey, B. L., 1989, Pillow basalts of the Angayucham Terrane; Oceanic plateau and island crust accreted to the Brooks Range: Journal of Geophysical Research, v. 94, p. 15901–15923.

Patton, W. W., Jr., and Miller, T. P., 1966, Regional geologic map of the Hughes Quadrangle, Alaska: U.S. Geological Survey Miscellaneous Investigations Map I-459, 1 sheet, scale 1:250,000.

——, 1973, Bedrock geologic map of Bettles and southern Part of Wiseman

Quadrangles, Alaska: U.S. Geological Survey Miscellaneous Field Studies Map MF-492, 1 sheet, scale 1:250,000.
Patton, W. W., Tailleur, I. L., Brosgé, W. P., and Lanphere, M. A., 1977, Preliminary report on the ophiolites of northern and western Alaska: Oregon Department of Geology and Industries Bulletin 95, p. 51–57.
Pearce, J. A., 1982, Trace element characteristics of lavas from destructive plate boundaries, *in* Thorpe, R. S., ed., Andesites; Orogenic andesites and related rocks: Chichester, John Wiley and Sons, p. 525–548.
——, 1983, Role of subcontinental lithosphere in magma genesis at active continental margins, *in* Hawkesworth, C. J., and Norry, M. J., eds., Continental basalts and mantle xenoliths: Cheshire, United Kingdom, Shiva Publications, p. 230–249.
Pearce, J. A., and Cann, J. R., 1973, Tectonic setting of basic volcanic rocks determined using trace element analysis: Earth and Planetary Science Letters, v. 19, p. 290–300.
Plafker, G., 1987, Regional geology and petroleum potential of the northern Gulf of Alaska continental margin, *in* Scholl, D. W., Grantz, A., and Vedder, J. G., eds., Geology and resource potential of the continental margin of western North America and adjacent ocean basins: Houston, Texas, Circum-Pacific Council for Energy and Mineral Resources Earth Science Series, v. 6, p. 11-1–11-38.
Plafker, G., and Hudston, T., 1980, Regional implications of Upper Triassic metavolcanic and metasedimentary rocks on the Chilkat Peninsula, southeastern Alaska: Canadian Journal of Earth Sciences, v. 17, p. 681–689.
Plafker, G., Jones, D. L., and Pessagno, E. A., Jr., 1977, A Cretaceous accretionary flysch and mélange terrane along the Gulf of Alaska margin: U.S. Geological Survey Circular 751-B, p. B41–B43.
Plafker, G., Nokleberg, W. J., Roeske, S. M., and Winkler, G. R., 1986, Nature and timing of deformation along the Contact fault system in the Cordova, Bering Glacier, and Valdez Quadrangles, *in* Bartsch-Winkler, S., and Reed, K., eds., Geologic studies in Alaska by the U.S. Geological Survey during 1985: U.S. Geological Survey Circular 978, p. 74–77.
Plafker, G., Blome, C. D., and Silberling, N. J., 1989a, Reinterpretation of lower Mesozoic rocks on the Chilkat Peninsula, Alaska, as a displaced fragment of Wrangellia: Geology, v. 17, p. 3–6.
Plafker, G., Nokleberg, W. J., and Lull, J. S., 1989b, Bedrock geology and tectonic evolution of the Wrangellia, Peninsular, and Chugach terranes along the Trans-Alaskan Crustal Transect in the northern Chugach Mountains and southern Copper River basin, Alaska: Journal of Geophysical Research, v. 94, p. 4255–4295.
Redman, E., 1986, An unconformity with associated conglomeratic sediments in the Berners Bay area of southeast Alaska: Alaska Divison of Geological and Geophysical Surveys Professional Report 86, p. 1–4.
Richter, D. H., 1976, Geologic map of the Nabesna Quadrangle, Alaska: U.S. Geological Survey Miscellaneous Geologic Investigations Map I-932, 1 sheet, scale 1:250,000.
Richter, D. H., and Jones, D. L., 1973a, Structure and stratigraphy of eastern Alaska Range, Alaska, *in* Arctic geology: American Association of Petroleum Geologists Memoir 19, p. 408–420.
——, 1973b, Reconnaissance geologic map of the Nabesna A-2 Quadrangle, Alaska: U.S. Geological Survey Miscellaneous Geologic Invesigations Map I-749, scale 1:63,360.
Roeder, D., and Mull, C. G., 1978, Tectonics of Brooks Range ophiolites, Alaska: American Association of Petroleum Geologists Bulletin, v. 62, p. 1698–1713.
Roeske, S. M., 1986, Field relations and metamorphism of the Raspberry Schist, Kodiak Islands, Alaska, *in* Evans, B. W., and Brown, E. H., eds., Blueschists and eclogites: Geological Society of America Memoir 164, p. 169–184.
Rubin, C. M., and Saleeby, J. B., 1987a, The inner boundary zone of the Alexander terrane; A newly discovered thrust belt: Geological Society of America Abstracts with Programs, v. 19, p. 445.
——, 1987b, The inner boundary of the Alexander terrane in southern SE Alaska; Part 1, Cleveland Peninsula to Revillagigedo Island: Geological Society of America Abstracts with Programs, v. 19, p. 826.
Shervais, J. W., 1982, Ti-V plots and the petrogenesis of modern and ophiolitic lavas: Earth and Planetary Science Letters, v. 59, p. 101–118.
Sisson, V. B., and Onstott, T. C., 1986, Dating blueschist metamorphism; A combined $^{40}Ar/^{39}Ar$ and electron microphobe approach: Geochemica et Cosmochimica Acta, v. 50, p. 2111–2117.
Smith, J. G., and MacKevett, E. M., Jr., 1970, The Skolai Group on the McCarthy B-4, C-4, and C-5 Quadrangles, Wrangell Mountains, Alaska: U.S. Geological Survey Bulletin 1274-Q, p. Q1–Q26.
Sun, S. S., and McDonough, W. F., 1989, Chemical and isotopic systematics of oceanic basalts; Implications for mantle composition and processes, *in* Saunders, A. D., and Norry, M. J., eds., Magmatism in the ocean basins: Geological Society of London Special Publication 42, p. 313–345.
Sun, S. S., and Nesbitt, R. W., 1977, Chemical heterogeneity of the Archean mantle, composition of the Earth, and mantle evolution: Earth and Planetary Science Letters, v. 35, p. 429–448.
Sun, S. S., Nesbitt, R. W., and Sharaskin, A. Y., 1979, Geochemical characteristics of mid-ocean ridge basalts: Earth and Planetary Science Letters, v. 44, p. 119–138.
Thompson, R. N., Morrison, M. A., Dickin, A. P., and Hendry, G. L., 1983, Continental flood basalts; Arachnids rule OK?, *in* Hawkesworth, C. J., and Norry, M. J., eds., Continental basalts and mantle xenoliths: Cheshire, United Kingdom, Shiva Publications, p. 158–185.
Thompson, R. N., Morrison, M. A., Hendry, G. L., and Parry, S. J., 1984, An assessment of the relative roles of crust and mantle in magma gneiss; An elemental approach: Philosophical Transactions of the Royal Society of London, v. A310, p. 549–590.
Thurston, S. P., 1985, Structure, petrology, and metamorphic history of the Nome Group blueschist terrane, Salmon Lake area, Seward Peninsula, Alaska: Geological Society of America Bulletin, v. 96, p. 600–617.
Till, A. B., Dumoulin, J. A., Gamble, B. M., Kaufman, D. S., and Carroll, P. I., 1986, Preliminary geologic map and fossil data, Solomon, Bendeleben, and southern Kotzebue Quadrangles, Seward Peninsula, Alaska: U.S. Geological Survey Open-File Report 86–276, scale 1:250,000.
Tysdal, R. G., and Case, J. E., 1979, Geologic map of the Seward and Blying Sound Quadrangles, Alaska: U.S. Geological Survey Miscellaneous Investigations Map I-1150, 12 p., scale 1:250,000.
Tysdal, R. G., and Plafker, G., 1978, Age and continuity of the Valdez Group, southern Alaska, *in* Shol, N. F., and Wright, W. B., compilers, Changes in stratigraphic nomenclature by the U.S. Geological Survey, 1977: U.S. Geological Survey Bulletin 1457-A, p. 120–124.
Vallier, T. L., 1986, Tectonic implications of arc axis (Wallowa Terrane) and back-arc (Vancouver Island) volcanism in Triassic rocks of Wrangellia [abs.]: American Geophysical Union Transactions, v. 67, p. 1233.
Winkler, G. R., 1976, Deep-sea fan deposition of the lower Tertiary Orca Group, eastern Prince William Sound, Alaska, *in* Miller, T. P., ed., Recent and ancient sedimentary environments in Alaska: Alaska Geological Society, p. R1–R20.
Winkler, G. R., Silberman, M. L., Grantz, A., Miller, R. J., and MacKevett, E. M., Jr., 1981, Geologic map and summary geochronology of the Valdez Quadrangle, southern Alaska: U.S. Geological Survey Open-File Report 80-892A, 2 sheets, scale 1:250,000.
Wood, D. A., 1980, The application of a Th-Hf-Ta diagram to problems of tectonomagmatic classification and to establishing the nature of crustal contamination of basaltic lavas of the British Tertiary Volcanic Prvoince: Earth and Planetary Science Letters, v. 50, p. 11–30.
Zuffa, G. G., Nilsen, T. H., and Winkler, G. R., 1980, Rock-fragment petrography of Upper Cretaceous Chugach terrane, southern Alaska: U.S. Geological Survey Open-File Report 80–713, 28 p.

MANUSCRIPT ACCEPTED BY THE SOCIETY FEBRUARY 14, 1991

Printed in U.S.A.

The Geology of North America
Vol. G-1, The Geology of Alaska
The Geological Society of America, 1994

Chapter 18

Latest Cretaceous and Cenozoic magmatism in mainland Alaska

Elizabeth J. Moll-Stalcup
U.S. Geological Survey, National Center, Reston, Virginia 22092

INTRODUCTION

Continental Alaska has been the site of widespread magmatism throughout much of the late Mesozoic and Cenozoic, but until recently, most of this magmatism was unrecognized due to the lack of modern geologic maps or isotopic age data for large tracts of Alaska. Although parts remain unmapped, progress in reconnaissance mapping and dating have enabled workers to identify major late Mesozoic and Cenozoic magmatic provinces outside the well-known Aleutian arc and to speculate as to their tectonic implications and origin (Wallace and Engebretson, 1984).

This chapter defines major Late Cretaceous and Cenozoic magmatic provinces in Alaska outside the Aleutian arc (Kay and Kay, this volume; Vallier and others, this volume; Miller and Richer, this volume) and southeast Alaska (Brew, this volume), and discusses their distribution, age, petrology, and tectonic implications. The available data suggest that Late Cretaceous and Cenozoic magmatism in continental Alaska can be roughly divided into three periods: (1) latest Cretaceous and early Tertiary (76 to 50 Ma), (2) middle Tertiary (43 to 37 Ma), and (3) late Tertiary and Quaternary (6 Ma to the present). Late Cretaceous and early Tertiary calc-alkalic volcanism and plutonism were widespread over much of western, central, and southern Alaska and on the Bering Sea shelf. Middle Tertiary magmatism was characterized by the eruption of small volumes of calcalkalic rocks in interior Alaska, contemporaneous with the inception of a major pulse of magmatism in the Aleutian arc. Late Tertiary and Quaternary volcanism has been characterized by the eruption of voluminous basaltic magma at numerous sites along the western margin of Alaska and on the Bering Sea shelf.

This chapter is accompanied by a map showing Cenozoic volcanic and plutonic rocks for the entire state at a scale of 1:2.5 million (Moll-Stalcup and others, this volume, Plate 5). Place names used in this chapter appear on that map. Although only major belts of regional significance are discussed in this chapter, tables summarizing age, lithologic, and chemical data for all the Latest Cretaceous to Quaternary volcanic and plutonic rocks outside the Aleutian arc and southeast Alaska are found on Plate 5.

Rock nomenclature used in this chapter generally follows that of Streckeisen (1979), Gill (1981), and Morrison (1980). On an anhydrous basis, basalts have less than 53 percent SiO_2, andesites have 53 to 63 percent SiO_2, dacites have 63 to 70 percent SiO_2, and rhyolites have more than 70 percent SiO_2. Fe_2O_3/FeO was set to 0.15 for the late Cenozoic basalts in order to calculate Mg numbers and normative mineralogies. The late Cenozoic rocks are classified as follows: Basalt having normative nepheline is called alkali basalt or alkali olivine basalt if it contains more than 10 percent normative olivine. Basalt having normative hypersthene is called tholeiite or olivine tholeiite if it contains more than 10 percent normative olivine. Basanites have 10 to 20 percent normative nepheline; nephelinites have more than 20 percent normative nepheline. Hawaiites have more than 5 percent total alkalies ($Na_2O + K_2O$) and less than 5 percent MgO. The Late Cretaceous, early Tertiary, and middle Tertiary suites are classified using the Peacock index and are then divided into low-, moderate-, or high-K, after Gill (1981), or shoshonitic after Morrison (1980). An upper limit for Fe_2O_3 was set by the formulate $\%Fe_2O_3 = \%TiO_2 + 1.5$ (after Irvine and Baragar, 1971). All ages were obtained by K/Ar methods unless otherwise noted.

All the Late Cretaceous and early Tertiary volcanic and plutonic rocks and some of the middle Tertiary rocks are hydrothermally altered and weathered, and I therefore have relied heavily on trace elements for interpretation of the geochemical data. Typical plutonic samples have about 1 percent total H_2O and 0.2 percent CO_2, and typical volcanic rocks have 1 to 4 percent H_2O^t and 0.2 percent CO_2. Pyroxenes and feldspars are generally fresh, but olivine, biotite, and hornblende are altered in some samples. Alteration of rocks from five volcanic fields, which are typical of much of the magmatic province, are described in more detail in Moll-Stalcup (1987). The late Cenozoic volcanic rocks are fresh, and even olivine is well preserved in most samples.

LATEST CRETACEOUS AND EARLY TERTIARY MAGMATISM

Latest Cretaceous and early Tertiary magmatic activity occurred in a vast region of Alaska stretching from the southern continental margin north to the Arctic Circle and west to the

Moll-Stalcup, E. J., 1994, Latest Cretaceous and Cenozoic magmatism in mainland Alaska, *in* Plafker, G., and Berg, H. C., eds., The Geology of Alaska: Boulder, Colorado, Geological Society of America, The Geology of North America, v. G-1.

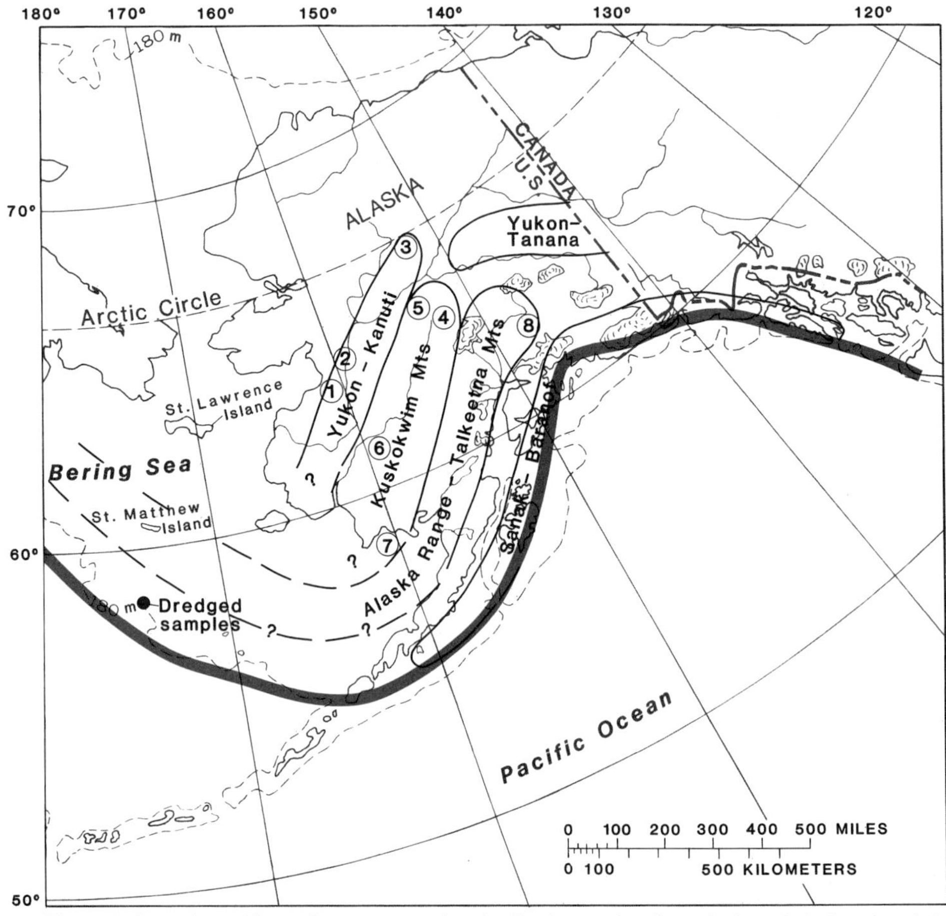

Figure 1. Location of Late Cretaceous and early Tertiary volcanic and plutonic belts of mainland Alaska. Proposed Paleocene continental margin shown in red. A few locations are: 1, Blackburn Hills volcanic field; 2, Yukon River area; 3, Kanuti volcanic field; 4, Sischu volcanic field; 5, Nowitna volcanic field; 6, Sleetmute area; 7, Bristol Bay; and 8, Talkeetna Montains. Dashed line marking 180-m water depth delineates edge of the Bering Sea shelf.

Bering Sea shelf (Fig. 1), possibly extending as far west as contemporaneous magmatic belts in eastern Siberia. Hudson (1979) and Wallace and Engebretson (1984) group the widespread volcanic and plutonic rocks in southern, western, and central Alaska into volcano-plutonic belts. From south to north, these are: The Sanak-Baranof belt, the Alaska Range–Talkeetna Mountains belt, and the Kuskokwim Mountains belt. An additional, previously unnamed, belt occurs farther to the northwest and is herein named the Yukon-Kanuti belt. Little-known volcanic and plutonic rocks of latest Cretaceous and early Tertiary age also occur in the Yukon-Tanana area of east-central Alaska (Foster and others, this volume), but their correlation and tectonic affinities are unknown. Data on the Late Cretaceous and early Tertiary rocks in the Yukon-Tanana upland are summarized on Table 1 of Plate 5, but those rocks are not discussed further in this chapter. The Sanak-Baranof belt, which consists of early Tertiary granitic plutons emplaced into the Gulf of Alaska accretionary wedge, extends along the southern margin of Alaska to Baranof Island in southeast Alaska and is described by Hudson (this volume).

Alaska Range–Talkeetna Mountains belt

The Alaska Range–Talkeetna Mountains belt consists of numerous coalescing plutons and subordinate volcanic rocks that extend in a broad belt, about 150 km wide, from the central Alaska Range, west and south to the Iliamna Lake region (Fig. 1). Most of the rocks occur south of the Denali fault, except for a few small bodies north of the fault near Farewell Lake and the Tonzona River (Reed and Nelson, 1980). Plutonic and volcanic rocks in the Alaska Range–Talkeetna Mountains belt intrude and overlie the Dillinger, Kahiltna, and Peninsular terranes, as well as a number of smaller terranes in the Mt. McKinley area (Silberling and others, this volume). Aeromagnetic anomalies on the Bering Sea shelf (Godson, 1984; Cooper and others, 1986) and the presence of compositionally similar contemporaneous volcanic rocks on St. Matthew Island (Patton and others, 1975) suggest that the belt may continue southwest of Iliamna Lake under Bristol Bay, curving west and north along the submerged continental shelf of Alaska (Fig. 1).

Plutonic activity in the Alaska Range–Talkeetna Mountains belt is divided into an early stage (75 to 60 Ma), which occurred chiefly on the south-southeast flank of the belt; and a late stage (65 to 50 Ma), which occurred chiefly on the north-northwest flank of the belt. The early stage consists of dominantly intermediate to felsic plutons and includes many of the rocks of Summit Lake (Reed and Lanphere, 1972), the Mount Susitna pluton (Magoon and others, 1976), and a large tonalite pluton in the southern Talkeetna Mountains (Csejtey, 1974). The late stage consists generally of felsic plutons and includes the quartz monzonite of Tired Pup, the Crystal Creek sequence, the McKinley sequence (Reed and Lanphere, 1972), and numerous granitic plutons in the northern Talkeetna Mountains and the adjacent southern Alaska Range (Csejtey and others, 1978, 1986).

Volcanic rocks in the Talkeetna Mountains consist of several small fields and one large field approximately 90 by 25 km that trends southeast perpendicular to the belt. The large volcanic field is more than 1,500 m thick and is composed of rhyolite and dacite stocks, irregular dikes, lenticular flows, and thick pyroclastic rocks at the base grading up into gently dipping interlayered basalt and andesite flows at the top (Csejtey and others, 1978). Three rocks from about midsection give ages of 56.5 to 50.4 Ma, indicating that the lower part of the section is Paleocene and Eocene in age. The stratigraphically high mafic and intermediate flows are thought to be equivalent in age to Miocene lava flows in the Wrangell Mountains (Csejtey and others, 1978).

The volcanic rocks of the Cantwell Formation crop out north of the Talkeetna Mountains volcanic rocks in the central Alaska Range, covering about 165 km^2 in the eastern part of the Mount McKinley National Park. They consist of at least 3,750 m of mostly andesite and rhyolite flows and subordinate basalt flows, felsic pyroclastic rocks, and related intrusive rocks (Gilbert and others, 1976). Gilbert and others (1976) considered the K/Ar ages of 60.6, 57.2, and 41.8 Ma to be minimum ages and interpret the formation as Paleocene in age.

Little-known volcanic rocks near Lake Clark consist of undivided Paleocene and Eocene volcanic and associated plutonic rocks that crop out discontinuously over more than 3,000 km^2 in the area between Lake Clark and the Mulchatna River (Nelson and others, 1983). The volcanic rocks are 62.7 to 56.2 Ma (Eakin and others, 1978) and 44.4 to 39.7 Ma (Thrupp and Coe, 1986); adjacent shallow plutons are 71.3 to 60.5 Ma (Nelson and others, 1983). The volcanic rocks are described as rhyolite breccia, lava flows and ash-flow tuffs, and subordinate mafic to intermediate flows; the intrusive rocks are described as granite, granodiorite, and diorite (Nelson and others, 1983).

The east Susitna batholith (Plate 5) occurs at the east end of the Alaska Range–Talkeetna Mountains belt on the east limb of the bend in the Denali fault. Although the batholith yields Late Cretaceous and early Tertiary minimum ages, it is not considered to be part of the Alaska Range–Talkeetna Mountains belt because it consists of regionally metamorphosed and penetratively deformed diorite, granodiorite, and quartz monzonite that give a wide range of K/Ar ages (Nokleberg and others, 1982; Table 1 on Plate 5), and because it is thought to have been 400 km from its present position at the time of its emplacement (Nokleberg and others, 1985). Tertiary displacements on regional strike-slip faults along the east side of the bend in the Denali fault, where the east Susitna batholith occurs, are thought to be much greater than displacements on the west side, where most of the Alaska Range–Talkeetna Mountains belt occurs. Because the batholith probably is allochthonous relative to the Alaska Range–Talkeetna Mountains belt and thus is not part of this belt, and because its age is ambiguous, it is not discussed further in this chapter.

Petrogenesis. The calc-alkalic plutons and volcanic rocks in the Alaska Range–Talkeetna Mountains belt are typical of continental-margin arc rocks, and are characterized chemically by low TiO_2, moderate K_2O and lack of Fe-enrichment (data from Reed and Lanphere, 1972, 1974a; Csejtey, 1974; Csejtey and others, 1978; Gilbert and others, 1976; Lanphere and Reed, 1985). The early-stage plutons in the Alaska Range are dominantly intermediate to felsic (54.5 to 70 percent SiO_2) and are compositionally equivalent to medium-K orogenic andesites and dacites. Plots of SiO_2 versus Na_2O and Ca-Na-K distinguish two suites of rocks: a diorite, tonalite, trondhjemite suite similar to the calc-alkalic-trondhjemite suite of southwest Finland (Arth and others, 1978) and a "normal" calc-alkalic suite. I was not able to determine from the published data (above references) whether the two suites are temporally or geographically distinct.

The late-stage plutons in the Alaska Range are generally more felsic and are divided into two groups on the basis of mineralogy and chemistry. One group is a normal calc-alkalic suite similar to the early-stage plutons and consists of the early Tertiary Crystal Creek pluton and numerous small granitic bodies in the northern Alaska Range. The other group is represented by the McKinley sequence and the quartz monzonite of Tired Pup (Lanphere and Reed, 1985; M. A. Lanphere, personal communication, 1984), which consist of siliceous peraluminous granites that plot at minimum-melt compositions on a Q-Ab-Or diagram and have moderately high strontium initial ratios ($^{87}Sr/^{86}Sr$ = 0.7054 to 0.7085; Lanphere and Reed, 1985). Lanphere and Reed (1985) interpret the chemical and isotopic data of the McKinley sequence as the result of mixing of mantle-derived magmas with upper Mesozoic flysch into which the plutons were intruded. They believe that this mixing took place in the early Tertiary during the collision between stable Alaska and southern accreted terranes. Paleomagnetic data, however, suggest that the terranes were accreted by Late Cretaceous time (Hillhouse and Coe, this volume). K/Ar ages (Lanphere and Reed, 1985; Reed and Lanphere, 1972, 1974a) suggest that these plutons were emplaced at the end of a long period of arc magmatism and that they may simply mark the end of this event. Chondrite-normalized multi-element diagrams for the McKinley sequence show large depletions in Nb and Ta, which are diagnostic of arc magmatism (Fig. 2; Perfit and others, 1980; Gill, 1981; Thompson and others, 1984).

Volcanic rocks having ages between 58 and 50 Ma crop out in the northern Alaska Range and Talkeetna Mountains at the east end of the belt. Limited petrologic data on the lower se-

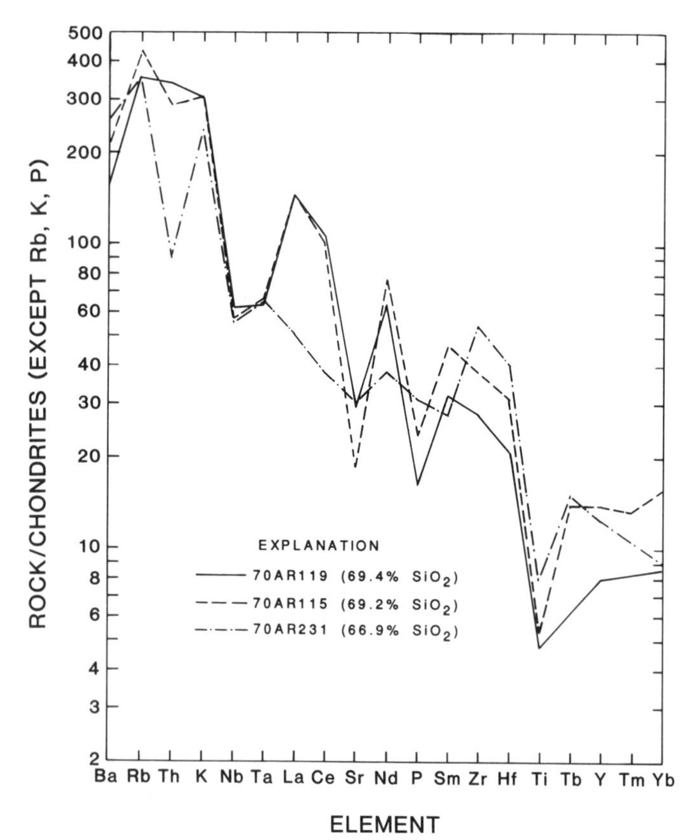

Figure 2. Chondrite-normalized spidergrams for rocks from the McKinley sequence. Data from Lanphere and Reed (1985); normalization factors from Thompson and others (1984). The rocks have deep Nb-Ta troughs characteristic of subduction- related magmas. Sharp spikes at Sr, P, and Ti suggest that the rocks are highly differentiated and have fractionated plagioclase, Fe-Ti oxides, and apatite.

quence of the Talkeetna Mountains volcanic rocks suggest they are typical of orogenic belts and are similar in composition to the older Summit Lake plutons. Analyzed rock samples from the Cantwell Formation have higher TiO_2 and lower Al_2O_3 (Gilbert and others, 1976)—both immobile elements—than the other clearly orogenic suites. Granitic plutons in the same region also have ages from 58 to 50 Ma, but their affinities are not known. The lack of muscovite and presence of rare hornblende suggest that they are similar to the Crystal Creek sequence, but the association of some plutons with tin mineralization suggests instead that they may be correlative with the peraluminous McKinley sequence.

The age and composition of the Talkeetna Mountains volcanic rocks suggest that arc volcanism occurred later in the eastern part of the belt than in the western or southern parts. Arc volcanism appears to have migrated gradually north between about 65 and 58 Ma, then east, until it shut off at about 50 Ma.

Kuskokwim Mountains belt

The Kuskokwim Mountains belt is a Late Cretaceous and early Tertiary volcanic-plutonic belt that extends along the entire length of the Kuskokwim Mountains for more than 800 km, from Bristol Bay to about latitude 64° (Fig. 1). Studies in the Medfra Quadrangle (Patton and others, 1980; Moll and others, 1981), in the Iditarod and McGrath Quadrangles (M. L. Miller, personal communication, 1986; Bundtzen and Laird, 1982, 1983a, b, c), in the Sleetmute Quadrangle area (Robinson and others, 1984; Decker and others, 1984, 1985, 1986; Reifenstuhl and others, 1984, 1985), and in the Tikchik Lakes–Bristol Bay region (Hoare and Coonrad, 1978b; Wilson, 1977; Globerman, 1985) have led to the recognition of this volcanic and plutonic belt of Late Cretaceous and early Tertiary age (Moll and Patton, 1982; Wallace and Engebretson, 1984). Compilation of more than 90 K/Ar ages from volcanic and plutonic rocks along the belt (E. J. Moll, unpublished data, 1985) suggests that the main magmatic pulse occurred between 72 and 60 Ma, contemporaneous with the older-stage plutonism in the Alaska Range. It is not yet clear if the Kuskokwim Mountains belt is separate from the Alaska Range–Talkeetna Mountains belt. Much of the area separating the two belts is covered by Quaternary surficial deposits on the north side of the Farewell fault and in the vicinity of the Mulchatna fault. Exposures of plutonic and volcanic rocks in low hills in the Farewell Lake area and on both sides of the Mulchatna fault suggest that the two belts may be continuous. The volcanic and intrusive rocks in the Kuskokwim Mountains belt are divided into six groups: (1) northern volcanic fields; (2) northern volcanoplutonic complexes; (3) northern plutons, dikes, and sills; (4) volcanic and plutonic rocks in the Sleetmute-Nyac area; (5) plutonic rocks of the southern Kuskokwim Mountains region; and (6) the Bristol Bay volcanic sequence. The northern part of the belt overlies the Ruby, Innoko, Nixon Fork, and Minchumina terranes; the southern part overlies the Dillinger, Tikchik Lake, Nyac, Kilbuck, Togiak, and Goodnews terranes of Jones and others (1987), and Silberling and others (this volume).

The northern volcanic fields include the Sischu, Nowitna, Dishna, and Yetna volcanic fields (Fig. 1; Plate 5). The Sischu volcanic field (71 to 66 Ma) consists of a narrow belt of poorly exposed rhyolite and dacite domes, flows, and tuff that extends from the Sischu Mountains to the southern Chitanatala Mountains (Moll and others, 1981; Moll-Stalcup and Arth, 1989). The main volcanic field covers an area of more than 725 km^2, is at least 500 m thick, and is fault bounded on the southeast side. A felsic pluton that crops out just east of the volcanic rocks has an age of 64 Ma (Silberman and others, 1979).

The Nowitna volcanic field consists of more than 1,500 m of chiefly andesitic flows preserved in a gently folded northeast-trending syncline that is fault-bounded on the southeast side. The field covers more than 2,700 km^2, overlapping the suture between the Innoko and Nixon Fork terranes. At least seven highly altered rhyolite domes overlie the andesite flows. K/Ar ages on three whole-rock andesite samples collected near the top of the section are 64 to 63 Ma (Silberman and others, 1979).

The Dishna volcanic field consists of calc-alkalic dacite, rhyolite, and minor andesite poorly exposed in a series of isolated ridges and hills that rise above the alluvium in the Innoko-Dishna

Rivers area (Chapman and others, 1985). These undated rocks are presumed to be Late Cretaceous or early Tertiary in age.

Volcanoplutonic complexes occur in the McGrath area at Page Mountain, Cloudy Mountain, Candle Mountain, Takotna Mountain, Mount Joaquin, the Beaver Mountains, and in the Lonesome Hills. These complexes have circular-shaped outcrop areas that consist of andesite flows and shallow hypabyssal rocks intruded by small granitic stocks. Most of the volcanic rocks are highly altered by the intrusions. The margins of many of the complexes appear to be fault-bounded and the complex at Page Mountain is down-faulted against the surrounding sedimentary rocks. The complexes are interpreted as being deeply eroded volcanic centers. Dates on volcanic and intrusive rocks from these complexes yield K/Ar ages ranging from 73 to 65 Ma (Moll and others, 1981; Bundtzen, and Laird, 1982, 1983a, b, c).

Widespread intrusive rocks, many too small to be shown on published maps, occur throughout the Kuskokwim Mountains belt. In the northern Kuskokwim Mountains, numerous dikes, sills, and small stocks, usually 1 to 9 km in diameter, give similar K/Ar ages (72 to 62 Ma) and are compositionally similar to the volcanic rocks. Most of the intrusive rocks are compositionally homogeneous monzonite, monzodiorite, quartz monzodiorite, quartz monzonite, or granite. Plutons at Von Frank and Stone Mountain, however, are compositionally zoned, and commonly grade inward from gabbro or monzogabbro at the margin, to quartz monzonite in the center of the pluton.

Volcanic and plutonic rocks in the Sleetmute-Nyac area have K/Ar ages ranging from 75 to 61.7 Ma (Robinson and others, 1984; Decker and others, 1985, 1986; Reifenstuhl and others, 1984). The volcanic sequence, named the Holokuk Basalt by Cady and others (1955), consists in fact of more than 1,000 m of chiefly andesite flows and lahars and minor rhyolite vitric tuff and breccia (Decker and others, 1986). Most of the andesites are older (74.5 to 64.3 Ma) than the rhyolites. Intrusive rocks include the Chuilnuk and Kiokluk granodiorite plutons dated at 68.7 to 67.5 Ma, intermediate stocks and dikes dated at 69.8 Ma, and a number of small rhyolite porphyries dated at 70.5 to 67.9 Ma and 61.5 Ma. Similar biotite and biotite-muscovite rhyolite porphyries occur to the north along the Nixon Fork–Iditarod fault and contain garnet (Bundtzen and Swanson, 1984).

More than 30 plutons, usually small stocks 3 to 15 km in diameter, occur in the southern Kuskokwim Mountains in the area between the Nushagak and Kuskokwim Rivers (Wilson, 1977; Hoare and Coonrad, 1978b). Hoare and Coonrad (1978b) describe them as monzonite, granodiorite, and quartz diorite stocks, "mafic" dikes and sills; and felsic dikes, sills, tuffs, and breccias. K/Ar ages for all the rock types in the Goodnews-Hagemeister Quadrangles range from 72.5 to 60.7 Ma. Poorly known granitic stocks to the north and east of the Goodnews-Hagemeister Quadrangles appear to be contemporaneous with, and compositionally similar to, those in the quadrangle (Wilson, 1977; J. M. Hoare, oral communication, 1980).

A thick sequence of volcanic rocks called the Bristol Bay volcanic sequence (Globerman, 1985) is exposed on Hagemeister, Walrus, and Summit Islands in Bristol Bay and on the adjacent mainland. The rocks are dated at 68.7 to 64.5 Ma (Globerman, 1985; Box, 1985). The volcanic rocks consist of andesitic lava flows interbedded with tuffs, breccias, and volcanogenic sedimentary rocks that are exposed in a section more than 2 km thick.

Petrogenesis. The Kuskokwim Mountains belt consists of moderate-K calc-alkalic to shoshonitic suites that range in composition from basalt to rhyolite. Present exposures suggest that andesite, followed by rhyolite, are the overwhelmingly dominant volcanic rocks types. Dacite and basalt are relatively uncommon, and rocks having less than 52 percent SiO_2 are rare. Most of the intrusive rocks have intermediate to felsic compositions, and many are compositionally equivalent to dacites, plotting in the silica gap (63 to 70 percent SiO_2) defined by the volcanic rocks. Mineralogies vary considerably according to rock type and K_2O content (Table 1, Moll-Stalcup and others, this volume).

Major-element data on the volcanic and plutonic rocks show trends typical of most igneous calc-alkalic suites: MgO, FeO*, TiO_2, Al_2O_3, and CaO decrease with increasing SiO_2; K_2O and Na_2O increase with increasing SiO_2. TiO_2 is low (less than 1.75 percent), and Al_2O_3 is moderate (12 to 17 percent). None of the suites shows Fe enrichment. K_2O varies from moderate (1.3 percent at 56 percent SiO_2) to very high values (4 percent at 56 percent SiO_2). Moderate to high-K suites plot in the subalkaline field of Irvine and Baragar (1971) on a total alkalies versus SiO_2 diagram and are calc-alkalic. Very high-K suites (Von Frank and Whirlwind on Fig. 3) plot in the alkalic field and are classified as shoshonitic (Morrison, 1980). Shoshonitic and calc-alkalic suites are similar in all major elements except K and P, which correlates with K. In the northern Kuskokwim Mountains the high-K calc-alkalic and shoshonitic suites tend to be older (71 to 65 Ma) than the moderate-K suites (68 to 62 Ma), although there is considerable overlap.

Major- and trace-element data suggest that the volcanic and plutonic rocks are highly enriched in Ba, Rb, Th, K, and Sr and depleted in Nb and Ta relative to La (Fig. 4). These features are characteristic of subduction-related arc rocks (Perfit and others, 1980; Gill, 1981; Thompson and others, 1984). All the rocks are LREE (light rare earth element) enriched, but the degree of enrichment varies, correlating with the abundance of K and other incompatible elements: shoshonitic rocks have La about 150; high-K rocks have La about 100; and moderate-K rocks have La about 75 × chondritic abundances. In contrast, andesites from the entire belt have similar HREE (heavy rare earth element) contents (6 to 13 × chondrites). There is also a rough correlation between geographic area and degree of incompatible-element enrichment. Andesites from the Bristol Bay volcanic sequence in the southernmost part of the belt have lower incompatible-element contents than andesites from Sleetmute, 360 km to the north (Table 1), which have lower incompatible-element contents than andesites and intermediate plutonic rocks from the northern Kuskokwim Mountains. LIL (large ion litho-

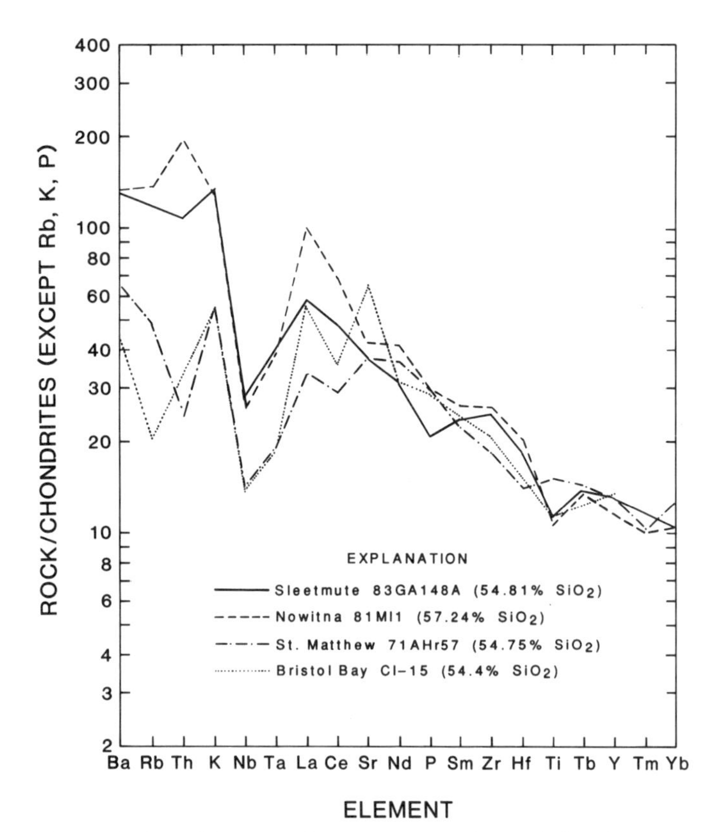

Figure 3. Chondrite-normalized spidergrams for andesites from the Kuskokwim Mountains belt and St. Matthew Island. Data for northern Kuskokwim Mountains (Nowitna) and St. Matthew Island from E. J. Moll-Stalcup and W. W. Patton Jr. (unpublished data, 1985). Data for Sleetmute from Decker and others (1986); for Bristol Bay from Globerman (1985).

phile) elements (K, Rb, Ba, Th, LREE) show the greatest increase from south to north; high-field-strength (HFS) elements (Zr, Hf, Nb, and Ta), which are also incompatible, increase to a lesser degree (Table 1). Alkali element contents in the highly altered volcanoplutonic complexes are high and variable, but REE data from the complexes are similar to data from the less altered Nowitna volcanic field, suggesting that the two suites are chemically similar (compare Cloudy Mountain and Nowitna on Fig. 5D). Trace-element ratios (Ba/Ta, Ba/La, La/Nb) in andesites from all three K-groups are similar to arc andesites (Gill, 1981) despite the higher contents of these elements in the high-K calc-alkalic and shoshonitic groups that are "typical" in arcs.

REE patterns for rhyolites, dacites, quartz monzonites, and granites from the Kuskokwim Mountains belt are highly variable. Most of the rhyolites and dacites from the Sischu volcanic field have patterns that have very high LREE and low HREE (3 to 5 × chondrites), but some have extremely high LREE and moderate HREE (Figs. 4A and 4B). The samples with low HREE contents show a weak correlation between silica and decreasing HREE, which Moll-Stalcup and Arth (1989) attribute to hornblende fractionation or formation by partial melting of a hornblende-bearing source (Moll-Stalcup, 1987). Rhyolites from the Sischu field that have extremely high LREE and very large negative Eu anomalies are probably highly fractionated high-silica rhyolites. Rhyolites and granites from the Sleetmute area have more moderate REE patterns (Fig. 4C) similar to those of the granitic intrusive rocks in the northern Kuskokwim Mountains.

Even the most mafic rocks in the Kuskokwim Mountains belt have compositions that suggest that they have undergone significant fractionation, and many show evidence for interaction with continental crust. In the northern Kuskokwim Mountains, where the basement is Precambrian and Paleozoic rocks, andesites from the Nowitna volcanic field have isotopic and trace-element variations that are best explained as assimilation of small amounts of continental crust during crystal fractionation (Moll and Arth, 1985; Moll-Stalcup and Arth, 1989). Rhyolites from the nearby Sischu volcanic field have high Sn, Be, U, W, and F contents and initial $^{87}Sr/^{86}Sr$ greater than 0.7080 (Moll and Arth, 1985; Moll and Patton, 1983), which suggests that they either were strongly contaminated by continental crust or were generated by partial melting of the crust.

In the Sleetmute area, initial $^{87}Sr/^{86}Sr$ varies from 0.704 to 0.706 (M. Robinson and R. Reifenstuhl, written communication, 1985). Initial $^{87}Sr/^{86}Sr$ does not correlate with SiO_2, Rb/Sr, or 1/Sr, indicating that crustal contamination of isotopically homogeneous magmas is not the only cause of this isotopic variation. Andesites, rhyolites, granites, and quartz monzonites have a similar range of initial $^{87}Sr/^{86}Sr$ (andesites: 0.7040 to 0.7060; felsic rocks: 0.7049 to 0.7063; M. Robinson and R. Reifen stuhl, written communication, 1985). The volcanic and plutonic rocks overlie sedimentary rocks of the Cretaceous Kuskokwim Group, but the nature and age of the older basement rocks is uncertain (S. Box, oral communication, 1986). The presence of andesites having initial $^{87}Sr/^{86}Sr$ as high as 0.706 suggests that old (Precambrian and Paleozoic) radiogenic continental crust or lithosphere, or sedimentary rocks derived from old continental crust occur in the Sleetmute area, although the mechanism for interaction with the crust is not yet understood. In contrast, initial $^{87}Sr/^{86}Sr$ on andesites from the Bristol Bay volcanic sequence are uniformly low (0.7037 to 0.7041; Globerman, 1985), indicating that old continental crust does not underlie this area.

St. Matthew Island, on the Bering Sea shelf, is composed entirely of Late Cretaceous and early Tertiary volcanic rocks ranging in composition from basalt to rhyolite (Patton and others, 1975). The island may be part of the Alaska Range–Talkeetna Mountains belt, part of the Kuskokwim Mountains belt, or neither, but the K_2O contents are most like those in volcanic rocks in the Alaska Range–Talkeetna Mountain belt (Fig. 5). No trace-element data on rocks from Alaska Range–Talkeetna Mountains belts have been published except for those on the anomalous McKinley sequence. Comparisons of incompatible-element contents between rocks from the Kuskokwim Mountains belt (Table 1) and rocks from St. Matthew Island show that the St. Matthew Island rocks have slightly lower abundances of incompatible elements than the Bristol Bay volcanic sequence, which has the lowest contents in the belt. REE data for St. Matthew andesites have lower LREE than andesites from either Sleetmute or the northern Kuskokwim Mountains (Fig. 4D), but have La contents similar to that of the Bristol Bay volcanic sequence (Table 1).

The unusually large range in incompatible element contents of rocks in the Kuskokwim Mountains belt is not well understood. K and incompatible elements vary geographically and temporally; along the strike of the belt, K increases from south to north, and in the northern Kuskokwim Mountains, the K content decreases with time from shoshonitic to moderate-K.

The volcanic and plutonic rocks in the Kuskokwim Mountains belt are compositionally similar to arc volcanic rocks that are thought to be generated by partial melting in a mantle wedge that has been modified by alkali-enriched fluids derived from the subducted slab and/or subducted sediments (Arculus and Powell, 1986). In arc magmas, K contents are separately influenced by the composition of the crust and the depth to the slab. K has been empirically observed to increase with distance from the trench, or depth to the slab. Additional factors that can influence the degree of K and other incompatible-element enrichment include greater degrees of crustal contamination or the presence of old, thicker, or more enriched crust or lithosphere. Estimates of crustal thickness from gravity studies (Barnes, 1977) suggest that the present crust is 30 to 35 km thick under the Kuskokwim Mountains, except in the southernmost part near Bristol Bay where it is 25 to 30 km thick. However, these present-day estimates are not well constrained because of the lack of seismic refraction data for the entire region. Furthermore, present-day crustal thickness may not be the same as crustal thicknesses in the early Tertiary.

The basement in the northern Kuskokwim Mountains consists of Precambrian and Paleozoic continental crust of the Ruby, Nixon Fork, and Minchumina terranes, which contrasts sharply with the late Paleozoic and Mesozoic mafic and intermediate oceanic rocks of the terranes in the south. Isotopic data suggest that the igneous rocks in the northern part of the belt have interacted with old continental crust. Thus, the increase in incompatible-element contents from south to north along the belt is probably related to the presence of old continental crust in the northern Kuskokwim Mountains.

Increasing incompatible-element contents along the strike of a continental arc has been documented by Hildreth and Moorbath (1988) in the Chilean Andes. They attribute the increase in incompatible elements from south to north to continental influence, because other factors, such as sediment subduction or depth to the Benioff zone, do not vary geographically, and the basement in the north is both thicker and older than the basement in the south.

K variation in rocks emplaced through old continental crust in the northern Kuskokwim Mountains cannot be attributed solely to the type of crust because all three suites overlie the same type of basement. Furthermore, there is no correlation between initial $^{87}Sr/^{86}Sr$ and K-type (E. J. Moll and M. L. Silberman, unpublished data, 1985), which suggests that the high K contents are neither the result of greater degrees of crustal contamination nor of local crustal inhomogeneities. There is, however, a weak correlation between age and K type, because the shoshonitic to high-K suites tend to be older than the moderate- to high-K suites. Perhaps the variation in K in the northern Kuskokwim Mountains is tectonically controlled by depth to the Benioff zone and the dip of the slab may have become shallower over time. This model would account for both the higher K content of the rocks and the occurrence of a narrower magmatic belt prior to 65 Ma. Alternatively, the older rocks may be more enriched in incompatible elements simply because those elements were concentrated in the first partial melts.

Yukon-Kanuti belt

The Yukon-Kanuti belt (Fig. 1) consists of an aligned group of early Tertiary volcanic fields that form a northeast-trending belt located north and west of the Kuskokwim Mountains belt. The belt extends from the Arctic Circle near the town of Bettles for more than 300 km southwest to the Kaltag fault and continues south of the fault on the west side of the Yukon River (Plate 5, Fig. 5). There is no clear boundary between the Kuskokwim Mountains and Yukon-Kanuti belts in the region south of the Kaltag fault. Late Cretaceous and early Tertiary volcanic rocks are divided into the two belts because the Yukon-Kanuti belt contains generally younger rocks (66 to 47 Ma) than are found in the Kuskokwim Mountains belt (76 to 60 Ma), and because the Yukon-Kanuti belt lies within the Yukon-Koyukuk province (Patton and others, this volume, chapter 7). The Yukon-Koyukuk province consists chiefly of late Paleozoic and Mesozoic mafic and intermediate volcanic rocks and associated sedimentary rocks, which contrast sharply with the Precambrian and lower Paleozoic schist and carbonate rocks that underlie the northern Kuskokwim Mountains belt. A further distinction is that the depth of post–early Tertiary erosion in the Yukon-Koyukuk province is shallower; as a result, there are more volcanic rocks and fewer plutonic rocks exposed in the Yukon-Kanuti belt than in the Kuskokwim Mountains belt.

Three areas in the Yukon-Kanuti belt have been studied in detail: The Kanuti volcanic field in the northern part, the Yukon River area located south of the Kaltag fault, and the Blackburn Hills volcanic field about 50 km farther southwest (Moll-Stalcup and Arth, 1989). Numerous small volcanic bodies, plutons, dikes, and sills, having K/Ar ages chiefly between 65 and 53 Ma, occur in the northern and central parts of the belt. Extensive volcanic and plutonic rocks of Tertiary age are shown on the geologic map of Alaska farther south (Beikman, 1974), but no detailed map, age, or petrologic data are available.

The Kanuti field lies within the Yukon-Koyukuk province near its southeast margin (Fig. 1). The field consists of dacite, andesite, and rhyodacite flows, domes, and tuffs exposed in a broad syncline that trends northeast and covers an area of more than 550 km^2. The base of the volcanic section is dated at 59.5 and 59.7 Ma, and the top at 55.9 Ma (Patton and Miller, 1973; Moll-Stalcup and Arth, 1989).

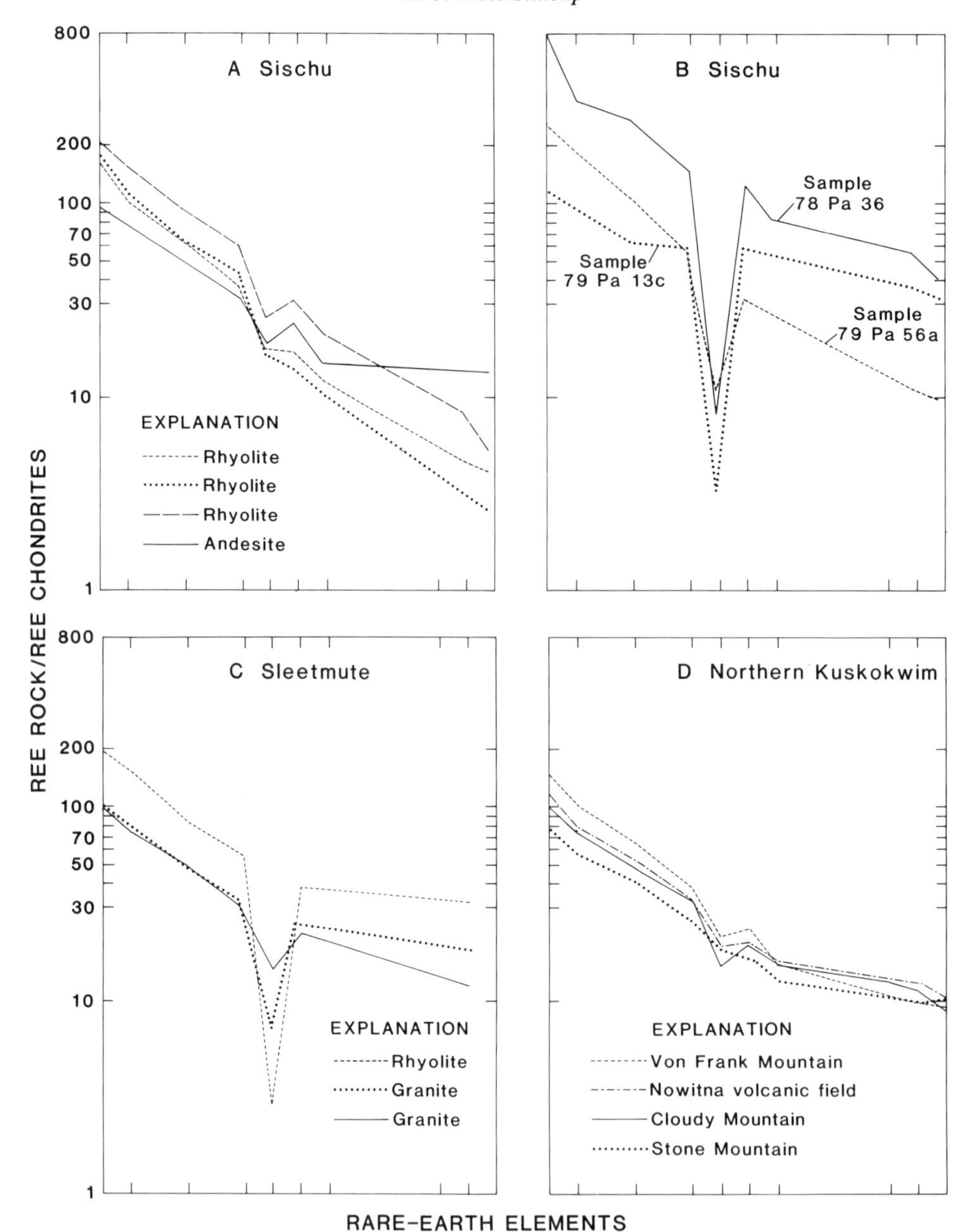

Figure 4. Eight chondrite-normalized rare-earth-element diagrams for rocks from five areas. A, andesite and rhyolites from the Sischu volcanic field having low HREE, which probably indicates hornblende or garnet fractionation. B, patterns for highly fractionated (high silica?) rhyolites from the Sischu volcanic field. C, rhyolite and granites from the Sleetmute area. D, REE data for andesites from the northern Kuskokwim Mountains: Von Frank = high K to shoshonitic; Nowitna = moderate to high K; Stone Mountain = moderate K; Cloudy Mountain = highly altered, but are thought to be shoshonitic on the basis of K_2O content. However, REE data suggest the Cloudy Mountain andesite may actually have moderate K contents. E, REE data for Sleetmute andesites. Chondrite-normalized La values for andesites of the Bristol Bay volcanic unit shown by a bar at La. F, REE data for andesites from St. Matthew Island. These rocks have very low K, Rb, Th, and LREE and may be correlative with the Alaska Range–Talkeetna Mountains belt for which we have no REE data. G and H, REE data for the Blackburn Hills volcanic field, showing the two andesite types that are distinguished on the basis of HREE: G, group 1 pyroxene andesites; H, group 2 pyroxene and hornblende andesites. Data for Sleetmute from Decker and others (1986); for Bristol Bay from Globerman (1985). All other data from E. J. Moll-Stalcup and W. W. Patton, Jr. (unpublished data, 1979–1986).

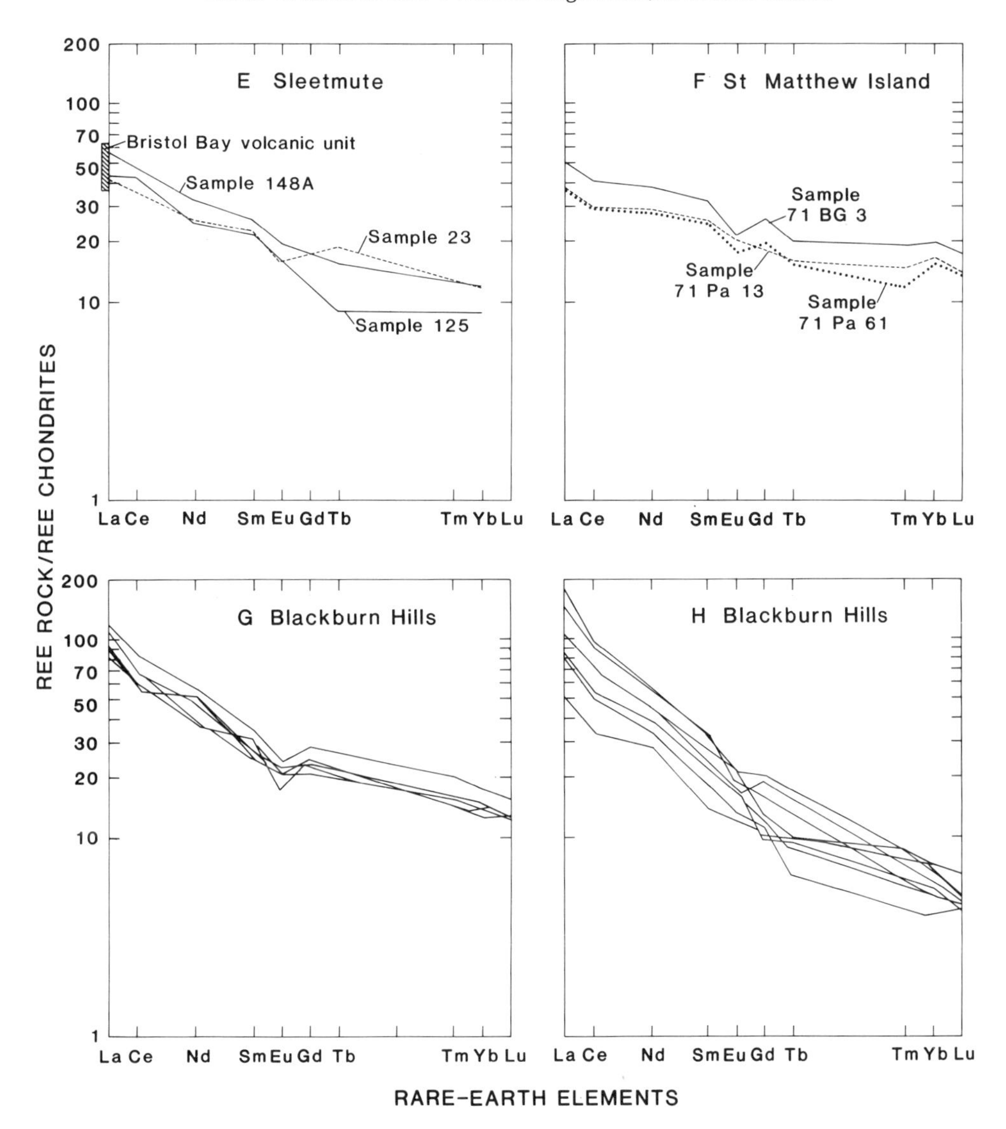

The Tokatjikh volcanic field, located about 75 km southwest of the Kanuti field near Tokatjikh Creek (Patton and others, 1978), consists of 275 km^2 of andesite and dacite flows and tuffs petrologically similar to those in the Kanuti field. A date of 59.6 ± 0.6 Ma on hornblende (W. W. Patton, Jr., written communication, 1989) indicates that the two fields are contemporaneous.

Another small volcanic field, located to the west, just south of the village of Huslia, near Roundabout Mountain, is undated and described as andesitic (Patton, 1966; Table 1, Moll-Stalcup and others, this volume). It is probably part of the early Tertiary volcanic belt, although it could be part of the younger mid-Tertiary (40 Ma) province discussed below.

Isolated domes, cones, and flows and one large volcanic field crop out in the dense vegetation along the west banks of the Yukon River south of the village of Kaltag (number 2 on Fig. 1). These rocks are the youngest volcanic rocks (53 to 47 Ma) in the Late Cretaceous and early Tertiary province of western Alaska and occur in three main locations. From north to south they are: (1) a large (400 km^2), poorly exposed volcanic field near Poisen Creek and Stink Creek that consists of basalt, andesite, dacite, and rhyolite (50.6 and 47.6 Ma, W. W. Patton, Jr. and E. J. Moll, unpublished data, 1981); (2) a large composite rhyolite dome and associated olivine basalt and pyroxene andesite flows at Eagle Slide (53.2 Ma, Patton and Moll, 1985); and (3) a morphologicaly well-preserved andesite cone located 6 km west of Bullfrog Island (53.8 Ma; Patton and Moll, 1985). Harris (1985) reports similar K-Ar ages on rocks from the same area.

A well-exposed volcanic field (150 km^2) located in the Blackburn Hills, about 10 km west of the Yukon River and 70

TABLE 1. INCOMPATIBLE ELEMENTS IN ANDESITES FROM THE KUSKOKWIM MOUNTAINS BELT*

	Von Frank (shoshonitic)	Nowitna (moderate to high-K)	Page Mountain (altered, probably high-K to shoshonitic)	Sleetemute (moderate-K)	Bristol Bay (moderate-K)	St. Matthew Island (low- to moderate-K)
	n = 5	n = 7	n = 7	n = 12	n = 6	n = 6
SiO_2	53.8 to 59.1	55.0 to 62.0	57.8 to 61.9	54.0 to 59.5	53.3 to 56.9	54.8 to 64.0
Nb	8 to 16	9 to 18	9 to 14	7.7 to 10.7	4 to 6	5 to 11
Zr	92 to 239	174 to 253	145 to 183	109 to 168	97 to 127	117 to 158
Y	17 to 25	22 to 33	14 to 25	17 to 26	21 to 32	23 to 26
Sr	911 to 1420	475 to 525	466 to 670	307 to 547	511 to 760	350 to 544
Rb	59 to 155	48 to 90	79 to 132	38 to 64	4 to 32	14 to 25
Ba	1792 to 2210	846 to 1350	1097 to 1870	349 to 910	376 to 786	318 to 470
La	49 to 57	30 to 44	25 to 37	12.8 to 28.8	12 to 21	1 to 16
Th	16.6 to 30.5	8.2 to 13.1	9.74 to 12.2	3.0 to 6.5	n.d.	1.0 to 2.0
Ta	0.902 to 1.18	0.79 to 1.13	0.664 to 0.85	0.43 to 1.06	n.d.	0.38 to 0.71
Hf	5.0 to 6.6	4.1 to 5.8	3.4 to 4.3	2.42 to 4.24	n.d.	2.5 to 4.1
$^{87}Sr/^{86}Sr$	0.7047 to 0.7051	0.70434 to 0.70508	0.7049 to 0.7059	0.70403 to 0.70601	0.70370 to 0.70414	n.d.

*Note: Von Frank, Nowitna, and Page Mountain are in the northern Kuskokwim Mountains, Sleetmute is in the central Kuskokwim Mountains, and Bristol Bay is at the southern end of the Kuskokwim Mountains. Incompatible elements increase from north to south, left to right. SiO_2 is in weight percent, all other elements are in parts per million.

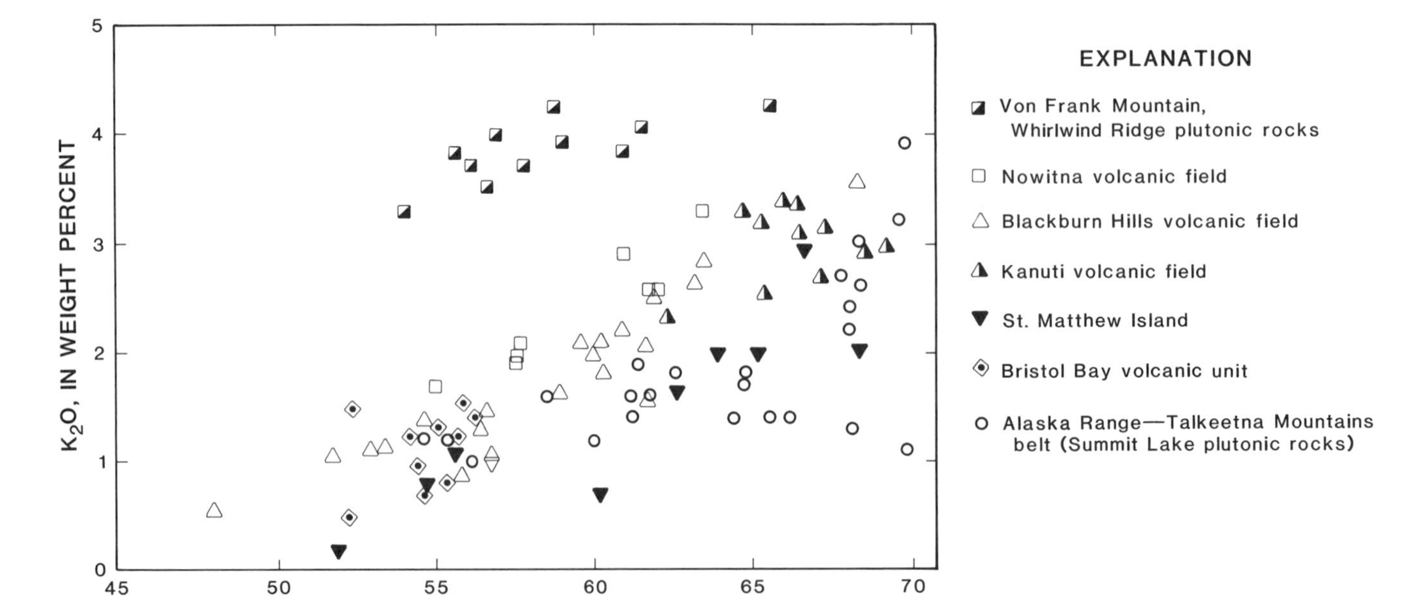

Figure 5. K_2O versus SiO_2 for Late Cretaceous and early Tertiary volcanic and plutonic suites. Alaska Range–Talkeetna Mountains belt (Summit Lake plutonic rocks: Reed and Lanphere, 1974a); St. Matthew Island; southern Kuskokwim Mountains (Bristol Bay volcanic unit); northern Kuskokwim Mountains (Nowitna volcanic field and Von Frank Mountain pluton–shoshonitic suite); Yukon-Kanuti belt (Blackburn Hills and Kanuti volcanic fields).

km south of the Kaltag fault (number 1 on Fig. 1), consists of a thick section of andesite flows exposed in a northeast-trending syncline that is bounded by the Thompson Creek fault on the west flank and by an unnamed fault on the northwest flank. The flows are interlayered with rhyolite domes and basalt flows near the top of the section. The core of the syncline consists of a thick section of highly altered green tuff intruded by a granodiorite pluton. The base of the volcanic pile is dated at 65 Ma; the rhyolite domes and granodiorite pluton at 56 Ma (Patton and Moll, 1985; Moll-Stalcup, 1987). The green altered tuff is interpreted as intracaldera tuff, and the granodiorite pluton is interpreted as an eroded resurgent dome (Moll-Stalcup, 1987).

A thick sequence of east-dipping andesite flows crop out in low hills about 20 km south of the Blackburn Hills on the opposite side of the Thompson Creek fault. Available data suggest that these flows are part of the west flank of the Blackburn Hills volcanic field, which in turn suggests that 20 km of left-lateral movement has taken place along the Thompson Creek fault since early Tertiary time.

Petrogenesis. The Yukon-Kanuti belt is generally similar to the Kuskokwim Mountain belt in that it consists of moderate- to high-K calc-alkalic suites of mafic to felsic compositions. Basalt, however, is more abundant in the Yukon-Kanuti belt, although still much less common than andesite. A further distinction between the two belts is that most rocks in the Yukon-Kanuti belt having 63 to 68 percent SiO_2 are volcanic rather than intrusive.

The Blackburn Hills volcanic field consists of a thick pile (1 km) of chiefly andesite flows at the base, and interlayered rhyolite, basalt, and andesite at the top. The lower andesite section is divided into two groups on the basis of REE patterns: One group consists of one- and two-pyroxene andesites and basalts that have REE patterns similar to the andesites in the northern Kuskokwim Mountains (Fig. 4D, G). The second group consists of hornblende and pyroxene andesites that have higher LREE and lower HREE than the first group (Fig. 4H). Rocks in the second group also have lower FeO* and TiO_2, and higher MgO at a given SiO_2, and higher initial $^{87}Sr/^{86}Sr$ than the first group, and they occur only in the northwest part of the volcanic field. The rhyolites at the top of the section have a wide variety of mineral assemblages; some are typical of calc-alkalic suites, and some of mildly alkalic rocks. Common mineral assemblages include anorthoclase+hedenbergite, oligoclase+biotite, and oligoclase+ orthopyroxene.

The occurrence of both calc-alkalic and mildly alkalic suites at the top of the Blackburn Hills section suggests that there was a chemical and mineralogical transition at about 56 Ma. Rocks that are older than 56 Ma are compositionally similar to those in the Kuskokwim Mountains belt in that the andesites are enriched in K, Rb, B, Th, U, and Sr and depleted in Nb and Ta, relative to the LREE (Fig. 6A). These andesites contain minerals typical of calc-alkalic rocks such as plagioclase, orthopyroxene, clinopyroxene, and hornblende. Rocks between 56 and 46 Ma constitute a mixed assemblage of calc-alkalic and mildly alkalic rocks, which occur at the top of the Blackburn Hills section and in the nearby Yukon River area. Some of the post–56 Ma rocks show the characteristic enrichments and depletions and have the typical calc-alkalic mineralogy (rhyolites: plag+biot; andesites: plag+ opx+cpx) of the earlier suite. The post–56 Ma assemblage differs, however, in that: (1) there is a higher proportion of basalt and these basalts have smaller Nb-Ta depletions and lower alkali/ LREE ratios than typical arc rocks (Fig. 6A; Thompson and others, 1982); and (2) some of the rhyolites have mildly alkalic mineral assemblages, such as anorthoclase+hedenbergite (Moll-Stalcup, 1987). This chemical and mineralogical transition is not strongly reflected in the major-element data, which indicate a calc-alkaline affinity for all the rocks.

The shift in mineralogy and chemistry upsection is found only in the Blackburn Hills field. The rocks in the Yukon River area (53 to 47 Ma) postdate the transition, whereas the rocks in the Kanuti field (59 to 56 Ma) predate the transition. The volcanic rocks in the Yukon River area include a mixed assemblage of typical calc-alkalic andesites and dacites (andesite: plag+opx+ cpx; dacite: plag+horn+biot) and mildly alkalic latites and basalts (latites: anorthoclase+plag+biot; basalt: ol+plag+cpx and groundmass ol+biot). Rocks containing calc-alkalic mineral assemblages have moderate Nb-Ta depletions (La/Nb greater than 2); rocks with mildly alkalic mineral assemblages have smaller Nb-Ta depletions. No correlation between age and rock types has been found in the Yukon River area, and it appears that both calc-alkalic "arc-type" and mildly alkalic "post–arc-type" rocks erupted between 53 and 47 Ma.

The Kanuti volcanic field is dated at 59 to 56 Ma and predates the transition noted at the southern end of the belt. All the analyzed rocks from the Kanuti field have a high alkali and LREE content and deep troughs at Nb and Ta on chondrite-normalized multi-element plots (Fig. 6C), similar to the older suite of rocks in the Blackburn Hills. The volcanic rocks in the Kanuti field range from high-silica andesite to rhyodacite. Most contain hornblende and have REE patterns similar to hornblende-bearing andesites in the Blackburn Hills (Fig. 5A).

Samples of andesite, basalt, and dacite that give K/Ar ages of 59.8 to 50.2 Ma have been dredged from the edge of the Bering Sea shelf (Fig. 1; Davis and others, 1987, 1989). These rocks are altered but appear to have typical calc-alkalic compositions and may be correlative with the Yukon-Kanuti belt. Whether they are part of the Yukon-Kanuti or some other belt, however, is not known. They may be rocks that were accreted to the continental margin after eruption.

Volcanic rocks that have K-Ar ages from 64 to 62 Ma also occur on St. Lawrence Island in the Bering Sea. Patton and Csejtey (1980) describe them as basalt, soda rhyolite, trachyandesite, and andesite. Chemical analyses (W. W. Patton, Jr., unpublished data, 1971) suggest the rocks are mildly alkalic and possibly a bimodal basalt-rhyolite assemblage (SiO_2 48 percent and 68 to 71 percent). The K-Ar ages and mildly alkaline compositions suggest that the rocks are probably not part of any of the Late Cretaceous and early Tertiary belts, but they could be part of the Yukon-Kanuti belt if the transition from subduction-related

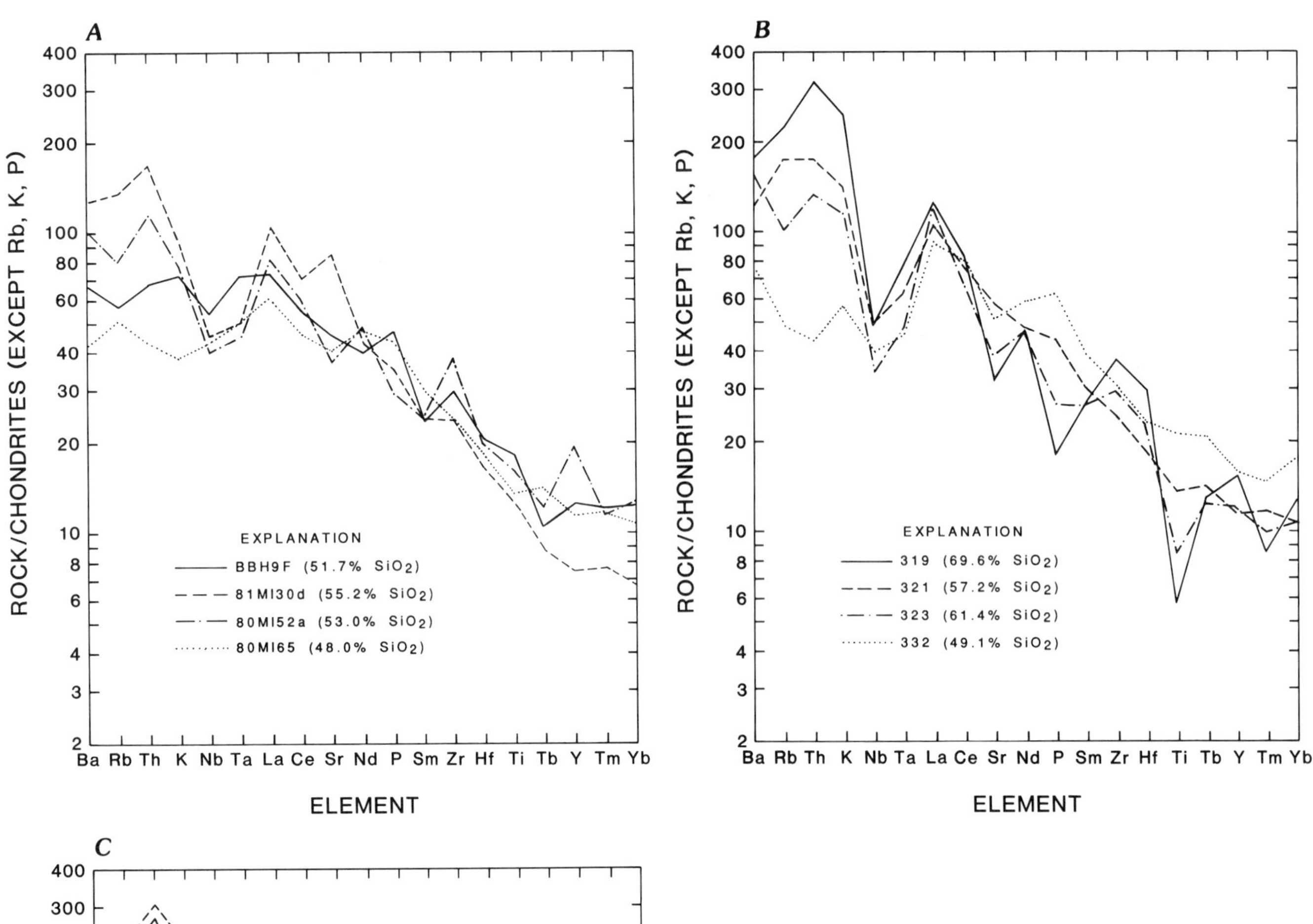

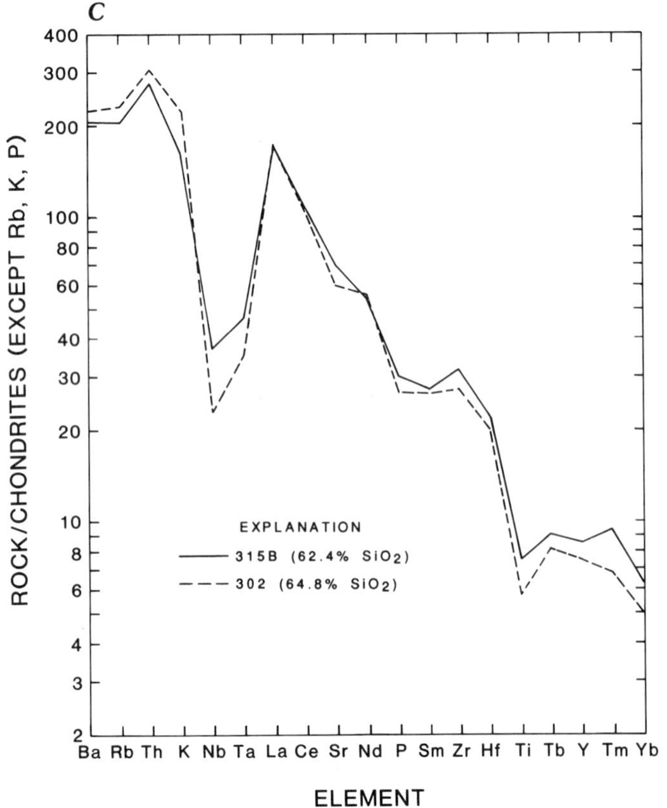

Figure 6. Chondrite-normalized spidergrams. A, samples from the Blackburn Hills in the Yukon–Kanuti belt. Samples 80ML52a and 81ML30d are from the older andesite section; samples 80ML9f and 80ML65 are basalts that overlie rhyolite domes or flows and are younger than 56 Ma. The younger rocks are more mafic, have lower alkalis relative to La and smaller Nb-Ta depletions. B, samples from Yukon River Eocene volcanic rocks. All data from E. J. Moll-Stalcup and W. W. Patton, Jr. (unpublished data, 1981). All the rocks except 81Pa332, a basalt dated at 53 Ma, have large Nb-Ta anomalies. The more siliceous rocks have negative spikes at Sr, P, and Ti due to fractionation. C, samples from the Kanuti volcanic field. All data from E. J. Moll-Stalcup and W. W. Patton, Jr. (unpublished data, 1980–1981).

calc-alkalic rocks to mildly alkalic post-subduction magmatism occurred earlier in that part of the Bering Sea.

Late Cretaceous and early Tertiary tectonic implications

The Alaska Range–Talkeetna Mountains, the Kuskokwim Mountains, and the Yukon-Kanuti belts constitute a Late Cretaceous and early Tertiary magmatic province that extends over a vast region of Alaska, from the Alaska Range north to the Arctic Circle and west to the Bering Sea shelf. The volcanic and plutonic rocks in this province overlap a number of tectonostratigraphic terranes that lie between the Alaska Range, Bristol Bay, and the Yukon-Koyukuk province, which suggests that these terranes were assembled by late Mesozoic time (Moll and Patton, 1982). Paleomagnetic studies of Late Cretaceous and early Tertiary volcanic rocks at Bristol Bay (Globerman, 1985), Lake Clark (Thrupp and Coe, 1986), the northern Kuskokwim Mountains (Nowitna field, Coe and others, 1985; Blackburn Hills field, Thrupp and Coe, 1986), the Talkeetna Mountains (Hillhouse and others, 1985), and the east-central Alaska Range (Cantwell Formation, Hillhouse and Gromme, 1982) indicate about 30 to 55° of counterclockwise rotation, but no major latitudinal displacement relative to North America since eruption (Hillhouse and Coe, this volume; Thrupp and Coe, 1986).

Igneous rocks erupted between 75 and 56 Ma from all three belts have chemical compositions typical of subduction-related magmatism. Dominant compositions are intermediate and felsic, and rocks from all three belts have high K, Ba, Sr, Rb, and Th, and low Nb, Ta, and Ti relative to LREE—the most diagnostic characteristics of arc magmatism. Most workers (e.g., Hudson, 1979; Reed and Lanphere, 1973) agree that the Alaska Range–Talkeetna Mountains belt is a continental volcanic arc related to subduction of the Kula plate under southern Alaska during the Paleocene. The tectonic environment of the Kuskokwim Mountains and Yukon-Kanuti belts has not been so well documented. Wallace and Engebretson (1984) outline three possible relations between the Alaska Range–Talkeetna Mountains and Kuskokwim Mountains belts: (1) the Kuskokwim Mountains belt formed in an extensional environment (or intracontinental back-arc setting) behind the Alaska Range–Talkeetna Mountains belt (Bundtzen and Gilbert, 1983; Gemuts and others, 1983); (2) the Alaska Range–Talkeetna Mountains and Kuskokwim Mountains belts constitute an unusually wide arc analogous to Oligocene volcanism in the western U.S. (Gill, 1981, p. 39, and references therein); or (3) the Alaska Range–Talkeetna Mountains and the Kuskokwim Mountains belts were two separate arcs juxtaposed by subsequent tectonic activity. Of the three possibilities, the third is the least likely because no major Tertiary suture has been mapped between the two belts, and geologic and paleomagnetic data suggest that the two belts were essentially in place during formation—not accreted at some later time. Both belts may have been brought into their current position by strike-slip faults, but Tertiary offsets along faults in western Alaska are thought to be relatively small, probably less than 150 km (Grantz, 1966). Finally, the abundance of high-K calc-alkalic and shoshonitic rocks in the Kuskokwim Mountains suggest that it was not a piece of the Alaska Range–Talkeetna Mountains belt that was originally along strike and was later strike-slip faulted to its present position.

The chemical composition of the plutonic and volcanic rocks in the Kuskokwim Mountains and Yukon-Kanuti belts strongly supports their formation in an arc rather than back-arc environment. Other workers (T. K. Bundtzen and S. E. Swanson, written communication, 1986) have argued that the Kuskokwim Mountains magmatism is the result of regional extension because many of the rocks in the northern Kuskokwim Mountains have high contents of total alkalis, and therefore plot in the alkalic field on total-alkalis-versus-silica diagrams, and the volcanic fields and intrusive rocks are associated with numerous normal/strike-slip faults. I distinguish between arc-related alkalic rocks (shoshonites) that are characterized by K_2O/Na_2O greater than, or equal to 1, low TiO_2 and high-field-strength-element contents, and high alkalis and Al_2O_3 contents (Morrison, 1980; Bloomer and others, 1989) and other types of alkalic rocks that have higher contents of Ti, Nb, Ta, and Zr. Although earlier workers believed most of the rocks were basalts and rhyolites and therefore constituted a bimodal assemblage, many of the chemical analyses published on earlier maps (Bundtzen and Laird, 1982, 1983a, b, c) have SiO_2 more than 53 percent, when the analyses are recalculated 100 percent anhydrous. Of 49 analyses having LOI less than 10 percent, 4 have less than 53 percent SiO_2, 36 have 53 to 63 percent SiO_2, 4 have 63 to 70 percent SiO_2 (all plutonic), and 4 have more than 70 percent SiO_2. Rocks having more than 53 percent SiO_2 are andesites even when they are olivine-bearing, because olivine is stable at higher silica contents in high-K suites than in low or moderate-K suites (Kushiro, 1975). Furthermore, the silica gap in the published data occurs in the dacite range (63 to 70 percent SiO_2), which is not uncommon in modern arc volcanoes (Grove and Donnelly, 1986) and is distinct from the Miocene bimodal basalt-rhyolite suite associated with extension in the western United States (Christiansen and Lipman, 1972). Bundtzen and Swanson's argument for extension based on the abundance of faults is discussed further below.

The chemical data strongly support the interpretation that the Alaska Range–Talkeetna Mountains, Kuskokwim Mountains, and Yukon-Kanuti belts constituted an anomalously wide magmatic arc in the Late Cretaceous and early Tertiary, analogous to the western United States in the Oligocene. In their present configuration the three belts span a width of 550 km, extending approximately 990 km measured orthogonal to the Aleutian trench near the Alaska Peninsula. In the conterminous states, Oligocene volcanism as far east as the Rocky Mountains has been attributed to convergence along the western continental margin (Lipman and others, 1972; Snyder and others, 1976). The compositions of andesites in the northern Kuskokwim Mountains are remarkably similar to those in the Oligocene San Juan volcanic field in Colorado (Table 2; Lipman and others, 1978), which also was located at least 900 km from the Oligocene continental margin.

TABLE 2. TRACE ELEMENT ABUNDANCES (ppm) AND RATIOS IN REPRESENTATIVE ANDESITE SAMPLES FROM THE SAN JUAN VOLCANIC FIELD (SUMMER COON), COLORADO, AND FROM THE NORTHERN KUSKOKWIM MOUNTAINS BELT (NOWITNA) AND THE YUKON–KANUTI BELT (BLACKBURN HILLS), ALASKA

	Summer Coon*	Nowitna†	Blackburn Hills†
(wt %)			
SiO_2	56	57	56.5
Rb (ppm)	65	52	35
Ba (ppm)	1,160	870	891
Sr (ppm)	930	505	493
K/Rb	344	298	299
La (ppm)	30	32	30
Ce (ppm)	79	66	52
Yb (ppm)	1.5	2.4	2.2
La/Yb	20	13.3	13.6
Th (ppm)	3.6	8.3	6.1
U (ppm)	1.1	3.1	2.2
Th/U	3.2	2.7	2.8

*Zielinski and Lipman (1976)
†Moll-Stalcup (1987)

Interpreting trends in K_2O within and among the three belts is complex because the three belts cut across numerous terranes of different age and lithology. K_2O contents are higher in the Kuskokwim Mountains and Yukon-Kanuti belt (Fig. 5) than in the Alaska Range–Talkeetna Mountains belt, as expected in an arc environment if the trench was located along the southern margin of Alaska. However, rocks in the northern Kuskokwim Mountains have higher K_2O-contents than those in either the southern Kuskokwim Mountains belt or the Yukon-Kanuti belt. As discussed earlier, high K_2O (and all incompatible element) contents in the northern Kuskokwim Mountains belt appears to be related to interaction of the magmas with old, possibly thicker, continental crust or lithospheric mantle. Even moderate-K rocks in the northern Kuskokwim Mountains have higher K and incompatible-element contents than rocks from any of the other areas or belts, suggesting that the magmas in the northern Kuskokwim Mountains have interacted with old continental crust or mantle, which has resulted in their enriched composition. However, K contents in Late Cretaceous and early Tertiary rocks from the northern Kuskokwim Mountains vary from moderate-K to high-K to shoshonitic, so not all of the K variation can be attributed to the age and thickness of the underlying lithosphere. Some must be the result of some other factor, such as changing depth to the subducting slab.

The age data suggest that between 75 and 65 Ma the magmatic arc was narrower, consisting chiefly of the Alaska Range–Talkeetna Mountains and Kuskokwim Mountains belts, and that the K gradient across the arc was steeper. During this period the K-content in the magmas increased across the arc, from moderate-K in the southern Alaska Range–Talkeetna Mountains belt, to high-K and shoshonitic in the northern Kuskokwim Mountains. Much of this steep gradient in K, across the Alaska Range–Talkeetna Mountains to the Kuskokwim Mountains belt, was probably tectonically controlled by increasing depth to the slab, but some K-enrichment in the northern Kuskokwim Mountains was due to lithospheric interaction.

During the period from 65 to 56 Ma, the arc broadened to include the Yukon-Kanuti belt, and the K gradient across the arc was more gradual, probably due to a decrease in dip of the subducting slab. During this time period, K_2O at a given silica content was lower in the Yukon-Kanuti belt than in the northern Kuskokwim Mountains, despite the fact that the Kuskokwim Mountains rocks were closer to the trench. This reversal in K trend across the arc is interpreted to be a result of interaction between old, enriched lithosphere and arc magmas in the northern Kuskokwim Mountains region. Variation in crustal thickness could be a factor in the K_2O gradient, but is not indicated by the gravity data. Gravity studies suggest that the present-day crust under both the Yukon-Kanuti belt, in the study area, and the northern Kuskokwim Mountains is between 30 and 35 km thick (Barnes, 1977). Although crustal thicknesses may have changed since the early Tertiary, they probably have not because the Tertiary has been characterized by mild deformation (open folds and strike-slip faulting) and low sedimentation rates.

The K_2O content of the arc magmas in the southern Kuskokwim Mountains belt (Bristol Bay volcanic rocks) is about the same as in the Yukon-Kanuti belt, both of which are underlain by late Paleozoic and Mesozoic oceanic crust and island-arc rocks of the Yukon-Koyukuk province and Togiak terrane (Fig. 5). These late Paleozoic and Mesozoic mafic and intermediate basement rocks would be expected to have much lower incompatible-element contents, lower $^{87}Sr/^{86}Sr$, and higher $^{143}Nd/^{144}Nd$ than the old sialic continental crustal rocks of the Ruby and Nixon Fork terranes, which underlie the northern Kuskokwim Mountains. The strong contrast between basement lithologies is evident in trace-element and isotope composition of rhyolites from each area. Rhyolites from the Blackburn Hills, in the Yukon-Kanuti belt, have lower LREE, Rb, Th, U, Sn, F, and initial $^{87}Sr/^{86}Sr$, and higher $^{143}Nd/^{144}Nd$ than rhyolites from the northern Kuskokwim Mountains (Moll and Patton, 1983; Moll and Arth, 1985; Moll-Stalcup and Arth, 1989). Andesites from the Nowitna volcanic field show evidence of interaction with enriched continental crust, whereas those from the Blackburn Hills do not (Moll and Arth, 1985; Moll-Stalcup and Arth, 1989). I attribute the very high K_2O and incompatible-element contents in the northern Kuskokwim Mountains magmas to interaction with old sialic crust in that area. Such old crust might also affect the K_2O gradient along the arc, from north to south, as well as cause the reversal in K_2O gradient across the arc from the northern Kuskokwim Mountains to the Yukon-Kanuti belt. Interaction with the crust seems especially common in subduction-related suites because these magmas rise slowly to the surface and have ample time to differentiate and assimilate country rock.

Bundtzen and Swanson (1984) note that many of the Late Cretaceous and early Tertiary volcanic and intrusive bodies are associated with normal or strike-slip faults. Little is known about the age and type of motion for most of these faults, but most appear to have significant right-lateral movement (Grantz, 1966). Some of the volcanic fields are cut by normal strike-slip faults, but it is uncertain whether the faults entirely postdate the volcanism, and thus control only the exposure of the volcanic fields, or if some of the volcanism was contemporaneous with, or postdates the faulting. Rhyolite domes dated at 58 Ma occur at Old Woman Mountain in the Kaltag fault zone just west of the Yukon River (W. W. Patton, Jr., and E. J. Moll, unpublished data, 1983), and similar domes dated at 61 Ma occur in the Nixon Fork–Iditarod fault zone near McGrath (Bundtzen and Laird, 1983c). Some of these rhyolite domes thus appear to have intruded into preexisting fault zones. In addition, some felsic dikes or sills of Late Cretaceous and early Tertiary age strike parallel to major faults (Patton and others, 1980; Bundtzen and Laird, 1983b), which suggests some of the faulting had occurred by the time the dikes and sills were emplaced. Most of the Late Cretaceous and early Tertiary volcanic piles are gently folded, with maximum dips of 20 to 45 degrees, but most of the Middle Tertiary volcanic piles are flat lying and undeformed (Harris, 1985; W. W. Patton, Jr., personal communication, 1986). These data demonstrate that at least some of the deformation postdates the Late Cretaceous and early Tertiary magmatism, and also suggests that some may have been contemporaneous with or predated it. I believe that these equivocal relations do not support the interpretation that the magmatism is the result of continental rifting. The chemical composition of rocks in both the Yukon-Kanuti and Kuskokwim Mountains belts was dominantly controlled by subduction, not continental rifting. Although the timing of movement along the faults is not well constrained, I find the idea that most of the movement along the faults took place in the Eocene and was related to the clockwise rotation or oroclinal bending of western Alaska (Globerman and Coe, 1984) very appealing. The implications of this idea are discussed further below.

In contrast to the 75- to 56-Ma rocks, many of the rocks in western Alaska that are 56 m.y. old or younger have chemistry and mineralogy that are not typical of subduction-related magmas. These rocks include the McKinley sequence granites and the quartz monzonite of Tired Pup in the western and northern Alaska Range, the rhyolites and basalts at the top of the Blackburn Hills, and the volcanic rocks in the Yukon River area in the Yukon-Kanuti belt. I believe subduction-related magmatism in western Alaska ceased at about 56 Ma, and the period from 56 to 50 Ma represents a transition from subduction-related magmatism to post-subduction, possibly intraplate, magmatism, during which rocks typical of both environments erupted. This transition period is marked by: (1) a marked decrease in the volume of magma erupted over the entire province; (2) an increase in the proportion of basalt and the eruption of mildly alkalic basalt, latite, and alkali rhyolite in the Yukon-Kanuti belt; and (3) the emplacement of peraluminous silicic granites of the McKinley sequence in the northern and western Alaska Range. Subduction-related volcanism younger than 56 Ma occurred in the Talkeetna Mountains at the eastern end of the Alaska Range–Talkeetna Mountains belt, but insufficient petrologic data are available on these rocks to determine whether a similar chemical or mineralogical transition marked the end of subduction-related magmatism there.

In summary, the age and composition data suggest that the Alaska Range–Talkeetna Mountains, Kuskokwim Mountains, and Yukon-Kanuti belts constituted an anomalously wide magmatic arc related to north-directed subduction at a trench along the Paleocene continental margin of Alaska (Fig. 1) between 75 and 56 Ma. The data also suggest that only the southeastern two-thirds of the province was active between 75 and 66 Ma, broadening to the entire province between 66 and 56 Ma. A sharp decrease in magma volume and a mineralogical and compositional change at 56 Ma mark the end of the subduction event in most of the province except the easternmost end of the province in the Talkeetna Mountains.

Plate-motion models (Engebretson and others, 1982) show rapid north-northeast–directed subduction of the Kula plate under southern Alaska between 74 and 56 Ma. In early Eocene time, the trench jumped out to its present position, and the Aleutian arc began to form (Rea and Duncan, 1986; Scholl and others, 1986). The convergence angle between the present position of the three belts and the Paleocene plate-motion vectors is near the minimum (25°) required for arc magmatism (Gill, 1981, p. 27; Wallace and Engebretson, 1984). Paleomagnetic data on the volcanic rocks indicate that western Alaska has rotated counter-clockwise 30 to 55° but has not been latitudinally displaced since the Paleocene (Globerman and Coe, 1984; Hillhouse and Coe, this volume). These data suggest that the magmatic belts may have had a convergence angle of 55 to 80° prior to rotation in the Eocene. Paleomagnetic data from the Ghost Rocks Formation on Kodiak Island in southern Alaska indicate it has moved 2,000 km to the north since the Paleocene (Plumley and others, 1982). These data suggest that a Tertiary suture associated with the accretion of southernmost Alaska should be located somewhere between southern Kodiak Island and the Alaska Range batholith (Thrupp and Coe, 1986). This hypothetical suture may represent a strike-slip fault (Moore and others, 1983) or the subduction zone associated with the wide Late Cretaceous and early Tertiary magmatic province.

The location of the magmatic belts relative to the trench is further obscured by numerous right-lateral strike-slip faults, which have apparently cut the wide magmatic arc since the Paleocene. The amount of offset since the Paleocene is not known. Most data suggest that the amount of offset along the strike-slip faults in western Alaska is considerably less than the amount of offset along the faults in eastern Alaska (Grantz, 1966). The

Yukon-Kanuti belt appears to be offset along the Kaltag fault, which Patton and Hoare (1968) suggest has had 60 to 130 km of right-lateral offset since the Late Cretaceous. If movement on faults is reconstructed and the trench is located along the northern part of Kodiak Island, the entire magmatic arc was within 750 km of the trench when it was active.

MIDDLE TERTIARY MAGMATISM

The period from 55 to 43 Ma was characterized by a hiatus in magmatic activity in most of mainland Alaska except along the hinge line of the Alaska orocline (Talkeetna Mountains, Arkose Ridge, Prince William Sound, and Yakutat terrane), the Sanak-Baranof belt (Hudson, this volume), and the Yukon River area. Small amounts of volcanic activity occurred in the Aleutian arc and Alaska Peninsula from 50 to 43 Ma (Scholl and others, 1986; Wilson, 1985; Vallier and others, this volume; Kay and Kay, this volume; Miller and Richter, this volume). At about 40 Ma (±3), a brief magmatic pulse occurred in western interior Alaska, and a major pulse of magmatic activity, which lasted 10 m.y., began in the Aleutian arc. The arc volcanism occurred in a narrow belt extending from the Aleutian Islands probably as far northeast as Sugar Loaf Mountain on the north side of the Denali fault in the central Alaska Range (Plate 5). Magmatism on the Aleutian Islands, on the west side of the Alaska Peninsula (the so-called Meshik arc of Wilson, 1985), on the north and west flank of the Alaska–Aleutian Range batholith (Merill Pass of Reed and Lanphere, 1973), at Mount Foraker and McGonagall (Reed and Lanphere, 1974b), at Mount Galen (Decker and Gilbert, 1978), and possibly as far east as Sugarloaf Mountain (Albanese, 1980; Albanese and Turner, 1980) was probably part of the mid-Tertiary Alaska-Aleutian arc.

The Talkeetna Mountains and Yukon River areas were discussed in the previous section and are interpreted as the last remnants of a widespread Late Cretaceous and early Tertiary subduction event in western and southern Alaska. Eocene dikes in Prince William Sound are listed in Table 1 on Plate 5 and are described in more detail by Hudson (this volume), who considers them part of the Sanak-Baranof belt of anatectic granites. The Eocene volcanic rocks in south-central and western interior Alaska are described below.

South-central Alaska

Although most of mainland Alaska experienced relative quiescence between 55 and 43 Ma, volcanic and plutonic rocks of this age occur in the hinge of the Alaska orocline in an area between the Border Ranges and Denali fault or in the southern accretionary wedge south of the Border Ranges fault. Rocks in the hinge area include the Arkose Ridge Formation, the Talkeetna Mountains volcanic rocks, and many unnamed plutons in the northern Talkeetna Mountains.

Basalt dikes, sills, and altered tholeiitic flows (55 to 43 Ma) occur in the Arkose Ridge Formation of southern Alaska, just south of the Talkeetna Mountains volcanic rocks (Silberman and Grantz, 1984). A. Grantz (personal communication, 1986) considers these rocks to be late Paleocene in age on the basis of plant fossils in interbedded sandstones. The rocks have smaller Nb-Ta depletions on multi-element diagrams than do contemporaneous arc rocks in the Talkeetna Mountains (Fig. 7), but their trace-element ratios do not unambiguously resolve whether they are arc tholeiites related to subduction or intraplate volcanic rocks possibly related to the Castle Mountain fault. Therefore, both their tectonic environment and their precise age are uncertain.

Early to middle Eocene basaltic flows, hyaloclastites, and flow breccias, along with interbedded clastic marine sedimentary rocks, occur in the Yakutat terrane south of the Border Ranges fault (Davis and Plafker, 1986). The basalts consist of LILE-depleted and LILE-enriched tholeiites, which Davis and Plafker (1986) interpret as normal mid-ocean ridge and oceanic island basalt. They suggest that these basalts are correlative with contemporaneous geochemically similar basalts, which occur in a linear belt extending from southern Vancouver Island to the southern Oregon Coast Range. They further suggest that the basalts originated as seamounts near the Kula-Farallon spreading center in the early to middle Eocene, and were accreted to the

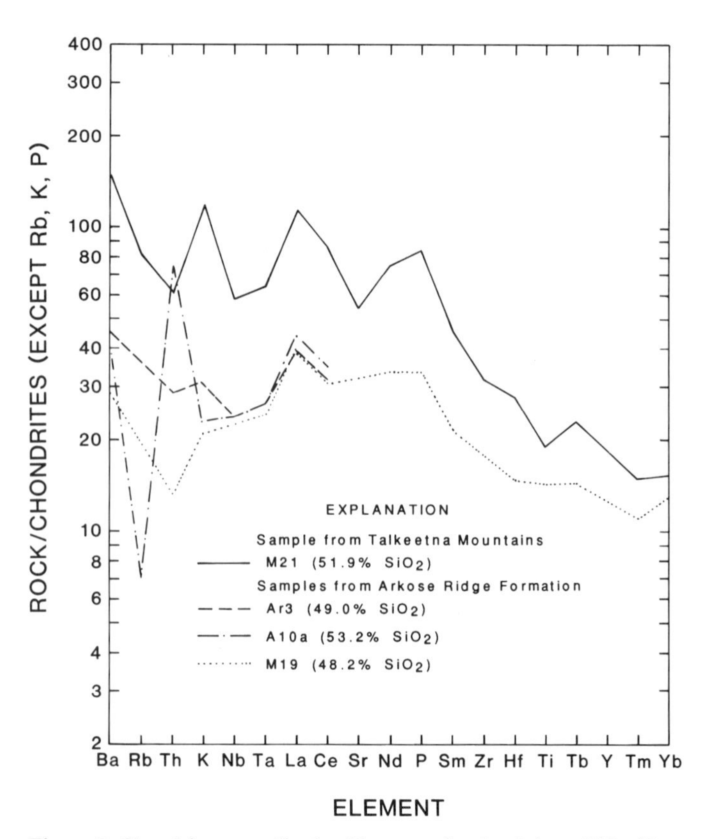

Figure 7. Chondrite-normalized spidergrams for the Arkose Ridge Formation and Talkeetna Mountains volcanic rocks. Data from A. Grantz and M. L. Silberman (unpublished data, 1981). The variation in the shape of the pattern for the Arkose Ridge samples for Ba, Rb, Th, and K suggests that these rocks are altered.

southern continental margin of Alaska at about 48 Ma during subduction of the Kula-Farallon ridge and Kula plate.

Western Interior Alaska

Volcanic rocks (37 to 43 Ma) occur on St. Lawrence Island, in the Melozitna area, in the Yukon River area, and in the Sleetmute area (Fig. 8). Limited data suggest that these constitute a bimodal assemblage of felsic volcanic rocks and associated basalt that have calc-alkalic to mildly alkalic affinities.

Patton and Csejtey (1980) describe volcanic rocks on St. Lawrence Island (39.3 Ma) as rhyolite and dacite tuff, tuff breccia, and flows. The unit is not well exposed but is thought to be flat-lying or gently dipping (Patton and Csejtey, 1980).

Three volcanic fields (40 ± 3 Ma) occur between the Melozitna and Koyukuk Rivers on the north side of the Kaltag fault at Indian Mountain, Takhakhdona Hills, and Dulbi River. All three consist chiefly of rhyolite tuff, flows, and breccia (Patton and others, 1978). Dark vesicular basalt flows (Patton and others, 1978) and andesite and dacite (Patton and Moll-Stalcup, unpublished data, 1989) also occur in the Takhakhdona Hills. Rhyolite obsidian (39.9 to 41.6 Ma) occurs at Indian Mountain (Miller and Lanphere, 1981) and is a possible source of the obsidian artifacts found in northwestern Alaska (Patton and Miller, 1970).

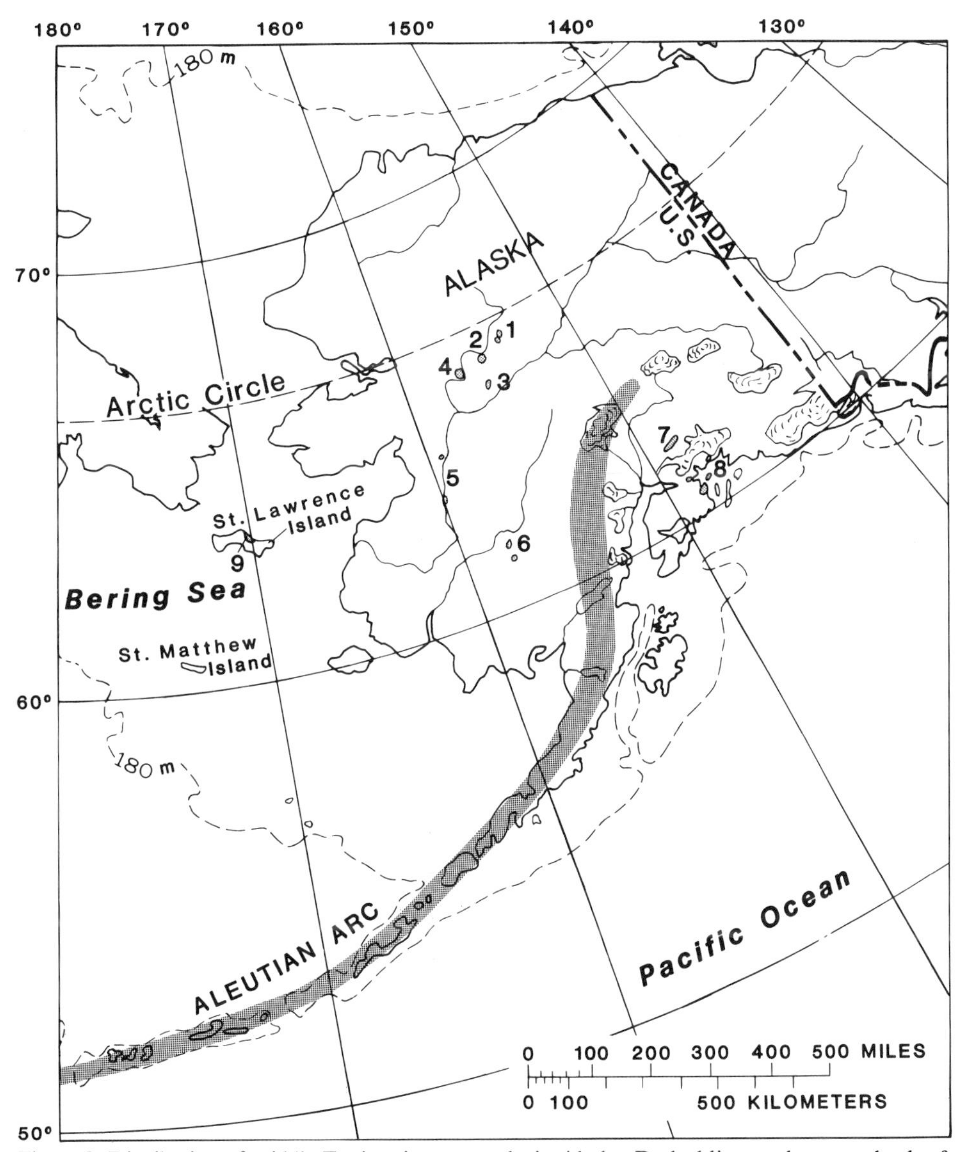

Figure 8. Distribution of middle Tertiary igneous rocks in Alaska. Dashed line marks water depth of 180 m. Middle Tertiary Alaska–Aleutian Range arc shown in shaded pattern. Middle Tertiary igneous localities shown in red. Locations are: 1, Indian Mountain; 2, Dulbi River; 3, Takhakhdona Hills; 4, Kateel River; 5, Yukon River; 6, Sleetmute area; 7, Matanuska Valley and Arkose Ridge; 8, Prince William Sound; and 9, St. Lawrence Island.

Harris (1985) reports K-Ar ages of 40 ± 3 Ma at two localities on the Yukon River, one north of Morgan Island in the southernmost Nulato Quadrangle (42.7 Ma) and one near the village of Grayling in the northernmost Holy Cross Quadrangle (42.4 Ma). He also reports that the Oligocene volcanic rocks are flat-lying and undeformed. No lithologic data are given, but at least some are basalt because the K-Ar dates were run on basalt whole-rock samples. Additional volcanic rocks of this age may occur farther south in the Holy Cross Quadrangle, where widespread undated volcanic rocks are exposed.

Little-known volcanic rocks in this age range occur in the Sleetmute area. Olivine basalt (38.2 Ma) is reported from the Chuilnuk River area, and a rhyolite sequence (43.8 Ma) is reported from Tang Mountain (Decker and others, 1986). The rhyolites at Tang Mountain are apparently interbedded(?) with trachyandesite near the base of the rhyolite section. Decker and others (1986) describe a black glassy rhyolite unit, probably a basal vitrophyre, overlain by a gray rhyolite unit.

Petrogenesis. Few petrologic data are available for the middle Tertiary volcanic rocks in western interior Alaska. Major-element data from St. Lawrence Island, Indian Mountain, the Takhakhdona Hills, and Tang Mountain suggest that the rocks are dominantly felsic (66 to 77 percent SiO_2), along with minor mildly alkalic basalt and andesite. Basalt occurs in the Takhakhdona Hills and Yukon River area; andesite occurs at Tang Mountain. No chemical or mineralogical data are available for the basalt on the Yukon River. Rocks in the Takhakhdona Hills range from basalt through rhyolite, and appear to have arc-like patterns on multi-element spidergrams (Fig. 9). Chemical analyses of the andesite at Tang Mountain (Decker and others, 1986) show that the rock has high TiO_2 (1.97 percent), K_2O (3.00 percent), Nb (49 ppm), Ta (3.85 ppm), and La (51 ppm) and low Ba (445 ppm). The composition of this rock, along with its trace-element ratios (Ba/Ta = 116, La/Nb = 1.03), suggests it is not a subduction-related orogenic andesite but a trachyandesite of some other affinity. Normalized multi-element data on trachyandesites from Tang Mountain show very high LREE and virtually no Nb-Ta depletion (Fig. 9). The initial $^{87}Sr/^{86}Sr$ on the same rock is 0.7033, much lower than initial $^{87}Sr/^{86}Sr$ on the Late Cretaceous and early Tertiary rocks from the same area.

Mineralogical data are available only for rocks from a few localities and are summarized in Table 1 on Plate 5. Most of the felsic rocks have minerals typical of calc-alkalic rocks, but some are more alkalic. Dacite having anorthoclase, plagioclase, biotite, and oxides, and rhyolite having anorthoclase, plagioclase, and biotite have been reported from the Takhakhdona Hills, and trachyandesite having sanidine in addition to plagioclase and clinopyroxene occurs at Tang Mountain.

The trachyandesite at Tang Mountain does not appear to be related to subduction. The volcanic fields in the Takhakhdona Hill, at Indian Mountain, the Dulbi River, and the Yukon River are dominantly composed of felsic tuff, flow, and breccia, but the Takhakhdona Hills field also contains subordinate basalt, andesite, and dacite. Mafic, intermediate, and felsic rocks from the

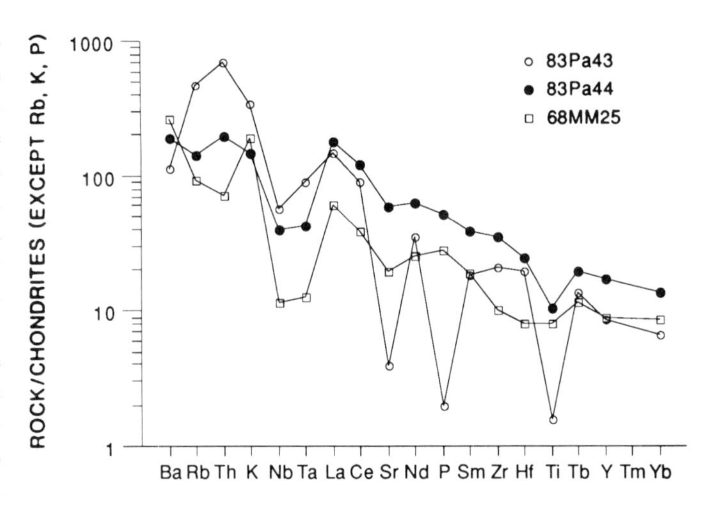

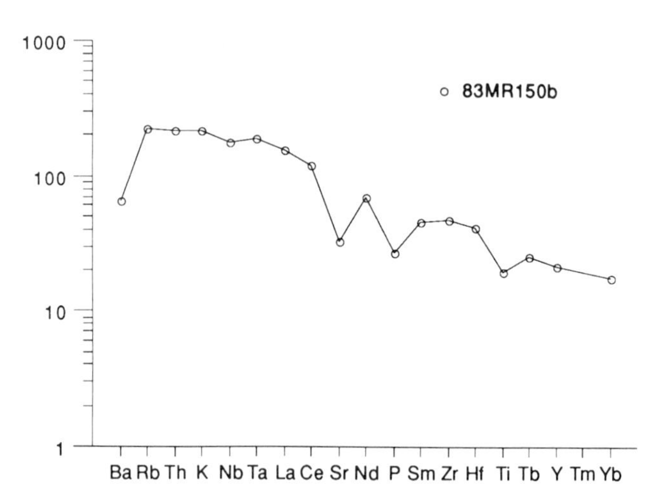

Figure 9. Chondrite-normalized spidergram of mid-Tertiary rocks from the Takhakhdona volcanic field in the Melozitna area and Tang Mountain in the Sleetmute area. Rocks from the Takhakhdona field include a rhyolite (77 percent SiO_2—sample 83Pa43), an andesite (58 percent SiO_2—83Pa44), and a basalt (52 percent SiO_2—68MM25). All three rocks appear to have low Nb and Ta relative to alkalis and LREE on chondrite-normalized spidergrams; patterns that are typical of arcs. In contrast the trachyandesite from Tang Mountain has higher Ta relative to La and only slightly less Nb relative to La, and is not considered an orogenic andesite. Data from the Takhakhdona field from Patton and Moll-Stalcup (unpublished data, 1983); data for Tang Mountain from Decker and others (1986). Spidergrams plotted using the program of Wheatley and Rock (1987).

Takhakhdona Hills have arc-like trace-element signatures (Fig. 9), but felsic compositions predominate and the rocks are more than 300 km from the most inland part of the active Aleutian–Alaska Range arc. The mid-Tertiary rocks in interior Alaska erupted contemporaneously with the start of a major pulse of volcanism in the Aleutian–Alaska Range arc and the bend in the Hawaiian-Emperor seamount chain (Clague and others, 1975).

Wallace and Engebretson (1984) also show a change from rapid northward-directed subduction of the Kula plate to slow northwest subduction of the Pacific plate at 43 Ma. The eruption of the volcanic rocks may be related to this change in the rate and angle of plate motion or to the switch from the Kula to the Pacific plate. Changes in angle of subduction may have resulted in movement along the many strike-slip faults in the area; however, the volcanic fields are not located on or near mapped faults, and their lack of deformation suggests that they have not experienced major movement. Also, paleomagnetic studies of Oligocene volcanic rocks in the Yukon River (Harris, 1985) and Lake Clark areas (Thrupp and Coe, 1986) suggest that the rocks have not rotated or moved latitudinally relative to North America since their formation.

Rocks of mid-Tertiary age may occur in at least two offshore basins. Eocene basalts having K/Ar ages of 40.7 ± 2 and 42.3 ± 10 Ma were found at the bottom of the Cape Espenberg and Nimiuk wells, drilled in the Kotzebue basin (Tolson, 1986). Poorly dated basalt flows or sills were also found near the bottom of wells drilled in Norton Sound basin and are thought to be middle to late Eocene (Kirschner, this volume). Both basins are extensional basins with horsts and grabens (Norton Sound) or graben and half-graben (Kotzebue) structures (Kirschner, this volume). Although no chemical data are available on these rocks, their occurrence at the bottom of offshore basins clearly indicates an origin in an extensional tectonic environment.

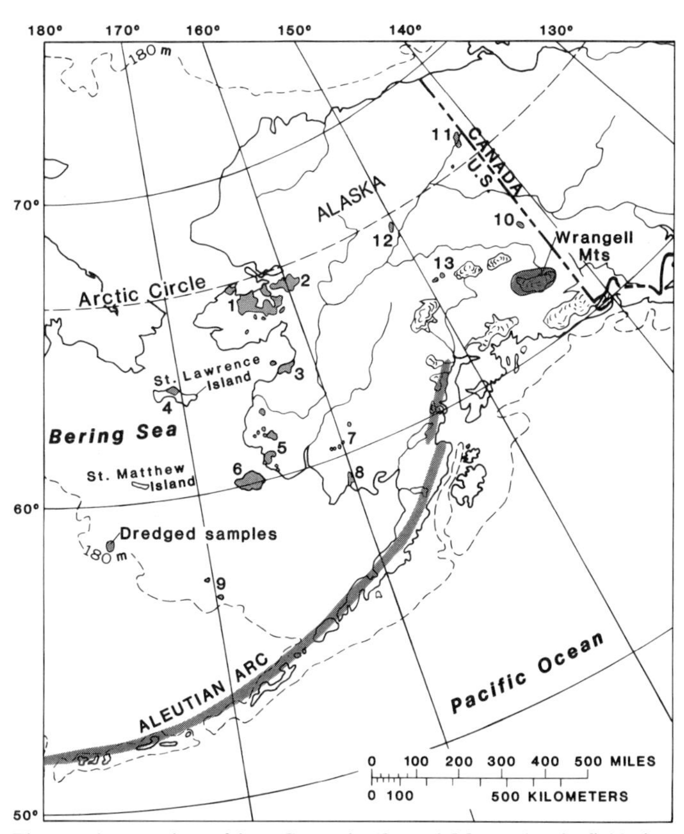

Figure 10. Location of late Cenozoic (0 to 6 Ma) volcanic fields in Alaska. 1, Imuruk Lake area; 2, Devil Mountain; 3, St. Michael volcanic field; 4, St. Lawrence Island; 5, Ingakslugwat volcanic field; 6, Nunivak Island; 7, Flat Top Mountain; 8, Togiak Basalt; 9, Pribilof Islands; 10, Prindle Volcano; 11, Porcupine-Black Rivers; 12, Ray Mountains; 13, Jumbo Dome (west) and Buzzard Creek (east). Red, chiefly basalt and basanite; shaded pattern, chiefly arc volcanic rocks. Dashed line marking 180-m water depth delineates the edge of the Bering Sea shelf.

LATE CENOZOIC VOLCANISM

Late Cenozoic volcanism in Alaska is dominated by calc-alkalic activity along the Aleutian arc and in the Wrangell Mountains and by contemporaneous alkalic and tholeiitic volcanism behind the arc in the Bering Sea region and easternmost interior Alaska (Fig. 10). The Aleutian arc has been active for at least the last 50 m.y. (Scholl and others, 1986), whereas most of the alkalic and tholeiitic Bering Sea basalts are restricted to the last 6 m.y.

Bering Sea region

The Bering Sea volcanic province consists of a number of large late Cenozoic basalt fields that occur in a vast region extending from St. Lawrence and the Pribilof Islands on the submerged Bering Sea shelf landward to the Seward Peninsula and Togiak River valley in western Alaska (Fig. 10). Most fields consist of a broad plain or shield composed of voluminous basalt flows overlain by steep cones and maars of undersaturated highly alkalic magma. Basaltic volcanism began as an isolated event about 28 to 26 m.y. ago with the eruption of the Kugruk Volcanics on the Seward Peninsula (Hopkins and others, 1963; Swanson and others, 1981). However, most of the volcanism in the Bering Sea region is much younger. Volcanism on Nunivak Island began 6 m.y. ago (Hoare and others, 1968), about the same time it resumed on the Seward Peninsula (5.8 Ma; Swanson and others, 1981). Most of the other fields began erupting in the Pliocene and Pleistocene.

On Nunivak Island, geologic mapping, paleomagnetic reversal stratigraphy, and K/Ar dating were used to determine the time and volume relations of volcanism (Hoare and others, 1968). The volcanic rocks range in age from 6 Ma to Holocene; the older series (3 to 6 Ma) underlies the western third of the island, and the younger series (0 to 1.7 Ma) covers the eastern two-thirds of the island.

Two suites occur on Nunivak Island: about 98 percent of the volcanic rocks form broad, thin pahoehoe flows of alkali basalt and subordinate tholeiite that make up 30 to 50 small shield volcanoes. The remaining 2 percent are basanites and nephelinites that form small, viscous flows, about 60 cinder cones, and ash deposits from four maar craters (J. M. Hoare, unpublished data, 1971). Subordinate eruptions of basanite or nephelinite commonly preceded and followed large eruptions of less alkalic basalt (J. M. Hoare, unpublished data, 1971). Analcime- or nepheline-bearing basanite or nephelinite overlie the Cretaceous sedimentary basement rocks on Nunivak Island and, in turn, are overlain by less alkalic basalts (J. M. Hoare, unpublished data, 1971). The youngest volcanism produced 40 to 50 basanite cones and four maar craters, which erupted in the last 0.5 m.y. in the south-central part of the island (Hoare and others, 1968). The cones and craters are aligned approximately east-west along several parallel fractures in a belt about 40 km long and 12 km wide. The basanite and nephelinite contain abundant inclusions of lherzolite, layered gabbro, chromite, conglomerate, sandstone, and basalt, and megacrysts as much as 10 cm long of anorthoclase, augite, and kaersutite.

The late Cenozoic (0.24 to 1.5 Ma) volcanic shield in the Kookooligit Mountains on St. Lawrence Island covers 42 by 33 km, is at least 500 m thick, and overlies volcanic, plutonic, and sedimentary rocks of early Paleozoic to Tertiary age (Patton and Csejtey, 1980). About 95 to 97 percent of the volcanic rocks are alkali-olivine basalt and subordinate amounts of olivine tholeiite, and 3 to 5 percent are basanite and minor nephelinite (J. M. Hoare, unpublished data, 1981). Large fluid flows of alkali-olivine and olivine-tholeiite basalt erupted from many small craters (20 to 60 m in diameter) and from one or more larger craters (100 to 150 m in diameter) located in the central part of the field (J. M. Hoare, unpublished data, 1981). Many small basanite flows and 60 to 80 cinder cones are aligned along an east-west belt across the north-central part of the field. Most of the basanite flows and cones contain xenoliths of deformed peridotite and/or gabbro. Sparse xenoliths of granitic rock and siliceous tuff in basanite cones and flows have been found at three localities.

Widespread late Cenozoic volcanic fields cover abut 10,000 km^2 of the Seward Peninsula (Till and Dumoulin, this volume). The Imuruk Lake area was originally mapped by Hopkins (1963), who distinguished five volcanic formations on the basis of weathering, degree of frost brecciation, and thickness of wind-blown silt. K/Ar studies in the Imuruk Lake region document the five eruptive episodes: (1) 28 to 26 Ma, (2) 5.8 to 2.2 Ma, (3) 0.9 to 0.8 Ma, (4) late Pleistocene, and (5) Holocene (Swanson and others, 1981). These data indicate that the earliest eruptive episode occurred 20 m.y. before the onset of mafic alkalic volcanism elsewhere in the Bering Sea region. The Imuruk Lake volcanic field is composed dominantly of alkali-olivine basalt and subordinate olivine tholeiite, quartz tholeiite, and basanite erupted from small shield volcanoes, cinder cones, plugs, and maar craters. Xenolithic inclusions, which occur in alkalic basalt and basanite, consist dominantly of lherzolite and subordinate harzburgite, chromite, schist, and granite. Additional basalt fields on the Seward Peninsula include basanite, tephrite, and alkali olivine basalt from a 2.5 to 2.9-Ma field north of Teller on the margin of the Imuruk basin and one at Devil Mountain in the northernmost part of the Peninsula (S. Swanson and D. Turner, unpublished data, 1984).

The Pribilof Islands, located near the edge of the Bering Sea shelf, were also the site of late Cenozoic volcanism. Volcanism occurred on St. George, the northern island, between 2.2 and 1.6 Ma, and on St. Paul Island, located 70 km to the south, from 0.374 Ma to the present (Cox and others, 1966; Lee-Wong and others, 1979). Volcanism was active in the intervening time on a submarine ridge located between St. George and St. Paul Islands, where dredged whole-rock basalt samples yield ages of 0.774 and 0.836 Ma (Simpson and others, 1979). St. George Island consists chiefly of deeply dissected olivine tholeiite flows. Its topography is controlled by numerous east-west–trending normal faults. St. Paul Island consists of coalescing small volcanoes, each composed of a central cinder cone and a surrounding shield of lava flows composed of alkali-olivine basalt and basanite. Dredged samples from the submarine ridge are chiefly olivine tholeiite, subordinate alkali-olivine basalt, minor basanite, and nephelinite (Lee-Wong and others, 1979).

The Togiak Basalt in southwestern Alaska (location 8 on Fig. 10) consists of about 9 km^3 of tholeiitic and alkali-olivine basalt flows (0.76 Ma) overlain by a tuya, or table mountain, formed by a subglacial basaltic eruption (Hoare and Coonrad, 1978a, 1980). The tuya consists of palagonitized glassy tuff and pillow lava capped by glassy subaerial flows of alkali-olivine basalt that probably erupted during glacial advances more than 39,000 yr ago.

The Yukon Delta contains numerous small volcanic fields that consist of alkali-olivine basalt flows, erupted from low saucer-shaped cones, overlain by young steep-sided cones composed of basanite and nephelinite (Hoare and Condon, 1971a; J. M. Hoare, unpubished data, 1971). Olivine nephelinite ash, perhaps erupted from a maar crater in the Ingakslugwat Hills, contains abundant olivine gabbro and lherzolite nodules (Hoare and Condon, 1971a). Some of the young steep-sided cones are aligned west-northwest, indicating a probable fracture control (Hoare and Condon, 1971a).

A large volcanic field covering about 2,000 km^2 near St. Michael Island in southeastern Norton Sound consists chiefly of voluminous olivine tholeiite and alkali-olivine basalt flows, and basanite tuffs, cones, and maar craters. Some of the young cones and short flows, such as those at Crater Mountain, consist of basanite with lherzolite nodules; others, such as a recent lava flow southeast of Crater Mountain, consist of olivine tholeiite. Flows from the base of the St. Michael volcanic field give whole-rock ages of 3.25 and 2.80 Ma (D. L. Turner, written communication, 1986). The approximate age of flows in the St. Michael volcanic field and on the Yukon delta were determined by Hoare and Condon (1966, 1968, 1971a, b) and Hoare and Coonrad (1959, 1978a) using magnetic polarity and physiographic expression of the rocks. Their data indicate that the flows were erupted during both the Brunhes Normal- (0.7 Ma to the present) and Matuyama Reversed- (2.4 to 0.7 Ma) Polarity Chrons. A few highly dissected flows in these areas may be as old as Pliocene.

A small olivine basalt field at Flat Top Mountain east of the Kuskokwim delta (4.62 to 4.72 Ma; Decker and others, 1986) is presumed herein to be part of the Bering Sea province.

Petrogenesis. Much confusion over the composition of the Bering Sea volcanic rocks exists because one of the few published papers describes the rocks on Nunivak Island as chiefly tholeiitic basalt and subordinate, highly undersaturated alkalic basalt (Hoare and others, 1968), although alkali olivine basalt is the most common rock type (J. M. Hoare, unpublished data, 1971). More recent studies in the province have focused on the peridotite inclusions from Nunivak (Francis, 1976a, b, 1978; Menzies and Murthy, 1980a, b; Roden and Murthy, 1985; Roden and others, 1984), the tectonic stress orientation of the province (Nakamura and others, 1977, 1980; Nakamura and Uyeda, 1980), and the Sr and Nd isotopic composition of the inclusions and volcanic rocks (Mark, 1971; Von Drach and others, 1986). Most authors describe the rocks as "back-arc" basalt because of their location behind the Aleutian arc.

For this chapter, I examined 200 analyses from St. Lawrence, Nunivak, St. Michael, and Ingakslugwat (J. M. Hoare, unpublished data, 1967–1971), 36 from the Pribilof Islands (Lee-Wong and others, 1979), 3 of the Togiak Basalt (Hoare and Coonrad, 1980), and 19 from the Seward Peninsula (Swanson and Turner, unpublished data, 1981). Thin sections from Hoare's collections show fresh phenocrysts and groundmass minerals, including olivine, and almost all of the analyses have less than 1 percent total H_2O.

Volcanic rocks in most fields range from nephelinites having more than 25 percent normative nepheline to tholeiites having 15 percent normative hypersthene. Most of the fields are compositionally similar to the volcanic field on Nunivak. Alkaline olivine basalt and olivine tholeiite represent at least 95 percent of the volcanic rocks present in all the volcanic fields (J. M. Hoare, unpublished data, 1967–1972) and form flat-lying flows and shield volcanoes. About 2 to 3 percent of the rocks are highly alkalic undersaturated basanite and nephelinite, which form short, viscous flows, cones, and ash. Eruptions of small volumes of highly alkalic undersaturated magma generally preceded and postdated voluminous outpouring of alkali and tholeiitic basalt (J. M. Hoare, unpublished data, 1971).

Most of the volcanic rocks have a 100 $Mg/[Mg + Fe^{2+}]$ greater than 65, and many contain ultramafic mantle xenoliths, which suggest that they may be primary or near-primary melts of mantle peridotite that have experienced little or no residence time in shallow magma chamgers and rose relatively rapidly to the surface. Data from both the basanite-nephelinite suite and the basalt suite cluster on AFM diagrams, further supporting the general lack of differentiation in the suite. Rare hawaiite (100 $Mg/[Mg + Fe^{2+}]$ = 55 to 64) has been reported on St. Lawrence Island, from the St. Michael volcanic field, and in dredgings from the ridge between St. Paul and St. George Islands, but has not been reported from the other areas.

Both the alkali-olivine and tholeiitic basalts have phenocrysts of olivine, plagioclase, clinopyroxene, magnetite, and ilmenite; they are characterized by diktytaxitic textures. Hawaiites have less olivine; basanites lack plagioclase phenocrysts. Basanites have nepheline and analcime, and nephelinites have sodic pyroxene and nepheline in the groundmass mesotasis (Hoare and others, 1968). None of the samples examined in thin section contained modal hypersthene, but it has been reported in the Imuruk Volcanics from the Imuruk Lake area (Hopkins, 1963).

The basanites and nephelinites commonly contain megacrysts and/or xenoliths of peridotite, gabbro, or bedrock. The megacrysts consist of unzoned anorthoclase, clinopyroxene, and kaersutite, as much as 10 cm long (Hoare and others, 1968). Most of the megacrysts are deformed, showing kink bands and undulatory extinction, and most show reaction relations to their host basalt. Aluminous clinopyroxene, kaersutite, and feldspar are common megacrysts in alkalic basalts around the world (Irving, 1974). Experimental studies of megacrysts and their host basalts suggest that megacrysts of clinopyroxene and kaersutite are near-liquidus phases of the host basalt at high pressures (10 to 20 kb). In contrast, experimental work suggests that anorthoclase is never a liquidus phase in a basalt or basanite at any pressure, and they therefore are generally thought to have precipitated from more evolved magmas, which later mixed with the host basalt (Irving and Frey, 1984). The lack of differentiated magmas in the Bering Sea basalt province, however, makes this mechanism unlikely in these magmas. H. Wilshire (oral communication, 1985) believes that anorthoclase megacrysts in the Mojave Desert, California, formed in mantle veins due to several generations of partial melting. The anorthoclase megacrysts in the Bering Sea basalts may have formed in a similar manner and were disaggregated during their rise to the surface.

The most abundant xenoliths, 75 percent of Nunivak's xenolith population (Francis, 1976a), are lherzolite nodules composed of olivine, enstatite, clinopyroxene, and spinel. Less than 1 percent of the lherzolite nodules on Nunivak contain chromian pargasitic amphibole, but about 50 percent contain zones of fine-grained diopside, olivine, spinel, and Al-rich glass, which Francis (1976a) interprets as relicts of melted amphibole. Red-brown chromian mica occurs between the amphibole and included spinel in some samples. Francis (1976a) believes that the spinel lherzolite xenoliths are accidental fragments of upper mantle and that amphibole formed during a mantle metasomatic event accompanied by infiltration of aqueous fluids enriched in alkalis and incompatible elements. He further suggests that the metasomatism predates entrainment of these nodules in the alkali basalt.

Other common xenoliths include corona-bearing pyroxene granulites (9 percent of Nunivak's xenolith population), which range from plagioclase to olivine dominated (Francis, 1976b). The reaction of olivine and plagioclase to clinopyroxene-spinel symplectite and aluminus orthopyroxene suggests that the xenoliths last equilibrated at 950°C under at least 9 kbar pressure. Francis (1976b) interprets the xenoliths as fragments of the base of the crust, and proposes that the reaction took place in the corona structures in response to crustal thickening of the Bering Sea shelf from thin oceanic crust (10 km) in the early Mesozoic to thicker crust (about 30 km) in the Quaternary.

Other reported xenoliths include bedrock of obvious crustal origin and dunite, harzburgite, chromite, gabbro, and amphibole-bearing pyroxenite of less certain origin (Hoare and others, 1968;

Francis, 1978). Lithologies of the bedrock xenoliths vary depending on the location of the volcanic field. At least some at each locality consist of fragments of underlying basalt flows.

The Bering Sea province is characterized by high alkali and low silica content and ranges in composition from nephelinite to basanite through alkali-olivine basalt to olivine tholeiite and tholeiite. Total alkalis decrease with increasing SiO_2 (Fig. 11) as in the Hawaiian suites (Clague and Frey, 1982; Frey and Clague, 1983). But, Bering Sea basalts have lower CaO than Hawaiian lavas, which result in greater silica saturation at a given total alkali content. Figure 11 shows the dividing line for alkalic and tholeiitic rocks in the Bering Sea province and the line of silica saturation, as defined by MacDonald and Katsura (1964) for the Hawaiian Islands.

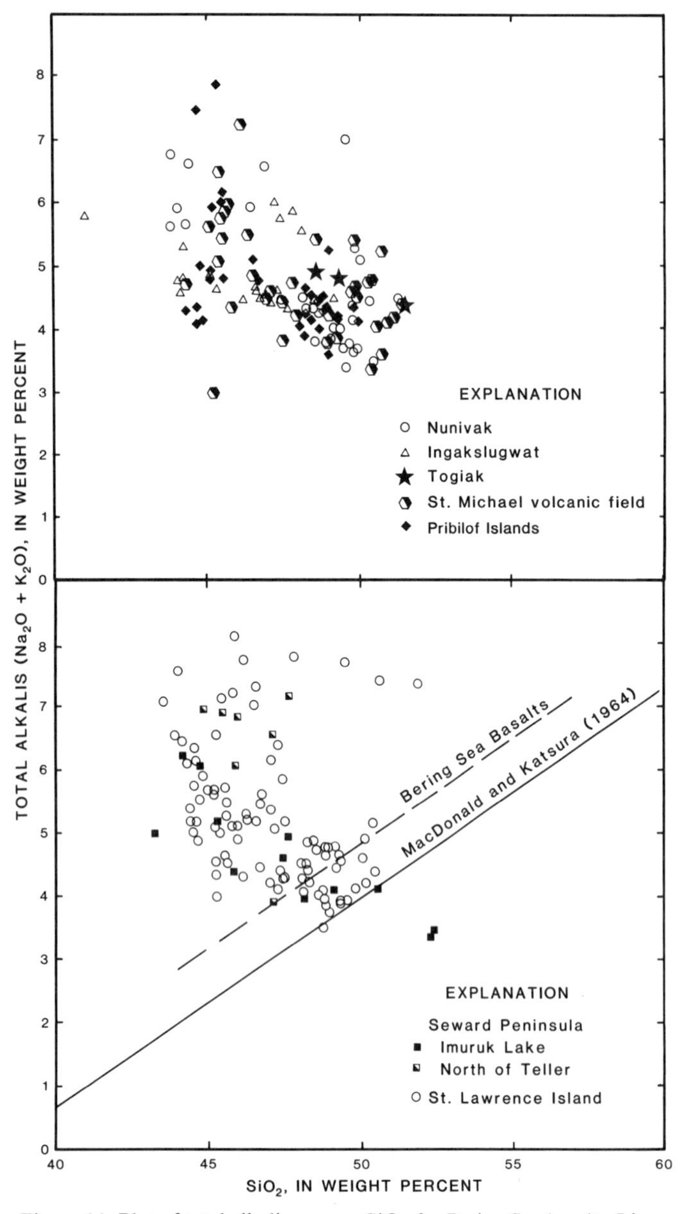

Figure 11. Plot of total alkalies versus SiO_2 for Bering Sea basalts. Lines divide silica-saturated and silica-undersaturated rocks on Hawaii and in Bering Sea basalts. Data from J. M. Hoare (unpublished data, 1971), S. Swanson and D. Turner (unpublished data, 1985), and Lee-Wong and others (1979).

REE data from Nunivak (Roden, 1982), the Pribilof Islands (Kay, 1977; Florence Lee-Wong, unpublished data, 1986), and St. Michael volcanic field (E. J. Moll and W. W. Patton, Jr., unpublished data, 1987) show that lavas from these fields are LREE enriched. LREE are correlated with alkalinity: nephelinites have 40 ppm La and tholeiites have about 10 ppm La.

P_2O_5, Rb, Sr, Ba, K_2O, and Na_2O all decrease with increasing silica and correlate positively with each other (J. M. Hoare, unpublished data, 1967–1971; Mark, 1971). These trends cannot be due to crystal fractionation of any phenocryst commonly found in the basanites or basalts. Studies of suites of similar composition (Frey and others, 1978; Clague and Frey, 1982) have shown that basalt to nephelinite to melilite suites formed by decreasing degrees of partial melting of a garnet peridotite source. By analogy, this suggests that Bering Sea nephelinites, which have the highest Rb, Sr, Ba, K, Na, and P and the lowest SiO_2, originated by the smallest degree of partial melting; and that tholeiites, which have the lowest alkalis and highest SiO_2, originated by the greatest degree of partial melting. Experimental petrologists have suggested that strongly silica-undersaturated magmas form from a peridotite mantle rich in carbon (Wyllie and Huang, 1976; Eggler and Holloway, 1977; Eggler, 1978). I examined P and K data from more than 200 major-element analyses of rocks from St. Lawrence Island to estimate the range of partial melting required by this volcanic field using the method of Clague and Frey (1982). For the olivine tholeiite to nephelinite series, K_2O varies from 0.4 to 2.2 percent and P_2O_5 varies from 0.22 to 0.71 percent. K content has the widest range, which suggests that it is the most incompatible major element. If K behaves as a totally incompatible element (D = 0) during partial melting, and no fractional crystallization occurs, its compositional range requires that the degree of partial melting vary by a factor of at least 5. Trace elements are probably more incompatible than major elements. Trace element data are not available for most of the Bering Sea basalt fields, but Rb data on a small suite of samples from Nunivak Island vary from 5.8 to 64.4 ppm (Mark, 1971), a factor of 11, over the range olivine tholeiite to nephelinite. A factor of 11 corresponds to a melting interval of at least 11 percent or 1 to 11 percent, 2 to 22 percent. However, as discussed below, Sr isotope data on the Bering Sea magmas suggest they have not originated in a mantle that has isotopically homogeneous $^{87}Sr/^{86}Sr$. The source for the nephelinites has lower $^{87}Sr/^{86}Sr$, and therefore probably lower Rb/Sr, than the source for the less alkalic basalts, which suggests that the melt interval is probably larger. Frey and others (1978) suggest that 4 to 25 percent melting produced a compositional range from olivine melilite to quartz tholeiite in the Tertiary to Holocene basalts of Victoria, Australia and Tasmania. Although I cannot rigorously constrain the degree of partial melting for the Nunivak Island suite, the data suggest that the nephelinites form by small amounts of partial melting, probably on the order of 1 to 3 percent, and the tholeiites formed by larger amounts of melting, on the order of 11 to 33 percent.

Three isotopic studies have been done on the Bering Sea province, and a summary of the reported data is shown in Figure 12. In the first study, Mark (1971) analyzed the Sr isotopic composition of more than 24 samples, ranging from nephelinite to tholeiite, from Nunivak Island and of a few samples from St. Lawrence and St. Michael Islands, Ingakslugwat Hills, and the Pribilof Islands. He found that $^{87}Sr/^{86}Sr$ = 0.7026 to 0.7033 and decreased with increasing alkalinity and decreasing silica saturation in the Nunivak samples. Nine basanites have $^{87}Sr/^{86}Sr$ of 0.70286 ± 0.00018, and 15 alkali-olivine and tholeiitic basalts have $^{87}Sr/^{86}Sr$ of 0.70311 ± 0.00013. Samples from the Pribilofs, Ingakslugwat, and St. Michael fall within the range of Sr isotope ratios given for Nunivak. $^{87}Sr/^{86}Sr$ on samples from St. Lawrence Island, which overlies Paleozoic sedimentary rocks, are higher, ranging from 0.7036 to 0.7039, and also decrease with increasing alkalinity.

Menzies and Murthy (1980a, b) studied the Sr and Nd isotopic composition of basalts, kaersutite megacrysts, and paragasite from lherzolite nodules from Nunivak in the second isotopic study. They report Sr isotope data for volcanic rocks, ranging from nephelinites to olivine tholeiites, similar to values reported by Mark in 1971 ($^{87}Sr/^{86}Sr$ = 0.70251 to 0.70330 and $^{143}Nd/^{144}Nd$ ranging from 0.51289 to 0.51304). In contrast to Sr, the Nd isotopes do not correlate with alkalinity. The nodules and megacrysts have $^{87}Sr/^{86}Sr$ ranging from 0.7027 to 0.7033, which suggests that the mantle under Nunivak is locally inhomogeneous in $^{87}Sr/^{86}Sr$. Coexisting pargasite and nodules have identical isotopic composition within analytical uncertainty. The Nunivak data plot on a $^{87}Sr/^{86}Sr$ versus $^{143}Nd/^{144}Nd$ diagram in the field where MORB (mid-ocean-ridge basalt) and OIB (oceanic-island basalt) overlap (Menzies and Murthy, 1980a, b; Roden and others, 1984) but are LREE enriched (Roden, 1982). Thus, the low-Nd isotopic composition of the basalts and nodules requires a time-integrated LREE—depleted source for the basalts, but all the basalts, even the tholeiites, are LREE enriched (Fig. 13). Menzies and Murthy (1980a) suggest that the LREE and other incompatible elements were enriched in the source region by relatively recent (within the last 200 m.y.) mantle metasomatism and that the range in Sr isotopic composition can be explained by local inhomogeneities in Rb/Sr that developed during the metasomatic event and resulted in small variations in $^{87}Sr/^{86}Sr$ over time.

The third isotope study (von Drach and others, 1986) focused on the Nd and Sr isotopic composition of the Aleutian Islands but also reported data from Nunivak, St. George, and St. Lawrence Islands. Data for Nunivak and St. George are identical to the previously published results on Nunivak (Menzies and Murthy, 1980b) and plot on the $^{87}Sr/^{86}Sr$–$^{143}Nd/^{144}Nd$ dia-

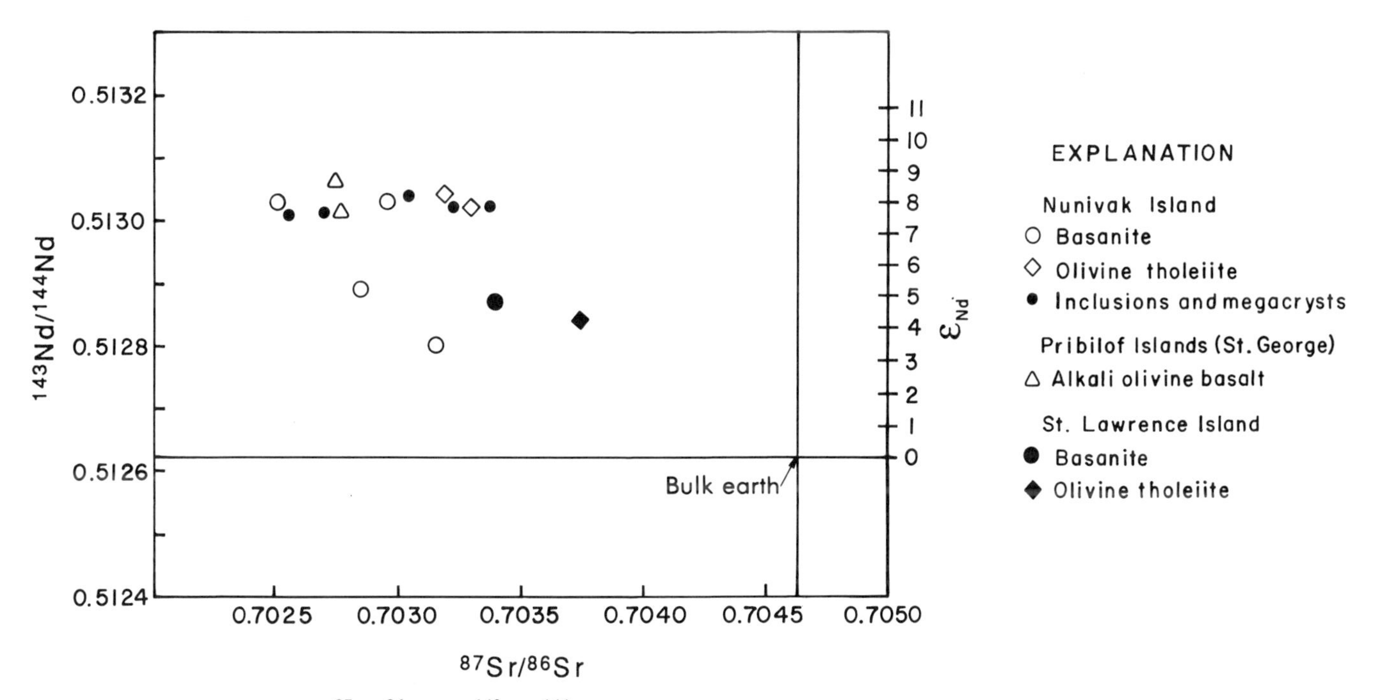

Figure 12. $^{87}Sr/^{86}Sr$ and $^{143}Nd/^{144}Nd$ data for Bering Sea basalts. Data from Menzies and Murthy (1980a), von Drach and others (1986), and Roden (1982). Rocks from St. George in the Pribilof Islands and Nunivak Island plot in the field where values for MORB (mid-ocean-ridge basalt) and oceanic island basalt overlap. Analyses of samples from St. Lawrence Island plot closer to bulk-earth compositions. Bulk-earth values from Allegre and others (1984).

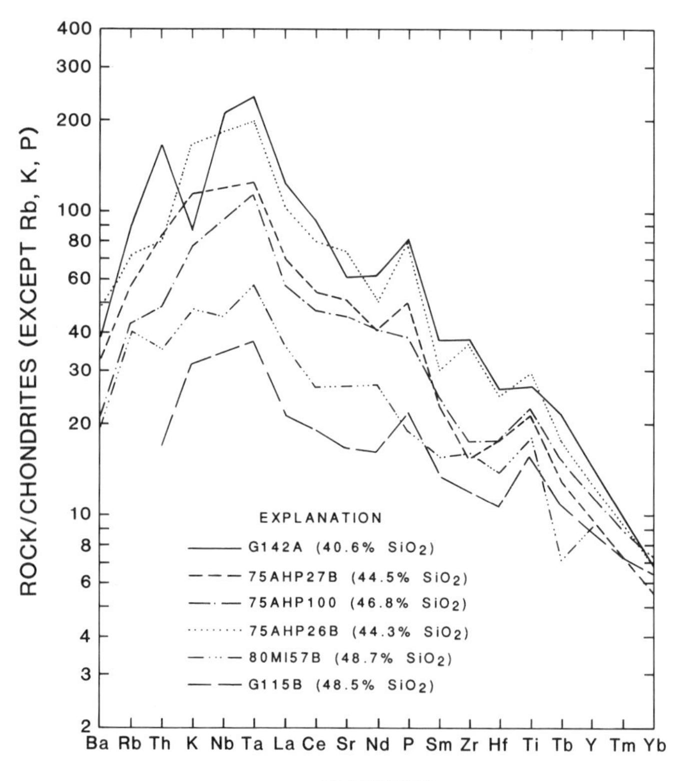

Figure 13. Chondrite-normalized spidergrams for volcanic rocks from the St. Michael volcanic field (80ML57B) and the Pribilof Islands (P26B, G142A, P27B, HP100, and G115B). Data from E. J. Moll-Stalcup and W. W. Patton, Jr. (unpublished data, 1980–1981), and F. Lee-Wong (unpublished data, 1981). The rocks having the highest alkalies and LREE are nephelinites; those having the lowest are tholeiites. Note the positive Nb-Ta anomaly.

gram in the field where MORB and OIB overlap. $^{143}Nd/^{144}Nd$ is 0.5133 ± 0.0002 and $^{87}Sr/^{86}Sr$ is 0.7025 to 0.7033 for six of the seven samples from both Nunivak and St George. Rocks from St. Lawrence Island have significantly higher $^{87}Sr/^{86}Sr$ and lower $^{143}Nd/^{144}Nd$ and plot along the mantle array within the field for oceanic basalts (Fig. 12).

Data from all three studies suggest that volcanic rocks from Nunivak, the Pribilofs, Ingakslugwat, and St. Michael are isotopically similar and plot in the field where MORB and OIB overlap. $^{143}Nd/^{144}Nd$ values are about 0.5132, and $^{87}Sr/^{86}Sr$ ranges from 0.7025 to 0.7033 for all the analyzed fields except St. Lawrence Island, which has a basement composed of rocks at least as old as middle Paleozoic, and has higher $^{87}Sr/^{86}Sr$ and lower $^{143}Nd/^{144}Nd$ than the other volcanic fields (Fig. 12). In all the fields where data are sufficient, $^{87}Sr/^{86}Sr$ appears to be negatively correlated with silica undersaturation and alkalinity. Hawaiian volcanic rocks on Oahu show a similar trend—the Honolulu Group, which is composed of undersaturated alkalic rocks, has lower $^{87}Sr/^{86}Sr$ (0.70331) than the underlying tholeiitic shield (0.70370) (Lanphere and others, 1980; Lanphere and Dalrymple, 1980; Clague and Frey, 1982). The source of the St. Lawrence Island magmas is either mantle that has higher $^{87}Sr/^{86}Sr$ and lower $^{143}Nd/^{144}Nd$ than the other volcanic fields (continental lithospheric mantle?) or crustally contaminated isotopically similar mantle. Inclusions of sialic rock have been reported in highly alkalic magmas on St. Lawrence Island (Patton and Csejtey, 1980). However, the primitive composition of the magmas, which suggests little differentiation and thus little or no residence time in shallow magma chambers, argues against the crustal contamination hypothesis. Although it is possible that crustal contamination may be responsible for the range in $^{87}Sr/^{86}Sr$ within individual volcanic fields, it is not required because a similar range in $^{87}Sr/^{86}Sr$ is found in the mantle, as evidenced from xenolith studies from Nunivak (Menzies and Murthy, 1980b). The proposed metasomatic event that enriched the mantle under the Bering Sea and western Alaska in K, LREE, and P probably occurred within the last 200 m.y., as suggested by Menzies and Murthy (1980b), but was not synchronous with the alkalic volcanism, as suggested by Roden and others (1984). Consequent metasomatism as suggested by Roden and others (1984) does not account for the range in $^{87}Sr/^{86}Sr$ in the basalts nor does it explain how large volumes of LREE-enriched magmas (including tholeiites) could form from a LREE-depleted mantle that had a small volume of LREE-enriched veins. It seems more likely that a separate metasomatic event, perhaps related to earlier subduction in the Neocomian or Late Cretaceous and early Tertiary, was responsible for the enrichment.

The Ingakslugwat Hills, Nunivak Island, and St. Michael Islands lie within the Yukon-Koyukuk basin, and the Pribilof Islands may lie within an offshore extension of this basin. The Yukon-Koyukuk basin is thought to be a Mesozoic island arc that collided with western Alaska in mid-Cretaceous time (Patton and Box, 1989). Studies of crustal xenoliths included in the basaltic flows indicate that sialic rocks of pre-Cretaceous age do not occur in the St. Michael, Nunivak, or Ingakslugwat volcanic fields (J. M. Hoare, unpublished data, 1981). The lack of pre-Cretaceous sialic inclusions suggests that Paleozoic strata are not present beneath these areas and further suggests that the lithosphere under the basin might have lower $^{87}Sr/^{86}Sr$ and higher $^{143}Nd/^{144}Nd$ than the lithosphere under older, long-lived continental areas such as St. Lawrence Island.

In summary, the Bering Sea basalt suites of nephelinite to tholeiite probably originated by increasing degrees of partial melting of a peridotite mantle rich in carbon. Most of the magmas rose quickly to the surface, and few, if any, were significantly differentiated. The basanites and nephelinites originated in a source having lower $^{87}Sr/^{86}Sr$ than the source of the less alkalic basalts. The small range in $^{87}Sr/^{86}Sr$ is probably due to mantle metasomatism that enriched the source area in K, LREE, Sr, Rb, and P within the last 200 m.y. (Menzies and Murthy, 1980b). Furthermore, this metasomatism may be related to previous subduction events in western Alaska during the Neocomian or Late Cretaceous and early Tertiary. The volcanic rocks on St. Lawrence Island have more radiogenic $^{87}Sr/^{86}Sr$ and less radiogenic

$^{143}Nd/^{144}Nd$ than volcanic rocks from the other fields and suggest the presence of more enriched mantle under St. Lawrence Island—possibly continental lithospheric mantle. The correlation between exposed crustal type and Sr and Nd isotope composition can be explained by tectonic models for western Alaska that require younger and more mafic crust beneath the Yukon-Koyukuk province than the crust under the surrounding metamorphic borderlands and St. Lawrence Island.

Eastern and central interior Alaska

Several small, isolated basaltic volcanoes occur in easternmost Alaska, between the Fortymile and Tanana Rivers. The cones and associated flows are undated but apparently are young, as evidenced by their well-preserved volcanic morphology. The best-preserved and probably youngest cone is Prindle volcano, a small cinder cone that was the source of a narrow basanite lava flow more than 10 km long (Foster and others, 1966). The cone and adjacent flow contain abundant inclusions of harzburgite, wehrlite, lherzolie, pyroxenite, and granulite-facies schist. Prindle volcano is undated but underlies the (informally named) White River ash bed, which was dated by ^{14}C methods at approximately 1,900 B.P. (Lerbekmo and Campbell, 1969).

Large volumes of olivine basalt flows, as much as 100 m thick, are exposed in fault blocks near the Porcupine and Black Rivers (Brosgé and Reiser, 1969). The flows are considered to be Tertiary or Quaternary because of their youthful appearance. One chemical analysis, of an alkali-olivine basalt, is available from flows along the Black River (Brabb and Hamachi, 1977).

Miscellaneous isolated volcanic rocks of late Cenozoic age also occur in central Alaska. A small, isolated maar volcano composed of olivine basalt erupted 3,000 years ago at Buzzard Creek on the north side of the Alaska Range (Albanese, 1980). Flat-lying olivine basalt flows cover approximately 100 km^2 of north-central Alaska between the Yukon and upper Koyukuk Rivers, about 125 km northeast of the town of Tanana. These rocks are undated, but their lack of deformation suggests they are late Tertiary or Quaternary in age (Patton and Miller, 1973).

The few available analyses of volcanic rocks in central and eastern Alaska indicate that the rocks are compositionally similar to Bering Sea basalts. Alkalic basalts in the Porcupine and Yukon-Tanana upland area are located north of arc volcanoes in the Wrangell Mountains (Fig. 10). By analogy with Bering Sea basalts, these magmas probably represent regional extension behind the present arc. Cones and flows of olivine basalt, some of which contain ultramafic inclusions, also occur in a south-trending regional belt that extends from eastern Alaska into the Yukon Territory and continues down along the western North American continental margin through British Columbia (Foster and others, 1966; Sinclair and others, 1978).

Jumbo Dome, the only occurrence of Quaternary (0.80 to 2.8 Ma) orogenic andesite in interior Alaska, occurs about 10 km southwest of the maar volcano at Buzzard Creek (Fig. 10; Albanese, 1980). The hornblende andesite dome is probably related to subduction under this area, making it the only Quaternary occurrence of arc volcanism in the 300-km-wide magmatic gap between Mount Spurr, at the northeast end of the Aleutian arc, and the Wrangell Mountains.

Late Cenozoic tectonic implications

Bering Sea basalts have compositions similar to suites found in a variety of tectonic environments, including oceanic islands (Hawaii: Clague and Frey, 1982; Frey and Clague, 1983), stable continents associated with regional faulting (southern and eastern Australia: Irving, 1974; Frey and others, 1978; and the western U.S.: Menzies and others, 1987), continental rifts (east Africa: King, 1970), and behind volcanic arcs a great distance from the arc (China and Korea: Nakamura and others, 1985). The locations of the various volcanic fields do not define a narrow volcanic belt, a rift axis or a hot-spot trend. Most recent authors (Nakamura and others, 1977; von Drach and others, 1986) have labeled the Bering Sea basalts as back-arc basalts, although they do not constitute a classic back arc characterized by a spreading rift axis or by high heat flow (Marshall, 1978; Smirnov and Sugrobov, 1979a, b), nor do they have typical back-arc compositions, which usually range from N-MORB to arc tholeiite (Saunders and Tarney, 1984; Hawkins and Melchoir, 1985). Turner and others (1981) suggest that the basalts on the Seward Peninsula are related to an interconnected system of rifts and transform faults. Major- and trace-element data suggest that the Bering Sea basalts came from a source similar to oceanic island basalts. Trace-element ratios and isotopic compositions of the Bering Sea basalts are similar to Hawaiian volcanic rocks and different from N-MORB (Table 3). The Bering Sea volcanic fields, however, are not aligned along the trend of a hot spot. Chondrite-normalized multi-element diagrams show that, unlike arc basalts, Bering Sea basalts have positive Nb and Ta anomalies (Fig. 13) and, therefore, have a different source than the Aleutian arc.

Many of the voluminous basalt fields are located on or near strike-slip or normal faults, and fault displacements suggest that at least some of the faulting began while volcanism was still active. The Togiak Basalt is located in a north-northeast–trending graben (Hoare and Coonrad, 1980; Globerman, 1985). Late Pliocene volcanic rocks on St. George Island in the Pribilofs are cut by numerous normal faults, most of which trend approximately east-northeast (Hopkins, 1976). Several of the Yukon delta volcanic fields are located along a trace of the Anvik fault, and the volcanic field near St. Michael Island is probably intersected by a trace of a Kaltag fault splay (W. W. Patton, Jr., oral communication, 1984). Young volcanic cones are aligned approximately east-west, apparently defining a fracture or fault in the St. Lawrence, Nunivak, and St. Michael volcanic fields, and in a small field north of Aropuk Lake on the Yukon delta. Late Cenozoic volcanism in the Seward Peninsula is associated with transform faulting, geothermal anomalies, large late Tertiary grabens, and high levels of seismicity in the central Seward Peninsula (Turner and Forbes, 1980).

TABLE 3. COMPARISON OF SELECTED TRACE-ELEMENT RATIOS FOR A BERING SEA THOLEIITE WITH A HAWAIIAN THOLEIITE AND N-MORB

	Bering Sea basalts*	Hawaii†	Average N-MORB§
P_2O_5/Ce	87.0	81.3	0.02
Rb/Sr	0.044	0.031	0.008
K/Rb	409	432	1,060
Zr/Hf	46.4	45 ± 4	33.5
Hf/Ta	2.44	1.35	15.45
Th/La	0.125	0.091	0.065
Th/Ce	0.065	0.047	0.021
Th/Sm	0.47	0.39	0.063
Th/Nd	0.088	0.094	0.023
Sr/Th	214	246	660
Ba/Th	88.7	165.0	60.0
Sr/Ba	2.4	1.5	11.0
Ba/La	11.1	14.9	3.9
Ba/Ce	5.8	7.6	1.3
Zr/Ta	113.0	60.0	518.0
Sr/Ce	14.0	11.3	13.9

*St. Michael volcanic field (80ML57b).
†Clague and Frey (1982).
§Wood (1979).

The distribution of active faults and monogenetic cones in late Quaternary volcanic fields on St. Lawrence, St. Michael, Ingakslugwat, Nunivak, the Pribilofs, and the Seward Peninsula were used by Nakamura (Nakamura and others, 1977; Nakamura and Uyeda, 1980; Nakamura and others, 1980) to define the tectonic stress field in the Bering Sea region in the late Quaternary. Cones on most of the volcanic fields are aligned east-west, corresponding to east-west maximum horizontal compression. Maximum horizontal compression in the Aleutian arc is oriented north-south, perpendicular to the trench. The axis of maximum horizontal compression (MHC) can represent either the intermediate stress axis or the maximum stress axis. Nakamura uses a presumed tectonic environment of the volcanism to interpret the MHC in the Bering Sea region as representing north-south extension and the MHC in the arc as representing north-south compression.

Holocene surface faults (Hudson and Plafker, 1978; Plafker and others, this volume, Plate 12) and focal-mechanism solutions for earthquakes on the Seward Peninsula and adjacent northwestern Alaska show dominantly normal fault movement with extension in the northwest or northeast directions (Biswas and others, 1986). Biswas and others (1986) classify all of western Alaska and the Bering Sea shelf as an area of "tensional stress regime." However, they provide no mechanisms for western Alaska south of the Seward Peninsula area or for the Bering Sea shelf.

In late Cenozoic time the Bering Sea shelf was located in the vicinity of the Eurasian, North American, and Pacific plates. The shelf was probably part of the North American plate, and the plate boundary between Eurasia and North America was located to the west of the Bering Sea shelf in eastern Siberia (Zonenshain and others, 1984). Harbert and others (1987) suggest slight convergence between Alaska and Eurasia in a north to northeast direction from 37 to 0 Ma in the Bering Sea region, based on spreading patterns in the North Atlantic. North to northeast convergence contradicts Nakamura and Biswas's studies and suggests that compressive stress from convergence between the North American and Eurasian plates was localized along the plate boundary in Siberia and did not affect the Bering Sea shelf. Therefore, plate motion between the North American and Eurasian plates does not appear to be responsible for volcanism on the Bering Sea shelf.

In the Bering Sea region, motion between the North American and Pacific plates was dominated by north-directed subduction of the Pacific plate along the Aleutian trench for at least the past 50 m.y. None of the Bering Sea basalts, not even those from the Pribilof Islands, only 550 km from the trench, have the Nb-Ta depletions characteristic of arc magmas. Neither do they occur along a rift axis, nor do they have typical back-arc compositions. However, they were erupted in a broad extensional environment located behind the Aleutian arc. Thus, although the rocks do not constitute a classic back arc, the occurrence of the tholeiitic and alkalic basalt in the Bering Sea region behind the Aleutian arc and in east-central Alaska behind the Wrangell Mountains volcanoes suggests that they represent a broad zone of regional extension behind the arc. Alkaline basalts that lack Nb-Ta anomalies occur behind the Japanese arc in Korea and China (Nakamura and others, 1985) and this occurrence may be similar to that of the Bering Sea basalts.

Voluminous eruptions of Bering Sea basalts began at 6 Ma, contemporaneous with small changes in Pacific plate motion (Barron, 1986; Cox and Engebretson, 1985) and the start of a major pulse in volcanic activity in the Aleutian arc that continues to the present. The timing of eruptions in the Bering Sea region may be related to the change in the angle of Pacific plate motion at 6 Ma, or it may possibly represent the time necessary to heat the back-arc region before volcanism began.

SUMMARY AND CONCLUSIONS

The Alaska Range–Talkeetna Mountains, Kuskokwim Mountains, and Yukon-Kanuti belts constitute an anomalously wide volcanic arc that was active during the Late Cretaceous and early Tertiary. The arc was narrower, consisting of the Alaska Range–Talkeetna Mountains and Kuskokwim Mountains belts from 75 to 66 Ma and broadened considerably to include the Yukon-Kanuti belt from 65 to 56 Ma. Plate motion models predict rapid north-northeast–directed subduction of the Kula plate under southern Alaska between 75 and 56 Ma (Engebretson and others, 1982). The angle of convergence between Paleocene plate motions and the present continental margin and three parallel magmatic belts is too small to generate arc magmatism (Wallace and Engebretson, 1984; Gill, 1981). This enigma is resolved by paleomagnetic models that suggest that western Alaska has been rotated 30 to 55 degrees counterclockwise since the Paleocene (Globerman and Coe, 1984; Hillhouse and Coe, this volume).

Assuming that models for counterclockwise rotation of western Alaska are correct, the continental margin of southern Alaska, which now has a tightly curved S-shape, may have had a more open S-shape in the Late Cretaceous and early Tertiary. This configuration places St. Matthew Island close to the trench in Paleocene time, which is consistent with its low K contents and tentative correlation with the Alaska Range–Talkeetna Mountains belt. Unrotating western Alaska also places the continental margin and three magmatic belts approximately east-west in Paleocene time—orthogonal to the direction of subduction. Compression between Alaska and Eurasia related to opening in the North Atlantic was probably responsible for flexure of the southern continental margin into its present tight S-curve. This bending is proably responsible for the post-Paleocene counterclockwise rotation of western Alaska.

At about 56 Ma, the trench jumped away from the continental margin to its present position, and formation of the Aleutian ridge began (Scholl and others, 1986). Paleomagnetic data on Paleocene and Oligocene volcanic rocks suggest that rotation of western Alaska occurred between 56 and 43 Ma (Thrupp and Coe, 1986; Harris, 1985). Between 56 and 43 Ma, Engebretson and others (1982) show rapid north-directed subduction under southern Alaska. Subduction-related volcanic rocks between 56 and 48 m.y. old are restricted to the hinge line of the oroclinal bend (Arkose Ridge Formation and Talkeetna Mountains volcanic rocks), which may have been more orthogonal to the plate motion than the rotated (or rotating?) southwestern continental margin in the Eocene.

Mid-Tertiary volcanism in interior Alaska is chiefly felsic and appears to be related to regional extension or movement along strike-slip faults. This volcanism occurred at 40 ± 3 Ma, coincident with a change in the angle of Pacific plate motion at 43 Ma and the start of a peak in magmatic activity in the Aleutian arc that occurred between 40 and 30 Ma.

Bering Sea basalts were erupted in a broad extensional environment behind the Aleutian arc, but are not a classic back arc. Eruptions of the basalts, which started at about 6 Ma, are contemporaneous with changes in Pacific plate motion and the beginning of a major eruptive pulse in the Aleutian arc. Bering Sea basalts originated in a mantle source similar to that for oceanic island basalts. The source of the magmas had been previously metasomatized by the addition of K, P, REE, and Ti. This metasomatic event occurred within the last 200 m.y. (Menzies and Murthy, 1980b), probably during the widespread Early Cretaceous, or Late Cretaceous and early Tertiary subduction events.

REFERENCES CITED

Albanese, M. D., 1980, The geology map of three extrusive bodies in the central Alaska Range [M.S. thesis]: Fairbanks, University of Alaska, 104 p.

Albanese, M. D., and Turner, D. L., 1980, ^{40}K–^{40}Ar ages from rhyolite of Sugar Loaf Mountain, central Alaska Range; Implications for offset along the Hines Creek Strand of the Denali fault system: Alaska State Division of Geological and Geophysical Surveys Short notes on Alaska Geology 1979–1980, p. 7–10.

Allegre, C. J., Hart, S. R., and Minster, J. F., 1984, Chemical structure and evolution of the mantle and continents determined by inversion of Nd and Sr isotopic data, II. Numerical experiments and discussion: Earth and Planetary Science Letters, v. 66, p. 191–213.

Arculus, R. J., and Powell, R., 1986, Source component mixing in the regions of arc magma generation: Journal of Geophysical Research: v. 19, p. 5913–5926.

Arth, J. G., Barker, F., Peterman, Z. E., and Friedman, I., 1978, Geochemistry of the gabbro-diorite-tonalite-trondhjemite suite of southwest Finland and its implications for the origin of tonalitic and trondhjemitic magmas: Journal of Petrology, v. 19, p. 289–316.

Barnes, D. F., 1977, Bouguer gravity map of Alaska: U.S. Geological Survey Geophysical Investigations Map GP-913, scale 1:2,500,000.

Barron, J. A., 1986, Paleoceanographic and tectonic controls of deposition of the Monterey Formation and related siliceous rocks in California: Palaeogeography, Palaeoclimatology, Palaeoecology, v. 53, p. 27–45.

Beikman, H. M., 1974, Preliminary geologic map of the southwest quadrant of Alaska: U.S. Geological Survey Miscellaneous Field Studies Map MF-611, 2 sheets, scale 1:1,000,000.

Biswas, N. N., Akim, K., Pulpan, H., and Tytgat, G., 1986, Characteristics of regional stresses in Alaska and neighboring areas: Geophysical Research Letters, v. 13, p. 177–180.

Bloomer, S. H., Stern, R. J., Fisk, E., and Geschwind, C. H., 1989, Shoshonitic volcanism in the northern Mariana arc, 1. Mineralogic and major trace element characteristics: Journal of Geophysical Research, v. 94, p. 4469–4496.

Box, S. E., 1985, Mesozoic tectonic evolution of the northern Bristol Bay region, southwestern Alaska [Ph.D. thesis]: Santa Cruz, University of California, 125 p.

Brabb, E. E., and Hamachi, B. R., 1977, Chemical composition of Precambrian, Paleozoic, Mesozoic, and Tertiary rocks from east-central Alaska: U.S. Geological Survey Open-File Report 77–631, p. 87.

Brosgé, W. P., and Reiser, H. N., 1969, Preliminary geologic map of the Coleen Quadrangle, Alaska: U.S. Geological Survey Open-File Report 69-25, scale 1:250,000.

Bundtzen, T. K., and Gilbert, W. G., 1983, Outline of geology and mineral resources of upper Kuskokwim region, Alaska: Journal of Alaska Geological Society, v. 3, p. 101–117.

Bundtzen, T. K., and Laird, G. M., 1982, Geologic map of the Iditarod D-2 and eastern D-3 Quadrangles, Alaska: Alaska Division of Geological and Geophysical Surveys Geologic Report 72, scale 1:63,360.

——, 1983a, Geologic map of the McGrath D-6 Quadrangle, Alaska: Alaska Geological and Geophysical Surveyes Professional Report 79, scale 1:63,360.

——, 1983b, Geologic map of the Iditarod D-1 Quadrangle, Alaska: Alaska Geological and Geophysical Survey Professional Report 78, scale 1:63,360.

——, 1983c, Preliminary geologic map of northeastern Iditarod C-3 Quadrangle, Alaska: Alaska Geological and Geophysical Surveys Report of Investigations 83-13, scale 1:63,360.

Bundtzen, T. K., and Swanson, S. E., 1984, Geology and petrology of igneous rocks in Innoko River area, Alaska: Geological Society of America Abstracts with Programs, v. 16, p. 273.

Cady, W. M., Wallace, R. E., Hoare, J. M., and Webber, E. J., 1955, The central Kuskokwim region, Alaska: U.S. Geological Survey Professional Paper 268, 132 p.

Chapman, R. M., Patton, W. W., Jr., and Moll, E. J., 1985, Reconnaissance geologic map of the Ophir Quadrangle, Alaska: U.S. Geological Survey Open-File Report 85–302, 17 p.

Christiansen, R. L., and Lipman, P. W., 1972, Cenozoic volcanism and plate-tectonic evolution of the western United States; Part 2 Late Cenozoic : Philosophical Transactions of the Royal Society of London series A, v. 271, p. 249–284.

Clague, D. A., and Frey, F. A., 1982, Petrology and trace element geochemistry of the Honolulu Volcanics, Oahu; Implications for the oceanic mantle below Hawaii: Journal of Petrology, v. 23, p. 447–504.

Clague, D. A., Dalrymple, G. B, and Moberly, R., 1975, Petrography and K-Ar ages of dredged volcanic rocks of the western Hawaiian Ridge and southern Emperor Seamount chain: Geological Society of America Bulletin, v. 86, p. 991–998.

Coe, R. S., Globerman, B. R., Plumley, P. W., and Thrupp, G. A., 1985, Paleomagnetic results from Alaska and their tectonic implications, *in* Howell, D. G., ed., Tectonostratigraphic terranes of the Circum-Pacific region: Houston, Texas, Circum-Pacific Council for Energy and Mineral Resources, v. 1, p. 85–108.

Cooper, A. K., Marlow, M. S., and Scholl, D. W., 1986, Geologic framework of the Bering Sea crust, *in* Scholl, D. W., ed., Geology and resource potential of the continental margin of western North America and adjacent ocean basins; Beaufort Sea to Baja California: Houston, Texas, Circum-Pacific Council for Energy and Mineral Resources Earth Science Series, v. 6, p. 73–102.

Cox, A., and Engebretson, D., 1985, Change in motion of Pacific plate at 5 m.y. B.P.: Nature, v. 313, p. 472–474.

Cox, A., Hopkins, D. M., and Dalrymple, G. B., 1966, Geomagnetic polarity epochs; Pribilof Islands, Alaska: Geological Society of America Bulletin, v. 77, p. 883–910.

Csejtey, B., Jr., 1974, Reconnaissance geologic investigations in the Talkeetna Mountains, Alaska: U.S. Geological Survey Open-File Report 74–147, 53 p., scale 1:63,360.

Csejtey, B., Jr., and 8 others, 1978, Reconnaissance geologic map and geochronology, Talkeetna Mountains Quadrangle, northern part of Anchorage Quadrangle, and southwest corner of Healy Quadrangle, Alaska: U.S. Geological Survey Open-File Report 78–558A, 60 p., 1 sheet, scale 1:250,000.

Csejtey, B., Jr., and 13 others, 1986, Geology and geochronology of the Healy Quadrangle: U.S. Geological Survey Open-File Report 86–396, 90 p., 4 sheets, scale 1:250,000.

Davis, A. S., and Plafker, G., 1986, Ecoene basalts from the Yakutat terrane; Evidence for the origin of an accreting terrane in southern Alaska: Geology, v. 14, p. 963–966.

Davis, A. S., Wong, F. L., Pickthorn, L.B.G., and Marlow, M. S., 1987, Petrology, geochemistry, and age of basanitoids dredged from the Bering Sea continental margin west of Navarian Basin: U.S. Geological Survey Open-File Report 87-407, 31 p.

Davis, A. S., Pickthorn, L.B.G., Vallier, T. L., and Marlow, M. S., 1989, Petrology and age of volcanic-arc rocks from the continental margin of the Bering Sea; Implications for Early Eocene relocation of plate boundaries: Canadian Journal of Earth Sciences, v. 26, p. 1474–1490.

Decker, J. E., and Gilbert, W. G., 1978, The Mount Galen volcanics; A new middle Tertiary volcanic formation in the central Alaska Range: Alaska Division of Geological and Geophysical Surveys Geologic Report 59, 11 p., scale 1:63,360.

Decker, J. E., Reifenstuhl, R. R., and Coonrad, W. L., 1984, Compilation of geologic data from the Russian Mission A-3 Quadrangle Alaska: Alaska Division of Geological and Geophysical Surveys Report of Investigations, 84-19, scale 1:63,360.

——, 1985, Compilation of geologic data from the Sleetmute A-7 Quadrangle, southwestern Alaska: Alaska Division of Geological and Geophysical Surveys Report of Investigations 85-1, scale 1:63,360.

Decker, J., Reifenstuhl, R. R., Robinson, M. F., and Waythomas, C. F., 1986, Geologic map of the Sleetmute A-5, A-6, B-5, and B-6 Quadrangles, Alaska: Alaska Divison of Geological and Geophysical Surveys Professional Report 93, 22 p., 1 sheet, scale 1:250,000.

Eakin, G. R., Gilbert, W. G., and Bundtzen, T. K., 1978, Preliminary bedrock geology and mineral resource potential of west-central Lake Clark Quadrangle, Alaska: Alaska Division of Geological and Geophysical Surveys Open-File Report AOF-118, 15 p., scale 1:125,000.

Eggler, D. H., 1978, The effect of CO_2 upon partial melting of peridotite in the system Na_2O-CaO-Al_2O_3-MgO-SiO_2-CO_2 to 35 kb, with an analysis of melting in a peridotite-H_2O-CO_2 system: American Journal of Science, v. 278, p. 305–343.

Eggler, D. H., and Holloway, J. R., 1977, Partial melting of peridotite in the presence of H_2O and CO_2; Principles and review, *in* Dick, H.J.B., ed., Magma genesis: Oregon Department of Mineral Industries Bulletin, v. 96, 169–183.

Engebretson, D. C., Cox, A., and Gordon, R. G., 1982, Relative motions between oceanic and continental plates in the Pacific Basin: Geological Society of America Special Paper 206, 59 p.

Foster, H. L., 1967, Geology of the Mount Fairplay area, Alaska: U.S. Geological Survey Bulletin 1241-B, p. 1–18.

——, 1970, Reconnaissance geologic map of the Tanacross Quadrangle, Alaska: U.S. Geological Survey Miscellaneous Investigations Series Map I-593, scale 1:250,000.

——, 1976, Geologic map of the Eagle Quadrangle, Alaska: U.S. Geological Survey Miscellaneous Investigations Series Map I-922, scale 1:250,000.

Foster, H. L., Forbes, R. B., and Ragan, D. M., 1966, Granulite and peridotite inclusions from Prindle Volcano, Yukon-Tanana upland, Alaska: U.S. Geological Survey Professional Paper 550-B, p. B115–B119.

Foster, H. L., Laird, J., Keith, T.E.C., Cushing, G. W., and Menzies, W. D., 1983, Preliminary geologic map of the Circle Quadrangle, Alaska: U.S. Geological Survey Open-File Report 83–170A, 29 p., scale 1:250,000.

Francis, D. M., 1976a, The origin of amphibole in lherzolite xenoliths from Nunivak Island, Alaska: Journal of Petrology, v. 17, p. 357-378.

——, 1976b, Corona-bearing pyroxene granulite xenoliths and the lower crust beneath Nunivak Island, Alaska: Canadian Mineralogist, v. 14, p. 291–298.

——, 1978, The implications of the compositional dependence of texture in spinel lherzolite xenoliths: Journal of Geology, v. 186, p. 473–486.

Frey, F. A., and Clague, D. A., 1983, Geochemistry of diverse basalt types from Loihi Seamount Hawaii; Petrogenic implications: Earth and Planetary Science Letters, v. 66, p. 337–355.

Frey, F. A., Green, D. H., and Roy, S. D., 1978, Integrated models of basalt petrogenesis; A study of quartz tholeiites to olivine melilites from southeastern Australia utilizing geochemical and experimental petrological data: Journal of petrology, v. 19, p. 463–513.

Gemuts, I., Puchner, C. C., and Steefel, C. I., 1983, Regional geology and tectonic history of western Alaska: Journal of Alaska Geological Society, v. 3, p. 67–85.

Gilbert, W. G., Ferrell, V. M., and Turner, D. L., 1976, The Teklanika Formation; A new Paleocene volcanic formation in the central Alaska Range: Alaska State Division of Geological and Geophysical Surveys Geologic Report 47, 16 p.

Gill, J. B., 1981, Orogenic andesites and plate tectonics: New York, Springer-Verlag, 390 p.

Globerman, B. R., 1985, A paleomagnetic and geochemical study of upper Cretaceous to lower Tertiary volcanic rocks from the Bristol Bay region, southwestern Alaska [Ph.D. thesis]: Santa Cruz, University of California, 292 p.

Globerman, B. R., and Coe, R. S., 1984, Paleomagnetic results from Upper Cretaceous volcanic rocks in northern Bristol Bay, SW Alaska, and tectonic implications, *in* Howell, D. G., Jones, D. L., Cox, A., and Nur, A., eds.,

Proceedings of the Circum-Pacific Terrane Conference: Stanford, California, Stanford University Publications in the Geological Sciences, v. 18, p 98–102.
Godson, R. H., 1984, Composite magnetic anomaly map of the United States; Part B, Alaska and Hawaii: U.S. Geological Survey Geophysical Investigations Map GP-954-B, scale 1:2,500,000.
Grantz, A., 1966, Strike-slip faults in Alaska: U.S. Geological Survey Open-File Report 66-53, 82 p.
Grove, T. L., and Donnelly-Nolan, J. M., 1986, The evolution of young silicic lavas at Medicine Lake Volcano, California; Implications for the origin of compositional gaps in calc-alkaline series lavas: Contributions to Mineralogy and Petrology, v. 92, p. 281–302.
Harbert, W. P., Frei, L. S, Cox, A., and Engebretson, D. C., 1987, Relative motions between Eurasia and North America in the Bering Sea region: Tectonophysics, v. 134, p. 239–261.
Harris, R. A., 1985, Paleomagnetism, geochronology, and paleotemperature of the Yukon-Koyukuk province, Alaska [M.S. thesis]: Fairbanks, University of Alaska, 143 p.
Hawkins, J. W., and Melchior, J. T., 1985, Petrology of Mariana Trough and Lau Basin Basalts: Journal of Geophysical Research, v. 90, p. 11431–11468.
Hildreth, W. E., and Moorbath, S., 1988, Crustal contributions to arc magmatism in the Andes of Central Chile: Contributions to Mineralogy and Petrology, v. 98, p. 455–489.
Hillhouse, J. W., and Grommé, C. S., 1982, Limits to northward drift of the Paleocene Cantwell Formation, central Alaska: Geology, v. 10, p. 552–556.
Hillhouse, J. W., Grommé, C. S., and Csejtey, B., Jr., 1985, Tectonic implications of paleomagnetic poles from early Tertiary volcanic rocks, south-central Alaska: Journal of Geophysical Research, v. 90, p. 12,523–12,535.
Hoare, J. M., and Condon, W. H., 1966, Geologic map of the Kwiguk and Black Quadrangles, western Alaska: U.S. Geological Survey Miscellaneous Geological Investigations Series Map I-469, scale 1:250,000.
—— , 1968, Geologic map of the Hooper Bay Quadrangle, Alaska: U.S. Geological Survey Miscellaneous Geological Investigations Series Map I-523, scale 1:250,000.
—— , 1971a, Geologic map of the Marshall Quadrangle, western Alaska, U.S. Geological Survey Miscellaneous Geological Investigations Series Map I-668, scale 1:250,000.
—— , 1971b, Geologic map of the St. Michael Quadrangle, Alaska: U.S. Geological Survey Miscellaneous Geologic Investigations Series Map I-682, scale 1:250,000.
Hoare, J. M., and Coonrad, W. L., 1959, Geology of the Russian Mission Quadrangle, Alaska: U.S. Geological Survey Miscellaneous Geologic Investigations Series Map I-292, scale 1:250,000.
—— , 1978a, A tuya in Togiak Valley, southwest Alaska: U.S. Geological Survey Journal of Research, v. 6, p. 193–201.
—— , 1978b, Geologic map of the Goodnews and Hagemeister Island Quadrangles region, southwestern Alaska, U.S. Geological Survey Open-File Report 78–9B, scale 1:250,000.
—— , 1980, The Togiak Basalt; A new formation in southwestern Alaska: U.S. Geological Survey Bulletin 1482-C, 11 p.
Hoare, J. M., Condon, W. H., Cox, A., and Dalrymple, G. B., 1968, Geology, paleomagnetism, and potassium-argon ages of basalts from Nunivak Island, Alaska, *in* Coats, R. R., Hay, R. L., and Anderson, C. A., eds., Studies in volcanology; A memoir in honor of Howell Williams: Geological Society of America Memoir 116, p. 377–414.
Hopkins, D. M., 1963, Geology of the Imuruk Lake area, Seward Peninsula, Alaska: U.S. Geological Survey Bulletin 1141-C, 101 p.
—— , 1976, Fault history of Pribilof Islands and its relevance to bottom stability in St. George Basin, *in* Environmental assessment of the Alaskan continental shelf: Boulder, Colorado, Environmental Research Laboratories, v. 13, p. 41–67.
Hudson, T., 1979, Mesozoic plutonic belts of southern Alaska: Geology, v. 7, p. 230–234.
Hudson, T., and Plafker, G., 1978, Kigluaik and Bendeleben faults, Seward Peninsula: U.S. Geological Survey Circular 772-B, p. B47–B50.
Irvine, T. N., and Baragar, W.R.A., 1971, A guide to the chemical classification of the common volcanic rocks: Canadian Journal of Earth Sciences, v. 8, p. 523–548.
Irving, A. J., 1974, Megacrysts from the Newer Basalts and other basaltic rocks of southeastern Australia: Geological Society of America Bulletin, v. 85, p. 1503–1514.
Irving, A. J., and Frey, F. A., 1984, Trace element abundances in megacrysts and their host basalts; Constraints on partition coefficients and megacryst genesis: Geochimica et Cosmochimica Acta, v. 48, p. 1201–1221.
Jones, D. L., Silberling, N. J., Coney, P., and Plafker, G., 1987, Lithotectonic terrane maps of Alaska (west of the 141st meridian): U.S. Geological Survey Miscellaneous Field Studies Map MF-1874-A, scale 1:2,500,000.
Kay, R. W., 1977, Geochemical constraints on the origin of Aleutian magmas, *in* Talwani, M., and Pittman, W., eds., Island arcs, deep sea trenches, and back-arc basins: American Geophysical Union Maurice Ewing Series 1, p. 229–242.
King, B. C., 1970, Volcanicity and rift tectonics in East Africa, *in* Clifford, T. N., and Gass, I. G., eds., African magmatism and tectonics: New York, Hofner, p. 263–283.
Kushiro, I., 1975, On the nature of silicate melt and its significance in magma genesis; Regularities in the shift of the liquidus boundaries involving olivine, pyroxene, and silica minerals: American Journal of Science, v. 275, p. 411–431.
Lanphere, M. A., and Dalrymple, G. B., 1980, Age and strontium isotopic composition of the Honolulu Volcanic Series, Oahu, Hawaii: American Journal of Science, v. 280-A, p. 736–751.
Lanphere, M. A., and Reed, B. L., 1985, The McKinley Sequence of granitic rocks; A key element in the accretionary history of southern Alaska: Journal of Geophysical Research, v. 90, p. 11413–11430.
Lanphere, M. A., Dalrymple, G. B., and Clague, D. A., 1980, Rb-Sr systematics of basalts from the Hawaiian-Emperor volcanic chain, *in* Shambach, J., and others, eds., Initial reports of the Deep Sea Drilling Project: Washington, D.C., U.S. Government Printing Office, v. 55, p. 695–706.
Lee-Wong, F., Vallier, T. L., Hopkins, D. M., and Silberman, M. L., 1979, Preliminary report on the petrography and geochemistry of basalts from the Pribilof Islands and vicinity, southern Bering Sea: U.S. Geological Survey Open-File Report 79–1556, 51 p.
Lerbekmo, T. F., and Campbell, F. A., 1969, Distribution, composition, and source of the White River ash, Yukon Territory: Canadian Journal of Earth Sciences, v. 6, p. 109–116.
Lipman, P., Prostka, H. J., and Christiansen, R. L., 1972, Cenozoic volcanism and plate tectonic evolution of western United States; Part 1, Early and middle Cenozoic: Philosophical Transactions of the Royal Society of London, series A, v. 271, p. 249–284.
Lipman, P. W., Doe, B. R., Hedge, C. E., and Steven, T. A., 1978, Petrologic evolution of the San Juan volcanic field, southwestern Colorado; Lead and strontium isotopic evidence: Geological Society of America Bulletin, v. 89, p. 59–82.
MacDonald, G. A., and Katsura, T., 1964, Chemical composition of Hawaiian lavas: Journal of Petrology, v. 6, p. 82–133.
Magoon, L. B., Adkison, W. L., and Egbert, R. M., 1976, Map showing geology, wildcat wells, Tertiary plant fossil localities, K-Ar age dates, and petroleum operations, Cook Inlet area, Alaska: U.S. Geological Survey Miscellaneous Investigations Series Map I-1019, scale 1:250,000.
Mark, R. K., 1971, Strontium isotopic study of basalts from Nunivak Island, Alaska [Ph.D. thesis]: Stanford, California, Stanford University, 50 p.
Marshall, M., 1978, The magnetic properties of some DSDP basalts from the North Pacific and inferences for Pacific plate tectonics: Journal of Geophysical Research, v. 83, no. B1, p. 289–308.
Menzies, M., and Murthy, V. R., 1980a, Mantle metasomatism as a precursor to the genesis of alkaline magmas; Isotopic evidence: American Journal of Science, v. 280-A, p. 622–638.
—— , 1980b, Nd and Sr isotope geochemistry of hydrous mantle nodules and their host alkali basalts; Implications for local heterogeneities in metasomatically veined mantle: Earth and Planetary Science Letters, v. 46, p. 323–334.

Menzies, M. A., and 7 others, 1987, A record of subduction processes and within-plate volcanism in lithospheric xenoliths of the southwestern USA, *in* Nixon, P. H., eds., Mantle xenoliths: New York, J. Wiley and Sons, p. 59–74.

Miller, T. P., and Lanphere, M. A., 1981, K-Ar age measurements on obsidian from the Little Indian River locality in interior Alaska, *in* Albert, N.R.D., and Hudson, T., eds., The United States Geological Survey in Alaska; Accomplishments during 1979: U.S. Geological Survey Circular 823-B, p. B39–B42.

Moll, E. J., and Arth, J. G., 1985, Sr and Nd isotopes from Late Cretaceous–early Tertiary volcanic fields in western Alaska; Evidence against old radiogenic continental crust under the Yukon-Koyukuk basin, EOS Transactions of the American Geophysical Union, v. 66, no. 46, p. 1102.

Moll, E. J., and Patton, W. W., Jr., 1982, Preliminary report on the Late Cretaceous–early Tertiary volcanic and related plutonic rocks in western Alaska, *in* Coonrad, W. L., ed., The United States Geological Survey in Alaska; Accomplishments during 1980: U.S. Geological Survey Circular 844, p. 73–76.

——, 1983, Late Cretaceous–early Tertiary calc-alkalic volcanic rocks of western Alaska: Geological Society of America Abstracts with Programs, v. 15, p. 406.

Moll, E. J., Silberman, M. L., and Patton, W. W., Jr., 1981, Chemistry, mineralogy, and K-Ar ages of igneous and metamorphic rocks of the Medfra Quadrangle, Alaska: U.S. Geological Survey Open-File Report 80–81C, 2 sheets, scale 1:250,000.

Moll-Stalcup, E. J., 1987, The petrology and Sr and Nd isotopic characteristics of five Late Cretaceous–early Tertiary volcanic fields in western Alaska [Ph.D. thesis]: Stanford, California, Stanford University, 310 p.

Moll-Stalcup, E. J., and Arth, J. G., 1989, The nature of the crust in the Yukon-Koyukuk province as inferred from the chemical and isotopic composition of five Late Cretaceous and early Tertiary volcanic fields in western Alaska: Journal of Geophysical Reserach, v. 94, p. 15989–16020.

Moll-Stalcup, E. J., and Arth, J. G., 1991, The petrology and Sr and Nd isotopic composition of the Blackburn Hills volcanic field, western Alaska: Geochimica et Cosmochimica Acta, v. 55, p. 3753–3776.

Moore, J. C., and 5 others, 1983, Paleogene evolution of the Kodiak Islands, Alaska; Consequences of ridge-trench interaction in a more southerly latitude: Tectonics, v. 2, p. 265–293.

Morrison, G. W., 1980, Characteristics and tectonic setting of the shoshonite rock association: Lithos, v. 13, p. 97–108.

Nakamura, E., Campbell, I. H., and Sun, S., 1985, The influence of subduction processes on the geochemistry of Japanese alkaline basalts: Nature, v. 316, p. 55–58.

Nakamura, K., and Uyeda, S., 1980, Stress gradient in arc–back arc regions and plate subduction: Journal of Geophysical Research, v. 85, p. 6419–6428.

Nakamura, K., Jacob, K. H., and Davies, J. H., 1977, Volcanoes as possible indicators of tectonic stress orientation; Aleutians and Alaska: Basel, Birkhauser Verlag, Pageoph, v. 115, p. 87–112.

Nakamura, K., Plafker, G., Jacob, K. H., and Davies, J. N., 1980, A tectonic stress trajectory map of Alaska using information from volcanoes and faults: Bulletin of the Earthquake Research Institute, v. 55, p. 89–100.

Nelson, W. H., Carlson, C., and Case, J. E., 1983, Geologic map of the Lake Clark Quadrangle, Alaska: U.S. Geological Survey Miscellaneous Field Studies Map MF-1114-A, scale 1:250,000.

Nokleberg, W. J., and 10 others, 1982, Geologic map of the southern Mount Hayes Quadrangle, Alaska: U.S. Geological Survey Open-File Report 82–52, 27 p., scale 1:250,000.

Nokleberg, W. J., Jones, D. L., and Silberling, N. J., 1985, Origin and tectonic evolution of the MacLaren and Wrangellia terranes, eastern Alaska Range, Alaska: Geological Society of America Bulletin, v. 96, p. 1251–1270.

Patton, W. W., Jr., 1966, Regional geology of the Kateel River Quadrangle, Alaska: U.S. Geological Survey Miscellaneous Geologic Investigations Series Map I-437, scale 1:250,000.

Patton, W. W., Jr., and Box, S. E., 1989, Tectonic setting of the Yukon-Koyukuk basin and its borderlands, western Alaska: Journal of Geophysical research, v. 94, p. 15,807–15,820.

Patton, W. W., Jr., and Csejtey, B., Jr., 1980, Geologic map of St. Lawrence Island, Alaska: U.S. Geological Survey Miscellaneous Geologic Investigations Series Map I-1203, scale 1:250,000.

Patton, W. W., Jr., and Hoare, J. M., 1968, The Kaltag fault, west-central Alaska, *in* Geological Survey research in 1968: U.S. Geological Survey Professional Paper 600D, p. D147–D153.

Patton, W. W., Jr., and Miller, T. P., 1970, A possible source for obsidian found in archeological sites in northwestern Alaska: Science, v. 169, p. 760–761.

——, 1973, Bedrock geologic m ap of Bettles and southern part of Wiseman Quadrangles, Alaska: U.S. Geological Survey Miscellaneous Field Studies Map MF-492, scale 1:250,000.

Patton, W. W., Jr., and Moll, E. J., 1985, Geologic map of northern and central parts of Unalakleet Quadrangle, Alaska: U.S. Geological Survey Miscellaneous Field Studies Map MF-1749, scale 1:250,000.

Patton, W. W., Jr., and 5 others, 1975, Reconnaissance geologic map of St. Matthew Island, Bering Sea, Alaska: U.S. Geological Survey Miscellaneous Field Studies Map MF-642, scale 1:125,000.

Patton, W. W., Jr., Miller, T. P., Chapman, R. M., and Yeend, W., 1978, Geologic map of the Melozitna Quadrangle: U.S. Geological Survey Miscellaneous Investigation Series Map I-1071, scale 1:250,000.

Patton, W. W., Jr., Moll, E. J., Dutro, J. T., Jr., Silberman, M. L., and Chapman, R. M., 1980, Preliminary geologic map of the Medfra Quadrangle: U.S. Geological Survey Open-File Report 80–811A, scale 1:250,000.

Perfit, M. R., Gust, D. A., Bence, A. E., Arculus, R. J., and Taylor, S. R., 1980, Chemical characteristics of island-arc basalts; Implications for mantle sources: Chemical Geology, v. 30, p. 227–256.

Plumley, P. W., Coe, R. S., Byrne, T., Reid, M. R., and Moore, J. C., 1982, Paleomagnetism of volcanic rocks of the Kodiak Islands indicates northward latitudinal displacement: Nature, v. 300, p. 50–52.

Rea, D. K., and Duncan, R. A., 1986, North Pacific plate convergence; A quantitative record of the past 140 m.y.: Geology, v. 14, p. 373–376.

Reed, B. L., and Lanphere, M. A., 1972, Generalized geologic map of the Alaska–Aleutian Range batholith showing potassium-argon ages of the plutonic rocks: U.S. Geological Survey Miscellaneous Field Studies Map MF-372. 2 sheets, scale 1:1,000,000.

——, 1973, The Alaska–Aleutian Range batholith; Geochronology, chemistry, and relation to circum-Pacific plutonism: Geological Society of America Bulletin, v. 84, p. 2583–2610.

——, 1974a, Chemical variations across the Alaska-Aleutian Range batholith: U.S. Geological Survey Journal of Research, v. 2, p. 343–352.

——, 1974b, Offset plutons and history of movement along the McKinley segment of the Denali fault system, Alaska: Geological Society of America Bulletin, v. 85, p. 1883–1892.

Reed, B. L., and Nelson, S. W., 1980, Geologic map of the Talkeetna Quadrangle, Alaska: U.S. Geological Survey Miscellaneous Investigations Series Map I-1174, scale 1:250,000.

Reifenstuhl, R. R., Robinson, M. S., Smith, T. E., Albanese, M. D., and Allegro, G. A., 1984, Geologic map of the Sleetemute B-6 Quadrangle, Alaska: Alaska Division of Geological and Geophysical Surveys Report of Investigations 84-12, scale 1:63,360.

Reifenstuhl, R. R., Decker, J., and Coonrad, W. L., 1985, Compilation of geologic data from the Taylor Mountains D-8 Quadrangle, southwestern Alaska: Alaska Division of Geological and Geophysical Surveys Report of Investigations 85-4, scale 1:63,360.

Robinson, M. S., Decker, J., Reifenstuhl, R. R., Murphy, J. M., and Box, S. E., 1984, Geologic map of the Sleetemute B-5 Quadrangle, Alaska: Alaska Division of Geological and Geophysical Surveys Report of Investigations 84-10, scale 1:63,360.

Roden, M. F., 1982, Geochemistry of the Earth's mantle, Nunivak Island, Alaska, and other areas; Evidence from xenolith studies [Ph.D. thesis]: Cambridge, Massachusetts Institute of Technology, 413 p.

Roden, M. F., and Murthy, V. R., 1985, Mantle metasomatism: Annual Review

of Earth and Planetary Sciences, v. 13, p. 269–296.
Roden, M. F., Frey, F. A., and Francis, D. M., 1984, An example of consequent mantle metasomatism in peridotite inclusions from Nunivak Island, Alaska: Journal of Petrology, v. 25, p. 546–577.
Saunders, A. D., and Tarney, J., 1984, Geochemical characteristics of basaltic volcanism within backarc basins, *in* Kolelaar, B. P., and Howells, M. F., eds., Marginal basin geology: Oxford University, p. 59–76.
Scholl, D. W., Vallier, T. L., and Stevenson, A. J., 1986, Geologic evolution and petroleum geology of the Aleutian ridge, *in* Scholl, D. W., ed., Geology and resource potential of the continental margin of western North America and adjacent ocean basins; Beaufort Sea to Baja California: Houston, Texas, Circum-Pacific Council for Energy and Mineral Resources Earth Science Series, v. 6, p. 59–72.
Silberman, M. L., and Grantz, A., 1984, Paleogene volcanic rocks of the Matanuska Valley area and the displacement history of the Castle Mountain fault, *in* Coonrad, W. L., and Elliott, R. L., eds., The U.S. Geological Survey in Alaska; Accomplishments during 1981: U.S. Geological Survey Circular 868, p. 82–86.
Silberman, M. L., Moll, E. J., Chapman, R. M., Patton, W. W., Jr., and Connor, C. L., 1979, Potassium-argon age of granitic and volcanic rocks from the Ruby, Medfra, and surrounding Quadrangles, west-central Alaska, *in* Johnson, K. M., and Williams, J. R., eds., The U.S. Geological Survey in Alaska; Accomplishments during 1978: U.S. Geological Survey Circular 804-B, p. B863–B866.
Simpson, G. L., Vallier, T. L., Pearl, J. E., and Lee-Wong, F., 1979, Potassium-argon ages and geochemistry of basalt dredged near Saint George Island, southern Bering Sea, *in* Johnson, K. M., and Williams, J. R., eds., The U.S. Geological Survey in Alaska; Accomplishments during 1978: U.S. Geological Survey Circular 804-B, p. B134–B136.
Sinclair, P. D., Templeman-Kluit, D. J., and Medaris, L. G., Jr., 1978, Lherzolite nodules from a Pleistocene cinder cone in central Yukon: Canadian Journal of Earth Sciences, v. 15, p. 220–226.
Smirnov, Y. B., and Sugrobov, V. M., 1979a, Heat flow in the northwest Pacific Ocean: Priroda, no. 8, p. 94–101.
——, 1979b, Terrestrial heat flow in the Kuril-Kamchatka, and Aleutian provinces; 1, Heat flow and tectonics: Moscow, Akademiya Nauk SSSR, Vulkanologiya i Seismologiya no. 1, p. 59–73.
Snyder, W. S., Dickinson, W. R., and Silberman, M. L., 1976, Tectonic implications of space-time patterns of Cenozoic magmatism in the western United States: Earth and Planetary Science Letters, v. 32, p. 91–106.
Streckeisen, A., 1979, Classification and nomenclature of volcanic rocks, lamprophyres, carbonatites, and melitic rocks; Recommendations and suggestions of the IUGS Subcommission on the Systematics of Igneous Rocks: Geology, v. 7, p. 331–335.
Swanson, S. E., Turner, D. L., and Fores, R. B., 1981, Petrology and geochemistry of Tertiary and Quaternary basalts from the Seward Peninsula, western Alaska: Geological Society of America Abstract with Programs, v. 13, p. 563.
Thompson, R. N., Morrison, M. A., Dickin, A. P., and Hendry, G. L., 1982, Continental flood basalts; Arachnids rule OK?, *in* Hawkesworth, C. J., and Norry, S J., eds., Continental basalts and mantle xenoliths: Nantwich, U.K., Shiva, p. 158–185.
Thompson, R. N., Morrison, M. A., Hendry, G. L., and Parry, S. J., 1984, An assessment of the relative roles of crust and mantle in magma genesis; An elemental approach: Philosophical Transactions of the Royal Society of London, series A, v. 310, p. 549–590.
Thrupp, G. A., and Coe, R. S., 1986, Early Tertiary paleomagnetic evidence and the displacement of southern Alaska: Geology, v. 14, p. 213–217.
Tolson, R. B., 1986, Structure and stratigraphy of the Hope Basin, *in* Scholl, D. W., ed., Geology and resource potential of the continental margin of western North America and adjacent ocean basins; Beaufort Sea to Baja California: Houston, Texas, Circum-Pacific Council for Energy and Mineral Resources Earth Science Series, v. 6, p. 59–72.
Turner, D. L., and Forbes, R. B., 1980, A geological and geophysical study of the geothermal energy potential of Pilgrim Springs, Alaska: Fairbanks, University of Alaska Geophysical Institute Report UAG R-271, 165 p.
Turner, D. L., Swanson, S. E., and Wescott, E., 1981, Continental rifting; A new tectonic model for geothermal exploration of the central Seward Peninsula, Alaska: Geothermal Resources Council Transactions, v. 5, p. 213–216.
von Drach, V., Marsh, B. D., and Wasserburg, G. J., 1986, Nd and Sr isotopes in the Aleutians; Multicomponent parenthood of island-arc magmas: Contributions to Mineralogy and Petrology, v. 92, p. 13–34.
Wallace, W. K., and Engebretson, D. C., 1984, Relationship between plate motions and Late Cretaceous to Paleogene magmatism in southwestern Alaska: Tectonics, v. 3, p. 295–315.
Wheatley, M. R., and Rock, N.M.S., 1987, Spider; A Macintosh program to generate normalized multi-element "spidergrams": American Mineralogist, v. 73, p. 919–921.
Wilson, F. H., 1977, Some plutonic rocks of southwestern Alaska, a data compilation: U.S. Geological Survey Open-File Report 77–501, 7 p., scale 1:1,000,000.
——, 1985, The Meshik arc; An Eocene to earliest Miocene magmatic arc on the Alaska Peninsula: Alaska Division of Geological and Geophysical Surveys Professional Report 88, 14 p.
Wood, D. A., 1979, A variably veined suboceanic upper mantle; Genetic significance for mid-ocean ridge basalts from geochemical evidence: Geology, v. 7, p. 499–503.
Wyllie, P. J., and Huang, W-L., 1976, Carbonation and melting reactions in the system CaO-MgO-SiO_2-CO_2 at mantle pressures with geophysical and petrological applications: Contributions to Mineralogy and Petrology, v. 54, p. 140–173.
Zielinski, R. A., and Lipman, P. W., 1976, Trace-element variations at Summer Coon volcano, San Juan Mountains, Colorado, and the origin of continental-interior andesite: Geological Society of America Bulletin, v. 87, p. 1477–1485.
Zonenshain, L. P., Savostin, L. A., and Sedov, A. P., 1984, Global paleogeodynamic reconstructions for the last 160 million years: Geotectonics, v. 18, p. 181–195.

MANUSCRIPT COMPLETED MAY 25, 1986
MANUSCRIPT ACCEPTED BY THE SOCIETY OCTOBER 24, 1990

ACKNOWLEDGMENTS

Many of the ideas contained in this chapter came from stimulating discussions with my colleagues at the U.S. Geological Survey, Stanford University, and the Alaska Division of Geological and Geophysical Surveys. I would especially like to thank Bill Patton, Steve Box, Gail Mahood, Dave Clague, and Howard Wilshire. I thank Sam Swanson, Don Turner, Michael Roden, Florence Lee-Wong, Alicia Davis, Gordon Thrupp, Brian Globerman, Art Grantz, Mark Robinson, John Decker, and Rocky Reifenstuhl for contributing unpublished data. Much of the section on the Bering Sea basalts is based on unpublished manuscripts and data from the late Joe Hoare. I thank Warren Coonrad for organizing Joe's field notes, chemical analyses, thin sections, and maps for my study. Comments by Steve Box, Gail Mahood, Alan Cox, Bill Patton, Elizabeth Miller, and Bob Coleman improved an early draft of the manuscript. The paper received helpful reviews from Tracy Vallier and Joe Arth.

Printed in U.S.A.

The Geology of North America
Vol. G-1, The Geology of Alaska
The Geological Society of America, 1994

Chapter 19

Latest Mesozoic and Cenozoic magmatism in southeastern Alaska

David A. Brew
U.S. Geological Survey, 345 Middlefield Road, Menlo Park, California 94025

INTRODUCTION

The most important latest Mesozoic and Cenozoic, post-accretionary geologic features of southeastern Alaska are those related to the magmatic activity that affected a large part of the region and to the resultant metamorphism and deformation. The metamorphic history is discussed elsewhere in this volume (Dusel-Bacon, this volume), and the magmatic activity is a continuation of the late Mesozoic activity discussed by Miller (this volume). Postaccretionary geologic history starts with the accumulation of the Gravina belt overlap assemblage of rocks in Late Jurassic and Early Cretaceous time (Berg and others, 1972). The locally voluminous volcanic rocks within that assemblage are probably the extrusive equivalents of island-arc intrusive rocks, which are preserved west of the Gravina belt over a large area in northern southeastern Alaska (Brew and Morrell, 1983). Neither the volcanics nor the granitoids are discussed in this chapter.

Previous syntheses concerned with the magmatic rocks of southeastern Alaska comprise a summary of post-Carboniferous volcanic activity (Brew, 1968), summaries of the distribution and general characteristics of the plutonic rocks (Brew and Morrell, 1980, 1983), a summary of the geochronologic data available (Wilson and Shew, 1982), and two reports concerned with the tectonic significance of major- and trace-element chemical data (Barker and Arth, 1984; Barker and others, 1986). Karl and Brew (1984) discussed migmatitic rocks associated with some of the intrusive rocks; that topic is not considered in this report.

In this chapter, the latest Mesozoic and Cenozoic magmatic rocks are grouped chronometrically (Table 1); the same time divisions are used elsewhere in this volume for the Cenozoic magmatic history of the rest of Alaska (Moll-Stalcup, this volume). The divisions are approximate, and several of the belts described in the region include rocks whose radiometric ages fall somewhat outside of the defined limits. Within each chronometric group, extrusive and intrusive rocks are identified compositionally and are separated into geographic belts. The general approach is similar to that of Brew and Morrell (1983). Table 2 provides summary information on modal and chemical compositions, chronometric data, and emplacement/eruptive environment for each of the chronometric groups. The chemical classifications of the rocks are those of Shand (1951) and Irvine and Baragar (1971). Figure 1 shows the general geographic distribution of the rocks of different ages and is an index map for the descriptions in the table.

Magmatic activity in southeastern Alaska ranges from early Paleozoic to Holocene in age but was most frequent in the late Mesozoic and Cenozoic. Currently available geochronologic data for southeastern Alaska are summarized in Figure 2, which shows the frequency distribution for 402 age determinations of all types of rocks from southeastern Alaska southeast of the Yakutat 1:250,000 scale Quadrangle. The relative recency of magmatic activity in the region is obvious, as are the dominance of mid-Tertiary events and the absence of any real break between Mesozoic and Cenozoic events.

DESCRIPTION OF THE TABLES

Table 1 links the major magmatic belts and areas of summary discussions in the text, shown on Figure 3, with the descriptions of the component belts and areas given in Table 2. The information presented in Tables 1 and 2 is derived from a report in preparation; that report contains more discussion of tectonic settings, emplacement situations, and extrusive activity than can be included here. Table 2 summarizes the data that support the conclusions of this chapter.

Table 2 is divided into columns for: (1) Figure 1 reference, which is the letter designation on those maps for the specific area; (2) area or belt name; (3) major and minor lithic types, with the latter shown in parentheses (granitic rock names are from Streckeisen, 1973); (4) chemical classification and chemical compositional types present, based on calculations using the PETCAL 4 program (Bingler and others, 1976) as revised by R. D. Koch (written communication, 1985); (5) SiO_2 range; (6) SiO_2 gap(s); (7) reference to map and diagram figures in this report: most figures include a Streckeisen (1973) QAP (quartz-alkali feldspar-plagioclase feldspar) classification diagram for granitic rocks, a silica-variation diagram, an AFM (alkaline element oxide–iron oxide–magnesium oxide) diagram, and a small map showing the area containing the rocks described; (8) age data; (9) discussion or remarks, focussed mainly on the environment of pluton emplacement or volcanic extrusion; and (10) references to the sources of the data.

Brew, D. A., 1994, Latest Mesozoic and Cenozoic magmatism in southeastern Alaska, *in* Plafker, G., and Berg, H. C., eds., The Geology of Alaska: Boulder, Colorado, Geological Society of America, The Geology of North America, v. G-1.

TABLE 1. MAJOR LATEST MESOZOIC AND CENOZOIC MAGMATIC BELTS AND AREAS OF SOUTHEASTERN ALASKA

Major Area or Belt Name (Age Division)	Components of the Major Area or Belt: Figure 1 and Table 2 Reference	Components of the Major Area or Belt: Individual Area or Belt Name
Great tonalite sill belt area (75–55 Ma)	GG	Juneau-Skagway area
	HH	Haines-Skagway area
	II	Juneau–Taku River area
	JJ	Sumdum area
	KK	Petersburg area
	LL	Bradfield Canal area
	MM	Ketchikan–Prince Rupert area
Coast Mountains belt (55–45 Ma)	AA	Haines-Skagway area
	BB	Juneau–Taku River area
	CC	Sumdum (Tracy Arm) area
	DD	Petersburg area
	EE	Bradfield Canal area
	FF	Ketchikan–Prince Rupert area
Fairweather-Baranof belt (45–35 Ma)	Y	Fairweather Range
	Z	Yakobi, Chichagof, and Baranof area
Glacier Bay region (45–35 Ma)	X	Glacier Bay region
Tkope-Portland Peninsula belt	F	Tkope volcanic-plutonic belt
(35–5 Ma)	J	William Henry Bay area
	K	Icy Strait volcanic-plutonic field
	M	Admiralty field
	N	Kuiu-Etolin volcanic-plutonic field
	P	Southern Etolin field
	S	Burroughs Bay area
	U	Ketchikan area
	V	Quartz Hill–Portland Peninsula area
Groundhog Basin–Cone Mountain (35–5 Ma)	O	Groundhog Basin area
	T	Cone Mountain area
Southern southeastern Alaska dike swarm (35–5 Ma)	W	Southern southeastern Alaska dike swarm
Kruzof–Kupreanof area (5–0 Ma)	A	Edgecumbe field
	B	Southern Kupreanof field
Behm Canal–Rudyerd Bay area	C	Blue River–Unuk River field
(5–0 Ma)	E	Behm Canal–Rudyerd Bay field

EVOLUTION OF MAGMATIC BELTS AND AREAS

The tectonic settings and compositional variations recorded in the several Cenozoic magmatic belts of southeastern Alaska indicate a varied and complicated evolutionary history. The older part of the record, from latest Cretaceous through about early Oligocene time, reflects the two main collisional events that dominate the Cenozoic history of the region. The younger part of the record, from the late Oligocene on, is the result of less well-understood events, ones that are probably related first to oblique subduction and then to extensional regimes associated with youngest Cenozoic strike-slip faulting.

The areas summarized in Tables 1 and 2 are grouped into nine major belts on Figure 3: one of latest Cretaceous and Paleocene age (75 to 55 Ma), one of early and middle Eocene age (55 to 45 Ma), two of middle and late Eocene and early Oligocene age (45 to 35 Ma), three of late Oligocene and Miocene age (35 to 5 Ma), and two of Pliocene and Quaternary age (5 to 0 Ma). Each of these belts is interpreted to record a specific magma-generating event (or series of events) and most have clear-cut chemical and/or modal compositional features that support the definition of the belts.

Figures 4 through 23 are keyed to Table 2 and are therefore not referred to specifically in the following summary discussion.

Great tonalite sill belt

The oldest belt discussed here, the latest Cretaceous and Paleocene "great tonalite sill" belt (Skagway/Ketchikan–Prince Rupert [75 to 55 Ma] on Fig. 3 and Tables 1 and 2), records only the youngest of a series of events that began in Early Cretaceous time in the "southeastern Alaska coincident zone" (Brew and Ford, 1985). The rocks of the tonalite sill belt are consistently calc-alkalic and dominantly metaluminous, locally have a prominent silica gap at 63 to 68 percent, and fall in the tonalite-granodiorite–quartz monzodiorite–quartz diorite fields of Streckeisen (1973). These plutons have emplacement ages that range from 67 to 55 Ma. The sill rocks with Paleocene emplacement ages of around 60 Ma are included with the older tonalite sill family because of their closely similar ages and habits. They are mostly granodiorite and have higher silica contents than the slightly older rocks.

The plutons of the great tonalite sill family are foliated and lineated tonalites that form a narrow belt. They have been localized along a profound, straight, structural discontinuity within a convergent setting in which the northeast side was moving upward over the southwest side (D.H.W. Hutton, personal communication, 1985, 1986). This discontinuity can be interpreted as either a within-plate rift margin (Brew and Ford, 1983) or as the boundary between two exotic terranes (Monger and others, 1982, 1983). The linear zone of compression persisted at least from 70 to 55 Ma, the tonalite period during which intrusions were emplaced. Metamorphism and major deformation occurred shortly before the emplacement of the intrusions.

The cause of the compression in this zone, whether it was

TABLE 2. DESCRIPTION OF LATEST MESOZOIC AND CENOZOIC MAGMATIC ROCKS OF SOUTHEASTERN ALASKA

Figure 1 and Table 1 references	Area or Belt Name	Major (and Minor) Lithic Types	Chemical Classification	SiO_2 Range (%)	SiO_2 Gap(s) (%)	Map and Diagrams on Figure	Age Data	Discussion	References
Pliocene and Quaternary Rocks (5–0 Ma)									
A	Edgecumbe field	Basalt, basaltic, andesite, andesite, dacite, (rhyolite)	Tholeiitic, calc–alkalic	47–72 (nontephra); 52–74 (tephra)	62–69 60–65	4, 5 (location)	Late Pleistocene and younger on K-Ar data (M. A. Lanphere, written communication, 1985)and microfossils (W.V. Slitter, written comm., 1985)	Basal tholeiitic basalt shield surmounted by calc-alkalic andesite cones and dacite plugs, basaltic to rhyolitic tephras all younger than 10,000 B.P.	Brew and others, 1969; Myers and others, 1984; Kosco, 1981; Riehle and Brew, 1984, unpublished data
B	Southern Kupreanof field	Olivine-bearing basalt	Mostly tholeiitic; aver. K content; some alkalic; sodic	45–53		5	Younger than 300 ka on K-Ar data	Pahoehoe and aa flows, some plugs	Brew and others, 1984, 1985; Douglass and others, 1989
C	Blue River–Unuk River field	Alkali olivine basalt	Mostly alkalic, sodic; some calc-alkalic; K-rich	46–48		5	As young as 360 ± 60 B.P. on radiocarbon	Valley-filling flows, small cinder cones	Elliott and others, 1981; Souther and others, 1984
D	Tlevak Strait-Suemez field	Olivine basalt	Alkalic; sodic	47			No data	Pahohoe surfaces, valley-filling flows	Eberlein and Churkin, 1970; Eberlein and others, 1983; G.D. Eberlein, written communication, 1986
E	Behm Canal–Rudyerd Bay field	Olivine basalt, basaltic breccia and tuff, andesite (trachyan-desite)	Alkalic; mostly potassic	43–61	46–59	5	Possibly two periods: 5 Ma and 1 Ma to 500 ka (Smith and Diggles, 1981)	Columnar flows, cinder cones	Wanek and Callahan, 1971; Berg and others, 1988; Smith and others, 1977; Ouderkirk, 1982; Doyle, 1983; Souther and others, 1984

TABLE 2. DESCRIPTION OF LATEST MESOZOIC AND CENOZOIC MAGMATIC ROCKS OF SOUTHEASTERN ALASKA (continued)

Figure 1 and Table 1 references	Area or Belt Name	Major (and Minor) Lithic Types	Chemical Classification	SiO_2 Range (%)	SiO_2 Gap(s) (%)	Map and Diagrams on Figure	Age Data	Discussion	References
Late Oligocene and Miocene Rocks (35–5 Ma)									
F	Tkope volcanic-plutonic belt	In Canada; granophyre, granite, quartz monzonite, granodiorite, quartz diorite gabbro. In U.S.: hornblende-biotite granite	Calc-alkalic except for gabbros, which are on calc-alkalic-tholeiitic boundary	49–77	51–59	N.A.	28–24 Ma on K-Ar, Rb-Sr, and fission track	Main expression is epizonal, composite Tkope River pluton in Canada; extension into U.S.A. consists of plugs, dikes, and small plutons	Jacobsen, 1979; Campbell and Dodds, 1983; D. A. Brew, unpublished data
G	Haines area	Biotite quartz monzonite, locally miarolitic	No data	No data	No data	N.A.	No data	Age inferred from lithic similarity to Kuiu-Etolin belt plutons	Redman and others, 1984
H	Fairweather Range	Garnet-muscovite-biotite granite and granodiorite	No data	No data	No data	N.A.	5.9 Ma on biotite and 16.6 Ma on muscovite (M. A. Lanphere, written communication, 1978)	May belong with nearby early Oligocene and late Eocene bodies	Brew and others, 1978; D. A. Brew, unpublished data
I	Lituya Bay area	Tuffs, flows of andesite and basaltic andesite	Calc-alkalic, K-poor, per Irvine and Baragar (1971)	54	N.A.	N.A.	Post–early Oligocene(?) to pre-middle Miocene (Miller, 1961)	Cenotaph volcanics unit of Miller (1961); nonmarine	Plafker, 1971; G. Plafker, written communication, 1986
J	William Henry Bay area	Biotite quartz monzonite, diorite	No data	No data	No data	N.A.	No data	Age inferred by Eakins (1975) on lithic grounds(?)	Eakins, 1975; Brew and Ford, [1985] 1986
K	Icy Strait volcanic-plutonic belt	Hornblende granite, hornblende quartz monzonite; breccia, flows, and tuff of dacite, andesite, and basalt	Volcanic rocks, mostly tholeiitic, aver. K content; also calc-alkalic, aver. K content; no data on granitoids	47–72	61–68	6, 7	Two episodes; 25 Ma and 16 Ma on whole-rock K-Ar (G. Plafker, written communication, 1986)	Linkage between plutons and volcanics is tenuous; REE diagram shows differentiated trend trend with higher SiO_2 rocks having negative Europium anomalies	Brew and Ford, 1986; D. A. Brew, unpublished data; G. Plafker, written communication, 1986; Fukuhara, 1986
L	Gut Bay area	Hornblende-biotite granodiorite, tonalite, tonalite, gabbro	No data	No data	No data	N.A.	24.3 Ma on biotite; 24.9 Ma and 31.5 Ma on coexisting biotite and hornblende, respectively	Heterogeneous intrusion	Loney and others, 1975

TABLE 2. DESCRIPTION OF LATEST MESOZOIC AND CENOZOIC MAGMATIC ROCKS OF SOUTHEASTERN ALASKA (continued)

Figure 1 and Table 1 references	Area or Belt Name	Major (and Minor) Lithic Types	Chemical Classification	SiO_2 Range (%)	SiO_2 Gap(s) (%)	Map and Diagrams on Figure	Age Data	Discussion	References
Late Oligocene and Miocene Rocks (35–5 Ma)									
M	Admiralty field	Andesite and basalt flows, (rhyolite tuff and breccia)	Mostly tholeiitic per MacDonald and Katsura (1964), but calc-alkalic (aver. K content) per Irvine and Baragar (1971)	47-58	N.A.	6, 7	Oligocene plant fossils (J. A. wolfe, written communication, 1985); 27 Ma whole-rock K-Ar (G. Plafker, written communication, 1986)	1,500–2,900 m thick alteration common	Loney, 1964; Lathram and others, 1965; G. Plafker, written communication, 1986
N	Kuiu-Etolin volcanic-plutonic belt	Basalt and andesite flows; rhyolite flows and tuffs, vent and other breccias; alkali granite, granite, quartz syenite (gabbro)	Volcanics mostly tholeiitic per MacDonald and Katsura (1964) and Miyashiro (1974), but calc-alkalic, (1974), aver. and low K-content per Irvine and Baragar (1971). Granitics are mostly peraluminous and metaluminous, calcalkalic, K-poor or K-aver., and only a few are peralkaline or alkalic	(volcanics) 46-76 (granitoids) 49-77	61-65 54-56, 61-65	8, 9, 10	Volcanics 22-20 Ma on whole-rock K-Ar; granitoids 19-24 Ma (Douglass and others, 1987)	Hetrogeneous volcanic and plutonic complex; gabbro and microgabbro low in section; siliceous volcaniclastic rocks associated with rhyolite; large, well-zoned granitoid body at east end of belt; basalts and andesites have no negative Europium anomaly, rhyolites a strong one, granitoids are in between	Brew and others, 1979, 1984; Hunt, 1984; Douglass and others, 1989
O	Groundhog Basin area	Rhyolite, biotite granite	Peraluminous per Shand (1951); tholeiitic per MacDonald and Katsura (1964), but calc-alkalic, K-poor per Irvine and Baragar (1971)	74–76		10	Sill 15 Ma on whole-rock K-Ar; plug 16 Ma on biotite K-Ar	Prominent rhyolite sill swarm apparently centered on granitic or felsic volcanic plugs	Brew and others, 1984; Douglass and others, 1989; R. P. Morrell, written communication, 1986
P	Southern Etolin field	Basalt flows, andesite breccias	No data	No data	No data	N.A.	No data	May be outlier of Kuiu-Etolin volcanics	Berg and others, 1976; Eberlein and others, 1983

TABLE 2. DESCRIPTION OF LATEST MESOZOIC AND CENOZOIC MAGMATIC ROCKS OF SOUTHEASTERN ALASKA (continued)

Figure 1 and Table 1 references	Area or Belt Name	Major (and Minor) Lithic Types	Chemical Classification	SiO_2 Range (%)	SiO_2 Gap(s) (%)	Map and Diagrams on Figure	Age Data	Discussion	References
Late Oligocene and Miocene Rocks (35–5 Ma)									
Q	East-central Prince of Wales field	Basalt and rhyolite breccia and tuff	No data	No data	No data	N.A.	No data	Very poorly known small isolated occurrences	Eberlein and others, 1983
R	Suemez field	Olivine basalt flows, basalt breccia, lapilli tuff, rhyolite and dacite flows	Peralkaline per Irvine and Baragar (1971); tholeiitic per MacDonald and Katsura (1964)	72	N.A.	N.A.	Associated with Tertiary(?) coal seams	Poorly known field, may be closely related to Tlevak field (D)	Eberlein and others, 1983; G. D. Eberlein, written communication, 1986
S	Burroughs Bay area	Biotite granite and biotite quartz monzonite	No data	No data	No data	N.A.	23 Ma K-Ar on biotite +chlorite (Hudson and others, 1979)	Quartz porphyry stock and dike swarm; explored for molybdenum	Hudson and others, 1979; Berg and others, 1988; R. L. Elliott and R. D. Koch, written communication, 1986
T	Cone Mountain area	Alkali-feldspar granite (rhyolite)	No data	No data	N.A.	N.A.	Miocene(?) reported by Koch and Elliott (1981)	May be similar to Groundhog Basin area (0)	Koch and Elliott, 1981; R. L. Elliott and R. D. Koch, written communication, 1986
U	Ketchikan area	Olivine-bearing pyroxene leucogabbro (gabbro, quartz diorite, ganodiorite)	No data	No data	N.A.	N.A.	24 Ma K-Ar on biotite, 25 Ma K-Ar on hornblende (Smith and Diggles, 1981)	May be distant member of Quartz Hill–Portland Peninsula group of plutons	Koch and Elliott, 1984; Berg and others, 1988
V	Quartz Hill–Portland Peninsula area	Olivine-hypersthene-augite gabbro, biotite granite, granite porphyry, biotite quartz monzonite	Granitoids: peraluminous, calc alcalic; gabbro: metaluminous, tholeiitic	47–78	48–73	11	Two episodes based on K-Ar: one at 30 Ma and one between 27 and 24 Ma	Four plutons in crude, E-trending belt–one contains major molybdenite deposit (Quartz Hill); strongly fractionated REE patterns and large Europium anomalies for granitoids, Sr initial ratios 0.747 to 0.7051 (Arth and others, 1986)	Elliott and others, 1976; Hudson and others, 1979

TABLE 2. DESCRIPTION OF LATEST MESOZOIC AND CENOZOIC MAGMATIC ROCKS OF SOUTHEASTERN ALASKA (continued)

Figure 1 and Table 1 references	Area or Belt Name	Major (and Minor) Lithic Types	Chemical Classification	SiO_2 Range (%)	SiO_2 Gap(s) (%)	Map and Diagrams on Figure	Age Data	Discussion	References
Late Oligocene and Miocene Rocks (35–5 Ma)									
W	Southern southeastern Alaska dike swarm	Lamprophyres, granitoids, basalt, dacite	Lamprophyres: alkalic, sodic; others: calc-alkalic	58–71	N.A.	11	7 Ma (Ouderkirk, 1982)	North-northeast striking swarm; probably deep expression of alkaline volcanic field to NE in British Columbia (Souther, 1970)	Smith, 1973; Elliott and others, 1976; Ouderkirk, 1982; Doyle, 1983
Late and Middle Eocene and Early Oligocene Rocks (45-35 Ma)									
X	Glacier Bay region	Biotite granite; alkali granite	Per- and metaluminous, dominantly calc-alkalic	58–76	71–75	12	42 to 31 Ma on K-Ar (12 determinations, M. A. Lanphere, oral communication, 1967, 1968, 1980)	Plutonic rocks slightly younger than in following two areas (Y and Z); body of Muir Inlet is associated with a Cu-Mo deposit	MacKevett and others, 1971, 1974; Brew and others, 1978; Brew, unpublished data; Himmelberg and Loney, 1981; Plafker and MacKevett, 1970
Y	Fairweather Range	Biotite granodiorite, biotite-hornblende quartz diorite and diorite; olivine gabbro, olivine norite, (olivine gabbronorite, anorthosite, wehrlite, dunite)	No data for granitoids but similar rocks not far NW are peraluminous, calc-alkalic; gabbroids are metaluminous, dominantly tholeiitic	Granitoids: 72–73 Gabbroids: 39–51	N.A.	N.A.	No data on granitoids; indirect dating of gabroids in that they are cut by felsic intrusives of above group (X)	This area and following area (Z) both contain a gabbroic and an intermediate to felsic suite; this area has dominant layered gabbros; area Z has dominant intermediate to felsic intrusives; LaPerouse layered gabbro is host of major magmatic Ni-Cu deposit	As above, plus Hudson and others, 1977; Loney and Himmelberg, 1983

TABLE 2. DESCRIPTION OF LATEST MESOZOIC AND CENOZOIC MAGMATIC ROCKS OF SOUTHEASTERN ALASKA (continued)

Figure 1 and Table 1 references	Area or Belt Name	Major (and Minor) Lithic Types	Chemical Classification	SiO_2 Range (%)	SiO_2 Gap(s) (%)	Map and Diagrams on Figure	Age Data	Discussion	References
Late and Middle Eocene and Early Oligocene Rocks (45–35 Ma)									
Z	Yakobi, Chichagof, and Baranof area	Garnet- or muscovite-bearing biotite-hornblende granodiorite and granite, biotite granodiorite, muscovite-hornblende-biotite tonalite, (hornblende quartz diorite), hornblende-pyroxene gabbronorite, hornblende pyroxenite, (quartz-bearing norite, leucogabbro)	Granitoids dominantly metaluminous, calc-alkalic–aver. K content; gabbroids metaluminous, calc-alkalic and tholeiitic, both aver. K content	Granitoids: 51–73 Gabbroids: 49–55	Granitoids: 58–65	12, 13	43–40 Ma K-Ar on hornblende and biotite from tonalite on Yakobi (M.A. Lanphere, written communication, 1982; F. H. Wilson, written commun., 1985); 49 Ma K-Ar on biotite from granodiorite on Kruzof (Loney and others, 1967); 47–42 Ma K-Ar on biotite and hornblende from granodiorite and tonalite on Baranof (Loney and others, 1967)	See above; gabbronorite on Yakobi is host of Cu deposit; REE diagram shows relatively undifferentiated trends. Kruzof pluton has $^{87}Sr/^{86}Sr$ values of about 0.70535; (Myers and others, 1984)	Loney and others, 1975; Johnson and Karl, 1985; Callahan, 1970; Wanek and Callahan, 1969; Himmelberg and others, 1987
Early and Middle Eocene Rocks (55–45 Ma)									
AA	Haines-Skagway area	K-feldspar-porphyritic hornblende quartz monzodiorite, K-feldspar-prophyritic hornblende-biotite quartz monzonite and granite, hornblende-biotite tonalite and granodiorite, biotite granite	Tonalite to W dominantly peraluminous (8:3), rest metaluminous, calk-alkalic, aver. and high K content; granodiorite and granite to E equally peraluminous and metaluminous, calc-	53–77	N.A.	14A, 15	Tonalite and granodiorite 54 Ma on Pb/U on zircon, 52 Ma on K-Ar on biotite; granite 52–51 Ma on Pb/U on zircon to E, 48 Ma to to W (Barker and others, 1986)	Assignment of rocks north of Haines based on lithic similarity to rocks to E and S; if assignment is correct, then this is the only known occurrence of an early Eocene body W of the "tonalite sill" family of plutons. Tonalite and granodiorite near Skagway REE diagram shows moderate to steep negative slopes and small negative EU anomalies, granites have flat slopes and distinct negative anomalies. Initial Sr ratios range from 0.70485 to 0.70770	Margaritz and Taylor, 1976; Redman and others, 1984; Lambert, 1974; Christie, 1957, 1959; Barker and others, 1986; D. A. Brew and A. B. Ford, unpublished data

TABLE 2. DESCRIPTION OF LATEST MESOZOIC AND CENOZOIC MAGMATIC ROCKS OF SOUTHEASTERN ALASKA (continued)

Figure 1 and Table 1 references	Area or Belt Name	Major (and Minor) Lithic Types	Chemical Classification	SiO_2 Range (%)	SiO_2 Gap(s) (%)	Map and Diagrams on Figure	Age Data	Discussion	References
Early and Middle Eocene Rocks (55–45 Ma)									
BB	Juneau-Taku River area	Homogeneous sphene-bearing biotite-hornblende granodiorite, tonalite, and granite; K-spar porphyritic hornblende-biotite granite, quartz monzodiorite, and granodiorite	Dominantly (2:1) metaluminous, rest peraluminous, calc-alkalic–aver. K content	60–76	N.A.	14B, 16	Published K-Ar of 53–50 Ma (Forbes and Engels, 1970) and Pb/U on zircon of 50 Ma (Gehrels and others, 1984) supported by abundant unpublished K-Ar data (J. G. Smith, written communication, 1976, 1986; F. H. Wilson, written communication, 1985, 1986)	Very large composite pluton; K-spar-porphyritic phase restricted to near International Boundary; associated with volcanic rocks of Sloko Group in same area (Souther, 1971)	Brew and Ford, 1986; Souther, 1971; D. A. Brew and A. B. Ford, unpublished data
CC	Sumdum (Tracy Arm) area	Sphene-hornblende biotite granodiorite and granite, sphene-biotite-hornblende granodiorite, hornblende granodiorite; locally porphyritic	Dominantly (2:1) metaluminous, rest peraluminous, calcalkalic–aver. K-content	59–75	N.A.	17	54–49 Ma K-Ar on numerous biotite and hornblende samples (J. G. Smith, written communications, 1976, 1986)	Southeastward continuation of very large composite pluton of Juneau–Taku River area	Brew and Grybeck, 1984; D. A. Brew and A. B. Ford, unpublished data
DD	Petersburg area	Locally K-spar-porphyritic sphene-bearing biotite-hornblende granodiorite, tonalite, and granite	Dominantly peraluminous, calc-alkalic–aver. K content	64–75	N.A.	17, 18	52–49 Ma K-Ar on biotite and hornblende (Douglass and others, 1987)	Large discrete pluton with complicated border phases	Brew and others, 1984; Webster, 1984; Douglass and others, 1989
EE	Bradfield Canal area	Locally K-spar-porphyritic sphene-bearing biotite-hornblende granodiorite and and granite, leucocratic quartz monzonite, alkali granite	Peraluminous, calc-alkalic–aver. K-content and K-poor	54–75	55–65	19	53–46 Ma on K-Ar from three different intrusive phases (Smith, 1977)	Several intrusive episodes probably represented	Koch and Elliott, 1981; Smith, 1977; R. D. Koch and R. L. Elliott, oral communication, 1986

TABLE 2. DESCRIPTION OF LATEST MESOZOIC AND CENOZOIC MAGMATIC ROCKS OF SOUTHEASTERN ALASKA (continued)

Figure 1 and Table 1 references	Area or Belt Name	Major (and Minor) Lithic Types	Chemical Classification	SiO_2 Range (%)	SiO_2 Gap(s) (%)	Map and Diagrams on Figure	Age Data	Discussion	References
Early and Middle Eocene Rocks (55–45 Ma)									
FF	Ketchikan–Prince Rupert area	Locally K-spar-porphyritic sphene-bearing biotite-hornblende granodiorite, granite, and tonalite	Peraluminous dominant over metaluminous 2:1; calc-alkalic–aver. K-content	58–72	60–65	19	55–45 Ma on K-Ar from biotite and hornblende (Smith and Diggles, 1981)	Major bodies are continuations of those in Bradfield Canal area; probably several close-spaced intrusive episodes represented; strongly fractionated REE patterns with small negative EU anomalies; Sr initial ratios 0.7046–0.7061; area connects to SE in British Columbia with Ponder pluton (Hutchison, 1982) and Mo-bearing Alice Arm intrusions (Woodcock and Carter, 1976; Christopher and Carter, 1976)	Berg and others, 1988; Smith, 1977; Smith and others, 1977; Woodcock and Carter, 1976; Hutchison, 1982; Arth and others, 1986
Paleocene Rocks (65–55 Ma)									
GG	Juneau-Skagway area	Foliated hornblende granodiorite and tonalite; biotite-hornblende granodiorite	Equal peraluminous and metaluminous; calc-alkalic–aver. K-content, some K-rich	57–76	N.A.	20	60 Ma on zircon (Gehrels and others, 1984) is supported by unpublished zircon age (G.R. Tilton, written communication, 1986) and K-Ar ages (F. H. Wilson, written communication, 1985, 1986)	Structurally more complicated than above (45–55 Ma) suite; modally and chemically in between the above suite and that below (65–75 Ma); plutons are generally stubby sills. Part of "great tonalite sill" family	Brew and Ford, 1986

TABLE 2. DESCRIPTION OF LATEST MESOZOIC AND CENOZOIC MAGMATIC ROCKS OF SOUTHEASTERN ALASKA (continued)

Figure 1 and Table 1 references	Area or Belt Name	Major (and Minor) Lithic Types	Chemical Classification	SiO_2 Range (%)	SiO_2 Gap(s) (%)	Map and Diagrams on Figure	Age Data	Discussion	References
Latest Cretaceous Rocks (75–65 Ma)									
HH	Haines-Skagway area	In Canada: Locally foliated K-spar porphyritic biotite-hornblende granite. In U.S.: well- to slightly-foliated locally lineated sphene-hornblende-biotite tonalite (hornblende quartz diorite, biotite-hornblende-granodiorite)	In Canada: Peraluminous, calcalkalic–aver. K-content and some K-poor. In U.S.: Equal per- and metaluminous, calc-alkalic–aver. K-content and some K-rich	In Canada 67–73. In U.S.: 57–73	N.A. N.A.	21, 22A	In Canada: U-Pb age of 72 Ma on zircon; In U.S.: U-Pb ages of 67–68 on zircons (Barker and others, 1986)	Pluton in Canada is unlike others in this suite structurally and and modally; it has Sr initial ratio of 0.70615; the plutons in the U.S. are part of the "great tonalite sill" family. REE diagram shows steep slopes with no or small negative Eu anomalies	Barker and others, 1986, Redman and others, 1984; D.A. Brew and A.B. Ford, unpublished data (1985)
II	Juneau–Taku River area	Well foliated, locally well lineated, locally sphene-bearing biotite-hornblende and hornblende-biotite tonalite, quartz diorite, and granodiorite	Dominantly metaluminous, calc-alkalic–aver. K-content slightly more common than K-rich	51–71	N.A.	21	Pb/U age of 67 Ma on zircon (Gehrels and others, 1984); unpublished K-Ar ages on biotite and hornblende suggest range may be 56–67 Ma (F. H. Wilson, written communication, 1985, 1986)	"Great tonalite sill" family consists of a single NW-digitating pluton to the NW and a group of narrow digitating plutons to the SE	Brew and Ford, 1986; Ford and Brew, 1981; D. A. Brew and A. B. Ford, unpublished data (1985)
JJ	Sumdum area	Generally well-foliated, locally well-lineated biotite-hornblende quartz diorite and tonalite, (biotite-hornblende granodiorite and quartz monzonite)	Meta- to peraluminous ratio 2.5:1; calcalkalic–aver. K-content and K-rich equally abundant	58–68	64–67?	23	K-Ar ages range from 66 Ma on hornblende to 50 Ma (J. G. Smith, written communication, 1977, 1978, 1986). Pb/U of 64 Ma reported by Gehrels and others (1984)	"Great tonalite sill" family consists of one narrow but continuous pluton plus several small sills of somewhat uncertain affinity	Brew and Grybeck, 1984; D. A. Brew and A. B. Ford, unpublished data
KK	Petersburg area	Generally well foliated, locally well-lineated biotite-hornblende and hornblende-biotite tonalite. quartz diorite (granodiorite)	Met- to peraluminous ratio is 10.5:1; calc-alkalic–dominantly K-rich, some aver. K-content	54–67	63–66?	22B, 23	No reliable ages available	"Great tonalite sill" family consists of four large homogeneous bodies; migmatic unit between two of them	Brew and others, 1984; Douglass and others, 1989

TABLE 2. DESCRIPTION OF LATEST MESOZOIC AND CENOZOIC MAGMATIC ROCKS OF SOUTHEASTERN ALASKA (continued)

Figure 1 and Table 1 references	Area or Belt Name	Major (and Minor) Lithic Types	Chemical Classification	SiO_2 Range (%)	SiO_2 Gap(s) (%)	Map and Diagrams on Figure	Age Data	Discussion	References
Latest Cretaceous Rocks (75–65 Ma)									
LL	Bradfield Canal area	Granodiorite and quartz diorite	No data	No data	No data	N.A.	No data	"Great tonalite sill" family consists of continuations from the Petersburg area (LL) that digitate or otherwise die out to the SE	Koch and Elliott, 1981; R. L. Elliott and R. D. Koch, oral communication, 1986
MM	Ketchikan–Prince Rupert area	Foliated biotite-hornblende quartz diorite and tonalite, granodiorite	Met- to per-aluminous ratio 5:1; calc-alkalic–aver. K-content and K-rich in equal amounts	56–62	N.A.	23	58 to 55 Ma on zircon (Berg and others, 1988)	Extension of "great tonalite sill" family into British Columbia is Quotoon pluton; in the U.S. adjacent to International Boundary that single homogeneous body resembles the family further N, but intervening area (to Bradfield Canal [LL]) is a poorly understood zone 25 km wide with several narrow sills; Arth and others (1986) report mildly fractionated REE patterns with small negative Eu anomalies and Sr initial ratios of 0.7063–0.7064	Berg and others, 1988; Smith and others, 1977; Hutchison, 1982; Arth and others, 1986

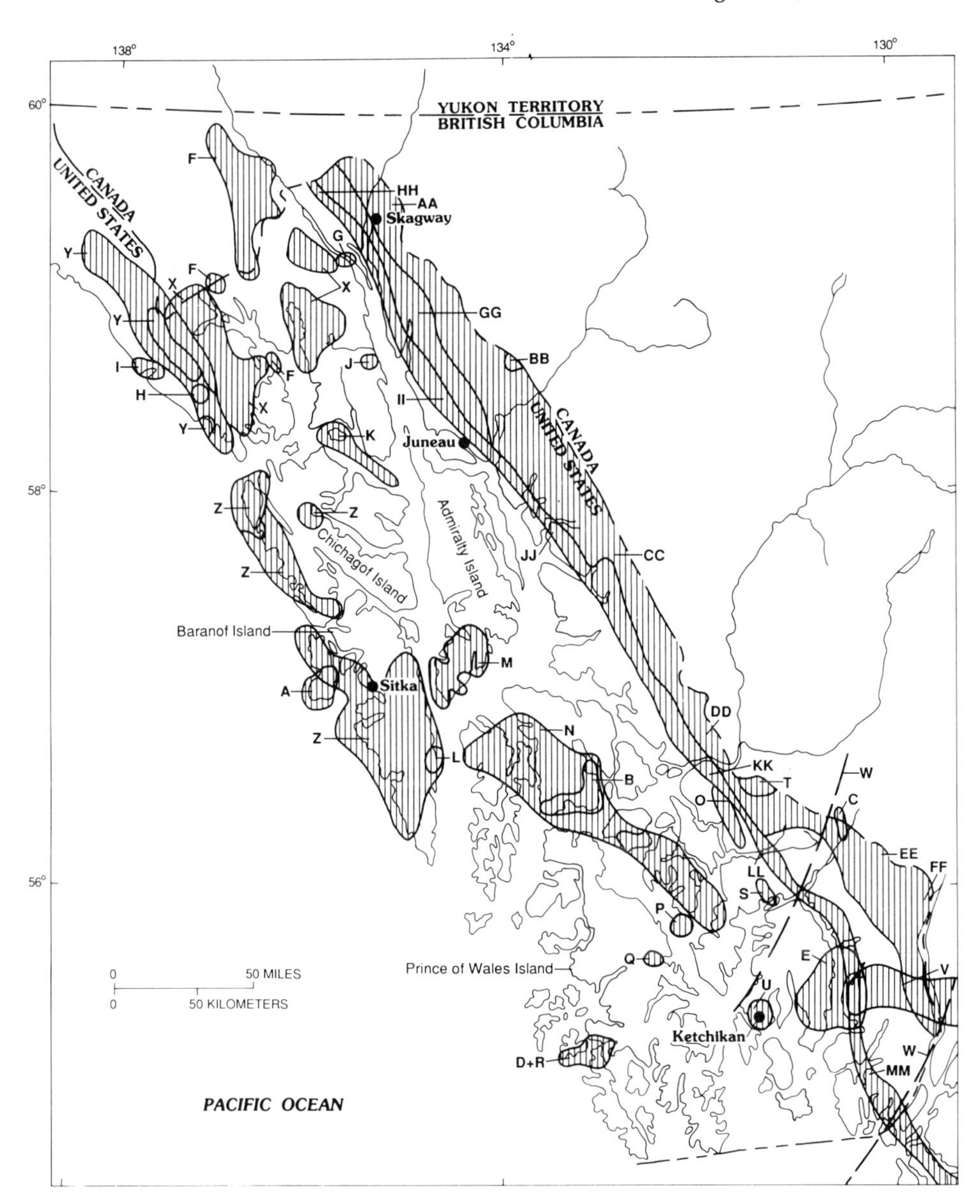

Figure 1. Cenozoic plutonic and volcanic rock localities in southeastern Alaska. Letters refer to areas described in Tables 1 and/or 2. Lined pattern indicates approximate extent of areas; boundaries between contiguous areas of the same age are omitted. Lines labelled "W" are the northwest and southeast boundaries of the southern southeastern Alaska dike swarm.

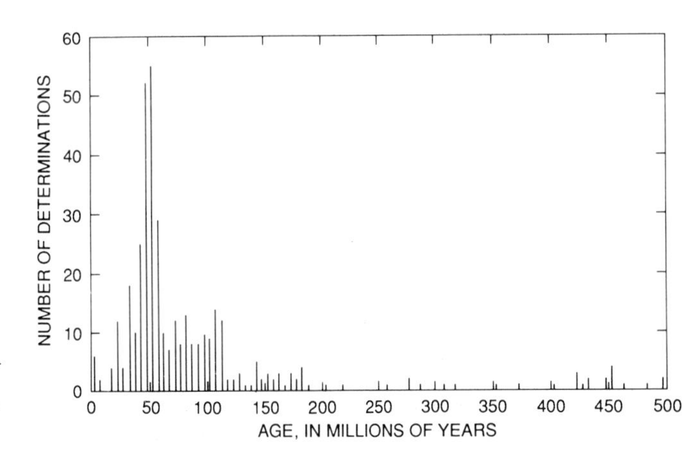

Figure 2. Histogram showing distribution of radiometric ages for southeastern Alaska.

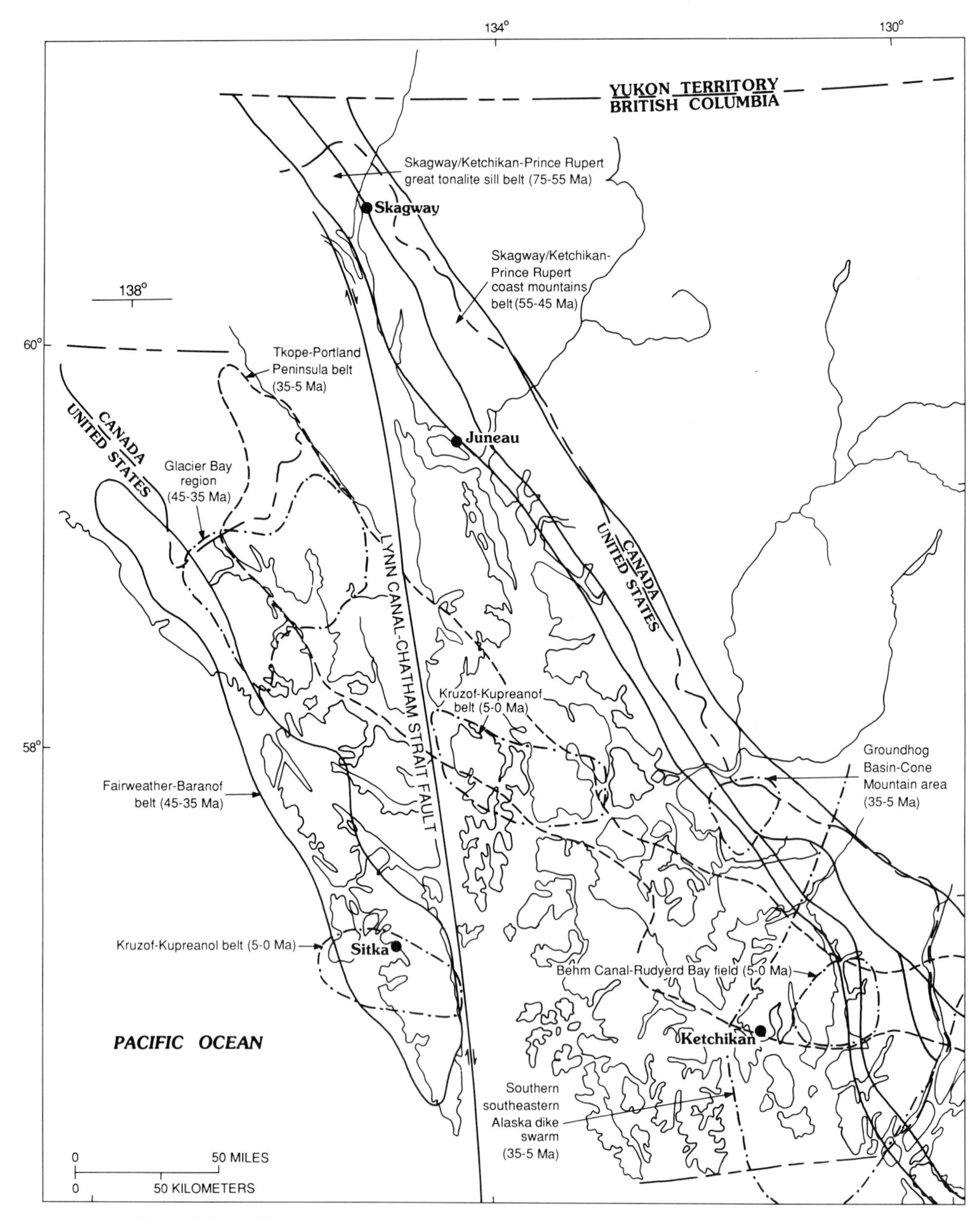

Figure 3. Latest Mesozoic and Cenozoic magmatic belts, fields, and areas in southeastern Alaska. Ages given in parentheses are those from the organization of the text and table and do not in every case reflect the full range of ages of the rocks in the belts. One hundred and twenty kilometers of right-lateral separation on the Lynn Canal–Chatham Strait fault has been removed. Different types of lines are used only to clarify distributions of overlapping belts.

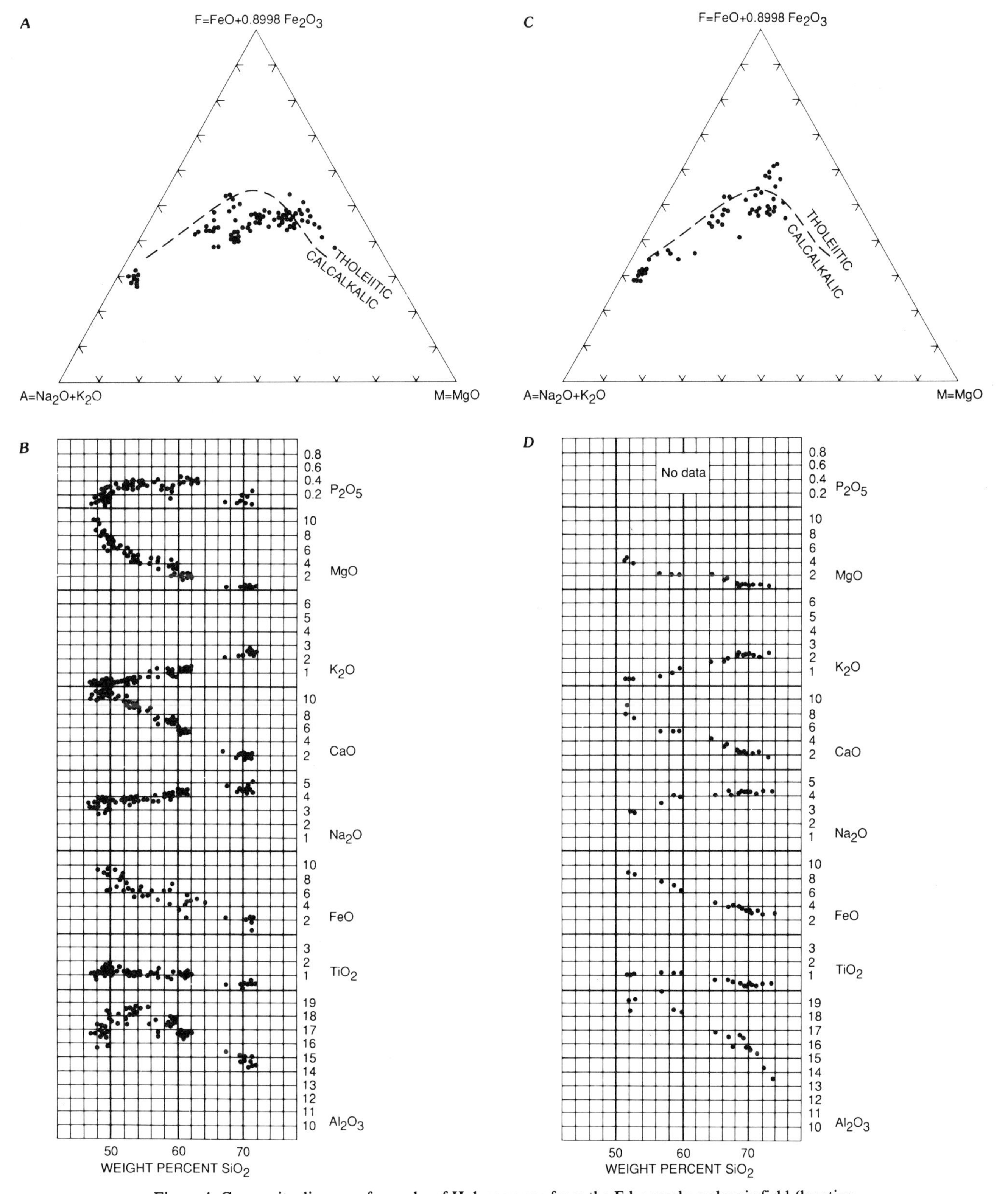

Figure 4. Composite diagrams for rocks of Holocene age from the Edgecumbe volcanic field (location shown on Fig. 5). *A,* AFM diagram after Irvine and Baragar (1971), non-tephra-deposit samples, data from J. R. Riehle (written communication, 1986). *B,* Silica-variation diagram, non-tephra-deposit samples, data from Brew and others (1969), Myers and Marsh (1981), and Kosco (1981). *C,* AFM diagram, tephra-deposit samples, data from J. R. Riehle (written communication, 1986). *D,* Silica-variation diagram, tephra-deposit samples, data from Riehle and Brew (1984).

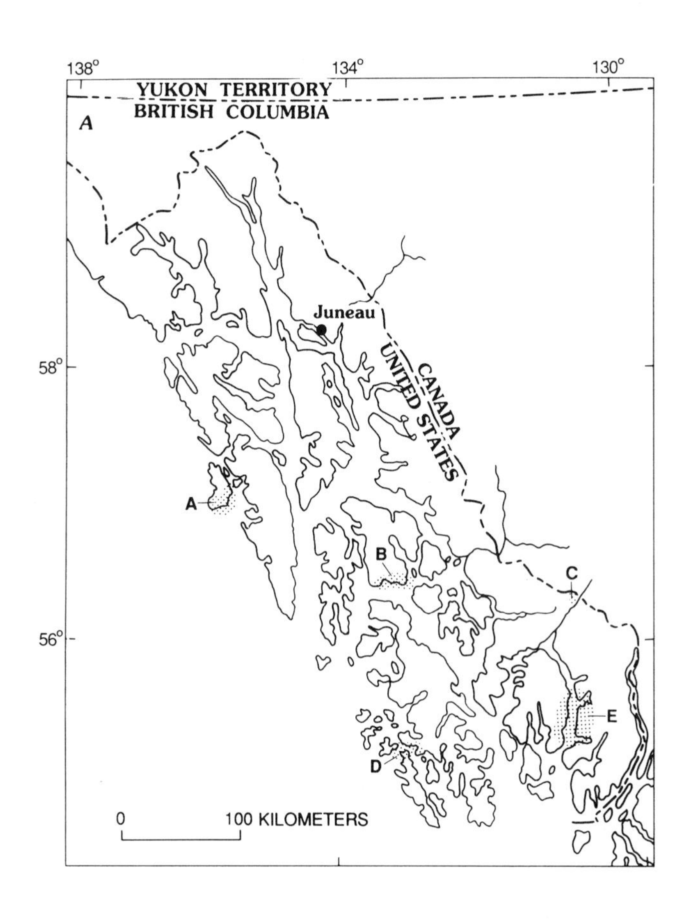

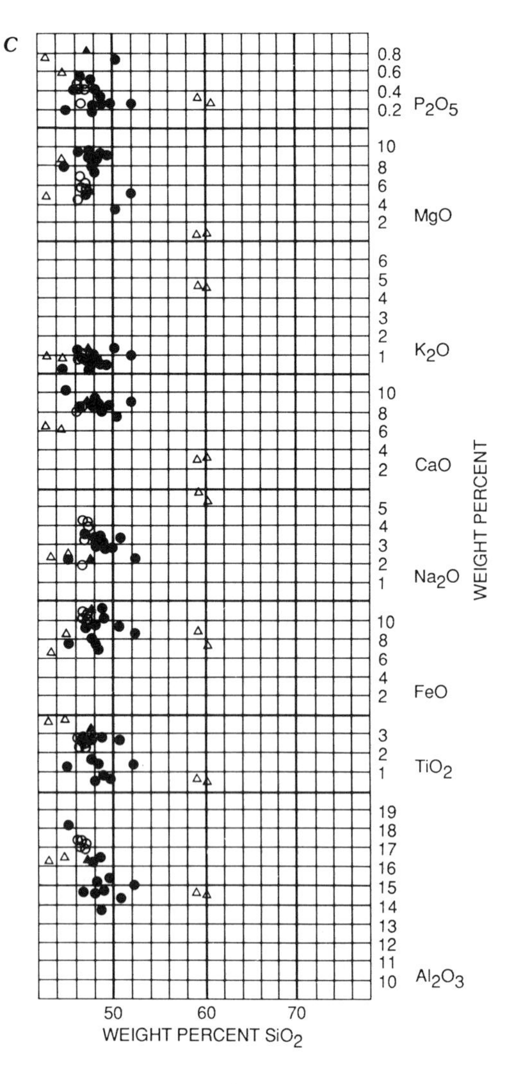

EXPLANATION

Rocks of Holocene age from

- ● Southern Kupreanof volcanic field
- ○ Blue River-Unuk River volcanic field
- ▲ Behm Canal-Rudyerd Bay volcanic field
- △ Tlevak Strait volcanic field

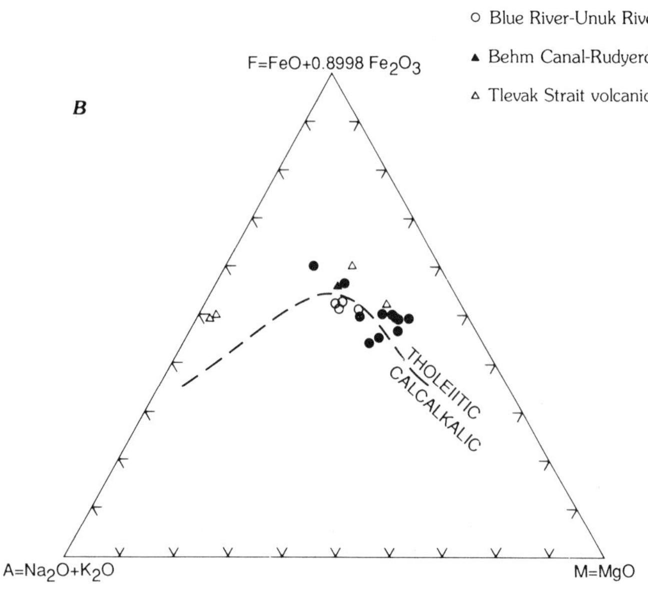

Figure 5. Location map and composition diagrams for rocks of Holocene age. *A,* the Edgecumbe volcanic field (A), southern Kupreanof volcanic field (B), Blue River–Unuk River volcanic field (C), Tlevak Strait field (D), and Behm Canal–Rudyerd Bay volcanic fields (E). *B,* AFM diagram (Irvine and Baragar, 1971). *C,* Silica-variation diagram. Data sources: Douglass and others (1989), R. L. Elliott (written communication, 1986), Wanek and Callahan (1971), and Ouderkirk (1982).

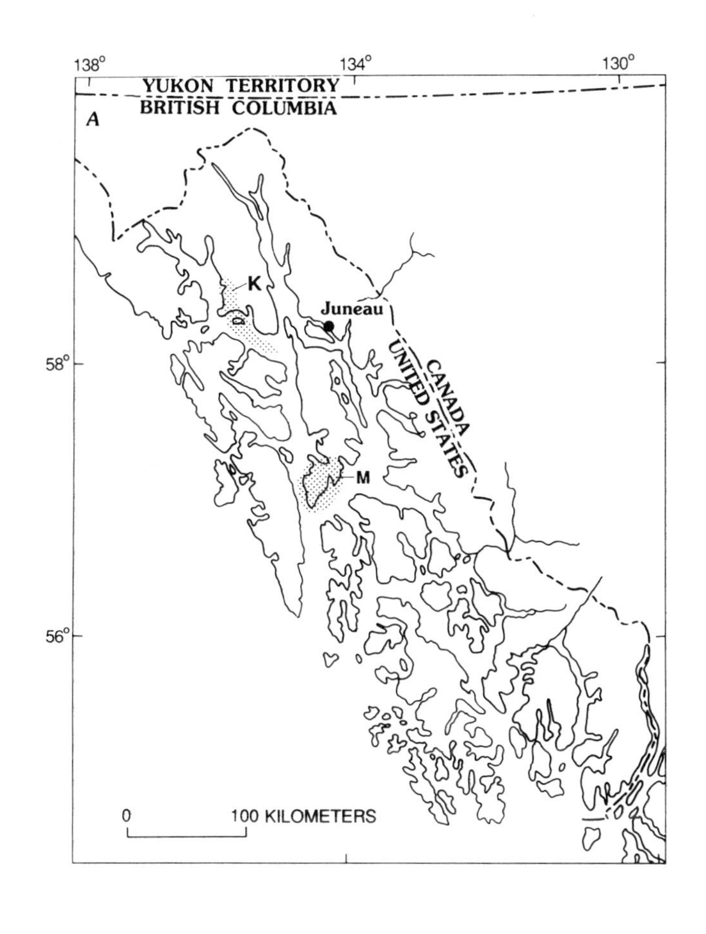

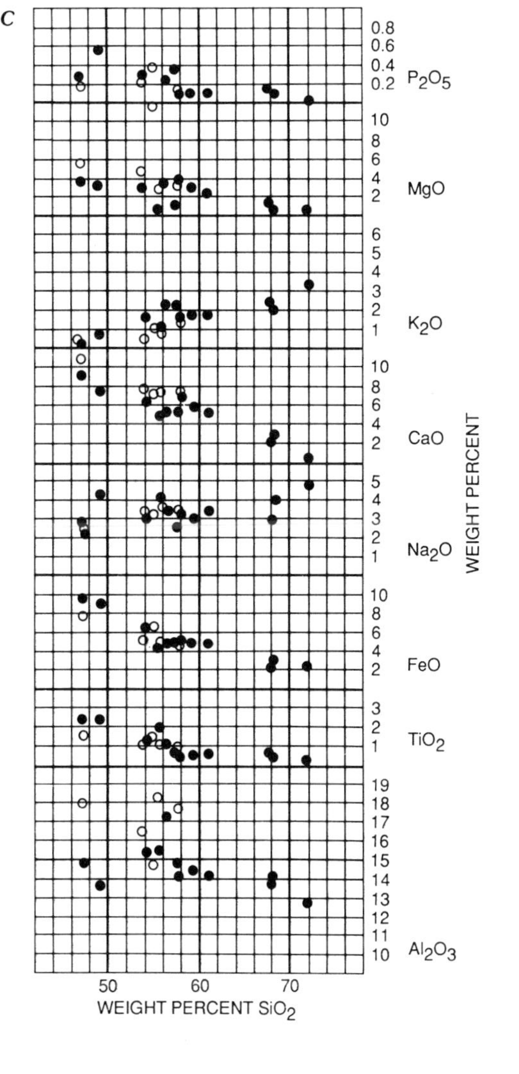

EXPLANATION

Rocks of late Oligocene and Miocene age from

- ● Icy Strait Belt
- ○ Admiralty Island volcanic field

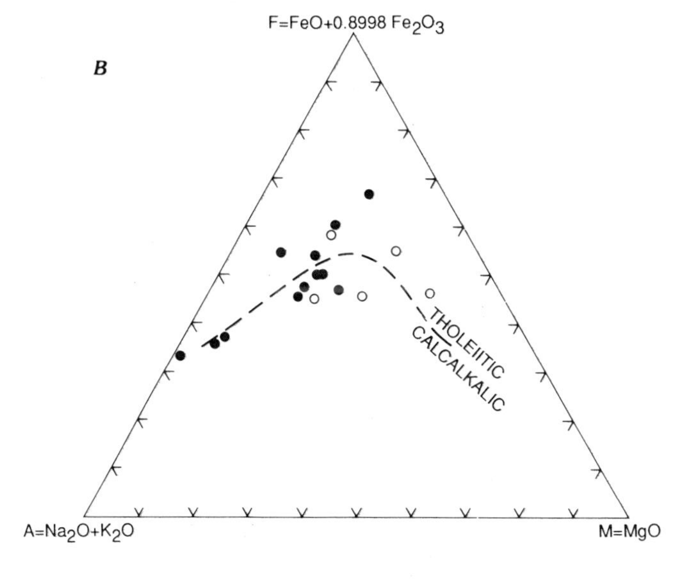

Figure 6. Location map and composition diagrams for rocks of late Oligocene and Miocene age. *A,* Icy Strait belt (K) and Admiralty Island volcanic field (M). *B,* AFM diagram (Irvine and Baragar, 1971). *C,* Silica-variation diagram. Data sources: D. A. Brew (unpublished data, 1985); Loney (1964), G. Plafker (written communication, 1986).

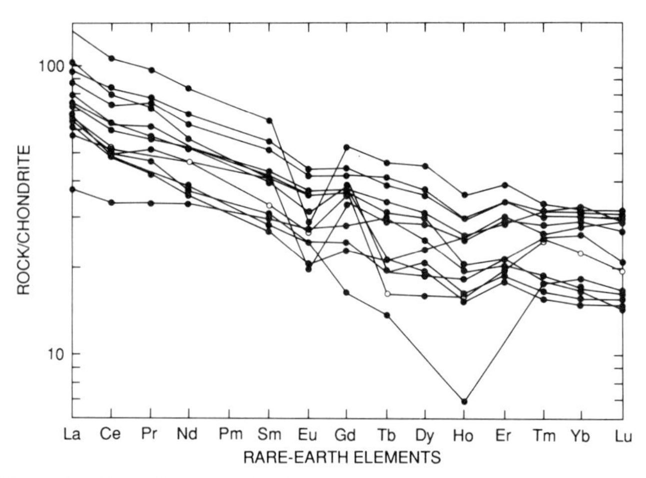

Figure 7. Chondrite-normalized rare-earth-element diagram for rocks of late Oligocene and Miocene age: Icy Strait belt (dots) and Admiralty Island volcanic field (circles). Data from D. A. Brew (unpublished data, 1985) and G. Plafker (written communication, 1986).

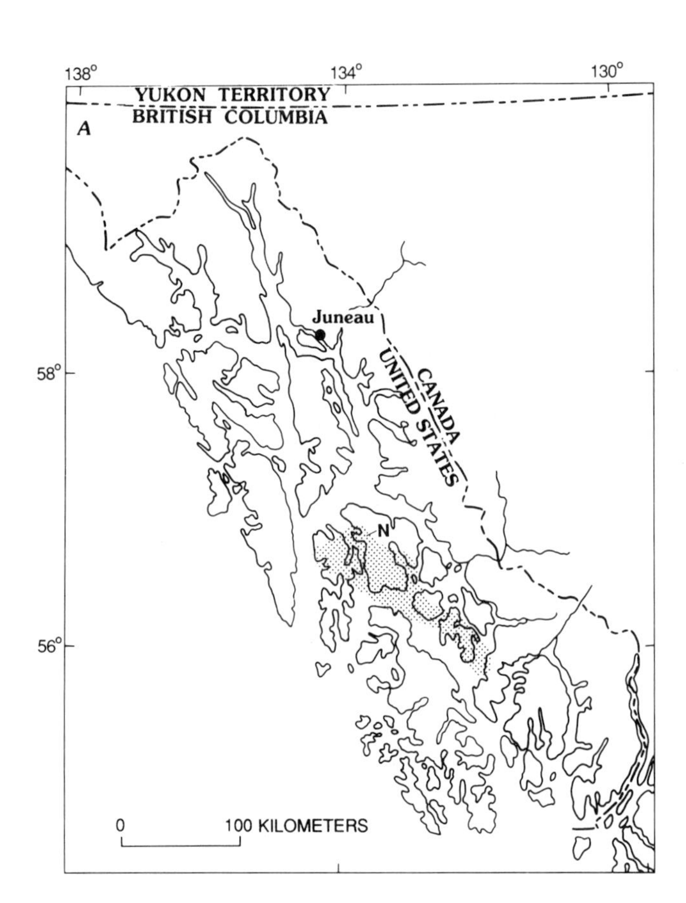

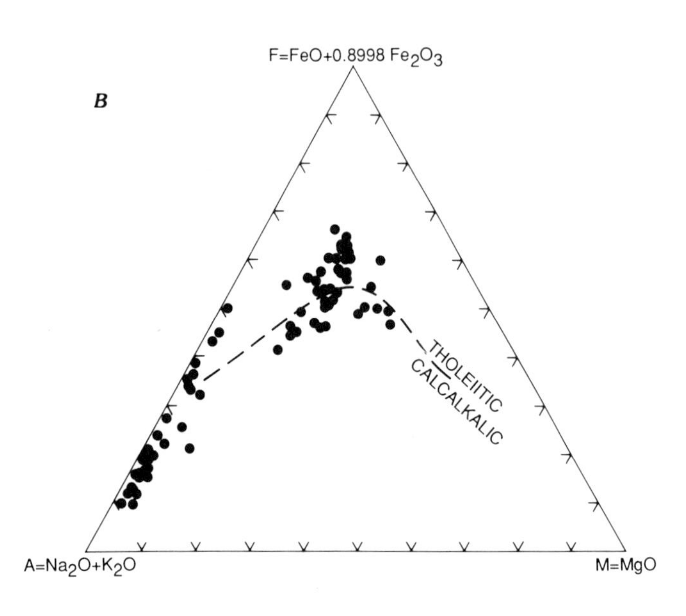

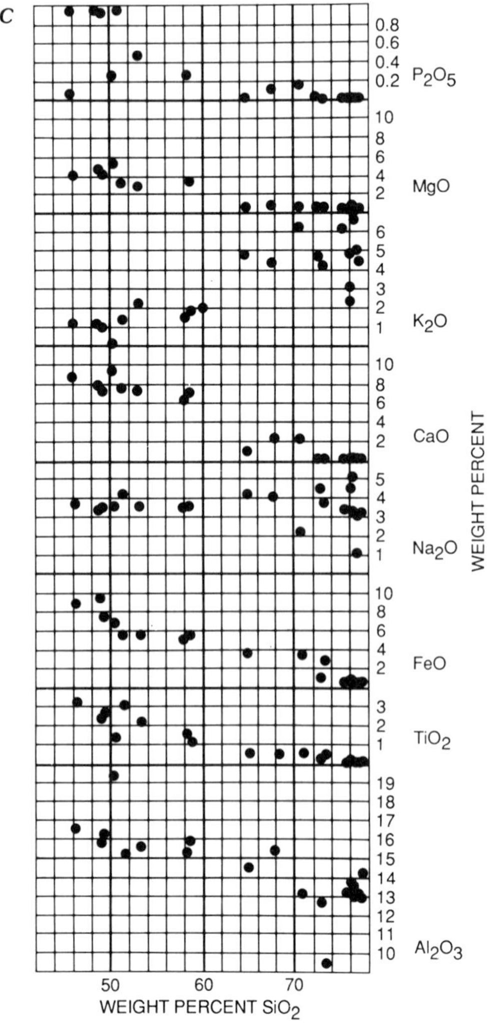

Figure 8. Location map and composition diagrams for volcanic rocks of late Oligocene and Miocene age. *A,* Kuiu-Etolin volcanic-plutonic belt (N). *B,* AFM diagram (Irvine and Baragar, 1971). *C,* Silica-variation diagram. Data source: Douglass and others (1989).

originally a rift or an ocean between two different terranes, was the movement of the outboard Alexander terrane (Silberling and others, this volume) toward the northeast (Monger and others, 1982). This movement is interpreted to have preceded the convergence of the Chugach terrane against the westward margin of the Alexander terrane (Plafker and others, 1977 and this volume, Chapter 12; Johnson and Karl, 1985). The consistent composition of the magmas argues for a deep and equilibrated source, even though the preliminary data of Arth and others (1986) on strontium initial ratios indicate possible derivation from continental source materials. This latter possibility can be used to support a within-plate-rift origin of the structural discontinuity that localized the tonalite sill belt and the other nearby parallel features of the southeastern Alaska coincident zone (Brew and Ford, 1985).

Coast Mountains belt

The linear Coast Mountains belt along the International Boundary (Skagway/Ketchikan–Prince Rupert Coast Mountains belt [55 to 45 Ma] on Fig. 3 and Tables 1 and 2) consists of a large volume of early and middle Eocene plutons that are probably a result of the convergence and crustal thickening associated with the compressive event just described. These rocks are consistently calc-alkalic, dominantly metaluminous north of the Sumdum area, and exclusively moderately peraluminous to the south. The overall range in silica content is 53 to 76 percent; the average silica content is about 67 percent to the north of the Sumdum area and 72 percent to the south. Modally, the rocks are dominantly sphene-hornblende-biotite granodiorite, granite, and tonalite. Available age determinations indicate that the plutons in the southern part of the Coast Mountains were emplaced between 55 and 45 Ma, and those in the northern part from 54 to 49 Ma. The Coast Mountains belt of large composite plutons parallels the great tonalite sill belt, commonly within a few kilometers, and in several places intrudes that belt.

The differences in age, structural habit, and composition of the rocks in this belt indicate a different origin from that of the great tonalite sill belt. The general absence of all structures but flow foliation, and the restricted thermal aureoles that are superposed on the earlier Barrovian-type metamorphism associated with the tonalite sill belt, indicate that these early and middle Eocene intrusions are post-tectonic and that their emplacement followed the abrupt uplift that accompanied and closely followed intrusion of the sill belt.

The composition of the plutons in the Coast Mountains belt, their location in relation to the highly deformed and presumably thickened crust near the tonalite sill belt, and the time-lag relations all indicate that the Coast Mountains belt is the result of the thickening that occurred during the latest Cretaceous and early Tertiary collision discussed above. I infer that the change from metaluminous to moderately peraluminous composition from north to south is related to the type of material conveyed to depth in the convergent zone. This may be a result of the greater thickness of older continental crust in the southern part of the Alexander terrane compared with the northern part.

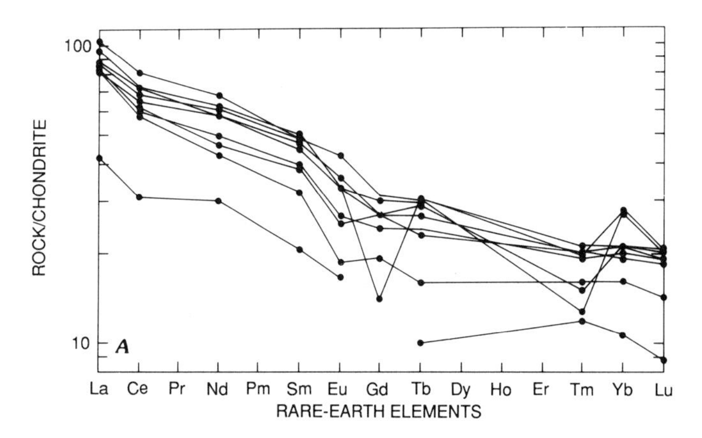

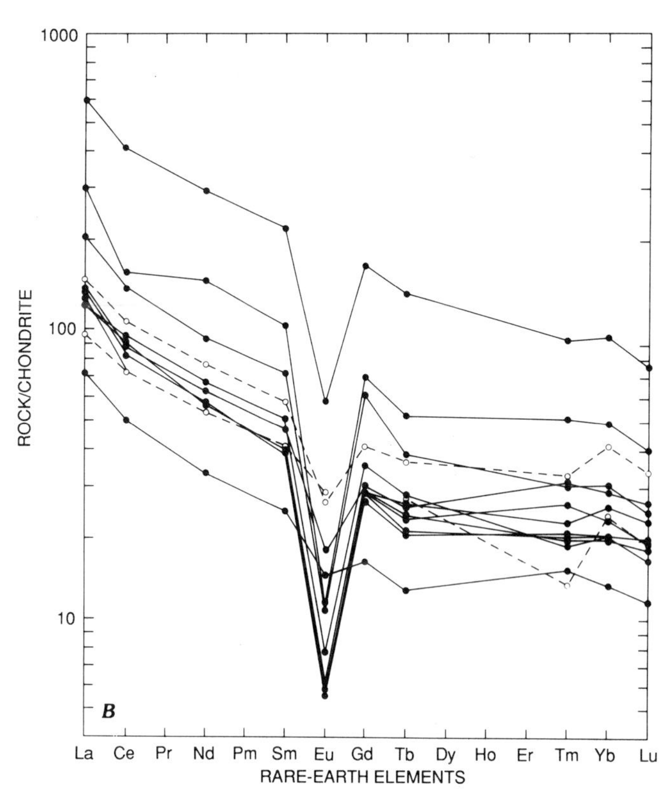

Figure 9. Chondrite-normalized rare-earth-element diagrams for rocks of late Oligocene and Miocene age from the Kuiu-Etolin volcanic-plutonic belt. *A,* Basalts and andesites. *B,* Rhyolites (dots) and granitic rocks (circles). Data source: D. A. Brew (unpublished data, 1985).

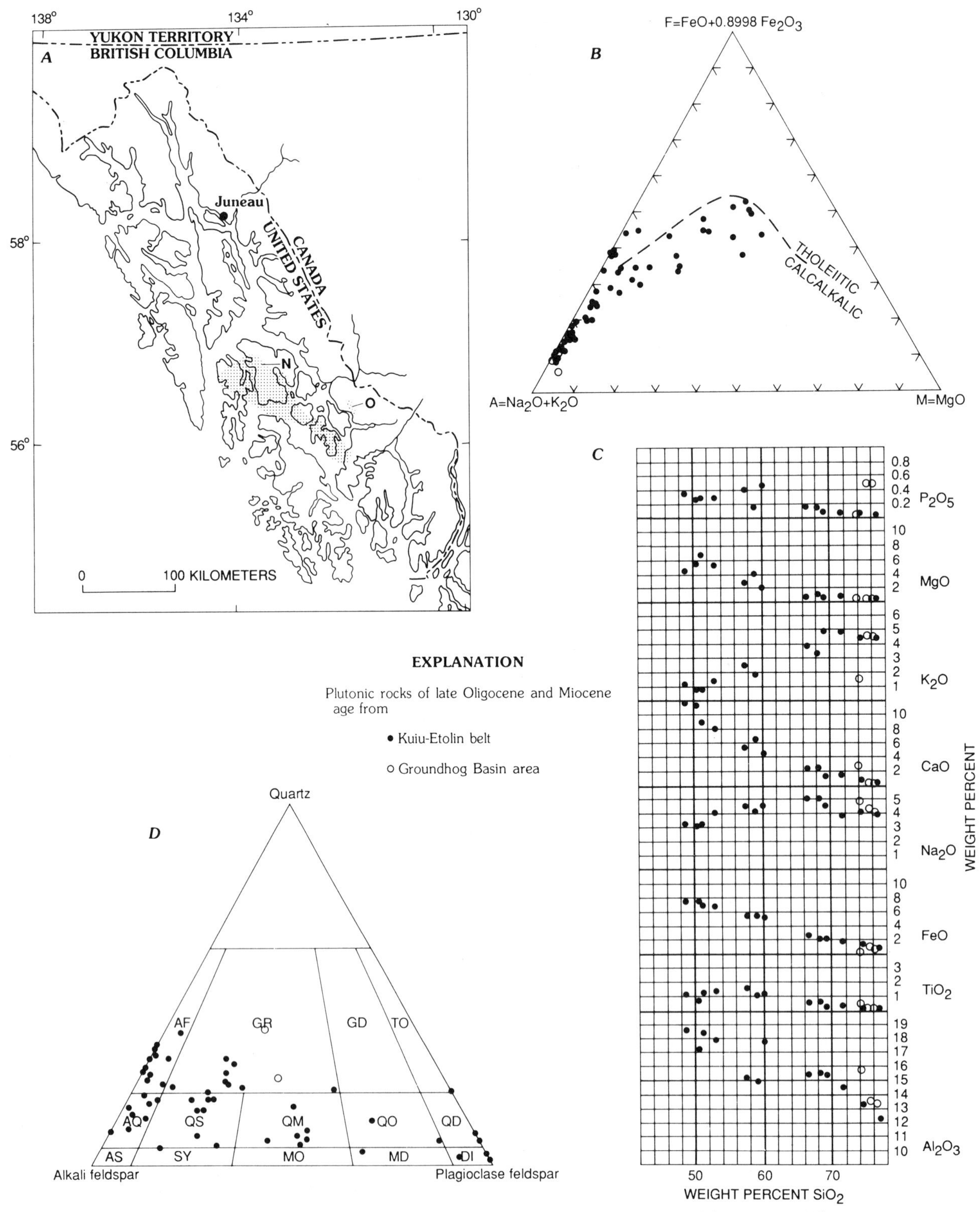

Figure 10. Location map and composition diagrams for plutonic rocks of late Oligocene and Miocene age. *A*, the Kuiu-Etolin volcanic-plutonic belt (N) and Groundhog Basin area (O). *B*, AFM diagram (Irvine and Baragar, 1971). *C*, Silica-variation diagram. *D*, General plutonic rock classification diagram (Streckeisen, 1973): AF, alkali-feldspar granite; AQ, alkali-feldspar quartz syenite; AS, alkali-feldspar syenite; DI, diorite; GR, granite; GD, granodiorite; MD, monzodiorite; MO, monzonite; QD, quartz diorite; QO, quartz monzodiorite; QM, quartz monzonite; QS, quartz syenite; SY, syenite; TO, tonalite. Data sources: Douglass and others (1989), and Hunt (1984).

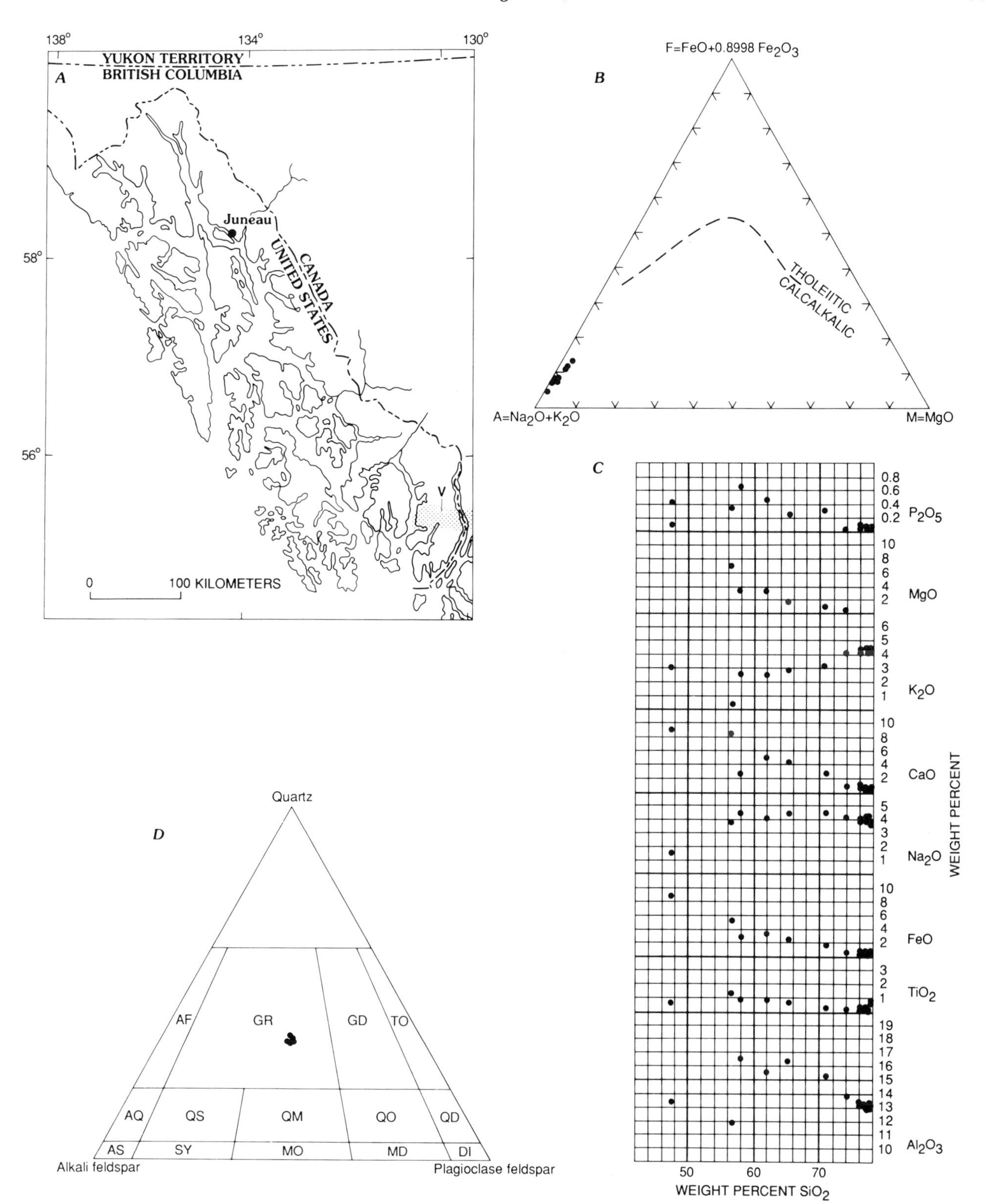

Figure 11. Location map and composition diagrams for plutonic rocks of late Oligocene and Miocene age. *A*, the Quartz Hill/Portland Peninsula area (V). *B*, AFM diagram (Irvine and Baragar, 1971). *C*, Silica-variation diagram; *D*, general plutonic rock classification diagram (Streckeisen, 1973): AF, alkali-feldspar granite; AQ, alkali-feldspar quartz syenite; AS, alkali-feldspar syenite; DI, diorite; GR, granite; GD, granodiorite; MD, monzondiorite; MO, monzonite; QD, quartz diorite; QO, quartz monzodiorite; QM, quartz monzonite; QS, quartz syenite; SY, syenite; TO, tonalite. Data sources: Smith and others (1977); Hudson and others (1979).

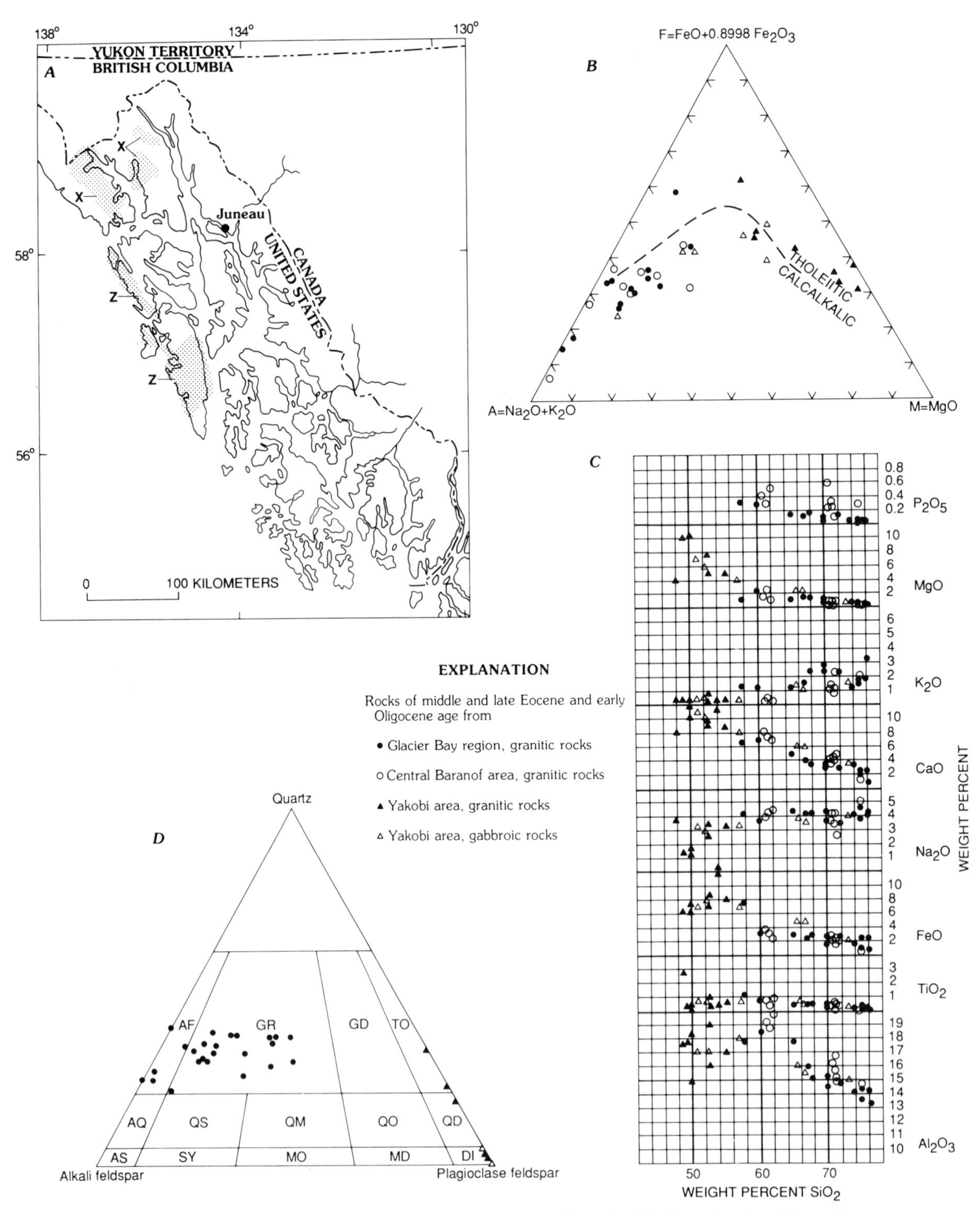

Figure 12. Location map and composition diagrams for granitic and gabbroic rocks of middle and late Eocene and early Oligocene age. *A,* the Glacier Bay region (X) and the Yakobi, Chichagof, and Baranof area (Z). *B,* AFM diagram (Irvine and Baragar, 1971). *C,* Silica-variation diagram. *D,* general plutonic rock classification diagram (Streckeisen, 1973): AF, alkali-feldspar granite; AQ, alkali-feldspar quartz syenite; AS, alkali-feldspar syenite; DI, diorite; GR, granite; GD, granodiorite; MD, monzondiorite; MO, monzonite; QD, quartz diorite; QO, quartz monzodiorite; QM, quartz monzonite; QS, quartz syenite; SY, syenite; TO, tonalite. Data sources: Himmelberg and Loney (1981), Himmelberg and others (1987), and Brew (unpublished data) for all but Baranof area; Wanek and Callahan (1969) and Callahan (1970) for Baranof area.

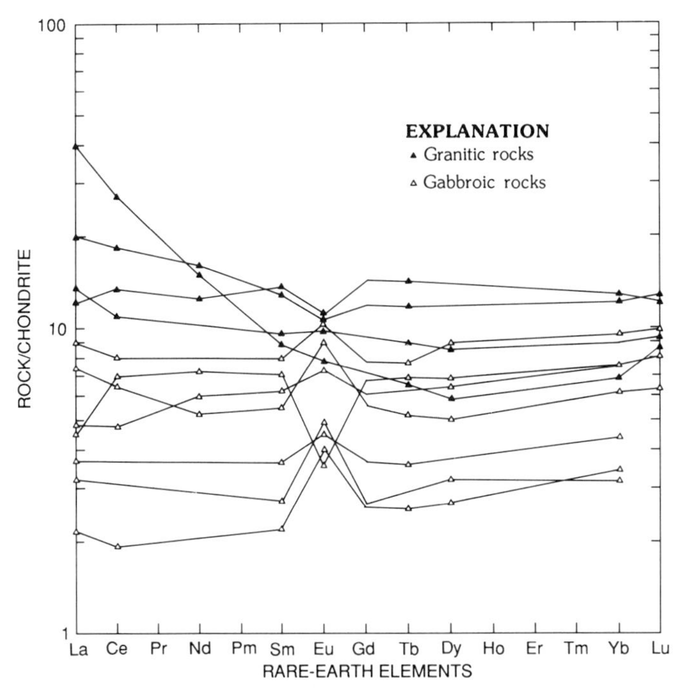

Figure 13. Chondrite-normalized rare-earth-element diagram for rocks of middle and late Eocene and early Oligocene age from the Yakobi Island area. Data source: Himmelberg and others (1987).

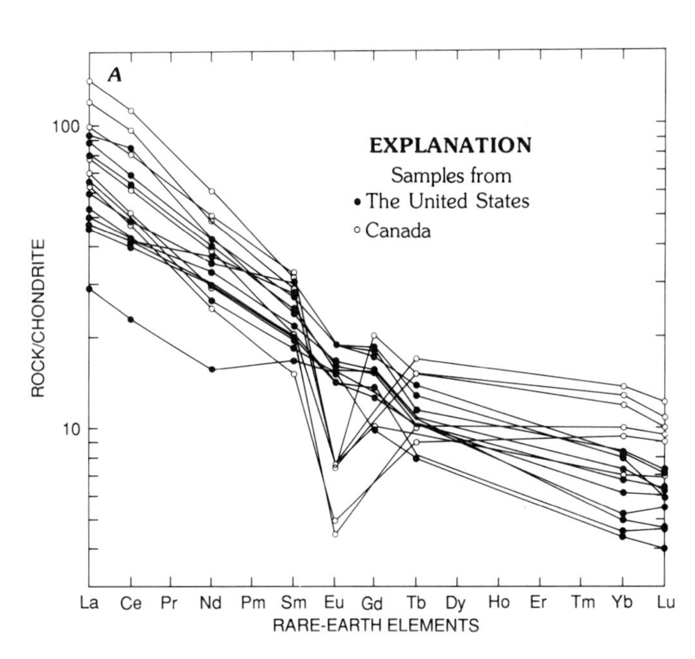

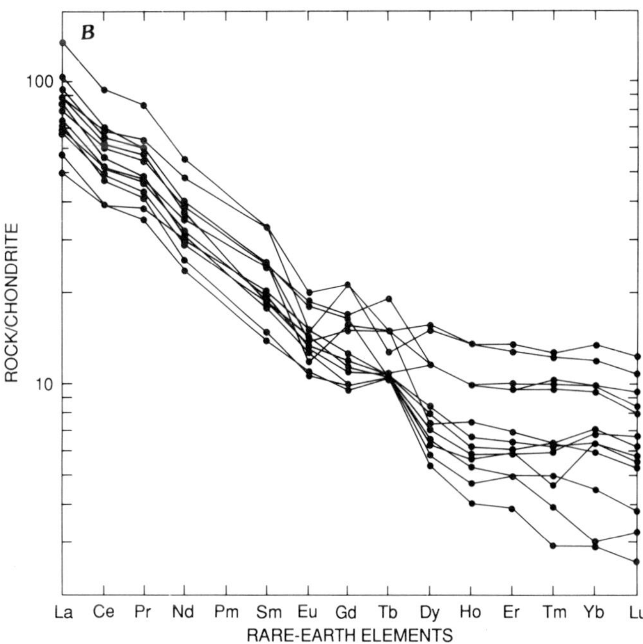

Figure 14. Chondrite-normalized rare-earth-element diagrams for granitic rocks of early and middle Eocene age. *A,* Haines-Skagway area; data source: Barker and others (1986). *B,* Juneau–Taku River area; data source: D. A. Brew and A. B. Ford (unpublished data, 1985).

Fairweather-Baranof belt

Plutons of the Fairweather-Baranof belt and the Glacier Bay region (Fig. 3, Tables 1 and 2) were emplaced in the time span of late Eocene to early Oligocene (45 to 35 Ma). The Fairweather-Baranof belt is parallel to the Coast Mountains belt and approximately 200 km to the southwest. Emplacement ages of 49 to 39 Ma indicate that the Fairweather-Baranof belt is younger than the Coast Mountains belt, although there is some overlap. Biotite-hornblende tonalite and granodiorite are the most common rock types in the southern part of the Fairweather-Baranof belt, and gabbronorite, pyroxenite, and other mafic and ultramafic rocks dominate the northern part. The granitic plutons are calc-alkalic, mostly peraluminous, and have silica contents that range from 60 to 73 percent. The belt occurs largely within the Chugach terrane (Silberling and others, this volume) and is interpreted to have formed as a result of the convergence and accretion of the Chugach terrane. Metamorphic mineral ages suggest that the convergence occurred in Late Cretaceous time, though before the time of the main deformation and metamorphism associated with the great tonalite sill belt.

In this interpretation, the Coast Mountains and the Fairweather-Baranof belts are not quite synchronous and are not directly related; either could have formed independently. The Coast Mountains belt is one result of the closure of the Gravina basin because of northeastward movement of the Alexander terrane, whereas the Fairweather-Baranof belt is one result of the

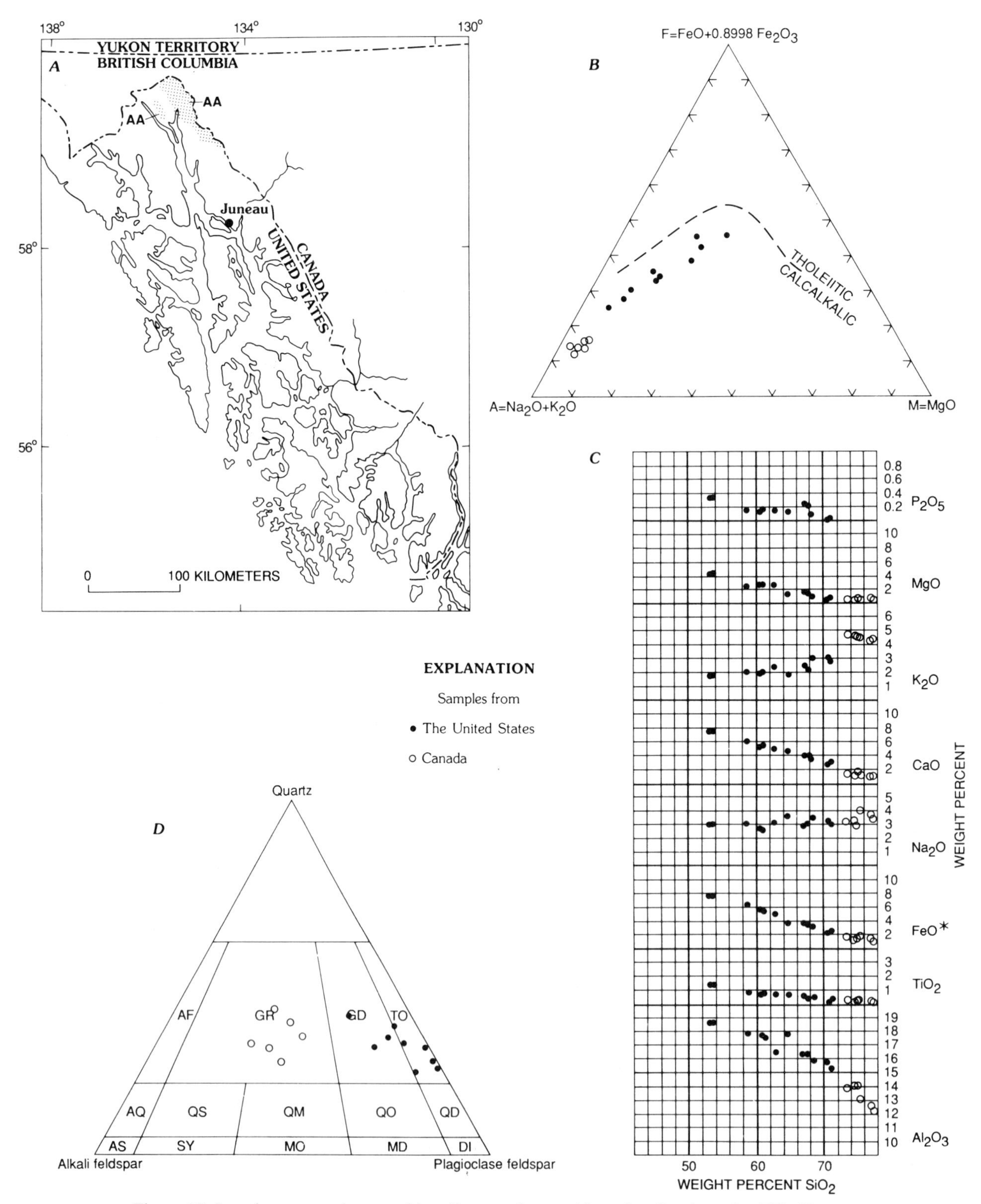

Figure 15. Location map and composition diagrams for granitic rocks of early and middle Eocene age. *A,* the Haines-Skagway area (AA). *B,* AFM diagram (Irvine and Baragar, 1971). *C,* Silica-variation diagram. *D,* General plutonic rock classification diagram (Streckeisen, 1973): AF, alkali-feldspar granite; AQ, alkali-feldspar quartz syenite; AS, alkali-feldspar syenite; DI, diorite, GR, granite; GD, granodiorite; MD, monzodiorite; MO, monzonite; QD, quartz diorite; QO, quartz monzodiorite; QM, quartz monzonite; QS, quartz syenite; SY, syenite; TO, tonalite. FeO* indicates total Fe as FeO. Data source: Barker and others (1986).

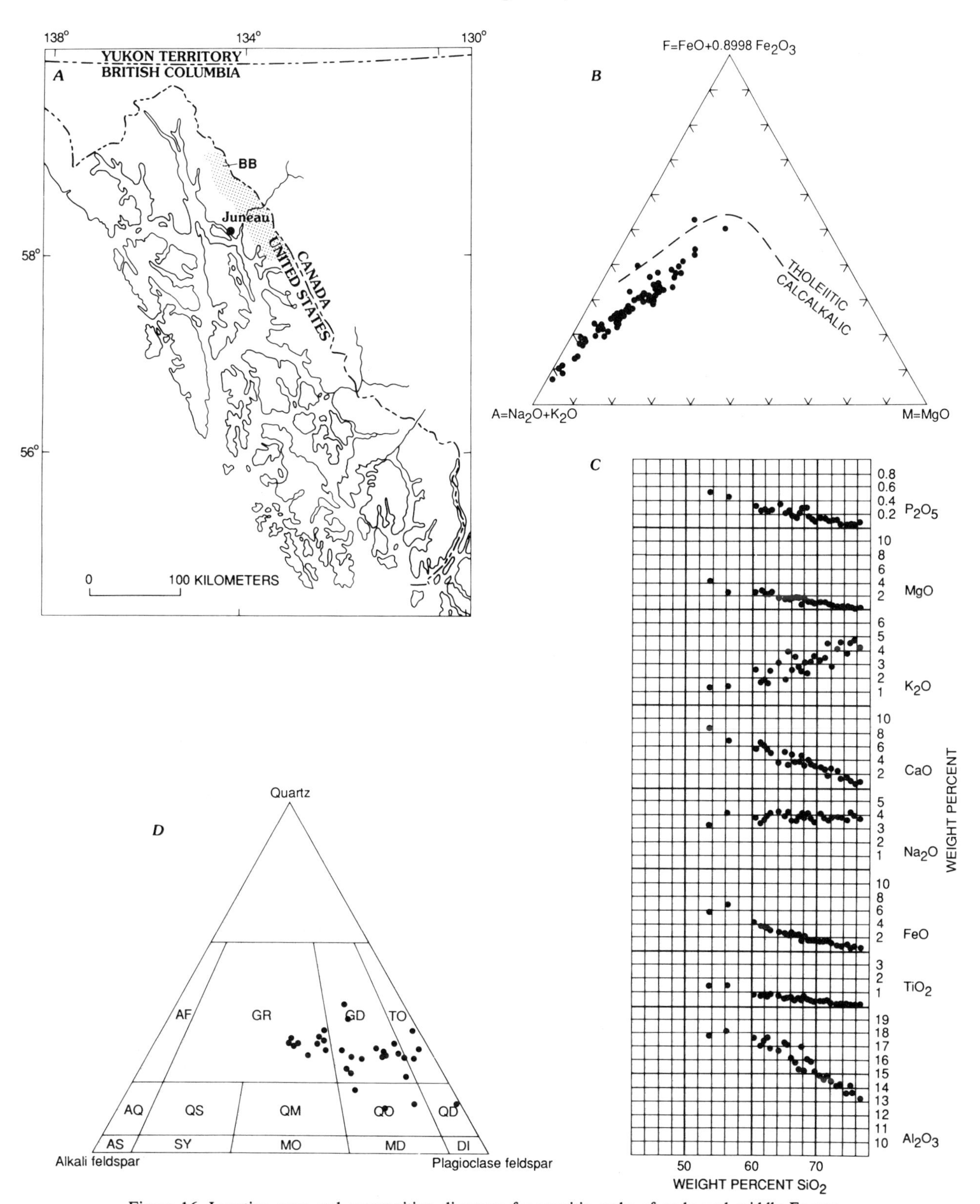

Figure 16. Location map and composition diagrams for granitic rocks of early and middle Eocene age. *A,* the Juneau-Taku River area (BB). *B,* AFM diagram (Irvine and Baragar, 1971). *C,* Silica-variation diagram. *D,* general plutonic rock classification diagram (Streckeisen, 1973): AF, Alkali-feldspar granite; AQ, alkali-feldspar quartz syenite; AS, alkali-feldspar syenite; DI, diorite; GR, granite; GD, granodiorite; MD, monzondiorite; MO, monzonite; QD, quartz diorite; QO, quartz monzodiorite; QM, quartz monzonite; QS, quartz syenite; SY, syenite; TO, tonalite. Data source: D. A. Brew and A. B. Ford (unpublished data, 1985).

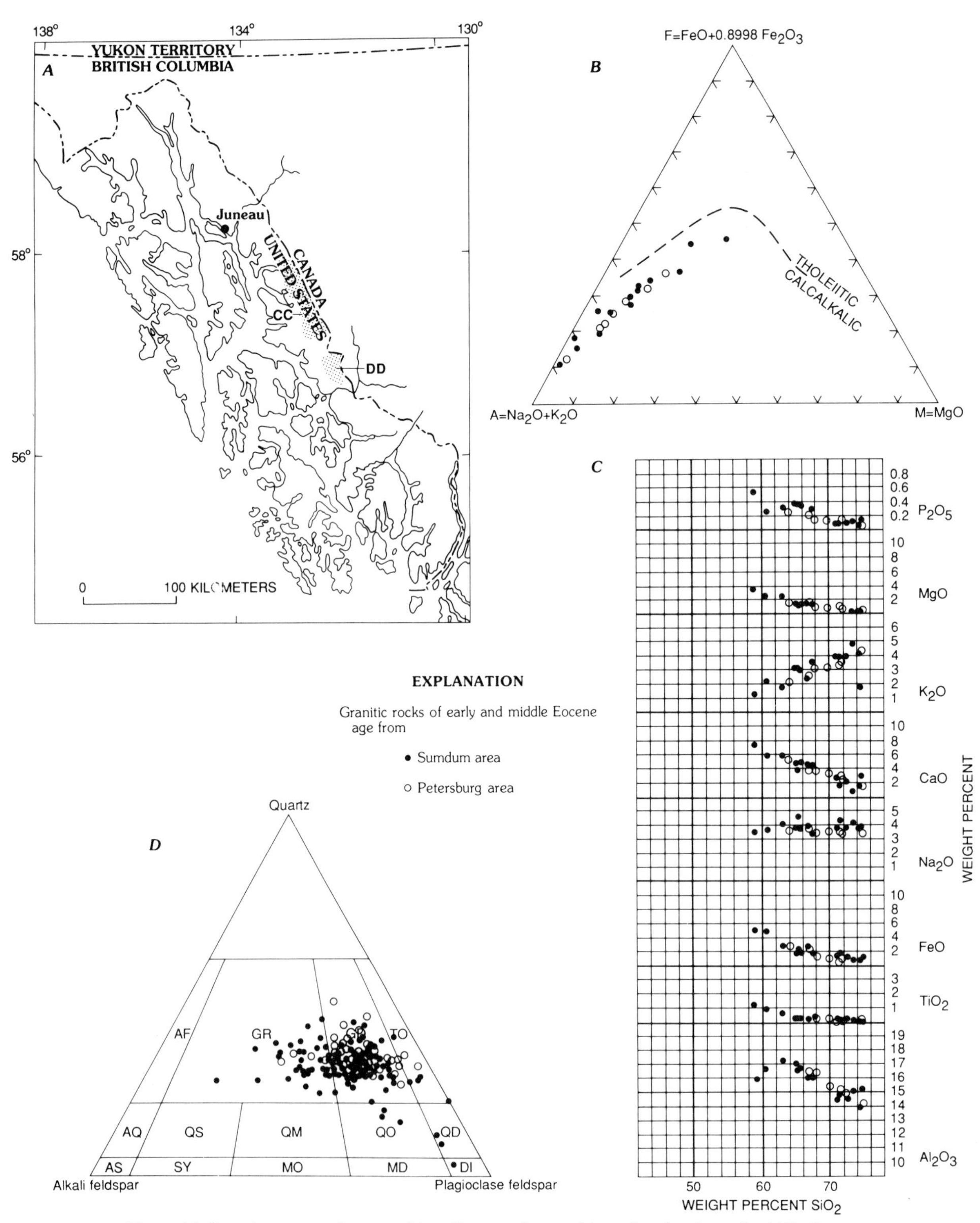

Figure 17. Location map and composition diagrams for granitic rocks of early and middle Eocene age. A, the Sumdum (CC) and Petersburg (DD) areas. *B,* AFM diagram (Irvine and Baragar, 1971). *C,*Silica-variation diagram. *D,* General plutonic rock classification diagram (Streckeisen, 1973): AF, alkali-feldspar granite; AQ, alkali-feldspar quartz syenite; AS, alkali-feldspar syenite; DI, diorite; GR, granite; GD, granodiorite; MD, monzondiorite; MO, monzonite; QD, quartz diorite; QO, quartz monzodiorite; QM, quartz monzonite; QS, quartz syenite; SY, syenite; TO, tonalite. Data sources: D. A. Brew and A. B. Ford (unpublished data, 1985) for Sumdum; Douglass and others (1989) for Petersburg.

accretion of the Chugach terrane to the west side of the Alexander terrane.

Glacier Bay region

The Glacier Bay region (Fig. 3, Tables 1 and 2) is, in contrast to the magmatic belts described, a northeast-trending, nearly rectangular area that slightly overlaps the northern part of the Fairweather-Baranof belt geographically and also in time, with ages ranging from about 42 to 30 Ma. The calc-alkalic plutons are dominantly nonfoliated to weakly foliated, metaluminous and moderately peraluminous biotite granite and alkali granite. Silica values range from 58 to 76 percent. This group of plutons is areally, compositionally, chemically, and structurally distinct from those in the Fairweather-Baranof belt, and their origin, though linked to the latter, must differ in some significant way. One possibility is that the plutons of the Glacier Bay region represent the silicic remnant of a magmatic system that produced the dominantly gabbroic plutons in the adjacent northern part of the Fairweather-Baranof belt and that the emplacement of the silicic portion was displaced to the northeast by the previously emplaced less fractionated mafic and ultramafic bodies. Another possibility is that they are an early manifestation of the younger (35 to 5 Ma) Tkope–Portland Peninsula belt, which itself is related to some obscure regime in the period after Chugach-terrane accretion and before transform faulting.

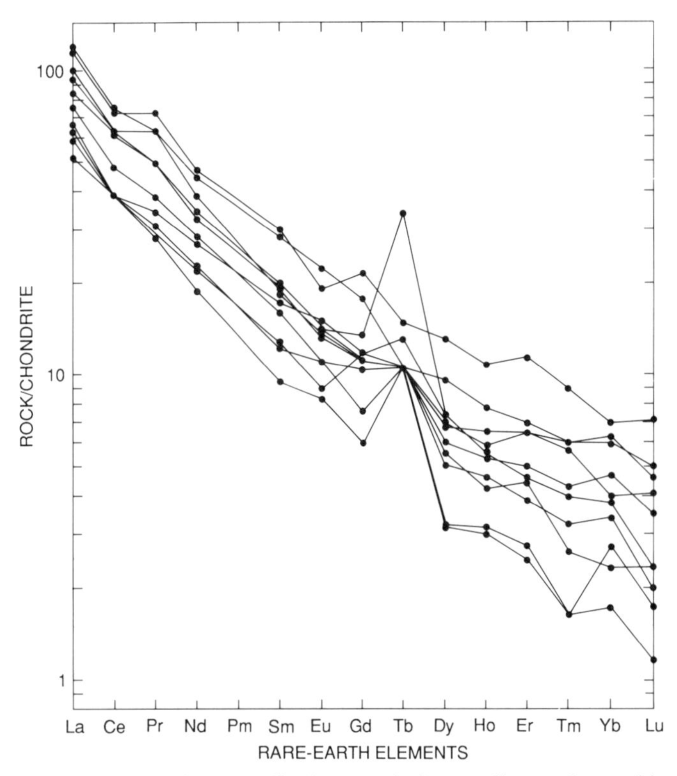

Figure 18. Chondrite-normalized rare-earth element diagram for granitic rocks of early and middle Eocene age from the Petersburg area. Data source: D. A. Brew (unpublished data, 1985).

Tkope–Portland Peninsula belt

The Tkope–Portland Peninsula belt, the Groundhog Basin–Cone Mountain area, and the southern southeastern Alaska dike swarm (Fig. 3, Tables 1 and 2) were emplaced within the time span of late Oligocene and Miocene (35 to 5 Ma). The three belts are clearly different from the collision-related belts just described; they each have distinct petrologic characteristics and represent different types of magma-generating events.

The Tkope–Portland Peninsula belt (Fig. 3, Tables 1 and 2) is the most prominent of the three belts or areas. It extends in a northwest-southeast direction for at least 560 km across all of southeastern Alaska, cutting across all tectonostratigraphic terranes except the Chugach terrane at an angle of about 15°. Both volcanic and plutonic rocks occur. The volcanic rocks are flows, tuff, and breccia of andesitic, basaltic, rhyolitic, and dacitic composition. All are calc-alkalic, and silica contents range from 47 to 77 percent, with a significant gap at 61 to 66 percent. Available age determinations indicate that the volcanics were erupted during the period from about 25 to 16 Ma. The granitics are both calc-alkalic and alkalic. Granite and granite porphyry are the most common rock types at the ends of the belt; most are moderately peraluminous. Alkali granite, granite, quartz syenite, and alkali quartz syenite are common types in the central part. Leucogabbro and gabbro occur locally. The plutonic rocks have a silica range of 49 to 77 percent with significant gaps at 54 to 56 and at 61 to 65 percent. The plutons were emplaced from 35 to 19 Ma: those at the northwest ends of the belt at about 28 to 24 Ma, those in the center at 24 to 19 Ma, and those at the southeast end at 30 Ma and 27 to 24 Ma.

The length and continuity of the Tkope–Portland Peninsula belt suggest that it could be the result of a significant collisional event of unusual orientation. However, no other evidence supporting such an origin has been preserved, and thus it is here considered unlikely. The composition of the plutons is unlike those in the other magmatic belts, probably because the magmas were generated at the base of or within the continental crust of the Alexander/Stikine terranes. The cause of the magmatic events is probably related to the change from convergence to oblique subduction to strike-slip movement between the Pacific and North American Plates, but the actual mechanism that caused the long belt to form is not clear. The axis of the belt coincides with the orientation of the tension planes that would be associated with the onset of differential strike-slip movement along the continental margin. The slight change in orientation of the belt near its southeast end could be related to differences in the thickness of the crust.

Groundhog Basin–Cone Mountain area

The Groundhog Basin–Cone Mountain area includes rhyolitic sills and biotite granite plugs (Fig. 3, Tables 1 and 2). Alkali

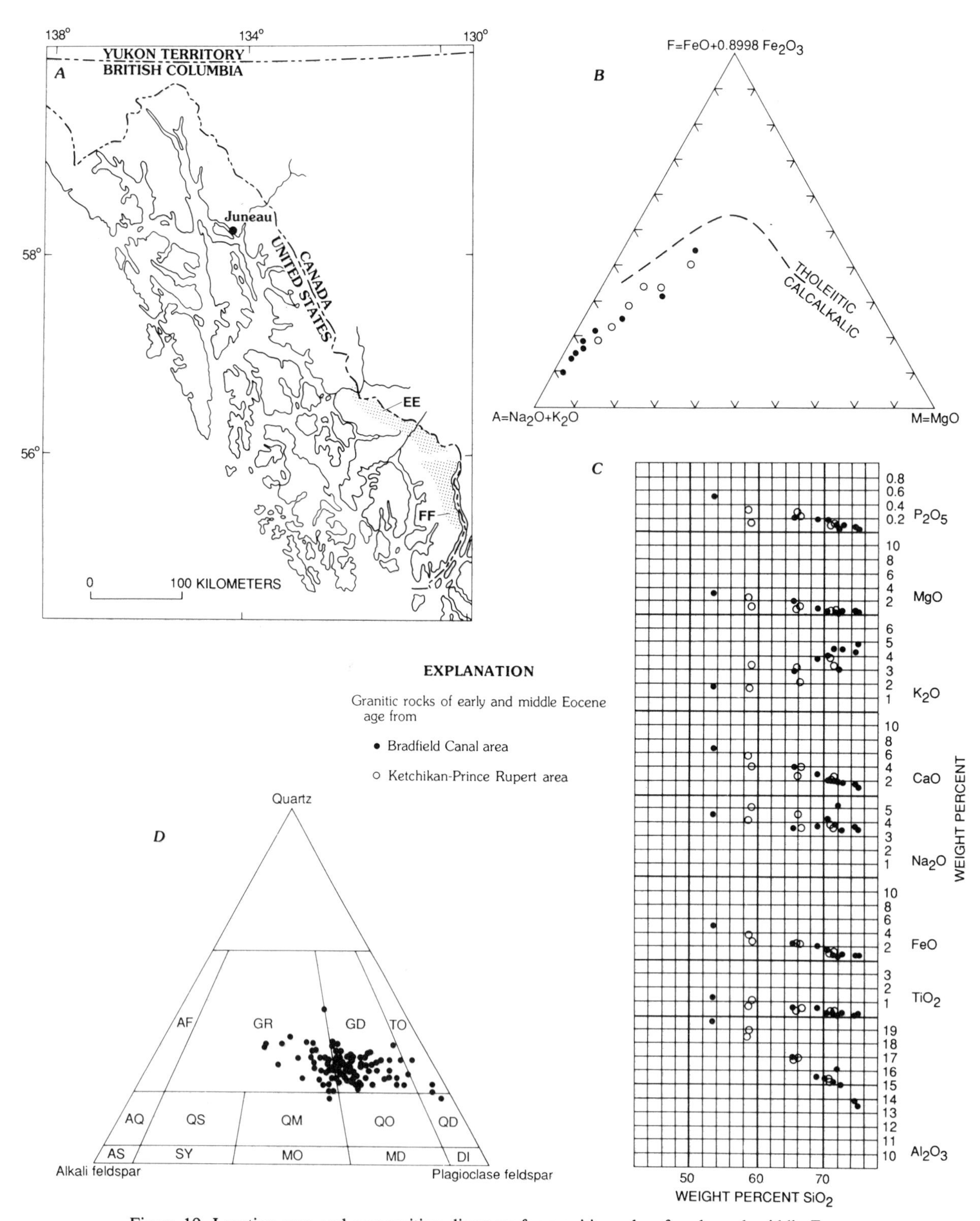

Figure 19. Location map and composition diagrams for granitic rocks of early and middle Eocene age. *A,* Bradfield Canal (EE) and Ketchikan (FF) areas. *B,* AFM diagram (Irvine and Baragar, 1971). *C,* Silica-variation diagram. *D,* General plutonic rock classification diagram (Streckeisen, 1973): AF, alkali-feldspar granite; AQ, alkali-feldspar quartz syenite; AS, alkali-feldspar syenite; DI, diorite; GR, granite; GD, granodiorite; MD, monzondiorite; MO, monzonite; QD, quartz diorite; QO, quartz monzodiorite; QM, quartz monzonite; QS, quartz syenite; SY, syenite; TO, tonalite. Data sources: Webster (1984) and Smith (1977) for Bradfield Canal area; Smith (1977) for Ketchikan–Prince Rupert area.

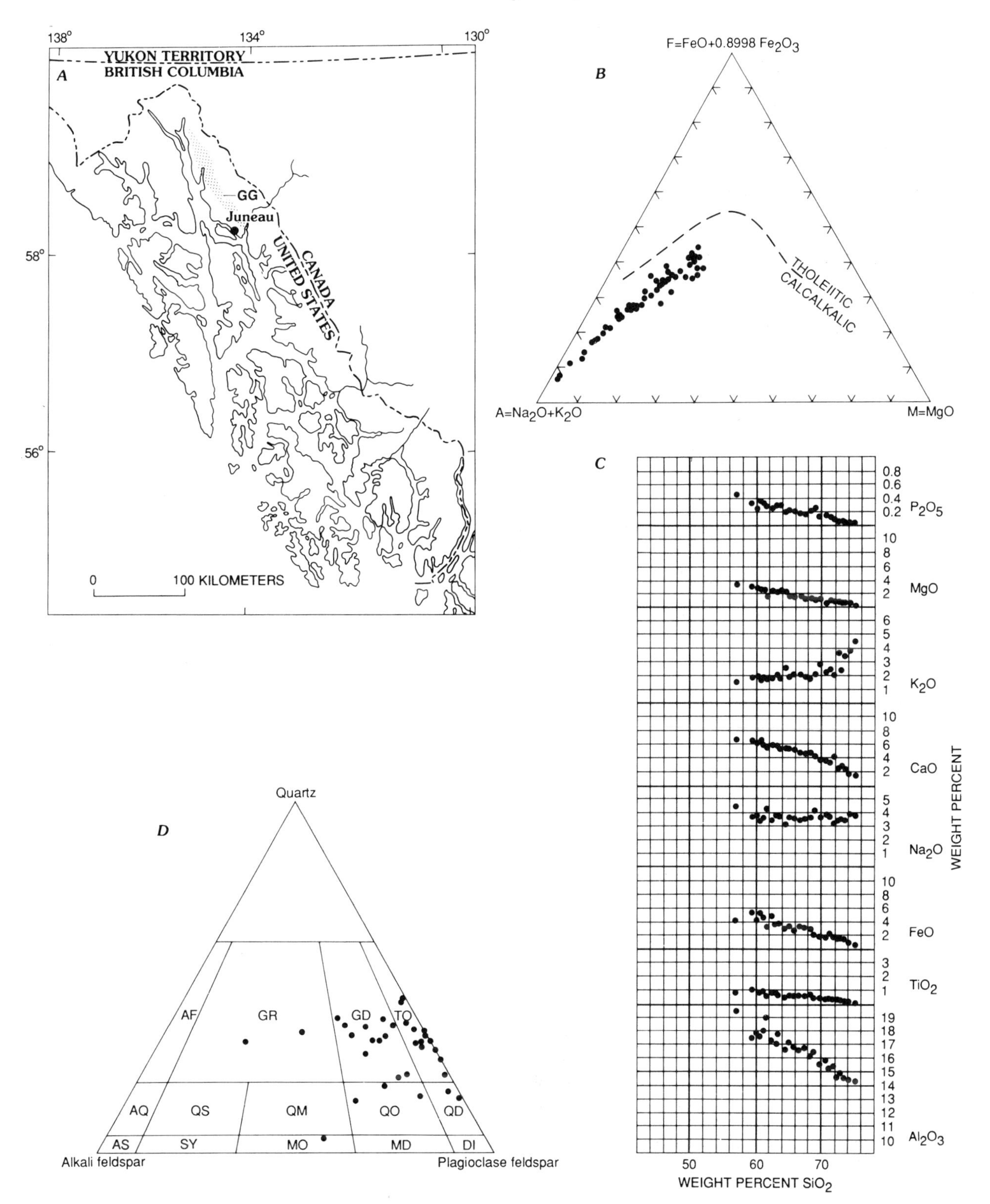

Figure 20. Location map and composition diagrams for granitic rocks of Paleocene age. *A,* the Juneau-Skagway area (GG). *B,* AFM diagram (Irvine and Baragar, 1971). *C,* Silica-variation diagram. *D,* General plutonic rock classification diagram (Streckeisen, 1973): AF, alkali-feldspar granite; AQ, alkali-feldspar quartz syenite; AS, alkali-feldspar syenite; DI, diorite; GR, granite; GD, granodiorite; MD, monzodiorite; MO, monzonite; QD, quartz diorite; QO, quartz monzodiorite; QM, quartz monzonite; QS, quartz syenite; SY, syenite; TO, tonalite. Data source: D. A. Brew and A. B. Ford (unpublished data, 1985).

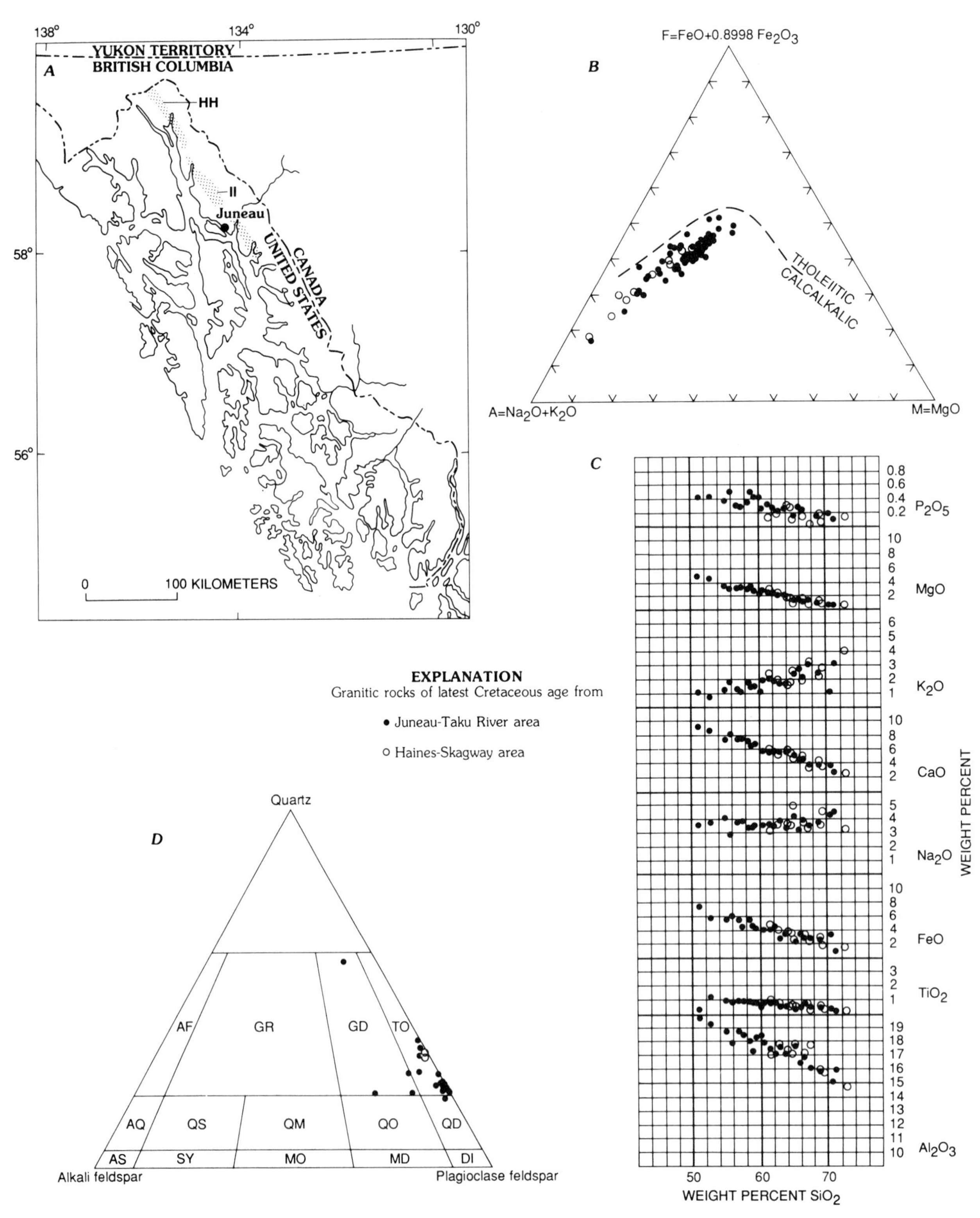

Figure 21. Location map and composition diagrams for granitic rocks of latest Cretaceous age. *A,* the Haines-Skagway (HH) and Juneau-Taku River (II) areas. *B,* AFM diagram (Irvine and Baragar, 1971). *C,* Silica-variation diagram. *D,* General plutonic rock classification diagram (Streickeisen, 1973): AF, alkali-feldspar granite; AQ, alkali-feldspar quartz syenite; AS, alkali-feldspar syenite; DI, diorite; GR, granite, GD, granodiorite; MD, monzodiorite; MO, monzonite; QD, quartz diorite; QO, quartz monzodiorite; QM, quartz monzonite; QS, quartz syenite; SY, syenite; TO, tonalite. Data sources: D. A. Brew and A. B. Ford (unpublished data, 1985) and Barker and others (1986).

granites may be present in the Cone Mountain area, but available data indicate that the rocks are calc-alkalic and moderately peraluminous and have a silica content from 74 to 76 percent. The granites were intruded at about 16 Ma, later than the rocks in the Tkope–Portland Peninsula belt. The plutons were intruded at a high crustal level under static conditions, but their relations to other belts and to possible localizing factors are obscure.

Southern southeastern Alaska dike swarm

The southern southeastern Alaska dike swarm (Fig. 3, Tables 1 and 2) consists mostly of lamprophyres that occupy a significant part of a northeast-trending belt about 100 km wide. Coeval granitic and volcanic rocks are also present. The swarm overlaps the southeastern end of the Tkope–Portland Peninsula belt, and at least some of the dikes are closely related to the plutons there. The lamprophyres are alkalic, and most are classified as alkali-olivine rocks. The non-lamprophyres are calc-alkalic and have a silica content ranging from 56 to 71 percent.

The age of intrusion of the lamprophyres is not well known; they cut plutons with ages of 27 to 24 Ma in the Tkope–Portland Peninsula belt but are not known to cut the plutons of the Grounding Basin area with ages of 17 to 14 Ma. Souther (1970) interprets them as the deeper expression of the dated Miocene alkalic volcanic fields to the northeast in British Columbia. Souther infers that those fields are localized in belts of large-scale crustal extension related to continental-margin transcurrent faulting; the dikes follow joints that are perpendicular to the foliation of the country rocks and resulted from the relaxation of the major stresses that affected the Coast Mountains belt in earlier Tertiary time.

Kruzof-Kupreanof and Behm Canal–Rudyerd Bay areas

Two areas or belts of Pliocene and Quaternary volcanic rocks are shown on Figure 3 and described in Tables 1 and 2. One, the Kruzof-Kupreanof area of Holocene rocks, appears on Figure 3 as two segments separated at the Chatham Strait fault because it postdates the major offset that was removed in constructing the palinspastic base for the figure. This area consists of two widely spaced volcanic fields of similar age and chemical composition. Those fields—Edgecumbe and southern Kupreanof—contain tholeiitic basalt, and the Edgecumbe field also has calc-alkalic younger flows and pyroclastic rocks. Most, but not all, of the flows are interpreted (J. R. Riehle and D. A. Brew, unpublished data) to be postglacial. Together, the two fields define an east-west–trending area similar in orientation to east-west Holocene volcanic belts in the west-central British Columbia region, which Souther (1970) relates to large-scale crustal extension.

The other area of Pliocene and Quaternary volcanic rocks is the Behm Canal–Rudyerd Bay volcanic field, most of which occurs within the area covered by the southern southeastern Alaska dike swarm. The small Blue River–Unuk River volcanic

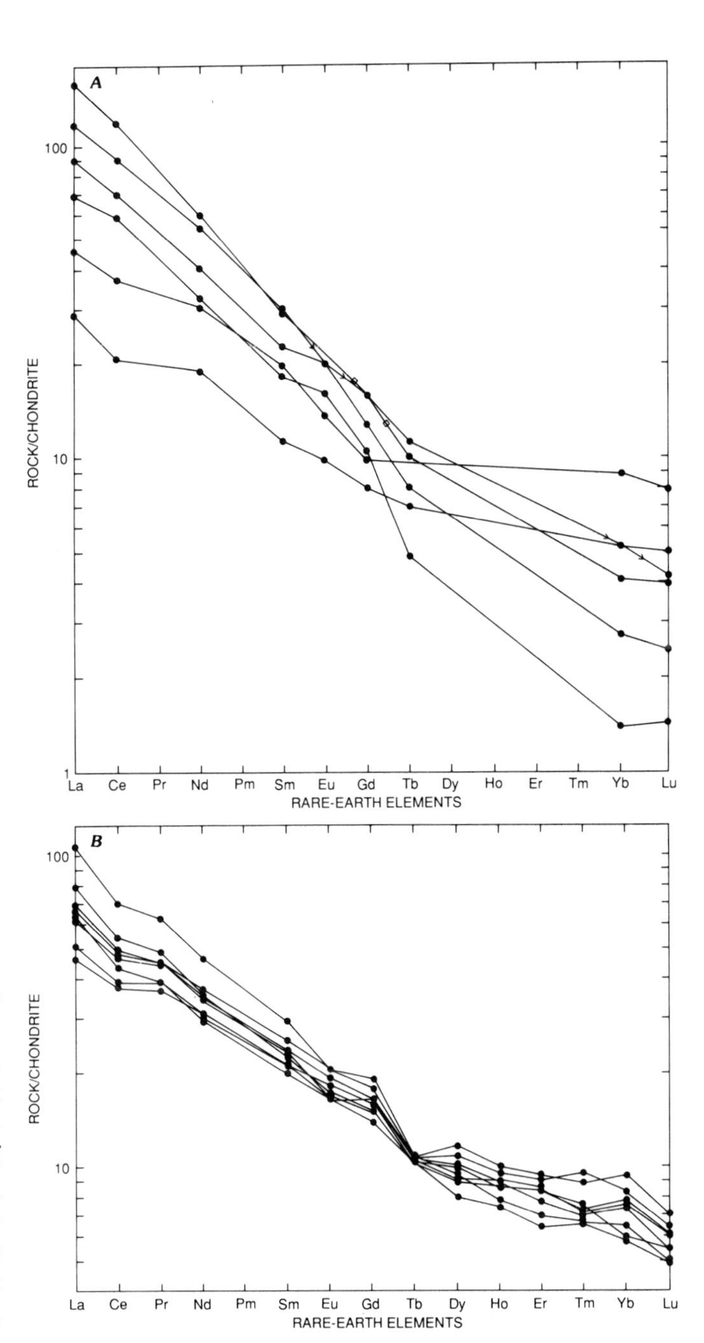

Figure 22. Chondrite-normalized rare-earth element diagrams for granitic rocks of latest Cretaceous age. *A*, Haines-Skagway area (data from Barker and others, 1986). *B*, Petersburg area (D. A. Brew, unpublished data, 1985).

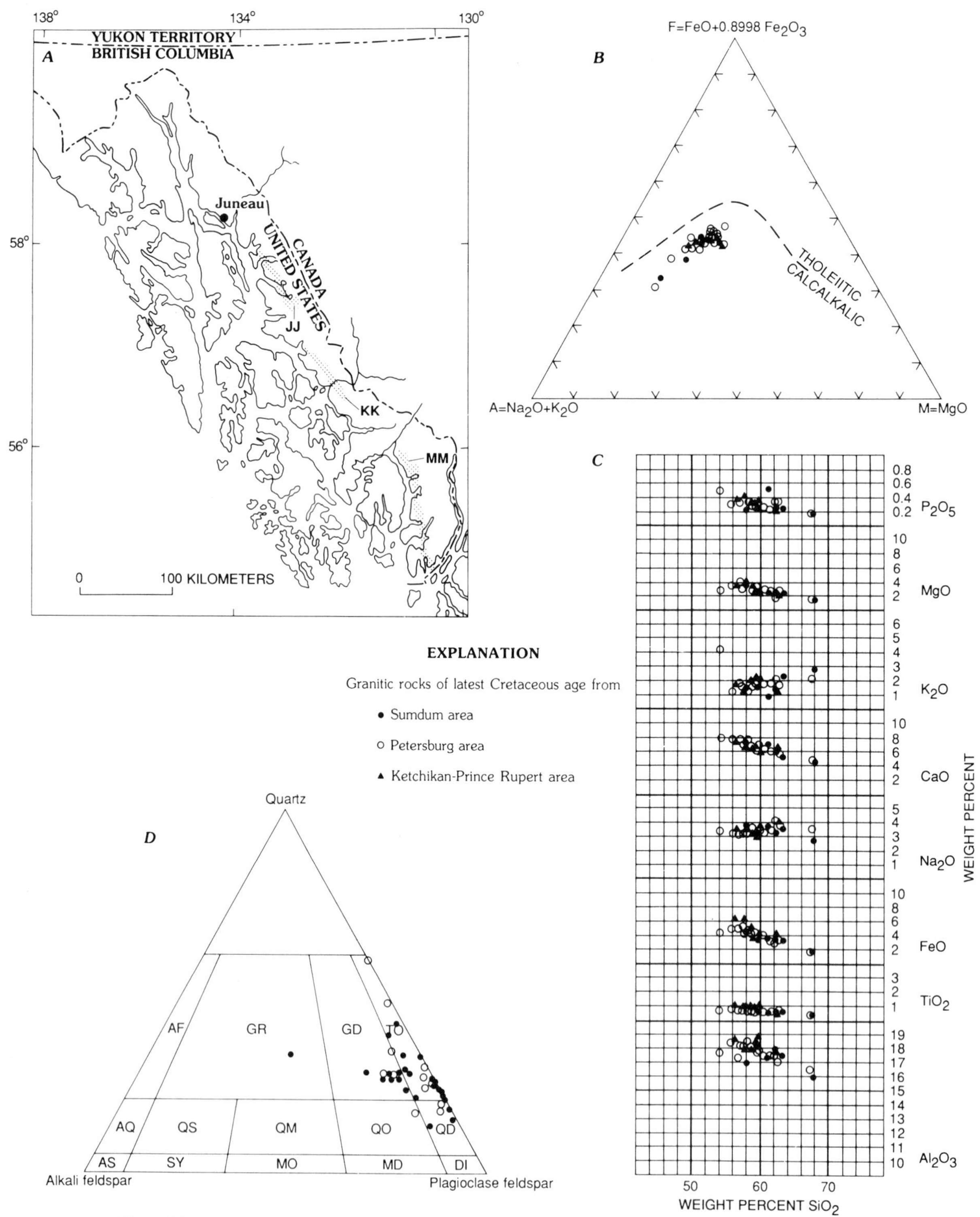

Figure 23. Location map and composition diagrams from granitic rocks of latest Cretaceous age. *A,* the Sumdum (JJ), Petersburg (KK), and Ketchikan–Prince Rupert (MM) areas. *B,* AFM diagram (Irvine and Baragar, 1971). *C,* Silica-variation diagram. *D,* General plutonic rock classification diagram (Streckeisen, 1973): AF, alkali-feldspar granite; AQ, alkali-feldspar quartz syenite; AS, alkali-feldspar quartz syenite; AS, alkali-feldspar syenite; DI, diorite; GR, granite; GD, granodiorite; MD, monzodiorite; MO, monzonite; QD, quartz diorite; QO, quartz monzodiorite; QM, quartz monzonite; QS, quartz syenite; SY, syenite; TO, tonalite. Data sources: D. A. Brew and A. B. Ford (unpublished data, 1985) and Brew and Grybeck (1984) for Sumdum; Douglass and others (1988) for Petersburg; and Smith (1977) for Ketchikan–Prince Rupert.

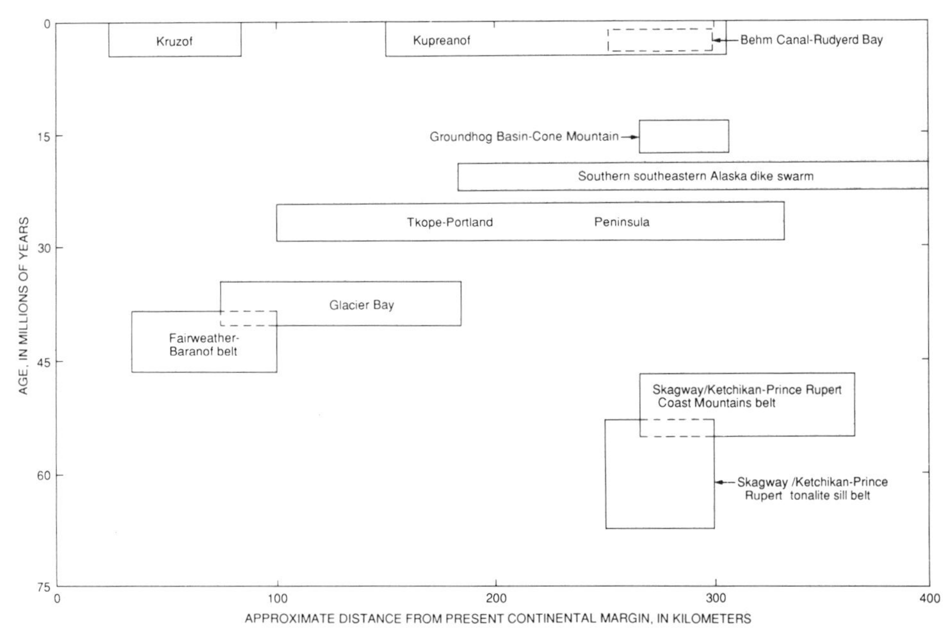

Figure 24. Time-space diagram summarizing latest Mesozoic and Cenozoic magmatism in southeastern Alaska.

field to the north (Fig. 1) is here considered an outlier of the Behm Canal–Rudyerd Bay field. Both fields contain alkali olivine basalts and other alkalic rocks that resemble the alkali to peralkaline basalts in the Mount Edziza field, which is located about 100 km to the north in Canada. Souther and Armstrong (1966), Souther (1970), and Souther and others (1984) relate the north-south orientation of the Mount Edziza field to large-scale crustal extension. It is likely that the Behm Canal–Rudyerd Bay field is an outlier of that large field.

SUMMARY

The Cenozoic volcanic and plutonic rocks of southeastern Alaska record a progression of events that are related to the tectonics of the northeastern Pacific margin. The progression started with the collisional/convergent events related to the accretion of the Alexander terrane to the Stikine terrane; the progression continued with the events related to the accretion and subduction of the Chugach terrane on the west side of the Alexander. These events along the northeastern Pacific margin occurred in Late Cretaceous and early Tertiary time; they were followed by events related first to oblique subduction and then to transition to dominantly transcurrent movements. The progression ended with magmatic events localized along extensional zones that may be related either to present-day right-lateral crustal displacements in the northeastern Pacific region or to residual stresses that originated in the late stages of convergence. Figure 24 summarizes these time and space relations.

REFERENCES CITED

Arth, J. G., Barker, F., Stern, T. W., and Zmuda, C., 1986, The Coast batholith near Ketchikan, southeast Alaska; Geochronology and geochemistry: Geological Society of America Abstracts with Programs, v. 18, p. 529.

Barker, F., and Arth, J. G., 1984, Preliminary results, Central Gneiss Complex of the Coast Range batholith, southeastern Alaska; The roots of a high-K, calc-alkaline arc?: Physics of the Earth and Planetary Interiors, v. 35, p. 191–198.

Barker, F., Arth, J. G., and Stern, T. W., 1986, Evolution of the coast batholith along the Skagway Traverse, Alaska and British Columbia: American Mineralogist, v. 71, no. 3-4, p. 632–643.

Berg, H. C., Jones, D. L., and Richter, D. H., 1972, Gravina-Nutzotin belt; Tectonic significance of an upper Mesozoic sedimentary and volcanic sequence in southern and southeastern Alaska, *in* Geological Survey research 1972: U.S. Geological Survey Professional Paper 800-D, p. D1–D24.

Berg, H. C., Elliott, R. L., Koch, R. D., Carten, R. B., and Wahl, F. A., 1976, Preliminary geologic map of the Craig D-1 and parts of the Craig C-1 and D-2 Quadrangles, Alaska: U.S. Geological Survey Open-File Report 76–430, 1 sheet, scale 1:63,360.

Berg, H. C., Elliott, R. L., and Koch, R. D., 1988, Geologic map of the Ketchikan and Prince Rupert Quadrangles, southeastern Alaska: U.S. Geological Survey Miscellaneous Investigations Series Map I-1807, 27 p., 1 sheet, scale 1:250,000.

Bingler, E. C., Trexler, D. T., Kemp, W. R., and Bonham, H. P., Jr., 1976, PETCAL; A *Basic* language computer program for petrologic calculations: Nevada Bureau of Mines and Geology Report 28, 27 p.

Brew, D. A., 1968, The role of volcanism in post-Carboniferous tectonics of southeastern Alaska and nearby regions, North America, *in* Proceedings, 23rd International Geological Congress, Prague 1968: Prague, Academia, p. 107–121.

Brew, D. A., and Ford, A. B., 1983, Comment *on* 'Tectonic accretion and the origin of the two major metamorphic and plutonic welts in the Canadian Cordillera': Geology, v. 11, p. 427–429.

——, 1985, Southeastern Alaska coincident zone, *in* Bartsch-Winkler, S., ed., The United States Geological Survey in Alaska; Accomplishments during 1984: U.S. Geological Survey Circular 967, p. 82–86.

——, 1986, Preliminary reconnaissance geologic map of the Juneau, Taku River, Atlin, and part of the Skagway Quadrangles, southeastern Alaska: U.S. Geological Survey Open-File Report 85–395, 23 p., scale 1:250,000.

Brew, D. A., and Grybeck, D., 1984, Geology of the Tracy Arm-Fords Terror Wilderness Study Area and vicinity, Alaska, *in* Mineral resources of the Tracy Arm-Fords Terror Wilderness Study Area and vicinity, Alaska: U.S. Geological Survey Bulletin 1525-A, p. 19-52, scale 1:125,000.

Brew, D. A., and Morrell, R. P., 1980, Preliminary map of intrusive rocks in southeastern Alaska: U.S. Geological Survey Miscellaneous Field Studies Map MF-1048, 1 sheet, scale 1:1,000,000.

——, 1983, Intrusive rocks and plutonic belts of southeastern Alaska, U.S.A., *in* Roddick, J. A., ed., Circum-Pacific plutonic terranes: Geological Society of America Memoir 159, p. 171–193.

Brew, D. A., Muffler, L.J.P., and Loney, R. A., 1969, Reconnaissance geology of the Mount Edgecumbe volcanic field, Kruzof Island, southeastern Alaska: U.S. Geological Survey Professional Paper 650-D, p. D1–D18.

Brew, D. A., and 7 others, 1978, Mineral resources of the Glacier Bay National Monument Wilderness Study Area, Alaska: U.S. Geological Survey Open-File Report 78–494, 692 p., scale 1:125,000.

Brew, D. A., and 5 others, 1979, The Tertiary Kuiu-Etolin volcanic-plutonic belt, southeastern Alaska, *in* Johnson, K. M., and Williams, J. R., eds., The United States Geological Survey in Alaska; Accomplishments during 1978: U.S. Geological Survey Circular 804-B, p. B129–B130.

Brew, D. A., Ovenshine, A. T., Karl, S. M., and Hunt, S. J., 1984, Preliminary reconnaissance geologic map of the Petersburg and parts of the Port Alexander and Sumdum Quadrangles, southeastern Alaska: U.S. Geological Survey Open-File Report 84–405, 43 p., scale 1:250,000.

Brew, D. A., Karl, S. M., and Tobey, E. F., 1985, Re-interpretation of the age of the Kuiu-Etolin belt volcanic rocks, Kupreanof Island, southeastern Alaska, *in* Bartsch-Winkler, S., and Reed, K. M., eds., The United States Geological Survey in Alaska; Accomplishments during 1983: U.S. Geological Survey Circular 945, p. 86–88.

Callahan, J. E., 1970, Geologic reconnaissance of a possible powersite at Takatz Creek, southeastern Alaska: U.S. Geological Survey Bulletin 1211-D, 18 p.

Campbell, R. B., and Dodds, C. J., 1983, Tatshenshini River map area: Geological Survey of Canada Open-File Report 926, 114 p.

Christie, R. L., 1957, Geology of the Bennett, Cassiar district, British Columbia: Canada Department of Mines and Technical Surveys Map 19-1957, 1 sheet, scale 1:253,440.

——, 1959, Geology of the plutonic rocks of the Coast Mountains in the vicinity of Bennett, British Columbia [Ph.D. thesis]: Toronto, Canada, University of Toronto, 182 p.

Christopher, P. A., and Carter, N. C., 1976, Metallogeny and metallogenic epochs for porphyry mineral deposits in the Canadian Cordillera: Canadian Institute of Mineralogy Special Volume 15, p. 64–71.

Douglass, S. L., Webster, J. H., Burrell, P. D., Lanphere, M. L., and Brew, D. A., 1989, Major element chemistry, radiometric ages, and locations of samples from the Petersburg and parts of the Port Alexander and Sumdum Quadrangles, southeastern Alaska: U.S. Geological Survey Open-File Report 89-527.

Doyle, M. C., 1983, A petrographic study of dike rocks in the Portland Peninsula area, southeastern Alaska and British Columbia [M.S. thesis]: Bryn Mawr, Pennsylvania, Bryn Mawr College, 29 p.

Eakins, G. R., 1975, Uranium investigations in southeastern Alaska: Alaska Division of Geological and Geophysical Surveys Geologic Report 44, 62 p.

Eberlein, G. D., and Churkin, M., Jr., 1970, Tlevak basalt, west coast of Prince of Wales Island, southeastern Alaska, *in* Cohee, G. V., Bates, R. G., and Wright, W. B., eds., Changes in stratigraphic nomenclature by the U.S. Geological Survey, 1968: U.S. Geological Survey Bulletin 1294-A, p. A25–A28.

Eberlein, G. D., Churkin, M., Jr., Carter, C., Berg, H. C., and Ovenshine, A. T., 1983, Geology of the Craig Quadrangle, Alaska: U.S. Geological Survey Open-File Report 83–91, 52 p., scale 1:250,000.

Elliott, R. L., Smith, J. G., and Hudson, T., 1976, Upper Tertiary high-level plutons of the Smeaton Bay area, southeastern Alaska: U.S. Geological Survey Open-File Report 76–507, 16 p.

Elliott, R. L., Koch, R. D., and Robinson, S. W., 1981, Age of basalt flows in the Blue River Valley, Bradfield Canal Quadrangle, *in* Albert, N.R.D., and Hudson, T., eds., The United States Geological Survey in Alaska; Accomplishments during 1979: U.S. Geological Survey Circular 823-B, p. B115–B116.

Forbes, R. B., and Engels, J. C., 1970, K^{40}/Ar^{40} age relations of the Coast Range batholith and related rocks of the Juneau Icefield area, Alaska: Geological Society of America Bulletin, v. 81, p. 579–584.

Ford, A. B., and Brew, D. A., 1981, Orthogneiss of Mount Juneau; An early phase of Coast Mountains plutonism involved in Barrovian regional metamorphism near Juneau, *in* Albert, N.R.D., and Hudson, T. L., eds., The United States Geological Survey in Alaska; Accomplishments during 1979: U.S. Geological Survey Circu;lar 823-B, p. B99–B102.

Fukuhara, C. R., 1986, Descriptions of plutons in the western part of the Juneau and parts of the adjacent Skagway Quadrangles, southeastern Alaska: U.S. Geological Survey Open-File Report 86–393, 81 p., scale 1:250,000.

Gehrels, G. E., Brew, D. A., and Saleeby, J. B., 1984, Progress report on U/Pb (zircon) geochronologic studies in the Coast plutonic-metamorphic complex east of Juneau, southeastern Alaska, *in* Reed, K. M., and Bartsch-Winkler, S., eds., The United States Geological Survey in Alaska; Accomplishments during 1982: U.S. Geological Survey Circular 939, p. 100–102.

Himmelberg, G. R., and Loney, R. A., 1981, Petrology of the ultramafic and gabbroic rocks of the Brady Glacier nickel-copper deposit, Fairweather Range, southeastern Alaska: U.S. Geological Survey Professional Paper 1195, 26 p.

Himmelberg, G. R., Loney, R. A., and Nabelek, P. I., 1987, Petrogenesis of gabbronorite at Yakobi and northwest Chichagof Islands, Alaska: Geological Society of America Bulletin, v. 98, p. 265–279.

Hudson, T., Plafker, G., and Lanphere, M. A., 1977, Intrusive rocks of the Yakutat–St. Elias area, south-central Alaska: U.S. Geological Survey Journal of Research, v. 5, no. 2, p. 155–172.

Hudson, T., Smith, J. G., and Elliott, R. L., 1979, Petrology, composition, and age of intrusive rocks associated with the Quartz Hill molybdenite deposit, southeastern Alaska: Canadian Journal of Earth Sciences, v. 16, no. 9, p. 1805–1822.

Hunt, S. J., 1984, Preliminary study of a zoned leucocratic-granite body on central Etolin Island, southeastern Alaska, *in* Coonrad, W. L., and Elliott, R. L., eds., The United States Geological Survey in Alaska; Accomplishments during 1981: U.S. Geological Survey Circular 868, p. 128–131.

Hutchison, W. W., 1982, Geology of the Prince Rupert–Skeena map area, British Columbia: Geological Survey of Canada Memoir 394, 116 p., scale 1:250,000.

Irvine, T. N., and Baragar, W. R., 1971, A guide to the chemical classification of the common volcanic rocks: Canadian Journal of Earth Sciences, v. 8, no. 5, p. 523–548.

Jacobson, B., 1979, Geochronology and petrology of the Tkope River batholith in St. Elias Mountains, northwestern British Columbia [B.Sc. thesis]: Vancouver, University of British Columbia, 47 p.

Johnson, B. R., and Karl, S. M., 1985, Geologic map of western Chichagof and Yakobi Islands, southeastern Alaska: U.S. Geological Survey Miscellaneous Investigations Series Map I-1506, 15 p., scale 1:125,000.

Karl, S. M., and Brew, D. A., 1984, Migmatites of the Coast plutonic-metamorphic complex, southeastern Alaska, *in* Reed, K. M., and Bartsch-Winkler, S., eds., The United States Geological Survey in Alaska; Accomplishments during 1982: U.S. Geological Survey Circular 939, p. 108–111.

Koch, R. D., and Elliott, R. L., 1981, Map showing distribution and abundance of gold and silver in geochemical samples from the Bradfield Canal Quadrangle, southeastern Alaska: U.S. Geological Survey Open-File Report 81–728-C, 2 sheets, scale 1:250,000.

——, 1984, Late Oligocene gabbro near Ketchikan, southeastern Alaska, *in* Coonrad, W. L., and Elliott, R. L., eds., The United States Geological Survey in Alaska; Accomplishments during 1981: U.S. Geological Survey Circular 868, p. 126–128.

Kosco, D. G., 1981, The Mount Edgecumbe volcanic field, Alaska; An example of tholeiitic and calc-alkaline volcanism: Journal of Geology, v. 89, no. 4, p. 459–477.

Lambert, M. B., 1974, The Bennett Lake cauldron subsidence complex, British Columbia and Yukon Territory: Geological Survey of Canada Bulletin 227, 213 p., scale 1:25,000.

Lathram, E. H., Pomeroy, J. S., Berg, H. C., and Loney, R. A., 1965, Reconnaissance geology of Admiralty Island, Alaska: U.S. Geological Survey Bulletin 1181-R, 48 p.

Loney, R. A., 1964, Stratigraphy and petrography of the Pybus-Gambier area, Admiralty Island, Alaska: U.S. Geological Survey Bulletin 1178, 103 p.

Loney, R. A., and Himmelberg, G. R., 1983, Structure and petrology of the La Perouse gabbro intrusion, Fairweather Range, southeastern Alaska: Journal of Petrology, v. 24, part 4, p. 377–423.

Loney, R. A., Brew, D. A., and Lanphere, M. A., 1967, Post-Paleozoic radiometric ages and their relevance to fault movements, northern southeastern Alaska: Geological Society of America Bulletin, v. 78, p. 511–526.

Loney, R. A., Brew, D. A., Muffler, L.J.P., and Pomeroy, J. S., 1975, Reconnaissance geology of Chichagof, Baranof, and Kruzof Islands, southeastern Alaska: U.S. Geological Survey Professional Paper 792, 105 p.

MacDonald, G. A., and Katsura, T., 1964, Chemical composition of Hawaiian lavas: Journal of Petrology, v. 5, no. 1, p. 82–133.

MacKevett, E. M., Jr., Brew, D. A., Hawley, C. C., Huff, L. C., and Smith, J. G., 1971, Mineral resources of Glacier Bay National Monument, Alaska: U.S. Geological Survey Professional Paper 632, 90 p.

MacKevett, E. M., Jr., Robertson, E. C., and Winkler, G. R., 1974, Geology of the Skagway B-3 and B-4 Quadrangles, southeastern Alaska: U.S. Geological Survey Professional Paper 832, 33 p.

Magaritz, M., and Taylor, H. P., Jr., 1976, Isotopic evidence for meteoric-hydrothermal alteration of plutonic igneous rocks in the Yakutat Bay and Skagway areas, Alaska: Earth and Planetary Science Letters, v. 30, p. 179–190.

Miller, D. J., 1961, Geology of the Lituya district, Gulf of Alaska Tertiary province: U.S. Geological Survey Open-File Report 61–100, 1 sheet.

Miyashiro, A., 1974, Volcanic rock series in island arcs and active continental margins: American Journal of Science, v. 274, no. 4, p. 321–355.

Monger, J.W.H., Price, R. A., Tempelman-Kluit, D. J., 1982, Tectonic accretion and the origin of two major metamorphic and plutonic welts in the Canadian Cordillera: Geology, v. 10, p. 70–75.

——, 1983, Reply *to* Comment *on* 'Tectonic accretion and the origin of two major metamorphic and plutonic welts in the Canadian Cordillera': Geology, v. 11, p. 428–429.

Myers, J. D., and Marsh, B. D., 1981, Geology and petrogenesis of the Edgecumbe volcanic field, southeastern Alaska; The interaction of basalt and sialic crust: Contributions to Mineralogy and Petrology, v. 77, p. 272–287.

Myers, J. D., Sinha, A. K., and Marsh, B. D., 1984, Assimilation of crustal material by basaltic magma; Strontium isotopic and trace element data from the Edgecumbe volcanic field, southeastern Alaska: Journal of Petrology, v. 25, no. 1, p. 1–26.

Ouderkirk, K. A., 1982, A petrographical study of dike and volcanic rocks in the Prince Rupert, British Columbia and Ketchikan area [Senior thesis]: Bryn Mawr, Pennsylvania, Bryn Mawr College, 31 p.

Plafker, G., 1971, Petroleum and coal, *in* MacKevett, E. M., Jr., Brew, D. A., Hawley, C. C., Huff, L. C., and Smith, J. G., eds., Mineral resources of Glacier Bay National Monument, Alaska: U.S. Geological Survey Professional Paper 632, p. 85–89.

Plafker, G., and MacKevett, E. M., 1970, Mafic and ultramafic rocks from a layered pluton at Mount Fairweather, Alaska: U.S. Geological Survey Professional Paper 700-B, p. 21–26.

Plafker, G., Jones, D. L., and Pessagno, E. A., Jr., 1977, A Cretaceous accretionary flysch and mélange terrane along the Gulf of Alaska margin, *in* Blean, K. M., ed., The United States Geological Survey in Alaska; Accomplishments during 1976: U.S. Geological Survey Circular 751-B, p. B41–B43.

Redman, E., Retherford, R. M., and Hickok, B. D., 1984, Geology and geochemistry of the Skagway B-2 Quadrangle, southeastern Alaska: Alaska Division of Geological and Geophysical Surveys Report of Investigations 84–31, 34 p.

Riehle, J. R., and Brew, D. A., 1984, Explosive latest Pleistocene(?) and Holocene activity of the Mount Edgecumbe volcanic field, Alaska, *in* Reed, K. M., and Bartsch-Winkler, S., eds., The United States Geological Survey in Alaska; Accomplishments during 1982: U.S. Geological Survey Circular 939, p. 111–115.

Shand, S. J., 1951, Eruptive rocks: New York, J. Wiley and Sons, Inc., 488 p.

Smith, J. G., 1973, A Tertiary lamprophyre dike province in southeastern Alaska: Canadian Journal of Earth Sciences, v. 10, no. 3, p. 408–420.

——, 1977, Geology of the Ketchikan D-1 and Bradfield Canal A-1 Quadrangles, southeastern Alaska: U.S. Geological Survey Bulletin 1425, 49 p.

Smith, J. G., and Diggles, M. F., 1981, Potassium-argon determinations in Ketchikan and Prince Rupert Quadrangles, southeastern Alaska: U.S. Geological Survey Open-File Report 78–73aN, 15 p., scale 1:250,000.

Smith, J. G., Elliott, R. L., Berg, H. C., and Wiggins, B. D., 1977, Map showing general geology and locations of chemically and radiometrically analyzed samples in parts of the Ketchikan, Bradfield Canal, and Prince Rupert Quadrangles, southeastern Alaska: U.S. Geological Survey Miscellaneous Field Studies Map MF-825, 2 sheets, scale 1:250,000.

Souther, J. G., 1970, Volcanism and its relationship to recent crustal movements in the Canadian Cordillera: Canadian Journal of Earth Sciences, v. 7, pt. 2 of no. 2, p. 553–568.

——, 1971, Geology and ore deposits of the Tulsequah map area, British Columbia: Geological Survey of Canada Memoir 362, 84 p., scale 1:250,000.

Souther, J. G., and Armstrong, J. E, 1966, North-central belt of the Cordillera of British Columbia: Canadian Institute of Mining and Metallurgy Special Volume 8, p. 171–184.

Souther, J. G., Armstrong, R. L., and Harakal, J., 1984, Chronology of the peralkaline, late Cenozoic Mount Edziza volcanic complex, northern British Columbia, Canada: Geological Society of America Bulletin, v. 95, p. 337–349.

Streckeisen, A. L., 1973, Plutonic rocks; Classification and nomenclature recommended by the I.U.G.S. subcommission on the systematics of igneous rocks: Geotimes, v. 18, no. 10, p. 26–30.

Wanek, A. A., and Callahan, J. E., 1969, Geology of proposed powersites at Deer Lake and Kasnyku Lake, Baranof Island, southeastern Alaska: U.S. Geological Survey Bulletin 1211-C, 25 p., scale 1:24,000.

——, 1971, Geologic reconnaissance of a proposed powersite at Lake Grace, Revillagigedo Island, southeastern Alaska: U.S. Geological Survey Bulletin 1211-E, 24 p.

Webster, J. H., 1984, Preliminary report on a large granitic body in the Coast Mountains, northeast Petersburg Quadrangle, southeastern Alaska, *in* Reed, K. M., and Bartsch-Winkler, S., eds., The United States Geological Survey in Alaska; Accomplishments during 1982: U.S. Geological Survey Circular 939, p. 116–118.

Wilson, F. H., and Shew, N., 1982, Apparent episodicity of magmatic activity deduced from radiometric age determinations, *in* Coonrad, W. L., ed., The United States Geological Survey in Alaska; Accomplishments during 1980: U.S. Geological Survey Circular 844, p. 13–15.

Woodcock, J. R., and Carter, N. C., 1976, Geology and geochemistry of the Alice Arm molybdenum deposits: Canadian Institute of Mineralogy Special Volume 15, p. 462–475.

MANUSCRIPT ACCEPTED BY THE SOCIETY MAY 21, 1990

NOTES ADDED IN PROOF

Most of the recent work that contributes to the subject of this chapter has been in the Coast Mountains and adjacent areas. That work consists mainly of the age-dating by G. E. Gehrels and coworkers, the detailed petrologic studies in the Juneau area by J. L. Drinkwater and coworkers, and the geochemical comparison of the Skagway and Ketchikan transects by F. Barker and J. G. Arth. In addition, D. A. Brew and coworkers are continuing studies of the Eocene plutons in the outer islands of southeastern Alaska and J. R. Riehle and coworkers have completed a series of reports on the Mount Edgecumbe volcanic field. References covering these and other studies follow.

ADDITIONAL REFERENCES

Barker, F., and Arth, J. G., 1990, Two traverses across the Coast batholith, southeastern Alaska, *in* Anderson, J. L., ed., The nature and origin of Cordilleran magmatism: Geological Society of America Memoir 174, p. 395–405.

Brew, D. A., Hammarstrom, J. M., Himmelberg, G. H., Wooden, J. L., Loney, R. A., and Karl, S. M., 1991, Crawfish Inlet pluton, Baranof Island, southeastern Alaska—A north-tilted Eocene body or an untilted enigma [abs.]: Geological Society of America, Abstracts with Programs, v. 23, no. 2, p. 8.

Brew, D. A., Karl, S. M., Loney, R. A., Ford, A. B., Himmelberg, G. R., and Hammarstrom, J. M., 1992, Jurassic and Cretaceous batholiths of southeastern Alaska: how many arcs? [abs.]: Geological Society of America, Abstracts with Programs, v. 24, no. 5, p. 9.

Burton, R., Brew, D. A., and Gray, S., 1992, Mount Edgecumbe volcanic field, Kruzof Island, southeastern Alaska (video): Rain Country Series Program 806, Juneau, Alaska, KTOO-TV Videos, 11 minutes.

Drinkwater, J. L., Brew, D. A., and Ford, A. B., 1989, Petrographic and chemical characteristics of the variably deformed Speel River pluton, south of Juneau, southeastern Alaska, *in* Galloway, J. P., and Dover, J. H., eds., Geologic studies in Alaska by the U.S. Geological Survey in 1988: U.S. Geological Survey Bulletin 1903, p. 104–112.

——, 1990, Petrographic and chemical data for the large Mesozoic and Cenozoic plutonic sills east of Juneau, southeastern Alaska: U.S. Geological Survey Bulletin 1918, 47 p.

Drinkwater, J. L., Ford, A. B., and Drew, D. A., 1992a, Magnetic susceptibilities and iron content of plutonic rocks across the Coast plutonic-metamorphic complex near Juneau, southeastern Alaska, *in* Bradley, D. W., and Dusel-Bacon, C., eds., Geologic studies in Alaska by the U.S. Geological Survey in 1991: U.S. Geological Survey Bulletin 2041, p. 125–139.

——, 1992b, Magnetic susceptibility measurements and sample locations of granitic rocks from along a transect of the Coast Mountains near Juneau, Alaska: U.S. Geological Survey Open-File Report 92-724, 22 p.

Ford, A. B., and Brew, D. A., 1987, The Wright Glacier volcanic plug and dike swarm, southeastern Alaska, *in* Hamilton, T. D., and Galloway, J. P., eds., Geologic studies in Alaska by the U.S. Geological Survey during 1986: U.S. Geological Survey Circular 998, p. 116–118.

Gehrels, G. E., McClelland, W. C., Samson, S. D., Patchett, P. J., and Brew, D. A., 1991, U-Pb geochronology and tectonic significance of Late Cretaceous–early Tertiary plutons in the northern Coast Mountains batholith: Canadian Journal of Earth Sciences, v. 28, no. 6, p. 899–911.

Gilbert, W. G., Clough, A. H., Burns, L. E., Kline, J. T., Redman, E., and Fogels, E. J., 1990, Reconnaissance geology and geochemistry of the northeast Skagway quadrangle, Alaska: Alaska Division of Geological and Geophysical Surveys, Report of Investigations 90-5, scale 1:125,000, 1 sheet.

Riehle, J. R., Brew, D. A., and Lanphere, M. A., 1989, Geologic map of the Mount Edgecumbe volcanic field, Kruzof Island, southeastern, Alaska: U.S. Geological Survey Map I-1983, scale 1:63,360.

Riehle, J. R., Champion, D. E., Brew, D. A., and Lanphere, M. A., 1992a, Pyroclastic deposits of the Mount Edgecumbe volcanic field, Alaska: Evolution of a stratified magma chamber: Journal of Volcanology and Geothermal Research, v. 53, p. 117–143.

Riehle, J. R., Mann, D. H., Peteet, D. M., Engstrom, D. M., Brew, D. A., and Meyer, C. E., 1992b, The Mount Edgecumbe tephra deposits, a marker horizon in southeastern Alaska near the Pleistocene-Holocene boundary: Quaternary Research, v. 37, p. 183–202.

Printed in U.S.A.

The Geology of North America
Vol. G-1, The Geology of Alaska
The Geological Society of America, 1994

Chapter 20

Crustal melting events in Alaska

Travis L. Hudson
ARCO Alaska, Inc., P.O. Box 100360, Anchorage, Alaska 99510

INTRODUCTION

Development of granitic magmas by melting of continental crust is documented by field studies in migmatite terranes (e.g., Kenah and Hollister, 1983; Johannes and Gupta, 1982), petrologic and mineralogic studies (e.g., White and Chappell, 1977; Speer, 1981; Wright and Haxel, 1982), geochemical and isotopic studies (e.g., Ben Othman and others, 1984; Kistler and others, 1981; Arth, this volume), and experimental studies (e.g., Clemens and Wall, 1981; Wyllie, 1983a). Although the processes that lead to melting of continental crust are complex (e.g., Leake, 1983), where present, these events are obviously an important element in the geologic evolution of a region. One expected result of such an event is the regional emplacement of felsic granitic rocks.

In this chapter, any granitic rocks whose composition can be inferred to be predominantly derived from sialic crustal materials—even juvenile crustal materials as is the case for some accretionary parts of southern Alaska—are considered to be the products of crustal melting. However, generally accepted criteria for recognizing such granitic rocks, such as high initial $^{87}Sr/^{86}Sr$, peraluminous composition, restite minerals or inclusions, and relations to metamorphism and migmatization, are generally not available for most granitic rocks in Alaska. For this reason, and because in some settings the distinctions between mantle and crustal processes are blurred (e.g., Leeman, 1983; Wyllie, 1983b; Ben Othman and others, 1984), a simplistic approach is taken here. The two principal criteria used to distinguish Alaskan granitic rocks whose origin may have been dominated by crustal melting processes are: (1) the compositional variation is dominantly restricted to granodiorite and granite, and (2) the granodiorite and granite represent temporally distinct, regionally distributed, plutonic events. These criteria can be supported by additional evidence, such as field relations, high initial $^{87}Sr/^{86}Sr$ ratios, and the presence of inherited zircons, in some cases. Based on the general criteria, widespread crustal melting in Alaska took place primarily in the Devonian, Cretaceous, and Paleogene. The distribution (all physiographic place names follow Wahrhaftig, this volume), age (all geologic time subdivisions follow Palmer, 1983), general character, and regional setting of the granitic rocks developed during each of these melting events is summarized separately below. These relations, combined with other aspects of the regional geology, enable the tectonic setting in which the crustal melting took place to be inferred. The inferred settings are different for the Devonian and Cretaceous events as well as for the two subdivisions of the Paleogene event.

DEVONIAN ORTHOGNEISSES

Granitic orthogneisses (Fig. 1) are exposed on Seward Peninsula (Till, 1983; Till and Dumoulin, this volume), in the Brooks Range (Nelson and Grybeck, 1980; Silberman and others, 1979a; Dillon and others, 1980; Sable, 1977; Moore and others, this volume), in the Ruby geanticline (Dover, 1984; W. W. Patton, oral communication, 1984; Patton and others, this volume, Chapter 7), and in the Yukon-Tanana Upland (Aleinikoff and others, 1981; Dusel-Bacon and Aleinikoff, 1985; Foster and others, this volume). Similar orthogneiss bodies are present in the Yukon Territory (Tempelman-Kluit and Wanless, 1980; Mortensen, 1983). The emplacement ages of the Alaska plutons as indicated by whole-rock Rb/Sr and zircon U/Pb ages, range from 344 ± 3 to 381 ± 6 Ma. They are primarily Devonian, although some in the Yukon-Tanana Upland are Mississippian (Dusel-Bacon and Aleinikoff, 1985).

The orthogneisses form subequant to irregular bodies (up to 35 km across) that are emplaced into metasedimentary country rocks of Precambrian(?) and early Paleozoic age. The orthogneisses have been variably metamorphosed to greenschist (Adams, 1983) and amphibolite facies (Dusel-Bacon and Aleinikoff, 1985; Dusel-Bacon, this volume), and the strong foliation developed within them is commonly concordant with the regional structural trends in their country rocks. Nevertheless, in places, sharp discordant contacts, apophyses and dikes into country rocks, and the presence of marginal hornfels and skarn, document the original intrusive character of these bodies (Adams, 1983; Sable, 1977). Original granitic features such as texture, schlieren, xenoliths, jointing, and aplitic and pegmatitic dikes are also preserved in some locations.

These metaplutonic rocks are medium- to coarse-grained, strongly foliate to gneissic, equigranular to blastoporphyritic, biotite (± muscovite) metagranite, and less abundant metagranodiorite and siliceous metatonalite. The accessory minerals now present are commonly zircon, apatite, garnet, sphene, ilmenite, and epidote-group minerals. The major-oxide data available for

Hudson, T. L., 1994, Crustal melting events in Alaska, *in* Plafker, G., and Berg, H. C., eds., The Geology of Alaska: Boulder, Colorado, Geological Society of America, The Geology of North America, v. G-1.

these rocks (Fig. 2) confirm their dominantly felsic character: silica varies from 67 to 78 percent, the combined Fe_2O_3 + FeO + MgO + CaO content varies from 0.7 to 7.9 percent, total alkalies vary from 6.1 to 8.5 percent, and the average K_2O/Na_2O ratio is about 1.6. The major-oxide variations primarily reflect decreasing mafic oxide contents with increasing silica; K_2O increases slightly with increasing silica. Rb/Sr isotopic data for the Arrigetch Peaks and Mount Igikpak plutons in the south-central Brooks Range indicate an initial $^{87}Sr/^{86}Sr$ ratio of 0.714 ± 0.003 (Silberman and others, 1979a), and similar data for augen gneiss bodies in the Yukon-Tanana Upland give initial $^{87}Sr/^{86}Sr$ ratios of 0.719 and 0.728 (Aleinikoff and others, 1986; Dusel-Bacon and Aleinikoff, 1985). Inherited zircons are locally present both in the Brooks Range and Yukon-Tanana Upland orthogneisses (Aleinikoff and others, 1981, 1986; J. T. Dillon, oral communication, 1984).

Thus, the available data support the interpretation that the Devonian orthogneisses of interior and northern Alaska represent anatectic melts of sialic crust. The regional relations listed below are important for interpreting the tectonic setting in which Devonian crustal melting took place.

1. The anatectic magmas intruded sialic crust upon which marine (meta)sedimentary sequences of early Paleozoic and, in places, possibly older age were deposited. Where the Yukon-Tanana Upland extends into Canada, these metasedimentary sequences probably represent the early Paleozoic or older continental margin of North America (Tempelman-Kluit, 1979).

2. Marine mafic and felsic metavolcanic rocks that have been documented as, or inferred to be of, Devonian age (Smith

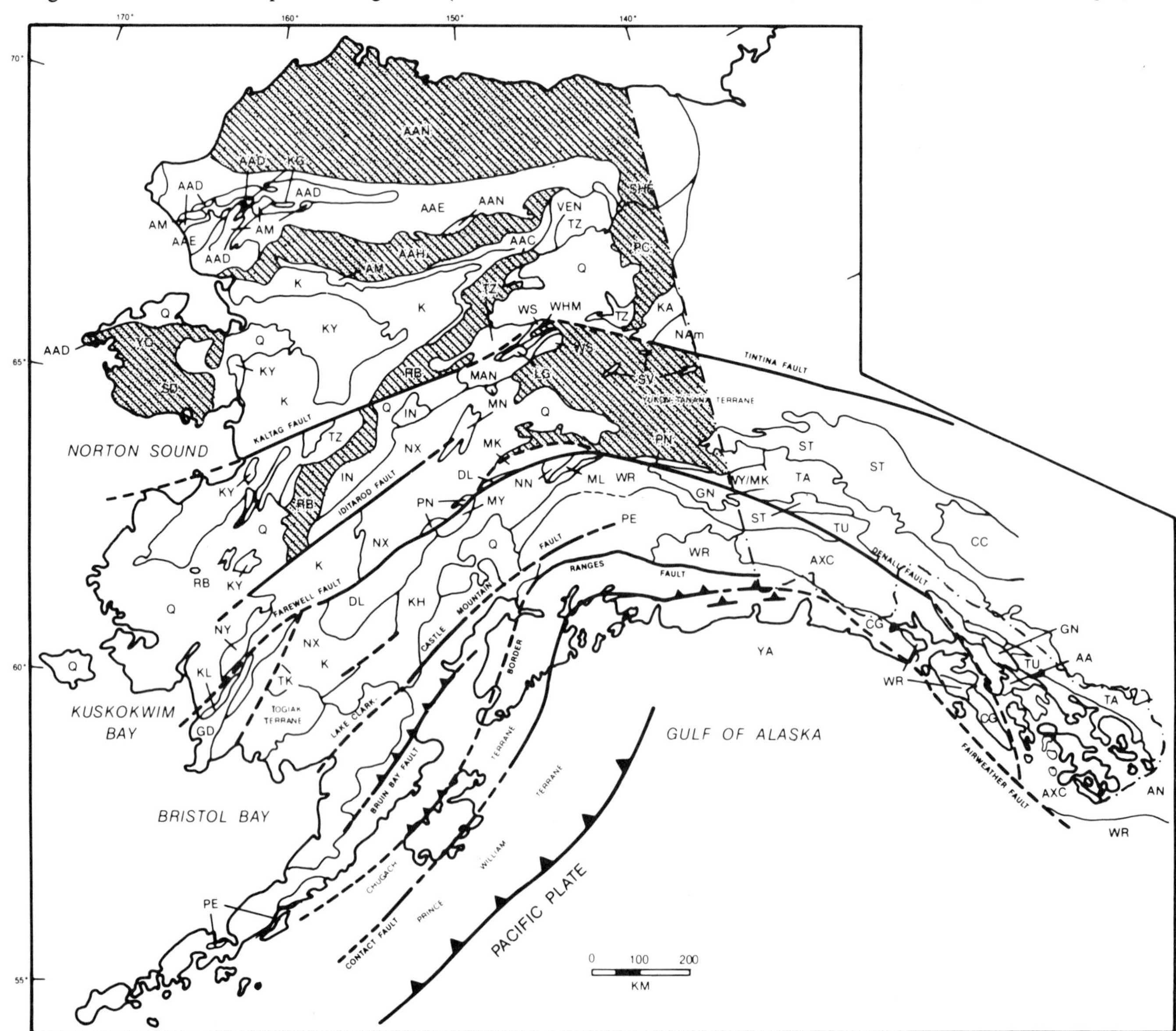

Figure 1. Map showing tectonostratigraphic terranes in Alaska (boundaries and symbols follow Jones and others, 1987; also see Silberling and others, this volume); shaded areas show the terranes that are known or suspected to contain Devonian and earliest Mississippian orthogneiss.

and others, 1978; Dillon and others, 1980; Aleinikoff, 1984; Aleinikoff and others, 1986; Duke and others, 1984; Hitzman, 1984) also developed in the regions where the Devonian orthogneisses were emplaced.

3. The metamorphic history of the regions containing the Devonian orthogneisses is not well known, but regional dynamothermal metamorphism appears to have taken place primarily in the Mesozoic and not the Devonian (Till, 1983; Forbes and others, 1984a, 1984b; Dillon and others, 1980; Dover, 1984; Cushing and others, 1984).

Previous interpretations of the tectonic setting for Devonian magmatism in Alaska fall into two categories: development of a convergent margin or island magmatic arc (Dillon and others, 1980; Nokleberg and others, 1983; Lange and Nokleberg, 1984; Dusel-Bacon and Aleinikoff, 1985) or development of extensional regions and marginal basins (Schmidt, 1984; Hitzman, 1984; Duke and others, 1984; Gemuts and others, 1983). In light of the regional relations listed above, and the observation that plutons developed in convergence-related magmatic arcs are more varied in composition than granodiorite and granite (e.g., Hudson, 1983), the interpretation preferred here is that the Devonian granites were emplaced in an extensional and/or transcurrent tectonic setting. If so, the extensional thinning of the continental crust, accompanied by emplacement of mafic magmas, could have led to high heat flow and anatectic melting.

CRETACEOUS GRANITIC ROCKS

The character of Cretaceous magmatic rocks in Alaska (Miller, this volume) is varied, but several suites of Cretaceous granodiorite and granite plutons (Fig. 3) may be the products of crustal melting. These include plutons on Seward Peninsula (Miller and Bunker, 1976; Hudson, 1977; Hudson and Arth, 1983), in the Ruby geanticline (Patton and Miller, 1973; Chapman and others, 1982; Silberman and others, 1979b; Puchner, 1984), and the Yukon-Tanana Upland and nearby parts of the Alaska Range (Forbes, 1982; Foster and others, 1978; Foster, 1976; Foster and others, 1979; Luthy and others, 1981; Foster and others, 1976; Foster, 1970; Richter and others, 1975; Richter, 1976; Blum, 1985). Plutons similar to those in the Yukon-Tanana Upland are present in the Yukon Territory of Canada (Tempelman-Kluit and Wanless, 1975). The reported K/Ar ages for these rocks (Wilson and others, this volume) range from 70 to 94 Ma on Seward Peninsula, from 91 to 111 Ma in the Ruby geanticline, and from 84 to 110 Ma in the Yukon-Tanana Upland and the Alaska Range. The similar plutons in the Yukon Territory give K/Ar ages in the 90 to 100 Ma range.

Several of the Cretaceous plutons are large (as much as 60 km across), subequant to elongate bodies that are emplaced into a regional assemblage of amphibolite-facies, metasedimentary and metavolcanic rocks of Precambrian(?) and Paleozoic age. These larger plutons display such mesozonal contact relations as grossly concordant foliated margins and broad irregular contact zones characterized by abundant dikes and sills in the country rocks. Smaller and more shallowly emplaced plutons, such as those on western Seward Peninsula, have sharp crosscutting contacts and distinct thermal aureoles in country rocks that are of low metamorphic grade or, in places, not metamorphosed.

The plutons are composite intrusions that are composed primarily of medium- to coarse-grained, equigranular to porphyritic biotite granite. Porphyritic granite containing large (several centimeters long) K-feldspar phenocrysts in a coarse-grained groundmass underlies large areas; porphyritic granite containing quartz, K-feldspar, and plagioclase phenocrysts in a fine-grained aplitic groundmass underlies smaller areas. Fine- to medium-grained equigranular granite occurs locally, both as marginal fa-

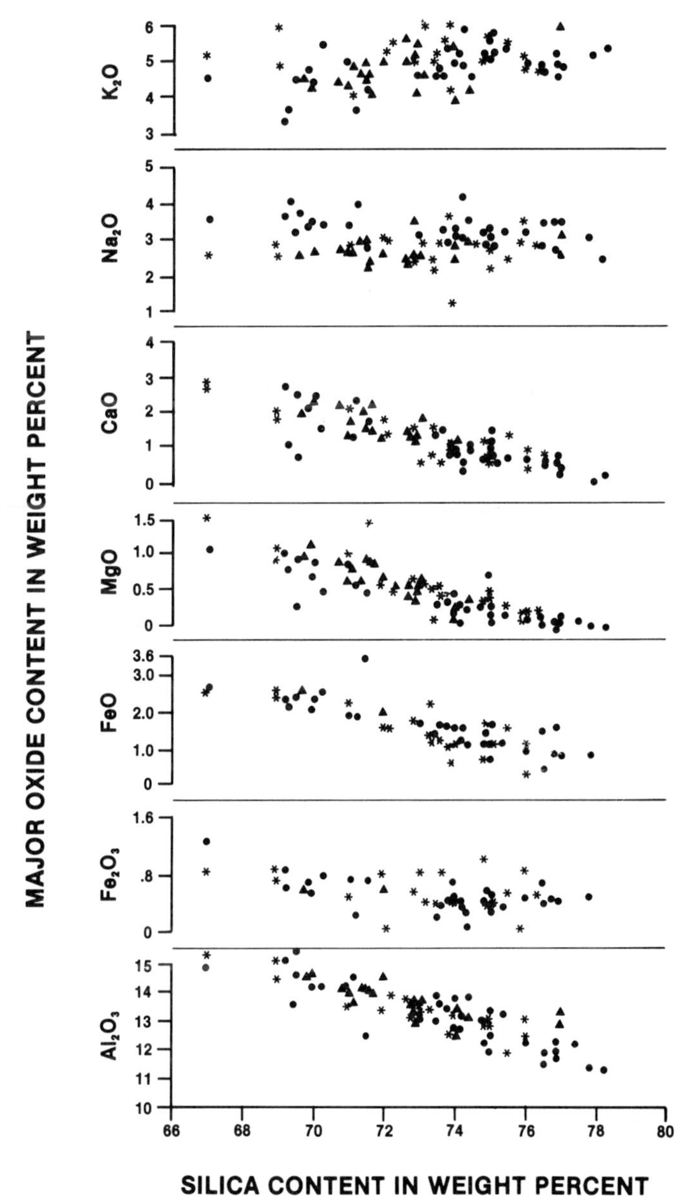

Figure 2. Variation of major oxides with silica content for orthogneisses from the Survey Pass Quadrangle in the south-central Brooks Range (●), the Okpilak batholith in the northeastern Brooks Range (*), and the Yukon-Tanana Upland (▲). See text for sources of data.

cies and as late intrusions. In general, biotite granodiorite and hornblende-biotite granodiorite are less common than granite, except for the plutons in the Yukon-Tanana Upland, where many of the samples described by Foster and others (1978) and Luthy and others (1981) are granodiorite. The accessory minerals commonly are zircon, apatite, sphene, allanite, and magnetite or ilmenite. The magnetite and ilmenite content is variable within some plutons (e.g., the Oonatut Granite Complex on Seward Peninsula: Hudson, 1979a; Hudson and Arth, 1983) and between batholiths (e.g., Darby and Bendeleben batholiths; Miller and Bunker, 1976).

Major-oxide contents (Fig. 4) have been determined for selected plutons on Seward Peninsula (Hudson and Arth, 1983; T. P. Miller, unpublished data), in the Ruby geanticline (C. Puchner, 1984, and unpublished data), and in the Yukon-Tanana Upland (Forbes, 1982; Foster and others, 1978; Luthy and others, 1981; Blum, 1985). In the Yukon-Tanana Upland, only Blum (1985) and Luthy and others (1981) have identified and described mid-Cretaceous plutons separately from other granitic rocks in this region. For this reason, only the petrologic and major-oxide data for two plutons in the Eagle Quadrangle (Foster and others, 1978) are included here. These plutons are located between the Shaw Creek and Mt. Harper lineaments (Wilson and others, 1985). They have been mapped as continuous intrusive bodies (Foster, 1976), and samples from two widely scattered localities within them give mid-Cretaceous K/Ar ages (Wilson and others, 1985).

On Seward Peninsula and in the Ray Mountains pluton of the Ruby geanticline (Fig. 4), silica varies from 68 to 77 percent,

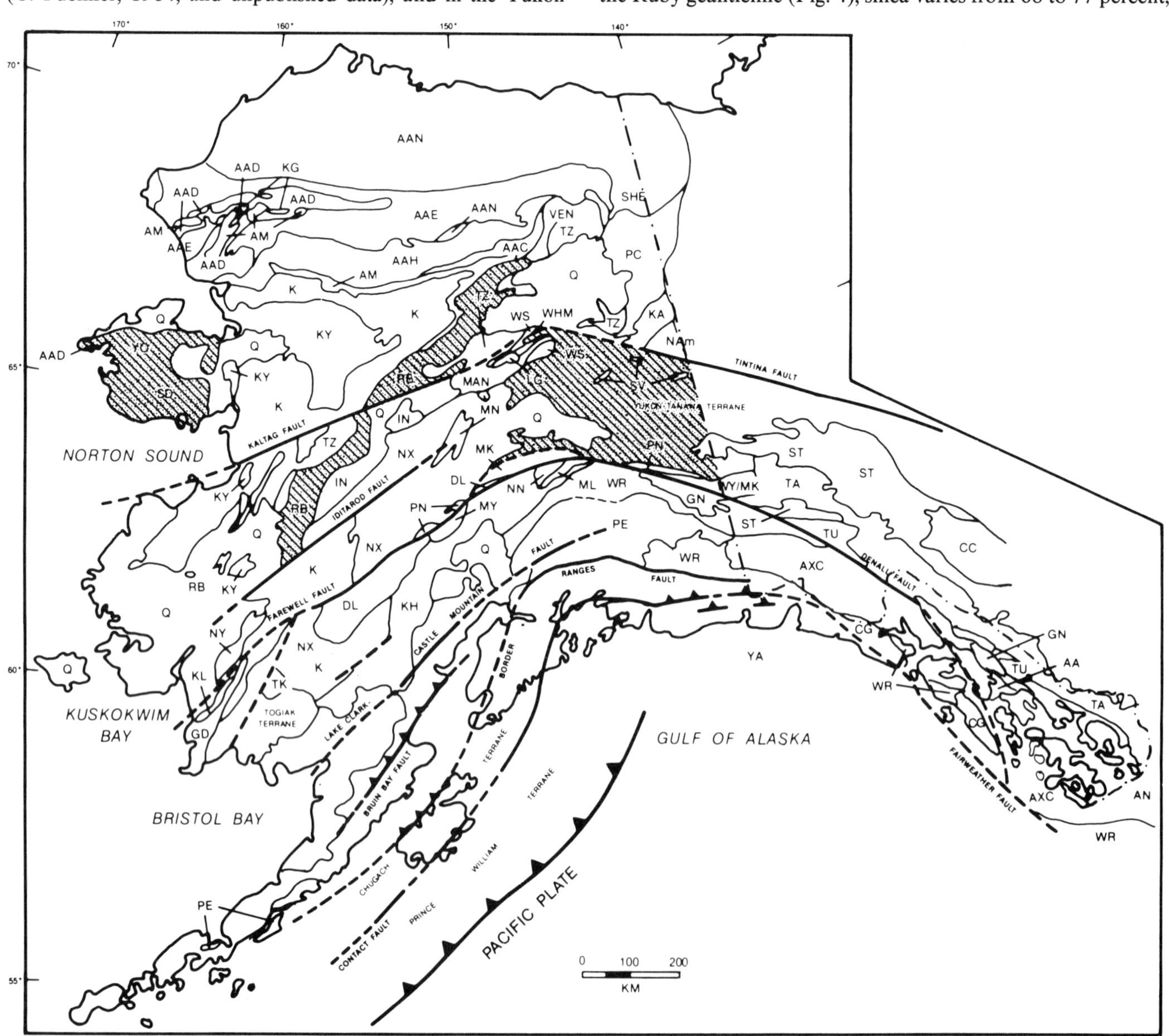

Figure 3. Map showing tectonostratigraphic terranes in Alaska (boundaries and symbols follow Jones and others, 1987; also see Silberling and others, this volume); shaded areas show the terranes that contain crustally derived Cretaceous granodiorite and granite.

total FeO + Fe_2O_3 + MgO + CaO varies from 0.5 to 3.5 percent, and total alkali content varies from 6.7 to 9.0 percent with a K_2O/Na_2O ratio of 1.1 to 1.8. The data for the Yukon-Tanana Upland indicate silica variation from 59 to 77 percent, total FeO + Fe_2O_3 + MgO + CaO content between 2.1 and 17.2 percent, and total alkalies ranging from 4.8 to 8.2 percent with a K_2O/Na_2O ratio of 0.7 to 2.0. Major-oxide variation within these plutons primarily reflects decreasing mafic oxide contents with increasing silica and alkalies. The Yukon-Tanana Upland plutons exhibit more compositional variation than the other Cretaceous plutons included in this study (Fig. 4), reflecting an increase of granodiorite in this region. The most mafic samples (<63 percent SiO_2) are from small plutons in the Fairbanks area that contain mafic xenoliths and minor amounts of pyroxene (Blum, 1985). Limited isotopic data (Hudson and Arth, 1983; Arth and others, 1984; Puchner, 1984; Blum, 1985) indicate initial $^{87}Sr/^{86}Sr$ ratios between 0.708 and 0.730 for the Cretaceous plutons; one lower value of 0.706 has a large uncertainty associated with it (Hudson and Arth, 1983). This isotopic variability suggests heterogeneous source regions for the magmas.

The regional relations important for interpreting the tectonic setting in which mid-Cretaceous crustal melting took place are listed below and summarized diagrammatically in Figure 5.

1. Jurassic and/or Early Cretaceous regional progressive metamorphism of high-pressure-facies series, and/or strong compressive deformation including thrust faulting, affected the Seward Peninsula (Till, 1983, 1984; Forbes and others, 1984a, b), the southern Brooks Range (Forbes and others, 1984a; Hitzman, 1982), the Ruby geanticline (Dover, 1984), and the Yukon-Tanana Upland (Brewer and others, 1984; Brown and Forbes, 1984; Cushing and others, 1984). The age of this deformation and metamorphism is not conclusively known, but it could be mostly Late Jurassic in age.

2. Late Early Cretaceous retrograde recrystallization of the high-pressure metamorphic rocks developed on Seward Peninsula (Forbes and others, 1984b) and the southern Brooks Range (Turner and others, 1979). Important regional recrystallization also took place in the Yukon-Tanana Upland (Wilson and others, 1985) and probably in the Ruby geanticline during the Early Cretaceous.

3. Early Cretaceous andesitic volcanism was widespread in the Yukon-Koyukuk province (Patton, 1973). An intermediate composition plutonic suite, associated with subsilicic and K_2O-rich intrusive rocks, was also emplaced in this region (97 to 108 Ma; Miller, 1972), while Cretaceous granodiorite and granite were emplaced elsewhere in Alaska.

4. Albian and younger Cretaceous sedimentation filled large successor basins of interior Alaska. These thick sections were deposited in deep marine to brackish-water environments.

5. Mid-Cretaceous (100 to 110 Ma) calc-alkaline magmatism, probably related to subduction, is present in the east-central Alaska Range (Richter and others, 1975; Hudson, 1979b).

6. Late Cretaceous magmatism occurred in the Alaska Range concurrently with fore-arc sedimentation in the Matanuska Valley and Cook Inlet and with accretion along the Gulf of Alaska margin (Hudson, 1979b). Late Cretaceous to earliest Tertiary calc-alkaline and related magmatism was widespread throughout interior Alaska and nearby parts of the Yukon Territory (Bergman and others, 1987).

The above events may reflect collisional, extensional, or subduction-related tectonic processes (Fig. 5). To understand the origins of mid-Cretaceous crustal melting, those processes that could emplace mantle-derived magma into the crust need to be considered first. The magmatism that may reflect these processes includes the Early Cretaceous andesitic volcanism and mid-Cretaceous plutonism of the Yukon-Koyukuk province and the

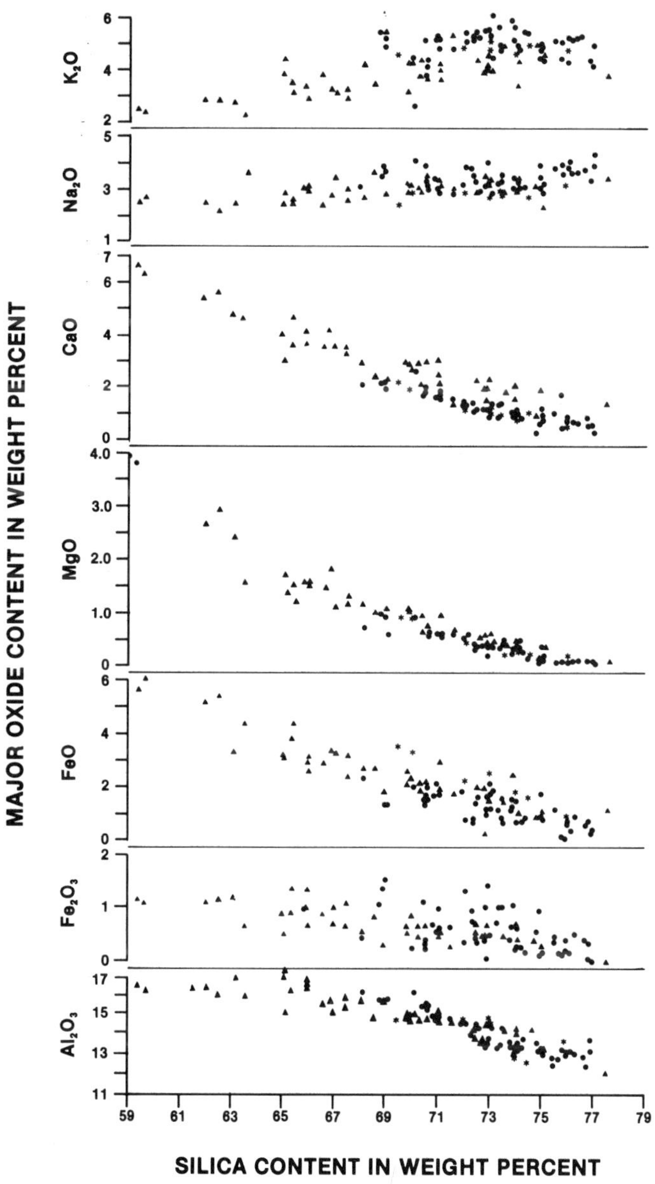

Figure 4. Variation of major oxides with silica content for Cretaceous granodiorite and granite from Seward Peninsula (●), the Ray Mountains pluton in the Ruby geanticline (*; FeO = total iron), and the Yukon-Tanana Upland (▲). See text for sources of data.

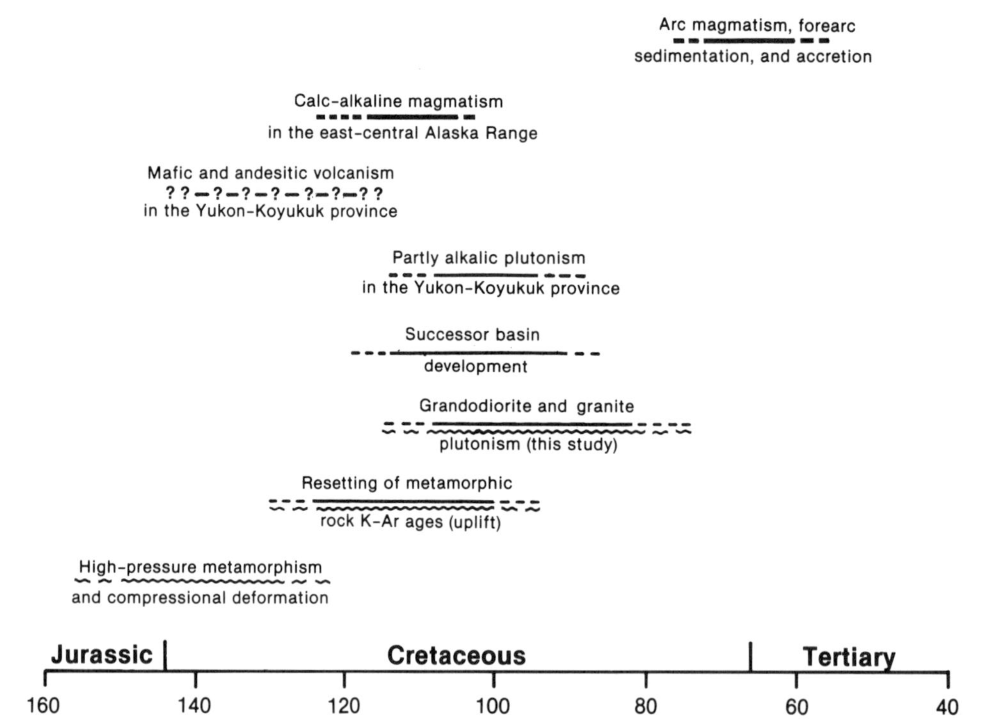

Figure 5. Diagrammatic summary of some major tectonic events bearing on the Cretaceous evolution of Alaska. These events are here linked to collisional (~), extensional (-), subduction (▪), or unknown (?-?) tectonic processes. The overlap in collisional and extensional events is interpreted here to represent an orogenic cycle of compression and subsequent decompression that affected all of interior Alaska in the Early to mid-Cretaceous. This interpretation implies that the Early Cretaceous volcanism of the Yukon-Koyukuk province may not be subduction related, although understanding its origin awaits further study.

mid-Cretaceous calc-alkaline magmatism of the east-central Alaska Range. The Late Cretaceous and Early Tertiary magmatism postdates the mid-Cretaceous granodiorite and granite plutonism of special concern here.

The Early Cretaceous volcanism and mid-Cretaceous plutonism of the Yukon-Koyukuk province are somewhat enigmatic in the regional geology of interior Alaska. The interpretation favored by some (e.g., Box, 1985) is that the volcanic assemblage represents an intraoceanic, subduction-related magmatic arc that was accreted to North America and is therefore allochthonous with respect to adjacent regions. If so, then this Early Cretaceous volcanism is not necessarily evidence of heat flux from the mantle in interior parts of Alaska. The mid-Cretaceous alkalic and related plutonism of the Yukon-Koyukuk province, however, suggests some mantle-related heat flux in interior Alaska, although as presently known, the plutons are restricted to the Yukon-Koyukuk province, nearby parts of Seward Peninsula, and St. Lawrence Island (Patton and Csejtey, 1980).

The calc-alkaline magmatism of the east-central Alaska Range is probably subduction related and could be the outer part of the magmatic arc that extended inland to include other mid-Cretaceous igneous rocks of the Canadian Cordillera (e.g., Le-Couteur and Tempelman-Kluit, 1976). If so, then mantle-derived magma could be responsible for some of the compositional variability of the Yukon-Tanana Upland granodiorites or for heating of deeper parts of the crust that contributed to abundant crustal melting in the Yukon-Tanana Upland (Blum, 1985). The influence of this possibly subduction-related magmatism appears to be restricted to east-central Alaska inasmuch as mid-Cretaceous magmatic rocks similar to those in the east-central Alaska Range have not been identified farther west. An adequately constrained interpretation of these regional relations will require careful reconstruction of the major strike-slip displacements on the Tintina and Denali fault systems.

Early and mid-Cretaceous mantle-related magmatism and an accompanying heat flux thus may have contributed to crustal melting in or near the Yukon-Koyukuk province and in east-central Alaska. However, it is the conclusion of this study that the most striking and widespread characteristics of the middle Mesozoic history of interior Alaska are (1) the Late Jurassic or Early Cretaceous high-pressure metamorphism, (2) mid-Cretaceous retrograde (uplift) recrystallization of the high-pressure metamorphic rocks, (3) mid-Cretaceous deep successor basin formation and filling, and (4) mid-Cretaceous voluminous granodiorite and granite emplacement (Fig. 5). The interpretation of these regional events that is preferred here is that a major Late Jurassic or Early Cretaceous period of compressional (collision-related?) deformation over large parts of interior Alaska significantly thickened the crust by structural imbrication, thus producing regionally distributed high-pressure metamorphic rocks. Crustal thickening and accompanying high-pressure metamorphism were followed by the reestablishment of thermal gradients in the lower crust and uplift as recorded by the widespread mid-Cretaceous retrograde recrystallization and resetting of K-Ar ages in metamorphic rocks. Melting in deeper parts of this uplifted crust probably took place at this time together with the extension necessary for development of the deep successor basins and for emplacement of any mantle-derived alkalic and associated magmas in the Yukon-Koyukuk province, Seward Peninsula, and St. Lawrence Island. The dominant tectonic processes leading to widespread mid-Cretaceous crustal melting in interior Alaska are therefore interpreted here to be the development of a compressional orogen and its subsequent extensional collapse. This interpretation implies that the Early Cretaceous mafic and andesitic volcanism of the Yukon-Koyukuk province may not reflect classic intraoceanic subduction, and that its origin deserves further study.

PALEOGENE GRANITIC ROCKS

Three belts of Paleogene granitic rocks (Fig. 6) appear to have an origin dominated by crustal melting. These include the McKinley sequence (Reed and Lanphere, 1973) and similar plutons elsewhere in the Alaska Range–Talkeetna Mountains belt (Hudson, 1983), the Sanak-Baranof belt along the Gulf of Alaska margin (Hudson and others, 1979; Hudson, 1983), and the Coast Mountains belt of southeastern Alaska (Coast Plutonic Complex belt I of Brew and Morrell, 1983; Brew, this volume). The K/Ar ages of these rocks vary from about 50 to 58 Ma in the Alaska Range–Talkeetna Mountains belt, from 58 to 60 Ma in the western segment and 45 to 52 Ma in the eastern segment of the Sanak-Baranof belt, and from about 45 to 54 Ma in the Coast Mountains belt.

The field and petrologic characteristics of the Paleogene granitic rocks in the Alaska Range–Talkeetna Mountains and Sanak-Baranof belts have been summarized by Hudson (1979b, 1983), Hudson and others (1979), and Hudson and Plafker (1982). The Paleogene plutons of the Coast Mountains belt form large subequant and coalescing plutons that are emplaced at shallow depths. The contacts are discordant to the deep-seated metamorphic and plutonic host rocks as well as to the low-grade metasedimentary and metavolcanic rocks that flank the Coast Mountains to the east. These plutons are dominantly massive, medium- to coarse-grained, equigranular biotite granite and granodiorite containing minor amounts of accessory hornblende, sphene, and, in some, garnet.

Major-oxide (Fig. 7) and other chemical data are available for the McKinley sequence in the Alaska Range (Reed and Lanphere, 1973, 1974a; Lanphere and Reed, 1985), some plutons of the Sanak-Baranof belt (Hill and others, 1981; Hudson and others, 1977), and for a number of plutons in the Coast Mountains belt (Barker and others, 1986; Smith and others, 1977). The analyses of the Paleogene granites of the Alaska Range–Talkeetna Mountains belt have an average SiO_2 content of 74.3 percent (range is 65.9 to 77.6 percent), an average CaO content of 1.1 percent (range is 0.34 to 2.9 percent), and an average $K_2O + Na_2O$ content of 8 percent. The average $K_2O + Na_2O$ ratio is 1:4. The analytical data for the Sanak-Baranof belt plutons show a wide range of compositional variation: SiO_2 ranges from 51 to 76 percent, Fe_2O_3* (total iron) + MgO + CaO from 1.9 to 10.9 percent and $K_2O + Na_2O$ from 5.9 to 8.6 percent. K_2O/Na_2O ratios are 0.6 to 2.0 (mostly about 1). The available data for the Coast Mountains belt (Barker and others, 1986; Smith and others, 1977) show a range of SiO_2 content from 53 to 77 percent (mostly 65 to 72 percent), a total FeO + Fe_2O_3 + MgO + CaO content of 2.8 to 16.0 percent (mostly <10 percent), and a $K_2O + Na_2O$ content of 6.3 to 8.5 percent; K_2O/Na_2O ratios are 0.4 to 1.6. The more mafic compositions in the Coast Mountains belt and the Sanak-Baranof belt are of sparse (only 11 percent of the samples have silica contents lower than 64 percent; Fig. 7), strongly foliated, hornblende-bearing rocks that in part include schlieren-like features. These more mafic rocks are present in places where mixing of amphibolitic country rocks with the granitic magma during emplacement can be inferred (e.g., the Mt. Draper and Mt. Stamy plutons in the Yukutat–St. Elias area; Hudson and others, 1977, p. 166). The available Sr isotope data (Lanphere and Reed, 1985; Hill and others, 1981; Hudson and others, 1979; Barker and others, 1986) all indicate initial $^{87}Sr/^{86}Sr$ ratios between 0.7048 and 0.7085. These values are consistent with those measured or inferred for the youthful, partly transitional continental margin crust (accreted deep sea sedimentary sequences and oceanic basalt) that underlies much of southern Alaska (Hill and others, 1981; Hudson and others, 1979).

The regional relations listed below are important for interpreting the tectonic setting in which Paleogene crustal melting took place (Plafker and Berg, this volume, Chapter 33).

1. A Late Cretaceous and earliest Tertiary calc-alkaline magmatic arc developed in the Alaska Range–Talkeetna Mountains region and in the Coast Mountains (Reed and Lanphere, 1973; Hudson, 1979b; Barker and others, 1986; Brew and Morrell, 1983).

2. Late Cretaceous and earliest Tertiary high-grade regional metamorphism and migmatization took place in the deeper parts of the Alaska Range and Talkeetna Mountains region (Smith, 1981, 1984) and the Coast Mountains (Kenah and Hollister, 1983; Brew and Ford, 1984). Such metamorphism and migmatization took place in the Gulf of Alaska region some time during the Paleocene or earliest Eocene (between 50 Ma and the time of accretion of the Upper Cretaceous flysch of the Valdez Group; Hudson and Plafker, 1982).

3. Right-lateral displacement of a few hundred kilometers along the principal strand of the Denali fault system may juxtapose similar Late Cretaceous and early Tertiary metamorphic and plutonic rocks of the Alaska Range and the Coast Mountains

(Forbes and others, 1974; Turner and others, 1974). Even if post–Late Cretaceous displacement on the Denali fault is tens of kilometers (Reed and Lanphere, 1974b; Csejtey and others, 1982), Figure 6 shows that the two northern plutonic belts are nearly connected and subparallel to the coeval Sanak-Baranof belt to the south.

The tectonic setting that immediately preceded emplacement of the Paleogene granitic rocks is well defined. In latest Cretaceous time, southern Alaska had a convergent subduction margin, as evidenced by development of a calc-alkaline magmatic arc, a fore-arc basin, and an accretionary complex seaward of the Border Ranges fault (Fig. 6).

The calc-alkaline magmatic front that was developed during this tectonic regime roughly coincides with the Alaska Range–Talkeetna Mountains and Coast Mountains belts of Paleogene plutons. The preferred interpretation here is that subduction-related magmatism had ceased by 60 Ma and that the 45 to 58-Ma granodiorite and granite plutons in these belts are post-subduction plutons whose origin is ultimately due to high-heat flow that accompanied the slightly older subduction-related magmatism. This origin for the Alaska Range–Talkeetna Mountains and Coast Mountains belts of Paleogene granodiorite and granite plutons seems much more likely than the accretionary origin for the McKinley sequence suggested by Lanphere and

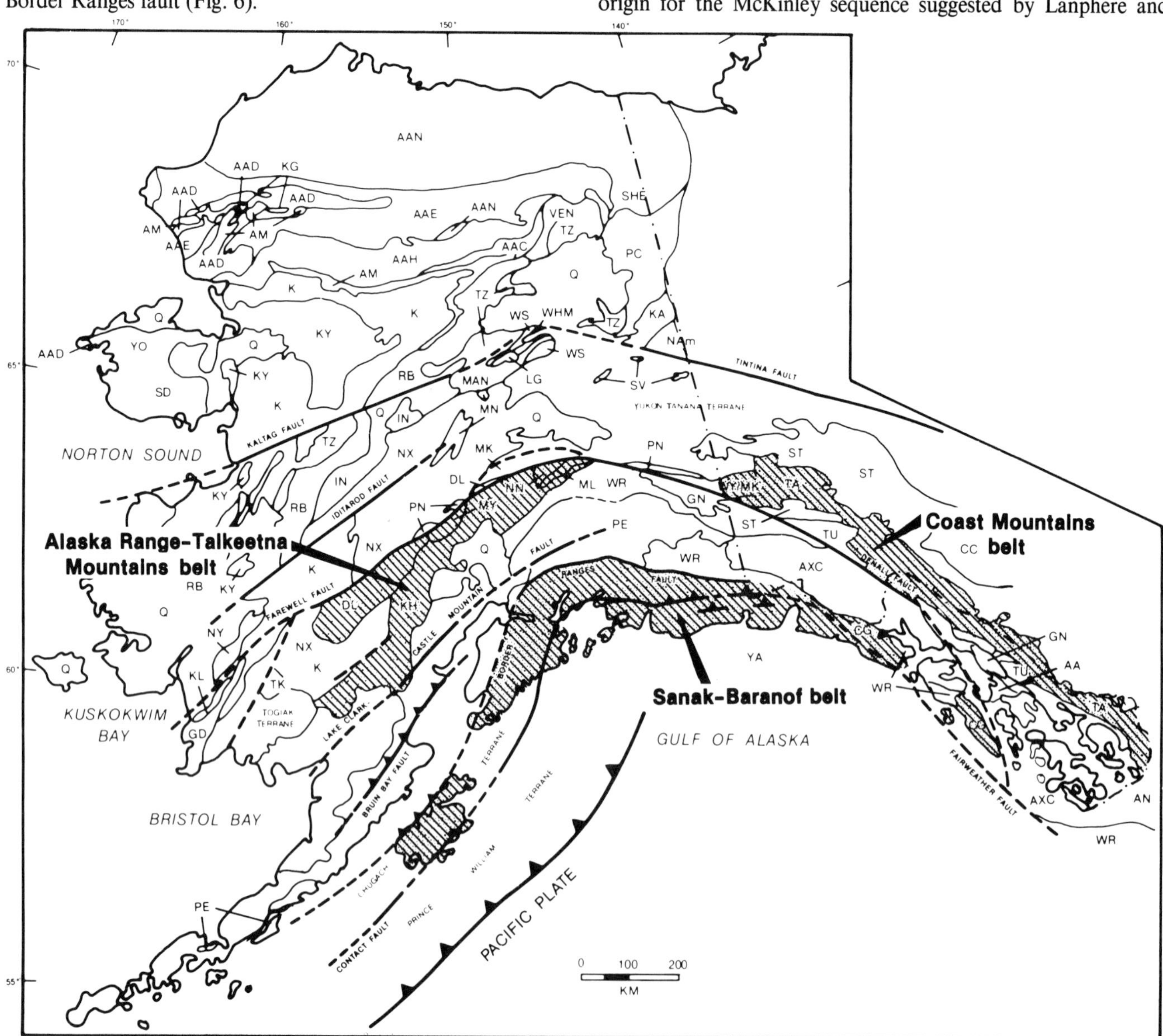

Figure 6. Map showing onshore extent of principal tectonostratigraphic terranes in Alaska (boundaries and symbols follow Jones and others, 1987; also see Silberling and others, this volume). Shaded areas show the terranes that contain crustally derived Paleogene granodiorite and granite. The Paleogene plutons in these terranes define three plutonic belts as indicated. Small dikes and plugs of similar granitic rocks are present in the upper plate of the Border Ranges fault (PE terrane) and locally within Wrangellia (WR terrane).

Reed (1985)—especially in light of the petrologic character of some composite plutons in the Alaska Range that show evidence of magma mixing. These composite 65-Ma plutons are mostly quartz monzonite, but they range in composition from peridotite to granite and contain such unusual components as K-feldspar-bearing ultramafic rocks and olivine-bearing pyroxene quartz monzonite (Reed and Nelson, 1980). These components strongly suggest that mantle-derived magmas were emplaced into the crust and incompletely mixed with felsic crustal melts. This interpretation also is consistent with the results of studies in the British Columbia part of the Coast Mountains belt just south of the area shown in Figure 6 (Kenah, 1979; Hollister, 1982; Crawford and Hollister, 1982; Kenah and Hollister, 1983), which show the importance of tonalitic and gabbric intrusions in elevating temperatures during early Tertiary granulite-facies metamorphism and anatexis.

The regional relations that constrain the interpretation of the tectonic setting for crustal melting in the Sanak-Baranof belt are listed below.

1. There is an apparent difference in age between the eastern (45 to 53 Ma) and western (58 to 60 Ma) segments of the belt.

2. The plutons postdate the deformation that accompanied accretion of the Upper Cretaceous and at least some of the Paleocene sedimentary and volcanic rocks in the Kodiak Island area and the Paleocene to early middle Eocene deformation of similar rocks in Prince William Sound and the eastern Chugach Mountains (Plafker and others, 1985). Mafic magmatism coeval with, or slightly older than, Sanak-Baranof granite emplacement, but also postdating accretionary deformation of Upper Cretaceous and Paleocene rocks, is recognized only locally in the eastern Prince William Sound area (G. Plafker, oral communication, 1986).

3. The plutons of the Sanak-Baranof belt postdate a major episode of subduction and are coeval with, and subparallel to, the Alaska Range–Talkeetna Mountains and Coast Mountains belts of crustally derived plutons (Fig. 7).

4. Paleogene high-grade metamorphism in the eastern Chugach Mountains may require a geothermal gradient of as much as 50°C/km at the time of metamorphism (Hudson and Plafker, 1982).

Three origins have been proposed to explain the plutons of the Sanak-Baranof belt: (1) reestablishment of geotherms within a thick accretionary prism upon slowdown or cessation of accretion (Hudson and others, 1979); (2) high heat flow associated with migration of a ridge-trench-trench triple junction along the continental margin (Marshak and Karig, 1977); and (3) high heat flow associated with emplacement of possibly ridge-related mafic magmas into the accretionary prism (Hill and others, 1981).

The second and third proposed origins are here considered unlikely because mafic magmatism, coeval with granitic plutonism and postdating accretionary deformation, is rare in the Gulf of Alaska region. Hill and others (1981), for example, argue that such mafic magmas explain some of the chemical and isotopic

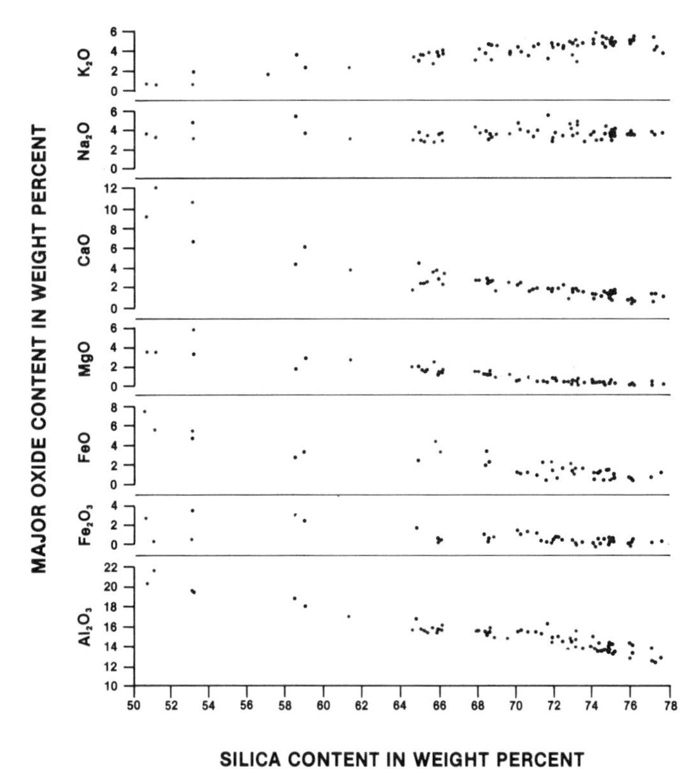

Figure 7. Variation of major oxides with silica for Paleogene granitic rocks in the Alaska Range–Talkeetna Mountains and Coast Mountains belts (*; does not include data for the tonalite of Skagway reported by Barker and others, 1986) and the Sanak-Baranof belt (●). See text for sources of data.

characteristics (e.g., $^{87}Sr/^{86}Sr_i$ and Sr content) of the Paleogene plutons in the western segment of the Sanak-Baranof belt. However, some of these compositional characteristics could be inherited from potential source rocks in the accretionary prism that include mafic volcanic assemblages along with the voluminous flysch sequences (e.g., Page and others, 1986).

The origin of the Sanak-Baranof belt, though uncertain, is probably closely linked to plate-tectonic processes. One additional hypothesis consistent with the foregoing constraints is presented here. This hypothesis assumes that young, relatively hot oceanic lithosphere was being subducted along the Gulf of Alaska margin during the Late Cretaceous and earliest Tertiary. When subduction slowed, this hotter lithosphere would have come to rest beneath the continental margin, producing the higher heat flow required throughout the region for Paleogene anatexis in the already tectonically thickened accretionary prism. Somewhat longer subduction on the eastern limb than on the western limb of the belt then explains the apparent age differences between the two limbs. This hypothesis provides for elevated heat flow on a regional scale and is consistent with the field relations that led to the first of the three proposed origins listed for the Sanak-Baranof belt (Hudson and others, 1979; Hudson and Plafker, 1982).

DISCUSSION

The crustal melting events indicated by the nature and timing of granitic plutonism provide some important insights into the regional geology and tectonic framework of Alaska. The key insight into the regional geology comes from the recognition that similar Devonian crustal melts evolved in all of the more deeply exposed cratonic blocks of interior and northern Alaska; the Seward Peninsula, Brooks Range, Ruby geanticline, and the Yukon-Tanana Upland. The setting in which these melts developed appears to be one of extensional or wrench tectonics, an interpretation that is consistent with some recent suggestions by others working elsewhere along the western Paleozoic margin of North America (Eisbacher, 1983; Miller and others, 1984; K. Ehman, written communication, 1984; Plafker and Berg, this volume, Chapter 33). This interpretation, with its inferred correlation, requires that the major cratonic blocks of interior and northern Alaska be related to the continental margin of North America and not exotic fragments that have been transported from some distant origin in the ancient Pacific (Coney and others, 1980). This conclusion does not preclude significant differential translation of these crustal blocks along the continental margin itself.

If these cratonic blocks are related, even indirectly, to one another and to the North American continental margin in Devonian time, then their subsequent regional evolution is likely to have some similarities. Such similarities seem to be present in areas affected by the middle Mesozoic metamorphism, deformation, and plutonism outlined above, with one possible exception. The Brooks Range apparently lacks evidence of mid-Cretaceous granitic plutonism (Dillon and others, 1980) such as that on Seward Peninsula, the Ruby geanticline, and in the Yukon-Tanana Upland (Fig. 5), although the Brooks Range underwent similar episodes of a Late Jurassic or Early Cretaceous high pressure metamorphism, compressional deformation, and subsequent uplift. This distinction suggests that vertical displacements are an important factor in the regional tectonics of interior Alaska.

All the areas characterized by voluminous mid-Cretaceous granodiorite and granite plutonism are deeply eroded, relatively high-grade metamorphic regions that have been uplifted with respect to nearby areas. These deeply eroded regions include the youthful, fault-bounded Kigluaik, Bendeleben, and Darby Mountains on Seward Peninsula, the northern part of the Ruby geanticline relative to adjacent areas in the southern Brooks Range and its continuations south of the Kaltag fault, and such subdivisions of the Yukon-Tanana Upland as the crustal block between the Shaw Creek and Mt. Harper lineaments. These relations suggest that most of the mid-Cretaceous granitic magmas were emplaced at intermediate (mesozonal) crustal levels. If such levels are not widely exposed, even in the regions where melting occurred, significant granite and granodiorite plutons will not crop out. The Brooks Range, for example, is a major fold and thrust belt that is not as deeply eroded as adjacent areas characterized by abundant mid-Cretaceous granitic rocks. This lack of deep erosion may explain why such plutons are not exposed in the southern Brooks Range although they may be present at depth. Vertical displacement and subsequent erosion are therefore considered to be important factors controlling the present exposures of mid-Cretaceous, crustally derived plutons in interior Alaska.

CONCLUSION

This study represents an attempt to identify the timing and setting of major crustal melting events in Alaska. In order to do so, it has focused on granodiorite and granite plutons that can reasonably be inferred to have an origin dominated by crustal melting. There probably are other less obvious plutonic and related volcanic rocks that also have a crustal origin (e.g., parts of the widespread, early Tertiary magmatic province of interior Alaska?). Because granitic plutonism in subduction-related plutonic belts may be closely linked to deeper crustal processes as well as to mantle processes (e.g., Leeman, 1983), the simplifying distinction originally made in this study seems somewhat arbitrary. However, the results show that this distinction—focusing as it does on regional plutonism of granodiorite and granite composition—helps identify important temporal and spatial regional relations.

This analysis suggests that crustal melting was an important process in the evolution of Alaska during the Devonian, the Cretaceous, and the Paleogene and that four different regional tectonic settings characterize the recognized times of crustal melting. These are: (1) an extensional and/or wrench setting during the Devonian in which high heat flow resulted in melting, (2) a post-compressive deformation setting during the mid-Cretaceous in which uplift of thickened crust resulted in melting, (3) a post-magmatic arc setting during the Paleogene in which high heat flow associated with arc magmatism resulted in melting, and (4) a post-accretion setting during the Paleogene in which crustal thickening combined with high heat flow resulted in melting. These four settings emphasize the diversity of conditions that can lead to crustal melting in the North America Cordillera.

ADDENDUM (11/89)

Since this report was first written in 1984, the opportunity to revise and update it has come about several times. At these times, editorial and technical review provided viewpoints and feedback that helped formulate revisions. However, throughout these revisions, adherence to the original available data, concepts, and conclusions has been maintained in order to correctly record the chronology of this work.

Considerable other work has been accomplished in Alaska since 1984, some of which is especially relevant to the main focus of this chapter. In the author's opinion, the recent contributions that have the most impact on the concepts and conclusions of this report are those clarifying the role of extension in the Cretaceous history of interior Alaska and those constraining interpretations of the origins of the Sanak-Baronof plutonic belt.

As noted in the original report, the regional distribution of metamorphic and plutonic rocks alone suggests major vertical tectonic displacements in interior Alaska during the mid-Cretaceous. However, the first work to place these vertical displacements and other aspects of Alaska Cretaceous history into a comprehensive extensional framework was that of Miller (personal communication, 1985; 1987). This work, together with the recognition of specific regional normal faults on the south side of the Brooks Range (Box, 1987; Oldow and others, 1987; Moore and others, this volume) has just started to clarify the character and extent of this extensional tectonic regime, but it has very important implications for all of the mid-Cretaceous regional elements in Alaska. For example, one implication might be that mid-Cretaceous and Late Cretaceous magmatism of the Yukon-Koyukuk province (112 to 80 Ma), represents mantle and crustal melting that accompanied extension—the variety of highly potassic, felsic, and intermediate composition plutonism in this province would primarily reflect the variety of source regions that underwent melting at this time. In this context, the mid-Cretaceous granodiorite and granite of regions peripheral to the Yukon-Koyukuk province would have formed in a similar tectonic regime but in regions with more uniform continental crustal substrata. Regardless, extension in the Cretaceous evolution of Alaska seems to be increasingly supported by recent studies and has the potential to unify many perplexing aspects of the tectonic history of this important time. The developing extensional concepts for Alaska have not required modification of the original conclusions of this report but they have clearly expanded the scope of implications that conclusions concerning the origin of mid-Cretaceous granodiorite and granite plutonism have; continuing efforts to place Alaska Cretaceous tectonic elements into a unifying tectonic framework are clearly justified.

The second area of recent contributions that are especially relevant to this report are those that help to constrain the origins of the Sanak-Baranof plutonic belt. These contributions include a new synthesis of Late Cretaceous and early Tertiary magmatism in southern Alaska (Bergman and others, 1987), thermal modeling (James and others, 1989), and petrologic and geochronologic studies (Sisson and others, 1989) of the Chugach Metamorphic Complex. The Chugach Metamorphic Complex (Hudson and Plafker, 1982) is a window into deeper structural levels of the Late Cretaceous accretionary prism that developed along the Gulf of Alaska continental margin. Metamorphism and anatexis in the Chugach Metamorphic Complex are considered representative of at least the shallower crustal melting environments that were the source of the plutons of the Sanak-Baranof plutonic belt (Hudson and others, 1979; Hudson and Plafker, 1982). The relations in the Chugach Metamorphic Complex require high geothermal gradients (40 to 60°C per km; Sisson and others, 1989) at the time of metamorphism and melting. The origin of the heat flow required to develop these thermal gradients was a topic of some discussion at the time this report was first written.

This author has emphasized that field relations around the Gulf of Alaska margin do not indicate significant influx of mafic magmas into the accretionary prism at the time of anatexis. This has been taken by the author as evidence that ridge subduction (Marshak and Karig, 1977; Hill and others, 1981) is not the cause of early Tertiary high heat flow. One alternative cause originally put forth in this report, the emplacement of youthful and warm oceanic crust beneath the accretionary prism, was tested by the thermal modeling studies of James and others (1989). These workers showed that it was possible to produce the required high heat flow if the oceanic crust emplaced beneath the accretionary prism was very young, on the order of 1 Ma or less. The youthfulness that is required seems to make the distinction between warm oceanic crust and a ridge somewhat academic. Therefore, modification of this hypothesis is in order—especially in light of a conclusion reached by Bergman and others (1987) concerning the tectonic framework of Late Cretaceous and early Tertiary magmatism throughout southern Alaska.

Bergman and others (1987) have concluded that a significant change in the tectonic framework of southern Alaska took place at about 63 Ma. In summary, this change marked the shift from rapid convergence along a shallow-dipping subduction zone in the Late Cretaceous to slower convergence along a more steeply dipping subduction zone like that of today in the early Tertiary (Eocene?). This shift suggests two important early Tertiary events: (1) the cessation or slowdown of subduction (Hudson and others, 1979), and (2) a breaking and sinking of the down-going oceanic plate upon resumption of subduction. The breaking and sinking of the oceanic plate could develop a tectonic regime that would allow mantle material to be emplaced near or subjacent to the accretionary prism. This presents another possible cause of high heat flow along the Gulf of Alaska margin at exactly the time it is needed (Paleocene and early Eocene). This new interpretation for the cause of high heat flow needs further evaluation, but it is consistent with several regional geologic relations in southern Alaska.

REFERENCES CITED

Adams, D. D., 1983, Geologic map of the northern contact area of the Arrigetch Peaks pluton, Brooks Range, Alaska: State of Alaska Department of Natural Resources Professional Report 83, scale 1:18,000.

Arth, J. G., Carlson, J. L., Foley, N. K., Friedman, I., Patton, W. W., Jr., and Miller, T. P., 1984, Crustal composition beneath the Yukon-Koyukuk Basin and Ruby geanticline as reflected in the isotopic composition of Cretaceous plutons: Geological Society of America Abstracts with Programs, v. 16, p. 267.

Aleinikoff, J. N., 1984, Age and origin of metaigneous rocks from terranes north and south of the Denali fault, Mt. Hayes Quadrangle, east-central Alaska: Geological Society of America Abstracts with Programs, v. 16, p. 266.

Aleinikoff, J. N., Dusel-Bacon, C., Foster, H. L., and Futa, K., 1981, Proterozoic zircon from augen gneiss, Yukon-Tanana Upland, east-central Alaska: Geology, v. 9, p. 469–473.

Aleinikoff, J. N., Dusel-Bacon, C., and Foster, H. L., 1986, Geochronology of augen gneiss and related rocks, Yukon-Tanana terrane, east-central Alaska: Geological Society of America Bulletin, v. 97, p. 626–637.

Barker, F., Arth, J. G., and Stern, T. W., 1986, Evolution of the Coast batholith along the Skagway traverse, Alaska and British Columbia: American Mineralogist, v. 71, p. 632–643.

Ben Othman, D., Fourcade, S., and Allegre, C. J., 1984, Recycling processes in granite-granodiorite complex genesis; The Querigut case studied by Nd-Sr isotope systematics: Earth and Planetary Science Letters, v. 69, p. 290–300.

Bergman, S. C., Hudson, T. L., and Doherty, D. J., 1987, Magmatic rock evidence for a Paleocene change in the tectonic setting of Alaska: Geological Society of America Abstracts with Programs, v. 19, p. 586.

Blum, J. D., 1985, A petrologic and Rb-Sr isotopic study of intrusive rocks near Fairbanks, Alaska: Canadian Journal of Earth Sciences, v. 22, p. 1314–1321.

Box, S. E., 1985, Early Cretaceous orogenic belt in northwestern Alaska; Internal organization, lateral extent, and tectonic interpretation, *in* Howell, D. A., ed., Tectonostratigraphic terranes of the Circum-Pacific region: Circum-Pacific Council for Energy and Mineral Resources, no. 1, p. 137–145.

—— , 1987, Lake Cretaceous or younger SW-directed extensional faulting, Cosmos Hills, Brooks Range, Alaska: Geological Society of America Abstracts with Programs, v. 19, p. 361.

Brew, D. A., and Ford, A. B., 1984, Timing of metamorphism and deformation in the Coast Plutonic-Metamorphic Complex, near Juneau, Alaska: Geological Society of America Abstracts with Program, v. 16, p. 272.

Brew, D. A., and Morrell, R. P., 1983, Intrusive rocks and plutonic belts of southeastern Alaska, U.S.A., *in* Roddick, J. A., Circum-Pacific plutonic terranes: Geological Society of America Memoir 159, p. 171–193.

Brewer, W., Sherwood, K., and Craddock, C., 1984, Metamorphic patterns in the central Alaska Range and the recognition of terrane bounding faults: Geological Society of America Abstracts with Programs, v. 16, p. 272.

Brown, E. H., and Forbes, R. B., 1984, Paragenesis and regional significance of eclogitic rocks from the Fairbanks district, Alaska: Geological Society of America Abstracts with Programs, v. 16, p. 272.

Chapman, R. M., Yeend, W., Brosgé, W. P., and Reiser, H. N., 1982, Reconnaissance geologic map of the Tanana Quadrangle, Alaska: U.S. Geological Survey Open-File Report 82–734, 18 p., scale 1:250,000.

Clemens, J. D., and Wall, V. J., 1981, Origin and crystallization of some peraluminous (S-type) granitic magmas: Canadian Mineralogist, v. 19, p. 111–131.

Coney, P. J., Jones, D. L., and Monger, J.W.H., 1980, Cordilleran suspect terranes: Nature, v. 288, p. 329–3333.

Crawford, M. L., and Hollister, L. S., 1982, Contrast of metamorphic and structural histories across the Work Channel lineament, Coast Plutonic Complex, British Columbia: Journal Geophysical Research, v. 87, p. 3849–3860.

Csejtey, B., Jr., Cox, D. P., Evarts, R. C., Stricker, G. D., and Foster, H. L., 1982, The Cenozoic Denali fault system and the Cretaceous accretionary development of southern Alaska: Journal of Geophysical Research, v. 86, no. 85, p. 3741–3754.

Cushing, G. W., Foster, H. L., and Harrison, T. M., 1984, Mesozoic age of metamorphism and thrusting in the eastern part of east-central Alaska: EOS Transactions of the American Geophysical Union, v. 65, no. 16, p. 290.

Dillon, J. T., Pessel, G. H., Chen, J. H., and Veach, N. C., 1980, Middle Paleozoic magmatism and orogenesis in the Brooks Range, Alaska: Geology, v. 8, p. 338–343.

Dover, J. H., 1984, Metamorphic rocks of the Ray Mountains, Tanana Quadrangle, central Alaska; Structure and regional tectonic significance: Geological Society of America, Abstracts and Programs, v. 16, no. 5, p. 279.

Duke, N. A., Nauman, C. R., and Newkirk, S. R., 1984, The Delta district, a mineralized lower Paleozoic marginal rift terrane in the east-central Alaska Range: Geological Society of America Abstracts and Programs, v. 16, p. 279.

Dusel-Bacon, C., and Aleinikoff, J. N., 1985, Petrology and tectonic significance of augen gneiss from a belt of Mississippian granitoids in the Yukon-Tanana terrane, east-central Alaska: Geological Society of America Bulletin, v. 96, p. 411–425.

Eisbacher, G. H., 1983, Devonian-Mississippian sinistral transcurrent faulting along the cratonic margin of western North America; A hypothesis: Geology, v. 11, p. 7–10.

Forbes, R. B., 1982, Bedrock geology and petrology of the Fairbanks mining district, Alaska: Alaska Division of Geological and Geophysical Surveys Open-File Report 169, 68 p.

Forbes, R. B., Smith, T. E., and Turner, D. C., 1974, Comparative petrology and structure of the McLaren, Rugy Range, and Coast Range belts; Implications for offset along the Denali fault system: Geological Society of America Abstracts with Programs, v. 6, p. 177.

Forbes, R. B., Evans, B. W., Thurston, S. P., and Armstrong, R. L., 1984a: Comparative Rb/Sr-K/Ar age data from the Seward Peninsula and Brooks Range, Alaska, blueschist terranes; A progress report: Geological Society of America Abstracts and Programs, v. 16, p. 284.

Forbes, R. B., Evans, B. W., and Thurston, S. P., 1984b, Regional progressive high-pressure metamorphism, Seward Peninsula, Alaska: Journal of Metamorphic Geology, v. 2, p. 43–54.

Foster, H. L., 1970, Reconnaissance geologic map of the Tanacross Quadrangle, Alaska: U.S. Geological Survey Miscellaneous Geologic Investigations Map I-593, scale 1:250,000.

—— , compiler, 1976, Geologic map of the Eagle Quadrangle, Alaska: U.S. Geological Survey Miscellaneous Geologic Investigations Map I-922, scale 1:250,000.

Foster, H. L., and 6 others, 1976, Background information to accompany folio of geologic and mineral resource maps of the Tanacross Quadrangle, Alaska: U.S. Geological Survey Circular 734, 23 p.

Foster, H. L., Donato, M. M., and Yount, M. E., 1978, Petrographic and chemical data on Mesozoic granitic rocks of the Eagle Quadrangle, Alaska: U.S. Geological Survey Open-File Report 78–253, 29 p.

Foster, H. L., and 6 others, 1979, The Alaskan Mineral Resource Assessment Program; Background information to accompany folio of geologic and mineral resource maps of the Big Delta Quadrangle, Alaska: U.S. Geological Survey Circular 783, 19 p.

Gemuts, I., Puchner, C. C., and Steffel, C. I., 1983, Regional geology and tectonic history of western Alaska, *in* Proceedings of 1982 Symposium on Western Alaska Resources and Geology: Alaska Geological Society Journal, p. 67–85.

Hill, M., Morris, J., and Whelan, J., 1981, Hybrid granodiorites intruding the accretionary prism, Kodiak, Shumagin, and Sanak Island, southwest Alaska: Journal of Geophysical Research, v. 86, no. B11, p. 10569–10590.

Hitzman, M. W., 1982, The metamorphic petrology of the southwestern Brooks Range, Alaska: Geological Society of America Abstracts with Programs, v. 14, p. 173.

—— , 1984, Geology of the Cosmos Hills; Constraints for Yukon-Koyukuk basin evolution: Geological Society of America Abstracts with Programs, v. 16, p. 290.

Hollister, L. S., 1982, Metamorphic evidence for rapid (2 mm/yr) uplift of a portion of the Central Gneiss Complex, Coast Mountains, British Columbia: Canadian Mineralogist, v. 20, p. 319–332.

Hudson, T., compiler, 1977, Geologic map of Seward Peninsula, Alaska: U.S. Geological Survey Open-File Report 77–796A, scale 1:1,000,000.

—— , 1979a, Igneous and metamorphic rocks of the Serpentine Hot Springs area, Seward Peninsula, Alaska: U.S. Geological Survey Professional Paper 1079, 27 p.

—— , 1979b, Mesozoic plutonic belts of southern Alaska: Geology, v. 7, p. 230–234.

—— , 1983, Calc-alkaline plutonism along the Pacific rim of southern Alaska, *in* Roddick, J. A., ed., Circum-Pacific plutonic terranes: Geological Society of America Memoir 159, p. 159–169.

Hudson, T., and Arth, J. G., 1983, Tin granites of Seward Peninsula, Alaska: Geological Society of America Bulletin, v. 94, p. 768–790.

Hudson, T., and Plafker, G., 1982, Paleogene metamorphism of an accretionary flysch terrane, eastern Gulf of Alaska: Geological Society of America Bulletin, v. 93, p. 1280–1290.

Hudson, T., Plafker, G., and Lanphere, M. A., 1977, Intrusive rocks of the Yakutat-St. Elias area, south-central Alaska: U.S. Geological Survey Journal of Research, v. 5, no. 2, p. 155–172.

Hudson, T., Plafker, G., and Peterman, Z. E., 1979, Paleogene anatexis along the Gulf of Alaska margin: Geology, v. 7, p. 573–577.

James, T. S., Hollister, L. S., and Morgan, W. J., 1989, Thermal modeling of the

Chugach Metamorphic Complex: Journal of Geophysical Research, v. 94, no. B4, p. 4411–4423.

Johannes, W., and Gupta, L. N., 1982, Origin and evolution of a migmatite: Contributions to Mineralogy and Petrology, v. 79, p. 114–123.

Jones, D. L., Silberling, N. J., Coney, P. J., and Plafker, G., 1987, Lithotectonic terrane map of Alaska (west of the 141st meridan), part A, *in* Silberling, N. J., and Jones, D. L., eds., Folio of the lithotectonic terrane maps of the North American Cordillera: U.S. Geological Survey Miscellaneous Field Studies Map MF-1874-A, 1 sheet, scale 1:2,500,000.

Kenah, C., 1979, Mechanism and physical conditions of emplacement of the Quottoon Pluton, British Columbia [Ph.D. thesis]: Princeton, New Jersey, Princeton University, 184 p.

Kenah, C., and Hollister, L. S., 1983, Anatexis in the Central Gneiss Complex, British Columbia, *in* Atherton, M. P. and Gribble, C. D., eds., Migmatites, melting, and metamorphism: Nantwich, Cheshire, Shiva Publishing Ltd., p. 142–162.

Kistler, R. W., Ghent, E. D., and O'Neil, J. R., 1981, Petrogenesis of garnet two-mica granites in the Ruby Mountains, Nevada: Journal of Geophysical Research, v. 86, no. B11, p. 10591–10606.

Lange, I. M., and Nokleberg, W. J., 1984, Massive sulfide deposits of the Jarvis Creek terrane, Mt. Hayes Quadrangle, eastern Alaskan Range: Geological Society of America Abstracts with Programs, v. 16, p. 294.

Lanphere, M. A., and Reed, B. L., 1985, The McKinley sequence of granitic rocks; A key element in the accretionary history of southern Alaska: Journal of Geophysical Research, v. 90, no. B13, p. 11413–11430.

Leake, B. E., 1983, Ultrametamorphism, migmatites, and granitoid formation, *in* Atherton, M. P., and Gribble, C. D., eds., Migmatites, melting, and metamorphism: Nantwich, Cheshire, Shiva Publishing, Ltd., p. 2–9.

LeCouteur, P. C., and Tempelman-Kluit, D. J., 1976, Rb/Sr ages and a profile of initial $^{87}Sr/^{86}Sr$ ratios for plutonic rocks across the Yukon-Crystalline terrane: Canadian Journal of Earth Sciences, v. 13, p. 319–330.

Leeman, W. P., 1983, The influence of crustal structure on compositions of subduction-related magmas: Journal of Volcanology and Geothermal Research, v. 18, p. 561–588.

Luthy, S. T., Foster, H. L., and Cushing, G. W., 1981, Petrographic and chemical data on Cretaceous granitic rocks of the Big Delta Quadrangle, Alaska: U.S. Geological Survey Open-File Report 81–398, 12 p.

Marshak, R. S., and Karig, D. E., 1977, Triple junctions as a cause for anomalously near-trench igneous activity between the trench and volcanic arc: Geology, v. 5, p. 233–236.

Miller, E. L., 1987, Dismemberment of the Brooks Range orogenic belt during middle Cretaceous extension: Geological Society of America Abstracts with Programs, v. 19, p. 432.

Miller, T. P., 1972, Potassium-rich alkaline intrusive rocks of western Alaska: Geological Society of America Bulletin, v. 83, p. 2111–2128.

Miller, T. P., and Bunker, C. M., 1976, A reconnaissance study of the uranium and thorium contents of plutonic rocks of the southeastern Seward Peninsula, Alaska: U.S. Geological Survey Journal of Research, v. 4, no. 3, p. 367–377.

Miller, E. L., Holdsworth, B. K., Whiteford, W. B., and Rodgers, D., 1984, Stratigraphy and structure of the Schoonover Sequence, northwestern Nevada; Implications for Paleozoic plate-margin tectonics: Geological Society of America Bulletin, v. 95, p. 1063–1076.

Mortensen, J. K., 1983, U-Pb and Rb-Sr evidence for the age and timing of tectonism of the Yukon-Tanana terrane in south-eastern Yukon Territory: Geological Association of Canada Program with Abstracts Joint Annual Meeting, p. A49.

Nelson, S. W., and Grybeck, D., 1980, Geologic map of the Survey Pass Quadrangle, Alaska: U.S. Geological Survey Map MF-1176-A, 2 sheets, scale 1:250,000.

Nokleberg, W. J., Aleinikoff, J. N., and Lange, I. M., 1983, Origin and accretion of Andean-type and island arc terranes of Paleozoic age juxtaposed along the Hines Creek fault, Mount Hayes Quadrangle, eastern Alaska Range, Alaska: Geological Society of America Abstracts with Programs, v. 15, p. 427.

Oldow, J. S., Gottschalk, R. R., and Ave Lallemant, H. G., 1987, Low-angle normal faults; southern Brooks Range fold and thrust belt, northern Alaska: Geological Society of America Abstracts with Programs, v. 19, no. 6, p. 438.

Page, R. A., and 6 others, 1986, Accretion and subduction tectonics in the Chugach Mountains and Copper River basin, Alaska; Initial results of TACT: Geology, v. 14, p. 501–505.

Palmer, A. R., 1983, The Decade of North American Geology geologic time scale: Geology, v. 11, p. 503–504.

Patton, W. W., Jr., 1973, Reconnaissance geology of the northern Yukon-Koyukuk province, Alaska: U.S. Geological Survey Professional Paper 774-A, 17 p.

Patton, W. W., Jr., and Csejtey, B., Jr., 1980, Geologic map of St. Lawrence Island, Alaska: U.S. Geological Survey Miscellaneous Investigations Series Map I-1203, scale 1:250,000.

Patton, W. W., Jr., and Miller, T. P., 1973, Bedrock geologic map of Bettles and southern part of Wiseman Quadrangles, Alaska: U.S. Geological Survey Miscellaneous Field Studies Map MF-492, scale 1:250,000.

Plafker, G., Keller, G., Barron, J. A., and Blueford, J. R., 1985, Paleontologic data on the age of the Orca Group, Alaska: U.S. Geological Survey Open-File Report 85–429, 24 p.

Puchner, C. C., 1984, Intrusive geology of the Ray Mountains batholith: Geological Society of America Abstracts with Programs, v. 16, no. 5, p. 329.

Reed, B. L., and Lanphere, M. A., 1973, Alaska-Aleutian Range batholith; Geochronology, chemistry, and relation to Circum-Pacific plutonism: Geological Society of America Bulletin, v. 84, p. 2583–2610.

—— , 1974a, Chemical variations across the Alaska-Aleutian Range batholith: U.S. Geological Survey Journal of Research, v. 2, no. 3, p. 343–352.

—— , 1974b, Offset plutons and history of movement along the McKinley segment of the Denali fault system, Alaska: Geological Society of America Bulletin, v. 85, p. 1883–1892.

Reed, B. L., and Nelson, S. W., 1980, Geologic map of the Talkeetna Quadrangle, Alaska: U.S. Geological Survey Miscellaneous Investigation Series Map I-1174, scale 1:250,000.

Richter, D. H., 1976, Geologic map of the Nabesna Quadrangle, Alaska: U.S. Geological Survey Miscellaneous Investigation Series Map I-932, scale 1:250,000.

Richter, D. H., Lanphere, M. A., and Matson, N. A., Jr., 1975, Granitic plutonism and metamorphism, eastern Alaska Range, Alaska: Geological Society of America Bulletin, v. 86, p. 819–829.

Sable, E. G., 1977, Geology of the western Romanzof Mountains, Brooks Range, northeastern Alaska: U.S. Geological Survey Professional Paper 897, 84 p.

Schmidt, J. M., 1984, Basalt-rhyolite volcanism of the Late Devonian Ambler Sequence, northwestern Alaska; Major element compositions and tectonic setting: Geological Society of America Abstracts with Programs, v. 16, p. 331.

Silberman, M. L., Brookins, D. G., Nelson, S. W., and Grybeck, D., 1979a, Rubidium-strontium and potassium-argon dating of emplacement and metamorphism of the Arrigetch Peaks and Mount Igikpak plutons, Survey Pass Quadrangle, Alaska, *in* Johnson, K. M., and Williams, J. R., eds., The United States Geological Survey in Alaska; Accomplishments during 1978: U.S. Geological Survey Circular 804-B, p. B18–B19.

Silberman, M. L., Moll, E. J., Chapman, R. M., Patton, W. W., Jr., and Connor, C. L., 1979b, Potassium-argon age of granitic and volcanic rocks from the Ruby, Medfra, and adjacent Quadrangles, west-central Alaska, *in* Johnson, K. M., and Williams, J. R., eds., The United States Geological Survey in Alaska; Accomplishments during 1978: U.S. Geological Survey Circular 804-B, p. B63–B66.

Sisson, V. B., Hollister, C. S., and Onstatt, T. C., 1989, Petrologic and age constraints on the origin of a low-pressure/high-temperature metamorphic complex, southern Alaska: Journal of Geophysical Research, v. 94, no. B4, p. 4392–4410.

Smith, J. G., Elliott, R. L., Berg, H. C., and Wiggins, B. D., 1977, Map showing general geology and location of chemically and radiometrically analyzed samples in parts of the Ketchikan, Bradfield Canal, and Prince Rupert Quad-

rangles, southeastern Alaska: U.S. Geological Survey Miscellaneous Field Studies Map MF-825, scale 1:250,000.
Smith, T. E., 1981, Geology of the western Clearwater Mountains, central Alaska: Alaska Division Geological and Geophysical Surveys Geologic Report 60, 72 p.
——, 1984, McClaren metamorphic belt of south-central Alaska revisited; The case for suprabatholithic recrystallization: Geological Society of America Abstracts with Programs, v. 16, p. 333.
Smith, T. E., and 5 others, 1978, Evidence for a mid-Paleozoic depositional age of volcanogenic base-metal massive sulfide occurrences and enclosing strata, Ambler district, northwest Alaska: Geological Society of America Abstracts with Program, v. 10, p. 148.
Speer, J. A., 1981, Petrology of cordierite and almandine-bearing granitoid plutons of the southern Appalachian Piedmont, U.S.A.: Canadian Mineralogist, v. 19, part 1, p. 35–46.
Tempelman-Kluit, D. J., 1979, Transported cataclasite, ophiolite, and granodiorite in Yukon; Evidence of arc-continent collision: Geological Survey of Canada Paper 79–14, 27 p.
Tempelman-Kluit, D. J., and Wanless, R. K., 1975, Potassium-argon age determinations of metamorphic and plutonic rocks in the Yukon Crystalline Terrane: Canadian Journal of Earth Sciences, v. 12, no. 11, p. 1895–1909.
——, 1980, Zircon ages for the Pelly Gneiss and Klotassin Granodiorite in western Yukon: Canadian Journal of Earth Sciences, v. 17, no. 3, p. 297–306.
Till, A. B., 1983, Granulite, peridotite, and blueschist; Pre-Cambrian to Mesozoic history of Seward Peninsula, *in* Proceedings of the 1982 Symposium on Western Alaska Resources and Geology: Alaska Geological Society Journal, p. 59–66.
——, 1984, Low-grade metamorphic rocks of Seward Peninsula, Alaska: Geological Society of America Abstracts with Programs, v. 16, p. 337.
Turner, D. L., Forbes, R. B., and Dillon, J. T., 1979, K-Ar geochronology of the southwestern Brooks Range, Alaska: Canadian Journal of Earth Sciences, v. 16, p. 1789–1804.
Turner, D. L., Smith, T. E., and Forbes, R. B., 1974, Geochronology of offset along the Denali fault system in Alaska: Geological Society of America Abstracts with Programs, v. 6, p. 268–269.
White, A.J.R., and Chappell, B. W., 1977, Ultrametamorphism and granitoid genesis: Tectonophysics, v. 43, p. 7–22.
Wilson, F. H., Weber, F. R., and Angleloni, L., 1984, Late Cretaceous thermal overprint and metamorphism, southeast Circle Quadrangle, Alaska: Geological Society of America Abstracts with Programs, v. 16, p. 340.
Wilson, F. H., Smith, J. G., and Shew, N., 1985, Review of radiometric data from the Yukon Crystalline Terrane, Alaska and Yukon Territory: Canadian Journal of Earth Sciences, v. 22, p. 525–537.
Wright, J. E., and Haxel, G., 1982, A garnet two-mica granite, Coyote Mountains, southern Arizona; Geologic setting, uranium-lead isotopic systematics of zircon, and nature of the granite source region: Geological Society of America Bulletin, v. 93, p. 1176–1188.
Wyllie, P. J., 1983a, Experimental studies on biotite- and muscovite-granites and some crustal magmatic sources, *in* Atherton, M. P., and Gribble, C. D., eds., Migmatites, melting, and metamorphism: Shiva Publishing, Ltd., p. 12–26.
——, 1983b, Experimental and thermal constraints on the deep-seated parentage of some granitoid magmas in subduction zones, *in* Atherton, M. P., and Gribble, C. D., eds., Migmatites, melting, and metamorphism: Nantwich, Cheshire, Shiva Publishing Ltd., p. 37–51.

MANUSCRIPT ACCEPTED BY THE SOCIETY MAY 18, 1990

NOTES ADDED IN PROOF

Work accomplished since 1989 that is pertinent to this report includes studies in the Sanak-Baranof plutonic belt, amplification of the mid-Cretaceous regional extension model for interior Alaska, and continuing discussion of the origin of the Devonian orthogneisses. None of the new data or discussions have required significant changes in the original interpretations of this report.

Geochemical and isotopic data on Eocene granodiorite in the eastern Gulf of Alaska (Barker and others, 1992), support an origin by partial melting of the accretionary prism intruded by the Sanak-Baranof belt plutons. Isotopic data on the accretionary rocks themselves has been recently reported by Farmer and others (1993).

The early thoughts concerning the role of extension in the mid-Cretaceous tectonic evolution of interior Alaska, and therefore the setting in which voluminous crustally derived granodiorite and granite was emplaced, evolved in a report by Miller and Hudson (1991). This report links several key geologic relations of interior and northern Alaska in a model requiring large-scale crustal extension during the mid-Cretaceous. This model and many interpretations that support it are the source of continuing controversy (Till and others, 1993).

The origin of the Devonian orthogneisses is further constrained by some new isotopic data (Nelson and others, 1993). Nelson and others (1993) favor a continental-margin volcanic arc setting for the origin of these metaigneous rocks as have Rubin and others (1990) and Plafker and Berg (this volume, Chapter 33). As originally discussed in this report, the compositional character of the Devonian orthogneisses and of the metavolcanic rocks that may be contemporaneous with them in Alaska is thought by the author to more likely have formed in an extensional or transcurrent tectonic setting rather than in a continental-margin volcanic arc setting.

ACKNOWLEDGMENTS

This study was possible because of willing consultations with Cynthia Dusel-Bacon, Helen Foster, Grant Cushing, John Dillon, Alison Till, Steven Nelson, and Joseph Arth. Thomas Miller and Christopher Puchner kindly provided unpublished analytical data. I thank George Plafker for encouraging me to contribute to this volume and the Atlantic Richfield Company for providing me the time to complete this synthesis while employed for other purposes. Catherine Campbell, Henry Berg, Paul C. Bateman, Joseph G. Arth, and Warren E. Yeend gave this manuscript constructive editorial and technical review.

ADDITIONAL REFERENCES

Barker, F., Farmer, G. L., Ayuso, R. A., Plafker, G., and Lull, J. S., 1992, The 50-MA granodiorite of the eastern Gulf of Alaska: Melting in an accretionary prism in the forearc: Journal of Geophysical Research, v. 97, no. B5, p. 6757–6778.
Farmer, G. L., Ayuso, R., and Plafker, G., 1993, A Coast Mountains provenance for the Valdez and Orca Groups, southern Alaska, based on Nd, Sr, and Pb isotopic evidence: Earth and Planetary Science Letters, v. 116, p. 9–12.
Miller, E. L., and Hudson, T. L., 1991, Mid-Cretaceous extensional fragmentation of a Jurassic–Early Cretaceous compressional origin, Alaska: Tectonics, v. 10, no. 4, p. 781–796.
Nelson, B. K., Nelson, S. W., and Till, A. B., 1993, Nd- and Sr-isotope evidence for Protozoic and Paleozoic crustal evolution in the Brooks Range, northern Alaska: Journal of Geology, v. 101, p. 435–450.
Rubin, C. M., Miller, M. M., and Smith, G. M., 1990, Tectonic development of Cordilleran mid-Paleozoic volcano-plutonic complexes: Evidence for convergent margin tectonism, *in* Harwood, D. S., and Miller, M. M., eds., Paleozoic and Early Mesozoic paleogeographic relations of the Sierra Nevada, Klamath Mountains, and related terranes: Geological Society of America Special Paper 255, p. 1–16.
Till, A. B., Box, S. E., Roeske, S. M., and Patton, W. W., Jr., 1993, Comment on Mid-Cretaceous extensional fragmentation of a Jurassic–Early Cretaceous compressional origin, Alaska, by Miller, E. L. and Hudson, T. L.: Tectonics, v. 12, no. 4, p. 1076–1081.

Printed in U.S.A.

The Geology of North America
Vol. G-1, The Geology of Alaska
The Geological Society of America, 1994

Chapter 21

Ophiolites and other mafic-ultramafic complexes in Alaska

William W. Patton, Jr., and Stephen. E. Box*
U.S. Geological Survey, 345 Middlefield Road, Menlo Park, California 94025
Donald J. Grybeck
U.S. Geological Survey, 4200 University Drive, Anchorage, Alaska 99508-4667

INTRODUCTION

Mafic and ultramafic complexes are widespread throughout Alaska, ranging in size from huge allochthonous masses several hundred square kilometers in area to tiny isolated blocks (Fig. 1). Some of these, such as the complexes in northern and western Alaska, clearly can be labeled ophiolites; others, such as the concentrically zoned bodies of southeastern Alaska, are not ophiolites; and still others, such as those in the Livengood belt of central Alaska, have uncertain affinities. All of the complexes discussed here, however, belong to well-defined belts that for the most part are confined to specific lithotectonic terranes or lie along terrane boundaries. Few of these complexes have been studied in detail, and the mode and time of emplacement of most are uncertain or controversial. In this chapter, we review available information on the structural setting and petrography of the complexes, and we describe the tectonic models that have been suggested to explain the mode of emplacement.

NORTHERN AND WEST-CENTRAL ALASKA

In the northern and west-central parts of Alaska, mafic-ultramafic complexes are exposed in three separate belts (Fig. 1): western Brooks Range, Angayucham, and Tozitna-Innoko. The complexes in the western Brooks Range and Angayucham belts are included in the Angayucham lithotectonic terrane as defined by Silberling and others (this volume); the complexes in the Tozitna-Innoko belt are included in the Tozitna and Innoko lithotectonic terranes as defined by Silberling and others (this volume). All three belts are composed of late Paleozoic and Mesozoic oceanic rocks that have been thrust onto Precambrian and Paleozoic miogeoclinal assemblages. Within each of these belts we recognize two separate thrust panels: a lower panel consisting of multiple thrust slices of pillow basalt, chert, nonlayered gabbro, and volcaniclastic rocks; and an upper panel composed of mafic-ultramafic complexes (Patton and others, 1977). Most workers agree that the mafic-ultramafic complexes of the upper panel represent the lower part of a classic ophiolite sequence (Roeder and Mull, 1978; Patton and others, 1977; Mayfield and others, 1983b; Loney and Himmelberg, 1984, 1985a, b). The composition of the upper panel and its close spatial relation to the higher-level oceanic rocks in the lower panel led to an earlier suggestion that the two panels together form a nearly complete dismembered ophiolite sequence (Patton and others, 1977; Roeder and Mull, 1978). However, recent dating by K-Ar and fossils has shown that the mafic-ultramafic complexes of the upper panel are of probable Jurassic age, whereas the pillow basalt, gabbro, chert, and volcaniclastic rocks of the lower panel range in age from Devonian to Jurassic.

Western Brooks Range

Scattered mafic-ultramafic complexes, including some of the largest in Alaska, extend in a broad arcuate belt across the western Brooks Range from Siniktanneyak Mountain to the Chukchi Sea, a distance of about 350 km. Five separate complexes have been mapped from east to west: Siniktanneyak Mountain, Misheguk Mountain, Asik Mountain, Avan Hills, and Iyikrok Mountain. The Avan Hills, Misheguk Mountain, and Asik Mountain complexes may coalesce beneath the alluviated lowlands of the lower Noatak River to form a single large body.

Recent workers in the western Brooks Range agree that these mafic-ultramafic complexes represent synformal remnants of a south-dipping allochthonous sheet that had its roots along the northern edge of the Yukon-Koyukuk basin (Mayfield and others, 1983b; Roeder and Mull, 1978; Patton and others, 1977). Both the synformal remnants in the western Brooks Range and the root zone along the northern edge of the Yukon-Koyukuk basin are included in the Angayucham terrane by Silberling and others (this volume). In the western Brooks Range, Mayfield and others (1983b) assign these rocks to the Misheguk Mountain allochthon, the highest of seven imbricated sequences that have overridden the western Brooks Range. They summarize the major features of the Misheguk Mountain sequence as follows:

This sequence is the lower part of a classical ophiolite sequence. The layers of the basalt and oceanic sediments above the gabbroic and ultra-

*Present address: U.S. Geological Survey, Room 656, West 920 Riverside Avenue, Spokane, Washington 99201.

Patton, W. W., Jr., Box, S. E., and Grybeck, D. J., 1994, Ophiolites and other mafic-ultramafic complexes in Alaska, *in* Plafker, G., and Berg, H. C., eds., The Geology of Alaska: Boulder, Colorado, Geological Society of America, The Geology of North America, v. G-1.

mafic rocks have not been reported and presumably were stripped away by erosion. The bottom of the sequence is a zone of olivine-rich peridotite and dunite with both tectonite and cumulate characteristics. Above this is a transitional zone that contains alternating layers of peridotite, pyroxenite, and gabbroic rocks. The top of the sequence mainly consists of gabbroic rock that tends to display cumulate layering of pyroxene-rich and plagioclase-rich layers in the lower part; in the upper part, it is not layered and is more leucocratic. Local occurrences of granitoid rocks are more prevalent in the upper part and seem to occur both as layers and as cross-cutting dikes in the gabbro (Roeder and Mull, 1978; Nelson and others, 1979; Zimmerman and others, 1981).

Mayfield and others (1983b) estimate that the Misheguk Mountain allochthon is no more than 1 to 2 km thick.

Description of complexes. *Siniktanneyak Mountain* (locality 1, Fig. 1). The easternmost complex of the western Brooks Range belt has a roughly oval-shaped exposed area of 150 km^2. Another small body of ultramafic rocks occurs 15 km to the southwest of the main mass, and if the covered area between is also underlain by ultramafic and mafic rocks, the total size of the complex may be as much as 300 km^2 (Inyo Ellersieck, oral communication, 1984). A geologic map by Nelson and Nelson (1982) indicates that the exposed body at Siniktanneyak Mountain is composed of about 20 percent ultramafic rocks—chiefly serpentinized dunite with cumulus layers of wehrlite, harzburgite, and olivine pyroxenite—and about 80 percent layered gabbro. The northern and western part of the complex is intruded by two small stocks ranging in composition from hornblende diorite to biotite-hornblende alaskite and the southeastern part by muscovite-biotite granite and alaskite dikes. A K-Ar age of 151 ± 15 Ma (Late Jurassic) was measured on hornblende from a hornblende pegmatite boulder in a stream draining Siniktanneyak Mountain (Patton and others, 1977). According to I. L. Tailleur

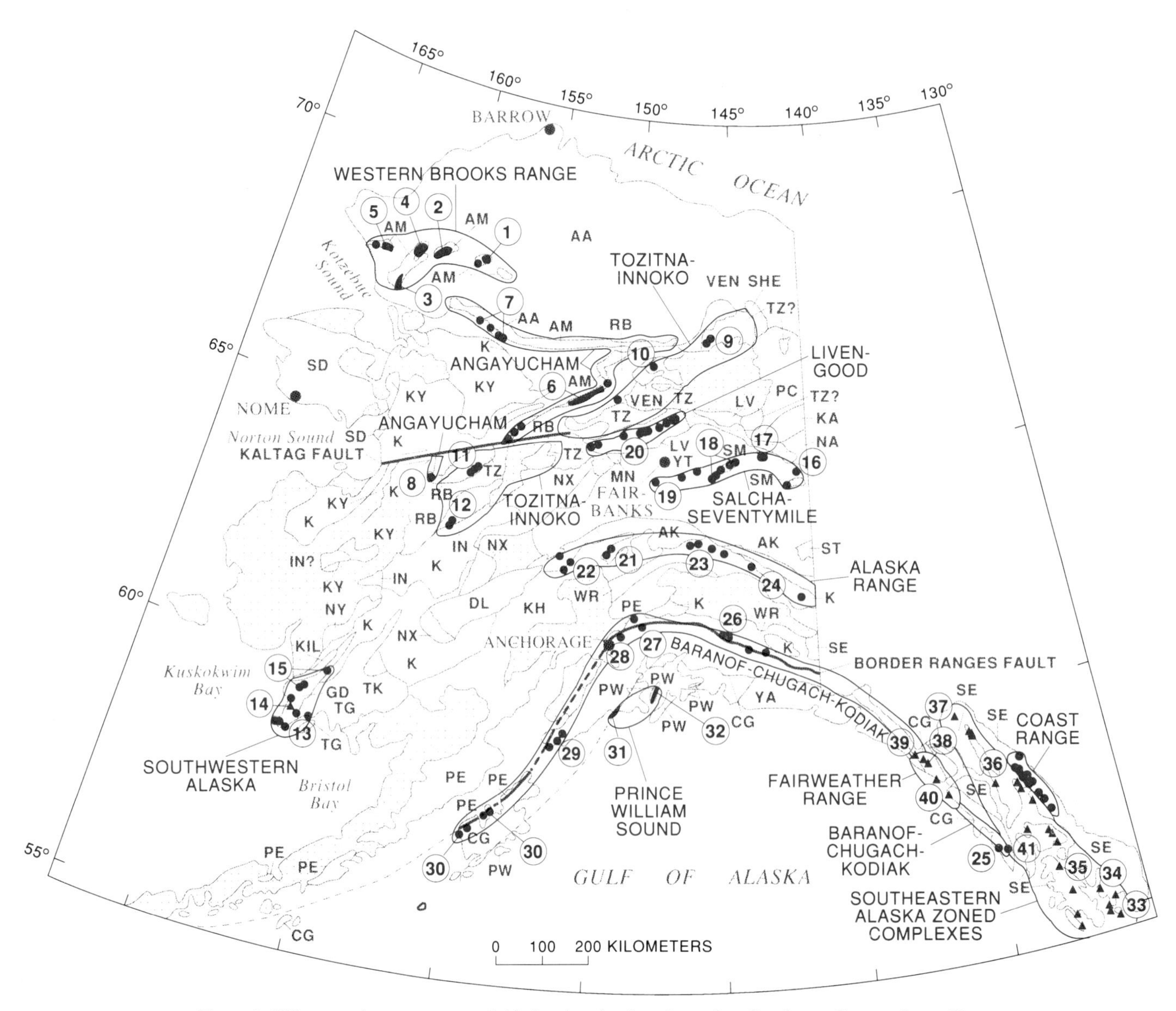

Figure 1. Lithotectonic terrane map of Alaska showing locations of mafic-ultramafic complexes. Terrane map modified from Jones and others (1987) and W. J. Nokleberg (written communication, 1989).

(oral communication, 1984), similar hornblende pegmatites are common within the complex. Potassium-argon ages of 162 ± 8 and 171 ± 9 Ma (Middle or Late Jurassic) were measured on a coexisting biotite-hornblende mineral pair from a quartz diorite intrusive body (C. F. Mayfield and Inyo Ellersieck, oral communication, 1984).

Misheguk Mountain (locality 2, Fig. 1). The largest of the western Brooks Range mafic-ultramafic complexes is 50 km long by 10 to 15 km wide and has a total exposed area of 460 km^2. Few petrographic details of the complex are known. Mapping by Ellersieck and others (1982) and Curtis and others (1982) indicates that only about 10 percent of the complex is made up of ultramafic rocks—chiefly peridotite, serpentinized dunite, and minor pyroxenite dikes. The main mass of the complex consists of medium- to coarse-grained layered gabbro intruded by dikes of peridotite, pyroxenite, and hornblende-plagioclase pegmatite. A small pluton of diorite, quartz diorite, and granite occurs in the northeastern part of the complex. Cross sections by Ellersieck and others (1982) show that the ultramafic rocks form a slablike body dipping to the southeast beneath medium- to coarse-grained layered gabbro, which in turn is succeeded along the southern margin of the complex by fine-grained gabbro. Locally, along the northwestern border of the complex, the ultramafic rocks are underlain by a wedge of pelitic schist and marble of uncertain age. K-Ar ages of 147.2 ± 4.4 and 155.8 ± 4.7 Ma have been obtained recently from hornblende gabbro and hornblende-bearing diorite from the upper part of the complex (Boak and others, 1985). The ages are thought to postdate crystallization of the mafic-ultramafic rocks and are attributed to thermal metamorphism during emplacement.

Avan Hills (locality 4, Fig. 1). This roughly oval-shaped complex has a total area of about 200 km^2 and probably extends another 100 km^2 beneath the heavily alluviated areas bordering the Avan River. The complex is divided into three separate zones:

EXPLANATION

Structurally bound mafic-ultramafic complexes (dot where complex too small to show at map scale)

Intrusive mafic-ultramafic complexes

Areas encompassing mafic-ultramafic complexes that are related by similar geologic settings—Name indicates belt or region

(8) Locality number discussed in text

MAP UNITS
[see Plate 3 for description of units]

POST-ACCRETION COVER DEPOSITS

Cenozoic deposits

K Cretaceous clastic deposits

ACCRETED TERRANES

AA Arctic Alaska
AK Alaska Range terranes, undivided—Broad Pass, Chulitna, Clearwater, Dillinger, Kahiltna, McKinley, Maclaren, Mystic, Nenana, Pingston, Susitna, West Fork, and Windy
AM Angayucham
CG Chugach
DL Dillinger
GD Goodnews
IN Innoko
KA Kandik River
KH Kahiltna
KIL Kilbuk
KY Koyukuk
LV Livengood area terranes, undivided—Baldry, Crazy Mountains, Livengood, Manley, Minook, White Mountains, and Wickersham
MN Minchumina
NX Nixon Fork
NY Nyac
PC Porcupine
PE Peninsular
PW Prince William
RB Ruby
SD Seward
SE Southeastern Alaska terranes undivided—Alexander, Stikinia, Tracy Arm, Taku, Wrangellia, and the post-accretionary Gravina-Nutzotin belt
SHE Sheenjek
SM Seventymile
ST Stikinia
TG Togiak
TK Tikchik
TZ Tozitna
VEN Venetie
WR Wrangellia
YA Yakutat
YT Yukon-Tanana

NONACCRETIONARY CONTINENTAL ROCKS

NA North America

an ultramafic tectonite zone of unknown thickness, a transitional zone about 1 km thick in which ultramafic and mafic rocks are interlayered, and a mafic zone about 2.5 km thick (Frank and Zimmerman, 1982; Zimmerman and others, 1981; Zimmerman and Soustek, 1979). The ultramafic tectonite zone is composed of interlayered dunite, harzburgite, and minor amounts of pyroxenite. The ultramafic rock is foliated nearly parallel to compositional layering and shows evidence of plastic strain and recrystallization. The mafic zone is composed of about 90 percent layered gabbro, chiefly leucogabbro, and subordinate amounts of troctolite and anorthosite. Small bodies of biotite granite, biotite tonalite, and tonalite are reported by Frank and Zimmerman (1982) in the western part of the complex, and a small body of "metabasite" is reported by Zimmerman and Frank (1982) at the base of the ultramafic zone. No age data for the Avan Hills complex have been published.

Iyikrok Mountain (locality 5, Fig. 1). This mafic-ultramafic complex is made up of two poorly exposed bodies, the larger of which is located at Iyikrok Mountain and the smaller about 20 km west of Iyikrok Mountain. The Iyikrok Mountain body has a total outcrop area of about 30 km^2, but Mayfield and others (1983a) suggest that it extends an additional 20 km^2 beneath an area of few exposures northwest of Iyikrok Mountain. The smaller body is about 15 km^2 in area, judging from the distribution of scattered exposures. Few details about the composition of either of these two bodies are known.

Asik Mountain (locality 3, Fig. 1). A roughly rectangular-shaped complex crops out in a 100-km^2 area at Asik Mountain on the east side of the lower Noatak River basin. Gravity and magnetic data suggest that it extends westward beneath the Noatak River for an additional 300 km^2 (Barnes and Tailleur, 1970; Mayfield and others, 1983b). No detailed study of the Asik complex has been carried out; based on reconnaissance mapping, Inyo Ellersieck (oral communication, 1984) estimates that the complex is made up of about 5 percent ultramafic rocks, exposed chiefly along the southern and western margins, and about 95 percent gabbro.

Geologic setting. The western Brooks Range mafic-ultramafic complexes belong to the highest of seven allochthonous sheets that constitute the stack of south-dipping imbricated thrust plates making up the western Brooks Range (Mayfield and others, 1983b). The mafic-ultramafic complexes of allochthon 7 and the pillow basalt and chert-forming allochthon 6 appear to have a close spatial, if not genetic, relation. They are preserved in small synformal remnants, which overlie Devonian to Early Cretaceous continental and continental-margin deposits composing allochthons 1 to 5. Allochthons 6 and 7 appear to be rooted along the northern margin of the Yukon-Koyukuk basin where they rest in thrust contact upon metamorphosed Proterozoic and early Paleozoic strata of the southern Brooks Range schist belt (Hammond subterrane of Silberling and others, this volume).

Thin slices of metamorphic rocks are reported locally along the sole of the mafic-ultramafic allochthon. On the northwest side of the Misheguk Mountain complex, a wedge of metamorphic rocks including quartz-muscovite-chlorite-garnet schist, actinolite-albite-chlorite schist, and marble separate the complex from the underlying allochthon of pillow basalt and chert (Ellersieck and others, 1982). The protoliths of these rocks are believed to be pelitic sedimentary rocks, mafic igneous rocks, and limestone, but their age and stratigraphic relations are uncertain. In the Avan Hills complex "metabasites" are reported by Zimmerman and Frank (1982) at the base of the mafic-ultramafic allochthon. At both localities, metamorphism has been attributed to emplacement of the mafic-ultramafic allochthons. From the zone of amphibolite and hornblende gneiss at the base of the Iyikrok Mountain complex, Boak and others (1985) obtained K-Ar ages of 154.2 ± 4.6 and 153.2 ± 4.6 Ma.

Time of emplacement. Several lines of evidence suggest that Brooks Range orogenic activity leading to emplacement of the mafic-ultramafic allochthon began in the Late Jurassic and continued into the Early Cretaceous. Orogenic flyschoid deposits rich in mafic debris derived from the mafic-ultramafic rocks and containing Late Jurassic and Early Cretaceous fossils are incorporated in thrust sheets that underlie the mafic-ultramafic allochthon (Mayfield and others, 1983b). Metamorphism and, in particular, the occurrence of "blueschist" (glaucophane) mineral assemblages in the southern Brooks Range are thought to be associated with overthrusting of the mafic-ultramafic allochthonous rocks. Greenschist-facies metamorphic mineral assemblages in this glaucophane-bearing schist belt yield predominantly Early Cretaceous potassium-argon cooling ages (Turner, 1984), which probably record the time of uplift following metamorphism. Added evidence on the timing of emplacement comes from the southeast side of the Yukon-Koyukuk basin, where garnet-amphibolite tectonites at the base of the mafic-ultramafic allochthon yield Late Jurassic metamorphic-mineral K-Ar ages (Patton and others, 1977). Emplacement of the mafic-ultramafic complexes must have occurred before late Early Cretaceous because: (1) Albian conglomerates, eroded from the Brooks Range and deposited along the northern margin of the Yukon-Koyukuk basin, contain debris from mafic-ultramafic complexes and from the underlying metamorphic rocks; and (2) granitic plutons of Albian and Cenomanian age on the southeast side of the basin intrude both the mafic-ultramafic complexes and the underlying metamorphic rocks (Patton, 1984).

Angayucham

The Cretaceous Yukon-Koyukuk basin of west-central Alaska (Patton and others, chapter 7, this volume) is bordered on the northern and southeastern sides by the Angayucham terrane (Fig. 1), a narrow but nearly continuous belt of oceanic rocks that dips inward beneath the basin and is thrust outward onto the Proterozoic and Paleozoic metamorphic rocks of the bordering Arctic Alaska and Ruby terranes. In this area we recognize three separate thrust panels in the Angayucham terrane (Patton and others, chapter 7, this volume): the lower or Slate Creek thrust panel, composed of phyllite and metagraywacke; the middle or Narvak thrust panel, composed of pillow basalt, chert, and gabbro; and the upper or Kanuti thrust panel, composed of mafic-ultramafic complexes. The middle and upper panels appear to correlate with allochthons 6 and 7 of the Angayucham terrane in the western Brooks Range. The Angayucham terrane in the Yukon-Koyukuk basin can be traced for about 500 km along the northern edge of the basin from the lower Kobuk River to the northeastern apex of the basin and for 400 km along the southeast side from the northeastern apex of the basin to the Kaltag fault. The Kaltag fault offsets the margin of the basin and the adjoining Angayucham terrane about 140 km to the west (Patton and Hoare, 1968; Patton and others, 1984). South of the fault the offset extension of the Angayucham terrane can be followed for an additional 50 km from near the village of Kaltag to the Innoko River lowlands (Fig. 1).

Description of complexes. Large mafic-ultramafic complexes composing the upper thrust panel of the Angayucham terrane are preserved in three separate localities bordering the Yukon-Koyukuk basin. Each of these localities is described below.

Kanuti (locality 6, Fig. 1). The largest volume of mafic-ultramafic complexes bordering the Yukon-Koyukuk basin is exposed along a 100-km segment of the Angayucham terrane on the southeastern limb of the basin in the Kanuti River drainage. The complexes vary in width from 1 to 10 km and have a total exposed area of about 300 km^2. According to Loney and Himmelberg (1985b), the complexes can be subdivided into a cumulus plutonic suite and a tectonite mantle suite. The cumulus suite, which makes up the upper part of the mafic-ultramafic thrust panel and lies generally northwest of the mantle suite, is composed of interlayered cumulus wehrlite, olivine clinopyroxenite, and gabbro. The mantle suite, which makes up the lower part of the thrust panel and lies southeast of the plutonic suite, is composed of harzburgite, dunite, and minor clinopyroxenite. Overall, the complexes are composed of about 65 percent cumulate plutonic suite and 35 percent mantle tectonite suite. The base of the mafic-ultramafic thrust panel commonly has a layer of amphibolite that is 10 to 25 m thick and consists of a highly tectonized aggregate of amphibole, plagioclase, and garnet. The basal thrust, where well exposed, dips from 15 to 26° to the northwest (Loney and Himmelberg, 1985b).

Two K-Ar ages of latest Jurassic (149 ± 9.6 Ma) and earliest Cretaceous (138 ± 8 Ma) age were obtained from amphibole-bearing dikes in the cumulate plutonic suite, and a K-Ar age of Late Jurassic (161 ± 4.9 Ma) was obtained from metamorphic amphibole from the thin layer of garnet amphibolite at the base of the mafic-ultramafic thrust panel (Patton and others, 1977). These ages are interpreted as cooling ages related to tectonic emplacement of the thrust panel.

Jade Mountains and Cosmos Hills (locality 7, Fig. 1). The mafic-ultramafic complexes of the upper thrust panel of the Angayucham terrane are preserved on the northern limb of the Yukon-Koyukuk basin in the Jade Mountains and Cosmos Hills. The main mass of the Jade Mountains is composed of pillow basalt belonging to the middle thrust panel. The pillow basalt is overlain on the southern flank of the mountains by the upper thrust panel consisting of a 500-m-thick slab of serpentinite that dips gently to the south. The total exposed area of the ultramafic rocks is about 9 km^2, but several small rubble patches of serpentinite in the adjoining Kobuk lowlands to the south suggest that the total area underlain by the serpentinite slab may be as much as 40 to 50 km^2. Although the ultramafic rock has been thoroughly serpentinized, the original fabric of the rock can be recognized. Remnant olivine and orthopyroxene are visible in thin section, indicating that the protolith rock was predominantly a harzburgite tectonite (Loney and Himmelberg, 1985a).

In the Cosmos Hills, the upper thrust panel of the Angayucham terrane is represented by tabular bodies of serpentinite, as much as 130 m thick and 8 km long, distributed around the metamorphic core of the Cosmos Hills antiform. Structurally, these highly tectonized bodies underlie sheared Cretaceous conglomerate of the Yukon-Koyukuk basin and overlie schist of the southern Brooks Range metamorphic belt (Hammond subterrane of Silberling and others, this volume). Highly altered and sheared bodies of mafic volcanic rocks are also found along this contact and may represent fragments of the middle thrust panel of the Angayucham terrane. Asbestos Mountain, a tabular body of serpentinite resting on the schist along the axis of the Cosmos Hills antiform, appears to be a klippe or remnant of the upper thrust panel that once covered the metamorphic core of the antiform.

No age data are available for the ultramafic complexes in the Jade Mountains or Cosmos Hills.

Magitchlie Range (locality 8, Fig. 1). Mafic-ultramafic complexes belonging to the upper thrust panel of the Angayucham terrane are also preserved on the southeastern limb of the Yukon-Koyukuk basin south of the Kaltag fault. This belt of complexes, which lies along the east side of the Magitchlie Range, can be traced by scattered patches of rubble and a strong positive aeromagnetic anomaly (Department of Energy, 1982) for about 50 km from the Kaltag fault to the Innoko River lowlands. Because of limited exposure, it is difficult to judge the size and continuity of the ultramafic bodies along this belt. The largest exposed complex, estimated to be about 10 km^2 in area, forms an unnamed isolated knob at benchmark "Peak No. 7" (Ophir A-5 Quadrangle). According to R. A. Loney and G. R. Himmelberg (oral communication, 1984), this complex consists chiefly of partly serpentinized cumulus wehrlite, dunite, and olivine gabbro cut by dikes of hornblende gabbro and hornblende pegmatite. Another smaller complex composed of highly altered, sheared, and locally talcose serpentinite occurs in the Innoko River lowlands 27 km southeast of "Peak No. 7."

The geologic setting of these ultramafic complexes is obscured by heavily vegetated lowlands that separate them from the surrounding geologic terranes. However, their location between the metamorphic rocks of the Ruby terrane in the Kaiyuh Mountains to the east and conglomerates of the Yukon-Koyukuk basin in the Magitchlie Range to the west, strongly supports the interpretation that they represent an extension of the Angayucham terrane.

A Late Jurassic K-Ar amphibole age of 151 ± 4.5 Ma was obtained from a gabbro pegmatite dike that cuts the peridotite at benchmark "Peak No. 7" (Patton and others, 1984).

Tectonic setting. The Angayucham terrane is composed of an upper thrust panel of mafic-ultramafic complexes; a middle thrust panel of pillow basalt, chert, and gabbro; and a lower thrust panel of phyllite and metagraywacke. It represents the root zone of huge nappes that overthrust the metamorphic cores of the southern Brooks Range (Arctic Alaska terrane) and the Ruby geanticline (Ruby terrane) in Early Cretaceous time (Patton and others, 1977; Roeder and Mull, 1978; Mayfield and others, 1983b). Emplacement of these three thrust panels occurred when a volcanic arc (Koyukuk terrane) rooted in oceanic crust collided with the Proterozoic and early Paleozoic continental rocks that border the basin (Roeder and Mull, 1978; Gealey, 1980; Box, 1985b; Patton and others, this volume, Chapter 7). Cretaceous K-Ar ages of metamorphic minerals from the metamorphosed continental borderlands probably are related to the Early Cretaceous overthrusting (Turner, 1984; Patton and others, 1984). Collision of the arc and thrusting of the oceanic rocks onto the continental rocks must have occurred before the late Early Cretaceous because: (1) Albian conglomerate, eroded from the borderlands and deposited along the margins of the basin, contains clasts of both oceanic and continental rocks; and (2) granitic plutons of Albian and Cenomanian (111 to 91 Ma) age intrude both the Ruby terrane and the Angayucham terrane along the margin of the Ruby geanticline (Patton, 1984). The Late Jurassic K-Ar ages obtained from metamorphic amphibole from garnet amphibolite tectonite on the sole of the upper thrust panel, and from amphibole-bearing dikes within the upper thrust panel are believed to reflect thermal events related to thrusting of the upper thrust panel over the middle thrust panel. These ages suggest that this overthrusting preceded thrusting of the Angayucham terrane over the metamorphic borderlands (Mayfield and others, 1983b; Patton and others, this volume, Chapter 7).

Although most workers agree that the oceanic rocks of the Angayucham terranes are allochthonous with respect to the metamorphic assemblages of the southern Brooks Range and Ruby genaticline, not all agree on the mode of emplacement. Coney (1984), for example, suggested that the oceanic rocks were rooted in an ocean basin that lay to the south and were thrust northward across the Ruby geanticline and Yukon-Koyukuk basin. In another interpretation, Gemuts and others (1983) envisaged generation of the oceanic rocks in rifts along the margins of the Yukon-Koyukuk basin and subsequent thrusting of the oceanic rocks onto the metamorphic borderlands during a later compressional event. Similarly, Dover (this volume) suggests that the mafic-ultramafic rocks were generated within the continental rocks of the Ruby geanticline and southern Brooks Range and were tectonically emplaced by nonaccretionary processes.

Tozitna-Innoko

The mafic-ultramafic complexes on the Ruby geanticline (Patton and others, this volume, Chapter 7) are shown on Figure 1 as the Tozitna-Innoko belt. They consist of allochthonous synformal masses of late Paleozoic and Mesozoic oceanic crustal and mantle rocks thrust onto the metamorphosed autochthonous and para-autochthonous Proterozoic(?) and early Paleozoic Ruby terrane in the core of the Ruby geanticline. They are roughly correlative in age and lithology with the Angayucham terrane and may represent remnants of the same overthrust sheets. Like the Angayucham terrane, they are composed of separate thrust panels: an upper thrust panel composed of mafic-ultramafic complexes and a lower thrust panel composed of multiple thrust sheets of pillow basalt, radiolarian chert, gabbro, volcaniclastic rocks, and argillite. A third thrust panel composed of phyllite and metagraywacke, similar to that found in the lower part of the Angayucham terrane in the Yukon-Koyukuk basin, is also present locally at the base of the Tozitna-Innoko complexes in the area north of the Kaltag fault.

Description of complexes. Four major mafic-ultramafic complexes located along the Tozitna-Innoko belt are described below:

Christian (locality 9, Fig. 1). Mafic-ultramafic complexes occur along the axis of a large synformal mass of Tozitna terrane located at the northeastern end of the Ruby geanticline where the geanticline merges with the Brooks Range. The bulk of the Tozitna terrane consists of nonlayered gabbro, basalt, diabase, chert, shale, and carbonate rock (Brosgé and Reiser, 1962). The structurally higher mafic-ultramafic complexes form two oblong bodies that have a total exposed area of about 170 km^2. They are composed largely of interlayered hornblende leucogabbro, anorthosite, hornblende gabbro, pyroxenite, and peridotite cumulates (Reiser and others, 1965; D. L. Jones, oral communication, 1984). The contact at the base of the complexes is poorly exposed, but we believe it is a thrust fault. A hornblende mineral separate from leucogabbro yielded a Middle Jurassic K-Ar age of 172 ± 8 Ma (Reiser and others, 1965; Patton and others, 1977).

Pitka (locality 10, Fig. 1). A mafic-ultramafic complex in the Tozitna terrane is located 140 km to the southwest of the Christian bodies in the upper drainages of the Hodzana River (Brosgé and others, 1973). This complex is exposed over an area of about 25 km^2 and is interpreted by Brosgé and others (1974) as a klippe resting on a structurally lower sheet of basalt, gabbro, and chert. Brosgé and others (1974) describe the complex as a synform of strongly foliated dunite and harzburgite, banded garnet amphibolite, and gneissic leucogabbro. At the base of the klippe a layer of garnet amphibolite, similar to the one that occurs on the sole of the mafic-ultramafic allochthon in the Kanuti segment of the Angayucham belt, yielded a Late Jurassic K-Ar metamorphic mineral age of 155 ± 4.6 Ma (Patton and others, 1977).

Yuki (locality 11, Fig. 1). Mafic-ultramafic complexes occur at the southwestern end of the Tozitna terrane in the Yuki River drainage of the Kaiyuh Mountains. Several large complexes that have a total exposed area of about 140 km^2 are distributed along a northeast-trending belt within a broad synformal mass of pillow basalt, radiolarian chert, and nonlayered gabbro (Patton and others, 1977). The complexes are composed of partly serpentinized harzburgite tectonite and of ultramafic cumulates that include dunite, wehrlite, and olivine clinopyroxenite (Loney and Himmelberg, 1984). Structural relations at the base of the complexes are obscured by a dense cover of vegetation, but scattered exposures of banded garnet amphibolite tectonite suggest that the contact is a thrust fault. A Middle Jurassic K-Ar age of 172 ± 5.2 Ma was measured on metamorphic amphibole from the amphibolite (Patton and others, 1984).

Mt. Hurst–Dishna (locality 12, Fig. 1). Mafic-ultramafic complexes within the Innoko terrane occur at Mt. Hurst and 35 km on strike to the southwest on the upper Dishna River (Chapman and others, 1985a; Miller and Angeloni, 1985). The contacts of the complexes are poorly exposed, and their precise relations to the subjacent upper Paleozoic and lower Mesozoic chert, argillite, basalt, and tuff of the Innoko terrane are uncertain. On the northwest side, the Mt. Hurst complex appears to be faulted against a narrow band of metamorphic rocks that presumably belongs to the Ruby terrane. The Mt. Hurst complex is well exposed over an area of about 30 km^2; the Dishna complex is less well exposed but appears to be of similar size. The Mt. Hurst complex is composed of about one-third harzburgite and two-thirds cumulate ultramafic rocks that include dunite, wehrlite, and olivine clinopyroxenite (Loney and Himmelberg, 1984). The Dishna complex consists of harzburgite, lherzolite, pyroxenite, dunite, and locally gabbro (Miller and Angeloni, 1985). All of the ultramafic rocks are partly to wholly serpentinized. Age data for the complexes are not available.

SOUTHWESTERN ALASKA

Mafic-ultramafic complexes are distributed along a belt extending 220 km north-northeast from Cape Newenham. Along this belt we recognize the following groups of complexes, each with a distinctive history and mode of occurrence.

Description of complexes. *Cape Newenham* (locality 13, Fig. 1). Mafic-ultramafic complexes in the vicinity of Cape Newenham are exposed over an area of about 90 km^2. They appear to represent thrust sheets of the disrupted base of the Togiak terrane resting in fault contact on the Goodnews terrane (Box, 1985a). The complexes consist of about 90 percent nonlayered gabbro and 10 percent serpentinized harzburgite and dunite. Pillow basalt and gabbro are closely associated with the mafic-ultramafic complexes in the Togiak terrane, and although contacts are everywhere faulted, we believe the pillow basalt and gabbro represent higher levels of the same ophiolite succession. The pillow basalt, which is Late Triassic in age, grades upward through a thick volcanic breccia section into Lower Jurassic shallow-marine andesitic volcanic and volcaniclastic strata. The structurally underlying Goodnews terrane is composed of schistose nappes, the highest of which consists of foliated basalt and gabbro. Lower nappes in the Goodnews terrane are composed of metasedimentary rocks. The schistose rocks are characterized by transitional greenschist-blueschist-facies metamorphic assemblages.

The mafic-ultramafic complexes at Cape Newenham have not been dated by isotopic methods. However, the associated pillow basalt, which we infer to be a stratigraphically higher part of this ophiolite succession, is intercalated with red radiolarian chert of Late Triassic age (J. M. Hoare, written communication, 1976). The Cape Newenham complexes were structurally emplaced along north-directed thrust faults onto the schistose rocks between Late Triassic and Middle Jurassic time. The older constraint on time of emplacement is the age of the radiolarian chert in the pillow basalt; the younger constraint is the K-Ar isotopic age of crosscutting intrusive rocks of the Platinum complexes (see below).

Platinum (locality 14, Fig. 1). The Platinum complexes underlie an area of about 25 km^2 near Goodnews Bay. They consist of two zoned bodies that have cores of weakly serpentinized dunite, intermediate shells of clinopyroxenite and hornblende gabbro-diorite, and outer rims of coarse pegmatitic hornblendite (Mertie, 1940; Bird and Clark, 1976; Southworth, 1984). Closely associated with these zoned bodies are scattered bodies of hornblende gabbro that yield K-Ar ages similar to those of the pegmatitic hornblendite. The zoned bodies intrude the schistose nappes of the Goodnews terrane; the hornblende gabbro bodies intrude both the schistose rocks of the Goodnews terrane and the overlying ophiolite and andesitic volcanic arc rocks of the allochthonous Togiak terrane. Hornblende pegmatites that rim the zoned bodies yield Middle Jurassic K-Ar ages (187 and 176 Ma); the associated hornblende gabbro bodies give Middle and Late Jurassic K-Ar ages (187 to 162 Ma).

Arolik (locality 15, Fig. 1). The Arolik mafic-ultramafic complexes have a total exposed area of about 70 km^2 in the drainage of the Arolik River north of Goodnews Bay. They also include scattered small serpentinite bodies extending for 120 km northeast from the Arolik River. The Arolik complexes consist of about 75 percent pyroxene gabbro and 25 percent harzburgite and minor amounts of dunite. Little is known about the petrography of these rocks, but generally they appear similar in composition to the Cape Newenham complexes. We interpret the Arolik complexes as klippen overlying transitional greenschist-blueschist-facies metamorphic rocks that are faulted on the northwest against the metamorphic rocks of the Kilbuck terrane. They appear to have been emplaced in Early Cretaceous time by northwestward thrusting associated with collision between the continental rocks of the Kilbuck terrane and the arc rocks of the amalgamated Togiak and Goodnews terranes. Cretaceous K-Ar and Rb-Sr metamorphic ages of 120 to 150 Ma (Turner and others, 1983) from the Kilbuck terrane may be related to this collision. The "root zone" of the Arolik bodies is uncertain, as are their igneous ages.

EAST-CENTRAL ALASKA

Salcha-Seventymile

A discontinuous belt of ultramafic complexes more than 200 km long is exposed in the Salcha and Seventymile Rivers area of the Yukon-Tanana Upland. This belt, which was named the Seventymile terrane by Jones and others (1987; Fig. 1), extends from the Alaska-Yukon boundary near Eagle west-southwest to the Wood River south of Fairbanks. The total exposed area of these complexes is approximately 200 km^2. From east to west they include: (1) the American River body south of Eagle (locality 16, Fig. 1); (2) the Mt. Sorenson body in the upper Seventymile River drainage (locality 17, Fig. 1); (3) a linear belt of smaller bodies north of the Salcha River (locality 18, Fig. 1); and (4) an isolated body in the Tanana-Kuskokwim Lowlands at Wood River Buttes (locality 19, Fig. 1). Aeromagnetic data (Decker and Karl, 1977) suggest that the body at Wood River Buttes occurs along a southwestward extension of the Salcha River belt. The ultramafic rocks consist of variably serpentinized bodies of harzburgite and subordinate amounts of dunite and clinopyroxenite. Most of these ultramafic bodies are closely associated with altered basalt ("greenstone") and minor amounts of gabbro and chert. The contact relations between those rocks and the ultramafic rocks are unclear, but their close spatial association suggests a possible genetic relation. Both the ultramafic rocks and the basalt, gabbro, and chert assemblages are surrounded by, and appear to rest structurally upon, greenschist-facies pelitic and psammitic schists of the Yukon-Tanana terrane. Glaucophane has been reported from one locality in the schists about 20 km east of the Mt. Sorenson body (Keith and others, 1981).

Direct information on the age of the ultramafic complexes is not available. Early Permian radiolarians and conodonts have been recovered from chert associated with an ultramafic body on the Salcha River (Weber and others, 1978). If the chert is part of a dismembered ophiolite succession that included the ultramafic rocks at the base, then the ultramafic rocks presumably are Permian or older.

All of the ultramafic complexes and the closely associated basalt, gabbro, and chert assemblages of the Seventymile terrane are allochthonous with respect to the underlying schist of the Yukon-Tanana terrane (H. L. Foster, oral communication, 1985). However, the direction of transport and location of the root zone of these thrust sheets are uncertain, and the time of emplacement cannot be fixed more closely than Permian or later.

To the southeast, in the adjoining Yukon Territory, the steep Teslin fault is considered the "root zone" for both the metamorphic and mafic-ultramafic nappes that lie to the northeast of the fault (Tempelman-Kluit, 1979). This fault zone separates rocks of the Stikinia and Cache Creek terranes on the southwest from rocks of the Yukon-Tanana metamorphic terrane and the Cassiar miogeoclinal terrane to the northeast. According to Tempelman-Kluit (1979), development of the Yukon-Tanana metamorphic terrane, including its "piggy-back" mafic-ultramafic nappes, is related to collision of an offshore island-arc complex (Stikinia, Cache Creek, and other terranes) with the North American miogeoclinal margin in Early Jurassic time (also see Monger and others, 1982). This interpretation suggests that the mafic-ultramafic nappes were rooted beneath the forearc of a former arc terrane, which presently lies south of the Salcha-Seventymile belt.

Livengood

Ultramafic and associated plutonic rocks crop out along a narrow but nearly continuous 95-km-long belt that trends southwestward across the Livengood and Tanana Quadrangles in east-central Alaska (locality 20, Fig. 1). These rocks extend along the boundary between the Cretaceous flysch deposits of the Manley terrane and the weakly metamorphosed Paleozoic rocks of the Livengood terrane. Sporadic outcrops continue on strike for another 100 km to the southwest where they have been included in the Manley terrane by Silberling and others (this volume) (Fig. 1). The complexes have a total outcrop area of about 150 km^2 and are composed predominantly of completely serpentinized ultramafic rocks. In the vicinity of Livengood, where they have been studied in greatest detail, Loney and Himmelberg (1988) report that the ultramafic rocks are largely serpentinized harzburgite and minor dunite. The serpentinite is locally intruded by lenticular bodies and dikes of altered diorite and gabbro.

The structural setting of the ultramafic complexes is unclear. In the Livengood Quadrangle, Chapman and others (1971, 1985b) interpret the contact with the Cretaceous flysch deposits of the Manley terrane as depositional, whereas Foster and others (1983) conclude that the contact, at least at the northeastern end of the belt, is a fault. Coney (1984) contends that the ultramafic rocks dip steeply northwestward and were emplaced against the flysch deposits by southeastward overthrusting in late Mesozoic time.

The age of the ultramafic complexes in the Livengood belt is uncertain, but probably pre-Ordovician. Near Livengood, diorite and gabbro intruding the ultramafic rocks yield K-Ar hornblende ages of 518 to 633 Ma (D. L. Turner, written communication, 1987). These K-Ar ages confirm a previous pre–Late Devonian age assignment by Foster (1966) based on the presence of serpentinite clasts in nearby conglomerates containing Late Devonian fossils. To the southwest, in the Tanana Quadrangle, Chapman and others (1982) assigned the ultramafic complexes a probable Late Cretaceous age on the basis of their apparent intrusive relation with the enclosing Cretaceous sedimentary rocks. However, Loney and Himmelberg (1988) argue that, although the ultramafic rocks in the Tanana Quadrangle may have undergone higher-temperature recrystallization, they appear to be an extension of the older Livengood ultramafic belt and should be assigned to the Livengood terrane rather than to the Manley terrane as indicated by Silberling and others (this volume) (Fig. 1).

SOUTHERN ALASKA

Alaska Range

Small ultramafic bodies are scattered along the central and eastern Alaska Range for at least 500 km, but a clear pattern of their age and origin is not apparent. Possibly the best known body in the central Alaska Range occurs in the Chulitna district (locality 21, Fig. 1), where an Upper Devonian ophiolite sequence of serpentinite, gabbro, pillow basalt, and radiolarian chert is well documented in the Chulitna terrane (Jones and others, 1980; Nokleberg and others, this volume, Chapter 10). Nearby in the Talkeetna Quadrangle (locality 22, Fig. 1), Reed and Nelson (1980) have identified several dunite and serpentinite bodies of uncertain age. The serpentinite occurs as a narrow fault sliver about 5 km in length along a prominent fault zone parallel to and just northwest of the Denali fault. In addition, "alpine-type" dunite and serpentinite are exposed discontinuously for 25 km in a thick, lithologically and structurally complex unit (chiefly marine, flyschoid, sedimentary rocks) of middle and late Paleozoic age.

Farther east in the central Alaska Range, at least 24 ultramafic bodies are exposed along a belt about 50 km long and 15 km wide that extends from the Maclaren Glacier to beyond Rainbow Mountain (locality 23, Fig. 1; Hanson, 1964; Rose, 1965, 1966; Stout, 1976; Nokleberg and others, 1982). The bodies consist mainly of locally serpentinized dunite and peridotite and of gabbro. Nokleberg and others (1979) suggest that these ultramafic bodies are the lower cumulus portion of a Triassic ophiolite. Subsequent work by Nokleberg and others (1981) suggests that the mafic-ultramafic rocks, together with adjoining belts of Nikolai Greenstone, represent rift-type magmatism along a Triassic spreading center. Farther to the east, near Slate Creek, Matteson (1973) has documented several small bodies of Cretaceous (122 Ma) biotite-hornblende pyroxenite, hornblendite, and gabbro that intrude Upper Jurassic and Lower Cretaceous Mentasta Argillite (Richter, 1967).

In the eastern Alaska Range (locality 24, Fig. 1; Richter, 1976), at least five alpine-type ultramafic bodies, chiefly serpentinite, serpentinized peridotite, and dunite, are scattered north of the Denali fault from the Canadian border to Mentasta Pass. The largest, near Mentasta Pass, is a thin body about 12 km long in Devonian phyllite. One of the bodies apparently intrudes a Cretaceous and Tertiary conglomerate, and on this basis, Richter (1976) has assigned these bodies to the Cretaceous. A nearby body of serpentinized peridotite and dunite in the Carden Hills is associated with gabbro and anorthosite in a metamorphosed Paleozoic volcanic-volcaniclastic sequence and has been assigned a probable Paleozoic age by Richter (1976).

Baranof-Chugach-Kodiak belt

A belt of ultramafic and related mafic complexes extends for more than 1,600 km in an arc from southern Baranof Island, along the north and west sides of the Chugach Range and then southwestward to Kodiak Island. The belt is generally coincident with the Border Ranges fault, a major structure along the north side of the Chugach terrane that can be traced from northern Baranof Island to Kodiak Island (Plafker and others, this volume, Chapter 12). Between Tonsina and Kodiak Island, where most of mafic-ultramafic complexes occur, the Border Ranges fault generally separates the subduction complexes of the Chugach terrane on the south and east from the island-arc assemblages of the Peninsular terrane to the north and west. The subduction complexes of the Chugach terrane include the Jurassic and Cretaceous McHugh complex of the northern Chugach Mountains (Clark, 1973; Winkler and others, 1981; Plafker and others, 1985), the Cretaceous Seldovia Bay complex of the southern Kenai Peninsula (Cowan and Boss, 1978), and the Cretaceous Uyak complex of Kodiak Island (Connelly, 1978). These complexes are heterogeneous assemblages of deformed and fragmented, weakly metamorphosed argillite, graywacke, radiolarian (meta)chert, greenstone, and local marble and tuff. They include rocks of diverse ages from Triassic to Cretaceous, but their age of emplacement is generally considered to be Cretaceous—probably slightly older than, but possibly partly contemporaneous with, the Late Cretaceous accretionary flysch that makes up most of the Chugach terrane. A series of thrust faults generally emplaces the subduction complexes onto the Late Cretaceous flysch. In the Peninsular terrane a volcanic arc is represented by the Lower Jurassic Talkeetna Formation, which consists of andesite, andesitic tuff, basalt and volcaniclastic sedimentary rocks and is extensively exposed just north of the Border Ranges fault from at least Tonsina to the southern Kenai Peninsula. The Jurassic calc-alkaline plutonic equivalents of the Talkeetna Formation are well documented on Kodiak Island (Kodiak-Kenai belt of Hudson, 1983) and from Tonsina southeastward to Baranof Island (Tonsina-Chichagof belt of Hudson, 1983); Hudson also indicated that the belts are probably continuous along the northern flank of the Chugach Range, where several Jurassic intermediate-composition plutons are known (Clark, 1971a, b; Magoon and others, 1976; Winkler and others, 1981; Pessel and others, 1981; Burns and others, 1983; Little and others, 1986; Pavlis, 1986).

The mafic-ultramafic complexes, which are distributed along the Border Ranges fault between Tonsina and Kodiak, were generally thought to be remnants of oceanic crust that formed the basement of the subduction complexes of the Chugach terrane (Beyer, 1980; Toth, 1981). However, recent investigations (Winkler and others, 1981; Pessel and others, 1981; Burns and others, 1983; Burns, 1985) suggest that they have arc affinities and are more closely associated with the Peninsular terrane. Several mafic-ultramafic complexes lie within the subduction complexes of the Chugach terrane, but their emplacement in that terrane is thought to be the result of Cretaceous and Tertiary tectonics (Burns, oral communication, 1986).

Near Tonsina, an ultramafic complex about 3 km by 20 km in outcrop extent is composed of layered dunite, harzburgitic dunite, wehrlite, websterite, and clinopyroxenite and occurs in a fault block along with Jurassic (about 175 Ma) layered gabbro and leucogabbro (locality 26, Fig. 1; Hoffman, 1974; Winkler and others, 1981; Coleman and Burns, 1985). Several smaller ultramafic bodies, including serpentinite, pyroxenite, and peridotite, are scattered for about 100 km east of Tonsina (Metz, 1975; MacKevett, 1978; MacKevett and Plafker, 1974; Herreid, 1970). Various ages ranging from Pennsylvanian to Jurassic have been suggested for these bodies, but most of them are fault-bounded, and their proposed ages are largely inferred. Page and others (1986) suggest that the ultramafic portion of the Tonsina complex may be part of a late Paleozoic arc upon which the Jurassic Talkeetna arc and associated plutonic rocks are superposed. Burns (1983, 1985), on the other hand, has suggested that some of the ultramafic rocks in the Tonsina complex are mantle rocks and thus may be older basement on which the Jurassic arc was built.

Near the Tonsina ultramafic complex, an extensive and probably cogenetic Jurassic (153 to 171 Ma) layered gabbro body is exposed along the foothills of the Chugach Range. It extends discontinuously westward to the Jurassic Wolverine complex, a body of layered dunite, clinopyroxenite, gabbro, and peridotite that crops out for a distance of about 8 km near Palmer (locality 27, Fig. 1; Clark, 1972; Burns, 1985). About 30 km southwest of Wolverine, near Anchorage, the Eklutna complex, which is mainly weakly serpentinized wehrlite, dunite and pyroxenite, occurs in an elongate, fault-bounded body about 3 km by 15 km in areal extent between a metamorphic complex to the northwest and the McHugh Complex to the southeast (locality 28, Fig. 1; G. R. Winkler, unpublished map, 1987; Rose, 1966; Clark and Bartsch, 1971a, b).

Several ultramafic bodies have long been known near Seldovia at the southern tip of the Kenai Peninsula (Guild, 1942; Toth, 1981). Red Mountain, the largest of these bodies, is a roughly oval-shaped area of layered dunite and clinopyroxenite (wehrlite) about 7 km in length. The body is faulted into the Cretaceous melange of the Seldovia Bay complex along a nearly vertical shell of serpentinite (locality 29, Fig. 1), between the Border Ranges fault and a prominent thrust fault that forms the eastern limit of the Seldovia Bay complex (Cowan and Boss, 1978; Magoon and others, 1976). Toth (1981) suggests that the Red Mountain body formed at an oceanic spreading center as part of an ophiolite sequence in a Late Cretaceous episode of subduction. More recently, Burns (1983, 1985) argued that the Red Mountain body formed in an island arc as a cumulate counterpart of volcanic rocks. In addition to the well-known ultramafic bodies in the Seldovia Bay complex, Forbes and Lanphere (1973) have identified serpentinite and associated chert and pillow basalt at the southeast side of a narrow fault block of quartz-sericite schist, marble, greenschist, and blueschist—the Seldovia schist terrane of Cowan and Boss (1978). They have dated the metamorphism within the Seldovia schist terrane as Late Triassic to Early Jurassic (about 190 Ma) and connect the event to an ophiolite-related subduction zone.

Several ultramafic bodies occur on western Kodiak Island near the Border Ranges fault (locality 30, Fig. 1). The ultramafic rocks, mainly dunite, ortho- and clinopyroxenite, and gabbro (Hill and Brandon, 1976), occur as kilometer-size "slabs" in the Uyak complex, a chaotic assemblage of chert, argillite, wacke, and greenstone (Connelly, 1978). The Uyak complex is thrust to the southeast over Upper Cretaceous turbidites of the Kodiak Formation. Connelly (1978) interpreted the Uyak complex and the ultramafic blocks within it as a subduction complex of probable Late Cretaceous age. Hill and Brandon (1976) suggested that the ultramafic rocks may represent the lower portion of a dismembered ophiolite; Burns (1983, 1985) interpreted the ultramafic bodies as a part of the Peninsular terrane that represents a portion of an island-arc ophiolite.

At the southeastern end of the Baranof-Chugach-Kodiak belt on Baranof Island, several small serpentinite bodies ("tectonic inclusions") occur in metamorphosed phyllitic rocks of the Kelp Bay Group (locality 25, Fig. 1; Loney and others, 1975). The association of these serpentinites with the greenstone, radiolarian metachert, phyllite, amphibolite, and metagabbro of the Kelp Bay Group suggests that they may represent part of an ophiolite assemblage (Berg and Jones, 1974). The nearby dunite-clinopyroxenite body at Red Bluff Bay (locality 40, Fig. 1; Guild and Balsley, 1942; Loney and others, 1975)—sometimes grouped with the zoned ultramafic complexes—is probably best included with the serpentinites on Baranof Island (Irvine, 1974). Although it is larger than the serpentinite bodies, the Red Bluff Bay body is similarly bounded by faults and is associated with phyllonite and greenschist of the Kelp Bay Group.

Prince William Sound

Ophiolites are exposed over extensive areas on the Resurrection Peninsula and on Knight Island in Prince William Sound (localities 31 and 32, Fig. 1). The complex on the Resurrection Peninsula near Seward consists of pillow basalt, a thick sequence of basaltic sheeted dikes and gabbro, and serpentinized ultramafic rocks that include clinopyroxenite, dunite, and harzburgite in a composite body about 8 km wide by 27 km long (Miller, 1984; Tysdal and others, 1977). The Resurrection Peninsula complex is faulted into Upper Cretaceous flysch and interbedded volcanic rocks of the Valdez Group in the Chugach terrane (Miller, 1984). The complex on Knight Island is strikingly similar to the one on Resurrection Peninsula and consists mainly of gabbro and pods of ultramafic rocks associated with a basalt sheeted-dike sequence. However, the Knight Island complex, which is exposed over an area about 13 km in width and 34 km in length, is interbedded with lower Tertiary Orca Group flysch of the Prince William terrane. Both complexes lack only basal tectonized peridotite and associated deep-marine sediments to be complete ophiolite sequences. The relation of the two complexes to each other is ambiguous. Miller (1984), in agreement with Tysdal and others (1977), assigned a Cretaceous age to the Resurrection Peninsula complex and a Tertiary age to the Knight Island complex but noted that the similar petrology suggests a similar origin.

SOUTHEASTERN ALASKA

A narrow belt of distinctive ultramafic bodies, variously called Alaska-type or (concentrically) zoned ultramafic complexes, extends for about 560 km along the length of southeastern Alaska, southwest of the Coast Range batholith. At least 35 such bodies occur (Taylor, 1967); the best known are on Duke Island (locality 33, Fig. 1; Irvine, 1974), at Union Bay (locality 34, Fig. 1; Ruckmick and Noble, 1959), on the Blashke Islands (locality

35, Fig. 1; Himmelberg and others, 1986), at Snettisham (locality 36, Fig. 1; Thorne and Wells, 1956), and near Klukwan (locality 37, Fig. 1; MacKevett and others, 1974). Several similar bodies are known elsewhere in the Americas, and an analogous belt occurs in the Ural Mountains of the U.S.S.R. (Taylor, 1967). The bodies in southeastern Alaska, however, are unique in their consistent petrology, chemistry, textures, and structures—especially remarkable because they are so scattered along the belt and are relatively small. Duke Island, the largest, is about 18 km in maximum outcrop dimension. Several of the bodies are concentrically zoned and have dunite cores; cumulus textures are common, notably at Duke Island, where sedimentary layering and structures are spectacularly displayed (Irvine, 1974). Where the complete zoned sequence is present, the complexes consist of cores of dunite surrounded by concentric shells of olivine pyroxenite, magnetite pyroxenite, and hornblende pyroxenite. Only partial sequences are exposed in most of the complexes, however, and shells commonly are discontinuous. Most of the bodies intrude gabbro. Lanphere and Eberlein (1966) have isotopically dated several of these ultramafic complexes by K-Ar methods as 100 to 110 Ma.

The pyroxene in the ultramafic rocks of the zoned bodies is consistently high-alumina diopsidic augite. Orthopyroxene and plagioclase do not occur in most of the complexes and occur only rarely in the Blashke Islands complex (Himmelberg and others, 1986). The olivine is highly magnesian. Magnetite is common, and magnetite deposits in the pyroxenite at Snettisham and Klukwan have been explored for commercial amounts of iron. Minor to trace amounts of chromite are ubiquitous in the dunite as disseminated grains, scattered lenses, and pods.

Two hypotheses have been proposed for the origin of these bodies. An earlier hypothesis, cited in numerous publications but put forth in detail by Taylor (1967), is that the complexes were formed from a fractionating series of increasingly ultramafic magmas, each of which intruded its precursor to form the zonal pattern. Taylor further suggested that the magmas may be derived by fusion within the mantle—the lowest-melting-temperature fraction developing first. Alternatively, Irvine (1974), in his detailed study of the complex at Duke Island, suggested that the various ultramafic rock types developed from one magma by differentiation and precipitation of cumulus minerals in a single magma chamber. The zonal pattern is due to subsequent diapiric remobilization of the layered complex in response to regional compression. Irvine also concluded that the ultramafic rocks crystallized from a single alkaline picritic magma, and he and others (Berg and others, 1972; Irvine, 1973; Ford and Brew, 1976) suggested that the zoned complexes are related to a coeval, generally coincident belt of Cretaceous alkaline volcanic rocks. Himmelberg and others (1986), on the other hand, present evidence that the Blashke Island body crystallized from a subalkaline magma.

Whatever the origin of these unusual bodies, the consistency of their petrology and chemistry argues for a noncontinental source for their parent magma(s), especially in view of the heterogeneity of the crust in southeastern Alaska. The most likely source is the mantle or, as indicated by Rb-Sr isotope data (Lanphere, 1968), the oceanic crust. Berg and others (1972) suggested that these complexes were emplaced during continental accretion and subduction in southeastern Alaska. Alternatively, Gehrels and Berg (this volume) propose that the zoned plutons may be related to a post-accretional tensional event. Although many workers believe that Cretaceous continental accretion played an important role in generating or localizing these bodies, the Cretaceous geology of southeastern Alaska suggests a more complicated tectonic history. For example, the region contains: (1) belts of 74- to 84-Ma calc-alkaline granodiorite, quartz diorite, and diorite plutons (Brew and Morrell, 1983); (2) the 55- to 65-Ma Coast plutonic complex (tonalite) sill belt just to the east of the belt of zoned ultramafic complexes; and (3) a belt of Upper Jurassic to mid-Cretaceous flysch and basaltic andesite that is nearly coincident with the belt of zoned ultramafic complexes (Berg and others, 1972).

In addition to the zoned ultramafic complexes, numerous other ultramafic bodies occur along the axial and eastern portions of southeastern Alaska. Grybeck and others (1976) and Brew and Grybeck (1984) described numerous ultramafic bodies scattered through the Coast Range batholith in the vicinity of Tracy and Endicott Arms. All of these bodies are only a few tens of meters in maximum dimension, and most are strongly foliated. They have a diverse lithology, including dunite, pyroxenite, peridotite, and abundant serpentinite. The bodies occur in several forms—as inclusions within the plutonic rocks of the Coast Range batholith, as fault blocks and slivers within the schist and gneiss roof pendants of the Coast Range batholith, and as intrafolial layers in the schist and gneiss. Several clearly pre-date the intrusion of the plutonic rocks of the Coast Range batholith; most appear to have undergone the same metamorphism that produced the schist and gneiss within the Coast Range batholith. The origin of these dismembered ultramafic bodies is obscure and possibly diverse. Their variety and distribution suggest that the protoliths of the Coast Range batholith were structurally and tectonically complex.

Numerous small ultramafic, hornblendite, and related(?) gabbro bodies occur along, or just west of the belt of zoned ultramafic complexes (Buddington and Chapin, 1929; Brew and Morrell, 1983). Some of them may be related to the zoned complexes, and others are probably mafic phases of larger intermediate plutons. Elsewhere in southeastern Alaska, an Ordovician (440 to 449 Ma) gabbro, pyroxenite, and hornblendite stock near southern Prince of Wales Island has been reported by Eberlein and others (1983), and on northern Admiralty Island, serpentinite bodies have been described by Berg and Jones (1974), who interpret them to be remnants of a dismembered upper Paleozoic or Mesozoic ophiolite suite.

A belt of distinctive Tertiary mafic(-ultramafic) plutons extends from northern Chicagof Island northward through the Fairweather Range. The best-known pluton in this belt is the layered gabbro body at Mt. La Perouse and Mt. Crillon (locality

38, Fig. 1; Rossman, 1963; Brew and others, 1978; Himmelberg and Loney, 1981; Loney and Himmelberg, 1983). At least 6,000 m of layered olivine gabbro and norite, lesser amounts of gabbronorite and troctolite, and rare anorthosite are spectacularly exposed, although nearly inaccessible, in the extremely mountainous terrane. The gabbro includes sparse, thin layers of ultramafic rocks and a basal cumulus ultramafic zone containing deposits of Cu, Ni, and Pt-group elements. The gabbro is a layered stratiform body formed in a single magma chamber by differentiation and cumulus processes. Himmelberg and Loney (1981) concluded that the unusually consistent composition of the body throughout its stratigraphic range resulted from repeated injections of similar magma during its cooling history. The body has a pronounced contact-metamorphic zone, in which the low-grade country rocks are upgraded to amphibolite and biotite schist. Hudson and Plafker (1981) bracketed the age of the gabbro between 19 and 41 Ma (with a probable age of 40 Ma) by K-Ar dating metamorphic minerals in the contact aureole; Loney and Himmelberg (1983) determined a K-Ar age of 28 Ma on minerals from the gabbro. Several other bodies of layered gabbro occur near the La Perouse body and as far north as Mt. Fairweather (locality 39, Fig. 1; Brew and others, 1978; Plafker and MacKevett, 1970). Tertiary gabbro or norite bodies, at least one of which is layered, and associated tonalite plutons that occur on Yakobi Island (locality 40, Fig. 1) and on the northern Chichagof Island probably are related to the layered gabbro pluton near Mt. La Perouse (Loney and others, 1975; Johnson and Karl, 1985; Rossman, 1959).

SUMMARY

The mafic-ultramafic complexes of Alaska are divided into four groups based on their geologic setting and tectonic history: (1) The large mafic-ultramafic complexes of northern and western Alaska are remnants of huge allochthonous sheets of oceanic and mantle rocks that overrode or were "obducted" onto the continental margin in Late Jurassic and Early Cretaceous time. These complexes, together with underlying allochthonous sheets of pillow basalt, gabbro, and chert, compose broad synformal masses or monoclinal slabs resting in thrust contact upon early Mesozoic and older continental and continental-margin deposits. (2) In the Alaska Range of central Alaska, the complexes are typically narrow, elongate bodies aligned along or near suture zones bounding diverse allochthonous terranes. The structural setting and common association with flyschoid deposits suggest that they represent fragments of former ocean basins that collapsed during accretion. (3) In the Baranof-Chugach-Kodiak belt of southern Alaska the mafic-ultramafic bodies are closely associated with subduction complexes along the Border Ranges fault and are bounded on the south by the Upper Cretaceous accretionary flysch of the Chugach terrane. Emplacement of these mafic-ultramafic bodies is generally attributed to subduction of the oceanic crust along the continental margin in Cretaceous time. (4) The concentrically zoned bodies of southeastern Alaska clearly are intrusive and therefore unlike the mafic-ultramafic complexes of the other three groups, most of which are bounded by faults. However, their Cretaceous age, their homogeneity of composition, and their parallelism with terrane boundaries for a distance of about 560 km strongly argue for a genetic relation with subduction and/or accretion.

The bulk of Alaskan ultramafic complexes appear to occur at the structural base of an adjacent volcanic arc terrane (western Brooks Range, Angayucham, and Tozitna-Innoko with the Koyukuk arc terrane; southwestern Alaska with the Togiak arc terrane; Baranof-Chugach-Kodiak belt with Peninsular arc terrane). These relations apparently resulted from the underthrusting of isostatically buoyant crust beneath the fore-arc limb of an adjacent volcanic arc terrane (a thick pile of continentally derived sediments beneath the Baranof-Chugach-Kodiak belt, and Paleozoic or older continental crust beneath the others). The other ultramafic-mafic complexes (Salcha-Seventymile, Livengood, Alaska Range, Prince William Sound, and in the Coast Range batholith) are either too complex or too poorly understood to speculate on their tectonic setting.

REFERENCES CITED (* = See Notes Added in Proof)

Barnes, D. F., and Tailleur, I. L., 1970, Preliminary interpretation of geophysical data from the lower Noatak River basin, Alaska: U.S. Geological Survey Open-File Report 70–18, 19 p.

Berg, H. C., and Jones, D. L., 1974, Ophiolite in southeastern Alaska: Geological Society of America Abstracts with Programs, v. 6, p. 144.

Berg, H. C., Jones, D. L., and Richter, D. H., 1972, Gravina–Nutzotin belt-tectonic significance of an upper Mesozoic sedimentary and volcanic sequence in southern and southeastern Alaska: U.S. Geological Survey Professional Paper 800-D, p. D1–D24.

Beyer, B. J., 1980, Petrology and geochemistry of ophiolite fragments in a tectonic melange, Kodiak Islands, Alaska [Ph.D. thesis]: Santa Cruz, University of California, 227 p.

Bird, M. L., and Clark, A. L., 1976, Microprobe study of olivine chromitites of the Goodnews Bay ultramafic complex, Alaska, and the occurrence of platinum: U.S. Geological Survey Journal of Research, v. 4, p. 717–725.

Boak, J. L., Turner, D. L., Wallace, W. K., and Moore, T. E., 1985, K-Ar ages of allochthonous mafic and ultramafic complexes and their metamorphic aureoles, western Brooks Range, Alaska [abs.]: American Association of Petroleum Geologists Bulletin, v. 69, no. 4, p. 656–657.

Box, S. E., 1985a, Mesozoic tectonic evolution of the northern Bristol Bay region, southwestern Alaska [Ph.D. thesis]: Santa Cruz, University of California, 163 p.

——, 1985b, Early Cretaceous orogenic belt in northwestern Alaska; Internal organization, lateral extent, and tectonic interpretation, *in* Howell, D. G., ed., Tectonostratigraphic terranes of the Circum-Pacific region: Circum-Pacific Council for Energy and Mineral Resources Earth Science Series, v. 1, p. 137–145.

Brew, D. A., and Grybeck, D., 1984, Geology of the Tracy Arm–Fords Terror wilderness study area and vicinity, Alaska: U.S. Geological Survey Bulletin 1525-A, 52 p.

Brew, D. A., and Morrell, R. P., 1983, Intrusive rocks and plutonic belts of southeastern Alaska, U.S.A., *in* Roddick, J. A., ed., Circum-Pacific plutonic terranes: Geological Society of America Memoir 159, p. 171–193.

Brew, D. A., Johnson, B. R., Grybeck, D., Griscom, A., and Barnes, D. F., 1978,

Mineral resources of the Glacier Bay National Monument wilderness study area, Alaska: U.S. Geological Survey Open-File Report 78–494, 670 p.

Brosgé, W. P., and Reiser, H. N., 1962, Preliminary geologic map of the Christian Quadrangle, Alaska: U.S. Geological Survey Open-File Report 62–15, 2 sheets, scale 1:250,000.

Brosgé, W. P., Reiser, H. N., and Yeend, W., 1973, Reconnaissance geologic map of the Beaver Quadrangle, Alaska: U.S. Geological Survey Miscellaneous Field Studies Map MF-525, scale 1:250,000.

Brosgé, W. P., Reiser, H. N., and Lanphere, M. A., 1974, Pitka ultramafic complex may be a klippe, *in* Carter, C., ed., U.S. Geological Survey Alaska Program, 1974: U.S. Geological Survey Circular 700, p. 42.

Buddington, A. F., and Chapin, T., 1929, Geology and mineral deposits of southeastern Alaska: U.S. Geological Survey Bulletin 800, 398 p.

Burns, L. E., 1983, The Border Ranges ultramafic and mafic complex; Plutonic core of an intraoceanic island arc [Ph.D. thesis]: Stanford, Cailfornia, Stanford University, 151 p., 2 sheets, scale 1:250,000.

—— , 1985, The Border Ranges ultramafic and mafic complex, south-central Alaska; Cumulate fractionates of island arc volcanics: Canadian Journal of Earth Sciences, v. 22, p. 1020–1038.

Burns, L. E., Little, T. A., Newberry, R. J., Decker, J. E., and Pessel, G. H., 1983, Preliminary geologic map of parts of the Anchorage C-2, C-3, D-2, and D-3 Quadrangles, Alaska: Alaska Division of Geological and Geophysical Surveys Report of Investigations 83–10, 3 sheets, scale 1:25,000.

Chapman, R. M., Weber, F. R., and Taber, B., 1971, Preliminary geologic map of the Livengood Quadrangle, Alaska: U.S. Geological Survey Open-File Report 71–66, 2 sheets, scale 1:250,000.

Chapman, R. M., Yeend, W., Brosgé, W. P., and Reiser, H. N., 1982, Reconnaissance geologic map of the Tanana Quadrangle, Alaska: U.S. Geological Survey Open-File Report 82–734, 18 p., scale 1:250,000.

Chapman, R. M., Patton, W. W., Jr., and Moll, E. J., 1985a, Reconnaissance geologic map of the Ophir Quadrangle, Alaska: U.S. Geological Survey Open-File Report 85–203, scale 1:250,000.

Chapman, R. M., Trexler, J. H., Jr., Churkin, M., Jr., and Weber, F. R., 1985b, New concepts of the Mesozoic flysch belt of east-central Alaska, *in* Bartsch-Winkler, S., ed., The United States Geological Survey in Alaska; Accomplishments in 1983: U.S. Geological Survey Circular 945, p. 29–32.

Clark, S.H.B., 1972, The Wolverine complex; A newly discovered layered ultramafic body in the western Chugach Mountains, Alaska: U.S. Geological Survey Open-File Report 72–70, 10 p.

—— , 1973, The McHugh complex of south-central Alaska: U.S. Geological Survey Bulletin 1372-D, p. D1–D11.

Clark, S.H.B., and Bartsch, S. R., 1971a, Reconnaissance geologic map and geochemical analyses of stream sediment and rock samples of the Anchorage B-6 Quadrangle, Alaska: U.S. Geological Survey Open-File Report 71–70, 63 p., scale 1:63,360.

—— , 1971b, Reconnaissance geologic map and geochemical analyses of stream sediment and rock samples of the Anchorage B-7 Quadrangle, Alaska: U.S. Geological Survey Open-File Report 71–71, 70 p., scale 1:63,360.

Coleman, R. G., and Burns, L. E., 1985, The Tonsina high-pressure mafic-ultramafic cumulate sequence, Chugach Mountains, Alaska: Geological Society of America Abstracts with Programs, v. 17, p. 348.

Coney, P. J., 1984, Structural and tectonic aspects of accretion in Alaska, *in* Howell, D. G., Jones, D. L., Cox, A., and Nur, A., eds., Proceedings of the Circum-Pacific Terrane Conference: Stanford University Publications in the Earth Sciences, v. 18, p. 68–70.

Connelly, W., 1978, Uyak complex, Kodiak Islands, Alaska; A Cretaceous subduction complex: Geological Society of America Bulletin, v. 89, p. 755–769.

Cowan, D. S., and Boss, R. F., 1978, Tectonic framework of the southwestern Kenai Peninsula, Alaska: Geological Society of America Bulletin, v. 89, p. 155–158.

*Curtis, S. M., Ellersieck, I., Mayfield, C. F., and Tailleur, I. L., 1982, Reconnaissance geologic map of southwestern Misheguk Mountain Quadrangle, Alaska: U.S. Geological Survey Open-File Report 82–611, 43 p., scale 1:63,360.

Decker, J. E., and Karl, S. M., compilers, 1977, Preliminary aeromagnetic map of the eastern part of southern Alaska: U.S. Geological Survey Open-File Report 77–169-E, scale 1:1,000,000.

Department of Energy, 1982, Total magnetic intensity anomaly map, Ophir Quadrangle: Department of Energy Open-File Map GJM–207(82), 4 plates, scale 1:250,000.

Eberlein, G. D., Churkin, M., Jr., Carter, C., Berg, H. C., and Ovenshine, A. T., 1983, Geology of the Craig Quadrangle, Alaska: U.S. Geological Survey Open-File Report 83–91, 52 p., scale 1:250,000.

*Ellersieck, I., Curtis, S. M., Mayfield, C. F., and Tailleur, I. L., 1982, Reconnaissance geologic map of south-central Misheguk Mountain Quadrangle, Alaska: U.S. Geological Survey Open-File Report 82–612, 38 p., scale 1:63,360.

Forbes, R. B., and Lanphere, M. A., 1973, Tectonic significance of mineral ages of blueschists near Seldovia, Alaska: Journal of Geophysical Research, v. 78, p. 1383–1386.

Ford, A. B., and Brew, D. A., 1976, Chemical nature of Cretaceous greenstone near Juneau, Alaska *in* Blean, K. M., ed., The United States Geological Survey in Alaska; Accomplishments during 1976: U.S. Geological Survey Circular 751-B, p. B88–B90.

Foster, H. L., Laird, J., Keith, T.E.C., Cushing, G. W., and Menzie, W. D., 1983, Preliminary geologic map of the Circle Quadrangle, Alaska: U.S. Geological Survey Open-File Report 83–170-A, 29 p., scale 1:250,000.

Foster, R. L., 1966, The petrology and structure of the Amy Dome area, Tolovana mining district, east-central Alaska [Ph.D. thesis]: Rolla, University of Missouri, 225 p.

Frank, C. O., and Zimmerman, J., 1982, Petrography of nonultramafic rocks from the Avan Hills complex, De Long Mountains, Alaska, *in* Coonrad, W. L., ed., The United States Geological Survey in Alaska; Accomplishments during 1980: U.S. Geological Survey Circular 844, p. 22–26.

Gealey, W. K., 1980, Ophiolite obduction mechanism, *in* Panayiotow, A., ed., Ophiolites; Proceedings International Ophiolite Symposium: Cyprus Geological Survey Department, p. 228–243.

Gemuts, I., Puchner, C. C., and Steffel, C. I., 1983, Regional geology and tectonic history of western Alaska, *in* Proceedings of the 1982 Symposium on Western Alaska Geology and Resource Potential: Alaska Geological Society, v. 3, p. 67–85.

Grybeck, D., Brew, D. A., Johnson, B. R., and Nutt, C. J., 1976, Ultramafic rocks in part of the Coast Range batholithic complex, southeastern Alaska, *in* Blean, K. M., ed., The United States Geological Survey in Alaska; Accomplishments during 1976: U.S. Geological Survey Circular 751-B, p. B82–B85.

Guild, P. W., 1942, Chromite deposits of Kenai Peninsula, Alaska: U.S. Geological Survey Bulletin 931-G, p. 139–175.

Guild, P. W., and Balsley, J. R., Jr., 1942, Chromite deposits of Red Bluff Bay and vicinity, Baranof Island, Alaska: U.S. Geological Survey Bulletin 936-G, p. G171–G187.

Hanson, L. G., 1964, Bedrock geology of the Rainbow Mountain area, Alaska Range, Alaska: Alaska Division of Geological Survey Report 2, 82 p.

Herreid, G., 1970, Geology of the Spirit Mountain nickel-copper prospect and surrounding area: Alaska Division of Mines and Geology Geologic Report 40, 19 p.

Hill, B. B., and Brandon, J., 1976, Layered basic and ultrabasic rocks, Kodiak Island, Alaska; The lower portion of a dismembered ophiolite?: EOS Transactions of the American Geophysical Union, v. 57, p. 1027.

Himmelberg, G. R., and Loney, R. A., 1981, Petrology of the ultramafic and gabbroic rocks of the Brady Glacier nickel-copper deposit, Fairweather Range, southeastern Alaska: U.S. Geological Survey Professional Paper 1195, 26 p.

Himmelberg, G. R., Loney, R. A., and Craig, J. T., 1986, Petrogenesis of the ultramafic complex at the Blashke Islands, southeastern Alaska: U.S. Geological Survey Bulletin 1662, 14 p.

Hoffman, B. L., 1974, Geology of the Bernard Mountain area, Tonsina, Alaska [M.S. thesis]: Fairbanks, University of Alaska, 68 p.

Hudson, T., 1983, Calc-alkaline plutonism along the Pacific rim of southern Alaska, *in* Roddick, J. A., ed., Circum-Pacific plutonic terranes: Geological Society of America Memoir 159, p. 159–169.

Hudson, T., and Plafker, G., 1981, Emplacement age of the Crillon–La Perouse pluton, Fairweather Range, *in* Albert, N.R.D., and Hudson, T., eds., The United States Geological Survey in Alaska; Accomplishments during 1979: U.S. Geological Survey Circular 823-B, p. B90–B93.

Irvine, T. N., 1973, Bridget Cove volcanics, Juneau area, Alaska; Possible parental magma of Alaska-type ultramafic complexes: Carnegie Institution of Washington Year Book 72, p. 478–491.

——, 1974, Petrology of the Duke Island ultramafic complex, southeastern Alaska: Geological Society of America Memoir 138, 240 p.

Johnson, B. R., and Karl, S. M., 1985, Geologic map of western Chichagof and Yakobi Islands, southeastern Alaska: U.S. Geological Survey Miscellaneous Field Studies Map I-1506, 15 p., 1 sheet, scale 1:250,000.

Jones, D. L., Silberling, N. J., Coney, P. J., and Plafker, G., 1987, Lithotectonic terrane maps of Alaska (west of the 141st meridian); Folio of the lithotectonic terrane maps of the North American Cordillera: U.S. Geological Survey Miscellaneous Field Studies Map MF-1874-A, scale 1:2,500,000.

Jones, D. L., Silberling, N. J., Csejtey, B., Jr., Nelson, W. H., and Blome, C. D., 1980, Age and structural significance of ophiolite and adjoining rocks in the upper Chulitna district, south-central Alaska: U.S. Geological Survey Professional Paper 1121-A, p. A1–A21.

Keith, T.E.C., Foster, H. L., Foster, R. L., Post, E. V., and Lehmbeck, W. L., 1981, Geology of an alpine-type peridotite in the Mount Sorenson area, east-central Alaska: U.S. Geological Survey Professional Paper 1170-A, p. A1–A9.

Lanphere, M. A., 1968, Sr-Rb-K and Sr isotopic relationships in ultramafic rocks, southeastern Alaska: Earth and Planetary Science Letters, v. 4, p. 185–190.

Lanphere, M. A., and Eberlein, G. D., 1966, Potassium-argon ages of magnetite-bearing ultramafic complexes in southeastern Alaska [abs.]: Geological Society of America Special Paper 87, p. 94.

Little, T. A., Pessel, G. H., Newberry, R. J., and Decker, J. E., 1986, Geologic map of parts of the Anchorage C-4, D-4 and C-5 quadrangles, Alaska: Alaska Division of Geological and Geophysical Surveys, Public Data File Map 86-28, scale 1:25,000.

Loney, R. A., and Himmelberg, G. R., 1983, Structure and petrology of the La Perouse gabbro intrusion, Fairweather Range, southeastern Alaska: Journal of Petrology, v. 24, p. 377–423.

——, 1984, Preliminary report on the ophiolites in the Yuki River and Mount Hurst areas, west-central Alaska, *in* Coonrad, W. L., and Elliott, R. L., eds., The United States Geological Survey in Alaska; Accomplishments during 1981: U.S. Geological Survey Circular 868, p. 27–30.

——, 1985a, Ophiolitic ultramafic rocks of the Jade Mountain–Cosmos Hills area, southwestern Brooks Range, Alaska, *in* Bartsch-Winkler, S., ed., The United States Geological Survey in Alaska; Accomplishments during 1984: U.S. Geological Survey Circular 967, p. 13–15.

*——, 1985b, Distribution and character of the peridotite-layered gabbro complex of the southeastern Yukon-Koyukuk ophiolite belt, Alaska, *in* Bartsch-Winkler, S., ed., The United States Geological Survey in Alaska; Accomplishments during 1983: U.S. Geological Survey Circular 945, p. 46–48.

——, 1988, Ultramafic rocks of the Livengood terrane, *in* Galloway, J. P. and Hamilton, T. D., eds., Geologic studies in Alaska by the U.S. Geological Survey during 1987: U.S. Geological Survey Circular 1016, p. 68–70.

Loney, R. A., Brew, D. A., Muffler, L.J.P., and Pomeroy, J. S., 1975, Reconnaissance geology of Chichagof, Baranof, and Kruzof Islands, southeastern Alaska: U.S. Geological Survey Professional Paper 792, 105 p.

MacKevett, E. M., Jr., 1978, Geologic map of the McCarthy Quadrangle, Alaska: U.S. Geological Survey Miscellaneous Investigations Series Map I-1032, 1 sheet, scale 1:250,000.

MacKevett, E. M., Jr., and Plafker, G., 1974, The Border Ranges fault in south-central Alaska: U.S. Geological Survey Journal of Research, v. 2, no. 3, p. 323–329.

MacKevett, E. M., Jr., Robertson, E. C., and Winkler, G. R., 1974, Geology of the Skagway B-3 and B-4 Quadrangles, southeastern Alaska: U.S. Geological Survey Professional Paper 832, 33 p.

Magoon, L. B., Adkison, W. L., and Egbert, R. M., 1976, Map showing geology, wildcat wells, Tertiary plant fossil localities, K-Ar age dates, and petroleum operations, Cook Inlet area, Alaska: U.S. Geological Survey Miscellaneous Investigations Series Map I-1019, 3 sheets, scale 1:250,000.

Matteson, C., 1973, Geology of the Slate Creek area, Mt. Hayes (A-2) Quadrangle, Alaska [M.S. thesis]: Fairbanks, University of Alaska, 66 p.

*Mayfield, C. F., Curtis, S. M., Ellersieck, I., and Tailleur, I. L., 1983a, Reconnaissance geologic map of the De Long Mountains A3, B3, and parts of A4, B4 Quadrangles, Alaska: U.S. Geological Survey Open-File Report 83–183, 2 sheets, scale 1:63,360.

*Mayfield, C. F., Tailleur, I. L., and Ellersieck, I., 1983b, Stratigraphy, structure, and palinspastic synthesis of the western Brooks Range, northwestern Alaska: U.S. Geological Survey Open-File Report 83–779, 58 p.

Mertie, J. B., Jr., 1940, The Goodnews platinum deposits, Alaska: U.S. Geological Survey Bulletin 918, 97 p.

Metz, P., 1975, Geology of the central portion of the Valdez C-2 Quadrangle, Alaska [M.S. thesis]: Fairbanks, University of Alaska, 124 p.

Miller, M. L., 1984, Geology of the Resurrection Peninsula, *in* Winkler, G. R., Miller, M. L., Hoekzema, R. B., and Dumoulin, J. A., Guide to the bedrock geology of a traverse of the Chugach Mountains from Anchorage to Cape Resurrection: Anchorage, Alaska Geological Society, p. 25–40.

Miller, M. L., and Angeloni, L. M., 1985, Ophiolitic rocks of the Iditarod Quadrangle, west-central Alaska [abs.]: American Association of Petroleum Geologists Bulletin, v. 69, no. 4, p. 669–670.

Monger, J.W.H., Price, R. A., and Tempelman-Kluit, D. J., 1982, Tectonic accretion and the origin of the two major metamorphic and plutonic welts in the Canadian Cordillera: Geology, v. 10, p. 70–75.

Nelson, S. W., and Nelson, W. H., 1982, Geology of the Siniktanneyak Mountain ophiolite, Howard Pass Quadrangle, Alaska: U.S. Geological Survey Miscellaneous Field Studies Map MF-1441, scale 1:63,360.

Nelson, S.W., Nokleberg, W. J., Miller-Hoare, M., and Mullen, M. W., 1979, Siniktanneyak Mountain ophiolite, *in* Johnson, K. M., and Williams, J. R., eds., The United States Geological Survey in Alaska; Accomplishments during 1978: U.S. Geological Survey Circular 804-B, p. B14–B16.

Nokleberg, W. J., Albert, N.R.D., and Zehner, R. E., 1979, The ophiolite of Tangle Lakes in the southern Mount Hayes Quadrangle, eastern Alaska Range; An accreted terrane?, *in* Johnson, K. M., and Williams, J. R., eds., The United States Geological Survey in Alaska; Accomplishments during 1978: U.S. Geological Survey Circular 804-B, p. B96–B98.

Nokleberg, W. J., Albert, N.R.D., Herzon, P. I., Miyaoka, R. T., and Zehner, R. E., 1981, Cross section showing accreted Andean-type arc and island arc terranes in southwestern Mount Hayes Quadrangle, Alaska, *in* Albert, N.R.D., and Hudson, T., eds., The United States Geological Survey in Alaska; Accomplishments during 1979: U.S. Geological Survey Circular 823-B, p. B66–B67.

Nokleberg, W. J., and 10 others, 1982, Geologic map of the southern Mount Hayes Quadrangle, Alaska: U.S. Geological Survey Open-File Report 82–52, 27 p., 1 sheet, scale 1:250,000.

Page, R. A., and 6 others, 1986, Accretion and subduction tectonics in the Chugach Mountains and Copper River basin, Alaska; Initial results of Trans-Alaska crustal transect: Geology, v. 14, p. 501–505.

Patton, W. W., Jr., 1984, Timing of arc collision and emplacement of oceanic crustal rocks on the margins of the Yukon-Koyukuk basin, western Alaska: Geological Society of America Abstracts with Programs, v. 16, p. 328.

Patton, W. W., Jr., and Hoare, J. M., 1968, The Kaltag fault, west-central Alaska, *in* Geological Survey Research 1968: U.S. Geological Survey Professional Paper 600-D, p. D147–D153.

Patton, W. W., Jr., Tailleur, I. L., Brosgé, W. P., and Lanphere, M. A., 1977, Preliminary report on the ophiolites of northern and western Alaska, *in* Coleman, R. G., and Irwin, W. P., eds., North American ophiolites: Oregon Department of Geology and Mineral Industries Bulletin 95, p. 51–57.

Patton, W. W., Jr., Moll, E. J., Lanphere, M. A., and Jones, D. L., 1984, New age data for the Kaiyuh Mountains, west-central Alaska, *in* Coonrad, W. L., and Elliott, R. L., eds., The United States Geological Survey in Alaska; Accomplishments during 1981: U.S. Geological Survey Circular 868, p. 30–32.

Pavlis, T. L., 1986, Geology of the Anchorage C-5 quadrangle, Alaska: Alaska Division of Geological and Geophysical Surveys, Public Data File 86-7, scale 1:63,360.

Pessel, G. H., Henning, M. W., and Burns, L. E., 1981, Preliminary geologic map of parts of the Anchorage C-1, C-2, D-1, and D-2 Quadrangles, Alaska: Alaska Division of Geological and Geophysical Surveys Open-File Report AOF–121, 1 sheet, scale 1:63,360.

Plafker, G., and MacKevett, E. M., Jr., 1970, Mafic and ultramafic rocks from a layered pluton at Mt. Fairweather, Alaska: U.S. Geological Survey Professional Paper 700-B, p. B21.

Plafker, G., Nokleberg, W. J., and Lull, J. S., 1985, Summary of 1984 TACT geologic studies in the northern Chugach Mountains and southern Copper River basin, *in* Bartsch-Winkler, S., ed., The United States Geological Survey in Alaska; Accomplishments during 1981: U.S. Geological Survey Circular 967, p. 76–79.

Reed, B. L., and Nelson, S. W., 1980, Geologic map of the Talkeetna Quadrangle, Alaska: U.S. Geological Survey Miscellaneous Investigations Series MF-1174, 1 sheet, 15 p., scale 1:250,000.

Reiser, H. N., Lanphere, M. A., and Brosgé, W. P., 1965, Jurassic age of mafic igneous complex, Christian Quadrangle, Alaska, *in* Geological Survey Research 1965: U.S. Geological Survey Professional Paper 525-C, p. C68–C71.

Richter, D. H., 1967, Geology of the upper Slana-Mentasta Pass area, south-central Alaska: Alaska Division of Geological and Geophysical Surveys Report 30, 25 p.

—— , 1976, Geologic map of the Nabesna Quadrangle, Alaska: U.S. Geological Survey Miscellaneous Investigations Map I-932, 1 sheet, scale 1:250,000.

Roeder, D., and Mull, C. G., 1978, Tectonics of Brooks Range ophiolites (Alaska): American Association of Petroleum Geologists Bulletin, v. 62, no. 9, p. 1696–1702.

Rose, A. W., 1965, Geology and mineral deposits of the Rainy Creek area, Mt. Hayes Quadrangle, Alaska: Alaska Division of Mines and Minerals Geologic Report 14, 51 p.

—— , 1966, Geology of chromite-bearing rocks near Eklutna, Anchorage Quadrangle, Alaska: Alaska Division of Mines and Minerals Geologic Report 18, 20 p.

Rossman, D. L., 1959, Geology and ore deposits of northwest Chichagof Island, Alaska: U.S. Geological Survey Bulletin 1058-E, p. 139–216.

—— , 1963, Geology and petrology of two stocks of layered gabbro in the Fairweather Range, Alaska: U.S. Geological Survey Bulletin 1121-F, p. F1–F50.

Ruckmick, J. C., and Noble, J. A., 1959, Origin of the ultramafic complex at Union Bay, southeastern Alaska: Geological Society of America Bulletin, v. 70, p. 981–1018.

Southworth, D. D., 1984, Red Mountain; A southeastern Alaska-type ultramafic complex in southwestern Alaska: Geological Society of America Abstracts with Programs, v. 16, p. 334.

Stout, J. H., 1976, Geology of the Eureka Creek area, east-central Alaska Range: Alaska Division of Geological and Geophysical Surveys Geologic Report 46, 32 p.

Taylor, H. P., Jr., 1967, The zoned ultramafic complexes of southeastern Alaska, *in* Wyllie, P. J., ed., Ultramafic and related rocks: New York, John Wiley and Sons, p. 96–118.

Tempelman-Kluit, D. J., 1979, Transported cataclastite, ophiolite, and granodiorite in Yukon; Evidence of arc-continental collision: Geological Survey of Canada Paper 79–14, 27 p.

Thorne, R. L., and Wells, R. R., 1956, Studies of the Snettisham magnetite deposits, southeastern Alaska: U.S. Bureau of Mines Report of Investigation 5195, 41 p.

Toth, M. I., 1981, Petrology, geochemistry, and origin of the Red Mountain ultramafic body near Seldovia, Alaska: U.S. Geological Survey Open-File Report 81–514, 86 p.

Turner, D. L., 1984, Tectonic implications of widespread Cretaceous overprinting of K-Ar ages in Alaskan metamorphic terranes: Geological Society of America Abstracts with Programs, v. 16, p. 338.

Turner, D. L., Forbes, R. B., Aleinikoff, J. N., Hedge, C. E., and McDougall, I., 1983, Geochronology of the Kilbuck terrane of southwestern Alaska: Geological Society of America Abstracts with Programs, v. 15, p. 407.

Tysdal, R. G., Case, J. E., Winkler, G. R., and Clark, S.H.B., 1977, Sheeted dikes, gabbro, and pillow basalt in flysch of coastal southern Alaska: Geology, v. 5, p. 377–383.

Weber, F. R., Foster, H. L., Keith, T.E.C., and Dusel-Bacon, C., 1978, Preliminary geologic map of the Big Delta Quadrangle, Alaska: U.S. Geological Survey Open-File Report 78–529-A, scale 1:250,000.

Winkler, G. R., and 5 others, 1981, Layered gabbroic belt of regional extent in the Valdez Quadrangle, *in* Albert, N.R.D., and Hudson, T., eds., The United States Geological Survey in Alaska; Accomplishments during 1979: U.S. Geological Survey Circular 823-B, p. B74–B76.

Zimmerman, J., and Frank, C. O., 1982, Possible obduction-related metamorphic rocks at the base of the ultramafic zone, Avan Hills complex, De Long Mountains, *in* Coonrad, W. L., ed., The United States Geological Survey in Alaska; Accomplishments during 1980: U.S. Geological Survey Circular 844, p. 27–28.

Zimmerman, J., and Soustek, P. G., 1979, The Avan Hills ultramafic complex, De Long Mountains, Alaska, *in* Johnson, K. M., and Williams, J. R., eds., The United States Geological Survey in Alaska; Accomplishments during 1978: U.S. Geological Survey Circular 804-B, p. B8–B11.

Zimmerman, J., Frank, C. O., and Bryn, S., 1981, Mafic rocks in the Avan Hills ultramafic complex, De Long Mountains, *in* Albert, N.R.D., and Hudson, T., eds., The United States Geological Survey in Alaska; Accomplishments during 1979: U.S. Geological Survey Circular 823-B, p. B14–B15.

MANUSCRIPT ACCEPTED BY THE SOCIETY MAY 17, 1990

ACKNOWLEDGMENTS

We wish to thank Henry Berg, Laurel Burns, Porter Irwin, and Robert Loney for their reviews and helpful suggestions for improving our manuscript.

NOTES ADDED IN PROOF

(1) The following preliminary reports listed in 'References Cited' have been superseded by later reports.

Curtis, S. M., Ellersieck, I., Mayfield, C. F., and Tailleur, I. L., 1982.
superseded by:
Curtis, S. M., Ellersieck, I., Mayfield, C. F., and Tailleur, I. L., 1984, Reconnaissance geologic map of southwestern Misheguk Mountain quadrangle, Alaska: U.S. Geological Survey Miscellaneous Investigation Series Map I-1502, scale 1:63,360.

Ellersieck, I., Curtis, S. M., Mayfield, C. F., and Tailleur, I. L., 1982.
superseded by:
Ellersieck, I., Curtis, S. M., Mayfield, C. F., and Tailleur, I. L., 1984, Reconnaissance geologic map of south-central Misheguk Mountain quadrangle, Alaska: U.S. Geological Survey Miscellaneous Investigation Series Map I-1504, scale 1:63,360.

Loney, R. A., and Himmelberg, G. R., 1985b
superseded by:
Loney, R. A., and Himmelberg, G. R., 1989, The Kanuti ophiolite, Alaska: Journal of Geophysical Research, v. 94, no. B11, p. 15,869–15,900.

Mayfield, C. F., Curtis, S. M., Ellersieck, I., and Tailleur, I. L., 1983a.
superseded by:
Mayfield, C. F., Curtis, S. M., Ellersieck, I., and Tailleur, I. L., 1990, Reconnaissance geologic map of the De Long Mountains A3, B3, and parts of A4, B4 quadrangles, Alaska: U.S. Geological Survey Miscellaneous Investigation

Series Map I-1929, scale 1:63,360.
Mayfield, C. F., Tailleur, I. L., and Ellersieck, I., 1983b.
superseded by:
Mayfield, C. F., Tailleur, I. L., and Ellersieck, I., 1988, Stratigraphy, structure, and palinspastic synthesis of the western Brooks Range, northwestern Alaska, *in* Gryc, G., ed., Geology and exploration of the National Petroleum Reserve in Alaska, 1974 to 1982: U.S. Geological Survey Professional Paper 1399, p. 143–186.

(2) The following reports, open-filed after this report was typeset, provide much additional information on the distribution, lithology, age, geochemistry, geologic setting, and associated mineral deposits of the major ophiolite assemblages in mainland Alaska.

Burns, L. E., 1992, Ophiolitic complexes and associated rocks near the Border Ranges fault zone, south-central Alaska: U.S. Geological Survey Open-File Report OF 92-20E, 7 p.
Foley, J. Y., 1992, Ophiolitic and other mafic-ultramafic metallogenic provinces in Alaska (west of the 141st meridian): U.S. Geological Survey Open-File Report OF 92-20B, 55 p.
Nelson, S. W., 1992, Ophiolitic complexes of the Gulf of Alaska: U.S. Geological Survey Open-File Report OF 92-20C, 9 p.
Patton, W. W., Jr., 1992a, Ophiolitic terrane of the western Brooks Range, Alaska: U.S. Geological Survey Open-File Report OF 92-20D, 7 p.
Patton, W. W., Jr., 1992b, Ophiolitic terrane bordering the Yukon-Koyukuk basin, Alaska: U.S. Geological Survey Open-File Report OF 92-20F, 7 p.
Patton, W. W., Jr., and Box, S. E., 1992, Ophiolitic terranes of east-central and southwestern Alaska: U.S. Geological Survey Open-File Report OF 92-20G, 13 p.
Patton, W. W., Jr., Murphy, J. M., Burns, L. E., Nelson, S. W., and Box, S. E., 1992, Geologic map of ophiolitic and associated volcanic arc and metamorphic terranes of Alaska (west of the 141st meridian): U.S. Geological Survey Open-File Report OF 92-20A, scale 1:2,500,000.

Printed in U.S.A.

The Geology of North America
Vol. G-1, The Geology of Alaska
The Geological Society of America, 1994

Chapter 22

Aleutian magmas in space and time

Suzanne Mahlburg Kay
Institute for the Study of the Continents (INSTOC), Snee Hall, Cornell University, Ithaca, New York 14853
Robert W. Kay
Department of Geological Sciences and Institute for the Study of the Continents (INSTOC), Snee Hall, Cornell University, Ithaca, New York 14853

INTRODUCTION

The Aleutian arc provided the setting for a proposal by Coats (1962) that arc magmas are related to a subducted oceanic plate in a convergent tectonic setting—a proposal that predated the theory of plate tectonics. Since that time, the study of Aleutian arc magmas has generated controversies over their origin and evolution that are fundamental to understanding the origin of magmas in all island arcs.

The first and longest controversy has been over the composition of the most important parental lava (precursor to most arc magmas) derived from the mantle. The question is whether this magma is a high-Al basalt generated by partial melting of the subducting plate (e.g., Marsh, 1982) or a more Mg-rich basalt derived from the peridotite overlying the plate that has been fluxed by a component from the plate (Kay, 1977; Perfit and others, 1980b).

The second major controversy is over the origin of the arc-type trace-element and isotopic characteristics of Aleutian magmas. The models considered for the magmatic source include (1) a mixture of subducted oceanic crustal and sedimentary components with a depleted mid-oceanic ridge-type mantle (Kay and others, 1978; Kay, 1980); (2) a subducted oceanic crustal source (Marsh, 1976) that includes a subducted sedimentary component (Brophy and Marsh, 1986); and (3) an upper mantle source composed of mid-oceanic ridge and oceanic island-type components (Morris and Hart, 1983).

The third major controversy is over the origin of the petrologic and geochemical diversity along the arc and within individual centers, and the relation of this diversity to differences in tectonic setting along the arc. The petrologic questions here involve the depth of fractionation in the crust, the role of magma mixing, the role of water, and the importance of crustal melting, assimilation, and incorporation of xenolithic fragments of the preexisting crust and mantle.

The aim of this chapter is to review the petrology and geochemistry of Aleutian arc magmas in order to present an overview of the regional variations in the arc and to identify the processes responsible for the origin and evolution of the magmatic rocks that constitute the arc crust. This overview leads us to a coherent model for the origin of the Aleutian crust. The petrologic and geochemical changes through time in the central part of the arc will also be briefly described.

DISTRIBUTION AND GENERAL CHARACTERISTICS OF ALEUTIAN ARC MAGMATISM

The geologic framework of the Aleutian arc, with emphasis on the offshore regions, is discussed by Vallier and others (this volume), and the general distribution and physical characteristics of Aleutian volcanoes have been summarized by Wood and Kienle (1990), Fournelle and others (this volume), and Kay and others (1982). Consequently, these aspects of the arc will not be discussed here. Instead, the purpose of this section is to review the aspects of the tectonic framework and physical characteristics of the volcanoes that are needed to understand the discussion of the petrology and geochemistry that follows.

As discussed by Marsh (1979), the main front of the Aleutian arc volcanoes can be grouped into linear segments based on geographic distribution. In the oceanic and western continental part of the arc, the segmentation scheme of Marsh (1979) was modified by Kay and others (1982) who defined the Rat, Andreanof, Four Mountains, and Cold Bay segments (Fig. 1). The ends of these linear segments correlate with major tectonic breaks in the overlying and underthrust plates and coincide with the terminations of rupture zones of major earthquakes (Kay and others, 1982). From west to east, the boundary between the Rat and the Andreanof segments coincides with the intersection of the Bowers ridge with the arc, the boundary between the Andreanof and Four Mountains segments coincides with the Amlia fracture zone on the lower plate, and the transition zone between the Four Mountains and Cold Bay segments coincides with the transition

Kay, S. Mahlburg, and Kay, R. W., 1994, Aleutian magmas in space and time, *in* Plafker, G., and Berg, H. C., eds., The Geology of Alaska: Boulder, Colorado, Geological Society of America, The Geology of North America, v. G-1.

from oceanic to continental crust. See Vallier and others (this volume) for further discussion of these features, and Kienle and Swanson (1983) for discussion of segmentation in the eastern Aleutian arc.

The region from the central part of the Cold Bay segment through the Four Mountains segment (Fig. 1) has several other complexities. First, for some 200 km between the Four Mountains and Cold Bay segments, the volcanoes do not form coherent lines. Second, volcanoes of the Four Mountains segment are closer to the trench than those of other segments. Finally, a secondary volcanic front consisting of small centers (such as Bogoslof, Amak, Uliaga, Kagamil, Carlisle, and smaller unnamed submarine centers) is developed behind the main arc in this region.

Variations in size, morphology, and petrologic characteristics of Aleutian arc volcanoes are related to position within the arc. As discussed by Marsh (1979), volcanic centers in the eastern part of the arc tend to be larger than those in the western part.

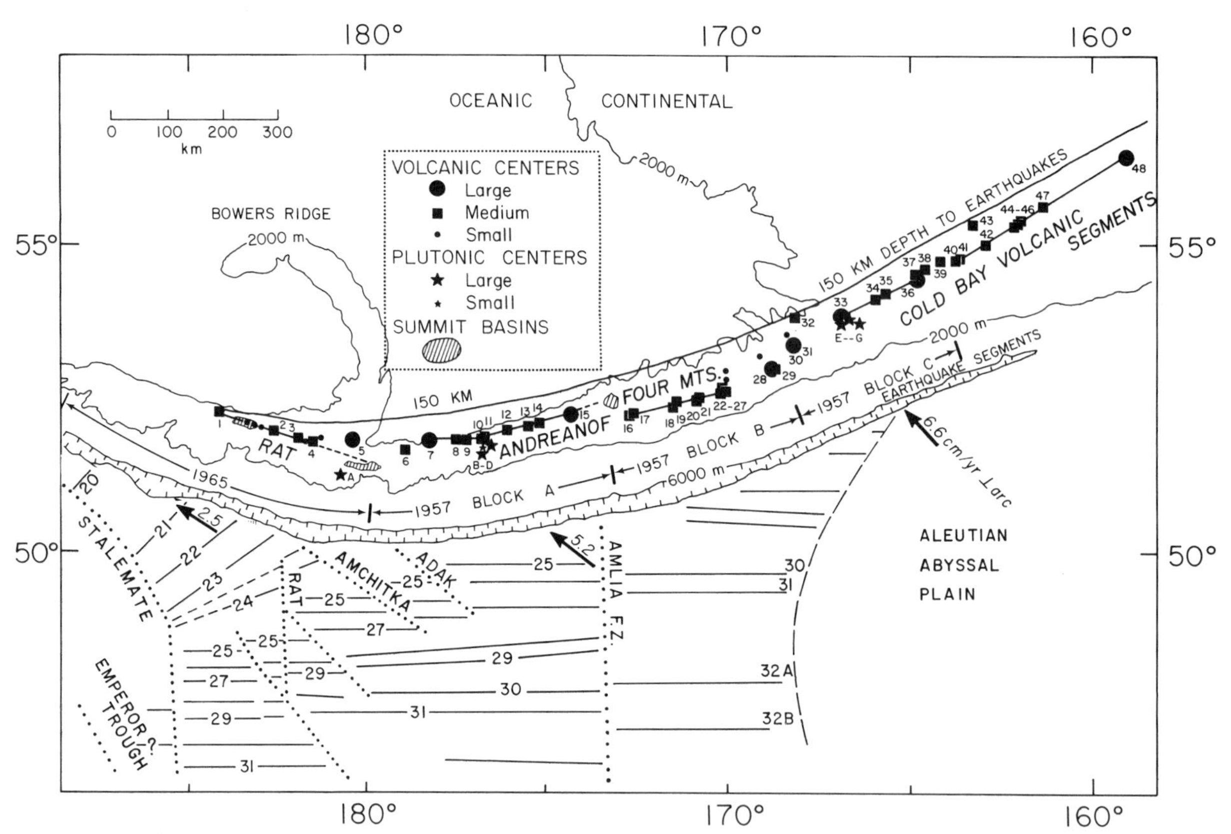

Figure 1. Bathymetric, tectonic, and magmatic features of the volcanically active part of the Aleutian Island arc. The figure is modified from Figure 1 of Kay and others (1982). Modifications include changes in the magnetic anomaly pattern (lines with numbers) on the Pacific Plate (from Lonsdale, 1988) and in the perpendicular convergence rates (from Engebretson and others, 1985). Volcanic (numbered) and plutonic centers (lettered) are listed below and are labeled as calc-alkaline (CA), tholeiitic (TH), or unknown or ambiguous (?) based on whole-rock composition, and mineralogy. Numbers in parentheses indicate number of whole-rock analyses available. Volcanoes: 1, Buldir (CA) (8); 2, Kiska (CA) (4); 3, Segula (TH?) 2; 4, Little Sitkin (CA) (13); 5, Semisopochni (TH) (15); 6, Gareloi (?); 7, Tanaga (?); 8, Bobrof (CA) (7); 9, Kanaga (CA) (20); 10, Moffett (Adak I.) (CA) (>20); 11, Adagdak (Adak I.) (CA) (10); 12, Great Sitkin (CA) (26); 13, Kasatochi (CA) (3); 14, Koniuji (CA); 15, Atka (TH) (>20); 16, Unnamed (Seguam I.) (?); 17, Unnamed (Seguam I.) (?); 18, Amutka (CA) (1); 19, Chagulak (TH?) (1); 20, Yunalaska (TH) (1); 21, Unnamed (Yunalaska I.); 22, Herbert (TH) (1); 23, Carlisle (CA) (1); 24, Cleveland; 25, Chuginadak (?) (1); 26, Kagamil (?) (1); 27, Uliaga (CA) (1); 28, Vsividof (Umnak I.) (TH) (4); 29, Recheschnoi (Umnak I.) (?) (4); 30, Okmok (Umnak I.) (TH) (>20); 31, Tulik (Umnak I.) (TH); 32, Bogoslof (CA) (3); 33, Makushin (Unalaska I.) (TH-CA) (14); 34, Akutan (TH) (10); 35, Akun (?); 36, Faris-Westdahl (Unimak I.) (TH) (7); 37, Pogromni (?); 38, Fisher (?); 39, Shishaldin (TH); 40, Isanotski (?); 41, Unnamed (?); 42, Frosty and Mt. Simeon (CA) (>10); 43, Amak (CA) (7); 44, Emmons (?); 45, Pavlof (TH): 46, Pavlof Sister (?); 47, Dana (?); 48, Veniaminof (TH) (4). Volcanoes to the east, including Augustine, are classified by Kienle and Swanson (1983). Plutons: A, East Cape (Amchitka I.) (3) (CA); B, Finger Bay (Adak I.) (TH) (26); C, Hidden Bay (Adak I.) (CA) (22); E, Shaler (Unalaska I.) (CA) (7); F, Captains Bay (Unalaska I.) (CA) (29); G, Beaver Inlet (Sedanka I.) (CA) (8).

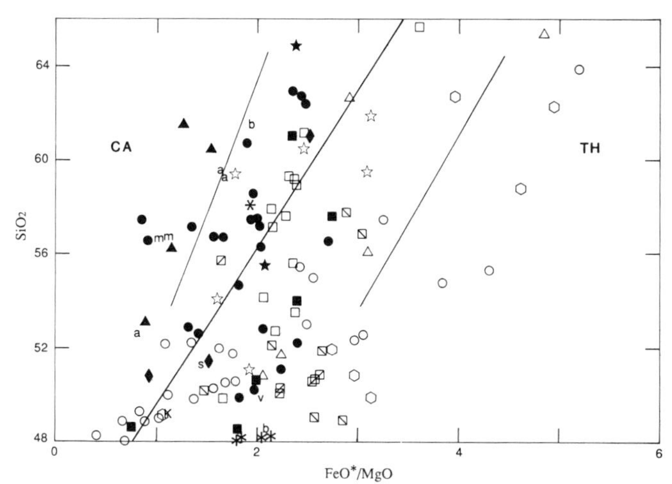

Figure 2. Plot of SiO_2 versus FeO* (total Fe as FeO)/MgO ratio for Pliocene-Recent Aleutian arc volcanic rocks for which trace-element data exist. A heavy line (from Miyashiro, 1974) separates the calc-alkaline (CA) from the tholeiitic (TH) field. Other lines separate the calc-alkaline from the transitional calc-alkaline field and the transitional tholeiitic from the tholeiitic field (following S. Kay and Kay, 1985a). Symbols: solid triangles, Buldir; solid stars, Little Sitkin; s, Segula; asterisks, Kanaga; solid circles, Moffett (m - high-Mg andesites of Kay, 1978); solid squares, Adagdak; K, Kasatochi; solid diamonds, Makushin; open squares, Great Sitkin; open stars, Islands of Four Mountains; open squares (NE-SW slash), Atka; open squares (SE-NW slash), Akutan; open triangles, Semisopochnoi; v, Veniaminof; open circles, Okmok; open hexagons, Westdahl; b, Bogoslof; a, Augustine. Data from Kay (1977), Kay and others (1982), Arculus and others (1977), Byers (1961), McCulloch and Perfit (1981), Delong and others (1985), Myers and others (1986b), Neuweld (1987), Nye and Reid (1986), and Appendix (see microfiche accompanying this volume).

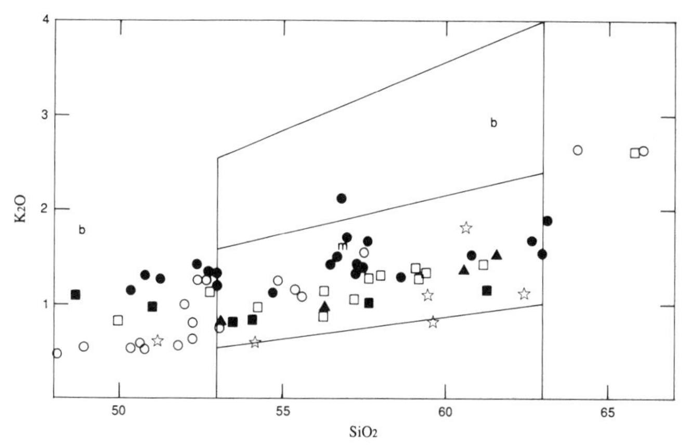

Figure 3. Plot of K_2O versus SiO_2 for selected Aleutian volcanic centers. Low, medium-, and high-K andesite fields of Gill (1981) are delineated. Symbols are the same as in Figure 2. Main arc tholeiitic andesites, and especially dacites, tend to have higher K_2O (Okmok, open circles; Westdahl, open squares) than main arc calc-alkaline andesites and dacites (solid circles, Moffett; solid triangles, Buldir; open squares, Great Sitkin; solid squares, Adagdak). Four Mountains samples (stars), except Uliaga, have low K; Bogoslof samples (b) have high K.

These size differences correlate with, and probably are due to, the decreasing perpendicular convergence rates to the west resulting from the oblique convergence of the Pacific Plate (Fig. 1). Superimposed on these east to west variations are size variations related to the location of volcanic centers within the segments. In general, volcanic centers at the ends of segments and between segments are larger than centers in the middle of segments. These variations are particularly well developed in the central and western part of the arc (Fig. 1). Compositionally, the largest volcanoes are predominantly basaltic with lesser amounts of nearly aphyric andesite and dacite; the smallest volcanoes have fewer basalts and much higher percentages of porphyritic andesite and dacite.

Chemical trends differ in volcanic rock series from large and small volcanic centers. By the criteria of Miyashiro (1974), lavas from the large centers are largely tholeiitic (that is, their FeO/MgO ratios increase with SiO_2), whereas those in the small centers are calc-alkaline (their FeO/MgO ratios are relatively constant with increasing SiO_2; Fig. 2). Based on this classification, volcanic rocks of the arc form a compositional continuum between tholeiitic and calc-alkaline characteristics. Following S. Kay and Kay (1985a), samples within this continuum will be designated tholeiitic, transitional tholeiitic, transitional calc-alkaline, and calc-alkaline (Fig. 2). The centers used as references will be: Buldir, a calc-alkaline center; Okmok and Westdahl, tholeiitic centers; and Moffett, Great Sitkin, and Semisopochnoi (also Atka), a sequence of centers that progress from transitional calc-alkaline through transitional tholeiitic centers. As shown in Figure 2, lavas within a single center can fall within several of these classifications (e.g., Moffett), but most often lavas above 54 percent SiO_2 fall within one group. The aim of this classification is to draw attention to the diversity of magmatic characteristics that require a diversity of processes.

Chemical differences in Aleutian arc magmas are also observed perpendicular to the arc. As in many arcs, these differences are shown by an increase in K_2O content at a given SiO_2 content with distance from the trench (see Figs. 1 and 3). When comparing the K_2O content of several samples, Kay and others (1982) showed that the Mg number (molar ratio of Mg to Mg + Fe) has to be considered, since Aleutian tholeiitic andesites have a higher K_2O content than calc-alkaline andesites with the same SiO_2, because of different fractionating mineralogies (see Fig. 3). For example, Bogoslof volcano, which is located north of Okmok volcano and is the largest of the centers behind the arc and the farthest from the trench, has the highest K_2O content of any center in the arc (Byers, 1961; Arculus and others, 1977). Similarly, lavas from Amak, north of the Cold Bay volcanic center, have a higher K_2O content than Cold Bay lavas (Marsh and Leitz, 1979). In contrast, Four Mountains volcanic centers, which are the closest to the trench (and presumably the closest to the seismic zone), have lavas with the lowest K_2O content. Four Mountain segment centers located behind the main volcanic front have higher K_2O content. The widening of the area of arc volcanism and the chemical diversity in this region are yet to be explained in any comprehensive Aleutian magmatic model.

Although the K_2O content of volcanoes decreases trenchward across the Aleutian volcanic arc, no low-K island arc tholeiites, such as those occurring in the frontal arc of the western Pacific island arcs (Gill, 1981), have been found in the Aleutian arc. One other important, and possibly related difference between the western Pacific arcs and the Aleutians is that the back-arc region in the Aleutians is a trapped piece of Mesozoic oceanic crust (Vallier and others, this volume) rather than a region where back-arc spreading is occurring or has occurred.

MODELS FOR THE ORIGIN OF MAGMATIC DIVERSITY IN THE ALEUTIAN ARC

Two basic models have been presented to explain the magmatic diversity in Aleutian volcanic rocks. In one model, Kay and others (1982) and S. Kay and Kay (1985a, b) have argued that the most common parental magmas are olivine tholeiitic (Mg-rich) basalts derived from mantle peridotite that has been modified by material from the subducted plate. In this model, a common process in all Aleutian volcanoes is the formation of high-Al basalt from parental olivine tholeiite at the base of the crust. Major-element and compatible trace-element variations and mineralogic differences between Aleutian magma series then result from fractionation and mixing processes in the Aleutian crust that are controlled by the tectonic setting of the volcano. Magmas that feed the high-volume tholeiitic centers pass quickly through crust that is under relative extension, resulting in shallow-level magma chambers where crystallization occurs at low pressure and there is little mixing or interaction with the arc crust. In contrast, magmas that feed the lower volume calc-alkaline centers pass slowly through crust under relative compression, and undergo crystallization and mixing at greater crustal depths. Important components of this model are correlation of volcanic volume with center type, and the observation that the smaller calc-alkaline centers are located in the middle of tectonically controlled arc segments, whereas the larger tholeiitic centers are located at the ends of these segments or in regions where segments are not well defined. Exceptions occur in the Cold Bay segment, where transitional tholeiitic lavas are common in intra-segment volcanoes such as at Akutan (Romick, 1982). However, in this region the largest volcanoes with the most extreme tholeiitic characteristics still occur at the end of or between segments.

In the second model, developed by Marsh and his co-workers (e.g., Myers and others, 1985, 1986b), differences along the arc are related to the relative age of the centers. In this model, melting of the down-going oceanic crust plus subducted sediment yields parental high-Al basalts, and crustal differentiation of high-Al basalts is a factor in creating the compositional spectrum observed in the volcanic centers. Calc-alkaline centers ("dirty" centers) form in an early immature stage of conduit development when significant lithospheric mantle debris is incorporated in the magma during transit through the lithosphere. In time, the volcanic center transforms into a tholeiitic center ("clean" center) as the conduit is thermally and chemically preconditioned and no longer contaminates the magmas. Duration of activity at a center explains the size differences between the tholeiitic and calc-alkaline centers. A problem with this model is that its evolutionary sequence is based on a comparison of volcanoes along the arc, since no volcano that has been studied shows a progression from calc-alkaline to tholeiitic lavas. On the contrary, the earliest (2 Ma) lavas from the large tholeiitic Okmok volcano (Bingham and Stone, 1972) are not calc-alkaline (Byers, 1961; Kay and others, 1982).

Clear distinctions between these two models involve different views of the evolution of Aleutian magmas in the arc crust and mantle, and different views of mantle dynamics beneath the arc. Differences in views on mantle dynamics are reflected in the relation between the arc lithosphere and asthenosphere in the two models. In the second (Marsh) model, the lithosphere includes much of the mantle above the down-going slab as well as the arc crust. In the first (Kay) model, most of the mantle above the down-going slab is undergoing ductile flow and is referred to as asthenosphere. DeBari and others (1987) suggest that this asthenospheric flow extends almost to the base of the crust, because of the penetratively deformed cumulate dunitic and wehrlitic xenoliths from Adagdak volcano (Adak Island) that are thought to have formed just below the crust.

Origin of Aleutian primitive and parental magmas

Fundamental to resolving the origin of the diversity in Aleutian arc volcanic rocks and the origin of island-arc crust, is an understanding of the formation of primitive and parental arc magmas. In this context, primitive basalts are those whose compositions are in equilibrium with the upper mantle. Primitive basalts lack evidence of crustal mixing, and those that have been at equilibrium with mantle peridotite have high Mg numbers and Cr and Ni content, indicating no (or more realistically, minimal) crustal-level fractionation. Mg-rich basalts (>8.5 percent MgO) that satisfy these criteria erupt infrequently in the Aleutians, but occur in both the calc-alkaline and tholeiitic centers. In contrast, important parental magmas are those magmas that evolve to produce the majority of the arc lavas. In this context, the most abundant basalts in the Aleutians, high-Al basalts (17 to 21 percent Al_2O_3), are important parental lavas in both models outlined above. Representative analyses of both Mg-rich and high-Al basalts are given in Tables 1 and 2 and plotted in Figure 4. The interrelationship of the major-element compositions of these basalts is critical to understanding the origin of the high-Al basalt type. Variations in incompatible trace-element ratios of both Mg- and Al-rich basalts reflect spatial and temporal differences in the magmatic source region that are largely decoupled from variations of major-element composition.

Mg-rich basalts (see Table 1) have been described from the tholeiitic center of Okmok (Byers, 1959, 1961; Kay and others, 1982; Nye and Reid, 1986), the calc-alkaline centers of Makushin (Perfit and others, 1980a), Kasatochi (S. Kay and Kay, 1985a) and Adagdak (DeBari and others, 1987), and from Ter-

TABLE 1. REPRESENTATIVE ANALYSES OF Mg-RICH AND HIGH-Mg ANDESITE FROM THE ALEUTIAN ISLANDS, ALASKA

Sample	OK1	OK1A	OK4	MK15	ADH	KAN2	KAS7A	MOF53A
SiO_2	45.70	48.07	48.97	51.20	48.46	50.28	49.07	55.50
TiO_2	0.51	0.73	0.72	0.75	0.69	0.75	0.66	0.86
Al_2O_3	12.90	15.64	16.27	15.69	15.14	16.10	15.75	15.50
Fe_2O_3	3.60	2.34	2.83					
FeO	5.60	6.98	6.24	9.21	9.03	8.74	8.82	6.21
MnO	0.03	0.15	0.01	0.18	0.14	0.13	0.14	0.10
MgO	18.20	12.68	9.62	9.64	11.83	11.19	8.77	5.58
CaO	9.60	10.76	12.86	10.12	11.37	9.99	12.52	9.51
Na_2O	1.83	2.14	2.11	2.77	2.12	2.57	2.24	2.98
K_2O	0.52	0.48	0.54	0.92	0.74	0.58	0.76	1.47
P_2O_5	0.10	0.12	0.13	0.21			0.15	0.32
Total	98.59	100.09	100.30	100.69	99.52	100.33	98.88	98.03
La	3.93	7.30	4.29	7.32	5.85	5.66	5.43	29.15
Ce	9.45	18.15	11.03	18.30	13.28	14.91	13.61	64.11
Nd	6.41	11.38	6.40	12.80	9.03	9.05	7.74	34.06
Sm	1.73	2.69	2.00	3.04	2.21	2.55	2.34	5.77
Eu	0.577	0.834	0.696	0.870	0.720	0.841	0.652	1.638
Tb	0.324	0.447	0.374	0.460	0.370	0.477	0.420	0.495
Yb	1.23	1.52	1.29	1.55	1.31	1.78	1.49	0.94
Lu	0.195	0.223	0.199		0.207	0.257	0.219	0.137
Rb				11			12	
Sr				533			482	
Ba	132	241	164	265	222	207	228	536
Cs	0.37	0.06	0.32	0.82	1.05	0.25	0.56	0.24
U	0.60	0.70	0.45		1.08	0.43	0.65	1.29
Th	0.66	1.42	0.92		1.99	0.84	1.60	3.00
Hf	1.10	1.48	1.04		2.09	1.75	1.30	4.03
Ta	0.05	0.09	0.08		0.08	0.13	0.10	
Sc	41.3	37.9	49.4		44.5	40.3	44.7	20.9
Cr	1230	698	608	255	614	946	338	377
Ni	360	272	98	66	219	222	61	153
Co	60	51	45		41	43	41	
FeO/MgO	0.49	0.72	0.91	0.96	0.76	0.78	1.01	1.11
Ba/La	33.5	33.1	38.1	36.2	38.0	36.6	42.0	18.4
La/Sm	2.3	2.7	2.1	2.4	2.7	2.2	2.3	5.1
La/Yb	3.2	4.8	3.3	4.7	4.5	3.2	3.7	30.9
Eu/Eu*	0.98	0.95	1.03	0.91	1.00	0.97	0.83	1.09
Th/Hf	0.60	0.96	0.88		0.95	0.48	1.24	0.74
Ba/Hf	119	163	158		106	119	176	133
Ba/Th	200	170	179		112	246	143	179
La/Hf	3.56	4.93	4.13		2.79	3.24	4.19	7.23
Th/U	1.1	2.0	2.0		1.8	1.9	2.5	2.3
$^{87}Sr/^{86}Sr$		0.70302		0.70333				0.7028
ε_{Nd}		7.54		6.17				11.3
$\delta^{18}O$	4.9	5.8	5.2		5.2	7.9		7.5

*OK1, OK1A, and OK4 are from the flank of Okmok volcano, Umnak Island (Byers, 1961). Major-element analyses are mostly from Byers (1961) and isotopic analyses are from McCulloch and Perfit (1981). MK15 is from Makushin volcano on Akutan Island (analyses from McCulloch and Perfit, 1981). ADH is xenolith host rock from Adagdak volcano, Adak Island (DeBari and others, 1987). KAN2 is Tertiary sill on Kanaga Island (Swanson and others, 1987). KAS7A is from Kasatochi volcano (major element analyses from Kay and Kay, 1985a). MOF53A is high-Mg andesite from below Moffett volcano (ADK53 *in* Kay, 1978). Partial analyses and isotopic data included in Kay (1978). New major-element analyses presented here were done by electron microprobe on glasses made from rock powders, and new trace element studies were done by INAA (instrumental neutron activation; Sr and Rb by XRF) at Cornell University. Analytical techniques and standard data are from Rubenstone (1984) and Kay and others (1987). Normalization factors for trace-element plots are Cs (3.77), K (116), Ba (3.77), U (0.015), Th (0.05), Hf (0.22), Ta (0.022), La (0.378), Ce (0.976), Nd (0.716), Sm (0.23), Eu (0.0866), Tb (0.589), Yb (0.249), and Lu (0.0387).

TABLE 2. REPRESENTATIVE ANALYSIS OF HIGH-Al BASALTS FROM THE ALEUTIAN ISLANDS, ALASKA*

Sample	OK7	CHAG	GS727	ADG14	MOF15	MOFA7	KAN508
SiO_2	50.60	51.04	49.68	48.62	49.35	52.40	47.66
TiO_2	1.10	0.98	0.85	0.99	0.86	0.98	1.18
Al_2O_3	18.06	21.12	19.39	18.71	18.57	20.43	18.58
Fe_2O_3	2.99						
FeO	6.75	7.73	9.33	10.27	9.52	8.44	10.28
MnO	0.12	0.12	0.14	0.15	0.15	0.15	0.16
MgO	5.58	3.99	5.60	5.68	4.84	3.52	6.06
CaO	11.46	10.90	10.97	11.52	10.80	9.57	11.41
Na_2O	2.74	3.21	2.66	2.75	2.85	3.27	2.68
K_2O	0.53	0.59	0.82	1.05	1.11	1.39	0.87
P_2O_5	0.28	0.13		0.21	0.15		
Total	100.21	99.81	99.44	99.95	98.20	100.15	98.88
La	5.35	5.88	5.99	8.44	7.76	5.22	7.12
Ce	13.57	13.69	13.86	18.48	18.99	11.28	16.93
Nd		10.76	10.66	11.64	11.67	8.20	11.80
Sm	2.55	2.75	2.77	3.22	3.28	2.09	3.47
Eu	0.929	0.978	0.911	1.040	1.032	0.982	1.085
Tb	0.577	0.519	0.500	0.556	0.469	0.336	0.578
Yb	1.89	1.90	1.91	1.94	1.66	1.40	1.90
Lu	0.280	0.278	0.259	0.270	2.58	0.192	0.266
Ba	191	238	261	441	317	489	352
Cs	0.85	0.40	0.52	0.36	0.32	0.30	0.39
U	0.43	0.29	0.54	0.97	0.72	1.01	0.81
Th	0.88	1.08	1.33	2.08	1.89	2.00	1.55
Hf	1.61	1.58	1.45	1.82	1.78	1.90	1.83
Ta	0.18	0.14	0.19	0.40	0.44	0.00	0.39
Sc	44.2	28.6	35.7	37.6	34.9	19.7	46.5
Cr	72	27	49	41	18	8	35
Ni	9	16	22	18	8	14	18
Co	32	26	32	37	41	16	47
FeO/MgO	1.69	1.94	1.67	1.81	1.97	2.40	1.70
Ba/La	35.7	40.5	43.5	52.3	40.8	93.7	49.4
La/Sm	2.1	2.1	2.2	2.6	2.4	2.5	2.1
La/Yb	2.8	3.1	3.1	4.3	4.7	3.7	3.8
Eu/Eu*	1.01	1.04	0.98	0.96	1.01	1.29	0.96
Th/Hf	0.54	0.68	0.92	1.15	1.06	1.05	0.85
Ba/Hf	119	150	181	243	178	257	192
Ba/Th	218	220	196	212	168	245	227
La/Hf	3.33	3.72	4.15	4.65	4.36	2.75	3.89
Th/U	2.0	3.7	2.5	2.1	2.6	2.0	1.9
$\delta^{18}O$	4.9	6.0			5.8	5.5	

*OK7 is from Okmok volcano, Umnak Island (major-element analyses from Byers, 1961); CHAG is from Chagulak volcano; GS727 is from Great Sitkin volcano (analyses from Neuweld, 1986); ADG-14 is from Adagdak volcano, Adak Island (basalt unit in Coats, 1952); MOF15 and MOFA7 (ADK7 *in* Kay and others, 1978) are from Moffett volcano, Adak Island (main cone, see Coats, 1952); KAN508 is from Round Head, Kanaga Island (Coats, 1952).

tiary sills on Kanaga Island (Pope, 1983; Swanson and others, 1987). Nye and Reid (1986) argue that some of the most Mg-rich (17 to 18 percent MgO, Table 1) of these basalts—those that occur on the flank of Okmok volcano—represent one type of primitive lava. These basalts have only olivine phenocrysts (cores are F092.9), which contain spinel inclusions. Several other Mg-rich basalts (12 to 13 percent MgO) from the same locality have predominantly olivine and minor clinopyroxene phenocrysts (Byers, 1959) and have been proposed as a second type of primitive lava by Nye and Reid (1986). Additional Mg-rich basalts (8.5 to 10 percent Mg), from Okmok and other centers, contain plagioclase as well as olivine and clinopyroxene phenocrysts. Perfit and others (1980a) proposed that a basalt from Makushin (MK15) was parental to the arc suite, and Kay and others (1982) and Conrad and Kay (1984) used one of the Mg-rich Okmok basalts (OK4; Byers, 1961) as a parent in crystal fractionation models that derived more evolved lavas from both the calc-alkaline and tholeiitic centers.

High-Al basalts (17 to 21 percent Al_2O_3) are the most common basalt type in the arc and have been proposed as the primary arc lava by Marsh and coworkers (Marsh, 1981, 1982; Myers and others, 1985, 1986a, b; Brophy and Marsh, 1986; Brophy, 1986). These basalts have high FeO/MgO ratios (>1.8), and low MgO (5 percent), and Ni and Cr content (Table 2); they are characterized by abundant plagioclase (AN80 to 90) and olivine phenocrysts (FO70 to 79) and less common clinopyroxene phenocrysts (Marsh, 1982; S. Kay and Kay, 1985a; Myers and others, 1985; Brophy, 1986). Titanomagnetite is ubiquitous. Although plagioclase-rich, the high Al content of most of these basalts appears not to be related to plagioclase accumulation, since Eu anomalies (Eu/Eu*; Eu* is interpolated from the REE pattern and equals Eu if there is no Eu anomaly) are both positive and negative (Eu/Eu* = 0.85 to 1.1 in 33 samples; Kay, 1977; Myers and others, 1986b; Table 2 and Figs. 4B and 5). Basalts with large positive Eu anomalies (Eu/Eu* > 1.1) and a very high Al_2O_3 content (>21 percent) are much more rare and may have excess plagioclase (Fig. 5). Despite arguments to the contrary by Marsh (1981), plagioclase-rich xenoliths whose minerals are similar to those in high-Al basalts provide evidence that some plagioclase accumulation can occur (S. Kay and Kay, 1985a).

Two end-member proposals have been offered to relate the Mg-rich and high-Al basalts. In one proposal, the Mg-rich lavas are primarily derived from above the subducting plate by melting of mantle peridotite that has been modified by the addition of a component from the plate below (i.e., Kay, 1977, 1980; Perfit and others, 1980a; Kay and others, 1982). In this model, the high-Al basalts are considered to be derivative magmas that form at the crust-mantle boundary or in the lower crust as the result of crystallization of olivine and clinopyroxene, but not plagioclase, from the high-Mg lavas (Conrad and Kay, 1984; S. Kay and Kay, 1985a, b; DeBari and others, 1987; Gust and Perfit, 1987).

In the other proposal, the high-Al basalts result from melting of the eclogitic oceanic crust that is near the top of the down-

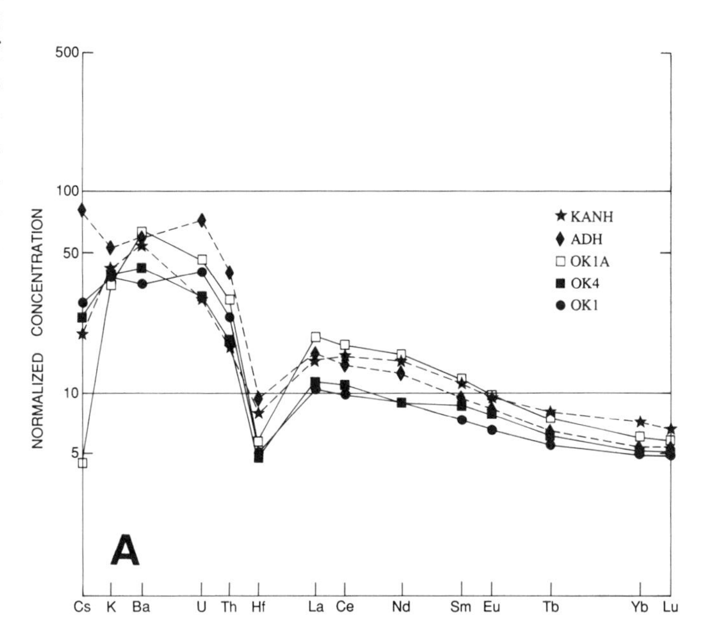

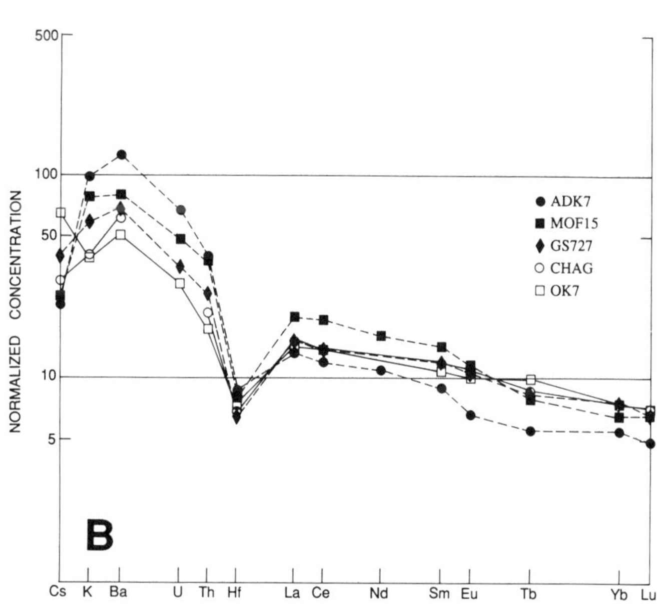

Figure 4. Trace elements of Aleutian basalts (data in Table 1) normalized to chondrites and to ocean-ridge basalts (Cs and K only). See Table 1 for normalization values. A. Mg-rich basalts. OK1, OK1a, OK4 (described in Byers, 1961) from outside the Okmok shield volcano (tholeiitic); ADH is the post-glacial host rock of the ultramafic xenoliths from Adagdak volcano (transitional calc-alkaline; DeBari and others, 1987); and KANH is a Tertiary sill hosting xenoliths on Kanaga Island (Swanson and others, 1987). B. High-Al basalts. ADK7 and MOF15 are from Moffett volcano (calc-alkaline to transitional calc-alkaline); GS727 (data from Neuweld, 1987) is from Great Sitkin volcano (transitional calc-alkaline); CHAG is from Chagulak volcano in the Four Mountains segment; and OK7 is from Okmok caldera (tholeiitic).

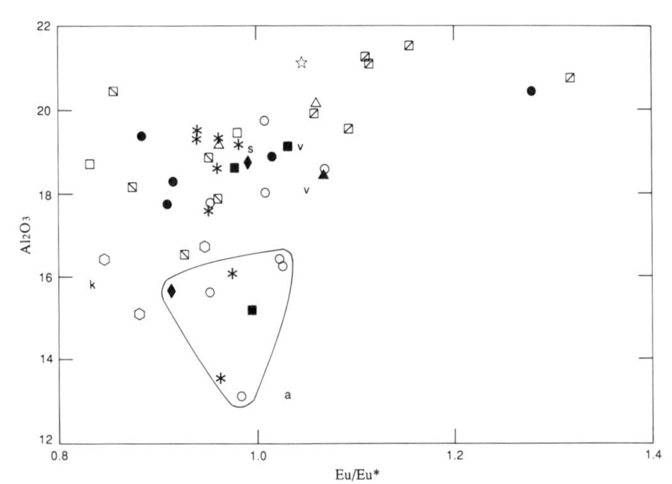

Figure 5. Plot of Al_2O_3 (wt. %) versus the Eu anomaly (Eu/Eu* where Eu* is interpolated from the REE pattern and equals Eu if there is no Eu anomaly) for Aleutian basalts. Symbols are the same as in Figure 2. Open symbols are for tholeiitic and transitional centers (except Great Sitkin, open squares), while closed symbols are for calc-alkaline centers. High-Al basalts with 18 to 20 percent Al_2O_3 commonly have little or no Eu anomaly, whereas basalts with >20 percent Al_2O_3 tend to have positive Eu anomalies. Data are from Kay (1977, 1978), Kay and Kay (unpublished data), McCulloch and Perfit (1981), Myers and others (1986b), and Neuweld (1987). Data from Nye and Reid (1986) are not included because we believe, based on analyses of similar samples from Umnak Island, that there are systematic errors in their Eu anomalies.

going plate. These melts pass through the overlying mantle lithosphere with little or no modification to be parental to Aleutian arc magmas (Marsh, 1982; Myers and others, 1985, 1986a, b; Brophy and Marsh, 1986; Brophy, 1986). In this model, the Mg-rich magmas are considered to be mixtures of high-Al basalts and upper mantle material that has been entrained in them (Myers and others, 1985; Brophy, 1986) or has been assimilated into them (Myers and Marsh, 1987) as they pass through the lithosphere.

The question of whether the Mg-rich lavas are mixtures of xenocrystic mantle material and primary high-Al basalts, or are primarily melts of slab-modified mantle, centers on their isotopic and mineralogic characteristics. Brophy (1986) argued (from a compilation of isotopic data) that the Mg-rich basalts have less radiogenic Sr and more radiogenic Nd than high-Al basalts, which is consistent with contamination of high-Al basalt by MORB-like mantle. However, as Brophy admits, the range of isotopic values is small and the correlation is poor. He supports his conclusion by quoting Myers and others (1985) who suggest that xenoliths described from Moffett volcano by Conrad and Kay (1984) provide evidence of this type of lithospheric contamination. However, these Moffett ultramafic and mafic xenoliths, as well as the dunitic and wehrlitic xenoliths from Adagdak volcano (DeBari and others, 1987) have Nd and Sr isotopic characteristics like those of the arc lavas (R. Kay and others, 1986, see below). Contamination of Moffett lavas by these xenoliths by mixing or assimilation does not change the Nd and Sr isotopic characteristics of the lavas. Furthermore, Nye and Reid (1986) have argued that olivine in the Okmok Mg-rich lavas is in equilibrium with the Mg-rich liquid with which it is associated, and is thus phenocrystic, not xenocrystic. The same is true for olivine in Mg-rich basalts from the Adagdak and Tertiary Mg-rich basalts from Kanaga (S. Kay, unpublished data).

The formation of high-Al basalts from Mg-rich basalts is supported by mass balance models that use observed rock types and are consistent with experimental phase equilibrium studies (Gust and Perfit, 1987). Major-element modeling by Conrad and Kay (1984), Brophy (1986), and Gust and Perfit (1987) shows that fractionation of approximately 20 percent olivine and clinopyroxene from Mg-basalts, such as OK4 and MK15, results in high-Al basalt liquids. The most detailed calculations are those of Conrad and Kay (1984) who combine major-element modeling with the analyses of partitioning of Cr and Ni in zoned clinopyroxene and olivine in cumulate-textured xenoliths to suggest that open-system mixing and fractionation produces a steady-state high-Al basalt. The dunite and wehrlite residuum predicted from these models are observed as xenoliths in the Mg-rich Adagdak basalt (Swanson and others, 1987; DeBari and others, 1987). These xenoliths are isotopically similar to the arc lavas (R. Kay and others, 1986).

Myers and others (1986b) object to these models by arguing that a compilation of trace-element data shows that concentration levels of incompatible elements in high-Al basalts are lower than those in high-Mg basalt, which is opposite to the trend expected if the two are related by crystal fractionation. However, their compilation consists of somewhat evolved tholeiitic basalts (data from Kay, 1977; DeLong and others, 1985) and does not include the high-Mg basalts considered to be parental to high-Al basalt in our discussion. Some of these Mg-rich basalts do have incompatible element concentrations that are lower than the high-Al basalts (Fig. 4A and B). Furthermore, major- and trace-element modeling by Neuweld (1986) shows that some fractionation models are successful: Great Sitkin high-Al basalts (e.g., GS727) can be derived from Mg-basalts like the Okmok sample OK4 (Tables 1 and 2). However, this model is only an approximation, because the parental basalt is from another center (no known Mg-rich basalts occur at Great Sitkin volcano). Because of trace-element differences between and within centers, not all high-Al basalts can be derived from all Mg-rich basalts (see below).

In further support of a derivative origin for Aleutian high-Al basalts, S. Kay and Kay (1985a) have suggested that the phenocryst assemblage common to the high-Al basalts (plagioclase and relatively Mg-, Ni-, and Cr-poor olivine and clinopyroxene) is the result of relatively low-pressure crystallization of high-Al basalt liquids that were produced by high-pressure fractionation of olivine and clinopyroxene in the lower crust. Marsh (1982), Brophy (1986), and Myers and others (1986a) argue that existing experimental evidence does not support this process, but does support the coexistence at equilibrium of anhydrous high-Al basalt liquids and clinopyroxene-garnet-bearing assemblages at 27 kb (Green and Ringwood, 1968; Johnston, 1986) in the subducted plate under the volcanoes. However, the experimental work of Gust and Perfit (1987) on Mg-rich basalt MK15 shows that, at pressures >5 kb, olivine and clinopyroxene crystallize before plagioclase. Interstitial glasses produced in these experiments approach the compositions of high-Al basalts.

Many workers have rejected the hypothesis that arc andesites are primary melts produced by low-percentage melting of garnet-bearing down-going plate, based on models that predict REE that are more fractionated (have high La/Yb ratios) than those in most andesites (see Gill, 1981). However, as pointed out by Apted (1981) and Brophy and Marsh (1986), REE patterns of high-Al basalts (e.g., 37 Aleutian samples have La/Yb ratios = 3 to 5: Table 2, Kay, 1977; Myers and others, 1986a; Brophy and Marsh, 1986) are consistent with a quartz eclogite melting model in which melt segregates at lower pressure and at higher degrees of melting, thereby reducing the amount of garnet (garnet retains the heavy REEs, i.e., Yb) in the solid residual assemblages. In a model of this process, Brophy and Marsh (1986) propose that this low pressure is achieved by decompression of eclogite diapirs. To generate a high-Al basalt (5 percent MgO), Figure 5 of Brophy and Marsh (1986) shows that ~55 percent melting is required if the melt has 5 percent H_2O, and ~65 percent if the melt is dry.

R. Kay and Kay (1985) outline reasons for doubting the validity of the eclogite diapir model. They say that a fundamental question is: at what percentage of melting does the magma segregate from the diapir as it rises after detaching from the subducting plate? Since unmelted eclogite is denser than peridotite, considerable melting is necessary for buoyancy (Brophy and Marsh, 1986, estimate 20 percent). Additional melting occurs during decompression. A key parameter is the segregation velocity at buoyancy (20 percent melt), which Brophy and Marsh (1986) calculate to be 3.2×10^{-4} to 3.2×10^{-8} m/yr, using the porosity-permeability relation of McKenzie (1984), melt viscosities of 10 to 10^5 Pas (10^2 to 10^6 poise), and a grain size of 0.1 mm. This velocity is obviously less than the diapir ascent velocity (presumed to be in the range of 10^{-2} to 10^{-1} m/yr). In contrast, use of the porosity-permeability relation from VonBargen and Waff (1986) and other parameters from McKenzie (1984), including a 1-mm grain size, over the same viscosity range yields segregation velocities of 20 to 2×10^{-3} m/yr. In this alternative calculation, only the most viscous magmas avoid segregation at 20 percent melting. For reference, Crater Lake andesite with 4 percent MgO has a viscosity of slightly less than 10^2 Pas (10^3 poise) at 20 kb and 1,350°C (Kushiro and others, 1976). Experiments on basalt at 10 Kb (Kushiro, 1986) indicate that 1 percent water reduces viscosity by a factor of 3, which is about the effect (in the opposite sense) of a 100°C temperature drop. Therefore, if water is present, equivalent viscosities are reached at lower temperatures. These alternative calculations suggest that, with or without water, andesite should segregate from eclogite at melting percentages below 20 percent, and that high-Al basalt-eclogite equilibrium, which occurs at higher melting percentage, will not be achieved. Note that segregation velocities in clinopyroxene-rich residues will be less than those in the olivine-rich residues assumed in our calculations, although Waff (1986) argues that the effect is small.

To summarize, although melts of silicic composition could be retained in ascending eclogite diapirs at low percentages of melting, melt that approaches high-Al basalt in composition (especially if it were hydrous) would require melt fractions in excess of 50 percent. However, melts segregate quickly at porosities well below 50 percent, halting the diapir. As at mid-oceanic ridges, melt migration by porous flow of basalt through peridotite and finally by magma-fracturing through the overlying lithosphere seems to be a workable alternative. Because this process produces olivine tholeiite (Fujii and Scarfe, 1985), the high-Al basalt is considered to be a derivative melt.

Rare high-Mg andesites (Table 1) from the base of Moffett volcano do have the type of isotopic and chemical signature expected from small degrees of melting of the oceanic crust (Kay, 1978). The steep REE patterns in these rocks suggest that small degrees of partial melt (3 percent melt) could segregate from eclogite in the subducted plate, in contrast to the large degree of melt required for segregation by Brophy and Marsh (1986). These high-Mg andesites further suggest that some magmas may pass through the mantle with relatively little interaction. Although the isotopic data would permit derivation of these andesites ($\epsilon_{Nd} = 11.3$, $^{87}Sr/^{86}Sr = 0.7028$, $^{206}Pb/^{204}Pb = 18.36$; Kay, 1978; R. Kay and others, 1986) from the melting of oceanic basalt buried in the arc crust, their steep REE patterns would be difficult to explain by using such models. Saunders and others (1987) propose that high-Mg andesites from Baja, California, similar to these Aleutian samples, are associated with the eclogite melting that accompanies ridge subduction. However, there is no evidence of ridge subduction contemporaneous with the formation of the Aleutian high-Mg andesites. Their origin remains a problem.

The Mg-rich lavas discussed by Nye and Reid (1986) and Gust and Perfit (1987), raise a further question concerning the composition of the mantle-derived magmas beneath the arc. The question is similar to that posed for ocean-ridge basalts: is the melting product of the mantle a picritic (around 15 percent MgO) basalt that fractionates olivine while still in the mantle, or an olivine tholeiite basalt more like MK15 or OK4 (9 to 10 percent MgO), or depending on conditions, are both types generated? Gust and Perfit (1987) conclude that the lack of three-phase multiple saturation (olivine, clinopyroxene, orthopyroxene) in their experimental studies of olivine tholeiite MK15 shows that MK15 is not a primary magma derived from two-pyroxene peridotite, but must have been derived from a more Mg-rich liquid. Mg-rich magmas have important implications for magma generation in the arc as they have liquidus temperatures above 1,280°C and as high as 1,400°C (Nye and Reid, 1986; Gust and Perfit, 1987).

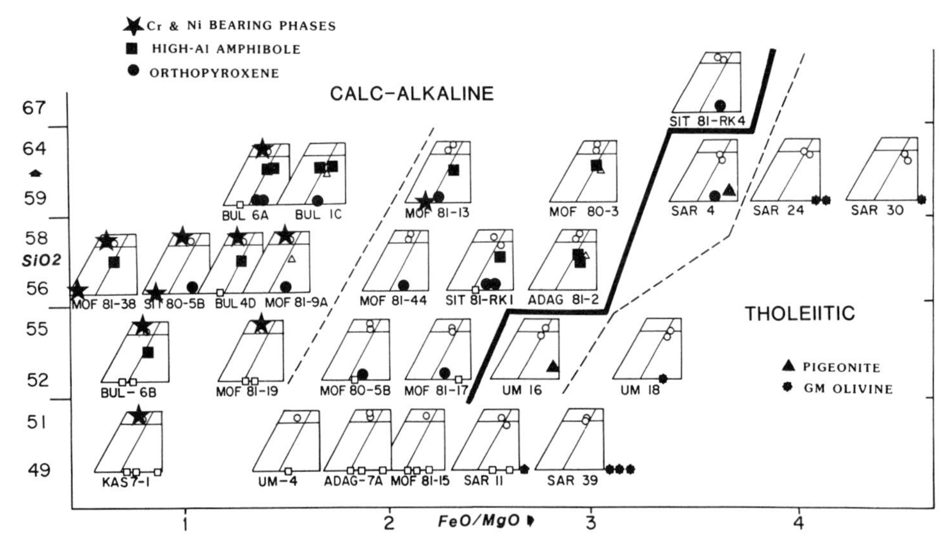

Figure 6. Mafic phases in Aleutian volcanic rocks plotted on the lower half of the Fe* (total Fe as Fe^{+2})-Mg-Ca (molar quantities) triangular diagram (pyroxene quadrilateral). Samples are arranged with respect to their whole-rock SiO_2 content and FeO/MgO ratio. The solid line separates the calc-alkaline and tholeiitic series and dashed lines separate groups within the series. Important mineral phases in distinguishing the series are identified on the figure. Other phases indicated are clinopyroxene (open circles), low-Al amphibole (open triangles) and Ni-poor olivine (open squares). In general, groundmass and phenocryst phases are not distinguished. Symbols represent the phase composition and are based on as many as ten analyses. Adapted from Figure 3 of S. Kay and Kay (1985a).

Mineralogic evidence for the evolution of Aleutian basaltic andesites, andesites, and dacites

The mineralogy of Aleutian volcanic rocks and their xenoliths is an important guide to the crystallization conditions, water content, extent of mixing, and compositions of mixing end members, and thus to the origin of the diversity in Aleutian magmas. Much petrographic information and some optically determined compositions are included in USGS Bulletin 1028 (see references in Kay and others, 1982) and in papers by Coats (1952) and Byers (1961). These studies, along with later microprobe studies by Marsh (1976), S. Kay and Kay (1985a), Brophy (1986), Romick (1982), Neuweld (1987), S. Kay and others (1986b) and Wolf (1987) show that Aleutian lavas are mineralogically diverse (Figs. 6 to 8) and formed under a range of conditions. This mineralogical diversity led S. Kay and Kay, 1985a to suggest that variable amounts of mixing of less fractionated and more fractionated magmas at variable crustal depths can account for much of the magmatic diversity in the arc (Fig. 9). The evidence for these processes is best seen in the basaltic andesites, andesites, and dacites.

Mineral assemblages in the extreme tholeiitic series (i.e., Okmok and Westdahl volcanoes, Figs. 1 and 2) are simple and suggest shallow-level closed-system crystal-fractionation processes (Fig. 9; S. Kay and Kay, 1985a). Evolved basalts (FeO/MgO >1.7) are often porphyritic, but andesites and dacites are generally microphyric to aphyric. Important phenocryst and groundmass phases include olivine, clinopyroxene, and plagioclase (Figs. 6 and 7). Generally, titanomagnetite (Fig. 8) is a groundmass phase. The FO content of olivine and the AN content of plagioclase decrease with increasing whole-rock FeO/MgO ratio and SiO_2 content (Figs. 6 and 7). Olivine (near FO_{50}) is a groundmass phase in the andesites and dacites. The TiO_2 content of titanomagnetite increases with increasing whole-rock TiO_2 content in the basalts and remains relatively constant in the more Si-rich rocks (Fig. 8). As expected in relatively simple fractionation processes, most phases are normally zoned.

Volcanic rocks in the transitional tholeiitic series are more phenocrystic than those in the tholeiitic series, and their mineral assemblages are slightly more complicated. The major mineralogic contrast with the tholeiitic series is that orthopyroxene occurs as a phenocryst in the transitional tholeiitic basaltic andesites, andesites, and dacites, and pigeonite occurs in the groundmass of some transitional tholeiitic andesites (Fig. 6). Normal zoning patterns in plagioclase, olivine, and pyroxene are most common, although some reverse zoning occurs. The mineralogic data for the transitional tholeiitic series are consistent with relatively shallow-level fractionation and some mixing of evolved lavas (Fig. 9; S. Kay and Kay, 1985a). Trace-element studies by DeLong and others (1985) on transitional tholeiitic rocks from Semisopochnoi also suggest that magma mixing occurs in the petrogenesis of the transitional tholeiitic series.

The mineralogy of the calc-alkaline series rocks, which are phenocryst-rich, suggests crystallization at higher pressure and water content and requires more mixing of evolved and unevolved magmas than in the tholeiitic series (Fig. 9). Variations in

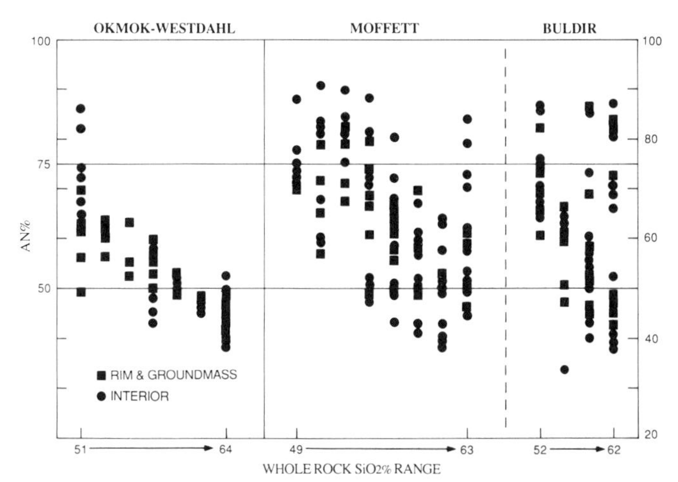

Figure 7. Compositions of Aleutian volcanic plagioclases from two tholeiitic centers (Okmok and Westdahl), a transitional calc-alkaline center (Moffett), and a calc-alkaline center (Buldir) plotted against whole-rock SiO_2 content. AN content of plagioclase (generally microphenocrysts and groundmass phases) from the tholeiitic centers decreases fairly regularly with increasing whole-rock SiO_2 content. In contrast, plagioclases in the calc-alkaline series show a broader composition range whose upper limit in AN content does not correlate with whole-rock SiO_2 content. AN85 (or higher) plagioclase occurs in almost every sample.

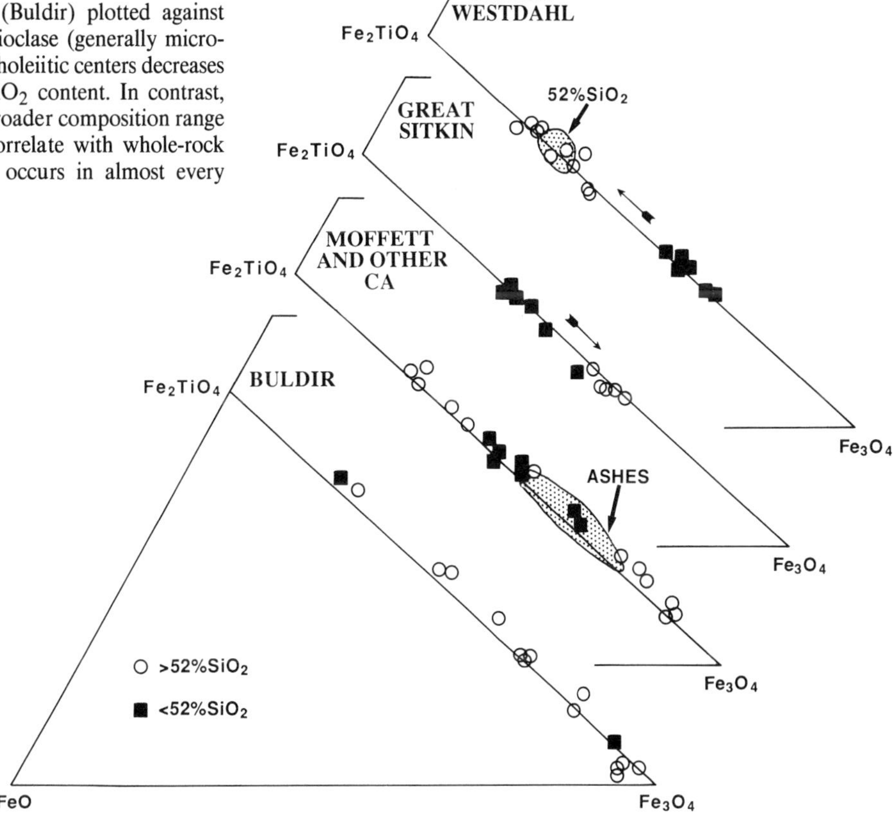

Figure 8. Compositions of magnetites from Aleutian volcanic rocks plotted on the ulvöspinel join (Fe_2TiO_4 - Fe_3O_4) of the system FeO - Fe_2O_3 - TiO_2. Magnetites in volcanic rocks with <52 percent SiO_2 are shown as solid squares, and those in rocks with >52 percent SiO_2 are shown as open circles. In general, magnetite composition correlates with whole-rock TiO_2 content. The magnetites in lavas from tholeiitic volcanoes (Okmok and Westdahl) increase in TiO_2 content as whole-rock SiO_2 increases to 55 percent and TiO_2 increases to 2.7 percent (magnetite compositions in lavas with 52 percent SiO_2 lie in the circled field). In more silicic rocks, TiO_2 content remains relatively constant or decreases as whole-rock TiO_2 decreases. Magnetites in the transitional calc-alkaline Great Sitkin show a decrease in TiO_2 content as whole-rock SiO_2 (50 to 61 percent) increases and TiO_2 (0.85 to 0.65 percent) decreases. Magnetites in the transitional calc-alkaline Moffett lavas are more variable, but the majority follow the same pattern as the Great Sitkin lavas. Most magnetites in calc-alkaline dacites (including ashes) plot in the field labeled ashes. Magnetites in calc-alkaline Buldir lavas are extremely variable, and some are almost pure magnetite. Magnetite compositions are consistent with higher oxygen fugacities in the calc-alkaline series. Data are from S. Kay and J. Romick (unpublished data) and Neuweld (1987).

crystallization conditions and in the composition of mixing end members distinguishes the transitional calc-alkaline and calc-alkaline series rocks (Fig. 9). In the calc-alkaline lavas, the crystallization sequence cannot be determined from plotting percentages of crystals against percentage of groundmass in the manner suggested by Marsh (1981), because magma mixing has resulted in nonequilibrium mineral assemblages.

Volcanic rocks in the transitional calc-alkaline series are characterized by complex phenocryst assemblages with minerals that have complex and often reversed zoning patterns. These characteristics suggest that mixing of high-Al basalts with derivative andesites and dacites is an important process in this series (Fig. 9; S. Kay and Kay, 1985a; Neuweld, 1987; Wolf, 1987). The series is characterized by the presence of either low-Al amphibole (7 to 9 percent Al_2O_3), orthopyroxene, or both as phenocrysts in the andesites and dacites (Fig. 6). Titanomagnetite is an important phenocryst phase (Fig. 8). Both normal and reversed zoning patterns occur in plagioclase, and in clinopyroxene and other mafic phases (Figs. 6 and 7). Compared to the tholeiitic series, the compositions of all mineral phases show more diversity in a single sample and often do not vary regularly with increasing whole-rock SiO_2 content. Plagioclase with $>AN_{80}$ occurs in almost all lavas regardless of their SiO_2 content (Fig. 7). Titanomagnetite compositions are more variable than in the tholeiitic series, and some are high in the magnetite component (Fig. 8). The mineralogy of the andesites and dacites, particularly their amphibole and magnetite phenocrysts, suggests that the transitional calc-alkaline series crystallized under more H_2O-rich conditions at higher fO_2, higher pressure, and lower temperature than the tholeiitic lavas (Fig. 9).

Volcanic rocks of the extreme calc-alkaline series show disequilibrium mineral assemblages and wide compositional variations in complexly zoned minerals. These lavas are modally richer in mafic phenocrysts than other Aleutian magmas (Neuweld, 1987) and are characterized by Ni-rich forsteritic olivine (up to FO_{92} with inclusions of Cr-spinel) and Cr-rich diopside, both of which are frequently partially replaced by high-Al amphibole (Fig. 6). Other phenocryst phases include hypersthene, salitic augite, magnetite, and very complexly zoned plagioclase. Reverse zoning is common in all phenocryst phases. Plagioclase compositions are variable and extend to $>AN_{85}$ over a large range of whole-rock SiO_2 (Fig. 7). Magnetite phenocrysts show a broader range of compositions extending to a higher magnetite component than in other Aleutian magmas (Fig. 8).

S. Kay and Kay (1985a) interpreted the disequilibrium mineral assemblages in the calc-alkaline series as representing mixing of Mg-rich basalt with more evolved andesite or dacite. The lack of plagioclase inclusions in the Mg-rich clinopyroxene and olivine phenocrysts suggests that plagioclase appeared on the liquidus after olivine and clinopyroxene in the mafic mixing end-member (Neuweld, 1987) and supports the crystallization sequence for the Mg-rich basalts discussed above. The replacement of the Mg-rich phases by high-Al amphibole and the occurrence of amphibole phenocrysts in some samples suggests that the mafic end member in these cases had a high H_2O content and that crystallization took place at a relatively high pressure and low temperature within the stability field of amphibole in basalts. These samples also contain salitic augite and Ca-rich plagioclase, which have been correlated with relatively H_2O-rich conditions (Conrad and Kay, 1984). One calc-alkaline andesite (SITR5B) does not contain amphibole phenocrysts but shows convincing evidence for magma mixing: Mg-rich clinopyroxene forms rims on orthopyroxene and occurs as microphenocrysts in the groundmass (S. Kay and Kay, 1985a).

Support for an important role for magma mixing and for the early crystallization of amphibole and titanomagnetite in the evolution of the Aleutian calc-alkaline series also comes from the

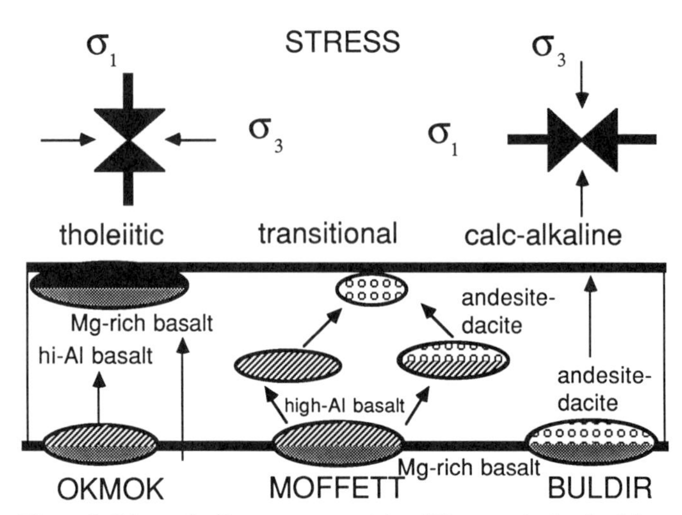

Figure 9. Schematic diagram summarizing differences in depth of fractionation and mixing in the Aleutian volcanic series. Boxed area represents the Aleutian crust and ellipses are magma chambers. In the tholeiitic series (e.g., Okmok), Mg-rich basalts occasionally erupt, but in general they fractionate olivine and clinopyroxene at the crust-mantle boundary, producing high-Al basalts, which then penetrate the crust in a relatively tensional (dike-like) environment. Mafic magmas pond beneath magma chambers causing reheating of partially crystallized magmas and melting of roof rocks. In the transitional tholeiitic (Semisopochnoi) and transitional calc-alkaline series (Moffett), parental Mg-rich magmas also fractionate olivine and clinopyroxene at the crust-mantle boundary producing high-Al basalts. These high-Al basalts can then migrate to magma chambers at various levels higher in the crust, where they can mix with andesites and dacites that have fractionated from high-Al basalts at higher levels, or themselves fractionate to andesite and dacite. Depending on the depth, amphibole may or may not fractionate from these lavas. Melting and assimilation of wall rocks can occur at depth. Because the stress system is mixed, eruption can take place from various levels, but in many cases, final eruption following storage at relatively shallow levels is suggested by low-pressure phenocryst assemblages in the erupted rocks. Calc-alkaline lavas (Buldir) penetrate the crust in relatively compressional (sill-like) environments. Mg-rich basalts fractionate to form high-Al basalts that in turn fractionate to andesites and dacites at depths near the crust-mantle boundary. Mixing of new batches of Mg-rich basalts with these fractionated andesites and dacites occurs at these depths, and the lavas are erupted to the surface relatively quickly as indicated by the presence of amphibole and mineral phases that are not at equilibrium. Remelting and assimilation of older crustal rocks probably alters the final character of these calc-alkaline lavas.

textural characteristics and mineralogy of nondeformed amphibole-bearing cumulate xenoliths found in a pyroclastic andesite from Moffett volcano on Adak Island (Conrad and Kay, 1984). These xenoliths suggest that olivine, clinopyroxene, and in some cases amphibole, crystallized before plagioclase. In particular, plagioclase-free, olivine-hornblende-clinopyroxene rocks (such as MM102; Conrad and Kay, 1984), have subhedral to euhedral Cr-rich amphibole replacing Cr-rich clinopyroxene and Ni-rich olivine (FO_{92}). Conrad and Kay (1984) also showed that mineral proportions of hornblende-bearing gabbro xenoliths are consistent with the residual mineralogy necessary to produce Moffett andesites from high-Al basalt.

The mineralogy of the calc-alkaline lavas and the Moffett xenoliths thus provides evidence for a spectrum of Mg-rich and high-Al basalts as mixing end members in calc-alkaline andesites. Furthermore, the existence of Ni-rich FO_{92} olivine in some of these lavas and xenoliths provides evidence that at least some magmas as mafic as the picritic basalts of Byers (1961) and Nye and Reid (1986) exist beneath both the calc-alkaline and tholeiitic centers.

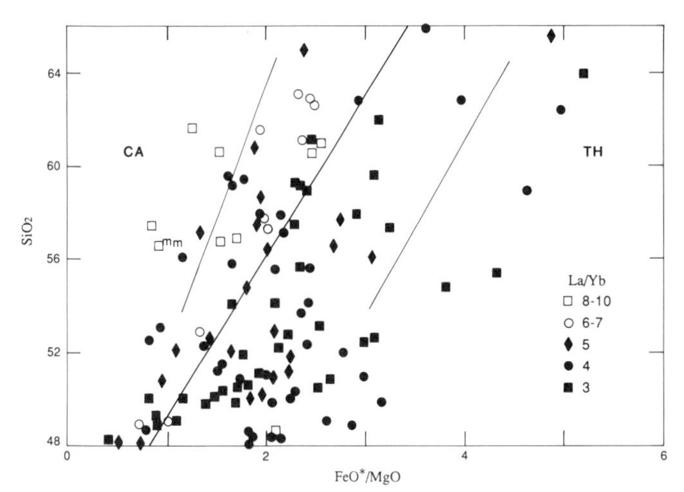

Figure 10. La/Yb ratios of Aleutian volcanic rocks shown superimposed on the plot of SiO_2 content versus FeO*/MgO ratio from Figure 2. The plot shows that lavas in the tholeiitic field and most of the lavas in the calc-alkaline field with <56 percent SiO_2 have La/Yb ratios of 5 or less. On the other hand, most lavas with La/Yb ratios from 6 to 8 and all lavas with La/Yb ratios >8 plot in the calc-alkaline field and have >56 percent SiO_2. The only exception is the Bogoslof basalt (FeO*/MgO = 2) from behind the main arc. All of the rocks with La/Yb ratios >6 contain amphibole, and fractionation of amphibole is an important factor in producing these high ratios. Data sources as in Figure 2.

Major- and trace-element evidence for the evolution of Aleutian basaltic andesites, andesites, and dacites

The mixing and fractionation trends deduced for the various Aleutian magma series from the phenocryst mineralogy are consistent with models based on major- and trace-element chemistry (Coats, 1956, 1959; Kay and others, 1982; S. Kay and Kay, 1985a; Marsh, 1982; Myers and others, 1985; DeLong and others, 1985; Neuweld, 1987; Wolf, 1987). Rare-earth element data are especially useful in supporting the role of amphibole fractionation in the calc-alkaline lavas. In contrast, incompatible trace-element differences (such as U, Th, Ba) do not vary systematically between different magma series and are established in the source region (S. Kay and others, 1986a).

Major-element characteristics of the Aleutian volcanic series are summarized by Kay and others (1982). In general, at the same Si content, tholeiitic series volcanic rocks have higher Fe, Ti, and K and lower Mg, Ca, and Al content than calc-alkaline series volcanic rocks. Differences are minimal in the basalts and increase with increasing Si content. The higher K and Ti content in the tholeiitic centers correlates with higher degrees of fractionation at the same Si content, and the late crystallization of K- and Ti-bearing phases such as titanomagnetite and amphibole. Higher Fe content in the tholeiitic samples can also be partially attributed to the late crystallization of oxide phases. Higher Mg content in the calc-alkaline lavas reflects the greater abundance and higher Mg content of mafic phases, and the higher Ca and Al content correlates with the suppression of plagioclase fractionation and the addition of calcic plagioclase to evolved melts by mixing. Compared with tholeiitic samples, the more primitive character of the major-element compositions (e.g., lower FeO/MgO ratio, higher Mg and Ca content) of the calc-alkaline samples is compatible with the mixing of more mafic lavas with differentiated lavas.

The REE pattern as indicated by the La/Yb ratios (Fig. 10) is an important trace-element parameter in Aleutian volcanic rocks that correlates with mineral fractionation (Kay, 1977; Kay and others, 1982). As with the major elements, differences are most pronounced in samples with >54 percent SiO_2. As shown in Figure 10, samples with the steepest REE patterns (La/Yb >6) have low FeO/MgO ratios and belong to the calc-alkaline series. In contrast, samples with high FeO/MgO ratios generally have La/Yb ratios <4, while those with intermediate FeO/MgO ratios have La/Yb ratios from 3 to 5. Variations in La/Yb ratio from 3 to 5 among centers can have as much to do with source area differences as with fractionating mineralogy (see below).

The observed differences in La/Yb ratio between Aleutian lavas are consistent with important amphibole fractionation in the evolution of the calc-alkaline, but not the tholeiitic, series. Contrasting trace-element patterns of Aleutian silicic andesites shown in Figure 11 illustrate this point. Amphibole-free tholeiitic andesites like SAR30 have relatively flat REE patterns, with larger Eu anomalies and higher REE levels, than do amphibole-bearing calc-alkaline andesites like MOF3A and BUL6A, which have steeper REE patterns that are more depleted in the middle REE range. REE patterns of low-Al amphibole-bearing calc-alkaline dacitic ashes are even more extreme than these calc-alkaline andesites (Romick and others, 1987). Because amphibole has higher REE distribution coefficients, particularly for the middle and heavy REE, than do olivine, the pyroxenes, the feld-

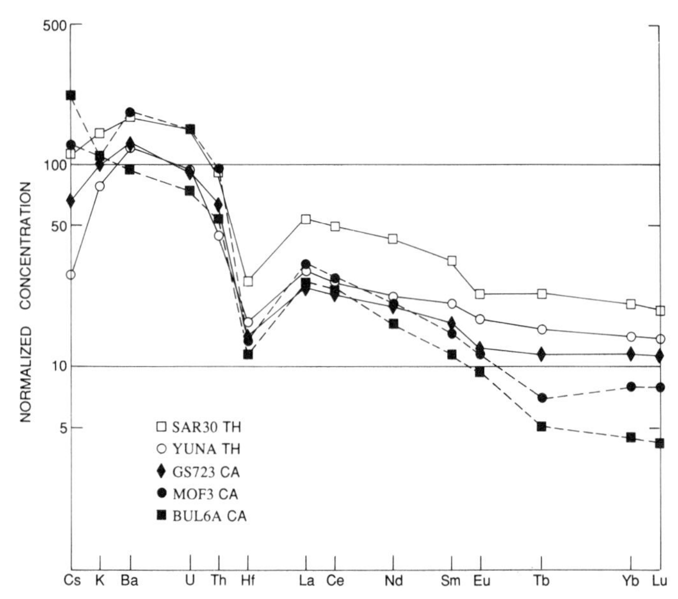

Figure 11. Trace elements for Aleutian silicic andesites normalization to chondrites and ocean-ridge basalts (see Table 1 for normalization values). Samples are from the tholeiitic Westdahl center (SAR30, 62 percent SiO_2, FeO*/MgO = 5.0), the transitional tholeiitic Four Mountains segment center of Yunalaska (YUNA, 63 percent SiO_2, FeO*/MgO = 3.2), the transitional calc-alkaline Great Sitkin center (GS723, 61 percent SiO_2, FeO*/MgO = 2.5), the transitional calc-alkaline Moffett center (MOF3, 63 percent SiO_2, FeO*/MgO = 2.4, data in Table 3), and the calc-alkaline Buldir center (BUL6a, 62 percent SiO_2, FeO*/MgO = 1.3). The tholeiitic dacites have higher REE levels and flatter REE patterns with larger Eu anomalies than the calc-alkaline dacites. The differences are particularly obvious in the levels of the heavy REE. Data are from the Appendix (see microfiche accompanying this volume) and Neuweld (1987).

spars, or the opaques (Arth, 1976), these differences are most easily explained by differences in the relative amount of amphibole fractionation.

The changes in trace-element characteristics in calc-alkaline samples with increasing SiO_2 content also support the idea of amphibole fractionation. For example, high-Al amphibole-bearing samples from Buldir show a progressive increase in the steepness of their REE patterns with increasing whole-rock SiO_2 content (Table 3; Fig. 12A), as would be expected with amphibole fractionation. The relatively constant Hf and Ta concentrations in these Buldir samples may also reflect partitioning of these elements into amphibole. On the other hand, highly incompatible elements, such as U and Th, that are present in very low concentrations in the phenocrysts increase with SiO_2 content, as would be expected with fractionation. The steep REE patterns in these Buldir samples show that the high-Al amphibole in them is not xenocrystic, for if it were, the excess amphibole would cause the REE patterns to become shallower (lower La/Yb ratio), rather than steeper.

Simple crystal fractionation will not explain the origin of the calc-alkaline lavas. For instance, Moffett andesite MOF38 (Table 3; Fig. 12B), which is similar in many respects to the Buldir samples, cannot be derived from Moffett high-Al basalt MOF15 (Table 1; Fig. 4B) even with amphibole fractionation, because Ni and Cr concentrations in MOF38 are much higher than those predicted by the models. As suggested by the mineralogy and major-element chemistry of calc-alkaline samples like MOF38, their trace-element chemistry also points to mixing of Mg-rich basalts, which are fractionating amphibole with calc-alkaline dacites (S. Kay and Kay, 1985a; Neuweld, 1987). Mixing of Mg-rich basalts and dacites, with no amphibole fractionation, can explain the trace-element characteristics of calc-alkaline andesites like SITR5b (56 percent and SiO_2, FeO/MgO = 1.2), which has a flatter REE pattern (La/Yb = 4.2) than MOF38 and lacks modal amphibole (Neuweld, 1987).

Comparison of the trace-element characteristics of Moffett andesite MOF44 and high-Al basalt MOF15 illustrates another important general point about amphibole in the Aleutian transitional calc-alkaline andesites (Table 3, Figs. 4B and 12B). Unlike the Buldir samples, transitional calc-alkaline andesite MOF44 lacks amphibole phenocrysts and has a REE pattern that is only slightly steeper than that of high-Al basalt MOF15. Although amphibole is not present in either rock, the more concave shape and slightly steeper REE pattern of MOF44 relative to MOF15 and other Moffett high-Al basalts is easiest to explain if some amphibole fractionation occurred. A similar argument has been made by Neuweld (1987) to explain the steeper REE patterns of transitional calc-alkaline andesites relative to high-Al basalts at Great Sitkin volcano. The REE patterns of the Aleutian andesites like these suggest a multistage fractionation process, with amphibole being fractionated before the crystallization of the final phenocryst assemblage.

In contrast to the calc-alkaline series, the trace-element characteristics of tholeiitic series samples are inconsistent with amphibole fractionation and support simple anhydrous crystal fractionation. As shown by the parallel trace-element patterns in Figure 12C for samples from the central region of Okmok volcano (Table 4), trace-element levels including the REE (except Eu which is fractionated by plagioclase) increase with increasing SiO_2 and FeO/MgO ratio. Fractionation models involving plagioclase, clinopyroxene and olivine—phases that have relatively low distribution coefficients for the REE—reproduce these trace-element patterns. The increasing Eu anomaly and lack of increase in the Sr content are consistent with plagioclase fractionation.

Temperature, pressure water content, and oxygen fugacity conditions for the evolution of Aleutian magmatic series

The physical conditions under which the Aleutian magma series formed can be partially constrained by comparing them with experimental analogues. Particularly relevant are the experiments of Baker and Eggler (1983, 1987) on a series of transitional tholeiitic high-Al basalts and andesites from Atka Volcano. The experimental results show that Atka andesites can evolve from Atka basaltic andesites by fractionation of plagioclase, oli-

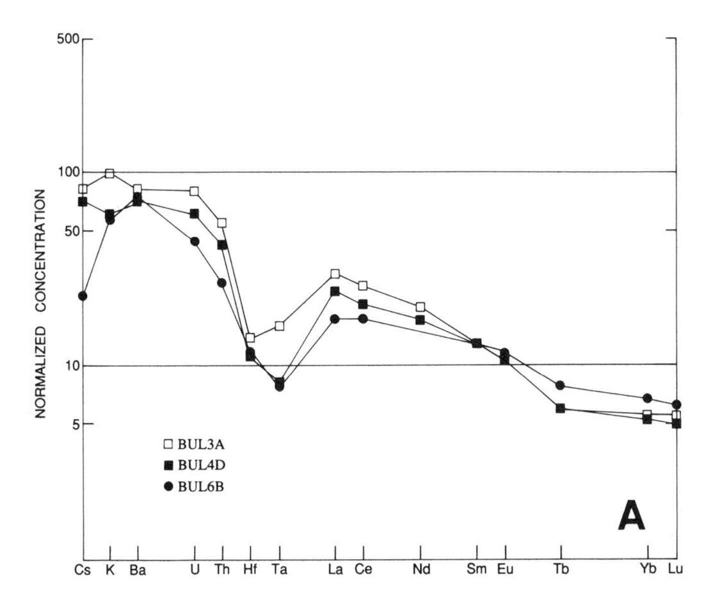

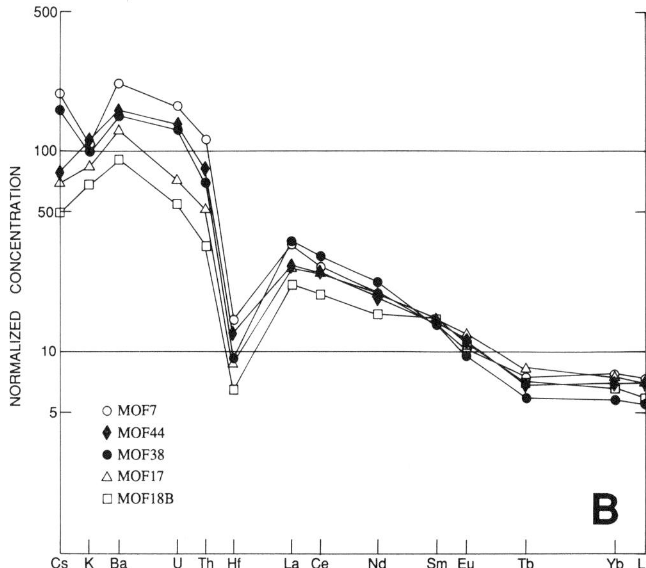

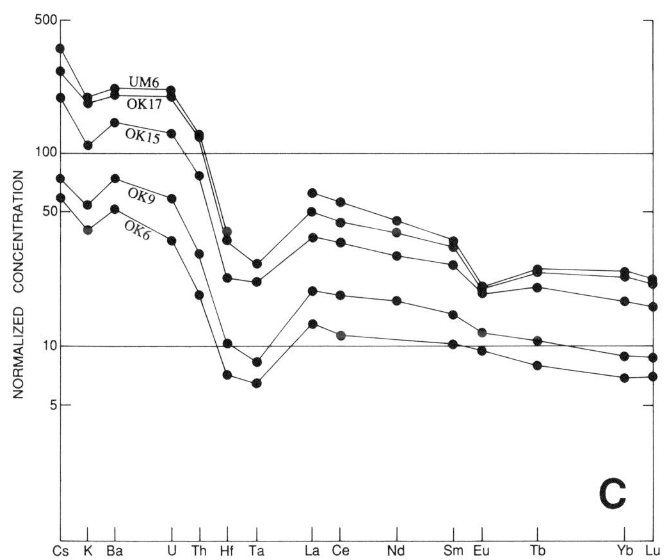

Figure 12. Trace-element patterns of Aleutian volcanic rocks (basalt to dacite) normalized to chondrites and ocean ridge basalts (see Table 1 for normalization values), showing contrasts between the tholeiitic and calc-alkaline series. A. Trace elements for calc-alkaline lavas from Buldir. Samples plotted are basaltic andesite BUL6b (52.7 percent SiO_2, FeO*/MgO = 0.9) and andesites BUL4d (55.9 percent SiO_2, FeO*/MgO = 1.16) and BUL3A (60.8 percent SiO_2, FeO*/MgO = 1.6). Data are in Table 3. B. Trace elements for transitional calc-alkaline and calc-alkaline lavas from Moffett. Samples plotted are high-Al basalt MOF18A (51 percent SiO_2, FeO*/MgO = 2.2), basaltic andesite MOF17 (53 percent SiO_2, FeO*/MgO = 2.1), andesite MOF44 (56.4 percente SiO_2, FeO*/MgO = 1.9), silicic andesite MOF7 (62.8 percent SiO_2, FeO*/MgO = 2.3). Also plotted is calc-alkaline andesite MOF38 (57.3 percent SiO_2, FeO*/MgO = 0.9). Data are from Table 3 and from Appendix (see microfiche accompanying this volume). C. Trace elements for tholeiitic lavas from Okmok. Samples plotted are in order of increasing La: basalt OK6 (50.5 percent SiO_2, FeO*/MgO = 1.8), basaltic andesites OK9 (52.2 percent SiO_2, FeO*/MgO = 1.71) and OK15 (54.8 percent SiO_2, FeO*/MgO = 3.9), and dacites OK17 (63.5 percent SiO_2, FeO*/MgO = 5.2) and UM6 (66.0 percent SiO_2, FeO*/MgO = 5.6). Data are given in Table 4.

vine, and augite between 2 and 5 kb with a magmatic H_2O content of 2 percent, and that Atka dacites can be produced from Atka andesites by fractionation of plagioclase, augite, and orthopyroxene at the same pressure and slightly higher H_2O content. Although complete experimental data are lacking, Baker and Eggler (1987) further predict that Atka basaltic andesites can fractionate from more primitive high-Al basalts under hydrous conditions at 5 to 8 kb.

From these results, Baker and Eggler (1987) conclude that crystallization of Atka lavas is a multi-stage, multi-level process and that the observed phenocryst mineralogy indicates only the final stage. This conclusion is consistent with the multistage crystallization process needed to explain the trace-element chemistry of the transitional calc-alkaline volcanic rocks from Great Sitkin (Neuweld, 1987) and Moffett volcanoes.

Baker and Eggler (1987) have proposed that phase boundaries on the diopside-olivine-quartz and orthoclase (SIOR) pseudoternary projection from plagioclase and magnetite are insensitive to water content, but that pressure expands the olivine phase volume. Projection of compositions of Aleutian samples on this diagram (Fig. 13A) supports the relative crystallization pressures that were suggested by S. Kay and Kay (1985a) on the basis of phenocryst mineralogy. As shown, the projected compositions indicate that the mixing components of the evolved calc-alkaline lavas (Buldir) crystallized at higher pressures than the evolved tholeiitic samples (Okmok anad Westdahl), which plot very near the 1 atmosphere liquid lines of multiple saturation (LLMS). Samples from transitional calc-alkaline to transitional tholeiitic centers (Adagdak, Great Sitkin, and Moffett) plot between these extremes, suggesting that they crystallized at intermediate pres-

TABLE 3. REPRESENTATIVE ANALYSES OF CALC-ALKALINE VOLCANIC ROCKS FROM BULDIR AND MOFFETT VOLCANOS, ALEUTIAN ISLANDS, ALASKA*

Sample	BUL6B	BUL4D	BUL6A	MOF17	MOFA54	MOF38	MOF44	MOF3
SiO_2	52.66	55.90	61.49	52.96	56.49	57.25	57.36	62.88
TiO_2	0.65	0.74	0.49	0.74	0.62	0.44	0.62	0.55
Al_2O_3	18.31	17.57	17.53	18.83	17.27	14.92	17.50	17.49
Fe_2O_3								
FeO	6.78	6.69	4.75	8.68	6.92	6.44	6.91	4.92
MnO	0.11	0.10	0.07	0.13	0.12	0.10	0.11	0.08
MgO	7.35	5.75	3.72	4.19	4.23	7.46	3.58	2.01
CaO	8.78	8.31	6.46	9.94	8.76	8.44	8.42	6.23
Na_2O	3.46	3.18	3.66	3.30	3.37	3.08	3.55	4.18
K_2O	0.81	0.96	1.50	1.17	1.65	1.42	1.57	1.51
P_2O_5	0.14	0.07	0.13	0.22		0.11	0.15	0.16
Total	99.05	99.27	99.80	100.14	99.43	99.66	99.77	100.01
La	6.67	8.99	9.88	10.04	8.99	13.68	10.06	12.07
Ce	17.16	20.54	22.70	24.01	18.68	29.48	24.35	26.93
Nd	7.54	12.40	11.87	14.10	8.23	15.71	13.19	14.13
Sm	2.97	3.04	2.66	3.41	2.66	3.13	3.23	3.31
Eu	1.000	0.956	0.794	1.078	0.881	0.823	0.983	0.994
Tb	0.448	0.365	0.308	0.499	0.349	0.356	0.457	0.421
Yb	1.63	1.34	1.15	1.89	1.25	1.46	2.00	1.97
Lu	0.245	0.194	0.169	0.278	0.196	0.216	0.311	0.302
Rb	18	18	35	28	33		42	49
Sr	377	517	497	572	747		449	474
Ba	256	262	358	474	574	548	581	685
Cs	0.30	0.92	2.84	0.88	0.84	2.08	1.03	1.60
U	0.83	0.82	1.12	1.08	1.80	1.86	1.96	2.22
Th	1.33	2.13	2.77	2.63	3.37	3.96	4.06	4.80
Hf	2.50	2.43	2.53	1.91	2.30	2.04	2.71	2.97
Ta	0.21	0.18	0.18	0.52	0.00	0.28	0.38	0.48
Sc	26.1	30.9	17.8	22.5	23.8	24.8	21.0	10.3
Cr	264	111	86	11	55	447	23	6
Ni	100	36	31	12	24	102	12	2
Co	35	30	55			29	25	20
FeO/MgO	0.94	1.16	1.28	2.08	1.64	0.86	1.93	2.45
Rb/Sr	0.05	0.03	0.07	0.05	0.04		0.09	0.10
Ba/La	38.4	29.1	36.3	47.2	63.8	40.1	57.7	56.8
La/Sm	2.2	3.0	3.7	2.9	3.4	4.4	3.1	3.6
La/Yb	4.1	6.7	8.6	5.3	7.2	9.4	5.0	6.1
Eu/Eu*	1.05	1.08	1.05	1.02	1.11	0.94	1.00	1.00
Th/Hf	0.53	0.88	1.09	1.38	1.47	1.94	1.50	1.62
Ba/Hf	102	108	142	248	250	268	214	231
Ba/Th	193	123	130	180	170	138	143	143
La/Hf	2.67	3.70	3.91	5.25	3.91	6.69	3.71	4.07
Th/U	1.61	2.58	2.46	2.43	1.87	2.13	2.07	2.16
$^{87}Sr/^{86}Sr$					0.70299			
ε_{Nd}					6.2			
$\delta^{18}O$	5.7	7.4		6.1	6.0	3.9	6.0	5.6

*BUL6B, BUL4D, and BUL6A are from Buldir Volcano; MOF17 (parasitic cone, see Coats, 1956); MOFA54, MOF38, MOF44, and MOF3 (main cone, see Coats, 1956) are from Moffett volcano. Major-element analyses largely from Kay and Kay (1985) and isotopic analyses from Kay and others (1978).

TABLE 4. REPRESENTATIVE ANALYSES OF THOLEIITIC LAVAS FROM OKMOK VOLCANO, ALEUTIAN ISLANDS, ALASKA*

Sample	OK6	OK9	OK15	OK16	OK17	OKUM6
SiO_2	50.52	52.23	54.80	57.40	63.50	66.03
TiO_2	0.97	1.24	2.48	1.54	0.81	0.76
Al_2O_3	19.71	17.04	14.90	15.49	16.10	14.81
Fe_2O_3	1.67	3.22	2.88	1.97	1.40	
FeO	6.82	6.60	9.00	7.86	5.00	6.06
MnO	0.12	0.14	0.17	0.15	0.10	0.08
MgO	4.64	5.54	3.00	2.94	1.20	1.08
CaO	11.96	10.39	7.15	6.74	3.80	3.66
Na_2O	2.74	2.90	3.94	4.11	4.50	4.66
K_2O	0.58	0.78	1.23	1.53	2.60	2.62
P_2O_5	0.19	0.19	0.65	0.32	0.23	0.23
Total	99.92	100.27	100.20	100.05	99.24	99.99
La	4.96	7.19	14.40	14.21	18.84	23.65
Ce	11.44	17.63	33.70	33.41	43.82	54.77
Nd	0.00	12.63	22.88	21.24	28.76	32.12
Sm	2.36	3.29	6.69	6.18	7.89	8.10
Eu	0.832	1.037	1.889	1.660	1.660	1.770
Tb	0.483	0.627	1.297	1.210	1.450	1.460
Yb	1.73	2.20	4.21	4.37	5.83	6.29
Lu	0.273	0.337	0.646	0.639	0.851	0.888
Rb						68
Sr						280
Ba	196	284	458	549	780	806
Cs	0.78	0.27	1.91	2.57	4.70	3.52
U	0.55	0.88	1.44	1.90	2.97	3.17
Th	0.92	1.55	2.81	3.84	6.38	6.27
Hf	1.606	2.249	4.180	5.130	7.980	8.320
Ta	0.14	0.18	0.41	0.48	0.61	
Sc	37.0	41.1	40.4	31.1	17.2	17.2
Cr	57	99	5	9	5	
Ni	18	29	2		1	
Co	28	32	25	20	6	
FeO/MgO	1.79	1.71	3.87	3.28	5.22	5.61
Rb/Sr`						0.243
Ba/La	39.4	39.5	31.8	38.6	41.4	34.1
La/Sm	2.1	2.2	2.2	2.3	2.4	2.9
La/Yb	2.9	3.3	3.4	3.3	3.2	3.8
Eu/Eu*	1.00	0.92	0.82	0.79	0.64	0.66
Th/Hf	251	1056	240	214	166	229
Ba/Hf	122	126	110	107	98	97
Ba/Th	213	183	163	143	122	129
La/Hf	3.09	3.20	3.44	2.77	2.36	2.84
Th/U	1.7	1.8	2.0	2.0	2.1	2.0
$\delta^{18}O$	4.2	5.6	4.8	4.2	4.3	5.4

*Samples OK6, OK9, OK15, OK16, and OK17 are from Okmok volcano, Umnak Island (see Byers, 1961, for location and most major elements). OKUM6 is from recent airfall on southern flank of Okmok volcano.

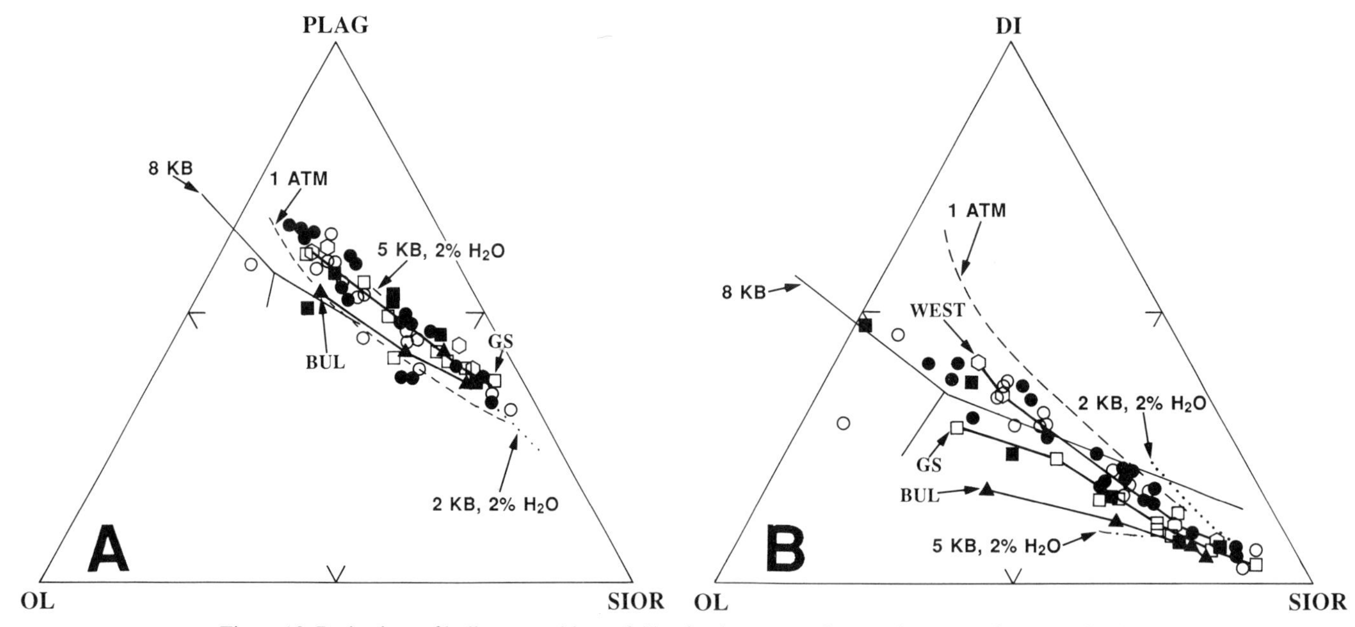

Figure 13. Projections of bulk compositions of Aleutian lavas onto the pseudoternary diagrams of Baker and Eggler (1983). Samples plotted are from the tholeiitic Okmok (open circles) and Westdahl (open hexagons, WEST); transitional calc-alkaline Great-Sitkin (open squares, GS), Moffett (solid circles), and Adagdak (solid squares); and calc-alkaline Buldir (solid triangles, BUL). Solid lines labeled BUL, GS, and WEST are sketched in from the Aleutian data. Labeled lines refer to liquid lines of multiple saturation (LLMS) determined experimentally under the conditions indicated by Baker and Eggler (1983, 1987). A. Plagioclase and magnetite projection onto olivine (OL)-diopside (DI)-silica-orthoclase (SIOR). Pressure moves the LLMS toward the OL corner (both for hydrous and anhydrous conditions). Diagram generally suggests that the tholeiitic centers (WEST) crystallized at lower pressures than the calc-alkaline centers (GS and BUL) in agreement with other criteria discussed in the text. Calc-alkaline samples from Moffett plot surprisingly near the tholeiitic samples. B. Diopside and magnetite projection onto OL-plagioclase (PLAG)-SIOR. Increasing water content shifts the LLMS toward the PLAG corner. The tholeiitic Aleutian data generally plot at lower water content than the calc-alkaline data (GS). Moffett and Buldir samples (BUL) plotting along the anhydrous LLMS are mixtures, and the diagram does not indicate their water content.

sures. Like the Atka transitional tholeiitic samples, nearly all of the transitional calc-alkaline samples plot between the 2 and 5 kb LLMS for 2 percent H_2O.

Baker and Eggler (1987) further propose that phase boundaries on the plagioclase-olivine-SIOR pseudoternary projection from diopside and magnetite are insensitive to pressure, but that increasing water content expands the plagioclase phase volume. In general, the projection of Aleutian samples on this diagram (Fig. 13B) is consistent with tholeiitic samples (Okmok, Westdahl) crystallizing at lower water content than transitional calc-alkaline samples (Great Sitkin, Adagdak, and Moffett). In detail, the tholeiitic samples plot between the 1 atmosphere anhydrous and the 5kb 2 percent H_2O LLMS, and the transitional calc-alkaline samples plot along or above the 5kb 2 percent H_2O LLMS with the Atka samples.

High-Al amphibole-bearing calc-alkaline basaltic andesites (e.g., Buldir) also plot along the anhydrous LLMS in Figure 13B. Their position on this diagram is inconsistent with the higher water content suggested by the occurrence of amphibole in these samples. Instead, the position of these samples could be an artifact—the result of the mixing of Mg-rich basalts and more evolved dacites that were crystallizing under different conditions (S. Kay and Kay, 1985a).

Oxygen fugacity estimates are difficult to make because of the lack of coexisting oxides in almost all Aleutian lavas. However, several lines of reasoning suggest that f_{O_2} is lower in the tholeiitic than the calc-alkaline lavas. For example, titanomagnetite is an important phenocryst in the calc-alkaline lavas, but is rare in mafic tholeiitic lavas and is only present sporadically in the more evolved lavas. The tholeiitic Okmok volcanic lavas have been suggested to form near the fayalite-magnetite-quartz (FMQ) oxygen buffer by Anderson (1976) and Nye and Reid (1986). Anderson's suggestion is based on poorly developed coexisting oxides in an evolved tholeiitic basalt from Okmok, whereas Nye and Reid's suggestion is based on the compositions of spinels included in olivine in Mg-rich lavas from Okmok. In contrast, Brophy (1986) suggests that transitional calc-alkaline Cold Bay lavas are 1 to 2 log units above the nickel-nickel oxide (NNO) oxygen buffer. Brophy's suggestion is based on fugacities calculated from whole-rock compositions based on the method of

Sack and others (1980); it is constrained by the experiments of Baker and Eggler (1983) on the stability of coexisting olivine and magnetite phenocrysts. Additional evidence for f_{O_2} above NNO is suggested by the experimental work of Rutherford and Devine (1986) who found that amphibole was stable in dacitic melts with 4.7 wt% H_2O (H_2O saturated at P = 0.5Gpa) at f_{O_2} of –10.0 (1.5 log units above NNO) at 920°C, but not at a higher f_{O_2}. In addition, Gust and Perfit (1987) show that spinel is a liquidus phase in the Mg-rich basalts from Makushin at fugacities above the NNO buffer but is a subsolidus phase at FMQ.

Differences in f_{O_2} may also be indicated by the generally larger Eu anomalies in the tholeiitic than in the calc-alkaline lavas. As shown in Figures 12C and 14, evolved tholeiitic lavas have moderate to large negative Eu anomalies (Okmok-Westdahl), whereas evolved calc-alkaline lavas (Buldir, Moffett) have small Eu anomalies. Relatively greater decreases in the Sr/Eu* and Eu/Eu* ratios (Eu* is interpolated Eu concentration, based on the concentration of other REE) in the tholeiitic center samples suggest more plagioclase fractionation, which agrees with the larger plagioclase phase volumes indicated in Figure 13B. Some have suggested that the small Eu anomalies in amphibole-bearing rocks occur because amphibole has a negative Eu anomaly that balances the positive Eu anomaly in plagioclase. However, the size of the Eu anomaly depends on the Eu^{+3}/Eu^{+2} ratio in addition to the minerals fractionated; therefore, the anomaly also decreases as conditions become more oxidizing, in agreement with the proposed differences in oxygen fugacities. The slopes (M) in Figure 14 may also suggest that tholeiitic centers (M = 3.6) like MORB (M = 5.0) have f_{O_2} near FMQ. Differences in Sr/Eu* ratios at the same Eu/Eu* suggest regional differences in parental lavas.

The data suggest that amphibole is stabilized in the calc-alkaline centers because of higher pressure and lower temperature crystallization conditions, combined with higher water content and f_{O_2} relative to the tholeiitic centers. The reason that calc-alkaline rocks may have a higher f_{O_2} than the tholeiitic rocks is not known. The differences may reflect differences in mantle history or alteration in the arc crust. Perhaps the mantle is more metasomatized by fluids from the subducted plate in the center of the segments, thus increasing H_2O content and f_{O_2} in the calc-alkaline lavas. The predominance of end-member tholeiitic centers in the eastern part of the arc also may reflect regional differences in mantle (perhaps also crustal) conditions. The f_{O_2} in the calc-alkaline series appears to be higher than that of most oceanic ridge basalts but is similar to that of some oceanic alkaline basalts and included xenoliths (Mattioli and Wood, 1986).

Further evidence for the depths at which Aleutian volcanic rocks evolve; Oxygen isotopes

Preliminary whole-rock oxygen isotope data from fresh Aleutian volcanic rocks (see Appendix on microfiche, at back of this volume) are consistent with chemical and mineralogical arguments for fractionation in shallow-level magma chambers under the tholeiitic centers. Analyses of volcanic rocks over the compositional range basalt to dacite from 15 centers show that while most Aleutian samples are isotopically similar ($\delta^{18}O$ = 5.5 to 6.5) to MORB, the six basalts and andesites with the lowest values ($\delta^{18}O$ = 4.7 to 5.2) are from the tholeiitic centers (Westdahl, Okmok). One explanation for the low values is that magma chamber roof rocks that have been altered at high temperature by

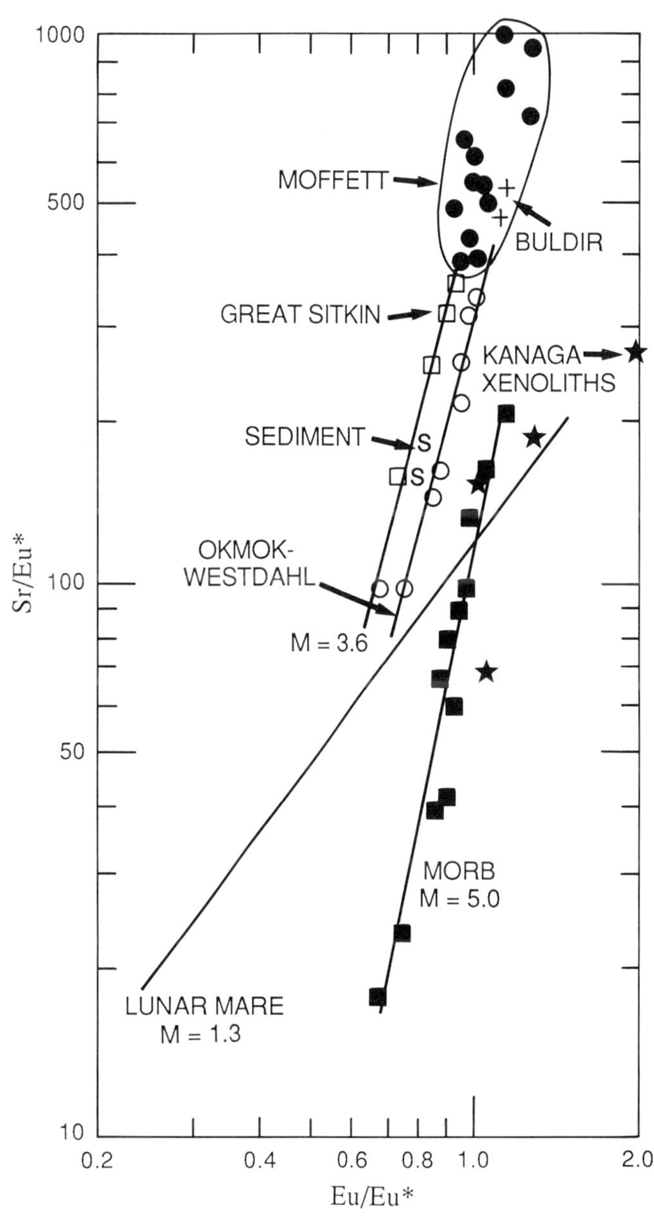

Figure 14. Plot of Sr/Eu* versus Eu/Eu* (Eu* is interpolated from REE pattern) for Aleutian volcanic rocks, xenoliths (from Kanaga Tertiary sill, stars) and sediments (s); MORB (closed squares) and lunar Mare (line represents trend). Slopes (M) of lines through points are suggested to be proportional to the relative fO_2 of the series; that is, Lunar Mare <MORB <Aleutians. The plot also shows the relatively larger Eu anomalies and lower Sr content in the tholeiitic (Okmok-Westdahl, open circles) and extreme transitional calc-alkaline lavas (Great Sitkin, open squares) compared to the more calc-alkaline lavas (circled field; Moffett, solid circles; and Buldir, solid field).

isotopically light meteoric water, founder into the shallow tholeiitic magma chambers underlying the calderas (such as Okmok, 10 km radius), contaminating the magmas with light oxygen (see model of Condomines and others, 1983). The presence of ^{3}He-enriched He in Aleutian hot springs (Poreda and others, 1981) attests to the penetration of light meteoric water into the volcanic pile (and also indicates a high $^{3}He/^{4}He$ ratio in the volcanic rocks).

Although the oxygen isotope ratios of the Aleutian calc-alkaline series volcanic rocks are not well characterized, 8 of 10 samples from Adak and Buldir volcanoes ($\delta^{18}O$ = 5.2 to 6.0) fall within or near the MORB range ($\delta^{18}O$ = 5.5 to 6.1). These data are consistent with mid to deep crustal fractionation and mixing processes, and little or no incorporation of isotopically light country rocks from shallow levels into the magmas. Two Buldir andesites have heavier oxygen ($\delta^{18}O$ = 7.1, 7.4), suggesting either incorporation of wall rock altered at low temperatures as in some central Aleutian plutonic and volcanic rocks (Perfit and Lawrence, 1979; Kay and others, 1983; Rubenstone, 1984), or more incorporation of oceanic sediment in the source region.

FRACTIONATION-INDEPENDENT TRACE ELEMENTS AND ISOTOPES; THE QUESTION OF SOURCE REGION DIVERSITY IN ALEUTIAN LAVAS

A successful model for the origin of Aleutian arc magmas must identify the mechanism for generating the geochemical characteristics that distinguish Aleutian and other volcanic rocks from the oceanic island basalts (OIB) and mid-ocean ridge basalts (MORB) that occur both north and south of the Aleutian island arc. The model must also explain the diversity of magmatic rocks within the Aleutian arc. The trace-element characteristics that Aleutian magmatic rocks have in common with other arcs include a depletion of some of the high field-strength elements (HFSE; such as Hf, Ta, Ti, Zr), an enrichment of the alkali and alkaline earths (Cs, Rb, K, Ba, Sr), and enrichment of U and Th relative to the light REE elements (Figs. 4, 11, 12A to C, 14, 15A to D). Also, compared to MORB, Aleutian and other arc volcanic rocks are characterized by higher ratios of $^{206}Pb/^{204}Pb$, $^{207}Pb/^{204}Pb$, $^{208}Pb/^{204}Pb$, and $^{87}Sr/^{86}Sr$ and lower ratios of $^{143}Nd/^{144}Nd$ (lower ϵ_{Nd}; Figs. 16 and 17). Another important trace-element characteristic of Aleutian and many other arc volcanic rocks (but not island-arc tholeiites, which are absent from the Aleutian arc) is enrichment (relative to chondritic meteorites) in light REE over heavy REE.

Several classes of models have been proposed to explain the geochemical differences between Aleutian lavas and oceanic lavas of the Pacific Ocean crust to the south of the arc. The first class of models involves the addition of sedimentary and hydrated oceanic crustal components from the subducted slab to the source of the parental magmas. In the models of Kay and others (1978), Kay (1980), McCulloch and Perfit (1981), and Perfit and Kay (1986), these components are added from the slab to the overlying peridotitic wedge, which then melts to produce the parental magma. In the models of Brophy and Marsh (1986), Myers and others (1986a), VonDrach and others (1986), and Myers and Marsh (1987), the sedimentary component is melted along with the basaltic oceanic crust in the slab to form the parental high-Al basalts. The high-Al basalts may be contaminated by the assimilation or mixing of material from the lithospheric peridotite wedge and, in extreme cases, become high-Mg basalts. In the second class of models, no sediment is involved, and the parental magma is derived from melting of a mixed MORB- and OIB-type source in the mantle wedge (Morris and Hart, 1983). In this model, a HFSE-bearing residual phase is required to explain the HFSE depletion of the arc magmas.

Examination of the chemistry of Aleutian magmatic rocks shows diversities in incompatible trace elements and isotopic ratios, which seem to suggest both spatial and temporal variability in the source regions of Aleutian arc magmas. Understanding this chemical diversity can help to distinguish and define processes occurring in the mantle wedge and the down-going plate. Although the isotopic ratios (except for oxygen) are independent of fractionation, trace-element ratios may not be; an important requirement in understanding the trace elements is to identify which elements are incompatible and reflect the source region, and which are compatible and controlled by mineral fractionation processes. Thus, U and Th appear to be virtually incompatible, since U-series data suggest that differentiation processes have not changed the Th/U ratio of six historic Aleutian (both calc-alkaline and tholeiitic) lavas by more than about 10 percent (Newman and others, 1984). Determining the compatibility of other elements is not so straightforward (Ryan and Langmuir, 1987), since differences in fractionating mineralogy may reflect differences in source composition, particularly with regard to volatile content and oxygen fugacity.

Trace-element evidence for source region diversity in the Aleutian arc

Of Aleutian magma types, Mg-rich basalts (Table 1; Fig. 4A) have the fewest fractionating liquidus phases (spinel, olivine, clinopyroxene, and in some cases, plagioclase) and should have the largest number of elements that behave incompatibly. Ratios of the most incompatible elements in these basalts are all of arc-type character (see above), but vary in magnitude—an observation most easily explained by source-region diversity. For example, peridotite and gabbro xenolith-bearing basalt from a Tertiary sill on Kanaga Island (KANH) has higher Ta and Hf, and lower Th and U levels relative to La, than other Mg-rich basalts (see Fig. 4A). These characteristics, combined with a relatively low La/Yb ratio (3.2), suggest a less extreme arc signature in this sample. In contrast, a peridotite xenolith-bearing basalt from Adagdak volcano (ADH) has the most extreme arc-type characteristics, since it has the highest Ba, Th, and U and lowest Ta relative to La of any of the Mg-rich basalts. In addition, it has a high La/Yb ratio (4.8). As shown in Figure 4A, trace elements distinguish these two basalts from three Mg-rich basalts from Umnak Island. Of the Umnak group, OK1A is distinct from the other two because of its higher La/Yb ratio (4.8; see Table 1).

High-Al basalts show the same types of incompatible-element ratio variations as the Mg-rich basalts (Table 2; Fig. 4B). Although the high-Al basalts have been involved in more extensive fractionation, mixing, and perhaps assimilation processes than the high-Mg basalts, much of their trace-element diversity is also believed to be due to variations in the source region.

The range and spatial pattern of variations in trace-element ratios in the arc can be illustrated by plotting trace-element ratios for basalts and basaltic andesites (<54 percent SiO_2) against distance along the arc. Representative plots are shown in Figure 15A to D for potentially incompatible element ratios that include the La/Yb (light REE/heavy REE), Ba/La (alkali/light REE), Th/Hf (Th/HFSE), and Ba/Hf (alkaline earth/HFSE) ratios. These plots suggest that while intracenter variations can be substantial, the majority of contemporaneous samples from single centers and centers within a region, have similar ratios, further suggesting regional variation of trace-element characteristics of the magmatic source region. Although the data for many centers are limited and could be argued to be nonrepresentative, a strong case can be made that the more extensive data for Moffett (and Adagdak), Great Sitkin, and Okmok are representative of the range that exists in these volcanoes. An understanding of the chemical diversity in the arc is just beginning, and no model has yet explained the observed intra- and intercenter variations.

Inspection of Figures 15A and B shows that there is little correlation between Ba/La and La/Yb ratios, indicating that the

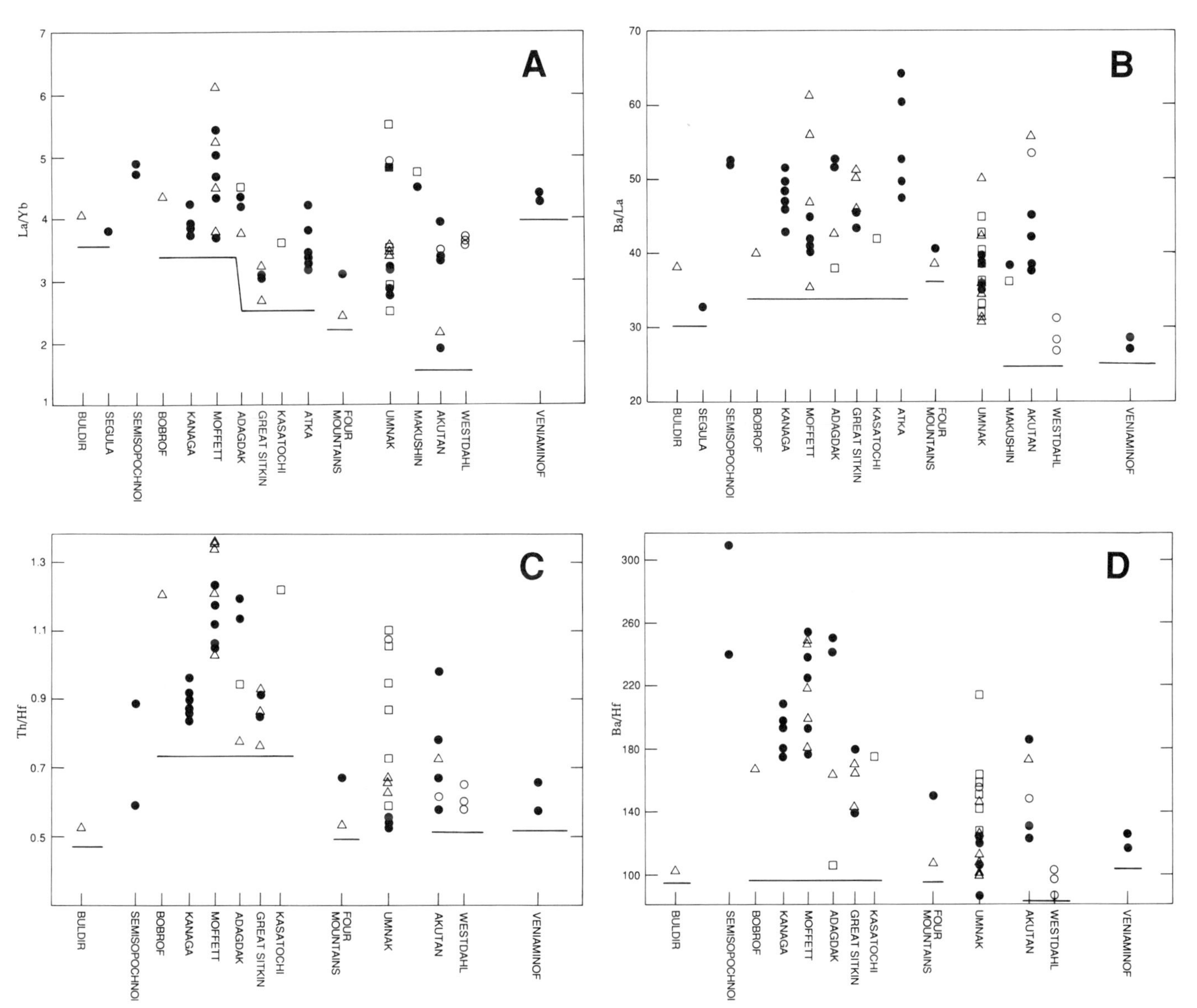

Figure 15. Plots of incompatible element ratios (A, La/Yb, B, Ba/La; C, Th/Hf; D, Ba/Hf) in Aleutian Mg-rich basalts (open squares), high-Al basalts (solid circles), other basalts with <52 percent SiO_2 (open circles), and basaltic andesites (52 to 55 percent SiO_2) (triangles) from centers arranged from west (Buldir) to east (Veniaminof) along the arc. Heavy lines indicate coherent volcanic segments which from west to east area: the Rat, the Andreanof, the Four Mountains, and the Cold Bay segments (see Fig. 1).

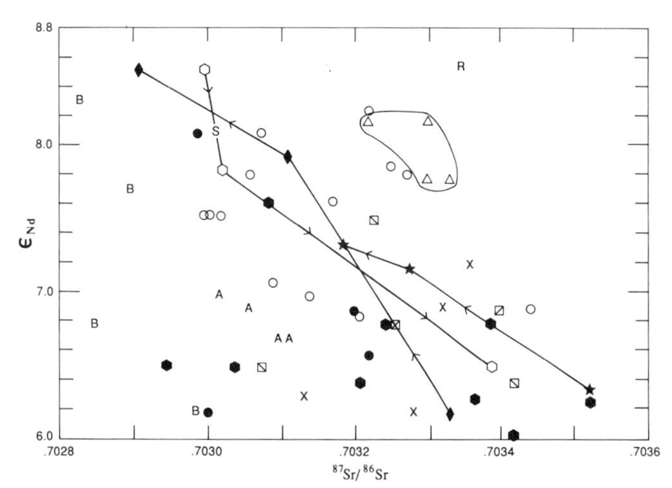

Figure 16. Plot of ϵ_{Nd} versus $^{87}Sr/^{86}Sr$ in Aleutian volcanic rocks. Symbols are the same as in Figure 2 with the addition of samples from Amak (A), Cold Bay (Mt. Simeon and Frosty, filled polygons), and Recheschnoi (R), and xenoliths from Adagdak and Kanaga (X). The Adak Mg-andesite (Kay, 1978) plots off-scale ($^{87}Sr/^{86}Sr$ = 0.7028, ϵ_{Nd} = 11.3). Arrows on lines connecting points (diamonds, Makushin; stars, Little Sitkin; hexagons, Westdahl) indicate direction of isotopic change with increasing SiO_2 content. Samples from Semisopochnoi are represented by triangles in circled field. Data are from DeLong and others (1985), Goldstein (personal communication, 1988), R. Kay and others (1978, 1986), McCulloch and Perfit (1981). Morris and Hart (1983), Nye and Reid (1986), VonDrach and others (1986), and White and Patchett (1984). All ϵ_{Nd} data are standardized to BCR = –0.4. $^{87}Sr/^{86}Sr$ ratios are reported as they were in original reference.

REE are decoupled from the alkalis and alkaline earths. This effect can be seen within the Andreanof segment volcanoes (Fig. 1) by noting that Ba/La ratios are in the same range from Atka to Kanaga, while La/Yb ratios are lower at Atka and Great Sitkin than they are at Adagdak and Kanaga. In contrast, both La/Yb and Ba/La ratios are generally lower at Okmok than they are at Moffett. These data provide evidence against the assumption that the range in Ba/La ratios is controlled by mineral fractionation because fractionation of hornblende, the only mineral that can significantly decrease the Ba/La ratio, increases the La/Yb ratio (see trace-element patterns of amphibole-rich xenoliths from Moffett volcano in DeBari and others, 1987). Regional differences in these ratios are thus independent of the chemical and mineralogical criteria that define the Aleutian calc-alkaline and tholeiitic series, and appear to be related to differences in the source regions of the parental lavas (Kay and others, 1986a).

Examination of La/Yb ratios for the arc as a whole shows that samples from both the eastern (Westdahl to Veniaminof) and the western (Buldir to Adagdak) part of the arc have ratios >3.5, whereas many samples from the central arc (Great Sitkin to Akutan) have ratios <3.5 (Fig. 15A: Note that Umnak samples with ratios >3.5 may be older than the main center). Some centers on the Alaska Peninsula that are underlain by continental crust have even higher La/Yb ratios, more like those found in continental-margin arcs (e.g., the Andes; Hickey and others, 1986). It is also worth noting that an important change in the age of the subducting oceanic crust occurs at the Amlia fracture zone (Fig. 1) in the vicinity of an apparent regional break in La/Yb ratios.

Regionally, the Ba/La ratios are lowest in the Alaska Peninsula, generally higher through the central part of the arc, and low again in the west (Segula and Buldir). The lower ratios (20 to 30) in the eastern arc are similar to those in continental arc segments that lack old basement, such as the Andean southern volcanic zone (Hickey and others, 1984, 1986). The Ba/La ratios in the central part of the arc (>40), particularly in the Andreanof segment, are higher than those in most other arcs (Hickey and others, 1984).

Th/Hf (Fig. 15C) and Ba/Hf (Fig. 15D) ratios, which monitor the behavior of Th, U, the alkaline earths, and alkali elements relative to the HFS elements, also show regional variations that correlate with Ba/La. Volcanoes in the Andreanof segment (Kanaga, Moffett, Adagdak, Great Sitkin, and Kasatochi) show the highest Ba/La, Th/Hf, and Ba/Hf ratios and thus have a more extreme arc-type signature than other centers in the arc. Ratios of Ba/Hf are particularly low from Westdahl volcano through the Four Mountains segment.

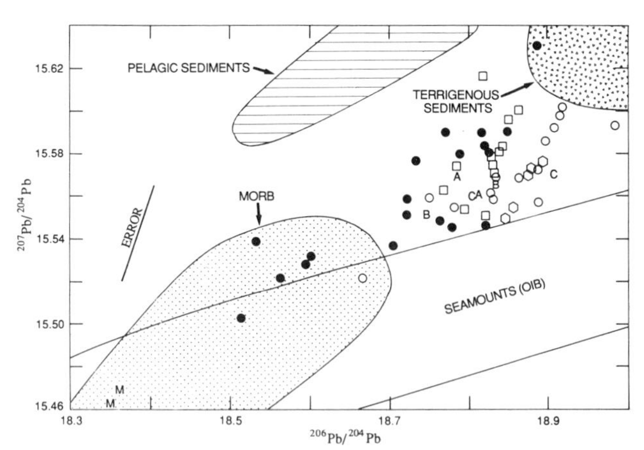

Figure 17. Plot of $^{207}Pb/^{204}Pb$ versus $^{206}Pb/^{204}Pb$ for Aleutian arc and other northern Pacific volcanic rocks and for Pacific oceanic sediments. Sediments in the pelagic field (horizontal stripes) are from the equatorial and northern Pacific, and those in the terrigenous field (dotted) are northeast Pacific oceanic sediments composed of continental detritus. OIB lavas (field containing seamount data are between parallel lines) are from the northeast Pacific, and MORB lavas (patterned field) are from the Gorda and Juan de Fuca ridges. Data trends can be produced by mixing oceanic type magmas with sediments. Aleutian volcanics from the western Andreanof segment (Adak Island, solid circles; high Mg andesites, M; Atka Island, open squares) have lower $^{206}Pb/^{204}Pb$, and so trend toward pelagic sediments, whereas those from closer to the continental Alaskan Peninsula (Westdahl Volcano, hexagons; Umnak Island, open circles; Cold Bay region, C; Amak Volcano, A; Bogoslof Volcano, B) have higher $^{206}Pb/^{204}Pb$ and trend toward terrigenous sediments. Bar indicates amount and direction of error in the measurements. Sediment and oceanic lava fields are from sources listed in Kay and others (1978). Aleutian data from Kay and others (1978), Morris and Hart (1983), Nye and Reid (1986), and Myers and Marsh (1987). Values are reported as in papers cited.

Although some regional trends emerge, trace-element variations within single centers can be large, suggesting that wide ranges in source components also occur beneath single centers. For example, La/Yb ratios are between 2.5 and 5.5 for most mafic samples along the arc, whereas ratios from Umnak Island samples alone are from 2.8 to 5.0. As pointed out by Nye and Reid (1986), these variations in the Umnak samples can be related to the source region because the Mg-rich basalts have ratios that cover almost this entire range. Looking more closely at the Umnak data, two groups emerge. The first group consists of samples from the main center of Okmok volcano, which have La/Yb ratios from 2.8 to 3.6, and can be interrelated by fractional crystallization (see Fig. 12C). The second group consists of samples distant from the main center that have La/Yb ratios ranging from 2.5 to 5.0 (OK1, OK1A, and OK4 in Fig. 4A). Ages of samples in this second group are poorly constrained, but they probably predate the main center (Byers, 1961), suggesting that part of the variation may depend on time.

To summarize, although some regions are inadequately represented, variations in trace-element ratios that are independent of fractionation imply both regional and local variations in the sources of Aleutian magmas. Volcanoes on the Alaska Peninsula have the most "continental" signatures. Those to the west fall into groups that have La/Yb ratios decoupled from the other ratios. The lowest La/Yb ratios occur in samples from Akutan to Great Sitkin volcanoes, whereas samples from the Andreanof segment have the highest Ba/Th, Ba/La, and Ba/Hf ratios. Before considering the reasons for these variations in more detail, we will discuss the isotopic evidence.

Radiogenic isotopic variations in Aleutian volcanic rocks

The range in the ratios of radiogenic to non-radiogenic isotopes in Aleutian Plio-Pleistocene to Recent volcanic rocks is small, and the ratios are relatively MORB-like compared to many arcs. If the high-Mg andesites of Kay (1978) are excluded, the overall range in ϵ_{Nd} is from +6 to +8.5, in $^{87}Sr/^{86}Sr$ from 0.70295 to 0.70345, in $^{206}Pb/^{204}Pb$ from 18.51 to 19.00, in $^{207}Pb/^{204}Pb$ from 15.50 to 15.63, and in $^{208}Pb/^{204}Pb$ from 38.0 to 38.6 (Figs. 16 and 17; data from Kay and others, 1978; McCulloch and Perfit, 1981; White and Patchett, 1984; Morris and Hart, 1983; Myers and others, 1985; DeLong and others, 1985; VonDrach and others, 1986; Nye and Reid, 1986; R. Kay and others, 1986; Myers and Marsh, 1987; Goldstein, personal communication, 1988).

As with incompatible trace elements, the range of isotopic ratios in the basalts is almost as great as that of samples covering the entire compositional range. For example, we note the isotopic differences among Mg-rich basalts from Umnak (OK1A, ID1, ID16, ID18, ID25) and Makushin (MK15) and high-Al basalts from Adagdak (AD14), Semisopochnoi, Akutan, and Umnak (Table 1 and 2; Figs. 16 and 17). As Nye and Reid (1986) point out, the Umnak samples show that considerable variation can occur in a restricted geographic region. However, since the ages of the older of these samples are unknown, the extent of variation with time is uncertain. The data show that the cause of variation is not simply the result of crustal contamination (which should be minimal in the Mg-rich basalts). Thus, inhomogeneities resulting from subcrustal processes must be instrumental in controlling these variations.

Nd and Sr isotopic ratios for all analyzed Aleutian lavas and xenoliths show only a rough correlation (Fig. 16) and are not easy to interpret. As pointed out by VonDrach and others (1986), samples from the secondary front islands (Amak and Bogoslof) plot below the trend of main arc. Some amphibole-bearing xenoliths and lavas from both Adak and Cold Bay volcanic centers (calc-alkaline centers on the main front) also plot below the trend of the main arc. Isotopic variations in these Moffett and Cold Bay samples show no simple correlation with major- or trace-element chemistry. All samples from Semisopochnoi and four samples from Umnak Island plot above the main trend. Samples that plot in the main Aleutian trend vary inconsistently with SiO_2 content. For example, three samples from calc-alkaline Makushin volcano show a decrease in $^{87}Sr/^{86}Sr$ and an increase in ϵ_{Nd} with increasing SiO_2 (McCulloch and Perfit, 1981), as do three samples from calc-alkaline Little Sitkin (White and Patchett, 1984). In contrast, the reverse pattern can be found with increasing SiO_2 in samples from tholeiitic Okmok and Westdahl (Goldstein, personal communication, 1988). At the transitional-tholeiitic Atka volcanoes center, samples with relatively small variations in ϵ_{Nd} (VonDrach and others, 1986) also show an increase in $^{87}Sr/^{86}Sr$ (0.70319 to 0.70345, in Myers and others, 1985) with increasing SiO_2 (Frost and Myers, 1987).

Regionally, isotopic and trace-element data show some correlations. VonDrach and others (1986) noticed that ϵ_{Nd} is lower in the Andreanof segment and on the Alaska Peninsula than it is in the western arc volcanoes and in the Okmok region. Although exceptions occur when the data of Goldstein (personal communication, 1988) are added, the pattern generally holds (Fig. 16). The generally lower ϵ_{Nd} in the volcanic rocks of the Andreanof segment correlates with their greater enrichment in the alkali and alkaline earth elements. However, lower ϵ_{Nd} and the trace-element enrichment-depletion patterns observed in the Cold Bay and Rat segments do not correlate, nor is a one-to-one correlation found in samples of the Andreanof segment.

Within the Andreanof region, Myers and others (1985) have suggested that volcanic samples from Atka show a smaller range in Sr isotopic ratios than samples from the Adak volcanoes. Additional data from Adak (R. Kay and others, 1986; Goldstein, personal communication, 1988) fall in the same range. Excluding a Mg-rich andesite from Adak (Kay, 1978), ϵ_{Nd} data ranges from 7.2 to 8.3 on Adak and from 6.7 to 8.0 on Atka (value of BCR standard not given by Frost and Myers, 1987).

Myers and others (1985) have used the range of Sr isotopic values within a center to support their contention that centers such as Atka represent "clean" (mature) systems, whereas centers such as Moffett represent "dirty" (immature) systems. In this model, chemical diversity occurs in the immature (calc-alkaline) centers because lithospheric mantle debris is picked up by the magma. In contrast, the mature (tholeiitic) centers have well-established magma conduits that are protected from lithospheric debris by thermal-chemical boundary layers. Although the model could explain the lower Sr isotopic values of some of the Moffett samples, samples like ADK54 which has both low Sr and low Nd isotopic ratios (Goldstein, personal communication, 1988) are difficult to explain by lithospheric debris.

In a somewhat similar model, the apparently greater isotopic variability in the calc-alkaline centers than in the tholeiitic centers—like their greater whole-rock chemical and mineralogic diversity (Kay and others, 1982)—can be explained by magma mixing. It is a plausible assumption that each mantle-derived magma batch is isotopically distinct because it has different amounts of sediment incorporated and mantle inhomogeneities can be variably modified by crustal contamination. The inefficient mixing of compositionally heterogenous magmas combined with assimilation and fractional crystallization (AFC) at intermediate to deep crustal levels in the calc-alkaline centers would result in some homogenization of incompatible element and isotopic signatures, but each mixed lava would retain evidence of its previous history. In contrast, in higher volume tholeiitic centers where larger magma batches more easily reach higher level magma chambers, isotopic signatures can be more uniform.

Sediment subduction in the Aleutian arc; Evidence from Pb and O isotopes and ^{10}Be data

Both of the models discussed above assume that subducted sediment plays a role in shaping the geochemical character of the source of Aleutian lavas. The available Pb isotopic data can be used to support the idea of sediment contamination in Aleutian lavas (Kay and others, 1978; Sun, 1980). On a plot of ^{207}Pb/^{204}Pb versus ^{206}Pb/^{204}Pb (Fig. 17), the Aleutian data fall in a field that diverges from the regional MORB and OIB fields toward the field of oceanic sediments (Kay and others, 1978; Nye and Reid, 1986). The Andreanof samples (Adak and Atka) have lower ^{206}Pb/^{204}Pb ratios suggesting a slightly larger pelagic sediment component than samples from the eastern arc (Umnak, Westdahl, Cold Bay). Pb isotopic values from Amak and Bogoslof, volcanoes behind the main arc, are indistinguishable from the other Aleutian samples.

In an alternative explanation, Morris and Hart (1983) have argued that the Aleutian Pb data lie within the global oceanic island basalt (OIB) field and could have an OIB mantle source. However, Perfit and Kay (1986) have pointed out that most of the Aleutian Pb values do not overlap the OIB field if the ^{207}Pb-rich Southern Hemisphere (DUPAL—type) OIBs are excluded. For instance, the least radiogenic of the Umnak Mg basalts (ID1, ID16) fall close to the Northern Hemisphere seamount (OIB) field and can be explained by an OIB mantle source without another component (Nye and Reid, 1986, their Fig. 5). But Pb data on other Aleutian samples are difficult to explain without a contribution from a source with more radiogenic ^{207}Pb.

^{10}Be data from young Aleutian volcanic rocks (Tera and others, 1986) also strongly suggest a sediment component (all rocks measured have ^{10}Be well above detection limits). As a cosmic-ray-produced nuclide with a 1.5-m.y. half-life, ^{10}Be is present only at the Earth's surface or at the top of oceanic sediment columns. Thus, ^{10}Be in the Aleutian volcanic rocks must be derived from subducted sediments or contamination at the Earth's surface, because ^{10}Be that was originally present in rocks buried in the crust of the main Aleutian arc would have completely decayed. The positive correlation of ^{10}Be and other incompatible elements in the Bogoslof 1927 basaltic and 1796 andesitic eruptions suggests that ^{10}Be is not the result of surface contamination. Recent work by Monaghan and others (1988) supports this suggestion because both whole rock and mineral separates from two Bogoslof eruptions have the same ^{10}Be/^{9}Be ratio, showing that the ^{10}Be was incorporated into the magma before the phenocrysts crystallized. We infer crystallization to be relatively deep, because the Bogoslof rocks are amphibole bearing. The high ^{10}Be content of the Bogoslof samples, which come from behind the main arc (Fig. 1), suggest that some sediment component reaches a depth of at least 150 km in the subducting slab (Morris and others, 1989).

The general correlation between the ^{10}Be and Ba content in volcanoes from both the eastern and central Aleutians (Fig. 18) suggests that enrichments in Ba, like those in ^{10}Be, are at least partially related to a subducted sedimentary component. Although the data are extremely limited, the somewhat higher ^{10}Be/Ba ratios in flows from the eastern volcanoes may correlate with the generally higher ^{206}Pb/^{204}Pb (Fig. 17) and lower Ba/La ratios (Fig. 15b) in the volcanic rocks of this region, suggesting that there are regional differences in the character of the sedimentary component being subducted along the arc.

Although a sedimentary component appears to be present in the source region of Aleutian volcanic rocks, how this component is mixed into the magma source remains a problem. Because the compositional variability within the sedimentary column is extreme (especially for trace elements) compared with the variability of the volcanic rocks, large-scale mixing is required (VonDrach and others, 1986; Kay and Kay, 1988). Calculations based on trace-element and isotopic data by Kay (1980, 1984) suggest that the amount of sediment relative to the total mass of mantle peridotite source is less than 1 percent. Even at this low percentage, a substantial fraction of elements such as Pb and Ba are furnished by the sediment. This model contrasts with the extreme view of Myers and others (1986b) that 8 to 13 percent carbonate and 21 to 44 percent of pelagic sediment are added to an eclogite melt to create parental high-Al basalt magmas. The high percentage of carbonate in this model is difficult to reconcile with the observed ^{87}Sr/^{86}Sr in the Aleutian arc lavas, and if applied to other arcs with higher ^{87}Sr/^{86}Sr, suggests extremely high percentages of carbonate in those arcs.

The oxygen isotopic composition of Aleutian lavas also argues against the large quantities of sediment in the model proposed by Myers and others (1986a). Nye and Reid (1986) observed that δ^{18}O values in Umnak lavas ranged from 5.2 to 6.5, which is within the range of mantle-derived tholeiitic rocks on

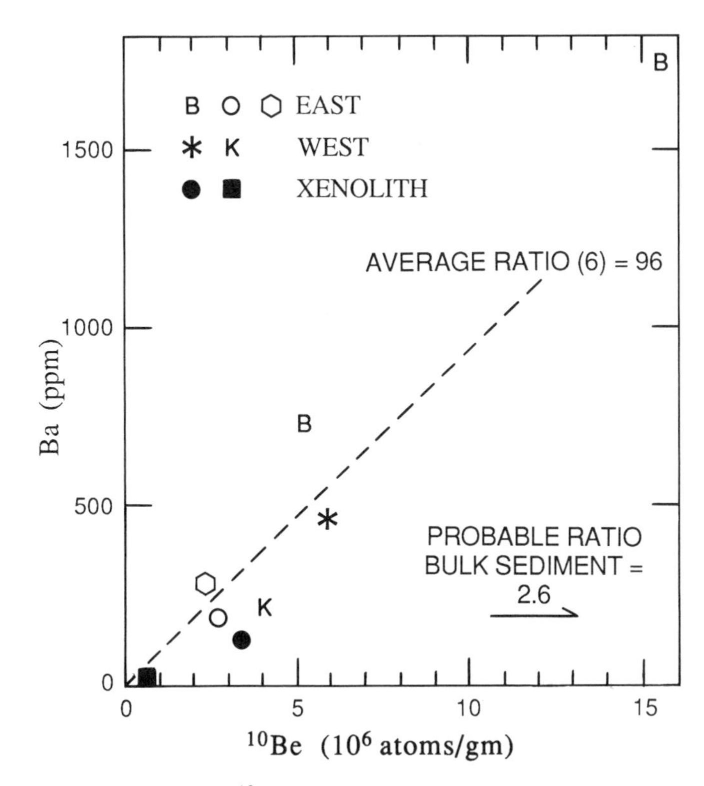

Figure 18. Ba versus ^{10}Be content for recent Aleutian volcanic rocks from Kanaga, Kasatochi, Okmok, Bogoslof, and Westdahl volcanoes (symbols are the same as in Fig. 2). Also plotted are ultramafic xenoliths from Moffett and Adagdak volcanoes. ^{10}Be data are from Tera and others (1986). Probable ratio in units of the figure) of the bulk sediment secton south of the Aleutian trench is estimated to be 2.6 by Tera and others (1986).

Hawaii (4.9 to 6.0) and in MORB (5.8 ± 0.3) as noted by Kyser and others (1981). A survey of lavas from all along the arc (see Tables 1–5 and Appendix in microfiche) shows that almost all samples, ranging from basalts to dacites. fall in only a slightly larger range ($\delta^{18}O$ = 4.8 to 6.5; 64 samples). Simple mass balance shows that only a small proportion of oxygen from isotopically heavy sediments ($\delta^{18}O$ >+20) can exist in these Aleutian lavas. Although more work is needed, it appears that primary Aleutian olivine tholeiites may have lower $\delta^{18}O$ than MORB (by ~one-half unit); it thus remains to explain light rather than heavy oxygen. The $\delta^{18}O$ data are consistent with the model of Morris and Hart (1983), because OIB have lower $\delta^{18}O$ than MORB (Kyser and others, 1981), but that model alone cannot explain the ^{10}Be data.

Some of the isotopic heterogeneity in Aleutian volcanic rocks could be generated as the lavas traverse the basal arc crust, as well as in the source region, as postulated for some Aleutian plutons (R. Kay and others, 1986; Kay and others, 1990). However, it is evident from the variable isotopic and trace-element signatures within single volcanoes, and especially within the most mafic arc magmas, that the mantle source beneath the arc is inhomogeneous.

GEOGRAPHIC AND TECTONIC CORRELATIONS

A goal of Aleutian studies is to correlate regional chemical and isotopic variability with regional tectonics. As a control on magmatic evolution in the crust, Kay and others (1982) postulated that tholeiitic magmas evolve in shallow magma chambers in regions of the arc that are under extension, and that calc-alkaline magmas evolve in deeper magma chambers in regions of the arc that are under compression. Regional tectonics exerts control on magma formation in two other ways: (1) by determining source and lithospheric mineralogy, and (2) by determining the proportion of components in the source. The basis for these controls can be outlined, but the details of their applicability remain obscure.

Mineralogic differences in the mantle source region could be related to the age of the oceanic crust that is being subducted. The intersection of the Amlia fracture zone with the arc (Fig. 1) between the Andreanof and Four Mountains segments is a case in point. The subducted crust to the east of this zone (Four Mountains side) is older than the subducted crust to the west (Andreanof side). Although the seismic data are not definitive (K. Jacob, oral communication, 1987), the older age of the subducting crust to the east (Four Mountains side) suggests that the slab would be thicker and heavier, and thus could be subducting at a greater angle in the east than to the west. A higher subduction angle would result in a thicker and hotter asthenosphere over the subducted plate, leading to more spinel peridotite relative to garnet peridotite in the wedge and higher degrees of melting in the peridotite source. Either less garnet in the source region or higher degrees of melting would lead to less retention of heavy REE in garnet and lower La/Yb ratios. Trace elements not fractionated by garnet are unaffected. However, the regional change in La/Yb ratio does not coincide exactly with the Amlia fracture zone, and the model does not explain why La/Yb ratios are high in the Rat segment, where the subducted oceanic crust is also younger.

Speculating further, differences in the subduction angle might also explain why there are relatively fewer calc-alkaline rocks in the region east of the Amlia fracture zone. If the peridotite source overlying the mantle is hotter in the east, the percentage of melting and thus the magma production rate are higher, leading to large shallow magma chambers. To the west the mantle is cooler, leading to a lower magma production rate and a greater chance that magmas will crystallize in the crust rather than erupt. If water is released from the subducted plate at equal rates east and west of the Amlia fracture zone, higher water content in the calc-alkaline centers (e.g., Moffett) might simply reflect the lower melting percentages and lower regional magma production rates.

Similarly, for secondary arc centers such as Bogoslof (north of Okmok), low production rates and higher water content (magmas are amphibole-bearing) correspond to a thicker lithosphere under the volcano. Thus, the higher K_2O (and trace element) content in the Bogoslof lavas could reflect less partial melting in the source.

Regional differences in trace-element content and isotopic ratios of arc magmas are related to differences in the proportions of the components that make up the magmas. Identification of the exact components remains a problem (except for ^{10}Be, which is a tracer of top sediments). The trace elements and isotopes that represent geochemically coherent "excess components" (e.g., non-OIB or MORB, see Kay and Kay, 1988) are a valuable tracing tool, and provide a constraint on the cause of regional differences in the arc.

A promising hypothesis to explain the regional differences in the arc involves differences in sediments that are subducted at the trench. Modern lavas in the Andreanof segment tend to have higher concentrations of Ba, Th, U, Rb, and Sr relative to La, higher ^{10}Be/Ba ratios, lower ϵ_{Nd}, and lower ^{206}Pb/^{204}Pb ratios than lavas from the Four Mountains segment and the adjacent region to the east (e.g., Umnak Island). Since the Andreanof segment is in the central part of the arc, far from both the Aleutian abyssal plain to the east and the Meiji sediment tongue to the west (Vallier and others, this volume), the subducting sediments might be more pelagic and less terrigenous. The closer trend toward the pelagic sediment field of the Andreanof segment samples on the Pb isotope diagram in Figure 17 supports this hypothesis. Proof, however, requires more data on the chemical character and distribution of sedimentary types in the Aleutian arc region. Interestingly, the boundary between the Rat and Andreanof segments corresponds to one of the trench sedimentation changes noted by Mogi (1969), which might explain the change in the range of Ba/La, Ba/Hf, and Th/Hf ratios that occur there.

A second hypothesis to explain the chemical variations in magmatic rocks along the arc involves the recycling of older Aleutian crust. This older crust includes both oceanic and older arc crust (S. Kay and Kay, 1985b) in the oceanic part of the arc, as well as continental crust in the Alaska Peninsula. Broadly speaking, these crustal differences are not reflected in the isotopic data (e.g., Kay and others, 1978; VonDrach and others, 1986), suggesting that crustal contamination is a second-order effect. Nonetheless, since the arc crust is relatively young, involvement of the preexisting crust is difficult to determine from isotopic data, and the extent of intracrustal recycling remains an important question. Intracenter isotopic variations that correlate with the stage of crustal differentiation (e.g., volcanic trends, Fig. 16; pluton trends, R. Kay and others, 1986) are probably due to crustal-level processes involving assimilation and mixing with older rocks. Trends toward higher ϵ_{Nd} in more silicic rocks could result from melting or assimilation of old oceanic crust buried in the arc. However, some important temporal chemical changes in Adak Island magmatic rocks described in the next section cannot easily be attributed to recycling of the preexisting crust.

Clear chemical differences exist between Tertiary arc-type basement rocks on Adak Island in the central Aleutians and spreading center–like basement rocks on Attu Island (173°E) in the volcanically inactive western Aleutians (Fig. 19; Table 5; Rubenstone, 1984; R. Kay and others, 1986). Since little work has been done on basement rocks in the region between the two islands, the effect of these differences on modern volcanic rocks is hard to assess. One logical place for the basement to change character is at the intersection of Bowers ridge with the arc—a change that could explain some of the trace-element differences between the rocks west and east of the western end of the Andreanof segment (Fig. 15).

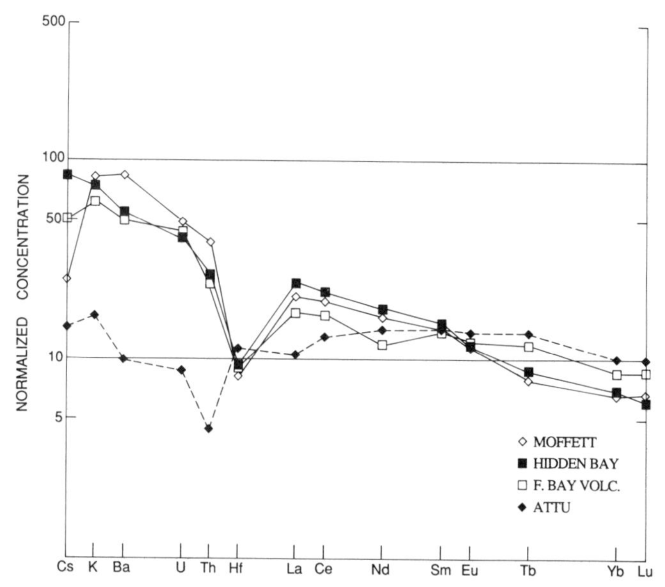

Figure 19. Trace-element patterns of Tertiary to Recent Aleutian mafic magmatic rocks normalized to chondrites and ocean ridge basalts (see Table 1 for normalization values). Trace-element patterns of Adak Island samples, which include a basalt from the Eocene Finger Bay volcanics (CMB-23), a gabbro from the Oligocene Hidden Bay Pluton (HB7-10), and a high-Al basalt from Moffett Volcano (MOF15), show the arc character of all samples studied from the central Aleutians. Note the increase in U, Th, and Ba relative to La in the Moffett lava compared to the older samples. In contrast, the trace-element pattern of a Tertiary pillow lava from Attu Island (SB80-1A) in the western Aleutians shows a fundamentally different pattern, which is related to a spreading center origin (Rubenstone, 1984; R. Kay and others, 1986). Data are given in Tables 1 and 4.

GEOCHEMICAL CHANGES THROUGH TIME IN CENTRAL ALEUTIAN MAGMATIC ROCKS

Temporal changes in the geochemistry of Aleutian magmas indicate processes occurring in the arc crust and mantle. During the existence of the Aleutian arc, the lithosphere (mainly crust) has changed because of the continued intrusion of arc magmas, and the exchange between these arc magmas and the preexisting lithosphere. If disruption of the crust by major folding and thrust faulting is minor (as suggested by field data), the crust that was under a section of arc at its inception is there today, and contains a record of the chemical growth and evolution of the arc. In contrast, the down-going slab and the asthenospheric mantle are continually renewed, and record only a transient situation. With this in mind, we will discuss the magmatic evolution of Adak Island in the middle of the modern Andreanof segment (Fig. 1).

TABLE 5. REPRESENTATIVE ANALYSES OF TERTIARY MAFIC MAGMATIC ROCKS FROM ATTU AND ADAK ISLANDS, ALEUTIAN ISLANDS, ALASKA*

Sample	SB80-1A	CM8-23A	BW8-R36	FB53	FB97	HB7-10
SiO_2	49.70	49.20	48.58	48.49	50.72	51.17
TiO_2	1.46	1.05	0.93	0.91	0.95	0.94
Al_2O_3	15.49	17.90	16.56	19.86	19.10	17.70
Fe_2O_3						
FeO	8.94	9.73	10.40	8.48	9.54	8.42
MnO	0.12	0.21	0.24	0.13	0.22	0.14
MgO	8.52	7.65	6.16	4.86	4.24	6.81
CaO	11.47	10.38	11.73	12.82	9.87	10.42
Na_2O	3.82	2.51	3.60	2.66	3.79	3.16
K_2O	0.23	0.87	0.76	0.81	1.38	1.04
P_2O_5	0.13	0.34	0.29	0.18	0.30	0.20
Total	99.88	99.84	99.25	99.20	100.11	100.00
La	3.90	6.46	9.60	7.66	10.10	8.99
Ce	12.70	16.30	22.70	18.10	23.30	21.28
Nd	10.10	8.66	14.60	15.30	16.50	12.82
Sm	3.22	2.64	3.81	3.01	3.81	3.43
Eu	1.190	1.030	1.140	0.894	1.310	1.023
Tb	0.811	0.692	0.616	0.449	0.563	0.512
Yb	2.47	2.15	1.65	1.30	1.60	1.74
Lu	0.385	0.330	0.260	0.197	0.262	0.239
Rb	5	17	14	18	25	28
Sr	183	432	900	728	753	736
Ba	39	188	148	233	329	214
Cs	0.19	0.67	0.31	0.32	0.73	1.09
U	0.12	0.65	0.49	0.76	0.95	0.61
Th	0.22	1.16	0.98	1.47	1.60	1.37
Hf	2.47	1.99	1.64	1.96	2.59	1.92
Sc	36.2	41.4	43.9	29.9	25.9	30.7
Cr	408	59	255	72	22	225
Ni	185	60	114	36	21	75
Co	408	44	45	40		34
FeO/MgO	1.05	1.27	1.69	1.74	2.25	1.24
Ba/La	10.00	29.10	15.42	30.42	32.57	23.80
La/Sm	1.2	2.4	2.5	2.5	2.7	2.6
La/Yb	1.6	3.0	5.8	5.9	6.3	5.2
Eu/Eu*	0.99	0.99	0.99	0.99	0.99	0.99
Th/Hf	0.09	0.58	0.59	0.75	0.62	0.71
Ba/Hf	16	94	90	119	127	111
Ba/Th	176	162	152	159	206	156
La/Hf	1.6	3.2	5.9	3.9	3.9	4.7
Th/U	1.84	1.80	2.00	1.93	1.68	2.25
$^{87}Sr/^{86}Sr$	0.70357		0.70306	0.70303	0.70326	0.70305
ε_{Nd}	10.10		7.60	9.30	8.40	9.40

*SB80-1A is a pillow lava from the Sarana-Chichagof region, Attu (analyses from Rubenstone, 1984); CM8-23A and BW8-R36 are volcanic clasts from the Finger Bay volcanics, Adak (analyses from Rubenstone, 1984); FB53 and FB97 are gabbros from the Finger Bay pluton, Adak (partial analyses in Kay and others, 1983); HB7-10 is a gabbro from the Hidden Bay pluton, Adak (see Citron and others, 1980). Isotopic analyses are from R. Kay and others (1986).

This discussion is representative of the central Aleutians near Adak Island, not as a model for the entire arc.

Geochemical studies have been made on the Tertiary to Recent magmatic units that crop out on Adak. The units studied in detail include the pre-Oligocene Finger Bay volcanics (Coats, 1956; Rubenstone, 1984) and Finger Bay pluton (Coats, 1956; Kay and others, 1983), the calc-alkaline Oligocene Hidden Bay pluton (Fraser and Snyder, 1959; Citron, 1980; Citron and others, 1980; Kay and others, 1990), and the Pliocene-Recent calc-alkaline volcanoes, Moffett and Adagdak. The distribution of these units is given in Kay and others (1983, 1990), and their geologic setting and history are summarized by Vallier and others (this volume).

The chemistry of the older Adak magmatic rocks is compared to those of the younger Moffett volcano in Table 5 and Figures 19 to 21. The Finger Bay volcanics unit, which shows variable zeolite- to greenschist-facies metamorphism, consists of dominantly mafic to intermediate composition pyroclastic rocks with some flows, dikes, and sills. Analyzed samples include volcanic clasts, flows, and dikes (Rubenstone, 1984). The small (3 km across) tholeiitic Finger Bay pluton is composed mainly of cumulate and noncumulate gabbro with lesser amounts of quartz monzodiorite and other intermediate units (Kay and others, 1983). In contrast, the larger (15 km across), calc-alkaline Hidden Bay pluton is composed dominantly of quartz monzodiorite and granodiorite (58 to 63 percent SiO_2) with lesser amounts of gabbro and leucogranodiorite (Citron, 1980).

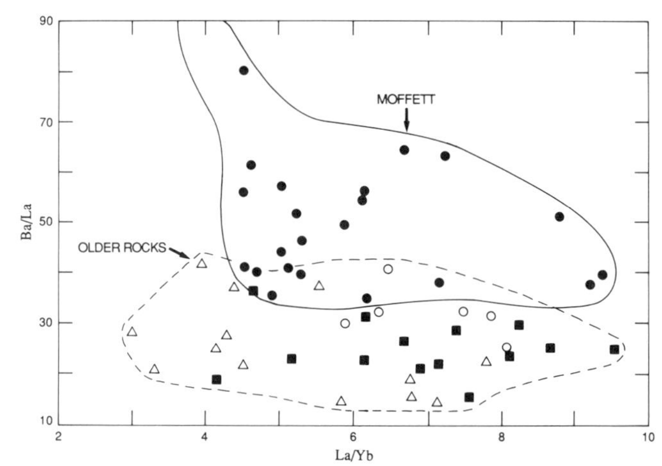

Figure 20. Plot of Ba/La versus La/Yb for Adak Island magmatic rocks from the Eocene Finger Bay volcanics (triangles), the Eocene(?) Finger Bay pluton (open circles), the Oligocene Hidden Bay pluton (solid squares), and the Plio-Pleistocene Moffett Volcano (solid circles). High-Mg andesites of Kay (1978) plot off-scale at Ba/La ~10 and La/Yb ~50. Although La/Yb ratios of the Moffett volcanics are similar to those of the older rocks, the Ba/La ratios of the Moffett volcanics are distinctly higher, suggesting a change in the source region beneath Adak Island over this time period. Except for the Finger Bay volcanic samples that have had La/Yb ratios >6 combined with Ba/La ratios <20, the older samples have Ba/La and La/Yb ratios that overlap those of Plio-Pleistocene to Recent volcanic rocks in other parts of the arc (compare with Figs. 15A and B).

In a broad sense, the petrology and geochemistry of the older Adak magmatic rocks are similar to those of the younger Aleutian arc volcanic rocks. The Tertiary calc-alkaline plutons (Fig. 1) on Unalaska (Perfit, 1977; Perfit and others, 1980a) and Kagalaska Islands (Citron, 1980; Kay and others, 1982) to the east, and Amchitka Island (Kay, 1980) to the west are also geochemically similar to the Adak rocks, particularly to the rocks of the Hidden Bay pluton (Citron, 1980; Citron and others, 1980; Kay and others, 1982; Kay and others, 1990). Like the younger arc volcanic rocks, none of these older magmatic rocks is chemically similar to the island-arc tholeiites of the western Pacific (Kay and others, 1982, 1983; Rubenstone, 1984).

The early mid-Tertiary magmatic rocks in the central and eastern arc are distinct from the early Tertiary volcanic and plutonic rocks on Attu Island in the volcanically interactive western Aleutian arc (Rubenstone, 1984; Shelton, 1986; R. Kay and others, 1986). These old Attu magmatic rocks appear to come from a spreading ridge environment and are geochemically like ocean-ridge rocks, as shown by their flat or light REE depleted REE patterns, lack of alkali and alkaline earth enrichment, lack of HFSE depletion relative to the REE (Fig. 19), and MORB-like ϵ_{Nd} (Table 5). Younger Tertiary volcanic rocks on Attu Island are more arc-like in character (Rubenstone, 1984; R. Kay and others, 1986).

Specific differences in the geochemistry and mineralogy between the Adak andesitic to dacitic composition volcanic and plutonic rocks can be explained by differences in the conditions of their crystallization. The plutons appear to represent the last unerupted stages of the volcanoes as suggested by mineral assemblages, which indicate more water-rich, cooler crystallization conditions (low-Al amphibole, biotite, quartz, and potassium feldspar; S. Kay and others, 1986b, 1990). Some plutonic rocks have lower Ba/La ratios at higher La/Yb ratios than observed in any of the young Aleutian volcanic rocks (Fig. 20), a trend that is consistent with an increased role for amphibole fractionation in the plutons, leading to higher La/Yb ratios and lower Ba/La ratios in the plutonic rocks.

In contrast, differences in incompatible and only slightly compatible element ratios (e.g., Ba, the light REE) in the Adak rocks record the evolution of magmas added to the crust on Adak. Some of these differences are illustrated in Figure 20, a plot of Ba/La (incompatible elements) versus La/Yb ratio (where Yb is more compatible than La). For example, the Ba/La ratios of most of the Tertiary samples are lower than those of samples from Moffett volcano, but are within the range of other recent Aleutian arc volcanic rocks (see Fig. 15B and recall that Moffett samples have relatively high Ba/La ratios). Even if amphibole fractionation has altered the Ba/La and La/Yb ratios in some of the plutonic rocks, the magnitude of the Ba/La ratios in the plutonic rocks as a whole still suggests that their source region, like that of the Finger Bay volcanic rocks, had a lower Ba/La ratio than the source region of the Moffett volcanic rocks.

The oldest Adak samples, those from the basaltic Finger Bay

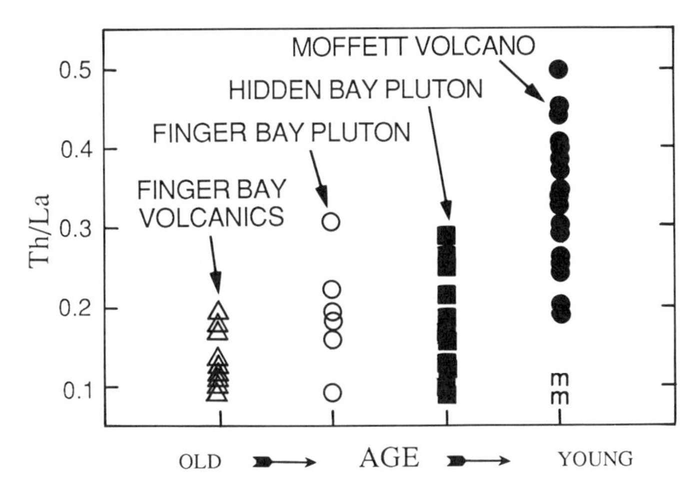

Figure 21. Plot of Th/La ratios versus relative age for Adak magmatic rocks. In addition to the symbols used in Figure 18, m represents the high-Mg andesites. The Th/La plot is representative of the high ratios of Th, U, Ba, Sr, and Rb to the other incompatible elements in many of the younger Moffett rocks relative to the older units.

volcanics unit, fall in two general groups as shown in Figure 20 (data from Rubenstone, 1984). The first group has Ba/La and La/Yb ratios that plot outside the Moffett field, but within the modern arc field. The isotopic characteristics of the one analyzed sample (ϵ_{Nd} = 6.8 and $^{87}Sr/^{86}Sr$ = 0.70332) from this group are like modern arc lavas (R. Kay and others, 1986). The second group has lower Ba/La (most are 15 to 20) at higher La/Yb ratios (~6 to 8) than observed in modern Aleutian basalts. Although the isotopic characteristics of this second group (3 samples, ϵ_{Nd} = 7.6 to 7.7 and $^{87}Sr/^{86}Sr$ = 0.70296 to 0.70316, R. Kay and others, 1986) are also within the range of modern Aleutian arc lavas, they are more oceanic-like than the first group. The subtle, somewhat more OIB-like character (lower Ba/La ratios combined with higher La/Yb ratios) of some of the second group of two samples could be consistent with their eruption in the early stages of the arc, especially if the initiation of the modern arc is related to an old fracture zone (Vallier and others, this volume). However, the similar characteristics of the Finger Bay volcanics and the modern arc lavas is much more striking than their differences (Rubenstone, 1984).

Comparison of other incompatible trace-element ratios between the older Adak rocks and those from Moffett Volcano shows that the older rocks have a lower U, Th, and Ba content relative to the other incompatible elements. Thus, ratios such as Th/La (see Fig. 21), Th/Hf, Ba/La, and Ba/Hf are generally lower in the older rocks than in the Moffett lavas, whereas ratios like Ba/Th show no obvious age correlations. The most notable exceptions are a few Finger Bay volcanic samples that have high Ba/Th ratios. These samples also have the highest Ba/La and Ba/Hf ratios, suggesting that they have higher Ba than the other Finger Bay volcanic samples. Isotopic data are not available on any of these samples.

Examination of the available isotopic data on Adak and Kagalaska plutonic rocks shows that ϵ_{Nd} in the plutonic samples (ϵ_{Nd} = +7.1 to +9.3) is, in some cases, higher than in any of the modern arc volcanic rocks (ϵ_{Nd} = +7.3 ± 1.25 for 56 samples; R. Kay and others, 1986). For example, samples of the Hidden Bay gabbro and the Finger Bay plutons have ϵ_{Nd} ranging from 8.4 to 9.4, and R. Kay and others (1986) suggest that these high ϵ_{Nd} values may result from assimilation of old oceanic crust by the plutonic magmas in the basement of the Adak crust. However, the process must be more complex, because ϵ_{Nd} decreases (9.4 to 7.1) and $^{87}Sr/^{86}Sr$ increases (0.70305 to 0.70328) with increasing SiO_2 in the three samples from the Hidden Bay pluton that have been analyzed. Although a systematic trend with SiO_2 content is observed in the Hidden Bay pluton data, no pattern emerges from the plutonic rock data as a whole (Kay and others, 1990).

As observed regionally for the young volcanic rocks, the higher ϵ_{Nd} in the older rocks appears to correlate in a general way with lower Ba/La, Ba/Hf, and Th/Hf ratios, suggesting that a component with relatively high Ba, Th, and U, and low ϵ_{Nd}, contributes to the distinctive trace-element signature of the younger rocks of the Adak region (e.g., Moffett Volcano). Contamination of the Moffett lavas with older Adak crustal rocks cannot easily account for the characteristics of this component, suggesting that the change is the result of change or addition of a component in the subcrustal source. The data on the older Adak rocks suggest that this component (perhaps a distinctive sediment type) became important after the mid-Tertiary and could be related to the formation of a well-defined accretionary prism in this part of the arc at about 5 Ma (Scholl and others, 1987).

ALEUTIAN MAGMATIC PROCESSES: AN INTEGRATED VIEW

The preceding sections have emphasized that magmatic and tectonic processes can be inferred from the mineralogy and the chemical and isotopic compositions of Aleutian igneous rocks. We have integrated these processes on a regional scale, to derive a tectonic and lithologic cross section of the Aleutian arc, as shown in Figure 22. This section is drawn across the middle of a coherent arc segment, such as near Adak Island. The discussion proceeds from deep to shallow tectonic levels.

Mantle level

We call our hypothesis for the formation of Aleutian arc magmas the convective limb process. In this process, a melt or solute-rich fluid (Eggler, 1989), which includes components from sediment and hydrothermally altered oceanic crust, is transferred from the subducted plate into the overlying mantle peridotite, causing the mantle to partially melt and rise as a broad convective limb (Fig. 22). The melt percolates continuously through the rising peridotite over a depth range from 100 to perhaps 30 km, where it segregates and ponds under the lithosphere, which is largely crust. A porous flow mechanism (Turcotte, 1987) appears to operate in favor of hydrofracture, because Aleutian mantle xenoliths have textures indicating penetrative deformation. Diapirism also appears to be ruled out, because the penetration velocity of hypothetical diapirs is slower than the segregation velocity of the interstitial melt required for buoyancy. Broadly speaking, the convective-limb hypothesis satisfies the observational demands of the distribution of seismic and velocity minima in the mantle over the plate (Engdahl and Billington, 1986, and Engdahl, oral communication, 1987) and the compositional coincidence of the least evolved arc magmas (olivine tholeiites) and partial melts of the spinel lherzolite mantle at the 30-km depth (e.g., Fujii and Scarfe, 1985).

Evidence that subducted sediment and hydrothermally altered oceanic crust is involved in the source region of Aleutian magmas comes from trace elements, radiogenic isotopes, and ^{10}Be. However, oxygen isotopic values of Aleutian arc lavas are close to those of MORB, indicating that the bulk of the Aleutian magmas comes from the mantle. Regional variations in incompatible trace-element and isotopic abundances in Aleutian arc volcanoes, as well as some of the heterogeneity within single centers, probably results from variability in the subducted component. This interpretation, however, requires a more extensive evaluation of the regional differences in the crust of the Pacific Plate and of the arc basement.

In major-element composition (except K_2O), the most mafic Aleutian basalts are similar to MORB, and the proposed convective-limb is similar to the process for magma generation and migration thought to occur at the mid-oceanic ridges (e.g., Turcotte, 1987). However, the driving force in the arc—H_2O-CO_2 flux-created buoyancy—is distinctive. The observation that the distance between volcanoes both along arcs and at mid-oceanic ridges is about the same supports our speculative model that these characteristic volcanic spacing result from transverse convective rolls. In this model, following the release of the basaltic fraction from the upwelling convecting limb, residual mantle largely descends both away from the trench and between the volcanic centers (not shown in Fig. 22). A quite different model commonly cited (e.g., Marsh, 1979) is that the spacing of arc volcanoes is due to Rayleigh-Taylor instabilities within a magma ribbon over the subducted plate at 100 km.

An important difference between arc and MORB tectonic regimes is that the magma-production rate (mass of magma per length of arc or ridge per time) is generally lower in arcs than at mid-oceanic ridges. However, the magma production rates for the least productive mid-oceanic ridges and the most productive arcs approach each other, suggesting a continuum of pressure-release melting mechanisms. For example, an Aleutian magma-production rate averaged over 60 m.y. is 35 km^3/km/m.y. (calculated from the crustal section of Grow, 1973, as modified by S. Kay and Kay, 1985b), which compares closely with magma production in the slowest mid-ocean ridges (southwest Indian Ridge: 40 km^3/km/m.y., calculated from an 8 mm/yr spreading rate and a 5-km-thick crust; Fisher and Sclater, 1983).

The role of volatiles released from the subducted plate is crucial in the production-rate differences of arc and MORB lavas. The higher volatile content of arc magmas means that the same percentage of melting is achieved at lower temperature in arcs

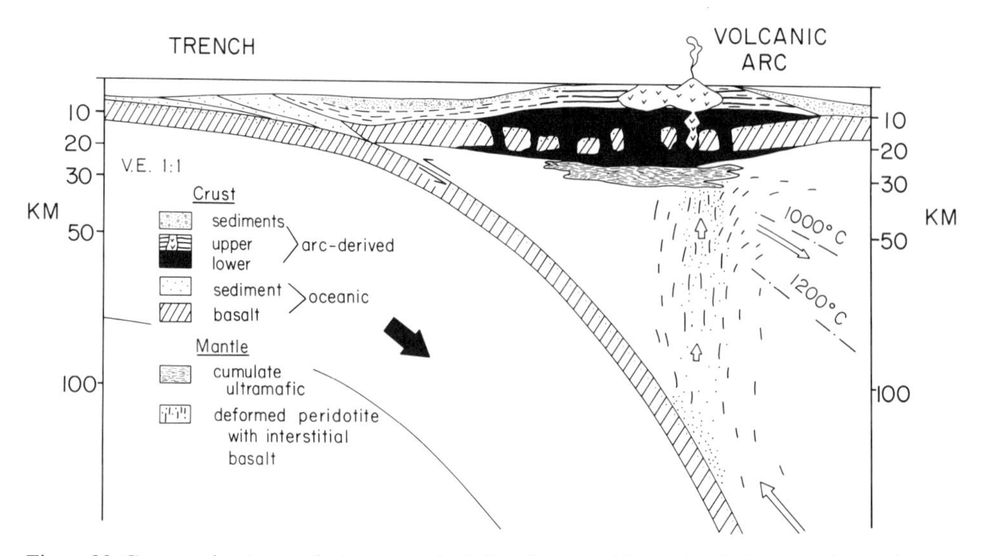

Figure 22. Cross section (no vertical exaggeration) showing crustal formation in the central Aleutian arc (modified from S. Kay and Kay, 1985b; R. Kay and others, 1986). Configuration of subducted plate, crustal thickness, and forearc lithologies are constrained by geophysical and seismological data (Grow, 1973; Engdahl and Billington, 1986). Melting (represented by dots) occurs in peridotite that lies above the subducting plate. Convective rise of the buoyant peridotite causes furhter melting (by decompression) accompanied by percolation of melt through the denser peridotite. Segregated mafic minerals from ponded olivine tholeiite form the cumulative ultramafic rocks that occur at the crust-mantle boundary. A basic lower crust consists of unassimilated residues of crystalline oceanic crust and intrusive hi-Al basalt and its mafic crystalline fractionates. Intermediate composition upper crust consists of igneous intrusions and extrusions of mafic to intermediate composition.

than at mid-oceanic ridges. For arcs, in the extreme cases of no released volatiles or of temperatures that are too low in the hanging wall, no melting of overlying peridotite occurs. In cases where the peridotite is hot enough to melt, volatile release drives buoyancy. If the magma-production rate does not vary directly with the release rate of volatiles into the mantle at the depth required for melting (100 km), then a variable volatile content of magmas could result (as suggested above for the Aleutian tholeiitic versus calc-alkaline magmas). Volatiles may also be stored in the hanging wall over the subducted plate.

Magmatic segregation processes in the mantle under the Aleutian arc have an important bearing on mantle evolution. Assuming that some continental crustal components in sediment and altered oceanic crust are recycled from the mantle back to the crust as components of arc magmas, any inefficiency of the recycling process will leave crustal components in the mantle. This inefficiency probably occurs at three sites: (1) during extraction of crustal components from the subducted plate, (2) during extraction of primitive basaltic melt from peridotite, and (3) during fractionation of basalt in the mantle or more likely at the crust-mantle boundary. Returned to the mantle at all three sites are mafic or ultramafic residues or fractionated liquidus minerals that are at equilibrium with arc magmas and therefore share their isotopic signatures. Some trace-element ratios in these arc-related mantle rocks are dominated by the recycled component. The production rate of these residues is large—perhaps one-fifth of the production rate of MORB residues, which is about four times the ratio of magma production rates reflecting lower percentages of melting at arcs (Kay, 1980). The return mechanism of the residues to deeper levels in the mantle is obvious for the subducted plate, but less apparent for the mantle residues and for the cumulates that pond immediately below the crust-mantle boundary. A delamination mechanism may apply to the cumulates.

Crustal level

The crust-mantle boundary is considered to be magmatic, and coincides with a series of short-lived magma ponds whose bases are formed by the accumulation of olivine and clinopyroxene—the liquidus phases of olivine tholeiite. Gabbros, together with crystallized high-Al basalts and their accumulated liquidus phases, constitute the lowermost crust. Because the present asthenospheric mantle reaches the crust-mantle boundary, oceanic mantle lithosphere that was under the arc at its inception has flowed away from under the volcanic line. Similarly, ultramafic cumulates from earlier stages of arc magmatism have also flowed away.

If a continuous elongated pool of magma accumulates at the crust/mantle boundary under the volcanic line, Rayleigh-Taylor instability could result in volcanic spacing. However, the magmatic production rate is probably too low to create a long-lived pool at such a shallow depth, because it would crystallize—a process that could be accelerated by crustal assimilation. Also, chemical and isotopic heterogeneity at the highest volume centers such as Okmok indicates incomplete mixing of magma batches—an observation that is difficult to reconcile with a magma chamber extending along the arc at the crust-mantle boundary.

Within the present mid- to lower crust beneath the arc, some of the original oceanic crustal basement probably remains (S. Kay and Kay, 1985b). This oceanic crust is most likely to be preserved in the large (about 500 km) coherent crustal segments. Pieces of this oceanic crust, metasomatized by arc intrusive bodies, occur as xenoliths in Tertiary dikes on Kanaga Island (Swanson and others, 1987). The hydrated upper layers of this oceanic crust, as well as sediments on top of it, should undergo dehydration and melting in response to the heating that accompanies magmatic intrusion, and can be recycled upward within the arc crust. Arc volcanic flows associated with the early stages of arc formation may also overlie the oceanic crust in the deeper parts of the arc crust.

Within coherent arc segments (the region shown in Fig. 22), the mid- to lower crust is the site of magma chambers in which the mixing and fractionation processes occur that yield the calc-alkaline series of rocks (Fig. 9). These processes imply that an important component of the crust at this level is the crystal residue from the fractionation processes that produpce the erupted calc-alkaline lavas. These magma chambers probably did not exist until the arc crust was built to some minimum thickness in the Tertiary (Kay and others, 1990). The upper crust in these regions consists of the plutons, which represent the last stages of calc-alkaline volcanoes, and the erupted volcanic rocks.

In the regions between coherent crustal blocks, the large volcanoes that are not in linear belts are attributed to extensional stress regimes (Fig. 9). If crustal dilation occurs in these regions, the preexisting oceanic crust may not be present, and either the crust is thinner or there is a higher intrusion rate. In either case, the arc crust probably consists of a large proportion of arc-derived magmas. Differentiation occurs along a tholeiitic trend, because fractionation and any mixing that occurs are generally happening at high temperatures at shallow levels. Remelted, hydrothermally altered roof rocks may be important in the genesis of some members of the tholeiitic series. Melting at shallow levels is facilitated by the high-temperature Mg-rich magmas, which can reach the surface in these regions.

Growth of the continental crust

As implied in this paper, the mean composition of the Aleutian arc crust is high-Al basalt (or more mafic basalt, depending on the proportion of preexisting oceanic crust). In contrast, the bulk composition of the continental crust is usually considered to be andesitic (e.g., Taylor and McLennan, 1985), similar to the composition of only the upper crust of the Aleutian arc. This creates a compositional dilemma if continental crust is composed of accreted arcs. As a solution to this problem, we have proposed (S. Kay and Kay, 1985b; and R. Kay and Kay, 1986, 1988) that lower crustal delamination accompanies structural suturing of oceanic arc terranes such as the Aleutians to existing continental margins, and that the mafic lower crust of arcs is recycled back into the mantle.

REFERENCES CITED

Apted, M. J., 1981, Rare earth element systematics of hydrous liquids from partial melting of basaltic eclogites; A re-evaluation: Earth and Planetary Science Letters, v. 52, p. 172–182.

Anderson, A. T., 1976, Magma mixing; Petrological process and volcanological tool: Journal of Volcanology and Geothermal Research, v. 1, p. 3–33.

Arculus, R. J., DeLong, S. E., Kay, R. W., and Sun, S. S., 1977, The alkalic rock suite of Bogoslof Island, eastern Aleutian arc, Alaska: Journal of Geology, v. 85, p. 177–186.

Arth, J. G., 1976, Behavior of trace elements during magmatic processes; A summary of theoretical models and their applications: U.S. Geological Survey Journal of Research, v. 4, p. 41–47.

Baker, D. R., and Eggler, D. H., 1983, Fractionation paths of Atka (Aleutians) high-alumina basaltsa; Constraints from phase relations: Journal of Volcanology and Geothermal Research, v. 18, p. 387–404.

——, 1987, Compositions of anhydrous and hydrous melts coexisting with plagioclase, augite, and olivine or low-Ca pyroxene from 1 atm to 8 kbar; Application to the Aleutian volcanic center of Atka: American Mineralogist, v. 72, p. 12–28.

Bingham, D. K., and Stone, D. B., 1972, Paleosecular variation of the geomagnetic field in the Aleutian Islands, Alaska: Geophysical Journal of the Royal Astronomical Society, v. 28, p. 317–335.

Brophy, J. G., 1986, The Cold Bay volcanic center, Aleutian volcanic arc I; Implications for the origin of high-alumina arc basalt: Contributions to Mineralogy and Petrology, v. 93, p. 368–380.

Brophy, J. G., and Marsh, B. D., 1986, On the origin of high-Alumina arc basalt and the mechanics of melt extraction: Journal of Petrology, v. 27, p. 763–789.

Byers, F. M., Jr., 1959, Geology of Umnak and Bogoslof Islands, Aleutian Islands: U.S. Geological Survey Bulletin 1028L, p. 267–369.

——, 1961, Petrology of three volcanic suites, Umnak and Bogoslof Islands, Aleutian Islands, Alaska: Geological Society of America Bulletin, v. 79, p. 93–128.

Citron, G. P., 1980, The Hidden Bay Pluton, Alaska; Geochemistry, origin, and tectonic significance of Oligocene magmatic activity in the Aleutian Island arc [Ph.D. thesis]: Ithaca, New York, Cornell University, 240 p.

Citron, G. P., Kay, R. W., Kay, S., Snee, L., and Sutter, J., 1980, Tectonic significance of early Oligocene plutonism on Adak Island, Central Aleutian Islands, Alaska: Geology, v. 8, p. 375–379.

Coats, R. R., 1952, Magmatic differentiation in Tertiary and Quaternary volcanic rocks from Adak and Kanaga Islands, Aleutian Islands, Alaska: Bulletin of the Geological Society of America, v. 63, p. 486–514.

——, 1956, Geology of northern Adak Island, Alaska: U.S. Geological Survey Bulletin 1028C, p. 45–66.

——, 1959, Geological reconnaissance of Semisopochnoi Island, western Aleutian Islands, Alaska: U.S. Geological Survey Bulletin 10280, p. 477–519.

——, 1962, Magma type and crustal structure in the Aleutian arc: American Geophysical Union Geophysical Monograph 6, p. 92–109.

Condomines, M., and 6 others, 1983, Helium, oxygen, strontium, and neodymium isotopic relationships in Icelandic volcanics: Earth and Planetary Science Letters, v. 66, p. 125–136.

Conrad, W. K., and Kay, R. W., 1984, Ultramafic and mafic inclusions from Adak Island; Crystallization history and implications for the nature of primary magmas and crustal evolution in the Aleutian arc: Journal of Petrology, v. 25, p. 88–125.

DeBari, S., Kay, S. M., and Kay, R. W., 1987, Ultramafic xenoliths from Adagdak Volcano, Adak, Aleutian Islands, Alaska; Deformed igneous cumulates from the Moho of an island arc: Journal of Geology, v. 95, p. 329–341.

DeLong, S. E., Perfit, M. R., McCulloch, M. T., and Ach, J., 1985, Magmatic evolution of Semisopochnoi Island, Alaska; Trace element and isotopic constraints: Journal of Geology, v. 93, p. 609–618.

Eggler, D. H., 1989, Influence of H_2O and CO_2 on melt and fluid chemistry in subduction, *in* Hart, S. R. and Gülen, L., eds., Crust/mantle recycling at subduction zones: Dordrecht, Kluwer Academic Publishers, NATO Advanced Study Institutes Series C, Mathematical and Physical Sciences, v. 258, p. 97–104.

Engdahl, E. R. and Billington, S., 1986, Focal depth determination of central Aleutian earthquakes: Bulletin of the Seismological Society of America, v. 76, p. 77–93.

Engebretson, D. C., Cox, A., and Gordon, R. G., 1985, Relative motions between oceanic and continental plates in the Pacific Basin: Geological Society of America Special Paper 206, 59 p.

Fisher, R. L. and Sclater, J. G., 1983, Tectonic evolution of the southwest Indian Ocean since the mid-Cretaceous; Plate motions and stability of the pole of Antarctica/Africa for at least 80 Myr.: Geophysical Journal of the Royal Astronomical Society, v. 73, p. 553–576.

Fraser, G. D., and Snyder, G. L., 1959, Geology of southern Adak Island and Kagalaska Island, Alaska: U.S. Geological Survey Bulletin 1028M, p. 371–408.

Frost, C. D., and Myers, J. D., 1987, Nd isotope systematics of Aleutian Island arc magma generation and ascent: EOS Transactions of the American Geophysical Union, v. 68, p. 462.

Fujii, T., and Scarfe, C. M., 1985, Composition of liquids coexisting with spinel lherzilite at 10 kbar and genesis of MORBs: Contributions to Mineralogy and Petrology, v. 90, p. 18–28.

Gill, J., 1981, Orogenic andesites and plate tectonics: New York, Springer-Verlag, 390 p.

Green, T. H., and Ringwood, A. E., 1968, The genesis of the calc-alkaline igneous rock suite: Contributions to Mineralogy and Petrology, v. 18, p. 105–162.

Grow, J. A., 1973, Crustal and upper mantle structure of the central Aleutian arc: Geological Society of America Bulletin, v. 84, p. 2169–2192.

Gust, D. A. and Perfit, M. R., 1987, Phase relations of a high-Mg basalt from the Aleutian island arc; Implications for primary island arc basalts and high-Al basalts: Contributions to Mineralogy and Petrology, v. 97, p. 7–18.

Hickey, R. L., Gerlach, D. C., and Frey, F. A., 1984, Geochemical variations in volcanic rocks from central-south Chile (33–42°S), *in* Harmon, R. S., and Barreiro, B. A., eds., Andean magmatism; Chemical and isotopic constraints: Cheshire, United Kingdom, Shiva, p. 72–95.

Hickey, R. L., Frey, F. A., Gerlach, D. C., and Lopez-Escobar, L., 1986, Multiple sources for basaltic arc rocks from the southern volcanic zone of the Andes (34°–41°S); Trace element and isotopic evidence for contributions from subducted oceanic crust, mantle, and continental crust: Journal of Geophysical Research, v. 91, p. 5963–5983.

Johnston, A. D., 1986, Anhydrous P-T phase relations of near primary high-Al basalt from the South Sandwich Islands; Implications for the origin of island arcs: Contributions to Mineralogy and Petrology, v. 92, p. 368–382.

Kay, R. W., 1977, Geochemical constraints on the origin of Aleutian magmas, *in* Talwani, M., and Pitman, W. C., eds., Island arcs, deep sea trenches, and back-arc basins: American Geophysical Union Ewing Series 1, p. 229–242.

——, 1978, Aleutian magnesian andesites; Melts from subducted Pacific ocean crust: Journal of Volcanology and Geothermal Research, v. 4, p. 117–132.

——, 1980, Volcanic arc magmas; Implications of a melting-mixing model for element recycling in the crust-upper mantle system: Journal of Geology, v. 88, p. 497–522.

——, 1984, Elemental abundances relevant to identification of magma sources: Philosophical Transactions of the Royal Society of London, Series A, v. 310, p. 535–547.

Kay, R. W., and Kay, S. M., 1985, Eclogite model and primary magmas of the Aleutian arc: EOS Transactions of the American Geophysical Union, v. 66, p. 422.

——, 1986, Petrology and geochemistry of the lower continental crust; An overview: Geological Society of London Special Publication 24, p. 147–159.

——, 1988, Crustal recycling and the Aleutian arc: Geochimica et Cosmochimica Acta, v. 52, p. 1351–1359.

——, 1989, Recycled continental crustal components in Aleutian Arc magmas, *in* Hart, S. R. and Gülen, L., eds., Crust/mantle recycling at subduction zones: Dordrecht, Kluwer Academic Publishers, NATO Advanced Study Institutes Series C, Mathematical and Physical Sciences, v. 258, p. 145–162.

Kay, R. W., Sun, S.-S., and Lee-Hu, C.-H., 1978, Pb and Sr isotopes in volcanic rocks from the Aleutian Islands and Pribilof Islands, Alaska: Geochimica et Cosmochimica Acta, v. 42, p. 263–273.

Kay, R. W., Rubenstone, J. L., and Kay, S. M., 1986, Aleutian terranes from Nd isotopes: Nature, v. 322, p. 605–609.

Kay, S. M., and Kay, R. W., 1985a, Aleutian tholeiitic and calc-alkaline magma series 1: The mafic phenocrysts: Contributions to Mineralogy and Petrology, v. 90, p. 276–290.

——, 1985b, Role of crystal cumulates and the oceanic crust in the formation of the lower crust of the Aleutian arc: Geology, v. 13, p. 461–464.

Kay, S. M., Kay, R. W., and Citron, G. P., 1982, Tectonic controls on tholeiitic and calc-alkaline magmatism in the Aleutian arc: Journal of Geophysical Research, v. 87, p. 4051–4072.

Kay, S. M., Kay, R. W., Brueckner, H. K., and Rubenstone, J. L., 1983, Tholeiitic Aleutian arc plutonism; The Finger Bay Pluton, Adak, Alaska: Contributions to Mineralogy and Petrology, v. 82, p. 99–116.

Kay, S. M., Kay, R. W., Romick, J., and Yogodzinski, G., 1986a, Spatial variations in trace element ratios in the Aleutian arc: Geological Society of America Abstracts with Programs, v. 18, p. 651.

Kay, S. M., Romick, J., and Kay, R. W., 1986b, Mineralogy of volcanic and plutonic rocks as a guide to processes in the formation of the Aleutian crust, *in* Abstracts, 13th Meeting, International Mineralogical Association, Stanford, California: International Mineralogical Association, p. 139.

Kay, S. M., Maksaev, V., Mpodozis, C., Moscoso R., and Nasi, C., 1987, Probing the evolving Andean Lithosphere; Middle to late Tertiary magmatic rocks in Chile over the modern zone of subhorizontal subduction (29–31.5°S): Journal of Geophysical Research, v. 92, p. 6173–6189.

Kay, S. M., Kay, R. W., Citron, G. P., and Perfit, M. R., 1990, Calc-alkaline pluton in an oceanic island arc; The Aleutian arc, Alaska, *in* Kay, S. M., and Rapela, C. W., eds., Plutonism from Antarctica to Alaska: Geological Society of America Special Paper 241, p. 233–256.

Kienle, J., and Swanson, S. E., 1983, Volcanism in the eastern Aleutian arc; Late Quaternary and Holocene centers, tectonic setting, and petrology: Journal of Volcanology and Geothermal Research, v. 17, p. 393–432.

Kushiro, I., 1986, Viscosity of partial melts in the upper mantle: Journal of Geophysical Research, v. 91, p. 9343–9350.

Kushiro, I., Yoder, H. S., and Mysen, B. O., 1976, Viscosities of basalt and andesite melts at high pressures: Journal of Geophysical Research, v. 81, p. 6351–6356.

Kyser, T. K., O'Neil, J. R., and Carmichael, I.S.E., 1981, Oxygen isotope thermometry of basic lavas and mantle nodules: Contributions to Mineralogy and Petrology, v. 77, p. 11–23.

Lonsdale, P., 1988, Paleogene history of the Kula Plate; Off-shore evidence and on-shore implications: Geological Society of America Bulletin, v. 100, p. 733–754.

Marsh, B. D., 1976, Some Aleutian andesites; Their nature and source: Journal of Geology, v. 84, p. 27–45.

——, 1979, Island arc development; Some observations, experiments, and speculations: Journal of Geology, v. 87, p. 687–713.

——, 1981, On the crystallinity, probability of occurrence, and rheology of lavas and magmas: Contributions to Mineralogy and Petrology, v. 78, p. 85–98.

——, 1982, The Aleutians, *in* Thorpe, R. S., ed., Andesites; Orogenic andesites and related rocks: New York, John Wiley, p. 99–115.

Marsh, B. D., and Leitz, R. E., 1979, Geology of Amak Island, Aleutian Islands, Alaska: Journal of Geology, v. 87, p. 715–723.

Mattioli, G. S., and Wood, B. J., 1986, Upper mantle oxygen fugacity recorded by spinel lherzolites: Nature, v. 322, p. 626–628.

McCulloch, M. T., and Perfit, M. R., 1981, $^{143}Nd/^{144}Nd$, $^{87}Sr/^{86}Sr$, and trace element constraints on the petrogenesis of Aleutian island arc magmas: Earth and Planetary Science Letters, v. 56, p. 167–179.

McKenzie, D., 1984, The generation and compaction of partially molten rock: Journal of Petrology, v. 25, p. 713–765.

Miyashiro, A., 1974, Volcanic rock series in island arcs and active continental margins: American Journal of Science, v. 274, p. 321–355.

Mogi, K., 1969, Relationship between the occurrence of great earthquakes and tectonic structures: Institute of the University of Tokyo Bulletin of Earthquake Research, v. 47, p. 429–451.

Monaghan, M. C., Klein, J., and Measures, C. I., 1988, The origin of ^{10}Be in island-arc volcanic rocks: Earth and Planetary Science Letters, v. 89, p. 288–298.

Morris, J., and Hart, S. R., 1983, Isotopic and incompatible element constraints on the genesis of island arc volcanics from Cold Bay and Amak Island, Aleutians, and implications for mantle structure: Geochimica et Cosmochimica Acta, v. 47, p. 2015–2030.

Morris, J., and 5 others, 1989, Sediment recycling at convergent margins; Constraints from the cosmogenic isotope ^{10}Be, *in* Hart, S. R. and Gülen, L., eds., Crust/mantle recycling at subduction zones: Dordrecht, Kluwer Academic Publishers, NATO Advanced Study Institutes Series C, Mathematical and Physical Sciences, v. 258, p. 81–88.

Myers, J. D., and Marsh, B. D., 1987, Aleutian lead isotopic data; Additional evidence for the evolution of lithospheric plumbing systems: Geochimica et Cosmochimica Acta, v. 51, p. 1833–1842.

Myers, J. D., Marsh, B. D., and Sinha, A. K., 1985, Strontium isotopic and selected trace element variations between two Aleutian volcanic centers (Adak and Atka); Implications for the development of arc volcanic plumbing systems: Contributions to Mineralogy and Petrology, v. 91, p. 221–234.

Myers, J. D., Frost, C. D., and Angevine, C. L., 1986a, A test of a quartz eclogite source for parental Aleutian magmas; A mass balance approach: Journal of Geology, v. 94, p. 811–828.

Myers, J. D., Marsh, B. D., and Sinha, A. K., 1986b, Geochemical and strontium isotopic characteristics of parental Aleutian arc magmas; Evidence from the basaltic lavas of Atka: Contributions to Mineralogy and Petrology, v. 94, p. 1–11.

Neuweld, M. A., 1987, The petrology and geochemistry of the Great Sitkin suite; Implications for the genesis of calc-alkaline magmas [M.S. thesis]: Ithaca, New York, Cornell University, 174 p.

Newman, S., Macdougall, J. D., and Finkel, R. C., 1984, $^{230}Th/^{238}U$ disequilibrium in island arcs; Evidence from the Aleutians and the Marianas: Nature, v. 308, p. 268–270.

Nye, C. J., and Reid, M. R., 1986, Geochemistry of primary and least fractionated lavas from Okmok volcano, central Aleutians; Implications for arc magmagenesis: Journal of Geophysical Research, v. 91, p. 10271–10287.

Perfit, M. R., 1977, The petrochemistry of igneous rocks from the Cayman Trench and the Captains Bay Pluton, Unalaska Island; Their relation to tectonic processes at plate margins [Ph.D. thesis]: New York, Columbia University, 273 p.

Perfit, M. R., and Kay, R. W., 1986, Comment *on* 'Isotopic and incompatible element constraints on the genesis of island arc volcanics from Cold Bay and Amak Island, Aleutians, and implications for mantle structure': Geochimica et Cosmochimica Acta, v. 50, p. 477–481.

Perfit, M. R., and Lawrence, J. R., 1979, Oxygen isotopic evidence for meteoric water interaction with the Captains Bay Pluton, Aleutian Islands: Earth and Planetary Science Letters, v. 45, p. 16–22.

Perfit, M. R., Brueckner, H., Lawrence, J. R., and Kay, R. W., 1980a, Trace element and isotopic variations in a zoned pluton and associated volcanic rocks, Unalaska Island, Alaska; Model for fractionation in the Aleutian calc-alkaline suite: Contributions to Mineralogy and Petrology, v. 73, p. 69–87.

Perfit, M. R., Gust, D. A., Bence, A. E., Arculus, R. J., and Taylor, S. R., 1980b, Chemical characteristics of island arc basalts: Implications for mantle sources: Chemical Geology, v. 30, p. 227–256.

Pope, R., 1983, The petrology of ultramafic and mafic xenoliths from Kanaga Island, the central Aleutians [M.S. thesis]: Ithaca, New York, Cornell University, 121 p.

Poreda, R., Craig, H., and Motyka, R., 1981, Helium isotope variations along the Alaskan-Aleutian arc: EOS Transactions of the American Geophysical Union, v. 62, p. 1092.

Romick, J. D., 1982, The igneous petrology and geochemistry of northern Akutan Island, Alaska [M.S. thesis]: Fairbanks, University of Alaska, 151 p.

Romick, J. D., Kay, S. M., and Kay, R. W., 1987, Amphibole fractionation and magma mixing in andesites and dacites from central Aleutians, Alaska: EOS Transactions of the American Geophysical Union, v. 68, p. 461.

Rubenstone, J. L., 1984, Geology and geochemistry of early Tertiary submarine volcanic rocks of the Aleutian Islands, and their bearing on the development of the Aleutian arc [Ph.D. thesis]: Ithaca, New York, Cornell University, 350 p.

Rutherford, M. J., and Devine, J., 1986, Experimental petrology of recent Mount St. Helens dacites; Amphibole, Fe-Ti oxides, and magma chamber conditions: Geological Society of America Abstracts with Programs, v. 18, p. 736.

Ryan, J., and Langmuir, C. H., 1987, The systematics of lithium abundances in young volcanic rocks: Geochimica et Cosmochimica Acta, v. 51, p. 1727–1741.

Sack, R. O., Carmichael, I.S.E., Rivers, M., and Ghiorso, M. S., 1980, Ferric-ferrous equilibria in natural silicate liquids at 1 bar: Contributions to Mineralogy and Petrology, v. 79, p. 169–186.

Saunders, A. D., Rogers, G., Marriner, G. R., Terrell, D. J., and Verma, S. P., 1987, Geochemistry of Cenozoic volcanic rocks, Baja California, Mexico; Implications for the petrogenesis of post-subduction magmas: Journal of Volcanology and Geothermal Research, v. 32, p. 233–245.

Scholl, D. W., Vallier, T. L., and Stevenson, A. J., 1987, Geologic evolution and petroleum geology of the Aleutian ridge, *in* Scholl, D. W., Grantz, A., and Vedder, J., eds., Geology and resource potential of the continental margin of western North America and adjacent ocean basins Beaufort Sea to Baja California: Circum-Pacific Council for Energy and Mineral Resources Earth Science Series, v. 6, p. 123–155.

Shelton, D. H., 1986, The geochemistry and petrogenesis of gabbroic rocks from Attu Island, Aleutian Islands, Alaska [M.S. thesis]: Ithaca, New York, Cornell University, 177 p.

Sun, S.-S., 1980, Lead isotope study of young volcanic rocks from mid-ocean ridges, oceanic islands, and island arcs: Philosophical Transactions of the Royal Society of London, Series A, v. 297, p. 409–445.

Swanson, S. E., Kay, S. M., Brearley, M., and Scarfe, C. M., 1987, Arc and back-arc xenoliths in Kurile-Kamchatka and western Alaska, *in* Nixon, P. H., ed., Mantle xenoliths: New York, John Wiley, p. 303–318.

Taylor, S. R., and McLennan, S. M., 1985, The continental crust; Its composition and evolution: Boston, Blackwell Scientific Publications, 312 p.

Tera, F., and 5 others, 1986, Sediment incorporation in island-arc magmas; Inferences from ^{10}Be: Geochimica et Cosmochimica Acta, v. 50, p. 535–550.

Turcotte, D. L., 1987, Physics of magma segregation processes: The Geochemical Society Special Publication 1, p. 69–74.

VonBargen, N., and Waff, H. S., 1986, Permeabilities, interfacial areas, and curvatures of partially molten systems; Results of numerical computations of equilibrium microstructures: Journal of Geophysical Research, v. 91, p. 9261–9276.

VonDrach, V., Marsh, B. D., and Wasserburg, J. G., 1986, Nd and Sr isotopes in the Aleutians; Multicomponent parenthood of island-arc magmas: Contributions to Mineralogy and Petrology, v. 92, p. 13–34.

Waff, H. S., 1986, Introduction to special section on partial melting phenomena in Earth and planetary evolution: Journal of Geophysical Research, v. 91, p. 9217–9222.

White, W. M., and Patchett, P. J., 1984, Hf-Nd-Sr and incompatible-element abundances in island arcs; Implications for magma origins and crust-mantle evolution: Earth and Planetary Science Letters, v. 67, p. 167–185.

Wolf, D. A., 1987, Identification of endmembers for magma mixing in Little Sitkin volcano, Alaska [M.S. thesis]: Albany, State University of New York, 210 p.

Wood, C., and Kienle, J., 1990, Volcanoes of North America: London, Cambridge University Press, 354 p.

Manuscript Accepted by the Society May 21, 1990

ACKNOWLEDGMENTS

Among many geological colleagues whose input has been indispensable, we acknowledge the work of many Cornell students over the last decade. They have spurred our own thinking and expanded the analytical data base that supports models of Aleutian magmatism. These students were G. Citron, W. Conrad, J. Rubenstone, J. Romick, G. Yogodzinski, R. Pope, D. Shelton, M. Neuweld, S. DeBari, G. Dilisio, L. Johnston, P. Kirk, L. Hardy and E. Fregeau. We also wish to thank former Cornell students who served as field assistants. They were J. Reynolds, T. Snedden, K. Dodd, S. Squyres, D. Coehlo, J. Lee, B. Hausback (UCLA student), and D. Gardner. Preceding this Cornell group were an equally indispensable group from Columbia, led by M. Perfit, and including A. Walker and Y. Chia. We thank D. White (USGS, Menlo Park, California) for the oxygen-isotope analyses. Cooperation and assistance from the USGS, US Navy, US Coast Guard, NOAA, US Fish and Wildlife Service, US Fish and Game Service, University of Washington Fisheries program, Reeve Aleutian Airways, and CIRES (University of Colorado) are gratefully acknowledged. Financial support has been provided by grants from NSF; student grants from GSA, AAPG, and NASA; and miscellaneous funds from Cornell University. A review by W. Hildreth (USGS) helped to improve the quality of the manuscript. The manuscript was written in 1987. This is INSTOC publication number 90.

NOTES ADDED IN PROOF

Our chapter, written in 1987, and slightly updated in 1989, was organized around three current controversies. What are the parental magmas in the Aleutians? What accounts for the trace element and isotope characteristics of the magmas? What are the underlying causes of petrologic and geochemical diversity of the magmas? The topics we covered did not include some found in Fournelle and others (this volume), which was written in 1991. The most important information in this update is a list of additional references. As the three controversies of 1987 continue at the time of this update (9/93), we will summarize our current views, which often contrast with those in Fournelle and others. We have not read their update.

Aleutian parental magmas

We recognized two types of parental (mantle-derived) magma types for which there was tangible evidence: a common basalt and an Mg-andesite. Alternatives for the origin of the basalt were: (1) melting of subducted mafic oceanic crust (with eclogite mineralogy) yielding high-Al basalt or (2) melting of peridotite of the mantle wedge, yielding high-Mg basalt. Our preference for the second alternative has been strengthened by recent experimental work (e.g., Rapp, 1990; Draper and Johnston, 1992). A critical result from the new experimental studies is the confirmation that at lower crustal conditions, high-Al basalt can result from fractionation of high-Mg basalt. For the overwhelming majority of experimental petrologists, arc basalt genesis by peridotite melting is now the choice, as it was for us in 1987.

The mechanics of the melting has been a topic of great interest, with a polybaric extraction of melt from a rising peridotite mass ("melting column") being popular (e.g., Plank and Langmuir, 1988). Our Figure 22 shows the "melting column." The along-arc variations within the mantle in a cross section like Figure 22 were described by us as "transverse rolls." Even given the popularity of models like that in Figure 22 (and alternatives), the problem of the thermal and flow structure of the subarc mantle remains unresolved. Progress is impeded by doctrinaire positions like that of Fournelle and others, who dismiss our model with the observation that our "mantle flow is opposite that of all geophysical models. . . . " This is not true, as evidenced by a careful reading of Davies and Stevenson (1992) and earlier references cited by them. Ultimately, tomographic work like that of Hasegawa and others (1991) in Japan will be needed to constrain any spatial patterns.

Despite its rarity, the Mg-andesite first reported in Kay (1978) has enjoyed a resurgent popularity due to the efforts of Defant and Drummond (1990). These authors coined the name "adakite" for the locality on Mt. Moffett, Adak Island, and concurred with our interpretation that it is a melt of the subducted oceanic crust. This remarkable magma, with the Nd, Pb, and Sr isotope ratios of MORB, is quite distinct from the magmas of the main Aleutian volcanoes, a fact not appreciated by Fournelle and others.

Finally, over the last 3 years we have investigated a third magma type, with chemical similarities toward boninites (with higher SiO_2 at the same Mg number compared to the Mg-basalt). We have concentrated on the locality at Piip volcano (Seliverstov and others, 1990; Volynets and others, 1993a; Yogodzinski and others, 1994). Piip is an active submarine volcano situated within a graben immediately behind the Komandorsky Islands, the westernmost islands of the Aleutian chain. The unique chemistry of the primary magma is attributed to shallow melting of hydrous peridotite. Recycled crustal components are much less abundant in these magmas than those of the Aleutian arc to the east. A subducted oceanic plate moving nearly parallel to the arc appears to lie under Piip (Boyd and Creager, 1991).

Trace element and isotopic characteristics

The nature of various components and their proportions in the mantle wedge is the most popular topic in arc magma genesis. In the Aleutians, a three-component model (slab igneous rock, sediment, depleted mantle) continues to be popular (e.g., Miller and others, 1992, which we look upon as an update to Kay, 1980, although this reference is not cited). The parallel between these two papers (which both contained models of Okmok basalt) extends to both using, as a component added to the mantle wedge, a melt of the basaltic part of the subducted slab. The weight of present-day opinion (except for the "adakite" mentioned above) comes down against this slab-melt component, in favor of a fluid component.

Mass balance calculations of ^{10}Be in arc magmas have spurred continued quantitative evolution of element inputs and outputs (recycling) in arcs, through correlations with elements like Ba (see our Fig. 18) and B (Morris and others, 1990). Sediment analyses relevant to the recycling problem (e.g., oceanic sediments south of the Aleutian trench) have been reported by Plank and Langmuir (1993) following the extensive analyses of Kay and Kay (1988).

Finally, the discovery of a geochemically distinct region in the westernmost Aleutian and Komandorski Islands (Yogodzinski and others, 1993) has caused a reassessment of common assumptions regarding the MORB-like and depleted mantle components. The extremely nonradiogenic Pb isotope ratios in "adakite" and boninitic magmas of the region strongly imply that mantle as well as crustal components vary along the arc.

Magmatic diversity

We recognized Aleutian magma series of two end member types: calcalkaline (CA) and tholeiitic (TH). We also recognized the full range of transitional types. We put great emphasis on the crustal environment as controlling the development of these magma series, as ultimately we related all magmas to mantle-derived parental olivine tholeiite. This was even true with CA plutons (Kay and others, 1990) although we held that the "relation" involved melting, assimilation, mixing, and fractional crystallization processes. Singer and others (1992) exploit the idea of extensional crustal environment (due to block rotations, see Geist and others, 1988) in their paper on the TH Seguam volcanic center. Romick and others (1992) show by trace element modelling of central Aleutian CA ashes that fractionation of a mid-crustal (amphibole-bearing) mineral assemblage was detectable even for magmas with upper crustal (amphibole-free) phenocryst populations. Romick and others (1992) found that some Aleutian dacites and their phenocrysts are quite similar to amphibole-bearing Mt. St. Helen's magmas. Not to see amphibole in Aleutian magmas is to practice selective vision.

Miller and others (1992) evaluate alternatives to tectonic controls of Aleutian magma series. They review the application of melt-peridotite interaction (e.g., Kelemen, 1990) in producing distinct olivine tholeiite parental magmas. They favor the view that "differences between CA and TH volcanoes originate in the slab, rather than being controlled by the stress regime in the crust." Specifically, the TH parent magmas are higher percentage partial melts associated with volatile-rich subducted oceanic crust of fracture zones. This contrasts with our view of the along-arc geochemical variability (see our Fig. 15) in which both CA and TH centers from a region have common parental magmas. The geochemical contrasts between the Islands of Four Mountains segment and the Andreanof segment are particularly obvious. Both these segments contain both CA and TH centers.

Finally, new data on Aleutian back-arc volcanism is reported by Volynets and others (1993b) for volcanic rocks dredged from seamounts north of the eastern Aleutian Islands. A cross-arc traverse that includes low-K magmas of the Islands of Four Mountains (see our Appendix on microfiche [back of the volume] and Singer and others, 1992) and high-K magmas erupted over deeper parts of the same subducted slab has been accomplished.

ADDITIONAL REFERENCES

Boyd, T. M., and Creager, K. C., 1991, The geometry of Aleutian subduction: Three-dimensional seismic imaging: Journal of Geophysical Research, v. 96, p. 2267–2291.

Brophy, J. G., 1989, Can high-alumina arc basalt be derived from low-alumina arc basalt? Evidence from Kanaga Island, Aleutian Arc, Alaska: Geology, v. 17, p. 333–336.

—— , 1991, Composition gaps, critical crystallinity, and fractional crystallization in orogenic (calc-alkaline) magmatic systems: Contributions to Mineralogy and Petrology, v. 109, p. 173–182.

Conrad, W. K., Kay, S. M., and Kay, R. W., 1983, Magma mixing in the Aleutian Arc: Evidence from cognate inclusions and composite xenoliths: Journal of Volcanology and Geothermal Research, v. 18, p. 279–295.

Creager, K. C., and Boyd, T. M., 1991, The geometry of Aleutian subduction: Three dimensional kinematic flow model: Journal of Geophysical Research, v. 96, p. 2293–2307.

Davies, J. H., and Stevenson, D. J., 1992, Physical model of source region of subduction zone volcanics: Journal of Geophysical Research, v. 97, p. 2037–2070.

Defant, M. J., and Drummond, M. S., 1990, Derivation of some modern arc magmas by melting of young subducted lithosphere: Nature, v. 347, p. 662–665.

Draper, D. S., and Johnston, A. D., 1992, Anhydrous PT phase relations of an Aleutian high-MgO basalt: An investigation of the role of olivine–liquid reaction in the generation of arc high-alumina basalts: Contributions to Mineralogy and Petrology, v. 112, p. 501–519.

Geist, E. L., Childs, J. R., and Scholl, D. W., 1988, The origin of summit basins of the Aleutian Ridge: Implications for block rotation of an arc massif: Tectonics, v. 7, p. 327–341.

Hasegawa, A., Zhao, D., Hori, S., Yamamoto, A., and Horiuchi, S., 1991, Deep structure of the northeastern Japan arc and its relationship to seismic and volcanic activity; Nature, v. 352, p. 683–689.

Helfrich, G. R., Stein, S., and Wood, B. J., 1989, Subduction zone thermal structure and mineralogy and their relationship to seismic wave reflections and conversions at the slab/mantle interface: Journal of Geophysical Research, v. 94, p. 753–763.

Kay, R. W., and Kay, S. M., 1991, Creation and destruction of lower continental crust: Geologische Rundschau, v. 80, p. 259–278.

Kay, S. M., Kay, R. W., Citron, G. P., and Perfit, M. R., 1990, Calc-alkaline plutonism in the intra-oceanic Aleutian arc, Alaska: Geological Society of America Special Paper 241, p. 233–255.

Kelemen, P. B., 1990, Reaction between ultramafic rock and fractionating basaltic magma I. Phase relations, the origin of calc-alkaline magma series, and the formation of discordant dunite, Journal of Petrology, v. 31, p. 51–98.

Miller, D. M., Langmuir, C. H., Goldstein, S. L., and Franks, A. L., 1992, The importance of parental magma composition to calc-alkaline and tholeiitic evolution: Evidence from Umnak Island in the Aleutians: Journal of Geophysical Research, v. 97, p. 321–343.

Morris, J. D., Leeman, W. P., and Tera, F., 1990, The subducted component in island arc lavas: Constraints from Be isotopes and B-Be systematics: Nature, v. 344, p. 31–35.

Plank, T., and Langmuir, C. H., 1988, An evaluation of the global variations in the major element chemistry of arc basalts: Earth and Planetary Science Letters, v. 90, p. 349–370.

—— , 1993, Tracing trace elements from sediment input to volcanic output at subduction zones: Nature, v. 362, p. 739–742.

Rapp, R. P., 1990, Vapor-absent partial melting of amphibolite/eclogite at 8–32 kbar: Implications for the origin and growth of the continental crust [Ph.D. Thesis]: Troy, N.Y., Rensselaer Polytechnic Institute.

Romick, J. D., Perfit, M. P., Swanson, S. E., and Shuster, R. D., 1990, Magmatism in the eastern Aleutian Arc: Temporal characteristics of igneous activity of Akutan Island: Contributions to Mineralogy and Petrology, v. 104, p. 700–721.

Romick, J. D., Kay, S. M., and Kay, R. W., 1992, The influence of amphibole fractionation on the evolution of calc-alkaline andesite and dacite tephra from the Central Aleutians, Alaska: Contributions to Mineralogy and Petrology, v. 112, p. 101–118.

Seliverstov, N. I., Avdieko, G. P., Ivanenko, A. N., Shrira, V. A., and Khubunaya, S. A., 1990, A new submarine volcano in the west of the Aleutian Island arc: Volcanology and Seismology, v. 8, p. 473–495.

Singer, B. S., and Myers, J. D., 1992, Intra-arc extension and magmatic evolution in the central Aleutian arc, Alaska: Geology, v. 18, p. 1050–1053.

Singer, B. S., Myers, J. D., and Frost, C. D., 1992, Mid-Pleistocene basalt from Seguam Volcanic Center, Central Aleutian Arc, Alaska: Local lithospheric structures and source variability in the Aleutian Arc: Journal of Geophysical Research, v. 97, p. 4561–4578.

Tsvetkov, A. A., 1991, Magmatism of the westernmost (Komandorsky) segment of the Aleutian Island Arc: Tectonophysics, v. 199, p. 289–317.

Volynets, O. N., Koloskov, A. V., Yogodzinski, G. M., Seliverstov, N. I., Igorov, Y. O., and Shkira, V. A., 1993a, Boninitic tendencies in lavas of the submarine Piip Volcano and surrounding area (far Western Aleutians): Geology, petrochemistry, and mineralogy: Volcanology and Seismology, v. 14, p. 1–22.

Volynets, O. N., and 6 others, 1993b, New data on volcanism in the rear zone of the eastern Aleutian arc (in Russian): Volcanology and Seismology, v. 4, p. 54–78.

Yogodzinski, G. M., Rubenstone, J. L., Kay, S. M., and Kay, R. W., 1993, Magmatic and tectonic development of the Western Aleutians: An oceanic arc in a strike-slip setting: Journal of Geophysical Research, v. 98, p. 11,807–11,834.

Yogodzinski, G. M., Volynets, O. N., Koloskov, A. V., and Seliverstov, N. I., 1994, Magnesian andesites and the subduction component in a strongly calcalkaline series at Piip Volcano in the far Western Aleutian arc: Journal of Petrology, v. 34 (in press).

Printed in U.S.A.

The Geology of North America
Vol. G-1, The Geology of Alaska
The Geological Society of America, 1994

Chapter 23

Age, character, and significance of Aleutian arc volcanism

John H. Fournelle* and Bruce D. Marsh
Department of Earth and Planetary Sciences, The Johns Hopkins University, Baltimore, Maryland 21218
James D. Myers
Department of Geology and Geophysics, University of Wyoming, Laramie, Wyoming 82071

INTRODUCTION

The Aleutian volcanic arc stretches nearly 3,000 km, from the Commander Islands off Kamchatka (U.S.S.R.), along the southern Bering Sea margin, and across the continental margin onto the Alaskan landmass. Intimately associated with the arc is the subduction of the Pacific Plate beneath the North American Plate. To the east, subduction is nearly orthogonal but becomes increasingly oblique westward. Near Buldir Island, motion between the two plates becomes strike-slip and volcanism ceases. The Aleutian volcanic front, which has in many places remained nearly fixed for at least several tens of millions of years, contains about 80 major volcanic vents, half of which have been historically active. These vents have yielded a spectrum of rock types, from basalt through andesite to dacite and rhyolite. This diversity of rock types is present throughout the history of nearly all the volcanic centers, almost regardless of volcano size and age.

Several features of the Aleutian volcanic arc—a long history of fixed volcanism extending from continental to oceanic crust, the focusing of large amounts of thermal energy on small areas of crust for long periods of time, the pattern of changing convergence, as well as the diversity of rock types—present an excellent opportunity to study the connection between global tectonics, magmatism, and continent evolution. The study of Aleutian volcanism can shed light both on deep-seated magmatic processes and on the interplay of the chemistry and physics of magma evolution, and particularly on the near-surface behavior of magma in various local tectonic and thermal regimes.

This chapter focuses primarily on the latest episode (the last 1 to 2 m.y.) of subaerial volcanism of the oceanic segment of the Aleutian arc, that is, the western and central parts of the volcanic arc, with some discussion of relevant features of volcanism on the Alaska Peninsula. Locations, physical characteristics, and historic eruptions of these volcanoes are given in Table 1.

*Present address: Department of Geology and Geophysics, University of Wisconsin, 1215 W. Dayton Street, Madison, Wisconsin, 53706-1692.

HISTORY OF GEOLOGICAL EXPLORATION

The harsh climate of the Aleutian Islands and the treacherous nature of the surrounding sea have made exploration there difficult, even today. In the 20 years following the discovery voyage (1741) by Bering and Chirikof, promyshleniki (frontiersmen) found every Aleutian island in their relentless harvest of sea otters and foxes. Beginning in 1760 with the Russian occupation and the appearance of missionaries, there were written accounts of volcanic eruptions (e.g., Veniaminof, 1840). Black (1981) chronicles some Aleut oral accounts of volcanic activity.

T. Jaggar visited Bogoslof Island one year after it erupted in 1906 (Jagger, 1908). He also visited Atka, Umnak, Unalaska, and Akutan Islands, as well as Pavlof volcano on the Alaska Peninsula. In the 1930s, R. H. Finch of the Hawaiian Volcano Observatory explored some of the eastern volcanoes (1934, Shishaldin; 1935, Akutan). At about the same time, Bernard Hubbard, the so-called "Glacier Priest", ascended Shishaldin and Veniaminof Volcanoes and explored Katmai and the giant Aniakchak Caldera (Hubbard, 1932a, 1932b, 1936).

The most comprehensive geologic field program in the Aleutians was that of the U.S. Geological Survey between 1945 and 1954. From this project, geologic reports were published on 13 of the central and western Aleutian Islands, and on two volcanoes of the Alaska Peninsula. These reconnaissance studies were published as *U.S. Geological Survey Bulletin 1028.* Much of our knowledge of Aleutian geology stems from that project. The most comprehensive study of any single volcanic event in the entire arc has been of the 1912 Katmai eruption (e.g., Griggs, 1922; Fenner, 1923, 1926; Curtis, 1968; Hildreth, 1983, 1987; an issue of Geophysical Research Letters, August 1991; Miller and Richter, this volume).

With the recognition of the relation between subduction and volcanism at convergent margins, the Aleutian island arc became the focus of renewed interest in the 1970's. Since then, six previously unmapped volcanic centers have been investigated. These centers include Amak Island (Marsh and Leitz, 1979), northern Akutan Island (Romick, 1982; Romick and others, 1990), northern Atka Island (Marsh, 1980), Seguam Island (Myers and Sing-

Fournelle, J. H., Marsh, B. D., and Myers, J. D., 1994, Age, character, and significance of Aleutian arc volcanism, *in* Plafker, G., and Berg, H. C., eds., The Geology of Alaska: Boulder, Colorado, Geological Society of America, The Geology of North America, v. G-1.

TABLE 1: LOCATIONS, PHYSICAL CHARACTERISTICS, AND DATES OF HISTORIC ERUPTIONS OF ALEUTIAN VOLCANOES WEST OF 162°W*

Island/Center	Position	Elevation (m)	Dimensions (km)	Historic Activity
Amak Island	163.15W, 55.42N	513	5.0 x 4.2	1700-10
Cold Bay Volc. Complex			31.3 x 23.4	
Frosty Peak	162.77W, 55.08N	1920	17.4	Holocene
Mt. Simeon	162.75W, 55.19N	287	3.0 x 2.0	
N. Walrus Peak	162.81W, 54.98N	893		
S. Walrus Peak	162.18W, 54.98N	892	15.9 x 10.9	
Unimak Island			110.7 x 56.5	
Roundtop	163.60W, 54.80N	1871	7.5	Holocene
Isanotski Peaks	163.73W, 54.75N	2446	11.2	1795, 1825, 1830, 1831, 1845
Shishaldin Volc.	163.97W, 54.75N	2857	17.4	1775-78,1790,1819,1824,1825,1826,1827-29,1830-31,
				1838,1842,1865,1880-81,1883,1897,1898,1899,1901,
				1912,1922,1925,1928,1929,1929-32,1932,1946,1946-47,
				1948,1951,1953,1955,1963,1967,1975,1976,1978, 1979
Fisher Caldera			17.4 x 11.2	1825,1826-27,1830
Mt. Finch	164.36W, 54.67N	478	2.8	
Eickelberg Peak	164.46W, 54.68N	1114	2.8	
Westdahl-Pogromni Volc. Complex			22.5 x 18.8	
Westdahl Peak	164.62W, 54.52N	1560		1964, 1978, 1979
Faris Peak	164.66W, 54.52N	1654		
Pogromni	164.67W, 54.57N	2002		1795, 1796, 1820, 1827-30
Akun Island			21.4 x 17.7	
Mt. Gilbert	165.65W, 54.25N	818		ca. 1834 (fumarole)
Akutan Island	166.00W, 54.13N	1303	29.9 x 21.6	1790,1828,1838,1845,1848,1852,1865,1867, 1883,
				1887,1892,1896,1907,1908,1911,1912,1928,1929,
				1931,1946-47,1948,1951,1953,1972-73,
				1974,1976-77, 1978,1980
Unalaska Island			128.9 x 53.5	
Makushin Volc.	166.93W, 53.90N	2036	38.6 x 17.4	1768-69,1790-92,1802,1818,1826-38,1844,1865,
				1867, 1883,1907,1912,1926,1938,1951,1852, 1980
Wide Bay Cone	166.60W, 53.97N	640		
Table Top Mtn.	166.66W, 53.97N	800		
Sugarloaf	166.73W, 53.94N	611		
Pakushin Cone	166.95W, 53.83N	1035		
Bogoslof Island	169.03W, 53.93N	46	1.5 x 0.5	1796,1804,1814,1820,1884,1890,1891,1906,1907,1909,
				1910.1913.1926. 1917.1931.1951
Umnak Island			118.7 x 27.9	
Okmok Caldera	168.13W, 53.42N	1072	44.8 x 27.4	1805,1817-20,1824-29,1830,1878,1899,1931,1936,
				1938, 1943, 1945, 1983; ca 1986 (Cone A)
Mt. Idak		585		
Jag Peak		900		
Mt. Tulik	168.05W, 53.37N	877		
Mt. Recheshnoi	168.55W, 63.15N	1984	10.0	Holocene
Mt. Vsevidof	168.68W, 53.13N	2076	12.4	1784, 1790, 1817, 1830, 1878, 1880, 1957
Kagamil Island	169.72W, 52.98N	893	10.2 x 6.2	1929
Uliaga Island	169.77W, 53.07N	888	3.7	Holocene
Chuginadak Island			23.4 x 13.9	
eastern	169.75W, 52.80N	1170	14.2 x 13.9	
Mt. Cleveland	169.95W, 52.82N	1730	8.7 x 8.2	1893,1897,1929,1932,1938,1944, 1951,1975, 1987
Carlisle Island	170.05W, 52.90N	1620	8.5 x 7.2	1774, 1828, 1838, 1987
Herbert Island	170.12W, 52.75N	1290	9.2 x 8.5	Holocene

TABLE 1: LOCATIONS, PHYSICAL CHARACTERISTICS, AND DATES OF HISTORIC ERUPTIONS OF ALEUTIAN VOLCANOES WEST OF 162°W* (continued)

Island/Center	Position	Elevation (m)	Dimensions (km)	Historic Activity
Yunaska Island			22.9 x 12.2	
northeast	170.70W, 52.63N	457	12.7 x 11.9	1817, 1824, 1825, 1830, 1873, 1920s?,1937
southwest	170.76W, 52.56N	915	11.9 x 7.0	
Chagulak Island	171.13W, 52.57N	1141	3.2 x 3.2	Holocene
Amukta Island	171.25W, 52.50N	1064	9.5 x 8.0	1786-91, 1876, 1963, 1987
Seguam Island				1786-90, 1827, 1891, 1892, 1902, 1927
· Pyre Peak	172.52W, 52.31N	1054		1977
eastern caldera	172.37W, 52.33N	847		
eastern volcano	172.33W, 52.35N	587		
western center	172.59W, 52.27N	713		
seafloor, N of Amlia	173.50W, 52.00N			1966-67 (submarine: hydrophonic)
Atka Island			98.3 x 34.6	
Korovin	174.15W, 52.38N	1533		1829, 1830, 1844, 1907, 1951, 1977
Klicheuf	174.14W, 52.33N	1451		
Sarichef	174.03W, 52.38N	1056		1812
Konia	174.13W, 52.36N	1125		
Koniuji Island	175.13W, 52.22N	266	1.2	
Kasatochi Island	175.50W, 52.18N	314	2.7 x 2.2	1760, 1827, 1828, 1899?
Great Sitkin Island	176.13W, 52.07N	1740	17.7 x 16.7	1760,1792,1829,1904,1933,1945,1949-50,1950, 1953, 1974
Adak Island			52.2 x 39.1	
Mt. Moffett	176.86W, 51.99N	1196	10.0 x 10.0	
Mt. Adagdak	176.60W, 51.94N	621	4.0 x 3.5	
Kanaga Island			49.8 x 12.4	
Kanaga Volcano	177.15W, 51.93N	1307	10.4 x 6.2	1768,1783-87,1790,1791,1827,1829,1904,1906, 1933, 1942
Bobrof Island	177.43W, 51.90N	738	3.7 x 2.2	
Tanaga Island			42.0 x 37.8	
Gash Bay	177.95W, 51.89N	650		
Takawagha	178.02W, 51.87N	1448	13.7 x 8.7	Holocene
Tanaga	178.15W, 51.88N	1777	5.8 x 5.0	1763-70, 1791, 1829, 1914
Sajaka	178.20W, 51.78N	1304	2.7	
Gareloi Island	178.80W, 51.80N	1573	10.7 x 9.2	1760,1790,1791,1792,1828-29,1873,1922,1929, 1930, 1950-51, 1952, 1980
Semisopochnoi Is.			20.6 x 17.4	
Ragged Top	179.67E, 51.82N	904	3.2	
Anvil Peak	179.60E, 51.99N	1223	3.7	
Mt. Cerberus	179.59E, 51.93N	774	3.7	1772, 1790, 1792, 1828, 1830, 1873, 1922, 1929, 1930
Sugarloaf	179.63E, 51.90N	856	3.5	
Perret Ridge	179.66E, 51.97N	889	5.0	
Little Sitkin Island	178.53E, 51.95N	1202	10.7 x 10.0	1776, 1828
Davidof Island	178.33E, 51.97N	1159	3.5 x 1.2	Holocene
Khvostof Island	178.29E, 51.98N	260	2.7 x 1.5	
Segula Island	178.13E, 52.02N	1140	7.7 x 7.0	Holocene
Kiska Island			41.0 x 15.7	
Kiska Volcano	177.60E, 51.10N	1220	9.0 x 8.2	1907, 1927, 1962-64, 1969
Buldir Island	175.92E, 52.35N	656	7.2 x 4.2	Holocene

***Dimensions are of either island, volcanic center/complex, or volcano; single dimension is diameter, otherwise maximum x minimum dimension.**

er, 1987; Singer and others, 1991), Tanaga (Coats and Marsh, 1984), and Shishaldin Volcano on Unimak Island (Fournelle, 1988; Fournelle and Marsh, 1991). Additional mapping of Makushin Volcano on Unalaska Island (Nye and others, 1986) has also materially improved our geologic knowledge of this large and important volcanic center. Initial geologic and geochemical results from Fisher caldera have been reported (Fournelle, 1990b). Sampling for geochemical studies has also provided additional data for some previously mapped centers. This work includes that on northern Kanaga by Brophy (1990), on northern Adak by R. W. and S. M. Kay and coworkers (Kay and others, 1978, 1982; Kay and Kay, this volume) as well as Marsh (1976), and at Okmok Caldera on Umnak Island by Nye (1983) and Nye and Reid (1986).

HISTORY OF VOLCANISM

Episodes of volcanism and plutonism

Aleutian arc volcanism is not continuous, but episodic. Recent volcanism is recorded by ash layers in the adjacent sea-floor sediment (Hein and others, 1978; Scheidegger and others, 1980), as well as tephra layers in coal beds on the Kenai Peninsula (Reinink-Smith, 1990a). Hein and others (1978) showed that Aleutian volcanism has waxed every 2.5 m.y. for the past 10 m.y., and about every 5 m.y. for the previous 10 to 20 m.y. The present episode of strong volcanism began about 1 Ma, and an equally strong episode occurred about 7.5 Ma and perhaps another near 16 Ma. The chemistry of deep-sea ash samples (see section on Chemistry and Petrography) is skewed toward andesitic-dacitic compositions, so it is likely that basaltic volcanic activity is under represented.

The Kenai coal tephra layers preserve a more detailed (but less complete) eruptive record for the Alaska Peninsula segment of the arc. Reinink-Smith (1990a), using mineral radiometric dates and estimated peat accumulation rates, found a strong episode of volcanism more than 10.5 m.y. ago, with eruptions every 125 to 500 yr. A reduced level of activity (one eruption every 9,000 yr) occurred between 10.5 and 7.5 Ma, with a pulse at 7.5 Ma. Volcanic activity dramatically increased about 5 Ma, with events every 1,700 to 2,400 yr.

These magmatic episodes are also apparent in the subaerial geology, especially in the central Aleutian Islands, where erosion has been deep enough to expose plutons within the volcanic pile. On Adak, the initial products of the present phase of volcanism occur as erosional residuals of fairly fresh lava flows and domes respectively near Heart Lake and along Kuluk Bay (Coats, 1956). These units are ~3 to 3.5 m.y. old (Marsh, 1976), and, in contrast to younger volcanic flows, commonly contain small amounts of carbonate and zeolite. On Atka, quartz diorite (8 to 9 Ma) crops out at the headland between Crescent Bay and Bechevin Bay, whereas basaltic lava flows (~11 Ma) occur at Martin Harbor and Bechevin Bay (Hein and others, 1984). Plutons near Sergief Bay as well as on Kugalaska have ages of 12 to 14 Ma. Vallier and others (this volume) summarize the setting of these and other plutons and consider the rocks into which they intrude.

This long history of volcanism has produced three main sequences of sedimentary-volcanogenic rocks, which have been most extensively studied and defined on Adak, Atka, and Amlia Islands. The Finger Bay volcanics, defined by Coats (1947), represent the lower series (~55 Ma) on Adak. The middle series (~35 to 10 Ma) apparently is not common in the central arc, and the upper series (<10 Ma) represents the more recent volcanism.

Sedimentary and igneous rocks of the entire arc result from volcanic episodes that have yielded plutons, lavas, pyroclastics, and volcaniclastics. Former emplacement and depositional environments were similar to today's, having local, quiet, and deep basins in the oceanic sector and shallow, swampy tidal flats in the continental sector.

Historic volcanic activity

Recorded Aleutian volcanic activity dates from the beginning of the eighteenth century, according to sparse written accounts (Grewingk, 1850; Dall, 1870; Veniaminov, 1840; Becker, 1898). Dall (1918), various U.S. Geological Survey geologists (in Coats, 1950), and the *Volcano Letter* of the Hawaiian Volcano Observatory (e.g., Jagger, 1932) also provide fragmentary information. The historical record, however, is far from complete, owing to a small native population concentrated in a few villages on a handful of islands. Dumond (1979) and Black (1981) have described the effects of volcanism on the indigenous peoples of the Aleutian Islands and Alaska Peninsula.

Many volcanoes are obscured by cloud cover for much of the year, hindering observation. Today most information about eruptions comes from aircraft pilots and satellite observations (e.g., McClelland and others, 1989). Much of the uncertainty in the early reports of volcanic activity stems from observations that simply record "smoke," some of which could be summit clouds. Although reporting of volcanic activity in the arc has materially improved in recent decades, a significant amount of volcanic activity may still go unreported.

Coats (1950), Simkin and others (1981), and McClelland and others (1989) document the record of volcanic eruptions over the past 300 years. Reported eruptions since 1700 are shown in Figure 1. The most active volcanoes have been Pavlof (42 eruptions), Shishaldin (33), and Akutan (30). The most active volcanoes are those on the western end of the Alaska Peninsula and the easternmost islands of the Aleutian arc.

DISTRIBUTION OF VOLCANISM

Volcanic front: Segmentation and spacing

The most distinctive spatial feature of present Aleutian volcanism is that "In detail, the volcanic line does not form a perfectly simple arc, but consists of segments of different lengths . . ." (Coats, 1950, p. 45). Coats recognized that arc vol-

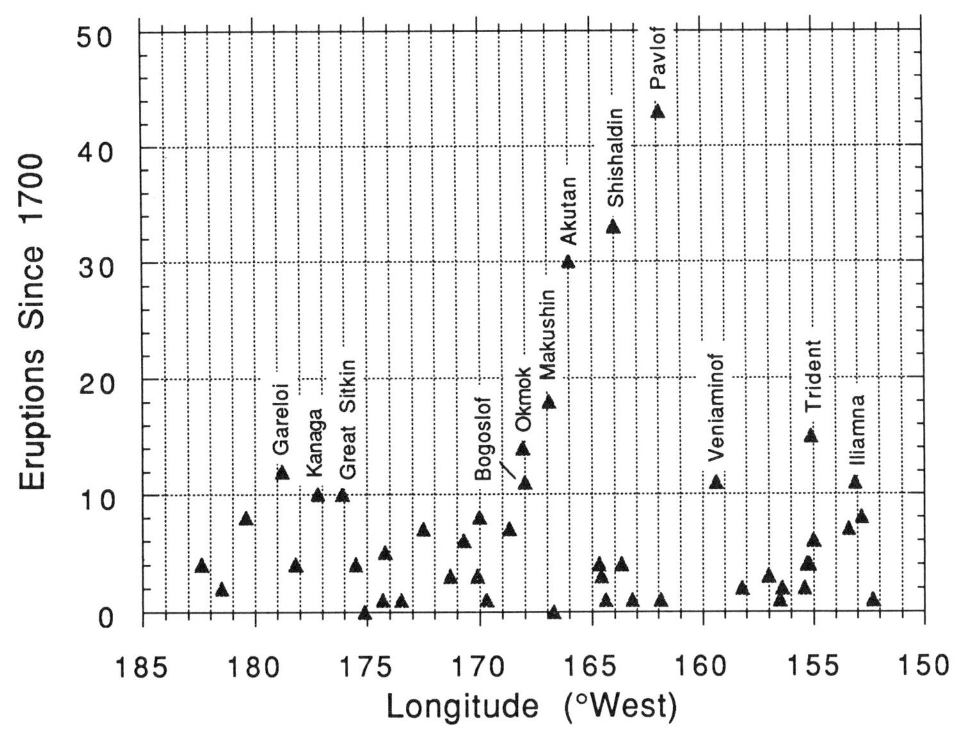

Figure 1. Plot of reported eruptions of Aleutian volcanoes since 1700 versus longitude.

canism produced a linear, segmented volcanic front. Marsh (1979a,b) used the active vents (i.e., summit craters) to more precisely define this segmentation, which is impressive when viewed from a volcanic summit, and holds best for the youngest volcanoes. Bogoslof Island, 35 km north of Umnak, and Amak Island, 50 km northwest of Cold Bay, are the only two places where subaerial volcanoes occur significantly off the volcanic front; they suggest a weak secondary volcanic front (Marsh, 1979a,b).

The intensity of the volcanic front, as measured by the magmatic flux (i.e., estimated subaerial volcano volume) over the most recent episode of volcanism, decreases strongly from east to west (Fig. 2). Also, the most active volcanoes over the past 300 years are located where convergence is the greatest (Fig. 1). The westernmost sign of recent subaerial volcanism is tiny Buldir Island near 176°E. This diminution of volcanism coincides with the systematic westward reduction in the rate of subduction of the Pacific Plate. The subduction rate in the far eastern part of the arc (152°W) is about 6.4 cm/yr (Minster and others, 1974), increases to about 7.1 cm/yr at 165°W, and then decreases continuously westward until 176°E, where the relative plate motion becomes strike-slip with no component of underthrusting.

The curvature of the arc is not the same as that of the earth itself, and so the subducting Pacific Plate is segmented into platelets or locally strongly bent. This deformation is indicated by variations in the record of seismicity (e.g., Spence, 1977; Kienle and Swanson, 1983). Major irregularities and breaks in the subducting plate mainly occur along original fracture zones (e.g., Amlia and Adak fracture zones) formed at a spreading center. The distinct change in the volcanic front east of Katmai (near Douglas at 153.5°W) correlates with a bend or kink in the subducting plate. There also may be some segmentation of the arc plate (e.g., Geist and others, 1988).

The position at depth of the subducting plate is fairly well known (e.g., Engdahl, 1973; Fujita and others, 1981; Plafker and others, Plate 12, this volume). As noted originally by Coats (1962) in his seminal introduction of the concept of subduction, the volcanic front is ~110 km above the seismic zone. The dip of the plate at these depths is generally about 45°, but at the shallowest part the dip decreases systematically eastward as the arc-trench gap widens; Jacob and others (1977) suggest that the Aleutian Trench in the Gulf of Alaska has migrated ~200 km seaward due to sediment accretion over the last 15 m.y. or so.

In the Aleutian oceanic sector, the long episodic history of Tertiary volcanism has produced extensive volcaniclastic deposits. These form the Aleutian Ridge, a large submarine basement structure. Most of the present landmass is thus related to much earlier phases of arc activity; the origin of the ridge itself is discussed in Vallier and others (this volume). An allochthonous terrane, the Umnak Plateau, forms the basement from west of Unimak Island to the Islands of Four Mountains (Howell and others, 1985).

The Alaska Peninsula has more complex basement of Early Permian and Mesozoic age, which is part of the Peninsular Terrane of Silberling and others (this volume). Wilson and others (1985) recognize two subterranes based on the age of the

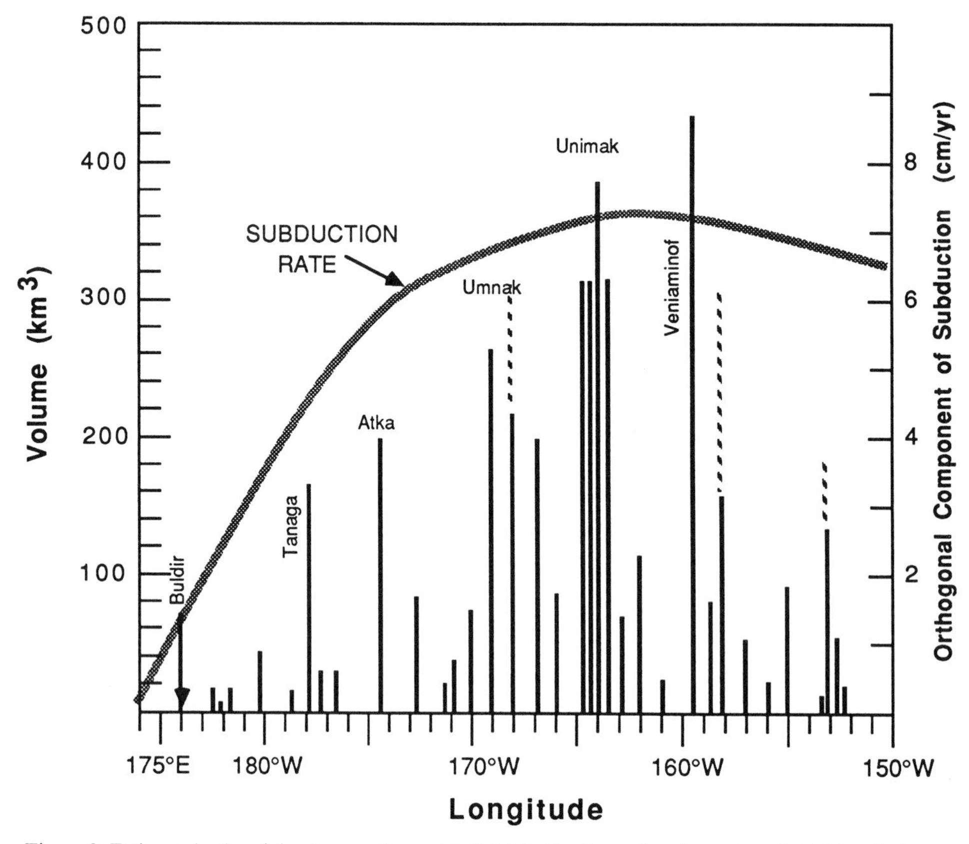

Figure 2. Estimated subaerial volumes of recent (≤2 Ma) Aleutian volcanic centers plotted in relation to their position along the arc. Volumes are calculated based upon the subaerial volcanic structures; they represent minimal volumes of extrusive material and are given to show the relative differences in sizes of volcanic centers. Dashed lines indicate uncertainties in calculating volumes. Also plotted is the orthogonal component of the rate of subduction of the North Pacific Plate along the arc (based on velocities of Minster and others, 1974). Note the correlation between volcano size (volume) and rate of subduction. (From Marsh, 1987.)

overlying Cenozoic volcanic rocks. The oldest subterrane occurs in the central Alaska Peninsula, formed during Eocene to early Miocene (48 to 22 Ma) volcanism (the Meshik arc of Wilson, 1985), is subparallel to the current one, and consists of basaltic to dacitic volcanic and intrusive rocks and intercalated volcaniclastic deposits. These rocks form the "transitional" crust upon which the presently active volcanic arc southwest of 157°W is built.

The present volcanic front in most instances coincides with earlier fronts. Most active centers in the volcanic front rest on the broad basement structure of the Aleutian Ridge, and Bogoslof and Amak Islands of the secondary front lie northward of this structure. Semisopochnoi Island appears to be an exception, where no older island-arc deposits exist.

Judging from the locations of plutons and the eroded remnants of earlier volcanoes, the volcanic front has moved systematically northward over the past 5 m.y. This migration is greatest in the centtral and western islands and lessens eastward. Midway up the Alaska Peninsula little, if any, migration is apparent. On Kiska, Tanaga, Kanaga, Adak, and Atka, volcanism currently occurs on the northernmost edges of the islands. The migration has not always been northward for, judging from the positions of the oldest plutons, volcanism may have first shifted slightly southward and then northward over the past 15 to 20 m.y. In some areas, like Tanaga, Seguam, and eastern Unimak, the migration may have been from east to west, parallel to the volcanic front. The magnitude of the northward motion on Adak and at Cold Bay may have been about 8 to 9 km over the past 3 to 5 m.y. Some of the largest volcanic centers (Okmok on Umnak and Veniaminof, Aniakchak, and others on the peninsula) and perhaps the Islands of Four Mountains may have not moved much during the past 1 m.y.

The continental sector (east of about 163°W) consists more commonly of large single stratocones, whereas in the oceanic sector the volcanic front is defined mainly by clusters of volcanoes, with each cluster defining a volcanic center. Regardless of size, active volcanic centers are usually spaced at intervals of about 60 to 70 km throughout the arc (Marsh and Carmichael, 1974), although the Katmai group of volcanoes appears to be an exception, having ~13 km intragroup spacing (Kienle and Swanson, 1983). Marsh (1979b) noted a rough correlation between spacing and volume, with the largest volcanic centers having a somewhat regular spacing. Some of these centers are presumably older, and younger centers have subdivided the original intervals. This process continues, as exemplified by the small and very young volcanoes of Bobrof between Tanaga and Kanaga, and by Kasatochi and Koniuji between Great Sitkin and Atka.

The segmentation of the front interrupts the regional spacing. Spacing within each segment may be independent of the next segment, making the intervals at the ends of segments irregular, for example, in the Islands of Four Mountains, near Douglas and Fourpeaked (~154°W), and the large interval between Atka and Seguam. There is no correlation between estimated subaerial volume and segmentation. Slightly more than half of the large centers lie within, rather than at the end, of their respective segments.

Tectonics and volcanism

S. M. Kay and others (1982) and Kay and Kay (this volume) proposed an alternative model of arc segmentation, in which local tectonic environment is the dominant control on magma chemistry. Their model (1) segmented the arc on a coarser basis, using islands rather than summit vents as the data points to be connected; (2) used the pattern of recent seismicity and tectonic (fracture zone) boundaries; and (3) gave special attention to 1957 and 1965 earthquake aftershock zones. The basis for assuming that these aftershock zones have been fixed throughout the current 1 to 2 m.y. of volcanism, however, is unclear. Based on these criteria, they divided the arc into four major tectonic blocks. They suggested that the size and composition of a volcanic center is determined by its position within a block. At the edges of blocks (extensional stress regimes), magmas ascend rapidly, experience little heat loss, and fractionate mostly at shallow levels producing large, tholeiitic centers. In the central parts of blocks (compressive stress regimes), magma ascent is inhibited, so that magmas ascend slowly, lose large amounts of heat rapidly, and fractionate at depth. By this model, small, calc-alkaline centers dominated by andesite are produced.

An alternative model for the generation of these two types of centers was suggested by Myers and others (1985). Based on Sr isotopic data from Atka and Adak, these authors suggested that the volumetric and geochemical differences (i.e., tholeiitic versus calc-alkaline) betwen the two volcanic centers could be explained by variations in the degree of development of the lithospheric magmatic conduits supplying each center.

The Aleutian Ridge in the central and western arc is fragmented into several structural blocks (Geist and others, 1987, 1988). At approximately 5 Ma, clockwise rotation of these blocks began, producing extended portions of the arc crust with summit basins. The modern volcanic centers of Kanaga, Adak, and Great Sitkin developed on unextended arc crust and Atka on incipiently extended crust, whereas Seguam grew atop strongly extended crust. Singer and Myers (1990) suggest that these contrasting crustal conditions may have had an effect on the different magmatic evolutionary histories of these volcanic centers.

EVOLUTION OF VOLCANIC CENTERS

Larger centers

Oceanic sector. Major volcanic centers along the arc show a remarkably similar sequence of morphologic development that varies significantly between the oceanic and continental parts (see Wood and Kienle, 1990, for brief descriptions with maps/photos of each Aleutian volcano). Myers and others (1985) classified Aleutian volcanic centers by their sizes (volumes), with large centers greater than 130 km^3 and small with less than 10 km^3.

In the oceanic sector, the lowest exposed subaerial lavas are mainly basaltic and form thin (3 to 5 m) flows in a gently dipping stack. These lavas are interspersed with coarse tephra and distributed around a central vent. The number of flows, which can vary from two or three to more than 20, are proportional to the size of the volcano. With continued eruption, the proportion of pyroclastic materials, especially lahars, increases greatly until only an occasional lava flow is found. Many units begin at the central vent, which is in most instances the locus of all eruptions. These flows soon encounter snow and ice or muddy debris, promoting autobrecciation, and rush down the volcano, sweeping the slopes as hot debris flows. These lahars, which may be as thick as 30 to 40 m, often show a multitude of detailed internal sedimentary structures, and fill glacial and river valleys, coming to rest with flat tops.

The deposition of coarse pyroclastic material, with a higher angle of repose than lava, results in the buildup of a steep stratocone upon the earlier shield. Abrupt changes in the style of the cone may reflect a change in the lava composition and/or viscosity of lavas that are erupted from a central vent amidst fields of ice, snow, and muddy debris.

During construction of the stratocone, the third major phase of volcanism begins with the formation of one or more new vents at some distance from the central vent on the flanks of the original shield. These smaller satellite cones, which may number as many as four or five, build simultaneously with the central cone and invariably show intercalated units. With continued buildup the compositions of the lavas of the main stratocone eventually become andesitic, then dacitic, whereas that of the satellite cones remain basaltic. Dome formation occurs in the dacitic stage of the main stratocone, but late plugs of the satellite vents remain basaltic. Near all the vents, small dikes and sometimes sills (10 to 100 cm wide) cut the volcanic pile.

Regardless of volcano size, the final cone-building stage is often culminated by caldera formation and eruption of a dacitic ash flow (Semisophochnoi), large dome (Atka), or rhyodacitic pumice beds (Okmok). The central-satellite vent systems cease eruptions and further activity is generally associated with the caldera itself. New activity consists of eruptions on the floor of the caldera, if the caldera floor is large and at low elevation (like Okmok, and in the continental sector, Aniakchak), or of growth of a new cone near the caldera rim, if the caldera is smaller and the floor is at high elevation (like Atka and Cold Bay). When eruptions within the caldera cease, activity shifts to a new location a few kilometers away where a new stratocone is built. Once established, eruptive activity in most instances stays within a relatively restricted circular zone about 10 to 20 km in diameter. Exceptions to this pattern occur on Tanaga, Seguam, and eastern Unimak, where the volcanism has migrated westward parallel to the arc some 10, 5 to 10, and 30 km, respectively, and in the continental sector at Pavlof where it has migrated about 20 km northeast.

Although high rates of erosion may occur during buildup, once dormancy or extinction comes, craters and high-level calderas become catchment basins for snow and ice builds. Basins are soon breached and become highly active cirques, excavating downward through the throat of the volcano and exposing the internal details of the vent.

Continental sector. In the continental part of the arc, major volcanic centers build with little sign of developing satellite vents, and there is more of a tendency to establish large solitary stratocones. Some have an array of substantial subsidiary cones, such as at Pavlof and adjoining Pavlof Sister, Little Pavlof, Double Crater, Mount Hague, and Mount Emmons, and perhaps at Katmai and Mageik. There is a tendency also for successive cones in any one region to form a linear series rather than a cluster as in the oceanic part of the arc. Also, the largest cones, beginning with Shishaldin, lie on or landward of the continental slope (~165°W). The chemical composition of the lavas and pyroclastics also changes eastward along the arc. Basalts and basaltic andesites volumetrically dominate the oceanic sector of the arc, whereas on the Alaska Peninsula basalts are far less common, or even absent from centers such as Katmai. Instead, they give way to more siliceous compositions that occur as voluminous and abundant ash flows. This transition in composition, which probably occurs near Veniaminof (159°W), may also be reflected in postcaldera eruptions. Miller and Smith (1984) found that these eruptions east of Veniaminof continue to be silicic, whereas Veniaminof and centers westward return to more mafic compositions.

Smaller centers

There are many small (<10 km^3) volcanic centers, especially in the oceanic sector, that show another style of development. Some evolve much as a large center, but never establish satellite vents, and generally in the last stages of activity produce summit and flank domes. The summit domes are andesitic to dacitic and the flank domes more basaltic. Others are more akin to large domal centers and although they produce lavas, they never develop a well-defined central eruptive crater. Their extrusive materials are for the most part highly crystalline. In many ways these centers (such as Adagdak and Moffett on Adak) may represent a critical point: for any less magma, no volcano would occur, whereas for more magma, a full stratovolcano develops.

Cinder cones

Cinder or monogenetic cones are not common throughout the arc, but where they are found, they are plentiful. On Shishaldin, two dozen cinder cones are scattered over the northwesterly flank at low elevations (<1,000 m), and some have issued substantial lava flows (Fournelle, 1988). Westdahl and Pogromni are surrounded by at least 22 cinder cones (Neal and Swanson, 1983). Eleven recent cinder cones occur in the vicinity of Pavlof Volcano (Kennedy and Waldron, 1955). These are small with a maximum height of about 100 m, and are similar to ones that occur on the flanks of other volcanoes, especially near vents that have issued lavas. They have a subparallel alignment to the six major stratocones (McNutt and Jacob, 1986). In addition, a group of cinder cones stretches in a line northwesterly from the northern flank of Veniaminof (Burk, 1965). The sizes of these postglacial cones seem to diminish away from the volcano. Cinder cones are also found on the floors of large calderas at Okmok, Aniakchak, and Fisher. Several of these have issued substantial amounts of tephra and lava.

The paucity of cinder cones within the arc may reflect the dominantly central-vent style of eruption characteristic of stratocones. Only when the central vent becomes high and narrow, or otherwise congested, is it easier for magma to be dispersed by dikes. On the southwest flank of Atka Volcano, a regional dike that likely formed accompanying caldera formation has been glacially excavated; apparently it was the feeder dike for a cinder cone. Cinder cones suggest fissure-style eruptions. Historic fissure eruptions are known in the Aleutians: Gareloi was split from summit to flank in April of 1929 (Coats, 1959a) with the creation of 13 explosion craters; three fissures formed on the north slope of Shishaldin in 1830 (Veniaminov, 1840); and Seguam had a fissure eruption in 1977.

In proposing a method to map regional stress patterns, Nakamura (1977) and Nakamura and others (1980) used cinder cone distributions and dikes at several Aleutian centers as indicators of the direction of the maximum horizontal stress. The results mainly seem to reflect a smoothly varying stress field dictated by the direction of convergence. However, direct evidence of regional stress is scarce because there are few cinder cones and fissures. Areas with major indicators of stress that Nakamura and associates did not consider include the alignments of active vents on Tanaga and at Pavlof, fissures and vents on Seguam, a cinder cone field (and adjacent dikes) on Shishaldin, and a set of regional dikes on Atka. Tanaga and Pavlof indicate a northeast-southwest principal stress, whereas Atka, Seguam, and Shishaldin show a northwest-southeast stress. Together they suggest that the regional stress field is not uniform, but is influenced by local stresses (e.g., McNutt and Jacob, 1986).

GEOCHEMICAL AND PETROGRAPHIC FEATURES OF THE ALEUTIANS

Chemistry and petrography

The early works of Fenner (1926) in the Katmai region and Coats (1952) in the central Aleutian Islands suggested a calc-alkaline character of the modern volcanic rocks. As the number of volcanic centers sampled has increased, however, the Aleutian oceanic suite has been shown to have greater compositional diversity than originally believed.

A large and varied geochemical and isotopic database exists for the Aleutian arc. The number and types of analytical data available for individual volcanic centers, however, varies considerably. Of the 39 Aleutian island volcanic centers, geochemical and/or isotopic data are available for only 23 (Appendix 1). A significant number of these are characterized by only a few analyses of a few elements. The analytical data set is also skewed toward the large volcanic centers of Shishaldin, Makushin, Okmok, and Atka, plus smaller Seguam, which have large and comprehensive geochemical databases. For these centers, many samples have been fully characterized, having been analyzed for REE and isotopes as well as major and trace elements. Other volcanic centers have been analyzed for only a single geochemical variable (e.g., REE or isotopes), and so their geochemical characterization is fragmentary.

Many of the 30 islands with recent volcanic activity have multiple centers, and therefore the number of major individual volcanoes is much larger than the number of islands. The islands range in size from Unimak (~4,000 km^2) to Koniuji (<1 km^2). These geologic differences suggest that volcanoes are supplied by magmatic systems that vary considerably in degree of development and complexity. Large volcanic centers with many eruptive stages and satellite vents, for example, may have noncommunicating *crustal* reservoirs and long-lived magma chambers (e.g., Myers and Frost, 1989). Lavas from such plumbing systems probably record the influence of a variety of magmatic processes, for example, crystal fractionation/accumulation, magma mixing, and crustal assimilation. Smaller centers, on the other hand, may be supplied by smaller and simpler systems. The geologic complexity of Aleutian volcanic centers may also reflect differences in *subcrustal* magmatic processes. In view of probable lithospheric and crustal complexities, analysis of chemical compositions without corresponding geologic information is of limited use in interpreting the processes of arc-magma generation and evolution. Geochemical data should be compared and contrasted between volcanic centers whose geologic setting and evolution are similar. The geology, geochemistry, and petrology of individual Aleutian volcanic centers is reviewed by Myers (in preparation).

The following section summarizes the general chemical features of Aleutian volcanic rocks. Various representative compositions from the volcanic centers of Cold Bay, Shishaldin, and Atka are given in Table 2. Silica-oxide variation diagrams for these three centers are presented in Figures 3 to 5.

Rocks are divided by SiO_2 content: basalts, $SiO_2 < 53$ wt. %; andesites, 53 to <63 wt. %; dacites, 63 to 70 wt. %. There are only a handful of published analyses of rhyolite, reflecting mainly the general scarcity of this composition, but also perhaps the tendency to ignore ash layers in soil when sampling. Maximum SiO_2 content is 76 wt. % at Vsevidof (n = 1) and Kanaga (n = 1); values of 72 to 75 wt. % are reported for Unimak (n = 2) and Umnak (n = 3), and 70 to 71 wt. % for Seguam (n = 7). On the Alaska mainland, Novarupta has several analyses of 76 to 78 wt. % SiO_2, but otherwise, between Mount Spurr and Cold Bay, there are no other volcanic rocks known to contain between 68 and 76 wt. % SiO_2 (see Miller and Richter, this volume).

Average compositions for the entire arc have also been calculated (Table 3), to provide a rough measure of chemical composition of Aleutian basalts, andesites (further divided into basaltic and silicic andesites), and dacites. The sources of all data are given in Appendix 2.

Basalts. Myers (1988) evaluated published chemical data for Aleutian basalts (<52 wt. % SiO_2; n = 205), dividing them into three major classes; high-MgO (>9 wt. %), low-MgO (<6 wt. %), and transitional (6 to 9 wt. % MgO) basalts. The low-MgO basalts were subdivided into low- and high-alumina (HAB) groups; the first subgroup has <18 wt. % Al_2O_3, <450 ppm Sr, >1.0 wt. % TiO_2, and >9 wt. % FeO*. Myers (1988) found that high-MgO basalts constitute 10% of the analyzed Aleutian basalts and occur at only a few centers; transitional basalts constitute 22%, and low-MgO basalts constitute 67% of the basaltic analyses. All basalts have intermediate levels of K_2O (0.5 to 1.0 wt. %).

The data set reviewed here includes more basalts (n = 346), with the upper SiO_2 limit at 53 wt. %. These basalts are very similar to those discussed by Myers (1988), with a few differences. For example, Al_2O_3 shows a bimodal distribution, with maxima at 16 to 17, and 18.5 to 19.5 wt %; to some extent, this reflects the inclusion of the Shishaldin data with its distinct FeTi-enriched, low-Alumina (low-Al) basalt. Similarly, whereas most basalts have TiO_2 in the 0.6 to 1.2 wt. % range, there is a distinct subset with values between 2.0 and 2.5 wt. %. Also, Sr shows a single maximum of 400 to 550 ppm (albeit with a rather large deviation), rather than a bimodal distribution as suggested by Myers (1988). Ba and Rb are present at, respectively, 200 to 300 ppm and 10 ppm levels; most Zr contents are in the 40 to 90 ppm range, whereas the range of Y is restricted to 15 to 25 ppm. The major element with the widest range of abundances is MgO (Cr and Ni have similar large ranges).

A common characteristic of island-arc lavas is their low (relative to MORB) concentration of base metals such as Ni, Co, and Cr. Ni contents in the basalts are in most instances less than about 25 ppm, although 100 to 200 ppm Ni is present in high-MgO basalts. Similarly, Cr is normally less than 50 ppm, although it too can reach 300 to 600 ppm.

High-MgO basalts contain moderately abundant (<16-vol. %) Mg-rich ($>Fo_{88}$) olivine phenocrysts with small chromite inclusions, plus occasional clinopyroxene of diopside-salite composition (plagioclase phenocrysts are for the most part absent or else show disequilibrium features). These phases are set in a fine-grained groundmass of olivine, clinopyroxene, plagioclase, magnetite, and sparse glass. Almost invariably, rocks with elevated whole-rock Ni (>40 ppm) and Cr (>50 ppm) contents contain Mg-rich olivine with Cr-spinel inclusions.

Transitional basalts have generally higher crystallinities (17 to 57 vol. % phenocrysts); $An_{70\text{-}90}$ plagioclase phenocrysts dominate (Myers, 1988). Also present are lesser amounts of $Fo_{50\text{-}70}$

TABLE 2: REPRESENTATIVE MAJOR- AND TRACE-ELEMENT CHEMISTRY OF BASALTS, ANDESITES, AND DACITES FROM THE ALEUTIAN VOLCANIC CENTERS OF COLD BAY, SHISHALDIN, AND ATKA*

Volc. Center	Cold Bay	Cold Bay	Cold Bay	Cold Bay	Cold Bay	Shishaldin	Shishaldin	Shishaldin	Shishaldin	Shishaldin
Description	Morzhovoi Hi-Al Bas.[†]	Morzhovoi Bas. And.[†]	Morzhovoi Andesite	Frosty Peak Andesite	Frosty Peak Andesite	Recent Lava Hi-Al Bas.[†]	Glac. Lava Hi-Mg Bas.[†]	Recent Lava FeTi Bas.[†]	Somma Andesite	Dacitic Pumice
Sample No.	CB42	CB30	CB39	CB74	CB51	SH5	SH101	SH9A	SH134	SH141
SiO_2	51.40	53.43	59.40	58.03	61.54	49.83	48.99	50.89	59.84	64.14
TiO_2	0.87	0.94	0.83	0.76	0.48	1.67	1.24	2.21	1.07	0.72
Al_2O_3	20.40	19.25	17.53	18.37	17.66	20.29	15.11	15.73	16.74	15.97
Fe_2O_3	3.03	3.45	2.87	3.24	n.d.	3.32	2.69	3.04	2.21	0.70
FeO	5.52	5.14	4.38	4.07	n.d.	6.71	7.94	8.64	5.07	4.24
FeO*	8.25	8.25	6.96	6.99	5.90	9.70	10.36	11.38	7.06	4.87
MnO	0.17	0.18	0.16	0.16	0.12	0.18	0.18	0.23	0.18	0.22
MgO	4.04	3.26	2.94	2.71	2.26	3.62	10.56	3.99	1.93	0.99
CaO	9.95	8.41	6.76	7.51	6.46	10.49	10.75	7.83	5.30	2.73
Na_2O	2.67	3.34	3.53	3.60	3.73	3.20	2.16	3.77	4.51	5.54
K_2O	0.60	1.06	1.58	1.26	1.08	0.56	0.74	1.14	1.63	2.17
P_2O_5	0.28	0.30	0.30	0.24	n.d.	0.27	0.19	0.59	0.37	0.19
LOI	0.43	0.55	0.01	0.11	n.d.	0.72	0.80	0.91	1.40	n.d.
Ba	220	472	485	475	480	231	246	456	550	878
Co	28	n.d.	n.d.	n.d.	n.d.	27	53	27	23	n.d.
Cr	6	n.d.	n.d.	n.d.	3	22	434	17	5	23
Ni	10	n.d.	n.d.	n.d.	3	12	136	10	7	11
Rb	10	24	41	32	32	17	17	41	42	56
Sc	n.d.	n.d.	n.d.	n.d.	n.d.	27	41.1	30.7	18	n.d.
Sr	384	507	328	400	400	629	530	470	429	255
V	192	n.d.	n.d.	n.d.	76	240	316	181	64	28
Y	16	n.d.	n.d.	n.d.	19	23	17	43	40	46
Zr	68	125	179	142	127	76	85	215	246	263

Volc. Center	Atka	Atka	Atka	Atka	Atka	Atka	Atka
Description	Four Flows Hi-Al Bas.[†]	Potainkof Bas. And.[†]	Milky Valley Andesite	Konia Andesite	Korovin Andesite	Kliuchef Dacite	Glassy Dacite
Sample No.	AT 1	AT 33	AT 34	AT119	AT 117	AT118	AT 20
SiO_2	49.80	53.29	58.24	61.39	62.49	66.03	66.88
TiO_2	0.96	0.95	0.69	0.87	0.62	0.46	0.46
Al_2O_3	20.12	18.31	17.33	16.75	16.37	16.12	15.57
Fe_2O_3	4.36	3.83	3.58	2.71	1.99	1.69	1.25
FeO	5.51	4.71	3.02	4.13	3.95	2.64	2.52
FeO*	9.43	8.16	6.24	6.57	5.74	4.16	3.64
MnO	0.18	0.20	0.16	0.18	0.15	0.13	0.10
MgO	4.72	3.92	3.24	1.64	1.69	0.64	0.74
CaO	9.66	8.82	7.10	4.37	4.23	2.36	2.22
Na_2O	3.71	3.29	3.74	4.59	4.53	4.72	4.96
K_2O	0.78	1.11	3.74	2.79	3.00	4.60	4.09
P_2O_5	0.23	0.26	0.19	0.40	0.19	0.15	0.12
LOI	0.24	1.26	0.44	0.59	0.21	1.05	0.94
Ba	273	530	745	750	n.d.	890	862
Co	29	29	24	n.d.	n.d.	n.d.	38
Cr	42	29	50	3	n.d.	<1	18
Ni	20	7	11	<1	n.d.	<1	3
Rb	10	21	40	64	n.d.	92	116
Sc	n.d.	n.d.	18	20	n.d.	13	11
Sr	630	575	487	343	n.d.	178	155
V	275	n.d.	136	39	n.d.	21	20
Y	22	23	22	40	n.d.	44	40
Zr	64	99	128	257	n.d.	344	342

*Data sources: Cold Bay: Brophy, 1984, 1986a, 1987; Shishaldin: Fournelle, 1988; Fournelle and Marsh, 1991; Atka: Marsh, 1982a, unpublished data.

Oxide and LOI values are in weight percent (wt %); trace elements are in parts per million (ppm); FeO* = all iron as FeO.

†Hi-Al, high alumina; Bas., basalt; And., andesite; Hi-Mg, high magnesium; n.d., no data.

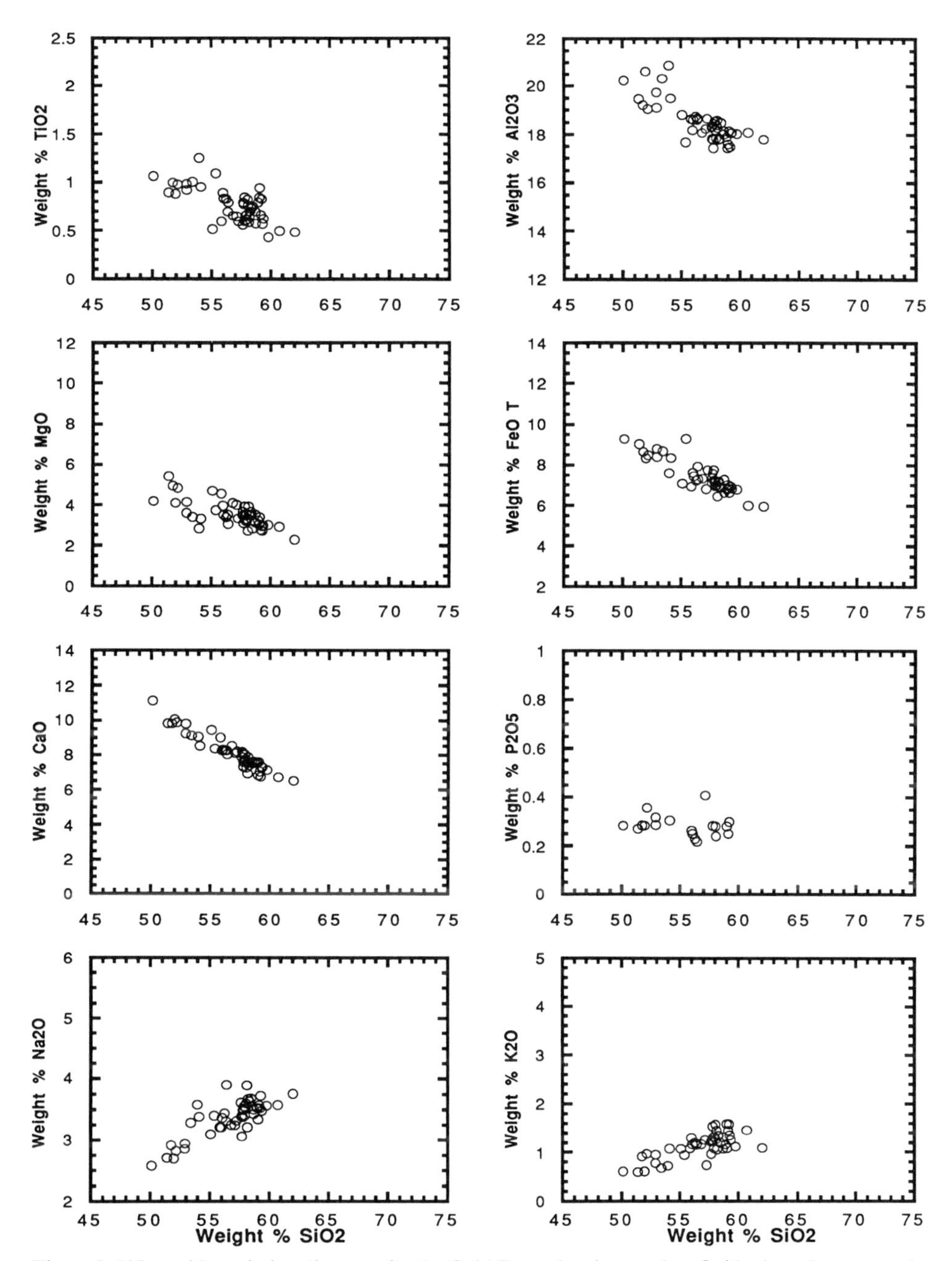

Figure 3. SiO_2-oxide variation diagrams for the Cold Bay volcanic complex. Oxides have been normalized to 100 wt. % "water-free" (samples with H_2O or LOI >2.5 wt. % not used). FeO T = all iron as FeO. Data sources given in Appendix 2.

olivine (3 to 7 vol. %), clinopyroxene (2 to 16 vol. %), and titanomagnetite.

High-alumina basalts (HABs) are highly porphyritic (25 to 73 vol. %), having abundant An_{70-90} plagioclase, lesser amounts of iron-rich (<Fo_{70}) olivine and titanomagnetite, and sporadic clinopyroxene (Myers, 1988). Groundmass phases in the low-MgO basalts include plagioclase, olivine, clinopyroxene, titanomagnetite, and rare interstitial glass. Pigeonite and hydrous phase phenocrysts are absent.

Little has been published on Aleutian low-alumina, low-Mg basalts. Myers (1988) recognized them as volumetrically minor. The FeTi-enriched, low-Al basalts from Shishaldin have, for the most part, less than 20% phenocrysts and in many cases, only microphenocrysts (Fournelle, 1988). Plagioclase dominates, with

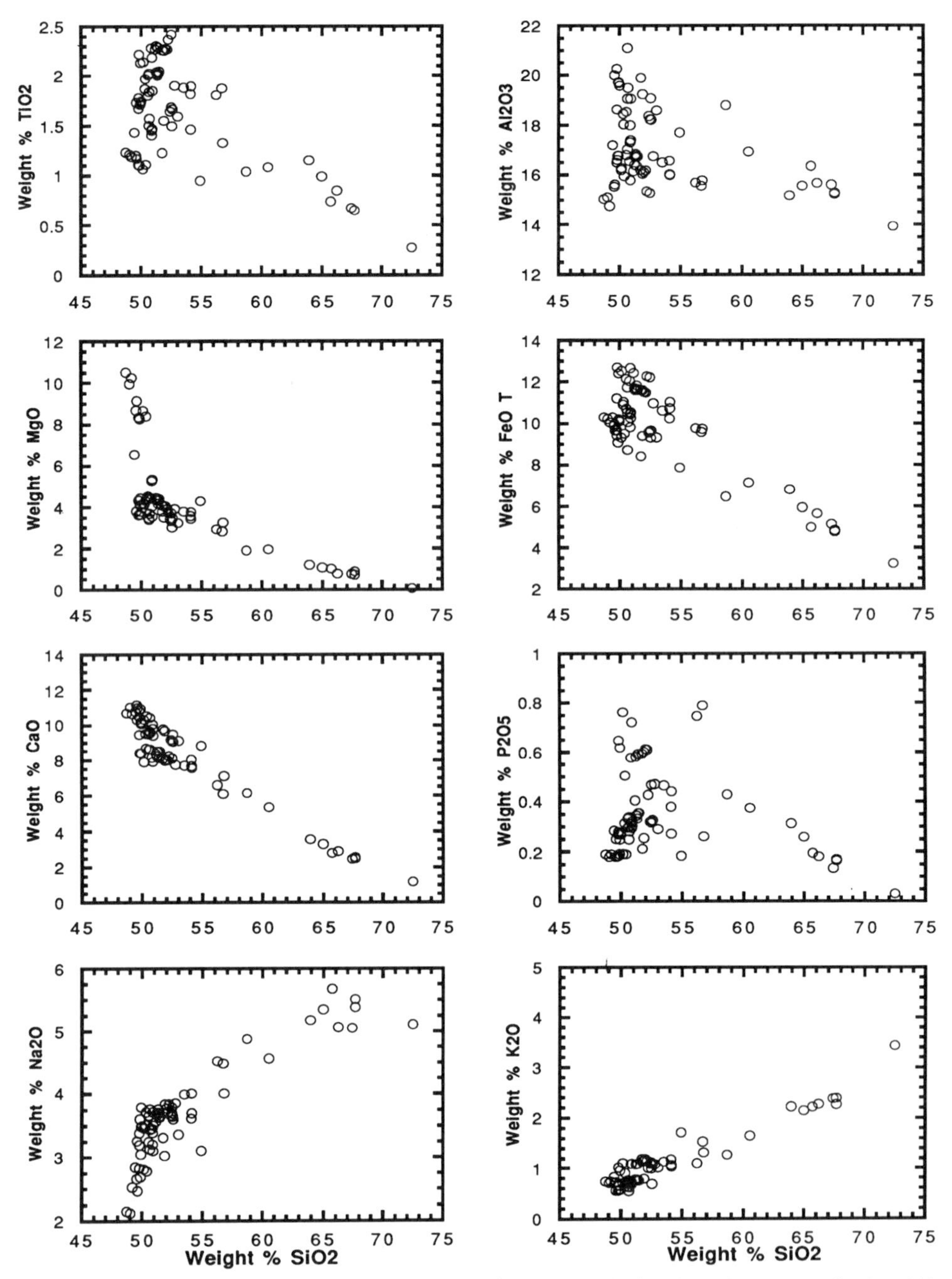

Figure 4. SiO_2-oxide variation diagrams for Shishaldin Volcano. Oxides have been normalized to 100 wt. % "water-free" (samples with H_2O or LOI > 2.5 wt. % not used). FeO T = all iron as FeO. Data sources given in Appendix 2.

An_{50-70} microphenocrysts, and less common phenocrysts as calcic as An_{90}. Olivine (Fo_{70}) and titanomagnetite (ti-mt) are scarce (<1 vol. %), and clinopyroxene (cpx) is commonly absent; plagioclase, cpx, and ti-mt compose the groundmass. A few quench crystals of plagioclase and olivine are present.

Andesites. The Aleutian data set contains 594 samples with SiO_2 contents between 53 and 63 wt. %. Relative to the basalts, the andesites contain decreased levels of FeO* (6 to 8.5 wt. %), MgO (2.5 to 4 wt. %), and CaO (6 to 9 wt. %); higher Na_2O (3 to 4 wt. %) and K_2O (0.5 to 2 wt. %); and similar TiO_2 (0.6 to 1.2 wt. %) and Al_2O_3 (16.5 to 19 wt. %).

Most trace elements vary, with increasing silica, in a consistent manner. Ni and Cr are less abundant (<15 ppm and <30 ppm, respectively), although there are a few high values (~100 and ~300 ppm). Ba increases (300 to 500 ppm) as does Rb (15 to 45 ppm). Sr, however, remains relatively constant (300 to 600 ppm) within the basalt to andesite range.

The andesites have been further broken down into basaltic

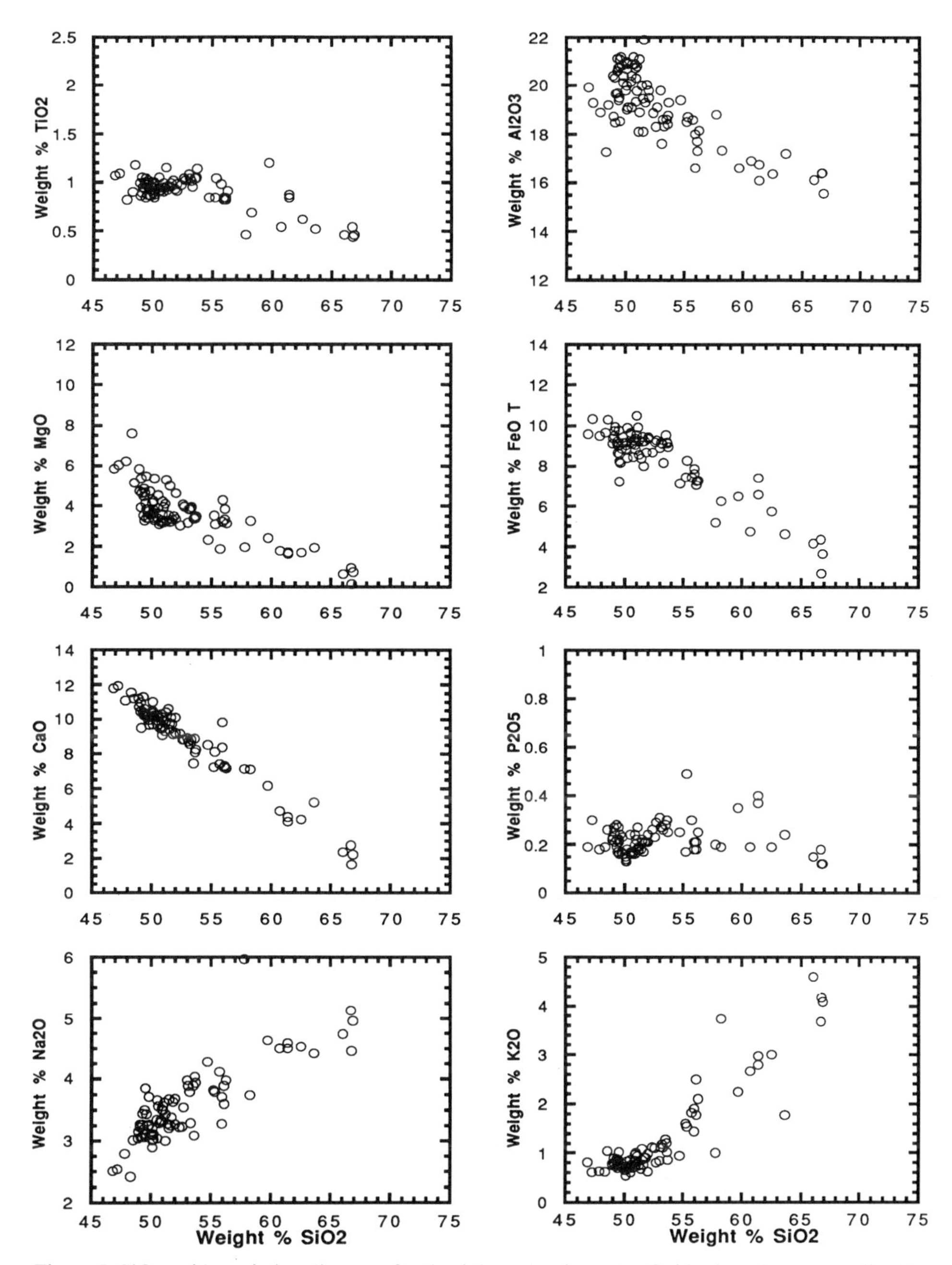

Figure 5. SiO_2-oxide variation diagrams for the Atka volcanic center. Oxides have been normalized to 100 wt. % "water-free" (sample with H_2O or LOI > 2.5 wt. % not used). FeO T = all iron as FeO. Data sources given in Appendix 2.

andesites (53 to 57 wt. % SiO_2) and silicic (acid) andesites (57 to 63 wt. % SiO_2). Calculated mean compositions are given in Table 3. The mean andesite composition is similar to the average worldwide andesite composition (n = 2,500) found by Gill (1981).

Andesites have the same high crystallinity as the basalts, approximately 35 to 50 vol. %. Plagioclase dominates, with 20 to 40 vol. %, followed by 6 to 9 vol. % cpx. Up to 3 vol. % orthopyroxene may be present; olivine and amphibole are absent or present in only trace amounts. Opaque minerals (mainly titano-magnetite) are more abundant (1 to 2 vol. %), at the volcanic centers such as Cold Bay, Akutan, Atka, Adak, and Kanaga, whereas at Okmok, Makushin, and Shishaldin they are generally absent.

Studies of andesites from the volcanic centers of Cold Bay (Brophy, 1987) and Kanaga (Brophy, 1990) show the following features: Kanaga plagioclase crystals range from fresh to corroded and sieve-textured, and the latter are more albitic than the former; most phenocryst cores are An_{50-60}. Cold Bay plagioclase show a bimodal distribution (An_{50-60}, An_{80-90}) and often complex zonation. Unzoned olivine (Fo_{62} to Fo_{81}, with ragged edges) is present in only one Cold Bay lava series; at Kanaga, Fo_{70-88} grains are corroded. Kanaga orthopyroxene compositions are in the range $Wo_{2-4}En_{58-71}Fs_{26-38}$, with the Mg-rich varieties showing normal zoning and the Fe-rich crystals reverse zoning (clinopyroxenes show the same patterns). Cold Bay clinopyroxene is in the range $Wo_{38-46}En_{40-47}Fs_{13-19}$. Amphibole is absent at Cold Bay, whereas trace amounts are present in one-third of all published Aleutian modes.

Dacites. A smaller number (n = 130) of dacites (i.e., 63 to 70 wt. % SiO_2) have been chemically analyzed. Compared to the andesites, there are significant decreases in FeO* (4.5 to 6 wt. %), MgO (0.5 to 2.5 wt. %), and CaO (2 to 6 wt. %), slight declines in TiO_2 (0.4 to 1 wt. %) and Al_2O_3 (15.5 to 16 wt. %), and increases in Na_2O (3.5 to 5 wt. %) and K_2O (1 to 2.5 wt. %). The major element with the largest spread of values is CaO (Sr has a similar wide range). The average Aleutian dacite composition is given in Table 3.

Ni and Cr (both <10 ppm) continue to decrease in concentration with increasing silica content. Ba and Rb increase respectively to 800 to 900 ppm and 30 to 60 ppm, whereas Sr decreases to 150 to 350 ppm. Zr shows elevated levels (125 to 275 ppm), and Y increases to 20 to 60 ppm.

Modal abundances have been determined for only a handful of samples (n = 8). These dacites have lower crystallinities, normally less than 20 vol. %. Plagioclase (≤15 vol. %) is the main phase, followed by cpx (<2 vol. %) and the opaques (≤1 vol. %). Small amounts (<1 vol. %) of olivine, orthopyroxene, and amphibole have been reported in individual samples but otherwise these phases appear to be rare. Plagioclase in Shishaldin dacites (Fournelle, 1988) ranges from An_{30} to An_{40}, green clinopyroxene is $Wo_{38}En_{36}Fs_{26}$, and golden-brown orthopyroxene is $Wo_4En_{49}Fs_{48}$. Kay and Kay (1985) report Great Sitkin clinopyroxene as $Wo_{42}En_{42}Fs_{16}$ and Westdahl clinopyroxene as $Wo_{40-41}En_{36-42}Fs_{18-22}$.

Hildreth (1983) determined mineral abundances of the 1912 Katmai (Novarupta) andesites, dacites, and rhyolites by mineral separation. He found the rhyolites to have 0.5 to 2 wt. % phenocrysts, whereas andesites and dacites contain 30 to 45 % (with dacites more crystal-rich). Present, in order of appearance, were plagioclase, hypersthene, Ti-magnetite, ilmenite, apatite, and pyrrhotite. In the rhyolites, quartz was as abundant as plagioclase; in the intermediate rocks, augite was about half as abundant as hypersthene.

The Deep Sea Drilling Project's (DSDP) northern Pacific cores, from locations adjacent to the Aleutian arc, contain volcanic ashes (Scheidegger and others, 1980). The ashes, of bulk

TABLE 3: MEAN COMPOSITIONS OF PUBLISHED ALEUTIAN VOLCANIC SAMPLES, FOR CATEGORIES BASED ON SiO_2 WEIGHT PERCENT, "WATER FREE"*

	Basalt		Basaltic Andesite		High-Silica Andesite		Andesite		Dacite	
Range	<53 wt% SiO_2		53-57 wt% SiO_2		57-63 wt% SiO_2		53-63 wt% SiO_2		63-70 wt% SiO_2	
	Mean (S.D.)	n=	Mean (S.D.)	n=	Mean (S.D.)	n=	Mean(S.D.)	n=	Mean (S.D.)	n=
SiO_2	50.6 (1.5)	346	54.9 (1.1)	270	59.4 (1.7)	324	57.4 (2.7)	594	65.8 (2.1)	130
TiO_2	1.10 (.42)	346	0.98 (.28)	270	0.80 (.21)	324	0.88 (.26)	594	0.72 (.17)	130
Al_2O_3	18.3 (1.9)	346	18.0 (1.1)	270	17.3 (.88)	324	17.6 (1.1)	594	15.7 (.71)	130
FeO*	9.57 (1.22)	346	8.17 (1.02)	270	6.68 (.96)	324	7.36 (1.24)	594	4.95 (.90)	130
MnO	0.18 (.03)	346	0.17 (.03)	269	0.15 (.03)	323	0.16 (.01)	593	0.14 (.04)	128
MgO	5.59 (2.25)	346	4.19 (1.02)	270	3.09 (.90)	324	3.59 (1.10)	594	1.53 (.65)	130
CaO	10.3 (1.2)	346	8.56 (.89)	270	6.79 (.89)	324	7.59 (1.29)	594	4.07 (1.10)	130
Na_2O	2.98 (.53)	346	3.43 (.45)	270	3.80 (.57)	324	3.63 (.55)	594	4.72 (.77)	130
K_2O	0.80 (.28)	346	1.15 (.36)	270	1.53 (.46)	324	1.36 (.46)	594	2.00 (.61)	130
P_2O_5	0.23 (.14)	322	0.23 (.11)	253	0.21 (.08)	274	0.22 (.01)	594	0.18 (.06)	128
Ba	309 (114)	193	401 (148)	110	584 (161)	102	489 (179)	212	710 (135)	74
Co	34 (9)	62	30 (8)	46	25 (8)	37	28 (8)	83	23 (12)	53
Cr	71 (133)	212	43 (32)	97	33 (72)	81	38 (54)	182	17 (13)	67
Cu	121 (75)	56	80 (118)	41	45 (36)	30	65 (94)	71	31 (43)	9
Ni	34 (61)	206	21 (25)	110	15 (19)	82	19 (23)	192	7 (6)	23
Rb	15 (8)	248	25 (10)	174	38 (12)	163	31 (13)	337	51 (18)	87
Sc	30 (7)	53	28 (8)	15	17 (6)	26	21 (7)	40	16 (3)	8
Sr	511	262	473 (140)	179	431 (135)	164	453 (139)	343	263 (79)	88
V	278 (66)	158	218 (55)	56	137 (70)	57	177 (75)	113	51 (44)	18
Y	24 (10)	213	24 (9)	159	29 (11)	137	26 (11)	296	39 (11)	82
Zn	82 (8)	17	74 (17)	6	84 (23)	4	78 (19)	10	81 (4)	2
Zr	88 (64)	222	102 (40)	172	136 (43)	145	118 (45)	317	180 (51)	82

S.D., standard deviation; n, number of chemical analyses. Oxide and LOI values are in weight percent (wt.%); trace element values are in parts per million (ppm); FeO = all iron as FeO.

SiO_2 54 to 65 wt. %, have small amounts of biotite and amphibole, with amphibole abundance increasing significantly for ash SiO_2 contents >66 wt. %. Biotite is uncommon in Aleutian subaerial deposits, and the only occurrence reported is in a rhyolitic dome at Recheschnoi (Byers, 1959). This rhyolite (75.4 wt. % SiO_2) contains "sparse, small euhedral phenocrysts of bluish gray quartz, feldspar, biotite, hornblende, and rare hypersthene" (Byers, 1959, p. 301).

Reinink-Smith (1990b) found that unaltered volcanic glass in the Pliocene coal beds of the Kenai Peninsula ranged in composition from andesite (60.3 wt. % SiO_2) to rhyolite (72.3 wt. % SiO_2).

Marsh (1982a) suggested two distinct styles of K_2O versus SiO_2 variation, leading to either ~2 or ~4 wt. % K_2O in dacites. The high potash trend is principally defined by the lavas of Semisopochnoi and Atka; both show a similar development from a central-satellite cone system to caldera formation and eruption of K_2O-rich dacite. High potash trends have also been found in some Japanese lavas (Aramaki and Ui, 1983).

Aleutian basalts and andesites have whole-rock sulfur contents in the range from 200 to 400 ppm (Marsh, unpublished data). Some volcanoes have large sulfur deposits, such as Little Sitkin, with up to 200,000 tons (Snyder, 1959), and Makushin, with up to 77,000 tons (Maddren, 1919). Ash erupted from Mount Spurr in 1953 was found to contain 1.5 wt. % SO_3 (Wilcox, 1959). Gerlach and others (1990) examined products of the 12/15/89 eruption of Redoubt Volcano and found up to 870 ppm of sulfur in melt inclusions in hornblende, 400 to 60 ppm in plagioclase and orthopyroxene melt inclusions, and ~60 ppm in the pumice matrix glass.

Arcwide variations in chemistry

Lavas and pyroclastics from Aleutian volcanoes compositionally span the SiO_2 range from 45 to 77 wt. %. Marsh (1982a) suggested that the dominant rock type in the arc was basaltic (49 to 52 wt. % SiO_2), based upon examination of the volcanic center of Atka. Examination of the numbers of published analyses for Atka and seven other Aleutian volcanic centers (Fig. 6) suggests that the situation may be more complex. The three very large volcanic centers of Shishaldin (300 to 400 km^3), Okmok (200 to 300 km^3), and Atka (200 km^3) are dominated by basalt, with minor amounts of dacite and andesite. The somewhat smaller volcanic centers of Makushin (145 km^3) and Akutan (80 km^3) appear to have approximately equal amounts of andesite and basalt, whereas Seguam (75 km^3) has a distinctive bimodal population of basalt/basaltic-andesite and dacite/rhyodacite. The small centers of Kanaga (25 km^3) and Adak (25 km^3, not shown on Fig. 6) are dominantly andesite, as is the volcanic compex of Cold Bay (~150 km^3) on the Alaska Peninsula. In addition to variations due to geographic position (continental versus oceanic, plus along-arc position), the volume of the volcanic center may be an important factor in determining its general chemical composition.

Figure 7 shows the longitudal variation of SiO_2 content of analyzed samples, from Mount Spurr in the east to Buldir Volcano in the west. Few basalts are present in the continental sector (i.e., east of Cold Bay); andesites through dacites predominate. Date from the oceanic volcanoes are presented here principally for comparison with those of the Alaska Peninsula–Cook Inlet sector.

Four oxides (Al_2O_3, FeO^*, MgO, and K_2O) have been plotted against longitude, by silicia categories (basalts, basaltic andesites, silicic andesites, and dacites), in Figure 8a to d. Considering only the well-sampled centers (at a given SiO_2 level), the variability of chemical compositions at some centers is striking, compared to the rather restricted compositions at others. For basaltic compositions, Shishaldin and Okmok (and to some extent, Akutan and Makushin) show significant heterogeneity, particularly for TiO_2, Al_2O_3, FeO^*, MgO, Y, and Zr. Atka and Seguam, on the other hand, appear to be homogeneous.

The andesitic database is relatively large and includes analyses of a number of centers, from Spurr to Yantarni, in the Alaska Peninsula–Cook Island section. Andesites from these volcanoes show a more restricted range (albeit for a small number of samples), relative to the oceanic sector, for Al_2O_3, TiO_2, FeO^*, CaO, P_2O_5, Y, Zr, and, to some extent, K_2O.

Dacitic compositions are also shown in Figure 8. Dacites (and less common rhyolite, n = 12) are not restricted to the continental portion of the Aleutian arc. There appears to be a wider range of some oxides (e.g., FeO^*, MgO, K_2O) in the oceanic sector compared with the Alaska Peninsula (AP)–Cook Inlet (CI) sector. In addition, K_2O appears to be somewhat enriched in the oceanic dacites and rhyolites relative to those in the AP-CI sector. For four oceanic rhyolites (72.5 to 76.4 wt. % SiO_2), there is an average of 3.9 wt. % K_2O, versus an average of 3.2 wt. % for four Novarupta rhyolites (which are at a higher silica content, ~77.3 wt. % SiO_2).

This arcwide variation in major (and trace element) chemistry is striking in the degree of diversity at some centers relative to others. There also appears to be less variation in composition of Alaska Peninsula–Cook Inlet basaltic andesites and silicic andesites, particularly in Al_2O_3, FeO^*, and K_2O, compared to those present in the oceanic part of the arc.

Transverse to the arc the only measure of change is recorded by the volcanic rocks of Bogoslof and Amak. Bogoslof's heterogeneous rocks (basalts and silicic andesites) are more alkalic than those of the nearby volcanic front (Arculus and others, 1977; Marsh and Leitz, 1979). They are also lower in MgO, and their Sr isotopic values are, for the most part, lower.

Lavas from Amak consist of homogeneous andesite that is compositionally distinct (i.e., having lower MgO, and higher K_2O and Na_2O) from the lavas of the nearby Cold Bay volcanic center. Additionally, Amak Sr and Nd isotope values are distinct from those in Cold Bay lavas (Morris and Hart, 1983; Marsh and Lietz, 1979; von Drach and others, 1986).

The largest Aleutian isotopic database is for $^{87}Sr/^{86}Sr$ (n = 149). Values for the oceanic sector range from about 0.7028

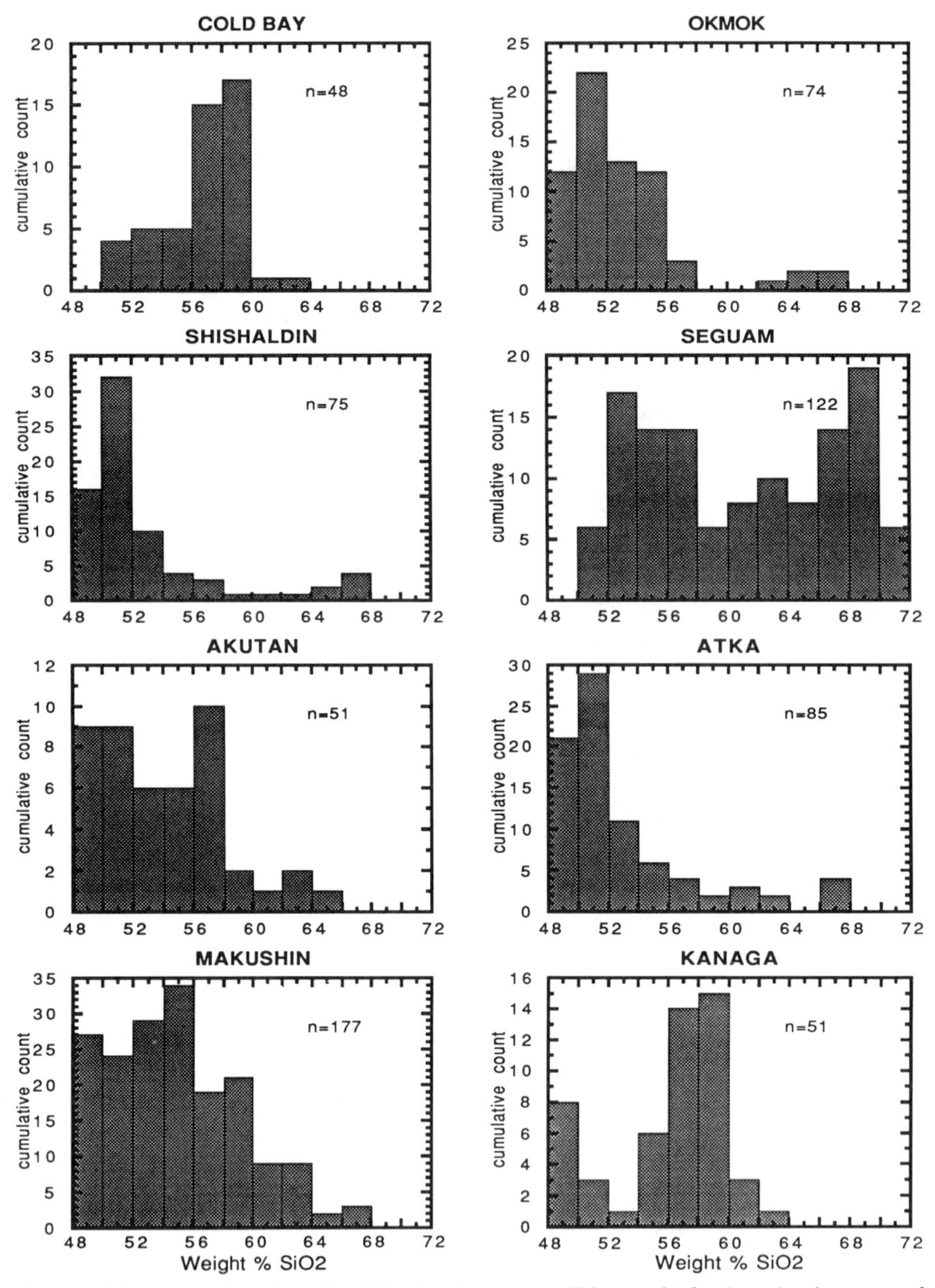

Figure 6. Histograms of number of published analyses versus SiO_2 (wt. %) for the volcanic centers of Cold Bay, Shishaldin, Akutan, Makushin, Okmok, Seguam, Atka, and Kanaga. Data sources given in Appendix 2. n = number of chemical analyses.

at Adak (sample ADK53 of Kay 1978, which we interpret as xenocryst-bearing) to 0.7037 at Seguam, with most data clustering between 0.70315 and 0.70335 (mean = 0.70331 ± 0.00024; Fig. 9). The coverage, however, is extremely variable, with one-third of the data from Adak and Atka. The secondary front volcanoes of Amak and Bogoslof have lower values, 0.7028 to 0.7032. With few data from the continental sector, there is a decrease in values from Spurr to Veniaminof, and continuing westward, average values are relatively constant although the least radiogenic values decrease to ~0.7029 through 167°W (Makushin). $^{87}Sr/^{86}Sr$ values show a distinct radiogenic enrichment at Seguam, with a westward decrease at Atka and then Adak.

Within this limited Sr isotopic range, a distinct feature is the

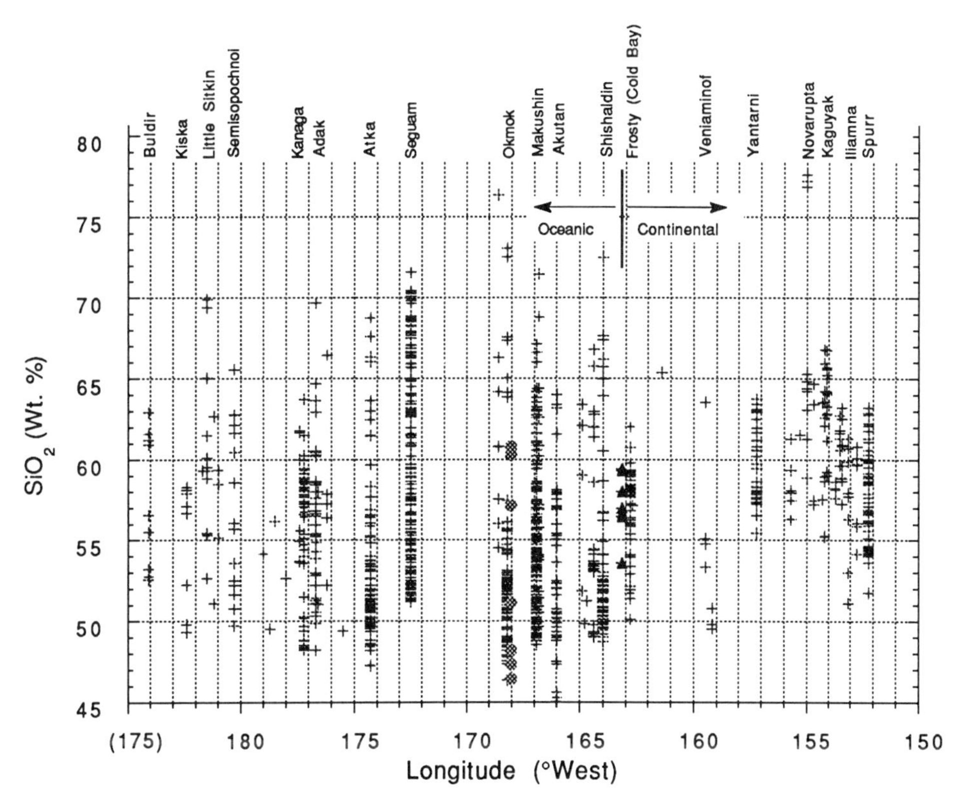

Figure 7. Plot of SiO_2 (wt. %, "water-free") versus longitude, from 152°W (Spurr) to 176°E (Buldir). Also plotted are the second-arc volcanoes of Amak (triangles; 163.15°W) and Bogoslof (hachured circles; 168.03°W). Data sources given in Appendix 2.

relatively large variation in some centers (for $n \geqslant 12$) such as Cold Bay, Okmok, and Adak, and the small variations at others, such as Seguam and Atka. Whereas this difference seems to reflect the diversity of lavas at some centers, such as Adak or Okmok, for some centers like Seguam, $^{87}Sr/^{86}Sr$ values show very little variation, despite the wide range of lava types present (basalt through rhyodacite).

The amount of $^{143}Nd/^{144}Nd$ data is limited (n = 68). Values range from 0.51294 to 0.51310 (Fig. 9), with an anomalous value of 0.51324 for Adak sample ADK53 (Kay, 1978). There is a range of data for each volcanic center, but there are only 7 to 9 data per center (12 for Seguam), making interpretation difficult. Seguam has the lowest range of values (0.51294 to 0.51299), similar to those for Cold Bay. There is a trend, from Seguam through Atka and Adak, of increasing $^{143}Nd/^{144}Nd$.

Lead isotopic data are limited (n = 80), with 75 percent from four centers (Okmok, Seguam, Atka, and Adak). There are differences in the spread of Pb isotopic values between these four centers (Fig. 9). Adak (n = 21) and Okmok (n = 15) show the greatest range of values, whereas Seguam (n = 12) and Atka (n = 12) have a restricted range, at more radiogenic values. There appear to be trends in the Pb isotopic data similar to those seen in $^{87}Sr/^{86}Sr$ between Seguam, Atka, and Adak, with radiogenic values decreasing westward.

Plots of $^{207}Pb/^{204}Pb$ and $^{208}Pb/^{204}Pb$ against $^{206}Pb/^{204}Pb$ show rather tight arrays (Myers and Marsh, 1987; Morris and Hart, 1983; Kay and others, 1978). The values of $^{207}Pb/^{204}Pb$ versus $^{206/204}Pb$ partially overlap the normal trend of oceanic ridge basalts and have a steeper slope than the oceanic trend; the Adak values have the greatest amount of scatter. $^{208}Pb/^{204}Pb$ versus $^{206}Pb/^{204}Pb$ overlaps the MORB array. The total range of values of $^{206}Pb/^{204}Pb$ is also much more restricted (~18.5 to 19.0) than for oceanic island and ridge basalts.

Calc-alkaline and tholeiitic classifications

Coats (1952) studied a small number of samples from Adak and Kanaga and classified them as calc-alkaline. Their alkali-lime (Peacock) index was 63, actually putting them in the "calcic" category (as are most of the subsequently published Aleutian suites). Analyses of these rocks, plotted on an AFM diagram, showed no iron enirchment. Byers (1961) noted an aphyric iron-enriched (12.32 wt. % FeO*) basalt at Okmok, which he compared with tholeiitic basalt of the Columbia River Basalt Group.

Kay (1977) and S. M. Kay and others (1982) applied the FeO*/MgO versus SiO_2 discrimination diagram of Miyashiro (1974) to Aleutian volcanic rocks. Suites having relatively low and constant FeO*/MgO were classified as calc-alkaline, where-

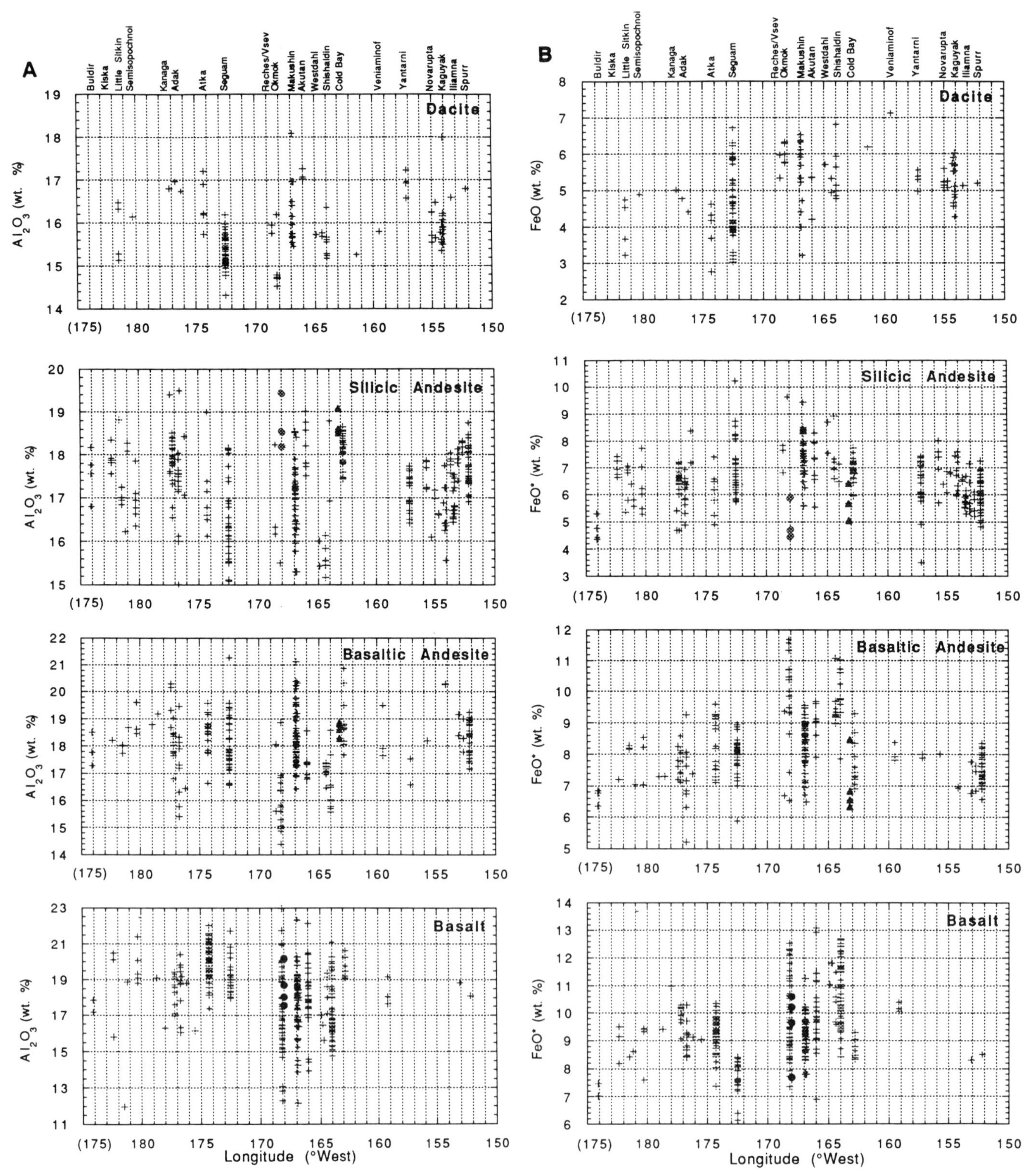

Figure 8. Plots of Al_2O_3 (A), FeO* (B), MgO (C), and K_2O versus longitude, from 152°W (Spurr) to 176°E (Buldir). Each oxide is given in wt. %, "water-free." Also plotted are the second-arc volcanoes of Amak (triangles; 163.15°W) and Bogoslof (solid and hachured circles; 168.03°W). For each oxide, there are separate plots for basalt (<53 wt. % SiO_2), basaltic andesites (53 to 57 wt. % SiO_2), silicic andesites (57 to 63 wt. % SiO_2), and dacites (63 to 70 wt. % SiO_2). Data sources given in Appendix 2.

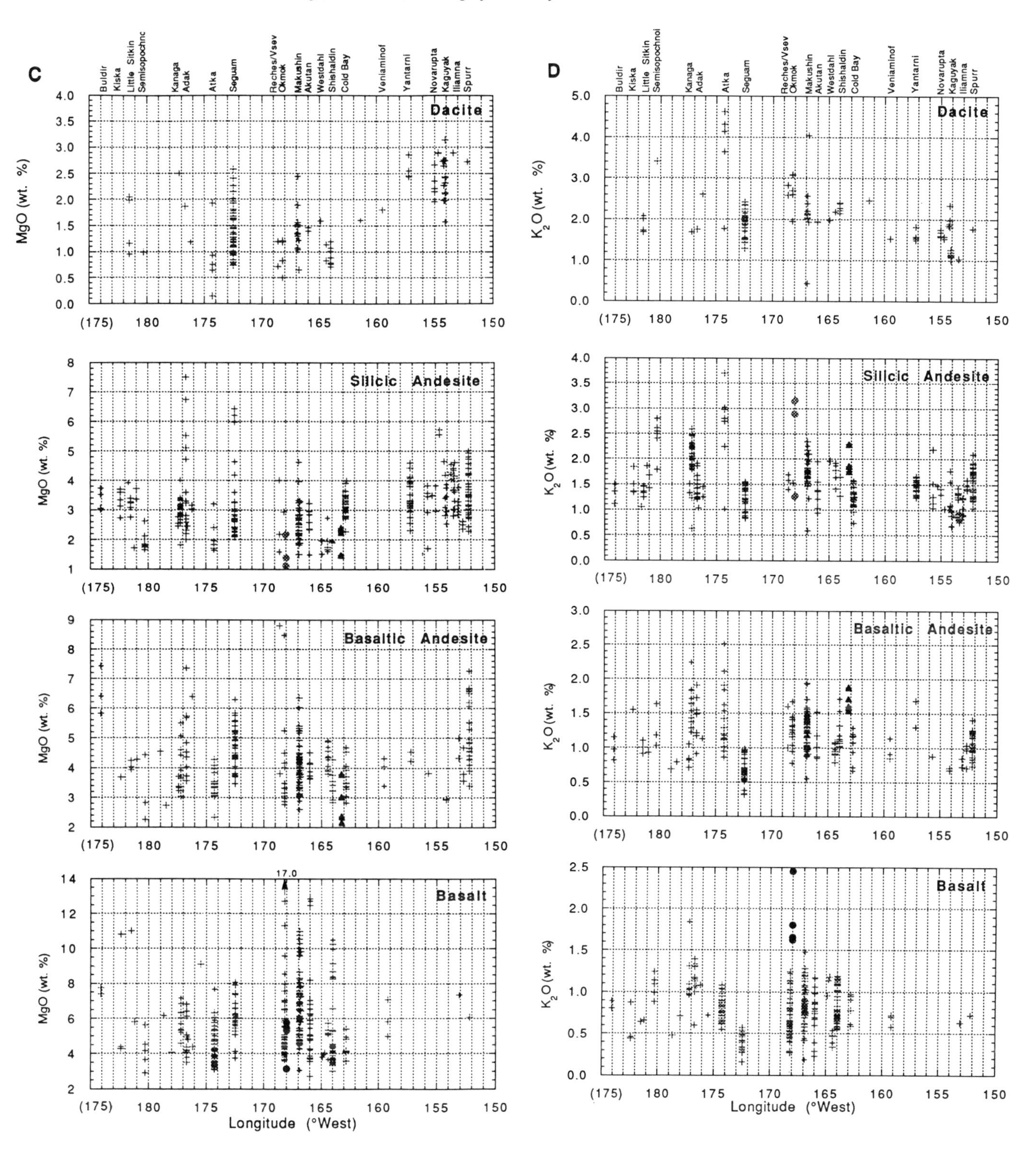
C
Buldir
Kiska
Little Sitkin
Semisopochnoi
Kanaga
Adak
Atka
Seguam
Reches/Vsev
Okmok
Makushin
Akutan
Westdahl
Shishaldin
Cold Bay
Veniaminof
Yantarni
Novarupta
Kaguyak
Iliamna
Spurr
MgO (wt. %)
Dacite
Silicic Andesite
Basaltic Andesite
Basalt
17.0
Longitude (°West)
D
K_2O (wt. %)
(175) 180 175 170 165 160 155 150

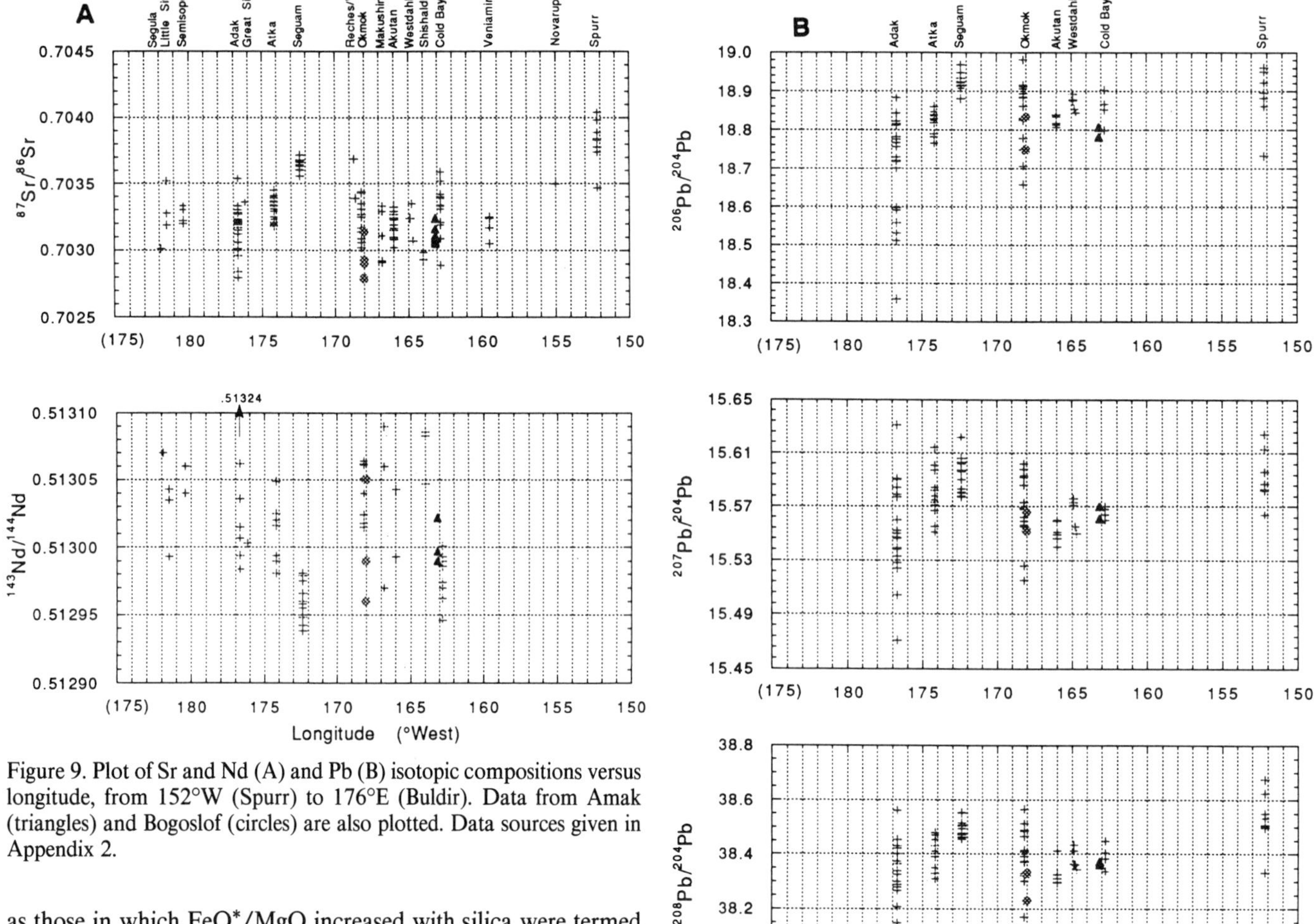

Figure 9. Plot of Sr and Nd (A) and Pb (B) isotopic compositions versus longitude, from 152°W (Spurr) to 176°E (Buldir). Data from Amak (triangles) and Bogoslof (circles) are also plotted. Data sources given in Appendix 2.

as those in which FeO*/MgO increased with silica were termed tholeiites (these rocks differ from MORB tholeiites in their higher K_2O content). S. M. Kay and others (1982) suggested petrographic, major- and trace-element, and volumetric differences among different centers. Large, mainly basaltic volcanic centers are tholeiitic; besides Fe-enrichment, they lack hydrous phenocrysts, have parallel REE patterns, and andesites or dacites are vitrophyric. Calc-alkaline centers are smaller, mainly andesitic, porphyritic with some amphibole, and have nonparallel REE trends. S. M. Kay and Kay (1985) further developed two intermediate classifications: transitional calc-alkaline and transitional tholeiite.

Several volcanic centers (Okmok, Shishaldin, Westdahl/Pogromni, and Seguam) have tholeiitic characteristics. Others such as Cold Bay, Kanaga, and Buldir are calc-alkaline. Myers and others (1985) also found these classifications to be useful for descriptive purposes. Many Aleutian suites, however, do not fit into either category. The dividing line is not clear, particularly in the continental sector of the arc, where Kienle and Swanson (1983) found that FeO*/MgO versus SiO_2 is not a useful discriminator.

Temporal chemical variations

Volcanism on Atka over the last ~7 m.y. shows that each volcanic system began anew with eruption of basalts, followed by andesite and perhaps some dacite. Over this period there has been no change in the fundamental character of the rocks (Marsh, 1980). Just south of Atka, on Amlia Island, there are rocks as old as ~40 Ma, some of which have been studied by Vallier and others (this volume). These rocks show the same diversity of types, but the basalts are virtually identical to those erupted from Korovin (on Atka) over the last 5,000 yrs.

Romick and others (1990) show that, over the past 5 m.y., there has been a decrease in the heterogeneity of Akutan lavas, reflected in bulk composition, mineralogy, and isotopes. The older (1.4 to 4.8 Ma) samples show a wide range in FeO/MgO, particularly in the basalts, plus extreme heterogeneity in plagioclase, olivine, and clinopyroxene compositions. The more recent (~1 Ma to present) volcanic materials show a more restricted range. Although there are less isotopic data for the younger rocks, there appears to be a significant decrease in $^{87}Sr/^{86}Sr$ heterogeneity from the older suite (n = 10, 0.703326 to 0.70302), through the 1-Ma suite (n = 5, 0.703190 to 0.703082), to the recent suite (n = 4, 0.703524 to 0.70342). The modern volcanic rocks show a distinct increase in $^{87}Sr/^{86}Sr$. Lead isotope ($^{207}Pb/^{204}Pb$ and $^{208}Pb/^{204}Pb$) data is less abundant (five old, two intermediate, one modern), but suggests—from the old to the intermediate stage—decreasing heterogeneity. Romick and others (1990), using incompatible element ratios, suggested that the three temporal groups of lavas were derived from different sources, and that older calc-alkaline or transitional tholeiitic lavas evolved to more depleted (in incompatible elements) and radiogenic (in Sr and Pb) modern tholeiitic lavas.

DSDP ash layers have provided information about Aleutian volcanic episodicity. The submarine ashes analyzed by Scheidegger and others (1980) provide some information, although they represent many combined sources and are likely biased toward more explosive (silicic) compositions. They indicate that volcanism may have become slightly more silicic over the last 1.5 m.y. Over the ~9 m.y. span of these ashes, silicic cycles have occurred every $\sim 10^5$ yr; however, the chemical nature of the volcanism has not changed systematically, and there have been no long-term changes.

Scheidegger and Kulm (1975) and Scheidegger and others (1980) determined chemical compositions for ashes from cores from near Kamchatka (DSDP Site 192), south of Unimak (Site 183), and east of Kodiak Island (Site 178). Because of the prevailing westerly winds, only ashes from the latter two cores probably have any significance for Aleutian volcanism. They conveniently and separately measure the regional volcanism of the oceanic (Site 183) and continental (Site 178) sectors of the arc. These data represent ashes as old as about 9 Ma, whereas the Aleutian volcanics described above are all less than about 3 Ma. Virtually all of the ashes have oxide totals ~96 wt. % (inferred ~4 wt. % H_2O), so some care must be taken when comparing them with subaerial data.

None of the submarine ashes has less than 54 wt. % silica, perhaps due to the lower explosivity of basaltic magma. As in the lavas, the average ash in the more oceanic sector is lower in silica (62 wt. %) than the average ash in the more continental sector (67.6 wt. %). There is significant scatter in Na_2O and CaO, particularly at silica contents above 60 wt. %, which may indicate the effects of alteration by seawater. As in some Aleutian volcanic rocks, there may be two trends in K_2O versus SiO_2.

There are some striking differences, mainly in the mafic (54 to 55 wt. % SiO_2) ashes from Site 183, which have higher TiO_2 and FeO* and lower Al_2O_3 than normally found in Aleutian subaerial volcanics. Ash and lapilli of similar composition were collected in September 1975, 90 km northeast of Shishaldin in the Bering Sea, by a NOAA fisheries research vessel. At that time, Shishaldin was observed to be vigorously erupting. Simkin (written communication, 1976) determined the chemical of this coarse basaltic (50 to 52 wt. % SiO_2) ash as highly enriched in TiO_2 (2.7 to 3.0 wt. %) and FeO* (13.5 to 14.8 wt. %) and depleted in Al_2O_3 (12.4 to 14.7). The apparently anomalous DSDP 183 ashes may actually be geochemical fingerprints of input from the large volcanic centers of Shishaldin and Westdahl on Unimak Island (Site 183 is southeast of Unimak).

PETROLOGY

Phase equilibria

Study of Aleutian lavas and their phenocrysts reveals several petrologic problems that may be resolved by phase equilibria considerations: the common presence of minor amounts of An_{90-100} plagioclase; of olivine either too Fo-rich or too Fo-poor for its host bulk composition; and of diopsidic-salitic clinopyroxene phenocrysts, distinct from normal augitic compositions.

Anorthite-rich plagioclase. Plagioclase dominates the phenocryst assemblage of most Aleutian basalts and andesites, with a wide range of compositions and textures. Cores are richer in An-content than adjacent mantles, although exceptions occur (e.g., Marsh and others, 1990). Drops of An_{15-30} are common at relatively narrow rims. Anorthitic ($>An_{90}$) plagioclase are present in minor amounts in many if not most Aleutian lavas. Such anorthite-rich plagioclase has sometimes been suggested to indicate high water content in arc magmas (e.g., Yoder, 1969). In water-saturated experiments in the simple albite-anorthite system, Yoder (1969) found that water lowers the plagioclase solidus at constant temperature, such that a more anorthitic crystal is produced (i.e., 150 bars H_2O pressure effects a change in the plagioclase from An_{60} to An_{74}).

The effect of water on multicomponent melts of basaltic composition, however, appears to be different. The thermodynamic melt model of Burnham (e.g., 1980) suggests that, for every 10 mole % (≈1 to 1.5 wt. %) increase in water in the melt, the plagioclase becomes only 1 mole % more anorthitic. An unrealistically large amount of water would thus be required to alter crystallization from, say, An_{80} to An_{90}. In fact, high H_2O contents would actually suppress plagioclase crystallization, with clinopyroxene consuming appreciable amounts of Ca; when plagioclase eventually crystallizes, it is of a lower An-content (Crawford and others, 1987). The silicate melt model of Ghiorso (1985) suggests a slight *decrease* in plagioclase An-conent with increasing H_2O. Regardless, both models indicate no major change in An-conent with addition of water to the silicate melt.

In addition, one-atmosphere melting experiments using Aleutian lavas suggest that the soda content of the melt largely dictates plagioclase composition (Marsh and others, 1990). Equilibrium plagioclase compositions of actual high-Al basalts vary, but mainly are An_{75-85}. The distinctive, large, unzoned anorthite crystals require equilibrium with a soda-poor melt, one which has not been identified in the Aleutians.

Olivine. Roeder and Emslie (1970) determined the Fe-Mg distribution between olivine and basaltic liquid and found K_D = 0.30 ± 0.03. The results of oxygen-buffered 1-atmosphere experiments on Aleutian lavas (Peterson and Marsh, unpublished data) similarly indicate K_D = 0.33 (± 0.01). Examination of the Aleutian olivines relative to their host lavas indicates *two* distinct disequilibrium conditions. The presence of olivines too Mg-rich (e.g., Fo_{90-93}) for their host liquid is commonly recognized; however, the second problem is that most olivine phenocrysts found in the Aleutians are generally *too Fe-rich* relative to the host lava (Marsh, unpublished data). Brophy (1984, 1986a) noted this at Cold Bay, where olivines in equilibrium with the host lava should be Fo_{78-80}. The olivines actually preent are Fo_{60-75}, which implies that the actual parental liquid would be more iron-rich (i.e., molar FeO/MgO of 1.1 to 1.8) than the lavas found.

The first disequilibrium condition could be caused by the interaction or contamination of low-Mg basalt by mafic wall rock (i.e., peridotite), which would elevate the bulk Mg content by the addition of Mg-rich olivine or clinopyroxene, with little time for reequilibration. If sufficient time were available, olivine-liquid Fe-Mg reequilibration would eliminate evidence of wall-rock interaction. Keleman (1986, 1990) and Keleman and Ghiorso (1986) have modelled reactions between peridotite and fractionating tholeiitic basalt.

The second disequilibrium condition may result from reequilibration of liquidus olivine with more Fe-rich differentiated residual liquids at higher degrees of crystallinity, such as observed in Hawaiian lava lakes (Moore and Evans, 1967). Alternatively, more Fe-rich magmas may be present at depth but fail to reach the surface.

Magmas having such low Mg-contents do occur at a few Aleutian volcanoes. Fournelle (1988) found that the majority of Shishaldin basalts are Fe-rich, with molar FeO/MgO of 1.1 to 1.4. They contain olivine phenocrysts of Fo_{70-75}, which are at equilibrium with the wallrock composition at the appropriate f_{O_2}. Byers (1961) and Nye and Reid (1986) observed low-Mg basalts at Okmok, and S. M. Kay and others (1982) at Westdahl. Thus, whereas these low-Mg and Fe-enriched basalts are relatively uncommon, they are present at some of the largest Aleutian volcanic centers.

Clinopyroxene. Distinctive green, nonaugitic clinopyroxene occur in lavas from several Aleutian volcanoes (Shishaldin, Okmok, Moffett), as well as volcanoes in Colombia, Nicaragua, and Japan (Fournelle, 1988). They are rich in a diopside or salite component, with elevated levels of Al_2O_3, TiO_2, Cr_2O_3, and Fe^{3+}/Fe^{2+}, and closely resemble those found in the southeastern Alaskan ultramafic layered intrusions. They have been referred to as the salitic trend by Conrad and Kay (1984). At Shishaldin, these clinopyroxene are $Wo_{45-48}En_{42-46}Fs_{8-11}$, with 2.5 to 5.5 wt. % Al_2O_3, 0.5 to 1.4 wt. % TiO_2, and up to 0.7 wt. % Cr_2O_3.

Conrad and Kay (1984) suggested that the presence of this clinopyroxene indicates a high H_2O content in parental Aleutian magmas. Following Irvine's (1974) suggestion that the Duke Island magmas were alkali basalts and thus of low silica activity, Conrad and Kay proposed that this mineral indicated high magmatic water content. This, however, is not the only possible way to cause low silica activity in the magma; a variety of reactions can be written to yield diopside-rich clinopyroxene. There is no need to invoke water, as shown by the following reactions (components in shorthand notation, phases in parentheses):

Di (cpx) + ½ SiO_2 (liq) + Al_2O_3 (liq) = An (pl) + ½ Fo (ol),
2 Di (cpx) = CaO (liq) + 3 SiO_2 (liq) + Fo (ol),
4 Di (cpx) + 3 Al_2O_3 (liq) = CaO (liq) + 3 An (liq) + 2 Fo (ol).

Diopsidic clinopyroxene is favored when activities of An and Fo are high, which would be the case if a high-alumina basalt eroded an olivine-rich body.

Silica-undersaturated liquids could be produced by fractionation of a high-Al basalt at pressures greater than a few kilobars (Marsh and others, 1990), for the silica content of fractionating plagioclase may be higher than that of the liquid. Or, a magma reacting with peridotite would be expected to develop a lower silica activity.

Diopsidic-salitic clinopyroxene may also be a reaction product. Manning and Bird (1986), suggested that Lower Zone Skaergaard clinopyroxene ($Wo_{40}En_{47}Fs_{13}$, with low Al_2O_3, TiO_2, and Cr_2O_3) is hydrothermal and formed by reaction of magmatic clinopyroxene with high-temperature aqueous fluids. Fournelle and Marsh (1987) suggested, based upon textural evidence, that Shishaldin diopsidic-salitic clinopyroxenes are products of a low-Mg basalt reacting with Fo-rich olivine bearing Cr-spinel inclusions.

Magmatic intensive parameters

Temperature and water content. Because the exact water content of arc magma is unknown, application of any geothermometer (i.e., plagioclase or olivine) that employs magmatic liquid in its defining reaction cannot be used. The characteristic low titania (<1.1 wt. %) content of arc magmas results in a scarcity of ilmenite, thus limiting the usefulness of the two-oxide (Buddington-Lindsley) geothermometer/oxygen barometer. The two-pyroxene geothermometer is applicable only to andesitic lavas. Direct estimation of magmatic temperatures in basaltic lavas thus is difficult.

The sequence of crystallization in most basalts, especially the stability of plagioclase, is strongly affected by water. At a few kilobars of pressure, saturation with water may cause plagioclase and all mafic phases to crystallize essentially together. A significant amount of water is, however, required for saturation: for a basaltic melt, $\sim$7 wt. % at 4 kbar, and $\sim$11 wt. % at 8 kbar (Burnham, 1979). The presence of dissolved water in the magma lowers the liquidus temperature to the point that amphibole is stable; this occurs at temperatures below about 1,050°C in a basalt, 950°C in an andesite (Eggler, 1972), and 900°C in a dacite (Merzbacher, 1983; Merzbacher and Eggler, 1984).

In most Aleutian-arc basalts, plagioclase appears on the liquidus well before any other phase. In addition, basaltic plugs containing 60 vol. % crystals show no sign of having entered the stability field of amphibole. These facts suggest that the amount of water in most Aleutian basalts is less than $\sim$2 wt. %, and probably less than 1%. Also, as the basalt is undersaturated with water at low pressure, it must be highly undersaturated at high pressures.

Basalts. As described above, pre-eruptive temperatures in basalts are not easy to estimate. Where a pair of pyroxenes or

Fe-Ti oxides occurs, the temperature commonly is in the range 1,100° to 1,200°C (e.g., Brophy, 1986a; Fournelle, 1988). Singer and others (1992b) found temperatures of 1,146° and 1,173°C in a Seguam basalt, using oxygen isotope thermometry on mineral separates. At Okmok, however, Nye and Reid (1986) suggest an unusually high temperature of 1,300° to 1,400°C (at >9 kbar), on the basis of the presumed existence of a (water-free) magma in equilibrium with Fo_{93} olivine. In general, if allowance is made for the crystal-rich nature of the common basalt, extrapolation to a crystal-free state puts the liquidus in the range of 1,200° to 1,250°C.

Andesites. Pre-eruptive temperatures range from about 950 to 1,050°C (Hildreth, 1983; Brophy, 1984, 1987; Fournelle, 1988; Romick and others, 1990; Singer and others, 1992a). Extrapolation to the crystal-free state suggests a liquidus temperature of near 1,200°C; this temperature, plus comparison with Merzbacher and Eggler's (1984) geohygrometer, suggests water contents of 3 to 5 wt. % (Brophy, 1984; Singer and others, 1992a).

Dacites and rhyolites. Crystallizing temperatures in these rocks fall in the range of 800° to 955°C. The dacites occupy the upper half of the range (Anderson, 1975; Hildreth, 1983; Fournelle, 1988; Singer and others, 1992a). Inferred water contents for dacites is 2 to 4 wt. % (Baker and Eggler, 1987; Fournelle, 1988; Singer and others, 1992a). Infrared spectroscopy indicates 2.7 wt. % H_2O in Katmai dacite and 4.5 wt. % HO in rhyolite (Lowenstern, 1990).

Pressure. There is no precise indicator of pressure of crystallization in these lavas. The best procedure is to compare the suite in question with experimentally determined crystallization sequences and coexisting melt compositions (i.e., cotectics) and work out the set of variables (P, H_2O content) most consistent with the rock compositions and mineral assemblages (Marsh, 1976).

The rare occurrence of alumina-rich orthopyroxene (at Cold Bay) suggests that some andesites contain a mineral assemblage that equilibrated at 8 to 10 kbar (Brophy, 1984). Pseudoternary projections suggest that some high-alumina basalts (bulk compositions) reflect a pressure of equilibration (i.e., separation from source) near 20 kbar (Fournelle and Marsh, 1991), whereas the modes and mineral compositions reflect lower pressure (<8 kbar) equilibria (Baker, 1987; Baker and Eggler, 1983, 1987; Gust and Perfit, 1987; Romick and others, 1990). The more silicic lavas appear to have crystallized at lower pressures; Singer and others (1992a), for example, suggest pressures of 3 to 4 kbar for Seguam andesites and 1 to 2 kbar for dacites.

Fugacity of oxygen (f_{O_2}). Oxygen fugacity is commonly calculated from magnetite and ilmenite pairs. Ilmenite is difficult to find in the low-TiO_2 Aleutian volcanic rocks, limiting application of this method for most basalts and andesites. Experimental studies (Sack and others, 1980; Kilinc and others, 1983) have found that the Fe^{3+}/Fe^{2+} of silicic glass is a function of bulk composition, temperature, and f_{O_2}. Oxygen fugacity can be inferred from minimum values of f_{O_2} determined from whole-rock Fe^{3+}/Fe^{2+} composition, assuming that they represent liquid compositions.

A distinctive chemical features of many island-arc lavas is their elevated f_{O_2} relative to ocean-ridge rocks. Whereas ocean-ridge rocks define a clear trend near the quartz-fayalite-magnetite (QFM) buffer, arc lavas show a trend nearly parallel to that buffer, but up to 2 log units higher, above the nickel-nickel oxide (NNO) buffer (Gill, 1981). (In the temperature range of interest, the NNO buffer curve is 0.7 log units above the QFM curve.)

Marsh (1980) suggested, based on the observed phenocryst assemblages, that Atka basalts crystallized about 0.5 to 1 log units above the NNO buffer. Hildreth (1983) found, from FeTi-oxide pairs in 65 of the 1912 Katmai samples, a trend with f_{O_2} above and at a slightly steeper slope than the NNO buffer curve (varying from high to low temperature samples—basalt to rhyolite).

Figure 10a shows the f_{O_2}–temperature relations, determined from FeTi oxides, for Shishaldin, Cold Bay, Okmok, Seguam, and Katmai lavas and pyroclastics.

Figure 10b shows the f_{O_2} for Shishaldin lavas calculated from whole-rock Fe^{3+}/Fe^{2+}; it indicates that f_{O_2} lies near the QFM and NNO buffers, consistent with the oxide data. Whereas no FeTi data exist for Atka, whole-rock Fe^{3+}/Fe^{2+} data indicates an f_{O_2} of approximately 1 to 1.5 units above the NNO buffer curve (Fig. 10c).

Andesites and dacites at Katmai (Hildreth, 1983), basalts and andesites at Cold Bay (Brophy, 1984, 1986a, 1987), and basalts at Atka (Marsh, 1980, 1982a) crystallized at an f_{O_2} of approximately NNO + 1 log unit, with it decreasing to NNO + 0.1 in rhyolite (Katmai: Hildreth, 1983).

Other Aleutian rocks, however, provide evidence for less oxidized preeruptive conditions. FeTi-oxides in volcanics (primarily basalts) from Shishaldin (Fournelle, 1988), Okmok (Anderson, 1975, written communication, 1987), and Seguam (Singer and others, 1992a) suggest crystallization at an f_{O_2} between the QFM and NNO buffers. It is probably no coincidence that lavas and tephra from these three volcanic centers have the tholeiitic characteristics of iron-enrichment, for the lower f_{O_2} would suppress magnetite (and ilmenite) crystallization.

The presence of sulfides has been documented at one Aleutian volcano, Novarupta. Hildreth (1983) found pyrrhotite (in apparent trace amounts) in mineral separates of the 1912 ejecta; inclusions of pyrrhotite in pyroxenes and oxides were also present. Reconnaissance study also indicates sulfides are present in some Shishaldin basalts (Fournelle, unpublished data).

The apparent absence of sulfides at most Aleutian volcanoes may be due to incomplete studies. However, if the absence is real, it may be a result of magmatic f_{O_2} being in the range of NNO to NNO+1 (or higher) at the particular volcano in question. Carroll and Rutherford (1987, 1988) found that magmatic sulfur speciation changes drastically, from sulfide to sulfate, over this f_{O_2} range. At higher f_{O_2}, anhydrite replaces pyrrhotite. Anhydrite has not been found yet in Aleutian volcanic rocks, but it is unstable in a wet environment. It has been found elsewhere, in freshly collected pumices (El Chichon: Luhr and others, 1984)

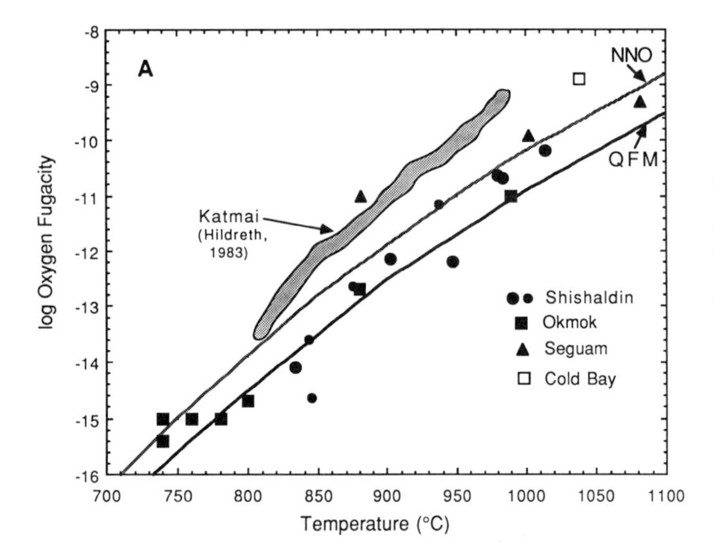

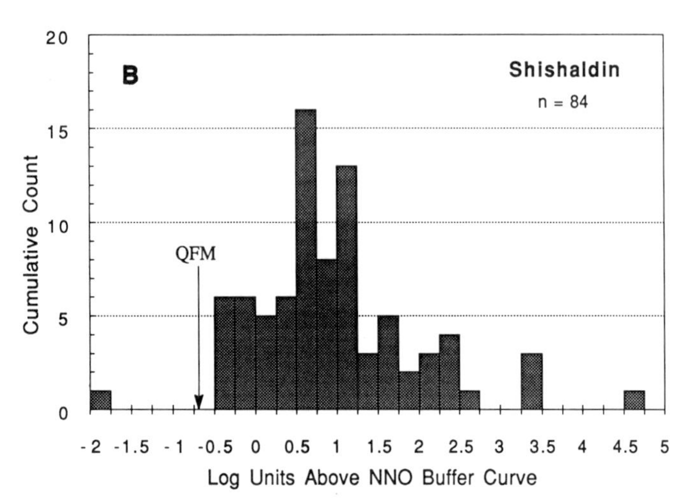

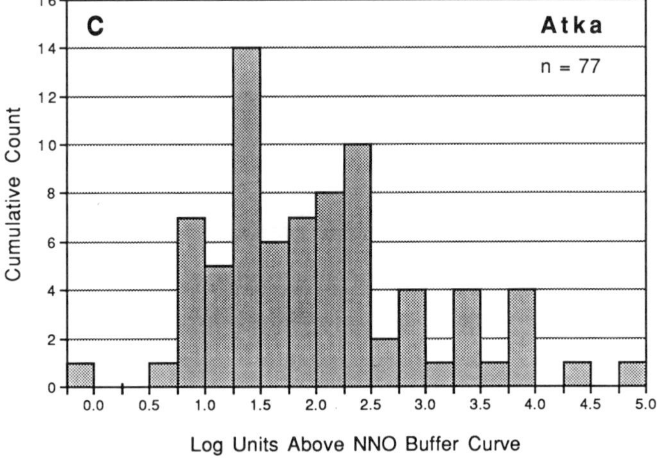

Figure 10A, Plot of log oxygen fugacity versus temperature (°C) determined from coexisting FeTi oxides from five Aleutian volcanoes. Data are from Cold Bay (Brophy, 1984), Shishaldin (Fournelle, 1988), Katmai (Hildreth, 1983), Okmok (Anderson, 1975, written communications, 1987), and Seguam (Singer and others, 1991). For Shishaldin, the larger symbol indicates grains >100 microns; smaller symbol, <100 microns. Also shown are the nickel-nickel oxide (NNO) and quartz-fayalite-magnetite (QFM) buffer curves. B, Calculated oxygen fugacities of Shishaldin volcanic rocks based upon whole-rock Fe^{3+}/Fe^{2+} following the experimental algorithm of Sack and others (1980) and Kilinc and others (1983). These values are consistent with f_{O_2} values between the QFM and NNO buffers, determined from ilmenite-magnetite pairs. (Three data above NNO+5 not included.) C, Calculated oxygen fugacities of Atka volcanic rocks, based upon whole-rock Fe^{3+}/Fe^{2+}, as for B. These values are consistent with modal mineralogy of Atka volcanic rocks, which suggests equilibrium 1 to 2 log units above the nickel-nickel oxide (NNO) buffer.

and in inclusions within pumice phenocrysts (Nevado de Ruiz: Fournelle, 1990a).

Like isotopic composition, f_{O_2} is inherited by a magma from its source rock. It is difficult to change f_{O_2}, except along the usual pseudobuffer or indicator curves by changing temperature, although it could be done by assimilating large amounts of highly reduced or oxidized wall rock, or by reaction of variable oxidation state sulfur phases in the magma (e.g., Whitney and Stormer, 1983).

High-alumina basalts and plagioclase accumulation

Many Aleutian high-alumina basalts (i.e., 17 to 22 wt. % Al_2O_3) contain 25 to 50 vol. % plagioclase. Crawford and other's (1987) global study concluded that such high-Al basalts are cumulates and they calculated that the groundmass of Aleutian high-Al basalt should be andesitic-dacitic. Later studies on Aleutian lavas have accepted these conclusions (Brophy, 1989a). Brophy (1989b) developed a model to explain plagioclase retention in crustal magma chambers.

In a study of plagioclase-rich recent lavas from Shishaldin, Fournelle and Marsh (1991) found no evidence of plagioclase accumulation. Groundmass and mineral separates were studied. The groundmass separates were basaltic, not andesitic or dacitic. Shishaldin high-Al basalts have positive Eu-anomalies, but they do not match those expected by plagioclase addition, and may instead be a signature of clinopyroxene in an eclogitic source (Brophy, 1986b). There is no evidence of disequilibrium between plagioclase and liquid/groundmass in the high-Al basalts. Plots of Al_2O_3, CaO, and Na_2O versus modal plagioclase do not correspond with plagioclase addition to high-Mg basalt or dacite. On the other hand, several older (~3 to 5 Ma) high-Al basalts from an adjacent glaciated unit show some evidence for plagioclase accumulation in a high-Mg basalt; Fournelle and Marsh (1991) suggested that this was a possible occurrence in the early development of the volcanic center.

Xenoliths and xenocrysts

True ultramafic (peridotite) xenoliths are uncommon in island arcs. Swanson and others (1987) examined the characteristics of Aleutian xenoliths from Kanaga and Adak. Those at Kanaga are in Tertiary rocks and not related to current volcanism; at Adak, ultramafic inclusions are present in two brecciated units (not lavas). Gabbroic xenoliths from Adak have been studied by Conrad and others (1983) and Conrad and Kay (1984), who suggested they were cumulate phases of early arc magmas.

More common, however, are individual crystals that stand out as foreign to their host lava.

Olivine. These olivine crystals are highly magnesian (Fo_{90-93}), large (0.1 to 0.2 cm), sometimes slightly strained and have a more fayalitic rim. Somewhat less magnesian (Fo_{85}), smaller, and highly corroded olivine crystals also sometimes occur in andesitic lavas that otherwise lack olivine (e.g., Adak: Marsh, 1976; Cold Bay: Brophy, 1984). These olivines are not in equilibrium with their host rocks, and appear to be from mantle peridotite or oceanic crust.

Based upon a global study of olivine minor element composition, Simkin and Smith (1970) suggested that xenocrysts had less than 0.10 wt. % Ca (elemental). Nye and Reid (1986) used this criterion to interpret Okmok Fo_{92-93} as nonxenocrystic. However, in a more recent study of xenocrystic olivine, Bodinier and others (1987) found 0.15 to 0.16 wt. % Ca in olivine cores in some garnet and spinel lherzolites. We suggest that the olivine Ca-content limit of 0.10 wt. % should not be used to preclude Fo-rich olivine from being characterized as xenocrystic.

Plagioclase. Plagioclase more calcic than An_{90}, and certainly above An_{95}, is xenocrystic, and could be from the underlying arc crust or from fragments of older decoupled slabs. Megacrysts of An_{90} have been found in a basalt dredged from the flank of a seamount just south of the Aleutian Trench in the Adak fracture zone (Fournelle, unpublished data).

Quartz. The presence of rounded ~1 cm quartz in high-Mg Shishaldin basalt indicates contamination (Fournelle, 1988).

Also, Marsh (unpublished data) found a Si-rich glassy bomb at Adak (sample AD97), with 88.6 wt. % SiO_2, 3.1 wt. % Al_2O_3, and 5.2 wt. % H_2O+CO_2, and having Sr and Pb isotopic compositions within the range of all Adak samples.

PETROGENESIS

Introduction

Discussions of arc petrogenesis have focused either on the deep source at the slab-mantle wedge interface, or upper level (crustal) processes. Three key source components have been hypothesized: the subducted oceanic crust, subducted sediments, and the overlying mantle wedge (Fig. 11). Upper level processes (e.g., fractionation, magma mixing, assimilation) are considered possible modifiers of the primary magma compositions. Presumably, by inverting the process and sorting out the upper level effects, the nature of the original source can be deciphered. This approach, however, assumes that the long ~100-km ascent through the lithosphere (i.e., upper mantle) will have a *negligible* effect on the composition of ascending magma. This assumption would be true only if the magma and the lithosphere were identical in composi-

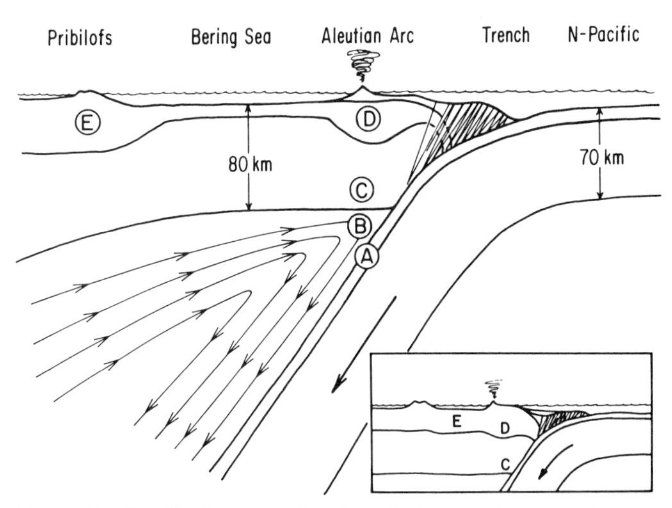

Figure 11. Simplified cross section through the oceanic part of the Aleutian arc. Possible regions of magma production and interaction are: A, subducted oceanic crust (with layer of sediments); B, asthenosphere ("mantle wedge"); C, arc lithosphere; D, subvolcanic front welt; E, "continental" crust. The insert shows a cross section through the Alaska Peninsula. Heavy arrow indicates motion of subducted plate. Streamlines indicate convective motion of the asthenosphere as it is coupled with downward motion of the subducted slab. (From von Drach and others, 1986.)

tion and at the same temperature, which is unlikely, as then there would be no buoyancy to force the magma to the surface. Instead, it appears most reasonable that chemical and thermal interactions between compositionally distinct magma and mantle may significantly alter the primary magma as it develops its ascent conduit (Marsh, 1978; Myers and others, 1985; Singer and others, 1989).

Diapir transport

How does magma travel upward from its source? There is no sign of earthquakes leading upward through the lithosphere (as at Hawaii) from near the subducting plate, even beneath the smallest, youngest volcanic centers. This suggests a passive mode of magma transfer such as diapirism. Enough heat must be carried upward to soften the wall sufficiently to allow the body to travel fast enough to reach the surface while still partially molten (Marsh, 1978, 1982b). To allow this, at least several bodies must successively travel essentially the same path through the lithosphere. As the initial mush begins ascent it undergoes increased melting as it tries to rise further away from its solidus. Once melting progresses to about 50 vol. %, the remaining solids can be repacked to free a magmatic liquid. The possibility that viscous compaction of the solids releases a liquid at very small amounts of melting seems much less likely (Brophy and Marsh, 1986).

The sharpness of the volcanic front and the fairly regular spacing of the volcanic centers imply that the magma may be transported to the surface via gravitational instability. The magmatic source "layer" produces a series of (ideally) evenly spaced diapirs, whose spacing and size is controlled by the thickness and viscosity of the layer and by the viscosity of the overlying material. An experimental and analytical study of this process, as applied to the Aleutians (Marsh, 1979b), suggests that the magmatic zone is thin ($<\sim$500 m), narrow ($\sim$10 to 30 km), highly viscous ($\sim 10^{12}$ to 10^{14} poise), and produces diapirs of 3 to 5 km in radius. Marsh (1979b) calculated the local mantle viscosity as $\sim 4 \times 10^{20}$ poise, which is in close accord with studies using glacial rebound.

Singer and others (1989) calculated that 3- to 6-km-radius diapirs, periodically ascending at 4×10^{-8} to 2×10^{-7} m/s, could establish magmatic conduits from 120 km depth to the crust in 60,000 to 900,000 years. In the process, large volumes of magma must solidify in the lithospheric mantle.

Evolution of volcanic plumbing systems

The rigors of fully penetrating the lithosphere almost guarantees that the first eruptions of a new volcanic center will be of magma that has interacted strongly, chemically and physically, with its wall rock (Marsh and Leitz, 1979). The first extrusives would be expected to have a heterogeneous isotopic, trace element, and bulk chemical signature, the net result of the original source material plus any lithospheric or asthenospheric wall rock it has encountered during ascent, as well as enhanced crystallization due to the initial cool state of the conduit walls. Bogoslof, appearing in 1796, is an excellent example.

Each magma traveling this general pathway will also react with the wall rock, but each time to a lesser degree as the conduit walls themselves are "contaminated" by solidified magma (Marsh and Kantha, 1978). As the conduit is gradually heated, successive diapirs may experience different crystallization sequences, which may in turn yield different liquid lines of descent (Singer and others, 1989). Passage of enough magma will produce a thermally and chemically insulated conduit, thus allowing magma to erupt that could carry the geochemical signature of its source (Fig. 12). Small volcanic centers, then, are likely to have "dirty" lithospheric plumbing systems, whereas larger, mature centers should have relatively "clean" systems (Myers and others, 1985).

An example of the foregoing model is Adak (Marsh, 1982a), a small to medium-sized volcanic center ($\sim$30 km^3) in the central Aleutians. Strontium and lead isotopic data, along with major- and trace-element data, show a good deal of scatter. One of the lowest exposed units—a breccia—that issued from Mount Moffett has an unusually low $^{87}Sr/^{86}Sr$ value of 0.7028; i.e., sample ADK-53 of Kay (1978), is andesitic (55 wt. % SiO_2), but contains high (5.58 wt. %) MgO, 150 ppm Ni, anomalously high Sr (1,783 ppm), and is dominated by large, unzoned fragments of Mg-rich clinopyroxene. Slightly higher in this same section is an andesitic flow containing abundant Fo-rich olivines (Marsh, 1976; sample AD-48); these features are consistent with strong interaction with lithospheric wall rocks.

A larger center, such as Atka ($\sim$200 km^3), on the other

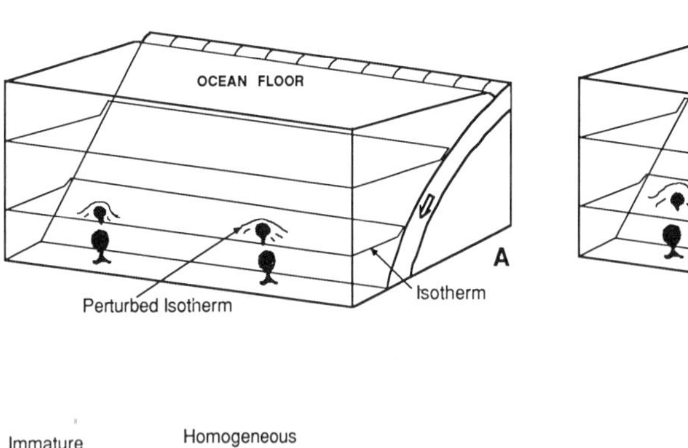

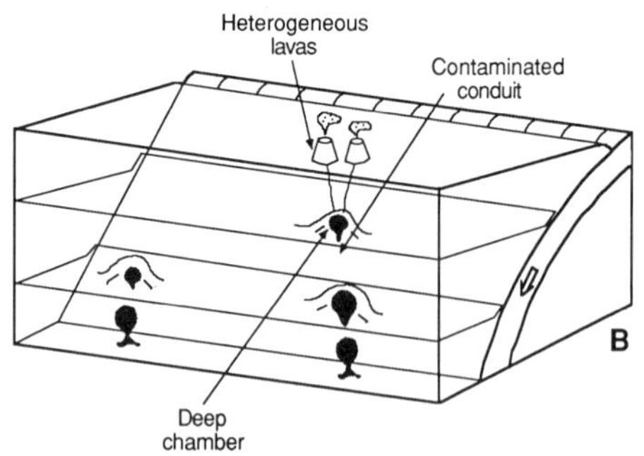

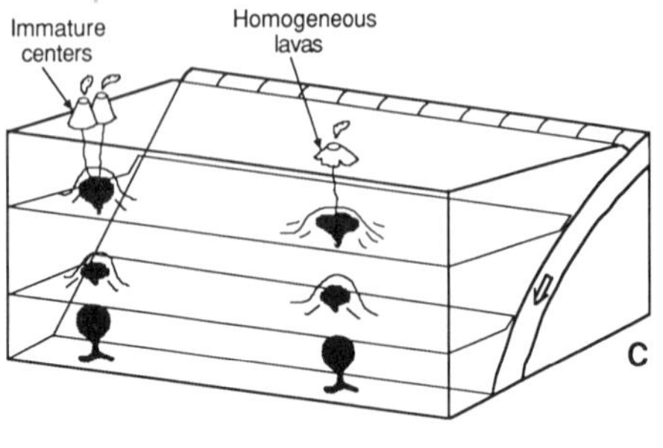

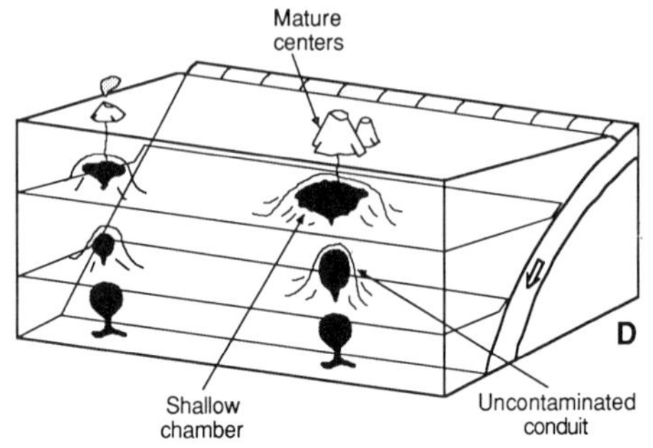

Figure 12. Schematic representation of the evolution of an Aleutian volcanic center. A, early stages of magma ascent with significant thermal and chemical interaction between magma and wall rock; B, eruption of the first highly contaminated lavas from deep-seated magma bodies; C, magma bodies are closer to the surface and erupted lavas are less contaminated; D, fully developed plumbing system. Eruptions are from well-established shallow magma chambers. Material in the plumbing system itself has been removed or chemically modified. Lavas erupted at this stage are petrographically and chemically homogeneous. (After Myers and others, 1985.)

hand shows a tight, restricted range of Sr and Pb isotopic data and smooth trends of major and trace elements. No xenoliths or xenocrysts (besides An-rich plagioclase) have been found in these lavas.

Volcanic centers that become too big (e.g., Okmok, 200 to 300 km^3; and Shishaldin, 300 to 400 km^3), may have so much magma fluxing the uppermost lithosphere that they erode and entrain fragments of ultramafic or mafic wall rock. Those lavas would show a distinct chemical signature that is indistinguishable from a peridotite source (e.g., Fo-rich olivine with chromite inclusions; high whole-rock Mg-content, Ni and Cr contents)—unless the magma were erupted so quickly that the intact wall rock (xenoliths or xenocrysts) were carried to the surface. This appears to have happened at Shishaldin, where chromite and clinopyroxene compositions in high-Mg basalts resemble those of the southeastern Alaskan zoned ultramafic complexes and of the Border Ranges ultramafic and mafic complex of Burns (1985). Based upon this evidence and geophysical studies of the Border Ranges fault (Fisher, 1981; Marlow and Cooper, 1983), Fournelle (1988) suggested that the Border Ranges ultramafic and mafic complex may be present below Shishdalin volcano. A similar scenario was suggested by Arculus and others (1983) at Mount Lamington (Papua New Guinea), where Mg-rich xenoliths were found in andesitic lavas. The Papuan Ultramafic Belt crops out within 10 km of the volcano, and geophysical evidence indicates it lies 12 to 16 km below the volcano.

The source of Aleutian magmas

The controversy over the source of Aleutian magmas is essentially the same today as when it was posed by Coats in 1962: whether the subducted slab yields an aqueous fluid (containing silica, Sr, Rb, Ba, and Pb) that causes melting of the mantle wedge; or whether the slab itself melts, yielding a silicate fluid (melt), containing some water and Sr, Rb, Ba, and Pb. This problem has been addressed using isotopes, trace elements, rare-earth elements, major elements, and magmatic intensive parameters, but seldom have all been considered simultaneously. McCulloch and Perfit (1981, p. 176) summed up the situation: "Although the ostensibly simple model of producing the Aleutian arc magmas by directly melting of a source consisting of partially altered MORB and several percent entapped sediment satisfies the Nd, Sr, and Pb isotopic constraints, it must also be compatible with the major and trace element data." The two main hypotheses are discussed below.

Slab dewatering–peridotite melting model. This hypothesis holds that an aqueous fluid from the subducted slab causes melting of the overlying mantle wedge, yielding a high-Mg basaltic melt that ascends to near (or above) the crust-mantle boundary, where fractionation can produce either high-Al basalts, or combined assimilation/fractional crystallization yields calc-alkaline andesites (Perfit and others, 1980a, 1980b; Gill, 1981; Nye and Reid, 1986; Gust and Perfit, 1987; Kay and Kay, this volume). Key to this model are the occasional arc basalts having elevated MgO (>10 wt. %), Ni (>100 ppm), and Cr (>100 ppm) and bearing Fo_{90-92} olivine. These unusual lavas are cited as evidence of an abundant but commonly untapped parental magma.

The presence of Fo-rich olivine in high-Mg arc basalts, however, is not conclusive evidence that these basalts were derived by partial melting of mantle peridotite (Myers, 1988). Such high-Mg basalts can be produced by low-Mg basalt interacting with ultramafic material in the lithosphere below the volcano (Arculus and others, 1983; Fournelle, 1988).

A major question of arc petrogenesis is how to melt the cold subducted slab. Mantle wedge metasomatism has been suggested (e.g., Tatsumi and others, 1986), which assumes that sufficient water is still present (having not been dehydrated earlier), and that the released aqueous phase travels vertically upward. Oxburgh and Turcotte (1976) suggest instead that at shallow depths, water may migrate downwards, deeper into the slab. Only at much greater depths (up to 700 km), could it be released upwards. Eggler (1989), on the other hand, suggests that any shallow (<75 km) slab-derived fluids will travel back up the slab, whereas fluids released at greater depths will be swept downward with the induced mantle flow.

If water-induced melting of peridotite is the process responsible for Aleutian arc magmatism, and if the commonly observed Aleutian high-alumina basalts are derived by fractionation from the high-magnesian basalt in equilibrium with this peridotite, then two consequences should be evident. First, the water content of the high-alumina basalts (the products of fractionation) should be unusually high if water-saturated melting of peridotite is invoked. If the peridotite melt is undersaturated with respect to water, then there should be noticeably high water contents of the high-alumina basalt. The common phenocryst assemblage in Aleutian basalts—lack of amphibole and abundance of plagioclase—suggests however that there is less than 1-2 wt. % H_2O in the basaltic magma at several kilobars of pressure, and therefore a parental magma must have much less than 0.5-1 wt. % H_2O at high pressures. Second, if water induces melting of peridotite by lowering the solidus temperature, then the temperatures recorded in the derivative high-alumina basalts should be even lower. Oxide, pyroxene, and oxygen isotope geothermometry show, however, that basalt temperatures are in the range 1,100° - 1,200°C (Brophy, 1986a; Fournelle, 1988; Singer and others, 1991, 1991b). The temperature of the parental magma should then be greater than 1,200°C.

A common belief is that the oxidized nature of arc lavas is a result of water fluxing the mantle. Studies of hydrothermal alteration of oceanic crust at mid-ocean ridge vents, coupled with the role of arc volcanoes in the global sulfur cycle (e.g., recent SO_2-rich eruptions of El Chichon, Nevado del Ruiz, and Mt. Pinatubo volcanoes), however, suggest that some iron in the oceanic crust may have been oxidized by the reduction of anhydrite (precipitated from seawater) to sulfide (Albarede and Michard, 1989) prior to subduction.

Kay and Kay (1989; this volume) have developed a model

of asthenospheric circulation, in which the mantle flow above the slab is opposite that of all geophysical models (e.g., Hsui and others, 1983; Tatsumi and others, 1986; see Fig. 11). Their model also is not consistent with the geophysical constraints of back-arc spreading, which occurs behind some arcs but not behind the Aleutian arc where the crust is too old, cold, and thick.

Eggler (1989) presented evidence that partial melts of the mantle wedge (with $H_2O + CO_2$), immediately above the slab at 75 to 125 km, are silica-poor and highly alkalic—carbonatitic, alkali carbonatitic, or melilitic. Increased melting would yield more olivine-rich melts approaching alkali picrites. These melts have little in common with arc-lava compositions.

Slab melting–lithospheric mining model. This model, which we prefer, holds that an upper portion of the subducted slab (oceanic crust plus sediment, metamorphosed to quartz eclogite) melts, and that high-alumina basalts represent primary magma compositions (Marsh, 1982a; Marsh and Carmichael, 1974; Myers, 1988; Johnston, 1986; Brophy and Marsh, 1986 [their Table 6 has a summary of the pros and cons of the peridotite versus quartz eclogite source models]).

Myers and others (1985) and Myers and Marsh (1987) suggested that slab-derived magmas may be compositionally modified by interaction with peridotite during ascent through the lithosphere (Fig. 12), producing high-Mg basalt. Assimilation of mafic material by less mafic magma may occur, as has been described by Arculus and others (1983), Kudo (1983), Kelemen and Sonnenfeld (1983), Evans (1985), Kelemen (1986), and Kelemen and Ghiorso (1986). The lack of olivine on (or near) the liquidus of Aleutian high-Al basalt suggests that it was never in equilibrium with peridotite.

Most Aleutian arc lavas display enrichment in alkalis with unfractionated REE. Brophy and Marsh (1986) explained the unfractionated (La/Yb) REE of Aleutian high-Al basalts lavas using a model of diapiric ascent and extraction of melt outside the garnet stability field. They calculated that the Rb and Ba levels (and most major elements) of the basalts were consistent with 40 to 60% melting of altered MORB plus 5 % Pacific pelagic sediment, although modelled Sr abundances are below those observed.

Hsui and others (1983) modelled the thermal regime above the cold subducting slab and suggested that enough hot mantle material could come in contact with the slab to begin to melt its upper surface at a depth of 100 to 150 km.

The crystal-rich nature of Aleutian basalts has generated some skepticism about the existence of primary high-Al basaltic liquids (e.g., Crawford and others, 1987). The study by Fournelle and Marsh (1991) of recent Shishaldin basalts suggests that these high-Al basalts are not plagioclase cumulates, and could be primary to ~20 kbar. On the other hand, older (~3 to 5 Ma) high-Al bsalts from an erosional remnant near Shishaldin have some features consistent with the addition of plagioclase to a high-Mg basalt. These features are consistent with the developmental history of the lithospheric conduit feeding the volcanic center; earlier magmas may have been more Mg-rich, owing to lithospheric interaction. It is possible that high-Al basalts developed from these magmas by fractionation of olivine and clinopyroxene, as in the model of Gust and Perfit (1987), with or without addition of plagioclase.

Production of island-arc andesite

Island arcs were once thought to consist dominantly of andesites. The Aleutian island arc, however, is dominated by high-Al basalts through andesites and only the Alaska Peninsula–Cook Inlet arc section is dominated by andesitic compositions (Fig. 6). Isotopically, the andesites (and dacites) are identical to the basalts, and could be derived from basalts by fractionation. Baker and Eggler (1987), however, found that at high pressure (8 kbar), Atka andesites and dacites can not be produced by fractionation of basalt owing to thermal divides (i.e., separation of minerals drives the residual liquid to lower silica content).

Gill (1981) concluded that calc-alkaline arc andesites are produced by a complex series of mechanisms: (1) deep interaction of slab-derived water or melt with the mantle wedge: melting, separation, and ascent; (2) intermediate level fractionation during ascent: stagnation at base of the crust or within it; (3) upper level fractionation. This general model has been applied to Aleutian andesites by Kay and others (1982). Brophy (1990) further developed it in relation to Kanaga volcanism by proposing that: A, low-Al (~high-Mg) basalt replenishes a magma chamber at 2 to 5 kbar and high-Al basalt is formed by fractionation; and B, significant amounts of crustal assimilation produce dacitic liquids. Eruption at this point, with conduit mixing of high-Al basalt and dacite, yields mixed andesites. This open-system model accounts for the eruption only of hybrid calc-alkaline basaltic andesites and andesites (54 to 63 wt. % SiO_2) and the absence of basalts or dacites.

There is another explanation for the abundance of calc-alkaline andesites, requiring no mixing of unseen end members. Kudo (1983) suggested that the textural evidence required to support magma mixing in Japanese calc-alkaline andesites could be explained adequately by assimilation processes. Mafic gabbroic inclusions, used as evidence that the magma has undergone fractionation, instead may be a result of the magma cooling as it attempted to assimilate contaminating xenoliths. This approach has gained further support by the thermodynamic modelling of Kelemen and Ghiorso (1986), who showed that combined fractional crystallization and assimilation of mafic wall rock (i.e., Fo_{90}) by a dioritic liquid with 2 wt. % H_2O could produce the calc-alkaline AFM trend. Kelemen (1990) suggested this as a viable mechanism for producing calc-alkaline suites at convergent margins worldwide.

Singer and others (1991) suggest a less complex, closed system fractionation model to explain Seguam tholeiitic andesites and dacites. Small batches of basalt or basaltic andesite separate from a larger ~3 to 4 kbar reservoir; by cooling, they can evolve to more silicic compositions, without crustal assimilation or mixing. Integral here is the extended nature of the crust at Seguam, compared to the unextended crust at Kanaga and Adak (Singer and Myers, 1990).

FUTURE WORK

Although the Aleutian island arc is one of the best studied oceanic arcs in the world, many parts of it have not been mapped or extensively sampled. For example, the recent volcanic products on Tanaga, the largest island in the western Aleutians, have been described only briefly (Coats and Marsh, 1984). In addition, a large segment of the central arc, from Amuktu to Kagamil, is virtually unknown geologically. Recent field mapping just to the west of this segment has shown that the volcanic centers of the central portion of the arc may exhibit geologic and geochemical characteristics unlike the rest of the arc (Myers and Singer, 1987; Singer and Myers, 1988). Owing to the ruggedness of the islands, their remoteness, the hardships of field work, and the short field seasons, geologic investigations to date have been mostly reconnaissance in nature. Consequently, even the "mapped" islands may hold surprises for future field geologists.

APPENDIX 1. SUMMARY OF THE NUMBERS OF CHEMICAL ANALYSES, BY INDIVIDUAL VOLCANIC CENTER (WEST OF 161°W), USED IN THIS CHAPTER

Volcanic Center	Major Elements	Trace Elements						Isotopes		
		Rb	Sr	Ba	Cr	Ni	REE	Sr	Nd	Pb
Cold Bay	48	29	29	21	9	8	1	13	8	4
Shishaldin	88	88	88	88	88	88	23	3	3	0
Fisher Caldera	24	24	24	24	24	24	0	0	0	0
Westdahl, Pogromni	6	5	5	2	0	0	2	3	0	5
Akutan	46	36	37	18	18	17	14	19	3	7
Makushin	177	106	93	23	17	18	4	6	3	0
Okmok	72	57	57	24	19	46	12	12	8	15
Recheshnoi	18	1	13	13	13	12	1	1	0	0
Vsevidof	4	2	2	2	0	0	2	1	0	0
Seguam	181	171	181	177	177	85	25	15	15	19
Atka	84	25	35	34	34	34	20	24	7	12
Kasatochi	1	0	0	0	0	0	0	0	0	0
Great Sitkin	5	1	1	0	0	0	0	1	1	0
Adak	47	31	32	30	24	26	29	23	13	21
Kanaga	52	44	44	44	33	1	34	0	0	0
Bobrof	7	0	0	0	0	0	0	0	0	0
Semisopochnoi	15	14	14	5	12	14	5	5	4	0
Little Sitkin	13	13	13	3	0	13	3	3	3	0
Segula	2	2	2	1	0	2	1	1	1	0
Kiska	7	5	5	0	0	5	0	0	0	0
Buldir	9	0	0	0	0	0	0	0	0	0
Bogoslof	9	5	5	5	2	2	3	4	3	2
Amak	7	7	7	4	4	1	0	7	3	2

Note: Specific trace elements and isotopes are indicated to show the relative proportions of chemical data between various volcanic centers.

APPENDIX 2. SOURCES OF GEOCHEMICAL DATA FOR ALEUTIAN VOLCANIC CENTERS

Spurr	Nye and Turner, 1990
Redoubt	Forbes and others, 1969; Kienle and others, 1983
Iliamna	Kienle and others, 1983
Augustine	Kienle and others, 1983
Douglas	Kienle and others, 1983
Fourpeaked	Kienle and others, 1983
Kaguyak	Kienle and others, 1983
Devils Desk	Kienle and others, 1983
Kukak	Kienle and others, 1983
Denison	Kienle and others, 1983
Snowy	Kienle and others, 1983
Novarupta	Hildreth, 1983, 1987
Trident	Forbes and others, 1969
Martin	Kienle and others, 1983
Kejulik	Kienle and others, 1983
Yantarni	Riehle and others, 1987
Veniaminof	Kay and others, 1982; Yount and others, 1985
Pavlof	Anderson, 1975
Cold Bay	Marsh, 1976; Kay, 1977, Kay and others, 1978; Marsh and Leitz, 1979; Kay and others, 1982; Brophy, 1984, 1986a, 1987
Shishaldin	Fournelle, 1988; Fournelle and Marsh, 1991
Fisher Caldera	Fournelle, 1990b, unpublished data; T. Miller, personal comm., 1985
Westdahl, Pogromni	Kay, 1977; Kay and others, 1978
Akutan	McCulloch and Perfit, 1981; Romick, 1982; Romick and others, 1990
Makushin	Drewes and others, 1961; DeLong, 1974; Perfit, 1977; Perfit and others,1980a; McCulloch and Perfit, 1981; Nye and others, 1986; Gust and Perfit, 1987
Okmok	Byers, 1959, 1961; Kay, 1977; Kay and others, 1978; McCulloch and Perfit, 1981; Nye, 1983; Kay and Kay, 1985; Nye and Reid, 1986, 1987; T. Miller, personal comm., 1985
Recheshnoi	Byers, 1959, 1961; Franks, 1981; Kay, 1977; Kay and others, 1978
Vsevidof	Byers, 1959, 1961; Kay, 1977; Kay and others, 1978
Seguam	Myers, unpublished data; Singer and others, 1992a
Atka	Marsh, 1980, 1982a, unpublished data; Myers and others, 1985, 1986; von Drach and others, 1986; Baker and Eggler, 1983, 1987; Myers and Marsh, 1987; Myers and Frost, unpublished data
Kasatochi	Kay and Kay, 1985
Great Sitkin	Simons and Mathewson, 1953; Marsh, 1976; Kay and Kay, 1985; von Drach and others, 1986
Adak	Coats, 1952; Marsh, 1976; Kay, 1977, 1978; Kay and others, 1978; Kay and others, 1982; Conrad and Kay, 1984; Kay and Kay, 1985; Myers and others, 1985; Kay and others, 1986; von Drach and others, 1986; Debari and others, 1987; Myers and Marsh, 1987; Myers and Frost, unpublished data
Kanaga	Coats, 1952; Fraser and Barrett, 1959; DeLong, 1974; Kay, 1977; Brophy, 1989a, 1990, unpublished data
Bobrof	Kay and others, 1982
Tanaga	Coats and Marsh, 1984
Semisopochnoi	Coats, 1959b; DeLong, 1974; DeLong and others, 1985
Little Sitkin	Snyder, 1959; DeLong, 1974; White and Patchett, 1984
Segula	Nelson, 1959; DeLong, 1974; McCulloch and Perfit, 1981
Kiska	Coats and others, 1961; DeLong, 1974
Buldir	Coats, 1953; Kay and Kay, 1985
Bogoslof	Fenner, 1926; Arculus and others, 1977; Kay, 1977; Kay and others, 1978; Marsh and Leitz, 1979; McCulloch and Perfit, 1981; von Drach and others, 1986
Amak	Marsh and Leitz, 1979; Morris and Hart, 1983; von Drach and others, 1986
submarine	Scholl and others, 1976

REFERENCES CITED

Albarede, F., and Michard, A., 1989, Hydrothermal alteration of the oceanic crust, *in* Hart, S. R., and Gulen, L., eds. Crust/mantle recycling at convergence zones: Dordrecht, Kluwer Academic, p. 29–36.

Anderson, A. T., 1975, Some basaltic and andesites gases: Reviews of Geophysics and Space Physics, v. 13, p. 37–55.

Aramaki, S., and Ui, T., 1983, Alkali mapping of the Japanese Quaternary volcanic rocks: Journal of Volcanology and Geothermal Research, v. 18, p. 549–560.

Arculus, R. J., DeLong, S. E., Kay, R. W., Brooks, C., and Sun, S. S., 1977, The alkalic rock suite of Bogoslof Island, eastern Aleutian arc, Alaska: Journal of Geology, v. 85, p. 177–186.

Arculus, R. J., Johnson, R. W., Chappell, B. W., McKee, C. O., and Sakai, H., 1983, Ophiolite-contaminated andesites, trachybasalts, and cognate inclusions of Mount Lamington, Papua New Guinea; Anhydrite-amphibole-bearing lavas and the 1951 cumulodome: Journal of Volcanology and Geothermal Research, v. 18, p. 215–247.

Baker, D. R., 1987, Depths and water content of magma chambers in the Aleutian and Mariana island arcs: Geology, v. 15, p. 496–499.

Baker, D. R. and Eggler, D. H., 1983, Fractionation paths of Atka (Aleutian) high-alumina basalts; Constraints from phase relations: Journal of Volcanology and Geothermal Research, v. 18, p. 387–404.

—— , 1987, Compositions of anhydrous and hydrous melts coexisting with plagioclase, augite, and olivine or low-Ca pyroxene from 1 atm to 8 kbar; Application to the Aleutian volcanic center of Atka: American Mineralogist, v. 72, p. 12–28.

Becker, G. F., 1898, Reconnaissance of the gold fields of southern Alaska, with some notes on general geology: U.S. Geological Survey 18th Annual Report, v. 3.

Black, L. T., 1981, Volcanism as a factor in human ecology; The Aleutian case: Ethnohistory, v. 28, p. 313–340.

Boudinier, J. -L., Dupuy, C., Dostal, J., and Meriet, C., 1987, Distribution of trace transition elements in olivine and pyroxenes from ultramafic xenoliths; Application of microprobe analysis: American Mineralogist, v. 72, p. 902–913.

Brophy, J. G., 1984, The chemistry and physics of Aleutian arc volcanism; The Cold Bay volcanic center, southwestern Alaska [Ph.D. thesis]: The Johns Hopkins University, 422 p.

—— , 1986a, The Cold Bay volcanic center, Aleutian volcanic arc; I, Implications for the origin of hi-alumina arc basalt: Contributions to Mineralogy and Petrology, v. 93, p. 368–380.

—— , 1986b, Source rock clinopyroxene, Eu anomalies, and the origin of hi-alumina basalt: EOS, v. 67, p. 405.

—— , 1987, The Cold Bay Volcanic Center, Aleutian volcanic arc; II, Implications for fractionation and mixing mechanism in calc-alkaline andesite genesis: Contributions to Mineralogy and Petrology, v. 97, p. 378–388.

—— , 1989a, Can high alumina basalt be derived from low alumina arc basalt? Evidence from Kanaga Island, Aleutian arc, Alaska: Geology, v. 17, p. 333–336.

—— , 1989b, Basalt convection and plagioclase retention, a model for the generation of high-alumina arc basalt: Journal of Geology, v. 97, p. 319–329.

—— , 1990, Andesites from northeastern Kanaga Island, Aleutians; Implications for calc-alkaline fractionation mechanisms and magma chamber development: Contributions to Mineralogy and Petrology, v. 104, p. 568–581.

Brophy, J. G., and Marsh, B. D., 1986, On the origin of high-alumina arc basalt and the mechanics of melt extraction: Journal of Petrology, v. 27, p. 763–789.

Burk, C. A., 1965, Geology of the Alaska Peninsula; Part 1, Island arc and continental margin: Geological Society of America Memoir 99, 250 p.

Burnham, C. W., 1979, The importance of volatile constituents, *in* Yoder, H. S., ed., The evolution of the igneous rocks: Princeton University Press, p. 439–482.

—— , 1980, The nature of multicomponent silica melts, *in,* Richard, D. T. and Wickman, F. E., eds., Chemistry and geochemistry of solutions at high temperatures and pressures: Pergamon Press, Physics and Chemistry of the Earth, v. 13 and 14, p. 197–229.

Byers, F. M., Jr., 1959, Geology of Umnak and Bogoslof Islands, Aleutian Islands, Alaska: U.S. Geological Survey Bulletin 1028-L, p. 267–369.

—— , 1961, Petrology of three volcanic suites, Umnak and Bogoslof Islands, Aleutian Islands, Alaska: Geological Society of America Bulletin, v. 72, p. 93–128.

Carroll, M. R., and Rutherford, M. J., 1987, The experimental stability of igneous anhydrite; Experimental results and implications for sulfur behavior in the 1982 El Chichon trachyandesite and other evolved magmas: Journal of Petrology, v. 28, p. 781–801.

—— , 1988, Sulfur speciation in hydrous experimental glasses of varying oxidation state; Results from measured wavelength shifts of sulfur X-rays: American Mineralogist, v. 73, p. 845–849.

Coats, R. R., 1947, Reconnaissance geology of some western Aleutian Islands: U.S. Geological Survey Alaskan Volcano Investigations, p. 95–105.

—— , 1950, Volcanic activity in the Aleutian arc: U.S. Geological Survey Bulletin 974-B, p. 35-49.

—— , 1952, Magmatic differentiation in Tertiary and Quaternary volcanic rocks from Adak and Kanaga Islands, Aleutian Islands, Alaska: Geological Society of America Bulletin, v. 63, p. 485–514.

—— , 1953, Geology of Buldir Island, Aleutian Islands, Alaska: U.S. Geological Survey Bulletin 989-A, p. 1–25.

—— , 1956, Geology of northern Adak Island, Alaska: U.S. Geological Survey Bulletin 1028-C, p. 45–67.

—— , 1959, Geologic reconnaissance of Semisopochnoi Island, western Aleutian Islands, Alaska: Geological Survey Bulletin 1028-O, p. 477–519.

—— , 1962, Magma type and crustal structure in the Aleutian arc, *in* The crust of the Pacific basin: American Geophysical Union Monograph 6, p. 92–109.

Coats, R. R., and Marsh, B. D., 1984, Reconnaissance geology and petrology of northern Tanaga, Aleutian Islands, Alaska: Geological Society of America Abstracts with Programs, v. 16, p. 474.

Coats, R. R., Nelson, W. H., Lewis, R. Q., and Powers, H. A., 1961, Geologic reconnaissance of Kiska Island, Aleutian Islands, Alaska: U.S. Geological Survey Bulletin 1028-R, p. 563–581.

Conrad, W. K., and Kay, R. W., 1984, Ultramafic and mafic inclusions from Adak Island; Crystallization history, and implications for the nature of primary magmas and crustal evolution of the Aleutian arc: Journal of Petrology, v. 25, p. 88–125.

Conrad, W. K., Kay, S. M., and Kay, R. W., 1983, Magma mixing in the Aleutian arc; Evidence from cognate inclusions and composite xenoliths: Journal of Volcanology and Geothermal Research, v. 8, p. 279–295.

Crawford, A. J., Falloon, T. J., and Eggins, S., 1987, The origin of island arc high-alumina basalts: Contributions to Mineralogy and Petrology, v. 97, p. 417–430.

Curtis, G. H., 1968, The stratigraphy of the ejecta from the 1912 eruption of Mount Katmai and Novarupta, Alaska, *in* Coats, R. R., Hay, R. L., and Anderson, C. A., eds., Studies in volcanology: Geological Society of America Memoir 166, p. 153–210.

Dall, W. H., 1870, Alaska and its resources: London, Lee and Shepard, 627 p.

—— , 1918, Reminiscences of Alaska volcanoes: Scientific Monthly, v. 7, p. 80–90.

DeBari, S., Kay, S. M., and Kay, R. W., 1987, Ultramafic xenoliths from Adagdak Volcano, Adak, Aleutian Islands, Alaska; Deformed igneous cumulates from the Moho of an island arc: Journal of Geology, v. 95, p. 329–341.

DeLong, S. E., 1974, Distribution of Rb, Sr and Ni in igneous rocks, central and western Aleutian Islands, Alaska: Geochimica et Cosmochimica Acta, v. 38, p. 245–266.

DeLong, S. E., Perfit, M. R., McCulloch, M. T., and Ach, J., 1985, Magmatic, evolution of Semisopochnoi Island, Alaska; Trace-element and isotopic constraints: Journal of Geology, v. 93, p. 609–618.

Drewes, H., Fraser, G. D., Synder, G. L., and Barnett, H. F., Jr., 1961, Geology of Unalaska Island and adjacent insular shelf, Aleutian Islands, Alaska: U.S. Geological Survey Bulletin 1028-S, p. 583–763.

Dumond, D. E., 1979, People and pumice on the Alaska Peninsula, *in* Sheets, P. D., and Grayson, D. K., eds., Volcanic activity and human ecology: New York, Academic Press, p. 373–392.

Eggler, D. H., 1972, Water-saturated and undersaturated melting relations in a Paricutin andesite and an estimate of water content in the natural magma: Contributions to Mineralogy and Petrology, v. 34, p. 261–171.

—— , 1989, Influence of H_2O and CO_2 on melt and fluid chemistry in subduction zones, *in* Hart, S. R., and Gulen, L., eds., Crust/mantle recycling at convergence zones: Dordrecht, Kluwer Academic, p. 97–104.

Engdahl, E. R., 1973, Relocation of intermediate depth earthquakes in the central Aleutians by seismic ray tracing: Nature, v. 245, p. 23–25.

Evans, C. A., 1985, Magmatic "metasomatism" in peridotites from the Zambales ophiolite: Geology, v. 13, p. 166–169.

Fenner, C. N., 1923, The origin and mode of emplacement of the great tuff deposit of the Valley of Ten Thousand Smokes: National Geographic Society, Contributed Technical Paper, Katmai Series, no. 1, 74 p.

—— , 1926, The Katmai magmatic province: Journal of Geology, v. 34, p. 673–772.

Finch, R. H., 1934, Shishaldin Volcano, Proceedings of the Fifth Pacific Science Conference, Canada: Toronto, University of Toronto Press, v. 3, p. 2369–2376.

—— , 1935, Akutan Volcano: Zeitschrift fur Vulkanologie, v. 16, p. 155–160.

Fisher, M. A., 1981, Location of the Border Ranges fault southwest of Kodiak Island, Alaska: Geological Society of America Bulletin, v. 92, p. 19–30.

Forbes, R. B., Ray, D. K., Katsura, T., Matsumoto, H., Haramura, H., and Furst, M. J., 1969, The comparative chemical compositions of continental vs. island arc andesites in Alaska, *in* McBirney, A. R., ed., Proceedings of the Andesite Conference: Oregon State Department of Geology and Mineral Industries, Bulletin 65, p. 111–120.

Fournelle, J. H., 1988, The geology and petrology of Shishaldin Volcano, Unimak Island, Aleutian arc, Alaska [Ph.D. thesis]: The Johns Hopkins University, 529 p.

—— , 1990a, Anhydrite in Nevado del Ruiz November 1985 pumice; Relevance to the sulfur problem: Journal of Volcanology and Geothermal Research, v. 42, p. 189–201.

—— , 1990b, Geology and geochemistry of Fisher Caldera, Unimak Island, Aleutians; Initial results: EOS, v. 71, p. 1,698.

Fournelle, J. H. and Marsh, B. D., 1987, Diopsidic cpx produced by assimilation of Fo-rich olivine by magmas at Shishaldin Volcano: Implications for Alaskan layered complexes: EOS Transactions of the American Geophysical Union, v. 68, p. 1525.

—— , 1991, Shishaldin Volcano; Aleutian high-alumina basalts and the plagioclase accumulation question: Geology, v. 19, p. 234–237.

Franks, A. L., 1981, A petrological study of Mt. Recheschnoi Volcano, Umnak Island, Alaska [M.S. thesis]: New York, Columbia University, 35 p.

Fraser, G. D., and Barnett, H. D., 1959, Geology of the Delarof and westernmost Andreanof Islands, Aleutian Islands, Alaska: U.S. Geologic Survey Survey Bulletin 1028-I, p. 211–245.

Fujita, K., Engdahl, E. R., and Sleep, N. H., 1981, Subduction zone calibration and teleseismic relocation of thrust zone events in the central Aleutian Islands: Bulletin of the Seismological Society of America, v. 71, p. 1805–1825.

Geist, E. L., Childs, J. R., and Scholl, D. W., 1987, The evolution and petroleum geology of the Amlia and Amukta intra-arc basins: Marine Petroleum Geology, v. 4, p. 334–352.

—— , 1988, The origin of summit basins of the Aleutian ridge; Implications for block rotation of an arc massif: Tectonophysics, v. 7, p. 327–341.

Gerlach, T. M., Westrich, H. R., and Casadevall, T. J., 1990, High sulfur and chlorine magma during the 1989–90 eruption of Redoubt Volcano, Alaska: EOS Transactions of the American Geophysical Union, v. 43, p. 1702.

Ghiorso, M. S., 1985, Chemical mass transfer in magmatic processes; I, Thermodynamic relations and numerical algorithms: Contributions to Mineralogy and Petrology, v. 90, p. 107–120.

Gill, J. B., 1981, Orogenic andesites and plate tectonics: Springer-Verlag, 390 p.

Grewingk, C., 1850, Beitrag zur Kenntniss der orographischen und geognostishen Beschaffenheit der nordwest-Kuste Amerikas mit den anliegenden Inseln, St. Petersburg: reprint from Mineralogisches Gesellschaft St. Petersburg Verhandlungen, 1848–49, 351 p.

Griggs, R. F., 1922, The Valley of Ten Thousand Smokes: Washington, D.C., National Geographic Society, 340 p.

Gust, D. A., and Perfit, M. R., 1987, Phase relations of a high-Mg basalt from the Aleutian island arc; Implications for primary island arc basalts and high-Al basalts: Contributions to Mineralogy and Petrology, v. 97, p. 7–18.

Hein, J. R., Scholl, D. W., and Miller, J., 1978, Episodes of Aleutian ridge explosive volcanism: Science, v. 199, p. 137–141.

Hein, J. R., McLean, H., and Vallier, T., 1984, Reconnaissance geology of Southern Atka Island, Aleutian Islands, Alaska: U.S. Geological Survey Bulletin 1609, 19 p.

Hildreth, W., 1983, The compositionally zoned eruption of 1912 in the Valley of Ten Thousand Smokes, Katmai National Park, Alaska: Journal of Volcanology and Geothermal Research, v. 18, p. 1–56.

—— , 1987, New perspectives on the eruption of 1912 in the Valley of Ten Thousand Smokes, Katmai National Park, Alaska: Bulletin of Volcanology, v. 49, p. 680–693.

Howell, D. G., Jones, D. L., and Schermer, E. R., 1985, Tectonostratigraphic terranes of the circum-Pacific region, *in* Howell, D. G., ed., Tectonostratigraphic terranes of the circum-Pacific region: Houston, Circum-Pacific Council for Energy and Mineral Resources, p. 3–30.

Hsui, A. T., Marsh, B. D., and Toksoz, M. N., 1983, On melting of the subducted oceanic crust; Effects of subduction induced mantle flow: Tectonophysics, v. 99, p. 207–220.

Hubbard, B., 1932a, Alaska's Shishaldin; The first ascent of "Smoky Moses": Saturday Evening Post, August 13, p. 8+.

—— , 1932b, Mush, you malemutes!: New York, The American Press, 179 p.

—— , 1936, Cradle of the storms: London, George G. Harrap and Co., 285 p.

Irvine, T. N., 1974, Petrology of the Duke Island ultramafic complex, southeastern Alaska: Geological Society of America Memoir, v. 138, 240 p.

Jacob, K. H., Nakamura, K., and Davies, J. N., 1977, Trench-volcano gap along the Alaska-Aleutian arc, facts and speculations on the role of terrigenous sediments for subduction, *in* Talwani, M., and Pitman, W. C., III, eds., Island arcs, deep sea trenches and back arc basins: American Geophysical Union, p. 243–258.

Jaggar, T. A., Jr., 1908, Journal of the technology expedition to the Aleutian Islands: The Technology Review, v. 10, p. 1–37.

—— , 1932, Aleutian eruptions, 1930–1932: Volcano Letter, no. 375, p. 1–4.

Johnston, A. D., 1986, Anhydrous P-T phase relations of nearly-primary high-alumina basalt from the South Sandwich Islands: Contributions to Mineralogy and Petrology, v. 92, p. 368–382.

Kay, R. W., 1977, Geochemical constraints on the origin of Aleutian magma, *in* Talwani, M., and Pitman, W. C., III, eds., Island arcs, deep sea trenches and back-arc basins: Washington, D.C., American Geophysical Union, p. 229–242.

—— , 1978, Aleutian magnesian andesites; Melts from subducted Pacific Ocean crust: Journal of Volcanology and Geothermal Research, v. 4, p. 117–132.

Kay, R. W., and Kay, S. M., 1989, Recycled continental crustal components in Aleutian arc magmas, implications for crustal growth and mantle heterogeneity, *in* Hart, S. R., and Gulen, L., eds., Crust/mantle recycling at convergence zones: Dordrecht, Kluwer Academic, p. 145–161.

Kay, R. W., Sun, S. S., and Lee-Hu, C. -N., 1978, Pb and Sr isotopes in volcanic rocks from the Aleutian Islands and Pribilof Islands, Alaska, Geochimica et Cosmochimica Acta, v. 42, p. 263–274.

Kay, R. W., Rubenstone, J. L., and Kay, S. M., 1986, Aleutian terranes from Nd isotopes: Nature, v. 322, p. 605–609.

Kay, S. M., and Kay, R. W., 1985, Aleutian tholeiitic and calc-alkaline magma series; I, The mafic phenocrysts: Contributions to Mineralogy and Petrology, v. 90, p. 276–290.

Kay, S. M., Kay, R. W., and Citron, G. P., 1982, Tectonic controls on tholeiitic and calc-alkalic magmatism in the Aleutian arc: Journal of Geophysical Research, v. 87, p. 4051–4072.

Kelemen, P. B., 1986, Assimilation of ultramafic rocks in subduction-related magmatic arcs: Journal of Geology, v. 94, p. 829–843.

—— , 1990, Reaction between ultramafic rock and fractionating basaltic magma; I, Phase relations, the origin of calc-alkaline magma series, and the formation of discordant dunite: Journal of Petrology, v. 31, p. 51–98.

Kelemen, P. B., and Ghiorso, M. S., 1986, Assimilation of peridotite in zoned calc-alkaline plutonic complexes; Evidence from the Big Jim complex, Washington Cascades: Contributions to Mineralogy and Petrology, v. 94, p. 12–28.

Kelemen, P. B., and Sonnfeld, M. D., 1983, Stratigraphy, structure, petrology and local tectonics, central Ladakh, NW Himalaya: Schweizer Mineralogische Petrographische Mitteilungen, v. 63, p. 267–287.

Kennedy, G. C., and Waldron, H. H., 1955, Geology of Pavlof Volcano and vicinity, Alaska: U.S. Geological Survey Bulletin 1028-A, p. 1–19.

Kienle, J., and Swanson, S. E., 1983, Volcanism in the eastern Aleutian arc; Late Quaternary and Holocene centers, tectonic setting and petrology: Journal of Volcanology and Geothermal Research, v. 17, p. 393–432.

Kienle, J., Swanson, S. E., and Pulpan, H., 1983, Magmatism and subduction in the eastern Aleutian arc, *in* Shimozuru, D., and Yokoyama, I., eds., Arc volcanism; Physics and tectonics: Tokyo, Terra Scientific Publishing Company, p. 191–224.

Kilinc, A., Carmichael, I.S.E., Rivers, M. L., and Sack, R. O., 1983, Ferric-ferrous ratio of natural silicate liquids equilibrated in air: Contributions to Mineralogy and Petrology, v. 83, p. 136–140.

Kudo, A. M., 1983, Origin of Calc-alkalic andesites, Nasu zone, northeastern Japan; Kuno revisited: Geochemical Journal, v. 17, p. 51–62.

Lowenstern, J. B., 1990, Pre-eruptive water content of high-silica rhyolite and dacite from the 1912 eruption at the Valley of Ten Thousand Smokes, Alaska: EOS Transactions of the American Geophysical Union, v. 71, p. 1690.

Maddren, A. G., 1919, Sulphur on Unalaska and Akun Islands and near Stepovak Bay, Alaska: U.S. Geological Survey Bulletin 692-E, p. 268–298.

Manning, C. E., and Bird, D. K., 1986, Hydrothermal clinopyroxenes of the Skaergaard intrusion: Contributions to Mineralogy and Petrology, v. 92, p. 437–447.

Marlow, M. S., and Cooper, A. K., 1983, Wandering terranes in southern Alaska; The Aleutia microplate and implications for the Bering Sea: Journal of Geophysical Research, v. 88, p. 3439–3446.

Marsh, B. D., 1976, Some Aleutian andesites, their nature and source: Journal of Geology, v. 84, p. 27–45.

—— , 1978, On the cooling of ascending andesitic magma: Philosophical Transaction Royal Society of London A, v. 288, p. 611–625.

—— , 1979a, Island arc volcanism: American Scientist, v. 67, p. 161.

—— , 1979b, Island arc development; Some observations, experiments, and speculations: Journal of Geology, v. 87, p. 687–713.

—— , 1980, Geology and petrology of northern Atka, Aleutian Islands, Alaska: Geological Society of America, Abstracts with Programs, v. 12, p. 476.

—— , 1982a, The Aleutians, *in* Thorpe, R. S., ed., Orogenic andesites: John Wiley & Sons, p. 99–114.

—— , 1982b, On the mechanics of igneous diapirism, stoping, and zone melting: American Journal of Science, v. 282, p. 808–855.

—— , 1987, Petrology and evolution of the N.E. Pacific including the Aleutians: Pacific Rim Congress 87, p. 309–315.

Marsh, B. D., and Carmichael, I.S.E., 1974, Benioff zone magmatism: Journal of Geophysical Research, v. 79, p. 1196–1206.

Marsh, B. D., and Kantha, L. H., 1978, On the heat and mass transfer from an ascending magma: Earth and Planetary Science Letters, v. 39, p. 435–443.

Marsh. B. D., and Leitz, R. E., 1979, Geology of Amak Island, Aleutian Islands, Alaska: Journal of Geology, v. 87, p. 715–723.

Marsh, B. D., Fournelle, J. H., Myers, J. D., and Chou, I. -M., 1990, On plagioclase thermometry in island-arc rocks; Experiments and theory, *in* Spencer, R. J., and Chou, I. -M., eds., Fluid-mineral interactions, A tribute to H. P. Eugster: Geochemical Society Special Publication, v. 2, p. 65–83.

McClelland, L., Simkin, T., Summers, M., Nielsen, E., and Stein, T. C., 1989, Global Volcanism, 1975–1985: Englewood Cliffs, New Jersey, Prentice Hall, 655 p.

McCulloch, M. T., and Perfit, M. R., 1981, $^{143}Nd/^{144}Nd$, $^{87}Sr/^{86}Sr$ and trace element constraints on the petrogenesis of Aleutian island arc magmas: Earth and Planetary Science Letters, v. 56, p. 167–179.

McNutt, S. R., and Jacob, K. H., 1986, Determination of large-scale velocity structure of the crust and upper mantle in the vicinity of Pavlof Volcano, Alaska: Journal of Geophysical Research, v. 91, p. 5013–5022.

Merrill, G. P., 1885, On hornblende andesites from the new volcano of Bogoslof Island in Bering Sea: Proceedings of U.S. National Museum, v. 8 (1886), p. 31–33.

Merzbacher, C., 1983, Water-saturated and -undersaturated phase relations of the Mount St. Helens dacite magma erupted on May 18, 1980, and an estimate of the pre-eruptive water content [M.S. thesis]: University Park, Pennsylvania State University, 54 p.

Merzbacher, C., and Eggler, D. H., 1984, A magmatic geohygrometer; Application to Mount St. Helens and other dacitic magmas: Geology, v. 12, p. 587–590.

Miller, T. P., and Smith, R. L., 1984, Calderas of the eastern Aleutian arc: Transactions of the American Geophysical Union, v. 64, no. 45, p. 87.

Minster, J. B., Jordan, T. H., Molnar, P., and Haines, E., 1974, Numerical modelling of instantaneous plate tectonics: Journal of Royal Astronomical Society, v. 36, p. 541–576.

Miyashiro, A., 1974, Volcanic rock series in island arcs and active continental margins: American Journal of Science, v. 274, p. 321–355.

Moore, J. G., and Evans, B. W., 1967, The role of olivine in the crystallization of the prehistoric Makaopuhi theoleiite lava lake, Hawaii: Contributions to Mineralogy and Petrology, v. 15, p. 202–223.

Morris, J. D., and Hart, S. R., 1983, Isotopic and incompatible element constraints on the genesis of island arc volcanics from Cold Bay and Amak Island, Aleutians, and implications for mantle structure: Geochimica et Cosmochimica Acta, v. 47, p. 2015–2030.

Myers, J. D., 1988, Possible petrogenetic relations between low- and high-MgO Aleutian basalts: Bulletin of Geological Society of America, v. 100, p. 1040–1053.

Myers, J. D., and Frost, C. D., 1989, Trace and rare earth element constraints on the origin and evolution of Aleutian arc magmas: EOS Transactions of the American Geophysical Union, v. 17, p. 721.

Myers, J. D., and Marsh, B. D., 1987, Aleutian lead isotopic data; Additional evidence for the evolution of lithospheric plumbing systems: Geochimica et Cosmochimica Acta, v. 51, p. 1833–1842.

Myers, J. D., and Singer, B. S., 1987, Seguam Island, central Aleutian Islands, Alaska; I, Geologic field relations: EOS Transactions of the American Geophysical Union, v. 68, p. 1525.

Myers, J. D., Marsh, B. D., and Sinha, A. K., 1985, Strontium isotopic and selected trace element variations between two Aleutian volcanic centers (Adak and Atka); Implications for the development of arc volcanic plumbing systems: Contributions to Mineralogy and Petrology, v. 91, p. 221–234.

—— , 1986, Geochemical and strontium isotopic characteristics of parental Aleutian arc magmas; Evidence from the basaltic lavas of Atka: Contributions to Mineralogy and Petrology, v. 94, p. 1–11.

Nakamura, K., 1977, Volcanoes as possible indicators of tectonic stress orientation; Principle and proposal: Journal of Volcanology and Geothermal Research, v. 2, p. 1–16.

Nakamura, K., Plafker, G., Jacob, K. H., and Davies, J. N., 1980, A tectonic stress trajectory map of Alaska using information from volcanoes and faults: Bulletin of the Earthquake Research Institute, v. 55, p. 89–100.

Neal, R. J., and Swanson, S. W., 1983, Petrology and geochemistry of Westdahl and Pogromni Volcanoes, Unimak Island, Alaska: EOS Transactions of the American Geophysical Union, v. 64, p. 893.

Nelson, W. H., 1959, Geology of Segula, Davidof and Khvostof Islands, Alaska: U.S. Geological Survey Bulletin 1028-K, p. 257–266.

Nye, C. J., 1983, Petrology and geochemistry of Okmok and Wrangell Volcanoes, Alaska [Ph.D. thesis]: University of California at Santa Cruz, 208 p.

Nye, C. J., and Reid, M. R., 1986, Geochemistry of primary and least fractionated lavas from Okmok Volcano, central Aleutians; Implications for magmagenesis: Journal of Geophysical Research, v. 91, p. 10271–10287.

——, 1987, Corrections to "Geochemistry of primary and least fractionated lavas from Okmok Volcano, central Aleutians; Implications for magmagenesis": Journal of Geophysical Research, v. 92, p. 8182.

Nye, C. J., and Turner, D. L., 1990, Petrology, geochemistry, and age of the Spurr volcanic complex, eastern Aleutian arc: Bulletin of Volcanology, v. 52, p. 205–226.

Nye, C. J., Swanson, S. E., and Reeder, J. W., 1986, Petrology and geochemistry of Quaternary volcanic rocks from Makushin Volcano, central Aleutian arc: Alaska Division of Geological and Geophysical Surveys, Open-File Report 86-80, p. 1–106.

Oxburgh, E. R., and Turcotte, D. L., 1976, The physico-chemical behavior of the descending lithosphere: Tectonophysics, v. 32, p. 107–128.

Perfit, M. R., 1977, The petrochemistry of igenous rocks from the Cayman Trench and the Captains Bay pluton, Unalaska Island; Their relation to tectonic processes at plate margins [Ph.D. thesis]: New York, Columbia University, 273 p.

Perfit, M. R., Brueckner, H., Lawrence, J. R., and Kay, R. W., 1980a, Trace element and isotopic variations in a zoned pluton and associated volcanic rocks, Unalaska Island, Alaska; A model for fractionation in the Aleutian calc-alkaline suite: Contributions to Mineralogy and Petrology, v. 73, p. 69–87.

Perfit, M. R., Gust, D. A., Bence, A. E., Arculus, R. J. and Taylor, S. R., 1980b, Chemical characteristics of island arc basalts; Implications for mantle sources: Chemical Geology, v. 30, p. 227–256.

Reinink-Smith, L. M., 1990a, Relative frequency of Neogene volcanic events as recorded in coal partings from the Kenai lowland, Alaska; A comparison with deep-sea core data: Geological Society of America Bulletin, v. 102, p. 830–840.

——, 1990b, Mineral assemblages of volcanic and detrital partings in Tertiary coal beds, Kenai Peninsula, Alaska: Clays and Clay Minerals, v. 38, p. 97–108.

Riehle, J. R., Yount, M. E., and Miller, T. P., 1987, Petrography, chemistry, and geologic history of Yantarni Volcano, Aleutian volcanic arc, Alaska: U.S. Geological Survey Bulletin 1761, 27 p.

Roeder, R. L., and Emslie, E. G., 1970, Olivine-liquid equilibrium: Contributions to Mineralogy and Petrology, v. 29, p. 275–289.

Romick, J. D., 1982, The igneous petrology and geochemistry of northern Akutan Island, Alaska [M.S. thesis]: Fairbanks, University of Alaska, 151 p.

Romick, J. D., Perfit, M. R., Swanson, S. E., and Shuster, R. D., 1990, Magmatism in the eastern Aleutian arc; Temporal characteristics of igneous activity on Akutan Island: Contributions to Mineralogy and Petrology, v. 104, p. 700–721.

Sack, R. O., Carmichael, I.S.E., Rivers, M., and Ghiorso, M. S., 1980, Ferric-ferrous equilibria in natural silicate liquids at 1 bar: Contributions to Mineralogy and Petrology, v. 75, p. 369–376.

Scheidegger, K. F., and Kulm, L. D., 1975, Late Cenozoic volcanism in the Aleutian arc; Information from ash layers in the northeastern Gulf of Alaska: Geological Society of America Bulletin, v. 86, p. 1407–1412.

Scheidegger, K. F., Corliss, J. B., Jazek, P. A., and Ninkovich, D., 1980, Compositions of deep-sea ash layers derived from North Pacific volcanic arcs; Variations in time and space: Journal of Volcanology and Geothermal Research, v. 7, p. 107–137.

Scholl, D. W., Marlow, M. S., MacLeod, N. S., and Buffington, E. C., 1976, Episodic Aleutian Ridge igneous activity; Implications of Miocene and younger submarine volcanism west of Buldir Island: Geological Society of America Bulletin, v. 87, p. 547–554.

Simkin, T., and Smith, J. V., 1970, Minor-element distribution in olivine: Journal of Geology, v. 78, p. 304–325.

Simkin, T., and 5 others, 1981, Volcanoes of the world: Stroudsburg, Pennsylvania, Hutchinson Ross, 233 p.

Simkin, T., Siebert, L., and McClelland, L., 1984, Volcanoes of the World: 1984 Supplement: Smithsonian Institution, Washington, D.C., 33 p.

Simons, F. S., and Mathewson, D. E., 1953, Geology of Great Sitkin Island, Alaska: U.S. Geological Survey Bulletin 1028-B, p. 21–42.

Singer, B. S., and Myers, J. D., 1988, Major and trace element characteristics of lavas from Seguam Island, central Aleutian Islands, Alaska: Geological Society of America Abstracts with Programs, v. 21, p. A56.

——, 1990, Intra-arc extension and magmatic evolution in the central Aleutian arc, Alaska: Geology, v. 18, p. 1050–1053.

Singer, B. S., Myers, J. D., Linneman, S. R., and Angevine, C., 1989, The thermal history of ascending magma diapirs and the thermal and physical evolution of magmatic conduits: Journal of Volcanology and Geothermal Research, v. 37, p. 273–289.

Singer, B. S., Myers, J. D., and Frost, C., 1992a, Mid-Pleistocene lavas from the Seguam volcanic center, central Aleutian arc; Closed-system fractional crystallization of a basalt to dacite eruptive suite: Contributions to Mineralogy and Petrology, v. 10, p. 87–112.

Singer, B. S., O'Neil, J. R., and Brophy, J. G., 1992b, Oxygen isotope constraints on the petrogenesis of Aleutian Arc magmas: Geology, v. 20, p. 367–370.

Snyder, G. L., 1959, Geology of Little Sitkin Island, Alaska: U.S. Geological Survey Bulletin 1028-H, p. 169–209.

Spence, W., 1977, The Aleutian arc; Tectonic blocks, episodic subduction, strain diffusion and magma generation: Journal of Geophysical Research, v. 82, no. 2, p. 213–230.

Swanson, S. E., Kay, S. M., Brearley, M., and Scarfe, C. M., 1987, Arc and back-arc xenoliths in Kurile-Kamchatka and western Alaska, *in* Nixon, P. H., ed., Mantle xenoliths: New York, John Wiley, p. 303–318.

Tatsumi, Y., Hamilton, D. L., and Nesbitt, R. W., 1986, Chemical characteristics of fluid phase released from a subducted lithosphere and origin of arc magmas; Evidence from high-pressure experiments and natural rocks: Journal of Volcanology and Geothermal Research, v. 29, p. 293–309.

Veniaminov, I., 1840, Notes on the Islands of the Unalashka District: (Black, L. T., and Geoghegan, R. H., translators; Pierce, R. A., ed.): 1984 Limestone Press Edition, Kingston, Ontario, 511 p.

von Drach, V., Marsh, B. D., and Wasserburg, G. J., 1986, Nd and Sr isotopes in the Aleutians; Multicomponent parenthood of island-arc magmas: Contributions to Mineralogy and Petrology, v. 92, p. 13–34.

White, W. M., and Patchett, J., 1984, Hf-Nd-Sr isotopes and incompatible element abundances in island arcs; Implications for magma origins and crust-mantle evolution: Earth and Planetary Science Letters, v. 67, p. 167–185.

Whitney, J. A., and Stormer, J. C., Jr., 1983, Igneous sulfides in the Fish Canyon Tuff and the role of sulfur in calc-alkaline magmas: Geology, v. 11, p. 99–102.

Wilcox, R. E., 1959, Some effects of recent volcanic ash falls with especial reference to Alaska: U.S. Geological Survey Bulletin 1028-N, p. 409–476.

Wilson, F. H., 1985, The Meshik arc; An Eocene to earliest Miocene magmatic arc on the Alaska Peninsula: Alaska Division of Geological and Geophysical Surveys Professional Report 88, 14 p.

Wilson, F. H., Detterman, R. L., and Case, J. E., 1985, The Alaska Peninsula terrane; A definition: U.S. Geological Survey Open-File Report 85-450, 17 p.

Wood, C. A., and Kienle, J., 1990, Volcanoes of North America: Cambridge, Cambridge University Press, 354 p.

Yoder, H. S., 1969, Calcalkalic andesites; Experimental data bearing on the origin of their assumed characteristics, *in* McBirney, A. R., ed., Proceedings of the Andesite Conference: Oregon Department of Geology, Mines and Industry Bulletin, v. 65, p. 77–89.

Yount, M. E., Miller, T. P., Emanuel, R. P., and Wilson, F. H., 1985, Eruption in an ice-filled caldera, Mount Veniaminof, Alaska Peninsula, *in* Bartsch-Winkler, S., and Reed, K. M., eds., The U.S. Geological Survey in Alaska; Accomplishments during 1983: U.S. Geological Survey Circular 945, p. 58–60.

Manuscript Accepted by the Society November 6, 1991

ACNOWLEDGMENTS

This chapter is dedicated to the memory of Rolf Werner Juhle of the U.S. Geological Survey, who died tragically in 1953 while conducting geologic studies in Katmai National Park. He had just received his Ph.D. from the Johns Hopkins University, having completed a dissertation on the geology of Iliamna Volcano.

Discussions with R. R. Coats and comments by W. Hildreth, E. Moll-Stalcup, G. Plafker, H. Berg, and two anonymous reviewers were helpful. S. McDuffie assisted with the figures. This work is supported by NSF grants EAR-8318240, EAR-8509005, EAR-8817394, and EAR-8916850 to the Johns Hopkins University (B.D.M.).

Printed in U.S.A.

The Geology of North America
Vol. G-1, The Geology of Alaska
The Geological Society of America, 1994

Chapter 24

Quaternary volcanism in the Alaska Peninsula and Wrangell Mountains, Alaska

Thomas P. Miller and Donald H. Richter
U.S. Geological Survey, 4200 University Drive, Anchorage, Alaska 99508-4667

INTRODUCTION

The numerous Quaternary volcanoes of the Alaska Peninsula, Cook Inlet area, and the Wrangell Mountains result from underthrusting of the Pacific Plate, or material coupled to the Pacific Plate, beneath the continental crust of North America. These volcanic centers are among the most prominent physiographic landforms in southern Alaska. They include some of the highest (>5,000 m), largest (>1,000 km^3), and most explosive (five Holocene eruptions with bulk volumes >50 km^3) volcanoes found along the entire circum-Pacific margin.

Edifices of the major Quaternary volcanoes dominate the Alaska Peninsula and Cook Inlet region (Fig. 1); numerous peaks rise 1,800 to 2,500 m above sea level. These volcanic centers, along with those of adjoining Unimak Island, constitute the eastern half of the Aleutian volcanic arc. This classic arc-trench system, equally divided between continental and oceanic segments, extends 2,600 km across the North Pacific. Separated from the northeast end of the Aleutian arc by 400 km is the subduction-related Wrangell volcanic field of Miocene to Holocene age, which underlies >10,000 km^2 of the Wrangell Mountains of south-central Alaska (Fig. 1).

Regional geologic mapping and topical volcanological studies since the early 1970s have resulted in an expanded understanding of the physical volcanology of the eastern Aleutian arc and Wrangell Mountains, including such parameters as size, stratigraphy, eruptive history, spacing, and geologic setting of the volcanic centers. Several volcanoes have now been mapped and studied in sufficient detail to clarify the physical and chemical processes associated with volcanic activity in this part of the circum-Pacific. This report incorporates new and previously unpublished information on the spatial distribution, volume, geologic setting, and major-element composition of the Quaternary volcanoes—information that was not available to authors of previous summary articles (Marsh, 1982; Kienle and Swanson, 1983).

EASTERN ALEUTIAN ARC: ALASKA PENINSULA-COOK INLET

Quaternary volcanism has resulted in a chain of subaerial stratocones on the Alaska Peninsula (including Unimak Island) and west side of Cook Inlet (Fig. 1) that constitutes the eastern Aleutian arc and includes the part of the entire arc built on continental crust. In this region, the Pacific Plate is impinging at close to a 90° angle on the North American Plate at rates of about 7.5 cm/year (Engebretson and others, 1985). The north-dipping Benioff zone lies about 100 km beneath the Cook Inlet portion of the arc and 75 to 100 km beneath the Alaska Peninsula centers (Kienle and Swanson, 1983). The arc-trench gap is about 300 km for much of the Alaska Peninsula but increases eastward to over 500 km in the Cook Inlet area (Jacob and others, 1977).

The recent completion of 1:250,000-scale geologic mapping for much of the region (Detterman and others, 1981, 1987; Riehle and others, 1987a; Wilson and others, 1994) has led to the identification of most major Quaternary volcanic centers. The term "volcanic center" is used in this report to describe a site of more or less continuous volcanism along the volcanic front during late Quaternary time. The actual vent area may have shifted 20 km or more during the life of the center, or there may be a cluster of active vents at an individual center. A center, therefore, may include overlapping lava flows and ejecta from more than one vent.

Using this arbitrary definition, 37 principal volcanic centers (Fig. 1; Table 1) have been identified in the eastern Aleutian arc, including Unimak Island. Centers included in this tabulation are generally aligned along the volcanic front parallel to the arc-trench system. The numerous closely spaced small volcanoes along the volcanic front in Katmai National Park are arbitrarily listed as separate centers. Several closely spaced vents on the lower Alaska Peninsula have been grouped together as the Kupreanof center (No. 28, Fig. 1). Pogromni and Westdahl volcanoes on the west end of Unimak Island have been included as a single

Miller, T. P., and Richter, D. H., 1994, Quaternary volcanism in the Alaska Peninsula and Wrangell Mountains, Alaska, *in* Plafker, G., and Berg, H. C., eds., The Geology of Alaska: Boulder, Colorado, Geological Society of America, The Geology of North America, v. G-1.

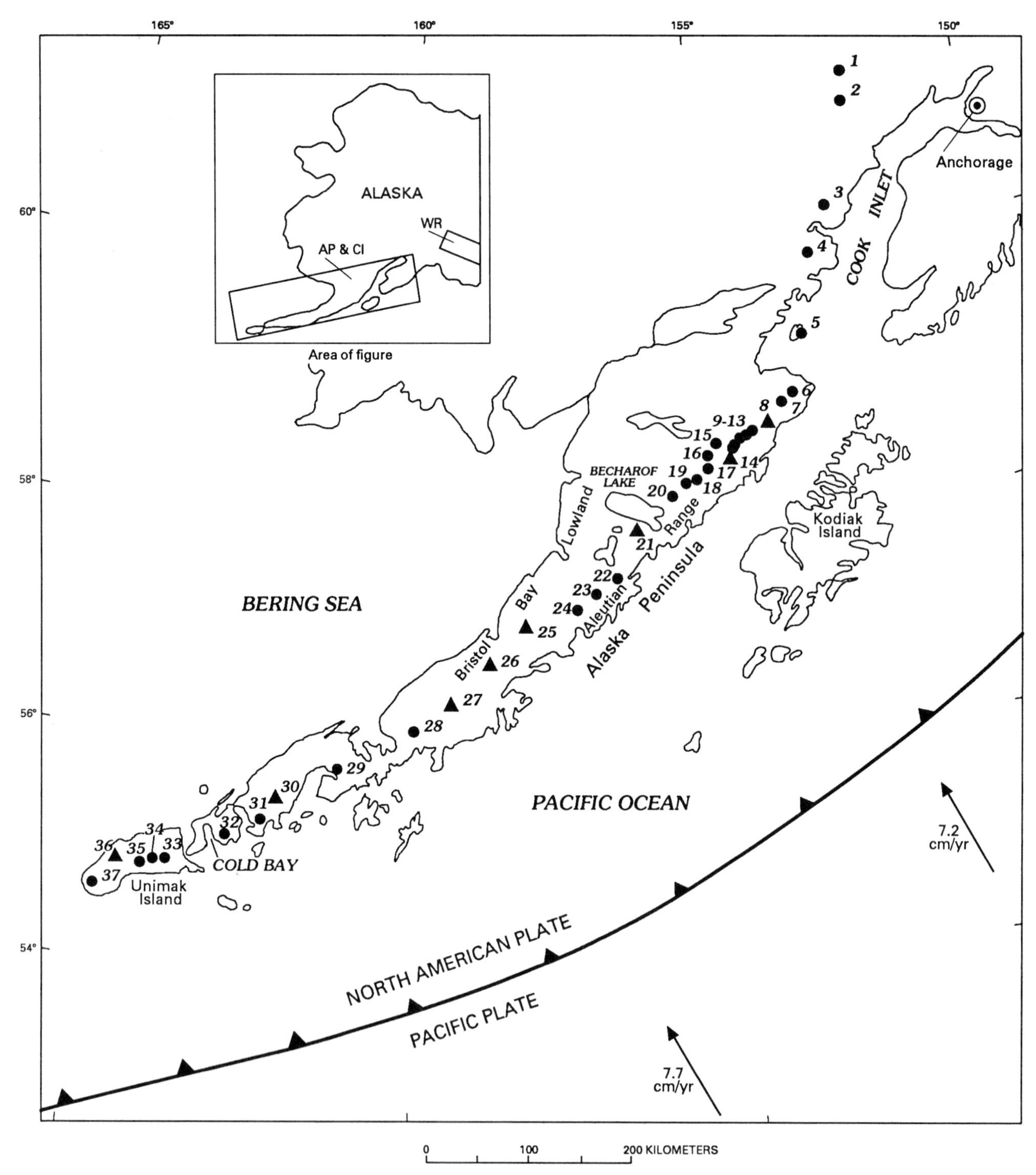

Figure 1. Major volcanic centers of the Alaska Peninsula (AP) and Cook Inlet (CI) region. Volcanic centers in the Wrangell Mountains (WR) are shown in Figure 6. ▲ denotes calderas; ● denotes stratavolcanoes. Numbers refer to centers as follows: 1, Hayes; 2, Spurr; 3, Redoubt; 4, Iliamna; 5, Augustine; 6, Douglas; 7, Fourpeaked; 8, Kaguyak; 9, Devils Desk; 10, Kukak; 11, Stellar; 12, Denison; 13, Snowy; 14, Katmai; 15, Griggs; 16, Novarupta; 17, Trident; 18, Mageik; 19, Martin; 20, Kejulik; 21, Peulik-Ugashik; 22, Kialagvik; 23, Chiginagak; 24, Yantarni; 25, Aniakchak; 26, Black Peak; 27, Veniaminof; 28, Kupreanof; 29, Dana; 30, Emmons Lake; 31, Dutton; 32, Frosty; 33, Round Top; 34, Isanotski; 35, Shishaldin; 36, Fisher; 37, Pogromni-Westdahl. Arrows and numbers adjacent to plate boundary denote direction and velocity of Pacific Plate movement relative to North American Plate (Engebretson and others, 1985). Sawteeth are on upper plate.

center (No. 37, Fig. 1) because the large topographic massif of which they are a part has not been mapped.

Not included in this tabulation is a series of widely separated domes and small monogenetic features such as Amak Island (Marsh and Leitz, 1978), Gas Rocks (Detterman and others, 1987), Ukinrek maars (Kienle and others, 1980), and three unnamed basaltic scoria cones in Katmai National Park (Riehle and others, 1987a); they are north of, and aligned parallel to, the Quaternary volcanic front. Many of these small volcanoes appear to be back-arc features, whose relation to nearby and larger volcanic centers of the Aleutian arc is uncertain.

Some relatively well-known volcanoes are now considered to be part of a much larger center. Pavlof volcano near Cold Bay (Fig. 1), for example, is one of the better known and most consistently active volcanoes in the Aleutian arc. Recent topical studies, however, have shown it to be only the most recently active of several small parasitic cones of Holocene age built on the flanks of the large Emmons Lake center (No. 30, Fig. 1; Table 1).

The Aleutian volcanic arc has been one of the least known volcanic provinces in the world because of its remoteness and the inherent logistical difficulties associated with any attempt to study it. Volcanic centers in the Aleutian Islands began to receive systematic attention from geologists only after World War II (Coats, 1950, 1962). More recently, many of these island volcanoes have been the subject of petrological studies, including trace-element and radiogenic isotope analyses (Kay and others, 1978, 1986; McCulloch and Perfit, 1981; Myers, 1988). Most recent discussions of the composition and petrogenesis of Aleutian arc volcanism are based on studies of the oceanic part of the arc.

The volcanic centers of the eastern Aleutian arc, however, received little attention until the 1960s. Notable exceptions were the studies by Griggs (1922), Fenner (1923, 1926), Allen and Zies (1923), and Zies (1929) of the 1912 eruption in the Valley of Ten Thousand Smokes in Katmai National Park. Regional syntheses of stratigraphy and structure of the basement rocks were lacking prior to the reconnaissance study by Burk (1965). The general structure of even the more noteworthy centers such as Katmai was little understood, and the existence of such large volcanic centers as Emmons Lake caldera (No. 30, Fig. 1) was virtually unknown.

Only within the past 20 years, starting with the classic study by Curtis (1968) of the ejecta of the 1912 Valley of Ten Thousand Smokes eruption have studies of eastern Aleutian arc volcanic centers and related phenomena begun (Hildreth, 1983, 1987; Riehle and others, 1987b; Miller and Smith, 1977; Nye, 1987). Previously unknown Quaternary volcanic centers such as those at Kialagvik, Hayes, and Yantarni (all of which exhibit Holocene activity) were identified during this period. Also, the existence of several of the more remote Katmai National Park volcanoes was confirmed by Kienle and Swanson (1983).

Geologic setting

The Alaska Peninsula–Cook Inlet centers are built entirely on continental crust (in contrast to the oceanic western arc) that is part of the Peninsular tectonostratigraphic terrane (Jones and others, 1987; Silberling and others, this volume). This terrane was formed far south of its present position and accreted onto mainland Alaska in Late Cretaceous and early Tertiary time (Stone and Packer, 1977; Vallier and others, this volume).

The eastern Aleutian volcanoes overlie continental and marine sedimentary rocks that range in age from mid-Paleozoic to Holocene (chiefly of Late Jurassic to early Tertiary age), volcanogenic rocks of late Mesozoic and mid- to late Tertiary age, and Jurassic to early Tertiary plutonic rocks of the Alaska-Aleutian Range batholith (Detterman and others, 1981, 1987; Riehle and others, 1987a; Reed and Lanphere, 1972; Wilson and others, 1991; Magoon and others, 1976). The volcanic centers rest on progressively younger basement rocks from east to west.

The regional structural grain is northeast-southwest, parallel to the axis of the Aleutian Range. The terrane is marked by normal, reverse, and thrust faults and numerous open folds. One of the major structural features of the region is the northeast-trending Bruin Bay fault, which has been mapped for 530 km from near Mt. Spurr (No. 2, Fig. 1) to Becharof Lake (Fig. 1) and may extend an additional 140 km to the southwest (Detterman and others, 1987).

Tertiary volcanism is represented by the Meshik and Aleutian volcanic arcs. The Meshik arc is predominantly of Eocene and Oligocene age, yielding K-Ar ages ranging from 48 to 22 Ma (Wilson, 1985), and is oriented subparallel to the Aleutian arc (Fig. 1). Aleutian arc volcanism includes the present-day Quaternary chain of volcanoes. It began in the middle Miocene and continued sporadically to the present. The locus of Aleutian arc magmatism has shifted from its Miocene position about 50 km northwestward to the present Pleistocene-Holocene volcanic front (Wilson, 1985).

Many of the volcanic centers served as sites of ice caps in Quaternary time; glacial deposits recording four major Pleistocene glacial episodes have been mapped in the region (Detterman, 1986). Volcanic edifices have been extensively glaciated, and many still host glaciers and extensive snow and ice fields. The volcanoes also provided much of the debris found in the morainal ridges and drift sheets of Pleistocene age that overlie most of the Bering Sea Lowland north of the Aleutian Range.

Age and eruptive activity

Potassium-argon ages from Veniaminof, Aniakchak, Yantarni, Mt. Spurr, and Redoubt volcanoes (Fig. 1) suggest that most of the Alaska Peninsula and Cook Inlet centers began forming within the last 10^6 years (Luedke and Smith, 1986; Riehle

TABLE 1. PHYSICAL CHARACTERISTICS OF PRINCIPAL QUATERNARY VOLCANIC CENTERS IN THE EASTERN ALEUTIAN ARC*

Volcanic Center	Location	Elevation (m)	Height (m)	Diameter (km)	Estimated Volume* (km^3)	Eruptive Activity: Holocene	Eruptive Activity: Historic	Eruptive Activity: Fumarolic	Remarks
1. Hayes	61.6°N 152.5°W	1,690	300 (?)	5 (?)	<20 (?)	◆			Volcano largely destroyed by catastrophic eruptions about 3,500 ^{14}C B.P. and by subsequent glacial activity. Remnants are snow- and ice-covered pyroclastic deposits adjacent to Hayes Glacier.
2. Spurr	61.3°N 152.3°W	3,374	>1,000	10 +	30–50	◆	◆	◆	Stratocone and summit dome; explosion amphitheater open to the south summit dome, and parasitic cone on south flank (Crater Peak); associated pyroclastic flows. Fumaroles on summit dome and at Crater Peak.
3. Redoubt	60.5°N 152.8°W	3,108	1,500–1,800	11	38	◆	◆	◆	Stratocone with widespread Holocene lahars; fumarolic area in ice-filled summit crater.
4. Iliamna	60.0°N 153.1°W	3,053	1,800	10	45	◆	?	◆	Deeply eroded, strongly altered andesitic stratocone. Slope failure is common with much landslide activity. Large fumarolic area near summit. Widespread Holocene pumice in quadrant northeast of volcano.
5. Augustine	59.4°N 153.4°W	1,259	950	6–8	15	◆	◆	◆	Nested summit domes surrounded by volcaniclastic apron composed of pyroclastic flow deposits, lahars, and debris avalanches; summit area breached to north. Fumaroles on and around summit.
6. Douglas	58.9°N 153.6°W	2,135	1,000	10	25	?	◆		Little studied, small, chiefly ice-covered stratocone; fumaroles and hot crater lake near summit.
7. Fourpeaked	58.8°N 153.7°W	2,104	900	7	10				Poorly exposed, ice-covered, strongly altered stratocone.
8. Kaguyak	58.6°N 154.1°W	901	600	5	4 *	◆			Stratocone with lake-filled caldera of Holocene age. At least one post-caldera dacitic dome. Surrounded by nonwelded pyroclastic flow deposit.
9. Devils Desk	58.5°N 154.3°W	1,955	430–735						
10. Kukak	58.5°N 154.4°W	1,040	500–800						
11. Stellar	58.4°N 154.4°W	2,272	400		60			◆	A group of small, closely spaced, largely ice-covered central vent volcanoes composed of andesitic flows, tuffs, and breccias. (9–12)
12. Denison	58.4°N 154.5°W	2,318	500						

TABLE 1. PHYSICAL CHARACTERISTICS OF PRINCIPAL QUATERNARY VOLCANIC CENTERS IN THE EASTERN ALEUTIAN ARC* (continued)

Volcanic Center	Location	Elevation (m)	Height (m)	Diameter (km)	Estimated Volume* (km^3)	Eruptive Activity: Holocene	Eruptive Activity: Historic	Eruptive Activity: Fumarolic	Remarks
13. Snowy Mountain	58.3°N 154.7°W	2,161	640	5	5	◆		◆	Largely ice-covered central vent volcano; two main vent areas about 3.5 km apart along ridge crest; fumarolic.
14. Katmai	58.3°N 155.0°W	2,047	800	7	10 *	◆	◆	◆	Glacier-clad stratocone with lake-filled caldera formed in 1912. Intracaldera dome activity; sub-lacustrine fumaroles.
15. Griggs	58.4°N 155.1°W	2,265	740	6	7	◆		◆	Stratocone with nested summit craters; fumaroles on south flank near summit.
16. Novarupta	58.3°N 155.2°W	841	?	1	1	◆	◆	◆	Post-caldera plug dome surrounded by 2-km-wide subsidence structure largely covered by caldera-fill tephra. Fumarolic. Source area for 1912 VTTS Pyroclastic flows.
17. Trident	58.2°N 155.1°W	2,070	850	5	5	◆	◆	◆	Deeply eroded multivent andesite-dacite stratocone consisting of three main peaks; fumarolic. Historic activity has been on south flank.
18. Mageik	58.2°N 155.3°W	2,210	1,000	6	10	◆	◆	◆	Multivent andesite-dacite stratocone: fumarolic.
19. Martin	58.2°N 155.4°W	1,844	550	6	5	◆	◆	◆	Stratocone with strongly fumarolic summit crater.
20. Kejulik	58.1°N 155.6°W	1,586	670	5	5				Small basaltic stratocone.
21. Peulik-Ugashik	57.7°N 156.3°W	921	300	8	15 *	◆	?		Ugashik is a deeply eroded stratocone with a small summit caldera containing a group of nested domes. Peulik is a Holocene satellitic andesitic stratocone built high on the north flank of Ugashik. The summit of Peulik contains a dome and is breached to the west; a large area of block-and-ash flows and avalanche deposits mantles the west flank of the volcano.
22. Kialagvik	57.2°N 156.7°W	+1,700	900	5	5–10	◆			Poorly exposed, ice-covered dome and ash-flow complex.
23. Chiginagak	57.1°N 157.0°W	2,135	1,300	6	15	◆	◆	◆	Symmetrical stratocone. Deep breach on south flank extends to summit crater; domes near summit. Fumarolic area high on north flank.
24. Yantarni	57.0°N 157.2°W	1,250	430	6	5	◆			Small andesitic stratocone with dacitic summit dome; cone is breached to the north. Extensive Holocene avalanche deposits and block-and-ash flows east and northeast of volcano.

TABLE 1. PHYSICAL CHARACTERISTICS OF PRINCIPAL QUATERNARY VOLCANIC CENTERS IN THE EASTERN ALEUTIAN ARC* (continued)

Volcanic Center	Location	Elevation (m)	Height (m)	Diameter (km)	Estimated Volume* (km^3)	Eruptive Activity: Holocene	Historic	Fumarolic	Remarks
25. Aniakchak	56.9°N 158.2°W	1,341	650	20	8 *	♦	♦	♦	Stratocone with large (10 km) ice-free summit caldera formed about 3,400 ^{14}C B.P. Intra-caldera domes, tuff cones, spatter cones, lava flows, and small lake. Hot (85°C) ground in northwest corner of caldera. Center is surrounded by pyroclastic flow deposits for distances of up to 50 km.
26. Black Peak	56.6°N 158.8°W	955	350	8	10 *	♦	♦		Small stratocone with summit caldera formed 4,100 to 4,700 ^{14}C BP. Intracaldera nested dacitic dome complex. Thick block-and-ash deposits in valleys north and west of caldera.
27. Veniaminof	56.2°N 159.4°W	2,507	1,900	35	400 *	♦	♦	♦	Large stratocone truncated by 10-km-wide summit caldera formed about 3,700 ^{14}C B.P. Caldera is occupied by an ice field that covers the south rim of the caldera. Alpine glaciers fill valleys on the north, east, and west flanks of the volcano. Active intracaldera basaltic cone. Lahars and pyroclastic flow deposits fill many of the valleys on the volcano's flanks. Northwest-trending belt of post-caldera cinder and scoria cones extends across the volcano from the Bering Sea to the Pacific Ocean side.
28. Kupreanof	55.9-56.1°N 159.8-160.1°W	1,320-1,890	400-900		70	♦	♦	♦	Easternmost of at least 5 known Quaternary vents occurring along a 30-km-long belt on the Aleutian Range crest; also called the Stepovak Bay Volcano Group (Wilson, 1989). Volcanic products from these vents are andesitic and commonly overlapping. Fumarolic activity at two localities.
29. Dana	55.6°N 161.2°W	1,354	1,000	5.5	8	♦	♦		Small central vent dome with lake-filled summit crater; surrounded by apron of volcaniclastic debris. Block-and-ash flow erupted 3,840 ^{14}C B.P. fills valleys north and south of volcano (Yount, 1990).

TABLE 1. PHYSICAL CHARACTERISTICS OF PRINCIPAL QUATERNARY VOLCANIC CENTERS IN THE EASTERN ALEUTIAN ARC* (continued)

Volcanic Center	Location	Elevation (m)	Height (m)	Diameter (km)	Estimated Volume* (km³)	Eruptive Activity: Holocene	Historic	Fumarolic	Remarks
30. Emmons Lake	55.4°W 162.0°W	2,519	850–975	30	350 *	◆	◆	◆	Large stratocone with oval-shaped 11 x 18 km caldera. At least two major caldera-forming eruptions in Pleistocene time. Post-caldera basaltic volcanism from Emmons, Hague, Pavlof Sister, and numerous other unnamed parasitic and flank volcanoes. Fumaroles on Pavlof and Hague volcanoes.
31. Dutton	55.2°N 162.3°W	1,506	760	6	5-10	◆			Small stratocone with Holocene domes in summit area and on northeast flank. Flanks are mantled by widespread debris flows and avalanche deposits.
32. Frosty	55.1°N 162.8°W	1,763	1,100	10	25	◆		?	Stratocone with central vent summit dome. Possible fumarolic area in ice-filled summit crater.
33. Round Top	54.8°N 163.6°W	1,871	600	8	10	◆			Ice-covered stratocone with summit and flank domes. Holocene pyroclastic flow deposits, lahars, and debris avalanches on south and southeast flanks.
34. Isanotski	54.8°N 163.7°W	2,446	1,800 (?)	10	45	◆	◆		Deeply eroded stratocone with possible domes.
35. Shishaldin	54.8°N 164.0°W	2,857	1,500 (?:)	10	500	◆	◆	◆	Symmetrical stratocone (upper 2,000 m ice-covered) built on older volcanic rocks (Fournelle and Marsh, 1986). Monogenetic cones and associated lava flows on northwest flanks; pyroclastic flow deposits; fumarolic summit area.
36. Fisher	54.7°N 164.4°W	1,094	960	21	300 *	◆	◆	◆	Large stratocone with oval-shaped (18 x 11 km) Holocene caldera (s); last major caldera-forming activity about 9,100 ^{14}C B.P. Thick intracaldera tephra fill; several post-caldera domes and cones. Fumarolic area near center of caldera.
37. Pogromni-Westdahl	54.5°N 164.7°W	2,000	1,700 (?)	20	200	◆	◆	◆	Large ice-covered volcanic massif with active vents at Pogromni and Westdahl volcanoes. Little information available.

*Pre-caldera volume.

and others, 1987b; Yount, 1984; Turner and Nye, 1986). Six K-Ar ages, ranging from 830 to 109 ka, have been obtained from Veniaminof (No. 27, Fig. 1)—the large volcanic center on the central Alaska Peninsula—and nearby Aniakchak volcano has yielded three K-Ar ages ranging from 661 to 242 ka. Turner and Nye (1986) conducted a detailed geochronologic study of Mt. Spurr volcano and concluded that most of the ancestral stratocone was built from 255 to 58 ka. In contrast, Augustine volcano, the small island center in lower Cook Inlet (No. 5, Fig. 1), is thought to have formed in the last 20,000 years (Johnston, 1978).

Historic eruptive activity has been reported from at least 19 of the 37 Alaska Peninsula–Cook Inlet volcanic centers (Table 1), more than 120 eruptions having taken place since 1760 (Simkin and others, 1981; T. P. Miller, unpublished data); more than half of the centers (21 of 37) show fumarolic activity. The number of historic eruptions is undoubtedly a minimum figure, since many eruptions probably went unreported prior to 1900 because of the remoteness of the region.

Holocene activity has been documented at 30 of the 37 principal volcanic centers (Table 1). Tephrochronologic studies in the region have only recently begun (Riehle, 1985) but they hold much promise for delineating the source and timing of major Holocene eruptions.

At least nine caldera-forming eruptions have occurred at eight of the volcanic centers on the Alaska Peninsula (Miller and Smith, 1987) during late Pleistocene and Holocene time. Five of these major eruptions may have had ejecta bulk volumes of >50 km^3, and all had estimated bulk volumes of >10 km^3; indeed 20 to 30 percent of all known Holocene eruptions of this magnitude worldwide appear to have occurred on the Alaska Peninsula (Miller and Smith, 1987). The 1912 eruption of about 35 km^3 of tephra from the Novarupta vent complex in Katmai National Park (No. 16, Fig. 1)—an event that resulted in the formation of calderas at Katmai and Novarupta—was the most voluminous rhyolitic eruption in the past 1,800 years and has been called the world's most important igneous event of this century (Hildreth, 1987).

Physical volcanology

Most Alaska Peninsula and Cook Inlet volcanoes are typical stratocones consisting of intercalated lava flows and volcaniclastic rocks and marked by summit craters and calderas. These volcanoes commonly have a basal unit that is predominantly volcaniclastic and an upper unit that is composed chiefly of flows. A few small centers (e.g., Augustine, Kialagvik, Dana, Mount Dutton; Nos. 5, 22, 29, 31 on Fig. 1) appear to consist chiefly of dacitic dome clusters surrounded by an apron of pyroclastic flow deposits.

Diameters of collapse calderas range from 3 km (Kaguyak) to 18 km (Emmons Lake; Miller, this volume), although the scalloped oval shape of larger calderas such as Emmons Lake and Fisher suggests more than one collapse event. The largest circular calderas (such as Aniakchak, Veniaminof; Table 1), each of which may represent a single collapse event, are about 10 km in diameter. Ash-flow tuffs, generally nonwelded, surround most calderas and, in the case of some larger calderas such as Aniakchak, originally covered areas of >2,000 km^2 (Miller and Smith, 1977) and extended as far as 80 km from their source volcano. At least six volcanic centers (Spurr, Augustine, Peulik, Kialagvik, Yantarni, Round Top; Nos. 2, 5, 21, 22, 24, 33, Fig. 1) have associated block-and-ash flows and avalanche and blast deposits indicative of smaller-scale explosive activity.

Attempts to correlate the spacing, size, compositional range, and tectonic setting of eastern Aleutian arc volcanic centers have been hampered by the lack of geological knowledge. When Marsh and Carmichael (1974) made a compilation of Aleutian arc volcanoes, for example, at least six Quaternary centers shown in Figure 1 were yet to be discovered, and few volcanoes of the Alaska Peninsula and Cook Inlet had been mapped in any detail. As additional mapping has become available, new volcanic centers have been identified, and some volcanoes previously considered separate entities have been identified as being part of a single eruptive center.

Volume. Obtaining meaningful figures for the volumes of volcanoes is difficult because of the erosion of deposits, the occurrence of calderas, the somewhat arbitrary classification into centers, and the cyclic eruptive character of many of the larger centers. However, approximate volumes have been estimated based on the regional geological mapping now available. Individual centers vary in size (Table 1) from small central-vent volcanoes, such as Yantarni with a volume of 6 km^3 (Riehle and others, 1987b), to large, long-lived, stratocone-caldera complexes such as Veniaminof, Emmons Lake, and Fisher (Table 1) that are probably cyclic in character and have volumes of 300 to 400 km^3. Most other volcanic centers have volumes generally less than 100 km^3. Volumes of individual volcanic centers in Katmai National Park are less meaningful, because the centers in this region are so close (in some cases overlapping) that it is difficult to determine what constitutes an individual center. Although Mount Spurr (3,375 m above sea level) and neighboring Redoubt (3,110 m) and Iliamna (3,075 m) are the highest volcanoes in the region, the actual height of each edifice is much less, since all three are built on a basement approximately 1,500 m high.

Marsh (1982) stated that the volume of the volcanic centers decreases from east to west along the Aleutian arc. He related this decrease to reduction in the rate of convergence between the Pacific and North American Plates. The largest volcanic centers in the arc, however, are actually in the 700-km-long segment between Umnak Island and Veniaminof volcano on the Alaska Peninsula (Fig. 2). Volcanic centers in the 770 km of arc east of Veniaminof and the 1,200 km of arc west of Umnak Island are smaller. The size of individual volcanic centers therefore seems not to be related in any simple way to plate convergence rates; rather, size is more apt to be due to a complex function of convergence angles and rates, crustal thickness and composition, and nature of the subducted material.

Spacing. Marsh and Carmichael (1974) and Marsh (1982) believed the Aleutian arc volcanoes had a rather regular spacing

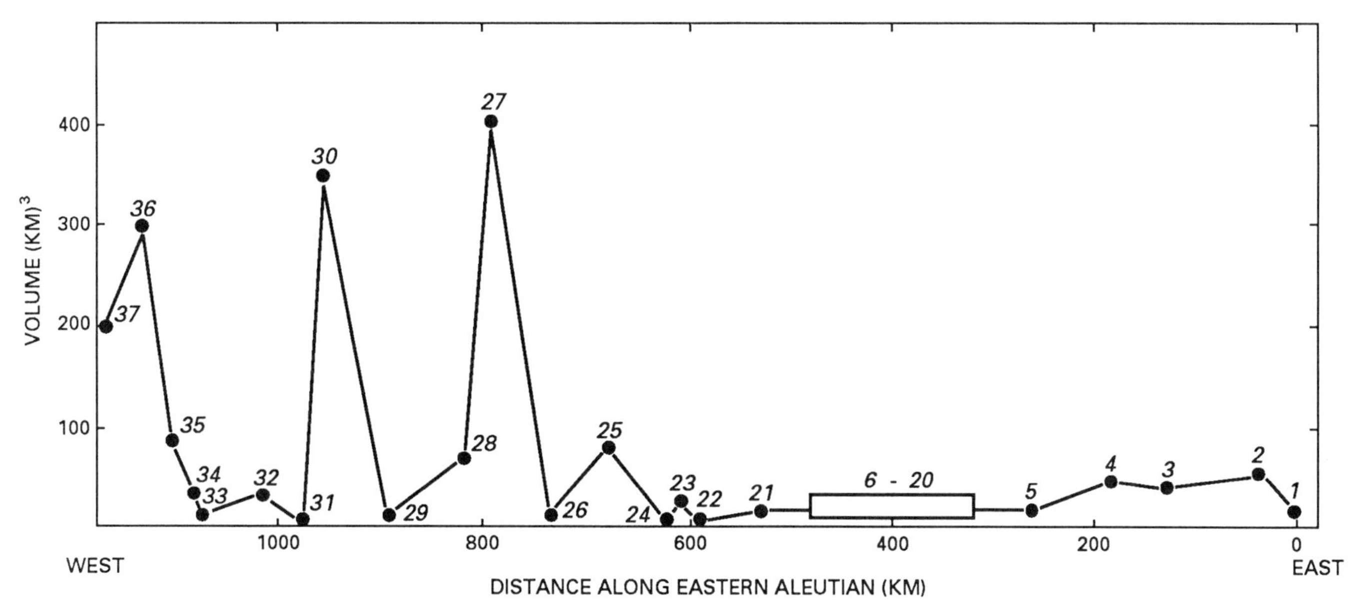

Figure 2. Approximate volumes of Alaska Peninsula–Cook Inlet volcanoes plotted against distance along Aleutian arc. Volcano numbers refer to Figure 1.

of 60 to 70 km, and Marsh (1979) commented on a segmentation of the arc based on the linear distribution of volcanoes. The spacing was thought to represent regularly spaced mantle plumes rising from melt developed at depths of 100 to 150 km along the strike of the subducted plate. Kay and others (1982) related segmentation of the arc to major tectonic breaks at the ends of rupture zones of major earthquakes. Kienle and Swanson (1983) recognized that segments of the easternmost part of the arc had different spacings. They noted a 70 ± 18-km spacing for Spurr-Redoubt-Iliamna volcanoes (excluding Hayes) in the Cook Inlet region, and a 13 ± 7-km spacing for volcanoes in the Katmai area.

Geological mapping in the area shows that spacing between centers varies considerably along this part of the Aleutian arc (Fig. 3). Some short "segments" of the arc indeed have a regular spacing between volcanic centers. The Kialagvik-Chiginagak-Yantarni volcanoes (Nos. 22 to 24, Fig. 1), for example, are almost exactly 18 km apart, and Aniakchak–Black Peak–Veniaminof centers (Nos. 25 to 27, Fig. 1) are almost exactly 54 km apart.

Distances between centers elsewhere in the region are much less regular. A 75-km-long arc segment in the Katmai area from Devil's Desk to Martin volcanoes (Nos. 9 and 19, Fig. 1) contains 11 closely spaced centers whose mutual relations are uncertain. S-wave shadowing beneath volcanoes in this region led Matumoto (1971) to infer two 10- to 20-km-wide, shallow (<10 km) magma chambers beneath volcanoes in the Katmai region and a third shallow chamber beneath Snowy volcano (No. 13, Fig. 1) 20 km to the northeast. Hildreth (1987, p. 692) has suggested that these chambers actually consist of dikes and sills that locally inflate into one or more pods of less than a few cubic kilometers of magma each. Some of the nine centers may therefore be hydraulically connected in a manner similar to that of Novarupta and Katmai. A 27-km-long arc segment extending southwest from Kupreanof volcano (No. 20, Fig. 1) includes a recently recognized group of at least five vents within a few kilometers of each other along the arc (Yount and others, 1985; Wilson, 1990); it is uncertain whether all these vents are part of the deeply eroded Kupreanof volcanic center, or represent additional centers. An "average" volcano spacing distance for the eastern Aleutian arc (and by inference the entire arc) would therefore appear to be relatively meaningless. It is more significant that various "segments" of the arc have quite different spacings between individual centers and that throughout the eastern Aleutian arc, the greatest distance between volcanic centers (the largest volcano-free segment) is the 94 km between Spurr and Redoubt (Nos. 2 and 3, Fig. 1).

Structural control. Many authors have suggested structural controls for the spacing of volcanoes. The Alaska Peninsula–Cook Inlet area is certainly a structurally complex and seismically active region characterized by broad open folds and normal, reverse, and thrust faults. Although crustal structure almost certainly influences where a volcanic center forms in a local area, recent 1:250,000-scale geological mapping (Detterman and others, 1981, 1987; Riehle and others, 1987a) indicates no single overriding crustal structural control of volcano spacing. Some individual volcanic centers, or clusters of centers, can be correlated with specific crustal structural features. Examples are the occurrence of Chiginagak and Kialagvik volcanoes on the Wide Bay anticline and Yantarni volcano on the trace of a prominent fault extension of the anticline (Detterman and others, 1987). On the other hand, the 560-km-long Bruin Bay fault is one of the

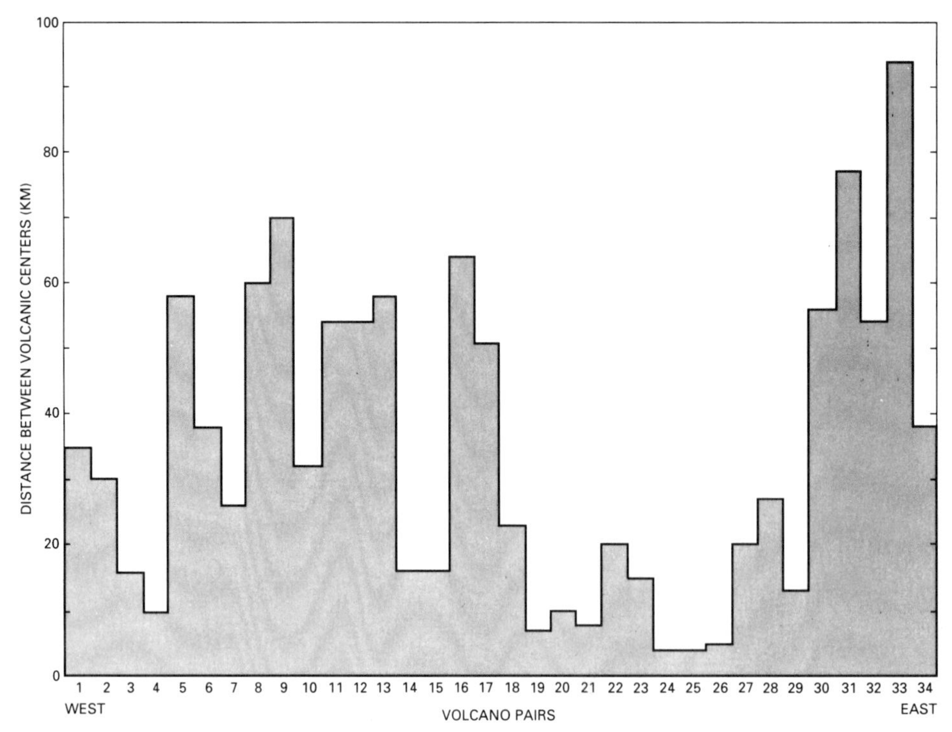

Figure 3. Distances between major volcanic centers in the Alaska Peninsula and Cook Inlet region. Volcano pairs and distance (km) are as follows: 1, Westdahl (Pogromni)-Fisher (35); 2, Fisher-Shishaldin (30); 3, Shishaldin-Isanotski (16); 4, Isanotski-Round Top (10); 5, Round Top–Frosty (58); 6, Frosty-Dutton (38); 7, Dutton–Emmons Lake (26); 8, Emmons Lake–Dana (60); 9, Dana-Kupreanof (70); 10, Kupreanof-Veniaminof (32); 11, Veniaminof–Black Peak (54); 12, Black Peak–Aniakchak (54); 13, Aniakchak-Yantarni (58); 14, Yantarni-Chiginagak (18); 15, Chiginagak-Kialagvik (18); 16, Kialagvik-Ugashik (64); 17, Ugashik-Kejulik (51); 18, Kejulik-Martin (23); 19, Martin-Mageik (7); 20, Mageik-Trident (10); 21, Trident-Katmai (8); 22, Katmai-Snowy (20); 23, Snowy-Denison (15); 24, Denison-Stellar (4); 25, Stellar-Kukak (4); 26, Kukak–Devils Desk (5); 27, Devils Desk–Kaguyak (20); 28, Kaguyak-Fourpeaked (27); 29, Fourpeaked-Douglas (13); 30, Douglas-Augustine (56); 31, Augustine-Iliamna (77); 32, Iliamna-Redoubt (54); 33, Redoubt-Spurr (94); 34, Spurr-Hayes (38). Note: Volcanoes such as Novarupta, Griggs, Ukinrek, and Amak (Simkin and others, 1981) lie north of the axis of the volcanic front and are not included in this pairing. The distance between Dana and Kureanof volcanoes is measured from the western most of the Holocene vents associated with the Kupreanof center.

major structural features of the region, and although a few volcanoes (Redoubt, Iliamna, Nos. 3 and 4, Fig. 1) are on or near the fault and its possible extension, the great majority of volcanic centers show little apparent relation to it.

Nakamura and others (1977) considered the small monogenetic volcanoes on the flank of the Veniaminof center to be aligned along regional stress fractures that developed normal to the trend of the volcanic arc as a result of regional compression in the direction of plate convergence. The volcanic activity marking this trend is chiefly Holocene in age, but a prominent dike swarm developed only in the older cone-building rocks of the center has the same orientation, suggesting that similar stress patterns extended further back in the Quaternary.

Composition

Over 500 major-element chemical analyses of the Quaternary volcanic rocks of the eastern Aleutian arc have recently been obtained. Many volcanic centers, however, have been sampled only in reconnaissance, and we have little detailed knowledge of their stratigraphy or compositional range. For this reason, approximately 330 analyses from ten of the better mapped volcanic centers in the area have been used for the summary presented in this report; at least 15 analyses were available from each of these centers. The following classification of volcanic rocks based on normalized SiO_2 content was used (similar to that adapted by Luedke and Smith, 1986): basalt <54 percent SiO_2, andesite 54

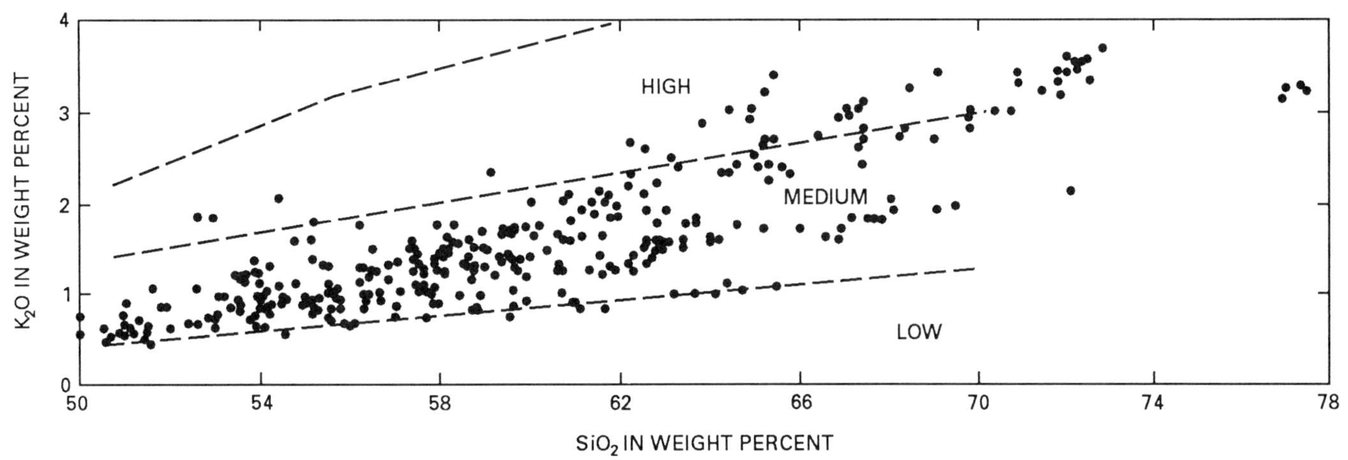

Figure 4. Plot of K_2O versus SiO_2 for 330 analyses from ten Alaska Peninsula and Cook Inlet (AP-CI) volcanoes. Data are from following volcanoes: Spurr, Redoubt, Iliamna, Augustine, Novarupta, Peulik-Ugashik, Yantarni, Aniakchak, Veniaminof, and Emmons Lake. Discriminant lines separating High, Medium, and Low are modified from Pecerrillo and Taylor (1976) and Ewart (1179).

to 62 percent SiO_2, dacite 62 to 70 percent SiO_2, and rhyolite >70 percent SiO_2.

The SiO_2 content ranges from 50 to 78 percent (Fig. 4). Most of the cone-building volcanic rocks, however, consist predominantly of andesite and low-SiO_2 dacite with a SiO_2 range of 55 to 64 percent; basalts (<54 percent SiO_2) are less common in this part of the arc. Rhyolite (>70 percent SiO_2) is known to be present only in ash-flow tuffs and post-caldera domes at the Ugashik-Peulik, Aniakchak, and Emmons Lake centers (Nos. 21, 25, and 30, Fig. 1) and in the Valley of Ten Thousand Smokes. High-SiO_2 (>75 percent) rhyolite is found in ejecta from the 1912 eruption and in the associated Novarupta dome, and nowhere else on the Alaska Peninsula. A composition gap of 72 to 77 percent SiO_2 between the high-silica rhyolite and volcanic rocks from other Alaska Peninsula–Cook Inlet centers is apparent in Figure 4 and attests to the uniqueness of the rhyolite. Hildreth (1983, 1987) has noted a SiO_2 compositional gap of 66 to 77 percent in the 1912 ejecta. Some small stratocones and dome complexes (Augustine, Black Peak, and Dana, Nos. 5, 26, and 29, Fig. 1) are composed almost entirely of andesite and dacite with small SiO_2 ranges of 57 to 63 percent.

Alkaline volcanic rocks appear to be restricted to a back-arc environment consisting of a belt of widely separated small monogenetic centers that subparallels the main andesitic arc for at least 800 km from the Ukinrek maars, near Becharof Lake on the Alaska Peninsula, west to Bogoslof Island in the Bering Sea (Fig. 1).

Radiogenic isotope data from Alaska Peninsula and Cook Inlet volcanic rocks is relatively sparse. Kay and others (1978) report $^{87}Sr/^{86}Sr$ ratios (SIR) ranging from 0.70305 to 0.70325 and from 0.70309 to 0.70334 from Veniaminof and Aniakchak volcanoes respectively. These ratios are similar to the 0.70368 SIR reported from Katmai volcanoes (Rubenstone and others, 1985; Hildreth, 1987) but lower than the 0.7040 to 0.7046 SIR range reported from Redoubt volcano by Bevier and Wheeler (1983). The SIR from Alaska Peninsula and Cook Inlet volcanic rocks are similar to the 0.70289 to 0.70342 values reported by McCulloch and Perfit (1981) from the western oceanic part of the Aleutian arc. As Hildreth (1987) has pointed out, however, the radiogenic isotopic data presently available are too limited to adequately determine the proportions of mantle, slab, and intracrustal contributions to andesitic magma, and more isotopic studies are needed.

Variations in potassium content of volcanic rocks, particularly andesites, are thought to have tectonic as well as petrogenetic significance (Gill, 1981). Andesites of the Alaska Peninsula–Cook Inlet region are medium-K (Fig. 4) on K_2O vs. SiO_2 plots (Taylor, 1969; Ewart, 1979), typical of many arc suites, and average 1.27 percent K_2O at the 57.5 percent SiO_2 reference level.

Kay and others (1982) recognized both tholeiitic and calc-alkaline differentiation trends in the oceanic part of the Aleutian arc on the basis of the change of FeO*/MgO ratio with SiO_2 (Miyashiro, 1974). This ratio increases with SiO_2 in tholeiitic centers but remains approximately constant in calc-alkaline centers.

Iron-enrichment plots of the Alaska Peninsula and Cook Inlet volcanoes (Fig. 5) exhibit calc-alkaline, tholeiitic, and transitional trends similar to those seen in the western part of the arc by Kay and others (1982). Only the westernmost two of the ten reference centers (Emmons Lake and Veniaminof), however, are tholeiitic; all the reference centers to the east are calc-alkaline, although Aniakchak is somewhat transitional in character (Fig. 5). The calc-alkaline character of eastern Aleutian arc volcanic rocks, first noted by Kienle and Swanson (1983), is thus extended through much of the Alaska Peninsula. The easternmost 30 percent (770 km) of the Aleutian arc therefore appears to be characterized by calc-alkaline magmatism, with few if any tholeiitic suites.

Conflicting views for the origin of calc-alkaline and tholeiitic magmas are discussed by Kay and Kay (this volume) and Myers and others (1985). Kay and Kay argue that these compositional differences result from processes in the Aleutian crust that are controlled by the tectonic setting of the volcano. Magma encountering crust that is under relative extension passes quickly through it and crystallizes at low pressures with little opportunity for intracrustal magma fractionation and mixing, resulting in tholeiitic centers. Magma encountering crust under relative compression, however, passes slowly through it and undergoes crystallization and mixing at greater crustal depths, resulting in calc-alkaline centers. Calc-alkaline centers would therefore form in the middle of tectonically controlled arc segments, and tholeiitic centers would form at segment boundaries, or in areas where segments are not well defined. The tectonic segments themselves are separated by rupture zones defined chiefly by the distribution of aftershocks following great earthquakes.

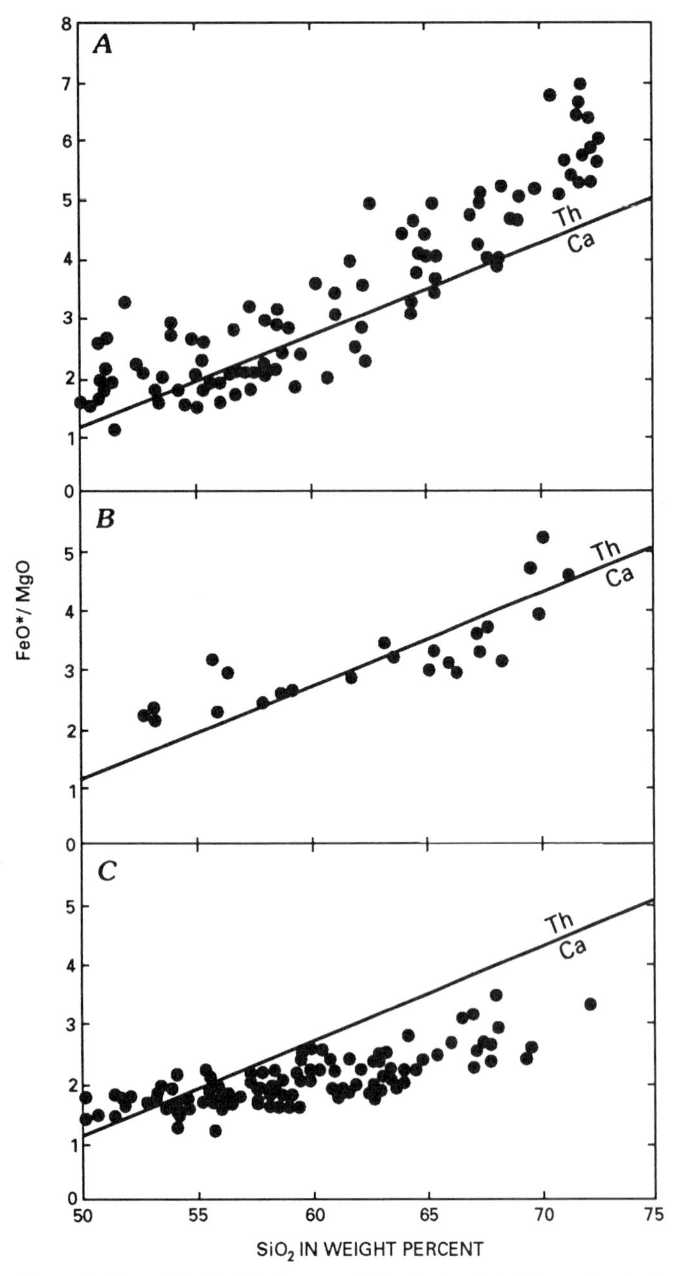

Figure 5. Plot of FeO*/MgO versus SiO_2 for the tholeiitic Veniaminof-Emmons Lake center, the transitional Aniakchak center, and the remaining calc-alkaline centers of the Alaska Peninsula–Cook Inlet (AP-CI). Data base is the same as Figure 2. Line separates tholeiitic/calc-alkaline plots from Miyashiro (1974).

Myers and others argue that calc-alkaline centers form in an early immature stage of conduit development, when mantle is incorporated in the magma during transit through the lithosphere. The volcanic center is transformed into a tholeiitic center over time as the conduit is thermally and chemically preconditioned and no longer contaminates the magmas.

Although the origin of calc-alkaline in contrast to tholeiitic magmatic trends is not the subject of this chapter, it is of interest that the conflicting theories were derived chiefly from studies of the western oceanic part of the Aleutian arc. The present study of the continental half of the arc indicates that the easternmost one-third of the entire Aleutian arc is characterized solely by calc-alkaline volcanic centers. Fisher and others (1981) and Kienle and Swanson (1983), however, consider this part of the arc to be segmented as is the oceanic part of the arc. Both parts show a variation in spacing and alignment of centers and aftershock rupture zones. Tectonic segmentation, therefore, does not appear to have influenced compositional trends (e.g., tholeiitic versus calc-alkaline) of individual centers, and the chemical diversity observed among volcanic centers in the oceanic part of the arc is lacking.

The change in compositional trends does not occur where the Aleutian arc intersects the continent west of Unimak Island but rather is 500 km to the east between Veniaminof and Black Peak volcanoes. This area is the same as the one in which a change in the character of post-caldera volcanism in the eastern Aleutian arc occurs (Miller and Smith, 1983). From Fisher to Veniaminof, post-caldera volcanism is mafic, whereas from Black Peak to Kaguyak, post-caldera volcanism is intermediate to silicic. The abrupt change in two different compositional trends in the same area suggests a common cause, which we believe is related to the nature and extent of continental crust. Kienle and Swanson (1983) suggested that the thickness of continental crust seems to control calc-alkaline versus tholeiitic differentiation trends in the Cook Inlet and Katmai regions—a suggestion we think may hold true for most of the Alaska Peninsula. The continental crust east of Veniaminof volcano may be thick and extensive enough to inhibit the easy passage of magma in spite of tectonic segmentation. Magma fractionation and mixing thus took place and resulted in calc-alkaline magmatic differentiation, similar to that proposed by Kay and Kay. However, crustal thickness figures are not well constrained beneath the Alaska

Peninsula. The thickness of the crust has been estimated to be 38.5 km in the vicinity of the Emmons Lake center on the lower Alaska Peninsula (see McNutt and Jacob, 1986, Fig. 3) and 30 to 35 km in the Katmai region (Kienle and Swanson, 1983). The exact nature of crustal changes (e.g., thickness) northeast of Veniaminof volcanic center remains to be worked out.

WRANGELL VOLCANIC FIELD

The Wrangell volcanic field of Miocene to Holocene age occurs chiefly in south-central Alaska, between 61°15′ and 62°40′N and 141° and 145°W (Fig. 6). Scattered remnants of the field extend into the Yukon Territory, as far south and east as 60°N and 136°W (Campbell and Dodds, 1985). In Alaska the Wrangell volcanic field covers as much as 10,000 km^2, including most of the high country of the Wrangell and St. Elias Mountains and parts of the contiguous eastern Alaska Range. Eight peaks are higher than 4,000 m, including Mt. Bona (5,005 m), the highest peak in the region. The central and western parts of the Wrangell volcanic field have received considerable attention in recent years, but the easternmost part has been mapped only in reconnaissance. Because of this general lack of information, short discussions of individual Quaternary volcanic centers are included in this report.

Wrangell volcanoes differ from those found elsewhere along the circum-Pacific margin because of their extremely large size (Table 2), their shield-like form and apparent nonexplosive summit calderas, and the highest eruptive rate yet reported for convergent plate margins (Nye, 1983).

Geologic setting

Almost all of the Wrangell volcanic field is on the Wrangellia terrane (Jones and others, 1987; Silberling and others, this volume), a major tectonostratigraphic terrane in south-central Alaska consisting of sedimentary and volcanic rocks that range in age from Pennsylvanian to Cretaceous (MacKevett, 1978; Richter, 1975). A small part of the extreme southeast end of the field overlies the Alexander terrane (Silberling and others, this volume), which is composed chiefly of lower to middle Paleozoic sedimentary rocks (MacKevett, 1978). Figure 6 shows the approximate boundary of the Wrangell volcanic field, the extent of perennial snow and ice cover, and the known major eruptive centers. Because of the extensive cover of snow and ice, geological field investigations have been extremely difficult over large parts of the area.

The Wrangell volcanic field and the Aleutian volcanic arc are separated by about 400 km (Fig. 7); the reason for the gap is not clear. Jacob and others (1977) pointed out that this is a region of relatively shallow subduction, and similar regions elsewhere in the circum-Pacific (e.g., the Peru-Chile subduction zone) are marked by gaps in active volcanism (Barazangi and Isacks, 1976). Cross and Pilger (1982) attributed the gap to low-angle subduction of relatively buoyant lithosphere, following subduction of an intraplate island-seamount chain.

Plafker (1969, 1987) has long believed that the Wrangell volcanic field is part of the Aleutian arc because of the seismicity and coseismic deformation associated with the 1964 Alaska earthquake in the area south of the 400-km gap, and because of similarities in composition and age between the Aleutian and Wrangell volcanoes. However, only recently has a Benioff zone associated with the Wrangell volcanic field been confirmed (Stephens and others, 1984). This seismically weak zone is subparallel to the general trend of the Wrangell volcanic field and is nearly normal to the trend of the Benioff zone under the northeasternmost segment of the Aleutian arc (Fig. 7; Stephens and others, 1984; Plafker, 1987). The gap between the Aleutian arc and the Wrangell volcanic field and the apparent misalignment of their Benioff zones may be due to some form of dextral shear along this complex plate boundary.

There is no evidence in marine seismic reflection data of any significant subduction of Pacific oceanic crust under the eastern Gulf of Alaska continental margin since the late Cenozoic (von Huene and others, 1979; Bruns, 1979), although weak seismicity between the Yakutat terrane and the Pacific Plate requires some relative motion along the boundary (Lahr and Plafker, 1980; Plafker, 1987). The Wrangell volcanic field lies north of the Gulf of Alaska on the Wrangellia terrane and inboard of the Yakutat terrane (Silberling and others, this volume), both of which are probably partially coupled to the Pacific Plate (Lahr and Plafker, 1980). In his model of the Cenozoic evolution of the Gulf of Alaska, Plafker (1987) suggests that Wrangell volcanism began about 20 to 25 Ma, following the subduction of about 225 km of the Yakutat terrane beneath the continental margin. Not fully understood, however, has been the continued voluminous outpouring of lava from the Wrangell volcanoes into Quaternary time.

Although the elongate Wrangell volcanic field roughly parallels the structural grain of the underlying basement rocks, no evidence suggests that individual eruptive centers of the field are structurally controlled. The generally older eastern part of the field may be offset 10 to 15 km by the Totschunda fault system (Figs. 6, 7)—a major Quaternary dextral strike-slip structure (Richter and Matson, 1971) related to the Denali fault. However, the fault system does not appear to have affected the location of contemporary volcanic activity. Moreover, there appears to be no pronounced alignment of volcanoes, as in the case of the Alaska Peninsula–Cook Inlet region, nor is there any obvious systematic spacing observed between volcanoes. The presently available data therefore suggest that the volcanic centers are randomly distributed throughout the Wrangell volcanic field.

Age and eruptive activity

Although only a scattering of K-Ar ages are available for the entire Wrangell volcanic field, the data indicate a general progression of eruptive activity from east to west across the field (Fig. 6). In the extreme western part, the known volcanoes are 1 m.y. old or younger (Richter and others, 1979; Nye, 1983); Mt. Wrangell,

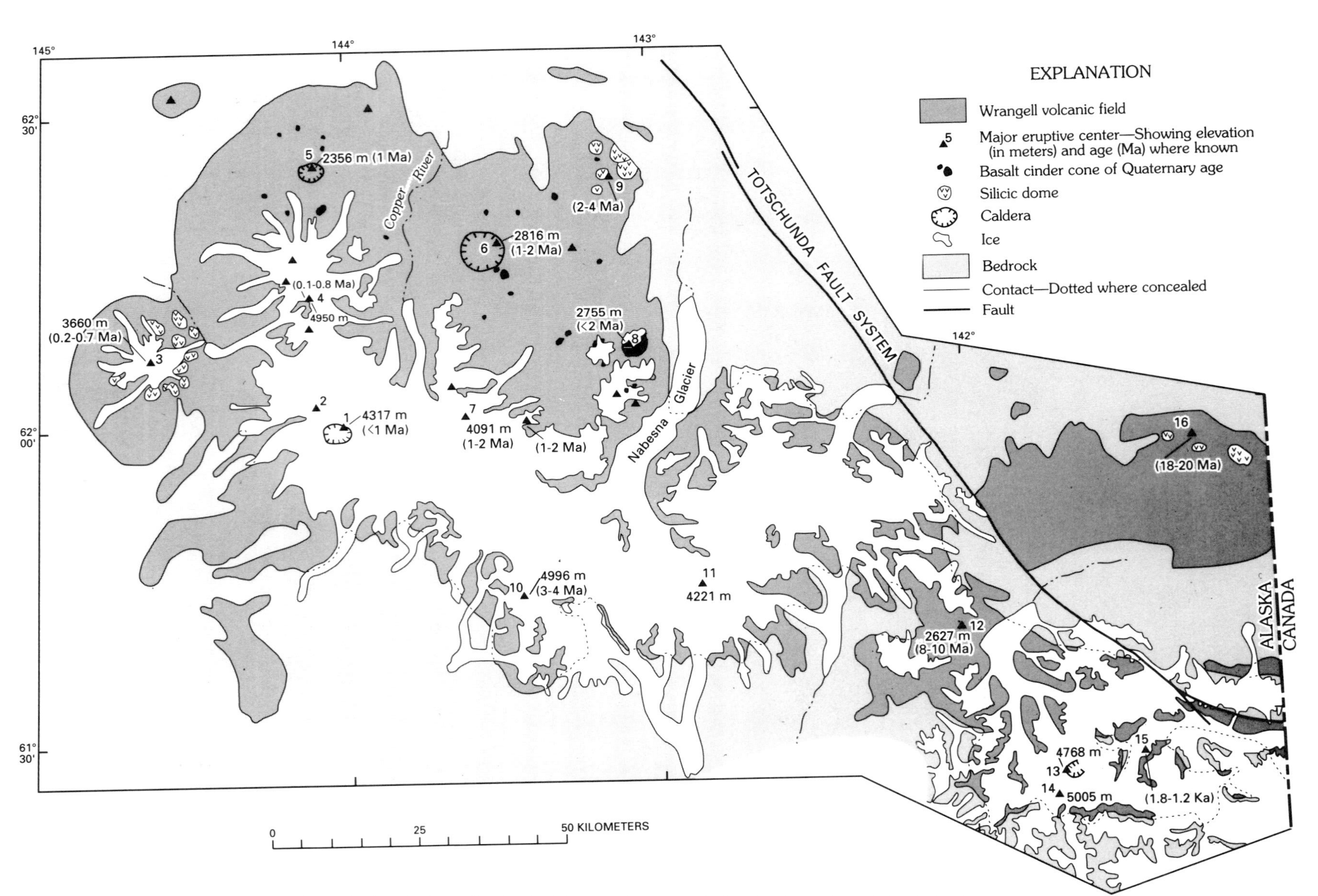

Figure 6. Map of Wrangell volcanic field in south-central Alaska. Numbers refer to individual volcanoes as follows: 1, Wrangell; 2, Zanetti; 3, Drum; 4, Sanford; 5, Capital; 6, Tanada; 7, Jarvis; 8, Gordon; 9, Skookum Creek; 10, Blackburn; 11, Regal; 12, Castle; 13, Churchill; 14, Bona; 15, White River Ash source; 16, Sonya Creek.

TABLE 2. PHYSICAL CHARACTERISTICS OF THE PRINCIPAL QUATERNARY VOLCANOES, WRANGELL VOLCANIC FIELD

Volcano	Location	Elevation (m)	Age (ma)	Height (km)	Diameter (km)	Estimated vol (km³)*
Wrangell	62.0°N 144.0°W	4,317	0–0.1	2.5	40	1,000
Drum	62.1°N 144.6°W	3,661	0.2–0.6	2.7	27	500
Sanford†	62.2°N 144.1°W	4,949	<1	3.1	34	900
Capital	62.4°N 144.1°W	2,356	1	1.3	18	100
Tanada	62.3°N 143.5°W	2,852	1–2	1.6	22	200
Jarvis	62.0°W 143.6°W	4,091	1.2	--------	Poorly defined	--------
Gordon	62.1°N 143.1°W	2,755	<2	0.6	5	4

*Volume estimated on basis of simple cone.
†Includes the north, west, and south Sanford eruptive centers.

in fact, has a large fumarolic area near its summit and is the only volcano in the field with reported historic activity. Through the central parts of the field, ages range between 1 and 5 Ma (Lowe and others, 1982; Richter and Smith, 1976), and in the east they range from 8 to 20 Ma (J. G. Smith, written communication, 1986). A significant exception to this progression of activity was the eruption of the White River Ash of Lerbekmo and others (1968) from a vent near the Alaska-Yukon border that occurred during two major events about 1,890 and 1,250 B.P. (Lerbekmo and others, 1975). This ash blanketed about 300,000 km^2 of eastern Alaska and western Canada. Other possible exceptions are Mt. Bona and Mt. Churchill near the east end of the Wrangell volcanic field, whose height and shape suggest young primary constructional forms.

Physical volcanology

The volcanic rocks of the Wrangell volcanic field are collectively referred to as the Wrangell Lava (Mendenhall, 1905). They consist predominantly of subaerial, or locally subglacial, lava flows and minor lahars, domes, pyroclastic rocks, dikes, and small subvolcanic intrusions. Thick and extensive piles of flat-lying to gently dipping lava flows, chiefly andesitic, characterize much of the field. Dacite to rhyolite domes are associated with a number of eruptive centers, such as Mt. Drum volcano and the Skookum Creek and Sonya Creek eruptive centers (Fig. 6), but are apparently absent from large areas of the Wrangell volcanic field. Bedded tephra, chiefly andesitic to basaltic, is relatively common and is interlayered with flows throughout the field, but pyroclastic flow deposits are rare. Poorly welded silicic ash-flow tuffs are present west of Mt. Wrangell interlayered with glacial deposits of the Copper River basin, and small volumes of pumiceous pyroclastic flow deposits are present in the Sonya Creek and Skookum Creek centers. With the exception of Mt. Drum—an apparent stratocone—and possibly Mt. Bona and Mt. Churchill—large snow-covered volcanic cones(?) at the east end of the field—the major volcanic structures appear to have been large shield volcanoes similar in form to the present shield of active Mt. Wrangell.

The three known shield volcanoes in the Wrangell volcanic field—Wrangell, Capital, and Tanada—have summit calderas that range in diameter from 4 to 8 km. The apparent absence of associated large-volume pyroclastic flow deposits suggests that these calderas had a nonexplosive origin. In addition to the larger volcanic constructional forms, more than 27 small (generally <2 km in diameter) cinder cones composed of basalt to basaltic andesite are found throughout the northwestern part of the field, especially between Mt. Drum and the Nabesna Glacier-River system (Fig. 6). These cones, many of which retain their original physiographic form, are probably all late Pleistocene and Holocene. They consist mostly of cinder, bombs, and spatter; short small-volume flows are associated with some of the larger cones. Mt. Gordon, largest of these young basaltic cones, is described below in more detail.

Volcano description

Some of the physical characteristics of the six principal Quaternary volcanoes and one large cinder cone in the Wrangell volcanic field (WVF) are shown in Table 2. In the following

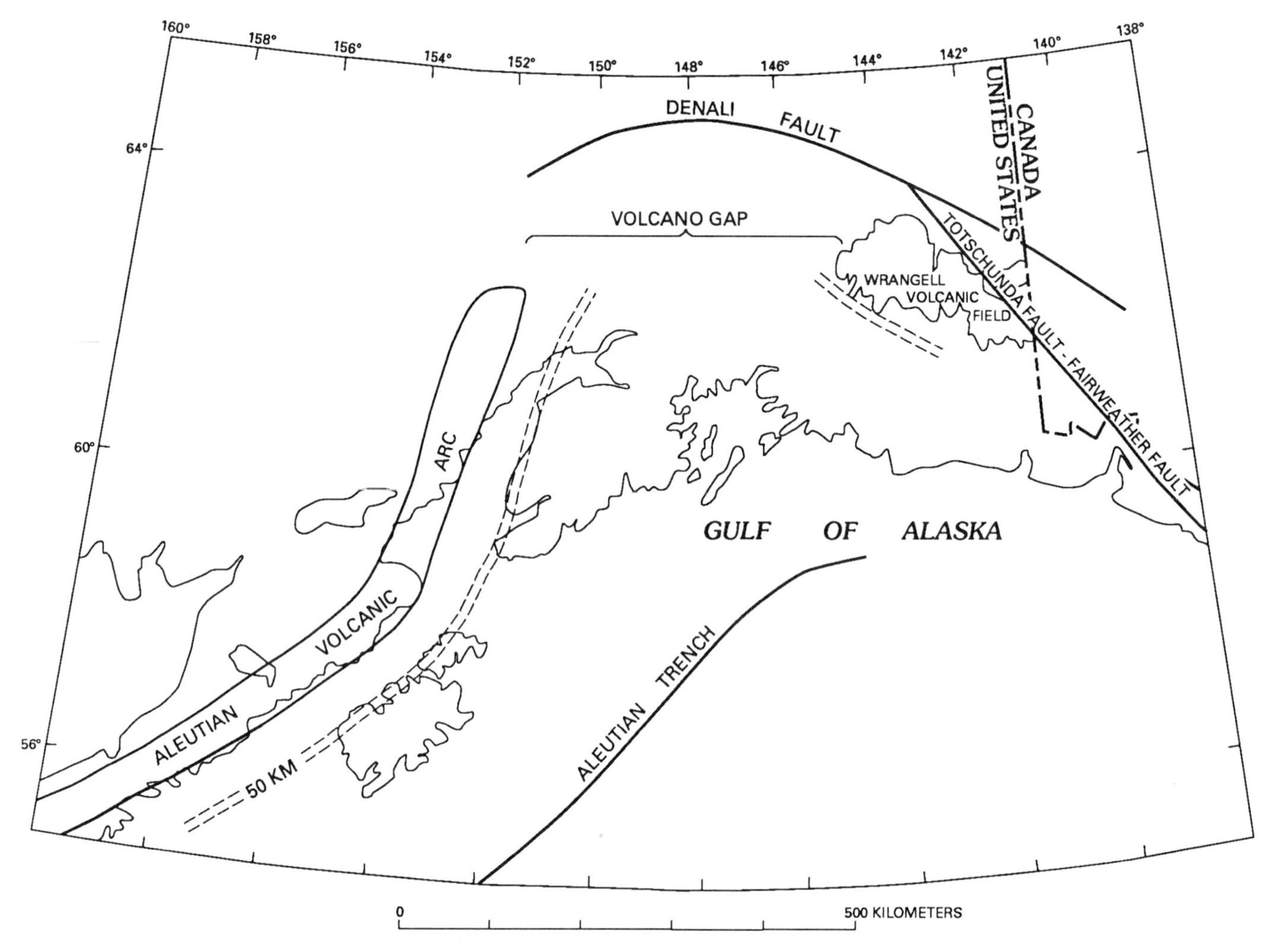

Figure 7. Map of southern Alaska showing location of Aleutian volcanic arc, Wrangell volcanic field, volcano gap, and principal regional tectonic features (modified from Lahr and Plafker, 1980). DF, Denali fault; TF, Totschunda fault; FF, Fairweather fault; AT, Aleutian trench; dashed double line, 50-km isobath of Benioff zone.

section, these volcanoes, together with a few eruptive centers that may be as young as 2 Ma, are described briefly in order of increasing age.

White River Ash. The White River Ash of Lerbekmo and others (1968) and Capps (1915) consists of two large rhyodacitic tephra lobes containing at least 25 km^3 of ejecta and covering more than 300,000 km^2 in eastern Alaska and the Yukon Territory (Lerbekmo and Campbell, 1969). The tephra was deposited during two major volcanic eruptions, 1,890 and 1,250 B.P (Lerbekmo and others, 1975). Source of the ash (Fig. 6) is believed to have been a vent now covered by the Klutlan Glacier about 22 km west of the Alaska–Yukon Territory border (Lerbekmo and Campbell, 1969) and within a few kilometers of a 200-m-high coarse pumice mound adjacent to the glacier. An alternative source may be snow-covered Mt. Churchill, about 10 km farther to the southwest, which has an apparent summit caldera 3 km in diameter that is filled with ice.

Mount Wrangell. This large, active, andesitic shielf volcano is indented by a 4- by 6-km ice-filled summit caldera. Three small (<1 km) post-caldera craters, all geothermally active, lie along the west and north rims of the caldera (Benson and Motyka, 1978). Mt. Zanetti—a large (450 m high) undissected parasitic cinder-spatter cone on the northwest flank of the shield, and possible source area for many flank flows—is probably younger than the latest glaciation. The bulk of Wrangell volcano was probably constructed between 0.3 and 0.6 Ma (D. H. Richter, unpublished data; Nye, 1983); however, some of the upper parts may be as young as 0.1 Ma. A lava flow, probably erupted from the summit area, has been dated at 80 ka (Nye, 1983), and the summit caldera may be as young as Holocene (Benson and Motyka, 1978). Nye (1983) believes that the modern Wrangell volcano is built on an older (1 to 2 Ma) dissected center which he refers to as the Chetaslina vent.

Mt. Drum. Mt. Drum is the westernmost volcano in the Wrangell volcanic field. Unlike its neighbors—Sanford, Wrangell, and Capital—it appears to have been more of a composite stratocone than a shield volcano (Richter and others, 1979). It was constructed between about 0.6 and 0.2 Ma (G. B. Dalrymple and M. A. Lanphere, unpublished data) during at least two cycles of cone building and ring dome extrusion (Richter and others, 1979). The first cycle began with the development of an andesitic to dacitic cone and culminated with the emplacement of a series of rhyolite ring domes. The second cycle of activity, followed without an apparent time break and continued to build the cone, but with more silicic lavas, and was accompanied by the emplacement of an early and late series of dacite ring domes. Following the second cycle, but with magma still available, violent explosive activity, apparently from the central vent area, destroyed a large part of the south half of the cone and deposited hot and cold volcanic avalanche debris over an area probably greater than 200 km^2.

Mt. Sanford. Perennial snow and ice almost completely cover Mt. Sanford, a bulbous-topped broad shield(?) volcano. Geological mapping (D. H. Richter, unpublished data) indicates that the upper parts of the volcano are built upon the coalescing flows from at least three main eruptive vents, referred to as the north, west, and south Sanford centers (Fig. 6). Aerial observations of the spectacular 2,500-m-high southwest wall of Mt. Sanford reveal that the bulbous top is composed of a massive 500-m-thick flow or dome that overlies a few hundred meters of flat-lying flows or pyroclastic deposits, which in turn overlie more than 1,500 m of andesitic flows and breccias cut by numerous dikes. Andesitic flows from the base of the west Sanford eruptive center have yielded K-Ar ages of 0.7 and 0.8 Ma (G. B. Dalrymple and M. A. Lanphere, unpublished data).

Capital Mountain. Capital volcano is a relatively small andesitic shield volcano with a roughly circular summit caldera 4 km in diameter. The shield consists of andesitic lava flows and lesser intercalated laharic rocks and pyroclastic deposits that dip gently away from the summit area. The caldera, with walls dipping 50° to 80° inward, is filled with flat-lying massive andesitic flows having an aggregate thickness of more than 450 m. A prominent andesite plug, 100 m high, intrudes the intracaldera lavas and forms the locus of a spectacular radial dike swarm. The hundreds of dikes are chiefly andesitic, but a few dikes are dacitic and rhyolitic. One large rhyolite dike extends westward about 10 km across the shield and caldera from a rhyolite laccolith on the east flank of the shield. K-Ar ages of the shield lavas and dikes are dated at about 1 Ma (Richter and others, 1984, 1989).

Tanada Peak. Tanada volcano, a somewhat larger and older version of neighboring Capital volcano, is an andesitic shield covering an area of more than 400 km^2. Much of the original andesitic shield has been stripped away, leaving the intracaldera flows as the high topographic features of the structure. Remaining shield lavas consist chiefly of andesitic lava flows and lesser laharic and pyroclastic deposits that dip gently away from the summit area. The summit caldera, approximately 8 km long by 6 km wide, is filled with massive flat-lying andesitic flows and dacitic agglutinates at least 900 m thick. The caldera walls consist mostly of pre-Tanada (2 to 3 Ma) flows, are less steep than at Capital volcano, and locally are mantled by thin pyroclastic air-fall beds. A few dikes intrude the structure, and a chain of younger andesitic to basaltic andesite cinder cones covers parts of the volcano's north flank. K-Ar ages indicate that the volcano was built between 1 and 2 Ma (Richter and others, 1984).

Mt. Gordon. This is the largest of the young ($<$1.5 Ma) basalt-basaltic andesite cinder cones that are common in the northwestern part of the Wrangell volcanic field. The cone of Mt. Gordon, about 5 km in diameter and 600 m high above its base of older Wrangell Lava, also erupted a significant volume of basaltic lava flows (Richter and Smith, 1976).

Mt. Jarvis. This mountain is the high point of a slightly curvilinear, north-trending, 10-km-long, 4,000-m-high ridge. The snow- and ice-covered ridge is composed of a thick sequence of dacitic and andesitic lava flows and capped by either a massive dacite flow or by a series of smaller dacite domes. One K-Ar age on basal (?) Jarvis flows suggests an age of about 1.6 Ma (Richter and Smith, 1976).

Skookum Creek. The Skookum Creek eruptive center is an erosionally dissected complex that may, in part, be as young as Quaternary. It consists principally of a series of rhyolite, rhyodacite, and andesite domes and their associated pyroclastic deposits, with an age of about 3.7 Ma, and an extensive sequence of relatively flat-lying andesite flows, some of which have been dated at 2.8 Ma (Lowe and others, 1982). Both the domes and the flows are intruded by a few rhyodacite and andesite dikes that appear to originate from a rhyodacite dome near the center of the complex. Relations between the volcanic units of the complex suggest that the flat-lying flows fill a caldera that is defined by the crude arcuate alignment of the domes.

Mt. Blackburn. The second highest peak in the Wrangell Mountains, Mt. Blackburn may be part of a large exhumed caldera (Winkler and Mackevett, 1981). The mountain is composed of 3- to 4-Ma plutonic rocks (J. G. Smith, unpublished data) that intrude, and possibly are overlain by, Wrangell Lava.

Mt. Bona and Mt. Churchill. Both Mt. Bona and Mt. Churchill are high snow- and ice-covered mountains in the eastern part of the Wrangell volcanic field. They appear to be relatively young constructional forms. Mt. Bona, the highest peak (5,005 m) in the Wrangell Mountains, may be a small stratocone built upon a high platform of Pennsylvanian and Lower Permian rocks (MacKevett, 1978). Mt. Churchill appears to have a 3-km-diameter caldera on its summit and, as mentioned earlier, may be an alternative source of the White River Ash.

Composition

Available major-element chemistry for the volcanic rocks of the Wrangell volcanic field indicates a range of 50 to 74 percent

SiO_2 (Fig. 8)—a compositional spread similar to that exhibited by the volcanic rocks in the Alaska Peninsula and Cook Inlet part of the Aleutian arc but lacking the high-silica rhyolites of the Valley of Ten Thousand Smokes. The volcanic rocks from both the Quaternary and older volcanoes in the Wrangell volcanic field have medium-K, calc-alkaline affinities (Fig. 8) typical of volcanoes associated with lithospheric plate interactions around the circum-Pacific tectonic belt. Unlike the volcanoes on the oceanic segment of the Aleutian arc and western Alaska Peninsula–Cook Inlet, no pronounced tholeiitic affinities are shown by the Wrangell volcanic rocks, although a few flows from Mt. Sanford appear to have tholeiitic tendencies (Fig. 9). At 57.5 percent SiO_2, the andesites contain an average of 1.44 percent K_2O (Fig. 8), which is somewhat, but probably not significantly, higher than the andesites of the Alaska Peninsula–Cook Inlet. Figure 9 also shows that, in general, Mt. Drum rocks have a low FeO*/MgO ratio, in contrast to other volcanic rocks in the Wrangell volcanic field—a feature that undoubtedly resul;ts from the abundance of olivine-rich (and hence MgO-rich) andesites and dacites from this volcano.

SUMMARY

The Aleutian arc and the Wrangell volcanic field, although separated by about 400 km, are both related to the underthrusting of the Pacific Plate beneath a collage of tectonostratigraphic terranes that make up the continental margin of southern Alaska. Volcanoes of both volcanic provinces are similar in composition and age but vary greatly in eruptive behavior and physical form. Quaternary volcanism in the region has resulted in some of the largest, most active, and most explosive volcanoes found along the entire circum-Pacific margin. The 1912 eruption of about 35 km^3 of tephra from the Novarupta vent complex in Katmai National Park, for example, was the most voluminous rhyolitic eruption in the past 1,800 years and has been called the most important igneous event of this century.

Aleutian arc Quaternary volcanism in the Alaska Peninsula–Cook Inlet region is defined by 37 major volcanic centers spread along 1,200 km of a seismically active arc-trench system. Over 100 eruptions have been reported from 19 centers since 1760, and more than half of the centers still show fumarolic activity. Eastern Aleutian centers were formed entirely on continental crust. Potassium-argon ages indicate that the centers were formed within the past million years.

Most Alaska Peninsula–Cook Inlet volcanoes are typical stratocones consisting of intercalated lava flows and volcaniclastic rocks, and are marked by summit craters and calderas. Caldera-forming eruptions with bulk ejecta volumes as large as 50 km^3 have occurred at eight of the volcanic centers on the Alaska Peninsula in late Pleistocene and Holocene time; 20 to 30 percent of known Holocene eruptions of this magnitude worldwide are located on the Alaska Peninsula and adjoining Unimak Island.

The centers generally have volumes of <100 km^3, although some large, long-lived stratocone-caldera complexes are as large as 300 to 400 km^3. The largest and most active volcanic centers in the eastern Aleutian arc (and the entire Aleutian arc) are found in the 700-km-long segment west of Veniaminof volcano on the Alaska Peninsula. The spacing between individual centers varies so much along the eastern Aleutian arc that an "average" volcano spacing figure is relatively meaningless. It is probably more significant for modeling purposes that various "segments" of the arc have quite different spacings between individual centers and that the longest volcano-free segment of the eastern Aleutian arc is less than 100 km.

The cone-building volcanic rocks consist predominantly of medium-K andesite and low-SiO_2 dacite; basalt is less common. Rhyolite is present in some ash-flow tuffs and intracaldera dome complexes. High-SiO_2 (>75 percent) rhyolite is restricted to the

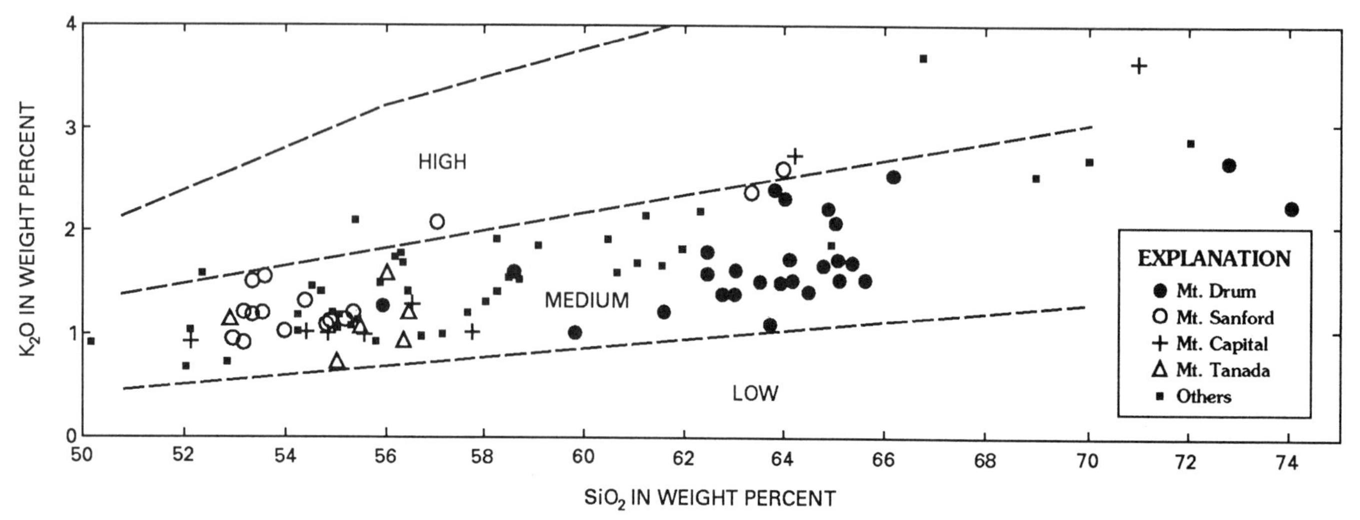

Figure 8. K_2O versus SiO_2 for volcanic rocks of the Wrangell volcanic field (WVF). Discriminant lines separating High, Medium, and Low are modified from Pecerrillo and Taylor (1976) and Ewart (1979).

ash-flow tuff of the Valley of Ten Thousand Smokes and the associated Novarupta dome.

Although the largest centers in the Alaska Peninsula–Cook Inlet region are tholeiitic, the easternmost 770 km of the Aleutian arc (one-third of the entire arc length) is characterized solely by calc-alkaline volcanic centers. Compositional trends of Quaternary volcanoes in the eastern Aleutian arc do not appear to have been influenced by tectonic segmentation such as has been proposed for the western oceanic part of the arc. Instead, continental crust east of Veniaminof volcano may be thick and extensive enough so as to inhibit the easy passage of magma. Magma fractionation and mixing could then lead to a calc-alkaline differentiation trend.

The Wrangell volcanic field, which covers more than 10,000 km^2 in south-central Alaska, contains at least six major volcanoes of Quaternary age. These volcanoes, located in the western part of the Wrangell volcanic field, as well as older volcanoes (5 to 20 Ma in the central and eastern parts of the field), are mostly large andesitic shield volcanoes with volumes as much as 1,000 km^3. Three of the Quaternary volcanoes contain recognizable near-circular summit calderas whose diameters range from 4 to 8 km. The absence of any known extensive pyroclastic flow deposits in the Wrangell volcanic field suggests that most of the calderas were of a nonexplosive origin. Neither the Quaternary volcanoes nor the older centers show any pronounced alignment or clustering that might be attributable to some form of structural control. The Wrangell volcanic field itself, however, does exhibit an east-west trend subparallel to the structural grain of southern Alaska.

The volcanic rocks of the Wrangell volcanic field are medium-K calc-alkaline rocks consisting predominantly of thick and extensive piles of flat-lying to gently dipping andesitic flows. Dacite to rhyolite domes, flows, and pyroclastic flow deposits are associated with some of the volcanoes. Bedded tephra is common, interlayered with flows throughout the field; small, young basalt to basaltic andesite cinder cones are scattered throughout the northwestern part of the field.

Regional mapping and topical studies of the Alaska Peninsula–Cook Inlet and Wrangell volcanic field regions, beginning in the early 1970s, have provided much new information on these previously little-understood volcanic provinces. The distribution, geologic setting, spacing, volume, and compositional trends of volcanic centers are now known in at least a general way. Future and more detailed topical studies including isotopic and trace-element analyses, geochronology, and large-scale geologic mapping of individual volcanic centers will undoubtedly contribute greatly to the understanding of these volcanic arcs in particular, and also to such arcs worldwide.

REFERENCES CITED

Allen, E. T., and Zies, E. G., 1923, A chemical study of the fumaroles of the Katmai Region: National Geographic Society Contributions Technical Papers, Katmai Series 2, p. 77–155.

Barazangi, M., and Isacks, B. L., 1976, Spatial distribution of earthquakes and subduction of the Nazca Plate beneath South America: Geology, v. 4, p. 686–692.

Benson, C. S., and Motyka, R. J., 1978, Glacier-volcano interactions on Mt. Wrangell, Alaska, *in* 1977–1978 Annual report of the Geophysical Institute:

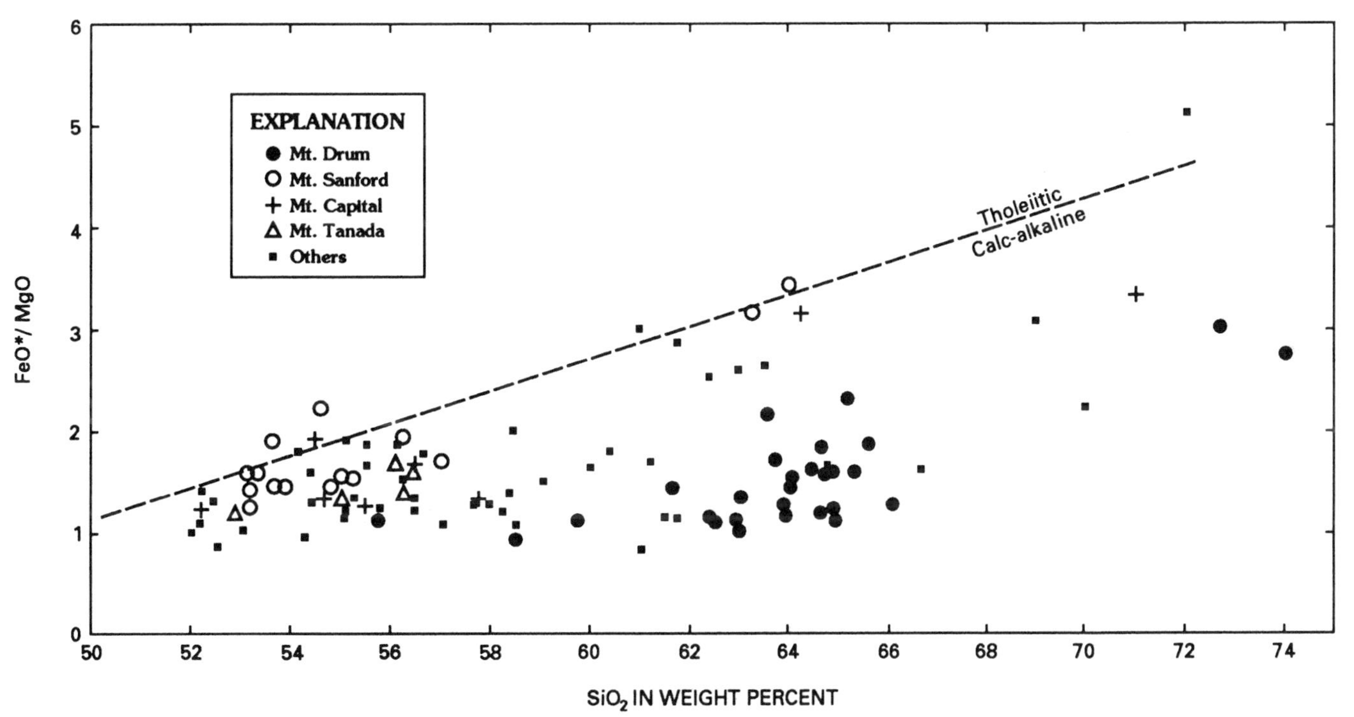

Figure 9. FeO*/MgO versus SiO_2 for volcanic rocks of the Wrangell volcanic field (WVF). Tholeiite–calc-alkaline line after Miyashiro (1974).

Fairbanks, University of Alaska, p. 1–25.
Bevier, M. L., and Wheller, K. R., 1983, Isotopic composition of lead and strontium in a suite of rocks from Redoubt volcano, Alaska: Geological Society of America Abstracts with Program, v. 15, p. 333.
Bruns, T. R., 1979, Late Cenozoic structure of the continental margin, northern Gulf of Alaska, *in* Sisson, A., ed., Proceedings of 6th Alaska Geological Society Symposium: Anchorage, Alaska Geological Society, p. J1–J33.
Burk, C. A., 1965, Geology of the Alaska Peninsula; Island arc and continental margin: Geological Society of America Memoir 99, 250 p.
Campbell, R. B., and Dodds, C. J., 1985, Geology of S. W. Kluane Lake map area (115G & FE. ½), Yukon Territory: Canadian Geological Survey Open-File 829, 2 sheets, scale 1:125,000.
Capps, S. R., 1915, An ancient volcanic eruption in the Upper Yukon Basin: U.S. Geological Survey Professional Paper 95-D, p. 59–64.
Coats, R. R., 1950, Volcanic activity in the Aleutian arc: U.S. Geological Survey Bulletin 974-B, p. 35–49.
—— , 1962, Magma type and crustal structure in the Aleutian arc, *in* Macdonald, G. A., and Kuno, H., eds., Crust of the Pacific Basin: American Geophysical Union Geophysical Monograph 6, p. 92–109.
Cross, T. A., and Pilger, R. H, Jr., 1982, Controls of subduction geometry, location of magmatic arcs, and tectonics of arc and back-arc regions: Geological Society of America Bulletin, v. 93, p. 545–562.
Curtis, G. H, 1968, The stratigraphy of the ejecta from the 1912 eruption of Mount Katmai and Novarupta, Alaska: Geological Society of America Memoir 116, p. 153–210.
Detterman, R. L., 1986, Glaciation of the Alaska Peninsula, *in* Hamilton, T. D., Reed, K. M., and Thorson, R. M., eds., Glaciation in Alaska; The geologic record: Alaska Geological Society, p. 151–170.
Detterman, R. L., Miller, T. P., Yount, M. E., and Wilson, F. H., 1981, Geologic map of the Chignik and Sutwik Island Quadrangles, Alaska: U.S. Geological Survey Miscellaneous Geologic Investigation Map I-1229, scale 1:250,000.
Detterman, R. L., Case, J. E., Wilson, F. H., and Yount, M. E., 1987, Geologic map of the Ugashik, Bristol Bay, and western part of Karluk Quadrangle, Alaska: U.S. Geological Survey Miscellaneous Investigation Series Map I-1685, scale 1:250,000.
Engebretson, D. C., Cox, A., and Gordon, R. B., 1985, Relative motions between oceanic and continental plates in the Pacific Ocean: Geological Society of America Special Paper 206, 59 p.
Ewart, A., 1979, A review of the mineralogy and chemistry of Tertiary-Recent dacitic, latitic, rhyolitic, and related salic volcanic rocks, *in* Barker, F., ed., Trondhjemites, dacites, and related rocks: New York, Elsevier Scientific Publishing Co., p. 13–22.
Fenner, C. N., 1923, The origin and mode of emplacement of the great tuff deposit of the Valley of Ten Thousand Smokes: National Geographic Society Contribution Technical Papers, Katmai Series, 74 p.
—— , 1926, The Katmai magmatic province: Journal of Geology, v. 34, p. 673–772.
Fournelle, J., and Marsh, B. D., 1986, Shishaldin volcano, Unimak Island, Aleutians; Unordinary arc lavas; 1. Chemistry, mineralogy, and petrology: EOS Transactions of the American Geophysical Union, v. 67, no. 44, p. 1276.
Fisher, M. A., Bruns, T. R., and von Huene, R., 1981, Transverse tectonic boundaries near Kodiak Island, Alaska: Geological Society of America Bulletin, v. 92, p. 10–18.
Gill, J. B., 1981, Orogenic andesites and plate tectonics: New York, Springer-Verlag, 390 p.
Griggs, R. F., 1922, The Valley of Ten Thousand Smokes: Washington, D.C., National Geographic Society, 340 p.
Hildreth, E. W., 1983, The compositionally zoned eruption of 1912 in the Valley of Ten Thousand Smokes, Katmai National Park, Alaska: Journal of Volcanology and Geothermal Research, v. 18, p. 1–56.
—— , 1987, New perspectives on the eruption of 1912 in the Valley of Ten Thousand Smokes, Katmai National Park, Alaska: Bulletin of Volcanology, v. 49, p. 680–693.
Jacob, K. H., Nakamura, K., and Davies, J. N., 1977, Trench-volcano gap along the Alaska-Aleutian arc; Facts and speculations on the role of terrigeneous sediments, *in* Talwani, M., and Pitman, W. C., III, eds., Island arcs, deep sea trenches, and back-arc basins: American Geophysical Union Maurice Ewing Series, p. 243–258.
Johnston, D. A., 1978, Volatiles, magma mixing, and the mechanism of eruption of Augustine volcano, Alaska [Ph.D. thesis]: Seattle, University of Washington, 177 p.
Jones, D. L., Silberling, N. J., Coney, P. J., and Plafker, G., 1987, Lithotectonic terrane map of Alaska (west of the 141st meridian): U.S. Geological Survey Map MF-1874-A, scale 1:2,500,000.
Kay, R. W., Sun, S.-S., and Lee-Hu, C.-H., 1978, Pb and Sr isotopes in volcanic rocks from the Aleutian Islands and Pribilof Islands, Alaska: Geochimica and Cosmochimica Acta, v. 42, p. 263–273.
Kay, S., Kay, R. W., and Citron, G. P., 1982, Tectonic controls on tholeiitic and calc-alkaline magmatism in the Aleutian arc: Journal of Geophysical Research, v. 87, no. B5, p. 4051–4072.
Kay, R. W., Rubenstone, J. L., and Kay, S. M., 1986, Aleutian terranes from Nd isotopes: Nature, v. 322, p. 605–609.
Kienle, J., and Swanson, S. E., 1983, Volcanism in the eastern Aleutian arc; Late Quaternary and Holocene centers, tectonic setting, and petrology: Journal of Volcanology and Geothermal Research, v. 17, p. 393–432.
Kienle, J., Kyle, P. R., Self, S., Motyka, R. J., and Lorenz, V., 1980, Ukinrek maars, Alaska; 1, April 1977 eruption sequence, petrology, and tectonic setting: Journal of Volcanology and Geothermal Research, v. 7, p. 11–37.
Lahr, J. C., and Plafker, G., 1980, Holocene Pacific–North American plate interaction in southern Alaska; Implications for the Yakataga seismic gap: Geology, v. 8, p. 483–486.
Lerbekmo, J. F., and Campbell, F. A., 1969, Distribution, composition, and source of the White River Ash, Yukon Territory: Canadian Journal of Earth Sciences, v. 6, p. 109–116.
Lerbekmo, J. F., Hanson, L. W., and Campbell, F. A., 1968, Application of particle size distribution to determination of source of a volcanic ash deposit: Proceedings of the 23rd International Geologic Congress, Prague, Section 2, p. 283–295.
Lerbekmo, J. F., Westgate, J. A., Smith, D.G.W., and Denton, G. H., 1975, New data on the character and history of the White River volcanic eruption, Alaska, *in* Suggage, R. P., Cresswell, M. M., eds., Quaternary studies: Wellington, Royal Society of New Zealand, p. 203–209.
Lowe, P. C., Richter, D. H., Smith, R. L., and Schmoll, H. R., 1982, Geologic map of the Nabesna B-5 Quadrangle, Alaska: U.S. Geological Survey Geologic Quadrangle Map GQ-1566, scale 1:63,360.
Luedke, R. G., and Smith, R. L., 1986, Map showing distribution, composition, and age of late Cenozoic volcanic centers in Alaska: U.S. Geological Survey Miscellaneous Investigation Series Map I-1091-F, scale 1:1,000,000.
MacKevett, E. M., Jr., 1978, Geologic map of the McCarthy Quadrangle, Alaska: U.S. Geological Survey Miscellaneous Investigations Map I-1032, scale 1:250,000.
Magoon, L. B., Adkinson, W. L., and Egbert, R. M., 1976, Map showing geology, wildcat wells, and Tertiary plant fossil localities, K-Ar age dates, and petroleum operations, Cook Inlet area, Alaska: U.S. Geological Survey Miscellaneous Investigations Series Map I-1019, 4 sheets, scale 1:250,000.
Marsh, B. D., 1979, Island arc development; Some observations, experiments, and speculations: Journal of Geology, v. 87, p. 687–714.
—— , 1982, The Aleutians, *in* Thorpe, R. S., ed., Andesites: New York, John Wiley and Sons, p. 99–114.
Marsh, B. D., and Carmichael, I.S.E., 1974, Benioff zone magmatism: Journal of Geophysical Research, v. 79, p. 1196–1206.
Marsh, B. D., and Leitz, R. E., 1978, Geology of Amak Island, Aleutian Islands, Alaska: Journal of Geology, v. 87, p. 715–723.
Matumoto, T., 1971, Seismic body waves observed in the vicinity of Mount Katmai, Alaska, and evidence for the existence of molten chambers: Geological Society of America Bulletin, v. 82, p. 2905–2920.
McCulloch, M. T., and Perfit, M. R., 1981, $^{143}Nd/^{144}Nd$, $^{87}Sr/^{86}Sr$, and trace element constraints on the petrogenesis of Aleutian island arc magmas: Earth

and Planetary Science Letters, v. 56, p. 167–179.

McNutt, S. R., and Jacob, K. H., 1986, Determination of large-scale velocity structure of the crust and upper mantle in the vicinity of Pavlof volcano, Alaska: Journal of Geophysical Research, v. 91, no. B5, p. 5013–5022.

Mendenhall, W. C., 1905, Geology of the central Copper River region, Alaska: U.S. Geological Survey Professional Paper 41, 133 p.

Miller, T. P., and Smith, R. L., 1977, Spectacular mobility of ash flows around Aniakchak and Fisher calderas, Alaska: Geology, v. 5, p. 173–176.

—— , 1983, Calderas of the eastern Aleutian arc [abs.]: EOS Transactions of the American Geophysical Union, v. 64, no. 45, p. 877.

—— , 1987, Late Quaternary caldera-forming eruptions in the eastern Aleutian arc, Alaska: Geology, v. 15, p. 434–438.

Miyashiro, A., 1974, Volcanic rock series in island arcs and active continental margins: American Journal of Science, v. 274, p. 321–355.

Myers, J. D., 1988, Possible petrogenetic relations between low- and high-MgO Aleutian basalts: Geological Society of America Bulletin, v. 100, p. 1040–1053.

Myers, J. D., Marsh, B. D., and Sinha, A. K., 1985, Strontium isotopic and selected trace element variations between two Aleutian volcanic centers (Adak and Atka); Implications for the development of arc volcanic plumbing systems: Contributions to Mineralogy and Petrology, v. 91, p. 221–234.

Nakamura, K., Jacob, K. H, and Davies, J. N., 1977, Volcanoes as possible indicators of tectonic stress orientation; Aleutians and Alaska: Pure and Applied Geophysics, v. 115, p. 87–112.

Nye, C. J., 1983, Petrology and geochemistry of Okmok and Wrangell volcanoes, Alaska [Ph.D. thesis]: Santa Cruz, University of California, 208 p.

—— , 1987, Stratigraphy, petrology, and geochemistry of the Spurr volcanic complex, eastern Aleutian arc, Alaska: University of Alaska Geophysical Institute Report UAG R-311, 135 p.

Peccerillo, A., and Taylor, S. R., 1976, Geochemistry of Eocene calc-alkaline volcanic rocks from the Kastamonu area, northern Turkey: Contributions to Mineralogy and Petrology, v. 58, p. 63–81.

Plafker, G., 1969, Tectonics of the March 27, 1964, Alaska earthquake: U.S. Geological Survey Professional Paper 543-I, 74 p.

—— , 1987, Regional geology and petroleum potential of the northern Gulf of Alaska continental margin, *in* Circum-Pacific Council for Energy and Mineral Resources Earth Science Series, v. 6, p. 229–267.

Reed, B. L., and Lanphere, M. A., 1972, Generalized geologic map of the Alaska-Aleutian Range batholith showing potassium-argon ages of the plutonic rocks: U.S. Geological Survey Miscellaneous Field Studies Map MF-372, 2 sheets, scale 1:1,000,000.

Richter, D. H., 1976, Geologic map of the Nabesna Quadrangle, Alaska: U.S. Geological Survey Miscellaneous Investigation Series Map I-932, scale 1:250,000.

Richter, D. H., and Matson, N. A., Jr., 1971, Quaternary faulting in the eastern Alaska Range, Alaska: Geological Society of America Bulletin, v. 82, p. 1529–1540.

Richter, D. H., and Smith, R. L., 1976, Geologic map of the Nabesna A-5 Quadrangle, Alaska: U.S. Geological Survey Geologic Quadrangle Map GQ-1292, scale 1:63,360.

Richter, D. H., Smith, R. L., Yehle, L. A., and Miller, T. P., 1979, Geologic map of the Gulkana A-2 Quadrangle, Alaska: U.S. Geological Survey Geologic Quadrangle Map GQ-1520, scale 1:63,360.

Richter, D. H., Smith, J. G., Ratte, J. C., and Leeman, W. P., 1984, Shield volcanoes in the Wrangell Mountains, *in* Reed, K. M., and Bartsch-Winkler, S., eds., The United States Geological Survey in Alaska; Accomplishments during 1982: U.S. Geological Survey Circular 939, p. 71–75.

Richter, D. H., and 5 others, 1989, Geologic map of the Gulkana B-1 Quadrangle, Alaska: U.S. Geological Survey Geologic Quadrangle Map GQ-1655, scale 1:63,360.

Riehle, J. R., 1985, A reconnaissance of the major Holocene tephra deposits in the upper Cook Inlet region, Alaska: Journal of Volcanology and Geothermal Research, v. 26, p. 37–74.

Riehle, J. R., Detterman, R. L., Yount, M. E., and Miller, J. W., 1987a, Preliminary geologic map of the Mt. Katmai Quadrangle and portions of the Afognak and Naknek Quadrangles, Alaska: U.S. Geological Survey Open-File Report 87-93, scale 1:250,000.

Riehle, J. R., Yount, M. E., and Miller, T. P., 1987b, Petrography, chemistry, and geologic history of Yantarni volcano, Aleutian volcanic arc, Alaska: U.S. Geological Survey Bulletin 1761, 27 p.

Rubenstone, J. L., Langmuir, L. H., and Hildreth, E. W., 1985, Isotope and trace element data bearing on the sources and evolution of magmas in the Katmai region, Alaska: Geological Society of America Abstracts with Program, v. 17, p. 704.

Simkin, T., and 5 others, 1981, Volcanoes of the world: Smithsonian Institution, Stroudsburg, Pennsylvania, Hutchinson Ross Publishing Co., 232 p.

Stephens, C. S., Fogelman, K. A., Lahr, J. C, and Page, R. A., 1984, Wrangell Benioff Zone, southern Alaska: Geology, v. 12, p. 373–376.

Stone, D. F., and Packer, D. R., 1977, Tectonic implications of Alaska Peninsula paleomagnetic data: Tectonophysics, v. 37, p. 183–201.

Taylor, S. R., 1969, Trace element chemistry of andesites and associated calc-alkaline rocks, *in* McBirney, A. R., ed., Proceedings of the Andesite Conference: Oregon Department of Geology and Mineral Resources Bulletin, v. 65, p. 43–64.

Turner, D. L., and Nye, C. J., 1986, Geochronology of eruptive events at Mt. Spurr, Alaska; Geothermal energy resource investigations at Mt. Spurr, Alaska: Fairbanks, University of Alaska Geophysical Institute Report UAG R-308, p. 2.1–2.8.

von Huene, R., Fisher, M. A., and Bruns, T. R., 1979, Continental margins of the Gulf of Alaska and late Cenozoic tectonic plate boundaries, *in* Sisson, A., ed., Proceedings of the 6th Alaska Geological Society Symposium: Anchorage, Alaska Geological Society, p. 11–130.

Winkler, G. R., and MacKevette, E. M., Jr., 1981, Geologic map of the McCarthy C-7 Quadrangle, Alaska: U.S. Geological Survey Geologic Quadrangle Map GQ-1533, scale 1:63,360.

Wilson, F. H., 1985, Meshik and Aleutian arcs; Tertiary volcanism on Alaska Peninsula, Alaska [abs.]: Pacific Section, American Association of Petroleum Geologists Program with Abstracts, p. 32.

—— , 1990, Stepovak Bay Group Volcanoes, *in* Wood, C. A., and Kienle, J., eds., Volcanoes of North America: Cambridge University Press, p. 55–56.

Wilson, F. H., Detterman, R. L., Miller, J. W., and Case, J. E., 1994, Geologic map of the Port Moller, Stepovak Bay, and Simeonof Island Quadrangles, Alaska Peninsula, Alaska: U.S. Geological Survey Miscellaneous Investigations Map I-2272, 1:250,000 scale (in press).

Yount, M. E., 1984, Redoubt Volcano, Cook Inlet, Alaska; Mineralogy and chemistry: Geological Society of America Abstracts with Programs, v. 16, p. 340–341.

—— , 1990, Dana volcano, *in* Wood, C. A., and Kienle, J., eds., Volcanoes of North America: Cambridge University Press, p. 54–55.

Yount, M. E., Wilson, F. H. and Miller, J. W., 1985, Newly discovered Holocene volcanic vents, Port Moller and Stepovak Bay Quadrangles, Alaska Peninsula, *in* Bartsch-Winkler, S., and Reed, K. M., eds., The U.S. Geological Survey in Alaska; Accomplishments in 1983: U.S. Geological Survey Circular 945, p. 60–62.

Zies, E. G., 1929, The Valley of Ten Thousand Smokes; 1, The fumarolic incrustations and their bearing on ore deposition; 2, The acid gases contributed to the sea during volcanic activity: National Geographic Society Contribution Technical Papers, Katmai Series 4, p. 1–79.

Manuscript Accepted by the Society May 23, 1990

Printed in U.S.A.

The Geology of North America
Vol. G-1, The Geology of Alaska
The Geological Society of America, 1994

Chapter 25

Isotopic composition of the igneous rocks of Alaska

Joseph G. Arth
U.S. Geological Survey, 345 Middlefield Road, Menlo Park, California 94025

INTRODUCTION

The isotopic-tracer data that are currently available on the igneous rocks of Alaska are inadequate to fully characterize the enormous areas of magmatic rock that are exposed on this subcontinent. Nevertheless, the existing data are quite useful for interpreting the origin and evolution of individual magmatic suites and their related ore deposits, the details of crustal growth in island arcs, the genesis of major continental-margin batholiths, and the nature of basement underlying vast areas of the remote interior.

This chapter is a summary of isotopic-tracer data on Mesozoic and Cenozoic igneous rocks of Alaska. Data published prior to the manuscript deadline of December 1986 are included, as are new data collected by the author from 1977 to 1986 in the course of collaborative studies with U.S. Geological Survey colleagues (references are provided if published prior to printing of this volume). The magmatic rocks are discussed in four sections: Mesozoic volcanic provinces, major Mesozoic to Cenozoic batholiths, magmatic suites of economic interest, and igneous rocks in Alaskan tectonostratigraphic terranes. Some of the longer subsections are followed by a one-paragraph summary that attempts a brief, generalized interpretation of the various isotopic results. The final section summarizes the Sr isotope data on a map of Alaska, and discusses the inferred nature of the deep continental crust.

Throughout the text, acronyms are used in place of the more cumbersome isotopic ratios, as follows:

Initial $^{87}Sr/^{86}Sr$ ratio(s) = SIR
Initial $^{143}Nd/^{144}Nd$ ratio(s) = NIR
Initial $^{176}Hf/^{177}Hf$ ratio(s) = HIR
Initial $^{206}Pb/^{204}Pb$ ratio(s) = PIR6
Initial $^{207}Pb/^{204}Pb$ ratio(s) = PIR7
Initial $^{208}Pb/^{204}Pb$ ratio(s) = PIR8

Locations of geologic, geographic/physiographic, and tectonostratigraphic features referred to in this chapter are shown respectively on Plate 1 (Beikman), Plate 2 (Wahrhaftig), and Plate 3 (Silberling and others) of this volume.

CENOZOIC VOLCANIC PROVINCES

The Aleutian magmatic province

Most of the published isotopic results in Alaska are on samples from the Aleutian Islands. The primary topic of interest has been the generation of island-arc magmas in an arc that is built on oceanic crust west of, and continental crust east of, the edge of the Bering Sea shelf. Various investigators have used isotope ratios to help evaluate the relative contribution of the many possible sources of arc magmas. These sources include rocks entrained on the subducted slab (ocean-ridge rocks, seamounts, pelagic and terrigenous deposits), the upper mantle overlying the slab, and the crust below the active volcanoes. A clearly written summary of the isotopic aspects of magmas erupted in arcs, including several examples from the Aleutians, is given by Hawkesworth (1982).

The earliest major study of Pb and Sr isotopes in Aleutian volcanic rocks was made by Kay and others (1978), who analyzed samples from nine volcanoes located both east and west of the edge of the Bering Sea shelf. The observed ratios do not vary along the strike of the arc (Table 1), suggesting that continental crust under the eastern part of the arc did not contribute significantly to the magmas. SIR of 0.7028 to 0.7037 are among the lowest observed in island arcs. PIR7 are intermediate between those of oceanic basalt and continent-derived sediment. Kay and Kay (this volume) proposed a source of arc magma that contained about 2 percent by mass of subducted sediment. The remaining melt could be derived from the subducted oceanic crust and the overlying mantle. In a related study, Kay (1978) described uncommon Aleutian magnesian andesites having distinct isotopic character and chemistry suggestive of an origin by equilibration of hydrous melt rich in large-ion-lithophile (LIL) elements from subducted oceanic basalt with LIL-element-poor mantle overlying the subduction zone.

The zoned Captains Bay pluton of gabbro to granite on Unalaska Island (eastern Aleutians) was studied by Perfit and Lawrence (1979) and Perfit and others (1980). They found SIR in the range 0.7030 to 0.7038, similar to that of Aleutian volcanic rocks. Delta ^{18}O values for whole-rock samples are –1.0 to 7.0

Arth, J. G., 1994, Isotopic composition of the igneous rocks of Alaska, *in* Plafker, G., and Berg, H. C., eds., The Geology of Alaska: Boulder, Colorado, Geological Society of America, The Geology of North America, v. G-1.

TABLE 1. ISOTOPIC RANGES FOR CENOZOIC VOLCANIC ROCKS OF ALASKA

Location	SIR*	NIR†	PIR6	PIR7	PIR8	HIR	References§
Aleutian Islands and Alaska Peninsula							
All except magnesian andesites	0.70284–0.70369	0.51295–0.51309	18.701–18.916	15.515–15.602	38.127–38.513	0.28317–0.28326	1, 3–11
Magnesian andesites	0.7027 –0.70285		17.856–18.359	15.416–15.471	37.313–37.815		2
Bering Sea							
Abyssal plain tholeiite	0.70278	0.51316					1, 11
Nunivak Island	0.70251–0.70330	0.51289–0.51315	18.628	15.457	38.112		12–16
Pribilof Islands	0.70274–0.70293	0.51303–0.51308	18.806–19.042	15.476–15.494	38.274–38.483		1, 11
Saint Lawrence Island	0.70340–0.70374	0.51285–0.51288					11
Gulf of Alaska area							
Kodiak and Hodgkins seamounts	0.70397		19.12–19.34	15.53–15.55	38.45–38.63		1
Mount Edgecumbe volcanic field	0.70291–0.70410		18.48–19.16	15.56–15.68	38.17–38.85		17–20
Mainland Alaska volcanic fields							
Blackburn Hills	0.7033–0.7052	0.51253–0.51290					21
Kanuti	0.7043–0.7048	0.51248–0.51267					21
Nowitna	0.7043–0.7051	0.51256–0.51257					21
Sischu	0.7075–0.7079	0.51244–0.51247					21
Yukon River	0.7037–0.7051	0.51266–0.51280					21

*All ratios normalized to $^{86}Sr/^{88}Sr = 0.11940$

†All ratios normalized to $^{146}Nd/^{144}Nd = 0.72190$

§1. Kay and others, 1978; 2. Kay, 1978; 3. Perfit and Lawrence, 1979; 4. Perfit and others, 1980; 5. McCulloch and Perfit, 1981; 6. Morris and Hart, 1983; 7. White and Patchett, 1984; 8. Myers and others, 1985; 9. Perfit and Kay, 1986; 10. Morris and Hart, 1986; 11. von Drach and others, 1986; 12. Mark, 1972; 13. Zartman and Tera, 1973; 14. Menzies and Murthy, 1980a; 15. Menzies and Murthy, 1980b; 16. Roden and others, 1984; 17. Myers and Marsh, 1981; 18. Kosco, 1981; 19. Myers and others, 1984; 20. Myers and Sinha, 1985; 21. Moll-Stalcup and Arth, 1989.

per mil. The lowest values, found in the intermediate rocks, reflect interaction of the pluton with large volumes of low-$\delta^{18}O$ meteoric water. No correlation of SIR and $\delta^{18}O$ was found; Sr was apparently unaffected by the exchange. Perfit and coworkers proposed that the intermediate rocks were the result of fractional crystallization of high-Al basaltic magma at crustal levels under conditions of increasing water fugacity. They suggested that aplitic granites crystallized from a late-stage, volatile-rich liquid.

A Nd and Sr isotopic study by McCulloch and Perfit (1981) included volcanic samples from islands farther to the west than the earlier studies discussed above, as well as two plutonic samples from Unalaska. SIR for volcanic and plutonic rocks were in the narrow range of 0.7029 to 0.7034. Combined Nd and Sr data led McCulloch and Perfit to suggest that 2 to 8 percent of ocean sediment was added to seawater-altered oceanic igneous crust to produce magmas related to the subducted slab. They suggested that the overlying mantle wedge may react with the slab-related magmas, but that overlying continental crust, if present, made little or no contribution to the magmas.

Morris and Hart (1983) compared SIR, NIR, and PIR between Cold Bay volcanics on the main Aleutian volcanic front near the tip of the Alaska Peninsula, and Amak Island volcanics located about 50 km behind the front. Both suites fell within the previously reported ranges, but Morris and Hart noted small secular increases in SIR and PIR at Cold Bay, which they correlated with increasing SiO_2 and decreasing CaO contents. For magmatic arcs in general, they noted the similarity of SIR and NIR values to those of ocean islands and suggested that the sources were the same. They thus favored derivation of Cold Bay and Amak suites by melting of a hybrid "plum-pudding" or metasomatically veined mantle composed of pieces of the lower mantle (of the type that yields some ocean-island basalts) intermixed with depleted upper mantle (of the type that is the source for normal ocean-ridge basalts). A discussion of the conflict between this and earlier models is given by Perfit and Kay (1986) and Morris and Hart (1986).

White and Patchett (1984) examined the isotopic ratios of Hf, Nd, and Sr in samples from a variety of arcs, including the Aleutians at Little Sitkin. They showed that HIR in island-arc magmas correlate positively with NIR and negatively with SIR. The HIR for several arcs, including the Aleutians, overlap the HIR of normal ocean-ridge basalts, but the HIR for other arcs have lower values. Of the arcs examined, the Aleutians have the lowest SIR, highest NIR, and highest HIR, and are closer in these values to ocean-ridge basalts than any other arc suite. White and Patchett acknowledge the similarity of the isotopic ratios for arc magmas to those of ocean-island magmas as noted by Morris and Hart (1983), but suggest that the mantle wedge above subduction zones is chemically like that beneath ocean ridges. They ascribed some of the isotopic differences between ridge basalts and island-arc magmas like those of the Aleutians to incorporation of 1 to 2 percent of subducted continent-derived or pelagic sediment in the sources.

Myers and others (1985) studied SIR and other chemical features of Adak and Atka lavas in order to compare a small, presumably immature volcanic center (Adak) with a larger, presumably more mature one (Atka). Although all data fell in the previously reported range of SIR, lavas at Adak are more variable in SIR and have a lower average value than those at Atka. The variability was ascribed to contamination of lavas by interaction with nonradiogenic ultramafic wall rock, in a "dirty plumbing system." Similar contamination would not occur in a "mature" edifice such as Atka where there is a "clean plumbing system" that provides lavas of more uniform isotopic composition.

Von Drach and others (1986) provided additional geographic coverage for Nd and Sr isotopic data. They noted the narrow range of NIR and larger spread of SIR in all Aleutian magmas, and reemphasized the role of seawater alteration because it has no substantial effect on NIR, but can influence SIR. After compiling average trace-element contents and isotopic ratios for arc volcanics and potential source materials, they estimated that Aleutian magmas are melts from mixtures of seawater-altered ocean-ridge basalt and about 4 percent of pelagic sediments.

Brown and others (1982) and Tera and others (1986) provided independent evidence for incorporation of sediments in Aleutian magma sources by measuring ^{10}Be contents in several Aleutian volcanic samples. ^{10}Be is formed in the atmosphere by reaction of oxygen and nitrogen with cosmic rays. It reaches the Earth's surface via precipitation and is incorporated into sediments and soils. Because ^{10}Be has a short half-life (1.5 m.y.), only magmas that include recently formed sediments in their source will show its presence. Sediments must be formed, subducted, and melted in less than about 10 m.y. Of 12 arcs that these authors measured, only the Aleutians and two others showed significant levels of ^{10}Be. Aleutian basalts and andesites range from 2.0 to 15.3×10^6 and average 4.1×10^6 atoms of ^{10}Be per gram of sample. These values compare with ^{10}Be concentrations of less than 1×10^6 atoms per gram (values indistinguishable from 0 at current analytical sensitivity) in lavas from ocean-ridge basalts, ocean islands, or flood basalts. Rocks from Bogoslof have the highest values (5.2 and 15.3×10^6), but otherwise, there is no geographic trend or polarity within the arc. Although Brown, Tera, and others find it a reasonable hypothesis to suggest that the ^{10}Be reflects incorporation of Miocene or younger sediments in the magma source, they note that two samples from Cold Bay that contain ^{10}Be do not have a sedimentary signature for Pb isotopes. Most workers rule out contamination by surface sources of ^{10}Be, as might occur during hydrothermal alteration, etc. The measured ^{10}Be levels are low enough, however, that they would be strongly influenced by contamination if it did occur.

Aleutian summary: Most isotopic studies indicate that the Aleutian-arc mafic magmas have a source composed of seawater-altered ocean-floor basalt combined with a few percent of sediment. Some interaction of subduction-produced magma with overlying mantle is probable. Interaction of magma with overlying continental crust, where present, is not required by the isotopic data, but cannot be precluded because the isotopic character of the crust is not much different from that of the Aleutian magmas (see section of this chapter on Peninsular terrane). Intermediate to silicic magmas appear to result primarily by fractional crystallization of mafic magmas. Evolution of highly evolved magmas may involve volatile-rich fluid.

Bering Sea magmatic rocks

The area north of the Aleutians includes the Bering Sea abyssal plain to the west and the Bering Sea shelf to the east. Igneous rocks have been isotopically studied primarily because they might reflect the nature of sources in a back-arc basin. Analyses are of samples from Deep Sea Drilling Project (DSDP) site 191 on the western side of the abyssal plain in the Kamchatka Basin; from the Pribilof Islands of St. George and St. Paul toward the edge of the Bering Sea Shelf and, respectively, about 350 and 425 km from the arc; and from Nunivak Island and St. Lawrence Island on the Bering Sea Shelf, respectively, about 550 and 950 km from the arc.

The abyssal plain tholeiite from DSDP site 191 showed NIR and SIR (Table 1) that are typical of ocean-ridge basalts (Kay and others, 1978; von Drach and others, 1986). Because only one site has been analyzed from the abyssal plain, no geographic variations on the plain are known.

The Pribilof Islands were examined by Kay and others (1978) for SIR and PIR, and by von Drach and others (1986) for SIR and NIR. Compared with Pacific ocean-ridge basalts, the Pribilof rocks have higher PIR6. Compared with Aleutian magmas they have lower PIR7 and generally lower SIR and higher NIR. All the ratios fall within the broad field of ocean-island basalts.

Samples from Nunivak Island were studied by Zartman and Tera (1973), who measured PIR on an inclusion of peridotite in basinite and on an alkali basalt sample. The PIR6, PIR7, and PIR8 are distinct for each sample, leading Zartman and Tera to suggest that the peridotite and basalt were not cogenetic. They noted that all ratios fall in the range of abyssal tholeiites and alkali-island basalts.

Mark (1972) measured SIR in young ($<$6 m.y.) basalt lavas on Nunivak Island and found a range of 0.7025 to 0.7033. The minor volume ($\approx$2 percent) of basanites had a mean SIR of 0.7027, whereas the more abundant tholeiites and alkali olivine basalts had a mean of 0.7031. Menzies and Murthy (1980a, b) reported NIR for these rocks of 0.51289 to 0.51304, and noted no correlation with chemistry or SIR. They also studied mantle inclusions of pargasite-lherzolite and kaersutite megacrysts and found the same range of SIR and NIR as in the lavas. However, pargasite, diopside, and mica in nodules have NIR near 0.51302, whereas kaersutite has NIR of 0.51281. Menzies and Murthy suggeted that recent ($<$200 m.y.) metasomatic event(s) produced an inhomogeneous mantle having parts that are "veined" by kaersutite, and parts that contain pargasite and mica. Melting of varying proportions of these components would produce the observed isotopic heterogeneity of the Nunivak lavas. They also noted the similarity of Nunivak values for SIR and NIR to those of ocean-island basalts. One sample measured by von Drach and others (1986) fell in the same field. Roden and others (1984) added SIR and NIR data for additional peridotite and pyroxenite inclusions at Nunivak, and also suggested that the mantle had been recently metasomatized. They proposed, however, that the peridotites were pieces of wall rock that had been infiltrated by a silicate melt or volatile-rich fluid derived from a parental magma, that this magma was probably related to the Nunivak volcanism, and that crystal segregations from it formed the pyroxenite inclusions. They concluded that the metasomatized peridotites could not be regarded as fragments of the source of Nunivak basalts, as had been concluded by Menzies and Murthy (1980a, b).

Two samples from St. Lawrence Island were analyzed for SIR and NIR by von Drach and others (1986). SIR are slightly higher and NIR slightly lower than those of Nunivak, but also fall in the field for ocean-island basalts.

The data of Table 1 show a small increase in SIR and decrease in NIR in the order Pribilofs, Nunivak, and St. Lawrence. This may correlate with distance from the Aleutian arc or from the edge of the Bering Shelf, or it may be fortuitous based on limited data.

Bering Sea summary: Bering Sea lavas of the abyssal plain (1 sample) show the isotopic character of ocean-ridge basalt. Rocks from the islands on the Bering Sea shelf show the isotopic character of ocean-island basalts and may be derived from a mantle that was zoned by metasomatism less than 200 m.y. ago. The relation of peridotite inclusions to magma sources is disputed. Small changes in the isotopic composition of the island magmas occur with increasing distance from the shelf edge or from the Aleutian arc, and with decreasing distance from the continental shorelines.

Mount Edgecumbe volcanic field

The Edgecumbe volcanic field of Kruzof Island lies on the Chugach terrane of the North American plate near its transform-fault boundary with the Pacific plate (Silberling and others, this volume). The rocks range in SiO_2 content from 47 to 72 weight percent, are dominantly basaltic (about 70 percent), and display a gap in SiO_2 content in the mid-60s. Myers and Marsh (1981) reported a few SIR and several whole-rock $\delta^{18}O$ values that showed a general increase from basalt to rhyodacite. They suggested that little fractional crystallization was apparent and that the variation from mafic to silicic comoposition was best explained by adding varying amounts of crustal melts to a basalt magma.

Kosco (1981) subdivided the Edgecumbe volcanics into tholeiitic and calc-alkaline suites. He suggested that crystal fractionation produced the variety of tholeiitic basalt having SIR near 0.703. He explained calc-alkalic basaltic-andesites, andesites, and dacites having SIR near 0.7037 as a second generation of magmas from a distinct mantle source, and thought these magmas evolved in an open system by crystal fractionation and by mixing with melts from the country rock.

Myers and others (1984) reported SIR of 34 Edgecumbe samples that have a range of 0.7029 to 0.7041. Most samples are in the range of 0.7032 to 0.7039, but plagioclase basalts have lower values (0.7029 to 0.7030) and some andesites have higher values (0.7040 to 0.7041). They suggested an origin for the suite by interaction of primary plagioclase-basalt magma with crustal contaminants that changed composition over time and were melts from older granitic rocks. They suggested that rhyodacites were mixtures of small amounts of basalt and the first, most siliceous, melts from the older granitic rocks, and that intermediate andesites would be produced by mixing of basalts with subsequent and more refractory melts from the older granitic rocks.

Myers and Sinha (1985) made 22 PIR analyses on the samples previously analyzed for SIR. As was observed for SIR, the PIR values were the lowest for plagioclase basalts, highest for some andesites, and intermediate for the remaining basalts through rhyodacites. Myers and Sinha suggested a revised model in which a hypothetical parent magma collects within the crust and interacts along its borders with a variety of country rocks to produce localized pools of hybrid intermediate to silicic magmas. The hypothetical parent magma pool is then contaminated by olivine basalt and mixed to produce the observed plagioclase basalt. Subsequent eruptions tap the localized pools and the principal magma chamber to produce the observed variety of lavas.

Mount Edgecumbe summary: Holocene lavas of Mount Edgecumbe include at least two types of basalt and a variety of more evolved magmas that were produced when mafic magmas interacted with crustal melts.

Late Cretaceous and Tertiary volcanism, mainland Alaska

Calc-alkalic to alkalic volcanism in a wide subduction-related arc occurred on mainland Alaska during the Late Cretaceous to Early Tertiary (75 to 48 Ma) in the Alaska Range, Talkeetna Mountains, Kuskokwim Mountains, and Yukon-Koyukuk basin (Moll-Stalcup, this volume). The isotopic composition of five volcanic fields in and near Yukon-Koyukuk basin was reported by Moll-Stalcup and Arth (1989). The objective of their study was to delimit the origin of the volcanic rocks and thereby identify the nature of the crust beneath the basin. Three of the volcanic fields (Blackburn Hills, Kanuti, and Yukon River) lie within the basin, and two (Sischu and Nowitna) overlie bordering Precambrian and Paleozoic metamorphic terranes to the southeast.

The volcanic fields within Yukon-Koyukuk basin have SIR of 0.7033 to 0.7052 and NIR of 0.5125 to 0.5129. Moll-Stalcup and Arth (1989) suggested that primary magma sources included oceanic mantle, and the mafic lower portions of the late Paleozoic and Mesozoic igneous assemblage of Koyukuk terrane. Based on these findings, they suggested that the Yukon-Koyukuk basin had no cratonal roots under its southeastern part. Blackburn Hills rocks (65 to 56 Ma) were subdivided into four groups, including Group 1 and Group 2 basalts and andesites, western rhyolites, and central and eastern granite intrusions. Moll-Stalcup and Arth found Group 1 compatible with an origin from ocean-island-type mantle and a small component of either ocean-floor sediment or the mafic lower crust of Koyukuk terrane. Group 2 rocks are interpreted as partial melts of the lower crust of Koyukuk terrane. Western rhyolites are proposed to be crystal fractionates of Group 1 and Group 2 magmas, or melts from the crust of Koyukuk terrane. Central and eastern granites may either be fractionates of Group 1 andesites, or melts of young mafic to intermediate crust. The Kanuti volcanic field (60 to 56 Ma) is largely dacite. Moll-Stalcup and Arth (1989) suggested an origin by melting of mafic to intermediate rocks in the lower crust of Koyukuk terrane or by fractionation of mafic magmas like those of Group 2 in Blackburn Hills. The Yukon River field (53 to 48 Ma) is a diverse collection of basalt through rhyolite and latite. Moll-Stalcup and Arth suggested that diverse processes led to the origin of these magmas from sources in the mantle and Mesozoic crust of both Koyukuk terrane and Angayucham-Tozitna terrane (an ophiolitic assemblage).

The Sischu volcanic field (71 to 66 Ma) overlies the Nixon Fork and Minchumina terranes (Silberling and others, this volume) where Precambrian to Paleozoic basement is exposed. The volcanic rocks are chiefly dacite and rhyolite, and there is minor andesite. They have SIR of 0.7075 to 0.708 and NIR of 0.5124 to 0.5125, reflecting involvement of old continental crust in their genesis (Moll-Stalcup and Arth, 1989).

The Nowitna volcanic field (66 to 63 Ma) is largely andesite and overlies three terranes: the Innoko terrane of Paleozoic to Mesozoic, oceanic sedimentary and igneous rocks; and the Ruby and Nixon Fork terranes that contain Precambrian to Paleozoic basement rocks. The Nowitna volcanic rocks have SIR of 0.7043 to 0.7051, and NIR are 0.5125 to 0.5126. Moll-Stalcup and Arth (1989) suggested a hybrid source consisting of oceanic and old continental rocks.

MAJOR MESOZOIC TO CENOZOIC BATHOLITHS

The Coast batholith of southeastern Alaska

The Coast batholith (or Coast Plutonic Complex of Canadian geologists) is the largest North American batholith and consists dominantly of plutons of quartz diorite to granite of Paleocene and Eocene age that intrude older ortho- and paragneisses. In the southeastern Alaska panhandle the plutons have been isotopically studied to the north in the Skagway area and to the south in the Ketchikan area (Table 2).

Skagway area plutons were analyzed for oxygen and hydrogen isotopes by Margaritz and Taylor (1976a). They found high values of $\delta^{18}O$ in quartz (9.7 to 11.5 per mil) in orthogneiss and tonalite of Skagway and suggested that the magmas either interacted with or assimilated high-$\delta^{18}O$ metasediments or weathered volcanic rocks. Some samples had low δD values and anomalous differences between the $\delta^{18}O$ of quartz and feldspar, indicating that meteoric water circulated through the intrusive rocks to depths of several kilometers after the plutons crystallized.

Barker and others (1986) reported SIR in the range 0.7048 to 0.7060 on five tonalitic to granitic plutons from the Skagway area. They note that these values are within the range of many orogenic andesitic suites, and also within the range expected in the country rocks surrounding the batholith, which consist of immature sedimentary rocks, island-arc volcanic rocks, and rift and intraplate basalts. The SIR values are low enough to suggest that the sources contained little, if any, radiogenic continental rocks of pre-Mesozoic age. Barker and others (1986) suggest that the magmas had a hybrid source consisting of subduction-related basalts and melts from the country rock.

Arth and others (1988) measured SIR and NIR in the major intrusive bodies of the Coast batholith near Ketchikan. The bodies include 55- to 57-Ma foliated tonalites on the west side of an Early Cretaceous or older migmatitic orthogneiss complex, and 52- to 54-Ma massive granodiorites and granites on the east side of the gneisses. The central orthogneisses have SIR in the range 0.7053 to 0.7066 (Arth and Barker, unpublished data), and are thought by Barker and Arth (1984) to be the roots of an Early Cretaceous continental-margin magmatic arc. The western tonalites have uniform SIR of 0.7063 to 0.7064 and NIR of 0.5123 to 0.5124 and are thought by Arth and others (1988) to have been generated from sources that included significant amounts of melt from mafic parts of the central orthogneisses. The eastern granodiorites and granites have SIR of 0.7046 to 0.7061 and NIR of 0.5124, and were probably generated by mixing and fractionation of subduction-related magmas with melts from Phanerozoic crustal rocks of intermediate to silicic composition.

West of Coast batholith, plutons of quartz diorite to tonalite are of distinct age and origin (Arth and others, 1988), and are discussed below in the section on Taku terrane. Arth and others (1988) suggest that the juxtaposition of the island-arc-like Taku terrane against the older magmatic arc (central gneisses) in early Tertiary time may have triggered the generation of the 57- to 52-Ma plutons of Coast batholith.

Several small mid-Tertiary granite bodies in the Ketchikan area intrude both Coast batholith and the Taku terrane. Based on SIR and NIR values, Arth and others (1988) suggested that these granites do not represent a further evolution of Coast batholith magmas, but instead reflect an origin from more primitive sources like those of Taku terrane. These observations provide evidence that Taku rocks underlay at least part of Coast batholith by mid-Tertiary time.

Coast batholith summary: The central gneisses of Coast batholith are probably the eroded remnant of an Early Cretaceous continental-margin magmatic arc. In early Tertiary time, foliated tonalites intruded along the western flank of the gneisses, and massive granodiorite and granite plutons intruded to the east. The tonalites formed from sources that included significant contributions of melt from mafic parts of the gneisses; whereas the granodiorites and granites probably formed by mixing of subduction-related mafic magmas with magmas derived by melting of metasedimentary, metavolcanic, and metaplutonic country rocks in which Precambrian rocks were absent or scarce. Some of the plutons experienced circulation of meteoric water after crystallization. The batholith probably formed at the time that Taku terrane was joined to North America at the site of an Early Cretaceous continental-margin arc.

The Alaska–Aleutian Ranges batholith of south-central Alaska

The Alaska–Aleutian Ranges batholith of south-central Alaska is one of the major circum-Pacific batholiths. Reed and Lanphere (1973) showed that it resulted from distinct intrusive episodes in Jurassic, Late Cretaceous to early Tertiary, and middle Tertiary time.

The southern and eastern part of the batholith is mostly Jurassic (158 to 174 Ma) quartz diorite and tonalite but varies from gabbro to granite. The rocks have SIR of 0.7033 to 0.7037 (M. A. Lanphere and B. L. Reed, unpublished data cited by Reed and others, 1983), values that are similar to those of oceanic island arcs. Reed and others (1983) suggest that these rocks formed the root of an oceanic island arc that was accreted to Alaska during late Mesozoic time.

The western and northern part of the batholith is composed principally of Late Cretaceous to early Tertiary (76 to 59 Ma) tonalite, granodiorite, and granite, as well as middle Tertiary (38 to 26 Ma) tonalite, granodiorite, and granite. Lanphere and Reed (1985) published SIR and $\delta^{18}O$ data on five McKinley-sequence, peraluminous, "minimum-melt" granite plutons of Paleocene age that constitute a small part of the Late Cretaceous to early Tertiary suite. These plutons largely intrude deformed upper Mesozoic flysch in the central Alaska Range. SIR range from 0.7054 to 0.7085 and $\delta^{18}O$ range from +11.2 to +14.6. The four plutons southeast of the Denali fault have SIR of 0.7054 to 0.7065, whereas a pluton northwest of the fault has SIR of 0.7071 to 0.7081. No polarity was observed in the oxygen ratios. Lanphere and Reed (1985) compared SIR in the granite with SIR in the flysch and found that simple melting of the flysch would produce granite of higher SIR than they observed. They suggested that mafic magma having lower SIR assimilated flysch to produce hybrid magmas that subsequently were fractionally crystallized. The pluton to the north of the Denali fault reflects assimilation of more radiogenic Precambrian or Paleozoic metasedimentary and metavolcanic rocks. They ascribed deformation of the schist belt and generation of magma to the tectonic event that produced accretion of the Talkeetna superterrane to stable Alaska in latest Cretaceous to early Tertiary time.

The Ruby batholith of central Alaska

A large percentage of outcrop in the Ruby geanticline in central Alaska is composed of Albian (~ 110 Ma) granitic intrusive rocks (Ruby batholith), which form a belt more than 350 km long (Patton and others, this volume, chapter 6). Most of these granitic plutons intrude Precambrian and Paleozoic schists and lesser orthogneisses, marble, quartzite, and amphibolite of the

Ruby terrane (Plate 3, this volume); a few intrude Paleozoic to Mesozoic oceanic rocks of the Angayucham terrane. The plutons consist dominantly of coarse-grained biotite and two-mica granites, most of which are feldspar porphyritic. Biotite-hornblende granodiorite occurs in the northeastern part of the batholith, as does a small pluton of monzonite and syenite near Jim River (Barker and others, this volume).

Arth and others (1984) determined SIR, PIR, and $\delta^{18}O$ values in four plutons in the center of the batholith near Kanuti, Hot Springs, and Sithylemenkat, and in the Ray Mountains. High values of SIR (0.706 to 0.730), $\delta^{18}O$ (>8.5), and PIR7 (15.54 to 15.70) led these authors to suggest that granite magmas of the geanticline either formed from—or incorporated large amounts of—continental crust of Paleozoic or greater age.

Blum and others (1987) reported SIR and $\delta^{18}O$ for Jim River and Hodzana plutons at the northeastern end of the batholith. They determined SIR of 0.7078 and 0.7079, respectively, in monzonitic and granitic rocks, and a range in $\delta^{18}O$ from +6.8 to +7.3 per mil. They suggested that monzonite was a primary magma formed in the lower crust or mantle, whereas granite magma may have formed when monzonite magma assimilated upper crustal melts.

Arth and others (1989a) reported SIR, NIR and $\delta^{18}O$ for all plutons of the batholith. They noted that SIR and NIR show a trend along strike from southwest to northeast, but that $\delta^{18}O$ has a constant range of +8.4 to +11.8 per mil. Rocks to the southwest have the highest SIR (0.7235) and lowest NIR (0.51150), whereas those to the northeast have the lowest SIR (0.7055) and highest NIR (0.51232). Some individual plutons show uniform isotopic values, whereas others show a range. Arth and others (1989a) suggested that isotopically uniform plutons underwent fractional crystallization; isotopically heterogeneous plutons were possibly affected by magma mixing and assimilation. The source for most of the magmas was within the deeper parts of the Proterozoic to Paleozoic crust that underlay both Ruby and Angayucham terranes at 110 Ma, and was probably lithologically similar to the presently exposed country rock of Ruby terrane. The source rocks were probably heterogeneous in age and composition, varying from Proterozoic to Paleozoic, and from silicic orthogneisses and schists to mafic to intermediate metaigneous rocks. The strong isotopic trends in the batholith may be related to changes in the proportions of the source rocks along strike under the batholith. Sources to the southwest were dominantly silicic and included both Paleozoic and Proterozoic rock, whereas those to the northeast were mostly Paleozoic and contained a higher proportion of mafic to intermediate rocks.

MAGMATIC SUITES OF ECONOMIC INTEREST

In addition to the major Mesozoic to Cenozoic volcanic provinces and batholiths discussed above, Alaska contains many other igneous suites of Mesozoic and Cenozoic age. Three suites that are of economic interest are discussed in this section. In the next section, many additional suites are discussed in relation to the tectonostratigraphic terrane in which they are located.

Duke Island-type ultramafic rocks, southeastern Alaska

More than 35 ultramafic complexes and associated gabbros form a belt about 50 km wide and 560 km long west of and parallel to the Coast batholith. These 102- to 113-Ma bodies, intruded after folding and metamorphism of the Paleozoic and early Mesozoic country rocks, are generally small and show a concentric zoning where dunite and peridotite in the core intrude successively surrounding zones of olivine pyroxenite, hornblende pyroxenite, and hornblendite. Lanphere (1968) determined SIR in four zoned bodies and found a range of 0.702 to 0.705. He noted the similarity of these values to those of oceanic volcanic rocks, and inferred that these largely cumulate rocks crystallized from magmas originating in oceanic-type mantle.

Taylor (1968) published $\delta^{18}O$ data for a variety of minerals from many of the ultramafic complexes of southeastern Alaska, and found them compatible with a magmatic origin for these rocks. $\delta^{18}O$ of clinopyroxene in various rock types has uniform values of +5.4 to +5.9 per mil. Magnetite in massive segregations has values of +3.3 to +4.4 per mil. Clinopyroxene-magnetite fractionation values of 5.1 to 2.1 per mil probably represent a close approach to equilibrium and are compatible with high-temperature magmatic crystallization (Taylor and Noble, 1969).

Tin granites of Seward Peninsula

Seven Late Cretaceous granite plutons that are spatially and genetically related to tin mineralization are exposed in a 170-km-long belt across the northwestern part of Seward Peninsula, Alaska. The plutons intrude a variety of terranes that include large areas of sedimentary and metasedimentary rocks of Precambrian and Paleozoic age. The field relations, ages, petrography, chemistry, and isotopic composition of these rocks are described by Hudson and Arth (1983), who determined SIR for five of the plutons. The data for individual plutons showed Sr isotopic homogeneity among the several textural facies that make up a composite body, and demonstrated that these facies are genetically linked. The initial ratios varied from pluton to pluton in the range 0.708 to 0.720. Hudson and Arth (1983) concluded that the high values reflect a dominantly crustal source for the parent magmas. They favored a multistage origin of the tin granites that involved melting of sialic crust to produce batholithic magmas, which subsequently fractionated to generate smaller volumes of highly evolved residual magma. Evolution of the individual plutons was largely by fractional crystallization, although the final stages involved transport of elements by a coexisting volatile phase.

Molybdenum-bearing granites of the Ketchikan area

Stockwork molybdenum deposits in the Ketchikan area are associated with shallow porphyritic biotite granite intrusions of late Oligocene to Miocene age. Magmatism took place in a postorogenic tectonic setting that was characterized by regional extension and possibly bimodal magmatism. Hudson and others (1981) reported SIR of 0.7051 for the Quartz Hill pluton, and

0.7049 to 0.7051 for the Burroughs Bay intrusion. The SIR of these bodies was lower than the late Oligocene to Miocene Sr isotopic ratios of the surrounding older plutons of Coast batholith and Taku terrane, or of the metamorphic country rocks. This distinction precludes a simple direct genetic relation between the molybdenum-related granites and the exposed rock around them. Hudson and others (1981) suggested a magma source region dominated by rocks of low Rb/Sr ratio, such as those in the mantle or lower crust or rapidly recycled, mantle-derived material at depths of less than 60 km. Arth and others (1988) suggested that the source was probably the deeper part of the Taku terrane, which probably underlay the whole area by mid-Tertiary time.

IGNEOUS ROCKS IN TECTONOSTRATIGRAPHIC TERRANES

All of the igneous rocks of Alaska intrude, overlie, or are otherwise part of one or more tectonostratigraphic (lithotectonic) terranes (Silberling and others, this volume). An igneous body that intrudes country rock of a given terrane may indicate the nature of the basement of that terrane at the time of intrusion. In this section, I review the isotopic composition (Table 2) and petrogenesis of Mesozoic and Cenozoic igneous rocks in the context of their host tectonostratigraphic terranes. In the next section, I discuss the implications of the data to the nature of the lower crust under Alaska.

Each of the tectonostratigraphic terranes of Alaska is bounded by faults and has a characteristic stratigraphy that records a different geologic history from that of adjacent terranes (Coney and others, 1980). Some of the terranes are "exotic" to North America in that they were transported as crustal blocks to their present location from elsewhere, and have been "accreted" to the continent in late Mesozoic or early Cenozoic time. Identifying the nature of the basement under some tectonostratigraphic terranes can be assisted by isotopic study of the igenous rocks. Many such rocks, particularly granitic intrusions, crystallize from magmas that originated, at least in part, by melting of substantial volumes of lower crust or subcrustal mantle. Their isotopic character thus can yield information about the age and composition of this deep source region. Because the deeper crust under a given terrane may change character during accretion, the igneous rocks that intrude at various times may reflect the crustal conditions before, during, or after the time of accretion.

The terranes discussed below are those for which at least some isotopic-tracer data are available for their igneous rocks. For most of the larger terranes there are at least some data. Many of the smaller terranes have not been studied. The terranes are described from east to west in southeastern Alaska, and from south to north in mainland Alaska.

Tracy Arm terrane

The Tracy Arm terrane consists largely of Paleozoic to Mesozoic schist, paragneiss, marble, amphibolite, and minor serpentinite that form roof pendants and wall rocks of the Coast batholith. These rocks may be the higher-metamorphic-grade equivalents of the Stikine terrane to the east, or of the Alexander or Taku terranes to the west. Early Tertiary plutons of the Coast batholith intrude the Tracy Arm terrane and have SIR of 0.7045 to 0.7064 and NIR of 0.5123 to 0.5124 (Barker and others, 1986, Arth and others, 1988). These plutons are discussed in the foregoing description of the Coast batholith. Arth and others (1988) suggested that the western foliated tonalites are largely the product of melting in the deeper mafic parts of the older central gneisses of the batholith, whereas the eastern granodiorites and granites are the product of mixing of subduction-related mafic magma with magmas derived by melting of metasedimentary and metaigneous country rock in which Precambrian rocks or metapelitic rocks were not abundant or absent. Thus the early Tertiary basement of this terrane was probably mostly Mesozoic orthogneisses and paragneisses. By mid Tertiary time, when small bodies of granite intruded, the basement source had a more primitive island-arc-like isotopic character similar to that found in plutons of Taku terrane (Arth and others, 1988). This change in magma sources in the basement beneath Tracy Arm terrane may reflect underthrusting by Taku terrane from the west during the early Tertiary.

Taku terrane

Taku terrane is immediately west of the Coast batholith in much of southeastern Alaska. On Revillagigedo Island and the mainland to the northwest, this terrane is intruded by tonalite and quartz diorite that constitute a suite of 90-Ma plutons and one body of possibly 135 Ma. These pre-accretionary rocks intrude high-grade mafic-to-intermediate metavolcanic rocks, metagraywacke, and metapelite, and have SIR of 0.704 to 0.705 and NIR of 0.5125 (Arth and others, 1988). The low SIR suggest that little, if any, radiogenic crust (Precambrian, granitic, or pelitic) was incorporated in or contributed melt to the magmas. The moderate NIR may reflect the incorporation or contribution of some Paleozoic and younger mafic to intermediate country rock. The tonalites probably solidified from magmas of hybrid origin, consisting of mixtures of subduction-related basalt and partial melts from Paleozoic or younger country rock of mafic to intermediate composition. The Taku, thus, may represent an accreted magmatic arc. Post-accretion Tertiary granites have SIR and NIR that are similar to those of the older intrusions (Arth and others, 1988) and may have originated from similar sources. Thus the basement beneath Taku terrane is probably of Phanerozoic oceanic to island-arc character.

Alexander terrane

Alexander terrane lies west of Taku terrane in much of southeastern Alaska and consists largely of Paleozoic schist and gneiss, felsic to mafic volcanic rocks, and minor pelitic rocks and carbonate (Plate 3, this volume.). On Prince of Wales Island the rocks were intruded by quartz monzonite and gabbro in late

TABLE 2. ISOTOPIC RANGES FOR MESOZOIC AND CENOZOIC PLUTONIC ROCKS OF ALASKA

Location	SIR*	NIR†	References§
Major Mesozoic to Cenozoic batholiths			
Coast batholith			
Western tonalites (Ketchikan area)	0.7063–0.7064	0.5123–0.5124	1
Central tonalite-granodiorite (Skagway area)	0.7054–0.7058		2
Central orthogneiss suite (Ketchikan area)	0.7053–0.7066		1
Eastern granodiorites and granites (Ketchikan and Skagway areas)	0.7046–0.7061	0.5124–0.5124	1, 2
Alaska-Aleutian Ranges batholith			
Southwestern Jurassic suite	0.7033–0.7037		3
McKinley series (Paleocene)			
southeast of Denali fault system	0.7054–0.7065		4
northwest of Denali fault system	0.7071–0.7081		4
Ruby batholith	0.7056–0.7294	0.5115–0.5124	5
Magmatic suites of economic interest			
Duke Island-type ultramafic rocks, SE Alaska	0.702–0.705		6
Tin granites of Seward Peninsula	0.708–0.720		7
Molybdenum-bearing granites, Ketchikan area	0.7049–0.7051		1, 8
Plutons in tectonostratigaphic terranes			
Chugach and Prince William terranes			
Sanak-Baranof belt pultons	0.7048–0.7063		9–12
Kahiltna terrane			
McKinley series plutons, Alaska Range	0.7054–0.7065		4
Koyukuk terrane			
eastern calc-alkalic suite	0.7038–0.7056	0.5124–0.5126	13
western alkalic suite	0.7069–0.7130	0.5121–0.5125	13
Peninsular terrane			
Kodiak and Afognak Islands-Jurassic	0.7035–0.7037		14
Aleutian Range Jurassic suite	0.7033–0.7037		3
Northern Chugach Jurassic suite	0.7032–0.7033		14
Talkeetna Mountains tonalites	0.7034–0.7037		14
Ruby terrane			
Ruby batholith (all)	0.7056–0.7294	0.5115–0.5124	5
Seward and York terranes			
Tin granites	0.708–0.720		7
Eastern Seward Peninsula calc-alkalic	0.7084–0.7109		15
Eastern Seward Peninsula alkalic	0.712–0.715		15
Tracy Arm terrane			
Coast batholith	0.7046–0.7064	0.5123–0.5124	1, 2
Taku terrane			
tonalitic suite	0.7041–0.7050		1
Wrangellia terrane			
Chitina Valley batholith	0.7038–0.7045		14
Yukon-Tanana terrane			
Gilmore Dome	0.7124		16

*All ratios normalized to $^{86}Sr/^{88}Sr = 0.11940$.
†All ratios normalized to $^{146}Nd/^{144}Nd = 0.72190$
§1. Arth and others, 1988; 2. Barker and others, 1986; 3. Reed and others, 1983; 4. Lanphere and Reed, 1985; 5. Arth and others, 1989a; 6. Lanphere, 1968; 7. Hudson and Arth, 1983; 8. Hudson and others, 1981; 9. Hudson and others, 1979; 10. Hill and others, 1981; 11. Myers and Marsh, 1981; 12. Kosco, 1981; 13. Arth and others, 1989b; 14. Arth and Hudson, unpublished data; 15, Arth, 1987; 16. Blum, 1985.

Ordovician to Early Silurian time (Lanphere and others, 1964). A Rb-Sr isochron study of these plutons by Armstrong (1985) gave an SIR of 0.7039, consistent with their emplacement in a Phanerozoic volcanic arc. No younger plutons are as yet isotopically characterized, so that the basement is tentatively considered to be of oceanic-arc character.

Chugach and Prince William terranes

Along the margin of the Gulf of Alaska, the Sanak-Baranof belt of early Tertiary granitic plutons intrudes a prism of flyschoid rocks that were accreted to the continental margin during Late Cretaceous or early Tertiary time (Hudson and others, 1979; Hudson, this volume). The plutons include biotite tonalite, granodiorite, and granite that are generally homogeneous, medium grained, and equigranular. Hornblende is generally absent, and white mica and garnet are present locally. Kyanite megacrysts occur in some plutons on Kodiak Island in the western part of the belt.

Hudson and others (1979) studied plutons and metamorphic country rocks of the Sanak-Baranof belt in the eastern Chugach and St. Elias Mountains. They showed that two 50-Ma granodiorites had SIR (0.7059 and 0.7063) that were close to the age-corrected Sr isotopic composition of two metagraywacke samples (0.7061 and 0.7062). Metasandstone and amphibolite, respectively, showed higher and lower values. Hudson and coworkers concluded that the granitic magmas are partial melts of the deeper parts of the accretionary prism and were generated after the prism was deformed against the continent.

Hill and others (1981) conducted an extensive geochemical study on plutons of about 60 Ma on Kodiak, Shumagin, and Sanak Islands in the western part of the Sanak-Baranof belt. The plutons are characterized by high $\delta^{18}O$ values of +10.9 to +13.2 per mil, which are similar to the values found in nearby graywacke, lower than those of nearby argillite, and presumably higher than those of local greenstone. SIR for the Shumagin batholith and Sanak pluton are about 0.7053, as estimated from mineral isochrons of about 60 Ma. At 60 Ma, the nearby flysch belt had $^{87}Sr/^{86}Sr$ ratios of 0.7046 to 0.7050 in graywacke and 0.7061 to 0.7089 in argillite. Hill and others (1981) calculated the proportion of sediment to mafic metavolcanic rocks needed to satisfy the Sr and O isotopic constraints, and concluded that the proportion of amphibolite required for melting was not present in the accreted prism. Therefore, they suggested that mantle-derived basalt magma was added to the prism to produce the observed compositions. The authors relied for their models on a $\delta^{18}O$ value of +10.6 per mil found in amphibolites of the Franciscan Complex in California (Margaritz and Taylor, 1976a). If nearby mafic volcanic rocks in the Sanak-Baranof belt have a lower $\delta^{18}O$ value, then much less mafic component would be required to model the granites, and the flysch belt alone may constitute an adequate source.

A granodiorite pluton on Kruzof Island near Sitka was intruded at about 50 Ma into Sitka graywacke of the Chugach terrane. The pluton is overlain by Holocene volcanic rocks of the Edgecumbe volcanic field. Myers and Marsh (1981) and Kosco (1981) measured SIR of about 0.7048 in the granodiorite. Myers and Marsh (1981) also measured $\delta^{18}O$ of 8.7 to 9.3 per mil in the granodiorite. The authors did not comment on the genesis of the pluton, but the values reported are compatible with an origin by melting of country rock or by mixing of country-rock melt and mafic magma, as hypothesized for other plutons in the Sanak-Baranof belt.

Chugach and Prince William summary: Tonalite-to-granite plutons of the early Tertiary Sanak-Baranof belt were probably derived by melting of deep parts of the accreted Chugach–Prince William terranes of Late Cretaceous or early Tertiary graywacke, argillite, and mafic volcanic rocks. An additional basalt component may have mixed with the crustal melts to produce the plutons.

Peninsular terrane

The Peninsular terrane of southwestern and south-central Alaska consists of Permian limestone; Upper Triassic limestone, argillite and volcanic rocks; Lower Jurassic andesitic volcanic rocks; and Middle Jurassic to Cretaceous clastic rocks. Jurassic plutons, largely tonalitic, are found in the Alaska–Aleutian Range batholith, on northern Kodiak Island, in the northern Chugach Mountains east of Anchorage, and in the Talkeetna Mountains.

Reed and others (1983) cite SIR of 0.7033 to 0.7037 in the southeastern part of the Alaska–Aleutian Range batholith (158 to 174 Ma). They consider this part of the batholith to be the root of an oceanic island arc.

Arth and Hudson (unpublished data) determined SIR values of 0.7033 to 0.7037 for 180- to 190-Ma plutons on Kodiak Island and in the northern Chugach Mountains, and values of 0.7034 and 0.7037 in tonalites of the Talkeetna Mountains. Hudson and others (1985) interpreted the Kodiak and Northern Chugach plutons to be a tholeiitic suite in a Jurassic island arc, and the Talkeetna Mountains tonalites to be part of a calc-alkalic suite in a Jurassic arc.

From the studies noted above, it appears that much of the Peninsular terrane is underlain by Mesozoic island-arc rocks. It should be noted that these rocks also underlay the eastern volcanoes of the Aleutian magmatic arc. Several studies of Aleutian volcanic rocks, discussed in the section of this chapter on Cenozoic volcanic provinces, did not reveal any "continental" component in the eastern Aleutian-arc magmas. However, no significant isotopic changes should accompany assimilation of "continental" rocks in the deeper parts of the Peninsular terrane, because the rocks are isotopically very similar to oceanic magma sources in the western Aleutians.

Wrangellia terrane

The Wrangellia terrane of east-central and southeastern Alaska is an upper Paleozoic volcanic-arc suite overlain by Permian limestone, pelite, and chert; Triassic subaerial basalt and overlying platform carbonate; and Mesozoic marine sedimentary rocks. Several Late Jurassic granite plutons of the Chitina Valley batholith, near the Canadian border, have SIR of 0.7038 to 0.7045 (Arth and Hudson, unpublished data), and may indicate a mafic source at depth. Aleinikoff (1984) reports that undated granites south of the Denali fault have PIR7 ratios of 15.52 to 15.57, which he finds suggestive of an oceanic source.

Yukon-Tanana terrane

The Yukon-Tanana terrane of east-central Alaska is composed mainly of Precambrian to Paleozoic polydeformed rocks, including quartz-mica schist and gneiss, quartzite, metavolcanic and metaplutonic rocks, and minor marble. Published isotopic data on Cretaceous plutons are limited. Blum (1985) determined a SIR of 0.7124 in a small 91-Ma porphyritic granodiorite-to-granite pluton of Gilmore Dome near Fairbanks. He suggested that the magma formed, at least in part, by anatectic melting of late Proterozoic or early Paleozoic crustal rocks.

Aleinikoff (1984) reported that Cretaceous plutons in several small terranes north of the Denali fault (Lake George, Macomb, Jarvis Creek, and Hayes Glacier) have PIR7 of 15.61 to 15.72, indicating a mixture of lead from both continental and oceanic sources. He also determined that Cretaceous plutons in the Maclaren terrane, just south of the Denali fault, have PIR7 of 15.56 to 15.60, suggesting derivation from an oceanic source with minor crustal input.

Ruby and Nixon Fork terranes

The Ruby and Nixon Fork terranes include Precambrian to Paleozoic pelitic and calcareous schists, orthogneisses, marble, quartzite, and metavolcanic rocks. Nixon Fork terrane contains a sequence of Paleozoic carbonates overlain by Permian, Triassic, and Cretaceous sedimentary rocks. The Ruby terrane is host to the Ruby batholith of Cretaceous granitic plutons, whose isotopic character is very radiogenic, reflecting a source in Proterozoic and Paleozoic continental crust (Arth and others, 1989a). The Ruby batholith is discussed in more detail in the section of this chapter on major batholiths.

The Nixon Fork terrane is host to the Sischu volcanic field, composed of rhyolite, dacite, and lesser andesite. The volcanic rocks have SIR of 0.7079 to 0.7141 and NIR of 0.5124 to 0.5125, reflecting magma generation involving old continental crust (Moll-Stalcup and Arth, 1989). Thus, both the Ruby and Nixon Fork terranes are probably underlain by continental rocks.

Koyukuk terrane and Yukon-Koyukuk basin

The Koyukuk terrane is an arcuate belt of Early Cretaceous volcanic rocks flanked by Upper Cretaceous flysch deposits in the vast Yukon-Koyukuk basin of west-central Alaska (Patton and Box, 1989). The basin is bordered by highlands of metamorphosed Precambrian and Paleozoic rocks of the Brooks Range, Seward Peninsula, and the Ruby geanticline. Two suites of plutonic rocks intrude the basin: a western alkalic suite of 100- to 113-Ma quartz monzonite, monzonite, syenite, and ultrapotassic subsilicic rocks; and an eastern calc-alkalic suite of 79- to 89-Ma tonalite, granodiorite, granite, and alaskite (Miller, 1989).

Arth and others (1984) reported SIR, NIR, PIR, and $\delta^{18}O$ on two granodiorite-to-tonalite plutons (Indian Mountain and Zane Hills). The plutons have low SIR (0.7038 to 0.7047), low PIR7 (15.50 to 15.53), and low $\delta^{18}O$ (<8.5) compared with Cretaceous granites in the adjacent borderlands (see descriptions of Ruby batholith and Seward terrane sections in this chapter). These values are characteristic of island-arc or convergent continental-margin magmatic suites in which tonalite to granodiorite may be generated by fractionation or melting of mafic volcanic rocks. Isotope data for these two plutons do not support the presence of Paleozoic or older continental crust below the eastern part of the Yukon-Koyukuk basin.

Arth (1985) reported the range of SIR and NIR for five plutons that form an east-to-west belt across the eastern half of the basin. From east to west (Indian Mountain to Purcell Mountains), the rocks become more potassic (from tonalite to granite), increase in SIR from 0.7038 to 0.7056, and have decreasing NIR from 0.51261 to 0.51245. The values were considered compatible with sources that may include oceanic mantle and Mesozoic supracrustal rocks. The isotopic variations would reflect an east-to-west increase in the degree of crustal involvement in magma generation.

In the western suite, the SIR and NIR of the potassic alkalic plutons display a continuation of the east-to-west trends of the plutons in the eastern suite. Arth (1987) reported SIR of 0.7069 to 0.7130 and NIR of 0.5125 to 0.5121. He ascribed the generation of the alkalic magmas to melting of Paleozoic and Precambrian subcontinental-type mantle, accompanied by interaction with rocks in the overlying lower crust.

Arth and others (1989b) reported isotopic and trace-element data for all plutons of the belt and noted remarkably continuous trends in isotopic ratios and concentration of several trace elements along the full 350-km length of the plutonic belt, from potassic alkalic plutons on the west through sodic calc-alkalic plutons in the east. From west to east, SIR vary from 0.712 to 0.704; NIR from 0.5121 to 0.5126; and $\delta^{18}O$ from +8.6 to +6.5. Arth and others (1989b) inferred from the chemical and isotopic features that the westernmost plutons originated by melting or fractionation in a Paleozoic or older continental section that included mantle and perhaps crustal rocks; that the easternmost plutons originated by melting or fractionation of mafic rocks of Mesozoic oceanic to island-arc affinity; and that plutons found between the eastern and western ends of the belt experienced individual fractionation histories, but their sources vary gradationally across the width of the basin in the proportion of older continental (mantle and crustal) and Mesozoic oceanic or island-arc rocks.

Moll-Stalcup and Arth (1989) reported SIR and NIR of three mid-Tertiary volcanic fields (Kanuti, Yukon River, and

Blackburn Hills) that lie along the southeastern margin of Yukon-Koyukuk basin. The low values for SIR (0.7037 to 0.7051) and high values for NIR (0.5125 to 0.5128) suggest sources within oceanic mantle or Mesozoic island-arc rocks.

From the foregoing studies, it appears that the Yukon-Koyukuk basin is underlain by continental lithosphere from its margin at Seward Peninsula eastward to about longitude 157°. East of this parallel and in a band at least 50 km wide along its southeastern margin, the basin is probably underlain by rocks of oceanic to island-arc character.

Seward and York terranes

The Seward Peninsula is largely composed of complex schists and gneisses of Precambrian to Devonian age (Seward terrane), and an assemblage of Precambrian(?) to Paleozoic metasedimentary rocks (York terrane). Igneous suites include Cretaceous tin-bearing granites, calc-alkalic intrusive rocks, and alkalic intrusive rocks.

The tin-granite suite intrudes the northwestern part of the peninsula and has high SIR, reflecting a source in the continental crust (Hudson and Arth, 1983). It is discussed separately in the economic section of this chapter.

Arth (1987) reported SIR of 0.708 to 0.711 in the 82- to 96-Ma calc-alkalic suite of eastern Seward Peninsula, including the Darby and Bendeleben plutons. He suggested generation within Precambrian to Paleozoic crustal rocks.

Arth (1987) reported that 99- to 113-Ma alkalic plutons of eastern Seward Peninsula (including Kachauik batholith and Dry Canyon Creek pluton) and the adjacent Yukon-Koyukuk basin have SIR of 0.707 to 0.713, and NIR of 0.5125 to 0.5121. He suggested an origin by melting in the Precambrian to Paleozoic continental mantle accompanied by interaction with rocks in the overlying crust.

From the studies cited above, it appears that most parts of Seward Peninsula are underlain by continental lithosphere of probable Precambrian to Paleozoic age.

A MAP OF THE DEEP CONTINENTAL CRUST BASED ON SIR OF IGNEOUS ROCKS

Figure 1 is a map of Alaska on which I infer the nature of the deep continental crust from the compositional and isotopic studies cited above. The magmas that formed many of the granitic and intermediate-to-silicic volcanic rocks were partly or totally produced within the deep crust, and therefore carry with them the isotopic signature of the materials that constitute that crust.

Among the isotopic data discussed in this chapter, only Sr ratios have been measured on enough samples to provide a geographically balanced data set that is adequate to characterize the general features of the deep Alaskan crust. In a manner analogous to the division of California plutons by Kistler and Peterman (1973), I divide Alaska plutons into three groups based on their SIR. The bounding SIR values for the groups are chosen on the basis of the genetic inferences made by the investigators who obtained and interpreted the petrologic, isotopic, and chemical data in each terrane. SIR values in the 0.702 to 0.705 range characterize igneous rocks formed largely from Paleozoic and younger rocks of generally oceanic character, such as the roots of island arcs. SIR values more than 0.708 characterize igneous rocks that formed largely from ancient continental lithosphere, such as Precambrian-to-Paleozoic crustal cratonic rocks and continental mantle. SIR values that fall between these two groups, in the range 0.705 to 0.708, characterize igneous rocks that formed either by melting of Paleozoic and Mesozoic flysch and melange or by mixing of oceanic and ancient continental sources.

Figure 1 depicts the three isotopic subdivisions and also shows the major faults and tectonostratigraphic terranes of Alaska. Areas of the figure shown in a dark stippled pattern have igneous rocks that have SIR higher than 0.708 and largely correspond to the parts of Alaska where Precambrian to Paleozoic rocks are exposed. I refer to the deep crust under these areas as Cratonia. Northern Cratonia, underlying Seward Peninsula and the Ruby Geanticline, is inferred from isotopic measurements on a large number of plutons. Central Cratonia, underlying the Nixon Fork and Yukon-Tanana terranes, is inferred from very limited data.

Areas shown in a light stipple pattern have igneous rocks that have SIR of less than 0.705, and largely correspond to parts of Alaska where Paleozoic to Mesozoic magmatic-arc and oceanic-type rocks are exposed. I refer to the deep crust under these areas as Volcania. Northern Volcania underlies the eastern and southern part of the Yukon-Koyukuk basin and is inferred from a large number of measurements. Southern Volcania underlies the rocks of Wrangellia and the Peninsular terranes, which have marked island-arc and oceanic affinities, and is inferred from a substantial number of measurements scattered throughout the area. Southeastern Volcania underlies the Taku terrane, which contains many volcanic and plutonic rocks that appear to be of magmatic-arc association. It is inferred from several measurements in the Ketchikan area.

Areas shown in a dot pattern have igneous rocks in the SIR range of 0.705 to 0.708 and largely correspond, with one exception, to parts of Alaska where Phanerozoic flysch belts are exposed. I refer to the basement under these areas as Detritia. Southern Detritia underlies the combined Chugach and Prince William terranes, which consist of Jurassic to Lower Tertiary flysch, melange, and minor mafic volcanic rocks. It is inferred from measurements on several plutons at widely separated localities along the length of the Sanak-Baranof plutonic belt. Central Detritia primarily underlies the Kahiltna terrane of Upper Jurassic to Cenomanian flysch. Eastern Detritia primarily underlies the Tracy Arm terrane and may include some areas under the Stikine terrane. It is inferred from measurements of the Coast batholith in the Ketchikan and Skagway areas and in Canada.

The only area of plutons having SIR of 0.705 to 0.708 that is probably not underlain by flysch is in the central part of the Yukon-Koyukuk basin. The intermediate SIR values in that region probably represent a transition from low-SIR calc-alkalic plutons of eastern Yukon-Koyukuk basin to alkalic high-SIR plutons of western Yukon-Koyukuk basin. The eastern plutons are probably related to melting of oceanic or island-arc-derived rocks, whereas the western Yukon-Koyukuk plutons may be generated in old continental lithosphere. Rocks in the central zone may reflect mixtures of these sources and may be underlain by deep crust and mantle of continental type (Cratonia) and deep crust of oceanic or island-arc type (Volcania).

The subdivision of the deep crust of Alaska into these three types leads to some interesting thoughts about the myriad tectonostratigraphic terranes that have been proposed in the upper crust (Plate 3, this volume). Most of the lower-crustal source rocks inferred from the isotopic and petrologic data are of the

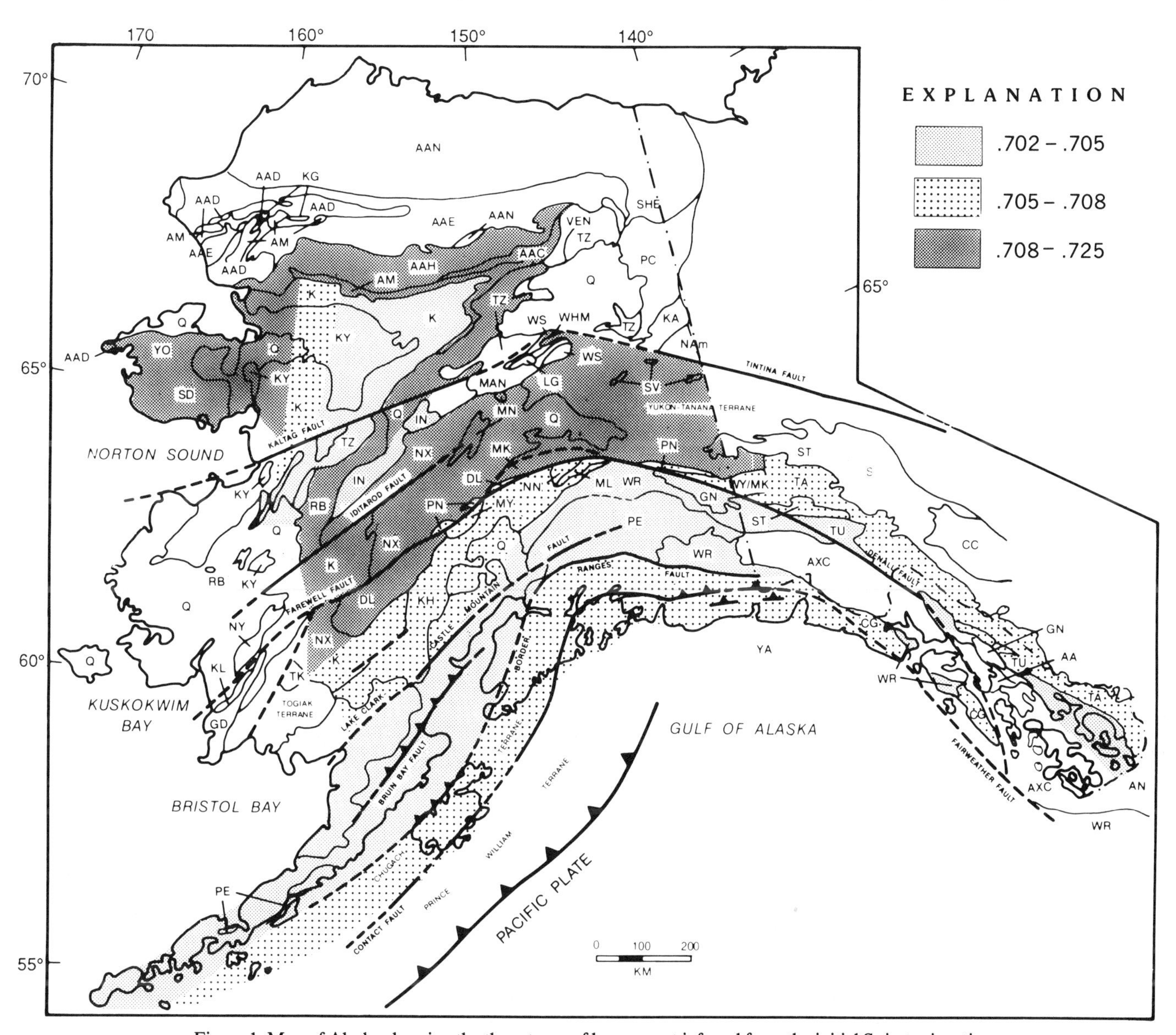

Figure 1. Map of Alaska showing the three types of lower crust inferred from the initial Sr isotopic ratios (SIR) and other geochemical data on the Jurassic to Tertiary igneous rocks. Areas shown in dark stipple have SIR of more than 0.708. They are underlain by Precambrian to Paleozoic continental lithosphere and are here called Cratonia. Areas shown in light stipple have SIR of 0.702 to 0.705 and are underlain by island-arc to oceanic lower crust, here called Volcania. Areas shown in a dot pattern have SIR of 0.705 to 0.708 and are underlain by infolded flysch belts, here called Detritia (except in the Yukon-Koyukuk basin where they are underlain by a mixture of Volcania and Cratonia). The major faults and tectonostratigraphic terranes are shown for reference. Many of the terranes are designated by abbreviations, including: Alexander—AXC, Angayucham—AM, Koyukuk—KY, Nixon Fork—NX, Peninsular—PE, Ruby—RB, Seward—SD, Taku—TU, Tracy Arm—TA, Wrangellia—WR, and York—YO.

same composition as the rock types that dominate the overlying, upper crustal terranes. The upper crust, thus, may be genetically linked to the lower crust in most cases, and I speculate that thrusting on a great scale has not decoupled most upper crustal terranes from their original substrate of lower crust.

On the other hand, some tectonostratigraphic terranes overlie lower crust of different character. For example, the Angayucham terrane is oceanic in character, but part of it lies along the northwestern flank of the Ruby geanticline, in which old cratonic rocks are exposed. An isolated granite body intrudes the Angayuchum terrane near the Ray Mountains, and has a high SIR, indicating the presence of cratonic rocks below the Angayucham rocks. Statewide, however, the surface area occupied by upper crustal terranes that are different from their basement appears to be quite small.

Finally, I note that many proposed tectonostratigraphic terranes have neighbors of similar type (cratonic, volcanic, or sedimentary), and these form coherent belts. These belts overlie large areas of basement that appear to be of matching type (Cratonia, Volcania, or Detritia). Thus, there may be much more genetic coherence in Alaskan crust than is implied by the myriad of proposed tectonostratigraphic terranes.

REFERENCES CITED

Aleinikoff, J. N., 1984, Age and origin of metaigneous rocks from terranes north and south of the Denali fault, Mt. Hayes Quadrangle, east-central Alaska: Geological Society of America Abstracts with Programs, v. 16, p. 266.

Armstrong, R. L., 1985, Rb-Sr dating of the Bokan Mountain granite complex and its country rocks: Canadian Journal Earth Sciences, v. 22, p. 1233–1236.

Arth, J. G., 1985, Neodymium and strontium isotopic composition of Cretaceous calc-alkaline plutons of the Yukon-Koyukuk Basin, Ruby geanticline, and Seward Peninsula, Alaska: EOS Transactions of the American Geophysical Union, v. 66, p. 1102.

—— , 1987, Regional isotopic variations in the Cretaceous plutons of northern Alaska: Geological Society of America Abstracts with Programs, v. 19, p. 355.

Arth, J. G., and 5 others, 1984, Crustal composition beneath the Yukon-Koyukuk basin and Ruby geanticline as reflected in the isotopic composition of Cretaceous plutons: Geological Society of America Abstracts with Programs, v. 16, p. 267.

Arth, J. G., Barker, F., and Stern, T. W., 1988, The Coast batholith near Ketchikan, Alaska; Petrography, geochronology, geochemistry, and isotopic character: American Journal of Science, v. 288A, p. 461–489.

Arth, J. G., and 5 others, 1989a, Isotopic and trace element variations in the Ruby batholith, Alaska, and the nature of the deep crust beneath the Ruby and Angayucham terranes: Journal of Geophysical Research, v. 94, p. 15941–15955.

Arth, J. G., and 5 others, 1989b, Remarkable isotopic and trace element trends in potassic through sodic Cretaceous plutons of the Yukon-Koyukuk basin, Alaska, and the nature of the lithosphere beneath the Koyukuk terrane: Journal of Geophysical Research, v. 94, p. 15957–15968.

Barker, F., and Arth, J. G., 1984, Preliminary results, central gneiss complex of the Coast Range batholith, southeastern Alaska; The roots of a high K, calc-alkaline arc?: Physics of the Earth and Planetary Interiors, v. 35, p. 191–198.

Barker, F., Arth, J. G., and Stern, T. W., 1986, Evolution of the Coast batholith along the Skagway Traverse, Alaska and British Columbia: American Mineralogist, v. 71, p. 632–643.

Blum, J. D., 1985, A petrologic and Rb-Sr isotopic study of intrusive rocks near Fairbanks, Alaska: Canadian Journal of Earth Sciences, v. 22, p. 1314–1321.

Blum, J. D., Blum, A. E., Davis, T. E., and Dillon, J. T., 1987, Petrology of cogenetic silica-saturated and -oversaturated plutonic rocks in the Ruby geanticline of north-central Alaska: Canadian Journal of Earth Sciences, v. 24, p. 159–169.

Brown, L., Klein, J., Middleton, R., Sacka, I. S., and Tera, F., 1982, ^{10}Be in island-arc volcanoes and implications for subduction: Nature, v. 299, p. 718–720.

Coney, P. J., Jones, D. L., and Monger, J.W.H., 1980, Cordilleran suspect terranes: Nature, v. 288, p. 329–333.

Hawkesworth, C. J., 1982, Isotope characteristics of magmas erupted along destructive plate margins, *in* Thorpe, R. S., ed., Andesites: New York, John Wiley and Sons, p. 549–571.

Hill, M., Morris, J., and Whelan, J., 1981, Hybrid granodiorites intruding the accretionary prism, Kodiak, Shumagin, and Sanak Islands, southwest Alaska: Journal of Geophysical Research, v. 86, p. 10569–10590.

Hudson, T., and Arth, J. G., 1983, Tin granites of Seward Peninsula, Alaska: Geological Society of America Bulletin, v. 94, p. 768–790.

Hudson, T., Plafker, G., and Peterman, Z. E., 1979, Paleogene anatexis along the Gulf of Alaska margin: Geology, v. 7, p. 573–577.

Hudson, T., Arth, J. G., and Muth, K. G., 1981, Geochemistry of intrusive rocks associated with molybdenite deposits, Ketchikan Quadrangle, southeastern Alaska: Economic Geology, v. 76, p. 1225–1232.

Hudson, T., Arth, J. G., Winkler, G. W., and Stern, T. W., 1985, Jurassic plutonism along the Gulf of Alaska: Geological Society of America Abstracts with Programs, v. 17, p. 362.

Kay, R. W., 1978, Aleutian magnesian andesites; Melts from subducted Pacific Ocean crust: Journal of Volcanology and Geothermal Research, v. 4, p. 117–132.

Kay, R. W., Sun, S. S., and Lee-Hu, C. N., 1978, Pb and Sr isotopes in volcanic rocks from the Aleutian Islands and Pribilof Islands, Alaska: Geochimica et Cosmochimica Acta, v. 42, p. 263–273.

Kistler, R. W., and Peterman, Z. E., 1973, Variations in Sr, Rb, K, Na, and initial ^{87}Sr/^{86}Sr in Mesozoic granitic rocks and intruded wall rocks in central California: Geological Society of America Bulletin, v. 84, p. 3489–3512.

Kosco, D. G., 1981, The Mount Edgecumbe volcanic field, Alaska; An example of tholeiitic and calc-alkaline volcanism: Journal of Geology, v. 89, p. 459–477.

Lanphere, M. A., 1968, Sr-Rb-K isotopic relationships in ultramafic rocks, southeastern Alaska: Earth and Planetary Science Letters, v. 4, p. 185–190.

Lanphere, M. A., and Reed, B. L., 1985, The McKinley sequence of granitic rocks: A key element in the accretionary history of southern Alaska: Journal of Geophysical Research, v. 90, p. 11413–11430.

Lanphere, M. A., MacKevett, E. M., Jr., and Stern, T. W., 1964, Potassium-argon and lead-alpha ages of plutonic rocks, Bokan Mountain area, Alaska: Science, v. 145, p. 705–707.

Margaritz, M., and Taylor, H. P., Jr., 1976a, Isotopic evidence for meteoric-hydrothermal alteration of plutonic igneous rocks in Yakutat Bay and Skagway areas, Alaska: Earth and Planetary Science Letters, v. 30, p. 179–190.

—— , 1976b, Oxygen, hydrogen, and carbon isotope studies on the Franciscan Formation, Coast Ranges, California: Geochimica et Cosmochimica Acta, v. 40, p. 215–234.

Mark, R. K., 1972, Strontium isotopic study of basalts from Nunivak Island, Alaska: Geological Society of America Abstracts with Programs, v. 4, p. 194–195.

McCulloch, M. T., and Perfit, M. R., 1981, ^{143}Nd/^{144}Nd, ^{87}Sr/^{86}Sr, and trace element constraints on the petrogenesis of Aleutian island arc magmas: Earth and Planetary Science Letters, v. 56, p. 167–179.

Menzies, M., and Murthy, V. R., 1980a, Mantle metasomatism as a precursor to the genesis of alkaline magmas; Isotopic evidence: American Journal of Science, v. 280A, p. 622–638.

——, 1980b, Nd and Sr isotope geochemistry of hydrous mantle nodules and their host alkali basalts; Implications for local heterogeneities in metasomatically veined mantle: Earth and Planetary Science Letters, v. 46, p. 323–334.

Miller, T. P., 1989, Contrasting plutonic rock suites of the Yukon-Koyukuk basin and the Ruby Geanticline, Alaska: Journal of Geophysical Research, v. 94, p. 15969–15987.

Moll-Stalcup, E. J., and Arth, J. G., 1989, The nature of the crust in the Yukon-Koyukuk province as inferred from the chemical and isotopic composition of five Late Cretaceous to early Tertiary volcanic fields in western Alaska: Journal of Geophysical Research, v. 94, p. 15989–16020.

Morris, J. D., and Hart, S. R., 1983, Isotopic and incompatible element constraints on the genesis of island arc volcanics from Cold Bay and Amak Island, Aleutians, and implications for mantle structure: Geochimica et Cosmochimica Acta, v. 47, p. 2015–2030.

——, 1986, Reply *to* Comment *on* 'Isotopic and incompatible element constraints on the genesis of island-arc volcanics from Cold Bay and Amak Island, Aleutians, and implications for mantle structure': Geochimica et Cosmochimica Acta, v. 50, p. 483–487.

Myers, J. D., and Marsh, B. D., 1981, Geology and petrogenesis of the Edgecumbe volcanic field, SE Alaska; The interaction of basalt and sialic crust: Contributions to Mineralogy and Petrology, v. 77, p. 272–287.

Myers, J. D., and Sinha, A. K., 1985, A detailed Pb isotopic study of crustal contamination/assimilation; The Edgecumbe volcanic field, SE Alaska: Geochimica et Cosmochimica Acta, v. 49, p. 1343–1355.

Myers, J. D., Sinha, A. K., and Marsh, B. D., 1984, Assimilation of crustal material by basaltic magma; Strontium isotopic and trace element data from the Edgecumbe volcanic field, SE Alaska: Journal of Petrology, v. 25, p. 1–26.

Myers, J. D., Marsh, B. D., and Sinha, A. K., 1985, Strontium isotopic and selected trace element variations between two Aleutian volcanic centers (Adak and Atka); Implications for the development of arc volcanic plumbing systems: Contributions to Mineralogy and Petrology, v. 91, p. 221–234.

Patton, W. W., Jr., and Box, S. E., 1989, Tectonic setting of the Yukon-Koyukuk basin and its borderlands, western Alaska: Journal of Geophysical Research, v. 94, p. 15807–15820.

Perfit, M. R., and Kay, R. W., 1986, Comment *on* 'Isotopic and incompatible element constraints on the genesis of island arc volcanics from Cold Bay and Amak Island, Aleutians, and implications for mantle structure': Geochimica et Cosmochimica Acta, v. 50, p. 477–481.

Perfit, M. R., and Lawrence, J. R., 1979, Oxygen isotopic evidence for meteoric water interaction with the Captains Bay pluton, Aleutian Islands: Earth and Planetary Science Letters, v. 45, p. 16–22.

Perfit, M. R., Brueckner, H., Lawrence, J. R., and Kay, R. W., 1980, Trace element and isotopic variations in a zoned pluton and associated volcanic rocks, Unalaska Island, Alaska; A model for fractionation in the Aleutian calcalkaline suite: Contributions to Mineralogy and Petrology, v. 73, p. 69–87.

Reed, B. L., and Lanphere, M. A., 1973, Alaska-Aleutian Range batholith; Geochronology, chemistry, and relation to Circum-Pacific plutonism: Geological Society of America Bulletin, v. 84, p. 2583–2610.

Reed, B. L., Miesch, A. T., and Lanphere, M. A., 1983, Plutonic rocks of Jurassic age in the Alaska-Aleutian Range batholith; Chemical variations and polarity: Geological Society of America Bulletin, v. 94, p. 1232–1240.

Roden, M. F., Frey, F. A., and Francis, D. M., 1984, An example of consequent mantle metasomatism in peridotite inclusions from Nunivak Island, Alaska: Journal of Petrology, v. 25, p. 546–577.

Taylor, H. P., 1968, The oxygen isotope geochemistry of igneous rocks: Contributions to Mineralogy and Petrology, v. 19, p. 1–71.

Taylor, H. P., and Noble, J. A., 1969, Origin of magnetite in the zoned ultramafic complexes of southeastern Alaska, *in* Wilson, H.D.B., ed., Magmatic ore deposits; A Symposium: Economic Geology Publishing Co., p. 209–230.

Tera, F., Brown, L., Morris, J., and Sacks, S., 1986, Sediment incorporation in island-arc magmas; Inference from ^{10}Be: Geochimica et Cosmochimica Acta, v. 50, p. 535–550.

von Drach, V., Marsh, B. D., and Wasserburg, G. J., 1986, Nd and Sr isotopes in the Aleutians; Multicomponent parenthood of island-arc magmas: Contributions to Mineralogy and Petrology, v. 92, p. 13–34.

White, W. M., and Patchett, J., 1984, Hf-Nd-Sr isotopes and incompatible element abundances in island arcs; Implications for magma origins and crust-mantle evolution: Earth and Planetary Science Letters, v. 67, p. 167–185.

Zartman, R. E., and Tera, F., 1973, Lead concentration and isotopic composition in five peridotite inclusions of probable mantle origin: Earth and Planetary Science Letters, v. 20, p. 54–66.

Manuscript Submitted January 1987

Manuscript Accepted by the Society October 25, 1990

Printed in U.S.A.

The Geology of North America
Vol. G-1, The Geology of Alaska
The Geological Society of America, 1994

Chapter 26

Paleomagnetic data from Alaska

John W. Hillhouse
U.S. Geological Survey, 345 Middlefield Road, Menlo Park, California 94025
Robert S. Coe
Earth Sciences Department, University of California, Santa Cruz, California 95064

INTRODUCTION

During the past decade, the study of paleomagnetism has provided compelling evidence for the displacement and accretion of Alaskan terranes. As indicated by paleomagnetic measurements of ancient latitudes, large areas of crust that now form part of Alaska were once located at lower latitudes with respect to the North American craton. Triassic volcanic rocks, for example, in the Wrangellia terrane of southern Alaska show a poleward shift in latitude greater than 27° (3,000 km). The large displacements that we infer from paleomagnetic data are consistent with the concept that most of Alaska is made up of displaced lithotectonic terranes, which Silberling and others (this volume) defined on the basis of paleontologic and stratigraphic differences. Paleomagnetic studies provide the basic data for measuring the rates of movement and for determining the arrival times of far-travelled terranes along the continental margin. Using this information, we can refine plate-tectonic models by comparing terrane movements with predicted relative motions between North America and the oceanic plates of the Pacific basin.

Paleomagnetic poles from the stable part of North America define a frame of reference for comparison with the paleolatitudes of Cordilleran terranes. This frame of reference, sometimes called the apparent polar wander (APW) path, has been compiled from numerous studies undertaken during the past three decades (McElhinny, 1973, p. 201–206; Irving, 1979; Irving and Irving, 1982). By 1970, the confirmation of sea-floor spreading and the refinement of cratonic reference poles strengthened the interpretation that paleomagnetic anomalies in the Cordillera were the result of translations and rotations of large-scale crustal blocks. In one of the early studies of paleomagnetism in British Columbia, Irving and Yole (1972) presented evidence for the northward movement of Vancouver Island, and proposed that the island is a remnant of a now-subducted oceanic plate. At about the same time, Packer and Stone (1972) presented paleomagnetic evidence for the northward movement of southern Alaska, and suggested that it was a continental fragment from the southwestern margin of North America. Since 1975, the pace of paleomagnetic research in Alaska has accelerated, and we are developing a clearer picture of the accretionary history of the region.

In this chapter, we review paleomagnetic evidence regarding accretionary tectonics in Alaska, paying particular attention to quality of the data according to reliability criteria, such as the fold test. We summarize the major tectonic events that are inferred from the paleomagnetic data, and then propose a plate-tectonic model for Alaska by integrating the displacement histories of terranes and of oceanic plates within the Pacific basin.

PALEOMAGNETIC METHODS

The central assumption of paleomagnetic methodology is that the geomagnetic field, when averaged over a sufficient interval of time, has a geocentric dipole source that is aligned with the Earth's rotational axis. When these conditions are met at a given site, the mean geomagnetic latitude equals the geographic latitude. Geomagnetic secular variation, which is caused by wobble of the dipole axis and by changes in the non-dipole component of the field, is roughly symmetrical about the rotational axis when sampled over a time interval of several thousand years. Sufficient time averaging of the geomagnetic signal is therefore an important requirement of the geocentric axial dipole (GAD) assumption and of paleolatitude determinations. The GAD assumption has been shown to be true to within a few degrees by worldwide compilations of paleomagnetic data from volcanic rocks younger than 5 Ma (McElhinny and Merrill, 1975). The assumption of a dipole field also holds for high-latitude sites in Alaska, as shown by a study of young basalts on Nunivak and the Pribilof Islands (Cox and Gordon, 1984).

As summarized by Merrill and McElhinny (1983), much research has been directed toward testing the validity of the GAD assumption for times prior to 5 m.y. ago. Late Mesozoic continental reconstructions of Pangea, which are independent of paleomagnetic pole determinations, confirm that the geomagnetic field had a geocentric dipole source since Jurassic time. This fact is demonstrated by the general agreement of continental APW paths when spreading in the Atlantic and Indian Oceans is back-tracked according to the pattern of magnetic stripes on the sea

Hillhouse, J. W., and Coe, R. S., 1994, Paleomagnetic data from Alaska, *in* Plafker, G., and Berg, H. C., eds., The Geology of Alaska: Boulder, Colorado, Geological Society of America, The Geology of North America, v. G-1.

floor. However, proving that the dipole source has always been aligned with the rotation axis requires independent evidence of ancient latitude, such as the distribution of coral, limestone, and other paleoenvironmental indicators. In studies of Proterozoic and younger rocks, these indicators generally form a more consistent pattern when plotted in terms of paleomagnetically derived latitudes as opposed to their present positions. This observation supports the GAD assumption, but lacks the resolution to prove the assumption conclusively.

Another approach to the GAD question uses oceanic hot spots (the Hawaiian Islands, for example) and the corresponding seamount chains to establish an absolute frame of reference for the lithosphere (Morgan, 1972; Duncan, 1981). Under the assumption that the hot spots have remained fixed relative to each other, the relative position of the rotation axis can be determined for each plate for periods back to about 150 m.y. ago. Then, for each plate, a comparison can be made between the position of the "hot spot" rotational pole and the time-averaged paleomagnetic pole (Jurdy, 1981; Harrison and Lindh, 1982; Gordon, 1983; Andrews, 1985). This comparison shows that the paleomagnetic pole has deviated as much as 20° from the "hot spot" reference pole during the Early Jurassic; the deviation decreased sharply during the early Tertiary. Any long-term separation of the two poles could be explained either by a shift of the entire lithosphere relative to the rotational axis (true polar wander), which would not necessarily violate the GAD assumption, or by prolonged wandering of the magnetic dipole axis away from the rotational axis. True polar wander can occur as a consequence of changes in the principal moment of inertia of the rotating, viscous mass of the Earth (Goldreich and Toomre, 1969). Long-term wandering of the dipole is less likely, given the positive paleoclimatic evidence and the known stability of the GAD field during the last 5 m.y.

The basic quantity that is measured in the paleomagnetic laboratory is the magnetic field direction of a rock, a vector specified by its inclination (plunge from horizontal) and declination (azimuth from true north). Given the assumption of a GAD field, the inclination (I) determines the paleolatitude (L) by the simple dipole formula: $2 \tan L = \tan I$. For bedded rocks, the magnetic inclination is easily determined relative to the ancient horizontal, so tilt of the rocks poses no problem in paleolatitude calculations. Magnetic field vectors are corrected for tilt of the beds by simply rotating the vector about the strike by the angle of dip. Tectonic rotations about a vertical axis are inferred from the paleomagnetic declination, when it is compared to the appropriate reference value. The simple tilt correction sometimes introduces errors in declination, especially in complexly deformed regions where there may be localized rotations about the vertical axis, oblique slip on faults, or multiple episodes of folding. Paleolatitudes, which are not affected by local structures, are therefore emphasized in the analysis of terrane movements.

The paleomagnetic pole is calculated from the mean paleomagnetic direction by use of the axial dipole formula. The inclination sets the angular distance (P) from the sampling site to the pole by the equation: $\cot P = \frac{1}{2} \tan I$, and the declination is the azimuth to the pole. Poles are particularly useful for combining paleomagnetic data from a large region, because the calculation corrects for the dipole field variation between distant sites. At a given site, the angular distance to the paleomagnetic pole of the site can be compared directly with the distance to the appropriate cratonic reference pole; a difference means that the site has moved in latitude relative to the craton. For simplicity a displacement to higher latitudes is termed "northward drift," although "poleward drift" is more correct because "north" changes as the craton moves.

Paleolatitudes are specified uniquely, whereas paleolongitudes are not. Consequently, the original position of a displaced terrane cannot be specified exactly by a single paleomagnetic pole, and displacements in latitude are only the minimum displacement that may have taken place. Also, tectonic rotations are not uniquely specified by a paleomagnetic pole unless the axis of rotation is constrained by independent geologic evidence.

Measuring the remanent magnetism of rocks will yield a valid paleomagnetic pole and paleolatitude if the following criteria are met: (1) the magnetization was acquired during or soon after deposition, (2) a sufficient number of samples were collected to establish a statistically valid magnetic direction, and (3) the collection represents thousands of years of geomagnetic secular variation to yield a geocentric axial dipole direction. Hence, the sampling scheme must be designed to meet these time and statistical constraints before the data can chart reliably the movements of terranes.

Metamorphism, chemical alteration, acquisition of viscous remanent magnetization, and lightning are the main sources of secondary magnetizations that mask the original magnetic signals in rocks. Practically all specimens, especially those from orogenic belts, must be treated in the laboratory to strip away magnetic overprints to reveal the underlying primary component. The common demagnetization techniques employ alternating fields (AF) or heating followed by cooling in a field-free furnace (thermal demagnetization). The change in magnetic direction during such treatments is analyzed either graphically on an orthogonal vector diagram (Zijderveld, 1967) or computationally by line-fitting methods (Kirschvink, 1980). These methods of magnetic vector analysis reveal the components that make up the total magnetization and ensure that a stable magnetic direction has been isolated in each rock specimen.

Once the stable component is identified, several tests help to ascertain whether or not the magnetization was acquired at the time of deposition. The fold test establishes the necessary condition that the magnetization pre-dates any folding of the strata. In applying the fold test, a comparison is made of the angular dispersion of magnetic directions before and after corrections are made for the tilt of the bedding, preferably from the opposite limbs of a single fold. The test is positive when application of the bedding corrections significantly decreases the angular dispersion of directions. Another useful test is termed the "consistency of reversals," which compares the mean directions of normal and reversed polarity specimens. If the two means are 180° apart, at least within the confidence limits of the means, the test is positive. A positive result reduces the probability that the data contain a secondary component that might have survived the demagnetization treatments. The "conglomerate test" can be applied when cobbles of the sampled strata occur in an overlying conglomerate bed. If the lower bed gives a coherent magnetic direction while the cobbles give a random distribution of directions, then the possibility of complete remagnetization of the strata can be ruled out. None of these tests prove that the age of magnetization equals the age of the rock; however, when all possible tests are positive, they virtually eliminate the possibility of an undetected secondary component due to heating or alteration.

Paleomagnetic vectors, given either as field directions or as poles, are usually assigned a confidence limit, the α_{95}, which encircles the mean vector. In practical terms this statistic means that the true mean of all vectors in the population will lie within the confidence circle of the sample population with a 95 percent probability. Because the α_{95} is a type of standard error, its radius decreases as the number of specimens increases; enough samples therefore must be collected to obtain a reasonably small confidence limit. The number of samples must also meet the criterion that the collection represents enough time to average out geomagnetic secular variation and thereby yields a geocentric axial dipole direction. Studies of the recent geomagnetic field indicate 2,000 to 10,000 yr as reasonable intervals for obtaining the dipole direction. Estimating the time span of a particular collection is very difficult, although inferences can be drawn from biostratigraphy, the presence of paleosols, the occurrence of magnetic reversals (a polarity transition requires about 1,000 yr), or isotopic dating. As a rough rule of thumb, 15 sites distributed over a reasonable stratigraphic interval is the minimum number to satisfy the statistical and time-representative requirements of a reliable paleomagnetic pole. The vector from each site should be averaged from 2 or more independently oriented specimens. Ideally, the sites should be distributed over a sizable geographic area that provides opportunity for a fold test.

PALEOMAGNETISM OF ALASKAN TERRANES

The following discussion of paleomagnetic results from the Alaskan terranes is keyed to Figure 1, which shows the distribution of paleomagnetic studies in pre–Late Cretaceous rocks, and to Figure 2, which shows results from Late Cretaceous and younger rocks. Hillhouse (1987) published tables of data that are depicted in Figure 1. Our discussion emphasizes the reliability of the results, as interpreted from the quality tests described above. Unless otherwise noted, locations of geologic, physiographic, and tectonostratigraphic features referred to in this chapter are respectively shown on Plates 1 (Beikman), 2 (Wahrhaftig), and 3 (Silberling and others) in this volume (also see Plafker and Berg, this volume, Chapter 33).

The Peninsular and Wrangellia terranes

Packer and Stone (1972, 1974) published the first paleomagnetic evidence that parts of Alaska have moved to higher latitudes relative to the North American craton. The original intent of their study was to use paleomagnetic poles from Jurassic rocks of the Alaska Peninsula to test for oroclinal bending of Alaska, as postulated by Carey (1955) and others. Packer and Stone collected their samples from Jurassic sedimentary rocks of the Naknek Formation, the Tuxedni Group, the Chinitna Formation, and from Jurassic quartz diorites on the western limb of the orocline (Burk, 1965). The surprising result was a paleolatitude for the Alaska Peninsula that was 19° lower than would be predicted for the region from the North American APW curve. Packer and Stone concluded that southern Alaska had drifted northward from an original position near the Oregon coast, and proposed a model similar to the present rifting and northward motion of Baja California. The "Baja Alaska" model was refined by the addition of Cretaceous and early Tertiary paleomagnetic results from the Alaska Peninsula (Stone and Packer, 1977, 1979). The Cretaceous poles fell into two groups, one defined by sites in the Upper Cretaceous Chignik Formation at Wide Bay and Chignik Lake, the other defined by sites in the Chignik at Herendeen Bay and the Shumagin Islands and in the Upper Cretaceous Hoodoo Formation at Canoe Bay. Paleolatitudes defined by these Cretaceous poles indicated that southern Alaska had moved approximately 25° south from its Jurassic position before beginning its rapid northward journey. Poles from the lower Tertiary Tolstoi Formation at Pavlof and Canoe Bays placed the Alaskan block near the latitude of Vancouver Island, requiring rapid northward drift of the block after 50 Ma.

These astonishing estimates of large-scale drift were attended by large uncertainties, not only in the error bars of the paleolatitude determinations but also in the overall reliability of the Peninsular terrane sedimentary rocks as clean recorders of the ancient geomagnetic field. For example, the generally mild deformation of the Naknek Formation precluded successful fold tests, because the angular dispersions of magnetic directions were not changed significantly by application of tilt corrections. All Jurassic results used in Packer and Stone's analysis were of one polarity, so they could not apply the test for secondary components that employs the antiparallelism of normal and reversed polarity magnetizations. They observed reversals in the Cretaceous and Tertiary rocks of the terrane, but the "consistency of reversals" test was not conclusive. Packer and Stone's experimental work did not demonstrate with certainty that the Mesozoic sedimentary rocks were free of secondary magnetic components, which might have been added to the primary remanence by heating during the extensive volcanism of the late Tertiary, or by chemical changes during weathering.

At about the same time as Packer and Stone's (1972) investigations, the concept that Alaska and British Columbia are composed of many far-travelled terranes was gaining acceptance following the stratigraphic studies of Monger and Ross (1971), Jones and others (1972), Monger and others (1972), and Berg and others (1972) (also see Nokleberg and others, this volume, Chapter 10). The recognition by Jones and others (1977) of Wrangellia as a large, displaced terrane was founded on its dis-

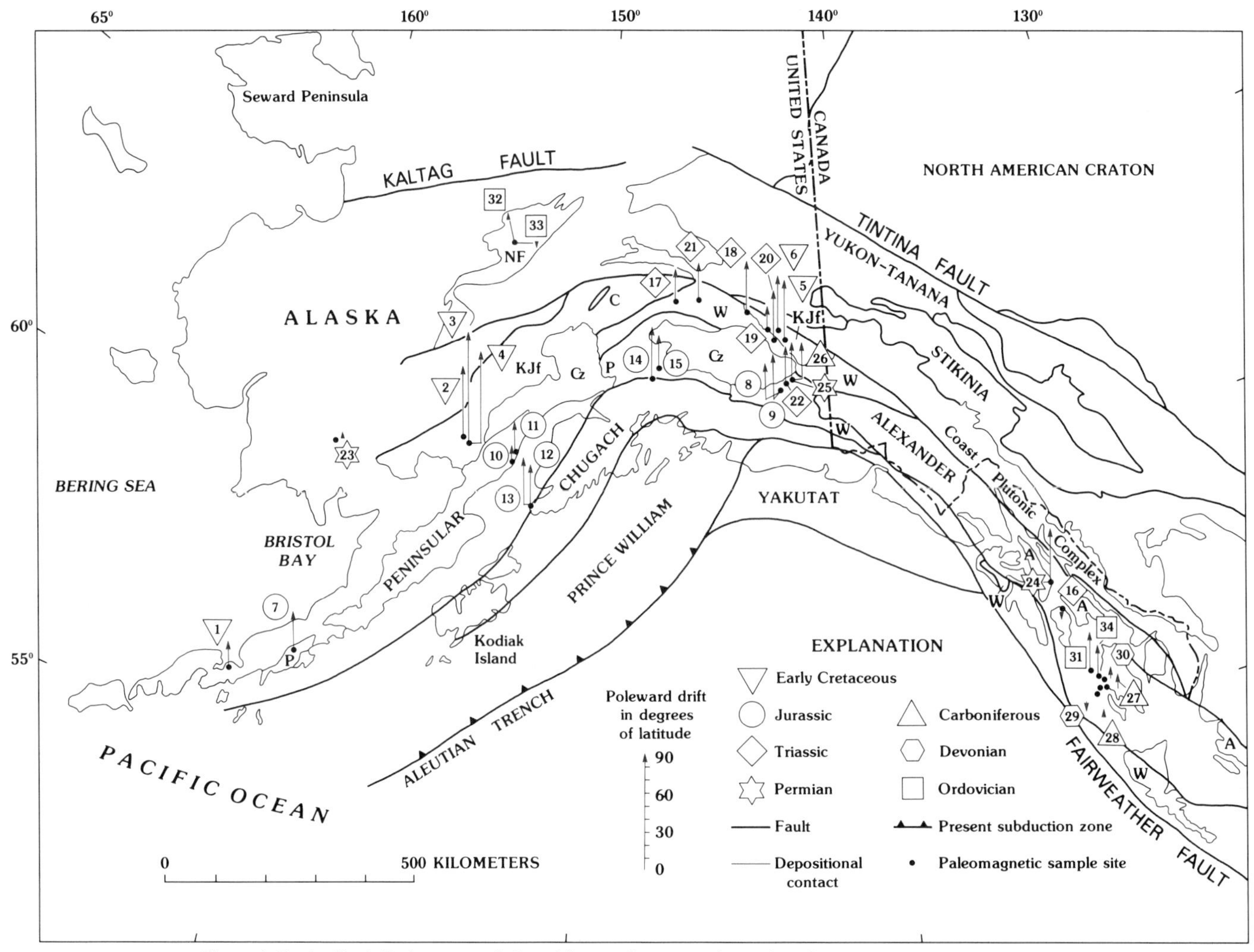

Figure 1. Latitudinal displacements inferred from paleomagnetic studies of pre–Late Cretaceous rocks in southern Alaska. Displacements are in degrees of latitude, assuming terranes originated in the northern hemisphere. Displacements to higher latitudes are indicated by upward arrows; downward arrows indicate displacements to lower latitudes relative to the North American craton. Terrane map modified from Jones and Silberling (1979) and Silberling and others (this volume). Selected terranes: A, Alexander; C, Chulitna; NF, Nixon Fork; P, Peninsular; W, Wrangellia. Sedimentary overlap assemblages: Cz, Cenozoic rocks, undifferentiated; KJf, Upper Jurassic and Lower Cretaceous flysch. See Table 1 of Hillhouse (1987) for references corresponding to the following paleomagnetic study areas: 1, Staniukovich Formation; 2–4, flysch near Lake Clark; 5, flysch near Nutzotin Mountain; 6, flysch near Nabesna; 7, Naknek Formation at Chignik Lake; 8–9, Nizina Mountain Formation; 10–11, Tuxedni Bay; 12–13, Talkeetna Formation at Seldovia Bay; 14–15, Talkeetna Formation in southern Talkeetna Mountains; 16, pillow basalt of Hound Island; 17, Nikolai Greenstone in Clearwater Mountains; 18, Nikolai Greenstone near Mentasta Pass; 19–20, Nikolai Greenstone near Nabesna; 21, Nikolai Greenstone in Mount Hayes Quadrangle; 22, Nikolai Greenstone near McCarthy; 23, basalt at Nuyukuk Lake; 24, Pybus Limestone; 25, Hasen Creek Formation; 26, Station Creek Formation; 27, Ladrones and Klawak Formations; 28, Peratrovich Formation; 29, Port Refugio Formation; 30, Wadleigh Formation; 31, lavas of the Descon Formation; 32, Telsitna Ridge; 33, Novi Mountain; 34, sedimentary rocks of the Descon Formation.

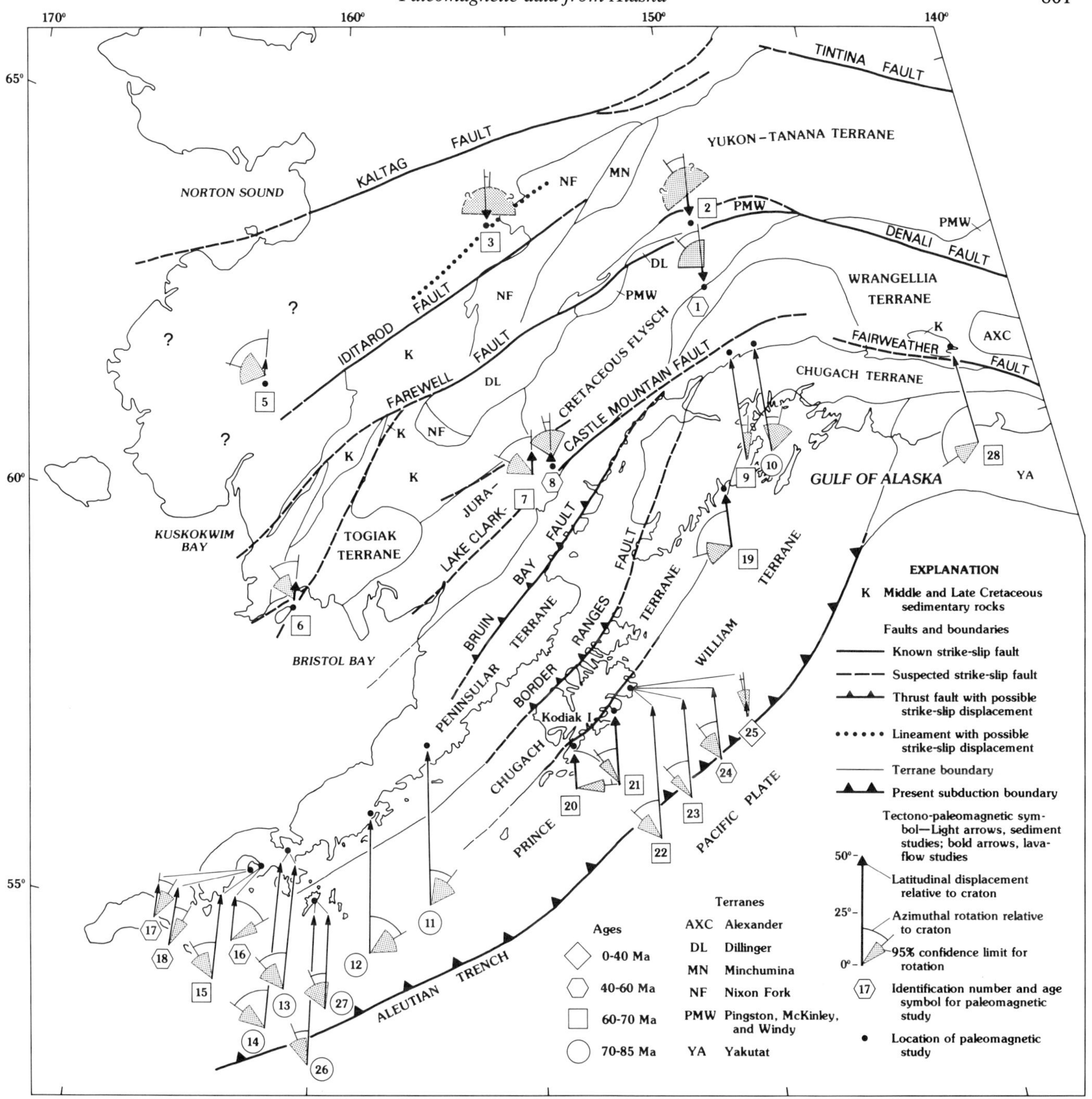

Figure 2. Latitudinal displacement and azimuthal rotation inferred from paleomagnetic studies of Late Cretaceous and younger rocks in Alaska, after Coe and others (1985). Bold arrows indicate results derived from volcanic rocks considered to be more reliable than sedimentary rocks for paleomagnetic study. See Table 2 of Coe and others (1985) for references. Sedimentary overlap assemblage: K, Cretaceous rocks. Identification numbers: 1, unnamed volcanic rocks, northern Talkeetna Mountains; 2, Cantwell Formation (lavas), Denali National Park; 3, Nowitna volcanic rocks, McGrath; 5, tuffaceous rocks, Ohogamiut Village; 6, unnamed volcanic rocks, northern Bristol Bay; 7–8, unnamed volcanic rocks, Lake Clark; 9, Chickaloon Formation, Matanuska Valley; 10, sedimentary rocks, Sheep Mountain; 11, Chignik Formation, Painter Creek; 12, Chignik Formation, Chignik Lagoon; 13–14, Chignik Formation, Herendeen Bay; 15, Hoodoo Formation, Canoe Bay; 16–17, Tolstoi Formation, Pavlof Bay; 18, Tolstoi Formation, Canoe Bay; 19, Valdez(?) Group, Resurrection Peninsula; 20, Ghost Rocks Formation (lavas), Alitak Bay, Kodiak Island; 21, Ghost Rocks Formation (lavas), Kiliuda Bay, Kodiak Island; 22, Kodiak Formation, Kodiak Island; 23, Ghost Rocks Formation (sedimentary rocks), Kodiak Island; 24, Sitkalidak Formation, Kodiak Island; 25, Narrow Cape Formation, Kodiak Island; 26–27, Shumagin Formation, Shumagin Islands; 28, MacColl Ridge Formation, McCarthy.

tinctive stratigraphy and its unique molluscan fauna compared to nearby terranes of similar age. Wrangellia contains a thick sequence of Triassic volcanic rocks that was the target of early paleomagnetic studies, providing clues to its exotic origin. In 1962, Richard R. Doell (U.S. Geological Survey) sampled the Triassic Nikolai Greenstone near McCarthy in the Wrangell Mountains; preliminary determinations gave magnetic directions that indicated anomalously low paleolatitudes, but this work was not pursued further for more than a decade. Before the terrane was recognized in British Columbia, Symons (1971) sampled the Triassic Karmutsen Formation on Vancouver Island and proposed northward displacement of the island as one of several possible explanations for the formation's unusual paleomagnetic pole. After careful thermal demagnetization experiments, Irving and Yole (1972) separated the Karmutsen magnetizations into a primary cooling component and an overprint, and postulated the northward displacement of Vancouver Island.

In 1976, the growing interest in the possible displacement of Wrangellia prompted more magnetic measurements of Doell's Nikolai collection (Hillhouse, 1977). The results were consistent with the low paleolatitudes that Canadian geologists had obtained from correlative volcanic rocks on Vancouver Island. At McCarthy, the paleolatitude derived from 50 lava flows was at least 27° south of the predicted Triassic latitude for the region, indicating that the terrane had moved 3,000 km north relative to continental North America. The magnetization of the Nikolai Greenstone was acquired before the rocks were folded, as indicated by a successful fold test, and the removal of secondary components was assured by the antiparallelism of normal and reversed directions. The magnetization had a high stability to thermal disturbances, as shown by thermal demagnetization experiments to 575°C.

Subsequent research on the Nikolai and its equivalents in Alaska, Oregon, and British Columbia confirmed the anomalously low paleolatitude of Wrangellia, adding credence to the concept of exotic terranes in the Cordillera (Yole and Irving, 1980; Schwarz and others, 1980; Stone, 1982; Hillhouse and others, 1982). In the Mt. Hayes Quadrangle of Alaska (Hillhouse and Grommé, 1984), some 250 km northwest of McCarthy, Triassic basalt of Wrangellia yielded a mean paleolatitude of 14°, similar to the latitude obtained at McCarthy (10°), on Vancouver Island (17°), and in Oregon (18°). All parts of Wrangellia apparently originated within 18° of the Triassic equator, after which fragments of the terrane were scattered along the continental margin from Oregon to Alaska. The fragment of Wrangellia in the Mount Hayes Quadrangle was carried the farthest north, at least 28° ± 6° or approximately 3,100 ± 660 km.

If the polarity of the dipole field is not known, it is difficult to determine whether a paleomagnetically-derived latitude should be assigned north or south of the paleoequator. This ambiguity arises for far-traveled terranes, because the paleodeclinations are no longer diagnostic after the terranes have undergone large vertical-axis rotations. Such is the case for Wrangellia, where the Nikolai basalts were magnetized mainly in a field characterized by upward inclinations during a polarity interval of unknown sign. The options for interpretation are as follows: (1) if the dominant polarity mode is normal, then the terrane originated in the southern hemisphere; (2) if the polarity is reversed, the terrane originated in the northern hemisphere. The displacements described in the foregoing paragraph assume a northern hemisphere origin; if the southern hemisphere option is correct, the possible displacements would be increased by about 30°.

To resolve the ambiguous original position of Wrangellia, Panuska and Stone (1981) investigated the paleomagnetism of Pennsylvanian and Permian volcaniclastic rocks, which were deposited during the long reversed-polarity interval of the Permo-Carboniferous. Their preliminary data showed that magnetic directions of the Paleozoic volcaniclastic rocks are similar to those in the Triassic Nikolai Greenstone, leading them to favor the tropical zone of the northern hemisphere (15°) as the original position of Wrangellia.

Tracking the late Mesozoic displacements of Wrangellia and the Peninsular terrane hinges on the accuracy of Panuska and Stone's conclusion concerning the polarity of the Nikolai paleomagnetic results. In Middle Jurassic and younger Mesozoic time, Wrangellia and the Peninsular terrane were overlapped by a sequence of marine sedimentary rocks, including the Chinitna and Nizina Mountain Formations, whose paleomagnetic poles should be the same regardless of their substrate (Jones and Silberling, 1979). Panuska and Stone (1981) combined the paleomagnetic data from both terranes to define an APW curve for southern Alaska, and chose the polarity of the Peninsular terrane Jurassic rocks to minimize the separation between the Jurassic pole of the Peninsular terrane and the Triassic pole of Wrangellia. Stone and others (1982) used this curve to reevaluate the Peninsular terrane data, arguing that it and Wrangellia moved southward from the northern hemisphere during the Jurassic to a position near 30°S before the final northward movement occurred. We believe that other scenarios that satisfy the APW curve are possible, including one that keeps Wrangellia and the Peninsular terrane in the northern hemisphere throughout the Mesozoic, by allowing northern Wrangellia to rotate relative to the Alaska Peninsula.

Although the exact Mesozoic movements of Wrangellia and the Peninsular terrane remain open to speculation, the paleomagnetic data reveal an overall northward drift until the beginning of the Tertiary. Early Tertiary and Late Cretaceous volcanic rocks consistently yield high paleolatitudes, indicating that Wrangellia and the Peninsular terrane had ceased drifting north relative to the North American continent by 65 to 55 Ma (Fig. 2). Hillhouse and Grommé (1982a) and Hillhouse and others (1985) studied the paleomagnetism of early Tertiary volcanic rocks along a 75-km transect from Denali National Park to the Talkeetna River. The transect crosses northern Wrangellia, deformed Jurassic and Cretaceous flysch, several smaller terranes within the flysch, and the Denali fault. In Denali National Park, Paleocene volcanic rocks of the Cantwell Formation yielded a paleolatitude of 81 ± 8°, which is 9 ± 8° higher than the latitude that would be predicted from cratonic poles. Panuska and Macicak (1986) combined those data with results of additional sampling of the Cantwell volcanic rocks, and determined a paleolatitude of 71 ± 10°. Their revision of the Cantwell paleolatitude brings it into close agreement with the reference paleolatitude for the region. Hillhouse and others (1985) obtained similar results from early Eocene andesites and dacites in the northern Talkeetna Mountains, where they determined a paleolatitude of 76 ± 10°. The reliability of the determinations is supported by successful fold tests, consistency of reversals, and measurements of the magnetization's thermal stability. In the southern Talkeetna Mountains, Eocene volcanic rocks along the southern boundary of the Peninsular terrane yielded a paleolatitude of 80 ± 9° (Panuska and Stone, 1985a).

In a transect from Lake Clark to Iliamna Lake, Coe and others (1985) and Thrupp and Coe (1984, 1986) determined paleolatitudes of early Tertiary volcanic rocks that overlap the Peninsular terrane. Near Lake Clark, one group of lava flows dated at about 66 Ma, gave a paleolatitude of 63 ± 9°, which is 9 ± 11° lower than predicted. Minor deformation of the Lake Clark area precluded a successful fold test, but the consistency of reversals test was met. Near Hagemeister Island in Bristol Bay, Globerman and Coe (1984a, b) determined a paleolatitude similar to the one near Lake Clark. The latest Cretaceous (68 Ma) volcanic rocks near Hagemeister Island yielded a paleolatitude of 65 ± 4° and a latitude anomaly of 9 ± 7°. The islands of northern Bristol Bay are not considered part of the Peninsular terrane, but the nearly identical paleomagnetic poles from coeval volcanic sequences of the two regions indicate that the two areas have not significantly moved relative to each other since 66 Ma.

Small northward displacements approaching the magnitudes of the confidence limits are implied by the Lake Clark and Bristol Bay results, in contrast to the small southward displacement that was determined in the Talkeetna Mountains transect. Because most of the studies yielded α_{95} confidence limits of about 9°, apparent differences in paleolatitude between those areas are probably not significant. In addition, the Cantwell Formation and the volcanic rocks of Lake Clark and the northern Talkeetna Mountains appear to be part of a Late Cretaceous and early Tertiary magmatic complex that overlaps the boundaries between Wrangellia, Jurassic and Cretaceous flysch, and the Peninsular terrane. Taken as a whole, the data indicate no significant change of latitude of southern Alaska relative to North America since 68 to 55 Ma, at least within current confidence limits of about 9°.

A significant counterclockwise rotation of southern Alaska can be inferred from the distribution of Late Cretaceous and early Tertiary paleomagnetic poles (Fig. 2). The declination anomalies vary from 29 to 54° in the counterclockwise sense when the Alaskan poles are compared with the appropriate reference poles (Coe and others, 1985; Hillhouse and others, 1985). Middle Eocene and Oligocene lavas near Lake Clark give no significant anomalies in declination, so the rotation occurred between 66 and 44 Ma (Thrupp and Coe, 1986).

In contrast to the paleomagnetic evidence provided by volcanic rocks, results from Upper Cretaceous (Maastrichtian) and lower Tertiary sedimentary rocks indicate that as late as 75 to 60 Ma the Peninsular terrane and Wrangellia were about 3,000 km south of their present positions relative to the craton. These results were obtained from the MacColl Ridge Formation near McCarthy (Panuska, 1985), the Tolstoi Formation on the western Alaska Peninsula (Stone and Packer, 1979; Stone and others, 1982), and the Chickaloon Formation in the Matanuska Valley (Stone, 1984). No consistent rotation of the region is discernible from the paleomagnetic poles. If, according to Panuska (1985), the major poleward movement of Wrangellia occurred after Maastrichtian time, then emplacement of the terrane by Eocene time would require relative plate velocities greater than 24 cm/yr. As an alternative to this unusually rapid velocity, he proposed that MacColl Ridge is separated from the Eocene volcanic rocks by an unrecognized suture.

In a review of results from Late Cretaceous and early Tertiary strata, Coe and others (1985) discussed the discrepancy between determinations of high paleolatitudes from the volcanic rocks and the unusually low paleolatitudes from the sedimentary rocks. They concluded that the magnetization of Tolstoi Formation is probably contaminated by strong secondary components of magnetization. The Chickaloon and MacColl Ridge Formations gave results that passed the fold test and the reversal test, so the paleomagnetic data appear to be free of magnetic overprints. However, systematically low inclinations are sometimes found in sedimentary rocks due to compaction and settling of the magnetic grains during deposition. This mechanism might explain the very low paleolatitudes from the Chickaloon and MacColl Ridge studies, because the anomalies are similar to inclination errors observed in redeposition experiments with sediments. Because the inclinations measured in Late Cretaceous and early Tertiary volcanic rocks of the Peninsular and Wrangellia terranes are consistently steeper than the inclinations of nearly coeval sedimentary rocks, and volcanic rocks are generally acknowledged as superior recorders of the ambient magnetic field, we are skeptical of the paleolatitude determinations from the Chickaloon and MacColl Ridge Formations.

Western Alaska

The few paleomagnetic studies made in the Kuskokwim Mountains region indicate that Paleozoic rocks of western Alaska have not moved great distances in latitude relative to the craton. Karl and Hoare (1979) reported a paleomagnetic pole from volcanic rocks of possible Permian age near Nuyakuk Lake, 100 km north of Dillingham. They collected a small number of specimens from nine sites in a gently dipping, fault-bounded block of basalt flows that did not permit a fold test. All sites gave field directions of reversed polarity. The Permian age of those volcanic rocks is not supported by direct evidence from fossils, and the rocks might be as young as Late Triassic. However, exclusively reversed field directions are to be expected if the magnetization was acquired during the long reversed-polarity interval of the Early Permian. The resultant paleomagnetic pole is indistinguishable from the Permian pole for stable North America. If the magnetization is actually as young as Late Triassic, then the results indicate a poorly resolved paleolatitude anomaly of 8 ± 9° and a counterclockwise rotation of about 40°. Regardless of the age uncertainty, post-Triassic northward drift of the region is minor compared to the displacements indicated for Wrangellia and the Peninsular terrane.

Upper Paleozoic rocks of the Nixon Fork terrane also reveal no statistically significant displacement in latitude relative to the craton (Plumley, 1984). In the Medfra Quadrangle, Plumley obtained useful paleomagnetic results from the Telsitna Formation (Middle and Upper Ordovician) and the Novi Mountain Formation (Lower Ordovician) after he applied thermal demagnetization treatments. The treatments revealed several components of magnetization: a high-temperature component that showed a slight decrease in angular dispersion when the bedding corrections were applied, and lower-temperature secondary components that were acquired after folding. The high-temperature component probably reflects the magnetic field at the time of deposition and indicates no significant paleolatitude anomaly compared to the expected value as calculated from a small set of cratonic poles. A large clockwise rotation is indicated by the anomalous declination of the mean paleomagnetic direction.

Cretaceous and early Tertiary paleomagnetic poles from western Alaska generally agree with poles from coeval volcanic sequences in the Peninsular terrane, the Jurassic and Cretaceous flysch belt, and Wrangellia. In the Nixon Fork terrane, Plumley's (1984) paleomagnetic study of the Paleocene Nowitna volcanic unit showed no significant displacement of the region, but his collection consisted of samples from only six cooling units. The confidence limit on this Paleocene pole encloses the sampling site, so no conclusion can be drawn regarding possible rotation of the region. Near Ohogamiut Village, about 80 km northwest of Bethel, Lower Cretaceous volcaniclastic rocks were sampled by Globerman and others (1983). The resultant magnetic directions yielded an inconclusive fold test, although application of the tilt corrections slightly increased the angular dispersion of directions. Although Globerman and his coauthors did not explicitly rule out an Early Cretaceous age of magnetization, they favored a Late Cretaceous remagnetization of the Ohogamiut tuffs. If the true age of the overprint is Late Cretaceous, then a paleolatitude anomaly of 7 ± 10° and a substantial counterclockwise rotation are implied (Coe and others, 1985).

From late Paleozoic paleomagnetic poles from western Alaska, we infer that Wrangellia and probably the Peninsular terrane were at latitudes at least 25° lower than those of the western Alaska terranes in Late Triassic time. Across western and south-central Alaska, the general concordance of paleomagnetic poles from Paleocene volcanic rocks indicates that the gap had closed and relative northward drift had ceased by 65 to 55 Ma. Block rotations occurred between the Paleozoic terranes of western Alaska before the Paleocene, as indicated by the discordant magnetic declinations of the Nixon Fork terrane and the Nuyakuk Lake volcanic rocks. After latest Cretaceous time (68 Ma), a consistent counterclockwise rotation of 30 to 50° affected possibly the entire region between the Kaltag fault and the southern margin of the Peninsular terrane.

Chugach and Prince William terranes

The Chugach terrane (Plafker and others, this volume, Chapter 12) makes up a large part of southern Alaska, but only a few pilot studies have been made of the terrane's paleomagnetism. Stone and Packer (1979) reported results from two sites in the Shumagin Formation of probable Late Cretaceous age at Nagai Island. Sedimentary rocks of similar age and character as the Shumagin Formation were sampled at the east shore of Kodiak Island in the Kodiak Formation (Stone and others, 1982). The mean paleolatitudes of the three sites range from 6°N to 32°N, substantially lower than the predicted value of 75°N from the North American reference. When the magnetic directions from these three sites are corrected for tilt of the bedding, the angular dispersion increases, and the spread in paleolatitude between the sites increases as well. Therefore, the magnetizations are probably contaminated by a post-deformational component, and the low paleolatitude determinations might not be valid.

The only other study of Late Cretaceous rocks in the Chugach terrane is on Resurrection Peninsula near Seward where Hillhouse and Grommé (1977) sampled one site in pillow basalt and two sections in the sheeted dike complex of the Valdez(?) Group. Correcting the magnetic directions for tilt of the pillow basalts and for tilt of pillow screens within the dike complex reduced the overall scatter of directions and gave a mean paleolatitude of 51 ± 10°. This value implies about 25° of northward drift relative to the craton since the Late Cretaceous. Grommé and Hillhouse (1981) obtained a paleolatitude anomaly of similar magnitude from a study of the La Perouse and Astrolabe gabbro plutons in the southeastern part of the Chugach terrane. Recent isotopic determinations from gabbro at La Perouse indicate that it is 28 m.y. old (Loney and Himmelberg, 1983), which reduces our confidence in the validity of Grommé and Hillhouse's (1981) paleolatitude determination. To reach its present position, the La Perouse body would have to have been translated northward at twice the Pacific plate rate (Plafker, 1984). Undetected tilt of the gabbro bodies might explain their apparently low paleolatitude values.

The Prince William terrane, which makes up the southern continental margin of Alaska, shows evidence of substantial northward drift in rocks as young as early Tertiary. The most convincing evidence comes from the southern shore of Kodiak Island, where the Paleocene Ghost Rocks Formation of Moore and others (1983) has been studied by Plumley and others (1982, 1983). Pillow basalts of the Ghost Rocks were sampled at Kiliuda and Alitak Bays. The results have high reliability, as indicated by positive fold tests at both sites and a positive reversal test at Kiliuda Bay. The paleomagnetic data indicate that in Paleocene time the Ghost Rocks were 25 ± 7° south of their present position relative to continental North America. Stone and others (1982) obtained larger paleolatitude anomalies from the sedimentary part of the Ghost Rocks Formation (45 ± 9°) and from the Eocene to Oligocene Sitkalidak Formation (32 ± 19°) on Kodiak

Island. They also reported results from the Narrow Cape Formation, showing that northward drift of the terrane has been only 6 ± 8° since Late Miocene time.

The paleomagnetic data from Paleocene and Late Cretaceous volcanic rocks indicate that at that time a latitude gap of about 25° separated the Ghost Rocks and volcanic rocks of the Resurrection Peninsula from the Wrangellia and Peninsular terranes. The gap subsequently might have been closed by strike-slip motion along the Border Ranges fault. Moore and others (1983), however, argued that plutons cross the fault and tie the Chugach and Peninsular terranes together after Paleocene time; if so, the Border Ranges fault is an unlikely mechanism for closing the latitude gap. Alternatively, Coe and others (1985) suggested that the emplacement of the far-travelled volcanic rocks might have occurred by cumulative slip along many faults in southern Alaska. We should stress that the Chugach and Prince William terranes have been sampled for paleomagnetism only in reconnaissance, primarily within fault-bounded igneous rocks near the southern margins of the terranes. Because these terranes are possibly melanges of off-scraped sediment and obducted oceanic crust, the large-scale translations may have involved only a few blocks now embedded within the terranes. This explanation eliminates the problem of requiring a major zone of convergence of Tertiary age between the Chugach terrane and the mainland.

Alexander terrane

The Alexander terrane of southeastern Alaska (Gehrels and Berg, this volume) has a distinctive Paleozoic stratigraphy that sets it apart from neighboring terranes in British Columbia (Berg and others, 1972; Jones and others, 1972). Early speculations regarding possible northward displacement of the Alexander terrane were stimulated by a study of Permian fusilinid faunas by Monger and Ross (1971), who recognized a pattern in which fusilinids of "Tethyan" affinity in the Cache Creek region are sandwiched between belts containing fauna that are more typical of North America. They postulated large-scale northward strike-slip motion of the outer coastal belt, which includes the Alexander terrane, as one of several explanations for the anomalous distribution of fusilinids. As suggested by Hamilton (1969), the presence of Tethyan faunas in western North America could be attributed to the accretion of exotic blocks from the proto-Pacific region. Jones and others (1972) favored strike-slip motion of the Alexander terrane on stratigraphic and lithologic grounds, noting similarities between the Alexander terrane and coeval rocks in northern California.

Paleomagnetic studies by Van der Voo and others (1980) further corroborated the evidence for northward displacement of the Alexander terrane. They sampled a thick section of sedimentary and volcanic rocks ranging in age from Late Ordovician to Late Carboniferous on the west side of Prince of Wales Island. After demagnetization treatments were applied, useful results were obtained from the Descon Formation (Middle and Upper Ordovician), Wadleigh Limestone (Devonian), Port Refugio Formation (Upper Devonian), Peratrovich Formation (Lower Carboniferous), and Ladrones Limestone and Klawak Formation (Upper Carboniferous). Application of tilt corrections significantly reduced the directional scatter of some of the Devonian and Ordovician units, indicating that the magnetization pre-dated folding. The Wadleigh Limestone exhibited consistently antiparallel directions, which passed the reversal test for the complete removal of secondary components. The dominant group of directions, which were inclined shallowly to the east, were chosen to be of reversed polarity.

When compared to coeval reference poles, the Alexander poles indicated movement to higher latitudes of 10 to 15° and about 25° of counterclockwise rotation relative to North America since the Late Carboniferous. The optimal fit of the Alexander poles to the North American apparent polar wander curve was obtained by having the original position of the Alexander terrane near northern California, in close agreement with the hypothesis of Jones and others (1972).

However, a controversy has arisen concerning the reference poles used by Van der Voo and others (1980) in reconstructing the paleoposition of the Alexander terrane. The controversy stems from the difference between late Paleozoic poles from the interior of North America and coeval poles from coastal New England and the Canadian Maritime provinces. Van der Voo and Scotese (1981) and Kent (1982) accounted for the difference by proposing northward drift of the New England block (Acadia) relative to the craton. Other workers argued that the difference in poles is due to Permian remagnetization of the cratonic rocks, because Late Carboniferous and Devonian poles lie close to the Permian pole (Roy, 1982; Roy and Morris, 1983). A recent study by Kent and Opdyke (1985) indicates that the Carboniferous Mauch Chunk Formation of Pennsylvania is contaminated by a Permian component of magnetization, and that when the overprint is removed, the Mauch Chunk paleolatitude is not significantly different from coeval paleolatitudes of Acadia. Irving and Strong (1984, 1985) also found no evidence of an Acadian displacement in their study of Lower Carboniferous rocks in Newfoundland. Revision of the Carboniferous and Devonian poles from the interior of North America is underway, and the paleomagnetic evidence for the displacement of the Alexander terrane must be reconsidered. If the polarity that Van der Voo and others (1980) assigned to the Alexander magnetic directions is correct, then the paleomagnetic evidence favors the origin of the Alexander terrane near its present latitude relative to the craton.

In Keku Strait between Kuiu and Kupreanof Islands, the Alexander terrane is capped by Upper Triassic rocks, including the Hound Island Volcanics. Hillhouse and Grommé (1980) reported paleomagnetic results from the Hound Island Volcanics, a formation consisting of basaltic pillow lava, pillow breccia, and aquagene tuff. Their results show good reliability, as indicated by successful fold and reversal tests. Thermal demagnetization experiments revealed magnetizations having high thermal stability. The paleolatitude of the Hound Island Volcanics does not differ significantly from the predicted latitude on the basis of Triassic and Early Jurassic poles from the interior of North America. However, the Hound Island paleolatitude is significantly higher than the values obtained from coeval strata in the nearby Stikine terrane of British Columbia. Interpretation of paleomagnetic results from Triassic volcanic rocks of the Stikine terrane suggests that the block has moved about 15° northward relative to the craton (Monger and Irving, 1980; Irving and others, 1980). As with the Paleozoic rocks, the validity of the reference poles in the Triassic rocks has been called into question. Gordon and others (1984) and May and Butler (1986) have argued that the apparent displacement of the Stikine terrane is an artifact due to improper positioning of the Late Triassic (Carnian-Norian) reference pole. Nevertheless, additional evidence of northward drift of coastal British Columbia comes from studies by Irving and others (1985) of middle Cretaceous plutons of the southern Coast Plutonic Complex. Their results imply that a composite block, including part of Wrangellia and the Stikine terrane, was displaced about 2,400 km northward and rotated clockwise after 90 Ma. Because the Stikine terrane lies between the craton and the Alexander terrane, the apparent lack of displacement of the Alexander terrane is puzzling.

As with Wrangellia, the polarity of the Alexander paleomagnetic poles cannot be selected with certainty, and it is possible that the terrane has moved from an original position in the southern hemisphere. For this option, normal polarity would be assigned to the shallow eastward directions from the Paleozoic rocks, and reversed polarity would be assigned to the steeply inclined, southwestward directions from the Triassic volcanic rocks. Under the southern option the terrane would move about 50° to the south following deposition of the Devonian Port Refugio Formation and then move rapidly northward to a position 85° higher in latitude in post-Triassic time. The southern option also requires a net rotation of about 160° about a vertical axis to be in accord with the Paleozoic declinations of the Alexander terrane. Moving the Alexander terrane to the southern hemisphere has the advantage of getting the terrane out of the way of the northward-moving Stikine terrane. Compared to the northern hemisphere option, however, the southern option requires much larger displacements and more changes in the direction of transport.

Recently, Panuska and Stone (1985b) obtained evidence in support of the southern hemisphere option from the Pybus Formation, which was deposited during the Permo-Carboniferous Reversed-Polarity Superchron. If the magnetization of the Pybus Formation is indeed of Permian age, then the results would require location of the Alexander terrane in the southern hemisphere. However, the possibility of remagnetization cannot be ruled out, because Panuska and Stone could not apply a conclusive fold test to their Pybus collection. More corroboration of the Permian pole from the Alexander terrane thus is needed to prove the validity of the southern hemisphere option.

Arctic Alaska

The oil boom in northern Alaska fostered several paleomagnetic investigations of the Brooks Range (Moore and others, this volume), mainly to test for large-scale counterclockwise rotation of the Arctic Alaska block. According to the rotation hypothesis, originally proposed by Carey (1955) and later advanced by Tailleur (1973) and Grantz and others (1979), northern Alaska is interpreted as a continental fragment that rifted from the Canadian Arctic islands. Presumably, sea-floor spreading created the Canada Basin during the late Mesozoic as Arctic Alaska rotated counterclockwise relative to North America. If the rotation hypothesis is correct, Paleozoic paleomagnetic poles of northern Alaska should be displaced southeast of the North American cratonic poles.

The first paleomagnetic test of the hypothesis, reported by Newman and others (1979), was attempted in the Brooks Range. Newman and coworkers sampled Devonian, Carboniferous, and Permian sedimentary rocks in thrust sheets from Cape Lisburne to the Arctic National Wildlife Refuge. Their preliminary analysis of the data apparently supported the argument for consistent counterclockwise rotation of the Brooks Range thrust sheets. Results of reliability tests, such as the fold test, were not presented. Hillhouse and Grommé's (1983) subsequent examination of Newman and others' unpublished data showed several problems with the original interpretation. In particular, they found that the distribution of Mississippian paleomagnetic poles, mainly from the Lisburne Group, clustered near the Cretaceous pole for central North America before corrections were made for tilt of the bedding. Application of the tilt corrections increased the overall dispersion of mean directions of the Brooks Range sites, indicating the retention of a post-deformational magnetic overprint. At most sites the angular dispersion of directions was so large that the "consistency of reversals" test could not be applied.

Additional sampling by Hillhouse and Grommé (1982b, 1983) confirmed the presence of a magnetic overprint of widespread extent in the central and eastern Brooks Range. They sampled the Kanayut Conglomerate (Upper Devonian and Lower Mississippian[?]) at 11 sites distributed from the Arctic Quadrangle to the Killik River (Fig. 3). The Kanayut site directions failed the fold test conclusively, despite attempts to remove secondary components by alternating field and thermal demagnetization methods. The overprint yielded a mean paleomagnetic pole that lies close to the Cretaceous reference pole for North America. Deep burial and subsequent uplift of the Brooks Range thrust sheets apparently reset the magnetization, precluding a successful test of the rotation hypothesis.

Oriented cores from exploratory wells on the North Slope are possibly free of the magnetic overprint that affects the Brooks Range. Hillhouse and Grommé (1983) tested a 10-m, fully oriented core from the East Simpson No. 2 test well near the Arctic coast in the National Petroleum Reserve. From a depth of 2,263 m, the drilling crew recovered a core of reddened argillite, believed to be part of the eroded surface on which Mississippian sandstone was deposited. The magnetization of the core is carried by hematite that probably formed by oxidation during deposition of the Mississippian sediments. The paleomagnetic pole from the reddened argillite is located east of Japan, in general agreement with Carey's predicted 30° rotation, which assumed a rotation

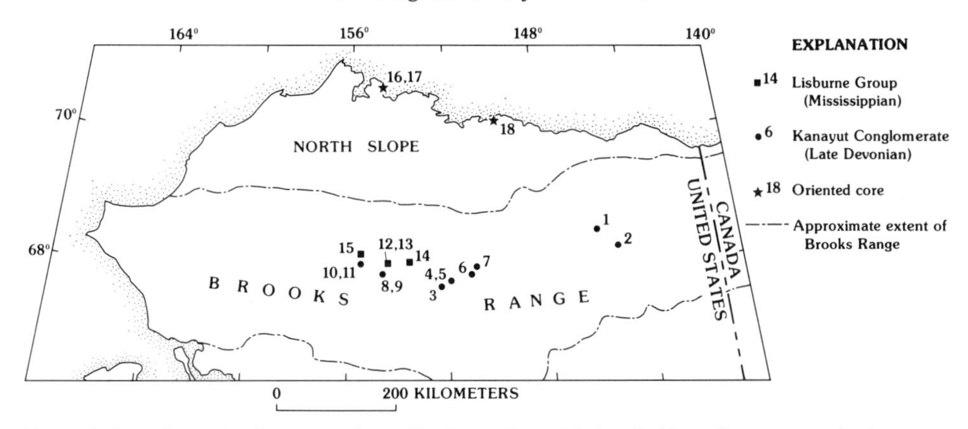

Figure 3. Locations of paleomagnetic studies in northern Alaska. 1–15, sedimentary rocks that were remagnetized after folding (Hillhouse and Grommé, 1983); 16–17, oriented cores from exploratory bore-holes in the National Petroleum Reserve (Hillhouse and Grommé, 1983); 18, two oriented cores from the Kuparuk River Formation near Prudhoe Bay (Halgedahl and Jarrard, 1987).

axis in central Alaska. Due to uncertainties in the age of magnetization and in the accuracy of the down-hole orientation procedure, Hillhouse and Grommé (1983) considered the conclusions of the study tentative.

Halgedahl and Jarrard (1987) conducted a more reliable test of the Arctic rotation hypothesis on oriented cores from two ARCO wells located in the Kuparuk River oil field on the North Slope. They obtained a paleomagnetic pole from Lower Cretaceous sedimentary rocks of the Kuparuk River Formation. The data are compatible with a model that calls for about 70° of counterclockwise rotation of Arctic Alaska about a pole of rotation in the Mackenzie River delta. The model places the North Slope region and Chukotka (U.S.S.R.) against the Canadian Arctic Islands prior to 130 Ma, when rifting began to form the Canada basin.

RELATIVE PLATE MOTIONS AND THE ACCRETION OF ALASKA

The literature concerning the paleomagnetism of Alaskan terranes has grown rapidly since 1975. Explaining every anomalous paleomagnetic result in terms of terrane movements leads to a bewildering sequence of displacements, often contradictory to other paleomagnetic studies and to geologic evidence (Plafker and Berg, this volume, Chapter 33). Our approach is to winnow the data, highlighting the better-substantiated paleomagnetic observations, mainly from studies of volcanic rocks, and to fit those observations with a plate-tectonic model for Alaska. We can distill three chief observations from the paleomagnetic literature of Alaska:

1. Wrangellia and the Peninsular terrane were separated from central Alaska by a gap in latitude of at least 25° in the early Mesozoic. By at least 55 Ma, the gap was closed by northward drift of Wrangellia and the Peninsular terrane.

2. In Late Cretaceous and Paleocene time, volcanic complexes in the southern parts of the Prince William and Chugach terranes were 25° south of the Alaskan margin.

3. After 68 Ma and before 44 Ma, central and western Alaska rotated 30 to 50° counterclockwise, presumably about a hinge line near 146°W.

Our goal is to fit these three observations from paleomagnetism into a plate-tectonic model that incorporates motions of oceanic plates in the Pacific and sea-floor spreading in the Arctic, which are the major tectonic forces that have shaped Alaska. Since at least 100 Ma, vast areas of sea floor have been subducted beneath the northern Pacific continental margin, as interpreted from the pattern of magnetic anomalies in the Gulf of Alaska. Hence, subduction and sea-floor spreading provide the mechanism for bringing terranes such as Wrangellia into Alaska. The past motions of the oceanic plates are difficult to reconstruct, because much of the older sea floor has been subducted and the early motion of the Pacific plate relative to North America is poorly known. Nevertheless, models of the now-vanished oceanic plates can be reconstructed from the magnetic anomalies of the Pacific basin, if it is assumed that sea-floor spreading has been symmetrical about the ridges. North America–Pacific plate motions can be estimated by following the circuit of spreading centers from the Atlantic Ocean through Antarctica to the Pacific Ocean (Atwater and Molnar, 1973), or by using the hot-spot reference frame (Engebretson, 1982; Engebretson and others, 1985). The assumptions behind both methods of reconstruction are subject to error, so that the positions of ancient spreading centers and the relative plate motions are imprecisely determined.

Nevertheless, oceanic plate reconstructions, such as the models presented by Stone and others (1982) or Wallace and Engebretson (1984) for the Alaska margin, offer a reasonable mechanism for the emplacement of Wrangellia, the Peninsular terrane, and the Prince William terrane. For example, prior to 150 Ma, Wrangellia might have been riding the Farallon plate in the southwest Pacific region (Engebretson, 1982). By middle Jurassic time, Wrangellia and the Peninsular terrane had joined together. The amalgamated terrane was carried northeastward toward the North American margin, possibly colliding with the continent near Oregon in the Early Cretaceous. We believe that part of Wrangellia was in Oregon by 130 Ma, because plutons of that age intrude the Seven Devils Group and yield concordant paleolatitudes with respect to the North American reference pole (Wilson and Cox, 1980; Hillhouse and others, 1982). Also, at about 130 Ma, Arctic Alaska rifted away from the Canadian Arctic Islands as the Canada basin began to form. Engebretson and others (1985) suggested that the Farallon plate was divided about 85 Ma by a spreading center, the northern part becoming the Kula plate. If the Kula-Farallon spreading center was south of Wrangellia, then the motion of the Kula plate relative to North America would be of the proper direction and velocity to bring Wrangellia into Alaska by 55 Ma. Transcurrent faults could detach Wrangellia from the mainland and slip it northward, leaving splinters of the terrane along the continental margin at Vancouver Island, Queen Charlotte Island, and southern Alaska. At that time, the zone of dextral shear probably extended east to the Tintina fault, causing northward drift of the Stikine terrane relative to the cratonic margin.

Our model for the accretion of Wrangellia and the Peninsular terrane is consistent with the timing of deformation and plutonism in the Jura-Cretaceous flysch terrane of south-central Alaska, which was the zone of convergence between Wrangellia and the mainland. Intense deformation of the flysch and obduction of smaller terranes such as Chulitna occurred after 100 Ma, because the youngest flysch beds are Cenomanian in age (Csejtey and others, 1982). Large-scale shortening of the flysch basin is indicated by the chaotic structure and penetrative deformation of the deposits. Relative motions between Wrangellia, the flysch, and the Nixon Fork terrane are constrained by the Alaska Range belt of plutons, which cuts across the terrane boundaries and the isoclinal folds within the flysch. The plutons were emplaced between 75 and 55 Ma (Reed and Lanphere, 1973; Lanphere and Reed, 1985).

Paleomagnetic evidence against post-Triassic displacement of the Alexander terrane remains an enigma, and cannot be simply explained by the oceanic plate model. The Tethyan fauna of the Cache Creek Group and the low paleolatitudes of the Coast Plutonic Complex imply that the Alexander terrane is out of place. The paleomagnetism of the Alexander terrane clearly warrants further study.

As preserved in volcanic rocks, the Paleocene and Early Eocene paleomagnetic record of Alaska constrains northward movement of the block between the Tintina fault and the Chugach terrane to have ended by 55 Ma, providing the backstop for the exotic southern blocks. The velocity and direction of the Kula plate are easily sufficient to close the 25° latitude gap between Paleocene volcanic rocks of the Prince William terrane and the previously accreted margin of Alaska, according to Wallace and Engebretson's (1984) model. Relative motion of the Kula plate accelerated in the interval 56 to 43 Ma, reaching a rate of 200 km/m.y. near the Kenai Peninsula. After the demise of the Pacific-Kula ridge at 43 Ma, when the Pacific plate began to subduct beneath Alaska, the rate of motion between North America and the Pacific plate was about 40 km/m.y. Therefore, most of the northward impetus was provided by the Kula plate in a northeast direction, with the Pacific plate adding a small component of slip along the British Columbia–Alaska margin after 43 Ma. Our model implies that the Paleocene volcanic rocks of the Prince William terrane originated in the Pacific Ocean far from the North American continental margin. The volcanic rocks were probably decoupled from the sedimentary rocks of the Prince William terrane, because the sediments were derived from a continental source (Plafker and others, this volume, Chapter 12). The volcanic rocks might be fragments of oceanic crust that collided with a sedimentary apron along the British Columbia–southeastern Alaska margin. This scenario solves the problem of the enigmatic suture between the Chugach and Peninsular terranes.

The third observation, that western Alaska rotated counterclockwise after 55 Ma, can be explained as a consequence of continental deformation at the time of sea-floor spreading in the North Atlantic and the Labrador Sea (Grantz, 1966; Patton and Tailleur, 1977). Spreading in the Labrador Sea between Greenland and North America was active from the Late Cretaceous to the early Oligocene (Srivastava, 1978), whereas spreading in the North Atlantic and Eurasian basin has occurred since 90 Ma (Pitman and Talwani, 1972). The opening of these basins requires overlap or compression of continental crust along the Eurasia–North America boundary from 70 to 50 Ma. The boundary is generally assumed to be in the northeast U.S.S.R., but it is not clearly defined. Continental overlap might have been accommodated by deformation in the Bering Sea region, which forced western Alaska to the southeast. If so, the motion would entail rotation about a hinge line near 146°W and dextral slip along the Beringian continental margin, a proposal that is consistent with the arcuate structural trends of southwestern Alaska. Restoration of western Alaska to its original position straightens out the arcuate trends and creates a geometry similar to the model of Freeland and Dietz (1973), although opening of the Canada Basin was the main rotational mechanism of their model. Rotation of western Alaska probably ceased by 50 Ma when a change in the pole of opening of the Arctic basin eased North America–Eurasia compression (Harbert and others, 1985). Also, at that time a fragment of the Kula plate was trapped behind the growing Aleutian volcanic arc, and motion ceased along the Beringian margin (Marlow and Cooper, 1980).

CONCLUSION

The accretion of southern Alaska was apparently a two-step process, according to our analysis of the paleomagnetic data. First, Wrangellia and the Peninsular terrane collided with the Nixon Fork and Yukon–Tanana terranes during the interval 100 to 55 Ma to make up the core of Alaska. Motion of the Kula plate probably provided the impetus to finally close the latitude gap between Wrangellia and the mainland. Secondly, volcanic complexes now embedded in the southern margins of the Prince William and Chugach terranes arrived in Alaska after 55 Ma, carried first by the Kula and then by the Pacific plate. The counterclockwise rotation of southwestern Alaska most likely occurred 68 to 44 m.y. ago as the latitude gap was closing between the volcanic complexes and the mainland. The rotation of western Alaska may be the result of deformation at the boundary between North America and Eurasia at the time of sea-floor spreading in the Eurasian basin, Labrador Sea, and North Atlantic.

REFERENCES CITED

Andrews, J. A., 1985, True polar wander; An analysis of Cenozoic and Mesozoic paleomagnetic poles: Journal of Geophysical Research, v. 90, p. 7737–7750.

Atwater, T., and Molnar, P., 1973, Relative motion of the Pacific and North American plates deduced from sea-floor spreading in the Atlantic, Indian, and South Pacific Oceans, *in* Proceedings of the Conference on Tectonic Problems of the San Andreas Fault System, Stanford University, California: Stanford, California, Stanford University Publications in the Geological Sciences, v. 13, p. 136–148.

Berg, H. C., Jones, D. L., and Richter, D. H., 1972, Gravina–Nutzotin belt; Tectonic significance of an upper Mesozoic sedimentary and volcanic sequence in southern and southeastern Alaska: U.S. Geological Survey Professional Paper 800-D, p. D1–D24.

Burk, C. A., 1965, Geology of the Alaska Peninsula; Island arc and continental margin: Geological Society of America Memoir 99, part 1, 250 pp.

Carey, S. W., 1955, The orocline hypothesis in geotectonics: Proceedings of the Royal Society of Tasmania, v. 89, p. 255–288.

Coe, R. S., Globerman, B. R., Plumley, P. W., and Thrupp, G. A., 1985, Paleomagnetic results from Alaska and their tectonic implications, *in* Howell, D. G., ed., Tectonostratigraphic terranes of the Circum-Pacific region: Circum-Pacific Council for Energy and Mineral Resources Earth Science Series, v. 1, p. 85–108.

Cox, A., and Gordon, R. G., 1984, Paleolatitudes determined from paleomagnetic data from vertical cores: Reviews of Geophysics and Space Physics, v. 22, p. 47–72.

Csejtey, B., Jr., Cox, D. P., Evarts, R. C., Stricker, G. D., and Foster, H. L., 1982, The Cenozoic Denali fault system and the Cretaceous accretionary development of southern Alaska: Journal of Geophysical Research, v. 87, p. 3741–3754.

Duncan, R. A., 1981, Hot spots in the southern oceans; An absolute frame of reference for motion of the Gondwana continents: Tectonophysics, v. 74, p. 29–42.

Engebretson, D. C., 1982, Relative motions between oceanic and continental plates in the Pacific Basin [Ph.D. thesis]: Stanford, California, Stanford University, 211 p.

Engebretson, D. C., Cox, A., and Gordon, R. G., 1985, Relative motions between oceanic plates in the Pacific basin: Geological Society of America Special Paper 206, 59 p.

Freeland, G. L., and Dietz, R. S., 1973, Rotation history of Alaskan tectonic blocks: Tectonophysics, v. 18, p. 379–389.

Globerman, B. R., and Coe, R. S., 1984a, Paleomagnetic results from Upper Cretaceous volcanic rocks in northern Bristol Bay, SW Alaska, and tectonic implications, *in* Proceedings of the Circum-Pacific Terrane Conference, Stanford University, California: Stanford, California, Stanford University Publications in the Geological Sciences, v. 18, p. 98–102.

——, 1984b, Discordant declinations from "in place" Upper Cretaceous volcanics, SW Alaska; Evidence for oroclinal bending?: Geological Society of America Abstracts with Programs, v. 16, p. 286.

Globerman, B. R., Coe, R. S., Hoare, J. M., and Decker, J., 1983, Palaeomagnetism of Lower Cretaceous tuffs from Yukon–Kuskokwim delta region, western Alaska: Nature, v. 305, p. 516–520.

Goldreich, P., and Toomre, A., 1969, Some remarks on polar wandering: Journal of Geophysical Research, v. 74, p. 2555–2567.

Gordon, R. G., 1983, Late Cretaceous apparent polar wander of the Pacific plate; Evidence for a rapid shift of the Pacific hot spots with respect to the spin axis: Geophysical Research Letters, v. 10, p. 709–712.

Gordon, R. G., Cox, A., and O'Hare, S., 1984, Paleomagnetic Euler poles and the apparent polar wander and absolute motion of North America since the Carboniferous: Tectonics, v. 3, p. 499–537.

Grantz, A., 1966, Strike-slip faults in Alaska: U.S. Geological Survey Open-File Report 66–53, 82 p., 4 sheets, scale 1:63,360.

Grantz, A., Eittrem, S., and Dinter, D. A., 1979, Geology and tectonic development of the continental margin of North America: Tectonophysics, v. 59, p. 263–291.

Grommé, S., and Hillhouse, J. W., 1981, Paleomagnetic evidence for northward movement of the Chugach terrane, southern and southeastern Alaska, *in* Albert, N.R.D., and Hudson, T., eds., The U.S. Geological Survey in Alaska: Accomplishments during 1979: U.S. Geological Survey Circular 823-B, p. B70–B72.

Halgedahl, S., and Jarrard, R., 1987, Paleomagnetism of the Kuparuk River Formation from oriented drill core; Evidence for rotation of the Arctic Alaska plate, *in* Tailleur, I., and Weiner, P., eds., Alaska North Slope, volume 2: Pacific Section, Society of Economic Paleontologists and Mineralogists, p. 581–617.

Hamilton, W., 1969, Mesozoic California and the underflow of Pacific mantle: Geological Society of America Bulletin, v. 80, p. 2409–2430.

Harbert, W. P., Frei, L. S., Cox, A., and Engebretson, D. C., 1985, Relative motions between Eurasia and North America in Bering Sea region: American Association of Petroleum Geologists Bulletin, v. 69, p. 665–666.

Harrison, C.G.A., and Lindh, T., 1982, Comparison between the hot spot and geomagnetic field reference frames: Nature, v. 300, p. 251–252.

Hillhouse, J. W., 1977, Paleomagnetism of the Triassic Nikokai Greenstone, McCarthy Quadrangle, Alaska: Canadian Journal of Earth Sciences, v. 14, p. 2578–2592.

——, 1987, Accretion of southern Alaska: Tectonophysics, v. 139, p. 107–122.

Hillhouse, J. W., and Grommé, C. S., 1977, Paleomagnetic poles from sheeted dikes and pillow basalt of the Valdez(?) and Orca Groups, southern Alaska: EOS Transactions of the American Geophysical Union, v. 58, p. 1127.

——, 1980, Paleomagnetism of the Triassic Hound Island volcanics, Alexander terrane, southeastern Alaska: Journal of Geophysical Research, v. 85, p. 2594–2602.

——, 1982a, Limits to northward drift of the Paleocene Cantwell Formation, central Alaska: Geology, v. 10, p. 552–556.

——, 1982b, Cretaceous overprint revealed by paleomagnetic study in the northern Brooks Range *in* Coonrad, W. L., ed., The U.S. Geological Survey in Alaska; Accomplishments during 1980: U.S. Geological Survey Circular 844, p. 43–46.

——, 1983, Paleomagnetic studies and the hypothetical rotation of Arctic Alaska: Alaska Geological Society Journal, v. 2, p. 27–39.

——, 1984, Northward displacement and accretion of Wrangellia; New paleomagnetic evidence from Alaska: Journal of Geophysical Research, v. 89, p. 4461–4477.

Hillhouse, J. W., Grommé, C. S., and Vallier, T. L., 1982, Paleomagnetism and Mesozoic tectonics of the Seven Devils volcanic arc in northeastern Oregon: Journal of Geophysical Research, v. 87, p. 3777–3794.

Hillhouse, J. W., Grommé, C. S., and Csejtey, B., Jr., 1985, Tectonic implications of paleomagnetic poles from lower Tertiary volcanic rocks, south central Alaska: Journal of Geophysical Research, v. 90, p. 12523–12536.

Irving, E., 1979, Paleopoles and paleolatitudes of North America and speculations about displaced terranes: Canadian Journal of Earth Sciences, v. 16, p. 669–694.

Irving, E., and Irving, G. A., 1982, Apparent polar wander paths Carboniferous through Cenozoic and the assembly of Gondwana: Geophysical Surveys, v. 5, p. 141–188.

Irving, E., and Strong, D. F., 1984, Evidence against large-scale Carboniferous strike-slip faulting in the Appalachian-Caledonian orogen: Nature, v. 310, p. 762–764.

——, 1985, Paleomagnetism of rocks from Burin Peninsula, Newfoundland; Hypothesis of late Paleozoic displacement of Acadia criticized: Journal of Geophysical Research, v. 90, p. 1949–1962.

Irving, E., and Yole, R. W., 1972, Paleomagnetism and kinematic history of mafic and ultramafic rock in fold mountain belts: Ottawa, Canada Department of Energy Mines and Resources, Publications of the Earth Physics Branch, v. 42, no. 3, p. 87–95.

Irving, E., Monger, J.W.H., and Yole, R. W., 1980, New paleomagnetic evidence for displaced terranes in British Columbia, *in* Strangway, D. W., ed., The continental crust and its mineral deposits: Geological Association of Canada Special Paper 20, p. 441–456.

Irving, E., Woodsworth, G. J., Wynne, P. J., and Morrison, A., 1985, Paleomagnetic evidence for displacement from south of the Coast Plutonic Complex, British Columbia: Canadian Journal of Earth Sciences, v. 22, p. 584–598.

Jones, D. L., and Silberling, N. J., 1979, Mesozoic stratigraphy; The key to tectonic analysis of southern and central Alaska: U.S. Geological Survey Open-File Report 79-1200, 37 p.

Jones, D. L., Irwin, W. P., and Ovenshine, A. T., 1972, Southeastern Alaska; A displaced continental fragment?: U.S. Geological Survey Professional Paper 800-B, p. B211–B217.

Jones, D. L., Silberling, N. J., and Hillhouse, J., 1977, Wrangellia; A displaced terrane in northwestern North America: Canadian Journal of Earth Sciences, v. 14, p. 2565–2577.

Jurdy, D. M., 1981, True polar wander: Tectonophysics, v. 74, p. 1–16.

Karl, S., and Hoare, J. M., 1979, Results of a preliminary paleomagnetic study of volcanic rocks from Nuyakuk Lake, southwestern Alaska *in* Johnson, K. M., and Williams, J. R., eds., The U.S. Geological Survey in Alaska; Accomplishments during 1978: U.S. Geological Survey Circular 804-B, p. B74–B78.

Kent, D. V., 1982, Paleomagnetic evidence for post-Devonian displacement of the Avalon platform (Newfoundland): Journal of Geophysical Research, v. 87, p. 8709–8716.

Kent, D. V., and Opdyke, N. D., 1985, Multicomponent magnetizations from the Mississippian Mauch Chunk Formation of the central Appalachians and their tectonic implications: Journal of Geophysical Research, v. 90, p. 5371–5383.

Kirschvink, J. L., 1980, The least-squares line and plane and the analysis of palaeomagnetic data: Geophysical Journal of the Royal Astronomical Society, v. 62, p. 699–718.

Lanphere, M. A., and Reed, B. L., 1985, The McKinley Sequence of granitic rocks; A key element in the accretionary history of southern Alaska: Journal of Geophysical Research, v. 90, p. 11413–11430.

Loney, R. A., and Himmelberg, G. R., 1983, Structure and petrology of the La Perouse gabbro intrusion, Fairweather Range, southeastern Alaska: Journal of Petrology, v. 24, part 4, p. 377–423.

Marlow, M. S., and Cooper, A. K., 1980, Mesozoic and Cenozoic structural trends under southern Bering Sea shelf: American Association of Petroleum Geologists Bulletin, v. 64, p. 2139–2155.

May, S. R., and Butler, R. F., 1986, North American Jurassic apparent polar wander; Implications for plate motion, paleogeography, and Cordilleran tectonics: Journal of Geophysical Research, v. 91, p. 11,519–11,544.

McElhinny, M. W., 1973, Palaeomagnetism and plate tectonics: Cambridge, Cambridge University Press, 358 p.

McElhinny, M. W., and Merrill, R. T., 1975, Geomagnetic secular variation over the past 5 m.y.: Reviews of Geophysics and Space Physics, v. 13, p. 687–708.

Merrill, R. T., and McElhinny, M. W., 1983, The earth's magnetic field; Its history, origin, and planetary perspective: New York, Academic Press, 401 p.

Monger, J.W.H., and Irving, E., 1980, Northward displacement of north-central British Columbia: Nature, v. 285, p. 289–294.

Monger, J.W.H., and Ross, C. A., 1971, Distribution of fusulinaceans in the western Canadian Cordillera: Canadian Journal of Earth Sciences, v. 8, p. 259–278.

Monger, J.W.H., Souther, J. G., and Gabrielse, H., 1972, Evolution of the Canadian Cordillera; A plate tectonic model: American Journal of Science, v. 272, p. 577–602.

Moore, J. C., and 5 others, 1983, Paleogene evolution of the Kodiak Islands, Alaska; Consequences of ridge-trench interaction in a more southerly latitude: Tectonics, v. 2, p. 265–293.

Morgan, W. J., 1972, Plate motions and deep mantle convection, *in* Shagam, R., and others, eds., Studies in Earth and space sciences: Geological Society of America Memoir 132, p. 7–22.

Newman, G. W., Mull, C. G., and Watkins, N. D., 1979, Northern Alaska paleomagnetism, plate rotation, and tectonics, *in* Sisson, A., ed., The relationship of plate tectonics to Alaskan geology and resources: Anchorage, Alaska Geological Society Symposium Proceedings, No. 6, p. C1–C7.

Packer, D. R., and Stone, D. B., 1972, An Alaskan Jurassic paleomagnetic pole and the Alaskan orocline: Nature, v. 237, p. 25–26.

——, 1974, Paleomagnetism of Jurassic rocks from southern Alaska, and the tectonic implications: Canadian Journal of Earth Sciences, v. 11, p. 976–997.

Panuska, B. C., 1985, Paleomagnetic evidence for a post-Cretaceous accretion of Wrangellia: Geology, v. 13, p. 880–883.

Panuska, B., and Macicak, M., 1986, A revised paleolatitude for the Paleocene Teklanika Volcanics, central Alaska Range, Alaska: EOS Transactions of the American Geophysical Union, v. 67, p. 921.

Panuska, B. C., and Stone, D. B., 1981, Late Palaeozoic palaeomagnetic data for Wrangellia; Resolution of the polarity ambiguity: Nature, v. 293, p. 561–563.

——, 1985a, Confirmation of the pre-50 m.y. accretion age of the Southern Alaska superterrane: EOS Transactions of the American Geophysical Union, v. 66, p. 863.

——, 1985b, Latitudinal motion of the Wrangellia and Alexander terranes and the Southern Alaska superterrane, *in* Howell, D. G., ed., Tectonostratigraphic terranes of the circum-Pacific region: Circum-Pacific Council for Energy and Mineral Resources Earth Science Series, v. 1, p. 109–120.

Patton, W. W., Jr., and Tailleur, I. L., 1977, Evidence in the Bering Strait region for differential movement beneath North America and Eurasia: Geological Society of America Bulletin, v. 88, p. 1298–1304.

Pitman, W. C., III, and Talwani, M., 1972, Sea-floor spreading in the North Atlantic: Geological Society of America Bulletin, v. 83, p. 619–646.

Plafker, G., 1984, Comment on 'Model for the origin of the Yakutat block, an accreting terrane in the northern Gulf of Alaska': Geology, v. 12, p. 563.

Plumley, P. W., 1984, A paleomagnetic study of the Prince William terrane and Nixon Fork terrane, Alaska [Ph.D. thesis]: Santa Cruz, University of California, 190 p.

Plumley, P. W., Coe, R. S., Byrne, T., Reid, M. R., and Moore, J. C., 1982, Palaeomagnetism of volcanic rocks of the Kodiak Islands indicates northward latitudinal displacement: Nature, v. 300, p. 50–52.

Plumley, P. W., Coe, R. S., and Byrne, T., 1983, Paleomagnetism of the Paleocene Ghost Rocks Formation, Prince William terrane, Alaska: Tectonics, v. 2, p. 295–314.

Reed, B. L., and Lanphere, M. A., 1973, Alaska-Aleutian Range batholith; Geochronology, chemistry, and relation to circum-Pacific plutonism: Geological Society of America Bulletin, v. 84, p. 2583–2610.

Roy, J. L., 1982, Paleomagnetism of Siluro-Devonian rocks from eastern Maine; Discussion: Canadian Journal of Earth Sciences, v. 19, p. 225–232.

Roy, J. L., and Morris, W. A., 1983, A review of paleomagnetic results from the Carboniferous of North America; The concept of Carboniferous geomagnetic field markers: Earth and Planetary Science Letters, v. 65, p. 167–181.

Schwarz, E. J., Muller, J. E., and Clark, K. R., 1980, Paleomagnetism of the Karmutsen basalts from southeast Vancouver Island: Canadian Journal of Earth Sciences, v. 17, p. 389–399.

Srivastava, S. P., 1978, Evolution of the Labrador Sea and its bearing on the early evolution of the North Atlantic: Geophysical Journal of the Royal Astronomical Society, v. 52, p. 313–357.

Stone, D. B., 1982, Triassic paleomagnetic data and paleolatitudes for Wrangellia, Alaska: Alaska Division of Geological and Geophysical Surveys Geologic Report 73, p. 55–62.

——, 1984, Concordance and conflicts in Alaskan paleomagnetic data, *in* Proceedings of the Circum-Pacific Terrane Conference: Stanford, California, Stanford University Publications in the Geological Sciences, v. 18, p. 186–189.

Stone, D. B., and Packer, D. R., 1977, Tectonic implications of Alaska Peninsula paleomagnetic data: Tectonophysics, v. 37, p. 183–201.

——, 1979, Paleomagnetic data from the Alaska Peninsula: Geological Society of America Bulletin, v. 90, p. 545–560.

Stone, D. B., Panuska, B. C., and Packer, D. R., 1982, Paleolatitudes versus time for southern Alaska: Journal of Geophysical Research, v. 87, p. 3697–3707.

Symons, D.T.A., 1971, Paleomagnetic notes on the Karmutsen Basalts, Vancouver Island, British Columbia: Canada Geological Survey Paper 71–24, p. 11–24.

Tailleur, I. L., 1973, Probable rift origin of Canada Basin: American Association of Petroleum Geologists Memoir 19, p. 526–535.

Thrupp, G. A., and Coe, R. S., 1984, Paleomagnetism of Tertiary volcanic rocks from the Alaska Peninsula; Is the Peninsular terrane a coherent tectonic entity?: Geological Society of America Abstracts with Programs, v. 16, p. 337.

——, 1986, Early Tertiary paleomagnetic evidence and the displacement of southern Alaska: Geology, v. 14, p. 213–217.

Van der Voo, R., and Scotese, C., 1981, Paleomagnetic evidence for a large (2,000 km) sinistral offset along the Great Glen fault during Carboniferous time: Geology, v. 9, p. 583–589.

Van der Voo, R., Jones, M., Grommé, C. S., Eberlein, G. D., and Churkin, M., Jr., 1980, Paleozoic paleomagnetism and northward drift of the Alexander terrane, southeastern Alaska: Journal of Geophysical Research, v. 85, p. 5281–5296.

Wallace, W. K., and Engebretson, D. C., 1984, Relationships between plate motions and Late Cretaceous to Paleogene magmatism in southwestern Alaska: Tectonics, v. 3, p. 295–315.

Wilson, D., and Cox, A., 1980, Paleomagnetic evidence for tectonic rotation of Jurassic plutons in Blue Mountains, eastern Oregon: Journal of Geophysical Research, v. 85, p. 3681–3689.

Yole, R. W., and Irving, E., 1980, Displacement of Vancouver Island; Paleomagnetic evidence from the Karmutsen Formation: Canadian Journal of Earth Sciences, v. 17, p. 1210–1228.

Zijderveld, J.D.A., 1967, A. C. demagnetization of rocks; Analysis of results, *in* Collinson, D. W., Creer, K. M., and Runcorn, S. K., eds., Methods in paleomagnetism: New York, Elsevier, p. 254–286.

MANUSCRIPT ACCEPTED BY THE SOCIETY MAY 17, 1990

ACKNOWLEDGMENTS

We thank the following people who contributed helpful comments and criticism of the early drafts of this manuscript: Henry C. Berg, Stephen E. Box (USGS), William P. Harbert (University of Pittsburgh), Steven R. May (Exxon Production Research), and George Plafker (USGS).

NOTES ADDED IN PROOF

Several new studies of paleomagnetism in Alaska were published during the interval 1989–1993 as this volume was being assembled and typeset. We wish to alert readers to the additional references cited below, and we highlight some new paleomagnetic results from the Peninsular, Chugach, and Alexander terranes.

Peninsular terrane

Stamatakos and others (1989) reported paleomagnetic results from Upper Cretaceous and lower Tertiary sedimentary rocks in Matanuska Valley, which lies within the Peninsular terrane of southern Alaska. Their results from four sites in the Matanuska Formation (Upper Cretaceous) north of the Castle Mountain fault indicated that the sediments were deposited 2,800 ± 2,000 km south of the formation's present latitudinal position with respect to North America. Furthermore, the Chickaloon and Arkose Ridge Formations of early Tertiary age yielded low magnetic inclinations corresponding to relative northward movement of 1,600 ± 1,200 km. On the basis of redeposition and compaction experiments, Stamatakos and others (1989) found no tendency for these sediments to develop systematically shallow inclinations relative to the ambient magnetic field during deposition. Therefore, the paleomagnetic record from sedimentary rocks continues to support the case for unusually rapid northward movement of the Peninsular terrane during the early Tertiary. In contrast, Matanuska Valley volcanic rocks that are slightly younger than the Chickaloon and Arkose Ridge Formations yield significantly higher paleolatitudes, implying no significant drift relative to the craton after the Eocene (38.8–53.6 Ma; Panuska and others, 1990). The disparity of paleomagnetic results from volcanic versus sedimentary rocks in southern Alaska poses a problem for future research.

Chugach terrane

The Resurrection Peninsula ophiolite was sampled extensively for paleomagnetism by Bol and others (1992). The new results imply a northward latitudinal displacement of the Chugach terrane of 13° ± 9° (1,400 ± 1,000 km), about half of the displacement inferred from previously reported results. Higher confidence is placed in these new results, owing to better age control (57 Ma from U/Pb), more extensive sampling, and better structural control. As inferred from the paleomagnetic evidence, the Resurrection Peninsula ophiolite moved northward on the order of 1,000 km relative to the lower Tertiary volcanic rocks of Matanuska Valley and the Talkeetna Mountains. The ophiolite may be part of the now-extinct Kula-Farallon ridge that accreted to the Alaska margin by 45 Ma.

Alexander terrane

The paradox posed by early work on the Triassic Hound Island Volcanics has been eliminated by a more extensive study (Haeussler and others, 1992a) of the paleomagnetism of the Alexander terrane. The paradox arose from the apparent lack of post-Triassic northward displacement of the Alexander terrane in contrast with substantial displacements inferred for coeval terranes in the mountains of western British Columbia. Careful separation of magnetic components by Haeussler and others (1992b) showed the previous results to be flawed by magnetic overprints. Elimination of the overprint yielded a paleolatitude of 19.2° ± 10.3° for the Triassic volcanic rocks, suggesting that the Alexander terrane and the neighboring Wrangellia terrane shared a common latitude during the Late Triassic. Analysis of paleomagnetic data from the Pybus Formation suggests that the Alexander terrane was in the northern hemisphere during the Permian.

ADDITIONAL REFERENCES

Bol, A. J., Coe, R. S., Gromme, C. S., and Hillhouse, J. W., 1992, Paleomagnetism of the Resurrection Peninsula, Alaska: Implications for the tectonics of southern Alaska and the Kula-Farallon ridge: Journal of Geophysical Research, v. 97, p. 17213–17232.

Haeussler, P., Coe, R. S., and Onstott, T. C., 1992a, Paleomagnetism of the Late Triassic Hound Island Volcanics: Revisited: Journal of Geophysical Research, v. 97, p. 19617–19639.

Haeussler, P.J., Coe, R. S., and Renne, P., 1992b, Paleomagnetism and geochronology of 23 Ma gabbroic intrusions in the Keku Strait, Alaska, and implications for the Alexander terrane: Journal of Geophysical Research, v. 97, p. 19641–19649.

Harbert, W., 1990, Paleomagnetic data from Alaska: Reliability, interpretation, and terrane trajectories: Tectonophysics, v. 184, p. 111–135.

Panuska, B. C., Stone, D. B., and Turner, D. L., 1990, Paleomagnetism of Eocene volcanic rocks, Talkeetna Mountains, Alaska: Journal of Geophysical Research, v. 95, p. 6737–6750.

Stamatakos, J. A., Kodama, K. P., Vittorio, L. F., and Pavlis, T. L., 1989, Paleomagnetism of Cretaceous and Paleocene sedimentary rocks across the Castle Mountain fault, south central Alaska: American Geophysical Union Geophysical Monograph Series, v. 50, p. 151–177.

Stone, D. B., and McWilliams, M. O., 1989, Paleomagnetic evidence for relative motion in western North America, *in* Ben-Avraham, Z., ed., Evolution of the Pacific Ocean Margins: Oxford, Oxford University Press, p. 53–72.

Wittbrodt, P. R., Stone, D. B., and Turner, D. L., 1989, Paleomagnetism and geochronology of St. Matthew Island, Bering Sea: Canadian Journal of Earth Sciences, v. 26, p. 2116–2129

Printed in U.S.A.

The Geology of North America
Vol. G-1, The Geology of Alaska
The Geological Society of America, 1994

Chapter 27

Late Cenozoic glaciation of Alaska

Thomas D. Hamilton
U.S. Geological Survey, 4200 University Drive, Anchorage, Alaska 99508

INTRODUCTION

Glaciers cover only about 5 percent of Alaska today, but they spread over as much as half of the state during the most widespread advances of the late Cenozoic (Fig. 1). Both modern and ancient glaciers have been most extensive in southern Alaska, where they were close to moisture sources in the North Pacific Ocean and the Gulf of Alaska (Fig. 2). Glaciers were smaller farther to the north because the nearest water bodies were ice-covered for much of the year and broad continental platforms were emergent during times of glacioeustatic sea-level lowering. The repeated glaciations of late Cenozoic time had great impact even on nonglaciated parts of Alaska, where they caused formation of proglacial lakes, construction of outwash terraces, loess deposition, and isostatic depression of coastal lowlands. Because the Alaskan glacial record interrelates with so many other physical processes and climate-related features, it provides the fundamental stratigraphic framework for the late Cenozoic history of much of the state.

The positions of readily accessible glaciers along the southeastern Alaskan coast were recorded during 18th century explorations of La Perouse, Cook, Vancouver, and others, and detailed studies of those glaciers and their fluctuations began near the close of the 19th century (Reid, 1896; Gilbert, 1904; Tarr and Martin, 1914). Compilations of the statewide glacial record have been published by Capps (1932), Péwé and others (1953), Karlstrom and others (1964), Coulter and others (1965), and Péwé (1975).

Ancient as well as modern glaciers in Alaska were primarily alpine in character. Glaciers that originated in mountains of northern and central Alaska were free to expand individually until they attained equilibrium, but the much larger glacier systems south of the crest of the Alaska Range coalesced to form a vast complex that constituted the northern part of the Cordilleran Ice Sheet (Hamilton and Thorson, 1983; Clague, 1989).

The Alaskan glacial record has been strongly influenced by tectonism. Volcanoes have been active throughout late Cenozoic time along zones of active plate convergence and in some intraplate areas, and lava flows and tephra layers are important means of dating and correlating Alaskan glacial advances. Tectonic deformation of drift sheets was especially active in southern parts of the state, severely disrupting glacial records but also providing a mechanism for correlating drifts on the basis of comparable amounts and/or types of deformation. Conversely, well-dated glacial successions can provide a framework for assessing long-term rates and recurrence intervals of faulting, volcanic eruptions, and other tectonic activities.

Glacial records from northern and southern Alaska differ in fundamental respects. The glacial successions in northern and central parts of the state extend back into late Tertiary time and commonly consist of five to seven distinct major glacial episodes. Farther to the south, glacial records tend to be shorter or more fragmentary because of (1) erosion or burial of older drifts during the latest Pleistocene ice advance, (2) dissipation of sediments where glaciers flowed into the sea, (3) obliteration of glacial features by tectonic activity, and (4) accelerated erosion and deposition owing to high relief caused by tectonism. The longer, climatically responsive records of northern and central Alaska therefore provide a baseline for interpreting less complete glacial records elsewhere in the state.

Most previous mapping of glacial deposits in Alaska has utilized formally named geologic-climate units, such as "glaciations" and "stades," as defined by the Code of Stratigraphic Nomenclature (American Commission on Stratigraphic Nomenclature, 1970). These units subsequently were abandoned by the North American Commission on Stratigraphic Nomenclature (1983, p. 849) because "inferences regarding climate are subjective and too tenuous a basis for the definition of formal geologic units." I have retained the term "glaciation" but use it informally in cases where a glaciation was formally named prior to the recent revision of the stratigraphic code. I also informally use the terms "drift sheet" (allostratigraphic units) and "glacial episode" and "glacial phase" (diachronic units). For a more thorough discussion of stratigraphic nomenclature, see Hamilton and others (1986b).

Much of this chapter is based on regional summaries in Hamilton and others (1986a). Emphasis is on studies carried out since the major statewide compilation by Péwé (1975). Space constraints prohibit discussion of the Holocene glacial record and of modern glaciation in Alaska, which have been reviewed recently by Calkin (1988) and Krimmel and Meyer (1989), respectively.

Time divisions employed here generally follow those of

Hamilton, T. D., 1994, Late Cenozoic glaciation of Alaska, *in* Plafker, G., and Berg, H. C., eds., The Geology of Alaska: Boulder, Colorado, Geological Society of America, The Geology of North America, v. G-1.

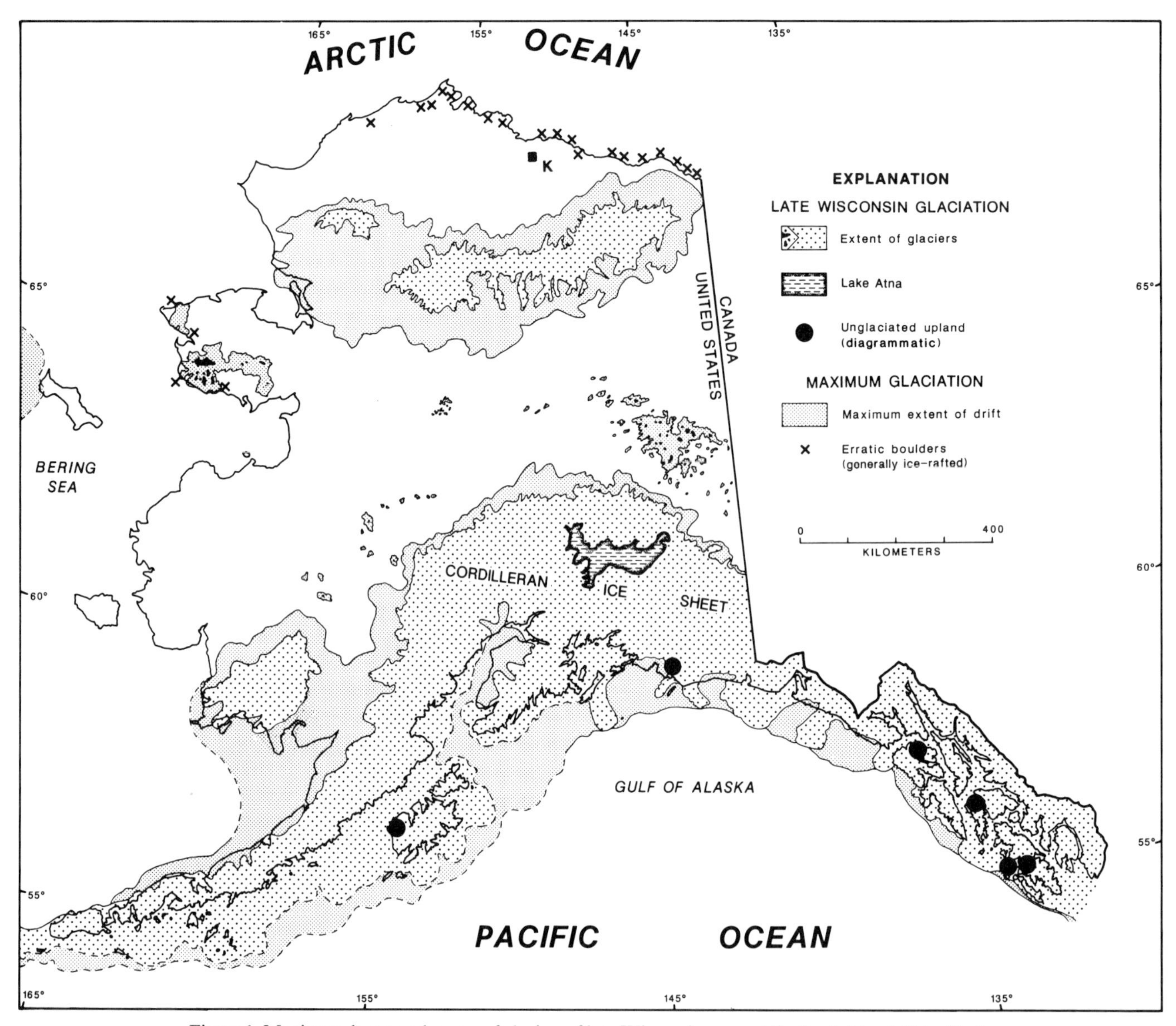

Figure 1. Maximum known advances of glaciers of late Wisconsin age and limits of older drift in Alaska (shown by dashed lines where inferred). Erratic boulders along Arctic Ocean margin represent the Flaxman Member of the Gubik Formation. K = Kuparuk gravel of Carter (1983).

Richmond and Fullerton (1986). The Pliocene-Pleistocene boundary is set at 1.65 Ma by international agreement. The early/middle Pleistocene boundary is the boundary between the Matuyama Reversed- and Brunhes Normal-Polarity Chrons, which has a generally accepted age of about 730 ka (Mankinen and Dalrymple, 1979), but which may be as old as 790 ka (Johnson, 1982). The boundary between the middle and late Pleistocene is the beginning of marine-oxygen-isotope stage 5, which Martinson and others (1987) have dated at about 125 ka. The age assigned to the Pleistocene-Holocene boundary, 10 ka, is the "nice round number" of Hopkins (1975).

For semantic simplicity, I use the general term "diamict" (Harland and others, 1966) for till-like deposits rather than the more familiar "diamicton" and "diamictite" for unlithified and indurated deposits, respectively. Till-like deposits in Alaska include abundant slightly to moderately lithified sediments of late Tertiary to early Pleistocene age, with properties intermediate between those of diamicton and diamictite.

TERTIARY GLACIATION

Glacial deposits of inferred Tertiary age occur widely in Alaska (Table 1). Many of these deposits have been faulted, tilted, uplifted, or otherwise tectonically deformed; other deposits are related to ancient valley or drainage systems that subsequently were abandoned. Depositional morphology seldom has been pre-

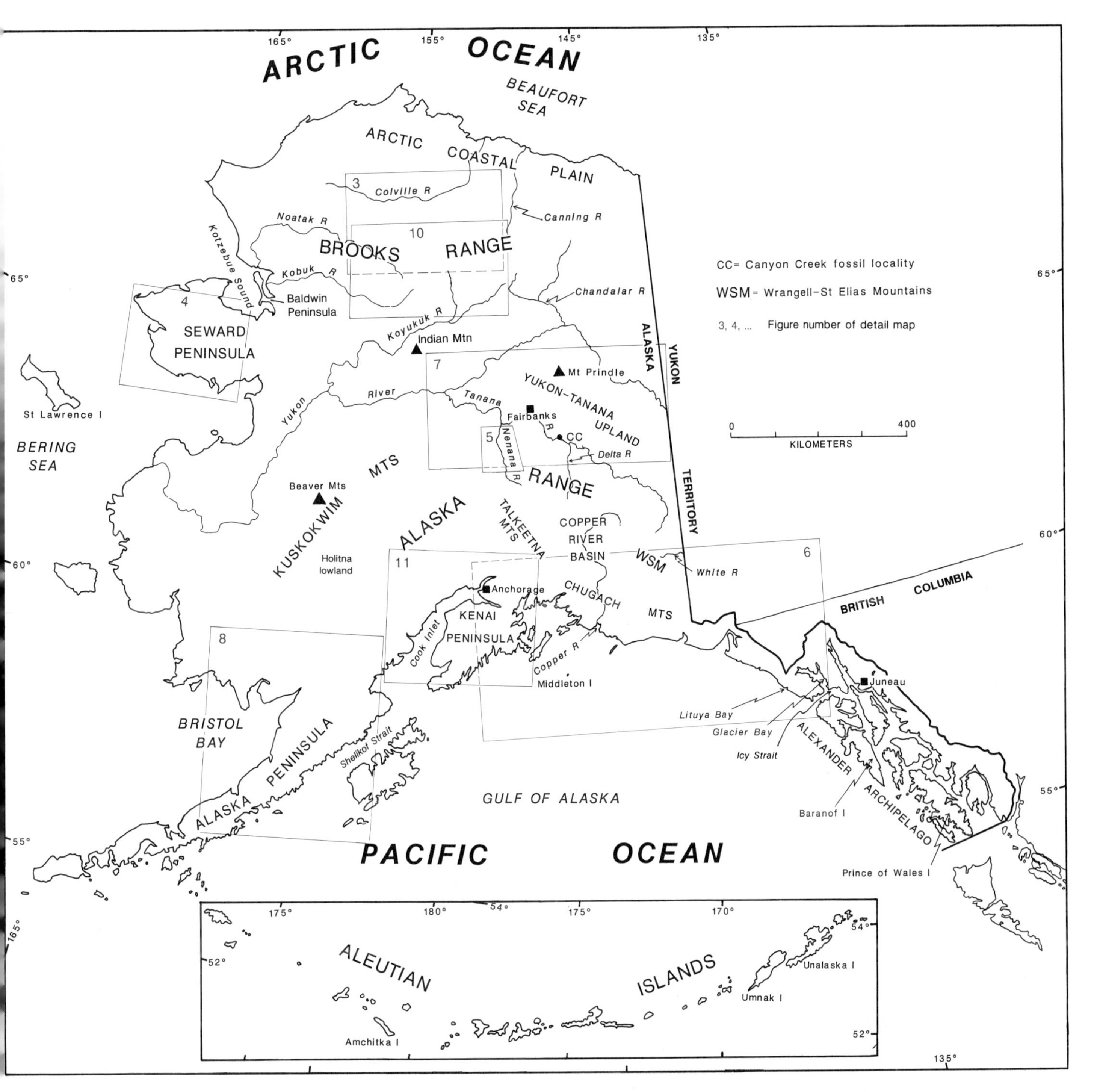

Figure 2. Index map of Alaska, showing locations of detailed maps and other localities mentioned in text.

TABLE 1. CORRELATION OF ALASKAN GLACIAL DEPOSITS

Age/Event	Arctic Coastal Plain[1]	Brooks Range[2]		Seward Peninsula[3]	Yukon-Tana Upland[4]	Kuskokwim Mountains[5]	North-Central Alaska Range[6]	
Late Wisconsin		Itkillik II	Late Phase 13-11.5 ka Main phase 24-13 ka	● 11.5 ka Mount Osborn (Esch Creek)	Salcha (Convert)	Tolstoi Lake	Riley Creek (Donnelly)	IV 10.5–9.5 ka III 12.8–11.8 ka II 15–13.5 ka I 26–17 ka
Penultimate Glaciation	Flaxman Member of Gubik Formation	Itkillik I	● >40 ka Phase B Phase A	● >40 ka Salmon Lake (Mint River) Stewart River*	Eagle (American Creek)	Bifurcation Creek	Healy (Delta)	(Late phase) (Early phase)
Middle Pleistocene		Sagavanirk-Tok River	Late phase Early phase	✧ 0.47 ± 0.05 Ma Nome River	Mount Harper (Little Champion)		Lignite Creek Bear Creek*	
			❑ 0.7 Ma	❑ 0.7 Ma				
Early Pleistocene		Anaktuvuk River (Sleepy Bear)		Sinuk	Charley River (Prindle)	Beaver Creek	Browne	
							Teklanika River	
Late Pliocene	Outwash	Gunsight Mountain		❑ 2.2 Ma Unnamed drift	Unnamed drift			
	2.4-3.5 Ma			▲ 2.8 Ma			▲ 2.8 Ma	
Older Tertiary	Kuparuk gravel						Nenana gravel	

served, but glacial origin of the deposits generally is indicated by (1) erratic boulders, (2) striated and faceted clasts, or (3) ice-rafted sand and stones in marine deposits.

Arctic Coastal Plain

The informally termed "Kuparuk gravel" (Carter, 1983), crops out east of the Colville River (see Fig. 1). It overlies Paleocene deposits and is truncated by a bluff that formed during a late Pliocene marine transgression (Carter and others, 1986b). It also is cut by fluvial terraces that are correlated with the late Tertiary Gunsight Mountain glacial interval of Hamilton (1979). Clasts as much as 1.5 m in diameter are common in the gravel. They are composed of resistant rock types common in nearby parts of the Brooks Range, suggesting that a Tertiary ice advance from that source must have reached to within 30 km of the modern coast (Carter, 1983).

Central Brooks Range

Intensely eroded drift of Gunsight Mountain age has been recognized beyond both flanks of the Brooks Range (Hamilton, 1979, 1986a, 1986b). Distinctive rock types such as the middle Paleozoic Kanayut Conglomerate (Nilsen and Moore, 1984) indicate sources deep within the range, and the lobate drift border demonstrates that some of the principal mountain valleys already were major conduits for alpine trunk glaciers by Gunsight Mountain time.

Drift remnants of Gunsight Mountain age are traceable for at least 330 km along the foothills north of the Brooks Range, where they define a series of large piedmont lobes that extended as much as 58 km beyond the range front (Fig. 3). Till remnants or erratic boulders occupy erosion surfaces that generally stand about 100 m above modern drainages and are associated with a regional terrace system along the Colville River. Rare arcuate

TABLE 1. CORRELATION OF ALASKAN GLACIAL DEPOSITS (continued)

Northwestern Alaska Range[7]	Wrangell and St. Elias Mountains[8]	Upper Cook Inlet[9]	Bristol Bay[10]	Gulf of Alaska[11]
Farewell 2	● 11.3 ka Macauley (Kluane) ● 29 ka	Elmendorf moraine 13.7-11.7 ka Naptowne	Brooks Lake: Iliuk Newhalen Iliamana (Alegnagik) Kvichak (Okstukuk)	Yakataga Formation: Eurhythmic ● 12,430 ± 100 Raven House
● 34,340 ± 1940 Farewell I (2 phases)	● >49 ka xxxxxxxxxxxxxxx Icefield	● >49 ka Knik	● >40 ka Mak Hill (Manokotak and Gnarled Mountain moraine belts)	❑ 40 ± 10 ka High terrace (drift unit) ● >72 ka
Selatna	Unnamed tills	Eklutna	Johnston Hill (Ekuk; Halfmoon Bay?)	Intense ice rafting
Lone Mountain		Caribou Hills* Mount Susitna Diamicts	Unnamed drift (Nichols Hill)	
Big Salmon Fork	▲ 2.7 ± 0.6 Ma Diamicts			❑ 2.5 Ma
	▲ 8.4 ± 0.7 Ma Diamicts	Kaloa deposits (late Miocene and younger)		Minor ice rafting

Place names below headings refer to informally named drift units. ( ) = local name used elsewhere in region; * = correlation uncertain; xxx = Old Crow tephra; ● = radiocarbon age; ✧ = Ar-Ar age; ▲ K-Ar age; ❑ = estimated age (see text for explanation).

Principal sources of data: 1. Carter, 1983; Carter and others, 1986a; 2. Hamilton, 1986a; 3. Kaufman and Hopkins, 1986; Kaufman and Calkin, 1988; 4. Weber, 1986; 5. Kline and Bundtzen, 1986; 6. Thorson, 1986; Ten Brink and Waythomas, 1985; 7. Kline and Bundtzen, 1986; 8. Denton, 1974; Denton and Armstrong, 1969; 9. Reger and Updike, 1983, 1989; Schmoll and Yehle, 1986; 10. Detterman, 1986; Lea and others, 1989; 11. Mann, 1986; Plafker, 1981; Armentrout, 1983.

morainal forms lack primary relief; they are defined principally by boulder concentrations and arcuate stream courses. Drift having nearly featureless topographic expression is more common and generally is associated with concentrations of relict boulders in the beds of streams. Elsewhere, the former drift sheet is represented only by scattered boulders of the highly resistant Kanayut Conglomerate, which are most common where loess cover of Quaternary age has been partly or completely stripped by erosion.

South of the Brooks Range, drift provisionally correlated with glacial advances of Gunsight Mountain age forms a broad (4 to 5 km), arcuate, probable end moraine that extends for about 70 km along the south margin of the Koyukuk River basin (Hamilton, 1986a). The moraine bears a thick cover of loess and lacustrine silt of Quaternary age, but its glacial origin is demonstrated by concentrations of erratic boulders along the beds and banks of rivers.

A broad U-shaped valley remnant in the west-central Brooks Range, which contains small nested end moraines of early Pleistocene and younger age, is interpreted as having been eroded during Gunsight Mountain time. Alluvial-fan deposits on the valley floor beneath the nested moraines contain cones of an extinct spruce that also occurs in the late Pliocene Beaufort Formation of arctic Canada (Matthews, 1990) but is not known to occur in Pleistocene deposits. The stratigraphic setting of the fossiliferous fan deposits indicates a probable late Tertiary age for the Gunsight Mountain glacial interval.

The Seward Peninsula

Glacial deposits older than the middle Pleistocene Nome River drift are poorly differentiated on the Seward Peninsula (Kaufman and Hopkins, 1986). Evidence for probable Tertiary glaciation is strongest in the California River area (Fig. 4), where surface erratics and erratic-bearing diamicts occur beyond the limits of moraines and drift assigned to the Sinuk glaciation of probable early Pleistocene age. Although not definitive, stratigraphic relations suggest that the pre-Sinuk diamict underlies marine transgressive deposits of late Pliocene age and is younger than a 2.8-Ma basaltic lava flow (Kaufman and Hopkins, 1986; D. M. Hopkins, personal communication, 1990); it therefore probably is a late Pliocene deposit.

North-central Alaska Range

Erratic boulders in the Nenana Gravel, a thick (1,000 to 1,300 m), weakly indurated conglomerate of late Miocene and early Pliocene age, have been described by Carter (1980) and Thorson (1986). Near the Delta River valley, boulder-bearing beds in the upper part of the Nenana Gravel dip northeast at 40 to 60° and contain clasts as much as 2 m in diameter, which include lithologies foreign to their present drainage basins (Carter, 1980). The Nenana Gravel also coarsens upward in exposures near the Nenana River valley; boulders and faceted cobbles are common near the top of the unit (Thorson, 1986). Other boulders are residually concentrated along ridge crests by erosional stripping of the Nenana Gravel. A borrow pit near the Nenana River exposes bouldery diamicton that contains grooved, striated, and faceted clasts and underlies lignite-bearing oxidized sand assignable to the Nenana Gravel (Thorson, 1986). The diamict is interpreted as lodgment till of probable late Tertiary age. A potassium-argon age determination of 2.79 ± 0.25 Ma provides a minimum age for the Nenana Gravel (Thorson, 1986).

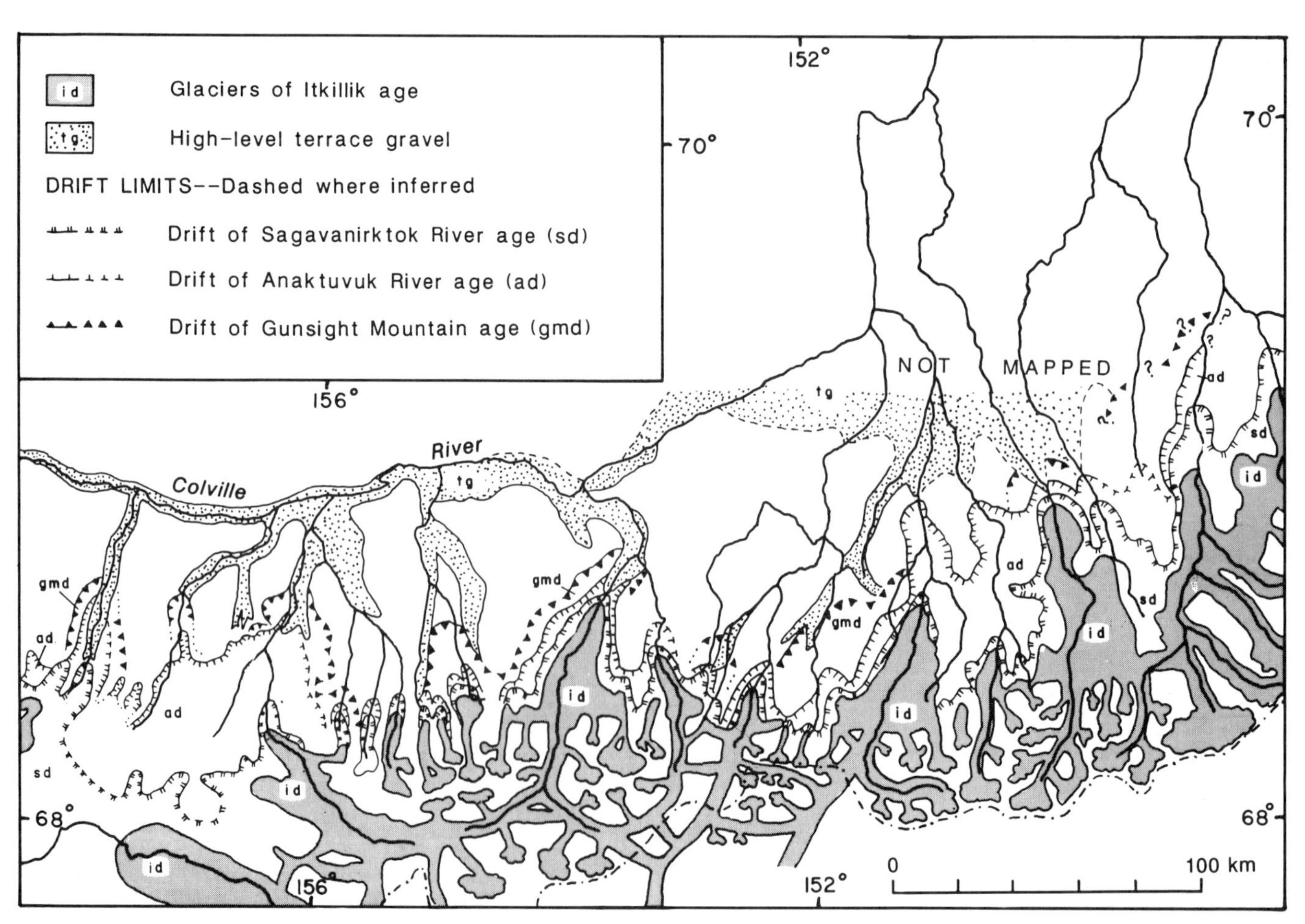

Figure 3. Glacier limits and high-level terrace gravel, northern Brooks Range and Arctic Foothills. Slightly modified from Hamilton (1986a).

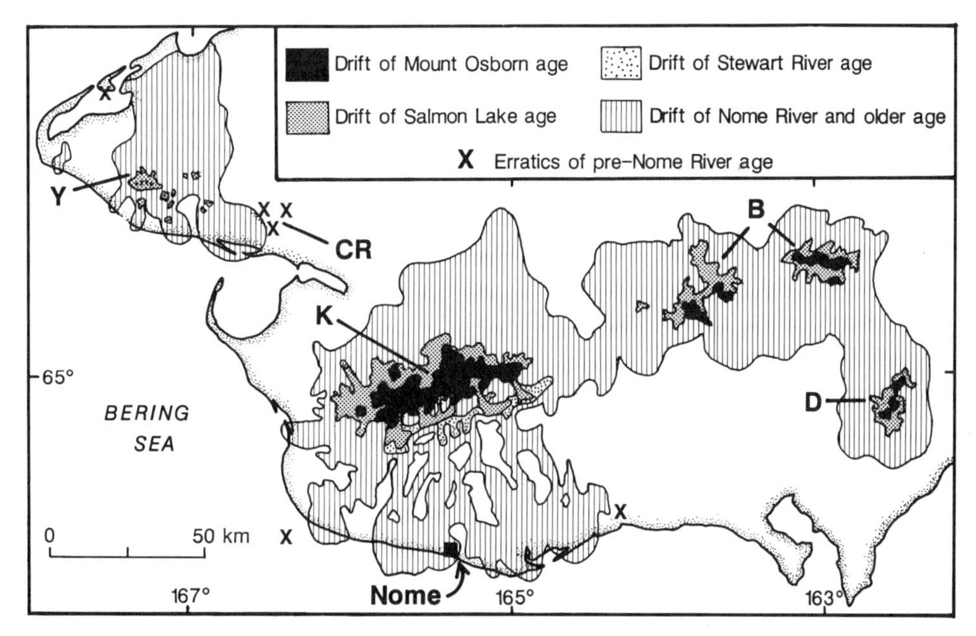

Figure 4. Seward Peninsula, showing ice advances of Mount Osborn, Stewart River, and greater age. B = Bendeleben Mountains, D = Darby Mountains, K = Kigluaik Mountains, Y = York Mountains. CR = California River locality. Modified from Kaufman and Hopkins (1986).

Teklanika River drift, a bouldery till of latest Tertiary or early Pleistocene age (Thorson, 1986), is exposed in a series of bluffs along the Teklanika River (Fig. 5). Diamict as much as 22 m thick contains striated and faceted boulders of Alaska Range lithologies in an oxidized sand-granule matrix and overlies the strongly oxidized, weakly indurated Nenana Gravel. The bluffs intersect the outer flank of a broad, gravelly morainal ridge with a thick cap of eolian sediments that defines a glacier lobe centered on the Nenana River valley and much larger than any younger glacier lobes. Because it generally was undeflected by foothills north of the Alaska Range, and because its deposits occur at altitudes much higher than those of younger drifts, the drift lobe probably was deposited before the Nenana River had incised its present valley and before erosion had etched the foothills into their present relief (Thorson, 1986).

Northwestern Alaska Range

Diamicts of the Big Salmon Fork glacial interval in foothills 15 to 25 km northwest of the Alaska Range consist of subrounded cobbles and boulders, many faceted and striated, in a silty sand matrix (Kline and Bundtzen, 1986). Associated conglomerates of poorly sorted, coarse cobble gravel may be outwash. Bedding dips northwest at 20 to 35° along the crests of hogbacks and homoclinal ridges. Clasts differ in composition from those of nearby younger glacial deposits, indicating significant changes in source areas since deposition of the diamicts and associated sediments.

Belts of lag boulders that lie beyond the limits of morainal topography in small creeks elsewhere in the foothills may also indicate very old and extensive glacier advances of late Tertiary age. Any moraines dating from these advances have either been destroyed by erosion or buried beneath younger sediments.

The Wrangell Mountains

Thick sequences of interbedded diamicts, fluvial sediments, and lava flows in fault-bounded blocks north of the Wrangell Mountains were initially described by Capps (1916) and later studied by Denton and Armstrong (1969). At least 12 separate diamicts in one of the structural blocks were identified as tillites of continental glaciers, because of (1) their massive, nonsorted, and nonstratified character; (2) presence of faceted, pentagonal, polished, and striated clasts that included allegedly exotic lithologies; and (3) identification of diagnostic glacial surface textures on sand grains (Denton and Armstrong, 1969). Potassium-argon ages on associated lavas indicated that the diamicts were deposited between 2.7 and at least 10 Ma. In one structural block, 10 diamicts underlie a flow dated at 8.4 ± 0.7 Ma, and flows dated at about 9.8 ± 0.3 to 10.2 ± 0.3 Ma are closely associated with diamicts in two other blocks.

Although Denton and Armstrong (1969) rejected alternative origins for the diamicts, Plafker and others (1977) argued that all or most of those units are actually lahars that originated on the Wrangell volcanoes. Rare glacially worked clasts in the deposits could have been shaped by alpine glaciers and later redeposited in lahars; Plafker and others (1977) found no clasts exotic to the upper parts of the contiguous Wrangell volcanoes. Recent studies by Eyles and Eyles (1989) have confirmed that the diamicts were probably emplaced as debris flows within alluvial-fan and fan-delta complexes in an actively subsiding fault-bounded basin. Eyles and Eyles (1989) concluded that the

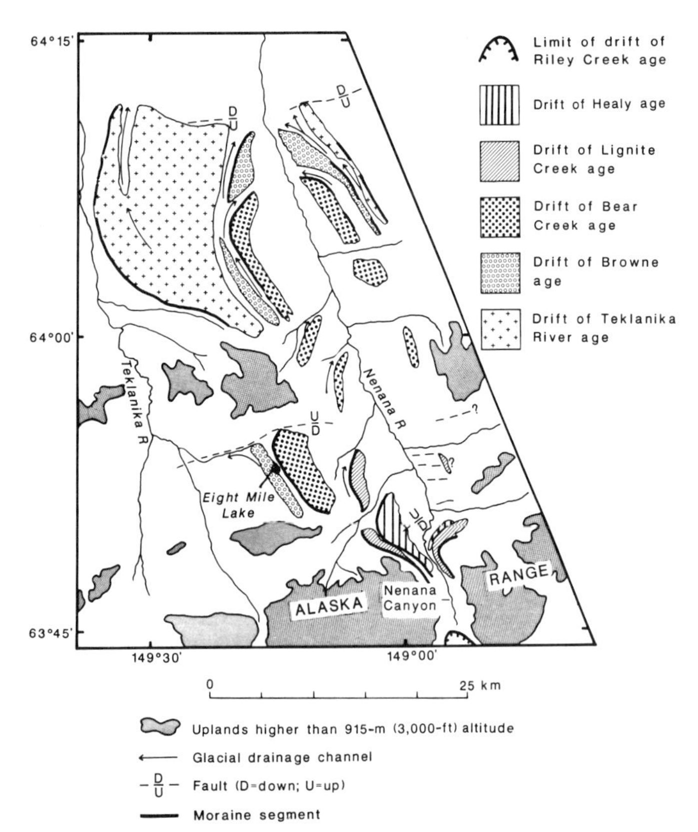

Figure 5. Major drift limits in the Nenana River valley, north-central Alaska Range. Modified from Thorson (1986).

glacially striated and faceted clasts within the diamicts were derived from deposits of local mountain glaciers rather than deposits of more extensive ice sheets, as proposed by Plafker and others (1977).

Upper Cook Inlet

Diamicts interpreted as tillites of late Miocene to Pliocene age occur in drill holes that penetrate the Beluga and Sterling Formations on the Kenai Peninsula (Boss and others, 1976, cited in Schmoll and others, 1984). Boss and others (1976) believe that tillite distribution was similar to that of late Pleistocene glacial deposits and, therefore, that the Kenai Mountains generated glaciers similar to the large piedmont lobes that covered most of the Kenai Peninsula during late Pleistocene time.

Diamicts are exposed over a distance of about 3.5 km at Granite Point on the west side of Cook Inlet (Schmoll and others, 1984, p. 15; Schmoll and Yehle, 1983). These diamicts, informally termed the "Kaloa deposits" (Schmoll and others, 1984), have a total stratigraphic thickness of about 110 m and are disconformably overlain by younger diamicts of Pleistocene age. The beds dip 1 to 6 degrees, and appear to occupy the limb of an anticline. Schmoll and others (1984) divided the Kaloa deposits into four major units: two thick massive diamicts with stone lines and two thinner units that contain interbeds of sandstone, siltstone, claystone, and (in the lower of the two units) coal. Pollen from the principal coal bed indicates that the diamicts probably are late Miocene in age.

The Mount Susitna glaciation (Karlstrom, 1964) is represented only by scattered erratics and by very old ice-scoured bedrock surfaces around upper Cook Inlet. Correlations with the central Alaska Range suggest a late Pliocene or early Pleistocene age for this glacial advance (Reger and Updike, 1989).

The Gulf of Alaska

Tertiary glaciation around the Gulf of Alaska is recorded in glaciomarine deposits exposed in the Yakataga Formation and recovered from marine piston cores. The Yakataga Formation, of Miocene through Holocene age, consists of interbedded marine, glaciomarine, and glaciofluvial deposits that locally exceed 5 km thickness (Plafker and Addicott, 1976; Plafker, 1981; Armentrout, 1983; Plafker and others, this volume, Chapter 12). It is exposed through an area of more than 30,000 km^2 along the Gulf of Alaska coast, including several islands on the continental shelf (Fig. 6). Glaciomarine deposition started either in the middle Miocene (Marincovich, 1989) or the late Miocene (Lagoe and others, 1989) in response to tectonic uplift of the Alaskan coast and to global climatic cooling, but its initiation was time transgressive owing to northwestward motion of the Yakutat terrane relative to the uplifted and glaciated coastal mountains (Plafker, 1987). Armentrout (1983) recognized three relatively cool paleoclimatic episodes associated with moderately abundant glacial sediments in lower parts of the Yakataga Formation; warmer conditions and normal marine sedimentation subsequently prevailed during most of the Pliocene. Diamicts later became abundant again at a time estimated by J. M. Armentrout (written communication, 1986) as about 2.5 Ma on the basis of correlations with onset of widespread glaciation in the North Pacific and North Atlantic regions (e.g., Carney and Krissek, 1986; Shackleton and others, 1984; Dowsett and Poore, 1990). Glaciomarine deposits continued to dominate the sediment record into the Holocene.

Piston cores taken by the Deep Sea Drilling Project (DSDP) on the deep sea floor in the Gulf of Alaska and adjoining parts of the North Pacific Ocean include abundant ice-rafted pebbles and sand. Recent revisions in diatom stratigraphy enable the onset of widespread ice rafting in the Gulf of Alaska to be dated at shortly after 2.48 Ma (Rea and Schrader, 1985). Rare erratic pebbles lower in the cores were deposited sometime prior to 3 Ma. Other DSDP cores taken north of the eastern Aleutian Islands record similar histories in which glaciers evidently first reached tidewater shortly after 2.48 Ma (Rea and Schrader, 1985).

Basal exposures of the Yakataga Formation on Middleton Island, which are described in a later section, are in part of latest Pliocene age (Mankinen and Plafker, 1987).

EARLY AND MIDDLE PLEISTOCENE GLACIATIONS

Glacial deposits of early to middle Pleistocene age are widespread in northern and central Alaska. High rates of Quater-

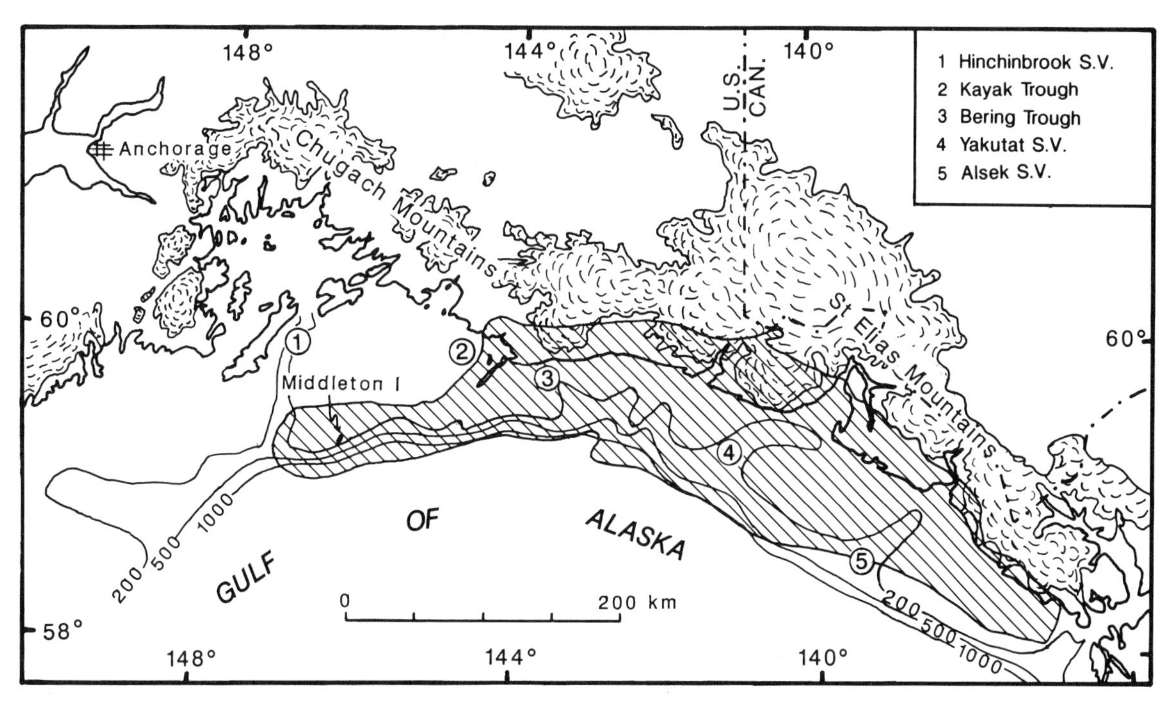

Figure 6. Extent of late Cenozoic Yakataga Formation across Gulf of Alaska continental shelf. Modern glaciers are shown by concentric dashed patterns. S.V. = Sea valleys crossing shelf. From Plafker and Addicott (1976).

nary tectonism farther to the south have caused comparable glacial deposits to be largely buried or obliterated by erosion, and drift successions older than late Pleistocene are uncommon in southern Alaska.

Two separate drift complexes are evident in many regions. The older, assigned to the early Pleistocene on the basis of paleomagnetic determinations, morphology, and position within regional drift successions, commonly was formed by glaciers that were much larger than any that formed subsequently. Early Pleistocene glaciation was followed by a long interval in which streams broadly eroded the drift sheets and incised glacial valleys to levels close to those of the present day. The younger drift complex, provisionally assigned to the middle Pleistocene, occupies modern valley floors. It forms extensive deposits in parts of northwestern Alaska, but elsewhere generally extends only short distances beyond the end moraines of late Pleistocene glaciers.

The central Brooks Range

Drift deposited during the Anaktuvuk River and Sagavanirktok River glaciations (Detterman and others, 1958; Keroher and others, 1966) is assigned respectively to the early and middle Pleistocene on the basis of limiting paleomagnetic determinations (Hamilton, 1986a) and recently revised correlations with the Seward Peninsula and Kotzebue Sound glacial sequences (Table 1).

During Anaktuvuk River time, valley glaciers in the central Brooks Range extended as much as 70 km beyond the margins of the range (Fig. 3). Glaciers flowing into the northern foothills formed paired lateral moraines; glaciers flowing south coalesced into broader piedmont lobes in the basins of the Kobuk, Koyukuk, and Chandalar Rivers. Large glaciers occupied the western Brooks Range and extended west into Kotzebue Sound. Some glacial valley systems of Anaktuvuk River age have been disrupted by tectonism, and deformation intensifies eastward toward the Canning River displacement zone of Grantz and others (1983). Drift of Anaktuvuk River age is distinguished from all younger glacial deposits by its association with ancient landscapes 50 m or more above the levels of modern streams and by the presence of tors, altiplanation terraces, and pediments on surfaces that were glaciated during Anaktuvuk River time but remained unglaciated thereafter.

In foothills north of the Brooks Range, drift of Anaktuvuk River age typically forms the outer sectors of massive compound lateral moraines several hundred meters thick (Hamilton, 1986a). End moraines typically have broad crests and gentle flanking slopes smoothly graded by solifluction (Table 2); they are covered by thick and continuous silt through which only a few large erratic boulders protrude. Drainage networks are well integrated. Kettle-like basins occur on some drift sheets, but they generally contain shallow thaw ponds that have developed in thick, postglacial, ice-rich silt. Drift sheets contain inset terraces 40 to 60 m above modern streams that continue downvalley into regional terrace systems 30 to 50 m high. Many kettles have developed on postglacial stream terraces, indicating that some glacier ice must have persisted for long periods after general deglaciation.

South of the Brooks Range, broad piedmont lobes of Anaktuvuk River age commonly terminated in proglacial lakes

TABLE 2. PHYSICAL CHARACTERISTICS OF PLEISTOCENE DRIFTS, NORTHERN AND CENTRAL ALASKA*

		Width of moraine crests (m)	Maximum flanking slopes (°)	Maximum boulder protrusion (cm)	Maximum depth of oxidation (m)
Central Brooks Range	Itkillik II	3–5 [2–3]	18–23 [16–22]	40–80 [20–30]	0.3
	Itkillik I	5–20 [5–15]	15–20 [14–20]	25–50 [≤20]	1.0–1.2
	Sagavanirktok River	75–200	2–3.5	100	>8
	Anaktuvuk River	500	1.2	10	>30
Seward Peninsula	Mount Osborn	24	14–16	70	0.2–0.3
	Salmon Lake	25	9–12	45	0.5
	Stewart River	50	6–7	30	
	Nome River	430	3.5	3	"Deep"
Yukon-Tanana Upland	Salcha	6	15–30		0.33
	Eagle	4–6	17–25		0.64
	Mount Harper	100	13–25		>0.53
	Charley River				>8
Nenana River Valley (Alaska Range)	Riley Creek	10–15	17–30		IV 0.2–0.3 III 0.3–0.5 II 0.7–0.8 I 1.0
	Healy	10–20	30		0.3
	Lignite Creek		15–20		15–20
	Bear Creek	<100	5–6		
	Browne	<50	15–25		

*Data from southern valleys of central Brooks Range are in brackets.
Note: Measurement techniques are not standardized and can vary between workers. See Kaufman and Calkin (1988) for recommended techniques.

(Hamilton, 1986a). Subdued, arcuate end moraines generally are covered by deposits of eolian and lacustrine silt more than 15 m thick. Knolls and ridges along many moraine crests were shaped by wave action; they contain washed gravel mixed with larger clasts derived from the original drift. River bluffs expose well-jointed, compact till and stony glaciolacustrine silt that are strongly oxidized along joint planes.

The Sleepy Bear glaciation, the oldest glacial advance recorded by Reger (1979) at Indian Mountain (see Fig. 2), probably is correlative with glacier advances of Anaktuvuk River age. End moraines having subdued morphology are overlapped by lacustrine sediments that formed when large glaciers from the Brooks Range crossed the Koyukuk River and created an extensive proglacial lake during Anaktuvuk River time (T. D. Hamilton, unpublished field mapping, 1987). The largest glacier of Sleepy Bear age occupied a valley floor that was incised 100 to 150 m prior to the next younger ice advance.

Glacier flow patterns of Sagavanirktok River age were similar to those of the Anaktuvuk River glaciation (Hamilton, 1986a), but glaciers were markedly smaller (Fig. 3). Paired lateral moraines in foothills north of the Brooks Range curve unbroken into valley centers, where they generally stand no more than 15 to

20 m above modern floodplains and extend only 3 to 8 km beyond the outermost terminal moraines of late Pleistocene age. Moraine crests are narrower than those of Anaktuvuk River drift, flanking slopes are steeper, and surface boulders are more abundant (Table 2). Kettles are rounded and partly filled by ice-rich silt; outwash terraces are capped by 1.5 to 2 m of frost-churned silty gravel. Outwash trains from late Pleistocene ice advances farther upvalley commonly contain kettles where they intersect end moraines of Sagavanirktok River age, indicating that subsurface glacier ice must still have been present when the younger glaciers advanced.

Glaciers of Sagavanirktok River age filled structural troughs along the south flank of the Brooks Range that had formed during a long interval of tectonism and nonglacial erosion that followed the Anaktuvuk River glaciation (Hamilton, 1989); they commonly terminated in lakes (Hamilton, 1986a). Drift is not as broadly eroded as is Anaktuvuk River drift, and it occurs close to modern valley floors. Some river bluffs expose compact, jointed, sandy to gravelly till, but silty glaciolacustrine diamict predominates elsewhere. Multiple end moraines and stratigraphic relations suggest two major ice advances in some valleys.

Till of Sagavanirktok River age, exposed in a bluff along the Kobuk River, overlies interglacial deposits that are magnetically reversed at the base but magnetically normal above (Hamilton, 1986a); the deposits probably span the boundary between the Matuyama Reversed- and Bruhnes Normal-Polarity Chrons. Because this part of the Kobuk River valley had been eroded previously during Anaktuvuk River time, the paleomagnetic data suggest that the Anaktuvuk River and Sagavanirktok River ice advances are of early and middle Pleistocene age, respectively.

The Seward Peninsula and Kotzebue Sound

Drift deposits of the Sinuk glacial interval, Nome River glaciation, and Stewart River glacial interval have been described by Kaufman and Hopkins (1986), who considered them to be of late Tertiary, early Pleistocene, and middle Pleistocene age, respectively. However, recent radiometric age determinations and amino-acid and paleomagnetic analyses suggest that the Sinuk drift may be of early Pleistocene age and that the succeeding Nome River advance probably took place during middle Pleistocene time (Table 1).

Sinuk drift consists of: (1) surface deposits with little primary glacial relief, (2) isolated erratic boulders, and (3) subsurface drift in excavations, natural exposures, boreholes, and marine seismic profiles. Surface exposures generally are limited to erratic boulders that are concentrated in the beds of streams, lie scattered on bedrock, or are exposed by placer mining. Tors and altiplanation terraces occur in upland areas glaciated during the Sinuk glacial interval, but these features have not been observed on surfaces covered by younger ice advances (D. M. Hopkins, oral communication, 1990). This relation is similar to that on surfaces glaciated during Anaktuvuk River time in the Brooks Range, suggesting that the Sinuk and Anaktuvuk River advances could be correlative.

Drift of the succeeding Nome River ice advance extends well beyond the major glaciated uplands of the Seward Peninsula (Kaufman and Hopkins, 1986; Kaufman and others, 1988). Glaciers were particularly extensive in the southwest part of the peninsula, where they reached the coast around Nome (Fig. 4). This distribution pattern suggests that glaciers received considerably more nourishment during Nome River time than during subsequent glaciations. Much of the Bering platform may have remained submerged during glacier growth, either due to tectonism (Hopkins, 1973) or because local glaciation began before onset of global sea-level lowering (J. Brigham-Grette, written communication, 1990). Morainal ridges of Nome River age generally are broad and smooth, with gentle slopes and thick, continuous loess cover (Kaufman and Calkin, 1988; Table 2).

The Baldwin Peninsula in Kotzebue Sound consists of marine, estuarine, glaciomarine, and glacial sediments that were deformed by glaciers flowing westward out of the Kobuk River valley and the Selawik River valley, which parallels the Kobuk valley to the south, and possibly southward out of the Noatak valley (Huston and others, 1990). The Baldwin Peninsula moraine is correlated with moraines of Nome River age on the Seward Peninsula by Brigham-Grette and Hopkins (1989) because of morphologic similarities and because both drift complexes represent the last extensive glaciation around the east margin of the Bering and Chukchi Seas. If the Baldwin Peninsula moraine were formed by glaciers that filled valleys of the western Brooks Range, it may correlate with the Anaktuvuk River glaciation as inferred by Hamilton (1986a, b) and by Kaufman and Hopkins (1986). However, if the moraine were formed by local ice caps in the lower Kobuk–Selawik River area, it more likely would correlate with the Sagavanirktok River glaciation of the central Brooks Range as concluded earlier by Hamilton and Hopkins (1982). Glaciers of Sagavanirktok River age in the central Kobuk River valley discharged into a large glacial lake (Hamilton, 1984) that may have formed behind an ice dam farther down the Kobuk River.

Age estimates for the Nome River glaciation are based on radiometric age determinations, paleomagnetic measurements, and amino-acid geochronology. Glacial deposits north of the Bendeleben Mountains on the Seward Peninsula are overlapped by a magnetically normal basaltic lava flow for which a potassium-argon age of 0.81 ± 0.08 Ma had been reported (Kaufman and Hopkins, 1986), but recent Ar-Ar micrograin analyses indicate contamination by an older phase of inherited argon (D. S. Kaufman, written communication, 1991). By excluding the older phase, D. S. Kaufman and R. C. Walter have determined a provisional age of 470 ± 190 ka for the basalt. Fossil molluscan shells from marine transgressive deposits that underlie Nome River till near Nome have amino-acid ratios that suggest a middle Pleistocene age (Kaufman and others, 1989b) of somewhere between 290 and 580 ka depending on the paleotemperature history of the site (D. S. Kaufman, written communication, 1990). Till from the type locality for the Nome River glaciation is normally magnetized, confirming that this glacial advance took place during the Brunhes Normal-Polarity Chron, and therefore is no older than about 730 ka (D. S. Kaufman, written communication, 1990). The glacially deformed marine and glaciomarine units on Baldwin Peninsula also have normal magnetic polarity, and amino-acid ratios on fossil molluscan shells suggest correlation with other marine deposits in northern Alaska that formed about 500 to 600 ka (Huston and others, 1990). Although the above data do not demonstrate an exact age for the Nome River glaciation, they show that it must be a middle Pleistocene event and that it most likely took place sometime between about 500 and 400 ka.

The Stewart River glacial interval is represented by smooth moraines with gravelly crests and thin loess 5 to 10 km beyond the limits of younger glacial deposits on the south side of the Kigluaik Mountains (Fig. 4; Kaufman and Hopkins, 1986; Kaufman and Calkin, 1988; Kaufman and others, 1988). Kettles generally are dry, stable, flat-bottomed depressions, but steep-sided and unstable dead-ice terrain on a Holocene terrace inset across one end moraine indicates that buried glacier ice may still be present locally. Relative-age criteria suggest that the Stewart River ice advance is closer in age to the drift of the penultimate glaciation than to the preceding, much more extensive ice advance of Nome River time. Its correlations with the Brooks Range glacial sequence are unclear.

Yukon-Tanana Upland

The glacial history of the Mount Prindle area, 85 km northeast of Fairbanks (Fig. 7), was described by Weber and Hamilton (1984). They correlate their oldest ice advance, the Prindle glaciation, with the Anaktuvuk River glaciation of the Brooks Range because of: (1) its great extent relative to all other ice advances, (2) major drainage changes and valley incisions since Prindle time, and (3) presence of tors on uplands last glaciated during Prindle time. They also correlate drift of the succeeding Little Champion glaciation, which occurs on modern valley floors, with drift of Sagavanirktok River age in the Brooks Range. Weber (1986) subsequently showed that this glacial succession extended throughout the Yukon-Tanana Upland. She correlated the oldest regional Pleistocene ice advance—the Charley River glacial episode—with the Prindle glaciation and the succeeding Mount Harper glacial episode with the Little Champion glaciation (Table 1).

Ice caps of Charley River age probably developed on all mountains above 900 m altitude. Outlet glaciers radiated to distances as great as 56 km and formed the most extensive glacial features preserved in the Yukon-Tanana Upland (Fig. 7). Evidence for the Charley River glacial episode consists primarily of concentrations of large erratic boulders at the downvalley limits of highly modified U-shaped troughs that radiate from the higher peaks. Possible moraines of Charley River age form solifluction-covered benches along some valley sides.

All mountains rising above 1,200-m altitude were glaciated during Mount Harper time (Weber, 1986). Valley glaciers extended as far as 29 km and reached altitudes of about 600 m (Fig. 7). Most end moraines of Mount Harper age occur in lower valleys and are largely vegetated (Weber, 1986). Their crests commonly are broad and flat, and they typically are fronted by broad outwash fans. Original moraine forms commonly are obscured by solifluction and talus deposits. Boulder surfaces are etched deeply, and many granite clasts have completely disintegrated. Multiple end moraines in many valleys indicate that the Mount Harper glacial episode probably consisted of at least two major ice advances (Weber, 1986).

The Kuskokwim Mountains

At least 13 isolated highlands within the Kuskokwim Mountains were sufficiently massive to generate alpine glaciers during Quaternary time (Kline and Bundtzen, 1986). Bundtzen (1980) recognized four major glacial advances in the Beaver Mountains. The oldest drift, assigned to the Beaver Creek glaciation, resembles early Pleistocene glacial deposits of the Seward Peninsula and the Yukon-Tanana Upland.

During Beaver Creek time, glaciers from a probable ice cap breached stream divides at high altitudes and formed long, planed upland surfaces (Kline and Bundtzen, 1986). Distribution of till and erratic boulders suggest that outlet glaciers advanced 14 to 26 km down each of the four major valleys that radiate from the center of the Beaver Mountains. Morainal landforms are absent, and evidence for former glaciation consists primarily of scattered patches of till, isolated erratic boulders, faint trim lines, and ice-marginal meltwater channels. The drift is covered by 1 to 3 m of loess that supports numerous thaw lakes.

North-central Alaska Range

Thorson (1986) recognized at least three glacial advances of probable early and middle Pleistocene age within the Nenana River valley (Fig. 5). From oldest to youngest, these are the Browne glaciation (Wahrhaftig, 1958) and the Bear Creek and Lignite Creek glacial episodes of Thorson (1986). The Browne glaciation probably is early Pleistocene in age, and the Lignite Creek event is late middle Pleistocene; the age assignment of the Bear Creek glacial episode is uncertain. The Teklanika River advance (discussed earlier in this chapter) could be either latest Tertiary or early Pleistocene in age.

The Browne glaciation was defined by Wahrhaftig (1958) on the basis of large, erratic boulders of granite on upland surfaces and in stream beds. Thorson (1986) later described probable paired end moraines of Browne age east and west of the Nenana River that are flanked by arcuate meltwater channels (Fig. 5). Original constructional forms are well defined, and ice-marginal channels and proximal outwash aprons also retain primary morphology (Table 2). A meltwater channel system east of the Nenana River grades into a broad fan of outwash gravel that is truncated either by a fault scarp or by a monocline about 100 m high.

The outer moraine of the Bear Creek drift sheet consists of arcuate, erratic-littered ridges east and west of the Nenana River that are flanked by marginal stream courses (Thorson, 1986). The ridges cross-cut moraines of Browne age and are inset within them. A correlative moraine segment near Eight Mile Lake contains bouldery gravel in which some clasts retain original glacial facets and grooves. Because of their cross-cutting and inset relations and because they appear to have higher surface relief and better internal drainage than drift of Browne age, the Bear Creek moraines probably represent a distinct younger glacial episode (Thorson, 1986). The Bear Creek glacier must have advanced down the Nenana valley when it was incised about 5 to 100 m below drift of Browne age and when the axis of the valley had shifted slightly westward.

The Lignite Creek glacial episode was named for a series of arcuate morainal ridges and intervening ice-marginal drainage channels close to the modern floor of the Nenana River valley (Thorson, 1986). These features are less subdued morphologically than drift of Bear Creek age. The ridges of pebble-cobble gravel mixed with sparse rounded boulders stand about 30 m high. Ice-marginal channels of Lignite Creek age are cut into gently sloping pediments and fans that developed after glacier retreat from the moraines of the Bear Creek advance.

Farther to the east, the early and middle Pleistocene glacial record in the Alaska Range is extremely fragmentary owing to uplift and erosional stripping within the mountains and to probable deep subsidence of glacial deposits that accumulated beyond their north flank in the Tanana River valley. An indirect record of glaciation in the Alaska Range is provided by the thick loess deposits that are exposed in mining excavations near Fairbanks (Péwé, 1989) and in river bluffs elsewhere in central Alaska (e.g., Edwards and McDowell, 1989; Begét and others, 1991). Paleomagnetic studies of the loesses (Begét and Hawkins, 1989; Begét and others, 1990) and dating of associated tephras (Stemper and others, 1989) promise to yield useful age estimates on major Pleistocene glacial and climatic fluctuations within the north-central Alaska Range.

Northwestern Alaska Range

Kline and Bundtzen (1986) describe two drift complexes that are younger than deformed Tertiary deposits but predate glacial deposits assigned to the late Pleistocene. The oldest drift,

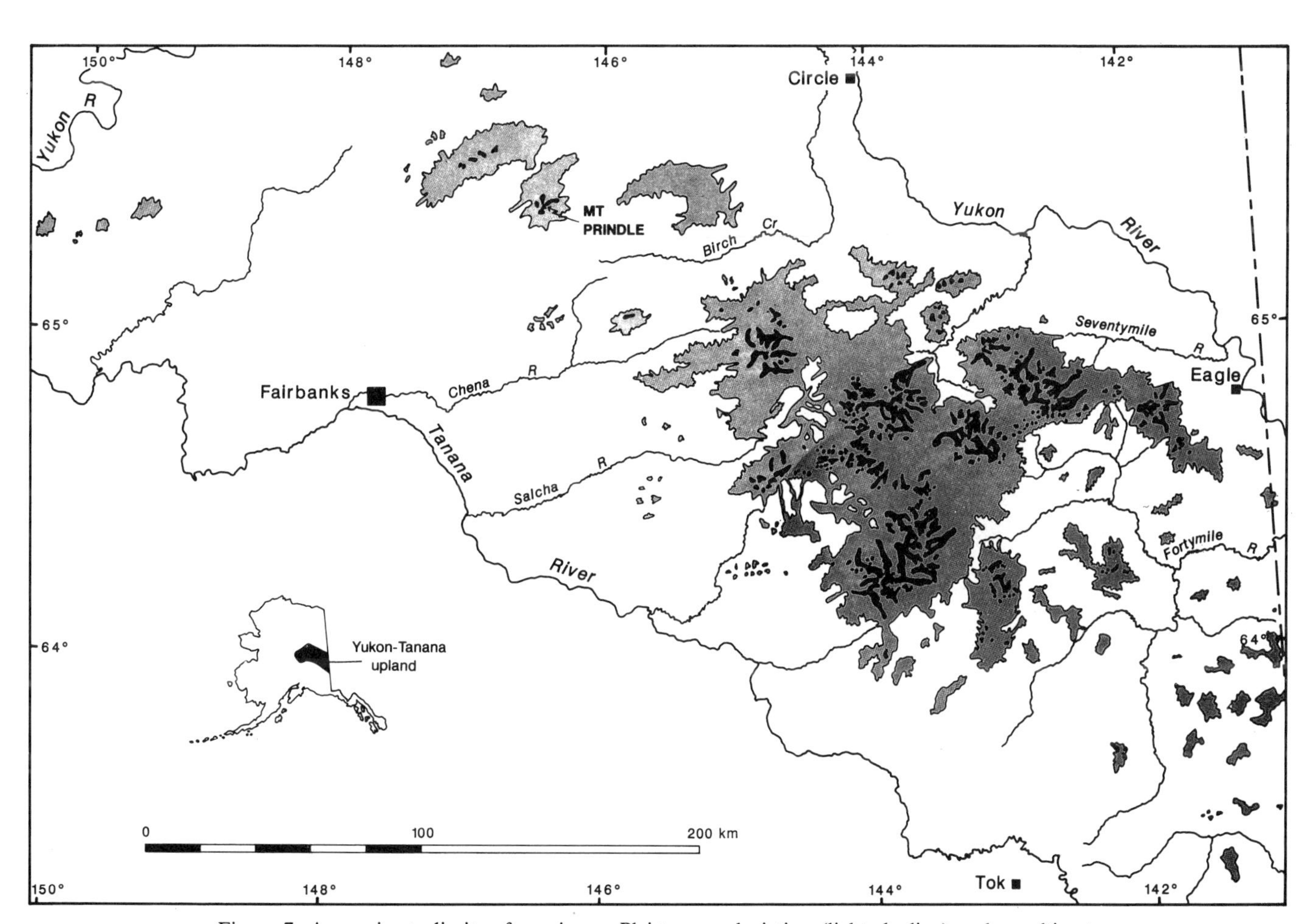

Figure 7. Approximate limits of maximum Pleistocene glaciation (light shading) and penultimate glaciation (Eagle glacial episode; dark shading) in the Yukon-Tanana Upland. Compiled by F. R. Weber, U.S. Geological Survey.

which formed during the Lone Mountain glaciation, is probably early Pleistocene in age. Younger glacial deposits, termed the Selatna drift by Fernald (1960), probably formed during the middle Pleistocene.

Glacial deposits that formed during Lone Mountain time are preserved only as erratic boulders and isolated patches of drift beyond the limits of younger drift sheets (Kline and Bundtzen, 1986). Drift remnants lack primary morainal morphology and have been smoothly graded by solifluction, but their glacial origin is indicated by erratic boulders derived from source areas deep within the Alaska Range. The deposits define several piedmont lobes that extended 40 to 50 km from the Alaska Range and terminated 5 to 10 km beyond the limtis of younger glacial advances.

Arcuate end moraines of Selatna age were deposited by piedmont lobes that extended as far as 40 km from the mountain front (Kline and Bundtzen, 1986). Subdued hummocky knob and kettle topography is common; kettles are partly or wholly drained, and most are integrated into postglacial drainage systems. Loess is more than 10 m deep locally. Vertical displacement along the Farewell fault system has offset some moraines by as much as 40 m. At least three nested moraine systems occur in some valleys, and morphologic differences suggest that they may span a long interval (Kline and Bundtzen, 1986).

Southern Alaska

The early and middle Quaternary glacial record of southern Alaska is extremely fragmentary because of late Quaternary tectonism and burial or destruction by the extensive glaciation of late Pleistocene age. Known drift deposits are described briefly here, but the attempted correlations in Table 1 are speculative.

Upper Cook Inlet. Glaciers from mountains to the north, east, and west flowed into upper Cook Inlet, where at various times they either coalesced or calved individually into tidewater (Hamilton and Thorson, 1983; Schmoll and Yehle, 1986; Reger and Updike, 1983, 1989). They generally did not leave clearly identifiable end moraines. Karlstrom (1964) defined five major Pleistocene ice advances in this region (Table 1). The oldest two advances, the Mount Susitna and Caribou Hills glaciations, are considered to be late Pliocene to early Pleistocene in overall age by Reger and Updike (1989), who believe that glaciers probably flowed down Cook Inlet to form an extensive ice shelf in the Gulf of Alaska during each of these events. Drift on Mount Susitna (see Fig. 11 for location) consists of scattered erratics on an abraded rock surface, and Reger and Updike (1983, 1989) describe similar ice-scoured upland surfaces elsewhere in upper Cook Inlet. The distribution of erratics and ice-modified surfaces indicates glacier flow patterns unrelated to modern valley systems.

The succeeding Caribou Hills glaciation was named for weathered drift in the southern Kenai lowland (see Fig. 11 for location), where glacial drainage channels and dissected lateral moraines and kame terraces still are recognizable (Reger and Updike, 1983, 1989). A possibly correlative deposit, the featureless, bouldery Mt. Magnificent ground moraine, occupies a glacially planed surface below the level of scattered mountain-top erratics near Anchorage (Schmoll and Yehle, 1986; Yehle and Schmoll, 1989). Throughout upper Cook Inlet, ice-scoured benches and truncated spurs occur at a distinct level between surfaces of Mount Susitna age and well-preserved features of younger ice advances (Reger and Updike, 1983, 1989).

Subdued moraines of the succeeding Eklutna glaciation, which occur within present drainage systems, were formed by the last Pleistocene ice masses to coalesce and entirely cover upper Cook Inlet (Reger and Updike, 1983). Ice may have overflowed westward across the Alaska Peninsula at that time (Hamilton and Thorson, 1983), forming moraines around the head of Bristol Bay that possibly correlate with Halfmoon Bay(?) drift described in the following section. Reger and Updike (1983, 1989) and Schmoll and Yehle (written communication, 1986) believe that glacier advances of Eklutna age must predate the last interglacial maximum (marine-isotope stage 5e) and therefore would be of middle Pleistocene age.

Bristol Bay region. Detterman (1986) described two drifts of possible early and middle Pleistocene age that are exposed on the Alaska Peninsula along the southeast shore of Bristol Bay. The oldest drift, which is unnamed, lacks glacial morphology and supports fully integrated drainage systems. Coastal bluffs expose moderately indurated till that unconformably underlies marine sediments and younger glacial deposits (Fig. 8). Upper surfaces of the drift and overlying marine deposits are oxidized, and both of those units are more indurated than overlying glacial sediments. The next-younger glacial unit, drift of the Johnston Hill glaciation, forms subdued morainal morphology (Detterman, 1986). Erratic lithologies suggest that glacial source areas were predominantly to the north and northeast of the Alaska Peninsula. Drift exposed in shore bluffs is weakly cemented and oxidized, and many clasts are partly disintegrated. Deposits within 40 m of present sea level have wave-cut scarps that formed when the sea stood higher against an isostatically depressed coast.

A possibly correlative glacial sequence in the Nushagak lowland along the north shore of Bristol Bay has been described by Lea and others (1989). The Nichols Hill drift unit, the oldest of the glacial deposits, was formed by a very extensive glacial advance from the Ahklun Mountains (Fig. 8). It is visible in coastal bluffs, but is nowhere exposed at the surface. The succeeding moraines, designated the Ekuk moraines by Lea and others (1989), were deposited by coalesced glacial lobes from the Ahklun Mountains (Fig. 8). Exposures in coastal bluffs show proglacial fluvial and intertidal deposits that have been strongly deformed by ice thrusting (Lea, 1990a). The Ekuk moraines may be correlative with the Halfmoon Bay(?) drift near the head of Bristol Bay, which was deposited from glaciers in the Alaska Range, and possibly also with Johnston Hill drift of the Alaska Peninsula (Table 1).

Copper River Basin. Glaciers have repeatedly flowed into the Copper River Basin from source areas in the Alaska and

Chugach Ranges to the north and south and in the Talkeetna and Wrangell Mountains to the west and east. The oldest recognized ice advances, several of which may have formed ice caps covering the entire basin (Nichols, 1989), probably date from the early and middle Pleistocene. A series of volcanic debris flows that originated in the Wrangell Mountains are interstratified with alluvium and glacial deposits in river bluffs along the Copper River and several of its tributaries (Nichols and Yehle, 1985; Richter and others, 1988). Radiometric dating of flow components or rocks from their source areas should provide a chronologic framework for much of the glacial record in the Copper River Basin.

Aleutian Islands. Extensive ice caps covered most of the Aleutian Islands during late Wisconsin time, and little evidence for earlier glaciation has been preserved (Black, 1983; Thorson and Hamilton, 1986). Glacial deposits older than late Pleistocene are recognized only on Amchitka Island, where indurated and faulted till is considered middle Pleistocene in age (Gard, 1980).

Gulf of Alaska. Exposures of the Yakataga Formation on Middleton Island document continental shelf sedimentation during latest Pliocene and early Pleistocene time—about 2.2 to 1.0 Ma (Mankinen and Plafker, 1987). Thick marine diamicts contain faceted and striated boulders that originated in the Chugach and St. Elias Mountains; their sandy mud matrix is similar to glacially derived nearshore mud in the modern Gulf of Alaska (Plafker and Addicott, 1976; Plafker, 1981; Eyles, 1988). The diamicts contain striated boulder pavements that represent repeated expansion of a partly floating ice shelf to the edge of the continental shelf (Eyles, 1988; Eyles and Lagoe, 1989). Submarine channel-fill deposits in the basal part of the unit probably formed when glaciers were less extensive. Foraminiferal biofacies indicate that sea level fluctuated by as much as 100 m during deposition of the Yakataga Formation, and that these fluctuations probably were caused by glacial eustacy (Lagoe and others, 1989).

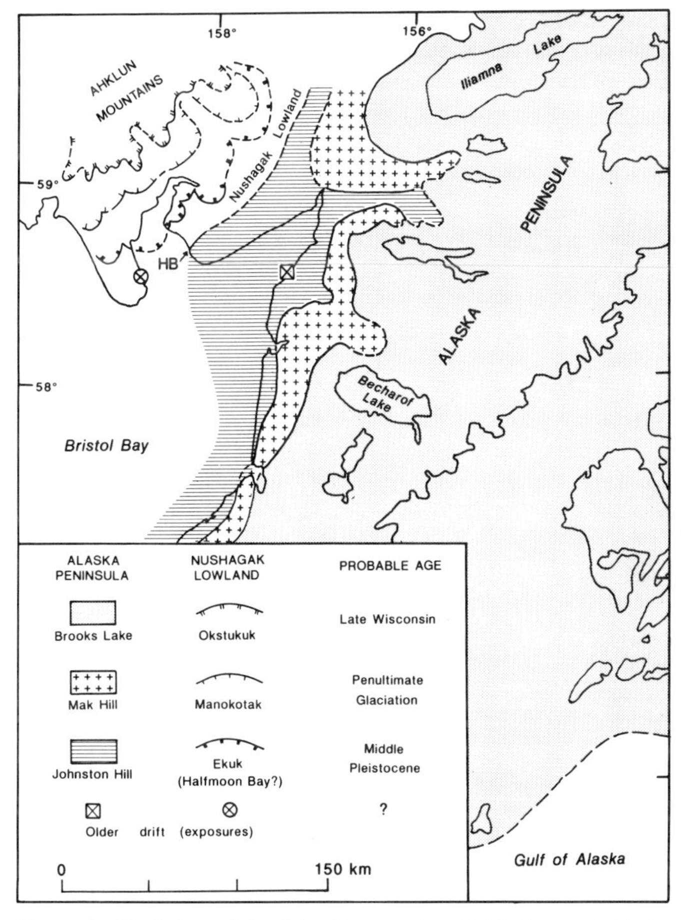

Figure 8. Glaciation of the Bristol Bay region. Drift sheets of Alaska Peninsula (shown by patterns) are from Detterman (1986) and Riehle and Detterman (1993); moraine belts of Nushagak lowland are from Lea and others (1989). Oldest drift of Nushagak lowland is the Nichols Hill drift of Lea and others (1989); oldest drift of Alaska Peninsula is unnamed. HB = outer limit of Halfmoon Bay(?) drift.

Abundant glacially derived sediments on the continental shelf around Middleton Island suggest that subsequent glacier advances must have reached this area, but the age of those advances is uncertain (Molnia, 1986). Marine sediment cores suggest nine ice-rafting maxima of middle Pleistocene age in the Gulf of Alaska (von Huene and others, 1976). The most intense ice rafting shown in these cores occurred about 400 ka, and ice rafting has been nearly continuous during the past 300 k.y.

LATE(?) PLEISTOCENE GLACIATIONS

Introduction

Glacial deposits of the younger ice advances in Alaska typically form rugged drift sheets, with primary depositional morphology only slightly to moderately altered. Two distinct components generally are recognized in northern and central Alaska. The younger drift is bracketed by radiocarbon ages of about 24 and 11.5 ka and therefore is equivalent in age to the late Wisconsin glaciation of midcontinental North America (Hamilton, 1982a, b) and to oxygen-isotope stage 2 of the marine sediment record. (For brevity, marine oxygen-isotope stages are termed "stage 1, stage 2," etc., through the remainder of this chapter.) Extensive ice advances of late Wisconsin age obliterated older glacial deposits in much of southern Alaska, where some areas may have remained glaciated throughout middle Wisconsin (stage 3) time.

Deposits of the next-older (penultimate), more extensive ice advance are beyond the range of radiocarbon dating, and their age assignment is controversial (Vincent, 1989; Fig. 9). Although generally inferred to be younger than the last interglacial maximum (stage 5e; e.g., Hamilton, 1986b; Heginbottom and Vincent, 1986), several recent studies have challenged that age assumption. Evidence against an early Wisconsin (stage 4) or late stage 5 age for the penultimate glaciation includes:

1. Redating of the Old Crow tephra, a widespread stratigraphic unit in Alaska and the Yukon (Westgate and others, 1983, 1985; Schweger and Matthews, 1985), from a formerly accepted age estimate of 86 ± 8 ka (Wintle and Westgate, 1986) to a present age estimate of 149 ± 13 ka (Westgate, 1988). The Old Crow tephra may have been deposited during the waning

phase of the penultimate glaciation in several parts of Alaska and the Yukon (Westgate and others, 1983; Hughes and others, 1989; Waythomas, 1990).

2. In the Yukon Territory, soils of apparent interglacial character developed on deposits of the Reid glaciation (Tarnoci, 1989), which generally is correlated with the penultimate glaciation of Alaska (e.g., Hamilton, 1986b; Heginbottom and Vincent, 1986).

3. Stratigraphic evidence suggests that the penultimate glaciation in parts of south-central and southwestern Alaska may be older than stage 5e. For example, drift of the penultimate Knik glaciation of the Upper Cook Inlet is overlain by the Goose Bay peat unit, which has an interglacial pollen flora (Ager and Brubaker, 1985) and a thermoluminescence age of about 175 ka (Reger and Updike, 1989).

4. In northern Alaska, morphologic "freshness" of drift of the penultimate glaciation, which has been used to argue for a post–stage 5e age, may be due in part to relict glacier ice that remained within these deposits and that continued to melt out during the Holocene (Hamilton, 1982b, 1986a).

5. "Early Wisconsin" deposits in several other regions of North America have been reassigned to greater ages following new geochronologic or stratigraphic studies (e.g., Curry, 1989; Easterbrook and others, 1981; Hughes and others, 1989; Pierce and others, 1976).

None of these lines of evidence, however, conclusively proves that penultimate glaciation through all of Alaska is of pre–stage 5e age, and equally strong arguments can be made for a younger age assignment. For example:

1. Deposits of the penultimate and late Wisconsin ice advances in the Brooks Range are components of a single drift sheet (Porter, 1964) whose outer limit is a more striking discontinuity than any within the drift complex (Hamilton, 1986a).

2. In areas of active tectonism, drift of the penultimate glaciation has been disrupted to a lesser degree than would be expected if it preceded stage 5e (i.e., if it formed prior to about 130 ka). For example, the Denali fault has undergone 50 to 60 m of right-lateral movement during the last 10 k.y. where it crosses the Delta River valley (Stout and others, 1973), yet during the interval that elapsed between the penultimate and late Wisconsin glaciations that glacial trough was not dislocated sufficiently to alter glacier flow patterns through it.

3. Little stratigraphic evidence is available for presence of interglacial deposits above drift of the penultimate glaciation. For example, the penultimate glaciation of the Nushagak lowland was followed by long intervals of thaw-lake and organic deposi-

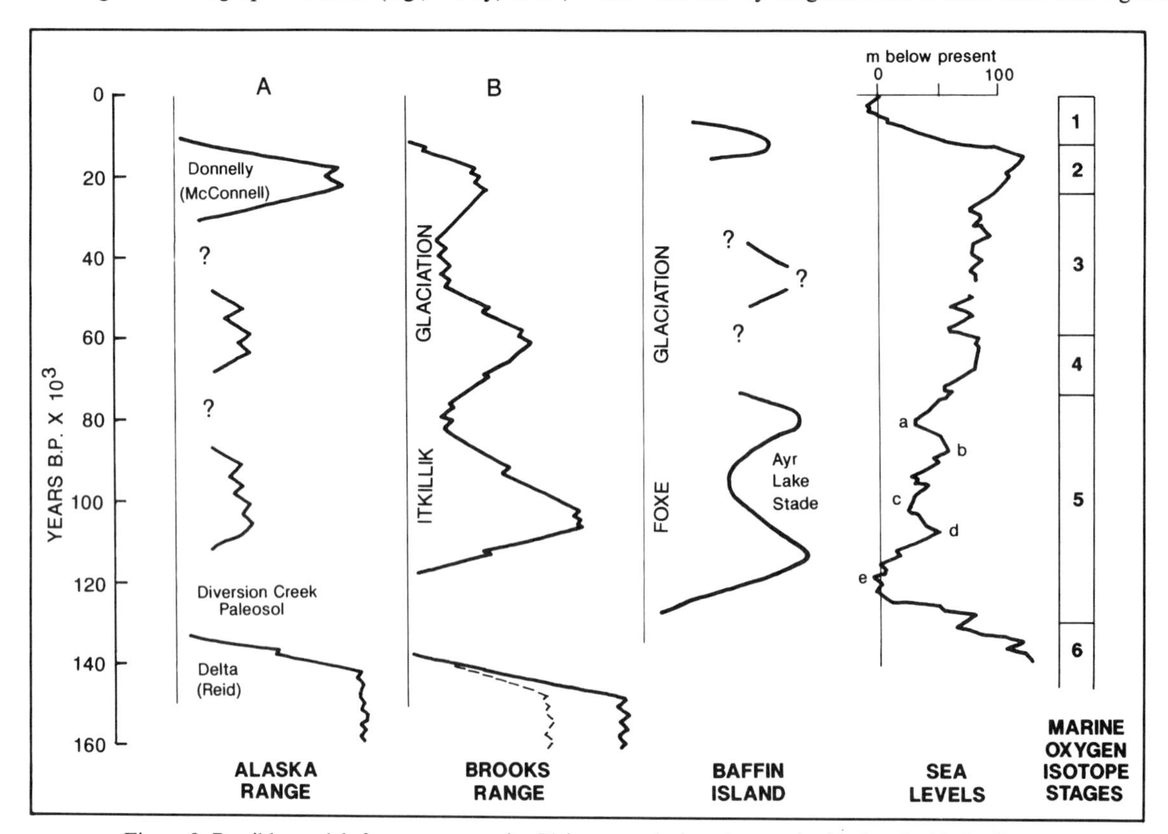

Figure 9. Possible models for youngest major Pleistocene glacier advances in Alaska. A, Alaska Range model. Penultimate glaciation preceded stage 5e; subsequent ice advances are obliterated by extensive late Wisconsin (stage 2) advance. B, Brooks Range model. Penultimate glaciation followed stage 5e and either partly overlapped or obliterated drift of late stage 6 (shown by solid and dashed lines, respectively). Baffin Island glacial record (from Andrews and Miller, 1984) and worldwide sea levels estimated from marine oxygen isotope record (from Shackleton, 1987) are shown for comparison.

tion under climates that were cooler than those of the Holocene (Lea and others, 1989).

4. On the arctic coast of Alaska, the Flaxman Member of the Gubik Formation, which records the breakup of an extensive ice shelf in the Canadian Arctic (see following section), is older than 52 ka but is younger than marine deposits believed to have formed during stage 5e (Carter and Ager, 1989; L. D. Carter, written communication, 1990).

5. Extensive glaciation in high latitudes late in stage 5 is demonstrated on Baffin Island, in the eastern Canadian Arctic, where stratified coastal exposures and marine sediment cores have been dated by radiocarbon and uranium-series methods and correlated by amino-acid ratios and pollen data (Andrews and Miller, 1984; Andrews and others, 1985). The most extensive glacier advance of the late Pleistocene on Baffin Island began shortly after stage 5e and terminated about 70 ka, and sediment cores from Baffin Bay show dominance of ice-rafted sediments from late stage 5 into stage 4. Uranium-series ages from Banks and Victoria Islands, in the western Canadian Arctic, suggest a similar glacial chronology (Causse and Vincent, 1989). Subsequent ice advances of late Wisconsin age are much more restricted in comparison.

Two models clearly are possible for the penultimate glaciation of Alaska (Fig. 9), and an additional possibility is that northern Alaska may follow a high-latitude model like that of Baffin Island, whereas southern Alaska may respond more like glaciers in more temperate regions of North America. According to Andrews and others (1985), circulation of warm water into Baffin Bay generated increased snowfall on Baffin Island during stages 5 and 4, causing glaciers to advance. Glaciers subsequently were so inhibited by absence of precipitation that maximum late Wisconsin advances did not occur until 12 to 8 ka, when retreat of the Laurentide ice sheet farther south allowed greater moisture to penetrate into the eastern Canadian Arctic (Andrews, 1987). If glaciers in northern Alaska were similarly controlled by moisture availability (Hamilton, 1981), in contrast to greater influence of temperature fluctuations farther south, there may be no single age for the "penultimate glaciation" of Alaska.

Arctic Coastal Plain

The Flaxman Member of the Gubik Formation consists of erratic-bearing glaciomarine sediments that occur locally along the Beaufort Sea coast to altitudes of about 7 m (Carter and others, 1986a, 1988; see Fig. 1). The erratic stones, of Canadian provenance, were transported to the Beaufort Sea coast by icebergs from an extensive ice sheet in the Canadian Arctic. Marine mollusk shells are enriched in ^{18}O relative to modern shells from the Beaufort Sea, confirming that more glacial ice was present in Flaxman time than occurs today. The Flaxman Member has been assumed to represent the distal glaciomarine facies of the Buckland advance of the northwestern sector of the Laurentide Ice Sheet (Vincent and Prest, 1987; Vincent and others, 1989), but the presence of spruce macrofossils suggests that it was deposited during an interstadial (Carter and Ager, 1989), and distinct erratic lithologies indicate a glacial source farther to the east (L. D. Carter, personal communication, 1990). Carter and Ager (1989) report that 11 thermoluminescence (TL) ages on sediments of the Flaxman Member range from 53 to 81 ka, and a uranium-series age on associated whale bone is 75 ka, but more recent unpublished TL determinations show wider age range and indicate that the TL method may be unreliable for the muddy glaciomarine sediments of the Alaskan arctic coast (L. D. Carter, personal communication, 1990).

A glaciomarine transgressive unit that locally exceeds 40 m thickness occurs on the Alaskan Beaufort shelf at water depths shallower than about 100 to 115 m (Dinter, 1985). It is composed of gravelly silt and sand derived from the Canadian Arctic and overlies a disconformity that probably represents the minimum sea-level stand of late Wisconsin time (Dinter, 1985). The transgressive unit evidently was deposited by sediment-laden icebergs that calved from the Laurentide Ice Sheet. Dinter (1985) suggested that the transgressive unit is late Wisconsin and Holocene in age, and that its outer limit marks the late Wisconsin minimum sea-level position. However, new core data and reinterpretation of seismic profiles indicate that the unit is probably older than the late Wisconsin (Dinter and others, 1990).

The central Brooks Range

The youngest major drift complex in the central Brooks Range is assigned to the Itkillik glaciation (Detterman and others, 1958), which is divisible into two major phases—Itkillik I and Itkillik II (Hamilton and Porter, 1975; Hamilton, 1986a). The outer border of Itkillik drift is the most striking discontinuity in the entire Brooks Range glacial succession; it separates subdued older glacial deposits from steeper, stonier, and more irregular drift with drainage anomalies and discontinuous vegetation.

During Itkillik I time, glaciers extended as far as 40 km north of the Brooks Range (Fig. 10), forming moraines having slightly flattened crests and irregular profiles (Table 2). Drainage nets are well integrated, but stream courses are largely controlled by original drift morphology. Streams cross moraines in bouldery riffles, and outwash terraces are 20 to 25 m high near moraine fronts. Nested sets of recessional moraines and ice-marginal drainage channels indicate that glaciers retreated dynamically from northern valleys rather than stagnating in place. Moraines in forested southern valleys generally have smooth, somewhat flattened, hummocky crests that support sparse vegetation; moist depressions are covered by dense muskeg.

The maximum advance of Itkillik I time (phase A) may have been followed by a younger major readvance (phase B) (Hamilton, 1986a, 1989); both advances are older than the effective range of radiocarbon dating.

Drift-mantled glacier ice persisted for long intervals in many northern valleys, and relict ice of Itkillik I age may still be widely present today (Hamilton, 1986a). Kettles that have turbid water, steep flanks of unstable gravel, and highly angular outlines remain active today in several valleys.

The largest glaciers of Itkillik II age extended about 25 km north of the Brooks Range (Fig. 10). Drift limits generally are marked by sharp lateral moraines, heads of conspicuously channeled outwash aprons, abrupt narrowing of stream incisions, and bulky end moraines and ice-stagnation deposits that nearly block valley centers. The deposits are slightly more irregular, steeper sided, stonier, and less vegetated than drift of Itkillik I age (Table 2), and surface drainage is poorly integrated. End moraines retain multiple crests and other minor relief features. Kettles are deep and steep sided; many have bare, unstable flanks, open tension cracks, turbid water, and active slumps indicative of continued enlargement. Solifluction occurs locally on lower slopes, but crests and upper slopes lack colluvium. Axial streams cross end moraines through steep, bouldery rapids, and outwash terraces are 10 to 20 m high. Recessional moraines are rare, indicating general downwastage rather than frontal retreat of glaciers. Rugged end moraines in southern valleys (Table 2) commonly are located at the margin of the Brooks Range and enclose elongate lakes or lacustrine plains. Drift of Itkillik II age formed between about 24 and 11.5 ka and correlates with the late Wisconsin fluctuations of the Laurentide Ice Sheet (Hamilton, 1982a, 1986a, b; Hamilton and others, 1987).

A major readvance of glaciers near the close of Itkillik II time formed distinctive end moraines as much as 15 to 20 km beyond the range front in the larger northern valleys (Hamilton, 1986a). Elongate lakes 20 to 60 km long formed in most northern valleys as glaciers retreated. Their upper limits rise southward by about 100 m, probably due to isostatic recovery following deglaciation. Radiocarbon ages demonstrate that the late Itkillik II readvance took place between 13 and 11.5 ka (Hamilton, 1982a, 1986a).

The Seward Peninsula

The Salmon Lake and Mount Osborn glaciations of the Seward Peninsula have relatively fresh morainal morphology (Table 2) and resemble Itkillik-age deposits of the Brooks Range (Kaufman and Hopkins, 1986). Salmon Lake drift, which is correlated with deposits of the Itkillik I phase, is older than 40 ka; the succeeding Mount Osborn advance probably is of late Wisconsin (Itkillik II) age.

Glaciers of Salmon Lake age averaged about 16 km long in the Kigluaik Mountains, where they filled trunk valleys and formed large lobate moraines beyond valley mouths (Kaufman

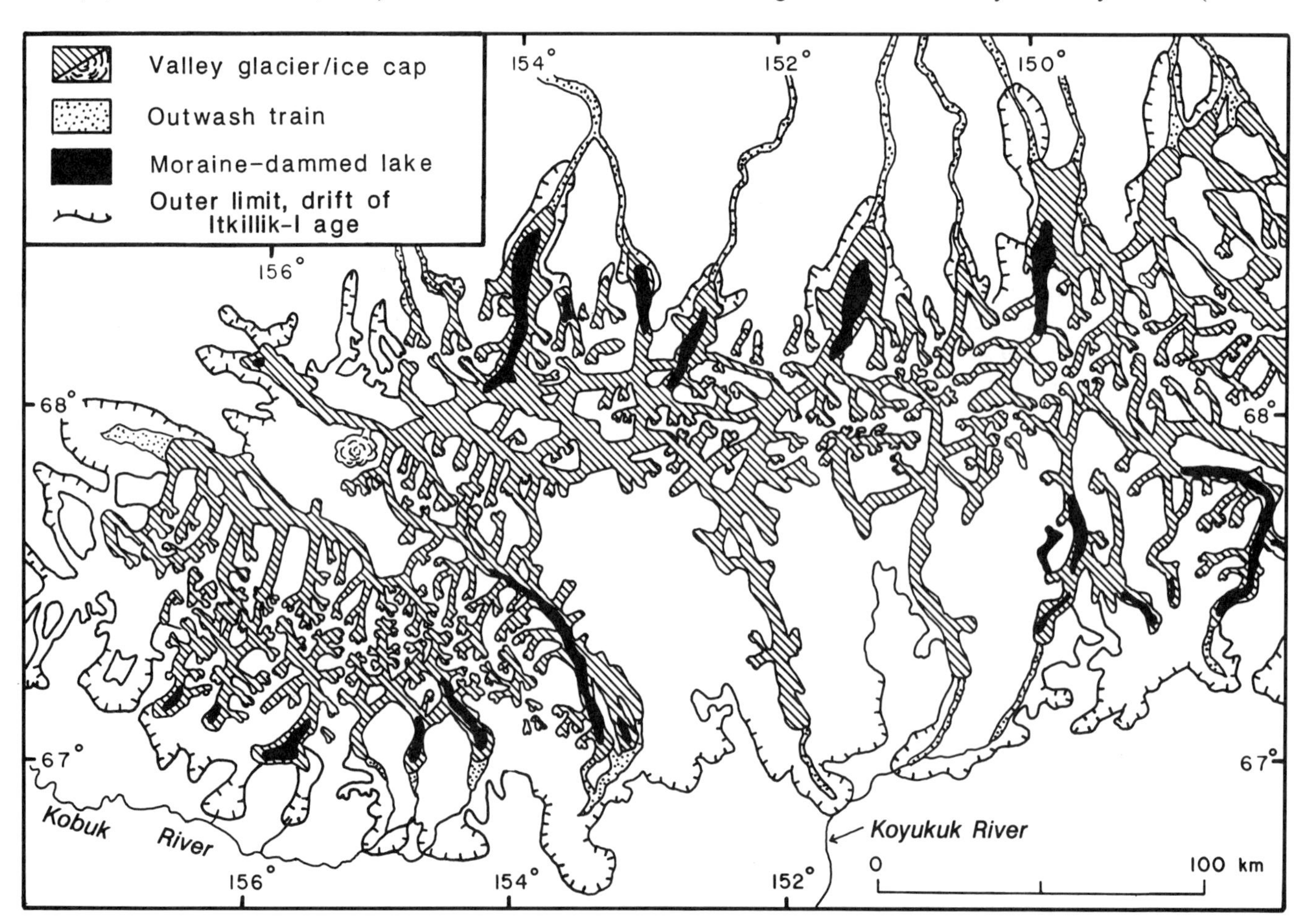

Figure 10. Distribution of glaciers, ice caps, and moraine-dammed lakes of Itkillik II age, central Brooks Range. Hachures show outer limits of preceding Itkillik I ice advance. From Hamilton (1986a).

and Hopkins, 1986; Kaufman and others, 1988); shorter glaciers in the Bendeleben and Darby Mountains were limited to tributary valleys and heads of trunk valleys (Fig. 4). Drift deposits have well-preserved glacial morphology, but have been significantly affected by periglacial processes (Kaufman and Calkin, 1988). In the York Mountains, drift of the Mint River glaciation (Sainsbury, 1967) is similar in appearance to drift of Salmon Lake age, and reconstructed equilibrium-line altitudes (ELAs) of both drifts are compatible (Kaufman and Hopkins, 1986). Morphologic criteria suggest that the Salmon Lake glaciation could be of either stage 4 or stage 6 age (Kaufman and Calkin, 1988).

During Mount Osborn time, small glaciers occupied high cirques and upper valleys in the Kigluaik Mountains, where they extended 2.5 to 3.0 km from cirque headwalls (Fig. 4). Moraines are sharply defined, with bouldery surfaces, and extend upvalley into prominent trimlines along valley walls (Kaufman and Calkin, 1988). Projection of reconstructed ELAs for Mount Osborn glaciers into the York Mountains indicates that snowline was at the appropriate altitude to generate the Esch Creek glaciation (Sainsbury, 1967), which therefore is considered to be correlative with the Mount Osborn advance (Kaufman and Hopkins, 1986). The highest north-facing cirques in the Kigluaik Mountains contain moraines that record a glacial stillstand or readvance late in Mount Osborn time, but a radiocarbon age from one cirque suggests that subsequent deglaciation was complete by 11.5 ka (Kaufman and others, 1988, 1989a).

The Yukon-Tanana Upland

Glacial deposits of Eagle and Salcha age in the Yukon-Tanana Upland were assigned to the late Pleistocene by Weber (1986). She correlated the Eagle glacial episode with ice advances of isotope stage 4 or late stage 5 age elsewhere in Alaska on the basis of tephrochronology, morphology (Table 2), and similar positions in glacial successions, and correlated the Salcha glacial episode with radiocarbon-dated glacial sequences of late Wisconsin age in the Alaska Range and the Brooks Range. Age-equivalent glacial advances in the Mount Prindle area were assigned by Weber and Hamilton (1984) to the American Creek and Convert glaciations, respectively.

Glaciers of Eagle age formed at and above altitudes of 1,460 m; they extended as much as 19 km through upper mountain valleys and generally terminated at about 900 m altitude (Fig. 7). End moraines generally occur above modern timberline; outwash trains form flat valley floors that are little dissected by modern streams. Almost all drift sheets contain multiple end moraines that indicate two distinct glacial advances.

Glaciers of Salcha age originated on the highest uplands, where they reoccupied the cirques and radiating valleys that had been shaped during the preceding glacial episode (Weber, 1986). Most glaciers were confined to upper valleys and terminated above modern timberline at altitudes above 1,200 m; a few of the longest ice tongues extended about 300 m lower. Moraine sequences in most valleys indicate four glacial advances. Moraines of the youngest two phases are weakly developed, and coalesce in some valleys.

The Kuskokwim Mountains

Drift deposits of Bifurcation Creek and Tolstoi Lake glaciations in the Kuskokwim Mountains have been described by Kline and Bundtzen (1986), who assigned them to the early Wisconsin and late Wisconsin, respectively.

Major cirque basins and U-shaped valleys in the Beaver Mountains were formed during ice advances of Bifurcation Creek age (Kline and Bundtzen, 1986). Glaciers extended about 10 km down valleys that radiated generally northward from the center of the massif and extended 4.5 to 6 km down south-trending valleys. Two and rarely three end moraines are present in almost all valleys. Cirque headwalls and valley flanks have been modified by stream incisions and by accumulation of colluvial rubble.

During Tolstoi Lake time, glaciers advanced as far as 6.5 km down northern valleys and 1.5 km or less down valleys that trend east or southeast (Kline and Bundtzen, 1986). South-facing cirques and valleys apparently did not contain ice at that time. Steep-fronted drift sheets were deposited in most major northern valleys, and multiple end moraines record three (rarely 4) distinct ice advances. Most cirque headwalls are unmodified except for talus cones, and cirque floors are undissected. A late Wisconsin age is inferred by Bundtzen (1980) for the drift on the basis of its topographic freshness, slight weathering of boulders, weak soil development, and unmodified cirques.

North-central Alaska Range

The youngest two major Pleistocene glacial advances formed conspicuous deposits in most valley systems of the north-central Alaska Range (Porter and others, 1983). The older of the two drifts is assigned to the Delta glaciation in the Delta River region (Péwé and Reger, 1989), whereas farther west in the Nenana River region, it is referred to the Healy glaciation (Wahrhaftig, 1958; Thorson, 1986). The Delta glaciation previously was believed to be of early Wisconsin age because Delta-age outwash at the Canyon Creek fossil locality underlies vertebrate fossils that had a uranium-series age estimate of about 80 ka (Weber and others, 1981; Hamilton and Bischoff, 1984; Weber, 1986). However, uranium-series ages on bones are now considered suspect (Bischoff and others, 1988; J. L. Bischoff, personal communication, 1990), and the age of the Delta glaciation is not longer constrained by the stratigraphy at Canyon Creek. The Healy terminal moraine of the Nenana River valley is a spatulate body of coarse, gravelly drift that consists of as many as ten individual ridges separated by undissected marginal channels (Thorson, 1986). Its sharply defined outer margin forms the limit of deposits with stony surfaces, original morainal morphology, and deranged drainage patterns. Drift of Healy age is older than the limit of radiocarbon dating; it was believed to postdate the last interglacial maximum (Ritter, 1982; Ten Brink, 1983;

Thorson, 1986) because it has not been severely weathered and because it is closer morphologically to drift of late Wisconsin age than it is to any older glacial deposits. Two main phases of Delta-age glaciation were suggested by Weber and others (1981) on the basis of drift morphology in the Delta River valley and stratigraphy of the Canyon Creek fossil locality, and two main phases of "early Wisconsin" glaciation also were reported by Ten Brink (1983) based on field studies in the Nenana valley region.

Expansion of succeeding glaciers of Donnelly or Riley Creek age to the north flank of the Alaska Range is closely constrained by two radiocarbon ages of 25.3 and 24.9 ka, and similar ages of about 26.4 and 23.9 ka provide maximum limits on the late Wisconsin McConnell glaciation in southwestern Yukon Territory (Hamilton and Fulton, 1994). Late Wisconsin glaciation in the north-central Alaska Range reportedly consists of four phases (Ten Brink and Waythomas, 1985), and outwash terraces formed during the oldest three phases are present in most valleys (Ritter and Ten Brink, 1986). During phase I, the maximum advance, glaciers constructed steep-fronted, bouldery, nested moraine systems in every major valley. They attained maximum positions at about 20 ka and began to retreat at about 17 ka. During stillstands and short readvances of phase II, dated at 15.0 to 13.5 ka, glaciers built a separate set of end moraines 2 to 10 km upvalley from the outermost moraines (Ten Brink and Waythomas, 1985). Phase II moraines generally are smaller than end moraines of phase I, and they contain more water-washed sediment. Phase II was followed by an episode of rapid glacier retreat and ice disintegration. Drift of phase III, a brief but strong glacier readvance at about 12.8 to 11.8 ka, includes abundant ice-disintegration features (Ten Brink and Waythomas, 1985). Moraines commonly cross-cut older moraines of the late Wisconsin succession, implying significant glacier retreat and readavance. This readvance is similar in age to the late Itkillik II readvance described previously in the central Brooks Range (Hamilton and Fulton, 1991). During Phase IV, the final late Wisconsin readvance, glaciers formed small moraines a few km beyond the limits of late Holocene ice advances in headward parts of the highest valleys. Ten Brink and Waythomas (1985) date phase IV at 10.5 to 9.5 ka, but this age assignment is based only on correlation with intervals of peat accumulation, which they believe represents wet and/or cool climate favorable for glacier growth.

Northwestern Alaska Range

The last major series of ice advances—the Farewell glaciation—in the western Alaska Range was subdivided by Fernald (1960) into two phases: Farewell 1 and Farewell 2. A radiocarbon age of 34,340 ± 1,940 B.P. from an organic layer near the base of overlying loess provides a minimum age limit on drift of the Farewell 1 advance, and Kline and Bundtzen (1986) believe that the Farewell 1 and Farewell 2 drifts probably are of early and late Wisconsin age, respectively.

During Farewell 1 time, glaciers formed piedmont lobes as much as 32 km beyond the Alaska Range (Kline and Bundtzen, 1986). Frozen peat and ice-rich silt as much as 10 m thick covers parts of the drift, and many ponds and lakes on its surface are thermokarst features. Multiple moraine systems in many valleys indicate two phases of glaciation.

Glaciers of Farewell 2 age advanced as much as 23 km from the range front where valleys drained ice fields south of the Alaska Range, but elsewhere they terminated close to the mouths of mountain valleys (Kline and Bundtzen, 1986). At least four glacial advances took place in most major valleys: the two earliest and most extensive formed arcuate moraines, the third was followed by widespread ice stagnation, and the fourth generally produced small moraines within mountain valleys. Correlations between these advances and the four glacial phases in the north-central Alaska Range are uncertain.

The St. Elias Mountains

Two drifts, assigned to pre-Macauley and Macauley glaciations, occur in stratigraphic sections and surface exposures along the White River valley. These units are correlative, respectively, with the Icefield and Kluane drift units in the northern St. Elias Mountains (Table 1). Both glacial events were assigned by Denton (1974) to the Wisconsin stage on the basis of radiocarbon ages, stratigraphic relations, and lack of significant weathering. However, drift of pre-Macauley age (Icefield glaciation) generally is correlated with drift of the Mirror Creek glaciation of southwest Yukon Territory, which is overlain by Old Crow tephra (Hughes and others, 1989). Therefore, it could be of pre-Wisconsin age. Deglaciation was followed by a nonglacial interval that extended from sometime prior to 49 ka until about 30 ka (Denton, 1974).

Macauley (Kluane) drift forms hummocky deposits having abundant kettles and sharp morphologic boundaries (Denton, 1974). Radiocarbon ages show that glaciers began to advance shortly after 29 ka and attained maxima sometime before about 13.7 ka. Deglaciation of major valleys was nearly complete by 12.5 ka, and valley heads within a few kilometers of modern large glaciers were ice free by 11.3 ka (Denton, 1974).

Upper Cook Inlet

The two youngest major glacial advances of Karlstrom (1964), the Knik and the Naptowne glaciations, entered upper Cook Inlet but probably did not entirely fill it (Fig. 11). During Knik time, ice lobes coalesced over the Anchorage area, but part of the Kenai lowland probably remained unglaciated. Moraines are slightly subdued and commonly are capped by as much as 1 m of loess (Reger and Updike, 1983). Subsurface diamict of probable Knik age in the Anchorage area underlies drift of early Naptowne age (Reger and Updike, 1989). Because it underlies peat with a probable interglacial flora (Ager and Brubaker, 1985), drift of the Knik glaciation appears to be older than stage 5e.

Deposits laid down during the Naptowne ice advance cover most lowlands of upper Cook Inlet (Reger and Updike, 1983,

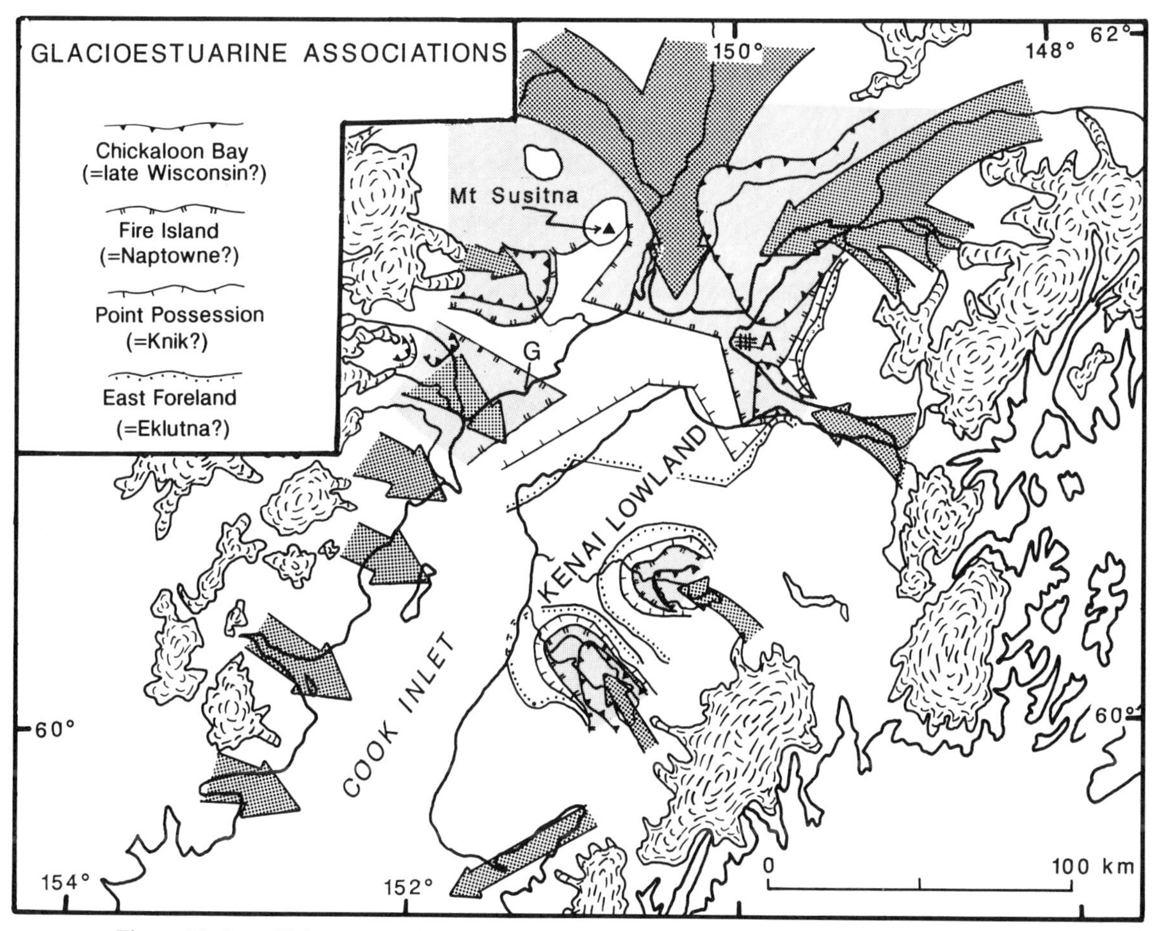

Figure 11. Late Pleistocene and present-day glaciation of Cook Inlet. Modern glaciers shown by concentric dashed patterns. Arrows show flow routes of principal glacier tongues; shaded areas are inferred extents of late Wisconsin glaciers in upper inlet. Principal glacioestuarine associations are shown for upper inlet only (Cairn Point GEA is not shown). A = Anchorage; G = Granite Point. Modified from Schmoll and Yehle (1986).

1989), Drift typically is little modified, with numerous kettles and poorly integrated stream systems; it has about 0.3 m of loess cover. Separate advances of early and late Wisconsin age may have taken place (Reger and Updike, 1989). Radiocarbon ages on early Naptowne drift are contradictory (Reger and Updike, 1983, p. 201), but they imply that the initial advance is older than the effective range of radiocarbon dating. No middle Wisconsin radiocarbon ages have been reported from upper Cook Inlet. Plant growth at that time evidently was severely inhibited by some combination of expanded glaciers and marine submergence (Schmoll and Yehle, 1986). Till and ice-stagnation deposits of early Naptowne age in the Anchorage area are overlain by the glaciomarine Bootlegger Cove Formation (Updike and others, 1984, p. 14–17). Deposition of the Bootlegger Cove Formation probably began sometime before about 18 ka, and radiocarbon ages of 12.2 to 11.5 ka on overlying peat provide a minimum age limit for this unit (Reger and Updike, 1983, 1989).

Most moraines in upper Cook Inlet become indistinct below about 120 m above present sea level, where glaciers probably reached tidewater in the isostatically depressed basin (Schmoll and Yehle, 1986). Schmoll and others (1984) have defined five glacioestuarine associations (GEAs) in upper Cook Inlet (Fig. 11). Each GEA is an assemblage of landforms and deposits that represents the inferred marginal positions of glaciers during a specific interval of time (Schmoll and Yehle, 1986). Some GEAs may have formed during major ice advances, but others probably represent only glacier stillstands or minor readvances. The East Foreland GEA, oldest of the succession, represents a glacier complex that filled all of upper Cook Inlet except for a small part of the Kenai Peninsula. Ice-margin positions are somewhat similar to those of the Eklutna glaciation, and they lie beyond the Knik ice limits mapped by Reger and Updike (1983). The glacier limits of the succeeding Point Possession GEA are close to the outermost Knik moraines. The Fire Island GEA may correspond to the early Naptowne ice limits mapped by Reger and Updike (1983) because reconstructions of both limits show that glacier ice last covered the Anchorage area at that time. The Chickaloon Bay GEA, which most closely coincides with deposition of the

Bootlegger Cove Formation at Anchorage, has a minimum limiting age of 13.8 ka. The Elmendorf moraine, which forms part of the succeeding Cairn Point GEA, was deposited between 13.7 and 11.7 ka (Schmoll and others, 1972).

Submarine diamicts in Shelikof Strait fill deep U-shaped troughs that Hampton (1985) believes formed during Naptowne time. The trough more likely formed during an earlier glacier advance out of upper Cook Inlet, and the thick (>800 m) diamicts may contain a long record of middle and late Pleistocene glaciomarine deposition.

Bristol Bay region

The youngest Pleistocene drift complexes of the Alaska Peninsula are assigned to the Mak Hill and Brooks Lake glaciations. Drift of Mak Hill age formed sometime prior to 40 ka (Detterman, 1986); the succeeding four-fold Brooks Lake drift sequence is assigned to the late Wisconsin.

Drift of Mak Hill age is widely exposed along the northwest side of the Alaska Peninsula and forms much of the Bristol Bay lowland (Fig. 8). Erratics are mainly rock types that originated from volcanoes of the Alaska Peninsula (Detterman, 1986). End moraines are distinct but somewhat subdued, and some segments have been modified by wave action. Drift of Mak Hill age is locally covered by up to 3 m of silt that contains tephra layers. Marine terraces 15 to 18 m above present sea level have been cut into the glacial deposits, and associated marine sediments also overlap Mak Hill drift (Detterman and others, 1987).

During Brooks Lake time, glaciers originating on volcanoes of the Alaska Peninsula coalesced with ice caps that formed south of the Alaska Peninsula (Weber, 1985) and expanded into a nearly continuous glacier complex. Four major glacial advances are termed the Kvichak, Iliamna, Newhalen, and Iliuk advances (Detterman, 1986). Drift of the Kvichak advance locally was deposited in water or was modified by subsequent wave action. The succeeding Iliamna advance was the last major glacial event on the Alaska Peninsula, and its end moraines enclose most of that region's large lakes. Drift morphology resembles that of the Kvichak advance, but moraine crests are a little sharper and flanks are somewhat steeper. The two youngest glacial advances, Newhalen and Iliuk, were minor events largely restricted to mountain valleys. Radiocarbon dates indicate that glaciers retreated from Newhalen moraines before 9 ka and probably prior to 10.6 ka (Detterman, 1986).

On the northwest shore of Bristol Bay, the youngest two Pleistocene glaciations formed distinctive moraines in the Nushagak lowland (Fig. 8). The penultimate glaciation may be represented by discontinuous segments of the Manokotak and Gnarled Mountain moraine belts, which exhibit subdued hummocky microrelief (Lea and others, 1989). The two moraine belts may correlate with the subdued end moraines of Mak Hill age on the Alaska Peninsula (Table 1).

The Okstukuk and Alegnagik moraine belts, which formed during the last glaciation, are within 10 km of range-front lakes and have little-modified hummocky topography (Lea and others, 1989). According to Lea and others (1989), these moraines probably are equivalent to the Kvichak and Iliamna moraines of the Brooks Lake glaciation on the Alaska Peninsula. Wind-deflated sediments from glacial outwash were redeposited as sand sheets and sand-loess intergrades; these deposits contain beds of slightly organic silt that probably formed during interstadial episodes of the last glaciation (Lea, 1989, 1990b). A distinctive silt bed near the base of the eolian sand has radiocarbon ages of 21.6 to about 18 ka (Lea, 1989). The silt bed represents a possible interstade that may correspond to the Hanging Lake thermal event of eastern Beringia, which is dated at about 22 to 18 ka by Matthews and others (1989) and 22 to 19 ka by Hamilton and Fulton (1994). A higher organic horizon marks another possible interstadial at about 14.3 ka. Eolian deposition ceased by about 12.5 ka, probably when glaciers receded into mountain valleys and their sediments became trapped by the range-front lakes (Lea, 1989).

Similar relations of eolian sand to glacial intervals 200 km farther north in the Holitna lowland are reported by Short and others (1990) and by Waythomas (1990). Eolian sand and silt that may correlate with the penultimate glaciation contains the Old Crow tephra near its upper contact, indicating that the tephra was deposited during waning stages of that glaciation. If the stratigraphic placement of the Old Crow tephra is valid, and if the 149 ka age estimate for this unit is correct, a stage 6 age would be indicated for the penultimate glaciation north of Bristol Bay.

The Copper River Basin

During late Pleistocene time, glaciers in the Chugach Range blocked the Copper River and caused a large lake to form in the basin enclosed between the Alaska Range and the Talkeetna, Chugach, and Wrangell Mountains (see Figs. 1 and 2). Large glaciers entered the basin from the Chugach Mountains; smaller glaciers from the other ranges intermittently blocked northern and western spillways at higher altitudes. Lava flows, volcanic debris flows, and tephra also entered the basin from active volcanoes of the Wrangell Mountains (Connor, 1984; Nichols and Yehle, 1985).

River bluffs contain multiple diamicts together with lacustrine, fluvial, and eolian deposits and volcanic detritus (Ferrians and others, 1983). Many of the diamicts are deposits of debris flows, turbidites, and sediment gravity flows that originated by slumping around the margins of the basin, perhaps triggered by earthquake shocks (Eyles, 1987; Ferrians, 1989). Glaciers probably covered much of the basin floor during the penultimate glacial advance (Hamilton and Thorson, 1983; Nichols, 1989); during later episodes, they calved into a deep lake that filled the central part of the basin (Williams and Galloway, 1986).

The highest recorded lake levels in the northeastern part of the basin are at 915 m altitude (Schmoll, 1984), and Williams and Galloway (1986) show lake deposits at altitudes as great as 975 m farther to the west. Discrepancies in shoreline altitudes may be due in part to isostatic tilting, but they also could result from different ages or differential preservation of the deposits.

The youngest major lake (glacial Lake Atna) began to fil the basin sometime before 40 ka (Williams and Galloway, 1986; Nichols, 1989) and perhaps before 50 ka (Ferrians, 1989); associated lacustrine deposits occur to altitudes as high as 775 m in the western part of the basin (Williams, 1989). An interstadial episode of lowered lake levels is dated at about 31 to 28 ka in the eastern sector of the basin (Ager, 1989; Ferrians, 1989; Nichols, 1989) and persisted from at least 31 ka until sometime after 22 ka in the northwestern part of the basin (Thorson and others, 1981). During this episode, lake level dropped below 655 m altitude across much of the basin (Ferrians, 1989).

The highest shorelines of the youngest lake stage occur at about 947 m asl throughout the Copper River Basin (Williams and Galloway, 1986), and indicate little or no local isostatic warping at this time. Moraines of Wisconsin age in southern and western parts of the basin are mantled with lacustrine silt to altitudes as high as 945 m, indicating that the lake persisted until glaciers had withdrawn into the Chugach Range (Williams and Galloway, 1986). Deglaciation was well advanced by 14 to 13 ka (Williams, 1986; Williams and Galloway, 1986; Sirkin and Tuthill 1987), and most of the basin became ice-free by 11.8 to 10.5 ka (Hamilton and Thorson, 1983).

The Aleutian Islands

Glaciers of late Wisconsin age covered nearly all of the inner Aleutian Islands and extended over much of the adjoining Aleutian platform (Black, 1983; Thorson and Hamilton, 1986; see Fig. 1). The axis of ice dispersal lay south of the inner islands, indicating that the Pacific Ocean was the primary moisture source for glaciers through this part of the Aleutian chain. Glaciers flowed north across the Aleutian platform and probably terminated as a floating ice shelf in the southern Bering Sea. Ice caps on Umnak and Unalaska Islands were confluent with the glacier complex on the Alaska Peninsula, but the western limit of completely coalesced ice is uncertain. Minimum limiting ages from postglacial sediments commonly are 8 to 6 ka (Black, 1983), but Thorson and Hamilton (1986) suggest that eustatically rising sea level, together with warming air and ocean temperatures, probably caused disintegration of the glacier complex sometime before 11 ka.

Gulf of Alaska

The Gulf of Alaska continental shelf consists of Tertiary and Pleistocene sedimentary rocks into which glacial troughs have been incised (Molnia, 1986). Overlying sediments include till, outwash, and glaciomarine deposits that thicken to more than 50 m in major troughs. The extent of late Pleistocene ice cover on the continental shelf is uncertain, but glaciers that reached the edge of the shelf may have been confined to the major troughs (see Fig. 1).

Carlson and others (1982) delineated eight major troughs and valleys that are incised into the continental shelf of the northeastern Gulf of Alaska (see Fig. 6). These features generally have U-shaped cross sections, concave longitudinal profiles, and sediment fillings whose seismic signatures are identical to those of known glacial deposits (Molnia, 1986). Their positions correspond to major onshore glacier systems or probable glacial flow routes during late Pleistocene time, and they probably were formed by major ice streams that flowed to the shelf edge (Carlson, 1989; Powell and Molnia, 1989).

The last major episode of ice retreat in the Gulf of Alaska region probably began at 12 to 15 ka (Heusser, 1985; Molnia, 1986). Radiocarbon ages from coastal areas suggest that glacier ice had retreated from the continental shelf by at least 10 ka and that glaciers draining the coastal mountains had reached positions comparable to those of the present day by about 9.6 to 9.3 ka.

Lituya Bay and Glacier Bay

Mann (1986) describes three informally named late Pleistocene drifts near Lituya Bay—the High terrace, Raven House, and Eurhythmic drift units. The High terrace drift unit, oldest of the succession, is preserved only on terraces more than 500 m above sea level (asl). Surface deposits are rare, and bogs blanket all glacial deposits. Weakly oxidized till and outwash in one exposure overlie peat that has a radiocarbon age of >72 ka and a pollen content that indicates interglacial vegetation. Because deep stream incisions are graded to a lower terrace, which probably formed between 50 and 30 ka, Mann (1986) assigned the High terrace drift unit to the early Wisconsin.

The Raven House drift unit includes subparallel moraines armored with wave-washed boulders that are distributed across a marine terrace at 180 to 210 m asl. Maximum limiting ages for Raven House drift were obtained from estimates of the age of formation of its associated terrace (Mann, 1986). Calculated rates of uplift and tilting suggest that the terrace system probably is older than 30 ka and that it formed about 60,000 yr after the last interglacial maximum. The terrace therefore probably formed during middle Wisconsin time, and its drift cover probably is late Wisconsin in age. The distribution of end moraines of Raven House age suggests that late Wisconsin glaciers in the Lituya Bay area formed separate piedmont lobes that extended as much as 16 km across the continental shelf from the present-day coast (Mann, 1986). Wave erosion of the Raven House drift unit probably resulted from isostatic depression due to crustal loading by glaciers, which was superimposed on slow tectonic uplift. Mann (1986) estimates that 60 to 100 m of isostatic depression would have been necessary to submerge and erode drift deposited at the time of the maximum late Wisconsin ice advance.

Till of the succeeding Eurhythmic readvance contains wood dated at 12.4 ka. Glaciers in the Lituya Bay area had receded to positions close to their present limits by that time (Mann, 1986).

Glaciers in Glacier Bay reached altitudes of 1,000 to 1,500 m during late Wisconsin time (McKenzie and Goldthwait, 1971);

they coalesced into a large ice stream that flowed through Glacier Bay and onto or across the continental shelf (Goldthwait, 1987). By about 13 ka, retreating glaciers had reached positions close to their present termini, and deglaciation was accompanied by marine transgression to as much as 60 m asl. The Brady Glacier, which drains into Icy Strait just west of Glacier Bay, reached altitudes of about 600 m asl along the coast during the late Wisconsin (Derksen, 1976, p. 13) and probably extended across the continental shelf at that time (Mann, 1986).

Raised beaches and bedrock terraces south of Glacier Bay record two marine incursions that probably resulted from ice loading (Ackerman and others, 1979). The highest terrace, at 12 to 15 m asl, was cut sometime between 13.4 and 9.2 ka; it records a halt or slowing in isostatic rebound.

Southeastern Alaska

The extent and history of glacial advances in most parts of southeastern Alaska are unknown. Regional morphology and limited field data suggest that the late Pleistocene glacier complex was formed by mountain ice sheets expanding out of icefields on the mainland and coalescing with small mountain ice caps on islands of the Alexander Archipelago (D. H. Mann, written communication, 1986). Outlet glaciers flowed through the fjord system that presently dissects the archipelago. Local valley glaciers and ice caps were generated on higher parts of the outer islands, but some local uplands and parts of the continental shelf probably remained ice free (see Fig. 1). This reconstruction is compatible with Mann's (1986) analysis of late Wisconsin glacier dynamics in the Lituya Bay area and with uplift curves and glacier flow patterns in coastal British Columbia (Warner and others, 1984; Barrie and Bornhold, 1989; Clague, 1989).

Peat beds beneath glaciomarine deposits near Juneau are older than 39 ka and may have been deposited during a middle Wisconsin interstade when glaciers had receded from inner fjords (Miller, 1973). The glaciomarine Gastineau Channel Formation subsequently was deposited during deglaciation and isostatic emergence about 13 to 9 ka. Miller's (1973) data show that emergent marine deposits as high as 230 m asl formed immediately after deglaciation and that subsequent isostatic recovery was interrupted by a stillstand or readvance of glaciers sometime after 13 ka.

The morphology of the Alexander Archipelago reflects intensive glacial erosion. Valleys connected by smoothly abraded passes cross most islands (Mann, 1986), and only a few areas that lie within precipitation shadows on large islands appear to have escaped inundation by regional ice streams or by locally generated alpine glaciers. Swanston (1969) describes two tills on Prince of Wales Island that differ in oxidation and therefore may represent two separate glacial advances. The younger till occurs only below 500 m asl and has a minimum limiting radiocarbon age of about 9.5 ka. Emergent marine deposits occur to heights as much as 150 to 210 m throughout this region (Mann, 1986).

SUMMARY AND DISCUSSION

Tertiary diamicts of probable glacial origin are widespread in Alaska, and they commonly represent several successive episodes of glaciation. The oldest glacial deposits in northern and central Alaska typically have been intensively eroded and occur only as erratic boulders redeposited in younger gravel; glacial deposits of younger Tertiary age more commonly occur as remnants of generally featureless drift. Tertiary glaciomarine deposits are more extensive around the Gulf of Alaska, where thick sections have been measured onshore (Plafker, 1981), and also are evident in seismic reflection profiles offshore (Carlson, 1989).

The general increase in glacial activity after about 2.5 Ma recorded in the Gulf of Alaska has possible counterparts elsewhere in Alaska, where glaciation of probable Miocene or early Pliocene age was followed by a second glacial episode late in the Pliocene. Tertiary glacial deposits in northern Alaska extend far beyond the most extensive drift sheets of Pleistocene age, but farther to the south they are exposed only locally and for short distances beyond the outer limits of younger drift. This distribution pattern strongly suggests that the Arctic Ocean was an important moisture source for glaciers in Tertiary time (Hamilton, 1986a, b), and it supports the arguments of Herman and Hopkins (1980) and Carter and others (1986b) that the Arctic Ocean probably lacked sea-ice cover until near the end of the Pliocene.

Glacial deposits of early Pleistocene age are identified by subdued morainal morphology, by association with ancient landscapes into which modern valley systems have been incised, and by the presence of alpine landforms that developed during a long postglacial interval of weathering and erosion. Few firm ages are available for these deposits. Glaciomarine sediments of the Yakataga Formation on Middleton Island were deposited during the interval from 2.2 to 1.0 Ma, and they record increasing intensity of glaciation upward in the section. Paleomagnetic data from the Kobuk River valley indicate that the Anaktuvuk River glaciation of the Brooks Range is older than 0.73 Ma.

Distribution of Anaktuvuk River drift and of possibly correlative deposits offshore in the Kotzebue Sound region (Decker and others, 1989) indicates that glaciers of early Pleistocene age were much more extensive throughout the Brooks Range than were glaciers of all succeeding advances. Similarly, extensive glacial deposits in the Yukon-Tanana Upland and the Kuskokwim Mountains were formed by ice caps that covered local highlands prior to incision of modern valley networks. The early Pleistocene glaciers and ice caps of northern and central Alaska probably required more abundant snowfall than at present.

The multiple glacial events in the Nenana River valley are difficult to correlate with the less complex early Pleistocene glacial records of tectonically stable areas farther to the north. Continued uplift of the Alaska Range, incision of the Nenana River valley, and segmentation of the valley by active faults evidently inhibited the obliterative overlap (Gibbons and others, 1984) that tends to simplify the glacial record in more stable regions.

Large kettles on broad river terraces that developed within drift of Anaktuvuk River age indicate that buried glacier ice remained for a long time after deglaciation of the foothills north of the Brooks Range. This persistent ice suggests the presence of continuous permafrost, which in turn generally is associated with perennial sea-ice cover. These relations suggest that the perennial ice cover of the Arctic Ocean could have developed during Anaktuvuk River time. However, deep weathering in drift deposits of northern and northwestern Alaska indicates that permafrost either disappeared or was significantly degraded during one or more interglacial intervals that followed the early Pleistocene.

Through much of northern and central Alaska, glaciers of middle Pleistocene age occupied valley floors close to modern levels and extended only short distances beyond the outermost limits of late Pleistocene ice advances. Glacier expansion followed a long interval of reduced glaciation during which mountain valleys were eroded to levels close to those of the present day. Two separate drifts having contrasting morphologies are present in some valleys, suggesting successive glacial advances spaced widely in time. Paleomagnetic determinations in the Kobuk River valley indicate that drift assigned to the middle Pleistocene formed during the Brunhes Normal-Polarity Chron (Hamilton, 1986a). Persistence of relict glacier ice in the Brooks Range and the Seward Peninsula suggests that much of the drift assigned to the middle Pleistocene in those regions could have been deposited as late as stage 6 and that continuous permafrost has persisted from that time until the present.

Unusually extensive glaciers or local ice caps on Seward Peninsula and in Kotzebue Sound are of middle Pleistocene age according to potassium-argon, amino-acid, and paleomagnetic age estimates. Glaciers also flowed across the Bering platform from the mountains of the Chukotsk Peninsula of Asia to St. Lawrence Island at about this time (Hopkins and others, 1972). Part of the Bering platform may have remained submerged and nourished locally large glaciers and ice caps, as advocated by Petrov (1967) and Hopkins (1973).

Little is known of the regional distribution of glaciers or moisture sources of middle Pleistocene age in southern Alaska. Drift around the head of Bristol Bay was derived from sources to the north and northeast (Detterman, 1986), a flow pattern that would be compatible with overflow of ice from upper Cook Inlet during either the Eklutna glaciation or an earlier ice advance that filled the inlet.

Drift of the penultimate glaciation is older than the age range of radiocarbon dating but, at least in northern Alaska, it appears to be younger than the maximum of the last interglaciation (stage 5e) because of its little-altered character, sharply defined outer boundaries, and poorly developed soil and weathering profiles. In addition, breakup of an ice sheet over the Canadian Arctic probably was responsible for deposition of the glaciomarine Flaxman Member of the Gubik Formation on the Arctic Coastal Plain sometime during stage 4 or late stage 5. Farther south, the penultimate glaciation may be older than stage 5e, but in some areas the distinction between late Wisconsin and older glaciations is blurred by the great extent of late Wisconsin ice advances and by persistence of glaciers through all or most of the Wisconsin stage. Two distinct penultimate-age glacial advances have been reported in many parts of Alaska (Table 1).

Alaska was partly deglaciated during the middle Wisconsin interstade, but parts of the state still remained ice covered. These areas incude some mountain valleys within the central Brooks Range (Hamilton, 1986a), parts of the St. Elias Mountains (Denton, 1974), upper Cook Inlet (Schmoll and Yehle, 1986), and parts of the Chugach Mountains near the lower course of the Copper River (Hamilton and Thorson, 1983).

The final major Pleistocene glacial episode in Alaska is correlated with late Wisconsin fluctuations of the Laurentide Ice Sheet, on the basis of numerous radiocarbon ages that generally show onset of glaciation at about 24 ka and deglaciation beginning at about 13.5 ka (Denton, 1974; Porter and others, 1983; Hamilton, 1982a, 1986a, b; Hamilton and Fulton, 1991). Throughout northern and central Alaska, the late Wisconsin glaciers were much less extensive than were those of the penultimate glaciation because of precipitation deficiencies resulting from emergence of the Bering platform, lower water temperatures in the North Pacific Ocean and Gulf of Alaska, and extensive seasonal and perennial sea-ice cover over the southern Bering Sea (Sancetta and others, 1985). Low precipitation and abundant sediment sources on barren glacial outwash and emergent marine shelves caused widespread deposition of eolian sand and silt (Hopkins, 1982; Hamilton and others, 1988; Lea and Waythomas, 1990). As many as four substages of late Wisconsin glaciation are reported from the Brooks Range, the Alaska Range, and the Bristol Bay region, but correlations among them are uncertain.

Glaciers of southern Alaska extended onto the continental shelf throughut the coastal area that stretches from the Aleutian Islands around the Gulf of Alaska into southeastern Alaska. Late Pleistocene ice caps may have covered much of the shelf south of the Alaska Peninsula, but farther to the east, glaciers probably reached the shelf edge only along a few U-shaped troughs. Most of those ice streams issued from major valleys that allowed passage of ice from the interior (Mann, 1986). Coastal areas probably were depressed isostatically 100 to 250 m, and records of deglaciation began at about 13 ka.

REFERENCES CITED

Ackerman, R. E., Hamilton, T. D., and Stuckenrath, R., 1979, Early culture complexes on the northern Northwest Coast: Canadian Journal of Archeology, v. 3, p. 195–209.

Ager, T. A., 1989, History of late Pleistocene and Holocene vegetation in the Copper River Basin, south-central Alaska, *in* Carter, L. D., Hamilton, T. D., and Galloway, J. P., eds., Late Cenozoic history of the interior basins of Alaska and the Yukon; Proceedings of a joint Canadian–American workshop: U.S. Geological Survey Circular 1026, p. 89–92.

Ager, T. A., and Brubaker, L., 1985, Quaternary palynology and vegetational history of Alaska, *in* Bryant, V., Jr., and Holloway, R., eds., Pollen records of Late Quaternary North American sediments: Dallas, Texas, American Association of Stratigraphic Palynologists Foundation, p. 353–384.

American Commission on Stratigraphic Nomenclature, 1970, Code of Stratigraphic Nomenclature, 2nd ed.: American Association of Petroleum Geologists, 45 p.

Andrews, J. T., 1987, The Late Wisconsin glaciation and deglaciation of the Laurentide Ice Sheet, *in* Ruddiman, W. F., and Wright, H. E., Jr., eds., North America and adjacent oceans during the last deglaciation: Boulder, Colorado, Geological Society of America, The Geology of North America, v. K-3, p. 13–37.

Andrews, J. T., and Miller, G. H., 1984, Quaternary glacial and nonglacial correlations for the eastern Canadian Arctic, *in* Fulton, R. J., ed., Quaternary stratigraphy of Canada; A Canadian contribution to IGCP Project 24: Geological Survey of Canada Paper 84-10, p. 101–116.

Andrews, J. T., and 6 others, 1985, Land/ocean correlations during the late interglacial/glacial transition, Baffin Bay, northwestern North Atlantic; A review: Quaternary Science Reviews, v. 4, p. 333–355.

Armentrout, J. M., 1983, Glacial lithofacies of the Neogene Yakataga Formation, Robinson Mountains, southern Alaska Coast Range, Alaska, *in* Molnia, B. F., ed., Glacial-marine sedimentation: New York, Plenum Press, p. 629–665.

Barrie, J. V., and Bornhold, B. D., 1989, Surficial geology of Hecate Strait, British Columbia continental shelf: Canadian Journal of Earth Sciences, v. 26, p. 1241–1254.

Begét, J. E., and Hawkins, D. B., 1989, Influence of orbital parameters on Pleistocene loess deposition in central Alaska: Nature, v. 337, p. 151–153.

Begét, J. E., Stone, D. B., and Hawkins, D. B., 1990, Paleoclimatic forcing of magnetic susceptibility variations in Alaskan loess during the late Quaternary: Geology, v. 18, p. 40–43.

Begét, J. E., Edwards, M., Hopkins, D. M., Keskinen, M. J., and Kukla, G., 1991, Old Crow tephra found at the Palisades of the Yukon, Alaska: Quaternary Research, v. 35, p. 291–297.

Bischoff, J. L., Rosenbauer, R. J., Tavoso, A., and de Lumley, H., 1988, A test of uranium-series dating of fossil tooth enamel; Results from Tournal Cave, France: Applied Geochemistry, v. 3, p 145–151.

Black, R. F., 1983, Glacial chronology of the Aleutian Islands, *in* Thorson, R. M., and Hamilton, T. D., eds., Glaciation in Alaska; Extended abstracts from a workshop: Fairbanks, University of Alaska Museum Occasional Paper 2, p. 5–10.

Boss, R. F., Lennon, R. B., and Wilson, B. W., 1976, Middle Ground Shoal oil field, Alaska, *in* Braunstein, J., ed., North American oil and gas fields: American Association of Petroleum Geologists Memoir 24, p. 1–22.

Brigham-Grette, J., and Hopkins, D. M., 1989, Extensive middle Pleistocene glaciation across central Beringia; Implications for climatic contrasts between mid and late Bruhnes: Geological Society of America Abstracts with Programs, v. 21, p. A53.

Bundtzen, T. K., 1980, Multiple glaciation in the Beaver Mountains, western interior Alaska, *in* Short notes on Alaskan geology, 1979–80: Alaska Division of Geological and Geophysical Surveys Geologic Report 63, p. 11–18.

Calkin, P. E., 1988, Holocene glaciation of Alaska (and adjoining Yukon Territory, Canada): Quaternary Science Reviews, v. 7, p. 159–184.

Capps, S. R., 1916, The Chisana–White River district, Alaska: U.S. Geological Survey Bulletin 630, 130 p.

——, 1932, Glaciation in Alaska, *in* Shorter contributions to general geology, 1931: U.S. Geological Survey Professional Paper 170, p. 1–8.

Carlson, P. R., 1989, Seismic reflection characteristics of glacial and glaciomarine sediment in the Gulf of Alaska and adjacent fjords: Marine Geology, v. 85, p. 391–416.

Carlson, P. R., Bruns, T. R., Molnia, B. F., and Schwab, W. C., 1982, Submarine valleys in the northeastern Gulf of Alaska; Characteristics and probable origin: Marine Geology, v. 47, p 217–242.

Carney, T. R., and Krissek, L. A., 1986, The late Pliocene and Pleistocene record of ice-rafting at DSDP site 580, northwest Pacific; A comparison of coarse-sand abundance and mass accumulation rate of ice-rafted detritus: Geological Society of America Abstracts with Programs, v. 18, p. 558.

Carter, L. D., 1980, Tertiary tillites(?) on the northeast flank of Granite Mountain, central Alaska Range, *in* Short notes on Alaska geology, 1979–80: Alaska Division of Geological and Geophysical Surveys Geologic Report 63, p. 23–27.

——, 1983, Cenozoic glacial and glaciomarine deposits of the central North Slope, Alaska, *in* Thorson, R. M., and Hamilton, T. D., eds., Glaciation in Alaska; Extended abstracts from a workshop: Fairbanks, University of Alaska Museum Occasional Paper 2, p. 17–21.

Carter, L. D., and Ager, T. A., 1989, Late Pleistocene spruce (*Picea*) in northern interior basins of Alaska and the Yukon; Evidence from marine deposits in northern Alaska, *in* Carter, L. D., Hamilton, T. D., and Galloway, J. P., eds., Late Cenozoic history of the interior basins of Canada and the Yukon; Proceedings of a joint Canadian-American workshop: U.S. Geological Survey Circular 1026, p. 11–14.

Carter, L. D., Brigham-Grette, J., and Hopkins, D. M., 1986a, Late Cenozoic marine transgressions of the Alaskan Arctic Coastal Plain, *in* Heginbottom, J. A., and Vincent, J.-S., eds., Correlation of Quaternary deposits and events around the margin of the Beaufort Sea: Geological Survey of Canada Open-File Report 1237, p. 21–26.

Carter, L. D., Brigham-Grette, J., Marincovich, L., Jr., and Hillhouse, J. W., 1986b, Late Cenozoic Arctic Ocean sea ice and terrestrial paleoclimate: Geology, v. 14, p. 675–678.

Carter, L. D., Brouwers, E. M., and Marincovich, L., Jr., 1988, Nearshore marine environments of the Alaskan Beaufort Sea during deposition of the Flaxman Member of the Gubik Formation, *in* Galloway, J. P., and Hamilton, T. D., eds., Geologic studies in Alaska by the U.S. Geological Survey during 1987: U.S. Geological Survey Circular 1016, p. 27–30.

Causse, C., and Vincent, J.-S., 1989, Th-U disequilibrium dating of Middle and Late Pleistocene wood and shells from Banks and Victoria Islands, Arctic Canada: Canadian Journal of Earth Sciences, v. 26, p. 2718–2723.

Clague, J. J., compiler, 1989, Quaternary geology of the Canadian Cordillera, *in* Fulton, R. J., ed., Quaternary geology of Canada and Greenland: Geological Survey of Canada, The Geology of North America, v. K-1, p. 15–96.

Connor, C. L., 1984, A middle Wisconsin pollen record from the Copper River Basin, Alaska, *in* Coonrad, W. L., and Ellicott, R. L., eds., The United States Geological Survey in Alaska; Accomplishments during 1981: U.S. Geological Survey Circular 868, p. 102–103.

Coulter, H. W., and others, 1965, Map showing extent of glaciations in Alaska: U.S. Geological Survey Miscellaneous Geologic Investigations Series Map I-415, scale 1:2,500,000.

Curry, B. B., 1989, Absence of Altonian glaciation in Illinois: Quaternary Research, v. 31, p. 1–13.

Decker, J., Robinson, M. S., Clough, J. G., and Lyle, W. M., 1989, Geology and petroleum potential of Hope and Selawik Basins, offshore northwestern Alaska: Marine Geology, v. 90, p. 1–18.

Denton, G. H., 1974, Quaternary glaciations of the White River valley, Alaska, with a regional synthesis for the northern St. Elias Mountains, Alaska and Yukon Territory: Geological Society of America Bulletin, v. 85, p. 871–892.

Denton, G. H., and Armstrong, R. L., 1969, Miocene-Pliocene glaciations in southern Alaska: American Journal of Science, v. 267, p. 1121–1142.

Derksen, S. J., 1976, Glacial geology of the Brady Glacier region, Alaska: Columbus, Ohio State University Institute of Polar Studies Report 60, 97 p.

Detterman, R. L., 1986, Glaciation of the Alaska Peninsula, *in* Hamilton, T. D., Reed, K. M., and Thorson, R. M., eds., Glaciation in Alaska; The geologic record: Anchorage, Alaska Geological Society, p. 151–170.

Detterman, R. L., Bowsher, A. L., and Dutro, J. T., Jr., 1958, Glaciation on the Arctic slope of the Brooks Range, northern Alaska: Arctic, v. 11, p. 43–61.

Detterman, R. L., Wilson, F. H., Yount, M. E., and Miller, T. P., 1987, Quaternary geologic map of the Ugashik, Bristol Bay, and western part of Karluk Quadrangles, Alaska: U.S. Geologic Survey Miscellaneous Investigations Series Map I-1801, scale 1:250,000.

Dinter, D. A., 1985, Quaternary sedimentation of the Alaska Beaufort shelf; Influence of regional tectonics, fluctuating sea levels, and glacial sediment sources: Tectonophysics, v. 114, p. 133–161.

Dinter, D. A., Carter, L. D., and Brigham-Grette, J., 1990, Late Cenozoic geologic evolution of the Alaskan North Slope and adjacent continental shelves, *in* Grantz, A., Johnson, L., and Sweeney, J., eds., The Arctic Ocean region: Boulder, Colorado, Geological Society of America, The Geology of North America, v. L, p. 459–490.

Dowsett, H. J., and Poore, R. Z., 1990, Pliocene paleoceanography of North Atlantic Deep Sea Drilling Project site 552; Application of planktonic foraminifera transfer function GSF18, *in* Gosnell, L. B., and Poore, R. Z., eds., Pliocene climates; Scenario for global warming: U.S. Geological Survey Open-File Report 90–64, p. 11–13.

Easterbrook, D. J., Briggs, N. D., Westgate, J. A., and Gorton, M. P., 1981, Age of the Salmon Springs glaciation in Washington: Geology, v. 9, p. 87–93.

Edwards, M. E., and McDowell, P. F., 1989, Quaternary deposits at Birch Creek, northeastern interior Alaska; The possibility of climatic reconstruction, *in* Carter, L. D., Hamilton, T. D., and Galloway, J. P., eds., Late Cenozoic history of the interior basins of Alaska and the Yukon; Proceedings of a joint Canadian-American workshop: U.S. Geological Survey Circular 1026, p. 48–50.

Eyles, C H., 1988, A model for striated boulder pavement formation on glaciated, shallow-marine shelves; An example from the Yakataga Formation, Alaska: Journal of Sedimentary Petrology, v. 58, p. 62–71.

Eyles, C. H., and Eyles, N., 1989, The upper Cenozoic White River "tillites" of southern Alaska; Subaerial slope and fan-delta deposits in a strike-slip setting: Geological Society of America Bulletin, v. 101, p. 1091–1102.

Eyles, N., 1987, Late Pleistocene debris-flow deposits in large glacial lakes in British Columbia and Alaska: Sedimentary Geology, v. 53, p. 33–71.

Eyles, N., and Lagoe, M. B., 1989, Sedimentology of shell-rich deposits (coquinas) in the glaciomarine upper Cenozoic Yakataga Formation, Middleton Island, Alaska: Geological Society of America Bulletin, v. 101, p. 129–142.

Fernald, A. T., 1960, Geomorphology of the upper Kuskokwim region, Alaska: U.S. Geological Survey Bulletin 1071-G, p. 191–279.

Ferrians, O. J., Jr., 1989, Glacial Lake Atna, Copper River Basin, Alaska, *in* Carter, L. D., Hamilton, T. D., and Galloway, J. P., eds., Late Cenozoic history of the interior basins of Alaska and the Yukon; Proceedings of a joint Canadian-American workshop: U.S. Geological Survey Circular 1026, p. 85–88.

Ferrians, O. J., Jr., Nichols, D. R., and Williams, J. R., 1983, Copper River Basin, *in* Péwé, T. L., and Reger, R. D., eds., Richardson and Glenn Highways, Alaska; Guidebook to permafrost and Quaternary geology: Fairbanks, Alaska Division of Geological and Geophysical Surveys Guidebok 1, p. 137–175.

Gard, L. M., Jr., 1980, The Pleistocene geology of Amchitka Island, Aleutian Islands, Alaska: U.S. Geological Survey Bulletin 1478, 38 p.

Gibbons, A. B., Megeath, J. D., and Pierce, K. L., 1984, Probability of moraine survival in a succession of glacial advances: Geology, v. 12, p. 327–330.

Gilbert, G. K., 1904, Glaciers and glaciation: New York, Doubleday, Page and Co., Harriman Alaska Series, v. 3, 231 p.

Grantz, A., Dinter, D. A., and Biswas, N. N., 1983, Map, cross sections, and chart showing late Quaternary faults, folds, and earthquake epicenters on the Alaskan Beaufort shelf: U.S. Geological Survey Miscellaneous Investigations Series Map I-1182-C, 3 sheets, 7 p., scale 1:500,000.

Goldthwait, R. P., 1987, Glacial history of Glacier Bay Park area, *in* Anderson, P. J., Goldthwait, R. P., and McKenzie, G. D., eds., Observed processes of glacial deposition in Glacier Bay, Alaska: Columbus, Ohio, Byrd Polar Research Center, Miscellaneous Publication No. 236, p. 5–16.

Hamilton, T. D., 1979, Late Cenozoic glaciations and erosion intervals, north-central Brooks Range, *in* Johnson, K. M., and Williams, J. R., eds., The United States Geological Survey in Alaska; Accomplishments during 1978: U.S. Geological Survey Circular 804-B, p. B27–B29.

——, 1981, Multiple moisture sources and the Brooks Range glacial record, *in* 10th Annual Arctic Workshop, March 12–14, 1981: Boulder, Colorado, Institute of Arctic and Alpine Research, p. 16–18.

——, 1982a, A late Pleistocene glacial chronology for the southern Brooks Range; Stratigraphic record and regional significance: Geological Society of America Bulletin, v. 93, p. 700–716.

——, 1982b, Relict Pleistocene glacier ice in northern Alaska, *in* Abstracts, 11th Annual Arctic Workshop, March 11–13, 1982: Boulder, Colorado, Institute of Arctic and Alpine Research, p. 25–26.

——, 1984, Surficial geologic map of the Ambler River Quadrangle, Alaska: U.S. Geological Survey Miscellaneous Field Studies Map MF-1678, 1:250,000.

——, 1986a, Late Cenozoic glaciation of the central Brooks Range, *in* Hamilton, T. D., Reed, K. M., and Thorson, R. M., eds., Glaciation in Alaska; The geologic record: Anchorage, Alaska Geological Society, p. 9–49.

——, 1986b, Correlation of Quaternary glacial deposits in Alaska, *in* Richmond, G. M., and Fullerton, D. S., eds., Quaternary glaciations in the United States of America: Quaternary Science Reviews, v. 5, Quaternary glaciations in the Northern Hemisphere, p. 171–180.

——, 1989, Upper Cenozoic deposits, Kanuti Flats and upper Kobuk Trench, northern Alaska, *in* Carter, L. D., Hamilton, T. D., and Galloway, J. P., eds., Late Cenozoic history of the interior basins of Alaska and the Yukon; Proceedings of a joint Canadian-American workshop: U.S. Geological Survey Circular 1026, p. 45–47.

Hamilton, T. D., and Bischoff, J. L., 1984, Uranium-series dating of fossil bones from the Canyon Creek vertebrate locality in central Alaska, *in* Reed, K. M., and Bartsch-Winkler, S., eds., The United States Geological Survey in Alaska; Accomplishments during 1982: U.S. Geological Survey Circular 939, p. 26–29.

Hamilton, T. D., and Fulton, R. J., 1994, Middle and late Wisconsin environments of eastern Beringia, *in* Bonnichsen, R., Frison, G. C., and Turnmire, K., eds., Ice-age peoples of North America: College Station, Texas A&M University Press (in press).

Hamilton, T. D., and Hopkins, D. M., 1982, Correlation of northern Alaskan glacial deposits; A provisional stratigraphic framework, *in* Coonrad, W., ed., The United States Geological Survey in Alaska; Accomplishments during 1980: U.S. Geological Survey Circular 844, p. 15–18.

Hamilton, T. D., and Porter, S. C., 1975, Itkillik glaciation in the Brooks Range, northern Alaska: Quaternary Research, v. 5, p. 471–497.

Hamilton, T. D., and Thorson, R. M., 1983, The Cordilleran Ice Sheet in Alaska, *in* Wright, H. E., Jr., and Porter, S. C., eds., Late Quaternary environments of the United States, v. 1: Minneapolis, University of Minnesota Press, p. 38–52.

Hamilton, T. D., Reed, K. M., and Thorson, R. M., eds., 1986a, Glaciation in Alaska; The geologic record: Anchorage, Alaska Geological Society, 265 p.

Hamilton, T. D., Reed, K. M., and Thorson, R. M., 1986b, Introduction and overview, *in* Hamilton, T. D., Reed, K. M., and Thorson, R. M., eds., Glaciation in Alaska; The geologic record: Anchorage, Alaska Geological Society, p. 1–8.

Hamilton, T. D., Lancaster, G. A., and Trimble, D. A., 1987, Glacial advance of late Wisconsin (Itkillik II) age in the upper Noatak River valley; A radiocarbon-dated stratigraphic record, *in* Hamilton, T. D., and Galloway, J. P., eds., Geologic studies in Alaska by the U.S. Geological Survey during 1986: U.S. Geological Survey Circular 998, p. 35–39.

Hamilton, T. D., Galloway, J. P., and Koster, E. A., 1988, Late Wisconsin eolian activity and related alluviation, central Kobuk River valley, *in* Galloway, J. P., and Hamilton, T. D., eds., Geologic studies in Alaska by the U.S. Geological Survey during 1987: U.S. Geological Survey Circular 1016, p. 39–43.

Hampton, M. A., 1985, Quaternary sedimentation in Shelikof Strait, Alaska: Marine Geology, v. 62, p. 213–253.

Harland, W. B., Herod, K. N., and Krinsley, D. H., 1966, The definition and identification of tills and tillites: Earth-Science Reviews, v. 2, p. 225–256.

Heginbottom, J. A., and Vincent, J.-S., eds., 1986, Correlation of Quaternary deposits and events around the margin of the Beaufort Sea; Contributions from a joint Canadian-American workshop, April 1984: Geological Survey of Canada Open File Report 1237, 60 p.

Herman, Y., and Hopkins, D. M., 1980, Arctic oceanic climate in late Cenozoic time: Science, v. 209, p. 557–562.

Heusser, C. J., 1985, Quaternary pollen records from the Pacific Northwest coast;

Aleutians to the Oregon-California boundary, *in* Bryant, V. M., and Holloway, R. G., eds., Pollen records of late Quaternary North American sediments: American Association of Stratigraphic Palynologists Foundation, p. 141–165.

Hopkins, D. M., 1973, Sea level history in Beringia during the past 250,000 years: Quaternary Research, v. 3, p. 520–540.

—— , 1975, Time-stratigraphic nomenclature for the Holocene epoch: Geology, v. 3, p. 10.

—— , 1982, Aspects of the paleogeography of Beringia during the late Pleistocene, *in* Hopkins, D. M., Matthews, J. V., Jr., Schweger, C. E., and Young, S. B., eds., Paleoecology of Beringia: New York, Academic Press, p. 3–28.

Hopkins, D. M., Rowland, R. W., and Patton, W. W., Jr., 1972, Middle Pleistocene mollusks from St. Lawrence Island and their significance for the paleo-oceanography of the Bering Sea: Quaternary Research, v. 2, p. 119–134.

Hughes, O. L., Rutter, N. W., and Clague, J. J., 1989, Quaternary stratigraphy and history; Yukon Territory, *in* Clague, J. J, compiler, Quaternary geology of the Canadian Cordillera, *in* Fulton, R. J., ed., Quaternary geology of Canada and Greenland: Geological Survey of Canada, The Geology of North America, v. K-1, p. 58–62.

Huston, M. M., Brigham-Grette, J., and Hopkins, D. M., 1990, Paleogeographic significance of middle Pleistocene glaciomarine deposits on Baldwin Peninsula, northwest Alaska: Annals of Glaciology, v. 14, p. 111–114.

Johnson, R. G., 1982, Bruhnes-Matuyama magnetic reversal dated at 790,000 yr B.P. by marine-astronomical correlations: Quaternary Research, v. 17, p. 135–147.

Karlstrom, T.N.V., 1964, Quaternary geology of the Kenai lowland and glacial history of the Cook Inlet region, Alaska: U.S. Geological Survey Professional Paper 443, 69 p.

Karlstrom, T.N.V., and others, 1964, Surficial geology of Alaska: U.S. Geological Survey Miscellaneous Geologic Investigations Series Map 1-357, scale 1:1,584,000.

Kaufman, D. S., and Calkin, P. E., 1988, Morphometric analysis of Pleistocene glacial deposits in the Kigluaik Mountains, northwestern Alaska, U.S.A.: Arctic and Alpine Reserach, v. 20, p. 273–284.

Kaufman, D. S., and Hopkins, D. M., 1986, Glacial history of the Seward Peninsula, *in* Hamilton, T. D., Reed, K. M., and Thorson, R. M., eds., Glaciation in Alaska; The geologic record: Anchorage, Alaska Geological Society, p. 51–77.

Kaufman, D. S., Hopkins, D. M., and Calkin, P. E., 1988, Glacial geologic history of the Salmon Lake area, Seward Peninsula, *in* Galloway, J. P., and Hamilton, T. D., eds., Geologic studies in Alaska by the U.S. Geological Survey during 1987: U.S. Geological Survey Circular 1016, p. 91–94.

Kaufman, D. S., and 6 others, 1989a, Surficial geologic map of the Kigluaik Mountains area, Seward Peninsula, Alaska: U.S. Geological Survey Miscellaneous Field Studies Map MF-2074, scale 1:63,360.

Kaufman, D. S., Hopkins, D. M., Brigham-Grette, J., and Miller, G. H., 1989b, Amino acid geochronology of marine and glacial events at Nome, Alaska: Geological Society of America Abstracts with Programs, v. 21, p. A210.

Keroher, G. C., and others, 1966, Lexicon of geologic names of the United States for 1936–1960: U.S. Geological Survey Bulletin 1200, 4431 p.

Kline, J. T., and Bundtzen, T. K., 1986, Two glacial records from west-central Alaska, *in* Hamilton, T. D., Reed, K. M., and Thorson, R. M., eds., Glaciation in Alaska; The geologic record: Anchorage, Alaska Geological Society, p. 123–150.

Krimmel, R. M., and Meier, M. F., 1989, Glaciers and glaciology of Alaska, *in* 28th International Geological Congress Field Trip Guidebook T301: American Geophysical Union, 61 p.

Lagoe, N. B., Eyles, C. H., and Eyles, Nicholas, 1989, Paleoenvironmental significance of foraminiferal biofacies in the glaciomarine Yakataga Formation, Middleton Island, Gulf of Alaska: Journal of Foraminiferal Research, v. 19, p. 194–209.

Lea, P. D., 1989, Last-glacial eolian deposits in the Nushagak lowland, southwestern Alaska; Distribution, lithostratigraphy, and radiocarbon chronology, *in* Lea, P. D., Quaternary environments and depositional systems of the Nushagak lowland, southwestern Alaska [Ph.D. thesis]: Boulder, University of Colorado, p. 217–294.

—— , 1990a, Pleistocene glacial tectonism and sedimentation on a macrotidal piedmont coast; Ekuk bluffs, southwestern Alaska: Geological Society of America Bulletin, v. 102, p. 1230–1245.

—— , 1990b, Pleistocene periglacial eolian deposits in southwestern Alaska; Sedimentary facies and depositional processes: Journal of Sedimentary Petrology, v. 60, p. 582–591.

Lea, P. D., and Waythomas, C. F., 1990, Late Pleistocene eolian sand sheets in Alaska: Quaternary Research, v. 34, p. 269–281.

Lea, P. D., Elias, S. A., and Short, S. K., 1989, Pleistocene stratigraphy and paleoenvironments of the southern Nushagak lowland, southwestern Alaska, *in* Lea, P. D., Quaternary environments and depositional systems of the Nushagak lowland, southwestern Alaska [Ph.D. thesis]: University of Colorado, p. 76–149.

Mankinen, E. A., and Dalrymple, G. B., 1979, Revised geomagnetic polarity time scale for the interval 0–5 m.y. B.P.: Journal of Geophysical Research, v. 84, p. 615–626.

Mankinen, E. A., and Plafker, G., 1987, Paleomagnetic evidence for a latest Pliocene and early Pleistocene age of the upper Yakataga Formation on Middleton Island, Alaska, *in* Hamilton, T. D., and Galloway, J. P., eds., Geologic studies in Alaska by the U.S. Geological Survey during 1986: U.S. Geological Survey Circular 998, p. 132–136.

Mann, D. H., 1986, Wisconsin and Holocene glaciation of southeast Alaska, *in* Hamilton, T. D., Reed, K. M., and Thorson, R. M., eds., Glaciation in Alaska; The geologic record: Anchorage, Alaska Geologic Society, p. 237–265.

Marincovich, L., Jr., 1989, Tectonically-induced early middle Miocene marine glaciation and molluscan faunas in the Gulf of Alaska: Geological Society of American Abstracts with Programs, v. 21, p. A115.

Martinson, D. G., and 5 others, 1987, Age dating and the orbital theory of the ice ages; Development of a high-resolution 0 to 300,000-year chronostratigraphy: Quaternary Research, v. 27, p. 1–29.

Matthews, J. V., Jr., 1990, New data on Pliocene floras/faunas from the Canadian Arctic and Greenland, *in* Gosnell, L. B., and Poore, R. Z., eds., Pliocene climates; Scenario for global warming: U.S. Geological Survey Open-File Report 90–64, p. 29–33.

Matthews, J. V., Jr., Schweger, C. E., and Hughes, O. L., 1989, Climatic change in eastern Beringia during oxygen isotope stages 2 and 3; Proposed thermal events, *in* Carter, L. D., Hamilton, T. D., and Galloway, J. P., eds., Late Cenozoic history of the interior basins of Alaska and the Yukon; Proceedings of a joint Canadian-American workshop: U.S. Geological Survey Circular 1026, p. 34–38.

McKenzie, G. D., and Goldthwait, R. P., 1971, Glacial history of the last eleven thousand years in Adams Inlet, southeastern Alaska: Geological Society of America Bulletin, v. 82, p. 1762–1782.

Miller, R. D., 1973, Gastineau Channel Formation; A composite glaciomarine deposit near Juneau, Alaska: U.S. Geological Survey Bulletin 1394-C, 20 p.

Molnia, B. F., 1986, Glacial history of the northeastern Gulf of Alaska; A synthesis, *in* Hamilton, T. D., Reed, K. M., and Thorson, R. M., eds., Glaciation in Alaska; The geologic record: Anchorage, Alaska Geological Society, p. 219–235.

Nichols, D. R., 1989, Pleistocene glacial events, southeastern Copper River Basin, Alaska, *in* Carter, L. D., Hamilton, T. D., and Galloway, J. P., eds., Late Cenozoic history of the interior basins of Alaska and the Yukon; Proceedings of a joint Canadian-American workshop: U.S. Geological Survey Circular 1026, p. 78–80.

Nichols, D. R., and Yehle, L. A., 1985, Volcanic debris flows, Copper River Basin, Alaska, *in* Proceedings, 4th International Conference and Field Workshop on Landslides, Tokyo, 1985: Tokyo, Japan Landslide Society, p. 365–372.

Nilsen, T. H., and Moore, T. E., 1984, Stratigraphic nomenclature for the Upper Devonian and Lower Mississippian(?) Kanayut Conglomerate, Brooks Range, Alaska: U.S. Geological Survey Bulletin 1529-A, 64 p.

North American Commission on Stratigraphic Nomenclature, 1983, North American Stratigraphic Code: American Association of Petroleum Geologists Bulletin, v. 67, p. 841–875.

Petrov, O. M., 1967, Paleogeography of Chukotka during late Neogene and Quaternary time, *in* Hopkins, D. M., ed., The Bering land bridge: Stanford, California, Stanford University Press, p. 144–171.

Péwé, T. L., 1975, Quaternary geology of Alaska: U.S. Geological Survey Professional Paper 835, 145 p.

——, 1989, Quaternary stratigraphy of the Fairbanks area, Alaska, *in* Carter, L. D., Hamilton, T. D., and Galloway, J. P., eds., Late Cenozoic history of the interior basins of Alaska and the Yukon; Proceedings of a joint Canadian-American workshop: U.S. Geological Survey Circular 1026, p. 72–77.

Péwé, T. L., and Reger, R. D., 1989, Middle Tanana River valley, *in* Péwé, T. L., and Reger, R. D., eds., Quaternary geology and permafrost along the Richardson and Glenn Highways between Fairbanks and Anchorage, Alaska; 28th International Geological Congress Field Trip Guidebook T102: American Geophysical Union, p. 17–24.

Péwé, T. L., and others, 1953, Multiple glaciation in Alaska: U.S. Geological Survey Circular 289, 13 p.

Pierce, K. L., Obradovich, J. D., and Friedman, I., 1976, Obsidian hydration dating and correlation of Bull Lake and Pinedale Glaciations near West Yellowstone, Montana: Geological Society of America Bulletin, v. 87, p. 703–710.

Plafker, G., 1981, Late Cenozoic glaciomarine deposits of the Yakataga Formation, Alaska, *in* Hambrey, M. J., and Harland, W. B., eds., Earth's pre-Pleistocene glacial record: New York, Cambridge University Press, p. 694–699.

——, 1987, Regional geology and petroleum potential of the northern Gulf of Alaska continental margin, *in* Scholl, D. W., ed., Geology and resource potential of the continental margin of western North America and adjacent ocean basins; Beaufort Sea to Baja California: Houston, Texas, Circum-Pacific Council for Energy and Mineral Resources Earth Science Series, v. 6, p. 229–268.

Plafker, G., and Addicott, W. O., 1976, Glaciomarine deposits of Miocene through Holocene age in the Yakataga Formation along the Gulf of Alaska margin, Alaska, *in* Miller, T. P., ed., Recent and ancient sedimentary environments in Alaska: Anchorage, Alaska Geological Society, p. Q1–Q23.

Plafker, G., Richter, D. H., and Hudson, T., 1977, Reinterpretation of the origin of inferred Tertiary tillite in the northern Wrangell Mountains, Alaska, *in* Blean, K. M., ed., The United States Geological Survey in Alaska; Accomplishments during 1976: U.S. Geological Survey Circular 751-B, p. B52–B54.

Porter, S. C., 1964, Late Pleistocene glacial chronology of north-central Brooks Range, Alaska: American Journal of Science, v. 262, p. 446–460.

Porter, S. C., Pierce, K. L., and Hamilton, T. D., 1983, Late Wisconsin mountain glaciation in the western United States, *in* Wright, H. E., Jr., and Porter, S. C., eds., Late Quaternary environments of the United States, v. 1: Minneapolis, University of Minnesota Press, p. 71–111.

Powell, R. D., and Molnia, B. F., 1989, Glacimarine and sedimentary processes, facies, and morphology of the south-southeast Alaska shelf and fjords: Marine Geology, v. 85, p. 359–390.

Rea, D. K., and Schrader, H., 1985, Late Pliocene onset of glaciation; Ice-rafting and diatom stratigraphy of North Pacific DSDP cores: Palaeogeography, Palaeoclimatology, Palaeoecology, v. 49, p. 313–325.

Reger, R. D., 1979, Glaciation of Indian Mountain, west-central Alaska: College, Alaska Division of Geological and Geophysical Surveys Geologic Report 61, p. 15–18.

Reger, R. D., and Updike, R. G., 1983, Upper Cook Inlet region and the Matanuska valley, *in* Péwé, T. L., and Reger, R. D., eds., Richardson and Glenn Highways, Alaska; Guidebook to permafrost and Quaternary geology: Fairbanks, Alaska Division of Geological and Geophysical Surveys Guidebook 1, p. 185–263.

——, 1989, Upper Cook Inlet region and Matanuska valley, *in* Péwé, T. L., and Reger, R. D., eds., Quaternary geology and permafrost along the Richardson and Glenn Highways between Fairbanks and Anchorage, Alaska, *in* 28th International Geological Congress Field Trip Guidebook T102: American Geophysical Union, p. 45–54.

Reid, H. F., 1896, Glacier Bay and its glaciers: U.S. Geological Survey Annual Report 1894–95, v. 16, part 1, p. 421–461.

Riehle, J. R., and Detterman, R. L., 1993, Quaternary geologic map of the Mount Katmai Quadrangle and adjacent parts of the Naknek and Afognak Quadrangles, Alaska: U.S. Geological Survey Miscellaneous Geologic Investigations Map I-2032, scale 1:250,000.

Richmond, G. M., and Fullerton, D. S., 1986, Introduction to Quaternary glaciations in the United States of America, *in* Richmond, G. M., and Fullerton, D. S., eds., Quaternary glaciations in the United States of America: Quaternary Science Reviews, v. 5, Quaternary glaciations in the Northern Hemisphere, p. 3–10.

Richter, D. H., Schmoll, H. R., and Bove, D. J., 1988, Source of the Sanford volcanic debris flow, south-central Alaska, *in* Galloway, J. P., and Hamilton, T. D., eds., Geologic studies in Alaska by the U.S. Geological Survey during 1987: U.S. Geological Survey Circular 1016, p. 114–116.

Ritter, D. F., 1982, Complex river terrace development in the Nenana Valley near Healy, Alaska: Geological Society of America Bulletin, v. 93, p. 346–356.

Ritter, D. F., and Ten Brink, N. W., 1986, Alluvial fan development and the glacial-glaciofluvial cycle, Nenana Valley, Alaska: Journal of Geology, v. 94, p. 613–625.

Sainsbury, C. L., 1967, Quaternary geology of western Seward Peninsula, Alaska, *in* Hopkins, D. M., ed., The Bering land bridge: Stanford, California, Stanford University Press, p. 121–143.

Sancetta, C., Heusser, L., Labeyrie, L., Naidu, A. S., and Robinson, S. W., 1985, Wisconsin-Holocene paleoenvironment of the Bering Sea; Evidence from diatoms, pollen, oxygen isotopes, and clay minerals: Marine Geology, v. 62, p. 55–68.

Schmoll, H. R., 1984, Late Pleistocene morainal and glaciolacustrine geology in the upper Copper River–Mentasta Pass area, Alaska: Geological Society of America Abstracts with Programs, v. 16, p. 332.

Schmoll, H. R., and Yehle, L. A., 1983, Glaciation in the upper Cook Inlet basin; A preliminary reexamination based on geologic mapping in progress, *in* Thorson, R. M., and Hamilton, T. D., eds., Glaciation in Alaska; Extended abstracts from a workshop: Fairbanks, University of Alaska Museum Occasional Paper 2, p. 75–81.

——, 1986, Pleistocene glaciation of the upper Cook Inlet basin, *in* Hamilton, T. D., Reed, K. M., and Thorson, R. M., eds., Glaciation in Alaska; The geologic record: Anchorage, Alaska Geological Society, p. 193–218.

Schmoll, H. R., Szabo, B. J., Rubin, M., and Dobrovolny, E., 1972, Radiometric dating of marine shells from the Bootlegger Cove Clay, Anchorage area, Alaska: Geological Society of America Bulletin, v. 83, p. 1107–1113.

Schmoll, H. R., Yehle, L. A., Gardner, C. A., and Odum, J. K., 1984, Guide to surficial geology and glacial stratigraphy in the upper Cook Inlet basin: Anchorage, Alaska Geological Society Guidebook, 89 p.

Schweger, C. E., and Matthews, J. V., Jr., 1985, Early and Middle Wisconsinan environments of eastern Beringia; Stratigraphic and paleoecological implications of the Old Crow tephra: Géographie physique et Quaternaire, v. 39, p. 275–290.

Shackleton, N. J., 1987, Oxygen isotopes, ice volume, and sea level: Quaternary Science Reviews, v. 6, p. 183–190.

Shackleton, N. J., and 16 others, 1984, Oxygen isotope calibration of the onset of ice-rafting and history of glaciation in the North Atlantic region: Nature, v. 307, p. 620–623.

Short, S. K., Waythomas, C. F., and Elias, S. A., 1990, Late Quaternary environments of the Holitna lowland, southwestern Alaska *in* 19th Arctic Workshop, March 8–10, Programs and Abstracts: Boulder, Colorado, Institute of Arctic and Alpine Research, p. 72–73.

Sirkin, L., and Tuthill, S. J., 1987, Late Pleistocene and Holocene deglaciation and environments of the southern Chugach Mountains, Alaska: Geological Society of America Bulletin, v. 99, p. 376–384.

Stemper, B. A., Westgate, J. A., and Péwé, T. L., 1989, Tephrochronology and paleomagnetism of loess in Fairbanks area of interior Alaska: 28th International Geological Congress, Washington, D.C., Abstracts, v. 2, p. 602.

Stout, J. H., Brady, J. B. Weber, F., and Page, R. A., 1973, Evidence for Quaternary movement on the McKinley strand of the Denali fault in the Delta River area, Alaska: Geological Society of America Bulletin, v. 84, p. 939–947.

Swanston, D. N., 1969, A late Pleistocene glacial sequence from Prince of Wales Island, Alaska: Arctic, v. 32, p. 25–33.

Tarnocai, C., 1989, Paleosols of northwestern Canada, *in* Carter, L. D., Hamilton, T. D., and Galloway, J. P., eds., Late Cenozoic history of the interior basins of Alaska and the Yukon; Proceedings of a joint Canadian-American workshop: U.S. Geological Survey Circular 1026, p. 39–44.

Tarr, R. S., and Martin, L., 1914, Alaskan glacier studies: Washington, D.C., National Geographic Society, 498 p.

Ten Brink, N. W., 1983, Glaciation of the northern Alaska Range, *in* Thorson, R. M., and Hamilton, T. D., eds., Glaciation in Alaska; Extended abstracts from a workshop: Fairbanks, University of Alaska Museum Occasional Paper 2, p. 82–91.

Ten Brink, N. W., and Waythomas, C. F., 1985, Late Wisconsin glacial chronology of the north-central Alaska Range; A regional synthesis and its implications for early human settlements, *in* Powers, W. R., and others, eds., North Alaska Range Early Man Project: National Geographic Society Research Reports, v. 19, p. 15–32.

Thorson, R. M., 1986, Late Cenozoic glaciation of the northern Nenana River valley, *in* Hamilton, T. D., Reed, K. M., and Thorson, R. M., eds., Glaciation in Alaska; The geologic record: Anchorage, Alaska Geological Society, p. 99–121.

Thorson, R. M., and Hamilton, T. D., 1986, Glacial geology of the Aleutian Islands, Based on the contributions of Robert F. Black, *in* Hamilton, T. D., Reed, K. M., and Thorson, R. M., eds., Glaciation in Alaska; The geologic record: Anchorage, Alaska Geological Society, p. 171–191.

Thorson, R. M., Dixon, E. J., Jr., Smith, G. S., and Batten, A. R., 1981, Interstadial proboscidean from south-central Alaska; Implications for biogeography, geology, and archeology: Quaternary Research, v. 16, p. 404–417.

Updike, R. G., Dearborne, L. L., Ulery, C. A., and Weir, J. L., 1984, Guide to the engineering geology of the Anchorage area: Anchorage, Alaska Geological Society, 72 p.

Vincent, J.-S., 1989, Late Pleistocene glacial advances, *in* Carter, L. D., Hamilton, T. D., and Galloway, J. P., eds., Late Cenozoic history of the interior basins of Alaska and the Yukon; Proceedings of a joint Canadian-American workshop: U.S. Geological Survey Circular 1026, p. 111–112.

Vincent, J.-S., and Prest, V. K., 1987, The Early Wisconsinan history of the Laurentide Ice Sheet: Géographie physique et Quaternaire, v. 41, p. 199–213.

Vincent, J.-S., Carter, L. D., Matthews, J. V., Jr., and Hopkins, D. M., 1989, Joint Canadian-American investigation of the Cenozoic geology of the lowlands bordering the Beaufort Sea, *in* Carter, L. D., Hamilton, T. D., and Galloway, J. P., eds., Late Cenozoic history of the interior basins of Alaska and the Yukon; Proceedings of a joint Canadian-American workshop: U.S. Geological Survey Circular 1026, p. 7–9.

von Huene, R., Crouch, J., and Larson, E., 1976, Glacial advance in the Gulf of Alaska area implied by ice-rafted material, *in* Cline, R. M., and Hayes, J. D., eds., Investigations of late Quaternary paleoceanography and paleoclimatology: Geological Society of America Memoir 145, p. 393–410.

Wahrhaftig, C., 1958, Quaternary geology of the Nenana River valley and adjacent parts of the Alaska Range: U.S. Geological Survey Professional Paper 293-A, p. 1–68.

Warner, B. G., Clague, J. J., and Mathewes, R. W., 1984, Geology and paleoecology of a mid-Wisconsin peat from the Queen Charlott Islands, British Columbia, Canada: Quaternary Research, v. 21, p. 337–350.

Waythomas, C. F., 1990, Late Quaternary stratigraphic framework of the Holitna lowland, southwest Alaska, *in* 19th Annual Arctic Workshop, March 8–10, Abstracts: Boulder, Colorado, Institute of Arctic and Alpine Research, p. 91.

Weber, F. R., 1985, Late Quaternary glaciation of the Pavlof Bay and Port Moller areas, Alaska Peninsula, *in* Bartsch-Winkler, S., ed., The United States Geological Survey in Alaska; Accomplishments during 1984: U.S. Geological Survey Circular 967, p. 42–44.

——, 1986, Glacial geology of the Yukon-Tanana Upland, *in* Hamilton, T. D., Reed, K. M., and Thorson, R. M., eds., Glaciation in Alaska; The geologic record: Anchorage, Alaska Geological Society, p. 79–98.

Weber, F. R., and Hamilton, T. D., 1984, Glacial geology of the Mt. Prindle area, Yukon-Tanana Upland, Alaska, *in* Short notes on Alaskan geology, 1982–83: Fairbanks, Alaska Division of Geological and Geophysical Surveys Professional Report 86, p. 42–48.

Weber, F. R., Hamilton, T. D., Hopkins, D. M., Repenning, C. A., and Haas, H., 1981, Canyon Creek; A late Pleistocene vertebrate locality in interior Alaska: Quaternary Research, v. 16, p 167–180.

Westgate, J., 1988, Isothermal plateau fission-track age of the late Pleistocene Old Crow tephra, Alaska: Geophysical Research Letters, v. 15, p. 376–379.

Westgate, J. A., Hamilton, T. D., and Gorton, M. P., 1983, Old Crow tephra; A new late Pleistocene stratigraphic marker across north-central Alaska and western Yukon Territory: Quaternary Research, v. 19, p. 38–54.

Westgate, J. A., Walter, R. C., Pearce, G. W, and Gorton, M. P., 1985, Distribution, stratigraphy, petrochemistry, and paleomagnetism of the late Pleistocene Old Crow tephra in Alaska and the Yukon: Canadian Journal of Earth Sciences, v. 22, p. 893–906.

Williams, J. R., 1986, New radiocarbon dates from the Matanuska Glacier bog section, *in* Bartsch-Winkler, S., and Reed, K. M., eds., Geological studies in Alaska by the U.S. Geological Survey during 1985: U.S. Geological Survey Circular 978, p. 85–88.

——, 1989, A working glacial chronology for the western Copper River Basin, Alaska, *in* Carter, L. D., Hamilton, T. D., and Galloway, J. P., eds., Late Cenozoic history of the interior basins of Alaska and the Yukon; Proceedings of a joint Canadian-America workshop: U.S. Geological Survey Circular 1026, p. 81–84.

Williams, J. R., and Galloway, J. P., 1986, Map of western Copper River Basin, Alaska, showing lake sediments and shorelines, glacial moraines, and location of stratigraphic sections and radiocarbon-dated samples: U.S. Geological Survey Open-File Report 86–390, 30 p., scale 1:250,000.

Wintle, A. G., and Westgate, J. A., 1986, Thermoluminescence age of Old Crow tephra in Alaska: Geology, v. 14, p 594–597.

Yehle, L. A., and Schmoll, H. R., 1989, Surficial geologic map of the Anchorage B-7 SW quadrangle, Alaska: U.S. Geological Survey Open-File Report 89-318, scale 1:25,000.

Manuscript Accepted by the Society October 24, 1990

ACKNOWLEDGMENTS

Parts of this chapter have been condensed from the volume *Glaciation in Alaska* (Hamilton and others, 1986a). I am grateful to the Alaska Geological Society for their support of that publication and for permission to adapt several illustrations from it for the present paper. Helpful comments on sections of this chapter were contributed by J. M. Armentrout, J. L. Bischoff, J. Brigham-Grette, L. D. Carter, R. L. Detterman, C. H. Eyles, D. M. Hopkins, D. S. Kaufman, D. H. Mann, B. F. Molnia, G. Plafker, R. D. Reger, H. R. Schmoll, R. M. Thorson, J. A. Westgate, and J. R. Williams. An initial draft was reviewed by O. J. Ferrians, Jr. and R. M. Thorson. Illustrations were drafted by G. A. Lancaster and Inyo Ellersieck.

NOTES ADDED IN PROOF

This chapter was last updated in 1989. Since that date, nearly 200 reports and abstracts have been published on Alaskan glaciation and related late Cenozoic climatic changes. The space allowed for this note permits discussion of only a few of those studies.

Tertiary glaciation

Drill cores from the Arctic Ocean basin and erratics in transgressive deposits around its margins indicate episodes of regional cold climate and glaciation. The oldest ice-rafted material in the Canada Basin is of late Miocene age (Grantz and others, 1990), and sediments on Alpha Ridge indicate a subsequent late Pliocene cold interval about 2.48 to 2.08 Ma (Scott and others, 1989). Erratic clasts in deposits of the Colvillian transgression (sometime between 2.7 and 2.48 Ma) and in basal beds of the Bigbendian transgression (about 2.48 Ma) indicate that glaciers were present somewhere around the margins of the Arctic Basin at those times (Brigham-Grette and Carter, 1992). Erratic boulders in late Tertiary deposits of the central Arctic Coastal Plain are reported by Rawlinson (1993), but ages of those deposits are uncertain.

Drill cores from the Gulf of Alaska and the North Pacific, seismic stratigraphy along the Gulf of Alaska shelf (Carlson, 1989), and outcrop studies of the Yakataga Formation provide additional insights into Tertiary glaciation in the Gulf of Alaska region. Sediment cores show that glaciation extended back into the latest Miocene (von Huene, 1989), and that a later major episode of glaciation began about 2.6 Ma (Morley and Dworetzky, 1991; Krissek and others, 1993). Glacial maxima are recorded at 2.5 to 1.8 Ma and during the last 1.3 m.y. (Morley and Dworetzky, 1991).

Initial widespread appearance of ice-rafted debris in the Yakataga Formation was assigned an early middle Miocene (15 to 16 Ma) age by Marincovich (1990) on the basis of mollusc faunas, but alternatively may be late Miocene (5 to 6 Ma) according to the foraminiferal record (Eyles and others, 1991; Zellers, 1990). Renewed glaciomarine deposition in the late Pliocene (Zellers, 1990; Zellers and others, 1992) followed a middle Pliocene cool-temperate episode (Zellers and others, 1992). Stratigraphy and sedimentology of the Yakataga Formation are reviewed by Eyles and Lagoe (1990), C. H. Eyles and others (1991), and N. Eyles and others (1992).

Early and middle Pleistocene glaciations

A general climatic reconstruction for the Arctic Ocean borderland by Repenning and Brouwers (1992) indicates mild conditions about 2 Ma followed by oscillating cool and warm periods 1.7 to 1.2 Ma and then by more severe cold and glacial conditions after 1.1 Ma. Sediment cores from the Arctic Ocean basin show repeated intense glaciations within the past 0.8 to 1.0 m.y. (Bischof and others, 1992; Phillips and others, 1992). Although an early Pleistocene age has been assumed for extensive, subdued drift sheets in northern Alaska, the age of these deposits is still poorly constrained. Correlative drift on Banks Island is now known to be magnetically reversed, hence older than 780 ka (Vincent, 1990). Moraines assigned to the early Pleistocene Anaktuvuk glaciation in the northeastern Brooks Range have been strongly deformed by tectonism (Rawlinson, 1993).

Cores taken from Alpha Ridge indicate that middle Pleistocene glaciation may have been interrupted by a warm interval 0.6 to 0.4 Ma (Scott and others, 1989), and insolation values suggest that major glaciation in northern Alaska would have been unlikely during 575 to 250 ka (Bartlein and others, 1991). The Anvilian marine transgression of the Nome area, which may date about 410 ka (Kaufman, 1992), probably occurred within this interval.

The middle Pleistocene Nome River glaciation had been dated as between 580 and 280 ka (Kaufman and others, 1991), and a correlative glacial advance built the Baldwin Peninsula moraine (Huston and others, 1990) sometime between 400 and 300 ka (Roof and Brigham-Grette, 1992). A probable correlative glacier from Chukotka flowed eastward across the Bering platform and encroached on St. Lawrence Island sometime prior to the last interglaciation (Brigham-Grette and others, 1992; Heiser and others, 1992).

The Old Crow tephra, which is widespread in Alaska and the Yukon Territory (Waythomas and others, 1993), generally occurs near the base of organic-rich deposits of interglacial character (Hamilton and Brigham-Grette, 1992). The fission-track age of 149 ± 13 ka for this deposit is supported by an age estimate of 135 ± 5 ka on stratigraphic grounds (Hamilton and Brigham-Grette, 1992) and by bracketing thermoluminescence ages of 110 ± 32 and 140 ± 30 ka (Berger and others, 1992). The Old Crow tephra is an important stratigraphic marker for the stage 6/5 glacial-interglacial transition (Begét and others, 1991; Hamilton and Brigham-Grette, 1992).

New studies of glacial sequences that include early and (or) middle Pleistocene as well as late Pleistocene ice advances include Peck and others (1990), Reger and Bundtzen (1990), Reger and Pinney (1993), and Waythomas (1990).

Late Pleistocene glaciation

Glaciation of probable isotope stage 4 or late stage 5 age was extensive in the Bering Strait region, where ice again advanced from Chukotka to St. Lawrence Island (Brigham-Grette and others, 1992; Hopkins and others, 1992). This advance may correlate with the glaciomarine Flaxman Formation of northern Alaska, which probably was deposited about 85 to 80 ka during isotope stage 5a (Carter and Whelan, 1991), and with extensive glaciation of the western Canadian Arctic (Vincent, 1992).

Multiple stades of Late Wisconsin glaciation are recognized and partly dated in the Cook Inlet region (Reger and Pinney, 1993) and on the Alaska Peninsula (Pinney and Begét, 1991). The most extensive advances ended by about 14 ka, when pollen records show a marked change from full-glacial to late-glacial environments (e.g., Anderson, 1991). Subsequent glacial readvances include the 12 to 10 ka Ukah Stade on the Alaska Peninsula (Pinney and Begét, 1991), a readvance about 12 ka at Valdez (Reger, 1991), and readvances about 13.4, 12 to 11.7, and 11 to 10 ka on Kodiak Island (Mann, 1992). Some of these readvances could coincide with Younger Dryas cooling, for which pollen evidence has been reported by Engstrom and others (1990).

ADDITIONAL REFERENCES

Anderson, P. M., 1991, Palynological data as tools for interpreting past climates—Some examples from northern North America, *in* Weller, G., Wilson, C. L., and Severin, B.A.B., eds., International Conference on the Role of the Polar Regions in Global Change, University of Alaska at Fairbanks, June 1990: Fairbanks, University of Alaska, v. II, p. 557–564.

Bartlein, P. J., Anderson, P. M., Edwards, M. E., and McDowell, P. F., 1991, A framework for interpreting paleoclimatic variations in eastern Beringia: Quaternary International, v. 10–12, p. 73–83.

Begét, J., Crumley, S., and Stone, D., 1991, The record of the last glacial cycle in Alaskan loess, *in* Program and Abstracts, 21st Arctic Workshop, University of Alaska, May 1991: Fairbanks, Alaska Quaternary Center, p. 51.

Berger, G. W., Pillans, B. J., and Palmer, A. S., 1992, Dating loess up to 800 ka by thermoluminescence: Geology, v. 20, p. 403–406.

Bischof, J. F., Clark, D. L., and Darby, D. A., 1992, Dropstone cycles in the western Arctic Ocean and their importance for continental glaciation in the North America Arctic region: Geological Society of America Abstracts with Programs, v. 24, p. 245.

Brigham-Grette, J., and Carter, L. D., 1992, Pliocene marine transgressions of northern Alaska—Circumarctic correlations and paleoclimatic interpretations: Arctic, v. 45, p. 74–89.

Brigham-Grette, J., Benson, S., Hopkins, D. M., Heiser, P., Ivanov, V. F., and Basilyan, A., 1992, Middle and late Pleistocene Russian glacial ice extent in the Bering Strait region—Results of recent field work: Geological Society of America Abstracts with Programs, v. 24, p. A346.

Carlson, P. R., 1989, Seismic reflection characteristics of glacial and glaciomarine sediment in the Gulf of Alaska and adjacent fjords: Marine Geology, v. 85, p. 391–416.

Carter, L. D., and Whelan, J. F., 1991, Isotopic evidence for restricted arctic sea ice during a late Pleistocene warm period—Implications for sea ice during future climatic warming at high latitudes: Geological Society of America Abstracts with Programs, v. 23, p. A237.

Engstrom, D. R., Hansen, B.C.S., and Wright, J. E., Jr., 1990, A possible Younger Dryas record in southeastern Alaska: Science, v. 250, p. 1383–1385.

Eyles, C. H., and Lagoe, M. B., 1990, Sedimentation patterns and facies geometries on a temperate glacially-influenced continental shelf—The Yakataga Formation, Middleton Island, Alaska, *in* Dowdeswell, J. A., and Scourse, J. D., eds., Glacimarine environments—Processes and sediments: Geological Society of London Special Publication 53, p. 363–386.

Eyles, C. H., Eyles, N., and Lagoe, M. B., 1991, The Yakataga Formation—A late Miocene to Pleistocene record of temperate glacial marine sedimentation in the Gulf of Alaska, *in* Anderson, J. B., and Ashley, G. M., eds., Glacial marine sedimentation: Geological Society of America Special Paper 261, p. 159–180.

Eyles, N., Vossler, S. M., and Lagoe, M. B., 1992, Ichnology of a glacially-influenced continental shelf and slope—the Late Cenozoic Gulf of Alaska (Yakataga Formation): Palaeogeography, Palaeoclimatology, Palaeoecology, v. 94, p. 193–221.

Grantz, A., May, S. D., Taylor, P. T., and Lawver, L. A., 1990, Canada Basin, *in* Grantz, A., Johnson, L., and Sweeney, J. F., eds., The Arctic Ocean region: Boulder, Colorado, Geological Society of America, The Geology of North America, v. L, p. 379–402.

Hamilton, T. D., and Brigham-Grette, J., 1992, The last interglaciation in Alaska—Stratigraphy and paleoecology of potential sites: Quaternary International, v. 10-12, p. 49–71.

Heiser, P. A., and 5 others, 1992, Pleistocene glacial geology of St. Lawrence Island, Alaska: Geological Society of America Abstracts with Programs, v. 24, p. A345–346.

Hopkins, D. M., Benson, S., Brigham-Grette, J., Heiser, P., and Ivanov, V., 1992, Ice sheets on the Bering shelf?, International Conference on Arctic Margins, Anchorage, Alaska, September 1992, Abstracts, p. 27.

Huston, M. M., Brigham-Grette, J., and Hopkins, D. M., 1990, Paleogeographic significance of middle Pleistocene glaciomarine deposits on Baldwin Peninsula, northwest Alaska: Annals of Glaciology, v. 14, p. 111–114.

Kaufman, D. S., 1992, Aminostratigraphy of Pliocene-Pleistocene high-sea-level deposits, Nome Coastal plain and adjacent nearshore area, Alaska: Geological Society of America Bulletin, v. 104, p. 40–52.

Kaufman, D. S., Walter, R. C., Brigham-Grette, J., and Hopkins, D. M., 1991: Middle Pleistocene age of the Nome River glaciation, northwestern Alaska: Quaternary Research, v. 36, p. 277–293.

Krissek, L. A., Rea, D. K., Basov, I., and Leg 145 Shipboard Scientific Party, 1993, Plio-Pleistocene paleoclimatic and paleoceanographic evolution of the North Pacific—Preliminary results of ODP Leg 145, *in* 23rd Annual Arctic Workshop, Columbus, Ohio, 1993, Abstracts: Columbus, Ohio State University Byrd Polar Research Center, Miscellaneous Series M-322, p. 49.

Mann, D. H., 1992, High resolution dating of moraines on Kodiak Island, Alaska, links Atlantic and North Pacific climatic changes during the late glacial: Geological Society of America Abstracts with Programs, v. 24, p. A346.

Marincovich, L., Jr., 1990, Molluscan evidence for early middle Miocene marine glaciation in southern Alaska: Geological Society of America Bulletin, v. 102, p. 1591–1599.

Morley, J. J., and Dworetzky, B. A., 1991, Evolving Pliocene-Pleistocene climate—A North Pacific perspective: Quaternary Science Reviews, v. 10, p. 225–237.

Peck, B. J., Kaufman, D. S., and Calkin, P. E., 1990, Relative dating of moraines using moraine morphometric and boulder weathering criteria, Kigluaik Mountains, Alaska: Boreas, v. 19, p. 227–239.

Phillips, R. L., Grantz, A., Mullen, M. W., Poore, R. Z., and Reick, H. G., 1992, Climatically controlled sedimentation—glacial-interglacial cycles from Northwind Ridge, Arctic Ocean, *in* 22nd Arctic Workshop, University of Colorado, March 1992, Program and Abstracts: Boulder, University of Colorado Institute of Arctic and Alpine Research, p. 116.

Pinney, D. S., and Begét, J. E., 1991, Deglaciation and latest Pleistocene and early Holocene glacier readvances on the Alaska Peninsula, *in* Weller, G., Wilson, C. L., and Severin, B.A.B., eds., International Conference on the Role of the Polar Regions in Global Change, University of Alaska, June 1990: Fairbanks, University of Alaska, v. II, p. 634–640.

Rawlinson, S. E., 1993, Surficial geology and morphology of the Alaskan central Arctic Coastal Plain: Fairbanks, Alaska Division of Geological and Geophysical Surveys, Report of Investigations 93-1, 172 p.

Reger, R. D., 1991, Deglaciation of the Allison-Sawmill Creeks area, southern shore of Port Valdez, Alaska, *in* Reger, R. D., ed., Short Notes on Alaskan Geology 1991: Fairbanks, Alaska Division of Geological and Geophysical Surveys, Professional Report 111, p. 55–61.

Reger, R. D., and Bundtzen, T. K., 1990, Multiple glaciation and gold-placer formation, Valdez Creek Valley, western Clearwater Mountains, Alaska: Fairbanks, Alaska Division of Geological and Geophysical Surveys, Professional Report 107, 29 p.

Reger, R. D., and Pinney, D. S., 1993, Late Wisconsin glaciation of the Cook Inlet region with emphasis on Kenai Lowland and implications for early peopling, *in* Davis, N., ed., Symposium on the Anthropology of Cook Inlet: Cook Inlet Historical Society (in press).

Repenning, C. A., and Brouwers, E. M., 1992, Late Pliocene–early Pleistocene ecological changes in the Arctic Ocean borderland: U.S. Geological Survey Bulletin 2036, 37 p.

Roof, S. R., and Brigham-Grette, J., 1992, Paleoclimatic significance of middle Pleistocene glacial deposits in the Kotzebue Sound region, northwest coastal Alaska: Geological Society of America Abstracts with Programs, v. 24, p. A314.

Scott, D. B., Mudie, P. J., Baki, V., MacKinnon, K. D., and Cole, F. E., 1989, Biostratigraphy and late Cenozoic paleoceanography of the Arctic Ocean—Foraminiferal, lithostratigraphic, and isotopic evidence: Geological Society of America Bulletin, v. 101, p. 260–277.

Vincent, J.-S., 1990, Late Tertiary and Early Pleistocene deposits and history of Banks Island, southwestern Canadian Arctic Archipelago: Arctic, v. 43, p. 339–363.

——, 1992, The Sangamonian and early Wisconsinan glacial record in the western Canadian Arctic, *in* Clark, P. U., and Lea, P. D., eds., The last interglacial-glacial transition in North America: Boulder, Colorado, Geological Society of America Special Paper 270, p. 233–252.

Von Huene, R., 1989, Continental margins around the Gulf of Alaska, *in* Winterer, E. L., Hussong, D. M., and Decker, R. W., eds., The eastern Pacific Ocean and Hawaii: Boulder, Colorado, Geological Society of America, The Geology of North America, v. N, p. 383–401.

Waythomas, C. F., 1990, Quaternary geology and late-Quaternary environments of the Holitna lowland, and Chuilnuk-Kiokluk Mountains region, interior southwestern Alaska [Ph.D. thesis]: Boulder, University of Colorado, 268 p.

Waythomas, C. F., Lea, P. D., and Walter, R. C., 1993, Stratigraphic context of Old Crow tephra, Holitna lowland, interior southwest Alaska: Quaternary Research, v. 40, p. 20–29.

Zellers, S. D., 1990, Foraminiferal biofacies analysis of the Yakataga formation, Icy Bay, Alaska—Insights into Pliocene glaciomarine paleoenvironments of the Gulf of Alaska: Palaios, v. 5, p. 273–296.

Zellers, S. D., Lagoe, M. B., and Ray, J. C., 1992, Paleoclimatic and depositional significance of an offshore Yakataga Fm. section, eastern Gulf of Alaska—Impact on a Neogene paleoclimatic framework for the far North Pacific: Geological Society of America Abstracts with Programs, v. 24, p. 93.

Printed in U.S.A.

The Geology of North America
Vol. G-1, The Geology of Alaska
The Geological Society of America, 1994

Chapter 28

Permafrost in Alaska

Oscar J. Ferrians, Jr.
U.S. Geological Survey, 4200 University Drive, Anchorage, Alaska 99508-4667

INTRODUCTION

Permafrost is defined as soil or rock material, with or without included moisture or organic matter, that has remained at or below 0°C for two or more years (Muller, 1945, p. 3). It is defined exclusively on the basis of temperature; however, one of the most important properties of permafrost is the amount of ice it contains. Permafrost with little or no ice generally does not cause engineering or environmental problems, but permafrost that is ice rich can cause extremely serious problems if allowed to thaw. Ice in active glaciers is not considered permafrost even though it fits the general definition of permafrost.

Permafrost underlies approximately 20 percent of the world's land mass (Muller, 1945, p. 3), and most of it is present in the Northern Hemisphere because the land area is much greater there than it is in the Southern Hemisphere (Fig. 1). However, permafrost is widespread in Antarctica. In addition to Alaska, where 85 percent of the land is within the permafrost region, extensive areas of permafrost are present in other countries in the Northern Hemisphere. These are Canada (50 percent), the U.S.S.R. (50 percent), and the People's Republic of China (20 percent).

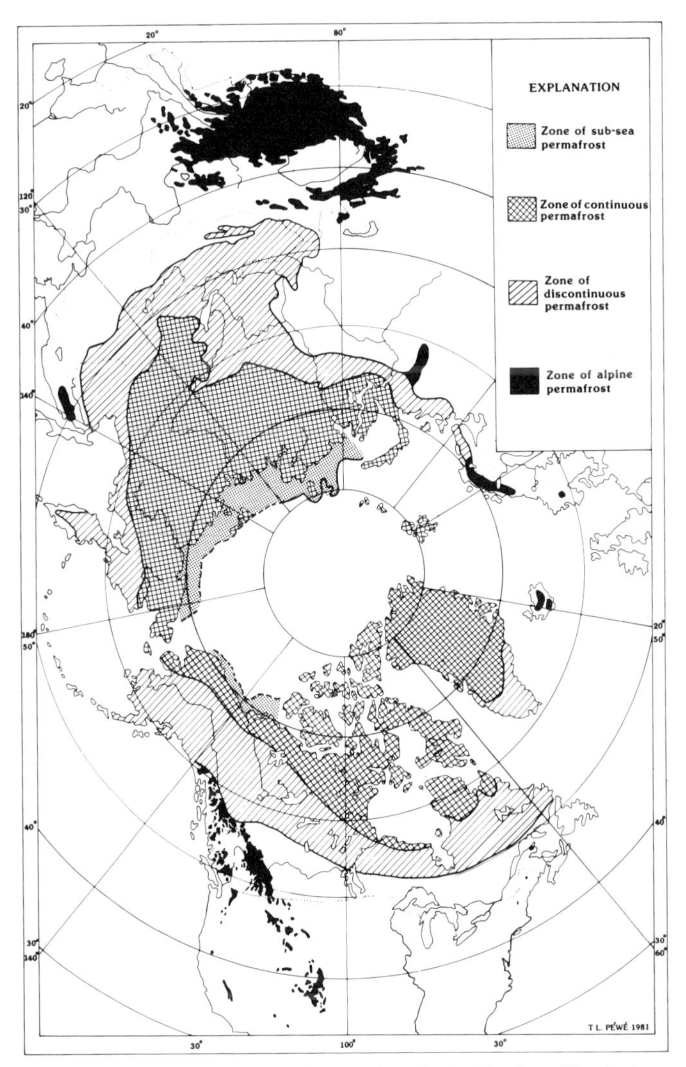

Figure 1. Map showing extent of permafrost in the Northern Hemisphere (from Péwé, 1982, Fig. 21).

ORIGIN AND THERMAL REGIME

Permafrost forms when the mean annual air temperature is low enough to maintain a mean annual near-surface ground temperature at or below 0°C. If climatic conditions in an area change in such a way as to reduce the average near-surface ground temperature to below 0°C, the depth of winter freezing will exceed the depth of summer thawing, and if the mean near-surface ground temperature remains below 0°C, a layer of frozen ground will be added each year to the base of the permafrost until the downward penetration (aggradation) of the frozen ground is balanced by heat flowing upward from the Earth's interior. Thus, the thickness to which permafrost can grow depends on the mean near-surface ground temperature and the geothermal gradient. The lower the temperature and gradient, the greater the thickness of the permafrost. In this manner, permafrost hundreds of meters thick can form during a period of several thousand years.

Because the formation of permafrost depends on ground

Ferrians, O. J., Jr., 1994, Permafrost in Alaska, *in* Plafker, G., and Berg, H. C., eds., The Geology of Alaska: Boulder, Colorado, Geological Society of America, The Geology of North America, v. G-1.

temperature near the surface, the thickness and areal distribution of permafrost are directly affected by the kind and size of natural surface features (bodies of water, topography, drainage, and vegetation) that act as a heat source or heat sink or as insulation. Changes in the surface environment—such as would be produced by a transgressing (or regressing) sea, or building of roads and other surface structures, draining of lakes, and clearing of vegetation—produce profound changes in the permafrost (Péwé, 1954, 1982; Lachenbruch, 1957a, b; Greene and others, 1960; Hok, 1969; Ferrians and others, 1969; Haugen and Brown, 1971; Brown, 1973; Lawson and Brown, 1978; Lawson and others, 1978; Brown and Grave, 1979a, b; Brown and Hemming, 1980; Nelson and Outcalt, 1982; Lawson, 1983, 1986; Walker, 1983, 1988; Walker, D.A., and others, 1987; Carter and others, 1987).

Although the permafrost table and the temperature in the upper part of permafrost may be quick to respond to surface changes (Lachenbruch and Marshall, 1986), centuries, or even tens of centuries, are required for surface changes to affect the bottom of thick permafrost. Therefore, the present thickness and distribution reflects former thermal environments (Lachenbruch and others, 1966).

In the absence of various modifying surface conditions such as those described previously, the top and bottom of the permafrost layer tend to parallel the ground surface, rising under hills and lowering beneath valleys.

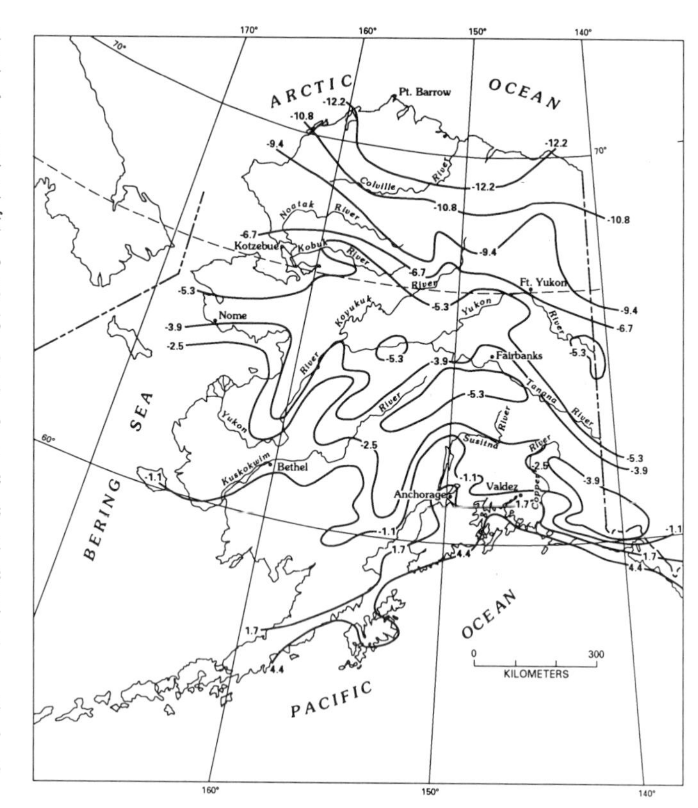

Figure 2. Map of Alaska showing mean annual air isotherms in degrees Fahrenheit (from Johnson and Hartman, 1969, Plate 35).

CLIMATE

Climate is the major factor that controls the regional (and global) distribution of permafrost. Within the continuous permafrost zone of Alaska, Barrow has the lowest recorded mean annual air temperature, −12.2°C, whereas Anchorage, just outside of the permafrost region, has a mean annual air temperature of 1.7°C (Fig. 2).

The climatic control is also reflected by the thickness of permafrost which ranges from more than 630 m at Prudhoe Bay to lenses less than a meter thick in the southernmost part of the permafrost region. Naturally, the thermal conductivity of the rock and soil material and the heat flow from the interior of the Earth also affect the depth to which permafrost can form.

Even though annual precipitation is low in most areas within permafrost regions, ground conditions generally are wet, especially in areas underlain by fine-grained sediments. This poor drainage is caused by the impervious permafrost layer, generally within a meter of the surface, and by the low evaporation rate at the prevailing low temperatures. Also, the low annual precipitation produces a thin snow cover, which limits the effectiveness of the snow cover as a ground insulator during the winter months when air temperatures are very cold.

In an area where the mean annual air temperature is about 0°C, microclimates, which are governed largely by local features, can be critical in determining whether or not permafrost is present. For example, the mean annual air temperature at a site underlain by unvegetated dry soils can be a few degrees higher than a nearby site underlain by vegetated wet soils. In addition, the great variability in the thermal conductivity of peaty soils between the summer when they are dry (low conductivity) and in the winter when they are wet and frozen (high conductivity) can cause permafrost to form and persist in areas where conditions are marginal for its formation. Slope exposure also controls microclimate in mountainous or hilly areas where south-facing slopes may receive considerably more solar heat than north-facing slopes. This is particularly important in southern parts of the permafrost region where south-facing slopes often are completely free of permafrost, whereas nearby north-facing slopes are underlain by permafrost.

EARLY INVESTIGATIONS IN ALASKA

In 1816, during an exploratory sailing mission to America to find a northwest passage, Otto von Kotzebue observed large masses of ground ice exposed in bluffs bordering Eschscholtz Bay of Kotzebue Sound in northwestern Alaska. His observations (von Kotzebue, 1821) are the earliest known published record of the presence of permafrost in Alaska. Subsequently, other explorers and scientists made observations and studies of frozen ground (Beechey, 1831; Richardson, 1841, 1854; Seemann, 1853; Dall,

1881; Hooper, 1881, 1884; Muir, 1883; Cantwell, 1887, 1896; Russell, 1890). Maddren (1905) prepared an excellent summary of this early work. He also described his own observations of frozen ground in Alaska and adjacent territory in this report. Many of these early observations and studies, limited to northwestern Alaska, were in conjunction with the study of rich Pleistocene fossil remains preserved in the frozen sediments exposed in bluffs bordering Eschscholtz Bay and in other areas. Even though most workers debated the origin of the ground ice, the existence of perennially frozen ground was clearly demonstrated. In fact, in the early 1880s, systematic ground temperature measurements were taken at the base of a 37.5-ft-deep (11.4 m) shaft at Point Barrow. A constant 12°F (–11°C) temperature was recorded (Ray, 1885). In addition, R. S. Woodward, U.S. Geological Survey, determined mathematically that over a period of a few thousand years, perennially frozen ground could reach depths of several hundred meters under the climatic conditions that exist today in northern Alaska (Russell, 1890, p. 130–132).

After the discovery of gold in the Circle district of Alaska in 1893, widespread prospecting and mining activity throughout Alaska and numerous geological studies by scientists provided considerable data about permafrost occurrences and factors that control its distribution. In interior Alaska the frozen ground posed major problems for placer miners (Purington, 1905). The organic-rich fine-grained sediments or "muck" deposits at the surface were frozen, and thus their removal was very difficult. The underlying gold-bearing gravel deposits were also frozen and, consequently, had to be thawed before they could be mined. With this new information about permafrost occurrences, it became possible to make estimates of the general distribution of permafrost.

DISTRIBUTION OF PERMAFROST

General distribution

One of the first efforts in the North American literature to show the distribution of permafrost in Alaska was by Nikiforoff (1928, Fig. 5). His page-size sketch map of the Northern Hemisphere shows only the southern boundary of "area of perpetually frozen ground," which approximately follows the 65th parallel across Alaska.

As part of a major report entitled "Areal Geology of Alaska," Smith (1939, Plate 14) prepared a map (scale 1:5,000,000) showing 72 areas of "Alaskan ground frost." These areas extend from Point Barrow in northernmost Alaska to as far south as the 60th parallel and from the Alaska/Yukon Territory border westward to the Bering Sea. No lines are shown delineating zones of permafrost or the southern boundary of permafrost; however, a 30°F (–1.1°C) isotherm shown on the map approximately coincides with the southernmost areas of permafrost.

The classic paper by Taber (1943) entitled "Perennially frozen ground in Alaska: Its origin and history" includes a page-size map of Alaska showing the "approximate boundary of perennially frozen ground." The boundary is largely similar to the southern boundary of permafrost shown on modern maps, particularly in the eastern half of the state.

Muller (1945, Fig. 1) and Terzaghi (1952, Fig. 9), on small-scale maps of the Northern Hemisphere, also show only the southern limit of the permafrost region; however, Black (1950, Fig. 1; 1953, Fig. 1; 1954, Fig. 1), Ives (1974, Fig. 4A.2), and Karte (1982, p. 16), on similar small-scale base maps, divided the permafrost region into three zones: continuous, discontinuous, and sporadic.

Benninghoff (1952, Fig. 1), Péwé (1954, Fig. 69; 1963, Fig. 32), and Hopkins and others (1955, Fig. 5), on small-scale index maps of Alaska, also divided the permafrost region into the continuous, discontinuous, and sporadic permafrost zones. The boundaries on three of these maps are compared on Figure 3.

This system of dividing the permafrost region into three zones was used first by the Soviets during the 1930s and 1940s. These permafrost zones generally are defined as follows. The continuous zone is underlain by permafrost almost everywhere except under large water bodies that are deep enough not to freeze to their bottoms during the winter. Also, most investigators consider that the temperature of continuous permafrost at a depth at which seasonal variation is barely detectable (10 to 20 m) is generally lower than –5°C. The discontinuous zone is mostly underlain by permafrost but includes numerous areas without permafrost, and the sporadic zone is mostly without permafrost but includes numerous areas of permafrost. Temperatures of permafrost also have been used to help define the discontinuous (from –5°C to –1°C) and sporadic (higher than –1°C) permafrost zones. Because of the small scale of these maps and the great difficulty in accurately delineating the boundary between the discontinuous and sporadic permafrost zones, most workers divided the onshore permafrost region into two zones: continuous and discontinuous (Brown, 1963, Fig. 3, 1970, Fig. 1; Williams, 1965, Plate 1, 1970, Figs. 1 and 10; Stearns, 1965, Fig. 6, 1966, Figs. 1 and 2a; Péwé, 1966, Fig. 1, 1969, Fig. 1, 1974a, Fig. 1, 1974b, Fig. 3.4, 1975, Figs. 22 and 23, 1976, Fig. 1, 1983b, Fig. 9.2; Ferrians and others, 1969, Fig. 1; Sater and others, 1971, Fig. 46; Mackay, 1972, Fig. 1; Price, 1972, Fig. 8; Brown and Péwé, 1973, Fig. 2; Jahn, 1975, Fig. 8; French, 1976, Fig. 4.3; Tedrow, 1977, Fig. 4-4; Linell and Tedrow, 1981, Fig. 3.2; Kreig and Reger, 1982, Fig. 1). On these maps, the continuous zone is defined essentially the same as it is above, but the discontinuous zone includes all of the general permafrost region south of the zone of continuous permafrost.

The most detailed and largest scale (1:2,500,000) map available showing the distribution and character of permafrost in Alaska was compiled by Ferrians (1965). On this map, the permafrost region is divided into two major units. One unit encompasses mountainous areas in which summits generally exceed 900 m in altitude, and which are underlain predominantly by bedrock at or near the surface. The other unit consists of lowland and upland areas, including hilly and mountainous areas in which summits are less than 900 m in altitude. The lowland and upland

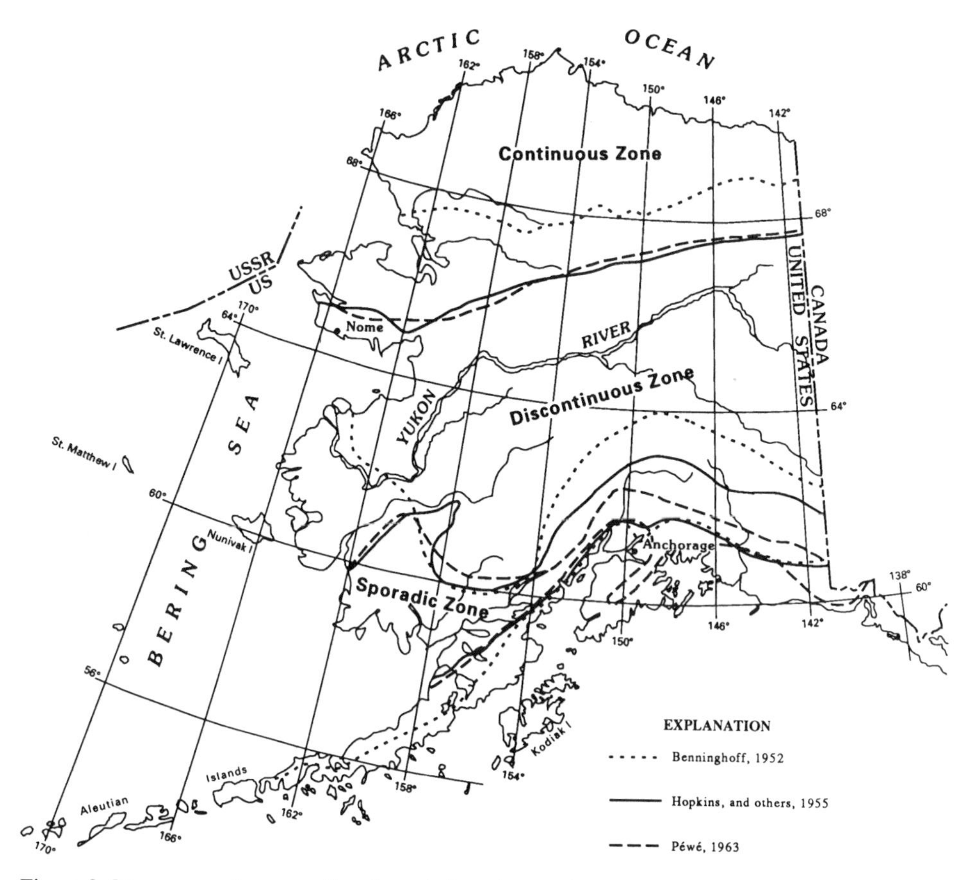

Figure 3. Map comparing permafrost boundaries of permafrost zones in Alaska (from Benninghoff, 1952, Fig. 1; Hopkins and others, 1955, Fig. 5; Péwé, 1963, Fig. 32).

areas are underlain predominantly by thick unconsolidated deposits, but locally they are underlain by bedrock at or near the surface.

In the mountainous areas, great differences in altitude, character of bedrock and surficial deposits, soil moisture, insolation received at ground surface, snow cover, and vegetative cover cause extreme variation in thickness and temperature of permafrost. Because of this variation, the mountainous areas are divided into three broad map units, primarily on the basis of differences in latitude and resultant climatic differences. From north to south, these map units are (1) areas generally underlain by permafrost, which include the Brooks Range and adjacent mountainous areas; (2) areas generally underlain by discontinuous permafrost, which include numerous mountainous areas in central Alaska (e.g., White, Ray, Bendeleben, Kigluaik, Wrangell, and Talkeetna Mountains and the Alaska Range); and (3) areas generally underlain by isolated masses of permafrost, which include the Chugach and Kilbuck Mountains and the northern part of the Aleutian Range.

The lowland and upland areas, where the thickness and temperature of permafrost are less variable than they are in the mountainous areas, are divided into six map units, which can be grouped into three broad zones: (1) the northern zone (largely north of the Brooks Range), which is generally underlain by thick permafrost in areas of both fine- and coarse-grained deposits; (2) the central zone (between the Brooks Range and the Alaska Range but including the Copper River Basin), which is generally underlain by moderately thick to thin permafrost in areas of fine-grained deposits and by discontinuous or isolated masses of permafrost in areas of coarse-grained deposits; and (3) the southern zone (including the Bristol Bay area and the eastern and western margins of the Susitna Lowland north of Anchorage), which is generally underlain by numerous isolated masses of permafrost in areas of fine-grained deposits, and which generally is free of permafrost in areas of coarse-grained deposits.

Outside the permafrost region, there are a few isolated occurrences of permafrost at high altitudes and in lowland areas where the ground is well insulated by peat and where solar radiation received at ground surface is low, especially near the border of the permafrost region.

The map also shows selected data from wells and borings concerning the presence or absence of permafrost and, if present, its thickness at 68 sites throughout the permafrost region.

Features shown on the map include the location of known thermal springs and active volcanoes. Permafrost is absent near these features, which are surface manifestations of high near-

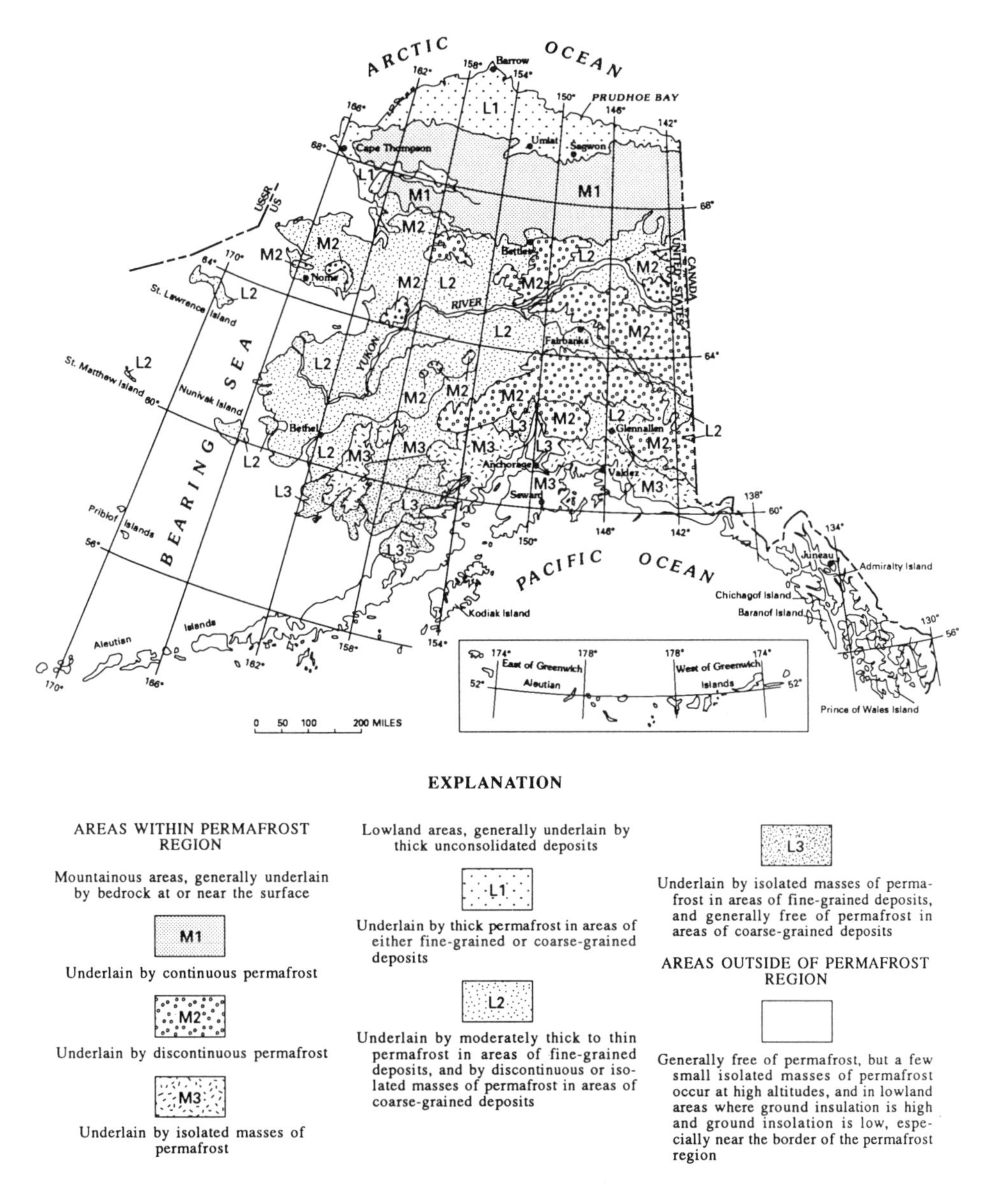

Figure 4. Generalized permafrost map of Alaska (from Ferrians and others, 1969, Fig. 2).

surface temperatures. Also, permafrost is either absent or at considerable depth beneath large rivers and large deep lakes throughout the permafrost region. The heat in these large water bodies, and also in the ocean waters, tends to decrease the thickness and to increase the temperature of permafrost in adjacent areas.

Corte (1969) included a large foldout map (no scale given) of the Northern Hemisphere in a review paper entitled "Geocryology and Engineering." The map shows the extent of perennially frozen ground, seasonal systematic freezing and thawing, and short duration and non-systematic freezing and thawing. The boundaries for these map units were taken from a permafrost map of the world by Baranov (1959, Fig. 22). Within Alaska, the southern boundary of the permafrost region is similar to that shown on most modern maps.

Permafrost maps of Alaska by Ferrians and others (1969, Fig. 2) and by Johnson and Hartman (1969, Plate 10) are generalizations of the larger scale permafrost map by Ferrians (1965). Figure 4 is a modified and generalized version of this map.

A 1:6,300,000-scale (approximately) map of Alaska by Williams (1970, Figs. 3 to 8) shows the location of pertinent shafts, borings, and wells and gives the depth of frozen ground and occurrence of ground water at these sites. No permafrost boundaries are shown on this map, but a large amount of permafrost-related data is presented.

A map of the Arctic region at a scale of 1:5,000,000 (American Geographical Society, 1975) shows the boundaries of the continuous and discontinuous permafrost zones. The boundaries within Alaska were taken from the "Permafrost Map of Alaska" (Ferrians, 1965).

A recent version of a page-size map showing the distribution of permafrost in the Northern Hemisphere (Fig. 1) compiled by Péwé (1982, Fig. 21, 1983a, Fig. 1, 1983b, Fig. 9.1) not only delineates the continuous and discontinuous zones, but it also delineates subsea and alpine permafrost. Subsea permafrost occurs under the sea, and alpine permafrost occurs at high altitudes in the middle and lower latitudes south of the southern boundary of the permafrost region. Subsea permafrost is present off the northern coast of Alaska under the continental shelf of the Beaufort Sea, and alpine permafrost is present locally in high mountainous areas in southern Alaska and the southeastern Alaska panhandle.

Another version of the Northern Hemisphere map, compiled by Heginbottom (1984, Fig. 1), has five map units: (1) continuous permafrost, (2) widespread discontinuous permafrost, (3) sporadic and alpine permafrost, (4) known subseabottom permafrost, and (5) Greenland ice-cap. Within Alaska, the extent of the continuous permafrost zone is similar to that shown on several other earlier maps and includes all of northern Alaska north of the southern flank of the Brooks Range; however, the discontinuous zone does not extend as far south as it does on the earlier maps, and sporadic and alpine permafrost are combined into one zone that covers the southern part of mainland Alaska and the higher mountainous areas of the southeastern Alaska panhandle. The distribution of subsea permafrost in Alaska is similar to that shown on the previously described map (Fig. 1) by Péwé.

A page-size map of North America by Harris (1986a, Fig. 7) shows permafrost zones similar to those compiled by Heginbottom (1984, p. 79) described above, but the boundaries between the zones are slightly different; moreover, unfortunately, on Harris' map a large part of the Arctic Coastal Plain in northernmost Alaska is erroneously shown as being underlain by "sub-seabottom" permafrost. Harris (1986b, Fig. 1.1) also compiled a permafrost map of the Northern Hemisphere.

The most recent version of a Northern Hemisphere permafrost map, prepared by P. J. Williams and M. W. Smith (1989, Fig. 1.7), uses intensity of shading to show increases in extent, both laterally and with depth. On their map the distribution of permafrost in Alaska was modified from Péwé (1983a).

Regional and local distribution

To help solve the serious permafrost-related engineering problems encountered in Alaska by the military during the early 1940s, special terrain investigations were carried out by the U.S. Geological Survey. As an outgrowth of these investigations, large-scale maps, some including profiles, showing permafrost distribution, were prepared for various areas including Galena (Elias and Vosburgh, 1946; Péwé, 1948b), Point Spencer (Black, 1946), Northway (Wallace, 1946), Umiat (Black and Barksdale, 1948), Fairbanks (Péwé, 1948a), and Dunbar (Péwé).

Using satellite imagery and available ground data, Anderson and others (1973, Fig. 25) prepared a 1:1,100,000-scale "permafrost terrain" map covering approximately 150,000 km^2 of north-central Alaska. Four different map units are differentiated. These are bedrock, alluvium-colluvium, active flood plain, and abandoned flood plain and terrace. According to the authors, the bedrock unit is characterized by a few scattered taliks, which are unfrozen zones, and a thaw depth of 0.3 to 1.0 m, except on south-facing slopes where thaw depths may exceed 2 m. The alluvium-colluvium unit has numerous taliks and a thaw depth of less than 0.5 m in areas of poor drainage, and from 0.5 to 2.0 m on moderately well-drained slopes. The active flood plain has numerous taliks and a thaw depth of more than 2.0 m. The abandoned flood plain and terrace unit has numerous taliks and many small thaw lakes; permafrost usually occurs at depths of less than 0.5 m.

Péwé and Bell (1975a, b, c, d, e), as part of a comprehensive engineering-geologic and hazards study of the Fairbanks area, prepared 1:24,000-scale maps showing the distribution of permafrost. They classified eight map units in order of increasing ice content, from permafrost free through permafrost with high ice content.

By using data from oil well logs, Osterkamp and Payne (1981, Fig. 2) and Osterkamp and others (1985, Figs. 1, 2, 3) prepared contour maps of the Prudhoe Bay and adjacent areas in northern Alaska showing estimated depth to base of ice-bearing permafrost. The ice-bearing permafrost is thickest (more than 600 m) in the Prudhoe Bay area and it thins rapidly to less than 350 m offshore to the north and to about 200 m to the south in the Northern Foothills of the Brooks Range.

Lachenbruch and others (1982, Fig. 1, 1987, Fig. 1, 1989, Fig. 28.1) prepared maps of northern Alaska showing generalized contours of long-term mean surface temperature, which is a major factor in controlling permafrost thickness. They concluded that even though these surface temperatures vary systematically from north to south (about −12.5°C to −4.6°C, respectively), there are large local variations and no conspicuous regional trends in permafrost thickness.

Collett and others (1989) prepared a map of the North Slope of Alaska that shows the depth to the base of the deepest ice-bearing permafrost as determined from well logs. They evaluated logs from 440 wells and summarized and tabulated data from 156 of them. Their map indicates a linear trend of maximum thickness of ice-bearing permafrost that parallels and is within 25 km of the coastline between the Colville and Canning Rivers. The ice-bearing permafrost ranges from less than 200 ft (61 m) in the west to greater than 2,000 ft (610 m) in the east, and it thins to the north and south.

Geophysical studies

Various types of geophysical sensors have been used to determine the distribution and thickness of ice-rich permafrost; however, because of the great vertical and horizontal variations in the character of permafrost and the limitations of the sensors used, remote sensing has not been successful in mapping large areas. Nevertheless, geophysical techniques have provided valuable information in certain local areas, and they have been especially helpful as a means of extrapolating between sites where permafrost conditions are known. Barnes (1966) described the geophysical methods used for delineating permafrost, and Ferrians and Hobson (1973) reviewed both traditional and geophysical methods used for mapping permafrost in North America.

Hoekstra and others (1975, Figs. 7, 12) made ground and airborne electrical resistivity surveys near Fairbanks and prepared resistivity contour maps as a means of delineating areas underlain by ice-rich permafrost. By using electromagnetic soundings, Daniels and others (1976, Fig. 16) prepared permafrost thickness and electrical resistivity contour maps in the Prudhoe Bay area of northern Alaska. Unfortunately, the data were proprietary, and consequently the exact location of the mapping could not be given. Olhoeft and others (1979) tested six different electromagnetic techniques to study permafrost in the National Petroleum Reserve in northwestern Alaska. Transient electromagnetic soundings were used by Ehrenbard and others (1983) in onshore and offshore areas west of Prudhoe Bay to map the bottom of thick permafrost, and Walker, G.G., and others (1987) also used transient electromagnetic techniques to detect subsea permafrost near Prudhoe Bay. In addition, seismic data have been used to delineate areas of ice-bonded subsea permafrost (high-velocity material) in the Beaufort Sea (Rogers and Morack, 1978, 1980, Fig. 9; Morack and Rogers, 1981a, b, 1982, 1984; Neave and Sellmann, 1983, Fig. 1, 1984, Fig. 1; Sellmann and Hopkins, 1984, Fig. 5). Kawasaki and Osterkamp (1989) discussed mapping shallow permafrost by using electromagnetic induction techniques.

PERMAFROST—RELATED LANDFORMS

Various geomorphic landforms in Alaska indicate the presence of permafrost; however, certain other landforms that are especially well developed in the permafrost regions also occur in nonpermafrost areas. Most types of patterned ground that develop owing to frost action fit in the latter category. Patterned ground includes sorted and nonsorted circles, polygons, nets, and stripes (Washburn, 1950, p. 8–9, 1956, p. 826, 1973, p. 103, 1980, p. 123). Solifluction features and string bogs or string moors also are common in permafrost regions, but they too occur in nonpermafrost areas.

The most significant landforms that indicate the presence of permafrost are the aggradational features, which include ice-wedge polygons, pingos, palsas (and peat plateaus), and frost blisters, and the various types of degradational features, which include thermokarst lakes, depressions, pits, gullies, and mounds, beaded drainage, and retrogressive thaw slumps. Aggradational features develop because of the formation of ground ice, whereas the degradational or thermokarst features develop because of the thawing of ground ice. Other significant landforms whose relations to permafrost are still being debated include cryoplanation terraces and active rock glaciers.

ENGINEERING PROBLEMS

In addition to the construction of roads, airstrips, buildings, and the various utilities in response to normal development, the critical need for finding and developing new and dependable sources of hydrocarbon and mineral resources has caused a tremendous increase in activities in the permafrost regions of Alaska. These regions pose special engineering problems and are environmentally "sensitive." Consequently, they require careful study and consideration to ensure the integrity of engineering structures and to minimize adverse environmental impacts.

There are two basic methods of constructing on permafrost: the active method and the passive method. The active method is used in areas where permafrost is thin and generally discontinuous, or where it contains relatively small amounts of ice. The object of this method is to thaw the permafrost and, if the thawed material has a satisfactory bearing strength, then proceed with construction in a normal manner. The object of the passive method is to keep the permafrost frozen so that it will provide a firm foundation for engineering structures. The passive method has wide application in interior and northern Alaska where permafrost is widespread, thick, and generally ice rich near the surface. When using this method, every effort should be made to minimize disturbing the ground surface, and when the ground surface is disturbed, to take carefully planned mitigative measures immediately. Various techniques used to help keep permafrost frozen include using different kinds of insulation (both natural and manmade), elevating heated structures above the ground surface, and using mechanical refrigeration.

Thawing of ice-rich permafrost is the most serious permafrost-related engineering problem. The thawing results in a loss of strength and a change in volume of the ice-rich soil. Under severe conditions, the ice-rich soil can liquefy and lose essentially all of its strength. A more common problem is differential settlement of the ground surface.

CONCLUSIONS

During the past 40 years, considerable progress has been made in determining the distribution and character of permafrost in Alaska; however, much remains to be done. The greatest shortcoming is the limited amount of borehole and ground temperature data; little or no information is available for most areas of Alaska. Because of the great expense, it is not practical to obtain permafrost data everywhere from boreholes. Nevertheless, as Alaska is developed, more borehole and ground temperature data in areas that represent various geologic and physiographic settings are essential to determine the temperature, thermal conductivity, thickness, and ice content of permafrost. Deep temperature measurements in permafrost are necessary to measure accurately the heat flow from the interior of the Earth, which limits the depth to which permafrost can form. Additional areas of research should include how climatic, geologic, hydrologic, topographic, and botanic conditions interact to control the distribution and character of permafrost. A better understanding in these and other subjects would make it possible to extrapolate more accurately between the widely scattered areas where permafrost conditions are known.

REFERENCES CITED

American Geographical Society, 1975, Map of the arctic region: New York, Cartographic Division, scale 1:5,000,000.

Anderson, D. M., and 6 others, 1973, An ERTS view of Alaska; Regional analysis of earth and water resources based on satellite imagery: U.S. Army Cold Regions Research and Engineering Laboratory Technical Report 241, 50 p.

Baranov, I. Ia., 1959, Geograficheskoe rasprostranenie sezonnopromerzayushchikh pochv i mnogoletnemerzlykh gornykh porod (Geographical distribution of seasonally frozen ground and permafrost), *in* Osnovy geokriologii (General geocryology): Moscow, Akademiya Nauk SSSR, pt. I, chap. 7, p. 193–218. (Translation by A. Nurklik, 1964, National Research Council of Canada Technical Translation TT-1121, 85 p.)

Barnes, D. F., 1966, Geophysical methods for delineating permafrost, *in* Proceedings, Permafrost, International Conference, Lafayette, Indiana, November 11–15, 1963: Washington, D.C., National Academy of Sciences, p. 159–164.

Beechey, F. W., 1831, Narrative of a voyage to the Pacific and Beering's [Bering] Strait: London, Henry Colburn and Richard Bently, v. 1 in 2 parts, pt. 1, p. 1–392, pt. 2, p. 393–742.

Benninghoff, W. S., 1952, Interaction of vegetation and soil frost phenomena: Arctic, v. 5, no. 1, p. 34–44.

Black, R. F., 1946, Permafrost investigations at Point Spencer, Alaska: U.S. Geological Survey Permafrost Program Progress Report 2, 20 p.

—— , 1950, Permafrost, *in* Trask, P. D., ed., Applied sedimentation: New York, John Wiley and Sons, p. 247–275.

—— , 1953, Permafrost; A review: New York Academy of Sciences, v. 15, no. 5, p. 126–131.

—— , 1954, Permafrost; A review: Geological Society of America Bulletin, v. 65, p. 839–855.

Black, R. F., and Barksdale, W. L., 1948, Terrain and permafrost, Umiat area, Alaska: U.S. Geological Survey Permafrost Program Progress Report 5, 23 p.

Brown, J., 1973, Environmental considerations for the utilization of permafrost terrain, *in* Permafrost, North American Contribution to Second International Conference, Yakutsk, U.S.S.R., July 13–28, 1973: Washington, D.C., National Academy of Sciences, p. 587–590.

Brown, J., and Grave, N. A., 1979a, Physical and thermal disturbance and protection of permafrost: U.S. Army Cold Regions Research and Engineering Laboratory Special Report 79-5, 42 p.

—— , 1979b, Physical and thermal disturbance and protection of permafrost, *in* Proceedings, Third International Conference on Permafrost, July 10–13, 1978, Edmonton, Alberta, V. 2: Ottawa, National Research Council of Canada, p. 51–91.

Brown, J., and Hemming, J. E., 1980, Workshop on environmental protection of permafrost terrain: The Northern Engineer, v. 12, no. 2, p. 30–36.

Brown, R.J.E., ed., 1963, Proceedings of the First Canadian Conference on Permafrost, April 17–18, 1962: National Research Council of Canada, Associate Committee on Soil and Snow Mechanics Technical Memorandum 76, 231 p.

—— , 1970, Permafrost as an ecological factor in the subarctic, *in* Ecology of the subarctic regions, v. 1 of Proceedings, Ecology and conservation; A symposium, Helsinki, 1966, Paris, UNESCO, p. 129–140.

Brown, R.J.E., and Péwé, T. L., 1973, Distribution of permafrost in North America and its relationship to the environment; A review, 1963–1973, *in* Permafrost, the North American contribution to the Second International Conference, Yakutsk, U.S.S.R., July 13–28, 1973: Washington, D.C., National Academy of Sciences, p. 71–100.

Cantwell, J. C., 1887, A narrative account of the exploration of the Kowak [Kobuk] River, Alaska, *in* Healy, Capt. M. A., Report of the cruise of the Revenue Marine Steamer *Corwin* in the Arctic Ocean in the year 1885: Washington, D.C., Treasury Department, p. 48–49.

—— , 1896, Ice-cliffs on the Kowak [Kobuk] River: National Geographic Magazine, v. 7, p. 345–346.

Carter, L. D., Heginbottom, J. A., and Woo, M., 1987, Arctic lowlands, *in* Graf, W. L., ed., Geomorphic systems of North America: Boulder, Colorado, Geological Society of America, Centennial Special Volume 2, p. 583–628.

Collett, T. S., Bird, K. J., Kvenvolden, K. A., and Magoon, L. B., 1989, Map showing the depth to the base of the deepest ice-bearing permafrost as determined from well logs, North Slope, Alaska: U.S. Geological Survey Oil and Gas Investigations Map OM-0222, scale 1:1,000,000.

Corte, A. E., 1969, Geocryology and engineering, *in* Varnes, D. J., and Kiersch, G., eds., Reviews in engineering geology: Boulder, Colorado, Geological Society of America, v. 2, p. 119–185.

Dall, W. H., 1881, Extract from a report to C. P. Patterson, Supt., Coast and Geodetic Survey: American Journal of Science, v. 21, nos. 121–126, p. 104–110.

Daniels, J. J., Keller, G. V., and Jacobson, J. J., 1976, Computer-assisted interpretation of electromagnetic soundings over a permafrost section: Geophysics, v. 41, no. 4, p. 752–765.

Ehrenbard, R. L., Hoekstra, P., and Rozenberg, G. S., 1983, Transient electromagnetic soundings for permafrost mapping, *in* Proceedings, Permafrost, Fourth International Conference, Fairbanks, Alaska, July 17–22., 1983: Washington, D.C., National Academy Press, p. 272–277.

Elias, M. M., and Vosburgh, R. M., 1946, Terrain and permafrost in the Galena area: U.S. Geological Survey Permafrost Program Progress Report 1, 25 p.

Ferrians, O. J., Jr., 1965, Permafrost map of Alaska: U.S. Geological Survey Miscellaneous Geologic Investigations Map I-445, scale 1:2,500,000.

Ferrians, O. J., Jr., and Hobson, G. D., 1973, Mapping and predicting permafrost in North America; A review, 1963–1973, *in* Permafrost, North American contribution to the Second International Conference, Yakutsk, U.S.S.R., July 13–28, 1973: Washington, D.C., National Academy of Sciences, p. 479–498.

Ferrians, O. J., Jr., Kachadoorian, R., and Greene, G. W., 1969, Permafrost and related engineering problems in Alaska: U.S. Geological Survey Professional Paper 678, 37 p.

French, H. M., 1976, The periglacial environment: London and New York, Longman, 309 p.

Greene, G. W., Lachenbruch, A. H., and Brewer, M. C., 1960, Some thermal effects of a roadway on permafrost, *in* Geological Survey research 1960: U.S. Geological Survey Professional Paper 400-B, p. B141–B144.

Harris, S. A., 1986a, Permafrost distribution, zonation, and stability along the eastern ranges of the Cordillera of North America: Arctic, v. 39, no. 1, p. 29–38.

—— , 1986b, The permafrost environment: Totowa, New Jersey, Barnes and Noble, 276 p.

Haugen, R. K., and Brown, J., 1971, Natural and man-induced disturbances of permafrost terrane, *in* Coates, D. R., ed., Environmental geomorphology: Binghamton, New York State University, p. 139–149.

Heginbottom, J. A., 1984, The mapping of permafrost: Canadian Geographer, v. 28, no. 1, p. 78–83.

Hoekstra, P., Sellmann, P. V., and Delaney, A. J., 1975, Ground and airborne resistivity surveys of permafrost near Fairbanks, Alaska: Geophysics, v. 40, no. 4, p. 641–656.

Hok, J. R., 1969, A reconnaissance of tractor trails and related phenomena on the North Slope of Alaska: U.S. Department of Interior, Bureau of Land Management, 66 p.

Hooper, C. L., 1881, Report of the cruise of the U.S. Treasury Department Steamer *Corwin* in the Arctic Ocean in 1880: Washington, D.C., U.S. Treasury Department Document 118, 71 p.

——, 1884, Report of the cruise of the U.S. Revenue Steamer *Corwin* in the Arctic Ocean in 1881: Washington, D.C., U.S. Treasury Department Document 601, 147 p.

Hopkins, D. M., and 6 others, 1955, Permafrost and ground water in Alaska: U.S. Geological Survey Professional Paper 264-F, p. 113–146.

Ives, J. D., 1974, Permafrost, *in* Ives, J. D., and Barry, R. G., eds., Arctic and alpine environments: London, Methuen Press, p. 159–194.

Jahn, A., 1975, Problems of the periglacial zone (Zagadnienia strefy perglackalnej): Warsaw, Panstwowe Wydawnictwo Naukowe, 223 p. (Translated from Polish for the National Science Foundation, Washington, D.C.)

Johnson, P. R., and Hartman, C. W., 1969, Environmental atlas of Alaska: College, University of Alaska, 111 p. (2nd edition, January 1971).

Karte, J., 1982, Development and present status of German periglacial research in the polar and subpolar regions: Polar Geography and Geology, v. 6, no. 1, 24 p.

Kawasaki, K., and Osterkamp, T. E., 1989, Mapping shallow permafrost by electromagnetic induction; Practical considerations: Cold Regions Science and Technology, v. 15, no. 3, p. 279–288.

Kreig, R. A., and Reger, R. D., 1982, Air-photo analysis and summary of landform soil properties along the route of the Trans-Alaska Pipeline System: Alaska Division of Geological and Geophysical Surveys Special Report 66, 149 p.

Lachenbruch, A. H., 1957a, Thermal effects of the ocean on permafrost: Geological Society of America Bulletin, v. 68, no. 11, p. 1515–1530.

——, 1957b, Three-dimensional heat conduction in permafrost beneath heated buildings: U.S. Geological Survey Bulletin 1052-B, p. 51–69.

Lachenbruch, A. H., and Marshall, B. V., 1986, Changing climate; Geothermal evidence from permafrost in the Alaskan Arctic: Science, v. 234, no. 4777, p. 689–696.

Lachenbruch, A. H., Greene, G. W., and Marshall, B. V., 1966, Permafrost and the geothermal regimes, *in* Wilimovsky, N. J., and Wolfe, J. N., eds., Environment of the Cape Thompson region, Alaska: U.S. Atomic Energy Commission Report PNE-481, p. 149–163.

Lachenbruch, A. H., Sass, J. H., Lawver, L. A., Brewer, M. C., and Moses, T. H., Jr., 1982, Depth and temperature of permafrost on the Alaskan Arctic Slope; Preliminary results: U.S. Geological Survey Open-File Report 82–1039, 20 p.

Lachenbruch, A. H., and 8 others, 1987, Temperature and depth of permafrost on the Alaskan North Slope, *in* Tailleur, I. L., and Weimer, P., eds., Alaskan North Slope geology: Bakersfield, California, Pacific Section of Society of Economic Paleontologists and Mineralogists and the Alaska Geological Society, v. 2, p. 545–558.

——, 1989, Temperature and depth of permafrost on the Arctic Slope of Alaska, *in* Gryc, G., ed., Geology and exploration of the National Petroleum Reserve in Alaska, 1974 to 1982: U.S. Geological Survey Professional Paper 1399, p. 645–656.

Lawson, D. E., 1983, Erosion of perennially frozen streambanks: U.S. Army Cold Regions Research and Engineering Laboratory Report 83-29, 22 p.

——, 1986, Response of permafrost terrain to disturbance; A synthesis of observations from northern Alaska, U.S.A.: Arctic Alpine Research, v. 18, p. 1–17.

Lawson, D. E., and Brown, J., 1978, Human-induced thermokarst at old drill sites in northern Alaska: The Northern Engineer, v. 10, no. 3, p. 16–23.

Lawson, D. E., and 7 others, 1978, Tundra disturbance and recovery following the 1949 exploratory drilling, Fish Creek, northern Alaska: U.S. Army Cold Regions Research and Engineering Laboratory Report 78-28, 91 p.

Linell, K. A., and Tedrow, J.C.F., 1981, Soil and permafrost surveys in the Arctic: New York, Oxford University Press, 279 p.

Mackay, J. R., 1972, The world of underground ice: Annals of the Association of American Geographers, v. 62, no. 1, p. 1–22.

Maddren, A. G., 1905, Smithsonian exploration in Alaska in 1904, *in* Search of mammoth and other fossil remains: Smithsonian Miscellaneous Collections, v. 49, no. 1584, 117 p.

Morack, J. L., and Rogers, J. C., 1981a, Beaufort and Chukchi Sea coast permafrost studies, *in* Environmental assessment of the Alaskan Continental Shelf, Annual reports of principal investigators for the year ending March 1981; Vol. 8, Hazards and data management: U.S. Department of Commerce and U.S. Department of the Interior, p. 293–332.

——, 1981b, Seismic evidence of shallow permafrost beneath islands in the Beaufort Sea, Alaska: Arctic, v. 34, no. 2, p. 169–174.

——, 1982, Marine seismic refraction measurements of near-shore subsea permafrost, *in* French, H. M., ed., Proceedings, Fourth Canadian Permafrost Conference, Calgary, Alberta, March 2–6, 1981: Ottawa, National Research Council of Canada, p. 249–255.

——, 1984, Acoustic velocities of nearshore materials in the Alaskan Beaufort and Chukchi Seas, *in* Barnes, P. W., Reimnitz, E., and Schell, D. M., eds., The Alaskan Beaufort Sea; Ecosystems and environments: Orlando, Florida, Academic Press, p. 259–274.

Muir, J., 1883, Botanical notes on Alaska, *in* Cruise of the Revenue Steamer *Corwin* in Alaska and northwestern Arctic Ocean in 1881: Washington, D.C., U.S. Treasury Department, p. 47–53.

Muller, S. W., 1945, Permafrost or permanently frozen ground and related engineering problems: U.S. Engineers Office, Strategic Engineering Study Special Report 62, 136 p.; Reprinted in 1947, Ann Arbor, Michigan, J. W. Edwards, Inc.

Neave, K. G., and Sellmann, P. V., 1983, Seismic velocities and subsea permafrost in the Beaufort Sea, Alaska, *in* Permafrost, Proceedings, Fourth International Conference, Fairbanks, Alaska, July 17–22, 1983: Washington, D.C., National Academy Press, p. 894–898.

——, 1984, Determining distribution patterns of ice-bonded permafrost in the U.S. Beaufort Sea from seismic data, *in* Barnes, P. W., Schell, D. M., and Reimnitz, E., eds., The Alaskan Beaufort Sea; Ecosystems and environments: Orlando, Florida, Academic Press, p. 237–258.

Nelson, F. E., and Outcalt, S. I., 1982, Anthropogenic geomorphology in northern Alaska: Physical Geography, v. 3, p. 17–48.

Nikiforoff, C., 1928, The perpetually frozen subsoil of Siberia: Soil Science, v. 26, no. 1, p. 61–77.

Olhoeft, G. R., Watts, R., Frischknecht, F., Bradley, F., and Dansereau, D., 1979, Electromagnetic geophysical exploration in the National Petroleum Reserve in Alaska, *in* Proceedings, Symposium on Permafrost Field Methods, Saskatoon, Saskatchewan, 1977: Ottawa, National Research Council of Canada Technical Memorandum 124, p. 184–190.

Osterkamp, T. E., and Payne, M. W., 1981, Estimates of permafrost thickness from well logs in northern Alaska: Cold Regions Science and Technology, v. 5, no. 1, p. 13–27.

Osterkamp, T. E., Pattersen, J. K., and Collett, T. S., 1985, Permafrost thickness in the Oliktok Point, Prudhoe Bay, and Mikkelsen Bay areas of Alaska: Cold Regions Science and Technology, v. 11, no. 2, p. 99–105.

Péwé, T. L., 1948a, Permafrost investigations, Fairbanks area, Alaska: U.S. Geological Survey Permafrost Program Preliminary Report, 16 p.

——, 1948b, Terrain and permafrost, Galena area, Alaska: U.S. Geological Survey Permafrost Program Progress Report 7, 52 p.

——, 1949, Preliminary report of permafrost investigations in the Dunbar area, Alaska: U.S. Geological Survey Circular 42, 3 p.

——, 1954, Effect of permafrost on cultivated fields, Fairbanks area, Alaska: U.S. Geological Survey Bulletin 989-F, p. 315–351.

——, 1963, Frost heaving of piles with an example from Fairbanks, Alaska: U.S. Geological Survey Bulletin 1111-I, p. 333–407.

——, 1966, Permafrost and its effect on life in the north, *in* Hansen, H. P., ed., Arctic biology, 2nd ed.: Proceedings, 18th Annual Biology Colloquium: Corvallis, Oregon State University Press (1957 edition revised by author); also republished as booklet, 40 p.

——, 1969, Terrasses d'altiplanation due Quaternaire ancien et moyen, Fairbanks, Alaska [abs.]: Proceedings, International Association for Quaternary Research, 8th Congress, Paris, France, 1969, p. 41.

——, 1974a, Permafrost: Encyclopedia Britannica, v. 14, p. 89–95.

——, 1974b, Geomorphic processes in polar deserts, *in* Smiley, T. L., and Zumberge, J. H., eds., Polar deserts: Tucson, University of Arizona Press, p. 33–52.

——, 1975, Quaternary geology of Alaska: U.S. Geological Survey Professional Paper 835, 145 p.

——, 1976, Permafrost, *in* 1976 Yearbook of Science and Technology: New York, McGraw Hill, p. 32–47.

——, 1982, Geologic hazards and their effect on man; Fairbanks area, Alaska: Alaska Division of Geological and Geophysical Surveys Special Report 15, 150 p.

——, 1983a, Alpine permafrost in the contiguous United States; A review: Arctic and Alpine Research, v. 15, no. 2, p. 145–156.

——, 1983b, The periglacial environment in North America during Wisconsin time, *in* Porter, S. C., ed., The late Pleistocene, v. 1 of Wright, H. E., Jr., ed., Late Quaternary environments of the United States: Minneapolis, University of Minnesota Press, p. 157–189.

Péwé, T. L., and Bell, J. W., 1975a, Map showing distribution of permafrost in the Fairbanks D-1 SW Quadrangle, Alaska: U.S. Geological Survey Miscellaneous Field Studies Map MF-671A, scale 1:24,000.

——, 1975b, Map showing distribution of permafrost in the Fairbanks D-2 NE Quadrangle, Alaska: U.S. Geological Survey Miscellaneous Field Studies Map MF-670A, scale 1:24,000.

——, 1975c, Map showing distribution of permafrost in the Fairbanks D-2 NW Quadrangle, Alaska: U.S. Geological Survey Miscellaneous Field Studies Map MF-668A, scale 1:24,000.

——, 1975d, Map showing distribution of permafrost in the Fairbanks D-2 SE Quadrangle, Alaska: U.S. Geological Survey Miscellaneous Field Studies Map MF-669A, scale 1:24,000.

——, 1975e, Map showing distribution of permafrost in the Fairbanks D-2 SW Quadrangle, Alaska: U.S. Geological Survey Miscellaneous Investigations Map I-829B, scale 1:24,000.

Price, L. W., 1972, The periglacial environment, permafrost, and man: Association of American Geographers, Commission on College Geography Resource Paper 14, 88 p.

Purington, C. W., 1905, Methods and costs of gravel and placer mining in Alaska: U.S. Geological Survey Bulletin 263, 273 p.

Ray, P. H., 1885, Report of the International Polar Expedition to Point Barrow, Alaska: Washington, D.C., U.S. Government Printing Office, 695 p.

Richardson, J., 1841, On the frozen soil of North America: Edinburgh, New Philosophical Journal, v. 30, p. 110–123.

——, 1854, The zoology of the voyage of H.M.S. *Herald,* under Captain Henry Kellet, during the years of 1845–51: London, p. 1–8, 61.

Rogers, J. C., and Morack, J. L., 1978, Geophysical investigations of offshore permafrost, Prudhoe Bay, Alaska, *in* Proceedings, Third International Conference on Permafrost, Edmonton, Alberta, July 10–13, 1978, v. 1: Ottawa, National Research Council of Canada, p. 560–566.

——, 1980, Geophysical evidence of shallow nearshore permafrost, Prudhoe Bay, Alaska: Journal of Geophysical Research, v. 85, no. B9, p. 4845–4853.

Russell, I. C., 1890, Notes on the surface geology of Alaska: Geological Society of America Bulletin, v. 1, p. 99–162.

Sater, J. E., Ronhovde, A. G., and van Allen, L. C., 1971, Arctic environment and resources: Washington, D.C., Arctic Institute of North America, 309 p.

Seemann, B., 1853, Narrative of the voyage of the H.M.S. *Herald* during the years 1845–51, under Captain Henry Kellett, R.N.: London, in 2 v.

Sellmann, P. V., and Hopkins, D. M., 1984, Subsea permafrost distribution on the Alaskan shelf, *in* Permafrost, Final Proceedings, Fourth International Conference, Fairbanks, Alaska, July 17–22, 1983: Washington, D.C., National Academy Press, p. 75–82.

Smith, P. S., 1939, Areal geology of Alaska: U.S. Geological Survey Professional Paper 192, 100 p.

Stearns, S. R., 1965, Selected aspects of geology and physiography of the cold regions: U.S. Army Cold Regions Research and Engineering Laboratory, Cold Regions Science and Engineering Monograph 1-A1, 40 p.

——, 1966, Permafrost (perennially frozen ground): U.S. Army Cold Regions Research and Engineering Laboratory, Cold Regions Science and Engineering Monograph 1-A2, 77 p.

Taber, S., 1943, Perennially frozen ground in Alaska; Its origin and history: Geological Society of America Bulletin, v. 54, p. 1433–1548.

Tedrow, J.C.F., 1977, Soils of the polar landscapes: New Brunswick, New Jersey, Rutgers University Press, 638 p.

Terzaghi, K., 1952, Permafrost: Boston Society of Civil Engineers Journal, v. 39, no. 1, p. 1–50.

von Kotzebue, O., 1821, A voyage of discovery into the South Sea and Behring's [Bering] Straits, for the purpose of exploring a northeast passage, undertaken in the years 1815–1818: London, Longman, Hurst, Rees, Orme, and Brown, v. 1 of 3 v., 358 p.

Walker, D. A., Cate, D., Brown, J., and Racine, C., 1987, Disturbance and recovery of arctic tundra terrain; A review of recent investigations: U.S. Army Cold Regions Research and Engineering Laboratory Report 87-11, 63 p.

Walker, G. G., Kawasaki, K., and Osterkamp, T. E., 1987, Transient electromagnetic detection of subsea permafrost near Prudhoe Bay, Alaska, *in* Tailleur, I. L., and Weimer, P., eds., Alaskan North Slope geology: Bakersfield, California, Pacific Section Society of Economic Paleontologists and Mineralogists and the Alaskan Geological Society, v. 2, p. 565–569.

Walker, H. J., 1983, Erosion in a permafrost-dominated delta, *in* Permafrost, Proceedings, Fourth International Conference, Fairbanks, Alaska, July 17–22, 1983: Washington, D.C., National Academy Press, p. 1344–1349.

——, 1988, Permafrost and coastal processes, *in* Permafrost, Proceedings, Fifth International Conference, Trondheim, Norway, Aug. 2–5, 1988: v. 33, Tapir, p. 35–42.

Wallace, R. E., 1946, Terrain analysis in the vicinity of Northway, Alaska, with special reference to permafrost: U.S. Geological Survey Permafrost Program Progress Report 3, 34 p.

Washburn, A. L., 1950, Patterned ground, Revue Canadienne Géographie, v. 4, no. 3–4, p. 59.

——, 1956, Classification of patterned ground and review of suggested origins: Geological Society of America Bulletin, v. 67, p. 823–865.

——, 1973, Periglacial processes and environments: New York, St. Martin's Press, 320 p.

——, 1980, Geocryology; A survey of periglacial processes and environments: New York, Halsted Press, 406 p.

Williams, J. R., 1965, Ground water in permafrost regions; An annotated bibliography: U.S. Geological Survey Water-Supply Paper 1792, 294 p.

——, 1970, Ground water in the permafrost regions of Alaska: U.S. Geological Survey Professional Paper 696, 83 p.

Williams, P. J., and Smith, M. W., 1989, The frozen earth; Fundamentals of geocryology: Cambridge, Cambridge University Press, 306 p.

Manuscript Accepted by the Society June 8, 1990

Printed in U.S.A.

The Geology of North America
Vol. G-1, The Geology of Alaska
The Geological Society of America, 1994

Chapter 29
Metallogeny and major mineral deposits of Alaska

Warren J. Nokleberg, David A. Brew, Donald Grybeck, and Warren Yeend
U.S. Geological Survey, 345 Middlefield Road, Menlo Park, California 94025
Thomas K. Bundtzen, Mark S. Robinson, and Thomas E. Smith
Alaska Division of Geological and Geophysical Surveys, 794 University Avenue, Fairbanks, Alaska 99709-4699
Henry C. Berg
115 Malvern Avenue, Fullerton, California 92632
With contributions by:
Gary L. Andersen, Edward R. Chipp, and David R. Gaard
Resource Associates of Alaska, Inc., Fairbanks, Alaska
P. Jeffery Burton
Jeffery Burton and Associates, Fairbanks, Alaska
John Dunbier, D. A. Scherkenbach
Noranda Exploration, Inc., Anchorage, Alaska
Jeffrey Y. Foley, Gregory Thurow, and J. Dean Warner
U.S. Bureau of Mines, Fairbanks, Alaska
Curtis J. Freeman
The F. E. Company, Inc., Fairbanks, Alaska
Bruce M. Gamble, Steven W. Nelson, and Jeanine M. Schmidt
U.S. Geological Survey, Anchorage, Alaska
Charles C. Hawley
Hawley Resource Group, Inc., Anchorage, Alaska
Murray W. Hitzman
Chevron Resources Company, San Francisco, California
Brian K. Jones
Bear Creek Mining, Kennecott Corporation, Anchorage, Alaska
Ian M. Lange
Department of Geology, University of Montana, Missoula, Montana
Christopher D. Maars, Christopher C. Puchner, and Carl I. Steefel
Anaconda Minerals Company, Anchorage, Alaska
W. David Menzie
U.S. Geological Survey, Menlo Park, California
Paul A. Metz
Mineral Industries Research Laboratory, University of Alaska, Fairbanks, Alaska
J. S. Modene, Joseph T. Plahuta, and Loren E. Young
Cominco Alaska, Inc., Anchorage, Alaska
Clint R. Nauman and Steven R. Newkirk
Research Associates of Alaska, Inc., Fairbanks, Alaska
Rainer J. Newberry
Geology/Geophysica Program, University of Alaska, Fairbanks, Alaska
Robert K. Rogers
WGM, Inc., Anchorage, Alaska
Charles M. Rubin
Anaconda Minerals Company, Denver, Colorado
Richard C. Swainbank
Geoprize, Limited, Anchorage, Alaska
P. R. Smith and Jackie E. Stephens
U.S. Borax and Chemical Corporation, Spokane, Washington

Nokleberg, W. J., Brew, D. A., Grybeck, D., Yeend, W., Bundtzen, T. K., Robinson, M. S., and Smith, T. E., , 1994, Metallogeny and major mineral deposits of Alaska, *in* Plafker, G., and Berg, H. C., eds., The Geology of Alaska: Boulder, Colorado, Geological Society of America, The Geology of North America, v. G-1.

INTRODUCTION AND PURPOSE

Alaska is commonly regarded as one of the frontiers of North America for the discovery of metalliferous mineral deposits. A recurring theme in the history of the state has been "rushes" or "stampedes" to sites of newly discovered deposits. Since about 1965, mining companies have undertaken much exploration for lode and placer mineral deposits. During the same period, because of the considerable interest in federal lands in Alaska and the establishment of new national parks, wildlife refuges, and native corporations, extensive studies of mineral deposits and of the mineral resource potential of Alaska have been conducted by the U.S. Geological Survey, the U.S. Bureau of Mines, and the Alaska Division of Geological and Geophysical Surveys. These studies have resulted in abundant new information on Alaskan mineral deposits. In the same period, substantial new geologic mapping has also been completed with the help of new logistical and technical tools. One result of the geologic mapping and associated geologic studies is the recognition of numerous fault-bounded assemblages of rocks designated as tectonostratigraphic (lithotectonic) terranes. This concept indicates that most of Alaska consists of a collage of such terranes (Silberling and others, this volume, Plate 3).

The purpose of this report is to summarize the local geology, geologic setting, and metallogenesis of the major metalliferous lode deposits and placer districts of Alaska. The term "major mineral deposit" is defined as a mine, mineral deposit with known reserve, prospect, or occurrence that the authors judged significant for any given geographic region. This report is based on published maps and reports and on unpublished data. The unpublished data were contributed by mineral-deposits geologists in private industry, universities, the U.S. Geological Survey, the Alaska Division of Geological and Geophysical Surveys, the U.S. Bureau of Mines, and by the authors. Contributors are listed in the Acknowledgments.

This chapter describes the regional metallogenesis and the major metalliferous lode deposits and placer districts. Classification of lode mineral and placer deposits is summarized in Appendix 1 at the end of the chapter. Locations of the lode deposits and placer districts are plotted on Plate 11 (Nokleberg and others, this volume), which also summarizes the characteristics of the lode deposits and placer districts (Tables 1 and 2 on Plate 11). The lithotectonic terrane map of Silberling and others (this volume) is used as the geologic underlay for Plate 11. More detailed, tabular descriptions of metalliferous lode deposits and placer districts of Alaska, and mineral deposit types are published in a U.S. Geological Survey Bulletin titled "Significant metalliferous lode deposits and placer districts of Alaska" by Nokleberg and others (1987). The following abbreviations or terms are used in this chapter: PGE = Platinum group elements; REE = rare-earth elements; and tonne = metric ton.

PREVIOUS STUDIES

Since 1964, several statewide and regional summaries of Alaskan metalliferous lode and placer deposits have been published by the U.S. Geological Survey (USGS) and the Alaska Division of Geological and Geophysical Surveys (ADGGS). In 1964, the USGS published a map of placer gold occurrences in Alaska (Cobb, 1964); subsequently, they published a statewide summary of metalliferous lode deposits (Berg and Cobb, 1967) and a summary of Alaskan placer deposits (Cobb, 1973). In 1976 and 1977, the USGS published a series of regional tables, maps, and references for metalliferous deposits. These regional reports cover the Brooks Range (Grybeck, 1977), the Seward Peninsula (Hudson and others, 1977), central Alaska (Eberlein and others, 1977), and southern Alaska (MacKevett and Holloway, 1977a, b). In 1981, the USGS published a report on all known mines, prospects, deposits, and occurrences in southeastern Alaska (Berg and others, 1981), and in 1982 a series of regional mineral-terrane maps of Alaska, prepared by C. C. Hawley and Associates, showing the location, size, and type of major metalliferous mineral deposits, was published by the Arctic Environmental Information and Data Center (AEIDC, 1982). In 1981 and 1984, the USGS published reports summarizing the regional geology, metallogenesis, and mineral resources of southeastern Alaska (Berg and others, 1981; Berg, 1984). In recent years, a yearly summary of Alaskan lode and placer deposits has been published by the Alaska Division of Geological and Geophysical Surveys; the summary for 1989 is by Bundtzen and others (1990). A preliminary version of this study was published by Nokleberg and others (1988).

CLASSIFICATION OF MINERAL DEPOSITS

Metalliferous deposits discussed in this report are classified into 29 lode deposit and 4 placer deposit models or types, listed below and described in Appendix 1. For simplicity, some types are grouped under one heading. This classification was derived mainly from the mineral-deposit models developed by Erickson (1982), Cox (1983a, b), Cox and Singer (1986), and other works listed in Appendix 1. Of the 29 types of lode and placer deposits identified in Alaska, four were newly formulated (Nokleberg and others, 1987): metamorphosed sulfide, Cu-Ag quartz vein, felsic-plutonic U lode deposits, and placer Sn deposits. The lode-deposit types are listed below, in order, from those at or near the surface, such as stratiform deposits, to those formed at deeper levels, such as zoned mafic-ultramafic and podiform chromite deposits. Each mineral deposit described in this chapter is classified into one of the types below.

Lode deposit types

Kuroko massive sulfide deposit.
Besshi massive sulfide deposit.
Cyprus massive sulfide deposit.
Sedimentary exhalative Zn-Pb deposit.
Kipushi Cu-Pb-Zn (carbonate-hosted Cu) deposit.
Metamorphosed sulfide deposit.
Bedded barite deposit.
Sandstone U deposit.
Basaltic Cu deposit.

Hot spring Hg deposit.
Epithermal vein deposit.
Low-sulfide Au quartz vein deposit (abbreviated to Au quartz vein deposit).
Cu-Ag quartz vein deposit.
Polymetallic vein deposit.
Sb-Au vein deposit.
Sn greisen, Sn vein, and Sn skarn deposits.
Cu-Zn-Pb (± Au, Ag), W, and Fe (± Au) skarn deposits.
Porphyry Cu-Mo, porphyry Cu, and porphyry Mo deposit.
Felsic plutonic U deposit.
Gabbroic Ni-Cu deposit.
Zoned mafic-ultramafic Cr-Pt (± Cu, Ni, Co, Ti or Fe) deposit.
Podiform chromite deposit.
Serpentine-hosted asbestos deposit.

Placer deposit types

Placer Au deposit.
Placer Sn deposit.
Placer PGE-Au deposit.
Shoreline placer Ti deposit.

LODE DEPOSITS, BROOKS RANGE

The northwestern Brooks Range contains several sedimentary exhalative Zn-Pb, one bedded barite, several podiform chromite, one Kipushi Cu-Pb-Zn, and various vein deposits. The southern flank of the central Brooks Range contains an extensive suite of major Kuroko massive sulfide deposits and one major carbonate-hosted Kipushi Cu-Pb-Zn deposit. The central Brooks Range contains a suite of moderate-size polymetallic vein, Au quartz vein, Sb-Au vein, porphyry Cu and Mo, and Cu-Pb-Zn and Sn skarn deposits. The northeastern Brooks Range contains a cluster of Pb-Zn skarn, polymetallic vein, and porphyry Mo deposits. Mineral deposits in the Brooks Range and Seward Peninsula are further summarized by Einaudi and Hitzman (1986).

Stratiform Zn-Pb-Ag and barite deposits, northwestern Brooks Range

The northwestern Brooks Range contains a belt of large sedimentary exhalative Zn-Pb-Ag and bedded barite deposits that extend along strike for more than 200 km (Table 1 on Plate 11, this volume). The larger Zn-Pb-Ag deposits are at Lik, which contains an estimated 25 million tonnes of ore, and at Red Dog Creek (described below), which contains an estimated 85 million tonnes of ore and ranks within the largest 20 percent of known deposits of this type. Both deposits have high values of Zn, Pb, and Ag. A somewhat similar deposit occurs at Drenchwater Creek (described below) in both sedimentary and volcaniclastic rocks. The Nimiuktuk bedded barite deposit contains an estimated 1.5 million tonnes of barite. Substantial potential exists for finding additional deposits in the northwestern Brooks Range (Nokleberg and Winkler, 1982; Lange and others, 1985).

The sedimentary exhalative Zn-Pb-Ag and bedded barite deposits occur in a tectonically disrupted and strongly folded assemblage of Mississippian and Pennsylvanian chert, shale, limestone turbidite, minor tuff, and sparse intermediate to silicic volcanic rocks, mainly keratophyres, named the Kuna Formation by Mull and others (1982). The Kuna Formation forms the basal unit of the Kagvik sequence of Churkin and others (1979) and the Kagvik terrane of Silberling and others (this volume). This unit, and younger late Paleozoic and early Mesozoic cherts and shales are interpreted either as a deep-water, allochthonous assemblage (Churkin and others, 1979; Nokleberg and Winkler, 1982; Lange and others, 1985) or as an assemblage deposited in an intracratonic basin (Mull and others, 1982; Mayfield and others, 1983). Depending on interpretation of stratigraphy, the sedimentary exhalative Zn-Pb-Ag and bedded barite deposits are interpreted to have formed either in an incipient submarine continental-margin arc or in the early stages of a long-lived, sediment-starved, epicontinental basin.

Red Dog Creek Zn-Pb-Ag deposit. *By J. T. Plahuta, L. E. Young, J. S. Modene.* The Red Dog Creek deposit is a major black-shale-hosted Zn-Pb-Ag deposit of submarine exhalative origin (Moore and others, 1986). The deposit occurs within a complexly deformed, northeast-trending belt of thrust slices. A Mesozoic compressional event greatly foreshortened the original basin and produced a thick stack of folded and internally imbricated thrust plates (Fig. 1). The deposit is hosted by black, fine-grained siliceous shales, by cherts, and by medium-grained limestone turbidites of the Mississippian and Pennsylvanian Kuna Fordation. The formation is informally subdivided into an upper, ore-bearing Ikalukrok unit and a lower, more calcareous Kivalina unit, which forms the stratigraphic footwall to the deposit (Fig. 1).

The sulfide deposits occur in two areas. The Main deposit is partially bisected by Red Dog Creek, and the smaller Hilltop deposit caps a hill about 800 m to the south. Drilling on the Main deposit indicates the presence of 85 million tonnes of ore grading 17.1% Zn, 5.0% Pb, and 82 g/t Ag. The main deposit trends north-northwest and forms a nearly flat-lying lens 1,600 m long and 150 to 975 m wide (Fig. 1). Depth to the top of the ore varies from zero to 60 m. The Hilltop deposit occurs in a flat-lying klippe as much as 90 m thick, 850 m long, and 600 m wide. Both deposits occur in five stacked thrust plates within a regionally telescoped sequence, and they are separated from the footwall units by a major fault zone. Two of the thrust plates contain significant sulfide deposits, and are separated by tectonic slivers of barren units. The deposits are structurally underlain by the Cretaceous Okpikruak Formation.

The deposits are intercalated, stratiform lenses composed of varying proportions of fine-grained sphalerite, galena, pyrite, marcasite, quartz, and barite. The dominant gangue is quartz. Local concentrations of fossil worm tubes, analogous to modern occurrences along the East Pacific Rise, are thought to represent vent-related life zones. Minor amounts of low-grade sulfides occur as beds and veins in the enclosing shales, especially near the base of the deposit, where the veins form a feeder system. In these areas, the veins are as much as 1 m thick and contain medium- to

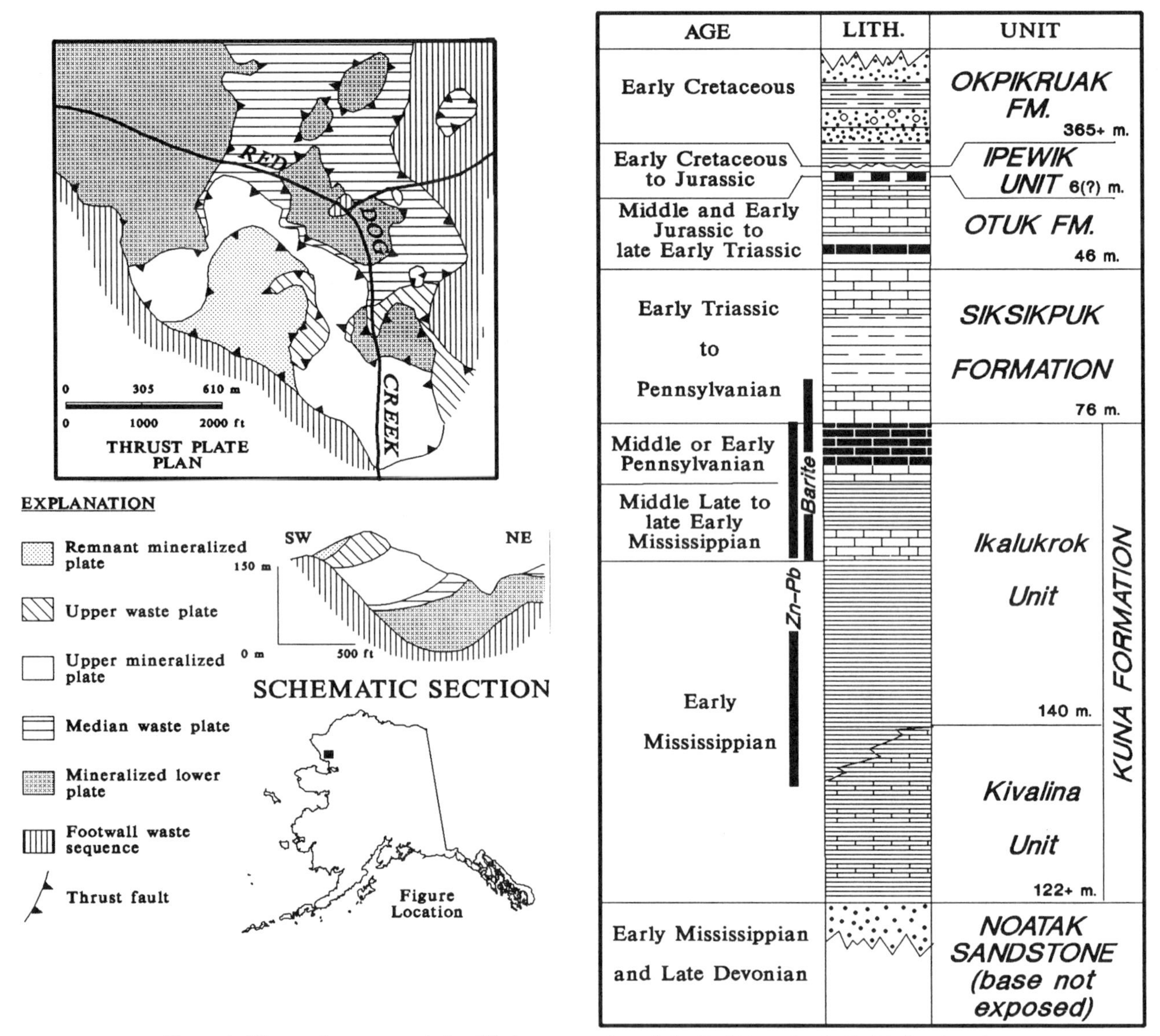

Figure 1. Thrust plate map and simplified cross section of the Red Dog Creek Zn-Pb-Ag deposit, Delong Mountains, northwestern Brooks Range. Base-metal sulfides and barite deposits occur in the Kuna Formation.

coarse-grained sulfides, as well as wall-rock inclusions. Barite-rich lenses, attaining thicknesses greater than 45 m, are most common as a cap to the deposit and locally extend upward into the Siksikpuk Formation.

The deposits are classified as submarine exhalative deposits that formed in the early stages of a long-lived starved epicontinental basin (Moore and others, 1986). Alternatively, Lange and others (1985) suggest that the deposit formed in an incipient submarine island-arc or submarine continental-margin-arc environment. The mineralizing event took place episodically for 25 to 30 m.y., concurrent with tectonic instability in the basin. Small, spatially associated igneous intrusions of possible Mississippian age provide evidence for locally elevated heat flow. Although the source of the metals is speculative, underlying Devonian fluvial-deltaic rocks may have provided a basin aquifer through which the mineralizing fluids moved before exhalation on the sea floor. Structurally adjacent parts of the Upper Devonian Noatak Sandstone contain vein and disseminated sulfides that may be coeval with the Red Dog Creek deposit.

Drenchwater Creek Zn-Pb-Ag deposit. *By W. J. Nokleberg.* The Drenchwater Creek deposit consists of sphalerite, galena, pyrite, marcasite, and sparse barite in a zone about 1,830 m long and as much as 45 m wide (Nokleberg and Winkler, 1982; Lange and others, 1985). The sulfides occur both in disseminations and in massive layers in deep-water marine chert, shale, tuff, tuffaceous sandstone, and keratophyre andesite flows and sills of Mississippian age. The rocks constitute the oldest part of the Kagvik sequence, which in this area occurs in the lowest structural plate in a belt of thrust faults thast strike east-west and dip gently south. The sulfides occur in hydrothermally altered chert and shale and in adjacent volcaniclastic rocks. Fragments of fine-grained feldspar, pumice lapilli, and mafic volcanic rocks are commonly replaced by aggregates of kaolinite, montmorillonite, sericite, chlorite, actinolite, barite, calcite, quartz, fluorite, and prehnite. Local sulfide-bearing quartz-rich exhalite is associated with the volcanic and volcaniclastic rocks. The sulfides and barite form disseminations and massive sphalerite-rich layers; more rarely, quartz-sulfide veins crosscut cleavage paralleling axial planes of south-dipping, north-verging folds. The veins represent minor, post-deformational transport and deposition of metals. Selected samples contain more than 1 percent Zn, 2 percent Pb, 150 g/t Ag, and 500 g/t Cd. The deposits probably formed from metal-laden hydrothermal fluids discharged onto a deep-ocean floor during submarine eruptions occurring in an incipient submarine continental margin-arc or island-arc environment (Nokleberg and Winkler, 1982).

Chromite deposits, northwestern Brooks Range

The southern flank of the northwestern Brooks Range contains podiform chromite deposits at Iyikrok Mountain, Avan, Misheguk Mountain, and Siniktanneyak Mountain (Table 1 on Plate 11, this volume). The deposits consist mainly of disseminated to fine-grained, discontinuous chromite layers and pods in complexly faulted, serpentinized dunite and harzburgite tectonite. The largest deposit, at Avan, contains an estimated 285,000 to 600,000 tonnes of chromite (Foley and others, 1991). The dunite and harzburgite tectonite are part of the Misheguk igneous sequence of pillow basalt (locally intensely serpentinized), gabbro, chert, and minor limestone that is interpreted as a dismembered ophiolite (Roeder and Mull, 1978; Zimmerman and Soustek, 1979; Nelson and Nelson, 1982). This sequence is named the Misheguk Mountain allochthon by Mayfield and others, 1983), which is part of the Angayucham terrane (Plate 3, this volume). The sedimentary rocks range in age from Mississippian to Jurassic; the mafic volcanic rocks range in age from Devonian to Triassic. The age of the ultramafic rocks is Jurassic or older. The ultramafic and mafic rocks occur in a series of klippen that are thrust over mainly Paleozoic and Mesozoic sedimentary rocks of the Arctic Alaska terrane to the north.

Sulfide vein deposits, northwestern Brooks Range

The eastern part of the northwestern Brooks Range contains a belt of sulfide vein deposits at Story Creek, Whoopee Creek, Frost, and Omar (Table 1 on Plate 11, this volume). They generally consist of sphalerite and galena with quartz and minor carbonate gangue in veins and fractures that are found in the Mississippian Kayak Shale of the Endicott Group at the Story Creek and Whoopee Creek deposits, and in dolomite and limestone of the Baird Group at Frost. The veins and fractures occur in linear zones from 1.5 to 3 km long and that cross tightly folded strata, indicating an epigenetic origin (I. F. Ellersieck and J. M. Schmidt, written communication, 1985). No tonnage and grade data are available. Insufficient data preclude assignment of these deposits to a specific mineral deposit type. The Omar deposit consists of disseminated to massive chalcopyrite and other sulfides in veinlets, stringers, and blebs in brecciated Ordovician to Devonian dolomite and in limestone of the Baird Group (Folger and Schmidt, 1986), and it is classified as a Kipushi Cu-Pb-Zn deposit. The Endicott Group, which hosts the Story Creek and Whoopee Creek deposits, forms part of the Brooks Range allochthon (Mayfield and others, 1983) and is part of the Endicott Mountains subterrane. The Baird Group, which hosts the Frost deposit, forms part of the Kelly River allochthon (Mayfield and others, 1983) and is part of the Delong Mountains subterrane.

Massive sulfide deposits, southern Brooks Range

An extensive belt of major Kuroko massive sulfide deposits and one Kipushi Cu-Pb-Zn deposit occurs along an east-west trend for about 260 km on the southern flank of the Brooks Range (Plate 11, this volume). The largest deposits are in the Ambler district (Hitzman and others, 1986) at Arctic, which contains an estimated 32 million tonnes of ore, and at Ruby Creek, which contains an estimated 91 million tonnes of ore averaging 1.2 percent Cu (Table 1 on Plate 11, this volume). Other Kuroko massive sulfide deposits in the belt are at Smucker, BT, Jerri Creek, Roosevelt Creek, and Michigan Creek (Hitzman and others, 1982). The Ann deposit in the southern Brooks Range may be either a polymetallic vein or a metamorphosed sulfide deposit.

The Kuroko massive sulfide deposits occur in or adjacent to submarine mafic and felsic metavolcanic rocks and associated carbonate, pelitic, and graphitic metasedimentary rocks of the Devonian and Mississippian Ambler sequence (Hitzman and others, 1982, 1986). The Ambler sequence, along with the Beaver Creek Phyllite, Bornite Marble, Mauneluk Schist, and Anirak Schist, forms the informally named "schist belt" of the southern Brooks Range (Hitzman and others, 1982). The Ambler sequence is generally multiply deformed and exhibits metamorphism of both greenschist and blueschist facies (Hitzman and others, 1982). The deposits occur within the Ambler sequence, which forms the southern part of the Hammond subterrane (Plate 3, this volume). Most workers in the southern Brooks range favor an origin of continental-margin rifting rather than an island arc for the Kuroko massive sulfide deposits. The Ruby Creek deposit (see below), classified as a Kipushi Cu-Pb-Zn deposit, is genetically related to Devonian submarine volcanism. Bernstein and Cox (1986) stress the significance of carrolite and the Cu-Ge sulfide renierite in the Ruby Creek deposit as a link to the dolomite-hosted deposits at Kipushi and Zaire (Cox and Singer, 1986).

Smucker Zn-Pb-Ag deposit. *By C. M. Rubin.* The Smucker deposit is the westernmost of the Kuroko massive sulfide deposits of the Ambler sequence. The deposit occurs on the limb of a recumbent, asymmetric antiform that plunges gently northwest. The stratiform sulfide horizon extends at least 1,000 m along strike and occurs structurally below quartz-muscovite schist and above quartz-graphite phyllite members of the Ambler sequence. The sulfide horizon consists of banded, fine- to medium-grained pyrite, sphalerite, galena, chalcopyrite, and minor owyheeite in a gangue of quartz, calcite, and pyrite. Hydrothermal alteration, sulfide stock-works, or sulfide veins are not observed. Grades of the major massive sulfide horizons range from 1 to 5 percent Pb, 5 to 10 percent Zn, and 103 to 343 g/t Ag with minor Au.

The deposit occurs in a Late Devonian, polydeformed sequence of mafic and felsic metavolcanic and metasedimentary rocks exhibiting mainly greenschist facies metamorphism. The mafic and felsic metavolcanic rocks consist of quartz-muscovite schist, quartz-feldspar-muscovite schist, quartz-chlorite-calcite phyllite, and porphyroblastic quartz-K-feldspar-muscovite schist. The interlayered metasedimentary rocks consist of quartz-muscovite-chlorite phyllite, quartz-graphite phyllite, calcite-mica schist, and marble. The host rocks probably are derived from a bimodal calcic to calc-alkalic volcanic suite, interlayered with impure clastic and carbonate sedimentary rocks. The host rocks strike west-northwest, dip moderately south, and have been deformed into tight to isoclinal folds overturned to the south. Interpretations of regional stratigraphy are obscured by intense deformation and transposition of bedding into foliation. The deposit is interpreted to have formed either along a Late Devonian epicratonic rift margin or in a pull-apart basin.

Arctic Zn-Cu-Ag deposit. *By J. M. Schmidt.* The Arctic deposit (Fig. 2) is the largest in the belt of volcanogenic (Kuroko) massive sulfide deposits in the Ambler sequence along the southern flank of the Brooks Range (Schmidt, 1983, 1986). The deposit occurs in a sequence of metamorphosed basaltic and rhyolitic volcanic rocks, volcaniclastic and minor plutonic rocks, and pelitic, carbonaceous, and calcareous sedimentary rocks. The main deposit is interlayered with graphitic schists between two metarhyolite porphyries that probably are submarine ash-flow tuffs (Schmidt, 1986).

The main deposit consists of semi-massive or less commonly massive sulfides in multiple lenses, each as much as 15 m thick, over a vertical interval of 6 to 80 m. The main deposit has an areal extent of about 900 by 1,050 m and is estimated to contain 32 million tonnes grading 4.0 percent Cu, 5.5 percent Zn, 1.0 percent Pb, 51.4 g/t Ag, and 0.65 g/t au (Sichermann and others, 1976). The sulfides are mainly chalcopyrite and sphalerite, with less pyrite, pyrrhotite, galena, tetrahedrite, and arsenopyrite and traces of bornite, magnetite, and hematite. The sulfides are slightly to strongly zoned laterally, with pyrrhotite and arsenopyrite more abundant in the central and northwestern parts of the deposit, and bornite more abundant in the southeast. Gangue minerals are also zoned, with calcite and dolomite dominant in the southeast, barite most abundant in the center, and quartz and minor phyllosilicates micas throughout. Minor sulfides occur at two horizons above the main deposit.

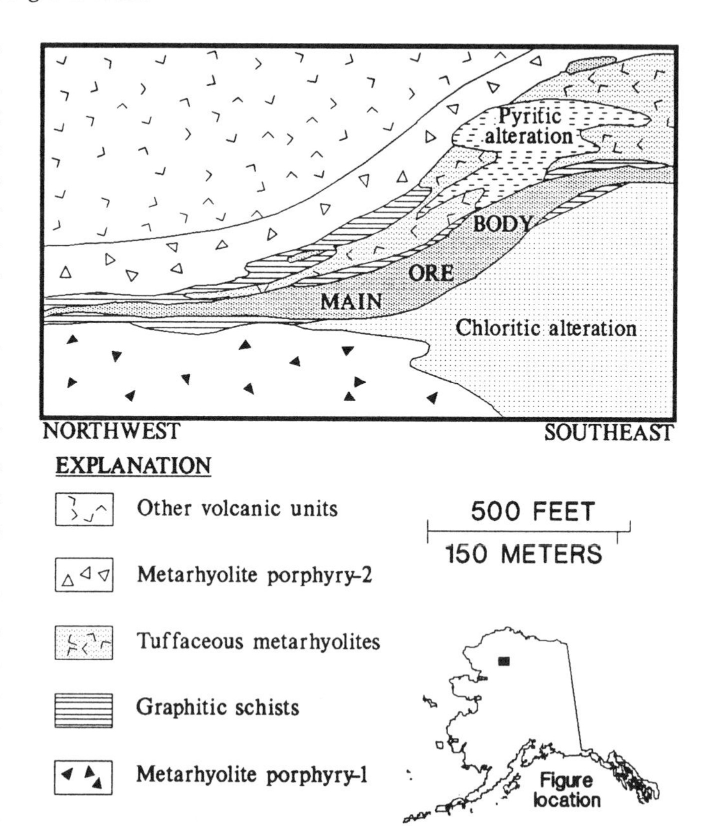

Figure 2. Schematic cross section of the Arctic Zn-Cu-Ag deposit, Ambler district, southern Brooks Range. The main ore body is overlain by a pyritic alteration zone and underlain by a zone of intense chloritic alteration.

Altered rocks are laterally and vertically zoned; they form a sequence that is thickest in the southeastern part of the deposit and thins rapidly westward. Magnesian chlorite-rich altered rock is limited to the footwall of the main deposit. Enveloping the sulfide lenses is an alteration zone containing barian fluorphlogopite, talc, barite, barian phengite, quartz, calcite, and magnesian chlorite. Irregularly overlying the main deposit are thinly laminated altered rocks containing of pyrite, calcite, and phengite. The altered rocks are thickest in a zone trending north-northeast on the southeast side of the deposit, coincident with its zone of high-Cu content.

Chemical changes in the host rocks adjacent to the main deposit include an overall decrease in Na, K, and Si, a decrease of Al in the talc-rich zones, and an increase in Mg, Ba, F, Fe, Mg/Fe, and total volatiles in the altered rocks. Most of the sulfides probably were deposited on surfaces having 0 to 450 m of relief from a topographically higher, linear vent area. An elongate zone containing high Cu, low Ag and Pb, and the most intense alteration, probably represent a fluid vent area that was fault controlled and active during short-lived Devonian and Mississippian rifting along a continental margin.

Ruby Creek Cu deposit. *By M. W. Hitzman.* The Ruby Creek deposit occurs on the north flank of the Cosmos Hills in

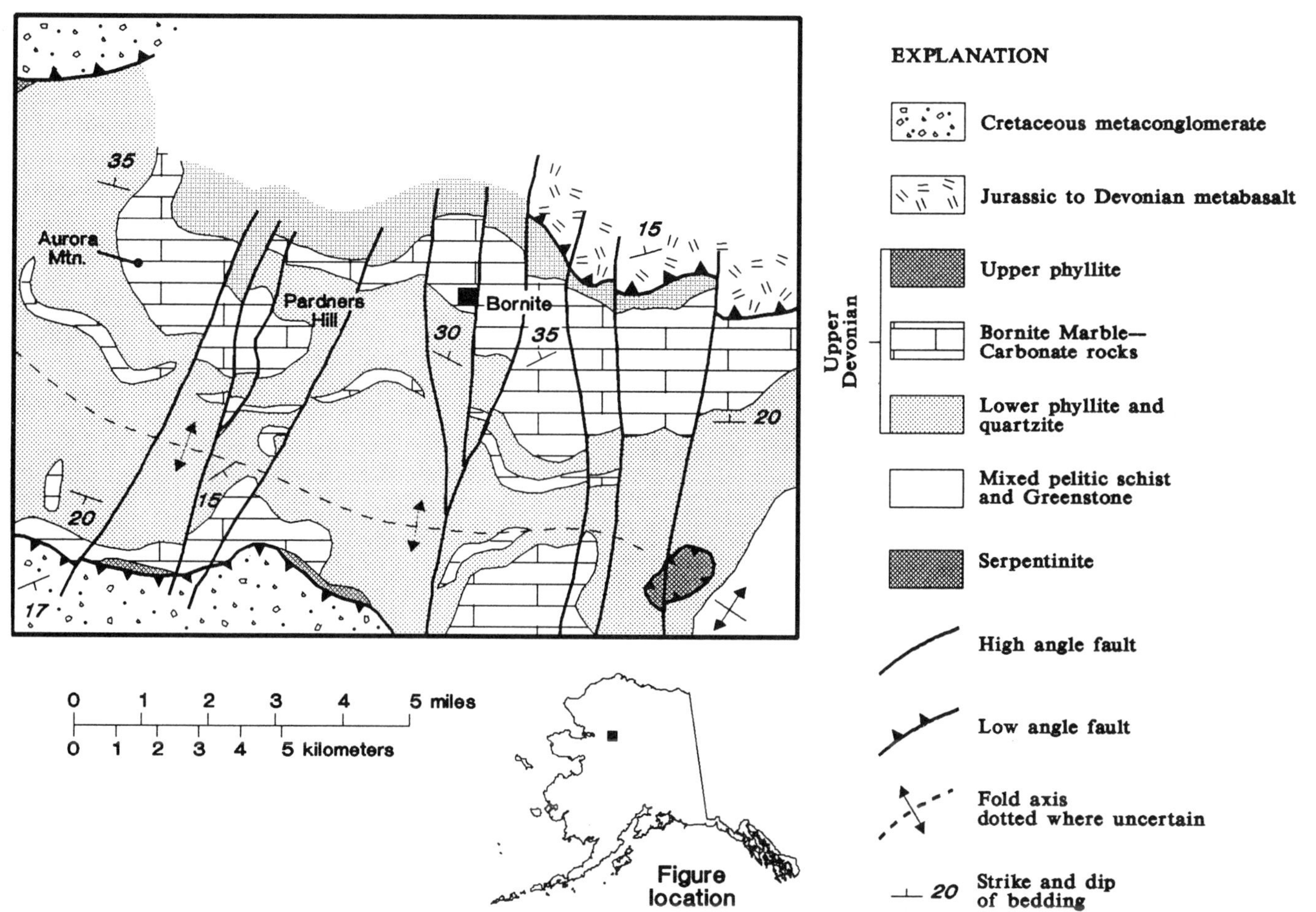

Figure 3. Generalized geologic map of the Cosmos Hills showing the Bornite (Ruby Creek) area, Ambler district, southern Brooks Range.

the west-central part of the Ambler district (Figs. 3, 4) (Hitzman and others, 1982). The deposit consists of disseminated to massive chalcopyrite, bornite, chalcocite, pyrite, and local sphalerite. Local sparse galena, pyrrhotite, marcasite, carrollite, cuprite, and renierite also are present. The deposit contains an estimated 91 million tonnes averaging 1.2 percent Cu, with 454,000 tonnes grading up to 4 percent Cu (Runnells, 1969). The deposit occurs in brecciated and intensely folded and faulted dolomite and limestone of the Devonian Bornite Marble (Hitzman and others, 1982), a unit that is about 1,000 m thick.

The deposit is interpreted to have formed along a rifted continental margin in the Late Devonian (Hitzman, 1986). It occurs along a fault(?)-controlled margin of a carbonate bank adjacent to a shale-filled graben. The carbonates were deposited as an intertidal bank with scattered organic mounds and a bioherm barrier at the basin edge; this carbonate bank is stratigraphically approximately equivalent to the bimodal metavolcanic rocks and metasedimentary rocks of the Ambler sequence, which contains volcanogenic massive sulfide deposits and which crops out about 15 km to the north. Paleogeographic reconstructions indicate that the Ruby Creek area and volcanogenic massive sulfide deposits in the Ambler sequence were located on opposite sides of a major shale-filled graben.

Mineralized hydrothermal dolostone bodies are recognized in biohermal and backreef facies of the Cosmos Hills along 10 km of the paleobasin margin (Hitzman and others, 1982). The overall trend of these bodies is parallel to the fault(?)-controlled margin of the basin, although several bodies formed along second-order structures. Clasts of hydrothermal dolostone are present in synsedimentary breccia, indicating mineralization occurred concurrently with sedimentation. Pb-Pb ages from galena in the deposit have yielded concordant Late Devonian ages (R. V. Kirkham, written communication, 1979).

The Ruby Creek hydrothermal dolostone (Fig. 4) displays evidence of three major hydrothermal dolomitizing events (Hitzman, 1986). The first two events, A and B, formed a roughly dome-shaped body about 450 m thick and 1,000 m by 1,500 m in lateral extent. Alteration was locally controlled by stratigraphy: pure clean limestone beds were most easily dolomitized. Less permeable argillaceous and highly carbonaceous limestones have

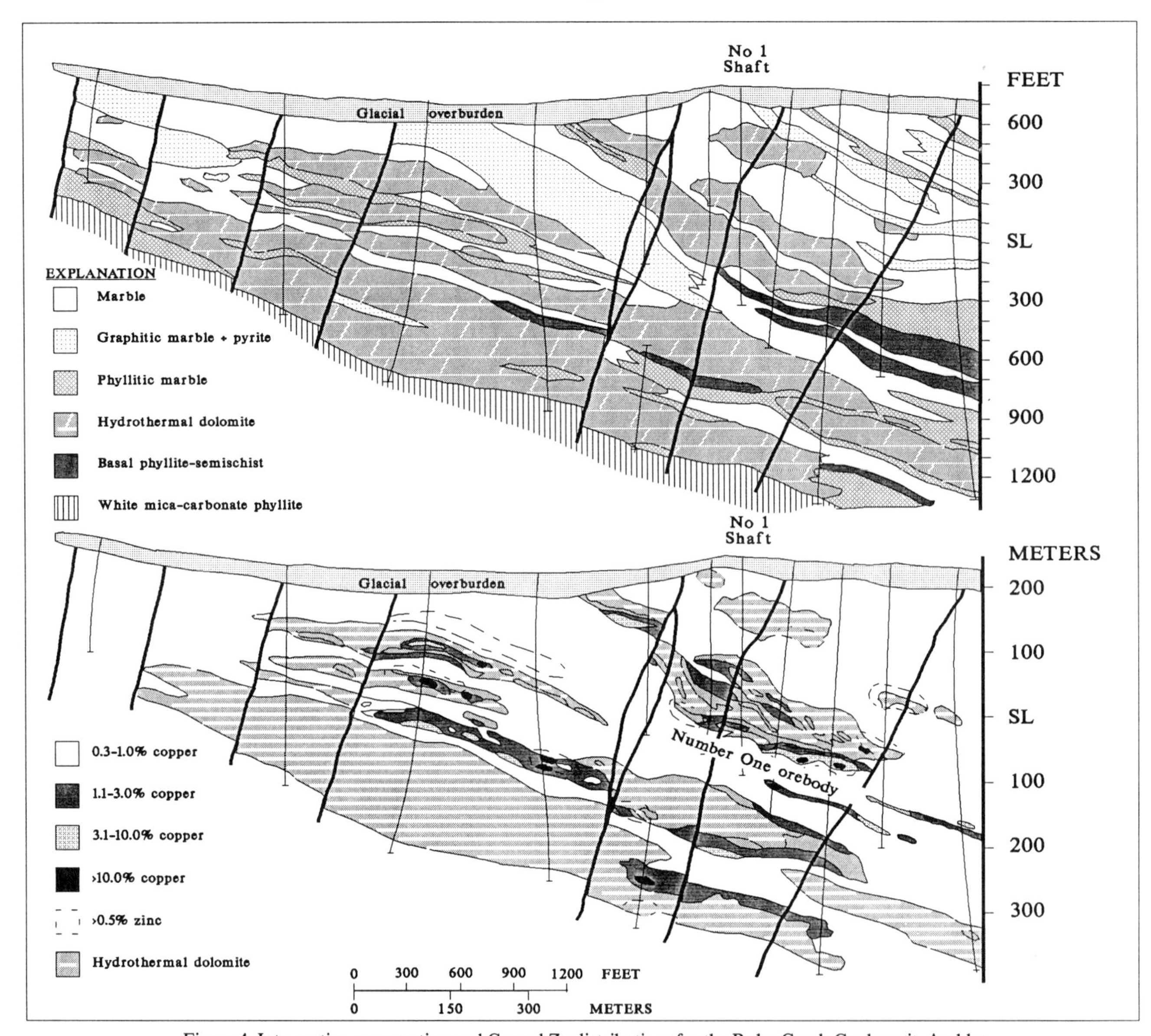

Figure 4. Interpretive cross section and Cu and Zn distributions for the Ruby Creek Cu deposit, Ambler district, southern Brooks Range. Note the relation between the sulfide deposits and the occurrence of hydrothermal dolomite.

been generally less altered. In the upper part of the system, these impermeable beds channeled fluids, while at the base, such beds were altered to zones of ferroan dolomite and/or siderite and chlorite.

The first alteration event produced the A-dolostone, consisting of ferroan dolomite near the base of the carbonate section, and grading upward to magnesian dolomite in the upper part of the section. The second hydrothermal event produced the B-dolostone, which cuts A-dolostone along irregular solution fronts and locally caused in-situ brecciation. Deep in the system, B-dolostone contains pyrite, chlorite, pyrrhotite, trace chalcopyrite and cymrite, and sphalerite. Late-B zones, defined by lenses of poorly ferroan dolomite and calcite in stratigraphic traps, commonly contain nearly massive pyrite and lesser sphalerite. The final hydrothermal event resulted in brittle fracturing of the dolostone body and production of C veins. The C veins are vertically zoned from mildly ferroan dolomite veins containing sparse pyrite deep in the system, to dolomite veins containing chalcopyrite and minor calcite in the central portion of the system, to dolomite-calcite veins containing subsidiary sphalerite and chalcopyrite on the outer fringes of the system. High-grade ore containing bornite, chalcocite, chalcopyrite, carrollite, and sphalerite formed where fracturing of the body allowed solutions access to the late B-massive pyrite zones. Copper sulfides replace the late-B pyrite; carrollite formed where the solutions forming the C veins interacted with the late-B cobaltiferous pyrite.

Sun Zn-Pb-Cu-Ag deposit. *By C. D. Maar.* The Sun deposit occurs in the easternmost part of the belt of Kuroko massive sulfide deposits in the Ambler district. The deposit consists of massive sulfides in at least three separate zoned horizons. The sulfides are sphalerite, chalcopyrite, galena, and argentiferous tetrahedrite. Gangue minerals are pyrite, arsenopyrite, and barite. The upper sulfide horizon is rich in Zn, Pb, and Ag, the middle horizon is rich in Cu, and the lower horizon is rich in Cu and Zn. Grades average 1 to 4 percent Pb, 6 to 12 percent, Zn, 0.5 to 2 percent Cu, and 685 to 1,030 g/t Ag. Individual quartz-barite beds contain as much as 690 to 1,030 g/t Ag.

The deposit is hosted in metarhyolite, muscovite-quartz-feldspar schist, micaceous calc-schist, marble, and greenstone of the Devonian and Mississippian Ambler sequence, which strikes northeast and dips moderately southeast. Most of the deposit is hosted in felsic or graphitic schist. Thin, concordant layers of sulfides occur in the metarhyolite, most of which is siliceous, light-colored, weakly schistose, and sparsely porphyritic. Small- and large-scale isoclinal folds are present both in host rocks and sulfide layers. The deposit probably formed during Devonian and Mississippian submarine volcanism.

Vein, Skarn, and Porphyry deposits, central Brooks Range

To the north of the Ambler sequence in the central Brooks Range is a long belt of vein, skarn, and porphyry deposits with an east-west strike length of about 240 km (Table 1 on Plate 11, this volume). At Mt. Igikpak and Arrigetch Peaks, the major deposits are polymetallic quartz veins with base-metal sulfides, Sn skarns with disseminated cassiterite and base-metal sulfides, and Cu-Pb-Zn skarns with disseminated Fe sulfides, and base-metal sulfides. At Sukapak Mountain, Sb-Au quartz vein deposits occur with sparse disseminated stibnite, cinnabar, and gold. At Victor and nearby areas, a porphyry Cu deposit contains veinlet and disseminated chalcopyrite and other base-metal sulfides in Devonian granodiorite porphyry, along with an adjacent Cu skarn deposit containing interstitial bornite, chalcopyrite, and other base metal sulfides. At Geroe Creek, veinlet, stockwork, and disseminated molybdenite occur in a Devonian granite, and are classified as a porphyry Cu-Mo deposit. In the Chandalar district, gold and sparse sulfides occur in Au quartz vein deposits in a zone about 4.0 km wide and 1.6 km long. In 1981, the Mikado Squaw mine in the Chandalar district contained an estimated 11,000 tonnes grading 75 g/t Au (Dillon, 1982). The Chandalar district contains an estimated 45,000 tonnes grading 80 g/t Au (Dillon, 1982).

These deposits are in a structurally complex and polymetamorphosed assemblage of Devonian or older carbonate rocks, including the Silurian and Devonian Skagit Limestone, calc-schist, quartz-mica schist, and quartzite, which is intruded by Proterozoic and Late Devonian gneissic granitic rocks and is part of the Hammond subterrane (Silberling and others, this volume). The Devonian gneissic granitic rocks and their related mineral deposits form an east-west-striking belt in the central part of the metasedimentary rocks. U-Pb zircon isotopic ages indicate that the Devonian gneissic granitic rocks intruded about 30 to 40 m.y. after the eruption of the submarine volcanic rocks that host the massive sulfide deposits in the Ambler district to the south (Newberry and others, 1986). The polymetallic vein deposits, Cu and Sn skarns, and porphyry deposits are interpreted as having formed during intrusion of the Devonian and older gneissic granitic rocks in a continental-margin subduction zone (Newberry and others, 1986). The Au quartz and Sb-Au quartz vein deposits are related to much later Mesozoic greenschist-to-amphibolite-facies regional metamorphism of the metasedimentary and gneissic granitic rocks.

Skarn, vein, and porphyry deposits, northeastern Brooks Range

A cluster of skarn, vein, and porphyry deposits is present in the northeastern Brooks Range (Table 1 on Plate 11, this volume) in an area that has mostly been withdrawn from mineral exploration. The major deposits are: (1) at Esotuk Glacier—a Pb-Zn skarn containing disseminated galena and sphalerite, and a fluorite vein; (2) at Porcupine Lake—tetrahedrite, enargite, and fluorite in veins and as replacements in a polymetallic vein(?); (3) in the Romanzof Mountains—disseminated galena, sphalerite, and chalcopyrite and base-metal sulfides in quartz veins in Devonian(?) granite (here classified as a superposed porphyry Cu and polymetallic vein deposit), and a Pb-Zn skarn with galena and sphalerite; (4) at Bear Mountain—a molybdenite- and wolframite-bearing Tertiary rhyolite porphyry stock, classified as a porphyry Mo deposit (Barker and Swainbank, 1986); and (5) at Galena Creek—disseminated galena and sphalerite in a polymetallic vein. No tonnage and grade estimates are available, and none of the deposits has been developed or productive. The paucity of deposits in the northeastern Brooks Range most likely reflects the limited geological exploration of the area.

The mineral deposits in the northeastern Brooks Range occur in a variety of host rocks. The skarn deposit at Esotuk Glacier is in Devonian or older marble and calc-schist intruded by Devonian gneissose granite. The polymetallic vein deposit at Porcupine Lake occurs in silicified tuffaceous limestone of the Mississippian and Pennsylvanian Lisburne(?) Group. The porphyry Cu, polymetallic vein, and skarn deposits at Romanzof Mountains are in Devonian(?) granite and Precambrian marble and calc-schist of the Neruokpuk Quartzite intruded by the Silurian or Early Devonian Okpilak (granite) batholith. The porphyry Mo deposit at Bear Mountain occurs in a molybdenite-wolframite-bearing Tertiary rhyolite porphyry stock intruding the Neroukpuk(?) Quartzite. The polymetallic vein deposit at Galena Creek is in metasedimentary rocks and greenstone of the Neruokpuk(?) Quartzite. The metasedimentary rocks of the Neruokpuk(?) Quartzite and the younger late Paleozoic and Mesozoic sedimentary rocks in the region are part of the North Slope subterrane (Plate 3, this volume).

LODE DEPOSITS, SEWARD PENINSULA

The Seward Peninsula contains a variety of lode deposits (Table 1 on Plate 11, this volume): (1) Sn-W vein, Sn skarn, and Sn greisen deposits; (2) polymetallic vein and porphyry deposits; (3) Au quartz vein deposits; (4) a felsic plutonic U and a sandstone U deposit; and (5) a metamorphosed sulfide deposit. The first three groups include most of the deposits in the region. The larger Sn lode deposits are mainly in the northwestern part of the peninsula; smaller Sn deposits occur elsewhere. The polymetallic and Au quartz vein deposits occur mainly in the southeastern and eastern parts of the peninsula. Felsic plutonic U, sandstone U, and metamorphosed sulfide deposits occur in the eastern part of the peninsula.

Sn vein, skarn, and greisen deposits

Sn lode deposits in the Seward Peninsula consist of Sn vein deposits at Cape Mountain and Potato Mountain, a Sn skarn deposit at Ear Mountain, Sn-W skarn and greisen deposits at Lost River, and a Sn greisen deposit at Kougarok. These deposits are commonly referred to as the Cretaceous tin province of the Seward Peninsula. The Sn vein deposits at Cape and Potato Mountains consist of cassiterite, pyrite, and a variety of other minerals, generally as disseminations in the margins or dikes of Cretaceous granite or in veins and veinlets in Precambrian or early Paleozoic metasedimentary rocks. The Sn skarn deposits at Lost River and Ear Mountain occur in Paleozoic limestone intruded by Late Cretaceous granite. The Sn greisen deposit at Kougarok occurs in steep pipes of greisenized Late Cretaceous granite. The larger Sn lode deposits occur at Lost River and Kougarok.

Lost River Sn-W deposits. The Lost River deposits occur in vein, skarn, greisen, and solution breccia near and along the upper margin of a Late Cretaceous granitic stock (Sainsbury, 1969; Dobson, 1982; D. Grybeck, written communication, 1984). The stock intrudes a thick sequence of argillaceous limestone of the Ordovician Port Clarence Limestone of former usage, which is part of the York terrane (Plate 3, this volume). Early-stage andradite-idocrase skarn, and later fluorite-magnetite-idocrase vein skarns are altered to chlorite-carbonate assemblages that formed contemporaneously with greisen formation and cassiterite deposition. Locally abundant beryllium concentrations occur in fluorite–white mica veins, some containing diaspore, chrysoberyl, and tourmaline. These veins are probably associated with the early stages of granite intrusion. The major ore minerals in the skarns and greisen are cassiterite and wolframite, with lesser stannite, galena, sphalerite, pyrite, chalcopyrite, arsenopyrite, and molybdenite and a wide variety of contact metasomatic and alteration minerals. Most of the production was from the Cassiterite Dike, a near-vertical granite dike extensively altered to greisen. Several small Sn deposits occur nearby, at or in the Tin Creek Granite; various polymetallic veins and skarns occur near the adjacent Brooks Mountain Granite.

Kougarok Sn deposit. *By Christopher C. Puchner.* The Kougarok deposit consists of tin and tantalum-niobium concentrations in granitic dikes, sills, and plugs and in schist adjacent to the granitic rocks (Puchner, 1986). Rb-Sr and K-Ar age determinations indicate a Late Cretaceous age for the granitic rocks, which probably are coeval with other tin-bearing granitic rocks of the Seward Peninsula (Hudson and Arth, 1983). The tin deposits occur in four geologic settings: (1) in steep cylindrical pipes of greisen formed in granite; (2) in greisen formed in dikes; (3) in greisen formed along the roof zone of granitic sills; and (4) as stockwork veinlets in adjacent schist. Tin occurs dominantly as disseminated cassiterite in quartz-tourmaline-topaz greisen. Sn grades range from 0.1 to 15 percent Sn and average approximately 0.5 percent Sn. Ta-Nb deposits are confined to the roof greisens. Tantalite-columbite occurs as disseminated grains in quartz-white mica greisen beneath and/or adjacent to tin-bearing quartz-tourmaline greisen. Grades range from 0.01 to 0.03 percent for both Ta and Nb. The wall rocks are part of the Nome Group (Sainsbury, 1972).

Polymetallic vein and porphyry deposits.

Sparse sulfide vein deposits, classified as polymetallic vein deposits, and one porphyry Mo deposit occur mainly on the eastern Seward Peninsula. The polymetallic vein deposits are at Serpentine Hot Springs, Omilak, Independence, and Quartz Creek, and the porphyry Mo deposit is at Windy Creek. The polymetallic vein deposits consist of veins, stringers, and disseminations of pyrite and base-metal sulfide. All but the Quartz Creek deposit occur in Paleozoic(?) marble and quartz-mica schists that are part of the Nome Group or the mixed unit of Till (1984) in an area intruded by Cretaceous granitic plutons. The Quartz Creek deposit occurs in altered andesite and granite of Jurassic or Cretaceous age. The porphyry Mo deposit at Windy Creek consists of veins and stringers of quartz, Fe sulfides, molybdenite, galena, and sphalerite in the hornblende granite of the Cretaceous(?) Windy Creek pluton. The pluton intrudes early Paleozoic mafic schist and marble of the Nome Group (or mixed unit of Till, 1984), which is part of the Seward terrane (Plate 3, this volume).

Au quartz vein deposits

The Seward Peninsula contains numerous Au quartz vein deposits, prospects, and one mine. The larger deposits are in the Nome district at Daniels Creek (Bluff), and at Big Hurrah (Gamble and others, 1985). Some of the deposits produced minor gold; only the Big Hurrah Mine in the Solomon River area has recorded production. Most of the deposits consist of gold and sparse sulfide minerals in thin quartz veins emplaced along fault zones in low-grade metamorphic rocks, mainly in the mixed unit of metasedimentary rocks and mafic schist of the Nome Group. The quartz veins cut the generally shallow-dipping metamorphic foliation. In addition to quartz, the veins typically contain minor carbonate and albite or oligoclase. Native gold is accompanied by sparse arsenopyrite and lesser pyrite. Total sulfide content is usually only a few percent.

According to Gamble and others (1985), the discordance of the veins to metamorphic foliations, the preliminary oxygen isotope and fluid inclusion data, similarities to other occurrences in metamorphic rocks, and the absence of known or suspected intrusives near the veins indicate that the gold deposits formed from fluids that equilibrated with the sedimentary and/or volcanic protoliths of the Nome Group under greenschist- or amphibolite-facies regional metamorphic conditions and then moved upward during a later, post-kinematic event to deposit the vein minerals. A Late Jurassic or Early Cretaceous age of metamorphism and vein formation seems likely (Gamble and others, 1985).

Big Hurrah Au deposit. *By B. M. Gamble.* The Big Hurrah Au deposit consists of four major quartz veins, and zones of ribbon quartz. The veins are 1 to 5 m thick and a few hundred meters long, and consist mainly of quartz with lesser amounts of plagioclase and carbonate minerals, and accessory gold, scheelite, arsenopyrite, and pyrite (Gamble and others, 1985, Read and Meinert, 1986). The veins are localized along faults that parallel major regional faults and strike northwest and mostly dip steeply southwest. The veins commonly have a ribbon structure caused by graphite- or carbon-coated fractures and/or inclusions of wall rock that parallel the veins. The veins occur in a graphitic quartz-rich schist that contains variable but small amounts of chlorite and/or sericite. An estimated 155,500 g of gold was recovered, mostly between 1903 and 1907, with some production in the 1940s and early 1950s (J. Orr, written communication, 1954). Several mining companies have drilled the deposit since 1954; recent assays show 25 to 65 g/t Au (Gamble and others, 1985). The schists hosting the veins are part of the Nome Group (mixed unit of Till, 1984), which forms part of the Seward terrane.

Smaller Au quartz vein deposits elsewhere on the Seward Peninsula share some similarities with the Big Hurrah deposit. The similarities are low sulfide-mineral concentration, fault localization, and confinement to low-grade, greenschist-facies metamorphic rocks. They differ, however, in being thinner and shorter, having lower grades and different host rocks, and containing a wider variety of minerals, including plagioclase, siderite, ferroan dolomite(?), arsenopyrite, minor pyrite, and locally stibnite.

U and metamorphosed sulfide deposits

A felsic plutonic U deposit occurs at Eagle Creek, and a sandstone U deposit occurs at Death Valley, both in the eastern part of the Seward Peninsula. The felsic plutonic deposit consists of disseminated U-, Th-, and REE-minerals along the margins of alkaline dikes intruded into a Cretaceous granite pluton and adjacent wall rocks (Miller, 1976; Miller and Bunker, 1976). The Death Valley sandstone U deposit consists mainly of metaautunite in Paleocene sandstone along the margin of a Tertiary sedimentary basin (Dickinson and Cunningham, 1984). The U in the sandstone probably was transported by groundwater from Cretaceous granitic plutons to the west (Dickinson and Cunningham, 1984).

A metamorphosed sulfide deposit occurs at Hannum Creek in the northeastern part of the Seward Peninsula. The deposit consists of blebs, stringers, massive boulders, and disseminations of base-metal sulfides and barite parallel to layering in Paleozoic quartz-mica schist and marble, part of the Nome Group (mixed unit in Till, 1984). The deposit is interpreted as a metamorphosed laminated exhalite that possibly originated as a sedimentary exhalative Zn-Pb deposit (J. A. Briskey, written communication, 1985).

LODE DEPOSITS, WEST-CENTRAL ALASKA

West-central Alaska contains a variety of lode deposits (Table 1 on Plate 11, this volume). The southwestern Kuskokwim Mountains contain an Sb-Hg vein deposit, a zoned mafic-ultramafic Fe-Ti deposit, and a hot-spring Hg deposit. The central Kuskokwim Mountains contain a complex and extensive suite of Au quartz vein, Sb-Au vein, polymetallic vein, epithermal vein, hot spring Hg, Cu and Fe skarn, and felsic plutonic U deposits, and a carbonate-hosted sulfide deposit. The northeastern Kuskokwim Mountains contains several podiform chromite deposits in thin discontinuous thrust sheets of ultramafic and related rocks. The west-central Yukon-Koyukuk basin contains a suite of polymetallic vein and porphyry Mo and Cu deposits. The northern and eastern Yukon-Koyukuk basin contains suites of felsic plutonic U, polymetallic and epithermal vein, and W skarn deposits and a suite of podiform chromite and serpentinite-hosted asbestos deposits in thin discontinuous thrust sheets of ultramafic and related rocks.

Vein, zoned mafic-ultramafic, and hot spring deposits, southwestern Kuskokwim Mountains

An Sb-Hg vein deposit at Kagati Lake, an Fe-Ti zoned mafic-ultramafic deposit at Kemuk Mountain, and a hot spring Hg deposit at Cinnabar Creek occur in the southwestern Kuskokwim Mountains. The Kagati Lake Sb-Hg deposit consists of stibnite cinnabar, and quartz veinlets along joint surfaces in a Late Cretaceous monzonite and granodiorite stock intruding Lower Cretaceous volcaniclastic rocks and andesite of the late Paleozoic and Mesozoic Gemuk Group (Sainsbury and MacKevett, 1965). The Kemuk Mountain Fe-Ti deposit consists of a buried titaniferous magnetite deposit in a crudely-zoned pyroxenite (Humble Oil and Refining Company, written communication, 1958), and is estimated to contain 2.2 billion tonnes grading 15 to 17 percent Fe and 2 to 3 percent TiO_2. The Cinnabar Creek Hg deposit consists of stibnite and cinnabar in shear zones, disseminations, and veinlets in or near silica-carbonate dikes in argillite and other clastic rocks of the Gemuk Group (Sainsbury and MacKevett, 1965), which is part of the Togiak terrane (Plate 3, this volume).

Vein, hot spring, skarn, and U deposits, central Kuskokwim Mountains region

A wide variety of magmatism-related lode deposits occur in the central Kuskokwim Mountains region. The major deposits are an epithermal vein deposit at Taylor Mountains; hot spring Hg deposits at Red Devil (described below), DeCoursey Mountain, and White Mountain (described below); an Sb-Au vein deposit at Snow Gulch; polymetallic vein deposits at Fortyseven Creek, Mission Creek, Owhat, Chicken Mountain, Golden Horn, Malemut, Granite, Broken Shovel, Cirque, Tolstoi, Independence, Candle, and Win-Won; a Cu-Au-Ag-Bi skarn deposit at Nixon Fork (described below); an Fe skarn deposit at Medfra; and a felsic plutonic U deposit at Sischu Creek. Most of the polymetallic vein and related deposits occur in the Flat and Innoko districts (described below).

The magmatism-related deposits are associated mainly with Late Cretaceous and early Tertiary granite, granodiorite, monzonite, and lesser gabbro that intrude an extensive suite of Cretaceous graywacke, argillite, basaltic to rhyolite volcanic flows, tuffs, and breccias of the Kuskokwim Group (Cady and others, 1955). Two unique deposits in the area are the Reef Ridge carbonate-hosted sulfide deposit and the Sischu Creek felsic plutonic U deposit. The Reef Ridge deposit consists mainly of stringers of sphalerite and minor galena in hydrothermal breccia in Silurian and Devonian carbonate rocks. The minimum strike length of the sulfides is 2,000 m, and the width is as much as 15 m. The felsic plutonic U deposit consists of strongly radioactive U- and Th-rich porphyritic sanidine rhyolite and quartz porphyry flows in belts as much as several km wide and long that are associated with Late Cretaceous volcanic piles and granitic stocks and plugs.

Red Devil Hg deposit. The Red Devil deposit consists of cinnabar and stibnite in about twenty plunging chimneylike bodies located along intersections of two altered basalt dikes in a wrench-fault zone (Herreid, 1962; MacKevett and Berg, 1963; H. R. Beckwith, written communication, 1965). The ore bodies are vertically zoned, with nearly pure cinnabar at the surface, and increasing proportions of stibnite at depth. At 200 m below the surface, mainly stibnite and quartz occur with a trace of cinnabar. The Red Devil mine has been the largest producer of mercury in Alaska; its production of 34,745 flasks accounts for about 80 percent of the state's production from 1942 through 1974 (Bundtzen and others, 1985). It is the largest and best exposed deposit of at least 15 known in the Kuskokwim mercury belt, which extends 400 km northeast from Dillingham to the upper Innoko River. The Red Devil deposit and others in the belt are hosted in the flysch of the Kuskokwim Group, a turbiditic to fluvial clastic sedimentary sequence interpreted by Bundtzen and Gilbert (1983) to have been deposited in an elongate structural trough in the mid- and Late Cretaceous.

The Red Devil and nearby deposits at Barometer, Parks, and Rhyolite are interpreted to have been deposited by ascending hydrothermal fluids, carrying Hg- and Sb-bearing solutions into epithermal conditions (Sainsbury and Mackevett, 1965). The sulfides were probably deposited when the fluids reached near-surface groundwater in hot-spring conduits. The Red Devil deposit is similar to other flysch-hosted deposits in California, Spain, and the U.S.S.R., where mercury was mobilized into high-level, structurally controlled deposits by igneous activity. Abundant 60- to 70-Ma plutons, dike swarms, and volcanic rocks prevalent in the central Kuskokwim area (Bundtzen and Gilbert, 1983) might have been heat sources that mobilized the mercury.

Flat and Innoko districts. *By T. K. Bundtzen.* The Flat district contains an extensive suite of polymetallic and lesser Sb-Au vein deposits at Chicken Mountain, Golden Horn, Broken Shovel, Granite, and Malemute, and the Innoko district contains an equally extensive suite of polymetallic vein deposits, locally rich in Sn, at Cirque, Tolstoi, Independence, and Win-Won (Bundtzen and Laird, 1982, 1983a, b, c; Bundtzen and Gilbert, 1983; Bundtzen and others, 1985).

The Chicken Mountain Au deposit consists of veinlets with arsenopyrite, scheelite, cinnabar, gold, and stibnite. The veinlets are hosted in an intensely hydrothermally altered zone in the southern part of the Chicken pluton (monzonitegabbro) of Chicken Mountain. Pervasive sericitic alteration occurs in a northeast-trending area of about 250 by 600 m.

The Golden Horn Au-Ag-Sb deposit consists of veins of stibnite, cinnabar, scheelite, sphalerite, Pb-Sb sulfosalts, and chalcopyrite in a gangue of quartz, tourmaline, and calcite in a shear zone 30 m wide and 3 km long on the eastern side of the Otter Creek pluton (monzonite). Stibnite and cinnabar locally crosscut other sulfides. The veins occur in irregularly distributed quartz-filled shear zones at or near the contact of the pluton with Cretaceous clastic rocks. Sb-Hg veins crosscut older sulfides but are not directly associated with Au-As-W zones. The deposit has produced 479 tonnes grading 174 g/t Au and 171 g/t Ag.

The Broken Shovel Ag-Pb-Sb deposit consists of tourmaline, quartz, arsenopyrite, and sulfosalts in veins in the central part of the Cretaceous Moose Creek pluton (monzonite). The veins vary from 1 to 3 m wide, and occur in a sericite-tourmaline alteration zone about 300 by 400 m. The deposit contains an estimated 13,600 tonnes grading 178 g/t Ag, 0.15 percent Pb, and 0.15 percent Sb.

The Cirque and Tolstoi Cu-Ag-Sn deposits consist of chalcopyrite, tetrahedrite, pyrite, arsenopyrite, and scheelite in a gangue of tourmaline, axinite, and quartz localized along faults or in tourmaline phyllic alteration in altered monzonite that is capped by altered olivine basalt. The two deposits, which exhibit characteristics both of polymetallic vein and porphyry Cu deposits, are associated with zoned, multiphase plutons ranging in composition from olivine gabbro to monzonite. The plutons and associated volcanic fields have alkalic affinities and show strong differentiation trends. The more promising deposits are in high cupolas or structural conduits of the most differentiated, felsic phases of the plutons.

The Independence Au deposit consists of quartz fissure veins containing gold, pyrite, and arsenopyrite in an altered dacite to rhyolite dike. The dike is part of the 60-km-long Yankee dike swarm. The deposit has been explored by several hundred meters of underground workings and has produced 1,773 g of Au from 113 tonnes of ore. The Candle deposit consists of cinnabar, arsenopyrite, and quartz in stockworks in a Late Cretaceous sericitized monzonite near the intrusive contact with overlying altered olivine basalt. The Win-Won Sn-Ag deposit consists of chalcopyrite in numerous en echelon quartz veinlets 10 to 20 cm thick, in a well-developed quartz stockwork in hornfels on the northeast margin of the Cloudy Mountains volcanic field and related plutonic complex.

White Mountain Hg deposit. *By B. K. Jones.* The White Mountain deposit consists of cinnabar in three zones between Ordovician limestone and shale of the Nixon Fork sequence (Sainsbury and MacKevett, 1965). The zones occur in a belt about 1 km wide and 3 km long on the northwest side of the Farewell fault. In the southern zone, cinnabar forms thin crystalline coatings in brecciated dolomite, coatings on breccia surfaces, and irregular veinlets. In the central zone, irregular lenses of cinnabar occur in silicified limestone and dolomite. In the northern zone, rich cinnabar lenses occur on both sides of a major fault between shale and limestone. The largest massive cinnabar lens is 35 m long and 10 to 15 cm thick. Local cinnabar also occurs in dolomitized limestone with small, karst-like solution caverns. The gangue minerals in the cinnabar lenses are mainly dolomite, chalcedony, calcite, dickite, and limonite. The deposit was mined mainly from 1964 to 1974 and produced about 3,500 flasks. Chip samples of the cinnabar zones contain from 5 to 30 percent cinnabar. The overall features of this deposit and the nearby Mary Margaret and Peggy Barbara deposits, including a sulfur-mercury spring, indicate low-temperature deposition of mercury in a hot-spring environment along structural conduits associated with the Farewell fault.

Nixon Fork–Medfra district. The Nixon Fork–Medfra district contains several Cu-Au skarn deposits from Nixon Fork to Medfra and a dolomitic Fe-skarn deposit at Medfra. The Cu-Au skarns consist of gold, chalcopyrite, pyrite, bornite, and Bi in skarn bodies that form irregular replacements in recrystallized Ordovician limestone of the Telsitna Formation near a Late Cretaceous monzonite (Martin, 1921; Brown, 1926; Jasper, 1961; Herreid, 1966; Patton and others, 1984) and in roof pendants overlying the pluton. Gangue minerals are diopside, andradite garnet, plagioclase, actinolite, epidote, and apatite. The skarns occur mainly in fractures from 1 to 4 m wide, 50 m long, and usually within 40 m of the pluton. A few skarns occur in roof pendants overlying the pluton, and several skarn bodies occur in fault-controlled veinlets as much as a few hundred m from the main trend of skarns. Local groundwater alteration produced extensive oxidized skarn containing limonite, quartz, malachite, pyrite, and gold. Phyllic or argillic clay mineral alteration, which occurs in the pluton, is typical of many Cu skarn deposits (Cox and Singer, 1986).

Skarns in the area have produced about 1.24 to 1.87 million g of Au and undisclosed amounts of Cu, Ag, and Bi. Individual skarn bodies in the area contain up to 113 g/t Au and 1.5 to 2 percent Cu. Most ore was mined from areas of secondary enrichment. Lower grade sulfide-rich ore occurs at depths greater than 60 m.

The Medfra Fe skarn deposit consists of magnetite and minor chalcopyrite and sphalerite in epidote and garnet skarn. The deposit contains an estimated 11,600 m^3 grading 85 percent Fe_2O_3, along with traces of Cu and Au (Patton and others, 1984).

Chromite deposits, Tozitna and Innoko areas

Several podiform chromite deposits occur in a series of intensely deformed ultramafic and associated rocks that form small, discontinuous thrust slices in the Tozitna and Innoko areas. The larger deposits are at Mount Hurst and Kaiyuh Hills. The Mount Hurst deposit consists of chrome spinel bands in dunite layers in wehrlite (Chapman and others, 1982; Roberts, 1984). The largest chromite band strikes north-south, and varies from 10 to 800 cm thickness over a strike length of 10 m. Grab samples contain 22 to 62.2 percent Cr_2O_3. The Kaiyuh Hills deposit consists of bands of chromite, 1 cm to 1 m thick, and disseminated chromite in fresh and serpentinized dunite (Loney and Himmelberg, 1984; Foley and others, 1985); it contains an estimated 15,400 to 33,500 tonnes Cr_2O_3. The ultramafic rocks at the two deposits consist of dunite, wehrlite, harzburgite, lherzolite, and clinopyroxenite. Structurally interlayered with the ultramafic rocks are chert, basalt, and carbonate. The ultramafic and associated rocks are interpreted as part of a complexly deformed and dismembered ophiolite of Jurassic(?) age (Loney and Himmelberg, 1984).

Vein and porphyry deposits, west-central Yukon-Koyukuk basin

The major deposits in the west-central Yukon-Koyukuk basin are a porphyry Mo deposit at McLeod, a combined polymetallic vein and porphyry Cu deposit at Illinois Creek, and polymetallic vein deposits at Perseverance (described below), Beaver Creek, and Quartz Creek. The porphyry Mo deposit at McLeod consists of aggregates of platy molybdenite in quartz veinlets in the altered core of a Late Cretaceous or early Tertiary granite porphyry stock that intrudes mid-Cretaceous graywacke (H. Noyes, written communication, 1984). At Illinois Creek, the combined polymetallic vein and porphyry Cu deposit consists of galena-sphalerite veins along the contact of altered Cretaceous granite porphyry with schist, and of propylitically altered Cretaceous granitic plutons containing chalcopyrite, galena, and precious metals (W. W. Patton, written communication, 1985). The granitic rocks intrude early Paleozoic or older greenschist, quartzite, and orthogneiss that are part of the Ruby terrane (Plate 3, this volume). The Quartz Creek deposit consists of disseminated sulfides in a zone as much as 8 km wide and more than 29 km long in altered andesite and granite of Jurassic or Cretaceous age.

Perseverance and Beaver Creek Pb-Ag-Zn deposits. *By B. K. Jones.* The Perseverance Pb-Ag-Sb deposit consists of veins of coarse-grained galena, tetrahedrite, and traces of jamesonite, in a gangue of dolomite and minor quartz, that crosscut bedding and schistosity of enclosing early Paleozoic or older chlorite mica schists of the Ruby terrane. Oxidized zones in the veins contain cerussite, azurite, malachite, and stibconite(?). The deposit has produced about 231 tonnes grading 73 percent Pb and 124 g/t Ag.

The Beaver Creek Ag-Pb-Zn deposit consists of a highly oxidized zone of limonite, goethite, argentiferous galena, quartz, and sphalerite. The zone contains local surface occurrences of massive galena and limonite-cerussite gossan. The zone is about 300 m long and varies from 2.5 to 5 m thick; it is parallel to or crosscuts layering in enclosing schists of the Ruby terrane. The Beaver Creek deposit contains an estimated 13,600 tonnes grading 103 g/t Ag, 0.8 percent Zn, and 0.5 percent Cu, and an additional 19,100 tonnes tons grading 26.1 g/t Ag, 4.2 percent Pb, 0.16 percent Zn, and 0.2 percent Cu.

U deposits, northern Yukon-Koyukuk basin

A suite of felsic plutonic U deposits occurs in Purcell district of the northwestern Yukon River Region at Wheeler Creek, Clear Creek, and Zanes Hills (Miller and Elliott, 1969; Miller, 1976; Jones, 1977). These deposits are part of a belt of U and Th deposits associated with a mid- and Late Cretaceous alkaline intrusive belt extending about 300 km from Hughes on the Koyukuk River to the Darby Mountains on the Seward Peninsula. The plutonic rocks vary from calc-alkaline to K-rich alkalic granitic rocks. They intrude a sequence of andesitic flows, tuffs, breccia, agglomerate, conglomerate, tuffaceous graywacke, and mudstone with local intercalations of Early Cretaceous limestone that are part of the Koyukuk terrane (Plate 3, this volume).

The Wheeler Creek deposit consists of uranothorite and gummite in small, smoky-quartz-rich veinlets, and in altered parts of a Late Cretaceous alaskite. The deposit is about 500 m long and 50 m wide. Grab samples contain as much as 0.0125 percent U. The Clear Creek deposit consists of uraniferous nepheline syenite and bostonite dikes that cut Early Cretaceous andesite. The dikes occur within the contact aureole of the Late Cretaceous monzonite to granodiorite of the Zane Hills and contain as much as 0.04 percent U, and 0.055 percent Th. The Zane Hills deposit consists of uranothorite, betafite, uraninite, thorite, and allanite in veinlets in a foliated monzonite border phase, locally grading to syenite, of the Late Cretaceous monzonite to granodiorite pluton of Zane Hills. Selected samples contain as much as 0.027 percent Th.

Asbestos and chromite deposits, northern and eastern flanks of Yukon-Koyukuk Basin

By J. Y. Foley

Several serpentinite-hosted asbestos deposits and a podiform chromite deposit occur along the eastern flanks of the Yukon-Koyukuk basin at Asbestos Mountain, Caribou Mountain, Lower Kanuti River, and Holonada. The Asbestos Mountain deposit consists of serpentinite with veins of cross- and slip-fiber tremolite and chrysotile in a klippe of ultramafic rocks. The larger podiform chromite deposits at Caribou Mountain and Holonada consist of bands of massive chromite and chromohercynite in layers as much as 3 m thick and 130 m long. The Caribou Mountain deposit contains an estimated 2,270 tonnes, and the Holonada deposit contains as much as 24,900 tonnes (Foley and McDermott, 1983; Foley and others, 1984). One layer at Caribou Mountain is at least 25 m long and contains 7.5 percnt Cr_2O_3. The average grade at Holonada is 20 percent Cr_2O_3. The asbestos and chromite deposits are in complexly faulted dunite layers associated with harzburgite, pillow basalt, locally intensely serpentinized gabbro, chert, and minor limestone of Permian through Jurassic age. The assemblage probably is a dismembered ophiolite (Zimmerman and Soustek, 1979; Nelson and Nelson, 1982; Loney and Himmelberg, 1985a, b) that is part of the Angayucham terrane (Plate 3, this volume).

Vein and skarn deposits, eastern Yukon-Koyukuk basin

A major polymetallic or epithermal vein deposit occurs at Upper Kanuti River, and a W skarn deposit occurs at Bonanza Creek. The Upper Kanuti River Pb-Ag deposit consists of disseminated pyrite, galena, and sphalerite in an extensive gossan zone in silicified rhyolite that may be a dike intruding a Cretaceous pluton (Patton and Miller, 1970). The Bonanza Creek W-Ag deposit consists of scheelite, chalcopyrite, and pyrrhotite in skarn adjacent to a Late Cretaceous granite pluton intruding early Paleozoic or older pelitic schist, quartzite, and marble (Clautice, 1980) of the Ruby terrane.

LODE DEPOSITS, EAST-CENTRAL ALASKA

East-central Alaska contains a variety of lode deposits (Table 1 on Plate 11, this volume). The Manley and Livengood area contains polymetallic vein, Sb-Au vein, Mn-Ag vein, Hg vein, felsic plutonic U, Sn greisen, and Sn vein deposits. The northern and central parts of the Yukon-Tanana Upland contains suites of Sb-Au vein, Au-quartz vein, polymetallic vein, W skarn, and porphyry Cu-Mo deposits, a Au-As vein deposit, and a Sn greisen deposit. The Manley and Livengood area also contains a serpentinite-hosted asbestos deposit and a minor Pt deposit in thrust sheets of ultramafic and associated rocks. The northern Alaska Range contains an extensive district of polymetallic and Sb-Au vein and Kuroko massive-sulfide deposits.

Vein and U deposits, Manley and Livengood region, northwest Yukon-Tanana Upland

The major lode deposits in the Manley area are a polymetallic vein deposit at Hot Springs Dome, a Mn-Ag vein at Avnet, and a Sb-Au vein deposit at Sawtooth Mountain. The Hot Springs Dome Au-Ag-Pb deposit consists of veins containing galena, limonite, siderite, copper, chalcopyrite, and Fe sulfides. The veins are in shear zones in Jurassic and Cretaceous flysch intruded by early Tertiary granite. The Avnet Mn-Ag deposit consists of masses of psilomelane in thin quartz veins in lower and middle Paleozoic chert, quartzite, limestone, dolomite, and greenstone. The Sawtooth Mountain Sb-Au deposit consists of a vertical cylindrical body about 3 m in diameter of massive stibnite in Jurassic and Cretaceous flysch near Cretaceous granite. The Jurassic and Cretaceous flysch consists of quartzite, graywacke, and argillite, and volcanic conglomerate that are complexly deformed and form part of the Manley terrane (Plate 3, this volume). The Paleozoic rocks at the Avnet deposit are part of the Baldry terrane.

Major lode deposits in the Livengood area include Sb-Au and Au quartz vein deposits at Gertrude Creek, Griffen, and Ruth Creek, and the Hudson Cinnabar Hg vein deposit. The Gertrude Creek, Griffen, and Ruth Creek Au deposits consist of quartz stringers as much as 8 cm wide containing pyrite, stibnite, and base metal sulfides in altered Cretaceous monzonite and in silica-carbonate rock. The Hudson Cinnabar deposit consists of cinnabar in disseminations and quartz veins in altered early Tertiary granite dikes and plutons. The wall rocks in the Livengood area consist of an intensely folded, weakly metamorphosed sequence of Ordovician Livengood Dome Chert and overlying Silurian and Devonian dolomite, chert, volcanic rocks, serpentinite, shale, sandstone, and minor limestone; these rocks form part of the Livengood terrane.

Northeast of Livengood, a felsic plutonic U deposit occurs at Roy Creek, and a Sn greisen and a Sn vein deposit occur at Lime Peak. The plutonic rocks hosting the deposits in both areas intrude a sequence of weakly deformed, quartz-rich sandstone, grit, shale, and slate, locally with probable Early Cambrian fossils. The sequence is informally named the Wickersham grit unit, and forms part of the Wickersham terrane.

Roy Creek U-Th deposit. *By P. J. Burton.* The Roy Creek (formerly Mount Prindle) deposit contains a varied suite of U, Th, and REE minerals, phosphates, carbonates, and oxides, including allanite, bastnaesite, britholite, monazite, neodymium phosphate(?), thorianite, thorite, uranite, and xenotime. These minerals occur in steeply dipping quartz fissure veins that locally pinch and swell. Hematitic alteration and leaching of magnetite from the wall rocks occurred during vein formation. The veins are hosted in a small alkaline intrusive complex consisting of Cretaceous syenite and granite. The major rock units are porphyritic biotite aegirine-augite syenite, aegirineaugite syenite, porphyritic biotite augite syenite, and alkali granite, along with minor magnetite-biotite-aegirine-augite lamprophyre dikes. The alkaline complex intrudes the Cambrian Wickersham grit unit.

Lime Peak Sn deposit. *By P. J. Burton and W. D. Menzie.* The Lime Peak deposit consists of veinlets, breccia zones, and pods of black tourmaline and of chlorite, sericite, green tourmaline, and quartz alteration in an early Tertiary hypabyssal, peraluminous, biotite granite pluton (Menzie and others, 1983; Burton and others, 1985). The pluton is cut by numerous felsic and minor intermediate dikes. The veinlets, breccia zones, and tourmaline pods suggest deuteric alteration, whereas the chlorite, sericite, and quartz are probably the result of hydrothermal alteration. Anomalously high geochemical values of Sn and associated Ag, B, Bi, Mo, Pb, and Zn occur in rock samples from and around the pluton. Rare fluorite, topaz, pyrite, chalcopyrite, and molybdenite occur in the altered rocks. Grab samples contain as much as 0.16 percent Sn, 0.1 percent Zn, 0.5 percent Cu, 0.2 percent Pb, and 14 g/t Ag. Cassiterite occurs in stream sediments in the surrounding area. The deposit is classified either as a Sn greisen or Sn vein deposit.

The granitic pluton varies from older, coarse-grained equigranular biotite granite to younger porphyritic biotite granite having a fine-grained groundmass. Local miarolitic cavities are present, and the pluton has a K-Ar age of 56.7 Ma. Epizonal emplacement of the pluton is implied by a wide contact metamorphic aureole, abundant miarolitic cavities, porphyritic textures, and abundant veins. The pluton intrudes the lower part of the Cambrian Wickersham grit unit.

Vein, skarn, Sn greisen, and porphyry deposits northern and eastern Yukon-Tanana Upland

The major lode mineral deposits in the northern and eastern Yukon-Tanana Upland are: Sb-Au vein deposits at Dempsey Pup and Scrafford; Au-quartz vein deposits at Table Mountain, Democrat, and Purdy; polymetallic vein deposits at Cleary Summit, Ester Dome, Blue Lead, Tibbs Creek, and Gray Lead; W skarn deposits at Salcha River and Gilmore Dome; porphyry Cu-Mo deposits at Mosquito, Asarco, Bluff, and Taurus (described below); a Au-As vein deposit at Miller House; a Sn greisen deposit at Ketchem Dome; a serpentinite-hosted asbestos at Slate Creek (described below); and a podiform chromite deposit at Eagle C3. The deposits at Scrafford, Cleary Summit, Gilmore Dome, Ester Dome, and Democrat are in the Fairbanks district, one of the major mining areas in Alaska.

The northern and eastern parts of the Yukon-Tanana Upland are underlain by multiply metamorphosed and penetratively deformed Devonian and older quartz-mica schist and gneiss, quartzite, quartz-rich grit, gneissic plutonic rocks, metavolcanic rocks, marble, and calc-schist that are intruded by Cretaceous and early Tertiary granitic plutons (Foster and others, this volume). The marble locally contains Devonian fossils, and parts of the gneissic plutonic rocks are dated as Devonian and Mississippian by U-Pb zircon isotopic studies. These metamorphic rocks are part of the Yukon-Tanana terrane (Plate 3, this volume).

The polymetallic and Sb-Au vein deposits are probably related either to greenschist-facies regional metamorphism and/or intrusion of Cretaceous and early Tertiary granitic plutons. The Au quartz vein deposits probably are related to a widespread regional metamorphic and deformational event that culminated with intrusion of Cretaceous granitic plutons, dikes, and sills. The porphyry and skarn deposits are related to an extensive suite of early Tertiary granitic plutons. The porphyry deposits occur at the western end of a broad belt of porphyry deposits extending from the Dawson Range in western Canada into eastern Alaska (Hollister, 1978).

Fairbanks district

By T. E. Smith and P. A. Metz

The Fairbanks district is one of the major mining areas in Alaska, with numerous lode and placer mines in an area of approximately 2,000 km^2. Since the discovery of gold placers in 1902, the district has produced 236 million g of placer gold and 7.8 million g of lode gold, nearly 25 percent of Alaska's production. The deposits were first described by Prindle and Katz (1913) and subsequently by Smith (1913), Chapin (1914, 1919), Mertie (1918), and Hill (1933). Hill (1933) first noted the close spatial relation of placer and lode deposits.

The Fairbanks district is underlain by three thrust-bounded metamorphosed stratigraphic sequences that are intruded by various granitic plutons, dikes, and sills (Fig. 5) (Smith and others, 1981; Forbes and Weber, 1982). The structurally lowest Fairbanks schist unit of late Precambrian or early Paleozoic age (Bundtzen, 1982), consists dominantly of quartzite and muscovite-quartz schist, regionally metamorphosed to greenschist facies. Within the Fairbanks schist unit is a succession known as the Cleary sequence, which is 120 to 240 m thick and consists of felsic schist, laminated white micaceous quartzite, chlorite actinolite greenstone, graphitic schist, minor metavolcanic rocks, calc-schist, and marble. The Cleary sequence hosts most of the lode deposits in the district and is present upstream from the major placer districts (Smith and others, 1981); it probably is of volcanogenic origin. Structurally above the Fairbanks schist unit is the Chena River sequence, consisting of banded amphibolite, tremolite marble, garnet-muscovite schist, biotite schist, calc-schist, and metachert, regionally metamorphosed to lower amphibolite facies (Forbes and Weber, 1982). In the northern part of the district, the Fairbanks schist is structurally overlain by the Chatanika sequence, consisting of garnet-pyroxene eclogite, garnet amphibolite, quartzite, marble, and pelitic schist.

Granitic rocks in the district include a hornblende granodiorite pluton near Pedro Dome, a younger, multiphase porphyritic quartz monzonite to granodiorite pluton at Gilmore Dome, and numerous small plutons or hypabyssal bodies and dikes of felsic or intermediate composition. Field relations indicate mesozonal emplacement; K-Ar and Rb-Sr isotopic ages range from 91 to 93 Ma (Late Cretaceous). Chemical, mineralogical, and isotopic criteria indicate that the porphyritic quartz monzonite may be an S-type granite, whereas the hornblende granodiorite displays features both of S- and I-type granite.

Two major deformations occurred in the district. The earlier resulted in synmetamorphic, overturned to recumbent, subisoclinal, northeast-verging folds with wavelengths of about 300 m. The later deformation refolded these structures into broad, northeast-trending open folds that control the distribution of major rock types (Section A-A′ on Fig. 5). Local minor structures, including shear and crush zones typically cluster in northeast- and north-northeast–trending sets, both of which have a close spatial and genetic relation to the discordant Au, Sb, and As lode deposits in the district. Northeast-trending faults typically show reverse displacement and southerly dips.

The district contains 188 lode gold deposits, of which 65 have produced an estimated 8.7 million g Au, with average grades ranging from 9.6 to 79 g/t (Thomas, 1973). The deposits are concentrated in the Cleary Summit, Ester Dome, Scrafford, and Gilmore Dome areas (Fig. 5). The lode deposits consist of five groups (Metz and Halls, 1981): (1) stratabound volcanogenic(?) deposits containing intergrown As, Zn, Sb, Pb, and Cu sulfides, gold, and scheelite in conformable lenses parallel to layering in metavolcanic rocks of the Cleary sequence; (2) Pb-sulfosalt quartz-sulfide veins containing argentiferous galena, sphalerite, chalcopyrite, stibnite, arsenopyrite, and gold in Cretaceous granitic plutons; (3) skarn deposits, mostly as replacements of calcareous layers of the Cleary sequence, containing scheelite in prograde hedenbergite pyroxene and subcalcic garnet skarn and in retrograde hornblende-quartz-calcite metasomatic mineral assemblages in calc-schist and marble adjacent to Cretaceous granitic plutons at Gilmore and Pedro Domes (Allegro, 1984); (4) polymetallic gold-sulfide quartz vein deposits within and crosscutting the Cleary sequence; and (5) stibnite gash veins and fracture fillings in axial plane shears in metavolcanic rocks of the Cleary sequence.

Field observations and chemical data lead to the following model of ore genesis: (1) Precambrian or early Paleozoic bimodal submarine volcanism in a rift environment, along with formation of volcaniclastic rocks and exhalites enriched in Au, Sb, As, and W; (2) regional polymetamorphism and deformation resulting in mobilization of metals into veins; (3) emplacement of post-tectonic Cretaceous granitic plutons, possibly during anatexis, with concurrent skarn formation and continued mobilization of metals in veins at favorable sites within the Cleary sequence (Smith and others, 1981).

Taurus Cu-Mo deposit

By E. R. Chipp

The Taurus porphyry Cu-Mo deposit is the best known of three porphyry deposits that are present in the Yukon-Tanana Upland about 11 to 22 km west of the Canadian border; the other deposits are at East Taurus and Bluff. The Taurus deposit consists of sparsely disseminated Cu and Mo sulfides and pervasive pyrite in altered parts of early Tertiary granite porphyry, and of disseminated pyrite in associated volcanic rocks. The sulfides occur in three settings: (1) in veinlets of quartz and sericite containing chalcopyrite, molybdenite, and pyrite; in veinlets of quartz and sericite with accessory biotite and orthoclase; in veinlets of quartz, magnetite and anhydrite; and in veinlets of clay, fluorite, and zeolite; (2) as sparse concentrations of Cu and Mo sulfides associated with potassic alteration in the magnetite-rich core of the

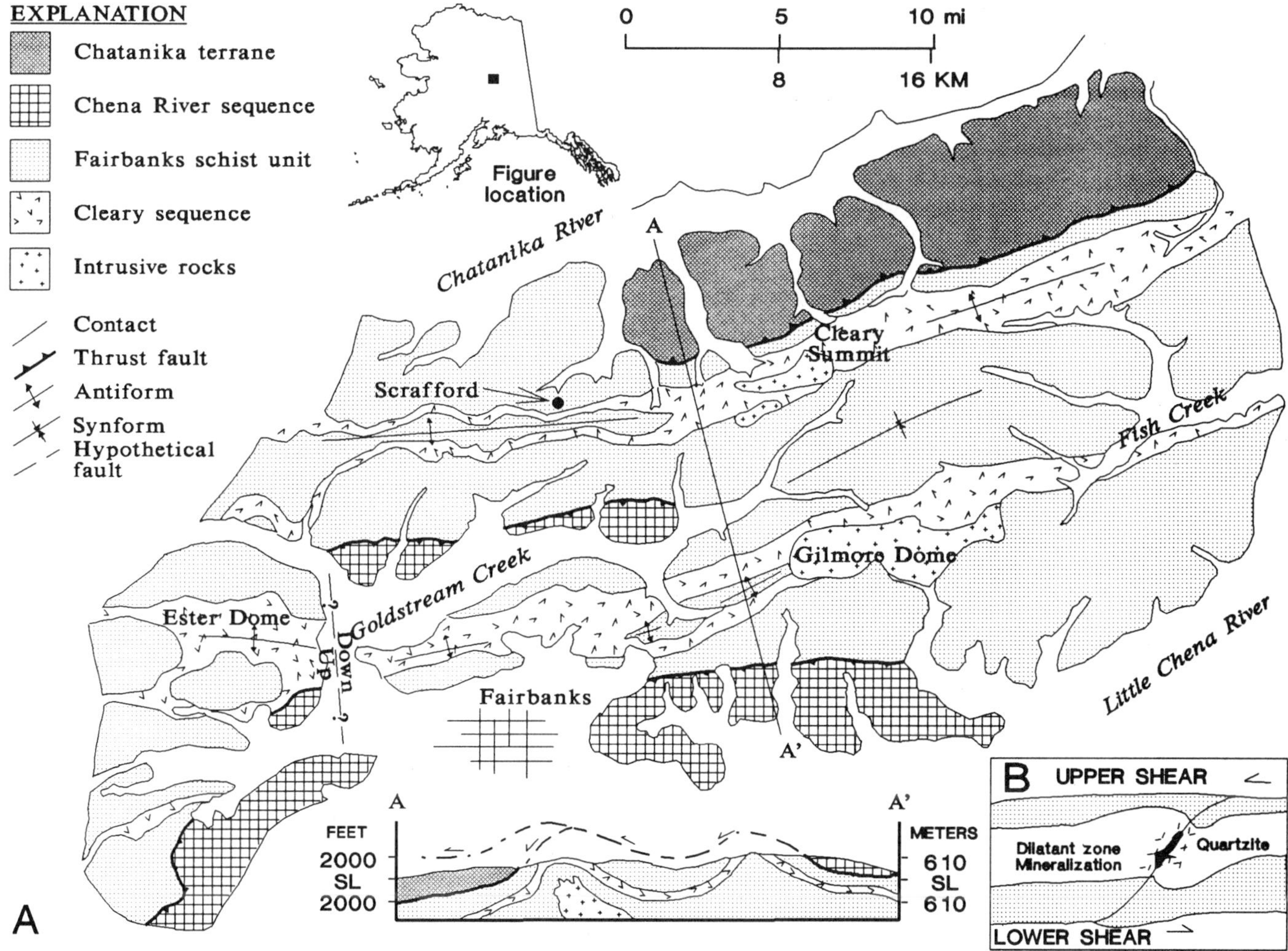

Figure 5. Fairbanks district, northern Yukon-Tanana Upland, east-central Alaska. (A) Generalized bedrock geologic map and cross section showing distribution of the Cleary sequence along anticlinal ridge crests and locations of four principal areas of mineral deposits. (B) Model of typical sulfide-bearing shear with crushed zone where shear intersects a competent quartzite horizon. The crush zones typically host gold-bearing quartz veins.

granite porphyry; and (3) as higher concentrations of Cu and Mo sulfides associated with phyllic alteration in the periphery of the granite porphyry. Propylitic alteration is minor in the core but is extensive in the periphery of the granite porphyry. Intense sericitic alteration occurs in gneiss along the southern and eastern contacts of the granite porphyry; these areas have very low concentrations of sulfides, which formed mostly by supergene processes.

Late hypogene alteration minerals such as montmorillonite fluorite, calcite, and zeolite occur locally in northeast-trending fractures, and may be contemporaneous with regional tourmaline-quartz alteration. Supergene alteration, resulting in abundant clay minerals and limonite, has occurred in all parts of the deposit to at least 30 m below the surface. Cu enrichment due to surficial oxidation and redeposition near the former water table is detectable but not significant. Chalcocite is the principal sulfide replacing chalcopyrite and coating pyrite for 30 m below the leached cap. Approximately 450 million tonnes grading 0.5 percent Cu nad 0.07 percent MoS_2 are present. One 120-m-long drill hole grades 0.104 percent MoS_2, indicating molybdenum may be more important at depth.

The granitic plutons at Taurus and at nearby porphyry deposits, and the associated felsic tuffs and breccias, are spatially related to the east-west–trending McCord Creek fault. Fault intersections and flexures apparently controlled emplacement of porphyries and intrusive breccias. The porphyries intrude multiply deformed and metamorphosed Devonian or older sedimentary and volcanic rocks and Devonian and Mississippian gneissic granitic rocks of the Yukon-Tanana Upland (Foster and others, this volume). Biotite from the granite porphyry at Taurus yields a K-Ar age of 57.0 Ma. Small stocks of Mesozoic(?) granodiorite also occur in the area. The porphyry deposits probably formed during hydrothermal alteration of magnetite-rich granite porphyry, probably within a back-arc environment.

Asbestos and Pt deposits, eastern Yukon-Tanana Upland

Asbestos and Pt deposits occur in the eastern Yukon-Tanana Upland, where they consist of a large serpentinite-hosted asbestos deposit at Fortymile (described below) and a Pt deposit in ultramafic rocks at Eagle C-3. The Eagle C-3 deposit contains relatively high PGE values in a small body of biotite pyroxenite. Both deposits are in discontinuous remnants of thrust sheets of ultramafic and associated rocks that are structurally above the Yukon-Tanana terrane (Foster and others, this volume). The thrust sheets consist of serpentinized harzburgite and associated ultramafic rocks, gabbro, pillow basalt, and local Permian chert, all of which may be part of a dismembered ophiolite, and which are part of the Seventymile terrane.

Fortymile Asbestos deposit. *By R. K. Rogers.* The Fortymile area contains numerous fairly small bodies of ultramafic rocks near the Tintina fault, eleven of which contain concentrations of chrysotile asbestos. The ultramafic rocks adjacent to the fault consist of partially serpentinized harzburgite and dunite, whereas those as much as 64 km south of the fault are completely serpentinized. The deposit in the Slate Creek area consists of antigorite with minor clinochrysotile, chrysotile, magnetite, brucite, and magnesite in completely serpentinized harzburgite and dunite. The serpentine probably replaced magnesium-rich olivine, minor orthopyroxene, and rare clinopyroxene. The chrysotile asbestos occurs in fracture zones near centers of thicker serpentinite bodies, primarily as cross-fiber chrysotile in randomly oriented veins about 0.5 to 1 cm thick. The veins contain alternating zones of chrysotile and magnetite and commonly exhibit magnetite selvages. Some chrysotile is altered to antigorite. The chrysotile veins appear to be the result of fracture filling from fluids migrating along fractures, or possibly from relatively immobile fluids locally dissolving and reprecipitating serpentine. Three of the ultramafic bodies in the Slate Creek area are estimated to contain a total of 58 million tonnes averaging 6.4 percent chrysotile fiber.

The harzburgite and dunite hosting the deposits form tabular tectonic lenses that range from 60 to 150 m thick and as much as 800 m long. The serpentinite is generally massive, whereas contacts of the ultramafic bodies commonly are zones of intense shearing. The serpentinite commonly is altered near fault zones and ultramafic contacts. Calcite, dolomite, magnesite, cryptocrystalline quartz, and limonite-goethite replace serpentine; these alteration minerals appear to have formed during reaction of serpentinite with CO_2-rich meteoric water. The Fortymile asbestos deposit probably is a low-temperature replacement deposit formed during alteration of the harzburgite.

Vein and massive sulfide deposits, northern Alaska Range region

The major lode deposits in the northern Alaska Range include: (1) an extensive district of polymetallic and Sb-Au vein deposits in the Kantishna District (described below) at Slate Creek, Eagles Den, Quigley Ridge, Banjo, Spruce Creek, and Stampede; and (2) an extensive suite of massive sulfide deposits at Liberty Bell, Sheep Creek, Anderson Mountain, WTF, Red Mountain, Miyaoka, Hayes Glacier, McGinnis Glacier, and in the Delta district. The massive sulfide deposits extend for 350 km along strike on the northern flank of the Alaska Range, and constitute one of the longer belts of massive sulfide deposits in Alaska. Deposits in this belt were discovered mainly in the period 1975 to 1985, and additional discoveries are likely.

Both the vein and massive sulfide deposits occur in a Devonian or older sequence of polymetamorphosed and polydeformed submarine metavolcanic rocks, pelitic schists, calcschist, and marble (Aleinikoff and Nokleberg, 1985; Nokleberg and Aleinikoff, 1985). This sequence is interpreted as the upper structural and stratigraphic level of the Yukon-Tanana terrane (Nokleberg and Aleinikoff, 1985). Metamorphic grade ranges from amphibolite facies at depth to greenschist facies at higher levels (Nokleberg and Aleinikoff, 1985; Nokleberg and others, 1986). Locally abundant Cretaceous(?) gabbro to diorite dikes and sills crosscut schistosity and foliation in the sequence. Structurally overlying these older rocks are the singly metamorphosed and deformed metasedimentary and metavolcanic rocks of the Precambrian or Paleozoic Keevy Peak Formation and Mississippian(?) Totlanika Schist (Wahrhafitg, 1968; Gilbert and Bundtzen, 1979). The massive sulfide deposits are classified as Kuroko massive sulfide deposits that formed during Devonian submarine volcanism. The polymetallic vein and Sb-Au deposits in northern Alaska Range probably formed during Cretaceous regional metamorphism and/or during intrusion of somewhat younger Late Cretaceous or early Tertiary dike swarms.

Kantishna district. *By T. K. Bundtzen.* The Kantishna district contains an extensive suite of polymetallic and Sb-Au vein deposits at Slate Creek, Quigley Ridge, Banjo, Spruce Creek, Eagle Den, and Stampede. Most of the deposits are in the middle Paleozoic or older metamorphosed volcanic and sedimentary rocks of the Spruce Creek sequence (Bundtzen, 1981), which is correlated by some workers with the Cleary sequence in the Fairbanks district.

Most of the vein deposits occur as crosscutting quartz-carbonate-sulfide veins and are confined to a 60-km long northeast-trending fault zone that extends from Slate Creek to Stampede. Mineralization occurred before, during, and after fault-zone movement, as illustrated by both crushed and undeformed ore shoots in the same vein system. The veins range from 30 to 500 m long and from a few cm to 9 m wide; they occur in various lithologies but are best developed in brittle rocks such as quartzite or metaigneous rocks. The vein deposits consist of Ag-Au-Sb-Pb-Zn quartz-carbonate-sulfide veins subdivided into the following three types: (1) polymetallic vein deposits composed of quartz, arsenopyrite, pyrite, gold, and scheelite; (2) polymetallic vein deposits composed of galena, sphalerite, tetrahedrite, pyrite, and chalcopyrite, often with silver, lead, and antimony sulfosalts; and (3) Sb-Au vein deposits composed of stibnite and quartz, largely free of other sulfides.

The Quigley Ridge deposit consists of type 2 veins and contains an estimated 381,000 tonnes grading 1,337 g/t Ag, 4.8 g/t Au, 6.4 percent Pb, and 2.3 percent Zn. The Banjo deposit consists of type 1 veins and contains an estimated 159,000 tonnes grading 13.4 g/t Au, 123 g/t Ag, and 1.5 percent combined Pb, Zn, and Sb. The Spruce Creek deposit also consists of type 1 veins and contains an estimated 77,000 tonnes grading 2.4 g/t Au, 276 g/t Ag, and 2.5 percent combined Pb, Zn, and Sb. The deposits at Slate Creek, Last Chance, and Stampede consist of type 3 veins and together contain an estimated 507,000 tonnes grading 11.9 percent Sb with minor Ag and Zn.

Textures indicate that arsenopyrite and pyrite formed early; sulfides such as sphalerite, chalcopyrite, galena, silver sulfides, and tetrahedrite formed next; and Sb minerals, such as boulangerite, jamesonite, and stibnite, formed late (Bundtzen, 1981). The highest Ag and Au values are in type 2 veins that contain tetrahedrite, polybasite, pyrargyrite, and pearceite. The Kantishna vein deposits were probably formed during hydraulic fracturing of the metalliferous host rocks of the Spruce Creek sequence. Metals were leached from the volcanic and sedimentary rocks and were transported by hydrothermal fluids into structural conduits. Heat probably was provided either by mid-Cretaceous intrusion or regional greeschist-facies metamorphism, or by emplacement of younger Late Cretaceous or early Tertiary dike swarms.

Liberty Bell and Sheep Creek massive sulfide deposits, Bonnifield district. *By T. K. Bundtzen.* Two small massive sulfide deposits occur at Liberty Bell and Sheep Creek in the Bonnifield district. Both are hosted by the volcanic and sedimentary rocks of the Precambrian or Paleozoic Keevy Peak Formation and the Mississippian(?) Totatlanika Schist. Both deposits illustrate diversity in texture, geometry, and metal content.

The Liberty Bell massive sulfide deposit consists of arsenopyrite, pyrite, pyrrhotite, chalcopyrite, and bismuthinite lenses and disseminations that occur parallel to layering in tuffaceous schist. The deposit has a maximum thickness of 10 m and a strike length of 200 m; it contains an estimated 91,000 tonnes grading 10 percent As, 2 percent Cu, and 34 g/t Au. The deposit is adjacent to a metamorphosed Paleozoic(?) plug that is probably coeval with the tuff protolith of the schist.

About 20 km south of the Liberty Bell deposit on the opposite limb of a major syncline, the Sheep Creek massive sulfide deposit consists of sphalerite, galena, pyrite and stannite in massive lenses in phyllite and metacon glomerate. The lenses are in a zone about 330 m long and are localized in the nose of an overturned anticline near the contact between the volcanic-rock-rich Totalanika Schist and the sedimentary-rock-rich Keevy Peak Formation. Selected samples average 11 percent combined Zn and Pb, 10 g/t Ag; local zones as much as 1 m thick average 1 percent Sn.

Anderson Mountain massive sulfide deposit, Bonnifield district. *By C. J. Freeman.* The Anderson Mountain massive sulfide deposit is in the Bonnifield district; it consists of massive sulfide layers with pyrite, chalcopyrite, galena, sphalerite, enargite, and arsenopyrite in a gangue of quartz, sericite, chlorite, calcite, barite, and siderite. The massive sulfides contain potentially recoverable Cu, Pb, Zn, and Ag, and geochemically anomalous Hg, As, Sb, W, Sn, and Ba. Thicknesses of the sulfide layers range from 0.6 to 3 m; grades range from 0.5 to 19 percent Cu, a trace to 5 percent Pb, a trace to 22 percent Zn, and a trace to 170 g/t Ag. The sulfide-rich layers occur in metamorphosed marine tuffaceous rhyolite interbedded with sedimentary rocks. The deposit is slightly discordant to the host horizon and appears to rest on an irregular paleosurface in the stratigraphic footwall. Metal contents progress gradually with time from early relatively low grades, through peak amounts, tapering again to late low grades. The lower contacts of the sulfide layers are sharp, whereas the upper contacts are irregular and have variable grade and geometry; the upper contacts are locally dome-shaped. Lateral and vertical metal-ratio trends indicate deposition near, but not at, an exhalative center.

The tuffaceous rhyolite and interbedded sedimentary rocks hosting the deposit are part of the Moose Creek Member of the Mississippian(?) Totatlanika Schist. The wall rocks beneath the deposit are mainly black carbonaceous shale and calcareous shale. The wall rocks above the deposit are mainly massive to pyroclastic basalt interbedded with lenses of thin black shale. Low-grade greenschist-facies metamorphism has altered the host rocks but has not destroyed relict sedimentary textures. Relict crossbedding, scours, and rare shelly fossils indicate marine deposition. The host units strike northeast, dip moderately southeast, and are dissected by numerous small high-angle faults.

WTF and Red Mountain Zn-Pb-Cu-Ag deposits. *By D. R. Gaard.* The WTF and Red Mountain deposits consist of massive pyrite, sphalerite, galena, and chalcopyrite in a quartz-rich gangue. Local alterations consist of intense silicification and talc formation. The deposits contain an estimated 1.12 million tonnes grading 0.15 percent Cu, 3.5 percent Pb, 7.9 percent Zn, 270 g/t Ag, and 1.9 g/t Au. The deposits occur on the limbs of a large east-west–trending asymmetric syncline: the Red Mountain deposit is on the south limb, and the WTF deposit is on the north limb (Fig. 6).

The deposits are in the upper part of the Mystic Creek Member of the Mississippian(?) Totatlanika Schist, near the contact with the overlying Sheep Creek Member. The Red Mountain deposit occurs in several silica-exhalite horizons in a sequence of metamorphosed dacitic to rhyolitic crystal tuff, lapilli tuff, minor flows, and metasedimentary rocks. The southern exhalite horizon at Red Mountain consists of sphalerite and coarse pyrite in black chlorite schist; the northern exhalite horizon at Red Mountain, about 90 to 120 m thick, contains pyrite-rich massive sulfide with Cu, Zn, Pb, and Ag and several massive pyrite horizons. The richest massive sulfides at Red Mountain are fine grained and finely to coarsely laminated. Local deformation of the deposit is illustrated by sparse sulfide augen in the massive sulfide layers.

The WTF deposit consists of an areally extensive pyrite-rich massive sulfide layer 3 m thick or less containing Ag, Zn, Pb, and Au. The sulfides are fine grained and finely to coarsely laminated.

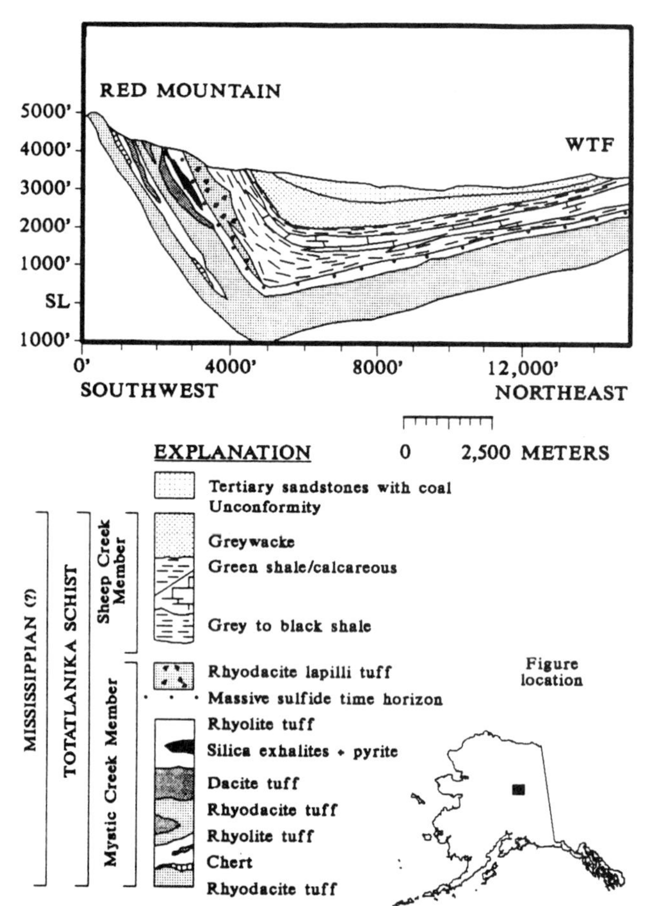

Figure 6. Generalized cross section through WTF and Red Mountain Zn-Pb-Cu-Ag deposits, northern Alaska Range, east-central Alaska. No vertical exaggeration. The deposits are localized near the top of the Mystic Creek Member of the Totatlanika Schist. The WTF and Red Mountain deposits appear to represent coeval proximal and distal volcanogenic sulfide deposits, respectively.

High Ag values are related to local tetrahedrite inclusions in galena. The quartz gangue content increases to the west, together with a decreasing Pb/Zn ratio and decrease in Ag. Synsedimentary pyrite gradually decreases from 10 to 20 percent to 2 to 5 percent in the black schist overlying the massive sulfide.

The thick massive sulfide deposit at Red Mountain and the thinner deposit at WTF probably are coeval proximal and distal deposits, respectively, and probably formed from a hydrothermal cell at a waning submarine volcanic center. Sulfidic exhalations precipitated the podiform massive sulfides at Red Mountain, whereas euxinic conditions in an extensive basin caused precipitation of the distal WTF deposit from fumarole-derived brines that originated near Red Mountain.

Miyaoka, Hayes Glacier, and McGinnis Glacier Cu-Pb-Ag-Au deposits. *By I. M. Lange and W. J. Nokleberg.* A suite of Kuroko massive sulfide deposits occurs at Miyaoka, Hayes Glacier, and McGinnis Glacier. The deposits consist of disseminated grains to massive lenses and pods of Fe-sulfides, chalcopyrite, and sphalerite in a gangue of quartz, chlorite, epidote, biotite, and actinolite (Nokleberg and Lange, 1985). The more extensive deposit at Miyaoka consists of sulfide pods and lenses as much as 1 m thick that occur discontinuously in a zone as much as 15 m thick and 2 km long. The sulfides are in an intensely deformed, interfoliated marine sequence of Devonian or older metavolcanic and metasedimentary rocks (Aleinikoff and Nokleberg, 1985). The host rocks show evidence of two periods of metamorphism: an older one of amphibolite facies and a younger one of greenschist facies. The deposits probably formed in a submarine island-arc setting.

Delta District. *By C. R. Nauman and S. R. Newkirk.* The Delta district deposits occur at the eastern end of the belt of massive sulfide deposits in eastern part of the northern Alaska range (Nauman and others, 1980). The district encompasses an area of approximately 1,000 km^2 and contains numerous stratiform, transported stratiform, and lesser replacement-type massive sulfide deposits in a thick sequence of metavolcanic and metasedimentary rocks metamorphosed at conditions of the greenschist facies.

The base metal deposits in the Delta district occur in four regional trends: the DD, DW, Trio, and PP-LZ trends. The central DD and DW trends contain massive sulfides in lenses and sheets, respectively. The DD South massive sulfide deposit, hosted in metavolcanic rocks, contains 1.5 million tonnes of brecciated and weakly banded pyrrhotite, pyrite, and Cu, Pb, and Zn sulfides with average grades of 1 percent Cu, 8 percent combined Pb and Zn, 62 g/t Ag, and 2g/t Au. The deposit forms a lens up to 545 m long, 212 m wide, and 15 m thick. The DD North deposit, another lens-like body, located about 1.6 km along strike from the DD south deposit, contains copper and gold grades similar to those in DD south, but is relatively depleted in Pb and Zn. To the northwest and southeast for several km along strike, thin beds of Pb-, Zn-, and Ag-rich massive sulfides crop out in pelitic and tuffaceous metasedimentary rock layers.

In contrast to the DD deposits, the tuff-hosted DW-LP deposit is composed of a laterally extensive, but structurally segmented, sheet-like massive pyrite bed containing more than 18 million tonnes of relatively low grade material. The bed is at least 606 m long, 3 to 15 m thick, and extends 1,500 m downdip. Typical grades range from 0.3 to 0.7 percent Cu, 1 to 3 percent Pb, 3 to 6 percent Zn, 34.3 to 109 g/t Ag, and 1 to 3.4 g/t Au. The deposits in the outlying Trio and PP-LZ trends are generally closely associated with calcareous and carbonaceous metasedimentary rocks that flank the central volcanic axis of the district, where the DD and DW trends occur. These deposits are relatively enriched in Pb, Zn, and Ag, and consist both of stratiform and replacement discontinuous massive sulfide deposits in a zone as much as 40 km long.

The sulfide deposits of the Delta district occur in the informally named Delta schist unit of the Yukon-Tanana terrane. This unit consists of a northwest-trending core of Devonian metavolcanic rocks (Aleinikoff and Nokleberg, 1985) flanked to the north by metamorphosed shallow marine sedimentary rocks, and to the south by metamorphosed deeper marine sedimentary rocks

(Nauman and others, 1980). The metavolcanic rocks, which host most of the major base and precious metal deposits, are derived from a volcanic suite varying in composition from spilite to keratophyre. Integral to this suite are numerous synvolcanic tholeiitic greenstone sills that are too thin to show on Figure 7. The greenstone sills probably are spatially related to the massive sulfide bodies and genetically related to the volcanic suite. Hydrothermal fluid flow probably radiated from the sills, producing overlapping stages of chloritization, silicification, sericitization, pyritization, and Pb-Ag-Au mineralization. The abundance and variety of sulfide deposits in the Delta district apparently resulted from evolving hydrothermal activity accompanying prolonged injection of syndepositional tholeiite sills into near-surface volcanic and sedimentary debris in a continental margin rift environment.

LODE DEPOSITS, ALEUTIAN ISLANDS AND ALASKA PENINSULA

The Aleutian Islands and Alaska Peninsula contain a limited variety of lode deposits (Table 1 on Plate 11, this volume). The Aleutian Islands and southwestern Alaska Peninsula contain an extensive suite of epithermal and polymetallic vein and porphyry Cu and Mo deposits. The northeastern Alaska Peninsula contains suites of Cu-Zn-Au and Fe skarn, polymetallic vein, and porphyry Cu deposits, and one epithermal vein deposit.

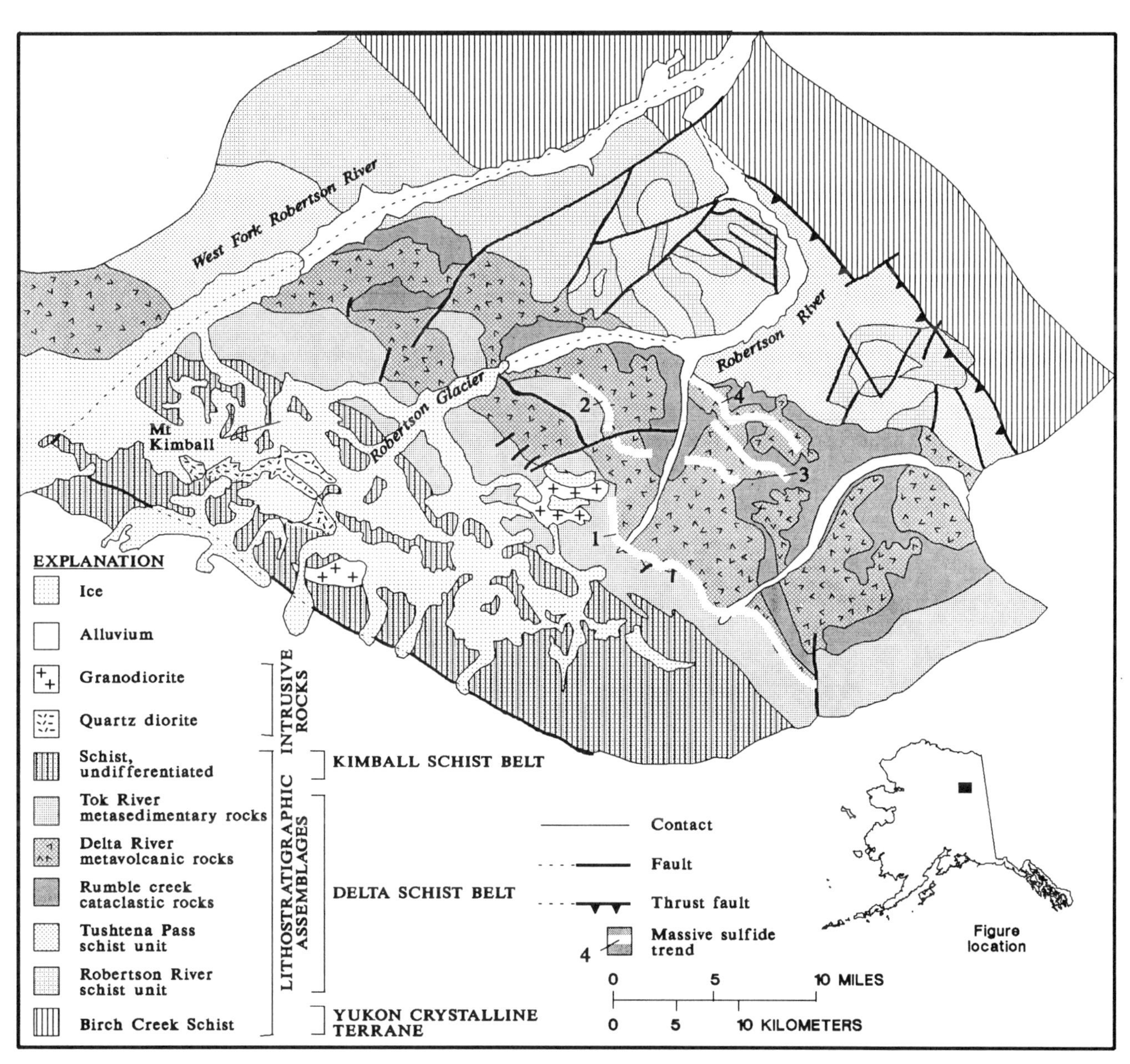

Figure 7. Generalized geologic map of the Pb-Zn-Cu-Ag deposits in the Delta district, northern Alaska Range, east-central Alaska.

Vein and porphyry deposits, Aleutian Islands, and southwestern Alaska Peninsula

Numerous epithermal and polymetallic vein and porphyry Cu and Cu-Mo deposits occur in the Aleutian Islands and southwestern Alaska Peninsula. They consist of: (1) epithermal vein deposits at Canoe Bay, Aquilla, Apollo-Sitka, Shumagin, San Diego Bay, Kuy, and Fog Lake; (2) polymetallic vein deposits at Sedanka, Warner Bay, Cathedral Creek, and Kilokak Creek; and (3) porphyry Cu and Mo deposits at Pyramid, Kawisgag, Mallard Duck Bay, Bee Creek, Rex, and Mike. The epithermal vein deposits generally consist of quartz-vein systems and silicified zones containing gold and sparse sulfides in Tertiary andesite and dacite and, to a lesser extent, in rhyodacite and rhyolite flows and breccias. The polymetallic vein deposits generally consist of base-metal sulfides in quartz veins and in disseminations in Tertiary diorite, granodiorite, and andesite and dacite stocks, in Tertiary andesite and dacite flows, and in volcanic sandstone intruded by stocks and dikes. The porphyry Cu and Mo deposits commonly consist of disseminated chalcopyrite and/or molybdenite and pyrite in altered zones, often along joints in stockworks in or near Tertiary or Quaternary andesite, dacite, and rhyodacite stocks.

The epithermal and polymetallic vein and porphyry deposits are along a linear belt more than 800 km long. This belt probably is related to hydrothermal and epithermal activity associated with the late-magmatic stages of Tertiary and Quaternary hypabyssal plutonic and associated volcanic centers. These centers are along part of the Aleutian arc, one of the classic igneous arcs along the rim of the Pacific Ocean. The arc is composed mainly of early Tertiary to Holocene andesite to dacite flows, tuff, and intrusive and extrusive breccia; hypabyssal diorite and quartz diorite and small silicic stocks, dikes, and sills; and volcanic graywacke, shale, and lahars (Burk, 1965; Wilson, 1985). Extensive late Tertiary and Quaternary volcanoes and associated volcanic and volcaniclastic rocks form major parts of the arc and dominate the landscape.

Underlying parts of the southwestern Alaska Peninsula, almost as far west as Cold Bay, is Mesozoic or Paleozoic bedrock, designated as part of the Peninsular terrane (Plate 3, this volume) (called Alaska Peninsula terrane by Wilson and others, 1985). This older bedrock is extensively intruded by the Jurassic, Cretaceous, and early Tertiary Alaska-Aleutian Range batholith (Reed and Lanphere, 1973). The Eocene and earliest Miocene volcanic and hypabyssal rocks deposited on, and intruded into this older bedrock, constitute part of the Meshik arc (Wilson, 1985). The major deposits in the southwestern Alaska Peninsula are the Apollo-Sitka, Shumagin, and Aquila Au-Ag epithermal vein deposits and the Pyramid porphyry Cu deposit.

Apollo-Sitka Au-Ag deposit. The Apollo-Sitka deposit consists of quartz-carbonate veins and silicified zones with gold, galena, sphalerite, chalcopyrite, tetrahedrite, native copper, and trace tellurides(?) (Brown, 1947; Alaska Mines and Geology, 1983; Eakins and others, 1985). Much of the gold is disseminated in sulfides. The veins and zones occur in a series of at least eight strongly-developed, northeast-striking fracture systems. The veins extend for several thousand meters along the surface and to a depth of at least 360 m; they range from a few centimeters to about 7 m thick. The higher-grade parts of the deposit occupy tensional flexures in the fracture systems. Abundant comb structure and euhedral crystal druses indicate that the veins formed at shallow depths. The fracture systems containing the veins are south of the Unga caldera system. The veins are hosted in propylitically altered shale, tuff, and intermediate to felsic volcanic rocks of probable late Tertiary age. From 1894 to 1906, the deposit produced about 3.33 million g of Au from 435,000 tonnes of ore grading 242 g/t Ag and 7.9 g/t Au. Most of the native gold ore was mined during this period. The gold in the remaining part of the deposit is associated with sphalerite and galena. Extensive exploration in the 1980's resulted in delineating an estimated additional 163,000 tonnes locally grading as much as 7.3 g/t Au, 240 g/t Ag, 15 percent Zn, and 1 percent Pb.

Pyramid Cu-Mo deposit. *By G. Anderson and T. K. Bundtzen.* The Pyramid deposit consists of disseminated molybdenite in Fe-stained dacite porphyry stock and dikes of late Tertiary age (Armstrong and others, 1976; Hollister, 1978; Wilcox and Cox, 1983). Zonal alteration is marked by a core of secondary biotite and about 3 to 10 percent magnetite grading outward to an envelope of quartz-sericite alteration. Fractures adjacent to the stock are filled with sericite. Local extensive oxidation and supergene enrichment by chalcocite and covellite occur in a blanket as much as 100 m thick. The stock intrudes Upper Cretaceous and lower Miocene fine-grained clastic rocks, which are contact metamorphosed adjacent to the stock. The deposit centers on a 3-km^2 outcrop area of the stock and contains an estimate of 113 million tonnes grading 0.4 percent Cu, 0.05 percent Mo, and a trace of Au.

The Pyramid deposit is the best known of a series of large-tonnage, low-grade porphyry Cu and Mo deposits in the Alaska Peninsula. The Pyramid Bee Creek, Rex, and Warner Bay porphyry Cu deposits occupy a transitional zone between the parts of the magmatic arc underlain by oceanic crust to the southwest and continental crust to the northeast. Some of the deposits are Mo-rich and contain anomalous concentrations of Bi, Sn, and W that may be characteristic of continental margin deposits (Wilson and Cox, 1983).

Vein, skarn, and porphyry deposits, northeastern Alaska Peninsula

A few skarn and porphyry deposits occur in the northeastern Alaska Peninsula. They consist of: (1) Cu-Au and Cu-Zn skarn deposits at Crevice Creek and Glacier Fork; (2) Fe skarn deposits at Kasna Creek and Magnetite Island; (3) an epithermal(?) vein deposit at the Johnson Prospect; and (4) polymetallic vein deposits at Kijik River and Bonanza Hills. The Cu-Au and Cu-Zn skarn deposits are in areas where Jurassic(?) quartz diorite and tonalite intrude the calcareous sedimentary rocks, and generally consist of epidote-garnet skarn in limestone or marble, with disseminations and layers of chalcopyrite, sphalerite, and pyrrhotite. The Fe skarn deposits are in dolomite or marble and generally consist of magnetite skarn with lesser garnet, amphibole, and local chalcopyrite. The Fe skarns occur in areas where Jurassic(?) quartz diorite and tonalite intrude calcareous sedimentary rocks. The polymetallic vein deposits generally consist of disseminated sulfides in altered Tertiary dacite porphyry or of base metal sulfides in quartz veins in metamorphosed dacite flows and sandstone near hypabyssal granite.

The foregoing deposits occur in marine sedimentary rocks of the Upper Triassic Kamishak Formation, Lower Triassic marble, and the volcanic and volcaniclastic rocks of the Early Jurassic Talkeetna Formation that are intruded by Jurassic, Cretaceous, and early Tertiary stocks and larger granitic plutons of the Alaska-Aleutian Range batholith.

Johnson Au-Zn deposit. *By C. I. Steefel.* The Johnson deposit consists of a quartz stockwork of quartz-sulfide veins with chalcopyrite, pyrite, sphalerite, galena, and gold. The veins also contain alteration minerals such as chlorite, sericite, anhydrite, and barite. Along a few meters of drill core, grades range from 20.6 to 41.2 g/t Au and 9.4 to 24.8 percent Zn, and average 2 percent Pb. The stockwork veins occur in a discordant pipelike body of silicified volcanic rocks. The deposit is hosted in volcaniclastic, pyroclastic, and volcanic rocks, part of the Portage Creek Agglomerate Member of the Lower Jurassic Talkeetna Formation. Near the deposits, the Talkeetna Formation is intruded by Late Jurassic quartz diorite and granite of the Alaska-Aleutian Range batholith. This deposit is an epithermal(?) vein deposit that probably formed during replacement and alteration associated with the late magmatic stage of nearby Jurassic plutons.

LODE DEPOSITS, SOUTHERN ALASKA

Southern Alaska contains a large variety of lode deposits (Table 1 on Plate 11, this volume). The southwestern Alaska Range contains a suite of Cu-Pb-Zn skarn, polymetallic vein, Sn greisen and vein, and porphyry Cu-Au and Mo vein deposits, a Besshi massive sulfide deposit, and a gabbroic Ni-Cu(?) deposit. The central and eastern Alaska Range and the Wrangell Mountains contain a suite of Cu-Ag and Fe skarn, polymetallic vein, and porphyry Cu and Cu-Mo deposits, and a suite of Cu-Ag quartz vein, basaltic Cu, and Besshi massive sulfide deposits. The Talkeetna Mountains contain a suite of Au quartz vein deposits, and a suite of podiform chromite deposits is present on Kodiak Island, the Kenai Peninsula, and the northern Chugach Mountains. The southern Chugach Mountains, southeast Kenai Peninsula, and Kodiak Island contain an extensive suite of Au quartz vein deposits, and the Prince William Sound district contains an extensive suite of Besshi and Cyprus massive sulfide deposits.

Skarn, vein, and massive sulfide deposits, southwestern Alaska Range

Major lode deposits in the southwestern and western Alaska Range consist of several Ag-Pb-Zn-Cu skarn deposits in the Farewell district at Bowser Creek, Rat Fork, Sheep Creek, and Tin Creek; a gabbroic Ni-Cu deposit at Chip Loy; and a Besshi massive sulfide deposit at Shellebarger Pass.

Farewell district. *By T. K. Bundtzen.* Major Cu-Ag-Pb-Zn skarn deposits occur at Bowser Creek, Rat Fork, Sheep Creek, and Tin Creek, and a Ni-Co deposit occurs at Chip Loy in the Farewell district, a 500 km^2 area of the southwestern Alaska Range. The Bowser Creek Ag-Pb skarn deposit consists of pyrrhotite, sphalerite, galena, and chalcopyrite in a hedenbergite-johannsenite endoskarn in marble adjacent to an early Tertiary felsic dike (Szumigala, 1985). Local fissures in marble adjacent to skarn contain Ag-rich galena and pyrrhotite. The deposit is estimated to contain as much as 272,000 tonnes with 10 percent Pb and Zn and 100 g/t Ag. The Tin Creek skarn deposit consists of pyroxene-rich skarn with abundant sphalerite and minor chalcopyrite, and of garnet skarn with chalcopyrite and minor sphalerite. Epidote and amphibole are locally abundant in the skarns (Szumigala, 1985). The pyroxene skarn is distal, and the garnet skarn is proximal, to a Tertiary granodiorite dike swarm. The Chip-Loy Ni-Co deposit, classified as a gabbroic Ni-Cu deposit, consists of massive to disseminated pyrrhotite, bravoite, and chalcopyrite along a steeply dipping contact between diabase and shale (Herreid, 1966; W. S. Roberts, oral communication, 1985).

Lode deposits in the Farewell district are generally in or near plutons 1 to 5 km^2 in outcrop area and in related igneous breccias, which are phases of early and/or middle Tertiary plutons of the Alaska-Aleutian Range batholith. The base-metal skarn deposits are typical of low-temperature, fracture-controlled zinc-lead skarns. The deposits occur either as skarns in lower and middle Paleozoic deep-water carbonate rocks or shale or as stockwork veinlet zones in fine-grained plutons. These stratified wall rocks are part of the Dillinger terrane (Plate 3, this volume).

Shellebarger Pass massive sulfide deposit. The Shellebarger Pass deposit consists of a very fine-grained mixture mainly of pyrite and marcasite with lesser sphalerite, chalcopyrite, galena, and pyrrhotite in a gangue of siderite, calcite, quartz, and dolomite (Reed and Eberlein, 1972). The sulfides occur in at least six individual bodies in carbonate-rich beds and as fracture fillings, mainly in chert and siltstone. The sulfides are hosted in Triassic or Jurassic chert, dolomite, siltstone, shale, volcanic graywacke, conglomerate, aquagene tuff, pillow basalt, agglomerate, and breccia. The highest chalcopyrite concentrations are in basal parts of the deposits. Minor sphalerite is present in zones in or near the hanging wall. The main sulfide bodies may be proximal to basaltic flow fronts. Extensive hydrothermal alteration occurs in the footwall but is rare to absent in the hanging wall. The deposit contains an estimated several hundred thousand tonnes of unknown grade. Selected samples contain as much as 5 percent Cu and average 2 percent Cu and 1 percent Zn. The host rocks are part of the Mystic terrane (Plate 3, this volume).

Polymetallic vein, sn greisen and vein, and porphyry deposits, western Alaska Range

The major lode deposits in the western Alaska Range are: Sn greisen(?) and vein deposits at Boulder Creek, Coal Creek (described below), and Ohio Creek; several polymetallic vein deposits at Partin Creek, Ready Cash, and Nim and Nimbus (described below); and porphyry Mo, Cu-Mo, and Cu-Au at Miss Molly, Treasure Creek, and Golden Zone.

The Sn greisen(?) and vein deposit at Boulder Creek (Purkeypile) consists of cassiterite and sulfides in fracture fillings in metasedimentary rocks near a Tertiary biotite granite (Conwell, 1977), and the Sn greisen and vein deposit at Ohio Creek consists of muscovite-tourmaline greisen and quartz-arsenopyrite veins in a Tertiary granite stock. The polymetallic vein deposits at Partin Creek and Ready Cash consist of Fe and base-metal sulfides in veinlets and disseminations in Triassic basalt and marble. The porphyry Mo deposit at Miss Molly consists of quartz veins with molybdenite, pyrite, and local fluorite in a Tertiary(?) granite stock intruding Jurassic and Cretaceous flsych (Fernette and Cleveland, 1984). The porphyry Cu-Mo deposits at Treasure Creek consist of disseminated molybdenite and other base-metal sulfides in a silicified and sheared Tertiary granite stock intruding Cretaceous flysch (Csejtey and Miller, 1978).

These magmatism-related deposits occur in the northeastern part of the Aleutian-Alaska Range batholith, mainly in the lower Tertiary McKinley sequence of granite and granodiorite plutons (Reed and Lanphere, 1973; Lanphere and Reed, 1985). In the western part of the area, the plutons intrude highly folded and thrusted Devonian mafic and ultramafic rocks, Devonian argillite and graywacke, Mississippian chert, Permian through Triassic volcanic and marine sedimentary rocks, and Jurassic argillite and sandstone, part of the Chulitna terrane (Plate 3, this volume). In the eastern part of the area, the plutons intrude highly deformed, mainly Late Jurassic and Early Cretaceous deep marine, partly volcaniclastic, flyschoid graywacke and argillite and minor amounts of chert, limestone, and conglomerate that are part of the Kahiltna terrane.

Coal Creek Sn deposit. *By G. Thurow and J. D. Warner.* The Coal Creek Sn greisen(?) and Sn vein system consists of (1) sporadic grains and local concentrations of cassiterite in a sheeted vein system and of minor disseminations of cassiterite in and above the apical dome of an early Tertiary granite intruding comagmatic older granite; and (2) cassiterite in thin quartz-topaz-sulfide veinlets, 1 to 3 mm wide, that postdate alteration and stockwork veinlets. The veins vary from a hairline to 1 cm in width, are nearly vertical, and attain a density of 10 veins per m in the most intensely fractured zones. Vein sulfides include arsenopyrite, pyrite, pyrrhotite, and sphalerite. Granite adjacent to the veins is pervasively altered to quartz, tourmaline, topaz, sericite, and minor fluorite. The granite intrudes and contact metamorphoses Devonian argillite, graywacke, and minor limestone. The deposit contains an estimated 5 million tonnes grading 0.28 percent Sn and 0.5 percent Cu.

Golden Zone Au deposit. *By C. C. Hawley.* The Golden Zone deposit consists of veins and mineralized shear zones, porphyry Au deposits, a siliceous breccia pipe, and skarn deposits, classified as parts of a complex polymetallic vein system and associated Au-Ag breccia pipe. The breccia pipe contains arsenopyrite. The pipe is in the center of a quartz diorite porphyry stock; both pipe and stock plunge steeply east-northeast and are barely unroofed. The pipe enlarges from about 75 m in diameter at the surface to about 100 m diameter at the 180-m level; it probably continues to enlarge at depth and splits into feeder zones. Veins in the breccia pipe vary from a few centimeters of massive sulfide to shear zones more than 15 meters across containing numerous sulfide veins. The breccia pipe may have formed during hydrothermal stoping and collapse of the quartz diorite, guided by northeast- and northwest-trending conjugate faults. The porphyry contains a network of hairline fractures and distinct fissures filled with arsenopyrite, pyrite, chalcopyrite, and quartz. The contact between the pipe and porphyry is sharp. The porphyry is dated at 68.0 Ma (Swainbank and others, 1977). The deposit has produced 49,000 g of Au, 268,000 g of Ag, and 19 tonnes of Cu, and still contains an estimated 5 million tonnes grading 4 g/t Au, along with minor Cu and Ag.

Nim and Nimbus Cu-Ag-Au Deposits. *By R. C. Swainbank.* The Nim deposit consists of: (1) veinlets and disseminations of pyrite, chalcopyrite, molybdenite, and arsenopyrite in contact metamorphic rocks and in intrusive breccia; (2) veins and disseminations of arsenopyrite, molybdenite, chalcopyrite, and chalcocite in an early Tertiary granite porphyry and in peripheral rhyolite dikes; and (3) disseminated arsenopyrite, molybdenite, chalcopyrite, and pyrite in rhyolite porphyry and quartz porphyry dikes. The country rocks are Jurassic(?) sedimentary rocks. Grab samples contain as much as 2 percent Cu, 137 g/t Ag, and 13 g/t Au. The deposit is in a zone about 0.5 km wide and 2 km long. The Nimbus deposit consists of a lens of massive arsenopyrite, pyrite, and sphalerite 1 to 2 m thick and 10 m long in a brecciated felsic dike in a strand of the Upper Chulitna fault.

Basaltic Cu, massive sulfide, vein, skarn, and porphyry deposits; central and eastern Alaska Range and Wrangell Mountains

The central and eastern Alaska Range and Wrangell Mountains contain a complex variety of large and small lode deposits. The largest and best known are the Cu deposits of the Kennecott district, which produced about 544 million kg Cu and 280 million g Ag from about 1913 to 1938, and the Nabesna mine, which produced about 1.66 million g Au from about 1931 to 1940. Major porphyry Cu-Mo deposits are at Orange Hill and Bond Creek, Horsfeld, and Carl Creek. Other major lode deposits are: (1) Basaltic Cu deposits at Westover, Nelson, and Erickson; (2) Cu-Au-Ag skarn deposits at Zackly, Rainy Creek, and Midas; (3) Fe skarn deposits at Nabesna and Rambler; (4) Cu-Ag quartz vein deposits at Kathleen-Margaret, Nugget Creek, and Nikolai; (5) porphyry Cu deposits at Rainbow Mountain, Slate Creek, Chistochina, Baultoff, and Carl Creek; (6) a porphyry Cu-Mo deposit at London and Cape; (7) a polymetallic vein deposit at Nabesna Glacier; (8) a Besshi massive sulfide deposit at Denali; and (9) a dunitic Ni-Cu deposit at Fish Lake.

The deposits occur in the Wrangellia terrane (Nokleberg and others, 1985; Plate 3, this volume), a complex stratigraphic assemblage of late Paleozoic island-arc volcanic and sedimentary rocks, metabasalt of the Triassic Nikolai Greenstone, Upper Triassic and Lower Jurassic limestone and calcareous argillite, and Upper Jurassic and Lower Cretaceous volcanic rocks and flysch of the Gravina-Nutzotin sequence. The older part of this assemblage is intruded by late Paleozoic hypabyssal plutons, and the entire assemblage is intruded by Jurassic and Cretaceous granitic plutons (Richter, 1975; MacKevett, 1978; Nokleberg and others, 1985, 1986). The metallogenesis and tectonic history of this part of the Wrangellia terrane is summarized by Nokleberg and others (1984) and by Nokleberg and Lange (1985).

Zackly Cu-Au deposit. *By R. J. Newberry and C. R. Nauman.* The Zackly Cu-Au skarn deposit consists of disseminated chalcopyrite, bornite, pyrite, and gold in a zone of andradite garnet-pyroxene skarn and sulfide bodies in Late Triassic marble adjacent to albitized Cretaceous quartz monzodiorite. The zone is about 650 m long and 30 m wide. Gold occurs only in the skarn, with higher Au grades mainly in a supergene(?) assemblage of malachite, limonite, chalcedony, and native Cu. A general skarn zonation outward from the pluton consists of: (1) brown garnet with chalcopyrite; (2) green garnet with bornite and chalcopyrite; and (3) clinopyroxene and wollastonite; and (4) marble with magnetite and bornite. The deposit contains an estimated 1.25 million tonnes grading 1.6 percent Cu and 6 g/t Au.

Denali Cu-Ag, Kathleen-Magaret Cu, Rainy Creek Cu-Ag, and Rainbow Mountain Cu deposits. The Denali Cu-Ag deposit consists of stratiform bodies of very fine-grained, thinly layered chalcopyrite and minor pyrite in thin-bedded carbonaceous, and calcareous argillite in a zone as much as 166 m long and 9 m wide in the Upper Triassic Nikolai Greenstone (Stevens, 1971; Seraphim, 1975). The sulfides typically are rhythmically layered. The argillite and greenstone locally are moderately folded and are metamorphosed to the lower greenschist facies. The deposit is classified as a Besshi (?) massive sulfide deposit, although it differs from Besshi deposits in having a low Fe sulfide content. The deposit most likely formed in a reducing or euxinic marine basin characterized by abundant organic matter and sulfate reducing bacteria.

The Kathleen-Margaret Cu-Ag vein deposit is in the Nikolai Greenstone and consists of a series of quartz veins, as much as 140 m long and 3 m wide, with disseminated to locally massive chalcopyrite, bornite, and malachite (Mackevett, 1965). The deposit probably formed during the waning stages of Cretaceous(?) greenschist-facies metamorphism and weak deformation of the Nilokai Greenstone (Nokleberg and others, 1984).

The Rainy Creek Cu-Ag skarn deposit consists of small masses and disseminations of chalcopyrite, bornite, minor sphalerite, galena, magnetite, secondary Cu-minerals, and sparse gold in a zone of garnet-pyroxene skarn. The deposit is part of a belt of skarns about 10 km long and as much as 5 km wide that are hosted by faulted lenses of marble adjacent to small hypabyssal intrusions, dikes, and sills of late Paleozoic(?) metagabbro, metadiabase, and meta-andesite to metadacite (Nokleberg and others, 1984).

The Rainbow Mountain porphyry Cu deposit is in a discontinuous zone of subvolcanic porphyry intrusions that contain disseminated grains and small masses of chalcopyrite and pyrite and minor sphalerite and galena. The zone of subvolcanic porphyry intrusions is about 6 km long and as much as 1 km wide. The plutons occur as small hypabyssal plutons, dikes, and sills, are hydrothermally altered, and intrude late Paleozoic submarine meta-andesite to metadacite and sedimentary rocks (Nokleberg and others, 1985). The Rainy Creek Cu-Ag skarn and the Rainbow Mountain porphyry Cu deposits are probably magmatism-related deposits formed during late Paleozoic island arc volcanism (Nokleberg and others, 1984).

Nabesna and Rambler Au deposits. *By R. J. Newberry and T. K. Bundtzen.* The Nabesna and Rambler Fe-Au skarn deposits formed in massive oxide and massive sulfide bodies that at Nabesna consist chiefly of pyrite and magnetite with minor chalcopyrite, galena, sphalerite, and arsenopyrite. At the Rambler deposit, the sulfide bodies consist of massive auriferous pyrrhotite and pyrite that crosscut previously formed skarn. The gold skarns are characteristically zoned into separate skarn, magnetite, and sulfide-silica bodies. The skarn and magnetite are usually poor in sulfide and gold. The gold-rich sulfide-silica bodies overlie the highest-level magnetite bodies in pipelike or manto-like replacements of marble between skarn and monzodiorite. In some cases, high-magnetite and high-sulfide bodies occur independently in marble near skarn. The Nabesna skarns are vertically zoned, with idocrase and pyroxene at depth and garnet, epidote, and magnetite toward the top. Crosscutting relations indicate that magnetite bodies are younger than the skarn and high silica-bodies. The deposits occur near the contact between monzodiorite and limestone. At Nabesna, the monzodiorite stock is exposed over a 2-km^2 area and contains sporadic albite-quartz-pyrite alteration. The monzodiorite has K-Ar hornblende and biotite ages respectively of 109 and 114 Ma. The Nabesna deposit produced about 1.66 million g Au and minor Cu and Ag (Wayland, 1943; Richter and others, 1975). The Rambler deposit contains an estimated 18,000 tonnes grading 34.3 g/t Au.

Orange Hill and Bond Creek Cu-Mo deposits. The Orange Hill and Bond Creek deposits consist of pyrite, chalcopyrite, and minor molybdenite in potassic and sericitic quartz veins and as disseminations (Richter and others, 1975). The deposits are hosted in the Cretaceous Nabesna pluton, a complex intrusion of quartz diorite and granodiorite intruded in turn by slightly younger granite porphyry. Most of the deposits consist of quartz-biotite-chalcopyrite-pyrite-anhydrite veinlets and quartz-sericite-pyrite veins that are localized in altered granite porphyry dikes (R. J. Newberry, written communication, 1985). Widespread, late-stage chlorite-sericite-epidote alteration is present within the Nabesna pluton. The main occurrences of altered rock occupy an area 1 by 3 km at Orange Hill and an area 2 by 3 km at the Bond Creek deposit. Associated skarn deposits consist of andradite garnet, pyroxene, pyrite, chalcopyrite, bornite, magnetite, massive pyrrhotite, pyrite, chalcopyrite, and sphalerite. The Nabesna pluton intrudes rocks as young as the Jurassic and Cretaceous flysch of the Gravina-Nutzotin belt (Plate 3, this volume). The Orange Hill deposit contains an estimated 320 million tonnes grading 0.35 percent Cu. The Bond Creek deposit contains an estimated 500 million tonnes grading 0.3 percent Cu.

Kennecott district. The Kennecott district includes the Bonanza, Jumbo, Erie, Mother Lode, and Green Butte mines. The deposits are localized in the lower, largely dolomitic parts of the Upper Triassic Chitistone Limestone, generally less than 100 m above the disconformably underlying Middle and (or) Upper Triassic Nikolai Greenstone (Fig. 8) (Bateman and McLaughlin, 1920; MacKevett, 1976; Armstrong and MacKevett, 1982). The major ore bodies are mainly irregular masses of Cu-sulfides. The largest known ore body at Jumbo was about 110 m high, as much as 18.5 m wide and extended 460 m along plunge. The Cu-sulfide minerals are chalcocite and covellite, subordinate enargite, bornite, chalcopyrite, luzonite and pyrite, and rare tennantite, galena, and sphalerite. Secondary malachite and azurite occur locally. From about 1913 to 1938, about 544 million kg Cu, and 279 million g Ag were produced from 4.4 million tonnes of ore. More than 96 km of underground workings were developed. Armstrong and Mackevett (1982) interpret the basaltic Cu deposits in the Kennecott district as having formed by derivation of Cu from the Nikolai Greenstone, followed by deposition from oxygenated groundwater in the lower part of the overlying Chitistone Limestone along dolomitic sabhka interfaces and as open-space fillings in fossil karsts. They interpret the age of deposition as Late Triassic, with possible later remobilization.

Willow Creek district, Talkeetna Mountains

Au quartz vein deposits in the Willow Creek district consist of a series of quartz vein with pyrite, chalcopyrite, magnetite, and gold, and minor arsenopyrite, sphalerite, tetrahedrite, and galena (Ray, 1954). Average grade ranges from about 17.2 to 68.6 g/t Au. About 18.4 million g of gold were produced from 1909 to 1950. The veins average about 0.3 to 1 mm thick, are locally as much as 2 m thick, and occupy east-northeast- and north-south-striking shear zones as much as 7 m wide. Wall-rock alteration along the veins consists of sericite, pyrite, carbonate, and chlorite. Clay-rich fault gouge is locally abundant along the margins of the veins and shear zones. The veins are in and along the margin of the early Tertiary granitic rocks of the Talkeetna Mountains batholith and locally also in adjacent mica schist. The main part of the district, which includes several mines and many prospects, occupies an area about 12.8 km long and 6.2 km wide along the southern margin of the batholith. Underground work-

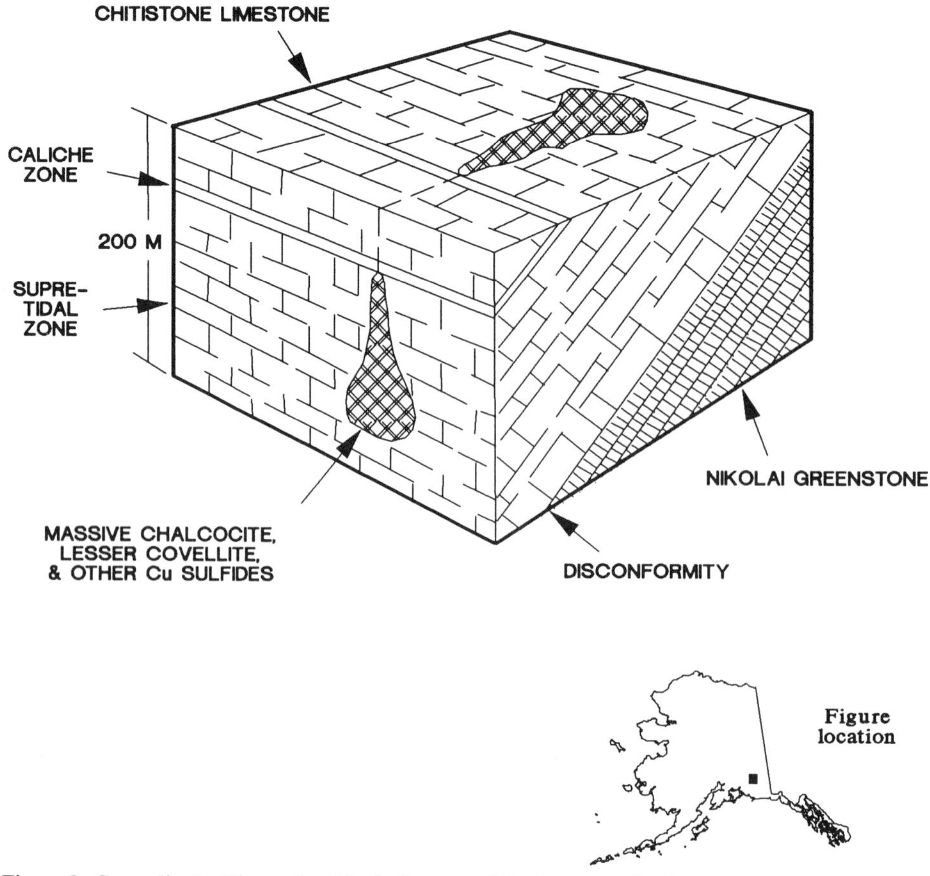

Figure 8. Generalized oblique view block diagram of the Bonanza Cu-Ag deposit in the Kennecott district, Wrangell Mountains, southern Alaska. Adapted from Bateman and McLaughlin (1920), and Armstrong and MacKevett (1982).

ings are estimated to total several thousand m. Nearly continuous mining and development has occurred from about 1909 to the present (1988).

Chromite and Ni-Cu deposits, Kodiak Island, Kenai Peninsula, and northern Chugach Mountains

A belt of podiform chromite deposits occurs in southern Alaska on northern Kodiak Island, on the Kenai Peninsula, and along the northern flank of the Chugach Mountains. The deposits are at Halibut Bay, Claim Point, Red Mountain (described below), and Bernard and Dust Mountains. A gabbroic Ni-Cu deposit occurs at Spirit Mountain. The podiform chromite deposits occur sporadically for a distance of more than 425 km in the Jurassic or older Border Ranges ultramafic and mafic complex (Burns, 1985), a belt of ultramafic tectonites and cumulate gabbros and norites that adjoins the Border Ranges fault system (MacKevett and Plafker, 1974; Burns, 1985) for 1,000 km in southern Alaska. The ultramafic and mafic rocks are interpreted as the roots of a Jurassic island arc (Burns, 1985; Plafker and others, 1985), and they form the southern margin of the Peninsular terrane (Plate 3, this volume). The Spirit Mountain deposit occurs at the eastern end of the belt of podiform chromite deposits. This deposit consists of Fe sulfides, pentlandite, chalcopyrite, and minor bravoite and sphalerite in small lenses and disseminations in serpentinized ultramafic rocks in gabbro sills that intrude late Paleozoic limestone, tuff, and chert. The ultramafic and mafic rocks at this deposit may be part of the distal, eastern end of the Border Ranges ultramafic and mafic complex.

Red Mountain chromite deposit. The Red Mountain deposit consists of layers and lenses of chromite as much as a few hundred m long and 60 m thick (Guild, 1942). The main chromite layer is about 190 m long and up to 0.3 m thick. The chromite layers are in Middle Jurassic or older Mesozoic layered dunite tectonite accompanied by minor wehrlite and clinopyroxenite; the ultramafic rocks locally are extensively serpentinized. The deposit contains an estimated 87,000 tonnes averaging 25 to 43 percent Cr_2O_3, including nearly 2.0 million tonnes in the "Turner stringer zone" (Foley and others, 1985). Approximately 36,700 tonnes of high-grade chromite ore was mined from six layers in the deposit. The ultramafic and associated rocks at Red Mountain may be cumulates that formed in the basal parts of an island arc, and then were subjected to penetrative deformation and high-grade metamorphism (Burns, 1985). Sporadic exploration and development has occurred from about 1919 to the present (1988). South of Red Mountain, the ultramafic rocks are faulted against metagraywacke and argillite of the Valdez Group of the Chugach terrane (Plate 3, this volume).

Au vein deposits, Kodiak Island, southeast Kenai Peninsula, and southern Chugach Mountains

An areally extensive suite of Au quartz vein deposits of small tonnage but locally high grade is present on Kodiak Island, on the southeast Kenai Peninsula, and in the southern Chugach Mountains. Major deposits include the Alaska Oracle, Chalet Mountain, Cliff, Crown Point, Gold King, Granite, Jewel, Kenai-Alaska, Lucky Strike, Mineral King, and Monarch.

On the mainland, the Au quartz vein and massive sulfide deposits occur in the Upper Cretaceous Valdez Group, which consists of complexly folded and weakly metamorphosed metagraywacke and argillite, locally interleaved with pillow basalt, basaltic tuff, and mafic plutons (Winkler and others, 1981a, b). Undeformed, narrow, early Tertiary granodiorite and diorite dikes and hypabyssal plutons locally intrude the intensely deformed Valdez Group. The lithologically similar, but less metamorphosed Kodiak Formation hosts Au quartz vein deposits on Kodiak Island. The Valdez Group and Kodiak Formation are parts of a flysch sequence deposited on oceanic crust (Plafker and others, 1985) and form the southern part of the Chugach terrane (Plate 3, this volume).

The Au quartz vein deposits are generally small, but high grade; most mines contain a maximum of a few hundred meters of underground workings. The largest deposit, at Cliff, produced about 1.6 million g Au. The gold typically occurs in quartz and minor carbonate fissure veins with minor pyrite, pyrrhotite, arsenopyrite, chalcopyrite, galena, stibnite, and sphalerite. Sulfides compose no more than a few percent of the veins. The veins range up to several hundred m long and a few m wide, with average grades from 34.3 to 64.8 g/t Au. the veins generally occur in metagraywacke and argillite and less often in early Tertiary diorite and granodiorite dikes.

The Au-bearing veins generally are the younger of two generations of quartz fissure veins in the Valdez Group (Richter, 1970; Goldfarb and others, 1986). The older and mostly barren veins are approximately parallel to the regional schistosity and parallel to axial planes of minor and major folds. Their strike varies from northwest in the east to northeast in the west. The younger veins locally carry gold, occur in a set of tensional cross joints or fractures, and are normal to the older quartz veins. The strike of the younger set of quartz veins also varies from northwest in the eastern part of the region to northeast in the western part of the region. Both sets of quartz veins generally dip steeply to vertically. The Au quartz vein deposits of the Chugach terrane probably formed during a widespread hydrothermal event that occurred in the waning stage of early Tertiary low-grade greenschist facies regional metamorphism, intense deformation, and granitic plutonism (Goldfarb and others, 1986).

Massive sulfide deposits, Prince William Sound district, Chugach Mountains

Besshi and Cyprus massive sulfide deposits are present in the Prince William Sound district along the eastern and northern margins of the Gulf of Alaska at Beatson, Copper Bullion, Ellamar, Fidalgo-Alaska, Knight Island, Latouche, Midas, Pandora, Standard Copper, and Threeman. The Midas deposit occurs in the southern part of the Valdez Group; the other deposits are in the Orca Group. Most of the deposits are classified as sediment-hosted Besshi massive sulfide deposits; the basalt-hosted deposits at Knight Island, Rua Cove, Standard Copper, and Threeman are classified as Cyprus massive sulfide deposits. The Orca Group, which hosts most of the deposits, consists of a strongly deformed, thick assemblage of Paleocene and Eocene (?) graywacke, argillite, minor conglomerate, pillow basalt, basaltic tuff, sills, and dikes (Winkler and Plafker, 1981). The assemblage is part of the Prince William terrane (Plate 3, this volume). A few gabbro plutons and locally abundant younger, early Tertiary diorite, granodiorite, and granite dikes and plutons intrude the Orca Group. Some of the plutonic rocks are intensely deformed.

Midas Cu-Ag-Au deposit. *By S. H. Nelson.* The stratiform Midas deposit consists of disseminated to massive chalcopyrite, pyrite, pyrrhotite, sphalerite, and minor galena in a folded, lens-shaped body as much as 7 m thick and 300 m long (Moffit and Fellows, 1950; Rose, 1965; Jansons and others, 1984). Margins of the ore body exhibit post-depositional shearing. Pillars in the main stope show sulfide layers and folds that are parallel to beds and folds in the host sedimentary rocks. The ore body occurs in highly deformed phyllite and metagraywacke of the Upper Cretaceous Valdez Group. Volcanic rocks have not been recognized in the mine, but they crop out in the footwall within a few hundred m of the ore body. Unmineralized to weakly mineralized quartz stockwork veins in the footwall could represent the feeder system for the main ore body. Pyrite is generally crystalline and subhedral, and is enclosed in a matrix of younger chalcopyrite, sphalerite, pyrrhotite, and quartz. Siliceous (chert?) beds are restricted to layers within the ore body. The deposit, classified as a Besshi massive sulfide deposit, produced 1.54 million kg Cu, 471,000 g Ag, and 78,900 g Au from 44,800 tonnes of ore, making it the fourth largest producer of Cu in the Prince William Sound district. The deposit still contains an estimated 56,200 tonnes of core grading 1.6 percent Cu.

Latouche and Beatson Cu-Ag deposits. The Latouche and Beatson mines worked two large and several small Besshi-type deposits in an extensive zone of massive sulfide lenses and sulfide disseminations (Johnson, 1915; Jansons and others, 1984). The sulfides are mainly pyrite and pyrrhotite accompanied by minor chalcopyrite, cubanite, sphalerite, galena, silver, and gold. The gangue minerals commonly are quartz, sericite, and ankerite. The two deposits collectively produced more than 84.4 million kg Cu from about 4.5 million tonnes of ore grading about 1.7 percent Cu and 9.3 g/t Ag. The deposits are part of as much as 120 m thick and 300 m long, and occur in a fault zone adjacent to metagraywacke and argillite.

LODE DEPOSITS, SOUTHEASTERN ALASKA

Southeastern Alaska contains varied and complex geology. Sedimentary and volcanic rocks range in age from Ordovician to Holocene and were intruded and deformed through a wide span of time. Most, but not all, of the intrusion, metamorphism, and deformation occurred in the Mesozoic and Cenozoic. In this paper, southeastern Alaska is divided into three north-northwest–trending regions, the Coast Mountains region, central southeastern Alaska, and coastal southeastern Alaska (Plate 11, this volume). The Coast Mountains region consists of the informally named Coast plutonic-metamorphic complex of Brew and Ford (1984a, b), which is approximately equivalent, from east to west, to part of the Stikinia terrane, all of the Tracy Arm and Taku terranes, and part of the Gravina-Nutzotin belt (Plate 3, this volume). Most of central southeastern Alaska consists of the Alexander terrane. Coastal southeastern Alaska consists of the Goon Dip Greenstone, Whitestripe Marble, and unnamed rocks that are part of the Wrangellia terrane, and the Kelp Bay Group, Sitka Graywacke and unnamed rocks that are part of the Chugach terrane.

Corresponding to the complex geology of the region are a complex variety of lode deposits. The Coast Mountains region contains extensive suites of Au quartz vein, metamorphosed sulfide, and zoned mafic-ultramafic deposits, a suite of Fe skarn and porphyry Mo deposits, and a Besshi massive sulfide deposit. Central southeastern Alaska contains extensive suites of Kuroko massive sulfide and bedded barite deposits, metamorphosed sulfide deposits, Cu-Zn-Au-Ag and Fe skarn and porphyry Cu deposits, Au quartz vein deposits, zoned mafic-ultramafic deposits, a gabbroic Ni-Cu deposit, and a felsic plutonic U and a sandstone U deposit. Coastal southeastern Alaska contains suites of Au quartz vein and gabbroic Ni-Cu deposits and a basaltic Cu deposit.

Coast Mountains region

The major lode deposits in the Coast Mountains region (Table 1 on Plate 11, this volume) are: (1) Au quartz vein deposits; (2) metamorphosed sulfide deposits; (3) a Besshi massive sulfide deposit; (4) an Fe skarn deposit; (5) a zoned mafic-ultramafic deposit; and (6) a porphyry Mo deposit.

Au Vein deposits. Au quartz vein deposits are present in the Coast Mountains region at Jualin, Kensington, Alaska-Juneau, Treadwell, Sumdum Chief, Riverside, Gold Standard, Sea Level, and Goldstream. These deposits are widespread and occur along a strike length of 300 km. Most deposits are in the Juneau gold belt (Spencer, 1906) in the northern part of the region, but a few are in an unnamed cluster in the southern part. In the Juneau gold belt, the deposits mostly occur in a metamorphic zone west of a large sill foliated tonalite (Brew, this volume). Host rocks are mainly metasedimentary and metavolcanic rocks of the Taku terrane (Jualin, Kensington, Alaska-Juneau, Sumdum Chief, and Sea Level deposits) and, to lesser extent, flysch of the Gravina-Nutzotin belt (Treadwell, Gold Standard, and Goldstream deposits). The deposit at Riverside is in the Stikinia terrane.

Substantial gold has been produced from these deposits: 108 million g from the Alaska-Juneau, 90.1 million g from the Treadwell, and 746,000 g each of Au and Ag from the Sumdum Chief. The Au quartz vein deposits in the western part of the Coast Mountains region, west of the foliated tonalite sill, probably formed during low-grade regional metamorphism and subsequent intrusion of intermediate and felsic postdeformational Tertiary plutons. Fluid inclusion studies at the Alaska-Juneau deposit indicate that the gold was deposited from deep-seated hydrothermal fluids in fault zones at temperatures greater than 230°C and pressures exceeding 1.5 kilobars, and that its deposition was accompanied by intense alteration and hydrofracturing of the host rocks (Goldfarb and others, 1986).

Alaska-Juneau Au deposit. The Alaska-Juneau deposit consists of quartz-calcite veins, a few centimeters to 1 m thick, containing sparse gold, pyrite, pyrrhotite, arsenopyrite, galena, sphalerite, chalcopyrite, and silver (Spencer, 1906; Twenhofel, 1952; Wayland, 1960; Herreid, 1962). The sulfide minerals also are present in adjacent, altered metamorphic rocks. The vein system is about 5.6 km long and as much as 600 m wide and consists of a series of semiparallel quartz-carbonate stringers in phyllite and schist near the contact of the Upper Triassic Perseverance Slate, and amphibolite derived from late (?) Mesozoic gabbro dikes and sills, with the Gastineau Volcanic Group of Permian and (or) Late Triassic age. The deposit produced about 108 million g Au, 59.1 million g Ag, and 21.8 million kg Pb from about 80.3 million tonnes of ore. The mine contains a few hundred km of underground workings.

Treadwell Au deposit. The Treadwell deposit consists of an extensive system of quartz and quartz-calcite replacements and veins with gold, pyrite, magnetite, molybdenite, chalcopyrite, galena, sphalerite, and tetrahedrite (Spencer, 1905; Buddington and Chapin, 1929; Twenhofel, 1952). The replacements and veins are in a shattered and altered granitic sill in slate and greenstone. Minor amounts of disseminated gold and sulfides occur in slate inclusions in the sill and in adjacent wall rock. The sill system is at least 1,100 m long and extends from a few hundred meters above sea level to almost 1,000 m below the surface of the Gastineau Channel. About 101 million g Au was produced from 25 million tonnes of ore.

Sumdum Chief Au deposit. The Sumdum Chief deposit consists of two quartz-calcite fissure veins with gold, auriferous pyrite, galena, sphalerite, chalcopyrite, and arsenopyrite (Spencer, 1906; Brew and Grybeck, 1984; Kimball and others, 1984). Gold is distributed unevenly and occurs mainly in pockets where small veins intersect large veins. The veins are as much as 6 m thick, and are in Paleozoic or Mesozoic graphitic slate and marble. About 746,000 g each of Au and Ag was produced.

Metamorphosed sulfide deposits. Metamorphosed sulfide deposits are present in the Coast Mountains region at Sweetheart Ridge, Sumdum (described below), Groundhog Basin, Alamo, Mahoney, Moth Bay, Reliance, and Red River. The deposits are widespread and occur along a strike length of about 300 km (Plate 11, this volume). The deposits consist of stratabound, massive to disseminated sulfides hosted in moderately to highly metamorphosed and deformed volcanic and sedimentary rocks. Original or primary features of the deposits have been so obscured by metamorphism and deformation as to preclude classification into a more specific deposit type. Some of the deposits may be metamorphosed Kuroko massive sulfide deposits, as indicated by high Pb and Ag values and the presence of metamorphosed felsic volcanic rocks. All but one of the deposits are in Taku terrane metamorphic rocks west of a persistent sill of foliated tonalite (Plate 3, this volume; Brew, this volume). The Red River deposit is in the central part of the Coast plutonic-metamorphic complex (Brew and Ford, 1984a, b) in the Tracy Arm terrane.

Substantial amounts of Cu, Pb, Zn, and Ag are present in these deposits. The Groundhog Basin deposit, which contains an estimated several hundred thousand tonnes grading 8 percent Zn, 1.5 percent Pb, and 51.5 g/t Ag, consists of disseminated to massive pyrrhotite, sphalerite, subordinate magnetite, galena, pyrite, and traces of chalcopyrite in several tabular or lenticular zones as much as 1 m thick. Host rocks are late Paleozoic or Mesozsoic calc-silicate gneiss, quartz-feldspar gneiss, and hornblende gneiss. The Moth Bay deposit contains an estimated 90,700 tonnes grading 7.5 percent Zn and 1 percent Cu and an additional estimated 181,400 tonnes grading 4.5 percent Zn and 0.75 percent Cu. This deposit consists of discontinuous lenses and layers of massive pyrite, pyrrhotite, minor chalcopyrite, and minor galena in late Paleozoic or Mesozoic muscovite-quartz-calcite schist, minor pelitic schist, and quartz-feldspar schist.

Sumdum Cu-Zn-Ag deposit. The Sumdum deposit consists of massive lenses and disseminations of pyrrhotite, pyrite, chalcopyrite, sphalerite, and lesser bornite, malachite, azurite, and galena in zones as much as 15 m wide (MacKevett and Blake, 1963; Brew and Grybeck, 1984; Kimball and others, 1984). The zones occur in metasedimentary schist and gneiss, mainly parallel to layering along the crest and flanks of isoclinal folds, but also in crosscutting veins and fault breccia. On the assumption that the deposit continues under the Sumdum Glacier, it is estimated to contain 24.2 million tonnes grading 0.57 percent Cu, 0.37 percent Zn, and 10.3 to 103 g/t Ag.

Massive sulfide, skarn, zoned mafic-ultramafic, and porphyry deposits. Four other lode deposit types occur in the Coast Mountains region: (1) A Besshi massive sulfide deposit at Yakima occurs in quartz-calcite-sericite schist of the Gravina-Nutzotin belt and consists of disseminated pyrite and minor galena and sphalerite in a zone 1,600 m long and 90 m wide. (2) An Fe skarn deposit at North Bradfield Canal occurs in marble and paragneiss intruded by Tertiary granite of the Coast plutonic-metamorphic complex (Brew and Ford, 1984a, b). This deposit consists of 11 magnetite-chalcopyrite skarn bodies that form crudely stratabound lenses in marble and paragneiss. The skarn bodies are as much as 106 m long and 12 m thick. (3) A Fe-Ti deposit is present at Union Bay (described below) in a zoned Cretaceous ultramafic pluton intruding the Gravina-Nutzotin belt. (4) A porphyry Mo deposit at Quartz Hill (described below) occurs in an Oligocene or Miocene granite porphyry intruding the central granitic belt of the Coast plutonic-metamorphic complex. The Quartz Hill deposit is regarded as a world-class Mo porphyry deposit (Eakins and others, 1985).

Union Bay Fe-Ti deposit. The Union Bay zoned mafic-ultramafic deposit is in dunite and consists of disseminated magnetite and chromite in small, discontinuous stringers up to a few centimeters long (Ruckmick and Noble, 1959). The dunite forms a pipe and lopolith in the center of the concentrically zoned Union Bay ultramafic pluton that intrudes the Late Jurassic and Early Cretaceous flysh of the Gravina-Nutzotin belt. A shell of peridotite encloses the dunite, and the peridotite in turn is enclosed by pyroxenite and hornblende pyroxenite that forms the periphery of the pluton. The deposit contains an estimated one billion tonnes grading 18 to 20 percent Fe. Selected samples average 0.093 g/t Pt and 0.20 g/t Pd. The ultramafic pluton at Union Bay is one of a series of 100-110 Ma mafic-ultramafic plutons, that intrude along the length of southeastern Alaska from Klukwan to Duke island (Taylor, 1967).

Quartz Hill Mo deposit. *By P. R. Smith and J. E. Stephens.* The Quartz Hill porphyry Mo deposit, 70 km east of Ketchikan, contains one of the world's largest concentrations of molybdenite. This large-tonnage deposit occurs in the hypabyssal late Oligocene or early Miocene intrusive complex of the informally named Quartz Hill stock. The stock is roughly ovoid in outcrop, approximately 5 km long by 3 km wide, and displays discordant contacts with the surrounding paragneiss and plutonic rocks of the Coast plutonic-metamorphic complex (Brew and Ford, 1984a, b) (Fig. 9). The stock is a complex suite of four distinct phases. The principal rock unit, the Quartz Hill granite body, is the oldest and most prominent phase, and makes up more than 75 percent of the outcrop area. The Quartz Hill granite body has been intruded by porphyritic quartz latite, younger granite, and dikes of quartz feldspar porphyry. Intrusive breccias are associated with some of the younger units. All of the rock units are similar in chemistry and mineralogy and consist of quartz, K-feldspar, sodic plagioclase, and minor biotite. Biotite from the Quartz Hill granite body has been dated at 26.9 Ma (Hudson and others, 1979).

The molybdenum deposit occurs predominantly in the Quartz Hill stock and is tabular to slightly concave upward. The surface dimensions are about 2,800 by 1,500 m, and the deposit extends to a depth of 370 to 500 m (Fig. 9). Two relatively high-grade zones occur: the Quartz Hill zone, south of the Stephens fault, and the Bear Meadow zone, north of the Stephens fault. The deposit, as determined from nearly 61,000 m of drill core, contains an estimated 1.7 billion tonnes grading 0.136 percent MoS_2, using a cutoff grade of 0.70 percent MoS_2. Within the deposit, a high-grade zone contains approximately 440 million tonnes grading 0.219 percent MoS_2 using a cutoff grade of 0.15 percent MoS_2.

Molybdenite and pyrite are the major sulfides and occur with or without quartz in randomly oriented veinlets forming a pervasive and well-developed stockwork. The molybdenite is in minute grains that range from 0.008 to 0.09 mm in diameter. Other sulfides, locally within or peripheral to the deposit include galena, sphalerite, and chalcopyrite, suggesting possibly recoverable byproduct Cu, Pb, Zn, Au, and Ag. Hydrothermal alteration of the stock is widespread and generally of weak to moderate intensity. Silicic, potassic, phyllic, argillic, and propylitic alterations are identified, but their recognition is complicated by subsequent effects of multiple intrusion and associated hydrothermal events.

Central southeastern Alaska

The major lode deposits in central southeastern Alaska (Table 1 on Plate 11, this volume) are: (1) Kuroko massive sulfide deposits at Glacier Creek, Orange Point, Greens Creek, Pyrola, Kupreanof Island, Helen S., Zarembo Island, Khayyam, Niblack, Barrier Islands, and Driest Point; (2) metamorphosed sulfide deposits at Cornwallis, Copper City, and Moonshine; (3) bedded barite deposits at Castle Island and Lime Point; (4) polymetallic vein deposits at Nunatak, Coronation Island, and Bay View; (5) Au quartz-vein deposits at Reid Inlet, Dawson, and Golden Fleece; (6) Cu-Zn-Au skarn deposits at Kupreanof Mountain and in the Jumbo district; (7) a Cu-Fe skarn deposit at Kasaan Peninsula; (8) a porphpyry Cu deposit at Margerie Glacier; (9) a gabbroic Ni-Cu deposit at Funter Bay; (10) zoned mafic-ultramafic Fe-Ti-V and Cr-PGE deposits at Klukwan and Duke Island, and an unclassified mafic-ultramafic Cu-Au-Ag-PGE deposit at Salt Chuck; (11) felsic plutonic U deposits at William Henry Bay, Salmon Bay, and Bokan Mountain; and (12) a sanstone U deposit at Port Camden.

These lode deposits are hosted in three main groups of rocks in central southeastern Alaska: (1) The Paleozoic and early Mesozoic sedimentary, volcanic, and plutonic rocks of the Alexander terrane (Brew and others, 1984); (2) various Mesozoic and early Tertiary plutonic rocks; and (3) Tertiary sandstone. The Alexander terrane consists mainly of: Paleozoic and Mesozoic carbonate rocks, carbonaceous flysch, chert, terrigenous and marine clastic rocks; Pre-Ordovician to Triassic metamorphosed basaltic to silicic flows and related volcaniclastic rocks; Ordovician and Silurian diorite and trondhjemite; and diverse Jurassic, Cretaceous, and Tertiary granitic rocks. Plutonic rocks hosting, or otherwise associated with some lode deposits in the Alexander terrane consist of Jurassic granite, Cretaceous granodiorite, Mesozoic (mainly Cretaceous) pyroxenite, gabbro-norite, and gabbro, and Tertiary granite, granite porphyry, and felsic dikes.

Massive sulfide and barite deposits. Kuroko massive sulfide deposits in central southeastern Alaska occur at Glacier Creek, Orange Point, Greens Creek (described below), Pyrola, Kupreanof Island, Helen S., Zarembo Island, Khayyam (described below), Niblack, Barrier Islands, and Driest Point. Most of the deposits consist of disseminated to massive Fe sulfides and base-metal sulfides in lenses and layers up to about 25 m wide and 170 m long. Host rocks are Ordovician, Silurian, Permian(?), and Triassic felsic to intermediate flows, tuff, and volcaniclastic rocks, interlayered with limestone, slate, chert, and lesser greenstone. These deposits are spread over 300 km along the strike length of the Alexander terrane.

Substantial amounts of Cu, Pb, Zn, Ag, and Au occur in the Kuroko massive sulfide deposits in central southeastern Alaska. The Glacier Creek deposit contains an estimated minimum

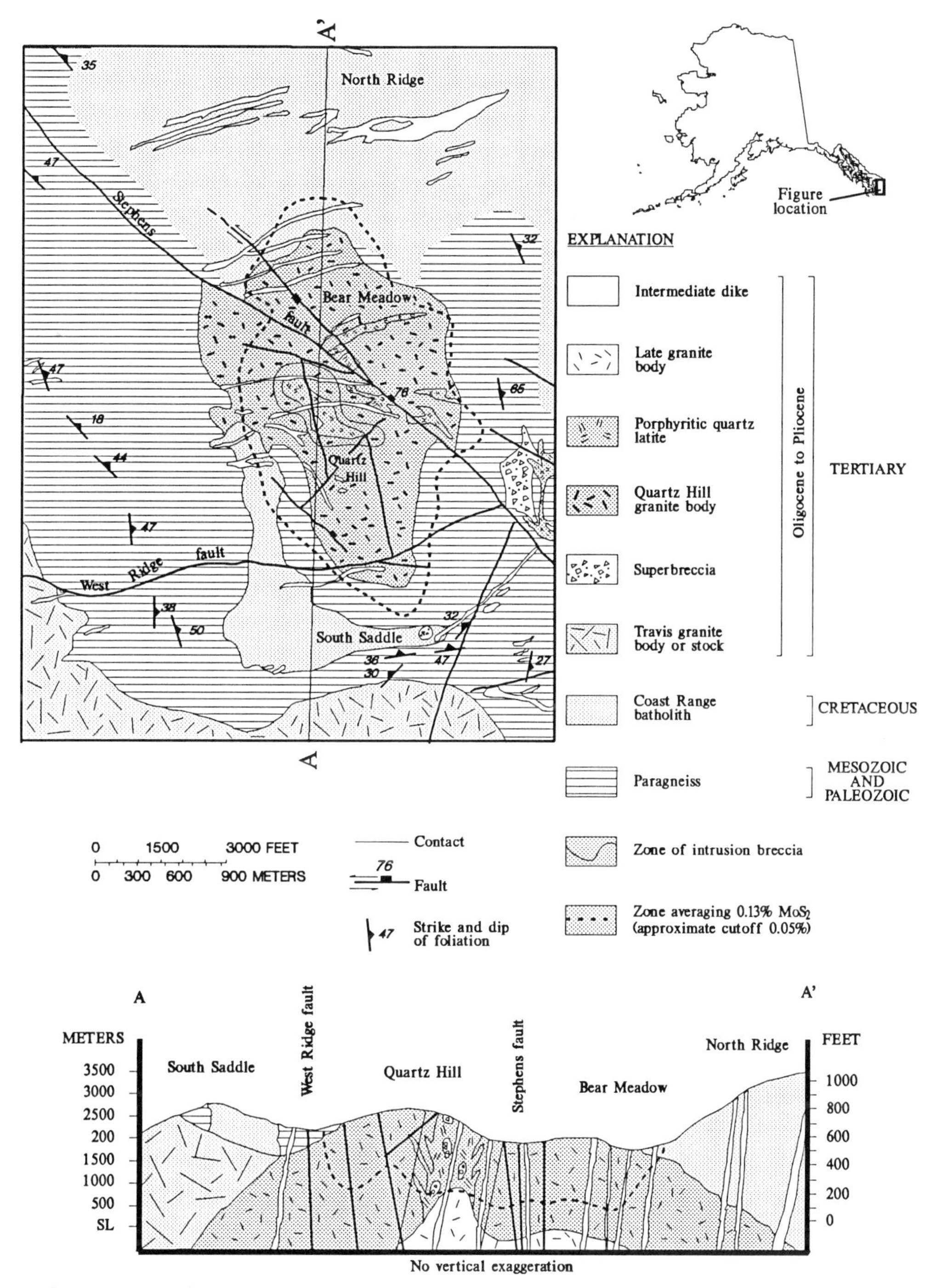

Figure 9. Generalized geologic map and cross section of the Quartz Hill porphyry Mo deposit, Coast Mountains region, southeastern Alaska. The Mo deposit is hosted in a hypabyssal Tertiary intrusive complex informally named the Quartz Hill stock.

680,000 tonnes grading as much as 3 percent combined Cu and Zn and as much as 45 percent $BaSO_4$. The Greens Creek deposit contains an estimated 3.6 million tonnes grading 8 percent Zn, 2.7 percent Pb, 0.4 percent Cu, 360 g/t Ag, and 3.4 g/t Au. The Khayyam deposit produced about 6.4 million kg Cu, 40,100 g Au, and 53,300 g Ag from 205,000 tonnes of ore. The Niblack deposit produced about 636,000 kg Cu, 34,200 g Au, and 466,500 g Ag.

Metamorphosed sulfide deposits in carbonate and metavolcanic host rocks are present at Cornwallis, Copper City, and

Moonshine. The Cornwallis Zn-Pb deposit consists of finely disseminated sphalerite, galena, and chalcopyrite in Carboniferous limestone breccia and is associated with pods, veins, and layers of barite as much as 2 m wide and 60 m long in Late Triassic felsic metavolcanic rocks. The Copper City Cu-Zn-Ag-Au deposit consists of massive chalcopyrite, pyrite, and sphalerite in layers and lenses about 1 m thick in metamorphosed early Paleozoic keratophyre and spilite. The Moonshine Ag-Pb deposit consists of galena, sphalerite, minor chalcopyrite and pyrite in fissure veins or pods as much as a few meters wide in dolomite veins cutting early Paleozoic marble. The Copper City deposit produced about 1,450 tonnes of ore, and the Moonshine deposit produced about 46,500 g Ag; no model is available to classify these two deposits.

Bedded barite deposits are present at Castle Island and Lime Point. The Castle Island deposit consists of lenses of massive barite interlayered with metamorphosed Triassic(?) limestone and calcareous and tuffaceous clastic rocks. The deposit produced 680,300 tonnes of ore grading 90 percent $BaSO_4$. The Lime Point deposit consists of interlayered lenses of barite and dolomite as much as 2 meters thick in lower Paleozoic marble. The deposit contains an estimated 4,500 tonnes grading 91 percent barite.

Greens Creek Zn-Pb-Cu-Ag-Au deposit. By J. Dunbier and D. A. Sherkenbach. The Greens Creek Zn-Pb-Cu-Ag-Au deposit consists of sulfide bands, laminations, and disseminations hosted in a sequence of chlorite-rich and sericite-rich metasedimentary rocks and of pyrite-chert-carbonate rocks that structurally overlie locally serpentinized mafic volcanic flows and tuffs (Dunbier and others, 1979; Drechsler and Dunbier, 1981). The mafic volcanic rocks crop out in the core of a large southeast-plunging antiform that is overturned to the northeast (Fig. 10); the metasedimentary rocks, exhalite, and associated sulfide bodies occur in the pinched nose and along the northeast limb of the structure several km from the rocks in the core. The sulfide content generally increases structurally upsection and culminates at the contact with overlying black carbonaceous argillite and graywacke. Deformation and lower greenschist-facies metamorphism characterize the host rocks.

The sulfide horizon has a structural hanging wall of finely bedded metasedimentary rocks and pyrite-carbonate-chert exhalite and a footwall of black graphitic argillite that overlies the metamorphosed tuff (Fig. 10). In the transitional contact zone, the sulfides occur in a series of south-plunging, elongate, massive pods as much as 25 m thick, with flanking units of black and white ore. The massive pods consist of layers, laminations, and disseminations of sphalerite, galena, chalcopyrite, and tetrahedrite in a pyrite-rich matrix. Black ore forms an extensive blanket in the deposit and consists of laminated, fine-grained pyrite, sphalerite, galena, and Ag-rich sulfosalt hosted in black carbonaceous exhalite and argillite. White ore is present along the edges of the massive pods and consists of minor amounts of tetrahedrite, pyrite, galena, and sphalerite in laminations, stringers, or disseminations that are hosted in massive chert, carbonate rocks, or sulfate-rich exhalite. Geopetal structures indicate that the ore horizon is overturned. Several vein assemblages are also present; the most interesting contain bornite, chalcopyrite, and gold. These veins are in chlorite-talc-carbonate alteration zones that are stratigraphically below the massive sulfide pods and may have been brine conduits. Extensive drilling has delineated a major mineral deposit, still open at depth, containing an estimated 3.6 million tonnes grading 0.4 percent Cu, 2.7 percent Pb, 7.9 percent Zn, 360 g/t Ag, and 3.4 g/t Au.

We infer that the Greens Creek deposit formed in a backarc or wrench-fault basin. Early deposition was dominated by arc- or continent-derived clastic and volcaniclastic sediments that intermixed with mafic flows and tuff. Late deposition was dominated by distal turbidites in a starved, euxinic basin. The basin remained tectonically active, with internal subbasins characterized by locally derived slump and debris breccias. Brines responsible for the deposit probably consisted of convective seawater that circulated in the lower basinal sequence. Brine flow was localized by structural rather than volcanic conduits, and the brines discharged into a dominantly sedimentary environment with local relief caused by active faulting. Ore deposition probably resulted from interaction between buoyant brine phases and seawater, in addition to

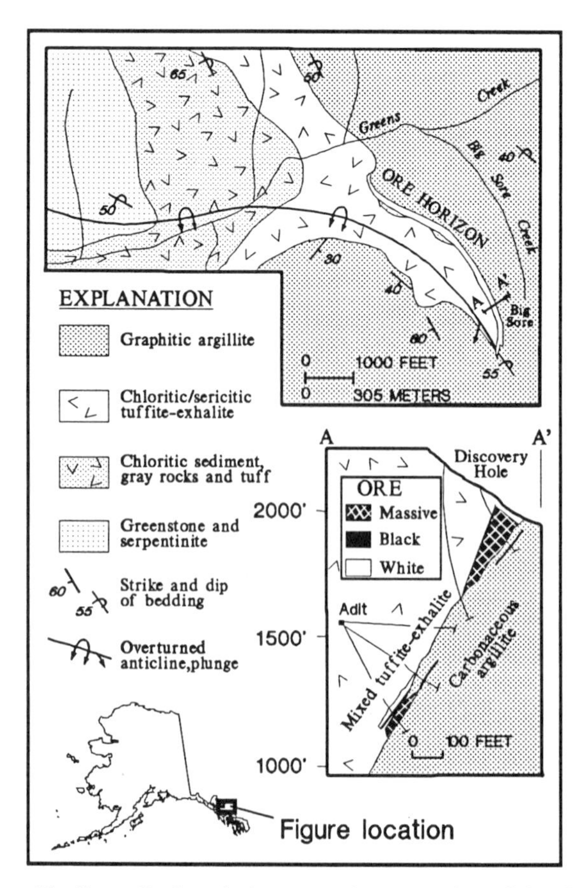

Figure 10. Generalized geologic map and cross section of the Greens Creek Zn-Pb-Ag-Au deposit, Admiralty Island, central southeastern Alaska. The location of sulfides is near the nose of an asymmetrical overturned anticline, between a stratigraphic footwall of mixed tuff and exhalite and a hanging wall of carbonaceous argillite.

precipitation in density-stratified pools. Unusually carbon-rich sedimentary rocks in the hanging wall may reflect blooming marine life associated with the brines. The Greens Creek deposit is an intriguing example of a massive sulfide deposit that shows some characteristics of Kuroko massive sulfide, sedimentary exhalative, and Cyprus massive sulfide deposits.

Khayyam Cu-Au-Ag deposit. The Khayyam deposit, classified as a Kuroko massive sulfide deposit, consists of irregular, elongate, nearly vertical lenses of massive pyrite, chalcopyrite, sphalerite, pyrrhotite, hematite, gahnite, and magnetite in a gangue of quartz, calcite, epidote, garnet, and chlorite (Fosse, 1946; Barrie, 1984a, b). The sulfides and associated minerals occur in about seven sulfide lenses as much as 70 m long and 6 m thick. The lenses are conformably enclosed in pre–Middle Ordovician felsic to mafic metavolcanic rocks of the Wales Group in the Alexander belt. The metavolcanic rocks show coarse fragmental textures, and intense chlorite alteration is present in the footwall below the sulfide lenses. Several hundred meters of underground workings exist. The deposit produced about 6.4 million kg Cu, 40,120 g Au, and 53,210 g Ag from about 205,000 tonnes of ore.

Polymetallic and Au quartz vein deposits. Polymetallic vein deposits occur at Nunatak (described below), Coronation Island, and Bay View. The Coronation Island Pb-Zn deposit consists of lenses of galena, sphalerite, and tetrahedrite in a clay-carbonate gangue in fault zones in Silurian(?) marble intruded by Tertiary(?) diorite. The deposit has produced more than 91 tonnes of ore. The Bay View Ag-Au deposit consists of quartz- and calcite-cemented fault breccia with disseminations and small masses of pyrite, chalcopyrite, and minor sphalerite and bornite. The host rock is a basalt dike that cuts fault-bounded Silurian trondhjemite. Selected samples contain as much as 10 g/t Ag and 0.1 g/t Au.

Au quartz vein deposits occur at Reid Inlet, Dawson, and Golden Fleece. The Reid Inlet deposit consists of narrow, discontinuous, steeply dipping quartz veins as much as a few hundred m long and 1.1 m thick in altered Cretaceous granodiorite, Permian(?) metamorphosed pelitic and volcanic rocks, and marble. The deposit has produced 220,000 to 250,000 g Au. The Dawson Au-Ag deposit consists of quartz stringers and veins as much as 1.8 m wide in Paleozoic black graphitic slate. The stringers and veins contain scattered pyrite and base metal sulfides. The deposit has probably produced at least several tens of thousands of g each of Au and Ag and minor amounts of Pb, and it contains an estimated 40,000 tonnes grading 34.3 g/t Au. The Golden Fleece Ag-Au deposit consists of irregular quartz fissure veins as much as 3 m thick containing pyrite, tetrahedrite, and gold in silicified and dolomitized marble cut by diabase dikes. The deposit has had considerable unrecorded production and contains several hundred meters of workings.

Nunatak Cu deposit. The Nunatak polymetallic vein deposit consists of abundant, closely spaced molybdenite-bearing quartz veins and minor disseminated molybdenite in hornfels, skarn, and a fault zone (Brew and others, 1984). The molybdenite-bearing veins, skarn, and fault zone are adjacent to a Tertiary(?) granite porphyry stock. Sulfides locally are disseminated in the porphyry. Besides molybdenite, sulfides include pyrite, pyrrhotite, chalcopyrite, and sparse tetrahedrite and bornite. The closely spaced vein stockwork contains an estimated 2.0 million tonnes grading 0.067 percent Mo and 0.16 percent Cu. The remaining stockwork has inferred resources of 118 thousand tonnes grading 0.026 percent Mo and 0.18 percent Cu. The granite porphyry intrudes tightly folded Paleozoic metasedimentary rocks. The deposit is classified either as a polymetallic vein or a porphyry Cu-Mo deposit; the polymetallic vein classification is more probable.

Skarn and porphyry deposits. A major Cu-Fe skarn deposit is present at Kasaan Peninsula (described below), and Cu-Zn-Au skarn deposits occur at Kupreanof Mountain and in the Jumbo district (described below). A combined porphyry Cu and lesser polymetallic vein deposit occurs at Margerie Glacier. The Kupreanof Mountains deposit consists of local massive pods, lenses, and disseminations of pyrrhotite, magnetite, and chalcopyrite, and minor sphalerite and pyrite in pyroxene-garnet skarn in Devonian (?) marble and in part in highly altered mafic igneous rocks. The deposit contains several hundred meters of underground workings.

Kasaan Peninsula Cu-Fe-Au-Ag deposits. The Kasaan Peninsula deposits consist of contorted tabular masses of magnetite, chalcopyrite, and pyrite in a gangue of calcite and calc-silicate minerals (Warner and Goddard, 1961). The masses generally are in conformable layers, mainly along contacts between calcareous metasedimentary rocks and mafic metavolcanic rocks adjacent to irregular dikes, sills, and plugs of Ordovician or Silurian diorite and quartz monzodiorite and mafic dikes. About 30 deposits are present on the 20-km-long peninsula. The largest deposit produced about 245,000 tonnes containing more than 5.8 million kg Cu, 216,000 g Au, and 1.74 million g Ag. This deposit contains an estimated 2.7 million tonnes averaging 53 to 59 percent Fe and 0.26 to 0.90 percent Cu. The deposit exhibits zoned calc-silicate minerals and sulfides and rather low Ag/Au and Zn/Cu ratios. The Kasaan Peninsula deposits are classified as Fe skarn deposits that probably formed during intrusion of Paleozoic plutonic rocks.

Jumbo district. The Jumbo district contains Cu-Au skarn deposits at Jumbo and smaller deposits at Magnetite Cliff, Copper Mountain and elsewhere in the area (Kennedy, 1953; Herreid and others, 1978). The skarns occur in early Paleozoic marble and pelitic metasedimentary rocks that are intruded by a mid-Cretaceous hornblende-biotite granodiorite having concordant hornblende and biotite K-Ar ages of 103 Ma. The Jumbo Cu-Au deposit consists of chalcopyrite, magnetite, sphalerite, and molybdenite in skarn at the contact between marble and an Early Cretaceous granodiorite stock. The gangue is mainly diopside and garnet. The Jumbo deposit, with more than 3.2 km of underground workings, produced 4.6 million kg Cu, 220,000 g Au, and 2.73 million g Ag from 112,000 tonnes of ore. The Magnetite Cliff deposit consists of a 25-m-thick shell of magnetite that mantles the mid-Cretaceous granodiorite in contact with garnet-diopside skarn. The skarn contains 2 to 3 percent chalcopyrite and an estimated 336,000 tonnes grading 45 percent Fe and 0.77 percent Cu. The Copper Mountain deposit is in granodiorite and consists of scasttered chalcopyrite and copper carbonate minerals in diopside endoskarn that comprises veins and masses of epidote, garnet, magnetite, and scapolite. The deposit has about 410 m of underground workings and produced 101,800 kg Cu, 321,000 g Ag, and 4,500 g Au.

Margerie Glacier Cu deposit. The Margerie Glacier deposit consists of chalcopyrite, arsenopyrite, sphalerite, molybdenite, and minor scheelite in quartz veins in shear zones, in massive sulfide bodies, and as disseminations (Brew and others, 1978). The veins, massive sulfides, and disseminations occur in a propylitically altered porphyritic Cretaceous (?) granite stock and in adjacent hornfels. The granite intrudes Permian (?) metamorphosed pelitic and volcanic rocks and minor marble. The deposit contains an estimated 145 million tonnes grading 0.02 percent Cu, 0.27 g/t Au, 4.5 g/t Ag, and 0.01 percent W, and is classified as a combined porphyry Cu and subordinate polymetallic vein deposit.

Gabbroic Ni-Cu and mafic-ultramafic deposits. A gabbroic Ni-Cu deposit is present at Funter Bay (described below), Fe, Ti, V, Cr, and PGE occur in zoned mafic-ultramafic deposits at Klukwan and Duke Island, and PGE has been recovered from an unclassified mafic-ultramafic body at Salt Chuck (described below). The zoned mafic-ultramafic deposits are part of a discontinuous belt of 100- to 110-Ma mafic-ultramafic plutons that extends the length of southeastern Alaska.

The Klukwan Fe-Ti-V deposit consists of titaniferous magnetite and minor chalcopyrite, hematite, and Fe sulfides in disseminations or in tabular zones in Cretaceous pyroxenite surrounded by diorite. The deposit contains an estimated 11.8 billion tonnes grading 0.2 percent V_2O_5, 13 percent magnetite, and 1.5 to 4.4 percent TiO_2. The Duke Island Cr-PGE deposit consists of disseminated to locally massive titaniferous magnetite and sparse chromite in hornblende and clinopyroxene zones in a Cretaceous zoned ultramafic pluton.

Funter Bay Ni-Cu deposit. The Funter Bay gabbroic Ni-Cu deposit consists of disseminated pyrrhotite, pentlandite, and chalcopyrite that occur in olivine-hornblende gabbro at the base of a gabbro-norite pipe of late(?) Mesozoic age (Barker, 1963; Noel, 1966). The pipe intrudes late Paleozoic or Triassic quartz-mica schist. The deposit contains an estimated 450 to 540 thousand tonnes grading 0.33 to 1.0 percent each of Cu and Ni and 0.05 to 0.32 percent Co.

Salt Chuck Cu-Au-Ag-PGE deposit. The Salt Chuck mafic-ultramafic deposit consists of irregularly and randomly distributed veinlets of bornite, minor chalcopyrite, and secondary chalcocite, covellite, native copper, and magnetite (Howard, 1935; Gault, 1945; Loney and others, 1987). The sulfides and oxides occur along cracks and fractures in a pipelike gabbro-pyroxenite stock of probable early Paleozoic (429 Ma) age that intrudes Ordovician and Silurian basalt and pyroclastic rocks. The deposit has produced about 296,000 tonnes grading 0.95 percent Cu, 1.2 g/t Au, 5.8 g/t Ag, and 2.2 g/t PGE.

U Deposits. Felsic plutonic U deposits occur at William Henry Bay, Salmon Bay, and Bokan Mountain (described below), and a sandstone U deposit occurs at Port Camden. The William Henry Bay deposit consists of veinlets carrying pyrite, chalcopyrite, galena, thorianite, and euxenite in a small Tertiary(?) granite pluton intruding Silurian(?) metavolcanic and metasedimentary rocks. The Salmon Bay deposit consists of carbonate fissure veins in Silurian metagraywacke near Tertiary felsic dikes. The veins contain a wide variety of minerals, including fluorite, hematite, magnetite, pyrite, chalcopyrite, thorite, monazite, zircon, parisite, and bastnaesite. The Port Camden sandstone U deposit consists of traces of U minerals in poorly sorted dolomitic sandstone of the Tertiary Kootznahoo Formation, which contains detritus derived from Tertiary or older granitic rocks.

Bokan Mountain U-Th-REE deposit. The Bokan Mountain felsic plutonic U deposit consists of disseminated accessory U-Th, REE, and niobate minerals, including uranothorite, uranoan thorianite, uraninite, xenotime, allanite, monazite, and accessory pyrite, galena, zircon, and fluorite (MacKevett, 1963; Lancelot and de Saint-Andre, 1982; Thompson and others, 1982; Armstrong, 1985). The U-Th, REE, and niobate minerals are hosted in an irregular, steeply dipping pipe of Jurassic peralkaline granite that grades outward into mostly barren granite. U-Th vein and pegmatite deposits occur in the outer parts of the granite and adjacent country rock. The deposit has produced about 109,000 tonnes grading about 1 percent U_3O_8. Equivalent grade Th was not recovered.

Coastal southeastern Alaska

Major gabbroic Ni-Cu, Au quartz vein, and basaltic Cu deposits are present in coastal southeastern Alaska (Table 1 on Plate 11, this volume). Gabbroic Ni-Cu deposits occur at Brady Glacier (described below), Bohemia Basin (described below), and Mirror Harbor. These deposits are in Tertiary mafic and ultramafic stocks that intrude Chugach terrane metagraywacke and phyllite of the Cretaceous Sitka Graywacke and metagraywacke, phyllite, and greenschist of the Cretaceous and Cretaceous(?) Kelp Bay Group. The mafic and ultramafic plutons are probably postdeformational and postmetamorphic.

Au quartz vein deposits occur at Apex, El Nido, Cobol, Chichagoff and Hirst-Chichagof (described below). These deposits generally consist of quartz fissure veins, as much as 4 m thick, and stockworks containing pyrite, arsenopyrite, chalcopyrite, galena, sphalerite, tetrahedrite, and gold. Minor sulfides locally occur in adjacent metasedimentary, metavolcanic, and granitic wall rocks. The Au quartz vein deposits at Apex, El Nido, and Cobol are in the Wrangellia terrane in late Paleozoic low-grade pelitic and metavolcanic rocks, including greenstone, quartzite, and siliceous limestone. The deposits at Chichagoff and Hirst-Chichagof are in the Cretaceous Sitka Graywacke of the Chugach terrane. The Au quartz vein deposits are probably formed during

Tertiary regional metamorphism, deformation, and subsequent granitic intrusion.

A possible basaltic Cu deposit occurs at Baker Creek. This deposit consists of small masses and disseminations of chalcopyrite and pyrite in zones as much as 4 m thick and 120 m long in metamorphosed subaerial basalt flows of the Triassic(?) Goon Dip Greenstone of the Wrangellia terrane.

Brady Glacier Ni-Cu deposit. The Brady Glacier deposit consists of disseminations and small masses of pentlandite, chalcopyrite, and sparse pyrite near the eastern edge and probable base of the La Perouse gabbro pluton, a layered Tertiary mafic-ultramafic pluton composed of gabbro and minor peridotite (Brew and others, 1978; Czamanske and Calk, 1981; Himmelberg and Loney, 1981). The deposit locally contains as much as 10 percent disseminated sulfides and has an estimated 82 to 91 million tonnes averaging 0.53 percent Ni, 0.33 percent cu;, 0.03 percent Co, and minor amounts of PGE. Selected samples with disseminated to massive sulfides contain 0.2 to 1.3 g/t PGE. The deposit occurs mainly beneath the Brady Glacier and is exposed only in three small nunataks. K-Ar ages of 25 to 30 Ma have been obtained for the La Perouse pluton, which intrudes metagraywacke and phyllite of the Cretaceous Sitka Graywacke.

Bohemia Basin Ni-Cu deposit. The Bohemia Basin deposit consists of magmatic segregations chiefly of pyrrhotite, pentlandite, and chalcopyrite (Kennedy and Walton, 1946; Johnson and others, 1982). The segregations occur in a trough-like body, about 45 m thick, near the base of a basin-shaped, composite norite stock of Tertiary age. The norite locally grades into gabbro and diorite. The stock intrudes metagraywacke, phyllite, and greenschist of the Cretaceous and Cretaceous(?) Kelp Bay Group. The deposit contains an estimated 19 million tonnes averaging 0.33 percent Ni, 0.21 percent Cu, and 0.04 percent Co.

Chichagoff and Hirst-Chichagof Au-Ag deposits. The Chichagoff and Hirst-Chichagof deposits consist of tabular to lenticular bodies of quartz carrying small masses of pyrite, arsenopyrite, galena, sphalerite, chalcopyrite, and local scheelite and tetrahedrite (Reed and Coats, 1941; Still and Weir, 1981; Johnson and others, 1982; Alaska Mines and Geology, 1985). The quartz bodies are mainly of ribbon quartz, as much as a few meters thick and a few thousand meters long. The main ore shoots are localized along intersections of various splays of the Hirst and Chichagoff faults and probably along warps in the faults. The deposit is in metagraywacke and argillite of the Cretaceous Sitka Graywacke. The deposits have produced about 25 million g Au, 1.24 million g Ag, and minor amounts of Pb and Cu.

PLACER DISTRICTS

More than 960 million g of gold have been produced from Alaskan mines since gold was discovered there in the late 1880's. Of this amount, more than 620 million g, or roughly two-thirds, has been obtained from placer deposits. Alaska is one of the few states where placer gold production has recently increased. Placers in Alaska have also yielded approximately 93 million g of silver, about 17 million g of PGE, 1.8 million kg of tin, and unspecified amounts of mercury and tungsten (Eakins and others, 1985). Placer mining in Alaska, principally for gold, is one of the major nonfuel mining industries on the basis of the value of mineral produced. It was second only to sand and gravel in 1982, when approximately $70,000,000 worth of gold was mined (Eakins and others, 1985). Silver in Alaska is produced primarily as a byproduct of placer gold. In 1982 Alaskan gold placer mines produced approximately 684,000 g of silver, as well as 6,160 kg of tin. Approximately 28,000 g of placer platinum was produced in 1981, the most recent year for which figures are available (Eakins and others, 1985). The fluctuating price of gold is a major factor in placer mining in Alaska, where operating costs are high. A number of mines are not economic when the price drops below about $10/g ($300/oz). Consequently, yearly production can vary greatly when the price fluctuates around this figure.

History of placer discovery and mining in Alaska

The native American in Alaska used native gold in ornamental jewelry and occasionally as decoration on pottery, utensils, and weapons. Gold was reported from Alaska as early as 1834 by a party of Russian-Americans exploring on the Russian River drainage of the Kenai Peninsula; however, gold was not actually mined until the late 1860s and early 1870s, initially near Sitka and near Juneau in the Silver Bow Basin area. The discovery of placer gold on tributaries of the Fortymile River in 1886 was instrumental in opening up the interior of Alaska to gold discovery and mining. Gold was found on Birch Creek, south of the Yukon River, leading to the development of the Circle Mining District in the 1890s, a district that continues to produce gold. The discovery of gold in the Klondike area of the Yukon Territory led to more discoveries in Alaska. Many of the Klondike prospectors who were not successful on the Dawson creeks drifted down the Yukon River into Alaska and eventually reached the beaches at Nome. A promising discovery was made there on Anvil Creek in 1898. The realization that the beaches around Nome could be worked sparked a major stampede to this new district.

The many prospectors in Alaska soon made other discoveries. A major discovery on tributaries of the Tanana River led to the founding of Fairbanks, in what would eventually become the richest gold mining region of the state. Gold in the upper Koyukuk River was discovered in 1898, and later discoveries in 1905 and 1908 resulted in establishing the gold mining towns of Coldfoot and Wiseman. The lower Yukon and Kuskokwim River basins witnessed minor rushes during 1909 to 1912; deposits were discovered in the Iditarod, Ruby, Flat, and Ophir districts. The remote Chandalar River district above the Arctic Circle was investigated in 1902, and a small rush ensued in 1906 with the discovery of deposits on Little Squaw Creek. During the same period, placer deposits in the Chistochina area in the eastern

Alaska Range were discovered. A general decline in new discoveries and production occurred about 1918. This trend continued until about 1928, when mechanization of mining resulted in increased production. In 1934 the price of gold was raised from $20 to $35 per ounce, and this resulted in a peak production of about 23 million g of gold in 1940.

War Production Board Order L-208 almost stopped gold mining by October 1942. Operating placer mines decreased from 554 to 142 by 1943. The recovery from 1944 to 1950 was slow because replacing equipment was costly. Lode mining suffered even more; as a result, 95 percent of the gold mined during this period came from placers. After World War II the price of gold remained at the 1934 standard of $35 per ounce, which caused a general decline in the industry. Almost all the great gold dredges had shut down by the early 1960's. Some small-scale placer mines survived by selectively mining the richer parts of deposits, but all-time low production figures were recorded in 1971 and 1972. The depressed price of gold, together with the high cost of labor and equipment, limited production to less than $500,000 from about a dozen gold-mining operations. The dramatic increase in the gold price in the late 1970s and early 1980s resulted in a corresponding increase in Alaskan gold production. By 1981 there were approximately 400 placer mines in the state employing about 3,000 miners, with annual production of about 4.2 million g of gold.

The major placer districts of Alaska are summarized in Table 2 on Plate 11 (this volume) using Cobb's (1973) regional divisions of Alaska. The districts are arranged by geographic region in the same order as for lode deposits on Table 1 (Plate 11). Data were compiled only for areas with production of more than 31,000 g (1,000 oz) of gold. The third column in Table 2 lists the major metals and other commodities in each placer district. Production figures are from Cobb (1973), and Robinson and Bundtzen (1979). Additional information on Alaskan placer deposits is provided by Cobb (1973), Robinson and Bundtzen (1979), and Cook (1983). Starting in 1980, annual conferences on Alaskan placer mining in Fairbanks have resulted in the yearly publication of a conference proceedings (for example, University of Alaska, 1989).

Placer districts, Brooks Range

Major placer districts in the Brooks Range are at Chandalar, Kiana, Noatak, Shungnak, and Wiseman in the Brooks Range. The largest gold-producing placer district is at Wiseman, which produced 9 million g of gold since its discovery in 1893. The next largest producer is the Chandalar district with 964,000 g of gold. Native gold and other heavy minerals in the Wiseman and Chandalar districts, including chalcopyrite, galena, magnetite, molybdenite, native bismuth, native copper, pyrite, scheelite, silver, and stibnite, are probably derived mainly from either volcanogenic massive sulfide deposits in the Ambler district or from Au quartz vein, Sb-Au vein, skarn, and porphyry deposits associated with Devonian or Mesozoic granitic plutons, or with Mesozoic metamorphism in the area. Gold and other heavy minerals in the Kiana, Shungnak, and Noatak districts may be derived from Au quartz vein deposits.

The Wiseman district contains perhaps the only year-round placer mine in Alaska. The operation has remained active by means of underground mining. From November to April, shafts and drifts are sunk and driven in frozen river gravel. This frozen gold-bearing gravel is brought to the surface and stacked by a self-dumping machine. From June until sometime in the fall, a three-man crew washes the thawed gravel to recover the contained gold.

Major placer districts, Seward Peninsula and western Yukon-Koyukuk, basin

Major placer districts on the Seward Peninsula are at Council, Kougarok, Nome, and Port Clarence on the Seward Peninsula. The Nome district placers are some of the larger producers of gold in Alaska and have produced as much as 140 million g of gold. Gold, cassiterite, cinnabar, columbite, scheelite-powellite, tantalite, wolframite, and other heavy minerals in these four major Seward Peninsula districts are derived mainly from Au quartz vein deposits occurring in regional metamorphic rocks and from Sn vein, Sn skarn, and Sn granite and polymetallic vein deposits associated with Cretaceous silicic granitic plutons.

The principal gold deposits in the Nome district are contained within the sand and gravel of five distinct emerged Pleistocene beaches. Several submerged beachlines are also known. The Alaska Gold Company currently operates two dredges at Nome which can process about 12,000 m^3 of gravel per day at maximum production. In addition, numerous small operators are working beaches and creeks with pans, rockers, sluice boxes, and suction dredges. In 1982, cold-water thawing continued ahead of the dredge on a 1,200-acre block of frozen ground that is estimated to contain as much as 31 million g of gold (Eakins and others, 1983). Exploration of offshore gold placers near Nome has been done by several companies.

Major placer districts in the western Yukon-Koyukuk basin are in the Fairhaven and Koyuk areas (Cobb, 1973) which are underlain by Late Cretaceous sedimentary, volcanic, and plutonic rocks. The largest placer gold producer is the Fairhaven district with production of 14 million g of gold. Gold, scheelite, magnetite, stibnite, uranothorianite, wolframite, and other heavy minerals in the districts are mainly derived from lode deposits associated with Cretaceous plutonic and volcanic centers.

Major placer districts, west-central Alaska

Major placer districts in west-central Alaska are at Aniak, Goodnews Bay, Hot Springs, Iditarod, Innoko, McGrath, Marshall, Melozitna, Rampart, Ruby, and Tolovana. The largest producer in the region is the Iditarod district with production of 41 million g of gold, followed by the Innoko district with 16.8 million g of gold, the Hot Springs district with 14 million g of gold, and the Ruby district with 12 million g of gold. The Goodnews Bay district has produced more than 16.8 million g of PGE and 0.9 million g of gold.

Gold, cassiterite, cinnabar, magnetite, native bismuth, pyrite, scheelite, stibnite, tourmaline, and other heavy minerals in the Aniak, Iditarod, Innoko, McGrath, Marshall, and Ruby districts are probably derived from polymetallic vein and porphyry deposits associated chiefly with Cretaceous and early Tertiary plutonic and volcanic centers and from Cretaceous sedimentary rocks. Chromite and platinum in the Innoko, McGrath, and Ruby districts are probably derived from Cr deposits in ultramafic rocks within thrust slices in the Tozitna and Innoko areas. Gold, cassiterite, cinnabar, magnetite, pyrite, REE minerals, scheelite, stibnite, tourmaline, and other heavy minerals in the Hot Springs, Melozitna, Rampart, and Tolovana districts are probably derived from vein deposits associated with Cretaceous or Tertiary granitic plutons and sedimentary rocks. PGE and gold in the Goodnews Bay district probably were derived from the Middle Jurassic Goodnews Bay mafic-ultramafic complex (Southworth and Foley, 1986).

Goodnews Bay district. The Goodnews Bay Pt deposit has been the largest producer of PGE in the United States (Mertie, 1976; Eakins and others, 1983; Southworth and Foley, 1986). From 1937 to 1975, approximately 16.8 million g of PGE was recovered. Large platinum nuggets are rare at Goodnews Bay; the largest recovered weighed about 124 g. Heavy-mineral concentrates, in addition to PGE minerals, include magnetite, ilmenite, chromite, and gold. Gold is a significant byproduct and makes up as much as 10 percent of the precious metal concentrate by volume. About 0.9 million g of gold has been produced. The pay streak on Salmon River is 105 to 140 m wide and reaches a maximum width of 180 m on Platinum Creek. The principal reserves remaining in this district are clay-rich parts of tailings and deep ground in the lower Salmon River drainage.

Innoko district. Placer gold has been mined intermittently in the Innoko district from modern stream and bench gravel since 1906 (Mertie, 1936; Bundtzen and Laird, 1980). Production has been about 17 million g of gold from approximately 25 placer mines over the last 75 years. The gravel is generally 2 to 6 m thick and is overlain by frozen muck 1 to 5 m thick. This muck layer must be thawed and stripped before mining the underlying gold-bearing gravel. The gold is concentrated in the lowest 1 m of gravel and in cracks in the uppermost 1 m of bedrock. Aplite and porphyry dikes intrude Cretaceous flysch bedrock; the dikes are more resistant to weathering and form ridges that act as barrier traps for the gold moving along streambeds. Gold also is found at the intersections of tributary streams with main streams where a gradient change occurs. The gold is generally fine grained and flattened, is occasionally iron stained, and includes adhering grains of quartz and magnetite. Yields of \$5.20 to \$10.50 per m^3 are common for the modern gravel; yields are as high as \$12.50 per m^3 for the bench gravels. Mineralized dikes, faults, and igneous rocks occur within or adjacent to creeks that are being mined in the area.

Most of the production in the district has been from the Ophir and Candle Hills deposits. The placer deposits in the Ophir area are downslope and downstream from basaltic to rhyolitic dike swarms and are concentrated along faults and dikes trending across stream channels (Bundtzen, 1980; Bundtzen and others, 1985). The dike swarms contain anomalous amounts of Au, Ni, Cr, and Zr. In the Candle Hills, fractures in plutons and hornfels locally contain anomalous amounts of base and precious metals and may be the lode source for the placer deposits.

Hot Springs district. About 14 million g of gold and 213,000 kg of tin have been produced from the Tofty area in the Hot Springs district; approximately 1.8 million kg of tin are estimated to remain (Wayland, 1961; Bundtzen, 1980; Robinson and others, 1982; Warner and Southworth, 1985). The heavy mineral concentrates include brown tourmaline, cassiterite (wood tin), chromite, aeschyite, tantalite, and monazite. Bedrock in the area consists of low-grade metamorphic rocks, serpentine, gabbro, quartz monzonite, and granite. The placers are in modern stream deposits and in bench deposits extending for a distance of 19 km. The lode source of Sn is probably related to granitic plutons in the area. Clasts of phyllite and quartz breccia are found with masses of cassiterite in the placers and indicate that some of the tin has been derived from Sn veins in metamorphic rocks.

Tolovana district. The Tolovana district deposits are in stream and bench gravel on a mature erosion surface largely buried by younger sediment. Gold also is present in buried bedrock benches that are not completely exhumed. Approximately 11.7 million g of gold has been produced. The Livengood gold placer deposit on Livengood Creek, the largest placer mine in the district, has been worked intermittently for 70 years. This deposit, which lies beneath a layer of frozen silt and barren gravel as much as 160 m thick, is estimated to contain 30 million m^3 of gravel averaging 1.4 g/m^3 Au. Gold in the district may be derived from polymetallic vein deposits associated with Cretaceous granitic plutons.

Major placer districts, east-central Alaska

Major placer districts in east-central Alaska are at Bonnfield, Circle, Eagle, Fairbanks, and Fortymile in the Yukon-Tanana Upland and at Kantishna. The largest producer of placer gold in Alaska is the Fairbanks district, which has produced about 238 million g of gold. Other major producers are the Fortymile district with 13 million g and the Circle district with 23 million g. Gold, base-metal sulfides, cinnabar, native silver, scheelite, and other heavy minerals in the placer deposits in these districts are probably derived from Au quartz vein, polymetallic vein, W skarn, and possible massive sulfide deposits in the region. Sparse chromite and PGE are probably derived from Cr deposits in mafic and ultramafic bodies in thrust slices.

Fairbanks district. The Fairbanks district, with production of 238 million g of gold, has produced more placer gold than any other district of Alaska (Cobb, 1973). The area also contains rich lode gold deposits. This region of Alaska has not been recently glaciated, which may account for the presence of well-developed and well-preserved deposits. The subdued topography reflects the long erosional cycles that have operated in the area, allowing ample time for the erosion of gold lode deposits and the development of placer deposits. Several cycles of alluviation during the Pleistocene have periodically concentrated and reconstructed the gold and associated heavy minerals. Late Tertiary and Quaternary alluviation, caused in part by tectonism and/or the rise of local base level, has resulted in the deposition of as much as 320 m of coarse gravel deposits. A later period of erosion has resulted in removal of much of the gravel, but basal paystreaks remain largely intact. The auriferous gravel, mainly of late Tertiary and Quaternary age, are now buried by frozen silt and other sediment, including windblown loess, which must be thawed before mining the underlying gravels.

Circle district. Placer gold has been mined in the Circle district since 1892, and approximately 23 million g of gold has been produced over the last 90 years (Yeend, 1982). Gold is concentrated in alluvial and colluvial deposits in the stream valleys draining into Birch and Crooked Creeks. In the North Fork of Harrison Creek, gold values range from $0.52 to $17.80 per m^3 but are most commonly $3.10 to $8.40 per m^3.

Substantial amounts of gold remain in the Circle district, and many placer deposits are still active. Many previously unmined gold-bearing stream channels have become attractive as a result of increases in the price of gold. A moderately large, low-grade, but as yet largely unevaluated gold resource may be contained in the extensive valley-fill deposits in the lower reaches of Crooked and Birch Creeks, as well as in the broad topographic trough on the south side of the Crazy Mountains.

Fortymile district. The Fortymile district is one of the oldest districts in Alaska; gold was discovered near the mouth of the Fortymile River in 1886 (Cobb, 1973; Eakins and others, 1983). From the time of discovery through 1961, placers in the Fortymile district were worked every year, yielding a total of about 13 million g of gold. The source of the gold in the placers is probably small polymetallic vein and Au quartz vein deposits in metamorphic rocks near contacts with Cretaceous and early Tertiary granitic plutons. Heavy minerals in the placer deposits consist of magnetite, ilmenite, hematite, barite, garnet, and pyrite and other sulfides. Small amounts of scheelite were reported from Chicken Creek and its tributaries. Both stream and bench placers have been mined in the Fortymile district. Gold nuggets as heavy as 780 g have been recovered from Jack Wade Creek, and commonly as much as 25 percent of the gold recovered is of jewelry size or larger. As recently as 1982 there were 26 active placer mines in the district.

Kantishna district. Gold placer deposits in the Kantishna district are present in modern streams and benches (Gilbert and Bundtzen, 1979; Bundtzen, 1981). Scheelite and native silver nuggets occur in the deposits. The gold and silver are probably derived mainly from polymetallic vein deposits that formed during Cretaceous regional metamorphism and plutonism. The Kantishna district contains a rich lode source for placer deposits, but placer production is currently modest; the disrict has produced about 1.4 million g Au.

Major placer districts, southern Alaska Range and Wrangell Mountains, southern Alaska

Major placer deposits in southern Alaska occur in the Chisana, Chistochina, Delta River, Nizina, Valdez Creek, and Yentna districts. The largest producer of placer gold in the area is the Chistochina district, with 4.4 million g of gold since discovery in 1898. Cassiterite, galena, magnetite, molybdenite, native copper, pyrite, silver, PGE, and other heavy minerals are probably derived from a variety of lode deposits, including Cu-Ag vein, polymetallic vein, skarn, and porphyry deposits, and from Late Jurassic and flysch, Late Jurassic and Early Cretaceous flysch of the Gravina-Nutzotin belt, and Tertiary sandstone. PGE and chromite are probably mainly derived from mafic and ultramafic rocks in the Chugach and Peninsula terranes (Plate 3, this volume). Native Cu is probably derived from the Nikolai Greenstone and associated basaltic Cu deposits.

Yentna district. Placer deposits in the Yentna district are in stream and bench deposits, Pleistocene glaciofluvial deposits, and Tertiary conglomerate and sandstone (Cobb, 1973). Placer mining in the Yentna district occurs mainly in the Petersville–Cache Creek area, which has had at least 12 separate mining operations. The largest mining operation uses two floating dredges supported by three large backhoes and has a capacity of about 3,800 m^3 per day. The district has produced approximately 3.58 million g of gold.

Valdez Creek district. Placer deposits in the Valdez Creek district are mainly in the buried gold-bearing gravel-filled Tammany channel in the Valdez Creek drainage. This channel has been mined by open-pit methods by as many as 70 persons (Bressler and others, 1985). Gold was originally discovered on Valdez Creek in 1902, and soon thereafter the buried channel was found to contain rich concentrations of placer gold. Approximately 1.2 million g of gold has been produced from the channel, and an estimated additional 2.1 million g of gold remains. The high-grade gravel averages more than 8 g gold per m^3 (Bressler and others, 1985). Exploration has identified multiple, superposed, gold-bearing paleochannels, indicating a history of successive downcutting and fluvial deposition. The placer gold mines in this district are some of the largest in Alaska.

Chistochina district. Placer gold deposits have been worked intermittently in the Chistochina district in the eastern Alaska Range since the early 1900s, with production of approximately 4.4 million g of gold (Yeend, 1981a, b). During the summer of 1985, about six deposits were being mined, three of which were in the Slate Creek area. The gold occurs in poorly sorted gravel that has diverse origins and includes alluvium, colluvium, and glaciofluvial deposits. Well-rounded boulders and cobbles derived from Tertiary(?) conglomerate are common in the deposits. Gold nuggets are rare and seldom exceed 6 mm in diameter. The bulk of the gold occurs as thin plates less than 1 mm in diameter, and large quantities of black sand make complete separation of the gold difficult. The source of the gold is probably Tertiary(?) conglomerate that is present in small isolated outcrops that commonly are small fault slivers. In the Slate Creek area, Tertiary(?) conglomerate caps the high hills to the north between Slate Creek and the Chistochina Glacier. The ultimate source of the gold and of the clasts in the Tertiary(?) conglomerate probably is rocks on the north side of the nearby Denali fault.

Major placer districts, Kodiak Island, Talkeetna and Chugach Mountains, southern Alaska

Major placer deposits occur in the Kodiak, Hope, Nelchina, and Willow Creek districts on Kodiak Island and in the Talkeetna and Chugach Mountains (Cobb, 1973). The largest producer is the Hope district with approximately 3.1 million g of gold. Gold, cinnabar, magnetite, native copper and silver, pyrite, and scheelite in the Hope, Kodiak, and Nelchina districts are probably derived mainly from Au quartz vein lode deposits in the Late Cretaceous metagraywacke and phyllite of the Valdez Group and possibly from early Tertiary granitic plutons intruding the Valdez Group. Chromite and PGE in the Kodiak district are probably derived from the Border Ranges ultramafic and mafic complex (Burns, 1985). Gold and chalcopyrite in the Willow Creek district are derived from Au quartz vein deposits occurring mainly in the Talkeetna Mountains batholith.

Placer Au and Ti in modern beach deposits occur in the Yakataga and Yakutat districts and include the Lituya Bay deposit (Thomas and Berryhill, 1962). The largest placer-gold producer is the Yakataga district, which has produced about 498,000 g of gold. Gold, chromite, magnetite, native copper, and other heavy minerals were probably derived from a combination of bedrock sources in eastern southern Alaska. These deposits are near the mouth of the Copper River, which drains parts of the Wrangellia, Peninsular, Chugach, and Prince William terranes (Plate 3, this volume).

Major placer districts, southeastern Alaska

Major placer deposits occur in the Junean and Porcupine Creek districts in southeastern Alaska (Wright, 1904; Cobb, 1973). The Juneau and Porcupine Creek district each has produced about 1.9 million g of gold. The Porcupine Creek placer deposits are in bench and stream gravel. The Juneau district deposits are in hill, residual, gulch, and creek placers. Alluvial gravel contains most of the placer gold; much of the gold, however, has been eroded and transported by glaciers and some is in submerged glacial deposits. Placer gold and associated heavy minerals in the Juneau district deposits are probably derived mainly from Au quartz vein deposits in the Juneau gold belt (Spencer, 1906).

SUMMARY

The local geology, classification, and metallogenesis of the major metalliferous lode and placer mineral deposits of Alaska are described for each of the state's seven regions. The deposits are classified into types by comparing the properties of each deposit with current mineral-deposit models. The mineral-deposit types in Alaska generally form specific suites for each geographic region. Within each region, the metalliferous lode mineral deposits are generally restricted to geologic units of narrow age ranges in major fold, thrust, and/or igneous belts. The origin and modification of the lode deposits in these belts is mainly related to specific sedimentary, magmatic, metamorphic, and/or deformational events such as deep marine, continental shelf, and epicontinental sedimentation; volcanism and plutonism in island-arc or in submerged continental-margin arc settings; arc and back-arc volcanism and plutonism along continental margins; oceanic rifting and continental rifting; regional metamorphism; and regional deformation.

The northwestern Brooks Range contains several sedimentary exhalative Zn-Pb, one bedded barite, several podiform chromite, one Kipushi Cu-Pb-Zn, and various vein deposits. The southern flank of the central Brooks Range contains an extensive suite of major Kuroko massive sulfide deposits and one major carbonate-hosted Kipushi Cu-Pb-Zn deposit. The central Brooks Range contains a suite of moderate-size polymetallic vein, Au quartz vein, Sb-Au vein, porphyry Cu and Mo, and Cu-Pb-Zn and Sn skarn deposits. The northeastern Brooks Range contains a cluster of Pb-Zn skarn, polymetallic vein, and porphyry Mo deposits. The Brooks Range contains five major districts of placer Au deposits.

The Seward Peninsula contains an extensive suite of Sn vein, Sn skarn, and Sn greisen deposits, a suite of Au quartz vein deposits, several polymetallic vein deposits, and individual porphyry Mo, felsic plutonic U, sandstone U, and metamorphosed sulfide deposits. The Seward Peninsula contains four major districts of placer Au deposits and two districts of combined placer Au and Sn deposits.

The southwestern Kuskokwim Mountains of west-central Alaska contain an Sb-Hg vein deposit, a zoned mafic-ultramafic Fe-Ti deposit, and a hot-spring Hg deposit. The central Kuskokwim Mountains contain a complex and extensive suite of Au quartz vein, Sb-Au vein, polymetallic vein, epithermal vein, hot spring Hg, Cu and Fe skarn, and felsic plutonic U deposits, and a carbonate-hosted sulfide deposit. The northeastern Kuskokwim Mountains contains several podiform chromite deposits in thin discontinuous thrust sheets of ultramafic and related rocks. The west-central Yukon-Koyukuk basin in west-central Alaska contains a suite of polymetallic vein and porphyry Mo and Cu deposits. The northern and eastern Yukon-Koyukuk basin contains suites of felsic plutonic U, polymetallic and epithermal vein, and W skarn deposits, and a suite of podiform chromite and serpentinite-hosted asbestos deposits in thin, discontinuous thrust sheets of ultramafic and related rocks. West-central Alaska contains seven major districts of placer Au deposits and two major districts of combined placer PGE and Au deposits.

The Manley and Livengood areas in east-central Alaska contain polymetallic vein, Sb-Au vein, Mn-Ag vein, Hg vein, felsic plutonic U, Sn greisen, and Sn vein deposits. The northern and central parts of the Yukon-Tanana Upland in east-central Alaska contain suites of Sb-Au vein, Au-quartz vein, polymetallic vein, W skarn, and porphyry Cu-Mo deposits, a Au-As vein deposit, and a Sn greisen deposit. The Manley and Livengood areas also contain a serpentinite-hosted asbestos deposit and a minor Pt deposit in thrust sheets of ultramafic and associated rocks. The northern Alaska Range in east-central Alaska contains an extensive district of polymetallic and Sb-Au vein and Kuroko massive sulfide deposits. East-central Alaska contains nine major placer Au districts.

The Aleutian Islands and southwest Alaska Peninsula contain an extensive suite of epithermal and polymetallic vein and porphyry Cu and Mo deposits. The northeast Alaska Peninsula contains suites of Cu-Zn-Au and Fe skarn, polymetallic vein, and porphyry Cu deposits, and one epithermal vein deposit. This region contains no major placer districts.

The southwestern Alaska Range in southern Alaska contains a suite of Cu-Pb-Zn skarn, polymetallic vein, Sn greisen and vein, and porphyry Cu-Au and Mo vein deposits, a Besshi massive sulfide deposit, and a gabbroic Ni-Cu(?) deposit. The central and eastern Alaska Range and the Wrangell Mountains in southern Alaska contain a suite of Cu-Ag and Fe skarn, polymetallic vein, and porphyry Cu and Cu-Mo deposits, and a suite of Cu-Ag quartz vein, basaltic Cu, and Besshi massive sulfide deposits. In southern Alaska, the Talkeetna Mountains contain a suite of Au quartz vein deposits, and a suite of podiform chromite deposits is present on Kodiak Island, the Kenai Peninsula, and in the northern Chugach Mountains. The southern Chugach Mountains, southeast Kenai Peninsula, and Kodiak Island contain an extensive suite of Au quartz vein deposits, and the Prince William Sound district contains an extensive suite of Besshi and Cyprus massive sulfide deposits. Southern Alaska contains eleven placer Au districts and one combined placer Au and Ti district.

The Coast Mountains region in southeastern Alaska contains extensive suites of Au quartz vein, metamorphosed sulfide, and zoned mafic-ultramafic deposits, a suite of Fe skarn and porphyry Mo deposits, and a Besshi massive sulfide deposit. Central southeastern Alaska contains extensive suites of Kuroko massive sulfide and bedded barite deposits, metamorphosed sulfide deposits, Cu-Zn-Au-Ag and Fe skarn and porphyry Cu deposits, Au quartz vein deposits, zoned mafic-ultramafic deposits, a gabbroic Ni-Cu deposit, and a felsic plutonic U and a sandstone U deposit. Coastal southeastern Alaska contains suites of Au quartz vein and gabbroic Ni-Cu deposits and a basaltic Cu deposit. Southeastern Alaska contains two major placer Au districts.

APPENDIX 1. MAJOR TYPES OF ALASKAN METALLIFEROUS LODE AND PLACER MINERAL DEPOSITS

CLASSIFICATION OF MINERAL DEPOSITS

Metalliferous lode deposits in this report are classified into 29 types, and placer deposits are classified into four types, described below. Types of placer deposits are listed last. In the following descriptions, some lode mineral deposits, such as various types of contact metasomatic or porphyry deposits, share a common origin and are grouped according to the classification of the dominant type.

The mineral-deposit models used in this report, and as described by various mineral-deposits geologists in Cox and Singer (1986), consist of both descriptive and genetic information that is systematically arranged to describe the essential properties of a class of mineral deposits. Some models are descriptive (empirical), in which case the various attributes are recognized as essential, even though their relations are unknown. An example is the basaltic Cu model, as adapted by Nokleberg and others (1987), in which the empirical geologic association of Cu sulfides with relatively Cu-rich metabasalt or greenstone is the essential attribute. Other models are genetic (theoretical), in which case the attributes are interrelated through some fundamental concept. An example is the W or Fe skarn (contact metasomatic) deposit model, in which the genetic process of contact metasomatism is the essential attribute. For additional information on the methodology of mineral deposit models, the reader is referred to the discussion by Cox and Singer (1986).

LODE DEPOSIT TYPES

Kuroko massive sulfide deposit

D. A. Singer* in *Cox and Singer, 1986

This deposit type consists of volcanogenic, massive to disseminated sulfides in felsic to intermediate marine volcanic and pyroclastic rocks and interbedded sedimentary rocks. The volcanic rocks are mainly rhyolite and dacite with subordinate basalt and andesite. The depositional environment is mainly hot springs related to marine volcanism in island-arc or extensional regimes. The deposit minerals include pyrite, chalcopyrite, sphalerite, and lesser galena, tetrahedrite, tennantite, and magnetite. Alteration products including zeolite, montmorillonite, silica, chlorite, and sericite may be present. Notable examples of Kuroko massive sulfide deposits in Alaska are the Arctic, Smucker, and Sun deposits in the Brooks Range, the WTF, Red Mountain, and Delta district deposits in east-central Alaska, and the Greens Creek, Glacier Creek, Khayyam, and Orange Point deposits in southeastern Alaska.

Besshi massive sulfide deposit

D. P. Box* in *Cox and Singer, 1986

This deposit type consists of thin, sheetlike bodies of massive to well-laminated pyrite, pyrrhotite, and chalcopyrite, and less abundant sulfide minerals, within thinly laminated clastic sedimentary rocks and mafic tuff. The rock types are mainly marine clastic sedimentary rocks, basaltic and lesser andesite tuff and breccia, and local black shale and red chert. The depositional environment is uncertain, but may possibly be submarine hot springs related to submarine basaltic volcanism. Associated minerals include sphalerite and lesser magnetite, galena, bornite, and tetrahedrite, with gangue quartz, carbonate, albite, white mica, and chlorite. Alteration is sometimes difficult to recognize because of metamorphism. Notable examples of Besshi massive sulfide deposits in Alaska are the Midas, Latouche, Beatson, Ellamar, and Fidalgo-Alaska mines in the Prince William Sound region of southern Alaska.

Cyprus massive sulfide deposit

D. A. Singer* in *Cox and Singer, 1986

This deposit type consists of massive sulfides in pillow basalt. The depositional environment consists of submarine hot springs along an axial graben in oceanic or backarc spreading ridges or of hot springs related to submarine volcanoes in seamounts. The deposit minerals consist mainly of pyrite, chalcopyrite, sphalerite, and lesser marcasite and pyrrhotite. The sulfides occur in pillow basalts that are associated with tectonized dunite, harzburgite, gabbro, sheeted diabase dikes, and fine-grained sedimentary rocks, all part of an ophiolite assemblage. Beneath the massive sulfides in places is stringer or stockwork pyrite, pyrrhotite, and minor chalcopyrite and sphalerite. The sulfide minerals are locally brecciated and recemented. Alteration in the stringer zone consists of abundant quartz, chalcedony, chlorite, and some illite and calcite. Some deposits are overlain by Fe-rich and Mn-poor ochre. Notable examples of Cyprus massive sulfide deposits in Alaska are the Knight Island and Threeman mines and the Copper Bullion deposit, all in coastal southern Alaska.

Sedimentary exhalative Zn-Pb deposit

J. A. Briskey* in *Cox and Singer, 1986

This deposit type consists of stratiform, massive to disseminated sulfides in sheet-like or lens-like tabular bodies that are interbedded with euxinic marine sedimentary rocks including dark shale, siltstone, limestone, chert, and sandstone. The depositional environment consists mainly of marine epicratonic embayments and intracratonic basins, with smaller local restricted basins. The deposit minerals include pyrite, pyrrhotite, sphalerite, galena, barite, and chalcopyrite. Extensive alteration may be present, including stockwork and disseminated sulfides, silica, albite, and chlorite. Notable examples of sedimentary exhalative Zn-Pb deposits in Alaska are the Lik and Red Dog Creek deposits in the northwestern Brooks Range.

Kipushi Cu-Pb-Zn (carbonate-hosted Cu) deposit

D. P. Cox* in *Cox and Singer, 1986

This deposit type consists of stratabound, massive sulfides hosted mainly in dolomitic breccia. The depositional environment consists mainly of strong fluid flow along faults or karst(?)-breccia zones. Generally no rocks of unequivocal igneous origin are related to the deposit. The deposit minerals include pyrite, bornite, chalcocite, chalcopyrite, carrollite, sphalerite, and tennantite with minor renierite and germanite. Local dolomite, siderite, and silica alteration may occur. Notable examples of carbonate-hosted Cu deposits in Alaska are the Ruby Creek and Omar deposits in the Brooks Range.

Metamorphosed sulfide deposit

Nokleberg and others, 1987

This deposit type consists of stratabound, massive to disseminated sulfides hosted in moderately to highly metamorphosed and deformed metavolcanic or metasedimentary rocks. Metamorphism and deformation have obscured protoliths of host rocks and deposits so as to preclude classification into more specific deposit types. The interpreted host rocks for these deposits are mainly felsic to mafic metavolcanic rocks and metasedimentary or metavolcanic schist and gneiss. The deposit minerals include chalcopyrite, sphalerite, galena, and bornite, sometimes with pyrite, magnetite, and hematite. Alteration is usually difficult to recognize because of metamorphism. These deposits occur mainly in the regional metamorphic rocks in southeastern Alaska in either the informally named Coast plutonic-metamorphic complex of Brew and Ford (1984a, b) or in the Alexander terrane. Notable examples of metamorphosed sulfide deposits are the Sweetheart Ridge, Sumdum, Groundhog Basin, and Moth Bay deposits, all in southeastern Alaska.

Bedded barite deposit

G. J. Orris* in *Cox and Singer, 1986

This deposit type consists of stratiform, massive barite interbedded either with marine cherty and calcareous sedimentary rocks, mainly dark chert, shale, mudstone, and dolomite, or with marine volc and volcaniclastic rocks. The depositional environment consists either of epicratonic marine basins or embayments, often with satellitic smaller local basins, or of volcanic arc or extensional regimes. Bedded barite deposits are often associated with sedimentary exhalative Zn-Pb or Kuroko massive sulfide deposits. Alteration consists of secondary barite veining and local, weak to moderate sericite replacement. Associated minerals include minor witherite, pyrite, galena, and sphalerite. Notable examples of bedded barite deposits in Alaska are the Nimiuktuk deposit in the northwestern Brooks Range and the Castle Island mine in southeastern Alaska.

Sandstone U deposit

C. T. Peterson and C. A. Hodges* in *Cox and Singer, 1986

This deposit type consists of concentrations of U oxides and related U minerals in localized, reduced environments in medium- to coarse-grained feldspathic or tuffaceous sandstone, arkose, mudstone, and conglomerate. The depositional environment is continental basin margins, fluvial channels, fluvial fans, or stable coastal plain, locally with nearby felsic plutons or felsic volcanic rocks. The deposit minerals include pitchblende, coffinite, carnotite, and pyrite. The notable example of a sandstone U deposit in Alaska is the Death Valley deposit in west-central Alaska.

Basaltic Cu deposit

Adapted from D. P. Cox* in *Cox and Singer, 1986

This deposit type consists of copper sulfides in large pipes and lenses in carbonate rocks within a few tens of meters of disconformably underlying subaerial basalt. The depositional environment consists of subaerial basalts overlain by mixed shallow marine and nearshore carbonate sedimentary rocks, including sabhka facies carbonate rocks; subsequent subaerial erosion, groundwater leaching and/or low-grade regional metamorphism may concentrate copper sulfides into pipes and lenses. The deposit minerals consist mainly of chalcocite and lesser bornite, chalcopyrite, and other Cu sulfides, pyrite, and oxidized Cu minerals such as malachite and azarite. Alteration minerals may include metamorphic chlorite, actinolite, epidote, albite, quartz, and zeolites, and secondary dolomite. Notable examples of basaltic Cu deposits in Alaska are the Kennecott, Westover, Nelson, and Erickson mines, all in southern Alaska. This deposit type may be transitional to Besshi massive sulfide deposits, particularly those that occur in pelitic sedimentary rocks interlayered with basalt and greenstone derived from basalt, such as at the Denali deposit.

Hot-spring Hg deposit

J. J. Rytuba* in *Cox and Singer, 1986

This deposit type consists of cinnabar, antimony, pyrite, and minor marcasite and native Hg in veins and in disseminations in graywacke, shale, andesite and basalt flows, andesite tuff and tuff breccia, and diabase dikes. The depositional environment is near the groundwater table in areas of former hot springs. Various alteration minerals such as kaolinite, alunite, Fe oxides, and natiave sulfur occur above the former ground-water table; pyrite, zeolites, potassium feldspar, chlorite, and quartz occur below it. Notable examples of hog-spring Hg deposits in Alaska are the Red Devil, De Coursey Mountain, and Cinnabar Creek mines in west-central Alaska.

Epithermal vein deposit

D. L. Mosier, T. Sato, N. J. Page, D. A. Singer, and B. R. Berger in *Cox and Singer, 1986*

This deposit type consists of quartz-carbonate-pyrite veins containing a variety of minerals, including gold, silver sulfosalts, chalcopyrite, argentite, galena, sphalerite, and arsenopyrite. The veins occur in felsic to intermediate volcanic rocks, sometimes overlying igneous intrusions or older volcanic sequences. One class of epithermal vein deposit, such as the one at Creede, Colorado, has high Pb, Zn, and Ag, sometimes high Cu, and low Au concentrations; another class, such as the one at Sado, Japan, has high Au, moderate to low Ag, locally high Cu, and generally low Pb and Zn concentrations. For both classes, the host volcanic rock composition ranges from andesite to rhyolite. The depositional environment is intermediate to felsic volcanic arcs and volcanic centers. Associated minerals include electrum, chalcopyrite, copper and silver sulfosalts, with lesser tellurides and bornite. Alteration minerals include quartz, kaolinite, montmorillonite, illite, and zeolites. Notable examples of epithermal deposits in Alaska are the Aquila and Shumagin deposits, and the Apollo-Sitka mine on the Alaska Peninsula.

Low-sulfide Au quartz vein deposit

B. R. Berger in *Cox and Singer, 1986*

This deposit type, abbreviated to Au quartz vein in this report, consists of gold in massive, persistent quartz veins in regionally metamorphosed volcanic rocks, and in metamorphosed graywacke, chert, and shale. The depositional environment is low-grade metamorphic belts. The veins are generally late metamorphic to postmetamorphic and locally cut granitic rocks. Associated minerals are minor pyrite, galena, sphalerite, chalcopyrite, arsenopyrite, and pyrrhotite. Alteration minerals include quartz, siderite, albite, and carbonate. Notable examples of Au quartz veins in Alaska are the Big Hurrah mine on the Seward Peninsula, the Chandalar district mines in the southern Brooks Range, the Willow Creek district mines, the Nuka Bay, Monarch, Jewel, Granite, and Cliff mines in southern Alaska, and the Alaska-Juneau, Jualin, Kensington, Sumdum Chief, Treadwell, Nido, and Chichagoff mines in southeastern Alaska.

Cu-Ag quartz vein deposit

Nokleberg and others, 1987

This deposit type consists of Cu sulfides and accessory Ag in quartz veins and disseminations in regionally metamorphosed mafic igneous rocks, mainly basalt, gabbro, and lesser andesite and dacite. The depositional environment is low-grade metamorphic belts. The veins are generally late-stage metamorphic. The deposit minerals include chalcopyrite, bornite, lesser chalcocite and pyrite, and rare native copper. Alteration minerals include epidote, chlorite, actinolite, albite, quartz, and zeolites. Notable examples of Cu-Ag quartz veins in Alaska are the Kathleen-Margaret and Nikolai mines in southern Alaska.

Polymetallic vein deposit

D. P. Cox in *Cox and Singer, 1986*

This deposit type consists of quartz-carbonate veins carrying Ag, Au, and base-metal sulfides. The veins are related to hypabyssal intrusions in sedimentary and metamorphic terranes or to metamorphic fluids forming during waning regional metamorphism. The igneous rocks range in composition from calcalkaline to alkaline and occur as dike swarms, shallow plugs or diapirs, and small to moderate-size, intermediate to felsic plutons that locally underlie coeval andesite to rhyolite flows. The depositional environment is near-surface fractures and breccias in thermal aureoles of the intrusions and also within the intrusions. The deposit minerals include native gold, electrum, pyrite, and sphalerite, locally with chalcopyrite, galena, arsenopyrite, tetrahedrite, Ag sulfosalts, and argentite. Alteration consists of wide propylitic zones and narrow sericitic and argillic zones. Notable examples of polymetallic veins in Alaska are the Independence and Golden Horn mines and the Broken Shovel and Beaver Creek deposits in west-central Alaska, the Quigley Ridge, Banjo, Spruce Creek, and Stampede deposits in the Kantishna district of east-central Alaska, the Cleary Summit and Ester dome mines in the Fairbanks district of east-central Alaska, the Sedanka and Bonanza Hills deposits of the Alaska Peninsula, and the Golden Zone deposit of southern Alaska.

Sb-Au vein deposit

Adapted from Sb deposit of J. D. Bliss and G. J. Orris in *Cox and Singer, 1986*

This deposit type consists of massive to disseminated stibnite and lesser gold in quartz-carbonate veins, pods, and stockworks in or adjacent to brecciated or sheared zones and faults; in sedimentary, volcanic, and metamorphic rocks adjacent to granitic plutons; in contact aureoles around granitic plutons; and in peripheries of granodiorite, granite, and monzonite stocks. The depositional environment is faults and shear zones that are epizonal fractures adjacent to, or within the margins of epizonal granitic plutons. Associated minerals include arsenopyrite, chalcopyrite, and tetrahedrite, locally with cinnabar and galena. This deposit type is locally associated with polymetallic vein deposits. Alteration consists mainly of silica, sericite, and argillite. Notable examples of Sb-Au veins in Alaska are the Slate Creek, Eagles Den, and Caribou Creek deposits in the Kantishna district of east-central Alaska and the Scrafford mine in east-central Alaska.

Sn greisen, Sn vein, and Sn skarn deposits

B. L. Reed and D. P. Cox in *Cox and Singer, 1986*

These three deposit types commonly occur in the same area and locally grade into one another. Sn greisen deposit type consists of disseminated cassiterite, cassiterite-bearing veinlets, and Sn sulfosalts in stockworks, lenses, pipes, and breccia in greisenized, mainly biotite and/or muscovite leucogranite emplaced in a mesozonal to deep volcanic environment. Sn greisens are generally postmagmatic and are associated with late-stage, fractionated granitic magma. Associated minerals include molybdenite, arsenopyrite, beryl, and wolframite. Alteration consists of incipient to massive greisen with quartz, muscovite, tourmaline, and fluorite replacement. Notable examples of Sn greisen deposits in Alaska are the Kougarok deposit on the Seward Peninsula and the Coal Creek deposit in southern Alaska.

Sn vein deposit type consists of simple to complex fissure fillings or replacement lodes in or near felsic, mainly mesozonal to hypabyssal plutons, often with dike swarms. The deposits tend to occur within or above the apices of granitic cusps and ridges. The deposit minerals are varied and include cassiterite, wolframite, arseonpyrite, molybdenite, scheelite, and beryl. Alteration minerals consist of sericite, tourmaline, quartz, chlorite, and hematite. The notable example of a Sn vein deposit in Alaska is the Lime Peak deposit in east-central Alaska.

Sn skarn deposit type consists of Sn, W, and Be minerals in skarns, veins, stockworks, and greisen near intrusive contacts between epizonal(?) granitic plutons and limestone. The deposit minerals include cassiter, locally accompanied by scheelite, sphalerite, chalcopyrite, pyrrhotite, magnetite, and fluorite. Alteration consists of greisen in granite margins and metasomatic development of andradite, idocrase, amphibole, chlorite, and mica in skarn. The notable example of a Sn skarn deposit in Alaska is the Lost River mine on the Seward Peninsula.

Cu-Zn-Pb (± Au, Ag), W, and Fe (± Au) skarn deposits

D. P. Cox and T. G. Theodore* in *Cox and Singer, 1986

Cu-Zn-Pb skarn deposit type consists of chalcopyrite, sphalerite, and galena in calc-silicate skarns that replace carbonate rocks along intrusive contacts with quartz diorite to granite and diorite to syenite plutons. Zn-Pb-rich skarns tend to occur farther from the intrusion; Cu-rich and Au-rich skarns tend to occur closer to the intrusion. The depositional environment is mainly calcareous sedimentary sequences intruded by felsic to intermediate granitic plutons. Associated minerals include pyrite, hematite, magnetite, bornite, arsenopyrite, and pyrrhotite. Metasomatic replacements consist of a wide variety of calc-silicate and related minerals. Notable examples of Cu-Zn-Pb skarn deposits in Alaska are the Bowser Creek, Rat Fork, Sheep Creek, and Tin Creek deposits. Notable examples of Cu-Au and Au skarn deposits in Alaska are the Nixon Fork-Medfra mine in west-central Alaska and the Jumbo mine in southeastern Alaska.

W skarn deposit type consists of scheelite in calc-silicate skarns that replace carbonate rocks along or near intrusive contacts with quartz diorite to granite plutons. The depositional environment is along contacts and in roof pendants of batholiths and in thermal aureoles of stocks that intrude carbonate rocks. Associated minerals are molybdenite, pyrrhotite, sphalerite, chalcopyrite, bornite, pyrite, and magnetite. Metasomatic replacements consist of a wide variety of calc-silicate and other contact metamorphic minerals. Notable examples of W skarns in Alaska are the deposits and mines in the Gilmore Dome area of the Fairbanks district in east-central Alaska.

Fe skarn deposit type consists of magnetite and/or Fe sulfides in calc-silicate skarns that replace carbonate rocks or calcareous clastic rocks along intrusive contacts with diorite, granodiorite, granite, and coeval volcanic rocks. The depositional environment is along intrusive contacts. The chief associated minerals are chalcopyrite and gold. Metasomatic replacements consist of a wide variety of calc-silicate and associated minerals. Notable examples of Fe skarns in Alaska are the Medfra deposit in west-central Alaska, and the Nabesna and Rambler mines in southern Alaska.

Porphyry Cu-Mo, porphyry Cu, and porphyry Mo deposit

D. P. Cox and T. G. Theodore* in *Cox and Singer, 1986

The porphyry Cu-Mo deposit type consists of stockwork veinlets of quartz, chalcopyrite, and molybdenite in or near porphyritic intermediate to felsic intrusions. The intrusions are mainly stocks and breccia pipes that intrude batholithic, volcanic, or sedimentary rocks. The depositional environment is high-level intrusive porphyries that are contemporaneous with abundant dikes, faults, and breccia pipes. Associated minerals include pyrite and peripheral sphalerite, galena, and gold. Alteration minerals consist of quartz, K-feldspar, and biotite or chlorite. Notable examples of porphyry Cu-Mo deposits in Alaska are the Taurus deposit in east-central Alaska, the Orange Hill, Bond Creek, Baultoff, Horsfeld, and Carl Creek deposits in southern Alaska, and the Pyramid deposit on the Alaska Peninsula.

Porphyry Cu deposit type consists of chalcopyrite in stockwork veinlets in hydrothermally altered porphyry and adjacent country rock. The porphyries range in composition from tonalite to monzogranite to syenitic porphyry. The depositional environment is epizonal intrusive rocks with abundant dikes, breccia pipes, cupolas of batholiths, and faults. Associated minerals include pyrite, molybdenite, magnetite, and bornite. Alteration consists of sodic, potassic, phyllic, argillic, and propylitic types. An example of a porphyry Cu deposit in Alaska is the Margerie deposit in southeastern Alaska.

Porphyry Mo deposit type consists of quartz-molybdenite stockwork veinlets in granitic porphyry and adjacent country rock. The porphyries range in composition from tonalite to granodiorite to monzogranite. The depositional environment is epizonal. Associated minerals are pyrite, scheelite, chalcopyrite, and tetrahedrite. Alteration is potassic grading outward to propylitic, sometimes with phyllic and argillic overprint. A notable example of a porphyry Mo deposit in Alaska is the Quartz Hill deposit in southeastern Alaska.

Felsic plutonic U deposit

Nokleberg and others, 1987

This deposit type consists of disseminated U, Th, and REE minerals in fissure veins and alkalic granite dikes in, or along the margins of alkalic and peralkalic granitic plutons or in granitic plutons including granite, alkalic granite, granodiorite, syenite, and monzonite. The depositional environment is mainly the margins of epizonal to mesozonal granitic plutons. The deposit minerals include allanite, thorite, uraninite, bastnaesite, monazite, uranothorianite, and xenotime, sometimes with galena and fluorite. Notable examples of felsic plutonic U deposits in Alaska are the Mount Prindle deposit in east-central Alaska and the Bokan Mountain deposits in southeastern Alaska.

Gabbroic Ni-Cu deposit

Adapted from synorogenic-synvolcanic Ni-Cu deposit of N. J. Page* in *Cox and Singer, 1986

This deposit type consists of massive sulfide lenses and disseminated sulfides in small to medium-size gabbroic intrusions in metamorphic belts of metasedimentary and metavolcanic rocks. In most areas of Alaska, the depositional environment consists of post-metamorphic and post-deformational, intermediate level intrusions of norite, gabbronorite, and ultramafic rocks. The deposit minerals include pyrrhotite, pentlandite, and chalcopyrite, sometimes with pyrite, Ti or Cr magnetite, and PGE minerals and alloys. Accessory Co also occurs in some deposits. Notable examples of gabbroic Ni-Cu deposits in Alaska are the Funter Bay, Brady Glacier, Bohemia Basin, and Mirror Harbor deposits, all in southeastern Alaska.

Zoned mafic-ultramafic Cr-Pt (± Cu, Ni, Co, Ti or Fe) deposit

Adapted from Alaskan PGE deposit type of N. J. Page and F. Gray* in *Cox and Singer, 1986

This deposit type consists of more or less concentrically zoned ultramafic to mafic intrusions that contain chromite, PGE (Pt-group elements), PGE minerals and alloys, and Ti-V magnetite. In most areas of Alaska, the depositional environment consists of post-metamorphic and post-deformational, intermediate-level intrusion of mafic and/or ultramafic plutons. The deposit minerals include combinations of chromite, PGE, pentlandite, pyrrhotite, Ti-V magnetite, bornite, and chalcopyrite. Notable examples of zoned mafic-ultramafic deposits in Alaska are the Kemuk Mountain deposit in west-central Alaska, and the Union Bay, Duke Island, and Klukwan deposits in southeastern Alaska. The Silurian gabbro-hosted Salt Chuck Au-Pd deposit in southeastern Alaska apparently in a unique type of deposit distinct from those hosted by the zoned mafic-ultramafic bodies. We include in it this section for convenience.

Podiform chromite deposit

J. P. Albers in *Cox and Singer, 1986*

This deposit type consists of podlike masses of chromite in the ultramafic parts of locally intensely faulted and dismembered ophiolite complexes. The host rock types are mainly dunite and harzburgite, commonly serpentinized. The depositional environment consists of magmatic cumulates in elongate magma pockets. The deposit minerals include chromite, magnetite, and PGE minerals and alloys. Notable examples of podiform chromite deposits in Alaska are the Iyikrok Mountain and Avan deposits in the northwestern Brooks Range, the Kaiyuh River deposit in west-central Alaska, and the Halibut Bay and Claim Point deposits and the Red Mountain mine in southern Alaska.

Serpentinite-hosted asbestos deposit

N. J. Page in *Cox and Singer, 1986*

This deposit type consists of chrysotile asbestos developed in stockworks in serpentinized ultramafic rocks. The depositional environment is usually an ophiolite sequence, locally with later deformation or igneous intrusion. Associated minerals are magnetite, brucite, talc, and tremolite. The notable example of a serpentinite-hosted asbestos deposit in Alaska is the Fortymile deposit in east-central Alaska.

PLACER DEPOSIT TYPES

Placer Au deposit

W. Yeend in *Cox and Singer, 1986*

This deposit type consists of elemental gold as grains and rarely as nuggets in gravel, sand, silt, and clay and their consolidated equivalents in alluvial, beach, aeolian, and glacial deposits. The depositional environment is high-energy alluvial where gradients flatten and river velocities lessen, as at the inside of meanders, below rapids and falls, and beneath boulders, and in shoreline and beaches where the winnowing action of surf causes Au concentrations. The major deposit minerals are gold, sometimes with attached quartz, magnetite, and/or ilmenite. Notable examples of placer Au deposits in Alaska are in the Wiseman district in the southern Brooks Range, the Nome, Council, and Fairhaven districts on the Seward Peninsula, the Marshall, Aniak, Iditarod, Innoko, McGrath, Ruby, Hughes, Hot Springs, and Tolovana districts in west-central Alaska, the Fairbanks, Circle, Fortymile, and Kantishna placer districts in east-central Alaska, the Valdez, Chistochina, Nizina, Hope, and Willow Creek districts in southern Alaska, and the Porcupine Creek and Juneau districts in southeastern Alaska.

Placer Sn deposit

Nokleberg and others, 1987

This deposit type consists mainly of cassiterite and elemental gold in grains in gravel, sand, silt, and clay and their consolidated equivalents mainly in alluvial deposits. The depositional environment is similar to that of placer Au deposits. Notable examples of placer Sn deposits in Alaska are those derived from Sn granites, such as in the Kougarok district on the Seward Peninsula and the Hot Springs district in west-central Alaska.

Placer PGE-Au deposit

W. Yeend and N. J. Page in *Cox and Singer, 1986*

This deposit type consists of PGE minerals and alloys in grains in gravel, sand, silt, and clay and their consolidated equivalents in alluvial, beach, aeolian, and glacial deposits. In some areas, placer Au and placer PGE deposits occur together. The depositional environment is high-energy alluvial where gradients flatten and river velocities lessen, as at the inside of meanders, below rapids and falls, and beneath boulders, and in shoreline areas where the winnowing action of surf causes PGE and Au concentrations in beaches. The major deposit minerals are Pt-group alloys, Os-Ir alloys, magnetite, chromite, and/or ilmenite. The notable example of a placer PGE deposit in Alaska is the Goodnews Bay placer district.

Shoreline placer Ti deposit

E. R. Force in *Cox and Singer, 1986*

This deposit type consists of ilmenite and other heavy minerals concentrated by beach processes and enriched by weathering. The hosting sediment types are medium- to fine-grained sand in dune, beach, and inlet deposits. The depositional environment is stable coastal region receiving sediment from bedrock regions. The major deposit minerals are low-Fe ilmenite, sometimes with rutile, zircon, and gold. Notable examples of shoreline placer Ti deposits in Alaska are the Yakutat (Lituya Bay) placer districts.

REFERENCES CITED

AEIDC (Arctic Environmental Information and Data Center), 1982, Mineral terranes of Alaska: AEIDC, Anchorage, Alaska, 1 p., 6 sheets, scale 1:1,000,000.

Alaska Mines and Geology, 1983, Shumagin Island gold mine shows promise of good returns, October, p. 13.

——, 1985, Firm wants to develop new gold mine at old (Chichagoff Mine) site, April, p. 7–8.

Allegro, G. L., 1984, Geology of the Old Smokey Prospect, Livengood C-4 quadrangle, Alaska: Alaska Division of Geological and Geophysical Surveys Report of Investigation ROI 84-1, 10 p.

Aleinikoff, J. N., and Nokleberg, W. J., 1985, Age of Devonian igneous arc terranes in the northern Mount Hayes quadrangle, eastern Alaska Range, Alaska, *in* Bartsch-Winkler, S., ed., The United States Geological Survey in Alaska: Accomplishments during 1984: U.S. Geological Survey Circular 967, p. 44–49.

Armstrong, A. K., and MacKevett, E. M., Jr., 1982, Stratigraphy and diagenetic history of the lower part of the Triassic Chitistone Limestone, Alaska: U.S. Geological Survey Professional Paper 1212-A, 26 p.

Amrstrong, R. L., 1985, Rb-Sr dating of the Bokan Mountain granite complex and its country rocks: Canadian Journal of Earth Sciences, v. 22, no. 8, p. 1233–1236.

Armstrong, R. L., Harakal, J. E., and Hollister, V. F., 1976, Age determinations of late Cenozoic copper deposits of the North American Cordillera: Institute of Mining and Metallurgical Engineers Transactions, Section B, v. 85, p. 239–244.

Barker, Fred, 1963, The Funter bay nickel-copper deposit, Admiralty Island, Alaska: U.S. Geological Survey Bulletin 1155, p. 1–10.

Barker, J. C., and Swainbank, R. C., 1986, A tungsten-rich porphyry molybdenum occurrence at Bear Mountain, northeast Alaska: Economic Geology, v. 81, p. 1753–1759.

Barrie, T.C.P., 1984a, The geology of the Khayyam and Stumble-On deposits, Prince of Wales Island, Alaska [M.A. thesis]: Austin, Texas, University of Texas, 172 p.

——, 1984b, Geology of the Khayyam and Stumble-On massive sulfide deposits, Prince of Wales Island, Alaska [abs.]: Geological Society of America Abstracts with Programs, v. 16, p. 268.

Bateman, A. M., and McLaughlin, D. H., 1920, Geology of the ore deposits of Kennocott, Alaska: Economic Geology, v. 15, p. 1–80.

Berg, H. C., 1984, Regional geologic summary, metallogenesis, and mineral resources of southeastern Alaska: U.S. Geological Survey Open-File Report

84-572, 298 p., 1 map sheet, scale 1:600,000.

Berg, H. C., and Cobb, E. H., 1967, Metalliferous lode deposits of Alaska: U.S. Geological Survey Bulletin 1256, 254 p.

Berg, H. C., Jones, D. L., and Richter, D. H., 1972, Gravina-Nutzotin belt—Tectonic significance of an upper Mesozoic sedimentary and volcanic sequence in southern and southeastern Alaska: U.S. Geological Survey Professional Paper 800-D, p. D1–D24.

Berg, H. C., Decker, J. E., and Abramson, B. S., 1981, Metallic mineral deposits of southeastern Alaska: U.S. Geological Survey Open-File Report 81-122, 136 p., 1 map sheet, scale 1:1,000,000.

Bernstein, L. R., and Cox, D. P., 1986, Geology and sulfide mineralogy of the Number One orebody, Ruby Creek copper deposit, Alaska: Economic Geology, v. 81, p. 1675–1689.

Bressler, J. R., Jones, W. C., and Cleveland, G., 1985, Geology of a buried channel system of the Denali Placer Gold Mine: Alaska Miner, January, 1985, p. 9.

Brew, D. A., and Ford, A. B., 1984a, The northern Coast plutonic-metamorphic complex, southeastern Alaska and northwestern British Columbia, *in* Coonrad, W. L., and Elliott, R. L., eds., The United States Geological Survey in Alaska: Accomplishments during 1981: U.S. Geological Survey Circular 868, p. 120–124.

——, 1984b, Tectonostratigraphic terranes in the Coast plutonic-metamorphic complex, *in* Reed, K. M., and Bartsch-Winkler, S., eds., The United States Geological Survey in Alaska: Accomplishments during 1982: U.S. Geological Survey Circular 939, p. 90–93.

Brew, D. A., and Grybeck, D., 1984, Geology of the Tracy Arm–Fords Terror Wilderness Study Area and Vicinity, Alaska: U.S. Geological Survey Bulletin 1525-A, 52 p.

Brew, D. A., Johnson, B. R., Grybeck, D., Griscom, A., and Barnes, D. F., 1978, Mineral resources of the Glacier Bay National Monument wilderness study area, Alaska: U.S. Geological Survey Open-File Report 78-494, 670 p.

Brew, D. A., Ovenshine, A. T., Karl, S. M., and Hunt, S. J., 1984, Preliminary reconnaissance geologic map of the Petersburg and parts of the Port Alexander and Sumdum 1:250,000 quadrangles, southeastern Alaska: U.S. Geological Survey Open-File Report 84-405, 43 p., 2 map sheets, scale 1:250,000.

Brown, F. R., 1947, Apollo Mine, Unga Island, Alaska: Alaska Division of Geological and Geophysical Surveys (Territorial Department of Mines) Report of Mineral Investigations MR-138-1, 33 p.

Browna, J. S., 1926, The Nixon Fork country: U.S. Geological Survey Bulletin 783-D, p. 97–144.

Buddington, A. F., and Chapin, T., 1929, Geology and mineral deposits of southeastern Alaska: U.S. Geological Survey Bulletin 800, 398 p.

Bundtzen, T. K., 1980, Geological guides to heavy mineral placers, *in* Second Annual Conference on Alaska Placer Mining: Mineral Industry Research Laboratory Report 46, p. 21–45.

——, 1981, Geology and mineral deposits of the Kantishna Hills, Mt. McKinley quadrangle, Alaska [M.S. thesis]: Fairbanks, University of Alaska, 237 p., 4 sheets, scale 1:63,360.

——, 1982, Bedrock geology of the Fairbanks mining district, western sector: Alaska Division of Geological and Geophysical Surveys Open-File Report 155, 2 map sheets, scale 1:24,000.

Bundtzen, T. K., and Gilbert, W. G., 1983, Outline of geology and mineral resources of upper Kuskokwim region, Alaska: Alaska Geological Society 1982 Symposium on Western Alaska, v. 3, p. 101–117.

Bundtzen, T. K., and Laird, G. M., 1980, Preliminary geology of the McGrath–Upper Innoko River area, western interior Alaska: Alaska Division of Mines and Geology Open-File Report 134, 36 p.

——, 1982, Geologic map of the Iditarod D-2 and eastern D-3 quadrangles, Alaska: Alaska Division of Geological and Geophysical Surveys Geologic Report 72, 26 p., 1 sheet, scale 1:63,360.

——, 1983a, Geologic map of the Iditarod D-1 quadrangle, Alaska: Alaska Division of Geological and Geophysical Surveys Geologic Report 78, 17 p., 1 map sheet, scale 1:63,360.

——, 1983b, Geologic map of the McGrath D-6 quadrangle, Alaska: Alaska Division of Geological and Geophysical Surveys Geological Report 79, 13 p., 1 map sheet, scale 1:63,360.

——, 1983c, Preliminary geologic map of the northeastern Iditarod C-3 quadrangle, Alaska: Alaska Division of Geological and Geophysical Surveys Report of Investigations 83-13, 6 p., 1 map sheet, scale 1:63,360.

Bundtzen, T. K., Miller, M. L., Laird, G. M., and Kline, J. T., 1985, Geology of heavy mineral placer deposits in the Iditarod and Innoko precincts, western Alaska, *in* Madonna, J. A., ed., 7th Annual Conference on Alaska Placer Mining: Alaska Prospectors Publication Company, p. 35–41.

Bundtzen, T. K., Swainbank, R. C., Deggen, J. R., and Moore, J. L., 1990, Alaska's Mineral Industry, 1989: Alaska Division of Geological and Geophysical Surveys Special Report 44, 100 p.

Burk, C. A .,1965, Geology of the Alaska Peninsula–island arc and continental margin: Geological Society of America Memoir 99, 250 p., 2 map sheets, scales 1:250,000 and 1:500,000.

Burns, L. E., 1985, The Border Ranges ultramafic and mafic complex, south-central Alaska: cumulate fractionates of island-arc volcanics: Canadian Journal of Earth Sciences, v. 22, p. 1020–1038.

Burton, P. J., Warner, J. D., and Barker, J. C., 1985, Reconnaissance investigation of tin occurrences at Rocky Mountain (Lime Peak), east-central Alaska: U.S. Bureau of Mines Open-File Report 31-85, 44 p.

Cady, W. M., Wallace, R. E., Hoare, J. M., and Webber, E. J., 1955, The central Kuskokwim region, Alaska: U.S. Geological Survey Professional Paper 268, 132 p.

Chapin, T., 1914, Lode mining near Fairbanks: U.S. Geological Survey Bulletin 592-J, p. 321–355.

——, 1919, Mining in the Fairbanks district: U.S. Geological Survey Bulletin 692-F, p. 321–327.

Chapman, R. M., Patton, W. W., Jr., and Moll, E. J., 1982, Preliminary summary of the geology of the eastern part of the Ophir quadrangle, Alaska: U.S. Geological Survey Circular 844, p. 70–73.

Churkin, M., Jr., Nokleberg, W. J., and Huie, C., 1979, Collision-deformed Paleozoic continental margin, western Brooks Range, Alaska: Geology, v. 7, no. 8, p. 379–383.

Clautice, K. H., 1980, Geological sampling and magnetic surveys of a tungsten occurrence, Bonanza Creek area, Hodzana Highlands, Alaska: U.S. Bureau of Mines Open-File Report 80-83, 80 p.

Cobb, E. H., 1964, Placer gold occurrences in Alaska: U.S. Geological Survey Mineral Investigation Resources Map MR-38, scale 1:2,500,000.

——, 1973, Placer deposits of Alaska: U.S. Geological Survey Bulletin 1324, 213 p.

Conwell, C. N., 1977, Boulder Creek tin lode deposit: Alaska Division of Geological and Geophysical surveys Geologic Report 55, p. 86–92.

Cook, D. L., 1983, Placer mining in Alaska: University of Alaska Mineral Industry Research Laboratory Report 65, 157 p.

Cox, D., 1983a, U.S. Geological Survey-INGEOMINAS Mineral Resource Assessment of Columbia: Ore deposit models: U.S. Geological Survey Open-File Report 83-423, 64 p.

——, 1983b, U.S. Geological Survey-INGEOMINAS Mineral Resource Assessment of Columbia: Additional ore deposit models: U.S. Geological Survey Open-File Report 83-901, 32 p.

Cox, D. P., and Singer, D. A., eds., 1986, Mineral deposit models: U.S. Geological Survey Bulletin 1693, 379 p.

Csejtey, B., and Miller, R. J., 1978, Map and table describing metalliferous and selected nonmetalliferous mineral deposits, Talkeetna Mountains quadrangle, Alaska: U.S. Geological Survey Open-File Report 78-558B, 20 p., 1 sheet, scale 1:250,000.

Czamanske, G. K., and Calk, L. C., 1981, Mineralogical records of cumulus processes, Brady Glacier Ni-Cu deposit, southeastern Alaska: Mining Geology, v. 31, p. 213–233.

Dickinson, K. A., and Cunningham, K.,1984, Death Valley, Alaska, uranium deposit [abs.]: Geological Society of America Abstracts with Programs, v. 16, p. 278.

Dillon, J. T., 1982, Source of lode- and placer-gold deposits of the Chandalar and upper Koyukuk districts, Alaska: Alaska Division of Geological and Geophysical Surveys Open-File Report 158, 22 p.

Dobson, D. D., 1982, Geology and alteration of the Lost River tin-tungsten-fluorite deposit, Alaska: Economic Geology, v. 77, p. 1033–1052.

Drechsler, J. S., Jr., and Dunbier, J., 1981, The Greens Creek ore deposit, Admiralty Island, Alaska [abs.]: Canadian Mining and Metallurgical Bulletin, v. 76, no. 833, p. 57.

Dunbier, J., Snow, G. G., and Butler, T. A., 1979, The Greens Creek project, Admiralty Island, Alaska, *in* Alaska's mineral and energy resources, economics and land status: Alaska Geological Society Symposium Program and Abstracts, p. 40.

Eakins, G. R., and 6 others, 1983, Alaska's mineral industry, 1982: Alaska Division of Geological and Geophysical Surveys Special Report 31, 63 p.

Eakins, G. R., Bundtzen, T. K., Lueck, L. L., Green, C. B., Gallagher, J. L., and Robinson, M. S., 1985, Alaska's mineral industry, 1984: Alaska Division of Geological and Geophysical Surveys Special Report 38, 57 p.

Eberlein, G. D., Chapman, R. M., Foster, H. L., and Gassaway, J. S., 1977, Map and table describing known metalliferous and selected nonmetalliferous mineral deposits in central Alaska: U.S. Geological Survey Open-File Report 77-1168D, 132 p., 1 map sheet, scale 1:1,000,000.

Einaudi, M. T., and Hitzman, M. W., 1986, Mineral deposits in northern Alaska: Introduction: Economic Geology, v. 81, p. 1583–1591.

Erickson, R. L., 1982, Characteristics of mineral deposit occurrences: U.S. Geological Survey Open-File Report 82-795, 248 p.

Fernette, G., and Cleveland, G., 1984, Geology of the Miss Molly molybdenum prospect, Tyonek C-6 quadrangle, Alaska: Alaska Division of Geological and Geophysical Surveys Professional Report 86, p. 35–41.

Foley, J. Y., and McDermott, M. M., 1983, Podiform chromite occurrences in the Caribou Mountain and lower Kanuti River areas, central Alaska: U.S. Bureau of Mines Information Circular IC-8915, 27 p.

Foley, J. Y., Hinderman, T., Kirby, D. E., and Mardock, C. L., 1984, Chromite occurrences in the Kaiyuh Hills, west-central Alaska: U.S. Bureau of Mines Open-File Report 178-84, 20 p.

Foley, J. Y., Barker, J. C., and Brown, L. L., 1985, Critical and strategic mineral investigation in Alaska: Chromium: U.S. Bureau of Mines Open File Report 97-85, 54 p.

Foley, J. Y., Dahlin, D. C., Barker, J. C., and Mardock, C. L., 1991, Chromite deposits in the western Brooks Range, Alaska: U.S. Bureau of Mines Information Circular.

Folger, P. F., and Schmidt, J. M., 1986, Geology of the carbonate-hosted Omar copper prospect, Baird Mountains, Alaska: Economic Geology, v. 81, p. 1690–1695.

Forbes, R. B., and Weber, F. L., 1982, Bedrock geologic map of the Fairbanks mining district: Alaska Division of Geological and Geophysical Surveys Open-File Report AOF-170, 2 map sheets, scale 1:63,360.

Fosse, E. L., 1946, Exploration of the copper-sulfur deposit, Khayyam and Stumble-On properties, Prince of Wales Island, Alaska: U.S. Bureau of Mines Report of Investigations 3942, 8 p.

Gamble, B. M., Ashley, R. P., and Pickthorn, W. J., 1985, Preliminary study of lode gold deposits, Seward Peninsula, *in* Bartsch-Winkler, S., ed., The United States Geological Survey in Alaska: Accomplishments during 1984: U.S. Geological Survey Circular 967, p. 27–29.

Gault, H. R., 1945, The Salt Chuck copper-palladium mine, Prince of Wales Island, southeastern Alaska: U.S. Geological Survey Open-File Report 45-25, 18 p.

Gilbert, W. G., and Bundtzen, T. K., 1979, Mid-Paleozoic tectonics, volcanism, and mineralization in north-central Alaska Range, *in* Sisson, A., ed., The relationship of plate tectonics to Alaskan geology and resources: Alaska Geological Society Symposium, 1977, p. F1–F21.

Goldfarb, R. J., Light, T. D., and Leach, D. L., 1986, Nature of the ore fluids at the Alaska-Juneau gold deposit, *in* Bartsch-Winkler, S., and Reed, K. M., eds., Geologic studies in Alaska by the U.S. Geological Survey during 1985: U.S. Geological Survey Circular 978, p. 92–95.

Grybeck, D., 1977, Known mineral deposits of the Brooks Range, Alaska: U.S. Geological Survey Open-File Report 77-166C, 45 p., 1 map sheet, scale 1:1,000,000.

Guild, P. W., 1942, Chromite deposits of Kenai Peninsula, Alaska: U.S. Geological Survey Bulletin 931-G, p. 139–175.

Herreid, G., 1962, Preliminary report on geologic mapping in the Coast Range mineral belt, *in* Alaska Division of Mines and Minerals Report for the year 1962, p. 44–59.

——, 1966, Geology and geochemistry of the Nixon Fork area, Medfra quadrangle, Alaska: Alaska Division of Mines and Minerals Geologic Report 22, 34 p.

Herreid, G., Bundtzen, T. K., and Turner, D. L., 1978, Geology and geochemistry of the Craig A-2 quadrangle and vicinity, Prince of Wales Island, southeastern Alaska: Alaska Division of Geological and Geophysical Surveys Geologic Report 48, 49 p., 2 plates, scale 1:40,000.

Hill, J. M., 1933, Lode deposits of the Fairbanks district, Alaska: U.S. Geological Survey Bulletin 849-B, p. 63–159.

Himmelberg, G. R., and Loney, R. A., 1981, Petrology of the ultramafic and gabbroic rocks of the Brady Glacier nickel-copper deposit, Fairweather Range, southeastern Alaska: U.S. Geological Survey Professional Paper 1195, 26 p.

Hitzman, M. W., 1986, Geology of the Ruby Creek copper deposit, southwestern Brooks Range, Alaska: Economic Geology, v. 81, p. 1644–1674.

Hitzman, M. W., Smith, T. E., and Proffett, J. M., 1982, Bedrock geology of the Ambler district, southwestern Brooks Range, Alaska: Alaska Division of Geological and Geophysical Surveys Geological Surveys Geologic Report 75, 2 map sheets, scale 1:125,000.

Hitzman, M. W., Proffett, J. M., Jr., Schmidt, J. M., and Smith, T. E., 1986, Geology and mineralization of the Ambler district, northwestern Alaska: Economic Geology, v. 81, p. 1592–1618.

Hollister, V. F., 1978, Geology of the porphyry copper deposits of the Western Hemisphere: Society of Mining Engineering, American Institute of Mining, Metallurgy, and Petroleum Engineers Incorporated, 218 p.

Howard, W. R., 1935, Salt Chuck copper-palladium mine: Alaska Territorial Department of Mines Report MR119-4, 22 p.

Hudson, T., and Arth, J. G., 1983, Tin granites of the Seward Peninsula, Alaska: Geological Society of America Bulletin, v. 94, no. 6, p. 768–790.

Hudson, T., Miller, M. L., and Pickthorn, W. J., 1977, Map showing metalliferous and selected nonmetalliferous mineral deposits, Seward Peninsula, Alaska: U.S. Geological Survey Open-File Report 77-796B, 46 p., 1 map sheet, scale 1:1,000,000.

Hudson, T., Smith, J. G., and Elliott, R. L., 1979, Petrology, composition, and age of intrusive rocks associated with the Quartz Hill molybdenite deposit, southeastern Alaska: Canadian Journal of Earth Sciences, v. 16, p. 1805–1822.

Jansons, U., Hoekzema, R. B., Kurtak, J. M., and Fechner, S. A., 1984, Mineral occurrences in the Chugach National Forest, south central Alaska: U.S. Bureau of Mines Open-File Report MLA 5-84, 43 p., 2 map sheets, scale 1:125,000.

Jasper, M. W., 1961, Mespelt mine, Medfra quadrangle: Alaska Division of Mines and Minerals 1961 Annual Report, p. 49–58.

Johnson, B. L., 1915, The gold and copper deposits of the Port Valdez district: U.S. Geological Survey Bulletin 622, p. 140–148.

Johnson, B. R., Kimball, A. L., and Still, J. C., 1982, Mineral resource potential of the Western Chichagof and Yakobi Islands wilderness study area, southeastern Alaska: U.S. Geological Survey Miscellaneous Field Map MF-1476-B, 10 p., 1 map sheet, scale 1:125,000.

Jones, B., 1977, Uranium-thorium bearing rocks of western Alaska [M.S. thesis]: Fairbanks, Alaska, University of Alaska, 80 p.

Kennedy, G. C., 1953, Geology and mineral deposits of Jumbo basin, southeastern Alaska: U.S. Geological Survey Professional Paper 251, 46 p.

Kennedy, G. C., and Walton, M. S., Jr., 1946, Geology and associated mineral deposits of some ultrabasic rock bodies in southeastern Alaska: U.S. Geological Survey Bulletin 947-D, p. 65–84.

Kimball, A. L., Still, J. C., and Rataj, J. L., 1984, Mineral deposits and occurrences in the Tracy Arm–Fords Terror wilderness study area and vicinity, Alaska: U.S. Geological Survey Bulletin 1525, p. 105–210.

Lancelot, J. R., and de Saint-Andre, B., 1982, U-Pb systematics and genesis of U deposits, Bokan Mountain (Alaska) and Lodeve (France) [abs.]: 5th International Conference on Geochronology, Cosmochronology, and Isotope Geology, Nikko National Park, Japan, June 27–July 2, Abstracts, p. 206–207.

Lange, I. M., Nokleberg, W. J., Plahuta, J. T., Krouse, H. R., and Doe, B. R., 1985, Geologic setting, petrology, and geochemistry of stratiform zinc-lead-barium deposits, Red Dog Creek and Drenchwater Creek areas, northwestern Brooks Range, Alaska: Economic Geology, v. 80, p. 1896–1926.

Lanphere, M. A., and Reed, B. L., 1985, The McKinley sequence of granitic rocks: A key element in the accretionary history of southern Alaska: Journal of Geophysical Research, v. 90, p. 11413–11430.

Loney, R. A., and Himmelberg, G. R., 1984, Preliminary report on ophiolites in the Yuki River and Mount Hurst areas, west-central Alaska, *in* Coonrad, W. L., and Elliott, R. L., eds., The United States Geological Survey in Alaska: Accomplishments during 1981: U.S. Geological Survey Circular 868, p. 27–30.

——, 1985a, Distribution and character of the peridotite-layered gabbro complex of the southeastern Yukon-Koyukuk ophiolite belt, *in* Bartsch-Winkler, S., and Reed, K. M., eds., The United States Geological Survey in Alaska: Accomplishments during 1983: U.S. Geological Survey Circular 945, p. 46–48.

——, 1985b, Ophiolitic ultramafic rocks of the Jade Mountains-Cosmos Hills area, southwestern Brooks Range, *in* Bartsch-Winkler, S., ed., The United States Geological Survey in Alaska: Accomplishments during 1984: U.S. Geological Survey Circular 967, p. 13–15.

Loney, R. A., Himmelberg, G. R., and Shew, N., 1987, Salt Chuck palladium-bearing ultramafic body, Prince of Wales Island, *in* Hamilton, T. D., and Galloway, J. P., eds., Geologic studies in Alaska by the U.S. Geologic Survey during 1986: U.S. Geological Survey Circular 998, p. 126–127.

MacKevett, E. M., Jr., 1963, Geology and ore deposits of the Bokan Mountain uranium-thorium area, southeastern Alaska: U.S. Geological Survey Bulletin 1154, 125 p.

——, 1965, Ore controls at the Kathleen-Margaret (Maclaren River) copper deposit, Alaska: U.S. Geological Survey Professional Paper 501C, p. C116–C120.

——, 1976, Mineral deposits and occurrences in the McCarthy quadrangle, Alaska: U.S. Geological Survey Miscellaneous Field Studies Map MF-773B, 2 map sheets, scale 1:250,000.

——, 1978, Geologic map of the McCarthy quadrangle, Alaska: U.S. Geological Survey Miscellaneous Investigations Series Map I-1032, scale 1:250,000.

MacKevett, E. M., Jr., and Berg, H. C., 1963, Geology of the Red Devil quicksilver mine, Alaska: U.S. Geological Survey Bulletin 1142-G, 16 p.

MacKevett, E. M., Jr., and Blake, M. C., Jr., 1963, Geology of the North Bradfield River iron prospect, southeastern Alaska: U.S. Geological Survey Bulletin 1108-D, p. D1–D21.

MacKevett, E. M., Jr., and Holloway, C. D., 1977a, Map showing metalliferous and selected nonmetalliferous mineral deposits in the eastern part of southern Alaska: U.S. Geological Survey Open-File Report 77-169A, 99 p., 1 map sheet, scale 1:1,000,000.

——, 1977b, Map showing metalliferous mineral deposits in the western part of southern Alaska: U.S. Geological Survey Open-File Report 77-169F, 39 p., 1 map sheet, scale 1:1,000,000.

MacKevett, E. M., Jr., and Plafker, G., 1974, The Border Ranges fault in south-central Alaska: U.S. Geological Survey Journal of Research, v. 2, no. 3, p. 323–329.

Martin, G. C., 1921, Gold lodes of the upper Kuskokwim region, Alaska: U.S. Geological Survey Bulletin 722, p. 149.

Mayfield, C. F., Tailleur, I. L., and Ellersieck, I., 1983, Stratigraphy, structure, and palinspastic synthesis of the western Brooks Range, northwestern Alaska: U.S. Geological Survey Open-File Report 83-779, 58 p., 5 map sheets, scale 1:1,000,000.

Menzie, W. D., Foster, H. L., Tripp, R. B., and Yeend, W. E., 1983, Mineral resource assessment of the Circle quadrangle, Alaska: U.S. Geological Survey Open-File Report 83-170B, 57 p., 1 map sheet, scale 1:250,000.

Mertie, J. B., Jr., 1918, Lode mining in the Fairbanks district, Alaska: U.S. Geological Survey Bulletin 662-H, p. 404–424.

——, 1936, Mineral deposits of the Ruby-Kuskokwim region, Alaska: U.S. Geological Survey Bulletin 864-C, p. 115–255.

——, 1976, Platinum deposits ofthe Goodnews Bay district, Alaska: U.S. Geological Survey Professional Paper 938, 42 p.

Metz, P. A., and Halls, C., 1981, Ore petrology of the Au-Ag-Sb-W-Hg mineralization of the Fairbanks mining district, Alaska [abs.]: Proceedings of Mineralization of the Precious Metals, Uranium, and Rare Earths, University College, Cardiff, Wales, 1981, p. 132.

Miller, T. P., 1976, Hardrock uranium potential in Alaska: U.S. Geological Survey Open-Report 76-246, 7 p.

Miller, T. P., and Bunker, C. M., 1976, A reconnaissance study of the uranium and thorium contents of plutonic rocks of the southeastern Seward Peninsula, Alaska: U.S. Geological Survey Journal of Research, v. 4, p. 367–377.

Miller, T. P., and Elliott, R. L., 1969, Metalliferous deposits near Granite Mountain, eastern Seward Peninsula, Alaska: U.S. Geological Survey Circular 614, 19 p.

Moffit, F. H., and Fellows, R. E., 1950, Copper deposits of the Prince William Sound district, Alaska: U.S. Geological Survey Bulletin 963-B, p. 47–80.

Moore, D. W., Young, L. E., Modene, J. S., and Plahuta, J. T., 1986, Geologic setting and genesis of the Red Dog zinc-lead-silver deposit, western Brooks Range, Alaska: Economic Geology, v. 81, p. 1696–1727.

Mull, C. G., Tailleur, I. L., Mayfield, C. F., Ellersieck, I., and Curtis, S., 1982, New upper Paleozoic and lower Mesozoic stratigraphic units, central and western Brooks Range, Alaska: American Association of Petroleum Geologists Bulletin, v. 66, no. 3, p. 348–362.

Nauman, C. R., Blakestad, R. A., Chipp, E. R., and Hoffman, B. L., 1980, The north flank of the Alaska Range, a newly discovered volcanogenic massive sulfide belt [abs.]: Geological Association of Canada Program with Abstracts, p. 73.

Nelson, S. W., and Nelson, W. H., 1982, Geology of the Siniktanneyak Mountain ophiolite, Howard Pass quadrangle, Alaska: U.S. Geological Survey Map MF-1441, 1 map sheet, scale 1,63,360.

Newberry, R. J., 1986, Mineral resources of the north-central Chugach Mountains, Alaska: Alaska Division of Geological and Geophysical Surveys Report of Investigation 86-23, 44 p.

Newberry, R. J., Dillon, J. T., and Adams, D. D., 1986, Skarn and skarn-like deposits of the Brooks Range, northern Alaska: Economic Geology, v. 81, p. 1728–1752.

Noel, G. A., 1966, The productive mineral deposits of southeastern Alaska: Canadian Institute of Mining and Metallurgy, v. 8, p. 215–229.

Nokleberg, W. J., and Aleinikoff, J. N., 1985, Summary of stratigraphy, structure, and metamorphism of Devonian igneous-arc terranes, northeastern Mount Hayes quadrangle, eastern Alaska Range, *in* Bartsch-Winkler, S., ed., The United States Geological Survey in Alaska: Accomplishments during 1984: U.S. Geological Survey Circular 967, p. 66–71.

Nokleberg, W. J., and Lange, I. M., 1985, Metallogenetic history of the Wrangellia terrane, eastern Alaska Range, Alaska [abs.]: U.S. Geological Survey Circular 949, p. 36–38.

Nokleberg, W. J., and Winkler, G. R., 1982, Stratiform zinc-lead deposits in the Drenchwater Creek area, Howard Pass quadrangle, northwestern Brooks Range, Alaska: U.S. Geological Survey Professional Paper 1209, 22 p., 2 map sheets, scale 1:20,000.

Nokleberg, W. J., Lange, I. M., and Roback, R. C., 1984, Preliminary accretionary terrane model for metallogenesis of the Wrangellia terrane, southern Mount Hayes quadrangle, eastern Alaska Range, Alaska, *in* Reed, K. M., and Bartsch-Winkler, S., eds., The United States Geological Survey in Alaska: Accomplishments during 1982: U.S. Geological Survey Circular 939, p. 60–65.

Nokleberg, W. J., Jones, D. L., and Silberling, N. J., 1985, Origin, migration, and accretion of the Maclaren and Wrangellia terranes, eastern Alaska Range, Alaska: Geological Society of America Bulletin, v. 96, p. 1251–1270.

Nokleberg, W. J., Aleinikoff, J. N., and Lange, I. M., 1986, Cretaceous deformation and metamorphism in the northeastern Mount Hayes quadrangle, eastern Alaska Range, *in* Bartsch-Winkler, S., and Reed, K. M., eds., Geologic Studies in Alaska by the U.S. Geological Survey during 1985: U.S. Geological Survey Circular 978, p. 64–69.

Nokleberg, W. J., and 7 others, 1987, Significant metalliferous lode deposits and placer districts of Alaska: U.S. Geological Survey Bulletin 1786, 104 p., 2 plates, scale 1:5,000,000.

—— , 1988, Metallogeny and major mineral deposits of Alaska: U.S. Geological Survey Open-File Report 88-73, 97 p., 2 plates, scale 1:5,000,000.

Patton, W. P., Jr., and Miller, T. P., 1970, Preliminary geologic investigations in the Kanuti River region, alaska: U.S. Geological Survey Buletin 1312-J, p. J1–J10.

Patton, W. W., Jr., Moll, E. J., and King, H. D., 1984, The Alaskan mineral resource assessment program: Guide to information contained in the folio of geologic and mineral resource maps of the Medfra quadrangle, Alaska: U.S. Geological Survey Circular 928, 11 p.

Plafker, G., Nokleberg, W. J., and Lull, J. S., 1985, Summary of 1985 TACT geologic studies in the northern Chugach Mountains and southern Copper River Basin, *in* Bartsch-Winkler, S., ed., The United States Geological Survey in Alaska: Accomplishments during 1984: U.S. Geological Survey Circular 967, p. 76–79.

Prindle, L. M., and Katz, F. J., 1913, Geology of the Fairbanks district, *in* Prindle, L. M., A geologic reconnaissance of the Fairbanks quadrangle, Alaska: U.S. Geological Survey Bulletin 525, p. 59–152.

Puchner, C. C., 1986, Geology alteration, and mineralization of the Kougarok Sn deposit, Seward Peninsula, Alaska: Economic Geology, v. 81, p. 1775–1794.

Ray, R. G., 1954, Geology and ore deposits of the Willow Creek mining district, Alaska: U.S. Geological Survey Bulletin 1004, 86 p.

Read, J. J., and Meinert, L. D., 1986, Gold-bearing quartz vein mineralization at the Big Hurrah mine, Seward Peninsula, Alaska: Economic Geology, v. 81, p. 1760–1774.

Reed, B. L., and Eberlein, G. D., 1972, Massive sulfide deposits near Shellebarger Pass, southern Alaska Range: U.S. Geological Survey Bulletin 1342, 45 p.

Reed, B. L., and Lanphere, M. A., 1973, Alaska-Aleutian Range batholith: Geochronology, chemistry, and relation to circum-Pacific plutonism: Geological Society of America Bulletin, v. 84, p. 2583–2610.

Reed, J. C., and Coats, R. R., 1941, Geology and ore deposits of the Chichagof mining district, Alaska: U.S. Geological Survey Bulletin 929, 148 p.

Richter, D. H., 1970, Geology and lode-gold deposits of the Nuka Bay area, Kenai Peninsula, Alaska: U.S. Geological Survey Professional Paper 625-B, p. B–B16.

—— , 1975, Geologic map of the Nabesna quadrangle, Alaska: U.S. Geological Survey Miscellaneous Investigations Series Map I-932, scale 1:250,000.

Richter, D. H., Singer, D. A., and Cox, D. P., 1975, Mineral resources map of the Nabesna quadrangle, Alaska: U.S. Geological Survey Miscellaneous Field Studies Map MF-655K, scale 1:250,000.

Roberts, W. S., 1984, Economic potential for chromium, platinum, and palladium in the Mount Hurst Ultramafics, west central area, Alaska: U.S. Bureau of Mines Open-File Report 22-84, 52 p.

Robinson, M. S., and Bundtzen, T. K., 1979, Historic gold production in Alaska—a minisummary: Alaska Division of Geological and Geophysical Surveys Mines and Geology Bulletin, v. 28, no. 3, p. 1–10.

Robinson, M. S., Smith, T. E., Bundtzen, T. K., and Albanese, M. D., 1982, Geology and metallogeny of the Livengood area, east-central Alaska [abs.]: Alaska Miners Association Annual Convention Program with Abstracts, p. 8.

Roeder, D., and Mull, C. G., 1978, Tectonics of Brooks Range ophiolites, Alaska: American Association of Petroleum Geologists Bulletin, v. 62, no. 9, p. 1696–1702.

Rose, A. W., 1965, Geology and mineralization of the Midas mine and Sulphide Gulch areas near Valdez: Alaska Division of Mines and Minerals Geologic Report 15, 21 p.

Ruckmick, J. C., and Noble, J. A., 1959, Origin of the ultramafic complex at Union Bay, southeastern Alaska: Geological Society of America Bulletin, v. 70, p. 981–1017.

Runnells, D. D., 1969, The mineralogy and sulfur isotopes of the Ruby Creek copper prospect, Bornite, Alaska: Economic Geology, v. 64, p. 75–90.

Sainsbury, C. L., 1969, Geology and ore deposits of the central York Mountains, western Seward Peninsula, Alaska: U.S. Geological Survey Bulletin 1287, 101 p.

—— , 1972, Geologic map of the Teller quadrangle, western Seward Peninsula, Alaska: U.S. Geological Survey Miscellaneous Geologic Investigations Map I-685, 1 map sheet, scale 1:250,000.

Sainsbury, C. L., and MacKevett, E. M., Jr., 1965, Quicksilver deposits of southwest Alaska: U.S. Geological Survey Bulletin 1187, 89 p.

Schmidt, J. M., 1983, Geology and geochemistry of the Arctic prospect, Ambler district, Alaska [Ph.D. dissertation]: Stanford, California, Stanford University, 253 p.

—— , 1986, Stratigraphic setting and mineralogy of the Arctic volcanogenic massive sulfide prospect, Ambler district, Alaska: Economic Geology, v. 81, p. 1619–1643.

Seraphim, R. H., 1975, Denali—A nonmetamorphosed stratiform sulfide deposit: Economic Geology, v. 70, p. 949–959.

Sichermann, H. A., Russell, R. H., and Fikkan, P. R., 1976, The geology and mineralization of the Ambler district, Alaska: Spokane, Washington, Bear Creek Mining Company, 22 p.

Smith, P. S., 1913, Lode mining near Fairbanks: U.S. Geological Survey Bulletin 542-F, p. 137–202.

Smith, T. E., Robinson, M. S., Bundtzen, T. K., and Metz, P. A., 1981, Fairbanks mining district in 1981: New look at an old mineral province [abs.]: Alaska Miners Association Convention Program with Abstracts, p. 12.

Southworth, D. D., and Foley, J. Y., 1986, Lode platinum-group metals potential of the Goodnews Bay ultramafic complex, Alaska: U.S. Bureau of Mines Open-File Report 51-86, 82 p.

Spencer, A. C., 1905, The Treadwell ore deposits, Douglas Island: U.S. Geological Survey Bulletin 259, p. 69–87.

—— , 1906, The Ju;neau gold belt, Alaska: U.S. Geological Survey Bulletin 287, p. 1–137.

Stevens, D. L., 1971, Geology and geochemistry of the Denali prospect, Clearwater Mountains, Alaska: [Ph. D. dissertation]: Fairbanks, University of Alaska, 81 p.

Still, J. C., and Weir, K. R., 1981, Mineral land assessment of the west portion of western Chichagof Island, southeastern Alaska: U.S. Bureau of Mines Open-File Report 89-81, 168 p.

Swainbank, R. C., Smith, T. E., and Turner, D. L., 1977, Geology and K-Ar age of mineralized intrusive rocks from the Chulitna mining district, central Alaska: Alaska Division of Geological and Geophysical Surveys Geologic Report 55, p. 23–28.

Szumigala, D. J., 1985, Geology of the Tin Creek zinc-lead skarn deposits, McGrath B-2 quadrangle, Alaska: Alaska Division of Geological and Geophysical Surveys Report of Investigations 85-50, 10 p.

Taylor, H. P., 1967, The zoned ultramafic intrusions of southeastern Alaska, *in* Wyllie, P. J., ed., ultramafic and related rocks: New York, John Wiley and Sons, p. 99–121.

Thomas, B. I., 1973, Gold-lode deposits, Fairbanks mining district, central Alaska: U.S. Bureau of Mines Information Circular 8604, 16 p.

Thomas, B. I., and Berryhill, R. V., 1962, Reconnaissance studies of Alaskan beach sands, eastern Gulf of Alaska: U.S. Bureau of Mines Report of Investigations 5986, 40 p.

Thompson, T. B., Pierson, J. R., and Lyttle, T., 1982, Petrology and petrogenesis of the Bokan mountain granite complex, southeastern Alaska: Geological Society of America Bulletin, v. 93, p. 898–908.

Till, A. B., 1984, Low-grade metamorphic rocks of Seward Peninsula, Alaska [abs.]: Geological Society of America Abstracts with Programs, v. 16, no. 5, p. 337.

Twenhofel, W. S., 1952, Geology of the Alaska-Juneau lode system, Alaska: U.S. Geological Survey Open-File Report 52-160, 170 p.

University of Alaska, 1989, Placer mining in todays world: Proceedings of the eleventh annual conference on placer mining: Fairbanks, Alaska, Polar Run Printing, 83 p.

Wahrhaftig, Clyde, 1968, Schists of the central Alaska Range: U.S. Geological Survey Bulletin 1254-E, 22 p.

Warner, J. D., and Southworth, D. D., 1985, Placer and lode sources of Niobium: Tofty, Alaska [abs.]: American Association of Petroleum Geologists, Pacific Section Programs with abstracts, p. 49.

Warner, L. A., and Goddard, E. N., 1961, Iron and copper deposits of Kasaan Peninsula, Prince of Wales Island, southeastern Alaska: U.S. Geological Survey Bulletin 1090, 136 p.

Wayland, R. G., 1943, Gold deposits near Nabesna, Alaska: U.S. Geological Survey Bulletin 933B, p. 175–199.

—— , 1960, The Alaska Juneau gold body: Neues Jahrbuch fur Mineralogie Abhandlungen, v. 94, p. 267–279.

—— , 1961, Tofty tin belt, Manley Hot Springs district, Alaska: U.S. Geological Survey Bulletin 1058-I, p. 363–414.

Wilson, F. H., 1985, The Meshik arc—an Eocene to earliest Miocene magmatic arc on the Alaska Peninsula: Alaska Division of Geological and Geophysical Surveys Professional Report 88, 14 p.

Wilson, F. H., and Cox, D. P., 1983, Geochronology, geochemistry, and tectonic environment of porphyry mineralization in the central Alaska Peninsula: U.S. Geological Survey Open-File Report 83-783, 24 p.

Wilson, F. H., Detterman, R. L., and Case, J. E., 1985, The Alaska Peninsula terrane; a definition: U.S. Geological Survey Open-File Report 85-450, 17 p.

Winkler, G. R., and Plafker, G., 1981, Geological map and cross sections of the Cordova and Middleton Island quadrangles, southern Alaska: U.S. Geological Survey Open-File Report 81-1164, 25 p., 1 map sheet, scale 1:250,000.

Winkler, G. R., Miller, R. J., MacKevett, E. M., Jr., and Holloway, C. D., 1981a, Map and summary table describing mineral deposits in the Valdez quadrangle, southern Alaska: U.S. Geological Survey Open-File Report 80-892-B, 2 map sheets, scale 1:250,000.

Winkler, G. R., Silberman, M. L., Grantz, A., Miller, R. J., and MacKevett, E. M., Jr., 1981b, Geologic map and summary geochronology of the Valdez quadrangle, southern Alaska: U.S. Geological Survey Open-File Report 80-892-A, 1 map sheet, scale 1:250,000.

Wright, C. W., 1904, The Porcupine placer district, Alaska: U.S. Geological Survey Bulletin 236, 35 p.

Yeend, W., 1981a, Placer gold deposits, Mt. Hayes quadrangle, Alaska, *in* Silberman, M. L. Field, C. W., and Berry, N. L., eds., Proceedings of the Symposium of Mineral Deposits of the Pacific Northwest: U.S. Geological Survey Open-File Report 81-355, p. 74–83.

—— , 1981b, Placer gold deposits, Mount Hayes quadrangle, Alaska, *in* Albert, N.R.D., and Hudson, T., eds., The United States Geological Survey in Alaska: Accomplishments during 1979: U.S. Geological Survey Circular 823-B, p. B68.

—— , 1982, Placers and placer mining, Circle District, Alaska, *in* Coonrad, W. L., ed., The United States Geological Survey in Alaska: Accomplishments during 1980: U.S. Geological Survey Circular 844, p. 64.

Zimmerman, J., and Soustek, P. G., 1979, The Avan Hills ultramafic complex, De Long Mountains, Alaska: U.S. Geological Survey Circular 804-B, p. B8–B11.

Manuscript Accepted by the Society February 28, 1991

ACKNOWLEDGMENTS

We thank numerous colleagues in private industry, universities, the U.S. Geological Survey, the Alaska Division of Geological and Geophysical Surveys, and the U.S. Bureau of Mines for contributing the data published in Nokleberg and others (1987), which is the major data base for this study, and for discussions of Alaskan mineral deposits. In particular, we thank Roger P. Ashley, James C. Barker, Joseph A. Briskey, William P. Brosge, Robert M. Chapman, Dennis P. Cox, Robert L. Detterman, John T. Dillon, Inyo F. Ellersieck, Peter F. Folger, Helen L. Foster, Wyatt G. Gilbert, Edward M. MacKevett, Jr., David W. Moore, William Morgan, Harold Noyes, William W. Patton, Jr., John Reed, Donald H. Richter, Alison B. Till, and Frederic H. Wilson for their contributions. This chapter was initiated mainly through the encouragement of the Geological Society of America and their plan to publish the Decade of North American Geology volumes on the geology of North America. We thank George Plafker for his encouragement, and Dennis P. Cox and Donald H. Richter for constructive and helpful reviews.

Printed in U.S.A.

The Geology of North America
Vol. G-1, The Geology of Alaska
The Geological Society of America, 1994

Chapter 30

Petroleum resources in Alaska

Leslie B. Magoon III
U.S. Geological Survey, MS 999, 345 Middlefield Road, Menlo Park, California 94025

INTRODUCTION

Alaska has the largest oil field in North America, and it has the nation's highest daily oil production of approximately 2 million barrels per day. This chapter summarizes the status of exploration and development for oil and gas resources of onshore Alaska and offshore state lands (Fig. 1). Emphasis is on geologic and geochemical evidence for the occurrence of oil and gas in three petroliferous areas and their evolution as studied by petroleum system models. Areas of primary interest are the North Slope, Cook Inlet, and Gulf of Alaska. The rest of Alaska has negligible petroleum potential and is discussed only briefly. For a more in-depth discussion of Alaskan geology, the reader is referred to other chapters on regional geology. Other chapters discuss onshore northern Alaska (Moore and others, this volume) and the offshore (Grantz and others, this volume), west-central Alaska (Patton and others, this volume), east-central Alaska (Dover, this volume; Foster and others, this volume), and south-central Alaska (Nokleberg and others, this volume, Chapter 10). The deformed flysch basins are discussed by Kirschner (this volume, Chapter 14), and the lithotectonic terranes are shown on Plate 3 (Silberling and others, this volume). The sedimentary basins and oil fields of Alaska are delineated on Plate 7 (Kirschner, this volume), and the petroleum potential of interior basins is discussed by Kirschner (this volume, Chapter 14). The petroleum potential of the Bering Sea shelf is considered by Marlow and others (this volume), and that of the southern Alaska shelves is in Plafker and others (this volume, Chapter 12), and Vallier and others (this volume). Miller and others (1959), Cram (1971), Miller and others (1975), Dolton and others (1981), and U.S. Geological Survey and Minerals Management Service (1988) are important earlier references that discuss the petroleum potential of the entire state.

Resources

The oil and gas resources of Alaska can be allocated to three categories: produced, discovered, and undiscovered (Table 1; U.S. Geological Survey and Minerals Management Service, 1988). As of December 31, 1987, 6.1 billion barrels of oil (bbo) and 3.8 trillion cubic feet of gas (tcfg) have been produced. In addition, a significant amount of gas has been reinjected back into the reservoirs to maintain pressure. Except for safety reasons, gas is seldom flared. Gas produced on the North Slope is consumed locally to run the facilities, the rest is reinjected into the Prudhoe Bay oil field. Cook Inlet gas is consumed locally, liquified and shipped overseas, or reinjected into other reservoirs. Reinjected gas is not included in the cumulative production.

Discovered hydrocarbons include measured reserves and inferred plus indicated reserves. Measured reserves is a known volume of petroleum remaining to be produced, whereas inferred plus indicated reserves is an expected volume of hydrocarbons from field extensions, infilling drilling, or enhanced recovery techniques.

Undiscovered recoverable resources, a quantity not yet found but based on geological information, is expected to exist; regardless of economics, it is technically and physically recoverable. For the undiscovered recoverable resources, two probability fractiles, F95 and F5, and the mean are shown for both oil and gas (Table 2). The F95, for example, indicates a 95-percent chance that at least 1.50 bbo is yet to be discovered under the Arctic Coastal Plain. If the numbers in Table 1 are correct, then Alaska could ultimately produce 32.6 bbo and 97.4 tcfg. Based on mean values for all of Alaska shown on Table 2, 95 percent (12.6 bbo) of the undiscovered oil and 93 percent (54.1 tcfg) of the undiscovered gas will be found on the North Slope. Areas of oil potential for the rest of Alaska, in descending order, are Cook Inlet, Gulf of Alaska, and Yukon-Kandik basin. Gas potential is highest for Cook Inlet, followed by the entire Alaska interior, Gulf of Alaska, and Bristol Basin.

Unconventional sources of hydrocarbons have been left out of all estimates. The West Sak and Ugnu sands of local usage above the Kuparuk River oil field on the North Slope contain billions of barrels of heavy oil (API oil gravity from 10 to 20°; Werner, 1987). On the North Slope, the volume of methane included in gas hydrate is estimated to be 8 to 10 tcfg (Collett and others, 1988).

Framework geology

The geology of Alaska is complex. Much of the state is made up of lithotectonic terranes that were assembled in their present

Magoon, L. B., III, 1994, Petroleum resources in Alaska, *in* Plafker, G., and Berg, H. C., eds., The Geology of Alaska: Boulder, Colorado, Geological Society of America, The Geology of North America, v. G-1.

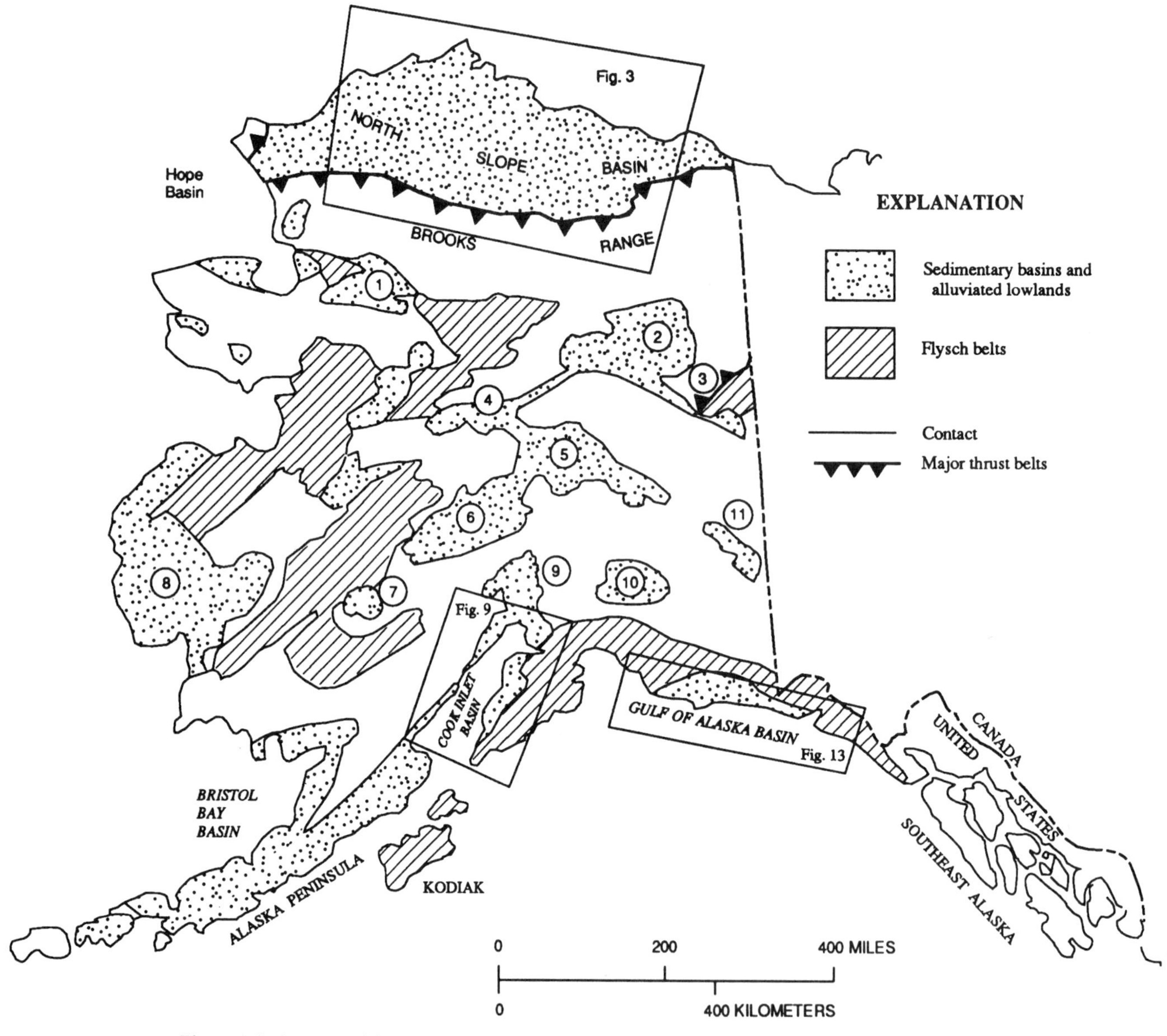

Figure 1. Index map of Alaska showing the three known onshore oil and gas provinces: North Slope basin, Cook Inlet basin, and Yakutat terrane in the Gulf of Alaska basin. The location of Figures 3, 9, and 13 are shown. Names of sedimentary basins are as follows: (1) Kobuk basin, (2) Yukon Flats basin, (3) Kandik basin, (4) Ruby-Rampart trough, (5) Nenana basin, (6) Minchumina basin, (7) Holitna basin, (8) Bethel basin, (9) Susitna basin, (10) Copper River basin, and (11) Northway lowlands. (Adapted from Kirschner, this volume, Chapter 14, Plate 7.)

positions relative to North America during the late Mesozoic and early Cenozoic (Silberling and others, this volume). These terranes contain myriad lithologies comprising sedimentary and igneous rocks, some of which were deformed and metamorphosed during accretion. South of the Brooks Range, deformed flysch deposits described by Kirschner (this volume, Chapter 14) have no petroleum potential and are considered economic basement. Throughout the Cenozoic within interior Alaska, significant thicknesses of nonmarine sedimentary rocks were deposited over these flysch basins (Kirschner, this volume, Plate 7 and Chapter 14). These sedimentary rocks contain coal, conglomerate, and sandstone where microbial gas might originate and accumulate.

The economic basement and geologic history of northern and southern Alaska differ from those of interior Alaska. Except for northeastern Alaska where pre-Mississippian rocks are petroleum reservoirs, basement rock is pre-Mississippian for the North Slope, pre–Middle Jurassic for Cook Inlet, and pre-Cenozoic for the Yakutat terrane. In northern Alaska, the hydrocarbon source-

TABLE 1. ULTIMATE OIL AND GAS RESOURCES FOR ONSHORE AND STATE OF ALASKA OFFSHORE WATERS*

	Produced Cumulative production	Discovered Measured reserves	Inf + ind reserves	Undiscovered Recoverable resources
Oil (bbo)	6.1	6.9	6.4	13.2
Gas (tcf)	3.8	32.7	3.0	57.9

*Inf + ind, inferred plus indicated; bbo, in billions of barrels of oil; tcf, in trillion cubic feet of gas. Undiscovered recoverable resources from U.S. Geological Survey and Minerals Management Service, 1988; cumulative production to 12/31/87; measured reserves, for 1/1/88 from Energy Information Administration, 1987, annual report; inferred + indicated, for 1/1/88 from D. H. Root (USGS), written communication.

rock interval ranges in age from Triassic through Cretaceous. The provenance for the primary quartz-rich reservoir rock is to the north and ranges in age from Mississippian through Early Creteceous, while the provenance for the secondary lithic-rich reservoir is to the south and ranges in age from Early Cretaceous through Tertiary. The north-to-south flip in provenance is marked by rifting in mid-Mesozoic time. In Cook Inlet, the Middle Jurassic marine-shale source rock was deposited in a back-arc basin, whereas the Cenozoic nonmarine conglomerate sandstone reservoirs were deposited in a fore-arc basin. A major unconformity separates the Mesozoic from the Cenozoic rocks. In the Gulf of Alaska, Paleogene marine source and reservoir rocks were deposited on the allochthonous Yakutat terrane in an unknown tectonic environment and were buried by thick glaciomarine sedimentary rocks deposited on the northward-moving Yakutat terrane. For these areas, each of these geologic events was important to the development of a petroleum system.

Petroleum system

The petroleum system model (Magoon, 1988a, b) is used to explain the occurrence of oil and gas in Alaska. As used herein, a petroleum system encompasses a hydrocarbon source rock and all generated oil and gas accumulation, and includes all those geologic elements and processes that are essential for a hydrocarbon deposit to exist in nature: namely, a petroleum source rock, migration path, reservoir rock, seal, trap, and the geologic processes that create each of these basic elements. All these elements must be correctly placed in time and space so that organic matter included in a source rock can be converted into a petroleum deposit. Or more simply, the petroleum system emphasizes the genetic relation between a source rock and an accumulation. A petroleum system exists wherever all the basic elements are known to occur, or are suspected to occur.

TABLE 2. ESTIMATES OF UNDISCOVERED RECOVERABLE RESOURCES FOR ONSHORE AND STATE OF ALASKA OFFSHORE WATERS AS OF 12/31/86*

	Oil (Billion barrels)			Gas (Trillion cubic ft)		
Area	F95	F5	Mean	F95	F5	Mean
North Slope						
Arctic Coastal Plain	1.50	14.80	6.00	4.66	58.24	22.11
Northern Foothills	0.67	5.12	2.24	4.03	24.31	11.49
Southern Foothills	0.58	13.18	4.35	2.85	61.56	20.49
Central Alaska						
Yukon-Kandik	0.00	0.49	0.11	0.00	0.49	0.11
Alaska Interior†	0.00	0.00	0.00	0.45	2.85	1.33
Bristol basin	0.00	0.00	0.00	0.11	0.67	0.32
Hope basin	....	....	....	....	....	....
Cook Inlet	0.09	0.64	0.29	0.35	3.91	1.53
Gulf of Alaska	0.03	0.58	0.19	0.03	2.00	0.56
Kodiak	....	....	....	....	....	....
Southeast Alaska	....	....	....	....	....	....
Total	3.6§	31.3§	13.2	15.6§	138.6§	57.9

*From U.S. Geological Survey and Minerals Management Service, 1988, Table IV.C.1, p. 218; ...., no information.
†Includes Alaska Peninsula, Bethel basin, Copper River basin, Holitna basin, Kobuk basin, Minchumina basin, Ruby-Rampart basin, Susitna basin, Nenana basin, Northway lowlands, and Yukon Flats basin.
§Fractile values are not additive.

The stratigraphic, areal, and temporal extent of each petroleum system is specific. Stratigraphically, the system is limited to the following rock units: a petroleum source rock, rock overburden required for maturity (time and temperature), rocks through which migration has occurred, and strata that make up the trap. The areal extent of the petroleum system is defined by a line that circumscribes the mature source rock and all oil and gas deposits, conventional and unconventional, originating from that source. The temporal extent of a petroleum system includes the duration or life-cycle of the system, and the preservation time indicates how long the hydrocarbons have been preserved in the geologic record.

A petroleum deposit includes high concentrations of any of the following substances: thermal and microbial natural gas, condensate, crude oil (including heavy oil), and solid bitumen. In any volume of hydrocarbons, these substances may be found in conventional siliciclastic and carbonate reservoirs as well as in gas hydrates, low-permeability rocks, fractured rocks, and coal beds.

A petroleum system (Table 3) can be identified in terms of a geochemical correlation of petroleum and source rocks at three levels of certainty: known, hypothetical, and speculative. In a known petroleum system, in the case of oil, a good geochemical match must exist between the source rock and oil accumulations; or, in the case of natural gas, the gas is produced from a gas source rock. In a hypothetical petroleum system, geochemical information is sufficient to identify a source rock, but no geochemical match exists between the identified source rock and a known petroleum deposit. In a speculative petroleum system, of which none are present in Alaska, the existence of source rocks and petroleum accumulations is postulated entirely on the basis of geologic or geophysical evidence. At the end of the system's name on Table 3, the level of certainty is indicated by a (!) for known, or a (.) for hypothetical.

The name of the petroleum system includes the name of the source rock followed by the name of the major reservoir rock and symbol expressing the level of certainty. For example, the Torok-Nanushuk(.) is a hypothetical system on the North Slope and is composed of the Cretaceous Torok Formation as the source rock and the sandstone of the Nanushuk Group as the major reservoir.

Seven petroleum systems are discussed. Their level of certainty is shown only on Table 3. For the North Slope, presently a foreland basin, there are three petroleum systems: Ellesmerian, Torok-Nanushuk, and Hue-Sagavanirktok. In Cook Inlet, presently a fore-arc basin, there are two systems: Tuxedni-Hemlock and Beluga-Sterling. The Stillwater-Kulthieth and Poul Creek petroleum systems explain the occurrence of petroleum in the Yakutat terrane along the Gulf of Alaska.

TABLE 3. PETROLEUM SYSTEMS IN ALASKA

Region	System Name
North Slope	Ellesmerian(!) Torok–Nanushuk(.) Hue–Sagavanirktok(!)
Cook Inlet	Tuxedni–Hemlock(.) Beluga–Sterling(.)
Yakutat terrane Gulf of Alaska	Stillwater–Kulthieth(.) Poul Creek(.)

NORTH SLOPE

The history of drilling exploratory and production wells on the North Slope commenced in 1944 with three wildcat wells (Fig. 2; Petroleum Information, 1988). Through 1987, 1,973 wells have been drilled, of which 283 were exploratory and 1,690 were field development wells (Petroleum Information, 1988). The first exploratory program was carried out by the U.S. Navy from 1944 through 1952 and included 69 wells drilled on the western half of the North Slope (Bird, 1981). Several small oil and gas fields were discovered, which prompted the drilling of 11 development wells (Figs. 2 and 3; Tables 4 and 5). In 1964, industry exploration shifted to the east-central North Slope, where in 1967, the Prudhoe Bay oil field was discovered. The

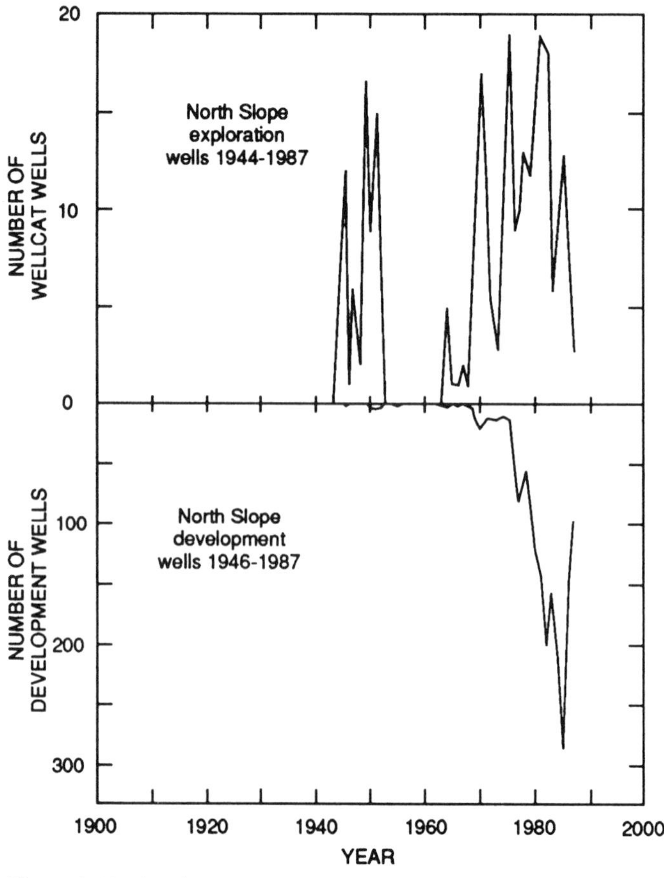

Figure 2. Exploration and development drilling history for the North Slope. Wells are plotted at year they reached total depth.

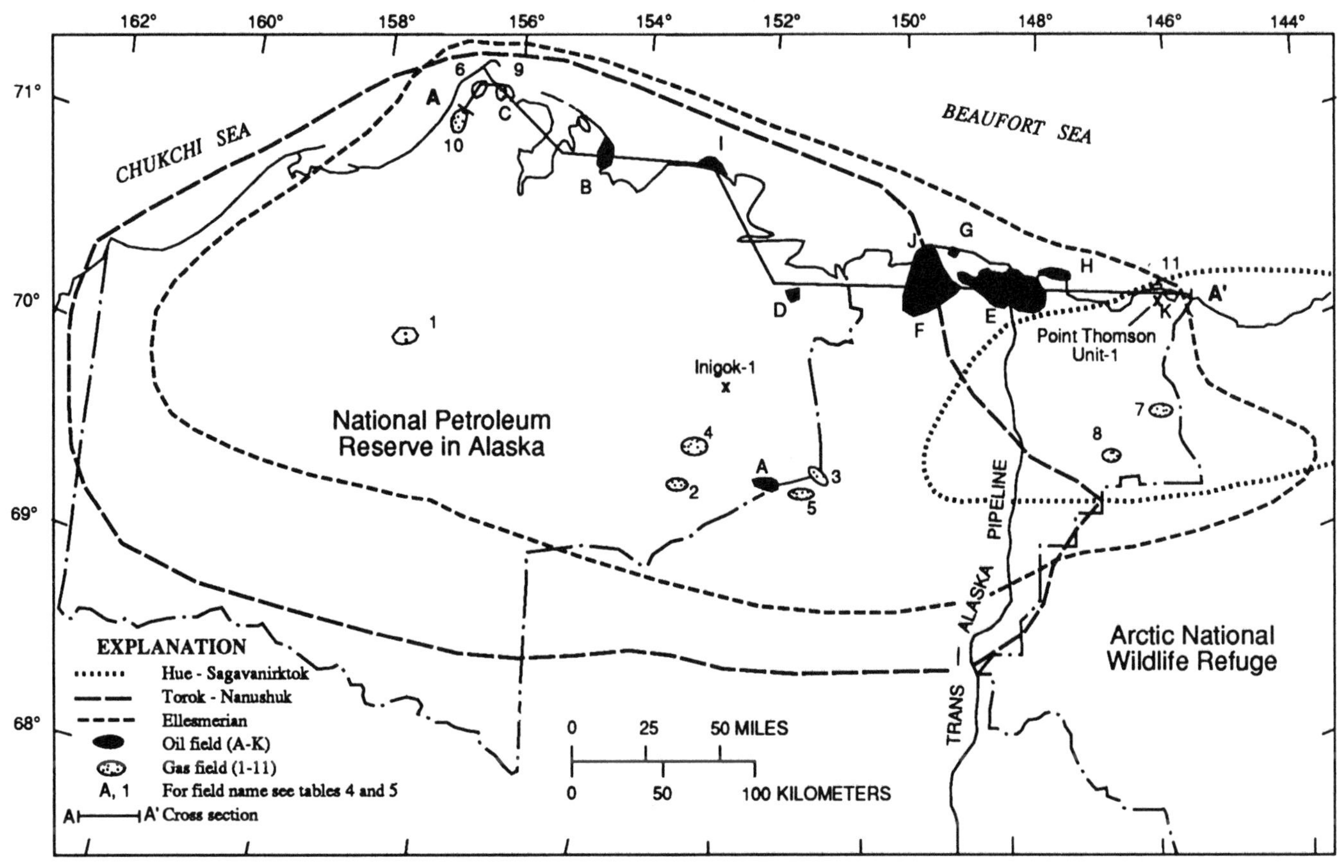

Figure 3. The North Slope showing oil and gas fields, National Petroleum Reserve in Alaska, Arctic National Wildlife Refuge, boundaries of petroleum systems, and location of cross section A–A′.

discovery of this supergiant oil field triggered a flurry of activity in exploratory drilling by industry and the federal government in the immediate area onshore and offshore as well as to the west, where there was hope of discovering another "Prudhoe" (Gryc, 1988). In addition, the trans-Alaska pipeline, completed in 1977, stimulated development drilling that peaked in 1985. Cumulative production (1988) on the North Slope is 5.9 bbo and 8.2 tcfg. This leaves 6.0 bbo and 25.4 tcfg left to be extracted (Tables 4 and 5). Much more gas is being produced than the cumulative production indicates because much of the gas is being reinjected into the Prudhoe Bay reservoirs to maintain pressure. Because there is no way to move the gas to market, gas consumption is restricted to production facilities on the North Slope. The exploration and petroleum geology of the North Slope are discussed by many workers (Miller and others, 1959; Collins and Robinson, 1967; Alaska Geological Society, 1972; Tailleur, 1973; Walker, 1973; Drummond, 1974; Bird, 1978, 1981, 1985, 1986a, b, 1988; Van Dyke, 1980; Sweeney, 1982; Huffman, 1985; McWhae, 1986; Molenaar and others, 1986; Bird and Magoon, 1987; Hubbard and others, 1987; Tailleur and Weimer, 1987; and Gryc, 1988).

The productive stratigraphic section on the North Slope can be divided into two sequences: the Ellesmerian and the Brookian (Fig. 4). Rocks older than the Ellesmerian, except in northeastern Alaska, are considered economic basement. The Ellesmerian sequence ranges in age from Mississippian to Early Cretaceous and includes important petroleum source, reservoir, and seal rocks. Except for the carbonate and coal deposits, the provenance of the quartz-rich Ellesmerian sediment is to the north, or the site of the present-day Arctic Ocean. The Colville basin, the site of deposition during this time, received many thousands of meters of sediment deposited in environments that ranged from nonmarine to deep marine. In Jurassic and Early Cretaceous time, rifting in the Arctic Basin reoriented the depositional pattern such that the Brooks Range was the provenance for lithic-rich sediment that prograded from the southwest to the northeast from Early Cretaceous to Holocene time (Ahlbrandt, 1979; Molenaar, 1983). Depositional environments ranged from nonmarine to deep marine and also include important source, reservoir, and seal rocks. Three petroleum systems, the Ellesmerian, Torok-Nanushuk, and Hue-Sagavanirktok, are present on the North Slope.

Ellesmerian system

The Ellesmerian petroleum system includes (Fig. 3; Table 4) the largest oil field in North America, Prudhoe Bay. Rocks involved in this petroleum system include both the Ellesmerian and

TABLE 4. OIL ACCUMULATIONS IN THE TOROK–NANUSHUK AND ELLESMERIAN PETROLEUM SYSTEMS BY DISCOVERY DATE, INDICATING CUMULATIVE PRODUCTION AS OF DECEMBER 31, 1987, REMAINING RESERVES, RESERVOIR CHARACTERISTICS, AND OIL CHEMISTRY*

Map symbol	Accumulation name	Year dis	Prod unit	Pool	Res lith	Prod stat	Trap type	Cum prod oil ($x10^3$bbl)	Cum prod gas (bcd)	Reserves oil ($x10^6$bbl)	Reserves gas (bcf)	Prod depth (m)	Orig press (kPa)	Sat press (kPa)
							Torok–Nanushuk							
A	Umiat	1946	Nanushuk	–	ss	shut-in	FA	–	–	70	–	75	–	–
B	Simpson	1950	Nanushuk	–	ss	abd	L	–	–	7	–	90	–	–
C	East Barrow	1980	pebble shale unit	–	ss	shut-in	FA	–	–	–	–	475	–	–
							Ellesmerian							
D	Fish Creek	1949	Nanushuk	–	ss	abd	L	–	–	–	–	915	–	–
E	Prudhoe Bay	1967	Kuparuk River	PBKR	ss	shut-in	C	–	–	–	–	1,890	22,130	20,550
E	Prudhoe Bay	1967	Sag River	Prudhoe	ss	prod	C	–	–	–	–	–	–	–
E	Prudhoe Bay	1967	Shublik	Prudhoe	co	prod	C	–	–	–	–	–	–	–
E	Prudhoe Bay	1967	Sadlerochit	Prudhoe	ss	prod	C	5,506,886	7,652	4,219	23,441	2,680	30,270	30,270
E	Prudhoe Bay	1967	Lisburne	PBL	co	prod	S	21,919	76	189	406	2,715	30,960	–
F	Kuparuk River	1969	Kuparuk River	Kuparuk	ss	prod	C	397,913	470	1,105	634	1,890	23,170	–
F	Kuparuk River	1969	Sagavanirktok	W. Sak	ss	shut-in	L	3	5	–	–	–	–	–
G	Milne Point	1969	Kuparuk River	Kuparuk	ss	shut-in	S	5,453	2	95	–	–	–	–
C	East Barrow	1974	Sag River	–	ss	shut-in	FA	–	–	–	–	670	6,895	–
H	Endicott	1978	Endicott	Endicott	ss	prod	C	8,807	8	366	907	3,050	33,580	–
I	Dalton	1979	Sadlerochit	–	ss	abd	L	–	–	–	–	2,400	–	–
I	Dalton	1979	Lisburne	–	co	abd	L	–	–	–	–	2,530	–	–
J	Niakuk	1981	–	–	–	shut-in	S	–	–	58	–	–	–	–
K	Point Thomson	1975	Canning	–	ss	shut-in	C	–	–	–	–	3,825	67,915	–
K	Point Thomson	1975	Thomson sand†	–	cg	shut-in	S	–	–	–	–	3,950	70,050	–

TABLE 4. OIL ACCUMULATIONS IN THE TOROK–NANUSHUK AND ELLESMERIAN PETROLEUM SYSTEMS BY DISCOVERY DATE, INDICATING CUMULATIVE PRODUCTION AS OF DECEMBER 31, 1987, REMAINING RESERVES, RESERVOIR CHARACTERISTICS, AND OIL CHEMISTRY* (continued)

Res temp (°C)	Net pay (m)	Por (%)	Perm (md)	Orig GOR (SCF/STB)	Water sat, S_{wi} (%)	Dev acres (ha)	Oil grav (API)	Sulfur (%)	$\delta^{34}S$ (‰)	Pr/Ph I	Pr/nC_{17}	$\delta^{13}C$ sat (‰)	$\delta^{13}C$ arom (‰)	Map symbol
							TOROK–NANUSHUK							
–	–	12	1-200	<100	–	2,430	36	0.1	-3.5	1.9	0.7	-28.0	-28.6	A
–	–	–	–	–	–	–	24	0.2	-5.2	B	B	-28.5	-28.1	B
–	–	–	–	–	–	–	31	0.2	+7.7	1.3	0.6	-29.5	-28.9	C
							ELLESMERIAN							
–	–	–	–	–	–	–	15	1.8	-2.3	B	B	-29.9	-29.6	D
66	9-25	23	3-200	450	28-47	–	23	–	–	–	–	–	–	E
–	–	–	–	–	–	–	–	–	–	–	–	–	–	E
–	–	–	–	–	–	–	–	–	–	–	–	–	–	E
93	0-135	22	265	730	21	61,110	28	0.9	-2.6	1.4	–	-29.8	-29.3	E
84	–	11	0.1-2.0	830	30	–	27	1.0	-2.7	–	–	-29.9	29.4	E
66	–	21	–	228-413	35	5,180	23	1.4	-0.4	–	–	-30.1	-29.7	F
–	–	–	–	–	–	–	–	–	–	–	–	–	–	F
–	–	–	–	–	–	–	–	–	–	–	–	–	–	G
14	–	22	44	–	–	730	18	1.5	+0.5	B	B	-29.5	-29.4	C
99	–	20	–	750	–	–	23	1.0	-0.2	–	–	-29.0	-28.7	H
–	–	–	–	–	–	–	–	1.9	+3.7	B	B	-29.6	-29.5	I
–	–	–	–	–	–	–	10	2.5	+4.4	B	B	-29.9	-30.1	I
–	–	–	–	–	–	–	–	–	–	–	–	–	–	J
91	–	–	–	864-934	–	–	23	–	–	–	–	–	–	K
96	–	–	–	5,830	–	–	18§	1.2	–	1.4	0.7	–	–	K

*Map symbols correspond to locations shown on Figure 3. References to complete this table include: Alaska Oil and Gas Conservation Commission, 1985, 1988; Bird and Magoon, 1987; R. P. Crandall, written communication, 1988; Hughes and Holba, 1988; Molenaar, 1982; Oil and Gas Journal, 1988; Petroleum Information, 1988. –, no information available; 1 ft = 0.3048 m; 1 psi = 6.895 kPa (kiloPascals); 1 acre = 0.4047 ha (hectare); 1 barrel = 0.1590 kiloliter; 1 cubic ft = 0.0283 cubic m; abd, abandoned; API, American Petroleum Institute; arom, aromatic hydrocarbons; B, biodegraded; bbl, barrel; bcf, billion cubic feet; C, combination trap; cg, conglomeratic sandstone; co, carbonate reservoir, cum prod, cumulative production through 12/31/87; dev, developed; dis, discovery; FA, faulted anticlinal trap; GOR, gas-to-oil ratio, in SCF/STB, standard cubic feet of gas per stock tank barrel of oil; grav, gravity; L, stratigraphic trap; PBKR, Prudhoe Bay–Kuparuk River; PBL, Prudhoe Bay Lisburne; perm, permeability; ph, phytane; por, porosity; pr, pristane; press, pressure; prod, producing; reserves, remaining reserves; res, reservoir; res lith, reservoir lithology; S, structural trap; sat, saturated hydrocarbons; ss, sandstone reservoir; stat, status; temp, temperature; W., west.

†Of local usage; includes oil and gas in pre-Mississippian reservoir rock

§Oil gravity ranges from 18°–45° API

the Brookian sequences. However, the source rocks and the major reservoir rocks are restricted to the Ellesmerian sequence (Fig. 4). Numerous authors have described the petroleum geology of the Ellesmerian sequence (Morgridge and Smith, 1972; Van Poollen and others, 1974; Jones and Speers, 1976; Bird and Jordan, 1977; Jamison and others, 1980; Seifert and others, 1980; Molenaar, 1981; Magoon and Claypool, 1981b, 1983, 1984, 1985, 1988; Carman and Hardwick, 1983; Alaska Oil and Gas Commission, 1984, 1985, 1988; Melvin and Knight, 1984; Oil and Gas Journal, 1984; Claypool and Magoon, 1985, 1988; Magoon and Bird, 1985, 1988; Magoon and Claypool, 1985; Specht and others, 1986; Whelan and others, 1986; Curiale, 1987; Sedivy and others, 1987; Werner, 1987; Farrington and others, 1988; and Hughes and Holba, 1988).

Both the Ellesmerian and Brookian sequences are required to complete this petroleum system. First, the Ellesmerian sequence includes three source rocks with mostly type II organic matter—the Shublik Formation, Kingak Shale, and to a lesser extent the pebble shale unit (Seifert and others, 1980; Magoon and Bird, 1985)—and three siliciclastic reservoir rocks—the Endicott Group, Sadlerochit Group (Ivishak Formation), and Kuparuk River Formation of Jamison and others (1980); it also includes a carbonate reservoir, the Lisburne Group (Bird, 1981, 1985; Figs. 4 and 5). The most important reservoir rock, in terms of volume of recoverable oil, is the Sadlerochit Group, a quartz-rich conglomeratic sandstone (Table 4). Other hydrocarbon accumulations in this system include both siliciclastic and carbonate reservoirs (Tables 4 and 5; Bird, 1981; Bird and Magoon, 1987). The Brookian, a northeasterly prograding deltaic and prodeltaic deposit, is the other sequence that provides the necessary overburden to mature the organic matter within the Ellesmerian source rocks. The areal distribution of the Ellesmerian system is controlled by the axis of the Colville trough and the present-day distribution of accumulations (Fig. 3).

Of the producing oil pools, the typical accumulation is situated in a combination structural-unconformity trap at a drill depth of 2,585 m (1,890 to 3,050 m) and covers 22,340 hectares (730 to 61,100 ha). The net pay ranges up to 135 m thick with an average porosity of 19 percent (11 to 22 percent) and permeabilities up to 265 millidarcies. The original pressure of the fluid in the reservoir is 29,500 kiloPascals (23,170 to 33,580 kPa); water saturation is 29 percent (21 to 35 percent), and temperature is 86 °C (66 to 99 °C). The API gravity of the oil is 25 degrees (23 to 28), and the oil has a gas-to-oil ratio of 635 (228 to 830) and a sulfur content of 1.1 percent (0.9 to 1.4 percent). When the oil is not biodegraded, the pristine/phytane ratio is 1.4, the carbon isotope value for the saturated hydrocarbons is –29.7 permil (–30.1 –29.0 permil), and for the aromatic hydrocarbons is –29.3 permil (–29.7 to –28.7 permil). Using carbon isotope and biological marker compounds, a source-rock-to-oil comparison is possible (Seifert and others, 1980; Sedivy and others, 1987). Therefore, the level of certainty for the Ellesmerian petroleum system is known.

The duration of the Ellesmerian petroleum system was 300

TABLE 5. GAS ACCUMULATIONS IN THE TOROK–NANUSHUK AND ELLESMERIAN PETROLEUM SYSTEMS BY DISCOVERY DATE, INDICATING ESTIMATED RESERVES, RESERVOIR CHARACTERISTICS, AND GAS CHEMISTRY*

Map No.	Accumulation name	Year dis	Prod unit	Pool	Res lith	Prod stat	Trap type	Cum prod oil ($x10^3$bbl)	Cum prod gas (bcd)	Reserves gas bcf	Prod depth (m)	Orig press (kPa)	Gas spec grav
						Torok–Nanushuk							
1	Meade	1950	Nanushuk	–	ss	abd	A	–	–	10-20	1,280	–	–
2	Wolf Creek	1951	Nanushuk	–	ss	abd	A	–	–	–	460	–	–
3	Gubik	1951	Colville	–	ss	abd	A	–	–	22-295	440	–	–
4	Square Lake	1952	Colville	–	ss	abd	A	–	–	–	500	–	–
5	East Umiat	1963	Nanushuk	–	ss	abd	A	–	–	–	550	5,170	0.600
						Ellesmerian							
6	South Barrow	1949	Kingak	Barrow ss†	ss	prod	FA	–	18	7	685	335	0.560
7	Kavik	1969	Sag River	–	ss	abd	FA	–	–	–	1,065	730	0.587
7	Kavik	1969	Sadlerochit	–	ss	abd	FA	–	–	–	–	730	0.588
8	Kemik	1972	Shublik	–	ss	abd	FA	–	–	–	2,620	815	0.600
9	East Barrow	1974	Kingak	Barrow ss†	ss	prod	FA	–	4	8	580	305	0.570
10	Walakpa	1980	pebble shale unit	–	ss	shut-in	L	–	–	–	635	315	–
11	Point Thomson	1975	Thomson sand†	–	cg	shut-in	C	–	–	–	3,930	3,050	–

TABLE 5. GAS ACCUMULATIONS IN THE TOROK–NANUSHUK AND ELLESMERIAN PETROLEUM SYSTEMS BY DISCOVERY DATE, INDICATING ESTIMATED RESERVES, RESERVOIR CHARACTERISTICS, AND GAS CHEMISTRY* (continued)

Res temp (°C)	Net pay (m)	Por (%)	Perm (md)	Water sat, S_{wi} (%)	Dev acres (ha)	Btu (Btu/ft^3)	$\delta^{13}C$ methane (‰)	C_1/C_{1-5}	Map No.
				Torok–Nanushuk					
–	–	–	–	–	–	–	–	–	1
–	–	–	–	–	–	–	–	–	2
–	–	–	–	–	–	–	–	–	3
–	–	–	–	–	–	–	–	–	4
10	20	15.4	15	32	–	–	–	–	5
				Ellesmerian					
17	8	20	30	52	1,415	–	-40.8	0.984	6
46	12	5	2	50	518	–	–	–	7
53	85	13	200	50	518	–	–	–	7
51	–	–	–	–	260	–	–	–	8
14	5	22	44	55	728	–	–	–	9
–	5	20	28	40	–	–	-37.8	0.997	10
96	–	–	–	–	–	–	–	–	11

*Map numbers correspond to locations shown on Figure 3. References to complete this table include: Alaska Oil and Gas Conservation Commission, 1985, 1988; Bird, 1981; R. P. Crandall, written communication, 1988; –, no information available; 1 ft = 0.3048 m; 1 psi = 6.897 kPa (kilo Pascals); 1 acre = 0.4047 ha (hectare); A, anticlinal trap; abd, abandoned; bbl, barrels; bcf, billion cubic feet; Btu, British thermal unit; cg, conglomeratic sandstone; cum prod, cumulative production through 12/31/87; dev, developed; dis, discovery; FA, faulted anticlinal trap; L, stratigraphic trap; orig press, original pressure; perm, permeability; por, porosity; prod, producing; res lith, reservoir lithology; res temp, reservoir temperature; sat, saturation; spec grav, specific gravity; ss, sandstone; stat, status.

†Of local usage.

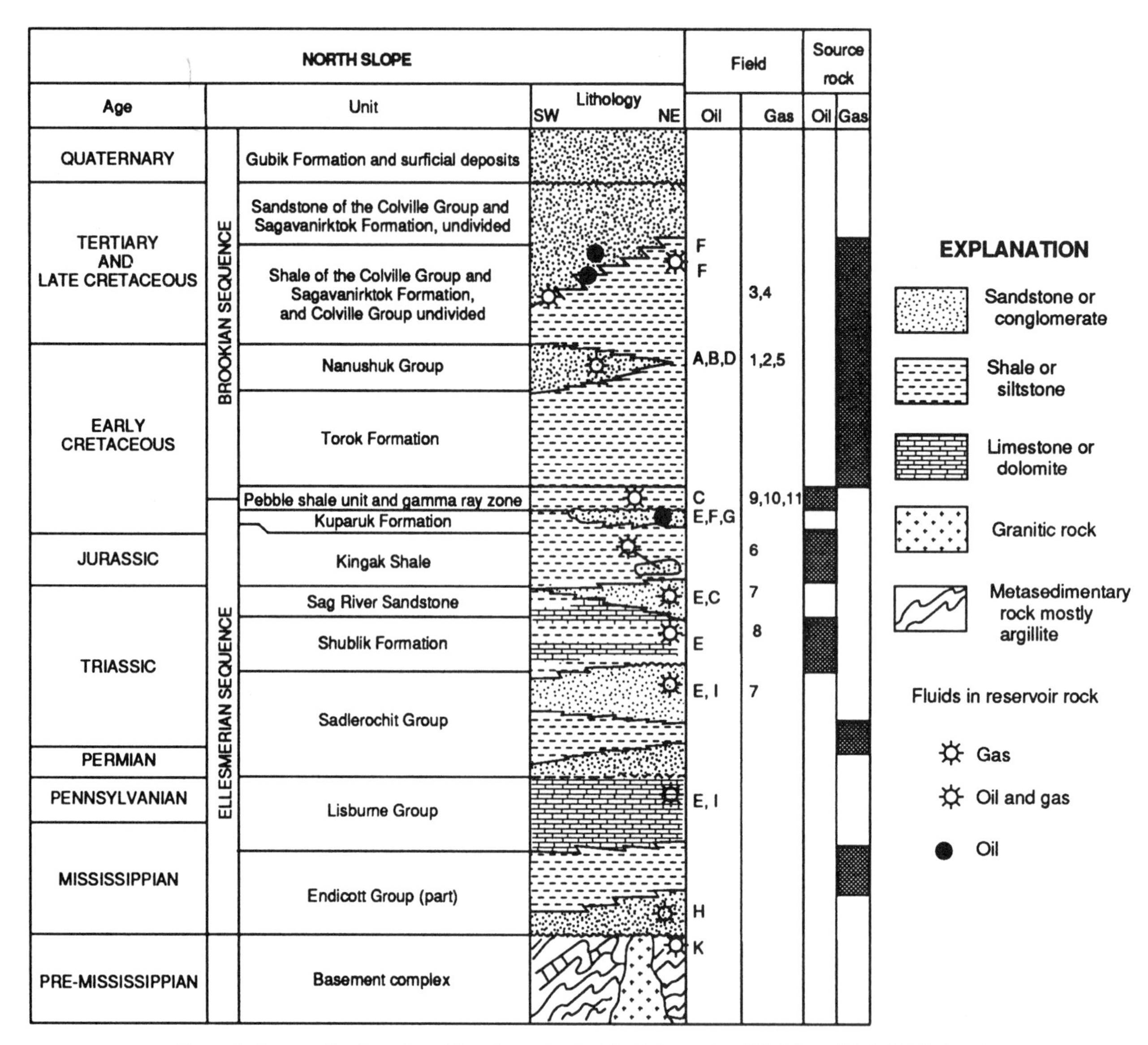

Figure 4. A generalized stratigraphic column for the North Slope (modified from Bird, 1985) showing petroleum in reservoir rocks and source rock intervals. See Tables 4 and 5 for oil and gas field names.

m.y.—from the Mississippian (360 Ma) to early Tertiary time (60 Ma; Fig. 6). Based on a Lopatin diagram for the Inigok No. 1 well in the eastern part of NPRA, the Shublik Formation and the Kingak Shale were well into the gas-generation phase (>160 TTI) by the end of the Cretaceous (Fig. 7). Extrapolating this model to the same rock units to the east in the Colville trough suggests that most oil near Prudhoe Bay migrated into the area by early Tertiary time. During the Tertiary some of the hydrocarbon generated was gas, but for the most part, already accumulated oil remigrated from the Prudhoe Bay pool into shallower reservoirs; the preservation time for this system is 60 m.y.

Torok-Nanushuk system

On the western part of the North Slope of Alaska, noncommercial quantities of hydrocarbons are associated with the Torok-Nanushuk petroleum system (Figs. 3, 4, and 5; Tables 4 and 5). Stratigraphic intervals involved include the Brookian se-

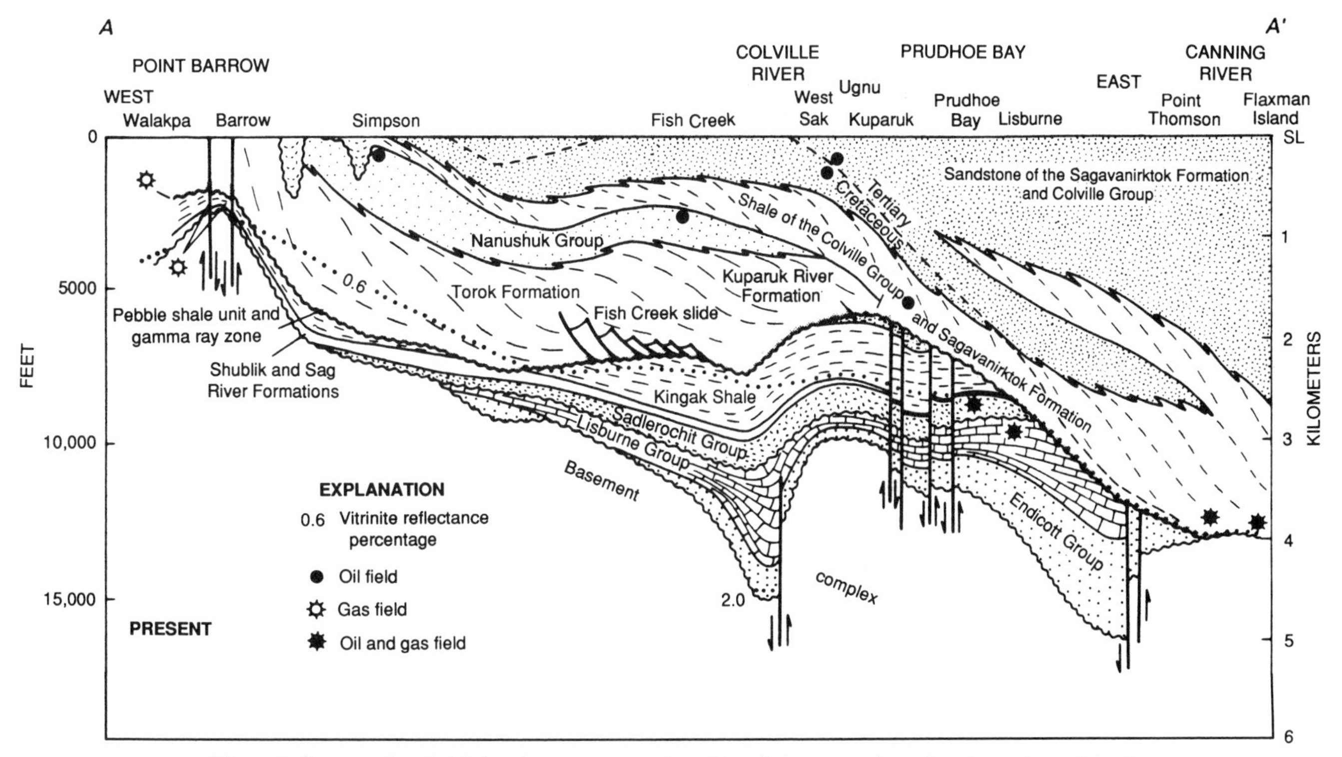

Figure 5. Cross section A-A′ showing present stratigraphic relations, petroleum-bearing units, and depth to the onset of petroleum generation (vitrinite reflectance = 0.6 percent; figure slightly modified from Bird, 1985). See Figure 3 for location of section. Rock symbols explained in Figure 4.

quence and the pebble shale unit; all are from the base of the unconformity below the pebble shale unit to the surface. References that describe the petroleum geology of this system are as follows: Ahlbrandt (1979); Magoon and Claypool (1981b, 1983, 1985, 1988); Molenaar (1981, 1982, 1988); Huffman (1985); and Magoon and Bird (1985, 1988).

The source rock for this system contains a type III kerogen and is most likely found in the lower part of the Torok Formation, and possibly in the underlying pebble shale unit (Magoon and Claypool, 1985). There are considerable geochemical data available on both the oils and source rocks, but a convincing geochemical correlation has not been demonstrated (Magoon and Bird, 1985; Magoon and Claypool, 1985). Therefore, the Torok-Nanushuk petroleum system is hypothetical. The geochemistry of these high-gravity, low-sulfur oils in this system separates them from the moderate-gravity, high-sulfur oils of the Ellesmerian system (Magoon and Claypool, 1981b, 1988). The composition of the Umiat and similar oils is consistent with their origin from the type III kerogen of the Torok Formation and pebble shale unit.

Sandstone of the Nanushuk Group of Early and Late Cretaceous age is the major reservoir rock (Bird, 1988). The reservoir rock is lithic-rich sandstone derived from the Brooks Range orogen (Ahlbrandt, 1979; Molenaar, 1981; Huffman, 1985). The undeveloped oil and gas accumulations are located in structural and stratigraphic traps (Collins and Robinson, 1967; Bird, 1985; Tables 4 and 5). The largest oil field is the Umiat, with 70 million barrels of recoverable oil in the Nanushuk Group (Molenaar, 1982; Table 4). It is unknown whether the gas in the Meade and Gubik fields is thermal or microbial. Since the oil and gas fields are, in many cases, relatively old, and have never been developed, comparatively little information is available for them.

The geologic setting, a foreland basin, remained the same during deposition of all the essential elements. The essential elements include the Torok Formation, and possibly the pebble shale unit, as the source rock; the Nanushuk Group as the reservoir rock; shale within the Nanushuk Group and shale of the Colville Group (Bird, 1988) as the regional seals; and the Brookian sequence, which includes the shale of the Colville Group and the Sagavanirktok Formation as the overburden necessary to mature the source rock. Although some hydrocarbons have been generated during Tertiary time, most of the oil has been preserved in the reservoir since early Tertiary time. Therefore, the duration of this petroleum system is estimated to be 105 m.y. and ranges from the Early Cretaceous (130 Ma) to the Paleogene (25 Ma); the preservation time is 25 m.y. The geographic distribution of

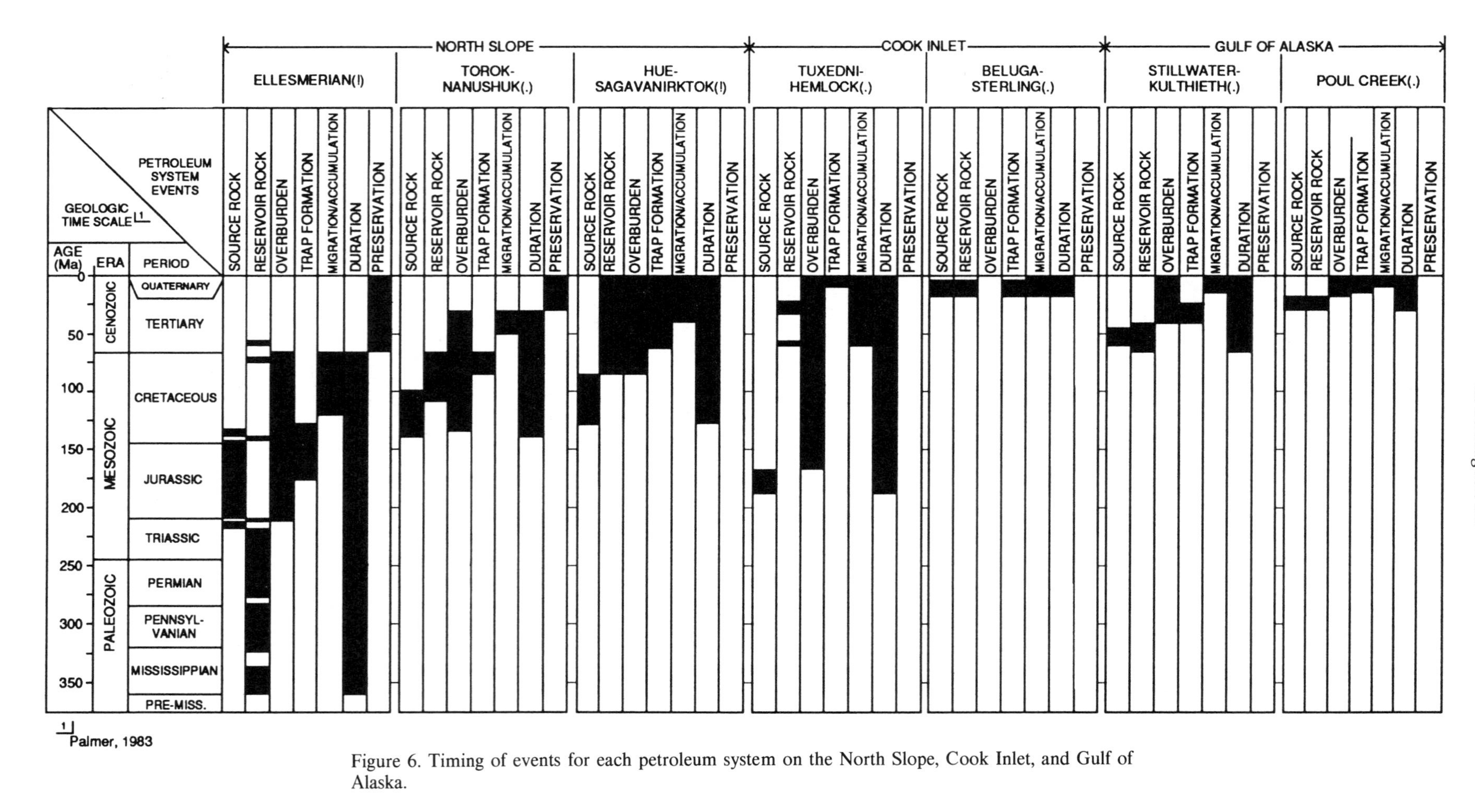

Figure 6. Timing of events for each petroleum system on the North Slope, Cook Inlet, and Gulf of Alaska.

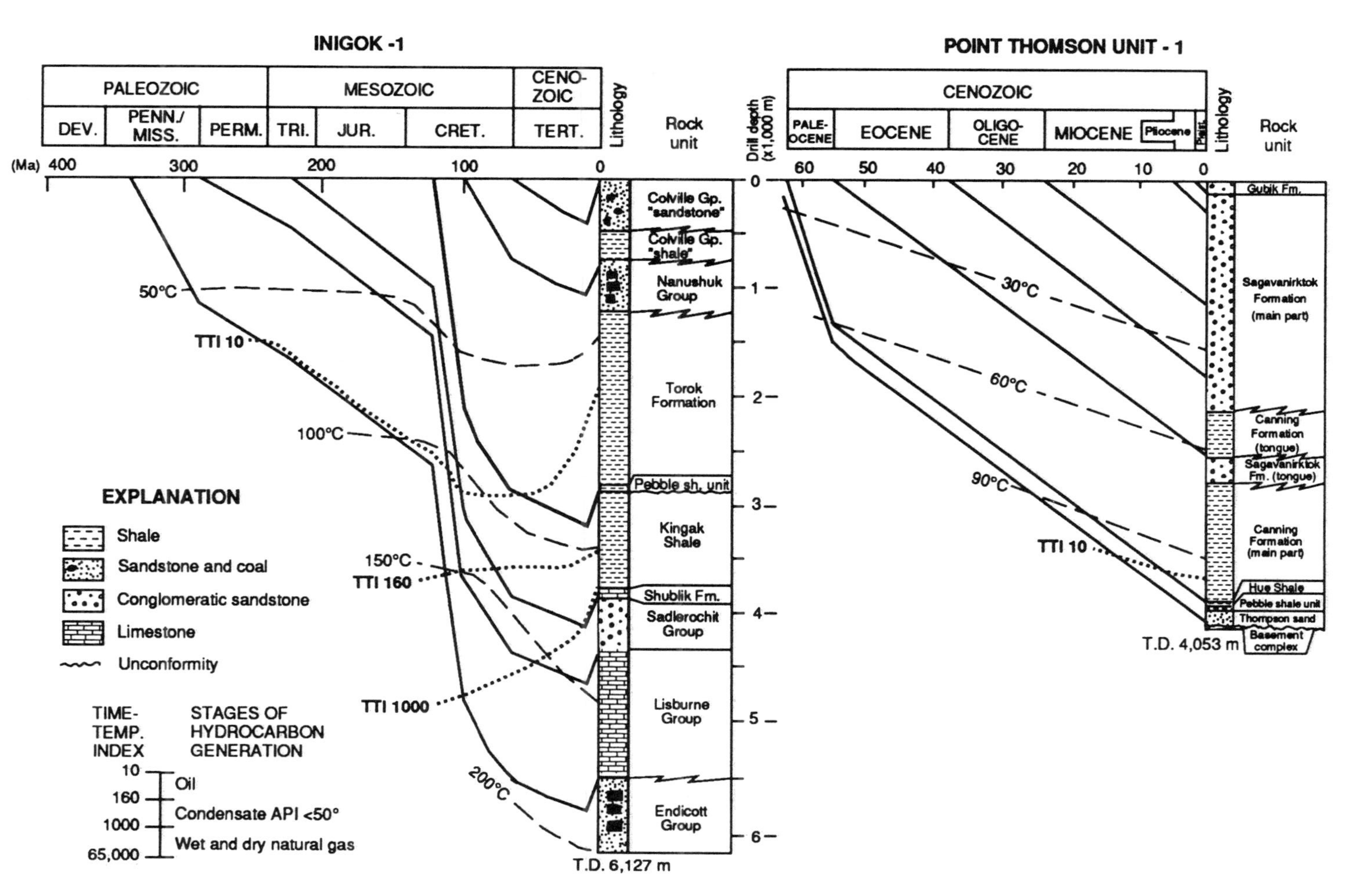

Figure 7. Lopatin diagrams indicating the time of oil and gas generation for the rocks penetrated in the Inigok No. 1 well and the Point Thompson unit No. 1 (from Magoon and Claypool, 1983; Magoon and others, 1987). Location shown on Figure 3.

the Torok-Nanushuk system is controlled on the east by the geographic extent of the Torok Formation and on the west by the known accumulations (Figs. 3 and 5).

Hue-Sagavanirktok system

On the eastern part of the North Slope in northeastern Alaska, an unknown quantity of hydrocarbons is associated with the Hue-Sagavanirktok petroleum system (Fig. 3). From the Point Thomson area east to the Canadian border, stratigraphic intervals involved in this system are the Hue Shale and Canning and Sagavanirktok Formations (Bird and Magoon, 1987). Additional references that relate to the petroleum geology of this system are relatively few because only recently has much subsurface information become available (Detterman and others, 1975; Wood and Armstrong, 1975; Mull and Kososki, 1977; Molenaar, 1983; Bader and Bird, 1986; Anders and others, 1987; Bird and Magoon, 1987; Magoon and others, 1987; and Molenaar and others, 1987).

The Hue Shale is the oil-source rock for this system. Other shale units (Shublik Formation, Kingak Shale, or pebble shale unit) that are oil prone west of the Canning River are either overmature or gas prone (type III kerogen) in northeastern Alaska. The Canning Formation and shale within the Sagavanirktok Formation are either immature or gas-prone source rocks. However, the Hue Shale contains large amounts of type II kerogen whose bitumen extract compares well with certain compounds and properties of the seep oil and oil that stains outcropping sandstones east of the Canning River. The major reservoir rock for this petroleum system probably is the Sagavanirktok Formation or turbidites in the Canning Formation because the oil seeps and stains commonly occur in these units. Because the Hue Shale unconformably overlies many of the Ellesmerian sequence and older units, future drilling may encounter oil pools that originated from this source rock in older reservoir rocks. The reservoir properties for most potential reservoirs in northeastern Alaska are discussed in Bird and Magoon (1987).

The Hue-Sagavanirktok petroleum system took approximately 95 m.y. to develop, from Late Cretaceous to Holocene time. Based on the Lopatin diagram in the Kavik area (field no. 7 in Fig. 3), the Hue Shale began generating oil 45 m.y. ago (Magoon and others, 1987), and in the Point Thompson No. 1 well (field K in Fig. 3), the Hue Shale began generating oil in Miocene time (10 Ma; Fig. 7). Trap formation occurred before, during, and after oil generation and migration. Since oil generation continues today, this petroleum system is incomplete, and the duration time is a minimum.

COOK INLET

Almost 60 years of exploration preceded the discovery of the first commercial oil field in the Cook Inlet area (Fig. 8). Exploration started on the Iniskin Peninsula in 1902, where seven wells were drilled. Between 1921 and 1957, only nine exploratory wells were drilled before the Swanson River oil field was discovered (Parkinson, 1962). Over the next 15 years, seven oil and 23 gas accumulations were discovered. Except for one oil pool (Redoubt Shoal), and 14 gas accumulations, all are still producing (Tables 6 and 7). By the end of 1987, almost 1.1 billion barrels of oil and 5.3 trillion cubic feet of gas had been produced, leaving 90 million barrels of oil and 3.3 tcf of gas yet to be extracted. The history of exploration and the framework and petroleum geology in this area are discussed by many workers (Kelly, 1963, 1968; Detterman and Hartsock, 1966; Crick, 1971; Blasko and others, 1972; Kirschner and Lyon, 1973; MacKevett and Plafker, 1974; Blasko, 1974, 1976a; Boss and others, 1976; Hite, 1976; Magoon and others, 1976; Fisher and Magoon, 1978; Claypool and others, 1980; Magoon and others, 1980; Magoon and Claypool, 1981a; Reed and others, 1983; Magoon, 1986; and Magoon and Egbert, 1986).

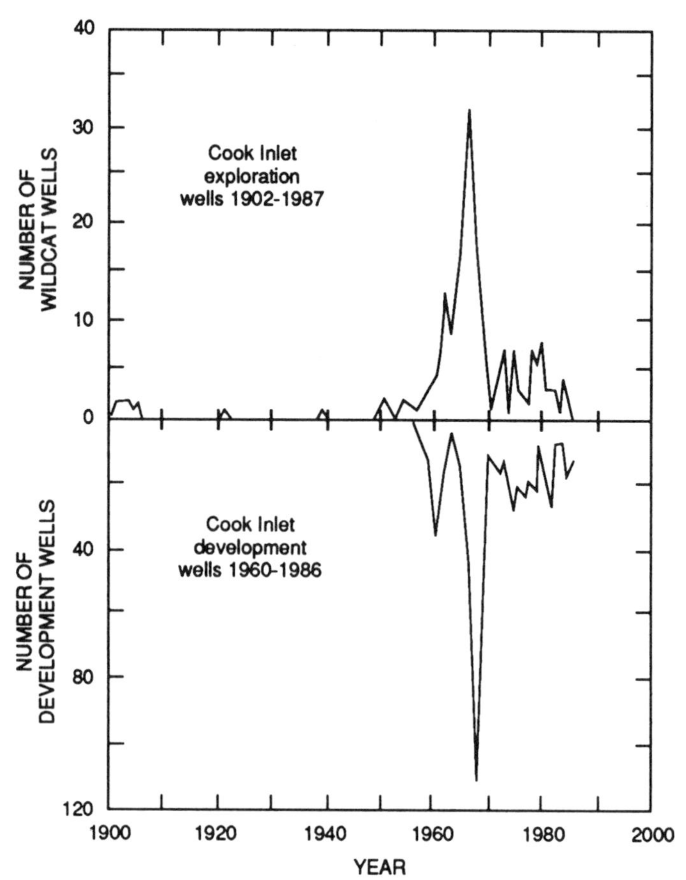

Figure 8. Exploration and development drilling history for Cook Inlet. Wells are plotted at year they reached total depth.

The tectonic evolution of the Cook Inlet area is complex because it is part of the northern Pacific margin, which has been the site of continuous convergence throughout the Mesozoic and Cenozoic (Coney and Jones, 1985). The Cook Inlet area is bounded on the northwest by the Alaska-Aleutian Range and the Talkeetna Mountains and on the southeast by the Kenai Mountains. The Jurassic and Lower Cretaceous rocks are included in the Peninsular terrane (Silberling and others, this volume); the Upper Cretaceous and Cenozoic rocks compose a post-amalgamation overlap sequence. Because all the commercial accumulations are associated with the Cenozoic rocks in the Cook Inlet area, the petroleum discussion is restricted to this area (Fig. 9).

In the Cook Inlet area, the pre–Late Cretaceous sequence and correlative plutonic rocks of the Alaska-Aleutian Range batholith constitute the Peninsular terrane (Silberling and others, this volume). The Border Ranges fault separates this terrane from the accreted Chugach terrane on the southeast. The Lower Jurassic Talkeetna Formation, a volcaniclastic sequence, and the Alaska-Aleutian batholith are economic basement for this petroleum province (Fig. 10). The Tuxedni Group of Middle Jurassic age is important because it contains rich source rocks and is the most likely source for all the oil and some gas in the area. The Upper Jurassic Naknek Formation contains a high percentage of feldspathic sandstone and conglomerate, but because of laumontite cementation, it is a poor reservoir (Franks and Hite, 1980; Bolm and McCulloh, 1986). The Cretaceous Matanuska Formation contains little organic matter, but does contain potential sandstone reservoirs. All these units in the Peninsular terrane were deposited in a coastal to deep marine environment and unconformably underlie the petroleum-bearing Cenozoic rocks.

The Cenozoic rocks in the Cook Inlet area overlap the Alaska-Aleutian batholith on the northwest and the Border Ranges fault on the southeast. Calderwood and Fackler (1972) defined and named the critical Cenozoic stratigraphic units—West Foreland Formation, Hemlock Conglomerate, Tyonek Formation, Beluga Formation, and Sterling Formation—that are regionally correlated (Alaska Geological Society, 1969a–d, 1970a, b) and mapped (Hartman and others, 1972). These rock units were all deposited in a nonmarine fore-arc basin setting. The provenance for the conglomerate, sandstone, siltstone, shale, and volcaniclastic debris was local highs flanking the basin and interior Alaska. Each of these rock units is a reservoir for oil or gas somewhere in the basin. Throughout the section, numerous large coal deposits formed and were preserved (Wahrhaftig and others, this volume; Barnes and Payne, 1956; Barnes and Cobb, 1959).

Tuxedni-Hemlock system

In the Cook Inlet area, all the large oil fields occur in reservoirs of early Tertiary age (Fig. 10; Table 6). Rock units that range in age from Middle Jurassic to Holocene are involved in the Tuxedni-Hemlock petroleum system.

Two different depositional settings, separated by a period of nondeposition and deformation, were required to complete this petroleum system. The first setting, a marine environment from Middle Jurassic to Late Cretaceous, accumulated sedimentary rocks deposited in both a fore-arc and back-arc basin associated with volcanic and plutonic rocks. The Middle Jurassic (187 to 163 Ma) Tuxedni Group, which contains a type III kerogen, is considered to be the source rock for the oil. The second setting, a fluvial environment, produced siliciclastic rocks deposited in a narrow fore-arc basin. Approximately 80 percent of the oil is contained in the Hemlock Conglomerate, a conglomeratic sandstone of Oligocene age (37 to 24 Ma; Magoon and others, 1976; Wolfe, 1981). The McArthur River field is the largest, with original reserves of almost 570 million barrels of oil, of which almost 500 million barrels are from the Hemlock Conglomerate (Table 6). The level of certainty for the Tuxedni-Hemlock petroleum system is hypothetical because geochemical information on the oils and rocks are insufficient to demonstrate a correlation (Magoon and Claypool, 1981a).

A typical oil pool in this petroleum system is located in a faulted anticline at a drill depth of 2,560 m (765 to 4,500 m). The pool covers 1,000 hectares (ranges from 195 to 5,000 ha), and has a net pay of 90 m (21 to 300 m). The reservoir rock is a conglomeratic sandstone with an average porosity of 17 percent (12 to 22 percent), an average permeability of 80 millidarcies (10 to 360 md), a reservoir pressure of 29,650 kiloPascals (14,045 to 52,071 kPa), and a temperature of 72 °C (44 to 102 °C). The typical oil in this reservoir has an API oil gravity of 34 $\pm$ 6°, a gas-oil ratio of 600 (175 to 3,850), a sulfur content of 0.1 percent, a pristine/phytane ratio of 2.7 (1.6 to 3.4), and carbon isotope values for saturated hydrocarbons of –30 permil (–30.4 to –29.6) and for aromatic hydrocarbons of –28 permil (–29.1 to –27.8). This information and biological marker data indicate that the oil originates from a marine-shale source rock (Peters and others, 1986).

Tertiary rocks include not only the reservoir rock but the overburden necessary to mature the Middle Jurassic marine source rock. Because the source rock is at maximum burial depth, the duration time for this petroleum system is 187 m.y., with no preservation time (Fig. 6). No Lopatin diagram is available for Cook Inlet; late Cenozoic oil generation is indicated by the thick Cenozoic sedimentary section east of the Middle Ground Shoal field (Figs. 10 and 11). Oil generation probably started within the last 5 m.y. and continues today; the petroleum system thus is still active, and the duration is lengthening. The geographic boundary for the Tuxedni-Hemlock petroleum system is restricted by known accumulations because the mature source rock is covered between the Swanson River and Middle Ground Shoal fields (Fig. 11).

Beluga-Sterling system

In Cook Inlet, a cumulative production of more than 6.1 tcf of dry gas is attributed to upper Tertiary rocks, which were deposited in a fore-arc basin (Magoon and Egbert, 1986; Fig. 5; Table 7). Three stratigraphic units are involved in the Beluga-Sterling petroleum system: the Tyonek, Beluga, and Sterling Formations (Fig. 10).

The siliciclastic Sterling Formation of late Miocene and Pliocene age is the major reservoir rock. Most of the microbial gas is in the Sterling Formation reservoirs, a medium-grained, well- to fairly well-sorted, and slightly conglomeratic sandstone (Crick,

TABLE 6. TUXEDNI–HEMLOCK OIL ACCUMULATIONS BY DISCOVERY DATE, INDICATING CUMULATIVE PRODUCTION AS OF DECEMBER 31, 1987, REMAINING RESERVES, RESERVOIR CHARACTERISTICS, AND OIL CHEMISTRY*

Map symbol	Accumulation name	Year dis	Prod unit	Pool	Res lith	Prod stat	Trap type	Cum prod oil ($x10^3$bbl)	Cum prod gas (bcf)	Reserves oil ($x10^6$bbl)	Reserves gas (bcf)	Prod depth (m)	Orig press (kPa)	Sat press (kPa)
A	Swanson River	–	Hemlock	34-10	cg	prod	A	–	–	–	–	3,285	39,300	7,240
A	Swanson River	–	Hemlock	Center	cg	prod	A	–	–	–	–	3,220	39,300	7,860
A	Swanson River	–	Hemlock	SCU	cg	prod	A	–	–	–	–	3,140	38,265	9,310
A	Swanson River	1957	**Field total**					208,469	1,752	10	250			
B	Mid Grd Shl	–	Tyonek	A	ss	prod	FA	2,000	4	–	–	765	–	–
B	Mid Grd Shl	–	Tyonek	B,C,D	ss	prod	FA	10,420	7	–	–	1,830	19,085	13,100
B	Mid Grd Shl	–	Ty-Hem	E,F,G	ss-cg	prod	FA	140,683	65	–	–	2,590	29,095	10,345
B	Mid Grd Shl	1962	**Field total**					153,103	77	11	7			
C	Granite Point	–	Tyonek	MGS	ss	prod	A	106,838	92	–	–	2,675	29,310	16,550
C	Granite Point	–	Hemlock	–	cg	prod	A	3	–	–	–	–	–	–
C	Granite Point	1965	**Field total**					106,841	92	19	15			
D	Trading Bay	–	Tyonek	C	ss	prod	FA	–	–	–	–	1,340	14,045	–
D	Trading Bay	–	Tyonek	D	ss	prod	FA	–	–	–	–	1,715	18,180	13,245
D	Trading Bay	–	Ty-Hem	–	ss-cg	prod	A	–	–	–	–	2,990	30,820	12,275
D	Trading Bay	–	Hemlock	–	cg	prod	FA	–	–	–	–	1,860	19,320	11,185
D	Trading Bay	1965	**Field total**					89,424	61	2	2			
E	McArthur River	–	Tyonek	MGS	ss	prod	A	36,769	18	–	–	2,695	27,640	12,590
E	McArthur River	–	Hemlock	–	cg	prod	A	474,421	171	–	–	2,820	29,305	12,320
E	McArthur River	–	W Foreland	–	ss	prod	A	19,317	6	–	–	2,940	30,730	8,170
E	McArthur River	1965	**Field total**					530,507	194	47	25			
F	Redoubt Shoal	1968	Hemlock	–	cg	abd	–	2	–	–	–	–	–	–
G	Beaver Creek	1972	Tyonek	MGS	ss	prod	–	3,521	1	1	1	4,510	52,070	–

TABLE 6. TUXEDNI–HEMLOCK OIL ACCUMULATIONS BY DISCOVERY DATE, INDICATING CUMULATIVE PRODUCTION AS OF DECEMBER 31, 1987, REMAINING RESERVES, RESERVOIR CHARACTERISTICS, AND OIL CHEMISTRY* (continued)

Res temp (°C)	Net pay (m)	Por (%)	Perm (md)	Orig GOR (SCF/STB)	Water sat, S_{wi} (%)	Dev acres (ha)	Oil grav (API)	Sulfur (%)	$\delta^{34}S$ (‰)	Pr/Ph I	Pr/nC_{17}	$\delta^{13}C$ sat (‰)	$\delta^{13}C$ arom (‰)	Map symbol
82	23	21	55	175	40	193	30	–	–	–	–	–	–	A
82	21	20	75	175	40	–	30	0.1	-1.9	1.6	0.8	-30.0	-29.4	A
82	67	22	360	350	40	1,077	37	–	–	–	–	–	–	A
														A
53	58	16	15	3,850	–	300	42	–	–	–	–	–	–	B
54	102	16	15	650	–	300	36	<0.1	+2.7	3.4	0.6	-30.0	-27.8	B
68	152	11	10	381	–	1,620	35	0.1	+1.9	2.1	0.4	-30.0	-28.2	B
														B
66	183	14	10	1,110	–	1,295	42	–	–	–	–	–	–	C
–	–	–	–	–	–	–	–	–	–	–	–	–	–	C
							41	0.1	0.0	2.6	0.5	-29.7	-28.0	C
–	–	–	–	–	–	–	28	0.1	-0.6	B	B	-29.6	-28.4	D
44	305	24	250	268	–	567	28	0.1	+1.7	2.7	2.2	-30.0	-28.1	D
82	66	12	12	275	36	202	36	0.1	+0.3	2.9	0.9	-30.2	-28.5	D
58	91	15	10	318	–	486	31	0.1	+2.5	3.1	3.6	-30.0	-28.7	D
														D
73	30	18	65	297	35	1,008	36	0.1	+0.4	2.4	0.9	-30.1	-28.7	E
82	88	11	53	404	35	5,018	35	0.1	+2.5	2.9	0.7	-30.1	-28.8	E
85	30	16	102	271	–	613	33	0.1	+2.4	2.7	0.7	-30.4	-29.1	E
														E
–	–	–	–	286	–	65	28	–	–	–	–	–	–	F
102	30	–	–	280	–	334	35	<0.1	+0.9	2.7	0.9	-30.1	-28.7	G

*Map symbols correspond to locations shown on Figure 9. References to complete this table include: Alaska Oil and Gas Conservation Commission, 1985, 1988; R. P. Crandall, written communication, 1988; Oil and Gas Journal, 1988. –, no information available; 1 ft + 0.3048 m; 1 psi = 6.895 kPa (kiloPascal); 1 acre = 0.4047 ha (hectare); 1 barrel = 0.1590 kiloliter; A, anticlinal trap; abd, abandoned; API, American Petroleum Institute; arom, aromatic hydrocarbons; B, biodegraded; bbl, barrel; bcf, billion cubic feet; cg, conglomeratic sandstone; cum prod, cumulative production through 12/31/87; dev, developed; dis, discovery; FA, faulted anticlinal trap; GOR, gas-to-oil ratio, in SCF/STB, standard cubic feet of gas per stock tank barrel of oil; grav, gravity; lith, lithology; MGS, Middle Ground Shoal Member of Debelius (1974); Mid Grd Shl, Middle Ground Shoal; perm, permeability; ph, phytane; pool A to G, industry pool designation; por, porosity; pr, pristane; press, pressure; prod, producing; res, reservoir, sat, saturated hydrocarbons; SCU, Soldatna Creek unit; ss, sandstone; stat, status; temp, temperature; Ty-Hem, Tyonek Formation and Hemlock Conglomerate, undivided; W, West; water sat, water saturation.

†Of Debelius (1974).

TABLE 7. BELUGA–STERLING GAS ACCUMULATIONS BY DISCOVERY DATE, INDICATING CUMULATIVE PRODUCTION AS OF DECEMBER 21, 1987, REMAINING RESERVES, RESERVOIR CHARACTERISTICS, AND GAS CHEMISTRY*

Map No.	Accumulation name	Year dis	Prod unit	Mbr or pool	Res lith	Prod stat	Trap type	Cum prod oil ($x10^3$bbl)	Cum prod gas (bcf)	Reserves gas (bcf)	Prod depth (m)	Orig press (kPa)	Gas spec grav
1	Kenai	1959	Sterling	3	ss	prod	D	–	263	–	1,130	12,840	0.577
1	Kenai	1959	Sterling	4	ss	prod	D	–	367	–	1,205	13,230	0.577
1	Kenai	1959	Sterling	5.1	ss	prod	D	–	429	–	1,225	13,225	0.577
1	Kenai	1959	Sterling	5.2	ss	shut-in	D	–	44	–	1,255	14,330	0.577
1	Kenai	1959	Sterling	6	ss	prod	D	–	397	–	1,390	12,270	0.557
1	Kenai	1959	Beluga	–	ss	prod	D	–	120	–	1,520	17,640	0.555
1	Kenai	1959	Tyonek	MGS	ss	prod	D	12	210	–	2,745	30,450	0.560
1	Kenai	1959	**Field total**					12	1,831	463			
2	Swanson River	1960	Sterling	B,D,E	ss	prod	A	–	13	–	875	9,205–31,030	0.600
3	West Fork	1960	Sterling	–	ss	shut-in	FA	–	2	6	1,520	14,045	0.560
4	Falls Creek	1961	Tyonek	MGS	ss	shut-in	–	–	–	13	2,145	23,470	0.600
5	Sterling	1961	Sterling	–	ss	shut-in	D	–	2	23	1,535	15,170	0.569
6	West Foreland	1962	Tyonek	MGS	ss	shut-in	–	–	–	20	–	29,410	0.600
7	North Cook Inlet	1962	Sterling	–	ss	prod	D	–	–	–	1,280	14,065	0.566
7	North Cook Inlet	1962	Beluga	–	ss	prod	D	–	–	–	1,555	17,085	0.566
7	North Cook Inlet	1962	**Field total**						820	680			
8	Beluga River	1962	Sterling	–	ss	prod	A	–	–	–	1,005	11,275	0.556
8	Beluga River	1962	Beluga	–	ss	prod	A	–	–	–	1,370	15,270	0.556
8	Beluga River	1962	**Field total**						253	604			
9	No Mid Grd Shl	1964	–	MGS	ss	shut-in	–	–	–	–	2,775	28,890	–
10	Birch Hill	1965	Tyonek	MGS	ss	shut-in	–	–	–	11	2,345	26,475	0.561
11	Moquawkie	1965	Tyonek	–	ss	shut-in	–	–	1	–	–	8,690–15,895	0.600
12	North Fork	1965	Tyonek	MGS	ss	shut-in	–	–	–	12	2,195	23,510	0.562
13	Nicolai Creek	1966	Ster-Bel	–	ss	shut-in	–	–	1	3	660	7,320–11,640	0.575
14	Ivan River	1966	Tyonek	Chuit	ss	shut-in	D	–	–	26	2,375	28,475	0.560
15	Beaver Creek	1967	Sterling	–	ss	prod	D	–	–	–	1,524	15,170	0.560
15	Beaver Creek	1967	Beluga	–	ss	prod	D	–	–	–	2,470	26,200	–
15	Beaver Creek	1967	**Field total**						63	177			
16	Albert Kaloa	1968	Tyonek	–	ss	shut-in	–	–	–	–	980	–	–

Map No.	Accumulation name	Year dis	Prod unit	Mbr or pool	Res lith	Prod stat	Trap type	Cum prod oil ($x10^3$bbl)	Cum prod gas (bcf)	Reserves gas (bcf)	Prod depth (m)	Orig press (kPa)	Gas spec grav
17	McArthur River	1968	Tyonek	MGS	ss	prod	A	–	97	–	–	11,955	0.564
17	McArthur River	1968	Tyonek	Chuit	ss	prod	A	–	36	–	–	–	–
17	McArthur River	1968	**Field total**						133	600			
18	Lewis River	1975	Beluga	–	ss	prod	–	–	4	18	1,435	19,030	–
19	Pretty Creek	1975	Beluga	–	ss	prod	–	–	1	25	1,830	–	–
20	Stump Lake	1978	Beluga	–	ss	shut-in	–	–	–	–	2,040	22,685–23,855	0.565
21	Theodore River	1979	Beluga	–	ss	shut-in	–	–	–	–	1,130	11,590–13,120	–
22	Cannery Loop	1979	Bel-Ty	–	ss	shut-in	–	–	–	300	–	27,580	0.560
23	Trading Bay	1979	Tyonek	MGS	ss	prod	FA	–	3	29	2,745	26,960	0.582
24	Mid Grd Shl	1982	Tyonek	MGS	ss	prod	FA	–	2	–	1,080	9,845	0.564

1971; Boss and others, 1976; Hayes and others, 1976; Claypool and others, 1980). When the microbial gas ($\delta^{13}C$ –57.6 permil) that was produced through 1987 is added to the remaining reserves, the Sterling Formation reservoirs account for 3 tcf, the Beluga Formation for 0.5 tcf, and the Tyonek Formation, after removal of the thermal gas ($\delta^{13}C$ –43.7 permil) from the McArthur River field, for 0.3 tcf. In the McArthur River field, the thermal gas in the Tyonek Formation is interpreted to have migrated from the underlying oil reservoirs in the Hemlock Conglomerate. In the Kenai field, the sandstones of the Sterling Formation contain almost 1.9 tcf or 82 percent of the gas. The siliciclastic Sterling Formation is the most important reservoir rock in the Beluga-Sterling petroleum system.

An average gas pool in this petroleum system has the following characteristics (Table 7). The pool is located in a domal structure that covers 1,050 hectares (20 to 3,360 ha) and has a net pay of 22 m (6 to 65 m). The sandstone reservoir has a water saturation of 40 percent (35 to 50 percent), a reservoir porosity of 27 percent (10 to 37 percent), and permeability of 1,100 millidarcies (3.5 to 4,400 md). The drill depth is 1,615 m (980 to 2,775 m) to a reservoir under 19,300 kiloPascals (7,320 to 31,000 kPa) at 50 °C (35 to 100 °C). The natural gas produced is 99 percent methane, with a specific gravity of 0.571 (0.555 to 0.600) and a heating value of 251 kilogram calories (250 to 256 kc).

The source for the gas is unclear, but the Beluga Formation and, to a lesser extent, the Sterling Formation have considerable coal and type III kerogen (Claypool and others, 1980). Most of the coal and type III kerogen are below the Sterling Formation, so this source is in a good position to charge overlying reservoirs with microbial gas (Claypool and others, 1980). Because this system requires no overburden to mature the source rocks, the duration time is short—from the late Miocene to Holocene, or about 12 m.y. (Palmer, 1983; Magoon and Egbert, 1986). The geographic boundary for the Beluga-Sterling petroleum system is presently restricted by known accumulations (Fig. 9).

GULF OF ALASKA

Oil and gas seeps in the northern Gulf of Alaska in the Katalla, Yakataga, and Samovar Hills areas (Fig. 13) indicate the possibility of commercial accumulations (Miller and others, 1959; Blasko, 1976b). From 1901 to 1932, 25 exploratory and 18 development wells were drilled (Fig. 12). The Katalla field, a fractured shale reservoir in the Poul Creek Formation, was discovered in 1902 and produced 153,922 barrels of 40° to 44° API gravity oil from 1904 to 1933 at a depth range of 110 and 460 m (Miller and others, 1959; Blasko, 1976b). Twenty-five exploratory wells and core holes were drilled onshore from 1954 to 1963 (Plafker, 1971; Plafker and others, 1975; Rau and others, 1983). In 1969, interest in the offshore began with the drilling of the Middleton Island State 1 well. From 1975 through 1978, one continental offshore stratigraphic test (COST) well and ten offshore exploratory wells were drilled and abandoned in the west-

TABLE 7. BELUGA–STERLING GAS ACCUMULATIONS BY DISCOVERY DATE, INDICATING CUMULATIVE PRODUCTION AS OF DECEMBER 21, 1987, REMAINING RESERVES, RESERVOIR CHARACTERISTICS, AND GAS CHEMISTRY* (continued)

Res temp (°C)	Net pay (m)	Por (%)	Perm (md)	Water sat, S_{wi} (%)	Dev acres (ha)	Btu (Btu/ft^3)	$\delta^{13}C$ methane (‰)	$\frac{C1}{C_{1-5}}$	Map No.
39	27	35.5	–	35	2,035	–	-57.0	0.999	1
41	18	36.5	–	35	3,060	–	-56.7	0.999	1
41	34	36.5	–	35	2,510	–	–	–	1
40	16	36.5	–	35	727	–	–	–	1
43	34	32	–	40	2,198	–	-56.5	–	1
46	65	15–20	–	40	518	–	–	–	1
62	30	18–22	–	40	1,150	–	-48.2	0.998	1
						1,005		0.989	1
51	–	30	650	35	260	1,002	-57.7	0.999	2
43	7	30	4,400	–	185	–	–	–	3
56	–	–	–	–	–	1,015	–	0.991	4
43	6	26	–	40	625	991	-60.9	0.998	5
77	8	–	–	–	260	929	–	0.921	6
43	40	28	178	40	3,360	–	–	–	7
48	9	28	175	40	1,010	–	–	–	7
–	–	–	–	–	–	993	-60.7	0.998	7
34	33	31	–	37	2,070	–	–	–	8
41	32	24	–	42	1,955	–	–	–	8
–	–	–	–	–	–	1,014	–	0.997	8
62	7	–	–	–	–	–	–	–	9
58	9	25	6	–	60	1,014	–	0.986	10
27–42	14-33	20-24	–	35-40	520	1,006	–	0.990	11
60	–	18	3.5	50	20	1,002	–	0.981	12
41–43	10	–	–	–	–	976	-62.2	0.995	13
53	11	20	1,600	45	980	1,004	–	0.989	14
42	34	30	2,000	40	1,280	–	-58.2	0.989	15
61	6	10	–	–	260	–	–	–	15
–	–	–	–	–	–	998	–	0.983	15
–	–	–	–	–	–	–	–	–	16

Res temp (°C)	Net pay (m)	Por (%)	Perm (md)	Water sat, S_{wi} (%)	Dev acres (ha)	Btu (Btu/ft^3)	$\delta^{13}C$ methane (‰)	$\frac{C_1}{C_{1-5}}$	Map No.
47	–	–	–	–	780	–	–	–	17
43	–	–	–	–	520	–	–	–	17
–	–	–	–	–	–	–	-43.7	0.994	17
44	26	–	–	–	415	–	–	–	18
–	–	–	–	–	–	–	–	–	19
40	–	–	–	–	–	–	–	–	20
–	–	–	–	–	–	–	–	–	21
52	–	–	–	–	520	–	–	–	22
79	18	–	–	–	260	–	–	–	23
54	–	–	–	–	–	–	–	–	24

*Map numbers correspond to locations shown on Figure 9. References to complete this table include: Alaska Oil and Gas Conservation Commission, 1985, 1988; Blasko, 1974; Claypool and others, 1980; R. P. Crandall, written communication, 1988. 1 ft = 0.3048 m; 1 psi = 6.895 kPa (kiloPascal); 1 acre = 0.4047 ha (hectare); 1 cubic ft = 0.0283 cubic meter; 1 Btu = 0.25198 kilogram calories; –, no information available; A, anticlinal trap; bbl, barrels; Bel-Ty, Beluga and Tyonek Formation, undivided; bcf, billion cubic feet; btu, British thermal unit; Chuit, Chuitna Member; cum prod, cumulative production through 12/31/87; D, domal trap; dev, developed; dis, discovery; FA, faulted anticlinal trap; let, letter; Mbr, member; MGS, Middle Ground Shoal Member of Debelius (1974); No Mid Grd Shl, North Middle Ground Shoal; orig press, original pressure; perm, permeability; pool B,D,E, industry pool designation; por, porosity; prod, producing; reserves, remaining reserves; res lith, reservoir lithology; res temp, reservoir temperature; sat, saturation; spec grav, specific gravity; ss, sandstone; sat, status; Ster-Bel, Sterling and Beluga Formations, undivided.

ern segment (Fig. 13). The last dry exploratory well was drilled in 1983 on a large structure in the central segment (Plafker, 1987). The history of exploration and the framework and petroleum geology in this area are discussed by several workers (Bruns, 1982a, b, 1983; Mull and Nelson, 1986; Palmer, 1976; Plafker, 1987; Plafker and others, 1980; Plafker and Claypool, 1979).

Because known oil and gas accumulations, seeps, and shows are associated with the Yakutat terrane, the petroleum potential in the Gulf of Alaska is restricted to this terrane (Fig. 13; Blasko, 1976b; Plafker, 1987; Plafker and others, this volume, Chapter 12). The Yakutat terrane is a large lanceolate slab pointing to the southeast that is being subducted base-first to the northwest. The base is located on the Kayak Island zone and Ragged Mountain fault; the northeast edge is marked by the Chugach–Saint Elias fault and the Fairweather fault, and the southwest edge by the Transition fault (Plafker, 1987). Structural deformation is maximum on the northwest and attenuates to the southeast.

The Pamplona and Dangerous River zones divide the Yakutat terrane into three segments: western, central, and eastern (Plafker and others, 1975; Bruns, 1983, 1985; Plafker, 1987). The basement complex for the eastern segment is Mesozoic; the basement for the central and western segments is Eocene and Paleocene(?) basalt. As much as 4,600 m of Paleogene sedimentary rocks overlie the basement complex within the central and western segments, and 4,000 to 5,000 m of Neogene sedimentary rocks is present over the entire terrane (Bruns, 1983, 1985; Plafker, 1987).

Three stratigraphic sequences in the Yakutat terrane include the essential elements for the present-day occurrence of petroleum. The oldest sequence ranges in age from early Eocene to as young as early Oligocene; it consists of the Kulthieth, Tokun, and Stillwater Formations and may be as thick as 4,600 m (Fig. 14; Miller and others, 1959; Stoneley, 1967; Plafker, 1967, 1971, 1987). The Kulthieth Formation is a nearshore paludal deltaic sequence that includes thick interfingering, coal-bearing units, sandstone, siltstone, and shale deposited in alluvial-plain, delta-plain, barrier-beach, and shallow-marine environments. The Tokun Formation is primarily a deltaic deposit, and the Stillwater Formation is a prodelta deposit. Offshore, rocks of this age were dredged from the continental slope; thus the sequence is presumed to underlie the continental shelf.

The intermediate sequence includes the Poul Creek Formation, which ranges in age from late Eocene to early Miocene (Fig. 14). This predominantly shaly unit was deposited in water depths that range from neritic to bathyal, includes glauconite and organic matter, and is intercalated with intrabasinal water-laid alkalic basaltic tuff, breccia, and pillow lava. This deep marine sequence may be as much as 2,000 m thick. This sequence was sampled in offshore wells (Lattanzi, 1981) and is also presumed to underlie the continental shelf.

The youngest sequence, the Yakataga Formation, ranges in age from late Oligocene(?) to Quaternary and unconformably overlies rock units of all three segments. This sequence includes an enormous thickness (<4,600 m) of siliciclastic strata, including

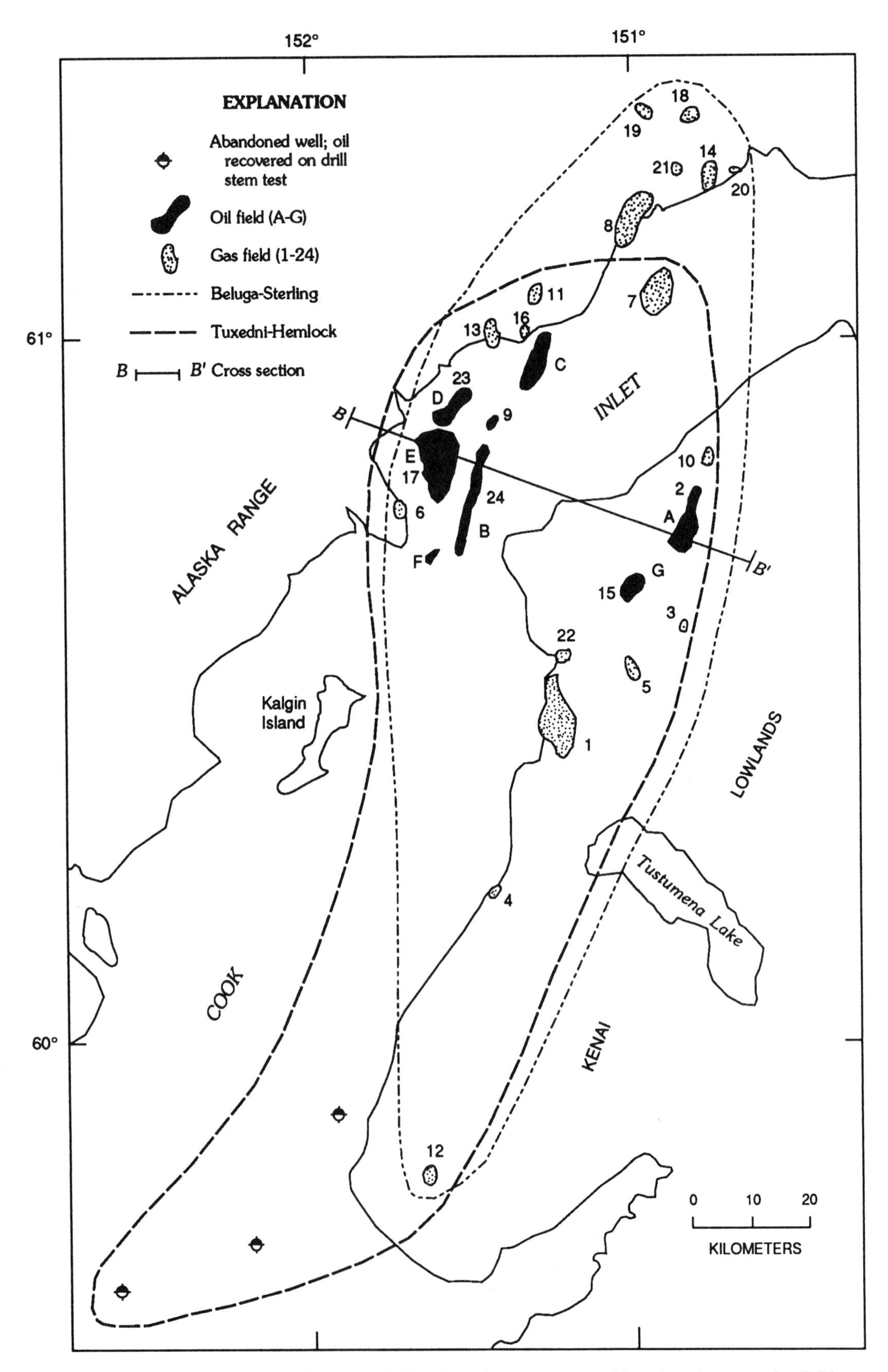

Figure 9. Cook Inlet oil and gas fields, boundaries of petroleum systems, and location of cross section B-B′.

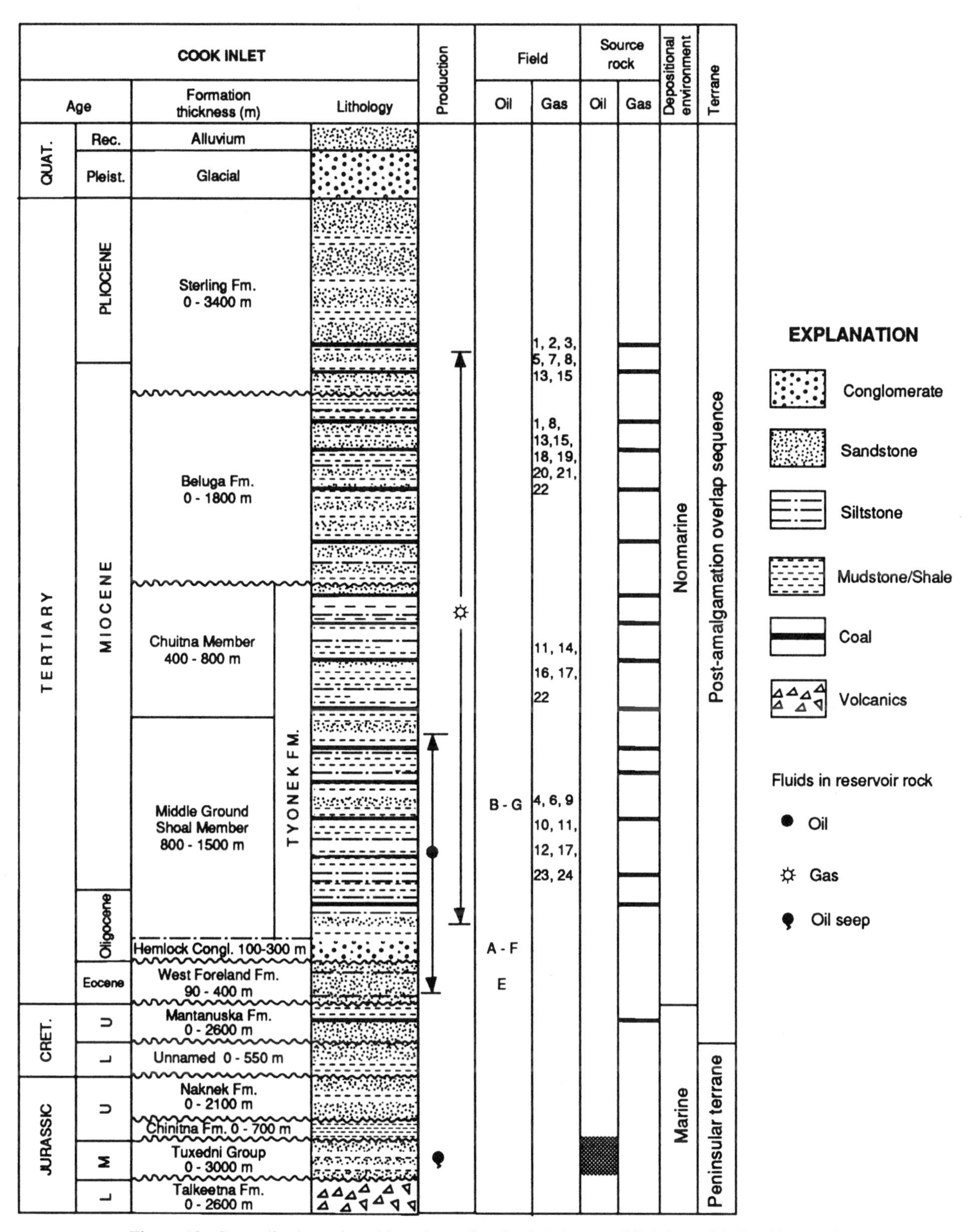

Figure 10. Generalized stratigraphic column for Cook Inlet (modified from Alaska Oil and Gas Conservation Commission, 1985, p. 182) showing petroleum in reservoir rocks and source rock intervals. See Tables 6 and 7 for oil and gas field names.

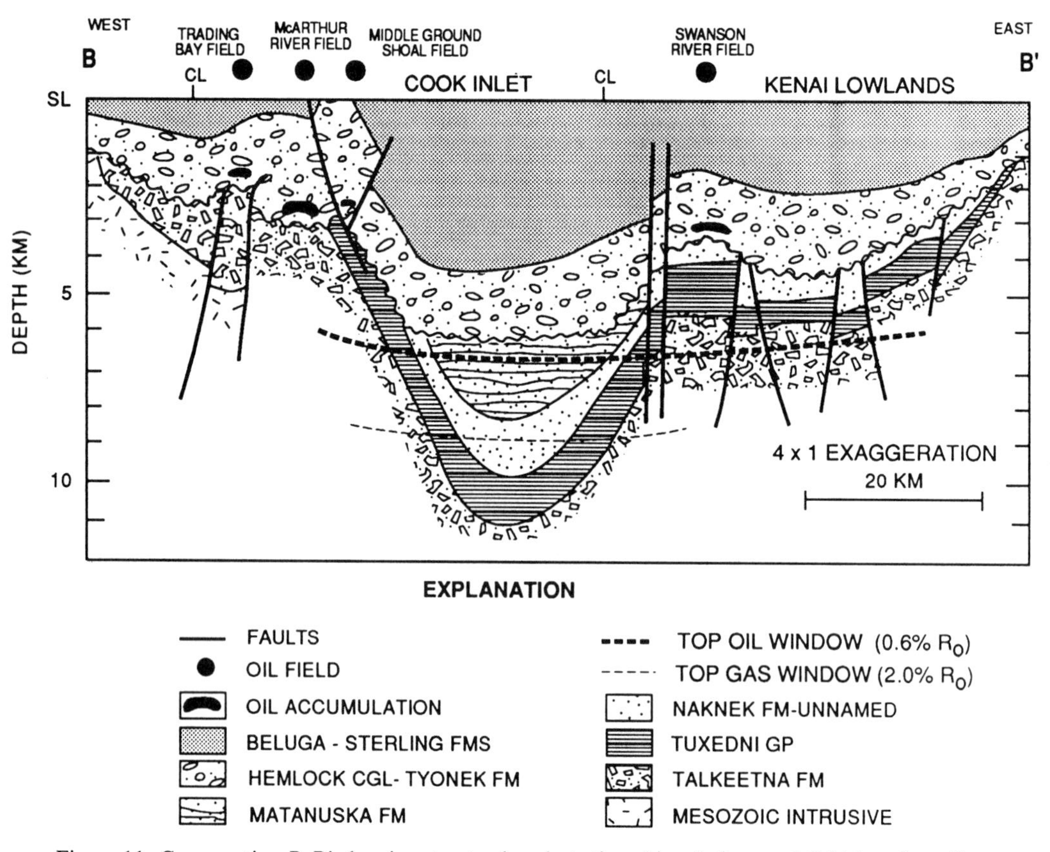

Figure 11. Cross section B–B′ showing structural and stratigraphic relations and field locations. See Figure 9 for location of section.

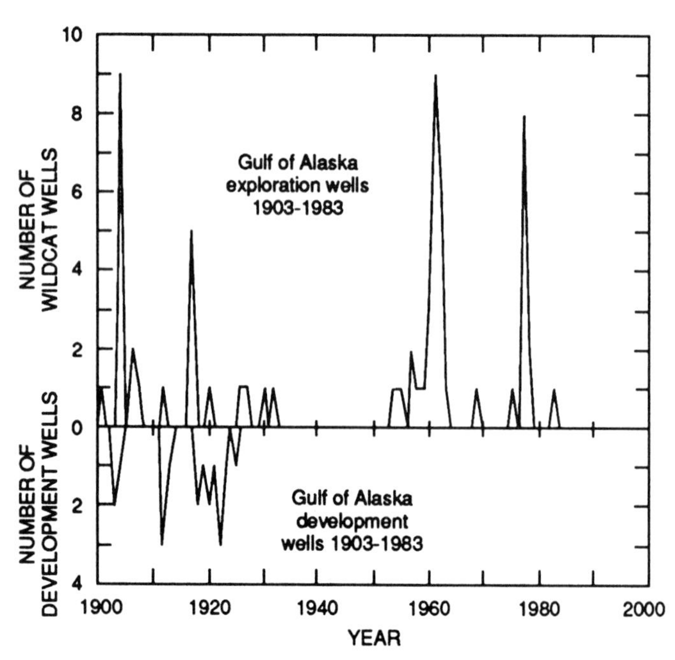

Figure 12. Exploration and development drilling history for the Yakutat terrane, Gulf of Alaska. Wells are plotted at year they reached total depth.

abundant glaciomarine sediment derived from the adjacent Fairweather Range and Chugach Mountains (Fig. 13). The Yakataga Formation is an essential element of the Gulf of Alaska petroleum systems because it is inferred to provide the burial depth necessary to generate the oil and gas.

Paleogene rocks in the Yakutat terrane increase in thermal maturity to the northwest (Plafker, 1987). The thermal maturity pattern is interpreted from coal beds, visual kerogen, Rock-Eval, and vitrinite reflectance data from outcrop, dredge, and well samples (Barnes, 1967; Palmer, 1976; Plafker and others, 1980; Mull and Nelson, 1986; Plafker, 1987). No thermal maturity information is available for the eastern segment. Coal rank indicates that the Paleogene is immature (0.4 to 0.6 percent R_o) west of Yakutat Bay (Plafker, 1987) to mature and overmature (0.6 to 2.5 percent R_o) in the Katalla area (Mull and Nelson, 1986). The Poul Creek Formation in the Katalla field is mature enough to have generated oil. Visual kerogen in two onshore wells (Socal, Riou Bay No. 1; Richfield, Duktoth No. 1) also indicates an increase in thermal maturity to the northwest in both the Paleogene and Neogene (Palmer, 1976). Thermal maturity of Paleogene dredge samples from the central segment along the Transition fault is immature (0.4 to 0.6 percent R_o; Plafker,

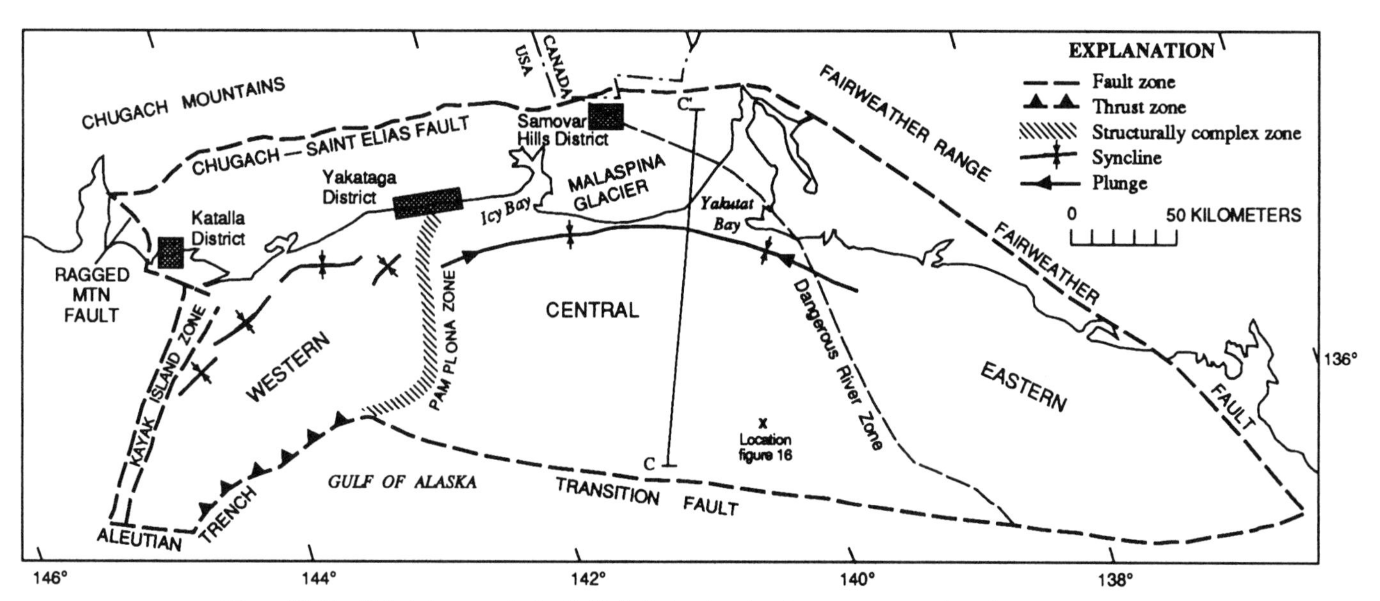

Figure 13. The Yakutat terrane in the Gulf of Alaska showing the western, central, and eastern segments. Also shown are locations of the Katalla, Yakataga, and Samovar Hills Districts, where oil and gas seeps are located, and the location of cross section C-C′.

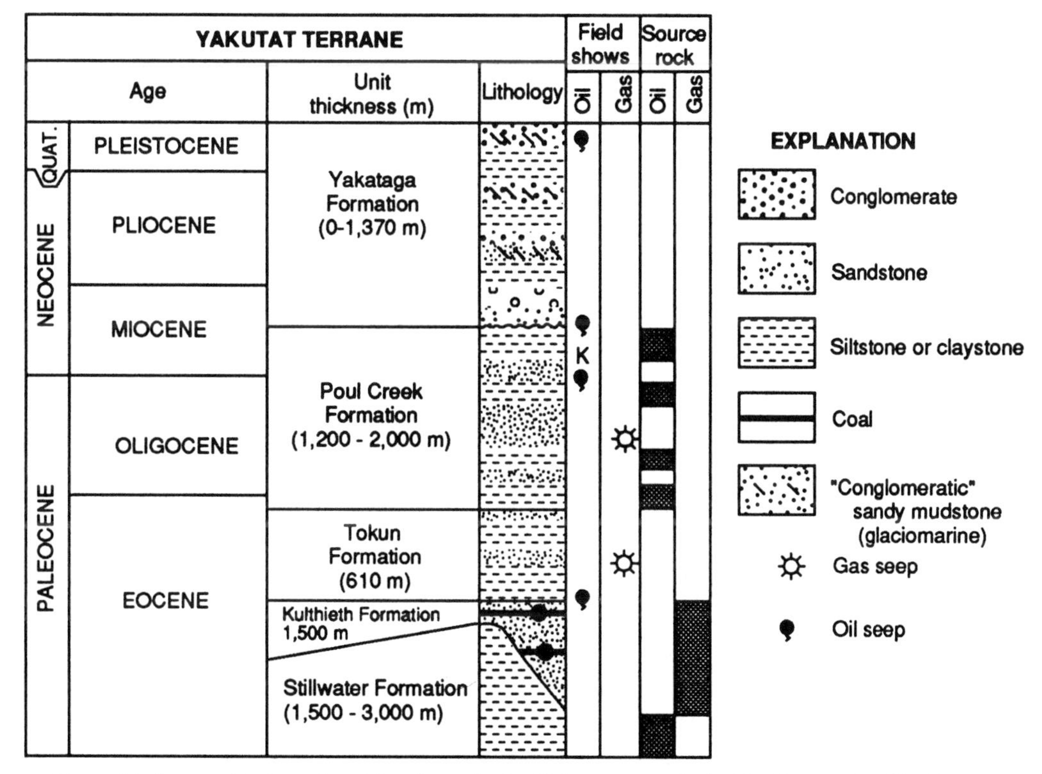

Figure 14. A generalized stratigraphic column for the Yakutat terrane in the Katalla area (modified from Plafker, 1987). Included are oil and gas shows (Plafker, 1987), and source rock intervals; K indicates oil production from the Katalla oil field.

1987), but the samples indicate an increase in thermal maturity to the northwest. The increase in thermal maturity to the northwest indicates that the uplift for the western segment is greater than the uplift for the central segment. Based on burial history calculations on the central segment (Bruns, 1983), the Neogene strata provided the overburden necessary to mature the Paleogene rocks in early Miocene time (~22 Ma) (Figs. 15 and 16).

Gas and oil seeps occur in three onshore areas—Katalla, Yakataga, and Malaspina Districts—in the northern part of the western and central segments (Miller and others, 1959; Blasko, 1976b; Plafker, 1971, 1987). The composition of the gas and oil indicates that the source rock is predominantly terrestrial or type III organic matter. Gas analyses from seeps in all three areas (Blasko, 1976b) show some ethane and higher hydrocarbons that strongly suggest that most of the gas is thermal, not microbial (Table 8). The unaltered oil from the Katalla oil field has high gravity (30 to 44° API), low sulfur (<0.1 wt percent), and a high pristane/phytane index (4.8; Table 9). Oils from the Yakataga and Samovar Hills Districts lack normal alkanes, have lower API gravities (13 to 37.4 °API), and higher sulfur content (0.1 to 0.3 percent; Table 9); these properties indicate that the oil is biodegraded.

On the basis of oil-source rock correlations, two petroleum systems, the Stillwater-Kulthieth and the Poul Creek, are present in the Yakutat terrane.

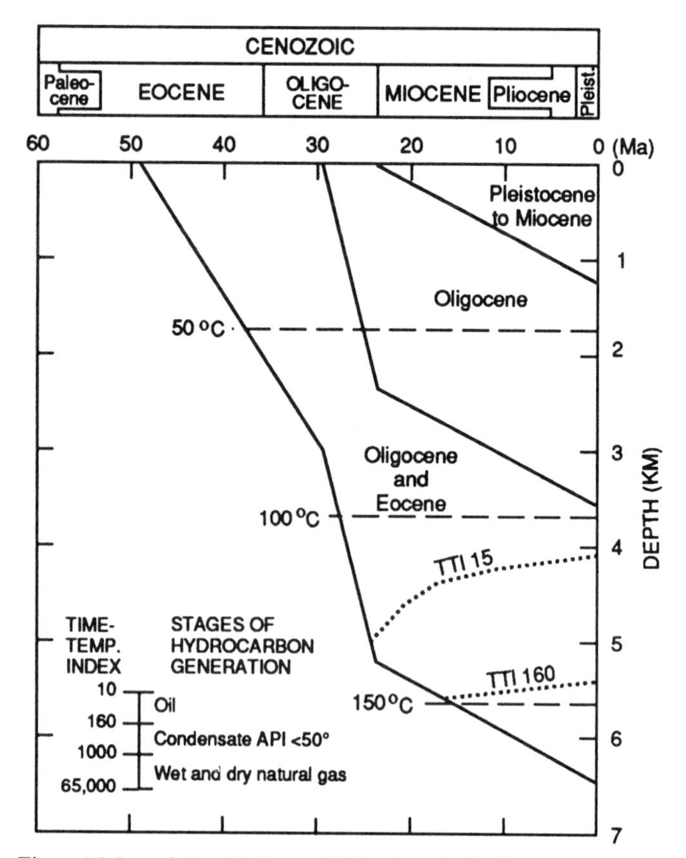

Figure 16. Lopatin burial plot and TTI values that indicate the time of oil generation in the central segment of the Yakutat terrane (Bruns, 1983). Location (x) shown on Figure 13.

Stillwater-Kulthieth system

The Stillwater-Kulthieth petroleum system is based on the interpreted facies relations between the Kulthieth, Tokun, and Stillwater Formations and on the oil–source rock correlation of Fuex (written communication, 1987). On the basis of the age of dredge samples from the continental slope, the dredged rocks are assumed to be the offshore equivalents of these onshore sequences (Plafker, 1987). The organic carbon content of the dredge samples ranges from 0.4 to 1.9 wt percent, averages over 1.0 wt percent, and has a high percentage of herbaceous organic matter (Plafker, 1987). Hydrogen and oxygen indices from Rock-Eval indicate that the dredge samples are type III kerogen (Plafker and Claypool, 1979; Bruns and Plafker, 1982), typical of deltaic depositional environments. In the central segment of the Yakutat terrane coal deposits onshore (Barnes, 1967) and vitrinite reflectance data from offshore dredge samples (Plafker, 1987) indicate that thermal maturity is immature but increases to the northwest. On the basis of Lopatin burial history (Bruns, 1983), these rocks should be mature in the deepest part of the central segment (Fig. 15).

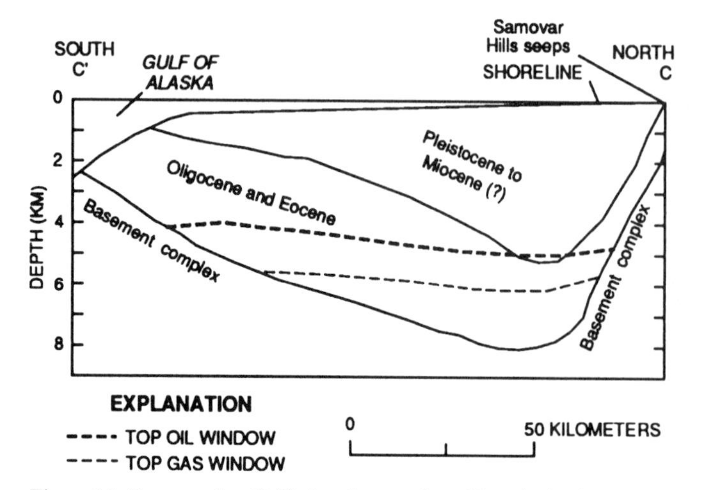

Figure 15. Cross section C-C′ showing stratigraphic units in the Yakutat terrane.

The oil attributed to this petroleum system is located in the Yakataga and Malaspina Districts, and on Wingham Island in the Katalla District (Fig. 13; Table 9). The results of the oil–source rock correlation are based on $\delta^{13}C$ of extract from a heated source rock (Kulthieth Formation) and on comparison with the topped oil (A. N. Fuex, written communication, 1987). The primary source rock for this system is interpreted to be the Stillwater Formation and its offshore equivalents that crop out on the continental slope and in the Samovar Hills; the Kulthieth Formation may be a secondary source. The Stillwater-Kulthieth is a hypothetical system because the correlation is based only on carbon isotopes and limited rock samples.

TABLE 8. U.S. BUREAU OF MINES GAS ANALYSES IN THE YAKUTAT TERRANE, GULF OF ALASKA*

Area	Sample from	C_1 (%)	C_2 (%)	C_3 (%)	nC_4 (%)	iso-C_4 (%)	nC_5 (%)	iso-C_5 (%)	Cyclo-C_5 (%)	C_{6+} (%)	Total (%)
					KATALLA AREA						
Katalla oil field	seep	67.3	14.0	11.0	3.2	2.9	0.6	0.4	0.1	0.3	99.9
Bering Lake	seep	99.9	tr	tr	0.0	0.0	0.0	0.0	0.0	0.0	99.9
Rathbun well	well	99.8	0.1	tr	tr	tr	tr	tr	tr	tr	99.9
					YAKTAGA AREA						
Crooked Creek	seep	97.2	2.1	0.5	tr	0.1	0.0	0.1	tr	tr	100
Munday Creek	seep	99.3	0.7	tr	0.0	0.0	0.0	0.0	0.0	0.0	100
Johnston Creek	seep	99.3	0.6	0.1	tr	tr	tr	0.0	tr	tr	100
Johnston Creek	seep	99.9	tr	tr	0.0	0.0	0.0	0.0	0.0	0.0	99.9
					SAMOVAR HILLS						
Oily Lake	seep	95.5	3.7	0.8	0.0	0.0	0.0	0.0	0.0	0.0	100
Oily Lake	seep	99.9	0.1	0.0	0.0	0.0	0.0	0.0	0.0	0.0	100

*From Blasko, 1976b; gas analyses normalized to include only hydrocarbons.

The reservoir rocks are included within the Kulthieth and Tokun Formations. Although there is a high percentage (60 percent) of thick (15 to 600 m), moderately porous (2 to 23 percent) sandstone, the sandstones have low permeability (0 to 43 md; Table 10). Induration and diagenesis are primarily responsible for the loss of good reservoir characteristics (Plafker, 1987).

The surface maturity pattern indicates that the western segment has undergone considerable uplift and the central segment is at maximum burial depth. Mature Paleogene rocks are exposed at the surface on the western segment. Obduction or uplift of the Yakutat terrane in late Cenozoic time would bring deeply buried Paleogene rocks to the surface in late Cenozoic time. An alternative interpretation requires high heat flow related to ridge subduction during Eocene time (Plafker, 1987) that may have resulted in early maturation in parts or all of this sequence (Plafker and Gilpin, this volume). In the central and western segments, the Eocene to Miocene Poul Creek Formation and its offshore equivalent is probably of insufficient thickness to mature the underlying regressive sequence. Therefore, the time of migration is based on the age of the post-Paleogene overburden, which is primarily the Yakataga Formation. The Yakataga Formation is sufficiently thick to provide the necessary overburden depending on the geothermal gradient (Bruns, 1983).

Migration of oil commenced in Neogene time and is presently moving up the northeast flank to the outcrop (Fig. 15). A major unconformity at the base of the Yakataga Formation and a change in depositional environment from deep marine to glaciomarine indicates a major tectonic reorientation. Because structures are the result of late Cenozoic subduction and hydrocarbon migration is presently taking place, petroleum accumulations should be present if reservoirs occur in a trapping position. However, exploratory drilling to date indicates a lack of commercial petroleum accumulations in Paleogene and younger strata.

Poul Creek system

The Poul Creek Formation is the source rock and the reservoir rock for the Poul Creek petroleum system. In the Katalla area, the Poul Creek Formation has an average organic carbon content of 1.8 wt percent; the organic matter is predominantly herbaceous (Mull and Nelson, 1986). On Kayak Island the organic carbon content of four Poul Creek samples averages 4.6 wt percent (Table 4 in Plafker, 1987), and is about 50 percent herbaceous and 50 percent amorphous. Vitrinite reflectance and thermal alteration index (TAI) data indicate that this unit is mature in the Katalla area (Mull and Nelson, 1986).

On the basis of carbon isotope data for the saturated and aromatic hydrocarbons, oil from the Katalla field is within 1.2 permil of the extract from a heated source rock (Poul Creek) from eastern Kayak Island (Table 9; A. N. Fuex, written communication, 1987). According to the same type of information, the oil from Wingham Island compares more favorably with the Kulthieth Formation from the Samovar Hills (Table 9). Since the oil-to-rock correlation is based only on carbon isotope data from very few oil and rock samples, the Poul Creek is a hypothetical petroleum system.

The reservoir is formed by fractures and possibly faults in the Poul Creek Formation that could have resulted from uplift as the Yakutat block collided with the Prince William terrane to the northwest. Although the Poul Creek Formation includes as much as 30 percent sandstone with porosities as high as 19 percent, the permeability is very low (<12 md; Table 10). As in the underlying units, excessive induration and diagenesis appear to have destroyed what reservoir properties existed (Plafker, 1987).

TABLE 9. OIL TO SOURCE-ROCK CORRELATION IN THE YAKUTAT TERRANE, GULF OF ALASKA*

Area	Rock unit or age	Field or API No.	Sample from	Sample type	API gravity	Pr/ph	Sulfur (wt %)	Heating program	$\delta^{13}C$ (permil) sat.	$\delta^{13}C$ (permil) arom.
					CONTINENTAL SLOPE (TRANSITION FAULT AREA)					
Central segment	Paleogene	78-22D3	dredge	rock	–	5.1	–	–	-27.7	-26.2
Central segment	Paleogene	79-39F	dredge	rock	–	3.4	–	–	-28.2	-27.6
Central segment	Paleogene	78-44D	dredge	rock	–	0.7	–	–	-30.6	-30.4
					KATALLA DISTRICT					
East Kyak Island	Poul Creek	AKA-S-577	outcrop	rock	–	–	–	unheated	-28.3	-26.8
								3/300	-27.6	-26.6
								6/330	-26.5	-25.5
								20/330	-26.4	-25.1
Katalla oil field†	Poul Creek	5006910026	well	oil	30.9	3.6	0.1		-25.4	-23.9
Wingham Island	Poul Creek	81APr51C	seep	oil	–	5.7	–	–	-27.4	-26.3
					YAKATAGA DISTRICT					
Johnston Creek	Poul Creek	AKA-O-58	seep	oil	15.7§	–	0.2	–	-28.3	-25.3
					MALASPINA DISTRICT (SAMOVAR)					
Samovar Hills	Kulthieth	AKA-S-80	outcrop	rock	–	–	–	3/300	-29.6	-27.4
								18/330	-28.5	-26.2
Samovar Hills	Kulthieth	AKA-O-56	seep	oil	17.0	–	0.2	-	-28.1	-25.3
Samovar Hills	Kulthieth	AKA-O-57	seep	oil	13.0	–	0.3	–	-28.3	-25.9
Hubbs Creek	Kulthieth	80APr127	seep	oil	37.4	6.2	0.1	–	-27.5	-25.8

*From A. N. Fuex, written communication, 1987; Continental slope, Wingham Island, Hubbs Creek results from Plafker, 1987; sat, saturate hydrocarbons; arom., aromatic hydrocarbons; Pr/ph, pristane/phytane ratio; 3/300, heated for 3 days at 300° F.

†Unpublished data.

§API gravity is low due to high water content of sample.

TABLE 10. RESERVOIR PROPERTIES FOR STRATIGRAPHIC UNITS IN THE YAKUTAT TERRANE, GULF OF ALASKA*

Stratigraphic unit	Percent sandstone (%)	Thickness range (m)	Porosity range (%)	Permeability range (md)
Yakataga and Redwood Formations	9–53	15–600	1.2–32.2	0–597
Poul Creek Formation	1–30	15–350	2.2–19.2	0–12
Kulthieth and Tokun Formations	60	15–490	1.8–22.7	0–43

*From Winkler and others, 1976; Lyle and Palmer, 1976; Plafker, 1987.

REFERENCES CITED

Ahlbrandt, T. S., ed., 1979, Preliminary geologic, petrologic, and paleontologic results of the study of Nanushuk Group rocks, North Slope, Alaska: U.S. Geological Survey Circulatory 794, 163 p.

Alaska Geological Society, 1969a, Northwest to southeast stratigraphic correlation section, Drift River to Anchor River, Cook Inlet basin, Alaska: Anchorage, Alaska Geological Society, vertical scale 1 inch = 500 feet

——, 1969b, South to north stratigraphic correlation section, Anchor Point to Campbell Point, Cook Inlet basin, Alaska: Anchorage, Alaska Geological Society, vertical scale 1 inch = 500 feet.

——, 1969c, South to north stratigraphic correlation section, Kalgin Island to Beluga River, Cook Inlet basin, Alaska: Anchorage, Alaska Geological Society, vertical scale 1 inch = 500 feet.

——, 1969d, West to east stratigraphic correlation section, West Foreland to Swan Lake, Cook Inlet basin, Alaska: Anchorage, Alaska Geological Society, vertical scale 1 inch = 500 feet.

——, 1970a, South to north stratigraphic correlation section, Campbell Point to Rosetta, Cook Inlet basin, Alaska: Anchorage, Alaska Geological Society, vertical scale 1 inch = 500 feet.

——, 1970b, West to east stratigraphic correlation section, Beluga River to Wasilla, Cook Inlet Basin, Alaska: Anchorage, Alaska Geological Society, vertical scale 1 inch = 500 feet.

——, 1972, Northwest to southeast stratigraphic correlation section, Prudhoe Bay to Ignek Valley, Arctic North Slope, Alaska: Anchorage, Alaska Geological Society, 1 sheet, vertical scale 1″ = 500′.

Alaska Oil and Gas Conservation Commission, 1984, Public hearing on Lisburne field rules, testimony by ARCO Alaska, Inc., November 29, 1984: Anchorage, Alaska Oil and Gas Conservation Commission, 187 p.

——, 1985, 1984 statistical report: Anchorage, Alaska Oil and Gas Conservation Commission, 187 p.

——, 1988, Alaska production summary by field and pool for 1987: Anchorage, Alaska Oil and Gas Convention Commission, 3 p.

Anders, D. E., Magoon, L. B., and Lubeck, S. C., 1987, Geochemistry of surface oil shows and potential source rocks, *in* Bird, K. J., and Magoon, L. B., eds., Petroleum geology of the northern part of the Arctic National Wildlife Refuge, northeastern Alaska: U.S. Geological Survey Bulletin 1778, p. 181–198.

Bader, J. W., and Bird, K. J., 1986, Geologic map of the Demarcation Point, Mt. Michelson, Flaxman Island, and Barter Island Quadrangles, Alaska: U.S. Geological Survey Miscellaneous Investigations Map I-1791, 1 sheet, scale 1:250,000.

Barnes, F. F., 1967, Coal resources of Alaska: U.S. Geological Survey Bulletin 1242-B, 36 p.

Barnes, F. F., and Cobb, E. H., 1959, Geology and coal resources of the Homer district, Kenai coal field, Alaska: U.S. Geological Survey Bulletin 1058-F, p. F217–F260.

Barnes, F. F. and Payne, T. G., 1956, The Wishbone district, Matanuska coal field, Alaska: U.S. Geological Survey Bulletin 1016, 88 p.

Bird, K. J., 1978, New information on the Lisburne Group (Carboniferous and Permian) in the National Petroleum Reserve in Alaska (NPRA) [abs.]: American Association of Petroleum Geologists Bulletin, v. 62, no. 5, p. 880.

——, 1981, Petroleum exploration of the North Slope, *in* Mason, J. F., ed., Petroleum geology in China; Principal lectures presented to the United Nations: Tulsa, Oklahoma, PennWell Publishing Co., p. 233–248.

——, 1985, The framework geology of the North Slope of Alaska as related to oil-source rock correlations, *in* Magoon, L. B., and Claypool, G. E., eds., Alaska North Slope oil/rock correlation study: American Association of Peroleum Geologists Studies in Geology 20, p. 2–39.

——, 1986a, A comparison of the play-analysis technique as applied in hydrocarbon resource assessments of the National Petroleum Reserve in Alaska and the Arctic National Wildlife Refuge, *in* Rice, D. D., ed., Oil and gas assessment; Methods and applications: American Association of Petroleum Geologists Studies in Geology 21, p. 133–142.

——, 1986b, The framework geology of the North Slope of Alaska as related to oil-source rock correlations, *in* Magoon, L. B., and Claypool, G. E., eds., Alaska North Slope oil/rock correlation study: American Association of Petroleum Geologists Studies in Geology 20, p. 3–29.

——, 1988, Alaskan North Slope stratigraphic nomenclature and data summary for government-drilled wells, *in* Gryc, G., ed., Geology of the National Petroleum Reserve in Alaska: U.S. Geological Survey Professional Paper 1399, p. 317–353.

Bird, K. J., and Jordan, C. F., 1977, Lisburne Group (Mississippian and Pennsylvanian), potential major hydrocarbon objective of Arctic Slope, Alaska: American Association of Petroleum Geologists Bulletin, v. 61, no. 9, p. 1493–1512.

Bird, K. J., and Magoon, L. B., eds., 1987, Petroleum geology of the northern part of the Arctic National Wildlife Refuge, northeastern Alaska: U.S. Geological Survey Bulletin 1778, 329 p.

Blasko, D. P., 1974, Natural gas fields; Cook Inlet basin, Alaska: U.S. Bureau of Mines Open-File Report 35-74, 29 p.

——, 1976a, Oil and gas seeps in Alaska; Alaska Peninsula, western Gulf of Alaska: U.S. Bureau of Mines Report of Investigations 8136, 123 p.

——, 1976b, Oil and gas seeps in Alaska; North-central Gulf of Alaska: U.S. Bureau of Mines Report of Investigations 8136, 123 p.

Blasko, D. P., Wenger, W. J., and Morris, J. C., 1972, Oilfields and crude oil characteristics, Cook Inlet basin, Alaska: U.S. Bureau of Mines Report of Investigations 7688, 44 p.

Bolm, J. G., and McCulloh, T. H., 1986, Sandstone diagenesis, *in* Magoon, L. B., ed., Geologic studies of the lower Cook Inlet COST No. 1 well, Alaska Outer Continental Shelf: U.S. Geological Survey Bulletin 1596, p. 51–53.

Boss, R. F., Lennon, R. B., and Wilson, B. W., 1976, Middle Ground Shoal oil field, Alaska, *in* Braunstein, J., ed., North American oil and gas fields: American Association of Petroleum Geologists Memoir 24, p. 1–22.

Bruns, T. R., 1982a, Hydrocarbon resource report for proposed OCS lease sale 88; Southeastern Alaska, northern Gulf of Alaska, Cook Inlet, and Shelikof Strait, Alaska: U.S. Geological Survey Open-File Report 82-928, 133 p.

——, 1982b, Structure and petroleum potential of the continental margin between Cross Sound and Icy Bay, northern Gulf of Alaska: U.S. Geological Survey Open-File Report 82-929, 63 p.

——, 1983, Structure and petroleum potential of the Yakutat segment of the northern Gulf of Alaska continental margin: U.S. Geological Survey Miscellaneous Field Studies Map MF-1480, 3 sheets, scale 1:500,000.

——, 1985, Tectonics of the Yakutat block, an allochthonous terrane in the northern Gulf of Alaska: U.S. Geological Survey Open-File Report 85-13, 112 p.

Bruns, T. R., and Plafker, G., 1982, Geology, structure, and petroleum potential of the southeastern Alaska and northern Gulf of Alaska continental margins, *in* Bruns, T. R., ed., Hydrocarbon resource report for proposed OCS leas sale 88; Southeastern Alaska, northern Gulf of Alaska, Cook Inlet, and Shelikof Strait, Alaska: U.S. Geological Survey Open-File Report 82-928, p. 11–52.

Calderwood, K. W., and Fackler, W. C., 1972, Proposed stratigraphic nomenclature for Kenai Broup, Cook Inlet basin, Alaska: American Association of Petroleum Geologists Bulletin, v. 56, no. 4, p. 739–754.

Carman, G. J., and Hardwick, P., 1983, Geology and regional setting of the Kuparuk oil field, Alaska: American Association of Petroleum Geologists Bulletin, v. 67, no. 6, p. 1014–1031.

Claypool, G. E., and Magoon, L. B., 1985, Comparison of oil-source rock correlation data for Alaskan North Slope; Techniques, results, and conclusions, *in* Magoon, L. B., and Claypool, G. E., eds., Alaska North Slope oil/rock correlation study: American Association of Petroleum Geologists Special Studies in Geology 20, p. 49–81.

——, 1988, Oil and gas source rocks in the National Petroleum Reserve in Alaska, *in* Gryc, G., ed., Geology of the National Petroleum Reserve in Alaska: U.S. Geological Survey Professional Paper 1399, 90 p.

Claypool, G. E., Threlkeld, C. N., and Magoon, L. B., 1980, Biogenic and thermogenic origins of natural gas in Cook Inlet basin, Alaska: American Association of Petroleum Geologists Bulletin, v. 64, no. 8, p. 1131–1139.

Collett, T. S., Bird, K. J., Kvenvolden, K. A., and Magoon, L. B., 1988, Geologic interrelations relative to gas hydrate within the North Slope of Alaska: U.S. Geological Survey Open-File Report 88-389, 150 p.

Collins, F. R., and Robinson, R. M., 1967, Subsurface stratigraphic, structural, and economic geology, northern Alaska: U.S. Geological Survey Open-File Report, 171 p.

Coney, P. J., and Jones, D. L., 1985, Accretion tectonics and crustal structure in Alaska: Tectonophysics, v. 119, p. 265–283.

Cram, I. H., ed., 1971, Future petroleum provinces of the United States; Their geology and potential: American Association of Petroleum Geologists Memoir 15, 803 p.

Crick, R. W., 1971, Potential petroleum reserves, Cook Inlet, Alaska, *in* Cram, I. H., ed., Future petroleum provinces of the United States: Their geology and potential: American Association of Petroleum Geologists Memoir 15, p. 109–119.

Curiale, J. A., 1987, Crude oil chemistry and classification, Alaska North Slope, *in* Tailleur, I. L., and Weimer, P., eds., Alaskan North Slope geology: Pacific Section, Society of Economic Paleontologists and Mineralogists, and Alaska Geological Society, p. 161–167.

Debelius, C. A., 1974, Environmental impact statement; Offshore oil and gas development in Cook Inlet, Alaska: U.S. Army Corps of Engineers, Alaska District, 446 p.

Detterman, R. L., and Hartsock, J. K., 1966, Geology of the Iniskin-Tuxedni region, Alaska: U.S. Geological Survey Professional Paper 512, 78 p.

Detterman, R. L., Reiser, H. N., Brosgé, W. P., and Dutro, J. T., Jr., 1975, Post-Carboniferous stratigraphy, northeastern Alaska: U.S. Geological Survey Professional Paper 86, 46 p.

Dolton, G. L., and 12 others, 1981, Estimates of undiscovered recoverable conventional resources of oil and gas in the United States: U.S. Geological Survey Circular 860, 87 p.

Drummond, K. J., 1974, Paleozoic Arctic margin of North America, *in* Burk, C. A., and Drake, C. L., eds., The geology of continental margins: New York, Springer-Verlag, p. 797–810.

Energy Information Administration, 1987, U.S. crude oil, natural gas, and natural gas liquids reserves—1986 annual report: U.S. Department of Energy, Energy Information Administration Report DOE/EIA-0216(86), 103 p.

Farrington, J. W. and 5 others, 1988, Bitumen molecular maturity parameters in the Ikpikpuk well, Alaskan North Slope, *in* Mattavelli, L., and Novelli, L., eds., Advances in organic chemistry 1987, Part I, Organic geochemistry in petroleum exploration: Organic Geochemistry, v. 13, p. 303–310.

Fisher, M. A., and Magoon, L. B., 1978, Geologic framework of lower Cook Inlet, Alaska: American Association of Petroleum Geologists Bulletin, v. 62, no. 3, p. 373–402.

Franks, S. G., and Hite, D. M., 1980, Controls of zeolite cementation in Upper Jurassic sandstones, lower Cook Inlet, Alaska [abs.]: American Association of Petroleum Geologists Bulletin, v. 64, no. 5, p. 708–709.

Gryc, G., ed., 1988, Geology of the National Petroleum Reserve in Alaska: U.S. Geological Survey Professional Paper 1399, 956 p.

Hartman, D. C., Pessel, G. H., and McGee, D. L., 1972, Preliminary report on stratigraphy of Kenai Group, upper Cook Inlet, Alaska: Alaska Department of Natural Resources, Division of Geological Survey Special Report 5, 4 p., 7 sheets, scale 1:500,000.

Hayes, J. B., Harms, J. C., and Wilson, T., Jr., 1976, Contrasts between braided and meandering stream deposits, Beluga and Sterling Formations (Tertiary), Cook Inlet, Alaska, *in* Miller, T. P., ed., Recent and ancient sedimentary environments in Alaska: Anchorage, Alaska Geological Society, p. J1–J27.

Hite, D. M., 1976, Some sedimentary aspects of the Kenai Group, Cook Inlet, Alaska, *in* Miller, T. P., ed., Recent and ancient sedimentary environments in Alaska: Anchorage, Alaska Geological Society, p. I1–I23.

Hubbard, R. J., Edrich, S. P., and Rattey, R. P., 1987, Geologic evolution and hydrocarbon habitat of the "Arctic Alaska microplate": Marine and Petroleum Geology, v. 4, p. 2–34.

Huffman, A. C., Jr., ed., 1985, Geology of the Nanushuk Group and related rocks, North Slope, Alaska: U.S. Geological Survey Bulletin 1614, 129 p.

Hughes, W. B. and Holba, A. G., 1988, Relationship between crude oil quality and biomarker patterns, *in* Mattavelli, L., and Novelli, L., eds., Advances in organic geochemistry in 1987, part I, Organic geochemistry in petroleum exploration: Organic Geochemistry, v. 13, p. 15–30.

Jamison, H. C., Brockett, L. D., and McIntosh, R. A., 1980, Prudhoe Bay; A 10-year perspective, *in* Halbouty, M. T., ed., Giant oil and gas fields of the decade 1968–1978: American Association of Petroleum Geologists Memoir 30, p. 289–314.

Jones, H. P., and Speers, R. G., 1976, Permo-Triassic reservoirs of Prudhoe Bay field, North Slope, Alaska, *in* Braunstein, J., ed., North American oil and gas fields: American Association of Petroleum Geologists Memoir 24, p. 23–50.

Kelly, T. E., 1963, Geology and hydrocarbons in Cook Inlet basin, Alaska, *in* Childs, O. E., and Beebe, B. W., eds., Backbone of the Americas: American Association of Petroleum Geologists Memoir 2, p. 278–296.

——, 1968, Gas accumulations in nonmarine strata, Cook Inlet basin, Alaska, *in* Beebe, B. W., and Curtis, B. F., eds., Natural gases of North America: American Association of Petroleum Geologists Memoir 9, v. 1, p. 49–64.

Kirschner, C. E., and Lyon, C. A., 1973, Stratigraphic and tectonic development of Cook Inlet petroleum province, *in* Pitcher, M. G., ed., Arctic geology: American Association of Petroleum Geologists Memoir 19, p. 396–407.

Lattanzi, R. D., 1981, Planktonic foraminiferal biostratigraphy of Exxon Company, U.S.A., wells drilled in the Gulf of Alaska during 1977 and 1978: Anchorage, Alaska Geological Society Journal, v. 1, p. 48–59.

Lyle, W. M., and Palmer, I. F., Jr., 1976, A stratigraphic study of the Gulf of

Alaska Tertiary Province, northern Gulf of Alaska area: Alaska Department of Natural Resources, Division of Geological and Geophysical Surveys Open-File Report 93, 24 p.

MacKevett, E. M., Jr., and Plafker, G., 1974, The Border Ranges fault in south-central Alaska: U.S. Geological Survey Journal of Research, v. 2, no. 3, p. 323–329.

Magoon, L. B., ed., 1986, Geologic studies of the lower Cook Inlet COST No. 1 well, Alaska Outer Continental Shelf: U.S. Geological Survey Bulletin 1596, 99 p.

——, 1988a, The petroleum system; A classification scheme for research, exploration, and resource assessment, *in* Magoon, L. B., ed., Petroleum systems of the United States: U.S. Geological Survey Bulletin 1870, p. 2–15.

——, 1988b, Petroleum systems of the United States: U.S. Geological Survey Bulletin 1870, 68 p.

Magoon, L. B., and Bird, K. J., 1985, Alaskan North Slope petroleum geochemistry for the Shublik Formation, Kingak Shale, pebble shale unit, and Torok Formation, *in* Magoon, L. B., and Claypool, G. E., eds., Alaska North Slope oil/rock correlation study: American Association of Petroleum Geologists Special Studies in Geology 20, p. 31–48.

——, 1988, Evaluation of petroleum source rocks in the National Petroleum Reserve in Alaska using carbon content, visual kerogen, and vitrinite reflectance, *in* Gryc, G., ed., Geology of the National Petroleum Reserve in Alaska: U.S. Geological Survey Professional Paper 1399, p. 381–450.

Magoon, L. B., and Claypool, G. E., 1981a, Petroleum geology of Cook Inlet basin, Alaska; An exploration model: American Association of Petroleum Geologists Bulletin, v. 65, no. 6, p. 1043–1061.

——, 1981b, Two oil types on the North Slope of Alaska; Implications for exploration: American Association of Petroleum Geologists Bulletin, v. 65, no. 4, p. 644–652.

——, 1983, Petroleum geology of the North Slope of Alaska; Time and degree of thermal maturity, *in* Bjorøy, M., ed., Proceedings of 10th International Meeting on Organic Chemistry; Advances in organic geochemistry, 1981: Chichester, John Wiley and Sons, p. 28–38.

——, 1984, The Kingak Shale of northern Alaska; Regional variations in organic geochemical properties and petroleum source rock quality, *in* Schenck, P. A., de Leeuw, J. W., and Lijmbach, G.W.M., eds., Advances in organic geochemistry 1983: Organic Geochemistry, v. 6, p. 533–542.

——, 1985, Alaska North Slope oil/rock correlation study: American Association of Petroleum Geologists Special Studies in Geology 20, 682 p.

——, 1988, Geochemistry of oil occurrences, National Petroleum Reserve in Alaska, *in* Gryc, G., ed., Geology of the National Petroleum Reserve in Alaska: U.S. Geological Survey Professional Paper 1399, p. 519–549.

Magoon, L. B., and Egbert, R. M., 1986, Framework geology and sandstone composition, *in* Magoon, L. B., ed., Geologic studies of the lower Cook Inlet COST no. 1 well, Alaska Outer Continental Shelf: U.S. Geological Survey Bulletin 1596, p. 65–90.

Magoon, L. B., Adkison, W. L., and Egbert, R. M., 1976, Map showing geology, wildcat wells, Tertiary plant-fossil localities, K-Ar age dates, and petroleum operations, Cook Inlet area, Alaska: U.S. Geological Survey Miscellaneous Investigations Map I-1019, 3 sheets, scale 1:250,000.

Magoon, L. B., Griesbach, F. R., and Egbert, R. M., 1980, Nonmarine Upper Cretaceous rocks, Cook Inlet, Alaska: American Association of Petroleum Geologists Bulletin, v. 64, no. 8, p. 1259–1266.

Magoon, L. B., Woodward, P. V., Banet, A. C., Griscom, S. B., and Daws, T. A., 1987, Thermal maturity, richness, and type of organic matter of source-rock units, *in* Bird, K. J., and Magoon, L. B., eds., Petroleum geology of the northern part of the Arctic National Wildlife Refuge, northeastern Alaska: U.S. Geological Survey Bulletin 1778, p. 127–179.

McWhae, J. R., 1986, Tectonic history of northern Alaska, Canadian Arctic, and Spitsbergen regions since Early Cretaceous: American Association of Petroleum Geologists Bulletin, v. 70, no. 4, p. 430–450.

Melvin, J., and Knight, A., 1984, Lithofacies, diagenesis, and porosity of the Ivishak Formation, Prudhoe Bay area, Alaska, *in* McDonald, D. A., and Surdam, R. C., eds., Clastic diagenesis: American Association of Petroleum Geologists Memoir 37, p. 347–365.

Miller, B. M., and 8 others, 1975, Geological estimates of undiscovered recoverable oil and gas resources in the United States: U.S. Geological Survey Circular 725, 78 p.

Miller, D. J., Payne, T. G., and Gryc, G., 1959, Geology and possible petroleum provinces in Alaska: U.S. Geological Survey Bulletin 1094, 127 p.

Molenaar, C. M., 1981, Depositional history and seismic stratigraphy of Lower Cretaceous rocks, National Petroleum Reserve in Alaska and adjacent areas: U.S. Geological Survey Open-File Report 81-1084, 42 p.

——, 1982, Umiat field; An oil accumulation in a thrust-faulted anticline, North Slope of Alaska, *in* Powers, R. B., ed., Geologic studies of the Cordilleran thrust belt: Denver, Colorado, Rocky Mountain Association of Geologists, p. 537–548.

——, 1983, Depositional relations of Cretaceous and lower Tertiary rocks, northeastern Alaska: American Association of Petroleum Geologists Bulletin, v. 67, no. 7, p. 1066–1080.

——, 1988, Depositional history and seismic stratigraphy of Lower Cretaceous rocks in the National Petroleum Reserve in Alaska and adjacent areas, *in* Gryc, G., ed., Geology of the National Petroleum Reserve in Alaska: U.S. Geological Survey Professional Paper 1399, p. 593–621.

Molenaar, C. M., Bird, K. J., and Collett, T. S., 1986, Regional correlation sections across the North Slope of Alaska: U.S. Geological Survey Miscellaneous Field Studies Map MF-1907, 1 sheet, scales 1″ = 2,000′; 1″ = 4,000′.

Molenaar, C. M., Bird, K. J., and Kirk, A. R., 1987, Cretaceous and Tertiary stratigraphy of northeastern Alaska, *in* Tailleur, I. L., and Weimer, P., eds., Alaskan North Slope geology: Pacific Section, Society of Economic Paleontologists and Mineralogists and Alaska Geological Society, v. 50, p. 513–528.

Morgridge, D. L., and Smith, W. B., Jr., 1972, Geology and discovery of Prudhoe Bay field, eastern Arctic Slope, Alaska, *in* King, R. E., ed., Stratigraphic oil and gas fields; Classification, exploration methods, and case histories: American Association of Petroleum Geologists Memoir 16, p. 489–501.

Mull, C. G., and Kososki, B. A., 1977, Hydrocarbon assessment of the Arctic National Wildlife Range, eastern Arctic Slope, Alaska, *in* Blean, K. M., ed., The United States Geological Survey in Alaska; Accomplishments during 1976: U.S. Geological Survey Circular 751-B, B20–B22.

Mull, C. G., and Nelson, S. W., 1986, Anomalous thermal maturity data from the Orca Group (Paleocene and Eocene), Katalla-Kayak Island area, *in* Bartsch-Winkler, S., and Reed, K. M., eds., Geologic studies in Alaska by the U.S. Geological Survey during 1985: U.S. Geological Survey Circular 978, p. 50–55.

Oil and Gas Journal, 1984, Exxon; N. Slope gas/condensate field is a giant: Oil and Gas Journal, v. 82, no. 11, p. 30.

——, 1988, U.S. fields with reserves exceeding 100 million bbl: Oil and Gas Journal, v. 86, no. 4, p. 60.

Palmer, A. R., 1983, The Decade of North American Geology 1983, geologic time scale: Geology, v. 11, p. 503–504.

Palmer, I. F., Jr., 1976, U.S.G.S. releases Alaskan well data: Oil and Gas Journal, v. 74, no. 12, p. 101–104.

Parkinson, L. J., 1962, One field, one giant; The story of Swanson River: Oil and Gas Journal, v. 60, no. 13, p. 180–183.

Peters, K. E., Moldowan, J. M., Schoell, M., and Hempkins, W. B., 1986, Petroleum isotopic and biomarker composition related to source rock organic matter and depositional environment, *in* Leythaeuser, D., and Rullkötter, J., eds., Advances in organic geochemistry; Part 1, Petroleum geochemistry: Organic Geochemistry, v. 10, no. 1-3, p. 17–27.

Petroleum Information, 1988, Alaska well history control system (WHCS): Petroleum Information Corporation digital file, proprietary.

Plafker, G., 1967, Geologic map of the Gulf of Alaska Tertiary Province, Alaska: U.S. Geological Survey Miscellaneous Geologic Investigations Map I-484, 1 sheet, scale 1:500,000.

——, 1971, Pacific margin Tertiary basin, *in* Cram, I. H., ed., Petroleum provinces of North America: American Petroleum of Petroleum Geologists Memoir 15, p. 120–135.

——, 1987, Regional geology and petroleum potential of the northern Gulf of

Alaska continental margin, *in* Scholl, D. W., Grantz, A., and Vedder, J. G., eds., Geology and resource potential of the continental margin of western North America and adjacent ocean basins; Beaufort Sea to Baja California: Houston, Texas, Circum-Pacific Council for Energy and Mineral Resources Earth Science Series 6, p. 229–267.

Plafker, G., and Claypool, G. E., 1979, Petroleum source rock potential or rocks dredged from the continental slope in the eastern Gulf of Alaska: U.S. Geological Survey Open-File Report 79-295, 24 p.

Plafker, G., Bruns, T. R., and Page, R. A., 1975, Interim report on petroleum resource potential and geologic hazards in the outer continental shelf of the Gulf of Alaska Tertiary province: U.S. Geological Survey Open-File Report 75-592, 74 p.

Plafker, G., Winkler, G., Coonrad, W., and Claypool, G. E., 1980, Preliminary geology of the continental slope adjacent to OCS lease sale 55, eastern Gulf of Alaska; Petroleum resource implications: U.S. Geological Survey Open-File Report 80-1089, 72 p.

Rau, W. W., Plafker, G., and Winkler, G. R., 1983, Foraminiferal biostratigraphy and correlations in the Gulf of Alaska Tertiary Province: U.S. Geological Survey Oil and Gas Investigations Chart OC-120.

Reed, B. L., Miesch, A. T., and Lanphere, M. A., 1983, Plutonic rocks of Jurassic age in the Alaska-Aleutian Range batholith; Chemical variations and polarity: Geological Society of America Bulletin, v. 94, p. 1232–1240.

Sedivy, R. A., and 5 others, 1987, Investigation of source rock-crude oil relationships in the northern Alaska hydrocarbon habitat, *in* Tailleur, I. L., and Weimer, P., eds., Alaskan North Slope geology: Pacific Section, Society of Economic Paleontologists and Mineralogists, p. 169–179.

Seifert, W. K., Moldowan, J. M., and Jones, J. W., 1980, Application of biological marker chemistry to petroleum exploration, *in* Proceedings 10th World Petroleum Congress: London, Heyden and Son, Ltd., p. 425–440.

Specht, R. N., Brown, A. E., Selman, C. H., and Carlisle, J. H., 1986, Geophysical case history, Prudhoe Bay field: Geophysics, v. 51, p. 1039–1049.

Stonely, R., 1967, The structural development of the Gulf of Alaska sedimentary province in southern Alaska: Quarterly Journal of the Geological Society of London, v. 123, p. 25–57.

Sweeney, J. F., 1982, Arctic-Alaska, a two-stage displaced terrane [abs.], *in* Proceedings, 33rd Alaska Science Conference, Fairbanks, Alaska: American Association for the Advancement of Science, p. 146.

Tailleur, I. L., 1973, Probable rift origin of Canada basin, Arctic Ocean, *in* Pitcher, M. G., ed., Arctic geology: American Association of Petroleum Geologists Memoir 19, p. 526–535.

Tailleur, I. L., and Weimer, P., eds., 1987, Alaskan North Slope geology: Pacific Section, Society of Economic Paleontologists and Mineralogists and Alaska Geological Society, v. 50, 874 p.

U.S. Geological Survey and Minerals Management Service, 1988, National Assessment of undiscovered conventional oil and gas resources; Working paper: U.S. Geological Survey Open-File Report 88-373, 511 p.

Van Dyke, W. D., 1980, Proven and probable oil and gas reserves, North Slope, Alaska: Alaska Department of Natural Resources, Division of Minerals and Energy Management, 11 p.

Van Poollen, and Associates and Alaska Division of Oil and Gas, 1974, in place volumetric determination of reservoir fluids, Sadlerochit Formation, Prudhoe Bay Field: Anchorage, Alaska Department of Natural Resources, Division of Oil and Gas, 41 p.

Walker, H. J., 1973, Morphology of the North Slope, *in* Britton, M. E., ed., Alaskan Arctic tundra: Arctic Institute of North America Technical Paper 25, p. 49–92.

Werner, M. R., 1987, Tertiary and Upper Cretaceous heavy-oil sands, Kuparuk River Unit area, Alaska North Slope, *in* Meyer, R. F., ed., Exploration for heavy crude oil and natural bitumen: American Association of Petroleum Geologists Studies in Geology 25, p. 537–547.

Whelan, J. K., Farrington, J. W., and Tarafa, M. E., 1986, Maturity of organic matter and migration of hydrocarbons in two Alaskan North Slope wells, *in* Leythaeuser, D., and Rullkötter, J., eds., Advances in organic chemistry; Part 1, Petroleum geochemistry: Organic Geochemistry, v. 10, no. 1-3, p. 207–219.

Winkler, G. R., McLean, H., and Plafker, G., 1976, Textural and mineralogical study of sandstones from the onshore Gulf of Alaska Tertiary Province, southern Alaska: U.S. Geological Survey Open-File Report 76-198, 48 p.

Wolfe, J. A., 1981, A chronologic framework for Cenozoic megafossil floras of northwestern North America and its relation to marine geochronology, *in* Armentrout, J. M., ed., Pacific Northwest Cenozoic biostratigraphy: Geological Society of America Special Paper 184, p. 39–42.

Wood, G. V., and Armstrong, A. K., 1975, Diagenesis and stratigraphy of the Lisburne Group limestones of the Sadlerochit Mountains and adjacent areas, northeastern Alaska: U.S. Geological Survey Professional Paper 857, 47 p.

Manuscript Accepted by the Society May 24, 1990

ACKNOWLEDGMENTS

In the early stages of preparation, K. J. Bird, T. R. Bruns, and G. Plafker made suggestions that were incorporated into the reviewed manuscript. I thank G. Gryc and C. M. Molenaar for their critical review of the final manuscript. A special thanks to R. P. Crandall of the Alaska Oil and Gas Conservation Commission who checked and added significant information to Tables 4 through 7. Zenon Valin carefully proofed the final manuscript and made helpful comments.

Printed in U.S.A.

The Geology of North America
Vol. G-1, The Geology of Alaska
The Geological Society of America, 1994

Chapter 31

Coal in Alaska

Clyde Wahrhaftig*
U.S. Geological Survey, MS-904, 345 Middlefield Road, Menlo Park, California 94025
Susan Bartsch-Winkler
U.S. Geological Survey, MS-905, Box 25046, Federal Center, Denver, Colorado 80225-0046
Gary D. Stricker
U.S. Geological Survey, MS-972, Box 25046, Federal Center, Denver, Colorado 80225-0046

INTRODUCTION

This chapter is an updated synthesis of information on an enormous, largely untapped natural resource—Alaskan coal. Herein we discuss the major fields as well as the smaller occurrences in the state, their geologic setting, depositional history, and, where known, the rank and chemistry of the coal and the size of the resources. Physiographic terminology is from Wahrhaftig (1965; this volume, Plate 2). Major basins of Alaska are delineated and discussed by Kirschner (this volume, Chapter 14 and Plate 7). Geographic or physiographic features referred to in this chapter that are not shown on the accompanying figures may be found in Plate 2 of this volume; geologic features not shown on the figures may be found in Plate 1 (Beikman, this volume).

Coal in Alaska was used by Inuit and Indian cultures before the advent of European explorers. The Beechy expedition of 1826–1827 reported the occurrence of coal in Alaska (Huish, 1836), and whaling ships mined coal from near Cape Beaufort north of the Arctic Circle before the turn of the twentieth century (Conwell and Triplehorn, 1976). The first coal mine, operated by the Russians at Port Graham on the southwest tip of the Kenai Peninsula, opened in 1855 and closed in 1867 (Martin, 1915) after the United States took possession of the Alaska Territory. Many mines were active after 1931, when Congress authorized the building of the Alaska Railroad, which created a market and transportation necessary for large-scale coal production. The first coal lease sale in the state in more than 17 years was held in 1983 in conjunction with an oil lease sale. In 1984, export of Alaskan coal began with shipments to South Korea. Other developments include construction of a coal terminal at the deep-water port at Seward, new loading facilities at the Usibelli coal mine, and upgrading of the Alaska Railroad, which hauls coal to Seward. In 1985, coal production increased by 60% over the previous year, with a gross value on production of 1.27×10^6 metric tons valued at $39.7 million (Bundtzen and others, 1986). An estimate of total coal in place is 9.4×10^{12} metric tons (Fig. 1), or about 50% of total United States resources.

Alaska's coal resources are enormous. Though little exploited, they will probably be a major component of industrial energy and hydrocarbon raw materials, at least for the nations bordering the Pacific, when the world's more easily won resources of oil and natural gas are depleted.

CONDITIONS OF COAL ACCUMULATION

Coal is defined as a "readily combustible rock containing more than 50 percent by weight and more than 70 percent by volume of carbonaceous material, formed from compaction or induration of variously altered plant remains similar to those of peaty deposits" (Schopf, 1956). The terrestrial vegetal remains that became coal accumulated under mainly anoxic conditions beneath stagnant or slowly moving swamp or marsh waters. The acidity of the water ultimately had to be high enough to kill off bacteria and fungi that would otherwise have digested and completely oxidized the peat.

Swamp peat appears to accumulate at rates of 0.1–2.3 m per thousand years (McCabe, 1984), and the most reliable estimates of the compaction ratio from peat to coal range from 7:1 to 20:1 (Ryer and Langer, 1980); thus 1 m of coal may have taken from 3,000 to 200,000 years to accumulate. Widespread, long-persisting swamps on extensive plains close to a stable base level (commonly sea level) were the sites of accumulation of the most extensive beds of coal.

The swamp vegetation has to be dense enough at the margins of the swamp to still any floodwater and filter out all inorganic sediment. It has been assumed that most deltaic or coastal-plain swamps do this; however, according to McCabe (1984), only modern swamp peats with convex surfaces where the water is introduced solely as precipitation are sufficiently low

*Deceased.

Wahrhaftig, C., Bartsch-Winkler, S., and Stricker, G. D., 1994, Coal in Alaska, *in* Plafker, G., and Berg, H. C., eds., The Geology of Alaska: Boulder, Colorado, Geological Society of America, The Geology of North America, v. G-1.

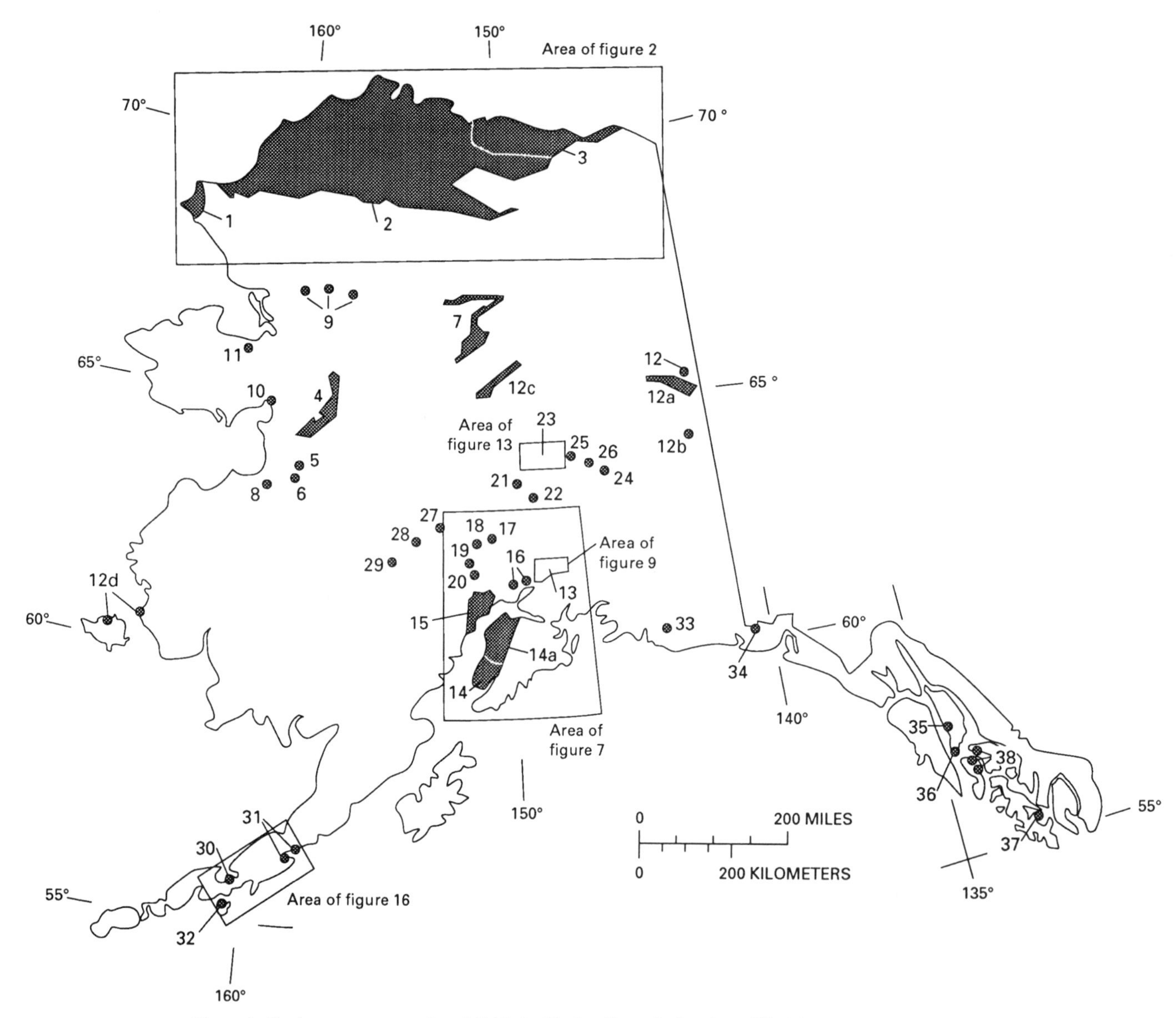

Figure 1. Coal occurrences and coal fields in Alaska. Compiled and modified from Merritt and Hawley (1986); Barnes (1967a, 1967b); Brew and others (1984); Magoon and others (1976); Plafker (1987). Base from Merritt and Hawley (1986). 1: Cape Lisburne field. 2: Cretaceous bituminous coal fields of the North Slope. 3: Cretaceous and Tertiary subbituminous coal and lignite fields of the North Slope. 4: Nulato field. 5: Williams Mine. 6: Coal Mine No. 1 (Blackburn). 7: Tramway Bar. 8: Anvik River coal occurrence. 9: Kobuk River coal occurrence. 10: Koyuk coal occurrence. 11: Chicago Creek coal deposit. 12: Nation River coal occurrence. 12a: Eagle-Circle district. 12b: Chicken district. 12c: Rampart district. 12d: Nunivak Island and Nelson Island occurrences. 13: Matanuska coal field. 14: Homer district, Kenai coal field. 14a: Kenai coal field. 15: Tyonek-Beluga coal field. 16: Little Susitna district. 17: Peters Hills district. 18: Fairview Mountain district. 19: Johnson Creek district. 20: Canyon Creek district. 21: Costello Creek coal field. 22: Broad Pass coal field. 23: Nenana coal field. 24: Jarvis Creek coal field. 25: Dry Creek– Newman Creek coal occurrence. 26: Delta Creek–Ptarmigan Creek coal occurrence. 27: Little Tonzona River coal occurrence. 28: Windy Fork coal occurrence. 29: Cheeneetnuk coal occurrence. 30: Herendeen Bay coal field. 31: Chignik coal field. 32: Unga Island coal occurrence. 33: Bering River coal field. 34: Samovar Hills coal occurrence. 35: Kootznahoo coal field. 36: Murder Cove coal occurrence. 37: Kasaan Bay coal occurrence. 38: Port Camden coal occurrence.

in ash to be precursors of commercial coal. Such peats have been found in deltaic and fluviatile environments in Borneo, Malaysia, and southwestern British Columbia (McCabe, 1984; Styan and Bustin, 1984).

Preservation of the swamp peat resulted from either (1) more rapid subsidence than peat accumulation, resulting in flooding of the swamp into a lake or marine embayment where shale or limestone would accumulate to cover the coal, or (2) burial by sheets of sand and gravel, such as crevasse splays or overbank and levee deposits, in migrating or avulsed channels.

Therefore, the geologic environments conducive to coal accumulation include delta plains, broad coastal plains with gentle slopes, and wide alluvial plains where swamps and flood basins can exist alongside naturally leveed rivers. Preservation of peat accumulations and their compaction and metamorphism into coal require input of sediment. A marine invasion, which brings peat accumulation to a stop, commonly leads to a roofrock of marine shale or limestone; in this setting, the sulfur content of the coal is likely to be high (Gluskoter and Simon, 1968; Gluskoter and Hopkins, 1970). Low-sulfur coals are preserved when burial is by crevasse splay or channel sand deposited by shifting fluvial or estuarine channels. Thus, the accumulation of thick sequences of sandstone and shale including low-sulfur coals, such as those that characterize the coal deposits in most of Alaska, require an enormous input of clastic sediment and a subsiding basin. Such large volumes of sediment may be derived from drainage of a huge tributary area, such as the Mississippi River system, or from streams that drain a tectonically active and mountainous hinterland, as has been characteristic of many Alaskan terranes since at least Early Cretaceous time.

CONSTITUTION AND METAMORPHISM OF COAL

The organic constituents of coal, comparable to minerals in an inorganic rock, are known as macerals. Three classes of macerals are recognized: (1) vitrinite, initially high in both oxygen and hydrogen and derived mainly from the cellulose and lignin of wood, leaves, roots, and their decay products; (2) exinite, high in hydrocarbons (fats and oils) and derived mainly from cuticles, resin, spores, and algae; and (3) inertinite, initially relatively high in carbon, the product of oxidation of organic matter (for example, fusinite, which is thought to be fossil charcoal) or originally sclerotinous tissue of fungi. Huminite is the maceral class in lignite and subbituminous coal equivalent to vitrinite, and liptinite is the maceral class equivalent to exinite (Stach, 1968; Neavel, 1981; Stach and others, 1982; McCabe, 1984). Macerals are identified by microscopic study of polished surfaces in reflected light on the basis of their form and reflectance—vitrinites have medium reflectance and appear gray, exinites have low reflectance and appear black, and inertinites have high reflectance. Many of the economic properties of coals depend on the proportions of macerals, and the classification into types is based on these proportions. Petrographic study of Alaskan coals is still in its early stages (Rao and Wolff, 1981; R. D. Merritt, 1986, written commun.).

The rank of a coal is a measure of the metamorphism it has undergone since the initiation of burial. The classification of coal by rank is listed in Table 1. Metamorphism of coal is due primarily to temperature, time, and, probably less significant, pressure (Teichmuller and Teichmuller, 1968). Temperature is that of Earth's geothermal gradient and increases with depth of burial; it probably has not exceeded 150–200 °C for most coals. Time plays an important role in determining coal rank. In general, coal buried for 50 m.y. at 50 °C will be subbituminous, whereas coal buried for 200 m.y. at 50 °C would be low-volatile bituminous. Tertiary coals are generally subbituminous, and Carboniferous coals are usually bituminous. This broad generalization is invalid in many geologic settings in Alaska at ancient plate margins, where heat produced by intrusions or by increased pressure brought on by tectonic activity can elevate coal rank, as in the Bering River field.

The grade of coal is its suitability for various economic uses, and it varies with the use. Grade depends on both type and rank, as well as on such deleterious factors as moisture content, ash content, and content of such undesirable elements as sulfur.

Terminology used in this report for coal classification and

TABLE 1. RANK CLASSIFICATION OF COAL*

Rank		Basis for classification†,§	Approximate carbon content (wt %)**
Lignite B	>31% volatile matter§	<3,500 cal/gm	72
Lignite A		3,500 to 4,600 cal/gm	
Subbituminous C		4,600 to 5,280 cal/gm	
Subbituminous B		5,280 to 5,840 cal/gm	76
Subbituminous A		5,840 to 6,400 cal/gm‡	
High-volatile			
bituminous C		6,400 to 7,230 cal/gm	
bituminous B		7,230 to 7,780 cal/gm	80
bituminous A			84
Medium-volatile bituminous	>7,780 cal/gm	22 to 31 wt % volatile matter	90
Low-volatile bituminous		14 to 22 wt % volatile matter	
Semianthracite		8 to 14 wt % volatile matter	
Anthracite		<8 wt % volatile matter	>90

*From Neavel (1981, p. 134).
†cal/gm calculated on moist, ash-free (wt%) basis. Note: Neavel states mineral-free, but with saturated pore-moisture content.
§Volatile matter calculated on dry, mineral-free (wt %) basis.
**Total carbon calculated on dry, ash-free (wt %) basis.
‡Bituminous if agglomerating, subbituminous if nonagglomerating.

reserve estimates, as well as supplementary terms, were defined by Wood and others (1983). Measured (or identified) reserves have the highest degree of geologic assurance; that is, they refer to coal deposits that have been studied in detail and for which there are many field and laboratory data. Indicated reserves have a moderate degree of geologic assurance; they are based on projecting thickness and other geologic data from nearby outcrops, trenches, workings, and drill holes, or from other measured reserves. Demonstrated reserves commonly refer to the sum of coal classified as measured and indicated resources. Inferred resources have a low degree of geologic assurance; estimates in this category are based on inferred continuity beyond measured and indicated reserves, for which there is geologic evidence, and estimates are computed by projection of coal data for a specified distance and depth beyond coal deposits classed as indicated. There are no sample or measurement sites in the area of inferred deposits. Hypothetical (or undiscovered) reserves have a low degree of geologic assurance; for these coal deposits, estimates of rank, thickness, and extent are based not on measurements at the coal site, but on assumed continuity of coal beyond inferred deposits.

Measurements are generally reported here in metric units, except where original measurements were in American units, in which cases the original units are given in parentheses following the metric units.

OCCURRENCE OF COAL IN ALASKA

The bulk of the known coal resources of Alaska are in two coal provinces: a province north of the Brooks Range, of mainly Cretaceous age (the Northern Alaska fields or North Slope province); and a province centered on Cook Inlet, of Tertiary age (the Southern Alaska–Cook Inlet province) (Fig. 1). The North Slope province is in the sedimentary wedge shed northeastward and northward from the Brooks Range during and after the Brookian orogeny, and eastward from now-collapsed highlands on the site of the present Chukchi basin, into a deep trough that lay between the Brooks Range and the Barrow arch. The coal-bearing section includes delta-plain and fluvial sediment that was deposited after the basin was filled to sea level and above, a filling that progressed mainly from west to east. These are the largest coal deposits in Alaska, but owing to their remoteness and formidable logistic and environmental problems inherent in Arctic exploitation, they are not yet of commercial value.

The Southern Alaska–Cook Inlet province is centered on the deep trough in the arc-trench gap between the Aleutian volcanic arc and the Aleutian Trench. Estuarine deposits in the southwestern part of this trough interfinger northeastward with deltaic and fluvial deposits of the lower end of a river system that had tributaries during middle Tertiary time as far north as the Yukon-Tanana Upland. Basins of coal accumulation that straddle these tributaries, such as the Nenana and Jarvis Creek coal fields, are included in this coal province. The Nenana coal field in the Southern Alaska–Cook Inlet province is the most commercially significant deposit in the state, followed by the Cook Inlet fields (a composite of several fields in the Cook Inlet region).

Outside these two large coal provinces, smaller deposits include, from north to south, scattered coal occurrences of Paleozoic and Tertiary age in northern and eastern Alaska, coal deposits of Cretaceous and Tertiary age in the Intermontane Plateaus and the Herendeen Bay, Chignik, and Bering River areas, and the coal deposits of southeastern Alaska (Fig. 1). Other minor coal occurrences in Alaska are shown in Figure 1 (Wood and Bour, 1988), but are not discussed herein.

On the basis of limited data and interpretations of onshore outcrops, onshore to offshore stratigraphic trends, wells located both onshore and offshore, and offshore seismic lines, 13 areas on the continental shelf of Alaska have potential of undiscovered coal. The resource potential of shelf areas in the Chukchi and Beaufort Seas and in the Norton, Bristol Bay, Navarin, and Cook Inlet basins is estimated to total 5.5×10^{12} metric tons (Affolter and Stricker, 1987b). Very meager information exists on these undiscovered offshore deposits, and they are not shown in entirety in Figure 1 (except those beneath the Beaufort Sea and Cook Inlet) or discussed further herein.

NORTH SLOPE PROVINCE

The northern Alaska coal fields are generally situated north of lat 69°N and include about 82,880 km^2 of coal-bearing rocks, both near the surface and in the deep subsurface (Fig. 2). Rocks on the North Slope range in age from Precambrian to Holocene; a representative columnar section is shown in Figure 3. The most important coal-bearing units of the North Slope are the Brookian sequence Nanushuk Group, but coal deposits of lesser quantity and quality occur on the western and eastern North Slope in the Ellesmerian Endicott Group and the Brookian Colville Group (the Fortress Mountain, Torok, and Sagavanirktok Formations) (Collier, 1906; Tailleur, 1965; Barnes, 1967b; Conwell and Triplehorn, 1976; Molenaar and others, 1984). Coal-bearing nonmarine sedimentary facies mainly of Cretaceous and Tertiary age prograde northward and northeastward from the Brooks Range to interfinger with coeval marine units.

As early as 1879, coal for fueling whaling ships was mined from Corwin Bluff on the Chukchi Sea north of Cape Lisburne (Schrader, 1904). Since 1944, various mining companies have carried out preliminary investigations, and during the Barrow fuel shortage in 1943–1944, at least one small mine was in operation on the Meade River (Clark, 1973). Operation of the Meade River coal mine demonstrated the technical feasibility of coal mining under arctic (permafrost) conditions, but no mining activity has taken place as of 1992. The coal deposits at Corwin Bluff were first described by Collier (1906). Later studies showed that coal is widespread in rocks that crop out in the foothills belt (Chapman and Sable, 1960; Moore and others, this volume), and is buried beneath the Arctic coastal plain (Tailleur and Brosgé, 1976; Moore and others, this volume); coal-bearing sequences are also likely to exist beneath the Chukchi Sea (Grantz and others, 1975; Grantz and others, this volume) and the Beaufort Sea (Affolter and Stricker, 1987b).

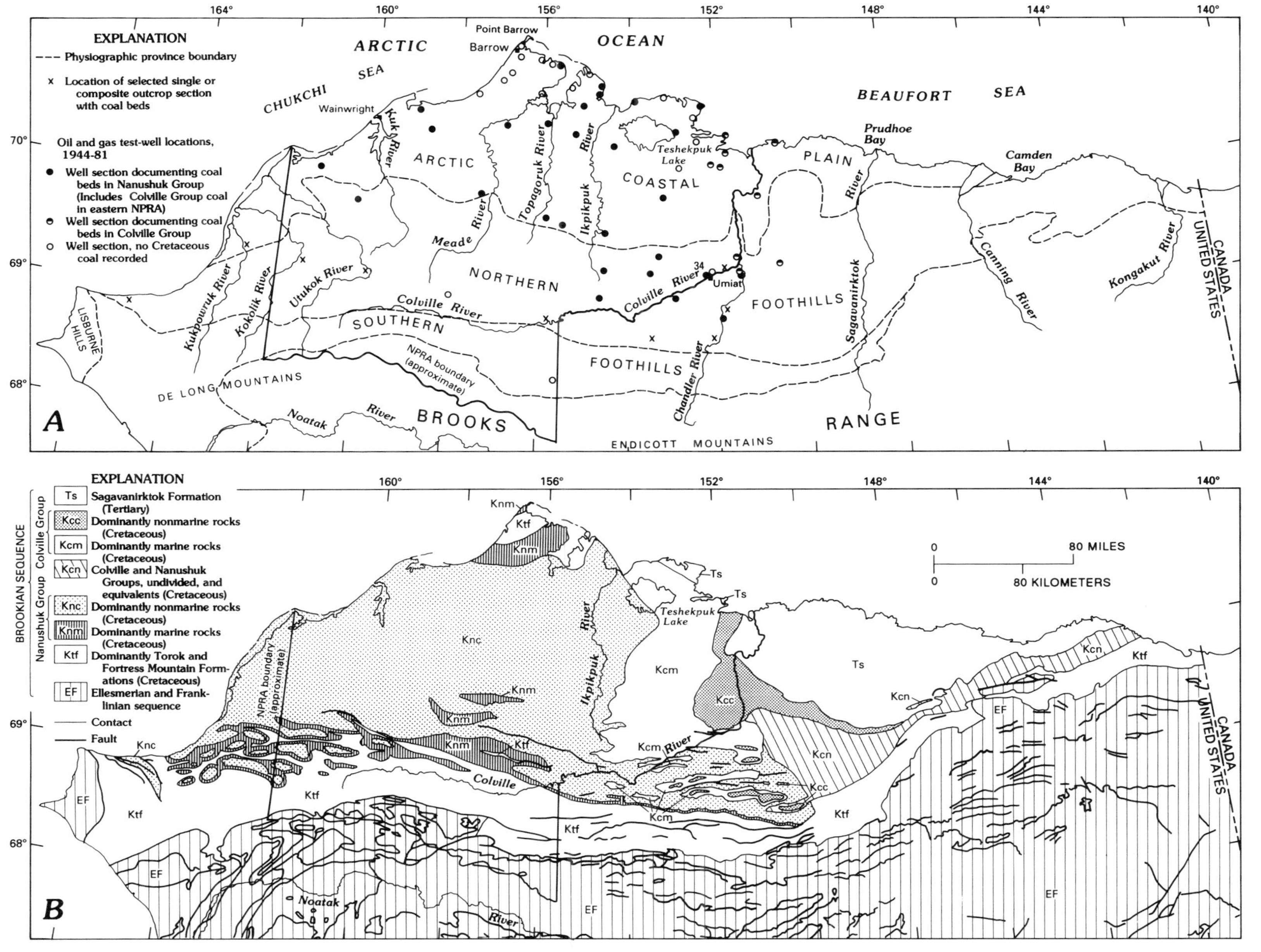

Figure 2. Maps of northern Alaska showing (A) physiographic provinces, National Petroleum Reserve Alaska (NPRA) boundary, and test well locations, and (B) generalized bedrock geology (from Sable and Stricker, 1987). Well location 34 near Umiat in (A) referred to in text.

Geologic units and setting

Details of the stratigraphy, structure, and tectonic setting of the North Slope and contiguous areas are presented by Grantz and others (this volume) and Moore and others (this volume). Three separate tectonic and genetic rock sequences exist on the North Slope (Figs. 3 and 4): the Franklinian, Ellesmerian, and Brookian sequences. The pre–Upper Devonian Franklinian sequence, of no importance for coal, is composed of weakly metamorphosed rocks; it is separated from the overlying Ellesmerian sequence by a regional erosional unconformity. The Upper Devonian to Lower Cretaceous Ellesmerian sequence consists of shallow-marine shelf and basinal clastic rocks, including the coal-bearing Upper Devonian to Lower Permian(?) Endicott

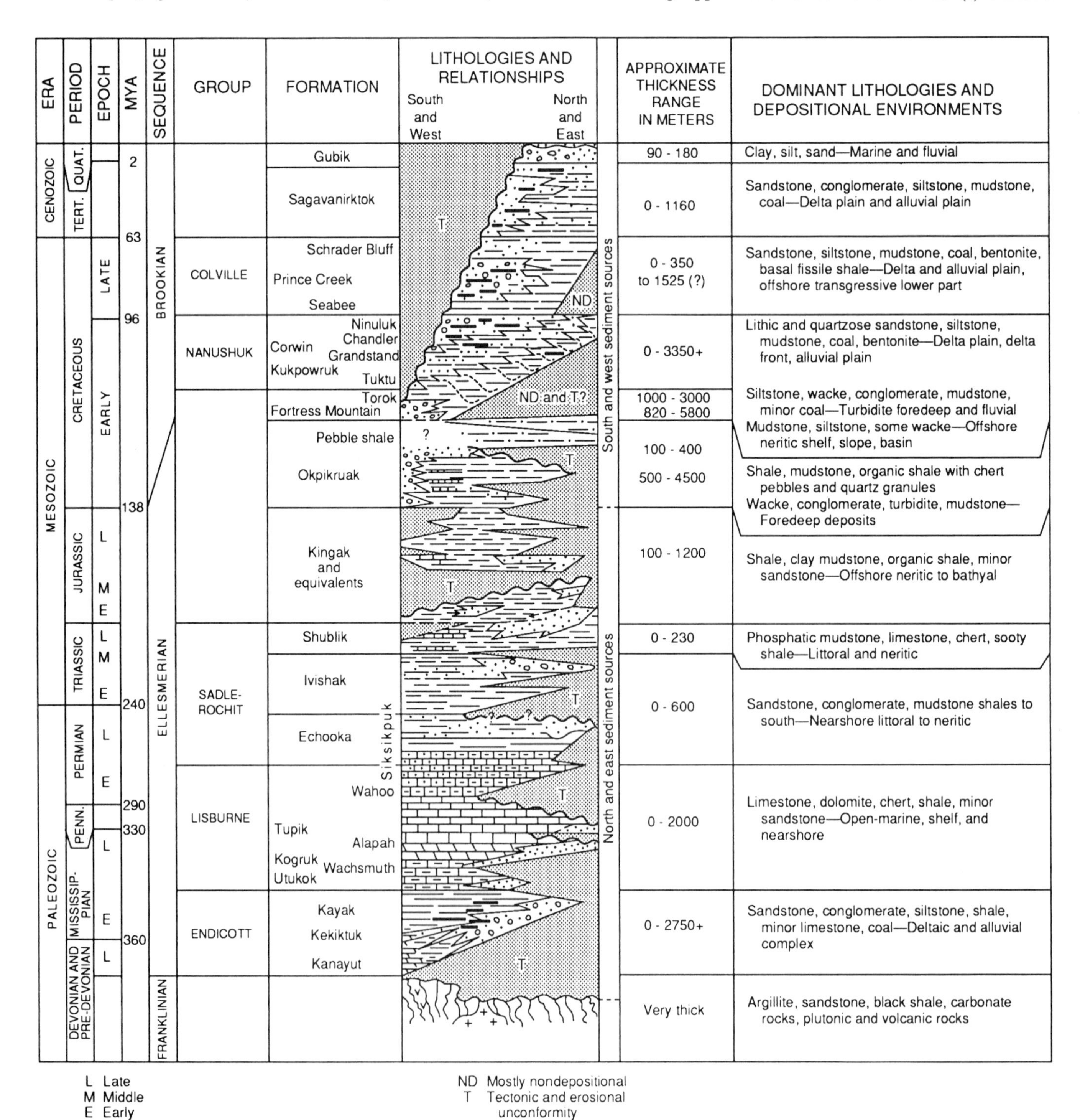

Figure 3. Generalized columnar section of rocks in the National Petroleum Reserve Alaska (from Sable and Stricker, 1987).

Group derived from northerly sources (Fig. 4). In the southern foothills belt, uppermost Ellesmerian rocks are represented by the Neocomian Okpikruak Formation (Fig. 4B), whereas farther north, uppermost Ellesmerian rocks are overlain by the pebble shale unit (Fig. 4B), which may be coeval with part of the Okpikruak Formation (Sable and Stricker, 1987). The lowermost Cretaceous to Quaternary Brookian sequence includes rocks deposited to the north and northeast into a foredeep north of the Brooks Range (Figs. 3 and 4). These rocks were deposited into the Cretaceous Colville basin and the Tertiary Camden basin (Fig. 5; Grantz and Eittreim, 1979). The Brookian sequence is broken by disconformities and unconformities or by nondeposition (Fig. 3; Sable and Stricker, 1987).

The Brooks Range, adjacent to and south of the North Slope

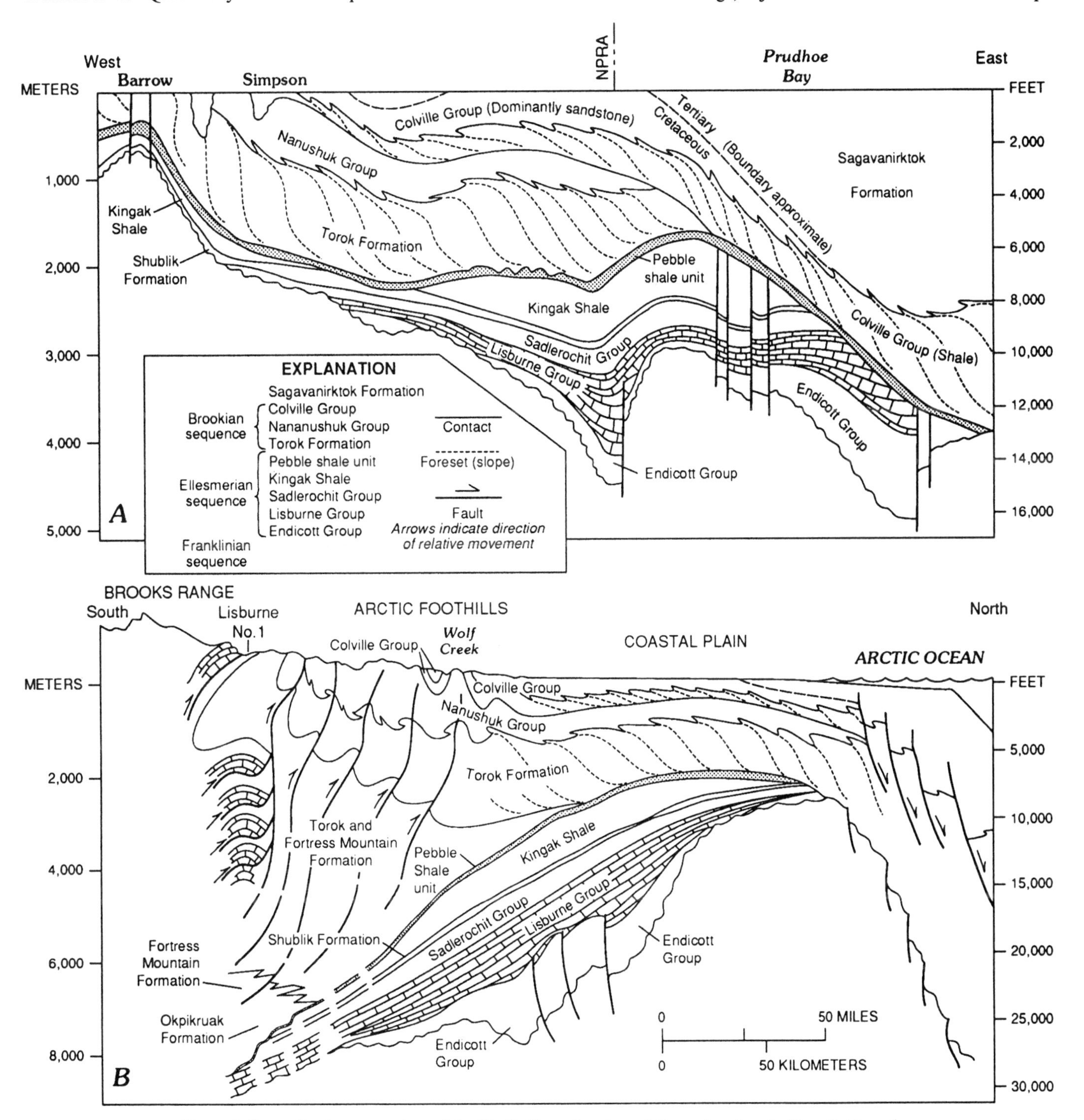

Figure 4. Generalized cross sections of the North Slope showing relations of rock units (from Sable and Stricker, 1987).

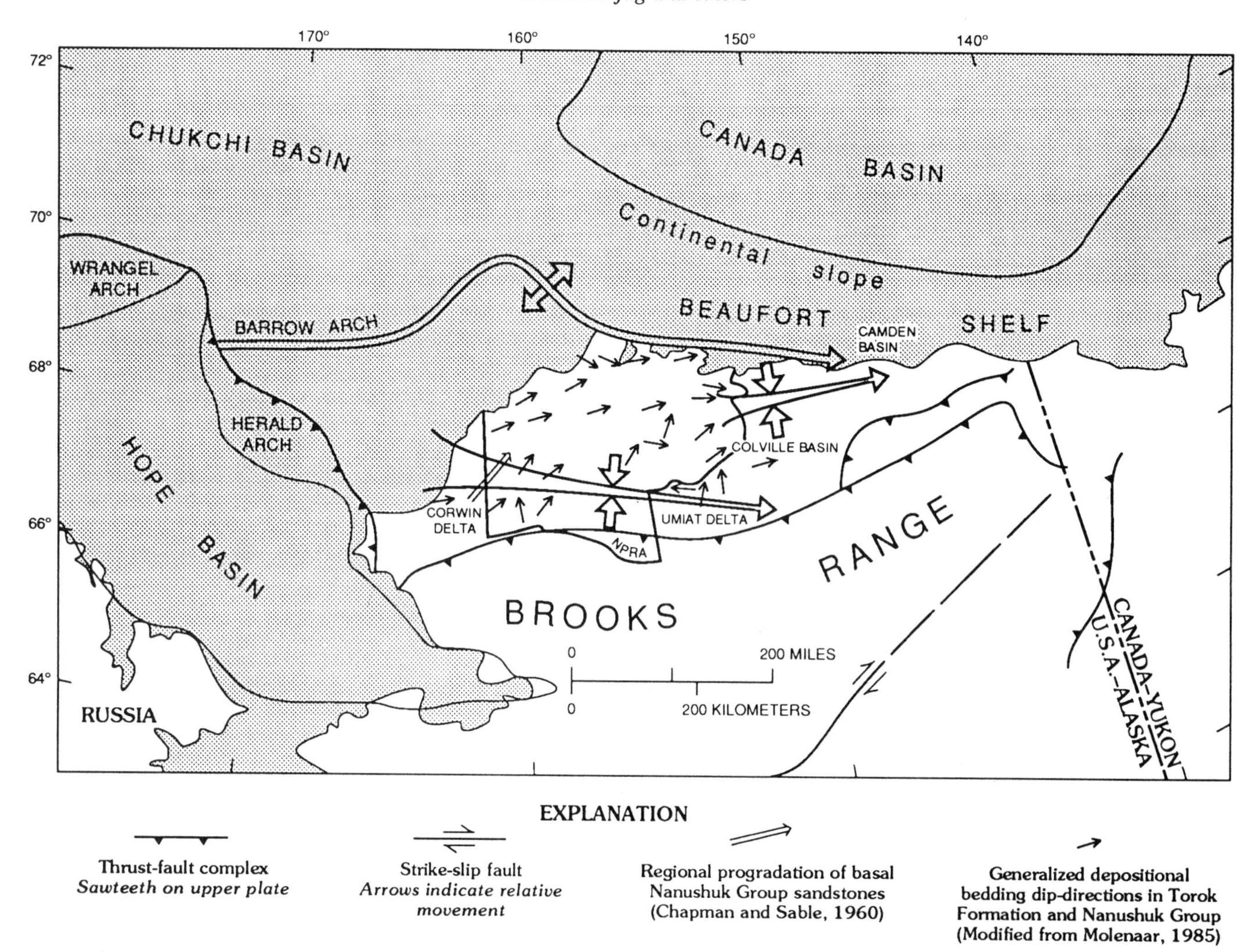

Figure 5. Structural elements of northern Alaska and adjoining regions pertinent to the depositional history of the Brookian sequence (from Sable and Stricker, 1987).

province (Fig. 5), is characterized by multiple, folded, generally east-west–striking and south-dipping, imbricate thrust sheets consisting mostly of Ellesmerian rocks (Mull, 1982; Mayfield and others, 1988; Moore and others, this volume). North of the Brooks Range, in the southern foothills belt of the North Slope province, these deformed strata include complexly overturned folds and thrust-sheet remnants of Ellesmerian and lower Brookian strata (Fig. 5). To the north in the northern foothills belt, Brookian rocks crop out; thrust faults and deformational features in anticlines in this area indicate that decollement thrust sheets also underlie the northern foothills, but die out to the north in the Coastal Plain province (Mayfield and others, 1988; Moore and others, this volume). The amount of deformation and tectonism is reflected in the rank of Nanushuk coal. Coal samples from the folded belt are generally subbituminous to bituminous, whereas those from undisturbed beds in the coastal plain are subbituminous to lignite A (Affolter and Stricker, 1987a; Sable and Stricker, 1987).

Paleozoic (Endicott Group) coal deposits

Paleozoic coal in the Kekiktuk Conglomerate occurs in wells on the Barrow arch east of Point Barrow (Sable and Stricker, 1987). Low-volatile bituminous coal beds, as thick as 3.3 m (11 ft), are exposed at several localities in highly folded and faulted Mississippian rocks beneath strata of the Lisburne Group in the Lisburne Hills between Cape Lisburne and Cape Thompson (Collier, 1906; Tailleur, 1965; Conwell and Triplehorn, 1976).

Endicott Group—the Kekiktuk Conglomerate. Coal beds of the Lower Mississippian Kekiktuk Conglomerate crop out 285 km west of the National Petroleum Reserve in Alaska (NPRA) in the Cape Lisburne region (location [loc.] 1, Fig. 1; Collier, 1906; Tailleur, 1965; Barnes, 1967b) and in the eastern Brooks Range (Sable and Stricker, 1987), and were penetrated in deep test wells (Husky Oil NPR Operations, Inc., 1982–1983). These Ellesmerian deposits, which have been described as anthracite to lower-rank coals, are as much as 1.5 m thick in wells south and southeast of Barrow (Sable and Stricker, 1987). They range in depth from 2,190 to 6,060 m and are of unknown lateral extent. Presumably coeval coal of Mississippian age that crops out at Cape Lisburne is low-volatile bituminous and low in sulfur (Tailleur, 1965). Tailleur reported 13 beds more than 0.7 m thick totaling 21 m in 366 m of section exposed in a seacliff south of Kapaloak Creek. Conwell and Triplehorn (1976) reported one bed 1.8 m thick of semianthracite on the Kukpuk River.

According to Moore and others (1984), this coal-bearing section, which they informally called the Kapaloak sequence, consists of nonmarine stream and levee sedimentary rocks as much as 70 m thick. The coal beds are associated with carbonaceous shale layers having detrital textures at the boundary between the marine and nonmarine units, interpreted to be interdistributary-bay deposits. The sequence is of Early Mississippian age. Moore and others (1984) regarded the Kapaloak sequence as most closely related to the autochthonous Endicott Group of the central Brooks Range and considered that the sequence can be regarded as a distal interfingering of elements of the fluviatile Kekiktuk Conglomerate with elements of the Kayak Shale.

Both the Kapaloak sequence of the Cape Lisburne area and the subsurface occurrences of coal in the Barrow arch appear to lie on the margins (possibly close to the shorelines) of a huge delta plain, defined by the Endicott Group of the Brooks Range, that was apparently fed from the Canadian shield before the counterclockwise rotation of Arctic Alaska away from the Canadian craton during the opening of the western Arctic Ocean (Canada basin) in Mesozoic time (Taylor and others, 1981; Mayfield and others, 1988; Grantz and others, this volume; Moore and others, this volume).

Coals of the Paleozoic Endicott Group are low in ash, with an apparent rank of low-volatile bituminous to semianthracite, and have a low moisture content (Conwell and Triplehorn, 1976) and a sulfur content of 0.5–1.1 wt% (Tailleur, 1965). No offshore trends are apparent for Mississippian coal deposits, and no assessments of offshore coal resources have been made (Affolter and Stricker, 1987b).

Cretaceous and Tertiary (Brookian) coal deposits

Fortress Mountain and Torok Formations. In the Fortress Mountain Formation, 12 beds less than 3 m thick of lignitic to subbituminous coal were described in the log of Seabee Test Well No. 1 at 3,260–3,960 m depth in the NPRA (well 34, Fig. 2; Husky Oil NPR Operations, Inc., 1982–1983). These beds occur with gilsonite-like hydrocarbons in the type area (Molenaar and others, 1987). The Torok Formation contains minor coal and carbonaceous shale beds in the southern NPRA and areas to the south (Sable and Stricker, 1987), and at 640–2,408 m depth in several NPRA wells.

Nanushuk Group. During Albian and Cenomanian time, equilibrium was apparently established between basin subsidence and sediment accumulation, resulting in very thick sequences of swamp and shallow-marine deposits that were subsequently compacted and preserved as coal of the Nanushuk Group (loc. 2, Fig. 1). These coal-bearing sequences are much thicker than the coal-bearing intervals in most basins in the conterminous Western United States (Sable and Stricker, 1987). The Cretaceous Corwin, Kukpowruk, Ninuluk, Chandler, Grandstand, and Tuktu Formations that compose the Nanushuk Group are about 3,000 m thick (Fig. 3; Smiley, 1969), and coal occurs in the middle and upper parts of the section (Callahan and Sloan, 1978). The sedimentary rocks are postorogenic molasse deposits representing offlap-delta and delta-plain environments (Huffman and others, 1988). The dominantly nonmarine coal-bearing rocks, principally within the Nanushuk Group (Martin and Callahan, 1978), are thickest and gently to moderately folded and faulted in the foothills belt (Fig. 6), but the bedding thickness and amplitude of folding apparently decrease in the subsurface to the north and east beneath the coastal plain (Bird and Andrews, 1979; Mull, 1985). Two major delta systems, the western (early Albian to Cenomanian) Corwin Delta and the eastern (middle Albian to Cenomanian) Umiat Delta of the Nanushuk Group (Ahlbrandt and others, 1979), prograded to the north and east (Fig. 5). The Corwin Delta, a large river-dominated system comparable to the Mississippi Delta in area and volume of sediment, contains numerous coal deposits as thick as 6 m (20 ft) in interdistributary-bay platforms and splay deposits within the middle (transitional) and upper delta plain (Roehler and Stricker, 1979). The Umiat Delta, which began as a high-constructional system higher in the section, had a lesser sediment volume and, presumably, a smaller source area than the Corwin Delta; it was progressively influenced by marine conditions (Huffman and others, 1985). Coal deposits become more sparse, thin, and discontinuous upward in the Umiat Delta section (Stricker, 1983). It is probable that most northern Alaska coal beds are lenticular and irregular, because they accumulated in interdistributary basins, infilled bays, or inland flood basins; some widespread tabular beds may have formed on broad, subsiding delta lobes. Nanushuk Group rocks exposed at Corwin Bluff include coal beds 1.7–2.7 m thick. At Cape Beaufort, these rocks contain coal beds 3.4 and 5.2 m thick, and on the Kukpowruk River, a coal bed 6.1 m thick has been described (Sanders, 1981).

Approximately 150 coal beds have been described in Nanushuk Group rocks; the coal is generally subbituminous A under the Arctic coastal plain and high-volatile bituminous in the folded foothills, is low in ash (less than 10 vol%) and sulfur (0.1–1.4 wt%) (Sanders, 1981; Affolter and Stricker, 1987a), and has environmentally safe contents of such elements as As, Be, Hg, Mo, Sb, and Se (Affolter and Stricker, 1987a).

Colville Group. The Upper Cretaceous Colville Group (Fig. 3) and the Tertiary rocks on the North Slope are of less economic potential than older units for oil and gas and coal deposits (loc. 3, Fig. 1), and these rocks have not been as well investigated as the older units. The coal beds in the Colville Group and in the Cretaceous and Tertiary Sagavanirktok Forma-

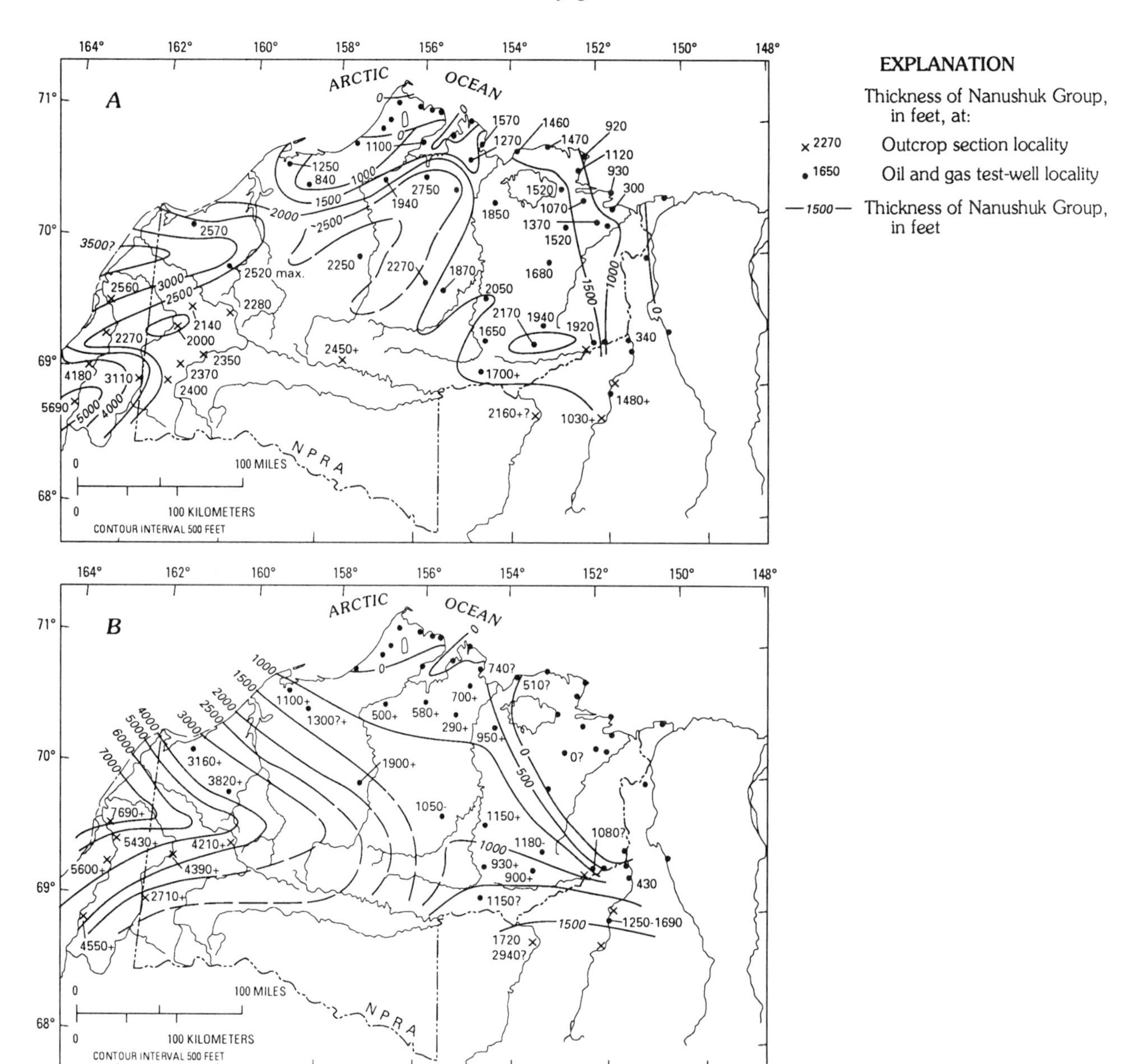

Figure 6. Map showing generalized thickness of Nanushuk Group rocks in and adjacent to National Petroleum Reserve Alaska (NPRA). A: Marine sequences including the Kukpowruk, Tuktu, and Grandstand Formations; contour interval 500 ft. B: nonmarine sequences including the Corwin Formation and the Killik Tongue of the Chandler Formation; contour interval 500 ft (152 m) (from Sable and Stricker, 1987).

tion are generally thinner and lower in rank than the Nanushuk Group coals.

Three formations make up the Colville Group, which is more than 1,525 m thick: the non-coal-bearing marine Seabee, the coal-bearing nonmarine Prince Creek, and the non-coal-bearing marine Schrader Bluff Formations. The sandstone, conglomerate, bentonitic shale, coal, bentonite, and tuff of the Prince Creek Formation probably represent delta-plain fluvial and delta-front foreshore environments of deposition (Fox and others, 1979). The uppermost nonmarine units of the Prince Creek Formation interfinger northward and eastward with marine strata of the 825-m-thick Schrader Bluff Formation, which also contains minor coal (Sable and Stricker, 1987).

In the vicinity of Umiat and Maybe Creek in the southeastern NPRA, coal beds as thick as 4 m occur, and some coaly zones are as much as 12 m thick (Brosgé and Whittington, 1966). On the lower Colville River, nine beds of coal and coaly beds as much as 4 m thick are exposed, and a 15-m-thick zone of interbedded shale and coal has been reported (Brosgé and Whittington, 1966). Subsurface coaly intervals are typically less than 15 m thick, but one 6-m-thick zone of coal interbedded with black shale and bentonite has been described from the Square Lake

Test Well No. 1 core (Collins, 1959). The coaly deposits in these units probably formed in lower delta-plain, alluvial-plain, and upper delta-plain environments (Sable and Stricker, 1987).

Sagavanirktok Formation. The Upper Cretaceous to Pliocene Sagavanirktok Formation underlies 19,400 km^2 in the eastern part of the North Slope in the Camden basin and under the Beaufort Sea shelf (Molenaar and others, 1987; Sable and Stricker, 1987). The coals accumulated in interdistributary bays and on platforms constructed by overbank splays. The Sagavanirktok consists of clastic deltaic rocks containing minor carbonaceous shale, coal, lignite, and bentonite, and sedimentary structures and facies indicate a continuation in Tertiary time of east-northeastward progradation. The formation has a 460-m-thick coal-bearing deltaic unit at its base (Roberts and others, 1991). Near Prudhoe Bay, a coal-bearing interval as thick as 400 m contains coal beds 0.6–6.7 m thick (Roberts, 1991); one 2-m-thick coal zone has been reported on the lower Shaviovik River (Roberts, 1991). Lignite and coaly shale as thick as 6 m occur in the lowermost part of the formation (Detterman and others, 1975; Molenaar and others, 1984). The coals accumulated in interdistributary bays on platforms constructed by overbank splays and alluvial-plain environments.

Coals of the Upper Cretaceous and Tertiary Colville Group and Sagavanirktok Formation are subbituminous C and low in sulfur, with a mean of 0.4 wt% (0.08–2.2 wt%) (Roberts and others, 1991).

Coal assessment of the North Slope Province

Barnes (1967a) calculated a total of 2.2×10^9 metric tons of demonstrated and 107×10^9 metric tons of undiscovered coal resources on the North Slope. Tailleur and Brosgé (1976) estimated the coal resources in northern Alaska by calculating the product of coal-bearing area and coal concentration. Using surface data and two oil and gas test wells, they estimated the coal resources on the North Slope at 109×10^9 metric tons of identified plus 104×10^9 to 34×10^{12} metric tons of hypothetical resources. A calculation based on the methodology of Sable and Stricker (1987) indicates that the Nanushuk Group contains an estimated 2.9×10^{12} metric tons of hypothetical coal resources on the surface and in the subsurface of onshore northern Alaska (Table 2); of this total, 1.2×10^{12} metric tons is subbituminous, and 1.7×10^{12} metric tons is bituminous (Stricker, 1991). In situ speculative Cretaceous Nanushuk coal under the Chukchi Sea has been estimated at 2.0×10^{12} tons of lignite A to high-volatile bituminous A (Affolter and Stricker, 1987b). Tertiary Sagavanirktok lignite coal beneath the Beaufort Sea is estimated to total 300×10^9 metric tons (Affolter and Stricker, 1987b).

A realistic assessment of the resources contained in the northern Alaska coal fields is probably not possible with the relatively meager geologic information available. The estimate by Martin and Callahan (1978) of 44×10^9 metric tons of demonstrated coal resources in an area of 65,000 km^2 of coal-bearing rocks illustrates the paucity of coal-data density. In comparison with the estimate of 2.9×10^{12} metric tons of hypothetical coal resources in this report, the estimate of identified coal resources does not give a true indication of the resource potential of the northern Alaska coal fields.

TABLE 2. COAL RESERVES FOR THE NANUSHUK GROUP, NORTH SLOPE PROVINCE

Rank	Attitude	Overburden (ft)	Resource estimate (tons)*
Subbituminous	Dips generally 15° or less	0 to 500	1,041
		500 to 1,000	18
		1,000 to 2,000	9
		>2,000	1
		Total	**1,068**
	Dips generally 15° or more	0 to 500	91
		500 to 1,000	5
		1,000 to 2,000	5
		>2,000	1
		Total	**102**
	Subbitimunious	**Total (Rounded)**	**1,170**
Bituminous	Dips generally 15° or less	0 to 500	1,216
		500 to 1,000	0
		Total	**1,216**
	Dips generally 15° or more	0 to 500	518
		500 to 1,000	0
		Total	**518**
	Bituminous	**Total (Rounded)**	**1,734**
North Slope		**Total (Rounded)**	**2,904**

*Reported in billions of metric tons.

YUKON-KOYUKUK PROVINCE

On the south side of the western Brooks Range is the Yukon-Koyukuk province, a large triangular area underlain by Cretaceous rocks containing some coal but with very little potential for coal resources. The Yukon-Koyukuk province was once an oceanic basin with an island arc that collided with and overrode terranes to the north, southeast, and west. The abrupt, almost catastrophic filling of this basin to sea level and above during mid-Cretaceous time gave little opportunity for long-lived paralic swamps in which the precursors of coal would accumulate, such as on the North Slope, and the intense episodes of deformation that followed accumulation greatly disturbed the small amount of coal that did form. In keeping with its age and deformation, the bulk of this coal is bituminous, and some is coking coal—a combustible, gray, hard, porous and coherent coal type, produced from bituminous coal that has undergone metamorphism in the absence of air (Wood and others, 1983). The geology of this region has been described by Patton (1973) and Patton and others (this volume, Chapter 7), and the paragraphs that follow are summarized from that work.

Geologic observations on the coal at Nulato (loc. 4, Fig. 1) were made in 1865 by W. H. Dall (Dall and Harris, 1892), who assigned it to the Neogene. Several small mines on the west bank of the Yukon River between the mouths of the Melozitna and Anvik Rivers provided coal for river steamers and blacksmithing during the gold rushes to the Klondike and other regions of Alaska at the turn of the twentieth century (Collier, 1906). Collier (1906) described the coal beds and mining operations; Martin (1926) wrote detailed lithologic descriptions of the sections containing the coals; and Patton and Bickel (1956) published detailed structure sections of the bluffs along this stretch of the Yukon River.

Geologic units and setting

Outcrops in the Yukon-Koyukuk province include a central section of Lower Cretaceous (Neocomian) andesitic, volcaniclastic, and associated plutonic rocks, extending eastward from the Seward Peninsula, through the Selawik drainage, into the adjacent Koyukuk drainage. These rocks represent the remains of a volcanic island arc built, presumably, on oceanic crust; they crop out in a belt that is convex to the north and east. Surrounding this central volcanic terrane are thick sedimentary-basin accumulations—to the north along the Kobuk drainage and to the east, southeast, and south along the Koyukuk and Yukon Rivers and into the Kuskokwim drainage. The sedimentary fill of these basins was derived partly from the volcanic island arc, but mostly from the surrounding highland provinces that include the Brooks Range to the north and the crystalline rocks of the Kokrines-Hodzana Highlands and Kaiyuh Mountains to the southeast. The crystalline highlands bordering the Yukon-Koyukuk province adjacent to its boundary contain overthrust sheets composed of an ophiolite sequence with Jurassic radiometric ages; these sheets are presumably parts of the original oceanic crust beneath the Yukon-Koyukuk province that were obducted onto the surrounding highlands. Thrusting of the ophiolite onto the highlands, between Neocomian and Albian time, provided much of the source for the Albian to Santonian basin fill.

The lower part of this basin fill is a turbidite sequence of mid-Cretaceous (Albian and Cenomanian) age, and the upper part, especially around the margins of the province, is a nonmarine to shallow-marine sequence of Late Cretaceous (Cenomanian to Santonian) age. The Cretaceous coal beds occur in the upper, continental part of the basin fill.

Cretaceous coal

At several localities, thin coal beds occur within continental and interbedded continental–shallow-marine sequences of mid-Cretaceous age. Barnes (1967a, p. B20) summarized the principal occurrences and thicknesses as follows.

"Twenty miles above Galena, one foot; ten miles above Nulato (Pickart mine), one and one-half to three feet; one mile above Nulato, six inches; four miles below Nulato (Bush mine), probably less than two feet; nine miles below Nulato (Blatchford mine), thickness unknown, sheared pockets eight feet across; 50 miles below Kaltag (Williams mine) [loc. 5, Fig. 1], three and one-half feet; and 16 miles above Blackburn (Coal Mine No. 1) [loc. 6, Fig. 1] two and one-half to three feet. The only localities where conditions are favorable for appreciable resources are the Williams and No. 1 mines, from each of which several hundred tons of coal was mined prior to 1903."

Additional occurrences of coal are (1) at Tramway Bar (loc. 7, Fig. 1) on the Middle Fork of the Koyukuk River, about 40 km (25 mi) downstream from Wiseman, in the extreme northeast corner of the Yukon-Koyukuk province, where Schrader (1900, p. 477, 485, 1904, p. 107, 114) reported a bed 3.5 m thick, the middle 3 m of relatively pure, high-volatile bituminous coal; (2) on the Anvik River, about 160 km upstream from its confluence with the Yukon River (loc. 8, Fig. 1), where Harrington (1918, p. 65) reported information from a local prospector that several beds as much as 3 m thick and other beds about 0.6 m thick are exposed along an 8-km-long stretch of the river; (3) on the Kobuk River above Kiana (loc. 9, Fig. 1), where small coal remnants occur as float (Patton and Miller, 1968); and (4) adjacent to the town of Koyuk on the beaches of Norton Sound (loc. 10, Fig. 1), where coal of Albian-Cenomanian age was mined until about 1970. This last deposit is now buried by beach gravel. The Tramway Bar locality was visited by members of the Alaska Division of Geological and Geophysical Surveys in 1985, who confirmed the presence of one coal bed at least 3.7 m thick and of two additional coal beds at least 1.2 m thick in a section of siltstone, sandstone, and quartz-rich conglomerate at least 90 m thick (Katherine Goff, 1985, written commun.; John Murphy, 1986, oral commun.). The Anvik River locality has not been visited by any geologist and is not shown on the reconnaissance geologic map of the Unalakleet quadrangle by Patton and Moll (1985).

Coal assessment of the Yukon-Koyukuk province

The overall poor coal prospects of the Yukon-Koyukuk province are exemplified in the measured section shown by Martin (1926, p. 404–406) at the Pickart mine, where 355 m of section included no more than 1.5 m of coal and only one bed as much as 0.6 m thick, giving a coal content of less than 0.5 vol%. Deposits in this province are of economic interest primarily for local use.

TERTIARY COAL OF THE SEWARD PENINSULA

Numerous small coal deposits occur on the Seward Peninsula (Fig. 1; Barnes, 1967a). Of these deposits, only one is of any significance—the Chicago Creek deposit of Tertiary age on the Kugruk River, about 40 km west of Candle (loc. 11, Fig. 1) (Henshaw, 1909; Toenges and Jolley, 1947). Two lignite beds have been mined: one is about 5.5 m thick, and the other at least 30 m thick, and both are known to extend at least 21 m below the surface (Barnes, 1967a). This lignite may be developed for use in Kotzebue (Sanders, 1981).

PALEOZOIC(?) COAL ON THE NATION RIVER

A bed of deformed and partially crushed coal, in pods as much as 2.4 m thick and 4 m long, was mined in 1897 and 1898 (Collier, 1906) from the south bank of the Nation River, about 1.5 km from its confluence with the Yukon River (loc. 12, Fig. 1). At the time of Collier's visit, the mine had already caved, and no exposures of coal were visible. The reported occurrence is in rocks of the Nation River Formation of latest Devonian age (Brabb and Churkin, 1967), a deep-sea turbidite fan deposit (Nilsen and others, 1976). The coal reported at the site by Collier (1903) may have occurred in a sliver of the coal-bearing continental rocks of Late Cretaceous to Pliocene age that were mapped by Brabb and Churkin (1969) in the Charley River south of the Yukon River. Subsequent mapping of the surrounding area disclosed no other coal exposures, and the original exposure has never again been found.

MINOR OCCURRENCES OF TERTIARY COAL IN THE YUKON BASIN

Coal of subbituminous or lignite rank occurs in association with rocks of apparent Tertiary age at the following localities (Barnes, 1967a): Eagle-Circle district (loc. 12a, Fig. 1), Chicken district (loc. 12b, Fig. 1), Rampart district (loc. 12c, Fig. 1), and Nunivak and Nelson Islands (loc. 12d, Fig. 1). Except for the Chicken district, which is reported to contain a coal bed at least 7 m thick, none of the coal reported in these occurrences is of commercial thickness or quality. Even in the Chicken district, where the bed was penetrated in a tunnel at the base of a 11-m-long shaft (Mertie, 1930) there are inconsistencies in the description, and the shaft is caved and inaccessible (Barnes, 1967a).

SOUTHERN ALASKA–COOK INLET COAL PROVINCE

Geologic setting

Many isolated occurrences of coal-bearing rocks of Tertiary age in southern and central Alaska appear to have been parts of a large river system that emptied into the Pacific Ocean through Cook Inlet (Fig. 1). The major coal resources of southern Alaska, beneath the Cook Inlet–Susitna Lowland and in the Matanuska Valley, appear to have accumulated in a basin of deep subsidence where this river system joined the Pacific (Kirschner, 1988). The two major forks of the trunk stream, the alluvial valley of which is now occupied by Cook Inlet, extended northward through the area now occupied by the Susitna Lowland and Broad Pass coal fields and eastward through the valley that contains the Matanuska coal field (Fig. 7). Three important coal fields along the north side of the Alaska Range—the Nenana, Jarvis Creek, and Tonzona (Farewell)—appear to have been places where tributaries of this river system flowed southward across areas of tectonic subsidence. The Yukon-Tanana Upland may have been in its headwaters. All the coal occurrences thought to have once been in this integrated drainage system are here considered parts of the Southern Alaska–Cook Inlet province (see Fig. 1). The Cook Inlet depression, the main basin of subsidence in this coal province, is a fore-arc basin that lies on the site of a middle Mesozoic open shelf between the volcanic arc represented by the Lower Jurassic Talkeetna Formation and the Middle Jurassic Talkeetna batholith on the north and an ancient Pacific oceanic crust at the site of the Kenai and Chugach Mountains on the south. A thick, mainly terrigenous, epiclastic sequence, ranging from Middle Jurassic to Late Cretaceous in age, accumulated on this shelf and lies, relatively undeformed, unconformably beneath the coal-bearing Tertiary deposits (Kirschner and Lyon, 1973; Fisher and Magoon, 1978; Magoon, this volume). The McHugh Complex and the Valdez Group, oceanic crust and deep-sea turbidite sequences, were accreted to southern Alaska during Late Cretaceous time to form the Chugach and Kenai Mountains, thus widening the arc-trench gap, which is now about 450 km wide, owing to accretion of Cenozoic oceanic terranes. The Cook Inlet basin lies in the northwesternmost part of this arc-trench gap (Fisher and Magoon, 1978). Irregular subsidence of some parts of the fore-arc basin began in latest Cretaceous time but continued sporadically throughout Cenozoic time, interrupted by mild uplift and erosion. Subsidence was greatest during Neogene time in a 250-km-long segment of Cook Inlet that became the trunk drainage system for much of central and southern Alaska and received sufficient sediment to remain mostly terrestrial during all of this period.

Paleocene precursors

The bulk of the coal in the Southern Alaska–Cook Inlet province is of Oligocene to early Pliocene age and is mostly subbituminous and lignite. However, during Paleocene and possibly early Eocene time, before the integration of this drainage system, coal-bearing continental sediment derived in large part from adjacent mountains accumulated in two narrow, approximately east-west–trending troughs. The coals in these Paleogene fields are of bituminous or higher rank.

Cantwell Formation. At the site of the present central Alaska Range, between Mount McKinley on the west and Mount Hayes on the east, at least 3,000 m of conglomerate, sandstone, shale, and rare thin coal beds accumulated to form the Paleocene Cantwell Formation (Wolfe and Wahrhaftig, 1970). An attempt was made to mine coal in the Cantwell Formation at mile 341 on the Alaska Railroad, but owing to structural complexity (the bed was displaced by faulting), the project was abandoned (Capps,

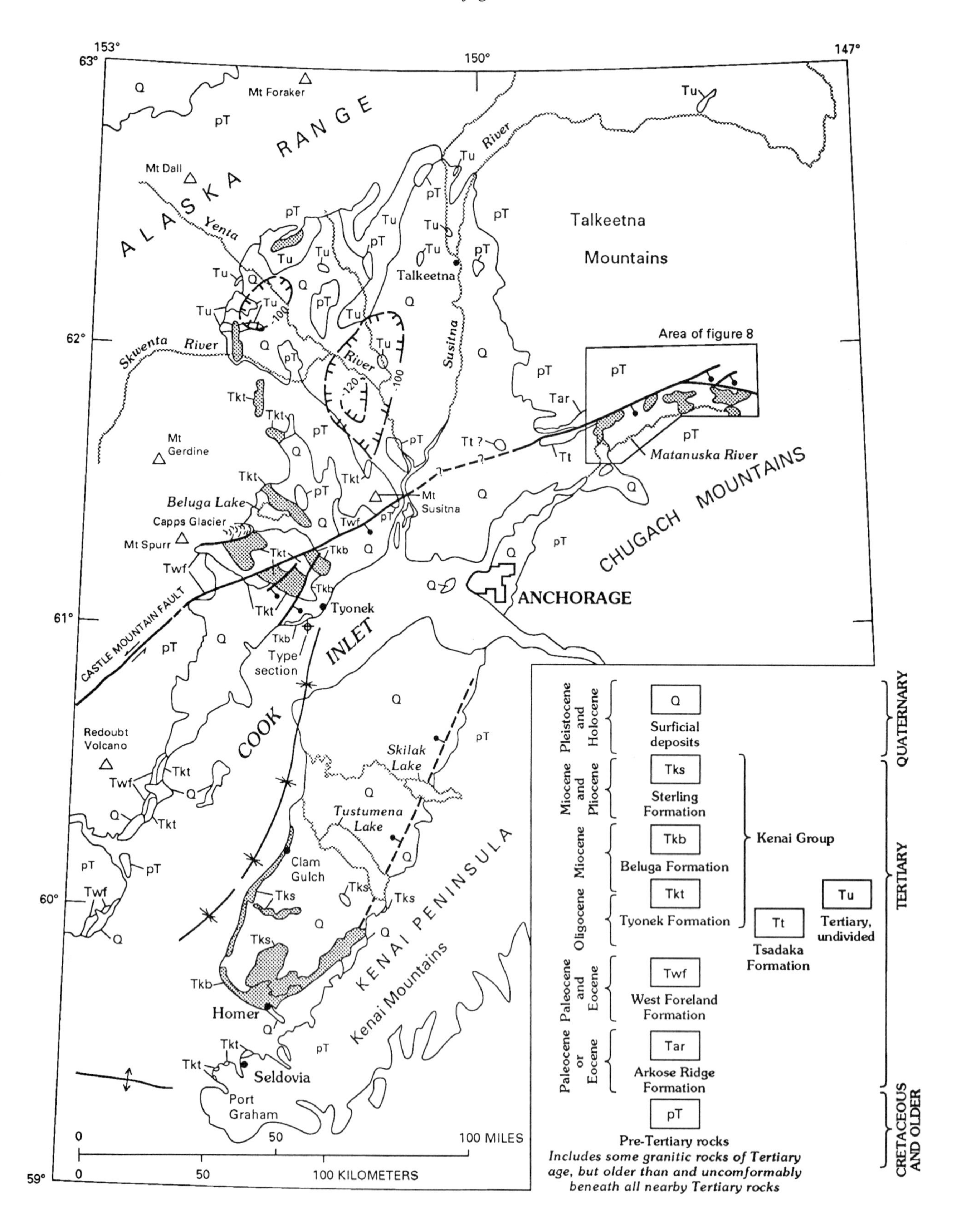

153°
150°
147°
63°
62°
61°
60°
59°
ALASKA RANGE
Mt Foraker
Mt Dall
Yenta
Skwenta River
Talkeetna
Talkeetna Mountains
Susitna
River
Area of figure 8
Matanuska River
CHUGACH MOUNTAINS
Mt Gerdine
Beluga Lake
Capps Glacier
Mt Spurr
Mt Susitna
ANCHORAGE
CASTLE MOUNTAIN FAULT
Tyonek
Type section
COOK INLET
Redoubt Volcano
Skilak Lake
Tustumena Lake
Clam Gulch
KENAI PENINSULA
Kenai Mountains
Homer
Seldovia
Port Graham
0 50 100 MILES
0 50 100 KILOMETERS
Pleistocene and Holocene
Q
Surficial deposits
QUATERNARY
Miocene and Pliocene
Tks
Sterling Formation
Miocene
Tkb
Beluga Formation
Kenai Group
Oligocene
Tkt
Tyonek Formation
Tt
Tsadaka Formation
Tu
Tertiary, undivided
TERTIARY
Paleocene and Eocene
Twf
West Foreland Formation
Paleocene or Eocene
Tar
Arkose Ridge Formation
pT
Pre-Tertiary rocks
Includes some granitic rocks of Tertiary age, but older than and uncomformably beneath all nearby Tertiary rocks
CRETACEOUS AND OLDER

EXPLANATION FOR FIGURE 7

Contact

Fault with vertical displacement—Ball and bar on downthrow side; dashed where uncertain or concealed

Near-vertical fault with horizontal displacement—Arrows indicate relative movement

Axis of tectonic arch or anticline

Axis of major basinal downwarp

-120 Isostatic gravity anomaly (in milligals) around major gravity lows in Susitna basin

Coal field and other area of significant coal outcrops

Figure 7. Generalized geologic map of the Cenozoic rocks in the Cook Inlet–Susitna lowland, showing coal fields and other areas of significant coal exposures (compiled from F. Barnes, 1966; D. Barnes, 1977; Barnes and Cobb, 1959; Barnes and Payne, 1956; Barnes and Sokol, 1959; Beikman, 1980; Capps, 1913, 1927, 1935, 1940; Detterman and others, 1976; Magoon and others, 1976; Reed and Nelson, 1980). For geology of inset, see Figure 9.

1940, p. 114). No study has been made of the coal, and reserves, if any, are unknown.

Matanuska Coal Field. The most important Paleocene coal field in Alaska is the Matanuska coal field (loc. 13, Fig. 1), which occupies the graben of the Matanuska Valley, between the Talkeetna Mountains on the north and the Chugach Mountains on the south (Figs. 7, 8). The part of the valley that contains coal is about 100 km long, from Moose Creek on the west to Anthracite Ridge on the east. The coal is found in the Chickaloon Formation, a 1,000–1,500-m-thick, Paleocene and lower Eocene sequence of claystone, siltstone, and sandstone, including minor conglomerate beds and many beds of coal (Triplehorn and others, 1984). The Chickaloon Formation rests with angular unconformity on the Lower and Upper Cretaceous Matanuska Formation, a sequence of marine sandstone and shale (Barnes and Payne, 1956; Grantz and Jones, 1960). The Chickaloon Formation is overlain by the Wishbone Formation (Fig. 9), a massive conglomerate about 900 m thick that contains clasts characteristic of the Talkeetna Mountains to the north. The Wishbone Formation, in the Talkeetna Mountains north of the east end of the coal field, is apparently unconformably overlain by nearly flat-lying Ter-

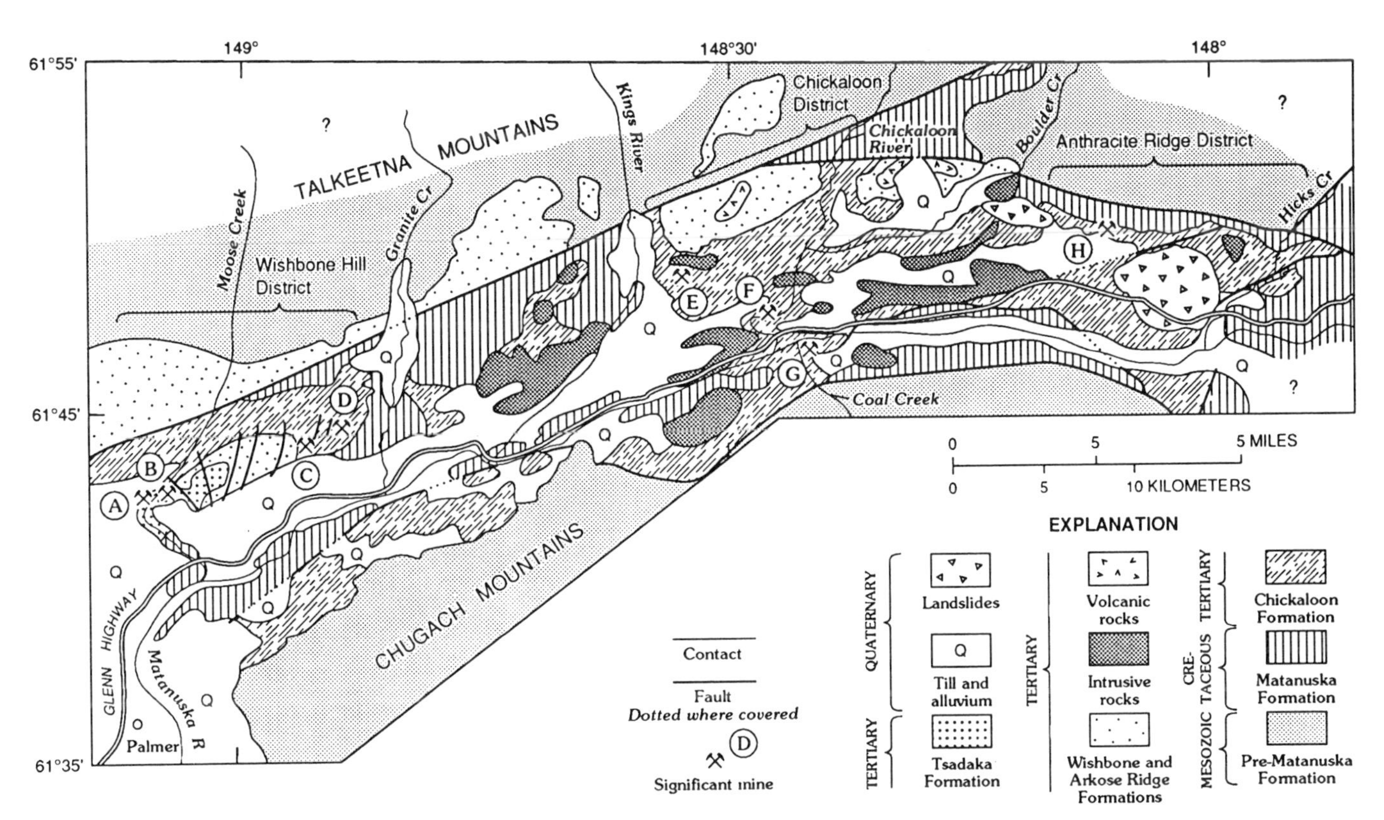

Figure 8. Simplified geologic map of the Matanuska Valley, showing location of coal districts and significant mines. A, Premier mine; B, Baxter mine; C, Evan Jones mine; D, Eska mine; E, Castle Mountain mine; F, Chickaloon mine; G, Coal Creek mine; H, Anthracite Ridge anthracite locality (compiled from Magoon and others, 1976; Detterman and others, 1976; Waring, 1936, plate 1). Blank areas, with question marks, geology not shown.

tiary basalt. Gabbro sills and dikes, some large and possibly related to the Tertiary volcanic rocks, intrude the Chickaloon Formation, most commonly east of Granite Creek.

Coal-bearing areas within the Matanuska Valley include (1) the Wishbone Hill district, an open faulted syncline between Moose and Granite Creeks at the west end of the valley, where there has been the most production (Barnes and Payne, 1956); (2) an area around Young Creek, just west of Kings River (Fig. 8) (Martin and Katz, 1912); (3) the Castle Mountain and Chickaloon districts, both on the north side of the Matanuska River and between the Kings and Chickaloon Rivers (Fig. 8) (Merritt and Belowich, 1984, plate 2); and (4) the Anthracite Ridge district, at the east end of the outcrop area of the Chickaloon Formation (Fig. 8; Waring, 1936; Merritt and Belowich, 1984). The coal in the Wishbone Hill district and around Young Creek is high-volatile bituminous; coal in the Castle Mountain and Chickaloon districts is predominantly low-volatile bituminous; and coal in the Anthracite Ridge district is mainly low-volatile bituminous, with patches of semianthracite and anthracite (Merritt and Belowich, 1984; Waring, 1936; Barnes, 1962).

The Matanuska coal field is important in the history of the settlement of Alaska. In 1914, the Alaska Railroad from the coast to the interior was routed to pass near the Matanuska Valley, owing to the availability of steaming coal from the coal field; until the early 1950s, when the railroad converted to oil, all coal for the locomotives came from the Wishbone Hill district.

The coal-bearing areas were divided into leasing units under the Federal Coal Leasing Act of 1915, and in 1917 the first mining began at the west end of the Wishbone Hill district. Because private coal mines were proving to be an uncertain source of coal for the Alaska Railroad, the government took over the Eska mine in 1917 and started developing a second coal mine, the Chickaloon, on the Chickaloon River. Although the Chickaloon mine was never a major producer, the Eska mine was kept on a standby basis throughout the first half of the twentieth century. Nine mines operated at one time or another in the Wishbone Hill district between 1917 and 1970, and three or four mines operated at one time or another in the Chickaloon and

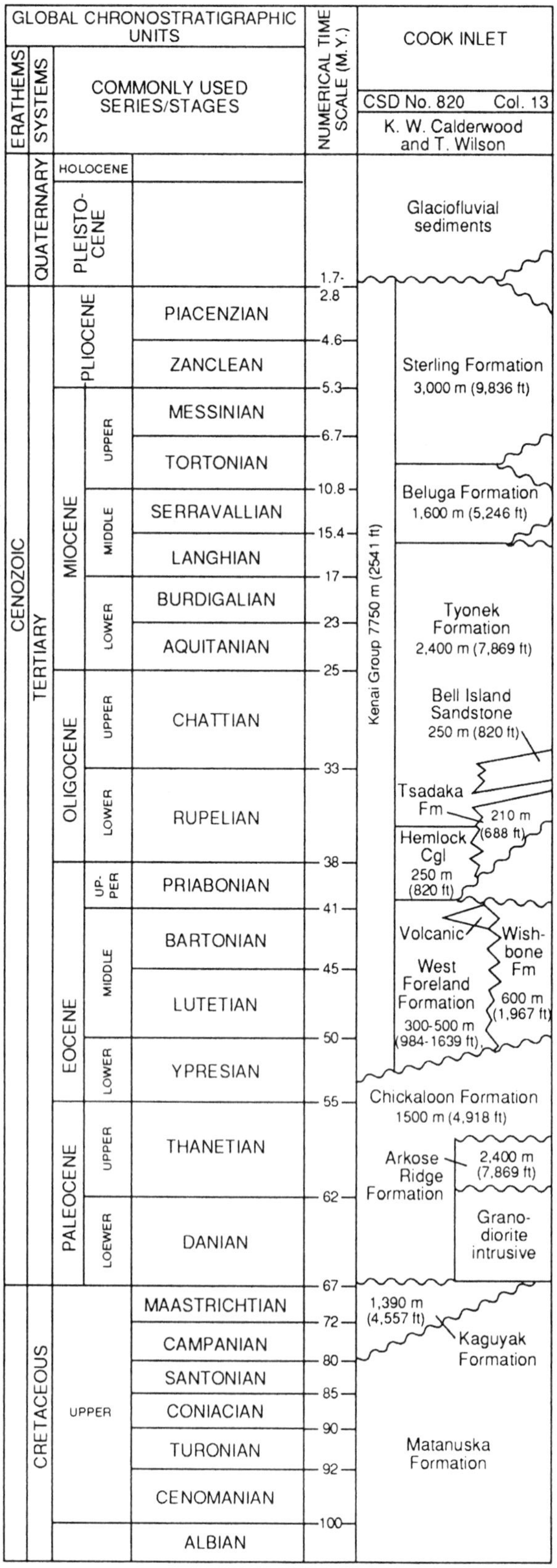

ERATHEMS	SYSTEMS	COMMONLY USED SERIES/STAGES			NUMERICAL TIME SCALE (M.Y.)	COOK INLET — CSD No. 820 Col. 13 — K. W. Calderwood and T. Wilson
	QUATERNARY	HOLOCENE				Glaciofluvial sediments
		PLEISTOCENE			1.7-2.8	
CENOZOIC	TERTIARY	PLIOCENE		PIACENZIAN	4.6	
				ZANCLEAN	5.3	Sterling Formation 3,000 m (9,836 ft)
		MIOCENE	UPPER	MESSINIAN	6.7	
				TORTONIAN	10.8	
			MIDDLE	SERRAVALLIAN	15.4	Beluga Formation 1,600 m (5,246 ft)
				LANGHIAN	17	
			LOWER	BURDIGALIAN	23	Tyonek Formation 2,400 m (7,869 ft)
				AQUITANIAN	25	
		OLIGOCENE	UPPER	CHATTIAN	33	Bell Island Sandstone 250 m (820 ft)
			LOWER	RUPELIAN	38	Tsadaka Fm 210 m (688 ft); Hemlock Cgl 250 m (820 ft)
		EOCENE	UPPER	PRIABONIAN	41	
			MIDDLE	BARTONIAN	45	Volcanic; West Foreland Formation 300-500 m (984-1639 ft); Wishbone Fm 600 m (1,967 ft)
				LUTETIAN	50	
			LOWER	YPRESIAN	55	Chickaloon Formation 1500 m (4,918 ft)
		PALEOCENE	UPPER	THANETIAN	62	Arkose Ridge Formation 2,400 m (7,869 ft)
			LOEWER	DANIAN	67	Grano-diorite intrusive
	CRETACEOUS	UPPER		MAASTRICHTIAN	72	1,390 m (4,557 ft) Kaguyak Formation
				CAMPANIAN	80	
				SANTONIAN	85	
				CONIACIAN	90	Matanuska Formation
				TURONIAN	92	
				CENOMANIAN	100	
				ALBIAN		

Kenai Group 7750 m (2541 ft)

A

Figure 9, A and B: Two interpretations of the Tertiary stratigraphy of the Cook Inlet–Susitna region. A: According to Calderwood and Wilson (cited *in* Schaff and Gilbert, 1987); continuously conformable sedimentation from Eocene through Pliocene in the center of the Cook Inlet lowland, with the West Foreland Formation treated as the basal member of the Kenai Group. B (on facing page): According to Magoon and others (1976), a major hiatus between the West Foreland Formation and the Hemlock Conglomerate; the West Foreland Formation not being a member of the Kenai Group. Following U.S. Geological Survey practice, in this paper the West Foreland Formation is not treated as a unit of the Kenai Group. C (on following page): Type section of the Tyonek Formation, the major coal-bearing unit of the Kenai Group in the Cook Inlet lowland (from Calderwood and Fackler, 1972, Fig. 7). The type section extends from a depth of 1.310 m to 3,430 m in Pan American Petroleum Corp. Tyonek State 17587 No. 2 well, sec. 30, T. 11 N., R. 11 W., S.M. (see Fig. 7 for location).

Castle Mountain districts. From 1917 to 1940, production was about 50,000 tons per year; from 1940 to 1951, production was about 160,000 tons per year; and from 1952 to 1970, production averaged 240,000 tons per year. A total of 3×10^6 tons was mined from open pits, and the rest from underground mines. Total coal production was about 7×10^6 metric tons between 1915 and 1970, when the availability of oil eliminated the market for coal (Merritt and Belowich, 1984, Fig. 5).

Geologic units and setting. Representative columnar sections of the Chickaloon Formation are shown in Figure 10; representative structural cross sections of coal-bearing rocks of the Matanuska coal field are shown in Figure 11. Coal beds within the Chickaloon Formation vary in thickness considerably or pinch out altogether within short distances. Correlation from outcrop to outcrop is difficult, and only in the Wishbone Hill district has a successful districtwide correlation been accomplished. In this area, four groups of minable coal beds, one to six beds in each group, were separated by 15–90 m of barren rock in a section

TERTIARY CORRELATION CHART SURFACE AND SUBSURFACE

AGE (In millions of years before present)	SYSTEMS		SERIES		FLORAL STAGE	GROUP	COOK INLET Lower: Copper Lake Cape Douglas	COOK INLET Upper: East Glacier Creek Homer area	COOK INLET Upper: Chuitna River Capps Glacier	COOK INLET Upper: Matanuska Valley
3–5	TERTIARY	NEOGENE	Pliocene	Upper, Lower	Clamgulchian	Kenai		Sterling Formation		Sterling (?) Formation
5–12	TERTIARY	NEOGENE	Miocene	Upper	Homerian	Kenai		Beluga Formation	Beluga Formation	
12–22.5	TERTIARY	NEOGENE	Miocene	Middle, Lower	Seldovian	Kenai	?	Tyonek Formation	Tyonek Formation	Tyonek Formation
22.5–25	TERTIARY	PALEOGENE	Oligocene	Upper	Angoonian	Kenai	Tyonek Formation	Tyonek Formation	Tyonek Formation	Tyonek Formation
25–30	TERTIARY	PALEOGENE	Oligocene	Upper	Angoonian	Kenai	Hemlock Conglomerate	Hemlock Conglomerate		Bell Island Sandstone of local usage[1] / Tsadaka Formation[1]
30–37	TERTIARY	PALEOGENE	Oligocene	Lower	Kummerian					
37–44	TERTIARY	PALEOGENE	Eocene	Upper	Ravenian					
44–48	TERTIARY	PALEOGENE	Eocene	Middle	Fultonian					
48–53	TERTIARY	PALEOGENE	Eocene	Lower	Franklinian		West Foreland Fm / Rocks near Copper Lake	West Foreland Formation	West Foreland Formation	Wishbone Formation
53–65	TERTIARY	PALEOGENE	Paleocene	Upper, Lower	Unnamed					Arkose Ridge Formation / Chickaloon Formation

[1]Not considered part of Kenai Group

B

360–460 m thick. In all, 12 minable beds totaled about 15 m in thickness of minable coal; the thickest bed had about 3.3 m of coal in a total thickness of 5 m (Barnes and Payne, 1956).

Six to ten coal beds were penetrated in the Chickaloon mine, most less than 1 m thick, but one more than 5 m thick, with 4.3 m of coal. The beds are lenticular and vary in thickness within 60–90 m laterally, making correlation, reserve calculations, and prospecting across transverse faults difficult.

The number of coal beds in the Anthracite Ridge district is uncertain, owing to poor exposures and complex structure (Fig. 11). The greatest thickness of coal reported in a single bed is 12 m, but the coal may be repeated by landslides; few beds in the

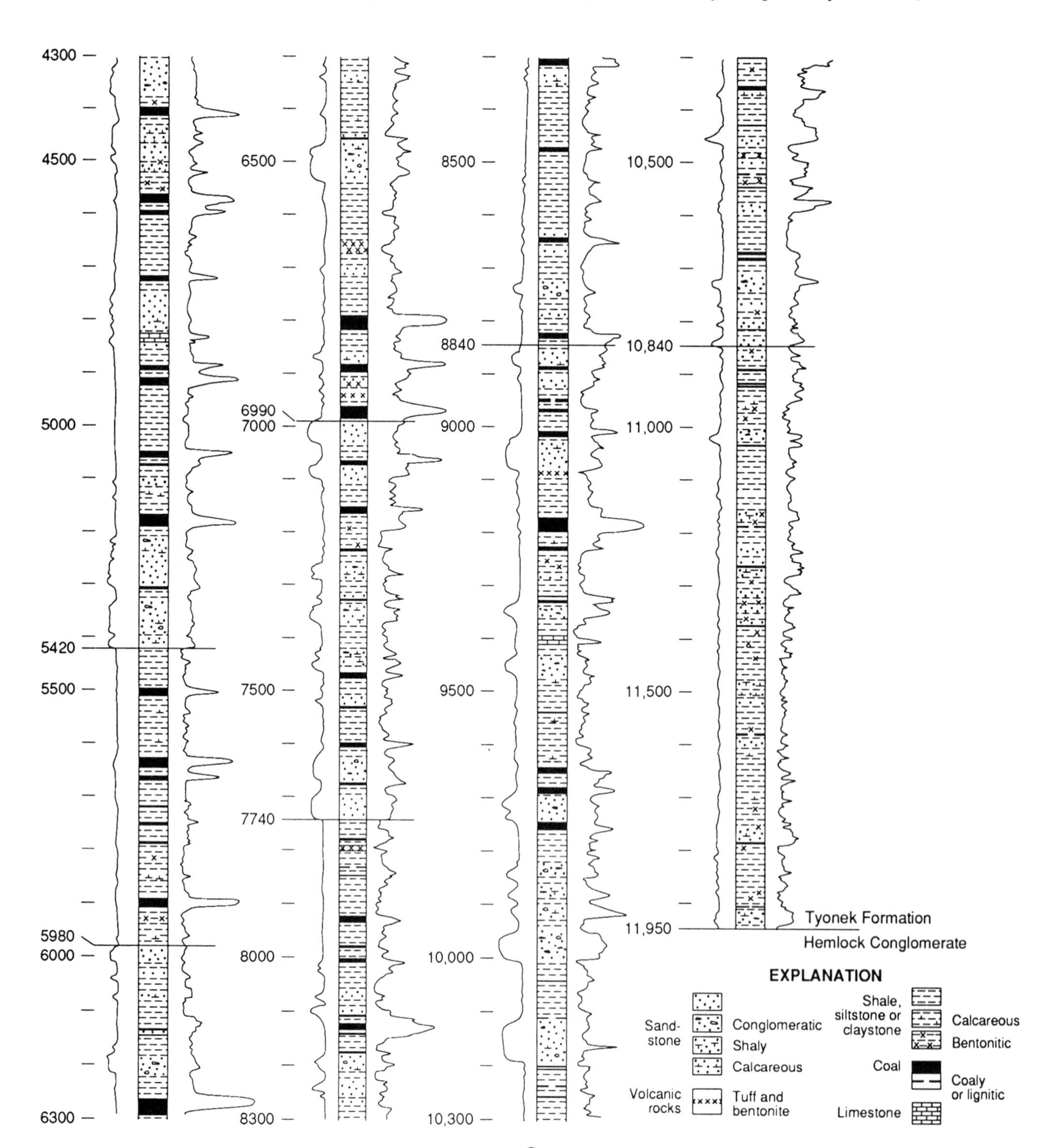

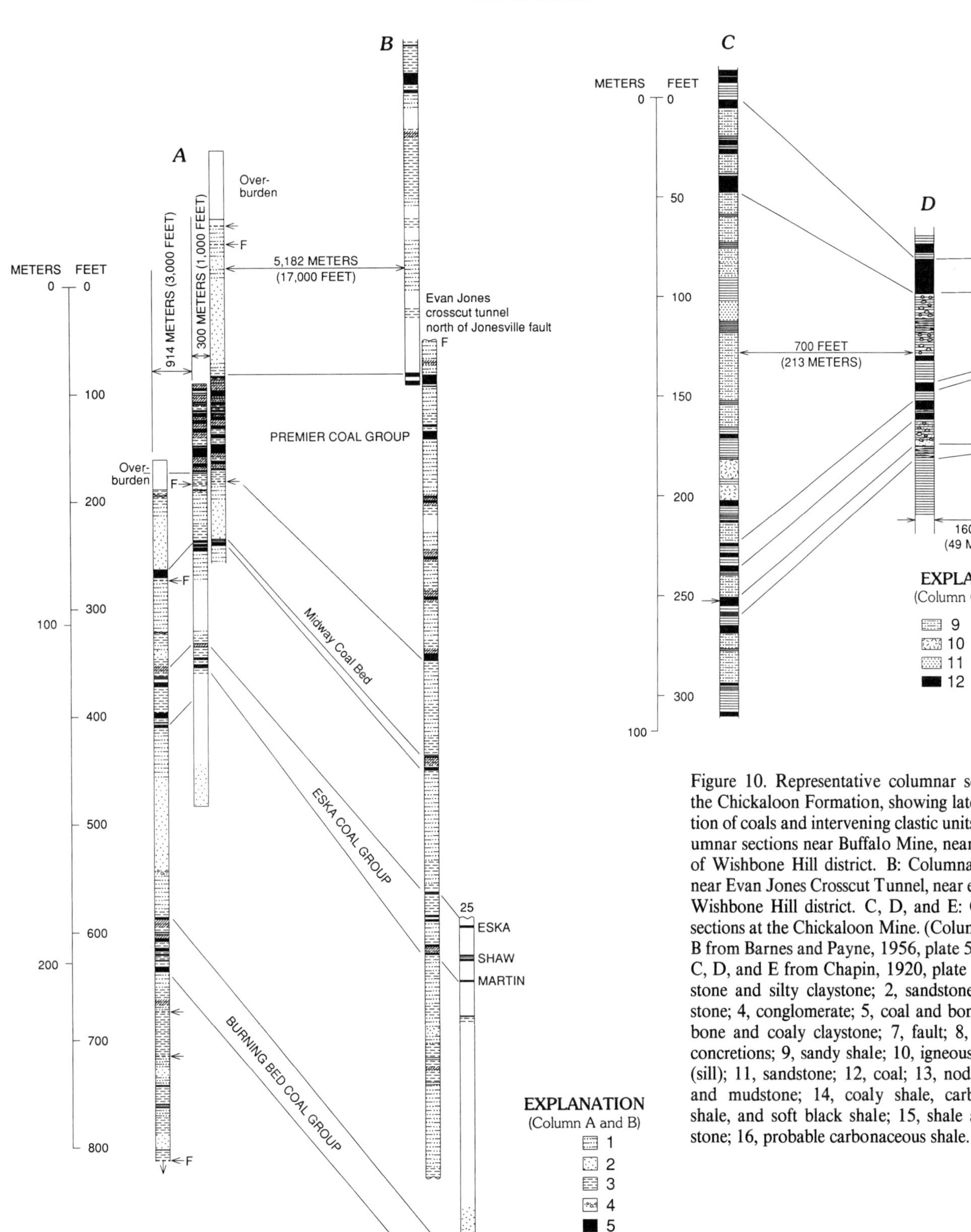

Figure 10. Representative columnar sections of the Chickaloon Formation, showing lateral variation of coals and intervening clastic units. A: Columnar sections near Buffalo Mine, near west end of Wishbone Hill district. B: Columnar sections near Evan Jones Crosscut Tunnel, near east end of Wishbone Hill district. C, D, and E: Columnar sections at the Chickaloon Mine. (Columns A and B from Barnes and Payne, 1956, plate 5; columns C, D, and E from Chapin, 1920, plate 5.) 1, siltstone and silty claystone; 2, sandstone; 3, claystone; 4, conglomerate; 5, coal and bony coal; 6, bone and coaly claystone; 7, fault; 8, ironstone concretions; 9, sandy shale; 10, igneous intrusion (sill); 11, sandstone; 12, coal; 13, nodular shale and mudstone; 14, coaly shale, carbonaceous shale, and soft black shale; 15, shale and mudstone; 16, probable carbonaceous shale.

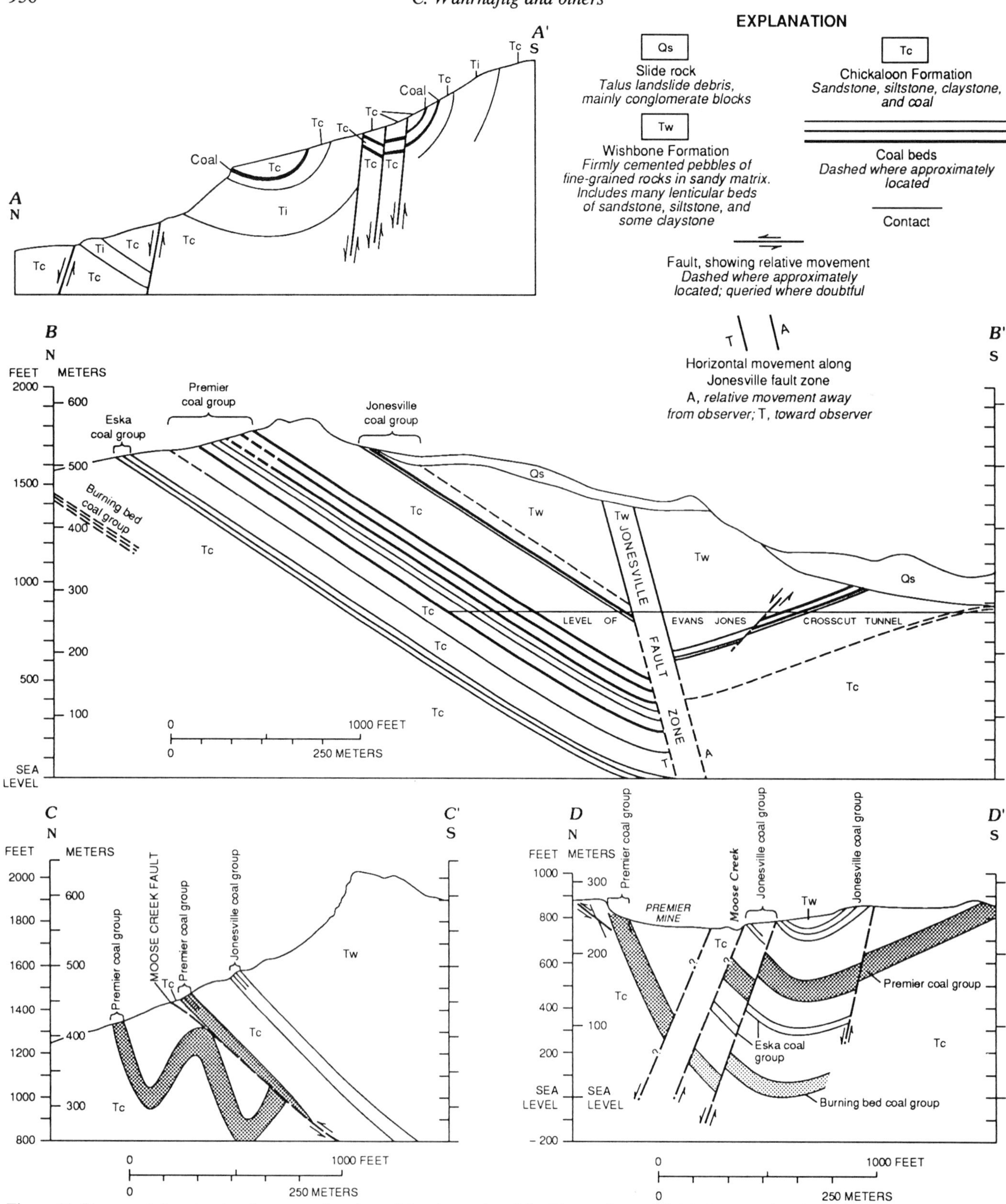

Figure 11. Representative structural cross sections in coal-bearing rocks of the Matanuska coal field. Cross section A-A′, at anthracite locality on Anthracite Ridge (from Waring, 1936, plate 8). Length of cross section A-A′ is 1.31 km; horizontal and vertical scales the same. Cross section B-B′ through Eska and Evan Jones mines, Wishbone Hill district. Cross sections C-C′ and D-D′, both near Premier Mine, Wishbone Hill district. (B-B′, C-C′, and D-D′ are slightly modified from Barnes and Payne, 1956, plates 7 and 9). Sections approximately north-south; north is to the left. Map units: Qs, Quaternary slide rock; Ti, Tertiary intrusive rocks; Tw, Paleocene and Eocene Wishbone Formation; Tc, Paleocene and Eocene Chickaloon Formation.

district are as thick as 1.2–2.0 m, and the coal beds are exceptionally lenticular.

The intensity of deformation and abundance of igneous dikes and sills increase eastward in the Chickaloon Formation of the Matanuska Valley. A few small dikes occur in the Wishbone Hill district, and thick sills are abundant in the Anthracite Ridge district. Heating induced by the igneous intrusions may be the main reason for the increase in average rank of the coal eastward in the coal field, although Barnes (1962) regarded heat generated by tectonic activity as more important. Merritt (1985a) described the natural coking of coal adjacent to an intrusive diabase sill and concluded that the temperature at the contact reached 550 °C. He also found that the bed that was locally coked had generally been raised in rank to semianthracite, whereas a coal bed about 50 m away was high-volatile bituminous A. The rank of coal in the Anthracite Ridge district also changes abruptly from low-volatile bituminous to semianthracite or anthracite within about 60 m, and the actual amount of anthracite coal appears to be quite small (Waring, 1936).

Structures in the Matanuska coal field are typically complex (Fig. 11). The doubly plunging Wishbone Hill syncline, a relatively simple structure in this field, has beds that dip 20°–40° on either flank; the structure is cut by two sets of transverse faults. Structural complications on its northwest flank make the coal beds in some structural blocks difficult to mine and reserves impossible to calculate (Barnes and Payne, 1956). With the possible exception of the Castle Mountain district, structural complexities increase eastward. In the Chickaloon district beds dip as much as 90°, and in the Chickaloon mine beds are overturned (Chapin, 1920) and abruptly faulted. Large areas of the Chickaloon Formation are covered by a thick mantle of glacial till and crop out only along stream bluffs (Capps, 1927). Anthracite occurrences on the south flank of Anthracite Ridge are bordered on the north by a high-angle fault of large displacement and are in tightly folded and locally overturned synclines cut by many faults.

Coal assessment of the Matanuska coal field. Estimates of coal resources reported by various workers for the coal field as a whole and for its various districts were tabulated by Merritt and Belowich (1984); they are as high as 200×10^6 tons of measured to inferred resources and as high as 2.2×10^9 metric tons of hypothetical resources. Considering the geology of the valley, the most reliable estimates appear to be those of Barnes (1967a), which are 113×10^6 metric tons of measured, indicated, and inferred resources, inclusive, and 250×10^6 metric tons of hypothetical resources. In the Anthracite Ridge district, the only identified minable bed of anthracite, 1.3–2.0 m thick, apparently underlies an area of no more than 0.01 km^2 and totals no more than 20,000 metric tons (Waring, 1936, locs. 24, 25, p. 34, 43, plate 8; Merritt and Belowich, 1984, locs. 31–32, p. 34, 55–56). One other reported anthracite occurrence (Merritt and Belowich, 1984, loc. 39), too thin to be mined, is on a large active landslide (Detterman and others, 1976).

West Foreland Formation. The West Foreland Formation, a type section of which is in a well on the West Foreland (Calderwood and Fackler, 1972), consists of tuffaceous claystone, sandstone, and conglomerate, including a few thin, probably unminable beds of coal. The formation underlies most of lower Cook Inlet southwest of the mouth of the Susitna River, and much of the lowland area of the northwestern Kenai Peninsula (Hartman and others, 1971). A maximum thickness of 1,000 m was reported at Cape Douglas (Fisher and Magoon, 1978, p. 380), but beneath most of Cook Inlet the unit has a maximum thickness of about 600 m (Hartman and others, 1971). Wolfe and Tanai (1980, p. 4) tentatively assigned it a latest Paleocene age; however, Magoon and others (1976) and Magoon and Egbert (1986) assigned it to the early Eocene (Fig. 9). It is probably correlative with the Wishbone Formation. Where exposed around the margins of the Cook Inlet lowland, the West Foreland Formation is unconformably overlain by rocks of the Kenai Group.

Calderwood and Fackler (1972) regarded the West Foreland as the basal formation of the Kenai Group, and Kirschner (1988; and 1989, written commun.) followed this assignment (see Fig. 9A). According to C. E. Kirschner (1989, written commun.), the West Foreland is conformable and probably continuous with the rest of the Kenai Group in the subsurface beneath Cook Inlet, and so it should be regarded as the basal formation of that group. He stated that the unconformity observed along the margins of the Cook Inlet lowland is between the West Foreland Formation and parts of the Tyonek Formation that are much younger than the rocks that rest on the West Foreland Formation in the subsurface.

Magoon and others (1976), however, regarded the unconformity at the top of the Wishbone Formation as regional, and so they removed the Wishbone from the Kenai Group (see Fig. 9B). Their argument for a regional unconformity stems from the observation that sections of the West Foreland Formation that have been dated are separated from rocks of the Kenai Group by a 20 m.y. hiatus. Because the assignment of Magoon and others (1976) has been adopted officially by the U.S. Geological Survey, it is followed here. However, the question of a regional unconformity at the top of the West Foreland remains undecided.

Conglomeratic rocks south of the Capps Glacier and along the north side of the Bruin Bay fault in the Beluga coal field have been correlated with the West Foreland Formation (Magoon and others, 1976). These conglomeratic rocks contain granitic boulders as much as 1 m across; the unit is about 365 m thick (Adkison and others, 1975).

The West Foreland Formation appears to have been deposited contemporaneously with the Wishbone Formation during downbowing of the Cook Inlet basin in early Tertiary time. If the 20 m.y. hiatus is real, then this period of subsidence probably was relatively short lived and was followed by little elevation change and only minor orogeny. The West Foreland Formation is not much more deformed than the overlying Kenai Group, nor is it significantly more consolidated.

Coal fields of the middle and upper Tertiary Kenai Group

The main coal deposits of the Cook Inlet–Susitna Lowland (see Fig. 7) are in the Kenai Group, which ranges in age from early Oligocene to late Pliocene. At its base is the Hemlock Conglomerate (correlated with the Tsadaka Formation and Bell Island Sandstone of Hartman and others, 1971), which is overlain successively by the Tyonek Formation, the Beluga Formation, and the Sterling Formation (Fig. 9; Calderwood and Fackler, 1972; Kirschner and Lyon, 1973; Magoon and others, 1976; Magoon, this volume). The Kenai Group contains the petroleum and natural gas reserves of the Cook Inlet petroleum province, and most information on the distribution and lithology of its formations, and on the amount of coal within them, comes from subsurface data obtained in drilling for oil and natural gas. The type localities of the formations established by Calderwood and Fackler (1972) are all subsurface well segments. Although excellent exposures of coal-bearing rocks exist in the Beluga coal field and in seacliffs near Tyonek and around the northwestern Kenai Peninsula between the head of Kachemak Bay and Kasilof (Barnes and Cobb, 1959; Barnes, 1966), those exposures are near the margins of the basin where the group is relatively thin and may contain hiatuses or unconformities.

The main basin of accumulation of the Kenai Group is between the Castle Mountain fault on the northwest and the presumably faulted northwest front of the Kenai Mountains on the southeast. The basin shallows over the transverse Augustine-Seldovia arch on the southwest and at the lower end of the Matanuska Valley on the northeast (Fig. 7). There was apparently no accumulation of sediment during Kenai time in the coal-bearing part of the Matanuska Valley. Total accumulation of sediment in the deepest part of this basin is more than 7,800 m (Hartman and others, 1971; Calderwood and Fackler, 1972). Kirschner and Lyon (1973) regarded the Kenai Group as intermittently estuarine in the central part of the basin, whereas Fisher and Magoon (1978) and Magoon and Egbert (1986) regarded the Kenai Group as continental throughout.

Hemlock Conglomerate. The Hemlock Conglomerate consists of conglomerate and conglomeratic sandstone containing quartz and chert, pebbles of metamorphic, volcanic, and plutonic rocks, and feldspathic sands with heavy minerals, predominantly epidote and garnet (Calderwood and Fackler, 1972; Magoon and Egbert, 1986). The Hemlock Conglomerate, which contains a few thin coal beds and many siltstone beds, is the main producing horizon for oil in the Cook Inlet province (Magoon, this volume); it is regarded as deltaic and estuarine in origin and apparently was derived from the north. Together with the Bell Island Sandstone and the Tsadaka Formation, temporal equivalents at the east end of the Cook Inlet basin, it makes a nearly uniform sheet 200 m thick, with a maximum thickness of about 450 m; it is Oligocene in age (Wolfe and Tanai, 1980; Magoon and Egbert, 1986).

Tyonek Formation. The lower Oligocene through middle Miocene Tyonek Formation (Wolfe and Tanai, 1980), which consists of as much as 2,330 m (Calderwood and Fackler, 1972) of sandstone, shale, and coal beds as much as 10 m thick, contains the bulk of the coal resources of the Cook Inlet basin and the Beluga coal field (see Fig. 9C). A sand/shale ratio map of the formation (Hartman and others, 1971) shows more than 50 vol% sandstone in an area along the west side of the basin between the Beluga coal field and West Foreland, and other areas of high sandstone content near Tuxedni Bay and in the extreme northeastern part of the basin. The locations of high sandstone content are thought to mark points of sediment input. An extensive clay band containing as little as 10 vol% sand extends south-southwestward along the western part of the Kenai Lowland and lies about 25 km southeast of the line of maximum thickness. Sand content increases southeastward of this zone of maximum clay content.

The cumulative thickness of coal penetrated in wells in the Tyonek Formation beneath Cook Inlet is shown in Figure 12. Coal in the Tyonek Formation is concentrated mainly along the northwest margin of the basin, from Kalgin Island northeastward along the west shore of the inlet to the Susitna River (Fig. 12). The coal concentration borders areas of highest sand input from the northwest. Secondary concentrations of coal occur beneath the lower Matanuska Valley and extend along the Kenai Lowland in a belt that coincides generally with the zone of minimum sand content. The thick Capps Glacier seam and seams along the Beluga and Chuitna Rivers (Barnes, 1966) in the Beluga coal field are in the Tyonek Formation (Magoon and others, 1976). Within a single oil field, correlation of the coal beds from well to well has proved difficult for distances of more than a few kilometers, suggesting considerable lenticularity of the coal seams and intervening sedimentary rocks. In outcrop, however, individual beds have been traced for as much as 10 km (Barnes, 1966; Ramsey, 1981).

Nearly flat-lying outliers of the Tyonek Formation along the southeast shore of Kachemak Bay near Seldovia and at Port Graham rest unconformably on metamorphic rocks of Triassic and Jurassic age, and appear in part to fill steep-sided valleys and in part to be downfaulted (Stone, 1906; Martin and others, 1915; Magoon and others, 1976). The occurrence on the northeast side of the entrance to Port Graham was the site of the plant fossils on which Oswald Heer in 1869 (cited *in* Hollick, 1936) established the "Arctic Miocene" flora of Alaska (see Stone, 1906), and the locality on which the name "Kenai Formation" (now Kenai Group) was established by Dall (1896). Coal was first reported there by Portlock in 1786 (see Stone, 1906, p. 54). Coal (chiefly lignite) was mined at this site by the Russians from 1855 to 1867, but it could not be produced at a profit, and operations ceased when Alaska was ceded to the United States in 1867 (Stone, 1906).

Beluga Formation. The middle and upper Miocene Beluga Formation (Wolfe and Tanai, 1980) has a maximum thickness of about 1,500 m (Hartman and others, 1971). The abundance of

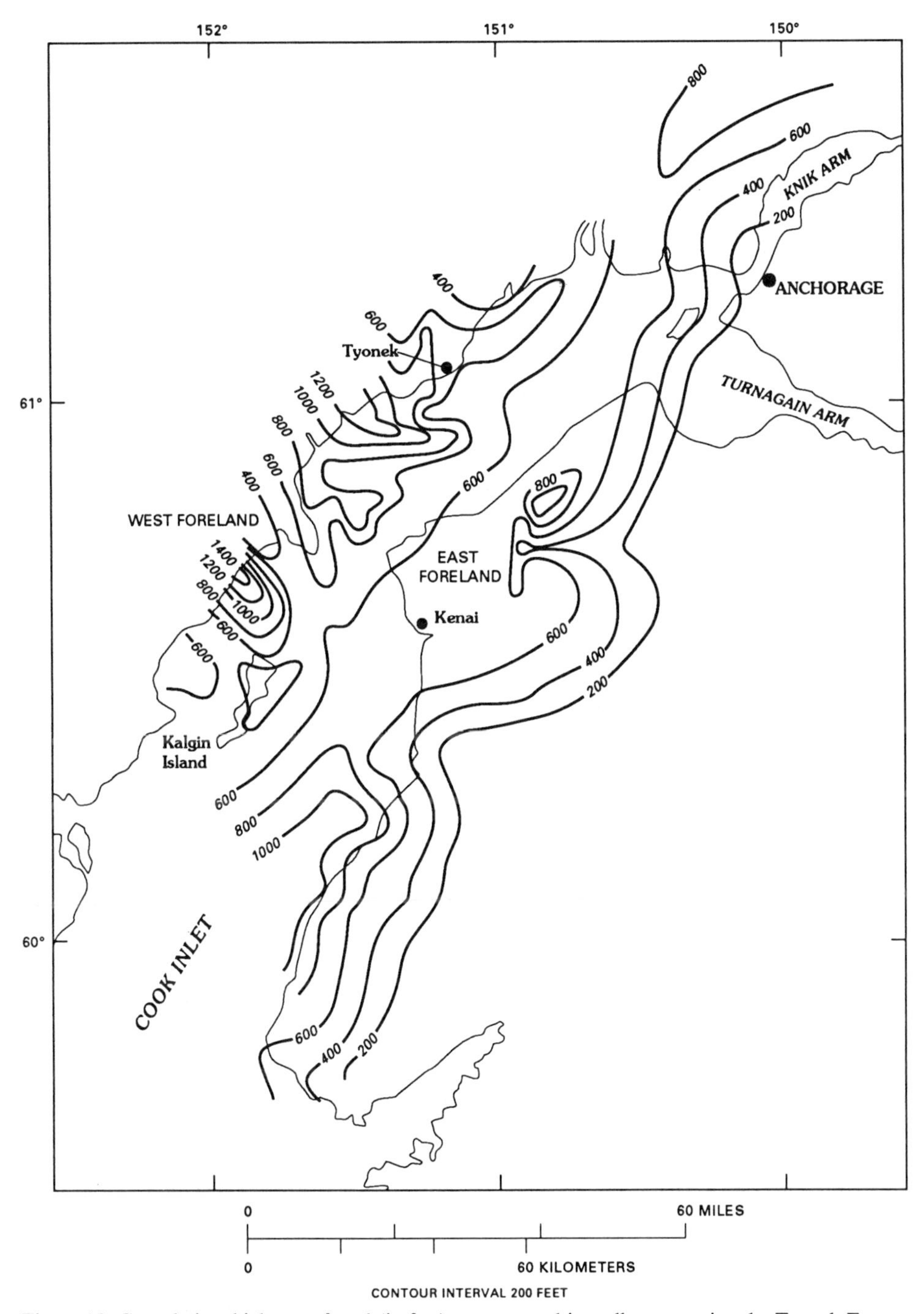

Figure 12. Cumulative thickness of coal (in feet) encountered in wells penetrating the Tyonek Formation beneath the Cook Inlet area (slightly modified from Hite, 1976).

epidote in the heavy-mineral suite (Kirschner and Lyon, 1973) and the relative abundance of metamorphic rock fragments in the locally pebbly sandstones led Hayes and others (1976) to interpret the Beluga Formation as an alluvial-fan complex derived mainly from the Kenai and Chugach Mountains. The Beluga Formation is exposed in beach bluffs along the northwest side of Kachemak Bay and the southwest end of the Kenai Peninsula south of Anchor Point (Merritt and others, 1987), and in the hills between Homer and the Anchor River (Magoon and others, 1976). It contains numerous coal beds as thick as 2 m (Barnes and Cobb, 1959).

Sterling Formation. The Sterling Formation of latest Miocene and Pliocene age (Wolfe and Tanai, 1980) consists of massive sandstone, conglomeratic sandstone, and interbedded claystone, and some thin coal and tuff beds (Kirschner and Lyon, 1973; Hayes and others, 1976; Hite, 1976), and is as thick as

3,350 m (Hartman and others, 1971; Calderwood and Fackler, 1972). Sandstone of the Sterling Formation contains abundant pumice grains and other fresh volcaniclastic detritus. Heavy minerals are predominantly hornblende and hypersthene, and volcanogenic hypersthene increases in abundance upsection; small amounts of garnet are present. These suggest that the Sterling Formation is apparently derived from the northwest, predominantly from the Alaska and Aleutian Ranges.

The Sterling Formation contains thin coal beds (generally no more than 1 m thick, but a few are as thick as 2.5 m [Barnes and Cobb, 1959; Calderwood and Fackler, 1972]). The coal is lignitic throughout much of the formation but is high-volatile subbituminous near the base. The Sterling and Beluga Formations have produced the bulk of the methane gas from the Cook Inlet petroleum province (Magoon, this volume).

Coal assessment of the Kenai Group. Coal resources of the Kenai Group in the Cook Inlet basin are large. Maps showing cumulative thicknesses of coal to depths of 610, 1,524, and 3,048 m in the subsurface, based on logs of oil wells throughout the basin, were prepared by McGee and O'Connor (1975), who calculated a speculative resource of 1.1×10^{12} metric tons of coal of apparent lignite rank to a depth of 3,048 m, and 100×10^{9} metric tons to a depth of 610 m. Affolter and Stricker (1987b) estimated that 0.7×10^{12} metric tons of this coal lie beneath the waters of Cook Inlet. In addition to the isopach map of McGee and O'Connor (1975), Hite (1976) showed isopachs on cumulative thickness of coal in the Hemlock and Tyonek Formations.

Extensive exploration for, and mapping of, coal has taken place within land areas of the Cook Inlet basin, and indicated reserves have been calculated for many areas.

For the Homer district (loc. 14, Fig. 1) of the Kenai coal field, Barnes and Cobb (1959) calculated (from surface exposures only) indicated reserves of 360×10^{6} metric tons of coal in beds 0.6 m thick or more, of which 45×10^{6} metric tons is in beds more than 1.5 m thick. Nearly all of these coal beds are covered by $<$300 m of overburden.

The Tyonek-Beluga coal field (loc. 15, Fig. 1), northwest of Cook Inlet, is cut by the northeast-trending Castle Mountain fault, and reserves have been calculated separately for areas south and north of the fault. From surface exposures, Barnes (1966) calculated a total of 1.5×10^{9} metric tons of coal in the part of the coal field south of the Castle Mountain fault. Most beds are more than 3 m thick, and some are almost 20 m thick. Ramsey (1981) identified six strippable beds averaging 3–10 m in thickness in a coal lease area of about 82 km^2 between the Chuitna and Beluga Rivers. Crude measurements from Ramsey's maps of this lease area suggest that it may contain 35 to 45×10^{9} metric tons to a depth of 300 m, provided the beds maintain their thickness laterally and downdip.

Between the Capps Glacier and Mount Susitna, the Tyonek-Beluga coal field extends northward across the Castle Mountain fault. The field includes an extensive area of coal-bearing rocks, assigned by Magoon and others (1976) to the Tyonek Formation, that are in fault contact along the Castle Mountain fault with the beds whose reserves were given in the foregoing paragraph. North of these coal-bearing rocks, Mesozoic basement rocks crop out in a continuous belt south of the Skwentna River, separating the Tyonek-Beluga coal field from other bodies of coal-bearing rocks farther north in the Susitna basin. For this part of the coal field, Barnes (1966) calculated 373×10^{6} metric tons of indicated reserves beneath the plateau south of Capps Glacier, 100×10^{6} metric tons of indicated reserves northeast of Beluga Lake, and 14×10^{6} metric tons of coal at the head of the Talachatna River, for a total of 487×10^{6} metric tons of indicated reserves. The reserves calculated beneath the plateau south of the Capps Glacier were mainly for the 15-m-thick Capps bed; Patsch (*in* Schmoll and others, 1981) reported a 7.5-m-thick coal bed beneath the Capps bed. Thus, the reserves in this area are probably greater than those reported by Barnes (1966).

Environmental and engineering problems connected with mining coal from the Tyonek-Beluga coal field, which were addressed by Schmoll and others (1981), include slope stability and landslide problems, earthquake hazards, volcanic hazards, and erosion and flood potential.

On the basis of surface exposures, drilling, and trenching, Barnes and Sokol (1959) concluded that the Little Susitna district (loc. 16, Fig. 1), which extends eastward from the Alaska Railroad at Houston for about 32 km along the south base of the Talkeetna Mountains, contains insufficient coal to warrant calculating reserves. The coal, in beds generally less than 0.6 m thick, is in the Tsadaka Formation, which is correlative with the Hemlock Conglomerate. About 10,000 tons are reported to have been produced from a mine at Houston.

Coal fields of the Susitna Basin

North of the bedrock barrier extending westward from Mount Susitna along the south side of the Yentna and Skwentna Rivers is the extensive Susitna Lowland, situated between the Talkeena Mountains and the Alaska Range. Most of this lowland is covered with glacial and alluvial deposits. Soft-coal-bearing continental sedimentary rocks, correlative with the Kenai Group of Cook Inlet, are exposed in a few places along the banks and tributaries of the Susitna and Yentna Rivers. Along the southeastern margin of the Alaska Range, these Kenai Group rocks generally dip toward the Susitna Lowland or lie in downfaulted or downwarped basins (Barnes, 1966; Magoon and others, 1976; Reed and Nelson, 1980). The Tertiary deposits of the Peters Hills–Cache Creek area (loc. 17, Fig. 1) and of the Mt. Fairview area (no. 18, Fig. 1) were recognized by Capps (1913) as the source for placer gold, and some coal from these deposits was used for fuel by gold prospectors. Most of the thick coal beds known from the southwestern part of the basin were not discovered until the early 1960s (Barnes, 1966).

The coal-bearing rocks were divided by Reed and Nelson (1980) into three units; from top to bottom these are: (1) an orange to light gray, massive pebble to boulder conglomerate, correlative with the Sterling Formation, as much as 770 m thick, deposited by streams flowing southeastward from the Alaska Range; (2) a predominantly sandstone unit about 170 m thick, consisting of repetitive cycles 7–23 m thick, grading from coarse-grained, pebbly sandstone at the base to silt and clay with coal or bony coal at the top; and (3) a basal member consisting of 40 vol% conglomerate, 20 vol% sandstone, and less than 40 vol% siltstone, claystone, and coal, the latter in beds as much as 17 m thick. The two lower units are correlative with the Tyonek Formation.

On the basis of surface exposures only, Barnes (1966) calculated indicated reserves of 4.1×10^6 metric tons of coal in beds less than 2 m thick in the Peters Hills, about 40×10^6 tons of coal mainly in beds more than 3 m thick in the Fairview Mountain area, 18×10^6 metric tons of coal mainly in beds more than 2 m thick in the Johnson Creek area (loc. 19, Fig. 1), and 100×10^6 tons of coal in the downfaulted half graben along Canyon Creek (loc. 20, Fig. 1). A drilling program by the Mobil Oil Corporation has demonstrated the existence of 450×10^6 metric tons of coal within 76 m of the surface in beds 3 to 15 m thick, in two leasing units totaling 9,300 ha. One unit includes the Canyon Creek coal basin and the other extends from the Skwentna River northward across Johnson Creek (Blumer, 1981).

Barnes (1977) showed two strongly negative gravity anomalies beneath the Susitna Lowland: one between the Johnson Creek coal outcrops and Yenlo Mountain and north of the Skwentna River, and the other between Yenlo Mountain and the Susitna River, centered at the junction of the Kahiltna and Yentna Rivers. He interpreted both anomalies as thick fill of low-density sediment, probably of the Kenai Group, beneath the lowland. There is a potential for large deposits of coal in the area of both of these gravity anomalies.

Coal in the Broad Pass depression

The Broad Pass depression is a narrow trough extending northeastward from the north end of the Cook Inlet–Susitna Lowland (Wahrhaftig, 1965); it is about 8 km wide, bordered by mountains that rise abruptly 1,000–2,500 m above the basin. Although most exposures consist of metamorphic and igneous rocks mainly of Mesozoic and older age, several patches and small basins of poorly consolidated, coal-bearing continental rocks, correlative with the Kenai Group, crop out on its floor. Only two of these areas have been investigated for coal resources: the Costello Creek basin (loc. 21, Fig. 1) and an area at Broad Pass station (loc. 22, Fig. 1). A detailed U.S. Bureau of Mines–U.S. Geological Survey drilling and trenching program of the Costello Creek basin during World War II (Wahrhaftig, 1944) disclosed a lower unit, 0–26 m thick, of predominantly gray sandstone, interbedded mudstone, and lenticular coal beds, overlain by an upper predominantly sandstone unit, as much as 150 m thick, lacking minable coal. Total reserves of the Costello Creek basin, calculated in 1944, were 317×10^3 metric tons of coal in two beds with a maximum thickness of 3 m. According to Barnes (1967a, p. B24), 58×10^3 metric tons of coal was mined in 1940–1954, and the rest was unminable. The coal is subbituminous, and the beds are tentatively correlated with the Tyonek Formation.

The coal at Broad Pass station, 13–16 km east of the Costello Creek basin, is lignite, interbedded with white to orange sand and gravel (Hopkins, 1951). Gently dipping Tertiary beds at this locality are correlated with the Sterling Formation of the Susitna Lowland. Hopkins estimated that at least 12.2×10^6 metric tons of coal might exist beneath the area of known exposures of Tertiary rocks at this locality, but only 270×10^3 metric tons of lignite with an ash content of 8–25 wt% were actually measured. Lignitic coal has been reported elsewhere in the Broad Pass depression, and orange to yellow gravels exposed in railroad cuts and streambanks resemble the Nenana Gravel of the north side of the Alaska Range and the Sterling Formation of the Susitna Lowland.

Nenana Coal Field

The Nenana coal field (loc. 23, Fig. 1) is the largest, most centrally located, and most thoroughly studied of the coal fields on the north side of the Alaska Range. It has produced more than half of the coal mined in Alaska. This coal field is located in the northern foothills of the Alaska Range, extending from about 80 km west to 80 km east of the Alaska Railroad (Fig. 13). It consists of several synclinal basins partly or wholly detached from each other by erosion of coal-bearing rocks from intervening structural highs.

Usibelli Group. Coal occurs in the Usibelli Group (Wahrhaftig, 1987), a poorly consolidated sequence of continental sedimentary rocks of Tertiary age, consisting of the Healy Creek, Sanctuary, Suntrana, Lignite Creek, and Grubstake Formations. The following discussions of the Usibelli Group are summarized from Wahrhaftig and others (1969), Wahrhaftig (1987), and C. Wahrhaftig (unpublished data, 1947–1963).

Healy Creek Formation. The oldest formation in the Usibelli Group is the Healy Creek Formation (Fig. 14), a sequence of lenticular beds of poorly sorted sandstone, conglomerate, siltstone, and claystone, including beds of bone and coal that abruptly thicken, split, or pinch out. The Healy Creek Formation was deposited on a surface having a few hundred meters of relief, and its thickness and number of coal beds vary markedly. As indicated by the mineral content and varying flow directions shown in crossbeds and other sedimentologic features, the clastic sediments were derived from nearby sources.

Locally, coal beds as thick as 15–20 m rest directly on deeply weathered crystalline basement rocks and persist laterally for less than 1 km. In some directions they pinch out against the buried land surface on which they were deposited, and in other directions they interfinger abruptly with clastic sedimentary deposits. They are probably ancient peat deposits that were laid down in reentrant valleys of surrounding hills on the margins of the alluvial plain of the Healy Creek Formation. The reentrants, the surrounding hillslopes of which were densely vegetated and

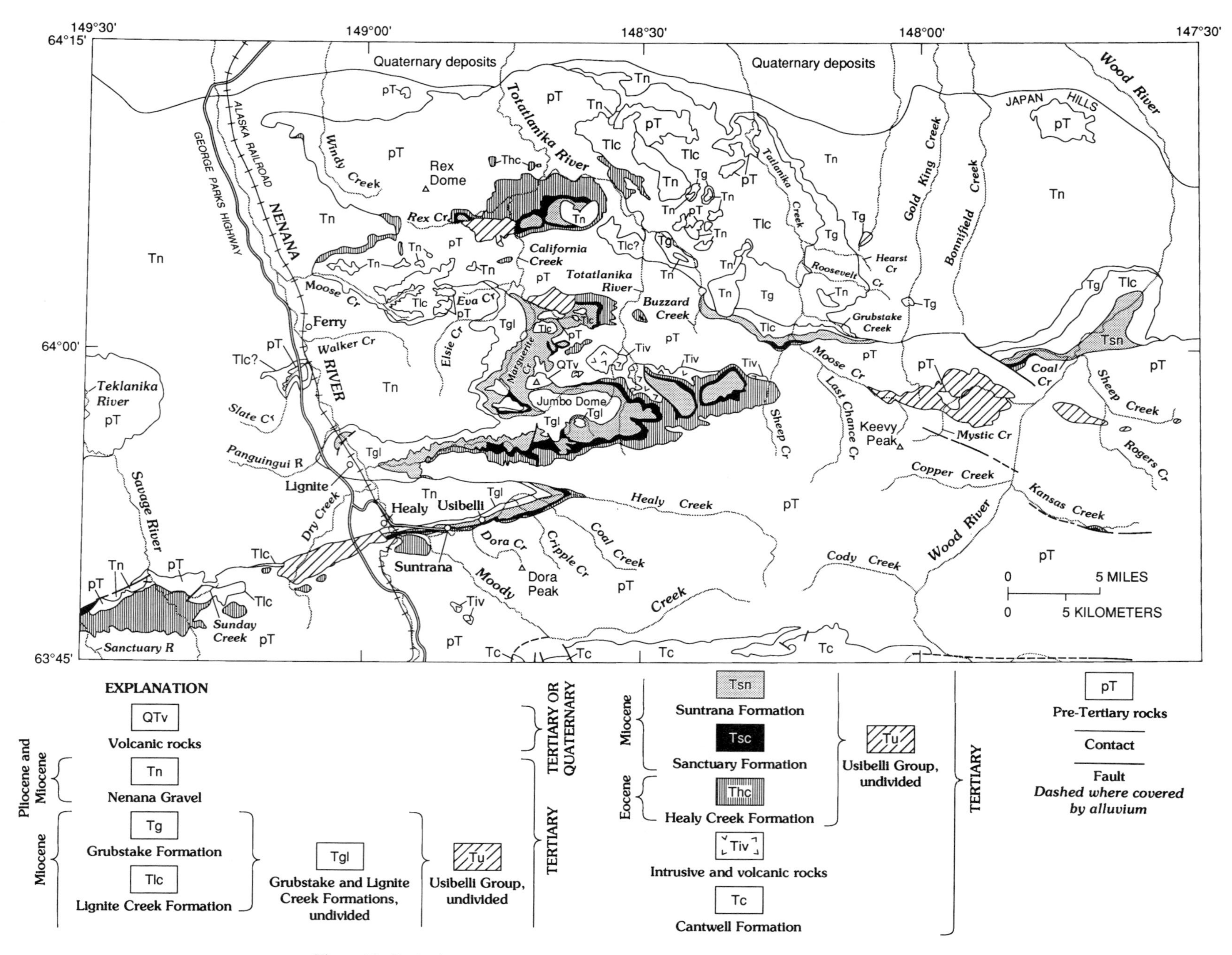

Figure 13. Geologic map of the Nenana coal field (from Wahrhaftig and others, 1969).

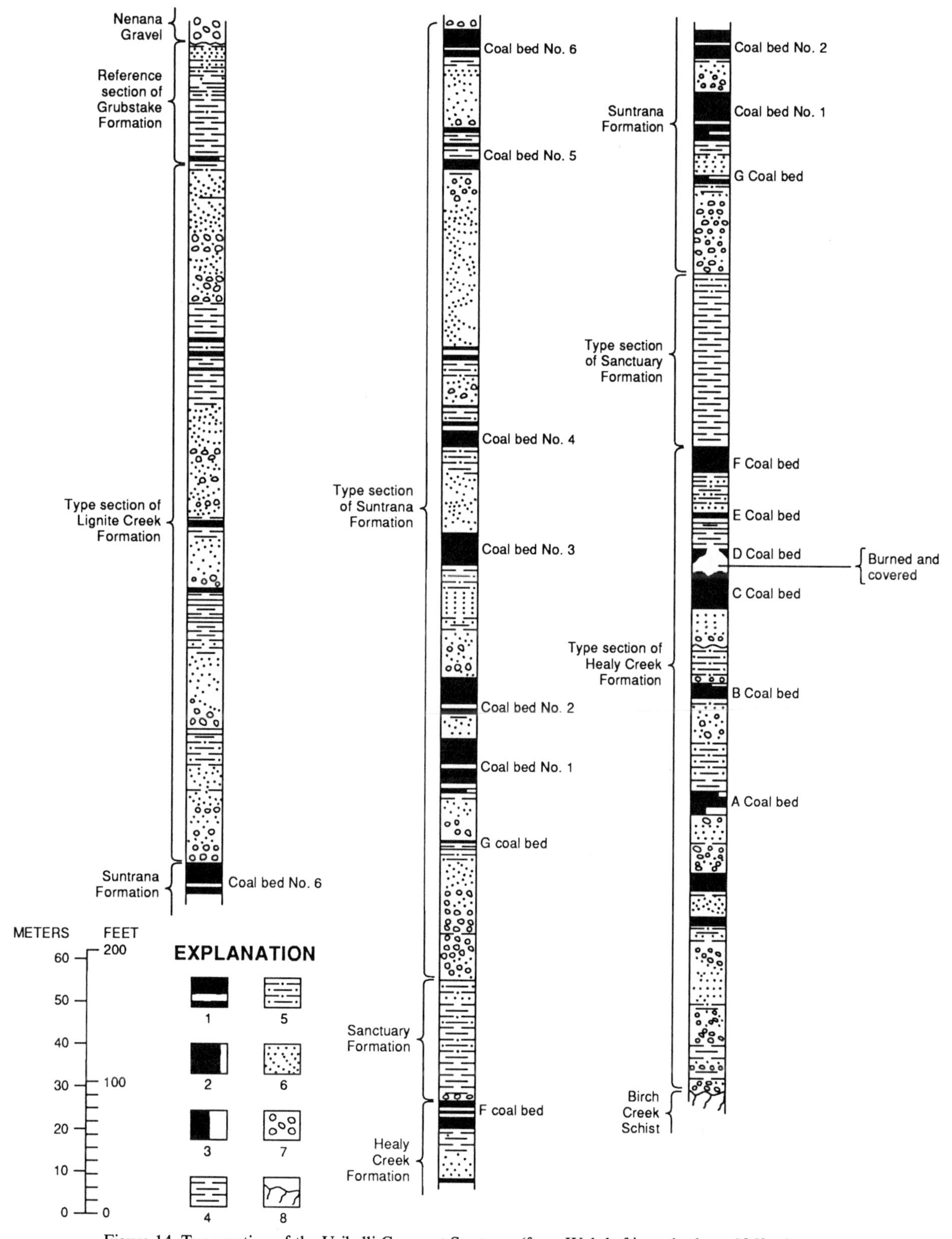

Figure 14. Type section of the Usibelli Group at Suntrana (from Wahrhaftig and others, 1969; also see Wahrhaftig, 1987). 1, coal, showing bone and clay partings; 2, bony coal; 3, bone; 4, claystone and shale; 5, siltstone; 6, sandstone, in part cross-bedded; 7, pebbles and conglomerate; 8, schist (unconformity at top).

shed no detritus, were protected from incursions of sediment from the alluvial plain by natural arboreal barriers or because they were raised bogs.

In most of the synclinal coal basins, the Healy Creek Formation is early and middle Miocene in age (Wolfe and Tanai, 1980; Wahrhaftig, 1987), but in the Rex Creek coal basin, where the formation was formerly thought to be as old as late Oligocene (Wolfe and Tanai, 1980), it is now regarded to be as old as late Eocene (Wolfe and Tanai, 1987).

Sanctuary Formation. The Sanctuary Formation is composed of 40 m of gray, thinly laminated (possibly varved) shale that weathers chocolate brown; the unit overlies the Healy Creek Formation and was assigned by Wolfe and Tanai (1980) to the middle Miocene. The Sanctuary Formation is interpreted to have accumulated in a large shallow lake. The highest coal bed of the Healy Creek Formation, the "F" bed (Fig. 14), which immediately underlies the Sanctuary Formation, is the only coal bed in the Healy Creek Formation of sufficiently continuous lateral extent to the analyzed for reserve estimates.

Suntrana Formation. The overlying Suntrana Formation is as thick as 400 m, and consists of 6–12 upward-fining cycles of well-sorted, crossbedded sandstone that is pebbly at the base. Each cycle has at its top a sequence of claystone, siltstone, and fine sandstone with one or more coal beds 0.5–20 m thick. Northward coarsening of sediment and current indicators in the crossbeds indicate that the clastic components were derived from sources to the north. Trough-crossbed sets are 1–2 m thick (C. Wahrhaftig, unpublished data, 1947–1963).

The middle Miocene Suntrana Formation (Wolfe and Tanai, 1980) contains the major coal resources of the Nenana coal field. The formation as a whole thickens gradually southeastward from a line of pinchout in the northwestern part of the coal field; most of the thick coal beds can be traced laterally over distances of as much as 25 km (Wahrhaftig, 1973). The bulk of the calculated reserves is in six coal beds 6–20 m thick. Isopachs on coal bed no. 6, the highest coal bed in the Suntrana Formation, are shown in Figure 15.

The lateral persistence of the coal beds and of the major intervening clastic units in a direction normal to the current direction poses problems in interpreting the conditions of deposition of the coal. The geometry suggests an alternation throughout the coal field of continuous peat swamps and sandy plains of fluvial sedimentation. One possible cause for this alternation would be that the clastic source area was a heavily vegetated region that underwent deep weathering beneath vegetation cover during long periods—represented by the coal beds—interrupted by short intervals of rapid erosion, possibly triggered by fires, catastrophic storms, or tectonic activity. Such a scenario, however, is not easily compatible with a system involving streams that have deposited trough crossbeds as much as 2 m thick and that flowed swiftly enough to transport pebble to cobble gravel across a plain flat enough to have acted as a peat bog at least 25 km wide in the direction of stream flow.

Lignite Creek Formation. Overlying the Suntrana Formation is the 150–240-m-thick, late middle to early late Miocene age Lignite Creek Formation (Fig. 14; Wolfe and Tanai, 1980), which is composed of repetitious sequences of well-sorted, crossbedded, pebbly and coarse-grained sandstone at the base, grading to interbedded clay and coal beds at the top. Grain mineralogy and pebble lithologies of the sandstone differ from those in the Suntrana Formation, and the coal beds are thin (generally less than 1.5 m), woody, and relatively lenticular. No coal reserves have been calculated for the Lignite Creek Formation. The coal beds pinch out northward. A non-coal-bearing conglomeratic facies, as much as 450 m thick, of the Lignite Creek Formation occurs along the north and west margins of the Nenana coal field.

Grubstake Formation. The highest formation assigned to the Usibelli Group is the Grubstake Formation, a dark gray laminated shale and claystone that is 180–300 m thick in the northeastern part of the Nenana coal field, but only 25–75 m thick in the southwestern part (Fig. 14). A K-Ar age on rhyolitic glass from an ash layer in the lower part of the Grubstake Formation is 8.3 ± 0.4 Ma, which coincides with a late Miocene age based on plant megafossils (Wahrhaftig and others, 1969; Wolfe and Tanai, 1980; Wahrhaftig, 1987). In the eastern part of the outcrop belt, the Grubstake Formation interfingers southward with coarse-grained dark sands similar to those in the overlying Nenana Gravel. The Grubstake Formation probably accumulated in the lake formed by the damming of formerly southward-directed drainage by the rising Alaska Range.

Nenana Gravel. The Nenana Gravel, a poorly consolidated, buff to red pebble to boulder conglomerate derived from the Alaska Range, rests on the Usibelli Group and ranges in thickness from 1,200 m at the south edge of the Nenana coal field to 300–400 m along the north edge of the foothills. The gravel detritus was shed northward from the rising Alaska Range that blocked the southward-flowing drainage tributary to the Cook Inlet–Susitna Lowland (Wahrhaftig, 1970). Its age is bracketed between 8.3 and 2.75 Ma, and so it is contemporaneous with the Sterling Formation. The Nenana Gravel is much more widely distributed than the Usibelli Group, which is primarily confined to synclinal basins deformed early in the orogeny that later deposited the Nenana Gravel. Along much of its outcrop length, the Nenana Gravel rests on rocks older than the Usibelli Group, and detritus from the Usibelli Group can be recognized in the Nenana Gravel.

Coal assessment of the Nenana coal field. The coal in the Nenana coal field ranges from lignite to subbituminous B but is mainly subbituminous C. It has an ash content of 5–20 wt%, a moisture content of 15–30 wt%, a heating value of 3,900–4,700 cal/gm (as received), and one of the lowest reported sulfur contents of any United States coal, reported to be 0.1–0.3 wt% (Rao and Wolfe, 1981; Affolter and others, 1981). Measured, indicated, and inferred resources of coal, in beds more than 0.75 m thick and within 1,000 m of the surface, are slightly less than 6.4×10^9 metric tons, of which about 5.4×10^9 metric tons is in beds more than 3 m thick (Barnes, 1967a). More than 5.4×10^9 metric tons of the resources are in the Healy Creek and Lignite

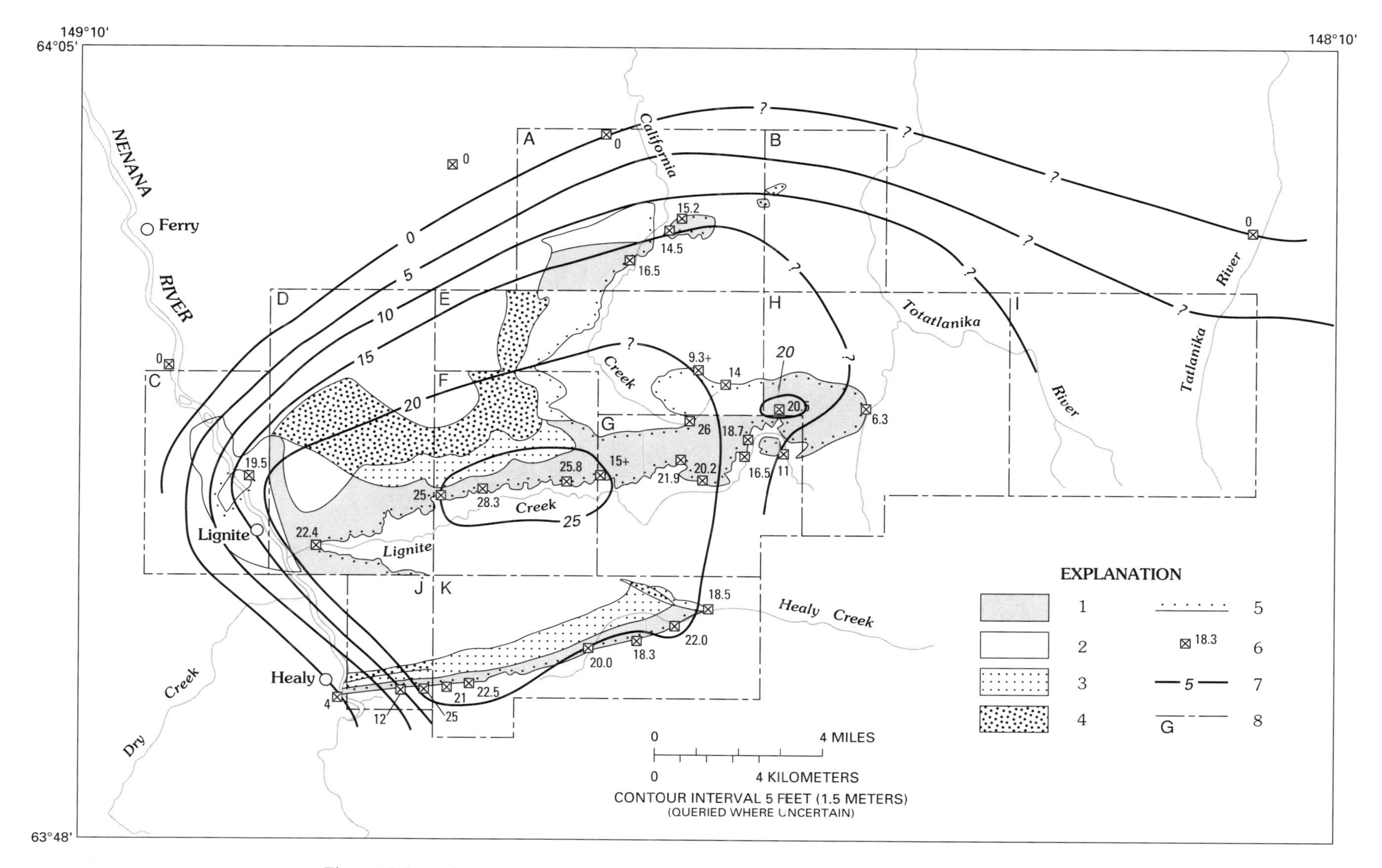

Figure 15. Isopachs and coal reserve areas of coal bed no. 6, Nenana coal field, Alaska. 1: Measured and indicated coal with less than 305 m (1,000 ft) of overburden. 2: Inferred coal with less than 305 m (1,000 ft) of overburden. 3: Measured and indicated coal with 305–914 m (1,000–3,000 ft) of overburden. 4: Inferred coal with 305–914 m (1,000–3,000 ft) of overburden. 5: Trace of outcrop of coal bed no. 6. 6: Location of measured section of coal bed no. 6, with thickness in feet (+ sign indicates measured thickness where either base or top is not exposed). 7: Isopach line for coal in bed no. 6, contour interval 1.5 m (5 ft), queried where position uncertain. 8: Boundary of coal reserve reporting unit in Table 3; letter designation in upper left or right corner of each unit (modified from Wahrhaftig, 1973).

Creek coal basins (Table 3; Wahrhaftig, 1973). Hypothetical resources are not calculated but are unlikely to be more than a fraction of the calculated reserves, because the coal-bearing rocks pinch out and are overlapped by the Nenana Gravel to the north, and because the coal beds thicken and increase in number southward, reaching their maximum thickness at the outcrops along the south margin of the coal field.

Coal was first mined when the Alaska Railroad reached the Nenana coal field in 1918. Mining was by underground methods, chiefly at the Suntrana mine, until the 1950s. Strip mining of coal, which began in 1944, is the only method used to mine coal today. Several 3–20-m-thick beds within the Suntrana Formation and Healy Creek Formation are separated by 10–60 m of poorly consolidated sandstone and are overlain at the surface by sand and gravel overburden with an overburden/coal ratio of less than 5:1. Essentially all of the stripping coal in the Healy Creek coal basin has been mined, but resources of as much as 360×10^6 metric tons may exist in the Lignite Creek coal basin. In 1985, Usibelli Coal Mine, Inc. produced about 1.4×10^6 metric tons of coal, 48% of which was consumed in Alaska; the rest was exported to South Korea (Green and Bundtzen, 1989).

Jarvis Creek coal field

The Jarvis Creek coal field (loc. 24, Fig. 1) is along the north flank of the Alaska Range on a plateau 1,000–1,300 m high, about 160 km east of the Nenana River and 3–10 km east of the Delta River and the Richardson Highway. Coal-bearing, poorly consolidated, continental sedimentary rocks totaling 600 m in thickness that underlie an area of about 40 km^2 of this plateau are correlated with the Healy Creek Formation of the Nenana coal field (Wahrhaftig and Hickcox, 1955). These coal-bearing sedimentary rocks rest on an erosional surface with at least 150 m of local relief that has been cut into crystalline rocks.

TABLE 3. RESERVES OF COAL IN THE LIGNITE CREEK AND HEALY CREEK COAL BASINS, NENANA COAL FIELD, ALASKA*

Reserve reporting unit	Average thickness of coal beds (m)	Total Number of Coal Beds	Reserves in millions of metric tons with overburden thickness of: Less than 30 m† inferred	30–91 m† inferred	Less than 305 m§ measured	Less than 305 m§ indicated	Less than 305 m§ inferred	305–610 m measured	305–610 m indicated	305–610 m inferred	610–915 m indicated	610–915 m inferred
A	>3	3	18.1	0.3	2	25	100	....	....	....	....	....
	1.5 to 3	3	14.3	5.2	4	16	13	....	....	....	....	....
	<1.5	2	1.9	0.2	....	....	9	....	....	....	....	....
B	>3	2	7.2	3.0	....	6.0	15	....	....	....	....	....
	<3	4	6.4	0.2	....	....	8	....	....	....	....	....
C	>3	2	0.9	....	....	....	96	....	....	....	....	....
D	>3	4	20.6	46.0	....	381	73	....	23	162	....	98
	1.5 to 3	2	....	....	....	....	31	....	....	17	....	....
	<1.5	1	....	....	....	....	22	....	....	....	....	....
E(NW)	>3	5	7.8	....	....	62	46	....	....	107	....	....
	<3	1	....	....	....	....	....	....	....	10	....	....
E(SE)	>3	7	4.4	5.7	....	94	80	....	5	3	....	....
	<3	1	....	....	....	8	....	....	1	....	....	....
F	>3	6	20.4	28.3	107	150	45	....	269	20	....	199
	1.5 to 3	1	....	....	....	7	13	....	....	30	....	....
	<1.5	1	....	....	....	....	3	....	....	3	....	....
G	>3	6	40.2	68.8**	113	414	145	....	123	22	....	....
	<3	3	....	....	14	....	16	....	6	....	....	....
H	>3	8	35.6	29.1	....	232	241	....	....	1	....	....
	<3	3	....	....	....	35	66	....	....	....	....	....
I	>3	2	3.3	....	....	....	99	....	....	....	....	....
	<3	4	6.2	....	....	....	110	....	....	....	....	....
J	<3	5	....	....	....	5.9	13	....	4	13	....	20
K	>3	11	‡	‡	272	79	100	248	53	96	222	79
	<3		....	....	....	1	15	....	1	12	....	2
	Totals		**187.3**	**186.5**	**512**	**1,514.9**	**1,356**	**248**	**485**	**496**	**222**	**398**
Totals by overburden category			**373.8**		**3,382.9**			**1,229**			**620**	
	Grand total											**5,605.7**

*C. Wahrhaftig, unpublished data, 1971. Reserve calculations are based on field work completed in 1960 and do not take into account coal mined since then. For location of reserve reporting units, see Figure 15.
†All classed as inferred because of possibility that beds are burned.
§Excluding reserves in first two reserve columns.
**Overburden 100 to 200 ft.
‡Extensive deposits of coal with less than 300 ft of overburden are mined in this reporting unit. Reserves have not been calculated because of difficulty of estimating the thickness of terrace gravel deposits.

The sedimentary rocks are gently deformed into an oval basin, the flanks of which dip generally less than 10°. The clastic components appear to be from two local sources: the lowermost beds are derived from the south, and the overlying beds are derived from the north and northeast. Although no connection between the Jarvis Creek coal field and the Cook Inlet area is known, the northern source of sediment implies that drainage from the coal field was to the south.

Thin, discontinuous coal beds are present throughout the section, but most of these beds are less than 0.75 m thick. Wahrhaftig and Hickcox (1955) calculated 12 × 10^6 metric tons of indicated and inferred resources of coal in seven beds exposed along the south and east sides of the coal field, and suggested that the field might contain as much as 57 × 10^6 metric tons of additional resources for which no outcrop evidence is available. Metz (1981) reported that drilling has blocked out about 1 × 10^6 metric tons of stripping coal in part of the coal field. The coal is subbituminous C, with a heating value of 4,400–5,000 cal/gm and an ash content of 5–13 wt%.

The north flank of the Alaska Range, between the east end of the Nenana coal field at the Wood River and the Jarvis Creek coal field, has broad, east-west–trending anticlinal and synclinal structures marked by arcuate cuestas of the Nenana Gravel. Coal-bearing rocks of the Usibelli Group are present discontinuously along the unconformable contact of the Nenana Gravel with underlying Paleozoic and Mesozoic crystalline rocks. The Usibelli Group has been recognized at several localities along Dry Creek (long 147°15′W; loc. 25, Fig.1); at the mouth of Newman Creek the group may be as much as 750 m thick. Two beds, each about 3 m thick, dipping 66°NE, are present on the west bank of Dry Creek at this locality (Inyo Ellersieck, 1987, oral and written commun. based on helicopter reconnaissance in 1977). Here the Usibelli Group extends from around the nose of a southeast-plunging anticline having a core of Paleozoic schist southward to about 6.5 km upstream in Newman Creek and, from there, about 8 km westward to Red Mountain Creek. In this vicinity, at least three coal beds averaging 1.4 m thick are exposed in a 45-m-thick section, and a borehole nearby penetrated at least two beds of minable coal, one 1.4 and the other 2.9 m thick (Merritt, 1985b). Patches of clinker indicate the presence of coal in the Usibelli Group on Dry Creek north of the mouth of Newman Creek. If the exposures of the Usibelli Group, which extend for about 24 km, have an average of at least 10 m of minable coal, then 60–70 × 10^6 metric tons of hypothetical resources may be present in this coal basin having an overburden depth of 500 m.

Coal has also been reported on the east bank of Delta Creek opposite the mouth of Ptarmigan Creek (lat 63°50′N, long 146°30′W, loc. 26, Fig. 1), where Ellersieck observed at least one coal bed more than 3 m thick in a complexly deformed exposure. Usibelli Group may also crop out in the core of an anticline that straddles the Little Delta River at lat 64°N; however, no coal has been reported there. Coal may be present elsewhere in the region between the Wood and Delta Rivers; however, a thick mantle of glacial and glaciofluvial deposits buries most of the lowland areas, and its extent, if any, is unknown.

Little Tonzona coal field

Exposures of coal of Tertiary age are present in a 145-km-long belt, locally as much as 3 km wide, along the Farewell fault on the northwest side of the Alaska Range between the Little Tonzona River on the northeast (lat 62°45′N, long 153°00′W) and the Cheeneetnuk River on the southwest (lat 62°00′N, long 155°15′W) (Barnes, 1967a; Sloan and others, 1979; Reed and Nelson, 1980; Solie and Dickey, 1982; Dickey, 1984; Gilbert and others, 1982). On the west bank of the Little Tonzona River, seven coal beds up to 9 m thick have a combined thickness of 35 m of subbituminous coal (loc. 27, Fig. 1). The beds are interbedded with ferruginous siltstone and silty shale; they strike N37°E and dip 47°–63°NW (Sloan and others, 1979; Reed and Nelson, 1980). The coal contains 6–11 wt% ash, 0.7–1.7 wt% sulfur, and is subbituminous C. Additional exposures of coal occur on Deepbank Creek, 11 km to the southwest.

Coal exposures are also on the west bank of Windy Fork of the Kuskokwim River. The section is 1,800 m thick and consists of about 90 vol% conglomerate; a 190-m-thick section of interbedded coal and shale is near the top and two or three 10-m-thick sequences of coal and shale are near the base of the section (loc. 28, Fig. 1; Solie and Dickey, 1982). According to Sloan and others (1979), who examined the 190-m-thick coal and shale section, the coal beds are mostly too bony to constitute a fuel resource. Only one bed has an ash content as low as 31 wt%; all other analyses give more than 40 wt% ash. According to Solie and Dickey (1982), within the 190-m-thick section are three beds totaling 20 m in thickness that contain 7–10 wt% ash and 0.5–0.6 wt% sulfur, and two other beds that total 6 m in thickness contain 10 and 18 wt% ash. The coal in both groups is high-volatile bituminous C or subbituminous A, and has moist ash-free heat content of 5,560–6,670 cal/cm. The beds of coal at this locality crop out on both limbs of a northeast-plunging syncline and dip 34°–70° (Gilbert and others, 1982).

About 18 km southwest of this locality, at the south edge of the alluvial plain of the Tanana-Kuskokwim lowland, an exposure of lignite was reported by Solie and Dickey (1982), apparently the same rock unit as the Windy Fork exposures. The marked difference in the rank of coal in these supposedly coeval rocks within the same structural context is puzzling. The structure of the exposure as described by Solie and Dickey (1982) is unlike that mapped by Gilbert and others (1982). About 5 km southwest of the lignite locality, where the Middle Fork of the Kuskokwim River emerges from the Alaska Range onto the Tanana-Kuskokwim Lowland, in a riverbank exposure of conglomerate, claystone, and coal striking N26°E and dipping 45°NW, Solie and Dickey (1982) observed 35 m of coal in six beds within the uppermost 100 m of the 600-m-thick section. At the southwest end of these coal occurrences, coal is exposed on the banks of the Cheeneetnuk River; the beds are within a block of Tertiary sedimentary rocks, about 1 km wide and a few kilometers long, that have been downfaulted into Paleozoic rocks (loc. 29, Fig. 1). The coal is intermediate in rank between subbituminous and high-volatile bituminous, and one bed was reported to be at least 5 m thick (Solie and Dickey, 1982).

Dickey (1984) reported that most of the clastic rocks in the Little Tonzona field were deposited by southward-flowing streams and that the abundant conglomerate in the section was probably deposited as alluvial fans along the base of a south-facing escarpment.

Both the southward-directed current indicators and the fact that the lower parts of the section are conglomeratic and resemble the section at Farewell Mountain in the Susitna Lowland suggest that the Little Tonzona field may have been one of the source regions for streams that flowed through the western Susitna Lowland to the Cook Inlet region.

CRETACEOUS COAL OF THE ALASKA PENINSULA

Coal occurs at Herendeen and Chignik Bays on the Alaska Peninsula in the Chignik Formation of Campanian (Late Cretaceous) age (Figs. 1 and 16). The coal accumulated in littoral and overbank swamps on a gentle shelf facing a major submarine trench to the south. To the north was a rolling upland underlain by plutonic, volcanic, and oceanic sedimentary rocks of Jurassic and older age.

The Chignik Formation is a four-times-repeated cyclic sequence, about 500 m thick, of shallow-water, nearshore marine, clastic sedimentary rocks overlain by continental sedimentary rocks (including coal beds). The clastic sedimentary rocks are typically calcareous and include fine- to medium-grained brown sandstone composed of quartz, chert, feldspar, and lithic fragments that are locally channeled and crossbedded; pebble and cobble conglomerate of black, gray, green, and red chert, white quartz, granitic clasts derived from the Jurassic Alaska–Aleutian Range batholith, and minor volcanic clasts; and dark, sandy micaceous siltstone (Detterman, 1978; Detterman and others, 1981). The Chignik rests with shallow angular unconformity on sedimentary rocks ranging in age from Late Jurassic to Early Cretaceous (late Neocomian) (Burk, 1965).

Rocks contemporaneous with the Chignik Formation that occur on offshore islands south of the Alaska Peninsula and beneath the Shelikof Strait are deep-marine volcaniclastic turbidites of a flysch sequence that accumulated in a deep oceanic trench (Plafker and others, this volume, Chapter 12). The shelf edge was apparently not far south of the depositional site of the Chignik Formation.

The Chignik Formation is thick and coal bearing only be-

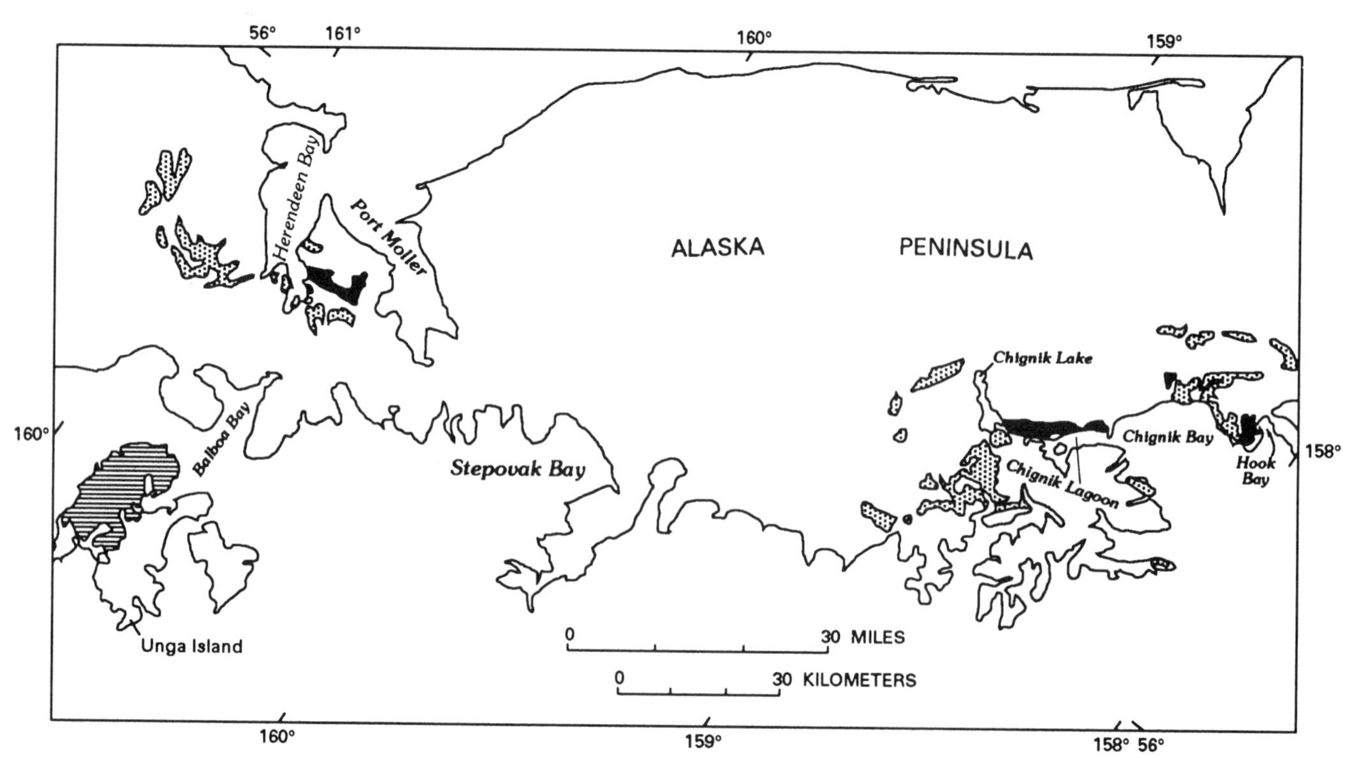

Figure 16. Map of part of the Alaska Peninsula and Unga Island, showing areas of outcrop of the Chignik and Bear Lake formations. Solid black is Chignik Formation where minable coal has been described; stippled area is Chignik Formation where it has not been examined for coal; lined area is Bear Lake Formation. North of latitude 56°00′ from Detterman and others (1981); south of latitude 56°00′ from Burk (1965).

tween Herendeen Bay and Hook Bay, a distance along the peninsula of about 200 km. Farther east (for example, just east of Aniakchak caldera) it is less than 30 m thick and contains no coal (R. L. Detterman, 1986, oral commun.). Presumably, a local highland immediately north of the Herendeen Bay–Chignik Bay area provided abundant sediment to a slowly subsiding marginal shelf, creating swampy flood plains with conditions amenable to coal accumulation.

According to F. H. Wilson and others (1991, written commun.), in the vicinity of Herendeen Bay, the Chignik Formation and other Mesozoic and lower Tertiary rocks form the upper plate of a nearly horizontal thrust fault, and the rocks in this thrust sheet are cut by numerous normal faults, resulting in almost chaotically jumbled kilometer-sized blocks.

The known coal exposures are located on the peninsula in the Herendeen Bay coal field (loc. 30, Fig. 1; also see Fig. 16) between Herendeen Bay and Port Moller, and in the Chignik Bay coal field (loc. 31, Fig. 1; also see Fig. 16) from Chignik Lake eastward along the north shore of Chignik Lagoon and Chignik Bay as far east as Hook Bay (Atwood, 1911; Conwell and Triplehorn, 1976). Several tunnels were driven along the coal seams in the syncline east of Mine Harbor on Herendeen Bay, but all penetrated faults, losing the coal within a few hundred meters of the portal. The coal mine on the Chignik River was operated between 1893 and 1911 to provide fuel for local fish canneries. The relatively high rank of the coal and its accessibility to ice-free harbors make it of economic interest.

Thicknesses of coal in the Herendeen Bay coal field have been reported for four localities (Paige, 1906; Atwood, 1911; Conwell and Triplehorn, 1976). Total thicknesses of coal in beds at least 0.8 m (2.5 ft) thick at these four localities range from 1.5 to 5 m (5–16 ft). A representative columnar section is shown in Figure 17A. All localities are in the western part of the synclinal body of the Chignik Formation that extends eastward on the peninsula from Mine Harbor in Herendeen Bay. This synclinal body covers an area of about 25 km^2, and if at least 1.2 m (4 ft) of minable coal is present throughout this area, it contains 35 × 10^6 metric tons of minable coal. The remaining areas of exposure of the Chignik Formation in the Herendeen coal field total about 85 km^2 (Burk, 1965, plate 2), and if 1.2 m (4 ft) of minable coal underlies 25% of that area, then 27–30 × 10^6 metric tons of additional coal may be present.

In view of the chaotic structure reported by F. H. Wilson and others (1991, written commun.) for the Chignik Formation in the Herendeen Bay coal field, the foregoing figures must be regarded as highly speculative, probably only an upper limit of what might prove to be minable. Thorough subsurface exploration will be necessary to establish any actual reserves.

Exposures of coal have been described at four localities between Chignik Lake and Hook Bay: on the Chignik River, north of Chignik Lagoon, north of the northeastern part of Chignik Bay, and at Hook Bay. A representative columnar section is shown in Figure 17B. The total thickness of coal in these occurrences is 1–4 m. If 1.2 m of minable coal is assumed to underlie the area north of Chignik Bay, Chignik Lagoon, and the area around Hook Bay, which together total about 39 km^2 ~55 × 10^6 metric tons of minable coal may exist beneath these areas. The area of exposure of the Chignik Formation in the Chignik coal field is 175 km^2 (Detterman and others, 1981); if 25% of this area is underlain by 1.2 m of minable coal, there could be an additional 62 × 10^6 metric tons of coal in this field. Thus, the two Cretaceous coal fields together have speculative resources amounting to slightly more than 180 × 10^6 metric tons.

According to Conwell and Triplehorn (1976), the coal is high-volatile bituminous B, with an ash content of 9.4–12.6 wt% in the Herendeen coal field and 20–40 wt% in the Chignik field. Sulfur content is less than 0.5 wt% in the Herendeen coal field and 0.3–1.4 wt% in the Chignik field. Coal beds in the Chignik field range from 0.3 to 1.5 m in thickness and are moderately folded and faulted.

TERTIARY COAL OF UNGA ISLAND

Tertiary strata, 7,600–9,140 m thick, are exposed in the western Alaska Peninsula and adjacent Shumagin Islands (Burk, 1965). A Paleogene sequence as thick as 6,100 m, largely derived from volcanic sources, is overlain unconformably by more than 1,500 m of marine and nonmarine sandstone and conglomerate of Miocene age (Bear Lake Formation). The sandstone consists of approximately equal amounts of volcanic clasts and chert, quartz, and granitic and other nonvolcanic clasts. Showings of thin lignite beds have been reported in these Tertiary rocks; however, the only minable occurrence of coal is in the northwestern part of Unga Island, on the west shore of Zachary Bay (loc. 32, Fig. 1; see Fig. 16).

Nilsen (1984) interpreted the bulk of the Bear Lake Formation to have accumulated within a broad shallow tidal-flat embayment in a back-arc setting on the north side of a middle Tertiary Aleutian arc, similar to, but larger than, the modern tidal flats at Port Heiden and Port Moller. The alluvial plain on the south margin of the tidal embayment included the site of Unga Island, accounting for the predominantly fluvial materials, abundant conglomerate, and scattered thin coal beds of Unga Island.

Northwestern Unga Island is a broad plateau, about 10 km on a side, underlain by the Oligocene(?) and lower and middle Miocene Unga Conglomerate Member of Burk (1965) of the Bear Lake Formation, which strikes northerly and dips 4°–5°W (Atwood, 1911; Burk, 1965, p. 92–93). A 10-m-thick section of fine-grained sandstone and shale containing two lignite beds, each including about 0.9 m (3 ft) of lignite, is exposed about 60 m above sea level near the south end of bluffs on the west side of Zachary Bay and is overlain by about 110 m of conglomerate and sandstone. According to Atwood (1911), the lignite was mined between 1882 and 1884 to provide fuel for coastal steamships, and a few hundred tons were reportedly shipped to San Francisco. At the time of Atwood's visit in 1908, the upper lignite bed was being developed with a crosscut tunnel, an

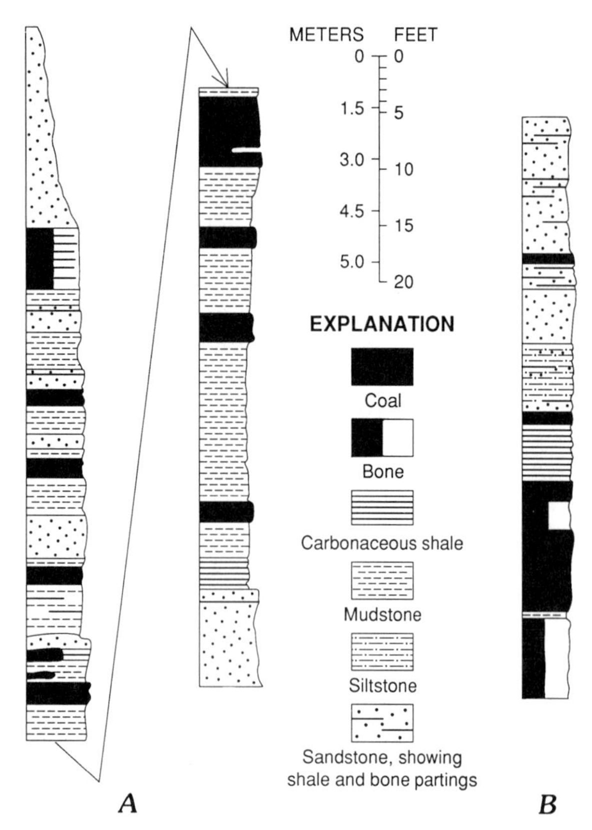

Figure 17. Measured sections of coal in the Chignik Formation. A: 2.4 km (1.5 mi) southeast of the village of Herendeen Bay. B: North bank of Chignik River near mouth (after Conwell and Triplehorn, 1978, Figs. 4 and 8).

aerial tramway, and bunkers for storing mined coal near the shore. A sample collected by Atwood contained 26 wt% ash and had a heating value of only 3,225 cal/gm. Lyle and others (1979) measured the same sequence about 1.6 km north of Atwood's locality and reported that the upper bed contained only 0.9 m of coal, 0.3 m of which is bony, and the lower bed contained only 0.5 m of coal, with a 0.2 m clay parting in the middle. There is no record of any coal being produced from this locality after Atwood's visit in 1908. Wilson and others (1991, written commun.) were unable to find the coal mine described by Atwood (1911), and have concluded that the coal will probably be used, if at all, for local purposes.

GULF OF ALASKA TERTIARY PROVINCE

Geologic setting

The Gulf of Alaska Tertiary province, a 485-km-long stretch of Tertiary rocks bordering the eastern Gulf of Alaska between Cross Sound and the Copper River, is the onland part of the Yakutat terrane, a nearly triangular, largely submarine crustal block south and west of the Chugach–St. Elias and Fairweather faults. The Yakutat terrane is still being accreted to the mainland of Alaska along the Aleutian Trench subduction zone (Plafker, 1987; Plafker and others, this volume, Chapter 12). Basement rocks that make up the east third of the Yakutat terrane consist of Mesozoic flysch and melange of the Yakutat Group, intruded by Eocene plutonic rocks; the basement of the offshore part of the rest of the Yakutat terrane is possibly Paleocene and lower Eocene oceanic crust. Along the north edge, between the shoreline and the Chugach–St. Elias fault, basement rocks are probably greenschist and zeolite facies, metamorphosed flysch and oceanic volcanic rocks of the Orca Group (Plafker and others, this volume, Chapter 12). Deposited on this basement are clastic sedimentary rocks that locally may exceed 12,000 m in thickness and range in age from late Eocene to late Pleistocene. These sedimentary rocks appear to have been derived from upland areas in present-day British Columbia and southeastern Alaska, as the Yakutat terrane migrated northwestward along the Fairweather transform fault. The bulk of this succession is marine—some of it is deep-water marine—but a thick continental and paralic sequence within it contains coal. Unlike coal elsewhere in Alaska, coal in the Gulf of Alaska Tertiary province accumulated when the Yakutat terrane was several hundred kilometers southeast of its present position relative to the rest of Alaska.

Kulthieth Formation—the Bering River coal field

The Bering River coal field (loc. 33, Fig. 1) is in the southern foothills of the Chugach Mountains, 30–65 km east of the mouth of the Copper River at the northwest corner of the Gulf of Alaska Tertiary province (Plafker, 1987). It underlies a northeast-trending area, about 35–40 km long and 3–10 km wide, between the Martin River Glacier and the Bering Glacier and River (Martin, 1908), and contains the great bulk of Alaska's known deposits of high-rank coal. The coals, which are low-volatile bituminous to anthracite, occur in the Kulthieth Formation of Eocene and Oligocene age (Fig. 18; Miller, 1957; Plafker, 1987). The Kulthieth Formation crops out in an east-west–trending belt, 3–19 km wide and 210 km long, from the Bering River coal field on the west to the Samovar Hills on the east. It lies in rugged, glacier-clad foothills at the base of the escarpment of the Chugach and St. Elias Mountains, 19–45 km north of the coast. Small outcrops of subbituminous coal on the west shore of Disenchantment Bay, 45 km north of Yakutat, are also assigned to the Kulthieth Formation. Although the Kulthieth Formation contains coal beds throughout its length of outcrop, its coal potential has been studied in detail in only two localities, the Bering River coal field and the Samovar Hills (loc. 34, Fig. 1). In most of the intervening belt, the formation is exposed on inaccessible cliffs, and long stretches of potential exposure in the belt are covered by vast glaciers.

The Kulthieth Formation is about 2,835 or more m thick in its type locality at the head of the Kulthieth River (Miller, 1957). It is overlain conformably by the 1,860-m-thick marine Poul Creek Formation, a unit composed of interbedded siltstone and sandstone of late Eocene to early Miocene age, in turn overlain

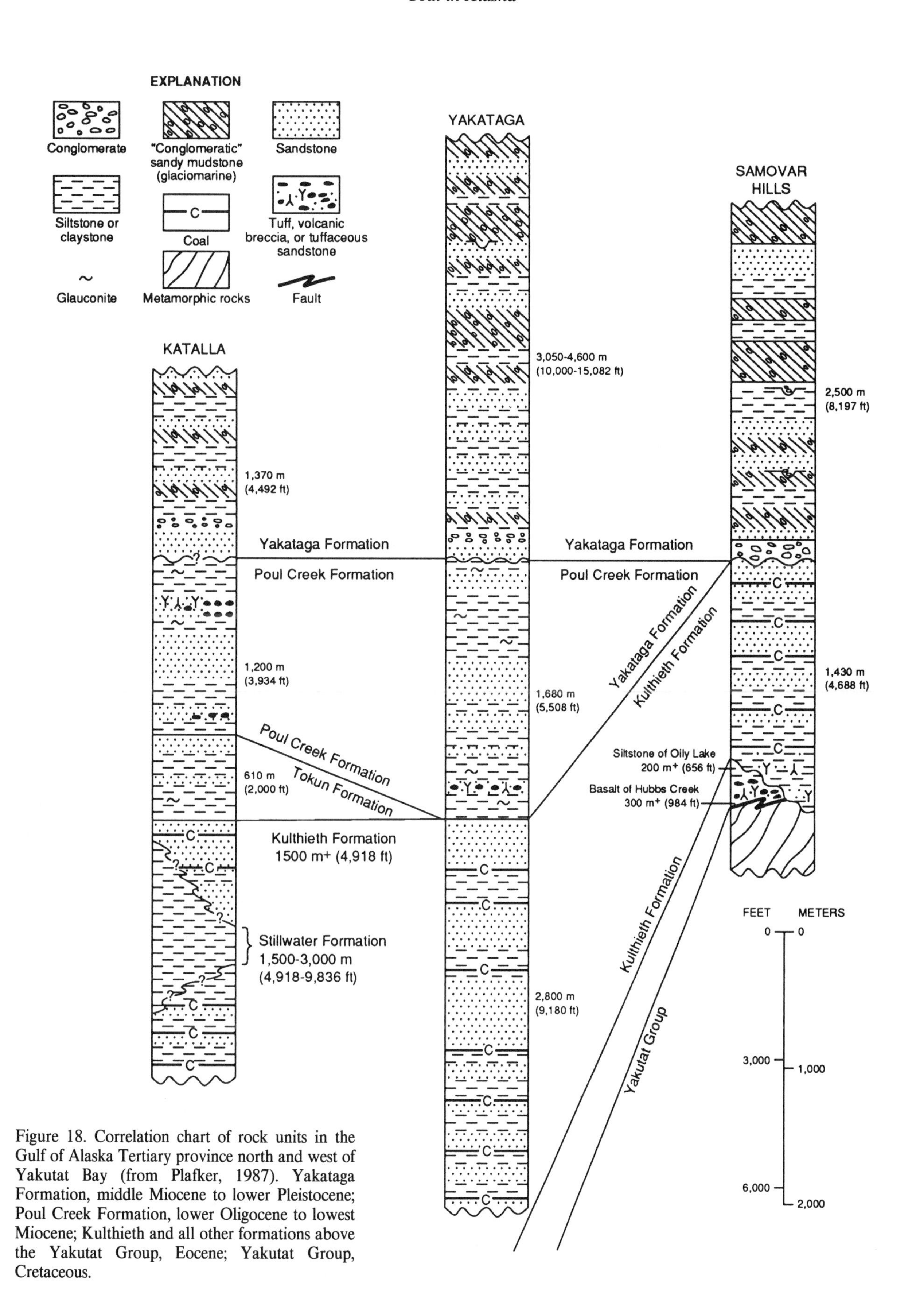

Figure 18. Correlation chart of rock units in the Gulf of Alaska Tertiary province north and west of Yakutat Bay (from Plafker, 1987). Yakataga Formation, middle Miocene to lower Pleistocene; Poul Creek Formation, lower Oligocene to lowest Miocene; Kulthieth and all other formations above the Yakutat Group, Eocene; Yakutat Group, Cretaceous.

by the 3,050–4,570-m-thick Yakataga Formation, a marine sequence that includes marine tillites of early Miocene to Pleistocene age (Fig. 18). According to Plafker and Miller (1957), The Kulthieth Formation just south of Mount St. Elias is about 3,050 m thick, but the section may be repeated by folds and thrusts. About 16–20 km east-southeast of this exposure, the Kulthieth of the Samovar Hills is more than 1,435 m thick, rests unconformably on the Yakutat Group, and is unconformably overlain by the Yakataga Formation. In the Bering River coal field, the Kulthieth Formation (Kushtaka Formation of Martin, 1908; Nelson and others, 1985) is variously reported to be 610–1,585 m thick; the unit interfingers with the underlying deep-water marine Stillwater Formation and the overlying shallow-marine Tokun Formation, which in part may be correlative with the basal part of the Poul Creek (Fig. 18).

The Kulthieth is a conspicuously banded sequence of sandstone, siltstone, shale, and coal. The sandstone is fine to medium grained, commonly crossbedded, well sorted, arkosic, cemented generally by calcite, and weathers orange. The siltstone and shale are dark gray and carbonaceous. Most of the coal beds apparently occur in the predominantly silt-shale sequences (Barnes, 1951; Miller, 1957; Plafker and Miller, 1957). Turner and Whateley (1989) regarded the lower two-thirds of the Kulthieth Formation, dominated by upward-coarsening sandstone sequences in the Bering River coal field, to be a deltaic and delta-plain sequence built onto the moderately deep sea floor on which shales of the Stillwater Formation accumulated. They interpreted the thin and discontinuous coals of this part of the formation to have accumulated as tidal-marsh deposits in the embayments between distributaries, and the upper, more coal-rich third of the deposits in the Bering River coal field to be of delta-plain and alluvial-plain origin.

The predominance of arkosic sandstones in the Kulthieth and other Paleogene formations indicates that the Yakutat terrane lay 600 km southeast of its present position, probably offshore of British Columbia or southern southeastern Alaska, when the coals accumulated (Plafker, 1987; Plafker and others, this volume, Chapter 12). Similarities between heavy-mineral assemblages of Eocene sandstones of the Yakutat terrane and the Kootznahoo Formation and related units in southeastern Alaska suggest that they both had a common source (Chisholm, 1985), possibly in the Coast Mountains of southeastern Alaska and British Columbia. Paleontologic evidence is equivocal, but is compatible with at least 600 km of displacement and, conceivably, much more (Bruns, 1983; see discussions by Plafker, 1983, 1984; Bruns, 1984; Bruns and Keller, 1984; Wolfe and McCoy, 1984; Chisholm, 1985; von Huene and others, 1985; Plafker, 1987; Marincovich, 1990). During the past 20–30 m.y., transcurrent motion along the Queen Charlotte–Fairweather transform fault has caused subduction of the leading edge of the Yakutat terrane beneath the Chugach terrane along the Chugach–St. Elias fault zone and accretion against the Chugach and Prince William terranes of the east-trending part of the Gulf of Alaska Tertiary province (Plafker and others, this volume, Chapter 12). To accommodate the change in direction of the coastline in the vicinity of Mount St. Elias, the province has been rotated counterclockwise and telescoped or shortened by progressively greater amounts of rotation to the west.

The Kulthieth Formation lies in an imbricate thrust zone characterized by south-verging overturned folds and numerous thrust faults. The intensity of deformation is greatest in the Bering River coal field, where the coal beds commonly are sheared, crushed, and squeezed from the flanks of the folds into pods along the hinges (Sanders, 1976).

Coal was discovered in the Bering River coal field in 1896 (Barnes, 1951), and was described in detail by Martin (1908). From 1912 to 1920, some experimental mines were opened, and the coal was tested by the U.S. Navy, but it was reported to be unsatisfactory for steaming, and there was essentially no interest in it until the early 1980s, when there was exploration for coal to be exported to the Far East (Eakins and others, 1983; Turner and Whateley, 1989). Most of the results of that exploration are proprietary however, and no production has yet resulted.

Assessment of the Bering River coal field. In the Bering River coal field, coal increases in rank from semibituminous at the west end of the field through semianthracite to anthracite in the Carbon Mountain area at the east end of the field (Cooper and others, 1946; Barnes, 1951). Martin (1908) described the coal as occurring in beds from a few centimeters to as much as 14 m in thickness; however, bed thickness changes abruptly, probably owing to tectonic squeezing of coal from flanks to noses of folds and to deformation along faults. According to Martin (1908) and Barnes (1951), the semibituminous coal (between bituminous and semianthracite) was found to be coking; however, Sanders (1976) stated that none of the coal is coking. U.S. Bureau of Mines analyses of 11 outcrop samples of coal collected from the Yakataga and Malaspina districts in 1952–1954 by D. J. Miller and George Plafker (1988, written commun.) show coal near the Chugach–St. Elias fault to be semianthracite, and coal in the Robinson Mountains, Samovar Hills, and Yakutat Bay to be bituminous.

The high rank of the Bering River coals appears to be due to heat provided from igneous intrusions at depth. Numerous basalt and diabase dikes and sills cut the coal-bearing rocks and are especially abundant in the east half of the coal field (Martin, 1908). Sills that intruded along coal seams are reported to have produced natural coke. The anomalously high thermal maturity of the coal-bearing and other Tertiary sedimentary rocks in comparison to that of the older Orca Group in the Chugach Mountains immediately north and west of the Yakutat block (Mull and Nelson, 1986) suggest that the thermal event that raised the rank of the coal (and presumably led to the intrusion of dikes and sills) occurred before accretion of the Yakutat block to the Alaskan continental margin.

The coal beds have been so intensely deformed that reserve estimates are highly speculative. Reserve estimates have been published only for the Bering River coal field, for which Eakins and others (1983) listed 56×10^6 metric tons of proved coal and an additional 32 to 900 × 10^6 metric tons of inferred coal. Sanders (1976) stated that the coal is mainly in the form of pods along fold axes and that coal reserves cannot be calculated in the normal way. By a technique that appears to compare the outcrop area of coal to that of the formation as a whole, Sanders (1976) calculated 3.3 × 10^9 metric tons of hypothetical coal to a depth of 914 m. It is doubtful whether mining of coal to that depth would ever be economically feasible.

A calculation made for this report (Plafker, 1987; George Plafker, 1988, unpublished data) of coal in the Samovar Hills implies that for a strike length of 4 km and a downdip distance of 1 km, the reserves are 135 × 10^6 metric tons of coal in beds more than 1 m thick, of which 100 × 10^6 metric tons is in beds more than 3 m thick. The greatest thickness reported is 12 m. The extent to which the beds have been tectonically thickened or thinned is unknown. No coal was reported in any of the 11 wells drilled for petroleum on the continental shelf southeast of the Bering River coal field (Bohm and others, 1976; Plafker and others, this volume, Chapter 12), and there is no estimate for offshore coal resources in this region (Affolter and Stricker, 1987b).

TERTIARY COAL OF SOUTHEASTERN ALASKA

Small coal deposits of little potential commercial value occur in southeastern Alaska (Fig. 1; Barnes, 1967a). All of these deposits, except the Kootznahoo deposit discussed below, are of lignite. The deposits are on Kootznahoo Inlet (loc. 35, Fig. 1) on the west side of Admiralty Island (Lathram and others, 1965); at Murder Cove (loc. 36, Fig. 1) on the south tip of Admiralty Island; at Kasaan Bay (loc. 37, Fig. 1) on Prince of Wales Island; on Zarembo Island; and at Port Camden on Kuiu Island and several nearby localities (loc. 38, Fig. 1).

The largest deposit at Kootznahoo Inlet is in rocks of Tertiary age and was mined by underground methods before 1929 for use at Juneau. The coal is bituminous rank, has a high sulfur content, and occurs only locally with many shale partings (Sanders, 1981). Coal beds range in thickness from 0.6 to 0.9 m (Affolter and Stricker, 1987a). No offshore extensions of these deposits are known (Affolter and Stricker, 1987b).

SUMMARY

The coal resources calculated for Alaska are enormous (Table 4), amounting to 3.4 × 10^{12} metric tons beneath onland Alaska, of which 210 × 10^9 metric tons are identified (measured, indicated, or inferred) and 3.2 × 10^{12} metric tons are hypothetical (based on coal penetrated in exploratory oil wells and judged to extend beneath likely basins). In addition, 5.5 × 10^{12} metric tons of hypothetical resources are thought to lie offshore of northern Alaska in the Beaufort and Chukchi Seas, and 0.7 × 10^{12} metric tons are thought to underlie the waters of Cook Inlet. These deposits amount to 50%–70% of the coal resources calculated for the entire United States (Wood and Bour, 1988).

TABLE 4. COAL RESOURCES OF ALASKA

Province or coal field	Identified resources* (refs)† (metric tons)	Hypothetical resources (refs)† (metric tons)
NORTH SLOPE PROVINCE		
On land	109 x 10^9 (A)	1.2 x 10^{12} (B, C) 1.7 x 10^{12} (B, D)
Offshore	44 x 10^9 (E)	5 x 10^{12} (F)
COOK INLET–SUSITNA PROVINCE		
Matanuska Valley	113 x 10^6 (G)	249 x 10^6 (G)
	18,000 anthracite (H)	
	180 x 10^6 bituminous (H)	22 x 10^9 (H)
KENAI GROUP IN THE COOK INLET–BELUGA LAKE LOWLAND		
Kenai Group as a whole		1.2 x 10^{12} of which 0.8 x 10^{12} is offshore (K)
Homer District	360 x 10^6 (I)	
Tyonek-Beluga District	36–45 x 10^9 (J)	
North of Castle Mountain fault	489 x 10^6 (L)	
KENAI GROUP IN SUSITNA VALLEY		
Peter Hills and Fairview Mountain Districts	40 x 10^6 (L)	
Johnson Creek and Canyon Creek Districts	450 x 10^6 (M)	
Broad Pass District	0.27 x 10^6 (N)	12.2 x 10^6 (N)
NORTH FLANK OF THE ALASKA RANGE		
Nenana Coal Field	6.4 x 10^9 (O)	58 x 10^6 (P)
North Side of the Alaska Range east of Nenana coal field	12 x 10^6 (P)	63–73 x 10^6 (Q)
Herendeen Bay Coal Field		70–75 x 10^6 (R)
Chignik Coal Field		128 x 10^6 (S)
GULF OF ALASKA TERTIARY PROVINCE		
Bering River Coal Field	56 x 10^6 (T)	3.3 x 10^9 (U)
Samovar Hills		150 x 10^6 (V)
Totals	**197–206 x 10^9**	**10.4 x 10^{12}**

*Identified resources include measured, indicated, and inferred resources of coal; resources were not calculated for coal of Paleozoic age, the Yukin-Koyukuk province, the Little Tonzona coal field, the Seward Peninsula, or southeastern Alaska.

†A = Tailleur and Brosgé (1976); B = New calculation based on method of Sable and Stricker (1987); C = Subbituminous coal; D = Bituminous coal; E = Martin and Callahan (1978); F = Affolter and Stricker (1987b); G = Barnes (1967a); H = Merritt and Belowich (1984); I = Barnes and Cobb (1959), overburden less than 300 m (1,000 ft); J = Ramsey (1981), overburden less than 300 m; K = Affolter and Stricker (1987a); L = Barnes (1966); M = Blumer (1981); N = Hopkins (1951); O = Barnes (1967a), overburden less than 1,000 m, see also Table 3; P = Wahrhaftig and Hickcox (1955); Q = Calculations based on oral and written communications (1985) from Inyo Ellersieck; R = Calculations based on data of Conwell and Triplehorn (1976) and Burk (1965); S = Calculations based on data of Detterman and others (1981); T = Eakins and others (1983) and Sanders (1976); U = Sanders (1976); V = Calculated from data of George Plafker, written communication (1988).

Nearly all of the coal is of Cretaceous and Tertiary age. A small amount of bituminous coal occurs in highly deformed rocks of Mississippian age south of Cape Lisburne, and coal of Mississippian age has been penetrated in deep wells near Point Barrow. Information on this coal is so fragmentary that resources have not been calculated.

The major coal provinces are the North Slope province, north of the Brooks Range, and a province centered on the Cook Inlet–Susitna Lowland on the south coast of Alaska. Coal in the North Slope province is mainly in fluviatile and deltaic sedimentary rocks of the Cretaceous Nanushuk Group, where it is mainly bituminous, and in the Upper Cretacous Colville Group and Upper Cretaceous and Tertiary Sagavanirktok Formation, where it is mainly subbituminous. Coal crops out along the seacoast at Corwin Bluff and elsewhere along riverbanks, but the enormous resources, amounting to 2.9×10^{12} metric tons, are based largely on records of exploratory wells for petroleum.

Coal in the Cook Inlet–Susitna province is Tertiary in age and occurs in several isolated basins that are thought to have once been connected by a Tertiary river system. The Matanuska Valley coal field, of Paleocene age, contains 295×10^6 metric tons of identified resources of bituminous coal and 18×10^3 metric tons of identified resources of anthracite, occuring where igneous intrusions cut the coal beds. The complex structure of this coal field hinders any development.

Nearly all the rest of the coal in the Cook Inlet–Susitna province is subbituminous and Neogene in age. The entire Kenai Group contains 1.1×10^{12} metric tons of hypothetical resources of coal within 1,000 m of the surface, 0.7×10^{12} metric tons of which are beneath Cook Inlet. A total of 360×10^6 metric tons of coal has been estimated for the Kenai Peninsula, and $35–45 \times 10^9$ metric tons for the Tyonek-Beluga district. About 500×10^6 metric tons have been calculated for basins in the Susitna Lowland north of Mount Susitna. Coal fields of this province along the north side of the Alaska Range include the Nenana, Jarvis Creek, and Little Tonsona fields. The bulk of the 6.4×10^9 metric tons of identified resources of coal is in the Nenana coal field.

About 180×10^6 metric tons of hypothetical resources of coal of Cretaceous age are calculated for the Herendeen Bay and Chignik coal fields of the Alaska Peninsula. As much as 3.3×10^9 metric tons of coal may occur in lower Tertiary beds of the Bering River coal field and Samovar Hills in the Gulf of Alaska Tertiary province, but no more than 55×10^6 metric tons is considered identified.

Minor deposits of coal, valuable only for local use, occur on the Seward Peninsula, in the Yukon-Koyukuk province, near Seldovia on the Kenai Peninsula, and at Kootznahoo on Admiralty Island in southeastern Alaska.

Deposits of clinker from burned coal show that it was set on fire by natural causes long before humans arrived in Alaska. The early natives probably used some of it for their campfires. Coal was first mined near Seldovia during the Russian occupation, but coal mining was not a significant component of Alaska's economy until the construction of the Alaska Railroad in the 1920s and 1930s. Bituminous coal from the Matanuska coal field powered the locomotives of the Alaska Railroad until they were converted to oil in the 1950s. At present (1992), the Usibelli mine, a large open-pit mine in the Nenana coal field, is the only active coal mine, producing between 1.3 and 1.8×10^6-metric tons per year, about half of which is used domestically; the other half is shipped to Korea. Considerable interest has been shown in the enormous, thick, tidewater-based coal reserves of the Tyonek-Beluga district.

The coals of Alaska have an unusually low sulfur content, averaging 0.2–0.4 wt%. They represent an enormous energy and organic-materials resource that, with proper care in mining and shipment, could be removed and transported with relatively little environment impact.

REFERENCES CITED

Adkison, W. L., Kelley, J. S., and Newman, K. R., 1975, Lithology and palynology of Tertiary rocks exposed near Capps Glacier and along Chuitna River, Tyonek quadrangle, southern Alaska: U.S. Geological Survey Open-File Report 75-71, 57 p., 1 plate.

Affolter, R. H., and Stricker, G. D., 1987a, Geochemistry of coal from the Cretaceous Corwin and Chandler formations, National Petroleum Reserve in Alaska (NPRA), *in* Tailleur, I. L., and Weimer, P., eds., Alaskan North Slope geology: Bakersfield, California, Pacific Section, Society of Economic Paleontologists and Mineralogists Special Publication 50, p. 217–224.

——, 1987b, Offshore Alaska coal, *in* Scholl, D. W., Grantz, A., and Vedder, J. G., Geology and resource potential of the continental margin of western North America and adjacent ocean basins—Beaufort Sea to Baja California: Houston, Texas, Circum-Pacific Council for Energy and Mineral Resources, Earth Science Series, v. 6, p. 639–647.

Affolter, R. H., Simon, F. H., and Stricker, G. D., 1981, Chemical analyses of coal from the Healy, Kenai, Seldovia, and Utukok River 1:250,000 quadrangles, Alaska: U.S. Geological Survey Open-File Report 81-654, 88 p.

Ahlbrandt, T. S., Huffman, A. C., Jr., Fox, J. E., and Pasternak, I., 1979, Depositional framework and reservoir quality studies of selected Nanushuk Group outcrops, North Slope, Alaska, *in* Ahlbrandt, T. S., ed., Preliminary geologic, petrologic, and paleontologic results of the study of Nanushuk Group rocks, North Slope, Alaska: U.S. Geological Survey Circular 794, p. 14–31.

Atwood, W. W., 1911, Geology and mineral resources of parts of the Alaska Peninsula: U.S. Geological Survey Bulletin 467, 137 p., 15 plates.

Barnes, D. F., 1977, Bouguer gravity map of Alaska: U.S. Geological Survey Geophysical Investigations Map GP-913, scale 1:2,500,000.

Barnes, F. F., 1951, A review of the geology and coal resources of the Bering River coal field, Alaska: U.S. Geological Survey Circular 146, 11 p.

——, 1962, Variation in rank of Tertiary coals in the Cook Inlet basin, Alaska: U.S. Geological Survey Professional Paper 450-C, p. C14–C16.

——, 1966, Geology and coal resources of the Beluga-Yentna region, Alaska: U.S. Geological Survey Bulletin 1202-C, p. C1–C34, plates 1–7.

——, 1967a, Coal resources of Alaska: U.S. Geological Survey Bulletin 1242-B, p. B1–B36, plate 1.

——, 1967b, Coal resources of the Cape Lisburne–Colville River region, Alaska: U.S. Geological Survey Bulletin 1242-E, p. E1–E37.

Barnes, F. F., and Cobb, E. H., 1959, Geology and coal resources of the Homer District, Kenai coal field, Alaska: U.S. Geological Survey Bulletin 1058-F, p. 217–258, plates 17–28.

Barnes, F. F., and Payne, T. G., 1956, The Wishbone Hill District, Matanuska coal field, Alaska: U.S. Geological Survey Bulletin 1016, 88 p., 20 plates.

Barnes, F. F., and Sokol, D., 1959, Geology and coal resources of the Little Susitna District, Matanuska coal field, Alaska: U.S. Geological Survey Bulletin 1058-D, p. 121–138, plates 7–11.

Beikman, H. M., 1980, Geologic map of Alaska: U.S. Geological Survey, scale 1:2,500,000.

Bird, K. J., and Andrews, J., 1979, Subsurface studies of the Nanushuk Group, North Slope, Alaska, *in* Ahlbrandt, T. S., ed., Preliminary geologic, petrologic, and paleontologic results of the study of Nanushuk Group rocks, North Slope, Alaska: U.S. Geological Survey Circular 794, p. 32–41.

Blumer, J. W., 1981, Review of Mobil coal leases—Yentna region, Alaska, *in* Rao, P. D., and Wolff, E. N., eds., Focus on Alaska's Coal '80 (Conference proceedings, University of Alaska, Fairbanks, October 21–23, 1980): University of Alaska Mineral Industries Research Laboratory Report no. 50, p. 122–126.

Bohm, J. G., Chmelik, F. B., Stewart, G. H., Turner, R. F., Waetjen, H. H., and Wills, J. C., 1976, Geological and operational summary, Atlantic Richfield Northern Gulf of Alaska COST Well No. 1: U.S. Geological Survey Open-File Report 76-635, 33 p.

Brabb, E. E., and Churkin, M., Jr., 1967, Stratigraphic evidence for the Late Devonian age of the Nation River Formation, east-central Alaska, *in* U.S. Geological Survey research 1967: U.S. Geological Survey Professional Paper 575-D, p. D4–D15.

—— , 1969, Geologic map of the Charley River quadrangle, east-central Alaska: U.S. Geological Survey Miscellaneous Geologic Investigations Series Map I-573, 1 sheet, scale 1:250,000.

Brew, D. A., Ovenshine, A. T., Karl, S. M., and Hunt, S. J., 1984, Preliminary reconnaissance geologic map of the Petersburg and parts of the Port Alexander and Sumdum 1:250,000 quadrangles, southeastern Alaska: U.S. Geological Survey Open-File Report 84-405, 2 sheets, 43 p.

Brosgé, W. P., and Whittington, C. L., 1966, Geology of the Umiat–Maybe Creek region, Alaska: U.S. Geological Survey Professional Paper 303-H, p. H501–H638.

Bruns, T. R., 1983, Model for the origin of the Yakutat block, an accreting terrane in the northern Gulf of Alaska: Geology, v. 11, p. 718–721.

—— , 1984, Reply *to* Comment *on* "Model for the origin of the Yakutat block, an accreting terrane in the northern Gulf of Alaska": Geology, v. 12, p. 563–564.

Bruns, T. R., and Keller, G., 1984, Reply *to* Comment *on* "Model for the origin of the Yakutat block, an accreting terrane in the northern Gulf of Alaska": Geology, v. 12, p. 565–567.

Bundtzen, T. K., Eakins, G. R., Green, C. B., and Lueck, L. L., 1986, Alaska's mineral industry, 1985: Alaska Division of Geological and Geophysical Surveys Special Report 39, 68 p.

Burk, C. A., 1965, Geology of the Alaska Peninsula—Island arc and continental margin: Geological Society of America Memoir 99, 250 p., geologic map, scale 1:250,000, tectonic map.

Calderwood, K. W., and Fackler, W. C., 1972, Proposed stratigraphic nomenclature for Kenai Group, Cook Inlet basin, Alaska: American Association of Petroleum Geologists Bulletin, v. 56, p. 739–754.

Callahan, J. E., and Sloan, E. G., 1978, Preliminary report on analyses of Cretaceous coals from northwestern Alaska: U.S. Geological Survey Open-File Report 78-319, 29 p, 1 plate.

Capps, S. R., 1913, The Yentna district, Alaska: U.S. Geological Survey Bulletin 534, 75 p., 13 plates.

—— , 1927, Geology of the Upper Matanuska Valley, Alaska: U.S. Geological Survey Bulletin 791, 92 p., 16 plates.

—— , 1935, The southern Alaska Range: U.S. Geological Survey Bulletin 862, 101 p., 8 plates.

—— , 1940, Geology of the Alaska Railroad region: U.S. Geological Survey Bulletin 907, 201 p., 9 plates.

Chapin, T., 1920, Mining developments in the Matanuska Coal Field: U.S. Geological Survey Bulletin 712, p. 131–167, plates 4–6.

Chapman, R. M., and Sable, E. G., 1960, Geology of the Utukok-Corwin region, northwestern Alaska: U.S. Geological Survey Professional Paper 303-C, 167 p.

Chisholm, W. A., 1985, Comment *on* "Model for the origin of the Yakutat block, an accreting terrane in the northern Gulf of Alaska": Geology, v. 13, p. 87.

Clark, P. R., 1973, Transportation economics of coal resources of northern slope coal fields, Alaska: University of Alaska Mineral Industry Research Laboratory Report 31, 134 p.

Collier, A. J., 1903, The coal resources of the Yukon, Alaska: U.S. Geological Survey Bulletin 218, 71 p.

—— , 1906, Geology and coal resources of the Cape Lisburne region, Alaska: U.S. Geological Survey Bulletin 278, 54 p.

Collins, F. R., 1959, Test wells, Square Lake and Wolf Creek areas, Alaska, *in* Exploration of Naval Petroleum Reserve No. 4 and adjacent areas, northern Alaska, part 5, Subsurface geology and engineering data: U.S. Geological Survey Professional Paper 305-H, p. 423–484.

Conwell, C. N., and Triplehorn, D. M., 1976, High-quality coal near Point Hope, northwestern Alaska, *in* Short notes on Alaskan geology, 1976: Alaska Division of Geological and Geophysical Surveys Geologic Report 51, p. 31–35.

—— , 1978, Herendeen Bay–Chignik coals, southern Alaska Peninsula: Alaska Division of Geological and Geophysical Surveys Special Report 8, 15 p., 2 plates.

Cooper, H. M., Snyder, N. H., Abernethy, R. F., Tarpley, E. C., and Swingle, R. J., 1946, Analyses of mine, tipple, and delivered samples, *in* Analyses of Alaska coals: U.S. Bureau of Mines Technical Paper 682, p. 19–69.

Dall, W. H., 1896, Coal and lignite in Alaska: U.S. Geological Survey Annual Report 17, p. 763–908.

Dall, W. H., and Harris, G. D., 1892, Correlation papers, Neocene: U.S. Geological Survey Bulletin 84, 349 p.

Detterman, R. L., 1978, Interpretation of depositional environments in the Chignik Formation, Alaska Peninsula: U.S. Geological Survey Circular 772-B, p. B62–B63.

Detterman, R. L., Reiser, H. N., Brosgé, W. P., and Dutro, J. T., Jr., 1975, Post-Carboniferous stratigraphy, northeastern Alaska: U.S. Geological Survey Professional Paper 886, 46 p.

Detterman, R. L., Plafker, G., Tysdal, R. G., and Hudson, T., 1976, Geology and surface features along part of the Talkeetna segment of the Castle Mountain–Caribou fault system, Alaska: U.S. Geological Survey Miscellaneous Field Studies Map MF-738, scale 1:63,360.

Detterman, R. L., Miller, T. P., Yount, M. E., and Wilson, F. H., 1981, Geologic map of the Chignik and Sutwik Island quadrangles, Alaska: U.S. Geological Survey Miscellaneous Investigations Series Map I-1229, scale 1:250,000.

Dickey, D. B., 1984, Cenozoic non-marine sedimentary rocks of the Farewell fault zone, McGrath quadrangle, Alaska: Sedimentary Geology, v. 38, p. 443–463.

Eakins, G. R., and 6 others, 1983, Alaska's mineral industry, 1982: Alaska Division of Geological and Geophysical Surveys Special Report 31, 63 p.

Fisher, M. A., and Magoon, L. B., 1978, Geologic framework of Lower Cook Inlet, Alaska: American Association of Petroleum Geologists Bulletin, v. 62, p. 373–402.

Fox, J. E., Lambert, P. W., Pitman, J. K., and Wu, C. H., 1979, A study of reservoir characteristics of the Nanushuk and Colville groups, Umiat test well 11, National Petroleum Reserve in Alaska: U.S. Geological Survey Circular 820, 47 p.

Gilbert, W. G., Solie, D. N., and Dickey, D. B., 1982, Preliminary bedrock geology of the McGrath B-3 quadrangle, Alaska: Alaska Division of Geological and Geophysical Surveys Alaska Open-File Report 148, 1:40,000

Gluskoter, H. J., and Hopkins, M. E., 1970, Distribution of sulfur in Illinois coals, *in* Smith, W. H., Nance, R. B., Hopkins, M. E., Johnson, R. G., and Shabica, C. W., eds., Depositional environments in parts of the Carbondale Formation—Western and northern Illinois: Illinois Geological Survey Guidebook Series 8, p. 89–95.

Gluskoter, H. J., and Simon, J. A., 1968, Sulfur in Illinois coals: Illinois State Geological Survey Circular 432, 28 p.

Grantz, A., and Eittreim, S., 1979, Geology and physiography of the continental margin north of Alaska and implications for the origin of the Canada basin: U.S. Geological Survey Open-File Report 79-288, 61 p.

Grantz, A., and Jones, D. L., 1960, Stratigraphy and age of the Matanuska Formation: U.S. Geological Survey Professional Paper 400-B, p. B347–B351.

Grantz, A., Holmes, M. L., and Kososki, B. A., 1975, Geologic framework of the Alaskan continental terrace in the Chukchi and Beaufort seas, *in* Yorath, C. J., Parker, E. R., and Glass, D. J., eds., Canada's continental margins and offshore petroleum exploration: Canada Society of Petroleum Geologists Memoir 4, p. 669–700.

Green, C. B., and Bundtzen, T. K., 1989, Summary of Alaska's mineral industry in 1988: Alaska Division of Geological and Geophysical Surveys Public Data File 89-7, 6 p.

Harrington, G. L., 1918, The Anvik-Andreafski region, Alaska: U.S. Geological Survey Bulletin 683, 70 p., 4 plates.

Hartman, D. C., Pessel, G. H., and McGee, D. L., 1971, Preliminary report, Kenai Group of Cook Inlet, Alaska: Alaska Division of Geological and Geophysical Surveys Special Report 5, 4 p., 11 plates.

Hayes, J. B., Harms, J. C., and Wilson, T., Jr., 1976, Contrasts between braided and meandering stream deposits, Beluga and Sterling formations (Tertiary), Cook Inlet, Alaska, *in* Miller, T. P., ed., Recent and ancient sedimentary environments in Alaska (Proceedings, Alaska Geological Society Symposium, April 2–4, 1975, Anchorage, Alaska): Anchorage, Alaska Geological Society, p. J1–J27.

Heer, Oswald, 1869, cited in Hollick, A., 1936, Tertiary floras of Alaska: U.S. Geological Survey Professional Paper 182, p. 2.

Henshaw, F. F., 1909, Mining in the Fairhaven precinct, *in* Mineral resources of Alaska: U.S. Geological Survey Bulletin 379, p. 355–369.

Hite, D. M., 1976, Some sedimentary aspects of the Kenai Group, Cook Inlet, Alaska, *in* Miller, T. P., ed., Recent and ancient sedimentary environments in Alaska (Proceedings, Alaska Geological Society Symposium, April 2–4, 1975, Anchorage, Alaska): Anchorage, Alaska Geological Society, p. I1–I23.

Hopkins, D. M., 1951, Lignite deposits near Broad Pass Station, Alaska: U.S. Geological Survey Bulletin 963-E, p. 187–191, plate 26.

Huffman, A. C., Jr., Ahlbrandt, T. S., Pasternack, I., Stricker, G. D., and Fox, J. E., 1985, Depositional and sedimentologic factors affecting the reservoir potential of the Cretaceous Nanushuk Group, central North Slope, Alaska, *in* Huffman, A. C., Jr., ed., Geology of the Nanushuk Group and related rocks, central North Slope, Alaska: U.S. Geological Survey Bulletin 1614, p. 61–74.

Huffman, A. C., Jr., Ahlbrandt, T. S., and Bartsch-Winkler, S., 1988, Sedimentology of the Nanushuk Group, North Slope, Alaska, *in* Gryc, G., ed., Geology and exploration of the National Petroleum Reserve in Alaska, 1974–1982: U.S. Geological Survey Professional Paper 1399, p. 281–298.

Huish, R., compiler, 1836, Voyages of Captain Beechey to the Pacific 1825–28 and of Captain Back to the Arctic Sea: London, William Wright, 704 p.

Husky Oil NPR Operations, Inc., 1982–1983, Geological reports of test wells in National Petroleum Reserve in Alaska: Unpublished reports of Husky Oil NPR Operations Inc. Copies of these reports are available for purchase from the National Geophysical and Solar-Terrestrial Data Center, NOAA, Boulder, Colorado 80303.

Kirschner, C. E., 1988, Map showing sedimentary basins of onshore and continental shelf areas, Alaska: U.S. Geological Survey Miscellaneous Investigations Series Map I-1873, scale 1:2,500,000.

Kirschner, C. E., and Lyon, C. A., 1973, Stratigraphic and tectonic development of Cook Inlet Petroleum Province, *in* Pitcher, M. G., ed., Arctic geology: American Association of Petroleum Geologists Memoir 19, p. 396–407.

Lathram, E. H., Pomeroy, J. S., Berg, H. C., and Loney, R. A., 1965, Reconnaissance geology of Admiralty Island, Alaska: U.S. Geological Survey Bulletin 1181-R, p. R1–R48.

Lyle, W. M., Morehouse, J. A., Palmer, I. F., Jr., and Bolm, J. G., 1979, Tertiary formations and associated Mesozoic rocks in the Alaska Peninsula area, Alaska, and their petroleum-reservoir and source-rock potential: Alaska Division of Geological and Geophysical Surveys Geologic Report 62, 65 p., 18 plates.

Magoon, L. B., and Egbert, R. M., 1986, Framework geology and sandstone composition, *in* Magoon, L. B., ed., Geologic studies of the Lower Cook Inlet COST No. 1 Well, Alaska outer continental shelf: U.S. Geological Survey Bulletin 1596, p. 65–90.

Magoon, L. B., Adkison, W. L., and Egbert, R. M., 1976, Map showing geology, wildcat wells, Tertiary plant fossil localities, K-Ar age dates, and petroleum operations, Cook Inlet Area, Alaska: U.S. Geological Survey Miscellaneous Investigations Series Map I-1019, 3 sheets, scale 1:250,000.

Marincovich, L., Jr., 1990, Molluscan evidence for early middle Miocene glaciation in southern Alaska: Geological Society of America Bulletin, v. 102, p. 1591–1599.

Martin, G. C., 1908, Geology and mineral resources of the Controller Bay region, Alaska: U.S. Geological Survey Bulletin 335, 141 p., 10 plates.

——, 1915, The western part of Kenai Peninsula, *in* Martin, G. C., Johnson, B. L., and Grant, U. S., Geology and mineral resources of Kenai Peninsula, Alaska: U.S. Geological Survey Bulletin 587, p. 41–112.

——, 1926, The Mesozoic stratigraphy of Alaska: U.S. Geological Survey Bulletin 776, 493 p.

Martin, G. C., and Callahan, J. E., 1978, Preliminary report on the coal resources of the National Petroleum Reserve in Alaska: U.S. Geological Survey Open-File Report 78-1033, 23 p., 2 plates.

Martin, G. C., and Katz, F. J., 1912, Geology and coal fields of the Lower Matanuska Valley, Alaska: U.S. Geological Survey Bulletin 500, 98 p., 19 plates.

Martin, G. C., Johnson, B. L., and Grant, U. S., 1915, Geology and mineral resources of Kenai Peninsula, Alaska: U.S. Geological Survey Bulletin 587, 243 p.

Mayfield, C. F., Tailleur, I. L., and Ellersieck, I., 1988, Stratigraphy, structure, and palinspastic synthesis of the western Brooks Range, northwestern Alaska, *in* Gryc, G., ed., Geology and exploration of the National Petroleum Reserve in Alaska, 1974–1982: U.S. Geological Survey Professional Paper 1399, p. 143–186.

McCabe, P., 1984, Depositional environments of coal and coal-bearing strata, *in* Rahmani, R. A., and Flores, R. M., eds., Sedimentology of coal and coal-bearing sequences (International Association of Sedimentologists Special Publication 7): Oxford, Blackwell Scientific Publications, p. 13–40.

McGee, D. L., and O'Connor, K. M., 1975, Cook Inlet Basin subsurface coal reserve study: Alaska Division of Geological and Geophysical Surveys Open-File Report 74, 24 p., 3 plates.

Merritt, R. D., 1985a, Review of coking phenomena in relation to an occurrence of prismatically fractured natural coke from the Castle Mountain Mine, Matanuska coal field, Alaska: International Journal of Coal Geology, v. 4, p. 281–298.

——, 1985b, Coal atlas of the Nenana basin, Alaska: Alaska Division of Geological and Geophysical Surveys Public-Data File 85-41, 197 p.

Merritt, R. D., and Belowich, M. A., 1984, Coal geology and resources of the Matanuska Valley, Alaska: Alaska Division of Geological and Geophysical Surveys Report of Investigations 84-24, 64 p., 3 plates.

Merritt, R. D., and Hawley, C. C., compilers, 1986, Map of Alaska's coal resources: Fairbanks, Alaska Division of Geological and Geophysical Surveys, scale 1:2,500,000.

Merritt, R. D., Lueck, L. L., Rawlinson, S. E., Belowich, M. A., Goff, K. M., Clough, J. G., and Reinick-Smith, L., 1987, Southern Kenai Peninsula (Homer District) coal resource assessment and mapping project—Final report: Alaska Division of Geological and Geophysical Surveys Public Data File 87-15, 125 p., 15 sheets.

Mertie, J. B., Jr., 1930, Mining in the Fortymile district, Alaska: U.S. Geological Survey Bulletin 813-C, p. 125–142.

Metz, P. A., 1981, Mining, processing, and marketing of coal from Jarvis Creek Field [abs.], *in* Rao, P. D., and Wolff, E. N., eds., Focus on Alaska's Coal '80 (Conference proceedings, University of Alaska, Fairbanks, October 21–23, 1980): University of Alaska Mineral Industries Research Laboratory Report no. 50, p. 171.

Miller, D. J., 1957, Geology of the southeastern part of the Robinson Mountains, Yakataga District, Alaska: U.S. Geological Survey Oil and Gas Investigations Map M-187, 2 sheets, scale 1:63,360.

Molenaar, C. M., 1985, Subsurface correlations and depositional history of the Nanushuk Group and related rocks, North Slope, Alaska, *in* Huffman, A. C., Jr., ed., Geology of the Nanushuk Group and related rocks, North Slope, Alaska: U.S. Geological Survey Bulletin 1614, p. 37–60.

Molenaar, C. M., Kirk, A. R., Magoon, L. B., and Huffman, A. C., 1984, Twenty-two measured sections of Cretaceous-Lower Tertiary rocks, eastern North Slope, Alaska: U.S. Geological Survey Open-File Report 84-695, 19 p.

Molenaar, C. M., Bird, K. J., and Kirk, A. R., 1987, Cretaceous and Tertiary stratigraphy of northern Alaska, *in* Tailleur, I. L., and Weimer, P., eds., Alaska North Slope geology: Bakersfield, California, Pacific Section, Society of Economic Paleontologists and Mineralogists Special Publication 50, p. 513–528.

Moore, T. E., Nilsen, T. H., Grantz, A., and Tailleur, I. L., 1984, Parautochthonous Mississippian marine and nonmarine strata, Lisburne Peninsula, Alaska, *in* Reed, K. M., and Bartsch-Winkler, S., eds., The United States Geological Survey in Alaska—Accomplishments during 1982: U.S. Geological Survey Circular 939, p. 17–21.

Mull, C. G., 1982, Tectonic evolution and structural style of the Brooks Range, Alaska—An illustrated summary, *in* Powers, R. B., ed., Geologic studies of the Cordilleran thrust belt: Denver, Rocky Mountain Association of Geologists, p. 1–46.

——, 1985, Cretaceous tectonics, depositional cycles, and the Nanushuk Group, Brooks Range and Arctic slope, Alaska, *in* Huffman, A. C., ed., Geology of the Nanushuk Group and related rocks, North Slope, Alaska: U.S. Geological Survey Bulletin 1614, p. 7–36.

Mull, C. G., and Nelson, S. W., 1986, Anomalous thermal maturity data from the Orca Group (Paleocene and Eocene), Katalla-Kayak Island area: U.S. Geological Survey Circular 978, p. 50–55.

Neavel, R. C., 1981, Origin, petrography, and classification of coal, *in* Elliott, M. A., ed., Chemistry of coal utilization (second supplementary volume): New York, Wiley-Interscience, p. 91–158.

Nelson, S. W., Dumoulin, J. A., and Miller, M. L., 1985, Geologic map of the Chugach National Forest, Alaska: U.S. Geological Survey Miscellaneous Field Studies Map MF-1645-B, scale 1:250,000, 16 p.

Nilsen, T. H., 1984, Miocene back-arc tidal deposits of the Bear Lake Formation, Alaska Peninsula, *in* Reed, K. M., and Bartsch-Winkler, S., eds., The United States Geological Survey in Alaska—Accomplishments during 1982: U.S. Geological Survey Circular 939, p. 85–88.

Nilsen, T. H., Brabb, E. E., and Simoni, T. R., 1976, Stratigraphy and sedimentology of the Nation River Formation, a Devonian deep-sea fan deposit in east-central Alaska, *in* Miller, T. P., ed., Recent and ancient sedimentary environments in Alaska: Anchorage, Alaska Geological Society, p. E1–E20.

Paige, S., 1906, The Herendeen Bay coal field: U.S. Geological Survey Bulletin 284, p. 101–108.

Patton, W. W., Jr., 1973, Reconnaissance geology of the northern Yukon-Koyukuk province, Alaska: U.S. Geological Survey Professional Paper 774-A, p. A1–A17.

Patton, W. W., Jr., and Bickel, R. S., 1956, Geologic map and structure sections along part of the lower Yukon River, Alaska: U.S. Geological Survey Miscellaneous Geologic Investigations Map I-197, scale 1:250,000.

Patton, W. W., Jr., and Miller, T. P., 1968, Regional geologic map of the Selawik and southeastern Baird Mountains quadrangles, Alaska: U.S. Geological Survey Miscellaneous Investigations Series Map I-530, scale 1:250,000.

Patton, W. W., Jr., and Moll, E. J., 1985, Geologic map of northern and central parts of Unalakleet quadrangle, Alaska: U.S. Geological Survey Miscellaneous Field Studies Map MF-1749, scale 1:250,000.

Plafker, G., 1983, The Yakutat block, an active tectonostratigraphic terrane in southern Alaska: Geological Society of America Abstracts with Programs, v. 15, p. 406.

——, 1984, Comment *on* "Model for the origin of the Yakutat block, an accreting terrane in the northern Gulf of Alaska": Geology, v. 12, p. 563.

——, 1987, Regional geology and petroleum potential of the northern Gulf of Alaska continental margin, *in* Scholl, D. W., Grantz, A., and Vedder, J. G., eds., Geology and resource potential of the continental margin of western North America and adjacent ocean basins: Houston, Texas, Circum-Pacific Council for Energy and Mineral Resources, Earth Science Series, v. 6, p. 229–268.

Plafker, G., and Miller, D. J., 1957, Reconnaissance geology of the Malaspina District, Alaska: U.S. Geological Survey Oil and Gas Investigations Map OM-189, scale 1:125,000.

Ramsey, J. P., 1981, Geology-coal resources and mining plan for the Chuitna River field, Alaska, *in* Rao, P. D., and Wolff, E. N., eds., Focus on Alaska's Coal '80 (Conference proceedings, University of Alaska, Fairbanks, October 21–23, 1980): University of Alaska, Mineral Industries Research Laboratory Report no. 50, p. 111–121.

Rao, P. D., and Wolff, E. N., 1981, Petrological, mineralogical, and chemical characterizations of certain Alaskan coals and washability products, *in* Rao, P. D., and Wolff, E. N., eds., Focus on Alaska's Coal '80 (Conference proceedings, University of Alaska, Fairbanks, October 21–23, 1980): University of Alaska, Mineral Industries Research Laboratory Report no. 50, p. 194–235.

Reed, B. L., and Nelson, S. W., 1980, Geologic map of the Talkeetna quadrangle, Alaska: U.S. Geological Survey Miscellaneous Investigations Series Map I-1174, scale 1:250,000, 15 p.

Roberts, S. B., 1991, Subsurface cross-section showing coal beds in the Sagavanirktok Formation, vicinity of Prudhoe Bay, east-central North Slope, Alaska: U.S. Geological Survey Coal Investigations Map 1C-139-A, 1 sheet.

Roberts, S. B., Stricker, G. D., and Affolter, R. H., 1991, Stratigraphy and chemical analyses of coal beds in the upper Cretaceous and Tertiary Sagavanirktok Formation, east-central North Slope, Alaska: U.S. Geological Survey Coal Investigations Map C-139-B, 1 sheet.

Roehler, H. W., and Stricker, G. D., 1979, Stratigraphy and sedimentation of the Torok, Kukpowruk, and Corwin formations in the Kokolik-Utukok River region, National Petroleum Reserve in Alaska: U.S. Geological Survey Open-File Report 79-995, 80 p.

Ryer, T. A., and Langer, A. W., 1980, Thickness change involved in the peat-to-coal transformation for a bituminous coal of Cretaceous age in central Utah: Journal of Sedimentary Petrology, v. 50, p. 987–992.

Sable, E. G., and Stricker, G. D., 1987, Coal in the National Petroleum Reserve in Alaska (NPRA)—Framework geology and resources, *in* Tailleur, I. L., and Weimer, P., eds., Alaskan North Slope geology: Bakersfield, California, Pacific Section, Society of Economic Paleontologist and Mineralogists Special Publication 50, p. 195–215.

Sanders, R. B., 1976, Geology and coal resources of the Bering River field, *in* Rao, P. D., and Wolff, E. N., eds., Focus on Alaska's Coal '75 (Conference proceedings, University of Alaska, Fairbanks, October 15–17, 1975): University of Alaska Mineral Industries Research Laboratory Report no. 37, p. 54–58.

——, 1981, Coal resources of Alaska, *in* Rao, P. D., and Wolff, E. N., eds., Focus on Alaska's Coal '80 (Conference proceedings, University of Alaska, Fairbanks, October 21–23, 1980): University of Alaska Mineral Industries Research Laboratory Report no. 50, p. 11–31.

Schaff, R. G., and Gilbert, W. G., coordinators, 1987, Southern Alaska region, Correlation of Stratigraphic Units of North America (COSUNA) Project: Tulsa, Oklahoma, American Association of Petroleum Geologists, 1 sheet.

Schmoll, H. R., Chleborad, A. F., Yehle, L. A., Gardner, C. A., and Patsch, A. D., 1981, Reconnaissance engineering geology of the Beluga coal resource area, south-central Alaska, *in* Rao, P. D., and Wolff, E. N., eds., Focus on Alaska's Coal '80 (Conference proceedings, University of Alaska, Fairbanks, October 21–23, 1980): University of Alaska Mineral Industries Research Laboratory Report no. 50, p. 92–110.

Schopf, J. M., 1956, A definition of coal: Economic Geology, v. 51, p. 521–527.

Schrader, F. C., 1900, Preliminary report on a reconnaissance along the Chandlar and Koyukuk rivers, Alaska, in 1899: U.S. Geological Survey Annual Report 21, Part 2, p. 441–486.

——, 1904, A reconnaissance in northern Alaska, across the Rocky Mountains, along Koyukuk, John, Anaktuvuk, and Colville rivers, and the Arctic Coast to Cape Lisburne, in 1901: U.S. Geological Survey Professional Paper 20, 139 p., map.

Sloan, E. G., Shearer, G. B., Eason, J. E., and Almquist, C. L., 1979, Reconnaissance survey for coal near Farewell, Alaska: U.S. Geological Survey Open-File Report 79-410, 18 p., 4 plates.

Smiley, C. J., 1969, Floral zones and correlations of Cretaceous Kukpowruk and Corwin formations, northwestern Alaska: American Association of Petroleum Geologists Bulletin, v. 53, p. 2079–2093.

Solie, D. N., and Dickey, D. B., 1982, Coal occurrences and analyses, Farewell–White Mountain area, southwest Alaska: Alaska Division of Geological and Geophysical Surveys Open-File Report 160, 17 p., 1 plate.

Stach, E., 1968, Basic principles of coal petrology: Macerals, microlithotypes, and some effects of coalification, *in* Murchison, D., and Westoll, T. S., eds., Coal and coal-bearing strata: New York, Elsevier, p. 3–17.

Stach, E., Mackowsky, M. T., Teichmueller, M., Taylor, G. H., Chandra, D., and Teichmueller, R., 1982, Stach's textbook of coal petrology (third edition): Berlin, Gebrueder Borntraeger, 535 p.

Stone, R. W., 1906, Coal fields of the Kachemak Bay region: U.S. Geological Survey Bulletin 277, p. 53–73.

Stricker, G. D., 1983, Coal occurrence, quality, and resource assessment, National Petroleum Reserve in Alaska [abs.], *in* U.S. Geological Survey polar research symposium—Abstracts with program: U.S. Geological Survey Circular 911, p. 32–33.

—— , 1991, Economic Alaskan coal deposits, *in* Gluskoter, H. J., Rice, D. D., and Taylor, R. B., eds., Economic geology, U.S.: Boulder, Colorado, Geological Society of America, The Geology of North America, v. P-2, p. 591–602.

Styan, W. B., and Bustin, R. M., 1984, Sedimentology of Fraser River delta peat deposits: A modern analogue for some deltaic coals, *in* Rahmani, R. A., and Flores, R. M., eds., Sedimentology of coal and coal-bearing sequences (International Association of Sedimentologists Special Publication 7): Oxford, Blackwell Scientific Publications, p. 241–271.

Tailleur, I. L., 1965, Low-volatile bituminous coal of Mississippian age on the Lisburne Peninsula, northwestern Alaska: U.S. Geological Survey Professional Paper 525-B, p. B34–B38.

Tailleur, I. L., and Brosgé, W. P., 1976, Coal resources of northern Alaska may be nation's largest, *in* Rao, P. D., and Wolff, E. N., eds., Focus on Alaska's Coal '75 (Conference proceedings, University of Alaska, Fairbanks, October 15–17, 1975): University of Alaska Mineral Industries Research Laboratory Report no. 37, p. 219–226.

Taylor, P. T., Kovacs, L. C., Vogt, P. R., and Johnson, G. L., 1981, Detailed aeromagnetic investigation off the Arctic basin, 2: Journal of Geophysical Research, v. 86, no. B7, p. 6323–6333.

Teichmueller, M., and Teichmueller, R., 1968, Geological aspects of coal metamorphism, *in* Murchison, D., and Westoll, T. S., eds., Coal and coal-bearing strata: New York, Elsevier, p. 233–267.

Toenges, A. L., and Jolley, T. R., 1947, Investigation of coal deposits for local use in the Arctic regions of Alaska and proposed mine development: U.S. Bureau of Mines Report of Investigations 4150, 19 p.

Triplehorn, D. M., Turner, D. L., and Naeser, C. W., 1984, Radiometric age of the Chickaloon Formation of south-central Alaska—Location of the Paleocene-Eocene boundary: Geological Society of America Bulletin, v. 95, p. 740–742.

Turner, B. R., and Whateley, M.K.C., 1989, Tidally influenced coal-bearing sediments in the Tertiary Bering River coalfield, south-central Alaska: Sedimentary Geology, v. 61, p. 111–123.

von Huene, R., Keller, G., Bruns, T. R., and McDougall, K., 1985, Cenozoic migration of Alaskan terranes indicated by paleontologic study, *in* Howell, D. G., ed., Tectonostratigraphic terranes of the Circumpacific region: Houston, Texas, Circumpacific Council for Energy and Mineral Resources, p. 121–136.

Wahrhaftig, C., 1944, Coal deposits of the Costello Creek basin, Alaska: U.S. Geological Survey Open-File Report 8, 8 p.

—— , 1965, Physiographic divisions of Alaska: U.S. Geological Survey Professional Paper 482, 52 p., 6 plates.

—— , 1970, Late Cenozoic orogeny in the Alaska Range: Geological Society of America Abstracts with Programs, v. 7, p. 713–714.

—— , 1973, Coal reserves of the Healy Creek and Lignite Creek coal basins, Nenana coal field, Alaska: U.S. Geological Survey Open-File Report 73-355, 6 p., 28 sheets.

—— , 1987, The Cenozoic section at Suntrana, Alaska, *in* Hill, M. L., ed., Cordilleran section of the Geological Society of America: Boulder, Colorado, Geological Society of America, Centennial Field Guide, v. 1, p. 445–450.

Wahrhaftig, C., and Hickcox, C. A., 1955, Geology and coal deposits, Jarvis Creek coal field, Alaska: U.S. Geological Survey Bulletin 989-G, p. 353–367, plates 10–12.

Wahrhaftig, C., Wolfe, J. A., Leopold, E. B., and Lanphere, M. A., 1969, The coal-bearing group in the Nenana coal field, Alaska: U.S. Geological Survey Bulletin 1274-D, p. D1–D30.

Waring, G. A., 1936, Geology of the Anthracite Ridge coal district, Alaska: U.S. Geological Survey Bulletin 861, 57 p., 14 plates.

Wolfe, J. A., and McCoy, S., Jr., 1984, Comment *on* "Model for the origin of the Yakutat block, an accreting terrane in the northern Gulf of Alaska": Geology, v. 12, p. 564–565.

Wolfe, J. A., and Tanai, T., 1980, The Miocene Seldovia Point flora from the Kenai Group, Alaska: U.S. Geological Survey Professional Paper 1105, 52 p., 25 plates.

—— , 1987, Systematics, phylogeny, and distribution of *Acer* (Maples) in the Cenozoic of western North America: Journal of the Faculty of Science of Hokkaido University, ser. 4, v. 22, p. 1–246.

Wolfe, J. A., and Wahrhaftig, C., 1970, The Cantwell Formation of the central Alaska Range: U.S. Geological Survey Bulletin 1294, p. A41–A46.

Wood, G. H., and Bour, W. V., III, 1988, Coal map of North America: U.S. Geological Survey Special Geologic Map, scale 1:5,000,000.

Wood, G. H., Jr., Kehn, T. M., Carter, M. D., and Culberston, W. C., 1983, Coal resource classification system of the U.S. Geological Survey: U.S. Geological Survey Circular 891, 65 p.

MANUSCRIPT ACCEPTED BY THE SOCIETY JULY 1, 1992

The Geology of North America
Vol. G-1, The Geology of Alaska
The Geological Society of America, 1994

Chapter 32

Geothermal resources of Alaska

Thomas P. Miller
U.S. Geological Survey, 4200 University Drive, Anchorage, Alaska 99508-4667

INTRODUCTION

Most of the major known geothermal systems of the world are associated in some manner with recent volcanism and thermal springs. Both phenomena occur in sufficient numbers in Alaska as to indicate potentially large geothermal resources. The volcanic areas of principal geothermal interest are (1) the Aleutian arc (Fig. 1), a seismically active volcanic arc-trench system extending some 2,500 km across the North Pacific and Alaska mainland; and (2) the Wrangell Mountains volcanic pile in east-central Alaska, which underlies an area of some 10,000 km^2 and ranges in age from Miocene to Holocene. Volcanism in these two areas is both tholeiitic and calc-alkaline in character and is related to the convergence of the North American and Pacific Plates. Both regions contain evidence of silicic and explosive volcanism, indicating high-level near-surface magma chambers with attendant large heat reservoirs. The distribution of volcanic rocks is shown on Plate 12 (Plafker and others, this volume).

Although other volcanic provinces occur in the state, they appear to be less important from a geothermal standpoint. The western Alaska volcanic province, consisting of Pliocene to Holocene olivine tholeiite and alkali basalt, underlies scattered areas totaling about 25,000 km^2 chiefly on the Seward Peninsula, the Norton Sound area (including St. Lawrence Island), the Yukon River delta, and the Pribilof Islands. Similar, widely scattered Quaternary volcanic rocks occur in the central and eastern interior of Alaska. These basaltic volcanic provinces appear to be related to an extensional tectonic regime because of their composition and mode of occurrence. Near-surface heat reservoirs of possible geothermal interest appear to be lacking. Quaternary volcanic rocks are also found in a few localities in southeastern Alaska, chiefly at Mount Edgecumbe near Sitka and on Kupreanof Island.

Of the more than 100 thermal springs known to occur in the state, approximately half are associated with the Aleutian volcanic arc. The remaining half are concentrated in interior and southeastern Alaska and have no apparent spatial or temporal association with recent volcanism. Many of these thermal springs have been sites of sporadic direct use as spas or small agriculture areas since the turn of the century. Thermal spring areas are shown on Plate 12 (Plafker and others, this volume).

PREVIOUS WORK

The first systematic description of geothermal areas in Alaska was a remarkably comprehensive report by Waring (1917), who at this early date, listed 75 of the presently known 108 hot springs in the state (Alaska Division of Geological and Geophysical Surveys, 1983). Little further study on either the thermal springs or igneous-related geothermal systems took place until the early 1970s when the U.S. Geological Survey began regional geothermal studies in Alaska as part of its Geothermal Resource Program. Miller (1973) compiled an updated summary of 94 known or suspected thermal springs, including a total of 34 chemical analyses of the thermal waters. Studies of igneous-related geothermal systems, principally in the Aleutian volcanic arc and the Wrangell Mountains were begun in 1973 (Miller and Barnes, 1976). Data gathered from this study formed the basis for the calculations of estimated thermal energy remaining in a particular volcanic system (Smith and Shaw, 1975, 1979). Miller and others (1975) conducted a study of the geologic setting and chemical characteristics of the thermal springs across 200,000 km^2 of interior Alaska. Their study confirmed the close association of thermal springs with the margins of granitic plutons and proposed a model for their origin whereby deeply circulating meteoric water gained access to the surface along the fractured contacts of massive plutonic and hornfelsic wall rocks.

The Geophysical Institute at the University of Alaska began studies of individual geothermal areas in the 1970s and subsequently was joined by the Alaska Division of Geological and Geophysical Surveys, with most of the studies supported by the U.S. Department of Energy. These studies were followed by regional and statewide summaries of available resources in the mid-1970s and 1980s. A partial listing of studies by these organizations, which included geological, geophysical (gravity, magnetics, electrical, He and Hg surveys, seismic refraction, and heat-flow investigations), and geochemical investigations, include reports by Motyka and others (1980b), Turner and Forbes (1980), Turner and others (1980), Motyka and others (1981), Wescott and Turner (1983), Turner and Wescott (1982), and East (1982).

Miller, T. P., 1994, Geothermal resources of Alaska, *in* Plafker, G., and Berg, H. C., eds., The Geology of Alaska: Boulder, Colorado, Geological Society of America, The Geology of North America, v. G-1.

A multidisciplinary geological, geochemical, and geophysical study of igneous-related systems on northern Adak Island, with particular emphasis on Mt. Adagdak volcano, was done by the U.S. Geological Survey at the request of the U.S. Navy (Miller and others, 1978). Detailed geological mapping, coupled with geochronological and major-element geochemical studies, was done on Mt. Drum volcano in the western Wrangell Mountains (Richter and others, 1978).

The most detailed geothermal exploration program in Alaska was conducted on the flanks of Makushin volcano on Unalaska Island (Fig. 1) as part of a cooperative study begun in 1981 by the Alaska Power Authority and the Alaska Division of Geological and Geophysical Surveys. The study included detailed geological mapping, geochemical and isotopic studies, and self-potential investigations, which were followed by the drilling of three 460-m-deep geothermal gradient holes and a 1,220-m-deep exploration well (Parmentier and others, 1983; Motyka, 1983; Matlick and Parmentier, 1983; Isselhardt and others, 1983).

GEOTHERMAL SITES

Aleutian volcanic arc

The area of Alaska with the greatest geothermal resource is the Aleutian volcanic arc, which extends some 2,500 km from Hayes volcano 130 km west of Anchorage to Buldir Island in the western Aleutians (Fig. 1). Over 60 major volcanic centers of Quaternary age, ranging in volume from 5 to more than 400 km^3, are included in this island-arc and continental margin system, which spans both ocean and continent; these centers include at least 37 volcanoes that have been active within the past 200 years (Kay and Kay, this volume; Miller and Richter, this volume). Of particular interest from a geothermal standpoint is the occurrence of more than 20 major calderas (indicative of near-surface magma chambers and heat reservoirs) associated with these centers (Table 1). Most of the calderas are Holocene in age (Miller and Smith, 1987) and range in diameter from less than 2 km (Great Sitkin) to over 18 km (Fisher). Calderas east of 159°W on the Alaska Peninsula (Fig. 1) are characterized by silicic post-caldera volcanism, suggesting the continued presence of high-level magma chambers and reservoirs (Miller and Smith, 1987). Calderas west of 159°W are generally characterized by more mafic post-caldera activity, suggesting a return to a more primitive system. Numerous other volcanoes (Augustine, Spurr, Dana, Kialagvik, Chiginagak, Yantarni, Frosty, Adagdak) exhibit young silicic domes, which along with the calderas, attest to the existence of high-level magma chambers and therefore probable associated near-surface heat sources.

Associated with these volcanic centers are many thermal areas consisting of fumaroles, mud pots, and more than 30 thermal springs. Unlike thermal springs elsewhere in Alaska, these

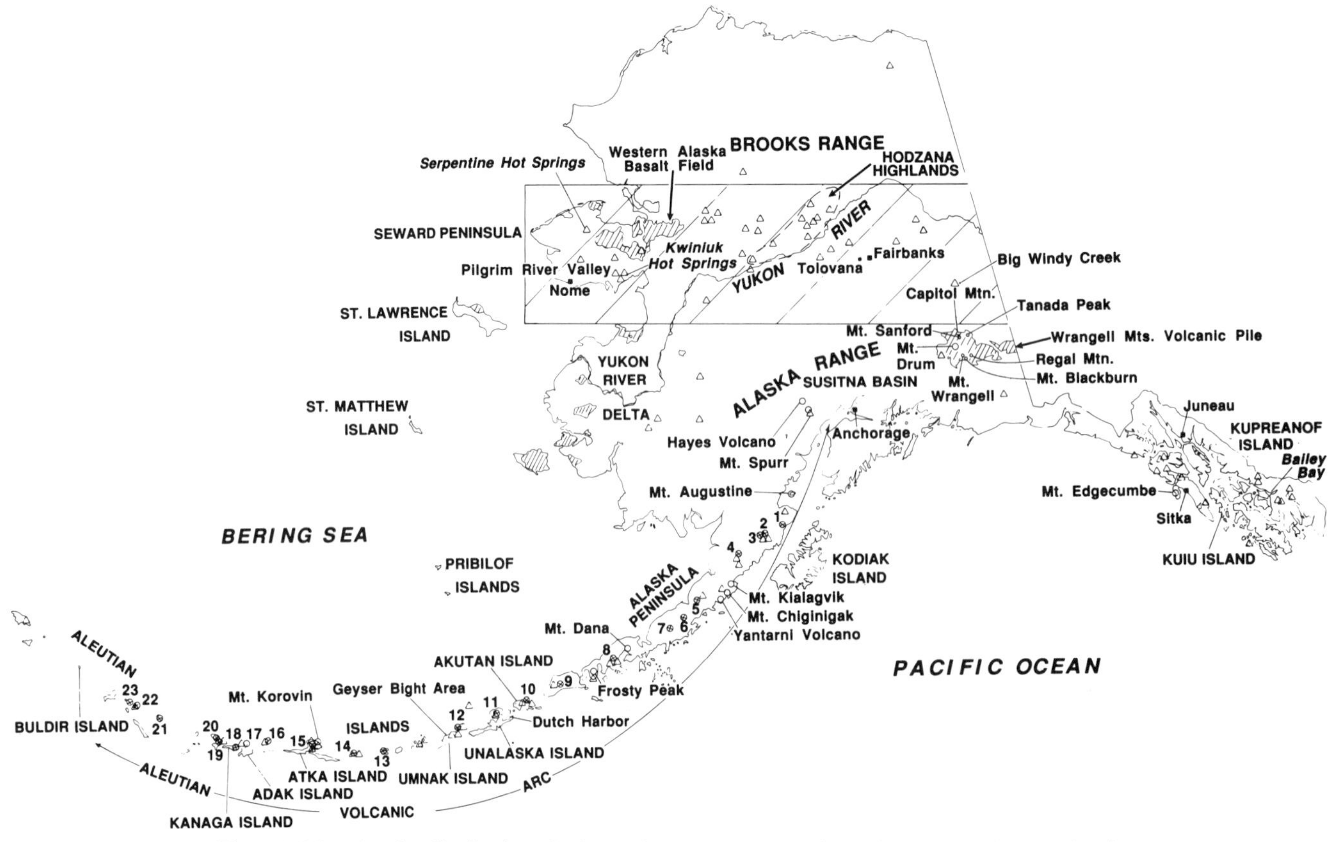

Figure 1. Map showing distribution of calderas, igneous-related geothermal systems, and thermal springs in Alaska. Pattern indicates areas of abundant thermal springs in interior Alaska. Triangle, thermal spring site; circle, volcano; circle with x, caldera; numbers refer to Table 1.

TABLE 1. CALDERAS AND RELATED IGNEOUS SYSTEMS OF THE ALEUTIAN ARC AND THEIR CONTAINED THERMAL ENERGY*

Name	Location	Approximate diameter (km)	Thermal energy remaining in system (10^{18} joules)
1. Kaguyak	Alaska Peninsula; 58°37'N,154°05'W	2.8	38
2. Katmai	Alaska Peninsula; 58°16'N,154°59'W	2.1 x 3.2	50
3. Novarupta	Alaska Peninsula; 58°16'N,155°09'W	3	120
4. Ugashik	Alaska Peninsula; 57°45'N,156°21'W	5	71
5. Aniakchak	Alaska Peninsula; 56°53'N,158°10'W	9.7 x 8.4	540
6. Black Peak	Alaska Peninsula; 56°32'N,158°37'W	2.7 x 2.6	50
7. Veniaminof	Alaska Peninsula; 56°10'N,159°23'W	8.4	481
8. Emmons	Alaska Peninsula; 55°20'N,162°01'W	18 x 10	1,440
9. Fisher	Unimak Island 54°38'N,164°25'W	10 x 11	1,440
10. Akutan	Akutan Island 54°08'N,166°00'W	2	25
11. Makushin	Unalaska Island 53°52'N,168°56'W	2.4 x 3.2	25
12. Okmok I, II	Umnak Island 53°39'N,168°03'W	10; 11	603
13. Yunaska	Yunaska Island 52°39'N,170°39'W	2.9 x 3.4	96
14. Sequam I, II	Sequam Island 52°19'N,172°29'W	8; 5.5	480
15. Kliuchef	Atka Island 53°19'N,174°09'W	4.5	?
16. Great Sitkin	Great Sitkin Island 52°04'N,176°07'W	1.6	>13
17. Adagdak†	Adak Island 51°59'N,176°35'W	....	50
18. Kanaton	Kanaga Island 51°55'N,177°10'W	4	180
19. Takawangha	Tanaga Island 51°52'N,178°00'W	3.2	54
20. Tanaga	Tanaga Island 51°53'N,178°07'W	11	960
21. Semispochnoi	Semispochnoi Island 51°56'N,179°35'W	6.1 x 7.5	360
22. Little Sitkin	Little Sitkin Island 51°57'N,178°32'W	4.5	180
23. Davidof	Davidof Island 51°58'N,178°20'W	2.4	29

*After Smith and Shaw, 1979.
†Not a caldera.

springs are closely associated with the areas of active volcanism, and this association is reflected in both the high surface temperatures (in some cases at or near boiling temperature) of the spring waters and in the high reservoir temperatures (commonly in excess of 200°C as estimated from geochemical thermometers). The Geyser Bight area on the north-central side of Umnak Island has the most impressive surface manifestation of high heat flux of any hot-water convection system in Alaska. Numerous thermal springs and pools over a 4-km^2 area are at, or in some cases above, boiling and are superheated with temperatures to 104°C; small geysers erupting to heights of 25 cm are not uncommon. Similar superheating and geyser activity has been noted at a thermal spring on the east side of Kanaga Island on the southeast flank of Kanaga volcano.

Smith and Shaw (1975, 1979) have evaluated the heat content of young igneous-related systems in the United States to a depth of 10 km. They estimated thermal energy in these systems on the basis of assumed high-level silicic magma chambers (i.e., calderas, silicic domes, etc.), chamber volume, and the time of their latest eruptions. Smith and Shaw's thermal energy calculations assume that a fixed volume of magma cooled from an initial temperature of 850°C to its present temperature solely by conduction in surrounding rocks, starting from a fixed time.

Smith and Shaw (1979) calculated that the fixed-volume estimate of thermal energy remaining in 58 igneous systems in the United States is approximately 101,000 × 10^{18} joules. These 58 systems include 22 of the Aleutian volcanic arc systems and are the ones judged by Smith and Shaw to have sufficient age and volume data to make meaningful calculations. Smith and Shaw (1979) caution that their estimate of 101,000 × 10^{18} joules is conservative and that the total igneous-related energy is at least an order of magnitude greater. The lack of sufficient geochronologic data, and the likelihood that very young systems with relatively small single-chamber volumes have subchamber support systems much greater in size and longevity than are inferred from the age and extent of volcanic products, make it impossible at present to give more quantitative estimates of undiscovered thermal energy.

The 22 volcanic centers in the Aleutian arc for which Smith and Shaw (1979) calculated remaining thermal energy (Table 1) total more than 7,000 × 10^{18} joules. Although this figure is only about 7 percent of the total U.S. estimate of 101,000 × 10^{18} joules, it should be realized that 75 percent of the entire U.S. estimate comes from five large intracontinental caldera systems: Yellowstone, Island Park, Valles, Rexburg, and Long Valley. Also, at least a dozen Aleutian volcanoes whose volcanic products and age suggest high-level magma chambers have not been included in the calculations, chiefly because of the relatively small size of their erupted products or because age and volume data are still too tentative.

Brook and others (1979), in an assessment of hydrothermal convection systems with reservoir temperatures of 90°C or more, estimated that the identified resource base for the Aleutian arc was a relatively low 10 × 10^{18} joules, whereas the undiscovered accessible resource base was estimated at as much as 580 × 10^{18} joules.

Although the Aleutian arc has significant potential for large geothermal resources, the remoteness of the area had resulted in little interest in development until 1976, when the U.S. Navy collaborated with the U.S. Geological Survey in an investigation of the geothermal potential, particularly for direct-use space heating purposes, of Adak Island. Three Quaternary volcanoes—Mt. Moffett, Andrew Bay, and Mt. Adagdak—occur on the north side of the island. As a result of a multidisciplinary study of the area that included geological mapping, geophysical studies, and isotopic age-dating, Miller and others (1978) concluded that a high-level thermal anomaly probably existed beneath Mt. Adagdak volcano. This volcano is the youngest of the three—its most recent dated eruption having been dated at 140 ka. Geophysical studies indicated a low-density, very conductive, northeast-trending mass under the south and east side of Adagdak; low resistivity measurements suggested this rock might be hot. In 1977, the U.S. Navy drilled three holes in Adagdak in search of a heat source for space heating purposes. Unfortunately, because of access problems, the holes were not located on the geophysical anomaly. The closest hole to the anomaly, however, yielded a bottom-hole (600 m) temperature of about 66°C.

Unalaska Island is located approximately 1,500 km southwest of Anchorage in the eastern Aleutian Islands. The northern part of the island is dominated by Makushin volcano, a large (200 km^3 in volume), historically active andesitic stratocone reaching an altitude of 2,037 m and having a summit caldera about 3 km in diameter. Although not one of the larger volcanic centers in the Aleutian arc in terms of present heat content (Smith and Shaw estimate 25 × 10^{18} joules thermal energy still remaining in the system as compared to more than 1,400 × 10^{18} joules for such large volcanic centers as Fisher caldera), the volcanic center is attractive as a geothermal resource because it is only 19 km from the major fishing port and population center of Dutch Harbor.

Detailed geological mapping, geochemical studies, and a self-potential geophysical survey conducted by the Alaska Division of Geological and Geophysical Surveys, and by private consultants in 1982 under contract to the state, led to the drilling of three primary temperature gradient holes to depths of about 440 m (Isselhardt and others, 1983). Subsurface temperatures as high as 195°C were obtained from these holes, and a water-dominated geothermal system was thought to exist on the eastern flank of the volcano.

A small-diameter resource-confirmation well was drilled in the area in 1983 to a depth of 594 m and yielded a bottom-hole static temperature of 193°C. Three potential geothermal resource zones were encountered in the hole. A long-term 34-day test of the well in 1984 suggested a shallow steam zone overlying a liquid-dominated reservoir in the fractured diorite basement rock (Economides and others, 1985). The reservoir (unflashed) fluid is a NaCl-type water with a total dissolved solid content of about 600 mg/l. A flowing temperature of 193°C was found at a depth of 594 m. The test suggested the reservoir could be highly productive. Sustained flow through a 3-in-diameter well bore of 63,000 lb/hr was achieved with less than 2 psi of pressure drawdown from an initial pressure of 494 psi. Economides and others (1985) calculated that a theoretical electricity reserve exists that would be sufficient for the needs of the local island populace for several hundred years at current consumption rates.

It seems clear that a large geothermal resource base exists beneath Makushin volcano. Whether it wil be exploited—given the remoteness of the region, the large initial costs of geothermal development, and the relatively small local population—is problematical.

Motyka and others (1981) have described large thermal areas on Atka and Akutan Islands in the central and eastern Aleutian Islands. On northeastern Atka Island, two large (50,000 m^2) thermal areas occur on the flanks of Mt. Kliuchef volcano (Kliuchef thermal area) and in a valley southeast of Korovin volcano (Korovin thermal area). These thermal areas consist of fumaroles, mud pots, boiling springs, warm ground, and hydrothermally altered zones. Thermal spring waters are of an acidic, low-chloride, high-sulfate variety. A combination of chemical and gas geothermometry using the techniques described by Fournier (1977) and Ellis and Mahon (1978) indicates reservoir temperatures of 239°C and >150°C for the two areas. Motyka and others (1981) suggest the possibility of a shallow vapor-dominated system associated with the Kliuchef thermal area.

The thermal area on Akutan Island occurs along a northeast-trending stream valley approximately 6 km east of Akutan volcano—a small andesitic stratocone with a small summit caldera. Hot-spring temperatures as high as 85°C, associated sinter deposits, and indicated reservoir temperatures of 180°C (Motyka and others, 1981) attest to the high heat flux of the area.

Geothermal resources on Atka and Akutan Islands are of some interest in that, unlike many other geothermal areas in the Aleutian arc, they occur relatively close to small fishing villages. Local use is therefore possible, particularly as (and if) the fishing industry expands.

Wrangell Mountains volcanic pile

A thick calc-alkaline pile of Neogene volcanic rocks underlies about 10,000 km^2 of the Wrangell Mountains in east-central Alaska. The rocks range in composition from basalt to rhyolite and in age from Miocene to Holocene (Miller and Richter, this volume). The Wrangell volcanic pile appears to be related to a northwest-trending, northeast-dipping subduction zone resulting from underthrusting of the Pacific Plate beneath the North American Plate (Stephens and others, 1984). The Wrangell volcanic pile is separated from the Aleutian arc by a gap of some 400 km; the relation between the two arcs, if any, is uncertain.

Few thermal springs are associated with the Wrangell volcanic rocks, and only Mt. Wrangell itself, a 4,320-m-high shield volcano with a summit caldera, has an active thermal area. Historic eruptions have been rare in the Wrangell volcanic pile. The presence of several large Quaternary stratovolcanoes, however, attests to the extent and nature of recent volcanism. These include Mount Drum (3,622 m), Mount Sanford (4,950 m), Mount Blackburn (5,037 m), Regal Mountain (4,210 m), and large shield volcanoes such as Tanada Peak and Capitol Mountain, which cover areas of 400 and 200 km^2, respectively (Richter and others, 1984). K-Ar ages of 200 to 2,000 ka have been obtained from these volcanoes; a volcanic center now largely under the Klutlan Glacier about 10 km from the Canadian border erupted explosively about 1,900 and 1,250 ^{14}C B.P. (Lerbekmo and others, 1975), depositing the White River ash over 300,000 km^2 of eastern Alaska and northwest Canada.

Mount Drum, the westernmost stratovolcanic complex in the Wrangell Mountains, has received a considerable amount of study as a possible geothermal resource (Richter and others, 1978). This volcanic center consists of a deeply dissected stratocone ringed by peripheral silicic domes. The presence of young silicic rocks initially suggested the possibility of a high-level magma chamber (Miller and Barnes, 1976) that might have geothermal potential. Subsequent mapping and petrologic and geochronologic studies showed at least two principal cycles of volcanism, each cycle consisting of andesitic and dacitic cone-building followed by peripheral emplacement of dacite, rhyodacite, and rhyolite domes. The younger cycle culminated in an explosive eruption resulting in destruction of part of the central stratocone and emplacement of volcanic avalanche deposits. K-Ar ages indicate that the volcano began to form about 800 ka and that the youngest dated event occurred about 240 ka; field relations indicate one or more still younger events.

The two complete cycles of volcanism consisted of similar episodes of cone building and dome emplacement. After the eruption of the first-cycle rhyolite domes, the magma chamber beneath Mount Drum was either depleted of much of its silicic differentiate or received an influx of more mafic magma, because the second-cycle cone-building lavas were predominantly dacite and andesite. The larger area outlined by the distribution of the ring domes suggests that the size of the magma chamber had increased considerably by the end of the second cycle. Stratocone rocks range from basaltic andesite (54.5 percent SiO_2) to dacite (63 percent SiO_2), whereas the silicic domes are composed of dacite, rhyodacite, and rhyolite (SiO_2 content as much as 72 percent).

The distribution and composition of the silicic-ring domes peripheral to the Mount Drum cone suggest the presence of a high-level magma chamber. Calculations based on the probable size, age, and composition of this magma chamber indicate the existence of a thermal anomaly of considerable magnitude, 26.5 to 114×10^{19} cal, one of the largest estimates for any volcanic system in Alaska.

Thermal spring areas, central and northern Alaska

More than 60 thermal springs are scattered throughout Alaska, and most of them occur in settings seemingly unrelated to Quaternary volcanism. In the vast area north of the Alaska Range

(~1,000,000 km^2), 36 thermal springs have been reported, 32 of which are located in a 200-km-wide east-west band extending across interior Alaska from the Seward Peninsula to within 160 km of the Canadian border (Fig. 1). More thermal springs may occur in this area, but geological studies detailed enough to reveal them have been completed for less than 10 percent of the area.

The 36 known thermal springs (Miller, 1973) show no temporal or spatial association with Tertiary or Quaternary volcanism (Moll-Stalcup and others, this volume; Plafker and others, this volume, Plate 12) but rather (Miller and Barnes, 1976) are closely associated with the margins of granitic plutons. Of the 36 thermal springs, 33 occur within 5 km of the margin of a granitic pluton. One of the few apparent exceptions to this empirical observation is Pilgrim Hot Springs on the Seward Peninsula (Fig. 1) where the springs appear to be related to a faulted margin of a Tertiary basin.

The occurrence of thermal springs in this large area is independent of the absolute age or composition of the associated plutonic rocks. Plutons with associated thermal springs range in age from 315 to 380 Ma in the Brooks Range (Dillon, 1980) to 60 Ma in the Yukon-Tanana Upland (DuBois and others, 1986). Most of the plutons, however, are Cretaceous, ranging in age from 110 to 66 Ma. The plutons include biotite granite, two-mica granite, granodiorite, monzonite, syenite, and nepheline syenite.

The distribution of thermal springs is independent of the age and lithology of the country rock enclosing the pluton. The country rocks, for example, range in age from Paleozoic and perhaps Precambrian(?) to Late Cretaceous, and include limestone, graywacke, andesite, mafic volcanic rocks, and regionally metamorphosed rocks of low and high grade. The hot springs occur in a number of different geologic provinces (Seward, Koyukuk, Ruby, Arctic Alaska, Angayucham, Yukon-Tanana, etc.) and tectonostratigraphic terranes that have a large variety of geologic and structural features; some of these features are confined to a single province or terrane, while others are found in two or more provinces and terranes. The only feature common to thermal springs and the terrane in which they occur is the existence of granitic plutons.

A prerequisite for the occurrence of a thermal spring throughout this large area appears to be a mass of competent, well-fractured rock near the margins of a massive, nonfoliated granitic body. All known thermal springs that occur outside the boundary of a pluton occur in nonfoliated rocks such as graywacke, mudstone, basalt, and andesite that may have been affected to some extent by contact metamorphism but not by regional metamorphism. Fracture systems apparently were not developed or are not sufficiently open in crystalline regionally metamorphosed rocks with a well-developed planar fabric to allow deeply circulating hot water to gain access to the surface. Where the wall rock surrounding a pluton consists of well-foliated, regionally metamorphosed rock, the thermal spring invariably occurs within the pluton.

Oxygen and hydrogen isotopic analysis of waters collected from ten thermal springs scattered across an area extending from the western Seward Peninsula to the Hodzana Highlands (Fig. 1) are similar to isotopic analyses of locally derived meteoric waters (Miller and others, 1975). The thermal spring waters are weakly alkaline and somewhat high in chloride. They appear to be deeply circulating meteoric water whose increased solvent action due to the increase in temperature and long flow path brought about leaching of the country rock. Six of 34 analyzed thermal springs contain dissolved solids in excess of 1,000 mg/l, ranging from 1,150 to more than 6,000 mg/l. These springs are also more saline than the other 28 springs and are characterized by higher concentrations of chloride, sodium, calcium, potassium, fluorine, lithium, bromine, and boron. Saline thermal springs occur in the same geologic provinces as nonsaline thermal springs and are found as far west as Serpentine Hot Springs on the Seward Peninsula and as far east as Tolovana and Big Windy Hot Springs near Fairbanks in central Alaska. The chemical and isotopic data indicate that both types of hot springs originate from deeply circulating meteoric water; their compositions probably result from a difference in the extent of leaching or in water-rock reactions from one locality to another (Miller and others, 1975).

The most studied geothermal area north of the Alaska Range is Pilgrim Springs in the west-central Seward Peninsula (Fig. 1). Unlike the other thermal springs in the region, its geologic setting is unclear because it lies in the center of the 2-km-wide, alluvium-filled Pilgrim River valley, a fault-bounded, graben-like structure where the basement rocks are not exposed. The valley is flanked by amphibolite-facies metamorphic rocks of probable Precambrian and Paleozoic age cut by Cretaceous plutons. The thermal springs are located in an oval-shaped 1.5-km^2 area of thawed ground surrounded by permafrost. Chemical geothermometry (Miller and others, 1975; Motyka and others, 1980a) indicates possible reservoir temperatures of 146 to 154°C. Seismic, gravity, and resistivity surveys (Turner and Forbes, 1980) suggest that the crystalline basement rocks that floor the Pilgrim River valley are at least 200 m deep beneath the thermal area and that the springs lie in a fault-bounded zone.

A total of six holes were drilled to depths of 45 to 305 m at Pilgrim Springs between 1979 and 1982. Siting of these holes was based on geophysical studies and a helium survey. The results of this drilling (Kunze and Lofgren, 1983) indicate that a perched hot-water reservoir (maximum temperature of 88°C) exists at a depth of 15 to 37 m and its underlain by a cold-water reservoir. Kunze and Lofgren (1983) noted that this reservoir represented less than 10 percent of the resource to be expected in the area and that a much larger and deeper source of hot water must exist.

The origin of Pilgrim Hot Springs is uncertain. Turner and Forbes (1980) have postulated that the Pilgrim River valley is part of a rift system extending in a general east-west direction across the Seward Peninsula and that the thermal springs may be related to active rifting. Motyka and others (1980b) suggest that because of the highly saline, alkali-chloride character of the thermal spring waters, and because the large Quaternary basalt field of the Imuruk Lake area is only 60 km away, a volcanic association cannot entirely be discounted.

Perhaps of greater significance, however, is the fact that Pilgrim Springs is only one of more than 30 thermal springs that occur in the 200-km-wide band that crosses several geologic provinces and terranes almost to the Canadian border. These springs show no relation to recent volcanism (i.e., no recent volcanic rocks exist within about 100 km of most of these springs). The saline character of Pilgrim Springs water is not unique; saline water is found both at Serpentine and Kwiniuk Hot Springs on the Seward Peninsula (both of which are unequivocally associated with margins of granitic plutons) as well as at Tolovana and Big Windy Creek (Keith and Foster, 1979; Keith and others, 1981) thermal springs in east-central Alaska. These characteristics suggest to me that Pilgrim Hot Springs has an origin similar to the other thermal springs in interior Alaska, that a granite body exists at depth beneath the Tertiary and surficial cover, and that the faulted margin of the Tertiary basin only controls the actual site of upwelling.

Southeastern Alaska thermal springs

Thermal springs in southeastern Alaska also appear to be related to the fractured margins of granitic masses (Waring, 1917; Miller and others, 1975; Motyka and others, 1980b). Of 18 known or reported thermal springs in southeastern Alaska, at least 12 occur within, or very close to, granitic plutons. In addition, Twenhofel and Sainsbury (1958) found that 14 of these hot springs are located on prominent lineaments. These springs appear to have no spatial association with Pleistocene-Holocene volcanic rocks at Mount Edgecumbe volcano near Sitka, nor with rocks in the Kuiu-Etolin volcanic belt (Fig. 1) on Kupreanof Island (Brew and others, 1985). Motyka and others (1980b), in a study of thermal spring sites in southeastern Alaska, concluded that the thermal waters are alkali-sulfate to alkali-chloride in character and are probably derived from the interaction of deeply circulating meteoric waters with granitic wall rock. Assuming a geothermal gradient of 30 to 50 °C/km, chemical geothermometers suggest subsurface temperatures of 55 to 151°C, indicating circulation to depths of 1.5 to 5 km.

Other geothermal sites

Turner and Wescott (1982) have speculated that zones of discontinuous hot water may exist at depth in the Susitna Basin in south-central Alaska some 60 km north of Anchorage (Fig. 1). Although no surface manifestations of anomalous heat flux have been found, four dry wildcat wells drilled for petroleum in the area encountered anomalously high temperatures with gradients of 44 to 123 °C/km in the underlying Tertiary sedimentary rocks. The source of the heat is unknown, but a helium survey of soil gas, and water, encountered elongate anomalies possibly controlled by Tertiary faults, along which hydrothermal convection could have occurred (Turner and Wescott, 1982). No hot water has been found, and only preliminary work has been done in searching for possible reservoirs. An obvious attraction for any geothermal resource development in this area is its proximity to over 60 percent of Alaska's population.

SUMMARY

Alaska, perhaps more than any other single region in North America, probably has the greatest number of potential geothermal energy sites. Regional and reconnaissance geological studies indicate that more than 25 igneous-related systems and thermal areas in the state have thermal anomalies of sufficient magnitude to be of interest for large-scale geothermal energy development. Exploration and research into Alaska's geothermal resources began about 1972 and is still in an early phase. Few of the large volcanic centers likely to have geothermal resources have been studied in sufficient detail to document their potential. Geophysical studies have been done only on Makushin and Adagdak volcanoes in the eastern and central Aleutian Islands and at Pilgrim Hot Springs on the Seward Peninsula, and those are the only sites in the state where extensive drilling has been conducted.

The remoteness of many of the geothermal areas with the greatest potential, the sparse population base, the difficulty in delivering the energy to distant markets, and the high front-end development costs are factors that affect the utilization of the resource. An additional complicating factor is that virtually all of the potentially large geothermal systems in the Aleutian volcanic arc and the Wrangell Mountains are located within the boundaries of national parks and monuments and wildlife refuges.

The Makushin geothermal site is probably typical of a volcanic system in the Aleutian volcanic arc that might constitute a geothermal resource. An exploration well drilled to about 600 m encountered a shallow steam zone at a temperature of 163°C overlying a liquid-dominated reservoir at a temperature of 193°C. It was estimated that a single production well from this site could supply about 9 MW of electrical power. The potential seems high elsewhere in the Aleutian arc and Wrangell volcanic pile for more such vapor- and hot-water-dominated convection systems to exist associated with volcanoes that have a high-level magma chamber and heat reservoir.

Geological, geophysical, and drilling information from Adagdak volcano on Adak Island in the Aleutians suggests that it might be typical of a high-level heat source relatively close to the surface large enough to provide direct use energy for space heating.

A prerequisite to development of any geothermal resource is the existence of a potential user. In addition to Makushin and Adagdak volcanoes with their nearby population centers of Unalaska and Adak, Mount Spurr (130 km west of Anchorage), Mount Drum in east-central Alaska (55 km from the main highway networks), and possibly Mount Edgecumbe in southeastern Alaska (25 km west of Sitka) all have potential since they are close to major population areas.

Thermal springs in Alaska outside of the Aleutian volcanic arc are characterized by relatively low surface temperatures (generally less than 60°C) and by low reservoir temperatures as indicated by geothermometry (usually less than 150°C). They appear to be associated with the fractured margins of granitic plutons and have low porosity. Indicated reservoir temperatures are typically below the 150 to 180°C thought necessary to generate electricity. The low porosity makes most of the thermal spring localities relatively poor candidates for large-scale direct-heat sources. Only Pilgrim Hot Springs on the Seward Peninsula may have sufficient porosity and volume to be a viable geothermal resource for development as a direct heat source.

All of the thermal springs represent hydrothermal convection systems, and, based on their high chloride and alkaline pH levels, are water- rather than vapor-dominated. There has been little detailed study of these thermal springs, and fewer than half a dozen such systems in Alaska have been drilled. Of this group, only Bailey Bay in southeastern Alaska, with a calculated reservoir temperature of 151°C, qualifies as a high-temperature hot-water system as defined by White and Williams (1975): that is, a system with a reservoir temperature >150°C. Of the remaining thermal springs, approximately 25 percent are intermediate-temperature systems with estimated reservoir temperatures of 90 to 150°C, and about 75 percent are low-temperature systems with estimated reservoir temperatures below 90°C (Miller, 1973; Turner and others, 1980).

The thermal springs appear to have little potential for large-scale production of electricity or for space heating. Their geologic setting and chemical geothermometry suggest limited reservoirs and fracture porosity, and relatively low reservoir temperatures. Many of them, however, have a long history of use for local recreational and agricultural purposes, and those uses, together with space heating for small numbers of houses in isolated villages and resorts, seem likely to continue indefinitely.

REFERENCES CITED

Alaska Division of Geological and Geophysical Surveys, 1983, Geothermal resources of Alaska: Alaska Department of Natural Resources, Division of Geological and Geophysical Surveys, scale, 1:2,500,000.

Brew, D. A., Karl, S. M., and Tobey, E. F., 1985, Reinterpretation of age of Kuiu-Etolin volcanic belt, Kupreanof Island, southeastern Alaska, *in* Bartsch-Winkler, S., and Reed, K. M., eds., The United States Geological Survey in Alaska; Accomplishments in 1983: U.S. Geological Survey Circular 945, p. 86–88.

Brook, C. A., and 5 others, 1979, Hydrothermal convection systems with reservoir temperatures >90°C, *in* Muffler, L.J.P., ed., Assessment of geothermal resources of the United States; 1978: U.S. Geological Survey Circular 790, p. 18–85.

Dillon, J. T., Pessel, G. H., and Veach, N. C., 1980, Middle Paleozoic magmatism and orogenesis in the Brooks Range, Alaska: Geology, v. 8, p. 338–343.

DuBois, G. D., Wilson, F. H., and Shew, N., 1986, Map and tables showing potassium-argon age determinations and selected major-element chemical analyses from the Circle Quadrangle, Alaska: U.S. Geological Survey Open-File Report 86–392, scale 1:250,000.

East, J., 1982, Preliminary geothermal investigations at Manley Hot Springs, Alaska: University of Alaska Geophysical Institute Report UAG R-290, 76 p.

Economides, M. J., Morris, C. W., and Campbell, D. A., 1985, Evaluation of the Makushin geothermal reservoir, Unalaska Island, *in* Proceedings of the 10th Workshop on Geothermal Reservoir Engineering: Stanford, California, Stanford Geothermal Program, Report TR-84, p. 227–232.

Elis, A. J., and Mahon, W.A.J., 1978, Chemistry and geothermal systems: New York, Academic Press, 342 p.

Fournier, R. O., 1977, Chemical geothermometers and mixing models for geothermal systems: Geothermics, v. 5, p. 41–50.

Isselhardt, C. F., Motyka, R. J., Matlick, J. S., Parmentier, P. P., and Huttrer, G. W., 1983, Geothermal resource model for the Makushin geothermal area, Unalaska Island, Alaska: Geothermal Resources Council Transactions, v. 7, p. 99–102.

Keith, T.E.C., and Foster, H. L., 1979, Big Windy Creek hot springs, Circle A-1 Quadrangle, Alaska, *in* Johnson, K. M., and Williams, J. R., eds., The United States Geological Survey in Alaska; Accomplishments during 1978: U.S. Geological Survey Circular 804-B, p. 55–57.

Keith, T.E.C., Presser, T. S., and Foster, H. L., 1981, New chemical and isotope data for the hot springs along Big Windy Creek, Circle A-1 Quadrangle, Alaska, *in* Albert, N.R.D., and Hudson, T., eds., The United States Geological Survey in Alaska; Accomplishments during 1978: U.S. Geological Survey Circular 823-B, p. 25–28.

Kunze, J. F., and Lofgren, B. E., 1983, Pilgrim Springs, Alaska, geothermal resource exploration, drilling, and testing: Geothermal Resources Council Transactions, v. 7, p. 301–304.

Lerbekmo, J. F., Westgate, J. A., Smith, D.G.W., and Denton, G. H., 1975, New data on the character and history of the White River volcanic eruptions, *in* Suggate, R. P., and Cresswell, M. M., eds., Quaternary studies: Wellington, Royal Society of New Zealand, p. 203–209.

Matlick, J. S., and Parmentier, P. P., 1983, Geothermal manifestations and results of a mercury soil survey in the Makushin geothermal area, Unalaska Island, Alaska: Geothermal Resources Council Transactions, v. 7, p. 305–309.

Miller, T. P., 1973, Distribution and chemical analyses of thermal springs in Alaska: U.S. Geological Survey Open-File Report 73–570, scale 1:2,500,000.

Miller, T. P., and Barnes, I., 1976, Potential for geothermal-energy development in Alaska; Summary, *in* Halbouty, M. T., Mahar, J. C., and Lian, H. M., eds., Circum-Pacific energy and mineral resources: American Association Petroleum Geologists Memoir 25, p. 149–153.

Miller, T. P., and Smith, R. L., 1987, Late Quaternary caldera-forming eruptions in the eastern Aleutian arc, Alaska: Geology, v. 15, p. 434–438.

Miller, T. P., Barnes, I., and Patton, W. W., Jr., 1975, Geologic setting and chemical characteristics of hot springs in west-central Alaska: U.S. Geological Survey Journal of Research, v. 3, no. 2, p. 149–162.

Miller, T. P., Hoover, D. B., Smith, R. L., and Long, C., 1978, Geothermal exploration on Adak Island, Alaska; A case history [abs.]: American Association Petroleum Geologists Bulletin, v. 62, no. 7, p. 1227.

Motyka, R. J., 1983, Geochemical and isotopic studies of waters and gases from the Makushin geothermal area, Unalaska Island, Alaska: Geothermal Research Council Transactions, v. 7, p. 103–108.

Motyka, R. J., Forbes, R. B., and Moorman, M., 1980a, Geochemistry of Pilgrim Springs thermal waters, *in* Turner, D. L., and Forbes, R. B., eds., A geological and geophysical study of the geothermal energy potential of Pilgrim Springs, Alaska: University of Alaska Geophysical Institute Report UAG R-271, p. 43–52.

Motyka, R. J., Moorman, M. A., and Reeder, J. W., 1980b, Assessment of thermal spring sites in southern southeastern Alaska; Preliminary results and evaluation: Alaska Division Geological and Geophysical Surveys Open-File Report 127, 72 p.

Motyka, R. J., Moorman, M. A., and Liss, S. A., 1981, Assessment of thermal spring sites, Aleutian arc, Atka Island to Becharof Lake; Preliminary results and evaluation: Alaska Division of Geological and Geophysical Surveys Open-File Report 144, 173 p.

Parmentier, P. P., Reeder, J. W., and Henning, M. W., 1983, Geology and hydrothermal alteration of Makushin geothermal area, Unalaska Island, Alaska: Geothermal Resources Council Transactions, v. 7, p. 181–185.

Richter, D. H., Smith, R. L., Yehle, L. A., and Miller, T. P., 1978, Geologic map of the Gulkana A-2 Quadrangle, Alaska: U.S. Geological Survey Geologic Quadrangle Map GQ-1520, scale 1:63,360.

Richter, D. H., Smith, J. G., Ratte, J. C, and Leeman, W. P., 1984, Shield volcanoes in the Wrangell Mountains, Alaska, *in* Reed, K. M., and Bartsch-Winkler, S., eds., United States Geological Survey in Alaska; Accomplishments during 1983: U.S. Geological Survey Circular 939, p. 71–75.

Smith, R. L., and Shaw, H. R., 1975, Igneous-related geothermal systems, *in* White, D. E., and Williams, D. L., eds., Assessment of geothermal resources of the United States; 1975: U.S. Geological Survey Circular 726, p. 58–83.

——, 1979, Igneous-related geothermal systems, *in* Muffler, L.J.P., ed., Assessment of geothermal resources of the United States; 1978: U.S. Geological Survey Circular 790, p. 12–17.

Stephens, C. D., Fogleman, K. A, Lahr, J. C., and Page, R. A., 1984, Wrangell Benioff zone, southern Alaska: Geology, v. 12, p. 373–376.

Turner, D. L., and Forbes, R. B., eds., 1980, A geological and geophysical study of the geothermal energy potential of Pilgrim Springs, Alaska: University of Alaska Geophysical Institute Report UAG R-271, 165 p.

Turner, D. L., and Wescott, E. M., 1982, A preliminary investigation of the geothermal energy resources of the lower Susitna Basin: University of Alaska Geophysical Institue Report UAG R-287, 49 p.

Turner, D. L., and 5 others, 1980, Geothermal energy resources of Alaska: University of Alaska Geophysical Institute Report UAG R-279, 19 p.

Twenhofel, W. S., and Sainsbury, C. L., 1958, Fault patterns in southeastern Alaska: Geological Society of American Bulletin, v. 69, p. 1431–1442.

Waring, G. A., 1917, Mineral springs of Alaska: U.S. Geological Survey Water-Supply Paper 418, 114 p.

Wescott, E. M., and Turner, D. L., 1983, Geothermal energy resource exploration of the eastern Copper River basin, Alaska: Geothermal Resources Council Transactions, v. 7, p. 211–213.

White, D. F., and Williams, D. L., 1975, Introduction, *in* White, D. F., and Williams, D. L., eds., Assessment of geothermal resources of the United States; 1975: U.S. Geological Survey Circular 726, p. 1–4.

Manuscript Accepted by the Society May 21, 1990

Printed in U.S.A.

The Geology of North America
Vol. G-1, The Geology of Alaska
The Geological Society of America, 1994

Chapter 33

Overview of the geology and tectonic evolution of Alaska

George Plafker
U.S. Geological Survey, MS-904, 345 Middlefield Road, Menlo Park, California 94025
Henry C. Berg
115 Malvern Avenue, Fullerton, California 92632

INTRODUCTION

In this chapter we present a brief overview of major aspects of the geology and tectonic evolution of Alaska. Our objective has been to incorporate the lithologic, structural, paleontologic, geophysical, and paleomagnetic data presented in this volume and elsewhere into a series of generalized interpretive maps that depict our version of the Phanerozoic tectonic evolution of Alaska. Where necessary, adjacent areas of Canada and the conterminous United States are discussed and shown on the maps. Many stratigraphic and tectonic relations germane to Alaska are present in those regions and some terranes now in Alaska were derived from regions to the south.

Our interpretation emphasizes evaluation of the data according to plate tectonics models and the concept that much of Alaska is a collage of tectonostratigraphic terranes that have been displaced to varying degrees relative to each other and to the North American craton. Many problems remain, however, and alternative interpretations have been proposed by others for virtually every major aspect of the model presented here. We hope that this synthesis will highlight crucial areas for future research in order to resolve ambiguities among the array of data sets and analog models.

Principal data sources

The first part of this chapter presents the major constraints used to develop the models of the Phanerozoic tectonic evolution of Alaska that follow. This synthesis is based on the geologic and geophysical literature of Alaska and adjacent regions of the North American continent and ocean basins, much of which is summarized in this volume. Space limitations preclude acknowledging all sources of data and ideas. The cited papers, many of which are syntheses of broad areas and topics, contain extensive bibliographic references to original data sources.

Base maps used in all figures are Albers equal-area projections. Geologic time terms used are those of Palmer (1983). Unless otherwise stated, orientations of all geologic and geographic features refer to present geographic coordinates.

LITHOTECTONIC TERRANES OF ALASKA

Figure 1 shows all but the smallest structurally bound lithotectonic terranes and subterranes that have been delineated in Alaska and adjacent parts of Canada as well as major areas of Cretaceous and Cenozoic basinal deposits. As defined by Jones and others (1983), Howell and others (1985), and Howell (1989), lithotectonic terranes (also called "suspect" or "tectonostratigraphic" terranes) are fault-bounded geologic packages of regional extent characterized by a geologic history different from that of neighboring terranes. They may or may not have had large displacements relative to nearby terranes. Terrane recognition, displacement histories, timing of accretion, and boundary conditions have been the subjects of an extensive literature (e.g., Coney and others, 1980; Coney and Jones, 1985; Jones and others, 1977, 1983, 1987; Monger and others, 1982, 1991; Monger and Berg, 1987; Saleeby, 1983; Plafker, 1990), and many of the terrane names and definitions in Alaska have continued to evolve since the terrane concept was introduced. Detailed studies have shown that some of these terranes may have undergone little or no displacement relative to adjacent terranes and that the original terrane definitions either have been modified or are no longer applicable. Except as otherwise noted, in this paper we adhere to the terrane nomenclature of Silberling and others (this volume, Plate 3) to describe these lithologically distinctive belts of rock.

Terranes shown in Figure 1 include displaced and/or rotated noncrystalline and crystalline fragments of the continental margin; probable continental margin magmatic arcs; dominantly oceanic crust and associated sedimentary rocks; intraoceanic magmatic arcs, oceanic plateaus, and rift-fill assemblages; and arc-related accretionary prisms. The main tectonic affinity of the larger terranes is indicated by map patterns. Several unique terranes too small to depict or name in Figure 1 are present primarily along major suture boundaries; these include fault slivers of adjacent terranes, fragments of exotic terranes that were probably carried to their present positions along the margins of larger terranes, and small crystalline terranes, the origins of which are obscured by plutonism and/or metamorphism.

Plafker, G., and Berg, H. C., 1994, Overview of the geology and tectonic evolution of Alaska, *in* Plafker, G., and Berg, H. C., eds., The Geology of Alaska: Boulder, Colorado, Geological Society of America, The Geology of North America, v. G-1.

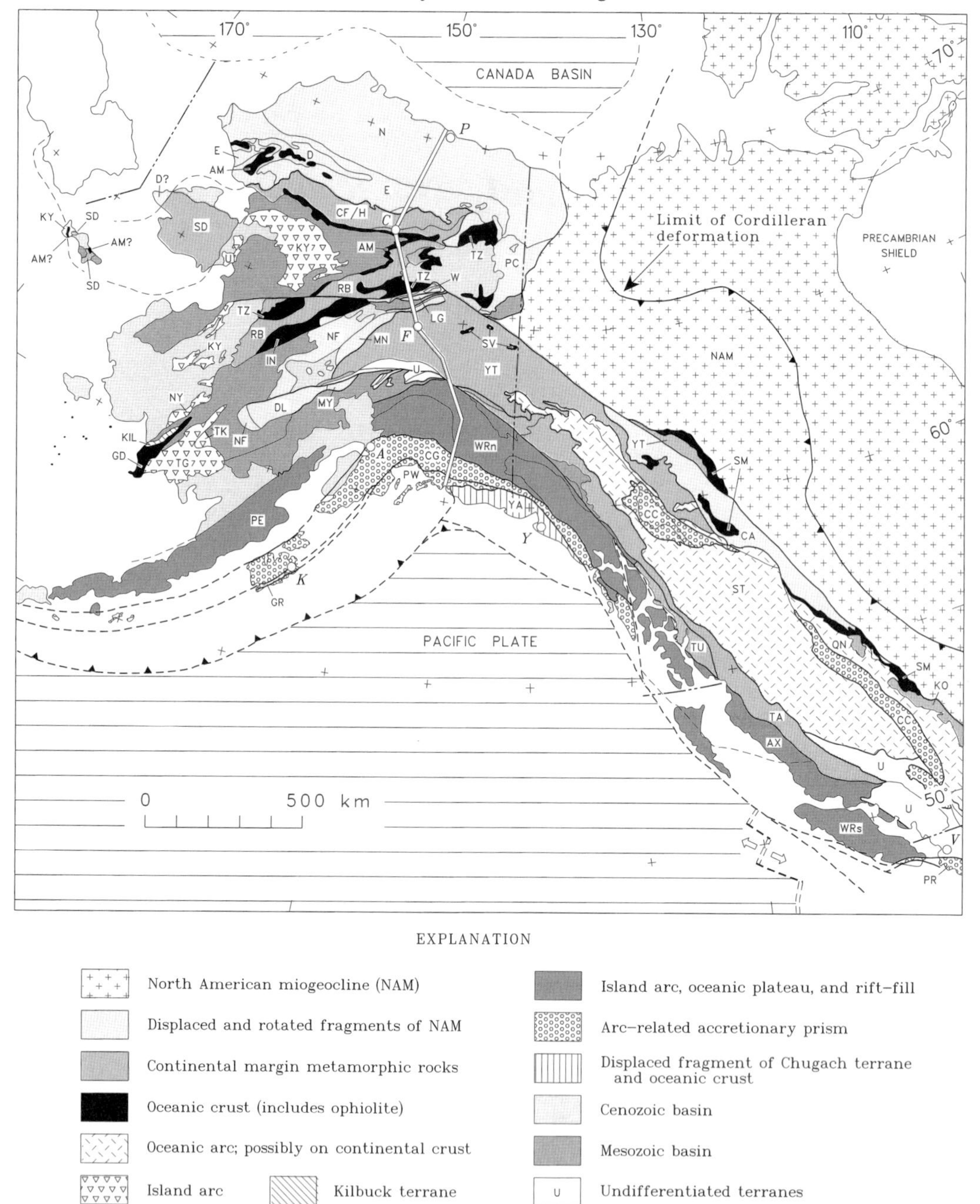

Figure 1. Generalized distribution and inferred tectonic affinity of selected terranes in Alaska and western Canada. North American craton: NAM: Displaced and/or rotated fragments of the North American craton margin: N, North Slope; E, Endicott; D, Delong Mountains; PC, Porcupine; MN, Minchumina; NF, Nixon Fork; DL, Dillinger; CA, Cassiar; MY, Mystic; W, undifferentiated Baldry, Minook, White Mountains, and Wickersham terranes. Dominantly metamorphic terranes of continental affinity: SD, Seward; CF/H, Coldfoot and Hammond; RB, Ruby; YT, Yukon-Tanana; KO, Kootenay; TA, Tracy Arm; TU, southern Taku. Precambrian metamorphic terrane of probable continental margin arc affinity: KIL, Kilbuck and related Idono Complex. Magmatic arc terrane of probable continental margin affinity: QN, Quesnellia; ST, Stikine. Dominantly oceanic terranes: AM, Angayucham; GD, Goodnews; IN, Innoko; LG, Livengood; SM, Slide Mountain; SV, Seventymile; TZ, Tozitna. Intraoceanic magmatic arc terranes: KY, Koyukuk; NY, Nyack; TG, Togiak; TK, Tikchik(?). Magmatic arc, oceanic plateau, and rift-fill assemblages: AX, Alexander; PE, Peninsular; WRn, Wrangellia (Alaska); WRs, Wrangellia (British Columbia). Arc-related accretionary prisms: CG, Chugach; GR, Ghost Rocks; PW, Prince William; CC, Cache Creek; PR, Pacific Rim. Displaced fragment of Chugach terrane and oceanic crust: YA, Yakutat terrane. Terranes generalized from Silberling and others (this volume, Plate 3). The double north-south line is the approximate route of the Trans-Alaska oil pipeline and of a major geological and geophysical study of the deep crust of Alaska (Trans-Alaska Crustal Transect). Locality abbreviations: A, Anchorage; C, Coldfoot; F, Fairbanks; K, Kodiak; P, Prudhoe Bay; V, Vancouver; Y, Yakutat.

TERRANES AND COMPOSITE TERRANES

In the tectonic reconstructions that follow, most larger terranes in the Cordillera of Alaska and Canada outboard of the North American craton are grouped into composite terranes (CTs) based on affinities in their inferred lithotectonic setting and tectonic evolution. These are the Arctic CT, Central CT, Yukon CT, Togiak-Koyukuk CT, Oceanic CT, Wrangellia CT, and Southern Margin CT (Fig. 2). As used in this chapter, a composite terrane is an aggregate of subordinate terranes which are grouped based on an interpretation of similar lithotectonic kindred or affinity. This is similar to Coney's (1989) definition for a superterrane. It differs from the Howell and others (1985) definition for a composite terrane in that there is no implication regarding whether the terranes came together before or after accretion to the continent. We adopt this definition because timing

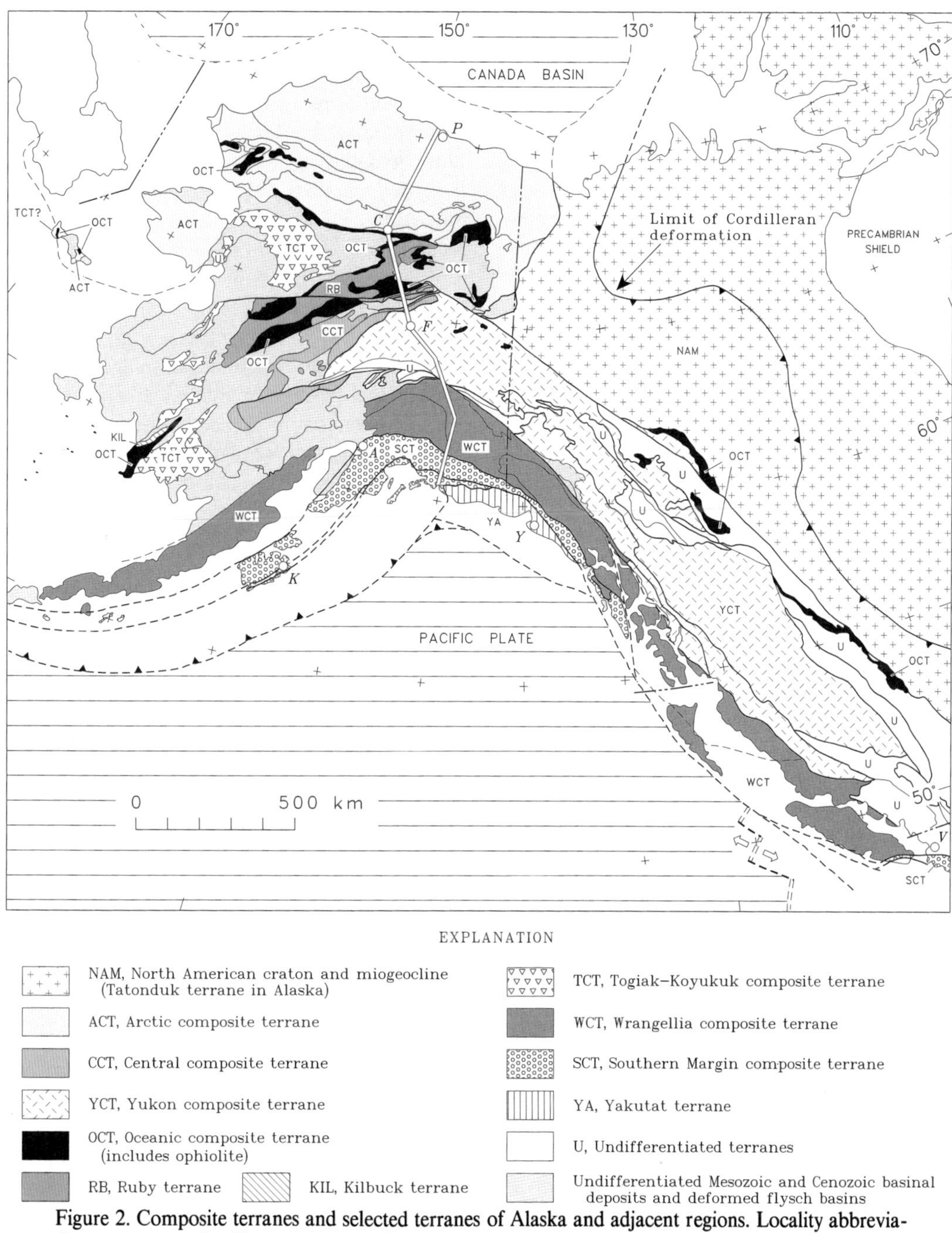

Figure 2. Composite terranes and selected terranes of Alaska and adjacent regions. Locality abbreviations are same as for Figure 1. See text for explanation and data sources.

of accretion is inapplicable in the case of composite terranes such as the Arctic CT, which rotated away from the craton but probably never was separated from it at the hinge (eastern) end.

In Alaska, the relatively large Ruby terrane in interior Alaska and the Yakutat terrane in southern Alaska (units RB and YT, Figs. 1 and 2) are not included with the CTs. Although it is likely that the Ruby terrane was originally part of the Arctic CT, we discuss it separately from the Arctic CT to emphasize differences in their Cretaceous and Tertiary rotation histories. The Yakutat terrane is discussed separately because its relation to nearby terranes is uncertain. In addition, there are small, isolated terranes whose relations to nearby terranes also are uncertain. Of these small terranes, the Kilbuck terrane in western Alaska is differentiated (KIL, Figs. 1 and 2) because it includes the only Early Proterozoic rocks in Alaska. Small terranes not discussed in this paper are mainly those along and near the Denali fault in south-central Alaska (unpatterned areas, Figs. 1 and 2).

All terranes west of the craton have moved to some extent, although many questions remain concerning the nature, timing, and magnitude of these displacements. The composite terranes and Ruby terrane were assembled into their present relative positions mainly from Middle Jurassic to Eocene time. The Yakutat terrane has been accreting to the southern margin of Alaska from Miocene time to the present. Variably metamorphosed Cretaceous flysch separates many of the internally deformed but coherent displaced terranes. Postaccretionary basin deposits, for which information on underlying terranes is sparse or absent, are present throughout Alaska and occupy extensive tracts of western Alaska (Kirschner, this volume).

North American craton margin

The small segment of the Cordilleran miogeocline in east-central Alaska (unit NAM, Figs. 1 and 2), which constitutes less than 1% of the total area of Alaskan terranes, is generally considered to be part of the autochthonous North American plate (Churkin and others, 1982; Jones and others, 1987). These strata display a well-preserved succession of Precambrian and Paleozoic miogeoclinal rocks that correlates closely with rocks to the east and southeast on the margin of the North American craton (Churkin and others, 1982; Dover, this volume). Howell and Wiley (1987) argued that strata in this area are an allochthonous part of the craton margin. The base of the sequence in Alaska consists of Late Proterozoic shelf to offshelf clastic rocks, including glacial diamictites, carbonate rocks, chert, and pillow basalt. Precambrian rocks are overlain by Cambrian to Devonian carbonate rocks, shale, and chert deposited during a major marine transgression that followed probable Late Proterozoic to early Paleozoic continental margin extension and subsidence (Howell and Wiley, 1987). The upper part of the sequence consists of Devonian clastic fan deposits derived from a western source terrain overlain by Upper Devonian and Mississippian siliceous shale and limestone. Permian and Triassic shallow-marine limestone and minor clastic rocks overlie the older sequence with angular unconformity.

Arctic composite terrane

The Arctic CT (unit ACT, Fig. 2) makes up about 25% of the area of Alaskan terranes. As used here the Arctic CT includes the Arctic Alaska terrane (Silberling and others, this volume, Plate 3; Moore and others, this volume) and its subterranes (North Slope, Endicott, Delong Mountains, Coldfoot, Hammond), the Porcupine and Seward Peninsula terranes, and probably parts of Saint Lawrence Island. The Arctic Alaska terrane and its constituent subterranes of Silberling and others (this volume) are equivalent to the Arctic Alaska superterrane and its constituent terranes as redefined by Moore (1992). In this volume, summaries of the regional geology of the Arctic CT are by Dover (southeastern Brooks Range), Grantz and others (Arctic basin and margins), Moore and others (Brooks Range and North Slope), and Till and Dumoulin (Seward Peninsula and St. Lawrence Island). Relevant topical studies include those by Barker (volcanic rocks), Dusel-Bacon (metamorphism), Hudson (anatectic granites), Kirschner (basins), Miller (plutonic rocks), and Patton (ophiolites). In this synthesis, we interpret the Arctic CT as a segment of the North American miogeocline that was displaced to its present position by rotation away from Arctic Canada.

The stratigraphy is characterized by deformed Late Proterozoic to Devonian rocks that record a history of continental margin and carbonate platform sedimentation, emplacement of Late Proterozoic and Cambrian tholeiitic basalt, and eruption of andesitic volcanic rocks of probable Ordovician age (Moore and others, this volume). A single occurrence of Cambrian trilobites of Siberian affinity suggests that at least part of the composite terrane is exotic to Alaska (Grantz and others, 1991) or that there was a marine connection with Eurasia at that time. In the metamorphic belt along the southern margin of the Arctic CT, metaclastic rocks are intruded by Proterozoic metagranitic rocks as old as 750 Ma and the rocks locally retain evidence of an amphibolite facies metamorphic event (Karl and others, 1989). The pre–Upper Devonian rocks (Franklinian sequence) were deformed, intruded by late Early to Middle Devonian granitic plutons, and locally (in the northeast Brooks Range and Arctic Slope subsurface) metamorphosed during the Late Devonian Ellesmerian orogeny. The Franklinian sequence is overlain by the thrust-imbricated Ellesmerian sequence of Upper Devonian to Lower Cretaceous clastic rocks, platform carbonate rocks, chert, and shale, in which the clastic rocks were derived from a source to the north. The Lower Cretaceous to Cenozoic Brookian sequence is made up of thrust-imbricated clastic foreland basin deposits derived from a southern source.

Major tectonic events recorded in the rocks are (1) Late Proterozoic deformation and metamorphism; (2) convergence and arc magmatism probably during the Ordovician; (3) probable continental-margin arc magmatism during the Early and

Middle Devonian followed by deposition of thick sequences of Upper Devonian coarse clastic rocks and regional deformation (Ellesmerian orogeny) (formation of rift basins may have accompanied or immediately followed the continental margin magmatic activity; (4) Late Jurassic to Early Cretaceous collision of the Togiak-Koyukuk CT along the southern margin of the Arctic CT that resulted in structural shortening of 200–500 km, obduction of oceanic crust, and high-pressure–high-temperature metamorphism of lower plate rocks; (5) synorogenic to postorogenic uplift and extension(?) in the southern Arctic CT accompanied by Early Cretaceous to mid-Cretaceous sedimentation northward into the Colville foreland basin and southward into the Yukon-Koyukuk basin; (6) incipient rifting in the Late Jurassic, followed by Barremian to Campanian rifting along the northern margin, large-scale counterclockwise rotation of the Arctic CT, and opening of the Arctic Ocean basin; and (7) Late Cretaceous to early Tertiary longitudinal compression, east-west shortening, and counterclockwise rotation of the Arctic CT.

Central composite terrane

The Central CT (unit CCT, Fig. 2) constitutes about 7% of the area of Alaskan terranes between the Brooks and Alaska ranges. The region is characterized by large areas of discontinuous and poor exposures, so that relations between and among terranes are generally not well known. Summaries in this volume of major aspects of the geology and tectonic evolution of the Central CT are by Decker and others (southwest), Dover and Foster and others (east-central), and Patton and others (west-central). Topical studies include those by Barker (volcanic rocks), Arth (crustal composition), Dusel-Bacon (metamorphism), Kirschner (basins), Miller (pre-Cenozoic plutonic rocks), Moll-Stalcup (Cretaceous and Cenozoic magmatic rocks), and Patton and others (ophiolites). The combined Central CT and Ruby terrane are equivalent to the Central CT of Plafker (1990). Although they are currently juxtaposed, they are discussed separately in this chapter to emphasize uncertainties in their original relative positions and possible differences in their kinematic histories.

The central CT consists of dominantly northeast-southwest–trending terranes interpreted as rifted, rotated, translated, and imbricated fragments of the northwest-trending miogeocline along the North American craton, and as having close stratigraphic ties to miogeoclinal rocks in adjacent parts of the Yukon Territory of Canada (Tempelman-Kluit, 1984). A possible exception is the belt of ultramafic and associated deep-marine bedded rocks (Livengood terrane) that is interleaved with the continental margin terranes north of Fairbanks (Fig. 1).

The Central CT consists of (1) Late Proterozoic and Cambrian(?) clastic deposits including abundant bimodal quartzite (Wickersham terrane) and (2) Late Proterozoic metamorphic basement overlain by shelf and offshelf Middle Cambrian to Upper Devonian carbonate rocks and mainly terrigenous clastic or cherty Permian, Triassic, and Lower Cretaceous marine sedimentary rocks (Nixon Fork terrane). A collection of Middle Cambrian trilobites from the Nixon Fork terrane suggests a connection with Siberia in the early Paleozoic (Palmer and others, 1985). Cordilleran Early Devonian and younger faunas indicate that by that time, the Nixon Fork and related shelf to basinal terranes of the southern Central CT were adjacent to the North American plate (Mamet and Plafker, 1982; Blodgett and Gilbert, 1992; Decker and others, this volume). The Central CT also consists of (3) Ordovician bimodal volcanic rocks, volcaniclastic rocks, and conglomerate overlain by Silurian and Devonian carbonate rocks and capped by undated clastic sedimentary rocks (White Mountains terrane); (4) a Paleozoic sequence mainly of complexly deformed Ordovician graptolitic shale, Ordovician and uppermost Devonian to Pennsylvanian radiolarian chert, quartzitic turbidites, pillow basalt, basinal marine carbonate rocks, Silurian and Upper Devonian reefal limestone, and Permian turbidites, chert, argillite, and conglomerate (Minchumina, Mystic, and Dillinger terranes); and (5) relatively small areas of Early Cambrian or older harzburgitic ultramafic rocks and associated gabbro that structurally underlie a sequence of deep-marine Ordovician radiolarian chert, interbedded undated mafic volcanic and carbonate rocks, and Devonian limestone (Livengood terrane).

The small Livengood terrane (unit LG, Fig. 1) consists of mantle peridotite intruded by Cambrian diorite and gabbroic rocks having K-Ar ages of 552–556 ± 17 Ma (Loney and Himmelberg, 1988; D. J. Turner, 1987, written commun.) and depositionally(?) overlying Upper Ordovician radiolarian chert and graptolitic shale. It is unique in that it is the only pre-Mesozoic ophiolite known in Alaska. The origin of the terrane is conjectural: it most likely represents an ultramafic ridge and overlying deep-sea sediments that formed during continental rifting in a manner analogous to the peridotite ridges formed along the west Iberia margin during opening of the North Atlantic basin (Boillot and Winterer, 1988). Extension along the continental margin at about 555–600 Ma is compatible with tectonic subsidence data for continental margin sequences of early Paleozoic age in Alaska (Howell and Wiley, 1987) and in the southern Canadian Rocky Mountains (Bond and Kominz, 1984). The ultramafic and mafic crystalline rocks of the Livengood terrane were later tectonically imbricated with continental margin sedimentary rocks, probably in Late Cretaceous to early Tertiary time.

Rocks in the Central CT record: (1) emplacement of ultramafic and deep-marine rocks (Livengood terrane) and continental margin clastic sedimentation and magmatism (White Mountain terrane) that is probably related to late Precambrian to Ordovician continental margin extension (Stewart, 1976; Aleinikoff and Plafker, 1989); (2) an originally south- to west-facing continental margin consisting of a dominantly shelf sequence (Nixon Fork and White Mountains terranes) that is inferred to have interfingered offshore with slope and basin facies, including mafic volcanic and local ultramafic rocks (Minchumina, Dillinger, Mystic, Wickersham, and Livengood terranes) (Decker and others, this volume; Patton and others, this volume, Chapter 7); and (3) Devonian(?) folding and northeast-verging thrusting toward the craton in the White Mountains terrane (Weber and others, 1988; Moore and Nokleberg, 1988).

Ruby terrane

The Ruby terrane (unit RB, Fig. 2) in interior Alaska constitutes about 3% of the area of Alaskan terranes. As with the Central CT, the Ruby terrane is an area of generally poor exposures, so relations between the Ruby terrane and adjacent terranes of the Central and Oceanic CTs are not well known. The geology and tectonic evolution of the Ruby terrane are summarized in this volume in the regional papers by Dover and by Patton and others (this volume, Chapter 7). Topical studies include those by Arth (crustal composition), Dusel-Bacon (metamorphism), and Miller (pre-Cenozoic plutonic rocks).

The Ruby terrane consists primarily of structurally complex quartz-mica schist, quartzite, calcareous schist, metabasalt, quartzofeldspathic schist and gneiss, and marble. The rocks probably originated as a continental margin assemblage, on the basis of both the quartz-rich compositions of the metamorphic rocks and the high silica and potassium compositions and high strontium initial ratios of crustal melt plutons that intrude the terrane (Patton and others, this volume, Chapter 7; Arth, this volume). The rocks contain sparse Ordovician, Silurian, and Devonian fossils, but could be as old as Proterozoic in age. Metamorphic grade is dominantly greenschist facies; high-pressure glaucophane-bearing greenschist facies rocks and intermediate- to high-pressure amphibolite facies rocks are present locally. The metamorphic assemblage is locally intruded by an Early Devonian metagranite (390 Ma) and by mid-Cretaceous granitic plutons that are especially voluminous in the part of the terrane north of the Kaltag fault.

Rocks in the Ruby terrane record: (1) early Paleozoic continental margin sedimentation followed by emplacement of granitic plutons, a history similar to that of the southern Arctic CT and Yukon CT, which suggests possible structural continuity of these terranes in Early Devonian time; (2) high-pressure–high-temperature metamorphism and penetrative deformation that is inferred to be related to Late Jurassic to Early Cretaceous ophiolite obduction and arc accretion (Patton and others, this volume, Chapter 7); and (3) widespread mid-Cretaceous plutonism and retrograde metamorphism related to arc magmatism and/or crustal melting (Arth, this volume; Miller, this volume, Chapter 16) that extended eastward into the Yukon CT and westward into the Arctic CT and stitched these terranes together by mid-Cretaceous time (Fig. 3).

Kilbuck terrane and Idono Complex

The Kilbuck terrane (Figs. 1 and 2) in southwestern Alaska and the tiny Idono Complex (located about 300 km northeast of the Kilbuck terrane) are the only Early Proterozoic rocks known in Alaska. They consist of fault-bounded slices of penetratively deformed gneiss and schist of upper amphibolite to granulite facies (Box and others, 1990; Miller and others, 1991). The metaplutonic rocks yield 2.06–2.07 Ga zircon U/Pb crystallization ages and show involvement of Archean crust with a model age of about 2.5 Ga. Isotopic systems were disturbed by younger metamorphic or plutonic processes at about 1.7–1.8 Ga, 190–180 Ma, and 140–120 Ma. The geochemical and isotopic data are most compatible with an origin of the early Precambrian rocks in a continental margin arc.

Structural relations of the Early Proterozoic rocks to nearby terranes are uncertain. Their location along strike with the metamorphic Late Proterozoic basement of the Ruby terrane suggests that, like the Ruby terrane, they may be displaced fragments of a continental margin arc built on the North American craton, as postulated by Miller and others (1991) and Decker and others (this volume). A North American craton origin, however, is problematic because (1) although rocks of this age are in the Wopmay orogen to the southeast in Canada (Hoffman, 1989), the Alaska rocks differ from analyzed Wopmay rocks in their Nd model ages (Box and others, 1990); and (2) known post-Jurassic dextral strike-slip displacements ($<$1000 km) on intracontinental faults in Alaska and Canada are insufficient to permit translation of fragments to southwestern Alaska from even the closest part of the present craton margin; an earlier episode of large-scale dextral displacement would be required. Permissible alternative interpretations are that the Kilbuck terrane and Idono Complex are part of an exotic terrane or that they are exotic continental margin fragments that were emplaced in an accretionary prism associated with the Togiak-Koyukuk arc and later dispersed to their present locations by strike-slip faulting.

Yukon composite terrane

The Yukon CT (unit YCT, Fig. 2) makes up about 10% of the area of Alaskan terranes. The composite terrane consists mainly of crystalline rocks (Yukon-Tanana terrane) and overlying arc-related rocks (Stikine terrane) in east-central and southeastern Alaska. Similar rocks underlie vast tracts in adjacent parts of Canada (Nisling-Kootenay terranes and the Pelly Gneiss) and probably also extend beneath parts of the Coast Mountains of southeastern Alaska, including all or parts of the Tracy Arm terrane and Taku terrane.

The Yukon CT as defined here differs from Composite terrane I of Monger and others (1982), the Stikinian superterrane of Saleeby (1983), and the Intermontane superterrane of Wheeler and McFeely (1991) mainly in that it does not include the complex of pericratonic, arc, and oceanic terranes (Cassiar, Quesnellia, Cache Creek, and Slide Mountain terranes) that lie east of the Yukon CT in Canada. Primary data sources in this volume for the following summary are Brew (late Mesozoic and Cenozoic magmatism), Dusel-Bacon (metamorphic rocks), Foster and others (Yukon-Tanana region), Gehrels and Berg (southeastern Alaska), Miller (pre-Cenozoic plutonic rocks), and Silberling and others (terranes). Data for the Canadian part of the composite terrane are mainly from Gabrielse and Yorath (1991); Hansen (1990), Monger and Berg (1987), Monger (1989), Monger and others, (1991), Mortensen (1992), Tempelman-Kluit (1979), and Wheeler and McFeely (1991).

Crystalline rocks of the Yukon CT all have strong affinities with rocks of the North America craton margin (Hansen, 1990; Samson and others, 1991; Mortensen, 1992; Foster and others, this volume). They consist mainly of (1) sparse orthogneiss of Late Proterozoic age; (2) polymetamorphosed and polydeformed quartz-rich schist and grit, pelitic schist, and sparse marble of Precambrian(?), Devonian, and Mississippian age; and (3) orthogneiss, augen gneiss, and intermediate to felsic volcanic rocks of Late Devonian to Early Mississippian age containing zircons with Proterozoic inheritance and Sm-Nd model ages consistent with values for North American cratonal basement. In metaclastic rocks of the Yukon-Tanana terrane, the presence of Early Proterozoic (about 2.1 to 2.3 Ga) zircons (Dusel-Bacon, this volume, Chapter 15; Foster and others, this volume), suggest a possible provenance in the Nahanni Province along the craton margin in northwest Canada (Hoffman, 1989).

Variably metamorphosed upper Paleozoic and Upper Triassic to Middle Jurassic carbonate and volcanic arc rocks (Stikine terrane) underlie small areas of east-central and southeastern Alaska, and they cover large tracts of the Yukon CT in contiguous areas of Canada. In east-central Alaska the Stikine terrane (Taylor Mountain terrane of Hansen, 1990) consists of intensely deformed epidote-amphibolite to amphibolite grade volcanic rocks, carbonate rocks, carbonate, chert, and quartz-rich schist and associated Upper Triassic to Lower Jurassic intrusive rocks. In Canada, Stikine terrane rocks as old as Upper Triassic (Mihalynuk and Mountjoy, 1989) locally depositionally overlie Yukon-Tanana terrane crystalline basement and are characterized by variably metamorphosed, stratigraphically stacked Mississippian through Jurassic marine and nonmarine volcanic and sedimentary strata and coeval intrusions (Monger and Berg, 1987). In southeastern Alaska the Stikine assemblage consists of upper Paleozoic carbonate and clastic strata and Upper Jurassic to Middle Jurassic arc-related rocks; contacts with adjacent terranes are obscured by faulting or younger plutonism (Gehrels and Berg, this volume). In this chapter we follow the interpretation, based on stratigraphic evidence cited by Mihalynuk and Mountjoy (1989) and others, that the Stikine terrane was originally depositional on the crystalline complex of the Yukon-Tanana terrane and that the original relations have been obscured in Alaska and in much of Canada by younger structural deformation and plutonism. Alternative interpretations, based mainly on Nd isotopic data from igneous rocks, are that the Stikine terrane has been structurally juxtaposed against the Yukon CT along major unidentified strike-slip faults or by overthrust faults involving hundreds of kilometers of displacement (Coney, 1989; Samson and others, 1991).

The rocks record (1) Late Devonian and mid-Cretaceous(?) continental margin magmatism inferred to be arc related; (2) arc magmatism in the late Paleozoic and in the Late(?) Triassic and Middle Jurassic (Stikine terrane); (3) closure of the Cache Creek Sea and accretion of the Yukon CT against the continental margin with intense deformation along a broad eastern suture zone in early Middle Jurassic time (Tempelman-Kluit, 1979; Hanson, 1990; Ricketts and others, 1992); (4) major episodes of regional plutonism and metamorphism to amphibolite facies mainly in Early Jurassic time in mainland Alaska (Dusel-Bacon, this volume, Chapter ??) and in pre–Late Triassic time in southeastern Alaska (Rubin and Saleeby, 1991); and (5) mid-Cretaceous plutonism and crustal extension accompanied by formation of subhorizontal metamorphic fabrics and detachments in east-central Alaska and adjacent areas (Pavlis and others, 1993; Hansen, 1990).

Paleomagnetic inclination data for Jurassic strata of the Stikine terrane have been cited as indicating post-Jurassic northward displacement of the Stikine terrane of about 13° relative to the craton (Monger and Irving, 1980); however, subsequent revisions in Jurassic North American reference poles appear to eliminate the need for latitudinal shift of the terrane (Irving and Wynne, 1990; Vandall and Palmer, 1990).

Togiak-Koyukuk composite terrane

The Togiak-Koyukuk CT (unit TCT, Fig. 2) in western Alaska constitutes about 5% of the area of terranes of Alaska. The geology of much of this region is imperfectly known because bedrock exposures commonly are poor and because large areas are covered with overlap deposits or water. Summaries of the geology in this volume are those of Decker and others (regional geology of southwest Alaska), Patton and others (regional geology of west-central Alaska), Barker (Mesozoic volcanic rocks), Dusel-Bacon (metamorphism), Miller (plutonic rocks), Moll-Stalcup (Late Cretaceous and Cenozoic magmatism), and Patton and others (ophiolite).

The Togiak-Koyukuk CT is a dismembered and rotated Late Triassic and older(?) through Early Cretaceous intraoceanic complex of arc-related volcanic and volcaniclastic rocks and their intrusive equivalents, that mainly comprises the Togiak, Koyukuk, and Nyac terranes but also includes Early Cretaceous rocks in parts of the Innoko terrane. Basement beneath the Togiak terrane consists of a younger assemblage of Late Triassic ophiolite and an older assemblage of Paleozoic and Triassic volcanic rocks, limestone, and chert. Arc rocks of the Koyukuk terrane are depositionally overlain by mid-Cretaceous (Albian-Cenomanian) marine and nonmarine sedimentary rocks in the Yukon-Koyukuk basin and by Upper Cretaceous and Paleogene volcanic rocks; in the Togiak terrane arc rocks are overlain by Upper Cretaceous volcanic rocks and marine sedimentary rocks.

The rocks record (1) imbrication, subduction, and local Middle Jurassic high-temperature–low-pressure metamorphism of Kobuk Sea oceanic crust of Devonian to Early Jurassic age in an accretionary prism; (2) Middle Jurassic through Early Cretaceous arc volcanism; (3) emplacement of the arc complex against an attenuated continental margin consisting of the Arctic and Central CTs and the Ruby and Kilbuck(?) terranes beginning in latest Jurassic or earliest Cretaceous time (about 144 Ma); (4) arc accretion that was preceded by at least 150 km subduction of the attenuated continental margin beneath Kobuk Sea oceanic crust (Oceanic CT) with attendant regional metamorphism and deformation of lower plate rocks; and (5) widespread mid-Cretaceous plutonism and Late Cretaceous and early Tertiary arc volcanism and related plutonism.

Oceanic composite terrane

The Oceanic CT makes up about 5% of the area of Alaskan terranes (unit OCT, Fig. 2). It comprises mainly the Angayucham, Tozitna, Innoko, Goodnews, and probably the Seventymile terranes, and defines a discontinuous belt that extends from western Alaska to the border with Canada. The geology and evolution of these oceanic terranes is summarized in this volume by Decker and others, Dover, Patton and others, Moore and others, and Foster and others; data on the composition of some of the volcanic rocks are presented by Barker and Barker and others (this volume). The terranes are similar in lithology to the Slide Mountain and related oceanic terranes located along structural strike to the southeast in adjacent parts of Canada (Fig. 1). They differ mainly in that available data suggest that the oceanic terranes in Canada and possibly the Seventymile terrane in eastern Alaska were accreted by early Middle Jurassic time (Mortensen, 1992), rather than in Late Jurassic and Early Cretaceous time (see below).

The Oceanic CT consists dominantly of oceanic basalt, volcanogenic and oceanic sedimentary rocks, and minor ultramafic rocks of Devonian to Early Jurassic age. In this chapter we follow the interpretation of Patton and others (1989; this volume, Chapter 21) that the oceanic terranes are dismembered fragments of the paleo-Pacific crust that were obducted as extensive thrust sheets onto the continental margin before and during accretion of magmatic arc rocks of the Togiak-Koyukuk CT in Late Jurassic and Early Cretaceous time. The closed ocean basin in Alaska has been referred to informally as the Kobuk Sea (Plafker, 1990). Lithologically equivalent rocks in contiguous parts of Canada, and possibly the Seventymile terrane in eastern Alaska, were emplaced against and over adjacent terranes during Middle Jurassic or earlier closure of the Cache Creek Sea. Obduction of the Oceanic CT was accompanied by widespread greenschist and blueschist metamorphism, by craton-verging compressional deformation, and by inverted structural stacking within much of the oceanic rocks. According to the model presented in this chapter, southeast vergence is predicted in the Ruby terrane. Results of a detailed study of the shear sense of mylonites along parts of the detachment zone in the northern Ruby terrane and the nearby Oceanic CT are ambiguous in that they indicate pre-mid-Cretaceous tectonic transport dominantly toward the northwest but locally toward the southwest (Miyaoka and Dover, 1990).

Our interpretation for the origin of the oceanic rocks differs from a more stabilistic model in which the rocks are considered to be disrupted remnants of local parautochthonous intracontinental rifts (Gemuts and others, 1983; Dover, this volume). In our judgment, the rift model does not account for the close association in time and space of the intraoceanic arc rocks of the Togiak-Koyukuk CT and thrust sheets of the Oceanic CT or for the long-lived dominantly pelagic sedimentary sequences that characterize much of the Oceanic CT.

Wrangellia composite terrane

The allochthonous Wrangellia composite terrane (unit WCT, Fig. 2) occurs mainly between the Denali and Border Ranges fault systems, where it constitutes about 20% of the area of Alaskan terranes; it also underlies large tracts of adjacent parts of Canada. The geology of the Wrangellia CT is summarized in this volume by Nokleberg and others (regional geology of south-central Alaska), Gehrels and Berg (regional geology of southeastern Alaska), Barker (pre-Cenozoic volcanism), Brew (late Mesozoic and Cenozoic magmatism in southeastern Alaska), Moll-Stalcup (Late Cretaceous and Tertiary magmatism in mainland Alaska), Miller and Richter (eastern Aleutian arc volcanism), Dusel-Bacon (metamorphism), and Patton and others (ophiolites).

The composite terrane is characterized by Late Proterozoic(?) and younger magmatic arc, oceanic plateau(?), and rift-fill assemblages (mainly the Wrangellia, Peninsular, Alexander, and northern Taku terranes), and the Late Jurassic to mid-Cretaceous magmatic arc and flysch deposits of the Gravina-Nutzotin belt. The northern part of the Taku terrane of Silberling and others (this volume, Plate 3) is included here as a fault-displaced fragment of Wrangellia (Plafker and others, 1989). As used here, the Wrangellia CT corresponds in part with Composite Terrane II of Monger and others (1982), the Wrangellia super-terrane of Saleeby (1983), and the Insular Composite Terrane of Wheeler and McFeely (1991). The Alexander terrane was together with Wrangellia and was the basement beneath at least part of Wrangellia by Early Pennsylvanian time (Gardner and others, 1988). The combined Wrangellia and Alexander terranes were together with the Peninsular terrane at least by Late Triassic time and possibly by the Early Permian (Plafker and others, 1989; Nokleberg and others, this volume, Chapter 10).

The central and southern segments of the Taku terrane of Silberling and others (this volume, Plate 3) have been reinterpreted as the exposed western margin of the Yukon-Tanana or comparable terranes of continental affinity (Rubin and Saleeby, 1992; McClelland and others, 1992a, 1992b). In addition, at least part of the dominantly crystalline Tracy Arm terrane in the Coast Mountains of Alaska and British Columbia was juxtaposed against the Wrangellia CT since mid-Cretaceous time (Crawford and others, 1987), and perhaps as early as the Middle Jurassic (McClelland and others, 1992a; Rubin and Saleeby, 1992).

The Alexander terrane consists of three subterranes that are characterized by a pre–Middle Ordovician volcanic-sedimentary-plutonic metamorphic complex overlain by Ordovician to Late Triassic variably metamorphosed mafic to felsic volcanic rocks, clastic and carbonate rocks, and by compositionally variable Ordovician and younger plutonic rocks.

The northern part of Wrangellia in Alaska (unit WRn, Fig. 1) is characterized by Early Pennsylvanian to Early Permian variably metamorphosed mafic to intermediate arc-related volcanic and intrusive rocks, overlying Middle to Late Triassic subaerial basalt and related intrusive rocks up to 6,000 m thick, and overlying Upper Triassic to Lower Jurassic shallow-marine clastic, siliceous, and calcareous rocks. In British Columbia, the southern part of Wrangellia (unit WRs, Fig. 1) shares the same Upper Triassic stratigraphy consisting of a sequence of dominantly marine basalt and marine sedimentary rocks; it differs from the Alaska part in that the oldest exposed units consist of arc volcanic rocks and associated intrusive rocks of Silurian and Devonian age, the Carboniferous and Early Permian are represented by chert and limestone, and the Paleozoic section is overlain by Upper Triassic and Jurassic andesitic volcaniclastic rocks (Monger and Berg, 1987; Brandon and others, 1986). Although the Wallowa terrane of Oregon has been considered to be part of Wrangellia (Jones and others, 1977), the calc-alkaline composition of the Triassic volcanic rocks (Sarewitz, 1983) makes such an association unlikely.

The Peninsular terrane is characterized by a well-stratified sequence of variably metamorphosed Paleozoic(?) volcanic rocks, Permian limestone, Upper Triassic (Norian) basalt, limestone, argillite, and tuff, Upper Triassic and Lower Jurassic andesitic flows, breccias, and volcaniclastic siltstone and sandstone, Middle Jurassic to Cretaceous fossiliferous clastic rocks and minor bioclastic limestone, and Jurassic batholithic granitic rocks (Detterman and Reed, 1980; Jones and others, 1987; Nokleberg and others, this volume, Chapter 10).

The rocks record (1) possible arc magmatism in the Alexander terrane during the Late Proterozoic and Cambrian(?) and arc magmatism during the Early Ordovician to Late Devonian and Late Jurassic to Early Cretaceous; (2) arc magmatism in the British Columbia part of the Wrangellia terrane during the Silurian, Devonian, and latest Triassic to earliest Middle Jurassic; (3) arc magmatism in the Alaska part of the Wrangellia terrane during the Early(?) Pennsylvanian to Early Permian, Late Jurassic to Early Cretaceous, and Cenozoic time; (4) arc magmatism on the Peninsular terrane during the latest Triassic and Early Jurassic and post-Eocene time intervals and local near-trench plutonism in mid-Cretaceous time (Gehrels and Berg, this volume; Nokleberg and others, this volume, Chapter 10). In addition, plutonism that may be arc related was widespread in the Wrangellia CT during the Middle and Late Jurassic and latest Cretaceous to early Tertiary (Hudson, 1983; Barker and Miller, this volume; Nokleberg and others, this volume, chapter 10); (5) Late Silurian to Early Devonian deformation in part of the Alexander terrane with development of a thick clastic wedge of redbeds (Klakas orogeny of Gehrels and Saleeby, 1987); (6) Middle to Late Triassic emplacement of voluminous sequences of tholeiitic basalt and gabbro in the Wrangellia, northern Taku, and Peninsular terranes and of bimodal basalt and felsic rocks in the Alexander terrane due either to rifting (Barker and others, 1989) or to initiation of a mantle plume (Richards and others, 1991); (7) paleomagnetic anomalies indicating large Late Triassic to mid-Cretaceous northward displacement (up to 25°) relative to the North American craton (Haeussler and others, 1992; Hillhouse and Coe, this volume; Plumley, 1990, oral commun.; Van der Voo and others, 1980); and (8) emplacement against the continental margin in Alaska and British Columbia by mid-Cretaceous time (Crawford and others, 1987; Rubin and Saleeby, 1992) and possibly as early as Middle Jurassic time (McClelland and others, 1992a), accompanied by intense contractional deformation and collapse of intervening flysch basins.

Southern Margin composite terrane

The Southern Margin CT (unit SCT, Fig. 2), which makes up about 20% of the area of the terranes of Alaska, includes the Chugach, Ghost Rocks, and Prince William terranes. The geology of this CT is described in this volume by Plafker and others (southern margin) and Gehrels and Berg (southeastern), Barker (volcanic rocks), Brew (Late Cretaceous and Tertiary magmatism in southeastern Alaska), Moll-Stalcup (Late Cretaceous and Tertiary magmatism in mainland Alaska), Dusel-Bacon (metamorphism), and Patton and others (ophiolite).

In Alaska the Southern Margin CT is a compound, complexly deformed, accretionary prism of Upper Triassic to Paleogene oceanic rocks, melange, and flysch. To the southeast, probably correlative rocks make up the Pacific Rim terrane of coastal Vancouver Island in British Columbia (Monger and Berg, 1987), and the various Mesozoic arc-related accretionary complexes in the conterminous United States and Mexico (Silberling and others, 1992).

The rocks of the Southern Margin CT record (1) intermittent offscraping of oceanic rocks and dominantly arc-related volcaniclastic and volcanic rocks against the Pacific margin of the Wrangellia CT during crustal convergence from Early Jurassic to middle Eocene time; (2) local blueschist facies metamorphism of accreted rocks in Early Jurassic and mid-Cretaceous time; (3) widespread high-temperature–low-pressure metamorphism and intrusion of anatectic tonalitic plutons in Eocene time; and (4) paleomagnetic anomalies indicating 16°–30° northward displacement relative to the craton of Late Cretaceous to middle Eocene volcanic rocks within the accretionary prism (Plumley and others, 1983; Hillhouse and Coe, this volume).

Yakutat terrane

The Yakutat terrane (unit YT, Fig. 2) along the northern Gulf of Alaska constitutes about 5% of the terranes of Alaska. The geology of this terrane is discussed in this volume by Plafker and others (regional), Moll-Stalcup (Late Cretaceous and Cenozoic magmatism) and Barker (accreted volcanic rocks). The Yakutat terrane is an allochthonous fragment of the continental margin that is inferred to have been displaced to its present position by dextral strike-slip faulting (Plafker and others, this volume, Chapter 12). The terrane has a composite basement of Paleocene(?) and Eocene oceanic crust in the western part and a fragment of the Southern Margin CT in the eastern part.

The rocks record (1) about 180 km offset of the Southern Margin CT along the Chatham Strait segment of the Denali fault in late Paleocene or early Eocene time; (2) deposition of a thick sequence of Eocene and younger siliciclastic rocks and minor mafic volcanic rocks and coal derived from the Coast Mountains, Wrangellia CT, and Southern Margin CT; (3) northwestward displacement of the terrane to its present position by about 600 km of post-Oligocene dextral slip on the Queen Charlotte–Fairweather transform fault system along its northeastern boundary and by a comparable amount of underthrusting of the northern and northwestern boundaries beneath the adjacent Southern Margin CT; and (4) collision-related deformation, mountain building, seismicity, and deposition of thick sequences of marine clastic and glaciomarine sedimentary rocks from middle Miocene time to the present.

MAGMATIC BELTS

The distribution, composition, and age of pre-Cenozoic plutonic and volcanic belts in Alaska help constrain timing of joining ("stitching") of one or more terranes and the petrotectonic setting of the terranes. The best-defined belts of plutonic rocks and arc-related volcanic rocks are depicted schematically in Figure 3; their salient features are outlined below.

Devonian to Early Mississippian magmatic belt

A magmatic belt of dominantly granitic intrusive rocks and local andesitic to felsic volcanic rocks of late Early to Middle Devonian age (mainly 380–390 Ma) extends along the southern

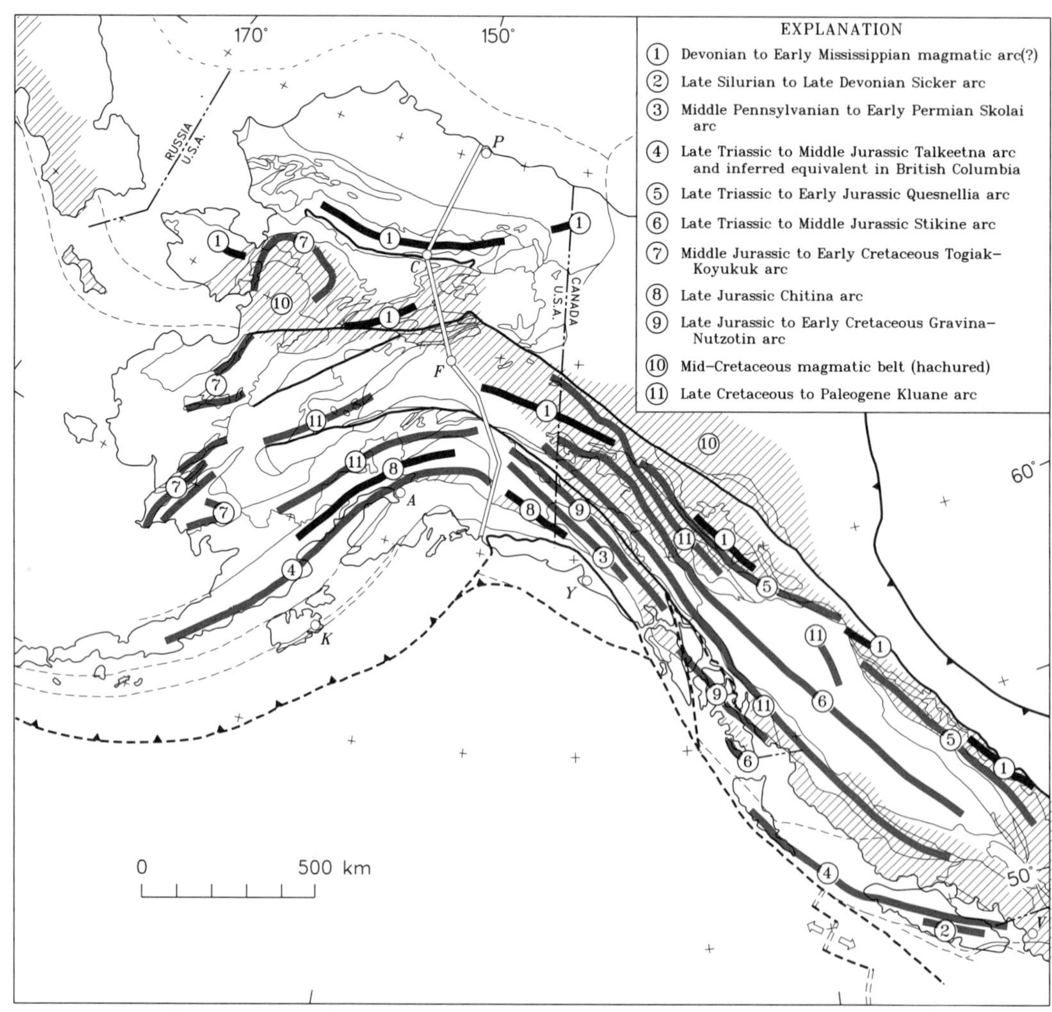

Figure 3. Major pre-Cenozoic magmatic belts of Alaska. Wide red lines indicate approximate axes of volcanic arc rocks; wide black lines indicate approximate axes of dominantly plutonic rocks interpreted as arc related; red hachures show approximate distribution of mid-Cretaceous plutonic and volcanic rocks. Locality abbreviations are same as for Figure 1. See text for explanation and data sources.

Arctic CT (Hammond and Coldfoot terranes) and occurs locally in the Ruby terrane (Fig. 3, no. 1). Rocks of similar composition, but Late Devonian to earliest Mississippian in age (350–370 Ma), underlie much of the Yukon CT in east-central Alaska. These crystalline rocks have been widely interpreted as part of a discontinuous magmatic belt that was emplaced in and on early Paleozoic or older rocks of the continental margin (Dillon and others, 1980; Nokleberg and others, 1989; Plafker, 1990; Mortensen, 1992). The magmatic belt is generally coeval with granitoid plutonism that extends southward through Yukon Territory in Canada and with plutonism and volcanism in the Klamath and Sierra Nevada mountains of Oregon and California, as summarized by Rubin and others (1991). An alternative explanation proposed for the belt is that the magmatism represents anatectic melting of continental crust along zones of extension (Hudson, this volume). However, this interpretation is less compatible with the local occurrences of comagmatic andesitic volcanic rocks and approximately coeval compressional deformation in parts of the belt.

Sicker arc

Metamorphosed volcanic and plutonic rocks of the Sicker arc (Fig. 3, no. 2) on Vancouver Island are of Late Silurian to Late Devonian age (370–420 Ma), although the oldest age for the base of the sequence is not known (Brandon and others, 1986). Possibly correlative rocks in the southern Alexander terrane make up a metamorphosed arc complex mainly of Early Ordovician to Early Devonian age that includes plutonic rocks of Early Silurian to Early Devonian age (430–405 Ma) (Gehrels and Saleeby, 1987). These data suggest a possible link between the southern Wrangellia and Alexander terranes as early as the Silurian.

Skolai arc

The Skolai arc (Fig. 3, no. 3) in the northern Wrangellia CT consists of andesitic volcanic rocks, shallow-marine sedimentary rocks, and shoshonitic plutonic rocks that range in age from at least Middle Pennsylvanian to Early Permian (Nokleberg and others, this volume, Chapter 10). Plutonic rocks having U-Pb zircon ages of 210 Ma are emplaced across the contact of the Wrangellia and Alexander terranes, indicating that they were together by Pennsylvanian time (Gardner and others, 1988).

Talkeetna arc

The Talkeetna arc (Fig. 4, no. 4) consists of intra-oceanic volcanic and volcaniclastic rocks of latest Triassic and Early Jurassic age in the Peninsular terrane of the northern Wrangellia CT and probably the coeval arc rocks of the latest Triassic to earliest Middle Jurassic Bonanza Group in the British Columbia part of the Wrangellia CT. The Talkeetna arc and the coeval arc rocks in the southern Wrangellia CT are inferred to have been part of the same arc that was disrupted by about 600 km of sinistral displacement (Plafker and others, 1989; Nokleberg and others, this volume, Chapter 10). The associated accretionary prism, the Chugach terrane, consists of volcaniclastic flysch, melange that contains exotic rocks, including sparse limestone blocks containing a Permian Tethyan fusulinid fauna, and minor blueschist facies rocks (Plafker and others, this volume, Chapter 12).

Quesnellia arc

The ocean-facing Quesnellia arc (Fig. 3, no. 5) and its associated accretionary prism complex, the Cache Creek terrane, are entirely in Yukon Territory and British Columbia, but their histories are a critical part of the evolution of outboard accreted terranes, including the Yukon-Tanana and Stikine terranes (Fig. 1). The Quesnellia arc (Quesnellia terrane) consists of Late Triassic and Early Jurassic calc-alkaline to alkaline volcanic rocks and comagmatic plutonic granitic rocks, along with volcaniclastic rocks, argillite, and minor limestone, that includes some upper Paleozoic carbonate rocks. The arc and associated rocks overlie late Precambrian to early Paleozoic metamorphic rocks that have continental crust affinity (Monger and Berg, 1987). The Cache Creek terrane along the ocean-facing side of the Quesnellia arc includes extensive tracts of carbonate rocks containing a distinctive Tethyan fusulinid and coral fauna of Permian age and blueschist facies metamorphic rocks of Late Triassic age (Monger and Berg, 1987). The Quesnellia arc can be interpreted as part of a long-lived fringing arc complex that evolved above an eastward-dipping subduction zone along the northeast Pacific margin (Saleeby, 1983; Miller, 1987). According to this interpretation, rocks of Tethyan affinity in the accretionary prism represent far-traveled guyots and their carbonate caps that were scraped off oceanic crust and incorporated as exotic blocks in the Cache Creek accretionary complex. In Alaska, the Quesnellia arc is broadly coeval with the Stikine and Talkeetna arcs and the Cache Creek accretionary prism may be partly coeval with the oldest rocks of the Southern Margin CT accretionary prism (Figs. 2 and 3).

Stikine arc

The Stikine arc (Fig. 3, no. 6) underlies much of the Cordillera in Yukon Territory and British Columbia, but locally underlies small areas of southeastern and east-central Alaska. In southeastern Alaska and contiguous parts of Canada it consists of a stacked sequence of Late Triassic to Middle Jurassic marine clastic and volcaniclastic rocks, volcanic rocks and comagmatic intrusive bodies, and minor carbonate rocks (Gehrels and Berg, this volume). In east-central Alaska it may be represented by an assemblage of variably metamorphosed volcanic rocks, carbonate rocks, chert, and quartz-rich pelitic rocks. Its age is uncertain, but it must be older than the Late Triassic to Early Jurassic granitic plutons that intrude it (Foster and others, this volume). The magmatic arc is built mainly on a thick upper Paleozoic arc sequence of the Stikine terrane (Monger and Berg, 1987) and probably in part on crystalline basement of the Yukon-Tanana terrane (Gehrels and Berg, this volume). Paleomagnetic data indicate latitudinal stability relative to the craton for Jurassic rocks in the Canadian part of the arc (Irving and Wynne, 1990; Vandall and Palmer, 1990).

Togiak-Koyukuk arc

The Togiak-Koyukak arc in western Alaska (Fig. 3, no. 7) consists of Middle Jurassic through Early Cretaceous calc-alkalic volcanic rocks, intrusive rocks, and volcaniclastic rocks that locally overlie oceanic rocks of Late Triassic and older(?) age (Barker, this volume; Decker and others, this volume; Patton and others, this volume, Chapter 7). The arc includes faulted and warped segments of the Togiak, Koyukuk, and Nyac terranes and parts of the Innoko terrane (Fig. 1). Together, these terranes compose the Togiak-Koyukuk CT described in a previous section.

Chitina arc

The Chitina arc (Fig. 3, no. 8) is a belt of calc-alkalic plutonic rocks of mainly Late Jurassic age (Tonsina-Chichagof belt of Hudson, 1983) that stitches across part of the northern Wrangellia and Alexander terranes (Plafker and others, 1989; Nokleberg and others, this volume, Chapter 10). It is interpreted as the roots of a magmatic arc, based on the dominantly tonalitic composition of the rocks and the linearity and continuity of the belt (Hudson, 1983). In Alaska the belt is locally truncated along its southern margin by the Border Ranges fault (MacKevett, 1978).

Gravina-Chisana arc

The Gravina-Chisana arc (Gravina-Nutzotin belt of Berg and others, 1972) (Fig. 3, no. 9) consists of a moderately to intensely deformed sequence of clastic and volcaniclastic sedimentary rocks with subordinate andesitic and basaltic volcanic rocks of Late Jurassic and Early Cretaceous age (Gehrels and Berg, this volume; Nokleberg and others, this volume, Chapter 10). The arc sequence, which is built on the inner margins of the Wrangellia and Alexander terranes, extends discontinuously from southeastern Alaska to the Alaska Range in south-central Alaska, where it links the Alexander and Wrangellia terrane. In southeastern and central Alaska it is thrust beneath the Yukon CT to the northeast (Csejtey and others, 1982; Gehrels and others, 1990; Nokleberg and others, this volume).

Mid-Cretaceous magmatic belt

Mid-Cretaceous (mainly late Early to early Late Cretaceous, or about 110–80 Ma) plutonism and associated thermal metamorphism occurred in a wide belt between the southern Brooks Range and the Alaska Range (Fig. 3, no. 10) and locally extends to the southern margin of the Wrangellia CT (Dusel-Bacon, this volume; Gehrels and Berg, this volume; Miller, this volume, Chapter 16). In west-central Alaska there were two episodes of magmatism—one from 113 to 99 Ma and a volumetrically lesser one from 89 to 79 Ma (Miller, this volume, Chapter 16; Patton and others, this volume, Chapter 7). The plutonic rocks have a wide compositional range, including ultrapotassic nepheline syenite, tonalite through granodiorite, and biotite granite and two-mica granite (Miller, this volume, Chapter 16). Granitic to tonalitic plutonism was accompanied in the Yukon CT in Alaska by volcanism (Bacon and others, 1990) and by extension along with local formation of subhorizontal fabrics and detachments (Pavlis and others, 1993). The magmatic belt in Alaska appears to be at least in part coextensive with major belts of plutonism and arc volcanism that extend southeastward into adjacent areas of Canada (Armstrong, 1988) and westward into the Russian far east (Albian to Turonian Ohkotsk-Chukotsk magmatic belt of Parfenov and Natal'in, 1986). Mid-Cretaceous magmatism is considered to be subduction related both in Canada and Siberia, and it is likely that at least some of the coeval magmatic activity in Alaska is also of subduction origin (Bacon and others, 1990; Miller, this volume, Chapter 16). An alternative interpretation is that the plutonic rocks of interior Alaska are crustal melts related to large-scale mid-Cretaceous extension (Miller and Hudson, 1991; Hudson, this volume), even though isotopic data indicate that some of the region in which the plutons occur is underlain by oceanic crust (Arth, this volume). In Alaska, this broad belt of mid-Cretaceous magmatism is broadly coeval with (1) extensive resetting of K-Ar and other isotopic ages (Dusel-Bacon, this volume); (2) greenschist facies metamorphism including widespread retrograde metamorphism of higher grade rocks; and (3) thermal overprinting of the preexisting remnant magnetization (Hillhouse and Coe, this volume) that appears to extend far beyond the magmatic belt.

Kluane arc

The Kluane arc (Fig. 3, no. 11) is represented by an arcuate belt of latest Cretaceous and Paleogene magmatism (75–56 Ma) that extends from southwestern to southeastern Alaska. Volcanic and plutonic rocks in a wide zone stitch across most major terranes in western, central, and east-central Alaska (Moll-Stalcup, this volume). Variably metamorphosed intermediate composition plutonic rocks that occur along the Wrangellia-Yukon CT boundary and underlie much of the Coast Mountains of southeastern Alaska and British Columbia (Armstrong, 1988; Plafker and others, 1989; Brew, this volume). A broad belt of coeval plutonic and volcanic rocks mainly within the Yukon CT in Yukon Territory (Armstrong, 1988) may be part of the arc complex, but is not shown in Figure 3. A distinct magmatic lull in Canada from 70 to 60 Ma (Armstrong, 1988) is not recognized in Alaska.

TRANSLATIONS AND ROTATIONS

The present complex pattern of lithotectonic terranes in Alaska reflects interactions of the oceanic plates with the North American plate and resulting large-scale translations and rotations. Figure 4 illustrates the most important assumptions regarding displacements and rotations in Alaska that were used in the reconstruction of the tectonic evolution model that follows. Much of the geologic data are summarized in the various regional papers in this volume. Paleomagnetic data are from well-dated volcanic and carbonate rocks for which the paleohorizontal can be determined (Hillhouse and Coe, this volume; Haeussler, 1992). The more conservative Northern Hemisphere option has been chosen in all cases where initial polarity can not be specified.

Counterclockwise rotation of the Arctic composite terrane

In northern Alaska, paleomagnetic data from clastic rocks in oriented well cores from near Prudhoe Bay indicate that the Arctic CT rotated some 66° counterclockwise away from the Arctic Canada margin between about 130–100 Ma (Halgedahl and Jarrard, 1987). In the Amerasian basin of Canada (not shown in Fig. 4), stratigraphic and structural data indicate that

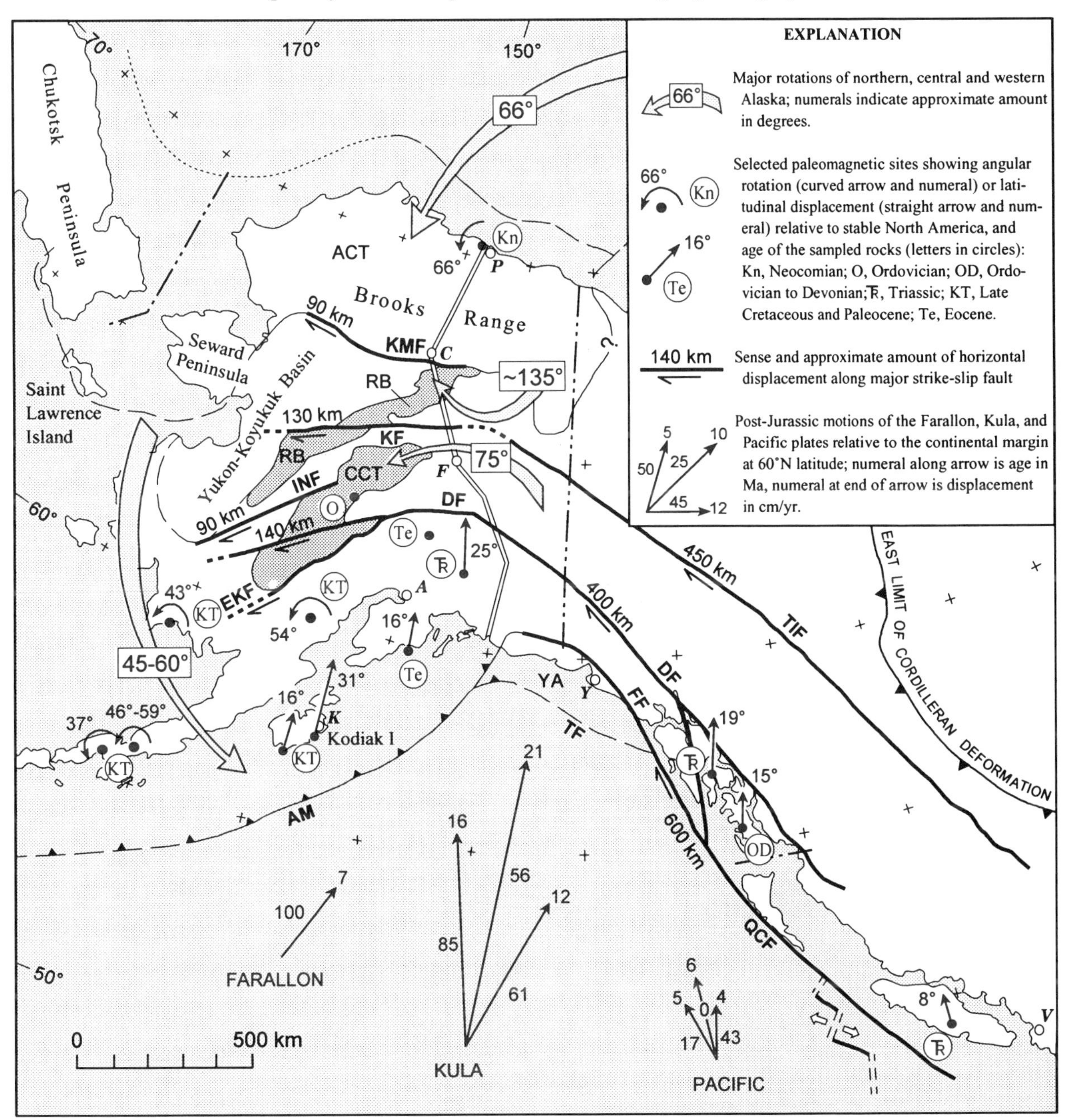

Figure 4. Significant rotations, translations, and plate motions in Alaska and adjacent parts of Canada. Open arrows indicate major rotations of northern and western Alaska. Fault abbreviations: AM, Aleutian megathrust; DF, Denali fault; EKF, East Kulukak fault; FF, Fairweather fault; KF, Kaltag fault; KMF, Kobuk-Malemute fault; INF, Iditarod-Nixon Fork fault; QCF, Queen Charlotte fault; TF, Transition fault; TIF, Tintina fault; ACT, Arctic composite terrane; CCT, Central composite terrane; RB, Ruby terrane. Locality abbreviations are same as for Figure 1. See text for explanation and data sources.

rifting was mainly during the late Albian to Cenomanian interval and that rotation of Arctic Alaska continued until about the end of the Cretaceous (Embry and Dixon, 1990).

Some of the more widely held alternative models for the evolution of the Arctic CT are reviewed by Nilsen (1981) and Moore and others (this volume). They include (1) a stabilist interpretation having the Arctic CT in its present position relative to North America since Paleozoic time (e.g., Bogdanov and Tilman, 1964; Churkin and Trexler, 1981); (2) northward translation from the northern Canadian part of the Cordilleran orogen in Yukon Territory by large-scale dextral slip along the Tintina fault followed by dextral offset along the postulated Porcupine lineament (Jones, 1982); and (3) westward translation from a position somewhere between the eastern Arctic Islands of Canada and northern Greenland by large-scale sinistral slip on a postulated transform fault along the present southern margin of the Canada basin (Oldow and others, 1987).

Counterclockwise oroclinal bend of western Alaska and western Arctic composite terrane

In western Alaska, geologic and paleomagnetic data indicate that at about 65–50 Ma, the region between the Kaltag fault and Kodiak Island was rotated 45°–60° counterclockwise about a vertical axial plane that coincides approximately with long 146°W. Before rotation of western Alaska, the continental margin is assumed to have been more linear with general northwest to north structural trends, except for the Ruby terrane, which may already have been rotated to a northeast trend during mid-Cretaceous time (see below). There are no reliable data for the paleolatitudes of rock units in western Alaska because all paleomagnetic sites are overprinted by younger chemical or thermal events.

Interpretation of oroclinal bending of western Alaska is based mainly on (1) abrupt bends in major intraplate transcurrent faults, most notably the Tintina and Denali faults (Grantz, 1966; Hickman and others, 1990); (2) the concave-oceanward distribution of Late Cretaceous arc-related rocks (Plafker and others, this volume, Chapter 12) and differences in the orientation of slip vectors in coeval rocks on opposite limbs of the bend (Moore and Connolly, 1979); (3) deformation of Late Cretaceous and Paleogene accreted rocks in the axis of the oroclinal bend in southern Alaska (Plafker and others, this volume, Chapter 12); and (4) paleomagnetic data from Late Cretaceous and early Tertiary rocks in western Alaska (Hillhouse and Coe, this volume; Plumley and others, 1983). A possible mechanism for oroclinal bending is Late Cretaceous and Paleogene convergence of about 2.5 to 6 cm/yr between the North American and Eurasian plates (Engebretsen and others, 1985).

Geologic data bracket the timing of the great southward-looping syntaxial bend, which extends from the western Brooks Range through the Seward Peninsula and Saint Lawrence Island to the Chukotsk Peninsula, as middle Late Cretaceous to the middle Tertiary (Patton and Tailleur, 1977).

Clockwise rotation of the Ruby terrane

The Ruby terrane and structurally overlying slabs of the Oceanic CT are inferred in our model to have rotated about 135° clockwise relative to the southern margin of the Arctic CT to form the southern boundary of the Yukon-Koyukuk basin (Fig. 4). This interpretation follows that of Tailleur (1980) and is based on correlation of the Ruby terrane with crystalline rocks of the southern Arctic CT and on the geometric relations among these terranes, which suggest rotation of the Ruby terrane about a hinge close to its present northeastern end. It differs from earlier models in that we suggest that rotation most likely occurred in two stages. The first stage, which was concurrent with opening of the Arctic basin in Early to mid-Cretaceous time, was accompanied by voluminous sedimentation in a proto-Koyukuk basin. The second stage, caused primarily by Late Cretaceous to early Tertiary northwestward displacement of the Yukon CT, resulted in final closure of the Yukon-Koyukuk basin and as much as 50% shortening by deformation of the enclosed sedimentary rocks.

A number of tectonic models have been proposed for emplacement of the Ruby and adjacent terranes, but no single model accounts adequately for the apparent continuity of crystalline and oceanic terranes around the eastern apex of the Yukon-Koyukuk basin, the distribution of overlying oceanic terranes, pre–Late Cretaceous filling and deformation of the Yukon-Koyukuk basin, and the timing and amount of strike-slip displacement available on the major intraplate faults of Alaska. Models for the northern group of terranes and Yukon-Koyukuk basin are (1) clockwise oroclinal bending of the Ruby terrane away from the southern Brooks Range in Late Cretaceous to early Tertiary time (Tailleur, 1980); (2) late Paleozoic rifting from the southern Brooks Range followed by pre–Early Cretaceous counterclockwise rotation to open the Yukon-Koyukuk basin (Patton, 1970); and (3) dextral offset of the Ruby and related terranes from the eastern end of the southern Brooks Range along a postulated major northeast-trending strike-slip fault named the Porcupine lineament by Churkin and Carter (1979).

Counterclockwise rotation and northwestward displacement of the Central composite terrane

Terranes that constitute the Central CT and structurally overlying slabs of the Oceanic CT are inferred in our model to have undergone as much as 75° counterclockwise rotation away from the margin of continental North America (Tailleur, 1980; Patton and others, this volume, chapter 7). The sense of rotation is compatible with early Paleozoic facies trends in the terranes, which indicate a shale-out toward the south and southeast in present coordinates (Churkin and others, 1982; Decker and others, this volume). Although poorly constrained, timing of this rotation corresponds to the Late Cretaceous to early Tertiary counterclockwise bending of western Alaska and northwestward displacement of the Yukon CT.

Paleomagnetic inclination data from Ordovician and

younger Paleozoic carbonate rocks in the Nixon Fork area of the Central CT (Figs. 1 and 4), indicate no significant difference (± 10°) between present and Ordovician paleolatitudes (Plumley, 1984). These data, together with stratigraphic similarities to lower Paleozoic strata of the North American miogeocline (Decker and others, this volume), are a key element in the tectonic reconstructions that follow because they constrain the Paleozoic location of the Nixon Fork and related terranes to nearby parts of the continental margin. The paleomagnetic data further suggest that the Nixon Fork terrane may have rotated about 75°–110° clockwise (Plumley, 1984); timing of the rotation is poorly constrained and we follow the interpretation of Patton and others (this volume, Chapter 7), who suggest that it could reflect post-Paleozoic rotation of the fault slices sampled, rather than rotation of the entire Central CT.

Tectonic models proposed for emplacement of all or parts of the Central CT and adjacent terranes include (1) emplacement of the Nixon Fork and related terranes by dextral strike-slip faulting (Tailleur, 1980; Patton and others, this volume, chapter 7); (2) large clockwise rotation of the Nixon Fork and related terranes away from the North American miogeocline, probably in the late Paleozoic or early Mesozoic (Churkin and others, 1982; Plumley, 1984; Plafker, 1990); and (3) a stabilistic model in which the Central CT and Ruby terrane were part of a southwesterly directed peninsular extension of the Paleozoic North American margin (Decker and others, this volume).

Northwestward movement of the Yukon composite terrane

The Yukon CT has been displaced northwest relative to the craton at least 450–900 km by dextral slip on the Tintina and related intraplate faults that make up its landward boundary (Gabrielse, 1985). Offsets on the Tintina fault and its splays are discussed below.

Mid-Cretaceous plutonic rocks of the Coast Plutonic Complex in British Columbia and the North Cascade Range in Washington have systematic inclination anomalies that have been interpreted to indicate (1) 1100–2400 km of northward translation (Irving and others, 1985; Beck and others, 1981; Umhoefer and others, 1989), (2) postemplacement tilt of the sampled rocks (Butler and others, 1989; Brown and Burmester, 1991), or (3) a combination of translation and tilt. Because of uncertainty regarding effects of regional tilt on the Cretaceous plutonic rocks, we have not incorporated these displacements in our tectonic model.

Northward movement of the Wrangellia composite terrane

Paleomagnetic data from Ordovician to Devonian limestone and Triassic basalt in the Alaska part of the Wrangellia CT (Figs. 1 and 4) indicate that it was 0°–15° south of its present position relative to the craton in the early Paleozoic (Van der Voo and others, 1980) and about 25° south in the Late Triassic (Hillhouse, 1987; Haeussler, 1992; Plumley, 1990, oral commun.). Paleomagnetic data from volcanic flow rocks indicate that the Wrangellia CT was at about its present latitude by late Paleocene or early Eocene time (Hillhouse, 1987). Paleomagnetic data from Vancouver Island in the southern part of the Wrangellia terrane indicate that it was at about the same paleolatitude as northern Wrangellia in Late Triassic and Jurassic time (Yole and Irving, 1980). At least 5° of the present difference in latitude of northern and southern Wrangellia may be due to postemplacement sinistral faulting (Plafker and others, 1989). The combined geologic data and plate tectonic reconstructions suggest that final suturing to southern Alaska was in the Late Cretaceous, most likely during the Cenomanian to Maastrichtian (Coney and Jones, 1985; Plafker and others, 1989; Plafker and others, this volume, Chapter 12). However, a strong case has been made for earlier accretion to North America in the mid-Cretaceous in Alaska (Csejtey and others, 1982; Nokleberg and others, 1985, 1989), by latest Albian time in southern British Columbia (Monger and others, 1982; Thorkelson and Smith, 1989), or as early as the Middle Jurassic in southeastern Alaska (McClelland and others, 1992a). In our model we follow McClelland and others (1992a) in assuming that the Wrangellia terrane moved northward to about its present position relative to inboard terranes by Late Jurassic time. Juxtaposition of the Wrangellia and Yukon CTs in their approximate present positions was primarily during the mid-Cretaceous. In central Alaska, at least local flysch deposition occurred in basins along the suture zone in Late Cretaceous time (Coney and Jones, 1985).

Northward displacement of oceanic crustal rocks in the Southern Margin composite terrane

Within the accretionary assemblage of the Southern Margin CT, paleomagnetic inclinations for Late Cretaceous to Paleocene volcanic rocks (Fig. 4) suggest 16°–31° northward translation relative to the craton (Plumley and others, 1983; Coe and others, 1985; Hillhouse, 1987; Hillhouse and Coe, this volume). We interpret these samples to be offscraped exotic oceanic fragments, and therefore they do not indicate the paleolatitude of the enclosing accretionary prism. In this respect they are similar to exotic rocks with large paleomagnetic anomalies that occur in accretionary prisms along the continental margin of the conterminous United States (Silberling and others, 1992). As noted in the following section, as much as 8° of the indicated northward displacement could occur by dextral slip on major inboard faults.

Displacement on major transcurrent faults

Geologic data indicate that the larger intraplate strike-slip faults in Alaska and adjacent parts of Canada have dextral offsets of tens to hundreds of kilometers (Fig. 4). Cumulative Late Cretaceous and early Tertiary dextral displacement is about 450 km on the northern Tintina fault and total post-Devonian displacement on the fault and its splays is about 900 km (Gabrielse, 1985; Price and Carmichael, 1986). Late Cretaceous to early Tertiary dextral slip on the Denali fault is about 400 km (Grantz, 1966; Nokleberg and others, 1985; Hickman and others, 1990). There has been about 600 km of late Cenozoic dextral displacement on the Fairweather fault (Plafker and others, this volume, Chapter 12). Cumulative displacement on these intraplate faults results in northward latitudinal shifts relative to the craton that can account for as much as 8° of the paleomagnetic inclination anomalies in terranes outboard of the Tintina fault.

About 450 km of post–Middle Jurassic to pre-Eocene offset has been documented along the northern Tintina fault in northwestern Yukon Territory (Gabrielse, 1985). The fault can be traced into Alaska discontinuously about 125 km to long 146°W, where it splays into a complex of west- to southwest-trending strike-slip and thrust faults (Dover, this volume). In east-central Alaska, the main strike-slip movement on the Tintina fault was post–Early Cretaceous in age and predates the deposition of poorly dated Tertiary(?) rocks (Dover, this volume). It is not known how slip is partitioned west of the mapped Tintina fault trace in Alaska. Approximately 310 km of dextral slip can be accounted for by offsets across known or inferred major splays of the Tintina fault in central and western Alaska (Fig. 4). Offset on the Kobuk-Malemute fault is poorly constrained at 90 km or more (Dillon and others, 1987; Dover, this volume). Offset is 160 on the Kaltag fault (Patton and Hoare, 1968), and about 90 km on the Iditarod–Nixon Fork fault (Miller and Bundtzen, 1988). The remainder of the Tintina fault slip may be accommodated by distributive shear on unidentified strike-slip faults between these structures (Dover, this volume), by compressional shortening during and after oroclinal bending (Chapman and others, 1985), or by a combination of these mechanisms.

Relative plate motions

Mid-Cretaceous and younger motions of the Farallon, Kula, and Pacific plates relative to a northwest-trending continental margin at lat 60°N are shown as vectors in Figure 4 (Engebretsen and others, 1985). Relative plate motions were about 7 cm/yr orthogonal convergence during subduction of the Farallon plate (100–85 Ma) and 12–21 cm/yr dextral-oblique convergence during subduction of the Kula plate (85–61 Ma). We assume that the demise of the Kula plate predated the onset of arc volcanism in the Alaska Peninsula (K-Ar ages 50 ± 2 Ma) and coincided approximately with the widespread early to middle Eocene thermal event along the eastern margin of the Gulf of Alaska (Plafker and others, this volume, Chapter 12). Since about 50 Ma, Pacific–North American plate relative motions were northwesterly at 4–6 cm/yr with dextral to dextral-oblique convergence on the northwest-trending transform margin and orthogonal convergence on the northeast-trending Aleutian Arc margin.

Poorly constrained ocean-floor magnetic anomalies for the Farralon plate from 180 to 100 Ma indicating sinistral oblique convergence of 4–12 cm/yr (Engebretsen and others, 1985) are not incorporated in our model or shown in Figure 4. If correct, an important consequence of these plate kinematic reconstructions is that significant post–Late Triassic northward transport of allochthonous terranes, such as the Wrangellia CT, was possible only after onset of Kula plate relative motions in the Late Cretaceous. This assumption was incorporated in previous models for the evolution of the southern Alaska margin (Plafker, 1990; Plafker and others, this volume, Chapter 12). We have not used the pre-100 Ma reconstructions in our model because the geologic data indicate that the Wrangellia CT in Alaska was probably close to its present position relative to inboard terranes by Late Jurassic time (McClelland and others, 1992a) and that it was juxtaposed against the Yukon CT by mid-Cretaceous time in Alaska (Gehrels and Berg, this volume; Nokleberg and others, this volume, Chapter 10) and British Columbia (Crawford and others, 1987; Monger and others, 1982; Thorkelson and Smith, 1989). Instead, we follow McClelland and others (1992a) in inferring a period of significant margin-parallel northward movement that may correspond with a lull in arc magmatism in the Wrangellia CT in the Jurassic from about Bajocian to Oxfordian time (183–156 Ma). Thus, the major part of the northward displacement or the Wrangellia CT could have taken place during this ~27 m.y. interval with the remainder accommodated by dextral transpression during Kula plate time.

TECTONIC EVOLUTION—A MODEL

In this section we present a model for the Phanerozoic tectonic evolution of the major elements of Alaska and contiguous areas based primarily on the geologic and geophysical constraints outlined above. The model illustrates the tectonic evolution of composite terranes and terranes in schematic plan view reconstructions for eight time intervals (Fig. 5). The figures do not incorporate internal deformation of the terranes and composite terranes: they do restore known displacements along major intraplate strike-slip faults.

Although large rotations and translations are indicated, the model is conservative in that it assumes Northern Hemisphere options for the paleomagnetic data, and probable North American affinities during the Phanerozoic for the Arctic, Central, Yukon, Wrangellia, and Southern Margin CTs as well as for the Ruby and Yakutat terranes. Much more mobilistic models for one or more of these terranes can be, and have been, postulated (e.g., Engebretsen and others, 1985; Gehrels and Saleeby, 1987; Stone and others, 1982).

We emphasize that no single model has yet been devised to accommodate all the geologic, paleontologic, geophysical, and paleomagnetic data, because many of these data are mutually contradictory and/or internally inconsistent. The main benefit of attempting to model the tectonic evolution of all of Alaska is that it provides a guide for future research by highlighting problems with the available data and interpretations.

Precambrian

The tectonic model in Figure 5 begins with the Cambrian because the Precambrian history is too fragmentary to permit a meaningful map reconstruction. Early Proterozoic metamorphic

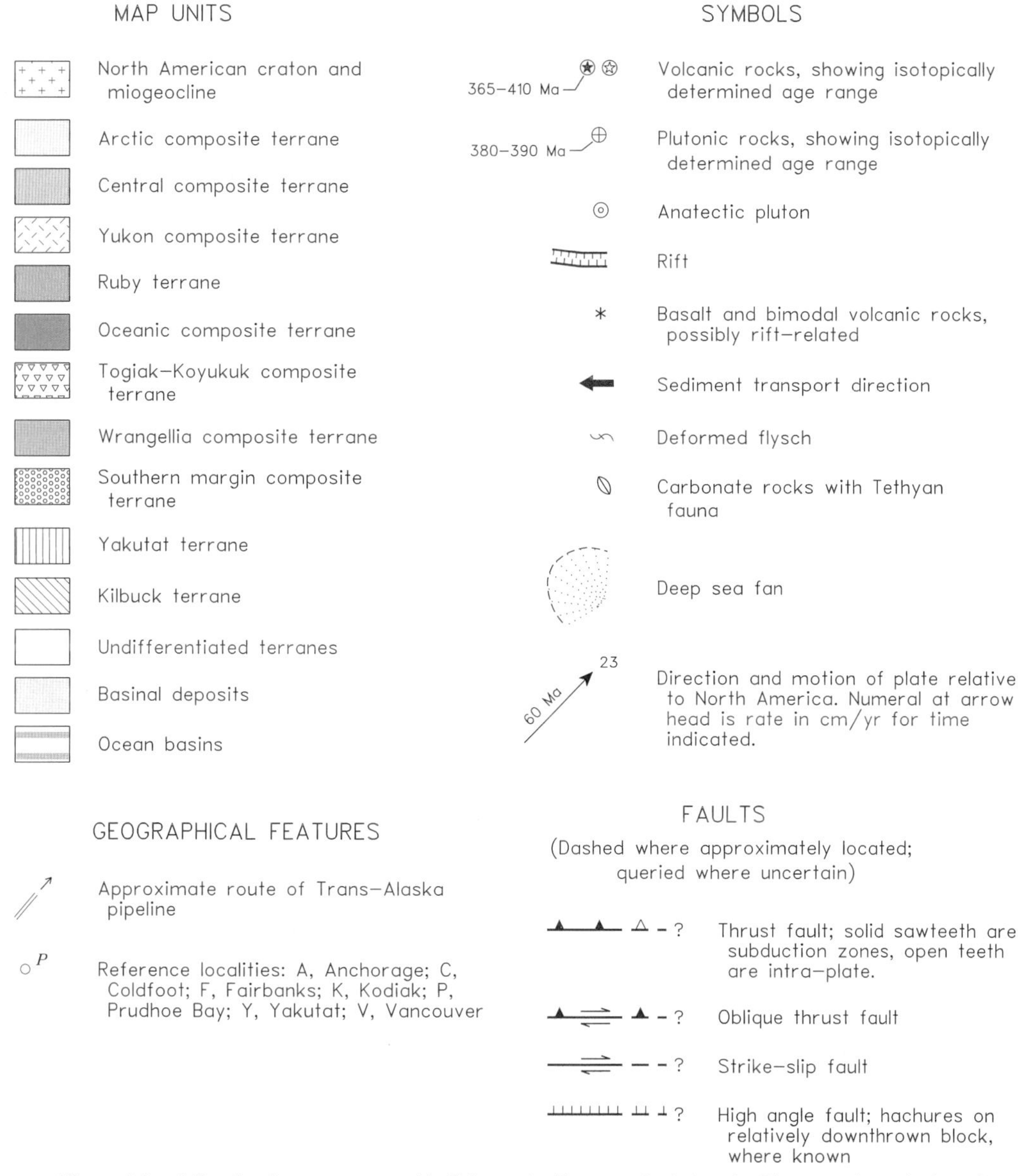

Figure 5 (on following four page spreads). Schematic diagrams depicting the Phanerozoic evolution of major tectonic elements of Alaska and adjacent areas of Canada. Terrane outlines are dashed where the presence or location are the most speculative. Orientations of geologic and geographic features are relative to present geographic coordinates. Terrane abbreviations are the same as in Figure 1. See text for explanation and data sources.

rocks (2.1 Ga) are known in Alaska only in the Kilbuck terrane and Idono Complex in western Alaska (Fig. 1). They have isotopic signatures that suggest an origin in a craton margin magmatic arc (Miller and others, 1991). Decker and others (this volume) suggest that they may be offset fragments of the Ruby terrane and that all of the Precambrian rocks in southwestern Alaska probably originated along the craton margin of North America. Although plausible, this interpretation can not be substantiated because the cratonic rocks have no close age and isotopic equivalents in the Ruby terrane or elsewhere in western North America.

The only other Early Proterozoic ages in Alaska are from detrital and inherited zircons in Late Proterozoic and Paleozoic rocks (Miller and others, 1991). They occur in (1) the western Brooks Range (2.06 and 2.26 Ga), (2) the eastern Alaska Range (2.1 Ga), and (3) the Yukon CT in east-central and southeastern Alaska (2.1–2.3 Ga). The provenance for the detrital zircons is unknown.

The western continental margin of North America, including what is now northern and eastern Alaska, was shaped by Late Proterozoic rifting beginning at about 850 Ma and possibly continuing into early Paleozoic time. Rifting was followed by gradual subsidence and initial deposition of thick sequences of clastic and carbonate rocks of the Cordilleran miogeocline (Stewart, 1976). Known and inferred Late Proterozoic rocks of the Arctic, Central, and Yukon CTs and the Ruby terrane are widely interpreted as having formed mainly, if not entirely, in the Cordilleran miogeocline (Dover, this volume; Foster and others, this volume; Moore and others, this volume; Patton and others, this volume, Chapter 7; Silberling and others, this volume).

In the Alexander terrane segment of the Wrangellia CT, magmatism began during the Late Proterozoic in a probable

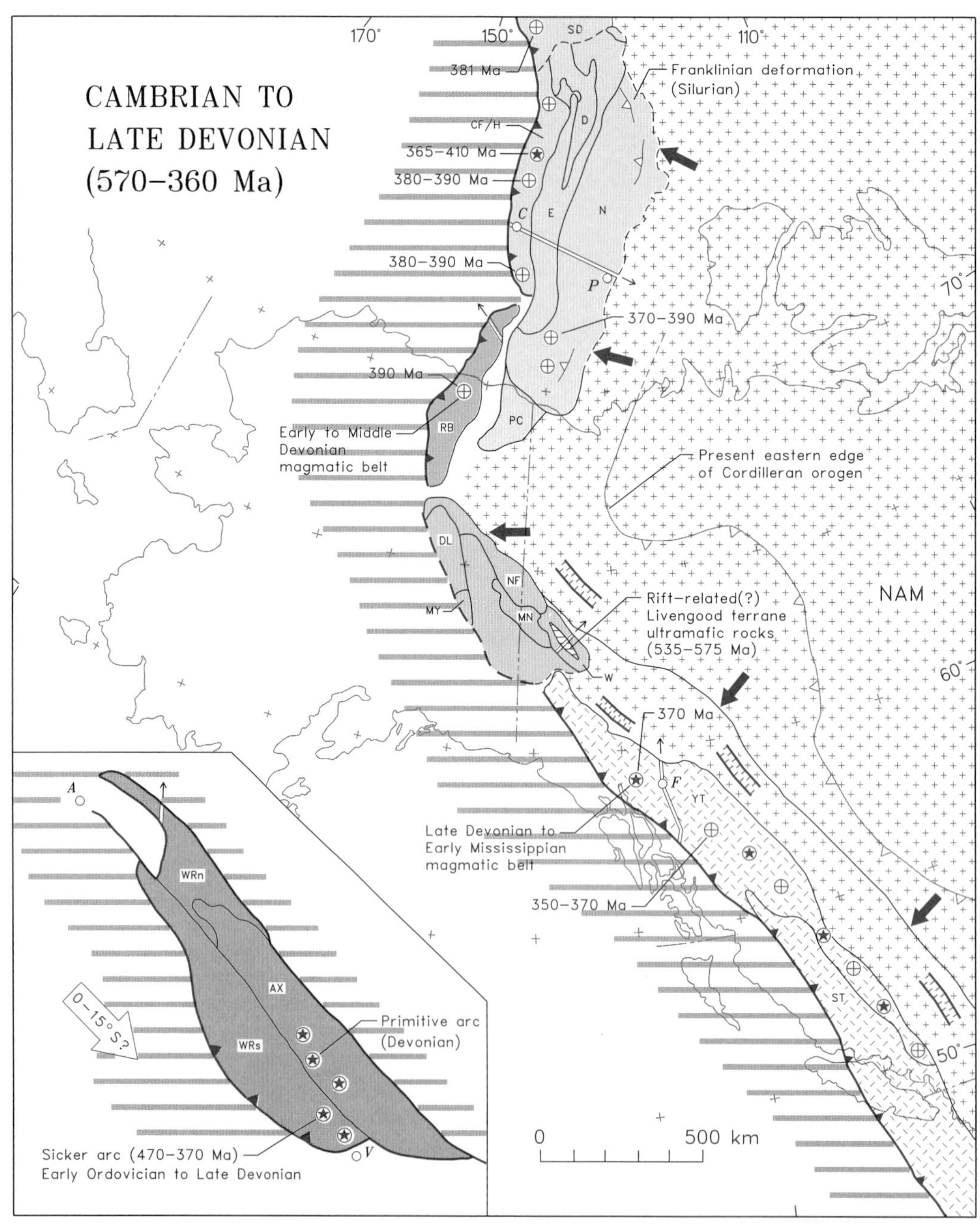

Figure 5a. Cambrian to Late Devonian: Inferred configuration of continental margin composite terranes and terranes showing sediment source direction, rift basins, Devonian to Early Mississippian arc-related(?) magmatic belt, and location of Cambrian or older rift-related(?) ultramafic rocks of the Livengood terrane. Box shows Wrangellia intraoceanic composite terrane and locations of Ordovician to Late Devonian arc magmatism; paleomagnetic data indicate a paleolatitude within 15° of its present latitude, assuming a Northern Hemisphere option.

intraoceanic environment; there are no data on the location of the composite terrane at that time.

Cambrian to Late Devonian (570–360 Ma)

The early Paleozoic western craton margin in what is now Alaska and contiguous parts of Canada can be established in a general way from the distribution of craton-related rocks, autochthonous sequences of the Cordilleran miogeocline, and by isotopic data. The approximate locations of the Arctic and Central CTs relative to the craton margin (Fig. 5A) are based on the paleomagnetic data indicating that the Nixon Fork terrane segment of the Central CT was at about its present latitude in Ordovician time, and by removing the rotation of the Arctic CT to restore it to a position contiguous with the Cordilleran miogeocline of Arctic Canada. The position of the Ruby terrane is based

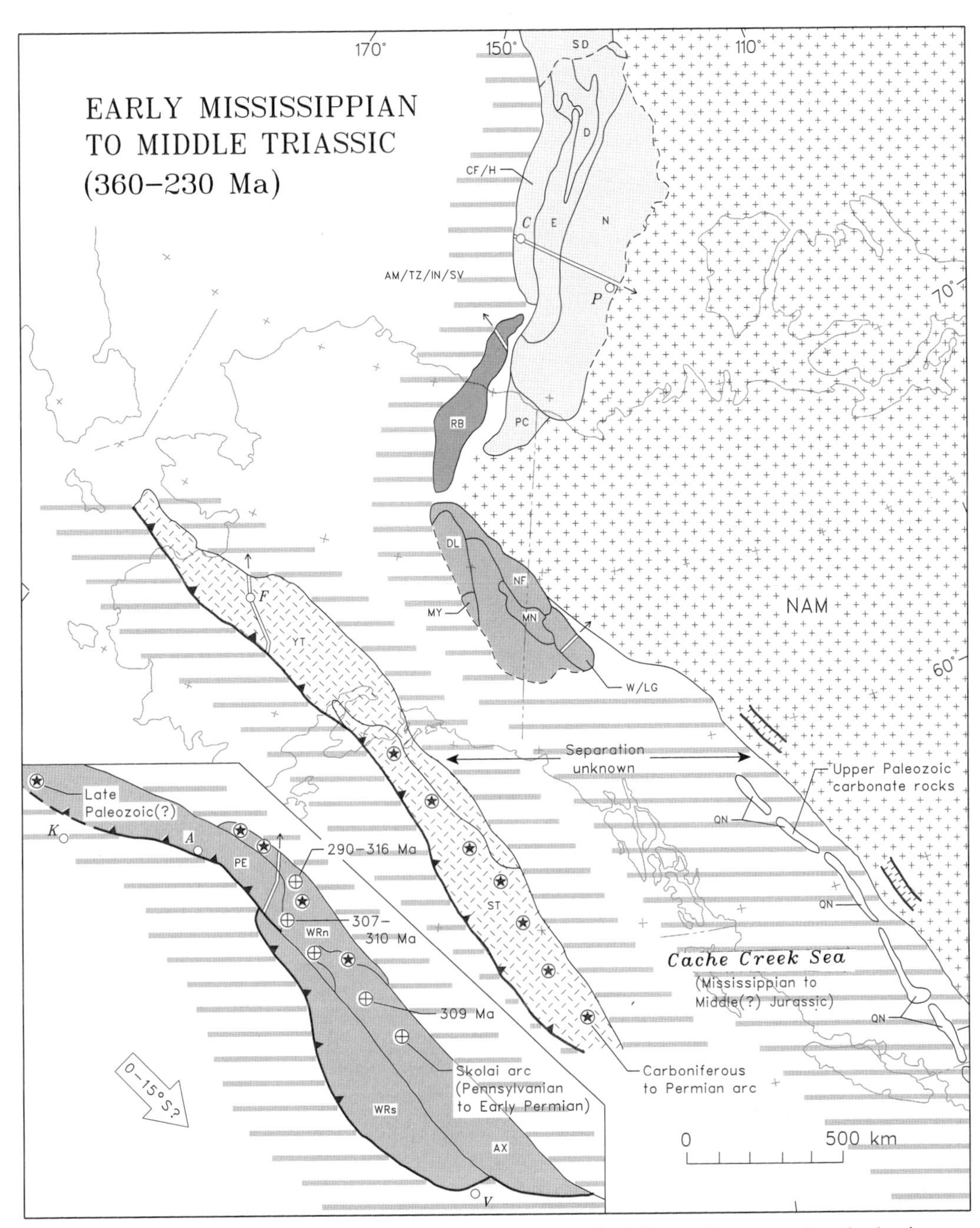

Figure 5B. Early Mississippian to Middle Triassic: Passive margin sedimentation occurred on the Arctic and Central CTs. The Cache Creek Sea opened in Mississippian time and the Yukon CT was a rifted continental fragment. Carboniferous to Early Permian magmatic arcs developed on the southern Yukon CT (Stikine terrane) and Wrangellia CT above east(?)-dipping subduction zones.

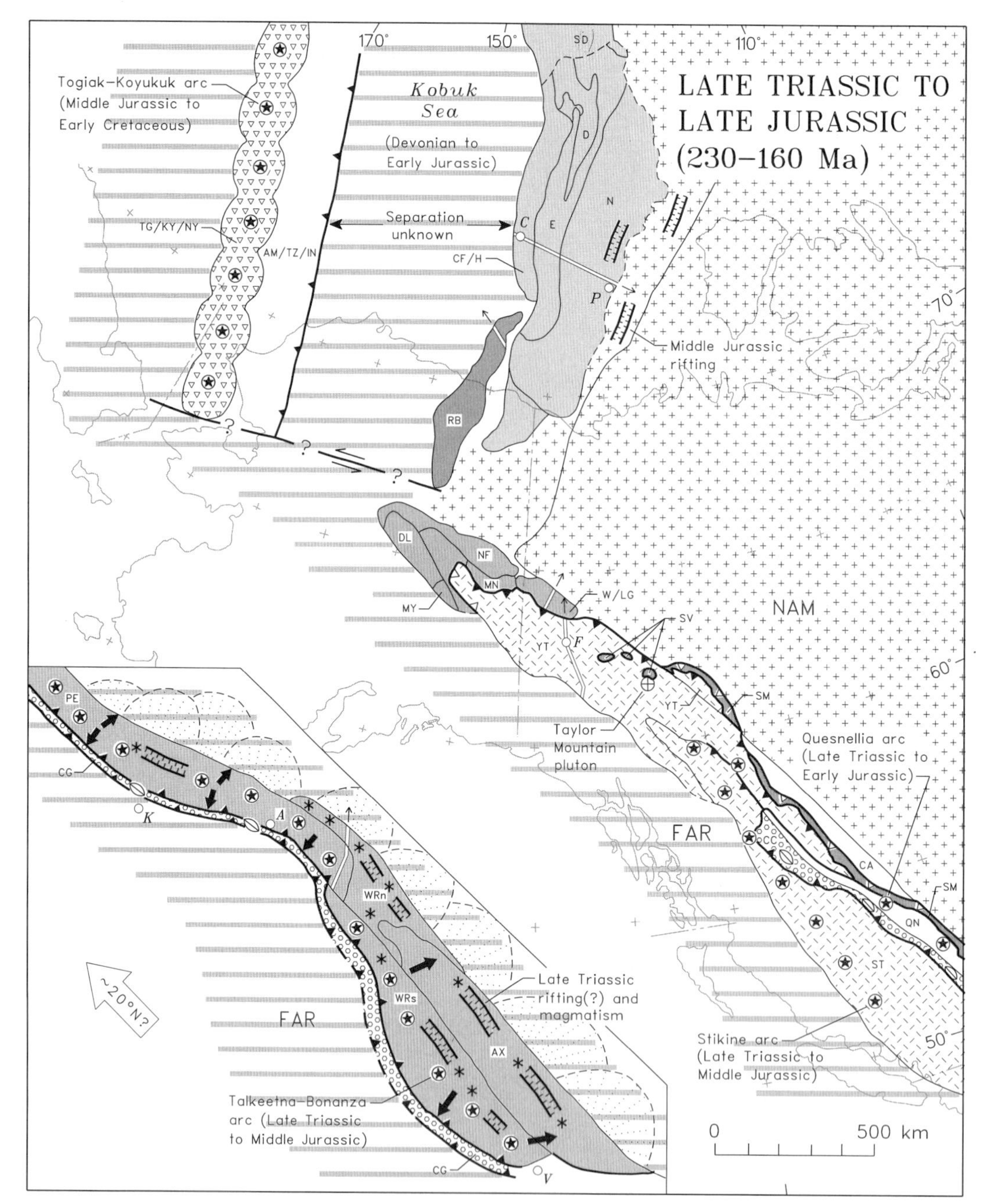

Figure 5C. Late Triassic to Late Jurassic: Passive margin sedimentation occurred on the Arctic CT. Arc volcanism and subduction took place on the intraoceanic Togiak-Koyukuk CT in the Middle Jurassic. The Late Triassic to Early Jurassic Quesnellia arc and its accretionary prism (Cache Creek terrane) were active along the continental margin in the Late Triassic and Early Jurassic above an east-dipping subduction zone. The Yukon CT closed the Cache Creek Sea from the Late Triassic to Early Jurassic, during which time Stikine arc was active above a probable west-dipping subduction zone. By Early Jurassic time, the Yukon CT collapsed against inboard terranes, overrode part of the Central CT and North American craton margin, and fragments of oceanic crust were obducted onto terranes along the eastern margin of the Yukon CT (Slide Mountain terrane) and were locally emplaced onto the Yukon CT in Alaska (Seventymile terrane). The Wrangellia CT was characterized by Middle and Late Triassic eruptions of tholeiitic basalt in the Wrangellia terrane, eruption of bimodal volcanic rocks in the Alexander and Peninsular terranes, and Late Triassic to early Middle Jurassic arc volcanism and formation of the accretionary prism of the Southern Margin composite terrane above an east-dipping subduction zone. Limestone with Tethyan faunas was offscraped into accretionary prisms of the Cache Creek terrane and the Southern Margin CT. There was probable major northward movement of Wrangellia CT between Late Triassic and Late Jurassic time. FAR, Farallon plate.

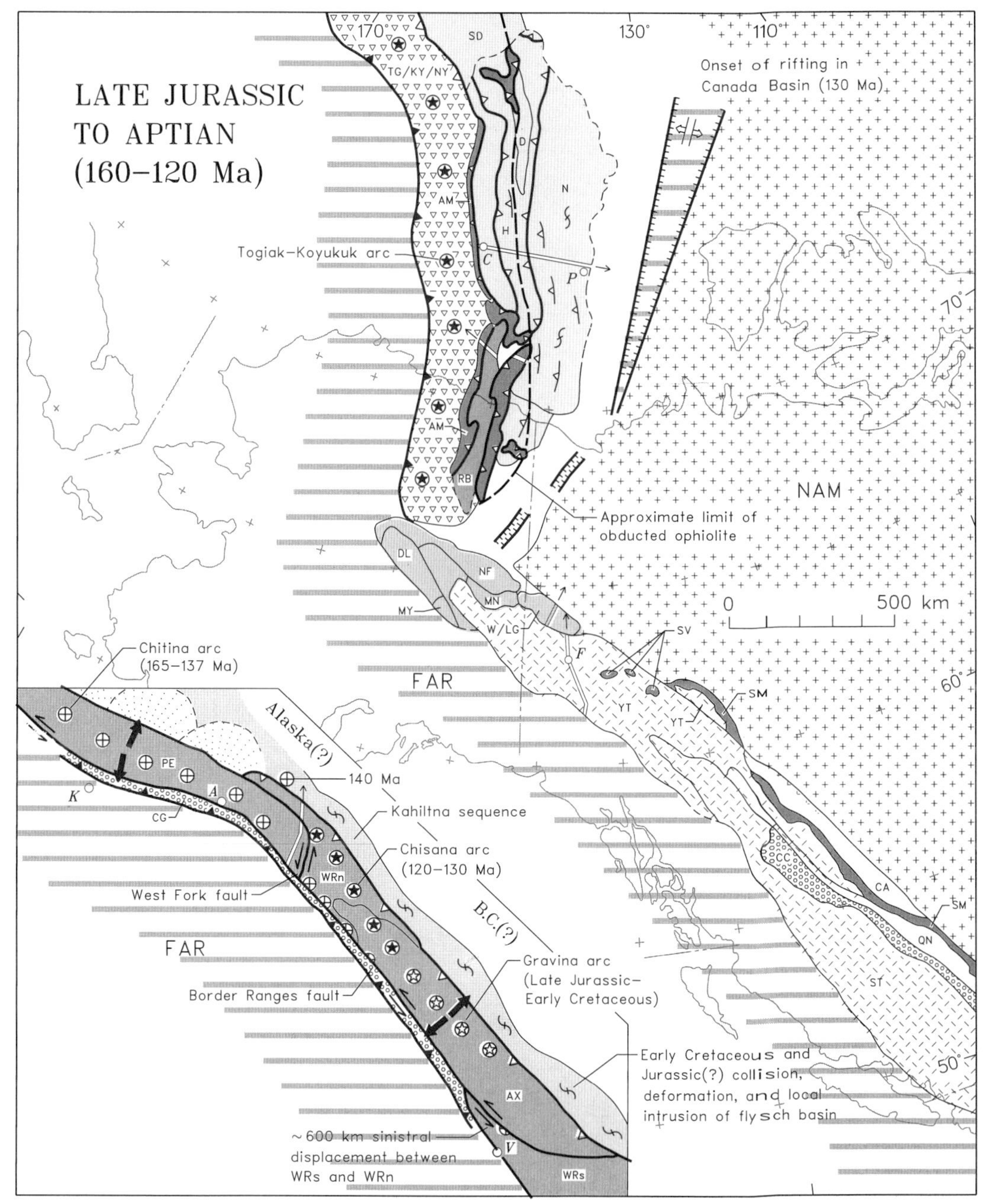

Figure 5D. Late Jurassic to Aptian: Incipient rifting occurred in the Canada basin. The Togiak-Koyukuk CT collided with the Arctic CT, resulting in 150+ km overriding of the Arctic CT by the oceanic crust of the closed Kobuk basin (Angayuchum, Tozitna, and Innoko terranes); there was major contractional deformation and regional high-pressure metamorphism in lower plate rocks and a probable flip in the subduction zone, above which arc magmatism continued through the Early Cretaceous. Arc magmatism and accretion occurred above the east-dipping subduction zone in the Wrangellia and Southern margin CTs. The Wrangellia terrane in Alaska was offset 600+ km from Wrangellia in British Columbia. There was deformation of flysch basins along the continentward margin of the Wrangellia CT by 140 Ma, probably by interaction with terranes along the continental margin. FAR, Farallon plate.

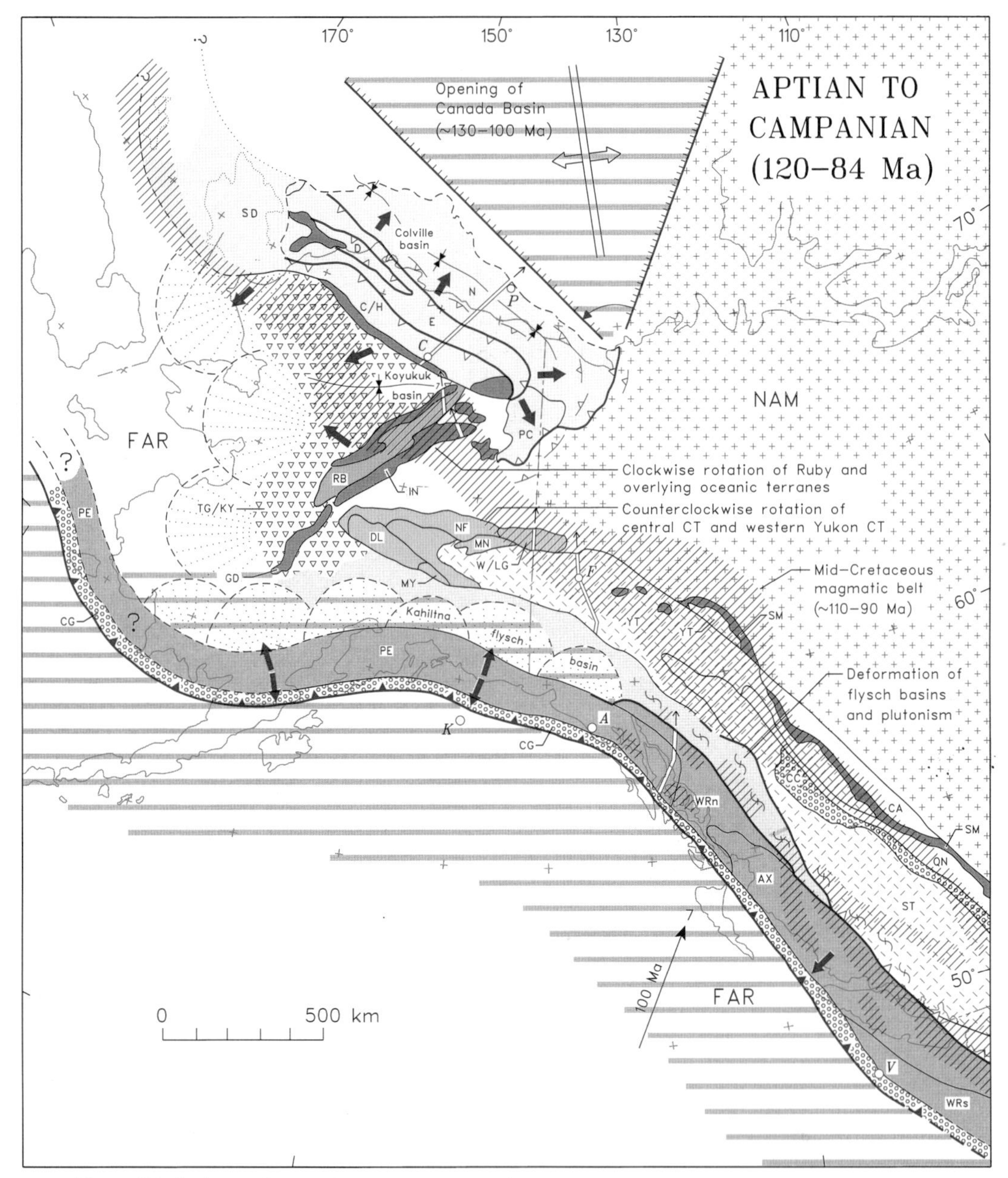

Figure 5E. Aptian to Campanian: The Canada basin opened, driving counterclockwise rotation of the Arctic CT and attached parts of the Togiak-Koyukuk and Oceanic CTs, and clockwise rotation of the Ruby terrane and attached parts of the Oceanic CT. The Wrangellia CT was emplaced against continental margin terranes with the collapse of flysch basins. A major magmatic belt (diagonal red lines) overprinted all major terranes in Alaska and adjacent parts of Canada; it is inferred to be related to east-dipping subduction along continental margin. Magmatism was accompanied by thermal metamorphism throughout much of Alaska and by at least local extension in the Yukon CT. FAR, Farallon plate.

on the inference that it was originally continuous with the southern part of the Arctic CT. There is no control on the latitudinal position of the Yukon CT; by Devonian time it was part of a continental margin magmatic arc.

The Cambrian through Early Devonian (Fig. 5A) continental margin in Alaska was characterized by (1) continued subsidence and local rifting in the miogeocline and deposition of carbonate rocks and of clastic sediments derived mainly from the craton; (2) local extension(?) with emplacement of ultramafic and mafic igneous rocks of the Livengood terrane on the ocean floor adjacent to the continental margin in latest Precambrian to Early Cambrian time; (3) local early Paleozoic extension and bimodal volcanism in parts of the Central and Yukon CTs; (4) possible early Paleozoic subduction and arc volcanism along the Arctic

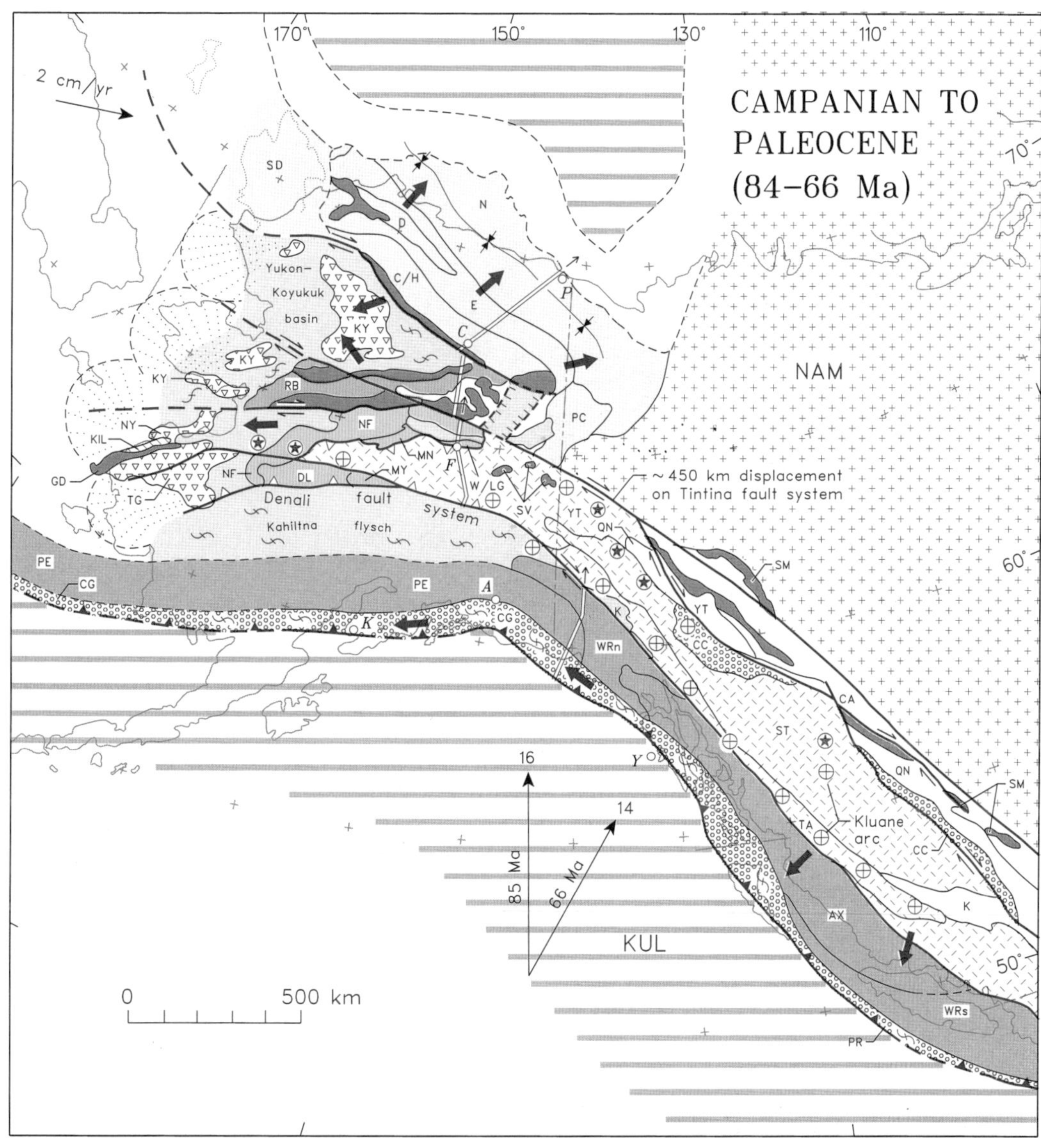

Figure 5F. Campanian to Paleocene: Dextral transpression along the continental margin resulted in northwestward movement of the Yukon CT along the Tintina and related faults, driving further clockwise rotation of the Ruby terrane and attached Oceanic CT to close the Yukon-Koyukuk basin. Kluane arc magmatism was active above the landward-dipping subduction zone, and there was large accretion of arc-derived sediments in the Southern Margin CT. KUL, Kula plate.

CT part of the continental margin or on an accreted continental fragment; and (5) Silurian (Franklinian orogeny) deformation of the inboard part of the Arctic CT.

The late Early Devonian to earliest Mississippian interval was characterized along the continental margin of Alaska and contiguous regions by compressional deformation and dominantly coarse clastic sedimentation (Ellesmerian orogeny of Arctic Alaska and Antler orogeny in the southern Cordillera). During this period, there was granitoid plutonism and local comagmatic(?) andesitic to felsic volcanism in the Arctic CT (Hammond, Coldfoot, and western North Slope subterranes), the Ruby terrane, and the Yukon CT (Yukon-Tanana terrane). Magmatism of this age has not been recognized in the Central CT. The Devonian to Early Mississippian magmatic belt is generally coeval and coextensive with granitoid plutonism that extends southward through Yukon Territory in Canada, and with plutonism and volcanism in the Klamath Mountains and Sierra Nevada of Oregon and California (see summary by Rubin and others, 1991).

The Wrangellia CT was a primitive oceanic arc or oceanic arc complex throughout much of the early Paleozoic (and probably the Late Proterozoic). In the southern segment of the Wrangellia CT on Vancouver Island, volcanism and plutonism of the Sicker arc are poorly dated in the interval from the Middle Silu-

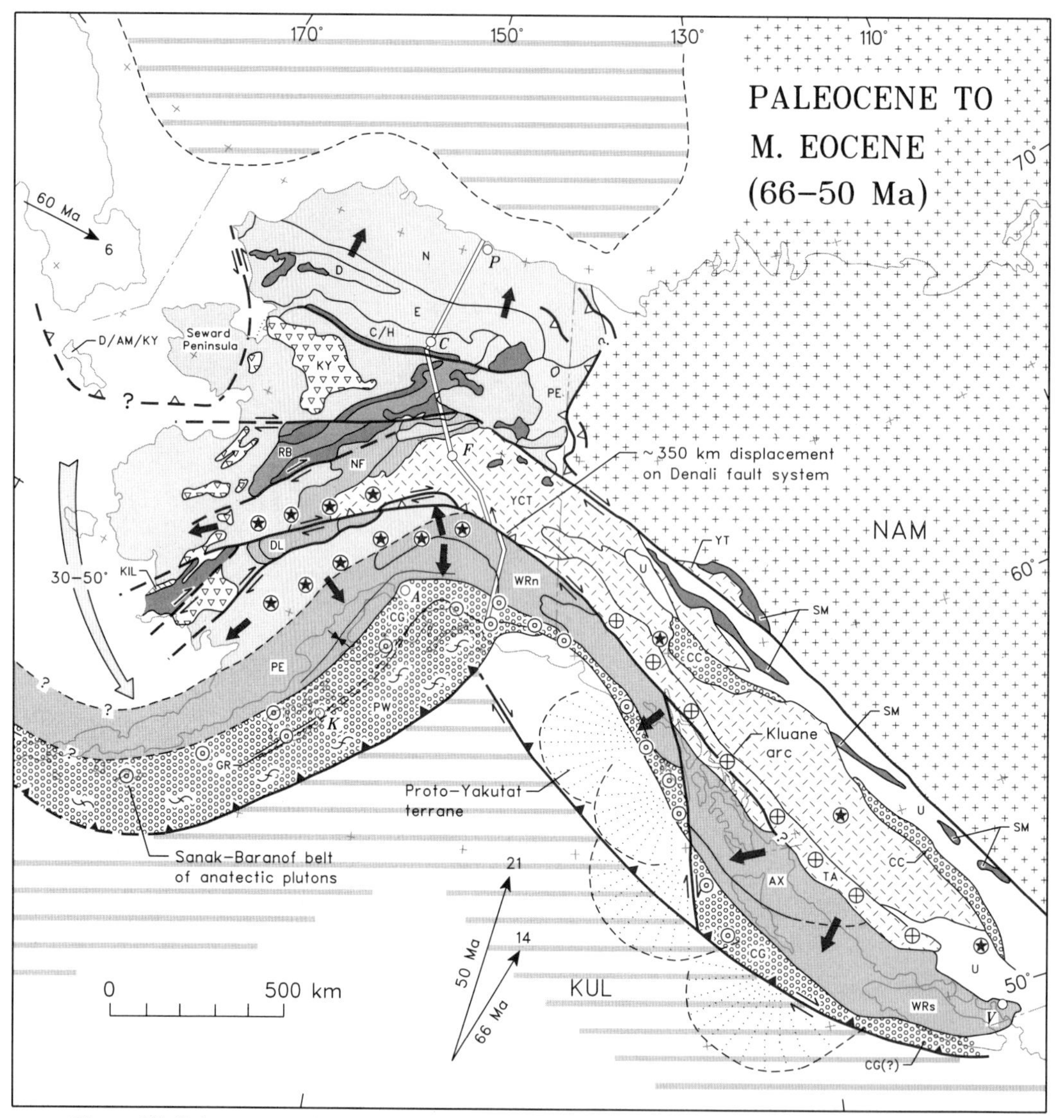

Figure 5G. Paleocene to middle Eocene: Large-scale counterclockwise rotation of western Alaska and displacement of the western Arctic CT (Seward Peninsula) was driven by compression between Eurasia and North America. Dextral displacement on the Denali fault system formed the proto-Yakutat terrane, and further dispersed terrane fragments in western Alaska. There was continued Kluane arc magmatism above landward-dipping subduction zones, and major volumes of arc-derived sediment were offscraped into western part of the Southern Margin composite terrane accretionary prism, followed by widespread anatectic magmatism along the southern Alaska margin. KUL, Kula plate.

rian to the Late Devonian (Brandon and others, 1986). The relation of this arc to the Late Devonian magmatism that characterized much of the western continental margin is unknown: one possibility is that it was an intraoceanic equivalent of the Devonian continental margin magmatic belt (Plafker, 1990).

Assuming the Northern Hemisphere option, the Ordovician to Devonian paleolatitude of the Wrangellia CT was 0°–15° south relative to the craton, depending upon the reference poles chosen (Van der Voo and others, 1980; Hillhouse and Coe, this volume). Its longitude relative to the craton is unknown. An intriguing alternative hypothesis assumes a Southern Hemisphere interpretation of the paleomagnetic data and suggests that the Alexander terrane part of the composite terrane was adjacent to eastern Australia until at least Middle Devonian time (Gehrels and Saleeby, 1987). This option is incompatible with Devonian faunal data from the Alexander terrane, which suggest North American affinities (Savage, 1988). Furthermore, magmatism of the Skolai and Talkeetna arcs indicates relative convergence between the Wrangellia CT and adjacent oceanic plates during most of the late Paleozoic and early Mesozoic. As a consequence, the Wrangellia CT could not have been transported passively on oceanic plates across the paleo-Pacific Ocean during this time

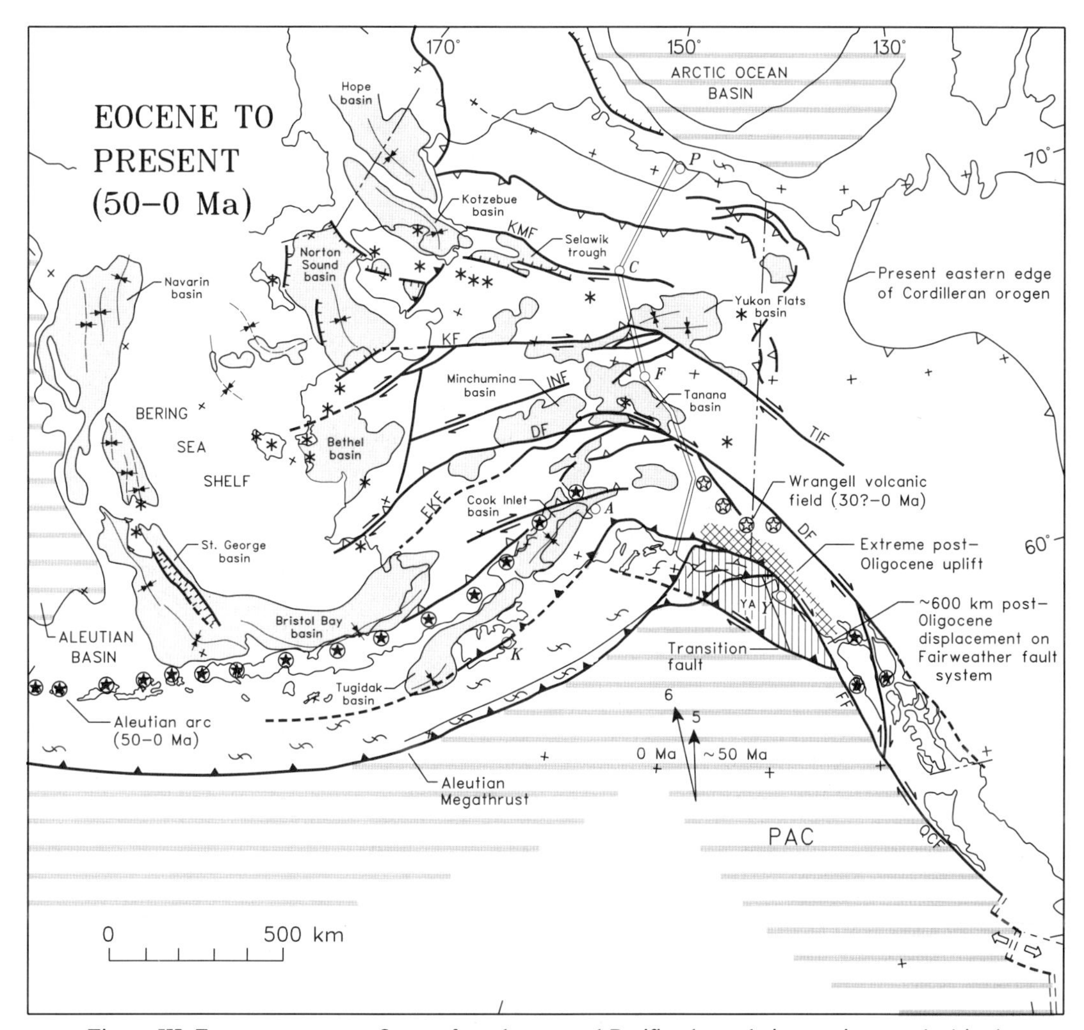

Figure 5H. Eocene to present: Onset of northwestward Pacific plate relative motions resulted in the formation of the present Aleutian–Alaska Peninsula arc, displacements on intraplate faults, and the formation of basins on land and on the contiguous continental shelves. At about 30 Ma, the Fairweather–Queen Charlotte transform boundary stepped northward to its present position, the Yakutat terrane began to move ~600 km northwestward along the transform boundary, and subduction of combined western Yakutat terrane and Pacific plate resulted in arc volcanism in the Wrangell Mountains and major uplift of eastern Chugach and Saint Elias mountains. Fault abbreviations are the same as for Figure 4. PAC, Pacific plate.

interval (Plafker and others, 1989; Nokleberg and others, this volume, Chapter 10).

Early Mississippian to Middle Triassic (360–230 Ma)

The Mississippian to Middle Triassic interval (Fig. 5B) in northern Alaska was characterized by deposition of the Ellesmerian sequence of continental clastic rocks, carbonate rocks, and chert on the subsiding ocean-facing shelf and rise of the Artic CT. Coeval rocks in the Central CT are scarce, but include Mississippian through Lower Jurassic carbonate rocks, chert, and volcaniclastic rocks (Innoko terrane) and Upper Permian and Upper Triassic sandy and spiculitic carbonate rocks (Nixon Fork terrane).

To the south, Devonian to earliest Mississippian orogeny in eastern Alaska, Yukon Territory, and British Columbia appears to have been immediately followed by extension and subsidence of the continental margin and rifting of fragments, including the Yukon CT, to open the Cache Creek Sea. Upper Paleozoic carbonate rocks were deposited on the Quesnellia terrane and the subsiding shelf. Relative motion between the seaward-moving Yukon CT and the oceanic plate resulted in Carboniferous and Permian arc magmatism in the southern part of the composite terrane (Stikine terrane); comparable magmatism has not been found in the Yukon-Tanana terrane to the north. The position of the Yukon CT relative to the craton during this time interval is unknown.

In the Wrangellia CT, late Paleozoic arc volcanic rocks and

comagmatic plutons (Skolai arc) resulted from an inferred craton-dipping subduction zone. The Skolai arc is roughly coeval with arc rocks in the collage of accreted terranes within the Blue Mountains, eastern Klamath Mountains, and northern Sierra Nevada of the conterminous United States (Saleeby, 1983; Miller, 1987) and may have been part of this same arc system.

Late Triassic to Middle Jurassic (about 230–160 Ma)

During most of the Late Triassic to Middle Jurassic interval (Fig. 5C), convergence prevailed between the Farallon plate and the continental margin. As a consequence, fringing intraoceanic arcs developed within the Farallon plate from Alaska to southern California. In what is now western Alaska, the southern part of the Togiak-Koyukuk CT (Togiak arc) became active in Middle Jurassic time above a west-dipping subduction zone. To the south, in Canada, the Late Triassic and Early Jurassic Quesnellia magmatic arc (Fig. 3) and its associated accretionary complex (Cache Creek terrane) developed along the continental margin above an east-dipping subduction zone. In the Yukon CT, the Late Triassic to Middle Jurassic Stikine arc was active above a probable west-dipping subduction zone.

Subduction beneath the Quesnellia and Stikine arcs ended with closure of the Cache Creek basin by Middle Jurassic time, after which these arc and oceanic assemblages were tied to terranes of continental margin affinities (Monger and Berg, 1987). Collision of the Yukon CT with the continental margin resulted in accretion and east-directed overthrusting of the Late Triassic to Middle Jurassic complex of arcs (Stikine and Quesnellia terranes), accretionary prisms (Cache Creek terrane), and oceanic crust (Seventymile and Slide Mountain terranes). Devonian to Late Triassic ages of the oceanic rocks and associated sedimentary cover suggest a very long-lived sea that may have been 1000–5000 km wide (Tempelman-Kluit, 1979). Accretion was accompanied by intense deformation and metamorphism that included local formation of blueschist and eclogite facies rocks (Hansen, 1990).

In the Wrangellia CT, the early part of the Late Triassic was characterized by rifting and the eruption of voluminous sequences of marine and subaerial dominantly tholeiitic basalt flows and pillow lavas and by local bimodal volcanic rocks (Nokleberg and others, this volume). Rift-related magmatism was followed by latest Triassic to early Middle Jurassic Talkeetna arc volcanism above an east-dipping subduction zone and by subduction and local blueschist facies metamorphism in the associated accretionary prism of the Southern Margin CT. Much of the northward movement of the Wrangellia CT may have occurred during the lull in arc volcanism from earliest Middle Jurassic to Late Jurassic time.

The occurrence of enigmatic Permian Tethyan faunas in rocks associated with these arcs has been the subject of considerable debate in tectonic reconstructions of the western Cordillera. In Alaska, Tethyan faunas occur mainly in Permian limestone blocks at localities within melange of the Southern Margin CT that is associated with the Talkeetna arc. At one locality near Anchorage, limestone containing Tethyan faunas occurs as a pod in schistose pelitic rocks, mafic volcanic rocks, and metachert that was interpreted as part of the Peninsular arc terrane (Clark, 1982). Late Triassic bivalve faunas in the Peninsular terrane, however, are endemic to the Americas and support an east Pacific origin (Newton, 1983), so autochthonous Tethyan faunas are unlikely. We suggest the alternative possibility that the limestone and associated rocks may be a metamorphosed facies of the same melange that contains limestone blocks of Tethyan affinity elsewhere in the Southern Margin CT. In adjacent parts of Canada, Permian limestone containing Tethyan faunas occurs as thick and extensive blocks in accretionary melange (Cache Creek terrane) associated with the Quesnellia arc. These blocks are interpreted to be disrupted limestone reefs capping guyots or plateaus that were rafted on oceanic crust from southern latitudes and were offscraped into the accretionary prism (Monger and Berg, 1987). In the conterminous United States, Tethyan faunas occur in coeval arc-related assemblages in a discontinuous belt that extends through the Blue Mountains, Klamath Mountains, and northern Sierra Nevada (Miller, 1987).

Late Jurassic to Aptian (about 160–120 Ma)

The Late Jurassic to Aptian interval (Fig. 5D) was characterized mainly by convergence accompanied by arc magmatism, accretion, and deformation along much of the western continental margin. In the Arctic, incipient rifting and sea-floor spreading of the Arctic Ocean basin is indicated by Late Jurassic to Early Cretaceous extensional structures along the present Canada basin margin of the Arctic CT (Grantz and others, this volume).

Beginning in the Late Jurassic and continuing through the Early Cretaceous, the Togiak-Koyukuk CT was accreted to the continental margin (Patton and others, this volume, chapter 7). This collision zone along ocean-facing parts of the Arctic and Central CTs was marked by subduction of the attenuated continental margin beneath the Togiak-Koyukuk accretionary prism and arc. As a consequence, Late Devonian to Early Jurassic oceanic basalt and ultramafic rocks derived from the closed Kobuk Sea (the Oceanic CT) were obducted onto the continental margin as extensive thrust sheets up to 150 km wide. As with the Cache Creek Sea, the considerable age span of the oceanic rocks and the scarcity of continent-derived detritus suggest that the width of the Kobuk Sea was on the scale of thousands of kilometers. Emplacement was accompanied by widespread metamorphism, including blueschist facies metamorphism in the lower continental plate, and by large-scale continent-verging thrusting and folding.

On the Wrangellia CT, magmatism occurred during much of Late Jurassic and Early Cretaceous time in the Chitina and Gravina-Nutzotin arcs above an east-dipping subduction zone. Magmatism in the Wrangellia CT could represent a southward continuation of the partly coeval Togiak-Koyukuk arc, which would imply that the Wrangellia CT was close to its present position at that time. Volcaniclastic flysch and melange, including olistostromal blocks derived from the Wrangellia and Peninsular terranes, were deposited in the trench and accreted to the Southern Margin CT at the same time that thick, dominantly volcaniclastic flysch sequences were deposited in basins along the backarc side of the composite terrane. Truncation and intense shearing of the southern margin of the Wrangellia CT in Alaska are inferred to have occurred during about 600 km of sinistral displacement of the British Columbia part of the Wrangellia CT relative to the Alaska part (Plafker and others, 1989; this volume, Chapter 12). Sinistral strike-slip offset along the boundary between the Wrangellia and Alexander terranes in British Columbia and along the Border Ranges fault or the West Fork fault in Alaska could explain abrupt truncation of both the Talkeetna and Chitina arcs in Alaska (Fig. 5D); however, structural evidence for the sense of movement along the boundary remains to be documented. The inferred displacement postdates Late Jurassic plutonism and predates Late Cretaceous accretion of the flysch and basalt assemblage of the Chugach terrane. Late Jurassic to Early Cretaceous sinistral displacements reported elsewhere in the Cordillera (Avé Lallemant and Oldow, 1988) are compatible with this hypothesis.

In what is now south-central Alaska, deformation of thick flysch basinal assemblages (Kahiltna sequence) along the inboard margin of the Wrangellia CT occurred before emplacement of postdeformational alkalic gabbro plutons that yield K-Ar dates of 140 Ma (Smith, 1981). McClelland and others (1992a) cited the presence of Late Jurassic to Early Cretaceous basinal assemblages along the inner margin of the Wrangellia CT from western Alaska to California, and the deformation within these basins, to postulate emplacement of the composite terrane against inboard terranes, mainly the Stikine and Yukon-Tanana terranes, by Middle Jurassic time. In Alaska and adjacent parts of Canada, however, the dominantly volcanogenic composition of the flysch in these basins and paleocurrent data reflecting westerly sources do not support close proximity to the continental margin (Berg and others, 1972; Eisbacher, 1976; Cohen and Lundberg, 1988; Plafker, 1985, unpublished data). Other models propose that the flysch basins were closed by a Late Jurassic and/or Early to mid-Cretaceous suture that migrated from south to north (Pavlis, 1989). Because there are no known direct geologic ties, the position of the Wrangellia CT relative to the continental margin has not been shown in Figure 5D.

Aptian to Campanian (about 120–84 Ma)

During the Aptian to Campanian interval (Fig. 5E), orthogonal to dextral oblique convergence characterized the boundary of the Farallon and North American plates, and most of northern, central, and western Alaska was affected by deformation and rotations related to opening of the Canada basin.

In northern Alaska counterclockwise rotation of the Arctic CT away from the craton from late Neocomian through Albian time resulted in the development of a passive margin along the Canada basin that persists to the present. Rotation was accompanied by continued thrusting and uplift of the southern part of the Arctic CT (O'Sullivan and others, 1993) and by continued clastic sedimentation in the Colville foredeep basin of the North Slope. Uplift of the Arctic CT may have been at least in part a consequence of postorogenic extension (Miller and Hudson, 1991). In the eastern Arctic CT, contractional deformation occurred in the hinge zone of rotation, and uplift is reflected in clastic sedimentation in the region (Howell and Wiley, 1987).

Rotation of the Arctic CT is inferred in our model to have driven clockwise rotation of the Ruby terrane, Togiak-Koyukuk CT, and contiguous oceanic terranes, and minor counterclockwise rotation of the Central and western Yukon CTs, to form the southern margin of the proto-Koyukuk basin. The basin was filled with synorogenic Lower to mid-Cretaceous marine flysch derived mainly from the uplifted Brooks Range orogen.

The Wrangellia CT was juxtaposed against the Yukon CT close to its present position during this interval with resulting intense deformation of intervening flysch basins and structural imbrication and continent-directed deep underthrusting, synorogenic plutonism, and metamorphism as high as kyanite grade in the tectonically thickened crust along the suture zone (Csejtey and others, 1982; Monger and others, 1982; Crawford and others, 1987; Rubin and Saleeby, 1992). In central Alaska, Stanley and others (1990) postulated large-scale mid-Cretaceous (about 115–95 Ma) underthrusting of flysch beneath the Yukon-Tanana terrane to the north to explain electrically conductive rocks at depth and lead isotope data (Aleinikoff and others, 1987) that suggest a flysch component in mid-Cretaceous plutons. In parts of the Yukon-Tanana terrane, crustal extension at about 130–110 Ma resulted in a "core-complex" style detachment, a regional subhorizontal metamorphic fabric, and elimination of as much as 10 km of crustal section (Pavlis and others, 1993). The tectonic mechanism for the extensional event is unknown. In western Alaska, the Wrangellia CT is inferred from dredge data to form the western boundary of what is now the Bering Sea shelf (Marlow and others, this volume). A major unknown is the nature of the crust between the outer Bering Sea shelf and the onshore terranes.

Mid-Cretaceous magmatism and associated thermal metamorphism occurred in a broad belt through much of Alaska (Miller, this volume, Chapter 16), the Canadian Cordillera (Armstrong, 1988), and the Russian far east (Parfenov and Natal'in, 1986). We believe that this magmatic belt is related mainly to continental margin arc magmatism above a shallow-dipping subduction zone, but it also includes many plutons probably formed by crustal melting (Crawford and others, 1987; Pavlis and others, 1993; Hudson, this volume; Miller, this volume, Chapter 16). Thermal metamorphism and magmatism of this age occurs in a belt as wide as 600 km that extends from the North American craton in Canada across all outboard terranes to the southern margin of the Wrangellia CT. This belt requires that the terranes were assembled in approximately their present relative positions by the end of this time interval, except for younger intraplate fault displacements.

Campanian to Paleocene (about 84–66 Ma)

From Campanian to Paleocene time (Fig. 5F), the terranes along the northeast Pacific Ocean margin assumed their present configuration. After formation of the Kula plate at about 85 Ma, relative motion north of the Kula–Farallon–North American plate triple junction was characterized by rapid dextral-oblique transpression.

Erosion of the north- and east-propagating thrust sheets in the Arctic CT continued to supply sediment to the Colville foreland basin (Howell and others, 1992; Dover, this volume; Moore and others, this volume). In central Alaska, northwestward displacement of the Yukon CT and the attached Central CT (Nixon Fork and related terranes) occurred mainly by dextral slip along the Tintina fault and its major splays in Alaska. This displacement resulted in additional clockwise rotation of the Ruby terrane and overlying oceanic terranes with resulting pervasive compressive deformation of the Yukon-Koyukuk basin fill. In southern Alaska, final suturing of the Wrangellia CT occurred upon closure of the intervening seaway, sometime after the Cenomanian. Closure was accompanied by deformation and metamorphism of the intervening Kahiltna flysch basin and associated miniterranes.

A continental margin magmatic arc (Kluane arc) was active during the Late Cretaceous to early Tertiary interval along and inboard of the Wrangellia CT–Yukon CT boundary and it extended westward into west-central Alaska. An enormous volume of volcaniclastic sediment was shed from the arc onto the ocean floor and this sediment was then offscraped to form a major part of the Southern Margin CT accretionary prism.

Although dextral offset along the Tintina fault is shown as occurring mainly in this interval, the actual age of offset in Alaska is uncertain. In contiguous areas of Canada, fault movement continued into late Eocene or Oligocene time (Gabrielse, 1985), and in Alaska a short segment was reactivated with vertical displacement as recently as Holocene time (Plafker and others, this volume, Plate 12). It is likely that at least some strike-slip movement occurred during this time interval on the Denali and other faults outboard of the Tintina fault, but there is no direct evidence for the timing or amount of these displacements.

Paleocene to Eocene (about 66–50 Ma)

During Paleocene to early Eocene time (Fig. 5F) large Kula–North American plate relative motions continued in Alaska and adjacent parts of Canada. At the same time there was up to 6 cm/yr compression between Eurasia and North America coincident with opening of the Labrador basin.

Compression between Eurasia and North America resulted in counterclockwise oroclinal bending, into about their present positions, of western Alaska and the Arctic CT. Possibly concurrent with rotation, the Seward Peninsula segment of the Arctic CT is inferred to have been offset to the south relative to the eastern Arctic CT; this offset is postulated as the result of faulting, rather than rotation, to maintain the subparallelism of coeval structures in the Seward Peninsula and nearby parts of the southern Brooks Range (Till and Dumoulin, this volume). Uplift in the eastern Arctic CT at about 60 Ma, indicated by fission-track data (O'Sullivan and others, 1993) and by thrusting of the eastern Arctic CT over the adjacent miogeocline, may be related to this compressional episode.

In interior Alaska, transpression along the southern continental margin resulted in major dextral strike slip on the Denali fault that preceded the main rotation of western Alaska, and resulted in continued minor dextral slip on other northwest-trending intraplate faults, such as the Tintina fault. The combination of dextral displacement on northwest-trending faults and oroclinal bending resulted in local development of extensional basins along these faults systems (Fig. 5H).

In southern Alaska, Kluane arc magmatism continued in southeastern Alaska and it extended into southwestern Alaska and probably the Bering Sea shelf, as a broad belt of arc-related volcanic rocks. Voluminous Paleocene and Eocene deep-sea fan and trench(?) deposits were derived from the Kluane arc and incorporated into the western limb of the Southern Margin CT. Accretion was followed by widespread early Eocene metamorphism and anatectic granitic plutonism in parts of the Southern Margin CT and adjacent Wrangellia CT inferred to have been related to subduction of the Kula-Farallon ridge. Along the southeastern margin of Alaska, the proto-Yakutat terrane was formed by early Eocene(?) oceanward stepping of the plate boundary following about 180 km of dextral displacement on the Chatham Strait segment of the Denali fault. During and immediately after faulting, lower to middle Eocene marine clastic deposits from a crystalline-complex source, carbonate reef detritus, and minor coal were deposited along the margins of the offset Chugach terrane and on oceanic crust to the west.

Eocene to present (about 50–0 Ma)

North to northwest motion of the Pacific plate relative to the continental margins of Alaska and most of Canada began by early middle Eocene time after subduction of the Kula plate and continues to the present. Figure 5G shows selected major structural features in Alaska and on the adjacent sea shelves; the present configuration of terranes and composite terranes is shown in Figures 1 and 2.

Nearly orthogonal to dextral-oblique convergence along most of the western Alaska margin resulted in development of the Aleutian arc as the Pacific–North American plate boundary from the Alaska Peninsula westward, thereby trapping a large segment of oceanic plate within the abyssal Bering Sea (Marlow and others, this volume; Vallier and others, this volume). Andesitic volcanism associated with the Aleutian arc (Fournelle and others, this volume; Kay and Kay, this volume) began on the Alaska Peninsula at about 50 Ma, a few million years after the onset of northwestward motion of the Pacific plate, and continues to the present. Subduction along the Aleutian megathrust beneath the arc was accompanied by accretion and underplating of deep-sea sediments and oceanic crust along the inner wall of the Aleutian Trench, which includes the Southern Margin CT. Along the western Aleutian ridge, a wide range of strike-slip, extensional, and rotational structures reflect the dextral-oblique to dextral relative motion across that part of the Aleutian arc (Vallier and others, this volume; Plafker and others, this volume, Plate 12). During this interval, some of the northeast- to east-trending intraplate strike-slip faults of western and central Alaska were reactivated as thrust or oblique thrust faults.

Dextral transpression characterized the transform plate margin off southeastern Alaska and British Columbia and resulted in dextral strike slip on northwest- to west-trending intraplate faults extending inland at least to the Kobuk and Tintina faults. Intermittent magmatism in southeastern Alaska and adjacent parts of British Columbia suggests a significant component of oblique underthrusting along that margin during Paleogene time (Brew, this volume). Throughout Alaska, continental sediment was deposited in interior basins formed as crustal sags, pull-aparts related to strike-slip faults, and half grabens (Kirschner, this volume, Chapter 14; Wahrhaftig and others, this volume), and dominantly marine sediment was deposited in shelf-margin and shelf basins (Kirschner, this volume, Plate 7). The larger basins of the Bering Sea shelf contain dominantly marine strata of post–middle Eocene age. Marlow and others (this volume) interpret the Norton and Bristol Bay basins as crustal sags that probably involve block faulting in basement rocks; they interpret the Navarin and St. George basins as grabens and half grabens formed by extensional collapse of the outer Bering shelf margin. An alternative view which emphasizes the scarcity of normal faults seen on seismic reflection profiles across many of these outer shelf basins, is that they are controlled by margin-parallel dextral shear zones along and near the continental shelf edge (Worrall, 1991).

At about 30 Ma, the transform margin stepped landward to its present position on the Queen Charlotte and Fairweather faults, and the Yakutat terrane began moving northwestward with the Pacific plate about 600 km to its present position (Plafker and others, this volume). Subduction of the leading edge of the Yakutat terrane beneath the Southern and Wrangellia CTs resulted in volcanism in the Wrangell arc beginning about 25 Ma and continuing to the present. Ongoing collision and subduction of the Yakutat terrane is marked by extreme uplift and topographic relief in the adjacent Chugach and Saint Elias mountains, by deposition of thick sequences of Miocene and younger clastic sediments (including abundant marine glacial diamictite) in continental shelf basins and the Aleutian Trench, by intense ocean-verging compressional deformation, and by high seismicity (Plafker and others, this volume, Chapter 12). Stress trajectories derived from neotectonic data (Plafker and others, this volume, Plate 12) suggest that late Cenozoic mountain-building deformation and seismicity throughout Alaska and adjacent parts of Canada and Neogene volcanism in the Aleutian arc and dominantly basaltic volcanism in large areas of interior and western Alaska, the Bering Sea shelf, and southeastern Alaska are driven mainly or entirely by interaction between the Pacific plate (and Yakutat terrane) with the southern margin of the continent.

Discussion

The tectonic model presented here is compatible with much of the geologic, geophysical, and paleomagnetic data for Alaska. There are two major differences between this model and previous interpretations of the tectonic evolution of Alaska (Plafker, 1990) and the southern Alaska margin (Plafker and others, this volume, Chapter 12). First, we have used geologic data to constrain the timing for northward movement and accretion of the Wrangellia CT to southern Alaska to Jurassic time, instead of the plate reconstructions of Engebretsen and others (1985), which preclude significant northward movement before Late Cretaceous time. Second, we postulate that clockwise rotation of the Central CT began in Early to mid-Cretaceous time, concurrent with opening of the Arctic basin, instead of late Paleozoic to early Mesozoic counterclockwise rotation indicated by paleomagnetic data (Plumley, 1984). Throughout this synthesis we have tried to highlight the inevitable problems that arise due to lack of data, to conflicting data from different studies, or to different interpretations of the same data. Resolution of these, and other major problems, should provide a fertile field for future geologic and geophysical research in Alaska.

ACKNOWLEDGMENTS

In an overview of this sort, it is not possible to cite all the many workers in Cordilleran tectonics who have contributed to the model we have presented, and we apologize for inevitable omissions. Many colleagues have provided helpful discussions and/or reviews of parts or all of early versions of this chapter. In particular we thank J. W. Hillhouse, D. G. Howell, J.W.H. Monger, C. S. Grommé, W. J. Nokleberg, W. W. Patton, and N. J. Silberling. We are especially grateful to Howell and Patton for their constructive technical reviews of this paper, to James Laney who prepared numerous versions of the illustrations, and to Judy Weathers, who provided invaluable editorial assistance.

REFERENCES CITED

Aleinikoff, J. N., and Plafker, G., 1989, In search of the provenance and paleogeographic location of the White Mountains terrane: Evidence from U-Pb data of granite boulders in the Fossil Creek Volcanics, *in* Dover, J. H., and Galloway, J. P., eds., Geological studies in Alaska by the U.S. Geological Survey, 1988: U.S. Geological Survey Bulletin 1903, p. 68–74.

Aleinikoff, J. N., Dusel-Bacon, C., Foster, H. L., and Nokleberg, W. J., 1987, Lead isotope fingerprinting of tectono-stratigraphic terranes, east-central Alaska: Canadian Journal of Earth Sciences, v. 24, p. 2089–2098.

Armstrong, R. L., 1988, Mesozoic and early Cenozoic magmatic evolution of the Canadian Cordillera, *in* Clark, S. P., Burchfiel, B. C., and Suppe, J., eds., Processes in continental lithospheric deformation: Geological Society America Special Paper 218, p. 55–91.

Avé Lallemant, H. G., and Oldow, J. S., 1988, Early Mesozoic southward migration of Cordilleran transpressional terranes: Tectonics, v. 7, p. 1057–1075.

Bacon, C. R., Foster, H. L., and Smith, J. G., 1990, Rhyolitic calderas of the Yukon-Tanana terrane, east-central Alaska: Volcanic remnants of a mid-Cretaceous magmatic arc: Journal of Geophysical Research, v. 95, p. 21,451–21,461.

Barker, F., Sutherland Brown, A., Budahn, J. R., and Plafker, G., 1989, Back-arc with frontal-arc component origin of Triassic Karmutsen basalt, British Columbia, Canada: Chemical Geology, v. 75, p. 81–102.

Beck, M. E., Burmester, R. F., and Shoonover, R., 1981, Paleomagnetism and tectonics of Cretaceous Mt. Stuart batholith of Washington: Translation or tilt?: Earth and Planetary Science Letters, v. 56, p. 336–342.

Berg, H. C., Jones, D. L., and Richter, D. H., 1972, Gravina-Nutzotin belt—Tectonic significance of an upper Mesozoic sedimentary and volcanic sequence in southern and southeastern Alaska, *in* Geological Survey research 1972: U.S. Geological Survey Professional Paper 800-D, p. D1–D24.

Blodgett, R. B., and Gilbert, W. C., 1992, Upper Devonian shallow-marine siliciclastic strata and associated fauna and flora, Lime Hills D-4 quadrangle, southwest Alaska *in* Bradley, D. C., and Dusel-Bacon, C., eds., Geological Studies in Alaska by the U.S. Geological Survey, 1991: U.S. Geological Survey Bulletin 2041, p. 106–1115.

Bogdanov, N. A., and Tilman, S. M., 1964, Similarities in the development of the Paleozoic structure of Wrangell Island and the western part of the Brooks Range (Alaska), *in* Soreshchanie po Problem Tektoniki: Moskva, Nauka, Skladchatye oblasti Evrazil, Materialy, p. 219–230.

Boillot, G., and Winterer, E. L., 1988, Drilling on the Galicia margin: Retrospect and prospect, *in* Boillot, G., Winterer, E. L., et al., eds., Proceedings of the Ocean Drilling Program, Scientific results, Volume 103: College Station, Texas, Ocean Drilling Program, p. 809–828.

Bond, G. C., and Kominz, M. A., 1984, Construction of tectonic subsidence curves for the early Paleozoic miogeocline, southern Canadian Rocky Mountains: Implications for subsidence mechanisms, age of breakup, and crustal thinning: Geological Society of America Bulletin, v. 95, p. 155–173.

Box, S. E., Moll-Stalcup, E. J., Wooden, J. L., and Bradshaw, J. Y., 1990, Kilbuck terrane: Oldest known rocks in Alaska: Geology, v. 18, p. 1219–1222.

Brandon, M. T., Orchard, M. J., Parrish, R. R., Sutherland Brown, A., and Yorath, C. J., 1986, Fossil ages and isotopic dates from Paleozoic Sicker Group and associated intrusive rocks, Vancouver Island, British Columbia, *in* Current research, Part A: Geological Survey of Canada Paper 86-1A, p. 683–696.

Brown, E. H., and Burmeister, R. F., 1991, Metamorphic evidence for tilt of the Spuzzum Pluton: Diminished basis for the "Baja British Columbia" concept: Tectonics, v. 10, p. 978–985.

Butler, R. F., Gehrels, G. E., McClelland, W. C., May, S. R., and Klepacki, D., 1989, Discordant paleomagnetic poles from the Canadian Coast Plutonic Complex: Regional tilt rather than large-scale displacement?: Geology, v. 17, p. 691–694.

Chapman, R. M., Trexler, J. H., Jr., Churkin, M., Jr., and Weber, F. R., 1985, New concepts of the Mesozoic flysch belt in east-central Alaska, *in* Bartsch-Winkler, S., and Reed, K. M., eds., The U.S. Geological Survey in Alaska: Accomplishments during 1983: U.S. Geological Survey Circular 945, p. 29–32.

Churkin, M., Jr., and Carter, C., 1979, Collision-deformed Paleozoic continental margin in Alaska—A foundation for microplate accretion: Geological Society of America Abstracts with Programs, v. 11, p. 72.

Churkin, M., Jr., and Trexler, J. H., Jr., 1981, Continental plates and accreted oceanic terranes in the Arctic, *in* Nairn, A.E.M., Churkin, M., Jr., and Stehli, F. G., eds., The ocean basin and margins, Volume 5, The Arctic Ocean: New York, Plenum Publishing, p. 1–20.

Churkin, M., Jr., Foster, H. L., Chapman, R. M., and Weber, F. R., 1982, Terranes and suture zones in east-central Alaska: Journal of Geophysical Research, v. 87, p. 3718–3730.

Clark, S.H.B., 1982, Reconnaissance bedrock geologic map of the Chugach Mountains near Anchorage, Alaska: U.S. Geological Survey Miscellaneous Field Studies Map MF-350, 70 p., scale 1:250,000.

Coe, R. S., Globerman, B. R., Plumley, P. W., and Thrupp, G. A., 1985, Paleomagnetic results from Alaska and their tectonic implications: American Association of Petroleum Geologists Circum-Pacific Earth Sciences Series, no. 1, p. 85–108.

Cohen, H. A., and Lundberg, N., 1988, Sandstone petrology of the Seymour Canal Formation (Gravina-Nutzotin belt): Implications for the accretion history of southeast Alaska: Geological Society of America Abstracts with Programs, v. 20, no. 7, p. A163.

Coney, P. J., 1989, Structural aspects of suspect terranes and accretionary tectonics in western North America: Journal of Structural Geology, v. 11, p. 107–125.

Coney, P. J., and Jones, D. L., 1985, Accretion tectonics and crustal structure in Alaska: Tectonophysics, v. 119, p. 265–283.

Coney, P. J., Jones, D. L., and Monger, J.W.H., 1980, Cordilleran suspect terranes: Nature, v. 288, p. 329–333.

Crawford, M. L., Hollister, L. S., and Woodsworth, G. J., 1987, Crustal deformation and regional metamorphism across a terrane boundary, Coast Plutonic Complex, British Columbia: Tectonics, v. 6, p. 343–361.

Csejtey, B., Jr., Cox, D. P., and Evarts, R. C., 1982, The Cenozoic Denali fault system and the Cretaceous accretionary development of southern Alaska: Journal of Geophysical Research, v. 87, p. 3741–3754.

Detterman, R. L., and Reed, B. L., 1980, Stratigraphy, structure, and economic geology of the Illiamna quadrangle, Alaska: U.S. Geological Survey Bulletin 1368-B, p. B28–B32, scale 1:250,000.

Dillon, J. T., Pessel, G. H., Chen, J. H., and Veach, N. C., 1980, Middle Paleozoic magmatism and orogenesis in the Brooks Range, Alaska: Geology, v. 8, p. 338–343.

Dillon, J. T., Tilton, G. R. Decker, J. E., and Kelly, M. J., 1987, Resource implications of magmatic and metamorphic ages for Devonian igneous rocks in the Brooks Range, *in* Tailleur, I. L., and Weimer, P., eds., Alaskan North Slope geology: Bakersfield, California, Pacific Section, Society of Economic Paleontologists and Mineralogists, Special Publication 50, p. 713–723.

Eisbacher, G. H., 1976, Sedimentology of the Dezadeash flysch and its implications for strike-slip faulting along the Denali fault, Yukon Territory and Alaska: Canadian Journal of Earth Sciences, v. 13, p. 1495–1513.

Embrey, A. F., and Dixon, J., 1990, The breakup unconformity of the Amerasia Basin, Arctic Ocean: Evidence from Arctic Canada: Geological Society of America Bulletin, v. 102, p. 1526–1534.

Engebretsen, D. C., Cox, A., and Gordon, R. G., 1985, Relative motions between oceanic and continental plates in the Pacific basin: Geological Society of America Special Paper 206, 59 p.

Gabrielse, H., 1985, Major dextral transcurrent displacements along the Northern Rocky Mountain Trench and related lineaments in north-central British Columbia: Geological Society of America Bulletin, v. 96, p. 1–14.

Gabrielse, H., and Yorath, C. J., 1991, Tectonic synthesis, *in* Gabrielse, H., and Yorath, C. J., eds., Geology of the Cordilleran orogen in Canada: Geological Survey of Canada, Geology of Canada, no. 4, p. 677–705.

Gardner, M. C., and eight others, 1988, Pennsylvanian pluton stitching of Wrangellia and the Alexander terrane, Wrangell Mountains, Alaska: Geology, v. 16, p. 967–971.

Gehrels, G. E., and Saleeby, J. B., 1987, Geologic framework, tectonic evolution, and displacement history of the Alexander terrane: Tectonics, v. 6, p. 151–173.

Gehrels, G. E., McClelland, W. C., Samson, S. D., Patchett, P. J., and Jackson, J. L., 1990, Ancient continental margin assemblage in the northern Coast Mountains, southeast Alaska and northwest Canada: Geology, v. 18, p. 208–211.

Gemuts, I., Puchner, C. C., and Steffel, C. I., 1983, Regional geology and tectonic history of western Alaska: Anchorage, Alaska Geological Society Journal, v. 3, p. 67–85.

Grantz, A., 1966, Strike-slip faults in Alaska: U.S. Geological Survey Open-File Report 267, 82 p.

Grantz, A., Moore, T. E., and Roeske, S. M., 1991, A-3 Gulf of Alaska to Arctic Ocean: Boulder, Colorado, Geological Society of America, Centennial Continent/Ocean Transect no. 15, 3 sheets, scale 1:500,000.

Halgedahl, S., and Jarrard, R., 1987, Paleomagnetism in the Kuparuk River Formation from oriented drill core: Evidence for rotation of the North Slope block, *in* Tailleur, I. L., and Weimer, P., eds., Alaskan North Slope geology: Bakersfield, California, Pacific Section, Society of Economic Paleontologists and Mineralogists Special Publication 50, p. 581–617.

Haeussler, P. J., 1992, Structural evolution of an arc-basin: The Gravina belt in central southeastern Alaska: Tectonics, v. 11, p. 1245–1265.

Haeussler, P. J., Coe, R. S., and Onstott, T. C., 1992, Paleomagnetism of the Late Triassic Hound Island Volcanics: Revisited: Journal of Geophysical Research, v. 97, p. 19,617–19,639.

Hansen, V. L., 1990, Yukon-Tanana terrane: A partial acquittal: Geology, v. 18, p. 365–369.

Hickman, R. G., Sherwood, K. W., and Craddock, C., 1990, Structural evolution of the early Tertiary Cantwell basin, south-central Alaska: Tectonics, v. 9, p. 1433–1449.

Hillhouse, J. W., 1987, Accretion of southern Alaska, *in* Kent, D. V., and Krs, M., eds., Laurasian paleomagnetism and tectonics: Tectonophysics, v. 139, p. 107–122.

Hoffman, P. F., 1989, Precambrian geology and tectonic history of North America, *in* Bally, A. W., and Palmer, A. R., eds., The geology of North America—An overview: Boulder, Colorado, Geological Society of America, Geology of North America, v. A, p. 447–512.

Howell, D. G., 1989, Tectonics of suspect terranes: London, New York, Chapman and Hall, 231 p.

Howell, D. G., and Wiley, T. J., 1987, Crustal evolution of northern Alaska inferred from sedimentology and structural relations in the Kandik area: Tectonics, v. 6, 619–631.

Howell, D. G., Jones, D. L., and Schermer, E. R., 1985, Tectonostratigraphic terranes of the circum-Pacific region, *in* Howell, D. G., ed., Tectonostratigraphic terranes of the Circum-Pacific region: Circum-Pacific Council for Energy and Mineral Resources Earth Science Series, no. 1, p. 3–30.

Howell, D. G., Johnsson, M. J., Underwood, M. B., Huafu, L., and Hillhouse, J. W., 1992, Tectonic evolution of the Kandik region, east-central Alaska: Preliminary interpretations, *in* Bradley, D. C., and Ford, A. B., eds., Geologic studies in Alaska by the U.S. Geological Survey, 1990: U.S. Geological Survey Bulletin 1999, p. 127–140.

Hudson, T. L., 1983, Calc-alkaline plutonism along the Pacific rim of southern Alaska, *in* Roddick, J. A., ed., Circum-Pacific plutonic terranes: Geological Society of America Memoir 159, p. 159–170.

Irving, E. M., and Wynne, P. J., 1990, Paleomagnetic evidence bearing on the evolution of the Canadian Cordillera: Royal Society of London Philosophical Transactions, v. 331, p. 487–509.

Irving, E. M., Woodsworth, G. J., Wynne, P. J., and Morrison, A., 1985, Paleomagnetic evidence for displacement from the south of the Coast Plutonic Complex, British Columbia: Canadian Journal of Earth Sciences, v. 22, p. 584–598.

Jones, D. L., Silberling, N. J., and Hillhouse, J. W., 1977, Wrangellia—A displaced terrane in northwestern North America: Canadian Journal of Earth Sciences, v. 14, p. 2565–2577.

Jones, D. L., Howell, D. G., Coney, P. J., and Monger, J.W.H., 1983, Recognition, character, and analysis of tectonostratigraphic terranes in western North America, *in* Hashimoto, M., and Uyeda, S., eds., Accretion tectonics in the circum-Pacific regions; proceedings of the Oji International Seminar on Accretion Tectonics, Japan, 1981: Advances in earth and planetary sciences: Tokyo, Terra Scientific Publishing Company, p. 21–35.

Jones, D. L., Silberling, N. J., Coney, P. J., and Plafker, G., 1987, Lithotectonic terrane map of Alaska (west of the 41st meridian): U.S. Geological Survey Miscellaneous Field Studies Map MF-874, scale 1:2,500,000.

Jones, P. B., 1982, Mesozoic rifting in the western Arctic Ocean basin and its relationship to Pacific seafloor spreading, *in* Embrey, A. F., and Balkwill, H. R., eds., Arctic geology and geophysics: Canadian Society of Petroleum Geologists Memoir 8, p. 83–99.

Karl, S. M., Aleinikoff, J. N., Dickey, C. F., and Dillon, J. T., 1989, Age and chemical composition of the Proterozoic intrusive complex at Mount Angayukaqsraq, western Brooks Range, Alaska, *in* Dover, J. H., and Galloway, J. P., eds., Geological studies in Alaska by the U.S. Geological Survey, 1988: U.S. Geological Survey Bulletin 1903, p. 10–19.

Loney, R. A., and Himmelberg, G. R., 1988, Ultramafic rocks of the Livengood terrane, *in* Galloway, J. P., and Hamilton, T. D., eds., U.S. Geological Survey in Alaska, accomplishments during 1987: U.S. Geological Survey Circular 1016, p. 68–70.

MacKevett, E. M., Jr., 1978, Geologic map of the McCarthy quadrangle, Alaska: U.S. Geological Survey Miscellaneous Investigations Series Map I-1032, scale 1:250,000.

Mamet, B., and Plafker, G., 1982, A Late Devonian (Frasnian) microbiota from the Farewell-Lyman Hills area, west-central Alaska: U.S. Geological Survey Professional Paper 1216-A, p. A1–A12.

McClelland, W. C., Gehrels, G. E., and Saleeby, J. B., 1992a, Upper Jurassic–Lower Cretaceous basinal strata along the Cordilleran margin: Implications for the accretionary history of the Alexander-Wrangellia-Peninsular terrane: Tectonics, v. 11, p. 823–835.

McClelland, W. C., Gehrels, G. E., Samson, S. D., and Patchett, P. J., 1992b, Protolith relations of the Gravina belt and Yukon-Tanana terrane in central southeastern Alaska: Journal of Geology, v. 100, p. 107–123.

Mihalynuk, M. G., and Mountjoy, K. J., 1989, Geology of the Tagish Lake area (104M/8, 9E): Ministry of Energy, Mines, and Petroleum Resources, British Columbia, Paper 1990-I, p. 181–196.

Miller, E. L., and Hudson, T. L., 1991, Mid-Cretaceous extensional fragmentation of a Jurassic–Early Cretaceous compressional orogen, Alaska: Tectonics, v. 10, p. 781–796.

Miller, M. L., and Bundtzen, T. K., 1988, Right-lateral offset solution for the Iditarod–Nixon Fork fault, western Alaska, *in* Galloway, J. P., and Hamilton, T. D., eds., Geologic studies in Alaska by the U.S. Geological Survey during 1987: U.S. Geological Survey Circular 1016, p. 99–103.

Miller, M. L., Bradshaw, J. Y., Kimbrough, T., Stern, T. W., and Bundtzen, T. K., 1991, Isotopic evidence for early Proterozoic age of the Idono Complex, west-central Alaska: Journal of Geology, v. 99, p. 209–223.

Miller, M. M., 1987, Dispersed remnants of a northeast Pacific fringing arc: Upper Paleozoic terranes of Permian McCloud faunal affinity, western U.S.: Tectonics, v. 6, p. 807–830.

Miyaoka, R. T., and Dover, J. H., 1990, Shear sense in mylonites, and implications for transport of the Rampart assemblage (Tozitna terrane), Tanana quadrangle, east-central Alaska, *in* Dover, J. H., and Galloway, J. P., eds., Geologic studies in Alaska by the U.S. Geological Survey, 1989: U.S. Geological Survey Bulletin 1946, p. 51–64.

Monger, J.W.H., 1989, Overview of Cordilleran geology, *in* Ricketts, B. D., ed., Western Canada sedimentary basin: A case history: Calgary, Canadian Society of Petroleum Geologists, p. 9–32.

Monger, J.W.H., and Berg, H. C., 1987, Lithotectonic terrane map of western Canada and southeastern Alaska: U.S. Geological Survey Miscellaneous Field Studies Map MF-1874-B, 12 p., 1 sheet, scale 1:2,500,000.

Monger, J.W.H., and Irving, E. M., 1980, Northward displacement of north-central British Columbia: Nature, v. 285, p. 289–294.

Monger, J.W.H., Price, R. A., and Tempelman-Kluit, D. J., 1982, Tectonic accretion and the origin of the two major metamorphic and plutonic welts in the Canadian Cordillera: Geology, v. 10, p. 70–75.

Monger, J.W.H., and nine others, 1991, Cordilleran terranes, *in* Gabrielse, H., and Yorath, C. J., eds., Geology of the Cordilleran orogen of Canada: Geological Survey of Canada, Geology of Canada, no. 4, p. 281–327.

Moore, J. C., and Connelly, W., 1979, Tectonic history of the continental margin of southwestern Alaska: Late Triassic to earliest Tertiary, *in* Proceeding of the Sixth Alaska Geological Symposium, Anchorage: Anchorage, Alaska Geological Society, p. H1–H29.

Moore, T. E., 1992, The Arctic Alaska superterrane, *in* Bradley, D. C., and Dusel-Bacon, C., eds., Geologic studies in Alaska by the U.S. Geological Survey, 1991: U.S. Geological Survey Bulletin 2041, p. 238–244.

Moore, T. E., and Nokleberg, W. J., 1988, Stratigraphy, sedimentology, and structure of the Wickersham terrane in the Cache Mountain area, east-central Alaska, *in* Galloway, J. P., and Hamilton, T. D., eds., Geologic studies in Alaska by the U.S. Geological Survey during 1987: U.S. Geological Survey Circular 1016, p. 75–80.

Mortensen, J. K., 1992, Pre-mid Mesozoic tectonic evolution of the Yukon-Tanana terrane, Yukon and Alaska: Tectonics, v. 11, p. 836–853.

Mull, C. G., Roeder, D. H., Tailleur, I. L., Pessel, G. H., Grantz, A., and May, S. D., 1987, Geologic sections and maps across Brooks Range and Arctic Slope to Beaufort Sea: Geological Society of America Map and Chart Series MC-28S, scale 1:500,000.

Newton, C. R., 1983, Paleozoogeographic affinities of Norian bivalves from the Wrangellian, Peninsular, and Alexander terranes, western North America, *in* Stevens, C. H., ed., Pre-Jurassic rocks in western North American suspect terranes: Los Angeles, California, Pacific Section, Society of Economic Paleontologists and Mineralogists, p. 37–48.

Nilsen, T. H., 1981, Upper Devonian and Lower Mississippian redbeds, Brooks Range, Alaska, *in* Miall, A. D., eds., Sedimentation and tectonics in alluvial basins: Geological Association of Canada Special Paper 23, p. 187–219.

Nokleberg, W. J., Jones, D. L., and Silberling, N. J., 1985, Origin and tectonic evolution of the Maclaren and Wrangellia terranes, eastern Alaska Range, Alaska: Geological Society of America Bulletin, v. 96, p. 1251–1270.

Nokleberg, W. J., Foster, H. L., and Aleinikoff, J. N., 1989, Geology of the northern Copper River basin, eastern Alaska Range, and southern Yukon-Tanana Upland, *in* Nokleberg, W. J., and Fisher, M. A., eds., Alaskan geological and geophysical transect, Field trip guidebook T104, 28th International Geological Congress: Washington, D.C., American Geophysical Union, p. 34–63.

Oldow, J. S., Avé Lallemant, H. G., Julian, F. E., and Seidensticker, C. M., 1987, Ellesmerian(?) and Brookian deformation in the Franklin Mountains, northeastern Brooks Range, Alaska, and its bearing on the origin of the Canada basin: Geology, v. 15, p. 37–41.

Oldow, J. S., Bally, A. W., Avé Lallemant, H. G., and Leeman, W. P., 1989, Phanerozoic evolution of the North American Cordillera; United States and Canada, *in* Bally, A. W., and Palmer, A. R., eds., The Geology of North America—An overview: Boulder, Colorado, Geological Society of America, Geology of North America, v. A, p. 139–232.

O'Sullivan, P. B., Murphy, J. M., Moore, T. E., and Howell, D. G., 1993, Results of 110 apatite fission track analyses from the Brooks Range and North Slope of northern Alaska, completed in cooperation with the Trans-Alaska Crustal Transect (TACT): U.S. Geological Survey Open-File Report 93-545, 104 p.

Palmer, A. R., compiler, 1983, The Decade of North American Geology 1983 geologic time scale: Geology, v. 11, p. 503–504.

Palmer, A. R., Dillon, J. T., and Dutro, J. T., Jr., 1985, Middle Cambrian trilobites with Siberian affinities from the central Brooks Range, northern Alaska: Geological Society of America Abstracts with Programs, v. 16, p. 327.

Parfenov, L. M., and Natal'in, B. A., 1986, Mesozoic tectonic evolution of northeastern Asia: Tectonophysics, v. 127, p. 291–304.

Patton, W. W., Jr., 1970, A discussion of tectonic history of northern Alaska, *in* Adkinson, W. L., and Brosgé, W. P., eds., Proceedings of the geological seminar on the North Slope of Alaska: Los Angeles, California, Pacific Section, American Association of Petroleum Geologists, p. E20.

Patton, W. W., Jr., and Hoare, J. M., 1968, The Kaltag fault, west-central Alaska: U.S. Geological Survey Professional Paper 600D, p. D147–D153.

Patton, W. W., Jr., and Tailleur, I. L., 1977, Evidence in the Bering Strait region for differential movement between North America and Eurasia: Geological Society of America Bulletin, v. 88, p. 1298–1304.

Patton, W. W., Jr., Box, S. E., and Grybeck, D. J., 1989, Ophiolites and other mafic-ultramafic complexes in Alaska: U.S. Geological Survey Open-File Report 89-648, 1 plate, 27 p.

Pavlis, T. L., 1989, Middle Cretaceous orogenesis in the northern Cordillera: A Mediterranean analog of collision-related extensional tectonics: Geology, v. 17, p. 947–950.

Pavlis, T. L., Sisson, V. B., Foster, H. L., Nokleberg, W. J., and Plafker, G., 1993, Mid Cretaceous extensional tectonics of the Yukon-Tanana terrane, Trans-Alaska Crustal Transect (TACT), east-central Alaska: Tectonics, v. 12, p. 103–122.

Plafker, G., 1990, Regional geology and tectonic evolution of Alaska and adjacent parts of the northeast Pacific Ocean margin: Proceedings of the Pacific Rim Congress 90: Queensland, Australia, Australasian Institute of Mining and Metallurgy, p. 841–853.

Plafker, G., Nokleberg, W. J., and Lull, J. S., 1989, Bedrock geology and tectonic evolution of the Wrangellia, Peninsular, and Chugach terranes along the Trans-Alaska Crustal Transect in the northern Chugach Mountains and southern Copper River basin, Alaska: Journal of Geophysical Research, v. 94, p. 4255–4295.

Plumley, P. W., 1984, A paleomagnetic study of the Prince William terrane and Nixon Fork terrane, Alaska [Ph.D. thesis]: Santa Cruz, University of California, 190 p.

Plumley, P. W., Coe, R. S., and Byrne, T., 1983, Paleomagnetism of the Paleocene Ghost Rocks Formation, Prince William terrane, Alaska: Tectonics, v. 2, p. 295–314.

Price, R. A., and Carmichael, D. M., 1986, Geometric test for Late Cretaceous–Paleogene intracontinental transform faulting in the Canadian Cordillera: Geology, v. 14, p. 468–471.

Richards, M. A., Jones, D. L., Duncan, R. A., and De Paolo, D. J., 1991, A mantle plume initiation model for the Wrangellia flood basalt and other oceanic plateaus: Science, v. 254, p. 263–267.

Ricketts, B. D., Evenchick, C. A., Anderson, R. G., and Murphy, D. C., 1992, Bowser basin, northern British Columbia: Constraints on the timing of initial subsidence and Stikinia-North America terrane interactions: Geology, v. 20, p. 1119–1122.

Rubin, C. M., and Saleeby, J. B., 1991, The Gravina sequence: Remnants of a mid-Mesozoic oceanic arc in southern southeast Alaska: Journal of Geophysical Research, v. 96, p. 14,551–14,568.

——, 1992, Tectonic history of the eastern edge of the Alexander terrane, southeast Alaska: Tectonics, v. 11, p. 586–602.

Rubin, C. M., Miller, M. M., and Smith, G. M., 1991, Tectonic development of Cordilleran mid-Paleozoic volcano-plutonic complexes: Evidence for convergent margin tectonism, *in* Harwood, D. S., and Miller, M. M., eds., Paleozoic and early Mesozoic paleogeographic relations of the Sierra Nevada, Klamath Mountains, and related terranes: Geological Society of America Special Paper 225, p. 1–16.

Saleeby, J. B., 1983, Accretionary tectonics of the North American Cordillera: Annual Review of Earth and Planetary Sciences, v. 11, p. 45–73.

Samson, S. D., Patchett, P. J., McClelland, W. C., and Gehrels, G. E., 1991, Nd isotopic characterization of metamorphic rocks in the Coast Mountains, Alaskan and Canadian Cordillera: Ancient crust bounded by juvenile terranes: Tectonics, v. 10, p. 770–780.

Sarewitz, D., 1983, Seven Devils terrane: Is it really a piece of Wrangellia?: Geology, v. 11, p. 634–637.

Savage, N. M., 1988, Devonian faunas and major depositional events in the southern Alexander terrane, southeastern Alaska, *in* McMillan, N. J., Embry, A. F., and Glass, D. J., eds., Devonian of the World, Volume I: Regional Syntheses: Canadian Society of Petroleum Geologists Memoir 14, p. 257–264.

Silberling, N. J., Jones, D. L., Monger, J.W.H., and Coney, P. J., 1992, Lithotectonic terrane map of the North American Cordillera: U.S. Geological Survey Miscellaneous Investigations Map I-2176, scale 1:5,000,000.

Smith, T. E., 1981, Geology of the Clearwater Mountains, south-central Alaska: Alaska Division of Geological and Geophysical Surveys Geologic Report 60, 72 p.

Stanley, W. D., Labson, V. F., Nokleberg, W. J., Csejtey, B., Jr., and Fisher, M. A., 1990, The Denali fault system and Alaska Range of Alaska: Evidence for suturing and thin-skinned tectonics from magnetotellurics: Geological Society of America Bulletin, v. 102, p. 160–173.

Stewart, J. H., 1976, Late Precambrian evolution of North America: Plate tectonics implications: Geology, v. 4, p. 11–15.

Stone, D. B., Panuska, B. C., and Packer, D. R., 1982, Paleolatitudes versus time for southern Alaska: Journal of Geophysical Research, v. 87, p. 3697–3708.

Tailleur, I. L., 1980, Rationalization of Koyukuk "crunch," northern and central Alaska [abs.]: American Association of Petroleum Geologists Bulletin, v. 64, p. 792.

Tempelman-Kluit, D. J., 1979, Transported cataclasite, ophiolite and granodiorite in Yukon: Evidence of arc-continent collision: Geological Survey of Canada Paper 79-14, 27 p.

—— , 1984, Counterparts of Alaska's terranes in Yukon, *in* Symposium of Cordilleran geology and mineral exploration: Status and future trends: Vancouver, Geological Association of Canada, p. 41–44.

Thorkelson, D. J., and Smith, A. D., 1989, Arc and intraplate volcanism in the Spences Bridge Group: Implications for Cretaceous tectonics in the Canadian Cordillera: Geology, v. 17, p. 1093–1096.

Umhoefer, P. J., Dragovich, J. C., and Engebretsen, D. C., 1989, Refinements of the "Baja British Columbia" plate-tectonic model for northward translation along the Late Cretaceous to Paleocene margin of western North America, *in* Hillhouse, J. W., ed., Deep structure and past kinematics of accreted terranes: International Union of Geodesy and Geophysics Geophysical Monograph 50, p. 101–111.

Van der Voo, R., Jones, M., Grommé, C. S., Eberlein, G. D., and Churkin, M., Jr., 1980, Paleozoic paleomagnetism and northward drift of the Alexander terrane, southeastern Alaska: Journal of Geophysical Research, v. 85, p. 5281–5296.

Vandall, T. A., and Palmer, H. C., 1990, Canadian Cordilleran displacement: Paleomagnetic results for the Early Jurassic Hazelton Group, Terrane I, British Columbia, Canada: Geophysical Journal International, v. 103, p. 609–619.

Weber, F. R., McCammon, R. B., Rinehart, C. D., Light, T. D., and Wheeler, K. L., 1988, Geology and mineral resources of the White Mountains National Recreational Area, east-central Alaska: U.S. Geological Survey Open-File Report 88-284, 120 p., 32 plates, scale 1:250,000.

Worrall, D. M., 1991, Tectonic history of the Bering Sea and the evolution of Tertiary strike-slip basins of the Bering shelf: Geological Society of America Special Paper 257, 120 p.

Yole, R. W., and Irving, E. M., 1980, Displacement of Vancouver Island; paleomagnetic evidence from the Karmutsen Formation: Canadian Journal of Earth Sciences, v. 17, p. 1210–1228.

MANUSCRIPT ACCEPTED BY THE SOCIETY DECEMBER 1, 1993

Printed in U.S.A.

Index

[Italic page numbers indicate major references]

B

C

D

E

G

H

I

J

K

N

O

P

Q

R

T

U

Z